ENVIRONMENTAL CONTAMINANT REFERENCE DATABOOK

VOLUME III

ENVIRONMENTAL CONTAMINANT REFERENCE DATABOOK

VOLUME III

JAN C. PRAGER

VAN NOSTRAND REINHOLD

I(T)P® an International Thomson Company

New York • Albany • Bonn • Boston • Detroit • London • Madrid • Melbourne
Mexico City • Paris • San Francisco • Singapore • Tokyo • Toronto

DISCLAIMER

Extreme care has been taken in preparation of this work.
However, neither the publisher nor the author shall be
held responsible or liable for any damages resulting in
connection with or arising from the use of any of the
information in this book.

Copyright © 1997 by Van Nostrand Reinhold

I(T)P® An International Thomson Publishing Company
The ITP logo is a registered trademark used herein under license.

Printed in the United States of America

For more information, contact:

Van Nostrand Reinhold
115 Fifth Avenue
New York, NY 10003

Chapman & Hall
2-6 Boundary Row
London
SE1 8HN
United Kingdom

Thomas Nelson Australia
102 Dodds Street
South Melbourne, 3205
Victoria, Australia

Nelson Canada
1120 Birchmount Road
Scarborough, Ontario
Canada, M1K 5G4

Champan & Hall GmbH
Pappelallee 3
69469 Weinheim
Germany

International Thomson Publishing Asia
221 Henderson Road #05-10
Henderson Building
Singapore 0315

International Thomson Publishing Japan
Hirakawacho Kyowa Building, 3F
2-2-1 Hirakawacho
Chiyoda-ku, 102 Tokyo
Japan

International Thompson Editores
Seneca 53
Col. Polanco
11560 Mexico D.F. Mexico

ISBN: 0-442-01971-8

1 2 3 4 5 6 7 8 9 10 HAM 01 00 99 98 97

Library of Congress Cataloging-in-Publication Data furnished upon request.

http://www.vnr.com
product discounts • free email newsletters
software demos • online resources
email: info@vnr.com

A service of I(T)P®

To my wife,
Anna,
and our newest granddaughter,
Sophie Lorraine O'Brien

Contents

Preface

The newest fad environmental regulation is to use ecosystem changes as a basis for setting permissible amounts of chemical contamination, a trend that alarms me because the science of ecology may not yet be up to the job. Its insufficiency is predicting ecological phenomena on the small-scale, cause-and-effect level. Although we can predict with acceptable certainty that a given large amount of toxic or oxygen-demanding substances will degrade the quality of air, land, or water, making it unsuitable for living creatures (ourselves included) and we even can predict qualitatively and quantitatively the degree of unsuitability, in many instances, we cannot tell the person who wants to use a small, specific amount of land, air, or water for disposal of a small amount of specific waste just how much damage that projected contamination will cause. It is the additive *de minimus* problems that undermine pollutant regulation at the specific permit level. Thus, for specific, local environmental protection decisions, we need return to the consideration of toxicological properties of pollutant chemicals—those "toxic materials in toxic amounts" that the 1970 Clean Water Act banned.

Prediction is a somewhat more straightforward matter in occupational health and safety because it involves only one species of air-breathing land animal (*Homo sapiens*) whose exposure to materials in the workplace can be or has been measured—and whose relationship to pollutants' effects on surrogate species is reasonably understood. Actual experiences are described by the subjects themselves. Granted, here too there is room for honest disagreement about exposure, dosages, and cause-and-effect, but no one is suggesting that we disregard much of our experience with the physical, chemical, and biological properties of pollutant materials in favor of a statistically supported abstraction predicting that a given amount of pollutant will degrade a given amount of environment to a given level of harm. Prediction at that level remains more a research-and-development goal than a reliable basis for regulation. Calls for applications of predictive ecology before its full verification in regulatory matters are certainly premature, as well-intentioned as they may be.

Environmental siting decisions that affect our lives most directly are made at the local government level. Although under the guidance of state and federal laws and regulations, local governments, except for the largest municipalities, are least well equipped to undertake highly technical evaluations. Siting, I believe, is the central issue of environmental protection. The geographic locations we choose to manufacture, transport, use and dispose of the substances and things we make govern who and what comes into contact with them. Zoning boards and planning commissions rarely have the scientific expertise to consider these matters in an environmental safety context, and the supply of competent and objective consultants to advise them is limited and unevenly distributed throughout the United States. Also, our adversarial system of adjudication and administrative procedures tends to erode the objectivity of technical experts, and this too needs study and modification to correct bias.

The current paradigm for environmental decision-making is risk analysis. Most environmental decisions now require some more-or-less formal analysis that follows the general pattern of first identifying the persons, places, or things at known risk in the proposed activity, then assessing the likelihood of exposure and effects combining to produce the identified adverse effect, and finally expressing that likelihood in some quantitative way that the public and those officials making the policy or permit decision can understand. Part of this expression is a statement of just how certain is the predicted result. As a general principle, the more systems involved in the analysis, the less certain will be the prediction. Uncertainty might be mitigated somewhat by integrating the prediction over a large number of systems taken as a whole (e.g., a physical model of an entire system such as the Army Corps of Engineers' models of various estuary flows or the University of Rhode Island's oceanographic miniature models of Narragansett Bay ecosystems), but this technique is in its infancy because there are few such models available to apply, and both physical and biological models of natural systems need further development, application, and verification.

Science is only as exact as its practitioners can make it—not as exact as we often need it to be for predicting risk. Sensitive dependence on initial conditions produces apparent uncertainty or even seemingly chaotic events, even in instances of exact mathematical certainty and Hamiltonian deterministic systems. Promulgating and compounding uncertainty additively or multiplicatively by combining a sequence of model results often diminishes the usefulness of not only the overall result, but, in the minds of the public, the component individual results as well. New techniques applied to modeling environmental consequences of contamination, such as fuzzy logic combined with artificial intelligence inferences, can reduce uncertainty, or at least manage to make it remain within acceptable limits in narrow domains, but here too much developmental work and even more verification are needed before public policy is based on large-scale models of these types. Throwing out the baby with the bath water—dropping attempts to model ecological processes to their lowest common denominator of certainty and prematurely using rough models of integrated processes—is a double mistake. Both model types are needed to make prudent decisions. Both types of models need specific information on the adverse effects of exposure to specific substances and compounds. It is that specific need that this book attempts to fill.

Introduction to Volume III

Volume III of this series is probably the last of my reference compendia of physical, chemical, and biological effects of environmental contaminants—primarily individual chemicals, but also some selected simple and complex mixtures. Each substance is summarized on a few pages. Summaries include information (as available) on CAS number; DOT and NIOSH numbers; synonyms; various detection limits; environmental transport, fate, and effects (selected narrative information on soil adsorption, volatilization, biodegradation, bioaccumulation, probable exposures, and effect types); water chemistry; metabolic pathways; a few structural formula drawings; molecular formulas; common uses; representative toxicities in nature (algae through mammals); odor thresholds; regulatory jurisdictions and authorities; standards (air and water); reactions; monitoring methods; international standards; and mitigation actions.

These data are intended for environmental regulators in federal, state, and local governments, as well as for conservation officers, emergency response personnel, public safety personnel, and chemical manufacturers, transporters, and sellers—a rather divergent group with convergent, sincere, and intense interest in the consequences of present, past, and future chemical contamination of the environment. In addition, the data may be used as a reference by students and teachers of environmental sciences, risk assessment, and disaster response. It also may be useful to media journalists for checking the accuracy of donated information. It is not a primary source of research information, however, and should not be sited as such. Scholars should instead seek information from peer-reviewed research literature, which is the scientific standard. This work is for quick reference only, and its value lies in its breadth of coverage. Its depth is limited. The series is intended to complement other VNR environmental data books, with somewhat more emphasis on ecological effects of regulated contaminant substances than on human health effects or occupational safety and health.

The basis for choice of chemicals to feature is still their frequency of appearance in several representative and important occupational safety and environmental databases. Databases chosen for chemical inclusion were CESARS, CHRIS, Hazardous Substances Data Bank, IRIS, and OHM/TADS. RTECS was used for information but, because the NIOSH mandate is to include all knowledge about all toxic substances in a single database, it was not used for choosing chemicals. All chemicals included in the chosen databases were listed, the resulting list was sorted, and the instances of each occurrence were counted. Those chemicals that appeared in all five databases were chosen for Volume I, and this volume continues from the list where Volume II ended. Some of the chemicals listed here appeared in as few as two databases. Sources of information used are all public domain databases. Among the databases listed above, CESARS was not used for textual material but all others were. RTECS is still a very good source of information. The disadvantage of using public domain databases as source materials for a data book is that they are poorly edited and formatted, requiring a great deal of attention to these matters. I have edited the materials to make them somewhat more readable, and have added content sparingly, as required to update or explain more clearly. Because this book is not intended as a primary source of research information, I chose to leave most of the references as they were found in the databases often incomplete and not given in a uniform acceptable style. There is enough information given for a good librarian to find a reference for you, but I felt no need to look up, correct, and complete each one myself partially because very few readers of data books check back to original research sources, in any case.

Most of the materials used in this volume are taken from databases available on compact disc from a variety of sources. My primary CD vendor was Silver Platter Information Inc., 100 River Ridge Drive, Norwood, MA 02062–5026 (phone 617–769–2599). Another good source of CDs is the Canadian Centre for Occupational Health and Safety, 250 Main Street, Hamilton, Ontario, Canada L8N 1H6 (phone 905–572–2981). For online searches, the National Library of Medicine's MD *Grateful Med* software and extensive collection of pertinent databases is my first choice. Online searching can be arranged through the U.S. Department of Commerce, National Technical Information Service, 5285 Port Royal Road, Springfield, VA 22161 (phone 703–321–8547). Chemical Information Systems operated now by Oxford Molecular Group Inc., 810 Gleneagles Court, Suite 300, Baltimore, MD 21286 (phone 410–821–5980), is another excellent source of information online.

This volume should be searched by means of its combined indices. They contain all the CAS numbers for each chemical described in all three volumes, although the text listings are alphabetical according to the chemical's primary name. This is not the most space-efficient method of listing, but it does have the advantage of allowing the reader to search by synonym—a luxury usually reserved for computer searches. Please let me know if you have suggestions for subsequent editions of this volume. Editors always like to hear from readers—particularly readers who offer constructive criticism.

—JCP

Acknowledgments

Once again I thank my wife, Anna, for her support, encouragement, and patience throughout the process of preparing Volume III. Joe Luszcz and Stephen Berard continue to be the best computer and editorial assistants that one could hope for, and their efforts are greatly appreciated. The crew at VNR, a constantly changing cast, remains constantly helpful however. Many and diverse thanks are due to Geraldine Albert, Peter Rocheleau, Barbara Mathieu, Louis Cabrera, Jesica Seacor, Caroline McCarra Sullivan, Renee Guilmette, Marianne Russell, and Nancy Olsen. Camillo Cimis copy edited this volume also, and again errors are much more likely mine than his as I do alter text after he has gone through it. I must thank, most of all, the many anonymous, dedicated, competent government scientists who cull research literature of many nations to create the public databases from which information in this volume comes.

■ ACEPHATE

CAS # 30560-19-1

SYNONYMS Ortho-124120 • orthene • ortran • o, s-dimethyl acetyl phosphoramidothioate • Chevron-orthene

MF $C_4H_{10}NO_3PS$

MW 183.16

COLOR AND FORM Colorless solid (technical grade 80-90% pure); white solid (technical); white crystals.

MELTING POINT 82-89°C (technical grade 80-90% purity).

FIREFIGHTING PROCEDURES Wear positive-pressure self-contained breathing apparatus when fighting fires involving this material. Do not extinguish fire unless flow can be stopped; or extinguish fire using agent suitable for type surrounding fire. Use water in flooding quantities as fog; solid streams of water may be ineffective; cool all affected containers with flooding quantities of water; apply water from as far a distance as possible. Use alcohol foam, carbon dioxide, or dry chemical. Organophosphorus pesticide, liquid, NOS (compound and preparation) (insecticides, other than agricultural, NEC); organophosphorus pesticide, liquid, NOS (compound and preparation) (agricultural insecticides, NEC, liquid); organophosphorus pesticide, solid, NOS (compound and preparation) (insecticides, other than agricultural, NEC); organophosphorus pesticide, solid, NOS (compound and preparation) (agricultural insecticides, NEC, other than liquid) [R12, p. 512-3].

TOXIC COMBUSTION PRODUCTS When heated to decomposition can emit highly toxic fumes of POx. (phosphates) [R3].

SOLUBILITY Solubility at room temperature: about 650 g/L water; >100 g/L acetone, ethanol; <50 g/L aromatic solvents (technical grade 80-90% purity); very soluble in water; moderately soluble in acetone, alcohol; low solubility in aromatic solvents; water solubility: 818,000 mg/L at 25°C.

VAPOR PRESSURE 1.7×10^{-6} mm Hg at 25°C.

OCTANOL/WATER PARTITION COEFFICIENT Log K_{ow} = -0.85 (measured).

STABILITY AND SHELF LIFE Relatively stable [R8, 4809].

OTHER CHEMICAL/PHYSICAL PROPERTIES Sp gr: 1.35; mp: 64-68°C (impure); Henry's Law constant for acephate can be estimated to be 5.0×10^{-13} atm-m³/mole (SRC).

PROTECTIVE EQUIPMENT Wear gas mask or respirator, goggles, and protective clothes and gloves [R2].

PREVENTATIVE MEASURES Keep out of reach of children. Avoid contact with mouth, skin, and eyes [R2].

DISPOSAL Group I Containers: Combustible containers from organic or metallo-organic pesticides (except organic mercury, lead, cadmium, or arsenic compounds) should be disposed of in pesticide incinerators or in specified landfill sites (organic or metallo-organic pesticides) [R4].

COMMON USES Contact and systemic insecticide. Effective against alfalfa looper, aphids, armyworms, bagworms, bean leafbeetle, bean leafroller, blackgrass bugs, bollworm, budworm, cabbage looper, cankerworms, corn earworm, cutworms, diamondback moth, European corn borer, fireworms, fleahopper, grasshoppers, green cloverworm, gypsy moth, hornworm, imported cabbage worm, imported fire ants, lace bugs, leafminers, leafhoppers, leafrollers, lygus, Mexican bean beetle, Mormon crickets, oak moth, saltmarsh caterpillar, soybean looper, spanworms, sparganothis, stinkbugs, tent caterpillars, three-cornered alfalfa hopper, thrips, tobacco hornworm, velvetbean caterpillar, webworms, and whitefly. For bell and non-bell peppers, brussel sprouts, cauliflower, celery, cotton, cranberries, dry beans, head lettuce, mint, peanuts, soybeans, and succulent beans. Cockroach control (spot treatment only) in residential and industrial buildings and insect control in forests, tobacco, and on ornamentals [R2].

NON-HUMAN TOXICITY LD_{50} rat male oral 945 mg/kg (technical grade) [R1]; LD_{50} rat female oral 866 mg/kg (technical grade) [R1]; LD_{50} rat, male acute, oral (technical) 866 mg/kg [R8, 4809]; LD_{50} rat, female acute, oral (technical) 945 mg/kg [R8, 4809]; LD_{50} mouse acute, oral (technical) 361 mg/kg [R8, 4809]; minimum lethal dose (MLD) dog 681 mg/kg [R8, 4809]; LD_{50} rabbit acute dermal greater than 2,000 mg/kg body weight [R8, 4810]; LD_{50} rabbit, male acute dermal >10,000 mg/kg [R9].

ECOTOXICITY LC_{50} *Gammarus pseudolimnaeus* >50 µg/L/96 hr at 12°C, mature. Static bioassay without aeration, pH 7.2–7.5, water hardness 40–50 mg/L as calcium carbonate, and alkalinity of 30–35 mg/L. Technical material, 94% [R10, 8]; LC_{50} *Salmo gairdneri* (rainbow trout) 1,100 µg/L/96 hr at 10°C, 1.5 g. Static bioassay without aeration, pH 7.2–7.5, water hardness 40–50 mg/L as calcium carbonate, and alkalinity of 30–35 mg/L. Technical material, 94% [R10, 8]; LC_{50} *Pimephales promelas* (fathead minnow) >1,000 µg/L/96 hr at 20°C, 1.0 g. Static bioassay without aeration, pH 7.2–7.5, water hardness 40–50 mg/L as calcium carbonate, and alkalinity of 30–35 mg/L. Technical material, 94% [R10, 9]; LC_{50} *Lepomis machrochirus* (bluegill) >1,000 µg/L/96 hr at 20°C, 0.4 g. Static bioassay without aeration, pH 7.2–7.5, water hardness 40–50 mg/L as calcium carbonate, and alkalinity of 30–35 mg/L. Technical material, 94% [R10, 9]; LC_{50} *Coturnix* 3,275 ppm/5 days (95% confidence interval: 2,691–3,986 ppm) [R11].

BIODEGRADATION River die-away studies determined that acephate degraded more rapidly in nonsterile creek water as compared to sterilized creek water (1); after 50 days of incubation, 54.8% of initial acephate was degraded in nonsterile water while only 23.6% had degraded in sterile water (1); in water plus sediment tests, 74.5% degraded in nonsterile media while only 45% degraded in sterile media (1) [R5].

BIOCONCENTRATION Based upon a measured log K_{ow} of −0.85 (1), a water solubility of 818,000 mg/L at 25°C (2), and regression-derived equations (3), the BCF for acephate can be estimated to be approximately 0.1–0.3, which indicates that it is not expected to bioconcentrate in aquatic organisms (SRC). In a model ecosystem study, acephate did not bioaccumulate in any of the organisms in the ecosystem that included algae, clam, crab, *Daphnia*, *Elodea*, fish, mosquito, and snail (4) [R6].

FIFRA REQUIREMENTS A registration standard was issued 9/87 for acephate used as an insecticide [R7].

REFERENCES

1. Spencer, E. Y. *Guide to the Chemicals Used in Crop Protection.* 7th ed. Publication 1,093. Research Institute, Agriculture Canada, Ottawa, Canada: Information Canada, 1982. 1.

2. *Farm Chemicals Handbook 1991.* Willoughby, OH: Meister, 1991, p. C6.

3. Sax, N. I. *Dangerous Properties of Industrial Materials.* 6th ed. New York, NY: Van Nostrand Reinhold, 1984. 2120.

4. 40 CFR 165.9 (a) (7/1/90).

5. Szeto, S. Y., et al. *J Environ Sci Health* B14: 635–54 (1979).

6. (1) Hansch, C., and A. J. Leo. *Medchem Project* Issue No 26. Claremont, CA: Pomona College (1985) (2) Wauchope R. D., et al. *Rev Environ Contam Toxicol* 123: 1–35 (1991) (3) Lyman, W. J., et al. *Handbook of Chemical Property Estimation Methods.* Washington, DC: American Chemical Soc pp. 5–4, 5–10 (1990) (4) Sanborn, J. R. *The Fate of Selected Pesticides in the USEPA.*

7. USEPA. *Report on the Status of Chemicals in the Special Review Program and Registration Standards in the Reregistration Program* p. 2–3 (1989).

8. Clayton, G.D., and F.E. Clayton (eds.). *Patty's Industrial Hygiene and Toxicology.* Volumes 2A, 2B, 2C: *Toxicology.* 3rd ed. New York: John Wiley and Sons, 1981–1982.

9. Purdue University, *National Pesticide*

Information Retrieval System, Acephate Fact Sheet No. 140 (1987).

10. U.S. Department of Interior, Fish and Wildlife Service. *Handbook of Acute Toxicity of Chemicals to Fish and Aquatic Invertebrates.* Resource Publication No. 137. Washington, DC: U.S. Government Printing Office, 1980.

11. Hill, E. F., and M. B. Camardese. *Lethal Dietary Toxicities of Environmental Contaminants and Pesticides to Coturnix.* Fish and Wildlife Technical Report 2. Washington, DC: United States Department of Interior, Fish and Wildlife Service, 1986. 21.

12. Association of American Railroads. *Emergency Handling of Hazardous Materials in Surface Transportation.* Washington, DC: Assoc. of American Railroads, Hazardous Materials Systems (BOE), 1987.

■ ACETANILIDE

NHCOCH₃

CAS # 103–84–4

SIC CODES 2830; 8000; 2815; 2850

SYNONYMS acetamide, N-phenyl • acetamidobenzene • acetanil • acetanilid • acetic acid anilide • acetic acid, amide, N-phenyl • acetoanilide • acetylaminobenzene • acetylaniline • N-acetylaniline • aniline, N-acetyl • antifebrin • benzenamine, N-acetyl • phenalgene • phenalgin • N-phenylacetamide • USAF EK-3

MF C₈H₉NO

MW 135.18

COLOR AND FORM Orthorhombic plates or scales from water; white shining crystalline scales.

ODOR Odorless.

TASTE Slightly burning.

DENSITY 1.2190 at 15°C.

DISSOCIATION CONSTANTS K_b: 1×10^{-13} at 28°C.

BOILING POINT 304°C at 760 mm Hg.

MELTING POINT 114.3°C.

AUTOIGNITION TEMPERATURE 985 − 10°F [R5].

SOLUBILITY 1 g soluble in 185 ml water, 3.4 ml alcohol, 20 ml boiling water, 3 ml methanol, 4 ml acetone, 0.6 ml boiling alcohol, in 3.7 ml chloroform, 5 ml glycerol, 8 ml dioxane, 47 ml benzene, 18 ml ether; very sparingly soluble in petroleum ether; chloral hydrate increases solubility in water; soluble in toluene; very soluble in hot toluene, in carbon tetrachloride.

VAPOR PRESSURE 1 mm Hg at 114.0°C.

STABILITY AND SHELF LIFE Rearranges under influence of UV light; acetyl group forms new bond on ring in ortho or para position [R6].

OTHER CHEMICAL/PHYSICAL PROPERTIES It forms eutectic mixture with phenol, resorcinol, and thymol.

DISPOSAL SRP: At the time of review, criteria for land treatment or burial (sanitary landfill) disposal practices are subject to significant revision. Prior to implementing land disposal of waste residue (including waste sludge), consult with environmental regulatory agencies for guidance on acceptable disposal practices.

COMMON USES Manufacture of medicinals, dyes; stabilizer for hydrogen peroxide; as addition to cellulose ester varnishes [R1]; medication: antipyretic, analgesic agent; (vet): analgesic, antipyretic agent [R1]; rubber accelerator [R3]; manufacture of intermediates; in synthetic camphor; precursor in penicillin manufacture [R2]; intermediate for sulfa drugs; hydrogen peroxide stabilizer; azo dye manufacture [R4].

PERSISTENCE Biodegrades at moderate rate.

DIRECT CONTACT Eczematous dermatitis may result—skin.

GENERAL SENSATION Can cause methemoglobinemia, cyanosis. Slightly burning taste; allergen.

PERSONAL SAFETY Inhalation of its dust should be avoided. Wear filter mask unless intense heat threatens high vapor concentrations, then employ canister or self-contained apparatus. Wear skin protection.

ACUTE HAZARD LEVEL Several grams taken orally can cause acute poisoning. Moderate ingestive and slight inhalation hazard. Allergen. Will produce BOD.

CHRONIC HAZARD LEVEL Allergen. Moderately toxic with chronic ingestion or inhalation. Chronic effects include methemoglobinemia and cyanosis. Repeated oral doses have caused damage to blood-forming organs.

DEGREE OF HAZARD TO PUBLIC HEALTH Allergen. Moderately toxic with acute or chronic ingestion. Acute inhalation threat is slight, but chronic hazard is moderate. Releases toxic vapors upon decomposition.

OCCUPATIONAL RECOMMENDATIONS None listed.

ACTION LEVEL Suppress suspension of dusts.

ON-SITE RESTORATION Dredge solids. Use carbon or peat on dissolved portion. Seek professional environmental engineering assistance through EPA's Environmental Response Team (ERT), Edison, NJ, 24-hour no., 908–548–8730.

BEACH OR SHORE RESTORATION Do not burn.

AVAILABILITY OF COUNTERMEASURE MATERIAL Carbon—water treatment plants, sugar refineries; peat—nurseries, floral shops.

DISPOSAL METHOD Add to a flammable solvent (alcohol or benzene). Pour into an iron pan in an open pit. Ignite or spray into an incinerator. Oxides of nitrogen may be scrubbed out with alkaline solution.

DISPOSAL NOTIFICATION Notify local air authority.

MAJOR WATER USE THREATENED Potable supply, recreation.

PROBABLE LOCATION AND STATE Orthorhombic plates, scales, colorless, and lustrous, but it can also be obtained as a white powder. Stable. It can be boiled without decomposition. Will sink and dissolve slowly.

WATER CHEMISTRY Acetanilide does not undergo appreciable chemical change when dissolved in natural waters. It is subject to biochemical oxidation, but concentrated amounts can be toxic to bacteria.

COLOR IN WATER Colorless.

BIOLOGICAL HALF-LIFE $T_{1/2}$ of acetanilide measured in populations of young (aged 20–35 yr) and elderly (aged over 65 yr) people (total = 93). $T_{1/2}$ were significantly longer in the elderly [R7].

REFERENCES

1. *The Merck Index.* 9th ed. Rahway, NJ: Merck and Co., Inc., 1976. 6.

2. Hawley, G. G. *The Condensed Chemical Dictionary.* 9th ed. New York: Van Nostrand Reinhold Co., 1977. 4.

3. Patty, F. (ed.). *Industrial Hygiene and Toxicology. Volume II: Toxicology.* 2nd ed. New York: Interscience Publishers, 1963. 1835.

4. *Kirk–Othmer Encyc Chemical Tech.* 3rd ed. 1978–present, V2, p. 318.

5. National Fire Protection Association. *Fire Protection Guide on Hazardous Materials.* 7th ed. Boston, MA: National Fire Protection Association, 1978 p. 325M–19.

6. Kearney, P. C., and D. D. Kaufman (eds.) *Herbicides: Chemistry, Degradation, and Mode of Action.* Volumes 1 and 2. 2nd ed. New York: Marcel Dekker, Inc., 1975. 863.

7. Farah, F., et al. "Hepatic Drug Acetylation and Oxidation: Effects of Aging in Man" *Br Med J* 2 (Jul 16), 155 (1977).

■ACETIC ACID, COBALT(2+) SALT

CAS # 71-48-7

SYNONYMS acetic acid, cobalt(2) salt • bis (acetato) cobalt • cobalt acetate • cobalt acetate (Co(OAc)₂) • cobalt diacetate • cobalt(2) acetate • cobalt(II) acetate • cobaltous diacetate

MF $C_4H_6O_4 \cdot Co$

MW 177.03

COLOR AND FORM Light-pink crystals.

TOXIC COMBUSTION PRODUCTS Cobalt oxide fumes may form in fire (tetrahydrate) [R7].

SOLUBILITY Readily soluble in water; 2.1 parts by wt (of the formula wt) /100 parts methanol by wt at 15°C.

STABILITY AND SHELF LIFE Solutions containing cobaltous ion, Co(2+), are relatively stable (cobaltous ion, Co(2+)) [R10, 246].

OTHER CHEMICAL/PHYSICAL PROPERTIES Intense red, monoclinic, prismatic crystals; soluble in alcohols, dilute acids, pentyl acetate; on heating becomes anhydr by 140°C; pH of 0.2-molar aqueous solution is 6.8 (tetrahydrate).

PROTECTIVE EQUIPMENT NIOSH-approved breathing apparatus, rubber gloves, goggles, or face shield, protective clothing (tetrahydrate) [R7].

PREVENTATIVE MEASURES SRP: Contaminated protective clothing should be segregated in such a manner that there is no direct personal contact by personnel who handle, dispose of, or clean the clothing. Quality assurance to ascertain the completeness of the cleaning procedures should be implemented before the decontaminated protective clothing is returned for reuse by the workers. Contaminated clothing should not be taken home at end of shift, but should remain at employee's place of work for cleaning.

CLEANUP Liquid material spills can be copiously flushed with water and channeled to a treatment system or holding tank for reclamation or proper disposal. Spills of dry material can be removed by vacuuming or wet mopping. Some spills can be removed by hosing, first with a mist of water to dampen the spilled material, and then with a more forceful stream that flushes it into a holding tank, or other facility for handling contaminated water. Work surfaces or contaminated clothing should never be cleaned by dry sweeping or blowing with pressurized hoses. Recovery systems used to reclaim waste metals should comply with federal, state, and local regulations. All waste materials generated in the handling of cobalt-containing substances should be disposed of in compliance with federal, state, and local regulations. (cobalt and cobalt salts) [R9].

DISPOSAL SRP: At the time of review, criteria for land treatment or burial (sanitary landfill) disposal practices are subject to significant revision. Prior to implementing land disposal of waste residue (including waste sludge), consult with environmental regulatory agencies for guidance on acceptable disposal practices.

COMMON USES Foam stabilizers for malt beverages (former use) [R1]; mineral supplement in cattle feed [R3]; bleaching agent and drier for lacquers, varnishes; anodizing agent; esterification catalyst [R1]; sympathetic inks; mineral supplement in feed additives (tetrahydrate) [R2]; oxidation catalyst [R4]; experimental use: cobalt edta, cobalt nitrate, and cobalt acetate ip injected simultaneously 3 min after, or 5 min after oral administration of potassium cyanide were effective antidotes for cyanide poisoning in albino mice [R5]; the only diseases in which the clinical use of cobalt is still advocated by some is the normochromic, normocytic anemia assoc with severe renal failure. (cobalt) [R6].

OCCUPATIONAL RECOMMENDATIONS
OSHA Permissible exposure limit (PEL): 8-hr TWA 0.1 mg/m³ (cobalt metal, fume, and dust (as Co)) [R8, 84].

THRESHOLD LIMIT 8-hr time-weighted average (TWA) 0.05 mg/m³ (1987) (cobalt, metal dust, and fume, as Co) [R14, 17].

INTERNATIONAL EXPOSURE LIMITS In the Federal Republic of Germany, industrial cobalt emission has been limited to 1 mg/m^3 for metallic cobalt and its slightly soluble compounds and 50 mg/m^3 for other compounds (cobalt and compounds) [R12, 260].

NON-HUMAN TOXICITY LD$_{50}$ mouse intravenous 28 mg/kg [R13, 286].

BIOLOGICAL HALF-LIFE It appears that independent of exposure route (inhalation, injection, or ingestion), most of the cobalt will be eliminated rapidly in humans. A small proportion is, however, eliminated slowly, having a biological half-time on the order of yr. (cobalt) [R11, 218].

BIOCONCENTRATION Food-chain concentration potential: bioconcentration of 200–1,000 fold only under constant exposure; not significant in spill condition (tetrahydrate) [R7].

GENERAL RESPONSE Stop discharge if possible. Keep people away. Avoid contact with solid and dust. Isolate and remove discharged material. Notify local health and pollution control agencies.

FIRE RESPONSE Not flammable. Poisonous gases may be produced in fire. Wear goggles and self-contained breathing apparatus.

EXPOSURE RESPONSE Call for medical aid. Dust irritating to eyes, nose, and throat. If inhaled will cause coughing or difficult breathing. If breathing has stopped, give artificial respiration. If breathing is difficult, give oxygen. Solid irritating to skin and eyes. If swallowed will cause nausea and vomiting. Remove contaminated clothing and shoes. Flush affected areas with plenty of water. If swallowed and victim is conscious, have victim drink water or milk, and have victim induce vomiting. If swallowed and victim is unconscious or having convulsions, do nothing except keep victim warm.

RESPONSE TO DISCHARGE Issue warning—water contaminant. Disperse and flush.

WATER RESPONSE Effect of low concentrations on aquatic life is unknown. May be dangerous if it enters water intakes. Notify local health and wildlife officials. Notify operators of nearby water intakes.

EXPOSURE SYMPTOMS Inhalation causes shortness of breath and coughing; permanent disability may occur. Ingestion causes pain and vomiting. Contact with eyes causes irritation. Contact with skin may cause dermatitis.

EXPOSURE TREATMENT Inhalation: move to fresh air; if breathing has stopped, begin artificial respiration. Ingestion: give large amounts of water; induce vomiting. Eyes: flush with water for at least 15 min. Skin: wash with soap and water.

REFERENCES

1. *The Merck Index.* 10th ed. Rahway, NJ: Merck Co., Inc., 1983. 346.

2. Sax, N. I., and R. J. Lewis, Sr. (eds.). *Hawley's Condensed Chemical Dictionary.* 11th ed. New York: Van Nostrand Reinhold Co., 1987. 293.

3. SRI.

4. *Kirk–Othmer Encyclopedia of Chemical Technology.* 3rd ed., Volumes 1–26. New York, NY: John Wiley and Sons, 1978–1984, p. 12 (80) 844.

5. El-Masry, Z., et al. *Ain Shams Med J* 26 (5–6): 710–7 (1975).

6. Gilman, A. G., L. S. Goodman, and A. Gilman (eds.). *Goodman and Gilman's The Pharmacological Basis of Therapeutics.* 7th ed. New York: Macmillan Publishing Co., Inc., 1985. 1,319.

7. U.S. Coast Guard, Department of Transportation. *CHRIS—Hazardous Chemical Data.* Volume II. Washington, DC: U.S. Government Printing Office, 1984–5.

8. NIOSH. *Pocket Guide to Chemical Hazards.* 2nd Printing. DHHS (NIOSH) Publ. No. 85–114. Washington, DC: U.S. Dept. of Health and Human Services, NIOSH/Supt. of Documents, GPO, February 1987.

9. NIOSH. *Criteria Document: Cobalt* p. 37 (1981) DHEW Pub. NIOSH 82–107.

10. National Research Council. *Drinking*

Water and Health. Volume 1. Washington, DC: National Academy Press, 1977.

11. Friberg, L., G. F. Nordberg, E. Kessler, and V. B. Vouk (eds.). *Handbook of the Toxicology of Metals.* 2nd ed. Vols. I, II. Amsterdam: Elsevier Science Publishers B. V., 1986, p. V2.

12. Seiler, H. G., H. Sigel, and A. Sigel (eds.). *Handbook on the Toxicity of Inorganic Compounds.* New York, NY: Marcel Dekker, Inc. 1988.

13. Venugopal, B., and T. D. Luckey. *Metal Toxicity in Mammals,* 2. New York: Plenum Press, 1978.

14. American Conference of Governmental Industrial Hygienists. *Threshold Limit Values for Chemical Substances and Physical Agents and Biological Exposure Indices for 1993–1994.* Cincinnati, OH: ACGIH, 1993.

■ ACETIC ACID, METHYL ESTER

CAS # 79–20–9

DOT # 1231

SIC CODES 2295; 2821

SYNONYMS acetic acid, methyl ester • methyl ester

MF $C_3H_6O_2$

MW 74.09

STANDARD CODES NFPA-1, 3, 0; ICC—flammable liquid, red label, 10 gallons in an outside container; USCG—class D combustible liquid; IATA—flammable liquid, red label, 1 liter passenger, 40 liter cargo.

PERSISTENCE Should biodegrade moderately fast.

INHALATION LIMIT 61 mg/m³.

DIRECT CONTACT Respiratory tract, mucous membranes.

GENERAL SENSATION Pleasant odor. Lacrimator. Recognition odor 200 ppm in air [R2]; skin irritation grade 1—no; narcotic in high-concentration effect, eye irritation grade 5—severe burns from 0.005 mL [R1]; irritant in air at 10,000 ppm. Lacrimator. Causes dyspnea, palpitation of the heart, depression, and dizziness.

PERSONAL SAFETY Wear skin protection and canister-type mask.

ACUTE HAZARD LEVEL Lethal concentration for rats in air, 35,000 ppm. Moderately toxic via all routes. May produce BOD.

CHRONIC HAZARD LEVEL Moderately toxic with chronic exposures by all routes.

DEGREE OF HAZARD TO PUBLIC HEALTH Moderately toxic via all routes with acute or chronic exposure. Mild irritant.

OCCUPATIONAL RECOMMENDATIONS None listed.

ACTION LEVEL Notify fire authority. If heat or flame is evident, evacuate area in case of explosion. Remove ignition sources.

ON-SITE RESTORATION Attempt treatment with carbon or peat. Seek professional environmental engineering assistance through EPA's Environmental Response Team (ERT), Edison, NJ, 24-hour no., 908-548-8730.

BEACH OR SHORE RESTORATION Burn off.

AVAILABILITY OF COUNTERMEASURE MATERIAL Carbon—water treatment plants, sugar refineries; peat—nurseries, floral shops.

DISPOSAL METHOD Spray into incinerator or burn in paper packaging. Additional flammable solvent may be added.

INDUSTRIAL FOULING POTENTIAL Volatility suggests rupture hazard when confined in boiler feed or cooling systems.

MAJOR WATER USE THREATENED Potable supply, fisheries, industrial.

PROBABLE LOCATION AND STATE Colorless liquid. Will dissolve.

WATER CHEMISTRY Subject to some biodegradation.

COLOR IN WATER Colorless.

GENERAL RESPONSE Shut off ignition sources, call fire department. Stop discharge if possible. Keep people away. Stay upwind, use water spray to knock down vapor. Isolate and remove discharged material. Notify local health and pollution control agencies.

FIRE RESPONSE Flammable. Containers may explode in fire. Flashback along vapor trail may occur. Vapor may explode if ignited in an enclosed area. Extinguish with dry chemicals, alcohol foam, or carbon dioxide. Water may be ineffective on fire. Cool exposed containers with water.

EXPOSURE RESPONSE Call for medical aid. Vapor irritating to eyes, nose, and throat. If inhaled will cause headache or dizziness. Move victim to fresh air. If breathing has stopped, give artificial respiration. If breathing is difficult, give oxygen. Liquid irritating to skin and eyes. Remove contaminated clothing and shoes. Flush affected areas with plenty of water. If swallowed and victim is conscious, have victim drink water or milk.

RESPONSE TO DISCHARGE Issue warning—high flammability. Restrict access. Disperse and flush.

WATER RESPONSE Effect of low concentrations on aquatic life is unknown. May be dangerous if it enters water intakes. Notify local health and wildlife officials. Notify operators of nearby water intakes.

EXPOSURE SYMPTOMS Very similar to those of methyl alcohol, which constitutes 20% of commercial grade. Inhalation causes headache, fatigue, and drowsiness; high concentrations can produce central nervous system depression and optic nerve damage. Liquid irritates eyes and may cause defatting and cracking of skin. Ingestion causes headache, dizziness, drowsiness, fatigue; may cause severe eye damage.

EXPOSURE TREATMENT Inhalation: remove victim from affected area; if breathing has ceased, apply artificial respiration; call doctor. Eyes: irrigate thoroughly with water for 15 min. and call doctor. Skin: wash affected area with water. Ingestion: get medical attention for methyl alcohol poisoning.

REFERENCES

1. (1) Smyth, H. F., Jr., C. P. Carpenter, C. S. Weil, U. C. Pozzani, J. A. Striegel, and J. S. Nycum, "Range-Finding Toxicity Data: List VII," *American Industrial Hygiene Association Journal*, 30: 470–476. 1969 (2) Smyth, H. F., C. P. Carpenter, and C. S. Weil, "Range-Finding Toxicity Data: List IV," *AMA Archives of Industrial Hygiene and Occupational Medicine*, 4: 119–122, 1951 (3) Smyth, H. F., C. P. Carpenter, C. S. Weil, U. C. Pozzani, and J. A. Striegel, "Range-Finding Toxicity Data: List VI," *American Industrial Hygiene Association Journal*, 23: 95–107, 1962 (4) Smyth, H. F., C. P. Carpenter, C. S. Weil, and U. C. Pozzani, "Range-Finding Toxicity Data: List V," *AMA Archives of Industrial Hygiene, and Occupational Medicine*, 10: 61–68, 1954 (5) Smyth, H. F., C. P. Carpenter, C. S. Weil, "Range-Finding Toxicity Data: List III," *Journal of Industrial Hygiene, and Toxicology*, 31: 60–62,1949 (6) Smyth, H. F., I. Seaton, L. Fischer, "The Single Dose Toxicity of Some Glycols, and Derivatives," *Journal of Industrial Hygiene and Toxicology*, 23 (6): 259–268, 1941 (7) Smyth, H. F., C. P. Carpenter, "Further Experience with the Range-Finding Test in the Industrial Toxicology Laboratory," *Journal of Industrial Hygiene and Toxicology*, 30: 63–68, 1948 (8) Smyth, H. F., C. P. Carpenter, "The Place of the Range-Finding Test in the Industrial Toxicology Laboratory," *Journal of Industrial Hygiene and Toxicology*, 26: 269–273, 1944.

2. Sullivan, R. J. *Air Pollution Aspects of Odorous Compounds*. NTIS PB 188 089, September 1969.

■ ACETIC ACID, THALLIUM(I) SALT

CAS # 563–68–8

DOT # 1707

SIC CODES 2800

SYNONYMS acetic acid, thallium(1) salt • thallium monoacetate • thallium(1) acetate • thallous acetate • thallium acetate

MF $C_2H_3O_2$ • Tl

MW 263.42

COLOR AND FORM White crystals.

ODOR Odorless.

DENSITY 3.765 at 137°C.

MELTING POINT 131°C.

FIREFIGHTING PROCEDURES Extinguish fire using agent suitable for type of surrounding fire (material itself does not burn or burns with difficulty). Use water in flooding quantities as fog; use alcohol foam, dry chemical, or carbon dioxide (thallium salt, solid, not otherwise specified) [R4]; water, carbon dioxide, foam, dry chemical. (soluble thallium compounds) [R5].

SOLUBILITY Soluble in alcohol; very soluble in chloroform; very soluble in cold water; insoluble in acetone.

OTHER CHEMICAL/PHYSICAL PROPERTIES Deliquescent.

PROTECTIVE EQUIPMENT Respirator selection: 0.5 mg/m³: any dust-and-mist respirator, except single-use respirators. 1 mg/m³: any dust-and-mist respirator, except single-use and quarter-mask respirators; any supplied-air respirator; any self-contained breathing apparatus. 2.5 mg/m³: any powered air-purifying respirator with a dust and mist filter; any supplied-air respirator operated in a continuous-flow mode. 5 mg/m³: any air-purifying full facepiece respirator with a high-efficiency particulate filter; any supplied-air respirator with a full facepiece; any self-contained breathing apparatus with a full facepiece; any powered air-purifying respirator with a tight-fitting facepiece and a high-efficiency particulate filter; any supplied-air respirator with a tight-fitting facepiece operated in a continuous-flow mode. 20 mg/m³: any supplied-air respirator with a full facepiece and operated in a pressure-demand or other positive-pressure mode. Emergency or planned entry in unknown concentration or IDLH conditions: any self-contained breathing apparatus with a full facepiece and operated in a pressure-demand or other positive-pressure mode; any supplied-air respirator with a full facepiece and operated in pressure-demand or other positive-pressure mode in combination with an auxiliary self-contained breathing apparatus operated in pressure-demand or other positive-pressure mode. Escape: any air-purifying full facepiece respirator with a high-efficiency particulate filter; any appropriate escape-type self-contained breathing apparatus (thallium and thallium compounds) [R6].

PREVENTATIVE MEASURES SRP: contaminated protective clothing should be segregated in such a manner so that there is no direct personal contact by personnel who handle, dispose of, or clean the clothing. Quality assurance to ascertain the completeness of the cleaning procedures should be implemented before the decontaminated protective clothing is returned for reuse by the workers.

CLEANUP 1. Ventilate area of spill. 2. For small quantities, sweep onto paper or other suitable material, place in sealed container for disposal. Large quantities may be reclaimed; however, if this is not practical, collect spilled material in the most convenient and safe manner and deposit in sealed containers for disposal. 3. Liquids containing soluble thallium compounds may be absorbed in vermiculite, dry sand, earth, or a similar material, and deposited in sealed containers for disposal. (thallium and thallium compounds) [R5].

DISPOSAL SRP: at the time of review, criteria for land treatment or burial (sanitary landfill) disposal practices are subject to significant revision. Prior to implementing land disposal of waste residue (including waste sludge), consult with environmental regulatory agencies for guidance on acceptable disposal practices.

COMMON USES High-specific-gravity solution used to separate ore constituents by flotation [R1]; formerly used as a depilating agent by dermatologists; also formerly as cosmetic depilatory cream [R2, 504]; thallium compounds are used in infrared spectrometers, in crystals, in other optical systems, and for coloring glass. (thallium compounds) [R3, 552].

STANDARD CODES IATA—poison B, poison label, 25 kg passenger, 95 kg cargo.

MAJOR SPECIES THREATENED 50 ml of a 400 mg/L solution caused death in birds 11 days; 10 mg thallium acetate in water caused death in birds 19 days; 0.05 mg at 2 days old caused death in chicks 11 weeks; 6 mg at 7 weeks old caused death in chicks 11 weeks; 341–2,730 mg/L Tl is toxic to plants and inhibits chlorophyll formation and seed generation; completely inhibits germination of cress and mustard seeds 18 days. [R13].

INHALATION LIMIT 0.1 mg/m^3.

GENERAL SENSATION Acute poisoning causes swelling of feet and legs, painful joints, vomiting, insomnia, hyperesthesia, and paraesthesia of hands and feet, mental confusion, polyneuritis, partial paralysis of the legs, angina-like pains, nephritis, wasting, and weakness, lymphocytosis, and eosinophilia, loss of hair. Fatal poisoning has occurred. Can be absorbed through skin. Can be absorbed through mucous membranes [R13].

PERSONAL SAFETY Wear industrial filter masks.

ACUTE HAZARD LEVEL Highly toxic when ingested. Moderately toxic when inhaled. Children have been known to tolerate 8 mg/kg body weight, whereas adults and adolescents have not. 100 ppm Tl in water is hazardous.

CHRONIC HAZARD LEVEL Highly toxic with repeated ingestion or inhalation. Marine waters should not exceed 1/20 of 96 hour LC$_{50}$. [R12].

DEGREE OF HAZARD TO PUBLIC HEALTH Cases of industrial poisoning are rare. Acute poisoning usually follows the ingestion of toxic quantities of a thallium-bearing depilatory or accidental or suicidal ingestion of rat poisoning. Highly toxic by ingestion or inhalation. Emits toxic fumes when heated.

OCCUPATIONAL RECOMMENDATIONS

OSHA 8-hr time-weighted average (TWA): skin: 0.1 mg/m^3 (thallium (soluble compounds) as Tl) [R9].

THRESHOLD LIMIT 8-hr time-weighted average (TWA) 0.1 mg/m^3, skin (1977) (thallium, soluble compounds, as Tl) [R10, 33].

ACTION LEVEL Suppress suspension of dusts. Isolate from heat.

ON-SITE RESTORATION Precipitate with sodium stearate or use cation exchanger. Seek professional environmental engineering assistance through EPA's Environmental Response Team (ERT), Edison, NJ, 24-hour no. 908–548–8730.

AVAILABILITY OF COUNTERMEASURE MATERIAL Sodium stearate—detergent manufacturers; cation exchangers—water softener suppliers.

DISPOSAL METHOD Thallium and its compounds should be dealt with separately. If the quantity justifies recovery, get in touch with the supplier for special instructions.

MAJOR WATER USE THREATENED Potable supply, recreation.

PROBABLE LOCATION AND STATE Silkwhite crystals. Will dissolve.

FOOD-CHAIN CONCENTRATION POTENTIAL Positive.

BIOLOGICAL HALF-LIFE Calculation of pharmacokinetic parameters for thallium in ethanol-intoxicated (alcohol-intoxicated) subjects led to extremely variable half-life values ranging from 1.7 to 30 days. [R7, 680].

BIOCONCENTRATION Bioconcentration factors for thallium: in freshwater plants, factor of 100,000; marine plants, factor of 100,000 (from table, total thallium) [R8, p. 18–5].

RCRA REQUIREMENTS As stipulated in 40 CFR 261.33, when thallium acetate, as a commercial chemical product or manufacturing chemical intermediate or an off-specification commercial chemical product or a manufacturing chemical intermediate, becomes a waste, it must be managed according to federal or state hazardous waste regulations. Also defined as a hazardous waste is any residue, contaminated soil, water, or other debris resulting from the cleanup of a spill, into water, or on dry land, of this waste. Generators of small quantities of this waste may qualify for partial exclusion from hazardous waste regulations (40 CFR 261.5) [R11].

GENERAL RESPONSE Avoid contact with solid and dust. Keep people away. Wear self-contained positive-pressure breathing apparatus and full protective clothing. Stop discharge if possible. Isolate and remove discharged material. Notify local health and pollution control agencies.

FIRE RESPONSE Nonflammable. Poisonous and irritating fumes are produced in a fire or when heated. Wear self-contained positive-pressure breathing apparatus and full protective clothing. Extinguish small fires: dry chemical, carbon dioxide, water spray, or foam; large fires: water spray, fog, or foam.

EXPOSURE RESPONSE Call for medical aid. Dust is poisonous. May be fatal if inhaled or absorbed through skin. Onset of symptoms may be delayed several hours. If in eyes or on skin, flush with running water for at least 15 minutes, holding eyelids open periodically, if appropriate. Remove and isolate contaminated clothing and shoes at the site. If breathing has stopped, give artificial respiration. If breathing is difficult, give oxygen. Solid is poisonous. May be fatal if swallowed or absorbed through skin. Onset of symptoms delayed 12 to 24 hours after ingestion. If swallowed, may cause nausea, vomiting, diarrhea, and abdominal pain. If in eyes, or on skin, flush with running water for at least 15 minutes, hold eyelids open periodically if appropriate. Remove and isolate contaminated clothing and shoes at the site. If swallowed and victim is conscious, have victim drink water, and induce vomiting by touching a finger to the back of the throat. If swallowed and victim is unconscious or having convulsions, do nothing except keep victim quiet and maintain body temperature.

RESPONSE TO DISCHARGE Issue warning—poison, water contaminant. Restrict access. Should be removed. Provide chemical and physical treatment.

WATER RESPONSE Harmful to aquatic life in very low concentrations. May be dangerous if it enters water intakes. Notify local health and wildlife officials. Notify operators of nearby water intakes.

EXPOSURE SYMPTOMS Thallium is one of the more toxic elements, both as an acute and a chronic poison. Effects of exposure are cumulative and onset of symptoms may be delayed 12 to 24 hours. May be fatal if inhaled, ingested, or absorbed through the skin. Irritating to skin and eyes. Readily absorbed through the skin and digestive tract. Ingestion of soluble thallium compounds has caused many deaths. Ingestion of sublethal quantities may cause nausea, vomiting, diarrhea, abdominal pain, and bleeding from the gut, accompanied or followed by drooping eyelids, crossed eyes, weakness, numbness, tingling of arms and legs, trembling, tightness, and pain in the chest. Loss of hair may occur in two to three weeks. Severe intoxication may cause prostration, rapid heartbeat, convulsions, and psychosis. Some effects may be permanent.

EXPOSURE TREATMENT Inhalation: move victim to fresh air; call emergency medical care. If breathing has stopped, give artificial respiration. Eyes or skin: immediately flush with running water for at least 15 minutes, lifting the upper and lower lids occasionally, if appropriate. Speed in removing material from skin is important. Remove and isolate contaminated clothing and shoes at the site. Ingestion: if conscious, have victim drink large quantities of water, and induce vomiting by touching the back of throat with a finger. If victim is unconscious or having convulsions, do nothing except keep victim warm.

REFERENCES

1. Sax, N. I., and R. J. Lewis, Sr. (eds.). *Hawley's Condensed Chemical Dictionary.* 11th ed. New York: Van Nostrand Reinhold Co., 1987. 1,142.

2. Doull, J., C. D. Klassen, and M. D. Amdur (eds.). *Casarett and Doull's Toxicology*. 3rd ed. New York: Macmillan Co., Inc., 1986.

3. Friberg, L., G. F. Nordberg, E. Kessler, and V. B. Vouk (eds). *Handbook of the Toxicology of Metals*. 2nd ed. Vols I, II. Amsterdam: Elsevier Science Publishers B. V., 1986, p. V2.

4. Association of American Railroads. *Emergency Handling of Hazardous Materials in Surface Transportation*. Washington, DC: Assoc. of American Railroads, Hazardous Materials Systems (BOE), 1987. 677.

5. Mackison, F. W., R. S. Stricoff, and L. J. Partridge, Jr. (eds.). *NIOSH/OSHA—Occupational Health Guidelines for Chemical Hazards*. DHHS (NIOSH) Publication No. 81–123 (3 vols). Washington, DC: U.S. Government Printing Office, Jan. 1981.

6. NIOSH. *Pocket Guide to Chemical Hazards*. 5th Printing/Revision. DHHS (NIOSH) Publ. No. 85–114. Washington, DC: U.S. Dept. of Health and Human Services, NIOSH/Supt. of Documents, GPO, Sept. 1985. 225.

7. Seiler, H. G., H. Sigel, and A. Sigel (eds.). *Handbook on the Toxicity of Inorganic Compounds*. New York, NY: Marcel Dekker, Inc. 1988.

8. Callahan, M. A., M. W. Slimak, N. W. Gabel, et al. *Water-Related Environmental Fate of 129 Priority Pollutants*. Volume I. EPA–440/4 79–029a. Washington, DC: U.S. Environmental Protection Agency, December 1979.

9. 29 CFR 1910.1000 (7/1/87).

10. American Conference of Governmental Industrial Hygienists. *Threshold Limit Values for Chemical Substances and Physical Agents and Biological Exposure Indices for 1993–1994*. Cincinnati, OH: ACGIH, 1993.

11. 40 CFR 261.33 (7/1/88).

12. *Proposed Criteria For Water Quality, Volume I*, U.S. Environmental Protection Agency. October 1973.

13. Zitko, V., "Toxicity, and Pollution Potential of Thallium," *The Science of the Total Environment*, 4, pp. 185–192, 1975.

ACETIC ACID, ZINC(II) SALT

CAS # 557–34–6

DOT # 9153

SIC CODES 3479

SYNONYMS acetic acid, zinc salt • dicarbomethoxyzinc • zinc diacetate • AI3-04465

MF $C_4H_6O_4 \cdot Zn$

MW 183.47

COLOR AND FORM White granules.

ODOR Faint vinegar odor.

TASTE Astringent taste.

DENSITY 1.735.

MELTING POINT 237°C.

FIREFIGHTING PROCEDURES If material on fire or involved in fire: Extinguish fire using agent suitable for type of surrounding fire (the material does not burn or burns with difficulty). [R8].

TOXIC COMBUSTION PRODUCTS When heated to decomposition it emits acrid smoke and fumes of zinc oxide. [R9].

SOLUBILITY 1 g soluble in 2.3 ml water; 1 g soluble in 1.6 ml boiling water; 1 g soluble in 30 ml alcohol; 1 g soluble in about 1 ml boiling alcohol; 29.000 lb/100 lb water at 68°F.

OTHER CHEMICAL/PHYSICAL PROPERTIES Neutral or slightly acid to litmus; pH about 5–6; faint, acetous odor; astringent taste; crystallizes from dilute acetic acid (dihydrate).

PROTECTIVE EQUIPMENT Bureau of Mines-approved respirator; rubber gloves; chemical goggles [R3].

PREVENTATIVE MEASURES In all cases where zinc is heated to the point where fume is produced, it is most important to ensure that adequate ventilation is provided. Individual protection is best ensured by education of the worker concerning metal-fume fever and the provision of local exhaust ventilation, or, in some situations, by wearing of supplied-air hood or mask. (zinc compounds) [R10, 2342].

CLEANUP Soil studied was from new lead belt of southeast Missouri, USA. Soil cationic exchange capacity of zinc acetate was high enough for disposal and would not represent danger to shallow groundwater table. [R11].

DISPOSAL SRP: at the time of review, criteria for land treatment or burial (sanitary landfill) disposal practices are subject to significant revision. Prior to implementing land disposal of waste residue (including waste sludge), consult with environmental regulatory agencies for guidance on acceptable disposal practices.

COMMON USES Preserving wood; as mordant in dyeing; manufacture glazes for painting on porcelain; protective (topical); as a reagent in testing for albumin, tannin, urobilin, phosphate, blood [R1]; feed additive; cross-linking agent for polymers; ingredient of dietary supplements (up to 1 mg daily) [R2]; in tobacco smoke filters and as a topical fungicide [R4]; medication: styptic, astringent, emetic [R1]; medication (vet): antiseptic, astringent, protective (topical); emetic [R1]; antihistamines were separated by TLC on metal-salt-impregnated silica gel plates with ethanol/DMF/ammonium hydroxide as developing solvent and visualization with Dragendorff reagent. Of the metal salts used to impregnate the TLC plates, zinc acetate (1%) gave the best results [R5]; it is a pharmaceutical necessity for zinc-eugenol dental cement for temporary fillings [R6]; zinc acetate is an indirect food additive for use only as a component of adhesives. [R7].

STANDARD CODES EPA 311; NFPA; no ICC; no USCG; Superfund designated (hazardous substances) list.

PERSISTENCE Zinc can persist indefinitely in water.

MAJOR SPECIES THREATENED Aquatic life.

DIRECT CONTACT Mildly irritating to skin and mucous membranes.

GENERAL SENSATION Faint acetous odor, astringent taste. Zn poisoning causes inflamed gills in fish.

PERSONAL SAFETY Filter mask may be required.

ACUTE HAZARD LEVEL THRESHOLD Con. for fish 0.1 ppm Zn [R13] toxic to fish. Zinc produces taste in water at low levels.

CHRONIC HAZARD LEVEL Daily dose of 10–15 mg for 4 months. In rats showed no effects. A low but significant mortality has been found among rainbow trout exposed; zinc is not believed to show chronic effects in man. Continuously for 4 months to constant concentrations of 0.2 of the 5-day LC_{50} and among rudd exposed for 8.5 months to 0.3 of the 7-day LC_{50}. Laboratory studies of avoidance reactions have shown that Atlantic salmon and rainbow trout may avoid concentrations of zinc in soft water that are 0.14–0.01 of the 7-day LC_{50}. Avoidance reactions have also been observed at 0.35–0.43 of the 7-day LC_{50} by migrating Atlantic salmon in a river polluted with copper and zinc. Carp and goldfish show avoidance of 0.3–0.45 of lethal concentrations under laboratory conditions [R15]. Maximum allowable toxicant concentration (MATC) for fathead minnow 0.032–0.18. Application factor for extrapolating 96 hour LC_{50} data 0.003–0.02 [R16]. Daphnids suffered decreased reproduction when exposed to 0.002 ppm Zn for 3 weeks, as did fathead minnows exposed to 0.18 ppm for 10 months [R17]. Freshwater should not exceed 0.003 of the 96 hour LC_{50} and marine waters 0.01 of the 96 hour LC_{50}. [R14].

DEGREE OF HAZARD TO PUBLIC HEALTH Mildly irritating to skin and mucous membranes.

OCCUPATIONAL RECOMMENDATIONS
OSHA Meets criteria for OSHA medical records rule. [R12].

ACTION LEVEL Suppress suspension of dusts.

ON-SITE RESTORATION Up to 4% Zn by weight can be removed on activated or ripe sludge. Use cation exchanger or lime to precipitate hydroxide. Seek professional environmental engineering assistance through EPA's Environmental Response Team (ERT), Edison, NJ, 24-hour no. 908–548–8730.

AVAILABILITY OF COUNTERMEASURE MATERIAL Cation exchanger—water softener suppliers; lime—cement plants.

DISPOSAL METHOD Add to large volume water. Stir in light excess soda ash (add slaked lime in presence of fluoride), decant and neutralize in second container with $6M$ HCl. Route to sewage plant. Landfill sludge.

DISPOSAL NOTIFICATION Local sewage and solid waste authority.

MAJOR WATER USE THREATENED Fisheries, potable supply.

PROBABLE LOCATION AND STATE Colorless crystals. Will dissolve.

WATER CHEMISTRY SOLUBILITY Dependent on pH, hardness, and alkalinity.

FOOD-CHAIN CONCENTRATION POTENTIAL Radioactive zinc (Zn-65) has been found to concentrate in plants, mold, and aquatic life. Positive; oysters concentrate it up to 200,000 times.

ECOTOXICITY TLm *Pimephales promelas* (fathead minnow) 0.88 ppm/96 hr (soft water) (conditions of bioassay not specified) [R3].

GENERAL RESPONSE Avoid contact with solid and dust. Keep people away. Stop discharge if possible. Isolate and remove discharged material. Notify local health and pollution control agencies.

FIRE RESPONSE Not flammable.

EXPOSURE RESPONSE Call for medical aid. Dust is irritating to eyes, nose, and throat. If inhaled will cause coughing or difficult breathing. If breathing has stopped, give artificial respiration. If breathing is difficult, give oxygen. Solid irritating to skin and eyes. If swallowed will cause nausea and vomiting. Remove contaminated clothing and shoes. Flush affected areas with plenty of water. If in eyes, hold eyelids open, and flush with plenty of water. If swallowed and victim is conscious, have victim drink water or milk, and have victim induce vomiting. If swallowed and victim is unconscious or having convulsions, do nothing except keep victim warm.

RESPONSE TO DISCHARGE Disperse and flush.

WATER RESPONSE Harmful to aquatic life in very low concentrations. May be dangerous if it enters water intakes. Notify local health and wildlife officials. Notify operators of nearby water intakes.

EXPOSURE SYMPTOMS Inhalation causes mild irritation of nose and throat, coughing, and sneezing. Ingestion can cause irritation or corrosion of the alimentary tract, resulting in vomiting. Contact with dust causes irritation of eyes and mild irritation of skin.

EXPOSURE TREATMENT Inhalation: move to fresh air; if exposure is severe, get medical attention. Ingestion: induce vomiting, followed by prompt and complete gastric lavage, cathartics, and demulcents. Eyes: flush with water for at least 10 min.; consult physician if irritation persists. Skin: wash with soap and water.

REFERENCES

1. *The Merck Index*. 10th ed. Rahway, NJ: Merck Co., Inc., 1983. 1,455.

2. Sax, N. I., and R. J. Lewis, Sr. (eds.). *Hawley's Condensed Chemical Dictionary*. 11th ed. New York: Van Nostrand Reinhold Co., 1987. 1,250.

3. U.S. Coast Guard, Department of Transportation. *CHRIS—Hazardous Chemical Data*. Volume II. Washington, DC: U.S. Government Printing Office, 1984–5.

4. *Kirk–Othmer Encyclopedia of Chemical Technology*. 3rd ed., Volumes 1–26. New York, NY: John Wiley and Sons, 1978–1984. p. 1 (78) 143.

5. Srivastava, S. P., and Reena. *Anal Lett* 15 (A5): 415–7 (1982).

6. Osol, A (ed.). *Remington's Pharmaceu-*

tical Sciences. 16th ed. Easton, PA: Mack Publishing Co., 1980. 718.

7. 21 CFR 175.105 (4/1/88).

8. Association of American Railroads. Emergency Handling of Hazardous Materials in Surface Transportation. Washington, DC: Assoc. of American Railroads, Hazardous Materials Systems (BOE), 1987. 730.

9. Sax, N. I. Dangerous Properties of Industrial Materials. 6th ed. New York, NY: Van Nostrand Reinhold, 1984. 2752.

10. International Labour Office. Encyclopedia of Occupational Health and Safety. Vols. I and II. Geneva, Switzerland: International Labour Office, 1983.

11. Jennett, J. C., S. M. Linnemann. J Water Pollut Control Fed 49: 1842–56 (1977).

12. 29 CFR 1910.20 (7/1/87).

13. Todd, D. K. The Water Encyclopedia, Maple Press, 1970.

14. Proposed Criteria For Water Quality, Volume I, U.S. Environmental Protection Agency. October 1973.

15. "Water Quality Criteria for European Freshwater Fish," Water Research EFAC Technical Paper No. 21, Vol. 8, pp. 683–684, 1974.

16. Pickering, Q. H., "Chronic Toxicity of Nickel to the Fathead Minnow," JWPCF, Vol. 46, No. 4, April 1974.

17. Environmental Protection Agency, "Report of the Pesticide Technical Committee to the Lake Michigan Enforcement Conference on Selected Trace Metals," NTIS PB-220 361, September 1972.

■ ACRYLIC ACID, ISOBUTYL ESTER

CAS # 106–63–8

DOT # 2527

SYNONYMS 2-methylpropyl acrylate • 2-propenoic acid, isobutyl ester • 2-propenoic acid, 2-methylpropyl ester • acrylic acid isobutyl ester • acrylic acid, isobutyl ester • isobutyl 2-propenoate • isobutyl propenoate • propenoic acid, isobutyl ester • isobutylacrylate

MF $C_7H_{12}O_2$

MW 128.19

COLOR AND FORM Clear liquid.

ODOR Sharp, fragrant odor.

ODOR THRESHOLD 2.00×10^{-3} ppm (odor detection in air, purity not specified) [R7].

DENSITY 0.8896 at 20 deg/4°C.

SURFACE TENSION 2.47 dynes/cm = 0.0247 N/m at 25°C.

VISCOSITY 0.822 cP at 70°F.

BOILING POINT 132°C.

HEAT OF COMBUSTION −13,500 Btu/lb = −7,500 cal/g = −314×10⁵ J/kg.

HEAT OF VAPORIZATION 130 Btu/lb = 71 cal/g = 3.0×10⁵ J/kg.

AUTOIGNITION TEMPERATURE 427°C [R5].

FLAMMABILITY LIMITS Lower 1.9%, upper 8.0% (in air) (% by vol) [R2].

FIREFIGHTING PROCEDURES Water may be ineffective [R5]. Do not extinguish fire unless flow can be stopped. Use water in flooding quantities as fog. Solid stream of water may spread fire. Cool all affected containers with flooding quantities of water. Apply water from as great a distance as possible. Use alcohol foam, dry chemical, or carbon dioxide [R6]. Evacuation: If fire becomes uncontrollable or container is exposed to direct flame—consider evacuation of one-third (1/3) mile radius [R6].

SOLUBILITY Soluble in methanol; soluble in alcohol, ether; water solubility of 1,800 mg/L.

VAPOR PRESSURE 10.7 mm Hg at 20°C.

OCTANOL/WATER PARTITION COEFFICIENT Log K_{ow} = 2.22.

OTHER CHEMICAL/PHYSICAL PROPERTIES Liquid water interfacial tension: 35 dynes/cm = 0.035 N/m at 27°C (est); ratio of specific heats of vapor: 1.044; heat of polymerization: -229 Btu/lb $= -127$ cal/g $= -5.32\times10^5$ J/kg.

PROTECTIVE EQUIPMENT Self-contained breathing apparatus, rubber gloves, chemical goggles [R2].

PREVENTATIVE MEASURES Awareness of the dangers and of good engineering design is essential to safety. Employees should be instructed about the necessity of cleansing the skin if it is contaminated by materials that are irritants or skin-absorbed. With careful design, however, and complete enclosure of those processes where toxic chemicals or intermediates occur, dangerous exposures can be avoided (acrylic acid and derivatives) [R8].

DISPOSAL SRP: At the time of review, criteria for land treatment or burial (sanitary landfill) disposal practices are subject to significant revision. Prior to implementing land disposal of waste residue (including waste sludge), consult with environmental regulatory agencies for guidance on acceptable disposal practices.

COMMON USES Chemical intermediate for polymers; comonomer in acrylic surface coatings [R1]; used as a monomer in synthetic resin manufacture [R4]; monomer for acrylate resins [R3].

STANDARD CODES No NFPA; no CPR; no USCG; no ICC.

MAJOR SPECIES THREATENED Waterfowl.

DIRECT CONTACT Skin irritant.

GENERAL SENSATION Skin irritation grade 3—capillary injection diluted; eye irritation grade 2—small burns from 0.5 mg [R14].

PERSONAL SAFETY Wear skin protection.

ACUTE HAZARD LEVEL Believed to be irritant and of moderate ingestive toxicity. May produce BOD. Lethal concentration for rats in air 2,000 ppm [R13].

CHRONIC HAZARD LEVEL Unknown.

DEGREE OF HAZARD TO PUBLIC HEALTH Similar compounds demonstrate irritant properties and moderate ingestive toxicity.

OCCUPATIONAL RECOMMENDATIONS None listed.

ACTION LEVEL Notify fire authority. Remove ignition source. Attempt to contain slick.

ON-SITE RESTORATION Oil-skimming equipment and sorbent foams can be applied to slick. Seek professional environmental engineering assistance through EPA's Environmental Response Team (ERT), Edison, NJ, 24-hour no. 908–548–8730.

BEACH OR SHORE RESTORATION Burn off.

AVAILABILITY OF COUNTERMEASURE MATERIAL Oil-skimming equipment—stored at major ports; sorbent foams (polyurethane)—upholstery shops.

DISPOSAL METHOD Spray into incinerator. Flammable solvent may be added.

INDUSTRIAL FOULING POTENTIAL Slick may reduce heat transfer or cause scaling and hot spots.

MAJOR WATER USE THREATENED Recreation, potable supply.

PROBABLE LOCATION AND STATE Colorless liquid. Will float in slick on surface.

ECOTOXICITY LC_{50} *Pimephales promelas* (fathead minnows) 29 days old 2.09 g/L/96 hr (95% confidence limit 1.92–2.28 g/L); >97% purity, water hardness 45.3 mg/L ($CaCO_3$), temperature 24.0°C, pH 7.61, dissolved oxygen 7.1 mg/L, alkalinity 49.8 mg/L ($CaCO_3$), tank volume 2.0 L, additions 18 vol/day (flow-through bioassay) [R9]; EC_{50} *Pimephales promelas* (fathead minnows) 29 days old 1.90 g/L/96 hr; >97% purity, water hardness 45.3 mg/L ($CaCO_3$), temperature 24.0°C, pH 7.61, dissolved oxygen 7.1 mg/L, alkalinity 49.8 mg/L ($CaCO_3$), tank volume 2.0 l, additions 18 vol/day (flow-through bioassay). Affected fish lost schooling behavior and swam near the tank surface in a corkscrew/spiral pattern. They were hyperactive and overreactive to external stimuli, had increased respiration and edema, and lost equilibrium prior to death [R9].

BIODEGRADATION By analogy to butyl acrylate, isobutyl acrylate may be susceptible to biodegradation (1). Butyl acrylate was confirmed to be significantly degradable according to the MITI test, the biodegradability screening test of the Japanese Ministry of International Trade and Industry (1). By analogy to ethyl acrylate, isobutyl acrylate may be susceptible to biodegradation using sewage seed (2). Ethyl acrylate readily biodegrades in screening tests using sewage seed in both fresh and salt water (2). The percent of theoretical BOD is 28 and 11%, respectively, after 5 days (2). After acclimation, the BOD after 5 days increased to 66% of theoretical (2) [R10].

BIOCONCENTRATION Using a reported log K_{ow} of 2.22 (1), a BCF of 29 has been calculated using a recommended regression-derived equation (2,SRC). This estimated BCF indicates that isobutyl acrylate will not be expected to bioconcentrate in aquatic organisms (SRC) [R11].

TSCA REQUIREMENTS Section 8 (a) of TSCA requires manufacturers of this chemical substance to report preliminary assessment information concerned with production, use, and exposure to EPA as cited in the preamble of the 51 FR 41329 [R12].

GENERAL RESPONSE Stop discharge if possible. Call fire department. Avoid contact with liquid and vapor. Isolate and remove discharged material. Notify local health and pollution control agencies.

FIRE RESPONSE Flammable. Flashback along vapor trail may occur. Vapor may explode if ignited in an enclosed area. Extinguish with dry chemical, foam, or carbon dioxide. Cool exposed containers with water.

EXPOSURE RESPONSE Call for medical aid. Vapor is irritating to eyes, nose, and throat. Move to fresh air. If breathing has stopped, give artificial respiration. If breathing is difficult, give oxygen. Liquid is irritating to skin and eyes. Harmful if swallowed. Remove contaminated clothing and shoes. Flush affected areas with plenty of water. If swallowed and victim is conscious, have victim induce vomiting. If swallowed, and victim is unconscious or having convulsions, do nothing except keep victim warm.

RESPONSE TO DISCHARGE Mechanical containment. Provide chemical and physical treatment.

WATER RESPONSE Effect of low concentrations on aquatic life is unknown. Fouling to shoreline. May be dangerous if it enters water intakes. Notify local health and pollution control officials. Notify operators of nearby water intakes.

EXPOSURE SYMPTOMS Moderate toxicity when swallowed. Contact with the eyes causes minor irritation no worse than that caused by hand soap.

EXPOSURE TREATMENT Inhalation: move victim to fresh air at once; give oxygen if breathing is difficult or artificial respiration if breathing has stopped; call a doctor. Ingestion: make victim vomit by sticking a finger down the throat or by giving strong, warm salt water to drink; get medical attention. Skin and eyes: remove chemical by flushing with plenty of clean, running water; remove contaminated clothing and wash exposed skin with soap and water.

REFERENCES

1. SRI.

2. U.S. Coast Guard, Department of Transportation. *CHRIS—Hazardous Chemical Data.* Volume II. Washington, DC: U.S. Government Printing Office, 1984–5.

3. Sax, N. I., and R. J. Lewis, Sr. (eds.). *Hawley's Condensed Chemical Dictionary.* 11th ed. New York: Van Nostrand Reinhold Co., 1987. 653.

4. Sittig, M. M. M. *Handbook of Toxic and Hazardous Chemicals and Carcinogens, 1985.* 2nd ed. Park Ridge, NJ: Noyes Data Corporation, 1985. 526.

5. National Fire Protection Association. *Fire Protection Guide on Hazardous Materials.* 9th ed. Boston, MA: National Fire Protection Association, 1986, p. 325M–61.

6. Association of American Railroads. *Emergency Handling of Hazardous Materials in Surface Transportation.* Washington, DC: Assoc. of American Railroads,

Hazardous Materials Systems (BOE), 1987. 392.

7. Fazzalari, F. A. (ed.). *Compilation of Odor and Taste Threshold Values Data.* ASTM Data Series DS 48A (Committee E-18). Philadelphia, PA: American Society for Testing and Materials, 1978. 26.

8. International Labour Office. *Encyclopedia of Occupational Health and Safety.* Vols. I and II. Geneva, Switzerland: International Labour Office, 1983. 53.

9. Geiger D. L., S. H. Poirier, L. T. Brooke, D. J. Call (eds.). *Acute Toxicities of Organic Chemicals to Fathead Minnows (Pimephales promelas).* Vol. III. Superior, Wisconsin: University of Wisconsin-Superior, 1986. 174.

10. (1) Sasaki, S., pp. 283–98 in *Aquatic Pollutants Transformation, and Biological Effects.* Pergamon Press (1978). (2) Price, K. S., et al. *J Water Pollut Contr Fed* 46: 63–77 (1974).

11. (1) Hansch, C., and A. J. Leo. *Medchem Project* Issue No. 26. Claremont, CA: Pomona College (1985). (2) Lyman, W. J., et al. *Handbook of Chemical Property Estimation Methods,* New York: McGraw-Hill, p. 5–5 (1982).

12. 40 CFR 712.30 (7/1/88).

13. Union Carbide Data Sheet, New York.

14. Carpenter, C. P., C. W. Weil, and H. F. Smyth, Jr., "Range-Finding Toxicity Data: List VIII," *Toxicology and Applied Pharmacology.* 28. DP. 313–319. 1974.

■ AFLATOXINS

CAS # 1402-68-2

SYNONYMS Aflatoxin

MF UVCB

SOLUBILITY Aflatoxins are soluble in methanol, acetone, and chloroform, but only slightly soluble in water and hydrocarbon solvents.

STABILITY AND SHELF LIFE Relatively unstable to light and air, particularly in solution in highly polar solvents; chloroform solutions are stable for years if kept in dark and cold.

OTHER CHEMICAL/PHYSICAL PROPERTIES Highly fluorescent, highly oxygenated, heterocyclic compounds characterized by dihydrodifurano or tetrahydrodifurano moieties fused to a substituted coumarin moiety. Lactone ring is susceptible to alkaline hydrolysis.

PROTECTIVE EQUIPMENT Latex gloves provide good protection against permeation by aflatoxins in dimethyl sulfoxide, but aflatoxins in chloroform permeate all types of laboratory gloves. If gloves are contaminated when aflatoxins in chloroform are handled, gloves should be changed immediately [R2].

PREVENTATIVE MEASURES Mycotoxins should be handled as very toxic substances. Perform manipulations under hood whenever possible, and take particular precautions, such as use of glove box, when toxins are in dry form because of electrostatic nature and resulting tendency to disperse in working areas. Swab accidental spills of toxin with 5% sodium hypochlorite bleach. Rinse all glassware exposed to aflatoxins with 1% sodium hypochlorite solution and then wash thoroughly (mycotoxins) [R3].

DISPOSAL Chemical treatment: Any chemical detoxification must reduce the mycotoxin level to within limits set by proper regulatory agencies, must have no toxic residues, and should not decrease the nutritive value of the treated commodity. These restrictions as well as simple economic considerations have narrowed the types of chemical agents likely to achieve significance in a commercially scaled detoxification process. Chemicals having good promise in this regard can be classified simply as oxidizing agents, acid, and bases. Of oxidizing systems that destroy all the aflatoxin only hydrogen peroxide has shown promise to detoxify foods and feeds. Aqueous suspensions of peanut meals containing 90 ppm aflatoxin were treated with 6% solution of hydrogen peroxide at pH 9.5 for 30 min at 80°C. The treatment destroyed 97% of the toxin, and the treated meals were shown to be nontoxic to ducklings. The use of acids as detoxifying agents has a definite potential in the recovery process of free fatty acids from commercial soapstock prepn. The use of inorganic and organic bases affords an efficient and relatively inexpensive means to achieve aflatoxin destruction or removal from large quantities of contaminated agricultural commodities. The refinement of edible oils is a classic example of the use of bases under conditions that would destroy or remove aflatoxins [R4].

COMMON USES Used for experimental purposes only [R1].

REFERENCES

1. IARC. *Monographs on the Evaluation of the Carcinogenic Risk of Chemicals to Man.* Geneva: World Health Organization, International Agency for Research on Cancer, 1972–present (multivolume work), p. V10 53 (1976).

2. Castegnaro, M. et al. *J Assoc of Anal Chem.* 65 (6): 1,520 (1982).

3. Association of Official Analytical Chemists. *Official Methods of Analysis.* 15th ed. and Supplements. Washington, DC: Association of Analytical Chemists, 1990, p. 15/1,184.

4. United Nations. *Treatment and Disposal Methods for Waste Chemicals* (IRPTC File). Data Profile Series No. 5. Geneva, Switzerland: United Nations Environmental Programme, Dec. 1985. 80.

■ ALDICARB

CAS # 116–06–3

SYNONYMS 2-methyl-2-(methylthio) propanal, o-((methylamino) carbonyl) oxime • 2-methyl-2-(methylthio) propionaldehyde o-(methylcarbamoyl) oxime • 2-Methyl-2-methylthio-propionaldehyd-o-(N-methyl-carbamoyl)-oxim (German) • 2-metil-2-tiometil-propionaldeid-o-(N-metil-carbamoil)-ossima (Italian) • aldecarb • aldicarbe (French) • carbamyl • propanal, 2-methyl-2-(methylthio)-,

o-((methylamino) carbonyl) oxime • propionaldehyde, 2-methyl-2-(methylthio)-, o-(methylcarbamoyl) oxime • sulfone-aldoxycarb • ent-27,093- • nci-c08640 • oms 771- • temic • temik • uc-21149 • Union Carbide-21149- • Union Carbide UC-21149 • temik-10-G

MF $C_7H_{14}N_2O_2S$

MW 190.29

COLOR AND FORM Crystals from isopropyl ether; white, crystalline solid.

ODOR Slightly sulfurous odor.

DENSITY 1.1950 at 25°C.

MELTING POINT 99–100°C.

FLAMMABILITY LIMITS Nonflammable [R21, 7].

FIREFIGHTING PROCEDURES If material is on fire or involved in a fire do not extinguish fire unless flow can be stopped. Use water in flooding quantities as fog. Solid streams of water may be ineffective. Cool all affected containers with flooding quantities of water. Apply water from as far a distance as possible. Use "alcohol" foam, dry chemical, or carbon dioxide (carbamate pesticide, liquid, nos, (compounds and preparations) (agricultural insecticides, NEC, liquid); carbamate pesticide, liquid, nos (compounds and preparations) (agricultural insecticides, nec, liquid); carbamate pesticide, liquid, nos (compounds and preparations) (insecticides, other than agricultural, nec)) [R23, 133]. Do not extinguish fire unless flow can be stopped. Use water in flooding quantities as fog. Cool all affected containers with flooding quantities of water. Apply water from as far a distance as possible. Solid streams of water may be ineffective. Use foam, dry chemical, or carbon dioxide (carbamate pesticide, liquid nos (compounds and preparations) (insecticides, other than agricultural, nec)) [R23, 133]. Extinguish fire using agent suitable for type of surrounding fire (material itself does not burn or burns with difficulty). Use water in flooding quantities as fog. Use "alcohol" foam, dry chemical, or carbon dioxide (carbamate pesticide, solid, nos (compounds and preparations) (insecticides, other than agricultural, nec)) [R23, 134]. Use water in flooding quantities as fog. Use foam, dry chemical, or carbon dioxide (carbamate pesticide, solid, nos (compounds and preparations) (agricultural insecticides, nec, other than liquid)) [R23, 134].

SOLUBILITY wt/wt at 25°C: in water 0.6%; acetone 35%; benzene 15%; xylene 5%; methylene chloride 30%; 17 µg/L of water at 25°C; 20% in ethyl ether; 10% in toluene; 15% in chlorobenzene; 5% in carbon tetrachloride, 24% in methyl isobutyl ketone (each at 30°C); soluble in ethanol; freely soluble in chloroform; slightly soluble in petroleum ether; insoluble in heptane; practically insoluble in hexane.

VAPOR PRESSURE 1×10^{-4} mm Hg at 25°C.

OCTANOL/WATER PARTITION COEFFICIENT Log K_{ow} = 1.13.

CORROSIVITY Noncorrosive to metal containers and equipment.

STABILITY AND SHELF LIFE Poor stability at 50°C; unstable in alkali [R5].

OTHER CHEMICAL/PHYSICAL PROPERTIES Decomposes above 100°C; vapor pressure: 3.5×10^{-5} mm Hg at 20°C.

PROTECTIVE EQUIPMENT Wear long-sleeved clothing and protective gloves when handling. [R22, p. C–281].

PREVENTATIVE MEASURES Change contaminated clothing daily and wash in strong washing soda solution and rinse thoroughly before re-use. Wash hands and face before eating or smoking. Bathe at the end of work day, washing entire body and hair with soap and water. [R22, p. C–281].

CLEANUP A system for removing pesticides from the wash water produced by pesticide applicators as they clean their equipment has been developed. The first step is the flocculation/coagulation and sedimentation of the pesticide-contaminated wash water. The supernatant from the first step is then passed through activated carbon columns. (pesticides) [R6].

DISPOSAL Generators of waste (equal to or greater than 100 kg/mo) containing this contaminant, EPA hazardous waste number P070, must conform with USEPA regulations in storage, transportation, treatment, and disposal of waste [R7].

COMMON USES Soil application for control of chewing and sucking insects (esp aphids, whiteflies, leaf miners, and soil-dwelling insects), spider mites, and nematodes in glasshouse and outdoor, ornamentals, sugar beet, fodder beet, strawberries, potatoes, onions, hops, vine nurseries, tree nurseries, groundnuts, soya beans, citrus fruit, bananas, coffee, sorghum, pecans, cotton, sweet potatoes, and other crops [R2]. Temik is used only as soil application to control certain insects, mites, and nematodes on citrus (grapefruit, lemons, limes, oranges only), cotton, dry beans; pecans (southeast only); sugarcane (Louisiana only) [R22, p. C–280]. Aldicarb is effective in reducing the foraging of skylarks in sugar-beet fields in the United Kingdom [R3]. Terrestrial non-food uses with rates (1 lb active ingredient): birch, 5.0–10.0; dahlias, 5.0–8.0; holly, 5.0–10.0; lilies (bulbs), 5.0–7.0; and roses, 7.0–10.0 [R1]. Commercial greenhouse uses with rates (1 lb active ingredient): carnations, 7.5–10.0; chrysanthemum, 7.5–10.0; Easter lilies, 5.0–7.5; gernera, 5.0–10.0; orchids, 7.5–10.0; poinsettia, 7.5–10.0; roses, 5.0–10.0; snapdragons, 5.0–10.0 [R1].

OCCUPATIONAL RECOMMENDATIONS None listed.

NON-HUMAN TOXICITY LD50 rat dermal 2,100–3,970 mg/kg (Temik 10G, dry) [R11]; LD50 rabbit dermal >5.0 mg/kg (technical aldicarb) [R11].

ECOTOXICITY LC50 *Phasianus colchicus* (ring-necked pheasant) oral greater than 300 ppm in 5-day diet *ad libitum* (no mortality to 300 ppm), age 10 days (technical material, 99.0%) [R12]; LC50 *Anas platyrhynchos* (mallard duck), oral less than 1,000 ppm in 5-day diet *ad libitum* (70% mortality at 1,000 ppm), age 10 days (technical material, 99.0%) [R12]; LC50 *Anas platyrhynchos* (mallard duck), oral 594 ppm in 5-day diet *ad libitum* (95% confidence limit 507–695 ppm), age 5 days (technical material, 99.0%) [R12]; LC50 *Coturnix japonica* (Japanese quail) oral 381 ppm in 5-day diet *ad libitum* (95% confidence limit 317–453 ppm), age 14 days (technical material, 99.0%) [R12]; LC$_{50}$ bobwhite quail, oral 2,400 mg/kg/7 days, age 56 days [R9]; LC$_{50}$ rainbow trout 8.8 mg/L/96 hr (conditions of bioassay not specified) [R9]; LC$_{50}$ *Paramecium multimicronucleatum* 145 ppm/9 hr; 122 ppm/13 hr; 104 ppm/17 hr; 93 ppm/24 hr (static bioassay) [R13]; LC$_{50}$ *Pimephales promelas* (fathead minnow) 1,370 µg/L/96 hr (flow-through bioassay) [R10]; LD$_{50}$ *Anas platyrhynchos* (mallard duck) oral 3.40 mg/kg (95% confidence limit 2.70–4.28 mg/kg), 3–4-mo-old females [R14]; LD$_{50}$ *Callipepla californica* (California quail) oral 2.58 mg/kg (95% confidence limit 1.96–3.40 mg/kg), 10-mo-old males [R14]; LD$_{50}$ *Callipepla californica* (California quail) oral 4.67 mg/kg (95% confidence limit 3.32–6.56 mg/kg), 10-mo-old females [R14]; LD$_{50}$ *Phasianus colchicus* (pheasant), oral 5.34 mg/kg (95% confidence limit 3.85–7.40 mg/kg), 3–4-mo-old females [R14]; LD$_{50}$ *Coturnix japonica* (Japanese quail) oral 387 ppm in 5-day diet *ad libitum* (95% confidence limit 336–445 ppm) (technical grade, 99% active ingredient) [R15].

IARC SUMMARY Evaluation: No data were available from studies in humans. There is inadequate evidence for the carcinogenicity of aldicarb in experimental animals. Overall evaluation: Aldicarb is not classifiable as to its carcinogenicity to humans (Group 3) [R8].

BIOLOGICAL HALF-LIFE In worms, aldicarb is rapidly converted to the sulfoxide which has a half-life in worms of 19 hr at 15°C, and 50 hr at 5°C [R16].

BIODEGRADATION Aldicarb has a half-life of 9 to 12 days in soil under laboratory conditions [R17].

BIOCONCENTRATION The accumulation of aldicarb in citrus leaves and fruit was studied. The effectiveness of the treatment was affected mainly by the rate of accumulation of the toxicant in the leaves. At 18 g/tree, the greatest residue found in the leaves was 106 µg/g fresh wt for aldicarb. The concentration in mature leaves was very similar to that in young leaves. The residue levels in the mature fruits were much lower than in the leaves. The main component of the residues in the leaves was aldicarb sulfoxide [R18].

CERCLA REPORTABLE QUANTITIES Persons in charge of vessels or facilities are required to notify the National Response Center (NRC) immediately when there is a release of this designated hazardous substance in an amount equal to or greater than its reportable quantity of 1.0 lb or 0.454 kg. The toll-free number of the NRC is (800) 424–8802; in the Washington, DC metropolitan area (202) 426–2675. The rule for determining when notification is required is stated in 40 CFR 302.4 (section IV. D. 3.b) [R19].

RCRA REQUIREMENTS As stipulated in 40 CFR 261.33, when aldicarb, as a commercial chemical product or manufacturing chemical intermediate or an off-specification commercial chemical product or a manufacturing chemical intermediate, becomes a waste, it must be managed according to federal or state hazardous waste regulations. Also defined as a hazardous waste is any container or inner liner used to hold this waste or any residue, contaminated soil, water, or other debris resulting from the cleanup of a spill, into water or on dry land, of this waste. Generators of small quantities of this waste may qualify for partial exclusion from hazardous waste regulations (40 CFR 261.5 (e)) [R20].

DOT *Health Hazards:* Poisonous if swallowed. Inhalation of dust poisonous. Fire may produce irritating or poisonous gases. Runoff from fire control or dilution water may cause pollution [R4].

Fire or Explosion: Some of these materials may burn, but none of them ignites readily [R4].

Emergency Action: Keep unnecessary people away; isolate hazard area and deny entry. Stay upwind; keep out of low areas. Self-contained breathing apparatus (SCBA) and structural firefighter's protective clothing will provide limited protection. Call CHEMTREC at 1–800–424–9300 for emergency assistance. If water pollution occurs, notify the appropriate authorities [R4].

Fire: Small Fires: Dry chemical, CO_2, halon, water spray, or standard foam. Large Fires: Water spray, fog, or standard foam is recommended. Move container from fire area if you can do so without risk [R4].

Spill or Leak: Do not touch spilled material; stop leak if you can do so without risk. Small Spills: Take up with sand or other noncombustible absorbent material and place into containers for later disposal. Small Dry Spills: With clean shovel place material into clean, dry container, and cover; move containers from spill area. Large Spills: Dike far ahead of liquid spill for later disposal [R4].

First Aid: Move victim to fresh air; call emergency medical care. Remove and isolate contaminated clothing and shoes at the site. In case of contact with material, immediately flush skin or eyes with running water for at least 15 minutes [R4].

REFERENCES

1. Purdue University, *National Pesticide Information Retrieval System* (1986).

2. Hartley, D., and H. Kidd (eds.). *The Agrochemicals Handbook.* 2nd ed. Lechworth, Herts, England: The Royal Society of Chemistry, 1987, p. A005.

3. *Kirk–Othmer Encyclopedia of Chemical Technology.* 3rd ed., Volumes 1–26. New York, NY: John Wiley and Sons, 1978–1984, p. 18 (82) 318.

4. Department of Transportation. *Emergency Response Guidebook 1987.* DOT P 5,800.4. Washington, DC: U.S. Government Printing Office, 1987, p. G–53.

5. Sunshine, I. (ed.). *CRC Handbook of Analytical Toxicology.* Cleveland: The Chemical Rubber Co., 1969. 533.

6. Nye, J. C. *ACS Symp Ser 259 (Treat Disposal Pestic Wastes):* 153–60 (1984).

7. 40 CFR 240–280, 300–306, 702–799 (7/1/89).

8. IARC. *Monographs on the Evaluation of the Carcinogenic Risk of Chemicals to Man.* Geneva: World Health Organization, International Agency for Research on Cancer, 1972–present (multivolume work), p. 53 108 (1991).

9. *American Hospital Formulary Service–Drug Information 85.* Bethesda, MD: American Society Hospital Pharmacists, 1985 (Plus supplements A and B, 1985). 4.

10. Pickering, Q. H., W. T. Gilliam. *Arch Environ Contam Toxicol* 11 (6): 699–702 (1982).

11. Hansch, C., A. Leo. *Substituent Constants for Correlation Analysis in Chemistry and Biology.* New York, NY: John Wiley and Sons, 1979. 168.

12. U.S. Department of the Interior, Fish and Wildlife Service, Bureau of Sports Fisheries and Wildlife. *Lethal Dietary Toxicities of Environmental Pollutants to Birds.* Special Scientific Report—Wildlife No. 191. Washington, DC: U.S. Government Printing Office, 1975. 8.

13. Edmiston, C. E., Jr., et al., *Environ Res* 36 (2): 338–350 (1985).

14. U.S. Department of the Interior, Fish and Wildlife Service. *Handbook of Toxicity of Pesticides to Wildlife.* Resource Publication 153. Washington, DC: U.S. Government Printing Office, 1984. 9.

15. Hill, E. F., and M. B. Camardese. *Lethal Dietary Toxicities of Environmental Contaminants and Pesticides to Coturnix.* Fish and Wildlife Technical Report 2. Washington, DC: United States Department of Interior Fish and Wildlife Service, 1986. 23.

16. Briggs, G. G., K. A. Lord. *Pestic Sci* 14 (4): 412–6 (1983).

17. DHEW/NCI. *Toxicology and Carcinogenesis Studies of Aldicarb.* p. 2 Report #136 NIH Pub #79–1391 (1979).

18. Neubauer, I., et al. *Pestic Sci* 13 (4): 387–94 (1982).

19. 54 FR 33419 (8/14/89).

20. 40 CFR 261.33 (7/1/88).

21. Worthing, C. R., and S. B. Walker (eds.). *The Pesticide Manual—A World Compendium.* 8th ed. Thornton Heath,

UK: The British Crop Protection Council, 1987.

22. *Farm Chemicals Handbook 1989.* Willoughby, OH: Meister Publishing Co., 1989.

23. Association of American Railroads. *Emergency Handling of Hazardous Materials in Surface Transportation.* Washington, DC: Assoc. of American Railroads, Hazardous Materials Systems (BOE), 1987.

■ ALLYLAMINE

CAS # 107-11-9

SIC CODES 2834; 2869; 2879; 2822

SYNONYMS 2-propen-1-amine • 2-propenylamine • 3-amino-1-propene • 3-aminopropene • 3-aminopropylene • monoallylamine

MF C_3H_7N

MW 57.11

COLOR AND FORM Colorless to light-yellow liquid.

ODOR Strong ammonia odor.

ODOR THRESHOLD Vapors are so unpleasant that it is unlikely human beings would voluntarily submit to systemically dangerous concentration (allyl amines) [R4].

TASTE Burning taste.

DENSITY 0.760 at 20°C.

BOILING POINT 55-58°C.

AUTOIGNITION TEMPERATURE 705°F [R6, p. 325M-23].

EXPLOSIVE LIMITS AND POTENTIAL Vapor forms explosive mixtures with air over wide range. [R6, p. 49-40].

FLAMMABILITY LIMITS 2.2%-22% [R6, p. 325M-23].

FIREFIGHTING PROCEDURES Personal protection: wear full protective clothing. Firefighting methods: use dry chemical, "alcohol" foam, or carbon dioxide; water may be used to keep fire-exposed containers cool. If leak or spill has ignited, use water spray to disperse vapors and to protect men attempting to stop leak. Water spray may be used to flush spills away from exposures and to dilute spills to nonflammable mixtures. [R6, p. 49-41].

SOLUBILITY Miscible with water, alcohol, chloroform, ether.

OTHER CHEMICAL/PHYSICAL PROPERTIES Conversion units: 1 mg/L = 428 ppm; 1 ppm = 2.23 mg/m³.

PROTECTIVE EQUIPMENT Personal protection: wear full protective clothing (fire conditions) [R6, p. 49-41].

PREVENTATIVE MEASURES Basic ventilation methods are local exhaust ventilation and dilution or general ventilation [R3].

COMMON USES Chemical intermediate for mercurial diuretics [R1]; pharmaceutical intermediate; organic synthesis [R2].

BINARY REACTANTS Highly reactive. Separate from oxidizing and combustible materials (ITII 80).

STANDARD CODES UN 2,334, ID no. 2334, Guide no. 28; NFPA-3, 3, 1—will travel to source of ignition, and flash back. Classified as flammable liquid, poison. Poison label required. IMCO—classified as nonflammable liquid.

DIRECT CONTACT Highly toxic via inhalation and dermal routes. Irritant to eyes, mucous membranes, and skin.

GENERAL SENSATION Highly toxic via oral, inhalation, and dermal routes. Animal experiments produced irritation to nose and mouth, congestion of the eyes. Extended exposure leads to irregular respiration, cyanosis, excitement, convulsions, and death. Can cause death or permanent injury.

PERSONAL SAFETY Butyl rubber gloves, coveralls, and face shield, or multipurpose gas mask required.

ACUTE HAZARD LEVEL High when exposed by inhalation during normal use.

OCCUPATIONAL RECOMMENDATIONS None listed.

ON-SITE RESTORATION Seek professional

environmental engineering assistance through EPA's Environmental Response Team (ERT), Edison, NJ, 24-hour no., 908–548–8730. Contain and isolate spill to limit spread. Construct clay/bentonite swale to divert uncontaminated portion of watershed around contaminated portion, bentonite or butyl-rubber-lined dams, interceptor trenches, or impoundments. Seek professional help to evaluate problem, implement containment measures, and conduct bench scale and pilot scale tests prior to full-scale decontamination program implementation. Treatment alternatives for contaminated water include powdered activated carbon sorption, granular activated carbon filtration, aeration, evaporation, and decomposition with an acid. Treatment alternatives for contaminated soils include well point collection and treatment of leachates as for contaminated water; bentonite/cement ground injection to immobilize spill. Contaminated soil residues may be packaged for disposal.

DISPOSAL METHOD Product residues and sorbent media may be packaged in 17h epoxy-lined drums and disposed of at an approved EPA disposal site. Destruction by high-temperature incineration or microwave plasma detoxification, if available. Encapsulation by organic polyester resin or silicate fixation. Confirm disposal procedures with responsible environmental engineer and regulatory officials.

PROBABLE LOCATION AND STATE Colorless to light-yellow liquid. Will dissolve.

COLOR IN WATER Colorless to light-yellow liquid.

DOT *Health Hazards:* Poisonous if inhaled or swallowed. Skin contact poisonous. Contact may cause burns to skin and eyes. Fire may produce irritating or poisonous gases. Runoff from fire control or dilution water may cause pollution. [R5, p. G–59].

Fire or Explosion: Some of these materials may burn, but none of them ignites readily. Some of these materials may ignite combustibles (wood, paper, oil, etc). [R5, p. G–59].

Emergency Action: Keep unnecessary people away; isolate hazard area and deny entry. Stay upwind; keep out of low areas. Self-contained breathing apparatus and chemical protective clothing that is specifically recommended by the shipper or producer may be worn but they do not provide thermal protection unless that is stated by the clothing manufacturer. Structural firefighter's protective clothing is not effective with these materials. Call CHEMTREC at 1–800–424–9300 as soon as possible, especially if there is no local hazardous-materials team available. [R5, p. G–59].

Fire: Some of these materials may react violently with water. Small Fires: Dry chemical, CO_2, halon, water spray, or alcohol foam. Large Fires: Water spray, fog, or standard foam is recommended. Move container from fire area if you can do so without risk. Cool containers that are exposed to flames with water from the side until well after fire is out. Stay away from ends of tanks. [R5, p. G–59].

Spill or Leak: Do not touch spilled material; stop leak if you can do so without risk. Use water spray to reduce vapors. Small Spills: Take up with sand or other noncombustible absorbent material and place into containers for later disposal. Large Spills: Dike liquid spill for later disposal. [R5, p. G–59].

First Aid: Move victim to fresh air, and call emergency medical care; if not breathing, give artificial respiration; if breathing is difficult, give oxygen. Remove and isolate contaminated clothing and shoes at the site. In case of contact with material, immediately flush skin or eyes with running water for at least 15 minutes. Keep victim quiet and maintain normal body temperature. Effects may be delayed; keep victim under observation. [R5, p. G–59].

Initial Isolation and Evacuation Distances: For a spill, or leak from a drum or small container (or small leak from tank): Isolate in all directions 75 feet. For a large spill from a tank (or from many containers or drums): First isolate in all directions 150 feet; then, evacuate in a downwind direction 0.2 mile wide and 0.4 mile long. [R5, p. Table].

FIRE POTENTIAL Moderate, when exposed to heat or flame [R3].

REFERENCES

1. SRI.

2. Hawley, G. G. *The Condensed Chemical Dictionary.* 9th ed. New York: Van Nostrand Reinhold Co., 1977. 29.

3. Sax, N. I. *Dangerous Properties of Industrial Materials,* 4th ed. New York: Van Nostrand Reinhold, 1975. 367.

4. Grant, W. M. *Toxicology of the Eye.* 2nd ed. Springfield, Illinois: Charles C. Thomas, 1974. 106.

5. Department of Transportation. *Emergency Response Guidebook 1987.* DOT P 5800.4. Washington, DC: U.S. Government Printing Office, 1987.

6. National Fire Protection Association. *Fire Protection Guide on Hazardous Materials.* 7th ed. Boston, MA: National Fire Protection Association, 1978.

■ ALLYL BROMIDE

CAS # 106-95-6

DOT # 1099

SYNONYMS 1-bromo-2-propene • 1-propene, 3-bromo • 2-propenyl bromide • 3-bromo-1-propene • 3-bromopropene • 3-bromopropylene • bromallylene • propene, 3-bromo

MF C_3H_5Br

MW 120.99

COLOR AND FORM Colorless to light-yellow liquid.

ODOR Unpleasant, pungent.

DENSITY 1.398 at 20°C.

SURFACE TENSION 26.9 dynes/cm = 0.0269 N/m at 20°C.

BOILING POINT 71.3°C at 760 mm Hg.

MELTING POINT −119°C.

HEAT OF COMBUSTION (est) 6,700 Btu/lb = 3,700 cal/g = 150×10^5 J/kg.

HEAT OF VAPORIZATION (est) 110 Btu/lb = 59 cal/g = 2.5×10^5 J/kg.

AUTOIGNITION TEMPERATURE 563°F (295°C) [R3].

EXPLOSIVE LIMITS AND POTENTIAL 4.4–7.3% [R7].

FLAMMABILITY LIMITS percent by vol: lower: 4.4%; upper: 7.3% [R9, p. 325M–23].

FIREFIGHTING PROCEDURES Water may be ineffective except as a blanket [R9, p. 325M–23]. Wear full protective clothing. Use water spray, dry chemical, alcohol foam, or carbon dioxide. Use water to keep fire-exposed containers cool. If a leak or spill has not ignited, use water spray to disperse vapors, and provide protection for men attempting to stop a leak. Water spray may be used to flush spills away from exposures. [R9, p. 49–42].

TOXIC COMBUSTION PRODUCTS At elevated temp, as in fire conditions, decomposition may occur to emit toxic fumes of hydrogen bromide. [R9, p. 49–41].

SOLUBILITY Slightly soluble in water; miscible with alcohol, chloroform, ether, carbon disulfide, carbon tetrachloride.

OTHER CHEMICAL/PHYSICAL PROPERTIES Liquid-water interfacial tension: (est) 40 dynes/cm = 0.040; N/m at 20°C; ratio of specific heats of vapor (gas): 1.1210.

PROTECTIVE EQUIPMENT Goggles and face shield; protective clothing; self-contained breathing apparatus for high vapor concentration [R4].

COMMON USES Manufacture of other allyl compounds [R2]; insecticidal fumigant [R5]; chemical intermediate in organic synthesis, for resins (copolymer with sulfur dioxide), and fragrances [R1]; contact poison [R6].

OCCUPATIONAL RECOMMENDATIONS
DOT *Health Hazards:* Poisonous; may be fatal if inhaled, swallowed, or absorbed through skin. Contact may cause burns to skin and eyes. Runoff from fire control or

dilution water may cause pollution. [R8, p. G–57].

Fire or Explosion: May be ignited by heat, sparks, or flames. Container may explode in heat of fire. Vapor explosion and poison hazard indoors, outdoors, or in sewers. [R8, p. G–57].

Emergency Action: Keep unnecessary people away; isolate hazard area and deny entry. Stay upwind, out of low areas, and ventilate closed spaces before entering. Self-contained breathing apparatus and chemical protective clothing that is specifically recommended by the shipper or producer may be worn, but they do not provide thermal protection unless that is stated by the clothing manufacturer. Structural firefighter's protective clothing is not effective with these materials. Call CHEMTREC at 1–800–424–9300 as soon as possible, especially if there is no local hazardous materials team available. [R8, p. G–57].

Fire: Small Fires: Dry chemical, CO_2, halon, water spray, or standard foam. Large Fires: Water spray, fog, or standard foam is recommended. Move container from fire area if you can do so without risk. Cool container with water using unmanned device until well after fire is out. Fight fire from maximum distance. Stay away from ends of tanks. Dike fire control water for later disposal; do not scatter the material. [R8, p. G–57].

Spill or Leak: Shut off ignition sources; no flares, smoking, or flames in hazard area. Do not touch spilled material; stop leak if you can do so without risk. Water spray may reduce vapor; but it may not prevent ignition in closed spaces. Small Spills: Take up with sand or other noncombustible absorbent material and place into containers for later disposal. Small Dry Spills: With clean shovel place material into clean, dry container, and cover; move containers from spill area. Large Spills: Dike far ahead of liquid spill for later disposal. [R8, p. G–57].

First Aid: Move victim to fresh air and call emergency medical care; if not breathing, give artificial respiration; if breathing is difficult, give oxygen. In case of contact with material, immediately flush skin or eyes with running water for at least 15 minutes. Speed in removing material from skin is of extreme importance. Remove and isolate contaminated clothing and shoes at the site. Keep victim quiet and maintain normal body temperature. Effects may be delayed; keep victim under observation. [R8, p. G–57].

Initial Isolation and Evacuation Distances: For a spill, or leak from a drum or small container (or small leak from tank): Isolate in all directions 50 feet. For a large spill from a tank (or from many containers or drums): First isolate in all directions 150 feet; then, evacuate in a downwind direction 0.2 mile wide and 0.2 mile long. [R8, p. Table].

FIRE POTENTIAL Vapor forms flammable mixtures with air. [R9, p. 49–41].

GENERAL RESPONSE Shut off ignition sources. Call fire department. Evacuate area in case of large discharge. Avoid contact with liquid and vapor. Isolate and remove discharged material. Notify local health and pollution control agencies.

FIRE RESPONSE Flammable. Poisonous gases may be produced in fire. Flashback along vapor trail may occur. Vapor may explode if ignited in an enclosed area. Wear goggles, self-contained breathing apparatus, and rubber overclothing (including gloves). Extinguish with dry chemicals, alcohol foam, or carbon dioxide. Water may be ineffective on fire. Cool exposed containers with water.

EXPOSURE RESPONSE Call for medical aid. Vapor irritating to eyes, nose, and throat. If inhaled will cause headache, dizziness, coughing, or difficult breathing. Move victim to fresh air. If breathing is difficult, give oxygen. Liquid irritating to skin and eyes. Remove contaminated clothing and shoes. Flush affected areas with plenty of water. If swallowed and victim is conscious, have victim drink water or milk.

RESPONSE TO DISCHARGE Issue warning—high flammability; restrict access; should be removed; chemical and physical treatment.

WATER RESPONSE Effect of low concentrations on aquatic life is unknown. May be dangerous if it enters water intakes. Notify local health and wildlife officials. Notify operators of nearby water intakes.

EXPOSURE SYMPTOMS Inhalation of vapor irritates mucous membranes and causes dizziness, headache, and lung irritation. Contact with liquid irritates eyes and skin. Ingestion causes irritation of mouth and stomach.

EXPOSURE TREATMENT Inhalation: remove from exposure; if breathing is difficult, give oxygen; call physician. Eyes: flush with water for at least 15 min. and call physician. Skin: flush with water; get medical attention for skin irritation. Ingestion: do not induce vomiting; get medical attention.

REFERENCES

1. SRI.

2. *The Merck Index.* 10th ed. Rahway, NJ: Merck Co., Inc., 1983. 44.

3. Hawley, G. G. *The Condensed Chemical Dictionary.* 10th ed. New York: Van Nostrand Reinhold Co., 1981. 35.

4. U.S. Coast Guard, Department of Transportation. *CHRIS—Hazardous Chemical Data.* Manual Two. Washington, DC: U.S. Government Printing Office, Oct., 1978.

5. Gosselin, R. E., H. C. Hodge, R. P. Smith, and M. N. Gleason. *Clinical Toxicology of Commercial Products.* 4th ed. Baltimore: Williams and Wilkins, 1976, p. II–113.

6. *Kirk–Othmer Encyc Chemical Tech.* 3rd. ed. 1978–present V4 p. 256.

7. International Labour Office. *Encyclopedia of Occupational Health and Safety.* Vols. I and II. Geneva, Switzerland: International Labour Office, 1983. 128.

8. Department of Transportation. *Emergency Response Guidebook 1987.* DOT P 5800.4. Washington, DC: U.S. Government Printing Office, 1987.

9. National Fire Protection Association. *Fire Protection Guide on Hazardous Materials.* 7th ed. Boston, MA: National Fire Protection Association, 1978.

■ ALLYL CHLOROFORMATE

CAS # 2937–50–0

DOT # 1722

SYNONYMS 2-propenyl chloroformate • allyl chlorocarbonate • carbonochloridic acid, 2-propenyl ester • formic acid, chloro-, allyl ester

MF $C_4H_5ClO_2$

MW 120.54

COLOR AND FORM Colorless liquid.

ODOR Pungent odor.

ODOR THRESHOLD 1.4 ppm [R4].

DENSITY 1.1394 at 20°C.

SURFACE TENSION 25 dynes/cm = 0.025 newtons/m at 20°C (estimated).

VISCOSITY 0.71 mPa–s(cP) at 20°C.

BOILING POINT 110°C at 760 mm Hg.

MELTING POINT Freezing point: −112°F = −80°C = 193K.

HEAT OF COMBUSTION −7,800 Btu/lb = −4,300 cal/g = −180×10⁵ J/kg (est).

HEAT OF VAPORIZATION 100 Btu/lb = 56 cal/g = 2.3×10⁵ J/kg (est).

FIREFIGHTING PROCEDURES Use dry chemical, foam, carbon dioxide, or water spray. Water may be ineffective. Use water spray to keep fire-exposed containers cool. Approach fire from upwind to avoid hazardous vapors and toxic decomposition products [R7, p. 49–23]. Do not extinguish fire unless flow can be stopped. Use "alcohol" foam, dry chemical, or carbon dioxide. Cool all affected containers with flooding quantities of water. Apply water from as far a distance as possible. Do not use water on material itself. If large quantities of combustibles are involved, use water in flooding quantities as spray and fog. Use water spray to knock down vapors [R3].

SOLUBILITY Insoluble in water.

VAPOR PRESSURE 20 mm Hg at 25°C.

OTHER CHEMICAL/PHYSICAL PROPERTIES Ratio of specific heats of vapor (gas) 1.0804; hydrolyze in water (chloroformic esters); when heated to decomposition it emits toxic fumes of phosgene; can react with oxidizing materials; flammable liquid; decomposes in water to form allyl alcohol and chloroformic acid. Chloroformates are reactive intermediates that combine acid chloride and ester functions. They undergo many reactions of acid chlorides, however, the rates are usually slower. Reactions of chloroformates, like other acid chlorides, proceed faster with better yields when alkali hydroxides or tertiary amines are present to react with the HCl as it forms. (chloroformates).

PROTECTIVE EQUIPMENT Vapor-proof protective goggles and face shield; plastic or rubber gloves, shoes, and clothing; gas mask or self-contained breathing apparatus [R4].

PREVENTATIVE MEASURES If material not on fire and not involved in fire: Keep sparks, flames, and other sources of ignition away. Keep material out of water sources and sewers. Build dikes to contain flow as necessary. Attempt to stop leak if without undue personnel hazard. Use water spray to knock down vapors. Do not use water on material itself. Neutralize spilled material with crushed limestone, soda ash, or lime [R3].

CLEANUP Eliminate all ignition sources. Keep water away from release. Stop or control the leak, if this can be done without undue risk. Use appropriate foam to blanket release and suppress vapors. Absorb in noncombustible material for proper disposal. [R7, p. 49–23].

COMMON USES Used as intermediates in synthesis of numerous compounds (chloroformic esters) [R1]. Can be used as an intermediate in production of diethylene glycol bis (allyl carbonate), a monomer used in the manufacture of break-resistant optical lenses [R1].

OCCUPATIONAL RECOMMENDATIONS None listed.

NON-HUMAN TOXICITY LD_{50} rats oral 244 mg/kg [R5]; LD_{50} mice oral 210 mg/kg [R5].

DOT *Health Hazards:* Poisonous; may be fatal if inhaled, swallowed, or absorbed through skin. Contact may cause burns to skin and eyes. Runoff from fire control or dilution water may cause pollution. [R6, p. G–57].

Fire or Explosion: May be ignited by heat, sparks, or flames. Container may explode in heat of fire. Vapor explosion and poison hazard indoors, outdoors, or in sewers. [R6, p. G–57].

Emergency Action: Keep unnecessary people away; isolate hazard area and deny entry. Stay upwind, out of low areas, and ventilate closed spaces before entering. Positive pressure self-contained breathing apparatus (SCBA) and chemical protective clothing which is specifically recommended by the shipper or manufacturer may be worn. It may provide little or no thermal protection. Structural firefighter's protective clothing is not effective for these materials. See the Table of Initial Isolation and Protective Action Distances. If you find the ID Number and the name of the material there, begin protective action. Call emergency response telephone number on shipping paper first. If shipping paper not available or no answer, call CHEMTREC at 1–800–424–9300. [R6, p. G–57].

Fire: Small Fires: Dry chemical, CO_2, water spray, or regular foam. Large Fires: Water spray, fog, or regular foam. Move container from fire area if you can do so without risk. Cool container with water using unmanned device until well after fire is out. Fight fire from maximum distance. Stay away from ends of tanks. Dike fire-control water for later disposal; do not scatter the material. [R6, p. G–57].

Spill or Leak: Shut off ignition sources; no flares, smoking, or flames in hazard area. Fully encapsulating, vapor-protective clothing should be worn for spills and leaks with no fire. Do not touch or walk through spilled material; stop leak if you can do so without risk. Water spray may reduce vapor; but it may not prevent ignition in closed spaces. Small Spills: Take up with sand or other noncombustible absorbent material and place into containers for later disposal. Small Dry Spills: With clean shovel place material into clean, dry container, and cover loosely; move containers from spill area. Large Spills: Dike far ahead of liquid spill for later disposal. [R6, p. G–57].

First Aid: Move victim to fresh air, and call emergency medical care; if not breathing, give artificial respiration; if breathing is difficult, give oxygen. In case of contact with material, immediately flush skin or eyes with running water for at least 15 minutes. Speed in removing material from skin is of extreme importance. Remove and isolate contaminated clothing and shoes at the site. Keep victim quiet and maintain normal body temperature. Effects may be delayed; keep victim under observation. [R6, p. G–57].

Initial Isolation and Protective Action Distances: Small Spills (from a a small package, or small leak from a large package): First, isolate in all directions 500 feet; then, protect persons downwind 0.1 mile (day), or 1.3 miles (night). Large spills (from a large package or from many small packages): First, isolate in all directions 500 feet; then, protect persons downwind 0.2 mile (day), or 1.6 miles (night). [R6, p. G–Table].

FIRE POTENTIAL Fire hazard: high when exposed to heat, open flame (or sparks), or powerful oxidizers [R2].

GENERAL RESPONSE Stop discharge if possible. Keep people away. Shut off ignition sources. Call fire department. Isolate and remove discharged material. Notify local health and pollution control agencies.

FIRE RESPONSE Flammable flashback along vapor trail may occur. Vapor may explode if ignited in an enclosed area. Wear goggles, self-contained breathing apparatus, and rubber overclothing (including gloves). Extinguish with dry chemicals, foam, or carbon dioxide. Water may

be ineffective on fire. Cool exposed containers with water.

EXPOSURE RESPONSE Call for medical aid. Vapor irritating to eyes, nose, and throat. If inhaled will cause difficult breathing. Move victim to fresh air. If breathing is difficult, give oxygen. Liquid irritating to skin and eyes. Harmful if swallowed. Remove contaminated clothing and shoes. Flush affected areas with plenty of water. If swallowed and victim is conscious, have victim drink water or milk.

RESPONSE TO DISCHARGE Issue warning—corrosive. Restrict access. Disperse and flush.

WATER RESPONSE Effect of low concentrations on aquatic life is unknown. May be dangerous if it enters water intakes. Notify local health and wildlife officials. Notify operators of nearby water intakes.

EXPOSURE SYMPTOMS Vapor irritates eyes and respiratory tract. Contact with liquid causes eye and skin irritation, and ingestion irritates mouth and stomach.

EXPOSURE TREATMENT Inhalation: remove from exposure; support respiration if necessary; call physician. Eyes: if irritated by either vapor or liquid, flush with water for at least 15 min. Skin: wash with large amounts of water for at least 15 min. Ingestion: do not induce vomiting; give water; call physician.

REFERENCES

1. Gerhartz, W (exec ed.). *Ullmann's Encyclopedia of Industrial Chemistry.* 5th ed. Vol A1: Deerfield Beach, FL: VCH Publishers, 1985 to Present, p. V5 87.

2. Sax, N. I. *Dangerous Properties of Industrial Materials,* 6th ed. New York, NY: Van Nostrand Reinhold, 1984. 163.

3. Association of American Railroads. *Emergency Handling of Hazardous Materials in Surface Transportation.* Washington, DC: Association of American Railroads, Bureau of Explosives, 1992. 37.

4. U.S. Coast Guard, Department of Transportation. *CHRIS—Hazardous Chemical Data.* Volume II. Washington, DC: U.S. Government Printing Office, 1984–5.

5. Oskerko, E. F., E. I. Klimova. *Toxic Effects of Allyl Chloroformate*; Gig Tr Prof Zabol 5: 51–2 (1984).

6. U.S. Department of Transportation. *Emergency Response Guidebook 1993.* DOT P 5800.6. Washington, DC: U.S. Government Printing Office, 1993.

7. National Fire Protection Guide. *Fire Protection Guide on Hazardous Materials.* 10th ed. Quincy, MA: National Fire Protection Association, 1991.

■ ALPHACYPERMETHRIN

CAS # 67375–30–8

SYNONYMS alfamethrin • dominex • (1R) cis (S) and (1S) cis (R) enantiomeric isomer pair of alpha-cyano-3- phenoxybenzyl-3-(2,2-dichlorovinyl)-2,2-dimethylcyclopropane carboxylate • [1alpha (S),3alpha]-(-)-cyano (3-phenoxyphenyl) methyl 3-(2,2-dichloroethenyl)— 2,2-dimethylcyclopropanecarboxylate • alphamethrin • fastac • concord • fendona • WL-85871 • bestox

MF $C_{22}H_{19}Cl_2NO_3$

MW 416.32

COLOR AND FORM Viscous yellowish-brown semisolid mass; colorless crystals.

BOILING POINT 200°C at 0.07 mm Hg.

MELTING POINT 80.5°C.

FIREFIGHTING PROCEDURES Use carbon dioxide, foam, or dry chemical (on fires involving pyrethroids) (pyrethrum) [R7, 2]. Self-contained breathing apparatus with a full facepiece operated in pressure-demand or other positive-pressure mode (pyrethrum) [R7, 5] Fire: Small Fires: Dry chemical, CO_2, water spray, or regular foam. Large Fires: Water spray, fog, or regular foam. Move container from fire area if you can do so without risk. Fight fire from maximum distance. Stay away from ends of tanks. Dike fire-control water for later disposal; do not scatter the material (pesticide, liquid, or solid, poisonous, not otherwise specified) [R5, p. G–55]. Extinguish fire using agent suitable for type of surrounding fire (pyrethrins). [R2].

SOLUBILITY Water soluble: 0.01–0.2 mg/L at 20°C; soluble in chloroform, cyclohexane, xylene, 337 g/L ethanol. In water at 25°C, ca 0.01 mg/L. In acetone 620, dichloromethane 550, cyclohexanone 515, ethyl acetate 440, chlorobenzene 420, acetophenone 390, o-xylene 315, n-hexane 7 (all in g/L at 25°C). In maize oil 19–20, ethylene glycol <1 (both in g/kg at 20°C).

VAPOR PRESSURE 170 nPa at 20°C.

STABILITY AND SHELF LIFE Very stable in neutral and acidic media. Hydrolyzed in strongly alkaline media. Thermally stable up to 220°C. Relatively stable to light [R1].

PROTECTIVE EQUIPMENT Employees should use dust and splashproof safety goggles where pyrethroids may contact the eyes (pyrethroids) [R7, 3].

PREVENTATIVE MEASURES Skin that becomes contaminated with pyrethrum should be promptly washed or showered with soap or mild detergent and water (pyrethrum) [R7, 3].

CLEANUP Spillages of pesticides at any stage of their storage or handling should be treated with great care. Liquid formulations may be reduced to solid phase by evaporation. Dry sweeping of solids is always hazardous: these should be removed by vacuum cleaning or by dissolving them in water or other solvent in the factory environment (pesticides) [R3].

DISPOSAL Incineration would be an effective disposal procedure where permitted. If an efficient incinerator is not available, the product should be mixed with large amounts of combustible material, and contact with the smoke should be avoided (pyrethrin products) [R4].

COMMON USES Control of a wide range of chewing and sucking insects (particularly Lepidotera, Coleoptera, and Hemiptera) in fruit (including citrus), vegetables, vines, cereals, maize, beet, oilseed rape, potatoes, cotton, rice, soya beans, forestry, and other crops. Control of cockroaches, mosquitoes, flies, and other insect pests in public health; and flies in animal houses. Also used as an animal ectoparasiticide [R1]. Insecticide (pyrethrins) [R4]. Control of cotton leaf perforator, cotton semi-looper, false codling moth, red bollworm, cotton stainers, spiny bollworm, cotton spotted bollworm, American bollworm, native budworm, tobacco budworm, cotton bollworm, pink bollworm, cotton leafworm, boll weevil in cotton. For major pests in coffee, maize, sweet corn, sorghum, flax, soybean, mung bean, navy bean, sunflower, tobacco, rice, bush, and trellis tomato, cruciferous crops, field peas, lupines, and pasture. Cutworm control in all row crops [R6, p. C–16].

NON-HUMAN TOXICITY LD_{50} rat oral 79–400 mg/kg (in corn oil, value depending on concentration) [R1].

ECOTOXICITY LC_{50} rainbow trout 0.0028 mg/L/96 hr (conditions of bioassay not specified) [R1].

EMERGENCY ACTION Keep unnecessary people away; isolate hazard area and deny entry. Stay upwind, out of low areas, and ventilate closed spaces before entering. Positive-pressure self-contained breathing apparatus (SCBA) and chemical protective clothing that is specifically recommended by the shipper or manufacturer may be worn. It may provide little or no thermal protection. Structural firefighters' protective clothing is not effective for these materials. Remove and isolate contaminated clothing at the site. Call CHEM-TREC at 1–800–424–9300 as soon as possible, especially if there is no local hazardous materials team available (pesticide, liquid, or solid, poisonous, not otherwise specified) [R5, p. G–55].

Spill or Leak: Do not touch or walk through spilled material; stop leak if you can do so without risk. Fully encapsulating, vapor-protective clothing should be worn for spills and leaks with no fire. Use water spray to reduce vapors. Small Spills: Take up with sand or other noncombustible absorbent material and place into containers for later disposal. Small Dry Spills: With clean shovel place material into clean, dry container, and cover loose-ly; move containers from spill area. Large Spills: Dike far ahead of liquid spill for later disposal. (pesticide, liquid, or solid, poisonous, not otherwise specified) [R5, p. G–55].

FIRST AID Move victim to fresh air, and call emergency medical care; if not breathing, give artificial respiration; if breathing is difficult, give oxygen. In case of contact with material, immediately flush skin or eyes with running water for at least 15 minutes. Speed in removing material from skin is of extreme importance. Remove and isolate contaminated clothing and shoes at the site. Keep victim quiet and maintain normal body temperature. Effects may be delayed; keep victim under observation. (pesticide, liquid, or solid, poisonous, not otherwise specified) [R5, p. G–55].

REFERENCES

1. Hartley, D., and H. Kidd (eds.). *The Agrochemicals Handbook.* 2nd ed. Lechworth, Herts, England: The Royal Society of Chemistry, 1987, p. A649/Aug 87.

2. Bureau of Explosives. *Emergency Handling of Haz Matl in Surface Trans* p. 434 (1981).

3. International Labour Office. *Encyclopedia of Occupational Health and Safety.* Vols. I and II. Geneva, Switzerland: International Labour Office, 1983. 1,619.

4. Sittig, M. M. M. *Handbook of Toxic and Hazardous Chemicals and Carcinogens, 1985.* 2nd ed. Park Ridge, NJ: Noyes Data Corporation, 1985. 762.

5. U.S. Department of Transportation. *Emergency Response Guidebook 1990.* DOT P 5800.5. Washington, DC: U.S. Government Printing Office, 1990.

6. *Farm Chemicals Handbook 1993.* Willoughby, OH: Meister Publishing Co., 1993.

7. Mackison, F. W., R. S. Stricoff, and L. J. Partridge, Jr. (eds.). *NIOSH/OSHA—Occupational Health Guidelines for Chemical Hazards.* DHHS (NIOSH) Publication No. 81–123 (3 vols). Washington, DC: U.S. Government Printing Office, Jan. 1981.

◼ ALPHA-PINENE

CAS # 80–56–8

SYNONYMS bicyclo (3.1.1) hept-2-ene, 2,6,6-trimethyl • 2,6,6-trimethylbicyclo (3.1.1) hept-2-ene • 2,6,6-trimethyl-bicyclo-(3,1,1)-2-heptene • 2-pinene • bicyclo (3.1.1) hept-2-ene, 2,6,6-trimethyl • pinene • FEMA number 2902

MF $C_{10}H_{16}$

MW 136.23

COLOR AND FORM Colorless, mobile liquid.

ODOR Characteristic odor of pine; odor of turpentine.

DENSITY 0.8590 at 20°C.

BOILING POINT 156°C at 2760 mm Hg.

MELTING POINT –62.5°C.

AUTOIGNITION TEMPERATURE 491°F (255°C) [R7].

FIREFIGHTING PROCEDURES Foam, carbon dioxide, dry chemical [R6]; do not extinguish fire unless flow can be stopped. Use water in flooding quantities as fog. Solid streams of water may spread fire. Cool all affected containers with flooding quantities of water. Apply water from as far a distance as possible. Use foam, dry chemical, or carbon dioxide [R8].

SOLUBILITY Almost insoluble in propylene glycol and glycerine; soluble in alcohol, chloroform, ether, glacial acetic acid.

VAPOR PRESSURE 4.9 mm Hg at 27°C.

OCTANOL/WATER PARTITION COEFFICIENT Log K_{ow} = 4.12 (est).

OTHER CHEMICAL/PHYSICAL PROPERTIES Max absorption (alcohol): 210 NM (log E = 3.64) (DL form); odor of turpentine (DL form); alpha-pinene in North American oils is dextrorotatory, and in most European oils it is levorotatory. Oxidized material has rosin-like odor. Structure would account for 4 optically active and 2 optically inactive isomers; only D-, L-, and DL-alpha-pinene are known. Melting point: –55°C; density: 0.8582 at 20°C; boiling point: 156.2°C (DL form); BP: 155–156°C; density: 0.8591 at 20°C; index of refraction: 1.4663 at 20°C/D; specific optical rotation: 51.14 at 20°C/D (D-form); oil; soluble in all proportions in alcohol, ether, chloroform; index of refraction: 1.4658 at 20°C/D (DL form); liquid; soluble in glacial acetic acid (DL-form); specific optical rotation: 33.52 at 20°C/D (alcohol) (D-form, hydrochloride); boiling point: 155–156°C (L-form); can react with oxidizing materials. Henry's Law constant = 0.107 atm m^3/mole at 25°C (calc); 0.8625 at 15°C.

PROTECTIVE EQUIPMENT Wear appropriate chemical protective gloves, boots, and goggles [R8].

PREVENTATIVE MEASURES If material not on fire and not involved in fire: Keep sparks, flames, and other sources of ignition away. Keep material out of water sources and sewers. Build dikes to contain flow as necessary. Attempt to stop leak if without undue personnel hazard. Use water spray to knock down vapors [R8].

COMMON USES Manufacture of camphor, insecticides, solvents, plasticizers, perfume bases [R1, 1182]; flavor ingredient [R1]; synthesis of geraniol, terpineol, terpene esters, lube oil additives [R2]; synthetic resins and their derivatives, odorant [R3]; chemical intermediate for synthetic pine oil, camphene, camphor, terpin hydrate, terpene resins, and insecticide, toxaphene [R4]; solvent for protective coatings, polishes, and waxes [R2]. Reported uses include 16–54 ppm in non-alcoholic beverages; 64 ppm in ice cream, ices, etc.; 48 ppm in candy; 160 ppm in baked goods; 2.6–150 ppm in condiments [R1].

OCCUPATIONAL RECOMMENDATIONS None listed.

NON-HUMAN TOXICITY LD_{50} rat oral 3,700 mg/kg [R6]; LC_{Lo} rat inhalation 625 µg/m^3 [R6]; LC_{Lo} guinea pig inhalation 572 µg/m^3 [R6]; LC_{Lo} mouse inhalation 364 µg/m^3 [R6].

BIODEGRADATION The concentration of alpha-pinene in seawater samples decreased from 0.41 ng/L to 0.25 ng/L when incubated with macrophytes for 6 hrs at

10°C (1). The concentration of alpha-pinene in the influent to a kraft mill-aerated stabilization basin with a 7–8-day retention time decreased from 0.20 ppm to 0.04 ppm (2). Pure cultures of *Pseudomonas putida* isolated from soil oxidized alpha-pinene to 3-isopropylbut-3-enoic acid, (Z)-2-methyl-5-isopropylhexa-2,5-dienoic acid, 2,4,5-trimethylhept-2-en-5-olide, and (E)-2-methyl-5-isopropylhexa-2,5-dienoic acid, 3,4-dimethylvaleric acid, 2,5,6-trimethyl-hept-3-enoic acid, and 2-methyl-5-isopropylhexa-2,5-dienoic acid under aerobic conditions (3) [R9].

BIOCONCENTRATION Based on an estimated log octanol/water partition coefficient of 4.12 (1,SRC), a bioconcentration factor of 799 can be calculated using an appropriate regression equation (2,SRC). This value indicates that alpha-pinene may bioconcentrate in fish and aquatic organisms (SRC) [R10].

DOT *Fire or Explosion:* Flammable/combustible material; may be ignited by heat, sparks, or flames. Vapors may travel to a source of ignition, and flash back. Container may explode in heat of fire. Vapor explosion hazard indoors, outdoors, or in sewers. Runoff to sewer may create fire or explosion hazard [R5].

Health Hazards: May be poisonous if inhaled or absorbed through skin. Vapors may cause dizziness or suffocation. Contact may irritate or burn skin and eyes. Fire may produce irritating or poisonous gases. Runoff from fire control or dilution water may cause pollution [R5].

Emergency Action: Keep unnecessary people away; isolate hazard area and deny entry. Stay upwind; keep out of low areas. Positive-pressure self-contained breathing apparatus (SCBA) and structural firefighters' protective clothing will provide limited protection. Isolate for 1/2 mile in all directions if tank, rail car, or tank truck is involved in fire. Call CHEMTREC at 1–800–424–9300 for emergency assistance. If water pollution occurs, notify the appropriate authorities [R5].

Fire: Small Fires: Dry chemical, CO_2, water spray, or alcohol-resistant foam.

Large Fires: Water spray, fog, or alcohol-resistant foam. Do not use dry chemical extinguishers to control fires involving nitromethane or nitroethane. Move container from fire area if you can do so without risk. Apply cooling water to sides of containers that are exposed to flames until well after fire is out. Stay away from ends of tanks. For massive fire in cargo area, use unmanned hose holder or monitor nozzles; if this is impossible, withdraw from area and let fire burn. Withdraw immediately in case of rising sound from venting safety device or any discoloration of tank due to fire [R5].

Spill or Leak: Shut off ignition sources; no flares, smoking, or flames in hazard area. Stop leak if you can do so without risk. Water spray may reduce vapor; but it may not prevent ignition in closed spaces. Small Spills: Take up with sand or other noncombustible absorbent material and place into containers for later disposal. Large Spills: Dike far ahead of liquid spill for later disposal [R5].

First Aid: Move victim to fresh air and call emergency medical care; if not breathing, give artificial respiration; if breathing is difficult, give oxygen. In case of contact with material, immediately flush eyes with running water for at least 15 minutes. Wash skin with soap and water. Remove and isolate contaminated clothing and shoes at the site. [R5].

GENERAL RESPONSE Stop discharge if possible. Keep people away. Shut off ignition sources. Evacuate area. Call fire department. Avoid contact with liquid and vapor. Isolate and remove discharged material. Notify local health and pollution control agencies.

FIRE RESPONSE Combustible. Flashback along vapor trail may occur. Vapor may explode if ignited in an enclosed area. Water may be ineffective on fire. Extinguish with dry chemical, alcohol foam, or CO_2. Cool exposed containers with water.

EXPOSURE RESPONSE Call for medical aid. Vapor irritating to eyes, nose, and throat. If inhaled, will cause nausea, vomiting, headache, difficult breathing, or loss of consciousness. Move to fresh air. If breathing has stopped, give artificial respiration. If breathing is difficult, give oxygen. Liquid poisonous if swallowed. Irritating to skin and eyes. Remove contaminated clothing and shoes. Flush affected areas with plenty of water.

RESPONSE TO DISCHARGE Evacuate area. Mechanical containment. Should be removed. Provide chemical and physical treatment.

WATER RESPONSE May be dangerous to aquatic life in high concentrations. Fouling to shoreline. May be dangerous if it enters water intakes. Notify local health and wildlife officials. Notify operators of nearby water intakes.

EXPOSURE SYMPTOMS Harmful if swallowed, inhaled, or absorbed through skin. High concentrations are extremely destructive to mucous membrane and upper respiratory tract, eyes, and skin. Symptoms of exposure may include burning sensation, coughing, wheezing, laryngitis, shortness of breath, headache, nausea, and vomiting.

EXPOSURE TREATMENT Inhalation: Call a physician. Remove the victim to fresh air. If not breathing, give artificial respiration. If breathing is difficult, give oxygen. Eyes or skin: immediately flush with copious amount of water for at least 15 minutes while removing contaminated clothing and shoes. Assure adequate flushing of the eyes by holding eyelids open with the fingers.

FIRE POTENTIAL Dangerous when exposed to heat or flame [R6].

FDA Alpha-pinene is a food additive permitted for direct addition to food for human consumption, as long as (1) the quantity added to food does not exceed the amount reasonably required to accomplish its intended physical, nutritive, or other technical effect in food, and (2) when intended for use in or on food it is of appropriate food grade and is prepared and handled as a food ingredient. Synthetic flavoring substances and adjuvants (includes alpha-pinene) may be safely used in foods [R11].

REFERENCES

1. *Fenaroli's Handbook of Flavor Ingredients*. Volume 2. Edited, translated, and revised by T. E. Furia and N. Bellanca. 2nd ed. Cleveland: The Chemical Rubber Co., 1975. 486.

2. Lewis, R. J., Sr. (Ed.). *Hawley's Condensed Chemical Dictionary*. 12th ed. New York, NY: Van Nostrand Reinhold Co., 1993. 918.

3. Hawley, G. G. *The Condensed Chemical Dictionary*. 9th ed. New York: Van Nostrand Reinhold Co., 1977. 687.

4. SRI.

5. U.S. Department of Transportation. *Emergency Response Guidebook 1990*. DOT P 5800.5. Washington, DC: U.S. Government Printing Office, 1990, p. G−26.

6. Sax, N. I. *Dangerous Properties of Industrial Materials*, 6th ed. New York, NY: Van Nostrand Reinhold, 1984. 2233.

7. National Fire Protection Guide. *Fire Protection Guide on Hazardous Materials*. 10th ed. Quincy, MA: National Fire Protection Association, 1991 p. 325M−80.

8. Association of American Railroads. *Emergency Handling of Hazardous Materials in Surface Transportation*. Washington, DC: Association of American Railroads, Bureau of Explosives, 1992. 765.

9. (1) Button, D. K., F. Juttner, *Marine Chemical* 26: 57−66 (1989) (2) Hrutfiord, B. F., et al., *Tappi* 58: 98−100 (1975) (3) Tudroszen, N. J., et al., *Biochemical J* 168: 315−8 (1977).

10. (1) GEMS. *Graphic Exposure Modeling System*. CLOGP USEPA (1987) (2) Lyman, W. J., et al., *Handbook of Chemical Property Estimation Methods*. New York: McGraw-Hill, pp. 5−1 to 5−30 (1982).

11. 21 CFR 172.515 (4/1/91).

12. Budavari, S. (ed.). *The Merck Index— Encyclopedia of Chemicals, Drugs, and Biologicals*. Rahway, NJ: Merck and Co., Inc., 1989.

■ ALUMINUM HYDROXIDE

CAS # 21645–51–2

SIC CODES 2815; 3569; 3211; 3255; 3269; 2893; 2841; 2261; 2834; 2833

SYNONYMS Af-260 • alcoa-331 • alcoa-c-30bf • alcoa-c-330 • Alcoa-c-333 • alpha-alumina trihydrate • alugel • alumigel • alumina hydrate • alumina hydrated • alumina trihydrate • aluminum hydrate • aluminum oxide hydrate • aluminum oxide trihydrate • aluminum oxide, trihydrate • aluminum oxide 3H₂O • aluminum trihydrate • aluminum (III) hydroxide • amphojel • baco af 260 • British aluminum af 260 • c-31 • c-31c • c-31f • c-33 • c-31-f • calmogastrin • ci-77002 • gha-331 • gha-332 • gha-431 • h-46 • higilite • higilite-h-31s • higilite-h 32 • higilite-h-42 • hychol-705 • hydrafil • hydral-705 • hydral-710 • hydrated alumina • hydrated aluminum oxide • martinal • martinal a • martinal-a/s • martinal f-a • p-30bf • reheis-f-1000 • trihydrated alumina • trihydroxyaluminum

MF AlH₃O₃

MW 78.01

COLOR AND FORM White monoclinic crystals; white powder, balls, or granules.

DENSITY 2.42.

MELTING POINT 300°C.

SOLUBILITY Insoluble in water, alcohol; soluble in hydrochloric acid, sulfuric acid, alkaline aqueous solution, in strong acids in presence of water.

STABILITY AND SHELF LIFE Forms gels on prolonged contact with water; absorbs acids, carbon dioxide [R2].

PREVENTATIVE MEASURES SRP: The scientific literature supports the wearing of contact lenses in industrial environments, as part of a program to protect the eye against chemical compounds and minerals causing eye irritation. However, there may be individual substances whose irritating or corrosive properties are such that the wearing of contact lenses would be harmful to the eye. In those specific cases contact lenses should not be worn.

DISPOSAL Recommended methods: discharge to sewer, chemical treatment, landfill. Peer-review: small amounts of aluminum hydroxide can be discharged to sewer. Neutralize to pH 6. Coagulant aid improves separation of Al(OH)₃ (aluminum hydroxide); supernatant to sewer after dilution and aluminum hydroxide to landfill (peer-review conclusions of an IRPTC expert consultation (May 1985)) [R7].

COMMON USES Desiccant powder [R4]; in packaging materials [R5]; chemical intermediate; filler in paper, plastics, rubber, and cosmetics; soft abrasive for brass and plastics; glass additive to increase mechanical strength and resistance to thermal shock, weathering, and chemicals; smoke suppressant for plastics and latex foams [R6]; medication [R3, p. II–128]; glass, ceramics, iron-free aluminum, and aluminum salts, manufactures of activated alumina, base for organic lakes, flame retardants, mattress batting; finely divided form (0.1–0.6 μm) used for rubber reinforcing agent, paper coating, filler, cosmetics [R1]; adsorbent; emulsifier; ion-exchanger, in chromatography; mordant in dyeing; filtering medium; manufacture of glass, fire clay, paper, pottery, printing inks, lubricating compositions, detergents; waterproofing fabrics; in antiperspirants, dentifrices; used in pharmacy as the gel or dried gel [R2].

BINARY REACTANTS Bismuth hydroxide.

STANDARD CODES NFPA—no; CFR—no; USCG—no; ICC—no; USCG—no.

PERSISTENCE Will remain in present form for extended periods of time.

MAJOR SPECIES THREATENED No real threat exists because of the limited solubility.

ACUTE HAZARD LEVEL Very low. No apparent aquatic effects from 0.05 ppm Al [R10]; powder has exchange capacity for phosphates and may remove some nutrients from natural waters.

CHRONIC HAZARD LEVEL Very low. Al has been shown to affect mortality and growth in rainbow trout (8 weeks) and daphnids (3 weeks) at 0.52 and 0.68 ppm respectively [R12]; aluminum below 0.05 is safe for trout [R13]; marine waters should not exceed 1/100 hour LC_{50}, 1/10 LD_{50} [R11].

DEGREE OF HAZARD TO PUBLIC HEALTH Very low.

OCCUPATIONAL RECOMMENDATIONS None listed.

ACTION LEVEL Although generally insoluble and innocuous, the solids can build up on the bottom of waterways, smothering benthos, and burying spawning ground. Try to block runoff that may carry the solid into any nearby surface waters.

ON-SITE RESTORATION Dredge the deposited solid. Seek professional environmental engineering assistance through EPA's Environmental Response Team (ERT), Edison, NJ, 24-hour no., 908–548–8730.

BEACH OR SHORE RESTORATION Will not harm the shoreline. Acid wash can be used to leach it, however, if so desired.

DISPOSAL Landfill, or return to supplier.

INDUSTRIAL FOULING POTENTIAL The solids, if in a dispersion, may settle out in piping and cause plugging; recommended (Al) for general use <3 ppm.

MAJOR WATER USE THREATENED Recreation due to unsightly dispersions.

PROBABLE LOCATION AND STATE White powder or granules. Will sink rapidly. Will be deposited on bottom of water course or be present as white dispersion.

WATER CHEMISTRY Aluminum solubility is highly pH-dependent, with the minimum occurring at pH 5.5. Under acid conditions, soluble forms are polymeric and cationic, whereas under alkaline conditions monomers eliminate water, leave a dimer sharing OH- ions on a common edge. These grow in chains and six-membered rings. Insoluble forms include gels and large, less soluble polymers.

COLOR IN WATER White dispersion; white precipitate.

FOOD-CHAIN CONCENTRATION POTENTIAL Negative.

BIOLOGICAL HALF-LIFE The mean plasma half-life of aluminum after iv administration in dogs is approximately 4.5 hr (aluminum) [R8, 1009].

FIFRA REQUIREMENTS Residues of aluminum hydroxide are exempted from the requirement of a tolerance when used as a diluent or carrier in accordance with good agricultural practices as inert (or occasionally active) ingredients in pesticide formulations applied to growing crops or to raw agricultural commodities after harvest [R9].

REFERENCES

1. Sax, N. I., and R. J. Lewis, Sr. (eds.). *Hawley's Condensed Chemical Dictionary*. 11th ed. New York: Van Nostrand Reinhold Co., 1987. 43.

2. *The Merck Index*. 10th ed. Rahway, NJ: Merck Co., Inc., 1983. 51.

3. Gosselin, R. E., R. P. Smith, H. C. Hodge. *Clinical Toxicology of Commercial Products*. 5th ed. Baltimore: Williams and Wilkins, 1984.

4. Osol, A., and J. E. Hoover, et al. (eds.). *Remington's Pharmaceutical Sciences*. 15th ed. Easton, PA: Mack Publishing Co., 1975. 752.

5. Sax, N. I. *Dangerous Properties of Industrial Materials*, 6th ed. New York, NY: Van Nostrand Reinhold, 1984. 177.

6. SRI.

7. United Nations. *Treatment and Disposal Methods for Waste Chemicals* (IRPTC File). Data Profile Series No. 5. Geneva, Switzerland: United Nations Environmental Programme, Dec. 1985. 82.

8. Ellenhorn, M. J., and D. G. Barceloux. *Medical Toxicology—Diagnosis and Treatment of Human Poisoning*. New York, NY: Elsevier Science Publishing Co., Inc. 1988.

9. 40 CFR 180.1001 (c) (7/1/91).

10. Turnbull–Kemp, P. St. J., "Trout in Southern Rhodesia. V. on the Toxicity of Copper Sulfate to Trout," *Rhodesia Agr. J.* 1958. 55 (6): 637–640.

11. *Proposed Criteria for Water Quality.*

Volume I. U.S. Environmental Protection Agency. October 1973.

12. Environmental Protection Agency, *Report of the Pesticide Technical Committee to the Lake Michigan Enforcement Conference on Selected Trace Metals*, NTIS PB–220 361, September 1972.

13. Everhart, W. H., and R. H. Freeman. *Effects of Chemical Variations in Aquatic Environments*. Volume 11, EPA R3–73–011 D, February 1973.

■ ALUMINUM SODIUM FLUORIDE

CAS # 15096–52–3

SYNONYMS Chiolite • cryolite ($AlNa_3F_6$) • cryolite ($Na_3(AlF_6)$) • ent-24,984 • Greenland spar • ice-spar • icetone • koyoside • kriolit • kryocide • Kryolith (German) • Natriumaluminiumfluorid (German) • Natriumhexafluoroaluminate (German) • sodium aluminofluoride • sodium aluminum fluoride • sodium fluoaluminate • sodium hexafluoroaluminate • villiaumite • kryolith • cryolite

MF $AlF_6 \cdot 3Na$

MW 209.97

COLOR AND FORM Monoclinic crystals (natural product); amorphous powder (synthetic product); snow-white, semi-opaque masses, vitreous fracture; natural form may be colored reddish or brown or even black but loses this discoloration on heating.

DENSITY 2.95.

MELTING POINT 1,000°C.

SOLUBILITY Soluble in dilute alkali; soluble in fused aluminum, ferric salts; insoluble in alcohol; soluble in concentrated sulfuric acid.

STABILITY AND SHELF LIFE More stable than fluosilicate with respect to hydrolysis [R11, p. II–113].

OTHER CHEMICAL/PHYSICAL PROPERTIES

Aluminum oxide is soluble and dissociable in molten cryolite, and electrodeposition of aluminum metal is thus possible; Mohs' hardness 2.5–3; fuses fairly easily.

PREVENTATIVE MEASURES SRP: The scientific literature supports the wearing of contact lenses in industrial environments, as part of a program to protect the eye against chemical compounds and minerals causing eye irritation. However, there may be individual substances whose irritating or corrosive properties are such that the wearing of contact lenses would be harmful to the eye. In those specific cases contact lenses should not be worn.

COMMON USES Stomach and contact insecticide [R1]; in aluminum and fluorine industry [R4]; principally an electrolyte in production and refining of aluminum; flux in production of various metals and alloys; flux and opacifier in manufacture of glass and enamels; coating for welding rods; filler for resin, rubber, ceramic-bonded grinding wheels [R6]; electric insulation; explosives; polishes [R8]; added to fireworks to produce yellow explosion [R9]; added to several gun propellants at 0.3% [R13].

OCCUPATIONAL RECOMMENDATIONS

OSHA 8-hr time-weighted average: 2.5 mg/m^3 (fluoride (as (F)) [R5].

NIOSH 10-hr time-weighted average: 2.5 mg/m^3 (fluorides (as (F)) [R3]; threshold-limit 8-hr time-weighted average (TWA) 2.5 mg/m^3 (1977) (fluorides (as (F)) [R15, 22].

INTERNATIONAL EXPOSURE LIMITS Former USSR: 1 mg/m^3 (fluorides, as hydrogen fluoride) [R12, 892].

NON-HUMAN TOXICITY LD$_{50}$ rat ip 59 mg/kg [R14, 1496].

ECOTOXICITY EC$_{50}$ *Simocephalus* 5.0 mg/L/48 hr at 15°C, first instar (95% confidence limit 3.6–6.8 mg/L) static bioassay without aeration, pH 7.2–7.5, water hardness 40–50 mg/L as calcium carbonate, and alkalinity of 30–35 mg/L (technical material, 96%) [R2]; EC$_{50}$ *Daphnia pulex* 10.0 mg/L/48 hr at 15°C, first instar (95% confidence limit 7.6–13.0

mg/L) static bioassay without aeration, pH 7.2–7.5, water hardness 40–50 mg/L as calcium carbonate, and alkalinity of 30–35 mg/L (technical material, 96%) [R10]; LC$_{50}$ rainbow trout 47.0 mg/L/96 hr at 12°C, wt 1.8 g static bioassay without aeration, pH 7.2–7.5, water hardness 40–50 mg/L as calcium carbonate, and alkalinity of 30–35 mg/L (technical material, 96%) [R10]; LC$_{50}$ bluegill more than 400 mg/L/96 hr at 24°C, wt 0.8 g; static bioassay without aeration, pH 7.2–7.5, water hardness 40–50 mg/L as calcium carbonate, and alkalinity of 30–35 mg/L (technical material, 96%) [R10].

BIOLOGICAL HALF-LIFE The mean plasma half-life of aluminum after iv admin in dogs is approximately 4.5 hr. (aluminum) [R13, 1009].

FIFRA REQUIREMENTS A tolerance of 7 ppm is established for combined residues of the insecticidal fluorine compound cryolite and synthetic cryolite (sodium aluminum fluoride) in or on the following agricultural commodities: apples; apricots; beans; beets (roots and tops); blackberries; blueberries (huckleberries); boysenberries; broccoli; brussels sprouts; cabbage; carrots; cauliflower; citrus fruits; collards; corn; cranberries; cucumbers; dewberries; eggplant; grapes; kale; kohlrabi; lettuce; loganberries; melons; mustard greens; nectarines; okra; peaches; peanuts; pears; peas; peppers; plums (fresh prunes); pumpkins; quinces; radish (roots and tops); raspberries; rutabaga (roots and tops); squash (winter and summer); strawberries; tomatoes; turnips (roots and tops); and youngberries [R7].

REFERENCES

1. Worthing, C. R. (ed.). *Pesticide Manual.* 6th ed. Worcestershire, England: British Crop Protection Council, 1979. 132.

2. U.S. Department of Interior, Fish and Wildlife Service. *Handbook of Acute Toxicity of Chemicals to Fish and Aquatic Invertebrates.* Resource Publication No. 137. Washington, DC: U.S. Government Printing Office, 1980. 23.

3. NIOSH. *NIOSH Pocket Guide to Chemical Hazards.* DHHS (NIOSH) Publication No. 90–117. Washington, DC: U.S. Government Printing Office, June 1990 116.

4. 40 CFR 180.145 (a) (7/1/91).

5. 54 FR 2920 (1/19/89).

6. SRI.

7. *The Merck Index.* 10th ed. Rahway, NJ: Merck Co., Inc., 1983. 374.

8. Sax, N. I., and R. J. Lewis, Sr. (eds.). *Hawley's Condensed Chemical Dictionary.* 11th ed. New York: Van Nostrand Reinhold Co., 1987. 325.

9. *Kirk–Othmer Encyclopedia of Chemical Technology.* 3rd ed., Volumes 1–26. New York, NY: John Wiley and Sons, 1978–1984, p. V19 487 (1982).

10. U.S. Department of Interior, Fish and Wildlife Service. *Handbook of Acute Toxicity of Chemicals to Fish and Aquatic Invertebrates.* Resource Publication No. 137. Washington, DC: U.S. Government Printing Office, 1980. 23.

11. Gosselin, R. E., R. P. Smith, H. C. Hodge. *Clinical Toxicology of Commercial Products.* 5th ed. Baltimore: Williams and Wilkins, 1984.

12. International Labour Office. *Encyclopedia of Occupational Health and Safety.* Vols. I & II. Geneva, Switzerland: International Labour Office, 1983.

13. Ellenhorn, M. J., and D. G. Barceloux. *Medical Toxicology—Diagnosis and Treatment of Human Poisoning.* New York, NY: Elsevier Science Publishing Co., Inc. 1988.

14. Clayton, G.D., and F.E. Clayton (eds.). *Patty's Industrial Hygiene and Toxicology.* Volumes 2A, 2B, 2C: Toxicology. 3rd ed. New York: John Wiley and Sons, 1981–1982.

15. American Conference of Governmental Industrial Hygienists. *Threshold Limit Values for Chemical Substances and Physical Agents and Biological Exposure Indices for 1994–1995.* Cincinnati, OH: ACGIH, 1994.

■ ALUMINUM SULFATE (2:3)

CAS # 10043–01–3

DOT # 9078

SIC CODES 2899; 2819

SYNONYMS alum • aluminum alum • aluminum sulfate (2:3) • aluminum sulfate ($Al_2(SO_4)_3$) • aluminum sulphate • aluminum trisulfate • cake alum • dialuminum sulphate • dialuminum sulfate • dialuminum trisulfate • sulfuric acid aluminum(3) salt (3: 2) • sulfuric acid, aluminum salt (3: 2) • pickle alum • filter alum • papermaker's alum • pearl alum

MF $O_{12}S_3 \cdot 2Al$

MW 342.14

pH Aqueous solution (1 g/1 ml water) not less than 2.9.

COLOR AND FORM White, lustrous crystals, pieces, granules, or powder.

ODOR Odorless.

TASTE Sweet, mildly astringent.

DENSITY 2.710 at 25°C.

MELTING POINT 770°C (with decomposition).

SOLUBILITY Soluble in 1 part water; practically insoluble in alcohol; soluble in dilute acid.

VAPOR PRESSURE Essentially zero.

CORROSIVITY Aluminum sulfate solution is corrosive to aluminum.

STABILITY AND SHELF LIFE Stable in air [R1].

OTHER CHEMICAL/PHYSICAL PROPERTIES Melts when gradually heated; at 250°C loses its water; on long boiling of aqueous solution, insoluble basic salt precipitates (aluminum sulfate octadecahydrate).

PROTECTIVE EQUIPMENT Personal protective equipment includes dust respirator; safety glasses or face shield; rubber gloves [R10].

PREVENTATIVE MEASURES SRP: The scientific literature supports the wearing of contact lenses in industrial environments, as part of a program to protect the eye against chemical compounds and minerals causing eye irritation. However, there may be individual substances whose irritating or corrosive properties are such that the wearing of contact lenses would be harmful to the eye. In those specific cases contact lenses should not be worn.

CLEANUP Methods are discussed for recovery of valuable components from or disposal of aluminum sulfate, formed during aluminum coagulation in water purification [R11].

DISPOSAL Pretreatment involves hydrolysis, followed by neutralization with sodium hydroxide [R12].

COMMON USES Tanning leather, sizing paper, mordant in dyeing; purifying water; manufacture of lakes, aluminum resinate; fireproofing and waterproofing cloth; clarifying oils and fats; treating sewage; waterproofing concrete; deodorizing and decolorizing petroleum; antiperspirants; agricultural pesticides; manufacture of aluminum salts [R3]; catalyst in manufacture of ethane; lubricating compositions; food additive [R2]; applied to soils, esp in West, to make them less alkaline; in East it is used to produce acid condition for such plants as rhododendrons, azaleas, camellias, blueberries (aluminum sulfate octadecahydrate) [R4]; medication: anti-infective [R3]; used to remove trihalomethanes (THMs) precursors from lake water [R5]. Alum spray is used for decontamination of radiocontaminated metal surfaces [R6]. Miscellaneous application in cosmetics and soap manufacture [R7]. Swimming-pool water is adjusted to pH 7.2–7.6 and sprinkled with alum 15.2 g/m³ (2 oz/1,000 gal). The trace metals, entrapped in the floc of aluminum hydroxide that forms, are removed by filtration or vacuuming (alum, aluminum sulfate tetradecahydrate) [R8]; an early step in the treatment of raw water: either aluminum sulfate or ferric chloride proved a good primary flocculant for virus removal. Polycationic coagulation aids were useful in removing virus, but anionic or nonionic polyelectrolytes probably would not be effective. Clays are sometimes added, to ensure formation of a visible floc to which virus may adsorb. Organic matter in the water interfered with virus removal by coagulation and settling; this effect may be prevented by prechlorination or preozonation. Finally, it should be noted that the virus removed had not been inactivated. Virus adsorbed to floc was quite capable of causing infection [R9].

STANDARD CODES EPA 311; no NFPA; no ICC; Superfund designated (hazardous substances) list.

PERSISTENCE Will slowly be precipitated as $Al(OH)_3$ due to natural alkalinity.

MAJOR SPECIES THREATENED Aquatic life.

PERSONAL SAFETY Guard against ingestion. Avoid breathing dust.

ACUTE HAZARD LEVEL 106 ppm was found to immobilize *Daphnia magna* in Lake Erie water.

CHRONIC HAZARD LEVEL Unknown but suspected of being very low. Al has been shown to affect mortality and growth in rainbow trout (8 weeks) and daphnids (3 weeks) at 0.52 and 0.68 ppm, respectively [R23]; aluminum below 0.05 is safe for trout [R9517]; marine waters should not exceed 1/100 96 hour LC_{50}, 1/10 LD_{50} [R22].

DEGREE OF HAZARD TO PUBLIC HEALTH Where lead service pipes are used, high concentrations (<5 ppm) may cause an increase in the lead content of drinking water. Can render water acid and hence capable of burns.

OCCUPATIONAL RECOMMENDATIONS
OSHA 8-hr time-weighted average: 2 mg/m^3. Final rule limits were achieved by any combination of engineering controls, work practices, and personal protective equipment during the phase-in period, Sept. 1, 1989, through Dec. 30, 1992. Final rule limits became effective Dec 31, 1992 (aluminum soluble salts, as Al) [R20]; threshold-limit 8-hr time-weighted average (TWA) 2 mg/m^3 (1979). (aluminum, soluble salts, as Al) [R21, 12].

ACTION LEVEL Spills into water will result in hydrolysis to sulfuric acid solution with the capability of producing burns. Keep people away from water until pH monitoring suggests acidity has subsided.

ON-SITE RESTORATION Treat with sodium bicarbonate to precipitate $Al(OH)_3$. Seek professional environmental engineering assistance through EPA's Environmental Response Team (ERT), Edison, NJ, 24-hour no. 908–548–8730.

AVAILABILITY OF COUNTERMEASURE MATERIAL Sodium bicarbonate—grocery distributors, large bakeries.

DISPOSAL METHOD Precipitate as $Al(OH)_3$ using lime, and landfill solids. Supernatant can be routed to sewage plant. Alum solutions are used by many water treatment facilities. These outlets may be anxious to use the waste solution rather than have it disposed of.

DISPOSAL NOTIFICATION Contact local water treatment authority.

INDUSTRIAL FOULING POTENTIAL Undesirable for paper and textile industry. Excessive quantities of alum in water used for paper have a deleterious effect on the strength of the paper product. In dyeing rayon a residual content of alum <0.4 ppm is desirable. Recommended (Al) <3 ppm. For general industrial use.

MAJOR WATER USE THREATENED Fisheries, irrigation.

PROBABLE LOCATION AND STATE Colorless crystals will dissolve in water.

WATER CHEMISTRY Aluminum solubility is highly pH dependent, with the minimum occurring at pH 5.5. Under acid conditions, soluble forms are polymeric and cationic, whereas under alkaline conditions monomeric anions are present. Polymer forms are a result of the hydrolysis of monomers eliminating water and leaving a dimer sharing OH^- ions on a common edge. These grow in chains and six-membered rings. Insoluble forms include gels and large, less soluble polymers.

FOOD-CHAIN CONCENTRATION POTENTIAL
Negative.

HUMAN TOXICITY The fatal adult dose of alum by mouth is about 30 g (form not stated) [R13, 169].

NON-HUMAN TOXICITY LD_{50} mouse, oral 4,210 mg/kg [R17, 109]; LD_{50} rat, oral 1,930 mg/kg [R17, 109]; LD_{50} mouse, oral 770 mg/kg [R10]; LD_{50} mouse ip 6.3 mg/kg [R16, 16]; LD_{50} mouse, oral 970 mg/kg [R16, 164]; LD_{50} mouse, oral about 6.1 g/kg [R15].

BIOLOGICAL HALF-LIFE The mean plasma half-life of aluminum after iv admin in dogs is approximately 4.5 hr. [R14, 1009].

BIOCONCENTRATION A study was undertaken to assess the possibility of aluminum bioaccumulation (in rainbow trout). Trout tissues, plankton, and water were analyzed for total aluminum concentration. Statistical comparisons of experimental and control tissues revealed few overall significant differences in the level of aluminum between alum-exposed and nonexposed fish, but significant differences existed between tissues within a given treatment and age class [R18].

FIFRA REQUIREMENTS Aluminum sulfate is exempted from the requirement of a tolerance when used as a Safener adjuvant in accordance with good agricultural practice as inert (or occasionally active) ingredients in pesticide formulations applied to growing crops only [R19].

GENERAL RESPONSE Avoid contact with liquid and vapor. Keep people away. Wear goggles, self-contained breathing apparatus, and rubber overclothing (including gloves). Shut off ignition sources. Call fire department. Stop discharge if possible. Isolate and remove discharged material. Notify local health and pollution control agencies.

FIRE RESPONSE Not flammable. Wear goggles, self-contained breathing apparatus, and rubber overclothing (including gloves). Extinguish with dry chemicals or carbon dioxide. Do not use water on fire.

EXPOSURE RESPONSE Call for medical aid. Dust is irritating to eyes, nose, and throat.

If inhaled will cause difficult breathing. If breathing has stopped, give artificial respiration. If breathing is difficult, give oxygen. Solid irritating to skin and eyes. If swallowed will cause nausea or vomiting. Remove contaminated clothing and shoes. Flush affected areas with plenty of water. If swallowed and victim is conscious, have victim drink water or milk. If swallowed, and victim is unconscious or having convulsions, do nothing except keep victim warm.

RESPONSE TO DISCHARGE Issue warning—water contaminant. Should be removed. Provide chemical and physical treatment.

WATER RESPONSE Harmful to aquatic life in very low concentrations. May be dangerous if it enters water intake. Notify local health and wildlife officials. Notify operators of nearby water intakes.

EXPOSURE SYMPTOMS Inhalation of dust irritates nose and mouth. Ingestion of large doses causes gastric irritation, nausea, vomiting, and purging. Dust irritates eyes and skin.

EXPOSURE TREATMENT Inhalation: rinse nose and mouth with water. Ingestion: give large amounts of water. Eyes: flush with water for at least 15 min. Skin: flush with water, wash with soap and water.

REFERENCES

1. Osol, A. (ed.). *Remington's Pharmaceutical Sciences.* 16th ed. Easton, PA: Mack Publishing Co., 1980. 721.

2. Sax, N. I., and R. J. Lewis, Sr. (eds.). *Hawley's Condensed Chemical Dictionary.* 11th ed. New York: Van Nostrand Reinhold Co., 1987. 49.

3. *The Merck Index.* 10th ed. Rahway, NJ: Merck Co., Inc., 1983. 54.

4. *Farm Chemicals Handbook 1984.* Willoughby, OH: Meister Publishing Co., 1984, p. B-18.

5. Hoehn, R. C., et al. Office of Water Research and Technology Rep. No. W83-01279 (1982).

6. *Decontamination of Radiocontaminated Metals*; Jpn Kokai Tokkyo Koho Patent

No. 84 27300 02/13/84 (Hitachi Plant Engineering and Construction Co. Ltd.)

7. *Chemical Products Synopsis: Aluminum Sulfate*, 1985.

8. *Kirk–Othmer Encyclopedia of Chemical Technology*. 3rd ed., Volumes 1–26. New York, NY: John Wiley and Sons, 1978– 1984, p. 24 (84) 434.

9. National Research Council. *Drinking Water and Health* Volume 1. Washington, DC: National Academy Press, 1977. 106.

10. U.S. Coast Guard, Department of Transportation. *CHRIS—Hazardous Chemical Data*. Volume II. Washington, DC: U.S. Government Printing Office, 1984-5.

11. Fulton, G. P. *Proc Awwa Semin Water Treat Waste Disposal*; paper no 3: 25 (1978).

12. Sittig, M. *Handbook of Toxic and Hazardous Chemicals*; p. 42 (1981).

13. Thienes, C., and T. J. Haley. *Clinical Toxicology*. 5th ed. Philadelphia: Lea and Febiger, 1972.

14. Ellenhorn, M. J., and D. G. Barceloux. *Medical Toxicology—Diagnosis and Treatment of Human Poisoning*. New York, NY: Elsevier Science Publishing Co., Inc. 1988.

15. Gosselin, R. E., R. P. Smith, H. C. Hodge. *Clinical Toxicology of Commercial Products*. 5th ed. Baltimore: Williams and Wilkins, 1984, p. II–128.

16. National Research Council. *Drinking Water, and Health*, Volume 4. Washington, DC: National Academy Press, 1981.

17. Venugopal, B., and T. D. Luckey. *Metal Toxicity in Mammals*, 2. New York: Plenum Press, 1978.

18. Buergel, P. M., R. A. Soltero. *J Freshwater Ecol* 2 (1): 37–44 (1983).

19. 40 CFR 180.1001 (d) (7/1/91).

20. 29 CFR 1910.1000 (7/1/91).

21. American Conference of Governmental Industrial Hygienists. *Threshold Limit Values for Chemical Substances and Physical Agents and Biological Exposure Indices for 1993–1994*. Cincinnati, OH: ACGIH, 1993.

22. *Proposed Criteria for Water Quality, Volume I*, U.S. Environmental Protection Agency. October 1973.

23. Environmental Protection Agency, "Report of the Pesticide Technical Committee to the Lake Michigan Enforcement Conference on Selected Trace Metals," NTIS PB–220 361, September 1972.

■ AMERICIUM-241

CAS # 14596–10–2

SYNONYMS americium

MF Am

ATOMIC WT 243.

STANDARD CODES Labels for packages of radioactive materials must be of diamond shape, in colors specified, with each side at least 4 inches long. Printing must be in black inside a black line border measuring at least 3½ inches on each side. "Radioactive white-I" label—label must be white. The single vertical bar on the lower half of the label must be bright red. Labels must be applied on two opposite sides of each package having a dose rate not exceeding 0.5 millirem per hour at any point on the external surface of the package. Not authorized for fissile class II packages. "Radioactive yellow-II" label—the upper half of the label must be bright yellow and the bottom half must be white. The two vertical bars on the lower half of the label must be bright red. Labels must be applied on two opposite sides of: (a) each package having a dose rate not exceeding 10 millirem per hour at any point on the external surface of the package and not exceeding 0.5 millirem per hour at 3 feet from the external surface of the package; or (b) each package for which the transport index does not exceed 0.5 at any time during transportation. "Radioactive yellow-III" label—the upper half of the label must be bright yellow and the bottom half must be white. The three vertical bars on the lower half of the label must be bright red. Labels must be applied on two opposite sides of: (a) each package having a surface dose rate exceeding 10 millirem per hour; (b) each fissile class III package; (c) each package containing a large quantity of radioactive material as: 20 curies of group I radionuclides, 20 curies of group II radionuclides, 200 curies of group III radionuclides, 200 curies of group IV radionuclides, 5,000 curies of group V radionuclides, 50,000 curies of group VI radionuclides, 500,000 curies of group VII radionuclides, or 5,000 curies of special form radioactive materials; or (d) each package transported under a special permit issued in response to a petition.

PERSISTENCE 458-year radioactive half-life; 20,000-day biological half-life in total body, 73,000 days in bone, 24,000 days in kidneys, and 3,000 days in liver.

COLOR AND FORM A silvery, somewhat malleable radioactive metal.

MELTING POINT 994°C.

BOILING POINT 2,607°C.

DENSITY 13.67 at 20°C.

INHALATION LIMIT $2.\times10^{-13}$ mg/m^3.

DIRECT CONTACT Varies with specific compounds.

PERSONAL SAFETY Alpha, gamma radiation. Do not allow contaminated water to come in contact with skin or personal clothing. Wear waterproof protection. If the radioactivity is also airborne, a mask with air filter may be required.

ACUTE HAZARD LEVEL Soluble: organ of reference maximum permissible 40-hour week burden in total UCI/cm^3 body (9), UCI mpc air, mpc water kidney (0.1) 6×10^{-12}/1×10^{-4} bone (0.05) $6.\times10^{-12}$/$1.\times10^{-4}$ liver (0.4) $9.\times10^{-12}$/$2.\times10^{-4}$ total body (0.3) $2.\times10^{-11}$/$2.\times10^{-4}$ gastrointestinal tract/ 0 $(2.\times10^{-7})$ $8.\times10^{-4}$ insoluble: organ of reference mpc/w mpc/a lung $(10^{-10}$ gastrointestinal tract) $8.\times10^{-4}$ /10^{-7}.

DEGREE OF HAZARD TO PUBLIC HEALTH High, due to exposure to radiation.

OCCUPATIONAL RECOMMENDATIONS Not listed.

ACTION LEVEL Notify local air authorities and the Nuclear Regulatory Commission. Do not enter area without radiation monitoring equipment.

ON-SITE RESTORATION 1. Cation exchange resin, 2. lime treatment plus coagulant. Seek professional environmental engineering assistance through EPA's Environmental Response Team (ERT), Edison, NJ, 24-hour no. 908–548–8730.

BEACH OR SHORE RESTORATION Remove the sand and bury at authorized burial site.

AVAILABILITY OF COUNTERMEASURE MATERIAL Cation exchange resin—water softening and conditioning suppliers, water treatment plants; lime sulfate—water treatment plants.

DISPOSAL METHOD Burial at an authorized radioactive disposal site.

DISPOSAL NOTIFICATION Contact the (Federal) Nuclear Regulatory Commission.

INDUSTRIAL FOULING POTENTIAL The safe radiation levels are below industrial fouling potential levels.

MAJOR WATER USE THREATENED All water uses.

PROBABLE LOCATION AND STATE 1. Pure element is silvery in appearance. 2. All compounds except AmF_3 are soluble in water and will dissolve.

REFERENCES
1. Lewis, R. J., Sr. *Sax's Dangerous Properties of Industrial Materials.* 9th Ed. New York: Van Nostrand Reinhold, 1995.

2. EPA's *OHMTADS* Database.

■ AMIBEN

$$COOH$$

(structure: benzoic acid ring with COOH at top, Cl at 2-position, NH$_2$ at 3-position, Cl at 5-position)

CAS # 133-90-4

SYNONYMS 2,5-dichloro-3-aminobenzoic acid • 3-amino-2,5-dichlorobenzoic acid • acp-m-728 • ambiben • amibin • amoben • benzoic acid, 3-amino-2,5-dichloro • chlorambed • chloramben • chlorambene • NCI-C00055 •, ornamental-weeder • vegiben

MF C$_7$H$_5$Cl$_2$NO$_2$

MW 206.03

COLOR AND FORM Colorless crystalline solid.

ODOR Odorless.

MELTING POINT 200-201°C.

SOLUBILITY Solubility in water: 700 mg/L at 25°C; soluble in alkali; solubility in acetone: 23.27 g/100 g at 29°C; solubility in benzene: 200 ppm at 24°C; insoluble in carbon tetrachloride; solubility in chloroform: 900 ppm at 25°C; solubility in ethanol: 172 g/L at 25°C; solubility in isopropanol: 11.29 g/100 g at 26°C; 354 g/100 g ether.

VAPOR PRESSURE 930 mPa at 100°C.

STABILITY AND SHELF LIFE Rapidly decomposed by light [R12, 553].

OTHER CHEMICAL/PHYSICAL PROPERTIES MP: above 195°C (technical product); forms water-soluble alkali metal and ammonium salts; there is no problem of precipitation with hard water; stable to heat, to oxidation, and to hydrolysis by acid or alkaline media but is decomposed by sodium hypochlorite solution; purplish-white powder (technical product).

PREVENTATIVE MEASURES Avoid skin and eye contact, inhalation of spray mists, and drifting [R2].

DISPOSAL Reuse: The use of chemical reagents as a means of disposal is not recommended at the present time. Amiben's shelf life is of sufficient duration to allow for total use of the product in accordance with label directions. On this basis, there should be few instances of small quantities of excess amiben requiring disposal. For the decontamination of amiben containers, the National Agricultural Chemical Association (USA) recommends triple rinse. "Triple rinse" means the flushing of containers three times, each time using a volume of the normal diluent equal to approximately ten percent of the container's capacity and adding the rinse liquid to the spray mixture or disposing of it by the method prescribed for disposing of the pesticide [R3]. Incinerate in a unit equipped with an effluent gas scrubber at a temperature above 1,000°C about 1-2 seconds. Recommendable method: Incineration [R3]. Wet Oxidation: Studies of wet oxidation process applied to amiben herbicide process wastes indicate 88 to 99.5% destruction of the active ingredient [R3].

COMMON USES Herbicide for grasses and broadleaf weeds on soybeans, dry beans, lima beans, asparagus, pumpkins, squash, corn, tomatoes, peppers, sweet potatoes [R1]; preplant incorporated, or pre-emergence weed control, applied at planting of corn, dry beans, lima beans, peanuts, pumpkins, seedling asparagus, soybeans, squash, sunflowers, established tomatoes and peppers, cucumbers, cantaloupes, snapbeans, and sweet potatoes. Post emergence in soybeans [R2].

OCCUPATIONAL RECOMMENDATIONS None listed.

NON-HUMAN TOXICITY LD$_{50}$ rat oral 3,500 to 5,620 mg/kg [R4]; LD$_{50}$ rabbit dermal 3,136 mg/kg [R4].

BIODEGRADATION Studies with soil microflora indicated that breakdown of amiben occurred and that the carboxyl was cleaved slowly but steadily [R5]. Moves readily in sandy soils or following heavy rains. Broken down by microorganisms. Some photochemical loss in aqueous solution. On moist soil loss reported, but soil must remain moist for loss to occur. Average persistence at recommended rates: approximately 6 to 8 weeks [R2, 93]. When a column of Ella Island soil type (pH

3.8, 1.6%, organic matter, 82.7% sand, 12.2% silt, and 5.2% clay) was perfused for 160 days with a microbial medium containing 5.7 and 26.9 ppm (14)-COOH-labelled Amiben, 56.2 and 39.1% biodegradation were observed for Amiben, respectively. When a column of Kewaunee clay soil type (pH 6.4, 3.8%, organic matter, 14% sand, 35.1% silt, and 48.7% clay) was perfused for 160 days with a microbial medium containing 5.7 and 26.9 ppm (14)-COOH-labelled Amiben, 42.7 and 27.7% biodegradation were observed for Amiben, respectively. When a column of Poygan silty clay soil type (pH 7.2, 10%, organic matter, 6% sand, 58.9% silt, and 33.6% clay) was perfused for 160 days with a microbial medium containing 5.7 and 26.9 ppm (14)-COOH-labelled Amiben, 40.3 and 22.6% biodegradation were observed for Amiben, respectively (1) [R8]. In an aerobic biodegradation study in a Honeoye silt loam at 25°C and pH 6.5, Amiben biodegraded slowly at rate of 4,500, 6,500, 11,500, and 74,000 cpm (cumulative (14)-CO_2 evolution) after 7, 14, 28, and 112 days, respectively (1). According to a bioassay using the soybean plant and 50 ppm Amiben in soil, there was no breakdown of Amiben over an 11-month incubation period in moist soil at 15°C; at incubation temperatures of 25 and 35°C, Amiben breakdown over an 11-month period was 42 and 46 percent, respectively (2). Amiben showed the greatest loss of activity in Staten Island peaty muck (42% dissolved organic carbon and pH 5.8), which suggests that the rate of degradation depends on the organic content (3) [R9].

BIOCONCENTRATION Based on a measured water solubility of 700 mg/L at 25°C (2) and a regression-derived equation (1), a BCF of 15.34 can be estimated for Amiben (SRC). This BCF value suggests that Amiben may not bioconcentrate in aquatic organisms (SRC) [R6].

FIFRA REQUIREMENTS Tolerances are established for residues of the herbicide chloramben in or on the following raw agricultural commodities: beans dried; beans, lima; beans, snap; beans, vines; cantaloupes; corn, field, fodder; corn,

field, forage; corn, field, grain; cucumbers; peanuts; forage; peas, pigeon; peas, pigeon, forage; peppers; pumpkins; soybeans; soybeans, forage; squash, summer; squash, winter; sunflower seed; sweet potatoes; and tomatoes [R10]. A registration standard was issued July 1981 for chloramben used as an herbicide. [R11].

FIRE POTENTIAL Nonflammable [R13, 93].

REFERENCES

1. SRI.

2. *Farm Chemicals Handbook 1991.* Willoughby, OH: Meister, 1991, p. C-19.

3. United Nations. *Treatment and Disposal Methods for Waste Chemicals* (IRPTC File). Data Profile Series No. 5. Geneva, Switzerland: United Nations Environmental Programme, Dec. 1985. 110.

4. National Research Council. *Drinking Water and Health,* Volume 1. Washington, DC: National Academy Press, 1977. 519.

5. Menzie, C. M. *Metabolism of Pesticides.* U.S. Department of the Interior, Bureau of Sport Fisheries and Wildlife, Publication 127. Washington, DC: U.S. Government Printing Office, 1969. 39.

6. (1) Lyman, W. J., et al. *Handbook of Chemical Property Estimation Methods.* Washington DC: Amer Chemical Soc, pp. 4–9, 5–4, 5–10, 7–4, 7–5, 15–15 to 15–32 (1990) (2) Worthing, C. R., S. B. Walker. eds. *The Pesticide Manual* 8th ed. Suffolk, England: Lavenham Press Ltd (1987).

7. Weed Science Society of America. *Herbicide Handbook.* 5th ed. Champaign, Illinois: Weed Science Society of America, 1983.

8. Widlung, R. E., et al. *Weed Research* 8: 213–25 (1968).

9. (1) Macrae, I. C., M. Alexander. *J Agric Food Chem* 13: 72–6 (1965) (2) Burnside, O. C. *Weeds* 13: 274–6 (1965) (3) Donaldson, T. W., C. L. Foy. *Weeds* 13: 195–202 (1965).

10. 40 CFR 180.266 (7/1/91).

11. USEPA. *Report on the Status of*

Chemicals in the Special Review Program and Reregistration Standards in the Reregistration Program. p. 2–4 (1989).

12. Kearney, P. C., and D. D. Kaufman (eds.). *Herbicides: Chemistry, Degradation, and Mode of Action.* Volumes 1 and 2. 2nd ed. New York: Marcel Dekker, Inc., 1975.

13. Weed Science Society of America. *Herbicide Handbook.* 5th ed. Champaign, Illinois: Weed Science Society of America, 1983.

■ 2-AMINO-5-AZOTOLUENE

CAS # 97–56–3

SYNONYMS 2′,3-dimethyl-4-aminoazobenzene • 2-methyl-4-((o-tolyl) azo) aniline • 4-(o-tolylazo)-o-toluidine • 4-amino-2′,3-dimethylazobenzene • 5-(o-tolylazo)-2-aminotoluene • aat • benzenamine, 2-methyl-4-((2-methylphenyl) azo)- • brasilazina oil yellow-r • butter-yellow • ci-11160 • ci-11160b • ci-solvent-yellow-3- • fast-yellow-at • fast-yellow-b • hidaco-oil-yellow • o-aminoazotoluene • o-at • oaat • oil-yellow-21 • oil-yellow-2681 • oil-yellow-2r • oil-yellow-at • oil-yellow-c • oil-yellow-i • organol-yellow-2t • ortho-tolueneazo-ortho-toluidine • Ortho-toluol-azo-ortho-toluidin (German) • somalia-yellow-r • sudan-yellow-rra • toluazotoluidine • tulabase fast garnet gb • tulabase fast garnet gbc • waxakol yellow nl

MF $C_{14}H_{15}N_3$

MW 225.32

COLOR AND FORM Golden crystals; reddish-brown to yellow crystals; yellow leaves from alcohol.

MELTING POINT 101–102°C.

SOLUBILITY Soluble in alcohol, ether, chloroform; soluble in oils and fats; soluble in acetone, cellosolve, and toluene; water solubility = 7.64 mg/L at 25°C (est).

VAPOR PRESSURE 7.5×10^{-7} mm Hg at 25°C (extrapolated).

OCTANOL/WATER PARTITION COEFFICIENT Log K_{ow} = 3.92 (est).

OTHER CHEMICAL/PHYSICAL PROPERTIES Henry's Law constant = 2.91×10^{-8} atm cu-m/mole at 25°C (est).

PROTECTIVE EQUIPMENT Dispensers of liquid detergent should be available. Safety pipettes should be used for all pipetting. In animal laboratory, personnel should wear protective suits (preferably disposable, one-piece, and close-fitting at ankles and wrists), gloves, hair covering, and overshoes. In chemical laboratory, gloves and gowns should always be worn; however, gloves should not be assumed to provide full protection. Carefully fitted masks or respirators may be necessary when working with particulates or gases, and disposable plastic aprons might provide addnl protection. Gowns should be of distinctive color, as a reminder that they are not to be worn outside the laboratory (chemical carcinogens) [R5, 8].

PREVENTATIVE MEASURES Smoking, drinking, eating, storage of food, or of food and beverage containers, or utensils, and the application of cosmetics should be prohibited in any laboratory. All personnel should remove gloves, if worn, after completion of procedures in which carcinogens have been used. They should wash hands, preferably using dispensers of liquid detergent, and rinse thoroughly. Consideration should be given to appropriate methods for cleaning the skin, depending on nature of the contaminant. No standard procedure can be recommended, but the use of organic solvents should be avoided. Safety pipettes should be used for all pipetting. (chemical carcinogens) [R5, 8].

CLEANUP A high-efficiency particulate arrestor (HEPA) or charcoal filters can be used to minimize amount of carcinogen in exhausted-air-ventilated safety cabinets, lab hoods, glove boxes, or animal rooms. Filter housing that is designed so that used filters can be transferred into plastic bag without contaminating maintenance staff is avail commercially. Filters should be placed in plastic bags immediately after removal. The plastic bag should be sealed immediately. The sealed bag should be labelled properly. Waste liquids should be placed or collected in proper containers for disposal. The lid should be secured and the bottles properly labelled. Once filled, bottles should be placed in plastic bag, so that outer surface is not contaminated. The plastic bag should also be sealed and labelled. Broken glassware should be decontaminated by solvent extraction, by chemical destruction, or in specially designed incinerators. (chemical carcinogens) [R5, 15].

DISPOSAL PRECAUTIONS There is no universal method of disposal that has been proved satisfactory for all carcinogenic compounds, and specific methods of chemical destruction published have not been tested on all kinds of carcinogen-containing waste. Summary of avail methods and recommendations (given) must be treated as guide only. (chemical carcinogens) [R5, 14].

COMMON USES Manufacture of dyes; medicine [R2]; dye for coloring oils, fats, and waxes; chemical intermediate for the dyes solvent red 24, acid red 115 [R1].

OCCUPATIONAL RECOMMENDATIONS

IARC SUMMARY No data are available for humans. Sufficient evidence of carcinogenicity in animals. Overall evaluation: Group 2B: The agent is possibly carcinogenic to humans [R3].

BIOCONCENTRATION Based on the estimated water solubility of 2-amino-5-azotoluene, 7.64 mg/L (1,SRC), and an estimated log octanol/water partition coefficient of 3.921 (1,SRC), bioconcentration factors ranging from 196–562 can be calculated using appropriate regression equations (2,SRC). These values indicate that 2-amino-5-azotoluene may bioconcentrate in fish and aquatic organisms (SRC) [R4].

REFERENCES

1. SRI.

2. Sax, N. I., and R. J. Lewis, Sr. (eds.). *Hawley's Condensed Chemical Dictio-nary*. 11th ed. New York: Van Nostrand Reinhold Co., 1987. 55.

3. IARC. *Monographs on the Evaluation of the Carcinogenic Risk of Chemicals to Man*. Geneva: World Health Organization, International Agency for Research on Cancer, 1972–present (multivolume work), p. S7 56 (1987).

4. (1) GEMS. *Graphic Exposure Modeling System PCCHEM and CLOGP USEPA* (1987). (2) Lyman, W. J., et al. *Handbook of Chemical Property Estimation Methods*. New York: McGraw-Hill, Chapt 5 (1982).

5. Montesano, R., H. Bartsch, E. Boyland, G. Della Porta, L. Fishbein, R. A. Griesemer, A.B. Swan, L. Tomatis, and W. Davis (eds.). *Handling Chemical Carcinogens in the Laboratory: Problems of Safety*. IARC Scientific Publications No. 33. Lyon, France: International Agency for Research on Cancer, 1979.

■ 3-AMINOTOLUENE

CAS # 108–44–1

SYNONYMS 1-amino-3-methylbenzene • 3-amino-1-methylbenzene • 3-aminophenylmethane • 3-aminotoluen (Czech) • 3-methylaniline • 3-methylbenzenamine • 3-toluidine • aniline, 3-methyl • benzenamine, 3-methyl • m-aminotoluene • m-methylaniline • m-methylbenzenamine • m-toluidin (Czech) • m-toluidine • m-tolylamine

MF C_7H_9N

MW 107.17

COLOR AND FORM Colorless liquid.

DENSITY 0.990 at 25°C.

DISSOCIATION CONSTANTS $pK_a = 4.70$.

BOILING POINT 203–204°C.

MELTING POINT −30.4°C.

FIREFIGHTING PROCEDURES Foam, carbon dioxide, dry chemical [R3]. Do not extinguish fire unless flow can be stopped. Use water in flooding quantities as fog. Solid streams of water may be ineffective. Cool all affected containers with flooding quantities of water. Apply water from as far a distance as possible (toluidine mixture) [R4]. Wear self-contained breathing apparatus when fighting fires involving this material (toluidine mixture) [R4].

TOXIC COMBUSTION PRODUCTS Dangerous when heated; it emits highly toxic fumes [R3].

SOLUBILITY Slightly soluble in water; soluble in alcohol, ether, dilute acids; infinitely soluble in carbon tetrachloride, heptane; soluble in acetone and benzene; water solubility = 17 g/L.

VAPOR PRESSURE 1 mm Hg at 41°C.

OCTANOL/WATER PARTITION COEFFICIENT Log K_{ow} = 1.40.

OTHER CHEMICAL/PHYSICAL PROPERTIES Solidifies at about −50°C.

PROTECTIVE EQUIPMENT Wear boots, protective gloves, and goggles (toluidine mixture) [R4].

PREVENTATIVE MEASURES If material not on fire and not involved in fire: Keep sparks, flames, and other sources of ignition away. Keep material out of water sources and sewers. Build dikes to contain flow as necessary. Use water spray to knock down vapors. (toluidine mixture) [R4].

DISPOSAL Toluidine is a waste chemical stream constituent that may be subjected to ultimate disposal by controlled incineration. Controlled incineration (oxides of nitrogen are removed from the effluent gas by scrubbers or thermal devices). (toluidine) [R5].

COMMON USES Manufacture of dyes and other organic chemicals [R1].

OCCUPATIONAL RECOMMENDATIONS

OSHA 8-hr time-weighted average: 2 ppm (9 mg/m³). Final rule limits were achieved by any combination of engineering controls, work practices, and personal protective equipment during the phase-in period, Sept 1, 1989 through Dec 30, 1992. Final rule limits became effective Dec 31, 1992. Skin absorption designation in effect as of Sept 1, 1989 [R9].

THRESHOLD LIMIT 8-hr time-weighted average (TWA) 2 ppm, 8.8 mg/m³, skin (1986) [R11, 34].

INTERNATIONAL EXPOSURE LIMITS Former USSR (1980) 0.7 ppm ceiling [R6].

BIODEGRADATION 3-Aminotoluene is readily biodegradable in screening tests (1–3). In one test, 74% of theoretical BOD was utilized in 7.5 days with an activated sludge inoculum acclimated to aniline (1). In another, 97.7% degradation occurred in 5 days with an activated sludge inoculum (2). Complete degradation of 3-aminotoluene was obtained within 8 days with a soil inoculum (3). The half-life of 3-aminotoluene in natural water from ponds and rivers in which the microbial populations were increased 10- to 100-fold by filtration and nutrients added was 4 hr (4) [R7].

BIOCONCENTRATION The log octanol/water partition coefficient for 3-aminotoluene is 1.40 from which one can estimate a BCF of 6.8 using a recommended regression equation (2,SRC). Therefore 3-aminotoluene would not be expected to bioconcentrate in aquatic organisms [R8].

ATMOSPHERIC STANDARDS This action promulgates standards of performance for equipment leaks of volatile organic compounds (VOC) in the synthetic organic chemical manufacturing industry (SOCMI). The intended effect of these standards is to require all newly constructed, modified, and reconstructed SOCMI process units to use the best demonstrated system of continuous emission reduction for equipment leaks of VOC, considering costs, non-air-quality health, and environmental impact and energy requirements. Toluidines are produced, as intermediates or final products, by process units covered under this subpart. (toluidines) [R10].

DOT *Health Hazards:* Poisonous; may be fatal if inhaled, swallowed, or absorbed through skin. Contact may cause burns to skin and eyes. Runoff from fire control or dilution water may give off poisonous gases and cause water pollution. Fire may produce irritating or poisonous gases. (toluidines (ortho-, meta-, and para-)) [R2].

Fire or Explosion: Some of these materials may burn, but none of them ignites readily. Container may explode violently in heat of fire. (toluidines (ortho-, meta-, and para-)) [R2].

Emergency Action: Keep unnecessary people away; isolate hazard area and deny entry. Stay upwind, out of low areas, and ventilate closed spaces before entering. Positive-pressure self-contained breathing apparatus (SCBA) and chemical protective clothing that is specifically recommended by the shipper or manufacturer may be worn. It may provide little or no thermal protection. Structural firefighters' protective clothing is not effective for these materials. Remove and isolate contaminated clothing at the site. Call CHEMTREC at 1–800–424–9300 as soon as possible, especially if there is no local hazardous materials team available. (toluidines (ortho-, meta-, and para-)) [R2].

Fire: Small Fires: Dry chemical, CO_2, water spray, or regular foam. Large Fires: Water spray, fog, or regular foam. Move container from fire area if you can do so without risk. Fight fire from maximum distance. Stay away from ends of tanks. Dike fire-control water for later disposal; do not scatter the material. (toluidines (ortho-, meta-, and para-)) [R2].

Spill or Leak: Do not touch or walk through spilled material; stop leak if you can do so without risk. Fully encapsulating, vapor-protective clothing should be worn for spills and leaks with no fire. Use water spray to reduce vapors. Small Spills: Take up with sand or other noncombustible absorbent material and place into containers for later disposal. Small Dry Spills: With clean shovel place material into clean, dry container, and cover loosely; move containers from spill area. Large Spills: Dike far ahead of liquid spill for later disposal. (toluidines (ortho-, meta-, and para-)) [R2].

First Aid: Move victim to fresh air and call emergency medical care; if not breathing, give artificial respiration; if breathing is difficult, give oxygen. In case of contact with material, immediately flush skin or eyes with running water for at least 15 minutes. Speed in removing material from skin is of extreme importance. Remove and isolate contaminated clothing and shoes at the site. Keep victim quiet and maintain normal body temperature. Effects may be delayed; keep victim under observation. (toluidines (ortho-, meta-, and para-)) [R2].

GENERAL RESPONSE Shut off ignition sources. Call fire department. Avoid contact with liquid. Keep people away. Stop discharge if possible. Isolate and remove discharged material. Notify local health and pollution control agencies.

FIRE RESPONSE Combustible. Poisonous gases may be produced in fire. Wear goggles and self-contained breathing apparatus. Extinguish with dry chemicals, alcohol or polymer foam, or carbon dioxide. Cool exposed containers with water.

EXPOSURE RESPONSE DATA Call for medical aid. Liquid irritating to skin and eyes. If swallowed will cause nausea, vomiting, or loss of consciousness. Remove contaminated clothing and shoes. Flush affected areas with plenty of water. If in eyes, hold eyelids open and flush with plenty of water. If swallowed and victim is conscious, have victim drink water or milk and induce vomiting. If swallowed and victim is unconscious or having convulsions, do nothing except keep victim warm.

RESPONSE TO DISCHARGE Issue warning—water contaminant. Restrict access. Mechanical containment. Should be removed. Chemical and physical treatment.

WATER RESPONSE Effect of low concentration on aquatic life is unknown. Fouling to shoreline. May be dangerous if it enters water intakes. Notify local health and wildlife officials. Notify operators of nearby water intakes.

EXPOSURE SYMPTOMS Absorption of toxic quantities by any route causes cyanosis (blue discoloration of lips, nails, skin); nausea, vomiting, and coma may follow. Repeated inhalation of low concentrations may cause pallor, low-grade secondary anemia, fatigability, and loss of appetite. Contact with eyes causes irritation.

EXPOSURE TREATMENT Get medical attention following all exposures to this compound. Inhalation: move to fresh air. Ingestion: if victim is conscious, promptly induce vomiting. Eyes: flush with copious amounts of water for at least 15 min., holding lids apart. Skin: remove all contaminated clothing; wash affected areas immediately and thoroughly with plenty of warm water and soap.

FIRE POTENTIAL Moderate, when exposed to heat or flame [R3].

REFERENCES

1. *The Merck Index.* 10th ed. Rahway, NJ: Merck Co., Inc., 1983. 1365.

2. U.S. Department of Transportation. *Emergency Response Guidebook 1990.* DOT P 5800.5. Washington, DC: U.S. Government Printing Office, 1990, p. G-55.

3. Sax, N. I. *Dangerous Properties of Industrial Materials,* 6th ed. New York, NY: Van Nostrand Reinhold, 1984. 2593.

4. Association of American Railroads. *Emergency Handling of Hazardous Materials in Surface Transportation.* Washington, DC: Assoc. of American Railroads, Hazardous Materials Systems (BOE), 1987. 687.

5. USEPA. *Engineering Handbook for Hazardous Waste Incineration,* p. 2-10 (1981) EPA 68-03-3025.

6. American Conference of Governmental Industrial Hygienists. *Documentation of the Threshold Limit Values and Biological Exposure Indices,* 5th ed. Cincinnati, OH: American Conference of Governmental Industrial Hygienists, 1986. 589.

7. (1) Malaney, G. W. *J Water Pollut Control Fed* 32: 1300-11 (1960). (2) Pitter, P. *Water Res* 10: 231-5 (1976). (3) Alexander, M., B. K. Lustigman, *J Agric Food Chem* 14: 410-3 (1966). (4) Paris, D.

F., N. L. Wolfe. *Appl Environ Microbiol* 53: 911-6 (1987).

8. (1) Hansch, C., and A. J. Leo. *MED Chemical Project.* Pomona College, Claremont, CA (1985). (2) Lyman, W. J., et al., *Handbook of Chemical Property Estimation Methods.* New York: McGraw-Hill, pp. 5.1-5.30 (1982).

9. 29 CFR 1910.1000 (7/1/91).

10. 40 CFR 60.489 (7/1/91).

11. American Conference of Governmental Industrial Hygienists. *Threshold Limit Values for Chemical Substances and Physical Agents and Biological Exposure Indices for 1994-1995.* Cincinnati, OH: ACGIH, 1994.

■ AMMONIUM TARTRATE

CAS # 14307-43-8

SIC CODES 2834

SYNONYMS tartaric acid, ammonium salt • butanedioic acid, 2, 3-dihydroxy[R-(R*, R*)]-, diammonium salt • diammonium tartrate • ammonium d-tartrate • l-tartaric acid, ammonium salt

MF $C_4H_{12}N_2O_6$

MW 184.14

STANDARD CODES EPA 311; TSCA; not listed IATA; not listed NFPA; not listed CFR 49; not listed AAR; CFR 14 CAB code 8; ICC STCC Tariff no. 1-C 28 191 81.

PERSISTENCE Nonpersistent.

DIRECT CONTACT Eye, skin, respiratory irritant.

GENERAL SENSATION Inhalation—coughing, vomiting, reddening of lips, mouth, nose, throat, and conjunctiva. Higher concentrations cause swelling of lips and conjunctiva, temporary blindness, restlessness, tightness in chest, frothy sputum indicating pulmonary edema, cyanosis, and rapid and weak pulse. Skin contact causes severe burning pain and corrosive damage. Eye contact causes immediate and severe pain followed by conjunctival edema and corneal clouding. Later cataract formation and atrophy of retina and iris may occur.

PERSONAL SAFETY Protect against both inhalation and contact with the skin. Must wear protective clothing including NIOSH-approved rubber gloves and boots, safety goggles, or face mask, and a respirator whose canister is specifically approved for this material. Decontaminate fully or dispose of all equipment after use.

COMMON USES Topical opthalmic; textile industry; medicine.

OCCUPATIONAL RECOMMENDATIONS None listed.

ACTION LEVEL Avoid contact with the spilled cargo. Stay upwind. Notify local air, water, and fire authorities of the accident.

ON-SITE RESTORATION Dam stream if possible to reduce the flow and prevent further dissipation by water movement. Apply activated carbon (or other sorbent) to absorb the dissolved material. Bottom pumps or underwater vacuum systems may be employed in small bodies of water; dredging may be effective in larger bodies to remove any undissolved material and adsorbent from the bottom. An alternative method is to pump water into a suitable container. Add hydrochloric acid (HCl) to ph 7. filter, use ion exchange. Carbon can be used to remove tartrate. For more information see ENVIREX Manual EPA 600/2–77–227. Seek professional environmental engineering assistance through EPA's Environmental Response Team (ERT), Edison, NJ, 24-hour no. 908–548–8730.

BEACH OR SHORE RESTORATION Close beach and shore to the public until material has been removed.

AVAILABILITY OF COUNTERMEASURE MATERIAL Bottom pumps, dredges—fire departments, U.S. Coast Guard, Army Corps of Engineers; vacuum systems—swimming-pool operators; activated carbon—water treatment plant, chemical companies.

DISPOSAL METHOD After the material has been contained, remove with contaminated soil and place in impervious containers. The material may be incinerated in a chemical incinerator at the specified temperature/dwell time combination. Any liquids, sludges, or solid residues generated should be disposed in accordance with all applicable federal, state, and local pollution control requirements. If appropriate incineration facilities are not available, material may be buried in a chemical waste landfill. An alternative method, when material is neutralized and diluted, is biological treatment at a municipal sewage treatment plant.

DISPOSAL NOTIFICATION Notify local and state health authorities, local solid waste disposal authorities, supplier, and shipper of material.

INDUSTRIAL FOULING POTENTIAL Low industrial water fouling potential.

MAJOR WATER USE THREATENED Recreational; fisheries.

PROBABLE LOCATION AND STATE White crystals, will dissolve.

WATER CHEMISTRY No reaction. Will form basic solution.

COLOR IN WATER Colorless.

FOOD-CHAIN CONCENTRATION POTENTIAL Negative. Ammonium compounds are biodegradable and will not accumulate in the food chain.

REFERENCES

1. *Hawley's Condensed Chemical Dictionary.*

2. EPA *OHMTADS* Database.

■ TERT-AMYL ACETATE

CAS # 625–16–1

DOT # 1104

SYNONYMS tert-pentyl acetate

SIC CODES 2851; 2869

STANDARD CODES EPA 311; not in TSCA;

IATA flammable liquid, flammable liquid label, 1 ltr passenger, 40 ltr cargo; NFPA 704 m system 1, 3, 0; CFR 49 flammable liquid, flammable liquid label, 1 qt passenger, 10 gal cargo, storage code 1, 2; cfr 14 cab code 8; AAR, Bureau of Explosives, STCC 4909111; Superfund Designated (Hazardous Substances) List.

PERSISTENCE Nonpersistent.

INHALATION LIMIT 525 mg/m³.

DIRECT CONTACT Can cause acute irritation to the skin and eyes but is not considered an ingestive toxin.

GENERAL SENSATION Irritation of the eyes, nose, and throat, followed by a relatively slow and gradual onset of narcosis, with slow recovery after exposure ceases. Tendency to acidosis, headache, cough, pain in chest, shortness of breath, dermatitis, dizziness, nausea, and fatigue.

PERSONAL SAFETY Protect against both inhalation and absorption through the skin. Wear protective clothing including NIOSH-approved rubber gloves and boots, safety goggles or face mask, and either a respirator whose canister is specifically approved for this material, or a self-contained breathing apparatus.

ACUTE HAZARD LEVEL The vapors are irritating to the skin, eyes, and mucous membranes. Inhalation of high concentrations can lead to narcosis in man. In guinea pigs, inhalation of air containing 2,000 ppm of sec-amyl acetate for 13 hours leads to eye and nose irritation, no narcosis, and full recovery. At 4 hours of 10,000 ppm, the guinea pigs experienced eye and nose irritation, narcosis, and eventual death. In man, a brief exposure to air containing 5,000–10,000 ppm of sec-amyl acetate gave marked eye and nose irritation. The lowest concentration that gave an irritating effect is 200 ppm of sec-amyl acetate. In highly turbid water, the toxicity to mosquitofish for a 96-hour tlm is about 65 ppm, based on sec-amyl acetate.

CHRONIC HAZARD LEVEL Tert-amyl acetate is believed to be of relatively low chronic toxicity. The threshold-limit-value–time-weighted average for a 40-hour work week is 525 mg/m³, and the short-term exposure limit (15 minutes) is 790 mg/m³ as N-amyl acetate. No skin sensitization and only minor dryness of the skin have been noted after prolonged contact.

DEGREE OF HAZARD TO PUBLIC HEALTH Toxicity chiefly concerned with inhalation of fumes; ingestion is relatively nontoxic or of slight toxicity.

OCCUPATIONAL RECOMMENDATIONS None listed.

ACTION LEVEL Avoid contact with the spilled cargo. Eliminate all ignition sources. Stay upwind. Notify local air, water, and fire authorities of the accident. Evacuate all people to a distance of at least 200 feet upwind and 1,000 feet downwind of the spill.

ON-SITE RESTORATION Dam stream if possible to reduce the flow and prevent further dissipation by water movement. Use of floating booms to contain the slick is advisable. Floating material can be skimmed or pumped out. Activated carbon can be applied to the water to adsorb the dissolved material. Bottom pumps, dredges, or vacuum systems can be used to remove the sorbent from the bottom. Under controlled conditions, an alternative method is to pump the water into a suitable container, transfer to a gravity separation tank, and pass through a filtration system with an activated-carbon filter. For more details, see Envirex manual EPA 600/2-77-227. Seek professional environmental engineering assistance through EPA's Environmental Response Team (ERT), Edison, NJ, 24-hour no. 908–548–8730.

BEACH OR SHORE RESTORATION Close beach and shore to the public until material has been removed. If spill results in significant amounts of unabsorbed material, an alternative method may be ignition by the local fire department, with notification of local air authorities.

AVAILABILITY OF COUNTERMEASURE MATERIAL Pumps and floating booms—Army Corps of Engineers; carbon—water treatment plants, chemical companies.

DISPOSAL METHOD After the material has been contained, apply a sorbent material (peat, straw, sawdust, etc.) to the contaminated area. Remove all contaminated sorbent and soil, and place in impervious containers. The material may be incinerated in an approved chemical incinerator at the specified temperature/dwell time combination. Any liquids, sludges, or solid residues generated should be disposed of in accordance with all applicable federal, state, and local pollution control requirements. If appropriate incineration facilities are not available, material may be buried in a chemical waste landfill. Amenable to biological treatment at a municipal sewage treatment plant when diluted.

DISPOSAL NOTIFICATION Notify local and state health authorities, local solid waste disposal authorities, supplier and shipper of material.

INDUSTRIAL FOULING POTENTIAL Potential explosion in boiler feed and cooling waters due to its flammability.

MAJOR WATER USE THREATENED Recreational; potable supply; industry.

PROBABLE LOCATION AND STATE Colorless liquid, will float and dissolve slowly.

WATER CHEMISTRY No reaction.

COLOR IN WATER Colorless.

GENERAL RESPONSE Shut off ignition sources and call fire department. Stop discharge if possible. Keep people away. Avoid contact with liquid and vapor. Stay upwind and use water spray to knock down vapor. Isolate and remove discharged material. Notify local health and pollution control agencies.

FIRE RESPONSE Flammable. Flashback along vapor trail may occur. Vapor may explode if ignited in an enclosed area. Wear goggles and self-contained breathing apparatus. Extinguish with dry chemical, alcohol foam, or carbon dioxide. Water may be ineffective on fire. Cool exposed containers with water.

EXPOSURE RESPONSE Call for medical aid. Vapor. Irritating to eyes, nose, and throat. If inhaled, will cause nausea, headache, or dizziness. Move to fresh air. If breathing has stopped, give artificial respiration. If breathing is difficult, give oxygen. Liquid. Irritating to skin and eyes. Remove contaminated clothing and shoes. Flush affected areas with plenty of water. If in eyes, hold eyelids open and flush with plenty of water.

RESPONSE TO DISCHARGE Issue warning—high flammability. Mechanical containment. Chemical and physical treatment.

WATER RESPONSE Harmful to aquatic life in very low concentrations. Fouling to shoreline. May be dangerous if it enters water intakes. Notify local health and pollution control officials. Notify operators of nearby water intakes.

EXPOSURE SYMPTOMS Inhalation and ingestion: Irritates the mucous membranes, depresses the central nervous system, and is narcotic. Damage to kidney, liver, and lung can occur. Ingestion may irritate gastrointestinal tract. Eyes: irritation. Skin: irritation.

EXPOSURE TREATMENT Call a physician. Inhalation: remove from exposure. Administer oxygen if needed. Eyes: flush with water for at least 15 min. Skin: remove contaminated clothing and shoes. Wash with soap and water. Subsequent treatment is symptomatic and supportive in nature.

REFERENCES
1. *CHRIS* database.

2. *OHMTADS* database.

■ AMYL MERCAPTAN

CAS # 110–66–7

DOT # 1111

SYNONYMS 1-pentanethiol • 2-methyl-1-butanethiol • 2-methylbutyl mercaptan • amyl-hydrosulfide • amyl-sulfhydrate • amyl-thioalcohol • mercaptan-amilique (French) • n-amyl mercaptan • n-pentylmercaptan • n-thioamyl alcohol

• pentanethiol • pentyl mercaptan • thioamyl alcohol

MF $C_5H_{12}S$

MW 104.23

COLOR AND FORM Liquid; water-white to light-yellow liquid (mixture of isomers).

ODOR Strong, offensive; garlic odor.

ODOR THRESHOLD The odor can be detected in the air at 3.0×10^{-7} mg/m^3 [R9].

DENSITY 0.8421 at 20°C.

SURFACE TENSION 26.8 dynes/cm = 0.0268 N/m at 20°C.

BOILING POINT 126.6°C at 460 mm Hg; 99.5°C at 10 mm Hg.

MELTING POINT −75.7°C.

HEAT OF COMBUSTION −17,070 Btu/lb = 9,480 cal/g = -397×10^5 J/kg.

HEAT OF VAPORIZATION 171 Btu/lb = 94.9 cal/g = 3.97×10^5 J/kg.

EXPLOSIVE LIMITS AND POTENTIAL Vapor forms explosive mixtures with air; readily ignitable. [R10, 42].

FIREFIGHTING PROCEDURES Foam, carbon dioxide, dry chemical [R5]. Do not extinguish fire unless flow can be stopped. Use water in flooding quantities as fog. Solid streams of water may be ineffective. Cool all affected containers with flooding quantities of water. Apply water from as far a distance as possible. Use foam, dry chemical, or carbon dioxide [R7].

TOXIC COMBUSTION PRODUCTS Sulfur dioxide gas is formed [R8].

SOLUBILITY Soluble in alcohol, ether; practically insoluble in water.

VAPOR PRESSURE 13.8 mm Hg at 25°C.

OTHER CHEMICAL/PHYSICAL PROPERTIES Bulk density = 6.99 lb/gal (mixture of isomers); liquid molar volume = 0.124405 m^3/kmol; heat of formation = -1.0979×10^8 J/kmol; heat of fusion = 1.7531×10^7 J/kmol at melting point; heat of combustion = -3.5657×10^9 J/kmol; Mercaptan content at least 90.0%; initial bp: not below 104.0°C; final bp not above 130°C; wt/gallon: 6.99 lb (mixture of isomers).

PROTECTIVE EQUIPMENT Respirators may be necessary to prevent pulmonary irritation and systemic effects. At low concentration (less than 5 ppm) a chemical cartridge respirator with a half-mask facepiece and organic vapor cartridges can be used. At high concentration, supplied air respirators with full facepiece are necessary. (thiols) [R11, 2173].

PREVENTATIVE MEASURES Control at source of exposure may involve enclosure of the operation or the use of local exhaust ventilation. (thiols) [R11, 2172].

CLEANUP Spills of thiols can be neutralized with a household bleach solution and flushed with an abundant flow of water (thiols) [R11, 2172]. Cover with a weak solution of calcium hypochlorite (up to 15%). Transfer into a large beaker. After 12 hours, neutralize with 6 MHCl, or 6M NH$_4$OH, if necessary. Drain into a sewer with abundant water. [R10, 43].

DISPOSAL SRP: At the time of review, criteria for land treatment or burial (sanitary landfill) disposal practices are subject to significant revision. Prior to implementing land disposal of waste residue (including waste sludge), consult with environmental regulatory agencies for guidance on acceptable disposal practices.

COMMON USES Synthesis of organic sulfur compound; chief constituent of odorant used in gas lines to locate leaks (mixture of isomers) [R2]; synthetic flavoring ingredient [R1].

OCCUPATIONAL RECOMMENDATIONS

DOT *Fire or Explosion:* Flammable/combustible material; may be ignited by heat, sparks, or flames. Vapors may travel to a source of ignition and flash back. Container may explode in heat of fire. Vapor explosion hazard indoors, outdoors, or in sewers. Runoff to sewer may create fire or explosion hazard. Material may be transported hot [R3].

Health Hazards: May be poisonous if inhaled or absorbed through skin. Vapors may cause dizziness or suffocation. Contact may irritate or burn skin and eyes. Fire may produce irritating or poisonous gases. Runoff from fire control or dilution water may cause pollution [R3].

Emergency Action: Keep unnecessary people away; isolate hazard area and deny entry. Stay upwind; keep out of low areas. Positive-pressure self-contained breathing apparatus (SCBA) and structural firefighters' protective clothing will provide limited protection. Isolate for 1/2 mile in all directions if tank, rail car, or tank truck is involved in fire. Call emergency response telephone number on shipping paper first. If shipping paper not available or no answer, call CHEMTREC at 1–800–424–9300. If water pollution occurs, notify the appropriate authorities [R3].

Fire: Small Fires: Dry chemical, CO_2, water spray, or regular foam. Large Fires: Water spray, fog, or regular foam. Move container from fire area if you can do so without risk. Apply cooling water to sides of containers that are exposed to flames until well after fire is out. Stay away from ends of tanks. For massive fire in cargo area, use unmanned hose holder or monitor nozzles; if this is impossible, withdraw from area and let fire burn. Withdraw immediately in case of rising sound from venting safety device or any discoloration of tank due to fire [R3].

Spill or Leak: Shut off ignition sources; no flares, smoking, or flames in hazard area. Stop leak if you can do so without risk. Water spray may reduce vapor; but it may not prevent ignition in closed spaces. Small Spills: Take up with sand or other noncombustible absorbent material and place into containers for later disposal. Large Spills: Dike far ahead of liquid spill for later disposal [R3].

First Aid: Move victim to fresh air and call emergency medical care; if not breathing, give artificial respiration; if breathing is difficult, give oxygen. In case of contact with material, immediately flush eyes with running water for at least 15 minutes. Wash skin with soap and water. Remove and isolate contaminated clothing and shoes at the site [R3].

FIRE POTENTIAL Flammable, dangerous fire risk (mixture of isomers) [R4]. Dangerous when exposed to heat or flame [R5]. Technical butanethiol (containing 28% of propane and 7% of pentanethiols) is hypergolic with 96% acid [R6]. Oxidation of several thiols to the sulfonic acids by addition to stirred concentrated acid had been effected normally, but when 2 new batches of pentanethiol were used, flame was observed in the vapor phase a few seconds after addition. No unusual impurities were detected [R6].

GENERAL RESPONSE Shut off ignition sources. Call fire department. Stop discharge if possible. Keep people away. Isolate and remove discharged material. Notify local health and pollution control agencies.

FIRE RESPONSE Flammable. Poisonous gases are produced in fire. Flashback along vapor trail may occur. Vapor may explode if ignited in an enclosed area. Containers may explode in fire. Wear goggles and self-contained breathing apparatus. Extinguish with dry chemicals, foam, or carbon dioxide. Water may be ineffective on fire. Cool exposed containers with water.

EXPOSURE RESPONSE Call for medical aid. Vapor: move victim to fresh air. Liquid: irritating to skin and eyes. Remove contaminated clothing and shoes. Flush affected areas with plenty of water. If swallowed and victim is conscious have victim drink water or milk.

RESPONSE TO DISCHARGE Issue warning—high flammability. Mechanical containment. Should be removed. Provide chemical and physical treatment.

WATER RESPONSE Effect of low concentrations on aquatic life is unknown. Fouling to shoreline. May be dangerous if it enters water intakes. Notify local health and wildlife officials. Notify operators of nearby water intakes.

EXPOSURE SYMPTOMS Inhalation may cause nausea because of offensive odor. Contact with eyes or skin causes slight irritation. Ingestion may cause vomiting.

EXPOSURE TREATMENT Inhalation: remove to fresh air; apply artificial respiration if required. Eyes: wash with water; see a physician. Skin: wash with soap and water, Ingestion: induce vomiting if it does not occur spontaneously.

REFERENCES

1. *Fenaroli's Handbook of Flavor Ingredients*. Volume 2. Edited, translated, and revised by T. E. Furia and N. Bellanca. 2nd ed. Cleveland: The Chemical Rubber Co., 1975. 351.

2. Sax, N. I., and R. J. Lewis, Sr. (eds.). *Hawley's Condensed Chemical Dictionary*. 11th ed. New York: Van Nostrand Reinhold Co., 1987. 883.

3. U.S. Department of Transportation. *Emergency Response Guidebook 1993*. DOT P 5800.6. Washington, DC: U.S. Government Printing Office, 1993, p. G–27.

4. Hawley, G. G. *The Condensed Chemical Dictionary*. 10th ed. New York: Van Nostrand Reinhold Co., 1981. 784.

5. Sax, N. I. *Dangerous Properties of Industrial Materials*. 6th ed. New York, NY: Van Nostrand Reinhold, 1984. 2134.

6. Bretherick, L. *Handbook of Reactive Chemical Hazards*. 4th ed. Boston, MA: Butterworth-Heinemann Ltd., 1990 1151.

7. Association of American Railroads. *Emergency Handling of Hazardous Materials in Surface Transportation*. Washington, DC: Association of American Railroads, Bureau of Explosives, 1992. 75.

8. U.S. Coast Guard, Department of Transportation. *CHRIS—Hazardous Chemical Data*. Volume II. Washington, DC: U.S. Government Printing Office, 1984–5.

9. Clayton, G.D., and F.E. Clayton (eds.). *Patty's Industrial Hygiene and Toxicology*. Volumes 2A, 2B, 2C: *Toxicology*. 3rd ed. New York: John Wiley and Sons, 1981–1982. 2076.

10. ITII. *Toxic and Hazardous Industrial Chemicals Safety Manual*. Tokyo, Japan: The International Technical Information Institute, 1988.

11. International Labour Office. *Encyclopedia of Occupational Health and Safety*. Vols. I&II. Geneva, Switzerland: International Labour Office, 1983.

■ ANETHOLE

CAS # 104–46–1

SYNONYMS anethol • anise camphor • anisole, p-propenyl • benzene, 1-methoxy-4-(1-propenyl)- • FEMA-number-2086 • isoestragole • p-methoxy-beta-methylstyrene • 1-methoxy-4-propenylbenzene • 1-methoxy-4-(1-propenyl) benzene • 4-methoxypropenylbenzene • nauli "gum" • oil of aniseed • propene,1-(p-methoxyphenyl)- • 1-propene, 1-(4-methoxyphenyl)- • p-propenylanisole • p-1-propenylanisole • 4-propenylanisole • p-propenylphenyl methyl ether

MF $C_{10}H_{12}O$

MW 148.22

COLOR AND FORM White crystals.

ODOR Anise oil odor.

TASTE Sweet taste.

DENSITY 0.9878 at 20°C.

VISCOSITY 2.45×10^{-3} Pa.

BOILING POINT 234°C.

MELTING POINT 21.3°C.

SOLUBILITY 1:8 in 80% alcohol; 1:1 in 90% ethanol; almost water insol; miscible with chloroform and ether; water solubility = 1.110×10^{-1} g/L at 25°C.

VAPOR PRESSURE 5.45 Pa at 294°K.

OTHER CHEMICAL/PHYSICAL PROPERTIES Max absorption (isooctane): 258 nm (log e = 4.25); melting point 21.35°C; bp: 234.5°C at 763 mm Hg (trans-isomer); crystalline mass at 20–21°C, liquid above

23°C (trans-isomer); index of refraction: 1.55455 at 20°C/D; max absorption (ethanol): 253 nm (e = 18,500); melting point −22.5°C; boiling point 79–79.5°C at 23 mm Hg (cis-isomer); distillation range: 234–237°C; liquid molar volume = 0.1507 m^3/kmol; heat of fusion at melting point = 1.60×10^7 J/kmol.

DISPOSAL SRP: At the time of review, criteria for land treatment or burial (sanitary landfill) disposal practices are subject to significant revision. Prior to implementing land disposal of waste residue (including waste sludge), consult with environmental regulatory agencies for guidance on acceptable disposal practices.

COMMON USES Manufacture of anisaldehyde; in perfumery, soap, and dentifrices; sensitizer in bleaching colors in color photography; as embedding material in microscopy; pharmaceutic aid (flavor); vet: has been used as a carminative and as a flavoring agent [R1]; as flavor ingredient [R9, 34]; in licorice candies [R2]; insect attractant [R3]; formerly used as an expectorant [R4].

OCCUPATIONAL RECOMMENDATIONS None listed.

BIOCONCENTRATION Based upon an experimental water solubility of 111 mg/L (1), the BCF of anethole can be estimated to be approximately 43.4 from a regression-derived equation (2). This estimated BCF value suggests that bioconcentration in aquatic organisms is not expected to be an important fate process (SRC) [R5].

FIFRA REQUIREMENTS In 1988, Congress amended FIFRA to strengthen and accelerate EPA's reregistration program. The nine-year reregistration scheme mandated by "FIFRA '88" applies to each registered pesticide product containing an active ingredient initially registered before November 1, 1984. Pesticides for which EPA had not issued Registration Standards prior to the effective date of FIFRA '88 were divided into three lists based upon their potential for exposure and other factors, with List B being of highest concern and D of least. List: C; Case: p-Anethole; Case No.: 3018; Pesticide type: Rodenticide; Case Status: Awaiting

Data/Data in Review: OPP awaits data from the pesticide's producer(s) regarding its human health and/or environmental effects, or OPP has received and is reviewing such data, in order to reach a decision about the pesticide's eligibility for reregistration. Active Ingredient (AI): anethole; AI Status: The producer(s) of the pesticide has made commitments to conduct the studies and pay the fees required for reregistration, and is meeting those commitments in a timely manner. [R8].

FDA Anethole used as a synthetic flavoring substance or adjuvant in food for human consumption is generally recognized as safe when used in accordance with good manufacturing practice [R6]. Anethole used as a synthetic flavoring substance or adjuvant in animal drugs, feeds, and related products is generally recognized as safe when used in accordance with good manufacturing or feeding practice [R7].

REFERENCES

1. Budavari, S. (ed.). *The Merck Index—Encyclopedia of Chemicals, Drugs, and Biologicals*. Rahway, NJ: Merck, and Co., Inc., 1989. 102.

2. Lewis, R. J., Sr. (Ed.). *Hawley's Condensed Chemical Dictionary*. 12th ed. New York, NY: Van Nostrand Reinhold Co., 1993 78.

3. *Farm Chemicals Handbook 1994*. Willoughby, OH: Meister, 1994, p. C–23.

4. *Kirk–Othmer Encyclopedia of Chemical Technology*. 4th ed. Volume 1: New York, NY. John Wiley and Sons, 1991–present. 1063.

5. (1) Yalkowsky, S. H. *Arizona Database of Aqueous Solubilities*. Univ of AZ, College of Pharmacy (1989). (2) Lyman, W. J., et al., *Handbook of Chemical Property Estimation Methods*. Washington, DC: Amer Chemical Soc pp. 4–9, 5–4, 5–10, 7–4, 7–5, 15–15 to 15–32 (1990).

6. 21 CFR 182.60 (4/1/93).

7. 21 CFR 582.60 (4/1/93).

8. USEPA/OPP. *Status of Pesticides in*

Reregistration and Special Review. p.177 (Mar, 1992) EPA 700-R-92-004.

9. *Fenaroli's Handbook of Flavor Ingredients*. Volume 2. Edited, translated, and revised by T.E. Furia and N. Bellanca. 2nd ed. Cleveland: The Chemical Rubber Co., 1975.

■ ANILINE, p-NITRO

CAS # 100-01-6

DOT # 1661

SYNONYMS p-nitroaniline • 1-amino-4-nitrobenzene

MF $C_6H_6N_2O_2$

MW 138.14

STANDARD CODES NFPA—3, 1, 1; ICC—poison B, poison label, 200 lbs in an outside container; USCG—poison B, poison label; IATA.

INHALATION LIMIT 0.006 mg/m³.

DIRECT CONTACT Not irritating—absorbable.

GENERAL SENSATION Odorless. Produces cyanosis on absorption into body, either by inhalation of dust or from skin contact. Inhalation of vapor at ambient temperature. Not hazardous because of low vapor pressure; at elevated exposures vapor inhalation is hazardous.

PERSONAL SAFETY Neoprene suits and eye protection are required. Respiratory apparatus is recommended only for high concentrations and poorly ventilated areas.

ACUTE HAZARD LEVEL Highly toxic via all routes. Emits toxic vapors when heated to decomposition. Toxic to aquatic life.

CHRONIC HAZARD LEVEL Low in humans. Can cause liver damage with repeated exposures. Solid may remain on bottom and provide equilibrium amounts of dissolved compound for prolonged periods.

DEGREE OF HAZARD TO PUBLIC HEALTH Highly toxic via all routes. Emits toxic vapors when heated to decomposition.

OCCUPATIONAL RECOMMENDATIONS None listed.

ACTION LEVEL Notify fire and air authority. Enter from upwind. Remove ignition source. Suppress suspension of dusts.

ON-SITE RESTORATION Dredge solid. Use carbon on dissolved portion. Seek professional environmental engineering assistance through EPA's Environmental Response Team (ERT), Edison, NJ, 24-hour no. 908-548-8730.

BEACH OR SHORE RESTORATION Do not burn.

AVAILABILITY OF COUNTERMEASURE MATERIAL Carbon—water treatment plants, sugar refineries.

DISPOSAL METHOD Can be buried in toxic landfills or burned under controlled conditions. (1) Pour onto sodium bicarbonate or a sand-soda-ash mixture (90/10). Mix in heavy paper cartons and burn in incinerator. May augment fire with wood or paper. (2) Burn packages of no. 1 in incinerator with afterburner and alkaline scrubber. (3) Dissolve in flammable solvent and burn in incinerator of no. 2.

DISPOSAL NOTIFICATION Landfill or local air authority.

MAJOR WATER USE THREATENED Potable supply, fisheries.

PROBABLE LOCATION AND STATE Yellow solid. Will sink and dissolve very very slowly. Most will remain there on the bottom.

WATER CHEMISTRY Presence of mineral acids sponsors formation of soluble salts.

COLOR IN WATER Yellow.

DOT *Health Hazards:* Poisonous; may be fatal if inhaled, swallowed, or absorbed through skin. Contact may cause burns to skin and eyes. Runoff from fire control or dilution water may give off poisonous gases and cause water pollution. Fire may produce irritating or poisonous gases (nitroaniline) [R1].

Fire or Explosion: Some of these materials may burn, but none of them ignites readily. Container may explode violently in heat of fire (nitroaniline) [R1].

Emergency Action: Keep unnecessary people away; isolate hazard area and deny entry. Stay upwind, out of low areas, and ventilate closed spaces before entering. Self-contained breathing apparatus and chemical protective clothing which is specifically recommended by the shipper or producer may be worn but they do not provide thermal protection unless it is stated by the clothing manufacturer. Structural firefighter's protective clothing is not effective with these materials. Remove and isolate contaminated clothing at the site. Call CHEMTREC at 1–800–424–9300 as soon as possible, especially if there is no local hazardous materials team available (nitroaniline) [R1].

Fire: Small Fires: Dry chemical, CO_2, halon, water spray or standard foam. Large Fires: Water spray, fog, or standard foam is recommended. Move container from fire area if you can do it without risk. Fight fire from maximum distance. Stay away from ends of tanks. Dike fire control water for later disposal; do not scatter the material (nitroaniline) [R1].

Spill or Leak: Do not touch spilled material; stop leak if you can do it without risk. Use water spray to reduce vapors. Small Spills: Take up with sand or other noncombustible absorbent material and place into containers for later disposal. Small Dry Spills: With clean shovel place material into clean, dry container and cover; move containers from spill area. Large Spills: Dike far ahead of liquid spill for later disposal (nitroaniline) [R1].

First Aid: Move victim to fresh air and call emergency medical care; if not breathing, give artificial respiration; if breathing is difficult, give oxygen. In case of contact with material, immediately flush skin or eyes with running water for at least 15 minutes. Speed in removing material from skin is of extreme importance. Remove and isolate contaminated clothing and shoes at the site. Keep victim quiet and maintain normal body temperature. Effects may be delayed; keep victim under observation. (nitroaniline) [R1].

GENERAL RESPONSE Avoid contact with solid and dust. Keep people away. Wear dust respirator. Stop discharge if possible. Call fire department. Isolate and remove discharged material. Notify local health and pollution control agencies.

FIRE RESPONSE Combustible. Poisonous gases may be produced in fire. Wear goggles and self-contained breathing apparatus. Extinguish with water, dry chemicals, foam, or carbon dioxide. Cool exposed containers with water.

EXPOSURE RESPONSE Call for medical aid. Dust poisonous if inhaled. If inhaled will cause headache, coughing, difficult breathing, or loss of consciousness. If in eyes, hold eyelids open and flush with plenty of water. If breathing has stopped, give artificial respiration. If breathing is difficult, give oxygen. Solid irritating to skin and eyes. If swallowed will cause headache, coughing, or loss of consciousness. Remove contaminated clothing and shoes. Flush affected areas with plenty of water. If in eyes, hold eyelids open and flush with plenty of water. If swallowed and victim is conscious, have victim drink water or milk and have victim induce vomiting. If swallowed and victim is unconscious or having convulsions, do nothing except keep victim warm.

RESPONSE TO DISCHARGE Issue warning—poison, water contaminant. Restrict access. Should be removed. Chemical and physical treatment.

WATER RESPONSE Harmful to aquatic life in very low concentrations. May be dangerous if it enters water intakes. Notify local health and wildlife officials. Notify operators of nearby water intakes.

EXPOSURE SYMPTOMS Inhalation or ingestion causes headache, drowsiness, shortness of breath, nausea, methemoglobinemia, and unconsciousness; fingernails, lips, and ears become bluish; prolonged and excessive exposures may also cause liver damage. Contact with eyes causes irritation and possible corneal damage. Contact with skin causes irritation; continued exposure may cause same symptoms as inhalation or ingestion.

EXPOSURE TREATMENT Inhalation: remove victim to fresh air; administer oxygen if required; get medical attention. Ingestion: induce vomiting; get medical attention. Eyes: flush with water for at least 15 min. Skin: flush with water, wash with soap and water; be sure that no compound remains in the hair or under the fingernails.

REFERENCES
1. Department of Transportation. *Emergency Response Guidebook 1987*. DOT P 5800.4. Washington, DC: U.S. Government Printing Office, 1987, p. G-55.

■ ANTIMONY-122

CAS # 14374-79-9

SYNONYMS radioactive antimony isotope 122 • ^{122}Sb • radioactive antimony

MF Sb

MW 122

EXPLOSIVENESS Nonfissionable.

MELTING POINT 630.5°C.

BOILING POINT 138°C.

SPECIFIC GRAVITY 6,684.

PERSISTENCE 2.80-day radioactive half-life; 38-day biological half-life in total body; 100-day biological half-life in bone and lung, 38 days in liver, and 4 days in thyroid.

BINARY REACTANTS Bromine; bromine trifluoride; bromoazide; chlorine; chlorine monoxide; fluorine; iodine; potassium; nitrate; potassium permanganate; potassium peroxide; sodium nitrate; sodium peroxide.

STANDARD CODES Labels for packages of radioactive materials must be of diamond shape, in colors specified, with each side at least 4 inches long. Printing must be in black inside a black line border measuring at least 3 1/2 inches on each side. "Radioactive white-I" Label. Label must be white. The single vertical bar on the lower half of the label must be bright red. Labels must be applied on two opposite sides of each package having a dose rate not exceeding 0.5 millirem per hour at any point on the external surface of the package. Not authorized for fissile class II packages. "Radioactive yellow II" label—the upper half of the label must be bright yellow and the bottom half must be white. The two vertical bars on the lower half of the label must be bright red. Labels must be applied on two opposite sides of: (a) each package having a dose rate not exceeding 10 millirem per hour at any point on the external surface of the package and not exceeding 0.5 millirem per hour at 3 feet from the external surface of the package; or (b) each package for which the transport index does not exceed 0.5 at any time during transportation. "Radioactive yellow-III" label—the upper half of the label must be bright yellow and the bottom half must be white. The three vertical bars on the lower half of the label must be bright red. Labels must be applied on two opposite sides of: (a) each package having a surface dose rate exceeding 10 millirem per hour; (b) each fissile class III package; (c) each package containing a large quantity of radioactive material as: 20 curies of group I radionuclides, 20 curies of group II radionuclides, 200 curies of group III radionuclides, 200 curies of group IV radionuclides, 5,000 curies of group V radionuclides, 50,000 curies of group VI radionuclides, 500,000 curies of group VII radionuclides, or 5,000 curies of special form radioactive materials, or (d) each package transported under a special permit issued in response to a petition.

MAJOR SPECIES THREATENED All species.

INHALATION LIMIT 0.6×10^{-8} mg/m^3.

DIRECT CONTACT Refer to specific compound.

PERSONAL SAFETY Beta and gamma radiation. Do not allow contaminated water to come in contact with skin or personal clothing. Wear waterproof protection. If radioactivity is also airborne, a mask with air filter may be required.

COMMON USES Tracer element; neutron source that may be reactivated in a nuclear reactor. [R2].

ACUTE HAZARD LEVEL Soluble: organ mpb in 40-hour week, 168-hour week of total body μC/cc μC/cc reference (μCi) mpc/w mpc/a mpc/w mpc/a gi / / $8.\times10^4$ / $2.\times10^7$ / $3.\times10^4$ / $6.\times10^8$ total body/ 20 / .3 / $4.\times10^6$ / .1 / 10^{-6} lung / 40 / .5 / $6.\times10^6$ / .2 / $2.\times10^6$ bone / 40 / .5 / $6.\times10^6$ / .2 / $2.\times10^6$ liver / 10^{-3}/ 10 / $2.\times10^4$/ 4 / $5.\times10^5$ thyroid / $3.\times10^3$ / 40 / $4.\times10^4$ / 10 / $2.\times10^4$ insoluble: gi / / $8.\times10^4$ / 10^{-7} / $3.\times10^4$ / $5.\times10^8$ lung / / / $4.\times10^7$ / / 10^{-7}.

DEGREE OF HAZARD TO PUBLIC HEALTH High, due to exposure to radiation; high, due to chemical toxicity.

OCCUPATIONAL RECOMMENDATIONS None listed.

ACTION LEVEL Notify local air authorities and the Nuclear Regulatory Commission. Do not enter area without radiation monitoring device.

ON-SITE RESTORATION 1. At neutral pH, use a coagulant such as aluminum sulfate. 2. At high pH, use anion exchange resin. 3. At low pH, use cation exchange resin. Seek professional environmental engineering assistance through EPA's Environmental Response Team (ERT), Edison, NJ; 24-hour phone number, 908-548-8730.

BEACH OR SHORE RESTORATION Remove the sand and bury at authorized burial site.

AVAILABILITY OF COUNTERMEASURE MATERIAL Coagulants such as aluminum sulfate or ferric sulfate—water-treatment plants; anion and cation exchange resins—water softening and conditioning suppliers, water-treatment plants.

DISPOSAL METHOD Burial at an authorized radioactive disposal site.

DISPOSAL NOTIFICATION Contact the Nuclear Regulatory Commission.

INDUSTRIAL FOULING POTENTIAL The safe radiation levels are below industrial fouling potential levels.

MAJOR WATER USE THREATENED All water uses.

PROBABLE LOCATION AND STATE 1. Pure element is silver-white metal. 2. Most compounds dissociate in water or are only very slightly soluble, $SbCl_3$ is quite soluble, will sink or dissolve and form precipitate.

FOOD-CHAIN CONCENTRATION POTENTIAL Can be concentrated by certain forms of marine life to over 300 times its concentration in the surrounding waters. Positive.

REFERENCES

1. EPA's *OHMTADS* Database.

2. Lewis, R.J., Sr. *Hawley's Condensed Chemical Dictionary*, 12th Ed. New York: Van Nostrand Reinhold Co., 1993.

■ ARAMITE

CAS # 140-57-8

SYNONYMS 2-(p-butylphenoxy)-1-methylethyl-2-chloroethyl sulfite • 2-(p-butylphenoxy) isopropyl-2-chloroethyl sulfite • 2-(p-t-butylphenoxy)-1-methylethyl sulphite of 2-chloroethanol • 2-(p-tert-butylphenoxy) isopropyl 2-chloroethyl sulfite • 2-(4-t-butylphenoxy) isopropyl-2-chloroethyl sulfite • 2-(p-t-butylphenoxy) isopropyl-2'-chloroethyl sulphite • 2-(p-t-butylphenoxy)-1-methylethyl 2-chloroethyl ester of sulphurous acid • 2-(p-butylphenoxy)-1-methylethyl 2-chloroethyl sulfite • 2-(p-t-butylphenoxy)-1-methylethyl 2'-chloroethyl sulphite • 2-chloroethyl 1-methyl-2-(p-t-butylphenoxy) ethyl sulphate • 2-chloroethyl sulphite of 1-(p-t-butylphenoxy)-2-propanol • 2-propanol, 1-(p-t-butylphenoxy)-, 2-chloroethyl sulfite • 88r • 88-r • aracide • aramit • aramite-15W •

aramiteararamite-15W • aratron • beta-chloroethyl-beta-(p-t-butylphenoxy)-alpha-methylethyl sulphite • beta-chloroethyl-beta'-(p-t-butylphenoxy)-alpha'-methylethyl sulfite • butylphenoxyisopropyl-chloroethyl-sulfite • ces • compound-88r • ent-16,519- • ethanol, 2-chloro-, 2-(p-t-butylphenoxy)-1-methylethyl sulfite • ethanol, 2-chloro-, ester with 2-(p-tert-butylphenoxy)-1-methylethyl sulfite • niagaramite • ortho-mite • sulfurous acid, 2-(p-t-butylphenoxy)-1-methylethyl-2-chloroethyl ester • sulfurous acid, 2-(p-tert-butylphenoxy)-1-methylethyl 2-chloroethyl ester • sulfurous acid, 2-chloroethyl 2-(4-(1,1-dimethylethyl) phenoxy)-1-methylethyl ester • beta-chloroethyl-beta-(p-tert-butylphenoxy)-alpha-methylethyl sulphite • 2-chloroethyl sulphite of 1-(p-tert-butylphenoxy)-2-propanol • ester of 2-chloroethanol with 2-(p-tert-butylphenoxy)-methyl sulphite • 2-(4-tert-butylphenoxy)-1-methylethyl 2-chloroethyl sulfite • 2-(p-tert-butylphenoxy)-1-methylethyl 2-chloroethyl sulfite

MF $C_{15}H_{23}ClO_4S$

MW 334.89

COLOR AND FORM Colorless liquid; clear light-colored oil.

DENSITY 1.145 at 20°C.

BOILING POINT 175°C at 0.1 mm Hg.

MELTING POINT −31.7°C.

SOLUBILITY Practically insoluble in water; miscible with many organic solvents; solubility in petroleum oils decr rapidly with decr temperatures; >10% in ethanol; soluble in dimethyl ketone; >10% in benzene; >10% in acetone; >10% in ether; water solubility of 0.1 ppm.

VAPOR PRESSURE <10 mm Hg at 25°C.

CORROSIVITY Noncorrosive.

STABILITY AND SHELF LIFE Stability in solution decreases in strong acid or alkali [R3].

OTHER CHEMICAL/PHYSICAL PROPERTIES Density: 1.145–1.162 (technical product); dark-amber liquid (technical product); incompatible with alkaline materials such as lime or Bordeaux mixture.

PROTECTIVE EQUIPMENT Dispensers of liquid detergent should be available. Safety pipettes should be used for all pipetting. In animal laboratory, personnel should wear protective suits (preferably disposable, one-piece, and close-fitting at ankles and wrists), gloves, hair covering, and overshoes. In chemical laboratory, gloves and gowns should always be worn; however, gloves should not be assumed to provide full protection. Carefully fitted masks or respirators may be necessary when working with particulates or gases, and disposable plastic aprons might provide addnl protection. Gowns should be of distinctive color, as a reminder that they are not to be worn outside the laboratory (chemical carcinogens) [R8, 8].

PREVENTATIVE MEASURES Smoking, drinking, eating, storage of food, or of food and beverage containers, or utensils, and the application of cosmetics should be prohibited in any laboratory. All personnel should remove gloves, if worn, after completion of procedures in which carcinogens have been used. They should wash hands, preferably using dispensers of liquid detergent, and rinse thoroughly. Consideration should be given to appropriate methods for cleaning the skin, depending on nature of the contaminant. No standard procedure can be recommended, but the use of organic solvents should be avoided. Safety pipettes should be used for all pipetting. (chemical carcinogens) [R8, 8].

CLEANUP A high-efficiency particulate arrestor (HEPA) or charcoal filters can be used to minimize amount of carcinogen in exhausted air ventilated safety cabinets, lab hoods, glove boxes, or animal rooms. Filter housing that is designed so that used filters can be transferred into plastic bag without contaminating maintenance staff is avail commercially. Filters should be placed in plastic bags immediately after removal. The plastic bag should be sealed immediately. The sealed bag should be labelled properly. Waste liquids should be placed or collected in proper containers for disposal. The lid should be secured and the bottles properly labelled. Once filled, bottles should be placed in plastic bag, so that outer surface is not contaminated. The plastic bag should also be sealed and labelled. Broken glassware should be decontaminated by solvent extraction, by chemical destruction, or in specially designed incinerators. (chemical carcinogens) [R8, 15].

DISPOSAL Mix aramite with excess CaO (calcium oxide) or NaOH (sodium hydroxide) and sand or other adsorbent in a pit or trench at least 0.5 m deep in a clay soil. NaOH (or Na_2CO_3, sodium carbonate) can also be added to the mixture to help speed the reactions when calcium oxide is used as the main alkali. The amount of calcium oxide or sodium hydroxide to use depends on the amount of pesticide to be disposed of and, to some extent, the concentration of active ingredient in the pesticide, and the actual chemical nature of the active ingredient. A practical guideline, in the absence of specific directions, is to use an approximate volume or wt of alkali from one-half of to the same as that of the pesticide. For dilute formulations, such as a 1% solution, or dust, the amount of calcium oxide or sodium hydroxide can be reduced by one-half. For very concentrated pesticides (over 80% active ingredient) the amount of calcium oxide or sodium hydroxide can be doubled, but the concentration should be mixed first with water (or soapy water) before reaction with the alkali. For safety, a preliminary test should be made in which very small amounts of the pesticide and alkali are mixed and observed briefly to make sure they do not react too vigorously. Sizeable quantities of pesticides can be disposed of in several smaller batches, rather than all at once, for added safety. Recommendable methods: Incineration, hydrolysis, landfill, and discharge to sewer. Peer review: Small amount when hydrolyzed can be flushed to sewer or landfilled. Large amount should be incinerated in a unit with effluent gas scrubbing. (peer-review conclusions of an IRPTC expert consultation (May 1985)) [R4].

COMMON USES Antimicrobicide agent (former use) [R9, 96]; miticide (former use) [R2].

OCCUPATIONAL RECOMMENDATIONS None listed.

NON-HUMAN TOXICITY LD_{50} rat oral 3,900 mg/kg [R10, 122]; LD_{50} guinea pig, oral 3.9 g/kg [R1]; LD_{50} mouse oral 2.0 g/kg [R1].

ECOTOXICITY LC_{50} *Coturnix japonica* (Japanese quail, 14-day-old), oral >5,000 ppm (5-day *ad libitum* diet) [R6].

IARC SUMMARY No data are available for humans. Sufficient evidence of carcinogenicity in animals. Overall evaluation: Group 2B: The agent is possibly carcinogenic to humans [R5].

BIOCONCENTRATION Based on a reported water solubility of 0.1 ppm (1), the BCF for aramite can be estimated to be 2,265 from a recommended regression-derived equation (2,SRC). This BCF value suggests a potential for significant bioconcentration in aquatic organisms (SRC) [R7].

REFERENCES

1. Spencer, E. Y. *Guide to the Chemicals Used in Crop Protection*. 6th ed. Publication 1093, Research Institute, Agriculture Canada, Ottawa, Canada: Information Canada, 1973. 20.

2. *The Merck Index*. 10th ed. Rahway, NJ: Merck Co., Inc., 1983. 112.

3. Sunshine, I. (ed.). *CRC Handbook of Analytical Toxicology*. Cleveland: The Chemical Rubber Co., 1969. 499.

4. United Nations. *Treatment and Disposal Methods for Waste Chemicals* (IRPTC File). Data Profile Series No. 5. Geneva, Switzerland: United Nations Environmental Programme, Dec. 1985. 287.

5. IARC. *Monographs on the Evaluation of the Carcinogenic Risk of Chemicals to Man*. Geneva: World Health Organization, International Agency for Research on Cancer, 1972–present (multivolume work), p. S7 57 (1987).

6. Hill, E. F., and M. B. Camardese. *Lethal Dietary Toxicities of Environmental Contaminants and Pesticides to Coturnix*. Fish and Wildlife Technical Report 2.

Washington, DC: United States Department of Interior, Fish and Wildlife Service, 1986. 27.

7. (1) Naishtein, S. Y. *Vopr Gigieny Naselen Mest*, Kiev, SB 5: 34-7 (1964). (2) Lyman, W. J., et al., *Handbook of Chemical Property Estimation Methods*. New York: McGraw Hill, p. 5-10 (1982).

8. Montesano, R., H. Bartsch, E.Boyland, G. Della Porta, L. Fishbein, R. A. Griesemer, A.B. Swan, L. Tomatis, and W. Davis (eds.). *Handling Chemical Carcinogens in the Laboratory: Problems of Safety*. IARC Scientific Publications No. 33. Lyon, France: International Agency for Research on Cancer, 1979.

9. Sax, N.I., and R.J. Lewis, Sr. (eds.). *Hawley's Condensed Chemical Dictionary*. 11th ed. New York: Van Nostrand Reinhold Co., 1987.

10. Sittig, M. (ed.). *Pesticide Manufacturing and Toxic Materials Control Encyclopedia*. Park Ridge, NJ: Noyes Data Corporation. 1980.

■ ARSENIC ACID

CAS # 1327-52-2

SIC CODES 2879; 2819; 3241; 2893

SYNONYMS arsenic pentoxide • arsenic oxide • orthoarsenic acid

MF AsH_3O_4

MW 141.95

STANDARD CODES USCG—poison B, poison label; NFPA-704M system health hazard 3; ICC—poison B, poison label, 200 lbs in an outside container.

PERSISTENCE Arsenic will persist. May cycle due to methylation.

MAJOR SPECIES THREATENED Beans and cucumbers are extremely sensitive to arsenic; most life forms.

INHALATION LIMIT 0.5 mg/m³.

DIRECT CONTACT Severe hazard on ingestion; skin—moderate hazard as an irritant or allergen.

GENERAL SENSATION Acute poisoning from ingestion shows marked irritation of the stomach and intestines with nausea, vomiting, and diarrhea. Vomitus and stools may be bloody. Patient goes into collapse and shock followed by coma and death in severe cases. Chronic poisoning may cause loss of appetite, cramps, nausea, constipation or diarrhea, liver damage, jaundice, disturbances of the blood, kidneys, and nervous system, and skin abnormalities.

PERSONAL SAFETY Emits highly toxic vapors when heated. Wear self-contained breathing apparatus if fire exists.

ACUTE HAZARD LEVEL 130 mg of arsenic ingested proves fatal in humans. Small doses may become fatal in time because arsenic accumulates in the body. Sublethal doses can affect liver. 10 ppm in water—acute hazard. Threshold concentration for fresh and salt water fish, 1 ppm [R1].

CHRONIC HAZARD LEVEL Marine waters should not exceed 0.01/96-hour LC_{50} [R3]. 1.4 ppm affects reproduction in daphnids over 3-week period [R4]. Application of 5 mg/L as in drinking water led to rapid die-off of rat and mouse breeding colonies [R2].

DEGREE OF HAZARD TO PUBLIC HEALTH Highly toxic when ingested, may be carcinogenic in water taken for long time, poisonous levels to humans have been reported as 0.21 ppm, 0.3-1 ppm, and 0.4-10 ppm.

OCCUPATIONAL RECOMMENDATIONS None listed.

ACTION LEVEL Clean immediate area. See tads file on arsenic also.

ON-SITE RESTORATION Alum floc ties up in insoluble form. Anion exchanges will pick up arsenates. Seek professional environmental engineering assistance through EPA's Environmental Response Team (ERT), Edison, NJ, 24-hour no., 908-548-8730.

AVAILABILITY OF COUNTERMEASURE MATERIAL In acute poisoning contents of intestinal tract should be evacuated. Bal is highly effective in both acute and chronic poisoning; alum—water treatment plant anion exchanges—water softener suppliers.

DISPOSAL METHOD Dissolve in minimum hydrochloric acid (concentrated reagent). Filter if necessary. Dilute with water until white precipitates form. Add just enough 6M hydrochloric acid to redissolve. Saturate with hydrogen sulfide. Filter, wash the precipitate, dry, package, and ship to the supplier.

INDUSTRIAL FOULING POTENTIAL Should not be present in food-processing water.

MAJOR WATER USE THREATENED Potable supply; fisheries; irrigation.

PROBABLE LOCATION AND STATE White crystals, hydroscopic. Is converted to As_2O_5 by heating above 300°C. Exists only in solution, very weak acid. Will dissolve, and some will precipitate as calcium arsenate.

WATER CHEMISTRY Arsenic exists in nature in +5, +3, −3 oxidation states. Based on available thermodynamic data, arsenate (+5) is the stable state in aerated water, but in very reducing sediments As (0) and AsH_3 (−3) can exist. In more moderately reducing sediments arsenite can exist. An EH—pH diagram has recently been constructed for the system involving arsenic, oxygen, water, and sulfur. Besides the inorganic forms of arsenic, organic derivatives numbering in the thousands have been prepared, some of which may contribute to the occurrence of arsenic in natural systems. The hydrolytic behavior of As (V) is simple since it forms an oxyacid, H_3AsO_4, whose properties resemble very closely those of H_3PO_4. Arsenate forms insoluble salts with many cations, and solubility products, probably good to ± 0.5 log unit, were reported for sixteen arsenates by Chukhlantsev. Arsenate is expected to be enriched in phosphate minerals by exchange with the phosphate ion, as was pointed out by Ferguson and Gavis. Arsenite[+3] is found in about equal amounts with arsenate in ocean water in dilute solutions As (III) exists as monomeric species believed to be As(OH)$_3$ and AsO(OH)$_2$ [−1]. There is some evidence for the existence of more basic anions, AsO_2 (OH)[−2] and AsO_3 [−3] in very concentrated base. The chemistry and ionization behavior of As (III) resemble more closely that of B (III) than D (III). The ionization constant for arsenious acid is known as a function of KCl concentration from the work of Antikainen and Rossi and as a function of temperature from the work of Antikainen and Tevanen. AsO_6 has a solubility of 0.103 M (As) at 25°C, and there is evidence for polymer formation in saturated solutions in the pH range 9 to 10; possible species are HAs_2O_4 [−1] and $H_2 As_3O_6$ [−1] (or As_2 (OH) (1 (mannitol, mannose, catechol, and pyrogallol) may be an important process in natural systems. A solution at pH 9 contains $10^{[−3]}$ M of the polyol.

COLOR IN WATER Colorless.

FOOD CHAIN CONCENTRATION POTENTIAL Modern—positive.

REFERENCES

1. Todd, D. K., *The Water Encyclopedia*. Maple Press, 1970.

2. Serkowitz, J. B., G. R. Schimke, and V. R. Valeri, *Water Pollution Potential of Manufactured Products*. Environmental Protection Agency, EPA–R2–73–179d, April 1973.

3. *Proposed Criteria for Water Quality*, Volume I, U.S. Environmental Protection Agency. October 1973.

4. Environmental Protection Agency, *Report of the Pesticide Technical Committee to the Lake Michigan Enforcement Conference on Selected Trace Metals*, NTIS PB-220 361, September 1972.

■ ARSENIC ACID, CALCIUM SALT (2:3)

CAS # 7778–44–1

DOT # 1573

SIC CODES 2842

SYNONYMS arseniate de calcium

(French) • arsenic acid (H_3AsO_4), calcium salt (2:3) • arsenic acid, calcium salt (2:3) • calcium arsenate ($Ca_3(AsO_4)_2$) • calcium orthoarsenate • calcium o-arsenate • tricalcium arsenate • tricalcium ortho-arsenate • tricalcium o-arsenate • Tricalciumarsenat (German) • chip cal • fencal • spra cal • chip cal granular • cucumber dust • flac • kalo • Kalziumarseniat (German) • kilmag • pencal • AI3-24838 • calciumarsenat • Caswell no. 137 • EPA pesticide chemical code 013501 • spracal

MF $As_2O_8•3Ca$

MW 398.08

COLOR AND FORM Colorless amorphous powder; white flocculent powder.

ODOR Odorless.

DENSITY 3.620.

MELTING POINT 1,455°C.

FIREFIGHTING PROCEDURES Extinguish fire using agent suitable for type of surrounding fire. Use water in flooding quantities as fog. Use foam, carbon dioxide, or dry chemical [R8]. Wear positive-pressure self-contained breathing apparatus when fighting fires involving calcium arsenate. [R8].

TOXIC COMBUSTION PRODUCTS Poisonous gases may be produced in fire. [R4].

SOLUBILITY 0.013 g/100 cc of water at 25°C; insoluble in organic solvents.

VAPOR PRESSURE 0 mm Hg.

CORROSIVITY Slight corrosive action on metals.

STABILITY AND SHELF LIFE Moisture and carbon dioxide cause slow decomposition to calcium carbonate and (phytotoxic) dicalcium hydrogen arsenate. In the presence of acids, water sol, strongly phytotoxic arsenic acid is produced. [R3].

OTHER CHEMICAL/PHYSICAL PROPERTIES Decomposition on heating.

PROTECTIVE EQUIPMENT Dust mask; goggles or face shield; protective gloves [R4].

PREVENTATIVE MEASURES Contact lenses should not be worn when working with this chemical [R10].

CLEANUP Both arsenate and arsenite can be removed from the water column by coprecipitation or adsorption onto iron oxides. Arsenate species can also be removed by adsorption onto aluminum hydroxide and clays, whereas arsenite is readily adsorbed onto metal sulfides. (arsenate and arsenite) [R11].

DISPOSAL SRP: At the time of review, criteria for land treatment or burial (sanitary landfill) disposal practices are subject to significant revision. Prior to implementing land disposal of waste residue (including waste sludge), consult with environmental regulatory agencies for guidance on acceptable disposal practices.

COMMON USES Insecticide; as molluscicide [R2]; used for pre-emergence treatment of turf and lawns to control crabgrass, bluegrass, chickweed, and certain soil insects, including Japanese beetle grubs [R6]; at one time widely used against leaf-eating insects, and still used on tobacco [R1]; herbicide—e.g., to control crabgrass in turf and lawns; insecticide—e.g., against Japanese beetle grubs; fortifier of Bordeaux mixture; insecticide for cotton (former use) [R7]; formerly used widely for boll weevil control on cotton, but USDA has cancelled this use [R6]. For control of *Poa annua*, crabgrass, and other annual grasses and weeds in turf. Also controls insect grubs. [R5].

STANDARD CODES EPA 311; ICC—(solid) class B poison, poison label, 200 lbs in an outside container; USCG—poison B, poison label; NFPA-3, 0, 0; IATA—poison B, poison label, 25 kg passenger, 95 kg cargo; Superfund designated (hazardous substances) list.

PERSISTENCE Will hydrolyze to arsenic acid over period of time.

MAJOR SPECIES THREATENED All species, especially molluscs.

DIRECT CONTACT Allergen, irritant to skin.

GENERAL SENSATION Ingestion results in

stomach irritation, vomiting, diarrhea, and intestinal discomfort.

PERSONAL SAFETY Wear full protective clothing and self-contained breathing apparatus.

ACUTE HAZARD LEVEL Highly toxic. Arsenic is toxic to Eurasion watermilfoil at 2.9–4.1 ppm as AsO_2 [R23]; allergen and irritant. Emits toxic vapors in disaster circumstances. Threshold concentration for fresh- and salt-water fish, 1 ppm As [R21].

CHRONIC HAZARD LEVEL Chronic allergen and irritant. 1.4 ppm As affects daphnid reproduction over a 3-week period [R24]; arsenic compounds are generally moderate toxicants with chronic dosing. Application of 5 mg/L in drinking water led to rapid die-off of rat and mouse breeding; arsenic itself is an accumulative poison. Breeding colonies [R22].

DEGREE OF HAZARD TO PUBLIC HEALTH Highly toxic. Allergen and irritant. Emits toxic vapors if contacted with acid or heated to decomposition.

OCCUPATIONAL RECOMMENDATIONS
OSHA 8-hr time-weighted average: 5 μg/m³ (inorganic arsenic) [R15].

NIOSH NIOSH recommends that the substance be treated as a potential human carcinogen with a 15-minute TWA: 2 μg As/m³. (inorganic arsenic) [R16].

THRESHOLD LIMIT 8-hr time-weighted average (TWA) 0.01 mg/m³ (1993) (arsenic, elemental, and inorganic compound (except arsine), as As) [R17, 13].

ACTION LEVEL Notify local air authority. Enter from upwind side. Suppress dust suspension in atmosphere.

ON-SITE RESTORATION Can precipitate arsenate out with iron. Alum floc ties up arsenate in insoluble form. Anion exchanges work well on arsenate. Seek professional environmental engineering assistance through EPA's Environmental Response Team (ERT), Edison, NJ, 24-hour no. 908–548–8730.

BEACH OR SHORE RESTORATION 1, As threshold [R21].

AVAILABILITY OF COUNTERMEASURE MATERIAL Iron salts—water treatment plants, photography shops; alum-water treatment plants; anion exchangers—water softener suppliers.

DISPOSAL METHOD Dissolve in minimum concentrated hydrocloric acid. Dilute with water until white precipitates form. Add just enough $6M$ HCl to redissolve. Saturate with H_2S. Filter, wash, dry, and ship to supplier.

INDUSTRIAL FOULING POTENTIAL May render water unacceptable for food processing uses.

MAJOR WATER USE THREATENED All uses.

PROBABLE LOCATION AND STATE White powder. Will sink or possibly form a white dispersion; will then slowly hydrolyze to arsenic acid.

WATER CHEMISTRY SOLUBILITY Increases with acidity.

FOOD-CHAIN CONCENTRATION POTENTIAL Not apparent for arsenates. Negative.

NON-HUMAN TOXICITY LD_{50} rat female oral 298 mg/kg [R2]; LD_{50} mouse oral 794 mg/kg [R9, 317].

BIOLOGICAL HALF-LIFE The half-life for the urinary excretion of arsenic (in man) is 3 to 5 days. (inorganic arsenic compounds) [R12, 1615].

BIODEGRADATION Microorganisms in sediments that contain arsenic convert arsenic into dimethyl arsine. A variety of arsenicals are converted into dimethyl arsine by methanobacteria. Methyl cobalamine serves as methyl donor. (inorg arsenic compounds) [R13].

BIOCONCENTRATION Biomagnification through the food chain does not occur with arsenicals. Brown algae contain about 30 ppm arsenic, and mollusks contain about 0.005 ppm As. Even plants grown in soils contaminated with As do not show higher concentrations of As than do plants grown on uncontaminated soil. (arsenical compounds) [R14].

ATMOSPHERIC STANDARDS Inorganic arsenic, pursuant to section 112 of the Clean Air Act, has been designated as a hazardous air pollutant. (inorganic arsenic) [R18].

RCRA REQUIREMENTS A solid waste containing calcium arsenate may become characterized as a hazardous waste when subjected to the toxicant extraction procedure listed in 40 CFR 261.24, and, if so characterized, must be managed as a hazardous waste [R19].

FIFRA REQUIREMENTS Under section 6 (c) (2) (b) of FIFRA, the Data Call-In Program, existing registrants are required to provide EPA with needed studies. For calcium arsenate, letters were issued 9/28/84 and 1/28/86 to indicate existing chronic data gaps [R20].

GENERAL RESPONSE Avoid contact with solid and dust. Keep people away. Wear a dust respirator. Stop discharge if possible. Isolate and remove discharged material. Notify local health and pollution control agencies.

FIRE RESPONSE Not flammable. Poisonous gases may be produced in fire. Wear goggles and self-contained breathing apparatus.

EXPOSURE RESPONSE Call for medical aid. Dust is poisonous if inhaled. If inhaled will cause coughing or difficult breathing. If breathing is difficult, give oxygen. Solid is poisonous if swallowed. If swallowed will cause nausea and vomiting. Remove contaminated clothing and shoes. Flush affected areas with plenty of water. If in eyes, hold eyelids open, and flush with plenty of water. If swallowed and victim is conscious, have victim drink water or milk, and have victim induce vomiting. If swallowed and victim is unconscious or having convulsions, do nothing except keep victim warm.

RESPONSE TO DISCHARGE Issue warning—poison, water contaminant. Restrict access. Should be removed. Provide chemical and physical treatment.

WATER RESPONSE Harmful to aquatic life in very low concentrations. May be dangerous if it enters water intakes. Notify local health and wildlife officials. Notify operators of nearby water intakes.

EXPOSURE SYMPTOMS Inhalation causes respiratory irritation. Ingestion causes irritation of mouth and stomach. Contact with eyes causes irritation.

EXPOSURE TREATMENT Inhalation: move to fresh air. Ingestion: give victim one tablespoonful of salt in glass of water; repeat until vomit is clear; then give 2 tablespoonfuls of Epsom salts or milk of magnesia and force fluids; call a physician in all cases of suspected poisoning. Eyes: flush with water for at least 15 min. Skin: flush with water; wash with soap and water.

REFERENCES

1. Worthing, C. R. (ed.). *Pesticide Manual.* 6th ed. Worcestershire, England: British Crop Protection Council, 1979. 74.

2. *The Merck Index.* 10th ed. Rahway, NJ: Merck Co., Inc., 1983. 227.

3. Hartley, D., and H. Kidd (eds.). *The Agrochemicals Handbook.* Old Woking, Surrey, United Kingdom: Royal Society of Chemistry/Unwin Brothers Ltd., 1983, p. A052/Oct 83.

4. U.S. Coast Guard, Department of Transportation. *CHRIS—Hazardous Chemical Data.* Volume II. Washington, DC: U.S. Government Printing Office, 1984–5.

5. *Farm Chemicals Handbook 1987.* Willoughby, OH: Meister Publishing Co., 1987, p. C–47.

6. *Farm Chemicals Handbook 1984.* Willoughby, OH: Meister Publishing Co., 1984, p. C–41.

7. SRI.

8. Association of American Railroads. *Emergency Handling of Hazardous Materials in Surface Transportation.* Washington, DC: Assoc. of American Railroads, Hazardous Materials Systems (BOE), 1987. 124.

9. Sax, N. I. *Dangerous Properties of Industrial Materials,.* 6th ed. New York, NY: Van Nostrand Reinhold, 1984.

10. NIOSH. *Pocket Guide to Chemical Hazards.* 2nd Printing. DHHS (NIOSH) Publ. No. 85–114. Washington, DC: U.S.

Dept. of Health and Human Services, NIOSH/Supt. of Documents, GPO, February 1987. 69.

11. USEPA. *Ambient Water Quality Criteria Doc; Arsenic* p. A–2 (1980) EPA 440/5–80–021.

12. Gilman, A. G., L. S. Goodman, and A. Gilman (eds.). *Goodman and Gilman's The Pharmacological Basis of Therapeutics.* 7th ed. New York: Macmillan Publishing Co., Inc., 1985.

13. Menzie, C. M. *Metabolism of Pesticides, an Update.* U. S. Department of the Interior, Fish and Wildlife Service, Special Scientific Report—*Wildlife* No. 184, Washington, DC: U.S. Government Printing Office, 1974. 37.

14. Brown, K. W., G. B. Evans, Jr., B. D. Frentrup (eds.). *Hazardous Waste Land Treatment.* Boston, MA: Butterworth Publishers, 1983. 242.

15. 29 CFR 1910.1018 (7/1/87).

16. NIOSH/CDC. NIOSH *Recommendations for Occupational Safety and Health Standards,* Sept. 1986 (Supplement to *Morbidity and Mortality Weekly Report* 35 No. 15, Sept. 26, 1986), p. 5S.

17. American Conference of Governmental Industrial Hygienists. *Threshold Limit Values for Chemical Substances and Physical Agents and Biological Exposure Indices for 1993–1994.* Cincinnati, OH: ACGIH, 1993.

18. 40 CFR 61.01 (7/1/87).

19. 40 CFR 261.24 (7/1/87).

20. USEPA/OPP; *Report on the Status of Chemicals in the Special Review Program, Registration Standards Program and the Data Call–In Program* p. 51 (1986).

21. Todd, D. K. *The Water Encyclopedia,* Maple Press, 1970.

22. Serkowitz, J. B., G. R. Schimke, and V. R. Valeri, *Water Pollution Potential of Manufactured Products,* Environmental Protection Agency, EPA–R2–73–179d, April 1973.

23. Stanley, R. A., "Toxicity of Heavy Metals and Salts to Eurasian Watermilfoil (*Myriophyllum spicatum L.*)," *Archives of Environmental Contamination and Toxicology,* Vol. 2, No. 4, 1974.

24. Mississippi–Alabama Sea Grant Consortium, *Effects of Mercury Compounds on Algae,* COM-75-10034, NTIS, May 1973.

■ ARSENIC ACID, LEAD(2+) SALT (2:3)

CAS # 3687–31–8

SIC CODES 2842

SYNONYMS arsenic acid, lead salt

MF As_2O_8 • $3Pb$

MW 899.41

STANDARD CODES EPA 311; NFPA-3, 0, 0; ICC—poison B, poison label, 200 lbs in an outside container; USCG—poison B, poison label; IATA—poison B, poison label, 25 kg passenger, 95 kg cargo; Superfund designated (Hazardous Substances) List.

PERSISTENCE Applications of 1,300 lb/acre have persisted at detectable levels for 15 years [R1].

MAJOR SPECIES THREATENED May smother benthic life.

INHALATION LIMIT 0.15 mg/m³.

DIRECT CONTACT Arsenic compounds are allergens, irritants to skin.

GENERAL SENSATION Can be absorbed through skin at chronically toxic levels. Lead poisoning symptoms include pica, anorexia, vomiting, malaise, and convulsions. Can cause permanent brain damage.

PERSONAL SAFETY Wear filter mask and skin protection.

ACUTE HAZARD LEVEL Threshold concentration for fresh- and salt-water fish—0.1 ppm Pb, 1 ppm As [R2]. Low solubility should negate most ingestive hazards. Highly toxic via inhalation. Toxic for fish.

CHRONIC HAZARD LEVEL Highly toxic via

all routes. Application factor to convert 96-hour LC_{50} to chronic safe limit—0.013 brook; 1.4 ppm affects daphnid reproduction over 3-week period [R5]. Application of irritant and allergen. Trout, 0.043 for rainbow trout [R6]. Deformity and sub-adult mortality: 5 mg/L in drinking water led to rapid die-off of rat and mouse breeding; chronic poisoning symptoms include weight loss and weakness, and anemia has been noted in trout exposed to 0.012–.14 ppm Pb for 19 months and 18 days colonies [R7], respectively. Freshwater should not exceed 0.03 ppm Pb and marine; waters 1/50 of 96-hour LC_{50}. Daphnid reproduction reduced 16% from 3 week; exposure to 0.030 ppm Pb [R8]. Administration of 25 mg/L Pb in drinking water led to rapid die-off of breeding colonies of mice and rats [R4].

DEGREE OF HAZARD TO PUBLIC HEALTH Acute or chronic exposure via ingestion or inhalation is highly hazardous. Can be absorbed through skin at chronically toxic levels. Arsenic compounds are irritants and allergens. Arsenates have mutagenic potential.

OCCUPATIONAL RECOMMENDATIONS None listed.

ACTION LEVEL Suppress suspension of dust. Remove from intense heat or flame.

ON-SITE RESTORATION Pump or vacuum undissolved portion from bottom. Arsenate can be absorbed on alum floc and lead can be precipitated with lime. Use mixed cation-anion exchangers. Seek professional environmental engineering assistance through EPA's Environmental Response Team (ERT), Edison, NJ, 24-hour no., 908–548–8730.

AVAILABILITY OF COUNTERMEASURE MATERIAL Pumps—fire department; vacuum—swimming-pool suppliers; lime—cement plants; alum—water treatment plants; mixed exchangers—water softener suppliers.

DISPOSAL METHOD Convert to nitrate with nitric acid. Evaporate in fume hood to thin paste. Add 500 mL water and saturate with H_2S, filter, wash, dry, and send back to supplier.

INDUSTRIAL FOULING POTENTIAL Should not be present in food-processing waters.

MAJOR WATER USE THREATENED Potable supply, fisheries, recreation.

PROBABLE LOCATION AND STATE Will sink in water; white crystals or powder. May appear as milky dispersion.

WATER CHEMISTRY Lead is stable in oxygenated water as the carbonate, hydroxide, or carbonate-hydroxide salts. Under reducing conditions, and in the presence of sulfur, lead sulfide will predominate. Lead is least soluble at pH 9–10 with carbon dioxide levels at $10^{-3} M$. At carbon dioxide levels of $10^{-2} M$ solubility is lowest at pH 8–10. At pH 7–8, solubility of total lead is 0.001–0.01 mg/L. [R3] See also file on arsenic for solution chemistry.

COLOR IN WATER White dispersion.

FOOD-CHAIN CONCENTRATION POTENTIAL Lead is concentrated by animals and fishes. Positive; barley has been found to concentrate 4–12 ppm in soil to 800–1,600 ppm in the roots and 10–17 ppm in the tops. Root crops generally show higher residues.

GENERAL RESPONSE Avoid contact with solid. Keep people away. Stop discharge if possible. Isolate and remove discharged material. Notify local health and pollution control agencies.

FIRE RESPONSE Not flammable.

EXPOSURE RESPONSE Call for medical aid. Solid poisonous if swallowed. If swallowed and victim is conscious, have victim drink water or milk and have victim induce vomiting. If swallowed and victim is unconscious or having convulsions, do nothing except keep victim warm.

RESPONSE TO DISCHARGE Issue warning—poison, water contaminant. Should be removed. Chemical and physical treatment.

WATER RESPONSE Harmful to aquatic life in very low concentrations. May be dangerous if it enters water intakes. Notify local health and wildlife officials. Notify operators of nearby water intakes.

EXPOSURE SYMPTOMS Inhalation or ingestion causes dizziness, headache, paralysis, cramps, constipation, collapse, coma. Subacute doses cause irritability, loss of weight, anemia, constipation. Blood and urine concentrations of lead increase.

EXPOSURE TREATMENT A specific medical treatment is used for exposure to this chemical; call a physician immediately. Give victim a tablespoon of salt in a glass of warm water and repeat until vomit is clear. Then give two tablespoons of epsom salt or milk of magnesia in water, and plenty of milk and water. Have victim lie down and keep quiet.

REFERENCES

1. Pimental, David, *Ecological Effects of Pesticides on Non-Target Species*, Presidential Report, Office of Science and Technology, June 1971.

2. Todd, D. K. *The Water Encyclopedia.* Maple Press, 1970.

3. Hem, J. D., and W. H. Durum, *Solubility and Occurrence of Lead in Surface Water*, MWWA, August 1973.

4. *Polychlorinated Biphenyls and the Environment.* Interdepartmental Task Force on PCBS, May 1972, published as Report No. ITF–PCB–72–1, Recipients Accession No. COM–72–10419, 20 March 1972, NTIS.

5. Environmental Protection Agency, *Report of the Pesticide Technical Committee to the Lake Michigan Enforcement Conference on Selected Trace Metals*, NTIS PB-220 361, September 1972.

6. Environmental Protection Agency, *Effects of Pesticides in Water*, NTIS PB-222 320, 1972.

7. Serkowitz, J. B., G. R. Schimke, and V. R. Valeri, *Water Pollution Potential of Manufactured Products*, Environmental Protection Agency, EPA-R2-73-179d, April 1973.

8. Biesinger, K. E. and G. M. Christensen, "Effects of Various Metals on Survival, Growth, Reproduction, and Metabolism of *Daphnia magna*," *J. Fish. Res. Bd. Can.*, Vol. 29, No. 12, 1972.

■ ARSENIC ACID, MONOPOTASSIUM SALT

CAS # 7784–41–0

DOT # 1677

SIC CODES 2819; 2879

SYNONYMS arsenic acid (H_3AsO_4), monopotassium salt • arsenic acid, monopotassium salt • Macquer's salt • monopotassium arsenate • monopotassium dihydrogen arsenate • potassium acid arsenate • potassium arsenate, monobasic • potassium dihydrogen arsenate • potassium dihydrogen arsenate (KH_2AsO_4) • potassium hydrogen arsenate • potassium hydrogen arsenate (KH_2AsO_4)

MF $AsH_2O_4 \cdot K$

MW 180.04

COLOR AND FORM Colorless crystals or white crystalline mass or powder; colorless tetragonal crystals.

ODOR Odorless.

DENSITY 2.867.

MELTING POINT 288°C.

FIREFIGHTING PROCEDURES Extinguish fire using agent suitable for type of surrounding fire (material itself does not burn or burns with difficulty). Use water in flooding quantities as fog. Use alcohol foam, dry chemical, or carbon dioxide. [R5].

SOLUBILITY Soluble in 5.5 parts cold water; very soluble in hot water; insoluble in alcohol; slowly soluble in 1.6 parts glycerol; soluble in ammonia.

OTHER CHEMICAL/PHYSICAL PROPERTIES Heat of solution: 49 Btu/lb = 27 cal/g = 1.1×10^5 J/kg.

PROTECTIVE EQUIPMENT Respirator selection: Upper-limit devices recommended by NIOSH: any detectable concentration: any self-contained breathing apparatus with a full facepiece and operated in a pressure-demand or other positive-pressure mode or any supplied-air respirator with a full facepiece and operated in a pressure-demand or other positive-pressure mode in combination with an auxiliary self-contained breathing apparatus operated in pressure-demand or other positive-pressure mode; escape: any air-purifying full facepiece respirator (gas mask) with a chin-style or front- or back-mounted acid gas canister having a high-efficiency particulate filter or any appropriate escape-type self-contained breathing apparatus. (arsenic and compounds (as As)) [R6].

PREVENTATIVE MEASURES Personnel protection: Avoid breathing dusts and fumes from burning material. Keep upwind. Avoid bodily contact with the material. Do not handle broken packages unless wearing appropriate personal protective equipment. Wash away any material that may have contacted the body with copious amounts of water or soap and water. If contact with the material is anticipated, wear appropriate chemical protective clothing. [R5].

CLEANUP Stay upwind. Use water spray to "knock down" dust. Isolate and remove discharged material. Notify local health and control agencies. [R4].

DISPOSAL SRP: at the time of review, criteria for land treatment or burial (sanitary landfill) disposal practices are subject to significant revision. Prior to implementing land disposal of waste residue (including waste sludge), consult with environmental regulatory agencies for guidance on acceptable disposal practices.

COMMON USES Used in textile, tanning, and paper industries; in insecticidal formulations (especially fly paper) [R1]; summer sprays of arsenic compounds were tried on some citrus varieties to determine whether the acid concentration of fruit could be reduced. Potassium arsenate is effective in lowering citrus fruit acidity [R2]. Insecticide—e.g., for fly bait and fly paper [R3]; laboratory reagent [R3].

STANDARD CODES EPA 311; TSCA; IATA poison class B, poison label, 25 kg passenger, 95 kg cargo; CFR 49 poison class B, poison label, 50 lb passenger, 200 lb cargo, storage code 1, 2; CFR 14 CAB code

8; not listed NFPA; AAR, Bureau of Explosives STCC 4923277; not listed ICC; Superfund designated (hazardous substances) list.

PERSISTENCE Persistent.

INHALATION LIMIT 10 mg/m³.

DIRECT CONTACT Eye, skin, respiratory irritant. Allergen.

GENERAL SENSATION Inhalation of arsenic dusts may cause acute nausea, pulmonary edema, restlessness, difficult breathing, cyanosis, cough with foamy sputum and rales. Skin contamination may produce burns, irritation, bronzing of skin, and loss of hair. Eye exposure may produce burns and ulcerations. May produce gastrointestinal upset. Absorption of arsenic may produce irritability, muscular paralysis, and liver and kidney dysfunction.

PERSONAL SAFETY Protect against both inhalation and absorption through the skin. Must wear protective clothing including NIOSH-approved rubber gloves and boots, safety goggles or face mask, hooded suit, and either a respirator whose canister is specifically approved for this material, or a self-contained breathing apparatus. Decontaminate fully or dispose of all equipment and clothing after use.

ACUTE HAZARD LEVEL This substance is highly toxic to humans by ingestion; between 7 drops and 1 teaspoonful may be fatal. The lowest toxic oral dose for a man is 300 mg/kg, ingested for 3 weeks intermittently. The oral LD_{50} for a rat is 14 mg/kg. Potassium arsenate is also a strong irritant to both the skin and respiratory tract. The aquatic toxicity to bluegill sunfish is approximately 50 ppm, based on 35 ppm 96-hour LC_{50} to bluegills for sodium arsenite. The threshold-limit-value time-weighted average for a 40-hour workweek is 0.01 mg/cu m as arsenic.

CHRONIC HAZARD LEVEL This material is an OSHA-recognized carcinogen. It has a neoplastic effect on rats when ingested. Chronic poisoning may lead to a loss of hair and nails and a loss of pigmentation to the skin. An abnormal incidence of carcinogenic effects involving both skin and respiratory tract has been reported among English factory employees exposed to arsenic compounds, but this has not been confirmed by experience in the U.S.

DEGREE OF HAZARD TO PUBLIC HEALTH
Very poisonous through inhalation or ingestion; an OSHA recognized carcinogen.

OCCUPATIONAL RECOMMENDATIONS
OSHA 8-hr time-weighted average: 5 µg/m³ (inorganic arsenic) [R11].

NIOSH NIOSH recommends that the substance be treated as a potential human carcinogen with a 15-minute TWA: 2 µg As/m³. (inorganic arsenic) [R12].

THRESHOLD LIMIT 8-hr time-weighted average (TWA) 0.01 mg/m³ (1993) (arsenic, elemental, and inorganic compounds (except arsine), as As) [R13, 13].

ACTION LEVEL Avoid contact with the spilled cargo. Stay upwind. Notify local air, water, and fire authorities of the accident. Evacuate all people to a distance of at least 200 feet upwind and 1,000 feet downwind of the spill.

ON-SITE RESTORATION Dam stream if possible to reduce the flow and prevent further dissipation due to water movement. Due to high solubility, dredging may not be useful. Pump water into a suitable container. Add lime to pH 10.5, then add ferric chloride ($FeCl_3$) to form floc. Settle, filter, and neutralize with sulfuric acid. For more details see Envirex manual EPA 600/2-77-227. Seek professional environmental engineering assistance through EPA's Environmental Response Team (ERT), Edison, NJ, 24-hour no. 908-548-8730.

BEACH OR SHORE RESTORATION Close beach and shore to the public until material has been removed. Avoid human contact. Do not burn material.

AVAILABILITY OF COUNTERMEASURE MATERIAL Pumps—fire stations, Army Corps of Engineers.

DISPOSAL METHOD After the material has been contained, remove it and the contaminated soil and place in impervious containers. If practical, transport material back to the supplier or chemical company to recover the heavy-metal content and for deactivation. If this is not practical, or if facilities are not available, the material should be encapsulated and buried in a specially designated chemical landfill. Not acceptable at a municipal sewage treatment plant.

DISPOSAL NOTIFICATION Notify local and state health authorities, local solid waste disposal authorities, supplier, and shipper.

INDUSTRIAL FOULING POTENTIAL Low industrial water-fouling potential.

MAJOR WATER USE THREATENED Recreational; fisheries; potable supply.

PROBABLE LOCATION AND STATE Colorless to white crystals, will sink and dissolve.

WATER CHEMISTRY Will form basic solution.

FOOD-CHAIN CONCENTRATION POTENTIAL Positive. Arsenic compounds tend to be accumulated by oysters and other molluscan shellfish.

NON-HUMAN TOXICITY LD$_{50}$ rat oral 14.0 mg/kg [R8].

BIOLOGICAL HALF-LIFE The biological half-life of ingested inorganic arsenic (in mouse) is about 10 hr and 50 to 80 percent is excreted in about 3 days. (inorganic arsenic compounds) [R7, 589].

BIODEGRADATION None [R9].

BIOCONCENTRATION Biomagnification through the food chain does not occur with arsenicals. Brown algae contain about 30 ppm arsenic and mollusks contain about 0.005 ppm As. Even plants grown in soils contaminated with As do not show higher concentrations of As than do plants grown on uncontaminated soil. (arsenical compounds) [R10].

ATMOSPHERIC STANDARDS Inorganic arsenic, pursuant to section 112 of the Clean Air Act, has been designated as a hazardous air pollutant. (inorganic arsenic) [R14].

RCRA REQUIREMENTS The Environmental

Protection Agency has amended regulations concerning ground-water monitoring with regard to screening suspected contamination at land-based hazardous waste treatment, storage, and disposal facilities. There are new requirements to analyze for a specified core list of chemicals plus those chemicals specified by the Regional Administrator on a site-specific basis. (Total arsenic (all species) is included on this list.) (total arsenic (all species)) [R15].

GENERAL RESPONSE Avoid contact with solid and dust. Keep people away. Wear goggles and dust respirator. Stay upwind. Use water spray to knock down dust. Isolate and remove discharged material. Notify local health and pollution control agencies.

FIRE RESPONSE Not flammable.

EXPOSURE RESPONSE Call for medical aid. Dust is poisonous if inhaled or swallowed. Irritating to eyes, nose, and throat. Move victim to fresh air. If breathing is difficult, give oxygen. Solid is poisonous if swallowed. Irritating to skin and eyes. Remove contaminated clothing and shoes. Flush affected areas with plenty of water. If in eyes, hold eyelids open, and flush with plenty of water. If swallowed and victim is conscious, have victim drink water or milk, and have victim induce vomiting. If swallowed and victim is unconscious or having convulsions, do nothing except keep victim warm.

RESPONSE TO DISCHARGE Issue warning—poison, water contaminant. Restrict access. Disperse and flush.

WATER RESPONSE Harmful to aquatic life in very low concentrations. May be dangerous if it enters water intakes. Notify local health and wildlife officials. Notify operators of nearby water intakes.

EXPOSURE SYMPTOMS Dust may irritate eyes. Ingestion or severe exposure by inhalation can cause burning of throat and mouth, abdominal pain, vomiting, diarrhea with hemorrhage, dehydration, jaundice, and collapse.

EXPOSURE TREATMENT Eyes: flush with water to remove dust. Ingestion: immediately induce evacuation of intestinal tract by inducing vomiting, giving gastric lavage, and saline cathartic; see physician at once; consider possible development of arsenic poisoning.

REFERENCES

1. *The Merck Index.* 10th ed. Rahway, NJ: Merck Co., Inc., 1983. 1098.

2. Procopiou, J., A. Wallace. *Alexandria J Agric Res* 27 (1): 93–8 (1979).

3. SRI.

4. U.S. Coast Guard, Department of Transportation. *CHRIS—Hazardous Chemical Data.* Volume II. Washington, DC: U.S. Government Printing Office, 1984–5.

5. Association of American Railroads. *Emergency Handling of Hazardous Materials in Surface Transportation.* Washington, DC: Assoc. of American Railroads, Hazardous Materials Systems (BOE), 1987. 570.

6. NIOSH. *Pocket Guide to Chemical Hazards.* 5th Printing/Revision. DHHS (NIOSH) Publ. No. 85–114. Washington, DC: U.S. Dept. of Health and Human Services, NIOSH/Supt. of Documents, GPO, Sept. 1985. 55.

7. Doull, J., C. D. Klassen, and M. D. Amdur (eds.). *Casarett and Doull's Toxicology.* 3rd ed., New York: Macmillan Co., Inc., 1986.

8. FAO/WHO. *Preliminary Working Paper on Calcium Arsenate and Lead Arsenate,* Joint Meeting on Pesticide Residues (1968) as cited in Nat'l Research Council Canada, *Effects of Arsenic in the Canadian Envir,* p. 219 (1978) NRCC No. 15391.

9. U.S. Coast Guard, Department of Transportation. *CHRIS—Hazardous Chemical Data.* Manual Two. Washington, DC: U.S. Government Printing Office, Oct., 1978.

10. Brown, K. W., G. B. Evans, Jr., B. D. Frentrup (eds.). *Hazardous Waste Land Treatment.* Boston, MA: Butterworth Publishers, 1983. 242.

11. 29 CFR 1910.1018 (7/1/87).

12. NIOSH/CDC. *NIOSH Recommenda-*

tions for Occupational Safety and Health Standards, Sept. 1986 (Supplement to *Morbidity and Mortality Weekly Report* 35, No. 15, Sept. 26, 1986), p. 5S.

13. American Conference of Governmental Industrial Hygienists. *Threshold Limit Values for Chemical Substances and Physical Agents and Biological Exposure Indices for 1993–1994*. Cincinnati, OH: ACGIH, 1993.

14. 40 CFR 61.01 (7/1/87).

15. 52 FR 25942 (7/9/87).

■ ARSENIC DISULFIDE

CAS # 1303-32-8

SIC CODES 2816; 2819; 2892

SYNONYMS arsenic bisulfide • ruby arsenic • red arsenic • red arsenic sulfide • realgar • red orpiment • arsenic sulfide • C.I. 77085 • arsenic sulfide red • c.i. pigment yellow 39 • red arsenic glass

STANDARD CODES EPA 311; TSCA; IATA poison class B, posion label, 25 kg passenger, 95 kg cargo; not listed NFPA; CFR 49 poison class B, poison label, 50 lb passenger, 200 lb cargo, storage code 1, 2; AAR, Bureau of Explosives STCC 4923211; CFR 14 cab code 8; IMCO/United Nations numerical designation 6.1/1557; ICC STCC tariff no 1–C 28 199 18.

PERSISTENCE Persistent.

INHALATION LIMIT 10 mg/m^3.

DIRECT CONTACT Contact with the eyes causes irritation. Irritates skin, especially where moist, and if not treated may cause ulceration. May also be an allergen.

GENERAL SENSATION Inhalation of arsenic dusts may cause acute nausea, pulmonary edema, restlessness, difficult breathing, cyanosis, cough with foamy sputum, and rales. Skin exposure may produce burns and and ulcerations. May produce gastrointestinal upset. Absorption of arsenic may produce irritability, muscular paralysis, and liver and kidney dysfunction.

PERSONAL SAFETY Protect against both inhalation and absorption through the skin. Must wear protective clothing including NIOSH-approved rubber gloves and boots, safety goggles or face mask, hooded suit, and either a respirator whose canister is specifically approved for this material or a self-contained breathing apparatus. Decontaminate fully or dispose of all equipment and clothing after use.

ACUTE HAZARD LEVEL Arsenic disulfide is an extremely toxic material to humans. Ingestion of between 7 drops and 1 teaspoon may be fatal. A related material, arsenic trioxide, has an oral LD$_{50}$ for man of 1.43 mg/kg, and the lowest toxic inhalation concentration is 0.7 mg/m^3. The arsenic disulfide compound will cause irritation to the eyes, skin, and mucous membranes.

CHRONIC HAZARD LEVEL This material is a recognized carcinogen of the lungs and skin. The ACGIH has designated 0.01 mg/m^3 of arsenic as the threshold-limit-value time-weighted average for a 40-hour workweek. OSHA has recommended the TLV-TWA be 10 μg/m^3 for a 40-hour workweek, and the ceiling limit to be 0.01 mg/m^3 for 15 minutes. Arsenic is retained in all tissues, bones, and especially in the hair.

DEGREE OF HAZARD TO PUBLIC HEALTH Material is highly toxic by ingestion and inhalation. Possible carcinogen of skin and lungs.

OCCUPATIONAL RECOMMENDATIONS None listed.

ACTION LEVEL Avoid contact with the spilled cargo. Stay upwind. Notify local air, water, and fire authorities of the accident. Evacuate all people to a distance of at least 200 feet upwind and 1,000 feet downwind of the spill.

ON-SITE RESTORATION Dam stream if possible to reduce the flow and prevent further dissipation by water movement. Bottom pumps or underwater vacuum systems may be employed in small bodies of water; dredging may be effective in larger bodies, to remove any undissolved material from the bottom. Alternatively, transfer to a suitable vessel, change pH to 6–7, and allow arsenic sulfide to precipitate, filter, and dilute if necessary. Dredging may be required because arsenic sulfide is quite insoluble. Add ferric chloride ($FeCl_3$) and alum to aid clarification. For more details refer to Envirex Manual EPA 600/277–227. Seek professional environmental engineering assistance through EPA's Environmental Response Team (ERT), Edison, NJ, 24-hour no., 908–548–8730.

BEACH OR SHORE RESTORATION Close beach and shore to the public until material has been removed. If tidal, scrape affected area at low tide with a mechanical scraper, and avoid human toxicity contact. Do not burn material.

AVAILABILITY OF COUNTERMEASURE MATERIAL Bottom pumps, dredges—fire departments, U.S. Coast Guard, Army Corps of Engineers; vacuum systems—swimming-pool operators.

DISPOSAL METHOD After the material has been contained, remove it and the contaminated soil and place in impervious containers. If practical, transport material back to the supplier or chemical company to recover the heavy-metal content and for deactivation. If this is not practical or facilities not available, the material should be encapsulated and buried in a specially designated chemical landfill. Not acceptable at a municipal sewage treatment plant.

DISPOSAL NOTIFICATION Notify local and state health authorities, local solid-waste disposal authorities, supplier, and shipper.

INDUSTRIAL FOULING POTENTIAL Low solubility, low industrial water-fouling potential.

MAJOR WATER USE THREATENED Recreational; fisheries; potable supply.

PROBABLE LOCATION AND STATE Orange-red powder, will sink.

WATER CHEMISTRY No reaction.

COLOR IN WATER Orange-red or colorless.

FOOD-CHAIN CONCENTRATION POTENTIAL Positive. Arsenic compounds tend to be accumulated by oysters and other molluscan shellfish.

GENERAL RESPONSE Avoid contact with solid and dust. Keep people away. Wear chemical protective suit with self-contained breathing apparatus. Isolate and remove discharged material. Notify local health and pollution control agencies.

FIRE RESPONSE Not flammable. Wear chemical protective suit with self-contained breathing apparatus.

EXPOSURE RESPONSE Call for medical aid. Dust poisonous if inhaled. Harmful to skin. Move to fresh air. If breathing has stopped, give artificial respiration. If breathing is difficult, give oxygen. Solid poisonous if swallowed. Will burn eyes and skin. Remove contaminated clothing and shoes. Flush affected areas with plenty of water. If in eyes, hold eyelids open and flush with plenty of water. If swallowed and victim is conscious, have victim drink water or milk and have victim induce vomiting. If swallowed and victim is unconscious or having convulsions, do nothing except keep victim warm.

RESPONSE TO DISCHARGE Issue warning—poison, water contaminant. Restrict access. Should be removed. Chemical and physical treatment.

WATER RESPONSES Harmful to aquatic life in very low concentrations. May be dangerous if it enters water intakes. Notify local health and wildlife officials. Notify operators of nearby water intakes.

EXPOSURE SYMPTOMS Acute and subacute poisoning are not common. Repeated inhalation causes irritation of nose, laryngitis, mild bronchitis. Ingestion causes weakness, loss of appetite, gastrointestinal disturbances, peripheral neuritis, occasional hepatitis. Contact with eyes causes irritation. Irritates skin, especially where moist; if not treated, may cause ulceration.

EXPOSURE TREATMENT Consult physician after all overexposure to this compound. Inhalation: move to fresh air. Ingestion: induce vomiting by giving warm salt water; repeat until vomit is clear. Eyes: flush with water for at least 15 min. Skin: wash well with water.

REFERENCES
1. *OHM/TADS* database.

■ 1H-AZEPINE-1-CARBOTHIOIC ACID, HEXAHYDRO-, S-ETHYL ESTER

CAS # 2212-67-1

SYNONYMS 1H-azepine-1-carbothioic acid, hexahydro-, S-ethyl ester • ethyl 1-hexamethyleneiminecarbothiolate • felan • jalan • molmate • ordram • r-4572 • S-Aethyl-N-hexahydro-1H-azepinthiolcarbamat (German) • S-ethyl 1-hexamethyleneiminothiocarbamate • S-ethyl ester hexahydro-1H-azepine-1-carbothioic acid • S-ethyl hexahydro-1H-azepine-1-carbothioate • S-ethyl hexahydroazepine-1-carbothioate • S-ethyl N,N-hexamethylenethiocarbamate • S-ethyl perhydroazepine-1-thiocarboxylate • Stauffer r-4,572 • yalan • yulan • S-ethyl-N-hexamethylenethiocarbamate • S-ethyl azepane-1-carbothioate • S-ethyl perhydroazepin-1-carbothioate • perhydroazepin-1-carbothioate • sakkimol • higalnate • malerbane-giavoni-l • ordam

MF $C_9H_{17}NOS$

MW 187.33

COLOR AND FORM Clear liquid; amber liquid.

ODOR Aromatic.

DENSITY 1.5156 at 30°C/D.

BOILING POINT 202°C at 10 mm Hg.

SOLUBILITY 880 mg/L water at 20°C; miscible with acetone, ethanol, kerosene, 4-methylpentan-2-one, xylene.

VAPOR PRESSURE 0.005 mm Hg at 25°C.

OCTANOL/WATER PARTITION COEFFICIENT Log K_{ow}= 3.53 (calc).

CORROSIVITY Noncorrosive.

STABILITY AND SHELF LIFE Stable at 100°C for 16 hours [R6].

OTHER CHEMICAL/PHYSICAL PROPERTIES Index of refraction: 1.5156 at 30°C/D (technical grade).

PROTECTIVE EQUIPMENT Wear protective clothing; wash it thoroughly after use [R4].

PREVENTATIVE MEASURES SRP: The scientific literature supports the wearing of contact lenses in industrial environments, as part of a program to protect the eye against chemical compounds and minerals causing eye irritation. However, there may be individual substances whose irritating or corrosive properties are such that the wearing of contact lenses would be harmful to the eye. In those specific cases contact lenses should not be worn.

DISPOSAL Incinerate in a unit equipped with an effluent gas scrubber to absorb polluting combustion product (sulfur dioxide). Incineration temperature about 1,000° for 1–2 sec. Recommended methods: Incineration (of treated residual) [R7].

COMMON USES Selective herbicide [R4]; registered for use on rice for the control of watergrass (*Echinochloa* species) and other weeds. [R3, 325]. Molinate is toxic to germinating broad-leaved and grassy weeds and is particularly useful for control of *Echinochloa* species in rice at 2–4 kg AI/ha. [R1, 578] (systemic herbicide) [R5].

OCCUPATIONAL RECOMMENDATIONS None listed.

NON-HUMAN TOXICITY LD_{50} rat oral 720 mg/kg [R2]; LD_{50} rabbit dermal 3,536 mg/kg [R2]; LD_{50} rat female oral 450 mg/kg [R1, 578]; LD_{50} rabbit dermal >10,000 mg/kg [R8, 539]; LD_{50} rabbit percutaneous >4,640 mg/kg [R1, 578]; LD_{50} rat male, oral 369 mg/kg [R1, 578].

ECOTOXICITY LC_{50} crayfish 14 mg/L/96 hr [R9]; LC_{50} *Gammarus fasciatus* 4.5 mg/L/96 hr at 21°C (95% confidence limit 3.5–5.8 mg/L), mature; static bioassay without aeration, pH 7.2–7.5, water hardness 40–50 mg/L as calcium carbonate, and alkalinity of 30–35 mg/L (technical, 98.6%) [R10]; LC$_{50}$ *Pteronarcys* 0.34 mg/L/96 hr at 15°C (95% confidence limit 0.24–0.47 mg/L), mature; static bioassay without aeration, pH 7.2–7.5, water hardness 40–50 mg/L as calcium carbonate, and alkalinity of 30–35 mg/L (technical, 98.6%) [R10]; LC$_{50}$ *Salmo gairdneri* (rainbow trout) 0.21 mg/L/96 hr at 12°C (95% confidence limit 0.16–0.29 mg/L), wt 1.6 g; static bioassay without aeration, pH 7.2–7.5, water hardness 40–50 mg/L as calcium carbonate, and alkalinity of 30–35 mg/L (technical, 98.6%) [R10]; LC$_{50}$ *Lepomis macrochirus* (bluegill) 0.32 mg/L/96 hr at 24°C (95% confidence limit 0.19–0.53 mg/L), wt 1.0 g; static bioassay without aeration, pH 7.2–7.5, water hardness 40–50 mg/L as calcium carbonate, and alkalinity of 30–35 mg/L (technical, 98.6%) [R10]; LC$_{50}$ *Coturnix*, oral >5,000 ppm for 5 days [R11]; LC$_{50}$ *Daphnia magna* 600 µg/L/96 hr (conditions of bioassay not specified) [R12]; TLm *Gambusia affinis* (mosquitofish) 16.4 ppm/96 hr (conditions of bioassay not specified) [R12]; LC$_{50}$ *Anas platyrhynchos* (mallard duckling) >9,300 for 5 days in diet [R12].

BIODEGRADATION In soil 50% is degraded in 14–35 days, mostly through microbial activity. [R1, 578].

BIOCONCENTRATION Using a continuous-flow water system and a 14-day exposure period, a molinate BCF of 26 was measured in a freshwater fish (topmouth gudgeon) (1). Average BCFs of 48 and 18 were calculated for pale chub and ayu sweetfish, respectively, collected from Japanese rivers by measuring the water concentration and the fish concentration (2). These measured BCF values indicate that bioconcentration in fish is not an important fate process (SRC) [R13].

FIFRA REQUIREMENTS Tolerances are established for negligible residues of the herbicide S-ethyl hexahydro-1H-azepine-1-carbothioate in or on the raw agricultural commodities rice and rice straw [R14].

REFERENCES

1. Worthing, C. R., and S. B. Walker (eds.). *The Pesticide Manual—A World Compendium.* 8th ed. Thornton Heath, UK: The British Crop Protection Council, 1987.

2. Hayes, W. J., Jr., E. R. Laws, Jr. (eds.). *Handbook of Pesticide Toxicology. Volume 3. Classes of Pesticides.* New York, NY: Academic Press, Inc., 1991. 1348.

3. Weed Science Society of America. *Herbicide Handbook.* 5th ed. Champaign, Illinois: Weed Science Society of America, 1983.

4. *Farm Chemicals Handbook 1992.* Willoughby, OH: Meister Publishing Co., 1992, p. C–230.

5. Hartley, D., and H. Kidd (eds.). *The Agrochemicals Handbook.* 2nd ed. Lechworth, Herts, England: The Royal Society of Chemistry, 1987, p. A282/Aug 87.

6. Sunshine, I. (ed.). *CRC Handbook of Analytical Toxicology.* Cleveland: The Chemical Rubber Co., 1969. 522.

7. United Nations. *Treatment and Disposal Methods for Waste Chemicals* (IRPTC File). Data Profile Series No. 5. Geneva, Switzerland: United Nations Environmental Programme, Dec. 1985. 97.

8. Hayes, Wayland J., Jr. *Pesticides Studied in Man.* Baltimore/London: Williams and Wilkins, 1982.

9. Cheah, M., et al. *Prog Fish Cult* 42 (3): 169–72 (1980).

10. U.S. Department of Interior, Fish and Wildlife Service. *Handbook of Acute Toxicity of Chemicals to Fish and Aquatic Invertebrates.* Resource Publication No. 137. Washington, DC: U.S. Government Printing Office, 1980. 53.

11. Hill, E. F., and M. B. Cammardese. *Lethal Dietary Toxicities of Environmental Contaminants and Pesticides to Coturnix.* Fish and Wildlife Technical Report 2. Washington, DC: United States Department of Interior Fish and Wildlife Service, 1986. 103.

12. Verschueren, K. *Handbook of Environmental Data of Organic Chemicals.* 2nd ed. New York, NY: Van Nostrand Reinhold Co., 1983. 883.

13. (1) Kanazawa, J. *Pestic Sci* 12:

417–24 (1981); (2) Tsuda, T., et al. *Toxicol Environchem* 34: 39–55 (1991).

14. 40 CFR 180.228 (7/1/91).

■ 2H-AZEPIN-2-ONE, HEXAHYDRO-

CAS # 105–60–2

SYNONYMS 1,6-hexolactam • 2-azacycloheptanone • 2-ketohexamethylenimine • 2-oxohexamethylenimine • 2-perhydroazepinone • 2H-azepin-2-one, hexahydro • 2H-azepin-7-one, hexahydro- • 6-caprolactan • 6-hexanelactam • aminocaproic-lactam • caprolattame (French) • cyclohexanone iso-oxime • e-kaprolaktam (Czech) • epsilon-caprolactam • epsylon-kaprolaktam (Polish) • hexahydro-2-azepinone • hexahydro-2h-azepin-2-one • hexamethylenimine, 2-oxo- • hexanoic acid, 6-amino-, cyclic lactam • hexanoic acid, 6-amino-, lactam • NCI-C50646 • omega-caprolactam

MF $C_6H_{11}NO$

MW 113.18

COLOR AND FORM Hygroscopic leaflets from petroleum ether; white crystals; flake; molten.

ODOR Unpleasant.

TASTE Unpleasant.

VISCOSITY At 78°C = 9 cP.

BOILING POINT 180°C at 50 mm Hg.

MELTING POINT 70°C.

HEAT OF VAPORIZATION 116 cal/g.

SOLUBILITY Soluble in water, alcohol, benzene, chloroform; freely soluble in methanol, ethanol, tetrahydrofurfuryl alcohol; ether, dimethylformamide, soluble in chlorinated hydrocarbons, cyclohexene, petroleum fractions.

VAPOR PRESSURE 1.9×10^{-3} mm Hg at 25°C.

OTHER CHEMICAL/PHYSICAL PROPERTIES Heat of fusion: 29 cal/g; hygroscopic.

PROTECTIVE EQUIPMENT Workers should be supplied with suitable protective clothing including gloves and eyewear; resp protective equipment may be necessary. [R2, 1094].

PREVENTATIVE MEASURES All vessels and piping should be regularly checked for leaks. Exhaust ventilation should be installed. [R2, 1094].

DISPOSAL Controlled incineration (oxides of nitrogen are removed from the effluent gas by scrubbers or thermal devices). Also caprolactam may be recovered from caprolactam still bottoms or nylon waste. Recommended method: Incineration [R3].

COMMON USES Manufacture of synthetic fibers of the polyamide type (perlon); solvent for high-mol-wt polymers [R1].

OCCUPATIONAL RECOMMENDATIONS
OSHA 8-hr time-weighted average: 1 mg/m³. Final rule limits were achieved by any combination of engineering controls, work practices, and personal protective equipment during the phase-in period, Sept 1, 1989 through Dec 30, 1992. Final rule limits became effective Dec 31, 1992. (dust) [R8].

THRESHOLD LIMIT 8-hr time-weighted average (TWA) 1 mg/m³; short-term exposure limit (STEL) 3 mg/m³ (1974) (dust) [R9].

NON-HUMAN TOXICITY LD_{50} mouse (male) oral 21 g/kg [R4].

IARC SUMMARY No data are available for humans. Evidence suggesting lack of carcinogenicity in animals. Overall evaluation: group 4: The agent is probably not carcinogenic to humans [R5].

BIODEGRADATION Grab sample, initial concentration 50 ppm, 21 days (% degradation): 50% in sterilized stream water (control); 75% in unsupplemented stream water; 100% in stream water plus sediment; 100% in stream water with yeast extract; 35% in sterilized lake water (control); 72% in unsupplemented lake water; 100% in lake water plus sediment;

and 100% in lake water with yeast extract. Grab sample, initial concentration 40.4 ppm, 21 days, % CO_2 evolution: <5% in sterilized stream water (control); <5% in unsupplemented stream water; 8% in stream water plus sediment; 36% in stream water with yeast extract; <5% in sterilized lake water (control); 5% in unsupplemented lake water; 10% in lake water plus sediment; and 50% in lake water with yeast extract. Grab sample, initial concentration 50 ppm, 21, 14, and 14 days—100% degradation in lake water with yeast extract at 10, 20, and 25°C, respectively. Grab sample, initial concentration 40.4 ppm, 21 days—8, 72, 79% CO_2 evolution in lake water with yeast extract at 10, 20, and 25°C, respectively. Grab sample, initial concentration 2,000, 1,000, and 100 ppm, 21 days—36, 85, and 100% degradation, respectively, in lake water supplemented with yeast extract. Grab sample, initial concentration 2,000, 1,000, and 100 ppm, 21 days—8, 32, and 80% CO_2 evolution, respectively, in lake water supplemented with yeast extract (1) [R6].

BIOCONCENTRATION A bioconcentration factor (BCF) of <1 was estimated for caprolactam based on a measured log K_{ow} of −0.19 (1,2,SRC). This BCF value and the complete water solubility of caprolactam suggest that this bioaccumulation in aquatic organisms will not be important (3,SRC) [R7].

ATMOSPHERIC STANDARDS This action promulgates standards of performance for equipment leaks of volatile organic compounds (VOC) in the synthetic organic chemical manufacturing industry (SOCMI). The intended effect of these standards is to require all newly constructed, modified, and reconstructed SOCMI process units to use the best demonstrated system of continuous emission reduction for equipment leaks of VOC, considering costs, non-air-quality health, and environmental impact and energy requirements. Caprolactam is produced, as an intermediate or a final product, by process units covered under this subpart [R10].

TSCA REQUIREMENTS Pursuant to section 8 (d) of TSCA, EPA promulgated a model health and safety data reporting rule. The section 8 (d) model rule requires manufacturers, importers, and processors of listed chemical substances and mixtures to submit to EPA copies and lists of unpublished health and safety studies. 2H-azepin-2-one, hexahydro- is included on this list [R11].

GENERAL RESPONSE Stop discharge if possible. Keep people away. Call fire department. Isolate and remove discharged material. Notify local health and pollution control agencies.

FIRE RESPONSE Combustible. Poisonous gases may be produced in fire. Extinguish with water, dry chemicals, foam, or carbon dioxide. Cool exposed containers with water.

EXPOSURE RESPONSE Call for medical aid. Liquid is irritating to skin and eyes. Harmful if swallowed. Remove contaminated clothing and shoes. Flush affected areas with plenty of water. If in eyes, hold eyelids open, and flush with plenty of water. If swallowed and victim is conscious, have victim drink water or milk.

RESPONSE TO DISCHARGE Issue warning—water contaminant. Disperse and flush.

WATER RESPONSE Effect of low concentrations on aquatic life is unknown. May be dangerous if it enters water intakes. Notify local health and wildlife officials. Notify operators of nearby water intakes.

EXPOSURE SYMPTOMS Inhalation causes coughing or mild irritation. Contact with hot liquid will burn eyes and skin.

EXPOSURE TREATMENT Inhalation: remove patient to fresh air. Eyes: wash with copious quantities of water for at least 15 min.; call physician. Skin: wash with water; call physician in case of thermal burn.

REFERENCES
1. Budavari, S. (ed.). *The Merck Index— Encyclopedia of Chemicals, Drugs, and Biologicals*. Rahway, NJ: Merck and Co., Inc., 1989. 266.

2. International Labour Office. *Encyclopedia of Occupational Health and Safety.* Volumes I and II. New York: McGraw-Hill Book Co., 1971.

3. United Nations. *Treatment and Disposal Methods for Waste Chemicals* (IRPTC File). Data Profile Series No. 5. Geneva, Switzerland: United Nations Environmental Programme, Dec. 1985. 98.

4. IARC. *Monographs on the Evaluation of the Carcinogenic Risk of Chemicals to Man.* Geneva: World Health Organization, International Agency for Research on Cancer 1972–present (multivolume work), p. V39 253 (1986).

5. IARC. *Monographs on the Evaluation of the Carcinogenic Risk of Chemicals to Man.* Geneva: World Health Organization, International Agency for Research on Cancer, 1972–present (multivolume work)., p. S7 59 (1987).

6. Fortman, L., A. Rosenberg. *Chemosphere* 13: 53–65 (1984).

7. (1) Hansch, C., and A. J. Leo. *Medchem Project* Issue no. 26 Claremont, CA: Pomona College (1985); (2) Lyman, W. J., et al. *Handbook of Chemical Property Estimation Methods* New York: McGraw-Hill p. 5–5 (1982); (3) Fischer, W. B., L. Crescentini. *Kirk–Othmer Ency Chemical Tech* 18: 425–36 (1982).

8. 29 CFR 1910.1000 (7/1/90).

9. American Conference of Governmental Industrial Hygienists. *Threshold Limit Values for Chemical Substances and Physical Agents and Biological Exposure Indices for 1993–1994.* Cincinnati, OH: ACGIH, 1993. 15.

10. 40 CFR 60.489 (7/1/90).

11. 40 CFR 716.120 (7/1/90).

■ BARIUM PEROXIDE

CAS # 1304–29–6

SYNONYMS bario (perossido di) (Italian) • barium binoxide • barium dioxide • barium superoxide • Bariumperoxid (German) • bariumperoxyde (Dutch) • dioxyde de baryum (French) • peroxyde de baryum (French) • barium peroxide (Ba(O₂)) • barium oxide, per-

MF BaO_2

MW 169.34

COLOR AND FORM White or grayish-white, heavy powder.

ODOR Odorless.

DENSITY 4.96.

BOILING POINT 800°C (loses oxygen).

MELTING POINT 450°C.

FIREFIGHTING PROCEDURES Use flooding amount of water in early stages of fire. Smother with suitable dry powder [R12, p. 49–19]. Cool all affected containers with flooding quantities of water. Apply water from as far a distance as possible. If fire becomes uncontrollable, consider evacuation of one-half (1/2) mile radius [R5]. Large amount of water should be used on adjacent fires (soluble barium compounds (as barium)) [R11, 2]. Self-contained breathing apparatus with a full facepiece operated in pressure-demand or other positive-pressure mode (soluble barium compounds (as barium)) [R11, 6].

SOLUBILITY Very slightly soluble in cold water; soluble in dilute acid; insoluble in acetone; 1.5 parts by wt of the formula wt/100 parts by wt of water at 0°C.

CORROSIVITY Corrodes metal slowly.

STABILITY AND SHELF LIFE Decomposes slowly in air [R1].

OTHER CHEMICAL/PHYSICAL PROPERTIES Powerful oxidizing material; heat of decomposition: −194 Btu/lb = −108 cal/g = −4.52×10⁵ J/kg; mol wt: 313.45; colorless hexagonal crystals; density or specific gravity: 2.292; MP: loses 8 H₂O at 100°C; solubility: 0.168 g/100 cc of water; soluble in dilute acid; insoluble in alcohol, ether, acetone (octahydrate). Solutions of barium salts yield a white precipitate with 2 *N* sulfuric acid. This precipitate is insoluble in hydrochloric acid and in nitric acid. These salts impart a yellowish-green color to a nonluminous flame, which appears blue when viewed through green glass. (soluble barium salts) [R3].

PROTECTIVE EQUIPMENT Wear appropriate chemical protective boots [R5].

PREVENTATIVE MEASURES If material not on fire and not involved in fire: Keep sparks, flames, and other sources of ignition away. Keep material out of water sources and sewers [R5].

CLEANUP If soluble barium compounds are spilled, the following steps should be taken: 1. Ventilate area of spill. 2. Collect spilled material in the most convenient and safe manner and deposit in sealed containers for reclamation or for disposal in a secured sanitary landfill. Liquid containing soluble barium compounds should be absorbed in vermiculite, dry sand, earth, or a similar material (soluble barium compounds (as barium)) [R11, 5]. The European Community guidelines regard soluble barium salts as moderately dangerous for water quality (LD₅₀ 5–50 mg/kg) and their removal may be accomplished by diluted sulfuric acid and subsequent neutralization. (sol barium salts) [R13, 100].

DISPOSAL SRP: At the time of review, criteria for land treatment or burial (sanitary landfill) disposal practices are subject to significant revision. Prior to implementing land disposal of waste residue (including waste sludge), consult with environmental regulatory agencies for guidance on acceptable disposal practices.

COMMON USES Bleaching animal substances, vegetable fibers, and straw; glass decolorizer; manufacture of hydrogen peroxide and oxygen; in cathodes; dyeing and printing textiles; in powdered aluminum in welding; in igniter compositions; oxidizing agent in organic synthesis [R1]; used for pyrotechnics and tracer-bullet formulations [R2]; as polysulfide curing agent [R3].

OCCUPATIONAL RECOMMENDATIONS

OSHA 8-hr time-weighted average: 0.5 mg/m³ (soluble barium compound, as Ba) [R7].

THRESHOLD LIMIT Time weighted average (TWA) 0.5 mg/m³ (1977) (barium, soluble compounds, as Ba) [R15, 13].

BIOLOGICAL HALF-LIFE The biologic half-life (in plasma) is short (less than 24 hours) (soluble barium salts) [R6].

BIOCONCENTRATION Marine animals concentrate the element 7–100 times, and marine plants 1,000 times from seawater. Soybeans and tomatoes also accumulate soil barium 2–20 times (barium compounds) [R14, 86].

RCRA REQUIREMENTS A solid waste containing barium (such as barium peroxide) may or may not become characterized as a hazardous waste when subjected to the toxicant extraction procedure listed in 40 CFR 261.24, and, if so characterized, must be managed as a hazardous waste [R8].

DOT *Health Hazards:* Poisonous if swallowed. Inhalation of dust poisonous. Contact may cause burns to skin and eyes. Fire may produce irritating or poisonous gases. Runoff from fire control or dilution water may cause pollution [R4].

Fire or Explosion: May burn rapidly. May ignite other combustible materials (wood, paper, oil, etc). Reaction with fuels may be violent [R4].

Emergency Action: Keep unnecessary people away; isolate hazard area and deny entry. Self-contained breathing apparatus (SCBA) and structural firefighter's protective clothing will provide limited protection. Call CHEMTREC at 1–800–424–9300 for emergency assistance. If water pollution occurs, notify the appropriate authorities [R4].

Fire: Small Fires: Dry chemical, CO₂, halon, or water spray. Large Fires: Water spray or fog. Move container from fire area if you can do so without risk. Cool contain-

ers that are exposed to flames with water from the side until well after fire is out. Stay away from ends of tanks. For massive fire in cargo area, use unmanned hose holder, or monitor nozzles [R4].

Spill or Leak: Do not touch spilled material. Keep combustibles (wood, paper, oil, etc) away from spilled material. Small Dry Spills: With clean shovel place material into clean, dry container, and cover; move containers from spill area. Large Spills: Dike far ahead of liquid spill for later disposal [R4].

First Aid: Move victim to fresh air; call emergency medical care. Remove and isolate contaminated clothing and shoes at the site. In case of contact with material, immediately flush skin or eyes with running water for at least 15 minutes [R4].

FIRE POTENTIAL This material itself is noncombustible [R5].

FDA Barium peroxide is an indirect food additive for use only as a component of adhesives [R9]. Bottled water shall, when a composite of analytical units of equal volume from a sample is examined by the methods described in paragraph (d) (1) (ii) of this section, meet the standards of chemical quality, and shall not contain barium in excess of 1.0 mg/L (total barium) [R10].

REFERENCES

1. *The Merck Index.* 10th ed. Rahway, NJ: Merck Co., Inc., 1983. 141.

2. *Kirk–Othmer Encyclopedia of Chemical Technology.* 3rd ed., Volumes 1–26. New York, NY: John Wiley and Sons, 1978–1984, p. 17 (82) 66.

3. *Kirk–Othmer Encyclopedia of Chemical Technology.* 3rd ed., Volumes 1–26. New York, NY: John Wiley and Sons, 1978–1984, p. 18 (82) 817.

4. Department of Transportation. *Emergency Response Guidebook 1987.* DOT P 5800.4. Washington, DC: U.S. Government Printing Office, 1987, p. G–42.

5. Association of American Railroads. *Emergency Handling of Hazardous Materials in Surface Transportation.* Wash-

ington, DC: Assoc. of American Railroads, Hazardous Materials Systems (BOE), 1987. 75.

6. Doull, J., C. D. Klaassen, and M. D. Amdur (eds.). *Casarett and Doull's Toxicology*. 2nd ed. New York: Macmillan Publishing Co., 1980. 438.

7. 29 CFR 1910.1000 (7/1/88).

8. 40 CFR 261.24 (7/1/88).

9. 21 CFR 175.105 (4/1/88).

10. 21 CFR 103.35 (4/1/88).

11. Mackison, F. W., R. S. Stricoff, and L. J. Partridge, Jr. (eds.). *NIOSH/OSHA— Occupational Health Guidelines for Chemical Hazards*. DHHS(NIOSH) Publication No. 81–123 (3 vols). Washington, DC: U.S. Government Printing Office, Jan. 1981.

12. National Fire Protection Association. *Fire Protection Guide on Hazardous Materials*. 9th ed. Boston, MA: National Fire Protection Association, 1986.

13. Seiler, H. G., H. Sigel, and A. Sigel, (eds.). *Handbook on the Toxicity of Inorganic Compounds*. New York, NY: Marcel Dekker, Inc. 1988.

14. Friberg, L., G. F. Nordberg, E. Kessler, and V. B. Vouk (eds). *Handbook of the Toxicology of Metals*. 2nd ed. Vols I, II. Amsterdam: Elsevier Science Publishers B.V., 1986, p. V2.

15. American Conference of Governmental Industrial Hygienists. *Threshold Limit Values for Chemical Substances and Physical Agents and Biological Exposure Indices for 1994–1995*. Cincinnati, OH: ACGIH, 1994.

■ BENZAL CHLORIDE

CAS # 98-87-3

SYNONYMS (dichloromethyl) benzene • alpha, alpha-dichlorotoluene • benzene, (dichloromethyl)- • benzyl dichloride • benzylene chloride • benzylidene chloride • chlorobenzal • chlorure de benzylidene (French) • dichlorophenylmethane • toluene, alpha, alpha-dichloro • ai-28,597 • cloruro de bencilideno (Spanish)

MF $C_7H_6Cl_2$

MW 161.03

COLOR AND FORM Colorless oily liquid.

ODOR Pungent odor; faint aromatic odor.

DENSITY 1.26.

SURFACE TENSION 20.20 mN/m at 203.5°C.

BOILING POINT 205°C.

MELTING POINT −16.4°C.

HEAT OF VAPORIZATION 11,075.9 cal/mole.

SOLUBILITY Insoluble in water; soluble in dilute alkali; >10% in ethanol; >10% in ether; soluble in most organic solvents.

VAPOR PRESSURE 1 mm Hg at 35.4°C.

OCTANOL/WATER PARTITION COEFFICIENT Log K_{ow} = 3.217 (calculated).

OTHER CHEMICAL/PHYSICAL PROPERTIES Undergoes reactions both at the side chain containing the chlorines and at the aromatic ring; conversion factor: 1 ppm = 0.152 mg/m³. Dipole moment: 6.9×10^{-30} coulomb m (in dilute solution).

PREVENTATIVE MEASURES SRP: Contact lenses should not be worn when working with this chemical.

DISPOSAL SRP: At the time of review, criteria for land treatment or burial (sanitary landfill) disposal practices are subject to significant revision. Prior to implementing land disposal of waste residue (including waste sludge), consult with environmental regulatory agencies for guidance on acceptable disposal practices.

COMMON USES In dyes (former use) [R1]; manufacture of benzaldehyde and cinnamic acid (former use) [R1]; to prepare benzoyl chloride [R2].

OCCUPATIONAL RECOMMENDATIONS None listed.

INTERNATIONAL EXPOSURE LIMITS Levels of benzal chloride in the working environment in former USSR may not exceed a maximum allowable concentration of 0.5 mg/m³ [R8].

NON-HUMAN TOXICITY LD_{50} rat oral 3,250 mg/kg [R4]; LD_{50} mouse oral 2,460 mg/kg [R5].

BIODEGRADATION Benzal chloride was found to readily biodegrade in water (1) [R6].

BIOCONCENTRATION Based on a computer-calculated log K_{ow} value of 3.217 (1), the BCF value for benzal chloride can be estimated to be 164 (2,SRC). Due to the relatively rapid hydrolysis of benzal chloride in water, bioconcentration in aquatic organisms is not likely to occur (SRC) [R7].

ATMOSPHERIC STANDARDS This action promulgates standards of performance for equipment leaks of volatile organic compounds (VOC) in the synthetic organic chemical manufacturing industry (SOCMI). The intended effect of these standards is to require all newly constructed, modified, and reconstructed SOCMI process units to use the best demonstrated system of continuous emission reduction for equipment leaks of VOC, considering costs, non-air-quality health, and environmental impact and energy requirements. Benzyl dichloride is produced, as an intermediate or final product, by process units covered under this subpart [R9].

CERCLA REPORTABLE QUANTITIES Persons in charge of vessels or facilities are required to notify the national response center (NRC) immediately, when there is a release of this designated hazardous substance, in an amount equal to or greater than its reportable quantity of 5,000 lb or 2,270 kg. The toll-free number of the NRC is (800) 424–8802; in the Washington DC metropolitan area (202) 426–2675. The rule for determining when notification is required is stated in 40 CFR 302.4 (section IV. D. 3.b) [R10].

TSCA REQUIREMENTS Section 8 (a) of TSCA requires manufacturers of this chemical substance to report preliminary assessment information concerned with production, use, and exposure to EPA [R11].

RCRA REQUIREMENTS As stipulated in 40 CFR 261.33, when dichloromethyl benzene as a commercial chemical product or manufacturing chemical intermediate or an off-specification commercial chemical product or a manufacturing chemical intermediate becomes a waste, it must be managed according to federal or state hazardous waste regulations. Also defined as a hazardous waste is any residue, contaminated soil, water, or other debris resulting from the cleanup of a spill, into water or on dry land, of this waste. Generators of small quantities of this waste may qualify for partial exclusion from hazardous waste regulations (40 CFR 261.5) [R12].

DOT *Health Hazards:* Poisonous; may be fatal if inhaled, swallowed, or absorbed through skin. Contact may cause burns to skin and eyes. Runoff from fire control or dilution water may give off poisonous gases and cause water pollution. Fire may produce irritating or poisonous gases [R3].

Fire or Explosion: Some of these materials may burn, but none of them ignites readily. Container may explode violently in heat of fire [R3].

Emergency Action: Keep unnecessary people away; isolate hazard area and deny entry. Stay upwind, out of low areas, and ventilate closed spaces before entering. Self-contained breathing apparatus and chemical protective clothing that is specifically recommended by the shipper or producer may be worn but they do not provide thermal protection unless that is stated by the clothing manufacturer. Structural firefighter's protective clothing is not effective with these materials. Remove and isolate contaminated clothing at the site. Call CHEMTREC at 1–800–424–9300 as soon as possible, especially if there is no local hazardous materials team available [R3].

Fire: Small Fires: Dry chemical, CO_2, halon, water spray, or standard foam. Large Fires: Water spray, fog, or standard foam is recommended. Move container from fire area if you can do so without risk. Fight fire from maximum distance. Stay away from ends of tanks. Dike fire control water for later disposal; do not scatter the material [R3].

Spill or Leak: Do not touch spilled material; stop leak if you can do so without risk. Use water spray to reduce vapors. Small Spills: Take up with sand or other noncombustible absorbent material and place into containers for later disposal. Small Dry Spills: With clean shovel place material into clean, dry container, and cover; move containers from spill area. Large Spills: Dike far ahead of liquid spill for later disposal [R3].

First Aid: Move victim to fresh air and call emergency medical care; if not breathing, give artificial respiration; if breathing is difficult, give oxygen. In case of contact with material, immediately flush skin or eyes with running water for at least 15 minutes. Speed in removing material from skin is of extreme importance. Remove and isolate contaminated clothing and shoes at the site. Keep victim quiet and maintain normal body temperature. Effects may be delayed; keep victim under observation [R3].

FIRE POTENTIAL Combustible [R1].

REFERENCES
1. Sax, N. I., and R. J. Lewis, Sr. (eds.). *Hawley's Condensed Chemical Dictionary.* 11th ed. New York: Van Nostrand Reinhold Co., 1987. 136.

2. Sittig, M. M. *Handbook of Toxic and Hazardous Chemicals and Carcinogens,* 1985. 2nd ed. Park Ridge, NJ: Noyes Data Corporation, 1985. 108.

3. Department of Transportation. *Emergency Response Guidebook 1987.* DOT P 5800.4. Washington, DC: U.S. Government Printing Office, 1987, p. G–55.

4. Vernot, E. H., et al., *Toxicol, and Appl Pharm* 42: 417–23 (1977).

5. Vernot, E. H., et al., *Toxicol and Appl Pharm* 42: 417–23 (1977) as cited in USEPA, Chemical Hazard Information Profile: Benzal chloride (draft), p. 6 (1982).

6. Steinhauser, K. G., et al., *Vom Wasser* 67: 147–54 (1986).

7. (1) GEMS. *Graphical Exposure Modeling System.* Fate of Atmospheric Pollutants (FAP) Data Base. Office of Toxic Substances. USEPA (1986). (2) Lyman, W. J., et al. *Handbook of Chemical Property Estimation Methods. Environmental Behavior of Organic Compounds.* New York: McGraw-Hill, p 5–4 (1982).

8. IARC. *Monographs on the Evaluation of the Carcinogenic Risk of Chemicals to Man.* Geneva: World Health Organization, International Agency for Research on Cancer, 1972–present (multivolume work), p. V28 67 (1982).

9. 40 CFR 60.489 (7/1/88).

10. 40 CFR 302.4 (7/1/88).

11. 40 CFR 712.30 (7/1/88).

12. 40 CFR 261.33 (7/1/88).

■ BENZAMIDE

CAS # 55–21–0

SYNONYMS ai3-01031 • benzoic-acid amide • benzoylamide • phenylcarboxyamide • amid-kyseliny-benzoove (Czech)

MF C_7H_7NO

MW 121.15

COLOR AND FORM Colorless crystals; monoclinic prisms or plates from water.

DENSITY 1.341.

SURFACE TENSION 47.26 dyn/cm.

BOILING POINT 288°C.

MELTING POINT 130°C.

HEAT OF COMBUSTION 847.6 kcal/g mol wt at 20°C.

SOLUBILITY >10% in benzene; 1 g in 6 ml ethanol; 1 g in 3.3 ml pyridine; slightly soluble in ether; 1 g in 74 ml water, more soluble in boiling water; soluble in ammonia.

VAPOR PRESSURE 1.65×10^{-4} mmHg (est).

OCTANOL/WATER PARTITION COEFFICIENT Log K_{ow} = 0.64.

OTHER CHEMICAL/PHYSICAL PROPERTIES
Dipole moment: 3.6.

DISPOSAL The following wastewater treatment technology has been investigated for benzamide: Concentration process: biological treatment [R2].

COMMON USES Organic synthesis [R1].

OCCUPATIONAL RECOMMENDATIONS None listed.

ECOTOXICITY LC_{50} *Pimephales promelas* (fathead minnow) 661 mg/L/96 hr (confidence limit 590–740 mg/L), flow-through bioassay with measured concentrations, 25.1°C, dissolved oxygen 7.2 mg/L, hardness 47.0 mg/L calcium carbonate, alkalinity 41.3 mg/L calcium carbonate, and pH 7.5 [R3].

BIODEGRADATION Benzamide is readily oxidized by activated sludge from a municipal wastewater plant (3). However, no information is available concerning biodegradation in aerobic natural waters. Benzamide is oxidized by soil bacteria adapted to phenol (4–6). In experiments in which 400 mg/L of benzamide was applied to soil columns, 96% degradation occurred within 3 days in a clay soil, and 98% degradation occurred within 13 days in an organic soil (2). Benzamide undergoes anaerobic biodegradation under both sulfate-reducing and methanogenic conditions. When incubated in aquifer slurries from a sulfate-reducing and methanogenic site, respectively, 45% and 40% of the benzamide was biodegraded after one month (1). All the benzamide disappeared within 8 months [R4].

BIOCONCENTRATION The BCF for benzamide determined from its octanol/water partition coefficient, 0.640 (1), is 1.804, using a recommended regression equation (2). Therefore benzamide would not be expected to bioconcentrate in aquatic organisms (SRC) [R5].

ATMOSPHERIC STANDARDS This action promulgates standards of performance for equipment leaks of volatile organic compounds (VOC) in the synthetic organic chemical manufacturing industry (SOCMI). The intended effect of these standards is to require all newly constructed, modified, and reconstructed SOCMI process units to use the best demonstrated system of continuous emission reduction for equipment leaks of VOC, considering costs, non-air-quality health, and environmental impact and energy requirements. Benzamide is produced, as an intermediate or a final product, by process units covered under this subpart [R6].

REFERENCES

1. Sax, N. I., and R. J. Lewis, Sr. (eds.). *Hawley's Condensed Chemical Dictionary*. 11th ed. New York: Van Nostrand Reinhold Co., 1987. 128.

2. USEPA. *Management of Hazardous Waste Leachate*, EPA Contract No. 68–03–2766 p. E–39 (1982).

3. Geiger D. L., S. H. Poirier, L. T. Brooke, D.J. Call (eds.). *Acute Toxicities of Organic Chemicals to Fathead Minnows (Pimephales promelas)*. Vol. II. Superior, Wisconsin: University of Wisconsin–Superior, 1985. 161.

4. (1) Kuhn, E. P., J. M. Suflita. *Haz Waste Haz Mater* 6: 121–33 (1989). (2) Fournier, J. C., J. Salle. *Chemosphere* 3: 77–82 (1974). (3) Lutin, P. A., et al. *Purdue Univ Eng Bull Ext Ser* 118: 131–45 (1965). (4) Chambers, C. W., et al. *J Water Pollut Contr Fed* 35: 1517–28 (1963). (5) Tabak, H. H., et al. *J Bacteriol* 87: 910 (1964). (6) Kramer, N., R. N. Doetsch. *Arch Biochemical Biophys* 26: 401–5 (1950).

5. (1) Hansch, C., and A. J. Leo. *Medchem Project* Issue No. 26, Claremont, CA: Pomona College (1985). (2) Lyman, W. J., et al. *Handbook of Chemical Property Estimation Methods*. New York: McGraw-Hill, Chapt 5 (1982).

6. 40 CFR 60.489 (7/1/91).

■ BENZ (a) ANTHRACENE, 7,12-DIMETHYL

CAS # 57-97-6

SYNONYMS benz (a) anthracene, 7,12-dimethyl • 1,4-dimethyl-2,3-benzphenanthrene • 7,12-dimethyl-1,2-benzanthracene • 7,12-dimethylbenzanthrascene • 7,12-dimethylbenzo (a) anthracene • 7,12-DMBA • dimethylbenz (a) anthracene • dimethylbenzanthracene • dmba • nci-c03918

MF $C_{20}H_{16}$

MW 256.36

COLOR AND FORM Plates, leaflets from acetone and alcohol, faint greenish-yellow tinge.

MELTING POINT 122-123°C.

TOXIC COMBUSTION PRODUCTS When heated to decomposition it emits acrid smoke, fumes [R2].

SOLUBILITY Insoluble in water; slightly soluble in alcohol; soluble in carbon disulfide, toluene; soluble in acetone, benzene; may be solubilized in water by purines such as caffeine, tetramethyluric acid; nucleosides, adenosine, and guanosine also show solvent action; 0.055 mg/L water at 24°C (99% purity).

OCTANOL/WATER PARTITION COEFFICIENT 5.65.

PREVENTATIVE MEASURES It is extremely important to avoid skin contact and release of anthracene vapor and dust into work atmosphere. Anthracene and its derivatives [R3].

DISPOSAL At the time of review, criteria for land treatment or burial (sanitary landfill) disposal practices are subject to significant revision. Prior to implementing land disposal of waste residue (including waste sludge), consult with environmental regulatory agencies for guidance on acceptable disposal practices [R4].

COMMON USES In experimental medicine to induce various malignant tumors in testing antineoplastic drugs [R1].

OCCUPATIONAL RECOMMENDATIONS OSHA meets criteria for proposed OSHA medical records rule [R7].

BIODEGRADATION 7,12-dimethylbenzanthracene was not degraded in 6 days in a standard BOD test using an activated sludge inoculum (1) [R5].

BIOCONCENTRATION When deposit feeding clams, *Macoma inquinata*, were exposed to detritus contaminated with Prudoe Bay crude oil to which an unspecified isomer of dimethylbenzanthracene had been added for 7 days, the log bioconcentration factor from seawater was 3.13 (1). The chemical associated with sediment was not available for uptake (1). From the log octanol/water partition coefficient, 5.65 (2), one would estimate the bioconcentration factor in fish to be 4.06 (3,SRC). When humic acids are added to water (2.0 mg dissolved organic carbon/L), the 6-hr log BCF in *Daphnia* was reduced only slightly from 2.99 to 2.82, demonstrating the presence of these adsorbing particulates does not significantly change the bioavailability of 7,12-dimethylbenzanthracene to *Daphnia* (4) [R6].

RCRA REQUIREMENTS As stipulated in 40 CFR 261.33, when 7,12-dimethylbenz (a) anthracene is a commercial chemical product or manufacturing chemical intermediate or an off-specification commercial chemical product, or a manufacturing chemical intermediate, it must be managed as a hazardous waste according to federal or state regulations. Also defined as a hazardous waste is any residue, contaminated soil, water, or other debris resulting from a cleanup of a spill, into water or on dry land, of this waste. Generators of small quantities of this waste may qualify for partial exclusion from hazardous waste regulations (see 40 CFR 261.5) [R8].

REFERENCES

1. Rossoff, I. S. *Handbook of Veterinary*

Drugs. New York: Springer Publishing Company, 1974. 179.

2. Sax, N. I. *Dangerous Properties of Industrial Materials,.* 6th ed. New York, NY: Van Nostrand Reinhold, 1984. 1136.

3. International Labour Office. *Encyclopedia of Occupational Health and Safety.* Vols. I and II. Geneva, Switzerland: International Labour Office, 1983. 163.

4. SRP.

5. Lutin, P. A., et al., Perdue Univ, *Eng Bull Ext Ser 118*: 131–45 (1965).

6. (1) Roesijadi, G., et al. *J Fish Res Board Canada 35*: 608–14 (1968). (2) Yalkowsky, S. H., et al. *Res Rev 89*: 43–55 (1983). (3) Lyman, W. J., et al., *Handbook of Chemical Property Estimation Methods. Environmental Behavior of Organic Compounds,* New York: McGraw Hill, pp. 5.1–5.30 (1982). (4) Leversee, G. J., et al., *Canadian J Fish Aquatic Sci 40* (suppl (2): 63–69 (1983).

7. 47 FR 30420 (7/13/82).

8. 40 CFR 261.33 (f) (7/1/85).

■ BENZENE, m-DINITRO-

CAS # 99–65–0

DOT # 1,597

SYNONYMS m-dinitrobenzene • 1,3-dinitrobenzol • 2,4-dinitrobenzene • benzene, 1,3-dinitro- • dinitrobenzene • dwunitrobenzen (Polish) • m-dnb • meta-dinitrobenzene • binitrobenzene • NSC-7189

MF $C_6H_4N_2O_4$

MW 168.12

COLOR AND FORM Yellowish crystals; rhombohedral plates from alcohol; pale-yellow solid.

DENSITY 1.575 at 18°C.

BOILING POINT 300–303°C.

MELTING POINT 89–90°C.

HEAT OF COMBUSTION −696.8 kcal at 20°C (solid).

EXPLOSIVE LIMITS AND POTENTIAL Severe explosion hazard when exposed to shock or flame [R9].

FIREFIGHTING PROCEDURES Firefighting respirator: self-contained breathing apparatus with a full facepiece operated in pressure-demand or other positive-pressure mode [R7, 5]. If dinitrobenzene, solid, is on fire or involved in fire: Use water in flooding quantities as fog, and cool all affected containers with flooding quantities of water. Apply water from as great a distance as possible, and use alcohol foam, carbon dioxide, or dry chemical. If fire is massive, back off, protect surroundings, and let burn (dinitrobenzene, solid) [R8]. For rescue purposes wear full protective clothing. Fight fire from an explosion-resistant location. In advanced or massive fires, area should be evacuated. If fire occurs in vicinity of this material, water should be used to keep containers cool [R6].

SOLUBILITY 1 g dissolves in 2,000 ml cold water, 320 ml boiling water; 1 g dissolves in 37 ml alcohol, 20 ml boiling alcohol; freely soluble in chloroform, ethyl acetate; very soluble in acetone, hot benzene, and pyridine; soluble in ether; water solubility: 2.2 mol/m³.

VAPOR PRESSURE 5.13×10^{-6} atm (est).

OCTANOL/WATER PARTITION COEFFICIENT Log K_{ow} = 1.49.

CORROSIVITY Liquid dinitrobenzene will attack some forms of plastics, rubber, and coatings. (dinitrobenzene, all isomers).

OTHER CHEMICAL/PHYSICAL PROPERTIES 1 mg/m³ = 0.14 ppm, 1 ppm = 6.99 mg/m³.

PROTECTIVE EQUIPMENT Employees should be provided with and required to use impervious clothing, gloves, face shields (eight-inch minimum), and other appropriate protective clothing necessary to prevent skin contact with dinitrobenzene or liquids containing dinitrobenzene. (SRP: Liquid dinitrobenzene will attack some forms of plastics, rubber, and coatings.) [R7, 3].

PREVENTATIVE MEASURES Eating and smoking should not be permitted in areas where dinitrobenzene or liquids containing dinitrobenzene are handled, processed, or stored. Employees who handle dinitrobenzene or liquids containing dinitrobenzene should wash their hands thoroughly with soap or mild detergent and water before eating, smoking, or using toilet facilities. [R7, 3].

CLEANUP 1. Remove all ignition sources. 2. Ventilate area of spill. 3. For small quantities, sweep onto paper, or other suitable material, and burn in a suitable combustion chamber that allows burning in an unconfined condition and is equipped with an appropriate effluent gas cleaning device. Large quantities may be reclaimed; however, if this is not practical, dissolve in fuel oil, and atomize in a suitable combustion chamber equipped with an appropriate effluent gas cleaning device. [R7, 4].

DISPOSAL 1. Make packages of dinitrobenzene in paper or other flammable material and burn in a suitable combustion chamber that allows burning in an unconfined condition and is equipped with an appropriate effluent gas cleaning device. 2. Dissolve dinitrobenzene in fuel oil and atomize in a suitable combustion chamber equipped with an appropriate effluent gas cleaning device. [R7, 4].

COMMON USES Chemical intermediate for m-phenylenediamine, used for aramid fibers, spandex fibers; dyes; explosives [R1]; used for the detection of 17-ketosteroid [R2]; used to prepare aminocresols by electrolytic reduction [R3]; used as an intermediate in the production of aromatic amines (by catalytic reduction with elemental iron in hydrochloric acid medium) for dye synthesis [R4]; 97% of nitrobenzene is used to produce aniline, which has wide application in the manufacture of dyes and medicines. (nitrobenzenes) [R5].

OCCUPATIONAL RECOMMENDATIONS
OSHA 8-hr time-weighted average: 1 mg/m³, skin absorption designation (dinitrobenzene, all isomers) [R13].

THRESHOLD LIMIT 8-hr time-weighted average (TWA) 0.15 ppm, 1.0 mg/m³, skin (1986) (dinitrobenzene, all isomers) [R14, 20].

INTERNATIONAL EXPOSURE LIMITS Former Czechoslovakia, Germany, Sweden, and former USSR: 0.15 ppm (no date) [R15].

NON-HUMAN TOXICITY LD_{50} rat oral as a 1% suspension in corn oil 83 mg/kg with fiducial limits 56–124 mg/kg [R10].

BIODEGRADATION Anaerobic and aerobic incubation of 1,3-dinitrobenzene with sewage effluent at 29°C resulted in about 85% and 40% degradation after 28 days, respectively, as determined by absorbance measurements (1). The products of the anaerobic incubations were aromatic amines, whereas, under aerobic conditions, ring cleavage and possible mineralization of the 1,3-dinitrobenzene occurred (1). More than 64 days were required to achieve a total loss of ultraviolet absorbancy of 1,3-dinitrobenzene when incubated with Niagara silt loam at 25°C (2). 1,3-Dinitrobenzene was incubated in Tennessee River water at 25°C (3). A second-order rate constant of 3.7×10^{-8} mL L/CFU (colony forming units) hr was obtained from the biodegradation data and was used to estimate a half-life of about 1 day (3). After 3 hr incubation at 30°C with phenol-adapted bacteria from soil and related environments, 75% of the initial amount of 1,3-dinitrobenzene was degraded (4) [R11].

BIOCONCENTRATION The log bioconcentration factor (BCF) of 1,3-dinitrobenzene in trout muscle is 0.93 (1). Using a log octanol/water partition coefficient of 1.49 (2), a log BCF of 0.9024 was estimated (3). Bioconcentration factors of these magnitudes suggest that bioconcentration of 1,3-dinitrobenzene in fish will not be significant (SRC) [R12].

GENERAL RESPONSE Avoid contact with solid and dust. Keep people away. Wear goggles, self-contained breathing apparatus, and rubber overclothing (including gloves). Call fire department. Isolate and remove discharged material. Notify local health and pollution control agencies.

FIRE RESPONSE Combustible. May explode

if exposed to heat or flames. Flood discharge area with water. Combat fires from behind barrier.

EXPOSURE RESPONSE Call for medical aid. Vapor or dust poisonous if inhaled or if skin is exposed. Move victim to fresh air. If breathing is difficult, give oxygen. Solid poisonous if swallowed or if skin is exposed. Remove contaminated clothing and shoes. Flush affected areas with plenty of water. If in eyes, hold eyelids open, and flush with plenty of water. If swallowed and victim is conscious, have victim drink water or milk, and have victim induce vomiting. If swallowed and victim is unconscious or having convulsions, do nothing except keep victim warm.

RESPONSE TO DISCHARGE Issue warning—poison, water contaminant. Restrict access. Should be removed. Provide chemical and physical treatment.

WATER RESPONSE Harmful to aquatic life in very low concentrations. May be dangerous if it enters water intakes. Notify local health and wildlife officials. Notify operators of nearby water intakes.

EXPOSURE SYMPTOMS Inhalation or ingestion causes loss of color, nausea, headache, dizziness, drowsiness, and collapse. Eyes are irritated by liquid. Stains skin yellow; if contact is prolonged, can be absorbed into blood and cause same symptoms as for inhalation.

EXPOSURE TREATMENT Inhalation: remove from exposure; get medical attention for methemoglobinemia. Eyes: flush with water for at least 15 min. Skin: wash well with soap and water. Ingestion: induce vomiting if conscious; give gastric lavage and saline cathartic; get medical attention.

REFERENCES

1. SRI.

2. *Kirk–Othmer Encyc Chemical Tech*, 3rd ed., 1978–present, V15, p. 85.

3. *Kirk–Othmer Encyc Chemical Tech*, 3rd ed., 1978–present, V2, p. 371.

4. *Kirk–Othmer Encyc Chemical Tech*, 3rd ed., 1978–present, V8, p. 186.

5. National Research Council. *Drinking Water and Health*, Volume 4. Washington, DC: National Academy Press, 1981. 222.

6. National Fire Protection Association. *Fire Protection Guide on Hazardous Materials*. 9th ed. Boston, MA: National Fire Protection Association, 1986, p. 44–49.

7. Mackison, F. W., R. S. Stricoff, and L. J. Partridge, Jr. (eds.). *NIOSH/OSHA—Occupational Health Guidelines for Chemical Hazards*. DHHS (NIOSH) Publication No. 81–123 (3 vols). Washington, DC: U.S. Government Printing Office, Jan. 1981.

8. Association of American Railroads. *Emergency Handling of Hazardous Materials in Surface Transportation*. Washington, DC: Assoc. of American Railroads, Hazardous Materials Systems (BOE), 1987. 269.

9. Sunshine, I. (ed.). *CRC Handbook of Analytical Toxicology*. Cleveland: The Chemical Rubber Co., 1969. 627.

10. Cody, T. E., et al. *J Toxicol Environ Health* 7 (5): 829–48 (1981).

11. (1) Hallas, L. E., M. Alexander. *Appl Environ Microbiol* 45: 1234–41 (1983); (2) Alexander, M., B. K. Lustigman. *J Agric Food Chem* 14: 410–3 (1966); (3) Mitchell, W.R., W. H. Dennis. *J Environ Sci Health* A17: 837–53 (1982); (4) Tabak, H. H., et al. *J Bact* 87: 910–9 (1964).

12. (1) Howard, P. H., et al. *Investigation of Selected Potential Environmental Contaminants*, USEPA 560/2–76–010 (1976); (2) Hansch, C., and A. J. Leo. *Medchem Project* Issue No. 19 (1981); (3) Lyman, W. J., et al. *Handbook of Chemical Property Estimation Methods. Environmental Behavior of Organic Compounds*. New York: McGraw-Hill, p. 5–5 (1982).

13. 29 CFR 1910.1000 (7/1/90).

14. American Conference of Governmental Industrial Hygienists. *Threshold Limit Values for Chemical Substances and Physical Agents and Biological Exposure Indices for 1993–1994*. Cincinnati, OH: ACGIH, 1993.

15. American Conference of Governmen-

tal Industrial Hygienists. *Documentation of the Threshold Limit Values and Biological Exposure Indices.* 5th ed. Cincinnati, OH: American Conference of Governmental Industrial Hygienists, 1986. 214.

■ BENZENE, 1,2,4-TRICHLORO-

CAS # 120-82-1

DOT # 2321

SYNONYMS 1,2,4-trichlorobenzol • benzene, 1,2,4-trichloro- • trojchlorobenzen (Polish) • unsym-trichlorobenzene • Hostetex l-pec • AI3-07775

MF $C_6H_3Cl_3$

MW 181.44

COLOR AND FORM Colorless liquid; rhombic crystals.

ODOR Aromatic.

ODOR THRESHOLD Industrial data report an odor threshold of approximately 3 ppm. [R7].

DENSITY 1.4542 at 20°C.

SURFACE TENSION 38.54 dyn/cm.

BOILING POINT 213.5°C at 760 mm Hg; 84.8°C at 10 mm Hg.

MELTING POINT 17°C.

HEAT OF VAPORIZATION 11,425.1 cal/mole.

AUTOIGNITION TEMPERATURE 571°C (1,060°C) [R6, p. 325M-88].

FLAMMABILITY LIMITS Lower flammable limit: 2.5% by volume; upper flammable limit: 6.6% by volume at 302°F (150°C) [R6, p. 325M-88].

FIREFIGHTING PROCEDURES Water, foam, carbon dioxide, dry chem [R6, 2,618].

Water may be used to blanket fire. [R6, p. 325M-88].

SOLUBILITY 19 ppm at 22°C in water; miscible with ether, benzene, petroleum ether, carbon disulfide; 0.000269 moles/L water at 20°C; water sol: 30 ppm at 20°C.

VAPOR PRESSURE 1 mm Hg at 38.4°C.

OCTANOL/WATER PARTITION COEFFICIENT Log K_{ow} = 4.02.

STABILITY AND SHELF LIFE Stable at room temperature [R7].

OTHER CHEMICAL/PHYSICAL PROPERTIES Heat capacity (liquid): 194.6 J/mole-K, at 1 atm (for calories/mol-K, multiply by 0.2390057).

PREVENTATIVE MEASURES SRP: The scientific literature supports the wearing of contact lenses in industrial environments, as part of a program to protect the eye against chemical compounds and minerals causing eye irritation. However, there may be individual substances whose irritating or corrosive properties are such that the wearing of contact lenses would be harmful to the eye. In those specific cases contact lenses should not be worn.

CLEANUP Absorb the spills with paper towels or like materials. Place in a hood to evaporate. Dispose by burning the towel [R8].

DISPOSAL Incineration, preferably after mixing with another combustible fuel. Assure complete combustion to prevent the formation of phosgene. An alkali scrubber is necessary to remove the halo acids produced. Recommended method: Incineration [R9].

COMMON USES Solvent in chemical manufacturing, dyes, and intermediates, dielectric fluid, synthetic transformer oils, lubricants, heat-transfer medium, insecticides [R1]; former use: as a soil treatment for termite control [R2]; used as a comonomer with p-dichlorobenzene in the production of arylene sulfide polymers [R3]; as a dye carrier, intermediate in the manufacture of herbicides, and higher chlorinated benzenes, dielectric fluid, solvent, heat-transfer medium [R4]; in degreasing agents, septic tank and drain cleaners, wood preservatives, and abrasive formulations [R5].

OCCUPATIONAL RECOMMENDATIONS

OSHA Ceiling value: 5 ppm (40 mg/m³) Final rule limits were achieved by any combination of engineering controls, work practices, and personal protective equipment during the phase-in period, Sept 1, 1989 through Dec 30, 1992. Final rule limits became effective Dec 31, 1992. [R16].

THRESHOLD LIMIT Ceiling limit 5 ppm, 37 mg/m³ (1978) [R17].

INTERNATIONAL EXPOSURE LIMITS Bulgaria, Poland, former USSR, Yugoslavia: 1.4 ppm [R7].

NON-HUMAN TOXICITY LD_{50} rat oral 756 mg/kg [R8]; LD_{50} mouse oral 766 mg/kg [R8].

ECOTOXICITY LC_{50} *Salmo gairdneri* (rainbow trout) 1.95 mg/L/48 hr at 15°C (conditions of bioassay not specified) [R10]; LC_{50} *Lepomis macrochirus* (bluegill sunfish) 109 mg/L/24 hr; 13.0 mg/L/48 hr; 3.36 mg/L/96 hr (conditions of bioassay not specified) [R10]; LC_{50} *Cyprinodon variegatus* (sheepshead minnow) >46.8 mg/L/24 hr; >46.8 mg/L/48 hr; 21.4 mg/L/96 hr (conditions of bioassay not specified) [R10]; LC_{50} *Poecilia reticulata* (guppy) 2.4 ppm/14 days (conditions of bioassay not specified) [R11]; LC_{50} *Pimephales promelas* (fathead minnow) 2,990 μg/L/96 hr (confidence limit 2,560–3,500 mg/L), flow-through bioassay with measured concentrations, 16.6°C, dissolved oxygen 9.5 mg/L, hardness 44.0 mg/L calcium carbonate, alkalinity 41.5 mg/L calcium carbonate, and pH 7.2 [R12].

BIOLOGICAL HALF-LIFE Approximate half-life is 5.5 days. (chlorinated benzenes) [R13].

BIODEGRADATION The formation rate of ¹⁴C-labeled carbon dioxide through the biodegradation of 1,2,4-trichlorobenzene by activated sludge was examined. After 5 days 13% of the 1,2,4-trichlorobenzene remained, 56% was converted to carbon dioxide, 23% to polar metabolites, and 7%

was volatilized. Approximately 80% of the 1,2,4-trichlorobenzene was adsorbed on solids, accounting for the low volatility from the system. [R14, p. 76–7].

BIOCONCENTRATION The bioconcentration factor for 1,2,4-trichlorobenzene by rainbow trout exposed over 119 days was studied (in a flow-through system). Fish exposed to water containing 3.2 ng/L showed a bioconcentration factor of 1,300 −320. Those exposed to 52 ng/L showed a bioconcentration factor of 3,200 −540. Bioconcentration values and information on concentration of 1,2,4-trichlorobenzene in Lake Ontario were used to predict residues in fish. There was excellent agreement between the predicted concentration and those determined from Lake Ontario fish [R15].

ATMOSPHERIC STANDARDS This action promulgates standards of performance for equipment leaks of volatile organic compounds (VOC) in the synthetic organic chemical manufacturing industry (SOCMI). The intended effect of these standards is to require all newly constructed, modified, and reconstructed SOCMI process units to use the best demonstrated system of continuous emission reduction for equipment leaks of VOC, considering costs, non-air-quality health, and environmental impact and energy requirements. 1,2,4-Trichlorobenzene is produced, as an intermediate or final product, by process units covered under this subpart [R18].

TSCA REQUIREMENTS Pursuant to section 8 (d) of TSCA, EPA promulgated a model health and safety data reporting rule. The section 8 (d) model rule requires manufacturers, importers, and processors of listed chemical substances and mixtures to submit to EPA copies and lists of unpublished health and safety studies. Benzene, 1,2,4-trichloro- is included on this list [R19].

GENERAL RESPONSE Avoid contact with liquid and vapor. Keep people away. Wear self-contained positive-pressure breathing apparatus and full protective clothing. Stop discharge if possible. Shut off ignition sources. Call fire department. Isolate and remove discharged material. Notify local health and pollution control agencies.

FIRE RESPONSE Combustible. Poisonous gases may be produced in fire. Wear self-contained positive-pressure breathing apparatus and full protective clothing. Extinguish small fires: dry chemical, CO_2, water spray or foam; large fires: water spray, fog, or foam.

EXPOSURE RESPONSE Call for medical aid. Vapor may be irritating to eyes, skin, and respiratory tract. Move to fresh air. If breathing has stopped, give artificial respiration. If breathing is difficult, give oxygen. Liquid may irritate skin and eyes. Poisonous if swallowed. If in eyes or on skin, flush with running water for at least 15 minutes; hold eyelids open if necessary. Remove and isolate contaminated clothing and shoes at the site. If swallowed and victim is conscious, have victim drink water or milk, and induce vomiting. If swallowed and victim is unconscious or having convulsions, do nothing except keep victim warm.

RESPONSE TO DISCHARGE Issue warning—poison, water contaminant. Should be removed. Provide chemical and physical treatment.

WATER RESPONSE Harmful to aquatic life in very low concentrations. May be dangerous if it enters water intakes. Notify local health and wildlife officials. Notify operators of nearby water intakes.

EXPOSURE SYMPTOMS Exposures to high concentrations via inhalation are potentially hazardous to the lungs, kidneys, and liver. Prolonged or repeated exposures or short exposure to high concentrations via inhalation are potentially hazardous to the lungs, kidneys, and liver. Prolonged or repeated exposure to the eyes is likely to result in moderate pain and transient irritation. Prolonged or repeated contact with the skin may result in moderate irritation and possible systemic effects. Ingestion: May cause kidney and liver damage.

EXPOSURE TREATMENT Inhalation: If breathing has stopped, give artificial respiration. If breathing is difficult, give oxy-

gen. Eyes or skin: flush with running water for at least 15 minutes; hold eyelids open if necessary. Wash skin with soap and water. Remove and isolate contaminated clothing and shoes at the site. Ingestion: If victim is conscious, have victim drink water or milk, and induce vomiting by touching a finger to the back of his throat.

REFERENCES

1. Sax, N. I., and R. J. Lewis, Sr. (eds.). *Hawley's Condensed Chemical Dictionary.* 11th ed. New York: Van Nostrand Reinhold Co., 1987. 1,175.

2. *Farm Chemicals Handbook 88.* Willoughby, OH: Meister Publishing Co., 1988, p. C-227.

3. *Kirk-Othmer Encyclopedia of Chemical Technology.* 3rd ed., Volumes 1-26. New York, NY: John Wiley and Sons, 1978-1984, p. V18 812 (1982).

4. USEPA. *Health Assess Document for Chlorinated Benzenes Part 1.* pp. 4-18 to 4-23 USEPA-600/8-84-015A (1984).

5. McNamara, P. W., et al. *Exposure and Risk Assess for 1,2,4-Trichlorobenzene,* USEPA-440/4-85-017 (1981).

6. National Fire Protection Guide. *Fire Protection Guide on Hazardous Materials.* 10th ed. Quincy, MA: National Fire Protection Association, 1991.

7. American Conference of Governmental Industrial Hygienists. *Documentation of the Threshold Limit Values and Biological Exposure Indices.* 5th ed. Cincinnati, OH: American Conference of Governmental Industrial Hygienists, 1986. 593.

8. ITII. *Toxic and Hazardous Industrial Chemicals Safety Manual.* Tokyo, Japan: The International Technical Information Institute, 1988. 535.

9. United Nations. *Treatment and Disposal Methods for Waste Chemicals (IRPTC File).* Data Profile Series No. 5. Geneva, Switzerland: United Nations Environmental Programme, Dec. 1985. 105.

10. USEPA. *Health Assessment Document* p. 6-5 (1980) EPA 600/8-84-015.

11. Verschueren, K. *Handbook of Environmental Data of Organic Chemicals.* 2nd ed. New York, NY: Van Nostrand Reinhold Co., 1983. 1123.

12. Geiger D. L., D. J. Call, L. T. Brooke (eds.). *Acute Toxicities of Organic Chemicals to Fathead Minnows (Pimephales promelas).* Vol. V. Superior, WI: University of Wisconsin–Superior, 1990. 91.

13. USEPA. *Ambient Water Quality Criteria Doc; Chlorinated Benzenes* p. C–38 (1980) EPA 440/5–80–028.

14. Callahan, M. A., M. W. Slimak, N. W. Gabel, et al. *Water-Related Environmental Fate of 129 Priority Pollutants.* Volume I. EPA–440/4 79–029a. Washington, DC: U.S. Environmental Protection Agency, December 1979.

15. Oliver, B.G., and A.J. Niimi, *Environ Sci Technol* 17 (5): 287–91 (1983).

16. 29 CFR 1910.1000 (7/1/90).

17. American Conference of Governmental Industrial Hygienists. *Threshold Limit Values for Chemical Substances and Physical Agents and Biological Exposure Indices for 1993–1994.* Cincinnati, OH: ACGIH, 1993. 34.

18. 40 CFR 60.489 (7/1/90).

19. 40 CFR 716.120 (7/1/90).

■ BENZENEACETIC ACID, 4-CHLORO-alpha-(1-METHYLETHYL)-, CYANO (3-PHENOXYPHENYL) METHYL ESTER

CAS # 51630–58–1

SYNONYMS pydrin • s-5602 • sanmarton • sumifly • sumipower • alpha-cyano-3-phenoxybenzyl 2-(4-chlorophenyl)-3-methylbutyrate • wl43775 • balmark • sumkidin • cyano (3-phenoxyphenyl) methyl 4-chloro-alpha-(1-methylethyl) benzene acetate • ectrin • alpha-cyano-3-phenoxybenzyl-alpha-(4-chlorophenyl) isovalerate • sd-43775 • sumicidin • alpha-cyano-3-phenoxy-benzyl alpha-isopropyl-4-chlorophenylacetate • alpha-cyano-3-phenoxy-benzyl alpha-(4-chlorophenyl) isovalerate • 4-chloro-alpha-(1-methylethyl) benzeneacetic acid cyano (3-phenoxyphenyl) methyl ester • cyano (3-phenoxyphenyl) methyl 4-chloro-alpha-(1-methylethyl) benzeneacetate • phenvalerate • belmark • pyridin • tirade • (cyano (3-phenoxyphenyl) methyl-4-chloro-alpha-(1-methylethyl) phenylacetate) • (rs)-alpha-cyano-3-phenoxybenzyl(rs)-2-(4-chlorophenyl)-3-methylbutyrate • alpha-cyano-3-phenoxybenzyl isopropyl-4-chlorophenylacetate • sumibac • sumifleece • sumitick • fenkill • EPA Shaughnessy code: 109301

MF $C_{25}H_{22}ClNO_3$

MW 419.93

COLOR AND FORM Clear, yellow, viscous liquid.

ODOR Mild chemical odor.

DENSITY 1.17 at 23°C.

FIREFIGHTING PROCEDURES Use carbon dioxide, foam, or dry chemical (on fires involving pyrethroids) (pyrethrum) [R8, 2]. Use self-contained breathing apparatus with a full facepiece operated in pressure-demand or other positive-pressure mode (pyrethrum) [R8, 5]. Extinguish fire using agent suitable for type of surrounding fire (pyrethrins) [R7].

TOXIC COMBUSTION PRODUCTS Hydrogen cyanide may be formed during thermal decomposition. [R2, p. II–261].

SOLUBILITY In water at 20°C, <1 mg/L. Readily soluble in most organic solvents, in acetone, ethanol, chloroform, cyclo-hexanone, xylene, all >1 kg/kg at 23°C.

VAPOR PRESSURE 1.1×10^{-8} mm Hg at 25°C.

OCTANOL/WATER PARTITION COEFFICIENT Log K_{ow}= 4.42.

CORROSIVITY Noncorrosive to metals.

STABILITY AND SHELF LIFE More stable in acidic solution than in alkaline solution [R4, 629].

OTHER CHEMICAL/PHYSICAL PROPERTIES Brown viscous liquid (technical grade, 90% min fenvalerate).

PROTECTIVE EQUIPMENT Protective gloves, goggles, or full face shield when handling. [R1, p. C139].

PREVENTATIVE MEASURES Avoid eye, skin, mouth contact. [R1, p. C139].

CLEANUP Spillages of pesticides at any stage of their storage or handling should be treated with great care. Liquid formulations may be reduced to solid phase by evaporation. Dry sweeping of solids is always hazardous: these should be removed by vacuum cleaning, or by dissolving them in water or other solvent in the factory environment. (pesticides) [R9].

DISPOSAL Incineration would be an effective disposal procedure where permitted. If an efficient incinerator is not available, the product should be mixed with large amounts of combustible material, and contact with the smoke should be avoided. (pyrethrin products) [R10].

COMMON USES Highly active contact insecticide effective against a wide range of pests, including strains resistant to organochlorine, organophosphorus, and carbamate insecticides. It controls insects that attack leaves or fruits on various crops, including cotton, fruit, vegetables, and vines at 25–250 g ai/ha, and is persistent under various field conditions. It is also used in public health and animal husbandry, controlling flies in cattle sheds for 60 days at 100 mg/sq m wall, and is effective against *Boophilus* at 200–300 mg/L [R5, 395]. A broad-spectrum insecticide for use on cotton and fruit [R2, p. II–261]. For treatment of animals, apply one or two tags per head as needed. Will aid in control of face flies. Will control ear ticks [R3]. Medication (vet): ectoparasiticide [R4, 629]. Uses include control of chewing, sucking, and boring insects (particularly Lepidoptera, Diptera, Orthoptera, Hemiptera, and Coleoptera) in fruit, vines, olives, hops, nuts, vegetables, cucurbits, cotton, oilseed, rape, sunflowers, lucerne, cereals, maize, sorghum, potatoes, beet, groundnuts, soya beans, tobacco, sugar cane, ornamentals, forestry, and on non-crop land. Also used for control of flying and crawling insects in public health situations and in animal houses [R6], insecticide. (pyrethrins) [R4, 1267].

OCCUPATIONAL RECOMMENDATIONS None listed.

NON-HUMAN TOXICITY LD$_{50}$ rat oral 451 mg/kg [R5, 395]; LD$_{50}$ rat percutaneous >5,000 mg/kg [R5, 395]; LD$_{50}$ rat, oral 3,200 mg/kg (technical pydrin suspended in water) [R2, p. II–261]; LD$_{50}$ rat, oral 1–3 g/kg (technical grade) [R14]; LD$_{50}$ rabbit dermal 1–3 g/kg (technical grade) [R14]; LC$_{50}$ rat inhalation >101 g/m^3/4 hr [R14]; LD$_{50}$ rat oral 200 mg/kg [R15]; LC$_{50}$ *Pimephales promelas* (fathead minnow) 5.14 mg/L/96 hr (confidence limit 4.89–5.40 mg/L), flow-through bioassay with measured concentrations, 25.5°C, dissolved oxygen 7.4 mg/L, hardness 45.5 mg/L calcium carbonate, alkalinity 41.6 mg/L calcium carbonate, and pH 7.3 [R16].

ECOTOXICITY LC$_{50}$ rainbow trout 0.0036 mg/L/96 hr (conditions of bioassay not specified) [R5, 395]; LC$_{50}$ *Salmo salar* (Atlantic salmon) 1 µg/L/96 hr, juvenile (conditions of bioassay not specified) [R17]; LC$_{50}$ *Pimephales promelas* (fathead minnow) 0.42 mg/L/96 hr (confidence limit 0.39–0.46 mg/L), flow-through bioassay with measured concentrations, 24.5°C, dissolved oxygen 7.3 mg/L, hardness 44.8 mg/L calcium carbonate, alkalinity 40.9 mg/L calcium carbonate, and pH 7.8 [R18]; EC$_{50}$ *Pimephales promelas* (fathead minnow) 0.04 mg/L/96 hr (confidence limit 0.36–0.44 mg/L), flow-through bioassay with measured concentrations, 24.5°C, dissolved oxygen 7.3 mg/L, hardness 44.8 mg/L calcium carbonate, alkalinity 40.0 mg/L calcium carbonate, and pH 7.8; effect: loss of equilibrium [R18]; LC$_{50}$ *Pimephales promelas* (fathead minnow) 5.14 mg/L/96 hr (confidence limit 4.89–5.40 mg/L), flow-through bioassay with measured concentrations, 25.5°C, dissolved oxygen 7.4 mg/L, hardness 45.5 mg/L calcium carbonate, alkalinity 41.6 mg/L calcium carbonate, and pH 7.3. Effect: Loss of equilibrium [R16]; LC_{50} rainbow trout technical grade 76.0 ppb active ingredient/24 hr (static test) [R19]; LC_{50} rainbow trout formulated product 21.0 ppb active ingredient/24 hr (static test) [R19]; LD_{50} mallard oral 9,932 mg/kg [R12]; LC_{50} bobwhite quail dietary >10,000 ppm [R12]; LC_{50} mallard dietary 5,500 ppm (conditions of bioassay not specified) [R12]; LC_{50} bluegill sunfish fish 0.42 ppb/96 hr (conditions of bioassay not specified) [R14].

IARC SUMMARY Evaluation: No data were available from studies in humans. There is inadequate evidence for the carcinogenicity of fenvalerate in experimental animals. Overall evaluation: fenvalerate is not classifiable as to its carcinogenicity to humans (group 3) [R11].

BIOLOGICAL HALF-LIFE Elimination from body fat is slow, with a half-life of 7–10 days; elimination from brain is less slow, with a half-life of 2 days (Marei et al., 1982), presumably due to the more effective perfusion of brain, and the presence of esterases in brain tissue [R13].

BIODEGRADATION In a laboratory study using sediment and seawater collected from a salt marsh near Escambia County, FL, fenvalerate was observed to have a half-life of about 34 days (1); however, when the medium was sterilized, fenvalerate showed no appreciable degradation after 28 days of incubation (1), thus suggesting that degradation was occurring through biotic (microbiological) means (SRC). In degradation tests using an activated sludge inoculum, the aerobic and anaerobic degradation rates of fenvalerate were 50–72% faster than in sterile controls (2); addition of a glucose medium to cometabolize the non-sterile flasks resulted in a 6-fold increase in the degradation rate under aerobic conditions (2). The aerobic (semi-open system, activated sludge inocula) and anaerobic (serum bottle technique) biodegradation rates of fenvalerate were determined in tests using

both inoculated and non-inoculated (control) experiments (3); the biodegradation half-life was determined to be 13 days under both aerobic and anaerobic conditions (3) [R20].

BIOCONCENTRATION In a 28-day laboratory study, steady-state BCFs of 4,700 and 570 were measured in eastern oysters (*Crassostrea virginica*) and sheepshead minnow (*Cyprinodon variegatus*), respectively (1). A BCF of 1,100 was measured for one isomer of fenvalerate (s, s-isomer) in carp in a 24-hr renewal exposure following 7 days of exposure (1). In a 30-day aquatic ecosystem study, fenvalerate BCFs of 100 for fish, 491 for snails, and 412 for algae were measured (1); relatively low residues in the organisms were attributed to metabolism, especially by the fish (1) [R21].

FIFRA REQUIREMENTS Tolerances are established for residues of the insecticide cyano (3-phenoxyphenyl) methyl-4-chloro-a-(1-methylethyl) benzeneacetate in or on the following raw agricultural commodities: almond hulls; almonds; apples; artichokes; beans, dried; beans, snap; broccoli; blueberries; cabbage; caneberries; cantaloupes; carrots; cattle (fat, meat byproducts, and meat); cauliflower; collards; corn, grain; corn, fodder; corn, forage; corn, sweet, kernels, and cobs; cottonseed; cucumbers; currants; eggplant; elderberries; English walnuts; filberts; goats (fat, meat by-products, and meat); gooseberries; hogs (fat, meat by-products, and meat); honeydew melons; horses (fat, meat by-products, and meat); huckleberries; milk; milk, fat; muskmelons; peanuts; peanut hulls; pears; peas; peas, dried; pecans; peppers; potatoes; pumpkins; radish, roots; radish, tops; sheep (fat, meat by-products, and meat); soybeans; stone fruits; sugarcane; summer squash; sunflower seed; tomatoes; turnip roots; turnip tops; watermelons; and winter squash [R22].

REFERENCES

1. *Farm Chemicals Handbook 1991.* Willoughby, OH: Meister, 1991.

2. Gosselin, R. E., R. P. Smith, H. C. Hodge. *Clinical Toxicology of Commercial*

Products. 5th ed. Baltimore: Williams and Wilkins, 1984.

3. Booth, N. H., L. E. McDonald (eds.). Veterinary Pharmacology and Therapeutics. 5th ed. Ames, Iowa: Iowa State University Press, 1982. 913.

4. Budavari, S. (ed.). The Merck Index—Encyclopedia of Chemicals, Drugs, and Biologicals. Rahway, NJ: Merck and Co., Inc., 1989.

5. Worthing, C. R., and S. B. Walker (eds.). The Pesticide Manual—A World Compendium. 8th ed. Thornton Heath, UK: The British Crop Protection Council, 1987.

6. Hartley, D., and H. Kidd (eds.). The Agrochemicals Handbook. 2nd ed. Lechworth, Herts, England: The Royal Society of Chemistry, 1987, p. A207/Aug 87.

7. Bureau of Explosives. Emergency Handling of Haz Matl in Surface Trans, p. 434 (1981).

8. Mackison, F. W., R. S. Stricoff, and L. J. Partridge, Jr. (eds.). NIOSH/OSHA—Occupational Health Guidelines for Chemical Hazards. DHHS (NIOSH) Publication No. 81–123 (3 vols). Washington, DC: U.S. Government Printing Office, Jan. 1981.

9. International Labour Office. Encyclopedia of Occupational Health and Safety. Vols. I and II. Geneva, Switzerland: International Labour Office, 1983. 1619.

10. Sittig, M. Handbook of Toxic and Hazardous Chemicals and Carcinogens, 1985. 2nd ed. Park Ridge, NJ: Noyes Data Corporation, 1985. 762.

11. IARC. Monographs on the Evaluation of the Carcinogenic Risk of Chemicals to Man. Geneva: World Health Organization, International Agency for Research on Cancer, 1972–present (multivolume work), p. 53 324 (1991).

12. Purdue University, National Pesticide Information Retrieval System, Fenvalerate Fact Sheet No. 145 (1987).

13. Hayes, W. J., E. R. Laws. (eds.). Handbook of Pesticide Toxicology V2 p. 597 (1991).

14. Purdue University, National Pesticide Information Retrieval System, Fenvalerate Fact Sheet No. 145 (1987).

15. Kirk–Othmer Encyclopedia of Chemical Technology. 3rd ed., Volumes 1–26. New York, NY: John Wiley and Sons, 1978–1984, p. V13 458 (1981).

16. Geiger D. L., D.J. Call, L. T. Brooke. (eds). Acute Toxicities of Organic Chemicals to Fathead Minnows (Pimephales promelas). Vol. IV. Superior, Wisconsin: University of Wisconsin–Superior, 1988. 335.

17. Murty, A. S. Toxicity of Pesticides to Fish. Volumes I, II. Boca Raton, FL: CRC Press Inc., 1986. 70.

18. Geiger D. L., D. J. Call, L. T. Brooke (eds.). Acute Toxicities of Organic Chemicals to Fathead Minnows (Pimephales promelas). Vol. V. Superior, WI: University of Wisconsin–Superior, 1990. 277.

19. Verschueren, K. Handbook of Environmental Data of Organic Chemicals. 2nd ed. New York, NY: Van Nostrand Reinhold Co., 1983. 670.

20. (1) Schimmel, S. C., et al. J Agric Food Chem 31: 104–13 (1983); (2) Kanazawa. J; Environ Monitor Assess 9: 57–70 (1987); (3) Kawamoto, K., K. Urano. Chemosphere 21: 1141–52 (1990).

21. Schimmel, S. C., et al. J Agric Food Chem 31: 104–13 (1983).

22. 40 CFR 180.379 (a) (7/1/91).

■ 1,3-BENZENEDIAMINE

CAS # 108–45–2

SYNONYMS 1,3-diaminobenzene • 1,3-phenylenediamine • 3-aminoaniline • benzene, 1,3-diamino • ci-76025- • ci-developer-11- • developer-11- • developer-c • developer-h • developer-m • direct-brown-br • direct-brown-gg •

m-aminoaniline • m-benzenediamine • m-diaminobenzene • m-fenylendiamin (Czech) • m-phenylenediamine • meta-aminoaniline • meta-benzenediamine • meta-diaminobenzene • metaphenylenediamine • phenylenediamine, meta, solid (DOT) • APCO-2330

MF $C_6H_8N_2$

MW 108.16

COLOR AND FORM White crystals becoming red on exposure to air; colorless needles; rhombic crystals from alcohol; colorless rhombic needles; dark oxidation products are present in compound exposed to moist air.

DENSITY 1.139.

BOILING POINT 284–287°C.

MELTING POINT 62–63°C.

TOXIC COMBUSTION PRODUCTS When heated to decomp, it emits toxic fumes of nitrogen compounds [R6].

SOLUBILITY Soluble in water, methanol, ethanol, chloroform, acetone, dimethylformamide, methyl ethyl ketone, dioxane; slightly soluble in ether, carbon tetrachloride, isopropanol, dibutyl phthalate; very slightly soluble in benzene, toluene, xylene, butanol.

VAPOR PRESSURE 1 mm Hg at 99.8°C.

OCTANOL/WATER PARTITION COEFFICIENT –1.23 (estimated).

STABILITY AND SHELF LIFE Unstable in air [R7].

OTHER CHEMICAL/PHYSICAL PROPERTIES Dipole moment, 1.79; specific gravity, 1.07 at 58/4°C.

DISPOSAL SRP: At the time of review, criteria for land treatment or burial (sanitary landfill) disposal practices are subject to significant revision. Prior to implementing land disposal of waste residue (including waste sludge), consult with environmental regulatory agencies for guidance on acceptable disposal practices.

COMMON USES Manufacture of dyes; rubber curing agents; ion exchange resins, decolorizing resins, formaldehyde condensates; resinous polyamides, block polymers, textile fibers, urethanes, petroleum additives, rubber chemicals; in corrosion inhibitors; in photography; as reagent for gold and bromine [R1]; detection of nitrous acid; lab reagent [R16, 800]; component of hair-dye formulations [R3]; it is employed as curing agent to harden epoxy resins used in casting and plastic tooling, laminating applications, and adhesives. It is added to isophthaloyl chloride to produce poly-meta-phenylene isophthalamide resin, a polyamide fiber (nomex) used in high-temp applications, e.g., protective clothing [R3]. Can be used for production of over 140 dyes, 37 of which are believed to have commercial significance. These dyes are used to color various textile fibers and other materials, e.g., leather, paper, polishes, spirit inks; also used as direct dye developer to obtain black, blue, and brown shades and it is included in hair-dye formulations to produce brown, golden-blond, blue, and gray shades on the hair [R3]. Chemical intermediate for dyes, e.g., direct black 22 and mordant brown 1 [R2]; captive monomer for nomex (aromatic polyamide fiber) [R2]; chain extender for spandex fibers [R2]; crosslinking agent for alkyltin polymers for marine paint [R2]; secondary coupler for hair-dye formulations [R2]; adhesion promotor for tire cords to rubber [R2]; analytical reagent for bromine and gold [R2]; Mg, Ca, Ba, and Sr can be determined by ion chromatography with m-phenylenediamine in perchloric acid as the eluent [R4].

OCCUPATIONAL RECOMMENDATIONS

OSHA Meets criteria for OSHA medical rules record [R14].

NIOSH NIOSH recommends that 2,4-diaminoanisole (4-methoxy-m-phenylenediamine) and its salts be handled in the workplace as if they were human carcinogens. Based primarily upon a preliminary analysis of National Cancer Institute data, laboratory rats and mice fed 2,4-diaminoanisole sulfate experienced a statistically significant excess of site-specific tumors. (2,4-diaminoanisole and its salts) [R15].

THRESHOLD LIMIT 8-hr time-weighted average (TWA) 0.1 mg/m³, skin (1991) [R18, 29].

NON-HUMAN TOXICITY LD_{Lo} mice intraperitoneal 400 mg/kg [R10]; Noael rat 6 mg/kg [R9].

IARC SUMMARY No data are available for humans. Inadequate evidence of carcinogenicity in animals. Overall evaluation: Group 3: The agent is not classifiable as to its carcinogenicity to humans [R8].

BIOLOGICAL HALF-LIFE Bacterial rate of degradation corresponding to $t_{1/2} = 100$ hr (estimated) [R11].

BIODEGRADATION 60% degradation of 1,3-benzenediamine at concentration levels of 25–30 ppm with acclimated activated sludge for 5 days (1). However, at 50 ppm it is reported to be toxic to 3 unacclimated activated sludges (2). No degradation occurred in 64 days when incubated with a soil inoculum (3). These results imply that 1,3-benzenediamine will degrade only in an acclimated or vigorous system (SRC) [R12].

BIOCONCENTRATION 1,3-Benzenediamine is very soluble (1) and partially ionized in water and therefore would not be expected to bioconcentrate in fish (SRC) [R13].

TSCA REQUIREMENTS Contained in section 8 (d) of TSCA, which requires manufacturers, importers, and processors to submit to EPA copies and lists of unpublished health and safety studies if they are involved in the manufacturing, importing, or processing of this chemical. [R17, p. IV–19].

DOT *Health Hazards:* Poisonous if swallowed. Inhalation of dust poisonous. Fire may produce irritating or poisonous gases. Runoff from fire control or dilution water may cause pollution [R5].

Fire or Explosion: This material may burn but will not ignite readily [R5].

Emergency Action: Keep unnecessary people away; isolate hazard area and deny entry. Stay upwind; keep out of low areas. Wear self-contained (positive-pressure if available) breathing apparatus and full protective clothing. If water pollution occurs, notify appropriate authorities. For emergency assistance call CHEMTREC (800) 424–9300 [R5].

Fire: Small Fires: Dry chemical, CO_2, water spray, or foam. Large Fires: Water spray, fog, or foam. Move container from fire area if you can do so without risk [R5].

Spill or Leak: Do not touch spilled material; stop leak if you can do so without risk. Small Spills: Take up with sand or other noncombustible absorbent material and place into containers for later disposal. Small Dry Spills: With clean shovel place material into clean, dry container, and cover; move container from spill area. Large Spills: Dike far ahead of spill for later disposal [R5].

First Aid: Move victim to fresh air; call emergency medical care. Remove and isolate contaminated clothing and shoes at the site. In case of contact with material, immediately flush skin or eyes with running water for at least 15 minutes [R5].

FIRE POTENTIAL Fire point: 175°C [R1].

REFERENCES

1. *The Merck Index.* 10th ed. Rahway, NJ: Merck Co., Inc., 1983. 1,050.

2. SRI.

3. IARC. *Monographs on the Evaluation of the Carcinogenic Risk of Chemicals to Man.* Geneva: World Health Organization, International Agency for Research on Cancer, 1972–present (multivolume work), p. V16 114 (1978).

4. *Kirk–Othmer Encyclopedia of Chemical Technology.* 3rd ed., Volumes 1–26. New York, NY: John Wiley and Sons, 1978–1984, p. 24: 318.

5. Department of Transportation. *Emergency Response Guidebook 1984* DOT P 5800.3 Washington, DC: U.S. Government Printing Office, 1984, p. G–53.

6. Sax, N. I. *Dangerous Properties of Industrial Materials,.* 6th ed. New York, NY: Van Nostrand Reinhold, 1984. 2184.

7. Hawley, G. G. *The Condensed Chemi-*

cal Dictionary. 9th ed. New York: Van Nostrand Reinhold Co., 1977. 672.

8. IARC. *Monographs on the Evaluation of the Carcinogenic Risk of Chemicals to Man*. Geneva: World Health Organization, International Agency for Research on Cancer, 1972–present (multivolume work), p. S7 70 (1987).

9. TRDB/USEPA; Technical Support Document (Draft), *Phenylenediamines*, p. 54 (1984).

10. IARC. *Monographs on the Evaluation of the Carcinogenic Risk of Chemicals to Man*. Geneva: World Health Organization, International Agency for Research on Cancer, 1972–present (multivolume work), p. 87/8,608.

11. Verschueren, K. *Hdbk of Environ Data on Organic Chem* (1977) as cited in ITC/USEPA; *Information Review #102* (Draft), m-Phenylenediamine, p. 12 (1979).

12. (1) Peter, P. *Water Res* 10: 231–5 (1976). (2) Marion, C. V., G. W. Malaney. *Eng Bull*, Purdue Univ Eng Ext Ser pp. 297–308 (1964) (3) Alexander, M., B. K. Lustigman, *J Agric Food Chem* 14: 410–3 (1966).

13. (1) Verschueren, K. *Handbook of Environmental Data on Organic Chemicals* 2nd ed. New York: Van Nostrand Reinhold Co., 1983. pp. 988–9.

14. 47 FR 30420 (7/13/82).

15. NIOSH. *Current Intelligence Bulletin* 2,4–Diaminoanisole #19, p. 1 (1978).

16. Hawley, G.G. *The Condensed Chemical Dictionary*. 10th ed. New York: Van Nostrand Reinhold Co., 1981.

17. Roytech/SOCMA. *Suspect Chemicals Source Book*. 4th ed. Burlingame, CA: Roytech Publications, 1985.

18. American Conference of Governmental Industrial Hygienists. *Threshold Limit Values for Chemical Substances and Physical Agents and Biological Exposure Indices for 1994–1995*. Cincinnati, OH: ACGIH, 1994.

■ BENZENESULFONIC ACID, DODECYL-, SODIUM SALT

CAS # 25155–30–0

SIC CODES 2841

SYNONYMS benzenesulfonic acid, dodecyl-, sodium salt • dodecylbenzene-sodium-sulfonate • dodecylbenzenesulfonic acid, sodium salt • dodecylbenzensulfonan-sodny (Czech) • p-1′,1′,4′,4′-tetramethyloktylbenzensulfonan sodny (Czech) • sodium dodecylbenzenesulphonate • sodium dodecylphenylsulfonate • sodium laurylbenzenesulfonate • sol-sodowa-kwasu-laurylobenzenosulfonowego (Polish)

MF $C_{18}H_{29}O_3S \cdot Na$

MW 348.52

COLOR AND FORM White to light-yellow flakes, granules, or powder.

DENSITY 1.0 at 20°C for 60% slurry.

FIREFIGHTING PROCEDURES Extinguish fire with water [R3]. Extinguish fire using agent suitable for type of surrounding fire (material itself does not burn or burns with difficulty). [R8].

SOLUBILITY Soluble in water.

OCTANOL/WATER PARTITION COEFFICIENT Log K_{ow} = 0.45.

PROTECTIVE EQUIPMENT Rubber gloves, safety glasses [R3].

PREVENTATIVE MEASURES If material not on fire and not involved in fire: Keep material out of water sources and sewers. Build dikes to contain flow as necessary [R8].

CLEANUP Land spill: dig a pit, pond, lagoon, or holding area to contain liquid or solid material. SRP: if time permits, pits, ponds, lagoons, soak holes, or holding areas should be sealed with an impermeable flexible-membrane liner. Cover solids with a plastic sheet to prevent dissolving in rain or firefighting water [R8].

DISPOSAL At the time of review, criteria for land treatment or burial (sanitary landfill) disposal practices are subject to significant revision. Prior to implementing land disposal of waste residue (including waste sludge), consult with environmental regulatory agencies for guidance on acceptable disposal practices [R9].

COMMON USES Anionic detergent [R2, 3904]; surface-active agent in home laundry and cleaning products and non-household formulated cleaners [R1]; nonionic and anionic surfactants are used in fertilizer manufacture (surfactants) [R4]; used in industrial, institutional, and chemical detergents and cleaners such as heavy-duty laundry products; car, truck, and bus cleaners; metal cleaning products; specialty cleaners and sanitation products; emulsifiers, suspension, or wetting agents, absorbants in pesticides and other agricultural chemicals; foaming and wetting agent in pulp and paper products; latex, textile, rubber, and polymer processing [R5]; dodocylbenzenesulfonate used in the foam separation of $Fe(OH)_3$. Flox is recovered from the collapsed foam by treatment with NaOH, Na_2CO_3 in surfactant recovery; Na_2SO_4 is less effective (dodecylbenzenesulfonate) [R6]; prune, plant growth regulator for lily, contains 25% of sodium dodecylbenzene sulfonate [R7].

STANDARD CODES EPA 311; no NFPA; Superfund designated (hazardous substances) list.

PERSISTENCE Half-life around 15 days.

MAJOR SPECIES THREATENED Waterfowl.

DIRECT CONTACT Skin irritant.

GENERAL SENSATION Causes vomiting when swallowed.

PERSONAL SAFETY Wear skin protection. Self-contained breathing apparatus is recommended if intense heat or flame is present.

ACUTE HAZARD LEVEL Threshold limit for fresh- and salt-water fish—2 ppm [R12]. Mild irritant. Moderately toxic with ingestion toxic to aquatic plants and can cause taste, odor, and other aesthetic nuisances at low levels. ABS dissolved oily coating on duck feathers, causing the ducks to drown. 0.5–1 ppm can cause frothing, especially at pH 3–9.

CHRONIC HAZARD LEVEL Limiting threshold for boating and fishing, 5 ppm [R12]. Doses greater than 0.2% over 6 months led to decreased weight gain and blood protein; long-term feeding experiments indicate no chronic hazard. 6 men fed 180 mg/day for 4 mo. showed no effects. 9 pigs fed water with 2,000 ppm for 180 days showed slight toxic effects and histopathological changes in kidney. Freshwater levels of ABS should never exceed 1/20 of 96-hour LC_{50}—0.2 mg/L maximum [R13].

DEGREE OF HAZARD TO PUBLIC HEALTH Mild irritant. Moderately toxic with ingestion. Emits toxic vapors when heated to decomposition.

OCCUPATIONAL RECOMMENDATIONS None listed.

ACTION LEVEL Notify air authority if intense heat threatens to decompose solid.

ON-SITE RESTORATION Seek professional environmental engineering assistance through EPA's Environmental Response Team (ERT), Edison, NJ, 24-hour no., 908–548–8730. Research has verified oxidation can be with hydrogen peroxide and a ferrous sulfate catalyst, but this is not advised for water with valued species in it. Foaming and carbon absorption can be effective.

AVAILABILITY OF COUNTERMEASURE MATERIAL For foaming use compressed air and skimmers—stored at major ports; carbon—water treatment plants, sugar refineries.

DISPOSAL METHOD Can be injected into red clay or high-iron-oxide-hydroxide soils. Carbon will adsorb abs also but is more expensive.

INDUSTRIAL FOULING POTENTIAL Detergent loosens iron and manganese from pipes; water softening systems hindered; coal washing waters cause a decrease in the water content of the filter cake and increase the time required for filtration.

MAJOR WATER USE THREATENED Recreation, fisheries, potable supply, industrial.

PROBABLE LOCATION AND STATE White-yellow flakes or powder. Will dissolve in water. Color in water colorless—may be evidenced by foaming.

FOOD-CHAIN CONCENTRATION POTENTIAL Levels up to 1.4–2.1 mg/L of anionic detergents have been found in marine life. These are below levels toxic to people eating the fish [R14].

BIODEGRADATION Only when massive aeration was employed was sodium dodecyl-benzenesulfonate degraded by some unspecified species of marine bacteria when present as sole carbon source [R10].

FIFRA REQUIREMENTS Unless designated as an active ingredient in accordance with paragraph (b) or (c) of this section, this substance, when used in antimicrobial products, is considered inert, having no independent pesticidal activity. The percentage of such an ingredient was included on the label in the total percentage of inert ingredients [R11].

REFERENCES

1. SRI.

2. Hawley, G. G. *The Condensed Chemical Dictionary.* 10th ed. New York: Van Nostrand Reinhold Co., 1981.

3. U.S. Coast Guard, Department of Transportation. *CHRIS—Hazardous Chemical Data.* Volume II. Washington, DC: U.S. Government Printing Office, 1984–5.

4. *Farm Chemicals Handbook 87.* Willoughby, OH: Meister Publishing Co., 1987, p. B–66.

5. *Chemical Products Synopsis: Alkyl Aryl Sulfonates,* (1982).

6. Huang Shang, D. A. *Sep Sci Technol* 18 (11): 1017–22 (1983).

7. Iyasaka, H., et al. Three newly registered agric chemicals; *Shokubutsu Boeki Plant Protect* 128 (4): 168–9 (1974).

8. Association of American Railroads. *Emergency Handling of Hazardous Materials in Surface Transportation.* Washington, DC: Assoc. of American Railroads,

Hazardous Materials Systems (BOE), 1987. 629.

9. SRP.

10. Castellvi, J., A. Ballester. *Rapp P–V Reun—Comm International Explor Sci Mer Mediterr* 25–6 (9): 121–2 (1979).

11. 40 CFR 162.60 (7/1/87).

12. Todd, D. K. *The Water Encyclopedia,* Maple Press, 1970.

13. *Proposed Criteria for Water Quality,* Volume I, U.S. Environmental Protection Agency. October 1973.

14. Banerji, S. K. "Detergents," *JWPCF, Annual Literature Review,* Vol. 47, No. 6, June 1975.

■ BENZENESULFONYL CHLORIDE

SO₂Cl

CAS # 98–09–9

SYNONYMS benzene sulfochloride • benzene sulfonechloride • benzenesulfonic (acid) chloride • benzenesulfonic chloride • benzenesulphonyl chloride • benzenosulfochlorek (Polish) • benzenosulphochloride • phenylsulfonyl chloride • benezenesulfochloride • benzenesulfon chloride

MF C₆H₅ClO₂S

MW 176.62

COLOR AND FORM Colorless, oily liquid.

DENSITY SP GR: 1.3842 at 15°C.

BOILING POINT 177°C at 100 mm Hg; 120°C at 10 mm Hg; decomposition 251–252°C at 760 mm Hg.

MELTING POINT 14.5°C (Merck, 11th Ed.).

SOLUBILITY Insoluble in water; soluble in ether, alcohol.

VAPOR PRESSURE 0.068 mm Hg at 25°C (est).

STABILITY AND SHELF LIFE Stable toward cold water [R1].

OTHER CHEMICAL/PHYSICAL PROPERTIES Solidifies at 0°C (Merck, 11th Ed.).

DISPOSAL Generators of waste (equal to or greater than 100 kg/mo) containing this contaminant, EPA hazardous waste number U020, must conform with USEPA regulations in storage, transportation, treatment, and disposal of waste [R6].

COMMON USES Chemical intermediate for benzene sulfonamides, thiophenol, glybuzole (hypoglycemic agent), N-2-chloroethyl amides, benzonitrile; for its esters—useful as insecticides, miticides; for fenson acaricide (former use) [R2]. Benzenesulfonyl chloride is a reagent for Friedel-Crafts sulfonylation [R3]. Benzenethiol is produced on a commercial scale by the red-phosphorus reduction of benzenesulfonyl chloride [R4].

OCCUPATIONAL RECOMMENDATIONS None listed.

INTERNATIONAL EXPOSURE LIMITS (In former USSR) max allowable concentration of 0.3 mg/m^3 air [R9].

NON-HUMAN TOXICITY LD$_{50}$ mouse acute oral 0.4 to 3.2 g/kg (benzenesulfonic acid) [R7].

ECOTOXICITY LC$_{50}$ brown trout yearlings 3 mg/L/48 hr (static bioassay) [R7].

BIODEGRADATION No data were located concerning the biodegradation of benzenesulfonyl chloride either in natural systems or in laboratory studies (SRC). Because benzenesulfonyl chloride rapidly hydrolyzes in water (1,SRC), biodegradation probably will not be an important process in the environment (SRC) [R8].

BIOCONCENTRATION Because benzenesulfonyl chloride rapidly hydrolyzes in water (1), bioconcentration in aquatic organisms is not expected to be a significant process (SRC) [R8].

CERCLA REPORTABLE QUANTITIES Persons in charge of vessels or facilities are required to notify the National Response Center (NRC) immediately when there is a release of this designated hazardous substance, in an amount equal to or greater than its reportable quantity of 100 lb or 45.4 kg. The toll-free number of the NRC is (800) 424–8802; in the Washington, DC, metropolitan area (202) 426–2675. The rule for determining when notification is required is stated in 40 CFR 302.4 (section IV. D. 3.b) [R10].

TSCA REQUIREMENTS Pursuant to section 8 (d) of TSCA, EPA promulgated a model health and safety data reporting rule. The section 8 (d) model rule requires manufacturers, importers, and processors of listed chemical substances and mixtures to submit to EPA copies and lists of unpublished health and safety studies. Benzenesulfonyl chloride is included on this list [R11].

RCRA REQUIREMENTS U020; as stipulated in 40 CFR 261.33, when benzenesulfonyl chloride, as a commercial chemical product or manufacturing chemical intermediate or an off-specification commercial chemical product or a manufacturing chemical intermediate, becomes a waste, it must be managed according to federal or state hazardous-waste regulations. Also defined as a hazardous waste is any residue, contaminated soil, water, or other debris resulting from the cleanup of a spill, into water or on dry land, of this waste. Generators of small quantities of this waste may qualify for partial exclusion from hazardous-waste regulations (40 CFR 261.5) [R12].

DOT *Health Hazards:* Poisonous if inhaled or swallowed. Skin contact poisonous. Contact may cause burns to skin and eyes. Fire may produce irritating or poisonous gases. Runoff from fire control or dilution water may cause pollution [R5].

Fire or Explosion: Some of these materials may burn, but none of them ignites readily. Some of these materials may ignite combustibles (wood, paper, oil, etc) [R5].

Emergency Action: Keep unnecessary people away; isolate hazard area and deny entry. Stay upwind; keep out of low areas. Positive-pressure self-contained breathing apparatus and chemical protective clothing that is specifically recommended by the shipper or manufacturer may be worn. It may provide little or no thermal protection. Structural firefighters' protective clothing is not effective with these materials. Call CHEMTREC at 1–800–424–9300 as soon as possible, especially if there is no local hazardous-materials team available [R5].

Fire: Some of these materials may react violently with water. Small Fires: Dry chemical, CO_2, water spray, or regular foam. Large Fires: Water spray, fog, or regular foam. Move container from fire area if you can do so without risk. Apply cooling water to sides of containers that are exposed to flames until well after fire is out. Stay away from ends of tanks [R5].

Spill or Leak: Do not touch or walk through spilled material; stop leak if you can do so without risk. Fully encapsulating, vapor-protective clothing should be worn for spills and leaks with no fire. Use water spray to reduce vapors. Small Spills: Take up with sand or other noncombustible absorbent material and place into containers for later disposal. Large Spills: Dike liquid spill for later disposal [R5].

First Aid: Move victim to fresh air and call emergency medical care; if not breathing, give artificial respiration; if breathing is difficult, give oxygen. In case of contact with material, immediately flush skin or eyes with running water for at least 15 minutes. Remove and isolate contaminated clothing and shoes at the site. Keep victim quiet and maintain normal body temperature. Effects may be delayed; keep victim under observation [R5].

REFERENCES

1. Budavari, S (ed.). *The Merck Index— Encyclopedia of Chemicals, Drugs, and Biologicals.* Rahway, NJ: Merck and Co., Inc., 1989. 167.

2. SRI.

3. *KIRK–Othmer Encyc Chemical Tech* 3rd ed., 1978–present V11 p. 287.

4. *KIRK–Othmer Encyc Chemical Tech* 3rd ed., 1978–present V22 p. 955.

5. U.S. Department of Transportation. *Emergency Response Guidebook 1990.* DOT P 5800.5. Washington, DC: U.S. Government Printing Office, 1990, p. G–59.

6. 40 CFR 240–280, 300–306, 702–799 (7/1/89).

7. Verschueren, K. *Handbook of Environmental Data of Organic Chemicals.* 2nd ed. New York, NY: Van Nostrand Reinhold Co., 1983. 234.

8. Haughton, A. R., et al., *J Chemical Soc Perkin Trans 2* 6: 637–43 (1975).

9. Ismagilova, A. K., V. G. Gilev, T. R. Ufim *Nauch–Issled Inst Gig Prof Zabol* 6: 151 (1971).

10. 54 FR 33419 (8/14/89).

11. 40 CFR 716.120 (7/1/88).

12. 40 CFR 261.33 (7/1/88).

■ BENZETHONIUM CHLORIDE

CAS # 121–54–0

SIC CODES 8000

SYNONYMS (2-(2-(4-diisobutylphenoxy) ethoxy) ethyl) dimethylbenzylammonium chloride • (diisobutylphenoxyethoxyethyl) dimethylbenzylammonium chloride • ammonium, benzyldimethyl (2-(2-(p-(1,1,3-tetramethylbutyl) phenoxy) ethoxy) ethyl)-, chloride • anti-germ 77 • antiseptol • banagerm • benzenemethanaminium, N, N-dimethyl-N-(2-(2-(4-(1,1,3,3-tetramethylbutyl) phenoxy) ethoxy) ethyl)-, chloride • benzethonium • benzethonium chloride 1622 • benzetonium chloride • benzyldimethyl (2-(2-(4-(1,1,3,3-tetramethylbutyl) phenoxy) ethoxy) ethyl) ammonium chloride • benzyldimethyl (2-(2-(p-(1,1,3,3-tetramethylbutyl) phenoxy) ethoxy) ethyl) ammonium chloride • benzyldimethyl-p-(1,1,3,3-tetramethylbutyl) phenoxyethoxy ethylammonium chloride • bzt • diapp • diisobutylphenoxyethoxyethyldimethyl benzyl ammonium-chloride • disilyn • formula-144 • hyamine • hyamine-1622 • inactisol • p-diisobutyl phenoxyethoxyethyl dimethyl benzylammonium chloride • p-tert-octylphenoxyethoxyethyldimethylbenzyl-ammonium chloride • phemeride • phemerol • phemerol chloride • phemersol chloride • phemithyn • polymine-d • qac • quatrachlor • sanizol • solamin • solamine • benzethoniumchloride • N, N-dimethyl-N-(2-(2-(4-(1,1,3,3-tetramethylbutyl) phenoxy) ethoxy) ethyl) benzenemethanaminium chloride • benzyldimethyl (2-(2-(p-1,1,3,3-tetramethylbutylphenoxy) ethoxy) ethyl) ammonium chloride • diisobutylphenoxyethoxyethyldimethyl benzyl ammonium chloride

MF $C_{27}H_{42}NO_2$ • Cl

MW 448.15

pH pH of 1% aqueous solution is between 4.8 and 5.5.

COLOR AND FORM Colorless crystals.

ODOR Mild.

TASTE Very bitter.

MELTING POINT 164–166°C (hot stage).

FIREFIGHTING PROCEDURES Alcohol foam, carbon dioxide, dry chemical, water fog (hyamine compounds) [R2].

SOLUBILITY Very soluble in water, giving foamy, soapy solution; soluble in alcohol, acetone, chloroform; slightly soluble in ether.

OTHER CHEMICAL/PHYSICAL PROPERTIES Splinters slightly at 120°C; optic sign; axial angle 2v = 48 deg; most fragments do not extinguish completely; hydrophilic part of molecule is positive rather than negative; detergents release anion (Cl- or Br-) (cationic detergents); eutectic temp: phenanthrene 86°C, benzanthrene 98°C. Mineral acids and many salt solutions ppt benzethonium chloride from solution more concentrated than 2% as oil that crystallizes on drying. Thin, hexagonal plates from chloroform ether (monohydrate); a synthetic quaternary ammonium compound; aqueous solution yields flocculent white precipitate with soap solutions.

PROTECTIVE EQUIPMENT Wear skin protection. Hot conditions may necessitate use of self-contained breathing apparatus [R8].

COMMON USES Medication: topical anti-infective; (vet) topical antiseptic [R4]; spermicide, astringent [R1]; antiseptic for wounds, infected skin surfaces, mild infections of the eye, and for nasal and oral mucosa; germicide for cleansing food and dairy utensils; for controlling swimming-pool algae; additive in hairdressing prepns to control dandruff; deodorant; preservative [R5]; applications (vet): topical anti-bacterial; 1:1,000 aqueous solution for cold sterilization of instruments; 1:3,000 for washing cow's teats and udder, kennels, and equipment; 1:5,000–1:10,000 in drinking water of chickens and turkeys; 1:500 topical powd insufflation for cervicitis and vaginitis in cattle; bacterial inactivator in vaccine manufacture [R6]; applications (vet): ringworm in calves has been treated with twice-weekly applications of 1:1,000 aqueous solution of phemerol applied to the area. Results have been variable [R7]; cationic detergent [R3].

MAJOR SPECIES THREATENED Detergents cause oils to be dissolved from waterfowl, rendering the birds subjected to becoming waterlogged or drowning. Fish can build up some resistance to detergents through acclimatization.

DIRECT CONTACT Mild.

GENERAL SENSATION Ingestion may cause vomiting, collapse, convulsions, coma. Forms alkaline solution.

PERSONAL SAFETY Wear skin protection. Heat or flame may necessitate self-contained breathing apparatus.

ACUTE HAZARD LEVEL Mild irritant. Moderate ingestive hazard. Ingestion of 1–3 grams can be fatal. May cause foaming. Will produce taste in water. >2,500 ppm in diet increased mortality in rats in 10–30 weeks (C1). Toxic to fish.

CHRONIC HAZARD LEVEL Dogs fed 500 ppm for 1 year showed no effect (C1).

Detergents inhibit the aeration of water bodies. Detergents are capable of solubilizing carcinogenic materials.

DEGREE OF HAZARD TO PUBLIC HEALTH Mild irritant. Moderate ingestive hazard. Emits toxic chlorides when heated to decomposition.

OCCUPATIONAL RECOMMENDATIONS None listed.

ACTION LEVEL Isolate from heat.

ON-SITE RESTORATION Treat spill with carbon or peat. Use cation exchange resin. Seek professional environmental engineering assistance through EPA's Environmental Response Team (ERT), Edison, NJ, 24-hour no., 908-548-8730.

AVAILABILITY OF COUNTERMEASURE MATERIAL Carbon—water treatment plants, sugar refineries; peat—nurseries, floral shops; cation exchangers—water softener suppliers.

DISPOSAL METHOD 1. Pour or sift onto sodium bicarbonate or a sand-soda-ash mixture (90–10). Mix and package in heavy paper cartons with plenty of paper packing to serve as fuel. Burn in an incinerator. Fire may be augmented with scrap wood. 2. The packages of #1 may be burned more effectively in an incinerator with afterburner and scrubber (alkaline). 3. The waste may be mixed with a flammable solvent (alcohol, benzene, etc.), and sprayed into the fire chamber of an incinerator with afterburners and scrubber. Mineral acids and salt solutions will precipitate as oil.

DISPOSAL NOTIFICATION Notify local air authority.

MAJOR WATER USE THREATENED Recreation, fisheries, potable supply.

PROBABLE LOCATION AND STATE Thin hexagonal plates; will be dissolved in water.

COLOR IN WATER Colorless, may produce foam.

ECOTOXICITY LC_{50} *Pimephales promelas* 1,600 µg/L/96 hr (conditions of bioassay not specified) [R9]; LC_{50} *Lepomis macrochirus* 1,400 µg/L/96 hr (conditions of bioassay not specified) [R9]; LC_{50} *Oncorhynchus kisutch* 53,000 µg/L/96 hr (conditions of bioassay not specified) [R9].

BIODEGRADATION Cationic surfactants had no bacteriostatic or bacteriolytic action on heterotrophic bacterial population isolated from sewage and river waters. Cyclical alkylammonium compounds are not degraded [R10].

FDA Manufacturers, packers, and distributors of drug and drug products for human use are responsible for complying with the labeling, certification, and usage requirements as prescribed by the Federal Food, Drug, and Cosmetic Act, as amended (secs 201–902, 52 Stat. 1,040 *et seq.*, as amended; 21 U.S.C. 321–392) [R11].

REFERENCES

1. Osol, A., and J. E. Hoover, et al. (eds.). *Remington's Pharmaceutical Sciences.* 15th ed. Easton, PA: Mack Publishing Co., 1975. 1089.

2. *Farm Chemicals Handbook 1993.* Willoughby, OH: Meister Publishing Co., 1993, p. C–185.

3. Lewis, R. J., Sr. (ed.). *Hawley's Condensed Chemical Dictionary.* 12th ed. New York, NY: Van Nostrand Reinhold Co., 1993 130.

4. Budavari, S (ed.). *The Merck Index—Encyclopedia of Chemicals, Drugs, and Biologicals.* Rahway, NJ: Merck and Co., Inc., 1989. 167.

5. SRI.

6. Rossoff, I. S. *Handbook of Veterinary Drugs.* New York: Springer Publishing Company, 1974. 41.

7. Jones, L. M., et al. *Veterinary Pharmacology and Therapeutics.* 4th ed. Ames: Iowa State University Press, 1977. 871.

8. Sittig, M. *Handbook of Toxic and Hazardous Chemicals and Carcinogens, 1985.* 2nd ed. Park Ridge, NJ: Noyes Data Corporation, 1985. 114.

9. Verschueren, K. *Handbook of Environmental Data of Organic Chemicals.* 2nd ed. New York, NY: Van Nostrand Reinhold Co., 1983. 244.

10. Baleux, B., P. Caumette. *Water Res* 11 (9): 833 (1977).

11. 21 CFR 200–299, 300–499, 820, and 860 (4/1/93).

▪ BENZILIC ACID, 4,4′-DICHLORO-, ETHYL ESTER

CAS # 510-15-6

SYNONYMS 4,4′-Dichlorbenzilsaeureaethylester (German) • 4,4′-dichlorobenzilic acid ethyl ester • benzeneacetic acid, 4-chloro-alpha-(4-chlorophenyl)-alpha-hydroxy-, ethyl ester • benzilic acid, 4,4′-dichloro-, ethyl ester • chlorbenzilate • chlorbenzylate • ent-18,596 • ethyl 2-hydroxy-2,2-bis (4-chlorophenyl) acetate • ethyl 4,4′-dichlorobenzilate • ethyl p, p′-dichlorobenzilate • ethyl-4,4′-diphenylglycollate • nci-c00408 • nci-c60413 • ethyl ester of 4,4′-dichlorobenzilic acid • ethyl 4-chloro-alpha-(4-chlorophenyl)-alpha-hydroxybenzeneacetate • ethyldichlorobenzilate • ethyl 2-hydroxy-2,2-di (p-chlorophenyl) acetate • chlorbenzilat • 4-chloro-alpha-(4-chlorophenyl)-alpha-hydroxybenzeneacetic acid ethyl ester • ethyl 4,4′-dichlorodiphenyl glycollate • ethyl di (p-chlorophenyl) glycollate • ethyl 4,4′-dichlorophenyl glycollate • kop-mite • folbex • g-338 • Geigy-338 • benzilan • acaraben • akar • akar-338 • benz-o-chlor • g23992 • acaraben-4e • acar • acarben-4e • akar-50 • Caswell no. 434 • EPA pesticide code-028801

MF $C_{16}H_{14}Cl_2O_3$

MW 325.20

COLOR AND FORM Colorless solid (pure).

DENSITY 1.2816 g/cm³ at 20°C.

BOILING POINT 146–148°C at 0.04 mm Hg.

MELTING POINT 36–37.3°C.

SOLUBILITY Solubility at 20°C: 10 mg/L water; 1 kg/kg acetone, dichloromethane, methanol, and toluene; 600 g/kg hexane; 700 g/kg 1-octanol; slightly soluble in benzene.

VAPOR PRESSURE 2.2×10^{-6} mm Hg at 20°C.

STABILITY AND SHELF LIFE Acaraben 4E shelf life of at least 3 to 5 years when stored in a dry place and minimum storage temperatures (above 32°F) are observed [R5].

OTHER CHEMICAL/PHYSICAL PROPERTIES Technical product is brownish liquid; density, 1.2816 at 20°C (about 93% pure) (technical grade).

PROTECTIVE EQUIPMENT It is particularly important to protect the skin from contamination through the proper use of protective clothing and gloves. Where indicated, respiratory protective equipment or NIOSH-approved breathing apparatus or combined respirators should be used. (pesticides, halogenated) [R8].

PREVENTATIVE MEASURES Wash hands with soap and water each time before eating, drinking, or smoking. At the end of each work day, bathe entire body with soap and plenty of water. Wear clean clothes each day, and launder before reusing [R9].

CLEANUP A high-efficiency particulate arrestor (HEPA) or charcoal filters can be used to minimize amount of carcinogen in exhausted-air-ventilated safety cabinets, lab hoods, glove boxes, or animal rooms. Filter housing that is designed so that used filters can be transferred into plastic bag without contaminating maintenance staff is available commercially. Filters should be placed in plastic bags immediately after removal. The plastic bag should be sealed immediately. The sealed bag should be labelled properly. Waste liquids should be placed or collected in proper containers for disposal. The lid should be secured and the bottles properly labelled. Once filled, bottles should be placed in plastic bag, so that outer surface is not contaminated. The plastic bag should also be sealed and labelled. Broken glassware should be decontaminated by solvent extraction, by chemical destruction, or in specially designed incinerators. (chemical carcinogens) [R10, 15].

DISPOSAL Generators of waste (equal to, or greater than 100 kg/mo) containing this contaminant, EPA hazardous waste number U038, must conform with USEPA regulations in storage, transportation, treatment, and disposal of waste [R11].

COMMON USES Acaricide in spider-mite control; used as a synergist for DDT (SRP: former use in USA) [R2]; on premises and for plant mite control. poor effectiveness versus *D. gallinae* on poultry (SRP: former use in USA) [R6]; chlorobenzilate was one of the several compounds which were effective against the citrus rust mite *Phyllocoptruta oleivora* (former use) [R7]; acaricide for citrus crops, ornamentals, cotton, (non-citrus) fruits, and nuts (SRP: former use in USA). [R1].

OCCUPATIONAL RECOMMENDATIONS None listed.

NON-HUMAN TOXICITY LD_{50} rat oral 2,784–3,880 mg/kg [R12]; LD_{50} rabbit percutaneous >10,000 mg/kg [R12]; LD_{50} rat male oral 1,040 mg/kg [R2]; LD_{50} rat female oral 1,220 mg/kg [R2]; LD_{50} rat 40 mg/kg diet [R4]; LD_{50} dog 500 mg/kg diet [R4].

ECOTOXICITY LC_{50} *Salmo gairdneri* (rainbow trout) 0.7 mg/L/96 hr at 13°C, wt 0.8 g, static bioassay [R13]; LC_{50} *Colinus virginianus* (bobwhite quail) 3,375 ppm/7 days (from table) [R14]; LC_{50} *Anas platyrhynchos* (mallard duck) >8,000 ppm/5 days (from table) [R14]; LC_{50} *Cyprinodon variegatus* (sheephead minnow) 1.0 mg/L/48 hr (from table) (conditions of bioassay not specified) [R14].

IARC SUMMARY No data are available for humans. Limited evidence of carcinogenicity in animals. Overall evaluation: group 3: The agent is not classifiable as to its carcinogenicity to humans [R15].

BIODEGRADATION The half-life of chlorobenzilate in two fine sandy soils was estimated to be 1.5–5 weeks following application of 0.5–1.0 ppm; removal was probably microbial (1). It is decarboxylated to 4,4'-dichlorobenzophenone by a yeast isolated from insecticide-treated soil under anaerobic conditions (2,3). In 22 days, 40, 29, and 39% of the ^{14}C-ring-labeled chlorobenzilate added to sediment-free water samples from 3 fresh-water lakes was converted to organic products; no $^{14}CO_2$ evolution was detected (4). Addition of sediment to the water samples from the three lakes gave $^{14}CO_2$ yields of 3.6, 0.0, and 18.3% (4). Chlorobenzilate was metabolized in water from another freshwater lake only when glucose and inorganic nutrients were added, and mineralized to $^{14}CO_2$ only when sediment was also added to the water (4) [R16].

BIOCONCENTRATION Using a reported water solubility of 13 ppm at 20°C (1), a BCF of 145 was calculated (2,SRC). Based on this estimated BCF, chlorobenzilate will not be expected to bioconcentrate in aquatic organisms (SRC). However its metabolite, dichlorobenzophenone, would be expected to bioconcentrate. Based in an estimated octanol water petroleum coefficient of 4.75 (3), its BCF is estimated to be 2,400 using a recommended regression relation (2) [R17].

RCRA REQUIREMENTS As stipulated in 40 CFR 261.33, when chlorobenzilate, as a commercial chemical product or manufacturing chemical intermediate or an off-specification commercial chemical product or a manufacturing chemical intermediate, becomes a waste, it must be managed according to federal or state hazardous waste regulations. Also defined as a hazardous waste is any residue, contaminated soil, water, or other debris resulting from the cleanup of a spill, into water or on dry land, of this waste. Generators of small quantities of this waste may qualify for partial exclusion from hazardous-waste regulations (40 CFR 261.5) [R18].

FIFRA REQUIREMENTS An interim 24-hour re-entry interval on citrus crops has been established until the Agency receives re-entry data. The Agency will make a determination as to the continued registerability of this chemical when the data requested in the Registration Standard Guidance Document are submitted and reviewed. The Agency determined in the Special Review process that chlorobenzilate end-use products be classified as Restricted Use to reduce exposure to loaders, mixers, and applicators. The chemical will continue to be classified for restricted use until the Agency receives data to re-evaluate its position [R3].

REFERENCES

1. SRI.

2. *The Merck Index.* 10th ed. Rahway, NJ: Merck Co., Inc., 1983. 298.

3. Purdue University. *National Pesticide Information Retrieval System* (1987).

4. Worthing, C. R., and S. B. Walker (eds.). *The Pesticide Manual—A World Compendium.* 8th ed. Thornton Heath, UK: The British Crop Protection Council, 1987. 162.

5. *Farm Chemicals Handbook 1989.* Willoughby, OH: Meister Publishing Co., 1989, p. C–69.

6. Rossoff, I. S. *Handbook of Veterinary Drugs.* New York: Springer Publishing Company, 1974. 104.

7. Mariconi F.A.M., et al. *Solo* 71 (2): 23–8 (1979).

8. International Labour Office. *Encyclopedia of Occupational Health and Safety.* Vols. I and II. Geneva, Switzerland: International Labour Office, 1983. 1636.

9. USEPA/OPP; *Chlorobenzilate: Position Document No 3* Appendix D, p. 1 (1979).

10. Montesano, R., H. Bartsch, E. Boyland, G. Della Porta, L. Fishbein, R. A. Griesemer, A. B. Swan, L. Tomatis, and W. Davis (eds.). *Handling Chemical Carcinogens in the Laboratory: Problems of Safety.* IARC Scientific Publications No. 33. Lyon, France: International Agency for Research on Cancer, 1979.

11. 40 CFR 240–280, 300–306, 702–799 (7/1/89).

12. Hartley, D., and H. Kidd (eds.). *The Agrochemicals Handbook.* 2nd ed. Lechworth, Herts, England: The Royal Society of Chemistry, 1987, p. A020/Aug 87.

13. U.S. Department of Interior, Fish and Wildlife Service. *Handbook of Acute Toxicity of Chemicals to Fish and Aquatic Invertebrates.* Resource Publication No. 137. Washington, DC: U.S. Government Printing Office, 1980. 82.

14. USEPA/OPP. *Chlorobenzilate: Position Document No. 3* Appendix B, p. 3 (1979).

15. IARC. *Monographs on the Evaluation of the Carcinogenic Risk of Chemicals to Man.* Geneva: World Health Organization, International Agency for Research on Cancer, 1972–present (multivolume work), p. S7 60 (1987).

16. (1) Wheeler, W. B., et al. *J Environ Qual* 2: 115–8 (1973); (2) Miyazaki, S., et al. *Appl Microbiol* 18: 972–6 (1969); (3) Miyazaki, S., et al. *J Agric Food Chem* 18: 87–91 (1970); (4) Wang, Y. S., et al. *J Agric Food Chem* 33: 495–9 (1985).

17. (1) Furer, R., M. Geiger. *Pestic Sci* 8: 337 (1977); (2) Lyman, W. J., et al. *Handbook of Chemical Property Estimation Methods* New York: McGraw-Hill, p. 5–5 (1982); (3) USEPA. *GEMS Database CLOGP3.* 90–8.

18. 40 CFR 261.33 (7/1/88).

■ BENZO (J) FLUORANTHENE

CAS # 205–82–3

SYNONYMS 10,11-benzofluoranthene • benzo-12,13-fluoranthene • dibenzo (a, jk) fluorene • 7,8-benzofluoranthene

MF $C_{20}H_{12}$

MW 252.32

COLOR AND FORM Yellow plates from alcohol, needles from acetic acid; orange needles from benzene and alcohol.

MELTING POINT 166°C.

SOLUBILITY Insoluble in water; slightly soluble in alcohol, acetic acid; soluble in hydrogen sulfide on heating.

VAPOR PRESSURE 1.50×10^{-8} mm Hg at 25°C.

OCTANOL/WATER PARTITION COEFFICIENT Estimated log K_{ow} of 6.12.

PROTECTIVE EQUIPMENT Dispensers of liquid detergent should be available. Safety pipettes should be used for all pipetting. In animal laboratory, personnel should wear protective suits (preferably disposable, one piece, and close fitting at ankles and wrists), gloves, hair covering, and overshoes. In chemical laboratory, gloves and gowns should always be worn; however, gloves should not be assumed to provide full protection. Carefully fitted masks or respirators may be necessary when working with particulates or gases, and disposable plastic aprons might provide addnl protection. Gowns should be of distinctive color, as a reminder that they are not to be worn outside the laboratory (chemical carcinogens) [R5, 8].

Operations connected with synth and purification should be carried out under well-ventilated hood. Analytical procedures should be carried out with care and vapors evolved during procedures should be removed. Expert advice should be obtained before existing fume cupboards are used and when new fume cupboards are installed. It is desirable that there be means for decreasing the rate of air extraction, so that carcinogenic powders can be handled without powder being blown around the hood. Glove boxes should be kept under negative air pressure. Air changes should be adequate, so that concn of vapors of volatile carcinogens will not occur (chemical carcinogens) [R5, 8].

Vertical-laminar-flow biological safety cabinets may be used for containment of in-vitro procedures, provided that the exhaust-air flow is sufficient to provide an inward air flow at the face opening of the cabinet, and that contaminated air plenums that are under positive pressure are leak tight. Horizontal laminar flow hoods or safety cabinets, where filtered air is blown across the working area toward the operator, should never be used. Each cabinet or fume cupboard to be used should be tested before work is begun (e.g., with fume bomb) and label fixed to it, giving date of test and avg air flow measured. This test should be repeated periodically and after any structural changes. (chemical carcinogens) [R5, 9].

Principles that apply to chem or biochem lab also apply to microbiological and cell culture labs. Special consideration should be given to route of admin. Safest method of administering volatile carcinogen is by injection of a soln. Admin by topical application, gavage, or intratracheal instillation should be performed under hood. If chem will be exhaled, animals should be kept under hood during this period. Inhalation exposure requires special equipment. Unless specifically required, routes of admin other than in the diet should be used. Mixing of carcinogen in diet should be carried out in sealed mixers under fume hood, from which the exhaust is fitted with an efficient particulate filter. Techniques for cleaning mixer and hood should be devised before expt is begun. When mixing diets, special protective clothing and, possibly, respirators may be required (chemical carcinogens) [R5, 9].

When the substance is admin in diet or applied to skin, animals should be kept in cages with solid bottoms and sides and fitted with a filter top. When volatile carcinogens are given, filter tops should not be used. Cages that have been used to house animals that received carcinogens should be decontaminated. Cage cleaning facilities should be installed in area in which carcinogens are being used, to avoid moving of contaminated cages. It is difficult to ensure that cages are decontaminated, and monitoring methods are necessary. Situations may exist in which the use of disposable cages should be recommended, depending on type and amount of carcinogen and efficiency with which it can be removed (chemical carcinogens) [R5, 10].

To eliminate risk that contamination in lab could build up during conduct of expt, periodic checks should be carried out on lab atmospheres, surfaces such as walls, floors, and benches, and interior of fume hoods and air ducts. As well as regular monitoring, check must be carried out after cleaning up of spillage. Sensitive methods are required when testing lab atmospheres. Methods should, where possible, be simple and sensitive (chemical carcinogens) [R5, 10].

Rooms in which obvious contamination has occurred, such as spillage, should be decontaminated by lab personnel engaged in expt. Design of expt should avoid contamination of permanent equipment. Procedures should ensure that maintenance workers are not exposed to carcinogens. Particular care should be taken to avoid contamination of drains or ventilation ducts. In cleaning labs, procedures should be used that do not produce aerosols or dispersal of dust; e.g., wet mop or vacuum cleaner equipped with high-efficiency particulate filter on exhaust, which are avail commercially, should be used. Sweeping, brushing, and use of dry dusters or mops should be prohibited. Grossly contaminated cleaning materials should not be reused. If gowns or towels are contaminated, they should not be sent to laundry, but decontaminated or burned, to avoid any hazard to laundry personnel (chemical carcinogens) [R4, 10].

Doors leading into areas where carcinogens are used should be marked distinctively with appropriate labels. Access should be limited to persons involved in expt. A prominently displayed notice should give the name of the scientific investigator or other person who can advise in an emergency and who can inform others (such as firefighters) on the handling of carcinogenic substances (chemical carcinogens) [R5, 11].

SRP: Contaminated protective clothing should be segregated in such a manner that there is no direct personal contact by personnel who handle, dispose of, or clean the clothing. Quality assurance to ascertain the completeness of the cleaning procedures should be implemented before the decontaminated protective clothing is returned for reuse by the workers. Contaminated clothing should not be taken home at end of shift, but should remain at employee's place of work for cleaning.

PREVENTATIVE MEASURES Smoking, drinking, eating, storage of food, or of food and beverage containers, or utensils, and the application of cosmetics should be prohibited in any laboratory. All personnel should remove gloves, if worn, after completion of procedures in which carcinogens have been used. They should wash hands, preferably using dispensers of liquid detergent, and rinse thoroughly. Consideration should be given to appropriate methods for cleaning the skin, depending on nature of the contaminant. No standard procedure can be recommended, but the use of organic solvents should be avoided. Safety pipettes should be used for all pipetting (chemical carcinogens) [R5, 8].

CLEANUP A high-efficiency particulate arrestor (HEPA) or charcoal filters can be used to minimize amount of carcinogen in exhausted-air-ventilated safety cabinets, lab hoods, glove boxes, or animal rooms Filter housing that is designed so that used filters can be transferred into plastic bag without contaminating maintenance staff is avail commercially. Filters should be placed in plastic bags immediately after removal. The plastic bag should be sealed immediately. The sealed bag should be labelled properly. Waste liquids should be placed or collected in proper containers for disposal. The lid should be secured and the bottles properly labelled. Once filled, bottles should be placed in plastic bag, so that outer surface is not contaminated. The plastic bag should also be sealed and labelled. Broken glassware should be decontaminated by solvent extraction, by chemical destruction, or in specially designed incinerators (chemical carcinogens) [R5, 15].

DISPOSAL SRP: At the time of review, criteria for land treatment or burial (sanitary landfill) disposal practices are subject to significant revision. Prior to implementing land disposal of waste residue (including waste sludge), consult with environmental regulatory agencies for guidance on acceptable disposal practices. There is no universal method of disposal that has been proved satisfactory for all carcinogenic compounds, and specific methods of chem destruction published have not been tested on all kinds of carcinogen-containing waste. Summary of avail methods and recommendations given must be treated as guide only (chemical carcinogens) [R5, 14].

Incineration may be only feasible method for disposal of contaminated laboratory waste from biological expt. However, not all incinerators are suitable for this purpose. The most efficient type is probably the gas-fired type, in which a first-stage combustion with a less-than-stoichiometric air-fuel ratio is followed by a second stage with excess air. Some are designed to accept aqueous and organic solvent solutions; otherwise, it is necessary to absorb soln onto suitable combustible material, such as sawdust. Alternatively, chem destruction may be used, especially when small quantities are to be destroyed in laboratory (chemical carcinogens) [R5, 15]. HEPA (high-efficiency particulate arrestor) filters can be disposed of by incineration. For spent charcoal filters, the adsorbed material can be stripped off at high temp and carcinogenic wastes generated by this treatment conducted to and burned in an incinerator. Liquid waste: Disposal should be carried out by incineration at temp that ensure complete combustion. Solid waste: Carcasses of lab animals, cage litter, and misc solid wastes should be disposed of by incineration at temp high enough to ensure destruction of chem carcinogens or their metabolites (chemical carcinogens) [R5, 15].

Small quantities of some carcinogens can be destroyed using chemical reactions but no general rules can be given. As a general technique, treatment with sodium dichromate in strong sulfuric acid can be used. The time necessary for destruction is seldom known but 1–2 days is generally considered sufficient when freshly prepd reagent is used. Carcinogens that are easily oxidizable can be destroyed with milder oxidative agents, such as saturated soln of potassium permanganate in acetone, which appears to be a suitable agent for destruction of hydrazines or of compounds containing isolated carbon–carbon double bonds. Concentrated or 50% aqueous sodium hypochlorite can also be used as an oxidizing agent (chemical carcinogens) [R5, 16].

Carcinogens that are alkylating, arylating, or acylating agents per se can be destroyed by reaction with appropriate nucleophiles, such as water, hydroxyl ions, ammonia, thiols, and thiosulfate. The reactivity of various alkylating agents varies greatly and is also influenced by sol of agent in the reaction medium. To facilitate the complete reaction, it is suggested that the agents be dissolved in ethanol or similar solvents. No method should be applied until it has been thoroughly tested for its effectiveness and safety on material to be inactivated. For example, in case of destruction of alkylating agents, it is possible to detect residual compounds by reaction with 4(4-nitrobenzyl)-pyridine. (chemical carcinogens) [R4, 17].

COMMON USES Experimental carcinogen; biochemical research [R1].

OCCUPATIONAL RECOMMENDATIONS

IARC SUMMARY No data are available for humans. Sufficient evidence of carcinogenicity in animals. Overall evaluation: Group 2B: The agent is possibly carcinogenic to humans [R2].

BIODEGRADATION After 1,280 days in soil treated with oil sludge, 79% of the original benzo (j) fluoranthene was recovered (1) [R3].

BIOCONCENTRATION Based on an estimated water solubility of 6.76×10^{-3} mg/L at 25°C (1) and an estimated log K_{ow} of 6.12 (2), the log BCF of benzo (j) fluoranthene has been calculated to range from 4.01 to 4.42 from various regression-derived equations (3,SRC). These log BCF values suggest benzo (j) fluoranthene has the potential to bioconcentrate in aquatic systems (SRC) [R4].

REFERENCES

1. SRI.

2. IARC. *Monographs on the Evaluation of the Carcinogenic Risk of Chemicals to Man.* Geneva: World Health Organiza-

tion, International Agency for Research on Cancer, 1972–present (multivolume work), p. S7 58 (1987).

3. Bossert, I. D., et al. *Applied Environ Microbiol* 47: 763–7 (1984).

4. (1) PCCHEM. *PCGEMS Graphical Exposure Modeling System* USEPA (1987). (2) CLOGP. *PCGEMS Graphical Exposure Modeling System* USEPA (1986). (3) Lyman, W. J., et al., *Handbook of Chemical Property Estimation Methods.* New York: McGraw-Hill, p. 5–4, 5–10 (1982).

5. Montesano, R., H. Bartsch, E. Boyland, G. Della Porta, L. Fishbein, R. A. Griesemer, A.B. Swan, L. Tomatis, and W. Davis (eds.). *Handling Chemical Carcinogens in the Laboratory: Problems of Safety.* IARC Scientific Publications No. 33. Lyon, France: International Agency for Research on Cancer, 1979.

■ BENZOPHENONE

CAS # 119–61–9

SIC CODES 2834; 2844; 2879; 2834; 3851

SYNONYMS benzoylbenzene • diphenyl ketone • phenyl ketone • biphenyl methanone • alpha-oxoditane •alpha-oxodiphenylmethane • diphenylmethanone • phenyl ketone • diphenyl ketone • benzoylbenzene • diphenylketone

MF $C_{13}H_{10}O$

MW 182.23

COLOR AND FORM Orthorhombic bisphenoidal prisms from alcohol or ether; white prisms.

ODOR Geranium-like odor; rose-like odor.

DENSITY 1.1108 at 18°C.

SURFACE TENSION 45.1 dyne/cm at 20°C.

BOILING POINT 305.4°C at 760 mm Hg.

MELTING POINT 48.5°C.

HEAT OF VAPORIZATION 23.86 cal/g = 99.83 J/g = 18,191 J/mol.

SOLUBILITY Insoluble in water; 1 g dissolves in 7.5 ml alcohol, 6 ml ether; soluble in chloroform; soluble in acetone and benzene.

VAPOR PRESSURE 1 mm Hg at 108.2°C.

COMMON USES Fixative for heavy perfumes, such as geranium, new-mown hay, especially when used in soaps. In the manufacture of antihistamines, hypnotics, insecticides [R1]. Organic synthesis; derivatives are used as ultraviolet absorbers; flavoring; polymerization inhibitor for styrene [R2].

DIRECT CONTACT Skin: toxicity unknown.

GENERAL SENSATION Persistent rose-like odor; geranium-like odor.

OCCUPATIONAL RECOMMENDATIONS None listed.

ACTION LEVEL Remove sources of spark or fire.

ON-SITE RESTORATION Dredge bottom. Seek professional environmental engineering assistance through EPA's Environmental Response Team (ERT), Edison, NJ, 24-hour no., 908–548–8730.

BEACH OR SHORE RESTORATION Burn off.

AVAILABILITY OF COUNTERMEASURE MATERIAL Vacuum or dredges—local swimming-pool suppliers.

DISPOSAL METHOD Incinerate.

DISPOSAL NOTIFICATION Local air authority.

PROBABLE LOCATION AND STATE Rhombic, white crystals. Will sink and remain on bottom.

COLOR IN WATER Colorless.

NON-HUMAN TOXICITY LC$_{50}$ *Pimephales promelas* (fathead minnow) 15.3 mg/L/96 hr (confidence limit 14.4–16.3 mg/L), flow-through bioassay with measured concentrations, 25.3°C, dissolved oxygen 6.9 mg/L, hardness 47.9 mg/L calcium carbonate, alkalinity 34.0 mg/L calcium carbonate, and pH 7.72 [R4]. EC$_{50}$ *Pimephales promelas* (fathead minnow) 15.3 mg/L/96 hr (confidence limit 14.4–16.3 mg/L), flow-through bioassay with measured concentrations, 25.3°C, dissolved oxygen 6.9 mg/L, hardness 47.9 mg/L calcium carbonate, alkalinity 34.0 mg/L calcium carbonate, and pH 7.72. Effect: loss of equilibrium [R4].

BIODEGRADATION Benzophenone exhibited 12% BODT over an incubation period of 5 days in an aerobic screening study using sewage inoculum (1). According to one study, microbial degradation was considered to be insignificant when feed solutions containing 5.8×10^{-5}, 1.3×10^{-4}, 1.1×10^{-4}, and 2×10^{-3} ppm benzophenone were treated in soil columns and 15, 41, 45, and 40% removal was observed, respectively (2). However, an increase in column effluent concentration of benzophenone was observed when mercuric chloride was added to the feed solution, indicating that some biodegradation took place in the soil column (2). In another soil column study, anaerobic conditions were simulated by flooding the column; anaerobic conditions inhibited biotic activity (3) [R5].

BIOCONCENTRATION Based on an experimental log K_{ow} of 3.18 (2) and two regression-derived equations (1), the BCF range for benzophenone can be estimated to be 70–90 (SRC). This BCF range suggests that benzophenone will not bioconcentrate in aquatic organisms (SRC) [R6].

ATMOSPHERIC STANDARDS This action promulgates standards of performance for equipment leaks of volatile organic compounds (VOC) in the synthetic organic chemical manufacturing industry (SOCMI). The intended effect of these standards is to require all newly constructed, modified, and reconstructed SOCMI process units to use the best demonstrated system of continuous emission reduction for equipment leaks of VOC, considering costs, non-air-quality health, and environmental impact and energy requirements. Benzophenone is produced, as an intermediate or a final product, by process units covered under this subpart [R7].

FIRE POTENTIAL Combustible [R2]; fire hazard: slight, when heated [R3].

FDA Benzophenone is a food additive permitted for direct addition to food for human consumption, as long as (1) the quantity added to food does not exceed the amount reasonably required to accomplish its intended physical, nutritive, or other technical effect in food, and (2) when intended for use in or on food it is of appropriate food grade and is prepared and handled as a food ingredient. Synthetic flavoring substances and adjuvants (includes benzophenone) may be safely used in foods [R8].

REFERENCES

1. Budavari, S. (ed.). *The Merck Index— Encyclopedia of Chemicals, Drugs, and Biologicals.* Rahway, NJ: Merck and Co., Inc., 1989. 171.

2. Sax, N. I., and R. J. Lewis, Sr (eds.). *Hawley's Condensed Chemical Dictionary.* 11th ed. New York: Van Nostrand Reinhold Co., 1987. 132.

3. Sax, N. I. *Dangerous Properties of Industrial Materials.* 6th ed. New York, NY: Van Nostrand Reinhold, 1984. 382.

4. Brooke, L. T., D. J. Call, D. T. Geiger, and C. E. Northcott (eds.). *Acute Toxicities of Organic Chemicals to Fathead Minnows (Pimephales promelas).* Superior, WI: Center for Lake Superior Environmental Studies, Univ. of Wisconsin-Superior, 1984. 392.

5. (1) Dore, M., et al., *Trib Cebedeau* 28: 3–11 (1975). (2) Hutchins, S. R., et al., *Environ Tox Chem* 2: 195–216 (1983). (3) Hutchins, S. R., et al., *Appl Environ Microb* 48: 1046–8 (1984).

6. (1) Lyman, W. J., et al., *Handbook of Chemical Property Estimation Methods.* Washington, DC: Amer Chemical Soc pp. 4–9, 5–4, 5–10, 7–4, 7–5, 15–15 to 15–32 (1990). (2) Hansch, C., and A. J. Leo. *Medchem Project,* Issue No. 26, Claremont, CA: Pomona College (1985).

7. 40 CFR 60.489 (7/1/91).

8. 21 CFR 172.515 (4/1/91).

■ BENZYL ACETATE

CAS # 140–11–4

SYNONYMS acetic acid, benzyl ester • acetic acid, phenylmethyl ester • alpha-acetoxytoluene • benzyl ethanoate • acetic acid benzyl ester • acetic acid phenylmethyl ester • phenylmethyl acetate • nci-c06508

MF $C_9H_{10}O_2$

MW 150.19

COLOR AND FORM Water-white liquid.

ODOR Pear-like odor; characteristic flowery (jasmine) odor.

TASTE Bitter, pungent taste.

DENSITY 1.050 at 25°C.

BOILING POINT 213°C.

MELTING POINT −51°C.

AUTOIGNITION TEMPERATURE 460°C [R5].

FIREFIGHTING PROCEDURES Alcohol foam, carbon dioxide [R4]. Extinguishing methods: Alcohol foam, water, or foam may cause frothing [R5].

SOLUBILITY Practically insoluble in water; miscible with alcohol, ether; soluble in benzene, chloroform; soluble in acetone.

VAPOR PRESSURE 1.9 mm Hg at 60°C.

OTHER CHEMICAL/PHYSICAL PROPERTIES Bp: 134°C at 102 mm Hg.

PREVENTATIVE MEASURES SRP: The scientific literature supports the wearing of contact lenses in industrial environments, as part of a program to protect the eye against chemical compounds and minerals causing eye irritation. However, there may be individual substances whose irritating or corrosive properties are such that the wearing of contact lenses would be harmful to the eye. In those specific cases contact lenses should not be worn.

COMMON USES Solvent for cellulose acetate and nitrate [R2]; varnish removers [R1]; fragrance ingredient in soaps, perfumes, and cosmetics; synthetic flavor in chewing gum, candy, gelatin, ice cream, beverages, baked goods; solvent for resins, oils, lacquers, polishes, inks [R3].

OCCUPATIONAL RECOMMENDATIONS

THRESHOLD LIMIT Notice of Intended Changes (1993–94): These substances, with their corresponding values, comprise those for which either a limit has been proposed for the first time, for which a change in the "Adopted" listing has been proposed, or for which retention on the Notice of Intended Changes has been proposed. In all cases, the proposed limits should be considered trial limits that will remain in the listing for a period of at least one year. If, after one year no evidence comes to light that questions the appropriateness of the values herein, the values will be reconsidered for the "Adopted" list. Time-weighted average (TWA) 10 ppm, 61 mg/m³ [R7]. A4. A4 = Not Classifiable as a Human Carcinogen [R7].

NON-HUMAN TOXICITY Rat (Osborne–Mendel) oral 2.49 g/kg [R8].

IARC SUMMARY No data are available for humans. Limited evidence of carcinogenicity in animals. Overall evaluation: Group 3: The agent is not classifiable as to its carcinogenicity to humans [R6].

FIRE POTENTIAL Slight, when exposed to heat or flame; can react with oxidizing materials [R4].

GENERAL RESPONSE Stop discharge if possible. Keep people away. Avoid contact with liquid and vapor. Call fire department. Isolate and remove discharged material. Notify local health and pollution control agencies.

FIRE RESPONSE Combustible. Wear self-contained breathing apparatus and protective clothing. Extinguish with water sprays, dry chemical, alcohol foam, or carbon dioxide. Cool exposed containers with water.

EXPOSURE RESPONSE Call for medical aid. Vapor irritating to eyes, nose, and throat. Move to fresh air. If breathing has stopped, give artificial respiration. If breathing is difficult, give oxygen. Liquid irritating to skin and eyes. Remove contaminated clothing and shoes. Flush affected areas with plenty of water. If in eyes, hold eyelids open and flush with plenty of water.

RESPONSE TO DISCHARGE Restrict access.

Mechanical containment. Should be removed. Chemical and physical treatment.

WATER RESPONSE Effect of low concentrations on aquatic life is unknown. May be dangerous if it enters water intakes. Notify local health and pollution control agencies. Notify operators of nearby water intakes.

EXPOSURE SYMPTOMS Harmful if inhaled. May be harmful if swallowed or absorbed through the skin. Vapor or mist is irritating to the eyes, mucous membranes, and upper respiratory tract.

EXPOSURE TREATMENT Inhalation: Call for medical aid. Remove the victim to fresh air. If not breathing, give artificial respiration. If breathing is difficult, give oxygen. Eyes or skin: immediately flush with copious amounts of water for at least 15 minutes.

FDA Benzyl acetate is a food additive permitted for direct addition to food for human consumption, as long as (1) the quantity added to food does not exceed the amount reasonably required to accomplish its intended physical, nutritive, or other technical effect in food, and (2) when intended for use in or on food it is of appropriate food grade and is prepared and handled as a food ingredient. Synthetic flavoring substances and adjuvants including benzyl acetate may be safely used in foods [R9].

REFERENCES

1. Lewis, R. J., Sr. (Ed.). *Hawley's Condensed Chemical Dictionary.* 12th ed. New York, NY: Van Nostrand Reinhold Co., 1993 134.

2. Budavari, S. (ed.). *The Merck Index— Encyclopedia of Chemicals, Drugs, and Biologicals.* Rahway, NJ: Merck, and Co., Inc., 1989. 176.

3. SRI.

4. Sax, N. I. *Dangerous Properties of Industrial Materials,.* 6th ed. New York, NY: Van Nostrand Reinhold, 1984. 399.

5. National Fire Protection Guide. *Fire Protection Guide on Hazardous Materials.* 10th ed. Quincy, MA: National Fire Protection Association, 1991, p. 325M–18.

6. IARC. *Monographs on the Evaluation of the Carcinogenic Risk of Chemicals to Man.* Geneva: World Health Organization, International Agency for Research on Cancer, 1972–present (multivolume work), p. S7 58 (1987).

7. American Conference of Governmental Industrial Hygienists. *Threshold Limit Values for Chemical Substances and Physical Agents and Biological Exposure Indices for 1994–1995.* Cincinnati, OH: ACGIH, 1994. 37.

8. DHHS/NTP. *Toxicology and Carcinogenesis Studies of Benzyl Acetate in F344/N Rats and B6C3F1 Mice* (Gavage Studies) p. 18 (1986) Technical Rpt Series No. 250 NIH Pub No. 86–2506.

9. 21 CFR 172.515 (4/1/91).

■ BENZYL BENZOATE

CAS # 120-51-4

SIC CODES 2869; 2844; 2879; 2861; 2261

SYNONYMS Ascabin • ascabiol • benylate • benzoic acid, benzyl ester • benzoic acid, phenylmethyl ester • benzyl alcohol benzoic ester • benzyl benzenecarboxylate • benzyl phenylformate • benzylets • colebenz • novoscabin • peruscabin • scabagen • scabanca • scabide • scabiozon • scobenol • vanzoate • venzonate • fema number 2138

MF $C_{14}H_{12}O_2$

MW 212.26

pH Solution in alcohol practically neutral to moistened litmus paper.

COLOR AND FORM Leaflets or oily liquid; clear, colorless liquid.

ODOR Faint, pleasant, aromatic odor.

TASTE Sharp burning taste.

DENSITY 1.118 at 25°C.

SURFACE TENSION 26.6 dyne/cm at 210.5°C.

VISCOSITY 8.292 cP at 25°C.

BOILING POINT 323–324°C.

MELTING POINT 21°C.

HEAT OF COMBUSTION -6.69×10^9 J/kmol.

HEAT OF VAPORIZATION 18.6 kcal/mol at 12–60°C.

AUTOIGNITION TEMPERATURE 480°C (896°F) [R5].

FIREFIGHTING PROCEDURES Water, foam, carbon dioxide, water spray, or mist, dry chemical [R10, 379].

SOLUBILITY Insoluble in water or glycerol; miscible with alcohol, chloroform, ether, oils; soluble in acetone, benzene.

VAPOR PRESSURE 0.000224 mm Hg at 25°C.

OCTANOL/WATER PARTITION COEFFICIENT Log K_{ow} = 3.97.

STABILITY AND SHELF LIFE A benzyl benzoate emulsion (20%) made with emulsifier OS 20 and wool wax alcohols was stable and retained its effectiveness for approximately 2 yr [R6].

OTHER CHEMICAL/PHYSICAL PROPERTIES Congeals at temperature not below 18°C; supercools easily; conversion factors: 8.66 mg/m³ is equivalent to 1 ppm (wt/vol); heat of formation = -1.91×10^8 J/kmol; liquid molar volume = 0.190348 m³/kmol.

PREVENTATIVE MEASURES SRP: The scientific literature for the use of contact lenses in industry is conflicting. The benefit or detrimental effects of wearing contact lenses depend not only upon the substance, but also on factors including the form of the substance, characteristics and duration of the exposure, the uses of other eye protection equipment, and the hygiene of the lenses. However, there may be individual substances whose irritating or corrosive properties are such that the wearing of contact lenses would be harmful to the eye. In those specific cases, contact lenses should not be worn. In any event, the usual eye protection equipment should be worn even when contact lenses are in place.

DISPOSAL SRP: At the time of review, criteria for land treatment or burial (sanitary landfill) disposal practices are subject to significant revision. Prior to implementing land disposal of waste residue (including waste sludge), consult with environmental regulatory agencies for guidance on acceptable disposal practices.

COMMON USES Camphor substitute in celluloid and plastic pyroxylin compd; pediculicide; acaricide [R2]; in synthetic musks, confectionery flavors, and chewing gum flavors; fixative; plasticizer for cellulose acetate and nitrocellulose; remedy for scabies; dye carrier; antispasmodic [R3]; repellant for chiggers, mosquitoes, and ticks on man [R1]; was used in the Vietnam War to eradicate and repel certain ticks and mites [R4].

STANDARD CODES NFPA 704m system 1, 1,-.

MAJOR SPECIES THREATENED Benthic life.

DIRECT CONTACT Vapors at high temperature may be irritating to respiratory tract. Slight hazard on ingestion.

GENERAL SENSATION May cause severe skin iritation. Faint, pleasant, aromatic odor, sharp burning taste.

PERSONAL SAFETY Avoid contact with eyes. Wear goggles and protect skin.

ACUTE HAZARD LEVEL Irritant and ingestive toxin.

DEGREE OF HAZARD TO PUBLIC HEALTH Irritant. May be toxic by dermal penetration.

OCCUPATIONAL RECOMMENDATIONS None listed.

ACTION LEVEL Clear downstream recreation areas.

ON-SITE RESTORATION Dredge or vacuum from bottom. Seek professional environmental engineering assistance through EPA's Environmental Response Team (ERT), Edison, NJ, 24-hour no., 908–548–8730.

BEACH OR SHORE RESTORATION Burn off.

AVAILABILITY OF COUNTERMEASURE MATERIAL Vacuum or dredges—swimming-pool suppliers.

DISPOSAL METHOD (1) A gas—pipe the gas into the incinerator. Or lower into a pit and allow it to burn away; (2) a liquid—atomize into an incinerator. Combustion may be improved by mixing with a more flammable solvent; (3) a solid—make up packages in paper or other flammable material. Burn in the incinerator. Or the solid may be dissolved in a flammable solvent and sprayed into the fire chamber.

DISPOSAL NOTIFICATION Local air and water authorities.

MAJOR WATER USE THREATENED Recreation.

PROBABLE LOCATION AND STATE Colorless leaflets or oily liquid; will sink and remain on bottom.

COLOR IN WATER Colorless.

NON-HUMAN TOXICITY LD_{50} rat oral 1,700 mg/kg [R10, 378]; LD_{50} rat dermal 4,000 mg/kg [R10, 378]; LD_{50} mouse, oral 1,400 mg/kg [R10, 378]; LD_{50} cat, oral 2,240 mg/kg [R10, 378]; LD_{50} rabbit, oral 1,680 mg/kg [R10, 378]; LD_{50} rabbit dermal 4,000 mg/kg [R10, 378]; LD_{50} guinea pig, oral 1,121 mg/kg [R11, 1,505]; LD_{50} dog, oral >22,440 mg/kg [R11, 1,505].

ATMOSPHERIC STANDARDS This action promulgates standards of performance for equipment leaks of volatile organic compounds (VOC) in the synthetic organic chemical manufacturing industry (SOCMI). The intended effect of these standards is to require all newly constructed, modified, and reconstructed SOCMI process units to use the best demonstrated system of continuous emission reduction for equipment leaks of VOC, considering costs, non-air-quality health, and environmental impact and energy requirements. Benzyl benzoate is produced, as an intermediate or a final product, by process units covered under this subpart [R7].

FIRE POTENTIAL Fire hazard slight when exposed to heat or flame [R10, 379].

FDA Benzyl benzoate is a food additive permitted for direct addition to food for human consumption as a synthetic flavoring substance and adjuvant in accordance with the following conditions: (1) the quantity added to food does not exceed the amount reasonably required to accomplish its intended physical, nutritive, or other technical effect in food, and (2) when intended for use in or on food it is of appropriate food grade and is prepared and handled as a food ingredient [R8]. Benzyl benzoate is an indirect food additive for use as a component of adhesives [R9].

REFERENCES

1. *Farm Chemicals Handbook 87.* Willoughby, OH: Meister Publishing Co., 1987. p. C–36.

2. Budavari, S. (ed.) *The Merck Index—Encyclopedia of Chemicals, Drugs, and Biologicals.* Rahway, NJ: Merck, and Co., Inc., 1989. 176.

3. SRI.

4. *Kirk–Othmer Encyclopedia of Chemical Technology.* 4th ed. Volumes 1: New York, NY. John Wiley and Sons, 1991–Present., p. V4 114.

5. National Fire Protection Guide. *Fire Protection Guide on Hazardous Materials.* 10th ed. Quincy, MA: National Fire Protection Association, 1991, p. 325M–18.

6. Bashura, G. S., et al. *Khim Farm ZH* 16 (1) 91 (1982).

7. 40 CFR 60.489 (7/1/92).

8. 21 CFR 172.515 (4/1/93).

9. 21 CFR 175.105 (4/1/93).

10. Sax, N.I. *Dangerous Properties of Industrial Materials.* 6th ed. New York, NY: Van Nostrand Reinhold, 1984.

11. Hayes, W.J., Jr., E.R. Laws, Jr. (eds.). *Handbook of Pesticide Toxicology.* Volume 3. Classes of Pesticides. New York, NY: Academic Press, Inc., 1991.

■ BENZYL CHLOROFORMATE

CH₂OCOCl

CAS # 501–53–1

DOT # 1739

SYNONYMS benzyl chlorocarbonate • benzylcarbonochloridate • benzylcarbonyl chloride • benzyloxycarbonyl chloride • carbobenzoxy chloride • carbobenzyloxy chloride • carbonochloridic acid, phenylmethyl ester • chloroformic acid benzyl ester • chloroformic acid, benzyl ester • formic acid, chloro-, benzyl ester • bzcf

MF C₈H₇ClO₂

MW 170.60

COLOR AND FORM Oily liquid; colorless to pale-yellow liquid.

ODOR Acrid odor; odor of phosgene; sharp, penetrating odor.

DENSITY 1.2166 20°C.

SURFACE TENSION 25 dynes/cm = 0.025 newtons/m at 20°C (estimated).

VISCOSITY 2.57 mPa-s (cP) at 20°C.

BOILING POINT 152°C at 760 mm Hg.

HEAT OF COMBUSTION −10.000 Btu/lb = −5,700 calories/g = −240×10⁵ J/kg (estimated).

HEAT OF VAPORIZATION 90 Btu/lb = 50 calories/g = 2.1×10⁵ J/kg (estimated).

EXPLOSIVE LIMITS AND POTENTIAL Containers may explode (during fire) [R2].

FIREFIGHTING PROCEDURES Extinguish with dry chemicals, foam, or carbon dioxide [R2]. Use dry chemical, dry sand, or carbon dioxide. Do not use water on material itself. If large quantities of combustibles are involved, use water in flooding quantities as spray and fog. Use water spray to knock down vapors. Cool all affected containers with flooding quanti-

ties of water. Apply water from as far a distance as possible [R5].

TOXIC COMBUSTION PRODUCTS Poisonous gases are produced in fire [R2].

SOLUBILITY Soluble in ether, acetone, benzene.

VAPOR PRESSURE 7 mm Hg at 85–87°C.

CORROSIVITY Corrosive.

OTHER CHEMICAL/PHYSICAL PROPERTIES Decomposition to CO₂ and benzyl chloride upon heating at 100–155°C; decomposition in hot ethanol and water; conversion factors: 6.96 mg/m³ = 1 ppm; reacts with water to form hydrochloric acid; flash point: 80.0°C (tag open cup), 107.9°C (tag closed cup).

PROTECTIVE EQUIPMENT Self-contained breathing apparatus or acid-type canister mask; goggles or face shield; rubber gloves; protective clothing [R2].

PREVENTATIVE MEASURES If material not on fire and not involved in fire: Keep sparks, flames, and other sources of ignition away. Keep material out of water sources and sewers. Build dikes to contain flow as necessary. Use water spray to knock down vapors. Do not use water on material itself. Neutralize spilled material with crushed limestone, soda ash, or lime [R5].

COMMON USES In peptide synthesis to block the amino group [R1]; reactive intermediate to pharmaceuticals and other chemicals [R3].

OCCUPATIONAL RECOMMENDATIONS
DOT *Health Hazards:* Poisonous if inhaled or swallowed. Contact causes severe burns to skin and eyes. Runoff from fire control or dilution water may cause pollution [R4].

Fire or Explosion: Some of these materials may burn, but none of them ignites readily. May ignite other combustible materials (wood, paper, oil, etc.). Violent reaction with water. Flammable/poisonous gases may accumulate in tanks and hopper cars. Runoff to sewer may create fire or explosion hazard. Material may be transported in a molten form [R4].

Emergency Action: Keep unnecessary people away; isolate hazard area and deny entry. Stay upwind, out of low areas, and ventilate closed spaces before entering. Positive-pressure self-contained breathing apparatus (SCBA) and chemical protective clothing that is specifically recommended by the shipper or manufacturer may be worn. It may provide little or no thermal protection. Structural firefighters' protective clothing is not effective for these materials. Isolate the leak or spill area immediately for at least 150 feet in all directions. See the Table of Initial Isolation and Protective Action Distances. If you find the ID number and the name of the material there, begin protective action. Call emergency response telephone number on shipping paper first. If shipping paper not available or no answer, call CHEMTREC at 1-800-424-9300 [R4].

Fire: Do not get water inside container. Small Fires: Dry chemical or CO₂. Large Fires: Flood fire area with water from a distance. Do not get solid stream of water on spilled material. Move container from fire area if you can do so without risk. Apply cooling water to sides of containers that are exposed to flames until well after fire is out. Stay away from ends of tanks [R4].

Spill or Leak: Do not touch or walk through spilled material; stop leak if you can do so without risk. Fully-encapsulating, vapor-protective clothing should be worn for spills and leaks with no fire. Use water spray to reduce vapor; do not put water directly on leak, spill area, or inside container. Keep combustibles (wood, paper, oil, etc.) away from spilled material. Spills: Dike for later disposal; do not apply water unless directed to do so. Cleanup only under supervision of an expert [R4].

First Aid: Move victim to fresh air and call emergency medical care; if not breathing, give artificial respiration; if breathing is difficult, give oxygen. In case of contact with material, immediately flush skin or eyes with running water for at least 15 minutes. Speed in removing material from skin is of extreme importance. Removal of solidified molten material from skin re-

quires medical assistance. Remove and isolate contaminated clothing and shoes at the site. Keep victim quiet and maintain normal body temperature [R4].

FIRE POTENTIAL Combustible [R2].

GENERAL RESPONSE Call fire department. Avoid contact with liquid. Keep people away. Stop discharge if possible. Isolate and remove discharged material. Notify local health and pollution control agencies.

FIRE RESPONSE Combustible. Poisonous gases are produced in fire. Containers may explode in fire. Wear chemical protective suit with self-contained breathing apparatus. Extinguish with dry chemicals, foam, or carbon dioxide.

EXPOSURE RESPONSE Call for medical aid. Liquid irritating to skin and eyes. Harmful if swallowed. Remove contaminated clothing and shoes. Flush affected areas with plenty of water. If swallowed and victim is conscious, have victim drink water or milk.

RESPONSE TO DISCHARGE Issue warning—corrosive. Restrict access. Should be removed. Provide chemical and physical treatment.

WATER RESPONSE Effect of low concentrations on aquatic life is unknown. May be dangerous if it enters water intakes. Notify local health and wildlife officials. Notify operators of nearby water intakes.

EXPOSURE SYMPTOMS Inhalation causes mucous membrane irritation. Eyes are irritated by excessive exposure to vapor. Liquid causes severe irritation of eyes and irritates skin. Ingestion causes irritation of mouth and stomach.

EXPOSURE TREATMENT Inhalation: remove from exposure; support respiration; call physician. Eyes: irrigate with copious amounts of water for 15 min. Skin: flush with large quantities of water; wash with soap and water. Ingestion: give large amounts of water; do not induce vomiting.

REFERENCES

1. Budavari, S. (ed.). *The Merck Index— Encyclopedia of Chemicals, Drugs, and*

Biologicals. Rahway, NJ: Merck, and Co., Inc., 1989. 273.

2. U.S. Coast Guard, Department of Transportation. *CHRIS—Hazardous Chemical Data*. Volume II. Washington, DC: U.S. Government Printing Office, 1984–5.

3. Kuneyy, J. H., J. M. Mullican (eds.). *Chemcyclopedia*. Washington, DC: American Chemical Society, 1994. 53.

4. U.S. Department of Transportation. *Emergency Response Guidebook 1993*. DOT P 5800.6. Washington, DC: U.S. Government Printing Office, 1993, p. G–39.

5. Association of American Railroads. *Emergency Handling of Hazardous Materials in Surface Transportation*. Washington, DC: Association of American Railroads, Bureau of Explosives, 1992. 120.

■ o-BENZYL-p-CHLOROPHENOL

CAS # 120–32–1

SYNONYMS 2-benzyl-4-chlorophenol • 4-chloro-2-(phenylmethyl) phenol • 4-chloro-2-benzylphenol • 4-chloro-alpha-phenyl-o-cresol • 4-chloro-alpha-phenyl-ortho-cresol • 5-chloro-2-hydroxydiphenylmethane • benzylchlorophenol • bio-clave • chlorophene • chlorophene, usan • clorofene • clorophene • ketolin-h • neosabenyl • o-cresol, 4-chloro-alpha-phenyl •, ortho-benzyl-para-chlorophenol • orthobenzyl-p-chlorophenol • orthobenzylparachlorophenol • p-chloro-o-benzylphenol • phenol, 4-chloro-2-(phenylmethyl)- • santophen • santophen-1 • santophen-l germicide • septiphene

MF $C_{13}H_{11}ClO$

MW 218.69

COLOR AND FORM Crystals; white to light-tan or pink flakes.

ODOR Slightly phenolic.

DENSITY 1.186–1.190 at 55 deg/15.5°C.

BOILING POINT 160–162°C at 3.5 mm Hg.

MELTING POINT 48.5°C.

SOLUBILITY Insoluble in water; highly soluble in alcohol and other organic solvent; water solubility = 149 mg/L at 25°C.

VAPOR PRESSURE 1.87×10^{-7} kPa.

OCTANOL/WATER PARTITION COEFFICIENT Log K_{ow} = 3.6.

CORROSIVITY Noncorrosive to most metals.

OTHER CHEMICAL/PHYSICAL PROPERTIES Crystallizing point 45°C min; dispersible in aqueous media with aid of soaps or synthesis dispersing agents.

DISPOSAL SRP: At the time of review, criteria for land treatment, or burial (sanitary landfill) disposal practices are subject to significant revision. Prior to implementing land disposal of waste residue (including waste sludge), consult with environmental regulatory agencies for guidance on acceptable disposal practices.

COMMON USES Active principle or enhancing agent for disinfectants [R3]; germicide in disinfectant solutions and soap formulations [R1]; o-benzyl-p-chlorophenol is produced as a fungicide [R2]; fungicide, herbicide, antimicrobial [R4].

OCCUPATIONAL RECOMMENDATIONS None listed.

BIOLOGICAL HALF-LIFE Male rats given an oral dose of 69 mg/kg or 206 mg/kg ^{14}C-labeled o-benzyl-p-chlorophenol. Excretion of o-benzyl-p-chlorophenol was biphasic with an initial rapid alpha phase with a half-life of 8 to 9 hr and slightly slower beta phase estimated to have a half-life of 52 to 140 hr [R5].

BIODEGRADATION Biodegradation rate constants and half-lives of 0.21 to 1.04 days and <1 to 3 days, respectively, were reported in river die-away studies using unacclimated river water containing 0.1

mg/L o-benzyl-p-chlorophenol (1). Sewage containing 0.5 mg/L o-benzyl-p-chlorophenol had a biodegradation rate and half-life for this compound of 0.42 to 2.07/days and 1 to 2 days, respectively. Activated sludge with an initial concentration of 0.5 mg/L o-benzyl-p-chlorophenol had a biodegradation rate constant of 3.88/days and a half-life of 0.18 days as measured by DOC loss (1). Ultimate biodegradation to CO_2 and H_2O, using acclimated sewage and 20–30 mg/L o-benzyl-p-chlorophenol, took slightly longer, with a half-life of 7 to 23 days (1). o-Benzyl-p-chlorophenol undergoes rapid biodegradation after a short acclimation period in river water; o-benzyl-p-chlorophenol at a concentration of 0.1 mg/L was 78% degraded aerobically after 8 days; first respike (0.1 mg/L) on day 8 with 90% removal after 2 days; second respike of 0.1 mg/L on day 10 with 100% removal by day 13 (2). o-Benzyl-p-chlorophenol, 0.5 mg/L, added to natural domestic sewage was substantially gone within one day; respiked (1.0 mg/L) with nearly 100% removal within one day (2). Acclimated activated sludge mixed liquor in a semicontinuous activated sludge unit (o-benzyl-p-chlorophenol concentration, 2.0 mg/L) gave complete removal of DOC (dissolved, organic carbon) in 24 hours. Sludge from a semicontinuous activated sludge unit was used as inoculum (o-benzyl-p-chlorophenol, 20 mg/L); there was a one-week lag period followed by CO_2 evolution at 60% of the theoretical within 27 days (2). CO_2 evolution was used as a test method in this experiment to ensure that degradation was due to microbial action (2) [R6].

BIOCONCENTRATION A BCF value of 75 was measured for bluegill sunfish (1). Bluegill sunfish (*Lepomis machrochirus*) were exposed to 0.057 mg/L o-benzyl-p-chlorophenol for a period of 96 hours followed by a depuration period of 96 hours to determine the ability of the fish to metabolize and eliminate o-benzyl-p-chlorophenol. Depuration rate constants were from 4.67 to 6.94/days with a half-life of 0.14 days (1). The rate constant of o-benzyl-p-chlorophenol uptake in fish ranged from 11.8 to 13.8/days with a half-life of

0.06 days. These BCF value suggests that bioconcentration in aquatic organisms is not an important fate process (SRC) [R7].

TSCA REQUIREMENTS Section 8 (a) of TSCA requires manufacturers of this chemical substance to report preliminary assessment information concerned with production, use, and exposure to EPA as cited in the preamble in 51 FR 41329 [R8].

REFERENCES

1. SRI.

2. SRI. 1994 *Directory of Chemical Producers—United States of America.* Menlo Park, CA: SRI International, 1994. 800.

3. Lewis, R. J., Sr. (Ed.). *Hawley's Condensed Chemical Dictionary.* 12th ed. New York, NY: Van Nostrand Reinhold Co., 1993 376.

4. USEPA/OPP. *Status of Pesticides in Reregistration and Special Review.* p. 121 (Mar, 1992) EPA 700–R–92–004.

5. Ridley, W. P., et al., *J Toxicol Environ Health* 18 (2): 267–83 (1986).

6. (1) Werner, F. A., et al., *Arch Environ Contam Toxicol* 12: 569–75 (1983). (2) Swisher, R. D., Gledhill, W. E. *Appl Microbiol* 26: 394–8 (1974).

7. Werner, F. A., et al., *Arch Environ Contam Toxicol* 12: 569–75 (1983).

8. 40 CFR 712.30 (7/1/94).

■ BERYLLIUM OXIDE

CAS # 1304–56–9

SYNONYMS beryllia • beryllium monoxide • beryllium oxide (BeO) • bromellete • natural bromellite • thermalox-995

MF BeO

MW 25.01

COLOR AND FORM White hexagonal crystals; light, amorphous powder.

ODOR Odorless.

DENSITY 3.0 at 20°C.

BOILING POINT Approximately 3,900°C.

MELTING POINT 2,530°C.

EXPLOSIVE LIMITS AND POTENTIAL Can react explosively with magnesium when heated [R4].

TOXIC COMBUSTION PRODUCTS Toxic beryllium oxide fume may form in fire [R3].

SOLUBILITY 2 µg/100 ml in water at 30°C; soluble in concentrated sulfuric acid, fused potassium hydroxide; slowly soluble in concentrated acids or fixed alkali hydroxides.

OTHER CHEMICAL/PHYSICAL PROPERTIES Amphoteric; after ignition it is almost insoluble in water, concentrated acids, and solutions of fixed alkali hydroxides; hardness: 9.

PROTECTIVE EQUIPMENT Wear goggles and self-contained breathing apparatus [R3].

PREVENTATIVE MEASURES In animal laboratory, personnel should remove their outdoor clothes and wear protective suits (preferably disposable, one-piece, and close-fitting at ankles and wrists), gloves, hair covering, and overshoes. Clothing should be changed daily but discarded immediately if obvious contamination occurs. Also, workers should shower immediately. In chemical laboratory, gloves, and gowns should always be worn; however, gloves should not be assumed to provide full protection. Carefully fitted masks or respirators may be necessary when working with particulates or gases, and disposable plastic aprons might provide addnl protection. Gowns should be of distinctive color, as a reminder that they should not be worn outside of lab. (chemical carcinogens) [R14, 8].

CLEANUP A high-efficiency particulate arrestor (HEPA) or charcoal filters can be used to minimize amount of carcinogen in exhausted-air-ventilated safety cabinets, lab hoods, glove boxes, or animal rooms. Filter housing that is designed so that used filters can be transferred into plastic bag without contaminating maintenance staff is avail commercially. Filters should be placed in plastic bags immediately after removal. The plastic bag should be sealed immediately. The sealed bag should be labelled properly. Waste liquids should be placed or collected in proper containers for disposal. The lid should be secured and the bottles properly labelled. Once filled, bottles should be placed in plastic bag, so that outer surface is not contaminated. The plastic bag should also be sealed and labelled. Broken glassware should be decontaminated by solvent extraction, by chemical destruction, or in specially designed incinerators. (chemical carcinogens) [R14, 15].

DISPOSAL SRP: At the time of review, criteria for land treatment or burial (sanitary landfill) disposal practices are subject to significant revision. Prior to implementing land disposal of waste residue (including waste sludge), consult with environmental regulatory agencies for guidance on acceptable disposal practices.

COMMON USES Agent in glass and ceramics MFR; component of nuclear fuels and moderators; catalyst in organic reactions; refractory material; component of electron and klystron tubes; component of resistor cores and transistor mountings; chemical intermediate for other beryllium compounds [R1]. Electric heat sinks; electrical insulators; microwave oven components; gyroscopes; military vehicle armor; rocket nozzles; crucibles; thermocouple tubing; laser structural components [R2].

OCCUPATIONAL RECOMMENDATIONS

OSHA During an 8-hr work shift, an employee may be exposed to a concentration of beryllium and beryllium compounds above 5 µg/m³ (but never above 25 µg/m³) only for a maximum period of 30 min. Such exposure must be compensated by exposures to concentrations less than 2 µg/m³ so that the cumulative exposure for the entire 8-hr work shift does not exceed a weighted average of 2 µg/m³. (beryllium and beryllium compounds) [R9].

NIOSH NIOSH recommends that the substance be treated as a potential human carcinogen with a 10-hr TWA: 0.5 µg/Be/m³. (beryllium) [R10].

THRESHOLD LIMIT 8-hr time-weighted average (TWA) 0.002 mg/m³ (1979) (beryllium and compounds, as Be) [R15, 13].

ECOTOXICITY LC$_{50}$ *Pimephales promelas* (fathead minnow) 150 µg/L/96 hr, soft water, 20,000 µg/L/96 hr, hard water (beryllium ion) (static bioassay) [R6]. LC$_{50}$ *Lepomis macrochirus* (bluegill) 1,300 µg/L/96 hr, soft water, 12,000 µg/L/96 hr, hard water (beryllium ion) (static bioassay) [R6]. EC$_{50}$ cladoceran (*Daphnia magna*) 2,500 µg/L/48 hr, effects on reproduction (beryllium ion) [R7]. LC$_{50}$ *Carassius auratus* (goldfish) 4,800 µg/Be/L, flow-through test, measured (beryllium ion) (time not specified) [R8]. LC$_{50}$ *Pimephales promelas* (fathead minnow) 3,250 µg/Be/L, flow-through test, measured (beryllium ion) (time not specified) [R8]. LC$_{50}$ *Jordanella floridae* (flagfish) 3,530 µg/Be/L, flow-through test, measured (beryllium ion) (time not specified) [R8].

IARC SUMMARY Evaluation: There is sufficient evidence in humans for the carcinogenicity of beryllium and beryllium compounds. There is sufficient evidence in experimental animals for the carcinogenicity of beryllium and beryllium compounds. Overall evaluation: Beryllium and beryllium compounds are carcinogenic to humans (Group (1) [R5].

BIOLOGICAL HALF-LIFE The biological half-life is long because absorption in the lung takes place rather slowly and a proportion of beryllium or its compound is stored in the liver and skeleton. Inhaled water-soluble beryllium salts are excreted mainly by the kidneys, with a half life in the range of 2–8 wk. (beryllium) [R13, 110].

BIOCONCENTRATION Bioconcentration of 100-fold can occur under constant exposure. Not significant in spill conditions [R3].

ATMOSPHERIC STANDARDS Beryllium ambient air concentration: 0.01 µg/m³ averaged over a 30-day period in the vicinity of the stationary source (total beryllium) [R11].

RCRA REQUIREMENTS The Environmental Protection Agency has amended its regulations concerning ground-water monitoring with regard to screening suspected contamination at land-based hazardous waste treatment, storage, and disposal facilities. There are new requirements to analyze for a specified core list of chemicals plus those chemicals specified by the Regional Administrator on a site-specific basis. (Total beryllium (all species) is included on this list.) (total beryllium (all species)) [R12].

GENERAL RESPONSE Avoid contact with solid and dust. Keep people away. Wear dust respirator and rubber overclothing (including gloves). Stop discharge if possible. Isolate and remove discharged material. Notify local health and pollution control agencies.

FIRE RESPONSE Not flammable. Poisonous gases may be produced in fire. Wear goggles and self-contained breathing apparatus.

EXPOSURE RESPONSE Call for medical aid. Dust poisonous if inhaled. If inhaled will cause coughing and difficult breathing. If in eyes, hold eyelids open and flush with plenty of water. If breathing has stopped, give artificial respiration. If breathing is difficult, give oxygen. Solid poisonous if swallowed. Irritating to skin and eyes. Remove contaminated clothing and shoes. Flush affected areas with plenty of water. If in eyes, hold eyelids open and flush with plenty of water. If swallowed and victim is conscious, have victim drink water or milk and have victim induce vomiting. If swallowed and victim is unconscious or having convulsions, do nothing except keep victim warm.

RESPONSE TO DISCHARGE Issue warning—poison, water contaminant. Restrict access. Should be removed. Chemical and physical treatment.

WATER RESPONSE Effect of low concentrations on aquatic life is unknown. May be dangerous if it enters water intakes. Notify local health and wildlife officials. Notify operators of nearby water intakes.

EXPOSURE SYMPTOMS Any dramatic, unexplained weight loss should be considered as possible first indication of beryllium disease. Other symptoms include anorexia, fatigue, weakness, malaise. Inhalation causes pneumonitis, nasopharyngitis, tracheobronchitis, dyspnea, chronic cough. Contact with dust causes conjunctival inflammation of eyes and irritation of skin.

EXPOSURE TREATMENT Inhalation: take chest x-ray immediately to check for pneumonitis. Ingestion: induce vomiting; get medical attention. Eyes: flush with water for at least 15 min.; get medical attention. Skin: cuts or puncture wounds in which beryllium may be embedded under the skin should be thoroughly cleansed immediately by a physician.

FIRE POTENTIAL Not flammable [R3].

REFERENCES

1. SRI.

2. USEPA. *Health Assessment Document for Beryllium* p. 3–4 (1978) EPA 600 8–84–026F.

3. U.S. Coast Guard, Department of Transportation. *CHRIS—Hazardous Chemical Data.* Volume II. Washington, DC: U.S. Government Printing Office, 1984–5.

4. National Fire Protection Association. *Fire Protection Guide on Hazardous Materials.* 9th ed. Boston, MA: National Fire Protection Association, 1986, p. 491M–21.

5. IARC. *Monographs on the Evaluation of the Carcinogenic Risk of Chemicals to Man.* Geneva: World Health Organization, International Agency for Research on Cancer, 1972–present (multivolume work), p. 58–103 (1993).

6. Tarzwell, C. M., C. Henderson. *Toxicity of Less Common Metals to Fishes, Ind Wastes* 5: 12 (1960) as cited in USEPA, *Ambient Water Quality Criteria Doc; Beryllium* p. B–2 (1980) EPA 440/5–80–024.

7. USEPA. *Ambient Water Quality Criteria Doc; Beryllium* p. B–3 (1980) EPA 440/5–80–024.

8. Cardwell, R. D., et al., *Acute Toxicity of Selected Toxicants to Six Species of Fish* (1976) EPA 600/3–76–008 as cited in USEPA, *Ambient Water Quality Criteria Doc; Beryllium* p. B–6 (1980) EPA 440/5–80–024.

9. 29 CFR 1910.1000 (7/1/87).

10. NIOSH/CDC. *NIOSH Recommendations for Occupational Safety and Health Standards* Sept. 1986 (Supplement to *Morbidity and Mortality Weekly Report* 35 No. 15, Sept. 26, 1986), p. 7S.

11. 40 CFR 61.32 (7/1/88).

12. 52 FR 25942 (7/9/87).

13. Seiler, H.G., H. Sigel, and A. Sigel (eds.). *Handbook on the Toxicity of Inorganic Compounds.* New York, NY: Marcel Dekker, Inc. 1988.

14. Montesano, R., H. Bartsch, E. Boyland, G. Della Porta, L. Fishbein, R. A. Griesemer, A.B. Swan, L. Tomatis, and W. Davis (eds.). *Handling Chemical Carcinogens in the Laboratory: Problems of Safety.* IARC Scientific Publications No. 33. Lyon, France: International Agency for Research on Cancer, 1979.

15. American Conference of Governmental Industrial Hygienists. *Threshold Limit Values for Chemical Substances and Physical Agents and Biological Exposure Indices for 1994–1995.* Cincinnati, OH: ACGIH, 1994.

■ BERYLLIUM SULFATE (1:1)

CAS # 13510–49–1

DOT # 1566

SYNONYMS beryllium sulfate (BeSO₄) • sulfuric acid, beryllium salt (1:1)

MF O₄S•Be

MW 105.07

COLOR AND FORM Colorless crystalline solid.

ODOR Odorless.

DENSITY 2.443.

SOLUBILITY Insoluble in cold water; soluble in water: 425 g/L at 25°C; 1 kg/L at 100°C.

OTHER CHEMICAL/PHYSICAL PROPERTIES
Decomposes to $BeSO_4 \cdot 4H_2O$.

PROTECTIVE EQUIPMENT Effective venting devices installed directly at the source of developing dusts or fumes are necessary. As an added precaution glove boxes or high-power fume cabinets are recommended, and, for high-risk assignments such as maintenance work on exhaust ducts or furnaces, respirators should be used. (beryllium) [R2, 107].

PREVENTATIVE MEASURES All preventive measures are focused on the necessity to prevent inhalation or skin contact. Adequate local exhaust ventilation and good housekeeping handling of beryllium within hoods for raised horizontal surfaces (either wet mopping or hosing vacuum cleaning), wet processing. (beryllium) [R3, 81].

CLEANUP Response to discharge: Issue warning: poison: corrosive; restrict access; should be removed; chemical and physical treatment [R4].

DISPOSAL SRP: At the time of review, criteria for land treatment or burial (sanitary landfill) disposal practices are subject to significant revision. Prior to implementing land disposal of waste residue (including waste sludge), consult with environmental regulatory agencies for guidance on acceptable disposal practices.

COMMON USES Chemical intermediate for beryllium hydroxide; intermediate for beryllium oxide [R1].

BINARY REACTANTS Aluminum.

STANDARD CODES ICC—poison B, poison label, 200 lb in an outside container; IATA—poison B poison label, 25 kg passenger, 95 kg cargo.

MAJOR SPECIES THREATENED At acid pH, Be is highly toxic to plants. Above pH 11.2, Be is beneficial to plants with Mg deficiency. 15–120 mg/L Be delays germination and retards growth of cress and mustard seeds in solution culture.

INHALATION LIMIT 0.002 mg/m³.

DIRECT CONTACT Strong irritant—skin.

GENERAL SENSATION Contact dermatitis. Dusts cause severe lung damage.

PERSONAL SAFETY Wear skin protection and industrial filter masks.

ACUTE HAZARD LEVEL Poisoning occurs at 1–100 µg/m³ in air. Toxic to fish and plants. A strong irritant. Highly toxic when inhaled. Poses no real ingestion threat and hence may not be highly hazardous to potable supply.

CHRONIC HAZARD LEVEL Rats healthy after 2 years on 18 mg Be/kg/day. Marine waters should not exceed 1/100 of 96-hour LC_{50} (1.5 ppm) [R15]; dogs healthy after 18 months on 10 mg $BeSO_4$/kg/day (C1). Mild irritant and allergen. Highly toxic with chronic inhalation.

DEGREE OF HAZARD TO PUBLIC HEALTH Strong irritant. Highly toxic with acute or chronic inhalation. Chronic allergen.

OCCUPATIONAL RECOMMENDATIONS
OSHA During an 8-hr work shift, an employee may be exposed to a concentration of beryllium and beryllium compounds above 5 µg/m³ (but never above 25 µg/m³) only for a maximum period of 30 min. Such exposure must be compensated by exposures to concentrations less than 2 µg/m³ so that the cumulative exposure for the entire 8-hr work shift does not exceed a weighted average of 2 µg/m³. (beryllium and beryllium compounds) [R10].

NIOSH NIOSH recommends that the substance be treated as a potential human carcinogen with a 10-hr TWA: 0.5 µg/Be/m³. (beryllium) [R11].

THRESHOLD LIMIT 8-hr time-weighted average (TWA) 0.002 mg/m³ (1979) (beryllium and compounds, as Be) [R12, 13].

ACTION LEVEL Restrict access to area. Suppress suspension of dusts.

ON-SITE RESTORATION Use cation exchangers or precipitate with CO_2 (carbonate solution) or lime. Seek professional environmental engineering assistance through EPA's Environmental Response Team (ERT); Edison, NJ, 24-hour no. 908–548–8730.

AVAILABILITY OF COUNTERMEASURE MA-

TERIAL Cation exchangers—water softener suppliers; CO_2 (carbonate solution)—soft-drink distributors; lime—cement plants.

DISPOSAL METHOD Dissolve in minimum of $6M$ HCl. Filter and treat filtrate with slight excess of $6M$ NH_4OH (use litmus). Boil and allow coagulated precipitate to settle for about 12 hours. Filter and dry. Package and ship to the supplier.

MAJOR WATER USE THREATENED Fisheries, recreation, irrigation.

PROBABLE LOCATION AND STATE Crystals—small, well-defined, colorless, tetragonal. Solid will sink and dissolve slowly.

WATER CHEMISTRY Forms strong acid solution.

FOOD-CHAIN CONCENTRATION POTENTIAL Potential.

NON-HUMAN TOXICITY LD_{50} rat (female) 0.51 mg/Be/kg, intravenously injected [R6].

ECOTOXICITY LC_{50} *Pimephales promelas* (fathead minnow) 150 µg/L/96 hr, soft water, 20,000 µg/L/96 hr, hard water (beryllium ion) (static bioassay) [R7]; LC_{50} *Lepomis macrochirus* (bluegill) 1,300 µg/L/96 hr, soft water, 12,000 µg/L/96 hr, hard water (beryllium ion) (static bioassay) [R7]; EC_{50} cladoceran (*Daphnia magna*) 2,500 µg/L/48 hr, effects on reproduction (beryllium ion) [R8]; LC_{50} *Carassius auratus* (goldfish) 4,800 µg/Be/L, flow-through test, measured (beryllium ion) (time not specified) [R9]; LC_{50} *Pimephales promelas* (fathead minnow) 3,250 µg/Be/L, flow-through test, measured (beryllium ion) (time not specified) [R9]; LC_{50} *Jordanella floridae* (flagfish) 3,530 µg/Be/L, flow-through test, measured (beryllium ion) (time not specified) [R9].

IARC SUMMARY Evaluation: There is sufficient evidence in humans for the carcinogenicity of beryllium and beryllium compounds. There is sufficient evidence in experimental animals for the carcinogenicity of beryllium and beryllium compounds. Overall evaluation: Beryllium and beryllium compounds are carcinogenic to humans (group (1) [R5].

BIOLOGICAL HALF-LIFE The biological half-life is long because absorption in the lung takes place rather slowly and a proportion of beryllium or its compounds is stored in the liver and skeleton. Inhaled water-soluble beryllium salts are excreted mainly by the kidneys, with a half-life in the range of 2–8 wk. (soluble beryllium salts) [R2, 110].

ATMOSPHERIC STANDARDS Beryllium ambient air concentration: 0.01 µg/m³ averaged over a 30-day period in the vicinity of the stationary source (beryllium and compounds) [R13].

RCRA REQUIREMENTS The Environmental Protection Agency has amended its regulations concerning groundwater monitoring with regard to screening suspected contamination at land-based hazardous waste treatment, storage, and disposal facilities. There are new requirements to analyze for a specified core list of chemicals plus those chemicals specified by the Regional Administrator on a site-specific basis. (Total beryllium (all species) is included on this list.) (total beryllium (all species)) [R14].

GENERAL RESPONSE Avoid contact with solid and dust; keep people away. Wear dust respirator and rubber overclothing (including gloves). Stop discharge if possible. Isolate and remove discharged material. Notify local health and pollution control agencies.

FIRE RESPONSE Not flammable. Poisonous gases may be produced in fire. Wear goggles and self-contained breathing apparatus.

EXPOSURE RESPONSE Call for medical aid. Dust poisonous if inhaled or if skin is exposed. If inhaled will cause coughing or difficult breathing. If breathing has stopped, give artifical respiration. If breathing is difficult, give oxygen. Solid irritating to skin and eyes. Harmful if swallowed. Remove contaminated clothing and shoes. Flush affected areas with plenty of water. If in eyes, hold eyelids open, and flush with plenty of water. If swallowed and victim is conscious, have victim drink water or milk, and have victim induce vomiting. If swallowed and victim is unconscious or having convulsions, do nothing except keep victim warm.

RESPONSE TO DISCHARGE Issue warning—poison, water contaminant. Restrict access. Should be removed. Provide chemical and physical treatment.

WATER RESPONSE Harmful to aquatic life in very low concentrations. May be dangerous if it enters water intakes. Notify local health and wildlife officials. Notify operators of nearby water intakes.

EXPOSURE SYMPTOMS Any dramatic, unexplained weight loss should be considered as possible first indication of beryllium disease. Other symptoms include anorexia, fatigue, weakness, malaise. Inhalation causes pneumonitis, nasopharyngitis, tracheobronchitis, dyspnea, chronic cough. Contact with eyes causes conjunctival inflammation. Contact with skin causes dermatitis of primary irritant or sensitization type; causes ulcer formation when in contact with cuts.

EXPOSURE TREATMENT Inhalation: take chest x-ray immediately to check for evidence of pneumonitis. Ingestion: induce vomiting; get medical attention. Eyes: flush with water for at least 15 min.; get medical attention. Skin: cuts or puncture wounds in which beryllium may be embedded under the skin should be thoroughly cleansed immediately by a physician.

REFERENCES

1. SRI.

2. Seiler, H. G., H. Sigel, and A. Sigel (eds.). *Handbook on the Toxicity of Inorganic Compounds.* New York, NY: Marcel Dekker, Inc. 1988.

3. Browning, E. *Toxicity of Industrial Metals.* 2nd ed. New York: Appleton-Century-Crofts, 1969.

4. U.S. Coast Guard, Department of Transportation. *CHRIS—Hazardous Chemical Data.* Volume II. Washington, DC: U.S. Government Printing Office, 1984–5.

5. IARC. *Monographs on the Evaluation of the Carcinogenic Risk of Chemicals to Man.* Geneva: World Health Organization, International Agency for Research on Cancer, 1972–present (multivolume work), p. 58–103 (1993).

6. Vacher, J., H. B. Stoner. *Biochemical Pharmacol* 17: 93 (1968) as cited in USEPA. *Ambient Water Quality Criteria Doc; Beryllium* p. C–8 (1980) EPA 440/5-80-024.

7. Tarzwell, C. M., and C. Henderson. *Toxicity of Less Common Metals to Fishes, Ind Wastes* 5: 12 (1960) as cited in USEPA, *Ambient Water Quality Criteria Doc; Beryllium,* p. B–2 (1980) EPA 440/5- 80-024.

8. USEPA. *Ambient Water Quality Criteria Doc; Beryllium,* p. B–3 (1980) EPA 440/5-80-024.

9. Cardwell, R. D. et al. *Acute Toxicity of Selected Toxicants to Six Species of Fish* (1976), EPA 600/3-76-008 as cited in USEPA, *Ambient Water Quality Criteria Doc; Beryllium* p. B–6 (1980) EPA 440/5-80-024.

10. 29 CFR 1910.1000 (7/1/87).

11. NIOSH/CDC. *NIOSH Recommendations for Occupational Safety and Health Standards* Sept. 1986 (Supplement to *Morbidity and Mortality Weekly Report* 35 No. 15, Sept. 26, 1986), p. 7S.

12. American Conference of Governmental Industrial Hygienists. *Threshold Limit Values for Chemical Substances and Physical Agents and Biological Exposure Indices for 1993–1994.* Cincinnati, OH: ACGIH, 1993.

13. 40 CFR 61.32 (7/1/88).

14. 52 FR 25942 (7/9/87).

15. *Proposed Criteria for Water Quality, Volume I,* U.S. Environmental Protection Agency. October 1973.

■ BETANAL

CAS # 13684–63–4

SYNONYMS 3-((methoxycarbonyl) amino) phenyl n-(3-methylphenyl) carbamate • 3-(carbomethoxyamino) phenyl 3-methylcarbanilate • 3-methoxycarbonylaminophenyl n-3′-methylphenylcarbamate • carbamic acid, (3-methylphenyl)-, 3-((methoxycarbonyl) amino) phenyl ester • carbanilic acid, m-hydroxy, methyl ester, m-methylcarbanilate (ester) • ep-452 • fenmedifam • methyl m-hydroxycarbanilate, m-methylcarbanilate • methyl n-(3-(n-(3-methylphenyl) carbamoyloxy) phenyl) carbamate • methyl-3-hydroxycarbanilate-3-methylcarbanilate • phenmedipham • Schering-38584 • sn-4075- • spin-aid • methyl-3-m-tolycarbamoloxyphenyl carbamate • sn-38584 • kemifam

MF $C_{16}H_{16}N_2O_4$

MW 300.34

COLOR AND FORM Colorless crystals.

ODOR Odorless.

DENSITY 0.25–0.30 g/cm³ at 20°C.

MELTING POINT 139–142°C (tech. 143–144°C).

SOLUBILITY 4.7 mg/L water; about 200 g/kg in acetone, cyclohexanone; about 50 g/kg in methanol; 20 g/kg in chloroform; 2.5 g/kg in benzene; about 500 mg/kg in hexene at room temperature.

VAPOR PRESSURE 1.3 nPa at 25°C.

CORROSIVITY Noncorrosive.

STABILITY AND SHELF LIFE Very low volatility; no changes were observed when held for 6 days at 50°C; hydrolysis at 22°C in buffer solution, 50% loss occurred in 70 days at pH 5, 24 hr at pH 7, 10 min at pH 9 [R1].

OTHER CHEMICAL/PHYSICAL PROPERTIES Colorless, odorless, crystalline solid; mp 143–144°C; solubility: 1 mg/L water at 20°C, acetone 20%, cyclohexanone 20%, methanol 5% (technical material 97% pure) L, Henry's Law constant = 8.41×10^{-13} atm–m³/mole (est).

PROTECTIVE EQUIPMENT Wear goggles to prevent splashing into eyes [R3].

PREVENTATIVE MEASURES Do not breathe spray mist. Do not eat, drink, or smoke while spraying. [R8, 376].

COMMON USES Post-emergent herbicide for control of certain broadleaf and grass weeds; recommended for use in sugar beets, strawberries, and sunflowers [R2].

NON-HUMAN TOXICITY LD_{50} dog oral 4,000 mg/kg [R1]; LD_{50} guinea pig oral 4,000 mg/kg [R1]; LD_{50} chicken oral 3,000 mg/kg [R1]; LD_{50} rat percutaneous 4,000 mg/kg [R1]; LD_{50} rat oral >8,000 mg/kg [R4].

ECOTOXICITY Harlequin fish 16.5 mg/L (conditions of bioassay not specified) (emulsifiable preparation) [R5].

BIODEGRADATION In slightly acid soil of low humus content, phenmedipham decomposed with a half-life of 28 to 55 days [R6].

BIOCONCENTRATION Based upon a water solubility of 4.7 mg/L at 25°C (1), the BCF of betanal can be estimated to be 260 from a regression-derived equation (2,SRC). This BCF value suggests that bioconcentration in aquatic organisms may occur (SRC) [R7].

REFERENCES

1. Worthing, C. R., and S. B. Walker (eds.). *The Pesticide Manual—A World Compendium.* 8th ed. Thornton Heath, UK: The British Crop Protection Council, 1987. 652.

2. Spencer, E. Y. *Guide to the Chemicals Used in Crop Protection.* 7th ed. Publication 1,093. Research Institute, Agriculture Canada, Ottawa, Canada: Information Canada, 1982. 448.

3. *Farm Chemicals Handbook 1992.* Willoughby, OH: Meister Publishing Co., 1992, p. C–259.

4. Sittig, M., (ed.) *Pesticide Manufacturing and Toxic Materials Control Encyclopedia.* Park Ridge, NJ: Noyes Data Corporation. 1980. 605.

5. Hartley, D., and H. Kidd (eds.). *The Agrochemicals Handbook.* 2nd ed. Lechworth, Herts, England: The Royal Society of Chemistry, 1987, p. A319/Oct 87.

6. Menzie, C. M. *Metabolism of Pesticides,* Update II. U.S. Department of the Interior, Fish and Wildlife Service, Special Scientific Report—*Wildlife* No. 212. Washington, DC: U.S. Government Printing Office, 1978. 222.

7. (1) Wauchope, R. D., et al., *Rev Environ Contam Toxicol* 123: 1–35 (1991) (2) Lyman, W. J., et al. *Handbook of Chemical Property Estimation Methods,* Washington, DC: Amer Chemical Soc, p. 5–10 (1990).

8. Weed Science Society of America. *Herbicide Handbook.* 5th ed. Champaign, Illinois: Weed Science Society of America, 1983.

■ BIFENTHRIN

CAS # 82657–04–3

SYNONYMS cyclopropanecarboxylic acid, 3-(2-chloro-3,3,3-trifluoro-1-propenyl)-2,2-dimethyl-, (2-methyl (1,1'-biphenyl)-3-yl) methyl ester, (z)- • biphenate • 2-methylbiphenyl-3-ylmethyl (z)-(1rs, 3rs)-3-(2-chloro-3,3,3-trifluoroprop-1-enyl)-2,2-dimethylcyclopropanecarboxylate • [1 alpha, 3 alpha (z)]-(-)-(2-methyl[1,1'-biphenyl]-3-yl) methyl 3-(2-chloro 3,3,3-trifluoro-1-propenyl)-2,2-dimethylcyclopropanecarboxylate • bifenthrine • FMC-54800 • [1alpha, 3alpha (z)]-(-)-3-(2-chloro-3,3,3-trifluoro-1-propenyl)-2,2-dimethylcyclopropanecarboxylic acid (2-methyl[1,1'-biphenyl]-3-yl) methyl ester • 2-methylbiphenyl-3-ylmethyl-(z)-(1rs)-cis-3-(2-chloro-3,3,3-trifluoroprop-1-enyl)-2,2-dimethylcyclopropanecarboxylate •

biphenthrin • biphentrin • fmc-54800 • brigade • talstar • capture

MF $C_{23}H_{22}ClF_3O_2$

MW 422.90

COLOR AND FORM Light-brown viscous oil; solid (pure); viscous oil hardens to a solid, light-brown mass.

DENSITY 1.212 g/ml at 25°C.

MELTING POINT 68–70°C.

FIREFIGHTING PROCEDURES Use carbon dioxide, foam, or dry chemical (on fires involving pyrethroids) (pyrethrum) [R7, 2]. Fire-fighting: Self-contained breathing apparatus with a full facepiece operated in pressure-demand or other positive-pressure mode (pyrethrum) [R7, 5]. Extinguish fire using agent suitable for type of surrounding fire (pyrethrins) [R2].

SOLUBILITY Soluble in methylene chloride, chloroform, acetone, ether, toluene; slightly soluble in heptane, methanol; soluble in water: 0.1 mg/L; soluble in water: <0.1 ppb.

VAPOR PRESSURE 1.81×10^{-7} torr at 25°C.

CORROSIVITY Noncorrosive.

STABILITY AND SHELF LIFE Stable for over one year at 25°C and 50°C (technical bifenthrin) [R1].

OTHER CHEMICAL/PHYSICAL PROPERTIES MP: 51–66°C (technical grade).

PROTECTIVE EQUIPMENT Employees should be provided with and required to use dust- and splash-proof safety goggles where pyrethroids may contact the eyes. (pyrethroids) [R7, 3].

PREVENTATIVE MEASURES Skin that becomes contaminated with pyrethrum should be promptly washed or showered with soap or mild detergent and water. (pyrethrum) [R7, 3].

CLEANUP Spillages of pesticides at any stage of their storage or handling should be treated with great care. Liquid formulations may be reduced to solid phase by evaporation. Dry sweeping of solids is always hazardous: these should be removed by vacuum cleaning or by dissolving them in water or other solvent in the factory environment. (pesticides) [R3].

DISPOSAL Incineration would be an effective disposal procedure where permitted. If an efficient incinerator is not available, the product should be mixed with large amounts of combustible material, and contact with the smoke should be avoided. (pyrethrin products) [R4].

COMMON USES Insecticide, acaricide [R6, 189].

NON-HUMAN TOXICITY LD$_{50}$ rat oral 54.5 mg/kg [R1].

ECOTOXICITY LD$_{50}$ bobwhite quail oral 1,800 mg/kg [R1].

FIFRA REQUIREMENTS This rule establishes time-limited tolerances (with an expiration date of 11/15/97) for residues of the synthetic pyrethroid bifenthrin and its 4'-hydroxy metabolite in or on the following agricultural commodities: corn (field, seed, and pop), foragem, fodder; milk, fat; meat of cattle, goats, hogs, horses, poultry, and sheep; meat by-products (mybp) of cattle, goats, hogs, horses, and sheep; poultry meat by-products; fat of cattle, goats, hogs, horses, and sheep; poultry fat; cotton seed; hops, dried; and eggs [R5].

REFERENCES

1. Hartley, D., and H. Kidd (eds.). *The Agrochemicals Handbook*. 2nd ed. Lechworth, Herts, England: The Royal Society of Chemistry, 1987, p. A0851/Jun 89.

2. Bureau of Explosives. *Emergency Handling of Haz Matl in Surface Trans*, p. 434 (1981).

3. International Labour Office. *Encyclopedia of Occupational Health and Safety*. Vols. I and II. Geneva, Switzerland: International Labour Office, 1983. 1619.

4. Sittig, M. *Handbook of Toxic and Hazardous Chemicals and Carcinogens*, 1985. 2nd ed. Park Ridge, NJ: Noyes Data Corporation, 1985. 762.

5. 59 FR 46190 (9/7/94).

6. Budavari, S. (ed.). *The Merck Index—Encyclopedia of Chemicals, Drugs and Biologicals*. Rahway, NJ: Merck and Co., Inc., 1989.

7. Mackison, F. W., R. S. Stricoff, and L. J. Partridge, Jr. (eds.). *NIOSH/OSHA—Occupational Health Guidelines for Chemical Hazards*. DHHS (NIOSH) Publication No. 81–123 (3 vols). Washington, DC: U.S. Government Printing Office, Jan. 1981.

■ BISPHENOL-A-DIGLYCIDYL-ETHER

CAS # 1675–54–3

SYNONYMS 2,2-bis (4'-glycidyloxyphenyl) propane • 2,2-bis (4-(2,3-epoxypropyloxy) phenyl) propane • 2,2-bis (4-glycidyloxyphenyl) propane • 2,2-bis (4-hydroxyphenyl) propane diglycidyl ether • 2,2-bis (p-(2,3-epoxypropoxy) phenyl) propane • 2,2-bis (p-glycidyloxyphenyl) propane • 2,2-bis (p-hydroxyphenyl) propane, diglycidyl ether • 4,4'-bis (2,3-epoxypropoxy) diphenyldimethylmethane • 4,4'-dihydroxydiphenyldimethylmethane diglycidyl ether • 4,4'-isopropylidenebis (1-(2,3-epoxypropoxy) benzene) • 4,4'-isopropylidenediphenol diglycidyl ether • bis (4-glycidyloxyphenyl) dimethylmethane • bis (4-hydroxyphenyl) dimethylmethane diglycidyl ether • d-e-r-332- • dian-diglycidyl ether • diglycidyl bisphenol-a • diglycidyl bisphenol-a ether • diglycidyl diphenylolpropane ether • diglycidyl ether of 2,2-bis (4-hydroxyphenyl) propane • diglycidyl ether of 2,2-bis (p-hydroxyphenyl) propane • diglycidyl ether of 4,4'-isopropylidenediphenol • diglycidyl ether of bisphenol-a • diomethane diglycidyl ether • epi-rez 510 • epoxide-a • erl-2774 • oxirane, 2,2'-((1-methylethylidene) bis (4,1-phenyleneoxymethylene))bis • p,p'-dihydroxydiphenyldimethylmethane diglycidyl ether • propane, 2,2-bis (p-(2,3-epoxypropoxy) phenyl)

MF $C_{21}H_{24}O_4$

MW 340.45

COLOR AND FORM Sticky and tacky.

ODOR Odorless.

FLAMMABILITY LIMITS Combustible [R2].

FIREFIGHTING PROCEDURES Because solvent curing agents of epoxy resins are flammable liquids, fire hydrants and control measures are required. Fire extinguishers should be located in area. (epoxy resins) [R10, 2218].

STABILITY AND SHELF LIFE Volatility of uncured epoxy resins is not great [R3].

PROTECTIVE EQUIPMENT Gloves should be used and skin contact avoided at all times (epoxy compounds) [R11, 469]. Adequate ventilation should be supplied to work area. In some operations, exhaust ventilation may be necessary to remove excessive concentration of vapors of curing agent or diluent (epoxy resins) [R12, 827]. SRP: The scientific literature supports the wearing of contact lenses in industrial environments, as part of a program to protect the eye against chemical compounds and minerals causing eye irritation. However, there may be individual substances whose irritating or corrosive properties are such that the wearing of contact lenses would be harmful to the eye. In those specific cases, contact lenses should not be worn. The most important measure for combating dermatitis is good personal hygiene. This requires supervisory instruction and good work habits, and provision of adequate facilities for removing material periodically (epoxy resins) [R10, 2217].

PREVENTATIVE MEASURES Mixing of volatile materials should always be done in a well-ventilated area and precautions should be taken to avoid splashing of materials on hands and face (epoxy compound) [R12, 469]. Adequate ventilation should be supplied to work area. In some operations, exhaust ventilation may be necessary to remove excessive concn of

vapors of curing agent or diluent (epoxy resins) [R7, 827]. SRP: The scientific literature supports the wearing of contact lenses in industrial environments, as part of a program to protect the eye against chemical compounds and minerals causing eye irritation. However, there may be individual substances whose irritating or corrosive properties are such that the wearing of contact lenses would be harmful to the eye. In those specific cases contact lenses should not be worn. The most important measure for combating dermatitis is good personal hygiene. This requires supervisory instruction and good work habits, and provision of adequate facilities for removing material periodically. (epoxy resins) [R10, 2217].

COMMON USES Used for sealing and encapsulating; making castings and pottings; and formulating light-weight foams (diglycidyl ethers of bisphenol A) [R10, 2217]. Used as binders in preparation of laminates of paper, polyester cloth, fiberglass cloth, and wood sheets (diglycidyl ethers of bisphenol A) [R10, 2217]. Major component of liquid epoxy resins [R1].

OCCUPATIONAL RECOMMENDATIONS None listed.

IARC SUMMARY Evaluation: There is limited evidence for the carcinogenicity of bisphenol A diglycidyl ether in experimental animals. No data were available from studies in humans on the carcinogenicity of glycidyl ethers. Overall evaluation: Bisphenol A diglycidyl ether is not classifiable as to its carcinogenicity to humans (Group 3) [R4].

TSCA REQUIREMENTS Pursuant to section 8 (d) of TSCA, EPA promulgated a model health and safety data reporting rule. The section 8 (d) model rule requires manufacturers, importers, and processors of listed chemical substances and mixtures to submit to EPA copies and lists of unpublished health and safety studies. Oxirane, 2,2′-((1-methylethylidene) bis (4,1-phenyleneoxymethylene))bis is included on this list [R5]. EPA has issued a Testing Consent Order that incorporates an Enforceable Consent Agreement (ECA) pursuant to the Toxic Substances Control Act (TSCA) with three chemical companies that have agreed to perform certain health effects tests and an exposure evaluation test with this compound [R9].

GENERAL RESPONSE Stop discharge if possible. Keep people away. Call fire department. Avoid contact with liquid. Isolate and remove discharged material. Notify local health and pollution control agencies.

FIRE RESPONSE Combustible. Extinguish with water, dry chemicals, foam, or carbon dioxide.

EXPOSURE RESPONSE Call for medical aid. Liquid irritating to skin and eyes. Harmful if swallowed. Remove contaminated clothing and shoes. Flush affected areas with plenty of water. If swallowed and victim is conscious, have victim drink water or milk. If swallowed, and victim is unconscious or having convulsions, do nothing except keep victim warm.

RESPONSE TO DISCHARGE Should be removed. Provide chemical and physical treatment.

WATER RESPONSE Effect of low concentration on aquatic life is unknown. May be dangerous if it enters water intakes. Notify local health and wildlife officials. Notify operators of nearby water intakes.

EXPOSURE SYMPTOMS Contact with liquid irritates eyes. Prolonged or repeated contact with skin causes irritation and dermatitis.

EXPOSURE TREATMENT Eyes: flush with water for at least 15 min. Skin: remove chemical with water or waterless skin cleaner.

REFERENCES

1. SRI.

2. U.S. Coast Guard, Department of Transportation. *CHRIS—Hazardous Chemical Data*. Volume II. Washington, DC: U.S. Government Printing Office, 1984–5.

3. International Labour Office. *Encyclopedia of Occupational Health and Safety*. Volumes I and II. New York: McGraw-Hill Book Co., 1971. 469.

4. IARC. *Monographs on the Evaluation of the Carcinogenic Risk of Chemicals to Man*. Geneva: World Health Organization, International Agency for Research on Cancer, 1972–present (multivolume work), p. 47 256 (1989).

5. 40 CFR 716.120 (7/1/90).

6. International Labour Office. *Encyclopedia of Occupational Health and Safety*. Vols. I & II. Geneva, Switzerland: International Labour Office, 1983.

7. Arena, J. M., and R. H. Drew (eds.). *Poisoning—Toxicology, Symptoms, Treatments*. 5th ed. Springfield, IL: Charles C. Thomas Publisher, 1986.

8. Clayton, G. D. and F. E. Clayton (eds.). *Patty's Industrial Hygiene and Toxicology*: Volume 2A, 2B, 2C: Toxicology. 3rd ed. New York: John Wiley Sons, 1981–1982.

9. 59 FR 38917 (8/1/94).

10. Clayton, G. D., and F. E. Clayton (eds.). *Patty's Industrial Hygiene and Toxicology*: Volume 2A, 2B, 2C: Toxicology. 3rd ed. New York: John Wiley Sons, 1981–1982.

11. Gosselin, R.E., R.P. Smith, H.C. Hodge. *Clinical Toxicology of Commercial Products*. 5th ed. Baltimore: Williams and Wilkins, 1984, p. II-411.

12. International Labour Office. *Encyclopedia of Occupational Health and Safety*. Vols. I & II. Geneva, Switzerland: International Labour Office, 1983.

13. 59 FR 38917 (8/1/94).

■ BORAX

CAS # 1303–96–4

SYNONYMS antipyonin • borascu • borax decahydrate • borax (Na_2 (B_4O_7) • $10H_2O$) • boric acid ($H_2B_4O_7$), disodium salt, decahydrate • boricin • bura • disodium tetraborate decahydrate • gerstley borate • neobor • polybor • sodium biborate decahydrate • sodium borate decahydrate ($Na_2B_4O_7$ • $10H_2O$) • sodium tetraborate decahydrate • solubor • boron sodium oxide ($B_4Na_2O_7$), decahydrate • sodium pyroborate decahydrate • sodium borate • jaikin • sodium biborate • sodium borate decahydrate • sodium pyroborate • polybor 3 • Caswell number 109 • EPA pesticide Code 011102

MF $B_4Na_2O_7$ • $10H_2O$

MW 381.4

pH Aqueous solution is alkaline to litmus and phenolphthalein.

COLOR AND FORM Colorless, monoclinic crystals; hard crystals, granules, or crystalline powder; white, gray, bluish, or greenish-white streak, vitreous or dull luster.

ODOR Odorless.

TASTE Alkaline.

DENSITY 1.73.

MELTING POINT 75°C when rapidly heated.

EXPLOSIVE LIMITS AND POTENTIAL A mixture of hydrated borax and zirconium explodes when heated [R7].

SOLUBILITY 1 g/16 ml water at 25°C; 1 g/about 1 ml glycerol at 25°C; insoluble in ethyl or isopropyl alcohol; insoluble in acid; 5.92 g/100 g of water at 25°C; 0.60 g/100 g acetone.

CORROSIVITY Solutions are not a corrosion hazard to ferrous metals.

STABILITY AND SHELF LIFE Stable [R14, 64].

OTHER CHEMICAL/PHYSICAL PROPERTIES At 100°C loses $5H_2O$; at 150°C loses $9H_2O$; becomes anhydrous at 320°C; efflorescent in dry air, crystals often being coated with white powder; dissolves many metallic oxides when fused with them. A 2.6% solution is iso-osmotic with serum.

PROTECTIVE EQUIPMENT Wear goggles or face shield when handling. [R14, 64].

PREVENTATIVE MEASURES Keep out of reach of children Wash thoroughly after handling. [R14, 64].

DISPOSAL SRP: At the time of review, criteria for land treatment or burial (sanitary landfill) disposal practices are subject to significant revision. Prior to implementing land disposal of waste residue (including waste sludge), consult with environmental regulatory agencies for guidance on acceptable disposal practices.

COMMON USES As preservative either alone or with other antiseptics, against wood fungus; artificial aging of wood; fireproofing fabrics and woods; soldering metals; manufacture of glazes and enamels; curing and preserving skins; tanning; pharmaceutic aid (alkalizing agent); vet: antiseptic, detergent, astringent for mucous membranes [R4]. Medication: Formerly used as a gargle or mouthwash in the treatment of aphthous ulcers and stomatitis [R3]. Component of insulation glass fiber to enhance durability; in borosilicate glasses to improve thermal properties; germicidal agent in household cleaning products; flux in enamels, frits, and glazes; nutrient in fertilizers; nonselective herbicide or soil sterilant; corrosion inhibitor in antifreeze; flux in nonferrous metallurgy; in production of ferrous and nonferrous boron alloys; flame retardant for cellulosic materials; chemical intermediate for perborates and numerous other boron derivatives [R1]. Powdered insecticide for crack and crevice treatment in food-handling areas [R5]. Formerly used as a lotion in bromidrosis and inflammatory conditions of the eye, and as a nasal douche [R3]. Used for prevention of mold on citrus; strongly phytotoxic and recommended at 7–10 lb/square rod for eradication of St. John's wort and poison ivy [R6]. As a larvicide for control of common house fly, and also for dog hookworm and swine kidney worm, spray applications of sodium borate are more effective [R15, p. C–46]. Borax and other borates are nonselective herbicides used to control weeds on the non-crop sites. They are used in combination with sodium chlorate to reduce the fire hazard of the latter [R2]. Nonselective vegetation control adapted to preventing emergence under asphalt. Dry application by hand or mechanical spreaders, 4 to 7 kg/10 sq m (9 to 15 kg/100 sq ft). Lighter rates may be used for annuals, heavier for deep-rooted perennials and paving treatment [R14, 64]. Vet: antiseptic, detergent, astringent for mucous membranes [R4]. Boric acid, borates, and perborates have been used as mild antiseptics or bacteriostats in eyewashes, mouthwashes, burn dressings, and diaper-rash powders; however, the effectiveness of boric acid has largely been discredited. [R16, 131].

OCCUPATIONAL RECOMMENDATIONS

OSHA 8-hr time-weighted average: 10 mg/m^3. Final rule limits were achieved by any combination of engineering controls, work practices, and personal protective equipment during the phase-in period, Sept 1, 1989 through Dec 30, 1992. Final rule limits became effective Dec 31, 1992 [R10].

THRESHOLD LIMIT 8-hr time-weighted average (TWA) 5 mg/m^3 (1977) [R19, 14].

INTERNATIONAL EXPOSURE LIMITS Both Belgium (1974) and the Netherlands (1976) have adopted the values of 5 mg/m^3 for decahydrate [R8].

NON-HUMAN TOXICITY LD$_{50}$ rat oral 396–689 mg boron/kg [R9]; LD$_{50}$ rat oral 5.66 g/kg [R4].

ECOTOXICITY LC$_{50}$ trout 27 ppm (soft water; exposure was initiated subsequent to fertilization and maintained through 4 days posthatching) (conditions of bioassay not specified) [R18, PB–267085]. LC$_{50}$ trout 54 ppm (hard water; exposure was initiated subsequent to fertilization and maintained through 4 days posthatching) (conditions of bioassay not specified) [R18, PB–267085]. LC$_{50}$ catfish 155 ppm (soft water; exposure was initiated subsequent to fertilization and maintained through 4 days posthatching) (conditions of bioassay not specified) [R18, PB–267085]. LC$_{50}$ catfish 71 ppm (hard water; exposure was initiated subsequent to fertilization and maintained through 4 days posthatching) (conditions of bioassay not specified) [R18, PB–267085]. LC$_{50}$ goldfish 65 ppm (soft water; exposure was initiated subsequent to fertilization and maintained through 4 days posthatching) (conditions of bioassay not specified) [R18, PB–267085]. LC$_{50}$ goldfish 59 ppm (hard water; exposure was initiated subsequent to fertilization

maintained through 4 days posthatching) (conditions of bioassay not specified) [R18, PB– 267085].

BIOLOGICAL HALF-LIFE Readily absorbed from the gastrointestinal tract and excreted in the urine with a half-life of about 24 hours. [R17, 3058].

BIOCONCENTRATION Accumulates in plants. [R14, 64].

FIRE POTENTIAL Nonflammable [R14, 64].

FDA Borax is an indirect food additive for use only as a component of adhesives [R11]. Textiles and textile fibers may be safely used as articles or components of articles intended for use in producing, manufacturing, packing, processing, preparing, treating, packaging, transporting, or holding food. Substances employed in the production of or added to textile and textile fibers may include borax for use as a preservative only [R12]. Borax is used in the manufacture of paper and paperboard products used in food packaging, for use in adhesives, sizes, and coatings [R13].

REFERENCES

1. SRI.

2. Worthing, C. R., and S. B. Walker (eds.). *The Pesticide Manual—A World Compendium*. 8th ed. Thornton Heath, UK: The British Crop Protection Council, 1987. 85.

3. Reynolds, J. E. F., and A. B. Prasad (eds.). *Martindale—The Extra Pharmacopoeia*. 28th ed. London: The Pharmaceutical Press, 1982. 337.

4. *The Merck Index*. 10th ed. Rahway, NJ: Merck Co., Inc., 1983. 1,231.

5. 38 FR 21685–6 (1973).

6. Spencer, E. Y. *Guide to the Chemicals Used in Crop Protection*. 7th ed. Publication 1093. Research Institute, Agriculture Canada, Ottawa, Canada: Information Canada, 1982. 48.

7. National Fire Protection Association. *Fire Protection Guide on Hazardous Materials*. 9th ed. Boston, MA: National Fire Protection Association, 1986, p. 491M–225.

8. American Conference of Governmental Industrial Hygienists. *Documentation of the Threshold Limit Values and Biological Exposure Indices.* 5th ed. Cincinnati, OH: American Conference of Governmental Industrial Hygienists, 1986. 60.

9. USEPA. *Health Advisory for Boron.* (Draft), p. 5 (1988).

10. 54 FR 2920 (1/19/89).

11. 21 CFR 175.105 (4/1/88).

12. 21 CFR 177.2800 (4/1/88).

13. 21 CFR 181.30 (4/1/88).

14. Weed Science Society of America. *Herbicide Handbook.* 5th ed. Champaign, Illinois: Weed Science Society of America, 1983.

15. *Farm Chemicals Handbook 1989.* Willoughby, OH: Meister Publishing Co., 1989.

16. Seiler, H. G., H. Sigel and A. Sigel (eds.). *Handbook on the Toxicity of Inorganic Compounds.* New York, NY: Marcel Dekker, Inc. 1988.

17. Clayton, G. D., and F. E. Clayton (eds.). *Patty's Industrial Hygiene and Toxicology:* Volume 2A, 2B, 2C: Toxicology. 3rd ed. New York: John Wiley Sons, 1981–1982.

18. Birge, W. J., J. A. Black. *Sensitivity of Vertebrate Embryos to Boron Compounds.* P.1–77 (1977) NTIS.

19. American Conference of Governmental Industrial Hygienists. *Threshold Limit Values for Chemical Substances and Physical Agents and Biological Exposure Indices for 1994–1995.* Cincinnati, OH: ACGIH, 1994.

■ BORON

CAS # 7440-42-8

MF B

MW 10.81

COLOR AND FORM polymorphic: alpha-rhombohedral form, clear red crystals; beta-rhombohedral form, black; alpha-tetragonal form, black, opaque crystals with metallic luster; amorphous form, black or dark-brown powder (other crystal forms known but not entirely characterized); yellow monoclinic or brown amorphous powder; filaments, powder, whiskers, single crystals.

DENSITY Amorphous, 2.3 g/cu cm; alpha-rhombohedral, 2.46 g/cu cm; alpha-tetragonal, 2.31 g/cu cm; beta-rhombohedral, 2.35 g/cu cm.

BOILING POINT 2,550°C.

MELTING POINT 2,300°C.

AUTOIGNITION TEMPERATURE Ignition temperature in air is 580°C. [R2, 64].

SOLUBILITY Insoluble in water; if finely divided, soluble in boiling sulfuric acid and in most molten metals, such as copper, iron, magnesium, aluminum, and calcium; insoluble in alcohol, ether.

VAPOR PRESSURE 1.56×10^{-5} atm at 2,140°C.

STABILITY AND SHELF LIFE Fairly stable at normal temperature [R8].

OTHER CHEMICAL/PHYSICAL PROPERTIES Atomic number 5; valence 3; naturally occurring isotopes: 10, 11; 3 short-lived artificial isotopes: 8, 12, 13.

DISPOSAL SRP: At the time of review, criteria for land treatment or burial (sanitary landfill) disposal practices are subject to significant revision. Prior to implementing land disposal of waste residue (including waste sludge), consult with environmental regulatory agencies for guidance on acceptable disposal practices.

COMMON USES In nuclear chemistry as neutron absorber; in ignitron rectifiers; in alloys, usually to harden other metals [R1]; [10]boron: as shield for nuclear radiation and in instruments used for detecting neutrons [R4]; deoxidizer in nonferrous metallurgy; grain refiner in aluminum; in delayed-action fuses; igniter in radio tubes; coating material in solar batteries [R5, 2979]; cementation of iron; oxygen scavenger for copper and other metals; in semiconductors, boron-coated tungsten wires; for fibers and filaments in composites with metals or ceramics; in high-temperature brazing alloys [R3]; amorphous: in pyrotechnic flares to provide distinctive green color, in rockets as igniter [R6]; as a catalyst for olefin polymerization and dehydration of alcohols. As an abrasive it imparts superior durability to cut-off wheels, effective as the major component of an ultra-high-pressure gasketing composition [R2, 66]; used in composite structural materials, in high-temperature abrasives, in special-purpose alloys, and in steel making [R7].

BINARY REACTANTS Ammonia, bromine tetrafluoride, cesium carbide, chlorine, fluorine, iodic acid, lead dioxide, nitric acid, nitric oxide, nitrosyl fluoride, nitrous oxide, potassium nitrite, rubidium carbide, silver fluoride.

STANDARD CODES NFPA—2, 2; ICC, USCG—no.

PERSISTENCE Natural calcium may slowly precipitate borates but not below levels toxic to plants.

MAJOR SPECIES THREATENED Plants.

GENERAL SENSATION Odorless.

PERSONAL SAFETY Protective clothing, goggles, and self-contained breathing apparatus are recommended for most boron compounds.

ACUTE HAZARD LEVEL Humans can tolerate several grams. High hazard to plants.

CHRONIC HAZARD LEVEL Moderate chronic toxicant via all routes; toxic to plants.

DEGREE OF HAZARD TO PUBLIC HEALTH Not considered highly toxic as a metal. Some compounds may be hazardous. Avoid dusts.

OCCUPATIONAL RECOMMENDATIONS None listed.

ACTION LEVEL Notify air authority if fire develops and threatens to spread toxic combustion fume. Prevent dust from suspending in air.

ON-SITE RESTORATION Add lime to precipitate calcium borates and neutralize with NaHCO₃. Seek professional environmental engineering assistance through EPA's En-

vironmental Response Team (ERT), Edison, NJ, 24-hour no. 908–548–8730.

AVAILABILITY OF COUNTERMEASURE MATERIAL Lime—cement plants; NaHCO₃—grocery distributors, large bakeries.

DISPOSAL METHOD Dump into landfill or release to air.

DISPOSAL NOTIFICATION Contact local solid waste or air authority.

MAJOR WATER USE THREATENED Irrigation.

PROBABLE LOCATION AND STATE Yellow or brown amorphous powder. Metal; will sink. Many salts will hydrolyze to boric acid.

REFERENCES

1. *The Merck Index*. 10th ed. Rahway, NJ: Merck Co., Inc., 1983. 186.

2. *Kirk–Othmer Encyclopedia of Chemical Technology*. 3rd ed., Volumes 1–26. New York, NY: John Wiley and Sons, 1978–1984, p. 4 (78).

3. Sax, N. I., and R. J. Lewis, Sr. (eds.). *Hawley's Condensed Chemical Dictionary*. 11th ed. New York: Van Nostrand Reinhold Co., 1987. 163.

4. Weast, R. C. (ed.) *Handbook of Chemistry and Physics*. 69th ed. Boca Raton, FL: CRC Press Inc., 1988–1989, p. B–10.

5. Clayton, G.D., and F.E. Clayton (eds.). *Patty's Industrial Hygiene and Toxicology*. Volume 2A, 2B, 2C: Toxicology. 3rd ed. New York: John Wiley and Sons, 1981–1982.

6. *The Merck Index*. 9th ed. Rahway, NJ: Merck and Co., Inc., 1976. 175.

7. Hawley, G. G. *Condensed Chemical Dictionary*, 10th ed New York: Van Nostrand Reinhold Company, p. 143–5 (1981).

8. International Labour Office. *Encyclopedia of Occupational Health and Safety*. Vols. I and II. Geneva, Switzerland: International Labour Office, 1983. 319.

■ BORON OXIDE

CAS # 1303–86–2

SYNONYMS Boric acid (HBO₂), anhydride; boric anhydride; boric oxide; boric oxide (B₂O₃); boron oxide (B₂O₃); boron sesquioxide; boron trioxide; diboron trioxide; fused boric acid; Caswell no. 109B

MF B_2O_3

MW 69.6

COLOR AND FORM Rhombic crystals; colorless, semitransparent lumps, or hard, white crystals.

ODOR Odorless.

TASTE Slightly bitter taste.

DENSITY 2.46 ± 0.01.

BOILING POINT 1860°C approximately.

MELTING POINT 450 ± 2°C.

SOLUBILITY Slowly soluble in 30 parts cold water; soluble in alcohol, glycerol; 2.77 g/100 g water at 20°C.; slowly soluble in 5 parts boiling water.

CORROSIVITY Corrosive to metals in presence of oxygen.

OTHER CHEMICAL/PHYSICAL PROPERTIES Brittle; hygroscopic; slowly reacts with water to form boric acid; colorless, vitreous crystals; solubility: in water at 0°C 1.1 g/100 CC, at 100°C 15.7 g/100 CC; soluble in acid, alcohol (oxide glass); index of refraction: 1.485; MP: approximately 450°C; density: 1.812 at 25°C (oxide glass); heat of vaporization at 298°K: 431.4 kJ/mol; viscosity at 350°C: 10.60; index of refraction at 14.4°C: 1.463; heat capacity (specific) at 298°K: 62.969 J/(kg–K) (vitreous boric oxide).

PROTECTIVE EQUIPMENT ACGIH-recommended respirator selection: For concentration up to 50 mg/m³ use any SRP: NIOSH-approved respirator. For concentration of 100 mg/m³ use any dust and mist respirator except single-use and quarter-mask respirators or any supplied-air respirator or any self-contained breathing apparatus. For concentration of 250 mg/m³ use any powered air-purifying respirator with a dust and mist filter or any supplied-air respirator operated in a continuous-flow mode. For 500 mg/m³ use any self-contained breathing apparatus with a full facepiece or any supplied-air respirator with a full face-piece or any air-purifying full facepiece respirator with a high-efficiency particulate filter or any powered air-purifying respirator with a tight-fitting facepiece and a high-efficiency particulate filter. For 7,500 mg/m³ use any supplied-air respirator with a full facepiece and operated in a pressure-demand or other positive-pressure mode. Emergency or planned entry in unknown concentration or IDLH conditions: use any self-contained breathing apparatus with a full facepiece and operated in a pressure-demand or other positive-pressure mode or any supplied-air respirator with a full facepiece and operated in a pressure-demand or other positive-pressure mode in combination with an auxiliary self-contained breathing apparatus operated in pressure-demand or other positive-pressure mode. For escape use any air-purifying full facepiece respirator with a high-efficiency particulate filter or any appropriate escape-type self-contained breathing apparatus [R8, 59]. Employees should be provided with and required to use impervious clothing, gloves, face shields (eight-inch minimum), and other appropriate protective clothing necessary to prevent repeated or prolonged skin contact with boron oxide [R9]. Non-impervious clothing that becomes contaminated with boron oxide should be removed promptly and not reworn until the boron oxide is removed from the clothing [R9]. Employees should be provided with and required to use dust and splashproof safety goggles where boron oxide or nonaqueous liquids containing boron oxide may contact the eyes. [R9].

PREVENTATIVE MEASURES Contact lenses should not be worn when working with this chemical [R8, 58]. Contact lens use in industry is controversial. A survey of 100 corporations resulted in the recommendation that each company establish its own contact lens use policy. One presumed hazard of contact lens use is possible chemical entrapment. It was found that contact lenses minimized injury or protected the eye. The eye was afforded more protection from liquid irritants. Soft contact lenses do not worsen corneal damage from strong chemicals and in some cases could actually protect the eye. Overall, the literature supports the wearing of contact lenses in industrial environments as part of the standard eye protection, e.g., face shields; however, more data are needed to establish the value of contact lenses [R10]. During production processes, atmospheric contamination by vapors and aerosols of boric anhydride should be minimized by use of local exhaust ventilation, process mechanization, and enclosure [R11, 321]. Skin that becomes contaminated with boron oxide should be promptly washed or showered with soap or mild detergent and water to remove any boron oxide. Employees who handle boron oxide or nonaqueous liquids containing boron oxide should wash their hands thoroughly with soap or mild detergent and water before eating, smoking, or using toilet facilities [R9]. SRP: Contaminated protective clothing should be segregated in such a manner so that there is no direct personal contact by personnel who handle, dispose of, or clean the clothing. Quality assurance to ascertain the completeness of the cleaning procedures should be implemented before the decontaminated protective clothing is returned for reuse by the workers. Contaminated clothing should not be taken home at end of shift, but should remain at employee's place of work for cleaning.

DISPOSAL SRP: At the time of review, criteria for land treatment or burial (sanitary landfill) disposal practices are subject to significant revision. Prior to implementing land disposal of waste residue (including waste sludge), consult with environmental regulatory agencies for guidance on acceptable disposal practices.

COMMON USES Used in herbicides; fire-resistant additive for paints, electronics; liquid encapsulation techniques [R2]. To determine silicon dioxide and alkalies in silicates; in metallurgy; in blowpipe analysis [R3]. Flux for enamels and glazes; additive for glass fibers; chemical intermediate for elemental boron, boron master alloys, borides, boron carbide, nitrides, and halides [R4]; as catalyst [R6, 71]. Boric acid, borates, and perborates have been used as mild antiseptics or bacteriostats in eyewashes, mouthwashes, burn dressings, and diaper-rash powders; however, the effectiveness of boric acid has largely been discredited. [R7, 131].

OCCUPATIONAL RECOMMENDATIONS

OSHA 8-hr time-weighted average: 15 mg/m^3. Transitional limits must continue to be achieved by any combination of engineering controls, work practices, and personal protective equipment during the phase-in period, Sept 1, 1989 through Dec 30, 1992. Final rule limits became effective Dec 31, 1992 (total dust) [R5]. 8-hr time-weighted avg: 10 mg/m^3. Final rule limits shall be achieved by any combination of engineering controls, work practices, and personal protective equipment during the phase-in period, Sept 1, 1989 through Dec 30, 1992. Final rule limits become effective Dec 31, 1992 (total dust) [R12]. Meets criteria for OSHA medical records rule (total dust) [R13].

THRESHOLD LIMIT Time-weighted average (TWA) 10 mg/m^3 (1986) [R14, 14]. Excursion limit Recommendation: Excursions in worker exposure levels may exceed three times the TLV-TWA for no more than a total of 30 min during a work day, and under no circumstances should they exceed five times the TLV-TWA. [R14, 5].

INTERNATIONAL EXPOSURE LIMITS 1–5 mg/m^3 (former USSR) (borates) [R7, 136].

FIRE POTENTIAL Noncombustible [R1].

REFERENCES

1. Sax, N. I., and R. J. Lewis, Sr. (eds.). *Hawley's Condensed Chemical Dictionary*. 11th ed. New York: Van Nostrand Reinhold Co., 1987. 162.

2. National Fire Protection Association. *Fire Protection Guide on Hazardous Materials*. 9th ed. Boston, MA: National Fire Protection Association, 1986. 61.

3. *The Merck Index*. 10th ed. Rahway, NJ: Merck Co., Inc., 1983. 185.

4. SRI.

5. 54 FR 2920 (1/19/89).

6. *Kirk–Othmer Encyclopedia of Chemical Technology*. 3rd ed., Volumes 1–26. New York, NY: John Wiley and Sons, 1978–1984.,p. 4(78).

7. Seiler, H.G., H. Sigel, and A. Sigel (eds.). *Handbook on the Toxicity of Inorganic Compounds*. New York, NY: Marcel Dekker, Inc. 1988.

8. NIOSH. *Pocket Guide to Chemical Hazards*. 2nd Printing. DHHS (NIOSH) Publ. No. 85–114. Washington, DC: U.S. Dept. of Health and Human Services, NIOSH/Supt.of Documents, GPO, February 1987.

9. Mackison, F. W., R. S. Stricoff, and L. J. Partridge, Jr. (eds.). *NIOSH/OSHA–Occupational Health Guidelines for Chemical Hazards*. DHHS(NIOSH) Publication No. 81–123 (3 VOLS). Washington, DC: U.S. Government Printing Office, Jan. 1981. 2.

10. Randolph, S. A., M. R. Zavon. *J Occup Med* 29: 237–42 (1987).

11. International Labour Office. *Encyclopedia of Occupational Health and Safety*. Vols. I & II. Geneva, Switzerland: International Labour Office, 1983.

12. 54 FR 2920 (1/19/89).

13. 29 CFR 1910.20 (7/1/88).

14. American Conference of Governmental Industrial Hygienists. *Threshold Limit Values for Chemical Substances and Physical Agents and Biological Exposure Indices for 1994–1995*. Cincinnati, OH: ACGIH, 1994.

■ 2-BROMO-2-CHLORO-1,1,1-TRIFLUOROETHANE

CAS # 151–67–7

SYNONYMS halothane • bromochlorotrifluoroethane • 1,1,1-trifluoro-2,2-chlorobromoethane • fluothane • rhodialothan • 1,1,1-trifluoro-2-bromo-2-chloroethane • 1,1,1-trifluoro-2-chloro-2-bromoethane • 2,2,2-trifluoro-1-chloro-1-bromoethane • ftorotan (Russian) • fluotane • alotano • anestan • freon-123B1 • ftorotan • halotano (Spanish) • halan • halothanum (Latin) • halsan • narcotan • narcotane • narcotann ne-spofa (Russian) • phthorothanum • chalothane • fluktan • fluorotane • halotan • halothan • narkotan • nsc-143490-

MF $C_2HBrClF_3$

MW 197.39

COLOR AND FORM Colorless, volatile liquid.

ODOR Characteristic, sweetish, not unpleasant odor.

DENSITY 1.871 at 20°C.

BOILING POINT 50.2°C.

TOXIC COMBUSTION PRODUCTS All fluorocarbons will undergo thermal decomposition when exposed to flame or red-hot metal. Decomposition products of the chlorofluorocarbons will include hydrofluoric and hydrochloric acid along with smaller amounts of phosgene and carbonyl fluoride. The last compound is very unstable to hydrolysis and quickly changes to hydrofluoric acid and carbon dioxide in the presence of moisture. (fluorocarbons) [R2].

SOLUBILITY Miscible with petroleum ether, other fat solvents; 3,900 mg/L at 25°C.

VAPOR PRESSURE 3.02×10^2 mm Hg at 25°C.

OCTANOL/WATER PARTITION COEFFICIENT Log K_{ow} = 2.30.

STABILITY AND SHELF LIFE Sensitive to light; may be stabilized with 0.01% thymol [R1].

OTHER CHEMICAL/PHYSICAL PROPERTIES Henry's Law constant: 0.0313 atm-m^3/mole at 25°C (est); vapor pressure: 243 mm Hg at 20°C.

PROTECTIVE EQUIPMENT Many of the fluorocarbons are good solvents of skin oil, so protective ointment should be used. (fluorocarbons) [R8, 544].

PREVENTATIVE MEASURES Sufficient exhaust and general ventilation should be provided to keep vapor concentration below recommended levels. (fluorocarbons) [R2].

CLEANUP Results of personal air sampling surveys of nitrous oxide and halothane in operating theaters and recovery areas at 27 hospitals performed between 1980 and 1984 were reported. Exposures for different groups of workers during general surgical procedures with and without anesthetic gas scavenging were compared. The survey covered 40 theaters and 18 recovery areas. Geometric-mean nitrous oxide exposures with consistent scavenging and without scavenging, respectively, were: anesthetists, 71 and 211 ppm; surgeons, 50 ppm and 150 ppm; other staff, 24 ppm and 70 ppm; and all staff, 32 ppm and 94 ppm. Geometric-mean halothane exposures with consistent scavenging and without scavenging, respectively, were: anesthesiologists, 1.1 ppm and 4.0 ppm; surgeons 0.6 ppm and 1.3 ppm; other staff, 0.7 ppm and 1.3 ppm; and all staff, 0.7 ppm and 1.7 ppm. Geometric-mean nitrous oxide and halothane exposures for recovery staff not entering the theater were 27 ppm and 0.6 ppm, respectively; no recovery areas had scavenging. Geometric-mean nitrous oxide exposures for anesthetists and for other staff excluding surgeons, respectively, were: 373 ppm and 203 ppm with zero air changes per hr, 202 ppm and 64 ppm with 10 to 15 air changes per hr, 124 ppm and 45 ppm with 15 to 20 air changes per hr, and 128 ppm and 40 ppm with more than 20 air changes per hr. Mean exposures to halothane and nitrous oxide were significantly greater with passive versus active scavenging for all staff and staff other than surgeons and anesthetists. With scavenging, 40 percent of nitrous oxide results were below 25 ppm, whereas 88% of halothane results were below 2 ppm for all staff. It was concluded that good general ventilation and effective gas scavenging systems are essential for control of exposure to inhalation anesthetics in operating theaters [R3].

DISPOSAL Because of recent discovery of potential ozone decomposition in the stratosphere by fluorotrichloromethane, this material should be released to the environment only as a last resort. Waste material should be recovered and returned to the vendor, or to licensed waste-disposal company [R4].

COMMON USES Medication: anesthetic (inhalation) [R1]; vet: anesthetic (inhalation) [R1].

OCCUPATIONAL RECOMMENDATIONS
THRESHOLD LIMIT 8-hr time-weighted average (TWA) 50 ppm, 404 mg/m^3, skin (1988) [R9, 22].

BIOCONCENTRATION Estimated bioconcentration factors ranging from 6 to 33 (SRC) can be calculated for 2-bromo-2-chloro-1,1,1-trifluoroethane based on its experimental water solubility, 3,900 mg/L at 25°C (1), and its experimental log octanol/water partition coefficient, 2.30 (2), using appropriate regression equations (3). These values indicate that 2-bromo-2-chloro-1,1,1-trifluoroethane will not bioconcentrate in fish and aquatic organisms (SRC) [R5].

FIRE POTENTIAL Nonflammable [R1].

FDA Manufacturers, packers, and distributors of drug and drug products for human use are responsible for complying with the labeling, certification, and usage requirements as prescribed by the Federal Food, Drug, and Cosmetic Act, as amended (secs 201–902, 52 Stat. 1040 et seq., as amended; 21 U.S.C. 321–392) [R6]. The Approved Drug Products with Therapeutic Equivalence Evaluations List identifies currently marketed drug products, including halothane, approved on the basis of safety and effectiveness by FDA under sections 505 and 507 of the Federal Food, Drug, and Cosmetic Act [R7].

REFERENCES
1. Budavari, S. (ed.). *The Merck Index— Encyclopedia of Chemicals, Drugs, and Biologicals*. Rahway, NJ: Merck and Co., Inc., 1989. 726.

2. International Labour Office. *Encyclopedia of Occupational Health and Safety.*

Vols. I and II. Geneva, Switzerland: International Labour Office, 1983. 897.

3. Gardner, R. J., *Annals of Occupational Hygiene* 33 (2): 159–73 (1989).

4. United Nations. *Treatment and Disposal Methods for Waste Chemicals* (IRPTC File). Data Profile Series No. 5. Geneva, Switzerland: United Nations Environmental Programme, Dec. 1985. 207.

5. (1) Suzuki, T. *J Comput–Aided Molec Design* 5: 149–66 (1991). (2) Hansch, C., and A. J. Leo. *Medchem Project* Issue No. 26, Claremont, CA: Pomona College (1985). (3) Lyman, W. J., et al., *Handbook of Chemical Property Estimation Methods*. New York: McGraw-Hill, Chapt. 4 (1982).

6. 21 CFR 200–299, 300–499, 820, and 860 (4/1/91).

7. DHHS/FDA. *Approved Drug Products with Therapeutic Equivalence Evaluations* 12th edition p. 3–138 (1992).

8. Zenz, C. *Occupational Medicine— Principles and Practical Applications*. 2nd ed. St. Louis, MO: Mosby-Yearbook, Inc, 1988.

9. American Conference of Governmental Industrial Hygienists. *Threshold Limit Values for Chemical Substances and Physical Agents and Biological Exposure Indices for 1994–1995*. Cincinnati, OH: ACGIH, 1994.

■ BROMOTRIFLUOROMETHANE

CAS # 75-63-8

SYNONYMS bromofluoroform • freon-13b1 • carbon monobromide trifluoride • f-13b1 • fc-13b1 • flugex-13b1 • fluorocarbon-1301 • halon-1301 • Khladon-13b1 • methane, bromotrifluoro- • monobromotrifluoromethane • r-13b1 • trifluorobromomethane • trifluoromethyl bromide • trifluoromonobromomethane • halocarbon-13B1 • refrigerant-13B1

MF $CBrF_3$

MW 148.92

COLOR AND FORM Colorless gas.

ODOR Slight ethereal odor.

DENSITY 1.538 g/ml at 25°C (liquid).

SURFACE TENSION 4 mN/m at 25°C.

VISCOSITY 0.157 cp at 25°C (liquid); 0.0154 cP at 25°C, 101.3 kPa (vapor).

BOILING POINT −57.86°C at 760 mm Hg.

MELTING POINT −166°C.

HEAT OF VAPORIZATION 118.7 kJ/kg at boiling point.

FIREFIGHTING PROCEDURES Extinguish fire using agent suitable for type of surrounding fire (material itself does not burn or burns with difficulty). Cool all affected containers with flooding quantities of water. Apply water from as far a distance as possible. [R8, 242].

TOXIC COMBUSTION PRODUCTS All fluorocarbons will undergo thermal decomposition when exposed to flame or red-hot metal. Decomposition products of the chlorofluorocarbons will include hydrofluoric and hydrochloric acid along with smaller amounts of phosgene and carbonyl fluoride. The last compound is very unstable to hydrolysis and quickly changes to hydrofluoric acid and carbon dioxide in the presence of moisture (fluorocarbons) [R2]. In contact with open flame or very hot surface, fluorocarbons may decompose into highly irritant and toxic gases: chlorine, hydrogen fluoride or chloride, and even phosgene (fluorocarbon refrigerant and propellants) [R9]. Under certain conditions, fluorocarbon vapors may decompose on contact with flames or hot surfaces, creating the potential hazard of inhalation of toxic decomposition products. (fluorocarbons) [R10, 3101].

SOLUBILITY Soluble in chloroform; 0.03% in water.

OCTANOL/WATER PARTITION COEFFICIENT log K_{ow} = 1.86.

CORROSIVITY Noncorrosive.

STABILITY AND SHELF LIFE Conditions contributing to instability: heat [R13, 2].

OTHER CHEMICAL/PHYSICAL PROPERTIES

(Air conc) 1 mg/L = 164 ppm; 1 ppm = 6.09 mg/m³ at 25°C, 760 mm Hg. Ionization potential 11.9 eV. Critical volume: 200 m³/mol; critical density: 0.745 g/cu cm; specific heat of liquid: 0.208 cal/g at 25°C; specific heat of vapor: 0.112 cal/g at 25°C and 1 atm. Dipole moment: 0.65 debye. Henry's Law constant = 0.4994 atm-m³/mole at 25°C.

PROTECTIVE EQUIPMENT

Recommendations for respirator selection, max concentration for use: 10,000 ppm: Respirator Classes: Any supplied-air respirator. Any self-contained breathing apparatus [R3]. Recommendations for respirator selection, max concentration for use: 25,000 ppm: Respirator Class: Any supplied-air respirator operated in a continuous-flow mode [R11]. Recommendations for respirator selection, max concentration for use: 50,000 ppm: Respirator Classes: Any self-contained breathing apparatus with a full facepiece. Any supplied-air respirator with a full facepiece. Any supplied-air respirator that has a tight-fitting facepiece and is operated in a continuous-flow mode [R11]. Recommendations for respirator selection condition: emergency or planned entry into unknown concentration or IDLH conditions: Respirator Classes: Any self-contained breathing apparatus that has a full facepiece and is operated in a pressure-demand or other positive-pressure mode. Any supplied-air respirator with a full facepiece and operated in pressure-demand or other positive-pressure mode in combination with an auxiliary self-contained breathing apparatus operated in pressure-demand or other positive-pressure mode [R11]. Recommendations for respirator selection, condition: escape from suddenly occurring respiratory hazards: Respirator Classes: Any air-purifying, full facepiece respirator (gas mask) with a chin-style, front- or back-mounted organic vapor canister. Any appropriate escape-type, self-contained breathing apparatus [R11]. Firefighters must wear self-contained breathing apparatus with a full facepiece operated in pressure-demand or other positive-pressure mode [R13, 4]. Many of the fluorocarbons are good sol-vents of skin oil, so protective ointment should be used (fluorocarbons) [R12, 544]. Neoprene gloves, protective clothing, and eye protection minimize risk of topical contact. Degreasing effect on skin can be treated with lanolin ointment (fluorocarbons) [R8, 3102]. Forced-air ventilation at level of vapor concentration, together with use of individual breathing devices with independent air supply, will minimize risk of inhalation. Lifelines should be worn when entering tanks or other confined spaces. (fluorocarbons) [R8, 3101].

PREVENTATIVE MEASURES

Persons not wearing protective equipment and clothing should be restricted from areas of leaks until cleanup has been completed [R13, 3]. If the use of respirators is necessary, the only respirators permitted are those that have been approved by the Mine Safety and Health Administration (formerly Mining Enforcement and Safety Administration) or by the National Institute for Occupational Safety and Health. In addition to respirator selection, a complete respiratory protection program should be instituted that includes regular training, maintenance, inspection, cleaning, and evaluation [R13, 2]. If material not involved in fire: Attempt to stop leak if without undue personnel hazard [R8, 242]. Personnel protection: Avoid breathing vapors. Keep upwind. Do not handle broken packages without protective equipment [R8, 242]. Sufficient exhaust and general ventilation should be provided to keep vapor concentration below recommended levels (fluorocarbons) [R2]. Inhalation of fluorocarbon vapors should be avoided (fluorocarbons) [R10, 3101]. Forced-air ventilation at the level of vapor concentration, together with the use of individual breathing devices with independent air supply, will minimize the risk of inhalation. Lifelines should be worn when entering tanks or other confined spaces (fluorocarbons) [R10, 3101]. Enclosure of process materials, and isolation of reaction vessels and proper design and operation of filling heads for packaging and shipping are administrative controls that may be instituted to limit occupational exposure to fluorocarbons during manufacture, packaging, and use (fluorocarbons) [R5, 3101]. If material not involved in fire: attempt to stop leak if without undue personnel hazard (refrigerants, NEC, gas, or liquid, nonflammable (refrigerant, gas, NOS, or dispersant gas, NOS) [R8, 599]. Sufficient exhaust and general ventilation should be provided to keep vapor concentration below recommended levels (fluorocarbons) [R2]. Filling areas should be monitored to ensure ambient concentration of fluorocarbons does not exceed 1000 ppm. Inhalation of fluorocarbon vapors should be avoided. If inhalation occurs, epinephrine or other sympathomimetic amines, and adrenergic activators should not be admin, because they will further sensitize heart to development of arrhythmias (fluorocarbons) [R10, 3101]. Appearance of toxic decomposition products serves as warning of occurrence of thermal decomposition, and detection of sharp acrid odor warns of presence. Halide lamps or electronic leak detectors may also be used. Adequate ventilation also avoids problem of toxic decomposition products. (fluorocarbons) [R10, 3101].

CLEANUP If trifluoromonobromomethane is leaked, the following steps should be taken: 1. Ventilate area of leak. 2. Stop flow of gas. [R13, 3].

DISPOSAL Incineration, preferably after mixing with another combustible fuel. Assure complete combustion to prevent the formation of toxic products. An acid scrubber is necessary to remove the halo acids produced [R4]. Because of recent discovery of potential ozone decomposition in the stratosphere by fluorotrichloromethane, this material should be released to the environment only as a last resort. Waste material should be recovered and returned to the vendor or to licensed waste-disposal company. [R14].

COMMON USES Low-temperature fire-extinguishing agent, refrigerant in environmental test chambers, metal and pharmaceutical processing [R1]. Chemical intermediate; metal hardening [R15]. As commercial and military fire extinguishant; as a refrigerant for food-processing

and storage; as a blowing agent to improve flame retardancy of rigid polyurethrane foams [R13, 2]. In organic synthesis in production of olefin resins; during manufacture of hydraulic fluids as an erosion inhibitor [R13, 2]. For special purposes in bubble chambers for ionization studies, in "quark" detection, and in radiation counters [R13, 2]. Mechanical vapor compression systems use fluorocarbons for refrigeration and air conditioning and account for majority of refrigeration capability in U.S. Fluorocarbons are used as refrigerants in home appliances, mobile air-conditioning units, retail food refrigeration systems, and chillers. (fluorocarbons) [R10, 3102].

OCCUPATIONAL RECOMMENDATIONS
OSHA 8-hr time-weighted average: 1,000 ppm (6,000 mg/m³ [R7].

NIOSH 10-hr time-weighted average: 1,000 ppm (6,100 mg/m³ [R3].

THRESHOLD LIMIT 8-hr time-weighted average (TWA) 1000 ppm, 6,090 mg/m³ (1986) [R16 35].

INTERNATIONAL EXPOSURE LIMITS Australia, Belgium, Finland, former West Germany, Netherlands, Yugoslavia, and Switzerland: 1,000 ppm; East Germany: 5,000 ppm average exposure, 10,000 ppm short-term exposure; Romania: 5,000 ppm average exposure, 7,000 max exposure [R5].

BIOCONCENTRATION Based on a reported log K_{ow} of 1.86 (1), the BCF for Freon 13B1 can be estimated to be 15.3 from a recommended regression equation (2,SRC). A BCF of 15.3 indicates that bioconcentration of Freon 13B1 in aquatic organisms will not be an important fate process (SRC) [R6].

REFERENCES
1. SRI.

2. International Labour Office. *Encyclopedia of Occupational Health and Safety.* Vols. I and II. Geneva, Switzerland: International Labour Office, 1983. 897.

3. NIOSH. *NIOSH Pocket Guide to Chemical Hazards.* DHHS (NIOSH) Publication No. 90–117. Washington, DC: U.S. Government Printing Office, June 1990, 220.

4. Sittig, M. *Handbook of Toxic and Hazardous Chemicals and Carcinogens*, 1985. 2nd ed. Park Ridge, NJ: Noyes Data Corporation, 1985. 893.

5. American Conference of Governmental Industrial Hygienists. *Documentation of the Threshold Limit Values and Biological Exposure Indices*. 5th ed. Cincinnati, OH: American Conference of Governmental Industrial Hygienists, 1986.

6. (1) Hansch, C., A. J. Leo. *Medchem Project Issue No. 26*, Claremont, CA: Pomona College (1985) (2) Lyman, W. J., et al., *Handbook of Chemical Property Estimation Methods*, New York: McGraw-Hill, pp. 5–10 (1982).

7. 29 CFR 1910.1000 (7/1/91).

8. Association of American Railroads. *Emergency Handling of Hazardous Materials in Surface Transportation*. Washington, DC Assoc. of American Railroads, Hazardous Materials Systems (BOE), 1987.

9. Gosselin, R.E., R.P. Smith, H.C. Hodge. *Clinical Toxicology of Commercial Products*. 5th ed. Baltimore: Williams and Wilkins, 1984, p. II-159.

10. Clayton, G. D., and F. E. Clayton (eds.). *Patty's Industrial Hygiene and Toxicology*: Volume 2A, 2B, 2C: Toxicology. 3rd ed. New York: John Wiley Sons, 1981–1982.

11. NIOSH. *NIOSH Pocket Guide to Chemical Hazards*. DHHS(NIOSH) Publication No. 90–117. Washington, DC: U.S. Government Printing Office, June 1990, 220.

12. Zenz, C. *Occupational Medicine—Principles and Practical Applications*. 2nd ed. St. Louis, MO: Mosby-Yearbook, Inc, 1988.

13. Mackison, F. W., R. S. Stricoff, and L. J. Partridge, Jr. (eds.). *NIOSH/OSHA – Occupational Health Guidelines for Chemical Hazards*. DHHS(NIOSH) Publication No. 81–123 (3 vols). Washington, DC: U.S. Government Printing Office, Jan. 1981.

14. United Nations. *Treatment and Disposal Methods for Waste Chemicals* (IRPTC File). Data Profile Series No. 5. Geneva, Switzerland: United Nations Environmental Programme, Dec. 1985. 207.

15. Sax, N.I., and R.J. Lewis, Sr. (eds.). *Hawley's Condensed Chemical Dictionary*. 11th ed. New York: Van Nostrand Reinhold Co., 1987. 174.

16. American Conference of Governmental Industrial Hygienists. *Threshold Limit Values for Chemical Substances and Physical Agents and Biological Exposure Indices for 1994–1995*. Cincinnati, OH: ACGIH, 1994.

■ BUTANE, 2,2-DIMETHYL-

CAS # 75–83–2

DOT # 1208

SYNONYMS butane, 2,2-dimethyl- • neohexane

MF C_6H_{14}

MW 86.20

COLOR AND FORM Colorless liquid.

DENSITY 0.6485 at 20°C.

BOILING POINT 49.7°C at 760 mm Hg.

MELTING POINT –99.9°C.

HEAT OF COMBUSTION 4,159.5 kJ/mol.

HEAT OF VAPORIZATION 27.68 kg/mol at 25°C.

AUTOIGNITION TEMPERATURE 761°F (405°C) [R3].

EXPLOSIVE LIMITS AND POTENTIAL Explosion limits 1.2–7% [R1].

FLAMMABILITY LIMITS Percent by vol: lower 1.2, upper 7.0 [R3].

FIREFIGHTING PROCEDURES Foam, carbon dioxide, dry chemical [R2].

SOLUBILITY Soluble in alcohol, diethyl ether, acetone, benzene; very soluble in petroleum ether, carbon tetrachloride.

VAPOR PRESSURE 400 mm Hg at 31.0°C.

OTHER CHEMICAL/PHYSICAL PROPERTIES

Octane rating 100. Heat capacity: 191.9 J/mol K at 25°C. Heat of fusion: 1.61 cal/g.

PROTECTIVE EQUIPMENT Air-supplied apparatus or organic vapor cartridge; goggles or face shield; rubber gloves [R4].

PREVENTATIVE MEASURES Ventilation control: the basic ventilation methods are local exhaust ventilation and dilution or general ventilation [R5].

COMMON USES Component of high-octane motor and aviation fuels; intermediate for agricultural chemicals [R1].

OCCUPATIONAL RECOMMENDATIONS None listed.

BIODEGRADATION Incubation with natural flora in the groundwater—in presence of the other components of high-octane gasoline (100 μL/L): biodegradation: 25% after 192 hr at 13°C (initial concentration 0.28 μL/L) [R6].

GENERAL RESPONSE Shut off ignition sources. Call fire department. Stop discharge if possible. Keep people away. Isolate and remove discharged material. Notify local health and pollution control agencies.

FIRE RESPONSE Flammable. Containers may explode in fire. Flashback along vapor trail may occur. Vapor may explode if ignited in an enclosed area. Extinguish with dry chemicals, foam, or carbon dioxide. Water may be ineffective on fire. Cool exposed containers with water.

EXPOSURE RESPONSE Call for medical aid. Vapor irritating to eyes, nose, and throat. If inhaled will cause dizziness, coughing, or difficult breathing. Move victim to fresh air. If breathing has stopped, give artificial respiration. If breathing is difficult, give oxygen. Liquid irritating to skin and eyes. If swallowed will cause nausea or vomiting. Remove contaminated clothing and shoes. Flush affected areas with plenty of water. If swallowed and victim is conscious, have victim drink water or milk. Do not induce vomiting.

RESPONSE TO DISCHARGE Issue warning—high flammability. Restrict access. Evacuate area. Disperse and flush.

WATER RESPONSE Effect of low concentrations on aquatic life is unknown. Fouling to shoreline. May be dangerous if it enters water intakes. Notify local health and wildlife officials. Notify operators of nearby water intakes.

EXPOSURE SYMPTOMS Inhalation causes dizziness, nausea, and vomiting; concentrated vapor may cause unconsciousness and collapse. Contact with liquid causes irritation of eyes; repeated contact may produce irritation of skin. Ingestion causes irritation of stomach. Aspiration causes severe lung irritation, rapidly developing pulmonary edema, and central nervous system excitement followed by depression.

EXPOSURE TREATMENT Inhalation: remove from exposure; if breathing has stopped, begin artificial respiration; call a physician. Eyes: flush with water for 15 min.; call physician if needed. Skin: flush well with water, then wash with soap and water. Ingestion: do not induce vomiting; guard against aspiration into lungs; call a doctor. Aspiration: enforce bed rest; give oxygen; get medical attention.

REFERENCES

1. Lewis, R. J., Sr. (ed.). *Hawley's Condensed Chemical Dictionary*. 12th ed. New York, NY: Van Nostrand Reinhold Co., 1993 812.

2. Sax, N. I. *Dangerous Properties of Industrial Materials*. 6th ed. New York, NY: Van Nostrand Reinhold, 1984. 1141.

3. National Fire Protection Guide. *Fire Protection Guide on Hazardous Materials*. 10th ed. Quincy, MA: National Fire Protection Association, 1991, p. 325M–42.

4. U.S. Coast Guard, Department of Transportation. *CHRIS—Hazardous Chemical Data. Manual Two*. Washington, DC: U.S. Government Printing Office, Oct., 1978.

5. Sax, N. I. *Dangerous Properties of Industrial Materials*. 4th ed. New York: Van Nostrand Reinhold, 1975. 673.

6. Verschueren, K. *Handbook of Environmental Data of Organic Chemicals*. 2nd

ed. New York, NY: Van Nostrand Reinhold Co., 1983. 549.

■ 1,3-BUTANEDIOL

CAS # 107–88–0

SYNONYMS 1,3-butylene glycol • 1,3-dihydroxybutane • 1-methyl-1,3-propanediol • beta-butylene glycol • butane-1,3-diol • methyltrimethylene glycol

MF $C_4H_{10}O_2$

MW 90.14

COLOR AND FORM Viscous liquid; pure compound is colorless.

ODOR Odorless when pure.

TASTE Sweet flavor with bitter aftertaste.

DENSITY 1.0059 at 20°C.

SURFACE TENSION 37.8 dynes/cm at 25°C.

VISCOSITY In centistokes: 24.6 at 50°C, 96 at 25°C, 590 at 0°C, 3,253 at −17.7°C, 6,059 at −23°C.

DISSOCIATION CONSTANTS $pK_a = 15.1$ at 25°C.

BOILING POINT 207.5°C at 760 mm Hg.

MELTING POINT FP: below 50°C.

HEAT OF COMBUSTION −2,488 kJ/mol at constant volume; −2,491 kJ/mol at constant pressure.

HEAT OF VAPORIZATION 585 J/g at boiling point.

AUTOIGNITION TEMPERATURE 741°F [R4].

FIREFIGHTING PROCEDURES Foam, CO_2, dry chemical [R6].

SOLUBILITY Practically insoluble in aliphatic hydrocarbons, benzene, toluene, carbon tetrachloride, ethanolamines, mineral and linseed oil; soluble in water, acetone, methyl ethyl ketone, ethanol, dibutyl phthalate, castor oil; slightly soluble in ether; miscible in water.

VAPOR PRESSURE 0.06 mm Hg at 20°C.

OTHER CHEMICAL/PHYSICAL PROPERTIES 1 gal weighs 8.398 lb at room temp; dielectric constant: 28.8 at 25°C; very hygroscopic; will absorb 38.5 wt % of water within 144 hr at 81% relative humidity. Mass: 183 (*Atlas of Mass Spectral Data*, John Wiley and Sons, New York) (1,3-butanediol); IR: 1260 (Coblentz Society Spectral Collection) (1,3-butanediol(dl)); UV: 5–22 (*Organic Electronic Spectral Data*, Phillips et al., John Wiley and Sons, New York) (1,3-butanediol(dl)); NMR: 86 (Varian Associates *NMR Spectra Catalogue*) (1,3-butanediol(dl)); Mass: 183 (*Atlas of Mass Spectral Data*, John Wiley and Sons, New York) (1,3-butanediol(dl)); Mass: 183 (*Atlas of Mass Spectral Data*, John Wiley and Sons, New York) (1,3-butanediol (l)); Mass: 54 (National Bureau of Standards *EPA–NIH Mass Spectra Data Base*, NSRDS–NBS–63) (1,3-butanediol (l)); vapor pressure = 0.0201 mm Hg at 25°C; liquid molar volume = 0.089913 m^3/kmol; heat of formation = -4.3460×10^8 J/kmol; heat of formation = −119.7 kcal/mol at 25°C (liquid); heat of formation = −103.5 kcal/mol at 25°C (gas); heat of vaporization = 16.2 kcal/mol at 25°C; specific optical rotation: −18.8°C at 25°C (ethanol) ((R)-(-) isomer).

PREVENTATIVE MEASURES SRP: The scientific literature for the use of contact lenses in industry is conflicting. The benefit or detrimental effects of wearing contact lenses depend not only upon the substance, but also on factors including the form of the substance, characteristics and duration of the exposure, the uses of other eye protection equipment, and the hygiene of the lenses. However, there may be individual substances whose irritating or corrosive properties are such that the wearing of contact lenses would be harmful to the eye. In those specific cases, contact lenses should not be worn. In any event, the usual eye protection equipment should be worn even when contact lenses are in place.

DISPOSAL SRP: At the time of review, criteria for land treatment or burial (sanitary landfill) disposal practices are subject to significant revision. Prior to implementing land disposal of waste residue (including waste sludge), consult with environmental regulatory agencies for guidance on acceptable disposal practices.

COMMON USES Intermediate in manufacture of polyester plasticizers; humectant for cellophane, tobacco [R1]; polyurethanes; surface active agents; coupling agent; solvent; food additive and flavoring [R4]; in cosmetic, pharmaceutical industry as glycerin substitute [R5]; esters of 1,3-butylene glycol with monocarboxylic acids have been used as plasticizer for cellulosics and polyvinyl chloride resins, as the terminal group in oil-free alkyds [R14, p. 567]; deicing of aircraft [R14, p. 568]; mainly used as a component of special polyester resins [R3]; 1,3-butylene glycol efficient antimicrobial agent, inhibiting Gram-neg and Gram-pos microorganisms, molds, and yeasts [R2].

STANDARD CODES NFPA—1, 1, 0; ICC—no; USCG—grade E combustible liquid.

GENERAL SENSATION Slightly toxic.

PERSONAL SAFETY Wear goggles and canister-type masks.

ACUTE HAZARD LEVEL May cause BOD problems.

CHRONIC HAZARD LEVEL 5,600 mg/kg in feed for 90 days had no effect on rats [R13].

DEGREE OF HAZARD TO PUBLIC HEALTH Slightly toxic by inhalation. Allowed as food additive.

OCCUPATIONAL RECOMMENDATIONS None listed.

ON-SITE RESTORATION Peat or carbon may be used to adsorb from water. Seek professional environmental engineering assistance through EPA's Environmental Response Team (ERT), Edison, NJ, 24-hour no., 908–548–8730.

AVAILABILITY OF COUNTERMEASURE MATERIAL Peat—nurseries, floral shops; carbon—water treatment plants, sugar refineries.

DISPOSAL METHOD Spray into incinerator.

INDUSTRIAL FOULING POTENTIAL No problems expected.

MAJOR WATER USE THREATENED Fisheries.

PROBABLE LOCATION AND STATE Viscous liquid; will dissolve in water.

COLOR IN WATER Colorless.

NON-HUMAN TOXICITY LD_{50} rat oral 22.8 g/kg [R1]; LD_{50} mouse subcutaneous 16.5 mL/kg [R7]; LD_{50} rat subcutaneous 20.1 mL/kg [R7]; LD_{50} guinea pig oral 11 g/kg [R6].

BIODEGRADATION Based primarily upon biodegradation screening studies with ethylene glycol (ethanediol) and propylene glycol (propanediol), the glycol chemical class is considered biodegradable by both acclimated and unacclimated soil, water, and sewage micro-organisms (1). Although several pure culture screening studies have demonstrated that 1,3-butanediol can be biodegraded (2–5), insufficient data are available from mixed culture screening studies to experimentally demonstrate environmental biodegradability (SRC) [R8].

BIOCONCENTRATION 1,3-butanediol is miscible in water (1); this suggests that bioconcentration in aquatic organisms will not be important environmentally (SRC) [R9].

ATMOSPHERIC STANDARDS This action promulgates standards of performance for equipment leaks of volatile organic compounds (VOC) in the synthetic organic chemical manufacturing industry (SOCMI). The intended effect of these standards is to require all newly constructed, modified, and reconstructed SOCMI process units to use the best demonstrated system of continuous emission reduction for equipment leaks of VOC, considering costs, non-air-quality health, and environmental impact and energy requirements. 1,3-butylene glycol is produced, as an intermediate or a final product, by process units covered under this subpart [R10].

FIRE POTENTIAL Combustible [R4]; slight, when exposed to heat or flame. Spontaneous heating: no [R6].

FDA 1,3-butanediol is an indirect food additive for use as a component of adhesives [R11]. The food additive 1,3-butylene glycol (1,3-butanediol) may be safely used in accordance with conditions prescribed in 21 CFR Part 573 [R12].

REFERENCES

1. Budavari, S. (ed.). *The Merck Index—Encyclopedia of Chemicals, Drugs, and Biologicals.* Rahway, NJ: Merck and Co., Inc., 1989. 239.

2. Harb, N. A., M. A. Toama. *Drug Cosmet Ind* 118 (5): 40–41, 136–137 (1976).

3. Gerhartz, W. (exec ed.). *Ullmann's Encyclopedia of Industrial Chemistry.* 5th ed. Vol A1: Deerfield Beach, FL: VCH Publishers, 1985 to present., p. VA4 461.

4. Lewis, R. J., Sr. (Ed.). *Hawley's Condensed Chemical Dictionary.* 12th ed. New York, NY: Van Nostrand Reinhold Co., 1993 185.

5. Browning, E. *Toxicity, and Metabolism of Industrial Solvents.* New York: American Elsevier, 1965. 664.

6. Sax, N. I. *Dangerous Properties of Industrial Materials,.* 6th ed. New York, NY: Van Nostrand Reinhold, 1984. 547.

7. Gosselin, R. E., R. P. Smith, H. C. Hodge. *Clinical Toxicology of Commercial Products.* 5th ed. Baltimore: Williams and Wilkins, 1984, p. II–179.

8. (1) Miller, L. M. *Investigation of Selected Potential Environmental Contaminants: Ethylene Glycol, Propylene Glycols, and Butylene Glycols.* USEPA–560/11–79–006 Washington, DC: USEPA, p. 67 (1979). (2) Daugherty, L.C. *Lubrication Engin* 36: 718–23 (1980). (3) Enomoto, K., et al., *Inst Phys Chemical Res* 53: 637–42 (1975). (4) Kersters, K., J. Deley.; *Biochim Biophys Acta* 71: 311–31 (1963). (5) Tsukamura, M. *Amer Rev Resp Dis* 394: 796–8 (1966).

9. (1) Riddick, J. A., et al., *Organic Solvents: Physical Properties and Methods of Purification. Techniques of Chemistry* 4th ed. New York: Wiley–Interscience, p. 268 (1986).

10. 40 CFR 60.489 (7/1/91).

11. 21 CFR 175.105 (4/1/91).

12. 21 CFR 573.225 (4/1/93).

13. Smyth, H. F., Jr., C. P. Carpenter, C. S. Weil, U. C. Pozzani, J. A. Striegel, and J. S. Nycum, "Range-Finding Toxicity Data: List Vll," *American Industrial Hygiene Association Journal*, 30: 470–476. 1969. Smyth, H. ·F., C. P. Carpenter, and C. S. Weil, "Range-Finding Toxicity Data: List IV," *AMA Archives of Industrial Hygiene and Occupational Medicine*, 4: 119–122, 1951. Smyth, H. F., C. P. Carpenter, C. S. Weil, U. C. Pozzani, and J. A. Striegel, "Range-Finding Toxicity Data: List Vl," *American Industrial Hygiene Association Journal*, 23: 95–107, 1962. Smyth, H. F., C. P. Carpenter, C. S. Weil, and U. C. Pozzani, "Range-Finding Toxicity Data: List V," *AMA Archives of Industrial Hygiene and Occupational Medicine*, 10: 61–68, 1954. Smyth, H. F., C. P. Carpenter, and C. S. Weil, "Range-Finding Toxicity Data: List 111," *Journal of Industrial Hygiene and Toxicology*, 31: 60–62,1949. Smyth, H. F., I. Seaton, and L. Fischer, "The Single Dose Toxicity of Some Glycols, and Derivatives," *Journal of Industrial Hygiene and Toxicology*, 23 (6): 259–268, 1941. Smyth, H. F., and C. P. Carpenter, "Further Experience with the Range-Finding Test In the Industrial Toxicology Laboratory," *Journal of Industrial Hygiene and Toxicology*, 30: 63–68, 1948. Smyth, H. F., and C. P. Carpenter, "The Place of the Range-Finding Test in the Industrial Toxicology Laboratory," *Journal of Industrial Hygiene and Toxicology*, 26: 269–273, 1 944.

14. Kirk-Othmer. *Concise Encyc Chem Tech.* 1985.

■ 1,4-BUTANEDIOL

CAS # 110–63–4

DOT # 1987

SYNONYMS 1,4-butylene glycol • 1,4-dihydroxybutane • 1,4-tetramethylene glycol • butane-1,4-diol • butanediol • diol-14b • sucol-b • tetramethylene glycol

MF $C_4H_{10}O_2$

MW 90.14

COLOR AND FORM Colorless, viscid liquid.

ODOR Nearly odorless.

DENSITY 1.0171 at 20°C.

SURFACE TENSION 44.6 dyne/cm at 20°C, 43.79 dyne/cm at 30°C.

VISCOSITY Viscosity coefficient = 89.24 at 20; and 71.5 cP 25°C.

DISSOCIATION CONSTANT pK_a = 14.5.

BOILING POINT 230°C at 760 mm Hg.

MELTING POINT 20.1°C.

HEAT OF COMBUSTION 2,585 kJ/mol.

HEAT OF VAPORIZATION 68.2 kJ/mol at 131.4°C; 59.4 kJ/mol at 193.2°C; 57.8 kJ/mol at 215.6°C; 56.5 kJ/mol at 230.5°C.

AUTOIGNITION TEMPERATURE 756°F [R6].

FIREFIGHTING PROCEDURES Alcohol foam, mist, foam, carbon dioxide, dry chemical [R4]. Water or foam may cause frothing. "Alcohol" foam [R5].

SOLUBILITY Soluble in water, alcohol; slightly soluble in ether; water solubility: infinite at 20°C.

VAPOR PRESSURE 0.0105 mm Hg at 25°C, estimated from experimentally derived coefficients.

CORROSIVITY Noncorrosive.

DISPOSAL SRP: At the time of review, criteria for land treatment or burial (sanitary landfill) disposal practices are subject to significant revision. Prior to implementing land disposal of waste residue (including waste sludge), consult with environmental regulatory agencies for guidance on acceptable disposal practices.

COMMON USES Chemical intermediate for tetrahydrofuran and other acetylenic chemicals (gamma butyrolactone); monomer for polybutylene terephthalate resins; chain extender for polyurethane resins [R1]. Intermediate in polyester resins [R10, 666]; in production of 1,4-butane-diol dimethanesulfonate (mylaran) [R2]; humectant, pharmaceuticals [R3].

OCCUPATIONAL RECOMMENDATIONS None listed.

HUMAN TOXICITY LD_{50} rat oral 1.78 g/kg [R9, 3878]; LD_{50} rat ip 1.37 g/kg [R9, 3878]; LD_{50} mouse, oral 2.18 g/kg [R9, 3878].

BIODEGRADATION Using an initial concentration of 100 mg/L 1,4-butanediol, 74–96 percent BOD was observed after a 2-week period in a biodegradation screening test using 30 mg/L sludge (1). Based on cell-free suspensions of several acetic acid bacteria, 1,4-butanediol is expected to oxidize to succinic acid (2) [R7].

BIOCONCENTRATION Based on an estimated log octanol-water partition coefficient of −0.22 (2) and a recommended regression-derived equation (1), the BCF for 1,4-butanediol can be estimated to be 0.4 (SRC). This BCF value indicates that 1,4-butanediol will not bioconcentrate in aquatic organisms (SRC) [R8].

FIRE POTENTIAL Low, when exposed to heat or flame [R4].

GENERAL RESPONSE Stop discharge if possible. Call fire department. Isolate and remove discharged material. Notify local health and pollution control agencies.

FIRE RESPONSE Combustible. Extinguish with dry chemical, alcohol foam, or carbon dioxide. Water may be ineffective on fire.

EXPOSURE RESPONSE Call for medical aid. Liquid or solid irritating to skin or eyes. Harmful if swallowed. Flush affected areas with plenty of water. If swallowed and victim is conscious have victim drink water or milk.

RESPONSE TO DISCHARGE Disperse and flush.

WATER RESPONSE Effect of low concentrations on aquatic life is unknown. May be dangerous if it enters water intakes. Notify local health and pollution control officials. Notify operators of nearby water intakes.

EXPOSURE SYMPTOMS Ingestion of large

amounts needed to produce any symptoms.

EXPOSURE TREATMENT Skin or eyes: wash off immediately with plenty of water.

REFERENCES

1. SRI.

2. IARC. *Monographs on the Evaluation of the Carcinogenic Risk of Chemicals to Man.* Geneva: World Health Organization, International Agency for Research on Cancer, 1972–present (multivolume work), p. V4 248.

3. Sax, N. I., and R. J. Lewis, Sr. (eds.). *Hawley's Condensed Chemical Dictionary.* 11th ed. New York: Van Nostrand Reinhold Co., 1987. 186.

4. Sax, N. I. *Dangerous Properties of Industrial Materials,.* 6th ed. New York, NY: Van Nostrand Reinhold, 1984. 547.

5. National Fire Protection Guide. *Fire Protection Guide on Hazardous Materials.* 10th ed. Quincy, MA: National Fire Protection Association, 1991, p. 325M–20.

6. U.S. Coast Guard, Department of Transportation. *CHRIS—Hazardous Chemical Data.* Volume II. Washington, DC: U.S. Government Printing Office, 1984–5.

7. (1) Chemicals Inspection and Testing Institute. Japan Chemical Industry Ecology—Toxicology and Information Center ISBN 4–89074– 101–1 (1992). (2) Kersters, K., J. De Ley. *Biochim Biophys Acta.* 71: 311–31 (1963).

8. (1) Lyman, W. J., et al., *Handbook of Chemical Property Estimation Methods.* Washington DC: Amer Chemical Soc, p. 5–4 (1990). (2) Meylan, W. M., P. H. Howard. *J Pharm Sci.* 84: 83–92 (1995).

9. Clayton, G. D. and F. E. Clayton (eds.). *Patty's Industrial Hygiene and Toxicology:* Volume 2A, 2B, 2C: Toxicology. 3rd ed. New York: John Wiley Sons, 1981–1982.

10. Browning, E. *Toxicity and Metabolism of Industrial Solvents.* New York: American Elsevier, 1965.

■ BUTANENITRILE

CAS # 109–74–0

SYNONYMS 1-cyanopropane • butane nitrile • butyric acid nitrile • butyronitrile • butyrylonitrile • n-butyronitrile • n-propyl cyanide • propyl cyanide • n-butyl nitrile • n-butanenitrile

MF C_4H_7N

MW 69.12

COLOR AND FORM Colorless liquid.

DENSITY 0.8091 at 0°C.

SURFACE TENSION 27.33 dyn/cm at 20°C.

VISCOSITY (Viscosity x 10^5) at 15°C = 624; at 30°C = 515.

BOILING POINT 117.5°C at 760 mm Hg.

MELTING POINT −112°C.

AUTOIGNITION TEMPERATURE 935°F (501°C) [R5].

FLAMMABILITY LIMITS Lower flammable limit: 1.65% [R5].

FIREFIGHTING PROCEDURES "Alcohol" foam [R5].

SOLUBILITY Miscible with alcohol, ether, dimethylformamide; soluble in benzene, 33,000 mg/L at 25°C (measured).

VAPOR PRESSURE Vapor pressure: 19.5 mm Hg at 25°C.

OTHER CHEMICAL/PHYSICAL PROPERTIES Density of saturated air: 1.07 at 38.4°C (air = 1); dipole moment 3.5; vapor pressure: 10 mm Hg at 15.4°C; Henry's Law constant = 5.23×10^{-5} atm m³/mole at 25°C (measured).

PROTECTIVE EQUIPMENT Wear appropriate equipment to prevent any possibility of skin contact (cyanides) [R6].

PREVENTATIVE MEASURES Workers should wash immediately when skin becomes contaminated (cyanides) [R6].

DISPOSAL Generators of waste (equal to or greater than 100 kg/mo) containing this contaminant, EPA hazardous waste numbers D003 and P030, must conform with USEPA regulations in storage, transportation, treatment, and disposal of waste. (cyanide compound and cyanides, not otherwise specified) [R7].

COMMON USES Basic material in industrial, chemical, and pharmaceutical intermediates and products; poultry medicines [R3]; chemical intermediate for butyric acid, butyramide, and pharmaceuticals, and other organics (e.g., ketones, esters) [R1].

OCCUPATIONAL RECOMMENDATIONS

NIOSH 10-hr time-weighted average: 8 ppm (22 mg/m^3) [R11].

NON-HUMAN TOXICITY Lethal ip dose of butyronitrile in rats was approximately 150 mg/kg [R8]; LD$_{50}$ rat oral 0.14 g/kg [R2]; LD$_{50}$ rat oral 135 mg/kg [R14, 343]; LD$_{50}$ rat sc 200 mg/kg [R14, 343]; LD$_{50}$ mouse ip 45.75 mg/kg [R14, 343]; LD$_{50}$ rabbit dermal 400 mg/kg [R14, 343]; LD$_{50}$ guinea pig dermal 0.1–0.5 mg/kg [R14, 343].

BIODEGRADATION Using a Warburg respirometer, a 500-ppm concentration of butanenitrile, and a 72-hr incubation period, butanenitrile was found to have a 10.5% theoretical BOD using an activated sludge seed from a Bordeaux, TN, waste treatment facility (1); with an activated sludge seed from a Nashville, TN, waste treatment facility, a 1.0% theoretical BOD was determined (1); butanenitrile was toxic to an activated sludge seed from a Franklin, TN, waste treatment facility (1). Butanenitrile was poorly oxidized during 24-hr incubations using a Warburg respirometer, 500-ppm concns, and activated sludge seeds from three different Ohio waste treatment facilities (2). These short-term-incubation and high-butanenitrile-concentration studies (1–2) may not represent environmental biodegradation, because the high concns may have been toxic to the microbes or the microbes may have been given inadequate time to acclimate to the butanenitrile (SRC) [R9].

BIOCONCENTRATION Based upon a measured water solubility of 33,000 mg/L at 25°C (1), the BCF for butanenitrile can be estimated to be about 2 from a recommended regression-derived equation (2,SRC), which suggests that bioconcentration in aquatic organisms is not important environmentally (SRC) [R10].

ATMOSPHERIC STANDARDS This action promulgates standards of performance for equipment leaks of volatile organic compounds (VOC) in the synthetic organic chemical manufacturing industry (SOCMI). The intended effect of these standards is to require all newly constructed, modified, and reconstructed SOCMI process units to use the best demonstrated system of continuous emission reduction for equipment leaks of VOC, considering costs, non-air-quality health, and environmental impact and energy requirements. Butyronitrile is produced, as an intermediate or a final product, by process units covered under this subpart [R12].

RCRA REQUIREMENTS D003; A solid waste containing a cyanide compound may become characterized as a hazardous waste when subjected to testing for reactivity as stipulated in 40 CFR 261.23, and, if so characterized, must be managed as a hazardous waste. (cyanide compounds) [R13].

DOT *Health Hazards:* Poisonous; may be fatal if inhaled, swallowed, or absorbed through skin. Contact may cause burns to skin and eyes. Runoff from fire control or dilution water may cause pollution [R4].

Fire or Explosion: Flammable/combustible material; may be ignited by heat, sparks, or flames. Vapors may travel to a source of ignition, and flash back. Container may explode in heat of fire. Vapor explosion and poison hazard indoors, outdoors, or in sewers. Runoff to sewer may create fire or explosion hazard [R4].

Emergency Action: Keep unnecessary people away; isolate hazard area and deny entry. Stay upwind; keep out of low areas. Positive-pressure self-contained breathing apparatus (SCBA) and chemical protective clothing that is specifically recommended by the shipper or manufacturer may be worn. It may provide little or no thermal protection. Structural firefighters' protective clothing is not effective for these materials. Isolate for 1/2 mile in all directions if tank, rail car, or tank truck is involved in fire. Call CHEMTREC at 1–800–424–9300 for emergency assistance [R4].

Fire: Small Fires: Dry chemical, CO_2, water spray, or alcohol-resistant foam. Large Fires: Water spray, fog, or alcohol-resistant foam. Move container from fire area if you can do so without risk. Dike fire-control water for later disposal; do not scatter the material. Apply cooling water to sides of containers that are exposed to flames until well after fire is out. Stay away from ends of tanks. Withdraw immediately in case of rising sound from venting safety device or any discoloration of tank due to fire [R4].

Spill or Leak: Shut off ignition sources; no flares, smoking, or flames in hazard area. Fully encapsulating, vapor-protective clothing should be worn for spills and leaks with no fire. Do not touch or walk through spilled material; stop leak if you can do so without risk. Water spray may reduce vapor, but it may not prevent ignition in closed spaces. Small Spills: Take up with sand or other noncombustible absorbent material and place into containers for later disposal. Large Spills: Dike far ahead of liquid spill for later disposal [R4].

First Aid: Move victim to fresh air and call emergency medical care; if not breathing, give artificial respiration; if breathing is difficult, give oxygen. In case of contact with material, immediately flush skin or eyes with running water for at least 15 minutes. Remove and isolate contaminated clothing and shoes at the site. Keep victim quiet and maintain normal body temperature. Effects may be delayed; keep victim under observation [R4].

GENERAL RESPONSE Avoid contact with liquid or vapor. Evacuate area. Shut off all sources of ignition. Call fire department. Wear self-contained breathing apparatus and protective clothing. Stop discharge if possible. Isolate and remove discharged material. Notify local health and pollution control agencies.

FIRE RESPONSE Flammable toxic fumes are released in fire. Flashback may occur along vapor trail. Containers may explode under fire conditions. Extinguish with dry chemical, alcohol foam, CO_2 Wear self-contained breathing apparatus and protective clothing.

EXPOSURE RESPONSE Call for medical aid. Vapor irritating to eyes, nose, throat, and skin. If inhaled: remove to fresh air. If not breathing, give artificial respiration. If breathing difficult, give oxygen. Liquid irritating to the skin. Remove contaminated clothing and shoes. Flush affected areas with plenty of water. If in eyes: hold eyelids open and flush with plenty of water. If swallowed and victim is conscious: have victim drink water or milk and induce vomiting. If swallowed and victim is unconscious: do nothing except keep victim warm.

RESPONSE TO DISCHARGE Issue warning—poison. Restrict access. Evacuate area. Mechanical containment. Should be removed. Chemical and physical treatment.

WATER RESPONSE Harmful to aquatic life in low concentrations. May be dangerous if it enters water intakes. Fouling to shoreline. Notify local health and wildlife officials. Notify operators of local water intakes.

EXPOSURE SYMPTOMS Dizziness, rapid respirations, headache, drowsiness, drop in blood pressure and pulse, delayed symptoms. May cause cyanosis (blue-grey coloring of skin and lips due to lack of oxygen).

EXPOSURE TREATMENT Inhalation: remove to fresh air; if breathing is stopped, give artificial respiration; if breathing is difficult, give oxygen. Eyes: hold eyelids open and flush with water for 15 minutes. Ingestion: if victim is conscious, have victim induce vomiting; if victim is unconscious, do nothing except keep victim warm. Skin: wash affected area with plenty of water.

FIRE POTENTIAL Flammable, dangerous fire risk [R3].

REFERENCES

1. SRI.

2. Budavari, S. (ed.). *The Merck Index—Encyclopedia of Chemicals, Drugs, and Biologicals.* Rahway, NJ: Merck and Co., Inc., 1989. 243.

3. Sax, N. I., and R. J. Lewis, Sr. (eds.). *Hawley's Condensed Chemical Dictionary.* 11th ed. New York: Van Nostrand Reinhold Co., 1987. 193.

4. U.S. Department of Transportation. *Emergency Response Guidebook 1990.* DOT P 5800.5. Washington, DC: U.S. Government Printing Office, 1990, p. G–28.

5. National Fire Protection Guide. *Fire Protection Guide on Hazardous Materials.* 10th ed. Quincy, MA: National Fire Protection Association, 1991, p. 325M–25.

6. NIOSH. *NIOSH Pocket Guide to Chemical Hazards.* DHHS (NIOSH) Publication No. 90–117. Washington, DC: U.S. Government Printing Office, June 1990, 76.

7. 40 CFR 240–280, 300–306, 702–799 (7/1/91).

8. Haguenoer, et al. *Bull Soc Pharm Lille Iss* (4): 161 (1974).

9. (1) Lutin, P. A. *J Water Pollut Control Fed* 42: 1632–42 (1970). (2) Malaney, G.W., R. M. Gerhold. Proc 17th Ind Waste Conf, Purdue Univ, Ext Ser 112: 249–57 (1962).

10. (1) Riddick, J. A., et al., *Organic Solvents: Physical Properties and Methods of Purification.* Techniques of Chemicals, 4th ed. New York: Wiley–Interscience, p. 587 (1986). (2) Lyman, W. J., et al., *Handbook of Chemical Property Estimation Methods,* Washington, DC: Amer Chemical Soc, p. 5–10 (1990).

11. NIOSH/CDC. *NIOSH Recommendations for Occupational Safety and Health Standards 1988,* Aug. 1988 (Suppl. to Morbidity and Mortality Wkly. Vol. 37 No. 5–7, Aug. 26, 1988). Atlanta, GA: National Institute for Occupational Safety and Health, CDC, 1988, p. V37 (S7) 21.

12. 40 CFR 60.489 (7/1/91).

13. 40 CFR 261.23 (7/1/91).

14. Snyder, R. (ed.). *Ethyl Browning's Toxicity and Metabolism of Industrial Solvents.* 2nd ed. Volume II: *Nitrogen and Phosphorus Solvents.* Amsterdam–New York–Oxford: Elsevier, 1990.

■ 2,4-D BUTOXYETHYL ESTER

CAS # 1929–73–3

SYNONYMS acetic acid, (2,4-dichlorophenoxy)-, 2-butoxyethyl ester • 2,4-d butoxyethyl ester • 2,4-d 2-butoxyethyl ester • 2,4-d-bee • 2,4-dbee • butoxyethanol ester of (2,4-dichlorophenoxy) acetic acid • 2,4-dichlorophenoxyacetic acid, butoxyethyl ester • weedone-638 • silvaprop-1 • weedone-100 emulsifiable • butoxy-d 3 • aqua-kleen • bladex-b • brush-killer-64 • planotox • weedone-lv-4 • esteron-99-concentrate • lo-estasol • weedone-lv4 • weedone lv-6 • weed-rhap lv-4d

MF $C_{14}H_{18}Cl_2O_4$

MW 311.2

COLOR AND FORM Amber liquid; colorless liquid.

ODOR Odorless when pure.

DENSITY 1.225 at 68°F.

BOILING POINT 185 to 190°C at 5.5 to 7 mm Hg.

FIREFIGHTING PROCEDURES Extinguish with dry chemicals, foam, or carbon dioxide. Water may be ineffective. Cool exposed containers with water (2,4-d esters) [R1].

TOXIC COMBUSTION PRODUCTS Irritating hydrogen chloride may form in fire. (2,4-d esters) [R1].

SOLUBILITY Soluble in most organic solvents; solubility in water: 12 mg/L.; soluble in oils (2,4-d esters).

VAPOR PRESSURE 1.70×10^{-3} at 25°C, method utilized: radioactive tracer (from table).

CORROSIVITY May attack some forms of plastics. (2,4-d esters).

STABILITY AND SHELF LIFE Shelf life of ester formulations varies, depending on the emulsifying system. Some retain satisfactory emulsifying properties after 3 yr (2,4-d) [R16, 132].

OTHER CHEMICAL/PHYSICAL PROPERTIES Viscous; decomposition temperature greater than 200°C at 760 mm Hg; vapor pressure: 4×10^{-6} at 25°C; method utilized: gas chromatography (from table); VAP: 40.5% and 60% ester had vap press of 3.64 and 2.67×10^{-6} mm Hg at 25°C; 2,4-d esters are generally immiscible or insoluble in water, but gradual hydrolysis (SRP) will occur in very alkaline waters (2,4-d esters); fuel-oil-like odor (2,4-d esters), SRP: technical product. Ester formulations have low solubility in water (2,4-d); SRP: they can be dispersed as aqueous emulsions; SRP: 2,4-d esters are soluble in nonpolar organic solvents such as hexane, benzene, acetone, and alcohols. (2,4-d esters).

PREVENTATIVE MEASURES Avoid contact with eyes, skin, and clothing. Avoid inhalation [R4].

CLEANUP Land spill: Dig a pit, pond, lagoon, holding area to contain liquid or solid material. (SRP: If time permits, pits, ponds, lagoons, soak holes, or holding areas should be sealed with an impermeable flexible-membrane liner. Cover solids with a plastic sheet to prevent dissolving in rain or firefighting water (2,4-dichlorophenoxyacetic acid ester) [R3]. Water spill: Use natural deep-water pockets, excavated lagoons, or sand-bag barriers to trap material at bottom. If dissolved in region of 10 ppm or greater concentration, apply activated carbon at ten times the spilled amount. Remove trapped material with suction hoses. Use mechanical dredges or lifts to remove immobilized masses of pollutants and precipitates. (2,4-dichlorophenoxyacetic acid ester) [R3].

DISPOSAL Generators of waste (equal to or greater than 100 kg/mo) containing this contaminant, EPA hazardous waste number U240, must conform with USEPA regulations in storage, transportation, treatment, and disposal of waste. (2,4-d, salts, and esters) [R5].

OCCUPATIONAL RECOMMENDATIONS
OSHA 8-hr time-weighted average: 10 mg/m³(2,4-D) [R13].

NON-HUMAN TOXICITY LD_{50} rat male oral 940 mg/kg [R17, 499]; LD_{50} rabbit percutaneous in the range of 4,000 mg/kg [R17, 499]; LD_{50} chicks 4-week-old, oral. Acid equivalent was 588 mg/kg [R7].

ECOTOXICITY LC_{50} *Gammarus lacustris* (scud) 440 µg/L/96 hr (conditions of bioassay not specified) [R17, 498]; LC_{50} *Gammarus fasciatus* (scud) 5,900 µg/L/96 hr (conditions of bioassay not specified) [R17, 498]; LC_{50} *Daphnia magna* (water flea) 5,600 µg/L/48 hr (conditions of bioassay not specified) [R17, 498]; LC_{50} *Cypridopsis vidua* (seed shrimp) 1,800 µg/L/48 hr (conditions of bioassay not specified) [R17, 499]; LC_{50} *Asellus brevicaudus* (sowbug) 3,200 µg/L/48 hr (conditions of bioassay not specified) [R17, 499]; LC_{50} *Palaemonetes kadiakensis* (glass shrimp) 1,400 µg/L/48 hr (conditions of bioassay not specified) [R17, 499]; LC_{50} *Pteronarcys californica* (stonefly) 1,600 µg/L/96 hr (conditions of bioassay not specified) [R17, 499]; LC_{50} *Pimephales promelas* (fathead minnow) 5,600 µg/L/96 hr (conditions of bioassay not specified) [R17, 499]; LC_{50} *Colinus virginianus* (bobwhite quail), oral greater than 5,000 ppm in 5-day diet (sample purity 69.3%) [R8]; LC_{50} *Coturnix japonica* (Japanese quail) oral greater than 5,000 ppm in 5-day diet (sample purity 69.3%) [R8]; LC_{50} *Phasianus colchicus* (ring-necked pheasant) oral greater than 5,000 ppm in 5-day diet (sample purity 69.3%) [R8]; LC_{50} *Anas platyrhynchos* (mallards) oral greater than 5,000 ppm in 5-day diet (sample purity 69.3%) [R8]; LC_{50} black bullhead 8.7–8.8 mg/L/24 hr (static bioassay) [R9]; LC_{50} goldfish 3.6–4.1 mg/L/24 hr (static bioassay) [R9]; LC_{50} rainbow trout 1.5–1.6 mg/L/24 hr (static bioassay) [R9]; LC_{50} *Pimephales promelas* (fathead minnow) 1,500 µg/L/48 hr (lethal to eggs) (conditions of bioassay not specified) [R17, 499].

IARC SUMMARY Classification of carcinogenicity: (1) evidence in humans: limited; overall summary evaluation of carcinogenic risk to humans is Group 2B: The agent is possibly carcinogenic to humans. (chlorophenoxy herbicides; from table) [R6].

BIOLOGICAL HALF-LIFE These herbicides do not accumulate in animals. They are not extensively metab but are actively excreted into the urine. Their plasma half-life in man is about 1 day. (chlorophenoxy compounds) [R10].

BIODEGRADATION The mean second-order rate constant for butoxyethyl ester of 2,4-D as determined in water samples from 31 sites around the USA was 5.4×10^{-10} L/organism-hr, with the coefficient of variation <65% over the sites (1). The half-life of the ester in a lake in northern Georgia during the summer was approximately 5 hr (1). The reaction pathway involves an initial hydrolysis to the free acid, 2,4-D (1). Half-lives determined in waters from 7 sites in western Washington ranged from 1.7 to 23 hr (4). Most of the variability in rates was due to differences in the microbial populations at different sites (4). The second-order rate contants ranged from 3.0×10^{-7} to 8.5×10^{-9} L/organism-hr but fell into two groups; the average rate in one group was an order of magnitude higher than the other, probably due to difference in the nature of microorganisms (4). Biodegradation half-lives measured for aufwuch colonies, microbial communities forming on surfaces, that were allowed to develop on teflon strips at four sampling sites (river surface, river bottom, pond surface, and pond bottom) ranged from 1 to 3.4 hr but decreased to as little as 38 min when the colonies were suspended (2). The second-order rate coefficients used to predict these rates were based on the ratio of colonized surface area to container volume and were approximately 17–22 L/m³-hr in one experiment (3,5). Half-lives obtained from batch cultures of grab samples taken from 4 field sites on vari-

ous dates ranged from 0.4 to 3 hr (3). The degradation rate by aufwuch colonies increases with mixing speed because the mass transport of the chemical to the surface is a controlling factor, and biotransformation rates are more detectable in quiescent waters than in faster-flowing waters (5,6). During the course of these experiments it was shown that the butoxyethyl ester of 2,4-D was rapidly transformed by fungi and green and blue-green algae (3) [R11].

BIOCONCENTRATION Channel catfish and bluegills exposed to radiolabeled 2,4-D butoxyethyl ester rapidly took up the chemical, showing maximum concentrations within 1 to 2 hr exposure for fed fish and 2 to 6 hr for fasted fish (1). The whole-body maximum BCF of 7–55 and 20–55 was obtained for fed and fasted fish, respectively (1). The fish convert the ester to the free acid, which is rapidly excreted as residues in most tissues and organs decline and exponentially approach negligible concentrations (1). Estimated BCF in aquatic organisms range from 162 to 408 (2) [R12].

CERCLA REPORTABLE QUANTITIES Persons in charge of vessels or facilities are required to notify the National Response Center (NRC) immediately, when there is a release of this designated hazardous substance, in an amount equal to or greater than its reportable quantity of 100 lb or 45.4 kg. The toll-free number of the NRC is (800) 424–8802; in the Washington DC metropolitan area (202) 426–2675. The rule for determining when notification is required is stated in 40 CFR 302.4 (section IV. D. 3.b). (2,4-D acid, salts, or esters) [R14].

RCRA REQUIREMENTS As stipulated in 40 CFR 261.33, when 2,4-D acid, salts, and esters, as commercial chemical products or manufacturing chemical intermediates or off-specification commercial chemical products or manufacturing chemical intermediates, become wastes, they must be managed according to federal or state hazardous-waste regulations. Also defined as a hazardous waste is any residue, contaminated soil, water, or other debris resulting from the cleanup of a spill, into water or on dry land, of these wastes. Generators of small quantities of these wastes may qualify for partial exclusion from hazardous-waste regulations (40 CFR 261.5) (2,4-d acid, salts, and esters) [R15].

DOT *Health Hazards:* Poisonous; may be fatal if inhaled, swallowed, or absorbed through skin. Contact may cause burns to skin and eyes. Runoff from fire control or dilution water may give off poisonous gases and cause water pollution. Fire may produce irritating or poisonous gases. (2,4-D ester) [R2]

Fire or Explosion: Some of these materials may burn, but none of them ignites readily. Container may explode violently in heat of fire (2,4-d ester) [R2].

Emergency Action: Keep unnecessary people away; isolate hazard area and deny entry. Stay upwind, out of low areas, and ventilate closed spaces before entering. Self-contained breathing apparatus and chemical protective clothing that are specifically recommended by the shipper or producer may be worn but they do not provide thermal protection unless that is stated by the clothing manufacturer. Structural firefighter's protective clothing is not effective with these materials. Remove and isolate contaminated clothing at the site. Call CHEMTREC at 1–800–424–9300 as soon as possible, especially if there is no local hazardous materials team available (2,4-d ester) [R2].

Fire: Small Fires: Dry chemical, carbon dioxide (CO_2), halon, water spray, or standard foam. Large Fires: Water spray, fog, or standard foam is recommended. Move container from fire area if you can do so without risk. Fight fire from maximum distance. Stay away from ends of tanks. Dike fire control water for later disposal; do not scatter the material (2,4-d ester) [R2].

Spill or Leak: Do not touch spilled material; stop leak if you can do so without risk. Use water spray to reduce vapors. Small Spills: Take up with sand or other noncombustible absorbent material and

place into containers for later disposal. Small Dry Spills: With clean shovel place material into clean, dry container, and cover; move containers from spill area. Large Spills: Dike far ahead of liquid spill for later disposal (2,4-D ester) [R2].

First Aid: Move victim to fresh air and call emergency medical care; if not breathing, give artificial respiration; if breathing is difficult, give oxygen. In case of contact with material, immediately flush skin or eyes with running water for at least 15 minutes. Speed in removing material from skin is of extreme importance. Remove and isolate contaminated clothing and shoes at the site. Keep victim quiet and maintain normal body temperature. Effects may be delayed; keep victim under observation (2,4-d ester) [R2].

FIRE POTENTIAL Material itself does not burn or burns with difficulty. (2,4-dichlorophenoxyacetic acid ester) [R3].

REFERENCES

1. U.S. Coast Guard, Department of Transportation. *CHRIS—Hazardous Chemical Data.* Volume II. Washington, DC: U.S. Government Printing Office, 1984–5.

2. Department of Transportation. *Emergency Response Guidebook 1987.* DOT P 5800.4. Washington, DC: U.S. Government Printing Office, 1987, p. G–55.

3. Association of American Railroads. *Emergency Handling of Hazardous Materials in Surface Transportation.* Washington, DC: Assoc. of American Railroads, Hazardous Materials Systems (BOE), 1987. 245.

4. *Farm Chemicals Handbook 1989.* Willoughby, OH: Meister Publishing Co., 1989, p. C–309.

5. 40 CFR 240–280, 300–306, 702–799 (7/1/89).

6. IARC. *Monographs on the Evaluation of the Carcinogenic Risk of Chemicals to Man.* Geneva: World Health Organization, International Agency for Research on Cancer, 1972–present (multivolume work), p. S7 60 (1987).

7. Whitehead, C. C., R. J. Pettigrew. *Toxi-*

col *Appl Pharmacol* 21: 348 (1972) as cited in *Who. Environ Health Criteria: 2,4–Dichlorophenoxyacetic Acid (2,4–D).* p. 66 (1984).

8. U.S. Department of the Interior, Fish and Wildlife Service, Bureau of Sports Fisheries and Wildlife. *Lethal Dietary Toxicities of Environmental Pollutants to Birds.* Special Scientific Report—*Wildlife* No. 191. Washington, DC: U.S. Government Printing Office, 1975. 15.

· 9. Inglis, A., E. L. Davis. *Bur Sport Fish Wild Tech Paper* 67: 1–22 (1972) as cited in *Nat'l Research Council Canada; Phenoxyherbicides.* p. 183 (1978) NRCC No. 16075.

10. Gilman, A. G., L. S. Goodman, and A. Gilman (eds.). *Goodman and Gilman's The Pharmacological Basis of Therapeutics.* 7th ed. New York: Macmillan Publishing Co., Inc., 1985. 1645.

11. (1) Paris, D. F., et al., *Appl Environ Microbiol* 41: 603–9 (1981). (2) Kollig, H. P. *Chemosphere* 14: 1779–87 (1985). (3) Lewis, D. L., et al., *Appl Environ Microbiol* 46: 146–51 (1983). (4) Rogers, J. E., et al. *Microbial Transformation Kinetics of Xenobioties in the Aquatic Environ.* USEPA–600/3–84–043 (1984). (5) Lewis, D. L., et al. *ASTM STP 865,* ASTM, Philadelphia, PA, pp. 3–13 (1985). (6) Kolling, H. P., et al. *Chemosphere* 16: 49–60 (1987).

12. (1) Rodgers, C. A., D. L. Stalling. *Weed Sci* 72: 101–5 (1972). (2) O'Kelly, J. C., T. R. Deason. *Degradation of Pesticides by Algae.* Environ Res Lab, pp. 51 USEPA–600/3–76–02 (1976).

13. 54 FR 2920 (1/19/89).

14. 54 FR 33419 (8/14/89). ·

15. 40 CFR 261.33 (7/1/88).

16. Weed Science Society of America. *Herbicide Handbook.* 5th ed. Champaign, Illinois: Weed Science Society of America, 1983.

17. Verschueren, K. *Handbook of Environmental Data of Organic Chemicals.* 2nd ed. New York, NY: Van Nostrand Reinhold Co., 1983.

■ sec–BUTYLAMINE

CAS # 13952–84–6

SIC CODES 2879.

SYNONYMS 2-butanamine • 2-ab • 2-aminobutane • 2-aminobutane base • propylamine, 1-methyl • 1-methylpropylamine • tutane • frucote • deccotane

MF $C_4H_{11}N$

MW 73.16

COLOR AND FORM Colorless liquid.

ODOR Ammoniacal.

DENSITY 0.724 at 20°C (dl-isomer).

SURFACE TENSION 22.42 dynes/cm = 0.02242 N/m at 20°C.

BOILING POINT 63°C.

MELTING POINT –104°C (freezing point).

HEAT OF COMBUSTION –17,600 Btu/Lb = –9,780 cal/g = -409×10^5 J/kg.

HEAT OF VAPORIZATION 178.09 Btu/Lb = 98.94 cal/g = 4.160×10^5 J/kg.

AUTOIGNITION TEMPERATURE 712°F [R15, 236].

EXPLOSIVE LIMITS AND POTENTIAL Containers may explode in fire [R2].

FLAMMABILITY LIMITS Lower 1.7%; Upper 9.8% [R8].

FIREFIGHTING PROCEDURES Alcohol foam, dry chemical, or carbon dioxide [R2].

TOXIC COMBUSTION PRODUCTS Toxic oxides of nitrogen may be formed in a fire [R2].

SOLUBILITY Miscible with water and most organic solvents.

VAPOR PRESSURE 135 mm Hg at 20°C.

OCTANOL/WATER PARTITION COEFFICIENT Log K_{ow} = 0.74.

CORROSIVITY Corrosive to tin, aluminum, and some steels.

STABILITY AND SHELF LIFE Stable [R6].

OTHER CHEMICAL/PHYSICAL PROPERTIES

Liquid; density 0.7308 at 15°C; index of refraction: 1.3963 at 15°C/D; specific optical rotation: 7.80 degrees (neat) at 15°C/D (sec-butylamine (d (S)); Liquid; density 0.728 at 19°C; specific optical rotation: −7.64°C (sec-butylamine (l)); it is a base, forming water-soluble salts with acids. Soluble in water, alcohol, ether, acetone (2–aminobutane (d), and (l)); soluble in alcohol, ether, acetone (2–aminobutane(dl)) Ratio of specific heats of vapor (gas): 1.073 at 20°C (est); heat of solution: −170 Btu/lb = −93 cal/g = −3.9×105 J/kg, wt/gal at 20°C = 6.0 lb; mass: 77 (*Atlas of Mass Spectral Data*, New York: John Wiley and Sons) (butane, 2–amino (d)); IR: 13484 (*Sadtler Research Laboratories Prism Collection*) (butane, 2–amino(dl)); NMR: 88 (Varian Associates *NMR Spectra Catalogue*) (butane, 2–amino(dl)); mass: 77 (*Atlas of Mass Spectral Data*, New York: John Wiley and Sons) (butane, 2–amino(dl)); mass: 77 (*Atlas of Mass Spectral Data*, New York: John Wiley and Sons) (butane, 2–amino (l)); mass: 21 (National Bureau of Standards *EPA–NIH Mass Spectra Data Base*, NSRDS–NBS–63) (butane, 2–amino (l)).

PROTECTIVE EQUIPMENT Chemical safety goggles, rubber gloves, and apron; respiratory protective equipment; non-sparking shoes [R2].

PREVENTATIVE MEASURES Wash thoroughly following contact [R4].

CLEANUP Small spill: Use absorbent paper to pick up spilled material. Follow by washing surfaces well with soap and water. Seal all wastes in vapor-tight plastic bags for eventual disposal. [R15, 237].

DISPOSAL After the material has been contained, apply a sorbent material (peat, straw, sawdust, etc.) to the contaminated area. Remove all contaminated sorbent and soil and place in impervious containers. The material may be incinerated in an approved chemical incinerator at the specified temperature/dwell-time combination. Any liquids, sludges, or solid residues generated should be disposed of in accordance with all applicable federal, state, and local pollution control requirements. If appropriate incineration facilities are not available, material may be buried in a chemical waste landfill. Amenable to biological treatment at a municipal sewage treatment plant when diluted.

SRP: At the time of review, criteria for land treatment or burial (sanitary landfill) disposal practices are subject to significant revision. Prior to implementing land disposal of waste residue (including waste sludge), consult with environmental regulatory agencies for guidance on acceptable disposal practices.

COMMON USES Fungistat [R3]; prevention of *Botrytis* on chrysanthemums and gladioli during storage and transit [R4].; Used as a post-harvest fungicide for citrus fruits. It effectively controls strains of *Penicillium digitatum* [R5]. It is a fungicide used to control many fruit-rotting fungi. Aqueous solutions of its salts, containing 5–20 g amine/L, are used as dips or sprays on harvested fruit to prevent decay in transport or storage. Harvested fruit may be fumigated at 317 mg/m³ for 4 hr, or the equivalent. Potatoes (seed or ware) are treated at 280 ml/ton for 2.5 hr to control *Phoma exigua var foveata* and *Polyscytalum pustulans*. In neutral aqueous solution, the decay of oranges due to *Penicillium digitatum* is largely controlled by the (R)–(−) isomer [R6].

STANDARD CODES EPA 311; TSCA; IATA flammable liquid, flammable liquid label, 1 L passenger, 40 L cargo; NFPA–3, 3,; CFR 49 flammable liquid, flammable liquid label, 1 quart passenger, 10 gallons cargo, storage code 1, 2; CFR 14 CAB code 8; not listed AAR; not listed ICC; Superfund designated (hazardous substances) list.

PERSISTENCE Nonpersistent.

DIRECT CONTACT Severe primary irritation and deep second-degree skin burns (blistering) in man, severe eye damage; ceiling TLV for skin is 5 ppm.

GENERAL SENSATION Burning and irritation of the eyes, nose, throat, and skin. Contact with the liquid can cause dermatitis, burns on skin, lacrimation, conjunctivitis, burns, corneal edema to eyes.

PERSONAL SAFETY Protect against both inhalation and absorption through the skin. Must wear protective clothing including NIOSH–approved rubber gloves and boots (non-sparking shoes), safety goggles or face mask, and either a respirator whose canister is specifically approved for this material, or a self-contained breathing apparatus.

ACUTE HAZARD LEVEL Sec-butylamine is very toxic to humans by ingestion; between 1 teaspoonful and 1 ounce may be fatal. The acute oral LD_{50} to rats for the tutane formulation is 380 mg/kg. When in contact with the skin, this material will cause severe second- and third-degree burns on short contact. It is extremely irritating to the eyes, causing corneal damage. Inhalation of concentrated fumes can cause asphyxiation and burns to the respiratory system (I37). The aquatic toxicity to bluegill sunfish in static water is 32 ppm for the 96 hour LC_{50}.

CHRONIC HAZARD LEVEL The ceiling-level threshold-limit value—time-weighted average for a 40–hour workweek for butylamine is 15 mg/m³.

DEGREE OF HAZARD TO PUBLIC HEALTH Liquid severely irritating to skin, eyes, respiratory and gastrointestinal tract. Very dangerous if ingested. Also liquid a great fire hazard.

OCCUPATIONAL RECOMMENDATIONS OSHA Meet criteria for proposed OSHA medical records rule [R11].
ACTION LEVEL Avoid contact with the spilled cargo. Stay upwind. Eliminate all ignition sources. Notify local air, water, and fire authorities of the accident. Evacuate all people to a distance of at least 200 feet upwind and 1,000 feet downwind of the spill.

ON-SITE RESTORATION Dam the stream if possible to reduce the flow and to prevent further spread of the slick. Use of floating booms is also advisable. Activated carbon can be applied to the water to adsorb the dissolved material. Bottom pumps, dredges, or vacuum systems can be used to remove the sorbent from the bottom. An alternative method is to transfer to a gravity separation tank, and pass through a filtration system with an activated carbon filter. For more details, see ENVIREX Manual, EPA 600/2–77–227. Seek professional environmental engineering assistance through EPA's Environmental Response Team (ERT), Edison, NJ, 24–hour no. 908–548–8730.

BEACH OR SHORE RESTORATION Close beach and shore to the public until material has been removed. If spill results in significant amounts of unabsorbed material, an alternative method may be ignition of the material by the local fire department, with notification of the local air authorities.

AVAILABILITY OF COUNTERMEASURE MATERIAL Bottom pumps, dredges—fire departments, U.S. Coast Guard, Army Corps of Engineers; vacuum systems—swimming-pool operators; activated carbon—water treatment plants, chemical companies; floating booms—Army Corps of Engineers, U.S. Coast Guard.

DISPOSAL NOTIFICATION Notify local and state health authorities, local solid waste disposal authorities, supplier, and shipper of material.

INDUSTRIAL FOULING POTENTIAL Low flash point, potential explosion in boiler feed and cooling water.

MAJOR WATER USE THREATENED Recreation; fisheries; potable supply; industry.

PROBABLE LOCATION AND STATE Colorless liquid is completely miscible in water.

WATER CHEMISTRY Will form basic solution when dissolved in water.

COLOR IN WATER Colorless.

FOOD-CHAIN CONCENTRATION POTENTIAL Negative. Not expected to accumulate in the food chain.

NON-HUMAN TOXICITY LD_{50} hen oral 250 mg/kg [R1]; LD_{50} rat oral 157.5 mg/kg (male); 146.8 mg/kg (female) [R9].

ECOTOXICITY LC_{50} bluegill (young) more than 50 mg/L [R10].

ATMOSPHERIC STANDARDS This action promulgates standards of performance for equipment leaks of volatile organic compounds (VOC) in the synthetic organic chemcial manufacturing industry (SOCMI). These standards implement Section 111 of the Clean Air Act and are based on the Administrator's determination that emissions from the synthetic organic, chemical manufacturing industry cause, or contribute significantly to, air pollution that may reasonably be anticipated to endanger public health or welfare. The intended effect of these standards is to require all newly constructed, modified, and reconstructed SOCMI process units to use the best demonstrated system of continuous emission reduction for equipment leaks of VOC, considering costs, non-air-quality health, and environmental impact and energy requirements. Sec-butylamine is covered under this rule. These standards of performance become effective upon promulgation but apply to affected facilities for which construction or modification commenced after January 5, 1981 [R12].

CERCLA REPORTABLE QUANTITIES Persons in charge of vessels or facilities are required to notify the National Response Center (NRC) immediately, when there is a release of this designated hazardous substance, in an amount equal to or greater than its reportable quantity of 1,000 lb or 454 kg. The toll-free telephone number of the NRC is (800) 424-8802; in the Washington metropolitan area (202) 426-2675. The rule for determining when notification is required is stated in 40 CFR 302.6 (section IV. D. 3.b) [R13].

DOT *Spill or Leak:* Shut off ignition sources; No flares, smoking, or flames in hazard area. Do not touch spilled material; Stop leak if you can do so without risk. Use water spray to reduce vapors. Small Spills: Take up with sand or other non-combustible absorbent material, and place into containers for later disposal. Large Spills: Dike far ahead of spill for later disposal (butylamine) [R7].

Fire: Small Fires: Dry chemical, CO_2, water spray, or foam. Large Fires: Water spray, fog, or foam. Move container from fire area if you can do so without risk. Stay away from ends of tanks. Cool containers that are exposed to flames with water from the side until well after fire is out. Withdraw immediately in case of rising sound from venting safety device or any discoloration of tank due to fire (butylamine) [R7].

Emergency Action: Keep unnecessary people away; Isolate hazard area and deny entry. Stay upwind; Keep out of low areas. Wear self-contained (positive-pressure if available) breathing apparatus and full protective clothing. Isolate for 1/2 mile in all directions, if tank car or truck is involved in fire. If water pollution occurs, notify appropriate authorities. For emergency assistance call CHEMTREC (800) 424-9300 (butylamine) [R7].

Fire or Explosion: Flammable/combustible material; May be ignited by heat, sparks, or flames. Vapors may travel to a source of ignition, and flash back. Container may explode in heat of fire. Vapor explosion hazard indoors, outdoors, or in sewers. Runoff to sewer may create fire or explosion hazard (butylamine) [R7].

Health Hazards: Poisonous if inhaled or swallowed. Contact may cause burns to skin and eyes. Fire may produce irritating or poisonous gases. Runoff from fire control or dilution water may cause pollution (butylamine) [R7].

First Aid: Move victim to fresh air. Call for emergency medical care. If not breathing, give artificial respiration. If breathing is difficult give oxygen. Remove and isolate contaminated clothing and shoes at the site. In case of contact with material, immediately flush skin and eyes with running water for at least 15 minutes. Keep victim quiet and maintain normal body temperature (butylamine) [R7].

FIRE POTENTIAL Flammable [R8].

FDA A tolerance of 90 ppm is established for residues of the fungicide sec-butylamine in citrus molasses and dried citrus pulp for cattle feed when present therein as a result of postharvest application of the fungicide to citrus fruit [R14].

REFERENCES

1. Spencer, E. Y. *Guide to the Chemicals Used in Crop Protection.* 7th ed. Publication 1093. Research Institute, Agriculture Canada, Ottawa, Canada: Information Canada, 1982. 73.

2. U.S. Coast Guard, Department of Transportation. *CHRIS—Hazardous Chemical Data.* Volume II. Washington, DC: U.S. Government Printing Office, 1984–5.

3. *The Merck Index.* 10th ed. Rahway, NJ: Merck Co., Inc., 1983. 215.

4. Hartley, D., and H. Kidd (eds.). *The Agrochemicals Handbook.* Old Woking, Surrey, United Kingdom: Royal Society of Chemistry/Unwin Brothers Ltd., 1983, p. A423/Oct 83.

5. Rippon, L.E.; *Australian Citrus News* 54: 8–9 (1978).

6. Worthing, C. R., S. B. Walker (eds.). *The Pesticide Manual—A World Compendium.* 7th ed. Lavenham, Suffolk, Great Britain: The Lavenham Press Limited, 1983. 80.

7. Department of Transportation. *Emergency Response Guidebook 1984* DOT P 5800.3 Washington, DC: U.S. Government Printing Office, 1984, p. G–68.

8. National Fire Protection Association. *Fire Protection Guide on Hazardous Materials.* 9th ed. Boston, MA: National Fire Protection Association, 1986, p. 325M–20.

9. Cheever, K.L., et al., *Toxicol Appl Pharmacol* 63 (1): 150–52 (1982).

10. Hartley, D., and H. Kidd (eds.). *The Agrochemicals Handbook.* Old Woking, Surrey, United Kingdom: Royal Society of Chemistry/Unwin Brothers Ltd., 1983, p. A432/Oct 83.

11. 47 FR 30420 (7/13/82).

12. 40 CFR 60.489 (7/1/85).

13. 50 FR 13456 (4/4/85).

14. 21 CFR 561.60 (4/1/86).

15. Keith, L.H., D.B. Walters (eds.). *Compendium of Safety Data Sheets for Research and Industrial Chemicals.* Parts I, II, and III. Deerfield Beach, FL: VCH Publishers, 1985.

◼ t-BUTYL MERCAPTAN

CAS # 75–66–1

SYNONYMS 1,1-dimethylethanethiol • 2-isobutanethiol • 2-methyl-2-propanethiol • 2-propanethiol, 2-methyl • t-butylmercaptan • tert-butanethiol • tert-butyl mercaptan • tert-butylmercaptan • tert-butylthiol

MF $C_4H_{10}S$

MW 90.20

COLOR AND FORM Mobile liquid; colorless liquid.

ODOR Heavy skunk odor.

DENSITY 0.79426 at 25°/4°C.

BOILING POINT –63.7–64.2°C at 760 mm Hg.

MELTING POINT –0.5°C.

EXPLOSIVE LIMITS AND POTENTIAL Lower members of the group (mercaptans), such as butyl mercaptan, form explosive mixtures with air. (mercaptans) [R4].

FIREFIGHTING PROCEDURES Alcohol foam, dry chemical, mist, fog [R3].

SOLUBILITY Slightly soluble in water; very soluble in alcohol, ether, liquid hydrogen sulfide; soluble in heptane.

STABILITY AND SHELF LIFE Remarkably stable to oxidizing agents [R2].

OTHER CHEMICAL/PHYSICAL PROPERTIES wt/gallon: 6.71 lb; distillation range: 62–67°C.

PROTECTIVE EQUIPMENT According to degree of possible exposure, persons working with mercaptans should wear personal protective equipment; contact with skin and mucous membranes of mercaptans in high or unknown concentration should be avoided. (mercaptans) [R4].

PREVENTATIVE MEASURES Ventilation control: the basic ventilation methods are local exhaust ventilation and dilution or general ventilation [R5]. Ventilation control: generally process from which mercaptan vapor is liable to be evolved should as far as practicable be enclosed, and exhaust ventilation should be provided to prevent vapors from diffusing into atmosphere of workroom. (mercaptans) [R4].

COMMON USES Odorant for natural gas [R1].

STANDARD CODES Codes for normal form: ICC—flammable liquid, red label, 10 gallons in an outside container; USCG—nonflammable liquid, red label; NFPA-2, 3, 0; IATA—flammable liquid, red label, 1 liter passenger, 40 liter cargo.

MAJOR SPECIES THREATENED Waterfowl.

INHALATION LIMIT 1.75 mg/m^3.

GENERAL SENSATION Skunk odor; recognition odor in air, 0.9×10^{-4} ppm [R6].

PERSONAL SAFETY Eye protection, protective clothing, and self-contained breathing apparatus required.

ACUTE HAZARD LEVEL Moderate ingestive and inhalative toxicant. Threatens aesthetics of all waters. Irritant. LC$_{50}$ for rats in air 4,020 mg/m^3.

CHRONIC HAZARD LEVEL Moderate chronic inhalative toxicant.

DEGREE OF HAZARD TO PUBLIC HEALTH Moderate ingestive and inhalative toxicant. Emits toxic fumes when heated to decomposition or when contacted with acid. Irritant.

OCCUPATIONAL RECOMMENDATIONS None listed.

ACTION LEVEL Notify fire and air authority. Remove all sources of ignition and acids. If fire develops, evacuate immediate area. Enter from upwind side.

ON-SITE RESTORATION Oil-skimming equipment and sorbent foams can be used on slicks. Peat or carbon will adsorb soluble portion. Seek professional environmental engineering assistance through EPA's Environmental Response Team (ERT), Edison, NJ, 24-hour no., 908-548-8730.

BEACH OR SHORE RESTORATION Should not be burned.

AVAILABILITY OF COUNTERMEASURE MATERIAL Oil-skimming equipment—stored at major ports; sorbent foams (polyurethane)—upholstery shops; peat—nurseries, floral shops; carbon—water treatment plants, sugar refineries.

DISPOSAL Dissolve in flammable solvent. Burn in incinerator with afterburner and SO$_2$ scrubber.

DISPOSAL NOTIFICATION Local air authority.

INDUSTRIAL FOULING POTENTIAL Volatile liquid, poses explosion danger when present in boiler feed or cooling water. Unacceptable in food-processing waters.

MAJOR WATER USE THREATENED Potable supply, recreation, industrial.

PROBABLE LOCATION AND STATE Colorless liquid. Will float in slick on surface and dissolve very slowly.

COLOR IN WATER Colorless.

REFERENCES

1. SRI.

2. *The Merck Index.* 9th ed. Rahway, NJ: Merck and Co., Inc., 1976. 201.

3. Sax, N. I. *Dangerous Properties of Industrial Materials.* 5th ed. New York: Van Nostrand Reinhold, 1979. 448.

4. International Labour Office. *Encyclopedia of Occupational Health and Safety.* Volumes I and II. New York: McGraw-Hill Book Co., 1971. 855.

5. Sax, N. I. *Dangerous Properties of Industrial Materials.* 4th ed. New York: Van Nostrand Reinhold, 1975. 494.

6. Sullivan, R. J., "Air Pollution Aspects of Odorous Compounds," *NTIS PB 188 089,* September 1969.

■ n-BUTYL METHACRYLATE

CAS # 97-88-1

DOT # 2227

SYNONYMS 2-methyl-butylacrylaat (Dutch) • 2-Methyl-butylacrylat (German) • 2-methyl-butylacrylate • 2-propenoic acid, 2-methyl-, butyl ester • butil-metacrilato (Italian) • butyl 2-methacrylate • butyl 2-methyl-2-propenoate • butylmethacrylaat (Dutch) • methacrylate de butyle (French) • methacrylic acid, butyl ester • Methacrylsaeurebutylester (German) • butyl methacrylate

MF $C_8H_{14}O_2$

MW 142.22

COLOR AND FORM Colorless liquid.

ODOR Faint characteristic odor; ester odor.

DENSITY 0.8936 at 20°C.

VISCOSITY 3.116 cP at 70°F.

BOILING POINT 160°C at 760 mm Hg.

MELTING POINT FP: below −75°C.

HEAT OF COMBUSTION −14,800 Btu/lb = −8,230 cal/g = −344×10⁵ J/kg (estimated).

AUTOIGNITION TEMPERATURE 562°F [R16, 1818].

EXPLOSIVE LIMITS AND POTENTIAL Lower: 2%; upper: 8%. [R16, 1818].

FLAMMABILITY LIMITS Upper: 2%, lower: 8% (in air, estimates) [R2].

FIREFIGHTING PROCEDURES Foam, dry chemical, carbon dioxide [R16, 1818]. Do not extinguish fire unless flow can be stopped. Use water in flooding quantities as fog. Solid streams of water may spread fire. Cool all affected containers with flooding quantities of water. Apply water from as far a distance as possible [R7].

SOLUBILITY Insoluble in water; >10% in ethanol; >10% in ether.

VAPOR PRESSURE 4.9 mm Hg at 20°C.

OCTANOL/WATER PARTITION COEFFICIENT Log K_{ow} = 2.88.

OTHER CHEMICAL/PHYSICAL PROPERTIES Polymerizes readily; liquid water interfacial tension: 35 dynes/cm = 0.035 N/m at 20°C; heat of polymerization: −180 Btu/lb = −100 cal/g = −4.2×10⁵ J/kg; Reid vapor

pressure: low; liquid heat capacity: 0.460 Btu/lb-F at 70°F; liquid thermal conductivity: 1.048 Btu-in./hr-sq ft-F at 70°F; saturated vapor density: 0.00203 lb/cu ft at 70°F; it weighs about 7.5 lb/gal.

PROTECTIVE EQUIPMENT Suitable protective clothing and self-contained resp protective apparatus should be available for use of those who may have to rescue persons overcome by fumes. (acrylic acid and derivatives) [R8].

PREVENTATIVE MEASURES Hazard is the generation of considerable exothermic heat in some of the reactions, with high pressures and temperature developing. This danger should be borne in mind when designing plant. Awareness of the dangers and of good engineering design are essential to safety. Employees should be instructed about the necessity of cleansing the skin if it is contaminated by materials that are irritants or skin-absorbed. With careful design, however, and complete enclosure of those processes where toxic chemicals or intermediates occur, dangerous exposures can be avoided. (acrylic acid, and derivatives) [R8].

COMMON USES Comonomer in acrylic surface coatings [R3]. Monomer for resins, solvent coatings, adhesives, oil additives; emulsions for textiles; leather, and paper finishing [R1]. In manufacture of contact lenses [R4]. Butyl methacrylate monomer and polymer are used in dental technology, as components in oil-dispersible pesticides, and as copolymers—for example, in paraffin embedding media. [R15, 2302].

OCCUPATIONAL RECOMMENDATIONS None listed.

NON-HUMAN TOXICITY LD_{50} mouse oral 15.8 g/kg [R9]; LD_{50} mouse subcutaneous 2.6 g/kg [R9]; LD_{50} rat oral 16.0 g/kg [R9].

BIOCONCENTRATION Based on a measured log K_{ow} of 2.88 (2), the BCF for n-butyl methacrylate can be estimated to be 91 using a recommended regression-derived equation (1,SRC). This BCF value suggests that n-butyl methacrylate would not bioconcentrate significantly in aquatic systems (SRC) [R10].

TSCA REQUIREMENTS Section 8 (a) of

TSCA requires manufacturers of this chemical substance to report preliminary assessment information concerned with production, use, and exposure to EPA as cited in the preamble of 51 FR 41329 [R11].

DOT *Fire or Explosion:* Flammable/combustible material; may be ignited by heat, sparks, or flames. Vapors may travel to a source of ignition, and flash back. Container may explode in heat of fire. Vapor explosion hazard indoors, outdoors, or in sewers. Runoff to sewer may create fire or explosion hazard [R5].

Health Hazards: May be poisonous if inhaled or absorbed through skin. Vapors may cause dizziness or suffocation. Contact may irritate or burn skin and eyes. Fire may produce irritating or poisonous gases. Runoff from fire control or dilution water may cause pollution [R5].

Emergency Action: Keep unnecessary people away; isolate hazard area and deny entry. Stay upwind; keep out of low areas. Self-contained breathing apparatus (SCBA) and structural firefighter's protective clothing will provide limited protection. Isolate for 1/2 mile in all directions if tank car or truck is involved in fire. Call CHEMTREC at 1–800–424–9300 for emergency assistance. If water pollution occurs, notify the appropriate authorities [R5].

Fire: Small Fires: Dry chemical, CO_2, halon, water spray, or alcohol foam. Large Fires: Water spray, fog, or alcohol foam is recommended. Move container from fire area if you can do so without risk. Cool containers that are exposed to flames with water from the side until well after fire is out. Stay away from ends of tanks. For massive fire in cargo area, use unmanned hose holder or monitor nozzles; if this is impossible, withdraw from area, and let fire burn. Withdraw immediately in case of rising sound from venting safety device or any discoloration of tank due to fire [R5].

Spill or Leak: Shut off ignition sources; no flares, smoking, or flames in hazard area. Stop leak if you can do so without

risk. Water spray may reduce vapor; but it may not prevent ignition in closed spaces. Small Spills: Take up with sand or other noncombustible absorbent material and place into containers for later disposal. Large Spills: Dike far ahead of liquid spill for later disposal [R5].

First Aid: Move victim to fresh air and call for emergency medical care; if not breathing, give artificial respiration; if breathing is difficult, give oxygen. In case of contact with material, immediately flush eyes with running water for at least 15 minutes. Wash skin with soap and water. Remove and isolate contaminated clothing and shoes at the site [R5].

FIRE POTENTIAL Flammable [R6].

FDA Butyl methacrylate is an indirect food additive for use only as a component of adhesives [R12]. Semirigid and rigid acrylic and modified acrylic plastics may be safely used as articles intended for use in contact with food, also as components of articles intended for use in contact with food. Optional substances that may be used include homopolymers and copolymers of the following monomer: n-Butyl methacrylate [R13]. Cross-linked polyester resins may be safely used as articles or components of articles intended for repeated use in contact with food. Cross-linked polyester resins are produced by the condensation of one or more acids with one or more alcohols or epoxides followed by copolymerization with one or more cross-linking agents including butyl methacrylate [R14].

GENERAL RESPONSE Stop discharge if possible. Keep people away. Call fire department. Avoid contact with liquid. Isolate and remove discharged material. Notify local health and pollution control agencies.

FIRE RESPONSE Combustible. Containers may explode in fire. Extinguish with dry chemicals, foam, or carbon dioxide. Water may be ineffective on fire. Cool exposed containers with water.

EXPOSURE RESPONSE Call for medical aid. Liquid irritating to skin and eyes. Harmful if swallowed. Remove contaminated clothing and shoes. Flush affected areas with plenty of water. If swallowed and victim is conscious, have victim drink water or milk, and have victim induce vomiting. If swallowed and victim is unconscious or having convulsions, do nothing except keep victim warm.

RESPONSE TO DISCHARGE Mechanical containment; should be removed; chemical and physical treatment.

WATER RESPONSE Effect of low concentrations on aquatic life is unknown. Fouling to shoreline. May be dangerous if it enters water intakes. Notify local health and wildlife officials. Notify operators of nearby water intakes.

EXPOSURE SYMPTOMS Inhalation may cause nausea because of offensive odor. Contact with liquid causes irritation of eyes and mild irritation of skin. Ingestion causes irritation of mouth and stomach.

EXPOSURE TREATMENT Inhalation: remove to fresh air; give oxygen or artificial respiration as required. Eyes: flush with copious amounts of water for 15 min. and consult physician. Skin: wash with soap and water. Ingestion: induce vomiting; call a physician.

REFERENCES

1. Sax, N. I., and R. J. Lewis, Sr. (eds.). *Hawley's Condensed Chemical Dictionary.* 11th ed. New York: Van Nostrand Reinhold Co., 1987. 188.

2. U.S. Coast Guard, Department of Transportation. *CHRIS—Hazardous Chemical Data.* Volume II. Washington, DC: U.S. Government Printing Office, 1984–5.

3. SRI.

4. *Kirk–Othmer Encyclopedia of Chemical Technology.* 3rd ed., Volumes 1–26. New York, NY: John Wiley and Sons, 1978–1984, p. 6 (79) 723.

5. Department of Transportation. *Emergency Response Guidebook 1987.* DOT P 5800.4. Washington, DC: U.S. Government Printing Office, 1987, p. G–26.

6. Commission of the European Communities. *Legislation on Dangerous Substances—Classification and Labelling in the European Communities.* Vol. II. London and Trotman Ltd., 1989. 358.

7. Association of American Railroads. *Emergency Handling of Hazardous Materials in Surface Transportation.* Washington, DC: Assoc. of American Railroads, Hazardous Materials Systems (BOE), 1987. 110.

8. International Labour Office. *Encyclopedia of Occupational Health and Safety.* Vols. I and II. Geneva, Switzerland: International Labour Office, 1983. 53.

9. Kustova, Z. R., et al. *Khim Prom–St, Ser; Toksikol Sanit Khim Plastmass* (3): 15–8 (1979).

10. (1) Lyman, W. J., et al., *Handbook of Chemical Estimation Methods.* New York: McGraw-Hill, p. 5–4 (1982). (2) Hansch, C., and A. J. Leo. *Medchem Project* Issue No. 26, Claremont, CA: Pomona College (1985).

11. 40 CFR 712.30 (7/1/88).

12. 21 CFR 175.105 (4/1/88).

13. 21 CFR 177.1010 (4/1/88).

14. 21 CFR 177.2420 (4/1/88).

15. Clayton, G. D. and F. E. Clayton (eds.). *Patty's Industrial Hygiene and Toxicology.* Volume 2A, 2B, 2C: Toxicology. 3rd ed. New York: John Wiley Sons, 1981–1982.

16. Sax, N.I. *Dangerous Properties of Industrial Materials.* 6th ed. New York, NY: Van Nostrand Reinhold, 1984.

■ n-BUTYL STEARATE

CAS # 123–95–5

SIC CODES 2844; 2869; 2841; 2893

SYNONYMS apex-4 • bs • butyl octadecanoate • butyl octadecylate • emerest-2325 • estrex-1b-54, -1b-55 • groco-5810 • kessco-bsc • kesscoflex-bs • n-butyl octadecanoate • octadecanoic acid, butyl ester • polycizer-332 • rc plasticizer b-17 • starfol bs-100 • stearic acid, butyl ester • tegester butyl stearate • wickenol-122 • wilmar butyl stearate • witcizer-200 • witcizer-201 • fema number 2214 • butyl stearate

MF $C_{22}H_{44}O_2$

MW 340.66

COLOR AND FORM Crystals from ethanol, propanol, or ether; waxy or oily (above 20°C); colorless or very-pale-yellow liquid above 20°C.

ODOR Odorless or faintly fatty odor; fatty fruity odor.

TASTE Fatty fruity taste.

DENSITY 0.855 at 20°C.

BOILING POINT 343°C.

MELTING POINT 27.5°C.

AUTOIGNITION TEMPERATURE 671°C [R16, p. 325M–24].

FIREFIGHTING PROCEDURES Carbon dioxide, dry chemical, fog, or mist [R10].

SOLUBILITY Soluble in ethanol, very soluble in acetone; soluble in ether; soluble in mineral or vegetable oils; solubility in water: 0.29% at 25°C.

OCTANOL/WATER PARTITION COEFFICIENT Log K_{ow} = 10.2 (est).

STABILITY AND SHELF LIFE Stable liquid [R2].

OTHER CHEMICAL/PHYSICAL PROPERTIES Wt/gal = 7.14 lb/gal at 20°C; hydrophile-lipophile balance value = 11 (in oil/water emulsion).

COMMON USES Direct food additive as synthetic flavor [R5]. Plasticizer for cellulose acetate butyrate, cellulose nitrate, ethyl cellulose, polystyrene; lubricant in extrusion and molding of polyvinyl chloride; emollient in creams, lotions, and lipsticks [R3]. Solvent for flavors—butter, banana [R6]. Spreading and softening agent in plastics, textiles, cosmetics, rubber industries [R1]. In manufacture of leather cloth and leather varnishes based on nitrocellulose and impact polystyrene [R4]. In lubricants for metals; in wax polishes as dye solvent; in carbon paper and inks; damp-proofer for concrete; plasticizer for laminated fiber products, rubber hydrochloride, chlorinated rubber,

and cable lacquers [R2]; in propellants [R7]; waterproofing agent [R8]; defoamer [R9].

BINARY REACTANTS Oxidizing materials.

STANDARD CODES NFPA-1, 1, 0.

PERSISTENCE Not expected to persist.

MAJOR SPECIES THREATENED Waterfowl.

ACUTE HAZARD LEVEL Probably slight.

DEGREE OF HAZARD TO PUBLIC HEALTH Slight toxic hazard; anesthetic.

OCCUPATIONAL RECOMMENDATIONS
THRESHOLD LIMIT 8-hr time weighted average (TWA) 10 mg/m³; the value is for total dust containing no asbestos and <1% free silica (1988). (stearates (does not include stearates of toxic metals)) [R17, 32].

ACTION LEVEL Isolate from sources of ignition.

ON-SITE RESTORATION Skim or boom. Use activated carbon or peat moss on dissolved portion. Seek professional environmental engineering assistance through EPA's Environmental Response Team (ERT), Edison, NJ, 24-hour no., 908-548-8730.

BEACH OR SHORE RESTORATION Burn off.

AVAILABILITY OF COUNTERMEASURE MATERIAL Activated carbon—water treatment plants, sugar refineries; peat moss—nurseries.

DISPOSAL Dilute and route to sewer.

DISPOSAL NOTIFICATION Local sewage authority.

MAJOR WATER USE THREATENED Recreation.

PROBABLE LOCATION AND STATE Clear liquid. Crystals from alcohol, propanol, or ether. Oily characteristics. Will float in slick on surface, and dissolve slowly.

COLOR IN WATER Colorless.

BIODEGRADATION Data specific to the biodegradation of butyl stearate were not located; however, aliphatic esters have been shown to biodegrade readily (1–2); therefore, butyl stearate is expected to biodegrade (SRC) [R11].

BIOCONCENTRATION Based upon an estimated log K_{ow} of 10.2 (1), the BCF for butyl stearate can be estimated to be 3.3×10^7 (SRC) from a regression-derived equation (2). Thus, bioconcentration should be very important in aquatic organisms that cannot metabolize butyl stearate (SRC) [R12].

FIRE POTENTIAL Slight, when exposed to heat or flame. It can react with oxidizing materials. [R16, 451].

FDA Butyl stearate is a food additive permitted for direct addition to food for human consumption, as long as (1) the quantity added to food does not exceed the amount reasonably required to accomplish its intended physical, nutritive, or other technical effect in food, and (2) when intended for use in or on food, it is of appropriate food grade and is prepared and handled as a food ingredient. Synthetic flavoring substances and adjuvants (includes butyl stearate) may be safely used in food [R13]. Butyl stearate is an indirect food additive for use only as a component of adhesives [R14]. Substances classified as plasticizers, when migrating from food-packaging material, shall including butyl stearate [R15].

REFERENCES

1. Budavari, S. (ed.). *The Merck Index—Encyclopedia of Chemicals, Drugs, and Biologicals*. Rahway, NJ: Merck and Co., Inc., 1989. 242.

2. Sax, N. I., and R. J. Lewis, Sr. (eds.). *Hawley's Condensed Chemical Dictionary*. 11th ed. New York: Van Nostrand Reinhold Co., 1987. 192.

3. SRI.

4. Lefaux, R. *Practical Toxicology of Plastics*. Cleveland: CRC Press Inc., 1968. 366.

5. Furia, T. E. (ed.). *CRC Handbook of Food Additives*. 2nd ed. Cleveland: The Chemical Rubber Co., 1972. 807.

6. Furia, T. E. (ed.). *CRC Handbook of Food Additives*. 2nd ed. Volume 2. Boca Raton, Florida: CRC Press, Inc., 1980. 262.

7. *Kirk–Othmer Encyclopedia of Chemical Technology*. 3rd ed., Volumes 1–26. New York, NY: John Wiley and Sons, 1978–1984, p. V9 640 (11980).

8. *Kirk–Othmer Encyclopedia of Chemical Technology*. 3rd ed., Volumes 1–26. New York, NY: John Wiley and Sons, 1978–1984, p. V9 335 (1980).

9. 40 CFR 180.1001 (d) (7/1/90).

10. Sax, N. I. *Dangerous Properties of Industrial Materials*. 5th ed. New York: Van Nostrand Reinhold, 1979. 451.

11. (1) Malaney, G.W., R. M. Gerhold. pp. 249–57 in *Proc 17th Ind Waste Conf*, Purdue Univ, Ext Ser 112 (1962). (2) Price, K. S., et al., *J Water Pollut Control Fed* 46: 63–77 (1974).

12. (1) USEPA. *Graphical Exposure Modeling System* (GEMS). CLOGP (1987). (2) Lyman, W. J., et al., *Handbook of Chemical Property Estimation Methods*. Washington, DC: Amer Chemical Soc, p. 5–4 (1990).

13. 21 CFR 172.515 (4/1/90).

14. 21 CFR 175.105 (4/1/91).

15. 21 CFR 181.27 (4/1/91).

16. National Fire Protection Guide. *Fire Protection Guide on Hazardous Materials*. 10th ed. Quincy, MA: National Fire Protection Association, 1991.

17. American Conference of Governmental Industrial Hygienists. *Threshold Limit Values for Chemical Substances and Physical Agents and Biological Exposure Indices for 1994–1995*. Cincinnati, OH: ACGIH, 1994.

■ 4-T-BUTYLTOLUENE

CAS # 98-51-1

SYNONYMS 1-methyl-4-tert-butylbenzene • 1-tert-butyl-4-methylbenzene • 4-methyl-tert-butylbenzene • 4-tert-butyl-1-methylbenzene • 4-tert-butyltoluene • benzene, 1-(1,1-dimethylethyl)-4-methyl • p-methyl-tert-butylbenzene • p-tert-butyltoluene • tbt • toluene, p-tert-butyl • P-tbt • 8-methylparacymene • ptbt • ai3-26435

MF $C_{11}H_{16}$

MW 148.27

COLOR AND FORM Clear, colorless liquid.

ODOR Distinct aromatic odor; gasoline-like odor.

ODOR THRESHOLD Odor low: 30 mg/m^3 [R4].

DENSITY 0.8612 at 20°C.

BOILING POINT 193°C at 760 mm Hg.

MELTING POINT −52°C.

SOLUBILITY Slightly soluble in alcohol; very soluble in ether; soluble in acetone, benzene; very soluble in chloroform; water solubility of 5.5 ppm at 25°C.

VAPOR PRESSURE 0.65 mm Hg at 25°C.

OTHER CHEMICAL/PHYSICAL PROPERTIES Henry's Law constant = 0.01535 atm-m^3/mole (est).

PROTECTIVE EQUIPMENT Recommendations for respirator selection: 250 ppm: Any supplied-air respirator operated in a continuous-flow mode. Any powered, air-purifying respirator with organic vapor cartridge(s). 500 ppm: Any chemical cartridge respirator with a full facepiece and organic vapor cartridge(s). Any air-purifying, full facepiece respirator (gas mask) with a chin-style, front- or back-mounted organic vapor canister. Any self-contained breathing apparatus with a full facepiece. Any supplied-air respirator with a full facepiece. 1000 ppm: Any supplied-air respirator with a full facepiece and operated in a pressure-demand or other positive-pressure mode. Emergency or planned entry into unknown concentrations or IDLH conditions: Any self-contained breathing apparatus that has a full facepiece and is operated in a pressure-demand or other positive-pressure mode. Any supplied-air respirator with a full facepiece and operated in pressure-demand or other positive-pressure mode in combination with an auxiliary self-contained breathing apparatus operated in pressure-demand or other positive-pressure mode. Escape: Any air-purifying, full facepiece respirator (gas mask) with a chin-style, front- or back-mounted organic vapor canister. Any appropriate escape-type, self-contained breathing apparatus. [R11, 55].

PREVENTATIVE MEASURES Contact lenses should not be worn when working with this chemical. [R11, 55].

CLEANUP Absorb on paper. Evaporate on a glass or an iron dish in hood. Burn the paper [R3].

COMMON USES As a solvent for resins; as a primary intermediate in chemical industry [R10, 107]; chemical intermediate for p-tert-butylbenzoic acid-resin raw material, pharmaceuticals; solvent for resin preparation [R1].

OCCUPATIONAL RECOMMENDATIONS
OSHA 8-hr time-weighted average: 10 ppm (60 mg/m^3) [R8].

NIOSH 10-hr time-weighted average: 10 ppm (60 mg/m^3); 15-min short-term exposure limit: 20 ppm (120 mg/m^3). [R11, 54].

THRESHOLD LIMIT 8-hr time-weighted average (TWA) 1 ppm, 6.1 mg/ (1993) [R12, 14].

NON-HUMAN TOXICITY LC_{50} rat inhalation 165 ppm/8 hr [R5]; LC_{50} rat inhalation 248 ppm/4 hr [R5]; LC_{50} rat inhalation 934 ppm/1 hr [R5]; LD_{50} rabbit percutaneous 13.8 to 27.8 ml/kg [R5].

ECOTOXICITY LD_{50} goldfish 3 mg/L/24 hr (modified ASTM-D1345) (conditions of bioassay not specified) [R5].

BIODEGRADATION Using a standard dilution method over a 5-day inoculation period and a seed from an effluent of a biological sanitary waste treatment plant, 4-t-butyltoluene was found to have a 2% theoretical BOD when the seed was not adapted and 6% theoretical BOD when the seed was adapted (1) [R6].

BIOCONCENTRATION Based upon a reported water solubility of 5.5 ppm at 25°C (1), the BCF for 4-t-butyltoluene can be estimated to be 236 from a recommended regression-derived equation (2,SRC). This estimated BCF indicates that some bio-concentration in aquatic organisms may occur (SRC) [R7].

TSCA REQUIREMENTS Pursuant to section 8 (d) of TSCA EPA promulgated a model health and safety data reporting rule. The section 8 (d) model rule requires manufacturers, importers, and processors of listed chemical substances and mixtures to submit to EPA copies and lists of unpublished health and safety studies. p-tert-Butyltoluene is included on this list [R9].

DOT *Fire or Explosion:* Flammable/combustible material; may be ignited by heat, sparks, or flames. Vapors may travel to a source of ignition, and flash back. Container may explode in heat of fire. Vapor explosion hazard indoors, outdoors, or in sewers. Runoff to sewer may create fire or explosion hazard (butyl toluene) [R2].

Health Hazards: May be poisonous if inhaled or absorbed through skin. Vapors may cause dizziness or suffocation. Contact may irritate or burn skin and eyes. Fire may produce irritating or poisonous gases. Runoff from fire control or dilution water may cause pollution (butyl toluene) [R2].

Emergency Action: Keep unnecessary people away; isolate hazard area and deny entry. Stay upwind; keep out of low areas. Positive-pressure self-contained breathing apparatus (SCBA) and structural firefighters' protective clothing will provide limited protection. Isolate for 1/2 mile in all directions if tank, rail car, or tank truck is involved in fire. Call CHEMTREC at 1–800–424–9300 for emergency assistance. If water pollution occurs, notify the appropriate authorities (butyl toluene) [R2].

Fire: Small Fires: Dry chemical, CO_2, water spray, or regular foam. Large Fires: Water spray, fog, or regular foam. Move container from fire area if you can do so without risk. Apply cooling water to sides of containers that are exposed to flames until well after fire is out. Stay away from ends of tanks. For massive fire in cargo area, use unmanned hose holder or monitor nozzles; if this is impossible, withdraw from area, and let fire burn. Withdraw immediately in case of rising sound from venting safety device or any discoloration of tank due to fire (butyl toluene) [R2].

Spill or Leak: Shut off ignition sources; no flares, smoking, or flames in hazard area. Stop leak if you can do so without risk. Water spray may reduce vapor; but it may not prevent ignition in closed spaces. Small Spills: Take up with sand or other noncombustible absorbent material and place into containers for later disposal. Large Spills: Dike far ahead of liquid spill for later disposal (butyl toluene) [R2].

First Aid: Move victim to fresh air and call for emergency medical care; if not breathing, give artificial respiration; if breathing is difficult, give oxygen. In case of contact with material, immediately flush eyes with running water for at least 15 minutes. Wash skin with soap and water. Remove and isolate contaminated clothing and shoes at the site (butyl toluene) [R2].

GENERAL RESPONSE Stop discharge if possible. Call fire department. Avoid contact with liquid. Isolate and remove discharges material. Notify local health and pollution control agencies.

FIRE RESPONSE Combustible. Water may be ineffective on fire. Wear self-contained breathing apparatus and protective clothing. Extinguish with dry chemical, alcohol foam, or CO_2. Cool exposed containers with water.

EXPOSURE RESPONSE Call for medical aid. Vapor irritating to skin, eyes, and respiratory tract. Move to fresh air. If breathing has stopped, give artificial respiration. If breathing is difficult, give oxygen. If in eyes, hold eyelids open and flush with plenty of water. Liquid irritating to skin, eyes, and respiratory tract. Remove contaminated clothing and shoes. Flush affected areas with plenty of water. If in eyes, hold eyelids open and flush with plenty of water. If swallowed, do not induce vomiting.

RESPONSE TO DISCHARGE Restrict access.

Mechanical containment. Should be removed. Chemical and physical treatment.

WATER RESPONSE Effect of low concentrations on aquatic life is unknown. Fouling to shoreline. May be dangerous if it enters water intakes. Notify local health and wildlife officials. Notify operators of nearby water intakes.

EXPOSURE SYMPTOMS May be harmful by inhalation, ingestion, or skin absorption. Vapor or mist is irritating to the eyes, mucous membranes, upper respiratory tract, and skin. May cause headache, tachycardia, abnormal cardiovascular system behavior, central nervous system depression, and hematopoietic depression.

EXPOSURE TREATMENT Inhalation: call for medical aid. Remove victim to fresh air. If not breathing, give artificial respiration. If breathing is difficult, give oxygen. Ingestion: do not induce vomiting. Eyes: flush with copious amounts of water for at least 15 minutes. Skin: wash with soap and copious amounts of water.

FIRE POTENTIAL Flammable. Moderately dangerous fire risk [R3].

REFERENCES

1. SRI.

2. U.S. Department of Transportation. *Emergency Response Guidebook 1990.* DOT P 5800.5. Washington, DC: U.S. Government Printing Office, 1990, p. G–27.

3. ITII. *Toxic and Hazardous Industrial Chemicals Safety Manual.* Tokyo, Japan: The International Technical Information Institute, 1988. 93.

4. Ruth, J.H., *Am Ind Hyg Assoc J* 47: A–142–51 (1986).

5. Verschueren, K. *Handbook of Environmental Data of Organic Chemicals.* 2nd ed. New York, NY: Van Nostrand Reinhold Co., 1983. 322.

6. Bridie, A. L., et al. *Water Res* 13: 627–30 (1976).

7. (1) Amoore, J. E., E. Hautala. *J Appl Toxicol* 3: 272–90 (1983) (2) Lyman, W.

J., et al., *Handbook of Chemical Property Estimation Methods.* Washington, DC: Amer Chemical Soc, p. 5–4 (1990).

8. 29 CFR 1910.1000 (7/1/90).

9. 40 CFR 716.120 (7/1/90).

10. Browning, E. *Toxicity and Metabolism of Industrial Solvents.* New York: American Elsevier, 1965.

11. NIOSH. *NIOSH Pocket Guide to Chemical Hazards.* DHHS(NIOSH) Publication No. 90–117. Washington, DC: U.S. Government Printing Office, June 1990.

12. American Conference of Governmental Industrial Hygienists. *Threshold Limit Values for Chemical Substances and Physical Agents and Biological Exposure Indices for 1994–1995.* Cincinnati, OH: ACGIH, 1994.

■ 1,4-BUTYNEDIOL

CAS # 110–65–6

SYNONYMS 1,4-dihydroxy-2-butyne • 2-butyne-1,4-diol • 2-butynediol • bis (hydroxymethyl) acetylene • butynediol • butynediol-1,4 (French) • 1,4-butinodiol (Spanish) • 2-butin-1,4-diol (Czechoslovakia)

MF $C_4H_6O_2$

MW 86.10

COLOR AND FORM Plates from benzene and ethyl acetate; white to light yellow.

BOILING POINT 238°C at 760 mm Hg.

MELTING POINT 58°C.

EXPLOSIVE LIMITS AND POTENTIAL Explosion hazard: moderate when exposed to heat or by spontaneous chemical reaction in contact with certain materials, such as mercury salts, strong acids, and alkaline-earth hydroxides and halides at high temp [R4].

FIREFIGHTING PROCEDURES Water, alcohol foam, dry chem, or carbon dioxide [R3].

SOLUBILITY In water, alcohol, acetone; soluble in aqueous acids.

OTHER CHEMICAL/PHYSICAL PROPERTIES Solution is straw to amber color.

PROTECTIVE EQUIPMENT Neoprene rubber gloves and safety goggles or face shield [R3].

PREVENTATIVE MEASURES SRP: The scientific literature supports the wearing of contact lenses in industrial environments, as part of a program to protect the eye against chemical compounds and minerals causing eye irritation. However, there may be individual substances whose irritating or corrosive properties are such that the wearing of contact lenses would be harmful to the eye. In those specific cases contact lenses should not be worn.

COMMON USES Intermediate; corrosion inhibitor; electroplating brightener; defoliant; polymerization accelerator; stabilizer for chlorinated hydrocarbons; cosolvent for paint and varnish removal [R1].

OCCUPATIONAL RECOMMENDATIONS None listed.

ECOTOXICITY LC_{50} *Pimephales promelas* (fathead minnow) 53.6 mg/L/96 hr (confidence limit 49.3–58.3 mg/L), flow-through bioassay with measured concentrations, 25.1°C, dissolved oxygen 6.8 mg/L, hardness 46.5 mg/L calcium carbonate, alkalinity 43.5 mg/L calcium carbonate, and pH 7.7 [R5]. EC_{50} *Pimephales promelas* (fathead minnow) 53.6 mg/L/96 hr (confidence limit 49.3–58.3 mg/L), flow-through bioassay with measured concentrations, 25.1°C, dissolved oxygen 6.8 mg/L, hardness 46.5 mg/L calcium carbonate, alkalinity 43.5 mg/L calcium carbonate, and pH 7.7. Effect: loss of equilibrium was not observed prior to death [R5].

DOT *Health Hazards:* Poisonous; may be fatal if inhaled, swallowed, or absorbed through skin. Contact may cause burns to skin and eyes. Runoff from fire control or dilution water may give off poisonous gases and cause water pollution. Fire may produce irritating or poisonous gases [R2].

Fire or Explosion: Some of these materials may burn, but none of them ignites readily. Container may explode violently in heat of fire [R2].

Emergency Action: Keep unnecessary people away; isolate hazard area and deny entry. Stay upwind, out of low areas, and ventilate closed spaces before entering. Positive-pressure self-contained breathing apparatus (SCBA) and chemical protective clothing that is specifically recommended by the shipper or manufacturer may be worn. It may provide little or no thermal protection. Structural firefighters' protective clothing is not effective for these materials. Remove and isolate contaminated clothing at the site. Call CHEMTREC at 1–800–424–9300 as soon as possible, especially if there is no local hazardous-materials team available [R2].

Fire: Small Fires: Dry chemical, CO_2, water spray, or regular foam. Large Fires: Water spray, fog, or regular foam. Move container from fire area if you can do so without risk. Fight fire from maximum distance. Stay away from ends of tanks. Dike fire-control water for later disposal; do not scatter the material [R2].

Spill or Leak: Do not touch spilled material; stop leak if you can do so without risk. Fully encapsulating, vapor-protective clothing should be worn for spills and leaks with no fire. Use water spray to reduce vapors. Small Spills: Take up with sand or other noncombustible absorbent material and place into containers for later disposal. Small Dry Spills: With clean shovel place material into clean, dry container, and cover loosely; move containers from spill area. Large Spills: Dike far ahead of liquid spill for later disposal [R2].

First Aid: Move victim to fresh air and call emergency medical care; if not breathing, give artificial respiration; if breathing is difficult, give oxygen. In case of contact with material, immediately flush skin or eyes with running water for at least 15 minutes. Speed in removing material from skin is of extreme importance. Remove and isolate contaminated clothing and shoes at the site. Keep victim quiet and maintain normal body temperature. Effects may be delayed; keep victim under observation [R2].

GENERAL RESPONSE Stop discharge if possible. Call fire department. Avoid contact with liquid and solid. Isolate and remove discharged material. Notify local health and pollution control agencies.

FIRE RESPONSE Combustible; extinguish with water, alcohol foam, dry chemical, or carbon dioxide.

EXPOSURE RESPONSE Call for medical aid. Liquid or solid irritating to skin and eyes. Harmful if swallowed. Remove contaminated clothing and shoes. Flush affected areas with plenty of water. If swallowed and victim is conscious, have victim drink water or milk.

RESPONSE TO DISCHARGE Disperse and flush.

WATER RESPONSE Effect of low concentrations on aquatic life is unknown. May be dangerous if it enters water intakes. Notify local health and pollution control officials. Notify operators of nearby water intakes.

EXPOSURE SYMPTOMS May cause dermatitis.

EXPOSURE TREATMENT Skin contact: wash affected skin area thoroughly with water. Eye contact: immediately wash with water for at least 15 minutes and get medical attention.

REFERENCES
1. Sax, N. I., and R. J. Lewis, Sr. (eds.). *Hawley's Condensed Chemical Dictionary.* 11th ed. New York: Van Nostrand Reinhold Co., 1987. 192.

2. U.S. Department of Transportation. *Emergency Response Guidebook 1990.* DOT P 5800.5. Washington, DC: U.S. Government Printing Office, 1990, p. G-55.

3. U.S. Coast Guard, Department of Transportation. *CHRIS—Hazardous Chemical Data.* Volume II. Washington, DC: U.S. Government Printing Office, 1984-5.

4. Sax, N. I. *Dangerous Properties of*

Industrial Materials. 6th ed. New York, NY: Van Nostrand Reinhold, 1984. 605.

5. Geiger D. L., D.J. Call, Brooke L. T. (eds). *Acute Toxicities of Organic Chemicals to Fathead Minnows (Pimephales promelas).* Vol. IV. Superior Wisconsin: University of Wisconsin–Superior, 1988. 55.

■ BUTYRIC ACID, 4-((4-CHLORO-O-TOLYL) OXY)-

CAS # 94–81–5

SYNONYMS (2,4-mcpb) • (4-chloro-o-tolyloxy) butyric acid • 2,4-mcpb • 2-methyl-4-chlorophenoxybutyric acid • 2M-4KhM • 2M-4✕M • 2M4KhM • 2M-4Kh-M • 4-(2-methyl-4-chlorophenoxy) butyric acid • 4-(4-Chlor-2-methyl-phenoxy)-buttersaeure [German] • 4-(4-Chlor-2-methylphenoxy)-buttersaeure [German] • 4-(4-chloro-2-methylphenoxy) butanoic acid • 4-(4-chloro-2-methylphenoxy) butyric acid • 4-(4-chloro-o-tolyl) oxy) butyric acid • 4-(MCB) • 4-chloro-2-methylphenoxybutyric acid • 4mcpb • bexane • bexone • butanoic acid, 4-(4-chloro-2-methylphenoxy)- • butyric acid, 4-((4-chloro-o-tolyl) oxy)- • gamma-(4-chloro-2-methylphenoxy) butyric acid • gamma-2-methyl-4-chlorophenoxybutyric acid • gamma-mcpb • legumex • MCPB • MCP-butyric • pdq • thitrol • trifolex • tritrol • trotox • u46-mcpb

MF $C_{11}H_{13}ClO_3$

MW 228.69

pH Acid.

COLOR AND FORM White solid; colorless crystals.

DISSOCIATION CONSTANTS pK_a = 4.84.

MELTING POINT 100–101°C.

SOLUBILITY Slightly soluble in carbon tetrachloride or benzene; soluble in ether; slightly soluble in petroleum oils; 44 mg/L water; 150 g/L ethanol; greater than 200 g/L acetone; in dichloromethane 16%, hexane 6.5%, toluene 0.8%.

CORROSIVITY Noncorrosive.

OTHER CHEMICAL/PHYSICAL PROPERTIES $pK_a = 4.84$, forming water-sol alkali metal and amine salts, though precipitation occurs with hard water; MP: 99–100°C (technical, about 90% pure); VP: Less than 1×10^{-7} mbar at 20°C (MCPB salts); noncorrosive to metals (MCPB salts); Henry's Law constant = 3.42×10^{-9} atm-m^3/mole at 25°C (est).

PREVENTATIVE MEASURES Handle carefully. Do not contaminate water, food, or feed by storage or disposal of this chemical [R2]. Cleaning glassware and spray equipment: Thoroughly flush with detergents, dilute alkali, and water [R6, 296]. SRP: The scientific literature for the use of contact lenses in industry is conflicting. The benefit or detrimental effects of wearing contact lenses depend not only upon the substance, but also on factors including the form of the substance, characteristics and duration of the exposure, the uses of other eye protection equipment, and the hygiene of the lenses. However, there may be individual substances whose irritating or corrosive properties are such that the wearing of contact lenses would be harmful to the eye. In those specific cases, contact lenses should not be worn. In any event, the usual eye protection equipment should be worn even when contact lenses are in place.

COMMON USES Herbicide [R6, 295].

OCCUPATIONAL RECOMMENDATIONS None listed.

NON-HUMAN TOXICITY LD$_{50}$ rat oral 680 mg active ingredient/kg [R1].

ECOTOXICITY LC$_{50}$ bobwhite quail oral greater than 5,000 ppm (no mortality at 2,236 ppm, 10% at 5,000 ppm), age 14 days [R3]; LC rainbow trout 75 mg/L/48 hr (conditions of bioassay not specified) [R2]; LC$_{50}$ ring-necked pheasant oral greater than 5,000 ppm (no mortality to 5,000 ppm), age 14 days [R3]; LC$_{50}$ mallard, oral greater than 5,000 ppm (no mortality at 5,000 ppm), age 10 days [R3]; LC$_{50}$ Japanese quail, oral greater than 5,000 ppm

(no mortality to 5,000 ppm), age 14 days [R3].

BIODEGRADATION Half-lives of less than 7 days were determined for MCPB at an initial concentration of 2 µg/g in a clay loam, heavy clay, and a sandy loam kept at about 85 percent field capacity moisture and 20°C; removal from air-dried soils was minimal (1). Furthermore, 2-methyl-4-chlorophenoxyacetic acid was formed in the moist soils as a result of beta-oxidation of the butyric acid sidechain of MCPB (1). Approximately 40% of 125 µg/L MCPB remained after an incubation period of 32 days in an anaerobic sludge digester at 37°C; about 15 percent remained after 32 days in the sterile control (2) [R4].

BIOCONCENTRATION Based on an estimated bioconcentration factor of 73 (1), MCPB is not expected to bioconcentrate in fish (SRC) [R5].

FIFRA REQUIREMENTS In 1988, Congress amended FIFRA to strengthen and accelerate EPA's reregistration program. The nine-year reregistration scheme mandated by "FIFRA 88" applies to each registered pesticide product containing an active ingredient initially registered before November 1, 1984. Pesticides for which EPA had not issued Registration Standards prior to the effective date of FIFRA '88 were divided into three lists based upon their potential for exposure and other factors, with List B being of highest concern and D of least. List: B; Case: MCPB, and salts; Case No.: 2365; Pesticide type: Herbicide; Case Status: Awaiting Data/Data in Review: OPP awaits data from the pesticide's producer(s) regarding its human health and/or environmental effects, or OPP has received and is reviewing such data, in order to reach a decision about the pesticide's eligibility for reregistration. Active Ingredient (AI): 4-(2-Methyl-4-chlorophenoxy) butyric acid; Data Call-in (DCI) Date: 07/10/91; AI Status: The producer of the pesticide has made commitments to conduct the studies and pay the fees required for reregistration, and is meeting those commitments in a timely manner [R7]. A tolerance is established for negligible residues of the herbicide 4-(2-methyl-4-chlorophenoxy) butyric acid in or on the raw agricultural commodity peas. [R8].

FIRE POTENTIAL Nonflammable [R6, 296].

REFERENCES

1. Worthing, C. R., and S. B. Walker (eds.). *The Pesticide Manual—A World Compendium*. 8th ed. Thornton Heath, UK: The British Crop Protection Council, 1987. 519.

2. *Farm Chemicals Handbook 1993*. Willoughby, OH: Meister Publishing Co., 1993, p. C–214.

3. U.S. Department of the Interior, Fish and Wildlife Service, Bureau of Sports Fisheries, and Wildlife. *Lethal Dietary Toxicities of Environmental Pollutants to Birds*. Special Scientific Report—*Wildlife* No. 191. Washington, DC: U.S. Government Printing Office, 1975. 25.

4. (1) Smith, A. E., B. J. Hayden. *Weed Res 21*: 179–83 (1981). (2) Kirk, P. W. W., J. N. Lester. *Environ Technol Let 10*: 405–14 (1989).

5. (1) Kenaga, E. E., *Ecotox Environ Safety* 4: 26–38 (1980).

6. Weed Science Society of America. Herbicide Handbook. 5th ed. Champaign, Illinois: Weed Science Society of America, 1983.

7. USEPA/OPP. *Status of Pesticides in Reregistration and Special Review.* p.141 (Mar, 1992) EPA 700-R-92–004.

8. 40 CFR 180.318 (7/1/91).

■ BUTYRIC ACID, 4-(2,4-DICHLOROPHENOXY)-

CAS # 94-82-6

SYNONYMS (2,4-dichlorophenoxy) butyric acid • 2,4-D butyric acid • 4-(2,4-dichlorophenoxy) butanoic acid • Butyrac-200 • Embutox-E • Butyrac ester • Butoxone ester • Butyrac • gamma-(2,4-dichlorophenoxy) butyric acid • Embutox • Embutone • Butoxone • Butirex • Butormone

MF $C_{10}H_{10}Cl_2O_3$

MW 349.09

COLOR AND FORM White crystals; white to light-brown crystals.

ODOR Slightly phenolic.

DISSOCIATION CONSTANT $pK_a = 4.95$ at 25°C.

MELTING POINT 118–120°C.

SOLUBILITY Water solubility: 46 ppm at 25°C; readily soluble in acetone, ethanol, and diethyl ether. Slightly soluble in benzene, toluene, and kerosene (Merck, 11th Ed.).

VAPOR PRESSURE 3.5×10^{-6} mm Hg at 25°C (est).

OCTANOL/WATER PARTITION COEFFICIENT Log $K_{ow} = 3.53$.

CORROSIVITY The acid is slightly corrosive to iron.

OTHER CHEMICAL/PHYSICAL PROPERTIES K_{ow} of 4-(2-4-dichlorophenoxy) butyric acid has been experimentally measured to be 370 at pH 7.9. Henry's Law constant estimated to be 2.29×10^{-9} atm-m³/mole at 25°C.

PREVENTATIVE MEASURES Avoid continuous exposure, even to small amounts [R4].

DISPOSAL Group I Containers: Combustible containers from organic or metallo-organic pesticides (except organic mercury, lead, cadmium, or arsenic compounds) should be disposed of in pesticide incinerators or in specified landfill sites (organic or metallo-organic pesticides) [R5].

COMMON USES Control of broad-leafed plants in row crops; defoliants, general brush control (chlorophenoxy compounds; from table) [R1]; post-emergence control of many annual and perennial broad-leafed weeds in lucerne, clovers, undersown cereals, grassland, forage legumes, soya beans, and groundnuts [R2]. Selective, translocated hormone-type herbicide [R3].

OCCUPATIONAL RECOMMENDATIONS None listed.

NON-HUMAN TOXICITY LD_{50} rat oral 700 mg/kg [R7]; LD_{50} mouse oral 400 mg/kg [R3].

IARC SUMMARY Classification of carcinogenicity: (1) evidence in humans: limited; overall summary evaluation of carcinogenic risk to humans is Group 2B: The agent is possibly carcinogenic to humans. (chlorophenoxy herbicides; from table) [R6].

BIODEGRADATION 4-(2,4-Dichlorophenoxy) butyric acid has been observed to undergo a beta-oxidation metabolism with soil microorganisms (1); the beta-oxidation yields (2,4-dichlorophenoxy) acetic acid (2,4-D) as the metabolite (1). At room temperatures, 4-(2,4-dichlorophenoxy) butyric acid was observed to degrade in a natural water die-away test (>12% in one week) (2); however, no degradation occurred in distilled water, or in natural water that had been sterilized by addition of acid (2) [R8].

BIOCONCENTRATION Based upon a measured log K_{ow} of 3.53 (1) and a water solubility of 46 mg/L at 25°C (2), the BCF for 4-(2,4-dichlorophenoxy) butyric acid can be estimated to be about 280 and 70, respectively, from linear regression-derived equations (3,SRC). These BCF values suggest that bioconcentration in aquatic organisms is not important (SRC) [R9].

DOT *Health Hazards:* Poisonous; may be fatal if inhaled, swallowed, or absorbed through skin. Contact may cause burns to skin and eyes. Runoff from fire control or dilution water may cause pollution (phenoxy pesticide, liquid, flammable, poisonous, not otherwise specified) [R10, p. G-28].

Fire or Explosion: Flammable/combustible material; may be ignited by heat, sparks, or flames. Vapors may travel to a source of ignition, and flash back. Container may explode in heat of fire. Vapor explosion and poison hazard indoors, outdoors, or in sewers. Runoff to sewer may create fire or explosion hazard (phenoxy pesticide, liquid, flammable, poisonous, not otherwise specified) [R10, p. G-28].

Emergency Action: Keep unnecessary people away; isolate hazard area, and deny entry. Stay upwind; keep out of low areas. Positive-pressure self-contained breathing apparatus (SCBA) and chemical protective clothing that is specifically recommended by the shipper or manufacturer may be worn. It may provide little or no thermal protection. Structural firefighters' protective clothing is not effective for these materials. Isolate for 1/2 mile in all directions if tank, rail car, or tank truck is involved in fire. Call CHEMTREC at 1-800-424-9300 for emergency assistance (phenoxy pesticide, liquid, flammable, poisonous, not otherwise specified) [R10, p. G-28].

Fire: Small Fires: Dry chemical, CO_2, water spray, or alcohol-resistant foam. Large Fires: Water spray, fog, or alcohol-resistant foam. Move container from fire area if you can do so without risk. Dike fire-control water for later disposal; do not scatter the material. Apply cooling water to sides of containers that are exposed to flames until well after fire is out. Stay away from ends of tanks. Withdraw immediately in case of rising sound from venting safety device or any discoloration of tank due to fire (phenoxy pesticide, liquid, flammable, poisonous, not otherwise specified) [R10, p. G-28].

Spill or Leak: Shut off ignition sources; no flares, smoking, or flames in hazard area. Fully encapsulating, vapor-protective clothing should be worn for spills and leaks with no fire. Do not touch, or walk through spilled material; stop leak if you can do so without risk. Water spray may reduce vapor; but it may not prevent ignition in closed spaces. Small Spills: Take up with sand or other noncombustible absorbent material and place into containers for later disposal. Large Spills: Dike far ahead of liquid spill for later disposal (phenoxy pesticide, liquid, flammable, poisonous, not otherwise specified) [R10, p. G-28].

First Aid: Move victim to fresh air, and

call for emergency medical care; if not breathing, give artificial respiration; if breathing is difficult, give oxygen. In case of contact with material, immediately flush skin or eyes with running water for at least 15 minutes. Remove and isolate contaminated clothing and shoes at the site. Keep victim quiet and maintain normal body temperature. Effects may be delayed; keep victim under observation (phenoxy pesticide, liquid, flammable, poisonous, not otherwise specified) [R10, p. G-28].

REFERENCES

1. Zenz, C. *Occupational Medicine— Principles and Practical Applications.* 2nd ed. St. Louis, MO: Mosby–Yearbook, Inc, 1988. 954.

2. Hartley, D., and H. Kidd (eds.). *The Agrochemicals Handbook.* 2nd ed. Lechworth, Herts, England: The Royal Society of Chemistry, 1987, p. A115.

3. Gosselin, R. E., R. P. Smith, H. C. Hodge. *Clinical Toxicology of Commercial Products.* 5th ed. Baltimore: Williams and Wilkins, 1984, p. II–323.

4. National Research Council. *Specifications and Criteria for Biochemical Compounds.* 3rd ed. Washington, DC: National Academy of Sciences, 1972, p. A115.

5. 40 CFR 165.9 (a) (7/1/90).

6. IARC. *Monographs on the Evaluation of the Carcinogenic Risk of Chemicals to Man.* Geneva: World Health Organization, International Agency for Research on Cancer, 1972–present (multivolume work), p. S7 60 (1987).

7. Worthing, C. R., and S. B. Walker (eds.). *The Pesticide Manual—A World Compendium.* 8th ed. Thornton Heath, UK: The British Crop Protection Council, 1987. 229.

8. (1) Gutenmann, W. H., et al., *Soil Sci Soc Amer Proc.* 28: 205–7 (1964). (2) Chau, A. S. Y., K. Thomson. *J Assoc of Anal Chemical* 61: 1481–5 (1978).

9. (1) Jafvert, C. T., et al., *Environ Sci Technol* 24: 1795–1803 (1990). (2) Smith, A. E. *Rev Weed Sci.* 4: 1–24 (1989). (3) Lyman, W. J., et al., *Handbook of Chemical Property Estimation Methods* Washington, DC: Amer Chemical Soc, pp. 5–4, 5–10 (1990).

10. U.S. Department of Transportation. *Emergency Response Guidebook 1990.* DOT P 5800.5. Washington, DC: U.S. Government Printing Office, 1990.

■ CARBON

CAS # 7440–44–0

SYNONYMS K-257 • acetylene black • acticarbone • activated carbon • adsorbit • ag-3 (adsorbent) • ag-5 (adsorbent) • ak (adsorbent) • anthrasorb • aqua-nuchar • ar-3 • art-2 • au-3 • bau • benzol-black • black-140 • black pearls • calcotone-black • carbolac • carbon-12 • carbopol-m • carbopol-z-4 • carbopol-extra • carbopol-z-extra • clf-II • cmb-50 • cmb-200 • coke-powder • columbia-lck • conductex • cwn-2 • filtrasorb • filtrasorb-200 • filtrasorb-400 • grosafe • irgalite-1104 • ma-100 [carbon] • norit • ou-b • pelikan-c-11/1431a • skg • skt • skt [adsorbent] • su2000 • suchar-681 • supersorbon-iv • supersorbon-s1 • witcarb-940 • xe-340 • xf-41751

MF C

MW 12.01115

COLOR AND FORM Usually soft, black scales (graphite); colorless mineral color can vary between pink, light blue, and even black due to incorporated impurities (diamond).

MELTING POINT 3,550°C (Graphite).

HEAT OF COMBUSTION -3.9351×10^8 J/kmol (graphite).

AUTOIGNITION TEMPERATURE Activated carbon showed an autoignition temperature in flowing air of 452–518°C. Presence of 5% of the base 'triethylenediamine' adsorbed on the carbon reduced the autoignition temperature to 230–260°C. [R5, 111].

EXPLOSIVE LIMITS AND POTENTIAL Explosion hazard, in the form of dust when exposed to heat, flame, or ammonium nitrate + heat, ammonium tetrachloride at 240°C, bromates, $Ca(OCl)_2$, chlorates, Cl_2, $(Cl_2 + Cr (OCl)_2)$, ClO, F_2, iodates, IO_5, $Pb(NO_3)_2$, $HgNO_3$, HNO_3, (oils + air), (potassium + air), Na_2S, $Zn(NO_3)_2$ [R2].

VAPOR PRESSURE 1 mm at 3,586°C.

OTHER CHEMICAL/PHYSICAL PROPERTIES Atomic number: 6; valence: 4; stable isotopes: 12 (98.892%); 13 (1.108%); radioactive isotopes: 9–11; 14–16. Nonmetallic element; divalent forms are known (carbenes); forms binary compound called carbides with many metals and some nonmetals, strong reducing agent, strong electrical conductivity. Atomic radius: 0.77 Å; ionic (crystal) radii: 2.60 Å (−4 oxidation state); 0.15 Å (4 oxidation state); orbital electrons: $[He]2s^22p^2$; electronegativity (Pauling scale): 2.5. Colorless, cubic crystals; insoluble in water, acid, alkali; index of refraction: 2.4173; density: 3.51; MP: greater than 3,550°C; BP: 4,827°C (diamond). Black hexagonal crystals; insoluble in water, acid, alkali; soluble in liquid iron; density: 2.25 at 20°C; sublimes: 3,652–3,697°C; BP: 4,827°C (graphite). Amorphous black crystals; insoluble in water, acid, alkali; density: 1.8–2.1; BP: 4,827°C; sublimes: 3,652–3,697°C (amorphous). Small transparent crystals formed during sublimation of pyrolytic graphite at low pressures and approximately 2,550 K ("white" carbon). The ^{12}C isotope, which comprises 99% of the element, is the standard to which atomic weights of all other elements are referred (C = 12.00 exactly). One mole of carbon atoms (6.02×10^{23}) is contained in 12 g of ^{12}C. Graphite is one of the softest known materials; diamond is one of the hardest. Graphite exists in two forms: alpha and beta. These have identical physical properties, except for their crystal structure. Naturally occurring graphites contain 30% rhombohedral (beta) form, whereas synthetic contain only alpha form. "White" carbon is a transparent birefringent material. Little information is currently avail about this allotrope ("white" carbon). The element carbon, due to its high heat capacity per unit weight, high energy of vaporization, and high temperature and pressure required for melting, has the highest "heat of ablation" (energy absorbed per mass lost) of any material, provided that mechanical removal of particulates does not occur. Liquid molar volume = 0.007455 m³/kmol (graphite). Heat of formation = 7.1668 $\times$ 10^8 J/kmol (graphite). Heat of fusion =

1.0460×10^8 J/kmol at melting point (graphite). Odorless, tasteless; fine black powder (activated charcoal). Insoluble in water or other known solvents (activated charcoal).

DISPOSAL SRP: At the time of review, criteria for land treatment or burial (sanitary landfill) disposal practices are subject to significant revision. Prior to implementing land disposal of waste residue (including waste sludge), consult with environmental regulatory agencies for guidance on acceptable disposal practices.

COMMON USES As strong reducing agent, and is used as such in purifying metals; in electrodes, electrical devices, and steel [R4, 218]. Jewelry; polishing, grinding, cutting glass, bearings for delicate instruments; manufacture of dies for tungsten wire, and similar hard wires; making styli for recorder heads; long-lasting phonograph needles; in semiconductor research (diamond) [R3, 470]. For "lead" pencils; refractory crucibles; stove polish; as pigment, lubricant, graphite cement; for matches and explosives; commutator brushes; anodes; arc-lamp carbons; electroplating; polishing compound, rust, and needle-paper; coating for cathode-ray tubes; moderator in nuclear piles (graphite) [R3, 470]. Medication: activated charcoal as antidote; adsorptive; vet: internally as an adsorptive in diarrhea; externally in foul wounds (amorphous) [R3, 274]. Used chiefly for clarifying, deodorizing, decolorizing, and filtering; pigment for rubber tires; for printing, stenciling, and drawing inks; for leather; stove polish; phonograph records; electrical insulating apparatus (amorphous, activated charcoal) [R3, 274]. Air purification, removal of sulfur dioxide from stack gasses, and "clean" rooms, deodorant, removal of jet fumes from airports, catalyst for natural-gas purification, brewing, chromium electroplating, air conditioning [R4, 219]. The carbonaceous residue of the destructive distillation of bituminous coal petroleum and coal-tar pitch: The principal type is that produced by heating bituminous coal in chemical recovery or beehive coke ovens (metallurgical coke), one ton of coal yielding approximately 0.7 ton of coke. Used

chiefly for the reduction of iron ore in blast furnaces, and as a source of synthetic gas. Coke from petroleum residues and coal-tar pitch is used for refractory furnace linings in the electrorefining of aluminum and other high-temperature services, and for electrodes in electrolytic reduction of Al$_2$O$_3$ to aluminum, as well as in electrothermal production of phosphorus, silicon carbide, and calcium carbide. (coke) [R4, 298].

OCCUPATIONAL RECOMMENDATIONS None listed.

NON-HUMAN TOXICITY LD$_{50}$ mouse iv 440 mg/kg [R2].

DOT *Fire or Explosion:* Flammable/combustible material; may be ignited by heat, sparks, or flames. May burn rapidly with flare-burning effect. Material may be transported in a molten form (carbon, activated; carbon, animal or vegetable origin) [R1].

Health Hazards: Fire may produce irritating or poisonous gases. Contact may cause burns to skin and eyes. Runoff from fire control or dilution water may cause pollution (carbon, activated; carbon, animal or vegetable origin) [R1].

Emergency Action: Keep unnecessary people away; isolate hazard area and deny entry. Stay upwind; keep out of low areas. Positive-pressure self-contained breathing apparatus (SCBA) and structural firefighters' protective clothing will provide limited protection. Call emergency response telephone number on shipping paper first. If shipping paper not available or no answer, call CHEMTREC at 1–800–424–9300. If water pollution occurs, notify the appropriate authorities (carbon, activated; carbon, animal or vegetable origin) [R1].

Fire: Small Fires: Dry chemical, sand, earth, water spray, or regular foam. Large Fires: Water spray, fog, or regular foam. Move container from fire area if you can do so without risk. Apply cooling water to sides of containers that are exposed to flames until well after fire is out. Stay away from ends of tanks. For massive fire in cargo area, use unmanned hose holder or monitor nozzles; if this is impossible, withdraw from area, and let fire burn. Magnesium Fires: Use dry sand, Met-L-X powder, or G-1 graphite powder (carbon, activated; carbon animal or vegetable origin) [R1].

Spill or Leak: Shut off ignition sources; no flares, smoking, or flames in hazard area. Do not touch or walk through spilled material. Small Dry Spills: With clean shovel, place material into clean, dry container, and cover loosely; move containers from spill area. Large Spills: Wet down with water and dike for later disposal (carbon, activated; carbon, animal or vegetable origin) [R1].

First Aid: Move victim to fresh air; call emergency medical care. In case of contact with material, immediately flush skin or eyes with running water for at least 15 minutes. Removal of solidified molten material from skin requires medical assistance. Remove and isolate contaminated clothing and shoes at the site (carbon, activated; carbon, animal or vegetable origin) [R1].

FIFRA REQUIREMENTS In 1988, Congress amended FIFRA to strengthen and accelerate EPA's reregistration program. The nine-year reregistration scheme mandated by "FIFRA 88" applies to each registered pesticide product containing an active ingredient initially registered before November 1, 1984. Pesticides for which EPA had not issued Registration Standards prior to the effective date of FIFRA '88 were divided into three lists based upon their potential for exposure and other factors, with List B being of highest concern, and D of least. List: D; Case: Carbon and Carbon Dioxide; Case No. 4019; Pesticide type: Insecticide, Herbicide; Case Status: RED Approved 09/91; OPP has reached a decision that some or all uses of the pesticide are eligible for reregistration, and issued a Reregistration Eligibility Document (RED). Active Ingredient(AI): Carbon; AI Status: OPP has completed a Reregistration Eligibility Document (RED) for the active ingredient/case [R7]. Residues of charcoal, activated (meets specifications in the Food Chemical Codex) are exempted from the requirement of a tolerance when used as a carrier in accordance with good agricultural practices as inert (or occasionally active) ingredients in pesticide formulations applied to growing crops or to raw agricultural commodities after harvest. [R6].

GENERAL RESPONSE Shut off ignition sources. Call fire department. Stay upwind. Use water spray to "knock down" dust. Isolate and remove discharged material. Notify local health and pollution control agencies.

FIRE RESPONSE Combustible. Poisonous gases may be produced in fire. Flood discharge area with water.

EXPOSURE RESPONSE Dust irritating to eyes, nose, and throat. Move victim to fresh air. Solid not harmful.

RESPONSE TO DISCHARGE Issue warning—high flammability. Disperse and flush.

WATER RESPONSE Effect of low concentrations on aquatic life is unknown. May be dangerous if it enters water intakes. Notify local health and wildlife officials. Notify operators of nearby water intakes.

EXPOSURE SYMPTOMS: No significant symptoms.

EXPOSURE TREATMENT: No treatment required.

FIRE POTENTIAL Activated carbon exposed to air is a potential fire hazard because of its very high surface area and adsorptive capacity. Freshly prepared material may heat spontaneously in air, and presence of water accelerates this. [R5, 111].

REFERENCES

1. U.S. Department of Transportation. *Emergency Response Guidebook 1993.* DOT P 5800.6. Washington, DC: U.S. Government Printing Office, 1993, p. G–32.

2. Sax, N. I. *Dangerous Properties of Industrial Materials.* 6th ed. New York, NY: Van Nostrand Reinhold, 1984. 640.

3. Budavari, S. (ed.). *The Merck Index— Encyclopedia of Chemicals, Drugs, and*

Biologicals. Rahway, NJ: Merck and Co., Inc., 1989.

4. Sax, N. I., and R. J. Lewis, Sr (eds.). *Hawley's Condensed Chemical Dictionary*. 11th ed. New York: Van Nostrand Reinhold Co., 1987.

5. Bretherick, L. *Handbook of Reactive Chemical Hazards*. 4th ed. Boston, MA: Butterworth–Heinemann Ltd., 1990.

6. 40 CFR 180.1001 (c) (7/1/94).

7. USEPA/OPP. *Status of Pesticides in Reregistration and Special Review*. p. 227 (Mar, 1992) EPA 700–R–92–004.

■ CARBON DIOXIDE

CAS # 124–38–9

SYNONYMS aer fixus • after-damp • anhydride carbonique (French) • carbon dioxide (CO_2) • Carbonica • carbonic acid gas • carbonic anhydride • carbon oxide (CO_2) • carbon oxide, di- • dioxido de carbono (Spanish) • dioxyde de carbone (French) • dry ice • Kohlendioxyd (German) • Kohlensaeure (German)

MF CO_2

MW 44.01

pH The pH of saturated CO_2 solutions varies from 3.7 at 101 kPa (1 atm) to 3.2 at 2,370 kPa (23.4 atm).

COLOR AND FORM Colorless gas or colorless liquid; solid: white snow-like flakes or cubes.

ODOR Odorless; faintly pungent odor.

ODOR THRESHOLD Odorless [R14].

TASTE Faint acid taste.

DENSITY 1.527 gas (air = 1); abs density: 0.1146 lb/ft³ at 25°C; volume at 25°C: 8.76 ft³/lb; density: (gas) 1.557 (N_2 = 1); (gas at 0°C) 1.976 g/L at 760 mm Hg; (liquid) 0.914 at 34.3 atm; (solid) at −56.6°C, 1.512; critical density: 0.464.

SURFACE TENSION 0.0162 newtons/meter at melting point.

VISCOSITY 20.3×10^{-6} Pa–s at 20°C (gas); 70.1×10^{-6} Pa–s at 20°C (liquid).

BOILING POINT −78.5°C (sublimation).

MELTING POINT −56.6°C at 5.2 atm.

HEAT OF COMBUSTION 0 J/kmol.

HEAT OF VAPORIZATION 83.12 cal/g.

EXPLOSIVE LIMITS AND POTENTIAL When carbon dioxide gas is passed over a mixture of powdered aluminum and sodium peroxide, the mixture explodes. Mixture of solid forms of potassium and carbon dioxide (as dry ice) explodes when subjected to shock. Solid carbon dioxide with sodium-potassium alloy will explode under slight impact. [R24, p. 491M–17].

FIREFIGHTING PROCEDURES Self-contained breathing apparatus with a full facepiece operated in pressure-demand or other positive-pressure mode [R12, 4]. Extinguish fire using agent suitable for type of surrounding fire. (Material itself does not burn or burns with difficulty.) Cool all affected containers with flooding quantities of water. Do not use water on material itself. Apply water from as far a distance as possible (carbon dioxide; carbon dioxide, refrigerated liquid) [R25, 190]. Extinguish fire using agent suitable for type of surrounding fire (material itself does not burn or burns with difficulty.) (Carbon dioxide, solid, or dry ice, or carbon ice) [R25, 192].

SOLUBILITY 0.145 g/100 mL water at 25°C; 0.097 g/100 ml water at 40°C; 0.058 g/100 mL water at 60°C; 0.348 g/100 mL water at 0°C; soluble in acetone; 90.1 cc/100 mL water at 20°C; 171.3 cc/100 mL water at 0°C; 31 cc/100 ml alcohol at 15°C; miscible with hydrocarbons and most organic liquids. Water solubility, amount dissolved, mL/kg (STP): 835 at 20°C, and 101 kPa; 634 at 30°C and 101 kPa; 1,137 at 10°C and 101 kPa; quantity dissolved, ml/g (STP) at 20°C: acetone: 8.2; ethanol: 3.6; benzene: 2.71; methanol: 4.1; toluene: 3.0; xylene: 2.31; heptane: 2.8; methyl acetate: 7.4; diethyl ether: 6.3.

VAPOR PRESSURE 5,728.9 kPa at 20°C.

STABILITY AND SHELF LIFE Gas is not affected by heat until temperature is about 2,000°C [R2].

OTHER CHEMICAL/PHYSICAL PROPERTIES Heat of formation: 94.05 kcal/mol; specific heat: 0.19 to 0.21 Btu/lb; absorbed by alkaline solution with formation of carbonates; 1 mg/m^3 is equivalent to 0.5 ppm; 1 ppm is equivalent to 2 mg/m^3; critical volume 2.137 cu dm/kg; critical compressibility factor 0.274; absolute density, gas at 101.325 kPa at 0°C 1.9770 kg/m^3; relative density, gas at 101.325 kPa at 0°C (air = 1) 1.53; viscosity: 10 μPa sec at 200 K; solid carbon dioxide sinks and boils in water; sublimes at −78.48°C at 760 mm Hg; ionization potential: 13.77 eV; air contains 0.030 vol/%; latent heat of fusion at −56.6°C; 518 kPa: 7.950 kJ/mol; 180.64 kJ/kg; 43.17 kcal/kg; triple point: −56.6°C at 5.11 atm; at atmospheric pressure solid form changes into gaseous phase without liquefaction; vapor pressure = 10.5 mm Hg at −120°C; 104.2 mm Hg at −100°C; 569.1 mm Hg at −82°C; all forms are noncombustible; viscosity of gas in micropascal seconds (μPa−s): 10.0 at 200 K; 15.0 at 300 K; 19.7 at 400 K; 24.0 at 500 K; 28.0 at 600 K; latent heat of vaporization = 353.4 J/g at the triple point; 231.3 J/g at 0°C; viscosity = 0.015 mPa−sec at 298 K and 101.3 kPa; gas density = 1.976 g/L at 273 K and 101.3 kPa.

PROTECTIVE EQUIPMENT Recommendations for respirator selection. Max concentration for use: 50,000 ppm: Respirator classes: Any supplied-air respirator. Any self-contained breathing apparatus [R13].

PREVENTATIVE MEASURES SRP: Local exhaust ventilation should be applied wherever there is an incidence of point-source emissions or dispersion of regulated contaminants in the work area. Ventilation control of the contaminant as close as possible to its point of generation is both the most economical and safest method to minimize personnel exposure to airborne contaminants.

CLEANUP (1) Ventilate area of leak to disperse gas (2) Stop flow of gas. If source of leak is cylinder and leak cannot be stopped in place, remove to safe place in open air, and repair leak, or allow cylinder to empty [R12]. Water spray may be used to convert any form of carbon dioxide to carbonic acid, which may then be neutralized with alkali [R15].

DISPOSAL Vent to atmosphere [R16].

COMMON USES Refrigerant; processing of foods; preserving foods; crusting of food; cryogenic freezing of food; production of urea, sodium carbonate (solvay process), methanol, carbonic acid, lead carbonate, potassium carbonate, potassium bicarbonate, ammonium carbonate, ammonium bicarbonate, sodium salicylate, carbonated petroleum, hydrocarbon products; provides an inert atmosphere for fire extinguishers, refinery products, petroleum products; displacing oxygen to prevent deterioration and flavor loss; in high-pressure applications; oil-well stimulation; in livestock slaughtering; as fertilizer; hardening of molds for metal castings [R1]; carbonation of beverages; for inerting flammable materials during mfr, handling, and transfer; as propellant in aerosols; to produce harmless smoke or fumes on stage [R2]; in shielded-arc welding; in market gardening; as an addition to atmosphere of greenhouses to incr growth rate; in cylinders for inflating life rafts; in dry ice form, it is used to embrittle the flash of molded rubber articles; to chill golf ball centers before winding; in solid or gaseous form to produce an inert atmosphere in vessels or plant where there would otherwise be an explosive risk [R23, 392]; used in the manufacture of aspirin [R4]; direct food additives, carbon dioxide, solid; MID (meat inspection division of USDA); limitations: for cooling during chopping and packaging of meat [R5]; the gas finds applications as resp stimulant in patient with nondepressed respiratory centers in whom it is desirable to increase respiratory minute volume. It is therefore used to hasten excretion of toxic gases and vapors from blood via lungs [R6]. Its most valuable use is with oxygen to avoid a reduction of carbon dioxide tension of blood. It is often used to relieve persistent hiccups; it only occasionally produces transient relief. It has been used to induce deep breathing and coughing to avoid postoperative atelectasis but is generally of little benefit [R7]; vet: wart destruction [R8]; carbon dioxide is used in the deactivation of soaps [R9]. A 10% concentration of ethylene oxide in carbon dioxide effectively kills most insect pests at all life stages [R10]. Mixtures of 10–30% ethylene oxide in carbon dioxide are used as a fumigant and sterilizing agent [R10]. Carbon dioxide has been used for a long time to render swine insensible prior to slaughter. Sheep and calves can be handled similarly [R22, 1061]. Chemical intermediate (carbonates, synthetic fiber, p-xylene, etc.), low-temperature testing, municipal water treatment, medicine, facturing, and acidizing of oil wells, mining (Cardox method), miscible pressure source, hardening of foundry molds and cores, shielding gas for welding, cloud seeding, moderator in some types of nuclear reactors, immobilization for humane animal killing, special lasers, blowing agent, as a demulsifier in tertiary oil recovery, possible source of methane, (liquid) carrier for powdered coal slurry [R3].

OCCUPATIONAL RECOMMENDATIONS None listed.

OSHA Table Z–1, 8-hour time-weighted average: 5,000 ppm (9,000 mg/m³) [R18].

NIOSH 10-hr time-weighted average: 5,000 ppm (9,000 mg/m³) [R13].

THRESHOLD LIMIT 8-hr time weighted average (TWA) 5,000 ppm, 9,000 mg/m³; short-term exposure limit (STEL) 30,000 ppm, 54,000 mg/m³ (1986) [R19].

INTERNATIONAL EXPOSURE LIMITS Australia: 5,000 ppm, STEL 30,000 ppm (1990); Federal Republic of Germany: 5,000 ppm, short-term level 10,000 ppm for 60 minutes, 3 times per shift (1989); Sweden: 5,000 ppm, 15-minute short-term level 10,000 ppm (1984); United Kingdom: 5,000 ppm, 10-minute STEL 15,000 ppm (1987). [R26, 223].

ECOTOXICITY Trout 240 mg/L/1 hour, toxic effect: lethal [R17]; rainbow trout 35 mg/L/96 hr, toxic effect: lethal [R17]; rainbow trout 60–240 mg/L/12 hr, toxic effect: lethal [R17]; harmful to some species of aquatic life in concentrations less than 20 mg/L [R14].

DOT *Fire or Explosion:* Cannot catch fire. Container may explode in heat of fire (carbon dioxide; carbon dioxide, refrigerated liquid (cryogenic liquid); carbon dioxide, solid) [R11].

Health Hazards: Vapors may cause dizziness or suffocation. Contact with liquid may cause frostbite (carbon dioxide; carbon dioxide, refrigerated liquid (cryogenic liquid); carbon dioxide, solid) [R11].

Emergency Action: Keep unnecessary people away; isolate hazard area and deny entry. Stay upwind, out of low areas, and ventilate closed spaces before entering. Positive-pressure self-contained breathing apparatus (SCBA) and structural firefighters' protective clothing will provide limited protection. Isolate for 1/2 mile in all directions if tank, rail car, or tank truck is involved in fire. Call emergency response telephone number on shipping paper first. If shipping paper not available or no answer, call CHEMTREC at 1–800–424–9300 (carbon dioxide; carbon dioxide, refrigerated liquid (cryogenic liquid); carbon dioxide, solid) [R11].

Fire: Move container from fire area if you can do so without risk. Apply cooling water to sides of containers that are exposed to flames until well after fire is out. Stay away from ends of tanks (carbon dioxide; carbon dioxide, refrigerated liquid (cryogenic liquid); carbon dioxide, solid) [R11].

Spill or Leak: Do not touch or walk through spilled material. Stop leak if you can do so without risk (carbon dioxide; carbon dioxide, refrigerated liquid (cryogenic liquid); carbon dioxide, solid) [R11].

First Aid: Move victim to fresh air, and call emergency medical care; if not breathing, give artificial respiration; if breathing is difficult, give oxygen. In case of frostbite, thaw frosted parts with water. Keep victim quiet, and maintain normal body temperature (carbon dioxide; carbon dioxide, refrigerated liquid (cryogenic liquid); carbon dioxide, solid) [R11].

FIRE POTENTIAL Noncombustible gas [R2].

FDA Substance added directly to human food affirmed as generally recognized as safe (GRAS) [R20]. Carbon dioxide used as a general-purpose food additive in animal drugs, feeds, and related products is generally recognized as safe when used in accordance with good manufacturing or feeding practice [R21].

REFERENCES

1. SRI.

2. Budavari, S. (ed.). *The Merck Index— Encyclopedia of Chemicals, Drugs, and Biologicals*. Rahway, NJ: Merck and Co., Inc., 1989. 274.

3. Lewis, R. J., Sr. (Ed.). *Hawley's Condensed Chemical Dictionary*. 12th ed. New York, NY: Van Nostrand Reinhold Co., 1993 219.

4. Environment Canada. *Tech Info for Problem Spills: Carbon Dioxide*, p. 13 (1984).

5. Furia, T. E (ed.). *CRC Handbook of Food Additives*. 2nd ed. Cleveland: The Chemical Rubber Co., 1972. 816.

6. Osol, A., and J. E. Hoover, et al. (eds.). *Remington's Pharmaceutical Sciences*. 15th ed. Easton, PA: Mack Publishing Co., 1975. 799.

7. Osol, A. (ed.). *Remington's Pharmaceutical Sciences*. 16th ed. Easton, PA: Mack Publishing Co., 1980. 804.

8. Rossoff, I. S. *Handbook of Veterinary Drugs*. New York: Springer Publishing Company, 1974. 82.

9. *Kirk–Othmer Encyclopedia of Chemical Technology*. 3rd ed., Volumes 1–26. New York, NY: John Wiley and Sons, 1978–1984, p. V11 94 (1980).

10. *Kirk–Othmer Encyclopedia of Chemical Technology*. 3rd ed., Volumes 1–26. New York, NY: John Wiley and Sons, 1978–1984, p. V9 464 (1980).

11. U.S. Department of Transportation. *Emergency Response Guidebook 1993*. DOT P 5800.6. Washington, DC: U.S. Government Printing Office, 1993, p. G–21.

12. Mackison, F. W., R. S. Stricoff, and L. J. Partridge, Jr. (eds.). *NIOSH/OSHA— Occupational Health Guidelines for Chemical Hazards*. DHHS (NIOSH) Publication No. 81–123 (3 vols). Washington, DC: U.S. Government Printing Office, Jan. 1981.

13. NIOSH. *NIOSH Pocket Guide to Chemical Hazards*. DHHS (NIOSH) Publication No. 90–117. Washington, DC: U.S. Government Printing Office, June 1990, 58.

14. Environment Canada. *Tech Info for Problem Spills: Carbon Dioxide*, p. 1 (1984).

15. Environment Canada, *Tech Info for Problem Spills: Carbon Dioxide*, p. 57 (1984).

16. Sittig, M. *Handbook of Toxic and Hazardous Chemicals*, p. 134 (1981).

17. Environment Canada. *Tech Info for Problem Spills: Carbon Dioxide*, p. 41 (1984).

18. 29 CFR 1910.1000 (58 FR 35338 (6/30/93)).

19. American Conference of Governmental Industrial Hygienists. *Threshold Limit Values for Chemical Substances and Physical Agents and Biological Exposure Indices for 1994–1995*. Cincinnati, OH: ACGIH, 1994. 15.

20. 21 CFR 184.1240 (4/1/94).

21. 21 CFR 582.1240 (4/1/94).

22. Booth, N.H., L.E. McDonald (eds.). *Veterinary Pharmacology and Therapeutics*. 5th ed. Ames, Iowa: Iowa State University Press, 1982.

23. International Labour Office. *Encyclopedia of Occupational Health and Safety*. Vols. I and II. Geneva, Switzerland: International Labour Office, 1983.

24. National Fire Protection Guide. *Fire Protection Guide on 'Hazardous Materials*. 10th ed. Quincy, MA: National Fire Protection Association, 1991.

25. Association of American Railroads. *Emergency Handling of Hazardous Materials in Surface Transportation*. Wash-

ington, DC: Association of American Railroads, Bureau of Explosives, 1992.

26. American Conference of Governmental Industrial Hygienists, Inc. *Documentation of the Threshold Limit Values and Biological Exposure Indices.* 6th ed. Volumes I, II, III. Cincinnati, OH: ACGIH, 1991.

■ CARBONYL DIFLUORIDE

CAS # 353–50–4

SYNONYMS carbon difluoride oxide • carbon fluoride oxide (COF_2) • carbon oxyfluoride • carbon oxyfluoride (COF_2) • carbonic difluoride • carbonyl fluoride • difluoroformaldehyde • fluophosgene • fluoroformyl fluoride • fluorophosgene

MF CF_2O

MW 66.01

COLOR AND FORM Colorless gas.

ODOR Nearly odorless.

DENSITY 1.139 at −114°C (liquid).

BOILING POINT −83°C at 760 mm Hg.

MELTING POINT −114°C.

STABILITY AND SHELF LIFE Unstable in presence of water, instantly hydrolyzed [R1].

OTHER CHEMICAL/PHYSICAL PROPERTIES Minimum purity: 97 mole percent; pungent, very hygroscopic gas; density (solid) at −190°C: 1.388; heat of formation: 166.6 kcal.

PREVENTATIVE MEASURES Contact lens use in industry is controversial. A survey of 100 corporations resulted in the recommendation that each company establish its own contact lens use policy. One presumed hazard of contact lens use is possible chemical entrapment. It was found that contact lenses minimized injury or protected the eye. The eye was afforded more protection from liquid irritants. Soft contact lenses did not worsen corneal damage from strong chemicals and in some cases could actually protect the eye.

Overall, the literature supports the wearing of contact lenses in industrial environments as part of the standard eye protection, e.g., face shields; however, more data are needed to establish the value of contact lenses [R5].

DISPOSAL Generators of waste (equal to or greater than 100 kg/mo) containing this contaminant, EPA hazardous waste number U033, must conform with USEPA regulations in storage, transportation, treatment, and disposal of waste [R6].

COMMON USES Chemical intermediate in organic synthesis, e.g., fluorinated alkyl isocyanates [R2]. Suggested for use as a military poison gas [R3].

OCCUPATIONAL RECOMMENDATIONS None listed.

OSHA 8-hr time-weighted average: 2 ppm (5 mg/m³). Final rule limits were achieved by any combination of engineering controls, work practices, and personal protective equipment during the phase-in period, Sept 1, 1989 through Dec 30, 1992. Final rule limits became effective Dec 31, 1992 [R8].

THRESHOLD LIMIT 8-hr time-weighted average (TWA) 2 ppm, 5.4 mg/m³; short-term exposure limit (STEL) 5 ppm, 13 mg/m³ (1986) [R9].

NON-HUMAN TOXICITY LC_{50} rat (inhalation) 360 ppm/1 hr [R7]; LC_{50} rat (inhalation) 90 ppm/4 hr [R7]; LD_{50} mouse ip 17.2 mg/kg [R12, (1982)]; LD_{50} mouse, oral 46.0 mg/kg [R12, (1982)]; LD_{50} mouse iv 23.0 mg/kg [R12, (1982)]; LD_{50} rat, oral 51.6 mg/kg (admin via stomach tube, under light ether anesthesia) [R12, (1982)]; LD_{50} rat, oral 32.0 mg/kg [R12, (1982)]; LD_{50} rat iv 11.8 mg/kg [R12, (1982)]; LD_{50} rat ip 24 mg/kg [R12, (1982)].

CERCLA REPORTABLE QUANTITIES Persons in charge of vessels or facilities are required to notify the National Response Center (NRC) immediately, when there is a release of this designated hazardous substance, in an amount equal to or greater than its reportable quantity of 1,000 lb or 454 kg. The toll-free number of the NRC is (800) 424–8802; In the Washington DC metropolitan area (202) 426–2675. The rule for determining when notification is required is stated in 40 CFR 302.4 (section IV. D. 3.b) [R10].

RCRA REQUIREMENTS U033; as stipulated in 40 CFR 261.33, when carbon oxyfluoride, as a commercial chemical product or manufacturing chemical intermediate or an off-specification commercial chemical product or a manufacturing chemical intermediate, becomes a waste, it must be managed according to federal or state hazardous-waste regulations. Also defined as a hazardous waste is any residue, contaminated soil, water, or other debris resulting from the cleanup of a spill, into water, or on dry land, of this waste. Generators of small quantities of this waste may qualify for partial exclusion from hazardous-waste regulations (40 CFR 261.5) [R11].

DOT *Health Hazards:* Poisonous; may be fatal if inhaled or absorbed through skin. Contact may cause burns to skin and eyes. Contact with liquid may cause frostbite. Clothing frozen to the skin should be thawed before being removed. Runoff from fire control or dilution water may cause pollution [R4].

Fire or Explosion: Some of these materials may burn, but none of them ignites readily. Cylinder may explode in heat of fire [R4].

Emergency Action: Keep unnecessary people away; isolate hazard area and deny entry. Stay upwind, out of low areas, and ventilate closed spaces before entering. Positive-pressure self-contained breathing apparatus (SCBA) and chemical protective clothing that is specifically recommended by the shipper or manufacturer may be worn. It may provide little or no thermal protection. Structural firefighters' protective clothing is not effective for these materials. Isolate the leak or spill area immediately for at least 150 feet in all directions. See the Table of Initial Isolation and Protective Action Distances. If you find the ID number and the name of the material there, begin protective action. Call CHEMTREC at 1–800–424–9300 as soon as possible, especially if

there is no local hazardous-materials team available [R4].

Fire: Small Fires: Dry chemical or CO_2. Large Fires: Water spray, fog, or regular foam. Do not get water inside container. Move container from fire area if you can do so without risk. Apply cooling water to sides of containers that are exposed to flames until well after fire is out. Stay away from ends of tanks. Isolate area until gas has dispersed [R4].

Spill or Leak: Stop leak if you can do so without risk. Fully encapsulating, vapor-protective clothing should be worn for spills and leaks with no fire. Use water spray to reduce vapor; do not put water directly on leak or spill area. Small Spills: Flush area with flooding amounts of water. Large Spills: Dike far ahead of liquid spill for later disposal. Do not get water inside container. Isolate area until gas has dispersed [R4].

First Aid: Move victim to fresh air and call emergency medical care; if not breathing, give artificial respiration; if breathing is difficult, give oxygen. In case of contact with material, immediately flush skin or eyes with running water for at least 15 minutes. Remove and isolate contaminated clothing and shoes at the site. Keep victim quiet and maintain normal body temperature. Effects may be delayed; keep victim under observation [R4].

GENERAL RESPONSE Avoid contact with liquid and vapor. Keep people away. Wear goggles and self-contained breathing apparatus. Stop discharge if possible. Evacuate area in case of large discharge. Stay upwind and use water spray to knock down vapor. Isolate and remove discharged material. Notify local health and pollution control agencies.

FIRE RESPONSE Not flammable. Poisonous gases are produced when heated. Wear goggles and self-contained breathing apparatus. Cool exposed containers and protect men effecting shutoff with water.

EXPOSURE RESPONSE Call for medical aid. Vapor poisonous if inhaled. Irritating to eyes, nose, and throat. Effects may be delayed. Move to fresh air. If breathing has stopped, give artificial respiration (but not mouth-to-mouth). If breathing is difficult, give oxygen. Maintain absolute rest until medical aid arrives.

RESPONSE TO DISCHARGE Issue warning—Poison. Restrict access. Evacuate area.

WATER RESPONSE Effect of low concentrations on aquatic life is unknown. May be dangerous if it enters water intakes. Notify local health and wildlife officials. Notify operators of local water intakes.

EXPOSURE SYMPTOMS Irritates lungs, causing delayed pulmonary edema. Slight gassing produces dryness or burning sensation in the throat, numbness, pain in the chest, bronchitis, and shortness of breath.

EXPOSURE TREATMENT Inhalation: remove victim from contaminated area; enforce absolute rest; call a doctor.

REFERENCES

1. Sax, N. I., and R. J. Lewis, Sr. (eds.). *Hawley's Condensed Chemical Dictionary.* 11th ed. New York: Van Nostrand Reinhold Co., 1987. 223.

2. SRI.

3. Sittig, M. *Hdbk Tox and Hazard Chemical and Carcinogens.* 2nd ed 1985 p. 196.

4. U.S. Department of Transportation. *Emergency Response Guidebook 1990.* DOT P 5800.5. Washington, DC: U.S. Government Printing Office, 1990, p. G-15.

5. Randolph, S. A., M. R. Zavon. *J Occup Med* 29: 237–42 (1987).

6. 40 CFR 240–280, 300–306, 702–799 (7/1/89).

7. American Conference of Governmental Industrial Hygienists. *Documentation of the Threshold Limit Values and Biological Exposure Indices.* 5th ed. Cincinnati, OH: American Conference of Governmental Industrial Hygienists, 1986. 111.

8. 54 FR 2920 (1/19/89).

9. American Conference of Governmental Industrial Hygienists. *Threshold Limit Values for Chemical Substances and Physical Agents and Biological Exposure Indices for 1994–1995.* Cincinnati, OH: ACGIH, 1994.

10. 54 FR 33419 (8/14/89).

11. 40 CFR 261.33 (7/1/88).

12. IARC. *Monographs on the Evaluation of the Carcinogenic Risk of Chemicals to Man.* Geneva: World Health Organization, International Agency for Research on Cancer, 1972–present (multivolume work), p. V27 273.

■ CARBONYL SULFIDE

CAS # 463–58–1

SYNONYMS Carbon monoxide monosulfide; carbon oxide sulfide; carbon oxysulfide

MF COS

MW 60.07

COLOR AND FORM Colorless gas.

ODOR Typical sulfide odor when pure.

DENSITY 1.028 at 17°C.

BOILING POINT –50°C at 760 mm Hg.

MELTING POINT –138°C.

EXPLOSIVE LIMITS AND POTENTIAL Explosive limits 12–29%; forms explosive mixtures with air [R1].

FLAMMABILITY LIMITS Lower: 12% by volume, upper 29% by volume [R4].

FIREFIGHTING PROCEDURES If fire becomes uncontrollable or container is exposed to direct flame, consider evacuation of one-third-mile radius [R9, 141]. If material is on fire or involved in a fire do not extinguish fire unless flow can be stopped. Do not apply water to point of leak in tank car or container. Use water in flooding quantities as fog. Cool all affected containers with flooding quantities of water. Apply water from as far a distance as possible. Use "alcohol" foam, dry chemical, carbon dioxide. [R9, 140].

SOLUBILITY Soluble in alcohol; soluble in toluene; very soluble in carbon disulfide; experimental water solubility: 1,220 mg/L at 25°C.

VAPOR PRESSURE 9,412 mm Hg at 25°C (extrapolated).

OCTANOL/WATER PARTITION COEFFICIENT Log K_{ow} = 0.8009 (estimated).

OTHER CHEMICAL/PHYSICAL PROPERTIES Burns with bluish flame; Henry's Law constant: 4.92×10^{-2} atm-m³/mol at 25°C (estimated).

PROTECTIVE EQUIPMENT Wear positive-pressure self-contained breathing apparatus. Wear appropriate chemical protective clothing. [R9, 141].

PREVENTATIVE MEASURES If material is not on fire and not involved in a fire keep sparks, flames, and other sources of ignition away. Keep material out of other water sources and sewers. Attempt to stop leak if without undue personnel hazard. Use spray water to knock down vapors. [R9, 141].

CLEANUP Cover with a weak solution of calcium hypochlorite (up to 15%). Transfer into a large breaker. After 12 hours, neutralize with 6M hydrocloric acid, or 6M ammonium hydroxide, if necessary. Drain into sewer with abundant water [R5].

DISPOSAL Generators of waste (equal to or greater than 100 kg/mo) containing this contaminant, EPA hazardous waste number D003, must conform with USEPA regulations in storage, transportation, treatment, and disposal of waste. (sulfide compounds) [R6].

COMMON USES Synthesis of thio organic compounds [R3]; chemical int, e.g., for alkyl carbonates and organic sulfur compounds [R2].

OCCUPATIONAL RECOMMENDATIONS None listed.

BIOCONCENTRATION Estimated bioconcentration factors for carbonyl sulfide ranging from 2 to 11 (1,SRC) can be calculated from an estimated log octanol/water partition coefficient of 0.8009 (2) and its experimental water solubility,

1,220 mg/L at 25°C (3). These values suggest that carbonyl sulfide will not bioconcentrate in fish and other aquatic organisms (SRC) [R7].

RCRA REQUIREMENTS D003; a solid waste containing sulfide compounds may become characterized as a hazardous waste when subjected to testing for reactivity as stipulated in 40 CFR 261.23, and, if so characterized, must be managed as a hazardous waste. (sulfide compounds) [R8].

DOT *Health Hazards:* Poisonous; may be fatal if inhaled, swallowed, or absorbed through skin. Contact causes burns to skin and eyes. Contact with liquid may cause frostbite. Runoff from fire control or dilution water may cause pollution. [R10, p. G–18].

Fire or Explosion: Extremely flammable; may be ignited by heat, sparks, or flames. Vapors may travel to a source of ignition, and flash back. Container may explode in heat of fire. Vapor explosion and poison hazard indoors, outdoors, or in sewers. [R10, p. G–18].

Emergency Action: Keep unnecessary people away; isolate hazard area, and deny entry. Stay upwind, out of low areas, and ventilate closed spaces before entering. Positive-pressure self-contained breathing apparatus (SCBA) and chemical protective clothing that is specifically recommended by the shipper or manufacturer may be worn. It may provide little or no thermal protection. Structural firefighters' protective clothing is not effective for these materials. Isolate the spill or leak area immediately for at least 150 feet in all directions. See the Table of Initial Isolation and Protective Action Distances. If you find the ID Number and the name of the material there, begin protective action. Isolate for 1/2 mile in all directions if tank, rail car, or tank truck is involved in fire. Call CHEMTREC at 1–800–424–9300 for emergency assistance. If water pollution occurs, notify the appropriate authorities. [R10, p. G–18].

Fire: Small Fires: Let burn unless leak can be stopped immediately. Large Fires: Water spray, fog, or regular foam. Move container from fire area if you can do so without risk. Apply cooling water to sides of containers that are exposed to flames until well after fire is out. Stay away from ends of tanks. For massive fire in cargo area, use unmanned hose holder or monitor nozzles; if this is impossible, withdraw from area, and let fire burn. Withdraw immediately in case of rising sound from venting safety device or any discoloration of tank due to fire. [R10, p. G–18].

Spill or Leak: Shut off ignition sources; no flares, smoking, or flames in hazard area. Fully encapsulating, vapor-protective clothing should be worn for spills and leaks with no fire. Stop leak if you can do so without risk. Use water spray to reduce vapors; isolate area until gas has dispersed. [R10, p. G–18].

First Aid: Move victim to fresh air, and call emergency medical care; if not breathing, give artificial respiration; if breathing is difficult, give oxygen. In case of contact with material, immediately flush skin or eyes with running water for at least 15 minutes. Remove and isolate contaminated clothing and shoes at the site. Keep victim quiet and maintain normal body temperature. Effects may be delayed; keep victim under observation. [R10, p. G–18].

Initial Isolation and Protective Action Distances: Small spills (leak, or spill from a small package or small leak from a large package): First, isolate in all directions 600 feet; then, protect those persons in the downwind direction 2 miles. Large spills (leak, or spill from a large package or spill from many small packages): First, isolate in all directions 600 feet; then, protect those persons in the downwind direction 2 miles. [R10, p. G–table].

REFERENCES

1. International Labour Office. *Encyclopedia of Occupational Health and Safety.* Vols. I and II. Geneva, Switzerland: International Labour Office, 1983. 2124.

2. SRI.

3. Hawley, G. G. *The Condensed Chemical Dictionary.* 9th ed. New York: Van Nostrand Reinhold Co., 1977. 165.

4. National Fire Protection Association. *Fire Protection Guide on Hazardous Materials.* 9th ed. Boston, MA: National Fire Protection Association, 1986, p. 325M–24.

5. ITII. *Toxic and Hazardous Industrial Chemicals Safety Manual.* Tokyo, Japan: The International Technical Information Institute, 1988. 111.

6. 40 CFR 240–280, 300–306, 702–799 (7/1/91).

7. (1) Lyman, W. J., et al., *Handbook of Chemical Property Estimation Methods.* New York: McGraw-Hill, Chapt 4 (1982). (2) Neely, W. B., G. E. Blau (eds). *Environmental Exposure from Chemicals,* Volume I, Boca Raton, FL: CRC Press, p. 207 (1985). (3) Macaluso, P. *Kirk–Othmer Encycl Chemical Tech.* 2nd ed. New York Wiley, 19: 371–424 (1969). (4) Swann, R. L., et al., *Res Rev* 85: 17–28 (1983).

8. 40 CFR 261.23 (7/1/91).

9. Association of American Railroads. *Emergency Handling of Hazardous Materials in Surface Transportation.* Washington, DC: Assoc. of American Railroads, Hazardous Materials Systems (BOE), 1987.

10. U.S. Department of Transportation. *Emergency Response Guidebook 1990.* DOT P 5800.5. Washington, DC: U.S. Government Printing Office, 1990.

■ CATECHOL

CAS # 120–80–9

SYNONYMS 1,2-benzene diol • 1,2-benzenediol • 1,2-dihydroxybenzene • 2-hydroxyphenol • benzene, o-dihydroxy • catechin • catechin (phenol) • catechol (phenol) • ci-76500 • ci-oxidation-base-26 • durafur developer c • fouramine-pch • fourrine-68 • o-benzenediol • o-dihydroxybenzene •

o-dioxybenzene • o-diphenol •
o-hydroquinone • o-hydroxyphenol •
o-phenylenediol • ortho-benzenediol •
ortho-dihydroxybenzene • ortho-
dioxybenzene • ortho-hydroquinone •
ortho-hydroxyphenol • ortho-
phenylenediol • oxyphenic acid • pelagol
grey-c • phthalhydroquinone •
pyrocatechin • pyrocatechine •
pyrocatechinic-acid • pyrocatechol •
pyrocatechuic acid • Katechol (Czech) •
Pyrokatechin (Czech) • Pyrokatechol
(Czech) • nsc-1573 • ai3-03995

MF $C_6H_6O_2$

MW 110.12

COLOR AND FORM Monoclinic tablets,
prisms from toluene; colorless crystals;
discolors to brown on exposure to air and
light, especially when moist.

ODOR Faint characteristic odor; phenolic
odor.

TASTE Sweet and bitter taste.

DENSITY 1.344.

DISSOCIATION CONSTANTS K at 18°C:
3.3×10^{-10} .

BOILING POINT 245.5°C at 760 mm Hg;
sublimes.

MELTING POINT 105°C.

AUTOIGNITION TEMPERATURE 510°C [R6].

FIREFIGHTING PROCEDURES Water, carbon dioxide, dry chemical [R5].

SOLUBILITY Soluble in 2.3 parts water;
soluble in chloroform, ether; very soluble
in pyridine, aqueous alkalis; slightly soluble in cold benzene; soluble in carbon
tetrachloride, and hot benzene; soluble in
alcohol, ether, acetate.

VAPOR PRESSURE 3×10^{-2} mm Hg at 20°C
(est).

OCTANOL/WATER PARTITION COEFFICIENT
Log K_{ow} = 0.88 (measured).

STABILITY AND SHELF LIFE Discolors in air
and light; aqueous solution soon turns
brown [R1].

OTHER CHEMICAL/PHYSICAL PROPERTIES
Equivalences: 1 mg/L = 222.3 ppm and 1

ppm = 0.00450 mg/L at 25°C, 760 mm Hg;
reduces ammoniacal silver nitrate and
Fehling's solution; forms a definite compound with boric acid; 10 mm Hg at
118.3°C; the rate constant has been estimated to be 24.5 cm^3/mole-sec at 25°C;
estimated Henry's Law Constant of
0.810×10^{-10}.

PROTECTIVE EQUIPMENT Provision of protective clothing and barrier creams [R7].

PREVENTATIVE MEASURES SRP: Local exhaust ventilation should be applied wherever there is an incidence of point-source
emissions or dispersion of regulated contaminants in the work area. Ventilation
control of the contaminant as close as
possible to its point of generation is both
the most economical and safest method to
minimize personnel exposure to airborne
contaminants.

DISPOSAL Incineration: It should be combined with paper or other flammable material. An alternative procedure is to dissolve the solid in a flammable solvent and
spray the solution into the fire chamber
[R8].

COMMON USES Medication [R1]; as an
antioxidant in the rubber, chemical, photographic, dye, fat, and oil industries; in
cosmetics and some pharmaceuticals
[R14, 2584]; anti-fungal preservative for
treating seed potato pieces (former use)
[R2]; photographic developer; developer in
fur dyes; intermediate for antioxidants in
rubber and lubricating oils [R3]; postharvest aid [R4].

OCCUPATIONAL RECOMMENDATIONS None
listed.

OSHA 8-hr time-weighted average: 5 ppm
(20 mg/m^3); final rule limits were achieved
by any combination of engineering controls, work practices, and personal protective equipment during the phase-in period, Sept 1, 1989 through Dec 30, 1992.
Final rule limits became effective Dec 31,
1992. Skin absorption designation in effect as of Sept 1, 1989 [R13].

THRESHOLD LIMIT 8-hr time-weighted average (TWA) 5 ppm, 23 mg/m^3, skin (1977)
[R15, 15].

NON-HUMAN TOXICITY LD_{50} mouse, oral 260 mg/kg [R1]; LD_{50} mouse ip 190 mg/kg [R1].

ECOTOXICITY EC_{50} *Pimephales promelas* (fathead minnow) 9.00 mg/L/96 hr (confidence limit 8.47–9.65 mg/L), flow-through bioassay with measured concentrations, 25.6°C, dissolved oxygen 6.4 mg/L, hardness 46.0 mg/L calcium carbonate, alkalinity 40.2 mg/L calcium carbonate, and pH 7.7. Effect: loss of equilibrium [R10]. LC_{50} *Pimephales promelas* (fathead minnow) 9.22 mg/L/96 hr (confidence limit 8.62–9.87 mg/L), flow-through bioassay with measured concentrations, 25.6°C, dissolved oxygen 6.4 mg/L, hardness 46.0 mg/L calcium carbonate, alkalinity 40.2 mg/L calcium carbonate, and pH 7.7 [R10].

IARC SUMMARY No data are available for humans. Inadequate evidence of carcinogenicity in animals. Overall evaluation: group 3: The agent is not classifiable as to its carcinogenicity to humans [R9].

BIODEGRADATION No data were located concerning the biodegradation of catechol in natural waters or sediments (SRC). Catechol is moderately to readily biodegraded in soils based on a residence time of 1 day for 500 mg of catechol in chernozem soil on hard carbonaceous woody loam (pH 7.1–7.5, 19°C) (1), and percent biodegradation (measured as percentage of recovered $^{14}CO_2$ activity) after 6 months at 23°C in Steinbeck loam (pH 5.0), Fallbrook sandy loam (pH 5.5), Greenfield sandy loam (pH 7.0), and Sorrento loam (pH 7.4) of 24, 50, 28, and 26%, respectively (2). Anaerobic biodegradation of catechol using digester sludge reported to be 67% (as CH_4/CO_2 produced) in 13 days following a 21-day lag period (3) and 98% biodegradation (as CH_4 produced) in 28 days (includes a 21-day lag period) (4) [R11].

BIOCONCENTRATION A bioconcentration factor (BCF) of 3 has been estimated (1,SRC) based on a measured log K_{ow} of 0.88 (2). Based on this estimated BCF, catechol is not expected to bioconcentrate in aquatic organisms (SRC) [R12].

FIRE POTENTIAL Slight when exposed to heat or flame; spontaneous heating: no [R5].

GENERAL RESPONSE Avoid contact with solid and dust. Keep people away. Wear rubber overclothing (including gloves). Stop discharge if possible. Call fire department. Isolate and remove discharged material. Notify local health and pollution control agencies.

FIRE RESPONSE Combustible. Poisonous gases may be produced when heated. Wear goggles and self-contained breathing apparatus. Extinguish with dry chemicals, alcohol foam, or carbon dioxide. Water may be ineffective on fire. Cool exposed containers with water.

EXPOSURE RESPONSE Call for medical aid. Dust irritating to eyes, nose, and throat. If inhaled will cause coughing or difficult breathing. If breathing has stopped, give artificial respiration. If breathing is difficult, give oxygen. Solid will burn skin and eyes. Harmful if swallowed. Remove contaminated clothing and shoes. Flush affected areas with plenty of water. If swallowed and victim is conscious, have victim drink water or milk and have victim induce vomiting. If swallowed and victim is unconscious or having convulsions, do nothing except keep victim warm.

RESPONSE TO DISCHARGE Issue warning—water contaminant; disperse and flush.

WATER RESPONSE Effect of low concentrations on aquatic life is unknown. May be dangerous if it enters water intakes. Notify local health and wildlife officials. Notify operators of nearby water intakes.

EXPOSURE SYMPTOMS Inhalation of dusts or mists may cause irritation of eyes, nose, and throat. Ingestion may cause convulsions and respiratory failure. Contact with eyes causes burns and possible permanent impairment of vision. Prolonged or repeated contact with skin may cause burn.

EXPOSURE TREATMENT Inhalation: if ill effects occur, get medical attention. Ingestion: promptly give milk or plenty of water, and induce vomiting; get medical attention promptly; no specific antidote known. Eyes and skin: immediately flush with plenty of water for at least 15 min; for eyes get medical attention promptly; remove and wash all contaminated clothing before reuse.

REFERENCES

1. Budavari, S. (ed.). *The Merck Index— Encyclopedia of Chemicals, Drugs, and Biologicals.* Rahway, NJ: Merck and Co., Inc., 1989. 1272.

2. *Farm Chemicals Handbook 1980.* Willoughby, OH: Meister, 1980, p. D-61.

3. SRI.

4. *Farm Chemicals Handbook 1991.* Willoughby, OH: Meister, 1991, p. C64.

5. Sax, N. I. *Dangerous Properties of Industrial Materials.* 6th ed. New York, NY: Van Nostrand Reinhold, 1984. 2339.

6. *Kirk–Othmer Encyclopedia of Chemical Technology.* 3rd ed., Volumes 1–26. New York, NY: John Wiley and Sons, 1978–1984, p. V13 41 (1981).

7. Sax, N. I. *Dangerous Properties of Industrial Materials.* 4th ed. New York: Van Nostrand Reinhold, 1975. 1070.

8. United Nations. Treatment and Disposal Methods for Waste Chemicals (IRPTC File). Data Profile Series No. 5. Geneva, Switzerland: United Nations Environmental Programme, Dec. 1985. 275.

9. IARC. *Monographs on the Evaluation of the Carcinogenic Risk of Chemicals to Man.* Geneva: World Health Organization, International Agency for Research on Cancer, 1972–present (multivolume work), p. S7 59 (1987).

10. Geiger D. L., D. J. Call, L. T. Brooke (eds.). *Acute Toxicities of Organic Chemicals to Fathead Minnows (Pimephales promelas).* Vol. V. Superior, WI: University of Wisconsin–Superior, 1990. 117.

11. (1) Medvedev, V.A., V. D. Davidov. pp. 245–54 in *Decomposition of Toxic and Nontoxic Organic Compounds in Soil.* Overcash, M. R. ed., Ann Arbor, MI: Ann Arbor Sci Publ (1981) (2) Stott, D. E., et al., *Soil Soc Am J* 47: 66–70 (1983) (3) Healy, J. B., Jr., L. Y. Young. *Appl Environ Microbiol* 38: 84–9 (1979). (4) Shelton, D. R., J. M. Tiedje. *Development of Tests for Determining Anaerobic Biodegradation Potential,* USEPA–560/5–81–013 (1981).

12. (1) Lyman, W. J., et al., *Handbook of Chemical Property Estimation Methods: Environ Behavior of Organic Compounds.* New York: McGraw-Hill, pp. 5–4 to 5–10 (1982) (2) Hansch, C., and A. J. Leo. *Medchem Project* Issue No. 26, Claremont, CA: Pomona College (1985).

13. 29 CFR 1910.1000 (7/1/91).

14. Clayton, G. D. and F. E. Clayton (eds.). *Patty's Industrial Hygiene and Toxicology.* Volume 2A, 2B, 2C: Toxicology. 3rd ed. New York: John Wiley Sons, 1981–1982.

15. American Conference of Governmental Industrial Hygienists. *Threshold Limit Values for Chemical Substances and Physical Agents and Biological Exposure Indices for 1994–1995.* Cincinnati, OH: ACGIH, 1994.

■ CETYL ALCOHOL

CAS # 36653–82–4

SYNONYMS adol-52 • adol-54 • adol-52nf • adol-52-nf • alcohol c-16 • aldol-54 • alfol-16 • atalco-c • cachalot c-51 • cetaffine • cetal • cetalol-ca • cetanol • cetostearyl alcohol • n-cetyl alcohol • cetylic alcohol • cetylol • crodacol-c • crodacol-cas • crodacol-cat • elfacos-c • ethal • ethol • FEMA number 2554 • hexadecanol • n-hexadecanol • n-1-hexadecanol • 1-hexadecanol • 1-hexadecyl alcohol • hexadecyl alcohol • hyfatol • lanol-c • lorol-24 • loxanol-k • loxanol-k-extra • loxanwachs-sk • normal primary hexadecyl alcohol • palmityl alcohol • product 308 • siponol-cc • siponol wax-a

MF $C_{16}H_{34}O$

MW 242.45

COLOR AND FORM FLakes from ethyl acetate; solid or leaf-like crystals; white crystals; liquid; unctuous, white flakes, granules, cubes, or castings; white, waxy solid.

ODOR Faint odor; odorless.

TASTE Bland, mild taste.

DENSITY 0.8187 at 50°C.

SURFACE TENSION 0.028449 newtons/m at melting point.

VISCOSITY 53 mPa–s (= cP) at 75°C.

BOILING POINT 334°C at 760 mm Hg.

MELTING POINT 49.3°C.

HEAT OF COMBUSTION -9.797×10^9 J/kmol.

HEAT OF VAPORIZATION 9.9829×10^7 J/kmol at boiling point.

FIREFIGHTING PROCEDURES Foam, carbon dioxide, dry chemical [R2].

SOLUBILITY Soluble in acetone; very soluble in ether; slightly soluble in alcohol; soluble in alcohol, chloroform, ether; water solubility = 1.34×10^{-5} g/L at 25°C.

VAPOR PRESSURE 3.06×10^{-6} mm Hg at 30°C.

OCTANOL/WATER PARTITION COEFFICIENT Log P = 6.65 (estimated from a reverse-phase high-pressure liquid chromatography/mass spectrometry method).

OTHER CHEMICAL/PHYSICAL PROPERTIES Congealing point: 46°C.

DISPOSAL SRP: At the time of review, criteria for land treatment or burial (sanitary landfill) disposal practices are subject to significant revision. Prior to implementing land disposal of waste residue (including waste sludge), consult with environmental regulatory agencies for guidance on acceptable disposal practices.

COMMON USES In cosmetics as emollient, emulsion modifier, coupling agent, pharmaceutical aid [R1]. Perfumery; foam stabilizer in detergents; face creams, lotions, lipsticks, toilet preparation; chemical intermediate; detergents; pharmaceuticals; cosmetics; base for making sulfonated fatty alcohols; to retard evaporation of water, when spread as film on reservoirs, or sprayed on growing plants [R7]. Cetyl alcohol and stearamides make dry powder dispersions of aluminum chlorhydroxide for deodorant-antiperspirant sticks [R8, 151]. Opacifying agent in shampoos [R8, 165]. Component of hair-straightening compounds [R10]. Opacifying agent in hair preparations [R11]. Used as a monomer lubricant in suspension polyermization [R12]. NF grade is high purity, used as primary structural agent in antiperspirant sticks; viscosity builder for creams, lotions, and other cosmetic emulsions [R9, 131].

NON-HUMAN TOXICITY LD_{50} mice oral 3.2–6.4 g/kg [R6, 4635]; LD_{50} mice intraperitoneal 1.6–3.2 g/kg [R6, 4635]; LD_{50} rat, oral 6.4–12.8 g/kg [R6, 4635]; LD_{50} rat intraperitoneal 1.6–3.2 g/kg [R6, 4635]; LD_{50} guinea pig skin absorption less than 10 g/kg [R6, 4635].

BIODEGRADATION In a 5-day incubation study using an activated sludge seed from a municipal sewage treatment plant, 28.0% of initial cetyl alcohol was mineralized (CO_2 measurement) (1). A theoretical BOD of 0% was observed using the AFNOR (the French norm procedure) screening test and a 5-day incubation period (2). In standard 5-day BOD tests using emulsified cetyl alcohol, 30–60% of initial cetyl alcohol was oxidized (3). In studies designed to examine the biodegradability of cetyl alcohol in thin films (monolayer) on water surfaces, it was found that biological destruction of the monolayer resulted in measurable consumption of the material with all substrates that were tested (1); substrates included 2% settled domestic sewage in BOD dilution water, 50% OH River water 50% BOD dilution water, water from a stock pond near San Antonio, TX, and other combinations of BOD dilution water and mineral supplements (3); oxidation rates varied with substrates (3); oxidation rates varied from 6.2 to 14.3% over incubation periods of 20 to 48 days (3). In Warburg respirometer tests using activated sludge and 500 mg/L of cetyl alcohol (well above its aqueous solubility), the theoretical oxygen demand was only 0.4% after a 12-hr incubation period (4). The anaerobic degradation of ^{14}C-labelled cetyl alcohol was studied in model sludge digester over a 28-day incubation period (5); 25.1% of radioactivity was recovered in methane gas and 72.0% was recovered in CO_2 gas (5) [R3].

BIOCONCENTRATION In a 3-day static exposure study using golden orfe fish (*Leuciscus idus melanotus*), a cetyl alcohol bioconcentration factor (BCF) of 56 was observed (1); a 24-hr BCF of 17,000 was observed in algae (*Chlorella fusca*) (1) [R4].

FIFRA REQUIREMENTS In 1988, Congress amended FIFRA to strengthen and accelerate EPA's reregistration program. The nine-year reregistration scheme mandated by "FIFRA 88" applies to each registered pesticide product containing an active ingredient initially registered before November 1, 1984. Pesticides for which EPA had not issued Registration Standards prior to the effective date of FIFRA '88 were divided into three lists based upon their potential for exposure and other factors, with List B being of highest concern, and D of least. List: D; Case: Aliphatic alcohols, C6–C16; Case No. 4004; Pesticide type: Insecticide, herbicide; Case Status: Awaiting Data/Data in Review: OPP awaits data from the pesticide's producer(s) regarding its human health or environmental effects, or OPP has received and is reviewing such data, in order to reach a decision about the pesticide's eligibility for reregistration. Active Ingredient (AI): Cetyl Alcohol; AI Status: The active ingredient is no longer contained in any registered products. Therefore, EPA is characterizing it as "cancelled." Under FIFRA, pesticide producers may voluntarily cancel their registered products. EPA also may cancel pesticide registrations if registrants fail to pay required fees, to make or meet certain reregistration commitments, or when the Agency reaches findings of unreasonable adverse effects [R5].

REFERENCES

1. Budavari, S. (ed.). *The Merck Index—Encyclopedia of Chemicals, Drugs, and Biologicals.* Rahway, NJ: Merck and Co., Inc., 1989. 311.

2. Sax, N. I. *Dangerous Properties of Industrial Materials.* 6th ed. New York, NY: Van Nostrand Reinhold, 1984. 1510.

3. (1) Freitag, D., et al. *Ecotox Environ Safety* 6: 60–81 (1982). (2) Dore, M., et al. *Trib Cebedeau* 28: 3–11 (1975). (3) Ludzack, F.J., M. B. Ettinger. *J Amer Water Works Assoc* 48: 849–58 (1957). (4) Gerhold, R. M., G. W. Malaney. *J Water Pollut Control Fed* 38: 562–79 (1966). (5) Steber, J., P. Wierich. *Water Res* 21: 661–7 (1987).

4. Freitag, D., et al. *Ecotox Environ Safety* 6: 60–81 (1982).

5. USEPA/OPP, *Status of Pesticides in Reregistration and Special Review*, p. 220 (Mar, 1992) EPA 700–R–92–004.

6. Clayton, G.D., and F.E. Clayton (eds.). *Patty's Industrial Hygiene and Toxicology*: Volume 2A, 2B, 2C: *Toxicology*. 3rd ed. New York: John Wiley and Sons, 1981–1982.

7. Lewis, R. J., Sr. (Ed.). *Hawley's Condensed Chemical Dictionary*. 12th ed. New York, NY: Van Nostrand Reinhold Co., 1993 246.

8. *Kirk–Othmer Encyclopedia of Chemical Technology*. 3rd ed., Volumes 1–26. New York, NY: John Wiley and Sons, 1978–1984, p. V7.

9. Kuney, J. H., J. M. Mullican (eds.). *Chemcyclopedia*. Washington, DC: American Chemical Society, 1994.

10. *Kirk–Othmer Encyclopedia of Chemical Technology*. 3rd ed., Volumes 1–26. New York, NY: John Wiley and Sons, 1978–1984, p. V17 113.

11. *Kirk–Othmer Encyclopedia of Chemical Technology*. 3rd ed., Volumes 1–26. New York, NY: John Wiley and Sons, 1978–1984, p. V12 91.

12. *Kirk–Othmer Encyclopedia of Chemical Technology*. 3rd ed., Volumes 1–26. New York, NY: John Wiley and Sons, 1978–1984, p. V1 400.

■ CETYLPYRIDINIUM CHLORIDE

CAS # 123–03–5

SYNONYMS 1-cetylpyridinium chloride • 1-hexadecylpyridinium chloride • acetoquat-cpc • aktivex • ammonyx-cpc • biosept • ceepryn chloride • cepacol chloride • ceprim • cetamium • dobendan • fixanol-c • hexadecylpyridinium chloride • intexsan-cpc • n-cetylpyridinium chloride • n-hexadecylpyridinium chloride • pristacin • pyrisept • quaternario-cpc • ceepryn • cepacol • medilave • merocet • cetyl pyridinium chloride

MF $C_{21}H_{38}N$ • Cl

MW 340.05

COLOR AND FORM White powder.

MELTING POINT 77 to 83°C.

OTHER CHEMICAL/PHYSICAL PROPERTIES Max absorption (H_2O): 259 nm (A = 121, 1%, 1 cm) (cetylpyridinium); white powder; MP: 77–83°C; freely soluble in water, alcohol, chloroform; very slightly soluble in benzene, ether; pH (1% aqueous solution): 6.0 to 7.0; surface tension (25°C): 43 dyn/cm (0.1% aqueous solution), 41 dyn/cm (1.0% aqueous solution), 38 dyn/cm (10% aqueous solution) (cetylpyridinium chloride monohydrate); index of refraction: 1.4842 at 25°C (monohydrate).

COMMON USES Germicide, fungicide, surfactant [R4, p. 52: 28]; antiseptic agent for skin and wound disinfection, and in mouthwashes; disinfectant for surgical instruments and other materials; bacteriostatic agent in medicated hairdressings [R1]; pharmaceutic aid (preservative); medication: topical anti-infective; medication (vet): topical antiseptic; disinfectant [R2].

MAJOR SPECIES THREATENED Detergents cause oils to be dissolved from waterfowl, rendering the birds subject to becoming waterlogged or drowning. Fish can build up some resistance to detergents through acclimatization.

DIRECT CONTACT Skin, mild irritant.

GENERAL SENSATION Large quantities may cause nausea, vomiting, collapse, convulsions, and coma.

PERSONAL SAFETY Wear skin protection. Heat or fire may necessitate use of self-contained breathing apparatus.

ACUTE HAZARD LEVEL Mild irritant; moderate ingestive hazard. May cause foaming. Will produce taste in water. Toxic to aquatic life.

CHRONIC HAZARD LEVEL Unknown. Detergents inhibit the aeration of water bodies. Detergents are capable of solubilizing carcinogenic materials.

DEGREE OF HAZARD TO PUBLIC HEALTH Mild irritant. Moderate ingestive hazard. Emits toxic chlorides when heated to decomposition.

OCCUPATIONAL RECOMMENDATIONS None listed.

ACTION LEVEL Isolate from heat.

ON-SITE RESTORATION Treat spill with carbon or peat. Can use cation exchanger. Seek professional environmental engineering assistance through EPA's Environmental Response Team (ERT), Edison, NJ, 24-hour no., 908–548–8730.

AVAILABILITY OF COUNTERMEASURE MATERIAL Carbon—water treatment plants, sugar refineries; peat—nurseries, floral shops; cation exchanger—water softener suppliers.

DISPOSAL 1. Pour or sift onto sodium bicarbonate or a sand-soda-ash mixture (90–10). Mix and package in heavy paper cartons with plenty of paper packing to serve as fuel. Burn in an incinerator. Fire may be augmented with scrap wood. 2. The packages of #1 may be burned more effectively in an incinerator with afterburner and scrubber alkaline. 3. The waste may be mixed with a flammable solvent (alcohol, benzene, etc.), and sprayed into the fire chamber of an incinerator with afterburners and scrubber.

DISPOSAL NOTIFICATION Notify local air authority.

MAJOR WATER USE THREATENED Recreation, fisheries, potable supply.

PROBABLE LOCATION AND STATE White powder; will dissolve.

COLOR IN WATER Colorless, may cause foaming.

BIODEGRADATION Alkylpyridinium derivatives are less biodegradable than mono-alkyltrimethyl and alkylbenzyl dimethyl ammonium chlorides. (alkylpyridinium derivatives) [R3].

FIFRA REQUIREMENTS In 1988, Congress amended FIFRA to strengthen and accelerate EPA's reregistration program. The nine-year reregistration scheme mandated by "FIFRA 88" applies to each registered pesticide product containing an active ingredient initially registered before November 1, 1984. Pesticides for which EPA had not issued Registration Standards prior to the effective date of FIFRA '88 were divided into three lists based upon their potential for exposure and other factors, with List B being of highest concern and D of least. List: C; Case: Alkyl pyridines, and pyridinium quaternaries; Case No.: 3013; Pesticide type: Insecticide, fungicide, rodenticide, and antimicrobial; Case Status: None of the active ingredients in the case are being supported for reregistration. All are unsupported, or some are unsupported and some are cancelled. Active Ingredient (AI): Cetyl pyridinium chloride; AI Status: The producer of the pesticide has not made or honored a commitment to seek reregistration, conduct the necessary studies, or pay the requisite fees. Unless some other interested party makes and meets such commitments, products containing the pesticide will be cancelled. [R5].

REFERENCES

1. SRI.

2. Budavari, S. (ed.). *The Merck Index—Encyclopedia of Chemicals, Drugs, and Biologicals.* Rahway, NJ: Merck and Co., Inc., 1989. 311.

3. Masuda. et al. *Studies on Biodegrad-*

ability of some Cationic Surfactants, TR.—Mezhdunar Kongr Poverkhn-AKT veshchestvam, 7th, 129–138 (1978).

4. *American Hospital Formulary Service.* Volumes I and II. Washington, DC: American Society of Hospital Pharmacists, to 1984.

5. USEPA/OPP. *Status of Pesticides in Reregistration and Special Review.* p.174 (Mar, 1992) EPA 700-R-92-004.

■ CHLORAL HYDRATE

$$Cl_3C—\underset{\underset{OH}{|}}{CHOH}$$

CAS # 302–17–0

SYNONYMS 1,1,1-trichloro-2,2-dihydroxyethane • 2,2,2-trichloro-1,1-ethanediol • aquachloral • bi-3411 • chloral,-monohydrate • Chloraldurat (German) • dormal • felsules • hs • hydral • kessodrate • lorinal • noctec • nycoton • nycton • phaldrone • rectules • somnos • sontec • tosyl • trawotox • Trichloracetaldehyd-hydrat (German) • trichloroacetaldehyde hydrate • trichloroacetaldehyde monohydrate • trichloroacetaldehyde, hydrated • trichloroethylidene glycol • escre • chloraldural • novochlorhydrate • chloralvan • chloralex • chlorali-hydras • kloralhydrat • 2,2,2-trichloroethane-1,1-diol • knockout drops

MF $C_2H_3Cl_3O_2$

MW 165.42

pH 3.5–4.4 (10% solution in water).

COLOR AND FORM Transparent, colorless crystals; large monoclinic plates; colorless or white crystals.

ODOR Aromatic, penetrating, and slightly acrid odor; pungent.

TASTE Slightly bitter, caustic taste.

DENSITY 1.908 at 20°C.

BOILING POINT 96.3°C at 764 mm Hg (decomp).

MELTING POINT 57°C.

SOLUBILITY Sparingly soluble in turpentine, petroleum ether, benzene, toluene, carbon tetrachloride; very soluble in pyridine; 2.4 g/ml water at 0°C; 1 g/68 g carbon disulfide; 14.3 g/ml water at 40°C; 1 G/1.3 ml alcohol; 1 g/1.4 ml olive oil; freely soluble in acetone, methyl ethyl ketone; 8.3 g/ml water at 25°C; 1 g/2 ml chloroform; 1 g/1.5 ml ether; 1 g/0.5 ml glycerol.

OCTANOL/WATER PARTITION COEFFICIENT Log K_{ow} = 0.99.

CORROSIVITY Corrosive to the skin and mucous membrane unless well diluted.

STABILITY AND SHELF LIFE Slowly volatilizes on exposure to air [R1].

OTHER CHEMICAL/PHYSICAL PROPERTIES Decomposed by sodium hydroxide into chloroform; reduces ammoniacal silver nitrate. Incompatible with alkaline substances. Incompatible with soluble barbiturates, tannin, oxidizing agents, and alcohol (chloral alcoholate may crystallize out), phenazone, phenol, thymol, and quinine salts. Pharmaceutical incompatibilities: iodide, cyanide, permanganate, borax, alkali hydroxides, and carbonates, lead acetate, monobromated camphor, diuretin, acetophenetidin, quinine sulfate, salol, theobromine sodiosalicylate, sodium phosphate, urea, urethane.

PREVENTATIVE MEASURES SRP: The scientific literature supports the wearing of contact lenses in industrial environments, as part of a program to protect the eye against chemical compounds and minerals causing eye irritation. However, there may be individual substances whose irritating or corrosive properties are such that the wearing of contact lenses would be harmful to the eye. In those specific cases contact lenses should not be worn [R2].

DISPOSAL Chloral hydrate is a waste chemical stream constituent that may be subjected to ultimate disposal by controlled incineration. Preferably after mixing with another combustible fuel, assure complete combustion to prevent the formation of phosgene; an acid scrubber is necessary to remove the halo acids produced [R5].

COMMON USES Medication: hypnotic, sedative [R1]; used as a rubifacient in topical preparations [R15, 1361]; medication (vet) [R16, p. 16/109]; as glue peptizing agent [R3].

OCCUPATIONAL RECOMMENDATIONS None listed.

NON-HUMAN TOXICITY LD_{50} rats oral 200–500 mg/kg [R17, 837]; LD_{50} horse, oral 100–150 g [R7].

IARC SUMMARY Evaluation: There is inadequate evidence in humans for the carcinogenicity of chloral and chloral hydrate. There is inadequate evidence in experimental animals for the carcinogenicity of chloral. There is limited evidence in experimental animals for the carcinogenicity of chloral hydrate. Overall evaluation: Chloral and chloral hydrate are not classifiable as to their carcinogenicity to humans (Group (3) [R6].

BIOLOGICAL HALF-LIFE The plasma half-life for therapeutic doses of chloral hydrate is 4 to 5 min, whereas for trichloroethanol (a metabolite) it is 8 to 12 hr, and for trichloroacetic acid (a metabolite), 67 hr. [R18, 586].

FIRE POTENTIAL Slight; when heated [R4].

FDA Manufacturers, packers, and distributors of drug and drug products for human use are responsible for complying with the labeling, certification, and usage requirements as prescribed by the Federal Food, Drug, and Cosmetic Act, as amended (secs 201–902, 52 Stat. 1,040 et seq., as amended; 21 U.S.C. 321–392) [R8]. Trichloroacetaldehyde hydrate is a chemical derivative of chloral, named in section 502 (d) of the Federal Food, Drug, and Cosmetic Act and is hereby designated as habit forming [R9]. Warning and caution statements for chloral hydrate are specifically required by law. Preparations containing habit-forming derivatives of substances named in section 502 (d) of the Federal Food, Drug, and Cosmetic act. The statement "Warning—May be habit forming" is required to appear on the labels of all drugs containing derivatives designated in 21 CFR 329.1 as habit forming [R10]. Schedules of controlled substances are established by section 202 of the Controlled Substances Act (21 U.S.C. 812). Schedule IV (c) includes the depressant chloral hydrate, its salts, isomers, and salts of isomers. DEA Code #2,465 [R11]. Chloral hydrate is an indirect food additive for use only as a component of adhesives [R12]. Animal drug specifications: Chloral hydrate, pentobarbital, and magnesium sulfate sterile aqueous solution contains 42.5 mg of chloral hydrate, 8.86 mg of pentobarbital, and 21.2 mg of magnesium sulfate in each ml of sterile aqueous solution containing water, 33.8% propylene glycol, and 14.25% ethyl alcohol. Conditions of use: For general anesthesia and as a sedative-relaxant in cattle and horses For intravenous use only: The drug is administered at a dosage level of 20 to 50 ml per 100 lb of body wt for general anesthesia until the desired effect is produced. Cattle usually require a lower dosage on the basis of body wt. When used as a sedative-relaxant, it is administered at a level of one-fourth to one-half of the anesthetic dosage level. Federal law restricts this drug to use by or on the order of a licensed veterinarian [R13]. The label when dispersed to or for a patient shall contain the following warning: "Caution: Federal law prohibits the transfer of this drug to any person other than the patient for whom it was prescribed." This statement is not required when dispensed for use in clinical investigations that are blind [R14].

REFERENCES

1. Budavari, S. (ed.). The Merck Index—Encyclopedia of Chemicals, Drugs, and Biologicals. Rahway, NJ: Merck and Co., Inc., 1989. 317.

2. Sax, N. I., and R. J. Lewis, Sr. (eds.). Hawley's Condensed Chemical Dictionary. 11th ed. New York: Van Nostrand Reinhold Co., 1987. 256.

3. Kirk–Othmer Encyclopedia of Chemical Technology. 3rd ed., Volumes 1–26. New York, NY: John Wiley and Sons, 1978–1984, p. V11 913 (1980).

4. Sax, N. I. Dangerous Properties of Industrial Materials. 6th ed. New York, NY: Van Nostrand Reinhold, 1984. 667.

5. USEPA. Engineering Handbook for Hazardous Waste Incineration, p. 2–5 (1981) EPA 68–03–3025.

6. IARC. Monographs on the Evaluation of the Carcinogenic Risk of Chemicals to Man. Geneva: World Health Organization, International Agency for Research on Cancer, 1972–present (multivolume work), p. 63 262 (1995).

7. Clarke, M. L., D. G. Harvey, and D. J. Humphreys. Veterinary Toxicology. 2nd ed. London: Bailliere Tindall, 1981. 107.

8. 21 CFR 200–299, 300–499, 820, and 860 (4/1/91.

9. 21 CFR 329.1 (4/1/91).

10. 21 CFR 369.22 (4/1/91).

11. 21 CFR 1308.14 (4/1/91).

12. 21 CFR 175.105 (4/1/91).

13. 21 CFR 522.380 (4/1/91).

14. 21 CFR 290.6 (4/1/91).

15. McEvoy, G.K. (ed.). American Hospital Formulary Service—Drug Information 92. Bethesda, MD: American Society of Hospital Pharmacists, Inc., 1992 (Plus Supplements 1992).

16. Aronson, C.E. (ed.). Veterinary Pharmaceuticals and Biologicals, 1982–1983. Edwardsville, Kansas: Veterinary Medicine Publishing Co., 1983.

17. Haddad, L.M., Clinical Management of Poisoning and Drug Overdose. 2nd ed. Philadelphia, PA: W.B. Saunders Co., 1990.

18. Ellenhorn, M.J., and D.G. Barceloux. Medical Toxicology—Diagnosis and Treatment of Human Poisoning. New York, NY: Elsevier Science Publishing Co., Inc. 1988.

■ CHLORENDIC ACID

CAS # 115–28–6

SYNONYMS 5-norbornene-2,3-dicarboxylic acid, 1,4,5,6,7,7-hexachloro • bicyclo (2.2.1) hept-5-ene-2,3-dicarboxylic acid, 1,4,5,6,7,7-hexachloro • het acid • hexachloroendomethylene-tetrahydrophthalic acid • kyselina 3,6-endomethylen-3,4,5,6,7,7-hexachlor-delta⁴-tetrahyroftalova (Czech) • kyselina-het (Czech)

MF $C_9H_4Cl_6O_4$

MW 388.84

COLOR AND FORM Crystalline solid.

OTHER CHEMICAL/PHYSICAL PROPERTIES Melting point: decomposes to the anhydride.

PROTECTIVE EQUIPMENT Dispensers of liquid detergent should be available. Safety pipettes should be used for all pipetting. In animal laboratory, personnel should wear protective suits (preferably disposable, one-piece, and close-fitting at ankles and wrists), gloves, hair covering, and overshoes. In chemical laboratory, gloves and gowns should always be worn; however, gloves should not be assumed to provide full protection. Carefully fitted masks or respirators may be necessary when working with particulates or gases, and disposable plastic aprons might provide addnl protection. Gowns should be of distinctive color, as a reminder that they are not to be worn outside the laboratory. (chemical carcinogens) [R10, 8].

PREVENTATIVE MEASURES Smoking, drinking, eating, storage of food, or of food and beverage containers, or utensils, and the application of cosmetics should be prohibited in any laboratory. All personnel should remove gloves, if worn, after completion of procedures in which carcinogens have been used. They should wash hands, preferably using dispensers of liquid detergent, and rinse thoroughly. Consideration should be given to appropriate methods for cleaning the skin, depending on nature of the contaminant. No stan-dard procedure can be recommended, but the use of organic solvents should be avoided. Safety pipettes should be used for all pipetting. (chemical carcinogens) [R10, 8].

CLEANUP A high-efficiency particulate arrestor (HEPA) or charcoal filters can be used to minimize amount of carcinogen in exhausted-air-ventilated safety cabinets, lab hoods, glove boxes, or animal rooms. Filter housing that is designed so that used filters can be transferred into plastic bag without contaminating maintenance staff is avail commercially. Filters should be placed in plastic bags immediately after removal. The plastic bag should be sealed immediately. The sealed bag should be labelled properly. Waste liquids should be placed or collected in proper containers for disposal. The lid should be secured and the bottles properly labelled. Once filled, bottles should be placed in plastic bag, so that outer surface is not contaminated. The plastic bag should also be sealed and labelled. Broken glassware should be decontaminated by solvent extraction, by chemical destruction, or in specially designed incinerators. (chemical carcinogens) [R10, 15].

DISPOSAL PRECAUTIONS There is no universal method of disposal that has been proved satisfactory for all carcinogenic compounds, and specific methods of chemical destruction published have not been tested on all kinds of carcinogen-containing waste. Summary of avail methods and recommendations given must be treated as guide only. (chemical carcinogens) [R10, 14].

COMMON USES Fire-retardant monomer [R1]; fire-retardant monomer in unsaturated polyester resins; fire retardant monomer in coatings, in epoxy resins, in polyurethane foams; extreme-pressure lubricant; chemical intermediate for dibutyl and dimethylchlorendate (plasticizers) [R2]. Chlorendic acid used primarily as a chemical intermediate in the manufacture of unsaturated polyester resins, and with special applications in electronic systems, paneling, engineering plastics, and paints. A major use is in fiberglass-reinforced resins for process equipment in chemical industries. Chlorendic acid is also used to impart flame resistance to polyurethane foams when reacted with nonhalogenated glycols to form halogenated polyols, and can be used in the manufacture of alkyd resins for special paints and inks [R3]. In Europe, 80% of the chlorendic acid produced is used in composites for flame-retardant building and transport materials. The remainder is used in composites for the manufacture of anti-corrosion equipment, such as tanks, piping, and scrubbers. In the USA, Latin America, and Far East, the usage pattern is reversed; 70–80% is used for anti-corrosion equipment, and 20–30% for flame-retardant applications [R3]. In the textile industry, the primary use for chlorendic acid is for flame retardant treatment of wool fabrics. The natural flame resistance of wool is enhanced by finishing treatments with chlorendic acid in dimethylformamide [R4].

OCCUPATIONAL RECOMMENDATIONS None listed.

NON-HUMAN TOXICITY LD$_{50}$ rat oral 1,170 mg/kg [R6].

IARC SUMMARY Classification of carcinogenicity: (1) No data are available for humans; (2) evidence in animals: sufficient. Overall summary evaluation of carcinogenic risk to humans is Group 2B: The agent is possibly carcinogenic to humans [R5].

BIODEGRADATION Resistant to hydrolytic dechlorination and may show considerable resistance to degradation [R7].

BIOCONCENTRATION Should have low potential for bioconcentration in organisms and food chains [R8].

TSCA REQUIREMENTS Section 8 (a) of TSCA requires manufacturers of this chemical substance to report preliminary assessment information concerned with production, use, and exposure to EPA as cited in the preamble of the 51 FR 41329 [R9].

REFERENCES

1. *Kirk–Othmer Encyclopedia of Chemical Technology.* 3rd ed., Volumes 1–26. New York, NY: John Wiley and Sons, 1978–1984, p. V10 373 (1980).

2. SRI.

3. IARC. *Monographs on the Evaluation of the Carcinogenic Risk of Chemicals to Man.* Geneva: World Health Organization, International Agency for Research on Cancer, 1972–present (multivolume work), p. V48 46 (1990).

4. IARC. *Monographs on the Evaluation of the Carcinogenic Risk of Chemicals to Man.* Geneva: World Health Organization, International Agency for Research on Cancer, 1972–present (multivolume work), p. V48 47 (1990).

5. IARC. *Monographs on the Evaluation of the Carcinogenic Risk of Chemicals to Man.* Geneva: World Health Organization, International Agency for Research on Cancer, 1972–present (multivolume work), p. V48 50 (1990).

6. Sax, N. I. *Dangerous Properties of Industrial Materials.* 6th ed. New York, NY: Van Nostrand Reinhold, 1984. 671.

7. Schuphan, I., K. Ballschmiter. *Metabolism of Polychlorinated Norborenes by Clostridium Butyricum.* Nature 237 (5350): 100–101 (1972).

8. TRDB/USEPA. *Final Technical Support Document (Draft) Chlorendic Acid*, p. 45 (1982).

9. 40 CFR 712.30 (7/1/91).

10. Montesano, R., H. Bartsch, E. Boyland, G. Della Porta, L. Fishbein, R. A. Griesemer, A.B. Swan, L. Tomatis, and W. Davis (eds.). *Handling Chemical Carcinogens in the Laboratory: Problems of Safety.* IARC Scientific Publications No. 33. Lyon, France: International Agency for Research on Cancer, 1979.

■ 3-CHLORO-2-METHYL-1-PROPENE

$$ClCH_2\overset{\overset{\displaystyle CH_3}{|}}{C}=CH_2$$

CAS # 563-47-3

DOT # 1993

SYNONYMS 2-methallyl chloride • 2-methyl-2-propenyl chloride • 2-methylallyl chloride • 3-Chlor-2-methyl-prop-1-en (German) • 3-chloro-2-methylpropene • 3-cloro-2-metil-prop-1-ene (Italian) • beta-methallyl chloride • beta-methylallyl chloride • chlorure de methallyle (French) • cloruro di metallile (Italian) • gamma-chloroisobutylene • isobutenyl chloride • methallyl chloride • methallylchloride • methylallyl chloride • propene, 3-chloro-2-methyl • 2-Methyl-allylchlorid (German)

MF C_4H_7Cl

MW 90.56

COLOR AND FORM Colorless to straw-colored liquid.

ODOR Sharp, penetrating.

DENSITY 0.9210 at 15°C.

BOILING POINT 71-72°C.

MELTING POINT -12°C.

EXPLOSIVE LIMITS AND POTENTIAL In air: 3.2% to 8.1% [R2].

FIREFIGHTING PROCEDURES Water may be ineffective. Alcohol foam [R6]. Do not extinguish fire unless flow can be stopped. Use water in flooding quantities as fog because solid streams of water may spread fire. Cool all affected containers with flooding quantities of water, being sure to apply water from as far a distance as possible. Use alcohol foam, carbon dioxide, or dry chemical [R7]. Personnel protection: Keep upwind of fire, and avoid breathing vapors. Do not handle broken packages without protective equipment and be sure to wash away any material that may have contacted the body with copious amounts of soap and water [R7]. Wear boots, protective gloves, goggles, and a self-contained breathing apparatus when fighting fires [R7].

SOLUBILITY Soluble in alcohol, ether, acetone, chloroform.

VAPOR PRESSURE 101.7 mm Hg (13.53 × 10^3 Pa) at 20°C.

STABILITY AND SHELF LIFE Volatile [R2].

PROTECTIVE EQUIPMENT Wear boots, protective gloves, goggles, as well as full protective clothing and self-contained breathing apparatus [R7].

PREVENTATIVE MEASURES Do not handle broken packages without protective equipment and be sure to wash away any material that may have contacted the body, with copious amounts of soap and water [R7].

COMMON USES Insecticide, fumigant; in, organic syntheses [R1]; intermediate for production of plastics, pharmaceuticals, and other chemicals [R2]; chemical intermediate for pesticides (e.g., carbofuran) [R3]. Seeds of a number of cultivars of cucumber, tomato, onion, and beetroot infested with certain insect pests may be fumigated with methyl bromide, methallyl chloride, and their combinations with carbon dioxide both at atmospheric pressure and 500 mm Hg within the temperature range from -1°C to 30°C; doses of fumigants may be 110-133 g/m³ giving actual concentration 110-130 g/m³ with various exposure times. These fumigants had no adverse effect on germination of seeds with moisture contents no more than 10-12%. It is not advisable to fumigate vegetable seeds with high initial moisture content of 20-24% in view of loss of germination of the order of 30-35% after 6 mo of storage. Seeds of high moisture content should be first dried to 10 or 12% and then fumigated according to the given schedules [R4].

OCCUPATIONAL RECOMMENDATIONS None listed.

NON-HUMAN TOXICITY LC50 mouse 57.0 mg/L (30 min) [R9]; LC50 rat 34.5 mg/L (30 min); *Salmonella typhimurium* mutagenicity: Number of revertents per µmol of test compound. With S-9 mix: 18 µmol; without S-9 mix: 44 µmol [R10].

ECOTOXICITY Median lethal concentration of methylallyl chloride for the small black beetle, *Tribolium destructor*, was 21.9–22.9 g/m^3 [R11]; LD_{50} goldfish 14 mg/L/24 hr [R12, 428].

IARC SUMMARY Evaluation: There is inadequate evidence in humans for the carcinogenicity of 3-chloro-2-methylpropene. There is limited evidence in experimental animals for the carcinogenicity of 3-chloro-2-methylchloropropene. Overall evaluation: 3-chloro-2-methylpropene is not classifiable as to its carcinogenicity to humans (Group 3) [R8].

BIODEGRADATION Oxidation parameters: biological oxygen demand (5-day test at 20°C): 0.81 NEN 3235–5.4 (NEN: Dutch Standard Test Method) [R12, 816].

DOT *Fire or Explosion:* Flammable/combustible material; may be ignited by heat, sparks, or flames. Vapors may travel to a source of ignition, and flash back. Container may explode in heat of fire. Vapor explosion hazard indoors, outdoors, or in sewers. Runoff to sewer may create fire or explosion hazard [R5].

Health Hazards: May be poisonous if inhaled or absorbed through skin. Vapors may cause dizziness or suffocation. Contact may irritate or burn skin and eyes. Fire may produce irritating or poisonous gases. Runoff from fire control or dilution water may cause pollution [R5].

Emergency Action: Keep unnecessary people away; isolate hazard area and deny entry. Stay upwind; keep out of low areas. Self-contained breathing apparatus (SCBA) and structural firefighter's protective clothing will provide limited protection. Isolate for 1/2 mile in all directions if tank car or truck is involved in fire. Call CHEMTREC at 1–800–424–9300 for emergency assistance. If water pollution occurs, notify the appropriate authorities [R5].

Fire: Small Fires: Dry chemical, CO$_2$, halon, water spray, or alcohol foam. Large Fires: Water spray, fog, or alcohol foam is recommended. Move container from fire area if you can do so without risk. Cool containers that are exposed to flames with water from the side until well after fire is out. Stay away from ends of tanks. For massive fire in cargo area, use unmanned hose holder, or monitor nozzles; if this is impossible, withdraw from area, and let fire burn. Withdraw immediately in case of rising sound from venting safety device or any discoloration of tank due to fire [R5].

Spill or Leak: Shut off ignition sources; no flares, smoking, or flames in hazard area. Stop leak if you can do so without risk. Water spray may reduce vapor; but it may not prevent ignition in closed spaces. Small Spills: Take up with sand or other noncombustible absorbent material and place into containers for later disposal. Large Spills: Dike far ahead of liquid spill for later disposal [R5].

First Aid: Move victim to fresh air and call emergency medical care; if not breathing, give artificial respiration; if breathing is difficult, give oxygen. In case of contact with material, immediately flush eyes with running water for at least 15 minutes. Wash skin with soap and water. Remove and isolate contaminated clothing and shoes at the site [R5].

FIRE POTENTIAL Flammable, dangerous fire risk [R2].

GENERAL RESPONSE Shut off ignition sources, call fire department. Stop discharge if possible. Keep people away. Stay upwind, use water spray to "knock down" vapor. Isolate, and remove discharged material. Notify local health and pollution control agencies.

FIRE RESPONSE Flammable. Irritating gases may be produced when heated. Containers may explode in fire. Flashback along vapor trail may occur. Vapor may explode if ignited in an enclosed area. Extinguish with dry chemicals, foam, or carbon dioxide. Water may be ineffective on fire. Cool exposed containers with water.

EXPOSURE RESPONSE Call for medical aid. Vapor harmful if inhaled. Move victim to fresh air. If breathing has stopped, give artificial respiration. If breathing is difficult, give oxygen. Liquid irritating to skin and eyes. Harmful if swallowed. Remove contaminated clothing and shoes. Flush affected areas with plenty of water. If swallowed and victim is conscious, have victim drink water or milk, and have victim induce vomiting. If swallowed and victim is unconscious or having convulsions, do nothing except keep victim warm.

RESPONSE TO DISCHARGE Issue warning—high flammability. Restrict access. Mechanical containment. Should be removed. Provide chemical and physical treatment.

WATER RESPONSE Effect of low concentrations on aquatic life is unknown. Fouling to shoreline. May be dangerous if it enters water intakes. Notify local health and wildlife officials. Notify operators of nearby water intakes.

EXPOSURE SYMPTOMS Inhalation causes irritation of nose and throat. Contact of vapor or liquid with eyes causes irritation. Liquid irritates skin. Ingestion causes irritation of mouth and stomach.

EXPOSURE TREATMENT Inhalation: remove victim to fresh air; if breathing stops, give artificial respiration and oxygen; subsequent treatment is symptomatic and supportive. Eyes: flush with water for at least 15 min.; get medical attention if exposure has been to liquid. Skin: flush with water; get medical attention if skin is burned. Ingestion: induce vomiting and follow with gastric lavage, demulcents, and saline cathartics; get medical attention.

REFERENCES

1. *The Merck Index.* 10th ed. Rahway, NJ: Merck Co., Inc., 1983. 302.

2. Hawley, G. G. *The Condensed Chemical Dictionary.* 10th ed. New York: Van Nostrand Reinhold Co., 1981. 667.

3. SRI.

4. Konokov, P. F., I. D. Kuznetsov. *Seed Sci Technol 10* (1): 95–103 (1982).

5. Department of Transportation. *Emergency Response Guidebook 1987.* DOT P 5800.4. Washington, DC: U.S. Government Printing Office, 1987, p. G–26.

6. National Fire Protection Association. *Fire Protection Guide on Hazardous Materials.* 7th ed. Boston, MA: National Fire Protection Association, 1978, p. 325M–130.

7. Bureau of Explosives. "Emergency Handling of Haz Matl" in *Surface Trans,* p. 332 (1981).

8. IARC. *Monographs on the Evaluation of the Carcinogenic Risk of Chemicals to Man.* Geneva: World Health Organization, International Agency for Research on Cancer, 1972–present (multivolume work), p. 63 332 (1995).

9. Bakhishev, G. N., et al. *Toxic Effect of Methallyl Chloride and Dichloroethane.* Fiziol Akt Veshchestva 6: 98–100 (1974).

10. Eder, E., et al., *Chemical Biol Interactions* 38: 303–15 (1982).

11. Cherkovskaya, A. Y., N. I. Anoskina. *Use of Fumigant Insecticides to Control the Small Black Beetle Tribolium Destructor* UYTT; TR Vses Nauchno–issled Inst Zerna Prod Ego Pererab 79: 119–22 (1974).

12. Verschueren, K. *Handbook of Environmental Data of Organic Chemicals.* 2nd ed. New York, NY: Van Nostrand Reinhold Co., 1983.

■ 1-CHLORO-2-NITROBENZENE

CAS # 88–73–3

SYNONYMS benzene, 1-chloro-2-nitro • o-chloronitrobenzene • 2-chloro-1-nitrobenzene • o-nitrochlorobenzene • oncb

MF $C_6H_4ClNO_2$

MW 157.56

COLOR AND FORM Yellow crystals; monoclinic needles.

DENSITY 1.368 g/L at 242°C.

SURFACE TENSION 4.37×10^{-2} N/m at 317.65 K.

VISCOSITY 2.09×10^{-3} Pa–s at 317.65°K.

BOILING POINT 245.5°C.

MELTING POINT 32°C.

HEAT OF VAPORIZATION 6.59×10^7 J/kmol at 306.14 K.

FIREFIGHTING PROCEDURES Use water spray, dry chemical, foam, or carbon dioxide [R6, p. 49–126]. Do not extinguish fire unless flow can be stopped. Use water in flooding quantities as fog. Solid streams of water may be ineffective. Cool all affected containers with flooding quantities of water. Apply water from as far a distance as possible. Use foam, dry chemical, or carbon dioxide. (nitrochlorobenzene, ortho, liquid) [R7, 690].

SOLUBILITY Insoluble in water; soluble in alcohol, benzene, ether; very soluble in acetone, pyridine; soluble in toluene, methanol, carbon tetrachloride; water solubility = 1.26×10^{-3} mol/L (198 mg/L); water solubility = 2,800 µM (440 mg/L) at 20°C.

OCTANOL/WATER PARTITION COEFFICIENT Log K_{ow} = 2.52.

OTHER CHEMICAL/PHYSICAL PROPERTIES Log K_{ow} = 2.24. The liquid molar volume is 0.116 m³/kmol. The ideal gas heat of formation is 3.72×10^7 J/kmol. Henry's Law constant = 4.45×10^{-5} atm m³/mole.

PROTECTIVE EQUIPMENT Job analysis to ensure proper handling procedures, adequate equipment design for both operating and maintenance and appropriate ventilation with air-pollution control are minimum requirements. The necessary protective measures in ascending order of effectiveness are respiratory protection, job rotation, limitation of exposure time, use of protective clothing, and whole-body protection. (nitro cmpds, aromatic) [R8, 1453].

PREVENTATIVE MEASURES SRP: Local exhaust ventilation should be applied wherever there is an incidence of point-source emissions or dispersion of regulated contaminants in the work area. Ventilation control of the contaminant as close as possible to its point of generation is both the most economical and safest method to minimize personnel exposure to airborne contaminants.

DISPOSAL SRP: At the time of review, criteria for land treatment or burial (sanitary landfill) disposal practices are subject to significant revision. Prior to implementing land disposal of waste residue (including waste sludge), consult with environmental regulatory agencies for guidance on acceptable disposal practices.

COMMON USES Chemical intermediate for carbofuran, a pesticide, other ints, e.g., o-nitrophenol, 2-chloroaniline, dyes (former use) [R1]. o-Chloronitrobenzene is used as an intermediate in the manufacture of o-aminophenol (used as a developer in the photography industry) [R2].

STANDARD CODES NFPA-3, 1, 1; ICC—(liquid) class B poison, poison label, 55 gallon in an outside container; USCG—poison B, poison label; IATA—poison B, poison label, 1 liter passenger, 220 liter cargo.

MAJOR SPECIES THREATENED May smother benthic life.

INHALATION LIMIT 0.001 mg/m³.

GENERAL SENSATION Causes cyanosis. Can be absorbed through skin.

PERSONAL SAFETY Wear skin protection and self-contained breathing apparatus.

ACUTE HAZARD LEVEL Highly toxic via all routes. Emits toxic vapors when heated to decomposition.

CHRONIC HAZARD LEVEL Highly toxic via all routes with chronic exposure.

DEGREE OF HAZARD TO PUBLIC HEALTH Highly toxic via all routes at acute or chronic exposure levels. Emits highly toxic vapors when heated to decomposition.

OCCUPATIONAL RECOMMENDATIONS None listed.

ACTION LEVEL Notify fire and air authority. If intense heat or flame is present, evacuate area. Enter from upwind. Remove ignition source. Restrict access to affected waters.

ON-SITE RESTORATION Dredge solids from bottom. Seek professional environmental engineering assistance through EPA's environmental response team (ERT), Edison, NJ, 24-hour no., 908–548–8730.

BEACH OR SHORE RESTORATION Do not burn.

DISPOSAL (1) Pour onto sodium bicarbonate or a sand-soda-ash mixture (90/10). Mix in heavy paper cartons and burn in incinerator. May augment fire with wood or paper. (2) Burn packages of no. 1 in incinerator with afterburner and alkaline scrubber. (3) Dissolve in flammable solvent and burn in incinerator of no. 2.

DISPOSAL NOTIFICATION Contact local air authority.

MAJOR WATER USE THREATENED Potable supply, recreation.

PROBABLE LOCATION AND STATE Yellow crystals may melt in warm weather. Will sink to bottom of water course.

BIODEGRADATION 21.1 ppm 1-chloro-2-nitrobenzene in OH River water inoculated weekly with settled sewage underwent no degradation in 175 days (1). 100 ppm 1-chloro-2-nitrobenzene inoculated with 30 ppm activated sludge at 25°C was less than 30% degraded after 2 weeks (2,3) [R4].

BIOCONCENTRATION Results of the Ministry of International Trade and Industry (MITI) test for bioaccumulation indicate that 1-chloro-2-nitrobenzene has a bioconcentration factor (BCF) of less than 100 (1). Using a recommended value for the log octanol-water partition coefficient of 2.24 (2) and a measured water solubility of 440 mg/L at 20°C (3), the BCF for 1-chloro-2-nitrobenzene has been estimated to be 30 and 20, respectively (4,SRC). Based on these BCF values, 1-chloro-2-nitrobenzene should not significantly bio-concentrate in aquatic organisms (SRC) [R5].

DOT *Health Hazards:* Poisonous; may be fatal if inhaled, swallowed, or absorbed through skin. Contact may cause burns to skin and eyes. Runoff from fire control or dilution water may give off poisonous gases and cause water pollution. Fire may produce irritating or poisonous gases (nitrochlorobenzene, solid, or liquid) [R3].

Fire or Explosion: Some of these materials may burn, but none of them ignites readily. Container may explode violently in heat of fire. Material may be transported in a molten form (nitrochlorobenzene, solid, or liquid) [R3].

Emergency Action: Keep unnecessary people away; isolate hazard area and deny entry. Stay upwind, out of low areas, and ventilate closed spaces before entering. Positive-pressure self-contained breathing apparatus (SCBA) and chemical protective clothing that is specifically recommended by the shipper or manufacturer may be worn. It may provide little or no thermal protection. Structural firefighters' protective clothing is not effective for these materials. See the Table of Initial Isolation and Protective Action Distances. If you find the ID number and the name of the material there, begin protective action. Remove and isolate contaminated clothing at the site. Call emergency response telephone number on shipping paper first. If shipping paper not available or no answer, call CHEMTREC at 1–800–424–9300 (nitrochlorobenzene, solid, or liquid) [R3].

Fire: Small Fires: Dry chemical, water spray, or regular foam. Large Fires: Water spray, fog, or regular foam. Move container from fire area if you can do so without risk. Fight fire from maximum distance. Stay away from ends of tanks. Dike fire control water for later disposal; do not scatter the material (nitrochlorobenzene, solid, or liquid) [R3].

Spill or Leak: Do not touch or walk through spilled material; stop leak if you can do so without risk. Fully encapsulating, vapor-protective clothing should be worn for spills and leaks with no fire. Use water spray to reduce vapors. Small Spills: Take up with sand or other noncombustible absorbent material and place into containers for later disposal. Small Dry Spills: With clean shovel place material into clean, dry container, and cover loosely; move containers from spill area. Large Spills: Dike far ahead of liquid spill for later disposal (nitrochlorobenzene, solid or liquid) [R3].

First Aid: Move victim to fresh air, and call emergency medical care; if not breathing, give artificial respiration; if breathing is difficult, give oxygen. In case of contact with material, immediately flush skin or eyes with running water for at least 15 minutes. Speed in removing material from skin is of extreme importance. Removal of solidified molten material from skin requires medical assistance. Remove and isolate contaminated clothing and shoes at the site. Keep victim quiet and maintain normal body temperature. Effects may be delayed; keep victim under observation (nitrochlorobenzene, solid, or liquid) [R3].

REFERENCES

1. SRI.

2. Jarvis, W., et al., *Health and Environmental Effects Document on Chloronitrobenzenes.* Syracuse Res Corp, Syracuse, NY, SRC# TR–92–009. p. 1–4 (1992).

3. U.S. Department of Transportation. *Emergency Response Guidebook 1993.* DOT P 5800.6. Washington, DC: U.S. Government Printing Office, 1993, p. G–55.

4. (1) Ludzack, F. J., M. B. Ettinger. *Eng Bull Ext Ser No 115*: 278–82 (1963). (2) Sasaki, S. pp. 283–98 in *Aquatic Pollutants Transformation and Biological Effects.* Hutzinger, O., et al. (eds.). Oxford: Pergamon Press (1978). (3) Kitano, M. *Biodegradation and Bioaccumulation Test on Chemical Substances.* OECD Tokyo Meeting Reference Book TSU–NO 3 (1978).

5. (1) Sasaki, S., pp. 283–98 in *Aquatic Pollutants Transformation and Biological Effects.* Hutzinger, O., et al. (eds.). Oxford: Pergamon Press (1978). (2)

Hansch, C., and A. J. Leo. *Medchem Project,* Issue No. 26, Pomona College Claremont, CA (1985). (3) Eckert, J. W., *Phytopathol 52*: 642–9 (1962). (4) Lyman, W. J., et al., *Handbook of Chemical Property Estimation Methods. Environ Behavior of Organic Compounds.* New York: McGraw-Hill, p. 5–5 (1982).

6. National Fire Protection Guide. *Fire Protection Guide on Hazardous Materials.* 10th ed. Quincy, MA: National Fire Protection Association, 1991.

7. Association of American Railroads. *Emergency Handling of Hazardous Materials in Surface Transportation.* Washington, DC: Association of American Railroads, Bureau of Explosives, 1992.

8. International Labour Office. *Encyclopedia of Occupational Health and Safety.* Vols. I & II. Geneva, Switzerland: International Labour Office, 1983.

■ 1-CHLORO-3-NITROBENZENE

CAS # 121–73–3

SYNONYMS benzene, 1-chloro-3-nitro • m-chloronitrobenzene • 3-chloro-1-nitrobenzene • metachloronitrobenzene • m-nitrochlorobenzene

MF $C_6H_4ClNO_2$

MW 157.56

COLOR AND FORM Pale-yellow orthorhombic prisms from alcohol.

DENSITY 1.534 at 20°C.

SURFACE TENSION 4.37×10^{-2} N/m at 317.65K.

BOILING POINT 236°C at 760 mm Hg.

MELTING POINT 46°C.

HEAT OF COMBUSTION -2.82×10^9 J/kmol.

HEAT OF VAPORIZATION 5.58×10^7 J/kmol at melting point at 317.65°K.

FIREFIGHTING PROCEDURES Use water spray, dry chemical, foam, or carbon dioxide [R7, p. 49–126]. Extinguish fire using agent suitable for type of surrounding fire (material itself does not burn or burns with difficulty). Use water in flooding quantities as fog. Use foam, dry chemical, or carbon dioxide. (nitrochlorobenzene, meta, solid) [R3].

SOLUBILITY Insoluble in water; sparingly soluble in cold and hot alcohol, chloroform, ether, carbon disulfide, glacial acetic acid; soluble in benzene; water solubility = 0.273 g/L at 20°C.

VAPOR PRESSURE 0.097 mm Hg at 25°C (extrapolated from experimentally derived coefficients).

OCTANOL/WATER PARTITION COEFFICIENT Log K_{ow} = 2.41.

OTHER CHEMICAL/PHYSICAL PROPERTIES Heat of fusion: 29.38 cal/g; can react with oxidizing materials; liquid molar volume = 0.117 m^3/kmol; IG heat of formation = 3.72×10^7 J/kmol; heat of fusion at melting point = 2.078×10^7 J/kmol.

PROTECTIVE EQUIPMENT Wear special protective clothing and positive-pressure self-contained breathing apparatus. [R7, p. 49–126].

PREVENTATIVE MEASURES Personnel protection: Avoid breathing dusts, and fumes from burning material. Avoid bodily contact with the material. Do not handle broken packages unless wearing appropriate personal protective equipment. Wash away any material that may have contacted the body with copious amounts of water or soap and water. (nitrochlorobenzene, meta, solid) [R3].

DISPOSAL SRP: At the time of review, criteria for land treatment or burial (sanitary landfill) disposal practices are subject to significant revision. Prior to implementing land disposal of waste residue (including waste sludge), consult with environmental regulatory agencies for guidance on acceptable disposal practices.

COMMON USES Chemical intermediate for meta-chloroaniline (dye and herbicide chemical int) [R1].

STANDARD CODES NFPA—3, 1, 1; ICC—Class B poison, poison label, 200 lbs in an outside container; USCG—poison B, poison label; IATA—poison B, poison label, 25 kg passenger, 95 kg cargo.

MAJOR SPECIES THREATENED May smother benthic life.

INHALATION LIMIT 0.001 mg/m^3.

GENERAL SENSATION Causes cyanosis. Can be absorbed through skin.

PERSONAL SAFETY Wear self-contained breathing apparatus and skin protection.

ACUTE HAZARD LEVEL Highly toxic via all routes. Emits toxic vapors when heated to decomposition.

CHRONIC HAZARD LEVEL Highly toxic via all routes with chronic exposure.

DEGREE OF HAZARD TO PUBLIC HEALTH Highly toxic via all routes at acute or chronic exposure levels. Emits highly toxic vapors when heated to decomposition.

OCCUPATIONAL RECOMMENDATIONS None listed.

ACTION LEVEL Notify fire and air authority. Warn civil defense of possible explosion. Evacuate area. Enter from upwind. Remove ignition source. Restrict access to affected waters.

ON-SITE RESTORATION Dredge solids. Seek professional environmental engineering assistance through EPA's Environmental Response Team (ERT), Edison, NJ, 24-hour no., 908–548–8730.

BEACH OR SHORE RESTORATION Do not burn.

DISPOSAL (1) Pour onto sodium bicarbonate or a sand-soda-ash mixture (90/10). Mix in heavy paper cartons and burn in incinerator. May augment fire with wood or paper. (2) Burn packages of no. 1 in incinerator with afterburner and alkaline scrubber. (3) Dissolve in flammable solvent and burn in incinerator of no. 2.

DISPOSAL NOTIFICATION Contact local air authority.

MAJOR WATER USE THREATENED Potable supply, recreation.

PROBABLE LOCATION AND STATE Yellow crystals. Will sink.

ECOTOXICITY LC_{50} *Pimephales promelas* (fathead minnow) 18.0 (17.7–18.3) mg/L 24, 48, 72, 96 hr, wt 148 mg, flow-through bioassay, dissolved oxygen 7.4 (4.6–8.8) mg/L, water hardness 44.9 (42.4–46.6) mg/L as $CaCO_3$, pH 6.9–7.7, alkalinity 42.9 (39.6–61.4) mg/L $CaCO_3$, temp: 26.4 ± 1.4°C, purity 98% [R4].

BIODEGRADATION m-Chloronitrobenzene was tested for biodegradability using the OECD, river die-away, and Pitter tests; the half-life for m-chloronitrobenzene was much greater than four weeks for these tests using both unadapted and adapted inoculum (1). A lack of significant ring cleavage by day 64, as measured by UV spectroscopy, showed that m-chloronitrobenzene at 10 µg/ml did not degrade readily in aqueous suspensions of Niagara silt loam (2). This analytical method was not selective enough, however, to determine whether biotransformations not involving aromatic ring cleavage may have taken place (2). Information from studies to determine the effectiveness of various drinking-water purification methods in the Netherlands suggests that m-chloronitrobenzene may be biodegraded in soil (3). The observed removal of m-chloronitrobenzene from bank-filtered water that was observed in these studies may have been due partly to biodegradation; however, the contribution of other processes such as adsorption was not determined. In an earlier study, m-chloronitrobenzene was measured in bank-filtered Rhine River water retained for a time of >1 year (4) [R5].

BIOCONCENTRATION Based on a water solubility of 0.273 g/L at 25°C (1), and an experimental log K_{ow} of 2.41 (2), respective BCFs of 26, and 40 were estimated for m-chloronitrobenzene from recommended regression-derived equations (3); these estimated BCF values suggest that bioconcentration in aquatic organisms is not an important environmental fate process. Rainbow trout (*Salmo gairdneri*) were fed a mixture of 14 different chloronitrobenzenes, including m-chloronitrobenzene, and the amount of m-chloronitrobenzene present in the fish was measured over 36 days (4). m-Chloronitrobenzene was present in trace amounts 3 days following exposure and was not detectable (<5 µg/kg fish) after 8 days. Fish in water containing 800 ng/L of m-chloronitrobenzene plus the same mixture of chloronitrobenzenes gave BCF values for m-chloronitrobenzene of 77 after 5 days to 91 after 36 days, with a mean BCF of 78 (4). The kinetic behavior, a short half-life, and low chemical retention of m-chloronitrobenzene in trout suggests that this chemical may not bioaccumulate appreciably in aquatic ecosystems (4) [R6].

DOT *Health Hazards:* Poisonous; may be fatal if inhaled, swallowed, or absorbed through skin. Contact may cause burns to skin and eyes. Runoff from fire control or dilution water may give off poisonous gases and cause water pollution. Fire may produce irritating or poisonous gases (nitrochlorobenzene, solid, or liquid) [R2].

Fire or Explosion: Some of these materials may burn, but none of them ignites readily. Container may explode violently in heat of fire. Material may be transported in a molten form (nitrochlorobenzene, solid, or liquid) [R2].

Emergency Action: Keep unnecessary people away; isolate hazard area and deny entry. Stay upwind, out of low areas, and ventilate closed spaces before entering. Positive-pressure self-contained breathing apparatus (SCBA) and chemical protective clothing that is specifically recommended by the shipper or manufacturer may be worn. It may provide little or no thermal protection. Structural firefighters' protective clothing is not effective for these materials. See the table of Initial isolation and protective action distances. If you find the ID number and the name of the material there, begin protective action. Remove, and isolate contaminated clothing at the site. Call emergency response telephone number on shipping paper first. If shipping paper not available or no answer, call CHEMTREC at 1–800–424–9300 (nitrochlorobenzene, solid, or liquid) [R2].

Fire: Small Fires: Dry chemical, water spray, or regular foam. Large Fires: Water spray, fog, or regular foam. Move container from fire area if you can do so without risk. Fight fire from maximum distance. Stay away from ends of tanks. Dike fire control water for later disposal; do not scatter the material (nitrochlorobenzene, solid, or liquid) [R2].

Spill or Leak: Do not touch or walk through spilled material; stop leak if you can do so without risk. Fully encapsulating, vapor-protective clothing should be worn for spills and leaks with no fire. Use water spray to reduce vapors. Small Spills: Take up with sand or other noncombustible absorbent material and place into containers for later disposal. Small Dry Spills: With clean shovel place material into clean, dry container, and cover loosely; move containers from spill area. Large Spills: Dike far ahead of liquid spill for later disposal (nitrochlorobenzene, solid, or liquid) [R2].

First Aid: Move victim to fresh air, and call emergency medical care; if not breathing, give artificial respiration; if breathing is difficult, give oxygen. In case of contact with material, immediately flush skin or eyes with running water for at least 15 minutes. Speed in removing material from skin is of extreme importance. Removal of solidified molten material from skin requires medical assistance. Remove and isolate contaminated clothing and shoes at the site. Keep victim quiet and maintain normal body temperature. Effects may be delayed; keep victim under observation (nitrochlorobenzene, solid, or liquid) [R2].

REFERENCES

1. SRI.

2. U.S. Department of Transportation. *Emergency Response Guidebook 1993.* DOT P 5800.6. Washington, DC: U.S. Government Printing Office, 1993, p. G–55.

3. Association of American Railroads. *Emergency Handling of Hazardous Materials in Surface Transportation.* Washington, DC: Association of American Railroads, Bureau of Explosives, 1992. 690.

4. Holcombe, G. W., et al., *Environ Pollut* 35 (Series (A): 367–81 (1984).

5. (1) Canton, J. H., et al., *Regulatory Toxicology and Pharmacology* 5: 123–131 (1985). (2) Alexander, M., B. Lustigman. *J Agr Food Chem* p. 410–413 (1966). (3) Piet, G. J., C. F. Morra. pp. 31–42 in *Water Resources Engineering* Ser. Huisman, L., T. N. Olsthorn, eds. Pitman Pub (1983). (4) Piet, G. J., et al., pp 69–80 in *Hydrocarbon Halo Hydrocarbon Aquatic Environ.* Afghan, B. K., D. Mackay, eds. New York, NY: Plenum Press (1980).

6. (1) Yalkowsky, S. H., R. M. Dannenfelser. *Arizona Database of Aqueous Solubilities.* Univ of AZ, College of Pharmacy (1992). (2) Hansch, C., and A. J. Leo. *Medchem Project* Issue No 26. Claremont, CA: Pomona College (1985). (3) Lyman, W. J., et al., *Handbook of Chemical Property Estimation Methods.* Washington, DC: Amer Chemical Soc, p. 5–4, 5–10 (1990). (4) Niimi, A. J., et al., *Environ Tox Chem* 8: 817–23 (1989).

7. National Fire Protection Guide. *Fire Protection Guide on Hazardous Materials.* 10th ed. Quincy, MA: National Fire Protection Association, 1991.

■ 1-CHLORO-4-NITROBENZENE

CAS # 100–00–5

SYNONYMS 1-chloor-4-nitrobenzeen (Dutch) • 1-Chlor-4-nitrobenzol (German) • 1-cloro-4-nitrobenzene (Italian) • 1-nitro-4-chlorobenzene • 4-chloro-1-nitrobenzene • 4-chloronitrobenzene • 4-nitro-1-chlorobenzene • 4-nitrochlorobenzene • benzene, 1-chloro-4-nitro • nitrochlorobenzene, para-, solid (DOT) • p-chloronitrobenzene • p-nitrochloorbenzeen (Dutch) • p-nitrochlorobenzene • p-Nitrochlorobenzol (German) • p-nitroclorobenzene (Italian) • p-nitrophenyl chloride • pncb • un-1578 (DOT)

MF $C_6H_4ClNO_2$

MW 157.56

COLOR AND FORM Monoclinic prisms; yellow crystals.

DENSITY 1.520.

SURFACE TENSION 3.71×10^{-2} N/m at 356.65 K.

VISCOSITY 1.07×10^{-3} Pa–s at 356.65°K.

BOILING POINT 242°C at 760 mm Hg.

MELTING POINT 82–84°C.

HEAT OF VAPORIZATION 6.21×10^7 J/kmol at 356.65 K.

FIREFIGHTING PROCEDURES Use water spray, dry chemical, foam, or carbon dioxide [R11, p. 49–126]. Carbon dioxide, dry chemical [R12, 858]. Extinguish fire using agent suitable for type of surrounding fire (material itself does not burn or burns with difficulty). Use water in flooding quantities as fog. Use foam, dry chemical, or carbon dioxide. (nitrochlorobenzene, para, solid) [R4].

SOLUBILITY The water solubility for p-chloronitrobenzene is 1.43×10^{-3} mol/L (225 mg/L). Water solubility is 2,877 µM (453 mg/L) at 20°C; sparingly soluble in cold alcohol and freely in boiling alcohol, ether, carbon disulfide.

VAPOR PRESSURE 0.094 mm Hg at 20°C.

OCTANOL/WATER PARTITION COEFFICIENT Log K_{ow} = 2.39.

OTHER CHEMICAL/PHYSICAL PROPERTIES The liquid molar volume is 0.121 m³/kmol. The IG heat of formation is 3.72×10^7 J/kmol. The heat fusion at the melting point is 1.41×10^7 J/kmol. Henry's Law constant = 5.44×10^{-5} atm m³/mol.

PROTECTIVE EQUIPMENT Wear special protective clothing and positive-pressure self-contained breathing apparatus. [R11, p. 49–126].

PREVENTATIVE MEASURES Contact lenses should not be worn when working with this chemical [R5].

CLEANUP 1. Ventilate area of spill. 2. For small quantities, sweep onto paper or other suitable material, place in an appropriate container, and burn in a safe place (such as a fume hood). Large quantities may be reclaimed; however, if this is not practical, dissolve in a flammable solvent (such as alcohol), and atomize in a suitable combustion chamber equipped with an appropriate effluent-gas-cleaning device [R6].

DISPOSAL SRP: At the time of review, criteria for land treatment or burial (sanitary landfill) disposal practices are subject to significant revision. Prior to implementing land disposal of waste residue (including waste sludge), consult with environmental regulatory agencies for guidance on acceptable disposal practices.

COMMON USES p-Chloronitrobenzene is used to manufacture p-nitrophenol, p-nitroaniline, p-aminophenol, phenacetin, acetominophen, parathion, and other agricultural chemicals, rubber chemicals, antioxidants, oil additives, and Dapsone (an antimalarial drug) [R2]. Chemical intermediate for ethyl and methyl parathion, acetaminophen, triclocarban, a bacteriostat, rubber-processing chemicals, and other ints, e.g., p-chloroaniline [R1].

STANDARD CODES NFPA-3, 1, 1; ICC—poison B, poison label, 200 lb in an outside container; USCG—poison B, poison label; IATA.

MAJOR SPECIES THREATENED May smother benthic life.

INHALATION LIMIT 0.001 mg/m³.

GENERAL SENSATION Causes cyanosis. Can be absorbed through skin.

PERSONAL SAFETY Wear skin protection and self-contained breathing apparatus.

ACUTE HAZARD LEVEL Highly toxic via all routes. emits toxic vapors when heated to decomposition.

CHRONIC HAZARD LEVEL Highly toxic via all routes at chronic exposure levels.

DEGREE OF HAZARD TO PUBLIC HEALTH Highly toxic via all routes at acute or chronic exposure levels. Emits highly toxic vapors when heated to decomposition.

OCCUPATIONAL RECOMMENDATIONS None listed.

OSHA Table Z-1 8-hour time-weighted average: 0.1 mg/m³. Skin designation [R9].

NIOSH Usually recommends that occupational exposures to carcinogens be limited to the lowest feasible concentration [R5].

THRESHOLD LIMIT 8-hr time-weighted average (TWA) 0.1 ppm, 0.64 mg/m³, skin (1988) [R14, 28].

ACTION LEVEL Notify fire and air authority. If intense heat or flame is present, evacuate surrounding area. Enter from upwind. Remove ignition source. Restrict access to affected waters.

ON-SITE RESTORATION Dredge bottom for solid. Use carbon or peat on dissolved portion. Seek professional environmental engineering assistance through EPA's environmental response team (ERT), Edison, NJ, 24-hour no., 908-548-8730.

BEACH OR SHORE RESTORATION Do not burn.

AVAILABILITY OF COUNTERMEASURE MATERIAL Carbon—water treatment plants, sugar refineries; peat—nurseries, floral shops.

DISPOSAL (1) Pour onto sodium bicarbonate or a sand-soda-ash mixture (90/10). Mix in heavy paper cartons and burn in incinerator. May augment fire with wood or paper. (2) Burn packages of no. 1 in incinerator with afterburner and alkaline scrubber. (3) Dissolve in flammable solvent and burn in incinerator of no. 2.

DISPOSAL NOTIFICATION Contact local air authority.

MAJOR WATER USE THREATENED Potable supply, recreation.

PROBABLE LOCATION AND STATE Yellow solid. Will sink and dissolve very slowly.

COLOR IN WATER Yellow.

NON-HUMAN TOXICITY LD$_{50}$ rat oral 530 mg/kg [R13, 1100]; LD$_{50}$ rabbit dermal >3,040 mg/kg [R13, 1100].

BIODEGRADATION 100 ppm p-chloronitrobenzene inoculated with 30 ppm activated sludge at 25°C was less than 30% degraded after 2 weeks (1,2). 10 µg/ml p-chloronitrobenzene inoculated with a mixed culture of microorganisms in soil was observed to be resistant to biodegradation (significant ring cleavage, as measured by UV absorbance, was not detected after 64 days) (3). 61–70% reduction of p-chloronitrobenzene after 8–13 days was observed when a mixture of p-chloronitrobenzene and 2,4-dinitrochlorobenzene under continuous-flow conditions involving feeding, aeration, settling, and reflux was inoculated with a species of *Arthrobacter simplex* isolated from industrial waste. When two aeration columns were used, one with *A. simplex*, and the other with *A. simplex*, *Streptomyces coelicolor*, Fusarium sp., and *Trichoderma viridis* isolated from soil, a 90% reduction of the nitro compounds was observed after 10 days. The reduction of p-chloronitrobenzene produced p-chloroaniline and some undefined products (4). The yeast *Rhodosporidiam* sp. reduced p-chloronitrobenzene under aerobic conditions to give 4-chloroacetanilide and 4-chloro-2-hydroxyacetanilide as final major metabolites (5). Whether these reductions in pure culture would occur in waste-water treatment plants is unknown (SRC) [R7].

BIOCONCENTRATION Results of the MITI test for bioaccumulation indicate p-chloronitrobenzene has a bioconcentration factor (BCF) of less than 100 (1). Using a recommended value for the log octanol-water partition coefficient of 2.39 (2) and a measured water solubility of 453 mg/L at 20°C (3), the BCF for p-chloronitrobenzene has been estimated to be 39 and 20, respectively (4,SRC). Based on these BCF values, p-chloronitrobenzene should not significantly bioconcentrate in aquatic organisms (SRC) [R8].

TSCA REQUIREMENTS Section 8 (a) of TSCA requires manufacturers of this chemical substance to report preliminary assessment information concerned with production, use, and exposure to EPA as cited in the preamble in 51 FR 41329 [R10].

DOT *Health Hazards:* Poisonous; may be fatal if inhaled, swallowed, or absorbed through skin. Contact may cause burns to skin and eyes. Runoff from fire control or dilution water may give off poisonous gases and cause water pollution. Fire may produce irritating or poisonous gases (nitrochlorobenzene, solid, or liquid) [R3].

Fire or Explosion: Some of these materials may burn, but none of them ignites readily. Container may explode violently in heat of fire. Material may be transported in a molten form (nitrochlorobenzene, solid, or liquid) [R3].

Emergency Action: Keep unnecessary people away; isolate hazard area and deny entry. Stay upwind, out of low areas, and ventilate closed spaces before entering. Positive pressure self-contained breathing apparatus (SCBA) and chemical protective clothing that is specifically recommended by the shipper or manufacturer may be worn. It may provide little or no thermal protection. Structural firefighters' protective clothing is not effective for these materials. See the Table of Initial Isolation and Protective Action Distances. If you find the ID Number and the name of the material there, begin protective action. Remove and isolate contaminated clothing at the site. Call emergency response telephone number on shipping paper first. If shipping paper not available or no answer, call CHEMTREC at 1–800–424–9300 (nitrochlorobenzene, solid, or liquid) [R3].

Fire: Small Fires: Dry chemical, water spray, or regular foam. Large Fires: Water spray, fog, or regular foam. Move container from fire area if you can do so without risk. Fight fire from maximum distance. Stay away from ends of tanks. Dike fire control water for later disposal; do not scatter the material (nitrochlorobenzene, solid, or liquid) [R3].

Spill or Leak: Do not touch or walk through spilled material; stop leak if you can do so without risk. Fully encapsulating, vapor-protective clothing should be worn for spills and leaks with no fire. Use water spray to reduce vapors. Small Spills: Take up with sand or other noncombusti-ble absorbent material and place into containers for later disposal. Small Dry Spills: With clean shovel place material into clean, dry container, and cover loosely; move containers from spill area. Large Spills: Dike far ahead of liquid spill for later disposal (nitrochlorobenzene, solid, or liquid) [R3].

First Aid: Move victim to fresh air, and call emergency medical care; if not breathing, give artificial respiration; if breathing is difficult, give oxygen. In case of contact with material, immediately flush skin or eyes with running water for at least 15 minutes. Speed in removing material from skin is of extreme importance. Removal of solidified molten material from skin requires medical assistance. Remove and isolate contaminated clothing and shoes at the site. Keep victim quiet and maintain normal body temperature. Effects may be delayed; keep victim under observation (nitrochlorobenzene, solid, or liquid) [R3].

FIRE POTENTIAL Fire hazard slight when exposed to heat or flame. [R12, 858].

REFERENCES

1. SRI.

2. Jarvis, W., et al., *Health and Environmental Effects Document on Chloronitrobenzenes.* Syracuse Res Corp, Syracuse, NY, SRC# TR–92–009. p. 1–4 (1992).

3. U.S. Department of Transportation. *Emergency Response Guidebook 1993.* DOT P 5800.6. Washington, DC: U.S. Government Printing Office, 1993, p. G–55.

4. Association of American Railroads. *Emergency Handling of Hazardous Materials in Surface Transportation.* Washington, DC: Association of American Railroads, Bureau of Explosives, 1992. 691.

5. NIOSH. *NIOSH Pocket Guide to Chemical Hazards.* DHHS (NIOSH) Publication No. 90–117. Washington, DC: U.S. Government Printing Office, June 1990, 164.

6. Mackison, F. W., R. S. Stricoff, and L. J. Partridge, Jr. (eds.). *NIOSH/OSHA—Occupational Health Guidelines for Chemical Hazards.* DHHS (NIOSH) Publication

No. 81–123 (3 vols). Washington, DC: U.S. Government Printing Office, Jan. 1981.

7. (1) Sasaki, S. p 283–98 in *Aquatic Pollutants: Transformation and Biological Effects.* Oxford: Pergamon Press (1978). (2) Kitano, M. *Biodegradation and Bioaccumulation Test on Chemical Substances* OECD Tokyo Meeting Reference Book TSU–No 3 (1978). (3) M. Alexander, B. K. Lustigman, *J Agric Food Chem* 14: 410–3 (1966) (4) Bielaszczyk, E., et al., *Acta Microbiol Pol* 16: 243–8 (1967) (5) Corbett, M. D., B. R. Corbett. *Appl Env Microbiol* 41: 942–9 (1981).

8. (1) Sasaki, S., p. 283–98 in *Aquatic Pollutants: Transformation and Biological Effects.* Oxford: Pergamon Press (1978). (2) Hansch, C., and A. J. Leo. *Medchem Project* Issue #26: Pomona College, Claremont, CA (1985). (3) Eckert, J. W. *Phytopathol* 52: 642–9 (1962). (4) Lyman, W. J., et al., *Handbook of Chemical Property Estimation Methods. Environmental Behavior of Organic Compounds.* New York: McGraw-Hill, p. 5–5 (1982).

9. 29 CFR 1910.1000 (58 FR 35338 (6/30/93)).

10. 40 CFR 712.30 (7/1/94).

11. National Fire Protection Guide. *Fire Protection Guide on Hazardous Materials.* 10th ed. Quincy, MA: National Fire Protection Association, 1991.

12. Sax, N.I. *Dangerous Properties of Industrial Materials.* 5th ed. New York: Van Nostrand Reinhold, 1979.

13. American Conference of Governmental Industrial Hygienists, Inc. *Documentation of the Threshold Limit Values and Biological Exposure Indices.* 6th ed. Volumes I,II, III. Cincinnati, OH: ACGIH, 1991.

14. American Conference of Governmental Industrial Hygienists. *Threshold Limit Values for Chemical Substances and Physical Agents and Biological Exposure Indices for 1994–1995.* Cincinnati, OH: ACGIH, 1994.

■ CHLOROACETYL CHLORIDE

CAS # 79–04–9

SYNONYMS acetyl chloride, chloro • chloracetyl chloride • chloroacetic acid chloride • chloroacetic chloride • chlorure de chloracetyle (French) • monochloroacetyl chloride

MF $C_2H_2Cl_2O$

MW 112.94

COLOR AND FORM Colorless to slightly yellow liquid; water-white liquid.

ODOR Sharp, pungent, irritating.

ODOR THRESHOLD An industrial hygienist was not able to detect odor at 0.011 ppm, found 0.023 ppm barely detectable, and 0.140 ppm strong. [R7, 269].

DENSITY 1.4202 at 20°C.

SURFACE TENSION 25 dynes/cm = 0.025 N/m at 20°C.

BOILING POINT 106°C.

MELTING POINT FP: −21.77°C.

HEAT OF COMBUSTION −4,000 Btu/lb = −2,000 cal/g = −90×10⁵ J/kg (est).

HEAT OF VAPORIZATION 166 Btu/lb = 92.0 cal/g = 3.85×10⁵ J/kg.

FIREFIGHTING PROCEDURES Do not use water on adjacent fires [R3]. Approach fire from upwind to avoid hazardous vapors and toxic decomposition products. Use water spray to keep fire-exposed containers cool. Extinguish fire, using agent suitable for surrounding fire [R2]. If material involved in fire: Use dry chemical or carbon dioxide. Do not use water on material itself. If large quantities of combustibles are involved, use water in flooding quantities as spray and fog. Use water spray to knock down vapors. Apply water from as far a distance as possible [R8, 217]. Wear positive-pressure self-contained breathing apparatus when fighting fires involving this material. [R8, 218].

TOXIC COMBUSTION PRODUCTS Heat of fire can cause decomposition with evolution of highly toxic and irritating hydrogen chloride and phosgene vapors [R3].

SOLUBILITY Soluble in ether and acetone.

VAPOR PRESSURE 19 mm Hg at 20°C.

CORROSIVITY Highly corrosive.

OTHER CHEMICAL/PHYSICAL PROPERTIES Ratio of specific heat of vapor (gas): 1.1191; heat of solution: −54 Btu/lb = −30 cal/g = −1.3×10⁵ J/kg (est).

PROTECTIVE EQUIPMENT Corrosion-resistant and impervious suits or overalls, foot protection, hand and arm protection, head protection, and eye and face protection; where corrosive gases may be expected, appropriate respiratory protective equipment is required, ranging from simple masks or respirators to air or oxygen lines or self-contained breathing apparatus, coupled with gas-tight goggles depending on type and quantity of the corrosive substance handled or liberated. (corrosive substances) [R6, 554].

PREVENTATIVE MEASURES SRP: The scientific literature for the use of contact lenses in industry is conflicting. The benefit or detrimental effects of wearing contact lenses depend not only upon the substance, but also on factors including the form of the substance, characteristics and duration of the exposure, the uses of other eye protection equipment, and the hygiene of the lenses. However, there may be individual substances whose irritating or corrosive properties are such that the wearing of contact lenses would be harmful to the eye. In those specific cases, contact lenses should not be worn. In any event, the usual eye protection equipment should be worn even when contact lenses are in place.

CLEANUP Approach release from upwind. Keep water away from release. Stop or control the leak, if this can be done without undue risk. Prompt cleanup and removal are necessary. Control runoff and isolate discharged material for proper disposal [R2].

DISPOSAL SRP: At the time of review, criteria for land treatment or burial (sanitary landfill) disposal practices are subject to significant revision. Prior to implementing land disposal of waste residue (including waste sludge), consult with environmental regulatory agencies for guidance on acceptable disposal practices.

COMMON USES Tear gas [R1].

OCCUPATIONAL RECOMMENDATIONS None listed.

THRESHOLD LIMIT 8-hr time-weighted average (TWA) 0.05 ppm, 0.23 mg/m³, skin; short-term exposure limit (STEL) 0.15 ppm, 0.69 mg/m³, skin [R5].

INTERNATIONAL EXPOSURE LIMITS Australia: 0.05 ppm (1990); United Kingdom: 0.05 ppm (1991). [R7, 269].

NON-HUMAN TOXICITY LD₅₀ rat, oral 120 mg/kg [R4].

GENERAL RESPONSE Avoid contact with liquid and vapor. Keep people away. Wear goggles and self-contained breathing apparatus. Stop discharge if possible. Isolate and remove discharged material. Notify local health and pollution control agencies.

FIRE RESPONSE Not flammable. Poisonous gases may be produced when heated. Do not use water on adjacent fires. Cool exposed containers with water.

EXPOSURE RESPONSE Call for medical aid. Vapor irritating to eyes, nose, and throat. Move victim to fresh air. If breathing is difficult, give oxygen. Liquid will burn skin and eyes. Harmful if swallowed. Remove contaminated clothing and shoes. Flush affected areas with plenty of water. If in eyes, hold eyelids open and flush with plenty of water. If swallowed and victim is conscious, have victim drink water or milk. Do not induce vomiting.

RESPONSE TO DISCHARGE Issue warning—air contaminant, corrosive. Restrict access. Disperse and flush.

WATER POLLUTION RESPONSES Effect of low concentrations on aquatic life is unknown. May be dangerous if it enters water intakes. Notify local health and wildlife officials. Notify operators of nearby water intakes.

EXPOSURE SYMPTOMS Inhalation causes

severe irritation of upper respiratory system. External contact causes severe irritation of eyes and skin. Ingestion causes severe irritation of mouth and stomach.

EXPOSURE TREATMENT Inhalation: remove from exposure; support respiration; call physician. Eyes: wash with copious amounts of water for 15 min. Call physician. Skin: wash with large amounts of water; treat burns as required. Ingestion: do not induce vomiting; give large amounts of water; call a physician.

REFERENCES

1. Sax, N. I., and R. J. Lewis, Sr. (eds.). *Hawley's Condensed Chemical Dictionary*. 11th ed. New York: Van Nostrand Reinhold Co., 1987. 261.

2. National Fire Protection Guide. *Fire Protection Guide on Hazardous Materials*. 10th ed. Quincy, MA: National Fire Protection Association, 1991, p. 49–51.

3. U.S. Coast Guard, Department of Transportation. *CHRIS—Hazardous Chemical Data*. Volume II. Washington, DC: U.S. Government Printing Office, 1984–5.

4. Sax, N. I. *Dangerous Properties of Industrial Materials*. 6th ed. New York, NY: Van Nostrand Reinhold, 1984. 678.

5. American Conference of Governmental Industrial Hygienists. *Threshold Limit Values for Chemical Substances and Physical Agents and Biological Exposure Indices for 1994–1995*. Cincinnati, OH: ACGIH, 1994. 16.

6. International Labour Office. *Encyclopedia of Occupational Health and Safety*. Vols. I & II. Geneva, Switzerland: International Labour Office, 1983.

7. American Conference of Governmental Industrial Hygienists, Inc. *Documentation of the Threshold Limit Values and Biological Exposure Indices*. 6th ed. Volumes I, II, III. Cincinnati, OH: ACGIH, 1991.

8. Association of American Railroads. *Emergency Handling of Hazardous Materials in Surface Transportation*. Washington, DC: Association of American Railroads, Bureau of Explosives, 1992.

■ 2-CHLOROANILINE

CAS # 95–51–2

SYNONYMS 1-amino-2-chlorobenzene • 2-chlorophenylamine • aniline, o-chloro • benzenamine, 2-chloro • fast-yellow-gc-base • o-chloraniline • o-chloroaniline

MF C_6H_6ClN

MW 127.58

COLOR AND FORM Amber liquid (SRP: technical grade).

ODOR Characteristic sweet odor; amine odor.

DENSITY 1.2114 at 22°C.

BOILING POINT 208.84°C.

MELTING POINT −14°C.

FIREFIGHTING PROCEDURES Water, dry chemical, foam, or carbon dioxide (4-chloroaniline) [R2].

TOXIC COMBUSTION PRODUCTS Irritating and toxic hydrogen chloride and oxides of nitrogen may form in fires. (4-chloroaniline) [R22].

SOLUBILITY Soluble in most organic solvents, also in acids.

VAPOR PRESSURE 0.17 mm Hg at 20°C.

OCTANOL/WATER PARTITION COEFFICIENT Log K_{ow} = 1.92.

STABILITY AND SHELF LIFE Darkens on exposure to air [R1].

OTHER CHEMICAL/PHYSICAL PROPERTIES MP: −1.9°C; BP: 208.84 at 760 mm Hg; density: 1.21266 at 20°C; index of refraction: 1.5889 at 20°C/D (beta); MP: −14°C; BP: 208.84 at 760 mm Hg; density: 1.21253 at 20°C; index of refraction: 1.58951 at 20°C/D (alpha); distillation range: 208–210°C; soluble in alcohol, ether, acetone (alpha, and beta); soluble in benzene (beta); max absorption alcohol: 236 nm (log E = 3.8); 291 nm (log E = 3.3) (beta); max absorption (ALC: 236 nm (log E = 3.8); 291 nm (log E =3.3) (alpha); heat of decomposition: 0.41 kJ/g.

PROTECTIVE EQUIPMENT Wear butyl rubber gloves, long-sleeve coveralls made of plastics, and self-contained breathing apparatus. [R8, 119].

PREVENTATIVE MEASURES Immediately wash contaminated areas of body with concentrated soap solution. Remove contaminated clothing, dry, then wash with concentrated soap solution, or dispose as waste. Contaminated shoes may be disposed of in an incinerator. [R8, 120].

CLEANUP Cover with a 9:1 mixture of sand and soda ash. After mixing, transfer into a paper carton stuffed with ruffled paper. [R8, 120].

DISPOSAL At the time of review, criteria for land treatment or burial (sanitary landfill) disposal practices are subject to significant revision. Prior to implementing land disposal of waste residue (including waste sludge), consult with environmental regulatory agencies for guidance on acceptable disposal practices [R3].

COMMON USES Dye intermediate; standards for colorimetric apparatus; manufacture of petroleum solvents and fungicides [R1].

OCCUPATIONAL RECOMMENDATIONS None listed.

ECOTOXICITY EC_{50} *Daphnia magna* (daphnids) 4.2 mg/L/24 hr (ability to swim) [R4]. EC_0 *Daphnia magna* (daphnids) 1.2 mg/L/24 hr (ability to swim) [R4]. EC_{100} *Daphnia magna* (Daphnids) 36 mg/L/24 hr (ability to swim) [R4]. EC_{50} *Daphnia magna* (daphnids) 1.8 mg/L/48 hr (ability to swim) [R4]. EC_0 *Daphnia magna* (daphnids) 0.3 mg/L/48 hr (ability to swim) [R4]. EC_{100} *Daphnia magna* (daphnids) 4.7 mg/L/48 hr (ability to swim) [R4].

BIODEGRADATION 2-Chloroaniline was found to be resistant to microbial degradation using the Japanese MITI protocol (1). A 36% BODT was measured over a 190 hr incubation period with a Warburg respirometer (2). 100% loss of UV absorbance in a mineral salts solution, with a soil inoculum, required in excess of 64 days (3). Using an acclimated activated sludge inoculum, 98% of initial 2-chloroaniline was degraded under the test conditions (4). Significant biological transformation

(31% degradation in 6 hr) was observed in an aqueous test system receiving activated sludge from two treatment plants (5). Half-lives greatly in excess of 4 weeks were observed using the modified OECD and Repetitive Die Away Test (6) [R5].

BIOCONCENTRATION Log BCF of 2-chloroaniline in fish were experimentally determined to be less than 2.0 using the Japanese MITI test procedures (1). 2-chloroaniline was found to have little or no bioconcentration in carp (2). The log BCF of 2-chloroaniline has been theoretically estimated to be 1.3 (3) [R6].

ATMOSPHERIC STANDARDS This action promulgates standards of performance for equipment leaks of volatile organic compounds (VOC) in the synthetic organic chemical manufacturing industry (SOCMI). These standards implement Section 111 of the Clean Air Act and are based on the administrator's determination that emissions from the SOCMI cause or contribute significantly to air pollution that may reasonably be anticipated to endanger public health or welfare. The intended effect of these standards is to require all newly constructed, modified, and reconstructed SOCMI process units to use the best demonstrated system of continuous emission reduction for equipment leaks of VOC, considering costs, non-airquality health, and environmental impact and energy requirements. o-Chloroaniline is produced, as an intermediate or final product, by process units covered under this subpart. These standards of performance become effective upon promulgation but apply to affected facilities for which construction or modification commenced after January 5, 1981 [R7].

DOT *Health Hazards:* Poisonous if swallowed. Inhalation of dust poisonous. Fire may produce irritating or poisonous gases. Runoff from fire control or dilution water may cause pollution (chloroaniline, solid) [R9, p. G–53].

Fire or Explosion: Some of these materials may burn, but none of them ignites readily (chloroaniline, solid) [R9, p. G–53].

Emergency Action: Keep unnecessary

people away; isolate hazard area, and deny entry. Stay upwind; keep out of low areas. Self-contained breathing apparatus (SCBA) and structural firefighter's protective clothing will provide limited protection. Call CHEMTREC at 1–800–424–9300 for emergency assistance. If water pollution occurs, notify the appropriate authorities (chloroaniline, solid) [R9, p. G–53].

Fire: Small Fires: Dry chemical, CO_2, Halon, water spray, or standard foam. Large Fires: Water spray, fog, or standard foam is recommended. Move container from fire area if you can do so without risk (chloroaniline, solid) [R9, p. G–53].

Spill or Leak: Do not touch spilled material; stop leak if you can do so without risk. Small Spills: Take up with sand or other noncombustible absorbent material and place into containers for later disposal. Small Dry Spills: With clean shovel place material into clean, dry container, and cover; move containers from spill area. Large Spills: Dike far ahead of liquid spill for later disposal (chloroaniline, solid) [R9, p. G–53].

First Aid: Move victim to fresh air; call for emergency medical care. Remove and isolate contaminated clothing and shoes at the site. In case of contact with material, immediately flush skin or eyes with running water for at least 15 minutes (chloroaniline, solid) [R9, p. G–53].

REFERENCES

1. Sax, N. I., and R. J. Lewis, Sr. (eds.). *Hawley's Condensed Chemical Dictionary.* 11th ed. New York: Van Nostrand Reinhold Co., 1987. 262.

2. U.S. Coast Guard, Department of Transportation. *CHRIS—Hazardous Chemical Data.* Volume II. Washington, DC: U.S. Government Printing Office, 1984–5.

3. SRP.

4. Kuhn, R., et al. *Water Res* 23 (4): 495–9 (1989).

5. (1) Kawasaki, M., *Ecotoxic Environ Safety* 4: 444–54 (1980). (2) Malaney, G. W. *J Water Pollut Control Fed* 32: 1300–11 (1960). (3) Alexander, M., B. K. Lustigman, *J Agric Food Chem* 14: 410–3 (1966). (4) Pitter, P. *Water Res* 10: 231–5 (1976). (5) Baird, R., et al. *J Water Pollut Control Fed* 49: 1609–15 (1977). (6) Canton, J. H., et al. *Regul Toxicol Pharmacol* 5: 123–31 (1985).

6. (1) Kawasaki, M., *Ecotoxic Environ Safety* 4: 444–54 (1980). (2) Sasaki, S. p. 283–98 in *Aquatic Pollutants: Transformations and Biological Effects*; Hutzinger, O., et al. (eds.). Oxford: Pergamon Press (1978). (3) Canton, J. H., et al., *Regul Toxicol Pharmacol* 5: 123–31 (1985).

7. 40 CFR 60.489 (7/1/87).

8. ITII. *Toxic and Hazarous Industrial Chemicals Safety Manual.* Tokyo, Japan: The International Technical Information Institute, 1982.

9. Department of Transportation. *Emergency Response Guidebook 1987.* DOT P 5800.4. Washington, DC: U.S. Government Printing Office, 1987.

■ 2-CHLOROBENZAL-MALONONITRILE

CAS # 2698–41–1

SYNONYMS (o-chlorobenzal) malononitrile • 2-chlorobenzylidene malononitrile • 2-chlorobmn • beta, beta-dicyano-o-chlorostyrene • cs • cs (lacrimator) • malononitrile, (o-chlorobenzylidene)- • nci-c55118 • o-chlorobenzylidene malonitrile • propanedinitrile, ((2-chlorophenyl) methylene)- • usaf kf-11 • ocbm • ortho-chlorobenzylidene malononitrile • ((2-chloro-phenyl) methylene) propanenitrile

MF $C_{10}H_5ClN_2$

MW 188.62

COLOR AND FORM White crystalline solid.

ODOR Odor of pepper.

BOILING POINT 310–315°C.

MELTING POINT 93–95°C.

SOLUBILITY Insoluble in water; soluble in acetone, dioxane, methylene chloride, ethyl acetate, benzene.

VAPOR PRESSURE 3.4×10^{-5} mm Hg at 20°C.

OCTANOL/WATER PARTITION COEFFICIENT Log K_{ow} = 1.849 (est).

OTHER CHEMICAL/PHYSICAL PROPERTIES Henry's law constant = 1.02×10^{-8} atm–m^3/mole (est).

PROTECTIVE EQUIPMENT Wear appropriate equipment to prevent repeated or prolonged skin contact [R3].

PREVENTATIVE MEASURES Workers should wash promptly when skin becomes contaminated and at the end of each work shift [R3].

CLEANUP 1. Ventilate area of spill. 2. For small quantities, sweep onto paper or other suitable material, place in appropriate container, and burn in safe place (such as fume hood). Large quantities may be reclaimed; however, if this is not practical, dissolve in flammable solvent (such as alcohol), and atomize in suitable combustion chamber equipped with appropriate effluent-gas-cleaning device. 3. Decontaminate area of spill: (A) by washing with a 5% solution of sodium hydroxide in 50/50 ethyl alcohol/water; or (B) by adding flake sodium hydroxide to a solution or slurry of the spill in isopropyl alcohol; or (C) by covering the spill with a 10% solution of sodium hydroxide in 50/50 isopropyl alcohol/water and letting stand 20 minutes before flushing with water [R4].

Chemical disposal method for cs was developed. The recommended reaction is aqueous alkaline hydrolysis of cs to o-chlorobenzaldehyde [R5].

DISPOSAL A poor candidate for incineration (cyanides) [R6].

COMMON USES It is used primarily as an incapacitating agent both by military and law enforcement personnel. It can be disseminated in burning grenades and weapon-fired projectiles, as an aerosol from the finely divided solid chemical, or from a solution of the chemical dissolved in methylene chloride or acetone [R1]. Used as a tear-gas and riot control agent [R2].

OCCUPATIONAL RECOMMENDATIONS None listed.

OSHA 8-hr time-weighted average: 0.05 ppm (0.4 mg/m³); transitional limits must continue to be achieved by any combination of engineering controls, work practices, and personal protective equipment during the phase-in period, Sept 1, 1989 through Dec 30, 1992. Final rule limits became effective Dec 31, 1992 [R12].

NIOSH Ceiling value: 0.05 ppm (0.4 mg/m³), skin [R3].

THRESHOLD LIMIT Ceiling limit 0.05 ppm, 0.39 mg/m³, skin (1983) [R13].

NON-HUMAN TOXICITY LD_{50} rat iv 28 mg/kg [R8]; LD_{50} rat ip 48 mg/kg [R8]; LD_{50} rat (male) oral 1,366 mg/kg [R9]; LD_{50} rat (female) oral 1,284 mg/kg [R9].

ECOTOXICITY LC_{50} rainbow trout 1.28 mg/L/12 hr (conditions of bioassay not specified) [R10]; LC_{50} rainbow trout >0.1 mg/L <1 wk [R7].

BIOLOGICAL HALF-LIFE Cats and rats were exposed to cs aerosol. Absorption of the compound occurred in the blood. The half-life in the blood was 5.5 seconds for cats [R1].

BIOCONCENTRATION Based upon an estimated log K_{ow} of 1.849 (1), the bioconcentration factor (BCF) for 2-chlorobenzalmalononitrile can be estimated to be 41.6 from a recommended regression-derived equation (2,SRC). This BCF value suggests that bioconcentration in aquatic organisms may not be important environmentally (SRC) [R11].

RCRA REQUIREMENTS D003; a solid waste containing a cyanide compound may become characterized as a hazardous waste when subjected to testing for reactivity as stipulated in 40 CFR 261.23, and, if so characterized, must be managed as a hazardous waste. (cyanide compounds) [R14].

REFERENCES

1. American Conference of Governmental

Industrial Hygienists. *Documentation of the Threshold Limit Values and Biological Exposure Indices.* 5th ed. Cincinnati, OH: American Conference of Governmental Industrial Hygienists, 1986. 124.

2. *Kirk–Othmer Encyclopedia of Chemical Technology.* 3rd ed., Volumes 1–26. New York, NY: John Wiley and Sons, 1978–1984, p. V5 401 (1979).

3. NIOSH. *NIOSH Pocket Guide to Chemical Hazards.* DHHS (NIOSH) Publication No. 90–117. Washington, DC: U.S. Government Printing Office, June 1990, 66.

4. Mackison, F. W., R. S. Stricoff, and L. J. Partridge, Jr. (eds.). *NIOSH/OSHA—Occupational Health Guidelines for Chemical Hazards.* DHHS (NIOSH) Publication No. 81–123 (3 vols). Washington, DC: U.S. Government Printing Office, Jan. 1981.

5. Brooks, M. E., et al. *US Ntis,* Ad Report Ad–A033469 (1976).

6. USEPA. *Engineering Handbook for Hazardous Waste Incineration.* p. 3–8 (1981) EPA 68–03–3025.

7. ABRAM, F. S. H., P. WILSON. *Water Res* 13 (7): 631 (1979).

8. Budavari, S. (ed.). *The Merck Index—Encyclopedia of Chemicals, Drugs, and Biologicals.* Rahway, NJ: Merck and Co., Inc., 1989. 328.

9. DHHS/NTP. *Toxicology and Carcinogenesis Studies of CS (94% o–Chlorobenzalmalononitrile) in F344/N Rats and B6C3F1 Mice.* p. 13 (1990) Technical Rpt Series No. 377, NIH Pub No. 90–2832.

10. Verschueren, K. *Handbook of Environmental Data of Organic Chemicals.* 2nd ed. New York, NY: Van Nostrand Reinhold Co., 1983. 361.

11. (1) USEPA. *PCGEMS* (Graphical Exposure Modeling System) CLOGP–Version April 1987 (1987). (2) Lyman, W. J., et al., *Handbook of Chemical Property Estimation Methods.* New York: McGraw-Hill, p. 5–4 (1982).

12. 29 CFR 1910.1000 (7/1/91).

13. American Conference of Governmental Industrial Hygienists. *Threshold Limit Values for Chemical Substances and Physical Agents and Biological Exposure Indices for 1994–1995.* Cincinnati, OH: ACGIH, 1994. 16.

14. 40 CFR 261.23 (7/1/91).

■ 2-CHLOROETHANOL

CAS # 107–07–3

DOT # 1135

SYNONYMS 2-chloorethanol (Dutch) • 2-Chloraethanol (German) • 2-chloro-1-ethanol • 2-chloroethyl alcohol • 2-cloroetanolo [Italian] • 2-hydroxyethyl chloride • 2-monochloroethanol • Aethylenechlorhydrin (German) • beta-chloroethanol • beta-chloroethyl alcohol • beta-hydroxyethyl chloride • chloroethanol • chloroethylowy-alkohol [Polish] • delta-chloroethanol • ethanol, 2-chloro • ethene, chlorohydrin • ethylchlorohydrin • ethyleenchloorhydrine [Dutch] • ethylene chlorhydrin • ethylene glycol, chlorohydrin • glicol monocloridrina [Italian] • glycol chlorohydrin • glycol monochlorohydrin • glycolmonochloorhydrine [Dutch] • glycomonochlorhydrin • monochlorhydrine du glycol [French] • nci-c50135

MF C_2H_5ClO

MW 80.52

COLOR AND FORM Colorless glycerine-like liquid.

ODOR Faint ethereal odor; sweet, pleasant.

DENSITY 1.197 at 20°C.

VISCOSITY 0.0343 poise at 20°C.

DISSOCIATION CONSTANTS $pK_a = 14.31$ at 25°C.

BOILING POINT 128–130°C at 760 mm Hg.

MELTING POINT −67.5°C.

HEAT OF COMBUSTION −6,487 btu/lb = −3,604 cal/g = -150.8×10^5 J/kg.

HEAT OF VAPORIZATION 10,740.6 cal/mole.

AUTOIGNITION TEMPERATURE 797°F; 425°C [R20, p. 325M–26].

EXPLOSIVE LIMITS AND POTENTIAL Upper 15.9%; lower 4.9% [R8].

FLAMMABILITY LIMITS Lower flammable limit: 4.9% by volume; upper flammable limit: 15.9% by volume [R20, p. 325M–26].

FIREFIGHTING PROCEDURES Water, alcohol foam, dry chemical, or carbon dioxide [R6].

TOXIC COMBUSTION PRODUCTS Toxic hydrogen chloride and phosgene fumes may be formed (during combustion) [R6].

SOLUBILITY Soluble in all proportions in water, alcohol, organic solvents; slightly soluble in ether; soluble in various resins; infinitely soluble in aqueous solution.

VAPOR PRESSURE 4.9 mm Hg at 20°C.

OCTANOL/WATER PARTITION COEFFICIENT Log K_{ow} = 0.03.

STABILITY AND SHELF LIFE It evaporates readily at room temperature [R9].

OTHER CHEMICAL/PHYSICAL PROPERTIES Coefficient of expansion 0.00089 at 20°C; wt/gal = 10.0 lb at 20°C C; autoignition temperature 425°C; conversion factors: 1 mg/L is equiv to 303.8 ppm and 1 ppm is equiv to 3.29 mg/m³ at 25°C, 760 mm Hg. When heated with water to 100°C it decomposes into glycol and aldehyde; when heated to 184°C, it decomposes into ethylene chloride and acetaldehyde; density of saturated air: 1.011 at 20°C; 1.022 at 30.3°C; percent in saturated air: 0.644 at 20°C; 1.32 at 30.3°C; vapor pressure = 7.18 mm Hg at 25°C; standard heat of fusion = −70.6 kcal/mol at 25°C; Henry's Law constant = 1.04×10^{-7} atm–m³/mole (est).

PROTECTIVE EQUIPMENT Organic canister mask or self-contained breathing apparatus; goggles or face shield [R6].

PREVENTATIVE MEASURES SRP: contaminated protective clothing should be segregated in such a manner that there is no direct personal contact by personnel who handle, dispose of, or clean the clothing. Quality assurance to ascertain the completeness of the cleaning procedures should be implemented before the decontaminated protective clothing is returned for reuse by the workers. Contaminated clothing should not be taken home at end of shift, but should remain at employee's place of work for cleaning.

CLEANUP 1. Remove all ignition sources. 2. Ventilate area of spill or leak. 3. For small quantities, absorb on paper towels. Evaporate in a safe place (such as a fume hood). Allow sufficient time for evaporating vapors to completely clear the hood ductwork. Burn the paper in a suitable location away from combustible materials. Large quantities can be collected and atomized in a suitable combustion chamber equipped with an appropriate effluent-gas-cleaning device. Ethylene chlorohydrin should not be allowed to enter a confined space, such as a sewer, because of the possibility of an explosion [R7].

DISPOSAL SRP: At the time of review, criteria for land treatment or burial (sanitary landfill) disposal practices are subject to significant revision. Prior to implementing land disposal of waste residue (including waste sludge), consult with environmental regulatory agencies for guidance on acceptable disposal practices.

COMMON USES Manufacture of insecticides; treating sweet potatoes before planting [R2]; introduction of hydroxyethyl group in organic synthesis; solvent for cellulose acetate, ethylcellulose; activate sprouting of dormant potatoes; manufacture of ethylene oxide and ethylene glycol, insecticides [R3]. Used to produce ethylene glycol and ethylene oxide; employed for sepn of butadiene from hydrocarbon mixt; in dewaxing and removing naphthenes from mineral oil, in refining of rosin; in extraction of pine lignin; solvent for cellulose ethers [R19, 4675]. Intermed for indigo, numerous other chemicals; intermed for thiodiethylene glycol (textile printing solvent); intermed for dichloroethyl formal for manufacture of polysulfides [R1]. 2-Chloroethanol may be used in the manufacture of dye intermediates, pharmaceuticals, plant-protecting agents, pesticides, and plasticizers [R4]. For removal of tar spots; cleaning agent for machines [R18, 397].

OCCUPATIONAL RECOMMENDATIONS None listed.

OSHA 8-hour time-weighted average: 5 ppm (16 mg/m³). Skin designation [R15].

NIOSH Ceiling value: 1 ppm (3 mg/m³), skin [R8].

THRESHOLD LIMIT Ceiling limit 1 ppm, 3.3 mg/m³, skin (1977) [R16].

INTERNATIONAL EXPOSURE LIMITS Australia: peak limitation, 1 ppm, skin (1990); former Federal Republic of Germany: 1 ppm, short-term level 5 ppm, 30 min, twice per shift, skin, pregnancy group C (no reason to fear a risk of damage to the developing embryo or fetus when MAK and BAT values are adhered to) (1990); Sweden: ceiling value 1 ppm, skin (1989); United Kingdom: 10-minute STEL 1 ppm, skin (1991). [R22, 601].

NON-HUMAN TOXICITY LD$_{50}$ rat, oral 58 mg/kg [R10]; LD$_{50}$ rat sc 72 mg/kg [R19, 4679]; LD$_{50}$ rabbit dermal 67.8 mg/kg [R19, 4679]; LD$_{50}$ rat dermal 84 mg/kg [R19, 4679].

ECOTOXICITY LC$_{50}$ *Pimephales promelas* (fathead minnow) 83.7 mg/L/96 hr (95% confidence limit 75.0–93.4 mg/L), flow-through bioassay with measured concentrations, 24.0°C, dissolved oxygen 7.1 mg/L, hardness 52.3 mg/L calcium carbonate, alkalinity 44.2 mg/L calcium carbonate, and pH 7.22 [R11].

BIOLOGICAL HALF-LIFE The elimination half-life and clearance value for 2-chloroethanol was 40.8 min following iv admin to beagle dogs [R12].

BIODEGRADATION 2-Chloroethanol is readily biodegradable in screening tests and biological treatment simulations using sewage and activated sludge inocula (1–6). Various investigators have obtained the following results in percent of theoretical BOD in screening tests using sewage inocula: 57% in 20 days (1); 50% in 10 days (2); and 87% in 10 days (3). The results of these screening tests indicate that acclimation is important in the biodegradation process [R13].

BIOCONCENTRATION Using the log octanol/water partition coefficient for 2-chloroethanol, 0.03 (1), one can estimate a BCF of 0.62 using a recommended regression equation (2,SRC). Therefore, 2-chloroethanol would not be expected to bioconcentrate in aquatic organisms [R14].

ATMOSPHERIC STANDARDS This action promulgates standards of performance for equipment leaks of volatile organic compounds (VOC) in the synthetic organic chemical manufacturing industry (SOCMI). The intended effect of these standards is to require all newly constructed, modified, and reconstructed SOCMI process units to use the best demonstrated system of continuous emission reduction for equipment leaks of VOC, considering costs, non-air-quality health, and environmental impact and energy requirements. Ethylene chlorohydrin is produced, as an intermediate or a final product, by process units covered under this subpart [R17].

DOT *Health Hazards:* Poisonous; may be fatal if inhaled, swallowed, or absorbed through skin. Contact may cause burns to skin and eyes. Runoff from fire control or dilution water may give off poisonous gases and cause water pollution. Fire may produce irritating or poisonous gases. [R21, p. G–55].

Fire or Explosion: Some of these materials may burn, but none of them ignites readily. Container may explode violently in heat of fire. Material may be transported in a molten form. [R21, p. G–55].

Emergency Action: Keep unnecessary people away; isolate hazard area, and deny entry. Stay upwind, out of low areas, and ventilate closed spaces before entering. Positive-pressure self-contained breathing apparatus (SCBA) and chemical protective clothing that is specifically recommended by the shipper or manufacturer may be worn. It may provide little or no thermal protection. Structural firefighters' protective clothing is not effective for these materials. See the table of initial isolation and protective action distances. If you find the ID number and the name of the material there, begin protective action. Remove, and isolate contaminated clothing at the site. Call emergency response telephone number on shipping paper first. If shipping paper not available or no answer, Call CHEMTREC at 1–800–424–9300. [R21, p. G–55].

Fire: Small Fires: Dry chemical, water spray, or regular foam. Large Fires: Water spray, fog, or regular foam. Move container from fire area if you can do so without risk. Fight fire from maximum distance. Stay away from ends of tanks. Dike fire control water for later disposal; do not scatter the material. [R21, p. G–55].

Spill or Leak: Do not touch or walk through spilled material; stop leak if you can do so without risk. Fully encapsulating, vapor-protective clothing should be worn for spills and leaks with no fire. Use water spray to reduce vapors. Small Spills: Take up with sand or other noncombustible absorbent material and place into containers for later disposal. Small Dry Spills: With clean shovel place material into clean, dry container, and cover loosely; move containers from spill area. Large Spills: Dike far ahead of liquid spill for later disposal. [R21, p. G–55].

First Aid: Move victim to fresh air and call for emergency medical care; if not breathing, give artificial respiration; if breathing is difficult, give oxygen. In case of contact with material, immediately flush skin or eyes with running water for at least 15 minutes. Speed in removing material from skin is of extreme importance. Removal of solidified molten material from skin requires medical assistance. Remove and isolate contaminated clothing and shoes at the site. Keep victim quiet and maintain normal body temperature. Effects may be delayed; keep victim under observation. [R21, p. G–55].

Initial Isolation and Protective Action Distances: Small Spills (from a small package or small leak from a large package): First, isolate in all directions 500 feet; then, protect persons downwind 0.1 mile (day), or 0.3 mile (night). Large spills (from a large package or from many small packages): First, isolate in all directions 500 feet; then, protect persons downwind 0.2 mile (day), or 1.6 miles (night). [R21, p. G–table].

FIRE POTENTIAL Moderate, when exposed to heat, flame, or oxidizers [R5].

GENERAL RESPONSE Stop discharge if possible. Keep people away. Stay upwind. Use water spray to "knock down" vapor. Call fire department. Isolate and remove discharged material. Notify local health and pollution control agencies.

FIRE RESPONSE Combustible. Poisonous gases may be produced in fire. Extinguish with water, dry chemicals, alcohol foam, or carbon dioxide. If breathing is difficult, give oxygen. Liquid irritating to skin and eyes. Harmful if swallowed. Remove contaminated clothing and shoes. Flush affected areas with plenty of water. If swallowed, and victim is conscious, have victim drink water or milk. Do not induce vomiting.

RESPONSE TO DISCHARGE Issue warning—water contaminant. Restrict access. Disperse and flush.

WATER RESPONSE Effect of low concentrations on aquatic life is unknown. May be dangerous if it enters water intakes. Notify local health and wildlife officials. Notify operators of nearby water intakes.

EXPOSURE SYMPTOMS Inhalation causes irritation of upper respiratory system, nausea, headache, delirium, coma, collapse. Liquid causes irritation of eyes and skin; prolonged contact with skin may allow penetration into body and cause same symptoms as following ingestion or inhalation. Ingestion causes nausea, headache, delirium, coma, and collapse.

EXPOSURE TREATMENT Inhalation: remove from exposure; give artificial respiration if breathing has stopped; call physician. Eyes: flush with water for at least 15 min. Get medical attention if irritation persists. Skin: wash off with copious amounts of water; call physician if contact has been prolonged. Ingestion: give large amounts of water; get medical attention.

REFERENCES

1. SRI.

2. Budavari, S. (ed.). *The Merck Index— Encyclopedia of Chemicals, Drugs, and Biologicals*. Rahway, NJ: Merck and Co., Inc., 1989. 598.

3. Sax, N. I., and R. J. Lewis, Sr. (eds.). *Hawley's Condensed Chemical Dictionary*. 11th ed. New York: Van Nostrand Reinhold Co., 1987. 485.

4. *Kirk–Othmer Encycl Chemical Technol*, 4th ed 6: 153 (1993).

5. Sax, N. I. *Dangerous Properties of Industrial Materials*. 6th ed. New York, NY: Van Nostrand Reinhold, 1984. 712.

6. U.S. Coast Guard, Department of Transportation. *CHRIS—Hazardous Chemical Data*. Volume II. Washington, DC: U.S. Government Printing Office, 1984–5.

7. Mackison, F. W., R. S. Stricoff, and L. J. Partridge, Jr (eds.). *NIOSH/OSHA—Occupational Health Guidelines for Chemical Hazards*. DHHS (NIOSH) Publication No. 81–123 (3 vols). Washington, DC: U.S. Government Printing Office, Jan. 1981.

8. NIOSH. *NIOSH Pocket Guide to Chemical Hazards*. DHHS (NIOSH) Publication No. 90–117. Washington, DC: U.S. Government Printing Office, June 1990, 108.

9. Arena, J. M. Poisoning: *Toxicology, Symptoms, Treatments*. Fourth Edition. Springfield, Illinois: Charles C. Thomas, Publisher, 1979. 208.

10. Gosselin, R. E., R. P. Smith, H. C. Hodge. *Clinical Toxicology of Commercial Products*. 5th ed. Baltimore: Williams and Wilkins, 1984, p. II–178.

11. Geiger D. L., Poirier S. H., Brooke L. T., D.J. Call (eds.). *Acute Toxicities of Organic Chemicals to Fathead Minnows (Pimephales promelas)*. Vol. III. Superior, Wisconsin: University of Wisconsin– Superior, 1986. 36.

12. Martis, L., et al., *J Toxicol Environ Health* 10 (4–5): 847–56 (1982).

13. (1) Conway, R. A., et al., *Environ Sci Technol* 17: 107–12 (1983). (2) Heukelekian, H., M. C. Rand. *J Water Pollut Contr Assoc* 29: 1040–53 (1955). (3) Lamb, C. B., G. F. Jenkins, pp. 326–9 in *Proc 8th Industrial Waste Conf*, Purdue Univ (1952). (4) Matsui, S., et al., *Prog Water Technol* 7: 645–59 (1975). (5) Mills, E. J., Jr., V. T. Stack, Jr., pp. 492–517 in *Proc 8th Indust Waste Conf*. Eng Bull, Purdue Univ, Eng Ext Ser (1954). (6) Klockner, D. *Chemical Ind* 39: 92–4 (1987).

14. (1) Hansch, C., and A. J. Leo. *Medchem Project* Issue No. 26 Claremont, CA: Pomona College (1985). (2) Lyman, W. J., et al., *Handbook of Chemical Property Estimation Methods*. New York: McGraw-Hill, Chapt 5 (1982).

15. 29 CFR 1910.1000 (58 FR 35338 (6/30/93)).

16. American Conference of Governmental Industrial Hygienists. *Threshold Limit Values for Chemical Substances and Physical Agents and Biological Exposure Indices for 1994–1995*. Cincinnati, OH: ACGIH, 1994. 21.

17. 40 CFR 60.489 (7/1/92).

18. Browning, E. *Toxicity and Metabolism of Industrial Solvents*. New York: American Elsevier, 1965.

19. Clayton, G. D. and F. E. Clayton (eds.). *Patty's Industrial Hygiene and Toxicology*. Volume 2A, 2B, 2C: Toxicology. 3rd ed. New York: John Wiley Sons, 1981–1982.

20. National Fire Protection Guide. *Fire Protection Guide on Hazardous Materials*. 10th ed. Quincy, MA: National Fire Protection Association, 1991.

21. U.S. Department of Transportation. *Emergency Response Guidebook 1993*. DOT P 5800.6. Washington, DC: U.S. Government Printing Office, 1993.

22. American Conference of Governmental Industrial Hygienists, Inc. *Documentation of the Threshold Limit Values and Biological Exposure Indices*. 6th ed. Volumes I, II, III. Cincinnati, OH: ACGIH, 1991.

■ 2-CHLOROETHYL VINYL ETHER

CAS # 110-75-8

SYNONYMS 2-vinyloxyethyl chloride • ethene, (2-chloroethoxy)- • ether, 2-chlorethyl vinyl • vinyl 2-chloroethyl ether • vinyl beta-chloroethyl ether

MF C_4H_7ClO

MW 106.56

COLOR AND FORM Colorless liquid.

DENSITY 1.0495 at 20°C.

BOILING POINT 109°C at 740 mm Hg.

MELTING POINT -70.3°C.

EXPLOSIVE LIMITS AND POTENTIAL Dangerous when heated or exposed to flame or sparks. Besides risk of explosion from air mixture of ether vapors, ethers tend to form peroxides upon standing. When ethers containing peroxides are heated they can detonate. (ethers) [R4].

FIREFIGHTING PROCEDURES Use foam, alcohol foam, carbon dioxide, dry chemical [R6].

TOXIC COMBUSTION PRODUCTS Dangerous; when heated to decomp, it emits highly toxic fumes of phosgene. (ethers) [R4].

SOLUBILITY Very soluble in alcohol and ether; soluble in 15,000 mg/L water at 25°C.

VAPOR PRESSURE Estimated vapor pressure is 30 torr at 25°C.

OCTANOL/WATER PARTITION COEFFICIENT Log K_{ow} = 0.99.

STABILITY AND SHELF LIFE Stable in caustic solution, hydrolyzes in acid solutions [R5].

OTHER CHEMICAL/PHYSICAL PROPERTIES 2-chloroethyl vinyl ether is hydrolyzed to 2-chloroethanol and acetaldehyde.

PREVENTATIVE MEASURES Only electrical equipment of explosion-proof type (group C classification) is permitted to be operated in ether areas. (ethers) [R4].

DISPOSAL Generators of waste (equal to or greater than 100 kg/mo) containing this contaminant, EPA hazardous waste number U042, must conform with USEPA regulations in storage, transportation, treatment, and disposal of waste [R7].

COMMON USES Used in the manufacture of anesthetics, sedatives, and cellulose ethers [R2]. Copolymer of 95% ethyl acrylate with 5% 2-chloroethyl vinyl ether has been used to produce acrylic elastomer [R3].

OCCUPATIONAL RECOMMENDATIONS None listed.

NON-HUMAN TOXICITY LD_{50} rat oral 250 mg/kg [R1]; LD_{50} rabbit dermal 3.2 ml/kg [R8].

ECOTOXICITY LC_{50} *Lepomis macrochirus* (bluegill) 194,000 µg/L/96 hr (static bioassay) [R9].

BIODEGRADATION 2-Chloroethyl vinyl ether gave a 76% and 52% (initial concentration 5 and 10 mg/L, respectively) theoretical biological oxygen demand in seven days using a settled domestic wastewater as a microbial inoculum. Complete biodegradation was obtained in seven days using the third subculture. 2-Chloroethyl vinyl ether was listed as showing significant degradation with rapid adaptation (the highest of a three-tier rating system) (1,2) [R10].

BIOCONCENTRATION Based on the log octanol/water partition coefficient (K_{ow}), 0.99 (1,SRC), and the water solubility, 15,000 mg/L (2), BCF values in the range 1-6 can be calculated for 2-chloroethyl vinyl ether (3,SRC), suggesting that bioconcentration in aquatic organisms should not be an important fate process (SRC) [R11].

CERCLA REPORTABLE QUANTITIES Persons in charge of vessels or facilities are required to notify the National Response Center (NRC) immediately, when there is a release of this designated hazardous substance, in an amount equal to or greater than its reportable quantity of 1,000 lb, or 454 kg. The toll-free number of the NRC is (800) 424-8802; in the Washington DC metropolitan area (202) 426-2675. The rule for determining when notification is required is stated in 40 CFR 302.4 (section IV. D. 3.b) [R12].

TSCA REQUIREMENTS Pursuant to section 8 (d) of TSCA, EPA promulgated a model health and safety data reporting rule. The section 8 (d) model rule requires manufacturers, importers, and processors of listed chemical substances and mixtures to submit to EPA copies and lists of unpublished health and safety studies. As cited in the preamble of 51 FR 41329, ethene, (2-chloroethoxy) is included on this list [R13].

RCRA REQUIREMENTS As stipulated in 40 CFR 261.33, when 2-chloroethyl vinyl ether, as a commercial chemical product or manufacturing chemical intermediate or an off-specification commercial chemical product or a manufacturing chemical intermediate, becomes a waste, it must be managed according to federal or state hazardous waste regulations. Also defined as a hazardous waste is any residue, contaminated soil, water, or other debris resulting from the cleanup of a spill, into water or on dry land, of this waste. Generators of small quantities of this waste may qualify for partial exclusion from hazardous waste regulations (40 CFR 261.5) [R14].

FIRE POTENTIAL Dangerous; shock or heat can cause gaseous ethers to escape from their containers (ethers) [R4]. Moderate fire hazard [R5].

REFERENCES

1. *The Merck Index.* 10th ed. Rahway, NJ: Merck Co., Inc., 1983. 300.

2. Sittig, M. *Handbook of Toxic and Hazardous Chemicals and Carcinogens, 1985.* 2nd ed. Park Ridge, NJ: Noyes Data Corporation, 1985. 231.

3. *Kirk–Othmer Encyclopedia of Chemical Technology.* 3rd ed., Volumes 1–26. New York, NY: John Wiley and Sons, 1978–1984, p. 8 (79) 463.

4. Sax, N. I. *Dangerous Properties of Industrial Materials.* 6th ed. New York, NY: Van Nostrand Reinhold, 1984. 1305.

5. Sax, N. I., and R. J. Lewis, Sr. (eds.). *Hawley's Condensed Chemical Dictio-*

nary. 11th ed. New York: Van Nostrand Reinhold Co., 1987. 266.

6. Sax, N. I. *Dangerous Properties of Industrial Materials.* 4th ed. New York: Van Nostrand Reinhold, 1975. 1235.

7. 40 CFR 240–280, 300–306, 702–799 (7/1/89).

8. USEPA. *Frequency of Organic Compounds Identified in Water* (1976) EPA 600/4–76–062 as cited in USEPA, *Health and Environmental Effects Profile for 2–Chloroethyl Vinyl Ether,* p. 46–6 (1980) ECAO–CIN–46.

9. USEPA. *Health and Environmental Effects Profile for 2–Chloroethyl Vinyl Ether,* p. 46–6 (1980) ECAO–CIN–46.

10. (1) Tabak, H. H., et al., *J Water Pollut Contr Fed* 53: 1503–18 (1981). (2) Tabak, H. H., et al., pp. 267–38 in *Test Protocols for Environmental Fate and Movement of Toxicants.* Proc Symp Assoc Official Analyt Chem, 94th Annual Mtg, Washington, DC (1981).

11. (1) GEMS. *Graphic Exposure Modeling System.* CLOGP USEPA (1987). (2) Callahan, M. A., et al., *Water Related Fate of 129 Priority Pollutants.* Volume II. USEPA–440/4–79–029B (1979). (3) Lyman, W. J., et al., *Handbook of Chemical Property Estimation Methods.* New York: McGraw-Hill, pp. 5–1 to 5–30 (1982).

12. 54 FR 33418 (8/14/89).

13. 40 CFR 712.30 (7/1/88).

14. 40 CFR 261.33 (7/1/88).

■ CHLOROPENTAFLUORO-ETHANE

CAS # 76–15–3

SYNONYMS ethane, chloropentafluoro • f-115 • fc-115 • fluorocarbon-115 • freon-115 • genetron-115 • monochloropentafluoroethane • propellant-115 • r-115 • refrigerant-115 • halocarbon-115 • chloropentafluorethane (French) • chloropentafluoretano (Spanish)

MF C_2ClF_5

MW 154.47

COLOR AND FORM Colorless gas.

ODOR Odorless.

DENSITY 1.526 kg/L (liquid at −20°C).

SURFACE TENSION 5 dynes/cm at 25°C.

VISCOSITY 0.193 cP at 25°C (liquid); 0.0125 cP at 25°C, 101.3 kPa (vapor).

BOILING POINT −38°C at 760 mm Hg.

MELTING POINT −106°C.

HEAT OF VAPORIZATION 126.0 kJ/kg at boiling point.

FIREFIGHTING PROCEDURES Extinguish fire using agent suitable for type of surrounding fire (material itself does not burn or burns with difficulty). Cool all affected containers with flooding quantities of water. Apply water from as far a distance as possible. [R12, 472].

TOXIC COMBUSTION PRODUCTS Under certain conditions, fluorocarbon vapors may decompose on contact with flames or hot surfaces, creating the potential hazard of inhalation of toxic decomposition products [R5, 3101]. All fluorocarbons will undergo thermal decomposition when exposed to flame or red-hot metal. Decomposition products of the chlorofluorocarbons will include hydrofluoric and hydrochloric acid along with smaller amounts of phosgene and carbonyl fluoride. The last compound is very unstable to hydrolysis and quickly changes to hydrofluoric acid and carbon dioxide in the presence of moisture (fluorocarbons) [R10]. In contact with open flame or very hot surface fluorocarbons may decompose into highly irritant and toxic gases: chlorine, hydrogen fluoride or chloride, and even phosgene (fluorocarbon refrigerant and propellants) [R11]. Under certain conditions, fluorocarbon vapors may decompose on contact with flames or hot surfaces, creating the potential hazard of inhalation of toxic decomposition products. (fluorocarbons) [R8, 3101].

SOLUBILITY Soluble in alcohol, ether; 0.006% in water at 25°C, 101.3 kPa.

VAPOR PRESSURE 804.6 kPa; 116.7 psia at 21.1°C.

CORROSIVITY Does not attack metals except at elevated temperatures.

STABILITY AND SHELF LIFE Has good thermal stability [R6, 269].

OTHER CHEMICAL/PHYSICAL PROPERTIES Dielectric constant 1.0035 at 27°C, 50.65 kPa (vapor). Critical volume: 259 cm³/mol; critical density: 0.596 g/cm³; specific heat: 0.285 cal/g at 25°C (liq). Henry's Law constant: 3.0 atm-m³/mole at 25°C(est). Dipole moment: 0.52 debye. Ozone depletion potential: 0.5. (Ozone depletion potential relative to R11 = 1.0. Scientific assessment of ozone: 1989.) (from table).

PROTECTIVE EQUIPMENT Forced-air ventilation at level of vapor concentration, together with use of individual breathing devices with independent air supply will minimize risk of inhalation. Lifelines should be worn when entering tanks or other confined spaces. [R5, 3101].

PREVENTATIVE MEASURES Filling areas should be monitored to ensure ambient concentration of fluorocarbons does not exceed 1,000 ppm. Inhalation of fluorocarbon vapors should be avoided. [R5, 3101].

DISPOSAL At the time of review, criteria for land treatment or burial (sanitary landfill) disposal practices are subject to significant revision. Prior to implementing land disposal of waste residue (including waste sludge), consult with environmental regulatory agencies for guidance on acceptable disposal practices [R2].

COMMON USES Refrigerant [R1]. SRP: Propellant in aerosol food preparations (former use) [R6]. Dielectric gas [R7, 269]. Mechanical vapor compression systems use fluorocarbons for refrigeration and air conditioning and account for majority of refrigeration capability in U.S. Fluorocarbons are used as refrigerants in home appliances, mobile air-conditioning units, retail food refrigeration systems and chillers. (fluorocarbons) [R8, 3102].

OCCUPATIONAL RECOMMENDATIONS None listed.

OSHA 8-hr time-weighted average: 1,000

ppm (6,320 mg/m³); final rule limits were achieved by any combination of engineering controls, work practices, and personal protective equipment during the phase-in period, Sept 1, 1989 through Dec 30, 1992. Final rule limits became effective Dec 31, 1992 [R4].

THRESHOLD LIMIT 8-hr time-weighted average (TWA) 1,000 ppm, 6,320 mg/m³ (1981) [R58, 16].

BIOCONCENTRATION Based on a water solubility of 60 ppm at 25°C (1), the bioconcentration factor for chloropentafluoroethane can be estimated to be 61 from a recommended regression equation (2,SRC). A bioconcentration factor of 61 does not indicate significant bioconcentration in aquatic organisms (SRC) [R3].

TSCA REQUIREMENTS (a) After October 15, 1978, no person may manufacture, except to import, any fully halogenated chlorofluoroalkane for any aerosol propellant use except for use in an article that is a food, food additive, drug, cosmetic, or device exempted under 15 U.S.C. 2602 (21 CFR 2.125), for those essential uses listed in 40 CFR 762.58, for exempted uses listed in 40 CFR 762.59. (b) After December 15, 1978, no person may import into the customs territory of the United States any fully halogenated chlorofluoroalkane, whether as a chemical substance, or as a component of a mixture or article, for any aerosol propellant use except for use in an article that is a food, food additive, drug, cosmetic, or device exempted under 15 U.S.C. 2602 (21 CFR 2.125), for those essential uses listed in 40 CFR 762.58, for exempted uses listed in 40 CFR 762.59. (c) Every person manufacturing fully halogenated chlorofluoroalkanes for aerosol propellant uses after October 15, 1978 must obtain a signed statement for aerosol propellant uses permitted under 40 CFR Part 762 or 21 CFR 2.125, or for other uses. (fully halogenated chlorofluoroalkanes) [R5].

FDA Essential uses of chlorofluorocarbons: (1) Metered-dose steroid human drugs for nasal inhalation; (2) metered-dose steroid human drugs for oral inhalation; (3) metered-dose adrenergic broncho-

dilator human drugs for oral inhalation; (4) contraceptive vaginal foams for human use; (5) metered-dose ergotamine tartrate drug products administered by oral inhalation for use in humans; (6) intrarectal hydrocortisone acetate for human use; (7) polymyxin B sulfate-bacitracin zinc-neomycin sulfate soluble antibiotic powder without excipients, for topical use on humans; (8) anesthetic drugs for topical use on accessible mucous membranes of humans where a cannula is used for application; (9) metered-dose nitroglycerin human drugs administered to the oral cavity; (10) metered-dose cromolyn sodium human drugs administered by oral inhalation; (11) metered-dose ipratropium bromide for oral inhalation; (12) metered-dose atropine sulfate aerosol human drugs administered by oral inhalation. (chlorofluorocarbons) [R12].

DOT *Fire or Explosion:* Some of these materials may burn, but none of them ignites readily. Cylinder may explode in heat of fire. [R9].

Health Hazards: Vapors may cause dizziness or suffocation. Contact with liquid may cause frostbite. Fire may produce irritating or poisonous gases. [R9].

Emergency Action: Keep unnecessary people away; isolate hazard area and deny entry. Stay upwind; keep out of low areas. Positive-pressure self-contained breathing apparatus (SCBA) and structural firefighters' protective clothing will provide limited protection. Isolate for 1/2 mile in all directions if tank, rail car, or tank truck is involved in fire. Call CHEMTREC at 1–800–424–9300 as soon as possible, especially if there is no local hazardous-materials team available. [R9].

Fire: Small Fires: Dry chemical or CO₂. Large Fires: Water spray, fog, or regular foam. Move container from fire area if you can do it without risk. Apply cooling water to sides of containers that are exposed to flames until well after fire is out. Stay away from ends of tanks. Withdraw immediately in case of rising sound from venting safety device or any discoloration of tank due to fire. Some of these materials, if spilled, may evaporate, leaving a flammable residue. [R9].

Spill or Leak: Stop leak if you can do it without risk. [R9].

First Aid: Move victim to fresh air and call emergency medical care; if not breathing, give artificial respiration; if breathing is difficult, give oxygen. [R9].

REFERENCES

1. Hawley, G. G. *The Condensed Chemical Dictionary.* 10th ed. New York: Van Nostrand Reinhold Co., 1981. 240.

2. SRP.

3. (1) Roy, W. R., R. A. Griffin. *Environ Geol Water Sci* 17: 241–7 (1985). (2) Lyman, W. J., et al., *Handbook of Chem Property Estimation Methods.* New York: McGraw-Hill, pp. 5–10 (1982).

4. 29 CFR 1910.1000 (7/1/91).

5. 40 CFR 762.45 (7/1/91).

6. Sittig, M. *Handbook of Toxic and Hazardous Chemicals and Carcinogens.* 1985. 2nd ed. Park Ridge, NJ: Noyes Data Corporation, 1985. 237.

7. Sax, N.I., and R.J. Lewis, Sr. (eds.). *Hawley's Condensed Chemical Dictionary.* 11th ed. New York: Van Nostrand Reinhold Co., 1987.

8. Clayton, G. D., and F. E. Clayton (eds.). *Patty's Industrial Hygiene and Toxicology:* Volume 2A, 2B, 2C: Toxicology. 3rd ed. New York: John Wiley and Sons, 1981–1982.

9. U.S. Department of Transportation. *Emergency Response Guidebook 1990.* DOT P 5800.5. Washington, DC: U.S. Government Printing Office, 1990, p. G-12.

10. International Labour Office. *Encyclopedia of Occupational Health and Safety.* Vols. I & II. Geneva, Switzerland: International Labour Office, 1983. 897.

11. Gosselin, R.E., R.P. Smith, H.C. Hodge. *Clinical Toxicology of Commercial Products.* 5th ed. Baltimore: Williams and Wilkins, 1984, p. II-159.

12. 21 CFR 2.125 (4/1/91).

■ CHLOROQUINE

CAS # 54–05–7

SIC CODES 2834

SYNONYMS 1,4-pentanediamine, N(4)-(7-chloro-4-quinolinyl)-N (1),N (1)-diethyl • amokin • aralen • arthrochin • artrichin • bemaphate • bipiquin • chingamin • chloraquine • chlorochin • delagil • gontochin • klorokin • mesylith • nivaquine-B • quinagamine • quinoline, 7-chloro-4-((4-(diethylamino)-1-methylbutyl) amino)- • resoquine • reumachlor • RP-3377 • sanoquin • sn-7,618 • tanakan • trochin • N4-(7-chloro-4-quinolinyl)-N1,N1-diethyl-1,4-pentanediamine • 7-chloro-4-(4-diethylamino-1-methylbutylamino) quinoline • SN-7618 • Capquin • (7-chloro-4-(4-diethylamino-1-methylbutylamino)-quinoline

MF $C_{18}H_{26}ClN_3$

MW 319.92

COLOR AND FORM White to slightly yellow, crystalline powder; colorless crystals.

ODOR Odorless.

TASTE Bitter taste.

MELTING POINT 87–92°C.

SOLUBILITY Very slightly soluble in water; soluble in dilute acids, chloroform, ether; insoluble in alcohol, benzene, chloroform, ether.

STABILITY AND SHELF LIFE Stable to heat in solns of pH 4.0 to 6.5 (chloroquine diphosphate) [R1].

OTHER CHEMICAL/PHYSICAL PROPERTIES Bitter colorless crystals; freely soluble in water, less soluble at neutral and alkaline pH; practically insoluble in alcohol, benzene, chloroform, ether; pH of 1% solution about 4.5; dimorphic, MP: 193–195°C, and 215–218°C (chloroquine diphosphate).

COMMON USES Medication: antimalarial; antiamebic. Antirheumatic. Lupus erythematosus suppressant [R1].

GENERAL SENSATION When used as a medicinal, may cause G.I. disturbances.

ACUTE HAZARD LEVEL Ingestion can cause gastric lesions at 50 mg/kg level. Toxicity may well be related to gastric activity [R3].

DEGREE OF HAZARD TO PUBLIC HEALTH Toxic vapors emitted upon burning. Internal injury may follow ingestion at low levels.

OCCUPATIONAL RECOMMENDATIONS None listed.

ACTION LEVEL Warn downstream water supply intakes.

ON-SITE RESTORATION Activated carbon or peat moss; seek professional environmental engineering assistance through EPA's Environmental Response Team (ERT), Edison, NJ, 24-hour no. 908–548–8730.

BEACH OR SHORE RESTORATION Do not burn.

AVAILABILITY OF COUNTERMEASURE MATERIAL Activated carbon—water treatment plants, sugar refineries; peat moss—nurseries.

MAJOR WATER USE THREATENED Potable supply.

PROBABLE LOCATION AND STATE As the diphosphate, $C_{18}H_{26}ClN_3 \cdot 2H_3PO_4$, colorless crystals. Dimorphic. Will dissolve in water column.

COLOR IN WATER Colorless.

HUMAN TOXICITY Fatalities have been reported following the accidental ingestion of relatively small doses of chloroquine (e.g., 750 mg, or 1 g of chloroquine phosphate in a 3-year-old child) [R4, 428].

IARC SUMMARY No data are available for humans. Inadequate evidence of carcinogenicity in animals. Overall evaluation, Group 3: The agent is not classifiable as to its carcinogenicity to humans [R2].

BIOLOGICAL HALF-LIFE The plasma half-life of chloroquine in healthy individuals is generally reported to be 72–120 hr. In one study, serum concentration of chloroquine appeared to decline in a biphasic manner, and the serum half-life of the terminal phase increased with higher dosage of the drug. In this study, the terminal half-life of chloroquine was 3.1 hr after a single 250-mg oral dose, 42.9 hr after a single 500-mg oral dose, and 312 hr after a single 1-g oral dose of the drug. [R4, 425].

FDA REQUIREMENTS Manufacturers, packers, and distributors of drug and drug products for human use are responsible for complying with the labeling, certification, and usage requirements as prescribed by the Federal Food, Drug, and Cosmetic Act, as amended (secs 201–902, 52 Stat. 1040 et seq., as amended; 21 U.S.C. 321–392). [R66].

REFERENCES

1. Budavari, S. (ed.). *The Merck Index—Encyclopedia of Chemicals, Drugs, and Biologicals.* Rahway, NJ: Merck and Co., Inc., 1989. 334.

2. IARC. *Monographs on the Evaluation of the Carcinogenic Risk of Chemicals to Man.* Geneva: World Health Organization, International Agency for Research on Cancer, 1972-present (multivolume work), p. S7 60 (1987).

3. Shriver, D. A., C. B. White, A. Sandor, and M. E. Rosenthale, "A Profile of the Rat Gastrointestinal Toxicity of Drugs Used to Treat Inflammatory Diseases," *Toxicology and Applied Pharmacology*, 32, pp. 73–83, 1975.

4. Hovious, I. C., G. T. Waggy, and R. A. Conway. *Identification and Control of Petrochemical Pollutants Inhibitory to Anaerobic Processes.* Environmental Protection Agency, EPA-R2-73-194, April 1973.

■ CHLOROTETRAFLUORO-ETHANE

CAS # 63938–10–3

DOT # 1021

SYNONYMS HCFC-124a

MF C_2HClF_4

MW 136.48

BOILING POINT: −12°C (1-chloro-1,2,2,2-tetrafluoroethane); −10.2 °C (HCFC-124a).

MELTING POINT: −199°C (1-chloro-1,2,2,2-tetrafluoroethane); −117°C (HCFC-124a). Also, 1.364 g/cm³ at 25°C (liquid density) (1-chloro-1,2,2,2-tetrafluoroethane); 1.379 g/cm³ at 20°C (liquid density) (HCFC-124a).

VISCOSITY 0.0138 mPa at 60°C (gas); 0.0314 mPa at 25°C (liquid) (1-chloro-1,2,2,2-tetrafluoroethane).

HEAT OF VAPORIZATION 167.9 kJ/kg (1-chloro-1,2,2,2-tetrafluoroethane).

FIREFIGHTING PROCEDURES If material involved in fire: Extinguish fire using agent suitable for type of surrounding fire (material itself does not burn or burns with difficulty). Cool all affected containers with flooding quantities of water. Apply water from as far a distance as possible. (refrigerants NEC, gas, or liquid, nonflammable; refrigerant, gas, NOS, or dispersant gas, NOS) [R2].

TOXIC COMBUSTION PRODUCTS All fluorocarbons will undergo thermal decomposition when exposed to flame or red-hot metal. Decomposition products of the chlorofluorocarbons will include hydrofluoric and hydrochloric acid along with smaller amounts of phosgene and carbonyl fluoride. The last compound is very unstable to hydrolysis and quickly changes to hydrofluoric acid and carbon dioxide in the presence of moisture. (fluorocarbons) [R3].

SOLUBILITY Solubility in water = 1.71 wt% at 24°C (1-chloro-1,2,2,2-tetrafluoroethane).

OTHER CHEMICAL/PHYSICAL PROPERTIES Heat capacity of liquid = 1.130 kJ/(kg–K) at 25°C; heat capacity of vapor = 0.741 kJ/(kg–K) at 101.3 kPa and 25°C; ozone depletion potential = 0.022; global warming potential = 0.11 (1-chloro-1,2,2-tetrafluoroethane).

PROTECTIVE EQUIPMENT Many of the fluorocarbons are good solvents of skin oil, so protective ointment should be used. (fluorocarbons) [R8, 544].

PREVENTATIVE MEASURES Sufficient exhaust and general ventilation should be provided to keep vapor concentration below recommended levels. (fluorocarbons) [R3].

COMMON USES The HCFCs are used as alternatives to CFCs in applications such as refrigerants, blowing agents, cleaning agents, and fire extinguishants (HCFCs) [R1]. HCFC–124 (1-chloro-1,2,2,2-tetrafluoroethane) is a potential substitute for CFC–11 and CFC–12 use as blowing agent in polyurethane foams [R5]. HCFC–124 (1-chloro 1,2,2,2-tetrafluoroethane) is a potential substitute for CFCs in aerosols, foams, and refrigerants [R6]. Proportional uses: Refrigeration/air conditioning, 43%; foam blowing agents, 20%; polymer precursors, 13%; solvent cleaning, 12%; aerosol propellants, 2%; medical equipment sterilization, 3%; other, 7%. (fluorocarbons) [R7].

BIOCONCENTRATION Based upon an estimated log K_{ow} of 1.86 (1), the BCF of chlorotetrafluoroethane can be estimated to be about 16 from a recommended regression-derived equation (2,SRC). This estimated BCF value suggests that bioconcentration in aquatic organisms is not an important fate process (SRC). [R4].

GENERAL RESPONSE Avoid contact with liquid or vapor. Keep people away. Stay upwind; keep out of low areas. Wear self-contained breathing apparatus and full protective clothing. Stop leak if possible.

FIRE RESPONSE Not flammable; container may explode in heat of fire. Fire may produce irritating or toxic gases. Move container from fire area if you can do so without risk. Stay away from ends of tanks. Cool containers that are exposed to flames with water from the side until well after fire is out. Withdraw immediately in case of rising sound from venting safety device or any discoloration of tanks due to fire.

EXPOSURE RESPONSE Call for medical aid. Vapor: vapors may cause dizziness or suffocation. Move victim to fresh air. If not breathing, give artificial respiration. If breathing is difficult, give oxygen. Liquid: contact with liquid may cause frostbite. Remove contaminated clothing and shoes. Flush affected areas with lukewarm water. Do not use hot water.

EXPOSURE SYMPTOMS Displaces air such that oxygen content may become too low to support life. Prolonged exposure can cause narcotic effect or rapid suffocation. Contact with liquid may cause frostbite.

EXPOSURE TREATMENT Call a physician. Inhalation: Move victim to fresh air. If breathing has stopped, give artificial respiration. If breathing is difficult, give oxygen. Skin or eyes: Remove contaminated clothing and shoes, flush affected area with lukewarm water. Do not use hot water.

REFERENCES

1. *Kirk–Othmer Encyclopedia of Chemical Technology.* 4th ed. Volumes 1: New York, NY. John Wiley and Sons, 1991–Present. 511.

2. Association of American Railroads. *Emergency Handling of Hazardous Materials in Surface Transportation.* Washington, DC: Assoc. of American Railroads, Hazardous Materials Systems (BOE), 1987. 599.

3. International Labour Office. *Encyclopedia of Occupational Health and Safety.* Vols. I and II. Geneva, Switzerland: International Labour Office, 1983. 897.

4. Meylan W.M., P.H. Howard, *J Pharm Sci*, Nov 1 (1994) (2) Lyman, W. J., et al., *Handbook of Chemical Property Estimation Methods* Washington, DC: Amer Chemical Soc, p. 5–4, 5–10 (1990).

5. Taverna, M., L. Hufnagel, pp. 41–7 in *Polyurethanes*, World Congr Proc, SPI/ISOPA 1991 (1991).

6. Denelzen, M.G.J., et al., *Role of CFC's, Substitutes and Other Halogenated Chemicals in Climatic Change.* RIVM-222901002 (NTIS PB91-164897). Bilthoven, Netherlands: Rijksinst, Volksgezond, Milieuhyg (1991).

7. Chemical Marketing Reporter; *Chemical Profile: Fluorocarbons*, Mar 16, 1992.

8. Zenz, C. *Occupational Medicine— Principles and Practical Applications.* 2nd ed. St. Louis, MO: Mosby-Yearbook, Inc, 1988.

■ CHLOROXURON

CAS # 1982–47–4

SYNONYMS 1-(4-(4-chloro-phenoxy) phenyl)-3,3-d'methyluré (French) • 3-(4-(4-chloor-fenoxy)-fenoxy)-fenyl)-1,1-dimethylureum (Dutch) • 3-(4-(4-Chlorphenoxy)-phenyl)-1,1-dimethylharnstoff (German) • 3-(4-(4-cloro-fenossil)-1,1-dimetil-urea (Italian) • 3-(p-(p-chlorophenoxy) phenyl)-1,1-dimethylurea • c-1983 • chloroxifenidim • ciba-1983 • N'-4-(4-chlorophenoxy) phenyl-N, N-dimethylurea • norex • tenoran • urea, 3-(p-(p-chlorophenoxy) phenyl)-1,1-dimethyl • urea, N'-(4-(4-chlorophenoxy)phenyl)-N, N-dimethyl • gesamoos • chlorphencarb • 3-(4-(4-chlorophenoxy) phenyl-1,1-dimethylurea • c-1933 • norex

MF $C_{15}H_{15}ClN_2O_2$

MW 290.75

COLOR AND FORM White crystals; colorless powder; colorless crystals.

ODOR Odorless.

DENSITY 1.34 g/cm^3 (20°C).

MELTING POINT 151–152°C.

SOLUBILITY at 20°C: 4 mg/L water; 44 g/kg acetone; 106 g/kg dichloromethane; 35 g/kg methanol; 4 g/kg toluene. Soluble in dimethylformamide and chloroform. Slightly soluble in benzene and diethyl ether.

VAPOR PRESSURE 239 nPa at 20°C.

CORROSIVITY Noncorrosive.

STABILITY AND SHELF LIFE Very stable under normal conditions. Sensitive to light [R5].

PROTECTIVE EQUIPMENT The use of personal protective equipment such as glasses, synthetic gloves, and face masks is important. (herbicides) [R4].

PREVENTATIVE MEASURES Only clean clothing and underwear should be worn and clothing should be changed daily for freshly washed clothes. Contaminated equipment should not be taken home. Adequate sanitary facilities and washing water should be provided for workers to wash before meals. Smoking and the consumption of alcoholic drinks before or during the handling of herbicides should be forbidden. Contaminated clothing should be removed immediately and a hot bath taken if possible. (herbicides) [R4].

DISPOSAL Incinerate in a unit with effluent gas scrubbing. Recommendable method: Incineration [R6].

COMMON USES Selective pre- and early post-emergent herbicide in soybeans, strawberries, various vegetable crops, and ornamentals (former use) [R1]. Selective herbicide, pre- and post-emergence control of annual broad-leaved weeds and some grasses in peas, beans, carrots, celery, celeriac, onions, leeks, garlic, chives, fennel, parsley, dill, tomatoes, cucurbits, soya beans, strawberries, ornamentals, fruit trees, and conifers. Control of mosses in ornamental and sports turf, on paths, and non-crop land, and in glasshouses (former use) [R2].

OCCUPATIONAL RECOMMENDATIONS None listed.

NON-HUMAN TOXICITY LD_{50} rat oral 3,000 mg/kg (technical chloroxuron) [R3]; LC_{50} rat inhalation >1.35 mg/L air/6 hr [R3]; LD_{50} rat oral 3,700 mg/kg, male [R11, 100]; LC_{50} rabbit, oral >10,000 mg/kg [R8].

ECOTOXICITY LC_{50} rainbow trout 0.43 mg/L/96 hr at 12°C (95% confidence limit 0.36–0.51 mg/L), wt 0.7 g. Static bioassay without aeration, pH 7.2–7.5, water hardness 40–50 mg/L as calcium carbonate, and alkalinity of 30–35 mg/L [R9]. LC_{50} channel catfish 0.43 mg/L/96 hr at 18°C (95% confidence limit 0.16–1.24 mg/L), wt 1.3 g. Static bioassay without aeration, pH 7.2–7.5, water hardness

40–50 mg/L as calcium carbonate, and alkalinity of 30–35 mg/L [R9]. LD_{50} mallard duck, oral greater than 2,000 mg/kg, 3–4-mo-old females (50% wettable powder) [R7]. LC_{50} killifish (*Oryzias latipes*) >15 ppm (conditions of bioassay not specified) [R11, 100]. LC_{50} crucian carp >150 mg/L (conditions of bioassay not specified) [R2]. LC_{50} bluegill 28 mg/L (conditions of bioassay not specified) [R3]. LC_{50} *Lepomis macrochirus* 25,000 µg/L/48 hr (conditions of bioassay not specified) [R8].

BIODEGRADATION Chloroxuron was N-demethylated by microorganisms from both Louisiana and Indiana soils. Metabolism was predominantly by direct hydrolysis to the aniline, which in turn underwent further transformations [R10].

FIRE POTENTIAL Nonflammable. [R11, 99].

REFERENCES

1. Spencer, E. Y. *Guide to the Chemicals Used in Crop Protection*. 7th ed. Publication 1093. Research Institute, Agriculture Canada, Ottawa, Canada: Information Canada, 1982. 119.

2. Hartley, D., and H. Kidd (eds.). *The Agrochemicals Handbook*. 2nd ed. Lechworth, Herts, England: The Royal Society of Chemistry, 1987, p. A083/Aug 87.

3. Worthing, C. R., and S. B. Walker (eds.). *The Pesticide Manual—A World Compendium*. 8th ed. Thornton Heath, UK: The British Crop Protection Council, 1987. 174.

4. International Labour Office. *Encyclopedia of Occupational Health and Safety*. Vols. I and II. Geneva, Switzerland: International Labour Office, 1983. 1039.

5. Hartley, D., and H. Kidd (eds.). *The Agrochemicals Handbook*. Old Woking, Surrey, United Kingdom: Royal Society of Chemistry/Unwin Brothers Ltd., 1983, p. A083/OCT 83.

6. United Nations. *Treatment and Disposal Methods for Waste Chemicals (IRPTC File)*. Data Profile Series No. 5. Geneva, Switzerland: United Nations Environmental Programme, Dec. 1985. 296.

7. U.S. Department of the Interior, Fish

and Wildlife Service. *Handbook of Toxicity of Pesticides to Wildlife.* Resource Publication 153. Washington, DC: U.S. Government Printing Office, 1984. 23.

8. Verschueren, K. *Handbook of Environmental Data of Organic Chemicals.* 2nd ed. New York, NY: Van Nostrand Reinhold Co., 1983. 389.

9. U.S. Department of Interior, Fish and Wildlife Service. *Handbook of Acute Toxicity of Chemicals to Fish and Aquatic Invertebrates.* Resource Publication No. 137. Washington, DC: U.S. Government Printing Office, 1980. 82.

10. Ross, J. A., B. G. Tweedy. *Soil Biol Biochemical* 5 (6): 739–46 (1973).

11. Weed Science Society of America. *Herbicide Handbook.* 5th ed. Champaign, Illinois: Weed Science Society of America, 1983.

■ CHROMIC OXIDE

CAS # 1308–38–9

SYNONYMS anhydride chromique (French) • chrome oxide • chromium oxide • chromium sesquioxide • chromium (3) oxide • chromium (3) trioxide • chromium (III) oxide (2: 3) • chromium oxide (Cr_2O_3) • dichromium trioxide • oxide of chromium • chromium oxide greens • c-Grun (German) • c-i-77288 • cosmetic hydrophobic green-9409 • cosmetic micro-blend chrome oxide 9229 • chromium oxide-X1134 • pure chromium oxide green-59 • green chrome oxide • green chromic oxide • chromic acid green • green chromium oxide • green cinnabar • green oxide of chromium • green oxide of chromium oc-31 • ci pigment green-17 • ci-77288 • chromium oxide green • chromium oxide pigment • chrome oxide green gn-m • chromia • casalis green • chrome green • ultramarine green • green rouge • leaf green • levanox green-ga • oil green • chrome ocher • 11661-green • anadonis green

MF Cr_2O_3

MW 152.00

COLOR AND FORM Light- to dark-green, fine, hexagonal crystals; green powder; bright-green crystals.

DENSITY 5.21.

BOILING POINT 4,000°C.

MELTING POINT 2,435°C.

SOLUBILITY Practically insoluble in water, alcohol, acetone.

STABILITY AND SHELF LIFE Exceptionally stable [R16, 469].

OTHER CHEMICAL/PHYSICAL PROPERTIES Turns brown on heating, but reverts to green color on cooling. Crystalline chromic oxide is extremely hard; will scratch quartz, topaz, zircon. Electrical resistivity: 1.3×10^9, 2.3×10^7, 6.8×10^3, and 4.5×10^3 microohm-cm at 350, 1,200,600, and 1,100°C, respectively. Calculated lattice energy: 15,276 kJ/mol; thermochemical cycle lattice energy: 114,957 kJ/mol. The 3 state is amphoteric (chromium(III)). Chromium(III) compounds are reduced to Cr(II) compounds by hypophosphites, electrolysis, or reducing metals, such as zinc, magnesium, and aluminum in acid solution (although it should be noted that Cr(II) compounds are stable only in the absence of air). In basic solution, Cr(III) is readily oxidized to chromium oxide (CrO_4^{-2}) by hypochlorite, hypobromite, peroxide, and oxygen under pressure at high temperature. In acid solution, Cr(III) is difficult to oxidize, and needs strong oxidizing agents, such as concentrated perchloric acid, sodium bismuthate, and permanganate (trivalent chromium). Cr(III) is most stable oxidation state of chromium and is always present as a coordination complex (trivalent chromium). Slightly soluble in acids, alkalies. Slightly soluble in water (trivalent chromium).

PROTECTIVE EQUIPMENT Employees should be provided with and required to use impervious clothing, gloves, face shields (eight-inch min), and other appropriate protective clothing necessary to prevent repeated or prolonged skin contact with solids or liquids containing insoluble chromium salts. Employees should be provided with and required to use dust- and splashproof safety goggles where solids or liquids containing insoluble chromium salts may contact the eyes. (chromium metal and insoluble chromium salts) [R15, 3].

PREVENTATIVE MEASURES Good industrial hygiene practices recommend that engineering controls be used to reduce environmental concentrations to the permissible exposure level. In addition to respirator selection, a complete respiratory protection program should be instituted that includes regular training sessions, maintenance, inspection, cleaning, and evaluation of the equipment. (chromium metal and insoluble chromium salts) [R15, 3].

CLEANUP Where possible, wet methods of cleaning should be used; at other sites, the only acceptable alternative is by vacuum cleaning. Spills of liquid or solid must be removed immediately to prevent dispersion as airborne dust (chromium compounds) [R16, 472]. If chromium metal or insoluble chromium salts are spilled, the following steps should be taken: 1. Remove all ignition sources where metallic chromium has been spilled. 2. Ventilate area of spill. 3. Collect spilled material in the most convenient and safe manner and deposit in sealed containers for reclamation or for disposal in a secured sanitary landfill. Liquid containing chromium metal or insoluble chromium salts should be absorbed in vermiculite, dry sand, earth, or a similar material. (chromium metal, and insoluble chromium salts) [R15, 4].

DISPOSAL SRP: At the time of review, criteria for land treatment or burial (sanitary landfill) disposal practices are subject to significant revision. Prior to implementing land disposal of waste residue (including waste sludge), consult with environmental regulatory agencies for guidance on acceptable disposal practices.

COMMON USES In abrasives and electric semiconductors; in alloys [R1]; in printing fabrics and banknotes [R1]; used in dyeing polymers; colorant for latex paints [R2]; in

manufacture of chromium metal and aluminum-chromium master alloys [R2]; agent in metallurgy [R3]; anhydrous chromic oxide is used in applications requiring heat, light, and chemical resistance (e.g., in glass and ceramics). It has special use in coloring portland cement, granules for asphalt roofing, and in camouflage paints [R4]. It is used as catalyst in preparation of methanol, butadiene, and high-density polyethylene. Used in refractory brick as minor component to improve performance. When used as mild abrasive for polishing jewelry and fine metal parts, it is known as "green rouge" [R4]. Hydrated chromic oxide is used as green pigment, esp for automotive finishes [R4]. Chromium and its compounds are used in metal alloys such as stainless steel; protective coatings on metal; magnetic tapes; and pigments for paints, cement, paper, rubber, composition floor covering, and other materials. Other uses include organic chemical synthesis, photochemical processing, and industrial water treatment. In medicine, chromium compounds are used in astringents, and antiseptics (chromium and its compounds) [R5]. Constituent of inorg pigments [R6]. Sensitizer in photographic industry; preparation of chromates. (total chromium) [R14, 120].

OCCUPATIONAL RECOMMENDATIONS None listed.

OSHA 8-hr time-weighted average: 1 mg/m^3(chromium, metal, and insoluble salts (as Cr)) [R8].

THRESHOLD LIMIT 8-hr time-weighted average (TWA) 0.5 mg/m^3 (1981) (chromium(III) compounds, as Cr) [R18, 16].

INTERNATIONAL EXPOSURE LIMITS Max allowable concentration (former USSR) 0.01 mg/m^3 [R16, 469].

BIOLOGICAL HALF-LIFE The elimination curve for chromium as measured by whole-body counting has an exponential form. In rats, 3 different components of the curve have been identified with the half-times of 0.5, 5.9, and 83.4 days, respectively. (chromium) [R17, 193].

BIOCONCENTRATION Generally, it can be stated that there is little tendency for chromium to accumulate along food chains in the trivalent inorganic form. (trivalent chromium) [R7].

ATMOSPHERIC STANDARDS Chromium has been designated as a hazardous air pollutant under section 112 of the Clean Air Act. (total chromium) [R9].

RCRA REQUIREMENTS A solid waste containing chromium (such as chromic oxide) may or may not become characterized as a hazardous waste when subjected to the toxicant extraction procedure listed in 40 CFR 261.24, and, if so characterized, must be managed as a hazardous waste [R10].

FDA Certification of the color additive (chromium oxide green) is not necessary for the protection of the public's health, and, therefore, batches thereof are exempt from certification pursuant to section 706 (C) of the act [R11]. Chromium oxide green is safe for use in coloring externally applied drugs, including those intended for use in the area of the eyes, in amounts consistent with good manufacturing practice [R11]. The color additive (chromium oxide green) and any mixture prepared therefrom intended solely, or in part, for coloring purposes shall bear, in addition to any information required by law, labeling in accordance with 21 CFR 70.25 [R11]. Bottled water shall, when a composite of analytical units of equal volume from a sample is examined by the methods described in paragraph (d) (1) (II) of this section, meet the standards of chemical quality, and shall not contain total chromium in excess of 0.05 mg/L (total chromium) [R12]. The color additives F, D, and C Green no. 3 were free from impurities other than those named (which include chromium, not more than 50 ppm) to the extent that such other impurities may be avoided by current manufacturing practice. (chromium (as Cr) salts) [R13].

REFERENCES

1. *The Merck Index.* 10th ed. Rahway, NJ: Merck Co., Inc., 1983. 316.

2. IARC. *Monographs on the Evaluation of the Carcinogenic Risk of Chemicals to Man.* Geneva: World Health Organization, International Agency for Research on Cancer, 1972-present (multivolume work), p. V2 104 (1973).

3. SRI.

4. IARC. *Monographs on the Evaluation of the Carcinogenic Risk of Chemicals to Man.* Geneva: World Health Organization, International Agency for Research on Cancer, 1972–present (multivolume work), p. V23 226 (1980).

5. DHHS/NTP. *Fourth Annual Report On Carcinogens,* p. 58 (1985) NTP 85–002.

6. Sax, N. I., and R. J. Lewis, Sr. (eds.). *Hawley's Condensed Chemical Dictionary.* 11th ed. New York: Van Nostrand Reinhold Co., 1987. 280.

7. Nat'l Research Council Canada. *Effects of Chromium in the Canadian Environment,* p. 49 (1976) NRCC No 15017.

8. 29 CFR 1910.1000 (7/1/88).

9. 40 CFR 61.01 (7/1/88).

10. 40 CFR 261.24 (7/1/88).

11. 21 CFR 73.1327 (4/1/88).

12. 21 CFR 103.35 (4/1/88).

13. 21 CFR 74.203 (4/1/88).

14. Browning, E. *Toxicity of Industrial Metals.* 2nd ed. New York: Appleton-Century-Crofts, 1969.

15. Mackison, F. W., R. S. Stricoff, and L. J. Partridge, Jr. (eds.). *NIOSH/OSHA – Occupational Health Guidelines for Chemical Hazards.* DHHS(NIOSH) Publication No. 81–123 (3 vols). Washington, DC: U.S. Government Printing Office, Jan. 1981.

16. International Labour Office. *Encyclopedia of Occupational Health and Safety.* Vols. I & II. Geneva, Switzerland: International Labour Office, 1983.

17. Friberg, L., G.F. Nordberg, E. Kessler, and V. B. Vouk (eds). *Handbook of the Toxicology of Metals.* 2nd ed. Vols I, II.: Amsterdam: Elsevier Science Publishers B.V., 1986, p. V2.

18. American Conference of Governmental Industrial Hygienists. *Threshold Limit*

Values for Chemical Substances and Physical Agents and Biological Exposure Indices for 1994–1995. Cincinnati, OH: ACGIH, 1994.

■ CIS-DIAMMINEDICHLORO-PLATINUM

CAS # 15663–27–1

SYNONYMS cacp • cddp • cis-ddp • cis-diaminodichloroplatinum • cis-diaminodichloroplatinum(II) • cis-diamminedichloroplatinum(II) • cis-diammineplatinum(II) chloride • cis-dichlorodiaminoplatinum • cis-dichlorodiaminoplatinum(II) • cis-dichlorodiammine platinum(II) • cis-dichlorodiammineplatinum • cis-dichlorodiammineplatinum(II) • cis-platinous diaminodichloride • cis-platinous diammine dichloride • cis-platinum • cis-platinum(II) • cis-platinum diaminodichloride • cis-platinum diamminedichloride • cis-platinum(II) diaminedichloride • cis-platinum(II) diaminodichloride • cis-platinum(II) diamminedichloride • cisplatin • cisplatino (Spanish) • cisplatyl • cispt(II) • cpdc • cpdd • ddp • nci-c55776 • neoplatin • nsc-119875 • pdd • Peyrone's chloride • platinol • platinum diamminodichloride • platinum, diamminedichlo-, cis • pt-01 • cis-platin • ai3-62048 • cisplatine • platiblastin • platinex

MF $Cl_2H_6N_2{\bullet}Pt$

MW 300.05

COLOR AND FORM Deep-yellow solid; yellow crystals; white powder.

DENSITY 3.738 g/m³.

FIREFIGHTING PROCEDURES In firefighting, the minimum respiratory protection required above 0.002 mg/m³ is a self-contained breathing apparatus with a full facepiece operated in pressure-demand or other positive-pressure mode. (soluble platinum salts) [R1].

TOXIC COMBUSTION PRODUCTS Toxic gases and vapors (such as chlorine) may be released when chlorine-containing soluble platinum salts decompose. (soluble platinum salts) [R1].

SOLUBILITY Solubility in water 0.253 g/100 g at 25°C; insoluble in most common solvents except dimethyl formamide; soluble 1 (part) in 42 (parts) of dimethyl-primanide.

STABILITY AND SHELF LIFE Slowly changes to trans-form in aqueous solution [R10, 329].

OTHER CHEMICAL/PHYSICAL PROPERTIES Slowly changes to trans-form in aqueous solution; decomposes at 270°C.

PROTECTIVE EQUIPMENT Dispensers of liquid detergent should be available. Safety pipettes should be used for all pipetting. In animal laboratory, personnel should wear protective suits (preferably disposable, one-piece, and close-fitting at ankles and wrists), gloves, hair covering, and overshoes. In chemical laboratory, gloves and gowns should always be worn; however, gloves should not be assumed to provide full protection. Carefully fitted masks or respirators may be necessary when working with particulates or gases, and disposable plastic aprons might provide addnl protection. Gowns should be of distinctive color as a reminder that they are not to be worn outside the laboratory. (chemical carcinogens) [R8, 8].

PREVENTATIVE MEASURES Smoking, drinking, eating, storage of food, or of food and beverage containers, or utensils, and the application of cosmetics should be prohibited in any laboratory. All personnel should remove gloves, if worn, after completion of procedures in which carcinogens have been used. They should wash hands, preferably using dispensers of liquid detergent, and rinse thoroughly. Consideration should be given to appropriate methods for cleaning the skin, depending on nature of the contaminant. No standard procedure can be recommended, but the use of organic solvents should be avoided. Safety pipettes should be used for all pipetting. (chemical carcinogens) [R8, 8].

CLEANUP A high-efficiency particulate arrestor (HEPA) or charcoal filters can be used to minimize amount of carcinogen in exhausted-air-ventilated safety cabinets, lab hoods, glove boxes, or animal rooms. Filter housing that is designed so that used filters can be transferred into plastic bag without contaminating maintenance staff is avail commercially. Filters should be placed in plastic bags immediately after removal. The plastic bag should be sealed immediately. The sealed bag should be labelled properly. Waste liquids should be placed or collected in proper containers for disposal. The lid should be secured and the bottles properly labelled. Once filled, bottles should be placed in plastic bag, so that outer surface is not contaminated. The plastic bag should also be sealed and labelled. Broken glassware should be decontaminated by solvent extraction, by chemical destruction, or in specially designed incinerators. (chemical carcinogens) [R8, 15].

DISPOSAL PRECAUTIONS There is no universal method of disposal that has been proved satisfactory for all carcinogenic compounds and specific methods of chemical destruction published have not been tested on all kinds of carcinogen-containing waste. Summary of avail methods and recommendations given must be treated as guide only. (chemical carcinogens) [R8, 14].

COMMON USES Medication: antineoplastic agent [R10, 361]; medication: treatment of a variety of malignancies [R7]; cisplatin has been shown to have trypanocidal effects. [R10, 361].

OCCUPATIONAL RECOMMENDATIONS None listed.

OSHA 8-hr time-weighted average: 0.002 mg/m^3 (platinum soluble salts, as Pt) [R4]; NIOSH 10-hr time-weighted average: 0.002 mg/m^3 (platinum soluble salts, as Pt) [R5].

THRESHOLD LIMIT 8-hr time-weighted average (TWA) 0.002 mg/m^3 (1970) (platinum soluble salts, as Pt) [R11, 30].

INTERNATIONAL EXPOSURE LIMITS Romania: 0.003 mg/m^3 (platinum soluble salts, as Pt) [R6].

NON-HUMAN TOXICITY LD$_{50}$ rat oral approximately 20 mg/kg [R3].

IARC SUMMARY Inadequate evidence of carcinogenicity in humans. Sufficient evidence of carcinogenicity in animals. Overall evaluation: Group 2A: The agent is probably carcinogenic to humans [R2].

BIOLOGICAL HALF-LIFE After rapid iv admin (to human patients) the drug has an initial half-life in plasma of 25 to 50 min; concentration declines subsequently with a half-life of 58 to 73 hr. [R9, 1250].

OSHA 8-hr time-weighted avg: 0.002 mg/m^3 (platinum sol salts, as Pt) [R12].

NIOSH 10-hr time-weighted avg: 0.002 mg/m^3 (platinum sol salts, as Pt) [R13].

REFERENCES

1. Mackison, F. W., R. S. Stricoff, and L. J. Partridge, Jr. (eds.). *NIOSH/OSHA—Occupational Health Guidelines for Chemical Hazards*. DHHS (NIOSH) Publication No. 81-123 (3 vols). Washington, DC: U.S. Government Printing Office, Jan. 1981.

2. IARC. *Monographs on the Evaluation of the Carcinogenic Risk of Chemicals to Man*. Geneva: World Health Organization, International Agency for Research on Cancer, 1972–present (multivolume work), p. S7 73 (1987).

3. Seiler, H. G., H. Sigel, and A. Sigel (eds.). *Handbook on the Toxicity of Inorganic Compounds*. New York, NY: Marcel Dekker, Inc. 1988. 535.

4. 29 CFR 1910.1000 (7/1/90).

5. NIOSH. *NIOSH Pocket Guide to Chemical Hazards*. DHHS (NIOSH) Publication No. 90-117. Washington, DC: U.S. Government Printing Office, June 1990, 186.

6. American Conference of Governmental Industrial Hygienists. *Documentation of the Threshold Limit Values and Biological Exposure Indices*. 5th ed. Cincinnati,

OH: American Conference of Governmental Industrial Hygienists, 1986. 492.

7. IARC; IARC *Monographs on the Evaluation of the Carcinogenic Risk of Chemicals to Humans*, IARC, Lyon, France 26: 151–64 (1981).

8. Montesano, R., H. Bartsch, E. Boyland, G. Della Porta, L. Fishbein, R. A. Griesemer, A.B. Swan, L. Tomatis, and W. Davis (eds.). *Handling Chemical Carcinogens in the Laboratory: Problems of Safety*. IARC Scientific Publications No. 33. Lyon, France: International Agency for Research on Cancer, 1979.

9. Gilman, A.G., T.W. Rall, A.S. Nies, and P. Taylor (eds.). *Goodman and Gilman's The Pharmacological Basis of Therapeutics*. 8th ed. New York, NY. Pergamon Press, 1990.

10. Budavari, S. (ed.). *The Merck Index – Encyclopedia of Chemicals, Drugs and Biologicals*. Rahway, NJ: Merck and Co., Inc., 1989.

11. American Conference of Governmental Industrial Hygienists. *Threshold Limit Values for Chemical Substances and Physical Agents and Biological Exposure Indices for 1994–1995*. Cincinnati, OH: ACGIH, 1994.

12. 29 CFR 1910.1000 (7/1/90).

13. NIOSH. *NIOSH Pocket Guide to Chemical Hazards*. DHHS (NIOSH) Publication No. 90–117. Washington, DC: U.S. Government Printing Office, June 1990 186.

■ CLOPYRALID

CAS # 1702–17–6

SYNONYMS 3,6-dichloro-2-pyridinecarboxylic acid • 3,6-dichloropicolinic acid • 3,6-dcp • dowco-290 • lontrel • shield • reclaim • acide dichloro-3,6 picolinique (French) • 3,6-dichloropyridine-2-carboxylic acid • format • lontrel-sf-100 • benzalox • campaign • crusader-s • escort • cirtoxin • matrigon • cyronal

MF $C_6H_3Cl_2NO_2$

MW 192.0

COLOR AND FORM White crystalline solid; colorless crystals.

ODOR Odorless.

DISSOCIATION CONSTANTS $pK_a = 2.3$.

MELTING POINT 151–152°C.

SOLUBILITY Approximately 1,000 ppm in water at 25°C; solubility at 25°C: >25/5 w/w in methanol, acetone, xylene; solubility (20°C): 9 g/kg water; 250 g/kg acetone, cyclohexanone; <11 g/kg xylene. Solubility at 25°C: acetone >25 g/100 ml; methanol >15 g/100 ml; xylene <0.5 g/100 ml.

VAPOR PRESSURE 1.2×10^{-5} mm Hg at 25°C.

CORROSIVITY Solutions are corrosive to aluminum, steel, and tin plate. (clopyralid salts).

STABILITY AND SHELF LIFE Stable in acidic media. Stable in UV irradiation [R1]. Storage stability: stable, excess of 2 yr. [R2].

OTHER CHEMICAL/PHYSICAL PROPERTIES All formulations are compatible with most types of hard water.

PROTECTIVE EQUIPMENT Wear goggles when handling concentrates [R2].

PREVENTATIVE MEASURES SRP: The scientific literature supports the wearing of contact lenses in industrial environments, as part of a program to protect the eye against chemical compounds and minerals causing eye irritation. However, there may be individual substances whose irritating or corrosive properties are such that the wearing of contact lenses would be harmful to the eye. In those specific cases contact lenses should not be worn.

COMMON USES Herbicide [R3]. Post-emergence control of many annual and perennial broad-leaved weeds of the families Polygonaceae, Compositae, Leguminosae, and Umbelliferae, in sugar beet, fodder beet, oilseed rape, maize, brassicas, onions, leeks, strawberries, flax, and grassland. Provides particularly good control of creeping thistle (*Cirsium arvense*), perennial sow-thistle, coltsfoot, mayweeds, and Polygonum species [R1].

OCCUPATIONAL RECOMMENDATIONS None listed.

NON-HUMAN TOXICITY LD_{50} rat male oral >5,000 mg/kg [R3]; LD_{50} rat female oral >4,300 mg/kg [R3]; LD_{50} rabbit percutaneous >2,000 mg/kg [R1]; LD_{50} mice oral >5,000 mg/kg [R1].

ECOTOXICITY LC_{50} rainbow trout 103.5 mg/L/96 hr (conditions of bioassay not specified) [R3]; LC_{50} bluegill sunfish 125.4 mg/L/96 hr (conditions of bioassay not specified) [R1]; LD_{50} bee oral >100 μg/bee/48 hr (conditions of bioassay not specified) [R1].

BIOLOGICAL HALF-LIFE 3,6-Dichloropicolinic (acid) degrades at a medium to fast rate with an average half-life range of 12 to 70 days in a wide range of soils across the U.S. [R2].

BIODEGRADATION Although some investigators found clopyralid to be biodegradable in field soils (2,3), others found that the herbicide was relatively persistent in field soil (4). Thus the biodegradation of clopyralid appears to be soil dependent and the rate of biodegradation in soil may be enhanced both by higher temperature and higher number of organisms that are capable of degrading the herbicide (1). The half-lives of clopyralid in a clay, clay loam, and sandy loam soil at 20°C and 85% field moisture capacity were estimated to be 38 days, 13 days, and 36 days, respectively (1) [R4].

BIOCONCENTRATION A bioconcentration factor of 13 estimated from the water solubility of 1,000 mg/L and a regression equation suggests that bioconcentration of clopyralid in aquatic organisms may not be important (1,SRC) [R5].

FIFRA REQUIREMENTS Tolerances are established for residues of the herbicide clopyralid (3,6-dichloro-2-pyridinecarboxylic acid) in or on the following raw agricultural commodities: barley, cattle, eggs: goats, grasses, hogs, horses, milk, oats, poultry, sheep, sugar beets, and wheat [R6]. Interim tolerances are established for residues of the herbicide clopyralid (3,6-dichloro-2-pyridinecarboxylic acid) in or on the following raw agricultural commodities (expiration date: 6/15/91): field corn (grain, fodder, and forage) [R7]. Tolerances are established for residues of the herbicide clopyralid (3,6-dichloro-2-pyridinecarboxylic acid) in or on the following foods: barley, milled fractions (except flour); oats, milled fractions (except flour); and wheat, milled fractions (except flour) [R8]. An interim tolerance is established for residues of the herbicide clopyralid (3,6-dichloro-2-pyridinecarboxylic acid) in or on the following foods (expiration date: 6/15/91): field corn, milling fractions [R8]. Tolerances are established for residues of the herbicide clopyralid (3,6-dichloro-2-pyridinecarboxylic acid) in or on the following feeds: barley, milled fractions (except flour); oats, milled fractions (except flour); wheat, milled fractions (except flour); and sugar beet molasses [R9]. An interim tolerance is established for residues of the herbicide clopyralid (3,6-dichloro-2-pyridinecarboxylic acid) in or on the following feeds (expiration date: 6/15/91): field corn, milling fractions. [R10].

REFERENCES

1. Hartley, D., and H. Kidd (eds.). *The Agrochemicals Handbook*. 2nd ed. Lechworth, Herts, England: The Royal Society of Chemistry, 1987, p. A433/AUG 87.

2. Weed Science Society of America. *Herbicide Handbook*. 5th ed. Champaign, Illinois: Weed Science Society of America, 1983. 165.

3. Budavari, S. (ed.). *The Merck Index— Encyclopedia of Chemicals, Drugs, and Biologicals*. Rahway, NJ: Merck and Co., Inc., 1989. 375.

4. (1) Smith, A. E., A. J. Aubin. *Bull Environ Contam Toxicol* 42: 670–5 (1989). (2) Bovey, R. W., C. W. Richardson. *J Environ Qual* 20: 528–31 (1991). (3) Galoux, M. P., et al., *J Agric Food Chem*

33: 965–8 (1985). (4) Pik, A. J., et al. *J Agric Food Chem* 25: 1054–61 (1977).

5. Kenaga, E. E., *Ecotoxicol Environ Safety* 4: 26–38 (1980).

6. 40 CFR 180.431(a) (7/1/91).

7. 40 CFR 180.431(b) (7/1/91).

8. 40 CFR 185.1100(a) (7/1/91).

9. 40 CFR 185.1100(b) (7/1/91).

10. (1) Smith, A. E., A. J. Aubin. *Bull Environ Contam Toxicol* 42: 670–5 (1989). (2) Bovey, R. W., C. W. Richardson. *J Environ Qual* 20: 528–31 (1991). (3) Bergstrom, L., et al. *Environ Toxicol Chem* 10: 563–71 (1991). (2) Lyman, W. J., et al. *Handbook of Chemical Property Estimation Methods*. Washington, DC: Amer Chem Soc, pp. 15–16 (1990).

■ COBALT MONOSULFIDE

CAS # 1317–42–6

SYNONYMS Cobalt sulfide • cobalt(2) sulfide • cobaltous sulfide • sycoporite

MF CoS

MW 91.01

COLOR AND FORM Reddish or silver-white, octahedral crystals; black amorphous powder (alpha); forms Co(OH)S in air (alpha).

DENSITY 5.45 at 18°C.

MELTING POINT Greater than 1,116°C.

SOLUBILITY 0.00038 g/100 cc H_2O at 18°C; slightly soluble in acid.

STABILITY AND SHELF LIFE Solutions containing cobaltous ion, Co(2+), are relatively stable (cobaltous ion, Co(2+)) [R7, 246].

OTHER CHEMICAL/PHYSICAL PROPERTIES Soluble in hydrochloric acid (alpha); grey powder or reddish-silver octahedral crystals; MP above 1,100°C; density 5.45; practically insoluble in water; soluble in acids (beta); exists in two forms: alpha and beta. In its compounds, cobalt occurs normally in the oxidation states 2 and 3, more seldom in the oxidation states 0, 1, and 4.

In normal salts the bivalent form is more stable than the trivalent one, the reverse is true for cobalt complexes. (cobalt compounds).

PROTECTIVE EQUIPMENT For temporary operations (which produce dust or fume) or when ventilation is not practicable, an air-line respirator should be worn. If ventilation not satisfactory, dust or fume respirator can be used (cobalt, alloys, and compounds) [R12, 495]. The maintenance worker should wear protective clothing, personal protection equipment, incl eye protection, and suitable respiratory protective equipment. When the catalyst takes the form of a harmful gas or vapor, exhaust ventilation, breathing apparatus, and protective clothing should be provided (catalysts, cobalt compounds) [R12, 426]. Respirator selection: Upper-limit devices recommended by NIOSH: 0.5 mg/m³: any dust and mist respirator except single-use respirators; 1 mg/m³: any dust and mist respirator except single-use and quarter-mask respirators or any dust, mist, and fume respirator with a full facepiece or any supplied-air respirator or any self-contained breathing apparatus; 2.5 mg/m³: any powered air-purifying respirator with a dust and mist filter or any supplied-air respirator operated in a continuous flow mode or any powered air-purifying respirator with a dust, mist, and fume filter; 5 mg/m³: any air-purifying full facepiece respirator with a high-efficiency particulate filter or any self-contained breathing apparatus with a full facepiece or any supplied-air respirator with a full facepiece; 20 mg/m³: any supplied-air respirator with a full facepiece and operated in a pressure-demand or other positive-pressure mode; emergency or planned entry in unknown concn or IDLH conditions: any self-contained breathing apparatus with a full facepiece and operated in a pressure-demand or other positive-pressure mode or any supplied-air respirator with a full facepiece and operated in a pressure-demand or other positive-pressure mode in combination with an auxiliary self-contained breathing apparatus operated in pressure-demand or other positive-pressure mode; escape: any air-purifying full facepiece respirator with a high-efficiency particulate filter or any appropriate escape-type self-contained breathing apparatus (cobalt metal, fume, and dust (as Co)) [R9, 85]. Employees should be provided with and required to use impervious clothing, gloves, face shields (eight-inch minimum), and other appropriate protective clothing necessary to prevent repeated or prolonged skin contact with cobalt dust (cobalt metal fume and dust) [R10, 2].

PREVENTATIVE MEASURES Processes that produce cobalt dust or fume should be provided with an effective local exhaust ventilation (cobalt, alloys, and compound) [R12, 495]. Avoiding skin contact may be difficult but barrier creams can be tried. Severely affected patient must be removed to occupations not involving cobalt exposure (cobalt, alloys and compound) [R12, 495]. Contact lenses should not be worn when working with these chemicals (cobalt metal, fumes and dust) [R13, 3]. Contact lens use in industry is controversial. A survey of 100 corporations resulted in the recommendation that each company establish their own contact lens use policy. One presumed hazard of contact lens use is possible chemical entrapment. It was found that contact lenses minimized injury or protected the eye. The eye was afforded more protection from liquid irritants. Soft contact lenses do not worsen corneal damage from strong chemicals and in some cases could actually protect the eye. Overall, the literature supports the wearing of contact lenses in industrial environments as part of the standard eye protection, e.g., face shields; however, more data are needed to establish the value of contact lenses [R14]. Employees should wash promptly when skin becomes contaminated; work clothing should be changed daily if it is reasonably probable that the clothing is contaminated; promptly remove non-impervious clothing that becomes contaminated (cobalt metal, fume, and dust (as Co)) [R9, 85]. If employees' clothing has become contaminated with cobalt dust, employees should change into uncontaminated clothing before leaving the work premises. Clothing contaminated with cobalt dust should be placed in closed containers for storage until it can be discarded or until provision is made for the removal of cobalt dust from the clothing. If the clothing is to be laundered or otherwise cleaned to remove the cobalt dust, the person performing the operation should be informed of cobalt dust's hazardous properties. Non-impervious clothing that becomes contaminated with cobalt dust should be removed promptly and not reworn until the cobalt dust is removed from the clothing (cobalt dust) [R13, 3]. Skin that becomes contaminated with cobalt dust should be promptly washed or showered with soap or mild detergent and water to remove any cobalt dust. Eating and smoking should not be permitted in areas where cobalt metal fume or dust are generated, handled, processed, or stored. Employees who handle cobalt metal fume or dust should wash their hands thoroughly with soap or mild detergent and water before eating, smoking, or using toilet facilities (cobalt metal fume and dust) [R13, 3]. Employers should institute programs that emphasize good personal hygiene to prevent skin and respiratory irritation caused by cobalt-containing dusts. After working with cobalt products, workers should thoroughly wash their hands and face before drinking, eating, or smoking. If skin contact with cobalt solutions occurs, the worker should wash the affected skin promptly. The employer should provide showers if workers have substantial contact with cobalt. These workers should be encouraged to wash or shower after each workshift. Employers should prohibit smoking or carrying of tobacco products, and should prohibit eating, food handling, or food storage within the work area (cobalt and cobalt salts) [R15]. General plant maintenance must be conducted regularly to prevent cobalt-containing dusts from accumulating in work areas. Cleaning should be performed with vacuum pickup or wet mopping to minimize the amount of dust dispersed into the air. A decontamination room should be available for cleaning equipment that is to receive major overhaul or maintenance. Spills of cobalt-containing material should be promptly cleaned up to minimize inhalation or dermal contact (cobalt and cobalt salts) [R16]. Special precautions are necessary when workers must enter tanks or vessels, such as reaction vessels containing cobalt catalysts or vessels used to prepare cobalt salts. Before any worker enters a vessel, all sources for transferring cobalt and other materials into or out of the vessel must be blanked to prevent their entry. The vessel interior must then be washed with water and purged with air. After purging, check the vessel's atmosphere with suitable instruments to ensure that no hazards from fire, explosion, oxygen deficiency, or dust inhalation exist. No one should enter a tank or vessel without first being equipped with an appropriate respirator and a secured lifeline or harness. Mechanical ventilation should be provided continuously when workers are inside the tank. At least one other worker similarly equipped with respiratory protection, lifeline, and harness should watch at all times from outside the vessel. Workers inside the tank must be able to communicate with those persons outside. Other workers must be available to assist in an emergency. Flame- or spark-generating operations, such as welding or cutting, should be performed only when an authorized representative of the employer has signed a permit based on a finding that all necessary safety precautions have been taken (cobalt and cobalt salts) [R17].

CLEANUP Liquid material spills can be copiously flushed with water and channeled to a treatment system or holding tank for reclamation or proper disposal. Spills of dry material can be removed by vacuuming or wet mopping. Some spills can be removed by hosing, first with a mist of water to dampen the spilled material, and then with a more forceful stream that flushes it into a holding tank, or other facility for handling contaminated water. Work surfaces or contaminated clothing should never be cleaned by dry-sweeping or blowing with pressurized hoses. Recovery systems used to reclaim waste metals should comply with federal, state, and local regulations. All waste materials generated in the handling of cobalt-contain-

ing substances should be disposed of in compliance with federal, state, and local regulations. (cobalt and cobalt salts) [R4].

DISPOSAL SRP: At the time of review, criteria for land treatment or burial (sanitary landfill) disposal practices are subject to significant revision. Prior to implementing land disposal of waste residue (including waste sludge), consult with environmental regulatory agencies for guidance on acceptable disposal practices. Proper mixing of the cobalt waste and the soil is essential to preventing excessive plant accumulation of cobalt (cobalt) [R18, 256]. Cobalt metal may be recovered from scrap as alternative to disposal [R7, 255]. Cobalt compound may be recovered from spent catalysts as alternative to disposal. [R7, 255].

COMMON USES Catalyst for hydrogenation or hydrodesulfurization [R6, 348]. Sulfur compounds and NOx are removed from waste gas (such as claus tail gas or stack gas) to give product gas suitable for discharge into atmosphere. Cobalt monosulfate was used as a catalyst [R1]. The only disease in which the clinical use of cobalt is still advocated by some is the normochromic, normocytic anemia assoc with severe renal failure. (cobalt) [R2].

OCCUPATIONAL RECOMMENDATIONS None listed.

OSHA Permissible exposure limit (PEL): 8-hr TWA 0.1 mg/m^3 (cobalt metal, fume, and dust (as Co)) [R9, 84].

THRESHOLD LIMIT 8-hr time-weighted average (TWA) 0.02 mg/m^3 (1994) (cobalt, elemental, and inorganic compound, as Co) [R11, 17]. Excursion Limit Recommendation: Excursions in worker exposure levels may exceed three times the TLV–TWA for no more than a total of 30 min during a work day, and under no circumstances should they exceed five times the TLV–TWA. (cobalt, elemental, and inorganic compound, as Co) [R11, 5]. A3. A3 = Animal carcinogen (cobalt, elemental, and inorganic compound, as Co) [R11, 17].

INTERNATIONAL EXPOSURE LIMITS In the Federal Republic of Germany, industrial cobalt emission has been limited to 1 mg/m³ for metallic cobalt and its slightly soluble compounds and 50 mg/m³ for other compounds. (cobalt and compounds) [R10, 260].

BIOLOGICAL HALF-LIFE It appears that independent of exposure route (inhalation, injection, or ingestion), most of the cobalt will be eliminated rapidly (in humans). A small proportion is, however, eliminated slowly, having a biological half-time on the order of years. (cobalt) [R8, 218].

BIOCONCENTRATION Only a few plant species accumulate cobalt above the 100 ppm that causes severe phytotoxicity. Hyperaccumulators of cobalt have been found that contain over 1% cobalt in dry leaves. (cobalt salts) [R5].

FIRE POTENTIAL Cobalt(II) sulfide dried at 300°C is pyrophoric [R3].

REFERENCES

1. Hass, R. H. U.S. Patent, 4123507, 10/31/78 (Union Oil Co of California).

2. Gilman, A. G., L. S.Goodman, and A. Gilman (eds.). *Goodman and Gilman's The Pharmacological Basis of Therapeutics.* 7th ed. New York: Macmillan Publishing Co., Inc., 1985. 1319.

3. Bretherick, L. *Handbook of Reactive Chemical Hazards.* 3rd ed. Boston, MA: Butterworths, 1985. 1016.

4. NIOSH. *Criteria Document: Cobalt.* p. 37 (1981) DHEW Pub. NIOSH 82–107.

5. Parr, J. F., P. B. Marsh, and J. M. Kla (eds.). *Land Treatment of Hazardous Wastes.* Park Ridge, NJ: Noyes Data Corporation, 1983. 176.

6. *The Merck Index.* 10th ed. Rahway, New Jersey: Merck Co., Inc., 1983.

7. National Research Council. *Drinking Water and Health.* Volume 1. Washington, DC: National Academy Press, 1977.

8. Friberg, L., G.F. Nordberg, E. Kessler, and V. B. Vouk (eds.). *Handbook of the Toxicology of Metals.* 2nd ed. Vols I, II.: Amsterdam: Elsevier Science Publishers B.V., 1986, p. V2.

9. NIOSH. *Pocket Guide to Chemical Hazards.* 2nd Printing. DHHS (NIOSH) Publ. No. 85–114. Washington, DC: U.S. Dept. of Health and Human Services, NIOSH/Supt.of Documents, GPO, February 1987.

10. Seiler, H.G., H. Sigel, and A. Sigel, (eds.). *Handbook on the Toxicity of Inorganic Compounds.* New York, NY: Marcel Dekker, Inc. 1988.

11. American Conference of Governmental Industrial Hygienists. *Threshold Limit Values for Chemical Substances and Physical Agents and Biological Exposure Indices for 1994–1995.* Cincinnati, OH: ACGIH, 1994.

12. International Labour Office. *Encyclopedia of Occupational Health and Safety.* Vols. I & II. Geneva, Switzerland: International Labour Office, 1983.

13. Mackison, F. W., R. S. Stricoff, and L. J. Partridge, Jr. (eds.). *NIOSH/OSHA – Occupational Health Guidelines for Chemical Hazards.* DHHS (NIOSH) Publication No. 81–123 (3 vols). Washington, DC: U.S. Government Printing Office, Jan. 1981.

14. Randolph, S. A., M. R. Zavon. *J Occup Med* 29: 237–42 (1987).

15. NIOSH. *Criteria Document: Cobalt.* p.36 (1981) DHEW Pub. NIOSH 82–107.

16. NIOSH. *Criteria Document: Cobalt.* p.37 (1981) DHEW Pub. NIOSH 82–107.

17. NIOSH. *Criteria Document: Cobalt.* p.38 (1981) DHEW Pub. NIOSH 82–107.

18. Brown, K.W., G. B. Evans, Jr., B.D. Frentrup (eds.). *Hazardous Waste Land Treatment.* Boston, MA: Butterworth Publishers, 1983.

19. Sittig, M. *Handbook of Toxic and Hazardous Chemicals and Carcinogens.*1985. 2nd ed. Park Ridge, NJ: Noyes Data Corporation, 1985.

■ COBALTOUS CARBONATE

CAS # 513–79–1

SYNONYMS carbonic acid, cobalt(2) salt (1: 1) • ci 77353 • cobalt carbonate (1: 1) • cobalt carbonate (CoCO₃) • cobalt(2) carbonate (CoCO₃) • sphaerocobaltite • cobalt spar

MF $CH_2O_3 \cdot Co$

MW 118.95

COLOR AND FORM Red powder or rhombohedral crystals; trigonal crystals.

DENSITY 4.13.

SOLUBILITY Almost insoluble in alcohol, methyl acetate; soluble in acids; insoluble in ammonia; 0.18 parts by wt (of formula wt)/100 parts of water by wt at 15°C.

STABILITY AND SHELF LIFE Oxidized by air to cobaltic carbonate [R1]; stable in air (hexahydrate) [R1]; solutions containing cobaltous ion Co(2+) are relatively stable. (cobaltous ion Co(2+)) [R16, 246].

OTHER CHEMICAL/PHYSICAL PROPERTIES Does not react with cold concentrated nitric acid or hydrochloric acid; when heated, dissolves with evolution of carbon dioxide; pink to violet-red crystalline needles; precipitated when excess carbon dioxide is present during prepn; on heating becomes anhydr by 140°C (hexahydrate); decomposition in hot water and before reaching MP (cobaltous carbonate, basic); pale-red powder, usually containing some water; practically insoluble in water; soluble in dilute acids and ammonia (cobaltous carbonate, basic); decomposition before reaching MP; molecular formula: $2CoCO_3 \cdot 3Co(OH)_2 \cdot H_2O$; mol wt: 534.74; violent-red prism; soluble in ammonium carbonate (cobalt(II) carbonate, basic); in its compound cobalt occurs normally in the oxidation states 2 and 3, more seldom in the oxidation states 0, 1, and 4. In normal salts the bivalent form is more stable than the trivalent one; the reverse is true for cobalt complexes (cobalt compounds).

PROTECTIVE EQUIPMENT For temporary operations (that produce dust or fume) or when ventilation is not practicable, an air-line respirator should be worn. If ventilation not satisfactory, dust or fume respirator can be used (cobalt, alloys, and compounds) [R7, 495]. The maintenance worker should wear protective clothing, personal protection equipment, including eye protection, and suitable respiratory protective equipment. When the catalyst takes the form of a harmful gas or vapor, exhaust ventilation, breathing apparatus, and protective clothing should be provided (catalysts, cobalt compounds) [R7, 426]. Respirator selection: Upper-limit devices recommended by NIOSH: 0.5 mg/m³: any dust and mist respirator except single-use respirators; 1 mg/m³: any dust and mist respirator except single-use and quarter-mask respirators or any dust, mist, and fume respirator with a full facepiece or any supplied-air respirator or any self-contained breathing apparatus; 2.5 mg/m³: any powered air-purifying respirator with a dust and mist filter or any supplied-air respirator operated in a continuous-flow mode or any powered air-purifying respirator with a dust, mist, and fume filter; 5 mg/m³: any air-purifying full facepiece respirator with a high-efficiency particulate filter or any self-contained breathing apparatus with a full facepiece or any supplied-air respirator with a full facepiece; 20 mg/m³: any supplied-air respirator with a full facepiece and operated in a pressure-demand or other positive-pressure mode; emergency or planned entry in unknown concn or IDLH conditions: any self-contained breathing apparatus with a full facepiece and operated in a pressure-demand or other positive-pressure mode or any supplied-air respirator with a full facepiece and operated in a pressure-demand or other positive-pressure mode in combination with an auxiliary self-contained breathing apparatus operated in pressure-demand or other positive-pressure mode; Escape: any air-purifying full facepiece respirator with a high-efficiency particulate filter or any appropriate escape-type self-contained breathing apparatus (cobalt metal, fume, and dust (as Co)) [R8, 85]. Employees should be provided with and required to use impervious clothing, gloves, face shields (eight-inch minimum), and other appropriate protective clothing necessary to prevent repeated or prolonged skin contact with cobalt dust. (cobalt metal, fume, and dust) [R11, 2].

PREVENTATIVE MEASURES Processes that produce cobalt dust or fume should be provided with an effective local exhaust ventilation (cobalt, alloys, and compounds) [R7, 495]. Avoiding skin contact may be difficult but barrier creams can be tried. Severely affected patient must be removed to occupations not involving cobalt exposure (cobalt, alloys and compounds) [R7, 495]. Contact lenses should not be worn when working with these chemicals (cobalt metal, fumes, and dust) [R11, 3]. Contact lens use in industry is controversial. A survey of 100 corporations resulted in the recommendation that each company establish their own contact lens use policy. One presumed hazard of contact lenses use is possible chemical entrapment. It was found that contact lenses minimized injury or protected the eye. The eye was afforded more protection from liquid irritants. Soft contact lenses do not worsen corneal damage from strong chemicals and in some cases could actually protect the eye. Overall, the literature supports the wearing of contact lenses in industrial environments as part of the standard eye protection, e.g., face shields; however, more data are needed to establish the value of contact lenses [R12]. Employees should wash promptly when skin becomes contaminated; work clothing should be changed daily if it is reasonably probable that the clothing is contaminated; promptly remove non-impervious clothing that becomes contaminated (cobalt metal, fume, and dust (as Co)) [R8, 85]. If employees' clothing has become contaminated with cobalt dust, employees should change into uncontaminated clothing before leaving the work premises. Clothing contaminated with cobalt dust should be placed in closed containers for storage until it can be discarded or until provision is made for the removal of cobalt dust from the clothing. If the clothing is to be laundered or otherwise cleaned to remove the cobalt dust, the person performing the operation should be informed of cobalt dust's hazardous properties. Non-impervious clothing that becomes contaminated with cobalt dust should be removed promptly and not reworn until the cobalt dust is removed from the clothing (cobalt dust) [R11, 3]. Skin that becomes contaminated with cobalt dust should be promptly washed or showered with soap or mild detergent and water to remove any cobalt dust. Eating and smoking should not be permitted in areas where cobalt metal fume or dust are generated, handled, processed, or stored. Employees who handle cobalt metal fume or dust should wash their hands thoroughly with soap or mild detergent and water before eating, smoking, or using toilet facilities (cobalt metal fume and dust) [R11, 3]. Employers should institute programs that emphasize good personal hygiene to prevent skin and respiratory irritation caused by cobalt-containing dusts. After working with cobalt products, workers should thoroughly wash their hands and face before drinking, eating, or smoking. If skin contact with cobalt solutions occurs, the worker should wash the affected skin promptly. The employer should provide showers if workers have substantial contact with cobalt. These workers should be encouraged to wash or shower after each workshift. Employers should prohibit smoking or carrying of tobacco products, and should prohibit eating, food handling, or food storage within the work area (cobalt and cobalt salts) [R13]. General plant maintenance must be conducted regularly to prevent cobalt-containing dusts from accumulating in work areas. Cleaning should be performed with vacuum pickup or wet mopping to minimize the amount of dust dispersed into the air. A decontamination room should be available for cleaning equipment that is to receive major overhaul or maintenance. Spills of cobalt-containing material should be promptly cleaned up to minimize inhalation or dermal contact (cobalt and cobalt salts) [R14]. Special precautions are necessary when workers must enter tanks or vessels, such as reaction vessels containing cobalt catalysts or vessels used to prepare cobalt salts. Before any worker enters a vessel, all sources for transferring cobalt and other materials into or out of the vessel must be blanked to prevent their entry. The vessel interior must then be washed with water and purged with air. After purging, analyze the vessel's atmosphere with suitable instruments to ensure that no hazards from fire, explosion, oxygen deficiency, or dust inhalation exist. No one should enter a tank or vessel without first being equipped with an appropriate respirator and a secured lifeline or harness. Mechanical ventilation should be provided continuously when workers are inside the tank. At least one other worker similarly equipped with respiratory protection, lifeline, and harness should watch at all times from outside the vessel. Workers inside the tank must be able to communicate with those persons outside. Other workers must be available to assist in an emergency. Flame- or spark-generating operations, such as welding or cutting, should be performed only when an authorized representative of the employer has signed a permit based on a finding that all necessary safety precautions have been taken. (cobalt and cobalt salts) [R15].

CLEANUP Liquid material spills can be copiously flushed with water and channeled to a treatment system or holding tank for reclamation or proper disposal. Spills of dry material can be removed by vacuuming or wet mopping. Some spills can be removed by hosing, first with a mist of water to dampen the spilled material, and then with a more forceful stream that flushes it into a holding tank, or other facility for handling contaminated water. Work surfaces or contaminated clothing should never be cleaned by dry-sweeping or blowing with pressurized hoses. Recovery systems used to reclaim waste metals should comply with federal, state, and local regulations. All waste materials generated in the handling of cobalt-containing substances should be disposed of in compliance with federal, state, and local regulations. (cobalt and cobalt salts) [R5].

DISPOSAL SRP: At the time of review, criteria for land treatment or burial (sanitary landfill) disposal practices are subject to significant revision. Prior to implementing land disposal of waste residue (including waste sludge), consult with environmental regulatory agencies for guidance on acceptable disposal practices. Proper mixing of the cobalt waste and the soil is essential to preventing excessive plant

accumulation of cobalt (cobalt) [R17, 256]. Cobalt metal may be recovered from scrap as alternative to disposal [R18, 255]. Cobalt compounds may be recovered from spent catalysts as alternative to disposal. [R18, 255].

COMMON USES In ceramics; manufacture of cobalt pigments; preparation of cobalt compounds; nutritional factor; used in cobalt deficiency in ruminants [R1]; trace element added to soils and animal feed; temperature indicator; catalyst; pigments [R2]; manufacture of cobaltous oxide; cobalt pigments; cobalt salts; intermediate (cobaltous carbonate, basic) [R2]; cobalt difluoride is prepared commercially by the reaction of cobalt carbonate with anhydrous hydrogen fluoride [R3]. The only diseases in which the clinical use of cobalt is still advocated by some is the normochromic, normocytic anemia assoc with severe renal failure. (cobalt) [R4].

OCCUPATIONAL RECOMMENDATIONS None listed.

OSHA Permissible exposure limit (PEL): 8-hr TWA 0.1 mg/m^3 (cobalt metal, fume, and dust (as Co)) [R8, 84].

INTERNATIONAL EXPOSURE LIMITS In the Federal Republic of Germany, industrial cobalt emission has been limited to 1 mg/m^3 for metallic cobalt and its slightly soluble compound and 50 mg/m^3 for other compounds. (cobalt and compounds) [R9, 260].

BIOLOGICAL HALF-LIFE It appears that independent of exposure route (inhalation, injection, or ingestion), most of the cobalt will be eliminated rapidly (in humans). A small proportion is, however, eliminated slowly, having a biological half-time on the order of years (cobalt) [R10, 218]. Behavior of (60)cobalt in man for up to 11 years after accidental inhalation was observed after fast clearance of main part of absorbed cobalt; the rest (about 10%), had a biological half-time in chest as well as in whole body of 5–15 years (cobalt) [R10, 219]. Accidentally swallowed (60)cobalt is eliminated with half-lives of 0.5, 2.7, and 59 days. Over 90% of cobalt taken up parenterally is eliminated within a few days. Only 10% of the dose shows a half-life of 2 yr. Following pulmonary uptake of (60)cobalt, a further very long half-life of 5–10 yr has been observed (cobalt) [R9, 254].

BIOCONCENTRATION Only a few plant species accumulate cobalt above the 100 ppm that causes severe phytotoxicity. Hyperaccumulators of cobalt have been found that contain over 1% cobalt in dry leaves. (cobalt salts) [R6].

REFERENCES

1. *The Merck Index.* 10th ed. Rahway, NJ: Merck Co., Inc., 1983. 346.

2. Sax, N. I., and R. J. Lewis, Sr. (eds.). *Hawley's Condensed Chemical Dictionary.* 11th ed. New York: Van Nostrand Reinhold Co., 1987. 293.

3. *Kirk–Othmer Encyclopedia of Chemical Technology.* 3rd ed., Volumes 1–26. New York, NY: John Wiley and Sons, 1978–1984, p. 10 (80) 717.

4. Gilman, A. G., L. S. Goodman, and A. Gilman (eds.). *Goodman and Gilman's The Pharmacological Basis of Therapeutics.* 7th ed. New York: Macmillan Publishing Co., Inc., 1985. 1319.

5. NIOSH. *Criteria Document: Cobalt* p. 37 (1981) DHEW Pub. NIOSH 82–107.

6. Parr, J. F., P. B. Marsh, and J. M. Kla (eds.). *Land Treatment of Hazardous Wastes.* Park Ridge, NJ: Noyes Data Corporation, 1983. 176.

7. International Labour Office. *Encyclopedia of Occupational Health and Safety.* Vols. I & II. Geneva, Switzerland: International Labour Office, 1983.

8. NIOSH. *Pocket Guide to Chemical Hazards.* 2nd Printing. DHHS (NIOSH) Publ. No. 85–114. Washington, DC: U.S. Dept. of Health and Human Services, NIOSH/Supt.of Documents, GPO, February 1987.

9. Seiler, H.G., H. Sigel, and A. Sigel, (eds.). *Handbook on the Toxicity of Inorganic Compounds.* New York, NY: Marcel Dekker, Inc. 1988.

10. Friberg, L., G.F. Nordberg, E. Kessler, and V.B. Vouk (eds.). *Handbook of the Toxicology of Metals.* 2nd ed. Vols I, II.: Amsterdam: Elsevier Science Publishers B.V., 1986, p. V2.

11. Mackison, F. W., R. S. Stricoff, and L. J. Partridge, Jr. (eds.). *NIOSH/OSHA – Occupational Health Guidelines for Chemical Hazards.* DHHS (NIOSH) Publication No. 81–123 (3 Vols). Washington, DC: U.S. Government Printing Office, Jan. 1981.

12. Randolph, S. A., M. R. Zavon. *J Occup Med* 29: 237–42 (1987).

13. NIOSH. *Criteria Document: Cobalt.* p.36 (1981) DHEW Pub. NIOSH 82–107.

14. NIOSH. *Criteria Document: Cobalt.* p.37 (1981) DHEW Pub. NIOSH 82–107.

15. NIOSH. *Criteria Document: Cobalt.* p.38 (1981) DHEW Pub. NIOSH 82–107.

16. National Research Council. *Drinking Water and Health.* Volume 1. Washington, DC: National Academy Press, 1977.

17. Brown, K.W., G. B. Evans, Jr., B.D. Frentrup (eds.). *Hazardous Waste Land Treatment.* Boston, MA: Butterworth Publishers, 1983.

18. Sittig, M. *Handbook of Toxic and Hazardous Chemicals and Carcinogens,* 1985. 2nd ed. Park Ridge, NJ: Noyes Data Corporation, 1985.

■ COBALTOUS FORMATE

CAS # 544–18–3

SIC CODES 2819

SYNONYMS cobalt formate • cobalt(II) formate • cobalt diformate • cobalt formate • formic acid, cobalt(2) salt

MF CH_2O_2 •1/2Co

MW 148.98

COLOR AND FORM Red crystalline solid.

FIREFIGHTING PROCEDURES Extinguish fire using agent suitable to type of surrounding fire [R5].

STABILITY AND SHELF LIFE Solutions containing cobaltous ion Co(2+) are relatively stable (cobaltous ion, Co(2+)) [R9, 246].

OTHER CHEMICAL/PHYSICAL PROPERTIES Dihydrate becomes anhydrous at 140°C; red crystalline powder; density: 2.13 at 22°C; almost insoluble in alcohol (dihydrate); decomposes at 175°C (dihydrate); solubility: 5.030 lb/100 lb at 68°F (dihydrate); in its compounds cobalt occurs normally in the oxidation states 2 and 3, more seldom in the oxidation states 0, 1, and 4. In normal salts the bivalent form is more stable than the trivalent one; the reverse is true for cobalt complexes. (cobalt compounds).

PROTECTIVE EQUIPMENT NIOSH-approved respirator if needed, rubberized fabric gloves, chemical dust goggles (dihydrate) [R6].

PREVENTATIVE MEASURES If material not involved in fire: Keep material out of water sources and sewers. Build dike to contain flow as necessary [R5].

CLEANUP Environmental consideration land spill: Dig a pit, pond, lagoon, holding area to contain liquid or solid material. (SRP: If time permits, pits, ponds, lagoons, soak holes, or holding areas should be sealed with an impermeable flexible-membrane liner.) Cover solids with a plastic sheet to prevent dissolving in rain or firefighting water [R5]. Environmental consideration, water spill: Neutralize with agricultural lime (CaO), crushed limestone (CaCO₃), or sodium bicarbonate (NaHCO₃). Adjust pH to neutral (pH = 7). Use mechanical dredges or lifts to remove immobilized masses of pollutants and precipitates [R5]. Liquid material spills can be copiously flushed with water and channeled to a treatment system or holding tank for reclamation or proper disposal. Spills of dry material can be removed by vacuuming or wet mopping. Some spills can be removed by hosing, first with a mist of water to dampen the spilled material, and then with a more forceful stream that flushes it into a holding tank, or other facility for handling contaminated water. Work surfaces or contaminated clothing should never be cleaned by dry-sweeping or blowing with pressurized hoses. Recov-

ery systems used to reclaim waste metals should comply with federal, state, and local regulations. All waste materials generated in the handling of cobalt-containing substances should be disposed of in compliance with federal, state, and local regulations. (cobalt and cobalt salts) [R7].

DISPOSAL SRP: At the time of review, criteria for land treatment or burial (sanitary landfill) disposal practices are subject to significant revision. Prior to implementing land disposal of waste residue (including waste sludge), consult with environmental regulatory agencies for guidance on acceptable disposal practices.

COMMON USES In preparation of cobalt catalysts [R1]. Manufacture of paint and varnish driers [R2]. The only disease in which the clinical use of cobalt is still advocated by some is the normochromic, normocytic anemia assoc with severe renal failure. (cobalt) [R3].

STANDARD CODES EPA 311; TSCA; not listed IATA; not listed CFR 49; not listed NFPA; not listed AAR; Superfund Designated (Hazardous Substances) List.

PERSISTENCE Persistent.

MAJOR SPECIES THREATENED Fish food organisms are quite sensitive to low concentrations.

INHALATION LIMIT 0.1 mg/m³.

DIRECT CONTACT Eye, skin, and respiratory irritant at fire temperatures.

GENERAL SENSATION At fire temperatures produces coughing, choking, difficult breathing, and cyanosis. May be irritating to eyes and skin.

PERSONAL SAFETY Protect against both inhalation and absorption through the skin. Must wear protective clothing including NIOSH-approved rubber gloves and boots, safety goggles, or face mask, and a respirator whose canister is specifically approved for this material. Decontaminate fully or dispose of all equipment and clothing after use.

ACUTE HAZARD LEVEL Cobalt salts are toxic to humans by ingestion; between 1 teaspoonful and 1 ounce may be fatal. In toxic doses this salt acts locally on the gastrointestinal tract, producing pain, vomiting, etc. This material seems to have a cumulative toxic action under conditions in which elimination cannot keep pace with absorption. On the other hand, there appears to be experimental evidence that a tolerance may develop to cobalt when initial doses are sufficiently low. The aquatic toxicity to fathead minnows is approximately 100 ppm, for a 96-hour LC_{50}.

CHRONIC HAZARD LEVEL Through chronic exposure, there may be an allergic-type sensitivity of the skin. The proposed threshold-limit-value time-weighted average for a 40-hour workweek is 0.05 mg/m³ as cobalt. Inhalation of cobalt dust may cause asthma-like symptoms with coughing and dyspnea. This may progress to interstitial pneumonia with marked fibrosis. Pneumoconiosis may develop that is believed to be reversible. Cobalt compounds are also suspected carcinogens of the connective tissues and lungs.

DEGREE OF HAZARD TO PUBLIC HEALTH Contact with the dust may be irritating to the eyes and skin.

OCCUPATIONAL RECOMMENDATIONS None listed.
OSHA Permissible exposure limit (PEL): 8-hr TWA 0.1 mg/m³ (cobalt metal, fume, and dust (as Co)) [R11, 84].

THRESHOLD LIMIT 8-hr time-weighted average (TWA) 0.02 mg/m³ (1994) (cobalt, elemental, and inorganic compound, as Co) [R10, 17].

INTERNATIONAL EXPOSURE LIMITS In the Federal Republic of Germany, industrial cobalt emission has been limited to 1 mg/m³ for metallic cobalt and its slightly soluble compounds and 50 mg/m³ for other compounds. (cobalt and compounds) [R13, 260].

ACTION LEVEL Avoid contact with the spilled cargo. Stay upwind. Notify local air, water, and fire authorities of the accident. Evacuate all people to a distance of at least 200 feet upwind and 1,000 feet downwind of the spill.

ON-SITE RESTORATION Dam stream if possible to reduce the flow and prevent further dispersion of the material by water movement. Due to solubility, dredging may be unproductive. Under controlled conditions, pump water into a suitable container. Add lime to pH 8–8.5, allow precipitate to settle, and neutralize to pH 7 with hydrochloric acid (HCl). For more information see Envirex Manual EPA 600/2–77–227. Seek professional environmental engineering assistance through EPA's Environmental Response Team (ERT), Edison, NJ, 24–hour no., 908–548–8730.

BEACH OR SHORE RESTORATION Close beach and shore to the public until material has been removed. Avoid human toxicity contact. Do not burn material.

AVAILABILITY OF COUNTERMEASURE MATERIAL Pumps—fire departments, Army Corps of Engineers.

DISPOSAL After the material has been contained, remove it and the contaminated soil and place in impervious containers. If practical, transport material back to the supplier for recovery. If this is not practical, the material should be encapsulated and buried in a specially designated chemical landfill. Not acceptable at a municipal sewage treatment plant.

DISPOSAL NOTIFICATION Notify local and state health authorities, local solid waste disposal authorities, supplier, and shipper.

INDUSTRIAL FOULING POTENTIAL May contaminate boiler feed water.

MAJOR WATER USE THREATENED Fisheries; potable water; industrial water.

PROBABLE LOCATION AND STATE Red crystals, will sink and dissolve.

COLOR IN WATER Red.

FOOD-CHAIN CONCENTRATION POTENTIAL Positive. Microorganisms tend to concentrate cobalt from the surrounding media.

BIOLOGICAL HALF-LIFE It appears that independent of exposure route (inhalation, injection, or ingestion), most of the cobalt will be eliminated rapidly (in humans). A small proportion is, however, eliminated slowly, having a biological half-time on the order of years. (cobalt) [R12, 218].

BIOCONCENTRATION Food-chain concentration potential: Microorganisms can concentrate cobaltous formate in water up to 1,070 to 1,500 times [R6].

CERCLA REPORTABLE QUANTITIES Persons in charge of vessels or facilities are required to notify the National Response Center (NRC) immediately, when there is a release of this designated hazardous substance, in an amount equal to or greater than its reportable quantity of 1,000 lb or 454 kg. The toll-free number of the NRC is (800) 424–8802; in the Washington DC metropolitan area (202) 426–2675. The rule for determining when notification is required is stated in 40 CFR 302.4 (section IV. D. 3.b) [R8].

DOT *Fire or Explosion:* Some of these materials may burn, but none of them ignites readily [R4].

Health Hazards: Contact may cause burns to skin and eyes. Fire may produce irritating or poisonous gases. Runoff from fire control or dilution water may cause pollution [R4].

Emergency Action: Keep unnecessary people away. Isolate hazard area and deny entry. Self-contained breathing apparatus (SCBA) and structural firefighter's protective clothing will provide limited protection. Call CHEMTREC at 1–800–424–9300 for emergency assistance. If water pollution occurs, notify the appropriate authorities [R4].

Fire: Small Fires: Dry chemical, carbon dioxide, halon, water spray, or standard foam. Large Fires: Water spray, fog, or standard foam is recommended. Move container from fire area if you can do so without risk. Do not scatter spilled material with high-pressure water streams. Dike fire-control water for later disposal [R4].

Spill or Leak: Stop leak if you can do so without risk. Small Dry Spills: With clean shovel place material into clean, dry container, and cover; move containers from spill area. Small Spills: Take up with sand or other noncombustible absorbent material and place into containers for later disposal. Large Spills: Dike far ahead of liquid spill for later disposal. Cover powder spill with plastic sheet or tarp to minimize spreading [R4].

First Aid: In case of contact with material, immediately flush eyes with running water for at least 15 minutes. Wash skin with soap and water. Remove and isolate contaminated clothing and shoes at the site [R4].

GENERAL RESPONSE Avoid contact with solid and dust. Keep people away. Wear goggles, self-contained breathing apparatus, and rubber overclothing (including gloves). Stop discharge if possible. Isolate and remove discharged material. Notify local health and pollution control agencies.

FIRE RESPONSE Fire data not available.

EXPOSURE RESPONSE Call for medical aid. Dust irritating to skin and eyes. Harmful if inhaled. Move to fresh air. If breathing has stopped, give artificial respiration. Solid will burn skin and eyes. Harmful if swallowed. Remove contaminated clothing and shoes. Flush affected area with plenty of water. If in eyes, hold eyelids open and flush with plenty of water. If swallowed and victim is conscious, have victim drink water or milk.

RESPONSE TO DISCHARGE Issue warning—water contaminant. Disperse and flush.

WATER RESPONSE Harmful to aquatic life in very low concentrations. May be dangerous if it enters water intakes. Notify local health and wildlife officials. Notify operators of nearby water intakes.

EXPOSURE SYMPTOMS Eyes: Causes burns. Skin: Can cause ulceration.

EXPOSURE TREATMENT Call a physician. Eyes: Flush with copious amounts of water. Skin: Wash thoroughly.

FIRE POTENTIAL Material itself does not burn or burns with difficulty [R5].

REFERENCES

1. *The Merck Index.* 10th ed. Rahway, NJ: Merck Co., Inc., 1983. 347.

2. Sax, N. I., ed. *Danger Props Indus Mater Report* 4 (1): 49–51 (1984).

3. Gilman, A. G., L. S. Goodman, and A. Gilman (eds.). *Goodman and Gilman's The Pharmacological Basis of Therapeutics.* 7th ed. New York: Macmillan Publishing Co., Inc., 1985. 1319.

4. Department of Transportation. *Emergency Response Guidebook 1987.* DOT P 5800.4. Washington, DC: U.S. Government Printing Office, 1987, p. G–31.

5. Association of American Railroads. *Emergency Handling of Hazardous Materials in Surface Transportation.* Washington, DC: Assoc. of American Railroads, Hazardous Materials Systems (BOE), 1987. 179.

6. U.S. Coast Guard, Department of Transportation. *CHRIS—Hazardous Chemical Data.* Volume II. Washington, DC: U.S. Government Printing Office, 1984–5.

7. NIOSH. *Criteria Document:* Cobalt, p. 37 (1981) DHEW Pub. NIOSH 82–107.

8. 54 FR 33419 (8/14/89).

9. National Research Council. *Drinking Water and Health Volume 1.* Washington, DC: National Academy Press, 1977.

10. American Conference of Governmental Industrial Hygienists. *Threshold Limit Values for Chemical Substances and Physical Agents and Biological Exposure Indices for 1994–1995.* Cincinnati, OH: ACGIH, 1994.

11. NIOSH. *Pocket Guide to Chemical Hazards.* 2nd Printing. DHHS (NIOSH) Publ. No. 85–114. Washington, DC: U.S. Dept. of Health and Human Services, NIOSH/Supt. of Documents, GPO, February 1987.

12. Friberg, L., G.F. Nordberg, E. Kessler, and V.B. Vouk (eds.). *Handbook of the Toxicology of Metals.* 2nd ed. Vols I, II.: Amsterdam: Elsevier Science Publishers B.V., 1986, p. V2.

13. Seiler, H.G., H. Sigel, and A. Sigel (eds.). *Handbook on the Toxicity of Inorganic Compounds.* New York, NY: Marcel Dekker, Inc. 1988.

■ COBALTOUS OXIDE

CAS # 1307–96–6

SYNONYMS ci-77322 • ci pigment black 13 • cobalt black • cobalt monooxide • cobalt monoxide • cobalt oxide (CoO) • cobalt(2) oxide • monocobalt oxide • zaffre • cobalt(II) oxide

MF CoO

MW 74.94

COLOR AND FORM Powder, or cubic or hexagonal crystals; color varies from olive green to red, depending on particle size, but commercial material is usually dark grey; pink cubic crystals; green-brown crystals.

DENSITY 5.7–6.7 (depending on preparation).

MELTING POINT About 1,935°C.

SOLUBILITY Practically insoluble in water; soluble in acids or alkalies; insoluble in alcohol, ammonium hydroxide.

OTHER CHEMICAL/PHYSICAL PROPERTIES Readily absorbs molecular oxygen even at room temp; easily reduced to cobalt by carbon or carbon monoxide. Reacts at high temperature with silica, alumina, zinc oxide to form pigments. When reduced with ammonia it contains 14–16% molecular oxygen and glows when exposed to air. In its compounds cobalt occurs normally in the oxidation states 2 and 3, more seldom in the oxidation states 0, 1, and 4. In normal salts the bivalent form is more stable than the trivalent one; the reverse is true for cobalt complexes. (cobalt compounds).

PROTECTIVE EQUIPMENT For temporary operations (that produce dust or fume) or when ventilation is not practicable, an air-line respirator should be worn. If ventilation not satisfactory, dust or fume respirator can be used (cobalt, alloys, and compounds) [R7, 495]. The maintenance worker should wear protective clothing, personal protection equipment, incl eye protection, and suitable respiratory protective equipment. When the catalyst takes the form of a harmful gas or vapor, exhaust ventilation, breathing apparatus, and protective clothing should be provided (catalysts, cobalt compounds) [R7, 426]. Respirator selection: Upper-limit devices recommended by NIOSH: 0.5 mg/m³: any dust and mist respirator except single-use respirators; 1 mg/m³: any dust and mist respirator except single-use and quarter-mask respirators or any dust, mist, and fume respirator with a full facepiece or any supplied-air respirator or any self-contained breathing apparatus; 2.5 mg/m³: any powered air-purifying respirator with a dust and mist filter or any supplied-air respirator operated in a continuous flow mode or any powered air-purifying respirator with a dust, mist, and fume filter; 5 mg/m³: any air-purifying full facepiece respirator with a high-efficiency particulate filter or any self-contained breathing apparatus with a full facepiece or any supplied-air respirator with a full facepiece; 20 mg/m³: any supplied-air respirator with a full facepiece and operated in a pressure-demand or other positive-pressure mode. Emergency or planned entry in unknown concn or IDLH conditions: any self-contained breathing apparatus with a full facepiece and operated in a pressure-demand or other positive-pressure mode or any supplied-air respirator with a full facepiece and operated in a pressure-demand or other positive-pressure mode in combination with an auxiliary self-contained breathing apparatus operated in pressure-demand or other positive-pressure mode; Escape: any air-purifying full facepiece respirator with a high-efficiency particulate filter or any appropriate escape-type self-contained breathing apparatus (cobalt metal, fume, and dust (as Co)) [R9, 85]. Employees should be provided with and required to use impervious clothing, gloves, face shields (eight-inch minimum), and other appropriate protective clothing necessary to prevent repeated or prolonged skin contact with cobalt dust (cobalt metal fume and dust) [R10, 2].

PREVENTATIVE MEASURES Processes that produce cobalt dust or fume should be provided with an effective local exhaust ventilation (cobalt, alloys, and compounds) [R7, 495]. Avoiding skin contact may be difficult but barrier creams can be tried. Severely affected patient must be removed to occupations not involving cobalt exposure (cobalt, alloys and compounds) [R7, 495]. Contact lenses should not be worn when working with these chemicals (cobalt metal, fumes and dust) [R10, 3]. Contact lens use in industry is controversial. A survey of 100 corporations resulted in the recommendation that each company establish their own contact lens use policy. One presumed hazard of contact lens use is possible chemical entrapment. It was found that contact lenses minimized injury or protected the eye. The eye was afforded more protection from liquid irritants. Soft contact lenses do not worsen corneal damage from strong chemicals and in some cases could actually protect the eye. Overall, the literature supports the wearing of contact lenses in industrial environments as part of the standard eye protection, e.g., face shields; however, more data are needed to establish the value of contact lenses [R14]. Employees should wash promptly when skin becomes contaminated; work clothing should be changed daily if it is reasonably probable that the clothing is contaminated; promptly remove non-impervious clothing that becomes contaminated (cobalt metal, fume, and dust (as Co)) [R9, 85]. If employees' clothing has become contaminated with cobalt dust, employees should change into uncontaminated clothing before leaving the work premises. Clothing contaminated with cobalt dust should be placed in closed containers for storage until it can be discarded or until provision is made for the removal of cobalt dust from the clothing. If the clothing is to be laundered or otherwise cleaned to remove the cobalt dust, the person performing the operation should be informed of cobalt dust's hazardous properties. Non-impervious clothing that becomes contaminated with cobalt dust should be removed promptly and not reworn until the cobalt dust is removed from the clothing (cobalt dust) [R14, 3]. Skin that becomes contaminated with cobalt dust should be promptly washed or showered with soap or mild detergent and water to remove any cobalt dust. Eating and smoking should not be permitted in areas where cobalt metal fume or dust is generated, handled, processed, or stored. Employees who handle cobalt metal fume or dust should wash their hands thoroughly with soap or mild detergent and water before eating, smoking, or using toilet facilities (cobalt metal fume and dust) [R14, 3]. Employers should institute programs that emphasize good personal hygiene to prevent skin and respiratory irritation caused by cobalt-containing dusts. After working with cobalt products, workers should thoroughly wash their hands and face before drinking, eating, or smoking. If skin contact with cobalt solutions occurs, the worker should wash the affected skin promptly. The employer should provide showers if workers have substantial contact with cobalt. These workers should be encouraged to wash or shower after each workshift. Employers should prohibit smoking or carrying of tobacco products, and should prohibit eating, food handling, or food storage within the work area (cobalt and cobalt salts) [R15]. General plant maintenance must be conducted regularly to prevent cobalt-containing dusts from accumulating in work areas. Cleaning should be performed with vacuum pickup or wet mopping to minimize the amount of dust dispersed into the air. A decontamination room should be available for cleaning equipment that is to receive major overhaul or maintenance. Spills of cobalt-containing material should be promptly cleaned up to minimize inhalation or dermal contact. (cobalt and cobalt salts) [R5]. Special precautions are necessary when workers must enter tanks or vessels, such as reaction vessels containing cobalt catalysts or vessels used to prepare cobalt salts. Before any worker enters a vessel, all sources for transferring cobalt and other materials into or out of the vessel must be blanked to prevent their entry.

The vessel interior must then be washed with water and purged with air. After purging, analyze the vessel's atmosphere with suitable instruments to ensure that no hazards from fire, explosion, oxygen deficiency, or dust inhalation exist. No one should enter a tank or vessel without first being equipped with an appropriate respirator and a secured lifeline or harness. Mechanical ventilation should be provided continuously when workers are inside the tank. At least one other worker similarly equipped with respiratory protection, lifeline, and harness should watch at all times from outside the vessel. Workers inside the tank must be able to communicate with those persons outside. Other workers must be available to assist in an emergency. Flame- or spark-generating operations, such as welding or cutting, should be performed only when an authorized representative of the employer has signed a permit based on a finding that all necessary safety precautions have been taken (cobalt and cobalt salts) [R16].

CLEANUP Liquid material spills can be copiously flushed with water and channeled to a treatment system or holding tank for reclamation or proper disposal. Spills of dry material can be removed by vacuuming or wet mopping. Some spills can be removed by hosing, first with a mist of water to dampen the spilled material, and then with a more forceful stream that flushes it into a holding tank or other facility for handling contaminated water. Work surfaces or contaminated clothing should never be cleaned by dry-sweeping or blowing with pressurized hoses. Recovery systems used to reclaim waste metals should comply with federal, state, and local regulations. All waste materials generated in the handling of cobalt-containing substances should be disposed of in compliance with federal, state, and local regulations. (cobalt and cobalt salts) [R5].

DISPOSAL SRP: At the time of review, criteria for land treatment or burial (sanitary landfill) disposal practices are subject to significant revision. Prior to implementing land disposal of waste residue (including waste sludge), consult with environmental regulatory agencies for guidance on acceptable disposal practices.

COMMON USES In pigments for ceramics; glass coloring and decolorization; preparation of cobalt metal catalysts; oxidation catalyst for drying oils, fast-drying paints, and varnishes; cobalt powder for binder in sintered tungsten carbide; in semiconductors [R1]. Used in enamel coatings on steel to improve the adherence of the enamel to the metal [R8, 1607]. Pigment in enamels; catalysts for afterburning of engine exhaust gases [R7, 493]. In petroleum industry, catalyzes conversion of thiophene to butane [R3]. Catalyst for removal of nitrogen, carbon oxides, and sulfur compounds from waste gases [R4]. Used as feed additive [R2].

OCCUPATIONAL RECOMMENDATIONS None listed.
OSHA Permissible exposure limit (PEL): 8-hr TWA 0.1 mg/m³ (cobalt metal, fume, and dust (as Co)) [R9, 84].

THRESHOLD LIMIT 8-hr time-weighted average (TWA) 0.02 mg/m³ (1994) (cobalt, elemental, and inorganic compound, as Co) [R10, 17].

INTERNATIONAL EXPOSURE LIMITS Max allowable concentration (former USSR) 0.5 mg/m³ [R7, 493].

NON-HUMAN TOXICITY LD_{50} rat, oral 1,700 mg/kg [R11, 286].

BIOLOGICAL HALF-LIFE Intratracheally instilled cobaltous oxide (1.5 µg) was retained in the lung for a relatively long period, with a half-time of about 15 days. [R12, 215].

BIOCONCENTRATION Only a few plant species accumulate cobalt above the 100 ppm that causes severe phytotoxicity. Hyperaccumulators of cobalt have been found that contain over 1% cobalt in dry leaves. (cobalt salts) [R6].

FIRE POTENTIAL Specially prepared (the form prepared by reducing the oxides in hydrogen) very fine cobalt dust will catch fire at room temperature in air. (cobalt metal fume and dust) [R14, 2].

REFERENCES
1. *The Merck Index.* 10th ed. Rahway, NJ: Merck Co., Inc., 1983. 347.

2. Sax, N. I., and R. J. Lewis, Sr. (eds.). *Hawley's Condensed Chemical Dictionary.* 11th ed. New York: Van Nostrand Reinhold Co., 1987. 295.

3. International Labour Office. *Encyclopedia of Occupational Health and Safety.* Volumes I and II. New York: McGraw-Hill Book Co., 1971. 271.

4. Osmanov, M. O., et al. former USSR patent no 521925 07/25/76 (Azerbaidzhan Institute of Petroleum and Chemistry).

5. NIOSH. *Criteria Document: Cobalt* p. 37 (1981) DHEW Pub. NIOSH 82–107.

6. Parr, J. F., P. B. Marsh, and J. M. Kla (eds.). *Land Treatment of Hazardous Wastes.* Park Ridge, NJ: Noyes Data Corporation, 1983. 176.

7. International Labour Office. *Encyclopedia of Occupational Health and Safety.* Vols. I & II. Geneva, Switzerland: International Labour Office, 1983.

8. Clayton, G. D., and F. E. Clayton (eds.). *Patty's Industrial Hygiene and Toxicology:* Volume 2A, 2B, 2C: Toxicology. 3rd ed. New York: John Wiley Sons, 1981–1982.

9. NIOSH. *Pocket Guide to Chemical Hazards.* 2nd Printing. DHHS (NIOSH) Publ. No. 85–114. Washington, DC: U.S. Dept. of Health and Human Services, NIOSH/Supt.of Documents, GPO, February 1987.

10. American Conference of Governmental Industrial Hygienists. *Threshold Limit Values for Chemical Substances and Physical Agents and Biological Exposure Indices for 1994–1995.* Cincinnati, OH: ACGIH, 1994.

11. Venugopal, B. and T.D. Luckey. *Metal Toxicity in Mammals,* 2. New York: Plenum Press, 1978.

12. Friberg, L., G.F. Nordberg, E. Kessler, and V.B. Vouk (eds.). *Handbook of the Toxicology of Metals.* 2nd ed. Vols I, II.:

Amsterdam: Elsevier Science Publishers B.V., 1986, p. V2.

13. Randolph, S. A., M. R. Zavon. *J Occup Med* 29: 237–42 (1987).

14. Mackison, F. W., R. S. Stricoff, and L. J. Partridge, Jr. (eds.). *NIOSH/OSHA—Occupational Health Guidelines for Chemical Hazards*. DHHS(NIOSH) Publication No. 81–123 (3 vols). Washington, DC: U.S. Government Printing Office, Jan. 1981.

15. NIOSH. *Criteria Document: Cobalt*. p.36 (1981) DHEW Pub. NIOSH 82–107.

16. NIOSH. *Criteria Document: Cobalt* p.38 (1981) DHEW Pub. NIOSH 82–107.

■ COBALTOUS SULFAMATE

CAS # 14017–41–5

SYNONYMS sulfamic acid, cobalt(2) salt (2:1)

MF $Co \cdot 2H_3NO_3S$

MW 253.09

COLOR AND FORM Reddish solid.

FIREFIGHTING PROCEDURES Extinguish fire using agent suitable for type of surrounding fire [R1].

SOLUBILITY Soluble in water.

STABILITY AND SHELF LIFE Solutions containing cobaltous ion, $Co(2+)$, are relatively stable. (cobaltous ion, $Co(2+)$) [R7, 246].

OTHER CHEMICAL/PHYSICAL PROPERTIES In its compounds cobalt occurs normally in the oxidation states 2 and 3, more seldom in the oxidation states 0, 1, and 4. In normal salts the bivalent form is more stable than the trivalent one; the reverse is true for cobalt complexes. (cobalt compound).

PROTECTIVE EQUIPMENT Wear appropriate chemical protective gloves, boots, and goggles [R1].

PREVENTATIVE MEASURES If material is not involved in fire: Keep material out of water sources and sewers. Build a dike to contain flow as necessary [R1].

CLEANUP Land spill: Dig a pit, pond, lagoon, holding area to contain liquid or solid material. (SRP: If time permits, pits, ponds, lagoons, soak holes, or holding areas should be sealed with an impermeable flexible-membrane liner.) Cover solids with a plastic sheet to prevent dissolving in rain or firefighting water [R1]. Water spill: Neutralize with agriculture lime (CaO), crushed limestone ($CaCO_3$), or sodium bicarbonate ($NaHCO_3$). Adjust pH to neutral (pH = 7). Use mechanical dredges or lifts to remove immobilized masses of pollutants and precipitates [R1]. Liquid-material spills can be flushed with water copiously and channeled to a treatment system or holding tank for reclamation or proper disposal. Spills of dry material can be removed by vacuuming or wet mopping. Some spills can be removed by hosing, first with a mist of water to dampen the spilled material, and then with a more forceful stream that flushes it into a holding tank, or other facility for handling contaminated water. Work surfaces or contaminated clothing should never be cleaned by dry-sweeping or blowing with pressurized hoses. Recovery systems used to reclaim waste metals should comply with federal, state, and local regulations. All waste materials generated in the handling of cobalt-containing substances should be disposed of in compliance with federal, state, and local regulations. (cobalt and cobalt salts) [R4].

DISPOSAL SRP: At the time of review, criteria for land treatment, or burial (sanitary landfill) disposal practices are subject to significant revision. Prior to implementing land disposal of waste residue (including waste sludge), consult with environmental regulatory agencies for guidance on acceptable disposal practices.

COMMON USES Used as a pigment and for electroplating metals [R1] The only diseases in which the clinical use of cobalt is still advocated by some is the normochromic, normocytic anemia assoc with severe renal failure. (cobalt) [R2].

STANDARD CODES EPA 311; TSCA; not

listed IATA; not listed CFR 49; CFR 14 CAB code 8; not listed NFPA; not listed AAR; ICC STCC tariff no 1–C 2819952; Superfund designated (hazardous substances) list.

PERSISTENCE Persistent.

MAJOR SPECIES THREATENED Fish food organisms are quite sensitive to low concentrations.

INHALATION LIMIT 0.1 mg/m³.

DIRECT CONTACT Eye, skin, and respiratory irritant at fire temperatures.

GENERAL SENSATION At fire temperatures produces coughing, choking, difficult breathing, and cyanosis. May be irritating to eyes and skin. May produce tearing.

PERSONAL SAFETY Protect against both inhalation and absorption through the skin. Must wear protective clothing including NIOSH-approved rubber gloves and boots, safety goggles, or face mask, and a respirator whose canister is specifically approved for this material. Decontaminate fully or dispose of all equipment after use.

ACUTE HAZARD LEVEL Cobalt salts may be very toxic to humans by ingestion; between 1 teaspoonful and 1 ounce may be fatal. This salt acts locally on the gastrointestinal tract, producing pain, vomiting, etc. This material has a cumulative toxic action under conditions in which elimination cannot keep pace with absorption. On the other hand, there appears to be experimental evidence that a tolerance may develop to cobalt when initial doses are sufficiently low. The aquatic toxicity to fathead minnows is approximately 100 ppm for a 96-hour LC₅₀.

CHRONIC HAZARD LEVEL Through chronic exposure, there might be a dermatitis of the allergic-sensitivity type to the skin. The proposed threshold-limit-value time-weighted average for a 40-hour workweek is 0.05 mg/m³ as cobalt. Inhalation of cobalt dust may cause an asthma-like disease with cough and dyspnea. This situation may progress to interstitial pneumonia with marked fibrosis. Pneumoconiosis may develop that is believed to

be reversible. Cobalt compounds are also suspected carcinogens of the lungs and connective tissue.

DEGREE OF HAZARD TO PUBLIC HEALTH Contact with the dust may be irritating to the eyes and skin.

OCCUPATIONAL RECOMMENDATIONS None listed.

OSHA Permissible exposure limit (PEL): 8-hr TWA 0.1 mg/m³ (cobalt metal, fume, and dust (as Co)) [R8, 84].

THRESHOLD LIMIT 8-hr time-weighted average (TWA) 0.02 mg/m³ (1994) (cobalt, elemental, and inorganic compound, as Co) [R8, 17].

INTERNATIONAL EXPOSURE LIMITS In the Federal Republic of Germany, industrial cobalt emission has been limited to 1 mg/m³ for metallic cobalt and its slightly soluble compounds and 50 mg/m³ for other compounds. (cobalt and compounds) [R10, 260].

ACTION LEVEL Avoid contact with the spilled cargo. Stay upwind. Notify local air, water, and fire authorities of the accident. Evacuate all people to a distance of at least 200 feet upwind and 1,000 feet downwind of the spill.

ON-SITE RESTORATION Dam stream if possible to reduce the flow and prevent further dissipation by water movement. Due to solubility, dredging may not be fruitful. Pump water to suitable tanks. Add hypochlorous acid (HOCl) to residual, then add calcium hydroxide (Ca(OH)₂) to pH 8.5. Allow to settle. Add calcium sulfate (CaSO₄) and hydrochloric acid (HCl) to pH 7. Filter. Ion exchange on weakly acidic resin, neutralize to pH 7 with calcium hydroxide. For more information see ENVIREX Manual EPA 600/2–77–227. Seek professional environmental engineering assistance through EPA's Environmental Response Team (ERT), Edison, NJ, 24-hour no. 908–548–8730.

BEACH OR SHORE RESTORATION Close beach and shore to the public until material has been removed. If tidal, scrape affected area at low tide with a mechanical scraper, and avoid human contact. Do not burn material.

AVAILABILITY OF COUNTERMEASURE MATERIAL Bottom pumps, dredges—fire departments, U.S. Coast Guard, Army Corps of Engineers.

DISPOSAL After the material has been contained, remove it and the contaminated soil and place in impervious containers. If practical, transport material back to the supplier for recovery. If this is not practical, the material should be encapsulated and buried in a specially designated chemical landfill. Not acceptable at a municipal sewage treatment plant.

DISPOSAL NOTIFICATION Notify local and state health authorities, local solid waste disposal authorities, supplier, and shipper.

INDUSTRIAL FOULING POTENTIAL May contaminate boiler feed water.

MAJOR WATER USE THREATENED Fisheries; potable water; industrial water.

FOOD-CHAIN CONCENTRATION POTENTIAL Positive. Microorganisms tend to concentrate cobalt from the surrounding media.

BIOLOGICAL HALF-LIFE It appears that independent of exposure route (inhalation, injection, or ingestion), most of the cobalt will be eliminated rapidly (in humans). A small proportion is, however, eliminated slowly, having a biological half-time on the order of years. (cobalt) [R9, 218].

BIOCONCENTRATION Only a few plant species accumulate cobalt above the 100-ppm level that causes severe phytotoxicity. Hyperaccumulators of cobalt have been found that contain over 1% cobalt in dry leaves. (cobalt salts) [R5].

CERCLA REPORTABLE QUANTITIES Persons in charge of vessels or facilities are required to notify the National Response Center (NRC) immediately, when there is a release of this designated hazardous substance, in an amount equal to or greater than its reportable quantity of 1,000 lb or 454 kg. The toll-free number of the NRC is (800) 424-8802; in the Washington DC metropolitan area, (202) 426-2675. The

rule for determining when notification is required is stated in 40 CFR 302.4 (section IV. D. 3.b) [R6].

DOT *Fire or Explosion:* Some of these materials may burn, but none of them ignites readily [R3].

Health Hazards: Contact may cause burns to skin and eyes. Fire may produce irritating or poisonous gases. Runoff from fire control or dilution water may cause pollution [R3].

Emergency Action: Keep unnecessary people away. Isolate hazard area and deny entry. Self-contained breathing apparatus (SCBA) and structural firefighter's protective clothing will provide limited protection. Call CHEMTREC at 1-800-424-9300 for emergency assistance. If water pollution occurs, notify the appropriate authorities [R3].

Fire: Small Fires: Dry chemical, CO_2, halon, water spray, or standard foam. Large Fires: Water spray, fog, or standard foam is recommended. Move container from fire area if you can do so without risk. Do not scatter spilled material with high-pressure water streams. Dike fire control water for later disposal [R3].

Spill or Leak: Stop leak if you can do so without risk. Small Dry Spills: With clean shovel place material into clean, dry container, and cover; move containers from spill area. Small Spills: Take up with sand or other noncombustible absorbent material and place into containers for later disposal. Large Spills: Dike far ahead of liquid spill for later disposal. Cover powder spill with plastic sheet or tarp to minimize spreading [R3].

First Aid: In case of contact with material, immediately flush eyes with running water for at least 15 minutes. Wash skin with soap and water. Remove and isolate contaminated clothing and shoes at the site [R3].

FIRE POTENTIAL Material itself does not burn or burns with difficulty [R1].

REFERENCES

1. Association of American Railroads. *Emergency Handling of Hazardous Ma-*

terials in Surface Transportation. Washington, DC: Assoc. of American Railroads, Hazardous Materials Systems (BOE), 1987. 179.

2. Gilman, A. G., L. S. Goodman, and A. Gilman (eds.). *Goodman and Gilman's The Pharmacological Basis of Therapeutics.* 7th ed. New York: Macmillan Publishing Co., Inc., 1985. 1319.

3. Department of Transportation. *Emergency Response Guidebook 1987.* DOT P 5800.4. Washington, DC: U.S. Government Printing Office, 1987, p. G–31.

4. NIOSH; *Criteria Document: Cobalt* p. 37 (1981) DHEW Pub. NIOSH 82–107.

5. Parr, J. F., P. B. Marsh, and J. M. Kla (eds.). *Land Treatment of Hazardous Wastes.* Park Ridge, NJ: Noyes Data Corporation, 1983. 176.

6. 54 FR 33419 (8/14/89).

7. National Research Council. *Drinking Water and Health,* Volume 1. Washington, DC: National Academy Press, 1977.

8. American Conference of Governmental Industrial Hygienists. *Threshold Limit Values for Chemical Substances and Physical Agents and Biological Exposure Indices for 1994–1995.* Cincinnati, OH: ACGIH, 1994.

9. Friberg, L., G.F. Nordberg, E. Kessler, and V.B. Vouk (eds.). *Handbook of the Toxicology of Metals.* 2nd ed. Vols I, II.: Amsterdam: Elsevier Science Publishers B.V., 1986, p. V2.

10. Seiler, H.G., H. Sigel, and A. Sigel (eds.). *Handbook on the Toxicity of Inorganic Compounds.* New York, NY: Marcel Dekker, Inc. 1988.

■ COPPER OXALATE

CAS # 5893–66–3

SIC CODES 2819; 2879

SYNONYMS oxalic acid, copper(2+) salt (1:1) • ethanedioic acid, copper(2+) salt (1:1) • copper oxalate

MF $CuC_2O_4 \cdot H_2O$

MW 160.6

STANDARD CODES EPA 311; TSCA; Coast Guard classification Poison B, Poison label; not listed in IATA; CFR 49 poison class B NOS, poison label, storage code 1, 2; CFR 14 Cab code 8; not listed ICC.

PERSISTENCE Persistent.

INHALATION LIMIT 0.2 mg/m^3.

GENERAL SENSATION Contact with skin may produce itching and an eruption on continued contact that may result in some degree of necrosis. Contact with the eye can cause severe conjunctivitis, edema of lids, and ulceration of the cornea. Inhalation may cause severe congestion of nasal mucosa and possibly ulceration. At fire temperatures may produce coughing and choking, burning of eyes, difficult breathing, and cyanosis.

PERSONAL SAFETY Protect against both inhalation and absorption through the skin. Must wear protective clothing, including NIOSH-approved rubber gloves and boots, safety goggles or face mask, hooded suit, and either a respirator whose canister is specifically approved for this material, or a self-contained breathing apparatus. Decontaminate fully or dispose of all equipment after use.

ACUTE HAZARD LEVEL Acute poisoning from the ingestion of copper salts is rarely severe, if the metal is removed promptly by emesis. Vomiting is provoked chiefly by the local irritant and astringent action of ionic copper on stomach and bowel. If vomiting fails to occur, gradual absorption from the bowel may cause systemic copper poisoning. The 96-hour TLM for fathead minnows is 1.02 ppm. The lethal dose of any copper salt varies widely, from less than 1 gram to several ounces in adults.

CHRONIC HAZARD LEVEL The OSHA standard for copper dust is 1 mg/m^3. The threshold-limit value for copper dust is 1 mg/m^3, and the short-term exposure limit is 2 mg/m^3. EPA has designated this material a category B, with an aquatic LC$_{50}$ value between 1 and 10 mg/L. Chronic copper poisoning due to excessive intake is rarely recognized in man. A type of chronic copper poisoning in man is recognized in the form of a metabolic disease called Wilson's disease, in which tissue copper levels are elevated, and this has been noted to precede the development of liver pathology.

DEGREE OF HAZARD TO PUBLIC HEALTH Material is an eye, skin, and respiratory irritant.

OCCUPATIONAL RECOMMENDATIONS None listed.

ACTION LEVEL Avoid contact and inhalation of the spilled cargo. Stay upwind. Notify local fire, air, and water authorities of accident. Evacuate all people to a distance of at least 200 feet upwind and 1,000 feet downwind of the spill.

ON-SITE RESTORATION Dam stream to reduce flow and to retard dissipation by water movement. Add activated carbon to the water. Bottom pumps, dredging, or underwater vacuum systems may be employed in small bodies of water; dredging may be effective in larger bodies to remove undissolved materials and sorbent. An alternative method, done under controlled conditions, is to pump water into a suitable container, pass water into a gravity separation tank, dual filtration system, and an activated carbon filter. For more details see Envirex Manual EPA 600/2-77-227. Seek professional environmental engineering assistance through EPA's Environmental Response Team (ERT), Edison, NJ, 24-hour no. 908-548-8730.

BEACH OR SHORE RESTORATION Close beach and shore to public until material has been removed. If tidal, scrape affected area at low tide with mechanical scraper (avoid human toxicity contact).

AVAILABILITY OF COUNTERMEASURE MATERIAL Bottom pumps—available through fire departments, EPA Regional Offices, U.S. Coast Guard, or Army Corps of Engineers; dredging—Army Corps of Engineers; underwater vacuum systems—swimming-pool operators; activated carbon—water treatment plants or chemical companies.

DISPOSAL After the material has been contained, remove it and the contaminated soil and place in impervious containers. If practical, transport material back to the supplier or chemical company to recover the heavy-metal content and for deactivation. If this is not practical, or facilities not available, the material should be encapsulated and buried in a specially designated chemical landfill. When diluted and neutralized, material is amenable to biological treatment at a municipal sewage treatment plant.

DISPOSAL NOTIFICATION Notify local and state health authorities, local solid waste disposal authorities, supplier, and shipper.

INDUSTRIAL FOULING POTENTIAL Potential for fouling of boilers is believed to be low.

MAJOR WATER USE THREATENED All users of downstream waters, monitor for limit of 1 mg/L as copper.

PROBABLE LOCATION AND STATE Blue-green or blue-white powder, will sink and can be almost completely recovered by physical means and would be evidenced as a layer on the bottom.

COLOR IN WATER Blue-green to blue-white.

FOOD-CHAIN CONCENTRATION POTENTIAL Positive. Copper has been shown to be accumulated by shellfish.

GENERAL RESPONSE Stop discharge if possible. Keep people away. Avoid contact with solid and dust. Isolate and remove discharged material. Notify local health and pollution control agencies.

FIRE RESPONSE Not flammable. Poisonous gases may be produced when heated.

EXPOSURE RESPONSE Call for medical aid. Dust irritating to eyes, nose, and throat. If inhaled will cause coughing or difficult breathing. If in eyes, hold eyelids open and flush with plenty of water. If breathing has stopped, give artificial respiration. If breathing is difficult, give oxygen. Solid irritating to skin and eyes. If swallowed will cause nausea, vomiting, or loss of consciousness. Remove contaminated clothing and shoes. Flush affected areas with plenty of water. If in eyes, hold eyelids open and flush with plenty of water. If swallowed and victim is conscious, have victim drink water or milk and have victim induce vomiting. If swallowed and victim is unconscious or having convulsions, do nothing except keep victim warm.

RESPONSE TO DISCHARGE: Should be removed; chemical and physical treatment.

WATER RESPONSE Effect of low concentrations on aquatic life is unknown. May be dangerous if it enters water intakes. Notify local health and wildlife officials. Notify operators of nearby water intakes.

EXPOSURE SYMPTOMS Inhalation causes irritation of nose and throat. Ingestion of very large amounts may produce symptoms of oxalate poisoning; watch for edema of the glottis and delayed constriction of esophagus. Contact with eyes causes irritation.

EXPOSURE TREATMENT Inhalation: remove to fresh air; if exposure has been prolonged, watch for symptoms of oxalate poisoning (nausea, shock, collapse, and convulsions). Ingestion: give large amount of water; induce vomiting; get medical attention. Eyes: flush with water for at least 15 min. Skin: flush with water.

REFERENCES
1. Chemical Hazard Response Information System. U.S. Dept. of Transportation, Coast Guard.

■ M-CRESOL

OH

CH₃

CAS # 108-39-4

DOT # 2076

SYNONYMS 3-cresol • m-cresol (ACGIH, DOT, OSHA) • m-cresole • m-cresylic acid • 1-hydroxy-3-methylbenzene • m-hydroxytoluene • 3-hydroxytoluene • m-Kresol • m-methylphenol •

3-methylphenol • m-oxytoluene • phenol, 3-methyl (9CI) • RCRA waste number U052 • m-toluol • UN2076 (DOT)

MF C_7H_8O

MW 108.15

SIC CODES 2842; 2892; 2861

SYNONYMS 3-methylphenol

BINARY REACTANTS Air.

STANDARD CODES EPA 311; NFPA—2, 1, 0; no ICC; USCG—grade D or E combustible liquid, depending on flash point cresol mixture; IATA—(liquid) poison B, poison label, 1 liter passenger, 220 liter cargo; Superfund designated (hazardous substances) list.

PERSISTENCE Biodegrades at moderate pace, but can alter aesthetics at very low levels.

MAJOR SPECIES THREATENED All species.

INHALATION LIMIT 0.022 mg/m³.

GENERAL SENSATION Skin, eye—dangerous; recognition odor in air 0.25 ppm [R2]; sweet tarry odor. May cause skin eruptions. Adsorption may lead to liver and kidney damage. Corrosive to body tissue. Toxic by inhalation, skin absorption, ingestion. TLV 5 ppm. Toxic hazard increases with increasing temperatures; toxic vapors prodiced upon heating.

PERSONAL SAFETY Eye and respiratory equipment, and rubberized clothing recommended. Protective creams are not adequate.

ACUTE HAZARD LEVEL 8 grams can be fatal to man. Toxic via all routes. Extremely corrosive irritant and allergen. Emits highly toxic fumes when heated to decomposition. Alters taste of water and fish flesh. Threshold concentration for fish 1 ppm as phenol; for salt water, 5 ppm as phenol [R1].

CHRONIC HAZARD LEVEL Dermatitis or liver and kidney injury. For boating and fishing, maximum concentration should be <10 ppm.

DEGREE OF HAZARD TO PUBLIC HEALTH

Toxic via all routes. Irritant and allergen. Highly corrosive to skin. Emits highly toxic fumes when heated to decomposition.

OCCUPATIONAL RECOMMENDATIONS None listed.

ACTION LEVEL Notify fire and air authority. Evacuate area. Warn civil defense of possible explosion. Enter from upwind side. Remove all ignition source. Restrict access to affected waters.

ON-SITE RESTORATION May pump some off bottom initially. Peat or carbon can be used on soluble portion. Seek professional environmental engineering assistance through EPA's Environmental Response Team (ERT), Edison, NJ, 24-hour no., 908–548–8730.

BEACH OR SHORE RESTORATION Should not be burned.

AVAILABILITY OF COUNTERMEASURE MATERIAL Pumps—fire department; peat refineries.

DISPOSAL Dilute concentrations can be routed through sewage treatment plants. Stripping, distillation, or adsorption are recommended for higher concentrations. Spray into incinerator or burn in paper packaging. Addition flammable solvent may be added.

DISPOSAL NOTIFICATION Contact local sewage authority.

INDUSTRIAL FOULING POTENTIAL Will taint flavors if present in food-processing water. May explode if present in boiler feed water.

MAJOR WATER USE THREATENED All uses. Can destroy fisheries by rendering flesh unpalatable.

PROBABLE LOCATION AND STATE Clear yellow liquid. Will sink initially and dissolve at moderate rate.

WATER CHEMISTRY Acts much like phenol. Forms weakly acidic solutions that are highly reactive in various situations. In the presence of acids, undergoes reactions; dilute nitric acid causes nitration. Phenols pick up chlorine readily, forming more objectionable compounds. Phenols readily oxidize (especially in alkaline solutions), forming a complex mixture of products including quinone and phenoquinone when the oxidant is air. Cresol darkens when exposed to light. Subject to biochemical degradation.

COLOR IN WATER Colorless.

GENERAL RESPONSE Avoid contact with liquid. Keep people away. Wear goggles, self-contained breathing apparatus, and rubber overclothing (including gloves). Stop discharge if possible. Call fire department. Isolate and remove discharged material. Notify local health and pollution control agencies.

FIRE RESPONSE Combustible. Poisonous flammable gases may be produced in fire. Wear goggles and self-contained breathing apparatus. Extinguish with water, dry chemical, foam, or carbon dioxide. Cool exposed containers with water.

EXPOSURE RESPONSE Call for medical aid. Liquid. Will burn skin and eyes. Poisonous if swallowed. Remove contaminated clothing and shoes. Flush affected areas with plenty of water. If swallowed and victim is conscious, have victim drink water or milk, and have victim induce vomiting.

RESPONSE TO DISCHARGE Issue warning—water contaminant, poison. Restrict access. Should be removed. Provide chemical and physical treatment.

WATER RESPONSE Harmful to aquatic life in very low concentrations. May be dangerous if it enters water intakes. Notify local health and wildlife officials. Notify operators of nearby water intakes.

EXPOSURE SYMPTOMS Inhalation: mucosal irritation and systemic poisoning. Eyes: intense irritation and pain, swelling of conjunctiva, and corneal damage may occur. Skin: intense burning, loss of feeling, wrinkling, white discoloration, and softening. Gangrene may occur. Ingestion: burning sensation in mouth and esophagus. Vomiting may result. Acute exposure by all routes may cause muscular weakness, gastroenteric disturbances, severe depression, collapse. Effects are primarily on central nervous system, and edema of lungs. Injury of spleen and pancreas may occur.

EXPOSURE TREATMENT Call a physician. Inhalation: move to fresh air. Irritation of nose or throat may be relieved to some extent by spraying or gargling with water until all odor disappears. For respiratory distress administer oxygen. Eyes: irrigate with copious quantities of running water for at least 15 min. Skin: remove contaminated clothing. Wash with soap and water until all cresol odor disappears. Follow with alcohol or glycerin (20% solution) wash. Follow with water. Ingestion: dilute with large quantities of liquid (salt water, weak sodium bicarbonate solution, milk, or gruel). Follow with demulcent such as raw egg white or corn starch paste. Induce vomiting.

REFERENCES

1. Todd, D. K., *The Water Encyclopedia*, Maple Press, 1970.

2. Sullivan, R. J., "Air Pollution Aspects of Odorous Compounds," *NTIS PB* 188 089, September 1969.

■ (E)-CROTONALDEHYDE

CAS # 123–73–9

SYNONYMS 2-butenal (trans) • 2-butenal, (e)- • aldehyde crotonique (French) • beta-methyl acrolein • crotenaldehyde • crotonaldehyde • crotonaldehyde, (e)- • e-2-butenal • nci-c56279 • trans-2-butenal • trans-crotonaldehyde • topanel • topanel-ca

MF C_4H_6O

MW 70.10

ODOR THRESHOLD Odor detection in air 5.25×10^2 ppb (gas chromatically pure) [R5].

DENSITY 0.869 at 20°C.

BOILING POINT 104°C.

MELTING POINT −74°C.

AUTOIGNITION TEMPERATURE +207°C [R17, 619].

EXPLOSIVE LIMITS AND POTENTIAL Vapor forms explosive mixtures with air. If polymerization takes place in container, there is possibility of violent rupture of container. Extremely violent polymerization reaction results when in contact with alkaline materials such as caustics, ammonia, or amines. [R16, p. 49–32].

FLAMMABILITY LIMITS In air % by vol—lower 2.1, upper 15.5 [R17, 619].

FIREFIGHTING PROCEDURES In advanced or massive fires, firefighting should be done from an explosion-resistant location. Use dry chemical, foam, or carbon dioxide. Water may be ineffective but should be used to keep fire-exposed containers cool. If leak or spill has not ignited, use water spray to disperse vapors. If it is necessary to stop a leak, use water spray to protect men attempting to do so. Water spray may be used to flush spills away from exposures [R16, p. 49–32]. Evacuation: If fire becomes uncontrollable, or container is exposed to direct flame, consider evacuation of one–third (1/3) mile radius [R4]. If material is on fire or involved in a fire: Do not extinguish fire unless flow can be stopped. Use water in flooding quantities as fog. Solid streams of water may spread fire. Cool all affected containers with flooding quantities of water. Apply water from as far a distance as possible. Use "alcohol" foam, dry chemical, or carbon dioxide [R4].

SOLUBILITY More than 50 g dissolve in 100 ml water; soluble in alcohol, ether, acetone, benzene. Water: solubility of 181,000 ppm at 20°C.

VAPOR PRESSURE 19 mm Hg at 20°C.

OCTANOL/WATER PARTITION COEFFICIENT Log K_{ow} = 0.63.

STABILITY AND SHELF LIFE May deteriorate in normal storage and cause hazard. [R17, 618].

PROTECTIVE EQUIPMENT Wear full protective clothing. [R16, p. 49–33].

PREVENTATIVE MEASURES Contact lenses should not be worn when working with this chemical [R6].

CLEANUP 1. Remove all ignition sources. 2. Ventilate area of spill or leak. 3. For small quantities, absorb on paper towels. Evaporate in a safe place (such as fume hood). Allow sufficient time for evaporating vapors to completely clear hood ductwork. Burn the paper in a suitable location away from other combustible materials. Large quantities can be reclaimed or collected and atomized in a suitable combustion chamber. Liquid crotonaldehyde should not be allowed to enter a confined space, such as a sewer, because of the possibility of an explosion (crotonaldehyde) [R15, 4]. Crotonaldehyde may be neutralized in chemical process waste water by adjusting the water to a pH of 8 with alkali hydroxide, such as NaOH, and heating for 15–30 min at 80–100°C [R18, 213]. Environmental considerations: land spill: Dig a pit, pond, lagoon, holding area to contain liquid or solid material. (SRP: If time permits, pits, ponds, lagoons, soak holes, or holding areas should be sealed with an impermeable flexible-membrane liner.) Dike surface flow using soil, sandbags, foamed polyurethane, or foamed concrete. Absorb bulk liquid with fly ash or cement powder. Apply appropriate foam to diminish vapor and fire hazard. Add sodium bisulfite (NaHSO$_3$) [R4]. Environmental considerations: water spill: Use natural barriers or oil spill control booms to limit spill travel. Use surface-active agent (detergent, soaps, alcohols), if approved by EPA. Inject "universal" gelling agent to solidify encircled spill, and increase effectiveness of booms. Remove trapped material with suction hoses. Use mechanical dredges or lifts to remove immobilized masses of pollutants and precipitates [R4]. Environmental considerations: air spill: Apply water spray or mist to knock down vapors [R4].

DISPOSAL At the time of review, criteria for land treatment or burial (sanitary landfill) disposal practices are subject to significant revision. Prior to implementing land disposal of waste residue (including waste sludge), consult with environmental regulatory agencies for guidance on acceptable disposal practices [R7].

COMMON USES Chemical intermediate for n-butyl alcohol, quinaldine, crotonic acid, surface-active agents, textile and paper sizes, insecticides, and flavoring agents; fuel-gas warning agent; solvent for polyvinyl chloride; alcohol denaturant; leather tanning agent [R1].

OCCUPATIONAL RECOMMENDATIONS None listed.

OSHA Meets criteria for OSHA medical records rule [R11].

THRESHOLD LIMIT 8-hr time-weighted average (TWA) 2 ppm, 5.7 mg/m^3 (1987) (crotonaldehyde) [R19, 17].

IARC SUMMARY Evaluation: There is inadequate evidence in humans for the carcinogenicity of crotonaldehyde. There is inadequate evidence in experimental animals for the carcinogenicity of crotonaldehyde. Overall evaluation: Crotonaldehyde is not classifiable as to its carcinogenicity to humans (Group (3) [R8].

BIODEGRADATION Crotonaldehyde (isomer not reported) was observed to have a 5-day BODT of 37% using the AFNOR T. 90 test protocol (1). Crotonaldehyde has been found to be degradable via anaerobic (methane fermentation) biotechnology (2,3) [R9].

BIOCONCENTRATION Based on a water solubility of 181,000 ppm at 20°C (1) using a log K_{ow} of 0.63 (2), the log BCF can be estimated to be −0.17, from a recommended regression-derived equation (3,SRC) [R10].

ATMOSPHERIC STANDARDS This action promulgates standards of performance for equipment leaks of volatile organic compounds (VOC) in the synthetic organic chemical manufacturing industry (SOCMI). These standards implement Section 111 of the Clean Air Act and are based on the Administrator's determination that emissions from the SOCMI cause, or contribute significantly to, air pollution that may reasonably be anticipated to endanger public health or welfare. The intended effect of these standards is to require all newly constructed, modified, and reconstructed SOCMI process units to use the best demonstrated system of continuous emission reduction for equipment leaks of VOC, considering costs, non-air-quality health and environmental impact, and energy requirements. Crotonaldehyde is produced, as an intermediate or final product, by process units covered under this subpart. These standards of performance become effective upon promulgation but apply to affected facilities for which construction or modification commenced after January 5, 1981. (crotonaldehyde) [R12].

TSCA REQUIREMENTS Pursuant to section 8 (d) of TSCA, EPA promulgated a model health and safety data reporting rule (40 CFR Part 716). The section 8 (d) model rule requires manufacturers, importers, and processors of listed chemical substances and mixtures to submit to EPA copies and lists of unpublished health and safety studies. Crotonaldehyde is included on this list. (crotonaldehyde) [R13].

RCRA REQUIREMENTS As stipulated in 40 CFR 261.33, when crotonaldehyde, as a commercial chemical product or manufacturing chemical intermediate or an off-specification commercial chemical product or a manufacturing chemical intermediate, becomes a waste, it must be managed according to federal or state hazardous waste regulations. Also defined as a hazardous waste is any residue, contaminated soil, water, or other debris resulting from the cleanup of a spill, into water or on dry land, of this waste. Generators of small quantities of this waste may qualify for partial exclusion from hazardous-waste regulations (see 40 CFR 261.5). (crotonaldehyde) [R14].

DOT *Health Hazards:* Poisonous; may be fatal if inhaled, swallowed, or absorbed through skin. Contact may cause burns to skin and eyes. Runoff from fire control or dilution water may cause pollution [R2].

Fire or Explosion: Flammable/combustible material; may be ignited by heat, sparks, or flames. Vapors may travel to a source of ignition, and flash back. Container may explode in heat of fire. Vapor explosion and poison hazard indoors, outdoors, or in sewers. Runoff to sewer may create fire or explosion hazard [R2].

Emergency Action: Keep unnecessary people away; isolate hazard area and deny entry. Stay upwind; keep out of low areas. Wear positive-pressure breathing apparatus and special protective clothing. Isolate for 1/2 mile in all directions if tank car or truck is involved in fire. If water pollution occurs, notify appropriate authorities. For emergency assistance call CHEMTREC (800) 424–9300 [R2].

Fire: Small Fires: Dry chemical, CO_2, water spray, and foam. Large Fires: Water spray, fog, or foam. Move container from fire area if you can do so without risk. Dike fire control water for later disposal; do not scatter the material. Cool containers that are exposed to flames with water from the side until well after fire is out. Withdraw immediately in case of rising sound from venting safety device or any discoloration of tank due to fire [R2].

Spill or Leak: Shut off ignition sources; no flares, smoking, or flames in hazard area. Do not touch spilled material; stop leak if you can do so without risk. Use water spray to reduce vapors. Small Spills: Take up with sand or other noncombustible absorbent material and place into containers for later disposal. Large Spills: Dike far ahead of spill for later disposal [R2].

First Aid: Move victim to fresh air; call emergency medical care. If not breathing, give artificial respiration. If breathing is difficult, give oxygen. Remove and isolate contaminated clothing and shoes at the site. In case of contact with material, immediately flush skin or eyes with running water for at least 15 minutes. Keep victim quiet and maintain normal body temperature. Effects may be delayed; keep victim under observation [R2].

FIRE POTENTIAL Flammable liquid [R16, p. 49–32].

FIRE HAZARD When exposed to heat or flame; can react with oxidizing materials [R3].

REFERENCES

1. SRI.

2. Department of Transportation. *Emer-*

gency Response Guidebook 1984 DOT P 5800.3 Washington, DC: U.S. Government Printing Office, 1984, p. G–28.

3. Sax, N. I. *Dangerous Properties of Industrial Materials.* 6th ed. New York, NY: Van Nostrand Reinhold, 1984. 817.

4. Association of American Railroads. *Emergency Handling of Hazardous Materials in Surface Transportation.* Washington, DC: Assoc. of American Railroads, Hazardous Materials Systems (BOE), 1987. 204.

5. Fazzalari, F. A. (ed.). *Compilation of Odor and Taste Threshold Values Data.* ASTM Data Series DS 48A (Committee E–18). Philadelphia, PA: American Society for Testing and Materials, 1978. 25.

6. NIOSH. *Pocket Guide to Chemical Hazards.* 2nd Printing. DHHS (NIOSH) Publ. No. 85–114. Washington, DC: U.S. Dept. of Health and Human Services, NIOSH/Supt. of Documents, GPO, February 1987. 89.

7. SRP.

8. IARC. *Monographs on the Evaluation of the Carcinogenic Risk of Chemicals to Man.* Geneva: World Health Organization, International Agency for Research on Cancer, 1972–present (multivolume work), p. 63 361 (1995).

9. (1) Dore, M., et al. *Trib Cebedeau* 28: 3–11 (1975). (2) Chou, W.L., et al. *Biotechnol Bioeng Symp* 8: 391–414 (1979). (3) Speece, R. E. *Environ Sci Technol* 17: 416A–27A (1983).

10. (1) Baxter, W. F., Jr. *Kirk–Othmer Encycl Chem Tech* 3rd ed. New York: Wiley, 7: 207–18 (1979). (2) GEMS. *Graphical Exposure Modeling System.* CLOGP. PC Version (1987). (3) Lyman, W. J., et al., *Handbook of Chemical Property Estimation Methods.* New York: McGraw-Hill, p. 5–4,5–10 (1982).

11. 29 CFR 1910.20 (7/1/87).

12. 40 CFR 60.489 (7/1/87).

13. 51 FR 32720 (7/1/87).

14. 53 FR 13382 (4/22/88).

15. Mackison, F. W., R. S. Stricoff, and L.

J. Partridge, Jr. (eds.). *NIOSH/OSHA— Occupational Health Guidelines for Chemical Hazards.* DHHS (NIOSH) Publication No. 81–123 (3 vols). Washington, DC: U.S. Government Printing Office, Jan. 1981.

16. National Fire Protection Association. *Fire Protection Guide on Hazardous Materials.* 9th ed. Boston, MA: National Fire Protection Association, 1986.

17. Sunshine, I. (ed.). *CRC Handbook of Analytical Toxicology.* Cleveland: The Chemical Rubber Co., 1969.

18. *Kirk-Othmer Encyclopedia of Chemical Technology.* 3rd ed., Volumes 1–26. New York, NY: John Wiley and Sons, 1978–1984, p. V7.

19. American Conference of Governmental Industrial Hygienists. *Threshold Limit Values for Chemical Substances and Physical Agents and Biological Exposure Indices for 1994–1995.* Cincinnati, OH: ACGIH, 1994.

■ CROTONIC ACID, 3-HYDROXY-, alpha-METHYLBENZYL ESTER, DIMETHYL PHOSPHATE, (e)-

CAS # 7700–17–6

SIC CODES 2879

SYNONYMS 1-methylbenzyl-3-(dimethoxyphosphinyloxo) isocrotonate • 1-phenylethyl (e)-3-[(dimethoxyphosphinyl) oxy]-2-butenoate • 2-butenoic acid, 3-((dimethoxyphosphinyl) oxy)-, 1-phenylethyl ester, (e)- • 3-hydroxycrotonic acid alpha-methylbenzyl ester dimethyl phosphate • 3-[(dimethoxyphosphinyl) oxy]-2-butenoic acid 1-phenylethyl ester •

a-methyl benzyl-3-hydroxy cis-crotonate dimethyl phosphate • alpha-methyl benzyl-3-(dimethoxy-phosphinyloxy)-cis-crotonate • alpha-methylbenzyl 3-hydroxycrotonate dimethyl phosphate • cis-2-(1-phenylethoxy) carbonyl-1-methylvinyl dimethylphosphate • crotonic acid, 3-hydroxy-, alpha-methylbenzyl ester, dimethyl phosphate, (e)- • crotoxyphos • cyodrin • dimethyl 2-(alpha-methylbenzyloxycarbonyl)-1-methylvinyl phosphate • dimethyl phosphate of alpha-methylbenzyl 3-hydroxy-cis-crotonate • dimethyl-cis-1-methyl-2-(1-phenylethoxycarbonyl) vinyl phosphate • ent-24,717 • kemdrin • o,o-dimethyl o-[1-methyl-2-(1-phenylcarbethoxy) vinyl] phosphate • pantozol-1 • sd-4294 • shell-sd-4294 • simax • volfazol • o,o-dimethyl o-(1-methyl-2-carboxy-alpha-phenylethyl) vinyl • cypona-ec • decrotox • duo-kill • duravos • crotoxyfos • alpha-methylbenzyl (e)-3-hydroxycrotonate ester with dimethyl phosphate

MF $C_{14}H_{19}O_6P$

MW 314.30

COLOR AND FORM Light-straw-colored liquid.

ODOR Mild ester.

DENSITY 1.19 at 25°C.

BOILING POINT 135°C at 0.03 mm Hg.

SOLUBILITY Solubility in water at room temp: 0.1%; slightly soluble in kerosene, saturated hydrocarbons; soluble in acetone, chloroform, ethanol, highly chlorinated hydrocarbons.

STABILITY AND SHELF LIFE In aqueous solution at 38°C, 50% is hydrolyzed in 87 hr at pH 1, 35 hr at pH 9. Will not attack fiberglass, reinforced polyester, rigid PVC, or the usual lacquers used for lining drums [R1].

OTHER CHEMICAL/PHYSICAL PROPERTIES Density: 1.2 at 15°C; index of refraction: 1.5005 at 25°C; slightly corrosive to mild steel, copper, lead, zinc, tin; soluble in ethanol, propan-2-ol, xylene; vapor pressure: 1.4×10^{-5} mm Hg at 20°C (technical).

PROTECTIVE EQUIPMENT Rubber gloves and respirator usually specified for commercial spraying [R4, p. C–55].

PREVENTATIVE MEASURES Avoid unnecessary exposure to spray operator or contamination of feedstuffs. [R3, 120].

COMMON USES Insecticide for external use on livestock [R2].

BINARY REACTANTS Incompatible with most mineral carriers, except synthetic silicas such as hisil 233, and colloidal silica k320. It is compatible with dichlorvos.

PERSISTENCE Half-life of ciodrin in aqueous environment ranges from 35 hr at pH 9 to 87 hr at pH 1 at 38°C. Persistence in Coachella fine sand as bioassayed by *Hippelates collusor* larvae is >6 months.

DIRECT CONTACT Nonreactive pupils, sweating.

GENERAL SENSATION Acute intoxication produces general weakness, headache, tightness in chest, blurred vision, nonreactive pupils, salivation, sweating, nausea, vomiting, diarrhea, abdominal cramps, or convulsions.

CHRONIC HAZARD LEVEL Rats fed for 90 days on a diet containing up to 900 ppm for males and up to 300 ppm for females were not affected in growth nor were histopathological changes observed. Dietary levels of 7 ppm do not affect whole-blood cholinesterase of rats. The effect on dogs is approximately the same.

OCCUPATIONAL RECOMMENDATIONS None listed.

ON-SITE RESTORATION Seek professional environmental engineering assistance through EPA's Environmental Response Team (ERT), Edison, NJ, 24-hour no., 908–548–8730. Contain and isolate spill by using clay/bentonite dams, interceptor trenches, or impoundments. Construct swale to divert uncontaminated portion of watershed around contaminated portion. Seek professional help to evaluate problem, implement containment measures, and conduct bench scale and pilot scale tests prior to full-scale decontamination program implementation. Treatment alternatives for contaminated water include powdered activated carbon, granular activated carbon, bentonite sorption, polypropylene sorption, aeration, evaporation, biodegradation, and alkaline hydrolysis. Treatment alternatives for contaminated soils include well point collection and treatment of leachates as for contaminated water, or bentonite/cement ground injection to immobilize spill with physical removal of immobilized residues, with pH adjustment to 8.0 and land application on an approved sanitary landfill. Residues may also be packaged for disposal.

DISPOSAL Product residues and sorbent media may be packaged in 17h epoxy-lined drums and disposed of at an approved EPA disposal site. Destruction by high-temperature incineration with gas-scrubbing equipment, if available. Encapsulation by organic polyester resin or silicate fixation. Confirm disposal procedures with responsible environmental engineer and regulatory officials.

PROBABLE LOCATION AND STATE Straw-colored liquid with a mild ester odor. Slightly soluble in H_2O. May dissolve and sink.

WATER CHEMISTRY Ciodrin decomposes moderately fast in water. Its half-life in aqueous solution at 38°C and pH 9 is 35 hr; at pH 1, 87 hr. It also undergoes chemical hydrolysis in basic media. It is considered to involve sorption-catalyzed hydrolysis. Color in water: light straw-colored liquid will be diluted and hydrolyzed.

FOOD-CHAIN CONCENTRATION POTENTIAL Residues in milk and meat are negligible.

ECOTOXICITY LD_{50} mallards oral 790 mg/kg (95% confidence limit 411–1,520 mg/kg), 3–4-mo-old males (technical, 85%) [R5].

REFERENCES

1. Worthing, C. R., S. B. Walker (eds.). *The Pesticide Manual—A World Compendium.* 7th ed. Lavenham, Suffolk, Great Britain: The Lavenham Press Limited, 1983. 138.

2. *The Merck Index.* 10th ed. Rahway, NJ: Merck Co., Inc., 1983. 373.

3. Rossoff, I. S. *Handbook of Veterinary Drugs.* New York: Springer Publishing Company, 1974.

4. *Farm Chemicals Handbook 1984.* Willoughby, OH: Meister Publishing Co., 1984.

5. U.S. Department of the Interior, Fish and Wildlife Service. *Handbook of Toxicity of Pesticides to Wildlife.* Resource Publication 153. Washington, DC: U.S. Government Printing Office, 1984. 25.

■ CUPRIC AMMONIA SULFATE

CAS # 10380–29–7

SIC CODES 2819; 2879

SYNONYMS Ammonio cupric • copper aminosulfate • copper(2+), tetraammine sulfate(1:1) monohydrate • ammoniated cupric sulfate • curpric tetraammine sulfate monohydrate • (tetraammine)copper sulfate hydrate • copper tetraammine sulfate monohydrate • copper(2+), tetraammine-, sulfate, monohydrate

MF $CuSO_3NH_4 \cdot H_2O$

MW 245.74

MELTING POINT Decomposes at 150°C.

BINARY REACTANTS Acids, soluble sulfides.

STANDARD CODES EPA 311; TSCA; Coast Guard Classification Poison B, poison label; not listed IATA; CFR 49 Poison class B NOS, Poison label, 50 lb; passenger 200 lb. cargo, storage code 1, 2; CFR 14 CAB code 8; ICC STCC tariff no 1–C 28 195 17.

PERSISTENCE Persistent.

INHALATION LIMIT 0.2 mg/m³.

GENERAL SENSATION Skin contact produces itching and an eruption on continued contact. Eye contact produces severe conjunctivitis, edema of lids, ulceration of cornea. Inhalation can lead to severe congestion of nasal mucosa and possible ulceration, coughing, vomiting, reddening of lips, mouth, nose, and throat. Also quite toxic by ingestion.

PERSONAL SAFETY Protect against both inhalation and skin contact. NIOSH-approved self-contained breathing apparatus, rubber gloves and boots, and hooded rubber suits must be worn.

ACUTE HAZARD LEVEL Acute poisoning from the ingestion of copper salts is rarely severe, if the metal is removed promptly by emesis. Vomiting is provoked chiefly by the local irritant and astringent action of ionic copper on stomach and bowel. If vomiting fails to occur, gradual absorption from the bowel may cause system copper poisoning. The 96-hour tlm for fathead minnows is 1.54 ppm. The lethal dose of any copper salt varies widely, from less than 1 g to several ounces in adults. The material is considerably less toxic than cupric sulfate.

CHRONIC HAZARD LEVEL The OSHA standard for copper dust is 1 mg/m³. The threshold-limit value for copper dust is 1 mg/m³ and the short-term exposure limit is 2 mg/m³. EPA has designated this material a category B, with an aquatic LC_{50} value between 1 and 10 mg/L. Chronic copper poisoning due to excessive intake is rarely recognized in man. A type of chronic copper poisoning in man is recognized in the form of a metabolic disease called Wilson's disease, where tissue copper levels are elevated, and this has been noted to precede the development of liver pathology.

DEGREE OF HAZARD TO PUBLIC HEALTH Eye, skin, and respiratory irritant. Highly toxic fumes at fire temperatures.

OCCUPATIONAL RECOMMENDATIONS None listed.

ACTION LEVEL Avoid contact and inhalation of the spilled cargo. Stay upwind. Notify local fire, air, and water authorities of the accident. Evacuate all people to a distance of at least 200 feet upwind and 1,000 feet downwind of the spill.

ON-SITE RESTORATION Contain body of water to retard dissipation by water movement. due to high solubility, check before dredging is attempted. If concentration in water is excessive, pump water into tank trucks or tank cars and return to supplier for reclamation or disposal. An alternative method, done under controlled conditions, is to pump water into suitable container, add hydrochloric acid (HCl) to pH 7, filter, ion exchange, add calcium hydroxide (Ca(OH)₂) to pH 9.5, settle, neutralize, and dilute. For more details see Envirex manual EPA 600/2–77–227. Seek professional environmental engineering assistance through EPA's Environmental Response Team (ERT), Edison, NJ, 24-hour no. 908–548–8730.

BEACH OR SHORE RESTORATION Remove by scraping at low tide. Avoid human toxicity contact.

AVAILABILITY OF COUNTERMEASURE MATERIAL Pumps—fire stations; carbon—water treatment plants, sugar refineries.

DISPOSAL After the material has been contained, remove it and the contaminated soil and place in impervious containers. If practical, transport material back to the supplier or chemical company to recover the heavy-metal content and for deactivation. If this is not practical or facilities not available, the material should be encapsulated and buried in a specially designated chemical landfill. When dilute and neutralized, the material is amenable to biological treatment at a sewage treatment plant.

DISPOSAL NOTIFICATION Notify local and state health authorities, local solid-waste disposal authorities, supplier, and shipper.

INDUSTRIAL FOULING POTENTIAL Corrosion greater than 0.05 inch/year in steel from copper sulfate solutions at less than 100°F and concentrations less than 30%.

MAJOR WATER USE THREATENED All users of downstream water, monitor for concentration in water.

PROBABLE LOCATION AND STATE Dark-blue crystals, will sink and will dissolve in the water by the time cleanup personnel arrive.

WATER CHEMISTRY Concentrated solution

will dissolve cellulose. Color in water, dark blue.

FOOD-CHAIN CONCENTRATION POTENTIAL Positive. Copper has been shown to be accumulated by shellfish.

REFERENCES

1. EPA database OHMTADS.

2. DOT database CHRIS.

■ CUPRIETHYLENEDIAMINE

CAS # 13426–91–0

SYNONYMS bis (ethylenediamine) copper ion • bis (ethylenediamine) copper(2) • bis (ethylenediamine) copper(2) ion • copper(2), bis (1,2-ethanediamine-N,N')- • copper(2), bis (1,2-ethanediamine-N,N')-, (sp-4-1)- • copper (2), bis (ethylenediamine)-, ion • copper-ethylenediamine complex • cupriethylene diamine • cupriethylenediamine, bis (ethylenediamine)-, ion • ethane, 1,2-diamino-, copper complex

MF $C_4H_{16}CuN_4$

MW 183.78

COLOR AND FORM Purple liquid.

ODOR Ammoniacal.

FIREFIGHTING PROCEDURES Do not extinguish fire unless flow can be stopped. Use water in flooding quantities as fog. Cool all affected containers with flooding quantities of water. Apply water from as far a distance as possible. Solid streams of water may be ineffective. Use alcohol foam, carbon dioxide, or dry chemical. [R7, 275].

OTHER CHEMICAL/PHYSICAL PROPERTIES Dissolves cellulose products.

PROTECTIVE EQUIPMENT Goggles or face shield; organic canister mask; rubber gloves; protective clothing (cupriethylenediamine hydroxide solution) [R5].

PREVENTATIVE MEASURES If material not on fire and not involved in fire: Keep sparks, flames, and other sources of ignition away. Keep material out of water

sources and sewers. Build dikes to contain flow as necessary. [R7, 276].

CLEANUP Experiments were made to determine the effectiveness of cation exchange resins in the removal of cupriethylenediamine from industrial wastewater. Resins ku 2–6, ku 2–8, and ku 2–tekh were effective [R6].

DISPOSAL SRP: At the time of review, criteria for land treatment or burial (sanitary landfill) disposal practices are subject to significant revision. Prior to implementing land disposal of waste residue (including waste sludge), consult with environmental regulatory agencies for guidance on acceptable disposal practices.

COMMON USES Cellulose solvent (cupriethylenediamine (CuEn) hydroxide) [R1]; cotton solvent (cupriethylenediamine hydroxide) [R2]; copper as elemental from copper—ethylenediamine complex [R3].

OCCUPATIONAL RECOMMENDATIONS None listed.

DOT *Health Hazards:* Vapors may cause dizziness or suffocation. Exposure in an enclosed area may be very harmful. Contact may irritate or burn skin and eyes. Fire may produce irritating or poisonous gases. Runoff from fire control or dilution water may cause pollution (cupriethylenediamine solution) [R4]. Fire or Explosion: Some of these materials may burn, but none of them ignites readily. Most vapors are heavier than air. Air/vapor mixtures may explode when ignited. Container may explode in heat of fire (cupriethylenediamine solution) [R4].

Emergency Action: Keep unnecessary people away; isolate hazard area and deny entry. Stay upwind, out of low areas, and ventilate closed spaces before entering. Positive-pressure self-contained breathing apparatus (SCBA) and structural firefighters' protective clothing will provide limited protection. Isolate for 1/2 mile in all directions if tank, rail car, or tank truck is involved in fire. Remove and isolate contaminated clothing at the site. Call emergency response telephone number on shipping paper first. If shipping paper not available or no answer, call CHEMTREC at 1–800–424–9300. If water pollution occurs, notify the appropriate authorities (cupriethylenediamine solution) [R4].

Fire: Small Fires: Dry chemical or CO_2. Large Fires: Water spray, fog, or regular foam. Apply cooling water to sides of containers that are exposed to flames until well after fire is out. Stay away from ends of tanks (cupriethylenediamine solution) [R4].

Spill or Leak: Shut off ignition sources; no flares, smoking, or flames in hazard area. Stop leak if you can do so without risk. Small Liquid Spills: Take up with sand, earth, or other noncombustible absorbent material. Large Spills: Dike far ahead of liquid spill for later disposal (cupriethylenediamine solution) [R4].

First Aid: Move victim to fresh air and call for emergency medical care; if not breathing, give artificial respiration; if breathing is difficult, give oxygen. In case of contact with material, immediately flush eyes with running water for at least 15 minutes. Wash skin with soap and water. Remove and isolate contaminated clothing and shoes at the site. Use first-aid treatment according to the nature of the injury (cupriethylenediamine solution) [R4].

FIRE POTENTIAL Nonflammable (cupriethylenediamine hydroxide solution) [R5].

REFERENCES

1. *Kirk–Othmer Encyclopedia of Chemical Technology.* 4th ed. Volumes 1: New York, NY. John Wiley and Sons, 1991–present. p. V5 491.

2. *Kirk–Othmer Encyclopedia of Chemical Technology.* 4th ed. Volumes 1: New York, NY. John Wiley and Sons, 1991–present. p. V7 633.

3. USEPA, *Status of Pesticides in Reregistration and Special Review.* Washington, DC: USEPA–738–R–94–008. p. 285 (1994).

4. U.S. Department of Transportation. *Emergency Response Guidebook 1993.* DOT P 5800.6. Washington, DC: U.S. Government Printing Office, 1993, p. G–74.

5. U.S. Coast Guard, Department of Transportation. *CHRIS—Hazardous Chemical Data.* Volume II. Washington, DC: U. S. Government Printing Office, 1984–5.

6. Makarova, V. A., et al., Izv Vyssh Uchebn Zaved, *Stroit Arkhit* 4: 119–22 (1980).

7. Association of American Railroads. *Emergency Handling of Hazardous Materials in Surface Transportation.* Washington, DC: Association of American Railroads, Bureau of Explosives, 1992.

■ CYANOPHOS

CAS # 2636–26–2

SYNONYMS o,o,-dimethyl o-(4-cyanophenyl) phosphorothioate • o,o-dimethyl o-(4-cyanophenyl) thionophosphate • phosphorothioic acid, o-(4-cyanophenyl) o,o-dimethyl ester • ai3-25675 • bay-34727 • Caswell no 268A • ciafos • cyanox • cyap • ent-25,675 • EPA pesticide chemical code 268200 • o-(4-cyanophenyl) o,o-dimethyl phosphorothioate • o,o-dimethyl phosphorothioate o-ester with p-hydroxybenzonitrile • o,o-Dimethyl-o-(4-cyano-phenyl)-monothiophosphat (German) • phosphorothioic acid, o,o-dimethyl ester, o-ester with p-hydroxybenzonitrile • s-4084 • sumitomo-s-4084 • 4-(dimethoxyphosphinothioyloxy) benzonitrile • phosphorothioic acid o-(4-cyanophenyl) o,o-dimethyl ester • phosphorothioic acid o,o-dimethyl ester, o-ester with p-hydroxybenzonitrile • cynock • (o-p-cyanophenyl-o,o-dimethylphosphorothioate)

MF $C_9H_{10}NO_3PS$

MW 243.21

COLOR AND FORM Yellow to reddish-yellow transparent liquid; clear amber liquid.

DENSITY 1.255–1.265 at 25°C.

BOILING POINT 119–120°C (decomp) at 0.09 mm Hg.

MELTING POINT 14–15°C.

FIREFIGHTING PROCEDURES Dry chemicals, carbon dioxide for small fires. Water spray, foam for large fires (malathion) [R4]. Area surrounding fire should be diked to prevent water runoff (malathion) [R5]. Wear self-contained breathing apparatus (or respirator for organophosphate pesticides) and rubber clothing while fighting fires of malathion with chlorine bleach solution. All clothing contaminated by fumes and vapors must be decontaminated (malathion) [R5]. Extinguish fire using agent suitable for type of surrounding fire. (malathion) [R6].

SOLUBILITY Very soluble in methanol, ethanol, acetone, and chloroform. Sparingly soluble in n-hexane, kerosene; slightly soluble in water; in water at 30°C, 46 mg/L. In n-hexane 27, methanol 1,000, xylene 1,000 (all in g/kg) at 20°C. Miscible in benzene, ketones, toluene, xylene.

VAPOR PRESSURE 105 mPa at 20°C.

STABILITY AND SHELF LIFE Stable to storage >2 yr under normal conditions. It is incompatible with alkaline materials [R1].

OTHER CHEMICAL/PHYSICAL PROPERTIES Vapor pressure: 105 mPa at 20°C (technical grade); amber liquid (technical grade). Solubility: 46 mg/L water at 30°C; 27 g/kg hexane at 20°C (technical grade).

PROTECTIVE EQUIPMENT Respirator selection: upper-limit devices recommended by NIOSH: At 150 mg/m^3: Any supplied-air respirator, or any self-contained breathing apparatus, or any chemical-cartridge respirator with organic vapor cartridge(s) in combination with a dust, mist, and fume filter. At 375 mg/m^3: Any supplied-air respirator operated in a continuous-flow mode, or any powered air-purifying respirator with organic vapor cartridge(s) in combination with a dust, mist, and fume filter. At 750 mg/m^3: Any supplied-air respirator with a full facepiece; or any self-contained breathing apparatus with a full facepiece; or any chemical-cartridge respirator with a full facepiece and organic vapor cartridge(s) in combination with a high-efficiency particulate filter, or any air-purifying full facepiece respirator (gas mask) with a chin-style or front- or back-mounted organic vapor canister having a high-efficiency particulate filter; or any powered air-purifying respirator with a tight-fitting facepiece and organic vapor cartridge(s) in combination with a high-efficiency particulate filter; or any supplied-air respirator with a tight-fitting facepiece operated in a continuous-flow mode. 5,000 mg/m^3: Any supplied-air respirator with a full facepiece and operated in a pressure-demand or other positive-pressure mode. Emergency or planned entry in unknown concentration or immediately dangerous to life or health (IDLH) conditions: Any self-contained breathing apparatus with a full facepiece and operated in a pressure-demand or other positive-pressure mode; or any supplied-air respirator with a full facepiece and operated in pressure-demand or other positive-pressure mode in combination with an auxiliary self-contained breathing apparatus operated in pressure-demand or other positive-pressure mode. Escape: Any air-purifying full facepiece respirator (gas mask) with a chin-style or front- or back-mounted organic vapor canister having a high-efficiency particulate filter; or any appropriate escape type self-contained breathing apparatus. (malathion) [R7].

CLEANUP 1. Ventilate area of spill or leak. 2. Collect for reclamation or absorb in vermiculite, dry sand, earth, or a similar material (malathion) [R13, 4]. Environmental consideration: Landspill: Dig a pit, pond, lagoon, or holding area to contain liquid or solid material. (SRP: If time permits, pits, ponds, lagoons, soak holes, or holding areas should be sealed with an impermeable flexible-membrane liner.) Dike surface flow using soil, sandbags, foamed polyurethane, or foamed concrete. Absorb bulk liquid with fly ash or cement powder (malathion) [R6]. Environmental consideration: Water spill: If dissolved, apply activated carbon at ten times the spilled amount (in region of 10 ppm or greater concentration). Remove trapped material with suction hoses. Use mechanical dredges or lifts to remove immobilized masses of pollutants, and precipitates (malathion) [R6]. Spills of malathion on floors were absorbed with absorbing clay. Sweeping compound may be utilized to facilitate the removal of all visible traces of malathion-contaminated clay. Equipment and fixtures contaminated with malathion were decontaminated with an alkaline solution (5% sodium hydroxide) (malathion) [R8]. Alternative treatment process for spent filter cake from manufacture of malathion requires the following steps: (1) hydrolysis; (2) steam stripping; (3) decantation; (4) composting; and (5) biological treatment (malathion) [R9]. Ultraviolet radiation in conjunction with ozone is a highly effective degradation technique for malathion (malathion) [R10].

DISPOSAL Malathion may be disposed of by absorbing in vermiculite, dry sand, earth, or a similar material, and disposing of so as to meet local, state, and federal regulations (malathion) [R13, 4]. Incineration together with flammable solvent in furnace equipped with afterburner and scrubber is recommended (malathion) [R14]. The following wastewater treatment technologies have been investigated for malathion: Biological treatment and reverse osmosis (malathion) [R15].

COMMON USES Insecticide [R3]. It is an insecticide used at 25–50 g ai/100 L to control lepidopterous pests and sucking insects on fruit, ornamentals, and vegetables. It is a locust-control insecticide and can be used for the control of household pests such as cockroaches, houseflies, and mosquitoes [R1]. Grain protectant [R2].

OCCUPATIONAL RECOMMENDATIONS None listed.

NON-HUMAN TOXICITY LD_{50} mouse oral 1,000 mg/kg [R3]; LD_{50} mouse ip 880 mg/kg [R3]; LD_{50} rat (male) oral 580 mg/kg [R1]; LD_{50} rat (female) oral 610 mg/kg [R1]; LD_{50} mouse percutaneous >2,500 mg/kg [R1].

ECOTOXICITY LC_{50} carp 5 mg/L (48 hr) [R1]; LC_{50} harlequin fish 36 mg/L (24 hr) [R11].

DOT *Health Hazards:* Poisonous; may be fatal if inhaled, swallowed, or absorbed through skin. Contact may cause burns to skin and eyes. Runoff from fire control or dilution water may cause pollution (organophosphorus pesticide, liquid, poisonous, flammable, not otherwise specified) [R12, p. G–28].

Fire or Explosion: Flammable/combustible material; may be ignited by heat, sparks, or flames. Vapors may travel to a source of ignition, and flash back. Container may explode in heat of fire. Vapor explosion and poison hazard indoors, outdoors, or in sewers. Runoff to sewer may create fire or explosion hazard (organophosphorus pesticide, liquid, poisonous, flammable, not otherwise specified) [R12, p. G–28].

Emergency Action: Keep unnecessary people away; isolate hazard area, and deny entry. Stay upwind; keep out of low areas. Positive-pressure self-contained breathing apparatus (SCBA) and chemical protective clothing that is specifically recommended by the shipper or manufacturer may be worn. It may provide little or no thermal protection. Structural firefighters' protective clothing is not effective for these materials. Isolate for 1/2 mile in all directions if tank, rail car, or tank truck is involved in fire. Call CHEMTREC at 1–800–424–9300 for emergency assistance (organophosphorus pesticide, liquid, poisonous, flammable, not otherwise specified) [R12, p. G–28].

Fire: Small Fires: Dry chemical, CO_2, water spray, or alcohol-resistant foam. Large Fires: Water spray, fog, or alcohol-resistant foam. Move container from fire area if you can do so without risk. Dike fire-control water for later disposal; do not scatter the material. Apply cooling water to sides of containers that are exposed to flames until well after fire is out. Stay away from ends of tanks. Withdraw immediately in case of rising sound from venting safety device or any discoloration of tank due to fire (organophosphorus pesticide, liquid, poisonous, flammable, not otherwise specified) [R12, p. G–28].

Spill or Leak: Shut off ignition sources; no flares, smoking, or flames in hazard area. Fully encapsulating, vapor protective clothing should be worn for spills and leaks with no fire. Do not touch or walk through spilled material; stop leak if you can do so without risk. Water spray may reduce vapor; but it may not prevent ignition in closed spaces. Small Spills: Take up with sand or other noncombustible absorbent material and place into containers for later disposal. Large Spills: Dike far ahead of liquid spill for later disposal (organophosphorus pesticide, liquid, poisonous, flammable, not otherwise specified) [R12, p. G–28].

First Aid: Move victim to fresh air, and call emergency medical care; if not breathing, give artificial respiration; if breathing is difficult, give oxygen. In case of contact with material, immediately flush skin or eyes with running water for at least 15 minutes. Remove and isolate contaminated clothing and shoes at the site. Keep victim quiet and maintain normal body temperature. Effects may be delayed; keep victim under observation (organophosphorus pesticide, liquid, poisonous, flammable, not otherwise specified) [R12, p. G–28].

REFERENCES

1. Worthing, C. R., and S. B. Walker (eds.). *The Pesticide Manual—A World Compendium.* 8th ed. Thornton Heath, UK: The British Crop Protection Council, 1987. 200.

2. *Farm Chemicals Handbook 1993.* Willoughby, OH: Meister Publishing Co., 1993, p. C 98.

3. Budavari, S. (ed.). *The Merck Index— Encyclopedia of Chemicals, Drugs, and Biologicals.* Rahway, NJ: Merck and Co., Inc., 1989. 352.

4. *Farm Chemicals Handbook 1989.* Willoughby, OH: Meister Publishing Co., 1989, p. C–180.

5. U.S. Coast Guard, Department of Transportation. *CHRIS—Hazardous Chemical Data.* Volume II. Washington, DC: U.S. Government Printing Office, 1984–5.

6. Asscociation of American Railroads. *Emergency Handling of Hazardous Materials in Surface Transportation.* Washington, DC: Assoc. of American Railroads, Hazardous Materials Systems (BOE), 1987. 425.

7. NIOSH. *Pocket Guide to Chemical Hazards.* 2nd Printing. DHHS (NIOSH) Publ. No. 85–114. Washington, DC: U.S. Dept. of Health and Human Services, NIOSH/Supt. of Documents, GPO, February 1987. 151.

8. NIOSH. *Criteria Document: Malathion.* p. 11 (1976) DHEW Pub. NIOSH 76–205.

9. Sittig, M. (ed.). *Pesticide Manufacturing and Toxic Materials Control Encyclopedia.* Park Ridge, NJ: Noyes Data Corporation. 1980. 476.

10. *Kirk–Othmer Encyclopedia of Chemical Technology.* 3rd ed., Volumes 1–26. New York, NY: John Wiley and Sons, 1978–1984, p. 24 (84) 310.

11. Verschueren, K. *Handbook of Environmental Data of Organic Chemicals.* 2nd ed. New York, NY: Van Nostrand Reinhold Co., 1983. 416.

12. U.S. Department of Transportation. *Emergency Response Guidebook 1990.* DOT P 5800.5. Washington, DC: U.S. Government Printing Office, 1990.

13. Mackison, F. W., R. S. Stricoff, and L. J. Partridge, Jr. (eds.). *NIOSH/OSHA— Occupational Health Guidelines for Chemical Hazards.* DHHS(NIOSH) Publication No. 81–123 (3 vols). Washington, DC: U.S. Government Printing Office, Jan. 1981.

14. Sittig, M. *Handbook of Toxic and Hazardous Chemicals and Carcinogens,* 1985. 2nd Ed. Park Ridge, NJ: Noyes Data Corporation, 1985. 555.

15. USEPA. *Management of Hazardous Waste Leachate,* EPA Contract No. 68-03-2766 p.E.56,90 (1982).

■ CYANURIC ACID

HO—N—OH structure (1,3,5-triazine ring)

CAS # 108–80–5

SYNONYMS 1,3,5-triazine-2,4,6 (1H, 3H, 5H)-trione • 2,4,6-trihydroxy-1,3,5-triazine • isocyanuric acid • kyselina-kyanurova (Czech) • pseudocyanuric acid • s-2,4,6-triazinetriol • s-triazine-2,4,6 (1H, 3H, 5H)-trione • s-triazine-2,4,6-triol • sym-triazinetriol • tricyanic-acid • trihydroxycyanidine • trihydroxytriazine

MF $C_3H_3N_3O_3$

MW 129.08

pH pH of saturated aqueous solution at room temperature = 4.8.

COLOR AND FORM Anhydrous crystals from concentration hydrochloric acid or sulfuric acid; crystalline powder.

ODOR Odorless.

TASTE Slightly bitter.

DENSITY 2.500 at 20°C.

DISSOCIATION CONSTANT pK_a = 7.2.

MELTING POINT 360°C.

SOLUBILITY 1 g dissolves in about 200 mL water; soluble in hot alcohols; soluble in pyridine, in concentrated hydrochloric acid without decomp; soluble in concentrated sulfuric acid without decomp; soluble in aqueous solution of sodium hydroxide; soluble in aqueous solution of potassium hydroxide; insoluble in cold methanol, acetone, benzene, ether, chloroform; water solubility = 2.00×10^3 mg/L at 25°C; water solubility = 2.969×10^3 mg/L at 2°C and 5.00×10^3 mg/L at 20°C.

OTHER CHEMICAL/PHYSICAL PROPERTIES Soluble in pyridine; insoluble in acetic acid; max absorption (water pH = 8): 214 nm (log e = 4.01); density: 1.768 at 0°C; Sadtler reference number: 7,027 (IR, PRISM) (dihydrate); 1 mg/L = 189 ppm; 1 ppm = 5.28 mg/m³; tends to form super-saturated solutions; exists in 2 tautomeric forms; the predominant form in the solid state is the keto form; in basic solution, the enol form is more stable; Ka_1 at 25°C = 6.3×10^{-8}; Ka_2 = 7.9×10^{-12}; efflorescent; water of crystallization is lost on exposure to air (dihydrate); monoclinic prisms (octahedra) from water (dihydrate).

CLEANUP Alkali and thermal treatment in filtration and precipitation was used to remove cyanuric acid from wastewaters [R3]. Dissolved cyanurate salts are removed from water by treatment with an aqueous solution of sodium hypochlorite whereby most of the cyanurate is converted to nitrogen, carbon dioxide, and sodium chloride. The treatment is carried out at pH 9.0–10.0 and 25°C to 55°C, the contact time being 2–6 hr. The mol ratio of sodium hypochlorite and cyanurate is kept between 6:1 and 8:1 [R4]. Sodium hypochlorite is used in removal of cyanurates from waste streams [R5].

DISPOSAL SRP: At the time of review, criteria for land treatment or burial (sanitary landfill) disposal practices are subject to significant revision. Prior to implementing land disposal of waste residue (including waste sludge), consult with environmental regulatory agencies for guidance on acceptable disposal practices.

COMMON USES Lab source of cyanic acid gas; selective herbicide; in preparation of melamine, sponge rubber [R1]. In chemical syntheses and as intermediate for chlorinated bleaches as whitening agent [R15, 2765]. Cyanuric acid is used in swimming pools to lower the rate of photochemical reduction of chlorine, hypochlorous acid, and hypochlorite ion [R2].

OCCUPATIONAL RECOMMENDATIONS None listed.

HUMAN TOXICITY LD_{50} rat oral 5 g/kg [R6].

BIODEGRADATION Studies were conducted in Greenfield sandy loam soil and in pure culture. 87% of labeled cyanuric acid had evolved as $^{14}CO_2$ after 16 days incr to 96% after 32 days [R7]. Cyanuric acid readily undergoes biodegradation under natural conditions, especially in systems of either low or zero dissolved-oxygen level, such as in anaerobic activated sludge and sewage, soils, muds, and river water [R9]. A facultative anaerobic bacterium that rapidly degrades cyanuric acid was isolated from the sediment of a stream that received industrial wastewater effluent. The bacterium used cyanuric acid or cysteine as a major, if not the sole, carbon and energy source under anaerobic conditions [R10]. A strain of *Sporothrix schenckii* was isolated that rapidly degraded cyanuric acid. Resting cells degraded the compd to carbon dioxide and ammonia [R11]. A study using a Warburg respirometer with acclimated sewage sludge and 100 mg cyanuric acid found that cyanuric acid was inhibitory to biodegradation(1). Another study using COD, TOC, and UV with activated sludge found no degradation with cyanuric acid, even with acclimated sludge (2). After 16 days, 87% of labeled cyanuric acid had evolved as $^{14}CO_2$, and after 192 days the percentage had increased to 99% in grab samples of aerobic soils (cyanuric acid conc 2.5 ppm)(3). In grab samples of saturated soil, 83% of labeled cyanuric acid had evolved as $14CO_2$ after 66 days and after 375 days the percentage increased to 99% (cyanuric acid conc 2.5 ppm)(3). Cyanuric acid completely degraded in a study using s-triazine herbicide production wastes, HPLC and mixed (aerobic and anaerobic) bacteria cultures(4) [R12]. Cyanuric acid biodegrades readily under a wide variety of natural conditions, and particularly well in systems of either low or zero dissolved-oxygen levels, such as anaerobic activated sludge and sewage, soils, muds, and muddy streams and river waters, as well as ordinary aerated activated sludge systems with typically low (1 to 3 ppm) dissolved-oxygen levels(1). For example: CO_2 evolution from ^{14}C-labeled cyanuric acid on soils, muds, and in natural waters ranges from 1% at 8 days to 116% at 23 days at room temperature(1). Biodegradation of cyanuric acid by aerated activated sludge ranges from 14% at 8.7 mg/L DO and 5 hours, to 100% at 2.5 mg/L DO and 10 hours(1). At a concentration of 10 µg/mL cyanuric acid in anaerobic sewage, cyanuric acid degraded by 25–50% in 48 hours, and 100% within 72–96 hours(1).

In an anaerobic mixed liquor, CO_2 evolution occurred at 4% within 7 hours, 11% in 17 hours, and 82% in 17 days(1). [R13].

BIOCONCENTRATION Based upon an experimental water solubility of 2,000 mg/L (1), the BCF of cyanuric acid can be estimated to be approximately 8.5 from a regression-derived equation (2). The BCF for cyanuric acid has also been experimentally determined to be <0.1 at 10 mg/L and <0.5 at 1 mg/L for a 6-week duration (3). According to these BCF values, bioconcentration in aquatic organisms is not expected to be an important fate process (SRC) [R8].

FIFRA REQUIREMENTS In 1988, Congress amended FIFRA to strengthen and accelerate EPA's reregistration program. The nine-year reregistration scheme mandated by "FIFRA 88" applies to each registered pesticide product containing an active ingredient initially registered before November 1, 1984. List A consists of the 194 chemical cases (or 350 individual active ingredients) for which EPA had issued Registration Standards prior to the effective date of FIFRA '88. List: A; Case: Chlorinated Isocyanurates Case No.: 0569; Pesticide type: antimicrobial; Registration Standard Date: 05/20/88; Case Status: Awaiting Data/Data in Review: OPP awaits data from the pesticide's producer(s) regarding its human health and/or environmental effects, or OPP has received and is reviewing such data, in order to reach a decision about the pesticide's eligibility for reregistration. Active Ingredient (AI): Cyanuric acid; AI Status: The active ingredient is no longer contained in any registered products. Therefore, EPA is characterizing it as "cancelled." Under FIFRA, pesticide producers may voluntarily cancel their registered products. EPA also may cancel pesticide registrations if registrants fail to pay required fees, to make or meet certain reregistration commitments, or when the agency reaches findings of unreasonable adverse effects. [R14].

REFERENCES

1. Budavari, S. (ed.). *The Merck Index— Encyclopedia of Chemicals, Drugs, and*

Biologicals. Rahway, NJ: Merck and Co., Inc., 1989. 420.

2. Gosselin, *CTCP* 5th ed 1984 p. II–333.

3. Salinkova, L. S., et al., *Purification of Wastewaters Containing Cyanuric Acid*; former USSR Patent Number 865822 09/23/81.

4. FMC Corp of the Netherlands. *Removal of Dissolved Cyanurate Compounds from Aqueous Waste Liquids*. Neth Appl Patent Number 78 00246 07/11/79 (FMC Corp).

5. Carlson R. H. *Sodium Hypochlorite Treatment for Removal of Cyanurate Compounds from Aqueous Waste Streams*, U.S. Patent Number 4075094, 2/21/78 (FMC CORP).

6. USEPA. *Chemical Hazard Information Profile (Draft)*: Cyanuric Acid and Chlorinated Derivatives (1981).

7. Wolf, D. C., J. P. Martin. "Microbial Decomposition of Ring–14C Atrazine, Cyanuric Acid and 2-Chloro-4,6-Diamino-s-Triazine," *J Environ Qual* 4 (1): 134–9 (1975).

8. (1) Burakevich, J. V., *Kirk–Othmer Encycl Chem Tech*. 3rd ed. New York, NY: Wiley Interscience 7: 397–410. (2) Lyman, W. J., et al., *Handbook of Chemical Property Estimation Methods*. Washington DC: Amer Chemical Soc, pp. 5–4, 5–10 (1990). (3) Chemicals Inspection and Testing Institute. *Biodegradation and Bioaccumulation Data of Existing Chemicals Based on the CSCL*. Japan p. 5–26 (1992).

9. Saldick J. "Biodegradation of Cyanuric Acid." *Appl Microbiol* 28 (6): 1004–8 (1974).

10. Jessee, J. A., et al. "Anaerobic Degradation of Cyanuric Acid, Cysteine and Atrazine by a Facultative Anaerobic Bacterium." *Appl Environ Microbiol* 45 (1): 97–102 (1983).

11. Zeyer, J., et al. "Rapid Degradation of Cyanuric Acid by *Sporothrix schenckii*." *Zentralbl Bakteriol Mikrobiol Hyg Abt 1, Orig C* 2 (2): 99–100 (1981).

12. (1) Helfgott, T. B., et al. *An Index of Refractory Organics*. Ada, OK: USEPA-600/2-77-174 (1977). (2) Pitter, P., J. Simanova. *Technol Vody Prostredi* 22: 93–113 (1978). (3) Wolf, D. C., J. P. Martin. *J Environ Qual* 4: 134–9 (1975). (4) Hogrefe, W., et al. *Biotechnology and Bioengineering* 27: 1291–6 (1985).

13. Saldick, J. *Appl Microbiol* 28: 1004–8 (1974).

14. USEPA/OPP. *Status of Pesticides in Reregistration and Special Review*. p.72 (Mar, 1992) EPA 700-R-92-004.

15. Clayton, G. D., and F. E. Clayton (eds.). *Patty's Industrial Hygiene and Toxicology*. Volume 2A, 2B, 2C: Toxicology. 3rd ed. New York: John Wiley Sons, 1981–1982.

■ CYCLONITE

CAS # 121-82-4

SYNONYMS 1,3,5-triaza-1,3,5-trinitrocyclohexane • 1,3,5-triazine, hexahydro-1,3,5-trinitro • 1,3,5-triazine, perhydro, 1,3,5-trinitro • 1,3,5-trinitro-1,3,5-triazacyclohexane • 1,3,5-trinitrohexahydro- 1,3,5-triazine • 1,3,5-trinitrohexahydro-s-triazine • 1,3,5-trinitroperhydro-1,3,5-triazine • cyclotrimethylenenitramine • cyclotrimethylenetrinitramine • esaidro-1,3,5-trinitro-1,3,5-triazina (Italian) • heksogen (Polish) • Hexahydro-1,3,5-trinitro-1,3,5-triazin (German) • hexahydro-1,3,5-trinitro-1,3,5-triazine • hexahydro-1,3,5-trinitro-s-triazine • hexogeen (Dutch) • hexogen • hexogen-5w • hexolite • pbx(af)-108 • rdx • s-triazine, hexahydro-1,3,5-trinitro • sym-trimethylenetrinitramine • t4- • trimethyleentrinitramine (Dutch) • trimethylenetrinitramine • sym-trimethylene trinitramine • trinitrotrimethylenetriamine

MF $C_3H_6N_6O_6$

MW 222.15

COLOR AND FORM White, crystalline powder.

DENSITY 1.82 at 20°C.

MELTING POINT 205–206°C.

EXPLOSIVE LIMITS AND POTENTIAL Cyclotrimethylene trinitramine is a Class A explosive. Class A explosives are explosives that decompose on detonation. This detonation occurs almost instantaneously and is violent. The explosion may be initiated by a sudden shock, high temperature, or a combination of the two [R3].

FIREFIGHTING PROCEDURES Fight fires from safe distance from explosion. In advanced or massive fires, the area should be evacuated. If fire occurs in the vicinity of this material, water should be used to keep containers cool. [R2]. Do not fight fires in a cargo of explosives. Evacuate area and let burn. Wear positive-pressure self-contained breathing apparatus when fighting fires involving this material. Evacuation: If the material is on fire or involved in fire consider evacuation of one (1) mile radius. (cyclotrimethylene trinitramine, desensitized; cyclotrimethylene trinitramine, wet with not less than 10% water) [R3].

SOLUBILITY Insoluble in water, alcohol, benzene, carbon disulfide; slightly soluble in ether, methanol, toluene; slightly soluble in ethyl acetate, glacial acetic acid; practically insoluble in carbon tetrachloride; readily soluble in hot aniline, phenol, warm nitric acid; 1 g dissolves in 25 mL acetone.

OCTANOL/WATER PARTITION COEFFICIENT Log K_{ow} = 0.87.

STABILITY AND SHELF LIFE Sensitive to heat. The stability of cyclonite is considerably superior to that of petn (pentaerythritol tetranitrate) and nearly equal to that of TNT. It withstands storage at 85°C for 10 months or at 100°C for 100 hr without measurable deterioration; hence from the viewpoint of stability, cyclonite must be considered highly satisfactory. [R13, 4196].

OTHER CHEMICAL/PHYSICAL PROPERTIES Orthorhombic crystals from acetone; impure military grades containing about 10% HMX (hexamethylenetetramine) have a melting point of about 190°C.

PROTECTIVE EQUIPMENT Personal protective equipment including eye protective equipment should be provided for normal process work as protection against eye splashes, acid burns, dermatitis, and skin absorption of toxic materials. (explosives industry) [R14, 809].

PREVENTATIVE MEASURES Floors of sheds and workrooms should be kept scrupulously clean. Special rules about acceptable types of footwear and clothing, and provision of shoe-cleaning mats prevent the bringing of exposed metal fasteners, loose metal, and gritty dirt into explosives buildings. Smoking and bringing of incendiary equipment into explosives area are forbidden. (explosive substances) [R14, 808].

CLEANUP After covering the spills with soda ash, mix, and spray with water. Scoop into a bucket of water and let it stand for 2 hr. Neutralize with 6-molar hydrochloric acid [R2]. The feasibility of treating the wastewaters from site x rdx/hmx manufacturing facilities in aerobic rotating biological contactor was evaluated. The results from the pilot-scale evaluation indicate that 95–100% soluble BOD removal can be achieved and maintained at soluble BOD loading rates of <2.5 lb/1,000 ft³/day [R4]. The feasibility of treating explosive-contaminated wastewater using an oxidant (hydrogen peroxide) in conjunction with short-wavelength uv light was evaluated. The process achieves high degree of decontamination without troublesome by-products [R5]. The use of ultraviolet-ozone and ultraviolet-oxidant to treat pink water from ammunition plants was evaluated. Dissolved rdx was removed to less than 1 mg/L with no by-products requiring disposal [R6]. Wet spilled material before picking it up. Do not attempt to sweep up dry material (cyclotrimethylene trinitramine, wet with not less than 10% water) [R3]. Wastewater containing rdx (cyclonite) was treated by adsorption on activated carbon; the carbon was regenerated by ozonization, which oxidizes the absorbed substance [R7].

COMMON USES High explosive [R12, 393]; used as base charge for detonators, and as an ingredient of bursting charges and plastic explosives by the military [R13, 4196]. Employed occasionally as a rodenticide [R1]. Because it is easily initiated by mercury fulminate, it may be used as a booster. [R15, 2668].

OCCUPATIONAL RECOMMENDATIONS None listed.

OSHA 8-hr time-weighted average: 1.5 mg/m^3. Skin absorption designation. Final rule limits must be achieved by any combination of engineering controls, work practices, and personal protective equipment—effective Sept 1, 1989 [R11].

THRESHOLD LIMIT 8-hr time-weighted average (TWA) 1.5 mg/m^3 (1990) [R16, 17].

INTERNATIONAL EXPOSURE LIMITS MAC former USSR 1 mg/m^3 [R14, 807].

ECOTOXICITY LC$_{50}$ bluegill 3.6 mg/L/96 hr, pH 6.0, static bioassay [R8].

BIODEGRADATION Found that nitro groups of RDX (cyclotrimethylene trinitramine) were reduced to amino groups during growth by anaerobic purple photosynthetic bacteria, although no actual RDX metabolism was found. Found (in other study) that neither RDX nor ammonium picrate could serve as a sole carbon source for Pseudomonads [R9].

BIOCONCENTRATION The whole-body BCF in bluegills (*Lepomis macrochirus*) was determined to be 24.8 (1), suggesting that bioaccumulation in aquatic organisms should not be an important fate process (SRC) [R10].

FIRE POTENTIAL Dangerous due to fire hazard (nitrates) [R15, 2002].

REFERENCES

1. Gosselin, R. E., R. P. Smith, H. C. Hodge. *Clinical Toxicology of Commercial Products.* 5th ed. Baltimore: Williams and Wilkins, 1984, p. II–212.

2. ITII. *Toxic and Hazarous Industrial Chemicals Safety Manual.* Tokyo, Japan: The International Technical Information Institute, 1982. 147.

3. Association of American Railroads. *Emergency Handling of Hazardous Materials in Surface Transportation.* Washington, DC: Assoc. of American Railroads, Hazardous Materials Systems (BOE), 1987. 214.

4. Kitchens, L. F., et al. 185 pp (1980) ARC–49–5766–1, ARCSL–CR–80028, AD–E410 256; order number AD–A084 657.

5. Andrews, C. C. 85 pp (1980) WQEC/C–80–137; order number AD–A084 684.

6. Roth, M., J. M. Murphy, Jr. 41 pp (1979), order number PB–300763, EPA/600/2–79/129.

7. Jain, K. K., A. J. Bryce. *Carbon Adsorption Handbook* 661–86 (1978).

8. Bentley, R. E., et al., 99 pp (1977); Iss order number AD–A061730.

9. Parr, J. F., P. B. Marsh, and J. M. Kla (eds.). *Land Treatment of Hazardous Wastes.* Park Ridge, NJ: Noyes Data Corporation, 1983. 302.

10. Ryon, M. G., et al., *Database Assessment of the Health and Environmental Effects of Munition Production Waste Water.* ORNL–6018 (NTIS DE84–016512) (1974).

11. 54 FR 2920 (1/19/89).

12. *The Merck Index.* 10th ed. Rahway, New Jersey: Merck Co., Inc., 1983.

13. Clayton, G. D. and F. E. Clayton (eds.). *Patty's Industrial Hygiene and Toxicology.* Volume 2A, 2B, 2C: Toxicology. 3rd ed. New York: John Wiley Sons, 1981–1982.

14. International Labour Office. *Encyclopedia of Occupational Health and Safety.* Vols. I & II. Geneva, Switzerland: International Labour Office, 1983.

15. Sax, N.I. *Dangerous Properties of Industrial Materials.* 6th ed. New York, NY: Van Nostrand Reinhold, 1984.

16. American Conference of Governmental Industrial Hygienists. *Threshold Limit Values for Chemical Substances and*

Physical Agents and Biological Exposure Indices for 1994–1995. Cincinnati, OH: ACGIH, 1994.

■ CYCLOPENTANE

CAS # 287–92–3

DOT # 1146

SYNONYMS pentamethylene

MF C_5H_{10}

MW 70.15

COLOR AND FORM Colorless liquid.

DENSITY 0.7457 at 20°C.

BOILING POINT 49.2°C at 760 mm Hg.

MELTING POINT –93.9°C.

AUTOIGNITION TEMPERATURE 682°F [R8, p. 325M–30].

SOLUBILITY Insoluble in water; miscible with hydrocarbon solvents; soluble in alcohol, benzene, acetone, ether; water solubility: 156 ppm at 25°C.

VAPOR PRESSURE 317.8 mm Hg at 25°C.

OCTANOL/WATER PARTITION COEFFICIENT Log K_{ow} = 3.00.

PROTECTIVE EQUIPMENT Protective clothing and barrier creams; medical control [R4].

PREVENTATIVE MEASURES Substitution of less irritating substances, redesign of operations, prevent contact, provision of a physical barrier against contact, proper washing facilities, work clothing, and storage facilities [R4].

COMMON USES As solvent; starting material for synthesis in chemical industry [R9, 3224]. Solvent for cellulose ethers; motor fuel; azeotropic distillation agent [R2]. Chemical intermediate in production of cyclopentadiene (insecticide) [R1].

OCCUPATIONAL RECOMMENDATIONS None listed.

OSHA 8-hr time-weighted average: 600 ppm (1,720 mg/m³). Final rule limits were achieved by any combination of engineering controls, work practices, and personal protective equipment during the phase-in period, Sept 1, 1989 through Dec 30, 1992. Final rule limits became effective Dec 31, 1992 [R7].

THRESHOLD LIMIT 8-hr time-weighted average (TWA) 600 ppm, 1,720 mg/m³ (1987) [R10, 17].

BIODEGRADATION Pure culture studies showed various isolated species of bacteria were unable to utilize cyclopentane as a single carbon source in soil (1,2). Mixed populations from groundwater contaminated with gasoline did not biodegrade cyclopentane at a initial concentration of 0.17 ppm in a gasoline mixture (3). After 192 hrs, the concentration was 0.04 ppm; however, the concentration in a sterilized control was 0.05 ppm (3) [R5].

BIOCONCENTRATION Based upon a water solubility of 156 ppm at 25°C (1) and log K_{ow} of 3.00 (2), bioconcentration factors (log BCF) for cyclopentane have been calculated to be 0.08 and 2.05, respectively, from recommended regression-derived equations (3). These BCF values indicate the potential for cyclopentane to bioconcentrate in aquatic organisms is low (SRC) [R6].

DOT *Fire or Explosion:* Flammable/combustible material; may be ignited by heat, sparks, or flames. Vapors may travel to a source of ignition, and flash back. Container may explode in heat of fire. Vapor explosion hazard indoors, outdoors, or in sewers. Runoff to sewer may create fire or explosion hazard [R3].

Health Hazards: May be poisonous if inhaled or absorbed through skin. Vapors may cause dizziness or suffocation. Contact may irritate or burn skin and eyes. Fire may produce irritating or poisonous gases. Runoff from fire control or dilution water may cause pollution [R3].

Emergency Action: Keep unnecessary people away; isolate hazard area and deny

entry. Stay upwind; keep out of low areas. Positive-pressure self-contained breathing apparatus (SCBA) and structural firefighters' protective clothing will provide limited protection. Isolate for 1/2 mile in all directions if tank, rail car, or tank truck is involved in fire. Call CHEMTREC at 1–800–424–9300 for emergency assistance. If water pollution occurs, notify the appropriate authorities [R3].

Fire: Small Fires: Dry chemical, CO_2, water spray, or regular foam. Large Fires: Water spray, fog, or regular foam. Move container from fire area if you can do so without risk. Apply cooling water to sides of containers that are exposed to flames until well after fire is out. Stay away from ends of tanks. For massive fire in cargo area, use unmanned hose holder or monitor nozzles; if this is impossible, withdraw from area, and let fire burn. Withdraw immediately in case of rising sound from venting safety device or any discoloration of tank due to fire [R3].

Spill or Leak: Shut off ignition sources; no flares, smoking, or flames in hazard area. Stop leak if you can do so without risk. Water spray may reduce vapor, but it may not prevent ignition in closed spaces. Small Spills: Take up with sand or other noncombustible absorbent material and place into containers for later disposal. Large Spills: Dike far ahead of liquid spill for later disposal [R3].

First Aid: Move victim to fresh air and call for emergency medical care; if not breathing, give artificial respiration; if breathing is difficult, give oxygen. In case of contact with material, immediately flush eyes with running water for at least 15 minutes. Wash skin with soap and water. Remove and isolate contaminated clothing and shoes at the site [R3].

FIRE POTENTIAL Flammable, dangerous fire risk [R2].

GENERAL RESPONSE Shut off ignition sources. Call fire department. Stop discharge if possible. Keep people away. Evacuate area. In case of large discharge. Isolate and remove discharged material.

Notify local health and pollution control agencies.

FIRE RESPONSE Flammable. Containers may explode in fire. Flashback along vapor trail may occur. Vapor may explode if ignited in an enclosed area. Extinguish with dry chemicals, foam, or carbon dioxide. Water may be ineffective on fire. Cool exposed containers with water.

EXPOSURE RESPONSE Call for medical aid. Vapor irritating to eyes, nose, and throat. If inhaled will cause dizziness, nausea, vomiting, difficult breathing, or loss of consciousness. Move victim to fresh air. If breathing has stopped, give artificial respiration. If breathing is difficult, give oxygen. Liquid irritating to skin and eyes. Harmful if swallowed. Remove contaminated clothing and shoes. Flush affected areas with plenty of water. If swallowed and victim is conscious, have victim drink water or milk. Do not induce vomiting.

RESPONSE TO DISCHARGE Issue warning—high flammability. Evacuate area. Disperse and flush.

WATER RESPONSE Effect of low concentrations on aquatic life is unknown. Fouling to shoreline. May be dangerous if it enters water intakes. Notify local health and wildlife officials. Notify operators of nearby water intakes.

EXPOSURE SYMPTOMS Inhalation causes dizziness, nausea, and vomiting; concentrated vapor may cause unconsciousness and collapse. Vapor causes slight smarting of eyes. Contact with liquid causes irritation of eyes and may irritate skin if allowed to remain. Ingestion causes irritation of stomach. Aspiration produces severe lung irritation and rapidly developing pulmonary edema; central nervous system excitement followed by depression.

EXPOSURE TREATMENT Inhalation: remove to fresh air; if breathing stops, apply artificial respiration, and administer oxygen. Eyes: flush with water for at least 15 min.; call a physician. Skin: flush well with water, then wash with soap and water. Ingestion: do not induce vomiting; guard against aspiration into lungs. Aspiration: enforced bed rest; give oxygen; get medical attention.

REFERENCES

1. SRI.

2. Sax, N. I., and R. J. Lewis, Sr. (eds.). *Hawley's Condensed Chemical Dictionary.* 11th ed. New York: Van Nostrand Reinhold Co., 1987. 339.

3. U.S. Department of Transportation. *Emergency Response Guidebook 1990.* DOT P 5800.5. Washington, DC: U.S. Government Printing Office, 1990, p. G–27.

4. Sax, N. I. *Dangerous Properties of Industrial Materials.* 6th ed. New York, NY: Van Nostrand Reinhold, 1984. 594.

5. (1) Beam, H. W., J. J. Perry. *J Gen Microb* 82: 163–9 (1974). (2) Perry, J.J. *The Role of Co-oxidation and Commensalism in the Biodegradation of Recalcitrant Molecules.* NTIS PB–80–28–034 US Army Res Off pp. 21 (1980). (3) Jamison, V. W., et al., pp. 187–96 in *Proc Intermediate Biodeg Symp* 3rd. Sharpley, J. M., A. M. Kapalan (eds.), Essex, Eng (1976).

6. (1) Yalkowsky, S. H., et al., *Arizona Data Base of Water Solubility* (1989). (2) Hansch, C., and A. J. Leo. *Medchem Project* Issue No 26. Claremont, CA: Pomona College (1985). (3) Lyman, W. J., et al., *Handbook of Chemical Property Estimation Methods.* New York: McGraw-Hill, pp. 5–4, 5–10 (1982).

7. 29 CFR 1910.1000 (7/1/90).

8. National Fire Protection Association. *Fire Protection Guide on Hazardous Materials.* 9th ed. Boston, MA: National Fire Protection Association, 1986.

9. Clayton, G. D., and F. E. Clayton (eds.). *Patty's Industrial Hygiene and Toxicology.* Volume 2A, 2B, 2C: Toxicology. 3rd ed. New York: John Wiley Sons, 1981–1982.

10. American Conference of Governmental Industrial Hygienists. *Threshold Limit Values for Chemical Substances and Physical Agents and Biological Exposure Indices for 1994–1995.* Cincinnati, OH: ACGIH, 1994.

■ CYCLOPROPANECARBOXYLIC ACID, 3-(2,2-DICHLOROVINYL)-2,2-DIMETHYL-, 3-PHENOXYBENZYL ESTER, (+n-)-, (cis, trans)-

CAS # 52645–53–1

SYNONYMS FMC-33297 • pynosect • ridect pour-on • (3-phenoxyphenyl) methyl-3-(2,2-dichlorovinyl)-2,2-dimethylcyclopropane carboxylate • coopex • corsair • pramex • permit • 3-phenoxybenzyl (1rs)-cis, trans-3-(2,2-dichlorvinyl)-2,2-dimethylcyclopropane carboxylate • (3-phenoxyphenyl) methyl 3-(2,2-dichloroethyl)-2,2-dimethylcyclopropane carboxylate • permethrine • perthrine • kafil • 3-(2,2-dichloroethenyl)-2,2-dimethylcyclopropanecarboxylic acid (3- phenoxyphenyl) methyl ester • 3-(phenoxyphenyl) methyl (-)-cis, trans-3-(2,2-dichloroethenyl)-2,2-dimethylcyclopropanecarboxylate • m-phenoxybenzyl (1)-cis, trans-3-(2,2-dichlorovinyl)-2,2-dimethylcyclopropane carboxylate • NIA-33297 • NRDC-143 • pp557 • sbp-1513 • S-3151 • ambush • cosair • dragnet • ectiban • eksmin • expar • nix • perigen • pounce • imperator • dragon • picket • pp-557 • permasec • talcord • outflank • stomoxin • quamlin

MF $C_{21}H_{20}Cl_2O_3$

MW 391.31

COLOR AND FORM Pale-brown liquid.

DENSITY 1.19–1.27 at 20°C.

BOILING POINT ca 200°C at 0.01 mm Hg.

MELTING POINT 34–35°C.

FIREFIGHTING PROCEDURES Use carbon dioxide, foam, or dry chemical on fires involving pyrethroids (pyrethrum) [R6, 2];

self-contained breathing apparatus with a full facepiece operated in pressure-demand or other positive-pressure mode (pyrethrum) [R6, 5].

SOLUBILITY Water solubility = 0.040 ppm at room temp; in water at 20°C, ca 0.2 mg/L. Soluble in most organic solvents except ethylene glycol. In xylene and hexane >1,000, methanol 258 (all in g/kg at 25°C); water solubility 0.2 mg/L at 30°C; soluble or miscible with organic solvents: acetone (450 g/L), hexane (>1 kg/kg), methanol (258 g/kg), xylene (>1 kg/kg).

VAPOR PRESSURE 0.045 mPa at 25°C.

OCTANOL/WATER PARTITION COEFFICIENT Log K_{ow}= 3.48.

CORROSIVITY Does not corrode aluminum.

STABILITY AND SHELF LIFE Stable at 40° for at least 2 years. Relatively stable to sunlight (when compared with natural pyrethrum), although some photochemical degradation does occur. More stable in acidic media than in alkaline media [R1].

OTHER CHEMICAL/PHYSICAL PROPERTIES Log K_{ow}= 6.5.

PROTECTIVE EQUIPMENT Employees should be provided with and required to use dust- and splash-proof safety goggles where pyrethroids may contact the eyes. (pyrethroids) [R6, 3].

PREVENTATIVE MEASURES Skin that becomes contaminated with pyrethrum should be promptly washed or showered with soap or mild detergent and water. (pyrethrum) [R6, 3].

CLEANUP Environmental consideration—land spill: dig a pit, pond, lagoon, or holding area to contain liquid or solid material. (SRP: if time permits, pits, ponds, lagoons, soak holes, or holding areas should be sealed with an impermeable flexible-membrane liner.) Dike surface flow using soil, sandbags, foamed polyurethane, or foamed concrete. Absorb bulk liquid with fly ash or cement powder. (pyrethrins) [R5].

DISPOSAL The following wastewater treatment technology has been investigated for chlorinated pesticides: concentration process: resin adsorption. (chlorinated pesticides) [R7].

COMMON USES Medication (vet): ectoparasiticide [R2, 1138]; it has a potential application for forest protection, and vector control against noxious insects in the household and on cattle, for the control of body lice, and in mosquito nets [R3]; nematocide, acaricide [R4]; control of larvae (and also adults and eggs) of chewing lepidopterous and coleopterous insect pests on pome fruit, stone fruit, berry fruit, citrus fruit, vines, olives, vegetables, cereals, maize, oilseed rape, cotton, tobacco, soya beans, and in conifer nurseries; whiteflies and other glasshouse pests on glasshouse cucumbers, tomatoes, and ornamentals; and sciarid flies and phorid flies on mushrooms; also used for control of crawling and flying insects (e.g., flies, ants, fleas, cockroaches, silverfish, etc.) in public health, and in agricultural premises including animal houses; and as an ectoparasiticide on animals [R1]; insecticide (pyrethrins) [R2, 1267].

OCCUPATIONAL RECOMMENDATIONS None listed.

NON-HUMAN TOXICITY LD_{50} rat oral 600 mg/kg [R10]; LD_{50} rat iv >270 mg/kg [R8, 81]; LD_{50} rat, oral 4,000 mg/kg (cis-trans-isomer ratio of 40:60) [R1]; rat oral LD_{50} 1.3 g/kg [R9, p. II–261]; LD_{50} rat, oral 6,000 mg/kg (cis-: trans-isomer ratio of 20:80) [R1]; LD_{50} rat percutaneous >4,000 mg/kg [R1]; LD_{50} rabbit percutaneous >2,000 mg/kg [R1]; LD_{50} chicken oral >3,000 mg/kg (cis-trans-isomer ratio of 40:60) [R1]; LD_{50} rat (female) 3,800 mg/kg (undiluted compound) [R8, 81]; LD_{50} rat (female) 410 mg/kg (ai dissolved in an unsaturated oil) [R8, 81]; LD_{50} rat (male) oral (in water) 2,949 mg/kg [R11]; LD_{50} rat (female) oral (in water) >4,000 mg/kg [R11]; LD_{50} rat (male) oral (in corn oil) 430 mg/kg [R11]; LD_{50} rat (female) oral (in corn oil) 470 mg/kg [R11]; LD_{50} rat (male) dermal (in water) >5,176 mg/kg [R11]; LD_{50} rat (male) dermal >25,000 mg/kg [R11]; LD_{50} rat (female) dermal >4,000 mg/kg [R11]; LD_{50} mouse (female) oral (in water) >4,000 mg/kg [R11]; LD_{50} mouse (male and female) oral (in DMSO) 250–500 mg/kg [R11]; LD$_{50}$ mouse (male) oral (in corn oil) 650 mg/kg [R11]; LD$_{50}$ mouse (female) oral (in corn oil) 540 mg/kg [R11]; LD$_{50}$ mouse dermal >2,500 mg/kg [R11]; LD$_{50}$ guinea pig oral (in water) >4,000 mg/kg [R11]; LD$_{50}$ rabbit (female) oral (in water) >4,000 mg/kg [R11]; LD$_{50}$ rabbit (female) dermal >2,000 mg/kg [R11]; LD$_{50}$ hen oral >1,500 mg/kg [R11].

ECOTOXICITY LC$_{50}$ *Pimephales promelas* (fathead minnow) 16.0 mg/L/96 hr (confidence limit 8.71–29.6 mg/L), flow-through bioassay with measured concentrations, 25.4°C, dissolved oxygen 7.5 mg/L, hardness 45.7 mg/L calcium carbonate, alkalinity 41.6 mg/L calcium carbonate, and pH 7.1 [R12]; LD$_{50}$ Japanese quail >13,500 mg/kg (cis-: trans-isomer ratio of 40: 60) [R1]; LC$_{50}$ bluegill sunfish 1.8 µg/L/48 hr (conditions of bioassay not specified) [R1]; LC$_{50}$ rainbow trout 5.4 µg/L/48 hr (conditions of bioassay not specified) [R1]; LC$_{50}$ brook trout (1.2 g) at 12°C 3.2 (2.2–4.8) µg/L/96 hr. Static bioassay without aeration, pH 7.2–7.5, water hardness 40–50 mg/L as calcium carbonate, and alkalinity of 30–35 mg/L (technical material 92.5%) [R13]; LC$_{50}$ brook trout (1.2 g) at 12°C. 5.2 (3.5–7.9) µg/L/96/hr. Static bioassay without aeration, pH 7.2–7.5, water hardness 40–50 mg/L as calcium carbonate, and alkalinity of 30–35 mg/L (liquid 5.7%) [R13]; LC$_{50}$ brook trout (1.2 g) at 12°C 2.3 (1.4–3.7) µg/L/96 hr. Static bioassay without aeration, pH 7.2–7.5, water hardness 40–50 mg/L as calcium carbonate, and alkalinity of 30–35 mg/L (emulsifiable concentrate 13.3%) [R13].

BIOLOGICAL HALF-LIFE A study was conducted to define permethrin toxicokinetics in Sprague Dawley rats after iv administration and to assess its oral bioavailability. Orally dosed rats received a single dose of 460 mg/kg by gastric intubation. Injected rats received 46 mg/kg intravenously. All animals were sacrificed 0.25, 0.5, 1, 2, 3, 4, 6, 8, 12, 24, or 48 hr after dosing. For permethrin the elimination half-life and the mean residence time from plasma were 8.67 and 11.19 hr after iv, and 12.37 and 17.77 hr after oral administration. The total plasma clearance was not influ-

enced by dose concentration or route and reached a value of 0.058 L/hr. After a single oral dose permethrin was absorbed slowly. The maximum plasma concentration was 49.46 µg/ml. The oral bioavailability of permethrin was 60.69%. The plasma concentration time data for permethrin metabolites as well as the tissue concentration time data for permethrin and its metabolites after an oral dose of permethrin were found to fit a one-compartment open model. The maximum amounts of permethrin in cerebellum, hippocampus, caudata putamen, frontal cortex, hypothalamus, and sciatic nerve were about 1.5, 2, 2, 2.7, 4.8, and 7.5 times higher than in plasma, respectively, suggesting an accumulation of pyrethroids by nervous tissue itself. The metabolites of permethrin, m-phenoxybenzyl alcohol and m-phenoxybenzoic acid, were detected in plasma, and in all selected tissues for 48 hr after dosing, suggesting that a combination of metabolism by the tissues, and diffusion into it from the blood may be present [R14].

BIODEGRADATION Pure cultures of *Bacillus cereus, Pseudomonas fluorescens*, and *Achromobacter sp.* transformed permethrin to 3-(2,2-dichlorovinyl)-2,2-dimethylcyclopropanecarboxylic acid, 3-phenoxybenzyl alcohol, 3-phenoxybenzoic acid, and 4-hydroxy-3-phenoxybenzoic acid; half-life of less than 5 days (1). The half-life of permethrin in aerobically incubated soil is less than 4 weeks, and the degradation of the trans-isomer is more rapid than that of the cis-isomer (2). Under anaerobic conditions in flooded silt loam soils, degradation half-lives were 32–34 days for ^{14}C-labeled trans-permethrin and greater than 64 days for ^{14}C-labeled cis-permethrin (2). The half-life in a sediment-seawater solution was measured to be less than 2.5 days; under sterile conditions there was no significant change in permethrin concentration (3) [R15].

BIOCONCENTRATION A BCF of about 480 was experimentally determined for sheepshead minnow (*Cyprinodon variegatus*) exposed to 1 µg/L permethrin for a

28-day test period (1). A BCF of 1,900 was also reported for oysters (1) [R16].

FIFRA REQUIREMENTS Tolerances are established for residues of the insecticide permethrin in or on the raw agricultural commodity cottonseed [R17].

REFERENCES

1. Hartley, D., and H. Kidd (eds.). *The Agrochemicals Handbook.* 2nd ed. Lechworth, Herts, England: The Royal Society of Chemistry, 1987, p. A316/Aug 87.

2. Budavari, S. (ed.). *The Merck Index—Encyclopedia of Chemicals, Drugs, and Biologicals.* Rahway, NJ: Merck and Co., Inc., 1989.

3. WHO. *Environmental Health Criteria 94: Permethrin,* p. 21 (1990).

4. Sax, N. I., and R. J. Lewis, Sr. (eds.). *Hawley's Condensed Chemical Dictionary.* 11th ed. New York: Van Nostrand Reinhold Co., 1987. 890.

5. Bureau of Explosives; *Emergency Handling of Haz Matl in Surface Trans* p. 434 (1981).

6. Mackison, F. W., R. S. Stricoff, and L. J. Partridge, Jr. (eds.). *NIOSH/OSHA—Occupational Health Guidelines for Chemical Hazards.* DHHS (NIOSH) Publication No. 81–123 (3 vols). Washington, DC: U.S. Government Printing Office, Jan. 1981.

7. USEPA. *Management of Hazardous Waste Leachate,* EPA Contract No. 68–03–2766 p. E–195 (1982).

8. Hayes, Wayland J., Jr. *Pesticides Studied in Man.* Baltimore/London: Williams and Wilkins, 1982.

9. Gosselin, R. E., R. P. Smith, H. C. Hodge. *Clinical Toxicology of Commercial Products.* 5th ed. Baltimore: Williams and Wilkins, 1984.

10. *Kirk–Othmer Encyclopedia of Chemical Technology.* 3rd ed., Volumes 1–26. New York, NY: John Wiley and Sons, 1978–1984, p. V13 458 (1981).

11. WHO. *Environmental Health Criteria 94: Permethrin* p. 64 (1990).

12. Geiger D. L., D.J. Call, L. T. Brooke. (eds). *Acute Toxicities of Organic Chemicals to Fathead Minnows (Pimephales promelas).* Vol. IV. Superior, Wisconsin: University of Wisconsin–Superior, 1988. 328.

13. U.S. Department of Interior, Fish and Wildlife Service. *Handbook of Acute Toxicity of Chemicals to Fish and Aquatic Invertebrates.* Resource Publication No. 137. Washington, DC: U.S. Government Printing Office, 1980. 58.

14. Anadon, A., et al. *Toxicol Appl Pharmacol* 110 (1): 1–8 (1991).

15. (1) Maloney, S. E., et al. *Appl Environ Microb* 54: 2874–6 (1988); (2) Jordan, E. G., and D. D. Kaufman. *J Agric Food Chem* 34: 880–4 (1986); (3) Schimmel, S. C., et al. *J Agric Food Chem* 31: 104–13 (1983).

16. Schimmel, S. C., et al. *J Agric Food Chem* 31: 104–13 (1983).

17. 40 CFR 180.378 (a) (7/1/91).

■ CYCLOPROPANECARBOXYLIC ACID, 2,2-DIMETHYL-3-(2-METHYLPROPENYL)-, (5-BENZYL-3-FURYL) METHYL ESTER

CAS # 10453–86–8

SYNONYMS (5-benzyl-3-furyl) methyl-2,2-dimethyl-3-(2-methylpropenyl)-cyclopropanecarboxylate • (5-benzyl-3-furyl) methyl chrysanthemate • 5-benzyl-3-furylmethyl (-)-cis, trans-chrysanthemate • 5-benzylfurfuryl chrysanthemate • benzofuroline • benzyfuroline • cyclopropanecarboxylic acid, 2,2-dimethyl-3-(2-methyl-1-propenyl)-, (5-(phenylmethyl)-3-furanyl) methyl ester • cyclopropanecarboxylic acid, 2,2-dimethyl-3-(2-methylpropenyl)-, (4-(2-benzyl) furyl) methyl ester • cyclopropanecarboxylic acid, 2,2-dimethyl-3-(2-methylpropenyl)-, (5-benzyl-3-furyl) methyl ester • dimethyl 3-(2-methyl-1-propenyl) cyclopropanecarboxylate • resmetrina

(Portuguese) • NSC-195022 • (5-(phenylmethyl)-3-furanyl) methyl 2,2-dimethyl-3-(2-methyl-1-propenyl) cyclopropanecarboxylate) • (5-(phenylmethyl)-3-furanyl) methyl 2,2-dimethyl-3-furylmethyl 2,2-dimethyl-3-(2-methylpropenyl) cyclopropanecarboxylate • 2,2-dimethyl-3-(2-methyl-1-propenyl) cyclopropanecarboxylic acid • (5-(phenylmethyl)-3-furanyl) methyl ester • ent-27474- • sb-pennick-1382- • oms-1206 • sbp-1382- • NIA-17370 • FMC-17370 • 5-benzyl-3-furymethyl(1rs)-cis, trans-chrysanthemate • resmethrine • chryson • synthrin • pynosect • isathrine • [5-phenylmethyl-3-furan]methyl-2,2-dimethyl-3-(2-methyl-1-propenyl) cyclopropane carboxylate. • 5-benzyl-3-furylmethyl (1rs, 3rs • 1rs, 3rs)-2,2-dimethyl-3-(2-methylprop-1-enyl) cyclopropanecarboxylate • 5-benzyl-3-furylmethyl (1rs)-cis-trans-2,2-dimethyl-3-(2-methylprop-1-enyl) cyclopropanecarboxylate • NRDC-104

MF $C_{22}H_{26}O_3$

MW 338.48

COLOR AND FORM Waxy off-white to tan solid; colorless crystals.

ODOR Chrysanthemate odor.

DENSITY 0.958–0.968 at 20°C.

BOILING POINT 180°C at 0.01 mm Hg.

MELTING POINT 43–48°C.

FIREFIGHTING PROCEDURES Use carbon dioxide, foam, or dry chemical (on fires involving pyrethroids) (pyrethrum) [R11, 2]; self-contained breathing apparatus with a full facepiece operated in pressure-demand or other positive-pressure mode (pyrethrum) [R11, 5]; extinguish fire using agent suitable for type of surrounding fire. (pyrethrins) [R10].

SOLUBILITY Very soluble in xylene and aromatic petroleum hydrocarbons; solubility in kerosene 10%; insoluble in water at 25°C; methylene chloride at 25°C: greater than 50% wt/wt; in acetone at 25°C: greater than 50% wt/wt; in ethanol

and isopropanol 8 and 7%, respectively; soluble at 30°C: 220 g/kg hexane; 81 g/kg methanol; in water at 30°C: <1 mg/L.

VAPOR PRESSURE 0.0015 mPa at 30°C.

CORROSIVITY Noncorrosive.

STABILITY AND SHELF LIFE Very stable in storage under dry conditions; decomposed by light (more slowly than pyrethrins), air, acids, and alkalis [R2].

OTHER CHEMICAL/PHYSICAL PROPERTIES Wax solid (technical resmethrin).

PROTECTIVE EQUIPMENT Employees should be provided with and required to use dust and splash-proof safety goggles where pyrethroids may contact the eyes. (pyrethroids) [R11, 3].

PREVENTATIVE MEASURES Skin that becomes contaminated with pyrethrum should be promptly washed or showered with soap or mild detergent and water. (pyrethrum) [R11, 3].

CLEANUP Spillages of pesticides at any stage of their storage or handling should be treated with great care. Liquid formulations may be reduced to solid phase by evaporation. Dry-sweeping of solids is always hazardous: these should be removed by vacuum cleaning, or by dissolving them in water or other solvent in the factory environment. (pesticides) [R12].

DISPOSAL SRP: At the time of review, criteria for land treatment or burial (sanitary landfill) disposal practices are subject to significant revision. Prior to implementing land disposal of waste residue (including waste sludge), consult with environmental regulatory agencies for guidance on acceptable disposal practices.

COMMON USES Potent contact insecticide effective against a wide range of insects [R1, 738]; highly active insecticide recommended for use against houseflies, German cockroaches [R5]; applied on insects found in household, greenhouse, indoor landscaping, mushroom houses, industrial, stored product, mosquito and insect control [R4, p. C-202]; pet sprays, pet shampoo, and application on horses and horse stables [R4, p. C-202]; cleared for use in aerosols, aqueous pressurized sprays, emulsifiable concentrates, transparent emulsions, and oil-base liquid, including ULV cleared for fabric protection [R4, p. C–202]; resmethrin (2%) has replaced pyrethrum-ddt formulation in aircraft disinfection [R6]; control on wood lice [R7]; used as a household and garden insecticide (especially for control of glasshouse whiteflies and aphids), and in agricultural premises [R2]; resmethrin is currently used for mosquito control (by aerial application) in the USA, and it can also be used for the control of whitefly in greenhouses [R8]; resmethrin is mainly used in aerosol formulations, but also in oil formulations and emulsifiable concentrates for the control of household and public-health insects. It is also used in combination with other insecticides (e.g., tetramethrin, malathion) [R8]; insecticide. (pyrethrins) [R9].

OCCUPATIONAL RECOMMENDATIONS None listed.

NON-HUMAN TOXICITY LD_{50} rat oral 1,400 mg/kg [R13]; LD_{50} rat oral >2,500 mg/kg [R2]; LD_{50} rat percutaneous >3,000 mg/kg [R1, 738]; LD_{50} rat (male) dermal 2,500 mg/kg [R14]; LD_{50} rat (female) dermal 2,500 mg/kg [R14]; LD_{50} rabbit dermal 2,500 mg/kg [R14]; LC_{50} rat inhalation >9,490 mg/m³/4 hr [R14]; LC_{50} rat inhalation >12,000 mg/m³/1 hr [R14]; LC_{50} rabbit inhalation >12,000 mg/m³/1 hr [R14]; LC_{50} dog inhalation >420 mg/m³/4 hr [R14]; LD_{50} rat (male) oral 1,244 mg/kg [R15]; LD_{50} rat (female) oral 1,721 mg/kg [R15]; LD_{50} rat (weanling, male), oral 1,987 mg/kg [R15]; LD_{50} mice (male) oral 690 mg/kg [R15]; LD_{50} mice (female) oral 940 mg/kg [R15]; LD_{50} rat dermal 3.0 g/kg [R3].

ECOTOXICITY LC_{50} *Oncorhynchus kisutch* (coho salmon) 1.8 µg/L/96 hr at 18°C (95% confidence limit 0.55–5.6 µg/L) wt 0.5 g (technical material, 84.5%) [R16]; LC_{50} *Salvelinus namaycush* (lake trout) 1.7 µg/L/96 hr at 12°C (95% confidence limit 1.1–2.5 µg/L) wt 0.7 g (technical material, 84.5%) [R16]; LC_{50} *Pimephales promelas* (fathead minnow) 3.0 µg/L/96 hr at 17°C (95% confidence limit 0.89–9.9 µg/L) wt

0.7 g (technical material, 84.5%) [R16]; LC_{50} *Ictalurus punctatus* (channel catfish) 16.6 µg/L/96 hr at 18°C (95% confidence limit 9.6–28.6 µg/L) wt 0.7 g (technical material, 84.5%) [R16]; LC_{50} *Lepomis macrochirus* (bluegill) 1.7 µg/L/96 hr at 18°C (95% confidence limit 0.31–9.3 µg/L) wt 0.6 g (technical material, 84.5%) [R16]; LD_{50} California quail, oral >2,000 mg/kg, 5–6 mo old male [R17]; LC_{50} coho salmon >150 µg/L/96 hr (static test); <0.277 µg/L/96 hr (flow-through test) [R18]; LC_{50} steelhead trout 0.450 µg/L/96 hr (static test); 0.275 µg/L/96 hr (flow-through test) [R18]; LC_{50} bluegill 2.62 µg/L/96 hr (static test); 0.750 µg/L/96 hr (flow-through test) [R18]; LC_{50} yellow perch 2.36 µg/L/96 hr (static test); 0.513 µg/L/96 hr (flow-through test) [R18]; LC_{50} *Daphnia pulex* 15,000 µg/L/3 hr (static at 25°C) (technical) [R19]; LC_{50} *Moina macrocopa* 14,000 µg/L/3 hr (static at 25°C) (technical) [R19]; LC_{50} *Sigara substriate* (size = 0.59 cm: 6.1 mg) 2 µg/L/48 hr (static at 25°C) (technical) [R19]; LC_{50} *Micronecta sedula* (size = 0.32 cm: 1.8 mg) 3.3 µg/L/48 hr (static at 25°C) (technical) [R19]; LC_{50} *Cloeon dipterum* (size = 0.93; 5.6 mg) 4.5 µg/L/48 hr (static at 25°C) (technical) [R19]; LC_{50} *Orthetrum albistylum speciosum* (size = 2.3 cm; 0.62 g) 7.3 µg/L/48 hr (static at 25°C) (technical) [R19]; LC_{50} *Eretes stricticus* (size = 1.5 cm; 0.2 g) 25 µg/L/48 hr (static at 25°C) (technical) [R19]; *LC_{50} Sympetrum frequens* (size = 2.1 cm; 0.56 g) 10 µg/L/48 hr (static at 25°C) (technical) [R19]; LC_{50} (*Cyprinus carpio*) carp 44 µg/L/48 hr (static system) (technical) [R20]; LC_{50} *Oryzias latipes* (killifish), adult 300 µg/L/48 hr (statis system) (technical) [R20].

BIOCONCENTRATION Based upon an estimated log K_{ow} of 6.16 (1), the BCF for resmethrin can be estimated to be log 4.45 from a recommended regression-derived equation (2,SRC). This BCF value suggests that bioconcentration in aquatic organisms (fish exhibit difficulty in rapidly degrading pyrethroids (3)) may be environmentally important in organisms that cannot metabolize resmethrin (SRC) [R21].

FIFRA REQUIREMENTS Tolerances are established for residues of the insecticide resmethrin in or on food items resulting from use of the insecticide in food handling and storage [R22].

REFERENCES

1. Worthing, C. R., and S. B. Walker (eds.). *The Pesticide Manual—A World Compendium.* 8th ed. Thornton Heath, UK: The British Crop Protection Council, 1987.

2. Hartley, D., and H. Kidd (eds.). *The Agrochemicals Handbook.* 2nd ed. Lechworth, Herts, England: The Royal Society of Chemistry, 1987, p. A362/Aug 87.

3. Gosselin, R. E., R. P. Smith, H. C. Hodge. *Clinical Toxicology of Commercial Products.* 5th ed. Baltimore: Williams and Wilkins, 1984, p. II-262.

4. *Farm Chemicals Handbook 1986.* Willoughby, OH: Meister Publishing Co., 1986.

5. Spencer, E. Y. *Guide to the Chemicals Used in Crop Protection.* 6th ed. Publication 1093, Research Institute, Agriculture Canada, Ottawa, Canada: Information Canada, 1973. 446.

6. Osol, A. (ed.). *Remington's Pharmaceutical Sciences.* 16th ed. Easton, PA: Mack Publishing Co., 1980. 1199.

7. *Jpn Kokai Tokkyo Koho* 82: 31601 (2/20/82) Nagaoka and Co. Ltd.

8. WHO. *Environmental Health Criteria 92: Resmethrins—Resmethrin, Bioresmethrin, Cisresmethrin* p. 25 (1989).

9. Budavari, S. (ed.). *The Merck Index—Encyclopedia of Chemicals, Drugs, and Biologicals.* Rahway, NJ: Merck and Co., Inc., 1989. 1267.

10. Bureau of Explosives, *Emergency Handling of Haz Matl in Surface Trans* p. 434 (1981).

11. Mackison, F. W., R. S. Stricoff, and L. J. Partridge, Jr. (eds.). *NIOSH/OSHA—Occupational Health Guidelines for Chemical Hazards.* DHHS (NIOSH) Publication No. 81–123 (3 vols). Washington, DC: U.S. Government Printing Office, Jan. 1981.

12. International Labour Office. *Encyclopedia of Occupational Health and Safety.* Vols. I and II. Geneva, Switzerland: International Labour Office, 1983. 1619.

13. *Kirk–Othmer Encyclopedia of Chemical Technology.* 3rd ed., Volumes 1–26. New York, NY: John Wiley and Sons, 1978–1984, p. V13 458 (1981).

14. WHO. *Environmental Health Criteria 92: Resmethrins—Resmethrin, Bioresmethrin, Cisresmethrin* p. 41 (1989).

15. WHO. *Environmental Health Criteria 92: Resmethrins—Resmethrin, Bioresmethrin, Cisresmethrin* p. 40 (1989).

16. U.S. Department of Interior, Fish and Wildlife Service. *Handbook of Acute Toxicity of Chemicals to Fish and Aquatic Invertebrates.* Resource Publication No. 137. Washington, DC: U.S. Government Printing Office, 1980. 71.

17. U.S. Department of the Interior, Fish and Wildlife Service. *Handbook of Toxicity of Pesticides to Wildlife.* Resource Publication 153. Washington, DC: U.S. Government Printing Office, 1984. 69.

18. Verschueren, K. *Handbook of Environmental Data of Organic Chemicals.* 2nd ed. New York, NY: Van Nostrand Reinhold Co., 1983. 1050.

19. WHO. *Environmental Health Criteria 92: Resmethrins—Resmethrin, Bioresmethrin, Cisresmethrin* p. 36 (1989).

20. WHO. *Environmental Health Criteria 92: Resmethrins—Resmethrin, Bioresmethrin, Cisresmethrin* p (1989).

21. (1) USEPA. *Graphical Exposure Modeling System (GEMS).* CLOGP (1987); (2) Lyman, W. J., et al. *Handbook of Chemical Property Estimation Methods.* Washington, DC: Amer Chemical Soc, p. 5–10 (1990); (3) Demoute, J. P. *Pestic Sci* 27: 375–85 (1989).

22. 40 CFR 185.5300 (7/1/91).

■ CYFLUTHRIN

CAS # 68359–37–5

SYNONYMS cyfoxylate • baythroid-h •
(rs)-alpha-cyano-4-fluoro-3-
phenoxybenzyl (1rs, 3rs: 1rs, 3sr)-3-
(2,2-dichlorovinyl)-2,2-
dimethylcyclopropanecarboxylate •
cyano (4-fluoro-3-phenoxyphenyl)
methyl 3-(2,2-dichloroethenyl-2,2-
dimethyl cyclopropanecarboxylate •
cyfluthrine • solfac • responsar • tempo
• 3-(2,2-dichloroethenyl)-2,2-
diethylcyclopropanecarboxylic acid
cyano (4-fluoro 3-phenoxyphenyl)
methyl ester • (r,s)-alpha-cyano-4-fluoro-
3-phenoxybenzyl-(1r,s)-cis, trans-3-(2,2-
dichlorovinyl)-2,2-
dimethylcyclopropanecarboxylate •
cyclopropanecarboxylic acid,
2-(2,2-dichlorovinyl)-3,3-dimethyl-, ester
with (4-fluoro-3-phenoxyphenyl)
hydroxyacetonitrile • fcr-1272 • Bay-fcr
1272 • baythroid

MF $C_{22}H_{18}Cl_2FNO_3$

MW 434.31

COLOR AND FORM Yellowish-brown oil,
viscous, amber, partly crystalline oil.

ODOR Aromatic solvent odor at room temperature.

MELTING POINT Ca 60°C.

FIREFIGHTING PROCEDURES Use carbon
dioxide, foam, or dry chemical (on fires
involving pyrethroids) (pyrethrum) [R7, 2].
Firefighting: Self-contained breathing apparatus with a full facepiece operated in
pressure-demand or other positive-pressure mode (pyrethrum) [R7, 5]. Extinguish
fire using agent suitable for type of surrounding fire. (pyrethrins) [R2].

SOLUBILITY Solubility in water (20°C) 2
μg/L.

VAPOR PRESSURE <1 mPa at 20°C.

STABILITY AND SHELF LIFE Pyrethrins are
stable for long periods in water-based
aerosols where emulsifiers give neutral
water systems. (pyrethrins) [R3].

OTHER CHEMICAL/PHYSICAL PROPERTIES
Colorless oil; specific optical rotation:
−15.0°C at 20°C/D (concentration by volume = 1.0 g in 100 ml chloroform)
((1r)(3r)(alpha r)-cyfluthrin). Pasty yellow

mass; contains 23–26% (r 1r)-cis- + (s 1s)-
cis- enantiomers (mp 57°C), 16–19% (s 1r-
cis-(MP 74°C), 33–36% (r 1r)-trans- + (s
1s)-trans-(MP 66°C), 22–25% (s 1r)- trans-
+ (r 1s)-trans-(MP 102°C) (technical cyfluthrin) Crystals from m-hexane; MP:
68–69°C; specific optical rotation: −2.1° at
20°C/D (concentration by volume = 1.0 g
in 100 ml chloroform) ((1r)(3s)(alpha s)-
cyfluthrin). Crystals; MP: 50–52°C; specific optical rotation: +24.5° at 20°C/D (concentration by volume = 1.0 g in 100 ml
chloroform) ((1r)(3r)(alpha s)-cyfluthrin).

PROTECTIVE EQUIPMENT Employees
should be provided with and required to
use dust- and splash-proof safety goggles
where pyrethroids may contact the eyes.
(pyrethroids) [R7, 3].

PREVENTATIVE MEASURES Skin that becomes contaminated with pyrethrum
should be promptly washed or showered
with soap or mild detergent and water
(pyrethrum) [R7, 3].

CLEANUP Spillages of pesticides at any
stage of their storage or handling should
be treated with great care. Liquid formulations may be reduced to solid phase by
evaporation. Dry sweeping of solids is
always hazardous: these should be removed by vacuum cleaning, or by dissolving them in water or other solvent in the
factory environment (pesticides) [R4].

DISPOSAL Incineration would be an effective disposal procedure where permitted. If
an efficient incinerator is not available,
the product should be mixed with large
amounts of combustible material, and
contact with the smoke should be avoided.
(pyrethrin products) [R5].

COMMON USES Agricultural insecticide
[R8, 432].

NON-HUMAN TOXICITY LD50 rat male oral
590 mg/kg [R1].

ECOTOXICITY LC50 golden orfe 0.0032
mg/L/96 hr (conditions of bioassay not
specified) [R1].

FIFRA REQUIREMENTS Tolerances are established for residues of the insecticide
cyfluthrin in or on the following raw agricultural commodities: Cattle (fat, meat,

and meat by-products), goats (fat, meat, and meat by-products), hogs (fat, meat, and meat by-products), horses (fat, meat, and meat by-products), sheep (fat, meat, and meat by-products), cottonseed, hops (fresh), and milk [R6].

REFERENCES

1. Hartley, D., and H. Kidd (eds.). *The Agrochemicals Handbook*. 2nd ed. Lechworth, Herts, England: The Royal Society of Chemistry, 1987, p. A799/Aug 87.

2. Bureau of Explosives, *Emergency Handling of Haz Matl in Surface Trans* p. 434 (1981).

3. *Farm Chemicals Handbook 1986*. Willoughby, OH: Meister Publishing Co., 1986, p. C-198.

4. International Labour Office. *Encyclopedia of Occupational Health and Safety*. Vols. I and II. Geneva, Switzerland: International Labour Office, 1983. 1619.

5. Sittig, M. *Handbook of Toxic and Hazardous Chemicals and Carcinogens, 1985*. 2nd ed. Park Ridge, NJ: Noyes Data Corporation, 1985. 762.

6. 40 CFR 180.436 (7/1/91).

7. Mackison, F. W., R. S. Stricoff, and L. J. Partridge, Jr. (eds.). *NIOSH/OSHA—Occupational Health Guidelines for Chemical Hazards*. DHHS (NIOSH) Publication No. 81-123 (3 vols). Washington, DC: U.S. Government Printing Office, Jan. 1981.

8. Budavari, S. (ed.). *The Merck Index— Encyclopedia of Chemicals, Drugs and Biologicals*. Rahway, NJ: Merck and Co., Inc., 1989.

■ CYPREX

CAS # 2439-10-3

SYNONYMS Aadodin • ac-5223 • American Cyanamid-5,223 • carpene • curitan • cyprex-65w • dodecylguanidine acetate • dodecylguanidine monoacetate • dodin • dodine • dodine acetate • dodine, mixture with glyodin • doguadine • doquadine • ent-16,436 • experimental fungicide 5223 • guanidine, dodecyl-, monoacetate • karpen • laurylguanidine acetate • melprex • melprex-65 • milprex • n-dodecylguanidine acetate • questuran • syllit • syllit-65 • tsitrex • venturol • vondodine • carpen

MF $C_{13}H_{29}N_3$ +m $C_2H_4O_2$.

MW 287.51

COLOR AND FORM Slightly waxy solid; colorless crystals.

MELTING POINT 136°C.

SOLUBILITY Soluble in alcohol, hot water; slightly soluble in other solvents. Insoluble in most organic solvents. In low-mol-wt alcohols ranges from about 7% to 23% at room temp; soluble in acids; 0.063% by wt in water at 25°C. Readily soluble in mineral acids. In distilled water 0.07%. Soluble in methanol, ethanol. Practically insoluble in most, organic solvents.

VAPOR PRESSURE Less than 0.01 mPa at 20°C.

CORROSIVITY Noncorrosive.

STABILITY AND SHELF LIFE At ordinary temperature compound is stable as solid or in solution; stable under moderately alkaline or acid conditions [R2].

OTHER CHEMICAL/PHYSICAL PROPERTIES Free base, reg number (112-25-2), is liberated by strong alkali. Compound is surface-active; salt of strong base and weak acid.

PROTECTIVE EQUIPMENT Protective clothing: Goggles, face shield, rubber gloves when handling [R7, p. C-128].

PREVENTATIVE MEASURES SRP: The scientific literature supports the wearing of contact lenses in industrial environments, as part of a program to protect the eye against chemical compounds and minerals causing eye irritation. However, there may be individual substances whose irritating or corrosive properties are such that the wearing of contact lenses would be harmful to the eye. In those specific cases contact lenses should not be worn. Transfer or mix (product) with adequate ventilation [R7, 1286].

DISPOSAL Recommendable methods: Incineration. Peer-review: Large amt: Incinerate in a unit equipped with effluent gas scrubbing. (peer-review conclusions of an IRPTC expert consultation (May 1985)) [R4].

COMMON USES Dodecylguanidine salts are highly effective for agricultural applications, particularly in the control of apple scab and cherry leaf spot. They are also effective algal growth inhibitors, useful in the manufacture of microorganism-stable oil- and water-emulsions and in treatment of saline process water [R3]. For scab on apples, pears, pecans. Leafspot on cherries. Foliar diseases of strawberries. Bacterial leafspot on peaches. Leaf blight of sycamores, black walnuts [R7, p. C–128].

OCCUPATIONAL RECOMMENDATIONS None listed.

NON-HUMAN TOXICITY LD_{50} rat (male) oral approximately 1,000 mg/kg [R1]; LD_{50} rabbit percutaneous >1,500 mg/kg [R5]; LD_{50} rat percutaneous >6,000 mg/kg [R5]; LD_{50} rat dermal >6,000 mg/kg [R7, p. C–128]; LD_{50} rabbit dermal >1,500 mg/kg (single 24-hr contact) [R7, p. C–128].

ECOTOXICITY LC_{50} Gammarus fasciatus 1.1 mg/L/96 hr. Static bioassay without aeration, pH 7.2–7.5, water hardness 40–50 mg/L as calcium carbonate, and alkalinity of 30–35 mg/L [R6]. LD_{50} Mallard duck, oral 1,142 mg/kg [R5].

FIFRA REQUIREMENTS In 1988, Congress amended FIFRA to strengthen and accelerate EPA's reregistration program. The nine-year reregistration scheme mandated by "FIFRA 88" applies to each registered pesticide product containing an active ingredient initially registered before November 1, 1984. List A consists of the 194 chemical cases (or 350 individual active ingredients) for which EPA had issued Registration Standards prior to the effective date of FIFRA '88. List: A; Case: Dodine; Case No.: 0161; Pesticide type: Fungicide, antimicrobiol; Registration Standard Date: 02/28/87; Case Status: Awaiting Data/Data in Review: OPP awaits data from the pesticide's producer(s) regarding its human health and/or environmental effects, or OPP has re-ceived and is reviewing such data, in order to reach a decision about the pesticide's eligibility for reregistration. Active Ingredient (AI): Dodine; Data Call-in (DCI) Date: 12/09/91; AI Status: The producer(s) of the pesticide has made commitments to conduct the studies and pay the fees required for reregistration, and is meeting those commitments in a timely manner [R10]. Tolerances are established for residues of the fungicide dodine (n-dodecyl-guanidine acetate) in or on the following raw agricultural commodities: apples; cherries (sour and sweet); meat; milk; peaches; pears; pecans; strawberries; and walnuts [R8]. Tolerances with regional registration, as defined in 180.1(n), are established for residues of dodine in or on the following raw agricultural commodities: spinach [R9].

REFERENCES

1. Worthing, C. R., and S. B. Walker (eds.). The Pesticide Manual—A World Compendium. 8th ed. Thornton Heath, UK: The British Crop Protection Council, 1987. 329.

2. Spencer, E. Y. Guide to the Chemicals Used in Crop Protection. 7th ed. Publication 1093. Research Institute, Agriculture Canada, Ottawa, Canada: Information Canada, 1982. 250.

3. Kirk–Othmer Encyclopedia of Chemical Technology. 3rd ed., Volumes 1–26. New York, NY: John Wiley and Sons, 1978–1984, p. S 520 (1984).

4. United Nations. Treatment and Disposal Methods for Waste Chemicals (IRPTC File). Data Profile Series No. 5. Geneva, Switzerland: United Nations Environmental Programme, Dec. 1985. 187.

5. Hartley, D., and H. Kidd (eds.). The Agrochemicals Handbook. 2nd ed. Lechworth, Herts, England: The Royal Society of Chemistry, 1987, p. A173/Aug 87.

6. U.S. Department of Interior, Fish and Wildlife Service. Handbook of Acute Toxicity of Chemicals to Fish and Aquatic Invertebrates. Resource Publication No. 137. Washington, DC: U.S. Government Printing Office, 1980. 82.

7. *Farm Chemicals Handbook 1993.* Willoughby, OH: Meister Publishing Co., 1993.

8. 40 CFR 180.172(a) (7/1/91).

9. 40 CFR 180.172(b) (7/1/91).

10. USEPA/OPP. *Status of Pesticides in Reregistration and Special Review.* p. 87 (Mar, 1992) EPA 700-R-92-004.

■ DAZOMET

CAS # 533–74–4

SYNONYMS 2-thio-3,5-dimethyl-tetrahydro-1,3,5-thiadiazine • 2h-1,3,5-thiadiazine-2-thione, tetrahydro-3,5-dimethyl • 3,5-dimethyl-1,2,3,5-tetrahydro-1,3,5-thiadiazinethione-2 • 3,5-dimethyl-1,3,5-2h-tetrahydrothiadiazine-2-thione • 3,5-dimethyl-2-thionotetrahydro-1,3,5-thiadiazine • 3,5-Dimethyl-perhydro-1,3,5-thiadiazin-2-thion [Czech, German] • 3,5-dimethyltetrahydro-1,3,5-2h-thiadiazine-2-thione • 3,5-dimethyl-tetrahydro-1,3,5-thiadiazine-2-thione • 3,5-dimethyltetrahydro-2h-1,3,5-thiadiazine-2-thione • 3,5-dimetil-peridro-1,3,5-tiadiazin-2-tione [Italian] • basamid • basamid-g • basamid-p • basamid-granular • basamid-puder • carbothialdin • carbothialdine • crag-85w • crag-974 • crag fungicide-974 • crag nemacide • dazomet powder basf • dimethylformocarbothialdine • dmtt • fennosan-b-100- • micofume • mylon [Czech] • mylone • mylone-85 • n-521 • nalcon-243 • nefusan • prezervit • stauffer-N-521- • tetrahydro-2h-3,5-dimethyl-1,3,5-thiadiazine-2-thione • tetrahydro-3,5-dimethyl-2h-1,3,5-thiadiazine-2-thione • thiadiazin [pesticide] • thiazon • thiazone • tiazon • ucc-974

MF $C_5H_{10}N_2S_2$

MW 162.27

COLOR AND FORM White crystals; needles from benzene; crystals from ethanol.

ODOR Nearly odorless; weakly pungent.

DENSITY 1.30 at 20°C.

MELTING POINT 106–107°C.

SOLUBILITY Soluble in water (0.12% at 25°C); insoluble in carbon tetrachloride; very soluble in hot alcohol, soluble in acids; carbon tetrachloride 0.0 g/100 ml, ether 0.0 g/100 ml, isopropanol 0.5 g/100 ml, xylene 1.1 g/100 ml, ethanol 3.0 g/100 ml, ethylene glycol 3.0 g/100 ml, ethylene glycol monomethyl ether 5.0 g/100 ml, dioxane 8.0 g/100 ml, acetone 13.1 g/100 ml, trichloroethylene 30.0 g/100 ml, chloroform 30 g/100 ml, dimethyl sulfoxide, 30 g/100 ml, dimethyl formamide 30 g/100 ml; water solubility 1.2 g/L at 25°C; solubility (20°C) 3 g/kg water; 173 g/kg acetone; 51 g/kg benzene; 391 g/kg chloroform; 400 g/kg cyclohexane; 15 g/kg ethanol; 6 g/kg ethyl ether.

VAPOR PRESSURE 370 micropascals at 20°C.

CORROSIVITY Noncorrosive in the dry state.

STABILITY AND SHELF LIFE Moderately stable but sensitive to heat above 35°C, and to moisture [R7].

OTHER CHEMICAL/PHYSICAL PROPERTIES Decomposes above 102°C. Sensitive to moisture and heat >35°C. Acid hydrolysis yields carbon disulfide. Vapor pressure = 2.8×10^{-6} mm Hg at 20°C.

PREVENTATIVE MEASURES Avoid contact with eyes, skin, or clothing [R6].

DISPOSAL SRP: At the time of review, criteria for land treatment or burial (sanitary landfill) disposal practices are subject to significant revision. Prior to implementing land disposal of waste residue (including waste sludge), consult with environmental regulatory agencies for guidance on acceptable disposal practices.

COMMON USES For soil fungi, nematodes, germinating weeds, soil insects. Preplant in seed beds for tobacco, nurseries, greenhouses, substrates for potted plants, turf, ornamentals. Antimicrobial in slimicide preparations [R1]. Antimicrobial agent for slimicide preparations and for adhesives; soil fungicide, nematocide and insecticide in seed beds for tobacco; used on turf and ornamentals [R2]. Paper-mill slimicide, paint, cooling water slimicide, adhesives [R3]; biocide in industrial wastewater [R4]; nematocide, fungicide [R5].

OCCUPATIONAL RECOMMENDATIONS None listed.

NON-HUMAN TOXICITY LD$_{50}$ rat oral 640 mg/kg [R8]. LD$_{50}$ mouse, male albino acute, oral 650 mg/kg [R11, 141]. LD$_{50}$ rabbit acute, oral 320–620 mg/kg [R11, 141].

BIODEGRADATION The rate of disappearance of dazomet from both an unamended and sterilized Williamson silt loam was identical, indicating that the initial phase of degradation in soil occurred predominately from nonenzymatic mechanisms (1). The rate of degradation of dazomet in soil, as measured by the release of methylisothiocyanate, was found to be independent of the soil microorganisms present (2) [R9].

BIOCONCENTRATION Based on an experimental water solubility of 1,200 mg/L at 25°C (1), the bioconcentration factor for dazomet can be estimated to be about 10 from a regression-derived equation (2,SRC). This estimated bioconcentration factor indicates that dazomet is not expected to bioconcentrate in fish and aquatic organisms [R10].

REFERENCES

1. *Farm Chemicals Handbook 1994.* Willoughby, OH: Meister, 1994, p. C–109.

2. SRI.

3. *Kirk–Othmer Encyclopedia of Chemical Technology.* 3rd ed., Volumes 1–26. New York, NY: John Wiley and Sons, 1978–1984, p. V13 244.

4. *Kirk–Othmer Encyclopedia of Chemical Technology.* 3rd ed., Volumes 1–26. New York, NY: John Wiley and Sons, 1978–1984, p. V24 381.

5. Gerhartz, W. (exec ed.). *Ullmann's Encyclopedia of Industrial Chemistry.* 5th ed. Vol A1: Deerfield Beach, FL: VCH Publishers, 1985 to present, p. A9 18.

6. *Farm Chemicals Handbook 1995.* Willoughby, OH: Meister, 1995, p. C–114.

7. Worthing, C. R., S. B. Walker (eds.). *The Pesticide Manual—A World Compendium.* 7th ed. Lavenham, Suffolk, Great Britain: The Lavenham Press Limited, 1983. 157.

8. Hayes, W. J., Jr., E. R. Laws, Jr. (eds.). *Handbook of Pesticide Toxicology. Volume 3. Classes of Pesticides.* New York, NY: Academic Press, Inc., 1991. 1389.

9. (1) Tate, R.L., and M. Alexander. *Soil Sci* 118: 317–21 (1974). (2) Munnecke, D. E., J. P. Martin. *Phytopathology* 54: 941–5 (1964).

10. (1) Yalkowsky, S. H., R. M. Dannenfelser. *Arizona Database of Aqueous Solubilities.* Univ of AZ, College of Pharmacy (1992). (2) Lyman, W. J., et al. (eds.); *Handbook of Chemical Property Estimation Methods*, Washington, DC: Amer Chemical Soc, Chapt 5 (1990).

11. Weed Science Society of America. *Herbicide Handbook.* 5th ed. Champaign, Illinois: Weed Science Society of America, 1983.

■ DECALDEHYDE

CAS # 112–31–2

SYNONYMS 1-decanal • 1-decyl aldehyde • aldehyde c-10 • aldehyde-c10 • c-10 aldehyde • capraldehyde • capric aldehyde • caprinaldehyde • caprinic aldehyde • caprylaldehyde • decanal • decanaldehyde • decyl aldehyde • decylic aldehyde • n-decaldehyde • n-decanal • n-decyl aldehyde

MF $C_{10}H_{20}O$

MW 156.30

COLOR AND FORM Colorless to light-yellow liquid.

ODOR Pronounced fatty odor that develops floral character on dil; strong orange-rose odor.

TASTE Sharp, orange flavor.

DENSITY 0.830 at 15°C.

BOILING POINT 208–209°C.

MELTING POINT –5°C.

SOLUBILITY Insoluble in water; soluble in

ethanol, ether, acetone; soluble in mineral oils, fixed oils, volatile oils; insoluble in glycerol; soluble in most organic solvents. Solubility in 80% alcohol: 1: 1.

STABILITY AND SHELF LIFE Fair [R3, 483].

OTHER CHEMICAL/PHYSICAL PROPERTIES Acid value: not more than 10.

COMMON USES Synthetic flavoring substance and adjuvant [R1].

STANDARD CODES NFPA, ICC, USCG—no.

PERSISTENCE Most aldehydes are readily degradable.

MAJOR SPECIES THREATENED Waterfowl.

DIRECT CONTACT Irritant to skin.

GENERAL SENSATION Floral, fatty odor skin irritation grade 5—necrosis with 0.005 ml; eye irritation grade 1—no effect [R2].

PERSONAL SAFETY Wear full protective clothing and self-contained breathing apparatus.

ACUTE HAZARD LEVEL Limited animal experiments suggest low toxicity. Irritant.

CHRONIC HAZARD LEVEL Unknown.

DEGREE OF HAZARD TO PUBLIC HEALTH Animal experiments suggest low toxicity. Irritant.

OCCUPATIONAL RECOMMENDATIONS None listed.

ACTION LEVEL Notify fire and air authority. Enter from upwind and attempt to contain slick. Remove ignition sources.

ON-SITE RESTORATION Oil-skimming equipment and sorbent foams can be applied to the slick. Seek professional environmental engineering assistance through EPA's Environmental Response Team (ERT), Edison, NJ, 24–hour no., 908-548-8730.

BEACH OR SHORE RESTORATION Should not be burned.

AVAILABILITY OF COUNTERMEASURE MATERIAL Oil-skimming equipment—major ports; sorbent foams (polyurethane) —upholstery shops.

DISPOSAL METHOD (1) Absorb on vermiculite and burn in open incinerator. (2) Dissolve in flammable solvent and spray in incinerator with afterburner.

DISPOSAL NOTIFICATION Contact local air authority.

INDUSTRIAL FOULING POTENTIAL Slick may reduce heat transfer or cause hot spots and scaling.

MAJOR WATER USE THREATENED Recreation.

PROBABLE LOCATION AND STATE Colorless to yellow liquid. Will float in slick on surface.

COLOR IN WATER Colorless to light yellow.

FDA 121.101, GRAS [R3, 830].

REFERENCES
1. Sax, N. I. *Dangerous Properties of Industrial Materials.* 5th ed. New York: Van Nostrand Reinhold, 1979. 536.

2. Smyth, H. F., Jr., C. P. Carpenter, C. S. Weil, U. C. Pozzani, J. A. Striegel, and J. S. Nycum, "Range-Finding Toxicity Data: List Vll," *American Industrial Hygiene Association Journal,* 30: 470–476. 1969. Smyth, H. F., C. P. Carpenter, and C. S. Weil, "Range-Finding Toxicity Data: List IV," *AMA Archives of Industrial Hygiene and Occupational Medicine,* 4: 119–122, 1951. Smyth, H. F., C. P. Carpenter, C. S. Weil, U. C. Pozzani, and J. A. Striegel, "Range-Finding Toxicity Data: List Vl," *American Industrial Hygiene Association Journal,* 23: 95–107, 1962. Smyth, H. F., C. P. Carpenter, C. S. Weil, and U. C. Pozzani, "Range-Finding Toxicity Data: List V," *AMA Archives of Industrial Hygiene and Occupational Medicine,* 10: 61–68, 1954. Smyth, H. F., C. P. Carpenter, and C. S. Weil, "Range-Finding Toxicity Data: List 111," *Journal of Industrial Hygiene and Toxicology,* 31: 60–62, 1949. Smyth, H. F., L. Seaton, and L. Fischer, "The Single Dose Toxicity of Some Glycols and Derivatives," *Journal of Industrial Hygiene and Toxicology,* 23 (6): 259–268, 1941. Smyth, H. F., and C. P. Carpenter, "Further Experience with the Range-Finding Test in the Industrial Toxicology Laboratory," *Journal of Industrial Hygiene and Toxicology,* 30: 63–68,

1948. Smyth, H. F., and C. P. Carpenter, "The Place of the Range-Finding Test in the Industrial Toxicology Laboratory," *Journal of Industrial Hygiene and Toxicology*, 26: 269–273, 1 944.

3. Furia, T.E. (ed.). *CRC Handbook of Food Additives*. 2nd ed. Cleveland: The Chemical Rubber Co., 1972.

■ DELTAMETHRIN

CAS # 52918–63–5

SYNONYMS Decamethrin • cislin • crackdown • k-othrine • (1r-(1-alpha (s),3-alpha))-cyano-(3-phenoxyphenyl) methyl-3-(2,2-dibromovinyl)-2,2-dimethylcyclopropanecarboxylate) • [1r-[1alpha (s),3alpha]]-cyano (3-phenoxyphenyl) methyl 3-[2,2-dibromoethenyl]-2, 2-dimethyl-cyclopropanecarboxylate • (s)-alpha-cyano-3-phenoxybenzyl (1r, 3r)-3-(2,2-dibromovinyl)-2,2-dimethyl cyclopropan-1-carboxylate • deltamethrine • k-othrin • butoss • 3-(2,2-dibromoethenyl)-2,2-dimethylcyclopropanecarboxylic acid cyano (3-phenoxy phenyl)-methyl ester • (s)-alpha-cyano-3-phenoxybenzyl-(1r)-cis-3-(2,2-dibromovinyl)-2,2-dimethyl cyclopropane carboxyate • esbecythrin • FMC-45498 • NRDC-161 • RU-22974 • butox • decis

MF $C_{22}H_{19}Br_2NO_3$

MW 505.22

COLOR AND FORM Crystals; colorless crystals; white or slightly beige powder.

ODOR Odorless.

MELTING POINT 98–101°C.

FIREFIGHTING PROCEDURES Use carbon dioxide, foam, or dry chemical (on fires involving pyrethroids) (pyrethrum) [R17, 2] Firefighting: Self-contained breathing apparatus with a full facepiece operated in pressure-demand or other positive-pressure mode (pyrethrum) [R17, 5]. Extinguish fire using agent suitable for type of surrounding fire. (pyrethrins) [R2].

SOLUBILITY Water soluble: 0.002 mg/L; acetone: 500 g/L; soluble in ethanol, acetone, dioxane; insoluble in water; water solubility ≤ 0.002 mg/L; acetone (500 g/L), ethanol (15 g/L), cyclohexanone (750 g/L), dioxan (900 g/L), xylene (250 g/L), ethyl acetate (soluble); solubility in cyclohexanone 750, dichloromethane 700, benzene 450, dimethyl sulphoxide 450, xylene 250, isopropanol 6 (all in g/L at 20°C).

VAPOR PRESSURE 0.002 mPa (25°C).

OCTANOL/WATER PARTITION COEFFICIENT Log K_{ow} = 5.43.

CORROSIVITY Noncorrosive to metals.

STABILITY AND SHELF LIFE Extremely stable on exposure to air. Under UV irradiation and in sunlight, a cis-trans isomerization, splitting of the ester bond, and loss of bromine occur. More stable in acidic media than in alkaline media [R1].

OTHER CHEMICAL/PHYSICAL PROPERTIES The technical material produced industrially contains greater than or equal to 98% deltamethrin m/m (proportion by mass) and is a colorless crystalline powder. (technical deltamethrin).

PROTECTIVE EQUIPMENT Employees should be provided with and required to use dust- and splashproof safety goggles where pyrethroids may contact the eyes. (pyrethroids) [R17, 3].

PREVENTATIVE MEASURES Skin that becomes contaminated with pyrethrum should be promptly washed or showered with soap or mild detergent and water. (pyrethrum) [R17, 3].

CLEANUP Spillages of pesticides at any stage of their storage or handling should be treated with great care. Liquid formulations may be reduced to solid phase by evaporation. Dry-sweeping of solids is always hazardous: these should be removed by vacuum cleaning, or by dissolving them in water or other solvent in the factory environment. (pesticides) [R3].

DISPOSAL Incineration would be an effective disposal procedure where permitted. If an efficient incinerator is not available, the product should be mixed with large amounts of combustible material, and contact with the smoke should be avoided. (pyrethrin products) [R4].

COMMON USES Insecticide [R18, 453].

HEALTH HAZARDS Poisonous; may be fatal if inhaled, swallowed, or absorbed through skin. Contact may cause burns to skin and eyes. Runoff from fire control or dilution water may give off poisonous gases and cause water pollution. Fire may produce irritating or poisonous gases. (pesticide, liquid, or solid, poisonous, not otherwise specified) [R8, p. G–55].

EMERGENCY ACTION Keep unnecessary people away; isolate hazard area, and deny entry. Stay upwind, out of low areas, and ventilate closed spaces before entering. Positive-pressure self-contained breathing apparatus (SCBA) and chemical protective clothing that is specifically recommended by the shipper or manufacturer may be worn. It may provide little or no thermal protection. Structural firefighters' protective clothing is not effective for these materials. Remove and isolate contaminated clothing at the site. Call CHEM-TREC at 1–800–424–9300 as soon as possible, especially if there is no local hazardous-materials team available (pesticide, liquid, or solid, poisonous, not otherwise specified) [R9, p. G–55].

Fire: Small Fires: Dry chemical, CO_2, water spray, or regular foam. Large Fires: Water spray, fog, or regular foam. Move container from fire area if you can do so without risk. Fight fire from maximum distance. Stay away from ends of tanks. Dike fire-control water for later disposal; do not scatter the material (pesticide, liquid, or solid, poisonous, not otherwise specified) [R9, p. G–55].

Spill or Leak: Do not touch or walk through spilled material; stop leak if you can do so without risk. Fully encapsulating, vapor-protective clothing should be worn for spills and leaks with no fire. Use water spray to reduce vapors. Small Spills: Take up with sand or other noncombustible absorbent material and place into containers for later disposal. Small Dry Spills: With clean shovel place material into clean, dry container, and cover loosely; move containers from spill area. Large Spills: Dike far ahead of liquid spill for later disposal (pesticide, liquid, or solid, poisonous, not otherwise specified) [R9, p. G–55].

First Aid: Move victim to fresh air, and call for emergency medical care; if not breathing, give artificial respiration; if breathing is difficult, give oxygen. In case of contact with material, immediately flush skin or eyes with running water for at least 15 minutes. Speed in removing material from skin is of extreme importance. Remove and isolate contaminated clothing and shoes at the site. Keep victim quiet and maintain normal body temperature. Effects may be delayed; keep victim under observation (pesticide, liquid, or solid, poisonous, not otherwise specified) [R9, p. G–55].

NON-HUMAN TOXICITY LD_{50} rat male oral 128 mg/kg (in vegetable oil) [R1]; LD_{50} dog (male and female), oral, in capsules >300 mg/kg (technical grade) [R10]; LD_{50} dog (male and female), oral, in PEG 200 (n.b., Sax number designation) 2 mg/kg (technical grade) [R10]; LD_{50} rabbit, dermal, in PEG 400 >2,000 mg/kg (technical grade) [R10]; LD_{50} rat (male), oral, in sesame oil 128 mg/kg (technical grade) [R11]; LD_{50} rat (female), oral, in sesame oil 139 mg/kg (technical grade) [R11]; LD_{50} rat (male), oral, in PEG 200 67 mg/kg (technical grade) [R11]; LD_{50} rat (female), oral, in PEG 200 86 mg/kg (technical grade) [R11]; LD_{50} rat (male adult), oral, in peanut oil 52 mg/kg (technical grade) [R11]; LD_{50} rat (female adult), oral, in peanut oil 31 mg/kg (technical grade) [R11]; LD_{50} rat (female weanling), oral, in peanut oil 50 mg/kg (technical grade) [R11]; LD_{50} rat dermal, 700 mg/kg (technical grade) [R11]; LD_{50} rat (male, and female), dermal, in methylcellulose (1%) >2,940 mg/kg (technical grade) [R11]; LD_{50} rat (female adult), dermal, in xylene >800 mg/kg (technical grade)

[R11]; LC$_{50}$ (male, and female), inhalation, dust 600 mg/m^3/6 hr (technical grade) [R11]; LD$_{50}$ mouse (male), oral, in sesame oil 33 mg/kg (technical grade) [R12]; LD$_{50}$ mouse (female), oral, in sesame oil 34 mg/kg (technical grade) [R12]; LD$_{50}$ mouse (male), oral, in PEG 200 21 mg/kg (technical grade) [R12]; LD$_{50}$ mouse (female), oral, in PEG 200 19 mg/kg (technical grade) [R12]; LD$_{50}$ mouse ip 33 mg/kg (technical grade) [R12]; LD$_{50}$ rat female oral 139 mg/kg (in vegetable oil) [R15]; LD$_{50}$ rat oral >5,000 mg/kg (in aqueous solution) [R15].

ECOTOXICITY LC$_{50}$ *Alburnus alburnus* (bleak) (static condition) 0.69 µg/L/96 hr (technical product) [R16]; LC$_{50}$ *Alburnus alburnus* (bleak) (static condition) 0.69 µg/L/96 hr (technical product) [R13]; LC$_{50}$ *Brachydanio rerio* (zebra fish) (flow system, static condition) 2.0 µg/L/96 hr (technical product) [R13]; *Cyrpinus carpio* (common carp) (flow system, static condition) 1.84 µg/L/96 hr, 0.86 µg/L/96 hr (technical product) [R13]; *Lctalurus nebulosus* (brown bullhead) (flow system, static conditions) µg/L/96 hr (technical product) [R13]; LC$_{50}$ *Lctalurus punctatus* (channel catfish) (flow system, static condition) 0.63 µg/L/96 hr (technical product) [R13]; LC$_{50}$ *Lepomis gibbosus* (pumpkinseed sunfish) (flow system, static condition) 0.58 µg/L/96 hr [R14]; LC$_{50}$ *Lepmois machrochirus* (bluegill sunfish) (flow system, static condition) 1.2 µg/L/96 hr [R14]; LC$_{50}$ *Rhodeus sericeus amarus* (static condition) 1.12 µg/L/96 hr [R14]; LC$_{50}$ *Salmo gairdneri* (rainbow trout) (flow system, static conditions) 0.39 µg/L/96 hr [R14]; LC$_{50}$ *Salmo salar* 1.97 µg/L/96 hr [R14]; LC$_{50}$ *Sarotherodon mossambicuse* (flow system, static conditions) 3.5 µg/L/96 hr [R14]; LD$_{50}$ mallard duck, oral >5,000 mg/kg [R15]; LC$_{50}$ quail dietary >10,000 mg/kg diet/8 day [R15].

IARC SUMMARY Evaluation: No data were available from studies in humans. There is inadequate evidence for the carcinogenicity of deltamethrin in experimental animals. Overall evaluation: Deltamethrin is not classifiable as to its carcinogenicity to humans (Group 3) [R5].

BIODEGRADATION In soil, undergoes mi-

crobial degradation within 1–2 weeks [R1].

BIOCONCENTRATION Based upon a measured log K_{ow} of 5.2 (1), the BCF for deltamethrin can be estimated to be log 3.7 from a recommended regression-derived equation (2,SRC); this BCF value suggests that bioconcentration may be an important fate process in aquatic organisms that cannot metabolize deltamethrin (SRC). In a static system for a 24-hr exposure period, deltamethrin BCFs of about 200 to 1,300 were measured for *Daphnia magna* (3); observed BCFs decreased with increases in dissolved organic carbon (3). BCF values of 39–303 were measured in larvae of the midge *Chronomus tentans* in sand, silt, or clay sediment water systems (1). In a pond study using radio-labelled C-14 deltamethrin, fathead minnows (*Pimephales promelas*) accumulated levels of extractable radioactivity 248–907 times higher than levels in the water at 24 hr after exposure although the nature of the radioactive compounds was not provided (4) [R7].

FIFRA REQUIREMENTS A tolerance is established for residues of the insecticide deltamethrin and its major metabolite, trans-deltamethrin, in or on the raw agricultural commodity tomatoes [R8].

REFERENCES

1. Hartley, D., and H. Kidd (eds.). *The Agrochemicals Handbook*. 2nd ed. Lechworth, Herts, England: The Royal Society of Chemistry, 1987, p. A122/Aug 87.

2. Bureau of Explosives; *Emergency Handling of Haz Matl in Surface Trans* p. 434 (1981).

3. International Labour Office. *Encyclopedia of Occupational Health and Safety.* Vols. I and II. Geneva, Switzerland: International Labour Office, 1983. 1619.

4. Sittig, M. *Handbook of Toxic and Hazardous Chemicals and Carcinogens,* 1985. 2nd ed. Park Ridge, NJ: Noyes Data Corporation, 1985. 762.

5. IARC. *Monographs on the Evaluation of the Carcinogenic Risk of Chemicals to Man.* Geneva: World Health Organiza-

tion, International Agency for Research on Cancer, 1972–present (multivolume work), p. 53 264 (1991).

6. WHO. *Environmental Health Criteria 97: Deltamethrin* p. 49 (1990).

7. (1) Muir, D.C.G., et al., *Environ Toxicol Chem* 4: 51–61 (1985). (2) Lyman, W. J., et al., *Handbook of Chemical Property Estimation Methods*. Washington, DC: Amer Chemical Soc, pp. 5–4, 5–10 (1990). (3) Day, K.E. *Environ Toxicol Chem* 10: 91–101 (1991) (4) Muir, D. C. G., et al., *J Agric Food Chem* 33: 603–9 (1985).

8. 40 CFR 180.435 (7/1/91).

9. U.S. Department of Transportation. *Emergency Response Guidebook 1990*. DOT P 5800.5. Washington, DC: U.S. Government Printing Office, 1990.

10. WHO. *Environmental Health Criteria 97: Deltamethrin* p. 63 (1990).

11. WHO. *Environmental Health Criteria 97: Deltamethrin* p. 61 (1990).

12. WHO. *Environmental Health Criteria 97: Deltamethrin* p. 62 (1990).

13. WHO. *Environmental Health Criteria 97: Deltamethrin* p. 49 (1990).

14. WHO. *Environmental Health Criteria 97: Deltamethrin* p. 50 (1990).

15. Hartley, D., and H. Kidd (eds.). *The Agrochemicals Handbook*. 2nd ed. Lechworth, Herts, England: The Royal Society of Chemistry, 1987, p. A122/Aug 87.

16. WHO. *Environmental Health Criteria 97: Deltamethrin* p. 86 (1990).

17. Mackison, F. W., R. S. Stricoff, and L. J. Partridge, Jr. (eds.). *NIOSH/OSHA—Occupational Health Guidelines for Chemical Hazards*. DHHS (NIOSH) Publication No. 81–123 (3 vols). Washington, DC: U.S. Government Printing Office, Jan. 1981.

18. Budavari, S. (ed.). *The Merck Index—Encyclopedia of Chemicals, Drugs and Biologicals*. Rahway, NJ: Merck and Co., Inc., 1989.

■ DIACETYL

CAS # 431–03–8

SYNONYMS 2,3-butadione • 2,3-butanedione • 2,3-diketobutane • biacetyl • butadione • butane-2,3-dione • dimethyl diketone • dimethylglyoxal • glyoxal, dimethyl

MF $C_4H_6O_2$

MW 86.09

COLOR AND FORM Yellowish-green, mobile liquid; yellow liquid.

ODOR Quinone odor; vapors have a chlorine-like odor; rancid-butter odor; strong odor.

ODOR THRESHOLD 8.6 ppb [R12].

TASTE Threshold taste detection in milk (gas chromatically pure): 1.40×10^{-2} ppm and 2.90×10^{-2} ppm; in skim milk: 1.00×10^{-1} ppm; threshold taste detection in water (gas chromatically pure): 5.40×10^{-3} ppm; in water: 1.00×10^{-8}% (volume/volume or wg/wg); with butter taste at 1 ppm.

DENSITY 0.990 at 15°C.

BOILING POINT 88°C at 760 mm Hg.

MELTING POINT −2.4°C.

FIREFIGHTING PROCEDURES Firefighting methods: alcohol foam, carbon dioxide, dry chemical [R10]. Do not extinguish fire unless flow can be stopped. Use water in flooding quantities as fog. Solid streams of water may be ineffective. Cool all affected containers with flooding quantities of water. Apply water from as far a distance as possible. Use "alcohol" foam, carbon dioxide, or dry chemical [R11].

SOLUBILITY Soluble in about 4 parts water; miscible with alcohol, ether; very soluble in acetone; soluble in carbitols; water solubility = 200 g/L at 15°C; readily soluble in all important organic solvents.

VAPOR PRESSURE 58.2 mm Hg at 25.1°C.

OCTANOL/WATER PARTITION COEFFICIENT Log P = −1.34.

OTHER CHEMICAL/PHYSICAL PROPERTIES

Enthalpy of formation (liquid) = −87.44 kcal/mol; enthalpy of sublimation (298 K) = 9.25 kcal/mol; Henry's Law constant = 0.0014 (dimensionless) at 298 K.

PREVENTATIVE MEASURES If material not on fire and not involved in fire: Keep sparks, flames, and other sources of ignition away. Keep material out of water sources and sewers. Build dikes to contain flow as necessary. Attempt to stop leak if without personnel hazard. Use water spray to disperse vapors and dilute standing pools of liquid [R11].

DISPOSAL SRP: At the time of review, criteria for land treatment or burial (sanitary landfill) disposal practices are subject to significant revision. Prior to implementing land disposal of waste residue (including waste sludge), consult with environmental regulatory agencies for guidance on acceptable disposal practices.

COMMON USES Carrier of aroma of butter, vinegar, coffee, and other foods [R1]. Synthetic flavoring substance and adjuvant [R5]; flavoring agent for oleomargarine [R2]; flavoring agent for candy, baked goods, and chewing gum [R2]; chemical intermediate for 2,3-butanedione 2-oxime [R2]. Diacetyl has been widely used as a chemical modifier of proteins, combining with arginine residues [R6]. Used as an electron-stabilizing compound; as a modifier of bacillus megaterium spore sensitivity to x-rays [R7]. 0.075% inhibited *E. coli* (except strain 0–128), *S. salivarius*, and *S. aureus*. 0.05% inhibited mycobacterium and 0.15% l arabinosus [R8]. Mainly important as a flavor component (buttery taste), it is used in low concentration in ice cream, baked goods, and margarine [R3]. Cyclocondensation with amines has been used to form triazine and pteridine ring systems; also used as a precursor to alpha-diones [R4].

OCCUPATIONAL RECOMMENDATIONS None listed.

BIODEGRADATION Only one reference could be located concerning the biodegradation of diacetyl. In this reference, the bacterium that was isolated from Amsterdam harbor water, using 2-butanol, and tentatively identified as a *Pseudomonas*, was able to degrade diacetyl. In fact diacetyl was identified as an intermediate in the microbial oxidation of 2-butanol (1). Because 2-butanol is biodegradable using river water or sewage inoculums with extensive mineralization (2–5), one would also predict diacetyl to be biodegradable if the mechanism is both correct and applicable to resident microbial populations (SRC) [R13].

BIOCONCENTRATION Because diacetyl is highly soluble in water (25% wt/wt) (1), it is not likely to bioconcentrate in fish (SRC) [R14].

DOT *Fire or Explosion:* Flammable/combustible material; may be ignited by heat, sparks, or flames. Vapors may travel to a source of ignition, and flash back. Container may explode in heat of fire. Vapor explosion hazard indoors, outdoors, or in sewers. Runoff to sewer may create fire or explosion hazard [R9].

Health Hazards: May be poisonous if inhaled or absorbed through skin. Vapors may cause dizziness or suffocation. Contact may irritate or burn skin and eyes. Fire may produce irritating or poisonous gases. Runoff from fire control or dilution water may give off poisonous gases and cause water pollution [R9].

Emergency Action: Keep unnecessary people away; isolate hazard area and deny entry. Stay upwind; keep out of low areas. Positive-pressure self-contained breathing apparatus (SCBA) and structural firefighters' protective clothing will provide limited protection. Isolate for 1/2 mile in all directions if tank, rail car, or tank truck is involved in fire. Call emergency response telephone number on shipping paper first. If shipping paper not available or no answer, call CHEMTREC at 1–800–424–9300. If water pollution occurs, notify the appropriate authorities [R9].

Fire: Small Fires: Dry chemical, CO_2, water spray, or alcohol-resistant foam. Large Fires: Water spray, fog, or alcohol-resistant foam. Move container from fire area if you can do so without risk. Apply cooling water to sides of containers that are exposed to flames until well after fire is out. Stay away from ends of tanks. For massive fire in cargo area, use unmanned hose holder, or monitor nozzles; if this is impossible, withdraw from area, and let fire burn. Withdraw immediately in case of rising sound from venting safety device or any discoloration of tank due to fire [R9].

Spill or Leak: Shut off ignition sources; no flares, smoking, or flames in hazard area. Stop leak if you can do so without risk. Water spray may reduce vapor; but it may not prevent ignition in closed spaces. Small Spills: Take up with sand or other noncombustible absorbent material and place into containers for later disposal. Large Spills: Dike far ahead of liquid spill for later disposal [R9].

First Aid: Move victim to fresh air and call emergency medical care; if not breathing, give artificial respiration; if breathing is difficult, give oxygen. In case of contact with material, immediately flush eyes with running water for at least 15 minutes. Wash skin with soap and water. Remove and isolate contaminated clothing and shoes at the site [R9].

FIRE POTENTIAL Fire hazard: dangerous when exposed to heat or flame [R10].

FDA Substance added directly to human food affirmed as generally recognized as safe (GRAS) [R15].

REFERENCES

1. Budavari, S. (ed.). *The Merck Index—Encyclopedia of Chemicals, Drugs, and Biologicals.* Rahway, NJ: Merck, and Co., Inc., 1989. 468.

2. SRI.

3. Gerhartz, W. (exec ed.). *Ullmann's Encyclopedia of Industrial Chemistry.* 5th ed. Vol A1: Deerfield Beach, FL: VCH Publishers, 1985 to present., p. VA$_{15}$ 90.

4. Aldrich. *Catalog/Handbook of Fine Chemicals 1994-95.* Aldrich Chemical Co, Milwaukee, WI, pp. 248 (1994).

5. Furia, T. E. (ed.). *CRC Handbook of Food Additives.* 2nd ed. Cleveland: The Chemical Rubber Co., 1972. 831.

6. Opdyke, D. L. *J. Food Cosmet Toxicol* 17: 765 (1979).

7. Tallentire, A. H., et al. *2,3-Butanedione, an Electron-stabilizing Compd, as a Modifier of Sensitivity of Bacillus Megaterium Spores to X-rays; Intermediate J Radiat Biol* 14: 394 (1968).

8. Gupta, K. G., et al. *Antibacterial Activity of Diacetyl and Its Influence on Keeping Quality of Milk, Zentralbl bakteriol Parasitenkd Infektionskr Hyg Erste Abt Orig Reihe B Hyg Praev Med* 158 (2) 202 (1973).

9. U.S. Department of Transportation. *Emergency Response Guidebook 1993.* DOT P 5800.6. Washington, DC: U.S. Government Printing Office, 1993, p. G-26.

10. Sax, N. I. *Dangerous Properties of Industrial Materials.* 6th ed. New York, NY: Van Nostrand Reinhold, 1984. 548.

11. Association of American Railroads. *Emergency Handling of Hazardous Materials in Surface Transportation.* Washington, DC: Association of American Railroads, Bureau of Explosives, 1992. 317.

12. Pietrzak, E., N. Barylko-Pikielna. *Acta Aliment Pol* 2: 207 (1976) as cited in Opdyke, D. L. *J. Food Cosmet Toxicol* 17: 765 (1979).

13. (1) Lijmbach, G. W. M., E. Brinkhuis. *Antonie von Leeuwenhoek* 39: 415-23 (1973). (2) Bridie, A. L., et al., *Water Res* 13: 627-30 (1979). (3) Dore, M., et al., *Trib Cebedeau* 28: 3-11 (1975) (4) Pitter, P., et al., *Water Res* 10: 231-5 (1976). (5) Hammerton, C. *J Appl Chemical* 5: 517-24 (1955).

14. *Merck Index.* 10th ed. p. 429 (1983).

15. 21 CFR 184.1278 (4/1/93).

■ DIACETYL PEROXIDE

CAS # 110-22-5

DOT # 2084

SYNONYMS acetyl peroxide • peroxide, diacetyl

MF $C_4H_6O_4$

MW 118.10

COLOR AND FORM Colorless crystals; needles from ether, leaves.

BOILING POINT 63°C at 21 mm Hg.

MELTING POINT 30°C.

EXPLOSIVE LIMITS AND POTENTIAL Pure (100%) material is severe explosion hazard; should not be stored after prepn, nor heated above 30°C [R1].

FIREFIGHTING PROCEDURES Fight with water from explosion-resistant location. In advanced or massive fires area should be evacuated. If fire occurs in vicinity of this material water should be used to keep containers cool. Cleanup and salvage operations should not be attempted until all cooled completely (25% solution) [R7, p. 49–111]. Firefighting method: water, dry chem, carbon dioxide (25% solution). [R3].

SOLUBILITY Slightly soluble in water; soluble in alcohol and hot ether; very soluble in carbon tetrachloride.

VAPOR PRESSURE 1.18 mm Hg at 20°C.

STABILITY AND SHELF LIFE Highly unstable (org peroxides) [R5].

OTHER CHEMICAL/PHYSICAL PROPERTIES Decomposition in sodium hydroxide; strong, pungent odor; shock-sensitive crystals may form below 17°F; vapor density: 4.07 (25% solution); liquid surface tension: (est) 30 dynes/cm = 0.030 N/m at 20°C (25% solution); liq-water interfacial tension: (est) 30 dynes/cm = 0.003 N/m at 20°C (25% solution); heat of decomposition: (est) −50 Btu/lb = −28 cal/g = −1.2×10⁵ J/kg (25% solution); density: 1.18 (25% solution); colorless liquid (25% solution); (est) −15,700 Btu/lb = −28 cal/g = −366×10⁵ J/kg (25% solution).

PROTECTIVE EQUIPMENT A face shield and rubber gloves should be worn when handling the substance, and a safety shield or hood door should be in front of apparatus containing it. [R4].

CLEANUP In event of spillage, spilled material should be absorbed with noncombustible absorbent, such as vermiculite. Sweep up, and place in plastic container for immediate disposal. Do not use spark-generating metals or cellulosic materials (paper, wood, etc.) for sweeping up or handling spilled material (25% solution) [R7, p. 49–111]. In event of spillage material should be absorbed with vermiculite. Dispose of absorbed peroxide by placing it on the ground in remote outdoor area and igniting with a long torch (25% solution). [R7, p. 49–111].

COMMON USES Initiator and catalyst for resins [R1]; catalyst employed to promote polymerization in manufacture of certain plastics [R8, 357].

OCCUPATIONAL RECOMMENDATIONS None listed.

DOT *Fire or Explosion:* May be ignited by heat, sparks, or flames. May burn rapidly with flare-burning effect. Container may explode in heat of fire. May explode from friction, heat, or contamination. Runoff to sewer may create fire or explosion hazard [R2].

Health Hazards: Contact may cause burns to skin and eyes. Fire may produce irritating or poisonous gases. Runoff from fire control or dilution water may cause pollution [R2].

Emergency Action: Keep unnecessary people away; isolate hazard area and deny entry. Stay upwind; keep out of low areas. Self-contained breathing apparatus (SCBA) and structural firefighter's protective clothing will provide limited protection. Call CHEMTREC at 1–800–424–9300 for emergency assistance. If water pollution occurs, notify the appropriate authorities [R2].

Fire: Small Fires: Dry chemical, CO₂, halon, water spray, or standard foam. Large Fires: Flood fire area with water. Do not move cargo or vehicle if cargo has been exposed to heat. If fire can be controlled, cool container with water from unmanned hose holder, or monitor nozzles until well after fire is out. If this is impossible, withdraw from area, and let fire burn [R2].

Spill or Leak: Shut off ignition sources; no flares, smoking, or flames in hazard area. Do not touch spilled material; stop leak if you can do so without risk. Small Spills: Take up with inert, damp, noncombustible material; move containers from spill area. Large Spills: Wet down with water and dike for later disposal [R2].

First Aid: Move victim to fresh air; call emergency medical care. In case of contact with material, immediately flush eyes with running water for at least 15 minutes. Wash skin with soap and water. Remove and isolate contaminated clothing and shoes at the site. Keep victim quiet and maintain normal body temperature [R2].

FIRE POTENTIAL Dangerous, by spontaneous chemical reaction. A powerful oxidizing agent; can cause ignition of organic materials on contact [R6, 357]. An oxidizing material. Sensitive to heat, and should not be subjected to temperature above 90°F; self-accelerating decomposition temp, 120°F (25% solution). [R7, p. 49–110].

GENERAL RESPONSE Stop discharge if possible. Keep people away. Call fire department. Avoid contact with liquid. Isolate and remove discharged material. Notify local health and pollution control agencies.

FIRE RESPONSE Combustible. May explode on contact with combustibles. Containers may explode in fire. Combat fires from safe distance or protected location. Flood discharge area with water. Cool exposed containers with water.

EXPOSURE RESPONSE Call for medical aid. Vapor irritating to eyes, nose, and throat. Move victim to fresh air. Liquid irritating to skin and eyes. Harmful if swallowed. Remove contaminated clothing and shoes. Flush affected areas with plenty of water. If swallowed, and victim is conscious, have victim drink water or milk, and have victim induce vomiting. If swallowed and victim is unconscious or having convulsions, do nothing except keep victim warm.

RESPONSE TO DISCHARGE Issue warning—oxidizing material, water contaminant; should be removed; chemical and physical treatment.

WATER RESPONSE Effect of low concentrations on aquatic life is unknown. May be dangerous if it enters water intakes. Notify local health and wildlife officials. Notify operators of nearby water intakes.

EXPOSURE SYMPTOMS Contact with liquid causes irritation of eyes and skin. If ingested, irritates mouth and stomach.

EXPOSURE TREATMENT Eyes: wash with plenty of water and get medical attention. Skin: wash with plenty of soap and water. Ingestion: induce vomiting and call a physician.

REFERENCES

1. Hawley, G. G. *The Condensed Chemical Dictionary.* 9th ed. New York: Van Nostrand Reinhold Co., 1977. 10.

2. Department of Transportation. *Emergency Response Guidebook 1987.* DOT P 5800.4. Washington, DC: U.S. Government Printing Office, 1987, p. G–49.

3. U.S. Coast Guard, Department of Transportation. *CHRIS—Hazardous Chemical Data.* Manual Two. Washington, DC: U.S. Government Printing Office, Oct., 1978.

4. National Research Council. *Prudent Practices for Handling Hazardous Chemicals in Laboratories.* Washington, DC: National Academy Press, 1981. 106.

5. International Labour Office. *Encyclopedia of Occupational Health and Safety.* Volumes I and II. New York: McGraw-Hill, Book Co., 1971. 1013.

6. Sax, N.I. *Dangerous Properties of Industrial Materials.* 4th ed. New York: Van Nostrand Reinhold, 1975.

7. National Fire Protection Association. *Fire Protection Guide on Hazardous Materials.* 7th ed. Boston, Mass.: National Fire Protection Association, 1978.

8. Grant, W. M. *Toxicology of the Eye.* 2nd ed. Springfield, Illinois: Charles C. Thomas, 1974.

■ 4,4'-DIAMINODIPHENYL-METHANE

H_2N—⬡—CH_2—⬡—NH_2

CAS # 101–77–9

SYNONYMS 4,4'-Diaminodiphenylmethan (German) • 4,4'-diamino ditan • 4,4'-diphenylmethanediamine • 4,4'-methylenebis (aniline) • 4,4'-methylenebisaniline • 4,4'-methylenedianiline • 4-(4-aminobenzyl) aniline • aniline, 4,4'-methylenedi • benzenamine, 4,4'-methylenebis • bis (4-aminophenyl) methane • bis (p-aminophenyl) methane • bis-p-Aminofenylmethan (Czech) • dadpm • dapm • di-(4-aminophenyl) methane • diaminodiphenylmethane • dianilinemethane • dianilinomethane • epicure-ddm • epikure-ddm • ht-972 • mda • methylenebis (aniline) • methylenedianiline • nci-c54604 • p, p'-Diaminodifenylmethan (Czech) • p,p'-diaminodiphenylmethane • p,p'-methylenedianiline • tonox

MF $C_{13}H_{14}N_2$

MW 198.29

pH Weak base.

COLOR AND FORM Crystals from water or benzene; tan flakes or lumps; plates or needles in water; plates in benzene; colorless to pale-yellow flakes.

ODOR Faint amine-like odor.

DENSITY 1.056 at 100 C/4 C.

BOILING POINT 398–399°C at 768 mm Hg.

MELTING POINT 91.5–92°C.

SOLUBILITY Very soluble in alcohol, benzene, and ether; solubility in acetone: 273 g/100 g; solubility in water: 0.1 g/100 g at 25°C.

VAPOR PRESSURE 1 mm Hg at 197°C.

OCTANOL/WATER PARTITION COEFFICIENT Log K_{ow} = 1.59 (measured).

STABILITY AND SHELF LIFE Oxidizes in air; pale-yellow crystals turn dark color when exposed to air [R1].

OTHER CHEMICAL/PHYSICAL PROPERTIES Has general characteristics of primary aromatic amines; Henry's Law Constant = 5.6×10^{-11} atm-m³/mole at 25°C (est).

PROTECTIVE EQUIPMENT Dispensers of liquid detergent should be available. Safety pipettes should be used for all pipetting. In animal laboratory, personnel should wear protective suits (preferably disposable, one-piece, and close-fitting at ankles and wrists), gloves, hair covering, and overshoes. In chemical laboratory, gloves and gowns should always be worn; however, gloves should not be assumed to provide full protection. Carefully fitted masks or respirators may be necessary when working with particulates or gases, and disposable plastic aprons might provide addnl protection. Gowns should be of distinctive color, as a reminder that they are not to be worn outside the laboratory. (chemical carcinogens) [R13, 8].

PREVENTATIVE MEASURES SRP: The scientific literature supports the wearing of contact lenses in industrial environments, as part of a program to protect the eye against chemical compounds and minerals causing eye irritation. However, there may be individual substances whose irritating or corrosive properties are such that the wearing of contact lenses would be harmful to the eye. In those specific cases contact lenses should not be worn.

CLEANUP Facility and process are discussed for removal of methylenedianiline [R4]. A high-efficiency particulate arrestor (HEPA) or charcoal filters can be used to minimize amount of carcinogen in exhausted-air-ventilated safety cabinets, lab hoods, glove boxes, or animal rooms. Filter housing that is designed so that used filters can be transferred into plastic bag without contaminating maintenance staff is avail commercially. Filters should be placed in plastic bags immediately after removal. The plastic bag should be sealed immediately. The sealed bag should be labelled properly. Waste liquids should be placed or collected in proper containers for disposal. The lid should be secured and the bottles properly labelled. Once filled, bottles should be placed in plastic bag, so that outer surface is not contaminated. The plastic bag should also be sealed and labelled. Broken glassware should be decontaminated by solvent extraction, by chemical destruction, or in specially designed incinerators. (chemical carcinogens) [R13, 15].

DISPOSAL PRECAUTIONS There is no universal method of disposal that has been proved satisfactory for all carcinogenic compounds and specific methods of chemical destruction published have not been tested on all kinds of carcinogen-containing waste. Summary of avail methods and recommendations given must be treated as guide only. (chemical carcinogens) [R13, 14].

COMMON USES Determination of tungsten and sulfates; as corrosion inhibitor; preparation of azo dyes [R12, 470]; as curing agent for polyurethane elastomers and polyurethane epoxy resins [R2]; chemical intermediate for polymethylene polyphenylisocyanate (captive), 4,4'-methylenediphenylisocyanate (captive); monomer for polyamide and polyimide resins and fibers; chemical intermediate for the dye pararosaniline; lab analytical reagent [R3].

OCCUPATIONAL RECOMMENDATIONS None listed.

NIOSH Recommends that 4,4'-methylenedianiline be regulated as a potential human carcinogen. Reduce exposure to lowest feasible concentration [R9].

THRESHOLD LIMIT 8-hr time-weighted average (TWA) 0.1 ppm (0.81 mg/m^3), skin (1986) [R14, 26].

NON-HUMAN TOXICITY LD$_{50}$ dog oral 300 ppm [R6]; LD$_{50}$ rat oral 120–250 mg/kg [R7]; LD$_{50}$ rat oral 597–830 mg/kg [R6].

IARC SUMMARY No data are available for humans. Sufficient evidence of carcinogenicity in animals. Overall evaluation: Group 2B: The agent is possibly carcinogenic to humans [R5].

BIOCONCENTRATION Based upon a water

solubility of 1,000 ppm at 25°C (1) and a measured log K$_{ow}$ of 1.59 (2), the bioconcentration factor (BCF) for MDA can be estimated to be 12.6 and 9.5, respectively, from recommended regression-derived equations (3,SRC). These BCF values are not indicative of significant bioaccumulation (SRC) [R8].

ATMOSPHERIC STANDARDS This action promulgates standards of performance for equipment leaks of volatile organic compounds (VOC) in the synthetic organic chemical manufacturing industry (SOCMI). The intended effect of these standards is to require all newly constructed, modified, and reconstructed SOCMI process units to use the best demonstrated system of continuous emission reduction for equipment leaks of VOC, considering costs, non-air-quality health, and environmental impact and energy requirements. Methylenedianiline is produced, as an intermediate or a final product, by process units covered under this subpart. [R10].

TSCA REQUIREMENTS Pursuant to section 8 (d) of TSCA, EPA promulgated a model health and safety data reporting rule. The section 8 (d) model rule requires manufacturers, importers, and processors of listed chemical substances and mixtures to submit to EPA copies and lists of unpublished health and safety studies. Benzenamine, 4,4'-methylenebis is included on this list [R11].

REFERENCES

1. IARC. *Monographs on the Evaluation of the Carcinogenic Risk of Chemicals to Man*. Geneva: World Health Organization, International Agency for Research on Cancer, 1972–present (multivolume work), p. V4 79 (1974).

2. IARC. *Monographs on the Evaluation of the Carcinogenic Risk of Chemicals to Man*. Geneva: World Health Organization, International Agency for Research on Cancer, 1972–present (multivolume work), p. V4 81 (1974).

3. SRI.

4. Young, D. A., B. G. Parker. *Removal of Methylenedianiline from Chemical Plant*

Wastewater; Report; Iss BDX–613–1981; Conf–780455–1, 1978, 29 PP.

5. IARC. *Monographs on the Evaluation of the Carcinogenic Risk of Chemicals to Man*. Geneva: World Health Organization, International Agency for Research on Cancer, 1972–present (multivolume work), p. S7 66 (1987).

6. American Conference of Governmental Industrial Hygienists. *Documentation of the Threshold Limit Values and Biological Exposure Indices*. 5th ed. Cincinnati, OH: American Conference of Governmental Industrial Hygienists, 1986, p. 393 (1986).

7. Verschueren, K. *Handbook of Environmental Data of Organic Chemicals*. 2nd ed. New York, NY: Van Nostrand Reinhold Co., 1983. 454.

8. (1) Moore, W. M., p. 338–48 in *Kirk–Othmer Encyclopedia of Chemical Technology* 3rd ed. New York: John Wiley and Sons (1978). (2) Hansch, C., and A. J. Leo. *Medchem Project* Issue No. 26, Clarmont CA: Pomona College (1985). (3) Lyman, W. J., et al., *Handbook of Chemical Property Estimation Methods*. New York: McGraw-Hill, p. 5–4, 5–10 (1982).

9. NIOSH/CDC. *NIOSH Recommendations for Occupational Safety and Health Standards*. 1988, Aug. 1988 (Suppl. to *Morbidity and Mortality Wkly*. Vol. 37 No. 5–7, Aug. 26, 1988). Atlanta, GA: National Institute for Occupational Safety, and Health, CDC, 1988. 19.

10. 40 CFR 60.489 (7/1/91).

11. 40 CFR 716.120 (7/1/91).

12. Budavari, S. (ed.). *The Merck Index— Encyclopedia of Chemicals, Drugs and Biologicals*. Rahway, NJ: Merck and Co., Inc., 1989.

13. Montesano, R., H. Bartsch, E.Boyland, G. Della Porta, L. Fishbein, R.A. Griesemer, A.B. Swan, L. Tomatis, and W. Davis (eds.). *Handling Chemical Carcinogens in the Laboratory: Problems of Safety*. IARC Scientific Publications No. 33. Lyon, France: International Agency for Research on Cancer, 1979.

14. American Conference of Governmental Industrial Hygienists. *Threshold Limit Values for Chemical Substances and Physical Agents and Biological Exposure Indices for 1994–1995*. Cincinnati, OH: ACGIH, 1994.

■ DIAZOMETHANE

CAS # 334–88–3

SYNONYMS azimethylene • acomethylene • diazirine • diazonium, methylide • methane, diazo

MF CH_2N_2

MW 42.05

COLOR AND FORM Yellow gas.

ODOR Musty.

DENSITY 1.45.

BOILING POINT –23°C.

MELTING POINT –145°C.

EXPLOSIVE LIMITS AND POTENTIAL Undiluted liquid and concentrated solution may explode violently, esp if impurities are present. Gaseous diazomethane may explode on heating to 100°C or on rough glass surfaces. Ground-glass apparatus and glass stirrers with glass sleeve bearings where grinding may occur should not be used. Alkali metals also produce explosions [R1]. May explode violently on heating or in contact with rough surfaces. In ether or benzene, in which diazomethane is soluble, the danger of explosion is much decreased [R7]. Diazomethane will undergo violent thermal decomposition above 200°C; vapor may explode violently. Explosions at low temperatures can occur if traces of organic matter are present [R8]. Calcium sulfate is unsuitable desiccant for drying tubes in diazomethane systems. Contact of diazomethane vapor and sulfate causes exotherm, which may lead to detonation [R9]. Severe, when shocked, exposed to heat or by chemical reaction [R10].

FIREFIGHTING PROCEDURES If material is on fire or involved in fire, use dry chemical, dry sand, or carbon dioxide. Do not use water on material itself. If large quantities of combustibles are involved, use water in flooding quantities as spray and fog. Use water spray to knock down vapors. Cool all affected containers with flooding quantities of water. Apply water from as far a distance as possible [R2].

SOLUBILITY Soluble in ether, dioxane; freely soluble in benzene; slightly soluble in ethyl alcohol, ethyl ether.

OTHER CHEMICAL/PHYSICAL PROPERTIES Solution of diazomethane and ether or dioxane decomposes only slowly at low temp; decomposes more rapid if alcohol or water present. Copper powder causes active decomposition with evolution of nitrogen and formation of insoluble white flakes of polymethylene, $(CH_2)_n$. Undiluted liquid and concentration solution may explode violently, particularly if impurities are present. Gaseous diazomethane may explode on heating to 100°C or on rough glass surfaces. Ground-glass apparatus and glass stirrers with glass sleeve bearings where grinding may occur should not be used. Alkali metals also produce explosions. May explode violently on heating or in contact with rough surfaces. In ether or benzene, in which diazomethane is sol, the danger of explosion is much decreased.

PROTECTIVE EQUIPMENT Eye protection, i.e., chemical safety goggles, should be worn as well as approved respirator [R3]. Wear appropriate equipment to prevent contamination or freezing of skin. Wear eye protection to prevent any possibility of eye contact. The following equipment should be available: eyewash. [R11, 83].

PREVENTATIVE MEASURES Contact lenses should not be worn when working with this chemical [R11, 83]. SRP: The scientific literature supports the wearing of contact lenses in industrial environments, as part of a program to protect the eye against chemical compounds and minerals causing eye irritation. However, there may be individual substances whose irritating or corrosive properties are such that the wearing of contact lenses would be harmful to the eye. In those specific cases contact lenses should not be worn.

Explosive (use safety screen), insidious poison (well-ventilated hood is absolutely necessary), avoid vapor [R1]. If material is not on fire and not involved in a fire, keep sparks, flames, and other sources of ignition away. Keep material out of water sources and sewers. Build dikes to contain flow as necessary. Use water spray to knock down vapors. Do not use water on material itself. Neutralize spilled material with crushed limestone, soda ash, or lime [R2]. Avoid breathing vapors. Keep upwind. Avoid bodily contact with the material. Do not handle broken packages unless wearing appropriate personal protective equipment. Wash away any material that may have contacted the body with copious amounts of water or soap and water. If contact with the material anticipated, wear appropriate chemical protective clothing [R2]. Workers should wash promptly when skin becomes contaminated. Remove clothing immediately if it becomes wet (to avoid flammability hazard) [R11, 83].

CLEANUP 1. Remove all ignition sources. 2. Ventilate area of leak. 3. Stop flow of gas. If source is cylinder and leak cannot be stopped in place, remove leaking cylinder to safe place in open air, and repair leak, or allow the cylinder to empty. 4. If in liquid form, allow to vaporize [R4].

DISPOSAL Diazomethane may be disposed of by burning at safe location or in suitable combustion chamber equipped with an appropriate effluent gas cleaning device [R4].

COMMON USES Powerful methylating agent for acidic compounds such as carboxylic acids, phenols, enols [R1].

OCCUPATIONAL RECOMMENDATIONS None listed.

OSHA 8-hr time-weighted average: 0.2 ppm (0.4 mg/m³) [R6].

NIOSH 10-hr time-weighted average: 0.2 ppm (0.4 mg/m³). [R11, 82].

THRESHOLD LIMIT 8-hr time-weighted average (TWA): 0.2 ppm (0.34 mg/m³) (1977) [R12, 18]. Excursion Limit Recommendation: Excursions in worker exposure levels may exceed three times the TLV–TWA for no more than a total of 30 min during a work day, and under no circumstances should they exceed five times the TLV–TWA [R12, 5].

IARC SUMMARY No data are available for humans. Limited evidence of carcinogenicity in animals. Overall evaluation: Group 3: The agent is not classifiable as to its carcinogenicity to humans [R5].

REFERENCES

1. Budavari, S. (ed.). *The Merck Index—Encyclopedia of Chemicals, Drugs, and Biologicals*. Rahway, NJ: Merck, and Co., Inc., 1989. 473.

2. Association of American Railroads. *Emergency Handling of Hazardous Materials in Surface Transportation*. Washington, DC: Assoc. of American Railroads, Hazardous Materials Systems (BOE), 1987. 253.

3. International Labour Office. *Encyclopedia of Occupational Health and Safety*. Volumes I and II. New York: McGraw-Hill Book Co., 1971. 384.

4. Mackison, F. W., R. S. Stricoff, and L. J. Partridge, Jr. (eds.). *NIOSH/OSHA—Occupational Health Guidelines for Chemical Hazards*. DHHS (NIOSH) Publication No. 81–123 (3 vols). Washington, DC: U.S. Government Printing Office, Jan. 1981.

5. IARC. *Monographs on the Evaluation of the Carcinogenic Risk of Chemicals to Man*. Geneva: World Health Organization, International Agency for Research on Cancer, 1972–present (multivolume work), p. S7 61 (1987).

6. 29 CFR 1910.1000 (7/1/90).

7. American Conference of Governmental Industrial Hygienists, Inc. *Documentation of the Threshold Limit Values*. 4th ed., 1980. Cincinnati, Ohio: American Conference of Governmental Industrial Hygienists, Inc., 1980. 122.

8. National Fire Protection Association. *Fire Protection Guide on Hazardous Materials*. 9th ed. Boston, MA: National Fire Protection Association, 1986, p. 491M-74.

9. Bretherick, L. *Handbook of Reactive Chemical Hazards*. 2nd ed. Boston MA: Butterworths, 1979. 296.

10. Sax, N.I. *Dangerous Properties of Industrial Materials*. 6th ed. New York, NY: Van Nostrand Reinhold, 1984. 898.

11. NIOSH. *NIOSH Pocket Guide to Chemical Hazards*. DHHS (NIOSH) Publication No. 90–117. Washington, DC: U.S. Government Printing Office, June 1990.

12. American Conference of Governmental Industrial Hygienists. *Threshold Limit Values for Chemical Substances and Physical Agents and Biological Exposure Indices for 1994–1995*. Cincinnati, OH: ACGIH, 1994.

■ DIBENZO (a,i) PYRENE

CAS # 189–55–9

SYNONYMS 1,2: 7,8-dibenzpyrene • 3,4: 9,10-dibenzopyrene • 3,4: 9,10-dibenzpyrene • benzo (rst) pentaphene • db (a,i) p • dibenzo (b,h) pyrene

MF $C_{24}H_{14}$

MW 302.38

COLOR AND FORM Greenish-yellow needles, prisms, or lamellae.

BOILING POINT 275°C at 0.05 mm Hg.

MELTING POINT 280°C.

SOLUBILITY Soluble in 1,4-dioxane, boiling glacial acetic acid (2 g/L), and boiling benzene (5 g/L); almost insoluble in diethyl ether and ethanol; 5.62×10^{-7} mol/m³ (estimated).

VAPOR PRESSURE 2.39×10^{-14} mm Hg (estimated).

OTHER CHEMICAL/PHYSICAL PROPERTIES Ionization potential: 7.06 eV; attacked by nitric acid and by other reagents, mainly at 5- and 8-positions; slowly oxidized by sulfuric acid to a quinone; oxidized by lead tetraacetate to the 5,8-diacetoxy derivative and by chromic oxide or selenium dioxide to the 5,8-quinone; concentrated H_2SO_4 solutions of dibenzo (a,i) pyrene) are blue with red fluorescence; it is almost insoluble in alcohol and ether; db (a,i) p can be sublimed unchanged.

PROTECTIVE EQUIPMENT Dispensers of liquid detergent. Safety pipettes for all pipetting. In animal laboratory protective suits (preferably disposable, one-piece, and close-fitting at ankles and wrists), gloves, hair covering, and overshoes. In chemical laboratory, gloves, and gowns should always be worn; however, gloves should not be assumed to provide full protection. Carefully fitted masks or respirators when working with particulates or gases, and disposable plastic aprons and gowns should be of distinctive color. (chemical carcinogens) [R11, 8].

PREVENTATIVE MEASURES Smoking, drinking, eating, storage of food or of food and beverage containers, or utensils, and the application of cosmetics should be prohibited in any laboratory. All personnel should remove gloves, if worn, after completion of procedures in which carcinogens have been used. They should wash hands, preferably using dispensers of liquid detergent, and rinse thoroughly. Consideration should be given to appropriate methods for cleaning the skin, depending on nature of the contaminant. No standard procedure can be recommended, but the use of organic solvents should be avoided. (chemical carcinogens) [R11, 8].

CLEANUP In surface waters, one-third of the total PAH is bound to larger suspended particles, a third is bound to finely dispersed particles, and the last third is present in dissolved form. The particle-bound portion of polycyclic aromatic hydrocarbons (PAH) can be removed by sedimentation, flocculation, and filtration processes. The remaining one-third dissolved PAH usually requires oxidation for partial removal/transformation (polynuclear aromatic hydrocarbons) [R3]. A high-efficiency particulate arrestor (HEPA) or charcoal filters can be used to minimize amount of carcinogen in exhausted-air-ventilated safety cabinets, lab hoods, glove boxes, or animal rooms. Filter housing that is designed so that used filters can be transferred into plastic bag without

contaminating maintenance staff is avail commercially. Filters should be placed in plastic bags immediately after removal. The plastic bag should be sealed immediately. The sealed bag should be labelled properly. Waste liquids should be placed or collected in proper containers for disposal. The lid should be secured and the bottles properly labelled. Once filled, bottles should be placed in plastic bag, so that outer surface is not contaminated. The plastic bag should also be sealed and labelled. Broken glassware should be decontaminated by solvent extraction, by chemical destruction, or in specially designed incinerators. (chemical carcinogens) [R11, 15].

DISPOSAL At the time of review, criteria for land treatment or burial (sanitary landfill) disposal practices are subject to significant revision. Prior to implementing land disposal of waste residue (including waste sludge), consult with environmental regulatory agencies for guidance on acceptable disposal practices [R4].

COMMON USES (SRP): experimental carcinogen; research chemical [R1].

OCCUPATIONAL RECOMMENDATIONS None listed.

OSHA Meets criteria for proposed OSHA medical records rule [R8].

IARC SUMMARY No data are available for humans. Sufficient evidence of carcinogenicity in animals. Overall evaluation: Group 2B: The agent is possibly carcinogenic to humans [R5].

BIODEGRADATION No information was found about dibenzo (a,i) pyrene biodegradation. In general, however, an increased number of rings in polynuclear aromatic hydrocarbons decreases the biodegradation rate (1). Thus, db (ai) p will probably biodegrade very slowly, if at all (1,SRC) [R6].

BIOCONCENTRATION Using an estimated log octanol/water partition coefficient of 7.298 (1), a BCF of 207,000 was estimated (2,SRC). Based on this estimated value dibenzo (a, i) pyrene would be expected to bioconcentrate in aquatic organisms. However, certain polynucleararomatic hydrocarbons are not likely to appreciably bioconcentrate in organisms that have microsomal oxidase, such as fish, as this enzyme enables the organism to metabolize them (3). db (ai) p is not expected to bioconcentrate for this reason [R7].

CERCLA REPORTABLE QUANTITIES Persons in charge of vessels or facilities are required to notify the National Response Center (NRC) immediately, when there is a release of this designated hazardous substance, in an amount equal to or greater than its reportable quantity of 1 lb or 0.454 kg. The toll-free telephone number of the NRC is (800) 424–8802; in the Washington metropolitan area, (202) 426–2675. The rule for determining when notification is required is stated in 40 CFR 302.6 (section IV. D. 3.b) [R9].

RCRA REQUIREMENTS When dibenzo (a,i) pyrene, as a commercial chemical product or manufacturing chemical intermediate or an off-specification commercial chemical product or a manufacturing chemical intermediate, becomes a waste, it must be managed according to federal or state hazardous waste regulations. Also defined as a hazardous waste is any residue, contaminated soil, water, or other debris resulting from the cleanup of a spill, into water, or on dry land, of this waste. Generators of small quantities of this waste may qualify for partial exclusion from hazardous waste regulations (see 40 CFR 261.5) [R10].

DOT *Fire or Explosion:* Flammable/combustible material; may be ignited by heat, sparks, or flames. May burn rapidly with flare-burning effect [R2].

Health Hazards: Fire may produce irritating or poisonous gases. Contact may cause burns to skin and eyes. Runoff from fire control or dilution water may cause pollution [R2].

Emergency Action: Keep unnecessary people away; isolate hazard area and deny entry. Stay upwind; Keep out of low areas. Wear self-contained (positive-pressure if available) breathing apparatus and full protective clothing. If water pollution occurs, notify appropriate authorities. for emergency assistance call CHEMTREC (800) 424–9300 [R2].

Spill or Leak: Shut off ignition sources; no flares, smoking, or flames in hazard area. Do not touch spilled material. Small Dry Spills: With clean shovel, place material into clean, dry container, and cover; move containers from spill area. Large Spills: Wet down with water and dike for later disposal [R2].

First Aid: Move victim to fresh air; Call emergency medical care. In case of contact with material, immediately flush skin or eyes with running water for at least 15 min. Remove and isolate contaminated clothing and shoes at the site [R2].

REFERENCES

1. SRI.

2. Department of Transportation. *Emergency Response Guidebook 1984* DOT P 5800.3 Washington, DC: U.S. Government Printing Office, 1984, p. G–32.

3. USEPA. *Ambient Water Quality Criteria Doc; Polynuclear Aromatic Hydrocarbons (Draft)* p. C–4 (1980).

4. SRP.

5. IARC. *Monographs on the Evaluation of the Carcinogenic Risk of Chemicals to Man.* Geneva: World Health Organization, International Agency for Research on Cancer, 1972–present (multivolume work), p. S7 62 (1987).

6. Santodonato, J., et al., *Health and Ecological Assessment of Polynuclear Aromatic Hydrocarbons.* Lee, S. D., L. Grant (eds.). Pathotox Publ Park Forest South IL (1981).

7. (1) GEMS. *Graphical Exposure Modeling System* CLOGP3 (1986). (2) Lyman, W. J., et al., *Handbook of Chemical Property Estimation Methods. Environmental Behavior of Organic Compounds.* New York: McGraw-Hill, p. 5–5 (1983). (3) Santodonato, J., et al., *Health and Ecological Assessment of Polynuclear Aromatic Hydrocarbons* Lee S. D., L. Grant (eds.). Pathotox Publ Park Forest South IL (1981).

8. 47 FR 30420 (7/13/82).

9. 50 FR 13456 (4/4/85).

10. 40 CFR 261.33 (7/1/87).

11. Montesano, R., H. Bartsch, E. Boyland, G. Della Porta, L. Fishbein, R.A. Griesemer, A.B. Swan, L. Tomatis, and W. Davis (eds.). *Handling Chemical Carcinogens in the Laboratory: Problems of Safety.* IARC Scientific Publications No. 33. Lyon, France: International Agency for Research on Cancer, 1979.

■ DIBUTYLAMINE

CAS # 111–92–2

DOT # 2248

SYNONYMS 1-butanamine, n-butyl • ai3-15329 • ai3-52649 • di-n-butylamine

MF $C_8H_{19}N$

MW 129.28

COLOR AND FORM Liquid; colorless.

ODOR Ammonia-like odor.

ODOR THRESHOLD Odor recognition in air = 2.7×10^1 ppm (chemically pure) [R5].

DENSITY 0.7601 at 20°C.

DISSOCIATION CONSTANT $pK_a = 11.31$.

BOILING POINT 159–160°C.

MELTING POINT −60 to −59°C.

EXPLOSIVE LIMITS AND POTENTIAL Vapor forms explosive mixtures with air. [R11, p. 49–37].

FLAMMABILITY LIMITS Lower, approximately 1.1%; upper, not established. [R11, p. 49–37].

FIREFIGHTING PROCEDURES "Alcohol" foam. Some flammable liquids, such as certain higher-molecular-wt alcohols and amines, will destroy "alcohol" foams even when applied at very high rates. Foam supplier consulted for recommendations regarding foam types and delivery rates [R11, p. 325M–32]. Use water spray, dry chem, "alcohol" foam, or carbon dioxide. Use water to keep fire-exposed containers cool. If leak or spill has not ignited, use water spray to disperse vapors and to protect men attempting to stop leak. Water spray may be used to flush spills away from exposures. [R11, p. 49–37].

SOLUBILITY Soluble in water, alcohol; soluble in ether, acetone, benzene; 3,100 mg/L in water.

VAPOR PRESSURE 1.9 mm Hg at 20°C.

OCTANOL/WATER PARTITION COEFFICIENT Log K_{ow} = 2.83.

OTHER CHEMICAL/PHYSICAL PROPERTIES Amines tends to be fat soluble. (amines).

PROTECTIVE EQUIPMENT Wear full protective clothing. [R11, p. 49–37].

PREVENTATIVE MEASURES Local exhaust ventilation should be applied wherever there is an incidence of point source emissions or dispersion of regulated contaminants in the work area. Ventilation control of the contaminant as close as possible to its point of generation is both the most economical and safest method to minimize personnel exposure to airborne contaminants [R6].

DISPOSAL At the time of review, criteria for land treatment or burial (sanitary landfill) disposal practices are subject to significant revision. Prior to implementing land disposal of waste residue (including waste sludge), consult with environmental regulatory agencies for guidance on acceptable disposal practices [R6].

COMMON USES Corrosion inhibitor; intermediate for emulsifiers, rubber accelerators, dyes, insecticides, flotation agent, inhibitor for butadiene [R2]; chemical intermediate for 2-dibutylaminoethanol, dibutyl ammonium oleate, rubber accelerators [R1].

OCCUPATIONAL RECOMMENDATIONS None listed.

OSHA Meets criteria for OSHA medical records rule [R9].

BIODEGRADATION In a screening study, di-n-butylamine completely degraded within 14 days at 10 ppm with both an activated sludge and freshwater sediment inoculum (1). BOD values obtained during this time period indicated that mineralization was essentially complete (1). River mud bacteria and activated sludge were inhibited by 50 and 100 ppm di-n-butylamine, respectively (1). In another study that utilized 100 ppm of di-n-butylamine, and an activated sludge inoculum, no oxygen consumption was observed until about three days, when the BOD increased sharply to about 30% of theoretical (2). A third screening test resulted in >90% degradation in 9 days including a 4-day lag period (3). Although low concns of the free diamine were degraded in 10 hr by acclimated mixed cultures, only 25% of the di-n-butylamine adsorbed on bentonite clay was degraded in this time (5). The sorbed diamine degraded in 2 days (5). The rate of degradation of the sorbed molecule does not depend on its desorption rate, but rather may be due to restricted access by microorganisms (5). Under anaerobic conditions with high nitrate loads (denitrification conditions), di-n-butylamine shows little tendenancy to form nitrosamines (4) [R7].

BIOCONCENTRATION Using a reported log octanol/water partition coefficient of 2.83 (1), an estimated BCF of 83 was calculated for di-n-butylamine using a recommended regression equation (2,SRC). Di-n-butylamine would therefore not be expected to bioconcentrate appreciably in aquatic organisms (SRC) [R8].

TSCA REQUIREMENTS Pursuant to section 8 (d) of TSCA, EPA promulagated a model health and safety data reporting rule. The section 8 (d) model rule requires manufacturers, importers, and processors of listed chemical substances and mixtures to submit to EPA copies and lists of unpublished health and safety studies. Dibutylamine is included on this list [R10].

DOT *Fire or Explosion:* Flammable/combustible material; may be ignited by heat, sparks, or flames. Vapors may travel to a source of ignition, and flash back. Container may explode in heat of fire. Vapor explosion hazard indoors, outdoors, or in sewers. Runoff to sewer may create fire or explosion hazard [R3].

Health Hazards: Poisonous if swallowed.

If inhaled, may be harmful. Contact may cause burns to skin and eyes. Fire may produce irritating or poisonous gases. Runoff from fire control or dilution water may cause pollution [R3].

Emergency Action: Keep unnecessary people away; isolate hazard area and deny entry. Stay upwind; keep out of low areas. Self-contained breathing apparatus (SCBA) and structural firefighter's protective clothing will provide limited protection. Isolate for 1/2 mile in all directions if tank car or truck is involved in fire. Call CHEMTREC at 1–800–424–9300 for emergency assistance. If water pollution occurs, notify the appropriate authorities [R3].

Fire: Small Fires: Dry chemical, CO_2, halon, water spray, or standard foam. Large Fires: Water spray, fog, or standard foam is recommended. Move container from fire area if you can do so without risk. Cool containers that are exposed to flames with water from the side until well after fire is out. Stay away from ends of tanks. Withdraw immediately in case of rising sound from venting safety device or any discoloration of tank due to fire [R3].

Spill or Leak: Shut off ignition sources; no flares, smoking, or flames in hazard area. Do not touch spilled material; stop leak if you can do so without risk. Water spray may reduce vapor; but it may not prevent ignition in closed spaces. Small Spills: Take up with sand or other noncombustible absorbent material and place into containers for later disposal. Large Spills: Dike far ahead of liquid spill for later disposal [R3].

First Aid: Move victim to fresh air and call for emergency medical care; if not breathing, give artificial respiration; if breathing is difficult, give oxygen. Remove and isolate contaminated clothing and shoes at the site. In case of contact with material, immediately flush skin or eyes with running water for at least 15 minutes. Keep victim quiet and maintain normal body temperature [R3].

FIRE POTENTIAL Moderate, when exposed to heat or flame [R4].

GENERAL RESPONSE Stop discharge if possible. Keep people away. Shut off ignition sources. Call fire department. Avoid contact with liquid and vapor. Isolate and remove discharged material. Notify local health and pollution control agencies.

FIRE RESPONSE Combustible. Poisonous gases may be produced in fire. Wear goggles and self-contained breathing apparatus. Extinguish with dry chemicals, alcohol foam, or carbon dioxide. Water may be ineffective on fire. Cool exposed containers with water.

EXPOSURE RESPONSE Call for medical aid. Vapor irritating to eyes, nose, and throat. If inhaled will cause headache, coughing, or difficult breathing. If breathing has stopped, give artificial respiration. If breathing is difficult, give oxygen. Liquid irritating to skin and eyes. If swallowed will cause nausea and vomiting. Remove contaminated clothing and shoes. Flush affected areas with plenty of water. If swallowed and victim is conscious, have victim drink water or milk. If swallowed, and victim is unconscious or having convulsions, do nothing except keep victim warm.

RESPONSE TO DISCHARGE Issue warning—water contaminant. Restrict access. Disperse and flush.

WATER RESPONSE Effect of low concentrations on aquatic life is unknown. May be dangerous if it enters water intakes. Notify local health and wildlife officials. Notify operators of nearby water intakes.

EXPOSURE SYMPTOMS Inhalation causes irritation of nose, throat, and lungs; coughing; nausea; headache. Ingestion causes irritation of mouth and stomach. Contact with eyes causes irritation. Contact with skin causes irritation and dermatitis.

EXPOSURE TREATMENT Inhalation: move from exposure; if breathing has stopped, start artificial respiration. Ingestion: give large amount of water. Eyes: irrigate with water for 15 min; get medical attention for possible eye damage. Skin: wash with large amounts of water for 15 min.

REFERENCES

1. SRI.

2. Sax, N. I., and R. J. Lewis, Sr. (eds.). *Hawley's Condensed Chemical Dictionary.* 11th ed. New York: Van Nostrand Reinhold Co., 1987. 370.

3. Department of Transportation. *Emergency Response Guidebook 1987.* DOT P 5800.4. Washington, DC: U.S. Government Printing Office, 1987, p. G–68.

4. Sax, N. I. *Dangerous Properties of Industrial Materials.* 6th ed. New York, NY: Van Nostrand Reinhold, 1984. 917.

5. Fazzalari, F. A. (ed.). *Compilation of Odor and Taste Threshold Values Data.* ASTM Data Series DS 48A (Committee E–18). Philadelphia, PA: American Society for Testing and Materials, 1978. 48.

6. SRP.

7. (1) Calamari, D., et al., *Chemosphere* 9: 753–62 (1980). (2) Yoshimura, K., et al. *J Amer Oil Chemical Soc* 57: 238–41 (1980). (3) Zahn, R., H. Wellens. *Wasser Abwasser Forsch* 13: 1–7 (1980). (4) Kaplan, D. L., et al., *Gov Rep Announce Index* 84: 67 (1984). (5) Wszolek, P. C., M. Alexander, *J Agric Food Chem* 27: 410–4 (1979).

8. (1) Hansch, C., and A. J. Leo. *Medchem Project,* Pomona, CA: Claremont College (1985). (2) Lyman, W. J., et al., *Handbook of Chemical Property Estimation Methods.* New York: McGraw-Hill, pp. 5–1 to 5–30 (1982).

9. 29 CFR 1910.20 (7/1/87).

10. 40 CFR 716.120 (7/1/87).

11. National Fire Protection Association. *Fire Protection Guide on Hazardous Materials.* 9th ed. Boston, MA: National Fire Protection Association, 1986.

■ DIBUTYL ETHER

CAS # 142–96–1

DOT # 1149

SYNONYMS 1-butoxybutane • butane,

1,1'-oxybis • butyl ether • butyl oxide • di-n-butyl ether • dibutyl oxide • ether butylique (French) • n-butyl ether • n-dibutyl ether

MF $C_8H_{18}O$

MW 130.26

COLOR AND FORM Colorless liquid.

ODOR Mild, ethereal odor.

ODOR THRESHOLD 0.37 mg/m³ (odor low) 2.50 mg/m³ (odor high) [R6].

DENSITY 0.7689 at 20°C.

SURFACE TENSION 23 dynes/cm = 0.023 N/m at 20°C.

VISCOSITY 0.0069 poise (20°C).

BOILING POINT 142°C at 760 mm Hg.

MELTING POINT −95.3°C.

HEAT OF COMBUSTION −17,670 Btu/lb = −9,820 cal/g = −411×10⁵ J/kg.

HEAT OF VAPORIZATION Latent heat of vaporization: 67.8 cal/g at 140.9°C.

AUTOIGNITION TEMPERATURE 382°F [R17, p. 325M–32].

EXPLOSIVE LIMITS AND POTENTIAL May form explosive peroxides, especially in anhydrous form [R2].

FLAMMABILITY LIMITS % by vol: lower, 1.5; upper, 7.6 [R17, p. 325M–32].

FIREFIGHTING PROCEDURES Use dry chemical, "alc" foam, or carbon dioxide. Water may be ineffective agent, but water should be used to keep fire-exposed containers cool. If a leak or spill has not ignited, use water spray to disperse the vapors and to protect men attempting to stop a leak. Water spray may be used to flush spills away from exposures. [R17, p. 49–37].

TOXIC COMBUSTION PRODUCTS Moderately dangerous; when heated, it emits acrid fumes. [R15, 584].

SOLUBILITY 0.03 to 0.05% by wt in water at 20°C; miscible with benzene and most organic solvents; soluble in all proportions in alcohol, ether; very soluble in acetone.

VAPOR PRESSURE 4.8 mm Hg at 20°C.

OCTANOL/WATER PARTITION COEFFICIENT Log K_{ow} = 3.08.

OTHER CHEMICAL/PHYSICAL PROPERTIES Percent in saturated air: 0.9 at 25°C, 760 mm Hg; density of saturated air: 1.1 at 25°C, 760 mm Hg (air = 1); 1 ppm = approximately 5.33 mg/m³ at 25°C, 760 mm Hg; 1 mg/L = approximately 188 ppm at 25°C, 760 mm Hg; liquid-water interfacial tension: (est) 30 dynes/cm = 0.030 N/m at 20°C; azeotrope 67 wt% at 92.9°C.

PROTECTIVE EQUIPMENT Wear goggles and self-contained breathing apparatus. [R17, p. 49–37].

PREVENTATIVE MEASURES Contact lens use in industry is controversial. A survey of 100 corporations resulted in the recommendation that each company establish its own contact lens use policy. One presumed hazard of contact lens use is possible chemical entrapment. It was found that contact lenses minimized injury or protected the eye. The eye was afforded more protection from liquid irritants. Soft contact lenses did not worsen corneal damage from strong chemicals and in some cases could actually protect the eye. Overall, the literature supports the wearing of contact lenses in industrial environments as part of the standard eye protection, e.g., face shields; however, more data are needed to establish the value of contact lenses [R7].

DISPOSAL The following wastewater treatment technologies have been investigated for butyl ether: concentration process: activated carbon [R8].

COMMON USES Extracting agent, used especially for separating metals; solvent purification;, organic synthesis (reaction medium) [R2]; solvent for esters, gums, hydrocarbons, alkaloids, oils, organic acids, and resins [R13, 1664]; used to produce butyl alcohol [R1]; has limited use as a solvent in the preparation of Grignard reagents [R3]; used to complex Ziegler catalysts to improve catalytic activity [R14, p. 584]; used in plutonium separation processes [R4].

OCCUPATIONAL RECOMMENDATIONS None listed.

NON-HUMAN TOXICITY LD_{50} rat oral 3.23–3.92 g/kg (from table) [R16, 2503]. LD_{50} rabbit percutaneous >20 ml/kg (from table) [R16, 2503]. LC_{50} rat inhalation 4,000 ppm/4 hr (from table) [R16, 2503]. Grade 4 primary irritation hazard to rabbit skin (from table) [R16, 2503]. Grade 1 hazard to rabbit eye (from table) [R16, 2503].

ECOTOXICITY LC_{50} *Cyprinodon variegatus* (sheepshead minnows) above 430 ppm/24, 48, 72, and 96 hr, juvenile 14–28 days old, seawater at 25–31°C, static bioassay [R9]. LC_{50} *Pimephales promelas* (fathead minnow) 32.5 mg/L 96-hr flow-through bioassay, wt 0.12 g, water hardness 45.5 mg/L $CaCO_3$, temp: 25 ± 1°C, pH 7.5, dissolved oxygen greater than 60% of saturation [R10].

BIODEGRADATION No data concerning the biodegradation of dibutyl ether in environmental media were located. The 5-day biological oxygen demand was 16% of theoretical in screening tests that utilized acclimated mixed microbial cultures (1). These data suggest that dibutyl ether may slowly be biodegraded in environmental media (SRC). Many ethers are known to be resistant to biodegradation (2) [R11].

BIOCONCENTRATION Based upon a reported log K_{ow} of 3.08 (1), a BCF of 129 has been estimated using a recommended regression equation (2). Based upon this estimated BCF, dibutyl ether will not be expected to bioconcentrate in aquatic organisms (SRC) [R12].

DOT *Fire or Explosion:* Flammable/combustible material; may be ignited by heat, sparks, or flames. Vapors may travel to a source of ignition, and flash back. Container may explode in heat of fire. Vapor explosion hazard indoors, outdoors, or in sewers. Runoff to sewer may create fire or explosion hazard [R5].

Health Hazards: May be poisonous if inhaled or absorbed through skin. Vapors may cause dizziness or suffocation. Contact may irritate or burn skin and eyes. Fire may produce irritating or poisonous gases. Runoff from fire control or dilution water may cause pollution [R5].

Emergency Action: Keep unnecessary people away; isolate hazard area and deny entry. Stay upwind; keep out of low areas. Self-contained breathing apparatus (SCBA) and structural firefighter's protective clothing will provide limited protection. Isolate for 1/2 mile in all directions if tank car or truck is involved in fire. Call CHEMTREC at 1–800–424–9300 for emergency assistance. If water pollution occurs, notify the appropriate authorities [R5].

Fire: Small Fires: Dry chemical, CO_2, halon, water spray, or alcohol foam. Large Fires: Water spray, fog, or alcohol foam is recommended. Move container from fire area if you can do so without risk. Cool containers that are exposed to flames with water from the side until well after fire is out. Stay away from ends of tanks. For massive fire in cargo area, use unmanned hose holder, or monitor nozzles; if this is impossible, withdraw from area, and let fire burn. Withdraw immediately in case of rising sound from venting safety device or any discoloration of tank due to fire [R5].

Spill or Leak: Shut off ignition sources; no flares, smoking, or flames in hazard area. Stop leak if you can do so without risk. Water spray may reduce vapor; but it may not prevent ignition in closed spaces. Small Spills: Take up with sand or other noncombustible absorbent material and place into containers for later disposal. Large Spills: Dike far ahead of liquid spill for later disposal [R5].

First Aid: Move victim to fresh air and call emergency medical care; if not breathing, give artificial respiration; if breathing is difficult, give oxygen. In case of contact with material, immediately flush eyes with running water for at least 15 minutes. Wash skin with soap and water. Remove and isolate contaminated clothing and shoes at the site [R5].

FIRE POTENTIAL Only electrical equipment of explosion-proof type (group C classification) is permitted to be operated in ether areas. Ether should not be stored near powerful oxidizers or in areas of high fire hazard. (ethers) [R15, 1305].

GENERAL RESPONSE Stop discharge if possible. Keep people away. Shut off ignition sources. Call fire department. Isolate and remove discharged material. Notify local health and pollution control agencies.

FIRE RESPONSE Flammable. Containers may explode in fire. Flashback along vapor trail may occur. Vapor may explode if ignited in an enclosed area. Extinguish with dry chemicals, alcohol foam, or carbon dioxide. Water may be ineffective on fire. Cool exposed containers with water.

EXPOSURE RESPONSE Call for medical aid. Vapor irritating to eyes, nose, and throat. Move victim to fresh air. If breathing is difficult, give oxygen. Liquid irritating to skin and eyes. Remove contaminated clothing and shoes. Flush affected areas with plenty of water. If swallowed and victim is conscious, have victim drink water or milk.

RESPONSE TO DISCHARGE Mechanical containment; should be removed; chemical and physical treatment.

WATER RESPONSE Effect of low concentrations on aquatic life is unknown. Fouling to shoreline. May be dangerous if it enters water intakes. Notify local health and wildlife officials. Notify operators of nearby water intakes.

EXPOSURE SYMPTOMS Inhalation causes irritation of nose and throat. Liquid irritates eyes and may irritate skin on prolonged contact. Ingestion causes irritation of mouth and stomach.

EXPOSURE TREATMENT Inhalation: remove to fresh air. Eyes: after contact with liquid, flush with water for at least 15 min. Skin: wipe off, wash well with soap and water. Ingestion: induce vomiting.

REFERENCES

1. *Kirk–Othmer Encyc Chemical Tech* 3rd ed. 1978–present, V4 p. 342.

2. Sax, N. I., and R. J. Lewis, Sr. (eds.). *Hawley's Condensed Chemical Dictio-* nary. 11th ed. New York: Van Nostrand Reinhold Co., 1987. 186.

3. *Kirk–Othmer Encyc Chemical Tech* 3rd ed. 1978–present, V12 p. 32.

4. *Kirk–Othmer Encyc Chemical Tech* 3rd ed. 1978–present, V18 p. 294.

5. Department of Transportation. *Emergency Response Guidebook 1987.* DOT P 5800.4. Washington, DC: U.S. Government Printing Office, 1987, p. G–26.

6. Ruth, J.H., *Am Ind Hyg Assoc J* 47: A–142–51 (1986).

7. Randolph, S. A., M. R. Zavon. *J Occup Med* 29: 237–42 (1987).

8. USEPA. *Management of Hazardous Waste Leachate*, EPA Contract No. 68–03–2766 p. E–151 (1982).

9. Heitmuller, P. T., et al. *Bull Environ Contam Toxicol* 27 (5): 596 (1981).

10. Vieth, G.D., et al., *Canadian J Fisheries Aquat Sci* 40 (6): 743–8 (1983).

11. (1) Babeu, L., D. D. Vaishnav. *J Indust Microbiol* 2: 107–15 (1987). (2) M. Alexander, *Biotechnol Bioeng* 15: 611–47 (1973).

12. (1) Abernethy, S. G., et al., *Environ Toxicol Chem* 7: 469–81 (1988). (2) Lyman, W. J., et al., *Handbook of Chemical Property Estimation Methods.* New York: McGraw-Hill, p. 5–5 (1982).

13. Patty, F. (ed.). *Industrial Hygiene and Toxicology.* Volume II: *Toxicology.* 2nd ed. New York: Interscience Publishers, 1963.

14. *Kirk-Othmer Encyc Chem Tech.* 3rd ed. 1978–present, V9.

15. Sax, N.I. *Dangerous Properties of Industrial Materials.* 6th ed. New York, NY: Van Nostrand Reinhold, 1984.

16. Clayton, G. D., and F. E. Clayton (eds.). *Patty's Industrial Hygiene and Toxicology.* Volume 2A, 2B, 2C: Toxicology. 3rd ed. New York: John Wiley Sons, 1981–1982.

17. National Fire Protection Association. *Fire Protection Guide on Hazardous Materials.* 9th ed. Boston, MA: National Fire Protection Association, 1986.

■ 2,6-DI-t-BUTYL-p-CRESOL

CAS # 128–37–0

SYNONYMS 2,6-bis (1,1-dimethylethyl)-4-methylpheno • 2,6-di-terc. butyl-p-kresol (Czech) • 2,6-di-tert-butyl-1-hydroxy-4-methylbenzene • 2,6-di-tert-butyl-4-methylphenol • 2,6-di-tert-butyl-p-cresol • 2,6-di-tert-butyl-p-methylphenol • 2,6-di-tert-butylcresol • 3,5-di-tert-butyl-4-hydroxytoluene • 4-hydroxy-3,5-di-tert-butyltoluene • 4-methyl-2,6-di-terc. butylfenol (Czech) • 4-methyl-2,6-di-tert-butylphenol • 4-methyl-2,6-tert butylphenol • antioxidant-4k • antioxidant-kb • ao-4k • bht • bht (food grade) • buks • butylated-hydroxytoluen • butylhydroxytoluene • cao-1 • cao-3 • dalpac • dbpc • dbpc (technical grade) • deenax • di-tert-butyl-p-creso • di-tert-butylcresol • dibunol • dibutylated-hydroxytoluene • impruval • impruvo • ionol • ionol-1 • ionole • methyldi-tert-butylphenol • nci-c03598 • nonox-tbc • o, o'-di-tert-butyl-p-cresol • parabar-441 • paranox-441 • phenol, 2,6-bis (1,1-dimethylethyl)-4-methyl • stavox • sustane-bht • tenamen-3 • tenox-bht • topanol • topanol-o • topanol-oc • vianol • antrancine • ionol-C • sustan • 2,6-tert-butyl-4-methylpheno • "Ionol" cp-antioxidan

MF $C_{15}H_{24}O$

MW 220.39

COLOR AND FORM White crystalline solid; pale-yellowish crystalline powder.

TASTE Tasteless.

DENSITY 1.048 at 20°C.

VISCOSITY 3.47 centistokes at 0°C, 1.54 centistokes at 120°C.

BOILING POINT 265°C.

MELTING POINT 70°C.

FIREFIGHTING PROCEDURES Carbon dioxide, dry chemial [R12, 427]; extinguishing method: water or foam may cause frothing [R6].

SOLUBILITY Insoluble in water; freely soluble in toluene; soluble in methanol, isopropanol, methyl ethyl ketone, acetone, cellosolve, benzene, most hydrocarbon solvents, ethanol, petroleum ether, liquid petrolatum (white oil): 0.5% wt/wt; more soluble in food oils and fats than butylated hydroxyanisol; good solubility in linseed oil; insoluble in aqueous alkal; soluble in naphtha; insoluble in 10% sodium hydroxide; 0.4 mg/L at 20°C in water.

STABILITY AND SHELF LIFE Stable in light and air [R13, 1221].

OTHER CHEMICAL/PHYSICAL PROPERTIES BP: 136°C at 10 mm Hg.

PREVENTATIVE MEASURES SRP: The scientific literature for the use of contact lenses in industry is conflicting. The benefit or detrimental effects of wearing contact lenses depend not only upon the substance, but also on factors including the form of the substance, characteristics and duration of the exposure, the uses of other eye protection equipment, and the hygiene of the lenses. However, there may be individual substances whose irritating or corrosive properties are such that the wearing of contact lenses would be harmful to the eye. In those specific cases, contact lenses should not be worn. In any event, the usual eye protection equipment should be worn even when contact lenses are in place.

COMMON USES Food additive [R12, 427]; antioxidant for synthetic rubbers, plastics, soaps, animal and vegetable oils; antiskinning agent in paints and ink [R1]; in chewing gum base, packaging materials, butter, candy, paraffin, and mineral oils [R14, 212]; employed to retard oxidative degradation of oils and fats in various cosmetics and pharmaceuticals [R13, 1222]; used by Shell Chemical Company as a stabilizer for monomers [R3]; BHT is used as an antioxidant in pyrethrum extract, an insecticide product [R4]; stabilizer in motor and aviation gasoline [R5]; antioxidant for petroleum products, jet fuels, food products, food packaging, animal feeds [R2]; satisfies ASTM D910–64T for use in aviation gasoline [R2].

OCCUPATIONAL RECOMMENDATIONS None listed.

OSHA 8-hr time-weighted average: 10 mg/m³ [R8].

THRESHOLD LIMIT 8-hr time-weighted average (TWA) 10 mg/m³ (1987) [R50, 20].

NON-HUMAN TOXICITY LD$_{50}$ rat oral 890 mg/kg [R12, 426]; LD$_{50}$ mouse, oral 1,040 mg/kg [R12, 426]; LD$_{50}$ guinea pig, oral 10,700 mg/kg [R12, 426].

IARC SUMMARY No data are available for humans. Limited evidence of carcinogenicity in animals. Overall evaluation: Group 3: The agent is not classifiable as to its carcinogenicity to humans [R7].

FIRE POTENTIAL Fire hazard: slight, when exposed to heat or flame; moderately dangerous; can react with oxidizing materials [R12, 427].

FDA Substances classified as antioxidants, when migrating from food-packaging material (limit of addition to food 0.005 percent) include butylated hydroxytoluene [R9]. Butylated hydroxytoluene used as a chemical preservative in food for human consumption is generally recognized as safe when the total content of antioxidants is not over 0.002 percent of fat or oil content, including essential (volatile) oil content of food, provided the substance is used in accordance with good manufacturing practice [R10]. Butylated hydroxytoluene used as a chemical preservative in animal drugs, feeds, and related products is generally recognized as safe when the total content of antioxidants is not over 0.002 percent of fat or oil content, including essential (volatile) oil content of food, provided the substance is used in accordance with good manufacturing or feeding practice [R11].

REFERENCES

1. Budavari, S. (ed.). *The Merck Index—Encyclopedia of Chemicals, Drugs, and Biologicals.* Rahway, NJ: Merck and Co., Inc., 1989. 238.

2. Lewis, R. J., Sr. (ed.). *Hawley's Condensed Chemical Dictionary.* 12th ed. New York, NY: Van Nostrand Reinhold Co., 1993. 372.

3. Chemical Mark Rep 224 (20): 13 (1983).

4. USEPA. *CHIP (DRAFT) Butylated Hydroxytoluene* p. 4 (1984).

5. *Kirk–Othmer Encyclopedia of Chemical Technology.* 3rd ed., Volumes 1–26. New York, NY: John Wiley and Sons, 1978–1984, p. V2 91 (1978).

6. National Fire Protection Guide. *Fire Protection Guide on Hazardous Materials.* 10th ed. Quincy, MA: National Fire Protection Association, 1991, p. 325M–33.

7. IARC. *Monographs on the Evaluation of the Carcinogenic Risk of Chemicals to Man.* Geneva: World Health Organization, International Agency for Research on Cancer, 1972–present (multivolume work), p. suppl 7 59.

8. 29 CFR 1910.1000 (7/1/91).

9. 21 CFR 181.24 (4/1/91).

10. 21 CFR 182.3173 (4/1/91).

11. 21 CFR 582.3173 (4/1/91).

12. Sax, N.I. *Dangerous Properties of Industrial Materials.* 6th ed. New York, NY: Van Nostrand Reinhold, 1984.

13. Osol, A. and J.E. Hoover, et al. (eds.). *Remington's Pharmaceutical Sciences.* 15th ed. Easton, Pennsylvania: Mack Publishing Co., 1975.

14. Furia, T.E. (ed.). *CRC Handbook of Food Additives.* 2nd ed. Cleveland: The Chemical Rubber Co., 1972.

■ 3,4-DICHLOROANILINE

CAS # 95–76–1

SYNONYMS 1-amino-3,4-dichlorobenzene • 3,4-dca • 3,4-dichloranilin • 3,4-dichloraniline • aniline, 3,4-dichloro

• benzenamine, 3,4-dichloro • dca • 3,4-dichlorobenzenamine

MF $C_6H_5Cl_2N$

MW 162.02

COLOR AND FORM Needles from petroleum ether; fine, light-tan crystals.

BOILING POINT 272°C.

MELTING POINT 71–72°C.

EXPLOSIVE LIMITS AND POTENTIAL Data are given for the prevention of explosion accidents during distillation of chlorine-substituted aromatic amines [R3].

FIREFIGHTING PROCEDURES Use dry chemical or carbon dioxide. Water or foam may cause frothing. Use water to keep fire-exposed containers cool. Water may be used to flush spills from exposures. [R10, p. 49–37].

SOLUBILITY Very soluble in alcohol, ether; slightly soluble in benzene; practically insoluble in water; soluble in most organic solvents; water solubility of 92 ppm at 20°C.

VAPOR PRESSURE Vapor pressure of 0.00975 mm Hg at 20°C.

OCTANOL/WATER PARTITION COEFFICIENT Log K_{ow} = 2.69 (measured).

PROTECTIVE EQUIPMENT Wear full protective clothing. [R10, p. 49–37].

PREVENTATIVE MEASURES SRP: Contaminated protective clothing should be segregated in such a manner so that there is no direct personal contact by personnel who handle, dispose of, or clean the clothing. Quality assurance to ascertain the completeness of the cleaning procedures should be implemented before the decontaminated protective clothing is returned for reuse by the workers.

DISPOSAL SRP: At the time of review, criteria for land treatment or burial (sanitary landfill) disposal practices are subject to significant revision. Prior to implementing land disposal of waste residue (including waste sludge), consult with environmental regulatory agencies for guidance on acceptable disposal practices.

COMMON USES Chemical intermediate for dyes and herbicides (e.g., diuron and propanil) [R1].

OCCUPATIONAL RECOMMENDATIONS None listed.

ECOTOXICITY LC_{50} *Ophryotrocha diadema* (adult) 15 mg/L/96 hr (conditions of bioassay not specified) [R4]; LC_{50} *Ophryotrocha diadema* (larvae) 4 mg/L/96 hr (conditions of bioassay not specified) [R4]; LC_{50} *Pimephales promelas* (fathead minnow) 34 days old 8.06 mg/L/96 hr (confidence limit: 7.26–8.95 mg/L) at 25°C (98% purity) (conditions of bioassay not specified) [R11, 146]; EC_{50} *Pimephales promelas* (fathead minnow) 34 days old 6.09 mg/L/96 hr (confidence limit: 5.71–6.49 mg/L) at 25°C (98% purity) (conditions of bioassay not specified) [R11, 146]; LC_{50} *Pimephales promelas* (fathead minnow) 28 days old 7.0 mg/L/96 hr (confidence limit: 6.6–7.5 mg/L) at 25.1°C (98% purity) (conditions of bioassay not specified) [R11, 143]; EC_{50} *Pimephales promelas* (fathead minnow) 28 days old 6.7 mg/L/96 hr (confidence limit: 6.4–6.9 mg/L) at 25.1°C (98% purity) (conditions of bioassay not specified) [R11, 143]; LC_{50} *Palaemonetes varians* (larvae) 2.0×10^{-5} mol/L/96 hr at 15°C, 100% sea water (conditions of bioassay not specified) [R5]; LC_{50} *Palaemonetes varians* (larvae) 1.1×10^{-5} mol/L/10 days at 15°C, 100% sea water (conditions of bioassay not specified) [R5]; LC_{50} *Palaemonetes variansu* (adult) 4.0×10^{-5} mol/L/96 hr at 15°C, 10% sea water (conditions of bioassay not specified) [R5]; LC_{50} *Neomysis integer* (adult) 9.1×10^{-6} mol/L/96 hr at 20°C, 10% sea water (conditions of bioassay not specified) [R5]; LC_{50} *Daphnia magna* 7.6×10^{-7} mol/L/96 hr at 20°C, 10% sea water (conditions of bioassay not specified) [R5].

BIODEGRADATION 3,4-Dichloroaniline (DCA), a biodegradation intermediate of several herbicides, is mineralized in soil only very slowly. In enrichment cultures, DCA failed to serve as the sole substrate, but analog enrichment yielded a *Pseudomonas putida* strain that, in presence of unchlorinated analog substrates, mineralized dca with release of $^{14}CO_2$, and Cl^-.

Mass spectrometric identification of the key biodegradation intermediates (3,4-dichloromuconate, 3-chlorobutenolide, and 3-chlorolevulinic acid) revealed that DCA biodegradation occurred through 4,5-dichlorocatechol, 3,4-dichloromuconate, 3-chlorobutenolide, 3-chloromaleylacetate, and 3-chloro-4-ketadipate to succinate plus acetate. Through the above pathway, dca was converted ultimately to inorganic end products. The slow mineralization of DCA in soil is not entirely explainable by the inherent recalcitrance of this compound but is explainable by the competing polymerization and binding reactions that decrease its availability [R6].

BIOCONCENTRATION Studies on the environmental hazards and fate of herbicides are reviewed. Herbicides do not accumulate by way of the trophic chains, except for diuron and 3,4-dichloroaniline, which accumulate in fish [R7].

ATMOSPHERIC STANDARDS This action promulgates standards of performance for equipment leaks of volatile organic compounds (VOC) in the synthetic organic chemical manufacturing industry (SOCMI). The intended effect of these standards is to require all newly constructed, modified, and reconstructed SOCMI process units to use the best demonstrated system of continuous emission reduction for equipment leaks of VOC, considering costs, non-air-quality health, and environmental impact and energy requirements. Dichloroaniline is produced, as an intermediate or final product, by process units covered under this subpart. (dichloroaniline) [R8].

TSCA REQUIREMENTS Section 8 (a) of TSCA requires manufacturers of this chemical substance to report preliminary assessment information concerned with production, use, and exposure to EPA as cited in the preamble of the 51 FR 41329 [R9].

DOT *Health Hazards:* Poisonous; may be fatal if inhaled, swallowed, or absorbed through skin. Contact may cause burns to skin and eyes. Runoff from fire control or dilution water may give off poisonous gases and cause water pollution. Fire may produce irritating or poisonous gases (dichloroaniline) [R2].

Fire or Explosion: Some of these materials may burn, but none of them ignites readily. Container may explode violently in heat of fire (dichloroaniline) [R2].

Emergency Action: Keep unnecessary people away; isolate hazard area and deny entry. Stay upwind, out of low areas, and ventilate closed spaces before entering. Self-contained breathing apparatus and chemical protective clothing that is specifically recommended by the shipper or producer may be worn but they do not provide thermal protection unless that is stated by the clothing manufacturer. Structural firefighter's protective clothing is not effective with these materials. Remove and isolate contaminated clothing at the site. Call CHEMTREC at 1–800–424–9300 as soon as possible, especially if there is no local hazardous-materials team available (dichloroaniline) [R2].

Fire: Small Fires: Dry chemical, CO_2, halon, water spray, or standard foam. Large Fires: Water spray, fog, or standard foam is recommended. Move container from fire area if you can do so without risk. Fight fire from maximum distance. Stay away from ends of tanks. Dike fire control water for later disposal; do not scatter the material (dichloroaniline) [R2].

Spill or Leak: Do not touch spilled material; stop leak if you can do so without risk. Use water spray to reduce vapors. Small Spills: Take up with sand or other noncombustible absorbent material and place into containers for later disposal. Small Dry Spills: With clean shovel place material into clean, dry container, and cover; move containers from spill area. Large Spills: Dike far ahead of liquid spill for later disposal (dichloroaniline) [R2].

First Aid: Move victim to fresh air and call for emergency medical care; if not breathing, give artificial respiration; if breathing is difficult, give oxygen. In case of contact with material, immediately flush skin or eyes with running water for at least 15 minutes. Speed in removing material from skin is of extreme importance. Remove and isolate contaminated clothing and shoes at the site. Keep victim quiet and maintain normal body temperature. Effects may be delayed; keep victim under observation (dichloroaniline) [R2].

FIRE POTENTIAL Combustible solid [R10, p. 49–37].

REFERENCES

1. SRI.

2. Department of Transportation. *Emergency Response Guidebook 1987*. DOT P 5800.4. Washington, DC: U.S. Government Printing Office, 1987, p. G–55.

3. Kotoyori, T. *Anzen Kogaku 21* (2): 107–14 (1982).

4. Hooftman, R. N., G. J. Vinke. *Ecotoxicol Environ Saf* 4 (3): 252–62 (1980).

5. Van der Meer, C., et al., *Bull Environ Contam Toxicol* 40 (2): 204–11 (1988).

6. You, I–S., R. Bartha. *J Agric Food Chem* 30 (2): 274–7 (1982).

7. Mathys, G. *Bull Organ Eur Mediterr Prot Plant* 5 (92): 87–100 (1975).

8. 40 CFR 60.489 (7/1/88).

9. 40 CFR 712.30 (7/1/88).

10. National Fire Protection Association. *Fire Protection Guide on Hazardous Materials*. 9th ed. Boston, MA: National Fire Protection Association, 1986.

11. Brooke, L.T., D.J. Call, D.T. Geiger, and C.E. Northcott (eds.). *Acute Toxicities of Organic Chemicals to Fathead Minnows (Pimephales promelas)*. Superior, WI: Center for Lake Superior Environmental Studies, Univ. of Wisconsin–Superior, 1984.

■ 1,3-DICHLOROBENZENE

Cl

Cl

CAS # 541–73–1

SYNONYMS benzene, 1,3-dichloro • benzene, m-dichloro • m-dichlorobenzene • m-dichlorobenzol • m-phenylene dichloride • meta-dichlorobenzene • m-dcb

MF $C_6H_4Cl_2$

MW 147.00

COLOR AND FORM Colorless liquid.

DENSITY 1.2884 at 20°C.

SURFACE TENSION 38.30 dynes/cm^{-1} at 20°C; 0.1147 dynes/cm^{-1} at 30°C.

VISCOSITY 1.045 mNXsXm^{-2} at 23°C; 0.955 mNXsXm^{-2} at 33°C.

BOILING POINT 173.53 at 10°C.

MELTING POINT −24.7°C.

HEAT OF VAPORIZATION 10,446.8.

FIREFIGHTING PROCEDURES Water, foam, carbon dioxide, dry chemical [R4].

SOLUBILITY Soluble in alcohol, ether, acetone, and benzene; 1,3-dichlorobenzene is soluble at 123 mg/L in water at 25°C.

VAPOR PRESSURE 2.3 mm Hg at 25°C.

OCTANOL/WATER PARTITION COEFFICIENT Log K_{ow} = 3.60.

OTHER CHEMICAL/PHYSICAL PROPERTIES Conversion factors: 1 mg/L = 166.3 ppm, 1 ppm = 6.01 mg/m³ at 25°C, and 760 mm Hg; partition coefficients at 37°C for 1,3-dichlorobenzene into blood = 201; into oil = 27,100; Dielectric constant = 5.04 at 25°C; 4.22 at 90°C; dipole moment: 1.68 (gas); 1.38 at 24°C, benzene as solvent.

PROTECTIVE EQUIPMENT The American Society for Testing and Materials (ASTM) cell was utilized to study permeation of chlorobenzene, o-dichlorobenzene, and m-dichlorobenzene, and o-, and p-chlorotoluenes through viton (unsupported) and nitrile (supported and unsupported) glove materials using isopropanol as collecting solvent, and FID (flame ionization detector)/gas chromatography for quantitation. Adequate mixing in the collection chamber was accomplished by externally agitating the ASTM cell at the required speed in a moving-tray water bath at 25°C. The viton glove did not show permeation even after 4 hr. The nitrile gloves showed breakthrough times of <1 hr. The steady-state molar flux rates for unsupported or supported nitrile gloves, or for the different challenge solvents were not statistically different. Breakthrough times were better indicators of permeation than steady-state molar flux rates. A mixed permeation mechanism was proposed, depending on swelling of the glove material [R5].

PREVENTATIVE MEASURES Contact lens use in industry is controversial. A survey of 100 corporations resulted in the recommendation that each company establish their own contact lens use policy. One presumed hazard of contact lens use is possible chemical entrapment. It was found that contact lenses minimized injury or protected the eye. The eye was afforded more protection from liquid irritants. Soft contact lenses do not worsen corneal damage from strong chemicals and in some cases could actually protect the eye. Overall, the literature supports the wearing of contact lenses in industrial environments as part of the standard eye protection, e.g., face shields; however, more data are needed to establish the value of contact lenses [R6].

CLEANUP Water spill: Use natural deep-water pockets, excavated lagoons, or sandbag barriers to trap material at bottom. If dissolved in region of 10 ppm or greater concentration, apply activated carbon at ten times the spilled amount. Remove trapped material with suction hoses. Use mechanical dredges or lifts to remove immobilized masses of pollutants and precipitates. Land spill: Dig a pit, pond, lagoon, or holding area to contain liquid or solid material. (SRP: If time permits, pits, ponds, lagoons, soak holes, or holding areas should be sealed with an impermeable flexible-membrane liner.) Cover solids with a plastic sheet to prevent dissolving in rain or firefighting water. (p-dichlorobenzene) [R7].

DISPOSAL SRP: At the time of review, criteria for land treatment or burial (sanitary landfill) disposal practices are subject to significant revision. Prior to implementing land disposal of waste residue (includ-

ing waste sludge), consult with environmental regulatory agencies for guidance on acceptable disposal practices.

COMMON USES Fumigant and insecticide [R1]; reacted with potassium hydroxide or sodium hydroxide to produce chlorophenols; used in the preparation of arylene sulfide polymers in the pps polymerization process [R2].

OCCUPATIONAL RECOMMENDATIONS None listed.

INTERNATIONAL EXPOSURE LIMITS MAC to skin former USSR, 20 mg/m^3 of air [R12].

ECOTOXICITY LC_{50} bluegill sunfish 21.8 mg/L/24 hr; 10.7 mg/L/48 hr; 5.02 mg/L/96 hr (static bioassay) [R8]; LC_{50} fathead minnow 12.7 mg/L/96 hr (static bioassay) [R8]; LC_{50} sheepshead minnow 8.46 mg/L/24 hr (static bioassay) [R8].

BIOLOGICAL HALF-LIFE Half-life in blood was 4.4 hr (oral administration, 200 mg/kg, male Wistar rats) [R9].

BIODEGRADATION Using a static-culture flask-screening procedure (5 or 10 mg/L test compound, a 7-day static incubation followed by three weekly subcultures, and a settled domestic wastewater as microbial inoculum), 1,3-dichlorobenzene was biodegraded 58–59%, 67–69%, 31–39%, and 33–35% after the original culture, first, second, and third subculture, respectively (1). In a continuous-flow activated-sludge system, nearly 100% of influent 1,3-dichlorobenzene was removed by an apparent combination of biodegradation and stripping (2). In a laboratory aquifer column simulating saturated-flow conditions typical of a river/groundwater infiltration system, the dichlorobenzenes were biotransformed under aerobic conditions but not under anaerobic conditions (3). 1,3-Dichlorobenzene was significantly removed in an aerobic biofilm column (9.8 ppb initial concentration) but was not removed in a methanogenic biofilm column (10 initial ppb) (4,5). The dichlorobenzenes are not expected to be biotransformed in anaerobic water conditions found in aquifers (6). An analysis of monitoring data from sediment cores indicated insufficient evidence to show occurrence of anaerobic dehalogenation of the chlorobenzenes in Lake Ontario sediment (7). The dichlorobenzenes were slowly biodegraded (6.3% of theoretical CO_2 evolution in 10 weeks) in an alkaline soil sample (8). Microbial decomposition of 1,3-dichlorobenzene by a *Pseudomonas* sp or by a mixed culture of soil bacteria yielded dichlorophenols as products (9). It has been suggested that the three dichlorobenzene isomers may undergo slow biodegradation in natural water (10) [R10].

BIOCONCENTRATION Mean 1,3-dichlorobenzene BCF values of 420–740 were experimentally determined for rainbow trout exposed up to 119 days in laboratory aquariums (1). A whole-body BCF of 89 was determined for bluegill sunfish exposed to 1,3-dichlorobenzene over a 28-day period in a continuous-flow system (2). Based on a water solubility of 133 ppm at 25°C (3) and a log K_{ow} of 3.60 (4), BCF values of 39 and 320 are estimated, respectively, from recommended equations (5,SRC) [R11].

ATMOSPHERIC STANDARDS This action promulgates standards of performance for equipment leaks of volatile organic compounds (VOC) in the synthetic organic chemical manufacturing industry (SOCMI). The intended effect of these standards is to require all newly constructed, modified, and reconstructed SOCMI process units to use the best demonstrated system of continuous emission reduction for equipment leaks of VOC, considering costs, non-air-quality health, and environmental impact and energy requirements. 1,3-Dichlorobenzene is produced, as an intermediate or final product, by process units covered under this subpart [R13].

CERCLA REPORTABLE QUANTITIES Persons in charge of vessels or facilities are required to notify the National Response Center (NRC) immediately, when there is a release of this designated hazardous substance, in an amount equal to or greater than its reportable quantity of 100 lb or 45.4 kg. The toll-free telephone number of the NRC is (800) 424–8802; in the Washington metropolitan area (202) 426–2675. The rule for determining when notifica-

tion is required is stated in 40 CFR 302.6 (section IV. D. 3.b) [R14].

TSCA REQUIREMENTS Pursuant to section 8 (d) of TSCA, EPA promulgated a model health and safety data reporting rule. The section 8 (d) model rule requires manufacturers, importers, and processors of listed chemical substances and mixtures to submit to EPA copies and lists of unpublished health and safety studies. 1,3-Dichlorobenzene is included on this list [R15].

RCRA REQUIREMENTS As stipulated in 40 CFR 261.33, when 1,3-dichlorobenzene, as a commercial chemical product or manufacturing chemical intermediate or an off-specification commercial chemical product or a manufacturing chemical intermediate, becomes a waste, it must be managed according to federal or state hazardous waste regulations. Also defined as a hazardous waste is any residue, contaminated soil, water, or other debris resulting from the cleanup of a spill, into water or on dry land, of this waste. Generators of small quantities of this waste may qualify for partial exclusion from hazardous waste regulations (see 40 CFR 261.5) [R16].

DOT *Fire:* Small Fires: Dry Chemical, CO_2, halon, water spray, or standard foam. Large Fires: Water spray, fog, or standard foam is recommended. Move container from fire area if you can do so with out risk. Fight fire from maximum distance. Stay away from ends of tanks. Dike fire control water for later disposal; do not scatter the material [R3].

Spill or Leak: Do not touch spilled material; stop leak if you can do so without risk. Small Spills: Take up with sand or other noncombustible absorbent material, and place into containers for later disposal. Large Spills: Dike far ahead of spill for later disposal [R3].

First Aid: Move victim to fresh air; call emergency medical care. If not breathing, give artificial respiration. If breathing is difficult, give oxygen. In case of contact with material, immediately flush skin or eyes with running water for at least 15 minutes. Remove and isolate contaminated clothing and at site. Effects should disappear after individual has been exposed to fresh air for approximately 10 minutes. Keep victim quiet and maintain normal body temperature. Effects may be delayed; keep victim under observation [R3].

Fire or Explosion: Some of these materials may burn, but none of them ignites readily [R3].

Emergency Action: Keep unnecessary people away; isolate hazard area and deny entry. Stay upwind; keep out of low areas. Ventilate closed spaces before entering them. Wear positive-pressure breathing apparatus and special protective clothing. For emergency assistance call CHEMTREC (800) 424–9300 [R3].

Health Hazards: Inhalation of vapor or dust is extremely irritating. May cause burning of eyes and flow of tears. May cause coughing, difficult breathing, and nausea. Brief-exposure effects last only a few minutes. Exposure in an enclosed area may be very harmful. Runoff from fire-control or dilution water may cause pollution [R3].

REFERENCES

1. Sax, N. I., and R. J. Lewis, Sr. (eds.). *Hawley's Condensed Chemical Dictionary.* 11th ed. New York: Van Nostrand Reinhold Co., 1987. 376.

2. *Kirk–Othmer Encyclopedia of Chemical Technology.* 3rd ed., Volumes 1–26. New York, NY: John Wiley and Sons, 1978–1984, p. 5 (79) 865.

3. Department of Transportation. *Emergency Response Guidebook 1987.* DOT P 5800.4. Washington, DC: U.S. Government Printing Office, 1987, p. G–55.

4. Sax, N. I. *Dangerous Properties of Industrial Materials.* 5th ed. New York: Van Nostrand Reinhold, 1979. 557.

5. Mikatavage, M., et al., *Am Ind Hyg Assoc J*, 45 (9): 617–21 (1984).

6. Randolph, S. A., M. R. Zavon. *J Occup Med* 29: 237–42 (1987).

7. Association of American Railroads. *Emergency Handling of Hazardous Materials in Surface Transportation.* Washington, DC: Assoc. of American Railroads, Hazardous Materials Systems (BOE), 1987. 240.

8. USEPA. *Health Assessment Document* p. 6–4 (1985) EPA 600/8–84–015.

9. Kimura, R., et al. *J Pharm Dyn* 7: 234–45 (1984).

10. (1) Tabak, H. H., et al., *J Water Pollut Control Fed* 53: 1503 (1981). (2) Kincannon, D. F., et al. *J Water Pollut Control Fed* 55: 157 (1983). (3) Kuhn, E. P., et al. *Environ Sci Technol* 19: 961 (1985). (4) Bouwer, E. J., *Environ Prog* 4: 43 (1985). (5) Bouwer, E. J., P. L. McCarty. *Ground Water* 22: 433 (1984). (6) Wilson, J. T., J. F. McNabb, "Bio Transformation of Organic Pollution in Groundwater EOS Transactions," Amer Geophys Union 64: 505–7 (1983). (7) Oliver, B. G., K. D. Nicol. *Environ Sci Technol* 17: 505 (1983). (8) Haider, K., et al., *Arch Microbiol* 96: 183 (1974). (9) Ballschmiter, K., et al. *Angew Chemical Intermediate* Ed English 16: 645 (1977). (10) Callahan, M. A., et al. *Water–Related Environ Fate of 129 Priority Pollut.* Vol. II USEPA–440/4–79–029B (1979A).

11. (1) Oliver, B.G.,and A.J. Niimi, *Environ Sci Technol* 17: 287 (1983). (2) Barrows, M. E., et al., pp. 379–92 in *Dyn Exposure Hazard Assess Toxic Chemical*, Ann Arbor, MI: Ann Arbor Sci (1980). (3) Banerjee, S., et al., *Environ Sci Technol* 14: 1227 (1980). (4) Hansch, C., and A. J. Leo. *Medchem Project* Issue No. 26, Claremont CA: Pomona College (1985). (5) Lyman, W. J., et al., *Handbook of Chemical Property Estimation Methods, Environ Behavior of Organic Compounds.* New York: McGraw-Hill, pp. 5–4, 5–10 (1982).

12. International Labour Office. *Encyclopedia of Occupational Health and Safety.* Vols. I and II. Geneva, Switzerland: International Labour Office, 1983. 459.

13. 40 CFR 60.489 (7/1/87).

14. 50 FR 13456 (4/4/85).

15. 40 CFR 716.120 (7/1/87).

16. 53 FR 13382 (4/22/88).

■ m-DICHLOROBENZENE

CAS # 25321–22–6

DOT # 1591

SIC CODE 2821

SYNONYMS 1,3-dichlorobenzene • metadichlorobenzene

MF $C_6H_4Cl_2$

MW 147.00

STANDARD CODES EPA 311; ICC, USCG—no; IATA—other restricted articles, class A, no label required, no limit passenger or cargo; NFPA 2, 2, 0; Superfund designated (hazardous substances) list.

PERSISTENCE Will undergo biochemical, chemical, and photochemical attack to form 2,5-dichlorophenol, dichloroquinol, and conjugates [R1].

MAJOR SPECIES THREATENED All animal life.

INHALATION LIMIT 300 mg/m³.

DIRECT CONTACT Eye, respiratory tract, skin.

GENERAL SENSATION Pleasant aromatic odor. Affects central nervous system; long exposures: central nervous system depressant. Chronic toxicity TLV 50 ppm. Good warning properties. Ortho more toxic than para. Can affect liver and kidneys. Ortho—strong irritant, para—allergen.

PERSONAL SAFETY Maximum 50 ppm. At spill sites of high vapor densities, eye and respiratory equipment and protective clothing are advisable. Use full face mask with self-contained breathing apparatus.

ACUTE HAZARD LEVEL Toxic via ingestion or inhalation. Strong irritant or allergen. Emits toxic vapors when heated to decomposition. Toxic to aquatic life.

CHRONIC HAZARD LEVEL A rat fed 0.003 mg/kg daily for 5 months showed no significant effects. Chronic irritant (ortho)

and allergen (para). Toxic via ingestion and inhalation on chronic basis.

DEGREE OF HAZARD TO PUBLIC HEALTH Ortho more toxic than para. Both toxic via ingestion or inhalation. Ortho strong irritant. Para is an allergen. Both emit toxic vapors when heated to decomposition. Para form shows potential mutagenic properties. Causes liver and kidney damage in humans.

OCCUPATIONAL RECOMMENDATIONS None listed.

ACTION LEVEL Notify fire and air authority. Enter from upwind. Remove ignition sources.

ON-SITE RESTORATION Pump or vacuum from bottom. Carbon or peat may be used on soluble portion. Seek professional environmental engineering assistance through EPA's Environmental Response Team (ERT), Edison, NJ, 24-hour no. 908–548–8730.

BEACH OR SHORE RESTORATION Do not burn.

AVAILABILITY OF COUNTERMEASURE MATERIAL Pumps—fire department; vacuum—swimming-pool suppliers; peat—nurseries, floral shops; carbon—water treatment plants, sugar refineries.

DISPOSAL (1) Pour onto sodium bicarbonate or a sand-soda-ash mixture (90/10). Mix in heavy paper cartons and burn in incinerator. May augment fire with wood or paper. (2) Burn packages of no. 1 in incinerator with afterburner and alkaline scrubber. (3) Dissolve in flammable solvent and burn in incinerator of no. 1.

DISPOSAL NOTIFICATION Contact air and fire authorities.

INDUSTRIAL FOULING POTENTIAL May reduce heat transfer or cause hot spots and scaling.

MAJOR WATER USE THREATENED Fisheries, potable supply, recreation.

PROBABLE LOCATION AND STATE Colorless liquid; will sink to bottom and stay there. Very little will dissolve.

WATER CHEMISTRY Subject to biochemical, photochemical, and chemical attack, but chlorinated by-products are quite persistent and toxic. The latter include dichlorophenol and dichloroquinol.

COLOR IN WATER Colorless.

GENERAL RESPONSE Avoid contact with liquid. Keep people away. Wear goggles and self-contained breathing apparatus. Stop discharge if possible. Call fire department. Isolate and remove discharged material. Notify local health and pollution control agencies.

FIRE RESPONSE Combustible. Poisonous gases are produced in fire. Wear goggles and self-contained breathing apparatus. Extinguish with water, dry chemical, foam, or carbon dioxide. Cool exposed containers with water.

EXPOSURE RESPONSE Call for medical aid. Liquid is irritating to skin and eyes. Harmful if swallowed. Remove contaminated clothing and shoes. Flush affected areas with plenty of water. If swallowed and victim is conscious, have victim drink water or milk, and have victim induce vomiting. If swallowed and victim is unconscious or having convulsions, do nothing except keep victim warm.

RESPONSE TO DISCHARGE Issue warning—water contaminant. Should be removed. Provide chemical and physical treatment.

WATER RESPONSE Harmful to aquatic life in very low concentrations. May be dangerous if it enters water intakes. Notify local health and pollution control officials. Notify operators of nearby water intakes.

EXPOSURE SYMPTOMS Inhalation: Causes headache, drowsiness, unsteadiness. Irritating to mucous membranes. Eyes: Severe irritation. Skin: Severe irritation. Ingestion: Irritation of gastric mucosa, nausea, vomiting, diarrhea, abdominal cramps, and cyanosis.

EXPOSURE TREATMENT Get medical aid. Inhalation: Remove from exposure. Keep quiet and warm. Eyes: Rinse with running water for 15 to 20 minutes. Skin: Wash with soap and water. Ingestion: Wash mouth, give emetic.

REFERENCES

1. Midwest Research Institute and RVR Consultants, *Production, Distribution, Use, and Environmental Impact Potential of Selected Pesticides*, TIS PB–238 795, March 15,1974.

2. EPA *OHMTADS* database.

3. DOT *CHRIS* database.

■ 3,3'-DICHLOROBENZIDINE DIHYDROCHLORIDE

CAS # 612–83–9

SYNONYMS (1,1'-biphenyl)-4,4'-diamine, 3,3'-dichloro-, dihydrochloride • benzidine, 3,3'-dichloro-, dihydrochloride

MF $C_{12}H_{10}Cl_2N_2$ • 2ClH

MW 326.06

COLOR AND FORM Needles; white crystals; white to light-gray powder.

ODOR Mild odor.

MELTING POINT 132–133°C.

SOLUBILITY 4 mg/L water at 22°C; readily soluble in alcohol; 3.99 ppm in water at pH 6.9 and 22°C; slightly soluble in dilute hydrochloric acid.

PROTECTIVE EQUIPMENT Full-body protective clothing and gloves should be employed in handling operations. Full-face supplied-air respirators of continuous-flow or pressure-demand type should also be used. [R8, 317].

PREVENTATIVE MEASURES Regulations in USA concerning 3,3'-dichlorobenzidine and its salts designate strict procedures to avoid worker contact: Mixtures containing 1.0% or more 3,3'-dichlorobenzidine and its salts must be maintained in isolated or closed systems. Employees must observe special personal hygiene rules, and certain procedures must be followed for movement of the material and in case of accidental spills and emergencies [R1].

CLEANUP Use of sodium hypochlorite bleach solution to decontaminate 3,3'-dichlorobenzidine (DCB) was partially effective. An aqueous solution of 5% tetrapotassium pyrophosphate and 10% sodium ethyl hexyl sulfate when blended in a jet sprayer effectively removed DCB from the work area (90–99% reduction). Once removed from the worksite and collected in a central location, it was then determined that the diazotization reaction (the addition of SO₄, ice, and NaNO₃) occurred to eliminate any detectable dcb from the washings (3,3′-dichlorobenzidine) [R2]. A high-efficiency particulate arrestor (HEPA) or charcoal filters can be used to minimize amount of carcinogen in exhausted-air-ventilated safety cabinets, lab hoods, glove boxes, or animal rooms. Filter housing that is designed so that used filters can be transferred into plastic bag without contaminating maintenance staff is avail commercially. Filters should be placed in plastic bags immediately after removal. The plastic bag should be sealed immediately. The sealed bag should be labelled properly. Waste liquids should be placed or collected in proper containers for disposal. The lid should be secured and the bottles properly labelled. Once filled, bottles should be placed in plastic bag, so that outer surface is not contaminated. The plastic bag should also be sealed and labelled. Broken glassware should be decontaminated by solvent extraction, by chemical destruction, or in specially designed incinerators. [R9, 15].

DISPOSAL At the time of review, criteria for land treatment or burial (sanitary landfill) disposal practices are subject to significant revision. Prior to implementing land disposal of waste residue (including waste sludge), consult with environmental regulatory agencies for guidance on acceptable disposal practices [R3].

COMMON USES Intermediate for organic pigments (15 organic dyes and pigments); 3,3′-dichlorobenzidine or its salts has also been reported to be used in a color test for presence of gold [R1].

OCCUPATIONAL RECOMMENDATIONS None listed.

THRESHOLD LIMIT A2, skin; A2 = suspect-

ed human carcinogen (1976) (3,3′-dichlorobenzidine) [R4].

NON-HUMAN TOXICITY LD₅₀ rat (Sprague–Dawley) oral 3.82 g/kg [R5].

BIOLOGICAL HALF-LIFE Male wistar rats and male beagles given 0.2 mg/kg body wt ¹⁴C-dichlorobenzidine iv (dissolved in 0.5% Tween-20 in water) displayed multiphasic blood clearances. The final phase, predominant by 24 hr after treatment, had a half-life of 68 hr in rats and 86 hr in dogs. [R10 (1982)].

BIODEGRADATION It appeared recalcitrant to degradation by naturally occurring aquatic microbial communities with only a minor loss of chemical detected over a 30-day incubation period [R7].

BIOCONCENTRATION The bioconcentration, elimination, and metabolism of 3,3′-dichlorobenzidine (DCB), an industrial agent and suspected human toxicity carcinogen, were investigated in bluegill sunfish (*Lepomis macrochirus*). ¹⁴C-DCB was rapidly accumulated by the fish from water containing 5 ppb or 0.1 ppm of the chemical. Based on total ¹⁴C residues, bioconcentration factors of 495–507 were observed in the whole fish with equilibria achieved in 96–168 hr. The ¹⁴C residues were distributed in both the edible and nonedible portions. ¹⁴C-DCB or its metabolites were not completely eliminated upon transfer of the fish to water free of dichlorobenzidine. The only metabolite detected in the fish was an acid-labile conjugate of DCB, which appears to be an N-glucuronide [R6].

REFERENCES
1. IARC. *Monographs on the Evaluation of the Carcinogenic Risk of Chemicals to Man.* Geneva: World Health Organization, International Agency for Research on Cancer, 1972–present (multivolume work), p. V29 242 (1982).

2. Hackman, R. J., T. Rust. "Removal and decontamination of residual 3,3′-dichlorobenzidine". *Am Ind Hyg Assoc J* 42 (5): 341–7 (1981).

3. SRP.

4. American Conference of Governmental

Industrial Hygienists. *Threshold Limit Values for Chemical Substances and Physical Agents and Biological Exposure Indices for 1994–1995.* Cincinnati, OH: ACGIH, 1994. 18.

5. USEPA. *Ambient Water Quality Criteria Doc; Dichlorobenzidine* p. C–9 (1980) EPA 440/5–80–040.

6. Appleton, H. T., H. C. Sikka. *Environ Sci Technol* 14 (1): 50–4 (1980).

7. Sikka, H. C., et al., *Fate of 3,3'-Dichlorobenzidine in Aquatic Environments* (1978) EPA 600/3–78/068.

8. Sittig, M. *Handbook of Toxic and Hazardous Chemicals and Carcinogens, 1985.* 2nd ed. Park Ridge, NJ: Noyes Data Corporation, 1985.

9. Montesano, R., H. Bartsch, E. Boyland, G. Della Porta, L. Fishbein, R. A. Griesemer, A.B. Swan, L. Tomatis, and W. Davis (eds.). *Handling Chemical Carcinogens in the Laboratory: Problems of Safety.* IARC Scientific Publications No. 33. Lyon, France: International Agency for Research on Cancer, 1979.

10. IARC. *Monographs on the Evaluation of the Carcinogenic Risk of Chemicals to Man.* Geneva: World Health Organization, International Agency for Research on Cancer, 1972–present (multivolume work), p. V29 249.

■ 3,4-DICHLORO-1-BUTENE

CAS # 760–23–6

SYNONYMS 1,2-dichloro-3-butene • 1-butene, 3,4-dichloro

MF $C_4H_6Cl_2$

MW 125.00

COLOR AND FORM Colorless liquid.

DENSITY 1.153 at 25°C.

SURFACE TENSION 0.044183 Newtons/m at 212K.

VISCOSITY 0.0050666 pascal-sec.

BOILING POINT 118.6°C.

MELTING POINT −61°C.

HEAT OF COMBUSTION −2.2×10⁹ J/kmol.

HEAT OF VAPORIZATION 4.3473×10⁷ J/kmol at 212K.

FIREFIGHTING PROCEDURES Use water in flooding quantities as fog. Cool all affected containers with flooding quantities of water. Apply water from as far a distance as possible. Solid streams of water may be ineffective. Use foam, dry chemical, or carbon dioxide. Use water spray to knock down vapors. (dichlorobutene) [R3].

SOLUBILITY 420 mg/L of H_2O; soluble in polar solvents; soluble in ethanol, ether, and benzene.

VAPOR PRESSURE 21.85 mm Hg at 25°C.

STABILITY AND SHELF LIFE Volatile liquid [R4].

OTHER CHEMICAL/PHYSICAL PROPERTIES Flash point = 83°F (28°C); liquid molar volume = 0.108919 m³/kmol; heat of formation = −6.44×10⁷ J/kmol; flash point = 301.15K; flammability limits = 2.4 to 13.3 volume percent.

PROTECTIVE EQUIPMENT Wear appropriate chemical protective gloves, boots, and goggles. (dichlorobutene) [R3].

PREVENTATIVE MEASURES If material not on fire and not involved in fire: Keep sparks, flames, and other sources of ignition away. Keep material out of water sources and sewers. Build dikes to contain flow as necessary. Use water spray to knock down vapors (dichlorobutene) [R3]. Personnel protection: Avoid breathing vapors. Keep upwind. Avoid bodily contact with the material. Do not handle broken packages unless wearing appropriate personal protective equipment. Wash away any material which may have contacted the body with copious amounts of water or soap and water. Wear positive-pressure self-contained breathing apparatus when fighting fires involving this material. If contact with the material anticipated, wear appropriate chemical protective clothing. (dichlorobutene) [R3].

DISPOSAL Wastewater from dichlorobutene manufacture and isomerization and waste brine from 3,4-dichlorobutene-1 dehydrogenation to chloroprene are mixed to form an aqueous solution containing approximately 1–5% NaCl. The solution, at pH less than 6, is extracted with liquid hydrocarbon, and the extract is incinerated. The treated water, which is not toxic to fish, is discharged [R5].

COMMON USES Chemical intermediate for chloroprene [R1]; chemical intermediate in the production of 1,4-butanediol in Japan [R2]; from 1951 to 1983 DuPont operated a butadiene chlorination process. The intermediate 1,4-dichloro-2-butene (obtained from 3,4-dichloro-1-butene) was converted to 3-hexenedinitrile and then hydrogenated to adiponitrile [R2].

OCCUPATIONAL RECOMMENDATIONS None listed.

NON-HUMAN TOXICITY LC$_{50}$ *Pimephales promelas* (fathead minnow) 490 mg/L/96 hr (95% confidence limit 8.51–10.2 mg/L), flow-through bioassay with measured concentrations, 25.1°C, dissolved oxygen 6.6 mg/L, hardness 44.1 mg/L calcium carbonate, alkalinity 42.3 mg/L calcium carbonate, and pH 7.52 [R6]. LC$_{50}$ *Pimephales promelas* (fathead minnow) 1,080 mg/L/96 hr (95% confidence limit not rel.), flow-through bioassay with measured concentrations, 25.1°C, dissolved oxygen 6.8 mg/L, hardness 46.5 mg/L calcium carbonate, alkalinity 43.5 mg/L calcium carbonate, and pH 7.7 [R7].

BIOCONCENTRATION In 8-week bioaccumulation studies using the Japanese MITI protocol (where the test fish are carp (*Cyprinus carpio*)), BCFs ranging from 0.59 to 13.34 have been observed for 3,4-dichloro-1-butene (1). These BCF values suggest that bioconcentration in aquatic organisms is not an important environmental fate process (SRC) [R8].

REFERENCES

1. *Kirk–Othmer Encyclopedia of Chemical Technology*. 4th ed. Volumes 1: New York, NY. John Wiley and Sons, 1991–present, p. V6 74.

2. Gerhartz, W. (exec ed.). *Ullmann's Encyclopedia of Industrial Chemistry.* 5th ed. Vol A1. Deerfield Beach, FL: VCH Publishers, 1985 to present, p. VA4 459.

3. Association of American Railroads. *Emergency Handling of Hazardous Materials in Surface Transportation.* Washington, DC: Association of American Railroads, Bureau of Explosives, 1992. 330.

4. Grant, W. M. *Toxicology of the Eye.* 3rd ed. Springfield, IL: Charles C. Thomas Publisher, 1986. 322.

5. Harris, A. T., et al. US Patent 4221659 9/9/80 (Du Pont de Nemours, EI, and Co).

6. Geiger D.L., S.H. Poirier L.T. Brooke, D.J. Call (eds.). *Acute Toxicities of Organic Chemicals to Fathead Minnows (Pimephales promelas).* Vol. II. Superior, Wisconsin: University of Wisconsin–Superior, 1985. 53.

7. Geiger D. L., D.J. Call, L.T. Brooke (eds). *Acute Toxicities of Organic Chemicals to Fathead Minnows (Pimephales promelas).* Vol. IV. Superior, Wisconsin: University of Wisconsin–Superior, 1988. 51.

8. Chemicals Inspection and Testing Institute. *Biodegradation and Bioaccumulation Data of Existing Chemicals Based on the CSCL Japan.* Japan Chemical Industry Ecology—Toxicology and Information Center. ISBN 4– 89074–101 –1, p 2–26 (1992).

■ 1,3-DICHLORO-5,5-DIMETHYLHYDANTOIN

CAS # 118–52–5

SYNONYMS dactin • daktin • dantoin • dcdmh • ddh • dichlorantin • dichloro-5,5-dimethylhydantoin • 1,3-dichloro-5,5-dimethyl-2,4-imidazolidinedione • halane • hydan • hydantoin, dichlorodimethyl • hydantoin, 1,3-dichloro-5,5-dimethyl •

2,4-imidazolidinedione, 1,3-dichloro-5,5-dimethyl • NCI-c03054 • omchlor

MF $C_5H_6Cl_2N_2O_2$

MW 197.03

pH Of aqueous solution, about 4.4.

COLOR AND FORM Four-sided, pointed prisms from chloroform; white powder.

ODOR Mild chlorine odor.

DENSITY 1.5 at 20°C.

MELTING POINT 132°C.

SOLUBILITY Solubility in water: 0.21% at 25°C, 0.60% at 60°C; solubility (at 25°C): 12.5% in carbon tetrachloride, 14% in chloroform, 30% in methylene chloride, 32.0% in ethylene dichloride, 17% in sym-tetrachlorethane, 9.2% in benzene; water solubility = 0.5 g/L at 20°C; 1.3 g/L at 40°C.

CORROSIVITY Presumably less corrosive than hypochlorite solutions with same concentration of available chlorine.

STABILITY AND SHELF LIFE Dry crystals [combined available chlorine 77.6% (theory)] may be stored without much loss of available chlorine [R5]. After 14 weeks at 60°C loss was 1.5% chlorine compared with a loss of 37.5% suffered by 70% calcium hypochlorite [R5].

OTHER CHEMICAL/PHYSICAL PROPERTIES Sublimes at 100°C; turns brown and burns at 212°C (after melting at 132°C); on contact with water and especially hot water, hypochlorous acid is liberated. At pH 9 nitrogen chloride is formed. At pH 9 it decomposes completely.

PROTECTIVE EQUIPMENT Wear appropriate equipment to prevent repeated or prolonged skin contact [R8, 86]. Wear eye protection to prevent any possibility of skin contact [R8, 86]. Remove clothing; eyewash [R8, 86]. Recommendations for respirator selection. Max concn for use: 2 mg/m³. Respirator Class(es): Any supplied-air respirator. Any self-contained breathing apparatus [R8, 86]. Recommendations for respirator selection. Max concn for use: 5 mg/m³. Respirator Class(es): Any supplied-air respirator op-erated in a continuous flow mode. Any self-contained breathing apparatus with a full facepiece. Any supplied-air respirator with a full facepiece [R8, 86]. Recommendations for respirator selection. Condition: Emergency or planned entry into unknown concn or IDLH conditions: Respirator Class(es): Any self-contained breathing apparatus that has a full facepiece and is operated in a pressure-demand or other positive-pressure mode. Any supplied-air respirator with a full facepiece and operated in pressure-demand or other positive-pressure mode in combination with an auxiliary self-contained breathing apparatus operated in pressure-demand or other positive-pressure mode [R8, 86]. Recommendations for respirator selection. Condition: Escape from suddenly occurring respiratory hazards: Respirator Class(es): Any air-purifying, full facepiece respirator (gas mask) with a chin-style, front- or back-mounted canister providing protection against the cmpd of concern and having a high-efficiency particulate filter. Any appropriate escape-type, self-contained breathing apparatus [R8, 86]. Wear butyl rubber gloves, protective clothing, self-contained breathing apparatus, protective shoes. [R9].

PREVENTATIVE MEASURES Contact lenses should not be worn when working with this chemical [R8, 66]. SRP: The scientific literature for the use of contact lenses in industry is conflicting. The benefit or detrimental effects of wearing contact lenses depend not only upon the substance, but also on factors including the form of the substance, characteristics and duration of the exposure, the uses of other eye protection equipment, and the hygiene of the lenses. However, there may be individual substances whose irritating or corrosive properties are such that the wearing of contact lenses would be harmful to the eye. In those specific cases, contact lenses should not be worn. In any event, the usual eye protection equipment should be worn even when contact lenses are in place. Work clothing should be changed daily if it is reasonably probable that the clothing may be contaminated [R10]. Workers should wash promptly when skin becomes contaminated [R8, 86]. Remove clothing promptly if it is non-impervious and becomes contaminated. [R8, 86].

CLEANUP Absorb the spills with paper towels or like materials. Place in hood to evaporate. Dispose by burning the towel [R4].

DISPOSAL SRP: At the time of review, criteria for land treatment or burial (sanitary landfill) disposal practices are subject to significant revision. Prior to implementing land disposal of waste residue (including waste sludge), consult with environmental regulatory agencies for guidance on acceptable disposal practices.

COMMON USES Chlorinating agent, disinfectant, industrial deodorant; in water treatment; active ingredient of powder laundry bleaches [R1]; intermediate for amino acids, drugs, insecticides; stabilizer for vinyl chloride polymers; polymerization catalyst [R1]; dichloro-5,5-dimethylhydantoin was once used as a chemical-warfare decontaminating agent [R3].

OCCUPATIONAL RECOMMENDATIONS None listed.

OSHA Table Z-1, 8-hour time-weighted average: 0.2 mg/m³ [R6].

NIOSH 10-hr time-weighted average: 0.2 mg/m³ [R8, 86]. 15-min short-term exposure limit: 0.4 mg/m³. [R8, 86].

THRESHOLD LIMIT 8-hr time-weighted average (TWA) 0.2 mg/m³ ; short-term exposure limit (STEL) 0.4 mg/m³ (1976) [R7].

FIFRA REQUIREMENTS In 1988, Congress amended FIFRA to strengthen and accelerate EPA's reregistration program. The nine-year reregistration scheme mandated by "FIFRA 88" applies to each registered pesticide product containing an active ingredient initially registered before November 1, 1984. Pesticides for which EPA had not issued Registration Standards prior to the effective date of FIFRA '88 were divided into three lists based upon their potential for exposure and other factors, with List B being of highest concern and D of least. List: C; Case: Dihalodialkylhydantoins; Case No.: 3055;

Pesticide type: Fungicide, herbicide, antimicrobial; Case Status: Awaiting Data/Data in Review: OPP awaits data from the pesticide's producer(s) regarding its human health and/or environmental effects, or OPP has received and is reviewing such data in order to reach a decision about the pesticide's eligibility for reregistration. Active Ingredient (AI): Dichloro-5,5-dimethylhydantoin; AI Status: The producer(s) of the pesticide has made commitments to conduct the studies and pay the fees required for reregistration, and is meeting those commitments in a timely manner. [R11].

FIRE POTENTIAL It is combustible, with evolution of chlorine at 210°C [R2].

REFERENCES

1. Budavari, S. (ed.). *The Merck Index— Encyclopedia of Chemicals, Drugs, and Biologicals.* Rahway, NJ: Merck and Co., Inc., 1989. 483.

2. American Conference of Governmental Industrial Hygienists, Inc. *Documentation of the Threshold Limit Values and Biological Exposure Indices.* 6th ed. Volumes I, II, III. Cincinnati, OH: ACGIH, 1991. 423.

3. *Kirk–Othmer Encyclopedia of Chemical Technology.* 4th ed. Volume 1: New York, NY. John Wiley and Sons, 1991–present, p. V5 923.

4. ITII. *Toxic and Hazardous Industrial Chemicals Safety Manual.* Tokyo, Japan: The International Technical Information Institute, 1988. 164.

5. *The Merck Index.* 9th ed. Rahway, NJ: Merck and Co., Inc., 1976. 404.

6. 29 CFR 1910.1000 (58 FR 35338 (6/30/93)).

7. American Conference of Governmental Industrial Hygienists. *Threshold Limit Values for Chemical Substances and Physical Agents and Biological Exposure Indices for 1994–1995.* Cincinnati, OH: ACGIH, 1994. 18.

8. NIOSH. *NIOSH Pocket Guide to Chemical Hazards.* DHHS (NIOSH) Publication

No. 90–117. Washington, DC: U.S. Government Printing Office, June 1990.

9. ITII. *Toxic and Hazardous Industrial Chemicals Safety Manual.* Tokyo, Japan: The International Technical Information Institute, 1988. 164.

10. NIOSH. *NIOSH Pocket Guide to Chemical Hazards.* DHHS (NIOSH) Publication No. 90–117. Washington, DC: U.S. Government Printing Office, June 1990, p. 86.

11. USEPA/OPP. *Status of Pesticides in Reregistration and Special Review.* p.191 (Mar, 1992) EPA 700-R-92-004.

■ 2,3-DICHLOROPHENOL

CAS # 576–24–9

SYNONYMS phenol, 2,3-dichloro

MF $C_6H_4Cl_2O$

MW 163.00

COLOR AND FORM Crystals from petroleum ether and benzene.

ODOR THRESHOLD In water 30 µg/L [R4].

TASTE Taste threshold in water: 0.04 µg/L.

DISSOCIATION CONSTANTS 2,3-Dichlorophenol has a pK_a of 7.70.

MELTING POINT 57–59°C.

EXPLOSIVE LIMITS AND POTENTIAL During vacuum fractionation of the mixed dichlorophenols produced by partial hydrolysis of trichlorobenzene, rapid admission of air to the receiver caused the column contents to be forced down into the boiler at 210°C, and a violent explosion ensued. (dichlorophenol mixed isomers) [R3].

SOLUBILITY Soluble in alcohol, ether, hot benzene, hot petroleum ether.

VAPOR PRESSURE 0.179 mm Hg at 25°C (calculated).

OCTANOL/WATER PARTITION COEFFICIENT Log K_{ow} of 2.84–3.15.

PROTECTIVE EQUIPMENT Wear boots, protective gloves, and goggles. (trichlorophenol) [R5].

PREVENTATIVE MEASURES SRP: Contaminated protective clothing should be segregated in such a manner so that there is no direct personal contact by personnel who handle, dispose of, or clean the clothing. Quality assurance to ascertain the completeness of the cleaning procedures should be implemented before the decontaminated protective clothing is returned for reuse by the workers.

CLEANUP Land Spill: Dig a pit, pond, lagoon, or holding area (SRP: If time permits, pits, ponds, lagoons, soak holes, or holding areas should be sealed with an impermeable flexible-membrane liner) to contain liquid or solid material. Cover solids with plastic sheet to prevent dissolving in rain or firefighting water (trichlorophenol) [R5]. Water Spill: Use natural deep-water pockets, excavated lagoons, or sandbag barriers to trap material at bottom. If dissolved, apply activated carbon at ten times the spilled amount in region of 10 ppm or greater concentration. Remove trapped material with suction hoses. Use mechanical dredges or lifts to remove immobilized masses of pollutants and precipitates (trichlorophenol) [R5]. Activated carbon is a good method for removing chlorophenols from water. Competitive adsorption occurs between chlorophenols and humic substances present in nearly all municipal water supplies. This competition decr the capacity of carbon for chlorophenols. (chlorophenols) [R6].

DISPOSAL SRP: At the time of review, criteria for land treatment or burial (sanitary landfill) disposal practices are subject to significant revision. Prior to implementing land disposal of waste residue (including waste sludge), consult with environmental regulatory agencies for guidance on acceptable disposal practices.

COMMON USES Research chemical [R1].

OCCUPATIONAL RECOMMENDATIONS None listed.

NON-HUMAN TOXICITY LD_{50} mouse (male CD-1 ICR) oral 2,585 mg/kg [R7]. LD_{50}

mouse (female CD-1 ICR) oral 2,376 mg/kg [R7].

BIODEGRADATION Approximately 100% of initially added 2,3-dichlorophenol had been degraded after 2–4 weeks of incubation in four freshwater-pond sediments (1); the chlorine at the 2-position was the most susceptible to the reductive dechlorination that occurred (1). After a lag period of approximately 2 weeks, 100% of the 2,3-dichlorophenol initially added to a freshwater-pond sediment (which had been contaminated with asphalt) was observed to degrade within two weeks (1). In anaerobic serum bottle tests using unacclimated sludge, 100% of added 2,3-dichlorophenol degraded within 6 wk, yielding 3-chlorophenol (2). In anaerobic serum bottle tests using sludge acclimated to 2-chlorophenol, only 4% of added 2,3-dichlorophenol degraded during a 30-day incubation (2); no degradation occurred over a 28-day period using sludge acclimated to 3-chlorophenol (2) [R8].

BIOCONCENTRATION Based on a measured log K_{ow} of 2.84–3.15 (1), the BCF of 2,3-dichlorophenol can be estimated to be 85–146 from a recommended regression-derived equation (2,SRC). This estimated BCF range suggests only a moderate potential for significant bioaccumulation in aquatic organisms (SRC) [R9].

ATMOSPHERIC STANDARDS This action promulgates standards of performance for equipment leaks of volatile organic compounds (VOC) in the synthetic organic chemical manufacturing industry (SOCMI). The intended effect of these standards is to require all newly constructed, modified, and reconstructed SOCMI process units to use the best demonstrated system of continuous emission reduction for equipment leaks of VOC, considering costs, non-air-quality health, and environmental impact and energy requirements. Chlorophenol is produced, as an intermediate or final product, by process units covered under this subpart. (chlorophenols) [R10].

DOT *Fire or Explosion:* Some of these materials may burn but none of them ignites readily (chlorophenol, solid) [R2].

Health Hazards: Poisonous if swallowed. Inhalation of dust poisonous. Fire may produce irritating or poisonous gases. Runoff from fire control or dilution water may cause pollution (chlorophenol, solid) [R2].

Fire: Small Fires: Dry chemical, CO_2, water spray, or standard foam. Large Fires: Water spray, fog, or standard foam is recommended. Move container from fire area if you can do so without risk (chlorophenol, solid) [R2].

Spill or Leak: Do not touch spilled material; stop leak if you can do so without risk. Small Spills: Take up with sand or other noncombustible absorbent material and place into containers for later disposal. Small Dry Spills: With clean shovel place material into clean, dry container, and cover; move container from spill area. Large Spills: Dike far ahead of liquid spill for later disposal (chlorophenol, solid) [R2].

First Aid: Move victim to fresh air; call for emergency medical care. Remove and isolate contaminated clothing and shoes at the site. In case of contact with material, immediately flush skin or eyes with running water for at least 15 minutes (chlorophenol, solid) [R2].

Emergency Action: Keep unnecessary people away; isolate hazard area and deny entry. Stay upwind; keep out of low areas. Self-contained breathing apparatus (SCBA) and structural firefighter's protective clothing will provide limited protection. If water pollution occurs, notify appropriate authorities. For emergency assistance call CHEMTREC (800) 424–9300 (chlorophenol, solid) [R2].

REFERENCES

1. SRI.

2. Department of Transportation. *Emergency Response Guidebook 1987.* DOT P 5800.4. Washington, DC: U.S. Government Printing Office, 1987, p. G–53.

3. Bretherick, L. *Handbook of Reactive Chemical Hazards.* 3rd ed. Boston, MA: Butterworths, 1985. 561.

4. Deitz, F., J. Traud. Gwf–Wasser/

Abwasser 199: 318 (1978) as cited in USEPA, *Ambient Water Quality Criteria Doc; Chlorinated Phenols* p. C–31 (1980) EPA 440/5–80–032.

5. Association of American Railroads. *Emergency Handling of Hazardous Materials in Surface Transportation*. Washington, DC: Assoc. of American Railroads, Hazardous Materials Systems (BOE), 1987. 694.

6. Murin, C.J., V. L. Snoeyink. *Environ Sci Technol* 13 (3): 305–11 (1979).

7. Borzelleca, J. F., et al. *Toxicol Lett* 29: 39–42 (1985).

8. (1) Rogers, J. E., D. D. Halet. *Am Chemical Soc Div Environ Chemical Preprints*, New Orleans, LA: 27: 699–701 (1987). (2) Boyd, S. A., D. R. Shelton. *Appl Environ Microbiol* 47: 272– (1984).

9. (1) Hansch, C., and A. J. Leo. *Medchem Project* Issue No 26. Claremont CA: Pomona College (1985). (2) Lyman, W. J., et al., *Handbook of Chemical Property Estimation Methods*. New York: McGraw-Hill, p. 5–4 (1982).

10. 40 CFR 60.489 (7/1/87).

■ 1,3-DICHLOROPROPANE

CAS # 142–28–9

DOT # 1279

SYNONYMS propane, 1,3-dichloro • trimethylene dichloride

MF $C_3H_6Cl_2$

MW 112.99

COLOR AND FORM Colorless liquid.

ODOR Sweet.

DENSITY 1.1876 at 20°C.

SURFACE TENSION 33.93 dynes/cm at 20°C.

BOILING POINT 120.4°C at 760 mm Hg.

MELTING POINT –99.5°C.

HEAT OF COMBUSTION –3,709 cal/g.

HEAT OF VAPORIZATION At boiling point 71.71 cal/g.

EXPLOSIVE LIMITS AND POTENTIAL Vapor may explode if ignited in an enclosed area [R3].

FLAMMABILITY LIMITS In air = 3.4%–14.5% (est) [R3].

TOXIC COMBUSTION PRODUCTS When heated to decomposition it emits highly toxic fumes of phosgene [R4].

SOLUBILITY Soluble in benzene, chloroform, alcohol, ether; sol: 2,870 mg/L.

VAPOR PRESSURE 18 torr.

OCTANOL/WATER PARTITION COEFFICIENT Log K_{ow} = 2.00.

OTHER CHEMICAL/PHYSICAL PROPERTIES IR: 1193 (Coblentz Society Spectral Collection) (propane, 2,2-dichloro); mass: 369 (*Atlas of Mass Spectral Data*, New York: John Wiley and Sons) (propane, 2,2-dichloro); miscible with water; Henry's Law constant: 9.76×10^{-4} atm-m³/mole.

PROTECTIVE EQUIPMENT Breakthrough times for dichloropropane on chlorinated polyethylene are less (usually significantly) than one hour as reported by two or more testers. (dichloropropane) [R5].

PREVENTATIVE MEASURES SRP: The scientific literature supports the wearing of contact lenses in industrial environments, as part of a program to protect the eye against chemical compounds and minerals causing eye irritation. However, there may be individual substances whose irritating or corrosive properties are such that the wearing of contact lenses would be harmful to the eye. In those specific cases contact lenses should not be worn.

COMMON USES Chemical intermediate for cyclopropane (anesthetic) [R1].

OCCUPATIONAL RECOMMENDATIONS None listed.

ECOTOXICITY LC_{50} *Poecilia reticulata* (guppy) 84 ppm/7 days [R6]; MATC *Pimephales promelas* 8–16 µg/L (est) [R7]; LC_{50} *Cyprinodon variegatus* (sheepshead minnow) 86,7000 µg/L 96 hr [R8]; EC_{50} *Selenastrum capricornutum* 72,200 µg/L/96

hr, toxic effect: cell numbers [R8]; LC_{50} *Daphnia magna* (cladoceran) 282,000 µg/L 96 hr [R8]; LC_{50} *Pimephales promelas* (fathead minnow) 131,000 µg/L 96 hr [R8]; LC_{50} *Mysidopsis bahia* (mysid shrimp) 10,300 µg/L 96 hr [R8].

BIODEGRADATION The BOD utilized by 1,3-dichloropropane in 5 days from a sewage seed inoculum was 16% of theoretical (1). No data could be found for biodegradation in environmental waters or soil [R9].

BIOCONCENTRATION Based on the experimental log octanol/water partition coefficient (2.00 (1)), one can estimate a log BCF of 1.06 (2). Therefore bioconcentration in fish would not be significant [R10].

CERCLA REPORTABLE QUANTITIES Persons in charge of vessels or facilities are required to notify the National Response Center (NRC) immediately when there is a release of this designated hazardous substance in an amount equal to or greater than its reportable quantity of 1,000 lb or 454 kg. The toll-free number of the NRC is (800) 424–8802; in the Washington DC metropolitan area (202) 426–2675. The rule for determining when notification is required is stated in 40 CFR 302.4 (section IV. D. 3.b) [R11].

TSCA REQUIREMENTS Pursuant to section 8 (d) of TSCA, EPA promulgated a model health and safety data reporting rule. The section 8 (d) model rule requires manufacturers, importers, and processors of listed chemical substances and mixtures to submit to EPA copies and lists of unpublished health and safety studies. Propane, 1,3-dichloro is included on this list [R12].

DOT *Fire or Explosion:* Flammable/combustible material; may be ignited by heat, sparks, or flames. Vapors may travel to a source of ignition, and flash back. Container may explode in heat of fire. Vapor explosion hazard indoors, outdoors, or in sewers. Runoff to sewer may create fire or explosion hazard (dichloropropane) [R2].

Health Hazards: May be poisonous if inhaled or absorbed through skin. Vapors may cause dizziness or suffocation. Contact may irritate or burn skin and eyes. Fire may produce irritating or poisonous gases. Runoff from fire control or dilution water may cause pollution (dichloropropane) [R2].

Emergency Action: Keep unnecessary people away; isolate hazard area and deny entry. Stay upwind; keep out of low areas. Positive-pressure self-contained breathing apparatus (SCBA) and structural firefighters' protective clothing will provide limited protection. Isolate for 1/2 mile in all directions if tank, rail car, or tank truck is involved in fire. Call CHEMTREC at 1–800–424–9300 for emergency assistance. If water pollution occurs, notify the appropriate authorities (dichloropropane) [R2].

Fire: Small Fires: Dry chemical, CO_2, water spray, or regular foam. Large Fires: Water spray, fog, or regular foam. Move container from fire area if you can do so without risk. Apply cooling water to sides of containers that are exposed to flames until well after fire is out. Stay away from ends of tanks. For massive fire in cargo area, use unmanned hose holder, or monitor nozzles; if this is impossible, withdraw from area, and let fire burn. Withdraw immediately in case of rising sound from venting safety device or any discoloration of tank due to fire (dichloropropane) [R2].

Spill or Leak: Shut off ignition sources; no flares, smoking, or flames in hazard area. Stop leak if you can do so without risk. Water spray may reduce vapor; but it may not prevent ignition in closed spaces. Small Spills: Take up with sand or other noncombustible absorbent material and place into containers for later disposal. Large Spills: Dike far ahead of liquid spill for later disposal (dichloropropane) [R2].

First Aid: Move victim to fresh air and call for emergency medical care; if not breathing, give artificial respiration; if breathing is difficult, give oxygen. In case of contact with material, immediately flush eyes with running water for at least 15 minutes. Wash skin with soap and water. Remove and isolate contaminated clothing and shoes at the site (dichloropropane) [R2].

FIRE POTENTIAL Flammable [R3].

GENERAL RESPONSE Stop discharge if possible. Keep people away. Shut off ignition sources and call fire department. Stay upwind and use water spray to "knock down" vapor. Avoid contact with liquid and vapor. Isolate and remove discharged material. Notify local health and pollution control agencies.

FIRE RESPONSE Flammable. Poisonous gases are produced in fire. Flashback along vapor trail may occur. Vapor may explode if ignited in an enclosed area. Wear goggles and self-contained breathing apparatus. Extinguish with foam, dry chemical, or carbon dioxide. Cool exposed containers with water.

EXPOSURE RESPONSE Call for medical aid. Vapor irritating to eyes, nose, and throat. Move to fresh air. If breathing has stopped, give artificial respiration. If breathing is difficult, give oxygen. Liquid irritating to skin and eyes. Harmful if swallowed. Remove contaminated clothing and shoes. Flush affected areas with plenty of water. If swallowed and victim is conscious, have victim drink water or milk.

RESPONSE TO DISCHARGE Issue warning—high flammability. Evacuate area.

WATER RESPONSE Effect of low concentrations on aquatic life is unknown. May be dangerous if it enters water intakes. Notify local health and wildlife officials. Notify operators of nearby water intakes.

EXPOSURE SYMPTOMS Inhalation: may cause some central nervous system depression. Eyes: may cause some pain and irritation. Skin: mild irritation.

EXPOSURE TREATMENT Call a doctor. Inhalation: remove to fresh air. If breathing has stopped, give artificial respiration. Eyes: flush with running water for 15 minutes. Skin: wash thoroughly with soap and water. Ingestion: gastric lavage or emesis and catharsis.

REFERENCES

1. SRI.

2. U.S. Department of Transportation. *Emergency Response Guidebook 1990.* DOT P 5800.5. Washington, DC: U.S. Government Printing Office, 1990, p. G–27.

3. U.S. Coast Guard, Department of Transportation. *CHRIS—Hazardous Chemical Data.* Volume II. Washington, DC: U.S. Government Printing Office, 1984–5.

4. Sax, N. I. *Dangerous Properties of Industrial Materials.* 6th ed. New York, NY: Van Nostrand Reinhold, 1984. 962.

5. ACGIH. *Guidelines Select of Chemical Protect Clothing.* Volume #1 Field Guide p. 48 (1983).

6. Verschueren, K. *Handbook of Environmental Data of Organic Chemicals.* 2nd ed. New York, NY: Van Nostrand Reinhold Co., 1983. 507.

7. Benoit, D. A., et al., *Environment and Pollution* 28: 189–97 (1982).

8. USEPA. *Ambient Water Quality Criteria Doc. Dichloropropanes and Dichloropropenes* p. B–6 (1980).

9. Bridie, A. L., et al., *Water Res* 13: 629–30 (1979).

10. (1) Hansch, C., and A. J. Leo. *Medchem Project* Pomona, CA (1985). (2) Lyman, W. J., et al., *Handbook of Chemical Property Estimation Methods.* New York: McGraw-Hill, pp. 5.1–5.30 (1982).

11. 40 CFR 302.4 (7/1/91).

12. 40 CFR 716.120 (7/1/91).

■ DICUMAROL

CAS # 66–76–2

SYNONYMS 2H-1-benzopyran-2-one, 3,3'-methylenebis (4-hydroxy • 3,3'-methyleen-bis (4-hydroxy-cumarine) (Dutch) • 3,3'-Methylen-bis (4-hydroxy-cumarin) (German) • 3,3'-methylene-bis (4-hydroxycoumarine) (French) • 3,3'-methylenebis (4-hydroxy-1,2-benzopyrone) • 3,3'-methylenebis (4-hydroxycoumarin) • 3,3'-metilen-bis (4-idrossi-cumarina) (Italian) • 4,4'-dihydroxy-3,3'-methylene bis coumarin • acadyl • acavyl •

antitrombosin • baracoumin • bhc • bis (4-hydroxycoumarin-3-yl) methane • bis-3,3'-(4-hydroxycoumarinyl) methane • bishydroxycoumarin • coumarin, 3,3'-methylenebis (4-hydroxy) • cuma • cumid • di-(4-hydroxy-3-coumarinyl) methane • di-4-hydroxy-3,3'-methylenedicoumarin • dicoumal • dicoumarin • dicoumarol • dicuman • dicumaol-r • dicumarine • dicumol • dufalone • kumoran • melitoxin • temparin • trombosan • 3,3'-methylenebis (4-hydroxy-2H-1-benzopyran-2-one)

MF $C_{19}H_{12}O_6$

MW 336.31

COLOR AND FORM White or creamy white, crystalline powder; minute crystals; needles.

ODOR Faint, pleasant odor.

TASTE Slightly bitter taste.

MELTING POINT 287–293°C.

SOLUBILITY Practically insoluble in water, alcohol, and ether. Insoluble in acetone. Soluble in aqueous alkaline solns, in pyridine, and similar organic bases. Slightly soluble in benzene and chloroform.

OCTANOL/WATER PARTITION COEFFICIENT 2.07.

OTHER CHEMICAL/PHYSICAL PROPERTIES Various esters of dicumarol (diacetate, dipivalate, and dinicotinate) were prepared, and their aqueous solubilities compared with that of dicumarol. The dinicotinate ester derivatives showed a five-fold greater solubility at pH 1–3 whereas the solubility of the diacetate ester was equal to that of the parent drug.

PROTECTIVE EQUIPMENT Adequate protective clothing should be worn at all times. In the lab this will consist of a lab coat, rubber or polyethylene gloves, and a NIOSH-approved respirator or respirator of a type applicable to the specific chemical being handled. (rodenticides) [R5, 1955].

PREVENTATIVE MEASURES Care should be taken not to contaminate foodstuffs; not to leave material within reach of children; to use prepared baits rather than scatter poison; to collect and destroy dead rodents; to bury baits and powder when operation is completed (coumarin derivatives) [R5, 561]. Wherever possible, toxic chemicals, concentrates and bait preparations should be handled in a fume cupboard. When bait mixing has been done in the field, operators should take care to remain sheltered from the wind (rodenticides) [R5, 1955]. Scrupulous personal hygiene must be adhered to when dealing with poisons. All cuts and abrasions on the hands and forearms must be covered with waterproof adhesive dressings before any operations are started. When the work is finished or when a break is taken in the middle of the day, protective clothing should be removed and hands washed thoroughly with soap and hot water. Contaminated protective clothing must not be taken into "clean" areas (rodenticides) [R5, 1955]. Smoking, eating, and drinking must be strictly prohibited in all rooms in which poisons are present (rodenticides) [R5, 1955]. In event of accidental poisoning in humans, it is important that proper medical help be enlisted at once. Local hospitals should be notified of the potential dangers that exist in places where rodenticides are present and be given precise details of the specific poisons that are used, with revelant information about antidotes, symptoms, etc. (rodenticides) [R5, 1955]. Normal first-aid facilities should be available and as many staff as possible should have proper first-aid training. (rodenticides) [R5, 1955].

COMMON USES Medication: anticoagulant [R1].

OCCUPATIONAL RECOMMENDATIONS None listed.

ECOTOXICITY LC_{50} Pimephales promelas (fathead minnow) 5.11 mg/L/96 hr (95% confidence limit 4.18–6.24 mg/L), flow-through bioassay with measured concentrations, 24.5°C, dissolved oxygen 7.2 mg/L, hardness 44.3 mg/L calcium carbonate, alkalinity 44.0 mg/L calcium carbonate, and pH 7.8 [R2].

BIOLOGICAL HALF-LIFE $T_{½}$ of dicumarol is

dose dependent, ranging from 10 hr at low dosage to 30 hr at high dosage. [R6, 1358].

BIOCONCENTRATION The log K_{ow} for dicumarol is 2.07 (1). Using this value of log K_{ow}, one can estimate a BCF of 22 using a recommended regression equation (2). This indicates that dicumarol will not bioconcentrate in aquatic organisms (SRC) [R3].

DOT *Health Hazards:* Poisonous; may be fatal if inhaled, swallowed, or absorbed through skin. Contact may cause burns to skin and eyes. Runoff from fire control or dilution water may cause pollution (coumarin derivative pesticides, liquid, flammable, toxic, NOS; coumarin derivative pesticides, liquid, toxic, flammable, NOS) [R4,p. G-28].

Fire or Explosion: Flammable/combustible material; may be ignited by heat, sparks, or flames. Vapors may travel to a source of ignition and flash back. Container may explode in heat of fire. Vapor explosion and poison hazard indoors, outdoors, or in sewers. Runoff to sewer may create fire or explosion hazard (coumarin derivative pesticides, liquid, flammable, toxic, NOS; coumarin derivative pesticides, liquid, toxic, flammable, NOS) [R4,p. G-28].

Emergency Action: Keep unnecessary people away; isolate hazard area and deny entry. Stay upwind; keep out of low areas. Positive-pressure self-contained breathing apparatus (SCBA) and chemical protective clothing that is specifically recommended by the shipper or manufacturer may be worn. It may provide little or no thermal protection. Structural firefighters' protective clothing is not effective for these materials. See the Table of Initial Isolation and Protective Action Distances. If you find the ID Number and the name of the material there, begin protective action. Isolate for 1/2 mile in all directions if tank, rail car, or tank truck is involved in fire. Call emergency response telephone number on shipping paper first. If shipping paper not available or no answer, call CHEMTREC at 1–800–424–9300 (coumarin derivative pesticides, liquid, flammable, toxic, NOS; coumarin derivative pesti-

cides, liquid, toxic, flammable, NOS) [R4,p. G-28].

Fire: Small Fires: Dry chemical, CO_2, water spray, or alcohol-resistant foam. Large Fires: Water spray, fog, or alcohol-resistant foam. Move container from fire area if you can do so without risk. Dike fire-control water for later disposal; do not scatter the material. Apply cooling water to sides of containers that are exposed to flames until well after fire is out. Stay away from ends of tanks. Withdraw immediately in case of rising sound from venting safety device or any discoloration of tank due to fire (coumarin derivative pesticides, liquid, flammable, toxic, NOS; coumarin derivative pesticides, liquid, toxic, flammable, NOS) [R4,p. G-28].

Spill or Leak: Shut off ignition sources; no flares, smoking, or flames in hazard area. Fully encapsulating, vapor-protective clothing should be worn for spills and leaks with no fire. Do not touch or walk through spilled material; stop leak if you can do so without risk. Water spray may reduce vapor, but it may not prevent ignition in closed spaces. Small Spills: Take up with sand or other noncombustible absorbent material and place into containers for later disposal. Large Spills: Dike far ahead of liquid spill for later disposal (coumarin derivative pesticides, liquid, flammable, toxic, NOS; coumarin derivative pesticides, liquid, toxic, flammable, NOS) [R4,p. G-28].

First Aid: Move victim to fresh air and call for emergency medical care; if not breathing, give artificial respiration; if breathing is difficult, give oxygen. In case of contact with material, immediately flush skin or eyes with running water for at least 15 minutes. Remove and isolate contaminated clothing and shoes at the site. Keep victim quiet and maintain normal body temperature. Effects may be delayed; keep victim under observation (coumarin derivative pesticides, liquid, flammable, toxic, NOS; coumarin derivative pesticides, liquid, toxic, flammable, NOS) [R4,p. G-28].

REFERENCES

1. Budavari, S. (ed.). *The Merck Index—*

Encyclopedia of Chemicals, Drugs, and Biologicals. Rahway, NJ: Merck and Co., Inc., 1989. 487.

2. Geiger, D. L., D. J. Call, L. T. Brooke, (eds.). *Acute Toxicities of Organic Chemicals to Fathead Minnows (Pimephales promelas).* Vol. V. Superior WI: University of Wisconsin–Superior, 1990. 262.

3. (1) Hansch, C., A. J. Leo. *Medchem Project Issue No 26.* Claremont, CA: Pomona College (1985). (2) Lyman, W. J., et al., *Handbook of Chemical Property Estimation Methods,* New York: McGraw-Hill, Chapt 5, Eqn 5.2 (1982).

4. U.S. Department of Transportation. *Emergency Response Guidebook 1993.* DOT P 5800.6. Washington, DC: U.S. Government Printing Office, 1993.

5. International Labour Office. *Encyclopedia of Occupational Health and Safety.* Vols. I and II. Geneva, Switzerland: International Labour Office, 1983.

6. Goodman, L.S., and A. Gilman (eds.). *The Pharmacological Basis of Therapeutics.* 5th ed. New York: Macmillan Publishing Co., Inc., 1975.

■ DIFLUBENZURON

CAS # 35367–38–5

SYNONYMS urea, 1-(p-chlorophenyl)-3-(2,6-difluorobenzoyl)- • dimilin • 1-(4-chlorophenyl)-3-(2,6-difluorobenzoyl) urea • N-(4-chlorophenylcarbamoyl-)2,6-difluorobenzamide • N-(((4-chlorophenyl) amino) carbonyl-)2,6-difluorobenzamide • 1-(p-chlorophenyl-)3-(2,6-difluorobenzoyl) urea • difluron • Thompson-Hayward-6040 • (N-((4-chlorophenyl) amino) carbonyl-)2,6-difluorobenzamide • N-((4-chlorophenyl) amino)-)2,6-difluorobenzamide • 1-(4-chlorophenyl-)3-(2,6-

difluorobenzoyl) urea • duphacid • micromite • dimilin wp-25 • dimilin-g1 • dimilin-g4 • dimilin odc-45 • larvakil • astonex • du-112307 • ent-29054 • oms-1804 • th 60-40 • ph-6040

MF $C_{14}H_9ClF_2N_2O_2$

MW 310.70

COLOR AND FORM Colorless crystals; white crystalline solid.

MELTING POINT 239°C.

SOLUBILITY Water solubility: values reported at: 0.00002%; soluble in water at approximately 0.3 ppm.; in water at 20°C, ca 0.14 mg/L. In acetone 6.5 g/L at 20°C. In dimethylformamide 104, dioxane 20 (both in g/L at 25°C). Moderately soluble in polar, organic solvents; very slightly soluble in nonpolar organic solvents.

VAPOR PRESSURE <0.033 mPa at 50°C.

CORROSIVITY Noncorrosive.

STABILITY AND SHELF LIFE Decomposition: <0.5% after 1-day storage at 100°C; <0.5% after 7 days at 50°C. The solid is stable to sunlight. Decomposition at 20°C in aqueous solution after 21 days in the dark is: 4% at pH 5.8, 8% at pH 7, 26% at pH 9. [R5, 287].

OTHER CHEMICAL/PHYSICAL PROPERTIES Technical-grade diflubenzuron (greater than or equal to 95% pure) is an off-white to yellow crystalline solid. (technical diflubenzuron). Decomposes on distillation. Melting point: 210–230°C (technical diflubenzuron).

COMMON USES Insect growth regulator; inhibits molting of larvae of mosquitoes, houseflies, stable flies, and black flies by interfering with chitin synthesis [R1].

NON-HUMAN TOXICITY LD50 rat oral >4,640 mg/kg [R2]; LD50 mice oral 4.64 g/kg (formulation with 50% kaolin) [R4]; LD50 rat oral >10 g/kg (formulation with 50% kaolin) [R4].

ECOTOXICITY LC50 coho salmon >150 mg/L/96 hr (conditions of bioassay not specified) [R1]; LC50 juvenile rainbow trout >150 mg/L/96 hr (conditions of bioassay not specified) [R3]; LC50 rainbow trout 250 mg/L/96 hr (static bioassay) [R3]; LC50 fathead minnow 430 mg/L/96 hr (static bioassay) [R3]; LC_{50} channel catfish 370 mg/L/96 hr (static bioassay) [R3]; LC_{50} bluegill 660 mg/L/96 hr (static bioassay) [R3]; LC_{50} *Mysidopsis bahia* 2.1 µg/L/96 hr (conditions of bioassay not specified) [R3].

BIODEGRADATION When applied to diflubenzuron, algae degraded 80% in one hour primarily to p-chlorophenylurea, and p-chloroaniline. [R6, 554].

BIOCONCENTRATION The BCF for diflubenzuron can be estimated to be 1,218 using the measured water solubility of 0.3 mg/L at 25°C (1) and a regression-derived equation (2). BCF values of 510 and 1,530 were also estimated for diflubenzuron (4). These BCF values suggest that bioconcentration may be important (SRC); however, one review suggests that diflubenzuron does not bioaccumulate in aquatic vertebrates (3) [R3].

FIFRA REQUIREMENTS Tolerances are established for residues of the insecticide diflubenzuron (N-((4-chlorophenyl) amino) carbonyl)-2,6-difluorobenzamide) in or on the following raw agricultural commodities: cottonseed; mushrooms; eggs; milk; soybeans; walnuts; cattle (fat, meat by-products, and meat); goats (fat, meat by-products, and meat); hogs (fat, meat by-products, and meat); horses (fat, meat by-products, and meat); poultry (fat, meat by-products, and meat; and sheep (fat, meat by-products, and meat) [R4].

REFERENCES

1. Verschueren, K. *Handbook of Environmental Data of Organic Chemicals.* 2nd ed. New York, NY: Van Nostrand Reinhold Co., 1983. 533.

2. *Kirk–Othmer Encyclopedia of Chemical Technology.* 3rd ed., Volumes 1–26. New York, NY: John Wiley and Sons, 1978–1984, p. V13 459 (1981).

3. (1) Carringer, R.D., et al., *J Agr Food Chem* 23: 568–72 (1975) (2) Lyman, W. J., et al., *Handbook of Chemical Property Estimation Methods.* Washington DC: Amer Chemical Soc pp. 4–9, 5–4, 5–10, 7–4, 7–5, 15–15 to 15–32 (1990) (3) Martinat, P.J., et al., *Bull Environ Contam Toxicol* 39: 142–9 (1987) (4) Kenaga, E. E., *Ecotox Environ Safety* 4: 26–38 (1980).

4. 40 CFR 180.377 (a) (7/1/91).

5. Worthing, C.R., and S.B. Walker (eds.). *The Pesticide Manual—A World Compendium.* 8th ed. Thornton Heath, UK: The British Crop Protection Council, 1987.

6. Menzie, C.M. *Metabolism of Pesticides—Update III.* Special Scientific Report—Wildlife No. 232. Washington, DC: U.S. Department of the Interior, Fish and Wildlife Service, 1980.

■ DIISODECYL PHTHALATE

CAS # 26761–40–0

SYNONYMS 1,2-benzenedicarboxylic acid, diisodecyl ester • bis (isodecyl) phthalate • didp • phthalic acid, bis (8-methylnonyl) ester • phthalic acid, diisodecyl ester • didp (plasticizer) • palatinol-z • plasticized-ddp • px-120 • sicol-184 • vestinol-dz

MF $C_{28}H_{46}O_4$

MW 446.74

DENSITY 0.966 at 20°C.

VISCOSITY 108 cP at 20°C.

BOILING POINT 250–257°C at 4 mm Hg.

MELTING POINT –58°F = –50°C = 223 K freezing point.

HEAT OF COMBUSTION Estimated at –16,600 Btu/lb = –9,200 cal/g = –3.86 × 10^5 J/kg.

AUTOIGNITION TEMPERATURE (755°F) 402°C [R2].

FLAMMABILITY LIMITS Lower: 0.3% by volume at 508°F [R2].

FIREFIGHTING PROCEDURES Extinguish with dry chemical, foam, or carbon dioxide. Water may be ineffective on fire [R3]. Alcohol foam. Water or foam may cause frothing [R2].

SOLUBILITY Insoluble in glycerol, glycols,

and some amines; soluble in most organic solvents; more soluble in sweet crude than in water and increasingly soluble with pH rise (K_{oc} = 1).

VAPOR PRESSURE 1.1 mm Hg at 200°C; 8.0 mm Hg at 250°C.

CORROSIVITY May attack some forms of plastics.

OTHER CHEMICAL/PHYSICAL PROPERTIES Will hydrolyze under acidic and basic conditions.

PROTECTIVE EQUIPMENT Goggles or face shield; rubber gloves [R3].

DISPOSAL At the time of review, criteria for land treatment or burial (sanitary landfill) disposal practices are subject to significant revision. Prior to implementing land disposal of waste residue (including waste sludge), consult with environmental regulatory agencies for guidance on acceptable disposal practices [R4].

COMMON USES Plasticizer for polyvinyl chloride for calendered film, sheet, coated fabrics, building wire jackets, and in wire and cable extrusion [R1]. Wire cable, and auto vinyl upholstery [R10]; used in non-plasticizer products, such as perfumes and cosmetics, and plasticized products, such as vinyl swimming pools, plasticized vinyl seats (on furniture and in cars), and clothing (jackets, raincoats, boots, etc.). (phthalates) [R9, p. 4–2].

OCCUPATIONAL RECOMMENDATIONS None listed.

BIOLOGICAL HALF-LIFE In fish, the half-life may be as short as 1.5 hr, yielding 99% clearance in 24 hr. (phthalate esters) [R5].

BIODEGRADATION It appears that the rate of phthalate ester biodegradation is relatively rapid for all biological systems investigated thus far. In fish, the half-life may be as short as 1.5 hr, yielding 99% clearance in 24 hr. (phthalate esters) [R5].

BIOCONCENTRATION The higher phthalate esters are fat-soluble materials with low water solubilities and therefore may be expected to bioconcentrate in aquatic organisms if not metabolized rapidly. The mean log BCF of diisodecyl phthalate in *Daphnia magna* as determined in a 21-day test using ring-labeled chemical is 2.06 (1). The mean log BCF in mussels (*Mytilus edulis*) was 3.54 between 14 and 28 days also using ring-labeled ester (2). However depuration was rapid in mussels, the half-life being 3.5 days (2). Diisodecyl phthalate is confirmed to be nonaccumulative or to have low bioconcentration according to test by the Japanese Ministry of International Trade and Industry (MITI) performed for 8 weeks in carp (3) [R6].

ATMOSPHERIC STANDARDS This action promulgates standards of performance for equipment leaks of volatile organic compounds (VOC) in the synthetic organic chemical manufacturing industry (SOCMI). These standards implement Section 111 of the Clean Air Act and are based on the Administrator's determination that emissions from the synthetic organic chemical manufacturing industry cause, or contribute significantly to, air pollution that may reasonably be anticipated to endanger public health or welfare. The intended effect of these standards is to require all newly constructed, modified, and reconstructed SOCMI process units to use the best demonstrated system of continuous emission reduction for equipment leaks of VOC, considering costs, non-air-quality health, and environmental impact and energy requirements. Diisodecyl phthalate is produced, as an intermediate or final product, by process units covered under this subpart. These standards of performance become effective upon promulgation but apply to affected facilities for which construction or modification commenced after January 5, 1981 [R7].

TSCA REQUIREMENTS Section 8 (a) of TSCA requires manufacturers of this chemical substance to report preliminary assessment information concerned with production, use, and exposure to EPA [R8].

GENERAL RESPONSE Stop discharge if possible. Isolate and remove discharged material. Notify local health and pollution control agencies.

FIRE RESPONSE Combustible. Extinguish

with dry chemicals, foam, or carbon dioxide. Water may be ineffective on fire.

EXPOSURE RESPONSE Liquid is not harmful.

RESPONSE TO DISCHARGE Mechanical containment. Should be removed. Provide chemical and physical treatment.

WATER RESPONSE Effect of low concentrations on aquatic life is unknown. Fouling to shoreline. May be dangerous if it enters water intakes. Notify local health and wildlife officials. Notify operators of nearby water intakes.

EXPOSURE SYMPTOMS No symptoms reported for any rate of exposure.

EXPOSURE TREATMENT Ingestion: call physician. Eyes: flush with water; call physician. Skin: wipe off; wash with soap and water.

REFERENCES

1. SRI.

2. National Fire Protection Association. *Fire Protection Guide on Hazardous Materials.* 9th ed. Boston, MA: National Fire Protection Association, 1986, p. 325M-40.

3. U.S. Coast Guard, Department of Transportation. *CHRIS—Hazardous Chemical Data.* Volume II. Washington, DC: U.S. Government Printing Office, 1984-5.

4. SRP.

5. Nat'l Research Council Canada; *Phthalate Esters* p. 20 (1980) NRCC No. 17583.

6. (1) Brown D., R. S. Thompson, *Chemosphere* 11: 417-26 (1982) (2) Brown D., R. S. Thompson, *Chemosphere* 11: 427-35 (1982) (3) Sasaki, S., pp. 283-98 in *Aquatic Pollut Transformation Bio Effects*, Hutzinger, O., L.H. Von Letyoeld, B.J.C. Zoeteman (eds.). Oxford: Pergamon Press (1978).

7. 40 CFR 60.489 (7/1/87).

8. 40 CFR 712.30 (7/1/87).

9. Kayser, R., D. Sterling, D. Viviani (eds.). *Intermedia Priority Pollutant Guidance Documents.* Washington, DC: U.S. Environmental Protection Agency, July 1982.

10. Nat'l Research Council Canada; *Phthalate Esters* p.49 (1980) NRCC No. 17583.

■ DIISOPROPYL METHYLPHOSPHONATE

CAS # 1445-75-6

SYNONYMS diisopropyl methanephosphonate • dimp • phosphonic acid, methyl-, bis (1-methylethyl) ester

MF $C_7H_{17}O_3P$

MW 180.21

DENSITY 0.976.

SURFACE TENSION 28.8 dynes/cm.

BOILING POINT 190°C (extrapolated).

SOLUBILITY 1-2 g/L at 25°C; water solubility = 160 g/L at 25°C.

VAPOR PRESSURE 0.267 kPa at 70°C.

OCTANOL/WATER PARTITION COEFFICIENT 1.03 (log).

OTHER CHEMICAL/PHYSICAL PROPERTIES Vapor pressure = 0.17 mm Hg at 25°C; boiling point = 121.05°C at 10 mm Hg.

DISPOSAL The IR and UV laser-induced photodestruction of diisopropyl methylphosphonate was examined. The excimer lasers ArF (193 nm), KrF (248 nm), XeCl (308 nm), quadrupled Nd:YAG (266 nm), and pulsed CO_2 laser were used. Samples were irradiated in the vapor and liquid phases. Photodissociation was observed at all irradiation wavelengths, being most efficient when ArF-irradiated diisopropyl methylphosphonate was in the vapor phase in the air or O_2. Pulsed, focused CO_2 radiation led to multiple photon dissociation and pyrolytic destruction. Light hydrocarbon gases were the principal decomposition products. H_2, CO, CO_2, and water were also detected. The residual liquid is likely to be a P-bearing acid [R1].

OCCUPATIONAL RECOMMENDATIONS None listed.

BIODEGRADATION No biodegradation was

observed when diisopropyl methylphosphonate was incubated in natural water for 12 weeks or in aqueous medium inoculated with soil microorganisms for 6 weeks (1). Additionally, diisopropyl methylphosphonate degradation was not observed when other carbon sources such as glucose, glycerol, and succinate were added to the medium. When diisopropyl methylphosphonate was incubated in soil at 25°C, slow biodegradation occurred, as was evidenced by the evolution of $14\text{-}CO_2$ (1). Approximately 1.5% and 5% of the carbon was released as CO_2 after 17 weeks in unacclimated and acclimated soil, respectively. The rate-limiting step is the enzymatic hydrolysis of diisopropyl methylphosphonate to isopropyl methylphosphonic acid. The estimated half-lives are 1 and 3 years in acclimated and unacclimated soil, respectively (1). When the soil temperature was reduced to 10°C, no biodegradation was observed. Another soil sample released 13.4% of its original activity as CO_2 after 34 weeks of incubation, indicating a half-life of 2 years (1) [R2].

BIOCONCENTRATION The BCF for diisopropyl methylphosphonate estimated from its log K_{ow}, 1.03 (1), using a recommended regression equation, is 5.1 (2). Therefore, diisopropyl methylphosphonate should not bioconcentrate in fish and other aquatic organisms (SRC). Experimental results with bluegill sunfish confirm that diisopropyl methylphosphonate does not bioconcentrate in fish (3). Diisopropyl methylphosphonate did not concentrate in the adipose tissue of ducks or quail (4). The lack of bioconcentration may be a result of metabolism that is known to occur in mammals and birds (4,5) [R3].

REFERENCES

1. Radziemski, L. J. *J Environ Sci Health Pt B, Pestic Food Contam Agric Wastes* 16 (3): 337–62 (1981).

2. Spanggord, R. J., et al., *Studies of Environmental Fate of DIMP and DCPD*. SRI International, Menlo Park, CA, U.S. NTIS AD–A078236 (1979).

3. (1) Krikorian, S. E., et al., *Quant Struct Act Relat* 6: 65–70 (1987). (2) Lyman, W. J., et al., *Handbook of Chemical Property Estimation Methods*, Chapt 5 Washington, DC: Amer Chemical Soc (1990). (3) Bentley, R.E., et al., *Acute Toxicity of Diisopropyl Methylphosphonate and Dicyclopentadiene to Aquatic Organisms*. U.S. NTIS AD–A037750 (1976). (4) Aulerich, R. J., et al., *Toxicological Study of Diisopropyl Methylphosphonate and Dicyclopentadiene in Mallard Ducks, Bobwhite Quail, and Mink*. U.S. NTIS AD–A087257 (1979). (5) Hart, Ec. R. *Mammalian Toxicological Evaluation of DIMP and DCPD*. U.S. NTIS AD–A082685 (1980).

■ 1,4:5,8-DIMETHANO-NAPHTHALENE, 1,2,3,4,10,10-HEXACHLORO-1,4,4a,5,8,8a-HEXAHYDRO-, endo, exo

CAS # 309–00–2

DOT # 1542

SIC CODES 2879; 2842

SYNONYMS 1,4:5,8-dimethanonaphthalene, 1,2,3,4,10,10-hexachloro-1,4,4a,5,8,8a-hexahydro-, (1alpha, 4alpha, 4abeta, 5alpha, 8alpha, 8abeta)- • 1,4:5,8-dimethanonaphthalene, 1,2,3,4,10,10-hexachloro-1,4,4a,5,8,8a-hexahydro-, endo, exo • hexachlorohexahydro-endo-exo-dimethanonaphthalene • NCI-C00044 • ent-15,949 • 1,2,3,4,10,10-hexachloro-1,4,4a,5,8,8a-hexahydro-1,4-endo, exo-5,8-dimethanonaphthalene • (1r, 4s, 4as, 5s, 8r, 8ar)-1,2,3,4,10,10-hexachloro-1,4,4a,5,8,8a-hexahydro-1,4:5,8-dimethanonaphthalene, not less than 95% • hhdn • compound-118 • sd-2794 • aldrosol • aldocit • aldrine (France) • kortofin • octalene • seedrin • tatuzinho • tipula • aldrex-40 • 1,2,3,4,10,10-hexachloro-1,4,4a,5,8,8a-hexahydro-1,4,5,8-dimethanonaphthalene • (1r, 4s, 5s, 8r)-1,2,3,4,10,10-

hexachloro-1,4,4a,5,8,8a-hexahydro-
1,4:5,8-dimethanonaphthalene •
1,2,3,4,10,10-hexachloro-1alpha,
4alpha, 4abeta, 5alpha, 8alpha, 8abeta
hexahydro-1,4:5,8-
dimethanonaphthalene • 1,2,3,4,10,10-
hexachloro-1,4,4a,5,8,8a-hexahydro-exo-
1,4-endo-5,8-dimethanonaphthalene •
aldrex • aldron • aldrec • algran •
soilgrin • 1,2,3,4,10,10-hexachloro-
1,4,4a,5,8,8a-hexahydro-1,4:5,8-
dimethanonaphthalene •
1,2,3,4,10,10-hexachloro-1,4,4a,5,8,8a-
hexahydro-exo-1,4-endo-5,8-
dimethanonaphthalene • (1alpha,
4alpha, 4abeta, 5alpha, 8alpha,
8abeta)-1,2,3,4,10,10-hexachloro
1,4,4a,5,8,8a-hexahydro-1,4:5,8-
dimethanonaphthalene • endo, exo-
1,2,3,4,10,10-hexachloro-1,4,4a,5,8,8a-
hexahydro-1,4:5,8-
dimethanonaphthalene

MF $C_{12}H_8Cl_6$

MW 364.90

COLOR AND FORM Colorless crystalline
solid; white crystalline substance; colorless needles; brown to white, crystalline
solid; tan to dark-brown solid.

ODOR Mild chemical odor.

DENSITY 1.6 at 20°C (solid).

BOILING POINT 145°C at 2 mm Hg.

MELTING POINT 104°C.

FIREFIGHTING PROCEDURES Firefighting:
self-contained breathing apparatus with a
full facepiece operated in pressure-demand or other positive-pressure mode (endrin) [R6, 5]. Use water to keep fire-exposed containers cool. If leak or spill has
not ignited, use water spray to disperse
the vapors, and to protect workmen repairing leak. Water spray to flush spills
away from exposures (endrin) [R5, p.
49–146]. Cool fire-exposed containers
with water [R7]. Fire extinguishing
agents: water spray, dry chemical, foam,
or carbon dioxide for fires involving solutions of aldrin in hydrocarbon solvents
[R7].

TOXIC COMBUSTION PRODUCTS Emits

highly toxic fumes of hydrogen chloride
and chlorinated breakdown products. [R5,
p. 49–37].

SOLUBILITY 0.027 mg/L in water at 27°C;
>600 g/L at 27°C in acetone, benzene,
xylene; soluble in aromatics, esters, ketones, paraffins, and halogenated solvents; 0.20 mg/L at 25°C; soluble in alcohol, ether; moderately soluble in
petroleum oils.

VAPOR PRESSURE 7.5×10^{-5} mm Hg at
20°C; 1.4×10^{-4} mm Hg at 25°C.

OCTANOL/WATER PARTITION COEFFICIENT
Log $K_{ow} = 3.01$.

CORROSIVITY Noncorrosive to steel, brass,
monel, copper, nickel, aluminum.

STABILITY AND SHELF LIFE Stable in presence of organic and inorg bases and stable
to action of hydrated metal chlorides and
mild acids [R9].

OTHER CHEMICAL/PHYSICAL PROPERTIES
Occurs as four isomers.

PROTECTIVE EQUIPMENT When opening
containers, mixing, or applying product,
wear protective rubber or pvc gloves, rubber boots, and clean overalls. Wear
NIOSH-approved dust mask when handling dust concentrates [R4].

PREVENTATIVE MEASURES Provide an eyewash station. Where there is any possibility of exposure of an employee's body to
endrin or liquids containing endrin, facilities for quick drenching of the body should
be provided within the immediate work
area for emergency use. (endrin) [R6, 3].

CLEANUP A process for removing pollutants from Du Pont's chambers works
plant in Deepwater, NJ is described. Process calls for treatment of wastes from
organic chemical manufacture by neutralization and settling, followed by a combined powdered-carbon-biological process
[R10].

DISPOSAL Potential candidate for rotary
kiln incineration with a temperature of
820–1,600°C with residence times for liquids and gases: seconds; solids: hours.
Also, a potential candidate for fluidized
bed incineration with a temperature range

of 450–980°C with residence times for liquids and gases: seconds; solids, longer [R11].

COMMON USES Insecticide (often for termites); insecticide against soil and cotton insects, turf pests, white grubs, and corn rootworms (SRP: former use) [R1]; insecticide highly effective against a range of soil-dwelling pests at 0.5 to 5.0 kg/hectare (SRP: former use) [R2]; aldrin has been used mainly against insects, primarily soil insects, that attack field, forage, vegetable, and fruit crops [R3, 824]; effective against termites, used for wood preservation, and to combat ant infestations [R2].

STANDARD CODES EPA 311; class B poison if more than 60% aldrin; NFPA—3, 1, 0; Superfund designated (hazardous substances) list.

PERSISTENCE Aldrin mixed with river water dropped to 20% of its original concentration in 8 weeks [R25]. Aldrin is oxidized to dieldrin under microbial activity. In soil, 26% may be present after 1 year, 5% in 3 years. Some inert diluents catalyze decomposition.

MAJOR SPECIES THREATENED Man, animals, and aquatic life.

INHALATION LIMIT 0.25 mg/m³.

DIRECT CONTACT Irritates skin.

PERSONAL SAFETY Protective clothing and filter mask.

ACUTE HAZARD LEVEL Fatal dose for man is 56 g/kg.

LETHAL THRESHOLD VALUES Minimum LC values trout—3.6; *Cambarus*—24; carp—4.8.

CHRONIC HAZARD LEVEL Adverse effects occurred at 5 mg/L fed to laboratory animals [R24].

DEGREE OF HAZARD TO PUBLIC HEALTH Poses threat to water supplies. May be carcinogenic.

OCCUPATIONAL RECOMMENDATIONS None listed.

OSHA 8-hr time-weighted average: 0.25 mg/m³. Skin absorption designation [R20].

NIOSH NIOSH recommends that aldrin be regulated as a potential human carcinogen [R8]. Threshold-limit–time-weighted average (TWA) 0.25 mg/m³, skin (1986) [R21, 12].

INTERNATIONAL EXPOSURE LIMITS FAO/WHO residue tolerance limit: 0.03–0.3 mg/kg/day [R17, 172].

ACTION LEVEL Evacuate area. Enter from upwind side. Notify local air authority and National Agricultural Chemicals Association.

ON-SITE RESTORATION Aldrin can be absorbed on activated carbon. Seek professional environmental engineering assistance through EPA's Environmental Response Team (ERT), Edison, NJ, 24-hour no. 908–548–8730.

BEACH OR SHORE RESTORATION Organic content will aid adsorption of aldrin. In these cases soil should be removed. Sand will hold little aldrin. Solvents can be burned off.

AVAILABILITY OF COUNTERMEASURE MATERIAL Activated carbon—water treatment plants, industrial waste treatment plants; carbon—sugar refineries.

DISPOSAL Mix in flammable solvent, spray in incinerator fire box (must have scrubber and afterburner). Or mix with vermiculite, sodium bicarbonate, or sand-soda mixture in paper boxes. Place in an open incinerator, cover with paper, and ignite with an excelsior train. Stay upwind.

DISPOSAL NOTIFICATION Contact local air and fire authorities.

MAJOR WATER USE THREATENED Fisheries, potable water, recreation.

PROBABLE LOCATION AND STATE Crystalline or emulsions. Crystals will not dissolve, but, as wettable powder or with emulsifiers, will quickly enter the water.

FOOD-CHAIN CONCENTRATION POTENTIAL High, as for all chlorinated hydrocarbons.

NON-HUMAN TOXICITY LD50 rat male oral 39 mg/kg [R3, 825]; LD50 rat female, oral 45 mg/kg [R3, 825]; LD50 rat dermal 98 mg/kg [R3, 825]; LD50 rabbit dermal 150 mg/kg [R3, 825].

ECOTOXICITY LD_{50} *Anas platyrhynchos* (mallard) oral female 520 mg/kg (95% confidence limit 229–1,210 mg/kg) 3–4 mo old [R13]; LD_{50} *Phasianus colchicus* (pheasant) female oral 16.8 mg/kg (95% confidence limit) 3–4 mo old [R13]; LD_{50} *Colinis virginianus* (bobwhite) female oral 6.59 mg/kg (95% confidence limit) 3–4 mo old [R13]; LD_{50} *Dendrocygna bicolor* (fulvous whistling duck) male oral 29.2 mg/kg (95% confidence limit) 3–6 mo old [R13]; LC_{50} *Gammarus fasciatus* (scud) 4,300 μg/L/96 hr, mature, at 21°C (95% confidence limit 3,500–5,300 μg/L) static bioassay without aeration, pH 7.2–7.5, water hardness 40–50 mg/L as calcium carbonate and alkalinity of 30–35 mg/L (technical material, 90%) [R14]; LC_{50} *Palaemonetes kadiakensis* (glass shrimp) 50 μg/L/96 hr, mature, at 21°C (95% confidence limit 38–65 μg/L) static bioassay without aeration, pH 7.2–7.5, water hardness 40–50 mg/L as calcium carbonate and alkalinity of 30–35 mg/L (technical material, 90%) [R14]; LC_{50} *Pteronarcys californica* (California stonefly) 1.3 μg/L/96 hr, second year class, at 15°C (95% confidence limit 0.8–2.2 μg/L) static bioassay without aeration, pH 7.2–7.5, water hardness 40–50 mg/L as calcium carbonate and alkalinity of 30–35 mg/L (technical material, 90%) [R14]; LC_{50} *Oncorhynchus tshawytscha* (chinook salmon) 14.3 μg/L/96 hr, wt 0.8 g, at 15°C static bioassay without aeration, pH 7.2–7.5, water hardness 40–50 mg/L as calcium carbonate and alkalinity of 30–35 mg/L (technical material, 90%) [R14]; LC_{50} *Salmo gairdneri* (rainbow trout) 2.6 μg/L/96 hr, wt 0.6 g, at 13°C (95% confidence limit 2.3–2.9 μg/L) static bioassay without aeration, pH 7.2–7.5, water hardness 40–50 mg/L as calcium carbonate and alkalinity of 30–35 mg/L (technical material, 90%) [R14]; LC_{50} *Pimephales promelas* (fathead minnow) 8.2 μg/L/96 hr, wt 0.6 g, at 18°C static bioassay without aeration, pH 7.2–7.5, water hardness 40–50 mg/L as calcium carbonate and alkalinity of 30–35 mg/L (technical material, 90%) [R14]; LC_{50} *Ictalurus melas* (black bullhead) 19 μg/L/96 hr, wt 1.5 g, at 24°C static bioassay without aeration, pH 7.2–7.5, water hardness

40–50 mg/L as calcium carbonate and alkalinity of 30–35 mg/L (technical material, 90%) [R14]; LC_{50} *Ictalurus punctatus* (channel catfish) 53 μg/L/96 hr, wt 5.2 g, at 18°C static bioassay without aeration, pH 7.2–7.5, water hardness 40–50 mg/L as calcium carbonate and alkalinity of 30–35 mg/L (technical material, 90%) [R14]; LC_{50} *Lepomis macrochirus* (bluegill) 6.2 μg/L/96 hr, wt 0.7 g, at 18°C (95% confidence limit 5.2–7.7 μg/L) static bioassay without aeration, pH 7.2–7.5, water hardness 40–50 mg/L as calcium carbonate and alkalinity of 30–35 mg/L (technical material, 90%) [R14]; LC_{50} *Micropterus salmoides* (largemouth bass) 5 μg/L/96 hr, wt 2.5 g, at 18°C static bioassay without aeration, pH 7.2–7.5, water hardness 40–50 mg/L as calcium carbonate and alkalinity of 30–35 mg/L (technical material, 90%) [R14]; EC_{50} *Simocephalus serrulatus* (daphnid) 23 μg/L/48 hr, first instar, at 15°C (95% confidence limit 17–30 μg/L) static bioassay without aeration, pH 7.2–7.5, water hardness 40–50 mg/L as calcium carbonate and alkalinity of 30–35 mg/L (technical material, 90%) [R14]; EC_{50} *Simocephalus serrulatus* (daphnid) 32 μg/L/48 hr, first instar, at 21°C (95% confidence limit 22–36 μg/L) static bioassay without aeration, pH 7.2–7.5, water hardness 40–50 mg/L as calcium carbonate and alkalinity of 30–35 mg/L (technical material, 90%) [R14]; EC_{50} *Daphnia pulex* (daphnid) 28 μg/L/48 hr, first instar, at 15°C (95% confidence limit 20–39 μg/L) static bioassay without aeration, pH 7.2–7.5, water hardness 40–50 mg/L as calcium carbonate and alkalinity of 30–35 mg/L (technical material, 90%) [R14]; EC_{50} *Cypridopsis vidua* (seed shrimp) 18 μg/L/48 hr, mature, at 21°C (95% confidence limit 15–21 μg/L) static bioassay without aeration, pH 7.2–7.5, water hardness 40–50 mg/L as calcium carbonate and alkalinity of 30–35 mg/L (technical material, 90%) [R14]; LC_{50} *Leiostomus xanthurus* (spot) 3.2 μg/L/2 days (conditions of bioassay not specified) [R15]; LC_{50} *Mugil curema* (white mullet) 2.8 μg/L/2 days (conditions of bioassay not specified) [R15]; LC_{50} *Mugil cephalus* (striped mullet) 2.0 μg/L/2 days (conditions of bioassay not specified) [R15]; LC_{50}

Aeroneuria pacifica (stonefly) 22 µg/L/30 days (conditions of bioassay not specified) [R16]; LC$_{50}$ *Anguilla rostrata* (American eel) 5 ppb/96 hr (static bioassay) [R17, 172]; TLm *Gasterosteus aculeatus* (threespine stickleback) 27.4 ppb/96 hr (static bioassay) [R17, 172]; LC$_{50}$ *Sphaeroides maculatus* (Northern puffer) 36 ppb/96 hr (static bioassay) [R17, 171].

IARC SUMMARY Inadequate evidence of carcinogenicity in humans. Limited evidence of carcinogenicity in animals. Overall evaluation: Group 3: The agent is not classifiable as to its carcinogenicity to humans [R12].

BIODEGRADATION Ability of fungi isolated from soils to degrade ^{14}C aldrin and its metabolites was assayed in culture growth medium. *Penicillium* metabolized parent compound or one of its metabolites. Field studies performed with soils packed into pvc tubes showed that added ^{14}C aldrin leached fastest in soil poor in organic matter [R18].

BIOCONCENTRATION Susceptible mosquitofish, *Gambusia affinis*, experienced greater accumulation of aldrin in tissues than resistant mosquitofish. Probable causes for resistance in decreased orders of importance: implied site insensitivity, barriers to penetration, and biotransformation [R19].

RCRA REQUIREMENTS P004; as stipulated in 40 CFR 261.33, when aldrin, as a commercial chemical product or manufacturing chemical intermediate or an off-specification commercial chemical product or a manufacturing chemical intermediate, becomes a waste, it must be managed according to federal or state hazardous waste regulations. Also defined as a hazardous waste is any container or inner liner used to hold this waste or any residue, contaminated soil, water, or other debris resulting from the cleanup of a spill, into water, or on dry land, of this waste. Generators of small quantities of this waste may qualify for partial exclusion from hazardous waste regulations (40 CFR 261.5 (e)) [R22].

FIFRA REQUIREMENTS In 1988, Congress amended FIFRA to strengthen and accelerate EPA's reregistration program. The nine-year reregistration scheme mandated by "FIFRA 88" applies to each registered pesticide product containing an active ingredient initially registered before November 1, 1984. List A consists of the 194 chemical cases (or 350 individual active ingredients) for which EPA had issued Registration Standards prior to the effective date of FIFRA '88. List: A; case: Aldrin; case No. 0172; pesticide type: Insecticide; registration Standard Date: 09/29/86; case status: The pesticide is no longer an active ingredient in any registered pesticide products. Therefore, EPA is characterizing it as cancelled. Under FIFRA, pesticide producers may voluntarily cancel their registered products. EPA also may cancel pesticide registrations if registrants fail to pay required fees, to make or meet certain reregistration commitments, or when the Agency reaches findings of unreasonable adverse effects; active ingredient (AI): aldrin [R23].

GENERAL RESPONSE Avoid contact with liquid or solid. Keep people away. Wear goggles, self-contained breathing apparatus, and rubber overclothing (including gloves). Stop discharge if possible. Call fire department if solution is discharged. Isolate and remove discharged material. Notify local health and pollution control agencies.

FIRE RESPONSE Solid is not flammable but usually is dissolved in a combustible liquid. Poisonous gases are produced when heated. Wear goggles, self-contained breathing apparatus, and rubber overclothing (including gloves). Extinguish with water, dry chemical, foam, or carbon dioxide. Cool exposed containers with water.

EXPOSURE RESPONSE Call for medical aid. Solid or solution poisonous if swallowed or if skin is exposed. Irritating to skin, eyes. Remove contaminated clothing and shoes. Flush affected areas with plenty of water. If in eyes, hold eyelids open, and flush with plenty of water. If swallowed and victim is conscious, have victim drink water or milk, and have victim induce vomiting. If swallowed and victim is unconscious or having convulsions, do nothing except keep victim warm.

RESPONSE TO DISCHARGE Issue warning—poison, water contaminant; liquid forms are flammable. Mechanical containment (of liquid form). Should be removed.

WATER RESPONSE Harmful to aquatic life in very low concentrations. May be dangerous if it enters water intakes. Notify local health and wildlife officials. Notify operators of nearby water intakes.

EXPOSURE SYMPTOMS Ingestion, inhalation, or skin absorption of a toxic dose will induce nausea, vomiting, hyperexcitability, tremors, epileptiform convulsions, and ventricular fibrillation. Aldrin may cause temporary reversible kidney and liver injury. Symptoms may be seen after ingestion of less than 1 gram in an adult; ingestion of 25 mg has caused death in children.

EXPOSURE TREATMENT Skin contact: wash with soap and running water. If material gets into eyes, wash immediately with running water for at least 15 minutes; get medical attention. Ingestion: call physician immediately; induce vomiting immediately. Repeat until vomit fluid is clear. Never give anything by mouth to an unconscious person. Keep patient prone and quiet. Physician: administer barbiturates as anti-convulsant therapy. Observe patient carefully because repeated treatment may be necessary.

REFERENCES

1. SRI.

2. Worthing, C. R., and S. B. Walker (eds.). *The Pesticide Manual—A World Compendium.* 8th ed. Thornton Heath, UK: The British Crop Protection Council, 1987. 11.

3. Hayes, W. J., Jr., E. R. Laws, Jr. (eds.). *Handbook of Pesticide Toxicology. Volume 2. Classes of Pesticides.* New York, NY: Academic Press, Inc., 1991.

4. *Farm Chemicals Handbook 1992.* Willoughby, OH: Meister Publishing Co., 1992, p. C–15.

5. National Fire Protection Association. *Fire Protection Guide on Hazardous Materials.* 7th ed. Boston, MA: National Fire Protection Association, 1978.

6. Mackison, F. W., R. S. Stricoff, and L. J. Partridge, Jr. (eds.). *NIOSH/OSHA—Occupational Health Guidelines for Chemical Hazards.* DHHS (NIOSH) Publication No. 81–123 (3 vols). Washington, DC: U.S. Government Printing Office, Jan. 1981.

7. U.S. Coast Guard, Department of Transportation. *CHRIS—Hazardous Chemical Data.* Volume II. Washington, DC: U.S. Government Printing Office, 1984–5.

8. NIOSH. *NIOSH Pocket Guide to Chemical Hazards.* DHHS (NIOSH) Publication No. 90–117. Washington, DC: U.S. Government Printing Office, June 1990, 34.

9. IARC. *Monographs on the Evaluation of the Carcinogenic Risk of Chemicals to Man.* Geneva: World Health Organization, International Agency for Research on Cancer, 1972–present (multivolume work), p. V5 26 (1974).

10. Hutton, D. G. *Ind Wastes* 26 (2): 22, 24, 26 (1980).

11. USEPA; *Engineering Handbook for Hazardous Waste Incineration* p. 3–8 (1981) EPA 68–03–3025.

12. IARC. *Monographs on the Evaluation of the Carcinogenic Risk of Chemicals to Man.* Geneva: World Health Organization, International Agency for Research on Cancer, 1972–present (multivolume work), p. S7 56 (1987).

13. U.S. Department of the Interior, Fish and Wildlife Service. *Handbook of Toxicity of Pesticides to Wildlife.* Resource Publication 153. Washington, DC: U.S. Government Printing Office, 1984. 10.

14. U.S. Department of Interior, Fish and Wildlife Service. *Handbook of Acute Toxicity of Chemicals to Fish and Aquatic Invertebrates.* Resource Publication No. 137. Washington, DC: U.S. Government Printing Office, 1980. 10.

15. USEPA. *Ambient Water Quality Criteria Doc; Aldrin/Dieldrin* p. B–42 (1980) EPA 440/5–80–019.

16. Jensen, L. D., A. R. Gaufin. *Jour Water Pollut Cont Fed* 38: 1273 (1966) as cited in USEPA, *Ambient Water Quality Criteria Doc; Aldrin/Dieldrin* p. B–39 (1980) EPA 440/5–80–019.

17. Verschueren, K. *Handbook of Environmental Data of Organic Chemicals.* 2nd ed. New York, NY: Van Nostrand Reinhold Co., 1983.

18. Flores–Ruegg, E. *Part of a Coordinated Program on Isotopic–Tracer–Aided Studies of Agrochemical Residue–Soil Biota Interactions.* Final report for the period 1 March 1978–30 June 1982; Report (IAEA–R–2161–F): 10 (1982).

19. Yarbrough, J. D., J. E. Chambers. *Acs Symp Ser* 99 (CH (9): 145 (1979).

20. 29 CFR 1910.1000 (7/1/91).

21. American Conference of Governmental Industrial Hygienists. *Threshold Limit Values for Chemical Substances and Physical Agents and Biological Exposure Indices for 1993–1994.* Cincinnati, OH: ACGIH, 1993.

22. 40 CFR 261.33 (7/1/91).

23. USEPA/OPP; *Status of Pesticides in Reregistration and Special Review* p. 61 (Mar, 1992) EPA 700–R–92–004.

24. A. D. Little Inc., "Relationship Between Organic Chemical Pollution of Fresh Water and Health," FWQA, 1970, 71632, Dec.

25. Eichelberger, J. W., J. J. Lichtenberg. "Persistence of Pesticides in River Water," *Environmental Science and Technology,* 1971, Vol. 5, No. 6. June.

■ 1,4:5,8-DIMETHANO-NAPHTHALENE, 1,2,3,4,10,10-HEXACHLORO-6,7-EPOXY-1,4,4a,5,6,7,8,8a OCTAHYDRO-, endo, endo-

CAS # 72–20–8

DOT # 2761

SIC CODES 2879; 2842

SYNONYMS
1,4:5,8-dimethanonaphthalene, 1,2,3,4,10,10-hexachloro-6,7-epoxy-1,4,4a,5,6,7,8,8a-octahydro endo, endo • 2,7:3,6-dimethanonaphth (2,3-b) oxirene, 3,4,5,6,9,9-hexachloro-1a,2,2a,3,6,6a, 7,7a-octahydro-, (1aalpha, 2beta, 2abeta, 3alpha, 6alpha, 6abeta, 7beta, 7aalpha)- • hexachloroepoxyoctahydro-endo, endo-dimethanonaphthalene • NCI-C00157 • ent-17,251 • nendrin • en-57 • endrex • endricol • experimental insecticide-269 • hexadrin • mendrin • oktanex • sd-3419 • endrine (French) • 3,4,5,6,9,9-hexachloro-1a,2,2a,3,6,6a,7,7a-octahydro-2,7:3,6-dimethanonaphth[2,3-b]oxirene • 1,2,3,4,10,10-hexachloro-6,7-epoxy-1,4,4a,5,6,7,8,8a-octahydro-endo, endo 1,4:5,8-dimethanonaphthalene • compound-269 • ent-17251 • 1,2,3,4,10,10-hexachloro-6,7-epoxy-1,4,4a,5,6,7,8,8a-octahydro-1,4-endo, endo 5,8-dimethanonaphthalene • (1r, 4s, 4as, 5s, 6s, 7r, 8r,8ar)-1,2,3,4,10,10-hexachloro-1,4,4a,5,6,7,8,8a octahydro-6,7-epoxy-1,4:5,8-dimethanonaphthalene • oms-197 • (1aalpha, 2beta, 2abeta, 3alpha, 6alpha, 6abeta, 7beta, 7aalpha)-3,4,5,6,9,9-hexachloro-1a,2,2a,3,6,6a,7,7a-octahydro-2,7:3,6-dimethanonaphth[2,3-b]oxirene • 1,2,3,4,10,10-hexachloro-1r, 4s, 4as, 5s, 6,7r, 8r, 8ar-octahydro-6,7-epoxy-1,4:5,8-dimethanonaphthalene • 3,4,5,6,9,9-hexachloro-1aalpha, 2beta, 2abeta, 3alpha, 6alpha, 6abeta, 7beta, 7aalpha octahydro-2,7:3,6-dimethanonaphth[2,3-b]oxirene

MF $C_{12}H_8Cl_6O$

MW 380.90

COLOR AND FORM White, crystalline solid.

ODOR Odorless.

ODOR THRESHOLD 1.80×10^{-2} ppm (perfume/flavor grade purity) [R13].

DENSITY Specific gravity 1.7 at 20°C.

MELTING POINT Below 392°F.

EXPLOSIVE LIMITS AND POTENTIAL During the blending of endrin and diethyl 4-nitrophenyl thionophosphate into a petroleum solvent, an unexpected exothermic reaction occurred that vaporized some solvent, and led to a vapor-air explosion. Faulty agitation was suspected [R11].

FLAMMABILITY LIMITS 1.1%–7% mixture in xylene (in air) [R1].

FIREFIGHTING PROCEDURES Fire extinguishing agents: dry chemical, foam, or carbon dioxide [R1]. Use water to keep fire-exposed containers cool. If leak or spill has not ignited, use water spray to disperse the vapors, and to protect workmen repairing leak. Water spray to flush spills away from exposures [R10]. Firefighting: Self-contained breathing apparatus with a full facepiece operated in pressure-demand or other positive-pressure mode. [R9, 5].

TOXIC COMBUSTION PRODUCTS Toxic hydrogen chloride and phosgene may be generated when solution burns [R1].

SOLUBILITY Solubility in g/100 ml solvent at 25°C: acetone 17, benzene 13.8, carbon tetrachloride 3.3, hexane 7.1, and xylene 18.3; solubility of endrin in fresh water: 200 µg/L; insoluble in methanol.

VAPOR PRESSURE 2×10^{-7} mm Hg at 25°C.

OCTANOL/WATER PARTITION COEFFICIENT Log K_{ow} = 5.6 (calculated).

CORROSIVITY Slightly corrosive to metals.

STABILITY AND SHELF LIFE Endrin is stable in the presence of most alkalis but rearranges to delta-keto endrin in the presence of strong acids, in sunlight, and when heated [R2].

OTHER CHEMICAL/PHYSICAL PROPERTIES An isomer of dieldrin.

PROTECTIVE EQUIPMENT Respirator for spray, fog, dust; rubber gloves and boots [R1].

PREVENTATIVE MEASURES Employees should wash immediately when skin is wet or contaminated. Work clothing should be changed daily if it is possible that clothing is contaminated. Provide emergency showers and eyewash [R14].

CLEANUP 1. Ventilate area of spill. 2. Collect spilled material in most convenient and safe manner and deposit in sealed containers for reclamation. Liquid containing endrin should be absorbed in vermiculite, dry sand, earth [R9].

DISPOSAL The reductive dechlorination of dieldrin and endrin was investigated as a possible procedure for field disposal of small quantities of these pesticides. The objective was to convert the parent compounds to environmentally less objectionable materials. Emulsifiable concentrate formulations of the pesticides in a soil slurry were mixed with powdered zinc, dilute acetic acid, and acetone to facilitate reaction. Analysis of the mixtures by GC–MS indicated essentially complete conversion of endrin and partial conversion of dieldrin to products probably formed by replacement of the bridge antichlorines with hydrogen [R15].

COMMON USES It is a nonsystemic and persistent insecticide used mainly on field crops. It is nonphytotoxic at insecticidal concentration, but is suspected of causing damage to maize (SRP: former use) [R5]. Insecticide used to control the army cutworm (*Euxoa aoxiliaris*), the pale western cutworm (*Agrotis orthogonia*), pine vole (*Microtus pinetorium*), meadow voles (*Microtus* species), and grasshoppers, but only when federal regulations are strictly followed (SRP: former use) [R6]. Insecticide for small grains, sugarcane, and apple orchards (SRP: former use) [R7]. The only known use of endrin is as an insecticide, as an avicide, and as a rodenticide (former use) [R8]. Endrin has been used mainly on field crops such as cotton and grains. It has also been used for grasshoppers in non-cropland and to control voles and mice in orchards (SRP: former use) [R3, 841]. Control of a wide range of insects (particularly lepidopterans) in cotton, maize, sugarcane, rice, cereals, ornamentals, and other crops (SRP: former use) [R4].

STANDARD CODES EPA 311; class B poison; NFPA—3, 1, 0; Superfund designated (hazardous substances) list.

PERSISTENCE Nonpersistent [R31]. Another report lists 100% remaining after 8 weeks in river water [R32]. Iyatomi claims toxicity persisted 1 month in a rice paddy.

MAJOR SPECIES THREATENED Fish are highly sensitive to endrin [R31]. Endrin is also especially poisonous to warmblooded animals.

INHALATION LIMIT 0.1 mg/m^3.

DIRECT CONTACT Irritates skin.

PERSONAL SAFETY Protective clothing and filter mask. Recommend approach from upwind side.

ACUTE HAZARD LEVEL Lethal threshold values: minimum LC values trout—0.096 µg/L; *Cambarus*—1.2 µg/L; carp—0.144 µg/L. Extremely toxic.

CHRONIC HAZARD LEVEL Endrin at 1 mg/kg in the diet of quail produced a 40% decrease in reproduction. Predator fish, frogs, etc. were fed mosquitofish reared in 2 ppm endrin (7 days). Mortality ranged from 72 to 100% in 24-hour period. Accumulative poison.

DEGREE OF HAZARD TO PUBLIC HEALTH Highly hazardous. Potential mutagenic effects.

OCCUPATIONAL RECOMMENDATIONS None listed.

OSHA 8-hr time-weighted average: 0.1 mg/m^3. Skin absorption designation [R27].

NIOSH 10-hr time-weighted average: 0.1 mg/m^3, skin [R12]; threshold-limit 8-hr time-weighted average (TWA) 0.1 mg/m^3, skin (1988) [R28, 20].

ACTION LEVEL Evacuate area. Enter from upwind side. Notify local air authority and National Agricultural Chemicals Association.

ON-SITE RESTORATION Carbon or peat may be used to adsorb spills. Seek professional environmental engineering assistance through EPA's Environmental Response Team (ERT), Edison, NJ, 24-hour no. 908–548–8730.

AVAILABILITY OF COUNTERMEASURE MATERIAL Carbon—water treatment plants, sugar refineries; peat—nurseries, floral shops.

DISPOSAL Rearranges above 200°C, rearranges in presence of acids, certain metal salts, and catalysts. (1) Mix with sodium bicarbonate or sand–soda-ash mixture in paper boxes. Burn in incinerator with scrap wood. (2) Dissolve in flammable solvent, spray into firebox of incinerator with scrubber and afterburner.

DISPOSAL NOTIFICATION Contact local air and fire authorities.

MAJOR WATER USE THREATENED Fisheries, recreation, potable supply.

FOOD-CHAIN CONCENTRATION POTENTIAL High, as for all chlorinated hydrocarbons. Birds of prey in the Netherlands contained up to 0.3 ppm in the liver and 15.7 ppm in mesenterial fat.

NON-HUMAN TOXICITY LD$_{50}$ guinea pig male oral 36 mg/kg [R3, 841]; LD$_{50}$ guinea pig female, oral 16 mg/kg [R3, 841]; LD$_{50}$ rat female dermal 15 mg/kg [R3, 841]; LD$_{50}$ rabbit female, oral 7–10 mg/kg [R3, 841]; LD$_{50}$ monkey, oral 3 mg/kg [R3, 841]; LD$_{50}$ rat, oral 3 mg/kg [R19]; LD$_{50}$ mouse oral 1.3 mg/kg [R19]; LD$_{50}$ mouse iv 2.3 mg/kg [R19]; LD$_{50}$ rat male dermal 18 mg/kg [R19].

ECOTOXICITY EC$_{50}$ *Simocephalus serrulatus lization* 45 µg/L/48 hr at 21°C (95% confidence limit 35–58 µg/L), first instar static bioassay without aeration, pH 7.2–7.5, water hardness 40–50 mg/L as calcium carbonate and alkalinity of 30–35 µg/L) first instar static bioassay without aeration, pH 7.2–7.5, water hardness 40–50 mg/L as calcium carbonate and alkalinity of 30–35 mg/L (technical material, 99%) [R20, 37]; EC$_{50}$ *Daphnia magna* (daphnid) 4.2 µg/L/48 hr at 21°C, first instar static bioassay without aeration, pH 7.2–7.5, water hardness 40–50 mg/L as calcium carbonate and alkalinity of 30–35 mg/L (technical material, 99%) [R20, 37]; EC$_{50}$ *Daphnia pulex* (daphnid) 20 µg/L/48 hr at 15°C (95% confidence limit 13–30 µg/L), first instar static bioassay without aeration, pH 7.2–7.5, water hardness 40–50 mg/L as calcium carbonate and alkalinity of 30–35 mg/L (technical material, 99%) [R20, 37]; EC$_{50}$ *Cypridopsis viuda* (seed shrimp) 1.8 µg/L/48 hr at 21°C, mature static bioassay without aeration, pH 7.2–7.5, water hardness 40–50 mg/L as calcium carbonate and alkalinity of 30–35 mg/L (technical material, 99%) [R20, 37]; LC$_{50}$ *Asellus brevicaudus* (sowbug) 1.5 µg/L/96 hr at 15°C (95% confidence limit 0.9–3.7 µg/L), mature. Static bioassay without aeration, pH 7.2–7.5, water hardness 40–50 mg/L as calcium carbonate and alkalinity of 30–35 mg/L (technical material, 99%) [R20, 37]; LC$_{50}$ *Gammarus lacustris* (scud) 3.0 µg/L/96 hr at 21°C (95% confidence limit 2.0–4.5 µg/L), mature. Static bioassay without aeration, pH 7.2–7.5, water hardness 40–50 mg/L as calcium carbonate and alkalinity of 30–35 mg/L (technical material, 99%) [R20, 37]; LC$_{50}$ *Gammarus fasciatus* (scud) 4.3 µg/L/96 hr at 21°C (95% confidence limit 3.5–5.2 µg/L), mature. Static bioassay without aeration, pH 7.2–7.5, water hardness 40–50 mg/L as calcium carbonate and alkalinity of 30–35 mg/L (technical material, 99%) [R20, 37]; LC$_{50}$ *Orconectes nais* (crayfish) 3.2 µg/L/96 hr at 21°C (95% confidence limit 1.6–7.5 µg/L), early instar. Static bioassay without aeration, pH 7.2–7.5, water hardness 40–50 mg/L as calcium carbonate and alkalinity of 30–35 mg/L (technical material, 99%) [R20, 37]; LC$_{50}$ *Palaemonetes kadiakensis* (glass shrimp) 3.2 µg/L/96 hr at 21°C (95% confidence limit 1.8–5.8 µg/L), mature. Static bioassay without aeration, pH 7.2–7.5, water hardness 40–50 mg/L as calcium carbonate and alkalinity of 30–35 mg/L (technical material, 99%) [R20, 37]; LC$_{50}$ *Pteronarcella badia* (stonefly) 0.54 µg/L/96 hr at 15°C (95% confidence limit 0.40–0.72 µg/L), naiad. Static bioassay without aeration, pH 7.2–7.5, water hardness 40–50 mg/L as calcium carbonate and alkalinity of 30–35 mg/L (technical material, 99%) [R20, 37]; LC$_{50}$ *Pteronarcys californica* (stonefly) 0.25 µg/L/96 hr at 15°C (95% confidence limit 0.20–0.31 µg/L), 2nd yr class. Static bioassay without aeration, pH 7.2–7.5, water hardness 40–50 mg/L as calcium carbonate and alkalinity of 30–35 mg/L (technical material, 99%) [R20, 37]; LC$_{50}$ *Claassenia sabulosa* (stonefly) 0.08 µg/L/96 hr at 15°C (95% confidence limit

0.06–0.09 µg/L), 2nd year class. Static bioassay without aeration, pH 7.2–7.5, water hardness 40–50 mg/L as calcium carbonate and alkalinity of 30–35 mg/L (technical material, 99%) [R20, 37]; LC$_{50}$ *Acroneuria* (stonefly) less than 0.18 µg/L/96 hr at 15°C, 2nd year class. Static bioassay without aeration, pH 7.2–7.5, water hardness 40–50 mg/L as calcium carbonate and alkalinity of 30–35 mg/L (technical material, 99%) [R20, 37]; LC$_{50}$ *Hexagenia bilineata* (mayfly) 62 µg/L/96 hr at 15°C (95% confidence limit 41–95 µg/L), first instar. Static bioassay without aeration, pH 7.2–7.5, water hardness 40–50 mg/L as calcium carbonate and alkalinity of 30–35 mg/L (technical material, 99%) [R20, 37]; LC$_{50}$ *Baetis* (mayfly) 0.90 µg/L/96 hr at 15°C, (95% confidence limit 0.57–1.4 µg/L), juvenile. Static bioassay without aeration, pH 7.2–7.5, water hardness 40–50 mg/L as calcium carbonate and alkalinity of 30–35 mg/L (technical material, 99%) [R20, 37]; LC$_{50}$ *Ischnura verticalis* (damselfly) 2.4 µg/L/96 hr at 21°C (95% confidence limit 1.5–3.8 µg/L), juvenile. Static bioassay without aeration, pH 7.2–7.5, water hardness 40–50 mg/L as calcium carbonate and alkalinity of 30–35 mg/L (technical material, 99%) [R20, 37]; LC$_{50}$ *Tipula* (crane fly) 12 µg/L/96 hr at 15°C (95% confidence limit 7.3–18 µg/L), juvenile. Static bioassay without aeration, pH 7.2–7.5, water hardness 40–50 mg/L as calcium carbonate and alkalinity of 30–35 mg/L (technical material, 99%) [R20, 37]; LC$_{50}$ *Atherix* (snipe fly) 4.6 µg/L/96 hr at 15°C (95% confidence limit 3.1–6.8 µg/L), juvenile. Static bioassay without aeration, pH 7.2–7.5, water hardness 40–50 mg/L as calcium carbonate and alkalinity of 30–35 mg/L (technical material, 99%) [R20, 37]; LC$_{50}$ *Salmo gairdneri* (rainbow trout) 0.75 µg/L/96 hr at 13°C (95% confidence limit 0.64–0.88 µg/L), wt 1.0 g. Static bioassay without aeration, pH 7.2–7.5, water hardness 40–50 mg/L as calcium carbonate and alkalinity of 30–35 mg/L (technical material, 99%) [R20, 37]; LC$_{50}$ *Carassius auratus* (goldfish) 0.44 µg/L/96 hr at 12°C (95% confidence limit 0.29–0.66 µg/L), fingerling (technical material, 99%) (flow-through toxicity test) [R20, 37]; LC$_{50}$

Cyprinus carpio (carp) 0.32 µg/L/96 hr at 12°C (95% confidence limit 0.25–0.41 µg/L), fingerling (technical material, 99%) (flow-through toxicity test) [R20, 37]; LC$_{50}$ *Pimephales promela* (fathead minnow) 1.8 µg/L/96 hr at 18°C (95% confidence limit 1.0–3.0 µg/L), wt 1.2 g. Static bioassay without aeration, pH 7.2–7.5, water hardness 40–50 mg/L as calcium carbonate and alkalinity of 30–35 mg/L (technical material, 99%) [R20, 37]; LC$_{50}$ *Ictalurus melas* (black bullhead) 1.1 µg/L/96 hr at 24°C (95% confidence limit 1.0–1.3 µg/L), wt 1.5 g. Static bioassay without aeration, pH 7.2–7.5, water hardness 40–50 mg/L as calcium carbonate and alkalinity of 30–35 mg/L (technical material, 99%) [R20, 37]; LC$_{50}$ *Ictalurus punctatus* (channel catfish) 0.32 µg/L/96 hr at 24°C (95% confidence limit 0.29–0.35 µg/L), wt 1.4 g. Static bioassay without aeration, pH 7.2–7.5, water hardness 40–50 mg/L as calcium carbonate and alkalinity of 30–35 mg/L (technical material, 99%) [R20, 37]; LC$_{50}$ *Gambusia affinis* (mosquitofish) 1.1 µg/L/96 hr at 17°C (95% confidence limit 0.4–3.4 µg/L), wt 0.6 g. Static bioassay without aeration, pH 7.2–7.5, water hardness 40–50 mg/L as calcium carbonate and alkalinity of 30–35 mg/L (technical material, 99%) [R20, 37]; LC$_{50}$ *Lepomis macrochirus* (bluegill) 0.61 µg/L/96 hr at 18°C (95% confidence limit 0.50–0.74 µg/L), wt 1.5 g. Static bioassay without aeration, pH 7.2–7.5, water hardness 40–50 mg/L as calcium carbonate and alkalinity of 30–35 mg/L (technical material, 99%) [R20, 38]; LC$_{50}$ *Micropterus salmoides* (largemouth bass) 0.31 µg/L/96 hr at 18°C (95% confidence limit 0.25–0.39 µg/L), wt 2.5 g. Static bioassay without aeration, pH 7.2–7.5, water hardness 40–50 mg/L as calcium carbonate and alkalinity of 30–35 mg/L (technical material, 99%) [R20, 38]; LC$_{50}$ *Perca flavescens* (yellow perch) 0.15 µg/L/96 hr at 12°C (95% confidence limit 0.12–0.18 µg/L) fingerling (technical material, 99%) (flow-through toxicity test) [R20, 38]; LC$_{50}$ bobwhite quail oral 14 ppm (95% confidence limit 11–24 ppm), age 17 days [R21]; LC$_{50}$ Japanese quail oral 18 ppm (95% confidence limit 15–20 ppm), age 14 days [R21]; LC$_{50}$ ring–necked pheasant

oral 14 ppm (95% confidence limit 11–17 ppm), age 22 days [R21]; LC$_{50}$ mallard oral 22 ppm (95% confidence limit 17–31 ppm), age 8 days [R21]; LC$_{50}$ mallard oral 18 ppm (95% confidence limit 15–21 ppm), age 5 days [R21]; LC$_{50}$ *Ophiocephalus punctatus* (fish) 0.033 ppm/96 hr (conditions of bioassay not specified) [R17]; LC$_{50}$ *Eupera singleyi* 60 µg/L/72 hr (conditions of bioassay not specified) [R22]; LD$_{50}$ *Anas platyrhynchos* (mallard) female oral 5.64 mg/kg (95% confidence limit 2.71–11.7 mg/kg) 12 mo old [R18]; LD$_{50}$ *Tympanuchus phasianellus* (sharp–tailed grouse) female oral 1.06 mg/kg (95% confidence limit) [R18]; LD$_{50}$ *Callipepla californica* (California quail) female oral 1.19 mg/kg (95% confidence limit 0.857–1.65 mg/kg) 9–10 mo old [R18]; LD$_{50}$ pheasant male oral 1.78 mg/kg (95% confidence limit 1.12–2.83 mg/kg) 3–4 mo old [R18]; LD$_{50}$ *Columba livia* (rock dove) oral 2.0–5.0 mg/kg [R18]; LD$_{50}$ *Odocoileus hemionus hemionus* (mule deer) female oral 6.25–12.5 mg/kg, 10 mo old [R18]; LC$_{50}$ *Pagurus longicarpus* (hermit crab) 12 ppb/96 hr, static lab bioassay [R23]; LC$_{50}$ *Coturonix* 17 ppm (95% confidence interval 25–20 ppm) [R24].

IARC SUMMARY No data are available for humans. Inadequate evidence of carcinogenicity in animals. Overall evaluation: Group 3: The agent is not classifiable as to its carcinogenicity to humans [R16].

BIOLOGICAL HALF-LIFE The half-life (in rats) is 2–3 days in males and 4 days in females after a dosage of 0.2 mg/kg. [R3, 842].

BIODEGRADATION Endrin appears to be resistant to biodegradation in natural waters and most soils. Greater than 80% of endrin recovered from sterilized and unsterilized natural water (drainage canal of Holland Marsh, Ontario, Canada), and distilled water after 16 weeks (1). No degradation of endrin observed—Little Miami River water, 8 weeks (2). Soil biodegradation half-lives of approximately 4 to 8 years (3), and 14 yrs (4) or more (5) have been reported. Considerable degradation (51%—4 days in Rutledge sand to essentially 100% in Magnolia sandy loam) has been reported (6); however, biodegradation was not conclusively proven to be the degradation route nor were the products identified (6). After 2 months endrin recovery in flooded Casiguran soil was only 8.4% while 88.24% was recovered in the upland soil (7); endrin persisted in all other soils for 2 months (7). Anaerobic conditions apparently stimulate biodegradation of endrin, although it remains persistent (8). The half-life in thick anaerobic sewage sludge is 5–14 days with 4 unidentified products observed (9) [R25].

BIOCONCENTRATION Stoneflies concentrated endrin 1,150 times above test water level; snails 8,600 times [R26].

RCRA REQUIREMENTS P051; as stipulated in 40 CFR 261.33, when endrin, as a commercial chemical product or manufacturing chemical intermediate or an off-specification commercial chemical product or a manufacturing chemical intermediate, becomes a waste, it must be managed according to federal or state hazardous waste regulations. Also defined as a hazardous waste is any container or inner liner used to hold this waste or any residue, contaminated soil, water, or other debris resulting from the cleanup of a spill, into water or on dry land, of this waste. Generators of small quantities of this waste may qualify for partial exclusion from hazardous waste regulations (40 CFR 261.5 (e)) [R29].

FIFRA REQUIREMENTS Classified for restricted use, limited to use by or under the direct supervision of a certified applicator. Formulations: All emulsions, dusts, wettable powders, pastes, and granular formulations 2% and above. Use patterns: All uses. Classification: Restricted. Criteria influencing restrictions: Acute dermal toxicity. Hazard to nontarget organisms [R30].

GENERAL RESPONSE Avoid contact with liquid, solid, and dust. Keep people away. Wear goggles, self-contained breathing apparatus, and rubber overclothing (including gloves). Stop discharge if possible. Call fire department. Isolate and remove discharged material. Notify local health and pollution control agencies.

FIRE RESPONSE Combustible solution or non-flammable solid. Poisonous gases are produced in fire. Extinguish with dry chemicals, foam, or carbon dioxide. Water may be ineffective on fire.

EXPOSURE RESPONSE Call for medical aid. Dust is poisonous if inhaled or if skin is exposed. Irritating to eyes, nose, and throat. Move victim to fresh air. If breathing is difficult, give oxygen. Liquid or solid poisonous if swallowed or if skin is exposed. Irritating to skin and eyes. Remove contaminated clothing and shoes. Flush affected areas with plenty of water. If in eyes, hold eyelids open, and flush with plenty of water. If swallowed and victim is conscious, have victim drink water or milk, and have victim induce vomiting. If swallowed and victim is unconscious or having convulsions, do nothing except keep victim warm.

RESPONSE TO DISCHARGE Issue warning—poison, water contaminant. Restrict access. Should be removed. Provide chemical and physical treatment.

WATER RESPONSE Harmful to aquatic life in very low concentrations. May be dangerous if it enters water intakes. Notify local health and wildlife officials. Notify operators of nearby water intakes.

EXPOSURE SYMPTOMS Inhalation causes moderate irritation of nose and throat; prolonged breathing may cause same toxic symptoms as for ingestion. Contact with liquid causes moderate irritation of eyes and skin. Prolonged contact with skin may cause same toxic symptoms as for ingestion. Ingestion causes frothing of the mouth, facial congestion, convulsions, violent muscular contractions, dizziness, weakness, nausea.

EXPOSURE TREATMENT Get medical attention after all exposures to this compound. Inhalation: remove from exposure. Eyes: flush with water for at least 15 min. Skin: wash with plenty of soap and water, but do not scrub. Ingestion: remove from the gastrointestinal tract, either by inducing vomiting (unless hydrocarbon solvents are involved, and the amount of insecticide is well below the toxic amount), or by gastric lavage with saline solution; saline cathartics may also be beneficial; fats and oils should be avoided; sedation with barbiturates is indicated if signs of central nervous system irritation are present; patient should have absolute quiet, expert nursing care, and a minimum of external stimuli to reduce danger of convulsions; epinephrine is contraindicated in view of the danger of precipitating ventricular fibrillation; if material ingested was dissolved in a hydrocarbon solvent, observe patient for possible development of hydrocarbon pneumonitis.

REFERENCES

1. U.S. Coast Guard, Department of Transportation. *CHRIS—Hazardous Chemical Data*. Volume II. Washington, DC: U.S. Government Printing Office, 1984–5.

2. IARC. *Monographs on the Evaluation of the Carcinogenic Risk of Chemicals to Man*. Geneva: World Health Organization, International Agency for Research on Cancer, 1972–present (multivolume work), p. V5 158 (1974).

3. Hayes, W. J., Jr., E. R. Laws, Jr. (eds.). *Handbook of Pesticide Toxicology*. Volume 2. *Classes of Pesticides*. New York, NY: Academic Press, Inc., 1991.

4. Hartley, D., and H. Kidd (eds.). *The Agrochemicals Handbook*. 2nd ed. Lechworth, Herts, England: The Royal Society of Chemistry, 1987, p. A177/Aug 87.

5. Worthing, C. R., S. B. Walker (eds.). *The Pesticide Manual—A World Compendium*. 7th ed. Lavenham, Suffolk, Great Britain: The Lavenham Press Limited, 1983. 235.

6. 44 FR 43635 (7/25/79).

7. SRI.

8. IARC. *Monographs on the Evaluation of the Carcinogenic Risk of Chemicals to Man*. Geneva: World Health Organization, International Agency for Research on Cancer, 1972–present (multivolume work), p. V5 159 (1974).

9. Mackison, F. W., R. S. Stricoff, and L. J. Partridge, Jr. (eds.). *NIOSH/OSHA—Occupational Health Guidelines for Chemi-cal Hazards*. DHHS (NIOSH) Publication No. 81–123 (3 vols). Washington, DC: U.S. Government Printing Office, Jan. 1981.

10. National Fire Protection Association. *Fire Protection Guide on Hazardous Materials*. 7th ed. Boston, MA: National Fire Protection Association, 1978, p. 49–146.

11. Bretherick, L. *Handbook of Reactive Chemical Hazards*. 3rd ed. Boston, MA: Butterworths, 1985. 776.

12. NIOSH. *NIOSH Pocket Guide to Chemical Hazards*. DHHS (NIOSH) Publication No. 90–117. Washington, DC: U.S. Government Printing Office, June 1990, 102.

13. Fazzalari, F. A. (ed.). *Compilation of Odor and Taste Threshold Values Data*. ASTM Data Series DS 48A (Committee E–18). Philadelphia, PA: American Society for Testing and Materials, 1978. 58.

14. Sittig, M. *Handbook of Toxic and Hazardous Chemicals and Carcinogens*, 1985. 2nd ed. Park Ridge, NJ: Noyes Data Corporation, 1985. 398.

15. Bulter, L. C., et al. *J Envir Sci Health* B16 (4): 395–408 (1981).

16. IARC. *Monographs on the Evaluation of the Carcinogenic Risk of Chemicals to Man*. Geneva: World Health Organization, International Agency for Research on Cancer, 1972–present (multivolume work), p. S7 63 (1987).

17. Sharma, S. K., et al. *Bull Envir Contam Toxicol* 23 (1–2): 153–7 (1979).

18. U.S. Department of the Interior, Fish and Wildlife Service. *Handbook of Toxicity of Pesticides to Wildlife*. Resource Publication 153. Washington, DC: U.S. Government Printing Office, 1984. 38.

19. American Conference of Governmental Industrial Hygienists. *Documentation of the Threshold Limit Values and Biological Exposure Indices*. 5th ed. Cincinnati, OH: American Conference of Governmental Industrial Hygienists, 1986. 231.

20. U.S. Department of Interior, Fish and Wildlife Service. *Handbook of Acute Toxicity of Chemicals to Fish and Aquatic Invertebrates*. Resource Publication No.

137. Washington, DC: U.S. Government Printing Office, 1980.

21. U.S. Department of the Interior, Fish and Wildlife Service, Bureau of Sports Fisheries and Wildlife. *Lethal Dietary Toxicities of Environmental Pollutants to Birds.* Special Scientific Report—*Wildlife* No. 191. Washington, DC: U.S. Government Printing Office, 1975. 21.

22. Nagvi, S. M., D. E. Ferguson. *Jour Miss Acad Sci* 14: 121 (1968) as cited in USEPA, *Ambient Water Quality Criteria Doc; Endrin* p. B–33 (1980) EPA 440/5–80–047.

23. Verschueren, K. *Handbook of Environmental Data of Organic Chemicals.* 2nd ed. New York, NY: Van Nostrand Reinhold Co., 1983. 608.

24. Hill, E. F., and M. B. Camardese. *Lethal Dietary Toxicities of Environmental Contaminants and Pesticides to Coturnix.* Fish and Wildlife Technical Report 2.Washington, DC: United States Department of Interior, Fish and Wildlife Service, 1986. 70.

25. (1) Sharom, M. S., et al. *Water Res* 14: 1089–93 (1980); (2) Eichelberger, J. W., J. J. Lichtenberg. *Environ Sci Technol* 5: 541–4 (1971); (3) Menzie, C. M. *Ann Rev Entomol* 17: 199–222 (1972); (4) Nash, R. G., E. A. Woolson. *Science* 157: 924–7 (1967); (5) Alexander, M. *Biotech Bioeng* 15: 611–47 (1973); (6) Bowman, M. C., et al. *J Agr Food Chem* 13: 360–5 (1965); (7) Castro, T. F., T. Yoshida. *J Agr Food Chem* 19: 1168–7 (1971); (8) Syracuse Research Corporation. *Hazard Assessment Report on Endrin*, First Draft, Syracuse, NY, pp. 124 TR69–119 (1980); (9) Hill, D. W., P. L. McCarthy. *J Water Pollut Control Fed* 39: 1259–77 (1967).

26. Anderson, R. L., et al. *Environ Pollut* SER A 22 (2): 111 (1980).

27. 29 CFR 1910.1000 (7/1/91).

28. American Conference of Governmental Industrial Hygienists. *Threshold Limit Values for Chemical Substances and Physical Agents and Biological Exposure Indices for 1993–1994.* Cincinnati, OH: ACGIH, 1993.

29. 40 CFR 261.33 (7/1/91).

30. 40 CFR 152.175 (7/1/91).

31. McKee, J. E., H. W. Wolf. *Water Quality Criteria*, California State Water Quality Control Board, 1963. 2nd Ed.

32. Eichelberger, J. W., J. J. Lichtenberg. Persistence of Pesticides in River Water, *Environmental Science and Technology*, 1971, Vol. 5, No. 6. June.

■ 1,4:5,8-DIMETHANONAPHTHALENE, 1,2,3,4,10,10-HEXACHLORO-6,7-EPOXY-1,4,4a,5,6,7,8,8a OCTAHYDRO, endo, exo

CAS # 60–57–1

DOT # 2761

SIC CODES 2879; 2842

SYNONYMS 1,2,3,4,10,10-hexachloro-6,7-epoxy-1,4,4a,5,6,7,8,8a-octa-hydro-endo, exo 1,4:5,8-dimethanonaphthalene • 1,2,3,4,10,10-hexachloro-exo-6,7-epoxy-1,4,4a,5,6,7,8,8a-octahydro-1,4-endo, exo-5,8-dimethanonaphthalene • 1,4:5,8-dimethanonaphthalene, 1,2,3,4,10,10-hexachloro-6,7-epoxy-1,4,4a,5,6,7,8,8a-octahydro-, endo, exo • 2,7:3,6-dimethanonaphth(2,3-b)oxirene, 3,4,5,6,9,9-hexachloro-1a, 2,2a,3,6,6a,7,7a-octahydro-, (1aalpha, 2beta, 2aalpha, 3beta, 6beta, 6aalpha, 7beta, 7aalpha), • dieldrine (French) • ent-16225 • exo-dieldrin • heod • hexachloroepoxyoctahydro-endo, exo-dimethanonaphthalene • NCI-C00124 • 1,8,9,10,11,11-hexachloro-4,5-exo epoxy-2,3-7,6-endo-2,1-7,8-exo-tetracyclo (6.2.1.1 3,6 0.0 2,7) dodec-9-ene • compound-497 • 3,4,5,6,9,9-hexachloro-1a,2,2a,3,6,6a,7,7a-octahydro-2,7:3,6-dimethanonaphth[2,3-b] oxirene • AI3-16225 • Caswell no. 333 • ENT-16,225 • EPA pesticide chemical code 045001 • 3,4,5,6,9,9-hexachloro-1a,2,2a,3,6,6a,7,7a-octahydro-2,7:3,6-dimethanonaphth (2,3-b) oxirene • sd-

3417 • endo, exo-3,4,5,6,9,9-hexachloro-1a,2,2a,3,6,6a,7,7a-octahydro-2,7:3,6-dimethenapth(2,3-b)oxirene • 1,2,3,4,10,10-hexachloro-6,7-epoxy-1,4,4a,5,6,7,8,8a-octahydro-1,4-endo-exo 5,8-dimethanonaphthalene • (1aalpha, 2beta, 2aalpha, 3beta, 6beta, 6aalpha, 7beta, 7aalpha)-3,4,5,6,9,9-hexachloro-1a,2,2a,3,6,6a,7,7a-octahydro-2,7:3,6-dimethanonaphth[2,3-b]oxirene • insecticide no. 497 • ent-16225 • octalox • panoram d-31 • (1r, 4s, 4as, 5r, 6r, 7s, 8s, 8ar)-1,2,3,4,10,10-hexachloro-1,4,4a,5,6,7,8,8a octahydro-6,7-epoxy-1,4:5,8-dimethanonaphthalene • 1,2,3,4,10,10-hexachloro-6,7-epoxy-1,4,4a,5,6,7,8,8a-octahydro-endo-1,4-exo 5,8-dimethanonaphthalene • endo, exo-1,2,3,4,10,10-hexachloro-6,7-epoxy-1,4,4a,5,6,7,8,8a-octahydro 1,4:5,8-dimethanonaphthalene • 1,2,3,4,10,10-hexachloro-1r, 4s, 4as, 5r, 6r, 7s, 8s, 8ar-octahydro-6,7-epoxy-1,4:5,8-dimethanonaphthalene • dieldrex • dieldrite • dielmoth • quintox • termitox • red shield • alvit • illoxol • dieldrex

MF $C_{12}H_8Cl_6O$

MW 380.9

COLOR AND FORM Colorless crystals; pale-tan flakes; white, crystalline substance.

ODOR Mild chemical odor.

ODOR THRESHOLD 0.041 ppm [R1].

DENSITY 1.75.

MELTING POINT 175–176°C.

FIREFIGHTING PROCEDURES Extinguish fire using an agent suitable for the type of surrounding fire. Material itself does not burn or burns with difficulty [R8].

TOXIC COMBUSTION PRODUCTS Toxic and irritating hydrogen chloride fumes may form in fire [R1].

SOLUBILITY Slightly soluble in mineral oils, moderately soluble in acetone; slightly soluble in petroleum ether, freely soluble in benzene; insoluble in methanol and aliphatic hydrocarbons; 186 µg/L water at 25°C; 48 g/100 ml ethylene dichloride at 20°C; in acetone 220, ethanol 40, dichlo-romethane 480, benzene 400, toluene 410, carbon tetrachloride 380, methanol 10 (all in g/L at 20°C).

VAPOR PRESSURE 7.78×10^{-7} mm Hg at 25°C.

CORROSIVITY Slightly corrosive to metals.

STABILITY AND SHELF LIFE Stable to light [R12, 203].

OTHER CHEMICAL/PHYSICAL PROPERTIES Buff to light-brown flakes (technical grade).

PROTECTIVE EQUIPMENT Wear rubber gloves, air-breathing apparatus, and overalls [R10].

PREVENTATIVE MEASURES Contact lenses should not be worn when working with this chemical [R9].

CLEANUP Absorb spills with paper towels [R10].

DISPOSAL Incineration (1,500°F, 0.5-second minimum for primary combustion; 3,200°F, 1.0 second for secondary combustion) with adequate scrubbing, and ash disposal facilities. [R11, 340].

COMMON USES Broad-spectrum insecticide used until 1974; EPA restricted its use to termite control by direct soil injection and non-food seed and plant treatment [R4]; wool processing industry [R5, 513]. Dieldrin was used in tropical countries as a residual spray on the inside walls and ceilings of homes for the control of vectors of diseases, mainly malaria [R6, 828]. Dieldrin is used to control locusts and tropical disease vectors, such as *Glossina* species. Industrial uses include timber preservation, termite-proofing of plastic, and rubber coverings of electrical and telecommunication cables, of plywood and building boards, and as a termite barrier in building construction [R3, 279]. Found in certain crops [R2]; control of public health insect pests, termites, locusts, and tropical disease vectors [R7].

STANDARD CODES EPA 311; Superfund designated (hazardous substances) list.

PERSISTENCE Dieldrin is said not to persist in soil, but one application of 1 lb/acre continued to kill aquatic beetles for over

10 months [R31]. Persistence studies showed 100% remaining in river water after 8 weeks [R32]. Some inert diluents catalyze decomposition. Edwards reports 75% left in soil after 1 year, 40% after 3 years.

MAJOR SPECIES THREATENED All forms of animal life. especially game fish.

INHALATION LIMIT 0.25 mg/m^3.

PERSONAL SAFETY May be absorbed through the skin. Protective clothing and filter mask are recommended.

ACUTE HAZARD LEVEL Lethal dose for man is estimated to be 71 mg/kg [R31]. Lethal threshold values, minimum LC values trout—0.24, *Cambarus* (crayfish)—2.4, carp—1.92 mg/kg.

CHRONIC HAZARD LEVEL Rats have shown imbalance in liver enzymes.

DEGREE OF HAZARD TO PUBLIC HEALTH High. Potential carcinogenic and mutagenic effects.

OCCUPATIONAL RECOMMENDATIONS None listed.

OSHA 8-hr time-weighted average: 0.25 mg/m^3). Transitional limits were achieved by any combination of engineering controls, work practices, and personal protective equipment during the phase-in period, Sept 1, 1989 through Dec 30, 1993. Final rule limits became effective Dec 31, 1993 [R26].

NIOSH NIOSH recommends that dieldrin be regulated as a potential human carcinogen [R9]. Threshold-limit 8-hr time-weighted average (TWA) 0.25 mg/m^3, skin (1986). [R27, 19].

INTERNATIONAL EXPOSURE LIMITS Other recommendations: former USSR (1977), Bulgaria (1971), Hungary (1974), and Poland (1976) 0.01 mg/m^3; Romania (1975) 0.2 mg/m^3 [R28].

ACTION LEVEL Evacuate area. Enter from upwind side. Notify local air authority and National Agricultural Chemicals Association.

ON-SITE RESTORATION Carbon or peat can be employed to adsorb spills. Seek profes-

sional environmental engineering assistance through EPA's Environmental Response Team (ERT), Edison, NJ, 24-hour no. 908–548–8730.

AVAILABILITY OF COUNTERMEASURE MATERIAL Carbon—water treatment plants, sugar refineries; peat—nurseries, floral shops.

DISPOSAL (1) Mix with vermiculite, sodium bicarbonate, or sand-soda-ash mixture, place in paper box in open incinerator, cover with paper and wood, and ignite with excelsior train. Stay on upwind side. (2) Dissolve in flammable solvent and spray in fire box of incinerator equipped with scrubber and afterburner.

DISPOSAL NOTIFICATION Contact local air and fire authorities.

MAJOR WATER USE THREATENED Fisheries, recreation water, potable supplies.

PROBABLE LOCATION AND STATE Will not dissolve unless accompanied by wetting agents.

FOOD-CHAIN CONCENTRATION POTENTIAL High, as for all chlorinated hydrocarbons; significant concentrations found in birds of prey in the Netherlands, both in the livers and mesenterial fat.

NON-HUMAN TOXICITY LD$_{50}$ rat oral 38.3 mg/kg [R14]; LD$_{50}$ sheep oral 50–75 mg/kg [R15]; LD$_{50}$ rat male dermal (20% emulsification concentration) 213.8 mg/kg [R15]; LD$_{50}$ rat female dermal (20% emulsification concentration) 119.9 mg/kg [R15]; LD$_{50}$ rat dermal (50% wettable powder) 213.4 mg/kg [R15]; LD$_{50}$ *Capra hircus* (domestic goat) male oral 100–200 mg/kg, 6–8 mo old [R13]; LD$_{50}$ rat male, oral 47 mg/kg [R6, 828]; LD$_{50}$ rat female, oral 38 mg/kg [R6, 828]; LD$_{50}$ rat newborn male, oral 167 mg/kg [R6, 828]; LD$_{50}$ rat preweanling male, oral 24 mg/kg [R6, 828]; LD$_{50}$ rat male dermal 90 mg/kg [R6, 828]; LD$_{50}$ rat female dermal 60 mg/kg [R6, 828]; LD$_{50}$ rabbit dermal <150 mg/kg [R6, 828].

ECOTOXICITY LC$_{50}$ *Tubifex* and *Limnodrilus* (mixed cultures) 6,700 µg/L/96 hr (conditions of bioassay not specified) [R16]; LC$_{50}$ *Cypretta kawati* (ostracod)

185 µg/L/24 hr; 12.3 µg/L/72 hr (conditions of bioassay not specified) [R17]; LC_{50} *Chironomus tentans* 0.9 µg/L/24 hr (conditions of bioassay not specified) [R18]; LC_{50} *Pseudacris triseriata* (frog, tadpoles) 100 µg/L/96 hr (conditions of bioassay not specified) [R19]; LD_{50} *Anas platyrhynchos* (mallard) female oral 381 mg/kg (95% confidence limit 141–1,030 mg/kg) 6–7 mo old [R13]; LD_{50} *Phasianus colchicus* (pheasant) male oral 79.0 mg/kg (95% confidence limit 33.3–187 mg/kg) 10–23 mo old [R13]; LD_{50} *Alectoris chukar* (chukar) oral 25.3 mg/kg (95% confidence limit 15.2–42.2 mg/kg) 8–11 mo old [R13]; LD_{50} *Passer domesticus* (house sparrow) female oral 47.6 mg/kg (95% confidence limit 34.3–66.0 mg/kg) [R13]; LD_{50} *Branta canadensis* (Canada goose) oral 50–150 mg/kg adult [R13]; LD_{50} *Dendrocygna bicolor* (fulvous whistling duck) female oral 100–200 mg/kg [R13]; LD_{50} *Perdix perdix* (gray partridge) female oral 8.84 mg/kg (95% confidence limit 3.32–23.6 mg/kg) 3–10 mo old [R13]; LD_{50} *Odocoileus hemionus hemionus* (mule deer) male oral 75–150 mg/kg (95% confidence limit) 8–18 mo old [R13]; LC_{50} *Salmo clarki* (cutthroat trout) 6.0 µg/L/96 hr (95% confidence limit 4.6–8.0 µg/L) wt 1.1 g, water temperature 9°C, static bioassay without aeration, pH 7.2–7.5, water hardness 40–50 mg/L as calcium carbonate and alkalinity of 30–35 mg/L (technical, 85% heod) [R20, 30]; LC_{50} *Salmo gairdneri* (rainbow trout) 1.2 µg/L/96 hr (95% confidence limit 0.9–1.7 µg/L) wt 1.4 g, water 13°C, static bioassay without aeration, pH 7.2–7.5, water hardness 40–50 mg/L as calcium carbonate and alkalinity of 30–35 mg/L (technical, 85% heod) [R20, 30]; LC_{50} *Carassius auratus* (goldfish) 1.8 µg/L/96 hr (95% confidence limit 1.2–2.8 µg/L) wt 1.0 g, water 18°C, static bioassay without aeration, pH 7.2–7.5, water hardness 40–50 mg/L as calcium carbonate and alkalinity of 30–35 mg/L (technical, 85% heod) [R20, 30]; LC_{50} *Pimephales promelas* (fathead minnow) 3.8 µg/L/96 hr (95% confidence limit 3.1–4.6 µg/L) wt 0.6 g, water 18°C, static bioassay without aeration, pH 7.2–7.5, water hardness 40–50 mg/L as calcium carbonate and alkalinity of 30–35 mg/L (technical, 85%

heod) [R20, 30]; LC_{50} *Ictalurus punctatus* (channel catfish) 4.5 µg/L/96 hr (95% confidence limit 2.5–7.9 µg/L) wt 1.4 g, water 18°C, static bioassay without aeration, pH 7.2–7.5, water hardness 40–50 mg/L as calcium carbonate and alkalinity of 30–35 mg/L (technical, 85% heod) [R20, 30]; LC_{50} *Lepomis macrochirus* (bluegill) 3.1 µg/L/96 hr (95% confidence limit 2.1–4.6 µg/L) wt 1.3 g, water 18°C, static bioassay without aeration, pH 7.2–7.5, water hardness 40–50 mg/L as calcium carbonate and alkalinity of 30–35 mg/L (technical, 85% heod) [R20, 30]; LC_{50} *Macropterus salmoides* (largemouth bass) 3.5 µg/L/96 hr (95% confidence limit 2.7–4.5 µg/L) wt 2.5 g, water 18°C, static bioassay without aeration, pH 7.2–7.5, water hardness 40–50 mg/L as calcium carbonate and alkalinity of 30–35 mg/L (technical, 85% heod) [R20, 30]; LC_{50} *Salmo clarki* (cutthroat trout) 12 µg/L/96 hr (95% confidence limit 11–14 µg/L) wt 1.3 g, hard water 8°C, static bioassay without aeration, pH 7.2–7.5, water hardness 40–50 mg/L as calcium carbonate and alkalinity of 30–35 mg/L. Static bioassay without aeration, pH 7.2–7.5, water hardness 40–50 mg/L as calcium carbonate and alkalinity of 30–35 mg/L (photodieldrin 98%) [R20, 30]; LC_{50} *Lepomis macrochirus* (bluegill) 11 µg/L/96 hr (95% confidence limit 9.3–13 µg/L) wt 1.4 g, water 18°C, static bioassay without aeration, pH 7.2–7.5, water hardness 40–50 mg/L as calcium carbonate and alkalinity of 30–35 mg/L (photodieldrin 98%) [R20, 30]; LC_{50} *Ictalurus punctatus* (channel catfish) 19 µg/L/96 hr (95% confidence limit 13–27 µg/L) wt 1.4 g, hard water 18°C, static bioassay without aeration, pH 7.2–7.5, water hardness 40–50 mg/L as calcium carbonate and alkalinity of 30–35 mg/L (photo-dieldrin 98%) [R20, 30]; LC_{50} *Asellus* 5.0 µg/L/96 hr (95% confidence limit 3.2–10.0 µg/L) mature, water temperature 21°C, static bioassay without aeration, pH 7.2–7.5, water hardness 40–50 mg/L as calcium carbonate and alkalinity of 30–35 mg/L (technical material 85% heod) [R20, 29]; LC_{50} *Gammarus fasciatus* 640 µg/L/96 hr (95% confidence limit 460–880 µg/L) mature, water temperature 21°C, static bioassay without aeration, pH 7.2–7.5, water hardness 40–50 mg/L as calcium carbonate and alkalinity of 30–35 mg/L (technical, 85% heod) [R20, 29]; LC_{50} *Orconectes* 740 μg/L/96 hr (95% confidence limit 680–1,200 μg/L) mature, water temperature 21°C, static bioassay without aeration, pH 7.2–7.5, water hardness 40–50 mg/L as calcium carbonate and alkalinity of 30–35 mg/L (technical, 85% heod) [R20, 30]; LC_{50} *Pteronarcys* 0.5 μg/L/96 hr (95% confidence limit 0.4–0.7 μg/L) 2nd yr class, water temperature 15°C, static bioassay without aeration, pH 7.2–7.5, water hardness 40–50 mg/L as calcium carbonate and alkalinity of 30–35 mg/L (technical, 85% heod) [R20, 30]; LC_{50} *Pteronarcella* 0.5 μg/L/96 hr (95% confidence limit 0.4–0.7 μg/L) 1st yr class, water temperature 15°C, static bioassay without aeration, pH 7.2–7.5, water hardness 40–50 mg/L as calcium carbonate and alkalinity of 30–35 mg/L (technical) [R20, 30]; LC_{50} *Claassemia* 0.6 μg/L/96 hr (95% confidence limit 0.4–0.8 μg/L) 2nd yr class, water temperature 15°C, static bioassay without aeration, pH 7.2–7.5, water hardness 40–50 mg/L as calcium carbonate and alkalinity of 30–35 mg/L (technical, 85% heod) [R20, 30]; LC_{50} *Ischnura* 12 μg/L/96 hr (95% confidence limit) juvenile, water temperature 24°C, static bioassay without aeration, pH 7.2–7.5, water hardness 40–50 mg/L as calcium carbonate and alkalinity of 30–35 mg/L (technical, 85% heod) [R20, 30]; LC_{50} Young *Coturnix* (Japanese quail) 60 ppm (95% confidence limit 57–63 ppm) 5-day diet [R21]; LC_{50} *Aedes aegypti* (mosquito) late 3rd instar larvae 6 ppb/24 hr (conditions of bioassay not specified) [R5, 516]; LD_{50} *Musca domestica* (housefly) female 9.8 μg/fly, 3 days old [R5, 516]; LC_{50} *Sphaeroides maculatus* (Northern puffer) 34 ppb/96 hr, static lab bioassay (100%) [R5, 517]; LC_{50} *Bufo woodhousi* (toad, tadpoles) 150 μg/L/96 hr (conditions of bioassay not specified) [R19].

IARC SUMMARY Inadequate evidence of carcinogenicity in humans. Limited evidence of carcinogenicity in animals. Overall evaluation: Group 3: The agent is not classifiable as to its carcinogenicity to humans [R22].

BIOLOGICAL HALF-LIFE Biological half-life of dieldrin in blood of humans ranges from 141 to 592 days with a mean of 369 days [R23].

BIODEGRADATION Dieldrin is not biodegraded in standard screening tests (1) and is extremely persistent in soils (2) under both aerobic and anerobic conditions (3). It took 7 yr for half of the dieldrin to disappear from soil field plots (4). No biodegradation in river waters has been noted (5,7). There is some evidence that microorganisms can form photodieldrin from dieldrin (4,6) [R24].

BIOCONCENTRATION Moderate to significant bioconcentration (100 to 10,000) in various aquatic species (1). BCF of 3–6,000 in fish (2–3) [R25].

RCRA REQUIREMENTS P037; as stipulated in 40 CFR 261.33, when dieldrin, as a commercial chemical product or manufacturing chemical intermediate or an off-specification commercial chemical product or a manufacturing chemical intermediate, becomes a waste, it must be managed according to federal or state hazardous waste regulations. Also defined as a hazardous waste is any container or inner liner used to hold this waste or any residue, contaminated soil, water or other debris resulting from the cleanup of a spill, into water or on dry land, of this waste. Generators of small quantities of this waste may qualify for partial exclusion from hazardous waste regulations (40 CFR 261.5 (e)) [R29].

FIFRA REQUIREMENTS Chemical: aldrin/dieldrin; pesticide and review criteria possibly met or exceeded: carcinogenicity, bio-accumulation, hazard to wildlife, and other chronic effects; final determination: 39 FR 37246 (10/18/74); cancellation for all but termiticide use [R30].

GENERAL RESPONSE Avoid contact with solid and dust. Keep people away. Wear goggles, dust respirator, and rubber overclothing (including gloves). Stop discharge if possible. Isolate and remove discharged material. Notify local health and pollution control agencies.

FIRE RESPONSE Not flammable. Poisonous gases may be produced when heated.

EXPOSURE RESPONSE Call for medical aid. Dust is poisonous if inhaled or if skin is exposed. If inhaled will cause headache, dizziness, or loss of consciousness. If breathing has stopped, give artificial respiration. If breathing is difficult, give oxygen. Solid is poisonous if swallowed or if skin is exposed. If swallowed will cause headache, nausea, dizziness, vomiting, or loss of consciousness. Remove contaminated clothing and shoes. Flush affected areas with plenty of water. If in eyes, hold eyelids open, and flush with plenty of water. If swallowed and victim is conscious, have victim drink water or milk, and have victim induce vomiting. If swallowed and victim is unconscious or having convulsions, do nothing except keep victim warm.

RESPONSE TO DISCHARGE Issue warning—water contaminant. Restrict access. Should be removed. Provide chemical and physical treatment.

WATER RESPONSE Harmful to aquatic life in very low concentrations. May be dangerous if it enters water intakes. Notify local health and wildlife officials. Notify operators of nearby water intakes.

EXPOSURE SYMPTOMS Inhalation, ingestion, or skin contact causes irritability, convulsions, or coma, nausea, vomiting, headache, fainting, tremors. Contact with eyes causes irritation.

EXPOSURE TREATMENT Inhalation: move to fresh air; give oxygen and artificial respiration as required. Ingestion: induce vomiting and get medical attention. Eyes: flush with plenty of water; get medical attention. Skin: flush with plenty of water.

REFERENCES

1. U.S. Coast Guard, Department of Transportation. *CHRIS—Hazardous Chemical Data*. Volume II. Washington, DC: U.S. Government Printing Office, 1984–5.

2. *Farm Chemicals Handbook 1992*. Willoughby, OH: Meister Publishing Co., 1992, p. C–115.

3. Worthing, C. R., and S. B. Walker (eds.). *The Pesticide Manual—A World Compendium*. 8th ed. Thornton Heath, UK: The British Crop Protection Council, 1987.

4. *USEPA; Ambient Water Quality Criteria Doc; Aldrin/Dieldrin* p. A–1 (1980) EPA 440/5–80–019.

5. Verschueren, K. *Handbook of Environmental Data of Organic Chemicals*. 2nd ed. New York, NY: Van Nostrand Reinhold Co., 1983.

6. Hayes, W. J., Jr., E. R. Laws, Jr. (eds.). *Handbook of Pesticide Toxicology. Volume 2. Classes of Pesticides*. New York, NY: Academic Preess, Inc., 1991.

7. Hartley, D., and H. Kidd (eds.). *The Agrochemicals Handbook*. 2nd ed. Lechworth, Herts, England: The Royal Society of Chemistry, 1987, p. A144/Aug 87.

8. Bureau of Explosives. *Emergency Handling of Haz Matl in Surface Trans*, p.191 (1981).

9. NIOSH. *NIOSH Pocket Guide to Chemical Hazards*. DHHS (NIOSH) Publication No. 90–117. Washington, DC: U.S. Government Printing Office, June 1990, 90.

10. ITII. *Toxic and Hazardous Industrial Chemicals Safety Manual*. Tokyo, Japan: The International Technical Information Institute, 1982. 172.

11. Sittig, M. *Handbook of Toxic and Hazardous Chemicals and Carcinogens*, 1985. 2nd ed. Park Ridge, NJ: Noyes Data Corporation, 1985.

12. Spencer, E. Y. *Guide to the Chemicals Used in Crop Protection*. 7th ed. Publication 1093. Research Institute, Agriculture Canada, Ottawa, Canada: Information Canada, 1982.

13. U.S. Department of the Interior, Fish and Wildlife Service. *Handbook of Toxicity of Pesticides to Wildlife*. Resource Publication 153. Washington, DC: U.S. Government Printing Office, 1984. 31.

14. Jager, K. W. *Aldrin, Dieldrin, Endrin,*

Telodrin: An Epidemiological and Toxicological Study of Long-Term Occupational Exposure (1970) as cited in NIOSH, *Special Occupational Hazard Review*: 78–201.

15. Jager, K. W. *Aldrin, Dieldrin, Endrin, Telodrin: An Epidemiological, and Toxicological Study of Long-Term Occupational Exposure* (1970) as cited in NIOSH, *Special Occupational Hazard Review: Aldrin/Dieldrin* p. 32 (1978) DHEW Pub. NIOSH 78–201.

16. Whitten, B. K., et al. *Jour Water Pollut Control Fed* 38: 227 (1966) as cited in USEPA. *Ambient Water Quality Criteria Doc; Aldrin/Dieldrin*, p. B–37 (1980) EPA 440/5–80–019.

17. Hansen, C. R., J. A. Kawatski. *J Fish Res Board Can* 33: 1198 (1976) as cited in USEPA. *Ambient Water Quality Criteria Doc; Aldrin/Dieldrin*, p. B–37 (1980) EPA 440/5–80–019.

18. Karnak, R. E., W. J. Collins. *Bull Environ Contam Toxicol* 12: 62 (1974) as cited in USEPA. *Ambient Water Quality Criteria Doc; Aldrin/Dieldrin*, p. B–37 (1980) EPA 440/5–80–019.

19. Sanders, H. O. *Copeia* 2: 246 (190) as cited in USEPA. *Ambient Water Quality Criteria Doc; Aldrin/Dieldrin* p. B–38 (1980) EPA 440/5–80–019.

20. U.S. Department of Interior, Fish and Wildlife Service. *Handbook of Acute Toxicity of Chemicals to Fish and Aquatic Invertebrates*. Resource Publication No. 137. Washington, DC: U.S. Government Printing Office, 1980.

21. Hill, E. F., and M. B. Camardese. *Lethal Dietary Toxicities of Environmental Contaminants and Pesticides to Coturnix. Fish and Wildlife Technical Report* 2. Washington, DC: United States Department of Interior, Fish and Wildlife Service, 1986. 60.

22. IARC. *Monographs on the Evaluation of the Carcinogenic Risk of Chemicals to Man*. Geneva: World Health Organization, International Agency for Research on Cancer, 1972–present (multivolume work), p. S7 62 (1987).

23. Hunter, C. G., et al. *Arch Environ Health* 18: 12 (1969) as cited in USEPA, *Hazard Profile: Dieldrin* p. 6 (1980).

24. (1) Tabak, H. H., et al. *J Water Poll Control Fed* 53: 1503–18 (1981); (2) Sanborn, J. R., et al. *The Degradation of Selected Pesticides in Soil: A Review of the Published Literature* USEPA 600/9–77–022 (1977); (3) Castro, T. F., T. Yoshida. *J Agric Food Chem* 19: 1168–70; (4) Nash, R. G., E. A. Woolson. *Science* 157: 924–7 (1967); (5) Eichelberger, J. W., J. J., Lichtenberg. *Environ Sci Technol* 5: 501–4 (1971); (6) Matsumura, F., et al. *Science* 170: 1206–7 (1970); (7) Sharom, M. S., et al. *Water Res* 14: 1089–93 (1980).

25. (1) Callahan, M. A., et al. *Water-Related Fate of 129 Priority Pollutants*, pp. 26–1 to 26–12 USEPA 440/4–79–029a (1979); (2) Sanborn, T. R., C.-C. Yu. *Bull Environ Cont Toxicol* 10: 340–6 (1973); (3) Metcalf, R. L., et al. *Environ Health Perspect.* 1973 No. 4: 35–44 (1973).

26. 29 CFR 1910.1000 (7/1/91).

27. American Conference of Governmental Industrial Hygienists. *Threshold Limit Values for Chemical Substances and Physical Agents and Biological Exposure Indices for 1993–1994.* Cincinnati, OH: ACGIH, 1993.

28. American Conference of Governmental Industrial Hygienists. *Documentation of the Threshold Limit Values and Biological Exposure Indices.* 5th ed. Cincinnati, OH: American Conference of Governmental Industrial Hygienists, 1986. 196.

29. 40 CFR 261.33 (7/1/91).

30. USEPA/OPP. *Status of Pesticides in Reregistration and Special Review* p. 43 (Mar, 1992) EPA 700–R–92–004.

31. McKee, J. E., H. W. Wolf. *Water Quality Criteria*, Caifornia State Water Quality Control Board, 1963. 2nd Ed.

32. Eichelberger, J. W., J. J.Lichtenberg. "Persistence of Pesticides in River Water," *Environmental Science and Technology*, 1971, Vol. 5, No. 6. June.

■ DIMETHOXANE

CAS # 828-00-2

SYNONYMS acetic acid, ester with
2,6-dimethyl-m-dioxan-4-ol • acetic acid,
2,6-dimethyl-m-dioxan-4-yl ester •
acetomethoxan • acetomethoxane •
6-acetoxy-2,4-dimethyl-meta-dioxane •
6-acetoxy-2,4-dimethyl-1,3-dioxane •
ddoa • 2,4-dimethyl-6-acetoxy-1,3-
dioxane • 2,6-dimethyl-meta-dioxan-4-ol
acetate • 2,6-dimethyl-1,3-dioxan-4-ol
acetate • 2,4-dimethyl-6-m-dioxanyl
acetate • 2,6-dimethyl-meta-dioxan-
4-yl acetate • m-dioxan-4-ol,
2,6-dimethyl-, acetate • 1,3-dioxan-4-ol,
2-6-dimethyl-, acetate • giv-gard-dxn •
giv gard dxn-co

MF $C_8H_{14}O_4$

MW 174.22

COLOR AND FORM Liquid; clear yellow to
light-amber liquid.

ODOR Mustard-like odor.

DENSITY 1.0655 at 20°C.

BOILING POINT 74-75°C at 6 mm Hg.

MELTING POINT FP: less than 25°C.

SOLUBILITY Miscible with water, many
organic solvents.

STABILITY AND SHELF LIFE Hydrolyzes in
aqueous solution to produce acetic acid
and corresponding free alcohol. [R6, 177].

DISPOSAL SRP: At the time of review,
criteria for land treatment or burial (sanitary landfill) disposal practices are subject
to significant revision. Prior to implementing land disposal of waste residue (including waste sludge), consult with environmental regulatory agencies for guidance
on acceptable disposal practices.

COMMON USES Preservative for cutting oil,
resin emulsions, water-based paints, cosmetics, inks. Effective range of concentration: 0.03-0.1%. Gasoline additive [R1].
Antimicrobial agent used to protect
against spoilage due to bacteria, fungi,
and yeasts in water-based paints; it can

be used in water-based cutting oils, specialty textile chemical emulsions, dyestuffs, fabric softeners, latex emulsions,
sizings, adhesives, antistatic lubricants,
and spinning emulsions at 500-1,500
mg/kg. [R6, 178].

OCCUPATIONAL RECOMMENDATIONS None
listed.

IARC SUMMARY No data are available for
humans. Limited evidence of carcinogenicity in animals. Overall evaluation:
Group 3: The agent is not classifiable as to
its carcinogenicity to humans [R2].

BIODEGRADATION In a 4-week biodegradation screening test (MITI test) using dimethoxane (100 ppm) and an activated
sludge inoculum, 76-83% of BOD was
removed (1) [R3].

BIOCONCENTRATION Using an estimated
log K_{ow} of 0.49 (1,SRC), one would estimate
a BCF of 1.4 for dimethoxane using a
recommended regression equation (2).
This would indicate that dimethoxane
would not bioconcentrate in aquatic organisms (2) [R4].

TSCA REQUIREMENTS Pursuant to section
8 (d) of TSCA, EPA promulgated a model
health and safety data reporting rule. The
section 8 (d) model rule requires manufacturers, importers, and processors of listed
chemical substances and mixtures to submit to EPA copies and lists of unpublished
health and safety studies. 1,3-Dioxin-4-ol,
2,6 dimethyl-, acetate is included on this
list [R5].

FIFRA REQUIREMENTS In 1988, Congress
amended FIFRA to strengthen and accelerate EPA's reregistration program. The
nine-year reregistration scheme mandated by "FIFRA 88" applies to each
registered pesticide product containing an
active ingredient initially registered before
November 1, 1984. Pesticides for which
EPA had not issued Registration Standards prior to the effective date of FIFRA
'88 were divided into three lists based
upon their potential for exposure and
other factors, with List B being of highest
concern and D of least. List: C; Case:
Dimethoxane; Case No.: 3064; Pesticide
type: Fungicide, antimicrobial; Case Sta-

tus: Awaiting Data/Data in Review: OPP awaits data from the pesticide's producer(s) regarding its human health and/or environmental effects, or OPP has received and is reviewing such data, in order to reach a decision about the pesticide's eligibility for reregistration. Active Ingredient (AI): Dimethyl-m-dioxan-4-ol acetate; AI Status: The producer of the pesticide has made commitments to conduct the studies and pay the fees required for reregistration, and is meeting those commitments in a timely manner. [R7].

REFERENCES

1. Budavari, S. (ed.). *The Merck Index—Encyclopedia of Chemicals, Drugs, and Biologicals*. Rahway, NJ: Merck and Co., Inc., 1989. 509.

2. IARC. *Monographs on the Evaluation of the Carcinogenic Risk of Chemicals to Man*. Geneva: World Health Organization, International Agency for Research on Cancer, 1972-present (multivolume work), p. S7 62 (1987).

3. Chemicals Inspection and Testing Institute. *Biodegradation and Bioaccumulation Data of Existing Chemicals Based on the CSCL Japan*. Japan Chemical Industry Ecology–Toxicology and Information Center. p. 5–21, ISBN 4–89074–101–1 (1992).

4. (1) Meylan, W. M., P. H. Howard. *J Pharm Sci* 84: 83–92 (1995). (2) Lyman, W. J., et al., *Handbook of Chemical Property Estimation Methods*. New York: McGraw-Hill, Chapt 5, Eqn 5–2 (1982).

5. 40 CFR 716.120 (7/1/92).

6. IARC. *Monographs on the Evaluation of the Carcinogenic Risk of Chemicals to Man*. Geneva: World Health Organization, International Agency for Research on Cancer, 1972–present. (multivolume work),p. V15.

7. USEPA/OPP. *Status of Pesticides in Reregistration and Special Review*. p.192 (Mar, 1992) EPA 700-R-92-004.

■ 3,3'-DIMETHYLBENZIDINE

CAS # 119–93–7

SYNONYMS (1,1'-biphenyl)-4,4'-diamine, 3,3'-dimethyl • 2-Tolidin (German) • 2-tolidina (Italian) • 2-tolidine • 3,3'-dimethyl-4,4'-biphenyldiamine • 3,3'-dimethyl-4,4'-diaminobiphenyl • 3,3'-dimethyl-4,4'-diphenyldiamine • 3,3'-dimethyl-(1,1'-biphenyl)-4,4'-diamine • 3,3'-dimethylbenzidin • 3,3'-dimethylbiphenyl-4,4'-diamine • 3,3'-dimethyldiphenyl-4,4'-diamine • 3,3'-tolidine • 4,4'-bi-o-toluidine • 4,4'-di-o-toluidine • 4,4'-diamino-3,3'-dimethylbiphenyl • 4,4'-diamino-3,3'-dimethyldiphenyl • benzidine, 3,3'-dimethyl • bianisidine • ci-37230- • ci azoic diazo component 113 • diaminoditolyl • fast dark-blue base-r • orthotolidine • o,o'-tolidine • o-tolidin • o-tolidine • diaminotolyl • dmb

MF $C_{14}H_{16}N_2$

MW 212.32

COLOR AND FORM White to reddish crystals or crystal powder.

BOILING POINT 300°C.

MELTING POINT 129–131°C.

HEAT OF COMBUSTION Liquid: 964.3 kgcal.

SOLUBILITY Slightly soluble in water; soluble in alcohol, ether, dilute acids.

OCTANOL/WATER PARTITION COEFFICIENT Log K_{ow} = 2.34.

STABILITY AND SHELF LIFE Affected by light [R2].

OTHER CHEMICAL/PHYSICAL PROPERTIES An estimated soil adsorption coefficient (K_{oc}) for 3,3'-dimethylbenzidine, using a measured log K_{ow} of 2.34 (1), and eqn 4–8 (2), is 447.

PROTECTIVE EQUIPMENT Wear clothing

and goggles to prevent skin and eye contact. [R13, 868].

PREVENTATIVE MEASURES SRP: The scientific literature supports the wearing of contact lenses in industrial environments, as part of a program to protect the eye against chemical compounds and minerals causing eye irritation. However, there may be individual substances whose irritating or corrosive properties are such that the wearing of contact lenses would be harmful to the eye. In those specific cases contact lenses should not be worn. [R14, 8].

CLEANUP A new method developed for the removal of carcinogenic aromatic amines from industrial aqueous effluents uses horseradish peroxidase and hydrogen peroxide, resulting in a nearly complete precipitation of carcinogenic aromatic amines from water due to enzymatic crosslinking [R4]. A high-efficiency particulate arrestor (HEPA) or charcoal filters can be used to minimize amount of carcinogen in exhausted-air-ventilated safety cabinets, lab hoods, glove boxes, or animal rooms. Filter housing that is designed so that used filters can be transferred into plastic bag without contaminating maintenance staff is avail commercially. Filters should be placed in plastic bags immediately after removal. The plastic bag should be sealed immediately. The sealed bag should be labelled properly. Waste liquids should be placed or collected in proper containers for disposal. The lid should be secured and the bottles properly labelled. Once filled, bottles should be placed in plastic bag, so that outer surface is not contaminated. The plastic bag should also be sealed and labelled. Broken glassware should be decontaminated by solvent extraction, by chemical destruction, or in specially designed incinerators. (chemical carcinogens) [R14, 15].

DISPOSAL Generators of waste (equal to, or greater than 100 kg/mo) containing this contaminant, EPA hazardous waste number U095, must conform with USEPA regulations in storage, transportation, treatment, and disposal of waste [R5].

COMMON USES Very sensitive reagent for detection of gold (1:10 million detectable) and free chlorine in water [R3]; chemical intermediate for azo dyes and 3,3′-dimethyl-4,4′-biphenylene diisocyanate [R1]; curing agent for urethane resins [R2].

OCCUPATIONAL RECOMMENDATIONS None listed.

NIOSH NIOSH recommends that 3,3′ dimethylbenzidine be regulated as a potential human carcinogen [R10].

THRESHOLD LIMIT A2, skin; A2 = suspected human carcinogen (1982) [R6].

IARC SUMMARY No data are available for humans. Sufficient evidence of carcinogenicity in animals. Overall evaluation: Group 2B: The agent is possibly carcinogenic to humans [R7].

BIODEGRADATION In a Warburg respirometer that used activated sludge from both domestic and industrial discharges, 100% of the 3,3′-dimethylbenzidine (initial concentration 20 mg/L) was depleted in 6 hours at 25°C (1) [R8].

BIOCONCENTRATION An estimated bioconcentration factor (BCF) for 3,3′-dimethylbenzidine, using a measured log K_{ow} of 2.34 (1) and eqn 5–2 (2), is 35. This indicates that 3,3′-dimethylbenzidine will not significantly bioconcentrate (SRC) [R9].

CERCLA REPORTABLE QUANTITIES Persons in charge of vessels or facilities are required to notify the National Response Center (NRC) immediately when there is a release of this designated hazardous substance, in an amount equal to or greater than its reportable quantity of 10 lb or 4.54 kg. The toll-free number of the NRC is (800) 424–8802; in the Washington DC metropolitan area (202) 426–2675. The rule for determining when notification is required is stated in 40 CFR 302.4 (section IV. D. 3.b) [R11].

RCRA REQUIREMENTS U095; as stipulated in 40 CFR 261.33, when 3,3′-dimethylbenzidine, as a commercial chemical product or manufacturing chemical intermediate or an off-specification commercial chemical product or a manufacturing chemical intermediate, becomes a waste, it must be managed according to federal or state hazardous waste regulations. Also defined as a hazardous waste is any residue, contaminated soil, water or other debris resulting from the cleanup of a spill, into water or on dry land, of this waste. Generators of small quantities of this waste may qualify for partial exclusion from hazardous waste regulations (40 CFR 261.5) [R12].

REFERENCES

1. SRI.

2. Sax, N. I., and R. J. Lewis, Sr. (eds.). *Hawley's Condensed Chemical Dictionary*. 11th ed. New York: Van Nostrand Reinhold Co., 1987. 1162.

3. Budavari, S. (ed.) *The Merck Index— Encyclopedia of Chemicals, Drugs, and Biologicals*. Rahway, NJ: Merck and Co., Inc., 1989. 1498.

4. Kilbanov, A. M., E. D. Morris. *Enzyme Microb Technol* 3 (2): 119–22 (1981).

5. 40 CFR 240–280, 300–306, 702–799 (7/1/90).

6. American Conference of Governmental Industrial Hygienists. *Threshold Limit Values for Chemical Substances and Physical Agents and Biological Exposure Indices for 1994–1995*. Cincinnati, OH: ACGIH, 1994. 34.

7. IARC. *Monographs on the Evaluation of the Carcinogenic Risk of Chemicals to Man*. Geneva: World Health Organization, International Agency for Research on Cancer, 1972–present (multivolume work), p. S7 62 (1987).

8. Baird, R., et al., *J Water Pollut Control Fed* 49: 1609–15 (1977).

9. (1) Hansch, C., and A. J. Leo. *Medchem Project* Issue No. 26, Claremont CA: Pomona College (1985). (2) Lyman, W. J., et al., *Handbook of Chemical Property Estimation Methods*. New York: McGraw-Hill, pp. 5–4 (1982).

10. NIOSH/CDC. *NIOSH Recommendations for Occupational Safety and Health Standards 1988*, Aug. 1988 (Suppl. to *Morbidity and Mortality Wkly. Vol. 37 No. 5–7*, Aug. 26, 1988). Atlanta, GA: National Institute for Occupational Safety and Health, CDC, 1988. 26.

11. 40 CFR 302.4 (7/1/91).

12. 40 CFR 261.33 (7/1/91).

13. Sittig, M. *Handbook of Toxic and Hazardous Chemicals and Carcinogens, 1985*. 2nd ed. Park Ridge, NJ: Noyes Data Corporation, 1985.

14. Montesano, R., H. Bartsch, E. Boyland, G. Della Porta, L. Fishbein, R.A. Griesemer, A.B. Swan, L. Tomatis, and W. Davis (eds.). *Handling Chemical Carcinogens in the Laboratory: Problems of Safety*. IARC Scientific Publications No. 33. Lyon, France: International Agency for Research on Cancer, 1979.

■ DIMETHYLDICHLOROSILANE

CAS # 75–78–5

DOT # 1162

SYNONYMS dichlorodimethylsilane • dichlorodimethylsilicon • dimethyl-dichlorsilan (Czech) • inerton dw-dmc • silane, dichlorodimethyl

MF $C_2H_6Cl_2Si$

MW 129.07

COLOR AND FORM Colorless liquid.

ODOR Sharp, like hydrochloric acid.

DENSITY 1.07.

SURFACE TENSION 20.1 dynes/cm = 0.0201 N/m at 20°C.

BOILING POINT 70.5°C.

MELTING POINT Lower than −70°C.

HEAT OF VAPORIZATION 100 Btu/lb = 58 cal/g = 2.4×10^5 joule/kg.

AUTOIGNITION TEMPERATURE Above 750°F [R1].

EXPLOSIVE LIMITS AND POTENTIAL Explosive limits: lower: 3.4 volume percent; upper: greater than 9.5 volume percent [R4, 942]. Vapor may explode if ignited in an enclosed area [R1]. Silanes form spontaneously explosive mixtures with air. (silanes) [R5, 3035].

FLAMMABILITY LIMITS Air: 1.4%–9.5% [R1].

FIREFIGHTING PROCEDURES Extinguish with dry chemical or carbon dioxide. Do not use water or foam on fire [R1].

TOXIC COMBUSTION PRODUCTS Hydrogen chloride and phosgene gases may form; both are toxic and irritating [R1].

SOLUBILITY Soluble in benzene and ether.

OTHER CHEMICAL/PHYSICAL PROPERTIES Reacts with water to form complex mixture of dimethylsiloxanes and liberates hydrogen chloride. Organo-functional silanes are noted for their ability to bond organic polymer systems to inorganic substrates (silane compounds). The reaction of organosilanes with halogens and halogen compound usually proceeds in good yield through cleavage of the Si—H bond and formation of the silicon-halogen bond. Reaction with fluorine, however, does not proceed satisfactorily because of cleavage of not only the Si—H but also C—Si and C—H bonds. Direct halogenation with chlorination, bromine, and iodine proceeds smoothly, however. (organosilanes).

PROTECTIVE EQUIPMENT Protective equipment should include corrosion-resistant and impervious suits, foot, hand and arm, head, and eye and face protection; where corrosive gases may be expected resp protective equipment is required; natural rubbers, synthetic rubbers, polyvinyl chloride, polypropylene, or polyethylene either in sheet form or with fabric backing are suitable (for personal protective equipment). (corrosive substances) [R6, 554].

PREVENTATIVE MEASURES Preventive measures should be directed primarily at preventing or minimizing contact between corrosive substances and skin, mucous membranes, and eyes. (corrosive substances) [R6, 553]. SRP: Local exhaust ventilation should be applied wherever there is an incidence of point-source emissions or dispersion of regulated contaminants in the work area. Ventilation control of the contaminant as close as

possible to its point of generation is both the most economical and safest method to minimize personnel exposure to airborne contaminants. The most satisfactory method of ensuring worker protection is to prevent contact with corrosive substances by utilizing only closed-circuit apparatus (corrosive substances) [R6, 438]. It is good practice to install emergency showers at all strategic locations; bath tubs filled with clean water can also provide valuable service in emergency (corrosive substances) [R6, 554]. SRP: The scientific literature supports the wearing of contact lenses in industrial environments, as part of a program to protect the eye against chemical compounds and minerals causing eye irritation. However, there may be individual substances whose irritating or corrosive properties are such that the wearing of contact lenses would be harmful to the eye. In those specific cases contact lenses should not be worn. SRP: Contaminated protective clothing should be segregated in such a manner so that there is no direct personal contact by personnel who handle, dispose of, or clean the clothing. Quality assurance to ascertain the completeness of the cleaning procedures should be implemented before the decontaminated protective clothing is returned for reuse by the workers. Contaminated clothing should not be taken home at end of shift, but should remain at employee's place of work for cleaning. Caution: silanes are toxic. Avoid contact with skin and eyes. Use effective fume removal device. (silanes) [R2, 43.321].

COMMON USES In ethchlorvynol assay [R2]. Intermediate for silicone products [R7]. High-purity derivatization reagent for gas chromatography. [R8].

OCCUPATIONAL RECOMMENDATIONS None listed.

NON-HUMAN TOXICITY LD_{50} rat oral 0.8 g/kg (from table) [R5, 2398]; LD_{50} rat ip approximately 0.06 g/kg (from table) [R5, 2398].

BIOCONCENTRATION Chlorosilanes are not expected to persist long enough to bioconcentrate. (chlorosilanes) [R3].

GENERAL RESPONSE Shut off ignition

sources. Call fire department. Avoid contact with liquid and vapor. Keep people away. Wear goggles and self-contained breathing apparatus. Stop discharge if possible. Isolate and remove discharged material. Notify local health and pollution control agencies.

FIRE RESPONSE Flammable. Poisonous gases may be produced in fire. Flashback along vapor trail may occur. Vapor may explode if ignited in an enclosed area. Extinguish with dry chemicals or carbon dioxide. Do not use water or foam on fire.

EXPOSURE RESPONSE Call for medical aid. Vapor irritating to eyes, nose, and throat. Move victim to fresh air. If breathing is difficult, give oxygen. Liquid will burn skin and eyes. Harmful if swallowed. Remove contaminated clothing and shoes. Flush affected areas with plenty of water. If swallowed and victim is conscious, have victim drink water or milk. Do not induce vomiting.

RESPONSE TO DISCHARGE Issue warning—high flammability; corrosive; restrict access; evacuate area. Disperse and flush with care.

WATER RESPONSE Effect of low concentrations on aquatic life is unknown. May be dangerous if it enters water intakes. Notify local health and wildlife officials. Notify operators of nearby water intakes.

EXPOSURE SYMPTOMS Inhalation irritates mucous membranes. Contact with liquid causes severe burns of eyes and skin. Ingestion causes severe burns of mouth and stomach.

EXPOSURE TREATMENT Inhalation: remove from exposure and support respiration; call physician if needed. Eyes: flush with water for 15 min; obtain medical attention immediately. Skin: flush with water; obtain medical attention immediately. Ingestion: if victim is conscious, give large amounts of water followed by milk or milk of magnesia.

REFERENCES

1. U.S. Coast Guard, Department of Transportation. *CHRIS—Hazardous Chemical Data. Volume II*. Washington, DC: U.S. Government Printing Office, 1984–5.

2. Association of Official Analytical Chemists. *Official Methods of Analysis*. 10th ed. and supplements. Washington, DC: Association of Official Analytical Chemists, 1965. New editions through 13th ed. plus supplements, 1982, p. 12/692.

3. ITC/USEPA. *Information Review #223 (Draft) Chlorosilanes* p. 10 (1981).

4. Sax, N. I. *Dangerous Properties of Industrial Materials*. 6th ed. New York, NY: Van Nostrand Reinhold, 1984.

5. Clayton, G.D., and F.E. Clayton (eds.). *Patty's Industrial Hygiene and Toxicology*: Volume 2A, 2B, 2C: *Toxicology*. 3rd ed. New York: John Wiley and Sons, 1981–1982.

6. International Labour Office. *Encyclopedia of Occupational Health and Safety*. Vols. I and II. Geneva, Switzerland: International Labour Office, 1983.

7. Sax, N. I., and R. J. Lewis, Sr. (eds.). *Hawley's Condensed Chemical Dictionary*. 11th ed. New York: Van Nostrand Reinhold Co., 1987. 413.

8. Kuney, J. H., and J. N. Nullican (eds.) *Chemcyclopedia*. Washington, DC: American Chemical Society, 1988. 283.

■ 2,4-DINITROANILINE

CAS # 97–02–9

DOT # 1596

SYNONYMS 1-amino-2,4-dinitrobenzene • 2,4-dinitraniline • 2,4-Dinitroanilin (German) • 2,4-dinitroanilina (Italian) • 2,4-dinitrobenzenamine • 2,4-dinitrophenylamine • aniline, 2,4-dinitro • benzenamine, 2,4-dinitro • NCI-c60753

MF $C_6H_5N_3O_4$

MW 183.14

COLOR AND FORM Yellow needles from dilute acetone, greenish-yellow plates from alcohol.

ODOR Musty odor.

DENSITY 1.615 g/ml at 14°C.

DISSOCIATION CONSTANTS pKa: 18.46.

BOILING POINT 56.7°C.

MELTING POINT 187.5–188°C.

EXPLOSIVE LIMITS AND POTENTIAL Can be detonated only by a very strong initiator. [R10, p. 49–45].

FIREFIGHTING PROCEDURES Water, carbon dioxide, dry chemical [R4]. Use extreme caution in approaching fire. Material may explode when exposed to heat or flame. No attempt should be made to fight advanced or massive fires except for remote activation of installed fire-extinguishing equipment or with unmanned fixed turrets. Area should be evacuated [R10, p. 49–45]. Water or foam may cause frothing. Personal protection: for rescue purposes wear full protective clothing. [R9, p. 325M–88].

SOLUBILITY Soluble in hot hydrochloric acid; practically insoluble in cold water; very sparingly soluble in boiling water; 5.8 parts soluble in 100 parts 88% alcohol at 18°C; 1 part soluble in 132.6 parts of 95% alcohol at 21°C.

VAPOR PRESSURE 5.94×10^{-7} mm Hg at 25°C (est).

OCTANOL/WATER PARTITION COEFFICIENT Log K_{ow} = 1.84 (est).

OTHER CHEMICAL/PHYSICAL PROPERTIES IR: 11388 (Sadtler Research Laboratories Prism Collection) (2,3-dinitroaniline); UV: 3120 (Sadtler Research Laboratories Spectral Collection) (2,3-dinitroaniline); IR: 17483 (Sadtler Research Laboratories Prism Collection) (2,6-dinitroaniline); UV: 5551 (Sadtler Research Laboratories Spectral Collection) (2,6-dinitroaniline); mass: 953 (National Bureau of Standards EPA–NIH Mass Spectra Data Base, NSRDS-NBS-63) (2,6-dinitroaniline); IR: 778 (Sadtler Research Laboratories IR Grating Collection) (3,5-dinitroaniline); UV: 9712 (Sadtler Research Laboratories Spectral Collection) (3,5-dinitroaniline); NMR: 1836 (Sadtler Research Laborato-ries Spectral Collection) (3,5-dinitroaniline).

PROTECTIVE EQUIPMENT Personal protection: for rescue purposes wear full protective clothing. [R10, p. 49–44].

PREVENTATIVE MEASURES SRP: Local exhaust ventilation should be applied wherever there is an incidence of point-source emissions or dispersion of regulated contaminants in the work area. Ventilation control of the contaminant as close as possible to its point of generation is both the most economical and safest method to minimize personnel exposure to airborne contaminants.

COMMON USES Preparation of azo dyes [R2]. Toner pigment in printing inks, corrosion inhibitor [R1].

OCCUPATIONAL RECOMMENDATIONS None listed.

ECOTOXICITY LC$_{50}$ *Pimephales promelas* (fathead minnow) 14.2 mg/L/96 hr (confidence limit 13.5 to 15.0 mg/L). Affected fish lost schooling behavior and swam near the tank surface, with half being hyperactive and half hypoactive. They had increased respiration and hemorrhaging, were darkly colored, and lost equilibrium prior to death [R5].

BIODEGRADATION 2,4-Dinitroaniline at an initial concentration of 100 ppm showed no biodegradation in river and sea water after 3 days using a cultivation test method (1) [R6].

BIOCONCENTRATION Based on an estimated log K_{ow} of 1.84 (2), the bioconcentration factor (BCF) for 2,4-dinitroaniline can be estimated to be about 15 from a recommended regression-derived equation (1,SRC). This BCF value suggests that bioconcentration in aquatic organisms may not be significant (SRC) [R7].

TSCA REQUIREMENTS Pursuant to section 8 (d) of TSCA, EPA promulgated a model health and safety data reporting rule. The section 8 (d) model rule requires manufacturers, importers, and processors of listed chemical substances and mixtures to submit to EPA copies and lists of unpublished health and safety studies. Benzenamine, 2,4-dinitro is included on this list [R8].

DOT *Health Hazards:* Poisonous; may be fatal if inhaled, swallowed, or absorbed through skin. Contact may cause burns to skin and eyes. Runoff from fire control or dilution water may cause pollution (dinitroaniline) [R3].

Fire or Explosion: Some of these materials may burn, but none of them ignites readily. May explode from friction, heat, or contamination (dinitroaniline) [R3].

Emergency Action: Keep unnecessary people away; isolate hazard area and deny entry. Stay upwind, out of low areas, and ventilate closed spaces before entering. Positive-pressure self-contained breathing apparatus (SCBA) and chemical protective clothing that is specifically recommended by the shipper or manufacturer may be worn. It may provide little or no thermal protection. Structural firefighters' protective clothing is not effective for these materials. Call CHEMTREC at 1–800–424– 9300 as soon as possible, especially if there is no local hazardous materials team available (dinitroaniline) [R3].

Fire: Small Fires: Dry chemical, CO_2, water spray, or regular foam. Large Fires: Water spray, fog, or regular foam. Move container from fire area if you can do so without risk. Apply cooling water to sides of containers that are exposed to flames until well after fire is out. Stay away from ends of tanks. For massive fire in cargo area, use unmanned hose holder or monitor nozzles; if this is impossible, withdraw from area, and let fire burn (dinitroaniline) [R3].

Spill or Leak: Do not touch or walk through spilled material; stop leak if you can do so without risk. Fully encapsulating, vapor-protective clothing should be worn for spills and leaks with no fire. Use water spray to reduce vapors. Small Spills: Take up with sand or other noncombustible absorbent material and place into containers for later disposal. Small Dry Spills: With clean shovel place material into clean, dry container, and cover; move

containers from spill area. Large Spills: Dike far ahead of liquid spill for later disposal (dinitroaniline) [R3].

First Aid: Move victim to fresh air and call for emergency medical care; if not breathing give artificial respiration; if breathing is difficult, give oxygen. In case of contact with material, immediately flush skin or eyes with running water for at least 15 minutes. Speed in removing material from skin is of extreme importance. Remove and isolate contaminated clothing and shoes at the site. Keep victim quiet and maintain normal body temperature. Effects may be delayed; keep victim under observation (dinitroaniline) [R3].

FIRE POTENTIAL Slight, when exposed to heat or flame [R4].

GENERAL RESPONSE Avoid contact with solid and dust. Keep people away. Wear goggles, self-contained breathing apparatus, and rubber overclothing (including gloves). Evacuate area. in case of large discharge. Call fire department. Isolate and remove discharged material. Notify local health and pollution control agencies.

FIRE RESPONSE Combustible. May explode if subjected to heat or flame. Poisonous gas is produced when heated. Evacuate surrounding area. Wear goggles, self-contained breathing apparatus, and rubber overclothing (including gloves). Combat fires from safe distance or protected location with unmanned hose holder or monitor nozzle.

EXPOSURE RESPONSE Call for medical aid. Dust poisonous if inhaled. Move to fresh air. Solids poisonous if swallowed or if skin is exposed. Irritating to eyes. Remove contaminated clothing and shoes. Flush affected areas with plenty of water. If swallowed and victim is conscious, have victim drink water or milk, and have victim induce vomiting. If swallowed and victim is unconscious or having convulsions, do nothing except keep victim warm.

RESPONSE TO DISCHARGE Should be removed; chemical and physical treatment.

WATER RESPONSE Effect of low concentrations on aquatic life is unknown. May be dangerous if it enters water intakes. Notify local health and wildlife officials. Notify operators of nearby water intakes.

EXPOSURE SYMPTOMS May cause headache, nausea, stupor. Irritating to skin and mucous membranes.

EXPOSURE TREATMENT Inhalation: artificial respiration if necessary. Ingestion: induce vomiting; give universal antidote; get prompt medical care. Contact with skin and eyes: remove victim from exposure; wash exposed skin with warm water and soap; flush eyes with water.

REFERENCES

1. Sax, N. I., and R. J. Lewis, Sr. (eds.). *Hawley's Condensed Chemical Dictionary.* 11th ed. New York: Van Nostrand Reinhold Co., 1987. 420.

2. Budavari, S. (ed.). *The Merck Index—Encyclopedia of Chemicals, Drugs, and Biologicals.* Rahway, NJ: Merck and Co., Inc., 1989. 516.

3. U.S. Department of Transportation. *Emergency Response Guidebook 1990.* DOT P 5800.5. Washington, DC: U.S. Government Printing Office, 1990, p. G–56.

4. Sax, N. I. *Dangerous Properties of Industrial Materials.* 6th ed. New York, NY: Van Nostrand Reinhold, 1984. 1209.

5. Geiger, D. L., S. H. Poirier, L. T. Brooke, D. J. Call (eds.). *Acute Toxicities of Organic Chemicals to Fathead Minnows (Pimephales promelas).* Vol. III. Superior, Wisconsin: University of Wisconsin–Superior, 1986. 119.

6. Kondo, M., et al., *Eisei Kagaku* 34: 188–95 (1988).

7. (1) Lyman, W. J., et al., *Handbook of Chemical Property Estimation Methods.* New York: McGraw-Hill, p. 5–4 (1990). (2) GEMS. *Graphical Exposure Modeling System PCGEMS* (1987).

8. 40 CFR 716.120 (7/1/90).

9. National Fire Protection Association. *Fire Protection Guide on Hazardous Materials.* 7th ed. Boston, Mass.: National Fire Protection Association, 1978.

10. National Fire Protection Association. *Fire Protection Guide on Hazardous Materials.* 9th ed. Boston, MA: National Fire Protection Association, 1986.

■ DINITROPHENOL

CAS # 25550–58–7

SYNONYMS phenol, dinitro • dnp

MF $C_6H_4N_2O_5$

MW 184.1

COLOR AND FORM Yellow crystals; yellow liquid when dissolved in a suitable solvent.

DENSITY 1.68.

EXPLOSIVE LIMITS AND POTENTIAL Severe explosion hazard when dry [R1].

FIREFIGHTING PROCEDURES Dangerously explosive. Flood with water. Cool all affected containers with flooding quantities of water. Apply water from as far a distance as possible (dinitrophenol solution (shipped water wet, with at least 15% water)) [R3]. Do not extinguish fire unless flow can be stopped. Use water in flooding quantities as fog. Solid streams of water may be ineffective. Cool all affected containers with flooding quantities of water. Apply water from as far a distance as possible. Use "alcohol" foam, carbon dioxide, or dry chemical. (dinitrophenol solution) [R3].

TOXIC COMBUSTION PRODUCTS Toxic oxides of nitrogen are produced in fires involving dinitrophenol. (dinitrophenol solution) [R3].

SOLUBILITY Slightly soluble in cold water, freely soluble in hot water, ethyl alcohol, ethyl ether, and benzene; soluble in chloroform.

VAPOR PRESSURE Vapor pressure of the dinitrophenols, on the order of 10–5 mm Hg or less (SRC).

OCTANOL/WATER PARTITION COEFFICIENT
Log K_{ow} = 1.54–2.36.

OTHER CHEMICAL/PHYSICAL PROPERTIES
IR: 19884 (Sadtler Research Laboratories IR Grating Collection) (2,3-dinitrophenol) UV: 17581 (Sadtler Research Laboratories Spectral Collection) (2,3-dinitrophenol) NMR: 12970 (Sadtler Research Laboratories Spectral Collection) (2,3-dinitrophenol).

PROTECTIVE EQUIPMENT People exposed to the solid or to a strong solution should wear protective clothing and hand protection and, in confined spaces, use respiratory protective equipment. [R13, 637].

PREVENTATIVE MEASURES During manufacture of dinitrophenol, dust must be kept to a minimum by the use of exhaust ventilation at points where containers are filled or centrifuges emptied. Contamination of clothing and the skin is readily detected by the yellow staining. [R13, 637].

CLEANUP Environmental considerations: Land spill: Dig a pit, pond, lagoon, or holding area to contain liquid or solid material. (SRP: If time permits, pits, ponds, lagoons, soak holes, or holding areas should be contained with a flexible impermeable-membrane liner.) Cover solids with a plastic sheet to prevent dissolving in rain or firefighting water (dinitrophenol solution (shipped water wet, with at least 15% water)) [R3]. Water spill: Use natural deep water pockets, excavated lagoons, or sandbag barrier to trap material at bottom. If dissolved, apply activated carbon at ten times the spilled amount in region of 10 ppm or greater concentration. Remove trapped material with suction hoses. Use mechanical dredges or lifts to remove immobilized masses of pollutants and precipitates of greater concentration. (dinitrophenol solution (shipped water wet, with at least 15% water)) [R12].

DISPOSAL At the time of review, criteria for land treatment or burial (sanitary landfill) disposal practices are subject to significant revision. Prior to implementing land disposal of waste residue (including waste sludge), consult with environmental regulatory agencies for guidance on acceptable disposal practices [R4]. Dinitrophenol: Incinerate (1800°F, 2.0 sec minimum) with adequate scrubbing equipment for the removal of NO_x. Dinitrophenol is a waste chemical stream constituent that may be subjected to ultimate disposal by controlled incineration [R7]. Group I Containers: Combustible containers from organic or metallo-organic pesticides (except organic mercury, lead, cadmium, or arsenic compounds) should be disposed of in pesticide incinerators or in specified landfill sites (organic or metallo-organic pesticides) [R8]. Group II Containers: Noncombustible containers from organic or metallo-organo pesticides (except organic mercury, lead, cadmium, or arsenic compounds) must first be triple-rinsed. Containers that are in good condition may be returned to the manufacturer or formulator of the pesticide product, or to a drum reconditioner for reuse with the same type of pesticide product, if such reuse is legal under Department of Transportation regulations (e.g., 49 CFR 173.28). Containers that are not to be reused should be punctured and transported to a scrap-metal facility for recycling, disposal, or burial in a designated landfill. (organic or metallo-organic pesticides) [R8].

COMMON USES Dyes, especially sulfur colors; picric acid; picramic acid; preservation of lumber; manufacture of the photographic developer diaminophenol hydrochloride; explosives manufacture; indicator; reagent for K and NH_4 ions [R2]. Used in weed control [R9]; manufacture of of dyes, diaminophenol, etc.; wood preservative; insecticide; indicator; as a reagent for the detection of potassium and ammonium ions (2,4-dinitrophenol) [R10]. Used as a diet aid [R11].

OCCUPATIONAL RECOMMENDATIONS None listed.

BIODEGRADATION Nitrophenols can inhibit aerobic microbial growth by uncoupling the metabolic process of oxidative phosphorylation (1). Static incubation of 5 and 10 mg/L of 2,4-dinitrophenol seeded with domestic wastewater resulted in 60 and 68% degradation, respectively, in 7 days

(2). Mixed cultures of phenol-adapted microorganisms exhibited some oxygen uptake in the presence of 2,4-dinitrophenol and little or no oxygen uptake in the presence of 2,5- and 2,6-dinitrophenol (3, 4,5), suggesting that the 2,4- isomer was slowly degraded under aerobic conditions, although the 2,5- or 2,6- isomers were not. Possible biotransformation processes of 2,4-dinitrophenol are: reduction of the nitro group, hydroxylation of the aromatic ring, and displacement of the nitro group by a hydroxyl group (6). A pure culture of the fungus *Fusarium oxysporum* was found to reduce 2,4-dinitrophenol to 2-amino-4-nitrophenol and 4-amino-2-nitrophenol (6). Nitrite release has been observed during the metabolism of 2,4-dinitrophenol by pure cultures of *Nocardia alba*, *Arthrobacter*, and *Cornebacterium simplex* (6) [R5].

BIOCONCENTRATION Based upon the observed relationship of the octanol/water partition coefficient and a compound's tendency to bioaccumulate in aquatic system, nitrophenols in general are not expected to bioaccumulate in aquatic organisms. (nitrophenols) [R14, p. 90–4].

CERCLA REPORTABLE QUANTITIES Persons in charge of vessels or facilities are required to notify the National Response Center (NRC) immediately, when there is a release of this designated hazardous substance, in an amount equal to or greater than its reportable quantity of 10 lb or 4.54 kg. The toll-free telephone number of the NRC is (800) 424–8802; in the Washington metropolitan area (202) 426–2675. The rule for determining when notification is required is stated in 40 CFR 302.6 (section IV. D. 3.b) [R6].

REFERENCES

1. Sax, N. I., and R. J. Lewis, Sr. (eds.). *Hawley's Condensed Chemical Dictionary*. 11th ed. New York: Van Nostrand Reinhold Co., 1987. 421.

2. Hawley, G. G. *The Condensed Chemical Dictionary*. 10th ed. New York: Van Nostrand Reinhold Co., 1981. 375.

3. Association of American Railroads. *Emergency Handling of Hazardous Ma-*

terials in Surface Transportation. Washington, DC: Assoc. of American Railroads, Hazardous Materials Systems (BOE), 1987. 270.

4. SRP.

5. (1) Callahan, M.A., et al., *Water-Related Fate 129 Priority Pollutants* Vol. II USEPA 440/4–79–029B (1979) (2) Tabak, H. H., et al., *J Wat Pollut Control Fed* 53: 1503 (1981) (3) Pitter, P. *Water Res* 10: 231 (1976) (4) Tabak, H. H., et al., *J Bacteriol* 87: 910 (1964) (5) Chambers, C.W., et al. *J Wat Pollut Control Fed* 35: 1517 (1963) (6) Overcash, M.R., et al., *Behavior of Organic Priority Pollut in the Terrestrial System, Di-n-Butyl Phthalate Ester, Toluene, and 2,4–Dinitrophenol* NTIS PB 82–224544 (1982).

6. 50 FR 13456 (4/4/85).

7. USEPA; *Engineering Handbook for Hazardous Waste Incineration* p. 2–6 (1981) EPA 68–03–3025.

8. 40 CFR 165 (7/1/87).

9. Gilman, A. G., L. S. Goodman, and A. Gilman (eds.). *Goodman and Gilman's The Pharmacological Basis of Therapeutics*. 7th ed. New York: Macmillan Publishing Co., Inc., 1985. 1645.

10. *The Merck Index*. 10th ed. Rahway, NJ: Merck Co., Inc., 1983. 479.

11. Horner WD; *Arch Ophthamal* 27: 1097 (1942).

12. Association of American Railroads. *Emergency Handling of Hazardous Materials in Surface Transportation*. Washington, DC: Assoc. of American Railroads, Hazardous Materials Systems (BOE), 1987. 270.

13. International Labour Office. *Encyclopedia of Occupational Health and Safety*. Vols. I and II. Geneva, Switzerland: International Labour Office, 1983.

14. Callahan, M.A., M.W. Slimak, N.W. Gabel, et al. *Water-Related Environmental Fate of 129 Priority Pollutants*. Volume I. EPA-440/4 79-029a. Washington, DC: U.S.Environmental Protection Agency, December 1979.

■ 2,5-DINITROPHENOL

CAS # 329–71–5

SYNONYMS phenol, 2,5-dinitro • phenol, gamma-dinitro • 2,5-dnp

MF $C_6H_4N_2O_5$

MW 184.12

pH Range: 4.0 colorless, 5.4 yellow.

COLOR AND FORM Yellow crystals; yellow monoclinic prisms or needles (diluted alcohol, water, ligroin).

ODOR THRESHOLD Detection: 2.4 mg/L [R3].

DISSOCIATION CONSTANTS $K_a = 0.7 \times 10^{-5}$ at 25°C.

MELTING POINT 108°C, also 104°C.

FIREFIGHTING PROCEDURES Dangerously explosive. Flood with water. Cool all affected containers with flooding quantities of water. Apply water from as far a distance as possible (dinitrophenol solution (shipped water wet, with at least 15% water)) [R2]. Do not extinguish fire unless flow can be stopped. Use water in flooding quantities as fog. Solid streams of water may be ineffective. Cool all affected containers with flooding quantities of water. Apply water from as far a distance as possible. Use "alcohol" foam, carbon dioxide, or dry chemical. (dinitrophenol solution) [R2].

TOXIC COMBUSTION PRODUCTS Toxic oxides of nitrogen are produced in fires involving dinitrophenol. (dinitrophenol solution) [R2].

SOLUBILITY Soluble in ether, benzene; slightly soluble in cold alcohol; soluble in hot alcohol, fixed alkali hydroxides; 0.68 g/L in water.

VAPOR PRESSURE 6.43×10^{-5} mm Hg at 25°C.

OCTANOL/WATER PARTITION COEFFICIENT Log K_{ow} = 1.75–2.00.

OTHER CHEMICAL/PHYSICAL PROPERTIES Heat of fusion: 30.80 cal/g = 128.87 J/g = 23,726 J/mol.

PROTECTIVE EQUIPMENT People exposed to the solid or to a strong solution should wear protective clothing and hand protection, and, in confined spaces, use respiratory protective equipment. (dinitrophenol) [R4].

PREVENTATIVE MEASURES During manufacture of dinitrophenol, dust must be kept to a minimum by the use of exhaust ventilation at points where containers are filled or centrifuges emptied. (dinitrophenol) [R4].

CLEANUP Land spill: Dig a pit, pond, lagoon, or holding area to contain liquid or solid material. (SRP: If time permits, pits, ponds, lagoons, soak holes, or holding areas should be contained with a flexible impermeable-membrane liner.) Cover solids with a plastic sheet to prevent dissolving in rain or firefighting water (dinitrophenol solution (shipped water wet, with at least 15% water)) [R2]. Water spill: Use natural deep-water pockets, excavated lagoons, or sandbag barriers to trap material at bottom. If dissolved, apply activated carbon at ten times the spilled amount in region of 10 ppm or greater concentration. Remove trapped material with suction hoses. Use mechanical dredges or lifts to remove immobilized masses of pollutants and precipitates of greater concentration (dinitrophenol solution (shipped water wet, with at least 15% water)) [R2]. Absorb bulk liquids with fly ash, cement powder, sawdust, or commercial sorbents (dinitrophenol solution) [R2]. Use natural barriers or oil-spill-control booms to limit spill motion. Use surface-active agent (e.g., detergent, soaps, alcohols) to compress, and thicken spilled material. Inject "universal" gelling agent to solidify encircled spill and increase effectiveness of booms.

DISPOSAL At the time of review, criteria for land treatment or burial (sanitary landfill) disposal practices are subject to significant revision. Prior to implementing land disposal of waste residue (including waste sludge), consult with environmental regulatory agencies for guidance on acceptable disposal practices [R5]. Dinitrophenol: Incinerate (1800°F, 2.0 sec minimum) with adequate scrubbing equipment for the removal of NO_x. Dinitrophenol is a waste chemical stream constituent which may be subjected to ultimate disposal by controlled incineration [R11]. Group I Containers: Combustible containers from organic or metallo-organic pesticides (except organic mercury, lead, cadmium, or arsenic compounds) should be disposed of in pesticide incinerators or in specified landfill sites (Organic or metallo-organic pesticides) [R12]. Group II Containers: Non-combustible containers from organic or metallo-organo pesticides (except organic mercury, lead, cadmium, or arsenic compounds) must first be triple-rinsed. Containers that are in good condition may be returned to the manufacturer or formulator of the pesticide product, or to a drum reconditioner for reuse with the same type of pesticide product, if such reuse is legal under Department of Transportation regulations (e.g., 49 CFR 173.28). Containers that are not to be reused should be punctured and transported to a scrap metal facility for recycling, disposal, or burial in a designated landfill (Organic or metallo-organic pesticides) [R12].

COMMON USES In manufacture of dyes and organic chemicals, as an indicator [R1]. As a reagent for the detection of potassium and ammonium ions [R1].

OCCUPATIONAL RECOMMENDATIONS None listed.

BIOLOGICAL HALF-LIFE The half-lives for elimination of 2,5-dinitrophenol from the blood of mice and rats following a single large dose given intraperitoneally are 3.3 min and 13.0 min, respectively (from table) [R6].

BIODEGRADATION 2,5-Dinitrophenol was stable to biochemical degradation by a mixed culture of phenol-adapted bacteria under aerobic conditions (0% COD chemical oxygen demand, removal after 20 days) (1) [R7].

BIOCONCENTRATION A bioconcentration factor (BCF) of 13 has been calculated for 2,5-dinitrophenol using recommended regression equations (1,SRC) and an experimentally determined log K_{ow} value of 1.75 (2). This BCF value suggests that 2,5-dini-

trophenol would not bioaccumulate significantly in aquatic organisms (SRC) [R8].

CERCLA REPORTABLE QUANTITIES Persons in charge of vessels or facilities are required to notify the National Response Center (NRC) immediately, when there is a release of this designated hazardous substance, in an amount equal to or greater than its reportable quantity of 10 lb or 4.54 kg. The toll-free telephone number of the NRC is (800)424–8802; in the Washington metropolitan area (202) 426–2675. The rule for determining when notification is required is stated in 40 CFR 302.6 (section IV. D. 3.b). (dinitrophenol) [R9].

DOT *Fire or Explosion:* May be ignited by heat, sparks, or flames. Container may explode in heat of fire. Vapor explosion and poison hazard indoors, outdoors, or in sewers (dinitrophenol solution) [R10, p. G–57].

Fire: Small Fires: Dry chemical, CO_2, water spray, or foam. Large Fires: Water spray, fog, or foam. Move container from fire area if you can do so without risk. Cool containers that are exposed to flames with water from the side until well after fire is out. Fight fire from maximum distance. Dike fire control water for later disposal; Do not scatter the material (dinitrophenol solution) [R10, p. G–57].

Health Hazards: Poisonous; May be fatal if inhaled, swallowed, or absorbed through skin. Contact may cause burns to skin and eyes. Runoff from fire control or dilution water may cause pollution (dinitrophenol solution) [R10, p. G–57].

Emergency Action: Keep unnecessary people away; isolate hazard area, and deny entry. Stay upwind; keep out of low areas. Ventilate closed spaces before entering them. Wear self-contained breathing apparatus and chemical protective clothing, but note that they do not provide thermal protection unless stated by the clothing manufacturer. Structural firefighter's protective clothing is not effective with these materials (dinitrophenol solutions) [R10, p. G–57].

Spill or Leak: Shut off ignition sources; no flares, smoking, or flames in hazard area. Do not touch spilled material; stop leak if you can do so without risk. Use water spray to reduce vapors. Small Spills: Take up with sand or other noncombustible absorbent material and place into containers for later disposal. Small Dry Spills: With clean shovel place material into clean, dry container, and cover; move containers from spill area. Large Spills: Dike far ahead of spill for later disposal (dinitrophenol solution) [R10, p. G–57].

First Aid: Move victim to fresh air; Call emergency medical care. In case of contact with material, immediately flush skin or eyes with running water for at least 15 minutes. Speed in removing material from skin is of extreme importance. Remove and isolate contaminated clothing and shoes at the site. Keep victim quiet and maintain normal body temperature. Effects may be delayed, keep victim under observation (dinitrophenol solution) [R10, p. G–57].

FIRE POTENTIAL Combustible, though it may require some effort to ignite. (dinitrophenol) [R2].

REFERENCES

1. *The Merck Index.* 10th ed. Rahway, NJ: Merck Co., Inc., 1983. 479.

2. Association of American Railroads. *Emergency Handling of Hazardous Materials in Surface Transportation.* Washington, DC: Assoc. of American Railroads, Hazardous Materials Systems (BOE), 1987. 270.

3. Verschueren, K. *Handbook of Environmental Data of Organic Chemicals.* 2nd ed. New York, NY: Van Nostrand Reinhold Co., 1983. 572.

4. International Labour Office. *Encyclopedia of Occupational Health and Safety.* Vols. I and II. Geneva, Switzerland: International Labour Office, 1983. 637.

5. SRP.

6. National Research Council. *Drinking Water and Health,* Volume 4. Washington, DC: National Academy Press, 1981. 234.

7. Pitter, P. *Water Res* 10: 231 (1976).

8. (1) Lyman, W. J., et al., *Handbook of Chemical Property Estimation Methods. Environmental Behavior of Organic Compounds.* New York: McGraw-Hill (1986). (2) Hansch, C., and A. J. Leo. *Medchem Project* Issue No. 26, Claremont, CA: Pomona College (1985).

9. 50 FR 13456 (4/4/85).

10. Department of Transportation. *Emergency Response Guidebook 1987.* DOT P 5800.4. Washington, DC: U.S. Government Printing Office, 1987.

11. USEPA. *Engineering Handbook for Hazardous Waste Incineration.* p.2–6 (1981) EPA 68-03-3025.

12. 40 CFR 165 (7/1/87).

■ DIPHOSGENE

CAS # 503–38–8

SYNONYMS perchloromethylformate • formic acid chlorotrichloromethyl ester • superpalite-phosgene • trichloromethyl chlorformate • carbonochloridic acid, trichloromethyl ester • diphosgen • formic acid, chloro-, trichloromethyl ester • perchloromethyl formate • superpalite • trichloromethyl chloroformate

MF $C_2Cl_4O_2$

MW 197.82

COLOR AND FORM Colorless liquid.

ODOR Odor similar to that of phosgene (newmown hay).

DENSITY 1.6525 at 14°C.

BOILING POINT 128°C at 760 mm Hg.

MELTING POINT –57°C.

SOLUBILITY Insoluble in water; soluble in ethanol, very soluble in ether; soluble in benzene, alcohol, and ether.

VAPOR PRESSURE 10 mm Hg at 20°C.

OTHER CHEMICAL/PHYSICAL PROPERTIES Decomposition by heat, porous substances, activated carbons (with evolution of phosgene), also by alkalies, hot water;

non-combustible; conversion factor: 1 ppm = 8.08 mg/m³; hydrolyze in water (chloroformates); suffocating liquid; boiling point = 49°C at 50 mm Hg; stable at room temp, decomposes to phosgene at approximately 300°C.

PROTECTIVE EQUIPMENT Air-supplied respirators should be provided [R4].

PREVENTATIVE MEASURES All materials should be handled so as to prevent all contact. Eyewash fountains should be provided. Also, special attention should be given to eye and respiratory tract medical examinations (chloroformates) [R4].

CLEANUP To decontaminate in enclosed spaces, use ammonia or steam [R3].

COMMON USES Organic synthesis; military poison gas [R2]; used in synthesis of isocyanides [R1].

BINARY REACTANTS Olefins.

STANDARD CODES NFPA, USCG—no; ICC—Class A poison, poison gas label, not accepted in an outside container.

PERSISTENCE Moderately persistent.

MAJOR SPECIES THREATENED All species.

DIRECT CONTACT Eyes, lungs.

GENERAL SENSATION Phosgene odor. Lacrimator. Recognition odor 8.8 mg/m³ in air [R5]; slightly lacrimatory. Physiological action delayed. Strong local irritant, very toxic with ingestion, and inhalation.

PERSONAL SAFETY Wear full protective clothing and self-contained breathing apparatus approved for warfare gases. Equipment should be capable of withstanding hydrochloric acid attack.

ACUTE HAZARD LEVEL Strong irritant, ingestive and inhalative toxin. Although insoluble, it will release corrosive hydrochloric acid into water, and create hazard. Emits highly dangerous vapors when heated to decomposition or in contact with water or steam. LC_{50} in air 344 mg/m³ for mice.

CHRONIC HAZARD LEVEL Unknown; if allowed to persist on bottom, will release toxic or corrosive amounts of hydrochloric acid for extended period.

DEGREE OF HAZARD TO PUBLIC HEALTH Strong irritant, ingestive and inhalative toxin. Emits highly dangerous vapors when heated to decomposition or in contact with water or steam.

OCCUPATIONAL RECOMMENDATIONS None listed.

ACTION LEVEL Evacuate area. Notify local air authority. Call Department of Defense—Army.

ON-SITE RESTORATION Pump or vacuum from bottom of water course. Neutralize water with $NaHCO_3$, as hydrochloric acid is produced. Seek professional environmental engineering assistance through EPA's Environmental Response Team (ERT), Edison, NJ, 24-hour no., 908–548–8730.

BEACH OR SHORE RESTORATION Do not burn.

AVAILABILITY OF COUNTERMEASURE MATERIAL Pump—fire department; vacuum—swimming-pool suppliers; $NaHCO_3$—grocery distributors, large bakeries.

DISPOSAL Sprinkle onto thick layer of mixed dry soda ash and slaked lime from behind body shield. Mix and spray with water mist, sift into large volume of water, neutralize, and route to sewage plant.

DISPOSAL NOTIFICATION Contact local sewage authority.

INDUSTRIAL FOULING POTENTIAL May produce corrosive HCl.

MAJOR WATER USE THREATENED Recreation, fisheries, irrigation.

PROBABLE LOCATION AND STATE Colorless liquid; will sink to bottom of water course.

COLOR IN WATER Colorless.

REFERENCES

1. Budavari, S. (ed.). *The Merck Index—Encyclopedia of Chemicals, Drugs, and Biologicals.* Rahway, NJ: Merck and Co., Inc., 1989. 527.

2. Sax, N. I., and R. J. Lewis, Sr. (eds.). *Hawley's Condensed Chemical Dictionary.* 11th ed. New York: Van Nostrand Reinhold Co., 1987. 1177.

3. Sax, N. I. *Dangerous Properties of Industrial Materials.* 5th ed. New York: Van Nostrand Reinhold, 1979. 626.

4. Clayton, G.D., and F.E. Clayton (eds.). *Patty's Industrial Hygiene and Toxicology*: Volume 2A, 2B, 2C: *Toxicology*. 3rd ed. New York: John Wiley and Sons, 1981–1982. 2387.

5. Sullivan, R. J., "Air Pollution Aspects of Odorous Compounds," *NTIS PB* 188 089, September 1969.

■ DIPROPYLAMINE

CAS # 142–84–7

DOT # 2383

SYNONYMS 1-propanamine, n-propyl • ai3-24037 • di-n-propylamine • n-dipropylamine • n-propyl-1-propanamine

MF $C_6H_{15}N$

MW 101.22

COLOR AND FORM Colorless liquid; water-white liquid.

ODOR Ammonia odor.

ODOR THRESHOLD Odor low = 0.4140 mg/m³; odor high = 0.8280 mg/m³ [R6].

DENSITY 0.738 at 20°C.

SURFACE TENSION 6.58 dynes/cm = 0.00658 newtons/m at 20°C.

VISCOSITY 0.528 cP at 70°F.

DISSOCIATION CONSTANT pK_b = 3; pK_a = 11.

BOILING POINT 109–110°C.

MELTING POINT –39.6°C.

HEAT OF COMBUSTION –18,750 Btu/lb = –10,420 cal/g = –436.0×10⁵ J/kg.

HEAT OF VAPORIZATION 143 Btu/lb = 79.5 cal/g = 3.33×10⁵ J/kg.

AUTOIGNITION TEMPERATURE 570°F (299°C) [R5].

FIREFIGHTING PROCEDURES Foam; dry chemical; carbon dioxide. Water may be ineffective [R2].

TOXIC COMBUSTION PRODUCTS Toxic oxides of nitrogen may form in fires [R2].

SOLUBILITY Soluble in alcohol; very soluble in acetone, benzene; soluble in all proportions in ether; 2.500 lb/100 lb water at 68°F.

VAPOR PRESSURE 30 mm Hg at 25°C.

OCTANOL/WATER PARTITION COEFFICIENT Log K_{ow} = 1.70.

OTHER CHEMICAL/PHYSICAL PROPERTIES Forms a hydrate with water; conversion factors: 1 mg/L = 242 ppm; 1 ppm = 4.14 mg/m³; wt/gal: 6.2 lb/L; amines tend to be fat soluble (amines); saturated liquid density = 46.010 lb/cu ft at 70°C; liquid heat capacity = 0.600 Btu/lb at 70°F; saturated vapor pressure = 0.433 lb/sq in. at 70°C; saturated vapor density = 0.00771 lb/ft³ at 70°C; ideal gas heat capacity = 0.390 Btu/lb at 75°C.

PROTECTIVE EQUIPMENT Self-contained breathing apparatus; butyl rubber gloves; butyl rubber apron; face shield [R2].

PREVENTATIVE MEASURES Local exhaust ventilation should be applied wherever there is an incidence of point-source emissions or dispersion of regulated contaminants in the work area. Ventilation control of the contaminant as close as possible to its point of generation is both the most economical and safest method to minimize personnel exposure to airborne contaminants [R7].

CLEANUP 1. Remove all ignition sources. 2. Ventilate area of spill or leak. 3. For small quantities, absorb on paper towels. Evaporate in safe place (such as fume hood). Allow sufficient time for evaporating vapors to completely clear the hood ductwork. Burn paper in suitable location away from combustible materials [R8].

DISPOSAL At the time of review, criteria for land treatment, or burial (sanitary landfill) disposal practices are subject to significant revision. Prior to implementing land disposal of waste residue (including waste sludge), consult with environmental regulatory agencies for guidance on acceptable disposal practices [R7].

COMMON USES Chemical intermediate for herbicide S-ethyl-di-n-propylthiocarbamate, S-propyl-di-n-propylthiocarbamate; in purification of perfluoro compounds [R1].

OCCUPATIONAL RECOMMENDATIONS None listed.

OSHA Meets criteria for OSHA medical records rule [R11].

BIODEGRADATION The Hoechst batch method was used to study the degree of biodegradation, degradation time, adaptation time, and degradation rate of 50 substances including dipropylamine [R9].

BIOCONCENTRATION Based on a log K_{ow} of 1.67 (1), the log BCF for dipropylamine can be estimated to be 1.04 from a recommended regression-derived equation (2,SRC), and therefore, dipropylamine will not be expected to bioconcentrate [R10].

TSCA REQUIREMENTS Pursuant to section 8 (d) of TSCA, EPA promulagated a model health and safety data reporting rule. The section 8 (d) model rule requires manufacturers, importers, and processors of listed chemical substances and mixtures to submit to EPA copies and lists of unpublished health and safety studies. Dipropylamine is included on this list [R12].

RCRA REQUIREMENTS As stipulated in 40 CFR 261.33, when dipropylamine, as a commercial chemical product or manufacturing chemical intermediate or an off-specification commercial chemical product or a manufacturing chemical intermediate, becomes a waste, it must be managed according to federal or state hazardous waste regulations. Also defined as a hazardous waste is any residue, contaminated soil, water, or other debris resulting from the cleanup of a spill, into water or on dry land, of this waste. Generators of small quantities of this waste may qualify for partial exclusion from hazardous waste regulations (see 40 CFR 261.5) [R13].

DOT *Fire or Explosion:* Flammable/combustible material; may be ignited by heat, sparks, or flames. Vapors may travel to a source of ignition, and flash back. Container may explode in heat of fire. Vapor explosion hazard indoors, outdoors, or in sewers. Runoff to sewer may create fire or explosion hazard [R3].

Health Hazards: Poisonous if swallowed. If inhaled, may be harmful. Contact may cause burns to skin and eyes. Fire may produce irritating or poisonous gases. Runoff from fire control or dilution water may cause pollution [R3].

Emergency Action: Keep unnecessary people away; isolate hazard area and deny entry. Stay upwind; keep out of low areas. Self-contained breathing apparatus (SCBA) and structural firefighter's protective clothing will provide limited protection. Isolate for 1/2 mile in all directions if tank car or truck is involved in fire. Call CHEMTREC at 1–800–424–9300 for emergency assistance. If water pollution occurs, notify the appropriate authorities [R3].

Fire: Small Fires: Dry chemical, CO_2, halon, water spray, or standard foam. Large Fires: Water spray, fog, or standard foam is recommended. Move container from fire area if you can do so without risk. Cool containers that are exposed to flames with water from the side until well after fire is out. Stay away from ends of tanks. Withdraw immediately in case of rising sound from venting safety device or any discoloration of tank due to fire [R3].

Spill or Leak: Shut off ignition sources; no flares, smoking, or flames in hazard area. Do not touch spilled material; stop leak if you can do so without risk. Water spray may reduce vapor; but it may not prevent ignition in closed spaces. Small Spills: Take up with sand or other noncombustible absorbent material and place into containers for later disposal. Large Spills: Dike far ahead of liquid spill for later disposal [R3].

First Aid: Move victim to fresh air and call emergency medical care; if not breathing, give artificial respiration; if breathing is difficult, give oxygen. Remove and isolate contaminated clothing and shoes at the site. In case of contact with material, immediately flush skin or eyes with running water for at least 15 minutes. Keep victim quiet and maintain normal body temperature [R3].

FIRE POTENTIAL Dangerous when exposed to heat or flame [R4].

GENERAL RESPONSE Shut off ignition sources. Call fire department. Avoid contact with liquid and vapor. Keep people away. Stop discharge if possible. Stay upwind. Use water spray to "knock down" vapor. Isolate and remove discharged material. Notify local health and pollution control agencies.

FIRE RESPONSE Flammable. Poisonous gases may be produced in fire. Containers may explode in fire. Flashback along vapor trail may occur. Vapor may explode if ignited in an enclosed area. Wear goggles and self-contained breathing apparatus. Extinguish with dry chemicals, foam, or carbon dioxide. Water may be ineffective on fire. Cool exposed containers with water.

EXPOSURE RESPONSE Call for medical aid. Vapor irritating to eyes, nose, and throat. If inhaled will cause headache, dizziness, coughing, or difficult breathing. If breathing has stopped, give artificial respiration. If breathing is difficult, give oxygen. Liquid will burn eyes. If swallowed will cause nausea and vomiting. Remove contaminated clothing and shoes. Flush affected areas with plenty of water. If swallowed and victim is conscious, have victim drink water or milk. If swallowed, and victim is unconscious or having convulsions, do nothing except keep victim warm.

RESPONSE TO DISCHARGE Issue warning—water contaminant., air contaminant, high flammability. Restrict access. Disperse and flush.

WATER RESPONSE Effect of low concentrations on aquatic life is unknown. Fouling to shoreline. May be dangerous if it enters water intakes. Notify local health and wildlife officials. Notify operators of nearby water intakes.

EXPOSURE SYMPTOMS Inhalation causes severe coughing and chest pain due to irritation of air passages; can cause lung edema; may also cause headache, nausea, faintness, and anxiety. Ingestion causes irritation and burning of mouth and stomach. Contact with eyes causes severe irritation and edema of the cornea. Contact with skin causes severe irritation.

EXPOSURE TREATMENT Inhalation: remove victim to fresh air; if he is not breathing, give artificial respiration; if breathing is difficult, give oxygen; call a physician. Ingestion: give large amount of water; get medical attention. Eyes: flush with water for 15 min; get medical attention for burns. Skin: flush with water for 15 min.

REFERENCES

1. SRI.

2. U.S. Coast Guard, Department of Transportation. CHRIS—*Hazardous Chemical Data.* Volume II. Washington, DC: U.S. Government Printing Office, 1984–5.

3. Department of Transportation. *Emergency Response Guidebook 1987.* DOT P 5800.4. Washington, DC: U.S. Government Printing Office, 1987, p. G–68.

4. Sax, N. I. *Dangerous Properties of Industrial Materials.* 6th ed. New York, NY: Van Nostrand Reinhold, 1984. 1244.

5. National Fire Protection Association. *Fire Protection Guide on Hazardous Materials.* 9th ed. Boston, MA: National Fire Protection Association, 1986, p. 325M–45.

6. Ruth, J.H., *Am Ind Hyg* Assoc J 47: A–142–51 (1986).

7. SRP.

8. Mackison, F. W., R. S. Stricoff, and L. J. Partridge, Jr. (eds.). *NIOSH/OSHA—Occupational Health Guidelines for Chemical Hazards.* DHHS (NIOSH) Publication No. 81–123 (3 vols). Washington, DC: U.S. Government Printing Office, Jan. 1981. 2.

9. Zahn, R., H. Wellens. *Z Wasser Abwasser Forsch* 13 (1): 1–7 (1980).

10. (1) Hansch, C., and A. J. Leo. *Medchem Project* Issue No. 26, Clare-

mont, CA: Pomona College (1985). (2) Lyman, W. J., et al., *Handbook of Chemical Property Estimation Methods.* New York: McGraw-Hill, p. 5–4 (1982).

11. 29 CFR 1910.20 (7/1/87).

12. 40 CFR 716.120 (7/1/87).

13. 53 FR 13382 (4/22/88).

■ DIPROPYLENE GLYCOL, MONOMETHYL ETHER

CAS # 34590–94–8

SYNONYMS dipropylene glycol, methyl ether • dipropylene glycol, monomethyl ether • dowanol-50b • dowanol-dpm • ppg-2 methyl ether • propanol, (2-methoxymethylethoxy)- • UCAR-solvent-2lm

MF $C_7H_{16}O_3$

MW 148.23

COLOR AND FORM Colorless liquid.

ODOR Mild ether odor in moderate concentration but a strong, objectionable odor at 1,000 ppm.

ODOR THRESHOLD 210 mg/m³ odor low; 6,000 mg/m³ odor high [R1].

TASTE Bitter taste.

DENSITY 0.95 at 25°C.

VISCOSITY 3.5 cP at 25°C.

BOILING POINT 374°F (190°C).

MELTING POINT (Freezing point) −80°C.

FIREFIGHTING PROCEDURES Dry chemical, carbon dioxide, mist, foam [R16, 1245].

SOLUBILITY Miscible with water and many organic solvents; completely miscible with acetone, ethanol, benzene, carbon tetrachloride, ether, methanol, monochlorobenzene, and petroleum ether.

VAPOR PRESSURE 0.4 mm Hg at 26°C.

STABILITY AND SHELF LIFE Volatile [R15, 657].

OTHER CHEMICAL/PHYSICAL PROPERTIES % in saturated air 0.047 at 25°C.

PREVENTATIVE MEASURES Contact lenses should not be worn when working with this chemical [R2]. If dipropylene glycol methyl ether saturates the clothing, promptly remove the clothing and wash or shower. [R14].

COMMON USES In manufacture of various cosmetics [R15, 658]; solvent in hard-surface liquid household cleaners, and water-based surface coatings; coupling agent, e.g., in water-based polishes [R8]; used as a solvent for nitrocellulose and other synthetic resins [R9]; solvent, hydraulic brake fluids [R10].

OCCUPATIONAL RECOMMENDATIONS None listed.

OSHA 8-hr time-weighted average: 100 ppm (600 mg/m³). Skin absorption designation [R6]. Threshold-limit 8-hr time-weighted average (TWA) 100 ppm, 606 mg/m³, skin; short-term exposure limit (STEL) 150 ppm, 909 mg/m³, skin (1976) [R7].

NON-HUMAN TOXICITY LD_{50} rat oral 5.35 g/kg [R3]; LD_{50} rabbit dermal 9.5 g/kg [R12]; LD_{50} rat oral 5.4 ml/kg [R11].

BIODEGRADATION Five-, ten-, and twenty-day BOD values for dipropylene glycol monomethyl ether were reported as 0, 0, and 31%, respectively (expressed as percentage of theoretical oxygen demand) (1). The type of inoculum, however, was not specified. This delayed oxygen demand suggests that an acclimation period is required in order for a dipropylene glycol monomethyl ether-degrading population to become established. Thus, intermittent releases of dipropylene glycol monomethyl ether to the environment or to wastewater treatment plants may also require an acclimation period before significant amounts of dipropylene glycol monomethyl ether are removed (SRC). No information was found on the biodegradation of dipropylene glycol monomethyl ether in soil or natural waters [R4].

BIOCONCENTRATION Because dipropylene glycol monomethyl ether is infinitely soluble in water, it will not be expected to bioconcentrate in aquatic organisms (SRC).

FIFRA REQUIREMENTS Dipropylene glycol monomethyl ether is exempted from the requirement of a tolerance when used as a stabilizer in accordance with good agricultural practice as inert (or occasionally active) ingredients in pesticide formulations applied to growing crops only [R5]. Dipropylene glycol monomethyl ether is exempted from the requirement of a tolerance when used as a surfactant or a related adjuvant of a surfactant in accordance with good agricultural practice as inert (or occasionally active) ingredients in pesticide formulations applied to animals [R12].

FDA Dipropylene glycol monomethyl ether is an indirect food additive for use only as a component of adhesives [R13].

REFERENCES

1. Ruth, J.H., *Am Ind Hyg Assoc J* 47: A–142–51 (1986).

2. NIOSH. *NIOSH Pocket Guide to Chemical Hazards*. DHHS (NIOSH) Publication No. 90–117. Washington, DC: U.S. Government Printing Office, June 1990, 102.

3. American Conference of Governmental Industrial Hygienists. *Documentation of the Threshold Limit Values and Biological Exposure Indices*. 5th ed. Cincinnati, OH: American Conference of Governmental Industrial Hygienists, 1986. 221.

4. Dow Chemical Company; *The Glycol Ethers Handbook* Form No. 110–363–81 (1981).

5. 40 CFR 180.1001 (d) (7/1/90).

6. 54 FR 2920 (1/19/89).

7. American Conference of Governmental Industrial Hygienists. *Threshold Limit Values for Chemical Substances and Physical Agents and Biological Exposure Indices for 1994–1995*. Cincinnati, OH: ACGIH, 1994. 20.

8. SRI.

9. Sittig, M. *Handbook of Toxicology and Hazardous Chemicals and Carcinogens* 2nd ed. 1985, p. 388.

10. Sax, N. I., and R. J. Lewis, Sr. (eds.). *Hawley's Condensed Chemical Dictionary*. 11th ed. New York: Van Nostrand Reinhold Co., 1987. 431.

11. American Conference of Governmental Industrial Hygienists. *Documentation of the Threshold Limit Values and Biological Exposure Indices*. 5th ed. Cincinnati, OH: American Conference of Governmental Industrial Hygienists, 1986. 221.

12. 40 CFR 180.1001 (e) (7/1/90).

13. 21 CFR 175.105 (4/1/90).

14. Mackison, F. W., R. S. Stricoff, and L. J. Partridge, Jr. (eds.). *NIOSH/OSHA— Occupational Health Guidelines for Chemical Hazards*. DHHS (NIOSH) Publication No. 81–123 (3 vols). Washington, DC: U.S. Government Printing Office, Jan. 1981.

15. Browning, E. *Toxicity and Metabolism of Industrial Solvents*. New York: American Elsevier, 1965.

16. Sax, N.I. *Dangerous Properties of Industrial Materials*. 6th ed. New York, NY: Van Nostrand Reinhold, 1984.

■ DISULFIRAM

$$(C_2H_5)_2N\overset{\overset{\displaystyle S}{\parallel}}{C}-SS-\overset{\overset{\displaystyle S}{\parallel}}{C}N(C_2H_5)_2$$

CAS # 97–77–8

SYNONYMS 1,1'-dithiobis (N,N-diethylthioformamide) • abstensil • abstinil • abstinyl • alcophobin • alkaubs • antabus • antabuse • antadix • antaenyl • antaethan • antaethyl • antaetil • antetan • antethyl • anteyl • anti-ethyl • antiaethan • antietanol • antietil • antikol • aversan • averzan • bis (N,N-diethylthiocarbamoyl) disulfide • bonibal • contrapot • disetil • disulfan • ephorran • ethyl thiudad • ethyl-tuads • ethyldithiourame • ethyldithiurame • exhoran • NCI-c02959 • nocbin • stopaethyl • stopethyl • tatd • tenurid • tenutex • tetidis • tetraethylthioperoxydicarbonic diamide • teturamin • thiosan • thioscabin •

thireranide • tillram • tiuram • thioperoxydicarbonic diamide (((H₂N)C(S))₂S₂), tetraethyl • bis (diethylthiocarbamoyl) disulfide • disulfuram • TETD • tetraethylthiuram disulfide • abstensyl (Argentina) • antivitium (Spain) • esperal (France) • Refusal (Netherlands) • ro-sulfram-500 (USA)

MF $C_{10}H_{20}N_2S_4$

MW 296.56

COLOR AND FORM White to off-white crystalline powder; light-gray crystalline powder.

ODOR Slight odor.

TASTE Slightly bitter taste.

DENSITY 1.30.

BOILING POINT 117°C at 17 mm Hg.

MELTING POINT 70°C.

SOLUBILITY Practically insoluble in water (0.02 g/100 ml); soluble in alcohol (3.82 g/100 ml), ether (7.14 g/100 ml), acetone, benzene, chloroform, carbon disulfide; slightly soluble in light petroleum.

OCTANOL/WATER PARTITION COEFFICIENT Log K_{ow} = 3.88.

OTHER CHEMICAL/PHYSICAL PROPERTIES Diethyldithiocarbamate is an avid chelator of copper and other metals and thereby inhibits the activity of dopamine beta-hydroxylase and alcohol dehydrogenase (diethyldithiocarbamate); wt/vol conversion: 12.10 mg/m³= 1 ppm;.

COMMON USES Rubber accelerator; vulcanizer; seed disinfectant; fungicide; medication: alcohol deterrent [R1]; isulfiram is used in the treatment of chronic alcoholism. It is not a cure and the treatment is likely to be of little value unless it is undertaken with the willing cooperation of the patient and is employed in conjunction with psychotherapy [R8, 579]; medication: (SRP: antidote for) severe trichloroethylene poisoning [R2].

OCCUPATIONAL RECOMMENDATIONS None listed.

OSHA 8-hr time-weighted average: 2 mg/ m³ (final rule limits) were achieved by any combination of engineering controls, work practices, and personal protective equipment during the phase-in period, Sept 1, 1989 through Dec 30, 1992. Final rule limits became effective Dec 31, 1992 [R5].

THRESHOLD LIMIT 8-hr time-weighted average (TWA) 2 mg/m³ (1986) [R7, 20]. Excursion Limit Recommendation: Excursions in worker exposure levels may exceed three times the TLV-TWA for no more than a total of 30 min during a work day, and under no circumstances should they exceed five times the TLV-TWA [R7, 5].

NON-HUMAN TOXICITY LD_{50} rat oral 8.6 g/kg [R2]; LD_{50} rabbit oral 2.05 g/kg [R2]; LD_{50} mouse intraperitoneal 75 mg/kg [R2].

IARC SUMMARY No data are available for humans. Inadequate evidence of carcinogenicity in animals. Overall evaluation, Group 3: The agent is not classifiable as to its carcinogenicity to humans [R3].

BIOLOGICAL HALF-LIFE The elimination half-life of disulfiram in plasma is 7.3 hr (from table) [R9, 424].

BIOCONCENTRATION The log octanol/water partition coefficient for disulfiram is 3.88 (1). Using this K_{ow}, one estimates a BCF of 136 and 523 using two recommended regression equations (2). Therefore, disulfiram would have low to moderate bioconcentration properties (SRC) [R4].

FDA Tetraethylthiuram disulfide is an indirect food additive for use only as a component of adhesives [R6].

REFERENCES

1. Budavari, S. (ed.). *The Merck Index—Encyclopedia of Chemicals, Drugs, and Biologicals.* Rahway, NJ: Merck and Co., Inc., 1989. 531.

2. American Conference of Governmental Industrial Hygienists. *Documentation of the Threshold Limit Values and Biological Exposure Indices.* 5th ed. Cincinnati, OH: American Conference of Governmental Industrial Hygienists, 1986. 225.

3. IARC. *Monographs on the Evaluation*

of the Carcinogenic Risk of Chemicals to Man. Geneva: World Health Organization, International Agency for Research on Cancer, 1972–present (multivolume work), p. S7 63 (1987).

4. (1) Hansch, C., and A. J. Leo. *Medchem Project* Issue No. 26, Claremont, CA: Pomona College (1985) (2) Lyman, W. J., et al., *Handbook of Chemical Property Estimation Methods*. New York: McGraw-Hill, Chapt 5 (1982).

5. 29 CFR 1910.1000 (7/1/90).

6. 21 CFR 175.105 (4/1/91).

7. American Conference of Governmental Industrial Hygienists. *Threshold Limit Values for Chemical Substances and Physical Agents and Biological Exposure Indices for 1994–1995.* Cincinnati, OH: ACGIH, 1994.

8. Reynolds, J.E.F., A. B. Prasad (eds.). *Martindale—The Extra Pharmacopoeia.* 28th ed. London: The Pharmaceutical Press, 1982.

9. Ellenhorn, M.J., and D.G. Barceloux. *Medical Toxicology—Diagnosis and Treatment of Human Poisoning.* New York, NY: Elsevier Science Publishing Co., Inc. 1988.

■ DIVANADIUM TRIOXIDE

CAS # 1314–34–7

SYNONYMS vanadic oxide • vanadium oxide • vanadium oxide (V_2O_3) • vanadium oxide, sesqui • vanadium sesquioxide • vanadium trioxide • vanadium(3) oxide

MF O_3V_2

MW 149.88

COLOR AND FORM Black powder.

DENSITY 4.87 at 18°C.

MELTING POINT 1,940°C.

SOLUBILITY Insoluble in water; soluble in nitric acid; hydrogen fluoride; alkali; slightly soluble in water; slightly soluble in cold water; soluble in hot water.

STABILITY AND SHELF LIFE On exposure to air it is gradually converted into indigo-blue crystals of vanadium tetraoxide (V_2O_4) [R1].

OTHER CHEMICAL/PHYSICAL PROPERTIES Has corundum (Al_2O_3) structure. The complex biochemistry of vanadium compounds depends on the oxidation states of the metal from −1 to 5. The vanadium compounds can easily change their oxidation states under physiological conditions, so that vanadium is mostly in the 5 oxidation state (except in the presence of reducing agents) (vanadium compounds). Valence states of −1 and 0 may occur in solid compounds, e.g., carbonyl and certain complexes. In oxidation state 5, vanadium is diamagnetic, and forms colorless or pale-yellow compounds. In lower oxidation states, the presence of one or more 3d electrons, usually unpaired, results in paramagnetic and colored compounds. All compounds of vanadium that have unpaired electrons are colored, but because the absorption spectra may be complex, a specific color does not necessarily correspond to a particular oxidation state (vanadium compounds). Coordination compounds of vanadium are mainly based on 6 coordination, in which vanadium has a pseudooctahedral structure. Coordination number 4 is typical of many vanadates. Coordination numbers 5 and 8 also are known for vanadium compounds, but numbers less than 4 have not been reported (vanadium compounds).

PROTECTIVE EQUIPMENT Because vanadium compounds cause irritation of the respiratory tract, it is recommended that protective equipment be worn while processing those compounds and that workers have periodic medical examinations. (vanadium compounds) [R4].

PREVENTATIVE MEASURES Safety showers and eyewash fountains should be located in or near areas where gross exposures to vanadium compounds are likely to occur and should be properly maintained. If vanadium compounds, especially the halide or oxyhalide liquids, come in contact with the skin, the affected area should be flushed promptly with water. The eyes, if splashed or otherwise contaminated with these reactive halides, should be flushed immediately and thoroughly with water at low pressure. The employee should then be taken promptly to the nearest medical facility to determine the need for further treatment. (vanadium compounds) [R5].

CLEANUP Land spill: Dig a pit, pond, lagoon, holding area to contain liquid or solid material. (SRP: If time permits, pits, ponds, lagoons, soak holes, or holding areas should be sealed with an impermeable flexible-membrane liner.) Cover solids with a plastic sheet to prevent dissolving in rain or firefighting water. (vanadium pentoxide) [R6].

DISPOSAL SRP: At the time of review, criteria for land treatment or burial (sanitary landfill) disposal practices are subject to significant revision. Prior to implementing land disposal of waste residue (including waste sludge), consult with environmental regulatory agencies for guidance on acceptable disposal practices.

COMMON USES Used as a catalyst such as in making ethanol from ethylene [R1]. Vanadium oxide is used as a catalyst in sulfuric and nitric acid manufacture [R9, 220]. Medication (vanadium compounds) [R9, 220].

OCCUPATIONAL RECOMMENDATIONS None listed.

OSHA An employee's exposure to vanadium fume (as V_2O_5) shall at no time exceed the ceiling value of 0.1 mg/m^3. Transitional limits must continue to be achieved by any combination of engineering controls, work practices, and personal protective equipment during the phase-in period, Sept 1 1989 through Dec 30, 1992. (vanadium fume (as V_2O_5)) [R7].

NIOSH Ceiling 0.05 mg vanadium/m^3 for 15 minutes for vanadium compounds (Aug 1977) (vanadium compounds) [R8].

INTERNATIONAL EXPOSURE LIMITS A European Community guideline recommends an every-4-month control at the workplace if a 50–μg vanadium/L urine level is reached and an annual control if the 5–μg vanadium/L urine level is reached. In case of levels of >50 μg vanadium/L urine, a temporary removal from risk should be done. (vanadium) [R11, 754].

NON-HUMAN TOXICITY LD$_{50}$ albino mouse oral 130 mg/kg [R12, 298].

BIOLOGICAL HALF-LIFE No adequate biokinetic data on the vanadium half-time in man are available. The ICRP (International Commission on Radiological Protection) estimate for the whole-body retention of vanadium in man is 42 days. Other studies on workers exposed to vanadium in workroom air report that blood and urinary values of vanadium drop to half the initial value within a few days after cessation of exposure. (vanadium compounds) [R10, 646].

DOT *Health Hazards:* Poisonous; may be fatal if inhaled, swallowed, or absorbed through skin. Contact may cause burns to skin and eyes. Runoff from fire control or dilution water may give off poisonous gases and cause water pollution. Fire may produce irritating or poisonous gases [R2].

Fire or Explosion: Some of these materials may burn, but none of them ignites readily. Container may explode violently in heat of fire [R2].

Emergency Action: Keep unnecessary people away; isolate hazard area and deny entry. Stay upwind, out of low areas, and ventilate closed spaces before entering. Self-contained breathing apparatus and chemical protective clothing that is specifically recommended by the shipper or producer may be worn but they do not provide thermal protection unless that is stated by the clothing manufacturer. Structural firefighter's protective clothing is not effective with these materials. Remove and isolate contaminated clothing at the site. Call CHEMTREC at 1–800–424–9300 as soon as possible, especially if there is no local hazardous materials team available [R2].

Fire: Small Fires: Dry chemical, carbon dioxide, halon, water spray, or standard foam. Large Fires: Water spray, fog, or standard foam is recommended. Move container from fire area if you can do so without risk. Fight fire from maximum distance. Stay away from ends of tanks.

Dike fire control water for later disposal; do not scatter the material [R2].

Spill or Leak: Do not touch spilled material; stop leak if you can do so without risk. Use water spray to reduce vapors. Small Spills: Take up with sand or other noncombustible absorbent material and place into containers for later disposal. Small Dry Spills: With clean shovel place material into clean, dry container, and cover; move containers from spill area. Large Spills: Dike far ahead of liquid spill for later disposal [R2].

First Aid: Move victim to fresh air and call for emergency medical care; if not breathing, give artificial respiration; if breathing is difficult, give oxygen. In case of contact with material, immediately flush skin or eyes with running water for at least 15 minutes. Speed in removing material from skin is of extreme importance. Remove and isolate contaminated clothing and shoes at the site. Keep victim quiet and maintain normal body temperature. Effects may be delayed; keep victim under observation [R2].

FIRE POTENTIAL Vanadium(III) oxide ignites on heating in air [R3].

REFERENCES

1. *The Merck Index.* 10th ed. Rahway, NJ: Merck Co., Inc., 1983. 1418.

2. Department of Transportation. *Emergency Response Guidebook 1987.* DOT P 5800.4. Washington, DC: U.S. Government Printing Office, 1987, p. G–55.

3. Bretherick, L. *Handbook of Reactive Chemical Hazards.* 3rd ed. Boston, MA: Butterworths, 1985. 1368.

4. Gul'ko, A.G. *Gig Saint* 21: 24–8 (1956) as cited in NIOSH. *Criteria Document: Vanadium* p. 26 (1977) DHEW Pub. NIOSH 77–222.

5. NIOSH. *Criteria Document: Vanadium* p. 91 (1977) DHEW Pub. NIOSH 77–222.

6. Association of American Railroads. *Emergency Handling of Hazardous Materials in Surface Transportation.* Washington, DC: Assoc. of American Railroads, Hazardous Materials Systems (BOE), 1987. 713.

7. 54 FR 2920 (1/19/89).

8. NIOSH/CDC. *NIOSH Recommendations for Occupational Safety and Health Standards 1988,* Aug. 1988 (Suppl. to *Morbidity and Mortality Wkly.* Vol. 37 No. 5–7, Aug. 26, 1988). Atlanta, GA: National Institute for Occupational Safety and Health, CDC, 1988. 28.

9. Venugopal, B., and T.D. Luckey. *Metal Toxicity in Mammals, 2.* New York: Plenum Press, 1978.

10. Friberg, L., G.F. Nordberg, E. Kessler, and V.B. Vouk (eds.). *Handbook of the Toxicology of Metals.* 2nd ed. Vols I, II: Amsterdam: Elsevier Science Publishers B.V., 1986, p. V2.

11. Seiler, H.G., H. Sigel, and A. Sigel (eds.). *Handbook on the Toxicity of Inorganic Compounds.* New York, NY: Marcel Dekker, Inc. 1988.

12. National Research Council. *Drinking Water and Health, Volume 1.* Washington, DC: National Academy Press, 1977.

■ DODECANE

CAS # 112–40–3

SYNONYMS n-Dodecan (German) • n-dodecane

MF $C_{12}H_{26}$

MW 170.38

COLOR AND FORM Colorless liquid.

ODOR THRESHOLD 37 mg/m^3 [R3].

DENSITY 0.7487 at 20°C.

VISCOSITY Less than 32 SUS (Saybolt universal seconds).

BOILING POINT 216.3°C at 760 mm Hg.

MELTING POINT –9.6°C.

HEAT OF VAPORIZATION 11,857.7 cal/mole.

AUTOIGNITION TEMPERATURE 397°F (203°C) [R4].

FLAMMABILITY LIMITS Lower 0.6% by volume [R4].

FIREFIGHTING PROCEDURES To fight fire: foam, CO_2, dry chemical [R5].

SOLUBILITY Very soluble in alcohol, ether, acetone, and chloroform, carbon tetrachloride; 0.005 mg/L at 20°C in salt water; 0.0029 mg/L at 25°C in sea water; 0.0037 mg/L at 25°C in distilled water.

VAPOR PRESSURE 0.3 mm Hg at 20°C; 1 mm Hg at 48°C.

OTHER CHEMICAL/PHYSICAL PROPERTIES Heat of fusion: 51.33 cal/g.

DISPOSAL This combustible material should be burned in a chemical incinerator equipped with an afterburner and scrubber [R6].

COMMON USES As solvent; in organic synthesis; as distillation chaser; in jet fuel research [R2]; chemical intermediate for n-dodecylbenzene [R1]; chemical intermediate for n-dodecanol [R1]; chemical intermediate for n-dodecyl chloride and bromide [R1]; chemical intermediate for chlorinated dodecanes, e.g., oil additives [R1]; component of paraffin mixture used as intermed for citric acid [R1]; solvent for printing inks and degreasing applications [R1]; chemical intermediate for n-dodecyl mercaptan [R1]; n-dodecane is used in the rubber industry and the paper processing industry [R3].

OCCUPATIONAL RECOMMENDATIONS None listed.

BIODEGRADATION Dodecane biodegrades in sewage, sediment, soil, and fresh and marine water, with the rate of degradation being strongly influenced by the acclimation of the degrading microorganisms (1–9). Thirty-seven percent of the dodecane was mineralized in a 5-day biodegradability test using activated sludge with most of the remaining radioactivity from the labelled substrate being bound to the sludge as unextractable residue (1). In other studies 74% of the theoretical BOD was achieved in 24 hr (2); 22 and 67% of the theoretical BOD was attained in 2 and 10 days, respectively, in a soil suspension (3); and 40 and 46% degradation occurred

when n-dodecane was exposed to microorganisms from polluted estuarial water and oil-rich sediment in Chesapeake Bay for an unspecified time period. Less degradation occurred in less-contaminated oil and water (4). In similar studies, 16 and 49% of dodecane in crude oil exposed to harbor water degraded in 5 and 15 days, respectively, whereas 21 and 87% exposed to harbor sediment degraded in the same time period (5), and 95.1% degradation occurred in 21 days in seawater inoculated with oil-oxidizing microorganisms (8). When anaerobically digested sewage sludge was amended on soil, all dodecane had disappeared from 0–1-cm and 6–7-cm core sections 1 yr after the last treatment (6). Anaerobic conditions tend to retard hydrocarbon degradation (6). Dodecane in jet fuels applied to soil cores and subjected to simulated rain had disappeared when tested for after 131 days (9). In a mesocosm experiment that simulated seasonal conditions in Narragansett Bay, RI, the half-life for dodecane was 1.1, 0.7, and 3.6 days under spring, summer, and winter conditions, respectively. Under summer conditions the half-life was tripled when mercury chloride was added to eliminate biodegradation (7). The degradation proceeded without any lag and mineralization was rapid (7) [R7].

BIOCONCENTRATION The log of the bioconcentration factor in static tests was 1.72 for golden orfes after 3 days and 3.80 for green algae after 24 hr (1). Only traces of dodecane were taken up by a marine diatom from crude oil (2) [R8].

FIRE POTENTIAL Combustible [R2].

REFERENCES

1. SRI.

2. Hawley, G. G. *The Condensed Chemical Dictionary*. 10th ed. New York: Van Nostrand Reinhold Co., 1981. 390.

3. Verschueren, K. *Handbook of Environmental Data of Organic Chemicals*. 2nd ed. New York, NY: Van Nostrand Reinhold Co., 1983. 595.

4. National Fire Protection Association. *Fire Protection Guide on Hazardous Materials*. 7th ed. Boston, MA: National Fire

Protection Association, 1978, p. 325M–91.

5. Sax, N. I. *Dangerous Properties of Industrial Materials.* 6th ed. New York, NY: Van Nostrand Reinhold, 1984. 1262.

6. Aldrich. *Catalog Hdbk Fine Chemical* p. 488 (1984).

7. (1) Freitag, D. *Ecotox Environ Safety* 6: 60–81 (1982). (2) Gerhold, R. M., E. W. Malaney. *J Water Pollut Control* 38: 562–79 (1966). (3) Haines, J. R., M. Alexander. *Appl Microbiol* 28: 1084–5 (1974). (4) Walker, J. D., R. R. Colwell. *Prog Water Technol* 7: 783–91 (1975). (5) Nagata, S., G. Kondo. *1977 Oil Spill Conf Amer Petrol Inst.* pp. 617–20 (1977). (6) Liu, D. *Bull Environ Contam Toxicol* 25: 616–22 (1980). (7) Wakeham, S.G., et al., *Environ Sci Technol* 17: 611–7 (1983). (8) Verschueren, K. *Handbook of Environmental Data on Organic Chemicals*; 2nd ed. New York: Van Nostrand Reinhold Co. pp. 595–6 (1983). (9) Ross, W. D., et al., *Environmental Fate and Biological Consequences of Chemicals Related to Air Force Activities.* p. 173 NTIS AD–A121 288/5 (1982).

8. (1) Freitag, D., et al., *Ecotox Environ Safety* 6: 60–81 (1982). (2) Karydis, M. *Microb Ecol* 5: 287–93 (1980).

■ 1-DODECANETHIOL

CAS # 112–55–0

SYNONYMS 1-dodecyl mercaptan • 1-mercaptododecane • dodecyl mercaptan • lauryl mercaptan • lauryl mercaptide • n-dodecanethiol • n-dodecyl mercaptan • n-lauryl mercaptan • NCI-c60935 • tert-dodecyl mercaptan • tert-dodecanthiol

MF $C_{12}H_{26}S$

MW 202.44

COLOR AND FORM Water-white to pale-yellow liquid; oily colorless liquid.

ODOR Mild, characteristic; mild skunk.

ODOR THRESHOLD 4 mg/m³ [R5].

DENSITY 0.8450 at 20°C.

SURFACE TENSION (est) 30 dynes/cm at 20°C.

BOILING POINT 142–145°C at 15 mm Hg.

MELTING POINT –7°C.

HEAT OF COMBUSTION (est) –10,100 cal/g.

HEAT OF VAPORIZATION (est) 60 cal/g.

EXPLOSIVE LIMITS AND POTENTIAL Explosion hazard: it produces explosive concentrations of vapor only at high temperatures. [R6, 855].

FIREFIGHTING PROCEDURES Dry chem or carbon dioxide. Water or foam may cause frothing. Water may be ineffective on fire. Cool exposed containers with water [R5]; "alcohol" foam [R4].

TOXIC COMBUSTION PRODUCTS Poisonous and irritating gases (e.g., sulfur dioxide) are generated in fires [R5].

SOLUBILITY Insoluble in water; soluble in ethanol, ether; soluble in methanol, acetone, benzene; soluble in gasoline, ethyl acetate.

VAPOR PRESSURE 154 mm Hg at 20°C (calc from exper. derived coeffic); Antoine constants: A = 7.0244, B = 1,817.8, C = 164.1; equation: $\log_{10} P = A - B/(Ct)$, in t–range 10–300°C, t is in°C, P is in mm Hg.

OTHER CHEMICAL/PHYSICAL PROPERTIES Conversion factors: 8.26 mg/m³ = 1 ppm; liquid-water interfacial tension: (est) 30 dynes/cm at 20°C; distillation range: 200–235°C; freezing point = –9.2°C.

PROTECTIVE EQUIPMENT Respirator when mist is present; rubber or vinyl gloves; chemical goggles; rubber shoes and apron [R5].

PREVENTATIVE MEASURES SRP: The scientific literature for the use of contact lenses in industry is conflicting. The benefit or detrimental effects of wearing contact lenses depend not only upon the substance, but also on factors including the form of the substance, characteristics and duration of the exposure, the uses of other eye protection equipment, and the hygiene of the lenses. However, there may be individual substances whose irritating or corrosive properties are such that the wearing of contact lenses would be harmful to the eye. In those specific cases, contact lenses should not be worn. In any event, the usual eye protection equipment should be worn even when contact lenses are in place.

DISPOSAL SRP: At the time of review, criteria for land treatment, or burial (sanitary landfill) disposal practices are subject to significant revision. Prior to implementing land disposal of waste residue (including waste sludge), consult with environmental regulatory agencies for guidance on acceptable disposal practices. Dissolve in flammable solvent. Burn in incinerator with afterburner and SO₂ scrubber.

COMMON USES In manufacture of synthetic rubber, plastics, pharmaceuticals, insecticides, fungicides [R6, 854]; polymerization regulator for styrene-butadiene rubber [R1]; in chemical syntheses of bactericides; as complexing agent, such as for the removal of metals from wastes [R7, 2078]; nonionic detergent [R2]; flotation reagent [R3].

STANDARD CODES NFPA—2, 1, 0; ICC, USCG—no; IATA—not exceeding 3 cc in hermetically sealed warning or odorizing devices; other hazardous article, class A, no limit passenger or cargo.

PERSISTENCE Chemical action causes release of sulfur oxides and sulfates.

MAJOR SPECIES THREATENED Waterfowl.

GENERAL SENSATION Offensive skunk odor. Cause nausea, headache, and unconsciousness, with cyanosis, cold extremities, and rapid pulse. Odor should provide good warning.

ACUTE HAZARD LEVEL As a mercaptan, may be strong irritant, and toxic via inhalation. Emits highly toxic fumes when heated to decomposition, or in contact with acid, water, or steam.

CHRONIC HAZARD LEVEL Unknown.

DEGREE OF HAZARD TO PUBLIC HEALTH Thought to be moderately toxic via inhalation and a strong irritant. Emits highly

toxic fumes when heated to decomposition or when in contact with acid, water, or steam.

OCCUPATIONAL RECOMMENDATIONS None listed.

ACTION LEVEL Notify fire and air authority. Enter from upwind. Remove ignition source. Isolate from oxidizing materials. Attempt to contain slick.

ON-SITE RESTORATION Oil-skimming equipment and sorbent foams can be used on slick. Seek professional environmental engineering assistance through EPA's Environmental Response Team (ERT), Edison, NJ, 24-hour no., 908–548–8730.

BEACH OR SHORE RESTORATION Do not burn.

AVAILABILITY OF COUNTERMEASURE MATERIAL Oil-skimming equipment—stored at major harbors; sorbent foams (polyurethane)—upholstery shops.

DISPOSAL NOTIFICATION Contact local air authority.

INDUSTRIAL FOULING POTENTIAL Slick may reduce heat transfer or cause hot spots and scaling. May render water unfit for food-processing purposes.

MAJOR WATER USE THREATENED Recreation, potable supply, industrial.

PROBABLE LOCATION AND STATE White–light-yellow liquid. Will float in slick on surface.

FIRE POTENTIAL Mercaptans will react with water, steam, or acids to produce flammable vapors. (mercaptans) [R8, 1742].

REFERENCES

1. SRI.

2. Sax, N. I., and R. J. Lewis, Sr. (eds.). *Hawley's Condensed Chemical Dictionary*. 11th ed. New York: Van Nostrand Reinhold Co., 1987. 440.

3. Kuney, J. H., J. M. Mullican (eds.). *Chemcyclopedia*. Washington, DC: American Chemical Society, 1994. 75.

4. National Fire Protection Guide. Fire Protection *Guide on Hazardous Materi-*

als. 10th ed. Quincy, MA: National Fire Protection Association, 1991, p. 325M–46.

5. U.S. Coast Guard, Department of Transportation. *CHRIS—Hazardous Chemical Data*. Volume II. Washington, DC: U.S. Government Printing Office, 1984–5.

6. International Labour Office. *Encyclopedia of Occupational Health and Safety*. Volumes I and II. New York: McGraw-Hill Book Co., 1971.

7. Clayton, G. D., and F. E. Clayton (eds.). *Patty's Industrial Hygiene and Toxicology*. Volume 2A, 2B, 2C: Toxicology. 3rd ed. New York: John Wiley Sons, 1981–1982.

8. Sax, N.I. *Dangerous Properties of Industrial Materials*. 6th ed. New York, NY: Van Nostrand Reinhold, 1984.

■ 1-DODECENE

CAS # 112–41–4

SYNONYMS adacene-12 • alpha-dodecene • alpha-dodecylene • n-dodec-1-ene

MF $C_{12}H_{24}$

MW 168.33

COLOR AND FORM Colorless liquid.

ODOR Mild, pleasant.

DENSITY 0.7584 at 20°C.

SURFACE TENSION 26.6 dynes/cm = 0.0256 N/m at 20°C.

VISCOSITY 1.72 sq mm/s at 20°C.

BOILING POINT 213.4°C at 760 mm Hg.

MELTING POINT –35.2°C.

HEAT OF COMBUSTION –18.911 Btu/lb.

HEAT OF VAPORIZATION 110 Btu/lb.

AUTOIGNITION TEMPERATURE 491°F (255°C) [R4].

FIREFIGHTING PROCEDURES Foam, carbon dioxide, dry chemical (dodecene) [R3].

SOLUBILITY Soluble in alcohol, benzene, ether, acetone.

VAPOR PRESSURE 0.0159 mm Hg at 25°C.

OCTANOL/WATER PARTITION COEFFICIENT Log K_{ow} = 6.5 (est).

OTHER CHEMICAL/PHYSICAL PROPERTIES Henry's Law Constant: 4.25 atm-m³/mole at 25°C.

PROTECTIVE EQUIPMENT Protective gloves; goggles or face shield [R5].

PREVENTATIVE MEASURES SRP: The scientific literature supports the wearing of contact lenses in industrial environments, as part of a program to protect the eye against chemical compounds and minerals causing eye irritation. However, there may be individual substances whose irritating or corrosive properties are such that the wearing of contact lenses would be harmful to the eye. In those specific cases contact lenses should not be worn.

COMMON USES Chemical intermediate for n-dodecyl dimethyl amine [R1]; 1-dodecene is used in flavors, perfumes, medicines, oils, dyes, and resins [R2].

STANDARD CODES NFPA; ICC, USCG—no.

PERSISTENCE Should degrade at low to moderate rate. Unsaturated bond is subject to photochemical attack.

MAJOR SPECIES THREATENED Waterfowl.

DIRECT CONTACT Irritant, skin.

GENERAL SENSATION Suspected of being irritant and narcotic at high concentrations.

PERSONAL SAFETY Wear skin protection and canister-type masks if vapor concentrations are high.

ACUTE HAZARD LEVEL Low toxicity. May pose BOD problem.

CHRONIC HAZARD LEVEL Unknown.

DEGREE OF HAZARD TO PUBLIC HEALTH Believed to be of low toxicity. May have irritant or narcotic properties at high concentrations.

OCCUPATIONAL RECOMMENDATIONS None listed.

ACTION LEVEL Notify fire authority. Remove ignition sources. Isolate from oxidizing material. Attempt to contain slick.

ON-SITE RESTORATION Oil-skimming equipment and sorbent foams can be used on slick. Seek professional environmental engineering assistance through EPA's Environmental Response Team (ERT), Edison, NJ, 24-hour no., 908–548–8730.

BEACH OR SHORE RESTORATION Burn off.

AVAILABILITY OF COUNTERMEASURE MATERIAL Oil-skimming equipment—stored at major ports; sorbent foams (polyurethane)—upholstery shops.

DISPOSAL Spray into incinerator. Additional flammable solvent may be added as required.

INDUSTRIAL FOULING POTENTIAL Slick may reduce heat transfer or cause hot spots and scaling.

MAJOR WATER USE THREATENED Recreation, industrial.

PROBABLE LOCATION AND STATE Colorless liquid; will float as a slick on surface of water.

COLOR IN WATER Colorless.

BIODEGRADATION Pure cultures of *Cornebacterium* sp (1), *C. resinae* (2,3), *Micrococcus cericans* (4), and yeast (5) were found to oxidize 1-dodecene. In pure-culture tests, 10 of 34 microorganism strains were found to grow on 1-dodecene (6) [R6].

BIOCONCENTRATION Based on an estimated log octanol/water partition coefficient of 6.5 (1,SRC), a bioconcentration factor of 51,000 can be calculated for 1-dodecene using an appropriate regression equation (2,SRC). The magnitude of this value indicate that 1-dodecene is expected to significantly bioconcentrate in fish and aquatic organisms (SRC) [R7].

ATMOSPHERIC STANDARDS This action promulgates standards of performance for equipment leaks of volatile organic compounds (VOC) in the synthetic organic chemical manufacturing industry (SOCMI). The intended effect of these standards is to require all newly constructed, modified, and reconstructed SOCMI process units to use the best demonstrated system of continuous emission reduction for equipment leaks of VOC, considering costs, non-air-quality health, and environmental impact and energy requirements. Dodecene is produced, as an intermediate or final product, by process units covered under this subpart. (dodecene) [R8].

FIRE POTENTIAL Low, when exposed to heat or flame (dodecene) [R3].

REFERENCES

1. SRI.

2. Sax, N. I., and R. J. Lewis, Sr. (eds.). *Hawley's Condensed Chemical Dictionary.* 11th ed. New York: Van Nostrand Reinhold Co., 1987. 439.

3. Sax, N. I. *Dangerous Properties of Industrial Materials.* 5th ed. New York: Van Nostrand Reinhold, 1979. 633.

4. National Fire Protection Association. *Fire Protection Guide on Hazardous Materials.* 9th ed. Boston, MA: National Fire Protection Association, 1986, p. 325M–46.

5. U.S. Coast Guard, Department of Transportation. *CHRIS—Hazardous Chemical Data.* Volume II. Washington, DC: U.S. Government Printing Office, 1984–5.

6. (1) Buswell, J. A., P. Jurshuk. *Arch Mikrobiol* 64: 215–22 (1969). (2) Cofone, L., et al. *J Gen Microbiol* 76: 243–6 (1973). (3) Walker, J. D., et al., pp. 821–73 in *Microbia Petroleum API/USEPA/USCG Conf of Prevention and Control of Oil Spills* (1973). (4) Makula, R., W. R. Finnerty. *J Bacteriol* 95: 2108–11 (1968). (5) Markovetz, A. J., R. E. Kallio. *J Bacteriol* 87: 96809 (1964). (6) Hoffman, B., H. J. Rehm. *European J Appl Microbiol* 3: 19–30 (1976).

7. (1) USEPA. *CLOGP–PCGEMS* (1988). (2) Lyman, W. J., et al., *Handbook of Chemical Property Estimation Methods.* New York: McGraw-Hill, Chapt 5 (1982).

8. 40 CFR 60.489 (7/1/90).

∎ n-DODECYL METHACRYLATE

CAS # 142–90–5

DOT # Not listed

SYNONYMS 2-propenoic acid, 2-methyl-, dodecyl ester • acrylic acid, 2-methyl-, dodecyl ester • dodecyl 2-methyl-2-propenoate • dodecyl methacrylate • laurylester-kyseliny-methakrylove (Czech) • metazene • methacrylic acid, dodecyl ester • methacrylic acid, lauryl ester • lauryl methacrylate • ai3-08765 • Caswell no. 521 • EPA pesticide code 053101

MF $C_{16}H_{30}O_2$

MW 254.46

DENSITY 0.868 at 20°C.

BOILING POINT 272–344°C.

MELTING POINT −20°C.

SOLUBILITY Insoluble in water.

PROTECTIVE EQUIPMENT Suitable protective clothing and self-contained resp protective apparatus should be available for use by those who may have to rescue persons overcome by fumes. (acrylic acid and derivatives) [R4].

PREVENTATIVE MEASURES The hazard is the generation of considerable exothermic heat in some of the reactions, in which high pressures and temperature may develop. This danger should be borne in mind when designing plant. Awareness of the dangers and of good engineering design is essential to safety. Employees should be instructed about the necessity of cleansing the skin if it is contaminated by materials that are irritants or skin-absorbed. With careful design, however, and complete enclosure of those processes where toxic chemicals or intermediates occur, dangerous exposures can be avoided (acrylic acid, and derivatives) [R4]. Contact lens use in industry is controversial. A survey of 100 corporations resulted in the recommendation that each company establish its own contact lens use policy. One presumed hazard of contact lens use is possible chemical entrapment. It was found that contact lenses minimized injury or protected the eye. The eye was afforded more protection from liquid irritants. Soft contact lenses did not worsen corneal damage from strong chemicals and in some cases could actually protect the eye. Overall, the literature supports the wearing of contact lenses in industrial environments as part of the standard eye protection, e.g., face shields; however, more data are needed to establish the value of contact lenses [R9]. SRP: Contaminated protective clothing should be segregated in such a manner so that there is no direct personal contact by personnel who handle, dispose of, or clean the clothing. Quality assurance to ascertain the completeness of the cleaning procedures should be implemented before the decontaminated protective clothing is returned for reuse by the workers. Contaminated clothing should not be taken home at end of shift, but should remain at employee's place of work for cleaning.

COMMON USES Polymerizable monomer for plastics, molding powders, solvent coatings, adhesives, oil additives; emulsions for textile, leather, and paper finishing [R2]. As a deodorant to mask methyl sulfide odors in industry [R11, 2084]. To delay volatilization of insecticides. Monomer for viscosity index improvers for lubricating oil [R1]. Monomer for pour-point depressants for distillate fuels [R1]. Used in dentistry as restorative materials, adhesives, prosthetic devices [R3].

OCCUPATIONAL RECOMMENDATIONS None listed.

NON-HUMAN TOXICITY LD$_{50}$ rats oral >5.0 g/kg [R12, 369]; LD$_{50}$ rabbits percutaneous >3.0 g/kg [R12, 369]; LD$_{50}$ rats intraperitoneal 21.6 g/kg [R12, 369]; LD$_{50}$ mice, oral greater than 87,250 mg/kg [R5].

TSCA REQUIREMENTS Section 8 (a) of TSCA requires manufacturers of this chemical substance to report preliminary assessment information concerned with production, use, and exposure to EPA as cited in the preamble of the 51 FR 41329 [R6].

FIFRA REQUIREMENTS Unless designated as an active ingredient as determined by EPA, this substance, when used in antimicrobial products as an emulsifier, is considered inert, having no independent pesticidal activity. The percentage of such an ingredient shall be included on the label in the total percentage of inert ingredients. [R10].

FIRE POTENTIAL Combustible [R2].

GENERAL RESPONSE Avoid contact with liquid and vapor. Keep people away. Wear self-contained positive-pressure breathing apparatus and full protective clothing. Stop leak if possible. Call fire department. Isolate and remove discharged material. Notify local health and pollution control agencies.

FIRE RESPONSE Combustible. Wear self-contained positive-pressure breathing apparatus and full protective clothing. Poisonous gases may be produced in fire. Containers may explode in fire. Extinguish with foam, CO_2, or dry chemicals. Cool exposed containers with water.

EXPOSURE RESPONSE Call for medical aid. Liquid irritating to skin and eyes. Remove and isolate contaminated clothing and shoes at the site. If in eyes or on skin, flush with running water for at least 15 min; hold eyelids open if necessary. Wash skin with soap and water. If swallowed and victim is conscious, have victim drink water or milk, and induce vomiting. If swallowed and victim is unconscious, or having convulsions, do nothing except keep victim warm.

RESPONSE TO DISCHARGE Mechanical containment; should be removed; chemical and physical treatment.

WATER RESPONSE Effect of low concentrations on aquatic life is unknown. Fouling to shoreline. May be dangerous if it enters water intakes. Notify local health and wildlife officials. Notify operators of nearby water intakes.

EXPOSURE SYMPTOMS Inhalation temporarily reduces blood pressure from 5 to 25%, increases respiratory rate, decreases heart rate, and causes some EKG changes. Liquid may cause irritation of eyes and skin. May be harmful if swallowed.

EXPOSURE TREATMENT Inhalation: move victim to fresh air. If breathing has stopped, give artificial respiration. If breathing is difficult, give oxygen. Eyes or skin: flush with running water for at least 15 min; hold eyelids open if necessary. Wash skin with soap and water. Ingestion: if victim is conscious, have victim drink milk or water, and induce vomiting. If victim is unconscious or having convulsions, do nothing except keep victim warm.

REFERENCES

1. SRI.

2. Sax, N. I., and R. J. Lewis, Sr. (eds.). *Hawley's Condensed Chemical Dictionary.* 11th ed. New York: Van Nostrand Reinhold Co., 1987. 686.

3. Fujisawa, S., et al., *J Biomed Mater Res* 18 (9): 1105–14 (1984).

4. International Labour Office. *Encyclopedia of Occupational Health and Safety.* Vols. I and II. Geneva, Switzerland: International Labour Office, 1983. 53.

5. Toida, S., H. Nishimura. *Toho Igakkai Zasshi* 18 (1): 179–80 (1971).

6. 40 CFR 712.30 (7/1/88).

7. Clayton, G. D., and F. E. Clayton (eds.). *Patty's Industrial Hygiene and Toxicology* Volume 2A, 2B, 2C: *Toxicology.* 3rd ed. New York: John Wiley Sons, 1981–1982.

8. *Kirk-Othmer Encyclopedia of Chemical Technology.* 3rd ed., Volumes 1–26. New York, NY: John Wiley and Sons, 1978–1984, p. 15(81).

9. Randolph, S. A., M. R. Zavon. *J Occup Med* 29: 237–42 (1987).

10. 40 CFR 162.60 (7/1/88).

11. Sax, N.I., and R.J. Lewis, Sr. (eds.). *Hawley's Condensed Chemical Dictionary.* 11th ed. New York: Van Nostrand Reinhold Co., 1987. 686.

12. Paulet, G., M. Vidal. *Arch Mal Prof Med Trac Secur Soc.* 36 (1–2): 58–60 (1975).

■ DODECYLTRICHLOROSILANE

CAS # 4484–72–4

DOT # 1771

SYNONYMS silane, dodecyltrichloro • silane, trichlorododecyl • trichlorododecylsilane

MF $C_{12}H_{25}Cl_3Si$

MW 303.81

COLOR AND FORM Colorless to yellow liquid.

ODOR Sharp odor, like hydrochloric acid; pungent.

DENSITY 1.026 at 25°C.

BOILING POINT 288°C.

HEAT OF COMBUSTION $-11,000$ Btu/lb $= -6,200$ cal/g $= -260 \times 10^5$ J/kg (estimated).

FIREFIGHTING PROCEDURES Dry chemical, carbon dioxide. Fire extinguishing agents not to be used: water, foam. Hydrochloric acid and phosgene fumes may form in fires. Behavior of fire: difficult to extinguish; re-ignition may occur. Contact with water applied to adjacent fires produces irritating hydrogen chloride fumes [R2].

TOXIC COMBUSTION PRODUCTS Hydrochloric acid and phosgene fumes may form in fires [R2].

CORROSIVITY Corrosive.

STABILITY AND SHELF LIFE Readily hydrolyzed by moisture, with production of hydrochloric acid [R5].

PROTECTIVE EQUIPMENT Acid-vapor type resp protection; rubber gloves; chemical worker's goggles; other protective equipment as necessary to protect eyes and skin [R2].

PREVENTATIVE MEASURES Preventive measures should be directed primarily at preventing or minimizing contact between corrosive substances and skin, mucous membranes, and eyes. Adequate ventilation and exhaust arrangements, whether general or local, should be provided whenever corrosive gases or dusts are present. (corrosive substances) [R6, 553].

COMMON USES Intermediate for silicones [R1].

OCCUPATIONAL RECOMMENDATIONS None listed.

DOT *Health Hazards:* Contact causes burns to skin and eyes. If inhaled, may be harmful. Fire may produce irritating or poisonous gases. Runoff from fire control or dilution water may cause pollution [R3].

Fire or Explosion: Some of these materials may burn, but none of them ignites readily. Flammable/poisonous gases may accumulate in tanks and hopper cars. Some of these materials may ignite combustibles (wood, paper, oil, etc.) [R3].

Emergency Action: Keep unnecessary people away; isolate hazard area and deny entry. Stay upwind; keep out of low areas. Self-contained breathing apparatus (SCBA) and structural firefighter's protective clothing will provide limited protection. Call CHEMTREC at 1–800–424–9300 for emergency assistance. If water pollution occurs, notify the appropriate authorities [R3].

Fire: Some of these materials may react violently with water. Small Fires: Dry chemical, CO_2, halon, water spray, or standard foam. Large Fires: Water spray, fog, or standard foam is recommended. Move container from fire area if you can do so without risk. Cool containers that are exposed to flames with water from the side until well after fire is out. Stay away from ends of tanks [R3].

Spill or Leak: Do not touch spilled material; stop leak if you can do so without risk. Small Spills: Take up with sand or other noncombustible absorbent material and place into containers for later disposal. Small Dry Spills: With clean shovel place material into clean, dry container, and cover; move containers from spill area. Large Spills: Dike far ahead of liquid spill for later disposal [R3].

First Aid: Move victim to fresh air; call for emergency medical care. Remove and isolate contaminated clothing and shoes at the site. In case of contact with material, immediately flush skin or eyes with running water for at least 15 minutes. Keep victim quiet and maintain normal body temperature [R3].

FIRE POTENTIAL Certain strong corrosives may, on contact with organic matter or other chemicals, cause fire. (corrosive substances) [R4].

GENERAL RESPONSE Avoid contact with liquid and vapor. Keep people away. Wear goggles and self-contained breathing apparatus. Stop discharge if possible. Call fire department. Isolate and remove discharged material. Notify local health and pollution control agencies.

FIRE RESPONSE Combustible. Poisonous gases may be produced in fire. Extinguish with dry chemicals or carbon dioxide. Do not use water or foam on fire.

EXPOSURE RESPONSE Call for medical aid. Vapor irritating to eyes, nose, and throat. Move victim to fresh air. If breathing is difficult, give oxygen. Liquid will burn skin and eyes. Harmful if swallowed. Remove contaminated clothing and shoes. Flush affected areas with plenty of water. If swallowed and victim is conscious, have victim drink water or milk. Do not induce vomiting.

RESPONSE TO DISCHARGE Issue warning—corrosive, water containment. Restrict access. Disperse and flush with care.

WATER RESPONSE Effect of low concentrations on aquatic life is unknown. May be dangerous if it enters water intakes. Notify local health and wildlife officials. Notify operators of nearby water intakes.

EXPOSURE SYMPTOMS Inhalation irritates mucous membranes. Contact with liquid causes severe burns of eyes and skin. Ingestion causes severe burns of mouth and stomach.

EXPOSURE TREATMENT Inhalation: remove from exposure; support respiration; call physician if needed. Eyes: flush with water for 15 min. Obtain medical attention immediately. Skin: flush with water; obtain medical attention if skin is burned. Ingestion: if victim is conscious, give large amounts of water, then milk or milk of magnesia.

REFERENCES

1. Hawley, G. G. *The Condensed Chemical Dictionary.* 9th ed. New York: Van Nostrand Reinhold Co., 1977. 326.

2. U.S. Coast Guard, Department of Transportation. *CHRIS—Hazardous Chemical Data.* Manual Two. Washington, DC: U.S. Government Printing Office, Oct., 1978.

3. Department of Transportation. *Emergency Response Guidebook 1987.* DOT P 5800.4. Washington, DC: U.S. Government Printing Office, 1987, p. G–60.

4. International Labour Office. *Encyclopedia of Occupational Health and Safety.* Volumes I and II. New York: McGraw-Hill Book Co., 1971. 338.

5. Sax, N. I. *Dangerous Properties of Industrial Materials.* 5th ed. New York: Van Nostrand Reinhold, 1979. 633.

6. International Labour Office. *Encyclopedia of Occupational Health and Safety.* Vols. I and II. Geneva, Switzerland: International Labour Office, 1983.

■ DI-n-PENTYLAMINE

CAS # 2050–92–2

SYNONYMS 1-pentanamine, n-pentyl • amine, dipentyl • di-n-amylamine • diamyl amine • diamylamine • dipentylamine • pentylamine, pentyl

MF $C_{10}H_{23}N$

MW 157.34

COLOR AND FORM Colorless to light-yellow liquid.

ODOR Ammoniacal odor.

DENSITY 0.7771 at 20°C.

BOILING POINT 202°C at 760 mm Hg.

EXPLOSIVE LIMITS AND POTENTIAL Vapor forms explosive mixtures with air [R2].

FLAMMABILITY LIMITS Flammable limits not recorded [R2].

FIREFIGHTING PROCEDURES Use carbon dioxide, alcohol foam, or dry chem. Water or foam may cause frothing. Use water to keep fire-exposed containers cool. If leak or spill has not ignited, use water spray to disperse vapors, and to provide protection for men attempting to stop leak. Water spray may be used to flush spills away from exposures [R2].

SOLUBILITY Soluble in acetone; very soluble in alcohol; miscible with ether; slightly soluble in water.

PROTECTIVE EQUIPMENT Wear full protective clothing [R2].

COMMON USES Rubber accelerators; flotation reagents; dyestuffs and corrosion inhibitors; solvent for oils, resins; solvent for some cellulose esters [R1].

STANDARD CODES NFPA—3, 2, 0; ICC, USCG—no.

PERSISTENCE Butylamines are degraded at moderate rate.

MAJOR SPECIES THREATENED All animal life. Waterfowl.

DIRECT CONTACT Irritant—may be sensitizer of skin.

GENERAL SENSATION High toxicity; suggested skin irritation grade 6—necrosis from undiluted; eye irritation grade 5—severe burns from 0.005 ml [R3].

PERSONAL SAFETY Wear full protective clothing and self-contained breathing apparatus.

ACUTE HAZARD LEVEL Toxic to fish at low levels. Appears toxic to animals. Irritant and potential sensitizer.

CHRONIC HAZARD LEVEL Unknown. May sensitize skin.

DEGREE OF HAZARD TO PUBLIC HEALTH Limited data suggest high toxicity. May be irritant or sensitizer.

OCCUPATIONAL RECOMMENDATIONS None listed.

ACTION LEVEL Notify fire and air authority. Enter from upwind. Remove oxidizing material and ignition source. Attempt to contain slick.

ON-SITE RESTORATION Oil-skimming equipment and sorbent foams may be effective initially. Peat or carbon can be used on the soluble portion. Cation exchangers will be effective under acid or neutral conditions. Seek professional environmental engineering assistance through EPA's Environmental Response Team (ERT), Edison, NJ, 24-hour no., 908–548–8730.

BEACH OR SHORE RESTORATION Do not burn.

AVAILABILITY OF COUNTERMEASURE MATERIAL Oil-skimming equipment—stored at major ports; sorbent foams (polyurethane)—upholstery shops; peat—nurseries, floral shops; carbon—water treatment plants, sugar refineries; cation exchangers—water softener suppliers.

DISPOSAL (1) Add contaminated amine to layer of sodium bisulfate in large evaporating dish. Spray with water. Neutralize and route to sewage plant. (2) Dissolve in flammable solvent and burn in open pit by use of excelsior train. Stay upwind. (3) Solution of no. 2 can be sprayed into incinerator with afterburner and scrubber.

DISPOSAL NOTIFICATION Contact local sewage or air authority.

MAJOR WATER USE THREATENED Potable supply, recreation, fisheries.

PROBABLE LOCATION AND STATE Colorless liquid. Will float in slick and dissolve very slowly.

WATER CHEMISTRY Forms slightly alkaline solution. Is biodegradable with time.

COLOR IN WATER Colorless.

DOT *Fire or Explosion:* Flammable/combustible material; may be ignited by heat, sparks, or flames. Vapors may travel to a source of ignition, and flash back. Container may explode in heat of fire. Vapor explosion hazard indoors, outdoors, or in sewers. Runoff to sewer may create fire or explosion hazard [R4, p. G–68].

Health Hazards: Poisonous if swallowed. If inhaled, may be harmful. Contact may cause burns to skin and eyes. Fire may produce irritating or poisonous gases. Runoff from fire control or dilution water may cause pollution [R4, p. G–68].

Emergency Action: Keep unnecessary people away; isolate hazard area, and deny entry. Stay upwind; keep out of low areas. Self-contained breathing apparatus (SCBA) and structural firefighter's protective clothing will provide limited protection. Isolate for 1/2 mile in all directions if tank car or truck is involved in fire. Call CHEMTREC at 1-800-424-9300 for emergency assistance. If water pollution occurs, notify the appropriate authorities [R4, p. G-68].

Fire: Small Fires: Dry chemical, CO_2, halon, water spray, or standard foam. Large Fires: Water spray, fog, or standard foam is recommended. Move container from fire area if you can do so without risk. Cool containers that are exposed to flames with water from the side until well after fire is out. Stay away from ends of tanks. Withdraw immediately in case of rising sound from venting safety device or any discoloration of tank due to fire [R4, p. G-68].

Spill or Leak: Shut off ignition sources; no flares, smoking, or flames in hazard area. Do not touch spilled material; stop leak if you can do so without risk. Water spray may reduce vapor; but it may not prevent ignition in closed spaces. Small Spills: Take up with sand or other non-combustible absorbent material and place into containers for later disposal. Large Spills: Dike far ahead of liquid spill for later disposal. [R4, p. G-68].

First Aid: Move victim to fresh air and call for emergency medical care; if not breathing, give artificial respiration; if breathing is difficult, give oxygen. Remove and isolate contaminated clothing and shoes at the site. In case of contact with material, immediately flush skin or eyes with running water for at least 15 minutes. Keep victim quiet and maintain normal body temperature [R4, p. G-68].

Initial Isolation and Evacuation Distances: For a spill or leak from a drum or small container (or small leak from tank): Isolate in all directions 50 feet. For a large spill from a tank (or from many containers or drums): First isolate in all directions 150 feet; then, evacuate in a downwind direction 0.2 mile wide, and 0.2 mile long. [R4, Table].

FIRE POTENTIAL Flammable liquid [R2].

REFERENCES

1. Hawley, G. G. *The Condensed Chemical Dictionary.* 9th ed. New York: Van Nostrand Reinhold Co., 1977. 267.

2. National Fire Protection Association. *Fire Protection Guide on Hazardous Materials.* 7th ed. Boston, MA: National Fire Protection Association, 1978, p. 49-112.

3. Smyth, H. F., Jr., C. P. Carpenter, C. S. Weil, U. C. Pozzani, J. A. Striegel, and J. S. Nycum, "Range-Finding Toxicity Data: List VII," *American Industrial Hygiene Association Journal,* 30: 470-476. 1969. Smyth, H. F., C. P. Carpenter, and C. S. Weil, "Range-Finding Toxicity Data: List IV," AMA *Archives of Industrial Hygiene and Occupational Medicine,* 4: 119-122, 1951. Smyth, H. F., C. P. Carpenter, C. S. Weil, U. C. Pozzani, and J. A. Striegel, "Range-Finding Toxicity Data: List VI," *American Industrial Hygiene Association Journal,* 23: 95-107, 1962. Smyth, H. F., C. P. Carpenter, C. S. Weil, and U. C. Pozzani, "Range-Finding Toxicity Data: List V," AMA *Archives of Industrial Hygiene and Occupational Medicine,* 10: 61-68, 1954. Smyth, H. F., C. P. Carpenter, and C. S. Weil, "Range-Finding Toxicity Data: List 111," *Journal of Industrial Hygiene and Toxicology,* 31: 60-62, 1949. Smyth, H. F., L. Seaton, and L. Fischer, "The Single Dose Toxicity of Some Glycols and Derivatives," *Journal of Industrial Hygiene and Toxicology,* 23 (6): 259-268, 1941. Smyth, H. F., and C. P. Carpenter, "Further Experience with the Range-Finding Test in the Industrial Toxicology Laboratory," *Journal of Industrial Hygiene and Toxicology,* 30: 63-68, 1948. Smyth, H. F., and C. P. Carpenter, "The Place of the Range-Finding Test in the Industrial Toxicology Laboratory," *Journal of Industrial Hygiene and Toxicology,* 26: 269-273, 1 944.

4. Department of Transportation. *Emergency Response Guidebook 1987.* DOT P 5800.4. Washington, DC: U.S. Government Printing Office, 1987.

■ENDOSULFAN SULFATE

CAS # 1031-07-8

SYNONYMS 6,7,8,9,10,10-hexachloro-1,5,5a,6,9,9a-hexahydro 6,9-methano-2,4,3-benzodioxathiepin-3,3-dioxide • 6,9-methano-2,4,3-benzodioxathiepin, 6,7,8,9,10,10-hexachloro-1,5,5a,6,9,9a hexahydro-, 3,3-dioxide • 5-norbornene-2,3-dimethanol, 1,4,5,6,7,7-hexachloro-, cyclic sulfate

MF $C_9H_6Cl_6O_4S$

MW 422.95

MELTING POINT 181°C.

SOLUBILITY 0.22 mg/L tap water (pH 7.2) at 22°C; water solubility of 0.117 ppm.

VAPOR PRESSURE 1×10^{-5} mm Hg at 25°C (estimated value).

OCTANOL/WATER PARTITION COEFFICIENT Log K_{ow} = 3.66 (calc).

OTHER CHEMICAL/PHYSICAL PROPERTIES Endosulfan sulfate did not degrade when thin films containing the compound were treated with radiation at 254 nm.

DISPOSAL SRP: At the time of review, criteria for land treatment or burial (sanitary landfill) disposal practices are subject to significant revision. Prior to implementing land disposal of waste residue (including waste sludge), consult with environmental regulatory agencies for guidance on acceptable disposal practices. Potential candidate for rotary kiln incineration with a temperature range of 820–1,600°C with a residence time for liquids and gases: seconds; solids: hours. (endosulfan) [R5].

COMMON USES SRP: Not commercially produced. Endosulfan is its precursor.

OCCUPATIONAL RECOMMENDATIONS None listed.

THRESHOLD LIMIT 8-hr time-weighted average (TWA) 0.1 mg/m³, skin (1986) (endosulfan) [R10, 21]. Excursion Limit Recommendation: Excursions in worker exposure levels may exceed three times the TLV-TWA for no more than a total of 30 min during a work day, and under no circumstances should they exceed five times the TLV-TWA. (endosulfan) [R10, 5].

ECOTOXICITY LC50 *Lebistes reticulatus* (guppy) 1.6 µg/L/48 hr (conditions of bioassay not specified) (from table) [R1]; LC50 *Carassius auratus* (goldfish) 17.5 µg/L/48 hr (conditions of bioassay not specified) (from table) [R1]; LC50 *Daphnia magna* (water flea) 140 µg/L/48 hr (conditions of bioassay not specified) (from table) [R1]; LC50 *Artemia salina* (brine shrimp) 750 µg/L/48 hr (conditions of bioassay not specified) (from table) [R1]; LC50 *Aedes aegypti* (mosquito) 150 µg/L/48 hr (from table) [R1]; LC50 *Limnea stagnalis* 6,000 µg/L/48 hr (conditions of bioassay not specified) (from table) [R1]; LC50 *Tubifex tubifex* (sludge worm) 2,500 µg/L/48 hr (from table) [R1].

BIODEGRADATION Endosulfan sulfate was reported to occur in the metabolic pathway of alpha- and beta-endosulfan exposed to mixed cultures from a sandy loam soil incubated at 20°C; endosulfan sulfate was subsequently converted to endosulfan diol; endosulfan diol was reported to be the final product of biodegradation (2). The half-life for endosulfan sulfate in the mixed cultures from a sandy loam soil was 11 weeks compared to >20 weeks in sterile inorganic salts nutrient control (2). Endosulfan sulfate at 5 and 10 ppm was not degraded in standard screening tests using a 7-day static incubation with settled domestic wastewater as inoculum, followed by 3 weekly subcultures (1). Alpha- and beta-endosulfan residues persisted for about 120 days in wet soils and about 100 days in dry soils (3), but it has been suggested that endosulfan sulfate is generally more persistent than these compounds, based on the limited information available (3) [R2].

BIOCONCENTRATION Algae: upper: 223, lower: 1,654; snail: upper: 5,457, lower: 29,430; mosquito: upper: 210, lower: 763; fish: upper: 935, lower: 1,741 [R3]. Endosulfan sulfate is more soluble in water and conversely less soluble in lipids than the organochlorine pesticides that usually associate with extensive accumulation through the food chain. Endosulfan sulfate is more readily degraded by living organisms. Based on these factors alone it would not tend to accumulate to as great an extent as some other organochlorines in the lipid-rich "pools" found within aquatic ecosystems [R6]. Using a reported solubility of 0.117 ppm (1) and a reported log octanol/water partition coefficient of 3.66 (1), BCFs of 2,070 and 356 were estimated, respectively (2,SRC). Based on these estimated BCFs, endosulfan sulfate may bioconcentrate in aquatic organisms (SRC). [R7].

CERCLA REPORTABLE QUANTITIES Persons in charge of vessels or facilities are required to notify the National Response Center (NRC) immediately, when there is a release of this designated hazardous substance, in an amount equal to or greater than its reportable quantity of 1 lb or 0.454 kg. The toll-free number of the NRC is (800) 424–8802; in the Washington DC metropolitan area (202) 426–2675. The rule for determining when notification is required is stated in 40 CFR 302.4 (section IV. D. 3.b) [R4].

FIFRA REQUIREMENTS Tolerances are established for the total residues of the insecticide endosulfan (6,7,8,9,10,10-hexachloro-1,5,5a,6,9,9a-hexahydro-6,9-methano-2, 4, 3-benzodioxathiephin-3-oxide) and its metabolite endosulfan sulfate (6,7,8,9,10,10-hexachloro-1,5,5a,6,9,9a-hexahydro-6,9-methano-2,4,3-benzodioxathiepin-3,3-dioxide) in or on the following raw agricultural commodities: apples, apricots, artichokes, beans, broccoli, brussels sprouts, cabbage, cauliflower, celery, cherries, collards, cucumbers, eggplants, grapes, kale, lettuce, melons, mustard greens, nectarines, peaches, pears, peas (succulent type), peppers, pineapples, plums, prunes, pumpkins, spinach, strawberries, summer squash, sunflower seeds, tomatoes, turnip greens, watercress, winter squash, alfalfa hay, almond hulls, and cottonseed milk fat (reflecting negligible residues in milk), sugarcane, alfalfa (fresh), carrots, sweet corn (kernels plus cob with husks removed), sweet potatoes, meat, fat, and meat by-products of cattle, goats, hogs, horses, and sheep, almonds, filberts, macadamia nut, mustard seed, pecans, potatoes, rape seed, safflower seed, straw of barley, oats, rye, and wheat, and walnuts, blueberries, raspberries, grain of barley, oats, rye, and wheat; and sugar beets (without tops) [R8]. A tolerance is established for combined residues of the insecticide endosulfan and its metabolite endosulfan sulfate in or on dried tea (reflecting less than 0.1 ppm residues in beverage tea) resulting from application of the insecticide to growing tea. [R9].

REFERENCES

1. Nat'l Research Council Canada. *Endosulfan* p. 49 (1975) NRCC No. 14098.

2. (1) Tabak, H. H., et al., *J Water Pollut Control Fed* 53: 1503–18 (1981). (2) Miles, J. R. W., P. Moy. *Bull Environ Contam Toxicol* 23: 13–9 (1979). (3) Callahan, M. A., et al., *Water-Related Environ Fate of 129 Priority Pollut* Volume 2, pp 27–1 to 27–16 USEPA 440/4–79–029a (1979).

3. Callahan, M. A., M. W. Slimak, N. W. Gabel, et al. *Water-Related Environmental Fate of 129 Priority Pollutants.* Volume I. EPA–440/4 79–029a. Washington, DC: U.S. Environmental Protection Agency, December 1979, p. 27–11.

4. 40 CFR 302.4 (7/1/88).

5. USEPA. *Engineering Handbook for Hazardous Waste Incineration.* p. 3–9 (1981) EPA 68-03-3025.

6. National Research Council Canada. *Endosulfan: Its Effects on Environmental Quality.* p. 39 (1975) NRCC No. 14098.

7. (1) Callahan, M. A., et al. *Water-Related Environ Fate of 129 Priority Pollut.* Vol 2, pp. 27–1 to 27–16 USEPA-440/4-79-029a (1979). (2) Lyman, W. J., et al. *Handbook of Chemical Property Estimation Methods. Environmental Behavior of Organic Compounds.* New York: McGraw-Hill, pp. 5–5 (1982).

8. 40 CFR 180.182 (7/1/88).

9. 40 CFR 185.2600 (7/1/88).

10. American Conference of Governmental Industrial Hygienists. *Threshold Limit Values for Chemical Substances and*

Physical Agents and Biological Exposure Indices for 1994–1995. Cincinnati, OH: ACGIH, 1994.

■ ESFENVALERATE

CAS # 66230-04-4

SYNONYMS sumi-alpha • a-alpha • oms-3023 • s-1844 • s-5602alpha • (s)-alpha-cyano-3-phenoxybenzyl (s)-2-(4-chlorophenyl)-3-methylbutyrate • s-(r,r)-cyano (3-phenoxyphenyl) methyl 4-chloro-2-(1-methylethyl) benzene-acetate • sumi-alfa • s-5620a alpha • asana

MF $C_{25}H_{22}ClNO_3$

MW 419.9

COLOR AND FORM Viscous yellow or brown liquid; white crystalline solid; amber liquid.

DENSITY 1.175 at 25°C.

BOILING POINT 151–167°C.

MELTING POINT 59–60.2°C.

FIREFIGHTING PROCEDURES Use carbon dioxide, foam, or dry chemical (on fires involving pyrethroids) (pyrethrum) [R8, 2]. Firefighting: Self-contained breathing apparatus with a full facepiece operated in pressure-demand or other positive-pressure mode (pyrethrum) [R8, 5]. Extinguish fire using agent suitable for type of surrounding fire. (pyrethrins) [R2].

SOLUBILITY Soluble in water: <1 mg/L at 20°C, chloroform; cyclohexanone; ethanol; xylene; 155 g/kg hexane; >1 kg/kg acetone. Solubility (1%): acetonitrile >60, chloroform >60, DMF >60, DMSO >60, ethyl acetate >60, acetone >60, ethyl cellosolve 40–50, n-hexane 1–5, kerosene <1, methanol 7–10, alpha-methylnaphthalene 50–60, xylene >60.

VAPOR PRESSURE 0.037 mPa (25°C).

STABILITY AND SHELF LIFE Pyrethrins are stable for long periods in water-based aerosols where emulsifiers produce neutral water systems. (pyrethrins) [R3].

OTHER CHEMICAL/PHYSICAL PROPERTIES Compatible with other herbicides and other insecticides.

PROTECTIVE EQUIPMENT Employees should use dust- and splash-proof safety goggles where pyrethroids may contact the eyes. (pyrethroids) [R8, 3].

PREVENTATIVE MEASURES Skin that becomes contaminated with pyrethrum should promptly be washed or showered with soap or mild detergent and water. (pyrethrum) [R8, 3].

CLEANUP Spillages of pesticides at any stage of their storage or handling should be treated with great care. Liquid formulations may be reduced to solid phase by evaporation. Dry-sweeping of solids is always hazardous: these should be removed by vacuum cleaning, or by dissolving them in water or other solvent in the factory environment (pesticides) [R4]. Land spill: Dig a pit, pond, lagoon, or holding area to contain liquid or solid material. (SRP: If time permits, pits, ponds, lagoons, soak holes, or holding areas should be sealed with an impermeable flexible-membrane liner.) Dike surface flow using soil, sandbags, foamed polyurethane, or foamed concrete. Absorb bulk liquid with fly ash or cement powder (pyrethrins) [R2]. Water spill: If pyrethrins are dissolved, apply activated carbon at ten times the spilled amount in the region of 10 ppm or greater concentration. Use mechanical dredges or lifts to remove immobilized masses of pollutants and precipitates (pyrethrin) [R2].

DISPOSAL Incineration would be an effective disposal procedure where permitted. If an efficient incinerator is not available, the product should be mixed with large amounts of combustible material, and contact with the smoke should be avoided. (pyrethrin products) [R5].

COMMON USES Control of a wide range of insect pests on cotton, maize, groundnuts, soya beans, sugar cane, sunflowers, sorghum, fruit trees, vegetables, ornamentals, and noncrop land [R1].

NON-HUMAN TOXICITY LD_{50} rat oral 75

mg/kg [R1]; LD$_{50}$ rabbit percutaneous >2,000 mg/kg [R1].

BIOCONCENTRATION The uptake of esfenvalerate by adult fathead minnow (*Pimephales promelas*) was studied in littoral enclosures in a field pond near Duluth, MN (1); esfenvalerate was applied to the surface of the pond as an insecticidal spray (1); the amount of esfenvalerate accumulated by the fish reached a maximum of 65.6 µg/kg (whole body basis) 1 day after application, at which time the aqueous concentration of esfenvalerate was 0.047 µg/L (1); based upon these concns, the BCF would be approximately 1,400 (SRC); 25 days after application, the esfenvalerate concentration in the fish had fallen to 12.5 µg/kg (1). Based upon an estimated log K$_{ow}$ of 6.77 (2), the BCF for esfenvalerate can be estimated to be 82,000 from a recommended regression-derived equation (3,SRC). These BCF values suggest that bioconcentration in aquatic organisms may be environmentally important (SRC) [R6].

FIFRA REQUIREMENTS A food-additive tolerance is established for residues of the insecticide cyano (3-phenoxyphenyl) methyl-4-chloro-alpha-(1-methylethyl) benzeneacetate and an isomer, (s)-cyano (3-phenoxyphenyl) methyl-(s)-4-chloro-alpha-(1-methylethyl)-benzeneacetate, as follows: In or on all food items (other than those already covered by a higher tolerance as a result of use on growing crops) in food-handling establishments where food products are held, processed, or prepared [R7].

REFERENCES

1. Hartley, D., and H. Kidd (eds.). *The Agrochemicals Handbook*. 2nd ed. Lechworth, Herts, England: The Royal Society of Chemistry, 1987, p. A990/Aug 87.

2. Bureau of Explosives. *Emergency Handling of Haz Matl in Surface Trans* p. 434 (1981).

3. *Farm Chemicals Handbook 1986*. Willoughby, OH: Meister Publishing Co., 1986, p. C–198.

4. International Labour Office. *Encyclopedia of Occupational Health and Safety.* Vols. I and II. Geneva, Switzerland: International Labour Office, 1983. 1619.

5. Sittig, M. *Handbook of Toxic and Hazardous Chemicals and Carcinogens, 1985.* 2nd ed. Park Ridge, NJ: Noyes Data Corporation, 1985. 762.

6. (1) Heinis, L.J., M.L. Knuth. *Environ Toxicol Chem* 11: 11–25 (1992) (2) USEPA, *Graphical Exposure Modeling System (GEMS).* CLOGP (1987) (3) Lyman, W. J., et al., *Handbook of Chemical Property Estimation Methods*. Washington, DC: Amer Chemical Soc, p. 5–4 (1990).

7. 40 CFR 185.1300 (a) (1) (7/1/91).

8. Mackison, F. W., R. S. Stricoff, and L. J. Partridge, Jr. (eds.). *NIOSH/OSHA—Occupational Health Guidelines for Chemical Hazards.* DHHS (NIOSH) Publication No. 81–123 (3 vols). Washington, DC: U.S. Government Printing Office, Jan. 1981.

■ ETHANE

CAS # 74–84–0

SYNONYMS bimethyl • dimethyl • ethyl hydride • methylmethane

MF C$_2$H$_6$

MW 30.08

COLOR AND FORM Colorless gas.

ODOR Odorless; mild gasoline-like odor.

ODOR THRESHOLD 8.99×10^2 ppm (detection in water, purity not specified) [R7].

DENSITY 0.446 at 0°C (liquid).

SURFACE TENSION 16 dynes/cm = 0.016 N/m at −88°C.

VISCOSITY 63.4 micropoises at −78.5°C.

BOILING POINT −88°C.

MELTING POINT −172°C.

HEAT OF COMBUSTION −1,727 Btu/ft^3 at 25°C.

HEAT OF VAPORIZATION 3,739.5 g cal/g mole.

AUTOIGNITION TEMPERATURE 882°F = 472°C [R4].

EXPLOSIVE LIMITS AND POTENTIAL Moderate, when exposed to flame [R3].

FLAMMABILITY LIMITS Lower: 3.0%; upper: 12.5% (% by volume) [R4].

FIREFIGHTING PROCEDURES Stop flow of gas [R3]. Do not extinguish fire unless flow can be stopped. Use water in flooding quantities as fog. Cool all affected containers with flooding quantities of water. Apply water from as great a distance as possible. If fire becomes uncontrollable or container is exposed to direct flame—consider evacuation of one-third (1/3) mile radius [R5].

SOLUBILITY Water solubility: 60.2 ppm at 25°C; 46 ml/100 ml alcohol at 4°C; >10% in benzene; 60.4 µg/ml water at 20°C; soluble in ether.

VAPOR PRESSURE 31,459 mm Hg at 25°C (calculated); saturated atmosphere.

OCTANOL/WATER PARTITION COEFFICIENT Log K_{ow} = 1.81.

OTHER CHEMICAL/PHYSICAL PROPERTIES 1 lb of ethane yields 20,420 Btu (net) at 15.56°C.

PROTECTIVE EQUIPMENT Self-contained breathing apparatus for high vapor concentration [R6].

PREVENTATIVE MEASURES If material not on fire and not involved in fire: keep sparks, flames, and other sources of ignition away. Keep material out of water sources and sewers. Attempt to stop leak if without undue personnel hazard. Use water spray to knock down vapors. If material leaking (not on fire) consider evacuation from downwind area based on amount of material spilled, location, and weather conditions. Avoid breathing vapors. Keep upwind [R5].

DISPOSAL SRP: At the time of review, criteria for land treatment or burial (sanitary landfill) disposal practices are subject to significant revision. Prior to implementing land disposal of waste residue (including waste sludge), consult with environmental regulatory agencies for guidance on acceptable disposal practices.

COMMON USES In the production of ethylene by high-temperature thermal cracking; as a feedstock in the production of vinyl chloride; in the synthesis of chlorinated hydrocarbons; as a refrigerant [R1]; as fuel gas (so-called "bottled gas" or "suburban propane" contains about 90% propane, 5% ethane, and 5% butane) [R2].

OCCUPATIONAL RECOMMENDATIONS None listed.

THRESHOLD LIMIT Simple asphyxiant inert gas or vapor. A TLV may not be recommended for each simple asphyxiant because the limiting factor is the available oxygen (1981) [R10].

BIOLOGICAL HALF-LIFE The elimination half-life of ethane in rats is 0.95 hr. [R11, 259].

BIODEGRADATION Within 24 hr, ethane was oxidized to its corresponding alcohol, ethanol, by cell suspensions of over 20 methyltrophic organisms isolated from lake water and soil samples (1,2). The average ethane utilization by microflora of 5 soils was 19 and 51% for single and mixed alkanes, respectively (3). The respective gas exchange and degradation rate constants were 0.83×10^{-5} cm^2 sec^{-1} and 0.027 day^{-1} for ethane contained in a model estuarine ecosystem at 10°C and a salinity of 30 parts per thousand. The corresponding biodegradation half-life was greater than 87 days (6). At 20°C and a salinity of 30, the respective gas exchange and degradation rate constants were 1.132×10^{-5} cm^2 sec^{-1} and 0.062 day^{-1}; the corresponding biodegradation half-life for *n*-ethane ranged from 21 to 33 days (6) [R8].

BIOCONCENTRATION Based upon a water solubility of 60.2 ppm (1) at 25°C and a log K_{ow} of 1.81 (2), the bioconcentration factor (log BCF) for ethane has been calculated, using recommended regression-derived equations, to be 1.15 and 1.79, respectively (3,SRC). These bioconcentration factors values do not indicate that bioconcentration in aquatic organisms is important (SRC) [R9].

GENERAL RESPONSE Stop discharge if possible. Keep people away. Shut off ignition sources and call fire department. Stay upwind and use water spray to knock down vapor. Avoid contact with liquid. Notify local health and pollution control agencies.

FIRE RESPONSE Flammable. Flashback along vapor trail may occur. Vapor may explode if ignited in an enclosed area. Stop flow of gas if possible. Cool exposed containers and protect men effecting shutoff with water. Let fire burn.

EXPOSURE RESPONSE Call for medical aid. Vapor: if inhaled will cause difficult breathing. Not irritating to eyes, nose, or throat. Move to fresh air. If breathing has stopped, give artificial respiration. If breathing is difficult, give oxygen. Liquid will cause frostbite. Flush affected areas with plenty of water. Do not rub affected areas.

RESPONSE TO DISCHARGE Issue warning—high flammability. Evacuate area.

WATER RESPONSE Not harmful.

EXPOSURE SYMPTOMS In high vapor concentrations, can act as simple asphyxiant. Liquid causes severe frostbite.

EXPOSURE TREATMENT Remove from exposure; support respiration.

REFERENCES

1. Snyder, R. (ed.) *Ethyl Browning's Toxicity and Metabolism of Industrial Solvents.* 2nd ed. Volume 1: *Hydrocarbons.* Amsterdam–New York–Oxford: Elsevier, 1987.

2. *The Merck Index.* 10th ed. Rahway, NJ: Merck Co., Inc., 1983. 540.

3. Sax, N. I. *Dangerous Properties of Industrial Materials,* 6th ed. New York, NY: Van Nostrand Reinhold, 1984. 1303.

4. National Fire Protection Association. *Fire Protection Guide on Hazardous Materials.* 9th ed. Boston, MA: National Fire Protection Association, 1986, p. 325M–46.

5. Association of American Railroads. *Emergency Handling of Hazardous Materials in Surface Transportation.* Washington, DC: Assoc. of American Railroads, Hazardous Materials Systems (BOE), 1987. 294.

6. U.S. Coast Guard, Department of Transportation. *CHRIS—Hazardous Chemical Data.* Volume II. Washington, DC: U.S. Government Printing Office, 1984–5.

7. Fazzallari, F. A. (ed.). *Compilation of Odor and Taste Threshold Values Data.* ASTM Data Series DS 48A (Committee E–18). Philadelphia, PA: American Society for Testing and Materials, 1978. 59.

8. (1) Patel, R. N., et al. *J Bacteriology* 139: 675–9 (1979); (2) Hou, C. T., et al. *Appl Environ Microbiol* 46: 178–84 (1983); (4) Jamison, V. W., et al. pp. 187–96 in *Proc International Biodeg Symp* 3rd Sharpley, J. M., and A. M. Kapalan (eds.), Essex, Eng (1976); (5) Brisbane, R. G., J. N. Ladd. *J Gen Appl Microbiol* 14: 447–50 (1968); (6) Bopp, R. F., et al. *Organic Geochemistry* 3: 9–14 (1981).

9. (1) McAuliffe, C. *J Phys Chem* 70: 1267–75 (1966); (2) Hansch, C., and A. J. Leo. *Medchem Project* Issue No. 26. Claremont, CA: Pomona College (1985); (3) Lyman, W. J., et al. *Handbook of Chemical Property Estimation Methods,* New York: McGraw-Hill, pp. 5– 4, 5–10 (1982).

10. American Conference of Governmental Industrial Hygienists. *Threshold Limit Values for Chemical Substances and Physical Agents and Biological Exposure Indices for 1993–1994.* Cincinnati, OH: ACGIH, 1993. 20.

11. Snyder, R. (ed.) *Ethyl Browning's Toxicity and Metabolism of Industrial Solvents.* 2nd ed. Volume 1: *Hydrocarbons.* Amsterdam–New York–Oxford: Elsevier, 1987.

■ETHANE, 1,1-DICHLORO-2,2-BIS (p-CHLOROPHENYL)-

CAS # 72–54–8

DOT # 2761

SIC CODES 2879; 2842

SYNONYMS 1,1-bis (4-chlorophenyl)-2,2-dichloroethane • 1,1-bis (p-chlorophenyl)-2,2-dichloroethane • 1,1-dichloor-2,2-bis (4-chloor fenyl)-ethaan (Dutch) • 1,1-Dichlor-2,2-bis (4-chlor-phenyl)-aethan (German) • 1,1-Dichlor-2,2-bis (4-chlorphenyl)-aethan (German) • 1,1-dichloro-2,2-bis (4-chlorophenyl)-ethane (French) • 1,1-dichloro-2,2-bis (4-chlorophenyl) ethane • 1,1-dichloro-2,2-bis (p-chlorophenyl) ethane • 1,1-dichloro-2,2-bis (parachlorophenyl) ethane • 1,1-dichloro-2,2-di(4-chloro-phenyl) ethane • 1,1-dicloro-2,2-bis (4-chloro-fenil)-etano (Italian) • 2,2-bis (4-chlorophenyl)-1,1-dichloroethane • 2,2-bis (p-chlorophenyl)-1,1-dichloroethane • 4,4'-ddd • benzene, 1,1'-(2,2-dichloroethylidene) bis (4-chloro • dichlorodiphenyl-dichloroethane • NCI-C00475 • p,p'-ddd • p,p'-dichloro-diphenyl-2,2-dichloroethylene • p,p'-dichlorodiphenyldichloroethane • p,p'-tde • tetrachlorodiphenylethane • Ent-4,225 • ethane, 1,1-dichloro-2,2-bis (p-chlorophenyl)- • TDE • dilene • Rhothane d-3 • Me-700 • Rhothane • Rothane • Caswell No. 307 • EPA pesticide code 029101

MF $C_{14}H_{10}Cl_4$

MW 320.04

COLOR AND FORM Crystals; colorless; white solid.

DENSITY 1.385.

BOILING POINT 193°C at 1 mm Hg.

MELTING POINT 109°C.

FIREFIGHTING PROCEDURES Fire extinguishing agents: water, foam, dry chemical, or carbon dioxide [R6].

TOXIC COMBUSTION PRODUCTS Irritating hydrogen chloride fumes may form in fires [R6].

SOLUBILITY In water: 0.005 ppm; 0.160 mg/L water at 24°C (99% purity); soluble in organic solvents.

VAPOR PRESSURE 10.2×10^{-7} torr at 30°C.

OCTANOL/WATER PARTITION COEFFICIENT Log $K_{ow} = 6.02$.

OTHER CHEMICAL/PHYSICAL PROPERTIES Specific gravity: 1.476 at 20°C (solid).

PROTECTIVE EQUIPMENT Wear goggles or face shield, dust mask, rubber gloves, and self-contained breathing apparatus [R6].

PREVENTATIVE MEASURES Stop discharge if possible. Keep people away. Avoid contact with solid and dust. Isolate and remove dischaged material. Notify local health and pollution control agencies [R6].

CLEANUP For land spill: Dig a pit, pond, lagoon, or holding area to contain liquid or solid material. (SRP: If time permits, pits, ponds, lagoons, soak holes, or holding areas should be sealed with an impermeable flexible-membrane liner.) Cover solids with a plastic sheet to prevent dissolving in rain or firefighting water [R7].

DISPOSAL Generators of waste (equal to or greater than 100 kg/mo) containing this contaminant, EPA hazardous waste number U060, must conform with USEPA regulations in storage, transportation, treatment, and disposal of waste [R8].

COMMON USES TDE less effective against moths than DDT, but it is superior against some other insects (e.g., tomato hornworms) (SRP: former use in USA) [R3]. Employed for controlling a number of pests on vegetables and tobacco. In medical entomology field, TDE is about equal to DDT against mosquito larvae but inferior against adults (former use in USA) [R4]; no longer any registered uses in USA [R5].

STANDARD CODES EPA 311; Superfund designated (hazardous substances) list.

PERSISTENCE Persistent. Very similar to DDT. Decomposes very slowly in soil. It hydrolyzes very slowly in water. Remains persistent in soil and water for eventual accumulation in the food chain. Tests in river water for 8 weeks produced no decrease in the concentration of TDE [R19].

MAJOR SPECIES THREATENED Marine life—shrimps.

DIRECT CONTACT Readily absorbed through the skin, causing hyperirritability, hyperexcitability, and other central nervous system disorders.

GENERAL SENSATION Following skin contamination or inhalation symptoms of hyperexcitability, tremors, ataxia, and convulsions followed by depression and respiratory failure may be seen. Kidney and liver damage may be seen. Irritating to skin and eyes.

PERSONAL SAFETY Self-contained breathing apparatus must be worn. Rubber gloves, hats, suits, and boots must be worn.

ACUTE HAZARD LEVEL This material is moderately toxic to humans by ingestion; between 1 ounce and 1 pound may be fatal. The lowest lethal dose for humans is 5,000 mg/kg and 600 mg/kg for mice.

CHRONIC HAZARD LEVEL Poisonings are slower in onset and longer in duration than for DDT. In chronic feeding experiments, TDE is stored in the body fat, but is mobilized and excreted faster than DDT when a normal diet is resumed. Chronic oral toxicity to rats fed daily for 2 years was 5 mg/kg.

DEGREE OF HAZARD TO PUBLIC HEALTH May cause kidney and lung damage and is a suspected carcinogen. Material is readily absorbed through the skin and is also quite irritating.

OCCUPATIONAL RECOMMENDATIONS None listed.

ACTION LEVEL Evacuate area. Enter from upwind side. Notify local air authority and National Agricultural Chemicals Association.

ON-SITE RESTORATION Dam the stream to reduce the flow and prevent further dissipation by water movement. Apply activated carbon to adsorb the dissolved material. Bottom pumps, dredges, or underwater vacuum systems may be employed to remove undissolved material and adsorbent from the bottom. An alternative method, done under controlled conditions, is to pump the water into a suitable container, and then pass through a filtration system and an activated carbon filter. For more details, see Envirex manual EPA 600/2–77–227. Seek professional environmental engineering assistance through EPA's Environmental Response Team (ERT), Edison, NJ, 24-hour no. 908–548–8730.

BEACH OR SHORE RESTORATION Close beach and shore to public until material has been removed. If tidal, scrape affected area at low tide with mechanical scraper. Avoid human contact.

AVAILABILITY OF COUNTERMEASURE MATERIAL Pumps, dredges—Army Corps of Engineers, fire department; vacuum systems—swimming-pool operator; activated carbon—water treatment plants, chemical companies.

DISPOSAL After the material has been contained, remove with contaminated soil and place in impervious containers. The material may be incinerated in a pesticide incinerator at the specified temperature/dwell-time combination. Any sludges or solid residues generated should be disposed of in accordance with all applicable federal, state, and local pollution control requirements. If appropriate incineration facilities are not available, material may be buried in a chemical waste landfill. Material is not amenable to biological treatment at a municipal sewage treatment plant.

DISPOSAL NOTIFICATION Notify local and state health authorities, local solid waste disposal authorities, supplier and shipper of material.

INDUSTRIAL FOULING POTENTIAL Not believed significant.

MAJOR WATER USE THREATENED Fisheries; potable supply; recreation water.

PROBABLE LOCATION AND STATE White crystalline solid. Will sink and associate with sediments.

WATER CHEMISTRY Persistent in water and sediments.

FOOD-CHAIN CONCENTRATION POTENTIAL High. As for all chlorinated hydrocarbons premolt penguins showed 16 ppb in liver and 2 ppb in fat. Crabeater seal showed 2 ppb in liver and 7 ppb in fat. Positive; birds of prey showed 0.08–8.6 ppm in liver and 5.1 ppm in mesenterial fat.

NON-HUMAN TOXICITY LD_{50} rat percutaneous >10,000 mg/kg [R1]; LD_{50} rat oral 3,400 mg/kg acute [R2].

ECOTOXICITY LC_{50} *Colinus virginianus* (bobwhite quail) oral 2,178 ppm, 5-day diet *ad libitum*, 23 days old (95% confidence limit 1,835–2,584 ppm) (technical grade, purity >95%) [R11]; LC_{50} *Coturnix japonica* (Japanese quail) oral 3,165 ppm, 5-day diet *ad libitum*, 7 days old (95% confidence limit 2,534–3,978 ppm) (technical grade, purity >95%) [R11]; LC_{50} *Coturnix japonica* (Japanese quail), oral 2,636 ppm in 5-day diet *ad libitum* (95% confidence limit 2,225–3,122 ppm) (technical grade, 100% AI) [R12]; LC_{50} *Phasianus colchicus* (ring-necked pheasant) oral 445 ppm, 5-day diet *ad libitum*, 10 days old (95% confidence limit 402–494 ppm) (technical grade, purity >95%) [R11]; LC_{50} *Anas platyrhynchos* (mallard duck) oral 4,814 ppm, 5-day diet *ad libitum*, 17 days old (95% confidence limit 3,451–7,054 ppm) (technical grade, purity >95%) [R11]; LC_{50} *Asellus brevicaudus* (sowbug) 16 µg/L/96 hr, mature; at 21°C, static bioassay (technical material, 99%) [R13]; LC_{50} *Gammarus fasciatus* (scud) 0.6 µg/L/96 hr, mature, at 21°C (95% confidence limit 0.1–1.2 µg/L), static bioassay (technical material, 99%) [R13]; LC_{50} *Palaemonetes kadiakensis* (glass shrimp) 2.4 µg/L/96 hr, mature, at 21°C, static bioassay (technical material, 99%) [R13]; LC_{50} *Pteronarcys californica* (stonefly) 380 µg/L/96 hr, second year class, at 15°C (95% confidence limit 280–520 µg/L), static bioassay (technical material, 99%) [R13]; LC_{50} *Ischnura venticalis* (damselfly) 34 µg/L/96 hr, juvenile, at 21°C, static bioassay (technical material, 99%) [R13]; LC_{50} *Salmo gairdneri* (rainbow trout) 70 µg/L/96 hr, wt 1.0 g, at 12°C (95% confidence limit 57–87 µg/L), static bioassay (technical material, 99%) [R13]; LC_{50} *Pimephales promelas* (fathead minnow) 4,400 µg/L/96 hr, wt 1.0 g, at 18°C (95% confidence limit 3,470–5,580 µg/L), static bioassay (technical material, 99%) [R13]; LC_{50} *Ictalarius punctatus* (channel catfish) 1,500 µg/L/96 hr, wt 0.8 g, at 18°C (95% confidence limit 1,180–1,910 µg/L), static bioassay (technical material, 99%) [R13]; LC_{50} *Micropterus salmoides* (largemouth bass) 42 µg/L/96 hr, wt 0.7 g, at 18°C (95% confidence limit 34–51 µg/L), static bioassay (technical material, 99%) [R13]; LC_{50} *Stizostedion vitreum vitreum* (walleye) 14 µg/L/96 hr, wt 1.0 g, at 18°C (95% confidence limit 11–19 µg/L), static bioassay (technical material, 99%) [R13]; EC_{50} *Simocephalus* (daphnid) 4.5 µg/L/48 hr, first instar, at 15°C (95% confidence limit 3.1–6.6 µg/L), static bioassay (technical material, 99%) [R13]; LD_{50} *Anas platyrhynchos* (mallard duck) female 3 months old oral >2,000 mg/kg (purity greater than 95%) [R10]; LD_{50} *Callipepla californica* (California quail) female 6 months old oral >760 mg/kg (purity greater than 95%) [R10]; LD_{50} *Phasianus colchicus* (ring-necked pheasant) female 3–4 months old oral 386 mg/kg (95% confidence limit 270–551 mg/kg) (purity greater than 95%) [R10]; LC_{50} *Gammarus lacustris* 0.64 µg/L/96 hr (conditions of bioassay not specified) [R14, 434]; LC_{50} *Gammarus fasciatus* 0.86 µg/L/96 hr (conditions of bioassay not specified) [R14, 434]; LC_{50} *Palaemonetes kadiakenis* 0.68 µg/L/96 hr (conditions of bioassay not specified) [R14, 434]; LC_{50} *Asellus breviacaudus* 10.0 µg/L/96 hr (conditions of bioassay not specified) [R14, 434]; LC_{50} *Daphnia pulex* (water flea) 3.2 µg/L/48 hr (conditions of bioassay not specified) [R14, 434]; TL_{50} *Palaemon marodactylus* (Korean shrimp) 2.5 ppb/96 hr, flow-through bioassay (range 1.6 to 4.0 ppb) [R14, 434]; TL_{50} *Palaemon marodactylus* (Korean shrimp) 8.3 ppb/96 hr, static bioassay (range 4.8 to 14.4 ppb) [R14, 434].

IARC SUMMARY Classification of carcinogenicity: (1) evidence in humans: inadequate; (2) evidence in animals: sufficient. Overall summary evaluation of carcinogenic risk to humans is Group 2B: The agent is possibly carcinogenic to humans (from table) [R15].

BIODEGRADATION No degradation of DDD exposed to ocean sediments in seawater under aerobic and anaerobic conditions observed after 12 months (1). p,p'-DDD was not significantly degraded in static screening tests using settled domestic wastewater inoculum with yeast extract, with 3 additional subcultures every 7 days (2). No degradation of DDD observed after 8 weeks of exposure of DDD in lab tests to raw water from a stream that received domestic and industrial wastes and farm runoff (3). DDD is very stable in Nixon sandy loam, where it had no effect on carbon dioxide production and nitrification rates (4) [R16].

BIOCONCENTRATION From 1949 through 1957, total of 54,800 kg of DDD (to control gnat population) was put directly into clear lake. Injury to grebes was attributed to chemical. In analysis of wide variety of samples, DDD was present: concentration in increasing order: plankton, small fish, big fish, and grebes. [R9, 490].

RCRA REQUIREMENTS As stipulated in 40 CFR 261.33, when DDD, as a commercial chemical product or manufacturing chemical intermediate or an off-specification commercial chemical product or a manufacturing chemical intermediate, becomes a waste, it must be managed according to federal or state hazardous waste regulations. Also defined as a hazardous waste is any residue, contaminated soil, water, or other debris resulting from the cleanup of a spill, into water or on dry land, of this waste. Generators of small quantities of this waste may qualify for partial exclusion from hazardous waste regulations (40 CFR 261.5) [R18].

FIFRA REQUIREMENTS Pesticide chemicals that cause related pharmacological effects will be regarded in the absence of evidence to the contrary, as having an additive deleterious action. Many pesticide chemicals within the following groups have

related pharmacological effects: chlorinated organic pesticides, arsenic-containing chemicals, metallic ditricarbamates, cholinesterase-inhibiting pesticides. Where residues from two or more chemicals in the same class are present in or on a raw agricultural commodity the tolerance for the total of such residues was the same as that for the chemical having the lowest numerical tolerance in this class, unless a higher tolerance level is specifically provided for the combined residues by a regulation in this part. DDD is a member of the class of chlorinated organic pesticides [R17].

GENERAL RESPONSE Stop discharge if possible. Keep people away. Avoid contact with solid and dust. Isolate and remove discharged material. Notify local health and pollution control agencies.

FIRE RESPONSE Combustible. Irritating gases may be produced when heated. Wear goggles and self-contained breathing apparatus. Extinguish with water, dry chemicals, foam, or carbon dioxide.

EXPOSURE RESPONSE Call for medical aid. Dust irritating to eyes, nose, and throat. Harmful if inhaled. If breathing has stopped, give artificial respiration. If breathing is difficult, give oxygen. Solid irritating to skin and eyes. Harmful if swallowed. Remove contaminated clothing and shoes. Flush affected areas with plenty of water. If in eyes, hold eyelids open, and flush with plenty of water. If swallowed and victim is conscious, have victim drink water or milk. If swallowed, and victim is unconscious or having convulsions, do nothing except keep victim warm.

RESPONSE TO DISCHARGE Issue warning—water contamination Should be removed. Provide chemical and physical treatment.

WATER RESPONSE Harmful to aquatic life in very low concentrations. May be dangerous if it enters water intakes. Notify local health and wildlife officials. Notify operators of nearby water intakes.

EXPOSURE SYMPTOMS Ingestion causes vomiting and delayed symptoms similar to those caused by DDT. Contact with eyes causes irritation.

EXPOSURE TREATMENT Ingestion: treatment should be given by a physician and is similar to that given following ingestion of DDT. Eyes: flush with water.

REFERENCES

1. *Kirk–Othmer Encyclopedia of Chemical Technology.* 3rd ed., Volumes 1–26. New York, NY: John Wiley and Sons, 1978–1984, p. 13 (81) 431.

2. *Farm Chemicals Handbook 1989.* Willoughby, OH: Meister Publishing Co., 1989, p. C–280.

3. IARC. *Monographs on the Evaluation of the Carcinogenic Risk of Chemicals to Man.* Geneva: World Health Organization, International Agency for Research on Cancer, 1972–present (multivolume work), p. V5 90 (1974).

4. Osol, A., J. E. Hoover, et al. (eds.). *Remington's Pharmaceutical Sciences.* 15th ed. Easton, PA: Mack Publishing Co., 1975. 1195.

5. PRC.

6. U.S. Coast Guard, Department of Transportation. *CHRIS—Hazardous Chemical Data.* Volume II. Washington, DC: U.S. Government Printing Office, 1984–5.

7. Bureau of Explosives. *Emergency Handling of Haz Matl in Surface Trans* p. 490 (1981).

8. 40 CFR 240–280, 300–306, 702–799 (7/1/89).

9. Hayes, W. J., Jr. *Toxicology of Pesticides.* Baltimore: Williams, and Wilkins, 1975.

10. U.S. Department of the Interior, Fish and Wildlife Service. *Handbook of Toxicity of Pesticides to Wildlife.* Resource Publication 153. Washington, DC: U.S. Government Printing Office, 1984. 77.

11. U.S. Department of the Interior, Fish and Wildlife Service, Bureau of Sports Fisheries and Wildlife. *Lethal Dietary Toxicities of Environmental Pollutants to Birds.* Special Scientific Report—*Wildlife* No. 191. Washington, DC: U.S. Government Printing Office, 1975. 34.

12. Hill, E. F., M. B. Camardese. *Lethal Dietary Toxicities of Environmental Contaminants and Pesticides to Coturnix.* Fish and Wildlife Technical Report 2. Washington, DC: United States Department of Interior, Fish, and Wildlife Service, 1986. 128.

13. U.S. Department of Interior, Fish and Wildlife Service. *Handbook of Acute Toxicity of Chemicals to Fish and Aquatic Invertebrates.* Resource Publication No. 137. Washington, DC: U.S. Government Printing Office, 1980. 24.

14. Verschueren, K. *Handbook of Environmental Data of Organic Chemicals.* 2nd ed. New York, NY: Van Nostrand Reinhold Co., 1983.

15. IARC. *Monographs on the Evaluation of the Carcinogenic Risk of Chemicals to Man.* Geneva: World Health Organization, International Agency for Research on Cancer, 1972–present (multivolume work), p. S7 186 (1987).

16. (1) Vind, H. P., et al. *Biodeterioration of Navy Insecticides in the Ocean,* p 14 NTIS AD–77310 (1973) (2) Tabak, H. H., et al. *J Water Pollut Control Fed* 53: 1503–18 (1981); (3) Eichelberger, J. W., J. J. Litchtenberg. *Environ Sci Technol* 5: 541–4 (1971); (4) Bartha, R., et al. *Appl Microbiol* 15: 67–75 (1967).

17. 40 CFR 180.3 (e) (7/1/88).

18. 40 CFR 261.33 (7/1/88).

19. Eichelberger, J. W., J. J. Lichtenberg, "Persistence of Pesticides in River Water," *Environmental Science and Technology,* 1971, Vol. 5, No. 6. June.

■ ETHANE, 1,2-DICHLORO-1,1,2,2-TETRAFLUORO-

CAS # 76–14–2

SYNONYMS freon-114 • ethane, 1,2-dichlorotetrafluoro- • f-114 • fc-114 • fluorocarbon-114 • propellant-114 •

r-114 • sym-dichlorotetrafluoroethane • 1,1,2,2-tetrafluoro-1,2-dichloroethane • dichlorotetrafluoroethane • ethane, 1,2-dichloro-1,1,2,2-tetrafluoro- • refrigerant-114 • halon-242 • Caswell No. 326a • EPA pesticide chemical code 326200 • ucon-114 • arcton-114 • arcton-33 • cryofluoran • cryofluorane • frigen-114 • frigiderm • genetron-114 • genetron-316 • ledon-114 • tetrafluorodichloroethane • 1,2-dichlorotetrafluoroethane • criofluorano (Spanish) • cryofluoranum (Latin)

MF $C_2Cl_2F_4$

MW 170.92

COLOR AND FORM Colorless liquid or gas; clear.

ODOR Very slight ethereal odor; odorless, but has a faint, ether-like odor at high concentrations.

DENSITY 1.5312 at 0°C (liquid).

SURFACE TENSION 13 dynes/cm at 77°C.

VISCOSITY Liquid: 0.386 cP at 70°F, 0.296 cP at 130°F.

BOILING POINT 4.1°C at 760 mm Hg.

MELTING POINT −94°C.

HEAT OF VAPORIZATION 6,134.6 cal/mole.

FIREFIGHTING PROCEDURES During firefighting wear self-contained breathing apparatus with full facepiece operated in pressure-demand or other positive-pressure mode [R2, 4]. Extinguish fire using agent suitable for type of surrounding fire (material itself does not burn or burns with difficulty). Cool all affected containers with flooding quantities of water. Apply water from as great a distance as possible. (dichlorodifluoromethane-dichlorotetrafluoroethane mixture, nonflammable gas) [R4, 242].

TOXIC COMBUSTION PRODUCTS Toxic substances may be formed on contact with a flame or hot metal surface [R5].

SOLUBILITY Soluble in alcohol, ether; 0.013% in water; water solubility = 130 mg/L at 25°C.

VAPOR PRESSURE Vapor pressure = 2,014 mm Hg at 25°C.

OCTANOL/WATER PARTITION COEFFICIENT Log K_{ow} = 2.82.

CORROSIVITY Noncorrosive.

STABILITY AND SHELF LIFE Conditions contributing to instability: heat [R2, 2].

OTHER CHEMICAL/PHYSICAL PROPERTIES Absorbs less than 0.0025% water.

PROTECTIVE EQUIPMENT Employees should be provided with and required to use impervious clothing, gloves, face shield (8-in. minimum), and other appropriate protective clothing necessary to prevent the skin from becoming wet with liquid. (refrigerant 114) [R2, 2].

PREVENTATIVE MEASURES SRP: The scientific literature supports the wearing of contact lenses in industrial environments, as part of a program to protect the eye against chemical compounds and minerals causing eye irritation. However, there may be individual substances whose irritating or corrosive properties are such that the wearing of contact lenses would be harmful to the eye. In those specific cases contact lenses should not be worn.

CLEANUP If spilled or leaked, the following steps should be taken: 1. Ventilate area of spill or leak. 2. If the gas is leaking, stop the flow. 3. If the liquid is spilled, or leaked, allow to vaporize. [R2, 3].

DISPOSAL Because of recent discovery of potential ozone decomposition in the stratosphere by fluorotrichloromethane, this material should be released to the environment only as a last resort. Waste material should be recovered and returned to the vendor or to licensed waste disposal company [R7].

COMMON USES Vet: in various "skin freezes" alone or with other agents by aerosol application. Recommended for spraying of snake and insect bites to retard absorption of venom [R1]; blowing agent for cellular polymers [R2, 2]; solvent and diluent in polymerization of fluoroolefins, cleaning, and degreasing printed circuit boards, preparation of explosives, and extraction of volatile substances [R2, 2]; foaming agent in fire extinguishing and aerosols [R2, 3]; inorganic synthesis in preparation of uranium tetrafluoride, freons, and polymer intermediates [R2, 3]; in aerosols with other freons to lower vapor pressure, and produce nonflammable aerosol propellants [R2, 2]; refrigerant in industrial cooling and air-conditioning systems [R2, 2]; used in inhibiting of metal erosion in hydraulic fluids; in strengthening glass bottles; in magnesium refining; and as a reflux liquid to assist heat removal [R2, 3]; mechanical vapor compression systems use fluorocarbons for refrigeration and air conditioning and account for majority of refrigeration capability in U.S. Fluorocarbons are used as refrigerants in home appliances, mobile air-conditioning units, retail food refrigeration systems, and chillers. (fluorocarbons) [R3, 3102].

OCCUPATIONAL RECOMMENDATIONS None listed.

OSHA 8-hr time-weighted average: 1,000 ppm (7,000 mg/m^3) [R11].

NIOSH 10-hr time-weighted average: 1,000 ppm (7,000 mg/m^3) [R6].

THRESHOLD LIMIT 8-hr time-weighted average (TWA) 1,000 ppm, 6,990 mg/m^3 (1986) [R12, 18].

INTERNATIONAL EXPOSURE LIMITS Germany (1971): 1,000 ppm [R8].

BIOCONCENTRATION Based on a water solubility of 130 mg/L at 25°C and a log K_{ow} value of 2.82, bioconcentration factors (BCF) of 10 and 82, respectively, were estimated for freon-114 (1–3,SRC). These BCF values suggest that freon 114 would not bioaccumulate significantly in aquatic organisms (SRC) [R9].

TSCA REQUIREMENTS (a) After October 15, 1978, no person may manufacture, except to import, any fully halogenated chlorofluoroalkane for any aerosol propellant use except for use in an article that is a food, food additive, drug, cosmetic, or device exempted under 15 U.S.C. 2602 (21 CFR 2.125) for those essential uses listed in 40 CFR 762.58 or for exempted uses listed in 40 CFR 762.59. (b) After December ber 15, 1978, no person may import into the customs territory of the United States any fully halogenated chlorofluoroalkane, whether as a chemical substance or as a component of a mixture or article, for any aerosol propellant use except for use in an article that is a food, food additive, drug, cosmetic, or device exempted under 15 U.S.C. 2602 (21 CFR 2.125) for those essential uses listed in 40 CFR 762.58 or for exempted uses listed in 40 CFR 762.59. (c) Every person manufacturing fully halogenated chlorofluoroalkanes for aerosol propellant uses after October 15, 1978 must obtain a signed statement for aerosol propellant uses permitted under 40 CFR Part 762 or 21 CFR 2.125, or for other uses. (fully halogenated chlorofluoroalkanes) [R13].

FIFRA REQUIREMENTS Residues of dichlorotetrafluoroethane are exempted from the requirement of a tolerance when used as a propellant in accordance with good agricultural practices as inert (or occasionally active) ingredients in pesticide formulations applied to growing crops or to raw agricultural commodities after harvest [R10].

REFERENCES

1. Rossoff, I. S. *Handbook of Veterinary Drugs.* New York: Springer Publishing Company, 1974. 165.

2. Mackison, F. W., R. S. Stricoff, and L. J. Partridge, Jr. (eds.). *NIOSH/OSHA—Occupational Health Guidelines for Chemical Hazards.* DHHS (NIOSH) Publication No. 81–123 (3 vols). Washington, DC: U.S. Government Printing Office, Jan. 1981.

3. Clayton, G.D., and F.E. Clayton (eds.). *Patty's Industrial Hygiene and Toxicology:* Volume 2A, 2B, 2C: *Toxicology.* 3rd ed. New York: John Wiley and Sons, 1981–1982.

4. Association of American Railroads. *Emergency Handling of Hazardous Materials in Surface Transportation.* Washington, DC: Assoc. of American Railroads, Hazardous Materials Systems (BOE), 1987.

5. Budavari, S. (ed.). *The Merck Index— Encyclopedia of Chemicals, Drugs, and*

Biologicals. Rahway, NJ: Merck and Co., Inc., 1989. 408.

6. NIOSH. *NIOSH Pocket Guide to Chemical Hazards.* DHHS (NIOSH) Publication No. 90–117. Washington, DC: U.S. Government Printing Office, June 1990, 90.

7. United Nations. *Treatment and Disposal Methods for Waste Chemicals* (IRPTC File). Data Profile Series No. 5. Geneva, Switzerland: United Nations Environmental Programme, Dec. 1985. 207.

8. American Conference of Governmental Industrial Hygienists. *Documentation of the Threshold Limit Values and Biological Exposure Indices.* 5th ed. Cincinnati, OH: American Conference of Governmental Industrial Hygienists, 1986. 191 9 (1) Riddick, J. A., et al. pp. 568–9 in *Organic Solvents: Physical Properties and Methods of Purification* 4th ed. New York: Wiley-Interscience, pp. 568–69 (1986); (2) Hansch, C., and A. J. Leo. *Medchem Project* Issue No. 26, Claremont, CA: Pomona College (1985); (3) Lyman, W. J., et al. *Handbook of Chemical Property Estimation Methods* New York: McGraw-Hill, pp. 5–5 (1982).

10. 40 CFR 180.1001 (c) (7/1/91).

11. 29 CFR 1910.1000 (7/1/91).

12. American Conference of Governmental Industrial Hygienists. *Threshold Limit Values for Chemical Substances and Physical Agents and Biological Exposure Indices for 1993–1994.* Cincinnati, OH: ACGIH, 1993.

13. 40 CFR 762.45 (7/1/91).

■ ETHANE, 1,1-DIFLUORO

CAS # 75–37–6

SYNONYMS algofrene-type-67 • difluoroethane • ethane, 1,1-difluoro • ethylene fluoride • ethylidene difluoride • ethylidene-fluoride • fc-152a • genetron-100 • genetron-152a • R-152a • refrigerant-152A • freon-152a

MF $C_2H_4F_2$

MW 66.06

COLOR AND FORM Colorless gas.

ODOR Odorless.

DENSITY 0.95 at 20°C.

SURFACE TENSION 11.25 dynes/cm = 0.01125 N/m at 20°C (liquid).

VISCOSITY 0.263 cP at 50°F.

BOILING POINT −24.7°C.

MELTING POINT −117°C.

HEAT OF COMBUSTION −7,950 Btu/lb = −4,420 cal/g = −1.85×10⁵ J/kg.

HEAT OF VAPORIZATION 140.5 Btu/lb = 78.03 cal/g = 3.265×10⁵ J/kg.

FLAMMABILITY LIMITS 3.7%–18% (in air) [R1].

FIREFIGHTING PROCEDURES Do not extinguish fire unless flow can be stopped. Use water in flooding quantities as fog. Cool all affected containers with flooding quantitites of water. Apply water from as far a distance as possible [R2]. Wear positive-pressure self-contained breathing apparatus when fighting fires involving this material. Approach fire with caution. If fire becomes uncontrollable or container is exposed to direct flame consider evacuation of one-half (1/2) mile radius [R2].

TOXIC COMBUSTION PRODUCTS Irritating hydrogen fluoride fumes may form in fire [R3].

SOLUBILITY Insoluble in water.

VAPOR PRESSURE 4,437.1 mm Hg at 25°C.

OCTANOL/WATER PARTITION COEFFICIENT Log K_{ow} = 0.75.

OTHER CHEMICAL/PHYSICAL PROPERTIES Ratio of specific heats of vapor (gas): 1.141; saturated vp: 83.520 psi at 75°F; saturated vapor density: 0.96110 lb/ft³ at 75°F; ideal gas heat capacity: 0.245 Btu/lb at 75°F.

PROTECTIVE EQUIPMENT Wear appropriate chemical protective gloves and goggles [R2].

PREVENTATIVE MEASURES If material not

on fire and not involved in fire: Keep sparks, flames, and other sources of ignition away. Keep material out of water sources and sewers. Attempt to stop leak if without undue personnel hazard. Use water spray to knock down vapors [R2].

COMMON USES Intermediate [R1].

OCCUPATIONAL RECOMMENDATIONS None listed.

BIOCONCENTRATION Based on a water solubility of 3,235 mg/L at 25°C (1), the bioconcentration factor (BCF) can be estimated for Freon 152A to be 6.5 from a recommended regression-derived equation (2,SRC). This BCF value suggests that Freon 152A would not bioconcentrate significantly in aquatic organisms (SRC) [R4].

ATMOSPHERIC STANDARDS This action promulgates standards of performance for equipment leaks of volatile organic compounds (VOC) in the Synthetic Organic Chemical Manufacturing Industry (SOCMI). The intended effect of these standards is to require all newly constructed, modified, and reconstructed SOCMI process units to use the best demonstrated system of continuous emission reduction for equipment leaks of VOC, considering costs, non-air-quality health, and environmental impact and energy requirements. Difluoroethane is produced, as an intermediate or final product, by process units covered under this subpart [R5].

REFERENCES

1. Sax, N. I., and R. J. Lewis, Sr. (eds.). *Hawley's Condensed Chemical Dictionary*. 11th ed. New York: Van Nostrand Reinhold Co., 1987. 397.

2. Association of American Railroads. *Emergency Handling of Hazardous Materials in Surface Transportation*. Washington, DC: Assoc. of American Railroads, Hazardous Materials Systems (BOE), 1987. 256.

3. U.S. Coast Guard, Department of Transportation. *CHRIS—Hazardous Chemical Data. Volume II*. Washington, DC: U.S. Government Printing Office, 1984-5.

4. (1) Hine, J., P. K. Mookerjee, *J Organic Chemical* 40: 292-8 (1975). (2) Lyman,

W. J., et al., *Handbook of Chemical Property Estimation Methods*, New York: McGraw-Hill, p. 5-10 (1982).

5. 40 CFR 60.489 (7/1/89).

■ ETHANE, 1,1'-OXYBIS

CAS # 60-29-7

DOT # 1155

SIC CODES 3833; 2834; 2892

SYNONYMS anaesthetic ether • anesthesia ether • anesthetic ether • diethyl ether • diethyl oxide • ether • ethoxyethane • solvent ether • 1,1'-oxybisethane • ethyl oxide • sulfuric ether

MF $C_4H_{10}O$

MW 74.14

BINARY REACTANTS Air or oxygen.

STANDARD CODES ICC—flammable liquid, red label, 10 gallon in an outside container; USCG—grade A flammable liquid, red label; NFPA—2, 4, 1; IATA—flammable liquid, red label, 1 liter passenger, 40 liter cargo.

PERSISTENCE Although ether does not degrade rapidly, it is volatile, and will disperse after short period of time.

INHALATION LIMIT 400 mg/m^3.

DIRECT CONTACT Mildly irritating to skin and mucous membranes. Low concentration of vapor in air causes unconsciousness. TLV, 400 ppm can be adsorbed via skin.

GENERAL SENSATION Sweet pungent odor. Skin irritation grade 1—no effect; burning taste. Eye irritation grade 2—small burns from 0.5 ml [R1]. Side effects include nausea, vomiting, and respiratory arrest. Death can come from respiratory paralysis.

PERSONAL SAFETY Chemical safety goggles, self-contained breathing apparatus, or canister masks (ether vapors), and protective clothing. PVC is considered poor for gloves [R2].

ACUTE HAZARD LEVEL Toxic via all routes; mild irritant; narcotic.

CHRONIC HAZARD LEVEL Unknown.

DEGREE OF HAZARD TO PUBLIC HEALTH Toxic via all routes; mild irritant; narcotic.

OCCUPATIONAL RECOMMENDATIONS None listed.

ACTION LEVEL Notify air and fire authority. Warn civil defense of possible explosion. Evacuate area. Enter from upwind and remove all ignition source. Use spark resistant equipment. Attempt to contain slick.

ON-SITE RESTORATION Skimming may pose a fire hazard. Carbon treatment, peat, or sorbent foams could lessen this hazard. Slick may be contained in booms.

BEACH OR SHORE RESTORATION Do not attempt to ignite. Seek professional environmental engineering assistance through EPA's environmental response team (ERT), Edison, NJ, 24-hour no. 201-321-6660.

AVAILABILITY OF COUNTERMEASURE MATERIAL Booms—stored at major ports; sorbent foams (polyurethane)—upholstery shops; carbon—water treatment plants, sugar refineries; peat—nurseries, floral shops.

DISPOSAL If ether has been standing in contact with oxygen for any protracted period of time, it should be treated with ferrous solution ($Fe_2(SO_4)_3$ 100 g, and 8 ml concentrated hydrochloric acid in 85 ml H_2O) or passed through an alumina column or shaker with 23% NaOH (1 part NaOH to 10 parts ether) to remove peroxides. Deperoxided ether can be burned on dry sand or ashes. Ignition should be made from remote location. Breathing equipment is advised.

DISPOSAL NOTIFICATION Contact fire authority.

INDUSTRIAL FOULING POTENTIAL Explosive hazard when confined in boiler feed or cooling-system water.

MAJOR WATER USE THREATENED Recreation, potable supply, industrial.

PROBABLE LOCATION AND STATE Colorless liquid. Will float on surface and dissolve very slowly. Most will volatilize.

WATER CHEMISTRY Volatilization will rapidly dissipate. Slick will autoxidize to form peroxides if allowed to stand for prolonged periods. Relatively inert to attack.

COLOR IN WATER Colorless.

GENERAL RESPONSE Avoid contact with liquid and vapor. Keep people away. Wear goggles and self-contained breathing apparatus. Shut off ignition sources and call fire department. Stop discharge if possible. Stay upwind and use water spray to "knock down" vapor. Isolate and remove discharged material. Notify local health and pollution control agencies.

FIRE RESPONSE Flammable. Flashback along vapor trail may occur. Vapor may explode if ignited in an enclosed area. Wear goggles and self-contained breathing apparatus. Extinguish with dry chemical, foam, or carbon dioxide. Water may be ineffective on fire. Cool exposed containers with water.

EXPOSURE RESPONSE Call for medical aid. Vapor irritating to eyes, nose, and throat. If inhaled, will cause nausea, vomiting, headache, or loss of consciousness. Move to fresh air. If breathing has stopped, give artificial respiration. If breathing is difficult, give oxygen. Liquid irritating to skin. Harmful if swallowed. Remove contaminated clothing and shoes. Flush affected areas with plenty of water. If swallowed, and victim is conscious, have victim drink water or milk.

RESPONSE TO DISCHARGE Issue warning—high flammability. Restrict access. Evacuate area.

WATER RESPONSE Effect of low concentrations on aquatic life is unknown. May be dangerous if it enters water intakes. Notify local health and wildlife officials. Notify operators of nearby water intakes.

EXPOSURE SYMPTOMS Vapor inhalation may cause headache, nausea, vomiting, and loss of consciousness. Contact with eyes will be irritating. Skin contact from clothing wet with the chemical may cause burns.

EXPOSURE TREATMENT Inhalation: remove victim to fresh air; if breathing has stopped, apply artificial respiration; if breathing is irregular, give oxygen; call a physician. Eyes: flush immediately with water for 15 min.

REFERENCES

1. (1) Smyth, H. F., Jr., C. P. Carpenter, C. S. Weil, U. C. Pozzani, J. A. Striegel, and J. S. Nycum, "Range-Finding Toxicity Data: List VII," *American Industrial Hygiene Association Journal*, 30: 470–476. 1969. (2) Smyth, H. F., C. P. Carpenter, and C. S. Weil, "Range-Finding Toxicity Data: List IV," *AMA Archives of Industrial Hygiene and Occupational Medicine*, 4: 119–122, 1951. (3) Smyth, H. F., C. P. Carpenter, C. S. Weil, U. C. Pozzani, and J. A. Striegel, "Range-Finding Toxicity Data: List VI," *American Industrial Hygiene Association Journal*, 23: 95–107, 1962. (4) Smyth, H. F., C. P. Carpenter, C. S. Weil, and U. C. Pozzani, "Range-Finding Toxicity Data: List V," *AMA Archives of Industrial Hygiene and Occupational Medicine*, 10: 61–68, 1954. (5) Smyth, H. F., C. P. Carpenter, and C. S. Weil, "Range-Finding Toxicity Data: List III," *Journal of Industrial Hygiene and Toxicology*, 31: 60–62,1949.(6) Smyth, H. F., L. Seaton, and L. Fischer, "The Single Dose Toxicity of Some Glycols and Derivatives," *Journal of Industrial Hygiene and Toxicology*, 23 (6): 259–268, 1941. (7) Smyth, H. F., C. P. Carpenter, "Further Experience with the Range-Finding Test in the Industrial Toxicology Laboratory," *Journal of Industrial Hygiene and Toxicology*, 30: 63–68, 1948. (8) Smyth, H. F. and C. P. Carpenter, "The Place of the Range-Finding Test in the Industrial Toxicology Laboratory," *Journal of Industrial Hygiene and Toxicology*, 26: 269–273, 1 944.

2. Gauerke, J. R., "Work Gloves to Meet OSHA Rules," *Chemical Engineering*, April 3, 1972.

■ETHANE, 1,1,2,2-TETRACHLORO-

CAS # 79–34–5

SIC CODES 2869

SYNONYMS 1,1,2,2-czterochloroetan (Polish) • 1,1,2,2-tetrachloorethaan (Dutch) • 1,1,2,2-Tetrachloraethan (German) • 1,1,2,2-tetrachlorethane (French) • 1,1,2,2-tetracloroetano (Italian) • 1,1-dichloro-2,2-dichloroethane • acetylene tetrachloride • dichloro-2,2-dichloroethane • ethane, 1,1,2,2-tetrachloro • s-tetrachloroethane • sym-tetrachloroethane • tetrachloroethane • tetrachlorure d'acetylene (French) • NCI-C03554 • a13-04597 • EPA pesticide chemical code 078601 • Westron • acetosal • acetylene tetrachloride • cellon • Bonoform

MF $C_2H_2Cl_4$

MW 167.84

COLOR AND FORM Colorless (pure) to pale-yellow liquid (technical); heavy, mobile liquid.

ODOR Sweetish, suffocating, chloroform-like odor; pungent odor.

ODOR THRESHOLD Less than 3 ppm [R4, 561].

DENSITY 1.58658 at 25°C.

SURFACE TENSION 37.85 dynes/cm = 0.03785 N/m at 20°C.

VISCOSITY 1.7 cP at 20°C.

BOILING POINT 146.5°C.

MELTING POINT −44°C.

HEAT OF VAPORIZATION 9,917.1 cal/mole.

EXPLOSIVE LIMITS AND POTENTIAL An explosion may occur when the solvent symmetrical tetrachloroethane is almost removed in the chlorinolysis of 2,4-dinitrophenyl disulfide. [R9, p. 491M–208].

TOXIC COMBUSTION PRODUCTS Irritating hydrogen chloride vapor may form in fire [R3].

SOLUBILITY 1.0 g/350 ml water at 25°C; soluble in alcohol, ether, acetone, and benzene; 2,900 mg/L water at 20°C; miscible with methanol, ethanol, petroleum ether, carbon tetrachloride, chloroform, carbon disulfide, dimethylformamide, and oils.

VAPOR PRESSURE 9 mm Hg at 30°C.

OCTANOL/WATER PARTITION COEFFICIENT Log K_{ow} = 2.39.

CORROSIVITY Corrosive liquid.

STABILITY AND SHELF LIFE Stable in absence of air, moisture, and light, even at high temperature [R10].

OTHER CHEMICAL/PHYSICAL PROPERTIES Average evaporative half-life of 55.2 min for concentration of 0.92 mg/L in water.

PROTECTIVE EQUIPMENT Chemical safety goggles; plastic face shield; air- or oxygen-supplied mask; safety hat with brim; solvent-proof apron; synthetic rubber gloves [R3].

PREVENTATIVE MEASURES Clothing contaminated with liquid 1,1,2,2-tetrachloroethane should be placed in closed containers for storage until it can be discarded or until provision is made for the removal of 1,1,2,2-tetrachloroethane from the clothing. If the clothing is to be laundered or otherwise cleaned to remove the 1,1,2,2-tetrachloroethane, the person performing the operation should be informed of 1,1,2,2-tetrachloroethane's hazardous properties. Skin that becomes contaminated with liquid 1,1,2,2-tetrachloroethane should be immediately washed or showered with soap or mild detergent and water to remove any 1,1,2,2-tetrachloroethane. Eating and smoking should not be permitted in areas where liquid 1,1,2,2-tetrachloroethane is handled, processed, or stored. Employees who handle liquid 1,1,2,2-tetrachloroethane should wash their hands thoroughly with soap or mild detergent and water before eating, smoking, or using toilet facilities. [R6, 3].

CLEANUP 1. Ventilate area of spill or leak. 2. Collect for reclamation or absorb in vermiculite, dry sand, earth, or a similar material. [R6, 4].

DISPOSAL Generators of waste (equal to, or greater than 100 kg/mo) containing this contaminant, EPA hazardous waste number U209, must conform with USEPA regulations in storage, transportation, treatment, and disposal of waste [R12]. Product residues and sorbent media may be packaged in 17h epoxy-lined drums and disposed of at an EPA-approved disposal site or destroyed with high temperatures, using incinerators with scrubbing equipment. Confirm disposal procedures with responsible environmental engineer and regulatory officials.

COMMON USES Nonflammable solvent for fats, oils, waxes, resins, cellulose acetate, copal, phosphorus, sulfur, rubber; solvent in certain types of Friedel–Crafts reactions, phthalic anhydride condensations; in manufacture of paints, varnish, rust removers; in soil sterilization and weed killer, insecticide formulations; in determination of theobromine in cacao; as immersion fluid in crystallography; in the biological laboratory to produce pathological changes in GI tract, liver, and kidneys. Intermediate in manufacture of of trichloroethylene and other chlorinated hydrocarbons having two carbon atoms [R1]; for cleansing and degreasing metals; paint removers, lacquers, photographic film; alcohol denaturant [R5]; chemical intermediate for trichloroethylene and tetrachloroethylene; chemical intermediate for 1,2- and 1,1-dichloroethylene; solvent, e.g., for polyesters and extractions; mothproofing agent for textiles [R2]; used in manufacture of cyanogen chloride, polymers, and tetrachloro-alkylphenol; use as a solvent in preparation of adhesives [R6, 3]. Bleach manufacturing [R7, 1075]; used in cement, lacquers; in manufacture of artificial silk and artificial pearls. Recently, it has also been used in the estimation of water content in tobacco, and in many drugs, and as a solvent for chromium chloride impregnation of furs [R8].

BINARY REACTANTS In the presence of moisture, the material gradually decomposes. As this breakdown occurs, hydrochloric acid is evolved, and metals in contact with the material may suffer corrosive damage. Other binary reactants include: 2,4-dinitrophenyl disulfide, potassium, sodium, potassium hydroxide, sodium-potassium alloy and bromoform, nitrogen tetroxide; mixtures of compound and potassium, sodium, nitrogen tetroxide may explode; compound mixed with solid potassium hydroxide and heated forms a spontaneously flammable gas.

STANDARD CODES Superfund designated hazardous substances list; U.N. no. 1702, Guide no. 55.

PERSISTENCE Persists in environment and is not degraded biologically. Undergoes hydrolysis in environment slowly, estimated half-life of several months to a few years. Photooxidation and photodissociation occur in stratosphere.

INHALATION LIMIT 5 mg/m^3.

DIRECT CONTACT Highly toxic by ingestion, inhalation, and skin absorption.

GENERAL SENSATION Symptoms of exposure: conjunctivitis, inflammation of the skin, irritation of respiratory tract, headache, dizziness, extreme exhaustion, enlarged liver and jaundice, oliguria, hematuria, albuminuria, mental instability and drowsiness, paralysis and coma, cardiac irregularity, tremors, lack of appetite, nausea, vomiting, epigastalgia, particularly polyneuritis. Initial symptoms of exposure are lacrimation, salivation, and irritation of nose and throat. Continued exposure to high concentrations results in restlessness, dizziness, nausea and vomiting, and narcosis. With less severe exposure, vague digestive and nervous-system complaints are common.

PERSONAL SAFETY Wear safety glasses, gas mask, rubber protective clothing. 50 ppm—chemical-cartridge respirator with an organic vapor cartridge or supplied-air respirator or self-contained breathing apparatus. 150 ppm—same as for 50 ppm but with a full facepiece on each or gas mask with an organic vapor canister or type-C supplied-air respirator operated in pressure-demand or other positive-pressure or continuous-flow mode.

OCCUPATIONAL RECOMMENDATIONS None listed.

OSHA Table Z–1 8-hour time-weighted average: 5 ppm (35 mg/m^3); skin designation [R27].

NIOSH NIOSH usually recommends that occupational exposures to carcinogens be limited to the lowest feasible concentration [R11, 230]; threshold-limit 8-hr time-weighted average (TWA) 1 ppm, 6.9 mg/m^3, skin (1986) [R28, 32].

INTERNATIONAL EXPOSURE LIMITS Former USSR (1967): 0.7 ppm; Federal Republic of Germany (1967): 1 ppm; Rumania (1967): 1.5 ppm; Yugoslavia (1967): 1 ppm [R4, 563].

ON-SITE RESTORATION Single-stage contactor dose of powdered carbon required to reduce the initial concentration (C.I.) of 1,1,2,2-tetrachloroethane at 1.0 mg/L to a final concentration (C.F.) of 0.001 mg/L at neutral pH is 2,200 mg/L; for C.I. 0.01 mg/L to C.F. 0.001 mg/L, use 20 mg carbon per liter. Granular carbon column doses for continuous operation until breakthrough (where effluent concentration equals influent concentration) are 95 mg/L to treat a C.I. of 1.0 mg/L or a dose of 5.3 mg/L to treat a C.I. of 0.01 mg/L. Multiplying the carbon dose given by 8,000 will convert it to the number of pounds required per one million gallons of water. Seek professional environmental engineering assistance through EPA's Environmental Response Team (ERT), Edison, NJ, 24-hour no. 908–548–8730. Contain and isolate spill to limit spread. Construct clay/bentonite swale to divert uncontaminated portion of watershed around contaminated portion. Isolation procedures include construction of bentonite or rubber-lined dams, interceptor trenches, or impoundments. Seek professional help to evaluate problem and implement containment procedures. Conduct bench-scale and pilot-scale tests prior to decontamination program implementation. Treatment alternatives for contaminated water include powdered activated-carbon sorption, granular activated-carbon filtration, acid hydrolysis, sorption by bentonite, peat moss, vermiculite, or polypropylene sorbent media, physical removal, and impoundment in a lined pit with leachate collection system. Treatment alternatives for contaminated soils include well-point injection to immobilize spill, physical removal and placement in a lined pit with leachate collection system, packaging for disposal. Confirm treatment procedures with responsible environmental engineer and regulatory officials.

WATER CHEMISTRY Rate of evaporation from rapidly stirred water at 25°C of 1 ppm solution—50% after 56 min, 90% after less than 120 min; heavy, very corrosive liquid with chloroform-like odor.

FOOD-CHAIN CONCENTRATION POTENTIAL The log octanol/water partition coefficient of 2.39 indicates bioaccumulation is unlikely.

NON-HUMAN TOXICITY LD_{50} rat, oral 200 mg/kg [R3]; LD_{50} rat oral 0.20 ml/kg [R1]; toxic dose dog oral 0.7 g/kg [R7, 1076].

ECOTOXICITY LC_{50} *Daphnia magna* 9,320 µg/L/48 hr, in a static unmeasured bioassay [R14]; LC_{50} *Pimephales promelas* (fathead minnow) 20,300 µg/L/96 hr, in a flow-through measured bioassay [R15]; LC_{50} *Lepomis macrochirus* (bluegill) 21,300 µg/L/96 hr, in a static unmeasured bioassay [R16]; LC_{50} *Mysiodopsis bahia* (mysid shrimp) 9,020 µg/L/96 hr, in a static unmeasured bioassay [R17]; LC_{50} *Cyprinodon variegatus* (sheepshead minnow) 12,300 µg/L/96 hr, in a static unmeasured bioassay [R17]; LC_{50} *Selenastrum capricornutum* 136,000–146,000 µg/L/96 hr (conditions of bioassay not specified) [R18, 440/5–89–0269]; LC_{50} *Skeletonema costatum* (alga) 6,230–6,440 µg/L/96 hr (conditions of bioassay not specified) [R18, 440/5–80–0269]; LC_{50} *Poecilia reticulata* (guppy) 37 ppm/7 day (conditions of bioassay not specified) [R7, 1076]; LC_{50} *Daphnia magna* 62 mg/L/48 hr (conditions of bioassay not specified) [R19]; EC_{50} *Daphnia magna* 14 mg/L/28 day; toxic effect: reproductive impairment (conditions of bioassay not specified) [R20]; LC_{50} Bluegill 21 mg/L/24 hr and 21 mg/L/96 hr (95% confidence limit 20–22 mg/L) at 21–23°C (conditions of bioassay not specified) [R21]; EC_{50} *Selenastrum*

capricornutum (alga) 136,000 µg/L/96 hr; toxic effect: chlorophyll a (conditions of bioassay not specified) [R22]; EC_{50} *Selenastrum capricornutum* (alga) 146,000 µg/L/96 hr; toxic effect: cell numbers (conditions of bioassay not specified) [R23]; EC_{50} *Skeletonema costatum* (alga) 6,440 µg/L/96 hr; toxic effect: chlorophyll a (conditions of bioassay not specified) [R23]; EC_{50} *Skeletonema costatum* (alga) 6,230 µg/L/96 hr; toxic effect: cell count (conditions of bioassay not specified) [R23]; LC_{50} *Pimephales promelas* (fathead minnow) 20.3 mg/L/96 hr (confidence limit 19.9–20.7 mg/L), flow–through bioassay with measured concentrations, 25.6°C, dissolved oxygen 7.8 mg/L, hardness 45.2 mg/L, calcium carbonate alkalinity 43.4 mg/L calcium carbonate, and pH 7.28 [R24].

IARC SUMMARY Inadequate evidence of carcinogenicity in humans. Limited evidence of carcinogenicity in animals. Overall evaluation: Group 3: The agent is not classifiable as to its carcinogenicity to humans [R13].

BIODEGRADATION One investigator who incubated the tetrachloroethane with sewage seed for 7 days and followed that with three successive 7-day subcultures found no significant degradation under these conditions (1). These results are in conflict with those of another investigator who obtained 41% degradation in 24 days in a modified shake-flask biodegradability test using an unacclimated inoculum and 19% degradation in a river die-away test, while 5 other chlorinated ethanes and ethenes tested were undegraded (2). A continuous-flow biofilm column operating under anaerobic conditions with a sewage inoculum achieved 97% steady-state removal during 4 months of operation (3). A product of the biodegradation was 1,1,2-trichloroethane (3). The most commonly found products of microbial degradation of these compounds evidently come from reductive dehalogenation, whereas nonmicrobial degradations tend to involve hydrolysis or oxidation (4) [R25].

BIOCONCENTRATION 1,1,2,2-Tetrachloroethane would not be expected to bioconcentrate in fish. The log of the bioconcentration factor in fish is reported to be 0.9–1 (1,2). After 14 days exposure to an average water concentration of 9.62 µg/L, the log bioconcentration factor of 1,1,2,2-tetrachloroethane in the tissue of bluegill sunfish (*Lepomis macrochirus*) was 0.9 (1–3) [R26].

ATMOSPHERIC STANDARDS This action promulgates standards of performance for equipment leaks of volatile organic compounds (VOC) in the synthetic organic chemical manufacturing industry (SOCMI). The intended effect of these standards is to require all newly constructed, modified, and reconstructed SOCMI process units to use the best demonstrated system of continuous emission reduction for equipment leaks of VOC, considering costs, non-air-quality health, and environmental impact and energy requirements. Tetrachloroethanes are produced, as intermediates or final products, by process units covered under this subpart. (tetrachloroethanes) [R29].

TSCA REQUIREMENTS EPA has issued a final rule, under section 4 of the Toxic Substances Control Act (TSCA), requiring manufacturers and processors to test this compound for certain health effects. Oral 14-day repeated dose and oral 90-day subchronic toxicity studies are required [R30].

RCRA REQUIREMENTS U209; as stipulated in 40 CFR 261.33, when 1,1,2,2-tetrachloroethane as a commercial chemical product or manufacturing chemical intermediate or an off-specification commercial chemical product or a manufacturing chemical intermediate, becomes a waste, it must be managed according to federal or state hazardous waste regulations. Also defined as a hazardous waste is any residue, contaminated soil, water, or other debris resulting from the cleanup of a spill, into water or on dry land, of this waste. Generators of small quantities of this waste may qualify for partial exclusion from hazardous waste regulations (40 CFR 261.5) [R31].

REFERENCES

1. Budavari, S. (ed.). *The Merck Index—Encyclopedia of Chemicals, Drugs, and Biologicals.* Rahway, NJ: Merck and Co., Inc., 1989. 1449.

2. SRI.

3. U.S. Coast Guard, Department of Transportation. *CHRIS—Hazardous Chemical Data.* Volume II. Washington, DC: U.S. Government Printing Office, 1984–5.

4. American Conference of Governmental Industrial Hygienists. *Documentation of the Threshold Limit Values and Biological Exposure Indices.* 5th ed. Cincinnati, OH: American Conference of Governmental Industrial Hygienists, 1986.

5. Sax, N. I., and R. J. Lewis, Sr. (eds.). *Hawley's Condensed Chemical Dictionary.* 11th ed. New York: Van Nostrand Reinhold Co., 1987. 1131.

6. Mackison, F. W., R. S. Stricoff, and L. J. Partridge, Jr. (eds.). *NIOSH/OSHA—Occupational Health Guidelines for Chemical Hazards.* DHHS (NIOSH) Publication No. 81–123 (3 vols). Washington, DC: U.S. Government Printing Office, Jan. 1981.

7. Verschueren, K. *Handbook of Environmental Data of Organic Chemicals.* 2nd ed. New York, NY: Van Nostrand Reinhold Co., 1983.

8. Sittig, M. *Handbook of Toxic and Hazardous Chemicals and Carcinogens,* 1985. 2nd ed. Park Ridge, N.J., Noyes Data Corporation, 1985. 637.

9. National Fire Protection Association. *Fire Protection Guide on Hazardous Materials.* 9th ed. Boston, MA: National Fire Protection Association, 1986.

10. IARC. *Monographs on the Evaluation of the Carcinogenic Risk of Chemicals to Man.* Geneva: World Health Organization, International Agency for Research on Cancer, 1972–present (multivolume work), p. V20 478 (1979).

11. NIOSH. *NIOSH Pocket Guide to Chemical Hazards.* DHHS (NIOSH) Publication No. 90–117. Washington, DC: U.S. Government Printing Office, June 1990.

12. 40 CFR 240–280, 300–306, 702–799 (7/1/91).

13. IARC. *Monographs on the Evaluation of the Carcinogenic Risk of Chemicals to Man.* Geneva: World Health Organization, International Agency for Research on Cancer, 1972–present (multivolume work), p. S7 72 (1987).

14. USEPA. *In-Depth Studies on Health and Environ Impacts of Selected Water Poll* (1978) Contract No. 68–01–4646, as cited in USEPA, *Ambient Water Quality Criteria Doc; Chlorinated Ethanes*, p. B–8 (1980) EPA 440/5–80–0269.

15. USEPA. *In-Depth Studies on Health and Environ Impacts of Selected Water Poll* (1978) Contract No. 68–01–4646 as cited in USEPA, *Ambient Water Quality Criteria Doc; Chlorinated Ethanes*, p. B–9 (1980) EPA 440/5–80–0269.

16. USEPA. *In-Depth Studies on Health and Environ Impacts of Selected Water Poll* (1978) Contract No. 68–01–4646 as cited in USEPA, *Ambient Water Quality Criteria Doc; Clorinated Ethanes*, p. B–9 (1980) EPA 440/5–80–0269.

17. USEPA. *In-Depth Studies on Health and Environ Impacts of Selected Water Poll* (1978) Contract No. 68–01–4646 as cited in USEPA, *Ambient Water Quality Criteria Doc; Chlorinated Ethanes*, p. B–10 (1980) EPA 440/5–80–0269.

18. USEPA. *In-Depth Studies on Health and Environ Impacts of Selected Water Poll* (1978) Contract No. 68–01–4646 as cited in USEPA, *Ambient Water Quality Criteria Doc; Chlorinated Ethanes*, p. B–13 (1980) EPA.

19. Richter, J. E., et al. *Arch Environ Contam Toxicol* 12 (6): 679–84 (1983).

20. Richter, J.E., et al. *Arch Environ Contam Toxicol* 12 (6): 679–84 (1983).

21. Buccafusco, R. J., et al. *Bull Environm Contam Toxicol* 26: 446 (1981).

22. USEPA. *In-Depth Studies on Health and Environ Impacts of Selected Water Poll* (1978) Contract No. 68–01–4646 as cited in USEPA, *Ambient Water Quality Criteria Doc; Chlorinated Ethanes*, p. B–13 (1980) EPA 440/5–80–0269.

23. USEPA. *In-Depth Studies on Health and Environ Impacts of Selected Water Poll* (1978) Contract No. 68–01–4646 as cited in USEPA, *Ambient Water Quality Criteria Doc; Chlorinated Ethanes*, p. B–13 (1980) EPA 440/5–80–0269.

24. Geiger, D. L., S. H. Poirier, L. T. Brooke, D.J. Call (eds.). *Acute Toxicities of Organic Chemicals to Fathead Minnows (Pimephales promelas).* Vol. II. Superior, Wisconsin: University of Wisconsin–Superior, 1985. 38.

25. (1) Tabak, H. H., et al. *J Water Pollut Control Fed* 53: 1503–18 (1981); (2) Mudder, T. I. Amer Chemical Soc Div Environ Chemical Presentation, Kansas City, MO, Sept p. 52–3 (1982); (3) Bouwer, E. J., P. L. McCarty. *Appl Environ Microbiol* 45: 1286–94 (1983); (4) Smith, L. R., J. Dragun. *Environ International* 10 (4): 291–8 (1985).

26. (1) Barrows, M. E., et al., 1980 *Dyn Exposure Hazard Assess Toxic Chemical,* Ann Arbor, MI: Ann Arbor Press, p. 379–392; (2) Kawaski, M. *Ecotox Environ Safety* 4: 444–454 (1980); (3) Vieth, G.D., et al. *An Evaluation of Using Partition Coefficients and Water Solubility to Estimate Bioconcentration Factors for Organic Chemicals in Fish,* ASTM STP 707, *Aquatic Toxicology* Easton, J. G., et al. (ed.) *Amer Soc Test Materials* p. 116–129 (1980).

27. 29 CFR 1910.1000 (58 FR 35338 (6/30/93)).

28. American Conference of Governmental Industrial Hygienists. *Threshold Limit Values for Chemical Substances and Physical Agents and Biological Exposure Indices for 1993–1994.* Cincinnati, OH: ACGIH, 1993.

29. 40 CFR 60.489 (7/1/92).

30. 58 FR 59667 (11/10/93).

31. 40 CFR 261.33 (7/1/92).

■ ETHANETHIOL

CAS # 75–08–1

SYNONYMS Aethanethiol (German) • Aethylmercaptan (German) • etantiolo (Italian) • ethaanthiol (Dutch) • ethanethiol • ethyl hydrosulfide • ethyl sulfhydrate • ethyl thioalcohol • ethylmercaptaan (Dutch) • ethylmerkaptan (Czech) • etilmercaptano (Italian) • lpg-ethyl-mercaptan 1010 • mercaptoethane • thioethanol • thioethyl alcohol

MF C_2H_6S

MW 62.14

COLOR AND FORM Colorless liquid.

ODOR Leek-like odor; penetrating garlic-like odor; strong skunk-like odor.

ODOR THRESHOLD For 50% response, 0.00047 ppm; 100% response, 0.001 ppm [R4].

DENSITY 0.83907 at 20°C.

SURFACE TENSION 23.63 dynes/cm at 2°C; 21.62 dynes/cm at 16.7°C.

VISCOSITY 0.003155 g/cm sec at 20°C.

BOILING POINT 35°C at 760 mm Hg.

MELTING POINT −144.4°C.

HEAT OF COMBUSTION −17.37 cal/mole at 25°C.

HEAT OF VAPORIZATION 6,728.7 cal/mol.

AUTOIGNITION TEMPERATURE 570°F [R4].

EXPLOSIVE LIMITS AND POTENTIAL Upper 18.0% and lower 2.8% by volume (at room temp) [R7].

FLAMMABILITY LIMITS Lower flammable limit: 2.8%; upper flammable limit: 18.0% [R3].

FIREFIGHTING PROCEDURES Do not extinguish fire unless flow can be stopped. Use water in flooding quantities as fog. Solid streams of water may spread fire. Cool all affected containers with flooding quantities of water. Apply water from as great a distance as possible. Use foam, dry chemical, or carbon dioxide [R5, 302]. Evacuation: if fire becomes uncontrollable or container is exposed to direct flame, consider evacuation of one-third (1/3) mile radius. If material leaking (not on fire) consider evacuation from downwind area based on amount of material spilled, location, and weather conditions. [R5, 303].

TOXIC COMBUSTION PRODUCTS Irritating fumes of sulfur dioxide are generated [R6].

SOLUBILITY In water at 20°C 6.76 g/L or 0.112 mole/L; soluble in alcohol, ether; soluble in acetone, dilute alkali.

VAPOR PRESSURE 442 mm Hg at 20°C.

STABILITY AND SHELF LIFE May deteriorate in normal storage and cause hazard. [R10, 626].

OTHER CHEMICAL/PHYSICAL PROPERTIES Azeotrope with n-pentane (51% ethanethiol) bp: 30.46°C; with ether (40% ethanethiol) bp: 31.50°C.

PROTECTIVE EQUIPMENT Wear appropriate equipment to prevent repeated or prolonged skin contact [R7].

PREVENTATIVE MEASURES Workers should wash promptly when skin becomes contaminated [R7].

CLEANUP Odorous gases, such as methylmercaptans, mainly present a cosmetic problem. Thus a variety of methods and procedures, and also patents, are available with methods to absorb odors in scrubbers, by use of catalytic oxidizers, or combined scrubber systems with oxidizing agents such as ozone or peroxides. [R9, 2070].

DISPOSAL Ethyl mercaptan is a waste chemical stream constituent that may be subjected to ultimate disposal by controlled incineration (2,000°F) followed by scrubbing with a caustic solution [R11].

COMMON USES Intermediate and starting material in manufacture of plastics, insecticides, antioxidants; odorant for natural gas [R2]; intermediate in production of acaricides, defoliants, pharmaceuticals, and adhesives [R1].

OCCUPATIONAL RECOMMENDATIONS None listed.

OSHA Ceiling value: 10 ppm (25 mg/m³). Transitional limits were achieved by any combination of engineering controls, work practices, and personal protective equipment during the phase-in period, Sept 1 1989 through Dec 30, 1992 [R13].

NIOSH 15-min ceiling value: 0.5 ppm (1.3 mg/m^3) [R7]. Threshold-limit 8-hr time-weighted average (TWA) 0.5 ppm, 1.3 mg/m^3 (1986) [R14, 22].

NON-HUMAN TOXICITY LD$_{50}$ rat ip 450 mg/kg [R4]; LD$_{50}$ rat oral 1,034 mg/kg [R8]; LC$_{50}$ rat inhalation 4,970 ppm/4 hr [R8]; LC$_{50}$ rat inhalation 2,770 ppm/4 hr [R4]; LC$_{50}$ mouse inhalation 4,420 ppm/4 hr [R4].

BIOCONCENTRATION Based upon a measured water solubility of 6,800 mg/L at 25°C (1), the BCF for ethyl mercaptan can be estimated to be approximately 4 from a linear regression-derived equation (2,SRC). This estimated BCF suggests that bioconcentration in aquatic organisms will not be important environmentally (SRC) [R12].

REFERENCES

1. SRI.

2. Budavari, S. (ed.). *The Merck Index—Encyclopedia of Chemicals, Drugs, and Biologicals.* Rahway, NJ: Merck and Co., Inc., 1989. 588.

3. National Fire Protection Guide. *Fire Protection Guide on Hazardous Materials.* 10th ed. Quincy, MA: National Fire Protection Association, 1991, p. 325M-53.

4. American Conference of Governmental Industrial Hygienists. *Documentation of the Threshold Limit Values and Biological Exposure Indices.* 5th ed. Cincinnati, OH: American Conference of Governmental Industrial Hygienists, 1986. 262.

5. Association of American Railroads. *Emergency Handling of Hazardous Materials in Surface Transportation.* Washington, DC: Assoc. of American Railroads, Hazardous Materials Systems (BOE), 1987.

6. U.S. Coast Guard, Department of Transportation. *CHRIS—Hazardous Chemical Data.* Volume II. Washington, DC: U.S. Government Printing Office, 1984-5.

7. NIOSH. *NIOSH Pocket Guide to Chemical Hazards.* DHHS (NIOSH) Publication No. 90-117. Washington, DC: U.S. Government Printing Office, June 1990 112.

8. *Kirk-Othmer Encyclopedia of Chemical Technology.* 3rd ed., Volumes 1-26. New York, NY: John Wiley and Sons, 1978-1984, p. V22 959 (1983).

9. Clayton, G.D., and F.E. Clayton (eds.). *Patty's Industrial Hygiene and Toxicology:* Volume 2A, 2B, 2C: *Toxicology.* 3rd ed. New York: John Wiley and Sons, 1981-1982.

10. Sunshine, I. (ed.). *CRC Handbook of Analytical Toxicology.* Cleveland: The Chemical Rubber Co., 1969.

11. USEPA. *Engineering Handbook for Hazardous Waste Incineration* p. 2-7 (1981) EPA 68-03-3025.

12. (1) Inst Natl de Recherche et de Securité. *CAH Notes Doc* 113: 597-600 (1983); (2) Lyman, W. J., et al. *Handbook of Chemical Property Estimation Methods,* Washington, DC: Amer Chemical Soc, p. 5-10 (1990).

13. 29 CFR 1910.1000 (7/1/91).

14. American Conference of Governmental Industrial Hygienists. *Threshold Limit Values for Chemical Substances and Physical Agents and Biological Exposure Indices for 1993-1994.* Cincinnati, OH: ACGIH, 1993.

■ ETHANOL, 2-(2-BUTOXYETHOXY)-

CAS # 112-34-5

SIC CODES 2818

SYNONYMS 2-(2-butoxyethoxy) ethanol • bucb • butoxydiethylene glycol • butoxydiglycol • butoxyethoxyethanol • butyl carbitol • butyl diglycol • butyl digol • butyl dioxitol • diethylene glycol, butyl ether • diethylene glycol, n-butyl ether • diglycol, monobutyl ether • dowanol-db • ethanol, 2,2'-oxybis-, monobutyl ether • ethanol, 2-(2-butoxyethoxy)- • o-butyl diethylene glycol • poly-solv db • butyl-ethyl-cellosolve • ektasolve-DB • Jeffersol-DB • diethylene glycol, monobutyl ether

MF $C_8H_{18}O_3$

MW 162.26

COLOR AND FORM Colorless liquid.

ODOR Faint, butyl odor.

DENSITY 0.9536 at 20°C.

VISCOSITY 0.0649 Poise at 20°C.

BOILING POINT 230.4°C.

MELTING POINT −68.1°C.

AUTOIGNITION TEMPERATURE 442°F [R4].

FIREFIGHTING PROCEDURES Alcohol foam, carbon dioxide, dry chemical [R4].

SOLUBILITY Miscible with water, oils; very soluble in ether, alcohol, acetone; soluble in benzene.

VAPOR PRESSURE 0.02 mm Hg at 20°C.

OCTANOL/WATER PARTITION COEFFICIENT Log K_{ow} = 0.91 (est).

OTHER CHEMICAL/PHYSICAL PROPERTIES Percent in saturated air: 0.003 at 25°C.

PREVENTATIVE MEASURES If materials are agitated or heated, or if skin contact is extensive and prolonged, it is advisable to protect personnel by enclosing process or providing local exhaust ventilation. Skin and eye protection should also be worn if possibility of such contact exists. (glycols and deriv) [R5].

COMMON USES Chemical intermediate for diethylene glycol monobutyl ether acetate; coalescing agent in latex paints; solvent for stamp pad inks; dye solvent; solvent in high-baked enamels; dispersant for vinyl chloride resins in organosols; diluent for hydraulic brake fluids; mutual solvent for soap, oil, water in household cleaners [R1]; in textile indust as wetting-out solution [R3]; solvent for nitrocellulose, oils, dyes, gums, soaps, polymers; plasticizer intermediate [R2].

STANDARD CODES NFPA—1,2, 0; ICC, USCG—no.

PERSISTENCE Should biodegrade at moderate rate.

DIRECT CONTACT Mild irritant to skin.

ACUTE HAZARD LEVEL Considered a moderate ingestive toxin; mild irritant. Generally more toxic than straight glycol.

CHRONIC HAZARD LEVEL Mild chronic irritant and ingestive toxin; 51 mg/kg is maximum level that produces no effects in rats when administered in drinking water for 30 days [R16].

DEGREE OF HAZARD TO PUBLIC HEALTH Ethers are considered more toxic than glycols. Moderate ingestive toxicant. Mild irritant.

OCCUPATIONAL RECOMMENDATIONS None listed.

NIOSH NIOSH recommends reducing exposure to lowest feasible concentration and preventing contact with the skin. (glycol ethers) [R13].

ACTION LEVEL Notify fire authority. Enter from upwind and remove ignition source. Isolate from oxidizing materials.

ON-SITE RESTORATION Carbon may be ineffective. Aeration may sponsor biodegradation. Seek professional environmental engineering assistance through EPA's Environmental Response Team (ERT), Edison, NJ, 24-hour no. 908–548–8730.

BEACH OR SHORE RESTORATION Burn off.

AVAILABILITY OF COUNTERMEASURE MATERIAL Carbon—water treatment plants, sugar refineries; aeration can be accomplished with compressors and piping.

DISPOSAL The following wastewater treatment technologies have been investigated for diethylene glycol monobutyl ether: Concentration process: Activated carbon [R6]. (1) Pour on ground in open air and ignite from distance or allow to evaporate. (2) Dissolve in higher alcohol, benzene, or petroleum ether; incinerate. (3) Ether of long standing may contain peroxides. Transport to isolated area in padded containers. Uncover and arrange excelsior train. From distance, puncture with rifle fire and ignite excelsior train.

DISPOSAL NOTIFICATION Contact local air and fire authority.

MAJOR WATER USE THREATENED Recreation, potable supply.

PROBABLE LOCATION AND STATE Colorless liquid; will dissolve.

HUMAN TOXICITY It has been estimated that the single oral dose of diethylene glycols lethal for humans is approximately 1 ml/kg. (diethylene glycols) [R7].

NON-HUMAN TOXICITY LD_{50} rat oral 6.56 g/kg [R8]; LD_{50} guinea pig oral 2.00 g/kg [R8].

ECOTOXICITY LC_{50} *Menidia beryllina* 2,000 ppm/96 hr (static bioassay in synthetic seawater at 23°C, mild aeration applied after 24 hr) [R9]; LC_{50} goldfish 2,700 mg/L/24 hr (modified ASTM-D-1345) (conditions of bioassay not specified) [R9]; LC_{50} *Lepomis macrochirus* 1,300 ppm/96 hr (static bioassay in fresh water at 23°C, mild aeration applied after 24 hr) [R9].

BIODEGRADATION Soil grab sample and river die-away test data pertaining to the biodegradation of diethylene glycol monobutyl ether in soil and natural waters were not located in the available literature. Although the rate cannot be determined, a few aerobic biological screening studies, which utilized settled waste water, sewage, or activated sludge for inocula, indicate that diethylene glycol monobutyl ether should biodegrade in the environment (1–3) [R10].

BIOCONCENTRATION Because diethylene glycol monobutyl ether is miscible in water (1), bioconcentration in aquatic systems is not expected to be an important fate process. Based upon an estimated log K_{ow} of 0.91 (2), a bioconcentration factor (log BCF) of 0.46 for diethylene glycol monobutyl ether has been calculated using a recommended regression-derived equation (3,SRC). This BCF value also indicates diethylene glycol monobutyl ether should not bioconcentrate in aquatic organisms (SRC) [R11].

ATMOSPHERIC STANDARDS This action promulgates standards of performance for equipment leaks of volatile organic compounds (VOC) in the synthetic organic chemical manufacturing industry (SOCMI). The intended effect of these standards is to require all newly constructed, modified, and reconstructed SOCMI process units to use the best demonstrated system of continuous emission reduction for equipment leaks of VOC, considering costs, non-air-quality health, and environmental impact and energy requirements. Diethylene glycol monobutyl ether is produced, as an intermediate or final product, by process units covered under this subpart [R14].

TSCA REQUIREMENTS Manufacturers and processors of diethylene glycol monobutyl ether are required to conduct subchronic toxicity, neurotoxicity/behavorial effects, developmental neurotoxicity, and pharmacokinetic tests under TSCA section 4 [R15].

FIFRA REQUIREMENTS Diethylene glycol monobutyl ether is exempted from the requirement of a tolerance when used as a deactivator for formulations used before crop emerges from soil or as a stabilizer in accordance with good agricultural practice as inert (or occasionally active) ingredients in pesticide formulations applied to growing crops only [R12].

GENERAL RESPONSE Stop discharge if possible. Keep people away. Call fire department. Isolate and remove discharged material. Notify local health and pollution control agencies.

FIRE RESPONSE Combustible. Extinguish with water, dry chemicals, alcohol foam, or carbon dioxide. Cool exposed containers with water.

EXPOSURE RESPONSE Call for medical aid. Liquid irritating to skin and eyes. Remove contaminated clothing and shoes. Flush affected areas with plenty of water. If in eyes, hold eyelids open, and flush with plenty of water. If swallowed and victim is conscious, have victim drink water or milk.

RESPONSE TO DISCHARGE Issue warning—water contaminant. Disperse and flush.

WATER RESPONSE Effect of low concentrations on aquatic life is unknown. May be dangerous if it enters water intakes. Notify local health and wildlife officials. Notify operators of nearby water intakes.

EXPOSURE SYMPTOMS Inhalation for brief periods has no significant effect. Contact with liquid causes moderate irritation of eyes and corneal injury. Prolonged contact with skin causes only minor irritation.

EXPOSURE TREATMENT Inhalation: remove to fresh air; if ill effects are observed, call a doctor. Eyes: Immediately flush with plenty of water for at least 15 min. Skin: Wash well with soap and water. Ingestion: Give large amounts of water.

REFERENCES

1. SRI.

2. Sax, N. I., and R. J. Lewis, Sr. (eds.). *Hawley's Condensed Chemical Dictionary.* 11th ed. New York: Van Nostrand Reinhold Co., 1987. 390.

3. Browning, E. *Toxicity and Metabolism of Industrial Solvents.* New York: American Elsevier, 1965. 635.

4. Sax, N. I. *Dangerous Properties of Industrial Materials.* 6th ed. New York, NY: Van Nostrand Reinhold, 1984. 1012.

5. International Labour Office. *Encyclopedia of Occupational Health and Safety.* Volumes I and II. New York: McGraw-Hill Book Co., 1971. 621.

6. USEPA *Management of Hazardous Waste Leachate*, EPA Contract No. 68–03–2766 p. E–152 (1982).

7. Doull, J., C. D. Klassen, and M. D. Amdur (eds.). *Casarett and Doull's Toxicology.* 3rd ed., New York: Macmillan Co., Inc., 1986. 656.

8. Budavari, S. (ed.). *The Merck Index— Encyclopedia of Chemicals, Drugs, and Biologicals.* Rahway, NJ: Merck and Co., Inc., 1989. 239.

9. Verschueren, K. *Handbook of Environmental Data of Organic Chemicals.* 2nd ed. New York, NY: Van Nostrand Reinhold Co., 1983. 524.

10. (1) Bridie, A. L., et al. *Water Res* 13: 627–30 (1979); (2) Dow Chemical Co. *The Glycol Ethers Handbook.* Midland, MI (1981); (3) Babeu, L., D. Vaishnav. *J Indust Microb* 2: 107–15 (1987).

11. (1) Dow Chemical Co. *The Glycol Ethers Handbook.* Midland, MI (1981); (2) CLOGP. *PCGEMS Graphical Exposure Modeling System* USEPA (1986); (3) Lyman, W. J., et al., *Handbook of Chemical Property Estimation Methods*, New York: McGraw-Hill, p. 5–4 (1982).

12. 40 CFR 180.1001 (d) (7/1/88).

13. NIOSH/CDC. *NIOSH Recommendations for Occupational Safety and Health Standards* 1988, Aug. 1988 (Suppl. to *Morbidity and Mortality Wkly.* Vol. 37 No. 5–7, Aug. 26, 1988). Atlanta, GA: National Institute for Occupational Safety and Health, CDC, 1988. 16.

14. 40 CFR 60.489 (7/1/90).

15. 40 CFR 799.1560 (7/1/88).

16. (1) Smyth, H. F., Jr., C. P. Carpenter, C. S. Weil, U. C. Pozzani, J. A. Striegel, and J. S. Nycum. "Range-Finding Toxicity Data: List VII," *American Industrial Hygiene Association Journal*, 30: 470–476. 1969. (2) Smyth, H. F., C. P. Carpenter, and C. S. Weil. "Range-Finding Toxicity Data: List IV," *AMA Archives of Industrial Hygiene and Occupational Medicine*, 4: 119–122, 1951. (3) Smyth, H. F., C. P. Carpenter, C. S. Weil, U. C. Pozzani, and J. A. Striegel. "Range-Finding Toxicity Data: List VI," *American Industrial Hygiene Association Journal*, 23: 95–107, 1962. (4) Smyth, H. F., C. P. Carpenter, C. S. Weil, and U. C. Pozzani. "Range-Finding Toxicity Data: List V," *AMA Archives of Industrial Hygiene and Occupational Medicine*, 10: 61–68, 1954. (5) Smyth, H. F., C. P. Carpenter, and C. S. Weil. "Range-Finding Toxicity Data: List III," *Journal of Industrial Hygiene and Toxicology*, 31: 60–62, 1949. (6) Smyth, H. F., I. Seaton, and L. Fischer. "The Single Dose Toxicity of Some Glycols and Derivatives," *Journal of Industrial Hygiene and Toxicology*, 23 (6): 259–268, 1941. (7) Smyth, H. F., and C. P. Carpenter. "Further Experience with the Range-Finding

Test in the Industrial Toxicology Laboratory," *Journal of Industrial Hygiene and Toxicology*, 30: 63–68, 1948. (8) Smyth, H. F., and C. P. Carpenter, "The Place of the Range-Finding Test in the Industrial Toxicology Laboratory," *Journal of Industrial Hygiene and Toxicology*, 26: 269–273, 1 944.

■ ETHANOL, 2-(2-BUTOXYETHOXY)-, ACETATE

CAS # 124–17–4

SYNONYMS 2-(2-butoxyethoxy) ethanol acetate • 2-(2-butoxyethoxy) ethyl acetate • butoxyethoxyethyl acetate • butyl diethylene glycol acetate • diethylene glycol butyl ether acetate • diethyleneglycol monobutyl ether acetate • diglycol monobutyl ether acetate

MF $C_{10}H_{20}O_4$

COLOR AND FORM Clear liquid.

ODOR Mild, not unpleasant odor.

TASTE Bitter taste.

DENSITY 0.985 at 20°C.

BOILING POINT 245°C at 760 mm Hg.

MELTING POINT −32°C.

FIREFIGHTING PROCEDURES Foam, carbon dioxide, dry chemical [R3].

SOLUBILITY Soluble in water; soluble in all proportions in alcohol, ether, acetone, organic solvents.

VAPOR PRESSURE Less than 0.01 mm Hg at 20°C.

OTHER CHEMICAL/PHYSICAL PROPERTIES 1 ppm = 8.34 mg/m³; 1 mg/L = 120 ppm.

COMMON USES Solvent for lacquers and other coatings, antibiotic extractions; coalescing aid for emulsion paints [R1]; solvent for cellulose acetate, ester gum, polyvinyl acetate; in paint and lacquer industry [R2, 639].

OCCUPATIONAL RECOMMENDATIONS None listed.

GENERAL RESPONSE Stop discharge if possible. Keep people away. Call fire department. Isolate and remove discharged material. Notify local health and pollution control agencies.

FIRE RESPONSE Combustible. Extinguish with water, dry chemicals, alcohol foam, or carbon dioxide. Cool exposed containers with water.

EXPOSURE RESPONSE Call for medical aid. Liquid irritating to skin and eyes. Remove contaminated clothing and shoes. Flush affected areas with plenty of water. If in eyes, hold eyelids open, and flush with plenty of water. If swallowed and victim is conscious, have victim drink water or milk.

RESPONSE TO DISCHARGE Issue warning—water contaminant. Disperse and flush.

WATER RESPONSE Effect of low concentrations on aquatic life is unknown. May be dangerous if it enters water intakes. Notify local health and wildlife officials. Notify operators of nearby water intakes.

EXPOSURE SYMPTOMS Prolonged breathing of vapor may cause irritation and nausea. Contact with liquid may cause mild irritation of eyes and skin. Can be absorbed through skin in toxic amounts.

EXPOSURE TREATMENT Inhalation: move victim to fresh air; if breathing has stopped, administer artificial respiration. Eyes: flush with water for at least 15 min. skin: wash skin with large amounts of water for 15 min; call physician if needed. Ingestion: induce vomiting; get medical attention.

REFERENCES

1. SRI.

2. Browning, E. *Toxicity, and Metabolism of Industrial Solvents*. New York: American Elsevier, 1965.

3. Sax, N. I. *Dangerous Properties of Industrial Materials*. 4th ed. New York: Van Nostrand Reinhold, 1975. 488.

■ ETHEPHON

CAS # 16672–87–0

SYNONYMS phosphonic acid, (2-chloroethyl)- • 2-(chloroethyl) phosphonic acid • 2-cepa • 2-Chloraethyl-phosphonsaeure (German) • 2-chloroethanephosphonic acid • 2-chloroethylphosphonic acid • amchemical 68–250 • camposan • cep • cepa • chlorethephon • chloroethylphosphonic acid • ethefon • ethel • etheverse • flordimex • g-996 • kamposan • phosphonic acid, (2-chloroethyl)- • roll-fruct • tomathrel • ethrel • prep • cepha • florel • bromoflor • (2-chloroethyl) phosphonic acid • ethrel-c • terpal

MF C$_2$H$_6$ClO$_3$P

MW 144.50

COLOR AND FORM Needles from benzene; white waxy solid.

DENSITY 1.58.

MELTING POINT 74–75°C.

SOLUBILITY Freely soluble in water, methanol, acetone, ethylene glycol, propylene glycol; slightly soluble in benzene, toluene; practically insoluble in petroleum ether; very slightly soluble in aromatic solvents; very soluble in alcohol; soluble in short-chain alcohols; sparingly soluble in nonpolar organic solvents; insoluble in kerosene; solubility: approximately 1 kg/L water; insoluble in diesel oil.

VAPOR PRESSURE Less than 0.01 mPa at 20°C.

CORROSIVITY Concentrate is corrosive.

STABILITY AND SHELF LIFE Aqueous solutions are stable below pH 3.5. Above pH 3.5 hydrolysis begins with the release of free ethylene [R2].

OTHER CHEMICAL/PHYSICAL PROPERTIES Very hygroscopic (needles: must be dried over P$_2$O$_5$).

PROTECTIVE EQUIPMENT Gloves and eye shields are advised. [R6, 219].

PREVENTATIVE MEASURES SRP: The scientific literature supports the wearing of contact lenses in industrial environments, as part of a program to protect the eye against chemical compounds and minerals causing eye irritation. However, there may be individual substances whose irritating or corrosive properties are such that the wearing of contact lenses would be harmful to the eye. In those specific cases contact lenses should not be worn.

DISPOSAL Recommendable methods: Incineration. Peer-review: Large amount should be incinerated in a unit with effluent gas scrubbing (peer-review conclusions of an IRPTC expert consultation (May 1985)) [R3].

COMMON USES Ethylene responses in plants (fruit ripening, abscission, flower induction, breaking apical dominance) [R6, 218]. Plant growth regulator on apples, cherries, walnuts, and filberts [R1]. Stimulator of latex flow in rubber, sugarcane ripener, fruit ripener, flowering agent for pineapple, color enhancer [R6]. Used to promote preharvest ripening in cranberries, Morello cherries, citrus fruit, figs, tomatoes, sugar beet and fodder beet seed crops, coffee, capsicums; to facilitate harvesting by loosening of the fruit in currants, gooseberries, cherries, and apples; to increase flower bud development in young apple trees; to prevent lodging in flax; to improve the sturdiness of onion seed crops [R7]. Plant growth regulator for barley, blackberries, blueberries, bromeliads, cantaloupes, coffee, cotton, cucumbers, figs, grapes, guava, macadamia nuts, ornamentals, peppers, rye, squash, tobacco, tomatoes, wheat. [R8].

OCCUPATIONAL RECOMMENDATIONS None listed.

INTERNATIONAL EXPOSURE LIMITS The maximum permissible aqueous concentration of ethephon was determined in large bodies of water. Based on toxicity data, the maximum permissible concentration of ethephon in water reservoirs was 3 mg/L [R5].

NON-HUMAN TOXICITY LD$_{50}$ rat oral 4,000 mg/kg [R1]; LD$_{50}$ rabbit percutaneous 5,730 mg/kg [R9, 346].

ECOTOXICITY LD$_{50}$ bobwhite quail oral 1,000 ppm [R6, 220]; LC$_{50}$ *Lepomis macrochirus* (bluegill) 300 mg/L/96 hr (conditions of bioassay not specified) [R9, 346]; LC$_{50}$ *Salmo gairdneri* (rainbow trout) 350 mg/L/96 hr (conditions of bioassay not specified) [R9, 346].

FIFRA REQUIREMENTS Tolerances are established for residues of the plant regulator ethephon [(2-chloroethyl) phosphonic acid] in or on the following raw agricultural commodities: apples; barley (grain); barley (straw); blackberries; blueberries; cantaloupes; cattle (fat); cattle (mbyp); cattle (meat); cherries; coffee beans; cottonseed; cranberries; cucumbers; figs; filberts; goats (fat); goats (mbyp); goats (meat); grapes; hogs (fat); hogs (mbyp); hogs (meat); horses (fat); horses (mbyp); horses (meat); lemons; macadamia nuts; milk; peppers; pineapples; pineapple (fodder); pineapple (forage); pumpkins; sheep (fat); sheep (mbyp); sheep (meal); tangerines; tangerine hybrids; tomatoes; walnuts; wheat (grain); and wheat (straw). [R4].

REFERENCES

1. *Kirk–Othmer Encyclopedia of Chemical Technology.* 3rd ed., Volumes 1–26. New York, NY: John Wiley, and Sons, 1978–1984, p. V18 2 (1982).

2. Budavari, S. (ed.). *The Merck Index— Encyclopedia of Chemicals, Drugs, and Biologicals.* Rahway, NJ: Merck and Co., Inc., 1989. 589.

3. United Nations. *Treatment and Disposal Methods for Waste Chemicals* (IRPTC File). Data Profile Series No. 5. Geneva, Switzerland: United Nations Environmental Programme, Dec. 1985. 231.

4. 40 CFR 180.300 (a) (7/1/91).

5. Ryskeldieva, E.F. *Gig Sanit* (8): 79 (1983).

6. *Kirk-Othmer Encyclopedia of Chemical Technology.* 3rd ed., Volumes 1–26. New York, NY: John Wiley and Sons, 1978–1984, p. V18 2 (1982).

7. Hartley, D., and H. Kidd (eds.). *The Agrochemicals Handbook.* 2nd ed. Lechworth, Herts, England: The Royal Society of Chemistry, 1987, p. A179/Aug 87.

8. *Farm Chemicals Handbook 1992.* Willoughby, OH: Meister Publishing Co., 1992, p. C-139.

9. Worthing, C.R., and S.B. Walker (eds.). *The Pesticide Manual—A World Compendium.* 8th ed. Thornton Heath, UK: The British Crop Protection Council, 1987.

■ ETHOPROP

$$CH_3CH_2CH_2S \diagdown \underset{CH_3CH_2CH_2S \diagup}{\overset{\overset{\displaystyle O}{\|}}{P}} \text{—} OCH_2CH_3$$

CAS # 13194–48–4

SIC CODES 2879

SYNONYMS Ent-27,318 • ethoprophos • jolt • o-ethyl s,s-dipropyl dithiophosphate • o-ethyl s,s-dipropyl phosphorodithioate • o-ethyl s,s-dipropylphosphorodithioate • phosphorodithioic acid, o-ethyl s,s-dipropyl ester • prophos • rovokil • v-c chemical v-c 9–104 • vc 9–104 • vc9–104 • Mobil v-c 9–104 • v-c 9–104 • Virginia-Carolina vc 9–104 • phosphorodithioic acid o-ethyl s,s-dipropyl ester • o-ethyl-s, s-dipropyl phosphorodithionate • mocap

MF C$_8$H$_{19}$O$_2$PS$_2$

MW 242.36

COLOR AND FORM Clear, pale-yellow liquid.

DENSITY 1.094 at 20°C.

BOILING POINT 86–91°C at 0.2 mm Hg.

MELTING POINT 20°C.

FIREFIGHTING PROCEDURES Do not extinguish fire unless flow can be stopped. Solid streams of water may be ineffective. Cool all affected containers with flooding quantities of water. Apply water from as far a distance as possible. Use "alcohol" foam, carbon dioxide, or dry chemical (organophosphorus pesticides, liquid, NOS) [R4]. Extinguish fire using agent suitable for type of surrounding fire (material itself does not burn or burns with difficulty). Use water in flooding quantities as fog. Use alcohol foam, carbon dioxide, or dry chemical. (organophosphorus pesticides, solid, NOS) [R4].

SOLUBILITY Readily soluble in most organic solvents; 750 mg/L of water; soluble in acetone, ethanol, hexane, kerosene, xylene; >300 g/kg in cyclohexane, 1,2-dichloroethane, diethyl ether, ethyl acetate, petroleum spirit.

VAPOR PRESSURE 3.8×10^{-4} mm Hg at 20–25°C (reported).

CORROSIVITY Noncorrosive to metals.

STABILITY AND SHELF LIFE Stable in water up to 100°C at pH 7, but rapidly hydrolyzed at 25°C, and pH 9 [R1].

OTHER CHEMICAL/PHYSICAL PROPERTIES 46.5 mPa at 26°C; Henry's Law constant = 1.62×10^{-7} atm-m^3/mole (est.).

PREVENTATIVE MEASURES Do not contaminate water, food, feed by storage, or disposal. [R10, p. C–142].

COMMON USES SRP (former use): wireworm and fleabeetle larvae control. It is recommended for corn rootworm, cutworm, and wireworm control on corn; wireworm control on white potato; and corn rootworm control on peanuts at pegging and root borer control on banana and plaintain [R2]. Control of plant-parasitic nematodes and soil insects in ornamentals, potatoes, sweet potatoes, tomatoes, vegetables, soya beans, groundnuts, cucurbits, strawberries, citrus, tobacco, pineapples, sugar cane, turf, and other crops [R3]. Nematicide, soil insecticide for bananas, beans (snap, lima), cabbage, corn, cucumber, flue-cured tobacco, peanuts, plantains, Bermuda, Zoysia, St. Augustine, Fescue, Kentucky blue, perennial rye, Bahia grasses in commercial turf. [R10, p. C–142] It is a non-systemic, nonfumigant nematicide, which is also effective against soil-dwelling insects, and used at 1.6–6.6 kg ai/hectare [R1].

PERSISTENCE Ethoprop is stable to hydrolysis in water at ambient temperatures. Its pesticidal action persists longer in dense than in loose soils. Ethoprop was effective for 2 to 4 weeks in preventing the up-to-24-hour-old larvae of *Diaprepes abbreviatus* from penetrating the soil when applied as 10 g or urea-foam-coated 10 g at the rate of 22.4 kg active ingredient per hectare. However, applications of 10 g at that rate did not hinder soil infestations after 3 months. Half-life in soil is 8 weeks. Half-life in fresh water at pH 11 to 12 at 25°C is 26 minutes.

DIRECT CONTACT Nonirritant to mallard ducks. Ethoprop is toxic via skin absorption. The dermal LD$_{50}$ in rabbits is approximately 26 mg/kg in rabbits and 10.6 mg/kg in ducks. 7 of 12 men exposed to 500 ppm ethoprop on the Maria Costa (a hull-damaged container ship) developed rashes. The rashes disappeared in a few hours. There were no other symptoms of exposure.

GENERAL SENSATION Symptoms of ethoprop poisoning are typical of cholinesterase inhibitors. They include: tightness across the chest, increased salivation, nausea, vomiting, abdominal cramps, diarrhea, abnormal heart rates, involuntary urination, constriction of eye pupils, and arm and leg weakness. If the poison enters the central nervous system, symptoms such as tension, insomnia, tremors, convulsions, and depression of circulatory and respiratory systems occur. Symptoms are generally localized. However, in large doses, the blood may absorb and spread the poison to other parts of the body. Cause of death is usually asphyxia caused by respiratory failure.

ACUTE HAZARD LEVEL Highly toxic via oral and dermal routes. Toxic to certain marine crabs and shrimp in the ppb range.

CHRONIC HAZARD LEVEL Dogs and rats fed a diet containing 100 ppm ethoprop for 90 days showed depression of cholinesterase levels but no other effects on pathology or histology.

OCCUPATIONAL RECOMMENDATIONS None listed.

ON-SITE RESTORATION Seek professional environmental engineering assistance through EPA's Environmental Response Team (ERT), Edison, NJ, 24-hour no. 908–548–8730. Contain and isolate spill by using clay/bentonite dams, interceptor trenches, or impoundments. Construct swale to divert uncontaminated portion of watershed around contaminated portion. Seek professional help to evaluate problem. Implement containment measures, and conduct bench scale and pilot scale tests prior to full-scale decontamination program implementation. Treatment alternatives for contaminated water include alkaline hydrolysis, powdered activated carbon, granular activated carbon, bentonite sorption, aeration, evaporation, biodegradation, and chemical oxidation. May be able to employ density stratification and impoundment removal techniques by removing product from bottom or top layers using skimming or polyethylene rope mop collection equipment. Treatment alternatives for contaminated soils include well-point collection and treatment of leachates as for contaminated waters, bentonite/cement ground injection to immobilize spill, physical removal of residues with pH adjustment to 8.0, and land application on an approved sanitary landfill. Residues may be packaged for disposal.

DISPOSAL Product residues and sorbent media may be packaged in 17h epoxy-lined drums and disposed of at an approved EPA disposal site. Destruction by high-temperature incineration or microwave plasma detoxification, if available. Encapsulation by organic polyester resin or silicate fixation. Confirm disposal procedures with responsible environmental engineer and regulatory officials.

PROBABLE LOCATION AND STATE Ethoprop is a pale-yellow liquid with a mild persistent odor. It is water-soluble and slightly denser than water so that it will tend to sink. It will solidify below 20°C.

WATER CHEMISTRY Very stable in water to 100°C, but is rapidly hydrolyzed in basic media at 25°C.

NON-HUMAN TOXICITY LD_{50} rats oral 62 mg/kg, albino [R1]; LD_{50} rabbits, oral 55 mg/kg [R1]; LD_{50} rabbit percutaneous 26 mg/kg [R3]; LD_{50} rat oral 216 mg/kg (technical imidan) [R11, p. II–298].

ECOTOXICITY LC_{50} *Colinis virginianus* (bobwhite) 33 ppm in 5-day diet (95% confidence limit 27–40 ppm), age 14 days (technical material, 95.8%) [R5]; LC_{50} *Coturnix japonica* (Japanese quail) 100 ppm in 5-day diet (95% confidence limit 85–117 ppm), age 14 days (technical material, 95.8%) [R5]; LC_{50} *Phasianus colchicus* (ring-necked pheasant) 118 ppm in 5-day diet (95% confidence limit 103–134 ppm), age 10 days (technical material, 95.8%) [R5]; LC_{50} *Anas platyrhynchos* (mallard duck) 287 ppm in 5-day diet (95% confidence limit 215–382 ppm), age 5 days (technical material, 95.8%) [R5]; LD_{50} *Anas platyrhynchos* (mallard duck), oral 61 mg/kg [R1]; LC_{50} *Carassius auratus* (goldfish) 13.6 mg/L/96 hr (conditions of bioassay not specified) [R1]; LC_{50} *Lepomis macrochirus* (bluegill) 2.07 mg/L/96 hr (conditions of bioassay not specified) [R1]; LC_{50} *Salmo gairdneri* (rainbow trout) 13.8 mg/L/96 hr (conditions of bioassay not specified) [R1]; LD_{50} hen oral 5.6 mg/kg [R1]; LC_{50} *Coturnix japonica* 89 ppm (95% confidence interval 72–109 ppm) [R6].

BIODEGRADATION During ethoprop metabolism studies, it was observed that ethoprop did not metabolize in sterilized soils, but did in nonsterile soils (1). Addition of ethoprop to soil has been observed to stimulate microbial oxygen consumption (2). During 4-week soil incubation studies using ^{14}C-labelled ethoprop, $^{14}CO_2$ evolution ranged from 23.4 to 50.9% in soils having no previous organophosphorus exposure and soils having prior exposure to isofenphos and fonophos (3); $^{14}CO_2$ from ethoprop was greater in soil having no prior exposure to the other pesticides (3). Mineralization of ethoprop was found to occur faster in soils that had been treated previously with ethoprop as compared to no prior treatment indicating microbial adaptation (4); during a 1-wk incubation, 32.7% mineralized in previously treated soil, and 19.9% mineralized in untreated soil (4). Faster biodegradation in previously ethoprop-treated soils was also noted in another study (5); sterilization of soils drastically reduced disappearance rates of ethoprop (5) [R7].

BIOCONCENTRATION Using an intermittent flow-through system, an ethoprop BCF range of 4 to 17 was measured in juvenile sheepshead minnow over a 28-day exposure period (1). This BCF value suggests that bioconcentration in aquatic organisms is not an important fate process (SRC) [R8].

DOT *Health Hazards:* Poisonous; may be fatal if inhaled, swallowed, or absorbed through skin. Contact may cause burns to skin and eyes. Runoff from fire control or dilution water may cause pollution (organophosphorus pesticide, liquid, flammable, poisonous, not otherwise specified) [R9, p. G–28].

Fire or Explosion: Flammable/combustible material; may be ignited by heat, sparks, or flames. Vapors may travel to a source of ignition, and flash back. Container may explode in heat of fire. Vapor explosion and poison hazard indoors, outdoors, or in sewers. Runoff to sewer may create fire or explosion hazard (organophosphorus pesticide, liquid, flammable, poisonous, not otherwise specified) [R9, p. G–28].

Emergency Action: Keep unnecessary people away; isolate hazard area, and deny entry. Stay upwind; keep out of low areas. Positive-pressure self-contained breathing apparatus (SCBA) and chemical protective clothing that is specifically recommended by the shipper or manufacturer may be worn. It may provide little or no thermal protection. Structural firefighters' protective clothing is not effective for these materials. Isolate for 1/2 mile in all directions if tank, rail car, or tank truck is involved in fire. Call CHEMTREC at 1–800–424–9300 for emergency assistance (organophosphorus pesticide, liquid, flammable, poisonous, not otherwise specified) [R9, p. G–28].

Fire: Small Fires: Dry chemical, CO_2, water spray, or alcohol-resistant foam. Large Fires: Water spray, fog, or alcohol-resistant foam. Move container from fire area if you can do so without risk. Dike fire-control water for later disposal; do not scatter the material. Apply cooling water to sides of containers that are exposed to flames until well after fire is out. Stay away from ends of tanks. Withdraw immediately in case of rising sound from venting safety device or any discoloration of tank due to fire (organophosphorus pesticide, liquid, flammable, poisonous, not otherwise specified) [R9, p. G–28].

Spill or Leak: Shut off ignition sources; no flares, smoking, or flames in hazard area. Fully encapsulating, vapor-protective clothing should be worn for spills and leaks with no fire. Do not touch or walk through spilled material; stop leak if you can do so without risk. Water spray may reduce vapor; but it may not prevent ignition in closed spaces. Small Spills: Take up with sand or other noncombustible absorbent material and place into containers for later disposal. Large Spills: Dike far ahead of liquid spill for later disposal (organophosphorus pesticide, liquid, flammable, poisonous, not otherwise specified) [R9, p. G–28].

First Aid: Move victim to fresh air, and call for emergency medical care; if not breathing, give artificial respiration; if breathing is difficult, give oxygen. In case of contact with material, immediately flush skin or eyes with running water for at least 15 minutes. Remove and isolate contaminated clothing and shoes at the site. Keep victim quiet and maintain normal body temperature. Effects may be delayed; keep victim under observation (organophosphorus pesticide, liquid, flammable, poisonous, not otherwise specified) [R9, p. G–28].

REFERENCES

1. Worthing, C. R., and S. B. Walker (eds.). *The Pesticide Manual—A World Compendium.* 8th ed. Thornton Heath, UK: The British Crop Protection Council, 1987. 355.

2. *Farm Chemicals Handbook 1984.* Willoughby, OH: Meister Publishing Co., 1984, p. C–155.

3. Hartley, D., and H. Kidd (eds.). *The Agrochemicals Handbook.* 2nd ed. Lech-

worth, Herts, England: The Royal Society of Chemistry, 1987, p. A185/Aug 87.

4. Association of American Railroads. *Emergency Handling of Hazardous Materials in Surface Transportation.* Washington, DC: Association of American Railroads, Bureau of Explosives, 1992. 713.

5. U.S. Department of the Interior, Fish, and Wildlife Service, Bureau of Sports Fisheries and Wildlife. *Lethal Dietary Toxicities of Environmental Pollutants to Birds.* Special Scientific Report—*Wildlife* No. 191. Washington, DC: U.S. Government Printing Office, 1975. 27.

6. Hill, E. F., and Camardese, M. B. *Lethal Dietary Toxicities of Environmental Contaminants and Pesticides to Coturnix.* Fish and Wildlife Technical Report 2. Washington, DC: United States Department of Interior, Fish and Wildlife Service, 1986. 72.

7. (1) Menzer, R.E., et al., *J Agric Food Chem* 19: 351–6 (1971) (2) Tu, C.M. *Can J Plant Sci* 53: 401–5 (1973) (3) Racke, K.D., J.R. Coats. *J Agric Food Chem* 36: 193–99 (1988) (4) Racke, K.D., J.R. Coats. In *Enhanced Biodegradation of Pesticides in the Environment.* ACS Symposium Series 426. pp. 68–81 (1990) (5) Smelt, J.H., et al. *Crop Protection* 6: 295–303 (1987).

8. USEPA; *Acephate, Aldicarb, Carbophenthion, DEF, EPN, Ethoprop, Methyl Parathion, and Phorate: Their Acute and Chronic Toxicity, Bioconcentration Potential and Persistence as Related to Marine Environments.* USEPA 600/4-81-041 (NTIS PB81-244477). Gulf Breeze, FL: USEPA pp. 47–48 (1981).

9. U.S. Department of Transportation. *Emergency Response Guidebook 1990.* DOT P 5800.5. Washington, DC: U.S. Government Printing Office, 1990.

10. *Farm Chemicals Handbook 1993.* Willoughby, OH: Meister Publishing Co., 1993.

11. Gosselin, R.E., R.P. Smith, H.C. Hodge. *Clinical Toxicology of Commercial Products.* 5th ed. Baltimore: Williams and Wilkins, 1984.

■ETHYL BROMIDE

CAS # 74–96–4

SYNONYMS bromic ether • bromoethane • bromure d'ethyle • ethane, bromo • etylu-bromek (Polish) • halon-2001 • hydrobromic ether • monobromoethane • NCI-554813 • NCI-C55481

MF C_2H_5Br

MW 108.98

COLOR AND FORM Colorless, liquid.

ODOR Ethereal.

ODOR THRESHOLD 890 mg/m^3 (odor low); 890 mg/m^3 (odor high) [R4].

TASTE Burning taste.

DENSITY 1.4612 at 20°C.

SURFACE TENSION 24.15 dyne/cm at 20°C.

VISCOSITY 0.379 cP at 25°C.

BOILING POINT 38.2°C.

MELTING POINT –119°C.

AUTOIGNITION TEMPERATURE 952°F (511°C) [R1].

EXPLOSIVE LIMITS AND POTENTIAL Lower 6.7%; upper 11.3% [R3].

FLAMMABILITY LIMITS Percent by vol: lower flammable limit: 6.8; upper flammable limit: 8.0 [R2].

TOXIC COMBUSTION PRODUCTS Readily decomposes into volatile toxic products hydrobromic acid and bromine, particularly in presence of hot surfaces or open flame [R3].

SOLUBILITY 1.067 g/100 g water at 0°C; 0.965 g/100 g water at 10°C; 0.914 g/100 g water at 20°C; 0.896 g/100 g water at 30°C; miscible in alcohol, ether, chloroform, organic solvents.

VAPOR PRESSURE 467 mm Hg at 25°C.

OCTANOL/WATER PARTITION COEFFICIENT log K_{ow} = 1.61 (measured).

CORROSIVITY Liquid ethyl bromide will attack some forms of plastics, rubber, and coatings.

STABILITY AND SHELF LIFE Conditions contributing to instability: heat [R10, 2].

OTHER CHEMICAL/PHYSICAL PROPERTIES Density in saturated air: 2.7 (air = 1); percent in saturated air: 62.5 at 25°C.

PROTECTIVE EQUIPMENT Wear safety glasses, rubber gloves, self-contained breathing apparatus, protective work gown. [R11, 221].

PREVENTATIVE MEASURES Skin that becomes wet with liquid ethyl bromide should be promptly washed or showered with soap or mild detergent and water to remove any ethyl bromide. [R10, 2].

CLEANUP Absorb the spills with paper towels or like materials. Place in hood to evaporate. Dispose by burning the towel [R11, 222].

DISPOSAL Dissolve in a combustible solvent. Scatter the spray of the solution into the furnace with afterburner and alkali scrubber [R11, 222].

COMMON USES Ethylating agent in organic synthesis, refrigerant [R1].

OCCUPATIONAL RECOMMENDATIONS None listed.

OSHA 8-hr time-weighted average: 200 ppm (890 mg/m³) [R7].

NIOSH NIOSH questioned whether the PEL proposed by OSHA for ethyl bromide was adequate to protect workers from recognized health hazards. [R12, 235].

THRESHOLD LIMIT 8-hr time-weighted average (TWA) 5 ppm, 22 mg/m³ (1992) [R13, 21].

INTERNATIONAL EXPOSURE LIMITS Romania: 90 ppm; former USSR: 1 ppm [R8].

IARC SUMMARY Evaluation: There is limited evidence for the carcinogenicity of bromoethane in experimental animals. No data were available from studies in humans on the carcinogenicity of bromoethane. Overall evaluation: Bromoethane is not classifiable as to its carcinogenicity to humans (Group 3) [R5].

BIOCONCENTRATION Based upon a water solubility of 8,939 mg/L at 25°C (1), the BCF for ethyl bromide can be estimated to be 3.7 from a regression-derived equation (2,SRC). Based upon a measured log K_{ow} of 1.61 (3), the K_{oc} for ethyl bromide can be estimated to be 9.9 from a regression-derived equation (2,SRC). These BCF values suggest that ethyl bromide will not bioconcentrate significantly in aquatic organisms (SRC) [R6].

ATMOSPHERIC STANDARDS This action promulgates standards of performance for equipment leaks of volatile organic compounds (VOC) in the synthetic organic chemical manufacturing industry (SOCMI). The intended effect of these standards is to require all newly constructed, modified, and reconstructed SOCMI process units to use the best demonstrated system of continuous emission reduction for equipment leaks of VOC, considering costs, non-air-quality health, and environmental impact and energy requirements. Ethyl bromide is produced, as an intermediate or a final product, by process units covered under this subpart [R9].

REFERENCES

1. Budavari, S. (ed.). *The Merck Index—Encyclopedia of Chemicals, Drugs, and Biologicals*. Rahway, NJ: Merck and Co., Inc., 1989. 596.

2. National Fire Protection Association. *Fire Protection Guide on Hazardous Materials*. 9th ed. Boston, MA: National Fire Protection Association, 1986, p. 325M–48.

3. Sax, N. I. *Dangerous Properties of Industrial Materials*. 6th ed. New York, NY: Van Nostrand Reinhold, 1984. 547.

4. Ruth, J.H., *Am Ind Hyg Assoc J* 47: A–142–51 (1986).

5. IARC. *Monographs on the Evaluation of the Carcinogenic Risk of Chemicals to Man*. Geneva: World Health Organization, International Agency for Research on Cancer, 1972–present (multivolume work), p. 52 311 (1991)

6. (1) Horvath, A. L., *Halogenated Hydrocarbons: Solubility–Miscibility with Water*. New York: Marcel Dekker p. 490 (1982) (2) Lyman, W. J., et al., *Handbook of Chemical Property Estimation Meth-*

ods New York: McGraw-Hill, p. 5–4, 5–10 (1982) (3) Hansch, C., and A. J. Leo. *Medchem Project* Issue No 26. Claremont, CA: Pomona College (1985).

7. 29 CFR 1910.1000 (7/1/91).

8. American Conference of Governmental Industrial Hygienists. *Documentation of the Threshold Limit Values and Biological Exposure Indices.* 5th ed. Cincinnati, OH: American Conference of Governmental Industrial Hygienists, 1986. 245.

9. 40 CFR 60.489 (7/1/91).

10. Mackison, F. W., R. S. Stricoff, and L. J. Partridge, Jr. (eds.). *NIOSH/OSHA— Occupational Health Guidelines for Chemical Hazards.* DHHS (NIOSH) Publication No. 81–123 (3 vols). Washington, DC: U.S. Government Printing Office, Jan. 1981.

11. ITII. *Toxic and Hazardous Industrial Chemicals Safety Manual.* Tokyo, Japan: The International Technical Information Institute, 1988.

12. NIOSH. *NIOSH Pocket Guide to Chemical Hazards.* DHHS (NIOSH) Publication No. 90–117. Washington, DC: U.S. Government Printing Office, June 1990.

13. American Conference of Governmental Industrial Hygienists. *Threshold Limit Values for Chemical Substances and Physical Agents and Biological Exposure Indices for 1994–1995.* Cincinnati, OH: ACGIH, 1994.

■ ETHYL CHLOROFORMATE

CAS # 541–41–3

DOT # 1182

SYNONYMS carbonochloridic acid, ethyl ester • cathyl chloride • Chlorameisensaeureaethylester (German) • chlorocarbonate d'ethyle (French) • chlorocarbonic acid ethyl ester • chloroformic acid ethyl ester • ethoxycarbonyl chloride • ethyl carbonochloridate • ethyl chlorocarbonate • ethyl chloromethanoate • ethylchloorformiaat (Dutch) • ethyle, chloroformiat d' (French) • etil clorocarbonato (Italian) • etil-cloroformiato (Italian) • formic acid, chloro-, ethyl ester • tl-423 • chloroformic acid ethyl ester • cloroformiato de etilo (Spanish)

MF $C_3H_5ClO_2$

MW 108.53

COLOR AND FORM Water-white liquid.

ODOR Sharp, like hydrochloric acid.

DENSITY 1.1403 at 20°C.

SURFACE TENSION 27.5 dynes/cm = 0.0275 N/m at 15°C.

BOILING POINT 95°C at 760 mm Hg.

MELTING POINT −80.6°C.

HEAT OF COMBUSTION −6,900 Btu/lb = −3,800 cal/g = −160×10⁵ J/kg (est).

HEAT OF VAPORIZATION 140 Btu/lb = 79 cal/g = 3.3×10⁵ J/kg (est).

AUTOIGNITION TEMPERATURE 932°F (500°C) [R9, p. 325M–50].

FIREFIGHTING PROCEDURES Use dry chemical, foam, carbon dioxide, or water spray. Water may be ineffective. Use water spray to keep fire-exposed containers cool. Approach fire from upwind to avoid hazardous vapors and toxic decomposition products. [R9, p. 49–82].

SOLUBILITY Practically insoluble, and gradually decomposes in water; miscible with alcohol, benzene, chloroform, ether.

CORROSIVITY Corrosive.

OTHER CHEMICAL/PHYSICAL PROPERTIES Decomposes in water and alcohol.

PROTECTIVE EQUIPMENT Air-line mask, self-contained breathing apparatus, or organic and acid canister mask; full protective clothing [R5].

PREVENTATIVE MEASURES Chloroformates should be handled so as to prevent all contact (with the worker). (chloroformates) [R8, 2387].

DISPOSAL Releases may require isolation or evacuation. Eliminate all ignition sources. Stop or control the leak, if this can be done without undue risk. Use water spray to cool and disperse vapors and protect personnel. Absorb in noncombustible material for proper disposal. [R9, p. 49–82].

COMMON USES Chemical intermediate for ore flotation agents; diethyl carbonate; isocyanates; polymers [R2]; organic synthesis [R1]; in the manufacture of ore flotation agents by reaction with various alkyl xanthates [R3].

OCCUPATIONAL RECOMMENDATIONS None listed.

THRESHOLD LIMIT Notice of Intended Changes (1993–94): These substances, with their corresponding values, comprise those for which either a limit has been proposed for the first time, for which a change in the "Adopted" listing has been proposed, or for which retention on the Notice of Intended Changes has been proposed. In all cases, the proposed limits should be considered trial limits that will remain in the listing for a period of at least one year. If, after one year no evidence comes to light that questions the appropriateness of the values herein, the values will be reconsidered for the "Adopted" list. Time-weighted average (TWA) 0.1 mg/m^3, skin [R6].

DOT *Health Hazards:* Poisonous; may be fatal if inhaled, swallowed, or absorbed through skin. Contact may cause burns to skin and eyes. Runoff from fire control or dilution water may cause pollution [R7, p. G–28].

Fire or Explosion: Flammable/combustible material; may be ignited by heat, sparks, or flames. Vapors may travel to a source of ignition, and flash back. Container may explode in heat of fire. Vapor explosion and poison hazard indoors, outdoors, or in sewers. Runoff to sewer may create fire or explosion hazard [R7, p. G–28].

Emergency Action: Keep unnecessary people away; isolate hazard area, and deny entry. Stay upwind; keep out of low areas. Positive-pressure self-contained breathing apparatus (SCBA) and chemical protective clothing that is specifically recommended by the shipper or manufacturer may be worn. It may provide little or no thermal protection. Structural firefighters' protective clothing is not effective for these materials. Isolate for 1/2 mile in all directions if tank, rail car, or tank truck is involved in fire. Call CHEMTREC at 1–800–424–9300 for emergency assistance [R7, p. G–28].

Fire: Small Fires: Dry chemical, CO$_2$, water spray, or alcohol-resistant foam. Large Fires: Water spray, fog, or alcohol-resistant foam. Move container from fire area if you can do so without risk. Dike fire-control water for later disposal; do not scatter the material. Apply cooling water to sides of containers that are exposed to flames until well after fire is out. Stay away from ends of tanks. Withdraw immediately in case of rising sound from venting safety device or any discoloration of tank due to fire [R7, p. G–28].

Spill or Leak: Shut off ignition sources; no flares, smoking, or flames in hazard area. Fully encapsulating, vapor-protective clothing should be worn for spills and leaks with no fire. Do not touch or walk through spilled material; stop leak if you can do so without risk. Water spray may reduce vapor; but it may not prevent ignition in closed spaces. Small Spills: Take up with sand or other noncombustible absorbent material and place into containers for later disposal. Large Spills: Dike far ahead of liquid spill for later disposal [R7, p. G–28].

First Aid: Move victim to fresh air and call for emergency medical care; if not breathing, give artificial respiration; if breathing is difficult, give oxygen. In case of contact with material, immediately flush skin or eyes with running water for at least 15 minutes. Remove and isolate contaminated clothing and shoes at the site. Keep victim quiet and maintain normal body temperature. Effects may be delayed; keep victim under observation [R7, p. G–28].

Initial Isolation and Protective Action Distances: Small Spills (leak or spill from a small package or small leak from a large package): First, isolate in all directions 150 feet; then, protect those persons in the downwind direction 0.2 mile. Large spills (leak or spill from a large package or spill from many small packages): First, isolate in all directions 150 feet; then, protect those persons in the downwind direction 0.2 mile. [R7, p. G–table].

FIRE POTENTIAL Dangerous when exposed to heat, flame, oxidizers [R4]. Flammable, dangerous fire risk [R1].

GENERAL RESPONSE Avoid contact with liquid and vapor. Keep people away. Wear goggles, self-contained breathing apparatus, and rubber overclothing (including gloves). Shut off ignition sources. Call fire department. Stop discharge if possible. Evacuate area in case of large discharge. Stay upwind. Use water spray to knock down vapor. Isolate and remove discharged material. Notify local health and pollution control agencies.

FIRE RESPONSE Flammable. Poisonous gases may be produced in fire. Flashback along vapor trail may occur. Vapor may explode if ignited in an enclosed area. Wear goggles, self-contained breathing apparatus, and rubber overclothing (including gloves). Extinguish with dry chemicals or carbon dioxide. Cool exposed containers with water.

EXPOSURE RESPONSE Call for medical aid. Vapor poisonous if inhaled or if skin is exposed. Irritating to eyes, nose, and throat. Move victim to fresh air. If breathing has stopped, give artificial respiration. If breathing is difficult, give oxygen. Liquid poisonous if swallowed or if skin is exposed. Remove contaminated clothing and shoes. Flush affected areas with plenty of water. If swallowed and victim is conscious, have victim drink water or milk. Do not induce vomiting.

RESPONSE TO DISCHARGE Issue warning—corrosive, high flammability, poison. Restrict access. Disperse and flush.

WATER RESPONSE Effect of low concentrations on aquatic life is unknown. May be dangerous if it enters water intakes. Notify local health and wildlife officials. Notify operators of nearby water intakes.

EXPOSURE SYMPTOMS Inhalation causes mucous membrane irritation, coughing, and sneezing. Vapor causes severe lachrymation; liquid causes acid-type burns of eyes and skin, like those of hydrochloric acid. Ingestion causes severe burns of mouth and stomach.

EXPOSURE TREATMENT Inhalation: remove to fresh air; use artificial respiration if breathing has stopped; call a doctor; keep victim quiet and administer oxygen if needed. Eyes: flush with water for at least 15 min; see a doctor. Skin: wash liberally with water for at least 15 min, then apply dilute solution of sodium bicarbonate or commercially prepared neutralizer. Ingestion: do not induce vomiting; give large amount of water; get medical attention.

REFERENCES

1. Sax, N. I., and R. J. Lewis, Sr. (eds.). *Hawley's Condensed Chemical Dictionary.* 11th ed. New York: Van Nostrand Reinhold Co., 1987. 483.

2. SRI.

3. *Kirk–Othmer Encyclopedia of Chemical Technology.* 3rd ed., Volumes 1–26. New York, NY: John Wiley, and Sons, 1978–1984, p. V4 765 (1978).

4. Sax, N. I. *Dangerous Properties of Industrial Materials.* 6th ed. New York, NY: Van Nostrand Reinhold, 1984. 1329.

5. U.S. Coast Guard, Department of Transportation. *CHRIS—Hazardous Chemical Data.* Volume II. Washington, DC: U.S. Government Printing Office, 1984–5.

6. American Conference of Governmental Industrial Hygienists. *Threshold Limit Values for Chemical Substances and Physical Agents and Biological Exposure Indices for 1993–1994.* Cincinnati, OH: ACGIH, 1993. 36.

7. U.S. Department of Transportation. *Emergency Response Guidebook 1990.* DOT P 5800.5. Washington, DC: U.S. Government Printing Office, 1990.

8. Clayton, G. D., and F. E. Clayton (eds.). *Patty's Industrial Hygiene and Toxicology.* Volume 2A, 2B, 2C: *Toxicology.* 3rd

ed. New York: John Wiley Sons, 1981–1982.

9. National Fire Protection Guide. *Fire Protection Guide on Hazardous Materials*. 10th ed. Quincy, MA: National Fire Protection Association, 1991.

■ ETHYL ISOCYANATE

CAS # 109–90–0

SYNONYMS ethane, isocyanato • isocyanatoethene • isocyanic acid, ethyl ester

MF C_3H_5NO

MW 71.09

DENSITY 0.9031 at 20°C.

BOILING POINT 60°C.

HEAT OF COMBUSTION 424.5 kcal (liquid).

FIREFIGHTING PROCEDURES Persons involved in fighting fires should wear a self-contained breathing apparatus with a full facepiece operated in a pressure-demand or other positive-pressure mode. Extinguishant: Carbon dioxide, dry chemical, or foam (methyl isocyanate) [R3, 2].

SOLUBILITY Soluble in alcohol, ether.

OTHER CHEMICAL/PHYSICAL PROPERTIES Primary alcohols react with isocyanates at room temperature, whereas secondary and tertiary alcohols react much more slowly at this same temperature. The usual reaction of an alcohol and an isocyanate leads to a urethane. In most reactions involving active hydrogen compounds, the aromatic isocyanates are more reactive than the aliphatic isocyanates. Substitution of electronegative groups on the aromatic ring enhance reactivity, whereas electropositive groups reduce the reactivity of the isocyanates. Steric hindrance on either the isocyanate or the active hydrogen compound retards the reaction. Reactions are typically catalyzed by acids and bases. The normal isocyanate reaction ultimately provides addition to the carbon–nitrogen double bond. In reactions involving compounds with an active hydrogen (i.e., one that can be replaced by sodium), the hydrogen becomes attached to the nitrogen of the isocyanate, and the remainder of the active hydrogen compound becomes attached to the carbonyl carbon. In addition to the usual reactions with active hydrogen, Friedel-Crafts and Grignard reactions may also be performed with isocyanates, to yield amides. Isocyanates also undergo self-condensation to form dimers, trimers, and polymers (isocyanates). Ethyl isocyanate was reduced with lithium aluminum deuteride and nitrosated with sodium nitrite to yield the carcinogen nitrosomethyl (d3) ethylamine.

PROTECTIVE EQUIPMENT Vapor concentration: 0.2 ppm or less: Any supplied-air respirator or any self-contained breathing apparatus. 1 ppm or less: Any supplied-air respirator with a full facepiece, helmet, or hood or any self-contained breathing apparatus with a full facepiece. A type C supplied-air respirator operated in a pressure-demand or other positive-pressure or continuous-flow mode. 20 ppm or less: A type C supplied-air respirator with a full facepiece operated in pressure-demand or other positive-pressure mode or with a full facepiece, helmet, or hood operated in continuous-flow mode. Greater than 20 ppm or entry and escape from unknown concentrations: Self-contained breathing apparatus with a full facepiece operated in a pressure-demand or other positive-pressure mode or a combination respirator that includes a type C supplied-air respirator with a full facepiece operated in pressure-demand or other positive-pressure or continuous-flow mode and an auxiliary self-contained breathing apparatus operated in pressure-demand or other positive-pressure mode. Escape: Any escape: self-contained breathing apparatus. (methyl isocyanate) [R3, 5].

PREVENTATIVE MEASURES Employees should change into uncontaminated clothing before leaving the work premises. Contaminated clothing should be placed in closed containers until it can either be discarded or until provision can be made for the removal of the contaminant. Employees should also be provided with emergency eyewash stations and notified that contact lenses should not be worn when working with this chemical. (methyl isocyanate) [R3, 3].

CLEANUP For small quantities absorb on paper towels. (methyl isocyanate) [R3, 3].

DISPOSAL For small quantities evaporate material in a fume hood. (methyl isocyanate) [R3, 3].

COMMON USES Pharmaceutical and pesticide intermediate [R1].

OCCUPATIONAL RECOMMENDATIONS None listed.

DOT *Health Hazards:* Poisonous; may be fatal if inhaled, swallowed, or absorbed through skin. Contact may cause burns to skin and eyes. Runoff from fire control or dilution water may cause pollution. [R2,p. G-57].

Fire or Explosion: May be ignited by heat, sparks, or flames. Container may explode in heat of fire. Vapor explosion and poison hazard indoors, outdoors or in sewers [R2,p. G-57].

Emergency Action: Keep unnecessary people away; isolate hazard area and deny entry. Stay upwind, out of low areas, and ventilate closed spaces before entering. Self-contained breathing apparatus and chemical protective clothing that is specifically recommended by the shipper or producer may be worn but they do not provide thermal protection unless that is stated by the clothing manufacturer. Structural firefighter's protective clothing is not effective with these materials. Call CHEMTREC at 1-800-424-9300 as soon as possible, especially if there is no local hazardous-materials team available [R2,p. G-57].

Fire: Small Fires: Dry chemical, CO_2, halon, water spray, or standard foam. Large Fires: Water spray, fog, or standard foam is recommended. Move container from fire area if you can do so without risk. Cool container with water using unmanned device until well after fire is out. Fight fire from maximum distance. Stay away from ends of tanks. Dike fire control water for later disposal; do not scatter the material [R2, p. G-57].

Spill or Leak: Shut off ignition sources; no flares, smoking, or flames in hazard area. Do not touch spilled material; stop leak if you can do so without risk. Water spray may reduce vapor, but it may not prevent ignition in closed spaces. Small Spills: Take up with sand or other noncombustible absorbent material and place into containers for later disposal. Small Dry Spills: With clean shovel place material into clean, dry container and cover; move containers from spill area. Large Spills: Dike far ahead of liquid spill for later disposal [R2,p. G-57].

First Aid: Move victim to fresh air and call emergency medical care; if not breathing, give artificial respiration; if breathing is difficult, give oxygen. In case of contact with material, immediately flush skin or eyes with running water for at least 15 minutes. Speed in removing material from skin is of extreme importance. Remove and isolate contaminated clothing and shoes at the site. Keep victim quiet and maintain normal body temperature. Effects may be delayed; keep victim under observation [R2,p. G-57].

Initial Isolation and Evacuation Distances: For a spill or leak from a drum or small container (or small leak from tank): Isolate in all directions 50 feet. For a large spill from a tank (or from many containers or drums): First isolate in all directions 150 feet; then, evacuate in a downwind direction 0.2 mile wide and 0.2 mile long [R2,p. TABLE].

REFERENCES

1. Sax. *Hawley's Condensed Chem Dictnry* 11th ed. 1987 p. 495.

2. Department of Transportation. *Emergency Response Guidebook 1987.* DOT P 5800.4. Washington, DC: U.S. Government Printing Office, 1987.

3. Mackison, F. W., R. S. Stricoff, and L. J. Partridge, Jr. (eds.). *NIOSH/OSHA—Occupational Health Guidelines for Chemical Hazards.* DHHS(NIOSH) Publication No. 81–123 (3 vols). Washington, DC: U.S. Government Printing Office, Jan. 1981.

■ ETHYL n-BUTYRATE

CAS # 105–54–4

DOT # 1180

SYNONYMS butanoic acid, ethyl ester • butyric acid, ethyl ester • butyric ether • ethyl butanoate • ethyl butyrate • FEMA number 2427

MF $C_6H_{12}O_2$

MW 116.18

pH Acid value: not more than 1 (0.06).

COLOR AND FORM Colorless liquid.

ODOR Fruity odor with pineapple undernote.

TASTE Sweet taste; can be tasted in water at a level of 0.450 ppm and at 0.015 ppm in milk.

DENSITY 0.879 at 20°C.

SURFACE TENSION 24.5 dynes/cm = 0.0245 N/m at 20°C.

VISCOSITY 0.711 cP at 15°C.

BOILING POINT 120–121°C.

MELTING POINT −100.8°C.

HEAT OF COMBUSTION 851.2 kcal at 20°C.

HEAT OF VAPORIZATION 9,468.5 cal/mol.

AUTOIGNITION TEMPERATURE 865°F, 463°C [R5].

FIREFIGHTING PROCEDURES Water may be ineffective. Alcohol foam [R5].

SOLUBILITY Soluble in 150 parts water, miscible with alcohol, ether; water solubility: 4.9×10^3 mg/L at 20°C.

VAPOR PRESSURE 12.8 mm Hg at 20°C.

OCTANOL/WATER PARTITION COEFFICIENT Log K_{ow} = 1.73 (est).

OTHER CHEMICAL/PHYSICAL PROPERTIES Density of saturated air: 1.08 (air = 1); may attack some forms of plastics.

PREVENTATIVE MEASURES SRP: The scientific literature supports the wearing of contact lenses in industrial environments, as part of a program to protect the eye against chemical compounds and minerals causing eye irritation. However, there may be individual substances whose irritating or corrosive properties are such that the wearing of contact lenses would be harmful to the eye. In those specific cases contact lenses should not be worn.

COMMON USES In manufacture of artificial rum; the alcohol solution constitutes the so-called "pineapple oil" [R3]; principally as flavor for foods, beverages, and chewing gums; fragrance ingredient for soap, perfume, creams, and lotions; solvent for cellulosic lacquers [R1].

OCCUPATIONAL RECOMMENDATIONS None listed.

NON-HUMAN TOXICITY LD_{50} rat oral 13,050 mg/kg [R6]; LD_{50} rabbit oral 5.2 g/kg [R7].

BIODEGRADATION No data regarding the biodegradation of ethyl butyrate in natural media or in laboratory screening studies were located (SRC).

BIOCONCENTRATION No data regarding the bioconcentration of ethyl butyrate were located (SRC). Based upon an estimated log K_{ow} of 1.73 (1), a BCF of 12 has been estimated using a recommended regression equation (2). Based upon an experimental water solubility of 4.9×10^3 mg/L (3), a BCF of 5 has been estimated using a recommended regression equation (2). Based upon these estimated BCF, ethyl butyrate will not be expected to bioconcentrate in aquatic organisms (SRC) [R8].

DOT Fire or Explosion: Flammable/combustible material; may be ignited by heat, sparks, or flames. Vapors may travel to a source of ignition, and flash back. Container may explode in heat of fire. Vapor explosion hazard indoors, outdoors, or in sewers. Runoff to sewer may create fire or explosion hazard [R4].

Health Hazards: May be poisonous if inhaled or absorbed through skin. Vapors may cause dizziness or suffocation. Contact may irritate or burn skin and eyes. Fire may produce irritating or poisonous gases. Runoff from fire control or dilution water may cause pollution [R4].

Emergency Action: Keep unnecessary people away; isolate hazard area and deny entry. Stay upwind; keep out of low areas. Positive-pressure self-contained breathing apparatus (SCBA) and structural firefighters' protective clothing will provide limited protection. Isolate for 1/2 mile in all directions if tank, rail car, or tank truck is involved in fire. Call CHEMTREC at 1–800–424–9300 for emergency assistance. If water pollution occurs, notify the appropriate authorities [R4].

Fire: Small Fires: Dry chemical, CO2, water spray, or alcohol-resistant foam. Large Fires: Water spray, fog, or alcohol-resistant foam. Do not use dry chemical extinguishers to control fires involving nitromethane or nitroethane. Move container from fire area if you can do so without risk. Apply cooling water to sides of containers that are exposed to flames until well after fire is out. Stay away from ends of tanks. For massive fire in cargo area, use unmanned hose holder, or monitor nozzles; if this is impossible, withdraw from area, and let fire burn. Withdraw immediately in case of rising sound from venting safety device or any discoloration of tank due to fire [R4].

Spill or Leak: Shut off ignition sources; no flares, smoking, or flames in hazard area. Stop leak if you can do so without risk. Water spray may reduce vapor; but it may not prevent ignition in closed spaces. Small Spills: Take up with sand or other noncombustible absorbent material and place into containers for later disposal. Large Spills: Dike far ahead of liquid spill for later disposal [R4].

First Aid: Move victim to fresh air and call emergency medical care; if not breathing, give artificial respiration; if breathing is difficult, give oxygen. In case of contact with material, immediately flush eyes with running water for at least 15 minutes. Wash skin with soap and water. Remove and isolate contaminated clothing and shoes at the site [R4].

FIRE POTENTIAL Flammable, dangerous fire risk [R2].

GENERAL RESPONSE Shut off ignition sources. Call fire department. Avoid contact with liquid and vapor. Keep people away. Stop discharge if possible. Stay upwind. Use water spray to "knock down" vapor. Isolate and remove discharged material. Notify local health and pollution control agencies.

FIRE RESPONSE Flammable. Containers may explode in fire. Flashback along vapor trail may occur. Vapor may explode if ignited in an enclosed area. Extinguish with dry chemicals, alcohol foam, or carbon dioxide. Water may be ineffective on fire. Cool exposed containers with water.

EXPOSURE RESPONSE Call for medical aid. Vapor irritating to eyes, nose, and throat. If inhaled will cause headache or dizziness. If breathing has stopped, give artificial respiration. If breathing is difficult, give oxygen. Liquid irritating to skin and eyes. If swallowed will cause nausea, vomiting, dizziness, or headache. Remove contaminated clothing and shoes. Flush affected areas with plenty of water. If swallowed and victim is conscious, have victim drink water or milk, and have victim induce vomiting. If swallowed and victim is unconscious or having convulsions, do nothing except keep victim warm.

RESPONSE TO DISCHARGE Issue warning—high flammability. Restrict access; mechanical containment; should be removed; chemical and physical treatment.

WATER RESPONSE Effect of low concentrations on aquatic life is unknown. Fouling to shoreline. May be dangerous if it enters water intakes. Notify local health and wildlife officials. Notify operators of nearby water intakes.

EXPOSURE SYMPTOMS Inhalation or ingestion causes headache, dizziness, nausea, vomiting, and narcosis. Contact with liquid irritates eyes.

EXPOSURE TREATMENT Inhalation: move victim to fresh air and call a physician; give artificial respiration if necessary. Ingestion: induce vomiting and call a physician. Eyes: flush with water for at least 15 min. Skin: flush with water; wash with soap and water.

REFERENCES

1. SRI.

2. Sax, N. I., and R. J. Lewis, Sr. (eds.). *Hawley's Condensed Chemical Dictionary.* 11th ed. New York: Van Nostrand Reinhold Co., 1987. 481.

3. *The Merck Index.* 10th ed. Rahway, NJ: Merck Co., Inc., 1983. 547.

4. U.S. Department of Transportation. *Emergency Response Guidebook 1990.* DOT P 5800.5. Washington, DC: U.S. Government Printing Office, 1990, p. G–26.

5. National Fire Protection Association. *Fire Protection Guide on Hazardous Materials.* 9th ed. Boston, MA: National Fire Protection Association, 1986, p. 325M–49.

6. Budavari, S. (ed.). *The Merck Index— Encyclopedia of Chemicals, Drugs, and Biologicals.* Rahway, NJ: Merck and Co., Inc., 1989. 596.

7. Gosselin, R. E., R. P. Smith, H. C. Hodge. *Clinical Toxicology of Commercial Products.* 5th ed. Baltimore: Williams and Wilkins, 1984, p. II–201.

8. (1) CLOGP. *PCGEMS Graphical Exposure Modeling System USEPA.* (1986). (2) Lyman, W. J., et al., *Handbook of Chemical Property Estimation Methods.* New York: McGraw-Hill, p. 5–5 (1982). (3) Riddick, J. A., et al., *Organic Solvents* New York: John Wiley and Sons Inc. (1984).

■ 5-ETHYLIDENE-2-NORBORNENE

CAS # 16219–75–3

DOT # Data not available

SYNONYMS 2-norbornene, 5-ethylidene • 5-ethylidenebicyclo (2,2,1) hep-2-ene • 5-ethylidenebicyclo (2.2.1) hept-2-ene • bicyclo (2.2.1) hept-2-ene, 5-ethylidene • enb • ethylidenenorbornene

MF C_9H_{12}

MW 120.21

COLOR AND FORM Colorless liquid; white liquid.

ODOR Turpentine-like.

ODOR THRESHOLD Between 0.007 and 0.014 ppm [R3].

DENSITY 0.8958 at 20°C.

BOILING POINT 67°C at 50 mm Hg.

MELTING POINT –80°C.

HEAT OF COMBUSTION (est) –10,450 cal/g.

FIREFIGHTING PROCEDURES Extinguish with dry chem, alcohol foam, or carbon dioxide. Water may be ineffective on fire. Cool exposed containers with water [R1].

VAPOR PRESSURE 4.2 mm Hg at 20°C.

OTHER CHEMICAL/PHYSICAL PROPERTIES Saturated air at 20°C and 1 atm contains estimated 5,526 ppm.

PROTECTIVE EQUIPMENT Organic canister or air-supplied mask; goggles or face shield; rubber gloves [R1].

CLEANUP Avoid contact with liquid and vapor. Keep people away. Shut off ignition sources. Call fire department. Stop discharge if possible. Stay upwind. Use water spray to "knock down" vapor. Isolate and remove discharged material [R1].

COMMON USES As third monomer in epdm (ethylene-propylene diene monomer) elastomers [R2].

OCCUPATIONAL RECOMMENDATIONS None listed.

THRESHOLD LIMIT Ceiling limit 5 ppm, 25 mg/m³ (1977) [R5].

NON-HUMAN TOXICITY LDLo rats oral 3,080 mg/kg [R4]; LCLo rats inhalation 963 ppm [R4]; LCLo mice inhalation 529 ppm [R4]; LDLo rabbits percutaneous 5,666 mg/kg [R4].

GENERAL RESPONSE Avoid contact with liquid and vapor. Keep people away. Shut off ignition sources. Call fire department. Stop discharge if possible. Stay upwind. Use water spray to "knock down" vapor. Isolate and remove discharged material. Notify local health and pollution control agencies.

FIRE RESPONSE Combustible. Extinguish with dry chemicals, alcohol foam, or carbon dioxide. Water may be ineffective on fire. Cool exposed containers with water.

EXPOSURE RESPONSE Call for medical aid. Vapor is irritating to eyes, nose, and throat. If inhaled will cause headache, coughing, or difficult breathing. If breathing has stopped, give artificial respiration. If breathing is difficult, give oxygen. Liquid is poisonous if swallowed. Irritating to skin and eyes. If swallowed will cause nausea and vomiting. Remove contaminated clothing and shoes. Flush affected areas with plenty of water. If in eyes, hold eyelids open, and flush with plenty of water. If swallowed and victim is conscious, have victim drink water or milk, and have victim induce vomiting. If swallowed and victim is unconscious or having convulsions, do nothing except keep victim warm.

RESPONSE TO DISCHARGE Mechanical containment. Should be removed. Provide chemical and physical treatment.

WATER RESPONSE Effect of low concentrations on aquatic life is unknown. Fouling to shoreline. May be dangerous if it enters water intakes. Notify local health and wildlife officials. Notify operators of nearby water intakes.

EXPOSURE SYMPTOMS Inhalation of vapors causes headache, confusion, and respiratory distress. Ingestion causes irritation of entire digestive system. Aspiration causes severe pneumonia. Contact with liquid causes irritation of eyes and skin.

EXPOSURE TREATMENT Inhalation: remove victim to fresh air; administer artificial respiration and oxygen if required; call a doctor. Ingestion: give large amount of water and induce vomiting; get medical attention at once. Eyes: flush with water for at least 15 min. Skin: wipe off, wash with soap and water.

REFERENCES

1. U.S. Coast Guard, Department of Transportation. *CHRIS—Hazardous Chemical Data*. Manual Two. Washington, DC: U.S. Government Printing Office, Oct., 1978.

2. Hawley, G. G. *The Condensed Chemical Dictionary*. 9th ed. New York: Van Nostrand Reinhold Co., 1977. 364.

3. American Conference of Governmental Industrial Hygienists, Inc. *Documentation of the Threshold Limit Values*, 4th ed., 1980. Cincinnati, OH: American Conference of Governmental Industrial Hygienists, Inc., 1980. 188.

4. Kincaid, E.R., et al., *Toxicol Appl Pharmacol* 20: 250 (1971).

5. American Conference of Governmental Industrial Hygienists. *Threshold Limit Values for Chemical Substances and Physical Agents and Biological Exposure Indices for 1994–1995*. Cincinnati, OH: ACGIH, 1994. 21.

■ ETHYLMERCURIC PHOSPHATE

CAS # 2235–25–8

SYNONYMS ethylmercury phosphate • lignasan • new improved ceresan • new improved granosan • ethyl mercury phosphate • ethyl mercuric phosphate

MF C₂H₇HgO₄P

MW 326.65

COLOR AND FORM White powder.

ODOR Garlic-like odor.

ODOR THRESHOLD The American National Standards Institute (ANSI) states that alkyl mercury compounds "are disagreeable in odor." (organo (alkyl) mercury) [R5, 3].

SOLUBILITY Soluble in water.

OTHER CHEMICAL/PHYSICAL PROPERTIES Ethylmercury radicals are strongly alkaline and form highly ionized salts that are generally water-soluble and appreciably volatile. These compounds are quantitatively decomposed by strong acids.

PROTECTIVE EQUIPMENT If the use of respirators is necessary, the only respirators permitted are those that have been approved by the Mine Safety and Health Admin (MSHA) (formerly Mining Enforcement and Safety Admin) or by the NIOSH (organo (alkyl) mercury) [R5, 3]. Respirator selection: upper-limit devices recommended by OSHA: For concn up to 0.1 mg/m^3, use any supplied-air respirator or any self-contained breathing apparatus. For concn up to 0.25 mg/m^3, use any supplied-air respirator operated in a continuous-flow mode. For concentration up to 0.5 mg/m^3, use any supplied-air respirator with a full facepiece, or any self-contained breathing apparatus with a full facepiece, or any supplied-air respirator with a tight-fitting facepiece operated in a continuous-flow mode. For concentration up to 10 mg/m^3, use any self-contained breathing apparatus with a full facepiece and operated in a pressure-demand or other positive-pressure mode. For emergency or planned entry in unknown concn or IDLH conditions, use any self-contained breathing apparatus with a full facepiece and operated in a pressure-demand or other positive-pressure mode, or any supplied-air respirator with a full facepiece and operated in a pressure-demand or other positive-pressure mode in combination with an auxiliary self-contained breathing apparatus operated in pressure-demand or other positive-pressure mode. Escape: use any appropriate escape-type self-contained breathing apparatus (mercury, (organo) alkyl compounds (as Hg)) [R6, 153]. Employees should be provided with and required to use impervious clothing, gloves, face shields (8-inch minimum), and other appropriate protective clothing necessary to prevent any possibility of skin contact with organo (alkyl) mercury or liquids containing organo (alkyl) mercury. Employees should be provided with and required to use dust- and splash-proof safety goggles where there is any possibility of organo (alkyl) mercury or liquids containing organo (alkyl) mercury contacting the eyes (organo (alkyl) mercury) [R5, 3].

PREVENTATIVE MEASURES Good industrial hygiene practices recommend that engineering controls be used to reduce environmental concentration to the permissible exposure level. However, there are some exceptions where respirators

may be used to control exposure. Respirators may be used when engineering and work practice are not technically feasible, when such controls are in the process of being installed, or when they fail and need to be supplemented. Respirators may also be used for operations which require entry into tanks or closed vessels, and in emergency situations. In addition to respirator selection, a complete respiratory program should be instituted which includes regular training, maintenance, inspection, cleaning, and evaluation (organo (alkyl) mercury) [R5, 3].

If employees' clothing has become contaminated with organo (alkyl) mercury or liquids containing organo (alkyl) mercury, employees should change into uncontaminated clothing before leaving the work premises. Clothing contaminated with organo (alkyl) mercury should be placed in closed containers for storage until it can be discarded or until provision is made for the removal of organo (alkyl) mercury from the clothing. If the clothing is to be laundered or otherwise cleaned to remove the organo (alkyl) mercury, the person performing the operation should be informed of organo (alkyl) mercury's hazardous properties. Non-impervious clothing that becomes contaminated should be removed immediately and not reworn until the organo (alkyl) mercury is removed (organo (alkyl) mercury) [R5, 3]. Where there is possibility of exposure of an employee's body to organo (alkyl) mercury or liquids containing organo (alkyl) mercury, facilities for quick drenching of the body and an eye-wash fountain should be provided within the immediate work area for emergency use (organo (alkyl) mercury) [R5, 3]. Eating and smoking should not be permitted in areas where organo (alkyl) mercury or liquids containing organo (alkyl) mercury are handled, processed, or stored. Employees should wash their hands thoroughly with soap or mild detergent and water before eating, smoking, or using toilet facilities (organo (alkyl) mercury) [R5, 4]. Contact lenses should not be worn when working with this chemical (mercury, (organo) alkyl compounds (as Hg)) [R6, 153]. Contact lens use in industry is controversial. A survey of 100 corporations resulted in the recommendation that each company establish their own contact lens use policy. One presumed hazard of contact lens use is possible chemical entrapment. Many authors found that contact lens minimized injury or protected the eye. The eye was afforded more protection from liquid irritants. It was found that soft contact lens do not worsen corneal damage from strong chemicals and in some cases could actually protect the eye. Overall, the literature supports the wearing of contact lenses in industrial environments as part of the standard eye protection, e.g. face shields; however, more data are needed to establish the value of contact lenses. [R7].

CLEANUP If organo (alkyl) mercury compounds are spilled or leaked, the following steps should be taken: (1) Remove all ignition sources. (2) Ventilate area of spill or leak. (3) If in the solid form, collect for reclamation, or disposal in sealed containers in a secured sanitary landfill. (4) If in the liquid form, for small quantities, absorb on paper towels. Evaporate in a safe place (such as a fume hood). Allow sufficient time for evaporating vapors to completely clear the hood ductwork. Burn the paper in a suitable location away from combustible materials. Large quantities can be collected and reclaimed or collected for reclamation or disposal in sealed containers in a secured sanitary landfill. (organo (alkyl) mercury) [R5, 4].

DISPOSAL SRP: At the time of review, criteria for land treatment or burial (sanitary landfill) disposal practices are subject to significant revision. Prior to implementing land disposal of waste residue (including waste sludge), consult with environmental regulatory agencies for guidance on acceptable disposal practices. Organo (alkyl) mercury may be disposed of: (1) If in the solid form, by collecting for reclamation or for disposal in sealed containers in a secured sanitary landfill. (2) If in the liquid form, for small quantities, absorb on paper towels. Evaporate in a safe place (such as a fume hood). Allow sufficient time for evaporating vapors to completely clear the hood ductwork. Burn the paper

in a suitable location away from combustible materials. Large quantities can be collected and reclaimed or collected for reclamation or disposal in sealed containers in a secured sanitary landfill (organo (alkyl) mercury) [R5, 5].

COMMON USES Seed fungicide (former use) and timber preservative [R1].

OCCUPATIONAL RECOMMENDATIONS None listed.

OSHA Permissible exposure limit (PEL) 0.01 mg/m³ for 8-hour time-weighted average (TWA). Ceiling concentration, 0.04 mg/m³. (mercury, (organo) alkyl compounds (as Hg)) [R6, 152].

THRESHOLD LIMIT 8–hr time-weighted average (TWA) 0.1 mg/m³, skin (1982) (mercury aryl compounds, as Hg) [R10, 24]. BEI (Biological Exposure Index): Total inorganic mercury in urine (preshift): 35 µg/g creatinine (mercury) [R10, 60]. BEI (Biological Exposure Index): Total inorganic mercury in blood (end of shift at end of workweek): 15 µg/L. (mercury) [R10, 60].

INTERNATIONAL EXPOSURE LIMITS East Germany reportedly has a limit of 0.05 mg/m³; other countries have generally adopted the 0.01 mg/m³ TLV. (mercury, alkyl compounds) [R3].

IARC SUMMARY Evaluation: There is inadequate evidence in humans for the carcinogenicity of mercury and mercury compounds. Overall evaluation: metallic mercury and inorganic mercury compounds are not classifiable as to their carcinogenicity to humans (Group 3) [R2].

BIOLOGICAL HALF-LIFE Half-lives may be different in different species. Initial half-lives of ethyl mercury were 4.4 days in mice and 7 to 10 days in rats of different ages. (organic mercury compounds) [R10, 16].

BIODEGRADATION A strain of *Pseudomonas*, isolated from soil, was found to be resistant to organic and inorganic mercurials. The organism adsorbed large amounts of mercury on the cell surface from culture media containing mercurials. Vaporization of the adsorbed mercury was induced. In addition to metallic mercury, ethane was produced when ethyl mercury phosphate was aerobically incubated with the organism. A cell-free extract of this organism gave similar results. The studies indicated that as sulfhydryl compound and NADH were required [R4]. In soil all the organomercury compounds are decomposed to mercury salts or to metallic mercury, which are the active fungicides. It has been suggested that this decomposition takes place through base exchange to form organomercury clays, which subsequently form mercury salts by further base exchange. These mercuric salts are then reduced to mercurous salts and to mercury. The metallic mercury liberated in the soil is ultimately converted to mercury sulfide by reaction with hydrogen sulfide liberated by soil microorganisms (mercury compounds) [R8, 22].

FIFRA REQUIREMENTS All uses of mercury are cancelled except the following: (1) as a fungicide in the treatment of textiles and fabrics intended for continuous outdoor use; (2) as a fungicide to control brown mold on freshly sawn lumber; (3) as a fungicide treatment to control Dutch elm disease; (4) as an in-can preservative in water based paints and coatings; (5) as a fungicide in water-based paints and coatings used for exterior application; (6) as a fungicide to control "winter turf diseases" such as *Sclerotinia boreales*, and gray and pink snow mold subject to the following: (a) the use of these products shall be prohibited within 25 feet of any water body where fish are taken for human consumption; (b) these products can be applied only by or under the direct supervision of golf course superintendents; (c) the products will be classified as restricted-use pesticides when they are reregistered and classified in accordance with section 4(c) of FEPCA. (mercury) [R9].

REFERENCES

1. Sax, N. I., and R. J. Lewis, Sr. (eds.). *Hawley's Condensed Chemical Dictionary.* 11th ed. New York: Van Nostrand Reinhold Co., 1987. 496.

2. IARC. *Monographs on the Evaluation of the Carcinogenic Risk of Chemicals to Man.* Geneva: World Health Organization, International Agency for Research on Cancer, 1972–present (multivolume work), p. 58 324 (1993).

3. American Conference of Governmental Industrial Hygienists. *Documentation of the Threshold Limit Values and Biological Exposure Indices.* 5th ed. Cincinnati, OH: American Conference of Governmental Industrial Hygienists, 1986. 360.

4. Menzie, C. M. *Metabolism of Pesticides, An Update.* U. S. Department of the Interior, Fish and Wildlife Service, Special Scientific Report—*Wildlife* No. 184, Washington, DC: U.S. GovernmentPrinting Office, 1974. 246.

5. Mackison, F. W., R. S. Stricoff, and L. J. Partridge, Jr. (eds.). *NIOSH/OSHA—Occupational Health Guidelines for Chemical Hazards.* DHHS (NIOSH) Publication No. 81–123 (3 vols). Washington, DC: U.S. Government Printing Office, Jan. 1981.

6. NIOSH. *Pocket Guide to Chemical Hazards.* 2nd Printing. DHHS (NIOSH) Publ. No. 85–114. Washington, DC: U.S. Dept. of Health and Human Services, NIOSH/Supt. of Documents, GPO, February 1987.

7. Randolph, S. A., M. R. Zavon. *J Occup Med* 29: 237–42 (1987).

8. White-Stevens, R. (ed.). *Pesticides in the Environment.*: Volume 1, Part 1, Part 2. New York: Marcel Dekker, Inc., 1971.

9. Environmental Protection Agency/OPTS. *Suspended, Cancelled and Restricted Pesticides.* 3rd Revision. Washington, DC: Environmental Protection Agency, January 1985. 16.

10. American Conference of Governmental Industrial Hygienists. *Threshold Limit Values for Chemical Substances and Physical Agents and Biological Exposure Indices for 1994–1995.* Cincinnati, OH: ACGIH, 1994.

ETHYL NITRITE

CAS # 109–95–5

DOT # 1194

SYNONYMS hyponitrous ether • nitrosyl ethoxide • nitrous acid ethyl ester • nitrous acid, ethyl ester • nitrous ethyl ether • spirit of ethyl nitrite • sweet spirit of niter • sweet spirit of nitre

MF $C_2H_5NO_2$

MW 75.08

COLOR AND FORM Colorless or yellowish, clear liquid.

ODOR Sweet, rum-like odor.

TASTE Fruity taste; burning, sweetish taste.

DENSITY 0.90 at 15°C.

SURFACE TENSION 30 dynes/cm = 0.030 N/m at 20°C.

BOILING POINT 17°C.

MELTING POINT FP: −58°F = −50°C = 233 K.

HEAT OF COMBUSTION −7,800 Btu/lb = −4,300 cal/g = −180×10⁵ J/kg.

HEAT OF VAPORIZATION 229 Btu/lb = 127 cal/g = 5.32×10⁵ J/kg.

AUTOIGNITION TEMPERATURE 194°F (decomposes) [R8, p. 325M–106].

EXPLOSIVE LIMITS AND POTENTIAL Severe, especially at greater than 90°C [R7].

FLAMMABILITY LIMITS Lower, 4.0%; upper, 50.0% by volume in air [R8, p. 325M–106].

FIREFIGHTING PROCEDURES Fight fires from an explosion-resistant location. In advanced or massive fires, the area should be evacuated. If a fire occurs in the vicinity of this material water should be used to keep containers cool. [R8, p. 49–161].

SOLUBILITY Slightly soluble in water; miscible with alcohol, ether; miscible with alcohol, ether.

STABILITY AND SHELF LIFE On keeping, it gradually decomposes, becoming acid, and oxides of nitrogen form. Decomposition is hastened by air, light, and moisture [R2].

OTHER CHEMICAL/PHYSICAL PROPERTIES Decomposition in water; decomposition spontaneously at 194°F (90°C); very mo-

bile; forms azeotropic mixture with isopentane (85%), amyl bromide (40%), and carbon disulfide (96%); floats on water; may boil on water; liq-water interfacial tension: 35 dynes/cm = 0.035 N/m at 20°C.

PROTECTIVE EQUIPMENT Self-contained breathing apparatus; goggles or face shield; rubber gloves [R3].

COMMON USES For preparing spirit nitrous ether [R2]; chemical intermediate; flavor in foods and beverages [R1]; organic reactions [R4]; flavors useful in rum, brandy, fruit flavors [R5].

OCCUPATIONAL RECOMMENDATIONS None listed.

DOT *Health Hazards:* Poisonous; may be fatal if inhaled, swallowed, or absorbed through skin. Contact may cause burns to skin and eyes. Runoff from fire control or dilution water may cause pollution [R6].

Fire or Explosion: Extremely flammable; may be ignited by heat, sparks, or flames. Vapors may travel to a source of ignition, and flash back. Container may explode violently in heat of fire. Vapor explosion and poison hazard indoors, outdoors, or in sewers. Runoff to sewer may create fire or explosion hazard [R6].

Emergency Action: Keep unnecessary people away; isolate hazard area and deny entry. Stay upwind; keep out of flow areas. Self-contained breathing apparatus and chemical protective clothing that is specifically recommended by the shipper or producer may be worn but they do not provide thermal protection unless it is stated by the clothing manufacturer. Structural firefighter's protective clothing is not effective with these materials. Evacuate the leak or spill area immediately for at least 50 feet in all directions. Isolate for 1/2 mile in all directions if tank car or truck is involved in fire. Call CHEMTREC at 1–800–424–9300 for emergency assistance. If water pollution occurs, notify the appropriate authorities [R6].

Fire: Small Fires: Dry chemical, CO_2, halon, water spray, or standard foam. Large Fires: Water spray, fog, or standard foam is recommended. Do not get water inside container. Cool containers that are exposed to flames with water from the side until well after fire is out. Stay away from ends of tanks. For massive fire in cargo area, use unmanned hose holder, or monitor nozzles; if this is impossible, withdraw from area, and let fire burn. Withdraw immediately in case of rising sound from venting safety device or any discoloration of tank due to fire [R6].

Spill or Leak: Shut off ignition sources; no flares, smoking, or flames in hazard area. Do not touch spilled material; stop leak if you can do so without risk. Use water spray to reduce vapor; do not get water inside container. Small Spills: Flush area with flooding amounts of water. Large Spills: Dike far ahead of liquid spill for later disposal [R6].

First Aid: Move victim to fresh air and call for emergency medical care; if not breathing, give artificial respiration; if breathing is difficult, give oxygen. Remove and isolate contaminated clothing and shoes at the site. In case of contact with material, immediately flush skin or eyes with running water for at least 15 minutes. Keep victim quiet and maintain normal body temperature. Effects may be delayed; keep victim under observation [R6].

FIRE POTENTIAL Highly dangerous when exposed to heat or flame. A powerful oxidizer [R7].

FDA 21 CFR 172.515: permitted synthetic flavoring additives and adjuvants for use in foods, under GMP. Ethyl nitrite is as synthetic flavor ingredient useful in rum, brandy, and fruit flavors [R5].

GENERAL RESPONSE Shut off ignition sources. Call fire department. Stop discharge if possible. Keep people away. Isolate and remove discharged material. Notify local health and pollution control agencies.

FIRE RESPONSE Flammable. Poisonous gases are produced in fire. Containers may explode in fire. Flashback along vapor trail may occur. Vapor may explode if ignited in an enclosed area. Wear goggles and self-contained breathing apparatus.

Combat fires from safe distance or protected location. Extinguish with water, dry chemicals, foam, or carbon dioxide. Cool exposed containers with water.

EXPOSURE RESPONSE Call for medical aid. Vapor if inhaled will cause headache, dizziness, or loss of consciousness. Move victim to fresh air. If breathing has stopped, give artificial respiration. If breathing is difficult, give oxygen. Liquid if swallowed will cause headache or loss of consciousness. If swallowed, and victim is conscious, have victim drink water or milk.

RESPONSE TO DISCHARGE Issue warning—high flammability. Restrict access. Evacuate area. Disperse and flush.

WATER RESPONSE Effect of low concentrations on aquatic life is unknown. May be dangerous if it enters water intakes. Notify local health and wildlife officials. Notify operators of nearby water intakes.

EXPOSURE SYMPTOMS Inhalation or ingestion causes headache, increased pulse rate, decreased blood pressure, and unconsciousness. Contact with liquid irritates eyes and skin.

EXPOSURE TREATMENT Inhalation: remove victim from exposure; if breathing has stopped, give artificial respiration; call physician. Eyes: flush with water for at least 15 min; get medical attention if irritation persists. Skin: flush with water, wash with soap and water. Ingestion: do not induce vomiting; call physician.

REFERENCES

1. SRI.

2. *The Merck Index.* 10th ed. Rahway, NJ: Merck Co., Inc., 1983. 554.

3. U.S. Coast Guard, Department of Transportation. *CHRIS—Hazardous Chemical Data. Manual Two.* Washington, DC: U.S. Government Printing Office, Oct., 1978.

4. Hawley, G. G. *The Condensed Chemical Dictionary.* 10th ed. New York: Van Nostrand Reinhold Co., 1981. 441.

5. Furia, T. E. (ed.). *CRC Handbook of Food Additives.* 2nd ed. Volume 2. Boca

Raton, Florida: CRC Press, Inc., 1980. 271.

6. Department of Transportation. *Emergency Response Guidebook 1987.* DOT P 5800.4. Washington, DC: U.S. Government Printing Office, 1987, p. G–30.

7. Sax, N. I. *Dangerous Properties of Industrial Materials.* 5th ed. New York: Van Nostrand Reinhold, 1979. 670.

8. National Fire Protection Association. *Fire Protection Guide on Hazardous Materials.* 7th ed. Boston, Mass.: National Fire Protection Association, 1978.

■ETHYL SILICATE

CAS # 78–10–4

DOT # 1292

SYNONYMS dynasil-a • es-100 • es-28 • es-28 (ester) • ethyl orthosilicate • etylu-krzemian (Polish) • orthosilicic acid, tetraethyl ester • silane, tetraethoxy • silicate d'ethyle (French) • silicate tetraethylique (French) • silicic acid, tetraethyl ester • silicon ethoxide • silicon tetraethoxide • silikan-l • teos • tetraethoxysilane • tetraethoxysilicon • tetraethyl silicate • tetraethyl-o-silicate

MF $C_8H_{20}O_4Si$

MW 208.37

COLOR AND FORM Colorless liquid; water-white.

ODOR Sharp, ester-like odor; faint odor.

ODOR THRESHOLD Water: decomposes in water; air: 17 µl/L; odor safety Class D; D = 10–50% of distracted persons can detect tlv concentration in the air [R6].

DENSITY 0.933 at 20°C.

SURFACE TENSION 22.8 dynes/cm = 0.0228 N/m at 20°C.

VISCOSITY 0.0179 poise at 20°C.

BOILING POINT 168.8°C at 760 mm Hg.

MELTING POINT −82.5°C.

HEAT OF COMBUSTION (est) −12,000 Btu/lb = −6,700 cal/g = −280×10⁵ J/kg.

HEAT OF VAPORIZATION 95 Btu/lb = 53 cal/g = 2.2×10⁵ J/kg.

EXPLOSIVE LIMITS AND POTENTIAL High temperatures may cause containers to explode [R4].

FLAMMABILITY LIMITS 1.3%–23% [R2].

FIREFIGHTING PROCEDURES Normal firefighting procedures may be used [R3]. Water, foam, dry chemical, carbon dioxide [R2].

TOXIC COMBUSTION PRODUCTS Toxic gases and vapors (such as carbon monoxide) may be released in a fire involving ethyl silicate [R4].

SOLUBILITY Practically insoluble in water; soluble in ether; slightly soluble in benzene; miscible with alcohol.

VAPOR PRESSURE 1.0 mm Hg at 20°C.

OTHER CHEMICAL/PHYSICAL PROPERTIES Hydrolyzed to an adhesive form of silica; wt/gal 7.8 lb at 20°C.

PROTECTIVE EQUIPMENT Rubber or polyethylene gloves; safety glasses or other form of eye protection; self-contained breathing apparatus or one that absorbs organic vapors [R2].

PREVENTATIVE MEASURES SRP: contaminated protective clothing should be segregated in such a manner so that there is no direct personal contact by personnel who handle, dispose of, or clean the clothing. Quality assurance to ascertain the completeness of the cleaning procedures should be implemented before the decontaminated protective clothing is returned for reuse by the workers. Contaminated clothing should not be taken home at end of shift, but should remain at employee's place of work for cleaning.

CLEANUP 1. Remove all ignition sources. 2. Ventilate area of spill or leak. 3. For small quantities, absorb on paper towels. Evaporate in safe place (such as a fume hood). Allow sufficient time for evaporating vapors to completely clear hood ductwork. Burn paper in suitable location away from combustible materials. Ethyl silicate should not be allowed to enter a confined space, such as a sewer, because of the possibility of an explosion. Sewers designed to preclude formation of explosive concentration of ethyl silicate vapors are permitted [R4].

DISPOSAL SRP: At the time of review, criteria for land treatment or burial (sanitary landfill) disposal practices are subject to significant revision. Prior to implementing land disposal of waste residue (including waste sludge), consult with environmental regulatory agencies for guidance on acceptable disposal practices.

COMMON USES In weatherproofing and for hardening stone, arresting decay, and disintegration; in manufacture of weatherproof and acid-proof mortars and cements; in the "lost wax" process for casting of high-melting alloys [R1].

OCCUPATIONAL RECOMMENDATIONS None listed.

OSHA 8-hour time-weighted average: 100 ppm (850 mg/m³) [R7].

NIOSH 10-hr time-weighted average: 10 ppm (85 mg/m³) [R5].

THRESHOLD LIMIT 8-hr time-weighted average (TWA) 10 ppm, 85 mg/m³ (1986) [R8, 21].

GENERAL RESPONSE Stop discharge if possible. Keep people away. Call fire department. Isolate and remove discharged material. Notify local health and pollution control agencies.

FIRE RESPONSE Combustible. Extinguish with water, dry chemicals, foam, or carbon dioxide.

EXPOSURE RESPONSE Call for medical aid. Liquid irritating to eyes. If swallowed will cause nausea and vomiting. Remove contaminated clothing and shoes. Flush affected areas with plenty of water. If swallowed and victim is conscious, have victim drink water or milk, and have victim induce vomiting. If swallowed and victim is unconscious or having convulsions, do nothing except keep victim warm.

RESPONSE TO DISCHARGE Disperse and flush.

WATER RESPONSE Effect of low concentrations on aquatic life is unknown. Fouling to shoreline. May be dangerous if it enters water intakes. Notify local health and wildlife officials. Notify operators of nearby water intakes.

EXPOSURE SYMPTOMS Inhalation of vapor causes eye and nose irritation, unsteadiness, tremors, salivation, respiratory difficulty, and unconsciousness. Contact with liquid irritates eyes and may cause dryness, cracking, and inflammation of skin. Ingestion may produce nausea, vomiting, and cramps.

EXPOSURE TREATMENT Inhalation: move patient from contaminated atmosphere; if breathing has ceased, start mouth-to-mouth artificial respiration; oxygen, if available, should be administered only by an experienced person when authorized by a physician; keep patient warm and comfortable; call physician immediately. Eyes: flush immediately with large quantities of running water for at least 15 min; obtain medical attention if irritation persists. Skin: immediately flush affected areas with large volumes of water; obtain medical attention if irritation persists. Ingestion: give large amounts of water or warm salty water and induce vomiting; milk, eggs, or olive oil may then be given; obtain medical attention if abdominal discomfort persists.

REFERENCES

1. Budavari, S. (ed.). *The Merck Index— Encyclopedia of Chemicals, Drugs, and Biologicals.* Rahway, NJ: Merck and Co., Inc., 1989. 605.

2. U.S. Coast Guard, Department of Transportation. *CHRIS—Hazardous Chemical Data. Volume II.* Washington, DC: U.S. Government Printing Office, 1984–5.

3. National Fire Protection Association. *Fire Protection Guide on Hazardous Materials.* 7th ed. Boston, MA: National Fire Protection Association, 1978, p. 325M–108.

4. Mackison, F. W., R. S. Stricoff, and L. J. Partridge, Jr. (eds.). *NIOSH/OSHA—Occupational Health Guidelines for Chemical Hazards.* DHHS (NIOSH) Publication

No. 81–123 (3 vols). Washington, DC: U.S. Government Printing Office, Jan. 1981.

5. NIOSH. *NIOSH Pocket Guide to Chemical Hazards*. DHHS (NIOSH) Publication No. 90–117. Washington, DC: U.S. Government Printing Office, June 1990, 114.

6. Amoore, J. E., E. Hautala. *J Appl Toxicol* 3 (6): 276 (1983).

7. 29 CFR 1910.1000 (58 FR 35338 (6/30/93)).

8. American Conference of Governmental Industrial Hygienists. *Threshold Limit Values for Chemical Substances and Physical Agents and Biological Exposure Indices for 1994–1995*. Cincinnati, OH: ACGIH, 1994.

∎FENBUTATIN OXIDE

CAS # 13356–08–6

SYNONYMS bendex • bis(tris(beta,beta-dimethylphenethyl)tin)oxide • bis(tris(2-methyl-2-phenylpropyl)tin)oxide • di(tri-(2,2-dimethyl-2-phenylethyl)tin)oxide • ent-27738 • fenbutatin oxide • fenbutatin-oxyde • fenylbutatin oxide • fenylbutylstannium oxide (Czech) • hexakis(beta,beta-dimethylphenethyl)-distannoxane • hexakis (2-methyl-2-phenylpropyl)-distannoxane • hexakis(2-methyl-2-phenylpropyl)distannoxane • osdaran • sd-14114 • shell sd-14114 • torque • vendex

MF $C_{60}H_{37}OSn_2$

MW 1,052.66

COLOR AND FORM White crystalline powder.

ODOR Mild odor.

BOILING POINT 235–240°C at 0.05 mm.

MELTING POINT 138–139°C.

FIREFIGHTING PROCEDURES If material is on fire or involved in fire, do not extinguish fire unless flow can be stopped. Use water in flooding quantities as fog. Solid streams of water may be ineffective. Cool all affected containers with flooding quantities of water. Apply water from as far a distance as possible. Use alcohol foam, dry chemical, or carbon dioxide (organotin pesticide, liquid) [R10, 514]. If material is on fire or involved in fire, extinguish fire using agent suitable for type of surrounding fire (material itself does not burn or burns with difficulty). Use water in flooding quantities as fog. Use "alcohol" foam, dry chemical, or carbon dioxide (organotin pesticide, solid) [R10, 515]. Wear positive-pressure self-contained breathing apparatus when fighting fires involving this material. (organotin pesticide, liquid, not otherwise specified (compounds and preparations)) (insecticides, not elsewhere classified) [R10, 514].

TOXIC COMBUSTION PRODUCTS Toxic gases and vapors may be released in a fire involving organic tin compounds. (organotin compounds) [R11, 3].

SOLUBILITY Solubility at 23°C: 0.005 mg/L water; 6 g/L acetone; 140 g/L benzene; 380 g/L dichloromethane. Very slightly soluble in aliphatic hydrocarbons and mineral oils; water solubility = 0.005 mg/L at 23°C.

OTHER CHEMICAL/PHYSICAL PROPERTIES Water causes conversion of fenbutatin oxide to tris (2-methyl-2-phenylpropyl) tin hydroxide, which is reconverted to parent compound slowly at room temperature and rapidly at 98°C. Triorganotin hydroxides behave not like alcohols, but more like inorganic bases, although strong bases remove the proton in certain triorganotin hydroxides because tin is amphoteric (triorganotin hydroxides). The bis (triorganotin) oxides are strong bases that react with inorganic and organic acids forming normal salt-like but nonconducting and water-insoluble compounds. They do not in the least resemble organic ethers, though they can occasionally form peroxides. (bis (triorganotin) oxides).

PROTECTIVE EQUIPMENT Wear appropriate chemical protective gloves, boots, and goggles. (organotin pesticide, liquid, not otherwise specified (compounds, and preparations) (insecticides, not elsewhere classified)) [R10, 514].

PREVENTATIVE MEASURES SRP: The scientific literature supports the wearing of contact lenses in industrial environments, as part of a program to protect the eye against chemical compounds and minerals causing eye irritation. However, there may be individual substances whose irritating or corrosive properties are such that the wearing of contact lenses would be harmful to the eye. In those specific cases contact lenses should not be worn.

CLEANUP If organic tin compounds are spilled or leaked: 1. Ventilate area of spill. 2. Collect spilled material in most convenient and safe manner and deposit in sealed containers for reclamation or for disposal. Liquid containing organic tin compounds should be absorbed in vermiculite, dry sand, earth, or similar material. (organic tin compounds (as tin)) [R2].

COMMON USES Control of the mobile stage of a wide range of phytophagus mites on citrus, glasshouse crops, ornamentally top fruit, vegetable, and vines [R1].

OCCUPATIONAL RECOMMENDATIONS None listed.

OSHA 8-hr time-weighted average: 0.1 mg/m^3. Final rule limits were achieved by any combination of engineering controls, work practices, and personal protective equipment during the phase-in period, Sept 1, 1989 through Dec 30, 1992. Final rule limits became effective Dec 31, 1992. Skin absorption designation in effect as of Sept 1, 1989. (tin, organic compounds, as Sn) [R7].

NIOSH Time-weighted average (TWA) recommended exposure limit (REL) 0.1 mg Sn/m^3 of air for exposures up to 10 hours (1976) (organotin compound) [R4]; threshold-limit 8-hr time-weighted average (TWA) 0.1 mg/m^3, skin; short-term exposure limit (STEL) 0.2 mg/m^3, skin (1992) (tin, organic compounds, as Sn) [R8].

INTERNATIONAL EXPOSURE LIMITS Threshold-limit values for hazardous substances in the air at the workplace: France 0.1 mg/m^3 (org tin compounds (as Sn)) [R12, 840].

NON-HUMAN TOXICITY LD$_{50}$ rat acute, oral 2,631 mg/kg [R3]; LD$_{50}$ mouse acute oral 1,450 mg/kg [R3]; LD$_{50}$ dog acute oral >1,500 mg/kg [R3]; LD$_{50}$ rabbit acute percutaneous >2,000 mg/kg [R3]; LD$_{50}$ rat acute percutaneous >1,000 mg/kg [R3].

ECOTOXICITY LC$_{50}$ rainbow trout 0.27 mg active ingredient [R3]; LD$_{50}$ bee acute oral >0.1 mg [R3].

BIODEGRADATION The aerobic and anaerobic biodegradation of fenbutatin oxide in soil and water should not be important (1) [R5].

BIOCONCENTRATION From the water solubility of 0.005 mg/L at 23°C (1) a log BCF of 4.09 has been estimated for fenbutatin oxide (2). The high BCF value indicates that bioconcentration of fenbutatin oxide

in aquatic organisms should be important (SRC) [R6].

DOT *Health Hazards:* Poisonous; may be fatal if inhaled, swallowed, or absorbed through skin. Contact may cause burns to skin and eyes. Runoff from fire control or dilution water may cause pollution (organotin pesticide, liquid, flammable, poisonous, not otherwise specified) [R9, p. G-28].

Fire or Explosion: Flammable/combustible material; may be ignited by heat, sparks, or flames. Vapors may travel to a source of ignition, and flash back. Container may explode in heat of fire. Vapor explosion and poison hazard indoors, outdoors, or in sewers. Runoff to sewer may create fire or explosion hazard (organotin pesticide, liquid, flammable, poisonous, not otherwise specified) [R9, p. G-28].

Emergency Action: Keep unnecessary people away; isolate hazard area, and deny entry. Stay upwind; keep out of low areas. Positive-pressure self-contained breathing apparatus (SCBA) and chemical protective clothing that is specifically recommended by the shipper or manufacturer may be worn. It may provide little or no thermal protection. Structural firefighters' protective clothing is not effective with these materials. Isolate for 1/2 mile in all directions if tank, rail car, or tank truck is involved in fire. Call CHEMTREC at 1-800-424-9300 for emergency assistance. If water pollution occurs, notify the appropriate authorities (organotin pesticide, liquid, flammable, poisonous, not otherwise specified) [R9, p. G-28].

Fire: Small Fires: Dry chemical, CO$_2$, water spray, or alcohol-resistant foam. Large Fires: Water spray, fog, or alcohol-resistant foam. Move container from fire area if you can do so without risk. Dike fire-control water for later disposal; do not scatter the material. Apply cooling water to sides of containers that are exposed to flames until well after fire is out. Stay away from ends of tanks. Withdraw immediately in case of rising sound from venting safety device or any discoloration of tank due to fire (organotin pesticide, liquid, flammable, poisonous, not otherwise specified) [R9, p. G-28].

Spill or Leak: Shut off ignition sources; no flares, smoking, or flames in hazard area. Fully encapsulating vapor-protective clothing should be worn for spills and leaks with no fire. Do not touch or walk through spilled material; stop leak if you can do so without risk. Water spray may reduce vapor, but it may not prevent ignition in closed spaces. Small Spills: Take up with sand or other noncombustible absorbent material and place into containers for later disposal. Large Spills: Dike far ahead of liquid spill for later disposal (organotin pesticide, liquid, flammable, poisonous, not otherwise specified) [R9, p. G–28].

First aid: Move victim to fresh air, and call for emergency medical care; if not breathing, give artificial respiration; if breathing is difficult, give oxygen. In case of contact with material, immediately flush skin or eyes with running water for at least 15 minutes. Remove and isolate contaminated clothing and shoes at the site. Keep victim quiet and maintain normal body temperature. Effects may be delayed; keep victim under observation. (organotin pesticide, liquid, flammable, poisonous, not otherwise specified) [R9, p. G–28].

REFERENCES

1. Worthing, C. R., and S. B. Walker (eds.). *The Pesticide Manual—A World Compendium.* 8th ed. Thornton Heath, UK: The British Crop Protection Council, 1987. 370.

2. Mackison, F. W., R. S. Stricoff, and L. J. Partridge, Jr. (eds.). *NIOSH/OSHA—Occupational Health Guidelines for Chemical Hazards.* DHHS (NIOSH) Publication No. 81–123 (3 vols). Washington, DC: U.S. Government Printing Office, Jan. 1981.

3. Hartley, D., and H. Kidd (eds.). *The Agrochemicals Handbook.* 2nd ed. Lechworth, Herts, England: The Royal Society of Chemistry, 1987, p. A 196/Aug 87.

4. NIOSH/CDC. *NIOSH Recommendations for Occupational Safety and Health Standards 1988,* Aug. 1988 (Suppl. to *Morbidity, and Mortality Wkly.* Vol. 37 No. 5-7, Aug. 26, 1988). Atlanta, GA:

National Institute for Occupational Safety and Health, CDC, 1988, p. V37 (S7) 22.

5. Lee, P. W., et al., Amer Chemical Soc Div Pest Chemical 188th ACS Natl Meet, Philadelphia, PA (1984).

6. (1) Worthing, C. R. *The Pesticide Manual,* 8th ed. The Lavenham Press Ltd, Lavenham, Suffolk, p. 370 (1987) (2) Lyman, W. J., et al., *Handbook of Chemical Property Estimation Methods,* New York: McGraw-Hill, p. 5–5 (1982).

7. 29 CFR 1910.1000 (7/1/90).

8. American Conference of Governmental Industrial Hygienists. *Threshold Limit Values for Chemical Substances and Physical Agents and Biological Exposure Indices for 1994–1995.* Cincinnati, OH: ACGIH, 1994. 34.

9. U.S. Department of Transportation. *Emergency Response Guidebook 1990.* DOT P 5800.5. Washington, DC: U.S. Government Printing Office, 1990.

10. Association of American Railroads. *Emergency Handling of Hazardous Materials in Surface Transportation.* Washington, DC: Assoc. of American Railroads, Hazardous Materials Systems (BOE), 1987.

11. Mackison, F. W., R. S. Stricoff, and L. J. Partridge, Jr. (eds.). *NIOSH/OSHA—Occupational Health Guidelines for Chemical Hazards.* DHHS(NIOSH) Publication No. 81–123 (3 vols). Washington, DC: U.S. Government Printing Office, Jan. 1981.

12. Seiler, H.G., H. Sigel, and A. Sigel (eds.). *Handbook on the Toxicity of Inorganic Compounds.* New York, NY: Marcel Dekker, Inc. 1988.

■ FENURON

CAS # 101–42–8

SIC CODES 2879

SYNONYMS N,N-dimethyl-N′-phenylurea • 1,2-dimethyl-3-phenylurea • PDU • 3 (phenyl)-1,1-dimethylurea • fenidin •

N,N-dimethyl-N-phenylurea • N-phenyl-N',N'-dimethylurea • dibar • 3-phenyl-1,1-dimethylurea • croptex-chrome • croptex-ruby • red • herbon-yellow • electrum • premalox • quintex • 1,1-dimethyl-3-phenylurea • fenulon • falisilvan • beet-kleen • dybar • phenyldimethylurea • fenidim • funulon • 1,1-dimethyl-3-phenylurea • N-phenyl-N-N-dimethylurea • urea, 1,1-dimethyl-3-phenyl

MF $C_9H_{12}N_2O$

MW 164.23

COLOR AND FORM White, crystalline solid; colorless crystalline solid; colorless crystals.

DENSITY 1.08 at 20°C.

MELTING POINT 133–134°C.

SOLUBILITY 3,850 ppm at 25°C; sparingly soluble in water (0.29% at 24°C); solubility (25°C): 3.85 g/L water; sparingly soluble in hydrocarbons. In ethanol 108.8, diethylether 5.5, acetone 80.2, benzene 3.1, chloroform 125, n-hexane 0.2, groundnut oil 1.0 (all in g/kg at 20–25°C); in water 3.85 g/L at 25°C.

VAPOR PRESSURE 1.6×10^{-4} mm Hg at 60°C.

CORROSIVITY Noncorrosive.

STABILITY AND SHELF LIFE Stable at neutral pH but hydrolyzed in acid and basic media. [R7, p. II–331].

OTHER CHEMICAL/PHYSICAL PROPERTIES Compatible with other wettable-powder herbicides; colorless crystals (fenuron trichloroacetate); molecular weight: 327.6 (fenuron trichloroacetate); melting point: 65–68°C (fenuron trichloroacetate); soluble in water 4.8 g/L at room temperature. Moderately soluble in acetone and aromatic solvents. sparingly soluble in petroleum oils (fenuron trichloroacetate); Henry's Law constant of 9.71×10^{-10} atm-m³/mole at 25°C (estimated).

DISPOSAL The products (of decomposition) are both toxic, thus precluding hydrolysis as a disposal method [R3]. Hydrolyzes at high temperature or in acid or alkali.

COMMON USES Weed and brush killer [R8, 987]; herbicide [R2]; control of woody plants and deep-rooted perennial weeds, particularly on non-crop land. Often used in combination with chloropropham to extend the weed spectrum and range of crops [R1].

PERSISTENCE Not persistent over 2 weeks in river water; persistent in soil for 8 months [R6].

MAJOR SPECIES THREATENED Marine life (shrimps). Livestock or people using water.

PERSONAL SAFETY Self-contained breathing apparatus must be worn. Rubber gloves, hats, suits, and boots must be worn.

ACUTE HAZARD LEVEL Extreme hazard to potable supplies and atmosphere when burned.

CHRONIC HAZARD LEVEL Not particularly high.

DEGREE OF HAZARD TO PUBLIC HEALTH Highly toxic in water or under disaster circumstances. May have mutagenic effects.

OCCUPATIONAL RECOMMENDATIONS None listed.

ACTION LEVEL Evacuate area. Enter from upwind side. Notify local air authority and National Agricultural Chemicals Association.

ON-SITE RESTORATION Carbon or peat may be used for adsorption. Seek professional environmental engineering assistance through EPA's Environmental Response team (ERT), Edison, NJ, 24-hour no., 908–548–8730.

AVAILABILITY OF COUNTERMEASURE MATERIAL Carbon—water treatment plants, sugar refineries; peat—nurseries, floral shops.

MAJOR WATER USE THREATENED Marine fisheries, potable supplies, farmstead water.

PROBABLE LOCATION AND STATE Will be dissolved in water. As time passes more will be dissolved as aniline than as original species. Will dissolve slowly.

FOOD-CHAIN CONCENTRATION POTENTIAL Negative.

NON-HUMAN TOXICITY LD_{50} rat acute oral 6,400 mg/kg [R8, 988]; LD_{50} rat, oral 7,500 mg/kg [R2]; LD_{50} rat oral 4,000–5,700 mg/kg (fenuron trichloroacetate) [R1].

ECOTOXICITY LC_{50} guppy 610 mg/L/48 hr (conditions of bioassay not specified) [R1].

BIODEGRADATION Based on extensive biological and chemical studies, the disappearance of fenuron from soil is due primarily to microbial degradation; however, no quantitative data were given (1). In a pure-culture study, fenuron was degraded 10% by the soil fungus *Rhizoctonia solani* after 6 days (2) [R4].

BIOCONCENTRATION Based on an estimated bioconcentration factor of 1–6 (1), fenuron is not expected to bioaccumulate in fish (SRC) [R5].

REFERENCES

1. Hartley, D., and H. Kidd (eds.). *The Agrochemicals Handbook*. 2nd ed. Lechworth, Herts, England: The Royal Society of Chemistry, 1987, p. A206 (1987).

2. Budavari, S. (ed.). *The Merck Index—Encyclopedia of Chemicals, Drugs, and Biologicals*. Rahway, NJ: Merck and Co., Inc., 1989. 629.

3. Sittig, M. (ed.) *Pesticide Manufacturing and Toxic Materials Control Encyclopedia*. Park Ridge, NJ: Noyes Data Corporation. 1980. 414.

4. (1) Abel, A. L., et al., *World Crops*. 328–30 (1957). (2) Weinberger, M., J. M. Bollag. *Appl Microbiol* 24: 750–4 (1972).

5. Kenaga, E. E., *Ecotox Environ Safety* 4: 26–38 (1980).

6. Eichelberger, J. W., J. J. Lichtenberg. "Persistence of Pesticides in River Water," *Environmental Science and Technology*, 1971, Vol. 5, No. 6. June.

7. Gosselin, R.E., R.P. Smith, H.C. Hodge. *Clinical Toxicology of Commercial Prod-*

ucts. 5th ed. Baltimore: Williams and Wilkins, 1984.

8. Verschueren, K. *Handbook of Environmental Data of Organic Chemicals*. 2nd ed. New York, NY: Van Nostrand Reinhold Co., 1983.

■ FERRIC AMMONIUM OXALATE

CAS # 2944–67–4

SIC CODES 2869

SYNONYMS ammonium ferric oxalate • ammonium trioxalate ferrate(III) • ammonium iron oxalate • ethanedioic acid, ammonium iron salt • oxalic acid, ammonium iron(3+) salt(3:3:1) • ethanedioic acid, ammonium iron(3+) salt (3:3:1) • ammonium oxalatoferrate(III)

MF $Fe(NH_4)_3(C_2O_4)_3 \cdot 3H_2O$

MW 428

BINARY REACTANTS Incompatible with strong oxidants.

STANDARD CODES EPA 311; TSCA; not listed IATA; not listed CFR 49; CFR 14 CAB code 8; not listed NFPA; not listed AAR; not listed ICC.

DIRECT CONTACT Eye, skin, and respiratory irritant.

GENERAL SENSATION Skin and eye contamination may be irritating. At fire temperatures produces coughing, choking, tearing, difficult breathing, and cyanosis. May produce kidney damage on prolonged contact.

PERSONAL SAFETY Protect against both inhalation and contact with the skin. Must wear protective clothing including NIOSH-approved rubber gloves and boots, safety goggles, or face mask, and either a respirator whose canister is specifically approved for this material, or a self-contained breathing apparatus in fire conditions. Decontaminate fully or dispose of all equipment and clothing after use.

ACUTE HAZARD LEVEL Orally, this iron salt is not strikingly toxic. When introduced directly into the bloodstream it is highly toxic. The bulk of the toxicological information on iron compounds is related to long-term exposures. Soluble iron salts such as this are cutaneous irritants, and their aerosols irritate the respiratory tract. The 96-hour LC_{50} aquatic toxicity to striped bass is 16 ppm, based on the toxicity of 4 ppm for the iron ion.

CHRONIC HAZARD LEVEL The inhalation of iron oxide fumes, a thermal decomposition product, may cause a benign pneumoconiosis in the lungs. The threshold-limit-value time-weighted average for a 40-hour workweek is 1 mg/m^3 (as iron) and the short-term exposure limit (15 minutes) is 2 mg/m^3. Iron oxide fumes can cause metal fume fever. Several iron compounds are suspected carcinogens of the lung, liver, connective tissue, and reticuloendothelial tissue.

DEGREE OF HAZARD TO PUBLIC HEALTH Moderately toxic, skin and mucous membrane irritant.

OCCUPATIONAL RECOMMENDATIONS None listed

ACTION LEVEL Avoid contact with the spilled cargo. Stay upwind. Notify local air, water, and fire authorities of the accident. Evacuate all people to a distance of at least 200 feet upwind and 1,000 feet downwind of the spill.

ON-SITE RESTORATION Dam stream if possible to reduce the flow and prevent further dispersion of the material by water movement. Pump water into a suitable container. Under controlled conditions add calcium hydroxide ($Ca(OH)_2$) to pH 9 and allow to settle. Reduce pH to 6–7 with hydrochloric acid (HCl) to precipitate iron. Filter, pass through an ion exchange, and neutralize with calcium hydroxide to pH 7. For more details see Envirex Manual EPA 600/2-77-227. Seek professional environmental engineering assistance through EPA's Environmental Response Team (ERT), Edison, NJ, 24-hour no., 908-548-8730.

BEACH OR SHORE RESTORATION Close beach and shore to the public until material has been removed. If tidal, scrape affected area at low tide with a mechanical scraper, and avoid human toxicity contact. Do not burn material.

DISPOSAL After the material has been contained, remove it and the contaminated soil and place in impervious containers. If practical, transport material back to the supplier for recovery. If this is not practical, the material should be buried in a specially designated chemical landfill. Amenable to biological treatment at a municipal sewage treatment plant.

DISPOSAL NOTIFICATION Notify local and state health authorities, local solid waste disposal authorities, supplier, and shipper.

INDUSTRIAL FOULING POTENTIAL May contaminate boiler feed water; hinders water softening.

MAJOR WATER USE THREATENED Fisheries; potable supply; industrial.

PROBABLE LOCATION AND STATE Bright-green granules, will sink to the bottom and dissolve.

COLOR IN WATER Greenish tint.

FOOD-CHAIN CONCENTRATION POTENTIAL Negative. Unlikely to accumulate in the food chain.

REFERENCES

1. DOT *CHRIS* database.

2. EPA *OHMTADS* database.

■ FLUCYTHRINATE

CAS # 70124–77–5

SYNONYMS cybolt • cythrin • 4-(difluoromethoxy)-alpha-(1-methylethyl) benzeneacetic acid cyano (3-phenoxy phenyl) methyl ester • (+ −)-cyano-(3-phenoxyphenyl) methyl (+)-4-(difluoromethoxy)-alpha-(methylethyl) benzeneacetate • ac-222705 • guardian • pay-off • stock-guard • (rs)-alpha-cyano-3-phenoxybenzyl (s)-2-(4-difluoromethoxyphenyl)-3-methylbutyrate • (rs)-cyano-(3-phenoxyphenyl) methyl (s)-4-(difluoromethoxy)-alpha-(1-methylethyl)-benzeneacetate

MF $C_{26}H_{23}F_2NO_4$

MW 451.48

COLOR AND FORM Viscous liquid.

DENSITY 1.189 at 22°C.

BOILING POINT 108°C at 0.35 mm Hg.

FIREFIGHTING PROCEDURES Use carbon dioxide, foam, or dry chemical (on fires involving pyrethroids) (pyrethrum) [R9, 2]. Firefighting: Self-contained breathing apparatus with a full facepiece operated in pressure-demand or other positive-pressure mode (pyrethrum) [R9, 5]. Extinguish fire using agent suitable for type of surrounding fire. (pyrethrins) [R2].

SOLUBILITY Soluble in acetone, xylene, 2-propanol; in water at 21°C, 0.5 mg/L. In corn oil 560, cottonseed oil 300, soy bean oil 300, hexane 90 (g/L at 21°C).

VAPOR PRESSURE 0.0012 mPa at 25°C.

STABILITY AND SHELF LIFE Stable to light [R1].

OTHER CHEMICAL/PHYSICAL PROPERTIES Dark-amber viscous liquid, with a faint ester-like odor (technical flucythrinate).

PROTECTIVE EQUIPMENT Employees should be provided with and required to use dust- and splash-proof safety goggles where pyrethroids may contact the eyes. (pyrethroids) [R9, 3].

PREVENTATIVE MEASURES Use with adequate ventilation, and wear goggles or a face shield when handling [R1].

CLEANUP Spillages of pesticides at any stage of their storage or handling should be treated with great care. Liquid formulations may be reduced to solid phase by evaporation. Dry sweeping of solids is always hazardous: These should be removed by vacuum cleaning, or by dissolving them in water or other solvent in the factory environment. (pesticides) [R3].

DISPOSAL Incineration would be an effective disposal procedure where permitted. If an efficient incinerator is not available, the product should be mixed with large amounts of combustible material, and contact with the smoke should be avoided. (pyrethrin products) [R4].

COMMON USES Insecticide; medication (vet): ectoparasiticide [R8, 645]; control of flies on cattle (Merck, 11th ed., p. 645).

NON-HUMAN TOXICITY LD_{50} rat male oral 81 mg/kg [R1].

ECOTOXICITY LD_{50} mallard duck oral >2,510 mg/kg [R1].

BIODEGRADATION In a laboratory study using sediment and seawater collected from a salt marsh near Escambia County, Florida, flucythrinate was observed to have a half-life of about 16 days (1); however, when the media were sterilized, flucythrinate showed no appreciable degradation after 28 days of incubation (1), thus suggesting that degradation was occurring through biotic means (SRC) [R5].

BIOCONCENTRATION Based upon a measured log K_{ow} of 6.20 (1) and a water solubility of 0.06 mg/L (taken from the USDA's evaluated database of pesticide properties) (2), the BCF for flucythrinate can be estimated to be 30,000 and 3,000, respectively, from recommended regression-derived equations (3,SRC). In a 28-day laboratory study, a steady-state BCF of 2,300 was measured in eastern oysters (*Crassostrea virginica*) (1) [R6].

FIFRA REQUIREMENTS Tolerances are established for residues of the insecticide flucythrinate in or on the following raw agricultural commodities: apples; cabbage; cattle, fat; cattle (meat by-products and meat); corn (fodder and forage); corn grain; cottonseed; goats, fat; goats (meat by-products and meat); hogs, fat; hogs (meat by-products and meat); horses, fat; horses (meat by-products, and meat; lettuce, head; milk; milk, fat; pears; sheep, fat; and sheep (meat by-products and meat) [R7].

REFERENCES

1. Hartley, D., and H. Kidd (eds.). *The Agrochemicals Handbook*. 2nd ed. Lech-

worth, Herts, England: The Royal Society of Chemistry, 1987, p. A550/Aug 87.

2. Bureau of Explosives; *Emergency Handling of Haz Matl in Surface Trans* p. 434 (1981).

3. International Labour Office. *Encyclopedia of Occupational Health and Safety.* Vols. I and II. Geneva, Switzerland: International Labour Office, 1983. 1619.

4. Sittig, M. *Handbook of Toxic and Hazardous Chemicals and Carcinogens, 1985.* 2nd ed. Park Ridge, NJ: Noyes Data Corporation, 1985. 762.

5. Schimmel S.C., et al., *J Agric Food Chem* 31: 104–113 (1983).

6. (1) Schimmel S.C., et al., *J Agric Food Chem* 31: 104–113 (1983) (2) Wauchope, R.D., et al., *Rev Environ Contam Toxicol* 123: 1–36 (1991) (3) Lyman, W. J., et al., *Handbook of Chemical Property Estimation Methods.* Washington, DC: Amer Chemical Soc pp. 5–4, 5–10 (1990).

7. 40 CFR 180.400 (7/1/91).

8. Budavari, S. (ed.). *The Merck Index—Encyclopedia of Chemicals, Drugs and Biologicals.* Rahway, NJ: Merck and Co., Inc., 1989.

9. Mackison, F. W., R. S. Stricoff, and L. J. Partridge, Jr. (eds.). *NIOSH/OSHA—Occupational Health Guidelines for Chemical Hazards.* DHHS(NIOSH) Publication No. 81–123 (3 vols). Washington, DC: U.S. Government Printing Office, Jan. 1981.

■ FLUOROBORIC ACID

CAS # 16872–11–0

SYNONYMS borate(1−), tetrafluoro-, hydrogen • borofluoric acid • fluoboric acid • fluoboric acid (H(BF$_4$)) • hydrofluoboric acid • hydrogen tetrafluoroborate • hydrogen tetrafluoroborate(1−) • tetrafluoroboric acid

MF BF$_4$•H

MW 87.82

pH Strong acid.

COLOR AND FORM Colorless liquid.

DENSITY Approximately 1.84.

DISSOCIATION CONSTANT pK (water): −4.9.

BOILING POINT Decomposition at bp of 130°C.

SOLUBILITY Soluble in hot water; miscible with water and alcohol.

OTHER CHEMICAL/PHYSICAL PROPERTIES Heat of formation: 388.5 kcal; forms crystalline salts with metals. Heat of formation (aqueous, 1 molal, at 25°C): −1,527 kJ/mol; specific gravity (48% solution): 1.37; specific gravity (42% solution): 1.32; specific gravity (30% solution): 1.20; surface tension (48% solution at 25°C): 65.3 mN/m (= dyn/cm). Undergoes limited hydrolysis in water to form hydroxyfluoborate ions. Major product is BF$_3$OH$^-$. Fluoroboric acid does not exist as a free, pure substance.

PROTECTIVE EQUIPMENT NIOSH: Respirator selection: 12.5 mg/m^3: any dust and mist respirator except single-use respirators. 25 mg/m^3: any dust and mist respirator except single-use and quarter-mask respirator or any supplied-air respirator or any self-contained breathing apparatus. 62.5 mg/m^3: any powered air-purifying respirator with a dust and mist filter (note: may need acid gas sorbent) or any supplied-air respirator operated in a continuous-flow mode. 125 mg/m^3: any self-contained breathing apparatus with a full facepiece or any air-purifying full facepiece respirator with a high-efficiency particulate filter or any supplied-air respirator with a full facepiece. 500 mg/m^3: any supplied-air respirator with a full facepiece and operated in a pressure-demand or other positive-pressure mode. Emergency or planned entry in unknown or IDLH conditions: any self-contained breathing apparatus with full facepiece and operated in a pressure-demand or other positive-pressure mode or any supplied-air respirator with a full facepiece and operated in pressure-demand or other positive-pressure mode in combination with auxiliary self-contained breathing apparatus operated in pressure-demand or other positive-pressure mode. Escape: any air-purifying full facepiece respirator (gas mask) with a chin-style or front- or back-mounted acid gas canister having a high-efficiency particulate filter or any appropriate escape-type self-contained breathing apparatus. (fluorides (as fluoride)) [R5, 127].

PREVENTATIVE MEASURES Contact lens use in industry is controversial. A survey of 100 corporations resulted in the recommendation that each company establish their own contact lens use policy. One presumed hazard of contact lens use is possible chemical entrapment. It was found that contact lenses minimized injury or protected the eye. The eye was afforded more protection from liquid irritants. Soft contact lenses do not worsen corneal damage from strong chemicals and in some cases could actually protect the eye. Overall, the literature supports the wearing of contact lenses in industrial environments as part of the standard eye protection, e.g., face shields; however, more data are needed to establish the value of contact lenses [R1].

HEALTH HAZARDS Contact causes burns to skin and eyes. If inhaled, may be harmful. Fire may produce irritating or poisonous gases. Runoff from fire control or dilution water may cause pollution [R4]. Fire or explosion: Some of these materials may burn, but none of them ignites readily. Flammable or poisonous gases may accumulate in tanks and hopper cars. Some of these materials may ignite combustibles (wood, paper, oil, etc.) [R4].

EMERGENCY ACTION Keep unnecessary people away; isolate hazard area, and deny entry. Stay upwind; keep out of low areas. Self-contained breathing apparatus (SCBA) and structural firefighter's protective clothing will provide limited protection. Call CHEMTREC at 1–800–424–9300 for emergency assistance. If water pollution occurs, notify the appropriate authorities [R4].

Fire: Some of these materials may react violently with water. Small Fires: Dry chemical, CO_2, Halon, water spray, or standard foam. Large Fires: Water spray, fog, or standard foam is recommended. Move container from fire area if you can do so without risk. Cool containers that are exposed to flames with water from the side until well after fire is out. Stay away from ends of tanks [R4].

Spill or Leak: Do not touch spilled material; stop leak if you can do so without risk. Small Spills: Take up with sand or other noncombustible absorbent material and place into containers for later disposal. Small Dry Spills: With clean shovel place material into clean, dry container, and cover; move containers from spill area. Large Spills: Dike far ahead of liquid spill for later disposal [R4].

First Aid: Move victim to fresh air; call emergency medical care. Remove and isolate contaminated clothing and shoes at the site. In case of contact with material, immediately flush skin or eyes with running water for at least 15 minutes. Keep victim quiet and maintain normal body temperature [R4].

CLEANUP Aqueous HBF_4 or its salt is treated by aluminum sulfate or ferric chloride and aqueous ammonia, ammonium sulfate, ammonium chloride, and urea, then with calcium hydroxide, and precipitate is filtered [R2].

DISPOSAL SRP: At the time of review, criteria for land treatment or burial (sanitary landfill) disposal practices are subject to significant revision. Prior to implementing land disposal of waste residue (including waste sludge), consult with environmental regulatory agencies for guidance on acceptable disposal practices.

COMMON USES As catalyst for preparing acetals, esterifying cellulose; to clean metal surfaces before welding; to brighten aluminum; reagent for sodium in presence of magnesium and potassium ions; as solute in electrolytes for plating metals such as chromium, iron, nickel, copper, silver, zinc, cadmium, indium, tin, and lead (has high throwing power); reagent for sodium in the presence of magnesium and potassium ions; for making stabilized diazo salts (diazonium and tetrazonium fluoborates); 0.1–0.5% solution retards fermentation [R6, 593].

THRESHOLD LIMIT 8-hr time-weighted average (TWA) 2.5 mg/m³ (1977) (fluorides, as F) [R8, 22]. Excursion Limit Recommendation: Excursions in worker exposure levels may exceed three times the TLV–TWA for no more than a total of 30 min during a work day, and under no circumstances should they exceed five times the TLV–TWA. (fluorides, as F) [R32, 5]. BEI (Biological Exposure Index): Fluoride in urine prior to shift is 3 mg/g creatinine. The determinant is usually present in a significant amt in biological specimens collected from subjects who have not been occupationally exposed. Such background levels are incl in the BEI value. The determinant is nonspecific, since it is observed after exposure to some other chemicals. These nonspecific tests are preferred because they are easy to use and usually offer a better correlation with exposure than specific tests. In such instances, a BEI for a specific, less quantitative biological determinant is recommended as a confirmatory test. (fluorides) [R32, 63].

OCCUPATIONAL RECOMMENDATIONS
OSHA 8-hr time-weighted average: 2.5 mg/m³ (fluorides, as F) [R3].

NIOSH 10-hr time-weighted average: 2.5 mg/m³ (fluorides, as F) [R5, 126]; threshold-limit 8-hr time-weighted average (TWA) 2.5 mg/m³ (1977) (fluorides, as F) [R7, 22].

REFERENCES
1. Randolph, S.A., M.R. Zavon. *J Occup Med* 29: 237–42 (1987).

2. *Treatment of Water Containing Fluoroboric Acid or Its Salt*; (CA/094/126846Q) Japan Kokai Tokkyo Koho Patent 80102490 08/05/80 (Showa Koji KK).

3. 54 FR 2920 (1/19/89).

4. Department of Transportation. *Emergency Response Guidebook 1987*. DOT P 5800.4. Washington, DC: U.S. Government Printing Office, 1987, p. G–60.

5. NIOSH. *Pocket Guide to Chemical Hazards*. 2nd Printing. DHHS (NIOSH) Publ. No. 85–114. Washington, DC: U.S. Dept. of Health and Human Services, NIOSH/Supt.of Documents, GPO, February 1987.

6. *The Merck Index*. 10th ed. Rahway, New Jersey: Merck Co., Inc., 1983.

7. American Conference of Governmental Industrial Hygienists. *Threshold Limit Values for Chemical Substances and Physical Agents and Biological Exposure Indices for 1994–1995*. Cincinnati, OH: ACGIH, 1994.

8. American Conference of Governmental Industrial Hygienists. *Threshold Limit Values (TLV's) for Chemical Substances and Physical Agents and Biological Exposure Indices (BEI's) for 1995–1996*. Cincinnati, OH: ACGIH, 1995.

◼ FLUVALINATE

CAS # 69409–94–5

SYNONYMS dl-valine, N-(2-chloro-4-(trifluoromethyl) phenyl)-, cyano (3-phenoxyphenyl) methyl ester • marvik • N-[2-chloro-4-(trifluoromethyl)-phenyl]-dl-valine cyano (3-phenoxyphenyl) methyl ester • zr-3210 • cyano (3-phenoxyphenyl) methyl N-[2-chloro-4-(trifluoromethyl) phenyl]-d-valinate • (rs)-alpha-cyano-3-phenoxybenzyl N-[2-chloro-alpha, alpha, alpha-trifluoro-p tolyl]-d-valinate. • (rs)-alpha-cyano-3-phenoxybenzyl (r)-2-[2-chloro-4-(trifluoromethyl) anilino-3-methyl butanoate] • spur • klartan

MF $C_{26}H_{22}ClF_3N_2O_3$

MW 502.95

COLOR AND FORM Yellow-amber liquid.

DENSITY 1.29 at 25°C.

BOILING POINT >450°C.

FIREFIGHTING PROCEDURES Use carbon dioxide, foam, or dry chemical (on fires involving pyrethroids) (pyrethrum) [R9, 2]. Firefighting: Self-contained breathing apparatus with a full facepiece operated in pressure-demand or other positive-pressure mode (pyrethrum) [R9, 5]. Extinguish fire using agent suitable for type of surrounding fire. (pyrethrins) [R2].

SOLUBILITY Solubility in water: 2.0 ppb; soluble in organic solvents; soluble in aromatic hydrocarbons, alcohols, diethylether, and dichloromethane.

VAPOR PRESSURE 1×10^{-7} mm Hg at 25°C.

CORROSIVITY Noncorrosive to slightly corrosive, depending upon the metal.

STABILITY AND SHELF LIFE Stable for at least 1 year at temperatures up to 50°C. Decreased stability under alkaline conditions. (technical fluvalinate) [R1].

OTHER CHEMICAL/PHYSICAL PROPERTIES Viscous yellow oil (technical fluvalinate).

PROTECTIVE EQUIPMENT Employees should be provided with and required to use dust- and splash-proof safety goggles where pyrethroids may contact the eyes. (pyrethroids) [R9, 3].

PREVENTATIVE MEASURES Wear respiratory protection and rubber gloves when opening drum and pouring (trace quantities of hydrogen cyanide may be present) [R1].

CLEANUP Spillages of pesticides at any stage of their storage or handling should be treated with great care. Liquid formulations may be reduced to solid phase by evaporation. Dry sweeping of solids is always hazardous: these should be removed by vacuum cleaning, or by dissolving them in water or other solvent in the factory environment. (pesticides) [R3].

DISPOSAL Incineration would be an effective disposal procedure where permitted. If an efficient incinerator is not available, the product should be mixed with large amounts of combustible material, and contact with the smoke should be avoided. (pyrethrin products) [R4].

COMMON USES Insecticide [R8, 659].

OCCUPATIONAL RECOMMENDATIONS None listed.

OSHA 8-hr time-weighted average: 5 mg/m³ (cyanides, as CN) [R6].

NON-HUMAN TOXICITY LD_{50} rat oral 261–282 mg/kg [R1].

ECOTOXICITY LD_{50} bobwhite quail oral >2,510 mg/kg [R1].

BIOCONCENTRATION Based upon a water solubility of 0.005 mg/L at 20–25°C (taken from the USDA's evaluated database of pesticide properties) (1), the BCF for fluvalinate can be estimated to be 12,000 from a recommended regression-derived equation (2,SRC). This BCF value suggests that bioconcentration may be important in aquatic organisms that cannot metabolize fluvalinate (SRC) [R5].

FIFRA REQUIREMENTS In 1988, Congress amended FIFRA to strengthen and accelerate EPA's reregistration program. The nine-year reregistration scheme mandated by "FIFRA 88" applies to each registered pesticide product containing an active ingredient initially registered before November 1, 1984. Pesticides for which EPA had not issued Registration Standards prior to the effective date of FIFRA '88 were divided into three lists based upon their potential for exposure and other factors, with List B being of highest concern, and D of least. List: B; Case: Fluvalinate; Case No. 2295; Pesticide type: Insecticide; Active Ingredient (AI): N-(2-Fluvinate; Data Call-in (DCI) Date: 4/12/91; Case Status: Awaiting Data/Data in Review: OPP awaits data from the pesticide's producer(s) regarding its human health or environmental effects, or OPP has received and is reviewing such data, in order to reach a decision about the pesticide's eligibility for reregistration [R7].

REFERENCES

1. Hartley, D., and H. Kidd (eds.). *The Agrochemicals Handbook*. 2nd ed. Lechworth, Herts, England: The Royal Society of Chemistry, 1987, p. A667/Aug 87.

2. Bureau of Explosives. *Emergency Handling of Haz Matl in Surface Trans* p. 434 (1981).

3. International Labour Office. *Encyclopedia of Occupational Health and Safety*. Vols. I and II. Geneva, Switzerland: International Labour Office, 1983. 1619.

4. Sittig, M. *Handbook of Toxic and Hazardous Chemicals and Carcinogens, 1985.* 2nd ed. Park Ridge, NJ: Noyes Data Corporation, 1985. 762.

5. (1) Wauchope, R.D., et al. *Rev Environ Contam Toxicol* 123: 1–36 (1991) (2) Lyman, W. J., et al., *Handbook of Chemical Property Estimmation Methods* Washington, DC: Amer Chemical Soc, p. 5–10 (1990).

6. 29 CCFR 1910.1000 (7/1/91).

7. USEPA/OPP. *Status of Pesticides in Reregistration and Special Review* p. 137 (Mar, 1992) EPA 700–R–92–004.

8. Budavari, S. (ed.). *The Merck Index— Encyclopedia of Chemicals, Drugs and Biologicals.* Rahway, NJ: Merck and Co., Inc., 1989.

9. Mackison, F. W., R. S. Stricoff, and L. J. Partridge, Jr. (eds.). *NIOSH/OSHA—Occupational Health Guidelines for Chemical Hazards.* DHHS (NIOSH) Publication No. 81–123 (3 vols). Washington, DC: U.S. Government Printing Office, Jan. 1981.

■ FOLPET

CAS # 133–07–3

SYNONYMS 1h-isoindole-1,3 (2h)-dione, 2-((trichloromethyl) thio) - • 2-((trichloromethyl) thio) -1h-isoindole-1,3 (2H)-dione • cosan-I • ent-26539 • faltan • folnit • folpan • folpel • ftalan • fungitrol-11 • intercide-tmp • N-((trichloromethyl) thio) phthalimide • N-(trichlormethylthio) phthalimide • N-(trichloromethylmercapto) phthalimide • orthophaltan • phaltane • phthalimide, N-((trichloromethyl) thio)- • phthaltan • spolacid • thiophal • trichloromethylthiophthalimide • vinicoll • N-(trichloromethylthio) phthalimide • N-(trichloromethylthio) phthalimide • fungitrol •

N-(trichloromethanesulphenyl) phthalimide • acryptan • phaltan • phaltane • faltex • ortho-phaltan-50w • Murphy's rose fungicide

MF $C_9H_4Cl_3NO_2S$

MW 296.55

COLOR AND FORM Crystals from benzene; white crystals; light-colored powder; colorless crystals.

MELTING POINT 177°C.

SOLUBILITY Slightly soluble in organic solvents; slightly soluble in ethanol and acetone; 1 mg/L water at room temp; 3–4% in aliphatic ketones and 0.1–1% in hydrocarbons; in chloroform 87, benzene 22, isopropanol 12.5 (all in g/L at 20°C).

VAPOR PRESSURE 1.3 mPa at 20°C.

OCTANOL/WATER PARTITION COEFFICIENT Log K_{ow} = 2.85.

CORROSIVITY Noncorrosive; decomposition products are corrosive.

STABILITY AND SHELF LIFE Virtually nonvolatile; stable when dry, but slowly hydrolyzes in water at ordinary temp, rapidly at high temperature, or under alkaline conditions [R1].

OTHER CHEMICAL/PHYSICAL PROPERTIES Yellow powder (technical folpet 90–95%); chemically related to captan; Henry's Law constant = 3.8×10^{-6} atm-m^3/mole at 20°C.

PROTECTIVE EQUIPMENT Goggles or a face shield [R9, p. C–161].

PREVENTATIVE MEASURES Avoid inhalation of dust and spray mist and keep away from eyes [R3].

CLEANUP A high-efficiency particulate arrestor (HEPA) or charcoal filters can be used to minimize amount of carcinogen in exhausted-air-ventilated safety cabinets, lab hoods, glove boxes, or animal rooms Filter housing that is designed so that used filters can be transferred into plastic bag without contaminating maintenance staff is avail commercially. Filters should be placed in plastic bags immediately after removal. The plastic bag should be sealed immediately. The sealed bag should be labelled properly. Waste liquids should be placed or collected in proper containers for disposal. The lid should be secured and the bottles properly labelled. Once filled, bottles should be placed in plastic bag, so that outer surface is not contaminated. The plastic bag should also be sealed and labelled. Broken glassware should be decontaminated by solvent extraction, by chemical destruction, or in specially designed incinerators. (chemical carcinogens) [R12, 15].

DISPOSAL SRP: At the time of review, criteria for land treatment or burial (sanitary landfill) disposal practices are subject to significant revision. Prior to implementing land disposal of waste residue (including waste sludge), consult with environmental regulatory agencies for guidance on acceptable disposal practices.

COMMON USES As fungicide in paints and plastics, and for treatment of structural internal and external surfaces of buildings [R3]. Broad-spectrum contact fungicide that is effective against a fairly wide range of fungi that cause plant diseases. It acts by inhibiting fungal mycelial growth, but does not eradicate established fungal infection [R10, 661]. Folpet is a protective fungicide used mainly for foliage application against *Venturia* species [R11, 426]. Control of downy mildews, powdery mildews, leaf spot diseases, scab, *Gloeosporium* rots, *Botrytis*, *Alternaria*, *Pythium*, and *Rhizoctonia* species in pome fruit, stone fruit, soft fruit, citrus fruit, vines, olives, hops, potatoes, lettuce, cucurbits, onions, leeks, celery, tomatoes, and ornamentals [R2]. For fruits, berries, vegetables, flowers, and ornamentals. Controls apple scab, cherry leaf spot, rose black spot, rose mildew. Seed and plant bed treatment [R9, p. C–160]. Fungicide-bactericide for vinyls, paints, and enamels [R4].

OCCUPATIONAL RECOMMENDATIONS None listed.

NON-HUMAN TOXICITY LC_{50} rat inhalation greater than 5.0 mg/L/2 hr (technical) [R5]; LC_{50} mouse inhalation greater than 6.0 mg/L/2 hr (technical) [R5]; LD_{50} rabbit percutaneous greater than 22,600 mg/kg [R11, 426]; LD_{50} rat, oral 10 g/kg [R10, 662]; LD_{50} rat ip 40 mg/kg [R10, 662].

ECOTOXICITY LC_{50} *Gammarus fasciatus* (scud) 2,500 µg/L/96 hr at 15°C, mature (95% confidence limit 1,994–3,134 µg/L). Static bioassay without aeration, pH 7.2–7.5, water hardness 40–50 mg/L as calcium carbonate, and alkalinity of 30–35 mg/L (technical material, 88–93%) [R6]; LC_{50} *Oncorhynchus kisutch* (coho salmon) 106 µg/L/96 hr at 12°C, wt 1.0 g (95% confidence limit 82–137 µg/L). Static bioassay without aeration, pH 7.2–7.5, water hardness 40–50 mg/L as calcium carbonate, and alkalinity of 30–35 mg/L (technical material, 88–93%) [R6]; LC_{50} *Salmo gairdneri* (rainbow trout) 39 µg/L/96 hr at 12°C, wt 1.5 g (95% confidence limit 18–85 µg/L). Static bioassay without aeration, pH 7.2–7.5, water hardness 40–50 mg/L as calcium carbonate, and alkalinity of 30–35 mg/L (technical material, 88–93%) [R6]. LC_{50} *Salmo trutta* (brown trout) 66 µg/L/96 hr at 12°C, wt 0.6 g (95% confidence limit 56–78 µg/L). Static bioassay without aeration, pH 7.2–7.5, water hardness 40–50 mg/L as calcium carbonate, and alkalinity of 30–35 mg/L (technical material, 88–93%) [R6]. LC_{50} *Salvelinus namaycush* (lake trout) 87 µg/L/96 hr at 12°C, wt 0.5 g (95% confidence limit 68–110 µg/L). Static bioassay without aeration, pH 7.2–7.5, water hardness 40–50 mg/L as calcium carbonate, and alkalinity of 30–35 mg/L (technical material, 88–93%) [R6]. LC_{50} *Pimephales promelas* (fathead minnow) 298 µg/L/96 hr at 12°C, wt 0.3 g (95% confidence limit 207–430 µg/L). Static bioassay without aeration, pH 7.2–7.5, water hardness 40–50 mg/L as calcium carbonate, and alkalinity of 30–35 mg/L (technical material, 88–93%) [R6]. LC_{50} *Ictalurus punctatus* (channel catfish) 108 µg/L/96 hr at 20°C, wt 1.2 g (95% confidence limit 59–201 µg/L). Static bioassay without aeration, pH 7.2–7.5, water hardness 40–50 mg/L as calcium carbonate, and alkalinity of 30–35 mg/L (technical material, 88–93%) [R6]. LC_{50} *Lepomis macrochirus* (bluegill) 72 µg/L/96 hr at 20°C, wt 0.6 g (95% confidence limit 58–89 µg/L). Static bioassay without aeration, pH 7.2–7.5, water hardness 40–50 mg/L as calcium carbonate, and alkalinity of 30–35 mg/L (technical material, 88–93%) [R6]. LC$_{50}$ *Micropterus dolomieui* (smallmouth bass) 91 µg/L/96 hr at 12°C, fingerling (95% confidence limit 73–113 µg/L). Static bioassay without aeration, pH 7.2–7.5, water hardness 40–50 mg/L as calcium carbonate, and alkalinity of 30–35 mg/L (technical material, 88–93%) [R6]. LC$_{50}$ *Perca flavescens* (yellow perch) 177 µg/L/96 hr at 12°C, wt 0.6 g (95% confidence limit 149–210 µg/L). Static bioassay without aeration, pH 7.2–7.5, water hardness 40–50 mg/L as calcium carbonate, and alkalinity of 30–35 mg/L (technical material, 88–93%) [R6]. LD$_{50}$ *Anas platyrhynchos* (mallard) male, oral more than 2,000 mg/kg, 3–4-month old (purity 92.4%) [R7].

BIOCONCENTRATION Based on a measured log K$_{ow}$ of 2.85 (1), the BCF for folpet can be estimated to be 86 from a regression-derived equation (2,SRC). This value of BCF indicates that the bioconcentration of folpet in aquatic organisms will be of minimum importance (SRC) [R8].

REFERENCES

1. Spencer, E. Y. *Guide to the Chemicals Used in Crop Protection.* 7th ed. Publication 1093. Research Institute, Agriculture Canada, Ottawa, Canada: Information Canada, 1982. 307.

2. Hartley, D., and H. Kidd (eds.). *The Agrochemicals Handbook.* 2nd ed. Lechworth, Herts, England: The Royal Society of Chemistry, 1987, p. A213/Aug 87.

3. Hartley, D., and H. Kidd (eds.). *The Agrochemicals Handbook.* Old Woking, Surrey, United Kingdom: Royal Society of Chemistry/Unwin Brothers Ltd., 1983, p. A213/Oct 83.

4. Lewis, R. J., Sr. (ed.). *Hawley's Condensed Chemical Dictionary.* 12th ed. New York, NY: Van Nostrand Reinhold Co., 1993 535.

5. Stevens, J. T., et al. *Toxicol Appl Pharmacol* 45 (1): 320 (1978).

6. U.S. Department of Interior, Fish and Wildlife Service. *Handbook of Acute Tox-*

icity of Chemicals to Fish and Aquatic Invertebrates. Resource Publication No. 137. Washington, DC: U.S. Government Printing Office, 1980. 42.

7. U.S. Department of the Interior, Fish and Wildlife Service. *Handbook of Toxicity of Pesticides to Wildlife.* Resource Publication 153. Washington, DC: U.S. Government Printing Office, 1984. 44.

8. (1) Hansch, C., and A. J. Leo. *Medchem Project* Issue No. 26, Claremont, CA; Pomona College (1985). (2) Lyman, W. J., et al., *Handbook of Chemical Property Estimation Methods* Washington, DC: Amer Chemical Soc pp. 5–4 (1990).

9. *Farm Chemicals Handbook 1993.* Willoughby, OH: Meister Publishing Co., 1993.

10. National Research Council. *Drinking Water and Health Volume 1.* Washington, DC: National Academy Press, 1977.

11. Worthing, C.R. and S.B. Walker (eds.). *The Pesticide Manual—A World Compendium.* 8th ed. Thornton Heath, UK: The British Crop Protection Council, 1987.

12. Montesano, R., H. Bartsch, E. Boyland, G. Della Porta, L. Fishbein, R. A. Griesemer, A.B. Swan, L. Tomatis, and W. Davis (eds.). *Handling Chemical Carcinogens in the Laboratory: Problems of Safety.* IARC Scientific Publications No. 33. Lyon, France: International Agency for Research on Cancer, 1979.

■ FORMETANATE HYDROCHLORIDE

CAS # 23422–53–9

SYNONYMS
3-dimethylaminomethyleneiminophenyl-N-methylcarbamate, hydrochloride • carbamic acid, methyl-, ester with N′-(m-hydroxyphenyl)-N,N-dimethylformamidine, monohydrochloride • carzol • carzol-sp • dicarzol • ent-27566 • ep-332 • formetanate monohydrochloride •

m-(((dimethylamino) methylene) amino) phenyl methylcarbamate hydrochloride • m-[((dimethylamino) methylene) amino] phenyl methylcarbamate hydrochloride • methanimidamide, N,N-dimethyl-N'-(3-(((methylamino) carbonyl) oxy) phenyl)-, monohydrochloride • morton-ep-332- • N,N-dimethyl-N'-(((methylamino) carbonyl) oxy) phenylmethanimidamide monohydrochloride • Schering-36056 • sn-36056

MF $C_{11}H_{15}N_3O_2 \cdot ClH$

MW 257.8

COLOR AND FORM Colorless crystals; white powder.

ODOR Faint odor.

SOLUBILITY In water 50%, methanol 20%; slightly soluble in organic solvents; benzene <0.1%, acetone <0.1%, chloroform 0.2%, methanol ca. 25%, hexane <0.1%; water solubility = >500,000 mg/L.

VAPOR PRESSURE Practically zero at room temperature.

STABILITY AND SHELF LIFE Non-volatile [R2].

OTHER CHEMICAL/PHYSICAL PROPERTIES Yellow crystalline solid; solubility: less than 1 g/L water; about 100 g/L acetone, chloroform; greater than 200 g/L methanol (formetanate).

DISPOSAL SRP: At the time of review, criteria for land treatment or burial (sanitary landfill) disposal practices are subject to significant revision. Prior to implementing land disposal of waste residue (including waste sludge), consult with environmental regulatory agencies for guidance on acceptable disposal practices.

COMMON USES Insecticide and acaricide on deciduous fruit and nuts and alfalfa [R1]. Active against the mobile stages of fruit tree spider mites and is recommended also for use in greenhouses. (formetanate) [R11].

OCCUPATIONAL RECOMMENDATIONS None listed.

NON-HUMAN TOXICITY LD_{50} rat oral 21 mg/kg [R3]; LD_{50} mouse oral 18 mg/kg [R7]; LD_{50} beagle dog oral 19.1 mg/kg [R7]; LD_{50} rat subcutaneous greater than 5,600 mg/kg [R7]; LD_{50} rabbit subcutaneous greater than 10,200 mg/kg [R7].

ECOTOXICITY LC_{50} Japanese quail (*Coturnix japonica*), 14 days old, oral (5-day *ad libitum* in diet) 993 ppm (95% confidence intervals 673–1,465 ppm) (technical grade, 93% active ingedient) [R4]; LC_{50} pheasant greater than 4,640 mg/kg in 10-day diet [R7]; LC_{50} bobwhite quail greater than 4,640 mg/kg in 10-day diet [R7]; LC_{50} duck 6,810 mg/kg in 10-day diet [R7]; LC_{50} rainbow trout 2.8 mg/L/96 hr [R7]; LC_{50} bluegill 20 mg/L/96 hr [R7]; LC_{50} black bullhead 75 mg/L/96 hr [R7].

BIODEGRADATION In river bottom soil of pH 8, the free amine decomposed rapidly, and the major identified degradation products were m-formaminophenyl-N-methylcarbamate, m-formaminophenol, and m-aminophenol (1). In soil, the free amine concentration decreased from 53.7% at 1 day to 10.4% after 16 days (1) [R5].

BIOCONCENTRATION Based upon an experimental water solubility of 500,000 mg/L (1), the BCF of formetanate hydrochloride can be estimated to be approximately 0.38 from a regression-derived equation (2). According to this BCF value, bioconcentration in aquatic organisms is not expected to be an important fate process (SRC) [R6].

FIFRA REQUIREMENTS In 1988, Congress amended FIFRA to strengthen and accelerate EPA's reregistration program. The nine-year reregistration scheme mandated by "FIFRA 88" applies to each registered pesticide product containing an active ingredient initially registered before November 1, 1984. List A consists of the 194 chemical cases (or 350 individual active ingredients) for which EPA had issued Registration Standards prior to the effective date of FIFRA '88. List: A; Case: Formetanate HCl; Case No. 0091; Pesticide type: Insecticide (acaricide); Registration Standard Date: 06/01/83; Case Status: Awaiting Data/Data in Review: OPP awaits data from the pesticide's producer(s) regarding its human health or environmental effects, or OPP has received and is reviewing such data, in order to reach a decision about the pesticide's eligibility for reregistration. Active Ingredient (AI): Formetanate Hydrochloride Data Call-in (DCI) Date: 09/11/90; AI Status: The producer(s) of the pesticide has made commitments to conduct the studies and pay the fees required for reregistration, and is meeting those commitments in a timely manner [R7]. Tolerances are established for residues of the insecticide formetanate hydrochloride in or on raw agricultural commodities as follows: peaches, grapefruit, lemons, limes, nectarines, oranges, tangerines, apples, pears, and plums (fresh prunes) [R9]. A tolerance is established for residues of the insecticide formetanate hydrochloride in dried prunes when present therein as a result of the application of the insecticide to growing plums (fresh prunes) [R10]. A tolerance is established for residues of the insecticide formetanate hydrochloride in citrus molasses resulting from the application of the insecticide to the growing raw agricultural commodities grapefruit, lemons, limes, oranges, and tangerines [R8].

REFERENCES

1. SRI.

2. Spencer, E. Y. *Guide to the Chemicals Used in Crop Protection*. 7th ed. Publication 1093. Research Institute, Agriculture Canada, Ottawa, Canada: Information Canada, 1982. 310.

3. Worthing, C. R. (ed.). *Pesticide Manual*. 6th ed. Worcestershire, England: British Crop Protection Council, 1979. 284.

4. Hill, E. F., and M. B. Camardese. *Lethal Dietary Toxicities of Environmental Contaminants and Pesticides to Coturnix*. Fish and Wildlife Technical Report 2. Washington, DC: United States Department of Interior, Fish and Wildlife Service, 1986. 81.

5. Arurkar S.K., C.O. Knowles. *Bull Env Contam Toxicol* 5: 324–28 (1970).

6. (1) Shiu, W.Y., et al. Rev *Environ Contam Toxicol* 116: 15–187 (1990) (2) Lyman, W. J., et al., *Handbook of Chemical Property Estimation Methods*. Washington DC: Amer Chemical Soc pp. 4–9, 5–4, 5–10, 7–44, 7– 5, 15–15 to 15–32 (1990).

7. USEPA/OPP. *Status of Pesticides in Reregistration and Special Review* p. 92 (Mar, 1992) EPA 700–R–92–004.

8. 40 CFR 186.3450 (7/1/94).

9. 40 CFR 180.276 (7/1/94).

10. 40 CFR 185.3450 (7/1/94).

11. Gerhartz, W. (exec ed.). *Ullmann's Encyclopedia of Industrial Chemistry*. 5th ed. Vol A1: Deerfield Beach, FL: VCH Publishers, 1985 to present, p. VA1 20.

■ FUEL OIL NO. 2

CAS # 68476–30–2

DOT # NA 1993

SYNONYMS diesel oil • API no. 2 fuel oil • fuel oil no. 2 (DOT) • gas oil • gas oil (DOT) • home heating oil no. 2 • #2 home heating oils • number 2 burner fuel • number 2 fuel oil

MF variable

MW variable

COLOR AND FORM Amber-colored distillate oil; brown, slightly viscous liquid.

DENSITY 0.8654 at 15°C.

VISCOSITY 268 cSt at 37.8°C.

FLASH POINT 100°F.

AUTOIGNITION TEMPERATURE 494°F [R2].

FIREFIGHTING PROCEDURES Do not extinguish fire unless flow can be stopped. Use water in flooding quantities as fog. Solid streams of water may be ineffective. Cool all affected containers with flooding quantities of water. Apply water from as far a distance as possible. Use foam, dry chemical, or carbon dioxide (Fuel Oil No. 1, 2, 4, or 5) [R6, 464].

TOXIC COMBUSTION PRODUCTS When heated to decomposition it emits acrid smoke and fumes [R2].

OTHER CHEMICAL/PHYSICAL PROPERTIES Pour point ≤ 18°C.

PROTECTIVE EQUIPMENT Wear appropriate chemical protective gloves, boots, and goggles. (fuel oil no. 1, 2, 4, or 5) [R6, 342].

PREVENTATIVE MEASURES If material not on fire and not involved in fire: Keep sparks, flames, and other sources of ignition away. Keep material out of water sources and sewers. Build dikes to contain flow as necessary. Use water spray to knock down vapors. (fuel oil no. 1, 2, 4, or 5) [R6, 342].

DISPOSAL Spray into incinerator. SRP: At the time of review criteria for land treatment or burial (sanitary landfill) disposal practices are subject to significant revision. Prior to implementing land disposal of waste residue (including waste sludge), consult with environmental regulatory agencies for guidance on acceptable disposal practices.

COMMON USES Most commonly used for domestic heating and power plant warm-up [R1].

STANDARD CODES NFPA—, 2, 0; USCG nonflammable liquid.

PERSISTENCE Loss of oil after 40 hr in bubbler apparatus—13% evaporated, 0.053% dissolved.

MAJOR SPECIES THREATENED Waterfowl and aquatic life.

DIRECT CONTACT Irritating to the skin in salt water.

GENERAL SENSATION Kerosene odor inhalation causes headache, stupor. Ingestion leads to irritation, nausea, and vomiting.

PERSONAL SAFETY Avoid sources of ignition.

ACUTE HAZARD LEVEL Coating action of oils can destroy water birds, plankton, algae, and fishes.

DEGREE OF HAZARD TO PUBLIC HEALTH Will cause taste and odor problems before reaching toxic levels.

OCCUPATIONAL RECOMMENDATIONS None listed.

ACTION LEVEL Call local fire authority. Remove all potential sources of ignition.

ON-SITE RESTORATION There are a wide variety of sorbents, sinking agents, gelling agents, combustion promoters, dispersants, and mechanical systems to treat oil spills. In addition, straw, polyurethane foam, activated carbon, and peat can be used to soak up oil. Seek professional environmental engineering asssistance through EPA's Environmental Response Team (ERT), Edison, NJ, 24-hour no. 908–548–8730.

BEACH OR SHORE RESTORATION Oil can be burned off beaches. May require additional fuel for complete combustion.

AVAILABILITY OF COUNTERMEASURE MATERIAL Oil treating agents are stocked in many major harbors. In addition, straw—farmstead stables; polyurethane foam—upholstery shops; activated carbon—water treatment plants, sugar refineries; and peat—nurseries, floral shops.

INDUSTRIAL FOULING POTENTIAL Boiler water feed should be limited to 7 ppm or less. Oil can result in poor heat transport, blistering, overheating, and foaming. In reused cooling water, no oil is acceptable. Oil causes tastes in food-processing water, and is especially detrimental to cement- and paper-making operations.

MAJOR WATER USE THREATENED Recreation, potable supply, fisheries, irrigation, industrial.

PROBABLE LOCATION AND STATE Brown viscous liquid; will float in slick.

COLOR IN WATER Irridescent; dark blue-purple slick.

FOOD-CHAIN CONCENTRATION POTENTIAL Negative.

NON-HUMAN TOXICITY LD_{50} rat oral 12.0 g/kg body weight [R3].

BIODEGRADATION The chief components of fuel oil no. 2 are expected to be C16–C19 unbranched alkanes (2,3,SRC). Microorganisms have the capacity to biodegrade normal alkanes in the range of

C16–C19 (1); BOD bottles incubated in the dark at 25°C using a soil inoculum exhibited 5.3 μg/ml O₂ consumption after a 5-day incubation period for hexadecane and 3.3 μg/ml O₂ consumption after 5 days for octadecane (1) [R4].

BIOCONCENTRATION To determine whether blue crabs can bioaccumulate and retain complex mixture of petroleum hydrocarbons, adult crabs were exposed for 14 days or 30–35 days in continuous flow-through seawater systems to 3 sublethal concentrations, nominally 0.00 (control), 0.01, or 1.0 ppm (mg/L), of the water-accommodated fraction of no. 2 fuel oils. Crabs exposed for 14 days were subsequently exposed to clean running seawater for 7 days whereas crabs exposed for 30–35 days were placed into clean running seawater for a 30-day depuration period. Gill, hepatopancreas, and muscle tissue samples were collected from control, water-accommodated fraction exposed, and depurated crabs and analyzed by GC, or GC/MS. No. 2 fuel oil compounds were not detected in any of the tissues collected from control crabs. Trace amounts of fuel oil compounds were detected in gill and hepatopancreas tissues collected from crabs exposed to the 0.01-ppm WAF, and no fuel oil compounds were detected in muscle tissues. All tissues of crabs exposed to the 1.0-ppm water-accommodated fraction accumulated no. 2 fuel oil compounds, and considerable amount remained in hepatopancreas and gill tissues following depuration for 30 days in clean seawater [R5].

REFERENCES

1. *Kirk–Othmer Encyclopedia of Chemical Technology.* 3rd ed., Volumes 1–26. New York, NY: John Wiley and Sons, 1978–1984, p. V11 571 (1980).

2. Sax, N. I. *Dangerous Properties of Industrial Materials.* 6th ed. New York, NY: Van Nostrand Reinhold, 1984. 1458.

3. IARC. *Monographs on the Evaluation of the Carcinogenic Risk of Chemicals to Man.* Geneva: World Health Organization, International Agency for Research on Cancer, 1972–present (multivolume work), p. V45 256 (1989).

4. (1) Haines, J.R., M. Alexander, *Appl Microbiol* 28: 1084–5 (1974), (2) Sax N.I., R.J. Lewis, Jr. *Hawley's Condensed Chemical Dictionary* 11th ed. New York, NY: Van Nostrand Reinhold Co p. 386 (1987) (3) Hoffman, HL; *Kirk–Othmer Encycl Chemical Tech* 3rd ed. 17: 268.

5. Melzian, B.D., J. Lake. *Oil Chemical Pollut* 3 (5): 367–99 (1987).

6. Association of American Railroads. *Emergency Handling of Hazardous Materials in Surface Transportation.* Washington, DC: Assoc. of American Railroads, Hazardous Materials Systems (BOE), 1987.

▪ GLUTARALDEHYDE

CAS # 111–30–8

SYNONYMS 1,5-pentanedial • 1,5-pentanedione • aldesen • cidex • glutaral • glutaraldehyd (Czech) • glutardialdehyde • glutaric acid dialdehyde • glutaric aldehyde • glutaric dialdehyde • hospex • nci-c55425 • pentanedial • sonacide • 1,3-diformylpropane

MF $C_5H_8O_2$

MW 100.13

COLOR AND FORM Colorless liquid.

ODOR Pungent.

DENSITY 0.72.

BOILING POINT 187–189°C (decomp).

MELTING POINT FP: −14°C.

SOLUBILITY Soluble in all proportions in water, alcohol; soluble in benzene; soluble in ether.

VAPOR PRESSURE 17 mm Hg at 20°C.

STABILITY AND SHELF LIFE Acid glutaraldehyde is more stable than alkaline glutaraldehyde [R15, 1103]. Stable in light, oxidizes in air, polymerizes in heat [R15, 1103]. Alkaline solution deposits polymeric film after few hr [R16, 994]. Glutaral loses activity within 2 wk after prepn [R17, 1620].

OTHER CHEMICAL/PHYSICAL PROPERTIES Conversion factors: 1 mg/L = 245 ppm; 1 ppm = 4.1 mg/m³; OH rate constant = 2.38×10^{-11} cm³/molecule-sec at 25°C.

PROTECTIVE EQUIPMENT Goggles or face shield; rubber gloves [R9]. Neoprene or butyl rubber gloves are protective. Latex rubber gloves are not as protective. [R14].

PREVENTATIVE MEASURES SRP: The scientific literature for the use of contact lenses in industry is conflicting. The benefit or detrimental effects of wearing contact lenses depend not only upon the substance, but also on factors including the form of the substance, characteristics and duration of the exposure, the uses of other eye protection equipment, and the hygiene of the lenses. However, there may be individual substances whose irritating or corrosive properties are such that the wearing of contact lenses would be harmful to the eye. In those specific cases, contact lenses should not be worn. In any event, the usual eye protection equipment should be worn even when contact lenses are in place. Containment of vapors and prevention of skin contact are important industrial hygiene principles to help avoid sensitization of the skin and respiratory irritation and/or asthma. Proper skin protection must be provided as well as ventilation controls. [R14].

DISPOSAL SRP: At the time of review, criteria for land treatment or burial (sanitary landfill) disposal practices are subject to significant revision. Prior to implementing land disposal of waste residue (including waste sludge), consult with environmental regulatory agencies for guidance on acceptable disposal practices.

COMMON USES Embalming fluid [R19, p. II–123]; tissue fixation; intermediate; cross-linking protein, and polyhydroxy materials; tanning of soft leathers [R2]; chemical intermediate for adhesives, sealants, electrical products [R4]; in sterilization of endoscopic instruments thermometers, rubber, or plastic equipment that cannot be heat sterilized [R5]; used as a biocide in the oil industry [R3]; the most popular enzyme cross-linking reagent; microbial cells are also cross-linked with glutaraldehyde to yield cell pellets [R6]; skin disinfectant [R7]; disinfectant that is very good not only against vegetative bacteria but also against spores. Its efficacy against fungi and viruses is good. Disinfectant of choice for cold sterilization of surgical instruments but is being displaced by ethylene oxide. Glutaraldehyde aerosols are also used to "sterilize" hospital rooms, operating areas, etc. Acid glutaraldehyde is more effective than alkaline glutaraldehyde [R15, 1103]. It possesses tuberculocidal action [R1].

OCCUPATIONAL RECOMMENDATIONS None listed.

THRESHOLD LIMIT Ceiling limit 0.2 ppm, 0.82 mg/m³ (1979) [R11].

NON-HUMAN TOXICITY LD_{50} rat oral 600 mg/kg [R8]; LD_{50} rabbit skin 2,560 mg/kg [R8]; LD_{50} rat oral 0.82 g/kg [R13, 2654]; LD_{50} rabbit skin 0.64 ml/kg [R13, 2654]; LC_{50} rat inhalation 5,000 ppm/4 hr exposure [R10].

FIFRA REQUIREMENTS In 1988, Congress amended FIFRA to strengthen and accelerate EPA's reregistration program. The nine-year reregistration scheme mandated by "FIFRA 88" applies to each registered pesticide product containing an active ingredient initially registered before November 1, 1984. Pesticides for which EPA had not issued Registration Standards prior to the effective date of FIFRA '88 were divided into three lists based upon their potential for exposure and other factors, with List B being of highest concern and D of least. List: B; Case: Glutaraldehyde; Case No.: 2315; Pesticide type: Fungicide, Herbicide, Antimicrobial; Case Status: Awaiting Data/Data in Review: OPP awaits data from the pesticide's producer(s) regarding its human health and/or environmental effects, or OPP has received and is reviewing such data, in order to reach a decision about the pesticide's eligibility for reregistration. Active Ingredient (AI): Glutaraldehyde; Data Call-in (DCI) Date: 06/10/91; AI Status: The producer(s) of the pesticide has made commitments to conduct the studies and pay the fees required for reregistration, and is meeting those commitments in a timely manner. [R18].

FDA Glutaraldehyde is an indirect food additive for use as a component of adhesives [R12].

GENERAL RESPONSE Stop discharge if possible. Keep people away. Isolate and remove discharged material. Notify local health and pollution control agencies.

FIRE RESPONSE Not flammable.

EXPOSURE RESPONSE Call for medical aid. Liquid irritating to skin and eyes. Harmful if swallowed. Remove contaminated clothing and shoes. Flush affected areas with plenty of water. If in eyes, hold eyelids open and flush with plenty of water. If swallowed and victim is conscious, have victim drink water or milk.

RESPONSE TO DISCHARGE Issue warning—water contaminant. Restrict access. Disperse and flush.

WATER RESPONSE Effect of low concentrations on aquatic life is unknown. May be dangerous if it enters water intakes. Notify local health and wildlife officials. Notify operators of nearby water intakes.

EXPOSURE SYMPTOMS Contact with liquid causes severe irritation of eyes and irritation of skin. Chemical readily penetrates skin in harmful amounts. Ingestion causes irritation of mouth and stomach.

EXPOSURE TREATMENT Eyes: immediately flush with plenty of water for at least 15 min; get medical attention. Skin: immediately flush with plenty of water for at least 15 min. Ingestion: give large amounts of water and induce vomiting; get medical attention.

REFERENCES

1. American Medical Association, AMA Department of Drugs. *AMA Drug Evaluations*. 3rd ed. Littleton, Massachusetts: PSG Publishing Co., Inc., 1977. 894.

2. Sax, N. I., and R. J. Lewis, Sr. (eds.). *Hawley's Condensed Chemical Dictionary*. 11th ed. New York: Van Nostrand Reinhold Co., 1987. 566.

3. Gerhartz, W. (exec ed.). *Ullmann's Encyclopedia of Industrial Chemistry*. 5th ed. Vol A1: Deerfield Beach, FL: VCH Publishers, 1985 to present, p. VA16 567.

4. SRI.

5. Budavari, S. (ed.). *The Merck Index—Encyclopedia of Chemicals, Drugs, and Biologicals*. Rahway, NJ: Merck and Co., Inc., 1989. 702.

6. Gerhartz, W. (exec ed.). *Ullmann's Encyclopedia of Industrial Chemistry*. 5th ed. Vol A1: Deerfield Beach, FL: VCH Publishers, 1985 to Present., p. VA9 386.

7. Gerhartz, W. (exec ed.). *Ullmann's Encyclopedia of Industrial Chemistry*. 5th ed. Vol A1: Deerfield Beach, FL: VCH Publishers, 1985 to present, p. VA8 555.

8. Sax, N. I. *Dangerous Properties of Industrial Materials.* 6th ed. New York, NY: Van Nostrand Reinhold, 1984. 1478.

9. U.S. Coast Guard, Department of Transportation. *CHRIS—Hazardous Chemical Data.* Volume II. Washington, DC: U.S. Government Printing Office, 1984–5.

10. American Conference of Governmental Industrial Hygienists, Inc. *Documentation of the Threshold Limit Values and Biological Exposure Indices.* 6th ed. Volumes I, II, III. Cincinnati, OH: ACGIH, 1991. 703.

11. American Conference of Governmental Industrial Hygienists. *Threshold Limit Values for Chemical Substances and Physical Agents and Biological Exposure Indices for 1994–1995.* Cincinnati, OH: ACGIH, 1994. 22.

12. 21 CFR 175.105 (4/1/93).

13. Clayton, G. D., and F. E. Clayton (eds.). *Patty's Industrial Hygiene and Toxicology:* Volume 2A, 2B, 2C: Toxicology. 3rd ed. New York: John Wiley Sons, 1981–1982.

14. Sullivan, J.B., Jr., G.R. Krieger (eds.). *Hazardous Materials Toxicology—Clinical Principles of Environmental Health.* Baltimore, MD: Williams and Wilkins, 1992. 983.

15. Osol, A., and J.E. Hoover, et al. (eds.). *Remington's Pharmaceutical Sciences.* 15th ed. Easton, Pennsylvania: Mack Publishing Co., 1975.

16. Goodman, L.S., and A. Gilman. (eds.) *The Pharmacological Basis of Therapeutics.* 5th ed. New York: Macmillan Publishing Co., Inc., 1975.

17. American Medical Association, Council on Drugs. *AMA Drug Evaluations Annual 1994.* Chicago, IL: American Medical Association, 1994.

18. USEPA/OPP. *Status of Pesticides in Reregistration and Special Review.* p. 138 (Mar, 1992) EPA 700-R-92-004.

19. Gosselin, R.E., H.C. Hodge, R.P. Smith, and M.N. Gleason. *Clinical Toxicology of Commercial Products.* 4th ed. Baltimore: Williams and Wilkins, 1976.

■ GLYCIDYL METHACRYLATE

CAS # 106–91–2

SYNONYMS 1-propanol, 2,3-epoxy-, methacrylate • 2,3-epoxypropyl methacrylate • 2-propenoic acid, 2-methyl-, oxiranylmethyl ester • cp-105 • glycidol-methacrylate • glycidyl alpha-methyl acrylate • methacrylic acid, 2,3-epoxypropyl ester

MF $C_7H_{10}O_3$

MW 142.17

COLOR AND FORM Colorless liquid.

ODOR Fruity odor.

DENSITY 1.073 g/m^3 at 25°C.

SURFACE TENSION 25 dynes/cm = 0.025 N/m at 20°C (est).

VISCOSITY 5.481 cP at 70°F.

BOILING POINT 189°C.

OTHER CHEMICAL/PHYSICAL PROPERTIES Liquid water interfacial tension: (est) 40 dynes/cm = 0.04 N/m at 20°C; ratio of specific heats of vapor: (est) 1.043; heat of polymerization: (est) −900 Btu/lb = −500 cal/g = −20×10^5 J/kg. Liquid heat capacity: 0.426 Btu/lb-F at 85°F; liquid thermal conductivity: 1.040 Btu-in./hr-sq ft-F at 70°F; ideal gas heat capacity: 0.336 Btu/lb-F at 90°F.

PROTECTIVE EQUIPMENT Suitable protective clothing and self-contained resp protective apparatus should be available for use of those who may have to rescue persons overcome by fumes (acrylic acid and derivatives) [R3]. Polyethylene-coated apron and gloves and close-fitting goggles. [R8].

PREVENTATIVE MEASURES One hazard is the generation of considerable exothermic heat in some of the reactions, so that high pressures and temperature may develop. This danger should be borne in mind when designing plant. Awareness of the dangers and of good engineering design is essential to safety. Employees should be instructed about the necessity of cleansing the skin if it is contaminated by materials that are irritants or skin-absorbed. With careful design, however, and complete enclosure of those processes where toxic chemicals or intermediates occur, dangerous exposures can be avoided (acrylic acid and derivatives) [R3]. Contact lens use in industry is controversial. A survey of 100 corporations resulted in the recommendation that each company establish its own contact lens use policy. One presumed hazard of contact lens use is possible chemical entrapment. It was found that contact lenses minimized injury or protected the eye. The eye was afforded more protection from liquid irritants. Soft contact lenses did not worsen corneal damage from strong chemicals and in some cases could actually protect the eye. Overall, the literature supports the wearing of contact lenses in industrial environments as part of the standard eye protection, e.g., face shields; however, more data are needed to establish the value of contact lenses [R9]. SRP: Contaminated protective clothing should be segregated in such a manner so that there is no direct personal contact by personnel who handle, dispose of, or clean the clothing. Quality assurance to ascertain the completeness of the cleaning procedures should be implemented before the decontaminated protective clothing is returned for reuse by the workers. Contaminated clothing should not be taken home at end of shift, but should remain at employee's place of work for cleaning.

COMMON USES Chemical intermediate for polymers; monomer and diluent in epoxy resin formulations [R1]; in hydrogel lenses [R7, 730]; in BIS–GMA dental resin [R2]; co-polymerization of methyl methacrylate (80–90%) with acrylic acid (10–20%), and diverse cross-linking agents, followed by neutralization of the polymerized acrylic acid with a basic substance, such as ammonium hydroxide or ethylenimine, produces hydrogels used for contact lenses. (acrylic acid) [R7, 735].

OCCUPATIONAL RECOMMENDATIONS None listed.

NON-HUMAN TOXICITY LD_{50} mouse oral 1.05 g/kg [R4]; LD_{50} rat oral 0.29 g/kg [R4];

LD_{50} mouse intraperitoneal 0.35 g/kg [R4]; LD_{50} rat intraperitoneal 0.29 g/kg [R4].

TSCA REQUIREMENTS Section 8 (a) of TSCA requires manufacturers of this chemical substance to report preliminary assessment information concerned with production, use, and exposure to EPA as cited in the preamble of the 51 FR 41329 [R5].

FDA Glycidyl methacrylate is an indirect food additive for use only as a component of adhesives. (glycidyl methacrylate polymers) [R6].

GENERAL RESPONSE Avoid contact with liquid. Keep people away. Wear rubber overclothing (including gloves). Call fire department. Stop discharge if possible. Isolate and remove discharged material. Notify local health and pollution control agencies.

FIRE RESPONSE Combustible. Wear goggles, self-contained breathing apparatus, and rubber overclothing (including gloves). Extinguish with foam, dry chemical, or carbon dioxide. Water may be ineffective on fire.

EXPOSURE RESPONSE Call for medical aid. Liquid will burn skin and eyes. Harmful if swallowed. Remove contaminated clothing and shoes. Flush affected areas with plenty of water. If swallowed and victim is conscious, have victim drink water or milk.

RESPONSE TO DISCHARGE Mechanical containment; should be removed; chemical and physical treatment.

WATER RESPONSE Effect of low concentrations on aquatic life is unknown. Fouling to shoreline. May be dangerous if it enters water intakes. Notify local health and wildlife officials. Notify operators of nearby water intakes.

EXPOSURE SYMPTOMS The liquid irritates eyes about as much as soap does. Prolonged contact with skin produces irritation and dermatitis.

EXPOSURE TREATMENT Skin: wash thoroughly with soap and water and treat as a chemical burn. Eyes: irrigate with clear and treat as a chemical burn. Eyes: irrigate with clear water for 15 min and get medical attention.

REFERENCES

1. SRI.

2. *Kirk–Othmer Encyclopedia of Chemical Technology.* 3rd ed., Volumes 1–26. New York, NY: John Wiley and Sons, 1978–1984, p. 7 (79) 5004.

3. International Labour Office. *Encyclopedia of Occupational Health and Safety.* Vols. I and II. Geneva, Switzerland: International Labour Office, 1983. 53.

4. Petrov, I. G., Mater, "Povolzh Konf Fiziol Uchastiem Biokhim", *Farmakol Morfol*, 6th (2): 49–50 (1973).

5. 40 CFR 712.30 (7/1/88).

6. 21 CFR 175.105 (4/1/88).

7. *Kirk-Othmer Encyclopedia of Chemical Technology.* 3rd ed., Volumes 1–26. New York, NY: John Wiley and Sons, 1978–1984.,p. 6(79).

8. U.S. Coast Guard, Department of Transportation. *CHRIS—Hazardous Chemical Data.* Volume II. Washington, DC: U.S. Government Printing Office, 1984–5.

9. Randolph, S. A., M. R. Zavon. *J Occup Med.* 29: 237–42 (1987).

■ GRISEOFULVIN

CAS # 126–07–8

SYNONYMS (2s-trans)-7-chloro-2′,4,6-trimethoxy-6′-methylspiro[benzofuran-2 (3h),1′-[2]cyclohexene]-3,4′-dione • 7-chloro-4,6,2′-trimethoxy-6′-methylgris-2′-en-3,4′-dione • 7-chloro-4,6-dimethoxycoumaran-3-one-2-spiro-1′-(2′-methoxy-6′-methylcyclohex-2′-en-4′-one) • amudane • biogrisin-fp • curling-factor • delmofulvina • fulcin • fulvicin • fulvicin-p/g • fulvicin-u/f • fulvina • fulvinil • fulvistatin • fungivin • greosin • gresfeed • gricin • grifulin • grifulvin • gris-peg • grisactin • griscofulvin • grisefuline • griseofulvin, +/- • griseofulvin-forte • griseofulvinum • grisetin • grisofulvin • grisovin •

grizeofulvin • grysio • guservin • lamoryl • likuden • murfulvin • neofulcin • nsc-34533- • poncyl • spiro (benzofuran-2 (3h),1′-(2) cyclohexene)-3,4′-dione, 7-chloro-2′,4,6-trimethoxy-6′-methyl-, (2s-trans)- • spiro (benzofuran-2 (3h),1′-(2)cyclohexene)-3,4′-dione, 7-chloro-2′,4,6-trimethoxy-6′-beta-methyl • spirofulvin • sporostatin • USAF sc-2

MF $C_{17}H_{17}ClO_6$

MW 352.77

COLOR AND FORM Stout octahedra or rhombs from benzene; white to creamy powder; colorless crystalline solid.

ODOR Odorless.

MELTING POINT 220°C.

SOLUBILITY Soluble in N,N-dimethylformamide at 25°C: 12–14 g/100 ml; slightly soluble in ethanol, chloroform, methanol, acetic acid, acetone, benzene, and ethyl acetate; practically insoluble in water and petroleum ether.

OCTANOL/WATER PARTITION COEFFICIENT Log P = 2.18.

STABILITY AND SHELF LIFE Preparations of griseofulvin have expiration dates of 2–5 yr following the date of manufacture. [R4, 83].

OTHER CHEMICAL/PHYSICAL PROPERTIES Sublimes without decomposition at 210°C.

PROTECTIVE EQUIPMENT Dispensers of liquid detergent should be available. Safety pipettes should be used for all pipetting. In animal laboratory, personnel should wear protective suits (preferably disposable, one-piece, and close-fitting at ankles and wrists), gloves, hair covering, and overshoes. In chemical laboratory, gloves and gowns should always be worn; however, gloves should not be assumed to provide full protection. Carefully fitted masks or respirators may be necessary when working with particulates or gases, and disposable plastic aprons might provide addnl protection. Gowns should be of distinctive color, as a reminder that they are not to be worn outside the laboratory. (chemical carcinogens) [R3, 8].

PREVENTATIVE MEASURES Smoking,

drinking, eating, storage of food, or of food and beverage containers, or utensils, and the application of cosmetics should be prohibited in any laboratory. All personnel should remove gloves, if worn, after completion of procedures in which carcinogens have been used. They should wash hands, preferably using dispensers of liquid detergent, and rinse thoroughly. Consideration should be given to appropriate methods for cleaning the skin, depending on nature of the contaminant. No standard procedure can be recommended, but the use of organic solvents should be avoided. Safety pipettes should be used for all pipetting. (chemical carcinogens) [R3, 8].

CLEANUP A high-efficiency particulate arrestor (HEPA) or charcoal filters can be used to minimize amount of carcinogen in exhausted-air-ventilated safety cabinets, lab hoods, glove boxes, or animal rooms. Filter housing that is designed so that used filters can be transferred into plastic bag without contaminating maintenance staff is avail commercially. Filters should be placed in plastic bags immediately after removal. The plastic bag should be sealed immediately. The sealed bag should be labelled properly. Waste liquids should be placed or collected in proper containers for disposal. The lid should be secured and the bottles properly labelled. Once filled, bottles should be placed in plastic bag, so that outer surface is not contaminated The plastic bag should also be sealed and labelled. Broken glassware should be decontaminated by solvent extraction, by chemical destruction, or in specially designed incinerators. (chemical carcinogens) [R3, 15].

DISPOSAL SRP: At the time of review, criteria for land treatment or burial (sanitary landfill) disposal practices are subject to significant revision. Prior to implementing land disposal of waste residue (including waste sludge), consult with environ-mental regulatory agencies for guidance on acceptable disposal practices.

COMMON USES Antifungal antibiotic in human toxicity mycotic diseases; antifungal antibiotic in vet medicine [R1].

OCCUPATIONAL RECOMMENDATIONS None listed.

BIOLOGICAL HALF-LIFE Drug has a half-life in plasma of about 1 day, and approximately 50% of oral dose can be detected in the urine within 5 days, mostly in the form of metabolites (SRP: 36% in the feces within 5 days). [R5, 1173].

FDA Manufacturers, packers, and distributors of drug and drug products for human use are responsible for complying with the labeling, certification, and usage requirements as prescribed by the Federal Food, Drug, and Cosmetic Act, as amended (secs 201–902, 52 Stat. 1040 et seq., as amended; 21 U.S.C. 321–392) [R2].

REFERENCES

1. SRI.

2. 21 CFR 200–299, 300–499, 820, and 860 (4/1/93).

3. Montesano, R., H. Bartsch, E. Boyland, G. Della Porta, L. Fishbein, R. A. Griesemer, A.B. Swan, L. Tomatis, and W. Davis (eds.). *Handling Chemical Carcinogens in the Laboratory: Problems of Safety.* IARC Scientific Publications No. 33. Lyon, France: International Agency for Research on Cancer, 1979.

4. McEvoy, G.K. (ed.). *American Hospital Formulary Service—Drug Information 94.* Bethesda, MD: American Society of Hospital Pharmacists, Inc. 1994 (Plus Supplements).

5. Gilman, A.G., T.W. Rall, A.S. Nies, and P. Taylor (eds.). *Goodman and Gilman's The Pharmacological Basis of Therapeutics.* 8th ed. New York, NY. Pergamon Press, 1990.

◼ HEPTYL ALCOHOL

CAS # 111–70–6

SYNONYMS 1-hydroxyheptane • alcohol c-7 • enanthic alcohol • gentanol • heptyl alcohol • hydroxy-heptane • n-heptan-1-ol • n-heptanol • n-heptyl alcohol • pri-n-heptyl alcohol • 1-heptanol

MF $C_7H_{16}O$

MW 116.23

COLOR AND FORM Colorless liquid.

ODOR Faint, aromatic, fatty.

TASTE Pungent, spicy taste.

DENSITY 0.8219 at 20°C.

BOILING POINT 176°C at 760 mm Hg.

MELTING POINT –34.6°C.

FIREFIGHTING PROCEDURES Use alcohol foam, carbon dioxide, dry chem [R2].

SOLUBILITY Miscible with alcohol and ether; 1.0 g/L of water at 18°C; 2.85 g/L of water at 100°C; 5.15 g/L of water at 130°C.

OTHER CHEMICAL/PHYSICAL PROPERTIES Water solutions are colloidal.

COMMON USES Food additive [R1]; used in non-alcoholic beverages 0.90 ppm; ice cream, ices, etc., 1.0–5.0 ppm; candy 3.0 ppm; baked goods 3.0 ppm [R1]. Found in a few essential oils: hyacinth, violet leaves, *Litsea zeylanica* [R1]. Organic intermediate, solvent, cosmetic formulations [R3]. Used as a solvent in the wet-process superphosphoric acid extraction by USS Agri-chemicals [R4].

NON-HUMAN TOXICITY Exposure to saturated vapor of heptanol for 4 hr did not produce mortality in rats. It was moderately irritant to eye and slightly to skin. Oral LD_{50} value was 6,200 and 5,500 mg/kg in male and female [R5]. Chronic exposure of rabbits to air containing max permissible concentration damaged structure and functional capacity of retina, optic nerve, chiasma, optic tract, anterior corpora quadrigemina, and cerebral optic lobe. Maximum permissible concentration apparently should be redefined [R6]. n-Hexyl alcohol was toxic to rats by aspiration route. Lower-mol-wt alcohols (C3–C10) caused death due to cardiac or respiratory arrest, or both [R7].

OCCUPATIONAL RECOMMENDATIONS OEL–Russia: STEL 10 mg/m³; skin Jan 1993.

GENERAL RESPONSE Stop discharge if possible. Call fire department. Isolate and remove discharged material. Notify local health and pollution control agencies.

FIRE RESPONSE Combustible. Extinguish with dry chemical, foam, or carbon dioxide. Cool exposed containers with water.

EXPOSURE RESPONSE Not harmful.

RESPONSE TO DISCHARGE Mechanical containment. Should be removed. Provide chemical and physical treatment.

WATER RESPONSE Effect of low concentrations on aquatic life is unknown. Fouling to shoreline. May be dangerous if it enters water intakes. Notify local health and wildlife officials. Notify operators of nearby water intakes.

EXPOSURE SYMPTOMS Low toxicity; liquid may irritate eyes.

EXPOSURE TREATMENT Flush all affected parts with plenty of water.

REFERENCES

1. *Fenaroli's Handbook of Flavor Ingredients*. Volume 2. Edited, translated, and revised by T. E. Furia and N. Bellanca. 2nd ed. Cleveland: The Chemical Rubber Co., 1975. 238.

2. Sax, N. I. *Dangerous Properties of Industrial Materials*. 5th ed. New York: Van Nostrand Reinhold, 1979. 722.

3. Hawley, G. G. *The Condensed Chemical Dictionary*. 9th ed. New York: Van Nostrand Reinhold Co., 1977. 435.

4. *Kirk–Othmer Encyc Chemical Tech* 3rd ed. 1978–present V10 p. 69.

5. Truhaut, R., et al., Toxicological study of primary n–heptyl alcohol; *Arch Mal Prof Med Trav Secur Soc* 35 (4–5) 501 (1974).

6. Feldman, N.G., et al. *Gig Tr Prof Zabol* (3): 55 (1973).

7. Gerarde, H. W., et al. Aspiration hazard and toxicity of homologous series of alcohols; *Arch Environ Health* 13 (4) 457 (1966).

■ HEXACHLORONAPHTHALENE

CAS # 1335-87-1

SYNONYMS Halowax 1014 • hexachlornaftalen (Czech) • hexachloronaphthalene (ACGIH: OSHA) • naphthalene, hexachloro

MF $C_{10}H_2Cl_6$

MW 334.82

COLOR AND FORM White solid.

ODOR Aromatic odor.

DENSITY 1.78 (water = 1).

BOILING POINT 344-388°C (approximately).

MELTING POINT 137°C (279°F).

FIREFIGHTING PROCEDURES Extinguishant: foam, carbon dioxide, dry chemical. Also wear a self-contained breathing apparatus (SCBA) with a full facepiece operated in the pressure-demand or other positive-pressure mode. [R8, 2].

TOXIC COMBUSTION PRODUCTS Toxic gases and vapors (such as hydrogen chloride, phosgene, and carbon monoxide) may be released in a fire involving hexachloronaphthalene. [R8, 2].

SOLUBILITY Insoluble in water; soluble in organic solvents.

VAPOR PRESSURE 3×10^{-8} mm Hg at 25°C (est).

OCTANOL/WATER PARTITION COEFFICIENT Log K_{ow} = 7.59 (est).

OTHER CHEMICAL/PHYSICAL PROPERTIES Henry's Law constant = 0.000087 atm-m³/mole (est); vapor pressure: less than 1 mm Hg at 20°C.

PROTECTIVE EQUIPMENT Persons should be provided with and required to use impervious clothing, gloves, face shields (eight-inch minimum), and other appropriate clothing to prevent skin contact with liquid, or molten hexachloronaphthalene or its fumes [R8, 3]. Employees should be provided with and required to use splashproof safety goggles where there is any possibility of molten Halowax 1014, solid Halowax 1014, or liquids containing Halowax 1014 contacting the eyes. [R8].

PREVENTATIVE MEASURES Clothing contaminated with hexachloronaphthalene should be removed immediately and placed in closed containers until it can be discarded or until provision is made for the removal of hexachloronaphthalene from the clothing [R8, 3]. SRP: The scientific literature supports the wearing of contact lenses in industrial environments, as part of a program to protect the eye against chemical compounds and minerals causing eye irritation. However, there may be individual substances whose irritating or corrosive properties are such that the wearing of contact lenses would be harmful to the eye. In those specific cases contact lenses should not be worn. Hexachloronaphthalene can affect the body if it is inhaled, comes in contact with the eyes or skin, or is swallowed. Every effort should be made to prevent skin, eye, oral, or inhalation contact with this material [R8]. Eating and smoking should not be permitted in areas where hexachloronaphthalene is handled, processed, or stored [R8]. Workers subject to skin contact with hexachloronaphthalene should wash with soap or mild detergent and water any areas of the body that may have contacted hexachloronaphthalene at the end of each work day [R8]. Employees who handle hexachloronaphthalene should wash their hands thoroughly with soap or mild detergent and water before eating, smoking, or using toilet facilities [R8]. Contact lenses should not be worn when working with this chemical [R9, 123]. Condenser impregnation and other operations involving melting of chloronaphthalene should be enclosed or provided with effective local exhaust ventilation (chloronaphthalenes) [R10]. SRP: Local exhaust ventilation should be applied wherever there is an incidence of point-source emissions or dispersion of regulated contaminants in the work area. Ventilation control of the contaminant as close as possible to its point of generation is both the most economical and safest method to minimize personnel exposure to airborne contaminants.

CLEANUP Collect spilled material in the most convenient and safe manner for reclamation or disposal in a secured sanitary landfill. Liquids containing hexachloronaphthalene should be absorbed in vermiculite, dry sand, or a similar material. [R8, 4].

DISPOSAL Incineration, preferably after mixing with another combustible fuel. Assure complete combustion to prevent the formation of phosgene. An acid scrubber is necessary to remove the halo acids produced. Recommendable method: Incineration. Peer-review: Ensure plentiful supply of hydrocarbon fuel (peer-review conclusions of an IRPTC expert consultation (May 1985)) [R1].

COMMON USES Used as an inert component of resins for coating or impregnating textiles, wood paper for flame, and waterproofing and fungicidal and insecticidal properties [R8, 1]. Used in electric wire insulation, and also as additives to special lubricants. [R13, 302].

OCCUPATIONAL RECOMMENDATIONS None listed.
OSHA 8-hr time-weighted average: (0.2 mg/m³). Skin absorption designation [R6].

THRESHOLD LIMIT 8-hr time-weighted average (TWA) 0.2 mg/m³, skin (1986) [R11, 22]. Excursion Limit Recommendation: Excursions in worker exposure levels may exceed three times the TLV–TWA for no more than a total of 30 min during a work day, and under no circumstances should they exceed five times the TLV–TWA. [R11, 5].

NON-HUMAN TOXICITY Swine, oral 176–220 mg/kg for 8–9 days. Death in 33–40 days [R2]. Sheep oral 117 mg/day penta-hexa for 106 days. Liver, heart, and renal disease (penta-hexachloronaphtha-lene) [R2]. Rabbit ip 15 mg/day penta-hexa for 12–26 days, liver disease. (penta-hexachloronaphthalene) [R3].

BIODEGRADATION Hexachloronaphthalene is reported to have "poor" biodegradability (1) [R4].

BIOCONCENTRATION Based upon an estimated log K_{ow} value of 7.59 (1), the BCF for hexachloronaphthalene can be estimated to be 3.45×10^5 using a recommended regression-derived equation (2,SRC), which suggests that bioconcentration in aquatic organisms may be an important fate process (SRC) [R5].

TSCA REQUIREMENTS Section 8 (a) of TSCA requires manufacturers of this chemical substance to report preliminary assessment information concerned with production, use, and exposure to EPA as cited in the preamble of the 51 FR 41329 [R7]. Pursuant to section 8(d) of TSCA, EPA promulgated a model Health and Safety Data Reporting Rule. The section 8(d) model rule requires manufacturers, importers, and processors of listed chemical substances and mixtures to submit to EPA copies and lists of unpublished health and safety studies. Hexachloronaphthalene is included on this list. [R12].

REFERENCES

1. United Nations. *Treatment and Disposal Methods for Waste Chemicals (IRPTC File)*. Data Profile Series No. 5. Geneva, Switzerland: United Nations Environmental Programme, Dec. 1985. 213.

2. Health and Welfare Canada. *Chloronaphthalenes: An Environmental-Health Perspective*. p. 79 (1982) 83–EHD–96.

3. Health and Welfare Canada. *Chloronaphthalenes: An Environmental-Health Perspective*. p. 79 (1982) 83–EHD–96.

4. Koda, Y., et al. *Nagoya Kogyo Gijutsu Shikensho Hokoku* 31: 299–305 (1982).

5. (1) GEMS. *Graphical Exposure Modeling System*. CLOGP. USEPA (1987). (2) Lyman, W. J., et al., *Handbook of Chemical Property Estimation Methods* Washington, DC: Amer Chemical Soc p. 5–4 (1990).

6. 29 CFR 1910.1000 (7/1/91).

7. 40 CFR 712.30 (7/1/91).

8. Mackison, F. W., R. S. Stricoff, and L. J. Partridge, Jr. (eds.). *NIOSH/OSHA— Occupational Health Guidelines for Chemical Hazards*. DHHS (NIOSH) Publication No. 81–123 (3 vols). Washington, DC: U.S. Government Printing Office, Jan. 1981.

9. NIOSH. *NIOSH Pocket Guide to Chemical Hazards*. DHHS (NIOSH) Publication No. 90–117. Washington, DC: U.S. Government Printing Office, June 1990.

10. Morgan, D. P., *Recognition and Management of Pesticide Poisonings*. 4th ed. p.138, EPA 540/9-88-001. Washington, DC: U.S. Government Printing Office, March 1989.

11. American Conference of Governmental Industrial Hygienists. *Threshold Limit Values for Chemical Substances and Physical Agents and Biological Exposure Indices for 1994–1995*. Cincinnati, OH: ACGIH, 1994.

12. 40 CFR 716.120 (7/1/91).

13. American Conference of Governmental Industrial Hygienists. *Documentation of the Threshold Limit Values and Biological Exposure Indices*. 5th ed. Cincinnati, OH: American Conference of Governmental Industrial Hygienists, 1986.

■ HEXAMETHYLENEIMINE

CAS # 111–49–9

DOT # 2493

SYNONYMS 1-azacycloheptane • 1h-azepine, hexahydro • azacycloheptane • cycloheptane, 1-aza • cyclohexamethylenimine • g-0 • hexahydro-1h-azepine • hexahydroazepine • hexamethylene-imine • hmi • homopiperidine • perhydroazepine

MF $C_6H_{13}N$

MW 99.20

COLOR AND FORM Clear colorless liquid.

ODOR Ammonia-like odor.

DENSITY 0.8643 at 22°C.

BOILING POINT 138.0°C at 749 mm Hg.

MELTING POINT −37°C.

FLAMMABILITY LIMITS 1.6%–2.3% (in air) [R1].

FIREFIGHTING PROCEDURES Dry chemical, alcohol foam, carbon dioxide. Water may be ineffective [R1].

TOXIC COMBUSTION PRODUCTS Toxic oxides of nitrogen may form in fire [R1].

SOLUBILITY Soluble in water; very soluble in alcohol, ether.

PROTECTIVE EQUIPMENT Self-contained breathing apparatus; impervious gloves; chemical safety goggles; impervious apron and boots [R1].

COMMON USES Chemical intermediate for pharmaceutical, agricultural, and rubber chemicals [R2]; in preparation of herbicides [R3].

OCCUPATIONAL RECOMMENDATIONS None listed.

DOT *Fire or Explosion:* Flammable/combustible material; may be ignited by heat, sparks, or flames. Vapors may travel to a source of ignition, and flash back. Container may explode in heat of fire. Vapor explosion hazard indoors, outdoors, or in sewers. Runoff to sewer may create fire or explosion hazard [R4].

Health Hazards: If inhaled, may be harmful; contact may cause burns to skin and eyes. Fire may produce irritating or poisonous gases. Runoff from fire control or dilution water may cause pollution [R4].

Emergency Action: Keep unnecessary people away; isolate hazard area and deny entry. Stay upwind; keep out of low areas. Self-contained breathing apparatus (SCBA) and structural firefighter's protective clothing will provide limited protection. Isolate for 1/2 mile in all directions if tank car or truck is involved in fire. Call CHEMTREC at 1–800–424–9300 for emergency assistance. If water pollution occurs, notify the appropriate authorities [R4].

Fire: Some of these materials may react violently with water. Small Fires: Dry chemical, CO_2, halon, water spray, or standard foam. Large Fires: Water spray, fog, or standard foam is recommended. Move container from fire area if you can do so without risk. Do not get water inside container. Cool containers that are exposed to flames with water from the side until well after fire is out. Stay away from ends of tanks. Withdraw immediately in case of rising sound from venting safety device or any discoloration of tank due to fire [R4].

Spill or Leak: Shut off ignition sources; no flares, smoking, or flames in hazard area. Do not touch spilled material; stop leak if you can do so without risk. Use water spray to reduce vapor; do not get water inside container. Small Spills: Take up with sand or other noncombustible absorbent material and place into containers for later disposal. Large Spills: Dike far ahead of liquid spill for later disposal [R4].

First Aid: Move victim to fresh air and call emergency medical care; if not breathing, give artificial respiration; if breathing is difficult, give oxygen. Remove and isolate contaminated clothing and shoes at the site. In case of contact with material, immediately flush skin or eyes with running water for at least 15 minutes. Keep victim quiet and maintain normal body temperature [R4].

GENERAL RESPONSE Avoid contact with liquid and vapor. Keep people away. Shut off ignition sources. Call fire department. Stop discharge if possible. Isolate and remove discharged material. Notify local health and pollution control agencies.

FIRE RESPONSE Combustible. Poisonous gases may be produced in fire. Flashback along vapor trail may occur. Vapor may explode if ignited in an enclosed area. Wear goggles and self-contained breathing apparatus. Extinguish with dry chemicals, alcohol foam, or carbon dioxide. Water may be ineffective on fire. Cool exposed containers with water.

EXPOSURE RESPONSE Call for medical aid.

Vapor irritating to eyes, nose, and throat. If inhaled will cause coughing, difficult breathing, or loss of consciousness. If breathing has stopped, give artificial respiration. If breathing is difficult, give oxygen. Liquid poisonous if swallowed. Will burn skin and eyes. If swallowed will cause nausea. Remove contaminated clothing and shoes. Flush affected areas with plenty of water. If swallowed and victim is conscious, have victim drink water or milk. If swallowed, and victim is unconscious or having convulsions, do nothing except keep victim warm. Do not induce vomiting.

RESPONSE TO DISCHARGE Issue warning—corrosive, air contaminant, water contaminant, restrict access, disperse, and flush.

WATER RESPONSE Effect of low concentrations on aquatic life is unknown. May be dangerous if it enters water intakes. Notify local health and wildlife officials. Notify operators of nearby water intakes.

EXPOSURE SYMPTOMS Inhalation of vapor irritates respiratory tract; high concentrations may cause disturbance of the central nervous system. Ingestion causes burns of mouth and stomach. Contact with concentrated vapor may cause severe eye injury. Contact with liquid causes burns of eyes and skin.

EXPOSURE TREATMENT Inhalation: remove victim to uncontaminated atmosphere; get medical attention. Ingestion: give large amount of water; do not induce vomiting; get medical attention if large amount was swallowed. Eyes: flush with water for 15 min and get medical attention. Skin: flush with water; wash with soap and water.

REFERENCES

1. U.S. Coast Guard, Department of Transportation. *CHRIS—Hazardous Chemical Data.* Manual Two. Washington, DC: U.S. Government Printing Office, Oct., 1978.

2. Hawley, G. G. *The Condensed Chemical Dictionary.* 9th ed. New York: Van Nostrand Reinhold Co., 1977. 439.

3. Shigematsu, S., et al; *Herbicides;* Ja-

pan Kokai Patent no 76 95133 08/20/76 (Kumiai Chemical Industry Co, Ltd).

4. Department of Transportation. *Emergency Response Guidebook 1987*. DOT P 5800.4. Washington, DC: U.S. Government Printing Office, 1987, p. G–29.

■ HEXANOIC ACID

CAS # 142–62–1

SYNONYMS caproic acid

MF $C_6H_{12}O_2$

MW 116.18

COLOR AND FORM Oily liquid.

ODOR Characteristic goat-like odor.

ODOR THRESHOLD Detection: 3.0 mg/kg [R7].

DENSITY 0.929 at 20°C.

SURFACE TENSION 23.4 mN/m at 70°C.

VISCOSITY 3.23 mPa. s at 20°C.

BOILING POINT 205.8°C.

MELTING POINT –3.4°C.

HEAT OF COMBUSTION –3,492.4 kJ/mol (liquid).

AUTOIGNITION TEMPERATURE 716°F [R5].

FIREFIGHTING PROCEDURES CO_2, dry chemical, fog, mist [R5]. Use water in flooding quantities as fog. Solid streams of water may be ineffective. Cool all affected containers with flooding quantities of water. Apply water from as far a distance as possible. Use "alcohol" foam, carbon dioxide, or dry chemical. Use water spray to knock down vapors [R6].

SOLUBILITY 11,000 mg/L (in water); slightly soluble in water (1.082 g/100 g); readily soluble in ethanol, ether.

VAPOR PRESSURE 0.2 mm at 20°C.

OCTANOL/WATER PARTITION COEFFICIENT Log K_{ow} = 1.88 to 1.92.

OTHER CHEMICAL/PHYSICAL PROPERTIES Heat of formation: –584.0 kJ/mol; specific heat 2.33 J/g; viscosity: 0.031 Poise at 20°C.

PROTECTIVE EQUIPMENT Strict precautions are necessary in handling; suitable protective equipment should be available. (saturated monocarboxylic acids) [R8].

PREVENTATIVE MEASURES If material not on fire and not involved in fire: Keep sparks, flames, and other sources of ignition away. Keep material out of water sources and sewers. Build dikes to contain flow as necessary. Use water spray to knock down vapors. Neutralize spilled material with crushed limestone, soda ash, or lime [R6].

DISPOSAL The following wastewater treatment technology has been investigated for caproic acid: concentration process: activated carbon [R9].

COMMON USES As insect attractant [R3]; manufacture of esters for artificial flavors and of hexyl derivatives, especially hexylphenols, and hexylresocinol [R2]; analytical chemistry, flavors, manufacture of rubber chemicals, varnish driers, resins, and pharmaceuticals [R1].

OCCUPATIONAL RECOMMENDATIONS None listed.

NON-HUMAN TOXICITY LD_{50} rat oral 6.44 g/kg [R7].

ECOTOXICITY TLm *Daphnia magna* 22 mg/L/24 hr [R7]; TLm *Lepomis macrochirus* 15–200 mg/L/24 hr [R7]; LC_{50} *Pimephales promelas* (fathead minnow), static bioassay in Lake Superior water at 18–22°C, 140 mg/L/1-hr [R7]; LC_{50} *Pimephales promelas* (fathead minnow), static bioassay in Lake Superior water at 18–22°C, 88 mg/L/24 hr [R7]; LC_{50} *Pimephales promelas* (fathead minnow), static bioassay in Lake Superior water at 18–22°C, 88 mg/L/48 hr [R7]; LC_{50} *Pimephales promelas* (fathead minnow), static bioassay in Lake Superior water at 18–22°C, 88 mg/L/72 hr [R7]; LC_{50} *Pimephales promelas* (fathead minnow), static bioassay in Lake Superior water at 18–22°C, 88 mg/L/96 hr [R7]; LC_{50} *Pimephales promelas* (fathead minnow) 320 mg/L/96 hr (confidence limit 306–334 mg/L), flow-through bioassay with measured concentrations, 25.3°C, dissolved oxygen 5.7 mg/L, hardness 43.0 mg/L calcium carbonate, alkalinity 46.0 mg/L calcium carbonate, and pH 7.58 [R10].

BIODEGRADATION A 5-day theoretical BOD of 44% was observed for hexanoic acid in an aerobic screening test using a sewage inoculum (1). Five- and 20-day theoretical BODs of 66 and 87% were observed in another aerobic screening test using a sewage inoculum (2). Using a Warburg respirometer, an adapted sewage inoculum, and 10,000 ppm concns of hexanoic acid, respective 5-, 10-, and 20-day theoretical BODs of 29, 66, and 69% were measured under aerobic conditions (3). One-day theoretical BODs of 26–54% were determined in a Warburg respirometer using various activated sludge inocula (4). Five-day theoretical BODs of 98–99% were achieved in an aerobic screening study using acclimated activated sludge inoculum (5). Respective 2-, 5-, 10-, and 30-day theoretical BODs of 42, 48, 54, and 65% were measured in an aerobic Warburg respirometer study using sewage inoculum (6). Using a Warburg respirometer and activated sludge inocula from three Tennessee municipal plants, theoretical BODs of 34.9–61.2% were measured over a 3-day inoculation period (7) [R11].

BIOCONCENTRATION Based upon a water solubility of 10,270 mg/L at 25°C (1) and an experimental log K_{ow} of 1.92 (2), respective BCFs of 3.4 and 17 can be estimated for hexanoic acid from recommended regression-derived equations (2,SRC); these estimated BCF values suggest that bioconcentration in aquatic organisms is not an important environmental fate process (SRC) [R12].

DOT *Health Hazards:* Contact causes burns to skin and eyes. If inhaled, may be harmful. Fire may produce irritating or poisonous gases. Runoff from fire control or dilution water may cause pollution [R4].

Fire or Explosion: Some of these materials may burn, but none of them ignites readily. Flammable/poisonous gases may accumulate in tanks and hopper cars. Some of these materials may ignite combustibles (wood, paper, oil, etc.) [R4].

Emergency Action: Keep unnecessary people away; isolate hazard area and deny entry. Stay upwind; keep out of low areas. Positive-pressure self-contained breathing apparatus (SCBA) and structural fire fighters' protective clothing will provide limited protection. Call CHEMTREC at 1–800–424–9300 for emergency assistance. If water pollution occurs, notify the appropriate authorities [R4].

Fire: Some of these materials may react violently with water. Small Fires: Dry chemical, CO_2, water spray, or regular foam. Large Fires: Water spray, fog, or regular foam. Move container from fire area if you can do so without risk. Apply cooling water to sides of containers that are exposed to flames until well after fire is out. Stay away from ends of tanks [R4].

Spill or Leak: Do not touch or walk through spilled material; stop leak if you can do so without risk. Small Spills: Take up with sand or other noncombustible absorbent material and place into containers for later disposal. Small Dry Spills: With clean shovel place material into clean, dry container, and cover loosely; move containers from spill area. Large Spills: Dike far ahead of liquid spill for later disposal [R4].

First Aid: Move victim to fresh air; call for emergency medical care. In case of contact with material, immediately flush skin or eyes with running water for at least 15 minutes. Remove and isolate contaminated clothing and shoes at the site. Keep victim quiet and maintain normal body temperature [R4].

FDA Hexanoic acid is a food additive permitted for direct addition to food for human consumption, as long as: (1) the quantity added to food does not exceed the amount reasonably required to accomplish its intended physical, nutritive, or other technical effect in food, and (2) when intended for use in or on food it is of appropriate food grade and is prepared and handled as a food ingredient. Synthetic flavoring substances and adjuvants (includes hexanoic acid) may be safely used in foods [R13].

REFERENCES

1. Sax, N. I., and R. J. Lewis, Sr. (eds.). *Hawley's Condensed Chemical Dictionary.* 11th ed. New York: Van Nostrand Reinhold Co., 1987. 214.

2. Budavari, S. (ed.). *The Merck Index—Encyclopedia of Chemicals, Drugs, and Biologicals.* Rahway, NJ: Merck and Co., Inc., 1989. 266.

3. *Kirk–Othmer Encyclopedia of Chemical Technology.* 3rd ed., Volumes 1–26. New York, NY: John Wiley and Sons, 1978–1984, p. V13 482 (1981).

4. U.S. Department of Transportation. *Emergency Response Guidebook 1990.* DOT P 5800.5. Washington, DC: U.S. Government Printing Office, 1990, p. G–60.

5. Sax, N. I. *Dangerous Properties of Industrial Materials.* 6th ed. New York, NY: Van Nostrand Reinhold, 1984. 1526.

6. Association of American Railroads. *Emergency Handling of Hazardous Materials in Surface Transportation.* Washington, DC: Assoc. of American Railroads, Hazardous Materials Systems (BOE), 1987. 367.

7. Verschueren, K. *Handbook of Environmental Data of Organic Chemicals.* 2nd ed. New York, NY: Van Nostrand Reinhold Co., 1983. 332.

8. International Labour Office. *Encyclopedia of Occupational Health and Safety.* Vols. I and II. Geneva, Switzerland: International Labour Office, 1983. 44.

9. USEPA. *Management of Hazardous Waste Leachate,* EPA Contract No. 68–03–2766 p. E–130 (1982).

10. Brooke, L. T., D. J. Call, D. T. Geiger, and C. E. Northcott (eds.). *Acute Toxicities of Organic Chemicals to Fathead Minnows (Pimephales promelas).* Superior, WI: Center for Lake Superior Environmental Studies, Univ. of Wisconsin–Superior, 1984. 175.

11. (1) Dore, M., et al., *Trib Cebedeau* 28: 3–11 (1975). (2) Gaffney, P. E., H. Heukelekian. *Sew Indust Wastes* 30: 673–9 (1958). (3) Gaffney, P. E., H. Heukelekian *J Water Pollut Control Fed* 33: 1169–83 (1961). (4) Malaney, G.W., R. M. Gerhold. *Proc 17th Ind Waste Conf, Purdue Univ, Ext Ser* 112: 249–57 (1962). (5) Engelbrecht, R. S., R. E. McKinney. *Sew Indust Wastes* 29: 1350 (1957). (6) Dias, F. F., M. Alexander, *Appl Microbial* 22: 1114–8 (1971). (7) Lutin, P. A. *J Water Pollut Control Fed* 42: 1632–42 (1970).

12. (1) Yalkowsky, S. H. *Arizona Database of Aqueous Solubilities.* Univ of AZ, College of Pharmacy (1989). (2) Hansch, C., and A. J. Leo. *Medchem Project* Issue No. 26, Claremont, CA: Pomona College (1985). (3) Lyman, W. J., et al., *Handbook of Chemical Property Estimation Methods* Washington, DC: Amer Chemical Soc pp. 5–4, 5–10 (1990).

13. 21 CFR 172.515 (4/1/91).

■ HEXAZINONE

CAS # 51235–04–2

SYNONYMS DPX-3674 • sha-107201 • 3-cyclohexyl-6-(dimethylamino)-1-methyl-1,3,5-triazine-2,4 (1h, 3h)-dione • 3-cyclohexyl-6-(dimethylamino)-1-methyl-s-triazine-2,4 (ih, 3h)-dione • velpar • velpar-k • velpar-l

MF $C_{12}H_{20}N_4O_2$

MW 252.36

CORROSIVITY Noncorrosive.

COLOR AND FORM White crystalline solid.

ODOR Odorless; negligible odor.

DENSITY Specific gravity = 1.25.

MELTING POINT 115–117°C.

SOLUBILITY Solubility in chloroform 388 g/100 g; solubility in methanol 265 g/100 g; solubility in benzene 94 g/100 g; solubility in dimethylformamide 83.6 g/100 g; solubility in acetone 79.2 g/100 g; solubility in toluene 38.6 g/100 g; sparingly soluble in hexane 0.3 g/100 g; 33 g/kg in water at 25°C.

VAPOR PRESSURE 3.0×10^{-7} mm Hg.

OCTANOL/WATER PARTITION COEFFICIENT Log K_{ow} = −4.40 (calculated).

CORROSIVITY Noncorrosive.

OTHER CHEMICAL/PHYSICAL PROPERTIES Octanol-water partition coefficient 4.01 (log, estimated).

PROTECTIVE EQUIPMENT Wear goggles or face shield and rubber gloves when handling this material [R1].

PREVENTATIVE MEASURES Keep away from heat, sparks, and open flame. Store above 32°C in a dry place. Keep container closed. Remove and wash contaminated clothing before re-use (velpar L) [R1]. Do not apply directly to water or wetlands. Do not apply where runoff is likely to occur. Do not contaminate water by cleaning of equipment or disposal of wastes (velpar L) [R1]. Avoid breathing dust or spray mist [R1]. Do not use on right-of-ways or other sites where marketable timber or other desirable plants are immediately adjacent to the treated area. Do not apply or drain or flush equipment on or near desirable trees or other plants, or on areas where their roots may extend, or in locations where the chemical may be washed or moved into contact with their roots [R8, p. 270–1]. Do not use on lawns, walks, driveways, tennis courts, or similar areas. Prevent drift of dry powder or spray to desirable plants. Do not contaminate any body of water. Do not reuse container [R8, 271].

COMMON USES Control of many annual, biennial, and perennial weeds and woody plants on non-cropland areas. Selective weed control in conifers, sugarcane, pineapple, rubber trees, and alfalfa [R1].

OCCUPATIONAL RECOMMENDATIONS None listed.

NON-HUMAN TOXICITY LD_{50} rat oral 1,690 mg/kg [R1]; LD_{50} rabbit > 5,278 mg/kg [R1]; LD_{50} guinea pig oral 860 mg/kg [R7]; LD_{50} dog oral 3,400 mg/kg [R7]; LC_{50} rat inhalation > 7.48 mg/L/1 hr exposure [R6, 272].

ECOTOXICITY LC_{50} *Salmo gairdneri* (rainbow trout) >100 mg/L/96 hr, static bioassay without aeration, pH 7.2–7.5, water hardness 40–50 mg/L as calcium carbonate, and alkalinity of 30–35 mg/L [R2]. LC_{50} *Lepomis macrochirus* (bluegill) > 100 mg/L/96 hr, static bioassay without aeration, pH 7.2–7.5, water hardness 40–50 mg/L as calcium carbonate and alkalinity of 30–35 mg/L [R2].

BIODEGRADATION The degradation of hexazinone in Taloka and Mountainberg soils incubated in the dark at 30°C and 10°C followed first-order kinetics (2). The respective half-lives were 77 and 76 days at 30°C and 502 and 426 days at 10°C (2). In contrast, when incubated in stream water, no significant degradation occurred at 10°C; the hexazinone concentration decreased about 2.2% in 200 days at 30°C, indicating a half-life of many years (2). Greenhouse soil degradation studies in both silt loam and sandy loam soil showed a half-life for primary degradation of <4 mo (1). Laboratory studies using C-14 labeled hexazinone with these soils showed that after an initial lag period of 10–20 days, the rate of CO_2 evolution increased rapidly with 45–75% of the applied radioactivity being evolved as CO_2 during the 80-day incubation period (1) [R3].

BIOCONCENTRATION The BCF in bluegill sunfish exposed to 0.01 and 1.0 ppm ^{14}C-radiolabeled hexazinone for 4 weeks was 2 and 1, respectively. Residue levels plateaued after 1–2 weeks of exposure. Maximum BCF were 2, 3–5, and 5–7 in the carcass, liver, and viscera, respectively (1). The hexazinone levels in the fish declined by over 90% after a 1-week depuration period and no residues were detected after 2 weeks (1) [R4].

FIFRA REQUIREMENTS In 1988, Congress amended FIFRA to strengthen and accelerate EPA's reregistration program. The nine-year reregistration scheme mandated by "FIFRA 88" applies to each registered pesticide product containing an active ingredient initially registered before November 1, 1984. List A consists of the 194 chemical cases (or 350 individual active ingredients) for which EPA had issued Registration Standards prior to the effective date of FIFRA '88. List: A; Case: Hexazinone; Case No. 0266; Pesticide type: Herbicide; Registration Standard Date: 12/31/86; Case Status: Awaiting Data/Data in Review: OPP awaits data from the pesticide's producer(s) regarding its human health or environmental effects, or OPP has received and is reviewing such data, in order to reach a decision about the pesticide's eligibility for reregistration. Active Ingredient (AI): Hexazinone; AI Status: The producer(s) of the pesticide has made commitments to conduct the studies and pay the fees required for reregistration, and is meeting those commitments in a timely manner [R5].

REFERENCES

1. *Farm Chemicals Handbook 1994.* Willoughby, OH: Meister, 1994, p. C 189.

2. U.S. Department of Interior, Fish and Wildlife Service. *Handbook of Acute Toxicity of Chemicals to Finfish and Aquatic Invertebrates.* Resource Publication No. 137. Washington, DC: U.S. Government Printing Office, 1980. 86.

3. (1) Rhodes, R.C. *J Agric Food Chem* 28: 311–5 (1980). (2) Bouchard, D.C., et al. *J Environ Qual* 14: 229–33 (1985).

4. Rhodes, R.C. *J Agric Food Chem* 28: 306–10 (1980).

5. USEPA/OPP. *Status of Pesticides in Reregistration and Special Review* p. 93 (Mar, 1992) EPA 700–R–92–004.

6. Weed Science Society of America. *Herbicide Handbook.* 5th ed. Champaign, Illinois: Weed Science Society of America, 1983.

7. Kennedy, G. L. *Fund Appl Toxicol.* 4 (4): 603–11 (1984).

8. USEPA. *Health Advisories for 50 Pesticides.* p.513 (1988) PB88-245931.

■ 1-HEXENE

CAS # 592-41-6

DOT # 2370

SYNONYMS hexene-1 • hexene • butylethylene • hexylene

MF C_6H_{12}

MW 84.16

COLOR AND FORM Colorless liquid.

BINARY REACTANTS Oxidizing materials.

SOLUBILITIES Insoluble in water; soluble in ethanol, ether, benzene, petroleum ether, chloroform.

VAPOR PRESSURE 310 mm Hg at 38°C.

COMMON USES In linear alpha-olefin mixt for production of primary alcohols; comonomer for high-density polyethylene [R1]. Synthesis of flavors, perfumes, dyes, resins, polymer modifier [R2].

STANDARD CODES NFPA—1, 3, 0; ICC, USCG—no.

PERSISTENCE Biodegrades at low rate. Photochemical action should break unsaturated bond.

MAJOR SPECIES THREATENED Waterfowl.

PERSONAL SAFETY Wear protective clothing and self-contained breathing apparatus.

ACUTE HAZARD LEVEL Moderate ingestive and inhalative toxicity. Irritant. Produces small BOD.

CHRONIC HAZARD LEVEL Unknown.

DEGREE OF HAZARD TO PUBLIC HEALTH Moderate ingestive and inhalative toxicity. Irritant.

OCCUPATIONAL RECOMMENDATIONS None listed.

ACTION LEVEL Notify fire and air authority. Enter from upwind and remove ignition source. Attempt to contain slick.

ON-SITE RESTORATION Oil-skimming equipment and sorbent foams can be applied to the slick. Seek professional environmental engineering assistance through EPA's Environmental Response Team (ERT), Edison, NJ, 24-hour no., 908–548–8730.

BEACH OR SHORE RESTORATION Burn off.

AVAILABILITY OF COUNTERMEASURE MATERIAL Oil-skimming equipment—stored at major ports; sorbent foams (polyurethane)—upholstery shops.

DISPOSAL Spray into incinerator or burn in paper packaging. Additional flammable solvent may be added.

INDUSTRIAL FOULING POTENTIAL Volatility suggests explosion hazard when confined in boiler feed or cooling-system water. Slick may reduce heat transfer or cause hot spots and scaling.

MAJOR WATER USE THREATENED Recreation, potable supply, industrial.

PROBABLE LOCATION AND STATE Colorless liquid. Will float as a slick on surface of water.

COLOR IN WATER Colorless.

BIODEGRADATION A Clark oxygen electrode was used to measure oxidation of hydrocarbons, using resting cell suspensions of corynebacterium strain that utilized n-octane as sole carbon and energy source. [R3].

GENERAL RESPONSE Stop discharge if possible. Keep people away. Shut off ignition sources and call fire department. Avoid contact with liquid and vapor. Stay upwind and use water spray to "knock down" vapor. Isolate and remove discharged material. Notify local health and pollution control agencies.

FIRE RESPONSE Flammable. Flashback along vapor trail may occur. Vapor may explode if ignited in an enclosed area. Extinguish with dry chemical, foam, or carbon dioxide. Water may be ineffective on fire. Cool exposed containers with water.

EXPOSURE RESPONSE Call for medical aid. Vapor, if inhaled, will cause dizziness, difficult breathing, or loss of consciousness. Move to fresh air. If breathing has stopped, give artificial respiration. If breathing is difficult, give oxygen. Liquid irritating to skin and eyes. Harmful if swallowed. Remove contaminated clothing and shoes. Flush affected areas with plenty of water. If swallowed and victim is conscious, have victim drink water or milk. Do not induce vomiting.

RESPONSE TO DISCHARGE Issue warning—high flammability. Evacuate area. Disperse and flush.

WATER RESPONSE Effect of low concentrations on aquatic life is unknown. Fouling to shoreline. May be dangerous if it enters water intakes. Notify local health and wildlife officials. Notify operators of nearby water intakes.

EXPOSURE SYMPTOMS Inhalation may cause giddiness or incoordination similar to that from gasoline vapor. Prolonged exposure to high concentrations may induce loss of consciousness or death.

EXPOSURE TREATMENT Skin or eyes: wash exposed skin areas with soap and water; thoroughly flush eyes with water to remove any splashes; launder contaminated clothing before reuse.

REFERENCES
1. SRI.

2. Hawley, G.G. *The Condensed Chemical Dictionary.* 9th ed. New York: Van Nostrand Reinhold Co., 1977. 440.

3. Buswell, J. A., et al. "Microbial Oxidation of Hydrocarbons Measured by Oxygraphy." *Arch Microbiol* 64(3) 215 (1969).

■ 2-HYDROXYETHYL ACRYLATE

CAS # 818–61–1

SYNONYMS 2-(acryloyloxy) ethanol • 2-propenoic acid, 2-hydroxyethyl ester • acrylic acid, 2-hydroxyethyl ester • beta-hydroxyethyl acrylate • bisomer-2hea • ethylene-glycol-monoacrylate • ethylene glycol acrylate • hea • hydroxyethyl acrylate

MF $C_5H_8O_3$

MW 116.13

DENSITY 1.1.

BOILING POINT 191°C.

EXPLOSIVE LIMITS AND POTENTIAL Containers may explode in fire (inhibited) [R2].

FLAMMABILITY LIMITS Lower 1.8% by volume [R5].

FIREFIGHTING PROCEDURES Extinguish with water, dry chemicals, alcohol foam, or carbon dioxide. Cool exposed containers with water [R2]. Use water in flooding quantities as fog. Solid stream of water may be ineffective. Cool all affected containers with flooding quantities of water. Apply water from as far a distance as possible. Use alcohol foam, dry chemical, or carbon dioxide [R6].

SOLUBILITY Soluble in water; estimated water solubility of 5.46×10^5 mg/L.

VAPOR PRESSURE 2.58×10^{-3} mm Hg at 25°C (calculated).

OCTANOL/WATER PARTITION COEFFICIENT Log K_{ow} = −0.21.

OTHER CHEMICAL/PHYSICAL PROPERTIES Colorless liquid; sweet pleasant odor; liquid surface tension (estimated): 28 dynes/cm = 0.028 N/m at 20°C; heat of polymerization (est): −218 Btu/lb = −121 cal/g = -5.06×10^5 J/kg; heat of combustion (est): −10,800 Btu/lb = −6,000 cal/g = -250×10^5 J/kg; liquid heat capacity = 0.491 Btu/lb-F at 70°F; liquid thermal conductivity = 1.048 Btu-in./hr-sq ft-F at 70°F; liquid viscosity: 5.438 cP at 70°C; saturated vapor pressure = 0.090 lb/sq in. at 177°F; saturated vapor density = 0.00153 lb/cu ft at 177°F; react readily with electrophilic, free-radical, and nucleophilic agent. (acrylic acid and esters).

PROTECTIVE EQUIPMENT Goggles or face shield; rubber gloves [R2].

PREVENTATIVE MEASURES Awareness of the dangers and of good engineering design is essential to safety. Employees should be instructed about the necessity of cleansing the skin if it is contaminated by materials that are irritants or skin-absorbed. With careful design, however, and complete enclosure of those processes where toxic chemicals or intermediates occur, dangerous exposures can be avoided. (acrylic acid and derivatives) [R7].

DISPOSAL SRP: At the time of review, criteria for land treatment or burial (sanitary landfill) disposal practices are subject to significant revision. Prior to implementing land disposal of waste residue (including waste sludge), consult with environmental regulatory agencies for guidance on acceptable disposal practices.

COMMON USES Chemical intermediate for polymers; comonomer in acrylic surface coatings [R1]; functional monomer for manufacture of thermosetting acrylic resins [R3]; polar monomer used in emulsion polymerization [R4].

OCCUPATIONAL RECOMMENDATIONS None listed.

ECOTOXICITY LC_{50} *Pimephales promelas* (fathead minnows) 4.8 g/L/96 hr; age 29 days old, water hardness 44.0 mg/L ($CaCO_3$), temperature 24.5°C, pH 7.69, dissolved oxygen 7.1 mg/L, alkalinity 49.8 mg/L ($CaCO_3$), tank volume 2.0 L, additions 18 vol/day, flow-through bioassay [R8].

BIODEGRADATION 2-Hydroxyethyl acrylate was readily degraded in screening tests using mixed microbial cultures isolated from sewage by an enrichment technique (1). After 5 days, 61% theoretical BOD was observed (1) [R9].

BIOCONCENTRATION Using a reported log K_{ow} of −0.21 (1), a BCF of 0.41 has been calculated using a recommended regression-derived equation (2,SRC). This estimated BCF indicates that 2-hydroxyethyl acrylate will not be expected to bioconcentrate in aquatic organisms (SRC) [R10].

TSCA REQUIREMENTS Section 8 (a) of TSCA requires manufacturers of this chemical substance to report preliminary assessment information concerned with production, use, and exposure to EPA as cited in the preamble of the 51 FR 41329 [R11].

GENERAL RESPONSE Avoid contact with liquid. Keep people away. Wear rubber overclothing (including gloves). Stop discharge if possible. Call fire department. Isolate and remove discharged material. Notify local health and pollution control agencies.

FIRE RESPONSE Combustible. Containers may explode in fire. Extinguish with water, dry chemicals, alcohol foam, or carbon dioxide. Cool exposed containers with water.

EXPOSURE RESPONSE Call for medical aid. Liquid will burn skin and eyes. Harmful if swallowed. Remove contaminated clothing and shoes. Flush affected areas with plenty of water. If swallowed and victim is conscious, have victim drink water or milk. Do not induce vomiting.

RESPONSE TO DISCHARGE Issue warning—water contaminant. Restrict access. Disperse and flush.

WATER RESPONSE Effect of low concentrations on aquatic life is unknown. May be dangerous if it enters water intakes. Notify local health and wildlife officials. Notify operators of nearby water intakes.

EXPOSURE SYMPTOMS Inhalation causes irritation of nose and throat. Contact with liquid irritates eyes and skin.

EXPOSURE TREATMENT Inhalation: remove victim from exposure; support respiration; call physician if needed. Eyes: wash with large amounts of water for 15 min; call physician. Skin: flush with water.

REFERENCES

1. SRI.

2. U.S. Coast Guard, Department of Transportation. *CHRIS—Hazardous Chemical Data*. Volume II. Washington, DC: U.S. Government Printing Office, 1984–5.

3. Sax, N. I., and R. J. Lewis, Sr. (eds.). *Hawley's Condensed Chemical Dictionary*. 11th ed. New York: Van Nostrand Reinhold Co., 1987. 622.

4. *Kirk–Othmer Encyclopedia of Chemical Technology*. 3rd ed., Volumes 1–26. New York, NY: John Wiley and Sons, 1978–1984, p. 14 (81) 87.

5. National Fire Protection Association. *Fire Protection Guide on Hazardous Materials*. 9th ed. Boston, MA: National Fire Protection Association, 1986, p. 325M–59.

6. Association of American Railroads. *Emergency Handling of Hazardous Materials in Surface Transportation*. Washington, DC: Assoc. of American Railroads, Hazardous Materials Systems (BOE), 1987. 379.

7. International Labour Office. *Encyclopedia of Occupational Health and Safety*. Vols. I and II. Geneva, Switzerland: International Labour Office, 1983. 53.

8. Geiger D.L., S.H. Poirier, L.T. Brooke, D.J. Call (eds.). *Acute Toxicities of Organic Chemicals to Fathead Minnows (Pimephales promelas)*. Vol. III. Superior, Wisconsin: University of Wisconsin–Superior, 1986. 90.

9. Babeau, L., D. D. Vaishnav. *J Indust Microbiol* 2: 107–15 (1987).

10. (1) Hansch, C., and A. J. Leo. *Medchem Project* Issue No 26. Claremont, CA: Pomona College (1985). (2) Lyman, W. J., et al., *Handbook of Chemical Property Estimation Methods*. New York: McGraw-Hill, p. 5–5 (1982).

11. 40 CFR 712.30 (7/1/88).

■ IRON

CAS # 7439–89–6

SYNONYMS Armco iron • eo5a • ferrovac-e • ferrum • loha • pzh-1m3 • pzh-2 • pzh1m1- • pzh2m • pzh2m1- • pzh2m2- • pzh3- • pzh3m • pzh4m • suy b-2

MF Fe

MW 55.847

COLOR AND FORM Silvery-white or gray, soft, ductile, malleable metal; in powder form it is black to gray; cubic; tenacious, lustrous metal.

DENSITY 7.86.

BOILING POINT 3,000°C.

MELTING POINT 1,535°C.

EXPLOSIVE LIMITS AND POTENTIAL Moderate, in form of dust when exposed to heat or flame [R2]. While steel piping that had held sulfuric acid in cold climates during winter was being refitted, 2 explosions occurred; these were possibly due to trapped hydrogen from acid-metal reaction. [R6, p. 491M-221].

FIREFIGHTING PROCEDURES Special mixtures of dry chemical [R2].

SOLUBILITY Insoluble in hot and cold water, alkali, alcohol, ether; soluble in acids.

VAPOR PRESSURE 1 mm Hg at 1,787°C.

STABILITY AND SHELF LIFE Stable in dry air but readily oxidizes in moist air, forming "rust" (chiefly oxide, hydrated) [R5, 670].

OTHER CHEMICAL/PHYSICAL PROPERTIES Readily attacked by dilute mineral acids and attacked or dissolved by organic acids. Not appreciably attacked by cold sulfuric acid or nitric acid, but is attacked by hot acids. 4 naturally occurring isotopes: (54)Fe, (56)Fe, (57)Fe, (58)Fe. Valences 2, 3; seldom 1, 4, 6. Density: cast 7.76; wrought 7.25 – 7.78; steel 7.6 – 7.78. MP: Cast 1,000 – 1,300°C; wrought 1,500°C; steel 1,300°C. Somewhat magnetic metal; holds magnetism only after hardening (as alloy steel, e.g., alnico).

Electrical resistivity (20°C): 9.71 microhm-cm.

PROTECTIVE EQUIPMENT Protective clothing and eye protection should be provided [R7, 741].

PREVENTATIVE MEASURES Toxic and flammable gases, vapors, and dusts are countered by local exhaust and general ventilation [R7, 741].

COMMON USES In manufacture of iron and steel castings (pig iron), of alloys with carbon, chromium, and other metals; as a material to increase density of oil-well drilling fluids [R1]. Alloyed with carbon, manganese, chromium, nickel, and other elements to form steels; (55)Fe and (59)Fe used in tracer studies; formerly in biological studies [R9, 670]. Catalysts [R7, 271]. The main uses for DRI (directly reduced iron ore) are as prime metallic iron units for electric furnace steelmaking and foundry operations, as a coolant in BOF (Bessemer–oxygen furnace) steelmaking, as a replacement for scrap in the open hearth, and for increasing productivity and decreasing coke ratio in blast furnaces and other smelting processes. Minor uses for DRI are as a chemical reagent for copper cementation and for the manufacture of iron powder for powder metallurgy applications. Recently, the solid-based direct reduction (DR) processes have been adapted to the reduction of agglomerates that are produced from waste-iron-bearing materials generated in steel plants primarily for recycling to the blast furnace [R7, p. 761].

BINARY REACTANTS Chlorine, chlorine trifluoride, fluorine, hydrogen peroxide, nitrogen dioxide, sulfuric acid.

STANDARD CODES ICC—iron mass spent, flammable solid, yellow label, not accepted in an outside container, iron sponge not oxidized, and iron sponge spent—same.

PERSISTENCE Low concentrations of iron can persist in water supplies indefinitely. Higher concentrations will be reduced by natural alkalinity.

MAJOR SPECIES THREATENED Fish.

INHALATION LIMIT 10 mg/m³.

GENERAL SENSATION Dust can cause conjunctivitis, chloroiditis, retinitis, and siderosis of tissues.

PERSONAL SAFETY Filter mask recommended.

ACUTE HAZARD LEVEL Iron hydroxide or oxide precipitates and coats the gills of fish, causing death. Iron is considered a slight inhalative hazard in dust form.

CHRONIC HAZARD LEVEL Continued exposure to 30 mg/m^3 of dust can cause chronic bronchitis.

DEGREE OF HAZARD TO PUBLIC HEALTH Slight inhalation hazard associated with dust or fume.

OCCUPATIONAL RECOMMENDATIONS None listed.

THRESHOLD LIMIT 8-hr time-weighted average (TWA) 5 mg/m^3 (1986) (iron oxide fume (Fe$_2$O$_3$) (1309–37–1), as Fe) [R8, 24]. Excursion Limit Recommendation: Excursions in worker exposure levels may exceed three times the TLV–TWA for no more than a total of 30 min during a work day, and under no circumstances should they exceed five times the TLV–TWA (iron oxide fume (Fe$_2$O$_3$) (1309–37–1), as Fe) [R8, 5]. B2. B2 = Welding fumes; total particulate (not otherwise classified). TLV–TWA 5 mg/m^3 (1986) (iron oxide fume (Fe$_2$O$_3$) (1309–37–1), as Fe) [R8, 24].

ACTION LEVEL Precautions should be taken to prevent suspension of metal dust. Soluble compounds may require specific procedures.

ON-SITE RESTORATION For soluble iron salts, treat with NaHCO$_3$ to precipitate hydroxide salt, or add cation exchanger. Metal can be removed with an electromagnet. Seek professional environmental engineering assistance through EPA's Environmental Response Team (ERT), Edison, NJ, 24-hour no., 908–548–8730.

AVAILABILITY OF COUNTERMEASURE MATERIAL Sodium bicarbonate—grocery distributor, bakeries; cation exchanger—water softener suppliers; electromagnet—salvage operators.

DISPOSAL Salvage.

INDUSTRIAL FOULING POTENTIAL Industrial use recommended level 80 ppm Fe; brewing—0.1 ppm; carbonated drinks—0.2 ppm; paper and pulp—0.3 ppm; canning—0.2 ppm; plastics—0.02 ppm; baking—0.2 ppm.

MAJOR WATER USE THREATENED Fisheries.

PROBABLE LOCATION AND STATE Metal, will sink to bottom. Many salts are soluble. May form insoluble phosphates or hydroxides in natural waters.

COLOR IN WATER Ferric—orange; ferrous—green.

FIRE POTENTIAL Hot iron (wire) burns in chlorine gas. Violent decomposition of hydrogen peroxide (52% by wt or greater) may be caused by contact with iron. Iron and hydrogen peroxide ignite immediately if trace of manganese dioxide is present [R6, p. 491M–220]. Moderate, in form of dust when exposed to heat or flame [R2]. Chlorine trifluoride reacts with iron with incandescence. Powdered iron reacts with fluorine below redness with incandescence. Reduced iron decomposes nitrogen dioxide at ordinary temperature with incandescence. Reacting mass formed by mixture of phosphorus and iron can become incandescent when heated. [R6, p. 491M–220].

FDA Iron, reduced: 121: 101; gras, nutrient, or dietary supplement [R3].

REFERENCES

1. SRI.

2. Sax, N. I. *Dangerous Properties of Industrial Materials*. 4th ed. New York: Van Nostrand Reinhold, 1975. 836.

3. Furia, T. E. (ed.). *CRC Handbook of Food Additives*. 2nd ed. Cleveland: The Chemical Rubber Co., 1972. 661.

4. *Kirk–Othmer Encyc Chemical Tech* 3rd ed. 1978–present v13.

5. *The Merck Index*. 9th ed. Rahway, NJ: Merck and Co., Inc., 1976.

6. National Fire Protection Association. *Fire Protection Guide on Hazardous Ma-*

erials. 6th ed. Boston, MA: National Fire Protection Association, 1975.

7. International Labour Office. *Encyclopedia of Occupational Health and Safety.* Volumes I and II. New York: McGraw-Hill Book Co., 1971.

8. American Conference of Governmental Industrial Hygienists. *Threshold Limit Values for Chemical Substances and Physical Agents and Biological Exposure Indices for 1994–1995.* Cincinnati, OH: ACGIH, 1994.

9. Bureau of Mines. *Mineral Commodity Summaries.* 1986 p. 78.

■ IRON, TRIS (DIMETHYL-DITHIOCARBAMATO)-

CAS # 14484–64–1

SIC CODES 2879

SYNONYMS aafertis • bercema-fertam-50 • dimethylcarbamodithioic acid, iron complex • dimethyldithiocarbamic acid, iron(3) salt • dimethyldithiocarbamic acid, iron salt • Eisen(III)-tris (N,N-dimethyldithiocarbamat) (German) • Eisendimethyldithiocarbamat (German) • ent-14,689 • ferbam-50 • ferbam, iron salt • ferbame • ferbeck • ferberk • fermate ferbam fungicide • fermocide • ferradow • ferric dimethyl-dithiocarbamate • fuklasin-ultra • fuklazin • hexaferb • hokmate • iron dimethyldithiocarbamate • iron, tris (dimethylcarbamodithioato-s,s')-, (oc-6-11)- • iron, tris (dimethyl-dithiocarbamato)- • karbam-black • knockmate • liromate • niacide • Stauffer-ferbam • sup'r-flo ferbam flowable • trifungol • tris (N,N-dimethyldithiocarbamato) iron(III) • tris (dimethyldithiocarbamato) iron • carbamic acid, dimethyldithio-, iron salt • dimethylcarbamodithioic acid, iron(3) salt • fuklasin • tris (dimethyl-carbamodithioato-s,s') iron • iron-tris (dimethyldithiocarbamate) • carbamate • iron(III)-dimethyldithio-carbamate • (oc-6-11)-tris (dimethylcarbamodithioato-s,s') iron • tris (dimethylcarbamodithioate-s,s') iron • vancide-Fe95 • aaferzimag • carbamic acid, dimethyldithio-, iron(3) salt • dimethyl carbamodithioic acid, iron(3) salt • cormate • dimethyldithiocarbamic acid, ferric salt • f-40 • ferban • fermacide • ferric dimethyl dithio-carbamate • ferric dimethyl-dithioate

MF $C_9H_{18}N_3S_6 \cdot Fe$

MW 416.51

pH 5.0.

COLOR AND FORM Black solid; fluffy powder; dark-brown powder; black powder.

ODOR Odorless.

MELTING POINT Above 180°C with decomposition.

SOLUBILITY 120 ppm in water; soluble in acetone, chloroform, pyridine, acetonitrile.

VAPOR PRESSURE Negligible (room temp).

CORROSIVITY Noncorrosive.

STABILITY AND SHELF LIFE Tends to decompose on prolonged storage or exposure to heat and moisture [R1].

OTHER CHEMICAL/PHYSICAL PROPERTIES Decomposition >180°C.

PROTECTIVE EQUIPMENT Wear appropriate equipment to prevent repeated or prolonged skin contact [R3].

PREVENTATIVE MEASURES SRP: The scientific literature supports the wearing of contact lenses in industrial environments, as part of a program to protect the eye against chemical compounds and minerals causing eye irritation. However, there may be individual substances whose irritating or corrosive properties are such that the wearing of contact lenses would be harmful to the eye. In those specific cases contact lenses should not be worn.

CLEANUP 1. Ventilate area of spill. 2. Collect spilled material in the most convenient and safe manner and deposit in sealed containers for reclamation or for disposal in a secured sanitary landfill. Liquid containing ferbam should be absorbed in vermiculite, dry sand, earth, or a similar material [R4].

DISPOSAL Ferbam may be disposed of in a secured sanitary landfill [R4]. Dissolve in flammable solvent and burn in incinerator with scrubber and afterburner.

COMMON USES Fungicide used on fruit, nuts, vegetables, ornamental crops, in household applications [R2].

STANDARD CODES NFPA 2, –, –.

PERSISTENCE Decomposes slightly with prolonged exposure to heat, air, and water. Persisted 28 days in soil; low pH and microbial life degrade quickly in soil.

MAJOR SPECIES THREATENED Fish.

INHALATION LIMIT 10 mg/m³.

DIRECT CONTACT May cause irritation of skin and mucous membranes and renal damage.

PERSONAL SAFETY Self-contained breathing apparatus must be worn. Rubber gloves, hats, suits, and boots must be worn.

ACUTE HAZARD LEVEL 1 to 4 mg/L toxic to fish.

CHRONIC HAZARD LEVEL Low, degrades fairly fast.

DEGREE OF HAZARD TO PUBLIC HEALTH Emits toxic fumes in disaster circumstances. Irritating. May be mutagenic. Metabolite may be carcinogenic.

OCCUPATIONAL RECOMMENDATIONS
OSHA 8-hr time-weighted average: 15 mg/m³. Transitional limits were achieved by any combination of engineering controls, work practices, and personal protective equipment during the phase-in period, Sept 1, 1989 through Dec 30, 1993. Final rule limits became effective Dec 31, 1993 [R7].

NIOSH 10-hr time-weighted average: 10 mg/m³ [R3]; threshold-limit 8-hr time-weighted average (TWA) 10 mg/m³ (1986) [R10, 22].

ACTION LEVEL Evacuate area. Enter from upwind side. Notify local air authority and National Agricultural Chemicals Association.

ON-SITE RESTORATION Carbon or peat may adsorb spilled ferbam. Seek professional environmental engineering assistance through EPA's Environmental Response Team (ERT), Edison, NJ, 24-hour no. 908–548–8730.

AVAILABILITY OF COUNTERMEASURE MATERIAL Carbon—water treatment plants, sugar refineries; peat—nurseries, floral shops.

DISPOSAL NOTIFICATION Contact local air authority.

MAJOR WATER USE THREATENED Fisheries.

PROBABLE LOCATION AND STATE Black wettable powder. Will dissolve.

FOOD-CHAIN CONCENTRATION POTENTIAL Negative.

NON-HUMAN TOXICITY LD_{50} rat oral 17 g/kg [R6]; LD_{50} rat oral >4,000 mg/kg [R9].

IARC SUMMARY No data are available for humans. Inadequate evidence of carcinogenicity in animals. Overall evaluation: group 3: The agent is not classifiable as to its carcinogenicity to humans [R5].

FIFRA REQUIREMENTS In 1988, Congress amended FIFRA to strengthen and accelerate EPA's reregistration program. The nine-year reregistration scheme mandated by "FIFRA 88" applies to each registered pesticide product containing an active ingredient initially registered before November 1, 1984. Pesticides for which EPA had not issued Registration Standards prior to the effective date of FIFRA '88 were divided into three lists based upon their potential for exposure and other factors, with List B being of highest concern, and D of least. List: B; Case: Dimethyldithiocarbamate salts; Case No. 2180; Pesticide type: Insecticide, fungicide, herbicide, rodenticide, and antimicrobial; Case Status: Awaiting Data/Data in Review: OPP awaits data from the pesticide's producer(s) regarding its human health or environmental effects, or

OPP has received and is reviewing such data, in order to reach a decision about the pesticide's eligibility for reregistration. Active Ingredient (AI): Ferric dimethyldithiocarbamate; Data Call-in (DCI) Date: 09/30/91; AI Status: The producer(s) of the pesticide has made commitments to conduct the studies and pay the fees required for reregistration, and is meeting those commitments in a timely manner [R8].

REFERENCES

1. Budavari, S. (ed.). *The Merck Index— Encyclopedia of Chemicals, Drugs, and Biologicals.* Rahway, NJ: Merck and Co., Inc., 1989. 630.

2. SRI.

3. NIOSH. *NIOSH Pocket Guide to Chemical Hazards.* DHHS (NIOSH) Publication No. 90–117. Washington, DC: U.S. Government Printing Office, June 1990, 114.

4. Mackison, F. W., R. S. Stricoff, and L. J. Partridge, Jr. (eds.). *NIOSH/OSHA—Occupational Health Guidelines for Chemical Hazards.* DHHS (NIOSH) Publication No. 81–123 (3 vols). Washington, DC: U.S. Government Printing Office, Jan. 1981.

5. IARC. *Monographs on the Evaluation of the Carcinogenic Risk of Chemicals to Man.* Geneva: World Health Organization, International Agency for Research on Cancer, 1972–present (multivolume work), p. S7 63 (1987).

6. American Conference of Governmental Industrial Hygienists. *Documentation of the Threshold Limit Values and Biological Exposure Indices.* 5th ed. Cincinnati, OH: American Conference of Governmental Industrial Hygienists, 1986. 269.

7. 29 CFR 1910.1000 (7/1/91).

8. USEPA/OPP. *Status of Pesticides in Reregistration and Special Review* p. 130 (Mar, 1992) EPA 700–R–92–004.

9. Worthing, C. R., and S. B. Walker (eds.). *The Pesticide Manual—A World Compendium.* 8th ed. Thornton Heath, UK: The British Crop Protection Council, 1987. 397.

10. American Conference of Governmental Industrial Hygienists. *Threshold Limit Values for Chemical Substances and Physical Agents and Biological Exposure Indices for 1994–1995.* Cincinnati, OH: ACGIH, 1994.

■ ISOBORNYL THIOCYANOACETATE

CAS # 115–31–1

SYNONYMS 1,7,7-trimethylbicyclo (2,2,1) hept-2-ylthiocyanatoacetate • acetic acid, thiocyanato-, 1,7,7-trimethylbicyclo (2.2.1) hept-2-yl ester, exo • acetic acid, thiocyanato-, isobornyl ester • bornate • cidalon • ent-92 • exo-1,7,7-trimethylbicyclo[2.2.1]hept-2-yl thiocyanatoacetate • isoborneol, thiocyanatoacetate • isobornyl thiocyanatoacetate • isobornyl thiocyanatoacetate, technical • terpinyl thiocyanoacetate • thanisol • thanite • thanite (herculespowder) • thiocyanatoacetic acid 1,7,7-trimethylbicyclo[2.2.1]hept-2-yl ester • thiocyanatoacetic acid, isobornyl ester

MF $C_{13}H_{19}NO_2S$

MW 253.39

COLOR AND FORM Clear, amber liquid; yellow, oily liquid; reddish-yellow oily liquid (technical product).

ODOR Terpene-like odor.

DENSITY 1.1465 at 25°C.

BOILING POINT 95°C at 0.06 mm Hg.

SOLUBILITY Very soluble in petroleum ether, most organic solvents, most oils, fats; very soluble in alcohol, benzene, chloroform, ether; practically insoluble in water.

VAPOR PRESSURE 0.06 mm Hg at 95°C.

CORROSIVITY Somewhat corrosive to galvanized iron.

STABILITY AND SHELF LIFE It is stable under normal storage conditions [R8].

OTHER CHEMICAL/PHYSICAL PROPERTIES
Stable under normal storage conditions.

PREVENTATIVE MEASURES Thiocyanates are somewhat volatile and should always be used with good ventilation. (thiocyanates) [R13, 82].

DISPOSAL SRP: At the time of review, criteria for land treatment or burial (sanitary landfill) disposal practices are subject to significant revision. Prior to implementing land disposal of waste residue (including waste sludge), consult with environmental regulatory agencies for guidance on acceptable disposal practices.

COMMON USES Used as fungicide to preserve industrial products such as glue, leather, paints, dyes, paper, and synthetic textile materials [R4]; immobilizing grass carp, green sunfish, and channel catfish for live capturing [R5]; pediculicide [R6]. It is a contact insecticide and knockdown agent used for fly, ant, silverfish, bedbug, cockroach, and mosquito control. Veterinary applications against fleas and lice on pets. Human toxicity louse control [R2]. It is used as a space spray (7%) in buildings [R3]. Isobornyl thiocyanoacetate is a non-systemic contact insecticide causing a rapid "knockdown" of flies and used mainly in domestic and livestock fly sprays [R1]. List: C; Case: Thanite, Case No. 3141; Pesticide type: Insecticide; Case Status: The pesticide is no longer an active ingredient in any registered pesticide products. Therefore, EPA is characterizing it as "cancelled." Under FIFRA, pesticide producers may voluntarily cancel their registered products. EPA also may cancel pesticide registrations if registrants fail to pay required fees, to make or meet certain reregistration commitments, or when the Agency reaches findings of unreasonable adverse effects [R7].

OCCUPATIONAL RECOMMENDATIONS None listed.

NON-HUMAN TOXICITY LD_{50} rat oral 1,000 mg/kg [R9]; LD_{50} rabbit oral 722 mg/kg [R9]; LD_{50} rabbit dermal 6,880 mg/kg [R9]; LD_{50} guinea pig oral 550 mg/kg [R10].

ECOTOXICITY LD_{50} mallard, male, 12 mo old, oral, more than 2,000 mg/kg [R11];

LD_{50} pheasant, male, 3 mo old, oral, more than 2,000 mg/kg [R11]; EC_{50} *Daphnia magna*, first-stage instar, 115 µg/L/48 hr (95% confidence limit 76–168 µg/L), static bioassay without aeration, pH 7.2–7.5, water hardness 40–50 mg/L as calcium carbonate, and alkalinity of 30–35 mg/L [R12]; LC_{50} *Gammarus fasciatus*, mature, 740 µg/L/96 hr at 15°C (95% confidence limit 451–1,214 µg/L), static bioassay without aeration, pH 7.2–7.5, water hardness 40–50 mg/L as calcium carbonate, and alkalinity of 30–35 mg/L [R12]. LC_{50} cutthroat trout, wt 0.3 kg, 160 µg/L/96 hr at 12°C (95% confidence limit 142–180 µg/L), tested in hard water, 162 ppm $CaCO_3$, static bioassay without aeration, pH 7.2–7.5, water hardness 40–50 mg/L as calcium carbonate, and alkalinity of 30–35 mg/L [R12]. LC_{50} lake trout, wt 0.3 kg, 109 µg/L/96 hr at 12°C (95% confidence limit 90–132 µg/L), static bioassay without aeration, pH 7.2–7.5, water hardness 40–50 mg/L as calcium carbonate, and alkalinity of 30–35 mg/L [R12].

REFERENCES

1. Martin, H., and C. R. Worthing (eds.). *Pesticide Manual.* 5th ed. Worcestershire, England: British Crop Protection Council, 1977. 309.

2. *Farm Chemicals Handbook 1981.* Willoughby, OH: Meister, 1981, p. C–331.

3. Rossoff, I. S. *Handbook of Veterinary Drugs.* New York: Springer Publishing Company, 1974. 290.

4. Umekawa, O., et al. Japanese patent number 75 30142 09/29/75 (Katayama Kagaku Kogyo Kenkyusho Co, Ltd).

5. Cumming, K. B., et al. *Prog Fish Cult.* 37 (2): 81 (1975).

6. Cutie, M. R. *Nard J* 102 (Jan): 41 (1980).

7. USEPA/OPP. *Status of Pesticides in Reregistration and Special Review.* p. 85 (Mar, 1992) EPA 700–R–92–004.

8. Hayes, Wayland J., Jr. *Pesticides Studied in Man.* Baltimore/London: Williams and Wilkins, 1982. 130.

9. Hayes, W. J., Jr., E. R. Laws, Jr. (eds.). *Handbook of Pesticide Toxicology. Volume 2. Classes of Pesticides.* New York, NY: Academic Press, Inc., 1991. 652.

10. Gosselin, R. E., R. P. Smith, H. C. Hodge. *Clinical Toxicology of Commercial Products.* 5th ed. Baltimore: Williams and Wilkins, 1984. p. II–2289.

11. U.S. Department of the Interior, Fish and Wildlife Service. *Handbook of Toxicity of Pesticides to Wildlife.* Resource Publication 153. Washington, DC: U.S. Government Printing Office, 1984. 81.

12. U.S. Department of Interior, Fish and Wildlife Service. *Handbook of Acute Toxicity of Chemicals to Fish and Aquatic Invertebrates.* Resource Publication No. 137. Washington, DC: U.S. Government Printing Office, 1980. 76.

13. White-Stevens, R. (ed.). *Pesticides in the Environment.* Volume 1, Part 1, Part 2. New York: Marcel Dekker, Inc., 1971.

■ ISOBUTYLAMINE

CAS # 78–81–9

DOT # 1214

SYNONYMS 1-amino-2-methylpropane • 1-propanamine, 2-methyl • 2-methylpropylamine • 3-methyl-2-propylamine • i-butylamine • iso-butylamine • monoisobutylamine

MF $C_4H_{11}N$

MW 73.16

COLOR AND FORM Colorless liquid.

ODOR Amine odor; fish-like odor.

DENSITY 0.724 at 25°C.

SURFACE TENSION 17.6 dynes/cm at 68°C (air).

BOILING POINT 68–69°C.

MELTING POINT –85°C.

HEAT OF COMBUSTION –713.6 kcal/mole at 20°C—liquid.

HEAT OF VAPORIZATION 478.3 cal/mole.

AUTOIGNITION TEMPERATURE 712°F [R3].

EXPLOSIVE LIMITS AND POTENTIAL 3.4% lower, 9% upper [R12, 1016].

FLAMMABILITY LIMITS Lower 3.4%; upper 9% [R12, 1016].

FIREFIGHTING PROCEDURES Dry chemical, foam, carbon dioxide, or alcohol foam [R2]. If isobutylamine is on fire or involved in fire: do not extinguish fire unless flow can be stopped. Use water in flooding quantities as fog. Solid streams of water may be ineffective. Cool all affected containers with flooding quantities of water. Apply water from as far a distance as possible. If fire becomes uncontrollable or container is exposed to direct flame, evacuate to a radius of 1,500 ft. [R4]. Wear self-contained breathing apparatus when fighting fires involving isobutylamine [R4].

TOXIC COMBUSTION PRODUCTS Can emit oxides of nitrogen on burning [R2].

SOLUBILITY Very soluble in water, alcohol, ether; soluble in acetone, benzene.

VAPOR PRESSURE 100 mm Hg at 18.8°C.

OCTANOL/WATER PARTITION COEFFICIENT Log K_{ow}-0.73–0.88.

CORROSIVITY Strongly caustic.

STABILITY AND SHELF LIFE Stable during shipment [R5].

OTHER CHEMICAL/PHYSICAL PROPERTIES Conversion units: 1 mg/L = 334 ppm; 1 ppm = 2.99 mg/m³.

PROTECTIVE EQUIPMENT Wear boots, protective gloves, and goggles. If contact with the material is anticipated, wear full protective clothing [R4].

PREVENTATIVE MEASURES Keep away from heat and open flame [R2].

CLEANUP Small Spill: use absorbent paper to pick up spilled material. Follow by washing surfaces well with soap and water. Seal all wastes in vapor-tight plastic bags for eventual disposal. [R12, 1017].

DISPOSAL At the time of review, criteria for land treatment or burial (sanitary landfill) disposal practices are subject to significant revision. Prior to implementing land disposal of waste residue (including waste sludge), consult with environmental regulatory agencies for guidance on acceptable disposal practices [R6].

COMMON USES Organic synthesis; insecticides [R1].

OCCUPATIONAL RECOMMENDATIONS

OSHA Meets criteria for proposed OSHA medical records rule [R10].

ECOTOXICITY LD0 creek chub 20 mg/L/24 hr in Detroit river water [R7].

BIODEGRADATION Isobutylamine was biologically oxidized by aniline-acclimated activated sludge using a Warburg respirometer (1). The bacteria *Alcaligenes faecalis*, isolated from activated sludge, was able to biodegrade isobutylamine using the Warburg technique (2) [R8].

BIOCONCENTRATION Based on an experimentally determined log K_{ow} of 0.73 (1), the BCF value for isobutylamine can be estimated to be 2.1 from a recommended regression-derived equation (2,SRC). The complete water solubility of isobutylamine (3) also suggests that isobutylamine will not bioconcentrate significantly (SRC) [R9].

CERCLA REPORTABLE QUANTITIES Persons in charge of vessels or facilities are required to notify the National Response Center (NRC) immediately, when there is a release of this designated hazardous substance, in an amount equal to or greater than its reportable quantity of 1,000 lb or 454 kg. The toll-free telephone number of the NRC is (800) 424–8802; in the Washington metropolitan area (202) 426–2675. The rule for determining when notification is required is stated in 40 CFR 302.6 (section IV. D. 3.b) [R11].

GENERAL RESPONSE Stop discharge if possible. Keep people away. Shut off ignition sources. Call fire department. Avoid contact with liquid and vapor. Isolate and remove discharged material. Notify local health and pollution control agencies.

FIRE RESPONSE Flammable poisonous gases may be produced in fire. Flashback along vapor trail may occur. Vapor may explode if ignited in an enclosed area.

Wear goggles and self-contained breathing apparatus. Extinguish with dry chemicals, alcohol foam, or carbon dioxide. Cool exposed containers with water.

EXPOSURE RESPONSE Call for medical aid. Vapor irritating to eyes, nose, and throat. If inhaled will cause coughing, difficult breathing, or loss of consciousness. If breathing has stopped, give artificial respiration. If breathing is difficult, give oxygen. Liquid will burn skin and eyes. If swallowed will cause nausea or loss of consciousness. Remove contaminated clothing and shoes. Flush affected areas with plenty of water. If swallowed and victim is conscious, have victim drink water or milk, and have victim induce vomiting.

RESPONSE TO DISCHARGE Issue warning—high flammability, air contaminant, water contaminant. Restrict access. Evacuate area. Disperse and flush.

WATER RESPONSE Effect of low concentrations on aquatic life is unknown. May be dangerous if it enters water intakes. Notify local health and wildlife officials. Notify operators of nearby water intakes.

EXPOSURE SYMPTOMS Inhalation causes severe coughing and chest pain due to irritation of air passages; can cause lung edema. Compound is sympathomimetic and is also a cardiac depressant and convulsant; ingestion causes nausea and profuse salivation. Contact with eyes causes severe irritation and edema of the cornea. Contact with skin causes severe irritation.

EXPOSURE TREATMENT Inhalation: remove victim to fresh air; if not breathing, give artificial respiration; if breathing is difficult, give oxygen; call a physician. Ingestion: give large amount of water followed by dilute vinegar or lemon juice; keep patient warm. Eyes: flush with water for 15 min. Skin: flush with water.

REFERENCES

1. Hawley, G. G. *The Condensed Chemical Dictionary.* 10th ed. New York: Van Nostrand Reinhold Co., 1981. 576.

2. Sax, N. I. *Dangerous Properties of*

Industrial Materials. 6th ed. New York, NY: Van Nostrand Reinhold, 1984. 1634.

3. National Fire Protection Association. *Fire Protection Guide on Hazardous Materials.* 9th ed. Boston, MA: National Fire Protection Association, 1986, p. 325M-61.

4. Bureau of Explosives. *Emergency Handling of Haz Matl in Surface Trans* p. 296 (1981).

5. U.S. Coast Guard, Department of Transportation. *CHRIS—Hazardous Chemical Data. Volume II.* Washington, DC: U.S. Government Printing Office, 1984-5.

6. SRP.

7. Verschueren, K. *Handbook of Environmental Data of Organic Chemicals.* 2nd ed. New York, NY: Van Nostrand Reinhold Co., 1983. 768.

8. (1) Malaney, G. W. *J Water Pollut Control Fed* 35: 1300 (1960). (2) Marion, C. V., G. W. Malaney. *J Water Pollut Control Fed* 35: 1269 (1963).

9. (1) Hansch, L. C., A. J. Leo. *Medchem Project* Issue No. 26, Claremont, CA: Pomona College (1985). (2) Lyman, W. J., et al., *Handbook of Chemical Property Estimation Methods. Environmental Behavior of Organic Compounds.* New York: McGraw-Hill, p. 5-4 (1982). (3) Perry, R. H., D. W. Green. *Perry's Chemical Engineers' Handbook.* 6th ed. pp. 3-28 (1984).

10. 47 FR 30420 (7/13/82).

11. 50 FR 13456 (4/4/85).

12. Keith, L.H., D.B. Walters (eds.). *Compendium of Safety Data Sheets for Research and Industrial Chemicals.* Parts I, II, and III. Deerfield Beach, FL: VCH Publishers, 1985.

■ ISODECYL ACRYLATE

CAS # 1330-61-6

SYNONYMS 2-propenoic acid, isodecyl ester • acrylic acid, isodecyl ester • isodecyl alcohol, acrylate • isodecyl propenoate

MF $C_{13}H_{24}O_2$

MW 212.37

COLOR AND FORM Colorless.

ODOR Weak acrylate.

DENSITY 0.885 at 20°C.

SURFACE TENSION Liquid surface tension: (est) 30 dynes/cm = 0.003 N/m at 20°C.

VISCOSITY Liquid 2.300 cP at 70°C.

HEAT OF COMBUSTION Heat of combustion: (est) -16,300 Btu/lb = -9,100 cal/g = -380×10^5 J/kg.

HEAT OF VAPORIZATION Latent heat of vaporization: 110 Btu/lb = 61 cal/g = 2.6×10^5 J/kg.

FIREFIGHTING PROCEDURES Extinguish with dry chem, foam, or carbon dioxide. Water may be ineffective on fire. Cool exposed containers with water [R1].

SOLUBILITY In water: 0.010 lb/100 lb at 68°C.

VAPOR PRESSURE Saturated vapor pressure = 0.792 lb/sq in at 304°F.

OTHER CHEMICAL/PHYSICAL PROPERTIES FP: -148°F = -100°C = 173 K; liq-water interfacial tension: (est) 30 dynes/cm = 0.003 N/m at 20°C; heat of polymerization: (est) -119 Btu/lb = -66 cal/g = -2.8×10^5 J/kg; liquid heat capacity = 0.461 Btu/lb-F at 70°F; liquid thermal conductivity = 1.048 Btu-in./hr-ft²-F at 70°F; react readily with electrophilic, free-radical, and nucleophilic agent (acrylic acid and esters).

PROTECTIVE EQUIPMENT Goggles or face shield; rubber gloves [R1]. Suitable protective clothing and self-contained resp protective apparatus should be available for use of those who may have to rescue persons overcome by fumes (acrylic acid and derivatives) [R2]. Protection required for safe handling of acrylic acid and esters commonly includes use of impervious gloves, shoe soles, and clothing; splash-proof goggles. (acrylic acid and derivatives) [R4, 351].

PREVENTATIVE MEASURES Awareness of the dangers and of good engineering design is essential to safety. Employees should be instructed about the necessity of cleansing the skin if it is contaminated by materials that are irritants or skin-absorbed. With careful design, however, and complete enclosure of those processes where toxic chemicals or intermediates occur, dangerous exposures can be avoided. (acrylic acid, and derivatives) [R2].

DISPOSAL SRP: At the time of review, criteria for land treatment or burial (sanitary landfill) disposal practices are subject to significant revision. Prior to implementing land disposal of waste residue (including waste sludge), consult with environmental regulatory agencies for guidance on acceptable disposal practices.

COMMON USES Reactive diluent in radiation curable coatings, adhesives, coatings, etc. [R5, 85]. Used for copolymerization [R5, 239].

OCCUPATIONAL RECOMMENDATIONS None listed.

TSCA REQUIREMENTS Section 8 (a) of TSCA requires manufacturers of this chemical substance to report preliminary assessment information concerned with production, use, and exposure to EPA as cited in the preamble of the 51 FR 41329 [R3].

GENERAL RESPONSE Stop discharge if possible. Keep people away. Call fire department. Isolate and remove discharged material. Notify local health and pollution control agencies.

FIRE RESPONSE Combustible. Extinguish with dry chemicals, foam, or carbon dioxide. Water may be ineffective on fire. Cool exposed containers with water.

EXPOSURE RESPONSE Call for medical aid. Liquid irritating to skin and eyes. Remove contaminated clothing and shoes. Flush affected areas with plenty of water. If swallowed and victim is conscious, have victim drink water or milk.

RESPONSE TO DISCHARGE Issue warning—water contaminant. Mechanical containment. Should be removed. Provide chemical and physical treatment.

WATER RESPONSE Effect of low concentrations on aquatic life is unknown. Fouling to shoreline. May be dangerous if it enters water intakes. Notify local health and wildlife officials. Notify operators of nearby water intakes.

EXPOSURE SYMPTOMS Inhalation causes mild irritation of nose and throat. Eyes are mildly irritated by vapor, more severely by liquid. Prolonged contact of liquid with skin may cause irritation.

EXPOSURE TREATMENT Inhalation: move to fresh air. Eyes: flush with water for at least 15 min after contact with liquid. Skin: wipe off, wash well with soap and water.

REFERENCES

1. U.S. Coast Guard, Department of Transportation. *CHRIS—Hazardous Chemical Data*. Volume II. Washington, DC: U.S. Government Printing Office, 1984–5.

2. International Labour Office. *Encyclopedia of Occupational Health and Safety*. Vols. I and II. Geneva, Switzerland: International Labour Office, 1983. 53.

3. 40 CFR 712.30 (7/1/88).

4. *Kirk-Othmer Encyclopedia of Chemical Technology*. 3rd ed., Volumes 1–26. New York, NY: John Wiley and Sons, 1978–1984, p. 1(78).

5. Kuney, J.H., and J.N. Nullican (eds.) *Chemcyclopedia*. Washington, DC: American Chemical Society, 1988.

■ ISOPENTANE

CAS # 78–78–4

DOT # 1265

SYNONYMS 1,1,2-trimethylethane • 2-methyl butane • 2-methylbutane • butane, 2-methyl • ethyldimethylmethane • iso-pentane • isoamyl hydride • isoamylhydride • dimethylethylmethane

MF C_5H_{12}

MW 72.17

COLOR AND FORM Colorless liquid.

ODOR Pleasant odor; gasoline-like odor.

DENSITY 0.6201 at 20°C.

SURFACE TENSION 13.72 dynes/cm at 20°C.

VISCOSITY 0.233 cP at 20°C.

BOILING POINT 27.8°C at 760 mm Hg.

MELTING POINT −159.9°C.

HEAT OF COMBUSTION Gas: −843.5 kcal/g at 20°C; liquid: −838.3 kcal/g at 20°C.

HEAT OF VAPORIZATION 146 Btu/lb.

AUTOIGNITION TEMPERATURE 788°F [R1].

EXPLOSIVE LIMITS AND POTENTIAL Explosive limits: lower 1.4%; upper 7.6% [R2].

FLAMMABILITY LIMITS Lower 1.4%; upper 7.6% [R3].

FIREFIGHTING PROCEDURES Foam, carbon dioxide, dry chemical [R2].

SOLUBILITY Miscible in alcohol and ether; soluble in hydrocarbons, oils; insoluble in water; water solubility: 48 mg/L.

VAPOR PRESSURE 595 mm Hg at 21.1°C.

OCTANOL/WATER PARTITION COEFFICIENT log K_{ow} = 2.30.

OTHER CHEMICAL/PHYSICAL PROPERTIES Vapor pressure: 689 mm Hg at 25°C.

PROTECTIVE EQUIPMENT Protective clothing, barrier creams; medical control [R4].

PREVENTATIVE MEASURES Substitution of less irritating substances, redesign of operations, prevent contact, provision of a physical barrier against contact, proper washing facilities, work clothing, and storage facilities [R4].

COMMON USES Solvent; manufacture of chlorinated derivatives; blowing agent for polystyrene [R1].

OCCUPATIONAL RECOMMENDATIONS None listed.

BIODEGRADATION Organisms isolated from groundwater contaminated by a gas-oline spill were found to grow when inoculated with pure samples of isopentane; however, no biodegradation of isopentane was observed when a mixed culture of these organisms was inoculated with gasoline (1) [R5].

BIOCONCENTRATION Estimated bioconcentration factors for isopentane ranging from 33 to 70 (1,SRC) can be calculated from its experimental log octanol/water partition coefficient, 2.30 (2), and its water solubility, 48 mg/L (3), respectively. These values indicated that isopentane is not expected to significantly bioconcentrate in fish and aquatic organisms (SRC) [R6].

ATMOSPHERIC STANDARDS This action promulgates standards of performance for equipment leaks of volatile organic compounds (VOC) in the synthetic organic chemical manufacturing industry (SOCMI). The intended effect of these standards is to require all newly constructed, modified, and reconstructed SOCMI process units to use the best demonstrated system of continuous emission reduction for equipment leaks of VOC, considering costs, non-air-quality health, and environmental impact and energy requirements. Isopentane is produced, as an intermediate or final product, by process units covered under this subpart [R7].

GENERAL RESPONSE Wear goggles and self-contained breathing apparatus. Stop discharge if possible. Keep people away. Shut off ignition sources and call fire department. Avoid contact with liquid and vapor. Stay upwind and use water spray to "knock down" vapor. Isolate and remove discharged material. Notify local health and pollution control agencies.

FIRE RESPONSE Flammable. Flashback along vapor trail may occur. Vapor may explode if ignited in an enclosed area. Extinguish with dry chemical, foam, or carbon dioxide. Water may be ineffective on fire. Cool exposed containers, and protect men effecting shutoff with water.

EXPOSURE RESPONSE Call for medical aid. Vapor irritating to nose and throat. If inhaled, will cause coughing, difficult breathing, or loss of consciousness. Move to fresh air. If breathing has stopped, give artificial respiration. If breathing is difficult, give oxygen. Liquid irritating to skin and eyes. If swallowed, will cause nausea or vomiting. Remove contaminated clothing and shoes. Flush affected areas with plenty of water. If swallowed and victim is conscious, have victim drink water or milk. Do not induce vomiting.

RESPONSE TO DISCHARGE Issue warning—high flammability. Evacuate area. Disperse and flush.

WATER RESPONSE Effect of low concentrations on aquatic life is unknown. May be dangerous if it enters water intakes. Notify local health and wildlife officials. Notify operators of nearby water intakes.

EXPOSURE SYMPTOMS Inhalation causes irritation of respiratory tract, cough, mild depression, irregular heartbeat. Aspiration causes severe lung irritation, coughing, pulmonary edema; excitement followed by depression. Ingestion causes nausea, vomiting, swelling of abdomen, headache, depression.

EXPOSURE TREATMENT Inhalation: maintain respiration, give oxygen if needed. Aspiration: enforce bed rest; give oxygen. Ingestion: do not induce vomiting; call a doctor. Eyes: wash with copious amounts of water. Skin: wipe off, wash with soap and water.

REFERENCES

1. Sax, N. I., R. J. Lewis, Sr. (eds.). *Hawley's Condensed Chemical Dictionary.* 11th ed. New York: Van Nostrand Reinhold Co., 1987. 659.

2. Sax, N. I. *Dangerous Properties of Industrial Materials.* 6th ed. New York, NY: Van Nostrand Reinhold, 1984. 1336.

3. National Fire Protection Association. *Fire Protection Guide on Hazardous Materials.* 9th ed. Boston, MA: National Fire Protection Association, 1986, p. 325M–62.

4. Sax, N. I. *Dangerous Properties of Industrial Materials.* 4th ed. New York: Van Nostrand Reinhold, 1975. 844.

5. Jamison, V. W., et al., pp. 187–96 in *Proc Intermediate Biodeg Symp 3rd.* Sharpley, J. M., A. M. Kapalan (eds.). Essex, England (1976).

6. (1) Lyman, W. J., et al., *Handbook of Chemical Property Estimation Methods.* New York: McGraw-Hill, Chapt 5 (1982). (2) Hansch, C., et al., *J Organic Chemical* 33: 347 (1968). (3) Riddick, J. A., et al., *Organic Solvents* 4th ed. New York, NY: Wiley (1986).

7. 40 CFR 60.489 (7/1/90).

■ KADETHRIN

CAS # 58769-20-3

SYNONYMS Spray-Tox • [1R-[1 alpha, 3 alpha (E)]]-[5-(phenylmethyl)-3-furanyl]methyl 3-[(dihydro-2-oxo-3 (2H)-thienylidene) methyl]-2,2-dimethylcyclopropanecarboxylate • 5-benzyl-3-furylmethyl(E)-(1R,3S)-2,2-dimethyl-3-(2-oxothiolan-3-ylidenemethyl)-cyclopropanecarboxylate • RU-15525 • 5-benzyl-3-furylmethyl(E)-(1R)-cis-2,2-dimethyl-3-(2-oxothiolan-3-ylidenemethyl) cyclopropanecarboxylate

MF $C_{23}H_{24}O_4S$

MW 396.51

FIREFIGHTING PROCEDURES Use carbon dioxide, foam, or dry chemical (on fires involving pyrethroids) (pyrethrum) [R6, 2]. Firefighting: Self-contained breathing apparatus with a full facepiece operated in pressure-demand or other positive-pressure mode (pyrethrum) [R6, 5]. Extinguish fire using agent suitable for type of surrounding fire. (pyrethrins) [R2].

SOLUBILITY Practically insoluble in water; soluble in dichloromethane, ethanol, benzene, toluene, xylene, acetone, piperonylbutoxide. Slightly soluble in kerosene.

VAPOR PRESSURE <0.1 mPa at 20°C.

STABILITY AND SHELF LIFE Hydrolyzed by aqueous alkalis. Rapidly decomposed by light (more slowly in mineral oils). Unstable to heat [R1].

OTHER CHEMICAL/PHYSICAL PROPERTIES Yellow-brown, viscous oil (technical kadethrin, 93%).

PROTECTIVE EQUIPMENT Employees should use dust- and splash-proof safety goggles where pyrethroids may contact the eyes (pyrethroids) [R6, 3]. Employees should use impervious clothing, gloves, and face shields (eight-inch minimum) (pyrethroids) [R6, 2]. Wear appropriate equipment to prevent repeated or prolonged skin contact (pyrethrum and pyrethrins) [R7]. Wear eye protection to prevent reasonable probability of eye contact. (pyrethrins) [R7].

PREVENTATIVE MEASURES Skin that becomes contaminated with pyrethrum should be promptly washed or showered with soap or mild detergent and water. (pyrethrum) [R6, 3].

CLEANUP Spillages of pesticides at any stage of their storage or handling should be treated with great care. Liquid formulations may be reduced to solid phase by evaporation. Dry sweeping of solids is always hazardous: these should be removed by vacuum cleaning, or by dissolving them in water or other solvent in the factory environment. (pesticides) [R3].

DISPOSAL Incineration would be an effective disposal procedure where permitted. If an efficient incinerator is not available, the product should be mixed with large amounts of combustible material, and contact with the smoke should be avoided. (pyrethrin products) [R4].

COMMON USES Control of household insect pests, particularly houseflies, mosquitoes, and cockroaches. Normally used in combination with other pyrethroid insecticides and synergists [R1].

NON-HUMAN TOXICITY LD50 rat male oral 1,324 mg/kg [R1]; LD50 rat female oral 650 mg/kg [R1]; LD50 dog oral >1,000 mg/kg [R1]; LD50 rat female percutaneous >3,200 mg/kg [R1].

FIFRA REQUIREMENTS In 1988, Congress amended FIFRA to strengthen and accelerate EPA's reregistration program. The nine-year reregistration scheme mandated by "FIFRA 88" applies to each registered pesticide product containing an active ingredient initially registered before November 1, 1984. Pesticides for which EPA had not issued Registration Standards prior to the effective date of FIFRA '88 were divided into three lists based upon their potential for exposure and other factors, with List B being of highest concern, and D of least. List: B; Case: Pyrethrin, and derivs; Case No. 2580; Pesticide type: Insecticide, fungicide, antimicrobial; Active Ingredient (AI): Pyrethrin coils; Case Status: The active ingredient is no longer contained in any registered products. Therefore, EPA is characterizing it as "cancelled." Under FIFRA, pesticide producers may voluntarily cancel their registered products. EPA also may cancel pesticide registrations if registrants fail to pay required fees, to make or meet certain reregistration commitments, or when EPA reaches findings of unreasonable adverse effects. (pyrethrin coils) [R5].

REFERENCES

1. Hartley, D., and H. Kidd (eds.). *The Agrochemicals Handbook.* 2nd ed. Lechworth, Herts, England: The Royal Society of Chemistry, 1987, p. A240/Aug 87.

2. Bureau of Explosives; *Emergency Handling of Haz Matl in Surface Trans* p. 434 (1981).

3. International Labour Office. *Encyclopedia of Occupational Health and Safety.* Vols. I and II. Geneva, Switzerland: International Labour Office, 1983. 1619.

4. Sittig, M. *Handbook of Toxic and Hazardous Chemicals and Carcinogens,* 1985. 2nd ed. Park Ridge, NJ: Noyes Data Corporation, 1985. 762.

5. USEPA/OPP. *Status of Pesticides in Reregistration, and Special Review* p. 156 (Mar, 1992) EPA 700–R–92–004.

6. Mackison, F. W., R. S. Stricoff, and L. J. Partridge, Jr. (eds.). *NIOSH/OSHA—Occupational Health Guidelines for Chemical Hazards.* DHHS (NIOSH) Publication No. 81–123 (3 vols). Washington, DC: U.S. Government Printing Office, Jan. 1981.

7. NIOSH. *NIOSH Pocket Guide to Chemical Hazards.* DHHS (NIOSH) Publication No. 90–117. Washington, DC: U.S. Government Printing Office, June 1990, 190.

■ LASIOCARPINE

CAS # 303-34-4

SYNONYMS ai3-51770 • (7alpha-
angelyloxy-5,6,7,8alpha-tetrahydro-3h-
pyrrolizin-1-yl) methyl 2,3-dihydroxy-2-
(1'-methoxyethyl)-3-methylbutyrate •
(z)-2-methyl-crotonic acid, 2,3-
dihydroxy-2-(1-methoxyethyl)-3-
methylbutyrate (ester) • (z)-2-
methylcrotonic acid, 2,3-dihydroxy-2-(1-
methoxyethyl)-3-methylbutyrate (ester) •
2,3,5,7alphabeta-tetrahydro-1-hydroxy-
1h-pyrrolizine-7-methanol 1-angelate-7-
(2,3-dihydroxy-2 (1-methoxyethyl))-3-
methyl-butyrate • 2-butenoic acid,
2-methyl-, 7-((2,3-dihydroxy-2-(1-
methoxyethyl)-3-methyl-1-oxobutoxy)
methyl)-2,3,5,7a-tetrahydro-1h-
pyrrolizin-1-yl ester, (1s-(1alpha (z),7
(2s, 3r),7aalpha))- • heliotridine ester
with lasiocarpum and angelic acid •
NCI-c01478 • NSC-30625-

MF $C_{21}H_{33}NO_7$

MW 411.5

COLOR AND FORM Colorless plates; colorless leaflets from petroleum ether.

MELTING POINT 96.4–97°C.

SOLUBILITY Soluble in most nonpolar organic solvents and ethanol; sparingly soluble in water (0.68%) and light petroleum;
soluble in ether, alcohol, and benzene.

STABILITY AND SHELF LIFE Decomposition
slowly on standing in air at room temperature [R1].

OTHER CHEMICAL/PHYSICAL PROPERTIES
Readily hydrolyzed with alkali. Reacts
readily with oxidizing agents (slowly with
atmospheric oxygen) to form dihydropyr-
rolizine derivative. Soluble in water. (lasiocarpine hydrochloride).

PROTECTIVE EQUIPMENT Dispensers of
liquid detergent should be available. Safety pipettes should be used for all pipetting.
In animal laboratory, personnel should
wear protective suits (preferably disposable, one-piece, and close-fitting at ankles
and wrists), gloves, hair covering, and
overshoes. In chemical laboratory, gloves

and gowns should always be worn; however, gloves should not be assumed to
provide full protection. Carefully fitted
masks or respirators may be necessary
when working with particulates or gases,
and disposable plastic aprons might provide addnl protection. Gowns should be of
distinctive color, as a reminder that they
are not to be worn outside the laboratory.
(chemical carcinogens) [R8, 8].

PREVENTATIVE MEASURES Smoking,
drinking, eating, storage of food, or of food
and beverage containers, or utensils, and
the application of cosmetics should be
prohibited in any laboratory. All personnel should remove gloves, if worn, after
completion of procedures in which carcinogens have been used. They should wash
hands, preferably using dispensers of liquid detergent, and rinse thoroughly. Consideration should be given to appropriate
methods for cleaning the skin, depending
on nature of the contaminant. No standard procedure can be recommended, but
the use of organic solvents should be
avoided. Safety pipettes should be used for
all pipetting. (chemical carcinogens) [R8,
8].

CLEANUP A high-efficiency particulate
arrestor (HEPA) or charcoal filters can be
used to minimize amount of carcinogen in
exhausted-air-ventilated safety cabinets,
lab hoods, glove boxes, or animal rooms
Filter housing that is designed so that
used filters can be transferred into plastic
bag without contaminating maintenance
staff is avail commercially. Filters should
be placed in plastic bags immediately after
removal. The plastic bag should be sealed
immediately. The sealed bag should be
labelled properly. Waste liquids should be
placed or collected in proper containers for
disposal. The lid should be secured and
the bottles properly labelled. Once filled,
bottles should be placed in plastic bag, so
that outer surface is not contaminated.
The plastic bag should also be sealed and
labelled. Broken glassware should be decontaminated by solvent extraction, by
chemical destruction, or in specially designed incinerators. (chemical carcinogens) [R8, 15].

DISPOSAL Generators of waste (equal to, or greater than 100 kg/mo) containing this contaminant, EPA hazardous waste number U143, must conform with USEPA regulations in storage, transportation, treatment, and disposal of waste [R4].

COMMON USES Research chemical [R2]; known liver carcinogen with antimitotic activity used in research.

OCCUPATIONAL RECOMMENDATIONS None listed.

NON-HUMAN TOXICITY LD_{50} rat oral 150 mg/kg [R3]; LD_{50} rat intraperitoneal 78 mg/kg [R3].

IARC SUMMARY Classification of carcinogenicity: (1) evidence in humans: not available; (2) evidence in animals: sufficient. Overall summary evaluation of carcinogenic risk to humans is Group 2B: The agent is possibly carcinogenic to humans. (from table) [R5].

CERCLA REPORTABLE QUANTITIES Persons in charge of vessels or facilities are required to notify the National Response Center (NRC) immediately when there is a release of this designated hazardous substance in an amount equal to or greater than its reportable quantity of 10 lb or 4.54 kg. The toll-free number of the NRC is (800) 424–8802; in the Washington, DC metropolitan area (202) 426–2675. The rule for determining when notification is required is stated in 40 CFR 302.4 (section IV. D. 3.b) [R6].

RCRA REQUIREMENTS U143; as stipulated in 40 CFR 261.33, when lasiocarpine, as a commercial chemical product or manufacturing chemical intermediate or an off-specification commercial chemical product or a manufacturing chemical intermediate, becomes a waste, it must be managed according to federal or state hazardous waste regulations. Also defined as a hazardous waste is any residue, contaminated soil, water, or other debris resulting from the cleanup of a spill, into water or on dry land, of this waste. Generators of small quantities of this waste may qualify for partial exclusion from hazardous waste regulations (40 CFR 261.5) [R7].

REFERENCES

1. IARC. *Monographs on the Evaluation of the Carcinogenic Risk of Chemicals to Man.* Geneva: World Health Organization, International Agency for Research on Cancer, 1972–present (multivolume work), p. V10 282 (1976).

2. IARC. *Monographs on the Evaluation of the Carcinogenic Risk of Chemicals to Man.* Geneva: World Health Organization, International Agency for Research on Cancer, 1972–present (multivolume work), p. V10 281 (1976).

3. Sax, N. I. *Dangerous Properties of Industrial Materials.* 6th ed. New York, NY: Van Nostrand Reinhold, 1984. 1685.

4. 40 CFR 240–280, 300–306, 702–799 (7/1/90).

5. IARC. *Monographs on the Evaluation of the Carcinogenic Risk of Chemicals to Man.* Geneva: World Health Organization, International Agency for Research on Cancer, 1972–present (multivolume work), p. S7 65 (1987).

6. 40 CFR 302.4 (7/1/90).

7. 40 CFR 261.33 (7/1/90).

8. Montesano, R., H. Bartsch, E. Boyland, G. Della Porta, L. Fishbein, R.A. Griesemer, A.B. Swan, L. Tomatis, and W. Davis (eds.). *Handling Chemical Carcinogens in the Laboratory: Problems of Safety.* IARC Scientific Publications No. 33. Lyon, France: International Agency for Research on Cancer, 1979.

■ LEAD(+2) FLUORIDE

CAS # 7783–46–2

DOT # 2811

SIC CODES 2819; 3832

SYNONYMS lead fluoride (PbF₂) • plomb fluorure (French) • lead difluoride • plumbous fluoride

MF F_2Pb

MW 245.21

COLOR AND FORM White to colorless crystals; rhombic, orthorhombic.

ODOR Odorless.

DENSITY 8.445 (orthorhombic); 7.750 (cubic).

BOILING POINT 1,293°C.

MELTING POINT 824°C.

FIREFIGHTING PROCEDURES Extinguishant: Dry sand, dry dolomite, or dry graphite (inorganic lead) [R5].

SOLUBILITY In water: 0.057 g/100 ml at 0°C; 0.065 g/100 ml at 20°C; soluble in nitric acid, insoluble in acetone, ammonia.

OTHER CHEMICAL/PHYSICAL PROPERTIES Dimorphous: orthorhombic, converted to cubic above 316°C; solubility incr in the presence of nitric acid or nitrates.

PROTECTIVE EQUIPMENT Respirators for heavy dust exposure; safety goggles [R1].

PREVENTATIVE MEASURES Workers should wash at the end of each work shift. (lead, as Pb) [R6].

CLEANUP A high-efficiency particulate arrestor (HEPA) or charcoal filters can be used to minimize amount of carcinogen in exhausted-air-ventilated safety cabinets, lab hoods, glove boxes, or animal rooms. Filter housing that is designed so that used filters can be transferred into plastic bag without contaminating maintenance staff is avail commercially. Filters should be placed in plastic bags immediately after removal. The plastic bag should be sealed immediately. The sealed bag should be labelled properly. Waste liquids should be placed or collected in proper containers for disposal. The lid should be secured and the bottles properly labelled. Once filled, bottles should be placed in plastic bag, so that outer surface is not contaminated. The plastic bag should also be sealed and labelled. Broken glassware should be decontaminated by solvent extraction, by chemical destruction, or in specially designed incinerators (chemical carcinogens) [R14, 15]. If inorganic lead is spilled or leaked, the following steps should be taken: 1. Remove all ignition sources. 2.

Ventilate area of spill or leak. 3. For small quantities of liquids containing inorganic lead, absorb on paper towels, and place in appropriate container. 4. Large quantities of liquid containing inorganic lead may be absorbed in vermiculite, dry sand, earth, or a similar material, and placed in appropriate container. 5. If in solid form, inorganic lead may be collected, and placed in appropriate container. 6. Inorganic lead may be collected by vacuuming with an appropriate system. (inorganic lead) [R5].

COMMON USES Electronic and optical applications; starting materials for growing single-crystal solid-state lasers; high-temperature dry-film lubricants in the form of ceramic-bonded coatings [R2]. A mild fluorinating agent, and is used in the conversion of molybdenum and tungsten oxide to oxyfluorides, as a raw material for lead tetrafluoride, in the manufacture of special grades of glass (qv), and an ingredient in underwater paints [R3].

STANDARD CODES EPA 311; TSCA; IATA poison Class B nos, poison label, 25 kg passenger, 95 kg cargo; CFR 49 poison Class B, poison label, 50 lb passenger, 200 lb cargo, storage code 1, 2; CFR 14 CAB Code 8; not listed NFPA; not listed AAR; ICC STCC Tariff no. 1–c 28 162 30; Superfund designated (hazardous substances) list.

PERSISTENCE Persistent.

MAJOR SPECIES THREATENED Fish species such as salmon, trout, minnows, and catfish are especially sensitive to lead. Toxicity intensified in soft water.

INHALATION LIMIT 0.15 mg/m^3.

DIRECT CONTACT Not irritating to skin or mucous membranes at low concentrations. The fluoride ion can cause protoplasmic poisoning and skin irritation at higher concentrations.

GENERAL SENSATION Inhalation or skin absorption may produce abdominal pain, loss of appetite, weight loss, constipation, apathy or irritability, vomiting, fatigue, headache, weakness, metallic taste, and muscle incoordination. At fire temperatures causes coughing, choking, headache, dizziness, weakness, difficult breathing, and cyanosis. Bloody sputum may be noted. Skin contact produces pain and severe burns. Eye contact produces corneal destruction.

PERSONAL SAFETY Protect against inhalation of the material. Must wear protective clothing including NIOSH-approved rubber gloves, safety goggles, or face mask, and a respirator whose canister is specifically approved for this material. Care must be exercised to decontaminate fully or dispose of all equipment and clothing after use.

ACUTE HAZARD LEVEL When lead salts are ingested, much passes through the body unabsorbed. Long periods of exposure are necessary to produce symptoms. Lead salts are easily absorbed in the respiratory tract when inhaled. Lead fluoride is not very irritating to the skin, but irritates the eyes. Ingestion of large amounts irritates the alimentary tract, with gastrointestinal distress. The lowest lethal oral dose for a guinea pig is 4 gm/kg. The aquatic toxicity to bluegill sunfish is approximately 82 ppm, based on 69 ppm of lead ion for a 96-hour TLM.

CHRONIC HAZARD LEVEL Lead is a cumulative poison. Repeated exposure through inhalation or ingestion will lead to damage of the liver, kidney, blood, and nervous system. Lead is a suspected carcinogen of the lungs and kidney. The threshold-limit-value, time-weighted average for a 40-hour workweek is 0.15 mg/m³, and the short-term exposure limit (15 minutes) is 0.45 mg/m³. The OSHA standard for an 8-hour workday is 0.2 mg/m³ for inorganic lead compounds.

DEGREE OF HAZARD TO PUBLIC HEALTH A suspected carcinogen of the lungs and kidney. Lead is accumulated in the body, and chronic poisoning is a serious hazard.

OCCUPATIONAL RECOMMENDATIONS

OSHA (1) The employer shall assure that no employee is exposed to lead at concentration greater than 50 µg/m³ averaged over an 8-hr period. (2) If an employee is exposed to lead for more than 8 hr in any work day, the permissible exposure limit,

as a TWA for that day, was reduced according to the following formula: Max permissible exposure limit (in µg/m³) = 400/hr worked in that day. (inorganic lead, as Pb) [R9].

NIOSH 10-hr time-weighted average: 0.100 mg/m³. Air concentration to be maintained so that worker blood lead remains <0.60 mg/100 g of whole blood. (lead, as Pb) [R6].

THRESHOLD LIMIT 8-hr time-weighted average (TWA) 0.15 mg/m³ (1986) (lead, inorganic dusts and fumes, as Pb) [R15, 24].

INTERNATIONAL EXPOSURE LIMITS Former USSR (1966) and Czechoslavakia (1969): 1 mg/m³; West Germany (1974) and Sweden (1975): 2.5 mg/m³ (fluorides, not otherwise classified, as F) [R17, 272].

ACTION LEVEL Avoid contact with the spilled cargo. Stay upwind. Notify local air, water, and fire authorities of the accident. Evacuate all people to a distance of at least 200 feet upwind and 1,000 feet downwind of the spill.

ON-SITE RESTORATION Dam stream if possible to reduce the flow and prevent further dispersion of the material by water movement. Bottom pumps or underwater vacuum systems may be employed in small bodies of water; dredging may be effective in larger bodies, to remove any undissolved material from the bottom. If treatment is to be attempted, pump water into a suitable container. Under controlled conditions add calcium hydroxide (Ca (OH)₂) to pH 8.5. Allow to settle and pump off clear water and add lime to pH 11. Let settle 24 hours, filter, and discharge. For more information see Envirex manual EPA 600/2–77–227. Seek professional environmental engineering assistance through EPA's Environmental Response Team (ERT), Edison, NJ, 24-hour no., 908–548–8730.

BEACH OR SHORE RESTORATION Close beach and shore to the public until material has been removed. If tidal, scrape affected area at low tide with a mechanical scraper, and avoid human toxicity contact. Do not burn material.

AVAILABILITY OF COUNTERMEASURE MATERIAL Bottom pumps, dredges—fire departments, U.S. Coast Guard, Army Corps of Engineers; vacuum systems—swimming-pool operators; activated carbon—water treatment plants, chemical companies.

DISPOSAL After the material has been contained, remove it and the contaminated soil and place in impervious containers. If practical, transport material back to the supplier for recovery. If this is not practical, the material should be buried in a specially designated chemical landfill. Not acceptable at a municipal sewage treatment plant. The following wastewater treatment technologies have been investigated for lead: Concentration process: Biological treatment. (lead) [R7].

DISPOSAL NOTIFICATION Notify local and state health authorities, local solid waste disposal authorities, supplier, and shipper.

INDUSTRIAL FOULING POTENTIAL In spite of low solubility, may corrode or foul industrial users.

MAJOR WATER USE THREATENED Fisheries; industrial.

PROBABLE LOCATION AND STATE White powder, will sink and very slowly dissolve.

WATER CHEMISTRY No reaction.

COLOR IN WATER White to colorless.

FOOD-CHAIN CONCENTRATION POTENTIAL Positive. Lead compounds are concentrated in the food chain.

NON-HUMAN TOXICITY LD_{Lo} guinea pig oral 4,000 mg/kg [R16, 1698].

IARC SUMMARY Classification of carcinogenicity: (1) evidence in humans: insufficient; (2) evidence in animals: sufficient. Overall summary evaluation of carcinogenic risk to humans is Group 2B: The agent is possibly carcinogenic to humans. (from table) [R8].

ATMOSPHERIC STANDARDS National primary and secondary ambient air-quality standards for lead and its compounds, measured as elemental lead are: 1 $\mu g/m^3$,

max arithmetic mean average over a calendar quarter. (lead compounds) [R10].

CERCLA REPORTABLE QUANTITIES Persons in charge of vessels or facilities are required to notify the National Response Center (NRC) immediately when there is a release of this designated hazardous substance in an amount equal to or greater than its reportable quantity of 10 lb or 4.54 kg. The toll-free telephone number of the NRC is (800) 424–8802; in the Washington metropolitan area (202) 426–2675. The rule for determining when notification is required is stated in the 40 CFR 302.6 (section IV. D. 3.b) [R11].

RCRA REQUIREMENTS D008; a solid waste containing lead (such as lead(2) fluoride) may or may not become characterized as a hazardous waste when subjected to the Toxicity Characteristic Leaching Procedure listed in 40 CFR 261.24, and, if so characterized, must be managed as a hazardous waste [R12].

DOT *Health Hazards:* Poisonous if swallowed. Inhalation of dust poisonous. Fire may produce irritating or poisonous gases. Runoff from fire control or dilution water may cause pollution [R4].

Fire or Explosion: Some of these materials may burn, but none of them ignites readily [R4].

Emergency Action: Keep unnecessary people away; isolate hazard area and deny entry. Stay upwind; keep out of low areas. Positive-pressure self-contained breathing apparatus (SCBA) and structural firefighters' protective clothing will provide limited protection. Call CHEMTREC at 1–800–424–9300 for emergency assistance. If water pollution occurs, notify the appropriate authorities [R4].

Fire: Small Fires: Dry chemical, CO_2, water spray, or regular foam. Large Fires: Water spray, fog, or regular foam. Move container from fire area if you can do so without risk [R4].

Spill or Leak: Do not touch or walk through spilled material; stop leak if you can do so without risk. Small Spills: Take up with sand or other noncombustible absorbent material and place into containers for later disposal. Small Dry Spills: With clean shovel place material into clean, dry container, and cover loosely; move containers from spill area. Large Spills: Dike far ahead of liquid spill for later disposal [R4].

First Aid: Move victim to fresh air; call for emergency medical care. In case of contact with material, immediately flush skin or eyes with running water for at least 15 minutes. Remove and isolate contaminated clothing and shoes at the site [R4].

FIRE POTENTIAL Not flammable [R1].

FDA Bottled water shall, when a composite of analytical units of equal volume from a sample is examined by the methods described in paragraph (d) (1) (II) of this section, meet the standards of chemical quality, and shall not contain lead in excess of 0.05 mg/L (lead) [R13]. Bottled water packaged in the USA to which no fluoride is added shall not contain fluoride in excess of 1.8 mg/L at 63.9–70.6°F. Bottled water packaged in the USA to which fluoride is added shall not contain fluoride in excess of 1.2 mg/L at 63.9–70.6°F. Imported bottled water to which no fluoride is added and imported bottled water to which fluoride is added shall not contain fluoride in excess of 1.4 mg/L and 0.8 mg/L, respectively. (fluoride) [R13].

GENERAL RESPONSE Avoid contact with solid and dust. Keep people away. Wear dust respirator. Stop discharge if possible. Isolate and remove discharged material. Notify local health and pollution control agencies.

FIRE RESPONSE Not flammable.

EXPOSURE RESPONSE Call for medical aid. Dust poisonous if inhaled. If inhaled will cause dizziness or loss of consciousness. If breathing has stopped, give artificial respiration. If breathing is difficult, give oxygen. Solid irritating to skin and eyes. If swallowed will cause nausea, vomiting, or loss of consciousness. Remove contaminated clothing and shoes. Flush affected areas with plenty of water. If swallowed

and victim is conscious, have victim drink water or milk and have victim induce vomiting. If swallowed and victim is unconscious or having convulsions, do nothing except keep victim warm.

RESPONSE TO DISCHARGE Issue warning—water contaminant. Restrict access. Should be removed. Provide chemical and physical treatment.

WATER RESPONSE Dangerous to aquatic life in high concentrations. May be dangerous if it enters water intakes. Notify local health and wildlife officials. Notify operators of nearby water intakes.

EXPOSURE SYMPTOMS Not irritating to skin or mucuous membranes; protect against chronic poisoning. Early symptoms of lead intoxication via inhalation or ingestion are most commonly gastrointestinal disorders, colic, constipation, etc.; weakness, which may go on to paralysis chiefly of the extensor muscles of the wrists, and less often the ankles, is noticeable in the most serious cases. Ingestion of a large amount causes local irritation of the alimentary tract; pain, leg cramps, muscle weakness, paresthesias, depression, coma, and death may follow in 1 or 2 days. Contact with eyes causes irritation.

EXPOSURE TREATMENT Remove at once all cases of lead intoxication from further exposure until the blood level is reduced to a safe value; immediately place the individual under medical care. Ingestion: give gastric lavage using 1% solution of sodium or magnesium sulfate; leave 15–30 g magnesium sulfate in 6–8 oz. of water in the stomach as antidote and cathartic; egg white, milk, and tannin are useful demulcents; atropine sulfate and other antispasmodics may relieve abdominal pain, but morphine may be necessary. Eyes or skin: flush with water.

REFERENCES

1. U.S. Coast Guard, Department of Transportation. *CHRIS—Hazardous Chemical Data.* Volume II. Washington, DC: U.S. Government Printing Office, 1984–5.

2. Sax, N. I., and R. J. Lewis, Sr. (eds.). *Hawley's Condensed Chemical Dictio-*

nary. 11th ed. New York: Van Nostrand Reinhold Co., 1987. 689.

3. *Kirk–Othmer Encyclopedia of Chemical Technology.* 3rd ed., Volumes 1–26. New York, NY: John Wiley and Sons, 1978–1984, p. V10 756 (1980).

4. U.S. Department of Transportation. *Emergency Response Guidebook 1990.* DOT P 5800.5. Washington, DC: U.S. Government Printing Office, 1990, p. G–53.

5. Mackison, F. W., R. S. Stricoff, and L. J. Partridge, Jr. (eds.). *NIOSH/OSHA—Occupational Health Guidelines for Chemical Hazards.* DHHS (NIOSH) Publication No. 81–123 (3 vols). Washington, DC: U.S. Government Printing Office, Jan. 1981.

6. NIOSH. *NIOSH Pocket Guide to Chemical Hazards.* DHHS (NIOSH) Publication No. 90–117. Washington, DC: U.S. Government Printing Office, June 1990, 136.

7. USEPA. *Management of Hazardous Waste Leachate.* EPA Contract No. 68–03–2766 p. E–53 (1982).

8. IARC. *Monographs on the Evaluation of the Carcinogenic Risk of Chemicals to Man.* Geneva: World Health Organization, International Agency for Research on Cancer, 1972–present (multivolume work), p. S7 (1987).

9. 29 CFR 1910.1025 (7/1/90).

10. 40 CFR 50.12 (7/1/90).

11. 58 FR 35328 (6/30/93).

12. 40 CFR 261.24 (7/1/91).

13. 21 CFR 103.35 (4/1/91).

14. Montesano, R., H. Bartsch, E. Boyland, G. Della Porta, L. Fishbein, R. A. Griesemer, A. B. Swan, L. Tomatis, and W. Davis (eds.). *Handling Chemical Carcinogens in the Laboratory: Problems of Safety.* IARC Scientific Publications No. 33. Lyon, France: International Agency for Research on Cancer, 1979.

15. American Conference of Governmental Industrial Hygienists. *Threshold Limit Values for Chemical Substances and Physical Agents and Biological Exposure Indices for 1994–1995.* Cincinnati, OH: ACGIH, 1994.

16. Clayton, G.D., and F.E. Clayton (eds.). *Patty's Industrial Hygiene and Toxicology:* Volume 2A, 2B, 2C: Toxicology. 3rd ed. New York: John Wiley and Sons, 1981–1982.

17. American Conference of Governmental Industrial Hygienists. *Documentation of the Threshold Limit Values and Biological Exposure Indices.* 5th ed. Cincinnati, OH: American Conference of Governmental Industrial Hygienists, 1986.

■LEAD FLUOROBORATE

CAS # 13814–96–5

SIC CODES 3691; 3714; 3519; 3429

SYNONYMS borate(1-), tetrafluoro-, lead(2) • borate(1-), tetrafluoro-, lead(2) (2:1) • lead borofluoride • lead boron fluoride • lead tetrafluoroborate • lead • plumbous fluoborate-tetrafluoroborate ($Pb(BF_4)_2$) • lead fluoroborate solution • lead fluoborate

MF B_2F_8•Pb

MW 380.81

COLOR AND FORM Crystalline powder; colorless solid.

FIREFIGHTING PROCEDURES Self-contained breathing apparatus with a full facepiece operated in pressure-demand or other positive-pressure mode (inorganic lead) [R4]. Extinguish fire using agent suitable for type of surrounding fire. If large quantities of combustibles are involved, use water in flooding quantities as spray fog [R22, 410]. Wear positive-pressure self-contained breathing apparatus when fighting fires involving this material. [R22, 410].

TOXIC COMBUSTION PRODUCTS Toxic and hazardous hydrogen fluoride gas may form in fire. (solution) [R2].

CORROSIVITY Corrosive to aluminum.

OTHER CHEMICAL/PHYSICAL PROPERTIES
Decomposes in water or alcohol. Colorless, faint-odored liquid; will corrode most metals; specific gravity: 1.75 at 20°C (liquid) (solution).

PROTECTIVE EQUIPMENT Not in excess of 0.5 mg/m³: Dust mask, except single use. Not in excess of 1 mg/m³: Dust and mist respirator, except single use, and quarter-mask respirators. Any fume respirator, or high-efficiency particulate filter respirator. Any supplied-air respirator or any self-contained breathing apparatus. Not in excess of 5 mg/m³: A high-efficiency particulate filter respirator with a full face-piece. Any supplied-air respirator with a full facepiece. Any self-contained breathing apparatus with a full facepiece. Not in excess of 100 mg/m³: Type C supplied-air respirator operated in pressure-demand or other positive-pressure or continuous-flow mode. A powered air-purifying respirator with high-efficiency particulate filter. Not in excess of 200 mg/m³: Type C supplied-air respirator with a full facepiece operated in pressure-demand or other positive-pressure mode, or with a full facepiece, helmet, or hood, operated in continuous-flow mode. Greater than 200 mg/m³, or entry and escape from unknown concentration: Self-contained breathing apparatus with a full facepiece operated in pressure-demand or other positive-pressure mode. A combination respirator, which includes a type C supplied-air respirator with a full facepiece, operated in pressure-demand or other positive-pressure or continuous-flow mode, and an auxiliary self-contained breathing apparatus operated in pressure-demand or other positive-pressure mode. (inorganic lead) [R5].

PREVENTATIVE MEASURES Avoid breathing dusts and fumes from burning material. Keep upwind. Avoid bodily contact with material. [R22, 410].

CLEANUP Land spills: Dig a pit, pond, lagoon, or holding area to contain liquid or solid material. (SRP: If time permits, pits, ponds, lagoons, soak holes, or holding areas should be contained with a flexible impermeable-membrane liner.) Cover solids with a plastic sheet to prevent dissolv-

ing in rain or firefighting water (lead fluoride) [R22, 410]. Wastewater at pH 1.3 containing 8,000 ppm fluorine as $Pb(BF)_2$ and $Sn(BF_4)_2$ is adjusted to pH 8 with sodium hydroxide, filtered, evaporated to 1/16 [R6]. Described treatment of wastewater from a tetraethyl lead manufacturing process: Two major categories of waste were inorganic lead wastewaters and organic lead wastewaters. After sedimentation in a holding basin to recover solid lead and lead oxide, the inorganic lead waste fraction (66.1 mg/L) was effectively treated by coagulation with ferric and ferrous sulfate (inorganic lead) [R7]. Water spill: Neutralize with agricultural lime (CaO), crushed limestone ($CaCO_3$), or sodium bicarbonate ($NaHCO_3$). Adjust pH to neutral (pH = 7). Use mechanical dredges or lifts to remove immobilized masses of pollutants and precipitates. [R22, 407].

DISPOSAL Route to metal salvage facility. SRP: At the time of review, criteria for land treatment or burial (sanitary landfill) disposal practices are subject to significant revision. Prior to implementing land disposal of waste residue (including waste sludge), consult with environmental regulatory agencies for guidance on acceptable disposal practices.

COMMON USES Electroplating solution for coating metal objects with lead; curing agent for epoxy resins; catalyst in production of linear polyesters [R1]; in metal finishing [R22, 410].

STANDARD CODES EPA 311; Superfund designated (hazardous substances) list.

PERSISTENCE Slowly neutralized and precipitated by natural carbonates.

MAJOR SPECIES THREATENED All life.

INHALATION LIMIT 0.15 mg/m³.

DIRECT CONTACT Solutions may have low pH—skin irritant.

GENERAL SENSATION Compounds can be absorbed through skin at toxic chronic levels. Symptoms include pica, anorexia, vomiting, malaise, and convulsions. May leave permanent brain damage.

ACUTE HAZARD LEVEL Threshold concentration for fresh and salt water fish, 0.1 ppm [R15]. Highly toxic when ingested or inhaled.

CHRONIC HAZARD LEVEL Application factor to derive chronic safe limit for Pb from 96-hour LC$_{50}$—0.013 for brook trout, 0.043 for rainbow trout [R21]. More of a chronic hazard than acute. Ingestive, inhalative toxin that can be absorbed through skin. Chronic poisoning symptoms include weight loss, weakness, and anemia. Deformity and subadult mortality have been noted in trout exposed to 0.012–0.14 ppm Pb for 19 months and 18 days, respectively [R17]. Freshwater should not exceed 0.03 ppm Pb, and marine waters 0.02 of 96-hour LC$_{50}$ [R18]. Reproduction in daphnids reduced 16% from 3-week exposure to 0.03 ppm Pb [R19]. Administration of 25 mg/L Pb in drinking water led to rapid die-off of breeding colonies of rats and mice [R16].

DEGREE OF HAZARD TO PUBLIC HEALTH Lead is an acute inhalative toxin and a chronic ingestive and inhalative toxin. Compounds are generally more toxic due to solubility.

OCCUPATIONAL RECOMMENDATIONS
OSHA OSHA permissible exposure limit: 50 μg/m³, 8-hr time-weighted average (fumes, and dusts, as Pb) [R23 (7/1/87)].

NIOSH Recommended exposure limit: <100 μg/m³ time-weighted average; air level to be maintained so that worker blood lead remains less than or equal to 60 μg/100 g. Recommendations are based on exposures up to 10 hr (inorganic lead) [R9]; threshold-limit 8-hr time-weighted average (TWA) 0.15 mg/m³ (1986). (lead, inorganic dusts and fumes, as Pb) [R24, 24].

ACTION LEVEL Clear downstream recreational areas.

ON-SITE RESTORATION Neutralize with lime or sodium hydrogen carbonate. Seek professional environmental engineering assistance through EPA's Environmental Response Team (ERT), Edison, NJ, 24-hour no. 908-548-8730.

BEACH OR SHORE RESTORATION Flush with mild alkaline solution.

AVAILABILITY OF COUNTERMEASURE MATERIAL Lime—cement plants; sodium hydrogen carbonate—grocery stores or bakeries.

INDUSTRIAL FOULING POTENTIAL Traces of lead in metal plating baths will affect the smoothness and brightness of deposits.

MAJOR WATER USE THREATENED Potable supply; fisheries; recreation.

PROBABLE LOCATION AND STATE Colorless liquid, will dissolve. Will be present in water as ionic lead and fluoboric acid.

COLOR IN WATER Colorless.

FOOD-CHAIN CONCENTRATION POTENTIAL Positive. Fishes and other animals will concentrate lead.

NON-HUMAN TOXICITY Minimum lethal dose (MLD) in rats after oral admin 50 mg (PbB$_2$F$_8$)/kg (27.2 mg Pb). [R25, 190].

BIOLOGICAL HALF-LIFE Biological half-life for lead in the bones of humans is 10 yr. (inorganic lead) [R8].

ATMOSPHERIC STANDARDS National primary and secondary ambient air-quality standard for lead and its compounds, measured as elemental lead, is: 1.5 μg/m³, maximum arithmetic mean averaged over a calendar quarter. (lead, and its compounds, as Pb) [R10].

CERCLA REPORTABLE QUANTITIES Persons in charge of vessels or facilities are required to notify the National Response Center (NRC) immediately when there is a release of this designated hazardous substance in an amount equal to or greater than its reportable quantity of 10 lb or 4.54 kg. The toll-free telephone number of the NRC is (800) 424-8802; in the Washington metropolitan area (202) 426-2675. The rule for determining when notification is required is stated in the 40 CFR 302.6 (section IV. D. 3.b) [R11].

RCRA REQUIREMENTS A solid waste containing lead fluoroborate may become characterized as a hazardous waste when subjected to the toxicant extraction procedure listed in 40 CFR 261.24, and, if so characterized, must be managed as a hazardous waste [R12].

DOT *Health Hazards:* Poisonous if swallowed. Inhalation of dust poisonous. Fire may produce irritating or poisonous gases. Runoff from fire control or dilution water may cause pollution [R3].

Fire or Explosion: Some of these materials may burn, but none of them ignites readily [R3].

Emergency Action: Keep unnecessary people away; isolate hazard area and deny entry. Stay upwind; keep out of low areas. Self-contained breathing apparatus (SCBA) and structural firefighter's protective clothing will provide limited protection. Call CHEMTREC at 1–800–424–9300 for emergency assistance. If water pollution occurs, notify the appropriate authorities [R3].

Fire: Small Fires: Dry chemical, CO_2, halon, water spray, or standard foam. Large Fires: Water spray, fog, or standard foam is recommended. Move container from fire area if you can do so without risk [R3].

Spill or Leak: Do not touch spilled material; stop leak if you can do so without risk. Small Spills: Take up with sand or other noncombustible absorbent material and place into containers for later disposal. Small Dry Spills: With clean shovel place material into clean, dry container, and cover; move containers from spill area. Large Spills: Dike far ahead of liquid spill for later disposal [R3].

First Aid: Move victim to fresh air; call emergency medical care. Remove and isolate contaminated clothing and shoes at the site. In case of contact with material, immediately flush skin or eyes with running water for at least 15 minutes [R3].

FDA The FDA action level of lead is 7.0 µg/ml of leaching solution for pottery (ceramics) flatware (avg of 6 units); 5.0 µg/ml of leaching solution for small hollowware (any one of 6 units); 2.5 µg/ml of leaching solution for large hollowware (any one of 6 units); 7.0 µg/ml of leaching solution if product intended for use by adults for silver-plated hollowware (avg of 6 units); and 0.5 µg/ml of leaching solution if product intended for use by infants and children for silver-plated hollowware (one or more of 6 units) (inorganic lead) [R13]. Bottled water shall, when a composite of analytical units of equal volume from a sample is examined by the methods described in paragraph (d) (1) (II) of this section, meet the standards of chemical quality, and shall not contain lead as Pb ion in excess of 0.05 mg/L. (lead as Pb ion) [R14].

REFERENCES

1. SRI.

2. U.S. Coast Guard, Department of Transportation. *CHRIS—Hazardous Chemical Data.* Volume II. Washington, DC: U.S. Government Printing Office, 1984–5.

3. Department of Transportation. *Emergency Response Guidebook 1987.* DOT P 5800.4. Washington, DC: U.S. Government Printing Office, 1987, p. G–53.

4. NIOSH. *Criteria Document: Inorganic Lead* p. I–10 (1978) DHEW Pub. NIOSH 78–158.

5. NIOSH. *Criteria Document: Inorganic Lead* p. I–7 (1978) DHEW Pub. NIOSH 78–158.

6. *Treatment for Wastewater Containing Fluoroborate*; Jpn Kokai Tokkyo Koho Patent No 80155785 12/04/80 (Hitachi, Ltd).

7. Patterson J. W., *Industrial Wastewater Treatment Technology* 2nd Edition p. 75 (1985).

8. USEPA. *The Health and Environmental Impacts of Lead:* p. 211 (1979) EPA 560/2-79-001.

9. NIOSH/CDC. NIOSH *Recommendations for Occupational Safety and Health Standards* Sept. 1986 (Supplement to *Morbidity and Mortality Weekly Report* 35 No. 15, Sept. 26, 1986), p. 21S.

10. 40 CFR 50.12 (7/1/87).

11. 58 FR 35328 (6/30/93).

12. 40 CFR 261.24 (7/1/87).

13. FDA. *Action Levels for Poisonous or Deleterious Substances in Human Food and Animal Feed* p. 9 (1982).

14. 21 CFR 103.35 (4/1/86).

15. Todd, D. K., *The Water Encyclopedia*, Maple Press, 1970.

16. Serkowitz, J. B., G. R. Schimke, and V. R. Valeri, *Water Pollution Potential of Manufactured Products*, Environmental Protection Agency, EPA–R2–73–179d, April 1973.

17. Environmental Protection Agency, *Report of the Pesticide Technical Committee to the Lake Michigan Enforcement Conference on Selected Trace Metals*, NTIS PB–220 361, September 1972.

18. *Proposed Criteria for Water Quality*, Volume I, U.S. Environmental Protection Agency. October 1973.

19. Biesinger, K. E., and G. M. Christensen, Effects of Various Metals on Survival, Growth, Reproduction, and Metabolism of *Daphnia magna*, *J. Fish. Res. Bd. Can.*, Vol. 29, No. 12, 1972.

20. Venugopal, B., and T. D. Luckey. *Metal Toxicity in Mammals*, 2. New York: Plenum Press, 1978.

21. Environmental Protection Agency. *Effects of Pesticides in Water*. NTIS PB-222 320, 1972.

22. Association of American Railroads. *Emergency Handling of Hazardous Materials in Surface Transportation*. Washington, DC: Assoc. of American Railroads, Hazardous Materials Systems (BOE), 1987.

23. 29 CFR 1910.1025.

24. American Conference of Governmental Industrial Hygienists. *Threshold Limit Values for Chemical Substances and Physical Agents and Biological Exposure Indices for 1994–1995*. Cincinnati, OH: ACGIH, 1994.

25. 52 FR 16482 (5/5/87).

■ LEAD NITRATE

CAS # 18256–98–9

SYNONYMS nitric acid, lead salt

MF $PbNO_3$

MW 331.2

BINARY REACTANTS Carbon, cyanides, phospham, phosphorus, sodium hyphosphite, stannous chloride, thiocyanates.

STANDARD CODES EPA 311; CFR—oxidizing material, yellow label, 100 lbs.; IATA—oxidizing material, yellow label, 12 kg passenger, 45 kg cargo; NFPA–1, 0, 1; Superfund designated (hazardous substances) list.

PERSISTENCE Slowly precipitated by natural alkalinity.

COMMON USES Manufacture of lead compounds; dye manufacture; matches; photography; textiles processing; synthetic rubber manufacture; paints; ceramics; engraving.

MAJOR SPECIES THREATENED Farm animals are poisoned by lead more frequently than by any other metallic poison. 0.18 mg/L Pb, animals, chronic poisoning soft water. >2.4 mg/L Pb, animals, chronic poisoning. 0.5 mg/L Pb, animals, maximum safe limit. 2 g Pb daily, calves have survived 2–3 years. 6–8 mg Pb/kg daily, calves tolerated several months to 3 years. 200–400 Pb/kg, a few days, calves death. Geese died from eating mine tailings. 100 mg/L Pb as $Pb(NO_3)_2$, calf died after 4 months. 2 mg Pb/kg dry soil, peas increased N content. 1.5–25 mg $Pb(NO_3)_2$/L, oats and potatoes stimulating effect. >50 mg $Pb(NO_3)_2$/L, oats and potatoes died in a week's time. 51.8 mg/L Pb in nutrient solution, sugar beets in sand culture slightly injurious. 2,760 mg/L Pb solution, cress and mustard seeds in solution, culture germination inhibited after 18 days. 345–1,380 mg/L Pb solution, cress and mustard seeds in solution, culture germination delayed and growth retarded. 0.005 mg Pb/kg, rats, chronic damage to central nervous system.

GENERAL SENSATION Can be absorbed

through skin at chronically toxic levels. 6 mg/m³/day, inhaled long-term, produces histological and pathological effects in man; 1.2 mg/day, ingested long-term, produces central nervous system disorders in man; lethal dose % bone <4.53 mg/100 g man, 17 mg/100 g child; advanced lead poisoning causes constipation, loss of appetite, anemia, abdominal pain, and gradual paralysis of the muscles, especially of the arms. Mild lead poisoning causes lethargy, moroseness, constipation, flatulence, and occasional abdominal pain.

PERSONAL SAFETY Wear approved filter mask.

ACUTE HAZARD LEVEL Highly toxic when ingested or inhaled. Toxic to fish and aquatic life.

CHRONIC HAZARD LEVEL Highly toxic by all routes. 6 mg/m³/day inhalation limit.

DEGREE OF HAZARD TO PUBLIC HEALTH 0.1 mg/L Pb in drinking water may cause chronic poisoning; highly toxic by acute and chronic routes.

OCCUPATIONAL RECOMMENDATIONS None listed.

ACTION LEVEL Suppress suspension of dusts. Isolate from heat.

ON-SITE RESTORATION Very difficult to recover—perhaps dithizone or EDTA followed by adsorption of the Pb complex with powdered C treatment. [R1]. Add lime to precipitate basic carbonate. Seek professional environmental engineering assistance through EPA's Environmental Response Team (ERT), Edison, NJ, 24-hour no. 908–548–8730.

AVAILABILITY OF COUNTERMEASURE MATERIAL Lime—cement plants; complexants—detergent manufacturers, analytical labs; carbon—water treatment plants, sugar refineries.

DISPOSAL Convert to nitrates with a minimum of nitric acid (concentrated, reagent). Evaporate in a fume hood to a thin paste. Add about 500 ml water, and saturate with hydrogen sulfide. Filter, wash, and dry the precipitate. Package and ship to the supplier.

INDUSTRIAL FOULING POTENTIAL Traces of lead in metal plating baths will affect the smoothness and brightness of deposits. Not acceptable in food-processing waters.

MAJOR WATER USE THREATENED All uses.

PROBABLE LOCATION AND STATE White or colorless translucent crystals. Will be dissolved in water.

COLOR IN WATER Colorless.

FOOD-CHAIN CONCENTRATION POTENTIAL Fish and other animals can concentrate lead.

GENERAL RESPONSE Avoid contact with solid and dust. Keep people away. Wear dust respirator. Stop discharge if possible. Isolate and remove discharged material. Notify local health and pollution control agencies.

FIRE RESPONSE Not flammable. Will increase the intensity of a fire. Poisonous gases may be produced when heated. Flood discharge area with water. Cool exposed containers with water.

EXPOSURE RESPONSE Call for medical aid. Dust is poisonous if inhaled. If inhaled will cause dizziness or loss of consciousness. If breathing has stopped, give artificial respiration. If breathing is difficult, give oxygen. Solid is irritating to skin and eyes. If swallowed will cause nausea, vomiting, or loss of consciousness. Remove contaminated clothing and shoes. Flush affected areas with plenty of water. If swallowed and victim is conscious, have victim drink water or milk, and have victim induce vomiting. If swallowed and victim is unconscious or having convulsions, do nothing except keep victim warm.

WATER POLLUTION RESPONSE Dangerous to aquatic life in high concentrations. May be dangerous if it enters water intakes. Notify local health and wildlife officials. Notify operators of nearby water intakes.

RESPONSE TO DISCHARGE Issue warning—water contaminant, oxidizing material. Restrict access. Disperse and flush.

PERSONAL PROTECTION Dust mask and protective gloves.

EXPOSURE SYMPTOMS: Early symptoms

of lead intoxication via inhalation or ingestion are most commonly gastrointestinal disorders, colic, constipation, etc.; weakness, which may go on to paralysis, chiefly of the extensor muscles of the wrists, and less often the ankles, is noticeable in the most serious cases. Ingestion of a large amount causes local irritation of the alimentary tract; pain, leg cramps, muscle weakness, paresthesias, depression, coma, and death may follow in 1 or 2 days. Contact with eyes causes irritation.

EXPOSURE TREATMENT: Remove at once all cases of lead intoxication from further exposure until the blood level is reduced to a safe value; immediately place the individual under medical care. Ingestion: give gastric lavage using 1% solution of sodium or magnesium sulfate; leave 15–30 g magnesium sulfate in 6–8 oz. of water in the stomach as antidote and cathartic; egg white, milk, and tannin are useful demulcents; atropine sulfate and other antispasmodics may relieve abdominal pain, but morphine may be necessary. Eyes or skin: flush with water.

REFERENCES

1. "Research Heightens Concern over PCB's," *Chemical and Engineering News*, pp. 27–28, April 17, 1972.

2. EPA *OHMTADS* database.

3. DOT *CHRIS* database.

■ LEAD OXIDE

CAS # 1317–36–8

SYNONYMS lead monooxide • lead monoxide • lead oxide yellow • lead protoxide • lead(2) oxide • lead(II) oxide • lead oxide (PbO) • ci-pigment-yellow-46 • CI-77577 • massicotite • plumbous oxide • yellow lead ocher • litharge • massicot • litharge-pure • litharge yellow L-28

MF OPb

MW 223.19

pH Strong base.

COLOR AND FORM Exists in 2 forms: red to reddish-yellow, tetragonal crystals, and yellow, orthorhombic crystals.

DENSITY 9.53.

MELTING POINT 888°C.

EXPLOSIVE LIMITS AND POTENTIAL Addition of 2–3 drops of about 90% peroxyformic acid (with lead oxide) causes an immediate violent explosion. [R21, 150].

FIREFIGHTING PROCEDURES Self-contained breathing apparatus with a full facepiece, operated in pressure-demand or other positive-pressure mode (inorganic lead) [R5].

TOXIC COMBUSTION PRODUCTS Dangerous; when heated it emits highly toxic fumes. (lead compounds) [R6].

SOLUBILITY Insoluble in alcohol; soluble in acetic acid, dilute nitric acid, warm solution of fixed alkali hydroxides; 0.017 g/L at 20°C; soluble in ammonium chloride, nitric acid; 0.0504 g/L at 20°C.

VAPOR PRESSURE 10 mm Hg at 1,085°C.

STABILITY AND SHELF LIFE Red to reddish-yellow, tetragonal crystals: stable at ordinary temperature [R1].

OTHER CHEMICAL/PHYSICAL PROPERTIES Litharge: density, 9.3; MP: 886°C; solubility: 0.001 g/100 cc water at 20°C, soluble in nitric acid, alkali, lead acetate, ammonium chloride, strontium chloride, calcium chloride; cofo: yellow, tetragonal crystals (litharge); commercial grades are yellow to reddish, depending on treatment, and purity. Massicot: an oxide of lead corresponding to same formula as litharge (PbO) but having different physical state; contains approximately 92.8% lead (massicot); massicot: index of refraction: 2.51, 2.61 (li), 2.71; solubility: 0.0023 g/100 cc water at 23°C, insoluble in cold water; soluble in alkali; density: 8.0; yellow, rhombic crystals (massicot). At 300–450°C in air, converted slowly into lead tetraoxide but at higher temperature reverts to lead oxide. Divalent lead has a strong affinity for inorganic ions containing oxygen (e.g., carbonate) or sulfur (sulfide). Lead can also complex with electron-rich ligands in many organic compounds, such as amino acids, proteins, and humic acid. (inorganic lead).

PROTECTIVE EQUIPMENT Avoid breathing dust. Wear dust mask approved by U.S. Bureau of Mines for this purpose [R1].

PREVENTATIVE MEASURES Cloakroom accommodation should be provided for this personal protective equipment with separate accommodation for clothing taken off during working hours. Washing accommodation, including bathing accommodation with warm water, should be provided and used. Time should be allowed for washing before eating. Arrangements should be made to prohibit eating and smoking in the vicinity of lead processes and suitable messrooms should be provided. (inorganic lead) [R22, 1204].

CLEANUP Described treatment of wastewater from a tetraethyllead manufacturing process. Two major categories of waste were inorganic lead wastewaters and organic lead wastewaters. After sedimentation in a holding basin to recover solid lead and lead oxide, the inorganic lead waste fraction (66.1 mg/L) was effectively treated by coagulation with ferric and ferrous sulfate (inorganic lead) [R7]. Lead oxide: Chemical conversion to the sulfide or carbonate, followed by collection of the precipitate, and recovery of lead by way of smelting operations. Landfilling of the oxide is also an acceptable procedure. Alternatively, it may be dissolved in nitric acid (HNO_3), precipitated as the sulfide, and returned to a supplier for reprocessing [R19, 545]. Environmental considerations for land spill: Dig a pit, pond, or lagoon holding area to contain liquid or solid material. (SRP: If time permits, pits, ponds, lagoons, soak holes, or holding areas should be sealed with an impermeable flexible-membrane liner.) Cover solids with a plastic sheet to prevent dissolving in rain or firefighting water (inorganic lead) [R8]. Environmental considerations for water spill: neutralize with agricultural lime (CaO), crushed limestone ($CaCO_3$), or sodium bicarbonate ($NaHCO_3$). Adjust pH to neutral (pH = 7). Use mechanical dredges or lifts to remove immobilized masses of pollutants and precipitates. (inorganic lead) [R8].

DISPOSAL SRP: At the time of review, criteria for land treatment or burial (sanitary landfill) disposal practices are subject to significant revision. Prior to implementing land disposal of waste residue (including waste sludge), consult with environmental regulatory agencies for guidance on acceptable disposal practices.

COMMON USES In ointments, plasters; preparing solution of lead subacetate; glazing pottery; glass flux for painting on porcelain and glass; lead glass; with glycerol as metal cement; producing iridescent colors on brass and bronze; in assay of gold and silver ores; coloring sulfur-containing substances, e.g., hair, nails, wool, horn; manufacture of artificial tortoise shell and horn; pigment for rubber; manufacture of boiled linseed oil [R1]. Manufacture of storage battery plates; chemical intermediate for pigments, basic lead carbonate, lead arsenate, red lead, lead soaps and greases, sodium plumbite, activator in rubber compounding; agent in manufacture of glass, glaze, and vitreous enamel; auxiliary drier in varnishes [R2]. Oil refining; in manufacture of paints, ink, acid-resisting compositions, and match-head compositions [R20, 706]. Lead oxide is used as a heat stabilizer for plastics [R3].

OCCUPATIONAL RECOMMENDATIONS

OSHA OSHA permissible exposure limit: 50 µg/m³, 8-hr time-weighted average (fumes, and dusts, as Pb) [R18 (7/1/87)].

NIOSH NIOSH-recommended exposure limit: <100 µg/m³ time-weighted average; air level to be maintained so that worker blood lead remains less than or equal to 60 µg/100 g. Recommendations are based on exposures up to 10 hr. (inorganic lead) [R13].

THRESHOLD LIMIT 8-hr time-weighted average (TWA) 0.15 mg/m³ (1986) (lead, inorganic dusts and fumes, as Pb) [R23, 24].

IARC SUMMARY No evaluation could be made of the carcinogenicity of lead oxide [R9].

BIOLOGICAL HALF-LIFE Biological half-life

for lead in the bones of humans is 10 yr. (inorganic lead compounds) [R10].

BIODEGRADATION Lake sediment microorganisms are able to directly methylate certain inorganic lead compounds (1). Although all lake sediments tested in one study were able to transform trimethyllead to tetramethyllead, only some could transform lead nitrate and chloride to tetramethyllead and no biotransformation occurred with lead oxide, hydroxide, bromide, cyanide, or palmitate (1). Under appropriate conditions, dissolution due to anaerobic microbial action may be significant in subsurface environments (2). In one investigation half of 40 intertidal sediments were found to release spasmodic and variable amounts of gaseous alkyllead over periods of 5–120 days, with a maximum release rate of 0.085 ng Pb/kg-hr (5). Indeed, the further from source areas, the more dominant was the fraction of alkyllead relative to total lead in air (up to 20%) (5). It is not known what forms of Pb in sediment may be amenable to alkylation (SRC). The mean percentage removal of lead during the activated sludge process was 82% and was almost entirely due to the removal of the insoluble fraction by adsorption onto the sludge floc and, to a much lesser extent, precipitation (3,4) [R11].

BIOCONCENTRATION Aquatic biota, both invertebrate and vertebrate, have been shown to bioconcentrate lead at levels greater than in water, and sometimes similar to those in sediments. However, concentration of lead tends to decrease with increasing trophic levels within aquatic systems. (inorganic lead) [R12].

ATMOSPHERIC STANDARDS National primary and secondary ambient air-quality standard for lead and its compounds, measured as elemental lead, is: 1.5 $\mu g/m^3$, maximum arithmetic mean averaged over a calendar quarter. (lead and its compounds, as Pb) [R14].

RCRA REQUIREMENTS A solid waste containing lead oxide may become characterized as a hazardous waste when subjected to the toxicant extraction procedure listed in 40 CFR 261.24, and, if so characterized, must be managed as a hazardous waste [R15].

DOT *Health Hazards:* Poisonous if swallowed. Inhalation of dust poisonous. Fire may produce irritating or poisonous gases. Runoff from fire control or dilution water may cause pollution (lead compounds, soluble, not otherwise specified) [R4].

Fire or Explosion: Some of these materials may burn, but none of them ignites readily (lead compounds, soluble, not otherwise specified) [R4].

Emergency Action: Keep unnecessary people away; isolate hazard area and deny entry. Stay upwind; keep out of low areas. Self-contained breathing apparatus (SCBA) and structural firefighter's protective clothing will provide limited protection. Call CHEMTREC at 1–800–424–9300 for emergency assistance. If water pollution occurs, notify the appropriate authorities (lead compounds, soluble, not otherwise specified) [R4].

Fire: Small Fires: Dry chemical, CO_2, halon, water spray, or standard foam. Large Fires: Water spray, fog, or standard foam is recommended. Move container from fire area if you can do so without risk (lead compounds, soluble, not otherwise specified) [R4].

Spill or Leak: Do not touch spilled material; stop leak if you can do so without risk. Small Spills: Take up with sand or other noncombustible absorbent material and place into containers for later disposal. Small Dry Spills: With clean shovel place material into clean, dry container, and cover; move containers from spill area. Large Spills: Dike far ahead of liquid spill for later disposal (lead compounds, soluble, not otherwise specified) [R4].

First Aid: Move victim to fresh air; call for emergency medical care. Remove and isolate contaminated clothing and shoes at the site. In case of contact with material, immediately flush skin or eyes with running water for at least 15 minutes (lead compounds, soluble, not otherwise specified) [R4].

FDA Bottled water shall, when a composite

of analytical units of equal volume from a sample is examined by the methods described in paragraph (d) (1) (II) of this section, meet the standards of chemical quality, and shall not contain lead as Pb ion in excess of 0.05 mg/L (lead as Pb ion) [R16]. The FDA action level of lead is 7.0 µg/ml of leaching solution for pottery (ceramics) flatware (avg of 6 units); 5.0 µg/ml of leaching solution for small hollowware (any one of 6 units); 2.5 µg/ml of leaching solution for large hollowware (any one of 6 units); 7.0 µg/ml of leaching solution if product intended for use by adults for silver-plated hollowware (avg of 6 units); and 0.5 µg/ml of leaching solution if product intended for use by infants and children for silver-plated hollowware (one or more of 6 units). (inorganic lead) [R17].

REFERENCES

1. *The Merck Index.* 10th ed. Rahway, NJ: Merck Co., Inc., 1983. 778.

2. SRI.

3. USEPA. *The Health and Environmental Impacts of Lead* p. 87 (1979) EPA 560/2–79–001.

4. Department of Transportation. *Emergency Response Guidebook 1987.* DOT P 5800.4. Washington, DC: U.S. Government Printing Office, 1987, p. G–53.

5. NIOSH. *Criteria Document: Inorganic Lead.* p. I–10 (1978) DHEW Pub. NIOSH 78–158.

6. Sax, N. I. *Dangerous Properties of Industrial Materials.* 4th ed. New York: Van Nostrand Reinhold, 1975. 866.

7. Patterson, J. W. *Industrial Wastewater Treatment Technolgy.* 2nd Edition p. 75 (1985).

8. Association of American Railroads. *Emergency Handling of Hazardous Materials in Surface Transportation.* Washington, DC: Assoc. of American Railroads, Hazardous Materials Systems (BOE), 1987. 407.

9. IARC. *Monographs on the Evaluation of the Carcinogenic Risk of Chemicals to Man.* Geneva: World Health Organization, International Agency for Research on Cancer, 1972–present (multivolume work), p. V23 387 (1980).

10. USEPA. *The Health and Environmental Impacts of Lead*: p. 211 (1979) EPA 560/2–79–001.

11. (1) USEPA. *Air Quality Criteria for Lead.* pp. 6–1 to 6–2 USEPA–600/8–77–017 (1977). (2) Frances, A. J. *Anaerobic Microbial Dissolution of Toxic Metals in Subsurface Environments.* BLN–36571, CONF–8540521–1 Upton, NY: Brookhaven National Lab (1985). (3) Stephenson, T., J. N. Lester. *Sci Tot Environ* 63: 199–214 (1987). (4) Stephenson, T., J. N. Lester. *Sci Tot Environ* 63: 215–30 (1987). (5) Hewitt, C. N., R. M. Harrison. *Environ Sci Technol* 21: 260–66 (1987).

12. USEPA. *The Health and Environmental Impacts of Lead*: p. 153 (1979) EPA 560/2–79–001.

13. NIOSH/CDC. *NIOSH Recommendations for Occupational Safety and Health Standards.* Sept. 1986 (Supplement to *Morbidity and Mortality Weekly Report* 35 No. 15, Sept. 26, 1986), p. 21S.

14. 40 CFR 50.12 (7/1/87).

15. 40 CFR 261.24 (7/1/87).

16. 21 CFR 103.35 (4/1/86) 17. FDA. *Action Levels for Poisonous or Deleterious Substances in Human Food and Animal Feed.* p. 9 (1982).

18. 29 CFR 1910.1025.

19. Sittig, M. *Handbook of Toxic and Hazardous Chemicals and Carcinogens, 1985.* 2nd ed. Park Ridge, NJ: Noyes Data Corporation, 1985.

20. Sax, N.I., and R.J. Lewis, Sr. (eds.). *Hawley's Condensed Chemical Dictionary.* 11th ed. New York: Van Nostrand Reinhold Co., 1987.

21. Bretherick, L. *Handbook of Reactive Chemical Hazards.* 3rd ed. Boston, MA: Butterworths, 1985.

22. International Labour Office. *Encyclopedia of Occupational Health and Safety.* Vols. I & II. Geneva, Switzerland: International Labour Office, 1983.

23. American Conference of Governmental Industrial Hygienists. *Threshold Limit Values for Chemical Substances and Physical Agents and Biological Exposure Indices for 1994–1995.* Cincinnati, OH: ACGIH, 1994.

■ LEAD(+2) SULFATE

CAS # 7446–14–2

SYNONYMS anglislite • sulfuric acid, lead(2) salt (1:1) • Bleisulfat (German) • ci-pigment-white-3 • ci-77630 • fast-white • Freemans white lead • lead bottoms • milk-white • mulhouse-white • sulfate de plomb (French) • natural anglesite • lead(II) sulfate (1:1)

MF O_4S • Pb

MW 303.25

COLOR AND FORM White, heavy crystal powder; white monoclinic or rhombic crystals.

DENSITY 6.2.

MELTING POINT 1,170°C.

FIREFIGHTING PROCEDURES Extinguish fire using agent suitable for type of surrounding fire (material itself does not burn or burns with difficulty). Cool all affected containers with flooding quantities of water. Apply water from as far a distance as possible. Keep run-off water out of sewers and water sources [R5].

TOXIC COMBUSTION PRODUCTS Special hazards of combustible products: toxic metal fumes [R4].

SOLUBILITY Soluble in cold water, hot water, or ammonium salts; slightly soluble in sulfuric acid; soluble in about 2,225 parts water; more soluble in nitric acid; soluble in sodium hydroxide, ammonium acetate, or tartrate solution; soluble in concentrated hydriodic acid; insoluble in alcohol.

CORROSIVITY Corrosive.

PROTECTIVE EQUIPMENT Wear appropriate equipment to prevent: Repeated or prolonged skin contact (lead, as Pb) [R6]. Wear eye protection to prevent: Reasonable probability of eye contact (lead, as Pb) [R6]. Recommendations for respirator selection. Max concentration for use: 0.5 mg/m³: Respirator Classes: Any supplied-air respirator. Any air-purifying respirator with a high-efficiency particulate filter (if an independent code); or a high-efficiency particulate filter. Any self-contained breathing apparatus (lead, as Pb) [R6]. Recommendations for respirator selection. Max concentration for use: 1.25 mg/m³: Respirator Classes: Any powered, air-purifying respirator with a high-efficiency particulate filter. Any supplied-air respirator operated in a continuous-flow mode (lead, as Pb) [R6]. Recommendations for respirator selection. Max concentration for use: 2.5 mg/m³: Respirator Classes: Any air-purifying, full facepiece respirator with a high-efficiency particulate filter. Any powered, air-purifying respirator with a tight-fitting facepiece, and a high-efficiency particulate filter. Any self-contained breathing apparatus with a full facepiece. Any supplied-air respirator with a full facepiece. Any supplied-air respirator that has a tight-fitting facepiece and is operated in a continuous-flow mode (lead, as Pb) [R6]. Recommendations for respirator selection. Max concentration for use: 50 mg/m³: Respirator Class: Any supplied-air respirator operated in a pressure-demand or other positive-pressure mode (lead, as Pb) [R6]. Recommendations for respirator selection. Max concentration for use: 100 mg/m³: Respirator Class: Any supplied-air respirator with a full facepiece and operated in a pressure-demand or other positive-pressure mode (lead, as Pb) [R6]. Recommendations for respirator selection. Condition: Emergency, or planned entry into unknown concentration or IDLH conditions: Respirator Classes: Any self-contained breathing apparatus that has a full facepiece and is operated in a pressure-demand or other positive-pressure mode. Any supplied-air respirator with a full facepiece and operated in pressure-demand or other positive-pressure mode in combination with an auxiliary self-contained breathing apparatus operated in pressure-demand or other positive-pressure mode (lead, as Pb) [R6]. Recommendations for respirator selection. Condition: Escape from suddenly occurring respiratory hazards: Respirator Classes: Any air-purifying, full facepiece respirator with a high-efficiency particulate filter. Any appropriate escape-type, self-contained breathing apparatus (lead, as Pb) [R6]. Wear goggles, self-contained breathing apparatus, and rubber gloves. [R4].

PREVENTATIVE MEASURES Workers should wash at the end of each work shift (lead, as Pb) [R6]. Remove clothing promptly if non-impervious clothing becomes contaminated (lead, as Pb) [R6]. SRP: The scientific literature supports the wearing of contact lenses in industrial environments, as part of a program to protect the eye against chemical compounds and minerals causing eye irritation. However, there may be individual substances whose irritating or corrosive properties are such that the wearing of contact lenses would be harmful to the eye. In those specific cases contact lenses should not be worn. If material not involved in fire keep sparks, flames, and other sources of ignition away. Keep material out of water sources and sewers [R5]. Personnel protection: Avoid breathing vapors or dusts. Avoid breathing fumes from burning material. Avoid bodily contact with the material. Do not handle broken packages unless wearing appropriate personal protective equipment. Wash away any material which may have contacted the body with copious amounts of water or soap and water. If contact with the material anticipated, wear appropriate chemical protective clothing [R5]. SRP: Contaminated protective clothing should be segregated in such a manner that there is no direct personal contact by personnel who handle, dispose of, or clean the clothing. Quality assurance to ascertain the completeness of the cleaning procedures should be implemented before the decontaminated protective clothing is returned for reuse by the workers. Contaminated clothing should not be taken home at end of shift, but should remain at employee's place of work for cleaning.

CLEANUP Environmental considerations, land spill: Cover solids with a plastic sheet to prevent dissolving in rain or firefighting water [R5]. Environmental considerations, water spill: Neutralize with agricultural lime, crushed limestone, or sodium bicarbonate. If dissolved, in region of 10 ppm, or greater concentration, apply activated carbon at ten times the spilled amount. Remove trapped material with suction hoses. Use mechanical dredges or lifts to remove immobilized masses of pollutants and precipitates [R5].

DISPOSAL The following wastewater treatment technologies have been investigated for lead: Concentration process: Biological treatment (lead) [R7]. The following wastewater treatment technologies have been investigated for lead: Concentration process: Chemical precipitation (lead) [R16]. The following wastewater treatment technologies have been investigated for lead: Concentration process: Reverse Osmosis (lead) [R17]. The following wastewater treatment technologies have been investigated for lead: Concentration process: Activated carbon (lead) [R18]. The following wastewater treatment technologies have been investigated for lead: Concentration process: Miscellaneous sorbents (lead) [R19].

COMMON USES Storage batteries; paint pigments [R1]. Used in manufacturing of alloys, in lithography; preparing rapidly drying oil varnishes; weighting fabrics [R2].

OCCUPATIONAL RECOMMENDATIONS
OSHA (1) The employer shall assure that no employee is exposed to lead at concentration greater than 50 $\mu g/m^3$ averaged over an 8-hr period. (2) If an employee is exposed to lead for more than 8 hr in any work day, the permissible exposure limit, as a TWA for that day, was reduced according to the following formula: Max permissible exposure limit (in $\mu g/m^3$) = 400/hr worked in that day (inorganic lead, as Pb) [R10].

NIOSH 10-hr time-weighted average: 0.100 mg/m^3. Air concentration to be maintained so that worker blood lead remains

<0.60 mg/100 g of whole blood (lead, as Pb) [R6].

THRESHOLD LIMIT 8-hr time-weighted average (TWA) 0.15 mg/m³ (1986) (lead, inorganic dusts, and fumes, as Pb) [R20, 24]. Excursion Limit Recommendation: Excursions in worker exposure levels may exceed three times the TLV–TWA for no more than a total of 30 min during a work day, and under no circumstances should they exceed five times the TLV–TWA (lead, inorganic dusts, and fumes, as Pb) [R20, 5]. BEI (Biological Exposure Index): Lead in blood (timing is not critical) is 50 µg/100 ml. The determinant is usually present in a significant amount in biological specimens collected from subjects who have not been occupationally exposed. Such background levels are included in the BEI value (1987–1988 adoption) (lead) [R20, 59]. BEI (Biological Exposure Index): Lead in urine (timing is not critical) is 150 µg/g creatinine. The determinant is usually present in a significant amount in biological specimens collected from subjects who have not been occupationally exposed. Such background levels are included in the BEI value (1987–1988 adoption) (lead) [R20, 59]. BEI (Biological Exposure Index): Zinc protoporphyrin in blood after 1-month exposure is 250 µg/100 ml erythrocytes or 100 µg/100 ml blood. The determinant is usually present in a significant amount in biological specimens collected from subjects who have not been occupationally exposed. Such background levels are included in the BEI value (1987–1988 adoption) (lead) [R20, 59]. Notice of Intended Changes (1993–1994): 8-hr time-weighted average (TWA) 0.05 mg/m³ (lead, elemental, and inorganic compounds, as Pb) [R20, 37]. Notice of Intended Changes (1993–1994): A3. A3 = Animal carcinogen (lead, elemental, and inorganic compound, as Pb) [R20, 37]. Notice of Intended Change (1994–1995): BEI (Biological Exposure Index): Lead in blood (timing is not critical) is 30 µg/100 ml. The determinant is usually present in a significant amount in biological specimens collected from subjects who have not been occupationally exposed. Such

background levels are included in the BEI value (lead) [R20, 63].

INTERNATIONAL EXPOSURE LIMITS former USSR (1977): 0.01 mg/m³; Hungary (1974): 0.02 mg/m³; former Czechoslovakia (1976), Poland (1976): 0.05 mg/m³; Romania (1975), Sweden (1975), and West Germany (1978): 0.1 mg/m³; East Germany (1973), Finland (1975), and Yugoslavia (1971): 0.15 mg/m³ (lead, inorganic compound, dust, and fume, as lead) [R11].

NON-HUMAN TOXICITY LD₅₀ guinea pig ip 300 mg/kg [R21, 190].

ECOTOXICITY LC₅₀ *Coturnix* >5,000 ppm/5 days (lead) [R9].

IARC SUMMARY Classification of carcinogenicity: (1) evidence in humans: insufficient; (2) evidence in animals: sufficient. Overall summary evaluation of carcinogenic risk to humans is Group 2B: The agent is possibly carcinogenic to humans. (from table) [R8].

ATMOSPHERIC STANDARDS National primary and secondary ambient air-quality standards for lead and its compounds, measured as elemental lead are: 1 µg/m³, max arithmetic mean averaged over a calendar quarter. (lead compounds) [R12].

CERCLA REPORTABLE QUANTITIES Persons in charge of vessels or facilities are required to notify the National Response Center (NRC) immediately when there is a release of this designated hazardous substance in an amount equal to or greater than its reportable quantity of 10 lb or 4.54 kg. The toll-free telephone number of the NRC is (800) 424–8802; in the Washington metropolitan area (202) 426–2675. The rule for determining when notification is required is stated in the 40 CFR IV. D. 3.b) [R13].

RCRA REQUIREMENTS D008; a solid waste containing lead (such as lead(2) sulfate) may or may not become characterized as a hazardous waste when subjected to the Toxicity Characteristic Leaching Procedure listed in 40 CFR 261.24, and, if so characterized, must be managed as a hazardous waste [R14].

DOT *Health Hazards:* Contact causes

burns to skin and eyes. If inhaled, may be harmful. Fire may produce irritating or poisonous gases. Runoff from fire control or dilution water may cause pollution (lead sulfate, with more than 3% free acid) [R3].

Fire or Explosion: Some of these materials may burn, but none of them ignites readily. Flammable/poisonous gases may accumulate in tanks and hopper cars. Some of these materials may ignite combustibles (wood, paper, oil, etc.) (lead sulfate, with more than 3% free acid) [R3].

Emergency Action: Keep unnecessary people away; isolate hazard area and deny entry. Stay upwind; keep out of low areas. Positive-pressure self-contained breathing apparatus (SCBA) and structural firefighters' protective clothing will provide limited protection. Call CHEMTREC at 1–800–424–9300 for emergency assistance. If water pollution occurs, notify the appropriate authorities (lead sulfate, with more than 3% free acid) [R3].

Fire: Some of these materials may react violently with water. Small Fires: Dry chemical, CO_2, water spray, or regular foam. Large Fires: Water spray, fog, or regular foam. Move container from fire area if you can do so without risk. Apply cooling water to sides of containers that are exposed to flames until well after fire is out. Stay away from ends of tanks (lead sulfate, with more than 3% free acid) [R3].

Spill or Leak: Do not touch or walk through spilled material; stop leak if you can do so without risk. Small Spills: Take up with sand or other noncombustible absorbent material and place into containers for later disposal. Small Dry Spills: With clean shovel place material into clean, dry container, and cover loosely; move containers from spill area. Large Spills: Dike far ahead of liquid spill for later disposal (lead sulfate, with more than 3% free acid) [R3].

First Aid: Move victim to fresh air; call emergency medical care. In case of contact with material, immediately flush skin or eyes with running water for at least 15 minutes. Remove and isolate contaminated clothing and shoes at the site. Keep victim quiet and maintain normal body temperature (lead sulfate, with more than 3% free acid) [R3].

FIRE RESPONSE Not flammable.

GENERAL RESPONSE Avoid contact with solid and dust. Keep people away. Wear goggles, self-contained breathing apparatus, and rubber gloves. Stop discharge if possible. Isolate and remove discharged material. Notify local health and pollution control agencies.

EXPOSURE RESPONSE Call for medical aid. Dust or solid poisonous if inhaled. Irritating to eyes. If swallowed will cause abdominal pain, nausea, vomiting, headache, and muscular weakness. Move to fresh air. Flush affected areas with plenty of water. If swallowed and victim is conscious, have victim drink water or milk, and induce vomiting.

RESPONSE TO DISCHARGE Issue warning—water contaminant, corrosive. Restrict access. Should be removed. Provide chemical and physical treatment.

WATER RESPONSE Harmful to aquatic life in very low concentrations. May be dangerous if it enters water intakes. Notify local health and wildlife officials. Notify operators of nearby water intakes.

EXPOSURE SYMPTOMS Inhalation: joint and muscle pains, headache, dizziness, and insomnia. Weakness, frequently of extensor muscles of hand and wrist (unilateral or bilateral). Heavy contamination: brain damage. Stupor progressing to coma—with or without convulsion, often death. Excitation, confusion, and mania less common. Cerebrospinal pressure may be increased. Eyes: caused a moderate purulent reaction and general inflammation of the rabbit eye. Ingestion: abdominal pain, diarrhea, constipation, loss of appetite, muscular weakness, headache, blue line on gums, metallic taste, nausea, and vomiting.

EXPOSURE TREATMENT Get medical aid. Inhalation: remove from source of exposure and keep quiet. Eyes: wash with running water. Skin: wash with soap and water. Ingestion: wash mouth, give emetic then epsom salts (30 g/250 ml hot water); get medical attention.

FIRE POTENTIAL Not flammable [R4].

FDA Bottled water shall, when a composite of analytical units of equal volume from a sample is examined by the methods described in paragraph (d) (1) (II) of this section, meet the standards of chemical quality, and shall not contain lead in excess of 0.05 mg/L (lead) [R15]. Bottled water shall, when a composite of analytical units of equal volume from a sample is examined by the methods described in paragraph (d) (1) (II) of this section, meet the standards of chemical quality, and shall not contain sulfate in excess of 250.0 mg/L. (sulfate) [R15].

REFERENCES

1. Sax, N. I., and R. J. Lewis, Sr. (eds.). *Hawley's Condensed Chemical Dictionary.* 11th ed. New York: Van Nostrand Reinhold Co., 1987. 693.

2. Budavari, S. (ed.). *The Merck Index—Encyclopedia of Chemicals, Drugs, and Biologicals.* Rahway, NJ: Merck and Co., Inc., 1989. 853.

3. U.S. Department of Transportation. *Emergency Response Guidebook 1990.* DOT P 5800.5. Washington, DC: U.S. Government Printing Office, 1990, p. G-60.

4. U.S. Coast Guard, Department of Transportation. *CHRIS—Hazardous Chemical Data.* Volume II. Washington, DC: U.S. Government Printing Office, 1984-5.

5. Assocciation of American Railroads. *Emergency Handling of Hazardous Materials in Surface Transportation.* Washington, DC: Assoc. of American Railroads, Hazardous Materials Systems (BOE), 1987. 413.

6. NIOSH. *NIOSH Pocket Guide to Chemical Hazards.* DHHS (NIOSH) Publication No. 90-117. Washington, DC: U.S. Government Printing Office, June 1990, 136.

7. USEPA. *Management of Hazardous Waste Leachate,* EPA Contract No. 68-03-2766 p. E-53 (1982).

8. IARC. *Monographs on the Evaluation of the Carcinogenic Risk of Chemicals to Man.* Geneva: World Health Organization, International Agency for Research on Cancer, 1972-present (multivolume work), p. S7 (1987).

9. Hill, E. F., and M. B. Camardese. "Lethal Dietary Toxicities of Environmental Contaminants and Pesticides to Coturnix." *Fish and Wildlife Technical Report.* 2.Washington, DC: United States Department of Interior Fish and Wildlife Service, 1986. 86.

10. 29 CFR 1910.1025 (7/1/90).

11. Arena, J. M., and R. H. Drew (eds.) *Poisoning—Toxicology, Symptoms, Treatments.* 5th ed. Springfield, IL: Charles C. Thomas Publisher, 1986. 344.

12. 40 CFR 50.12 (7/1/90).

13. 58 FR 35328 (6/30/93).

14. 40 CFR 261.24 (7/1/91.

15. 21 CFR 103.35 (4/1/91).

16. USEPA. *Management of Hazardous Waste Leachate.* EPA Contract No. 68-03-2766 p. E-71-72 (1982).

17. USEPA. *Management of Hazardous Waste Leachate.* EPA Contract No. 68-03-2766 p. E-88 (1982).

18. USEPA. *Management of Hazardous Waste Leachate.* EPA Contract No. 68-03-2766 p. E-164 (1982).

19. USEPA. *Management of Hazardous Waste Leachate.* EPA Contract No. 68-03-2766 p. E-202 (1982).

20. American Conference of Governmental Industrial Hygienists. *Threshold Limit Values for Chemical Substances and Physical Agents and Biological Exposure Indices for 1994-1995.* Cincinnati, OH: ACGIH, 1994.

21. Venugopal, B., and T. D. Luckey. *Metal Toxicity in Mammals,* 2. New York: Plenum Press, 1978.

■ LETHANE-384

CAS # 112-56-1

SIC CODES 2879

SYNONYMS 2-butoxy-2′-thiocyanodiethyl ether • 2-[2-(butoxy) ethoxy]ethyl ether of thiocyanic acid • butyl carbitol rhodanate • butyl carbitol thiocyanate • 1-butoxy-2-(2-thiocyanoethoxy) ethane • 1-butoxy-alpha-(2-thiocyanoethoxy) ethane • 2-(2-(butoxy) ethoxy) ethyl thiocyanic acid ester • 2-(2-butoxyethoxy) ethyl thiocyanate • 2-butoxy-2′-thiocyanodiethyl ether • 2-[2-(butoxy) ethoxy] ester of thiocyanic acid • beta-butoxy-beta′-thiocyanodiethyl ether • butoxyrhodanodiethyl-ether • ent-6 • ethane, 1-butoxy-2-(2-thiocyanatoethoxy)- • ethanol, 2-(2-butoxyethoxy)-, thiocyanate • thiocyanic acid, 2(2-butoxyethoxy) ethyl ester • thiocyanic acid 2-(2-butoxyethoxy) ethyl ester • 2-(2-butoxy) ethoxy) ethyl ester of thiocyanic acid • lethane-384-regular • butyl 'carbitol' thiocyanate • butyl 'carbitol' rhodanate

MF $C_9H_{17}NO_2S$

MW 203.33

COLOR AND FORM Liquid; clear brown oil.

DENSITY 0.915–0.930 at 25°C.

BOILING POINT 120–125°C at 0.25 mm Hg.

SOLUBILITY Practically insoluble in water; miscible with hydrocarbons and most organic solvents; soluble in petroleum oils.

STABILITY AND SHELF LIFE Stable at ordinary temperatures but tends to rearrange at higher temperatures [R1].

OTHER CHEMICAL/PHYSICAL PROPERTIES Brownish oil (commercial product).

COMMON USES Insecticide [R2].

BINARY REACTANTS Thiocyanate group reactivity: explodes with chlorates and nitrates including nitric acid. Explodes with oxidizing agents like peroxides. Reacts vigorously with boron triiodide (NFC 13, 80/NFPA).

STANDARD CODES UN 1615.

DIRECT CONTACT Moderately irritating to the skin and mucous membranes. Absorbed through intact skin. Prolonged absorption causes skin eruptions, runny nose, dizziness, cramps, nausea, and vomiting.

GENERAL SENSATION Skin irritant. Dermally absorbed: skin eruptions, runny nose, dizziness, cramps, nausea, and vomiting.

ACUTE HAZARD LEVEL Causes rapid knockdown; a narcotic at high concentrations. High concentrations cause central nervous system depression.

OCCUPATIONAL RECOMMENDATIONS None listed.

ON-SITE RESTORATION Seek professional environmental engineering assistance through EPA's Environmental Response Team (ERT), Edison, NJ, 24-hour no., 908–548–8730. Contain and isolate spill by using clay/bentonite dams, interceptor trenches, or impoundments. Construct swale to divert uncontaminated portion of watershed around contaminated portion. Seek professional help to evaluate problem, implement containment measures, and conduct bench scale and pilot scale tests prior to full-scale decontamination program implementation. Treatment alternatives for contaminated water include alkaline hydrolysis, powdered activated carbon, granular activated carbon, granular activated carbon, bentonite sorption, aeration, evaporation, biodegradation, and chemical oxidation. May be able to employ density stratification and impoundment removal techniques by removing product from bottom or top layers using skimming or polyethylene rope mop collection equipment. Treatment alternatives for contaminated soils include well-point collection and treatment of leachates as for contaminated waters, bentonite/cement ground injection to immobilize spill, physical removal of residues with pH adjustment to 8.0, and land application on an approved sanitary landfill. Residues may also be packaged for disposal.

DISPOSAL Product residues and sorbent media may be packaged in 17h epoxy-lined drums and disposed of at an approved EPA disposal site. Destruction by high-temperature incineration or microwave plasma detoxification, if available.

Encapsulation by organic polyester resin or silicate fixation. Confirm disposal procedures with responsible environmental engineer and regulatory officials.

PROBABLE LOCATION AND STATE Lethane is a brownish oil, which should float on water in case of spills.

NON-HUMAN TOXICITY LD_{50} rat oral 90 mg/kg [R3].

FIFRA REQUIREMENTS In 1988, Congress amended FIFRA to strengthen and accelerate EPA's reregistration program. The nine-year reregistration scheme mandated by "FIFRA 88" applies to each registered pesticide product containing an active ingredient initially registered before November 1, 1984. Pesticides for which EPA had not issued Registration Standards prior to the effective date of FIFRA '88 were divided into three lists based upon their potential for exposure and other factors, with List B being of highest concern and D of least. List: C; Case: Thiocyanoethyl derivs.; Case No.: 3142; Pesticide type: Insecticide; Case Status: The pesticide is no longer an active ingredient in any registered pesticide products. Therefore, EPA is characterizing it as "cancelled." Under FIFRA, pesticide producers may voluntarily cancel their registered products. EPA also may cancel pesticide registrations if registrants fail to pay required fees, to make or meet certain reregistration commitments, or when the Agency reaches findings of unreasonable adverse effects. Active Ingredient (AI): (Butoxyethoxy)ethyl thiocyanate; AI Status: The producer(s) of the pesticide has not made or honored a commitment to seek reregistration, conduct the necessary studies, or pay the requisite fees. Unless some other interested party makes and meets such commitments, products containing the pesticide will be cancelled. [R4].

REFERENCES

1. Spencer, E. Y. *Guide to the Chemicals Used in Crop Protection.* 6th ed. Publication 1093, Research Institute, Agriculture Canada, Ottawa, Canada: Information Canada, 1973. 59.

2. *Farm Chemicals Handbook 1993.* Willoughby, OH: Meister Publishing Co., 1993, p. C–203.

3. Matsumura, F. *Toxicology of Insecticides.* 2nd ed. New York, NY: Plenum Press, 1985. 86.

4. USEPA/OPP. *Status of Pesticides in Reregistration and Special Review.* p. 215 (Mar, 1992) EPA 700-R-92-004.

■ LITHIUM CHLORIDE

CAS # 7447–41–8

MF ClLi

MW 42.39

pH Aqueous solution neutral or slightly alkaline.

COLOR AND FORM Cubic crystals, granules, or crystalline powder; white.

TASTE Sharp saline taste.

DENSITY 2.07.

BOILING POINT 1,360°C.

MELTING POINT 613°C.

SOLUBILITY 1 g dissolves in 1.3 ml cold, 0.8 ml boiling water; soluble in amyl alcohol, pyridine; 25.10 g soluble in 100 cc alcohol at 30°C; 42.36 g soluble in 100 cc methanol at 25°C; 4.11 g soluble in 100 cc acetone at 25°C; 0.538 g soluble in 100 cc ammonium hydroxide at 33.9°C; soluble in ether, nitrobenzene.

OTHER CHEMICAL/PHYSICAL PROPERTIES Tendency to form hydrates.

COMMON USES Manufacture of mineral waters; in pyrotechnics; soldering aluminum; in refrigerating machines; medication: antidepressant [R3]. Brazing [R8, 1068]; medication: management of manic-depressive psychosis [R9, p. III–203]. Dehumidifying agent in air conditioning [R10, 3]. Welding, and soldering flux; dry batteries; heat-exchange media; salt baths; desiccant [R2]. Chemical intermediate for lithium metal and lithium borohydride; eutectic with potassium chloride, e.g., in metallurgy; additive to brazing fluxes and dry-cell batteries; dehumidifier in air-conditioning systems; static drying agent; antistatic finish for fabrics; additive to soft drinks and mineral waters [R1]. Catalyst for low-temperature reaction between epoxides and carboxyl groups to yield polyester product [R4]. Additive in low-freezing fire-extinguishing solution [R5].

PERSISTENCE Can persist indefinitely.

MAJOR SPECIES THREATENED 2–5 ppm lithium sulfate in water can harm citrus plants. 1, 2, and 4 ppm in soil has caused lithium poisoning symptoms. Low concentrations interfere with growth of eggs of aquatic organisms and cause monstrosities.

GENERAL SENSATION Sharp saline taste; prolonged absorption may cause disturbed electrolyte balance; impaired renal function, central nervous system disturbances. Cases have been reported where intake caused dizziness, ringing in ears, visual disturbances, tremors, and mental confusion. Some people use as salt substitute. Can damage kidneys if sodium intake is limited.

PERSONAL SAFETY Wear filter mask unless intense heat threatens release of chlorides, then self-contained devices are required.

ACUTE HAZARD LEVEL Moderate ingestive and inhalative hazard. Low toxicity to fish. Toxic to freshwater invertebrates. Can be damaging to irrigation water.

CHRONIC HAZARD LEVEL Slight hazard with chronic inhalation or ingestion. Marine waters should not exceed 0.01 LD_{50} 24-hour maximum [R7].

DEGREE OF HAZARD TO PUBLIC HEALTH Lithium compounds are moderately toxic when inhaled or ingested. Slight chronic hazard. Emits toxic vapors when heated to decomposition.

OCCUPATIONAL RECOMMENDATIONS None listed.

ON-SITE RESTORATION Use cation exchangers or precipitate with phosphates. Seek professional environmental engineering assistance through EPA's Environmental Response Team (ERT), Edison, NJ, 24-hour no., 908–548–8730.

AVAILABILITY OF COUNTERMEASURE MATERIAL Cation exchangers—water softener suppliers; phosphates—fertilizer industry, hardware stores.

DISPOSAL Sift or pour onto a dry layer of sodium bicarbonate in a large evaporating dish. After mixing thoroughly spray with $6M$ NH_4OH while stirring. Cover with a layer of crushed ice and stir. Dump this slurry into a large container. Repeat until all has been treated. Neutralize and slowly siphon the suspension into the drain with excess running water.

DISPOSAL NOTIFICATION Notify local sewage authority.

INDUSTRIAL FOULING POTENTIAL Cl tolerances: pulp and paper—75 ppm, textile, and brewing—100 ppm, carbonated drinks—100 ppm.

MAJOR WATER USE THREATENED Irrigation.

PROBABLE LOCATION AND STATE White deliquescent crystalline solid. Hygroscopic. Will dissolve.

WATER CHEMISTRY Phosphate salts insoluble (lithium).

COLOR IN WATER Colorless.

BIOLOGICAL HALF-LIFE Lithium chloride solution (24 mmol) admin to 7 healthy volunteers in single dose and multiple dose expt: mean biol $T_{1/2}$ was approximately 19.8 hr [R6].

REFERENCES

1. SRI.

2. Hawley, G. G. *The Condensed Chemical Dictionary.* 9th ed. New York: Van Nostrand Reinhold Co., 1977. 519.

3. *The Merck Index.* 9th ed. Rahway, NJ: Merck and Co., Inc., 1976. 722.

4. *Kirk–Othmer Encyc Chemical Tech.* 3rd ed. 1978–present, V18 p. 585.

5. Considine. *Chemical and Process Technol Encyc.* 1974 p. 699.

6. Nielsen–Kudsk, F., A. Amdisen. *Eur J Clin Pharmacol.* 16 (4) 271 (1979).

7. *Proposed Criteria for Water Quality*, Volume I, U.S. Environmental Protection Agency. October 1973.

8. Patty, F. (ed.). *Industrial Hygiene and Toxicology*. Volume II: *Toxicology*. 2nd ed. New York: Interscience Publishers, 1963.

9. Gosselin, R. E., H. C. Hodge, R. P. Smith, and M. N. Gleason. *Clinical Toxicology of Commercial Products*. 4th ed. Baltimore: Williams and Wilkins, 1976.

10. Venugopal, B., and T. D. Luckey. *Metal Toxicity in Mammals*, 2. New York: Plenum Press, 1978.

■ LITHIUM HYDRIDE

CAS # 7580–67–8

SYNONYMS hydrure de lithium [French] • lithium hydride (LiH) • lithium monohydride

MF HLi

MW 7.95

COLOR AND FORM Commercial product is usually gray; cubic crystals; white, translucent, crystalline mass or powder.

ODOR Odorless.

DENSITY 0.76–0.77.

MELTING POINT 680°C.

AUTOIGNITION TEMPERATURE 392°F [R7].

FIREFIGHTING PROCEDURES Use approved Class D extinguishers or smother with dry sand, dry clay, or dry ground limestone. Do not use carbon dioxide or halogenated extinguishing agents. Do not use water. Violent reaction may result [R15, p. 49–106]. A fire once started cannot be extinguished by ordinary methods; smothering by dolomite powder is recommended [R16, 1738]. Fires should be extinguished with dry chemicals; "Lith–X," the trademark for a graphite-base dry chemical extinguishing agent, is available for such fires [R16, 1739]. Use dry nitrogen, graphite, or lithium chloride. Never use water, foam, halogenated hydrocarbons, soda acid, dry chemical, or carbon dioxide [R7]. Do not use water. Do not use dry chemical or carbon dioxide. Use graphite, soda ash, or powdered sodium chloride (lithium hydride; lithium hydride in fused solid form) [R8].

TOXIC COMBUSTION PRODUCTS Combustion may produce irritants and toxic gases [R15, p. 49–106]. Irritating alkali fumes may form in a fire. [R7].

SOLUBILITY Very slightly soluble in acid. Decomposes in cold water. Slightly soluble in inert polar organic solvents such as ethers. Insoluble in benzene and toluene. Soluble in ether.

VAPOR PRESSURE 0 mm Hg at 20°C.

STABILITY AND SHELF LIFE Darkens rapidly on exposure to light [R10].

OTHER CHEMICAL/PHYSICAL PROPERTIES Decomposition pressure nil at 25°C; 0.7 mm at 500°C; 760 mm at approximately 850°C. Heat of solution: −7,200 Btu/lb. Hygroscopic and pyrophoric; darkens on exposure to light. Enthalpy of formation = −90.7 kJ/mol. Density = 0.82.

PROTECTIVE EQUIPMENT Employees should be provided with and required to use impervious clothing [R11]. Goggles or face shield; rubberized gloves; flameproof outer clothing; respirator; high boots or shoes [R7]. Wear appropriate personal protective clothing to prevent skin contact [R9]. Wear appropriate eye protection to prevent eye contact [R9]. Eyewash fountains should be provided in areas where there is any possibility that workers could be exposed to the substance; this is irrespective of the recommendation involving the wearing of eye protection (liquids containing >0.5 mg/m³ of contaminant) [R9]. Facilities for quickly drenching the body should be provided within the immediate work area for emergency use where there is a possibility of exposure. Note: It is intended that these facilities should provide a sufficient quantity or flow of water to quickly remove the substance from any body areas likely to be exposed. The actual determination of what constitutes an adequate quick drench facility depends on the specific circumstances. In certain instances, a deluge shower should be readily available, whereas in others, the availability of water from a sink or hose could be considered adequate (liquids containing >0.5 mg/m³ of contaminant) [R17]. Recommendations for respirator selection. Max concentration for use: 0.25 mg/m³. Respirator Class(es): Any supplied-air respirator. Any self-contained breathing apparatus. Any air-purifying respirator with a high-efficiency particulate filter (if an independent code); or a high-efficiency particulate filter [R9]. Recommendations for respirator selection. Max concentration for use: 0.625 mg/m³. Respirator Class(es): Any supplied-air respirator operated in a continuous-flow mode. Any powered air-purifying respirator with a high-efficiency particulate filter. May require eye protection [R9]. Recommendations for respirator selection. Max concentration for use: 1.25 mg/m³. Respirator Class(es): Any air-purifying, full facepiece respirator with a high-efficiency particulate filter. Any self-contained breathing apparatus with a full facepiece. Any supplied-air respirator with a full facepiece. Any powered air-purifying respirator with a tight-fitting facepiece, and a high-efficiency particulate filter. May require eye protection [R9]. Recommendations for respirator selection. Max concentration for use: 50 mg/m³. Respirator Class(es): Any supplied-air respirator with a full facepiece and operated in a pressure-demand or other positive-pressure mode [R9]. Recommendations for respirator selection. Condition: Emergency or planned entry into unknown concentration or IDLH conditions: Respirator Class(es): Any self-contained breathing apparatus that has a full facepiece and is operated in a pressure-demand or other positive-pressure mode. Any supplied-air respirator with a full facepiece and operated in pressure-demand or other positive-pressure mode in combination with an auxiliary self-contained breathing apparatus operated in pressure-demand or other positive-pressure mode [R9]. Recommendations for respirator selection. Condition: Escape from suddenly occurring respiratory hazards: Respirator Class(es): Any air-purifying, full facepiece respirator with a high-efficiency particulate filter. Any appropriate escape-type, self-contained breathing apparatus [R9]. Wear butyl rubber gloves, laboratory coat, and eye protection [R18, 210].

PREVENTATIVE MEASURES SRP: Contaminated protective clothing should be segregated in such a manner so that there is no direct personal contact by personnel who handle, dispose of, or clean the clothing. Quality assurance to ascertain the completeness of the cleaning procedures should be implemented before the decontaminated protective clothing is returned for reuse by the workers. Contaminated clothing should not be taken home at end of shift, but should remain at employee's place of work for cleaning. Exhaust ventilation should be provided for welding and brazing operations where there is a lithium hazard, and dust should be avoided by careful handling and enclosed processes (lithium) [R20]. Lithium hydride is pyrophoric and should be maintained and handled out of contact with air and moisture [R16, 1739]. Contact lenses should not be worn when working with this chemical [R19]. SRP: The scientific literature for the use of contact lenses in industry is conflicting. The benefit or detrimental effects of wearing contact lenses depend not only upon the substance, but also on factors including the form of the substance, characteristics and duration of the exposure, the uses of other eye protection equipment, and the hygiene of the lenses. However, there may be individual substances whose irritating or corrosive properties are such that the wearing of contact lenses would be harmful to the eye. In those specific cases, contact lenses should not be worn. In any event, the usual eye protection equipment should be worn even when contact lenses are in place. Workers should wash immediately when skin becomes contaminated [R19]. Work clothing should be changed daily if there is any possibility that the clothing may be contaminated [R19]. Remove clothing immediately if it becomes contaminated [R19]. Clothing that has had any possibility of being contaminated with lithium hydride or liquids containing lithium hydride should be placed in closed containers for storage until it can be discarded or until provision is made for the removal of lithium hydride from the clothing. If the clothing is to be laundered or otherwise cleaned to remove the lithium hydride, the person performing the operation should be informed of lithium hydride's hazardous properties [R19]. Skin that becomes contaminated with lithium hydride should be immediately brushed to remove any solid lithium hydride from the skin and washed or showered with copious quantities of water to remove any lithium hydride. Workers subject to skin contact with lithium hydride or liquids containing lithium hydride should wash any areas of the body that may have contacted lithium hydride at the end of each work day. Eating and smoking should not be permitted in areas where lithium hydride or liquids containing lithium hydride are handled, processed, or stored. Employees who handle lithium hydride or liquids containing lithium hydride should wash their hands thoroughly before eating, smoking, or using toilet facilities [R21]. Personnel protection: Avoid breathing dusts and fumes from burning material. Do not handle broken packages unless wearing appropriate personal protective equipment. Wash away any material that may have contacted the body with copious amounts of water or soap and water (lithium hydride; lithium hydride in fused solid form) [R8]. If material not on fire and not involved in fire: Do not use water. Keep sparks, flames, and other sources of ignition away. Keep material out of water sources and sewers. Keep material dry. (lithium hydride; lithium hydride in fused solid form) [R8].

CLEANUP 1. Remove all ignition sources. 2. Ventilate area of spill. 3. Collect spilled material in the most convenient and safe manner and deposit in sealed containers for reclamation or for disposal in a secured sanitary landfill. Avoid contact with water [R11]. Wearing butyl rubber gloves, fireproof clothing, face shield, and goggles, cover spill with sand. Transfer mixture into a dry plastic bag filled in advance with an inert gas. Carry outdoors for

incineration. After burning (if not in a proper incinerator), sprinkle water on the residue for complete destruction. Alternatively, in the fume hood, add butanol slowly to the solid mixture until the reaction ceases. Then carefully add water until all the hydride is destroyed. Let stand until solid settles. Decant liquid into drain with at least 50 times its volume of water. Discard solid residue with normal refuse [R16, 209]. Eliminate all ignition sources. Keep water away from release. Shovel into suitable dry container. Report any release in excess of 1 lb [R15, p. 49–106]. Transfer mixture into a dry plastic bag filled in advance with an inert gas. Carry outdoors for incineration. After burning, sprinkle water on the residue for complete destruction [R12].

DISPOSAL SRP: At the time of review, criteria for land treatment or burial (sanitary landfill) disposal practices are subject to significant revision. Prior to implementing land disposal of waste residue (including waste sludge), consult with environmental regulatory agencies for guidance on acceptable disposal practices.

COMMON USES Reducing agent, condensing agent with ketones and acetic esters; desiccant; in hydrogen generators [R1]. Prep of lithium amide and double hydrides; nuclear shielding material; reducing agent [R2]. In manufacture of electronic tubes, in ceramics, and in chemical synthesis [R3]. Most lithium hydride, which is produced in tonnage quantities, is used as a raw material in the production of lithium aluminum hydride and lithium amide; another important use is in the production of silane by the Sundermeyer process [R17, 203]. As a result of its high hydrogen content, it is used as a transportable hydrogen source, e.g., for inflatable rubber dinghies and balloons [R17, 203]. Vigorous reaction with water, producing large amounts of hydrogen, has been used in military applications and buoyancy devices [R4].

OCCUPATIONAL RECOMMENDATIONS
OSHA 8-hour time-weighted average: 0.025 mg/m³ [R13].

NIOSH 10-hr time-weighted average: 0.025 mg/m³ [R9].

THRESHOLD LIMIT 8-hr time-weighted average (TWA) 0.025 mg/m³ (1977) [R22, 23]. Excursion Limit Recommendation: Excursions in worker exposure levels may exceed three times the TLV–TWA for no more than a total of 30 min during a work day, and under no circumstances should they exceed five times the TLV–TWA. [R22, 5].

INTERNATIONAL EXPOSURE LIMITS Australia: 0.025 mg/m³ (1990); United Kingdom: 0.025 mg/m³ (1991). [R24, 863].

DOT *Fire or Explosion:* May ignite itself if exposed to air. May reignite after fire is extinguished. May ignite in presence of moisture. Violent reaction with water produces flammable gas. Runoff to sewer may create fire or explosion hazard (lithium hydride; lithium hydride, fused solid) [R5].

Health Hazards: May be poisonous if inhaled. Contact may cause burns to skin and eyes. Fire may produce irritating or poisonous gases (lithium hydride; lithium hydride, fused solid) [R5].

Emergency Action: Keep unnecessary people away; isolate hazard area and deny entry. Stay upwind; keep out of low areas. Positive-pressure self-contained breathing apparatus (SCBA) and structural firefighters' protective clothing may provide limited protection. See the Table of Initial Isolation and Protective Action Distances. If you find the ID Number and the name of the material there, begin protective action. Call emergency response telephone number on shipping paper first. If shipping paper not available or no answer, call CHEMTREC at 1–800–424–9300. If water pollution occurs, notify the appropriate authorities (lithium hydride; lithium hydride, fused solid) [R5].

Fire: Do not use water or foam. Small Fires: Dry chemical, soda ash, lime, or sand. Large Fires: Withdraw from area, and let fire burn. Lithium Fires: Use dry sand, sodium chloride powder, graphite powder, or copper powder. Move container from fire area if you can do so without risk

(lithium hydride; lithium hydride, fused solid) [R5].

Spill or Leak: Shut off ignition sources; no flares, smoking, or flames in hazard area. Do not touch or walk through spilled material; stop leak if you can do so without risk. No water on spilled material; do not get water inside container. Small Dry Spills: With clean shovel place material into clean, dry container, and cover loosely; move containers from spill area. Small Spills: Take up with sand or other noncombustible absorbent material and place into containers for later disposal. Large Spills: Dike liquid spill for later disposal. Cover powder spill with plastic sheet or tarp to minimize spreading (lithium hydride; lithium hydride, fused solid) [R5].

First Aid: Move victim to fresh air; call for emergency medical care. Wipe material from skin immediately; flush skin or eyes with running water for at least 15 minutes. Remove and isolate contaminated clothing and shoes at the site (lithium hydride; lithium hydride, fused solid) [R5].

GENERAL RESPONSE Avoid contact with solid and dust. Keep people away. Wear dust respirator and rubber overclothing (including gloves). Stop discharge if possible. Isolate and remove discharged material. Notify local health and pollution control agencies.

FIRE RESPONSE Combustible. Irritating flammable gas may be produced when heated. Extinguish with dry graphite, soda ash, or other inert powder. Do not use water, foam, carbon dioxide, or dry chemicals on fire or adjacent fires.

EXPOSURE RESPONSE Call for medical aid. Dust poisonous if inhaled. If inhaled will cause coughing or difficult breathing. If breathing has stopped, give artificial respiration. If breathing is difficult, give oxygen. Solid will burn skin and eyes. If swallowed will cause nausea or loss of consciousness. Remove contaminated clothing and shoes. Flush affected areas with plenty of water. If swallowed and victim is conscious, have victim drink water or milk. If swallowed, and victim is unconscious or having convulsions, do nothing except keep victim warm.

RESPONSE TO DISCHARGE Issue warning—high flammability, corrosive. Restrict access. Disperse, and flush with care.

WATER RESPONSE Effect of low concentrations on aquatic life is unknown. May be dangerous if it enters water intakes. Notify local health and wildlife officials. Notify operators of nearby water intakes.

EXPOSURE SYMPTOMS Inhalation of dust causes coughing, sneezing, and burning of nose and throat. Ingestion causes severe burns of mouth and stomach; symptoms of central nervous system damage may occur. Contact with eyes or skin causes severe caustic burns.

EXPOSURE TREATMENT Lithium hydride burns of the eyes, skin, or respiratory tract appear to be worse than those caused by an equivalent amount of sodium hydroxide. Inhalation: remove victim to fresh air; if irritation persists get medical attention at once. Ingestion: give large volumes of water and milk; gastric lavage may be indicated. Eyes: flush with copious quantities of running water for at least 15 min; get medical attention. Skin: flush with water; treat as a caustic burn.

FIRE POTENTIAL Lithium hydride is a flammable solid and is dangerous when wet [R6]. Spontaneous ignition occurs when nitrous oxide and lithium hydride or hydrazine are mixed. [R15, p. 491M–144].

REFERENCES

1. Budavari, S. (ed.). *The Merck Index— Encyclopedia of Chemicals, Drugs, and Biologicals*. Rahway, NJ: Merck and Co., Inc., 1989. 871.

2. Sax, N. I., and R. J. Lewis, Sr. (eds.). *Hawley's Condensed Chemical Dictionary*. 11th ed. New York: Van Nostrand Reinhold Co., 1987. 709.

3. Doull, J., C. D. Klaassen, and M. D. Amdur (eds.). *Casarett and Doull's Toxicology*. 2nd ed. New York: Macmillan Publishing Co., 1980. 448.

4. *Kirk–Othmer Encyclopedia of Chemical Technology*. 3rd ed., Volumes 1–26. New York, NY: John Wiley and Sons, 1978–1984, p. V14 463.

5. U.S. Department of Transportation. *Emergency Response Guidebook 1993*. DOT P 5800.6. Washington, DC: U.S. Government Printing Office, 1993, p. G–40.

6. Sullivan, J. B. Jr., G. R. Krieger (eds.). *Hazardous Materials Toxicology—Clinical Principles of Environmental Health*. Baltimore, MD: Williams and Wilkins, 1992. 1199.

7. U.S. Coast Guard, Department of Transportation. *CHRIS—Hazardous Chemical Data*. Volume II. Washington, DC: U.S. Government Printing Office, 1984–5.

8. Association of American Railroads. *Emergency Handling of Hazardous Materials in Surface Transportation*. Washington, DC: Association of American Railroads, Bureau of Explosives, 1992. 583.

9. NIOSH. *NIOSH Pocket Guide to Chemical Hazards*. DHHS (NIOSH) Publication No. 90–117. Washington, DC: U.S. Government Printing Office, June 1990, 136.

10. *The Merck Index*. 10th ed. Rahway, NJ: Merck Co., Inc., 1983. 793.

11. Mackison, F. W., R. S. Stricoff, and L. J. Partridge, Jr. (eds.). *NIOSH/OSHA— Occupational Health Guidelines for Chemical Hazards*. DHHS (NIOSH) Publication No. 81–123 (3 vols). Washington, DC: U.S. Government Printing Office, Jan. 1981.

12. ITII. *Toxic and Hazardous Industrial Chemicals Safety Manual*. Tokyo, Japan: The International Technical Information Institute, 1988. 304.

13. 29 CFR 1910.1000 (58 FR 35338 (6/30/93)).

14. Gerhartz, W. (exec ed.). *Ullmann's Encyclopedia of Industrial Chemistry*. 5th ed. Vol A1: Deerfield Beach, FL: VCH Publishers, 1985 to present, p. VA13.

15. National Fire Protection Guide. *Fire Protection Guide on Hazardous Materials*. 10th ed. Quincy, MA: National Fire Protection Association, 1991.

16. Clayton, G.D., and F.E. Clayton (eds.). *Patty's Industrial Hygiene and Toxicology*. Volume 2A, 2B, 2C: *Toxicology*. 3rd ed. New York: John Wiley and Sons, 1981–1982.

17. Association of American Railroads. *Emergency Handling of Hazardous Materials in Surface Transportation*. Washington, DC: Association of American Railroads, Bureau of Explosives, 1992. 583.

18. Armour, M. A. *Hazardous Laboratory Chemicals Disposal Guide*. Boca Raton, FL: CRC Press Inc., 1991.

19. NIOSH. *NIOSH Pocket Guide to Chemical Hazards*. DHHS (NIOSH) Publication No. 90–117. Washington, DC: U.S. Government Printing Office, June 1990, p. 136.

20. International Labour Office. *Encyclopedia of Occupational Health and Safety*. Vols. I and II. Geneva, Switzerland: International Labour Office, 1983. 1344.

21. Mackison, F. W., R. S. Stricoff, and L. J. Partridge, Jr. (eds.). *NIOSH/OSHA— Occupational Health Guidelines for Chemical Hazards*. DHHS (NIOSH) Publication No. 81–123 (3 vols). Washington, DC: U.S. Government Printing Office, Jan. 1981.

22. American Conference of Governmental Industrial Hygienists. *Threshold Limit Values for Chemical Substances and Physical Agents and Biological Exposure Indices for 1994–1995*. Cincinnati, OH: ACGIH, 1994.

23. American Conference of Governmental Industrial Hygienists, Inc. *Documentation of the Threshold Limit Values and Biological Exposure Indices*. 6th ed. Volumes I, II, III. Cincinnati, OH: ACGIH, 1991.

■ MANEB

CAS # 12427-38-2

SYNONYMS 1,2-ethanediylbis (carbamo-dithioato)(2-)manganese • 1,2-ethanediylbiscarbamodithioic acid, manganese complex • 1,2-ethanediylbis-maneb, manganese(2) salt (1:1) • 1,2-ethylenediylbis (carbamodithioato) manganese • carbamic acid, ethylenebis (dithio-), manganese salt • carbamodithioic acid, 1,2-ethanediylbis-, manganese(2) salt (1:1) • chem-neb • chloroble-m • cr-3029 • dithane-m-22 • ent-14,875 • ethylenebis (dithiocarbamato), manganese • ethylenebis (dithiocarbamic acid) manganous salt • ethylenebisdithiocarbamate manganese • f-10 (pesticide) • kypman-80 • lonocol-m • manam • maneba • manebe (French) • manebgan • manesan • mangaan(II) -(N,N'-ethyleen-bis(dithiocarbamaat)) (Dutch) • Mangan(II)-(N,N'-aethylen-bis (dithiocarbamate)) (German) • manganese(II)-ethylene-di (dithiocarbamate) • manganese ethylene-1,2-bisdithiocarbamate • manganese, ((1,2-ethanediylbis(carbamodithioato))(2-)) • manganese, (ethylenebis (dithiocarbamato))- • manzate • N,N'-ethylene bis (dithiocarbamate manganeux) (French) • N,N'-etilen-bis (ditiocarbammato) di manganese (Italian) • nereb • nespor • plantifog-160m • polyram-m • rhodianebe • sup-'r-flo • sopranebe • tersan-lsr • trimangol • tubothane • maneb-80 • manganous-ethylenebis (dithiocarbamate) • trimangol-80 • aamangan • ethylenebis (dithiocarbamicacid), manganese salt • 1,2-ethanediylbiscarbamodithioic acid, manganese(2) salt (1:1) • maneb-zl4 • manzate-200 • manzate-maneb fungicide • m-diphar • mnebd • meb • remasan-chloroble-m • griffin-manex • akzo-chemie-maneb • manganese ethylenebis-dithiocarbamate • maneb-r • dithane-m22 • manzeb • manzin • tersan-lsr • vancide-maneb-90 • manganeseethylene-bis-dithiocarbamate • farmaneb • manzate • vancide • dithane m-22 special • manzate-d • vancide-maneb-80 • granol-nm • granox-nm • granox-pfm • agrosol-s • delsene-m • curzate-m

MF $C_4H_6N_2S_4.Mn$

MW 265.3

COLOR AND FORM Yellow powder; crystals from alcohol; brown powder.

ODOR Faint.

DENSITY 1.92.

SOLUBILITY Moderately soluble in water; soluble in chloroform, pyridine; virtually insoluble in water and common organic solvents; dissolves in chelating agents.

VAPOR PRESSURE $<7.5\times10^{-8}$ mm Hg at 20°C.

STABILITY AND SHELF LIFE Stable under ordinary storage conditions but decomposes more or less rapidly when exposed to moisture or to acids. In presence of moisture, decomposition proceeds as in nabam, with formation of polymeric ethylene thiuram monosulfide. The biological activity of the product remains practically unvaried for 2 yr under environmental conditions, provided the product is stored in its unopened and undamaged original containers, in shaded and, if possible, well-aired places [R5].

OTHER CHEMICAL/PHYSICAL PROPERTIES Decomposes before melting; light-colored solid; slightly soluble in water, insoluble in most common organic solvents (technical); decomposes on prolonged exposure to air or moisture and rapidly on contact with acids; estem is one of the products formed on contact with moisture.

PREVENTATIVE MEASURES SRP: The scientific literature supports the wearing of contact lenses in industrial environments, as part of a program to protect the eye against chemical compounds and minerals causing eye irritation. However, there may be individual substances whose irritating or corrosive properties are such that the wearing of contact lenses would be harmful to the eye. In those specific cases contact lenses should not be worn.

DISPOSAL Generators of waste (equal to or

greater than 100 kg/mo) containing this contaminant, EPA hazardous waste number U114, must conform with USEPA regulations in storage, transportation, treatment, and disposal of waste. Ethylenebisdithiocarbamic acid, salts and esters) [R6].

COMMON USES Fungicide for vegetables, for seed treatment of vegetables, and field crops, for deciduous fruits and nuts [R1]; foliar fungicide for fruits and vegetables. Its water-sol analogs, zineb and maneb, have very wide spectra of activity as foliar sprays for fruits, vegetables, and ornamentals esp for control of blights, leafspots, blotches, and mildews. [R17, 14]; maneb is used exclusively as broad- spectrum contact fungicide. The principal diseases controlled by maneb are early and late blight of potato and tomato, downy mildew, and anthracnose on vegetables and so called "rot" diseases of fruits such as apricots, peaches, and grapes. It is also used for seed treatment of small grains, such as wheat [R3].

OCCUPATIONAL RECOMMENDATIONS

OSHA Ceiling value: 5 mg/m^3 (manganese compounds, as Mn) [R12].

NIOSH 10-hr time-weighted average: 1 mg/m^3 (manganese compounds, as Mn) [R13].

NON-HUMAN TOXICITY LD_{50} rat, oral 6,750 mg/kg [R9]; LD_{50} rat percutaneous >5,000 mg/kg [R2]; LD_{50} rabbit dermal >2 g/kg [R8]; LD_{50} rat oral 4,400 mg/kg [R8].

ECOTOXICITY LC_{50} carp 1.8 mg/L/48 hr (conditions of bioassay not specified) [R2]; LC_{50} mallard ducks and bobwhite quail oral >10,000 mg/kg (8-day dietary study) [R2].

IARC SUMMARY No data are available for humans. Inadequate evidence of carcinogenicity in animals. Overall evaluation: group 3: The agent is not classifiable as to its carcinogenicity to humans [R7].

BIODEGRADATION In a 5-day CO_2 evolution study using an activated sludge inoculicy, 0.6% of applied maneb (50 ppb; ^{14}C–radiolabeled) was detected as ^{14}C–labeled CO_2 (1). The effect of maneb on microorganisms in two agriculturally important soils in Puerto Rico was studied under laboratory conditions for an 18–month period (2); in general, bacterial numbers increased while actinomycetes and fungi decreased, especially at higher concns (up to 960 ppm) (2); it was postulated that bacteria are the principal microorganisms in soil that are responsible for maneb biodegradation (2) [R10].

BIOCONCENTRATION In a 3-day static system study, a BCF of <10 was measured in golden ide (*Leuciscus idus melanotus*) fish (1). This BCF value suggests that bioconcentration in aquatic organisms is not an important environmental fate process (SRC) [R11].

CERCLA REPORTABLE QUANTITIES Persons in charge of vessels or facilities are required to notify the National Response Center (NRC) immediately, when there is a release of this designated hazardous substance, in an amount equal to or greater than its reportable quantity of 5,000 lb, or 2,270 kg. The toll-free number of the NRC is (800) 424–8802; in the Washington DC metropolitan area (202) 426–2675. The rule for determining when notification is required is stated in 40 CFR 302.4 (section IV. D. 3.b). (ethylenebisdithiocarbamic acid, salts and esters) [R14].

RCRA REQUIREMENTS U114; as stipulated in 40 CFR 261.33, when ethylenebisdithiocarbamic acid, salts and esters, as a commercial chemical product or manufacturing chemical intermediate or an off-specification commercial chemical product or a manufacturing chemical intermediate, becomes a waste, it must be managed according to federal or state hazardous waste regulations. Also defined as a hazardous waste is any residue, contaminated soil, water, or other debris resulting from the cleanup of a spill, into water or on dry land, of this waste. Generators of small quantities of this waste may qualify for partial exclusion from hazardous waste regulations (40 CFR 261.5). (ethylenebisdithiocarbamic acid, salts and esters) [R15].

DOT *Fire or Explosion:* Flammable/combustible material. May ignite itself if exposed to air. May reignite after fire is extinguished. May burn rapidly with flare-burning effect. Runoff to sewer may create fire or explosion hazard (maneb or maneb prepn, stabilized against self-heating; maneb or maneb preparation with 50% or more maneb) [R4].

Health Hazards: If inhaled, may be harmful. Contact may cause burns to skin and eyes. Fire may produce irritating or poisonous gases. Runoff from fire control or dilution water may cause pollution (Maneb or maneb prepn, stabilized against self-heating; maneb or maneb preparation with 50% more maneb) [R4].

Emergency Action: Keep unnecessary people away; isolate hazard area and deny entry. Stay upwind; keep out of low areas. Positive-pressure self-contained breathing apparatus (SCBA) and structural firefighters' protective clothing will provide limited protection. Call CHEMTREC at 1–800–424–9300 for emergency assistance. If water pollution occurs, notify the appropriate authorities (maneb or maneb prepn, stabilized against self-heating; maneb or maneb preparation with 50% more maneb) [R4].

Fire: Some of these materials may react violently with water.

Hydrosulfite fires: use flooding quantities of water for any size fire.

Small fires: dry chemical, soda ash, lime, or sand. Large fires: flood fire area with water from a distance. Do not get water inside container. Move container from fire area if you can do so without risk. Apply cooling water to sides of containers that are exposed to flames until well after fire is out. Stay away from ends of tanks. For massive fire in cargo area, use unmanned hose holder, or monitor nozzles; if this is impossible, withdraw from area, and let fire burn (maneb or maneb prepn, stabilized against self-heating; maneb or maneb preparation with 50% or more maneb) [R4].

Spill or Leak: Do not touch or walk through spilled material; stop leak if you can do so without risk. Do not get water inside container.

Small Spills: Flush area with flooding amounts of water. Large Spills: Dike liquid spill for later disposal (maneb or maneb prepn, stabilized against self-heating; maneb or maneb preparation with 50% or more maneb) [R4].

First Aid: Move victim to fresh air; call emergency medical care. In case of contact with material, immediately flush skin or eyes with running water for at least 15 minutes. Remove and isolate contaminated clothing and shoes at the site (maneb or maneb prepn, stabilized against self-heating; maneb or maneb preparation with 50% or more maneb) [R4].

FDA Bottled water shall, when a composite of analytical units of equal volume from a sample is examined by the methods described in paragraph (d) (1) (II) of this section, meet the standards of chemical quality, and shall not contain manganese in excess of 0.05 mg/L. (manganese) [R16].

REFERENCES

1. SRI.

2. Hartley, D., and H. Kidd (eds.). *The Agrochemicals Handbook.* 2nd ed. Lechworth, Herts, England: The Royal Society of Chemistry, 1987, p. A252/AUG 87.

3. IARC. *Monographs on the Evaluation of the Carcinogenic Risk of Chemicals to Man.* Geneva: World Health Organization, International Agency for Research on Cancer, 1972–present (multivolume work), p. V12 139 (1976).

4. U.S. Department of Transportation. *Emergency Response Guidebook 1990.* DOT P 5800.5. Washington, DC: U.S. Government Printing Office, 1990, p. G–337.

5. Spencer, E. Y. *Guide to the Chemicals Used in Crop Protection.* 7th ed. Publication 1093. Research Institute, Agriculture Canada, Ottawa, Canada: Information Canada, 1982. 360.

6. 40 CFR 240–280, 300–306, 702–799 (7/1/91).

7. IARC. *Monographs on the Evaluation of the Carcinogenic Risk of Chemicals to Man.* Geneva: World Health Organiza-

tion, International Agency for Research on Cancer, 1972–present (multivolume work), p. S7 65 (1987).

8. Purdue University, *National Pesticide Information Retrieval System, Maneb Fact Sheet No. 182* (1988).

9. *Kirk–Othmer Encyclopedia of Chemical Technology.* 3rd ed., Volumes 1–26. New York, NY: John Wiley and Sons, 1978–1984, p. V14 885 (1981).

10. (1) Freitag, D., et al., *Chemosphere* 14: 1589–616 (1985) (2) Dubey. H.D., R.L. Rodriguez, *J Agr Univ PR* 58: 78–86 (1974).

11. (1) Freitag, D., et al., *Chemosphere* 14: 1589–616 (1985).

12. 29 CFR 1910.1000 (7/1/91).

13. NIOSH. *NIOSH Pocket Guide to Chemical Hazards.* DHHS (NIOSH) Publication No. 90–117. Washington, DC: U.S. Government Printing Office, June 1990, 138.

14. 40 CFR 302.4 (7/1/91).

15. 40 CFR 261.33 (7/1/91).

16. 21 CFR 103.35 (4/1/91).

17. White-Stevens, R. (ed.). *Pesticides in the Environment.* Volume 1, Part 1, Part 2. New York: Marcel Dekker, Inc., 1971.

■ MANGANESE

CAS # 7439-96-5

SYNONYMS colloidal manganese • cutaval • mangan (Polish)

MF Mn

MW 54.94

COLOR AND FORM Gray-pink metal, cubic or tetragonal; steel-gray, lustrous, hard, brittle metal; exists in 4 allotropic forms: alpha, beta, gamma, and delta; alpha is body-centered cubic; beta is cubic; delta is body-centered cubic; gamma, or electrolytic manganese, is face-centered cubic; when stabilized at room temperature it is face-centered tetragonal; manganese ob-tained by distillation of crude metal is whitish gray with silvery crystalline fracture; electrodeposited metal is silvery white, hard, and brittle.

BOILING POINT 1,962°C.

MELTING POINT 1,244°C or ±3°C.

AUTOIGNITION TEMPERATURE Manganese dust clouds have minimal ignition temperature of 450°C. The limiting oxygen (O_2) percentage preventing ignition of dust cloud is 15. [R2, 1767].

EXPLOSIVE LIMITS AND POTENTIAL Explosion hazard: moderate, in form of dust, when exposed to flame [R6].

FIREFIGHTING PROCEDURES Use dry chemical to extinguish fire [R3].

SOLUBILITY Readily dissolves in dilute mineral acids; soluble in aqueous solutions of sodium or potassium bicarbonate.

VAPOR PRESSURE 1 mm Hg at 1,292°C.

STABILITY AND SHELF LIFE Superficially oxidized on exposure to air [R1].

OTHER CHEMICAL/PHYSICAL PROPERTIES Density: alpha, 7.47 at 20°C; beta, 7.26 at 20°C; gamma, 6.37 at 1,100°C; delta, 6.28 at 1,143°C.

PROTECTIVE EQUIPMENT Respirator selection: 50 mg/m³: any dust and mist respirator, any supplied-air respirator, any self-contained breathing apparatus; 125 mg/m³: any powered air-purifying respirator with a dust and mist filter, any supplied-air respirator operated in a continuous-flow mode; 250 mg/m³: any air-purifying full facepiece respirator with a high-efficiency particulate filter, any powered air-purifying respirator with a tight-fitting facepiece and a high-efficiency particulate filter, any self-contained breathing apparatus with a full facepiece, any supplied-air respirator with a full facepiece, any supplied-air respirator with a tight-fitting facepiece operated in a continuous-flow mode; 5,000 mg/m³: any supplied-air respirator with a half-mask and operated in a pressure-demand or other positive-pressure mode; 7,500 mg/m³: any supplied-air respirator with a full facepiece and operated in a pressure-demand or other positive-pressure mode; emergency or planned entry in unknown concentration or IDLH conditions: any self-contained breathing apparatus with a full facepiece and operated in a pressure-demand or other positive-pressure mode, any supplied-air respirator with a full facepiece and operated in a pressure-demand or other positive-pressure mode in combination with an auxiliary self-contained breathing apparatus operated in pressure-demand or other positive-pressure mode; escape: any air-purifying full facepiece respirator with a high-efficiency particulate filter, any appropriate escape-type self-contained breathing apparatus (manganese and compounds (as Mn)) [R7].

PREVENTATIVE MEASURES SRP: Local exhaust ventilation should be applied wherever there is an incidence of point-source emissions or dispersion of regulated contaminants in the work area. Ventilation control of the contaminant as close as possible to its point of generation is both the most economical and safest method to minimize personnel exposure to airborne contaminants.

DISPOSAL SRP: At the time of review, criteria for land treatment or burial (sanitary landfill) disposal practices are subject to significant revision. Prior to implementing land disposal of waste residue (including waste sludge), consult with environmental regulatory agencies for guidance on acceptable disposal practices.

COMMON USES For rock crushers, railway points, and crossings, wagon buffers [R1]; manufacturing of ceramics, matches, glass, dyes, welding rods [R4]; component of steel, steel alloys, cast iron, superalloys, and nonferrous alloys; chemical intermediate for high-purity salts; purifying and scavenging agent in metal production [R5].

OCCUPATIONAL RECOMMENDATIONS
OSHA An employee's exposure to manganese shall at no time exceed the ceiling value of 5 mg/m^3 [R9]. Threshold-limit 8–hr time-weighted average (TWA) 1 mg/m^3; short-term exposure limit (STEL) 3 mg/m^3 (1979) (fume) [R10, 24].

BIOLOGICAL HALF-LIFE Experiments on rats have shown that elimination from brain is slower than from whole body. Whole body had a half-time of 14 days, whereas half-time for cerebrum was so long that it could not be determined accurately during an observation time of 34 days. (manganese and manganese compounds) [R8, 363].

ATMOSPHERIC STANDARDS Maximum permissible atmospheric concentration 5 mg/m^3; maximum permissible atmospheric concentration of fumes 1 mg/m^3 [R11].

REFERENCES
1. *The Merck Index*. 10th ed. Rahway, NJ: Merck Co., Inc., 1983. 816.

2. Clayton, G.D., and F.E. Clayton (eds.). *Patty's Industrial Hygiene and Toxicology*: Volume 2A, 2B, 2C: *Toxicology*. 3rd ed. New York: John Wiley and Sons, 1981–1982.

3. Worthing, C. R., and S. B. Walker (eds.). *The Pesticide Manual—A World Compendium*. 8th ed. Thornton Heath, UK: The British Crop Protection Council, 1987. 727.

4. Doull, J., C. D. Klassen, and M. D. Amdur (eds.). *Casarett and Doull's Toxicology*. 3rd ed. New York: Macmillan Co., Inc., 1986. 614.

5. SRI.

6. Sax, N. I. *Dangerous Properties of Industrial Materials*. 6th ed. New York, NY: Van Nostrand Reinhold, 1984. 1729.

7. NIOSH. *Pocket Guide to Chemical Hazards*. 2nd Printing. DHHS (NIOSH) Publ. No. 85–114. Washington, DC: U.S. Dept. of Health and Human Services, NIOSH/ Supt. of Documents, GPO, February 1987. 151.

8. Friberg, L., G. F. Nordberg, E. Kessler, and V. B. Vouk (eds.). *Handbook of the Toxicology of Metals*. 2nd ed. Vols I, II. Amsterdam: Elsevier Science Publishers B. V., 1986, p. V2.

9. 29 CFR 1910.1000 (7/1/88).

10. American Conference of Governmental Industrial Hygienists. *Threshold Limit*

Values for Chemical Substances and Physical Agents and Biological Exposure Indices for 1993–1994. Cincinnati, OH: ACGIH, 1993.

11. Reynolds, J.E.F., A. B. Prasad (eds.). *Martindale—The Extra Pharmacopoeia.* 28th ed. London: The Pharmaceutical Press, 1982. 936.

■ MEPHOSFOLAN

CAS # 950–10–7

SIC CODES 2879.

SYNONYMS diethyl 4-methyl-1, 3-dithiolan-2-ylidenephosphoramidate, 2-(diethoxyphosphinylimino)-4-methyl-1, 3-dithiolan, p,p-diethyl cyclic propylene ester of phosphonodithioimidocarbonic acid • phosphoramidic acid, (4-methyl-1,3-dithiolan-2-ylidene)-, diethyl ester • ac-47470 • ai3-25991 • cl-47470 • cyclic propylene p,p-diethyl phosphonodithioimidocarbonate • diethyl (4-methyl-1,3-dithiolan-2-ylidene) phosphoramidate • diethyl 4-methyl-1,3-dithiolan-2-ylidenephosphoramidate • ei-47470 • ent-25,991 • imidocarbonic acid, phosphonodithio-, cyclic propylene p,p-diethyl ester • mephospholan (French) • dithiolane iminophosphate • p,p-diethyl cyclic propylene ester of phosphonodithiomidocarbonic acid • cytrolane insecticide • 2-(diethoxy-phosphinylimino)-4-methyl-1,3-dithiolane • (40-methyl-1,3-dithiolan-2-ylid-ene) phosphoramidic acid diethyl ester • Phosphonodithioimidocarbonic acid cyclic propylene p,p-diethyl ester • 2-diethoxy-phosphinylimino-4-methyl-1,3-dithiolane • cyclic propylene (diethoxyphosphinyl) dithiomidocarbonate • ent-25991 • ent-25-991 • ei-47-470 • cytrolane • cytrolane • cyclic propylene (diethoxy-phosphinyl) dithioimidocarbonate • diethyl (4-methyl-1,3-dithiolan-2-ylidene)

phosphoroamidate • diethyl-4-methyl-1,3-dithiolan-2-ylidenephosphoramidate

MF $C_8H_{16}NO_3PS_2$

MW 269.34

COLOR AND FORM Colorless liquid; yellow to amber liquid

DENSITY 1.539 at 26°C.

BOILING POINT 120°C at 1×10^{-3} mm Hg.

SOLUBILITY Soluble in acetone, xylene, benzene, ethanol, and toluene, moderately soluble in water. Solubility in water at 25°C: 57 g/kg.

STABILITY AND SHELF LIFE It is stable in water under neutral, slightly acid or basic conditions. Photodegradation in an aqueous environment [R1].

COMMON USES Has shown significant activity against certain of the more important borer insects such as European corn borer, sugar cane borer, and fall armyworm [R1]. Insecticide, acaricide [R5, 5744]. It is a contact and stomach insecticide, and is used for the control of aphids, bollworms, mites, stem borers, and whiteflies on such major crops as cotton, maize, field crops, fruit, and vegetables [R2]. Control of thrips, jassids, mites, and leaf-eating larvae on hops, sorghum, rice, sugar cane, vegetables, tobacco, etc. [R3]. For cotton leafworm, *Spodoptera littoralis*, pink bollworm. Also aphids on hops [R4].

GENERAL SENSATION Symptoms are anorexia, nausea, vomiting, diarrhea, excessive salivation, pupillary constriction, bronchoconstriction, muscle twitching, convulsions, coma, and death by respiratory failure. Effects are cumulative.

OCCUPATIONAL RECOMMENDATIONS None listed.

ON-SITE RESTORATION Seek professional environmental engineering assistance through EPA's Environmental Response Team (ERT), Edison, NJ, 24–hour no., 908–548–8730. Contain and isolate spill to limit spread. Construct clay/bentonite swale to divert uncontaminated portion of watershed around contaminated portion. Isolation procedures include construction of bentonite or butyl-rubber-lined dams, interceptor trenches, or impoundments. Seek professional help to evaluate problem, and implement containment procedures. Conduct bench scale and pilot scale tests prior to decontamination program implementation. Treatment alternatives for contaminated water include powdered activated-carbon sorption, granular activated-carbon filtration, biodegradation activated-carbon filtration, biodegradation alkaline hydrolysis, and chemical precipitation using soda ash. Treatment alternatives for contaminated soils include well-point collection and treatment of leachates as for contaminated water, bentonite/cement ground injection to immobilize solids, physical removal of immobilized residues; package for disposal. Confirm treatment procedures with responsible environmental engineer and regulatory officials.

DISPOSAL Product residues and sorbent media may be packaged in 17h epoxy-lined drums and disposed of at an EPA–approved disposal site or be made to undergo high-temperature destruction using incinerators with scrubbing equipment. Confirm disposal procedures with responsible environmental engineer and regulatory officials.

PROBABLE LOCATION AND STATE Yellow to amber liquid; moderately soluble in water. Some may form a bottom layer in water.

WATER CHEMISTRY It is stable in water under neutral conditions but it is hydrolyzed by alkali at pH >9 or by acid at pH <2.

NON-HUMAN TOXICITY LD_{50} rat oral 8.9 mg/kg [R1]; LD_{50} rabbit albino dermal 9.7 mg/kg [R1]; LD_{50} mouse albino oral 11.3 mg/kg [R2]; LD_{50} (24 hr) albino rabbit male percutaneous 28.7 mg technical/kg [R2].

ECOTOXICITY LC_{50} *Salmo* spp (trout) 2.12 mg/L/96 hr (conditions of bioassay not specified) [R2]; LC_{50} *Cyprinus carpio* (carp) 54.5 mg/L/96 hr (conditions of bioassay not specified) [R2]; LD_{50} *Coturnix coturnix japonica* (Japanese quail) oral 12.8 mg/kg [R2]; TC_{50} *Apis mellifera* (ho-

neybee) topical application 0.0035 mg/bee [R2].

DOT *Health Hazards:* Poisonous; may be fatal if inhaled, swallowed, or absorbed through skin. Contact may cause burns to skin and eyes. Runoff from fire control or dilution water may cause pollution (organophosphorus pesticide, liquid, poisonous, flammable, not otherwise specified) [R6, p. G–28].

Fire or Explosion: Flammable/combustible material; may be ignited by heat, sparks, or flames. Vapors may travel to a source of ignition, and flash back. Container may explode in heat of fire. Vapor explosion and poison hazard indoors, outdoors, or in sewers. Runoff to sewer may create fire or explosion hazard (organophosphorus pesticide, liquid, poisonous, flammable, not otherwise specified) [R6, p. G–28].

Emergency Action: Keep unnecessary people away; isolate hazard area, and deny entry. Stay upwind; keep out of low areas. Positive-pressure self-contained breathing apparatus (SCBA) and chemical protective clothing that is specifically recommended by the shipper or manufacturer may be worn. It may provide little or no thermal protection. Structural firefighters' protective clothing is not effective for these materials. Isolate for 1/2 mile in all directions if tank, rail car, or tank truck is involved in fire. Call CHEMTREC at 1–800–424–9300 for emergency assistance (organophosphorus pesticide, liquid, poisonous, flammable, not otherwise specified) [R6, p. G–28].

Fire: Small fires: Dry chemical, CO_2, water spray, or alcohol-resistant foam. Large Fires: Water spray, fog, or alcohol-resistant foam. Move container from fire area if you can do so without risk. Dike fire-control water for later disposal; do not scatter the material. Apply cooling water to sides of containers that are exposed to flames until well after fire is out. Stay away from ends of tanks. Withdraw immediately in case of rising sound from venting safety device or any discoloration of tank due to fire (organophosphorus pesticide, liquid, poisonous, flammable, not otherwise specified) [R6, p. G–28].

Spill or Leak: Shut off ignition sources; no flares, smoking, or flames in hazard area. Fully encapsulating, vapor-protective clothing should be worn for spills and leaks with no fire. Do not touch or walk through spilled material; stop leak if you can do so without risk. Water spray may reduce vapor; but it may not prevent ignition in closed spaces.

Small Spills: Take up with sand or other noncombustible absorbent material and place into containers for later disposal. Large Spills: Dike far ahead of liquid spill for later disposal. (organophosphorus pesticide, liquid, poisonous, flammable, not otherwise specified) [R6, p. G–28].

First Aid: Move victim to fresh air, and call emergency medical care; if not breathing, give artificial respiration; if breathing is difficult, give oxygen. In case of contact with material, immediately flush skin or eyes with running water for at least 15 minutes. Remove and isolate contaminated clothing and shoes at the site. Keep victim quiet and maintain normal body temperature. Effects may be delayed; keep victim under observation (organophosphorus pesticide, liquid, poisonous, flammable, not otherwise specified) [R6, p. G–28].

REFERENCES

1. Spencer, E. Y. *Guide to the Chemicals Used in Crop Protection.* 7th ed. Publication 1093. Research Institute, Agriculture Canada, Ottawa, Canada: Information Canada, 1982. 367.

2. Worthing, C. R., and S. B. Walker (eds.). *The Pesticide Manual—A World Compendium.* 8th ed. Thornton Heath, UK: The British Crop Protection Council, 1987. 528.

3. Hartley, D., and H. Kidd (eds.). *The Agrochemicals Handbook.* 2nd ed. Lechworth, Herts, England: The Royal Society of Chemistry, 1987, p. A436/AUG 87.

4. *Farm Chemicals Handbook 1993.* Willoughby, OH: Meister Publishing Co., 1993, p. C 102.

5. Budavari, S. (ed.). *The Merck Index— Encyclopedia of Chemicals, Drugs and Biologicals.* Rahway, NJ: Merck and Co., Inc., 1989.

6. U.S. Department of Transportation. *Emergency Response Guidebook 1990.* DOT P 5800.5. Washington, DC: U.S. Government Printing Office, 1990.

■ MERCAPTOACETIC ACID

CAS # 68–11–1

SYNONYMS 2-mercaptoacetic acid • 2-thioglycolic acid • acetic acid, mercapto • acide thioglycolique (French) • alpha-mercaptoacetic acid • glycolic acid, 2-thio • glycolic acid, thio • thioglycolic acid • thioglycollic acid • thiovanic acid • USAF CB-35

MF $C_2H_4O_2S$

MW 92.12

COLOR AND FORM Colorless liquid.

ODOR Strong, unpleasant odor; odor has characteristic of sulfhydryl groups.

DENSITY 1.3253 at 20°C.

DISSOCIATION CONSTANTS $K_1 = 2.1 \times 10^{-4}$ at 25°C; $K_2 = 2.1 \times 10^{-11}$; pK_a: 3.68 at 148°C.

BOILING POINT 120°C at 20 mm Hg.

MELTING POINT –16.5°C.

SOLUBILITY Miscible with water, alcohol, ether, chloroform, benzene, and many other organic solvents.

VAPOR PRESSURE 10 mm Hg at 18°C.

STABILITY AND SHELF LIFE Readily oxidized by air [R2].

OTHER CHEMICAL/PHYSICAL PROPERTIES Vapor pressure: 1 mm Hg $(0.13 \times 10^3$ Pa) at 60°C.

PROTECTIVE EQUIPMENT When applying the solution of thioglycolic acid or its derivatives, the hairdresser should protect exposed areas of skin by the use of rubber or plastic gloves, and eye contact should be avoided. The solution should be neutralized as quickly as possible [R5, 2172].

PREVENTATIVE MEASURES Hairdressers should use thioglycolic acid or its derivatives in dilute solution with pH near neutral. Hairdressers using thioglycolic acid or its derivatives should be informed of hazards involved and be watchful for early signs of disorders (burning sensations, itching, etc.). They should not use these preparations when suffering from skin irritation. In hairdressing salons, effective ventilation should be provided to prevent the material from accumulating in the atmosphere in the form of mist [R5, 2172].

COMMON USES Chemical intermediate for thioglycolates (eg, calcium thioglycolate); ingredient of depilatories, permanent hair-wave solutions, and biological media for microorganism growth; reagent for detection of iron and other metal ions; chelating agent [R1].

STANDARD CODES NFPA—3, -, -; ICC, USCG—no.

PERSISTENCE Does not biodegrade well.

DIRECT CONTACT Severe burns and skin blisters.

GENERAL SENSATION Strong unpleasant odor. Releases toxic H_2S. Can be absorbed through skin.

PERSONAL SAFETY Wear full skin protection and self-contained breathing apparatus.

ACUTE HAZARD LEVEL Irritant. Highly toxic by all routes. Emits H_2S upon decomposition.

CHRONIC HAZARD LEVEL Highly toxic by all chronic exposure routes.

DEGREE OF HAZARD TO PUBLIC HEALTH Irritant; highly toxic by all routes with acute or chronic exposure. Emits H_2S upon decomposition.

OCCUPATIONAL RECOMMENDATIONS None listed.

THRESHOLD LIMIT 8–hr time-weighted average (TWA) 1 ppm, 3.8 mg/m³, skin (1978) [R4, 34]. Excursion Limit Recommendation: Excursions in worker exposure levels may exceed three times the TLV-TWA for no more than a total of 30 min during a work day, and under no circumstances should they exceed five times the TLV-TWA.

ACTION LEVEL Evacuate area. notify fire and air authority. Enter from upwind. Remove ingition sources and isolate material.

ON-SITE RESTORATION Neutralize with anion exchanger or $NaHCO_3$. Seek professional environmental engineering assistance through EPA's Environmental Response Team (ERT), Edison, NJ, 24–hour no., 908–548–8730.

BEACH OR SHORE RESTORATION Do not heat.

AVAILABILITY OF COUNTERMEASURE MATERIAL Anion exchanger—water softener supplier; $NaHCO_3$—grocery distributors, bakeries.

DISPOSAL Dissolve in flammable solvent. burn in incinerator with afterburner and SO_2 scrubber.

DISPOSAL NOTIFICATION Local air authority.

INDUSTRIAL FOULING POTENTIAL Can be hazardous when confined in boiler feed or cooling-system waters.

MAJOR WATER USE THREATENED Recreation, potable supply, industrial.

PROBABLE LOCATION AND STATE Liquid. Will dissolve.

WATER CHEMISTRY Acid is readily oxidized upon standing in air. Decomposition can lead to release of H_2S.

BIODEGRADATION Biological oxygen demand: 20°C, 1–5 days [R3].

REFERENCES
1. SRI.

2. *The Merck Index.* 10th ed. Rahway, NJ: Merck Co., Inc., 1983. 1337.

3. Verschueren, K. *Handbook of Environmental Data of Organic Chemicals.* 2nd ed. New York, NY: Van Nostrand Reinhold Co., 1983. 809.

4. American Conference of Governmental Industrial Hygienists. *Threshold Limit Values for Chemical Substances and Physical Agents and Biological Exposure*

Indices for 1994–1995. Cincinnati, OH: ACGIH, 1994.

5. International Labour Office. *Encyclopedia of Occupational Health and Safety.* Vols. I and II. Geneva, Switzerland: International Labour Office, 1983.

■ MERCAPTOBENZOTHIAZOLE

CAS # 149–30–4

SYNONYMS 2 (3h)-benzothiazolethione • 2-benzothiazolethiol • 2-benzothiazolethione • 2-benzothiazolinethione • 2-benzothiazolyl mercaptan • 2-mbt • 2-mercaptobenzothiazole • 2-mercaptobenzthiazole • 2-merkaptobenzotiazol (Polish) • accel-m • accelerator-m • benzothiazole-2-thione • captax • dermacid • ekagom-g • kaptaks • kaptax • mbt • mebetizole • mebithizol • mercaptobenzthiazole • mertax • nci-c56519 • nuodex-84 • pneumax-mbt • rotax • royal-mbt • soxinol-m • thiotax • USAF gy-3 • usaf xr-29 • vulkacit-m • vulkacit-mercapto • vulkacit-mercapto/c

MF $C_7H_5NS_2$

MW 167.25

COLOR AND FORM Pale-yellow monoclinic needles or leaflets; yellowish powder; needles from alcohol or dilute methanol; yellowish to tan crystalline powder.

ODOR Disagreeable odor; distinctive odor (depends on degree of purification).

DENSITY 1.42.

DISSOCIATION CONSTANT The dissociation constant for 2–mercaptobenzothiazole is between 6.7 and 7.2.

MELTING POINT 180.2–181.7°C.

EXPLOSIVE LIMITS AND POTENTIAL Combustible [R1].

SOLUBILITY Practically insoluble in water; solubility at 25°C (g/dl): alcohol 2.0; ether 1.0; acetone 10.0; benzene 1.0; carbon tetrachloride less than 0.2; naphtha less than 0.5; moderately soluble in glacial acetic acid; soluble in alkalies and alkali carbonate solution; soluble in dilute caustic soda, alcohol, chloroform; insoluble in gasoline; soluble in hot acetic acid; more soluble in crude sweat than in water.

OCTANOL/WATER PARTITION COEFFICIENT Log K_{ow} = 2.41.

OTHER CHEMICAL/PHYSICAL PROPERTIES Conversion factor: 6.83 mg/m^3 = 1 ppm; MP: 170–173°C (technical mercaptobenzothiazole); light-yellow powder; specific gravity: 1.70 at 25°C (zinc salt of mercaptobenzothiazole).

PROTECTIVE EQUIPMENT Respirators approved by the U.S. Bureau of Mines for nuisance dust and safety spectacles are recommended in the event of excessive dustiness in handling the compounds [R3].

PREVENTATIVE MEASURES SRP: The scientific literature for the use of contact lenses in industry is conflicting. The benefit or detrimental effects of wearing contact lenses depend not only upon the substance, but also on factors including the form of the substance, characteristics, and duration of the exposure, the uses of other eye protection equipment, and the hygiene of the lenses. However, there may be individual substances whose irritating or corrosive properties are such that the wearing of contact lenses would be harmful to the eye. In those specific cases, contact lenses should not be worn. In any event, the usual eye protection equipment should be worn even when contact lenses are in place.

DISPOSAL SRP: At the time of review, criteria for land treatment or burial (sanitary landfill) disposal practices are subject to significant revision. Prior to implementing land disposal of waste residue (including waste sludge), consult with environmental regulatory agencies for guidance on acceptable disposal practices.

COMMON USES Salts used as fungicides [R11, 923]; fungicide for preserving textiles (SRP: former use) [R2]; vulcanization accelerator for rubber (requires use of stearic acid for full activation), tire treads, and carcasses, mechanical specialities; corrosion inhibitor in cutting oils and petroleum products, extreme-pressure additive in greases [R1].

OCCUPATIONAL RECOMMENDATIONS None listed.

NON-HUMAN TOXICITY LD_{50} mice (Slc: ddY) male oral LD_{50} 1,558 (mg/kg) (suspension in 5% gum arabic solution) [R4]; LD_{50} mice (Slc: ddY) male oral LD_{50} 3,148 (mg/kg) (suspension in olive oil) [R4]; LD_{50} mice (Slc: ddY) female oral LD_{50} 1,490 (mg/kg) (suspension in 5% gum arabic solution) [R4].

ECOTOXICITY Approximate fatal concentration, *Carassius auratus* (goldfish) 2 mg/L/48 hr (conditions of bioassay not specified) [R5].

BIODEGRADATION Biodegradation studies with 2–mercaptobenzothiazole using an activated sludge seed indicate little or no biodegradation (2–4). Only 2.5% of the 2–mercaptobenzothiazole (100 ppm) degraded in a 2–week period in the standard biodegradability test of the Japanese Ministry International Trade and Industry (MITI), a BOD test utilizing a mixed inoculum of activated sludge, sewage, and surface water (6). On this basis, 2–mercaptobenzothiazole was judged to be difficult to biodegrade (5,6). It has been suggested that at the concns used in these studies, 56–100 ppm, 2–mercaptobenzothiazole is toxic to the microorganisms in the sludge (1). A more detailed study indicated that inhibitory effects occur between 20 and 50 ppm of 2–mercaptobenzothiazole (1). When the sludge is well acclimated and the concentration low enough, 2–mercaptobenzothiazole may completely biodegrade, forming 2–benzothiazolesulfonate or dibenzothiazole–2,2–disulfide, depending on conditions (1) [R6].

BIOCONCENTRATION In a 6–week test performed at two concentration levels, the BCF of 2–mercaptobenzothiazole in carp was <8 (1). Another investigator also reported little or no bioconcentration in fish (2). These experimental results are in agreement with the BCF of 41 estimated from the log K_{ow} of 2.41 (3) using a recommended regression equation (4). Because the dissociation constant (pK_a) for 2– mercaptobenzothiazole is between 6.7 and 7.2 (5,6), the percentage of ionized 2–mercaptobenzothiazole will increase with pH, and its K_{ow} and BCF may be pH sensitive (SRC) [R7].

TSCA REQUIREMENTS Pursuant to section 8 (d) of TSCA, EPA promulgated a model health and safety data reporting rule. The section 8 (d) model rule requires manufacturers, importers, and processors of listed chemical substances and mixtures to submit to EPA copies and lists of unpublished health and safety studies. Mercaptobenzothiazole is included on this list [R8].

FDA 2–Mercaptobenzothiazole is an indirect food additive for use only as a component of adhesives [R9]. Ophthalmic+n and topical-dosage-form new animal drugs not subject to certification. Specifications and conditions of use provided for dogs [R10].

REFERENCES

1. Lewis, R. J., Sr. (ed.). *Hawley's Condensed Chemical Dictionary.* 12th ed. New York, NY: Van Nostrand Reinhold Co., 1993 739.

2. SRI.

3. Sittig, M. *Handbook of Toxic and Hazardous Chemicals and Carcinogens,* 1985. 2nd ed. Park Ridge, NJ: Noyes Data Corporation, 1985. 566.

4. Ogawa, Y., et al., *Bull Natl Inst Hyg Sci* (Tokyo) (107): 44–50 (1989).

5. Verschueren, K. *Handbook of Environmental Data of Organic Chemicals.* 2nd ed. New York, NY: Van Nostrand Reinhold Co., 1983. 810.

6. (1) DeVos, D., et al., *Appl Macrobiol Biotechnol* 39: 622–6 (1993). (2) Mainprize, J., et al., *J Appl Bacteriol* 40: 285–91 (11976). (3) Chudoba, J., et al. *Act Hydrochemical Hydrobiol* 5: 495–8 (1977). (4) Kawasaki, M., *Ecotox Environ Safety* 4: 444–54 (1980). (5) Kondo, M., et al., *Eisei Kagaku* 34: 115–22 (1988). (6) Chemicals Inspection and Testing Institute. *Biodegradation and Bioaccumulation Data of Existing Chemicals Based on the CSCL Japan.* Japan Chemical

Industry Ecology–Toxicology and Information Center. ISBN 4–89074–101–1 p. 5–6 (1992).

7. (1) Chemicals Inspection and Testing Institute. *Biodegradation and Bioaccumulation Data of Existing Chemicals Based on the CSCL Japan.* Japan Chemical Industry Ecology–Toxicology and Information Center. ISBN 4–89074–101–1 p. 5–6 (1992). (2) Sasaki, S., pp. 283–98 in *Aquatic Pollutants*, Hutzinger, O., et al. (eds.), Oxford: Pergamon Press (1978). (3) Brownlee, B. G., et al., *Environ Toxicol Chemical* 11: 1153–68 (1992). (4) Lyman, W. J., et al. (eds.), *Handbook of Chemical Property Estimation Methods*. New York McGraw-Hill, Chapt 5, Eqn 5–2 (1982). (5) Danehy, J. P., K. N. Parameswaran. *J Chemical Eng Data* 13: 386–9 (1968). (6) Lomakina, L. N., E. K . Yakovskaya. *Vestn Mosk Univ, Khim* 24: 73–6 (1969).

8. 40 CFR 716.120 (7/1/94).

9. 21 CFR 175.105 (4/1/93).

10. 21 CFR 524.1376 (4/1/93).

11. Budavari, S. (ed.). *The Merck Index—Encyclopedia of Chemicals, Drugs and Biologicals*. Rahway, NJ: Merck and Co., Inc., 1989.

■ MERCURIC POTASSIUM IODIDE

CAS # 7783–33–7

SYNONYMS dipotassium tetraiodomercurate • dipotassium tetraiodomercurate(2-) • mercurate(2-), tetraiodo-, dipotassium • mercuric potassium iodide, solid • mercury potassium iodide • mercury potassium iodide (K_2HgI_4) • mercury(II) potassium iodide • potassium iodomercurate • potassium mercuric iodide • potassium mercuriiodide • potassium tetraiodomercurate(II) • solution potassium iodohydragyrate • Channing's solution • Mayer's reagent • Nessler reagent

MF HgI_4 • 2K

pH Neutral or alkaline to litmus.

COLOR AND FORM Sulfur-yellow crystals.

ODOR Odorless.

DENSITY 4.29.

FIREFIGHTING PROCEDURES Wear positive-pressure self-contained breathing apparatus when fighting fires involving this material [R3]. If material involved in *Fire:* Extinguish fire using agent suitable for type of surrounding fire. Use water in flooding quaantities as fog. Use "alcohol" foam, dry chemical, or carbon dioxide [R3].

TOXIC COMBUSTION PRODUCTS Fire may produce irritating or poisonous gases [R4].

SOLUBILITY Very soluble in cold water; soluble in alcohol, ether, acetone.

OTHER CHEMICAL/PHYSICAL PROPERTIES Deliquescent in moist air; crystallizes with 1–2 mol water. Mercury salts, when heated with Na_2CO_3, yield metallic Hg, and are reduced to metal by H_2O_2 in the presence of alkali hydroxide. Cu, Fe, Zn, and many other metals ppt metallic Hg from neutral or slightly acid solution of mercury salts (mercury salts). Soluble ionized mercuric salts give a yellow precipitate of HgO with NaOH and a red precipitate of HgI_2 with alkali iodide. (soluble mercuric salts).

PROTECTIVE EQUIPMENT (SRP: in the laboratory) Use skin and resp protection when dry mercuric salts are to be used. Use skin protection when concentrated aqueous solutions of mercuric salts are used. (mercuric salts) [R5].

PREVENTATIVE MEASURES Use disposable uniforms, so that a contaminated uniform is not a source of absorption through the skin. Preventative measure: adequate ventilation; careful attention to good housekeeping, e.g., avoidance of spills, and prompt and proper cleaning if a spill occurs; all containers of mercury and its compounds should be kept tightly closed; floors should be washed on a regular basis with dilute calcium sulfide solution or other suitable reactant; floors should be nonporous; all workers directly involved in the plant operation should shower thoroughly each day before leaving, (mercury compounds) [R6].

CLEANUP Mercury removal from waste water can be accomplished by the BMS process: Chlorine is added to the waste water, oxidizing any mercury present to the ionic state. The BMS adsorbent (an activated carbon concentrate with sulfur compound on its surface) is used to collect ionic mercury. The spent adsorbent is then distilled to recover the mercury, leaving a carbon residue for reuse or disposal. The TMR IMAC process can also be used: Waste water is fed into a reactor, whereby a slight excess of chlorine is maintained, oxidizing any mercury present to ionic mercury. The liquid is then passed through the TMR IMAC ion-exchange resin, where mercury ions are adsorbed. The mercury is then stripped from the spent resin with hydrochloric acid solution (mercury compound) [R7]. Spilled mercury compounds or solutions can be cleaned up by any method that does not cause excessive airborne contamination or skin contact. (mercury compounds) [R8].

DISPOSAL SRP: At the time of review, criteria for land treatment or burial (sanitary landfill) disposal practices are subject to significant revision. Prior to implementing land disposal of waste residue (including waste sludge), consult with environmental regulatory agencies for guidance on acceptable disposal practices.

COMMON USES Topical anti-infective, disinfectant; vet: topical anti-infective, disinfectant [R1]. A solution of potassium mercuriiodide in KOH is Nessler's reagent, which becomes yellow upon contact with traces of NH_3 (ammonia), making it a valuable reagent for detecting traces of NH_3. [R21, 1773].

OCCUPATIONAL RECOMMENDATIONS

OSHA During an 8–hr work shift, an employee may not be exposed to a concentration of mercury above 1 mg/m³. (mercury) [R13].

NIOSH 10–hr time-weighted average 0.05 mg/m³ (mercury and inorganic compounds (as Hg)) [R14].

THRESHOLD LIMIT 8-hr time-weighted average (TWA) 0.025 mg/m³, skin (1994) (mercury, in organic forms including metallic mercury) [R22, 25].

HUMAN TOXICITY Lethal blood level: The concentration of inorganic mercury present in the blood (serum or plasma) that has been reported to cause death in humans is: 0.04–2.2 mg%; 0.4–22 µg/mL. (inorganic mercury) [R10].

IARC SUMMARY Evaluation: There is inadequate evidence in humans for the carcinogenicity of mercury and mercury compounds. Overall evaluation: Metallic mercury and inorganic mercury compounds are not classifiable as to their carcinogenicity to humans (Group 3) [R9].

BIOLOGICAL HALF-LIFE The average biological half-time of a tracer dose of divalent inorganic mercury compound in man is 42 days for the whole body and 26 days for blood. (mercury compounds) [R11].

BIODEGRADATION Inorganic forms of Hg can be converted to organic forms by microbial action in the biosphere. (inorganic mercury) [R12].

BIOCONCENTRATION Mercuric salts are still widely employed in industry, and industrial discharge into rivers has polluted many parts of the world. Microorganisms convert inorg mercury to methyl mercury, which is taken up rapidly by planktonic algae and is concentrated in fish by way of food chain. (mercuric salts) [R23, 1611].

ATMOSPHERIC STANDARDS Emissions to the atmosphere from sludge incineration plants, sludge drying plants, or a combination of these sludge wastewater treatment plant processes shall not exceed 3,200 grams of mercury per 24-hour period. (total mercury) [R15].

RCRA REQUIREMENTS A solid waste containing mercuric potassium iodide may become characterized as a hazardous waste when subjected to the toxicant extraction procedure listed in 40 CFR 261.24, and, if so characterized, must be managed as a hazardous waste [R16].

DOT *Health Hazards:* Poisonous if swallowed. Inhalation of dust poisonous. Fire may produce irritating or poisonous gases. Runoff from fire control or dilution water may cause pollution [R2].

Fire or Explosion: Some of these materials may burn, but none of them ignites readily [R2].

Emergency Action: Keep unnecessary people away; isolate hazard area and deny entry. Stay upwind; keep out of low areas. Self-contained breathing apparatus (SCBA) and structural firefighter's protective clothing will provide limited protection. Call CHEMTREC at 1–800–424–9300 for emergency assistance. If water pollution occurs, notify the appropriate authorities [R2].

Fire: Small fires: Dry chemical, CO_2, halon, water spray, or standard foam. Large Fires: Water spray, fog, or standard foam is recommended. Move container from fire area if you can do so without risk [R2].

Spill or Leak: Do not touch spilled material; stop leak if you can do so without risk. Small Spills: Take up with sand or other noncombustible absorbent material and place into containers for later disposal.

Small Dry Spills: With clean shovel place material into clean, dry container, and cover; move containers from spill area. Large Spills: Dike far ahead of liquid spill for later disposal [R2].

First Aid: Move victim to fresh air; call for emergency medical care. Remove and isolate contaminated clothing and shoes at the site. In case of contact with material, immediately flush skin or eyes with running water for at least 15 minutes [R2].

FDA Bottled water shall, when a composite of analytical units of equal volume from a sample is examined by the methods described in paragraph (d) (1) (II) of this section, meet the standards of chemical quality, and shall not contain mercury in excess of 0.002 mg/L (total mercury) [R17]. The color additive F, D, and, C Blue Number 2 shall conform to the specifications in the CFR 74.102, and be free from impurities other than those named, including mercury (as Hg) in not more than 1 part per million, to the extent that such other impurities may be avoided by current good manufacturing practice (total mercury) [R18]. The color additive F, D, and, C Green Number 3 shall conform to the specifications in the CFR 74.203, and be free from impurities other than those named, including mercury (as Hg) in not more than 1 part per million, to the extent that such other impurities may be avoided by current good manufacturing practice (total mercury) [R19]. The color additive F, D, and, C Yellow Number 5 shall conform to the specifications in the CFR 74.705, and be free from impurities other than those named, including mercury (as Hg) in not more than 1 part per million, to the extent that such other impurities may be avoided by current good manufacturing practice (total mercury) [R20].

REFERENCES

1. *The Merck Index.* 10th ed. Rahway, NJ: Merck Co., Inc., 1983. 1105.

2. Department of Transportation. *Emergency Response Guidebook 1987.* DOT P 5800.4. Washington, DC: U.S. Government Printing Office, 1987, p. G–53

3. Association of American Railroads. *Emergency Handling of Hazardous Materials in Surface Transportation.* Washington, DC: Assoc. of American Railroads, Hazardous Materials Systems (BOE), 1987. 436

4. *Department of Transportation. Emergency Response Guidebook.* 1984 DOT P 5800.3 Washington, DC: U.S. Government Printing Office, 1984, p. G–53.

5. Association of Official Analytical Chemists. *Official Methods of Analysis.* 10th ed. and supplements. Washington, DC: Association of Official Analytical Chemists, 1965. New editions through 13th ed. plus supplements, 1982, p. 13/883 51.079.

6. *Kirk–Othmer Encyclopedia of Chemical Technology.* 3rd ed., Volumes 1–26. New York, NY: John Wiley and Sons, 1978–1984, p. V15 167.

7. Environment Canada. *Tech Info for Problem Spills: Mercury* (Draft), p. 59 (1982).

8. National Research Council. *Prudent Practices for Handling Hazardous Chemicals in Laboratories.* Washington, DC: National Academy Press, 1981. 53.

9. IARC. *Monographs on the Evaluation of the Carcinogenic Risk of Chemicals to Man.* Geneva: World Health Organization, International Agency for Research on Cancer, 1972–present (multivolume work), p. 58 324 (1993).

10. Winek, C. L. *Drug and Chemical Blood–Level Data 1985.* Pittsburgh, PA: Allied Fischer Scientific, 1985.

11. USEPA. *Mercury Health Effects Update.* p. 2–5 (1984) EPA 600/8–84–019F.

12. Schroeder, W. H. *Envir Sci Tech* 16 (7): 394A–400A (1982) as cited in *Environment Canada; Tech Info for Problem Spills: Inorganic Mercury* (Draft), p. 41 (1982).

13. 29 CFR 1910.1000 (7/1/87).

14. NIOSH. *Pocket Guide to Chemical Hazards.* 5th Printing/Revision. DHHS (NIOSH) Publ. No. 85–114. Washington, DC: U.S. Dept. of Health and Human Services, NIOSH/Supt. of Documents, GPO, Sept. 1985. 152.

15. 40 CFR 61.52 (b) (7/1/87).

16. 40 CFR 261.24 (7/1/87).

17. 21 CFR 103.35 (4/1/88).

18. 21 CFR 74.102 (4/1/88).

19. 21 CFR 74.203 (4/1/88).

20. 21 CFR 74.705 (4/1/88).

21. Clayton, G.D., and F.E. Clayton (eds.). *Patty's Industrial Hygiene and Toxicology:* Volume 2A, 2B, 2C: *Toxicology.* 3rd ed. New York: John Wiley and Sons, 1981–1982.

22. American Conference of Governmental Industrial Hygienists. *Threshold Limit Values for Chemical Substances and Physical Agents and Biological Exposure Indices for 1994–1995.* Cincinnati, OH: ACGIH, 1994.

23. Gilman, A. G., L. S. Goodman, and A. Gilman (eds.). *Goodman and Gilman's The Pharmacological Basis of Therapeutics.* 7th ed. New York: Macmillan Publishing Co., Inc., 1985.

■ MERCUROUS SULFATE

CAS # 7783–36–0

SYNONYMS dimercury sulfate • mercurous sulfate, solid (DOT) • sulfuric acid, dimercury(1) salt

MF Hg_2O_4S

MW 497.29

COLOR AND FORM White to slightly yellow crystalline powder; colorless monoclinic crystals.

DENSITY 7.56.

FIREFIGHTING PROCEDURES If mercurous sulfate involved in fire, extinguish fire using agent suitable for type of surrounding fire (material itself does not burn or burns with difficulty). Use water in flooding quantities as fog. Use foam, carbon dioxide, or dry chemical [R3]. Personnel protection: Wear positive-pressure self-contained breathing apparatus when fighting fires involving this material [R3].

SOLUBILITY 0.06 g/100 ml water at 25°C; 0.09 g/100 ml water at 100°C. Soluble in nitric acid and sulfuric acid. Insoluble in water (solid).

OTHER CHEMICAL/PHYSICAL PROPERTIES Becomes gray on exposure to light with production of mercury and mercuric sulfate. Mercury salts when heated with sodium carbonate yield metallic mercury (Hg) and are reduced to metal by hydrogen peroxide in the presence of alkali hydroxide. copper, iron, zinc, and many other metals precipitate metallic mercury (Hg) from neutral or slightly acidic solutions of mercury salts (mercury salts). Mercurous salts give a black precipitate with alkali hydroxides and a white precipitate of calomel with hydrogen chloride or soluble chlorides. (mercurous salts).

PROTECTIVE EQUIPMENT In areas where the exposures are excessive, respiratory protection was provided either by full-face canister-type mask or supplied air respirator, depending on the concentration of mercury fumes. Above 50 mg Hg/m³ requires supplied-air positive-pressure full-face respirators. Full-body work clothes including shoes or shoe covers and hats should be supplied and clean work clothes should be supplied daily. Work clohtes should not be stored with street clothes in the same locker. (inorganic mercury) [R4].

PREVENTATIVE MEASURES Use disposable uniforms, so that a contaminated uniform is not a source of absorption through the skin. Preventative measures: adequate ventilation; careful attention to good housekeeping, e.g., avoidance of spills, and prompt and proper cleaning if a spill occurs; all containers of mercury and its compounds should be kept tightly closed; floors should be washed on a regular basis with dilute calcium sulfide solution or other suitable reactant; floors should be nonporous; all workers directly involved in the plant operation should shower thoroughly each day before leaving. (mercury compounds) [R5].

CLEANUP Mercury removal from waste water can be accomplished by the BMS process: Chlorine is added to the waste water, oxidizing any mercury present to the ionic state. The BMS adsorbent (an activated carbon concentrate with sulfur compound on its surface) is used to collect ionic mercury. The spent adsorbent is then distilled to recover the mercury, leaving a carbon residue for reuse, or disposal. The TMR IMAC process can also be used: Waste water is fed into a reactor, whereby a slight excess of chlorine is maintained, oxidizing any mercury present to ionic mercury. The liquid is then passed through the TMR IMAC ion-exchange resin, where mercury ions are adsorbed. The mercury is then stripped from the spent resin with hydrochloric acid solution (mercury compounds) [R6]. Spilled mercury compounds or solutions can be cleaned up by any method that does not cause excessive airborne contamination or skin contact. (mercury compounds) [R7].

DISPOSAL SRP: At the time of review, criteria for land treatment or burial (sanitary landfill) disposal practices are subject to significant revision. Prior to implementing land disposal of waste residue (including waste sludge), consult with environmental regulatory agencies for guidance on acceptable disposal practices.

COMMON USES For making electric batteries; with zinc sulfate in standard Clark cell and with cadmium sulfate in std Weston cell [R1].

OCCUPATIONAL RECOMMENDATIONS
OSHA Meets criteria for OSHA medical records rule [R10].

NIOSH 10–hr time-weighted average 0.05 mg/m³ (mercury and inorganic compounds (as Hg)) [R11].

THRESHOLD LIMIT 8–hr time-weighted average (TWA) 0.025 mg/m³, skin (1994) (mercury, inorganic forms including metallic mercury) [R19, 25].

HUMAN TOXICITY Lethal blood level: The concentration of inorganic mercury present in blood (serum or plasma) that has been reported to cause death in humans is: 0.04–2.2 mg%; 0.4–22 µg/ml. (inorganic mercury) [R8].

BIOLOGICAL HALF-LIFE In general, body burden of mercury in man has a half-life of about 60 days. (mercury) [R18, 1612].

BIODEGRADATION Inorganic forms of mercury (Hg) can be converted to organic forms by microbial action in the biosphere. (inorganic mercury) [R9].

ATMOSPHERIC STANDARDS Emissions to the atmosphere from sludge incineration plants, sludge drying plants, or a combination of these that process wastewater treatment plant sludges shall not exceed 3,200 grams of mercury per 24–hour period. (total mercury) [R12].

RCRA REQUIREMENTS A solid waste containing mercurous sulfate may become characterized as a hazardous waste when subjected to the toxicant extraction procedure listed in 40 CFR 261.24, and, if so characterized, must be managed as a hazardous waste [R13].

DOT *Health Hazards:* Poisonous if swallowed. Inhalation of dust poisonous. Fire may produce irritating or poisonous gases. Runoff from fire control or dilution water may cause pollution [R2].

Fire or Explosion: Some of these materials may burn, but none of them ignites readily [R2].

Emergency Action: Keep unnecessary people away; isolate hazard area and deny entry. Stay upwind; keep out of low areas. Self-contained breathing apparatus (SCBA) and structural firefighter's protective clothing will provide limited protection. Call CHEMTREC at 1–800–424–9300 for emergency assistance. If water pollution occurs, notify the appropriate authorities [R2].

Fire: Small fires: Dry chemical, CO_2, halon, water spray, or standard foam. Large Fires: Water spray, fog, or standard foam is recommended. Move container from fire area if you can do so without risk [R2].

Spill or Leak: Do not touch spilled material; stop leak if you can do so without risk. Small Spills: Take up with sand or other noncombustible absorbent material and place into containers for later disposal.

Small Dry Spills: With clean shovel place material into clean, dry container, and cover; move containers from spill area. Large Spills: Dike far ahead of liquid spill for later disposal [R2].

First Aid: Move victim to fresh air; call for emergency medical care. Remove and isolate contaminated clothing and shoes at the site. In case of contact with material, immediately flush skin, or eyes with running water for at least 15 minutes [R2].

FDA Bottled water shall, when a composite of analytical units of equal volume from a sample is examined by the methods described in paragraph (d) (1) (II) of this section, meet the standards of chemical quality, and shall not contain mercury in excess of 0.002 mg/L (total mercury) [R14]. The color additive F, D, and C Blue Number 2 shall conform to the specifications in the CFR 74.102 and be free from impurities other than those named, including mercury (as Hg) in not more than 1 part per million, to the extent that such other impurities may be avoided by current good manufacturing practice (total mercury) [R15]. The color additive F, D, and C Green Number 3 shall conform to the specifications in the CFR 74.203, and be free from impurities other than those named, including mercury (as Hg) in not more than 1 part per million, to the extent that such other impurities may be avoided by current good manufacturing practice (total mercury) [R16]. The color additive F, D, and C Yellow Number 5 shall conform to the specifications in the CFR 74.705, and be free from impurities other than those named, including mercury (as Hg) in not more than 1 part per million, to the extent that such other impurities may be avoided by current good manufacturing practice. (total mercury) [R17].

REFERENCES

1. *The Merck Index.* 10th ed. Rahway, NJ: Merck Co., Inc., 1983. 842.

2. Department of Transportation. *Emergency Response Guidebook 1987.* DOT P 5800.4. Washington, DC: U.S. Government Printing Office, 1987, p. G–53.

3. Association of American Railroads. *Emergency Handling of Hazardous Materials in Surface Transportation.* Washington, DC: Assoc. of American Railroads, Hazardous Materials Systems (BOE), 1987. 438.

4. Sittig, M. M. *Handbook of Toxic and Hazardous Chemicals and Carcinogens.* 1985. 2nd ed. Park Ridge, NJ: Noyes Data Corporation, 1985. 571.

5. *Kirk–Othmer Encyclopedia of Chemical Technology.* 3rd ed., Volumes 1–26. New York, NY: John Wiley and Sons, 1978–1984, p. 15 (81) 167.

6. Environment Canada. *Tech Info for Problem Spills: Mercury.* (Draft), p. 59 (1982).

7. National Research Council. *Prudent Practices for Handling Hazardous Chem-*

icals in Laboratories. Washington, DC: National Academy Press, 1981. 53.

8. Winek, C. L. Drug and Chemical Blood- Level Data 1985. Pittsburgh, PA: Allied Fischer Scientific, 1985.

9. Schroeder, W. H. Envir Sci Tech 16 (7): 394A–400A (1982) as cited in Environment Canada. Tech Info for Problem Spills: Inorganic Mercury (Draft), p. 41 (1982).

10. 29 CFR 1910.20 (7/1/87).

11. NIOSH. Pocket Guide to Chemical Hazards. 5th Printing/Revision. DHHS (NIOSH) Publ. No. 85–114. Washington, DC: U.S. Dept. of Health and Human Services, NIOSH/Supt. of Documents, GPO, Sept. 1985. 152.

12. 40 CFR 61.52 (b) (7/1/87).

13. 40 CFR 261.24 (7/1/87).

14. 21 CFR 103.35 (4/1/88).

15. 21 CFR 74.102 (4/1/88).

16. 21 CFR 74.203 (4/1/88).

17. 21 CFR 74.705 (4/1/88).

18. Gilman, A.G., L.S. Goodman, and A. Gilman (eds.). Goodman and Gilman's The Pharmacological Basis of Therapeutics. 7th ed. New York: Macmillan Publishing Co., Inc., 1985.

19. American Conference of Governmental Industrial Hygienists. Threshold Limit Values for Chemical Substances and Physical Agents and Biological Exposure Indices for 1994–1995. Cincinnati, OH: ACGIH, 1994.

■ MERCURY FULMINATE

CAS # 628-86-4

SYNONYMS fulminate of mercury, dry • fulminate of mercury, wet • fulminic acid, mercury(2) salt • mercuric cyanate • mercury fulminate (dry) • mercury fulminate (wet) • mercury fulminate [Hg(ONC)$_2$] • mercury(II) fulminate • mercury, bis(fulminato)-

MF CHNO • 1/2Hg

MW 284.62

COLOR AND FORM White, cubic crystals; gray crystalline powder.

DENSITY 4.42.

EXPLOSIVE LIMITS AND POTENTIAL Detonation may be initiated when dry by flame, heat, impact, friction, or intense radiation. contact with sulfuric acid causes explosion [R5].

TOXIC COMBUSTION PRODUCTS When heated to decomposition it emits toxic fumes of mercury and oxides of nitrogen [R4].

SOLUBILITY Soluble in hot water, alcohol, ammonium hydroxide; slightly soluble in cold water.

PROTECTIVE EQUIPMENT Personal protective equipment, including eye protective equipment, should be provided for normal process work as protection against eye splashes, acid burns, dermatitis, and skin absorption of toxic materials. (explosives industry) [R6].

PREVENTATIVE MEASURES Workshops and magazines must be adequately protected against lightning; all safety devices should be inspected periodically and any defect remedied immediately. Operations should be separated and automated to the maximum possible extent; remote handling equipment and automatic safety devices should be installed. Tools should be made of non-sparking materials and protected from any contact with dust and grit that might produce sparks. Energy sources and methods of applying energy to machines should be designed, used, and maintained with due regard to the danger of fire and explosion. It may be that the use of electrical equipment is ruled out by the danger of sparking or flames. (explosives industry) [R6].

CLEANUP Mercury removal from waste water can be accomplished by the BMS process: Chlorine is added to the waste water, oxidizing any mercury present to the ionic state. The BMS adsorbent (an activated carbon concentrate with sulfur compound on its surface) is used to collect ionic mercury. The spent adsorbent is then distilled to recover the mercury, leaving a carbon residue for reuse or disposal. The TMR IMAC process may also be used: Waste water is fed into a reactor, whereby a slight excess of chlorine is maintained, oxidizing any mercury present to ionic mercury. The liquid is then passed through the TMR IMAC ion-exchange resin, where mercury ions are adsorbed. The mercury is then stripped from the spent resin with hydrochloric acid solution (mercury compounds) [R7]. Spilled mercury compounds or solutions can be cleaned up by any method that does not cause excessive airborne contamination or skin contact. (mercury compounds) [R8].

DISPOSAL SRP: At the time of review, criteria for land treatment or burial (sanitary landfill) disposal practices are subject to significant revision. Prior to implementing land disposal of waste residue (including waste sludge), consult with environmental regulatory agencies for guidance on acceptable disposal practices.

COMMON USES An initiating explosive; manufacture of caps and detonators for producing explosions of military, industr, and sporting purposes [R1]. Used as a catalyst in the oxynitration of benzene to nitrophenol [R2].

OCCUPATIONAL RECOMMENDATIONS .
OSHA During an 8–hr work shift, an employee may not be exposed to a concentration of mercury above 1 mg/m³. (mercury) [R13].

NIOSH 10–hr time-weighted average 0.05 mg/m³ (mercury and inorganic compound (as Hg)) [R14].

THRESHOLD LIMIT 8–hr time-weighted average (TWA) 0.025 mg/m³, skin (1994) (mercury, in organic forms including metallic mercury) [R19, 25].

HUMAN TOXICITY Lethal blood level: The concentration of inorganic mercury present in the blood (serum or plasma) that has been reported to cause death in humans is: 0.04–2.2 mg%; 0.4–22 µg/ml. (inorganic mercury) [R11].

IARC SUMMARY Evaluation: There is inadequate evidence in humans for the carcinogenicity of mercury and mercury compounds. Overall evaluation: Metallic mercury and inorganic mercury compounds are not classifiable as to their carcinogenicity to humans (Group 3) [R9].

BIOLOGICAL HALF-LIFE In general, body burden of mercury in man has a half-life of about 60 days. (mercury) [R10].

BIODEGRADATION Inorganic forms of Hg can be converted to organic forms by microbial action in the biosphere. (inorganic mercury) [R12].

ATMOSPHERIC STANDARDS Emissions to the atmosphere from sludge incineration plants, sludge drying plants, or a combination of these sludge wastewater treatment plant processes shall not exceed 3,200 grams of mercury per 24–hour period. (total mercury) [R15].

CERCLA REPORTABLE QUANTITIES Persons in charge of vessels or facilities are required to notify the National Response Center (NRC) immediately, when there is a release of this designated hazardous substance, in an amount equal to or greater than its reportable quantity of 10 lb or 4.54 kg. The toll-free telephone number of the NRC is (800) 424–8802; in the Washington metropolitan area (202) 426–2675. The rule for determining when notification is required is stated in 40 CFR 302.6 (section IV. D. 3.b) [R16].

RCRA REQUIREMENTS As stipulated in 40 CFR 261.33, when mercury fulminate, as a commercial chemical product or manufacturing chemical intermediate or as an off-specification commercial chemical product or a manufacturing chemical intermediate, becomes a waste, it must be managed as a hazardous waste according to federal or state regulations. Also defined as a hazardous waste is any container or inner liner used to hold this waste or any residue, contaminated soil, water, or other debris resulting from the cleanup of a spill, into water or on dry land, of this waste. Generators of small quantities of this waste may qualify for partial exlusion from hazardous waste regulations (40 CFR 261.5 (e)) [R17].

DOT *Health Hazards:* Poisonous if swallowed. Inhalation of dust poisonous. Fire may produce irritating or poisonous gases. Runoff from fire control or dilution water may cause pollution (mercury compounds, not otherwise specified) [R3].

Emergency Action: Keep unnecessary people away; isolate hazard area and deny entry. Stay upwind; keep out of low areas. Self-contained breathing apparatus (SCBA) and structural firefighter's protective clothing will provide limited protection. Call CHEMTREC at 1–800–424–9300 for emergency assistance. If water pollution occurs, notify the appropriate authorities (mercury compounds, not otherwise specified) [R3].

Fire: Small Fires: Dry chemical, CO_2, halon, water spray, or standard foam. Large Fires: Water spray, fog, or standard foam is recommended. Move container from fire area if you can do so without risk (mercury compounds, not otherwise specified) [R3].

Spill or Leak: Do not touch spilled material; stop leak if you can do so without risk. Small Spills: Take up with sand or other noncombustible absorbent material and place into containers for later disposal.

Small Dry Spills: With clean shovel place material into clean, dry container, and cover; move container from spill area. Large Spills: Dike far ahead of liquid spill for later disposal (mercury compounds, not otherwise specified) [R3].

First Aid: Move victim to fresh air; call for emergency medical care. Remove and isolate contaminated clothing and shoes at the site. In case of contact with material, immediately flush skin or eyes with running water for at least 15 minutes (mercury compounds, not otherwise specified) [R3].

FDA Bottled water shall, when a composite of analytical units of equal volume from a sample is examined by the methods described in paragraph (d) (1) (II) of this section, meet the standards of chemical quality, and shall not contain mercury fulminate in excess of 0.002 mg/L. (total mercury) [R18].

REFERENCES

1. Sax, N. I., and R. J. Lewis, Sr. (eds.). *Hawley's Condensed Chemical Dictionary.* 11th ed. New York: Van Nostrand Reinhold Co., 1987. 746.

2. *Kirk–Othmer Encyclopedia of Chemical Technology.* 3rd ed., Volumes 1–26. New York, NY: John Wiley and Sons, 1978–1984, p. V26 (85) 745.

3. Department of Transportation. *Emergency Response Guidebook 1987.* DOT P 5800.4. Washington, DC: U.S. Government Printing Office, 1987, p. G–53.

4. Sax, N. I. *Dangerous Properties of Industrial Materials.* 6th ed. New York, NY: Van Nostrand Reinhold, 1984. 1753.

5. Bretherick, L. *Handbook of Reactive Chemical Hazards.* 2nd ed. Boston, MA: Butterworths, 1979. 405.

6. International Labour Office. *Encyclopedia of Occupational Health and Safety.* Vols. I and II. Geneva, Switzerland: International Labour Office, 1983. 808.

7. Environment Canada. *Tech Info for Problem Spills: Mercury (Draft)*, p. 59 (1982).

8. National Research Council. *Prudent Practices for Handling Hazardous Chemicals in Laboratories.* Washington, DC: National Academy Press, 1981. 53.

9. IARC. *Monographs on the Evaluation of the Carcinogenic Risk of Chemicals to Man.* Geneva: World Health Organization, International Agency for Research on Cancer, 1972–present (multivolume work), p. 58 324 (1993).

10. Gilman, A. G., L. S. Goodman, and A. Gilman (eds.). *Goodman and Gilman's The Pharmacological Basis of Therapeutics.* 7th ed. New York: Macmillan Publishing Co., Inc., 1985. 1612.

11. Winek, C. L. *Drug and Chemical Blood–Level Data 1985.* Pittsburgh, PA: Allied Fischer Scientific, 1985.

12. Schroeder, W. H. *Envir Sci Tech* 16 (7): 394A–400A (1982) as cited in *Environment Canada; Tech Info for Problem Spills: Inorganic Mercury (Draft)*, p. 41 (1982).

13. 29 CFR 1910.1000 (7/1/87).

14. NIOSH. *Pocket Guide to Chemical Hazards*. 5th Printing/Revision. DHHS (NIOSH) Publ. No. 85–114. Washington, DC: U.S. Dept. of Health and Human Services, NIOSH/Supt. of Documents, GPO, Sept. 1985. 152.

15. 40 CFR 61.52 (b) (7/1/87).

16. 51 FR 34534 (9/29/86).

17. 51 FR 28296 (8/6/86).

18. 21 CFR 103.35 (4/1/88).

19. American Conference of Governmental Industrial Hygienists. *Threshold Limit Values for Chemical Substances and Physical Agents and Biological Exposure Indices for 1994–1995*. Cincinnati, OH: ACGIH, 1994.

■ MERCURIC BENZOATE

CAS # 583–15–3

SYNONYMS benzoic-acid, mercury(2) salt • beta-mercuribenzoate • mercury(II) benzoate

MF $C_7H_6O_2 \cdot 1/2Hg$

MW 442.83

COLOR AND FORM White crystalline powder.

MELTING POINT 165°C.

FIREFIGHTING PROCEDURES If mercuric benzoate involved.

Fire: Extinguish fire using agent suitable for type of surrounding fire (mercuric benzoate itself does not burn or burns with difficulty). Use water in flooding quantities as fog. Use alcohol foam, carbon dioxide, or dry chemical [R17, 432]. Wear positive-pressure self-contained breathing apparatus when fighting fires involving mercuric benzoate [R17, 432].

SOLUBILITY Soluble in ammonium chloride; benzene; soluble in solution of ammonium benzoate.

STABILITY AND SHELF LIFE Sensitive to light (monohydrate) [R3].

OTHER CHEMICAL/PHYSICAL PROPERTIES Freely soluble in sodium chloride solution; soluble in 90 parts cold, 40 parts boiling water; slightly soluble in alcohol; when boiled with water or alcohol it hydrolyzes to a basic salt and free benzoic acid; odorless (monohydrate). Mol wt 460.84 (mercury(II) benzoate monohydrate); mercury salts, when heated with sodium carbonate, yield metallic mercury, and are reduced to metal by hydrogen peroxide in the presence of alkali hydroxide. Copper, iron, zinc, and many other metals precipitate metallic mercury from neutral or slightly acid solution of mercury salts. (mercury salts).

PROTECTIVE EQUIPMENT Personnel protection: Wear boots, protective gloves, and goggles. Do not handle broken packages without protective equipment. If contact with the material anticipated, wear full chemical protective clothing. [R17, 433].

PREVENTATIVE MEASURES If material not involved in Fire: Keep material out of water sources and sewers. [R17, 432].

CLEANUP Mercury removal from waste water can be accomplished by the BMS process: chlorine is added to the waste water, oxidizing any mercury present to the ionic state. The BMS adsorbent (an activated carbon concentrate with sulfur compound on its surface) is used to collect ionic mercury. The spent adsorbent is then distilled to recover the mercury, leaving a carbon residue for reuse or disposal. The TMR IMAC process can also be used waste water is fed into a reactor, whereby a slight excess of chlorine is maintained, oxidizing any mercury present to ionic mercury. The liquid is then passed through the TMR IMAC ion-exchange resin, where mercury ions are adsorbed. The mercury is then stripped from the spent resin with hydrochloric acid solution (mercury compounds) [R4]. Spilled mercury compounds or solutions can be cleaned up by any method that does not cause excessive airborne contamination or skin contact. (mercury compounds) [R5].

DISPOSAL SRP: At the time of review, criteria for land treatment or burial (sanitary landfill) disposal practices are subject to significant revision. Prior to implementing land disposal of waste residue (including waste sludge), consult with environmental regulatory agencies for guidance on acceptable disposal practices.

COMMON USES Used formerly in therapeutics in treatment of syphilis and gonorrhea [R16, 86].

OCCUPATIONAL RECOMMENDATIONS

OSHA During an 8–hr work shift, an employee may not be exposed to a concentration of mercury above 1 mg/m³. (mercury) [R8].

NIOSH 10–hr time-weighted average 0.05 mg/m³ (mercury and inorganic compounds (as Hg)) [R9].

IARC SUMMARY Evaluation: There is inadequate evidence in humans for the carcinogenicity of mercury and mercury compounds. Overall evaluation: metallic mercury and inorganic mercury compounds are not classifiable as to their carcinogenicity to humans (Group 3) [R6].

BIOLOGICAL HALF-LIFE Except for excretion by saliva, mercury is excreted by liver through bile, and also by mucous membranes of small intestines and colon. In rats elimination curve is a multi-phasic exponential curve, having a rapid phase with half-life of about 5 days, another phase with half-life of 1 mo, and still another of about 3 mo. (mercuric compounds) [R18, 413].

BIODEGRADATION Certain bacteria are capable of transforming mercuric ion to volatile elemental mercury. (mercuric ion) [R7].

BIOCONCENTRATION Many organisms are capable of accumulating mercury from water. Bioconcentration up to 10,000-fold [R2].

ATMOSPHERIC STANDARDS Emissions to the atmosphere from sludge incineration

plants, sludge drying plants, or a combination of these sludge wastewater treatment plant processes shall not exceed 3,200 grams of mercury per 24-hour period. (total mercury) [R10].

RCRA REQUIREMENTS A solid waste containing mercuric benzoate may become characterized as a hazardous waste when subjected to the toxicant extraction procedure listed in 40 CFR 261.24 and, if so characterized, must be managed as a hazardous waste [R11].

DOT *Health Hazards:* Poisonous if swallowed. Inhalation of dust poisonous. Fire may produce irritating or poisonous gases. Runoff from fire control or dilution water may cause pollution [R1].

Fire or Explosion: Some of these materials may burn, but none of them ignites readily [R1].

Emergency Action: Keep unnecessary people away; isolate hazard area and deny entry. Stay upwind; keep out of low areas. Self-contained breathing apparatus (SCBA) and structural firefighter's protective clothing will provide limited protection. Call CHEMTREC at 1–800–424–9300 for emergency assistance. If water pollution occurs, notify the appropriate authorities [R1].

Fire: Small fires: Dry chemical, CO₂, halon, water spray, or standard foam. Large Fires: Water spray, fog, or standard foam is recommended. Move container from fire area if you can do so without risk [R1].

Spill or Leak: Do not touch spilled material; stop leak if you can do so without risk. Small Spills: Take up with sand or other noncombustible absorbent material and place into containers for later disposal.

Small Dry Spills: With clean shovel place material into clean, dry container, and cover; move containers from spill area. Large Spills: Dike far ahead of liquid spill for later disposal [R1].

First Aid: Move victim to fresh air; call for emergency medical care. Remove and isolate contaminated clothing and shoes at the site. In case of contact with material, immediately flush skin or eyes with running water for at least 15 minutes [R1].

FDA Bottled water shall, when a composite of analytical units of equal volume from a sample is examined by the methods described in paragraph (d) (1) (II) of this section, meet the standards of chemical quality, and shall not contain mercury in excess of 0.002 mg/L (total mercury) [R12]. The color additive F, D, and C Blue Number 2 shall conform to the specifications in the CFR 74.102 and be free from impurities other than those named, including mercury (as Hg) in not more than 1 part per million, to the extent that such other impurities may be avoided by current good manufacturing practice (total mercury) [R13]. The color additive F, D, and C Green Number 3 shall conform to the specifications in the CFR 74.203 and be free from impurities other than those named, including mercury (as Hg) in not more than 1 part per million, to the extent that such other impurities may be avoided by current good manufacturing practice (total mercury) [R14]. The color additive F, D, and C Yellow Number 5 shall conform to the specifications in the CFR 74.705, and be free from impurities other than those named, including mercury (as Hg) in not more than 1 part per million, to the extent that such other impurities may be avoided by current good manufacturing practice. (total mercury) [R15].

REFERENCES

1. Department of Transportation. *Emergency Response Guidebook 1987.* DOT P 5800.4. Washington, DC: U.S. Government Printing Office, 1987, p. G–53.

2. U.S. Coast Guard, Department of Transportation. *CHRIS—Hazardous Chemical Data.* Volume II. Washington, DC: U.S. Government Printing Office, 1984–5.

3. *The Merck Index.* 10th ed. Rahway, NJ: Merck Co., Inc., 1983. 839.

4. Environment Canada. *Tech Info for Problem Spills: Mercury* (Draft), p. 59 (1982).

5. National Research Council. *Prudent Practices for Handling Hazardous Chemicals in Laboratories.* Washington, DC: National Academy Press, 1981. 53.

6. IARC. *Monographs on the Evaluation of the Carcinogenic Risk of Chemicals to Man.* Geneva: World Health Organization, International Agency for Research on Cancer, 1972–present (multivolume work), p. 58 324 (1993).

7. Callahan, M. A., M. W. Slimak, N. W. Gabel, et al. *Water–Related Environmental Fate of 129 Priority Pollutants.* Volume I. EPA–440/4 79–029a. Washington, DC: U.S. Environmental Protection Agency, December 1979, p. 14–9.

8. 29 CFR 1910.1000 (7/1/87).

9. NIOSH. *Pocket Guide to Chemical Hazards.* 5th Printing/Revision. DHHS (NIOSH) Publ. No. 85–114. Washington, DC: U.S. Dept. of Health and Human Services, NIOSH/Supt. of Documents, GPO, Sept. 1985. 152.

10. 40 CFR 61.52 (b) (7/1/87).

11. 40 CFR 261.24 (7/1/87).

12. 21 CFR 103.35 (4/1/88).

13. 21 CFR 74.102 (4/1/88).

14. 21 CFR 74.203 (4/1/88).

15. 21 CFR 74.705 (4/1/88).

16. Venugopal, B., and T.D. Luckey. *Metal Toxicity in Mammals.* 2. New York: Plenum Press, 1978.

17. Association of American Railroads. *Emergency Handling of Hazardous Materials in Surface Transportation.* Washington, DC: Assoc. of American Railroads, Hazardous Materials Systems (BOE), 1987.

27. Friberg, L., G.F. Nordberg, E. Kessler, and V.B. Vouk (eds.). *Handbook of the Toxicology of Metals.* 2nd ed. Vols I, II.: Amsterdam: Elsevier Science Publishers B.V., 1986.

■ MERCURIC OLEATE

CAS # 1191–80–6

SYNONYMS 9-octadecenoic acid(z)-, mercury(2) salt • 9-octadecenoic acid, mercury salt • mercury oleate • oleate of mercury • oleic acid, mercury(2) salt • mercuric oleate, solid

MF $C_{18}H_{34}O_2$ • $1/2Hg$

MW 763.52

COLOR AND FORM Yellowish-brown, somewhat transparent, ointment-like mass.

ODOR Odor of oleic acid.

FIREFIGHTING PROCEDURES If mercuric oleate involved in fire: Extinguish fire using agent suitable for type of surrounding fire. Mercuric oleate itself does not burn or burns with difficulty. Use water in flooding quantities as in fog. Use foam, dry chemical, or carbon dioxide. [R17, 434].

SOLUBILITY Practically insoluble in water; slightly soluble in alcohol or ether, freely soluble in fixed oils.

OTHER CHEMICAL/PHYSICAL PROPERTIES Mercury salts, when heated with sodium carbonate, yield metallic mercury (Hg), and are reduced to metal by hydrogen peroxide in the presence of alkali hydroxide. Copper, iron, zinc, and many other metals precipitate metallic Hg from neutral or slightly acid solutions of mercury salts. (mercury salts).

PROTECTIVE EQUIPMENT Personnel protection: Avoid breathing dusts and fumes from burning mercuric oleate. Keep upwind. Avoid bodily contact with the mercuric oleate. Wear boots, protective gloves, and goggles. Do not handle broken packages without protective equipment. Wash away any material that may have contacted the body with copious amounts of water or soap and water. Wear self-contained breathing apparatus when fighting fires involving this material. If contact with the mercuric oleate is anticipated, wear full protective clothing. [R17, 434].

PREVENTATIVE MEASURES Use disposable uniforms, so that a contaminated uniform is not a source of absorption through the skin. Preventative measure: adequate ventilation; careful attention to good housekeeping, e.g., avoidance of spills, and prompt and proper cleaning if a spill occurs; all containers of mercury and its compounds should be kept tightly closed; floors should be washed on a regular basis with dilute calcium sulfide solution or other suitable reactant; floors should be nonporous; all workers directly involved in the plant operation should shower thoroughly each day before leaving. (mercury compounds) [R3].

CLEANUP Mercury removal from waste water can be accomplished by the BMS process: chlorine is added to the waste water, oxidizing any mercury present to the ionic state. The BMS adsorbent (an activated carbon concentrate with sulfur compound on its surface) is used to collect ionic mercury. The spent adsorbent is then distilled to recover the mercury, leaving a carbon residue for reuse or disposal. The TMR IMAC process may also be used: Waste water is fed into a reactor, whereby a slight excess of chlorine is maintained, oxidizing any mercury present to ionic mercury. The liquid is then passed through the TMR IMAC ion-exchange resin, where mercury ions are adsorbed. The mercury is then stripped from the spent resin with hydrochloric acid solution (mercury compounds) [R4]. Spilled mercury compounds or solutions can be cleaned up by any method that does not cause excessive airborne contamination or skin contact. (mercury compounds) [R5].

DISPOSAL SRP: At the time of review, criteria for land treatment or burial (sanitary landfill) disposal practices are subject to significant revision. Prior to implementing land disposal of waste residue (including waste sludge), consult with environmental regulatory agencies for guidance on acceptable disposal practices.

COMMON USES In antiseptics; antifouling paints [R1].

OCCUPATIONAL RECOMMENDATIONS
OSHA Meets criteria for OSHA medical records rule [R9].

NIOSH 10-hr time-weighted average 0.05 mg/m^3 (mercury and inorganic compounds (as Hg)) [R10].

HUMAN TOXICITY Lethal blood level: The

concentration of inorganic mercury present in the blood (serum or plasma) that has been reported to cause death in humans is: 0.04–2.2 mg%; 0.4–22 µg/ml. (inorganic mercury) [R7].

IARC SUMMARY Evaluation: There is inadequate evidence in humans for the carcinogenicity of mercury and mercury compounds. Overall evaluation: Metallic mercury and inorganic mercury compounds are not classifiable as to their carcinogenicity to humans (Group 3) [R6].

BIOLOGICAL HALF-LIFE In general, body burden of mercury in man has a half-life of about 60 days. (mercury) [R18, 1612].

BIODEGRADATION Certain bacteria are capable of transforming mercuric ion to volatile elemental mercury. (mercuric ion) [R8].

BIOCONCENTRATION Mercuric salts are still widely employed in industry, and industrial discharge into rivers has polluted many parts of the world. Microorganisms convert inorg mercury to methyl mercury, which is taken up rapidly by planktonic algae and is concentrated in fish by way of food chain. (mercuric salts) [R18, 1611].

ATMOSPHERIC STANDARDS Emissions to the atmosphere from sludge incineration plants, sludge drying plants, or a combination of these sludge wastewater treatment plant processes shall not exceed 3,200 grams of mercury per 24–hour period. (total mercury) [R11].

RCRA REQUIREMENTS A solid waste containing mercuric oleate may become characterized as a hazardous waste when subjected to the toxicant extraction procedure listed in 40 CFR 261.24 and, if so characterized, must be managed as a hazardous waste [R12].

DOT *Health Hazards:* Poisonous if swallowed. Inhalation of dust poisonous. Fire may produce irritating or poisonous gases. Runoff from fire control or dilution water may cause pollution [R2].

Fire or Explosion: Some of these materials may burn, but none of them ignites readily [R2].

Emergency Action: Keep unnecessary people away; isolate hazard area and deny entry. Stay upwind; keep out of low areas. Self-contained breathing apparatus (SCBA) and structural firefighter's protective clothing will provide limited protection. Call CHEMTREC at 1–800–424–9300 for emergency assistance. If water pollution occurs, notify the appropriate authorities [R2].

Fire: Small fires: Dry chemical, CO_2, halon, water spray, or standard foam. Large Fires: Water spray, fog, or standard foam is recommended. Move container from fire area if you can do so without risk [R2].

Spill or Leak: Do not touch spilled material; stop leak if you can do so without risk. Small Spills: Take up with sand or other noncombustible absorbent material and place into containers for later disposal.

Small Dry Spills: With clean shovel place material into clean, dry container, and cover; move containers from spill area. Large Spills: Dike far ahead of liquid spill for later disposal [R2].

First Aid: Move victim to fresh air; call for emergency medical care. Remove and isolate contaminated clothing and shoes at the site. In case of contact with material, immediately flush skin or eyes with running water for at least 15 minutes [R2].

FDA Bottled water shall, when a composite of analytical units of equal volume from a sample is examined by the methods described in paragraph (d) (1) (II) of this section, meet the standards of chemical quality, and shall not contain mercury in excess of 0.002 mg/L (total mercury) [R13]. The color additive F, D, and C Blue Number 2 shall conform to the specifications in the CFR 74.102 and be free from impurities other than those named, including mercury (as Hg) in not more than 1 part per million, to the extent that such other impurities may be avoided by current good manufacturing practice (total mercury) [R14]. The color additive F, D, and C Green Number 3 shall conform to the specifications in the CFR 74.203, and be free from impurities other than those named, including mercury (as Hg) in not more than 1 part per million, to the extent that such other impurities may be avoided by current good manufacturing practice (total mercury) [R15]. The color additive F, D, and C Yellow Number 5 shall conform to the specifications in the CFR 74.705, and be free from impurities other than those named, including mercury (as Hg) in not more than 1 part per million, to the extent that such other impurities may be avoided by current good manufacturing practice. (total mercury) [R16].

REFERENCES

1. Sax, N. I., and R. J. Lewis, Sr. (eds.). *Hawley's Condensed Chemical Dictionary.* 11th ed. New York: Van Nostrand Reinhold Co., 1987. 743.

2. Department of Transportation. *Emergency Response Guidebook 1987.* DOT P 5800.4. Washington, DC: U.S. Government Printing Office, 1987, p. G–53.

3. *Kirk-Othmer Encyclopedia of Chemical Technology.* 3rd ed. Volumes 1–26. New York, NY: John Wiley and Sons, 1978–1984, p. 15 (81) 167.

4. Environment Canada. *Tech Info for Problem Spills: Mercury (Draft)*, p. 59 (1982).

5. National Research Council. *Prudent Practices for Handling Hazardous Chemicals in Laboratories.* Washington, DC: National Academy Press, 1981. 53.

6. IARC. *Monographs on the Evaluation of the Carcinogenic Risk of Chemicals to Man.* Geneva: World Health Organization, International Agency for Research on Cancer, 1972–present (multivolume work), p. 58 324 (1993).

7. Winek, C. L. *Drug and Chemical Blood–Level Data 1985.* Pittsburgh, PA: Allied Fischer Scientific, 1985.

8. Callahan, M. A., M. W. Slimak, N. W. Gabel, et al. *Water–Related Environmental Fate of 129 Priority Pollutants.* Volume I. EPA–440/4 79–029a. Washington, DC: U.S. Environmental Protection Agency, December 1979, p. 14–9.

9. 29 CFR 1910.20 (7/1/87).

10. NIOSH. *Pocket Guide to Chemical Hazards.* 5th Printing/Revision. DHHS (NIOSH) Publ. No. 85–114. Washington, DC: U.S. Dept. of Health and Human Services, NIOSH/Supt. of Documents, GPO, Sept. 1985. 152.

11. 40 CFR 61.52 (b) (7/1/87).

12. 40 CFR 261.24 (7/1/87).

13. 21 CFR 103.35 (4/1/88).

14. 21 CFR 74.102 (4/1/88).

15. 21 CFR 74.203 (4/1/88).

16. 21 CFR 74.705 (4/1/88).

17. Association of American Railroads. *Emergency Handling of Hazardous Materials in Surface Transportation.* Washington, DC: Assoc. of American Railroads, Hazardous Materials Systems (BOE), 1987.

18. Gilman, A.G., L.S. Goodman, and A. Gilman. (eds.). *Goodman and Gilman's The Pharmacological Basis of Therapeutics.* 7th ed. New York: Macmillan Publishing Co., Inc., 1985.

■ MERCURIC POTASSIUM CYANIDE

CAS # 591–89–9

SYNONYMS dipotassium tetracyanomercurate(2-) • dipotassium tetrakis (cyano-c) mercurate(2-) • mercurate(2-), tetracyano-, dipotassium • mercurate(2-), tetrakis (cyano-c)-, dipotassium, (t-4)- • mercury potassium cyanide • potassium cyanomercurate • potassium tetracyanomercurate ($K_2Hg(CN)_4$) • potassium tetracyanomercurate(II)

MF C_4HgN_4 •2K

MW 382.87

COLOR AND FORM Colorless or white crystals.

FIREFIGHTING PROCEDURES If mercuric potassium cyanide involved in fire: Extinguish fire using agent suitable for type of surrounding fire. Mercuric potassium cyanide itself does not burn or burns with difficulty. Use alcohol foam, carbon dioxide, or dry chemical. Do not use water on mercuric potassium cyanide itself. If large quantities of combustibles are involved, use water in flooding quantities as spray and fog. Use water spray to absorb vapors [R3].

TOXIC COMBUSTION PRODUCTS Emit hydrocyanic acid. (cyanides) [R2].

SOLUBILITY Soluble in cold water and alcohol.

OTHER CHEMICAL/PHYSICAL PROPERTIES Mercury salts, when heated with sodium carbonate, yield metallic mercury (Hg), and are reduced to metal by hydrogen peroxide in the presence of alkali hydroxide. Copper, iron, zinc, and many other metals precipitate metallic Hg from neutral or slightly acid solutions of mercury salts (mercury salts). Soluble ionized mercuric salts give a yellow precipitate of mercuric oxide with sodium hydroxide and a red precipitate of mercuric iodide with alkali iodide. (soluble mercuric salts).

PROTECTIVE EQUIPMENT Personnel protection: Wear self-contained breathing apparatus. Wear boots, protective gloves, and goggles. If contact with the material is anticipated, wear full protective clothing [R3].

PREVENTATIVE MEASURES Use disposable uniforms, so that a contaminated uniform is not a source of absorption through the skin. Preventative measure: adequate ventilation; careful attention to good housekeeping, e.g., avoidance of spills, and prompt and proper cleaning if a spill occurs; all containers of mercury and its compounds should be kept tightly closed; floors should be washed on a regular basis with dilute calcium sulfide solution or other suitable reactant; floors should be nonporous; all workers directly involved in the plant operation should shower thoroughly each day before leaving the plant. (mercury compounds) [R4].

CLEANUP Mercury removal from waste water can be accomplished by the BMS

process: Chlorine is added to the waste water, oxidizing any mercury present to the ionic state. The BMS adsorbent (an activated carbon concentrate with sulfur compound on its surface) is used to collect ionic mercury. The spent adsorbent is then distilled to recover the mercury, leaving a carbon residue for reuse or disposal. The TMR IMAC process can also be used: Waste water is fed into a reactor, whereby a slight excess of chlorine is maintained, oxidizing any mercury present to ionic mercury. The liquid is then passed through the TMR IMAC ion-exchange resin, where mercury ions are adsorbed. The mercury is then stripped from the spent resin with hydrochloric acid solution (mercury compounds) [R5]. Spilled mercury compounds or solutions can be cleaned up by any method that does not cause excessive airborne contamination or skin contact (mercury compounds) [R6].

DISPOSAL SRP: At the time of review, criteria for land treatment or burial (sanitary landfill) disposal practices are subject to significant revision. Prior to implementing land disposal of waste residue (including waste sludge), consult with environmental regulatory agencies for guidance on acceptable disposal practices.

COMMON USES In manufacture of mirrors to prevent silver coating from yellowing; as reagent in testing for free acids [R16, 1105].

OCCUPATIONAL RECOMMENDATIONS
OSHA During an 8-hr work shift, an employee may not be exposed to a concentration of mercury above 1 mg/m^3. (mercury) [R11].

NIOSH 10hr time-weighted average 0.05 mg/m^3 (mercury and inorganic compounds (as Hg)) [R12].

THRESHOLD LIMIT 8-hr time-weighted average (TWA) 0.025 mg/m^3, skin (1994) (mercury, in organic forms including metallic mercury) [R17, 25].

HUMAN TOXICITY Lethal blood level: The concentration of inorganic mercury present in the blood (serum or plasma) that has been reported to cause death in humans is: 0.04–2.2 mg%; 0.4–22 µg/ml. (inorganic mercury) [R8].

IARC SUMMARY Evaluation: There is inadequate evidence in humans for the carcinogenicity of mercury and mercury compounds. Overall evaluation: Metallic mercury and inorganic mercury compounds are not classifiable as to their carcinogenicity to humans (Group 3) [R7].

BIOLOGICAL HALF-LIFE The average biological half-time of a tracer dose of divalent inorganic mercury compounds in man is 42 days for the whole-body and 26 days for blood. (mercury compounds) [R9].

BIODEGRADATION Inorganic forms of mercury can be converted to organic forms by microbial action in the biosphere. (inorganic mercury) [R10].

BIOCONCENTRATION Mercuric salts are still widely employed in industry, and industrial discharge into rivers has polluted many parts of the world. Microorganisms convert inorg mercury to methyl mercury, which is taken up rapidly by planktonic algae and is concentrated in fish by way of food chain. (mercuric salts) [R18, 1611].

ATMOSPHERIC STANDARDS Emissions to the atmosphere from sludge incineration plants, sludge drying plants, or a combination of these sludge wastewater treatment plant processes shall not exceed 3,200 grams of mercury per 24–hour period. (total mercury) [R13].

RCRA REQUIREMENTS A solid waste containing mercuric potassium cyanide may become characterized as a hazardous waste when subjected to the toxicant extraction procedure listed in 40 CFR 261.24, and, if so characterized, must be managed as a hazardous waste [R14].

DOT *Health Hazards:* Poisonous if swallowed. Inhalation of dust poisonous. Fire may produce irritating or poisonous gases. Runoff from fire control or dilution water may cause pollution [R1].

Fire or Explosion: Some of these materials may burn, but none of them ignites readily [R1].

Emergency Action: Keep unnecessary people away; isolate hazard area and deny entry. Stay upwind; keep out of low areas. Self-contained breathing apparatus (SCBA) and structural firefighter's protective clothing will provide limited protection. Call CHEMTREC at 1–800–424–9300 for emergency assistance. If water pollution occurs, notify the appropriate authorities [R1].

Fire: Small fires: Dry chemical, CO_2, halon, water spray, or standard foam. Large Fires: Water spray, fog, or standard foam is recommended. Move container from fire area if you can do so without risk [R1].

Spill or Leak: Do not touch spilled material; stop leak if you can do so without risk. Small Spills: Take up with sand or other noncombustible absorbent material and place into containers for later disposal.

Small Dry Spills: With clean shovel place material into clean, dry container, and cover; move containers from spill area. Large Spills: Dike far ahead of liquid spill for later disposal [R1].

First Aid: Move victim to fresh air; call for emergency medical care. Remove and isolate contaminated clothing and shoes at the site. In case of contact with material, immediately flush skin or eyes with running water for at least 15 minutes [R1].

FIRE POTENTIAL Moderate, by chemical reaction with heat, moisture, acid; emits hydrocyanic acid. (cyanides) [R2].

FDA Bottled water shall, when a composite of analytical units of equal volume from a sample is examined by the methods described in paragraph (d) (1) (II) of this section, meet the standards of chemical quality, and shall not contain mercury in excess of 0.002 mg/L. (total mercury) [R15].

REFERENCES

1. Department of Transportation. *Emergency Response Guidebook 1987.* DOT P 5800.4. Washington, DC: U.S. Government Printing Office, 1987, p. G–53.

2. Sax, N. I. *Dangerous Properties of Industrial Materials,*. 6th ed. New York, NY: Van Nostrand Reinhold, 1984. 2283.

3. Association of American Railroads. *Emergency Handling of Hazardous Materials in Surface Transportation.* Washington, DC: Assoc. of American Railroads, Hazardous Materials Systems (BOE), 1987. 436.

4. *Kirk–Othmer Encyclopedia of Chemical Technology.* 3rd ed., Volumes 1–26. New York, NY: John Wiley and Sons, 1978–1984, p. 15 (81) 167.

5. Environment Canada. *Tech Info for Problem Spills: Mercury (Draft),* p. 59 (1982).

6. National Research Council. *Prudent Practices for Handling Hazardous Chemicals in Laboratories.* Washington, DC: National Academy Press, 1981. 53.

7. IARC. *Monographs on the Evaluation of the Carcinogenic Risk of Chemicals to Man.* Geneva: World Health Organization, International Agency for Research on Cancer, 1972–present (multivolume work), p. 58 324 (1993).

8. Winek, C. L. *Drug and Chemical Blood–Level Data 1985.* Pittsburgh, PA: Allied Fischer Scientific, 1985.

9. USEPA. *Mercury Health Effects Update.* p. 2–5 (1984) EPA 600/8–84–019F.

10. Schroeder, W. H., *Envir Sci Tech* 16 (7): 394A–400A (1982) as cited in *Environment Canada. Tech Info for Problem Spills: Mercury (Draft),* p. 41 (1982).

11. 29 CFR 1910.1000 (7/1/87).

12. NIOSH. *Pocket Guide to Chemical Hazards.* 5th Printing/Revision. DHHS (NIOSH) Publ. No. 85–114. Washington, DC: U.S. Dept. of Health and Human Services, NIOSH/Supt. of Documents, GPO, Sept. 1985. 152.

13. 40 CFR 61.52 (b) (7/1/87).

14. 40 CFR 261.24 (7/1/87).

15. 21 CFR 103.35 (4/1/88).

16. *The Merck Index.* 10th ed. Rahway, New Jersey: Merck Co., Inc., 1983.

17. American Conference of Governmental Industrial Hygienists. *Threshold Limit Values for Chemical Substances and Physical Agents and Biological Exposure Indices for 1994–1995.* Cincinnati, OH: ACGIH, 1994.

18. Gilman, A.G., L.S. Goodman, and A. Gilman (eds.). *Goodman and Gilman's The Pharmacological Basis of Therapeutics.* 7th ed. New York: Macmillan Publishing Co., Inc., 1985.

■ MERCUROUS IODIDE

CAS # 15385–57–6

SYNONYMS dimercury diiodide • mercury diiodide • mercury iodide (Hg$_2$I$_2$) • mercury protoiodide • yellow mercury iodide

MF Hg$_2$I$_2$

MW 654.99

COLOR AND FORM Bright-yellow, amorphous powder; yellow tetragonal.

ODOR Odorless.

TASTE Tasteless.

DENSITY 7.70.

MELTING POINT 290°C.

FIREFIGHTING PROCEDURES If mercurous iodide involved in fire: Extinguish fire using agent suitable for type of surrounding fire (mercurous iodide itself does not burn or burns with difficulty). Use water in flooding quantities as fog. Use foam, carbon dioxide, or dry chemical. If mercurous iodide not involved in fire: Keep mercurous iodide out of water sources and sewers [R2].

SOLUBILITY Insoluble in alcohol, ether; soluble in solutions of mercurous or mercuric nitrates; very slightly soluble in cold water; soluble in ammonium hydroxide and potassium iodide; soluble in castor oil, liquid ammonia, aqua ammonia.

STABILITY AND SHELF LIFE Darkens or becomes greenish on exposure to light, mercuric iodide (HgI₂), and metallic mercury being formed [R1].

OTHER CHEMICAL/PHYSICAL PROPERTIES
Partial decomposition to mercury (Hg) and mercury iodide (HgI₂) at 290°C when rapidly heated; cold ammonia, its solution, or alkali iodide, decomposes it into mercury and mercuric iodide.

PROTECTIVE EQUIPMENT Wear boots, protective gloves, and goggles. Do not handle broken packages without protective equipment. Wear positive-pressure self-contained breathing apparatus when fighting fires involving this material. If contact with the material anticipated, wear full chemical protective clothing [R2].

PREVENTATIVE MEASURES Adequate ventilation; use of disposable uniforms, so that a contaminated uniform is not a source of absorption through the skin: use disposable mercury-vapor-absorbing masks; careful attention to good housekeeping, e.g., avoidance of spills, and prompt and proper cleaning if a spill occurs; all containers of mercury and its compounds should be kept tightly closed; floors should be washed on a regular basis with dilute calcium sulfide solution or other suitable reactant; floors should be nonporous; all workers directly involved in the plant operation should shower thoroughly each day before leaving. (mercury compounds) [R3].

CLEANUP (1) Ventilate area of spill. (2) Collect spilled material for reclamation using commercially available mercury vapor depressants or specialized vacuum cleaners. (inorganic mercury) [R4].

DISPOSAL SRP: At the time of review, criteria for land treatment or burial (sanitary landfill) disposal practices are subject to significant revision. Prior to implementing land disposal of waste residue (including waste sludge), consult with environmental regulatory agencies for guidance on acceptable disposal practices.

COMMON USES Antibacterial agent [R1]; mercurous iodide has been used in ointment in eye diseases [R12].

OCCUPATIONAL RECOMMENDATIONS
OSHA During an 8-hr work shift, an employee may be exposed to a concentration of mercury above 1 mg/m³. (mercury) [R8].

NIOSH 10–hr time-weighted average 0.05 mg/m³ (mercury and inorganic compounds (as Hg)) [R9]; threshold-limit 8–hr time-weighted average (TWA) 0.025 mg/m³, skin (1994) (mercury, in organic forms including metallic mercury) [R14, 25].

HUMAN TOXICITY Lethal blood level: the concentration of inorganic mercury present in blood (serum or plasma) that has been reported to cause death in humans is: 0.04–2.2 mg%; 0.4–22 μg/mL (inorganic mercury) [R6].

NON-HUMAN TOXICITY Discoloration of cornea is produced experimentally in animals by repeated systemic administration of mercury. Grayish ring in cornea just anterior to endothelium extending approximately 2 mm from limbus (inorganic mercury) [R12]. In animals mercury is a potent nephrotoxin and produces nephrosis and eventually uremia. At necropsy the firm, shrunken kidneys are obvious (mercury compounds) [R13, 717].

IARC SUMMARY Evaluation: There is inadequate evidence in humans for the carcinogenicity of mercury and mercury compounds. Overall evaluation: Metallic mercury and inorganic mercury compounds are not classifiable as to their carcinogenicity to humans (group 3) [R5].

BIOLOGICAL HALF-LIFE In general, body burden of mercury in man has a half-life of about 60 days. (mercury) [R15, 1612].

BIODEGRADATION Inorganic forms of mercury (Hg) can be converted to organic forms by microbial action in the atmosphere. (inorganic mercury) [R7].

ATMOSPHERIC STANDARDS Emissions to the atmosphere from sludge incineration plants, sludge drying plants, or a combination of these sludge wastewater treatment plant processes shall not exceed 3,200 grams of mercury per 24–hour period. (total mercury) [R10].

RCRA REQUIREMENTS A solid waste containing mercurou iodide may become characterized as a hazardous waste when subjected to the toxicant extraction procedure listed in 40 CFR 261.24, and, if so characterized, must be managed as a hazardous waste [R11].

REFERENCES

1. *The Merck Index.* 10th ed. Rahway, NJ: Merck Co., Inc., 1983. 842.

2. Association of American Railroads. *Emergency Handling of Hazardous Materials in Surface Transportation.* Washington, DC: Assoc. of American Railroads, Hazardous Materials Systems (BOE), 1987. 437.

3. *Kirk–Othmer Encyclopedia of Chemical Technology.* 3rd ed., Volumes 1–26. New York, NY: John Wiley and Sons, 1978–1984, p. 15 (81) 167.

4. Mackison, F. W., R. S. Stricoff, and L. J. Partridge, Jr. (eds.). *NIOSH/OSHA—Occupational Health Guidelines for Chemical Hazards.* DHHS (NIOSH) Publication No. 81–123 (3 vols). Washington, DC: U.S. Government Printing Office, Jan. 1981. 4.

5. IARC. *Monographs on the Evaluation of the Carcinogenic Risk of Chemicals to Man.* Geneva: World Health Organization, International Agency for Research on Cancer, 1972–present (multivolume work), p. 58 324 (1993).

6. Winek, C. L. *Drug and Chemical Blood–Level Data 1985.* Pittsburgh, PA: Allied Fischer Scientific, 1985.

7. Schroeder, W.H. *Envir Sci Tech* 16 (7): 394A–400A (1982) as cited in *Environment Canada; Tech Info for Problem Spills: Mercury* (Draft), p. 41 (1982).

8. 29 CFR 1910.1000 (7/1/87).

9. NIOSH. *Pocket Guide to Chemical Hazards.* 5th Printing/Revision. DHHS (NIOSH) Publ. No. 85–114. Washington, DC: U.S. Dept. of Health and Human Services, NIOSH/Supt. of Documents, GPO, Sept. 1985. 152.

10. 40 CFR 61.52 (b) (7/1/87).

11. 40 CFR 261.24 (7/1/87).

12. Grant, W. M. *Toxicology of the Eye.* 2nd ed. Springfield, Illinois: Charles C. Thomas, 1974. 653.

13. Casarett, L. J., and J. Doull. *Toxicology: The Basic Science of Poisons.* New York: Macmillan Publishing Co., 1975.

14. American Conference of Governmental Industrial Hygienists. *Threshold Limit Values for Chemical Substances and Physical Agents and Biological Exposure Indices for 1994–1995.* Cincinnati, OH: ACGIH, 1994.

15. Gilman, A.G., L.S. Goodman, and A. Gilman (eds.). *Goodman and Gilman's The Pharmacological Basis of Therapeutics.* 7th ed. New York: Macmillan Publishing Co., Inc., 1985.

■ MERCUROUS OXIDE

CAS # 15829–53–5

SYNONYMS mercurous oxide, black, solid (DOT) • mercury oxide (Hg₂O) • mercury(I) oxide

MF Hg_2O

MW 417.18

COLOR AND FORM Black or brownish-black powder.

DENSITY 9.8.

SURFACE TENSION 75.6 dynes/m² .

FIREFIGHTING PROCEDURES If material involved in fire: Extinguish fire using agent suitable for type of surrounding fire (material itself does not burn or burns with great difficulty). Use water in flooding quantities as fog. Use foam, dry chemical, or carbon dioxide [R1].

SOLUBILITY Insoluble in water; soluble in nitric acid.

OTHER CHEMICAL/PHYSICAL PROPERTIES Hydrochloric acid converts it into calomel.

PROTECTIVE EQUIPMENT Bureau of Mines-approved airline respirator; impervious suit; appropriate eye protection [R2].

PREVENTATIVE MEASURES Use disposable uniforms, so that a contaminated uniform is not a source of absorption through the skin. Preventative measure: adequate ventilation; careful attention to good housekeeping, e.g., avoidance of spills, and prompt and proper cleaning if a spill occurs; all containers of mercury and its compounds should be kept tightly closed; floors should be washed on a regular basis with dilute calcium sulfide solution or other suitable reactant; floors should be nonporous; all workers directly involved in the plant operation should shower thoroughly each day before leaving. (mercury compounds) [R3].

CLEANUP Mercury removal from waste water can be accomplished by the BMS process: Chlorine is added to the waste water, oxidizing any mercury present to the ionic state. The BMS adsorbent (an activated carbon concentrate with sulfur compound on its surface) is used to collect ionic mercury. The spent adsorbent is then distilled to recover the mercury, leaving a carbon residue for reuse or disposal. The TMR IMAC process may also be used: Waste water is fed into a reactor, whereby a slight excess of chlorine is maintained, oxidizing any mercury present to ionic mercury. The liquid is then passed through the TMR IMAC ion-exchange resin where mercury ions are adsorbed. The mercury is then stripped from the spent resin with hydrochloric acid solution. (mercury compounds) [R4].

DISPOSAL SRP: At the time of review, criteria for land treatment or burial (sanitary landfill) disposal practices are subject to significant revision. Prior to implementing land disposal of waste residue (including waste sludge), consult with environmental regulatory agencies for guidance on acceptable disposal practices.

OCCUPATIONAL RECOMMENDATIONS .

OSHA During an 8-hr work shift, an employee may not be exposed to a concentration of mercury above 1 mg/m³. (mercury) [R7].

NIOSH 10–hr time-weighted average 0.05 mg/m³ (mercury and inorganic compounds (as Hg)) [R8]; threshold-limit 8–hr time-weighted average (TWA) 0.025 mg/m³,

skin (1994) (mercury, in organic forms including metallic mercury) [R11, 25].

IARC SUMMARY Evaluation: There is inadequate evidence in humans for the carcinogenicity of mercury and mercury compounds. Overall evaluation: Metallic mercury and inorganic mercury compounds are not classifiable as to their carcinogenicity to humans (group 3) [R5].

BIODEGRADATION Inorganic forms of Hg can be converted to organic forms by microbial action in the biosphere. (inorganic mercury) [R6].

ATMOSPHERIC STANDARDS Emissions to the atmosphere from sludge incineration plants, sludge drying plants, or a combination of these sludge wastewater treatment plant processes shall not exceed 3,200 grams of mercury per 24–hour period. (total mercury) [R9].

RCRA REQUIREMENTS A solid waste containing mercurous oxide may become characterized as a hazardous waste when subjected to the toxicant extraction procedure listed in 40 CFR 261.24, and, if so characterized, must be managed as a hazardous waste [R10].

REFERENCES

1. Association of American Railroads. *Emergency Handling of Hazardous Materials in Surface Transportation.* Washington, DC: Assoc. of American Railroads, Hazardous Materials Systems (BOE), 1987. 438.

2. U.S. Coast Guard, Department of Transportation. *CHRIS—Hazardous Chemical Data. Volume II.* Washington, DC: U.S. Government Printing Office, 1984–1985.

3. *Kirk–Othmer Encyclopedia of Chemical Technology.* 3rd ed., Volumes 1–26. New York, NY: John Wiley and Sons, 1978–1984, p. 15 (81) 167.

4. Environment Canada. *Tech Info for Problem Spills: Mercury* (Draft), p. 59 (1982).

5. IARC. *Monographs on the Evaluation of the Carcinogenic Risk of Chemicals to Man.* Geneva: World Health Organization, International Agency for Research

on Cancer, 1972–present (multivolume work), p. 58 324 (1993).

6. Schroeder, W.H. *Envir Sci Tech* 16 (7): 394A–400A (1982) as cited in Environment Canada, *Tech Info for Problem Spills: Mercury* (Draft), p. 41 (1982).

7. 29 CFR 1910.1000 (7/1/87).

8. NIOSH. *Pocket Guide to Chemical Hazards.* 5th Printing/Revision. DHHS (NIOSH) Publ. No. 85–114. Washington, DC: U.S. Dept. of Health and Human Services, NIOSH/Supt. of Documents, GPO, Sept. 1985. 152.

9. 40 CFR 61.52 (b) (7/1/87).

10. 40 CFR 261.24 (7/1/87).

11. American Conference of Governmental Industrial Hygienists. *Threshold Limit Values for Chemical Substances and Physical Agents and Biological Exposure Indices for 1994–1995.* Cincinnati, OH: ACGIH, 1994.

■ MERCURY OXIDE SULFATE

CAS # 1312–03–4

SYNONYMS basic mercuric sulfate • mercuric basic sulfate • mercuric subsulfate • mercury oxide sulfate [$Hg_3O_2(SO_4)$] • mercury oxonium sulfate • mercury sulfate, basic • turpeth mineral

MF Hg_3O_6S

MW 729.83

COLOR AND FORM Lemon-yellow powder; bright-yellow scales.

ODOR Odorless.

DENSITY 6.44.

EXPLOSIVE LIMITS AND POTENTIAL Acetylides form in ammoniacal solution of mercury salts and acetylene. The dried acetylides are extremely sensitive and violent. (mercury salts) [R4].

FIREFIGHTING PROCEDURES If mercury is involved in fire: extinguish fire using agent suitable for type of surrounding fire (mercury oxide itself does not burn or burns with difficulty). Use water in flooding quantities as fog. Use foam, carbon dioxide, or dry chemical (mercury compounds, solid, not otherwise specified) [R3]. Wear self-contained breathing apparatus when fighting fires involving mercury. (mercury compounds, solid, not otherwise specified) [R3].

SOLUBILITY Solubility in water at 16°C: 0.003 g/100 cC; slightly soluble in hot water; insoluble in alcohol; soluble in acids.

OTHER CHEMICAL/PHYSICAL PROPERTIES Mercury salts, when heated with sodium carbonate, yield metallic mercury, and are reduced to metal by hydrogen peroxide in the presence of alkali hydroxide. Copper, iron, zinc, and many other metals precipitate metallic mercury from neutral or slightly acid solution of mercury salts. (mercury salts).

PROTECTIVE EQUIPMENT Personnel protection: Wear boots, protective gloves, and goggles. Do not handle broken packages without protective equipment. If contact with the material anticipated, wear full protective clothing. (mercury compounds, solid, not otherwise specified) [R3].

PREVENTATIVE MEASURES Preventive measure: adequate ventilation; use of disposable uniforms, so that a contaminated uniform is not a source of absorption through the skin; use of disposable mercury-vapor-absorbing masks; careful attention to good housekeeping, e.g., avoidance of spills, and prompt and proper cleaning if a spill occurs; all containers of mercury and its compounds should be kept tightly closed; floors should be washed on a regular basis with dilute calcium sulfide solution or other suitable reactant; floors should be nonporous; all workers directly involved in the plant operation should shower thoroughly each day before leaving. (mercury compound) [R5].

CLEANUP Spilled mercury compounds or solutions can be cleaned up by any method that does not cause excessive airborne contamination or skin contact. (mercury compounds) [R6].

DISPOSAL SRP: At the time of review, criteria for land treatment or burial (sanitary landfill) disposal practices are subject to significant revision. Prior to implementing land disposal of waste residue (including waste sludge), consult with environmental regulatory agencies for guidance on acceptable disposal practices.

COMMON USES Medication [R1].

OCCUPATIONAL RECOMMENDATIONS

OSHA During an 8–hr work shift, an employee may not be exposed to a concentration of mercury above 1 mg/m³. (mercury) [R11].

NIOSH 10–hr time-weighted average 0.05 mg/m³ (mercury and inorganic compounds (as Hg)) [R12].

THRESHOLD LIMIT 8–hr time-weighted average (TWA) 0.025 mg/m³, skin (1994) (mercury, in organic forms including metallic mercury) [R19, 25].

HUMAN TOXICITY Lethal blood level: The concentration of inorganic mercury present in the blood (serum or plasma) that has been reported to cause death in humans is: 0.04–2.2 mg%; 0.4–22 µg/ml. (inorganic mercury) [R8].

IARC SUMMARY Evaluation: There is inadequate evidence in humans for the carcinogenicity of mercury and mercury compounds. Overall evaluation: Metallic mercury and inorganic mercury compounds are not classifiable as to their carcinogenicity to humans (Group 3) [R7].

BIOLOGICAL HALF-LIFE The average biological half-time of a tracer dose of divalent inorganic mercury compound in man is 42 days for the whole-body and 26 days for blood. (mercury compound) [R9].

BIODEGRADATION Inorganic forms of Hg can be converted to organic forms by microbial action in the biosphere. (inorganic mercury) [R10].

BIOCONCENTRATION Mercuric salts are still widely employed in industry, and industrial discharge into rivers has polluted many parts of the world. Microorganisms convert inorg mercury to methyl mercury, which is taken up rapidly by planktonic algae and is concentrated in fish by way of food chain. (mercuric salts) [R20, 1611].

ATMOSPHERIC STANDARDS Emissions to the atmosphere from sludge incineration plants, sludge drying plants, or a combination of these sludge wastewater treatment plant processes shall not exceed 3,200 grams of mercury per 24–hour period. (total mercury) [R13].

RCRA REQUIREMENTS A solid waste containing mercury oxide sulfate may become characterized as a hazardous waste when subjected to the toxicant extraction procedure listed in 40 CFR 261.24, and, if so characterized, must be managed as a hazardous waste [R14].

DOT *Health Hazards:* Poisonous if swallowed. Inhalation of dust poisonous. Fire may produce irritating or poisonous gases. Runoff from fire control or dilution water may cause pollution (mercury compounds, liquid or solid) [R2].

Fire or Explosion: Some of these materials may burn, but none of them ignites readily (mercury compounds, liquid, or solid) [R2].

Emergency Action: Keep unnecessary people away; isolate hazard area and deny entry. Stay upwind; keep out of low areas. Self-contained breathing apparatus (SCBA) and structural firefighter's protective clothing will provide limited protection. Call CHEMTREC at 1–800–424–9300 for emergency assistance. If water pollution occurs, notify the appropriate authorities (mercury compounds, liquid, or solid) [R2].

Fire: Small fires: Dry chemical, CO₂, halon, water spray, or standard foam. Large Fires: Water spray, fog, or standard foam is recommended. Move container from fire area if you can do so without risk (mercury compounds, liquid or solid) [R2].

Spill or Leak: Do not touch spilled material; stop leak if you can do so without risk. Small Spills: Take up with sand or other noncombustible absorbent material and place into containers for later disposal.

Small Dry Spills: With clean shovel place material into clean, dry container, and cover; move container from spill area. Large Spills: Dike far ahead of liquid spill for later disposal (mercury compounds, liquid, or solid) [R2].

First Aid: Move victim to fresh air; call emergency medical care. Remove and isolate contaminated clothing and shoes at the site. In case of contact with material, immediately flush skin or eyes with running water for at least 15 minutes. (mercury compounds, liquid or solid) [R2].

FDA Bottled water shall, when a composite of analytical units of equal volume from a sample is examined by the methods described in paragraph (d) (1) (II) of this section, meet the standards of chemical quality, and shall not contain mercury in excess of 0.002 mg/L (total mercury) [R15]. The color additive F, D, and C Blue Number 2 shall conform to the specifications in the CFR 74.102 and be free from impurities other than those named, including mercury (as Hg) in not more than 1 part per million, to the extent that such other impurities may be avoided by current good manufacturing practice (total mercury) [R16]. The color additive F, D, and C Green Number 3 shall conform to the specifications in the CFR 74.203 and be free from impurities other than those named, including mercury (as Hg) in not more than 1 part per million, to the extent that such other impurities may be avoided by current good manufacturing practice (total mercury) [R17]. The color additive F, D, and C Yellow Number 5 shall conform to the specifications in the CFR 74.705 and be free from impurities other than those named, including mercury (as Hg) in not more than 1 part per million, to the extent that such other impurities may be avoided by current good manufacturing practice. (total mercury) [R18].

REFERENCES

1. Hawley, G. G. *The Condensed Chemical Dictionary.* 9th ed. New York: Van Nostrand Reinhold Co., 1977. 548.

2. Department of Transportation. *Emergency Response Guidebook 1987.* DOT P 5800.4. Washington, DC: U.S. Government Printing Office, 1987, p. G–53.

3. Association of American Railroads. *Emergency Handling of Hazardous Materials in Surface Transportation.* Washington, DC: Assoc. of American Railroads, Hazardous Materials Systems (BOE), 1987. 440.

4. National Fire Protection Association. *Fire Protection Guide on Hazardous Materials.* 7th ed. Boston, MA: National Fire Protection Association, 1978, p. 491M–253.

5. *Kirk–Othmer Encyclopedia of Chemical Technology.* 3rd ed., Volumes 1–26. New York, NY: John Wiley and Sons, 1978–1984, p. 15 (81) 1167.

6. National Research Council. *Prudent Practices for Handling Hazardous Chemicals in Laboratories.* Washington, DC: National Academy Press, 1981. 53.

7. IARC. *Monographs on the Evaluation of the Carcinogenic Risk of Chemicals to Man.* Geneva: World Health Organization, International Agency for Research on Cancer, 1972–present (multivolume work), p. 58 324 (1993).

8. Winek, C. L. *Drug and Chemical Blood–Level Data 1985.* Pittsburgh, PA: Allied Fischer Scientific, 1985.

9. USEPA. *Mercury Health Effects Update.* p. 2–5 (1984) EPA 600/8–84–019F.

10. Schroeder, W. H. *Envir Sci Tech* 16 (7): 394A–400A (1982) as cited in *Environment Canada: Tech Info for Problem Spills: Mercury* (Draft), p. 41 (1982).

11. 29 CFR 1910.1000 (7/1/87).

12. NIOSH. *Pocket Guide to Chemical Hazards.* 5th Printing/Revision. DHHS (NIOSH) Publ. No. 85–114. Washington, DC: U.S. Dept. of Health and Human Services, NIOSH/Supt. of Documents, GPO, Sept. 1985. 152.

13. 40 CFR 61.52 (b) (7/1/87).

14. 40 CFR 261.24 (7/1/87).

15. 21 CFR 103.35 (4/1/88).

16. 21 CFR 74.102 (4/1/88).

17. 21 CFR 74.203 (4/1/88).

18. 21 CFR 74.705 (4/1/88).

19. American Conference of Governmental Industrial Hygienists. *Threshold Limit Values for Chemical Substances and Physical Agents and Biological Exposure Indices for 1994–1995.* Cincinnati, OH: ACGIH, 1994.

20. Gilman, A.G., L.S. Goodman, and A. Gilman (eds.). *Goodman and Gilman's The Pharmacological Basis of Therapeutics.* 7th ed. New York: Macmillan Publishing Co., Inc., 1985.

■ MERCURY(1+), METHYL-, ION

CAS # 22967–92–6

SYNONYMS mercury(1+), methyl • mercury(1+), methyl-, ion • methylmercury ion • methylmercury ion (1+) • methylmercury(1+) • methylmercury(I)-cation

MF CH_3Hg

MW 215.63

OTHER CHEMICAL/PHYSICAL PROPERTIES In methylmercuric bromide, the covalent bonding of H_3CHg is more firm and stable than the ionic Hg–Br bond; CH_3Hg can exist free only in minute concentrations; the affinity of CH_3Hg for some ligands is high, particularly affinity to the thiol group.

PROTECTIVE EQUIPMENT Protective equipment: Wear appropriate clothing to prevent any possibility of skin contact. Wear eye protection to prevent any possibility of eye contact. Employees should wash immediately when skin is wet or contaminated. Work clothing should be changed daily if it is possible that clothing is contaminated. Remove non-impervious clothing immediately if wet or contaminated. Provide emergency showers and eyewash. (methylmercury compounds) [R3].

PREVENTATIVE MEASURES Keep material out of water sources and sewers. Keep sparks, flames, and other sources of ignition away. (mercury-based pesticides) [R4].

CLEANUP Experimental research on cleanup of wastewater containing methylmercury has been done using macroporous sulfhydryl resin [R5].

DISPOSAL SRP: At the time of review, criteria for land treatment or burial (sanitary landfill) disposal practices are subject to significant revision. Prior to implementing land disposal of waste residue (including waste sludge), consult with environmental regulatory agencies for guidance on acceptable disposal practices.

COMMON USES Formerly methylmercury compounds were extensively used in seed treatment as fungicides [R1]; used in treating seeds for fungi and seedborne diseases, as timber preservatives and disinfectants. (methylmercury compounds) [R2].

OCCUPATIONAL RECOMMENDATIONS
OSHA Meets criteria for OSHA medical records rule [R11]. Threshold-limit 8–hr time-weighted average (TWA) 0.01 mg/m^3, skin; short-term exposure limit (STEL) 0.03 mg/m^3, skin (1980). (mercury, alkyl compounds, as Hg) [R12, 24].

HUMAN TOXICITY Lethal blood level: the concentration of organic mercury present in blood (serum or plasma) that has been reported to cause death in humans is: >0.06 mg% (i.e., >0.6 µg/ml). (organic mercury) [R8].

IARC SUMMARY Evaluation: There is sufficient evidence in experimental animals for the carcinogenicity of methylmercury chloride. In making the overall evaluation, the Working Group took into account evidence that methylmercury compounds are similar with regard to absorption, distribution, metabolism, excretion, genotoxicity, and other forms of toxicity. Overall evaluation: methylmercury compounds are probably carcinogenic to humans (group 2B) [R6].

BIOLOGICAL HALF-LIFE Biologic half-life of CH_3Hg in blood is about 120 days. Relative to inorganic Hg the high half-life is attributable to reabsorption from intestines of any CH₃Hg that is excreted into the digestive tract with bile. [R7, 90].

BIODEGRADATION Growing bacterial cells can transform methylmercury to the volatile elemental mercury form, which is readily lost from the aquatic environment. However, living but non growing bacterial and algal cells cause the demethylation of methylmercury to inorganic mercury, and dead bacterial cells can lead to the methylation of inorganic mercury to methylmercury(I). Temperature, organic enrichment, and oxygen level also influence methylation activity. Low temperature has been found to limit methylation in sediment, although methylmercury production has been observed in river sediments at 4°C. Methylation activity is stimulated by the addition of organics and directly correlates with the organic content of the environment. Reports of the effects of oxygen level on mercury methylation are conflicting, although most investigators have observed higher methylation rates in anaerobic sediments than in aerobic sediments (2,3). Mercury freshly added and bound to river sediment was found to be available for methylation by microorganisms. Thus adsorption of inorganic mercury compounds may not render it biologically inactive (2). Observed methylation rates in anaerobic sediments with 0.1–50 mg/L Hg added, 0–1.9 µg Hg/day, volatile suspended solids; anaerobic sediments with 10,100 µg/g added, 1.9–9.7 mg Hg/day, volatile suspended solids; aerobic sediments with 10–100 µg/g added, 0–5.2 µg Hg/day, volatile suspended solids (3) [R9].

BIOCONCENTRATION Fish readily bioconcentrate methylmercury either directly through water or through components of the food chain. Factors that affect the observed levels of mercury in plants and animals at different trophic levels include age, surface area, metabolism, habitat, and activity (2). Loss appears to occur in two stages: first, methylmercury is distributed throughout the tissues over a period of a few weeks, and then is discharged very slowly from the established binding sites (1). The depurative half-life in fish varies between 1 and 3 years (2) [R10].

ATMOSPHERIC STANDARDS Emissions to the atmosphere from sludge incineration plants, sludge drying plants, or a combination of these sludge wastewater treatment plant processes shall not exceed 3,200 grams of mercury per 24–hour period. (total mercury) [R13].

RCRA REQUIREMENTS A solid waste containing methylmercury may become characterized as a hazardous waste when subjected to the toxicant extraction procedure listed in 40 CFR 261.24, and, if so characterized, must be managed as a hazardous waste [R14].

FIFRA REQUIREMENTS All uses of mercury are cancelled except the following: (1) as a fungicide in the treatment of textiles and fabrics intended for continuous outdoor use; (2) as a fungicide to control brown mold on freshly sawn lumber; (3) as a fungicide treatment to control Dutch elm disease; (4) as an in-can preservative in water-based paints and coatings; (5) as a fungicide in water-based paints and coatings used for exterior application; (6) as a fungicide to control "winter turf diseases" such as *Sclerotinia boreales*, and gray and pink snow mold subject to the following: a. the use of these products were prohibited within 25 feet of any water body where fish are taken for human consumption; b. these products can be applied only by or under the direct supervision of golf course superintendents; c. the products will be classified as restricted-use pesticides when they are reregistered and classified in accordance with section 4 (c) of FEPCA. (mercury) [R15].

REFERENCES

1. Bretherick, L. *Handbook of Reactive Chemical Hazards.* 3rd ed. Boston, MA: Butterworths, 1985. 393.

2. Sittig, M. *Handbook of Toxic and Hazardous Chemicals,* p. 420 (1981).

3. Sittig, M. *Handbook of Toxic and Hazardous Chemicals,* p. 422 (1981).

4. Association of American Railroads. *Emergency Handling of Hazardous Materials in Surface Transportation.* Washington, DC: Assoc. of American Rail-

roads, Hazardous Materials Systems (BOE), 1987. 440.

5. Yu, M., et al. *Huan Ching Ko Hsueh* 1 (4): 10 (1980).

6. IARC. *Monographs on the Evaluation of the Carcinogenic Risk of Chemicals to Man.* Geneva: World Health Organization, International Agency for Research on Cancer, 1972–present (multivolume work), p. 58 324 (1993).

7. Venugopal, B., and T. D. Luckey. *Metal Toxicity in Mammals,* 2. New York: Plenum Press, 1978.

8. Winek, C. L. *Drug and Chemical Blood-Level Data 1985.* Pittsburgh, PA: Allied Fischer Scientific, 1985.

9. (1) Ramamoorthy, S., et al. *Bull Environ Contam Tox* 29: 167–73 (1982); (2) Callister, S. M., M. R. Winfrey. *Water, Air, and Soil Poll* 29: 453–65 (1986); (3) Bisogni, J. J. *The Biogeochemistry of Mercury in the Environment*; Nriagu, J.O. (ed.), New York: Elsevier, pp. 221–230 (1974).

10. (1) Jenson, S., A. Jernelov. *Nature* 223: 753–4 (1969); (2)Callahan, M. A., et-al. *Water-Related Environmental Fate of 129 Priority Pollutants* Vol. 1 pp. 14–1 to 14–11 USEPA-44/4-79-029a (1979).

11. 29 CFR 1910.20 (7/1/87).

12. American Conference of Governmental Industrial Hygienists. *Threshold Limit Values for Chemical Substances and Physical Agents and Biological Exposure Indices for 1993-1994.* Cincinnati, OH: ACGIH, 1993.

13. 40 CFR 61.52 (b) (7/1/87).

14. 40 CFR 261.24 (7/1/87).

15. Environmental Protection Agency/OPTS. *Suspended, Cancelled, and Restricted Pesticides.* 3rd Revision. Washington, DC: Environmental Protection Agency, January 1985. 16.

■ MERCURY(2+) NTA

CAS # 53113-61-4

SYNONYMS mercuric nitrilotriacetic acid. mercury (2+) nitrilotriacetic acid

MF $C_6H_6HgNO_6 \cdot H$

MW 254.50

PREVENTATIVE MEASURES Use disposable uniforms, so that a contaminated uniform is not a source of absorption through the skin. Preventative measure: adequate ventilation; careful attention to good housekeeping, e.g., avoidance of spills, and prompt and proper cleaning if a spill occurs; all containers of mercury and its compounds should be kept tightly closed; floors should be washed on a regular basis with dilute calcium sulfide solution or other suitable reactant; floors should be nonporous; all workers directly involved in the plant operation should shower thoroughly each day before leaving. (mercury compounds) [R1].

CLEANUP Mercury removal from waste water can be accomplished by the BMS process: Chlorine is added to the waste water, oxidizing any mercury present to the ionic state. The BMS adsorbent (an activated carbon concentrate with sulfur compound on its surface) is used to collect ionic mercury. The spent adsorbent is then distilled to recover the mercury, leaving a carbon residue for reuse or disposal. The TMR IMAC process may also be used: Waste water is fed into a reactor, whereby a slight excess of chlorine is maintained, oxidizing any mercury present to ionic mercury. The liquid is then passed through the TMR IMAC ion-exchange resin where mercury ions are adsorbed. The mercury is then stripped from the spent resin with hydrochloric acid solution. (mercury compounds) [R2].

DISPOSAL SRP: At the time of review, criteria for land treatment or burial (sanitary landfill) disposal practices are subject to significant revision. Prior to implementing land disposal of waste residue (including waste sludge), consult with environmental regulatory agencies for guidance on acceptable disposal practices.

OCCUPATIONAL RECOMMENDATIONS

OSHA During an 8-hr work shift, an employee may be exposed to a concentration of mercury above 1 mg/m³ (mercury) [R6]. NIOSH: 10–hr time-weighted average 0.05 mg/m³. (mercury and inorganic compounds (as Hg)) [R7].

THRESHOLD LIMIT 8–hr time-weighted average (TWA) 0.1 mg/m³, skin (1982) (mercury aryl compounds, as Hg) [R10, 24]. ASTD: Emissions to the atmosphere from sludge incineration plants, sludge drying plants, or a combination of these sludge process wastewater treatment plant shall not exceed 3,200 grams of mercury per 24 hr period. (total mercury) [R11].

EMERGENCY ACTION Keep unnecessary people away; isolate hazard area and deny entry. Stay upwind; keep out of low areas. Self-contained breathing apparatus (SCBA) and structural firefighter's protective clothing will provide limited protection. Call CHEMTREC at 1–800–424–9300 for emergency assistance. If water pollution occurs, notify the appropriate authorities (mercury compounds, not otherwise specified) [R12]. *Fire:* Small fires: Dry chemical, CO₂, halon, water spray or standard foam. Large Fires: Water spray, fog, or standard foam is recommended. Move container from fire area if you can do it without risk (mercury compound, not otherwise specified) [R12]. Spill or Leak: Do not touch spilled material; stop leak if you can do it without risk. Small Spills: Take up with sand or other noncombustible absorbent material and place into containers for later disposal. Small Dry Spills: With clean shovel place material into clean, dry container and cover; move container from spill area. Large Spills: Dike far ahead of liquid spill for later disposal. (mercury compounds, not otherwise specified) [R12].

FIRST AID Move victim to fresh air; call for emergency medical care. Remove and isolate contaminated clothing and shoes at the site. In case of contact with material, immediately flush skin or eyes with running water for at least 15 minutes. (mercury compounds, not otherwise specified) [R1].

HUMAN TOXICITY Lethal blood level: The concentration of inorganic mercury present in the blood (serum or plasma) that

has been reported to cause death in humans is: 0.04–2.2 mg%; 0.4–22 µg/ml. (inorganic mercury) [R4].

IARC SUMMARY Evaluation: There is inadequate evidence in humans for the carcinogenicity of mercury and mercury compounds. Overall evaluation: Metallic mercury and inorganic mercury compounds are not classifiable as to their carcinogenicity to humans (group 3) [R3].

BIOLOGICAL HALF-LIFE In general, body burden of mercury in man has a half-life of about 60 days. (mercury) [R13, 1612].

BIODEGRADATION In general, NTA complexes of mercury can be expected to degrade much more slowly than those of nickel, zinc, and iron [R5].

ATMOSPHERIC STANDARDS Emissions to the atmosphere from sludge incineration plants, sludge drying plants, or a combination of these sludge-process wastewater treatment plants shall not exceed 3,200 grams of mercury per 24–hr period. (total mercury) [R8].

RCRA REQUIREMENTS A solid waste containing mercury(2) NTA may become characterized as a hazardous waste when subjected to the toxicant extraction procedure listed in 40 CFR 261.24, and, if so characterized, must be managed as a hazardous waste [R9].

REFERENCES

1. *Kirk–Othmer Encyclopedia of Chemical Technology.* 3rd ed., Volumes 1–26. New York, NY: John Wiley and Sons, 1978–1984, p. 15 (81) 167.

2. Environment Canada. *Tech Info for Problem Spills: Mercury (Draft)*, p. 59 (1982).

3. IARC. *Monographs on the Evaluation of the Carcinogenic Risk of Chemicals to Man.* Geneva: World Health Organization, International Agency for Research on Cancer, 1972–present (multivolume work), p. 58 324 (1993).

4. Winek, C. L. *Drug and Chemical Blood–Level Data 1985.* Pittsburgh, PA: Allied Fischer Scientific, 1985.

5. Nat'l Research Council Canada. *NTA*

(*Nitrilotriacetic Acid*) p. 13 (1976) NRCC No. 15023.

6. 29 CFR 1910.1000 (7/1/87).

7. NIOSH. *Pocket Guide to Chemical Hazards.* 5th Printing/Revision. DHHS (NIOSH) Publ. No. 85–114. Washington, DC: U.S. Dept. of Health and Human Services, NIOSH/Supt. of Documents, GPO, Sept. 1985. 152.

8. 40 CFR 61.52 (b) (7/1/87).

9. 40 CFR 261.24 (7/1/87).

10. American Conference of Governmental Industrial Hygienists. *Threshold Limit Values for Chemical Substances and Physical Agents and Biological Exposure Indices for 1994–1995.* Cincinnati, OH: ACGIH, 1994.

11. 40 CFR 61.52(b) (7/1/87).

12. Department of Transportation. *Emergency Response Guidebook 1987.* DOT P 5800.4. Washington, DC: U.S. Government Printing Office, 1987., p. G-53.

13. Gilman, A.G., L.S. Goodman, and A. Gilman (eds.). *Goodman and Gilman's The Pharmacological Basis of Therapeutics.* 7th ed. New York: Macmillan Publishing Co., Inc., 1985.

■ MERCURY(I) NITRATE (1:1)

CAS # 10415–75–5

DOT # 1627

SIC CODE 2842.

SYNONYMS mercury protonitrate • mercury(I) nitrate • nitrate mercureux (French) • nitric acid, mercury(1) salt • nitric acid, mercury(I) salt • mercury(1) nitrate

MF $NO_3 \cdot Hg$

MW 262.60

COLOR AND FORM Solid, white, monoclinic, colorless crystalline material.

ODOR Slight odor of nitric acid.

DENSITY 4.78 at 25°C.

MELTING POINT 70°C.

EXPLOSIVE LIMITS AND POTENTIAL Mixture of mercurous nitrate and phosphorus explodes violently when struck with hammer. [R6, p. 491M–161].

FIREFIGHTING PROCEDURES Flood the fire with water. Cool all affected containers with flooding quanities of water. Apply water from as great a distance as possible. [R5, 438].

TOXIC COMBUSTION PRODUCTS Smoke from fire may contain toxic mercury and oxides of nitrogen [R2].

SOLUBILITY Water-soluble.

CORROSIVITY Solution may corrode metals.

STABILITY AND SHELF LIFE Sensitive to light (mercurous nitrate hydrate) [R1].

OTHER CHEMICAL/PHYSICAL PROPERTIES MP: about 70°C with decomposition; density: 4.78; with water alone a basic salt is formed; blackened by ammonia, caustic alkali, and alkaline earth solutions (mercurous nitrate dihydrate).

PROTECTIVE EQUIPMENT In areas where the exposures are excessive, respiratory protection was provided either by full-face canister-type mask or supplied-air respirator, depending on the concentration of mercury fumes. Above 50 mg Hg/m^3 requires supplied-air positive-pressure full-face respirators. Full-body work clothes including shoes or shoe covers and hats should be supplied and clean work clothes should be supplied daily. Work clohtes should not be stored with street clothes in the same locker. (mercury, inorganic) [R7, 571].

PREVENTATIVE MEASURES Use disposable uniforms, so that a contaminated uniform is not a source of absorption through the skin: use disposable mercury vapor-absorbing masks; preventative measure: adequate ventilation; careful attention to good housekeeping, e.g., avoidance of spills, and prompt and proper cleaning if a spill occurs; all containers of mercury and its compounds should be kept tightly closed; floors should be washed on a regular basis with dilute calcium sulfide solution or other suitable reactant; floors should be nonporous; all workers directly involved in the plant operation should shower thoroughly each day before leaving. (mercury compounds) [R8].

CLEANUP Mercury removal from wastewater can be accomplished by the BMS process: Chlorine is added to the waste water, oxidizing any mercury present to the ionic state. The BMS adsorbent (an activated carbon concentrate with sulfur compound on its surface) is used to collect ionic mercury. The spent adsorbent is then distilled to recover the mercury, leaving a carbon residue for reuse or disposal. The TMR IMAC process may also be used: Waste water is fed into a reactor, whereby a slight excess of chlorine is maintained, oxidizing any mercury present to ionic mercury. The liquid is then passed through the TMR IMAC ion-exchange resin, where mercury ions are adsorbed. The mercury is then stripped from the spent resin with hydrochloric acid solution. (mercury compounds) [R9].

COMMON USES Fire gilding, blackening brass [R3]; reagent [R4].

BINARY REACTANTS Alkali; sulfides; ammonia; when heated in the presence of alcohol, forms explosive mercury fulminate.

STANDARD CODES EPA 311; TSCA; Coast Guard Classification poison B, poison label; AAR, Bureau of Explosives STCC 4918752; IATA poison class B, poison label 25-kg passenger, 95-kg cargo, CFR 49, oxidizer, oxidizer label, 50-lb passenger, 100-lb cargo, storage code 1, 2; CFR CAB code 8; not listed ICC; Superfund designated (hazardous substances) list.

PERSISTENCE Persistent. The major portion of the mercury will ultimately reside in the bottom sediments where, through microbial action, mono– and dimethylmercury can be formed.

MAJOR SPECIES THREATENED Fish and fish food organisms.

INHALATION LIMIT 0.01 mg/m^3.

DIRECT CONTACT May be absorbed through the skin. Will irritate skin and eyes on contact.

GENERAL SENSATION Metallic taste, diarrhea, difficult breathing, coughing, eye irritation, dizziness, clumsiness, slurred speech, irritation of mucous membranes, mental deterioration, tremors.

PERSONAL SAFETY Protect against both inhalation and absorption through the skin. NIOSH-approved self-contained breathing apparatus must be worn. Rubber gloves, hooded rubber suits, and rubber boots must be worn.

ACUTE HAZARD LEVEL Ingestion of inorganic mercury compounds can lead to acute systemic poisoning, and may be fatal in a few minutes. Acute poisoning has resulted from inhaling dust concentrations of 1.2–8.5 mg/m^3. The lowest published toxic oral dose for human toxicity is 10 mg/kg of mercury, with the toxic effect on the gastrointestinal tract. A fatal dose of mercury salts is 20 mg to 3 g, and an ingestion of 75 mg/day in drinking water is lethal [R19]. The oral LD$_{50}$ for rats is 297 mg/kg and the 96-hour LC$_{50}$ for fathead minnows is 0.2 ppm.

CHRONIC HAZARD LEVEL Mental and nervous symptoms are common to chronic mercury poisoning. The lowest published toxic concentration to the central nervous system is 0.169 mg/m^3. The threshold-limit-value-time-weighted average for a 40-hour work week is 0.1 mg/m^3, and the short-term exposure limit for 15 minutes is 0.15 mg/m^3. The EPA criterion for mercury in the domestic water supply is 2 µg/L; freshwater aquatic life 0.05 µg/L; and for marine aquatic life 0.10 µg/L . EPA has designated this material a category A, with an aquatic LC$_{50}$ value between 1 and 0.1 mg/L.

DEGREE OF HAZARD TO PUBLIC HEALTH Highly toxic through ingestion, skin contact, or inhalation. Can produce permanent damage to brain and kidney.

OCCUPATIONAL RECOMMENDATIONS
OSHA During an 8–hr work shift, an employee may not be exposed to a concentration of mercury above 1 mg/m^3. (mercury) [R14].

NIOSH 10–hr time-weighted average 0.05 mg/m³ (mercury and inorganic compounds (as Hg)) [R15]; threshold-limit 8–hr time-weighted average (TWA) 0.1 mg/m³, skin (1982) (mercury, aryl, and inorganic compounds, as Hg) [R16, 24].

ACTION LEVEL Avoid contact and inhalation of the spilled cargo. Stay upwind. Notify local fire, air, and water authorities of accident. Evacuate all persons to a distance of at least 200 feet upwind and 1,000 feet downwind of the spill.

ON-SITE RESTORATION Dam stream to contain flow and to retard dissipation by water movement. Vacuum or dredge from bottom sediments and transport back to the supplier in metal drums. If mercury concentrations are above 0.05 µg/L, pump out water into tank trucks or tank cars for transport back to the supplier for reclamation or disposal. An alternative method, done under controlled conditions, is to transfer to suitable container, add sodium hydroxide (NaOH) to pH 7 to 8, add sodium sulfide (Na₂S), and let stand until no mercurous ion is present. For more details see ENVIREX Manual EPA 600/2–77–227. Seek professional environmental engineering assistance through EPA's Environmental Response Team (ERT), Edison, NJ, 24-hour no. 908–548–8730.

BEACH OR SHORE RESTORATION Scrape material from soil at low tide.

AVAILABILITY OF COUNTERMEASURE MATERIAL Vacuum—swimming pool suppliers; dredge—Army Corp of Engineers.

DISPOSAL After the material has been contained, remove it and the contaminated soil and place in impervious containers. If practical, transport material back to the supplier or chemical company to recover the heavy-metal content and for deactivation. If this is not practical or facilities are not available, the material should be encapsulated and buried in a specially designated chemical landfill. Not amenable to biological treatment at sewage treatment plant. SRP: At the time of review, criteria for land treatment or burial (sanitary landfill) disposal practices are subject to significant revision. Prior to implementing land disposal of waste residue (including waste sludge), consult with environmental regulatory agencies for guidance on acceptable disposal practices.

DISPOSAL NOTIFICATION Notify local and state health authorities, local solid waste disposal authorities, supplier, and shipper.

INDUSTRIAL FOULING POTENTIAL Fouling potential believed low.

MAJOR WATER USE THREATENED All downstream water users, monitor water concentrations.

PROBABLE LOCATION AND STATE Colorless crystals will sink and decompose into a basic salt, which can be removed from the bottom sediments.

WATER CHEMISTRY Decomposes to a white insoluble precipitate. Color in water, white to yellow.

FOOD-CHAIN CONCENTRATION POTENTIAL Positive. Many organisms can accumulate mercury from water; bioaccumulative up to 10,000-fold. Readily accumulated in shellfish.

HUMAN TOXICITY Lethal blood level: the concentration of inorganic mercury present in blood (serum or plasma) that has been reported to cause death in humans is: 0.04–2.2 mg%; 0.4–22 µg/mL. (inorganic mercury) [R12].

IARC SUMMARY Evaluation: there is inadequate evidence in humans for the carcinogenicity of mercury and mercury compounds. Overall evaluation: metallic mercury and inorganic mercury compounds are not classifiable as to their carcinogenicity to humans (group 3). [R10].

BIOLOGICAL HALF-LIFE In general, the body burden of mercury in man has a half-life of about 60 days. (inorganic mercury salts) [R11, 1613].

BIODEGRADATION Inorganic forms of mercury (Hg) can be converted to organic forms by microbial action in the biosphere. (inorganic mercury) [R13].

ATMOSPHERIC STANDARDS Emissions to the atmosphere from sludge incineration plants, sludge drying plants, or a combination of sludge wastewater treatment plant processes shall not exceed 3,200 grams of mercury per 24-hour period. (total mercury) [R17].

RCRA REQUIREMENTS A solid waste containing mercurous nitrate may become characterized as a hazardous waste when subjected to the toxicant extraction procedure listed in 40 CFR 261.24, and, if so characterized, must be managed as a hazardous waste [R18].

GENERAL RESPONSE Avoid contact with solid and dust. Keep people away. Wear dust respirator and rubber overclothing (including gloves). Stop discharge if possible. Isolate and remove discharged material. Notify local health and pollution control agencies.

FIRE RESPONSE Not flammable. Will increase the intensity of a fire. Poisonous gases may be produced when heated.

EXPOSURE RESPONSE Call for medical aid. Dust is poisonous if inhaled or if skin is exposed. If inhaled it will cause coughing or difficult breathing. If breathing has stopped, give artificial respiration. If breathing is difficult, give oxygen. Solid is poisonous if swallowed or if skin is exposed. Irritating to skin and eyes. If swallowed will cause nausea and vomiting. Remove contaminated clothing and shoes. Flush affected areas with plenty of water. If in eyes, hold eyelids open, and flush with plenty of water. if swallowed and victim is conscious, have victim drink water or milk, and have victim induce vomiting. If swallowed and victim is unconscious or having convulsions, do nothing except keep victim warm.

RESPONSE TO DISCHARGE Issue warning—poison, water contaminant, oxidizing material. Restrict access. Disperse and flush.

WATER RESPONSE Effect of low concentrations on aquatic life is unknown. May be dangerous if it enters water intakes. Notify local health and wildlife officials. Notify operators of nearby water intakes.

EXPOSURE SYMPTOMS Acute systemic poisoning may be fatal within a few minutes; death by uremic poisoning is usually delayed 5–12 days. Acute poisoning has resulted from inhaling dust concentrations of 1.2–8.5 mg/m^3 of air; symptoms include tightness and pain in chest, coughing, and difficulty in breathing. Ingestion causes necrosis, pain, vomiting, and severe purging. Contact with eyes causes ulceration of conjunctiva and cornea. Contact with skin causes irritation and possible dermatitis; systemic poisoning can occur by absorption through skin.

EXPOSURE TREATMENT Inhalation: remove victim to fresh air; get medical attention. Ingestion: give egg whites, milk, or activated charcoal; induce vomiting; consult physician. Eyes: flush with water for at least 15 min. Skin: flush with water.

REFERENCES

1. Sax, N. I., and R. J. Lewis, Sr. (eds.). *Hawley's Condensed Chemical Dictionary.* 11th ed. New York: Van Nostrand Reinhold Co., 1987. 745.

2. U.S. Coast Guard, Department of Transportation. *CHRIS—Hazardous Chemical Data.* Volume II. Washington, DC: U.S. Government Printing Office, 1984–5.

3. *The Merck Index.* 10th ed. Rahway, NJ: Merck Co., Inc., 1983. 842.

4. Thienes, C., and T. J. Haley. *Clinical Toxicology.* 5th ed. Philadelphia: Lea and Febiger, 1972. 435.

5. Association of American Railroads. *Emergency Handling of Hazardous Materials in Surface Transportation.* Washington, DC: Assoc. of American Railroads, Hazardous Materials Systems (BOE), 1987.

6. National Fire Protection Association. *Fire Protection Guide on Hazardous Materials.* 9th ed. Boston, MA: National Fire Protection Association, 1986.

7. Sittig, M. *Handbook of Toxic and Hazardous Chemicals and Carcinogens,* 1985. 2nd ed. Park Ridge, NJ: Noyes Data Corporation, 1985.

8. *Kirk-Othmer Encyclopedia of Chemical Technology.* 3rd ed., Volumes 1–26.

New York, NY: John Wiley and Sons, 1978–1984, p. 15 (81) 167.

9. Environment Canada. *Tech Info for Problem Spills: Mercury* (Draft), p. 59 (1982).

10. IARC. *Monographs on the Evaluation of the Carcinogenic Risk of Chemicals to Man.* Geneva: World Health Organization, International Agency for Research on Cancer, 1972–present (multivolume work), p. 58 324 (1993).

11. Gilman, A. G., L. S. Goodman, and A. Gilman (eds.). *Goodman and Gilman's The Pharmacological Basis of Therapeutics.* 7th ed. New York: Macmillan Publishing Co., Inc., 1985.

12. Winek, C. L. *Drug and Chemical Blood–Level Data 1985.* Pittsburgh, PA: Allied Fischer Scientific, 1985.

13. Schroeder, W. H. *Envir Sci Tech* 16 (7): 394A–400A (1982) as cited in Environment Canada, *Tech Info for Problem Spills: Inorganic Mercury* (Draft), p. 41 (1982).

14. 29 CFR 1910.1000 (7/1/87).

15. NIOSH. *Pocket Guide to Chemical Hazards.* 5th Printing/Revision. DHHS (NIOSH) Publ. No. 85–114. Washington, DC: U.S. Dept. of Health and Human Services, NIOSH/Supt. of Documents, GPO, Sept. 1985. 152.

16. American Conference of Governmental Industrial Hygienists. *Threshold Limit Values for Chemical Substances and Physical Agents and Biological Exposure Indices for 1993-1994.* Cincinnati, OH: ACGIH, 1993.

17. 40 CFR 61.52 (b) (7/1/87).

18. 40 CFR 261.24 (7/1/87).

19. Shelford, V. E., *An Experimental Study of the Effects of Gas Waste Upon Fishes with Special Reference to Stream Pollution,* Bull. 111. State Lab. Nat. Hist., 1917, 11:381–412.

20. Zitko, V. Polychlorinated Biphenyls Solubilized in Waste by Non-Ionic Surfactants for Studies of Toxicity to Aquatic Animals, *Bull. Environ. Contam. & Toxicol.,* 1970, 5(2):279–285.

21. Shelford, V. E., *An Experimental Study of the Effects of Gas Waste upon Fishes with Special Reference to Stream Pollution,* Bull. 111. State Lab. Nat. Hist., 1917, 11:381–412.

■ MERPHOS

CAS # 150–50–5

SYNONYMS butyl phosphorotrithioite [(bus) 3p] • chemagro b-1776 • deleaf-defoliant • easy off-d • folex • phosphorotrithious acid, s,s,s-tributyl ester • phosphorotrithious-acid,-tributyl-ester • s,s,s-tributyl phosphorotrithioite • s,s,s-tributyl trithiophosphite • tributyl phosphorotrithioite • tributyl trithiophosphite

MF $C_{12}H_{27}PS_3$

MW 298.54

COLOR AND FORM Pale-amber liquid.

ODOR Mild characteristic odor.

DENSITY 0.99–1.01 at 20°C.

BOILING POINT 115–134°C at 0.08 mm Hg.

SOLUBILITY Sparingly soluble in water; soluble in acetone, ethyl alcohol, benzene, hexane, kerosene, diesel oil, heavy aromatic naphthas, xylene, methylated naphthalene.

PROTECTIVE EQUIPMENT For protection during application wear hat, long-sleeved shirt, trousers. In addition, mixer/loaders must wear rubber or neoprene gloves [R5].

DISPOSAL SRP: At the time of review, criteria for land treatment or burial (sanitary landfill) disposal practices are subject to significant revision. Prior to implementing land disposal of waste residue (including waste sludge), consult with environmental regulatory agencies for guidance on acceptable disposal practices.

COMMON USES Merphos is used to defoliate cotton at 1.1 to 2.2 kg active ingredient/acre and can induce leaf abscission in some other plants, such as roses and hydrangeas [R2]. Folex is particularly suited for total defoliation of cotton preparatory to machine harvesting because the natural action of folex causes the leaves to drop in a relatively green state with fresh weight adequate to cause them to fall to the ground [R1]. Case: Merphos; Case No. 2385; pesticide type: fungicide, herbicide; case status: The pesticide is no longer an active ingredient in any registered pesticide products. Therefore, EPA is characterizing it as cancelled [R3].

OCCUPATIONAL RECOMMENDATIONS None listed.

NON-HUMAN TOXICITY LD_{50} rabbit percutaneous 5–10 g/kg [R6]; LD_{50} rat oral male 1,475 mg/kg [R7]; LD_{50} rat oral female 910 mg/kg [R7]; LD_{50} rat dermal male 690 mg/kg [R7]; LD_{50} rat dermal female 615 mg/kg [R7]; LD_{50} rabbit dermal 4,600 mg/kg [R7]; LD_{50} rat oral 1.3 g/kg [R6].

ECOTOXICITY LC_{50} *Salmo gairdneri* wt 0.6 g, 33 mg/L/96 hr at 12°C (95% confidence limit: 20–53). Static bioassay without aeration, pH 7.2–7.5, water hardness 40–50 mg/L as calcium carbonate, and alkalinity of 30–35 mg/L [R8].

BIOCONCENTRATION Using an estimated log K_{ow} of 7.67 (1), one would estimate a BCF of 40,000 for merphos using a recommended regression equation (2). This would indicate that merphos would bioconcentrate in aquatic organisms. However, merphos' rapid oxidation in water would suggest little tendency to bioconcentrate (SRC) [R9].

FIFRA REQUIREMENTS In 1988, Congress amended FIFRA to strengthen and accelerate EPA's reregistration program. The nine-year reregistration scheme mandated by "FIFRA 88" applies to each registered pesticide product containing an active ingredient initially registered before November 1, 1984. Pesticides for which EPA had not issued Registration Standards prior to the effective date of FIFRA '88 were divided into three lists based upon their potential for exposure and other factors, with List B being of highest concern and D of least. List: B; Case: Merphos; Case No.: 2385; Pesticide type:

Fungicide, Herbicide; Case Status: The pesticide is no longer an active ingredient in any registered pesticide products. Therefore, EPA is characterizing it as "cancelled." [R3].

FIRE POTENTIAL Slight, when exposed to heat or flame. Can react vigorously with oxidizing materials [R4].

REFERENCES

1. *Farm Chemicals Handbook* 1994. Willoughby, OH: Meister, 1994, p. C–166.

2. *Farm Chemicals Handbook* 88. Willoughby, OH: Meister Publishing Co., 1988. 819.

3. USEPA/OPP. *Status of Pesticides in Reregistration and Special Review* p. 143 (Mar. 1992) EPA 700–R–92–004.

4. Sax, N. I. *Dangerous Properties of Industrial Materials,.* 5th ed. New York: Van Nostrand Reinhold, 1979. 1042.

5. *Farm Chemicals Handbook* 1983. Willoughby, OH: Meister Publishing Co., 1983, p. C–108.

6. Gosselin, R. E., R. P. Smith, H. C. Hodge. *Clinical Toxicology of Commercial Products.* 5th ed. Baltimore: Williams and Wilkins, 1984, p. II–3302.

7. Hayes, Wayland J., Jr. *Pesticides Studied in Man.* Baltimore/London: Williams and Wilkins, 1982. 408.

8. U.S. Department of Interior, Fish and Wildlife Service. *Handbook of Acute Toxicity of Chemicals to Fish and Aquatic Invertebrates.* Resource Publication No. 137. Washington, DC: U.S. Government Printing Office, 1980. 84.

9. (1) Meylan, W. M., P. H. Howard. *J Pharm Sci* 84: 83–92 (1995). (2) Lyman, W. J., et al., *Handbook of Chemical Property Estimation Methods.* New York: McGraw-Hill, Chapt 5, Eqn 5–2 (1982).

■ METHACRYLIC ACID, ETHYL ESTER

CAS # 97–63–2

DOT # 2277

SYNONYMS 2-propenoic acid, 2-methyl-, ethyl ester • ethyl 2-methacrylate • ethyl 2-methyl-2-propenoate • ethyl alpha-methyl acrylate • ethyl methacrylate • rhoplex ac-33

MF $C_6H_{10}O_2$

MW 114.16

COLOR AND FORM Colorless, liquid.

ODOR Acrid odor.

DENSITY 0.9135 at 20°C.

BOILING POINT 117°C at 760 mm Hg.

MELTING POINT > −75°C.

HEAT OF COMBUSTION −7.040 cal/g (−294×10⁵ J/kg; 12.670 Btu/lb).

HEAT OF VAPORIZATION 96 cal/g (4.0×10⁵ J/kg; 170 Btu/lb).

EXPLOSIVE LIMITS AND POTENTIAL 1.8% to saturation [R14, 2293].

FIREFIGHTING PROCEDURES Carbon dioxide, dry chemical [R5]. Do not extinguish fire unless flow can be stopped. Use water in flooding quantities as fog; solid streams of water may spread fire. Cool all affected containers with flooding quantities of water. Apply water from as far a distance as possible. If fire becomes uncontrollable or container is exposed to direct flame, consider evacuation of 1/3-mile radius [R6].

SOLUBILITY Soluble in alcohol, ether; water solubility, 5,600 ppm at 20°C.

VAPOR PRESSURE 14 torr at 20°C.

OCTANOL/WATER PARTITION COEFFICIENT Log K_{ow} = 1.94.

OTHER CHEMICAL/PHYSICAL PROPERTIES Conversion factors (wt/vol): 4.66 mg/m³ is equivalent to 1 ppm; heat capacity 1.9 J/(g°C (K) at 5.0 mm Hg; heat of polymerization 57.7 kJ/mol at 5.0 mm Hg.

PROTECTIVE EQUIPMENT Suitable protective clothing and self-contained resp protective apparatus should be available. (acrylic acid deriv) [R2].

PREVENTATIVE MEASURES Inhalation and skin contact should be avoided. (methacrylates) [R14, 2298].

DISPOSAL SRP: At the time of review, criteria for land treatment or burial (sanitary landfill) disposal practices are subject to significant revision. Prior to implementing land disposal of waste residue (including waste sludge), consult with environmental regulatory agencies for guidance on acceptable disposal practices.

COMMON USES Ethyl methacrylate is used to make polymers, which in turn are used for building, automotive, aerospace, and furniture industries [R15, 442]. Used as a co-monomer in acrylic polymers for surface coating resins, acrylic emulsion polymers for polishes, etc.; co-monomer in denture-base material; used as a chemical intermediate [R1]; butyl, or ethyl methacrylate copolymers with hydrophilic hydroxyethyl methacrylate and relatively large amounts of the cross-linking agent triethylene glycol dimethacrylate produce hard hydrophilic polymers that can be machined into hard contact lenses with wettable surfaces [R3].

OCCUPATIONAL RECOMMENDATIONS None listed.

HUMAN TOXICITY The estimated human fatal dose may be about 5.4 g/kg. [R14, 2299].

NON-HUMAN TOXICITY LD_{50} rabbits dermal >9.1 g/kg [R13, 369].

BIODEGRADATION Ethyl methacrylate has been listed as being "well biodegradable" by the Japanese for consideration in their chemical substance control law (1) [R7].

BIOCONCENTRATION Based on the log octanol/water partition coefficient of 1.94 (1), a BCF for ethyl methacrylate can be estimated at 18 (2,SRC); utilizing the water solubility, 5,600 ppm at 20°C (3,SRC), a value of 5 is calculated (2,SRC). The range of these values suggests that bioaccumulation in aquatic organisms should not be an important process (2,SRC) [R8].

CERCLA REPORTABLE QUANTITIES Persons in charge of vessels or facilities are required to notify the National Response Center (NRC) immediately, when there is a release of this designated hazardous substance, in an amount equal to or greater than its reportable quantity of 1,000 lb, or 454 kg. The toll-free telephone number of the NRC is (800) 424–8802; in the Washington metropolitan area (202) 426–2675. The rule for determining when notification is required is stated in 40 CFR 302.6 (see section IV. D. 3.b) [R9].

TSCA REQUIREMENTS Section 8 (a) of TSCA requires manufacturers of this chemical substance to report preliminary assessment information concerned with production, use, and exposure to EPA. (2–propanoic acid, 2 methyl ethyl ester) [R10].

RCRA REQUIREMENTS As stipulated in 40 CFR 261.33, when ethyl methacrylate, as a commercial chemical product or manufacturing chemical intermediate or an off-specification commercial chemical product or a manufacturing chemical intermediate, becomes a waste, it must be managed according to federal or state hazardous waste regulations. Also defined as a hazardous waste is any residue, contaminated soil, water, or other debris resulting from the cleanup of a spill, into water or on dry land, of this waste. Generators of small quantities of this waste may qualify for partial exclusion from hazardous waste regulations (40 CFR 261.5). (ethyl methacrylate) [R11].

DOT *Fire or Explosion:* Flammable/combustible material; may be ignited by heat, sparks, or flames. Vapors may travel to a source of ignition, and flash back. Container may explode in heat of fire. Vapor explosion hazard indoors, outdoors, or in sewers. Runoff to sewer may create fire or explosion hazard [R4].

Health Hazards: May be poisonous if inhaled or absorbed through skin. Vapors may cause dizziness or suffocation. Contact may irritate or burn skin and eyes. Fire may produce irritating or poisonous gases. Runoff from fire control or dilution water may cause pollution [R4].

Emergency Action: Keep unnecessary people away; isolate hazard area and deny entry. Stay upwind; keep out of low areas. Self-contained breathing apparatus (SCBA) and structural firefighter's protective clothing will provide limited protection. Isolate for 1/2 mile in all directions if tank car or truck is involved in fire. Call CHEMTREC at 1–800–424–9300 for emergency assistance. If water pollution occurs, notify the appropriate authorities [R4].

Fire: Small fires: Dry chemical, CO_2, halon, water spray, or alcohol foam. Large Fires: Water spray, fog, or alcohol foam is recommended. Move container from fire area if you can do so without risk. Cool containers that are exposed to flames with water from the side until well after fire is out. Stay away from ends of tanks. For massive fire in cargo area, use unmanned hose holder, or monitor nozzles; if this is impossible, withdraw from area, and let fire burn. Withdraw immediately in case of rising sound from venting safety device or any discoloration of tank due to fire [R4].

Spill or Leak: Shut off ignition sources; no flares, smoking, or flames in hazard area. Stop leak if you can do so without risk. Water spray may reduce vapor; but it may not prevent ignition in closed spaces.

Small Spills: Take up with sand or other noncombustible absorbent material and place into containers for later disposal. Large Spills: Dike far ahead of liquid spill for later disposal [R4].

First Aid: Move victim to fresh air and call emergency medical care; if not breathing, give artificial respiration; if breathing is difficult, give oxygen. In case of contact with material, immediately flush eyes with running water for at least 15 minutes. Wash skin with soap and water. Remove and isolate contaminated clothing and shoes at the site [R4].

FIRE POTENTIAL Dangerous when exposed to heat or flame; can react with oxidizing materials [R5]. The low flash points of the lower methacrylates create a fire hazard, and these compounds can form explosive mixtures with air. (methacrylates) [R13, 371].

FDA Ethyl methacrylate is an indirect

food additive polymer for use as a basic component of single and repeated-use food contact surfaces. (ethyl methacrylate) [R12].

GENERAL RESPONSE Shut off ignition sources. Call fire department. Avoid contact with liquid and vapor. Keep people away. Stop discharge if possible. Stay upwind. Use water spray to "knock down" vapor. Isolate and remove discharged material. Notify local health and pollution control agencies.

FIRE RESPONSE Flammable. Containers may explode in fire. Flashback along vapor trail may occur. Vapor may explode if ignited in an enclosed area. Extinguish with dry chemicals, foam, or carbon dioxide. Water may be ineffective on fire. Cool exposed containers with water.

EXPOSURE RESPONSE Call for medical aid. Vapor irritating to eyes, nose, and throat. If inhaled will cause coughing or difficult breathing. If breathing has stopped, give artificial respiration. If breathing is difficult, give oxygen. Liquid irritating to skin and eyes. If swallowed will cause nausea and vomiting. Remove contaminated clothing and shoes. Flush affected areas with plenty of water. If swallowed and victim is conscious, have victim drink water or milk, and have victim induce vomiting. If swallowed and victim is unconscious or having convulsions, do nothing except keep victim warm.

RESPONSE TO DISCHARGE Issue warning—high flammability. Restrict access, mechanical containment, should be removed, chemical and physical treatment.

WATER RESPONSE Effect of low concentration on aquatic life is unknown. Fouling to shoreline. May be dangerous if it enters water intakes. Notify local health and wildlife officials. Notify operators of nearby water intakes.

EXPOSURE SYMPTOMS Inhalation may cause irritation of the mucous membrane. Ingestion causes irritation of mouth and stomach. Contact with liquid irritates eyes and skin.

EXPOSURE TREATMENT Inhalation: re-move victim to fresh air; apply artificial respiration and oxygen if indicated. Ingestion: induce vomiting; call a physician. Eyes: wash with copious quantities of water for 15 min; call a physician. Skin: flush with water; wash with soap and water.

REFERENCES

1. SRI.

2. International Labour Office. *Encyclopedia of Occupational Health and Safety.* Vols. I and II. Geneva, Switzerland: International Labour Office, 1983. 53.

3. *Kirk–Othmer Encyclopedia of Chemical Technology.* 3rd ed., Volumes 1–26. New York, NY: John Wiley and Sons, 1978–1984, p. 6 (79) 723.

4. Department of Transportation. *Emergency Response Guidebook 1987.* DOT P 5800.4. Washington, DC: U.S. Government Printing Office, 1987, p. G–26.

5. Sax, N. I. *Dangerous Properties of Industrial Materials,.* 6th ed. New York, NY: Van Nostrand Reinhold, 1984. 1368.

6. Association of American Railroads. *Emergency Handling of Hazardous Materials in Surface Transportation.* Washington, DC: Assoc. of American Railroads, Hazardous Materials Systems (BOE), 1987. 303.

7. Sasaki, S. pp. 283–98 in *Aquatic Pollutants.* Hutzinger, O., et al. (ed.), Oxford: Pergamon Press (1978).

8. (1) Hansch, C., and A. J. Leo. *Medchemical Project Issue No. 26.* Claremont, CA: Pomona College (1985). (2) Lyman, W. J., et al., *Handbook of Chemical Property Estimation Methods.* New York: McGraw-Hill, pp. 5–1 to 5–30 (1982). (3) Brophy, M. O., et al., *Health and Environmental Effects Profile on Ethyl Methacrylate.* SRC TR–85–247. Syrcause, NY: Syracuse Res Corp (1986).

9. 40 CFR 302.4 (7/1/88).

10. 40 CFR 712.30 (7/1/88).

11. 40 CFR 261.33 (7/1/88).

12. 21 CFR 177.1010 (4/1/88).

13. *Kirk-Othmer Encyclopedia of Chemical Technology.* 3rd ed., Volumes 1–26. New York, NY: John Wiley and Sons, 1978–1984., p. 15(81).

14. Clayton, G. D., and F. E. Clayton (eds.). *Patty's Industrial Hygiene and Toxicology.* Volume 2A, 2B, 2C: Toxicology. 3rd ed. New York: John Wiley and Sons, 1981–1982.

15. Sittig, M. *Handbook of Toxic and Hazardous Chemicals and Carcinogens, 1985.* 2nd ed. Park Ridge, NJ: Noyes Data Corporation, 1985.

■ METHANE

CAS # 74–82–8

SYNONYMS fire-damp • marsh gas • methyl hydride • R-50 (refrigerant) • biogas

MF CH_4

MW 16.05

COLOR AND FORM Colorless gas.

ODOR Odorless; weak odor.

ODOR THRESHOLD 200 ppm [R2].

TASTE Tasteless.

DENSITY 0.7168 g/L.

SURFACE TENSION 14 dynes/cm = 0.014 N/m at −161°C.

VISCOSITY 34.8 µP at −181.6°C; 76.0 µP at −78.5°C; 102.6 µP at 0°C; 108.7 µP at 20°C; 133.1 µP at 100.0°C; 160.5 µP at 200.5°C; 181.3 µP at 284°C; 202.6 µP at 380°C; 226.4 µP at 499°C.

BOILING POINT −161.4°C.

MELTING POINT −182.6°C.

HEAT OF COMBUSTION 978 Btu/cu ft at 25°C (1 kg of methane yields 13,300 kcal).

HEAT OF VAPORIZATION 2,128.8 gcal/gmol.

AUTOIGNITION TEMPERATURE 999°F (537°C) [R13, p. 325M–65].

EXPLOSIVE LIMITS AND POTENTIAL Severe explosion hazard [R1, 752]. Methane, when emitted from the manure pit, tends to accumulate in the higher portions of the building. There have been several instances of explosions occurring when the methane accumulation was ignited by a pilot light or a workman's welding torch [R5, 1240]. Dangerous when exposed to heat or flame. [R10, 1762].

FLAMMABILITY LIMITS Lower: 5.0%; upper: 15% (% by volume) [R11, p. 325M–65].

FIREFIGHTING PROCEDURES Do not extinguish fire unless flow can be stopped. Use water in flooding quantities as fog. Cool all affected containers with flooding quantities of water. Apply water from as far a distance as possible. If fire becomes uncontrollable or container is exposed to direct flame consider evacuation of one-third (1/3) mile radius [R1]. Dry chemical is preferred as firefighting agent. [R12, 320].

SOLUBILITY 0.60 ml in 1 g ethyl alcohol at 20°C; 24.4 ppm at 25°C (measured). Soluble in benzene; 0.91 ml in 1 g ether at 20°C.

VAPOR PRESSURE 2 atm at −152.3°C; 5 atm at −138.3°C; 10 atm at −124.8°C; 20 atm at −108.5°C; 40 atm at −86.3°C.

OCTANOL/WATER PARTITION COEFFICIENT Log K_{ow} = 1.09.

OTHER CHEMICAL/PHYSICAL PROPERTIES Burns with a pale, faintly luminous flame; air containing more than 14% methane burns without noise.

PROTECTIVE EQUIPMENT Self-contained breathing apparatus for high concentrations; protective clothing if exposed to liquid [R2].

PREVENTATIVE MEASURES Good ventilation will prevent formation of harmful concentration in normal workplaces (hydrocarbons, aliphatic) [R5, 1072]. When a filling, storage, and dispatch depot is being selected, consideration must be given to the safety of both the site and the environment. Pump rooms and filling machinery must be located in fire-resistant buildings with roofs of light construction. Doors and other closures should open outwards from the building. The premises should be adequately ventilated and a system of lighting with flameproof electrical switches should be installed (gases and air, compressed) [R5, 947]. If material not on fire, and not involved in Fire: Keep sparks, flames, and other sources of ignition away. Keep material out of water sources and sewers. Attempt to stop leak if without undue personnel hazard. Use water spray to knock down vapors. If material leaking (not on fire) consider evacuation from downwind area based on amt of material spilled, location, and weather conditions [R6]. Do not handle broken packages unless wearing appropriate personal protective equipment. Avoid breathing vapors. Keep upwind. [R6].

CLEANUP By forced ventilation, maintain concentration of gas below the range of explosive mixture. Remove the tank or cylinder to an open area. Leave to bleed off in the atmosphere. [R12, 321].

DISPOSAL SRP: At the time of review, criteria for land treatment or burial (sanitary landfill) disposal practices are subject to significant revision. Prior to implementing land disposal of waste residue (including waste sludge), consult with environmental regulatory agencies for guidance on acceptable disposal practices.

COMMON USES In the production of methanol [R6, 256]. Constituent of illuminating and cooking gas; in manufacture of hydrogen, hydrogen cyanide, ammonia, acetylene, formaldehyde [R8, 853]. Chemical int (excluding fuel use); constituent of natural gas (about 85% methane); in manufacture of synthesis gas, halogenated methanes, carbon disulfide, sodium hydrosulfide, carbon black [R8]. Source of petrochemicals by conversion to hydrogen and carbon monoxide by steam cracking or partial oxidation; starting material for manufacture of synthetic proteins [R9, 752].

OCCUPATIONAL RECOMMENDATIONS None listed.

THRESHOLD LIMIT Simple asphyxiant inert gas or vapor. A TLV may not be recommended for each simple asphyxiant be-

cause the limiting factor is the available oxygen (1981) [R4].

BIODEGRADATION The utilization of methane by some cultures, such as *Methylococcus* and others, as a carbon source suggests that methane is biodegradable [R3].

GENERAL RESPONSE Stop discharge if possible. Keep people away. Shut off ignition sources and call fire department. Stay upwind and use water spray to "knock down" vapor. Evacuate area in case of large discharge. Avoid contact with liquid and vapor. Notify local health and pollution control agencies.

FIRE RESPONSE Flammable. Flashback along vapor trail may occur. May explode if ignited in an enclosed area. Stop discharge if possible. Cool exposed containers and protect men effecting shutoff with water. Let fire burn.

EXPOSURE RESPONSE Call for medical aid. Vapor not irritating to eyes, nose, or throat. If inhaled, will cause dizziness, difficult breathing, and loss of consciousness. Move to fresh air. If breathing has stopped, give artificial respiration. If breathing is difficult, give oxygen. Liquid will cause frostbite. Flush affected areas with plenty of water. Do not rub affected areas.

RESPONSE TO DISCHARGE Issue warning—high flammability. Restrict access. Evacuate area.

WATER RESPONSES Not harmful to aquatic life.

EXPOSURE SYMPTOMS High concentrations may cause asphyxiation. No systemic effects, even at 5% concentration in air.

EXPOSURE TREATMENT: Remove to fresh air. Support respiration.

REFERENCES

1. Association of American Railroads. *Emergency Handling of Hazardous Materials in Surface Transportation.* Washington, DC: Assoc. of American Railroads, Hazardous Materials Systems (BOE), 1987. 444.

2. U.S. Coast Guard, Department of Trans-

portation. *CHRIS—Hazardous Chemical Data.* Volume II. Washington, DC: U.S. Government Printing Office, 1984–5.

3. Clayton, G.D., and F.E. Clayton (eds.). *Patty's Industrial Hygiene and Toxicology:* Volume 2A, 2B, 2C: *Toxicology.* 3rd ed. New York: John Wiley and Sons, 1981–1982. 3180.

4. American Conference of Governmental Industrial Hygienists. *Threshold Limit Values for Chemical Substances and Physical Agents and Biological Exposure Indices for 1994–1995.* Cincinnati, OH: ACGIH, 1994. 24.

5. International Labour Office. *Encyclopedia of Occupational Health and Safety.* Vols. I and II. Geneva, Switzerland: International Labour Office, 1983.

6. Association of American Railroads. *Emergency Handling of Hazardous Materials in Surface Transportation.* Washington, DC: Assoc. of American Railroads, Hazardous Materials Systems (BOE), 1987. 444.

7. *The Merck Index.* 10th ed. Rahway, NJ: Merck Co., Inc., 1983.

8. Bureau of the Census. *Census of Mineral Industries.* 1982: Natural Gas Liquids.

9. Bureau of the Census. Foreign Trade Division—Trade Information Office.

10. Sax, N. I. *Dangerous Properties of Industrial Materials,.* 6th ed. New York, NY: Van Nostrand Reinhold, 1984.

11. National Fire Protection Association. *Fire Protection Guide on Hazardous Materials.* 9th ed. Boston, MA: National Fire Protection Association, 1986.

12. ITII. *Toxic and Hazarous Industrial Chemicals Safety Manual.* Tokyo, Japan: The International Technical Information Institute, 1982.

13. National Fire Protection Association. *Fire Protection Guide on Hazardous Materials.* 9th ed. Boston, MA: National Fire Protection Association, 1986.

■ METHANE, IODO-

CAS # 74–88–4

SYNONYMS iodometano (Italian) • iodure de methyle (French) • Jod-methan (German) • joodmethaan (Dutch) • methane, iodo- • methyl iodide • Methyljodid (German) • methyljodide (Dutch) • Metylu-jodek (Polish) • monoiodomethane • monoioduro di metile (Italian)

MF CH_3I

MW 141.94

COLOR AND FORM Colorless, transparent liquid.

ODOR Pungent; sweet, ethereal.

DENSITY 2.28 at 20°C.

SURFACE TENSION 25.8 dynes/cm at 43.5°C in contact with air.

VISCOSITY 0.606 cP at 0°C, 0.424 cP at 40°C.

BOILING POINT 42.5°C.

MELTING POINT −66.5°C.

HEAT OF COMBUSTION −194.7 kcal/mole.

HEAT OF VAPORIZATION 6,616.5 g cal/g mole.

EXPLOSIVE LIMITS AND POTENTIAL Iodomethane is not explosive [R8].

FIREFIGHTING PROCEDURES Self-contained breathing apparatus with a full facepiece operated in pressure-demand or other positive-pressure mode [R9, 4].

SOLUBILITY Miscible with alcohol, ether; soluble in alcohol, ethanol, benzene, and chloroform; soluble in carbon tetrachloride; solublity in water is 14 g/L at 20°C.

VAPOR PRESSURE 400 mm Hg at 25°C.

OCTANOL/WATER PARTITION COEFFICIENT Log K_{ow} = 1.69.

STABILITY AND SHELF LIFE Colorless liquid that turns yellow, red, or brown when exposed to light and moisture [R10].

OTHER CHEMICAL/PHYSICAL PROPERTIES % in saturated air: 53 at 25°C; conversion factors: 1 mg/L is equivalent (equiv) to 172 ppm and 1 ppm is equiv to 5.8 mg/m³ at 25°C, 760 torr.

PROTECTIVE EQUIPMENT Employees should be provided with and required to use impervious clothing, gloves, face shields (eight-inch min), and other appropriate protective clothing necessary to prevent repeated or prolonged skin contact with liquid [R9, 2].

PREVENTATIVE MEASURES Contact lenses should not be worn when working with this chemical [R11].

CLEANUP 1. Ventilate area of spill or leak. 2. Collect for reclamation or absorb in vermiculite, dry sand, earth, or similar material. [R9, 3].

DISPOSAL At the time of review, criteria for land treatment or burial (sanitary landfill) disposal practices are subject to significant revision. Prior to implementing land disposal of waste residue (including waste sludge), consult with environmental regulatory agencies for guidance on acceptable disposal practices [R12].

COMMON USES In methylations; in microscopy because of high refractive index; as imbedding material for examining diatoms; in testing for pyridine [R1]; used as methylating agent in preparation of pharmaceutical intermediates [R3]; chemical intermediate for methylamines and quaternary ammonium iodides [R2]; chemical intermediate for quaternary phosphonium iodides [R2]; chemical intermediate for organometallics, e.g., methyl mercuric iodide [R2]; used as a fire extinguisher [R4, 3446]; methyl iodide has been investigated in India for use as a fumigant to control internal fungi of grain sorghum [R5]; reacts with many compounds as alkylating agent [R6]; used as building block for radioactive tracers synthesis. Methyl iodide [¹⁴C] is used for carbon-14-labeled compounds; methyl iodide [³H] is used for tritium-labeled compounds [R7, p. 634]; reacts with dimethyl ethyl amine to yield quaternary ammonium compounds. [R7, p. 521].

OCCUPATIONAL RECOMMENDATIONS

OSHA Meets criteria for proposed OSHA medical rules record [R17]. Threshold-limit 8–hr time-weighted average (TWA) 2 ppm, 12 mg/m³, skin (1986) [R18, 26].

NON-HUMAN TOXICITY LC lo mouse inhalation 78,693 ppm/10 min [R14]; LC lo mouse inhalation 18,109 ppm/30 min [R14]; LD$_{50}$ of orally admin iodomethane in the rat was 76 mg/kg [R15]; LC$_{50}$ mouse inhalation 5 mg/L/57 min [R4, 3447].

IARC SUMMARY No data are available for humans. Limited evidence of carcinogenicity in animals. Overall evaluation: group 3: The agent is not classifiable as to its carcinogenicity to humans [R13].

BIOCONCENTRATION Methyl iodide has a low log octanol/water partition coefficient, 1.51 (1), and therefore would not be expected to bioconcentrate in fish (2). However, it has been found in marine fish and shellfish at concentrations that would indicate some bioaccumulation (2) [R16].

RCRA REQUIREMENTS As stipulated in 40 CFR 261.33, when methyl iodide, as a commercial chemical product or manufacturing chemical intermediate or as an off-specification commercial chemical product or a manufacturing chemical intermediate, becomes a waste, it must be managed as a hazardous waste according to federal or state regulations. Also defined as a hazardous waste is any residue, contaminated soil, water, or other debris resulting from the cleanup of a spill, into water or on dry land, of this waste. Generators of small quantities of this waste may qualify for partial exclusion from hazardous waste regulations (see 40 CFR 261.5) [R19].

REFERENCES

1. *The Merck Index.* 10th ed. Rahway, NJ: Merck Co., Inc., 1983. 872.

2. SRI.

3. IARC. *Monographs on the Evaluation of the Carcinogenic Risk of Chemicals to Man.* Geneva: World Health Organization, International Agency for Research on Cancer, 1972–present (multivolume work), p. V15 247 (1977).

4. Clayton, G.D., and F.E. Clayton (eds.). *Patty's Industrial Hygiene and Toxicolo-*

gy: Volume 2A, 2B, 2C: *Toxicology*. 3rd ed. New York: John Wiley and Sons, 1981–1982.

5. DHHS/NTP. *Fourth Annual Report on Carcinogens*, p. 132 (1985) NTP 85–002.

6. IARC. *Monographs on the Evaluation of the Carcinogenic Risk of Chemicals to Man*. Geneva: World Health Organization, International Agency for Research on Cancer, 1972–present (multivolume work), p. V15 245 (1977).

7. *Kirk-Othmer Encyc Chemical Tech*, 3rd ed. 1978–present, V19.

8. ITII. *Toxic and Hazarous Industrial Chemicals Safety Manual*. Tokyo, Japan: The International Technical Information Institute, 1982. 339.

9. Mackison, F. W., R. S. Stricoff, and L. J. Partridge, Jr. (eds.). *NIOSH/OSHA—Occupational Health Guidelines for Chemical Hazards*. DHHS (NIOSH) Publication No. 81–123 (3 vols). Washington, DC: U.S. Government Printing Office, Jan. 1981.

10. American Conference of Governmental Industrial Hygienists. *Documentation of the Threshold Limit Values and Biological Exposure Indices*. 5th ed. Cincinnati, OH: American Conference of Governmental Industrial Hygienists, 1986. 399.

11. NIOSH. *Pocket Guide to Chemical Hazards*. 2nd Printing. DHHS (NIOSH) Publ. No. 85–114. Washington, DC: U.S. Dept. of Health and Human Services, NIOSH/Supt. of Documents, GPO, February 1987. 165.

12. SRP.

13. IARC. *Monographs on the Evaluation of the Carcinogenic Risk of Chemicals to Man*. Geneva: World Health Organization, International Agency for Research on Cancer, 1972–present (multivolume work), p. S7 66 (1987).

14. Verschueren, K. *Handbook of Environmental Data of Organic Chemicals*. 2nd ed. New York, NY: Van Nostrand Reinhold Co., 1983. 855.

15. Johnson, M. K. "Metabolism of Iodomethane in the Rat," *Biochemical J* 98 (1): 38–43 (1966).

16. (1) Hansch, C., and A. J. Leo. *Substituent Constants for Correlation Analysis in Chemisty and Biology*; New York, NY: John Wiley and Sons, pp. 339 (1979); (2) Lyman, W. J., et al. *Handbook of Chemical Property Estimation Methods Environmental Behavior of Organic Commpounds*, New York: McGraw-Hill (1982); (3) Dickson, A. G., J. P. Riley. *Ma Pollut Bull* 7: 167–9 (1976).

17. 47 FR 30420 (7/13/82).

18. American Conference of Governmental Industrial Hygienists. *Threshold Limit Values for Chemical Substances and Physical Agents and Biological Exposure Indices for 1993–1994*. Cincinnati, OH: ACGIH, 1993.

19. 51 FR 28296 (8/6/86).

■ 1,3,4-METHENO-1H-CYCLOBUTA(cd) PENTALENE, 1,1a, 2,2,3,3a,4,5,5,5a, 5b, 6-DODECACHLORO-OCTAHYDRO-

CAS # 2385–85–5

SYNONYMS 1,1a, 2,2,3,3a, 4,5,5,5a, 5b, 6-dodecachlorooctahydro-1,3,4-metheno-1- H-cyclobuta(cd) decane • 1,1a, 2,2,3,3a, 4,5,5a, 5b, 6-dodecachloro-octahydro-1,3,4-metheno-1H-cyclobuta(cd) pentalene • 1,2,3,4,5,5-hexachloro-1,3-cyclopentadiene dimer • 1,3,4-metheno-1H-cyclobuta(cd) pentalene, 1,1a, 2,2,3,3a, 4,5,5,5a, 5b, 6-dodecachlorooctahydro- • 1,3-cyclopentadiene, 1,2,3,4,5,5-hexachloro-, dimer • bichlorendo • cyclopentadiene, hexachloro-, dimer • decane, perchloropentacyclo- • dechlorane • dechlorane-4070 • dechlorane-515 • dechlorane-plus-515 • dodeca-chlorooctahydro-1,3,4-metheno-2H-cyclobuta (c, d) pentalene • dodecachlorooctahydro-1,3,4-metheno-2h-cyclobuta(cd) pentalene • dodecachloropentacyclo (3.2.2.0

(sup2,6),0 (sup3,9),0 (sup5,10))decane • dodecachloropentacyclodecane • ent-25,719 • gc-1283 • hexachlorocyclopentadiene-dimer • hrs-1276 • NCI-C06428 • paramex • perchlorodihomocubane • perchloropentacyclo (5.2.1.0 (2,6).0 (3,9).0 (5,8))decane • perchloro-pentacyclo (5.2.1.0 (sup2,6).0 (sup3,9).0 (sup5,8))decane • perchloropentacyclodecane • ferriamicide • perchloropentacyclo (5.3.0.0 (2,6).0 (3,9).0 (4,8))decane • dodecachloro-pentacyclo (5.2.1.0 (2,6).0 (3,9).0 (5,8))decane • perchlordecone • fire-ant-bait

MF $C_{10}Cl_{12}$

MW 545.50

COLOR AND FORM Snow-white crystals from benzene.

ODOR Odorless.

ODOR THRESHOLD Odor low 5.0667 mg/m³; odor high 5.0667 mg/m³ [R8].

MELTING POINT 485°C.

SOLUBILITY Practically insoluble in water; 15.3% in dioxane at room temp; 14.3% in xylene at room temp; 12.2% in benzene at room temp; 7.2% in carbon tetrachloride at room temp; 5.6% in methyl ethyl ketone at room temp.

VAPOR PRESSURE 3×10^{-7} mm Hg at 25°C.

OCTANOL/WATER PARTITION COEFFICIENT Log K_{ow} = 5.28.

CORROSIVITY Practically noncorrosive to metals.

STABILITY AND SHELF LIFE Very stable at normal temperatures [R7].

OTHER CHEMICAL/PHYSICAL PROPERTIES Vapor specific gravity: 18.8 (calculated).

PROTECTIVE EQUIPMENT Dispensers of liquid detergent should be available. Safety pipettes should be used for all pipetting. In animal laboratory, personnel should wear protective suits (preferably disposable, one-piece, and close-fitting at ankles and wrists), gloves, hair covering, and overshoes. In chemical laboratory, gloves and gowns should always be worn; how-ever, gloves should not be assumed to provide full protection. Carefully fitted masks or respirators may be necessary when working with particulates or gases, and disposable plastic aprons might provide addnl protection. Gowns should be of distinctive color, as a reminder that they are not to be worn outside the laboratory. (chemical carcinogens) [R9, 8].

PREVENTATIVE MEASURES Smoking, drinking, eating, storage of food, or of food and beverage containers, or utensils, and the application of cosmetics should be prohibited in any laboratory. All person-nel should remove gloves, if worn, after completion of procedures in which carcin-ogens have been used. They should wash hands, preferably using dispensers of liq-uid detergent, and rinse thoroughly. Con-sideration should be given to appropriate methods for cleaning the skin, depending on nature of the contaminant. No stan-dard procedure can be recommended, but the use of organic solvents should be avoided. Safety pipettes should be used for all pipetting. (chemical carcinogens) [R9, 8].

CLEANUP A high-efficiency particulate arrestor (HEPA) or charcoal filters can be used to minimize amount of carcinogen in exhausted-air-ventilated safety cabinets, lab hoods, glove boxes, or animal rooms. Filter housing that is designed so that used filters can be transferred into plastic bag without contaminating maintenance staff is avail commercially. Filters should be placed in plastic bags immediately after removal. The plastic bag should be sealed immediately. The sealed bag should be labeled properly. Waste liquids should be placed or collected in proper containers for disposal. The lid should be secured and the bottles properly labeled. Once filled, bottles should be placed in plastic bag, so that outer surface is not contaminated The plastic bag should also be sealed and labeled. Broken glassware should be de-contaminated by solvent extraction, by chemical destruction, or in specially de-signed incinerators. (chemical carcino-gens) [R9, 15].

DISPOSAL SRP: At the time of review

criteria for land treatment or burial (sanitary landfill) disposal practices are subject to significant revision. Prior to implementing land disposal of waste residue (including waste sludge), consult with environmental regulatory agencies for guidance on acceptable disposal practices.

COMMON USES Fire retardant for plastics, rubber, paint, paper, electrical goods [R2]; insecticide used to control western harvester ants, yellow jackets, and imported fire ants (former use) [R4]; mirex is a stomach insecticide with little contact activity and has found its widest use against ants; for imported fire ant, baits containing 0.075% are used; for harvester ant bait is 0.15% [R5]; (former use) mirex is marketed under the tradename dechlorane for use in flame-retardant coatings for various materials. [R6, 879].

NON-HUMAN TOXICITY LD_{50} rat male oral 306 mg/kg [R3]; LD_{50} rabbit subcutaneous 800 mg/kg [R3]; LD_{50} rabbit dermal 800 mg/kg body wt (from table) [R13]; LD_{50} rat dermal 2,000 mg/kg body wt (from table) [R14]; LD_{50} dog male oral 1,000 mg/kg body wt (in corn oil) (from table) [R15]; LD_{50} hamster female oral 125 mg/kg body wt (from table) [R16]; LD_{50} hamster male oral 250 mg/kg body wt (from table) [R16]; LD_{50} rat male oral 740 mg/kg body wt (from table) [R14]; LD_{50} rat female oral 600 mg/kg body wt (in corn oil) (from table) [R14]; LD_{50} rat female intraperitoneal 365 mg/kg body wt (from table) [R17]; LD_{50} rabbit dermal 800 mg/kg body wt (from table) [R13].

ECOTOXICITY LC_{50} Colinus virginianus (bobwhite quail) oral 2,511 ppm in 5-day diet (95% confidence limit 2,160–2,908 ppm) (purity, 98%) [R18]; LC_{50} Japanese quail, oral greater than 5,000 ppm in 5-day diet (20% mortality at 5,000 ppm) (purity, 98%) [R18]; LC_{50} Phasianus colchicus (ring-necked pheasants), oral 1,540 ppm in 5-day diet (95% confidence limit 1,320–1,789 ppm) (purity, 98%) [R18]; LC_{50} Anas platyrhynchos (mallard ducks), oral greater than 5,000 ppm in 5-day diet (no mortality to 5,000 ppm) (purity 98%) [R18]; LD_{50} Anas platyrhynchos (mallard ducks), oral greater than 2,400

mg/kg, 3 mo-old-males (purity 98%) [R12]; LD_{50} Phasianus, (pheasants), oral greater than 2,000 mg/kg, 3 mo-old-females (purity 98%) [R12]; EC_{50} daphnid Simocephalus serrulatus, greater than 0.100 mg/L/48 hr at 16°C, 1st instar (technical material, 98%; static bioassay without aeration) [R19]; EC_{50} Daphnia pulex (daphnid), greater than 0.100 mg/L/48 hr at 16°C, 1st instar (technical material, 98%; static of bioassay without aeration) [R19]; EC_{50} Daphnia magna (daphnid), greater than 0.100 mg/L/48 hr at 17°C, 1st instar (technical material, 98%; static bioassay without aeration) [R19]; EC_{50} Chironomus plumosus (midge), greater than 1.0 mg/L/48 hr at 22°C, 4th instar (technical material, 98%; static of bioassay without aeration) [R19]; LC_{50} Gammarus pseudolimnaeus (scud), greater than 1.0 mg/L/96 hr at 17°C, mature (technical material, 98%; static of bioassay without aeration) [R19]; LC_{50} Salmo gairdneri (rainbow trout), greater than 100 mg/L/96 hr at 12°C, wt 1 g (technical material, 98%; static of bioassay without aeration) [R19]; LC_{50} Perca flavescens (yellow perch), greater than 100 mg/L/96 hr at 15°C, wt 2.6 g (technical material, 98%; static bioassay without aeration) [R19]; LC_{50} Pimephales promelas (fathead minnow), greater than 100 mg/L/96 hr at 18°C, wt 1.3 g (wettable powder, 50%; static bioassay without aeration) [R19]; LC_{50} Lepomis macrochirus (bluegill sunfish), greater than 100 mg/L/96 hr at 18°C, wt 1.1 g (wettable powder, 50%; static of bioassay without aeration) [R19]; LC_{50} Stizostedion vitreum (walleye), greater than 100 mg/L/96 hr at 18°C, wt 1.4 g (wettable powder, 50%; static of bioassay without aeration) [R19]; LD_{50} shrimp 1.01 ppm/72 hr (conditions of bioassay not specified) [R1]; a concentration of 0.1 µg/L technical grade mirex in flowing seawater was lethal to juvenile pink shrimp, Panaeus durorarum, in a 3-wk exposure [R20]; in static tests with larval stages (megalopal) of the mud crab, Rhithropanopeus harrisii, reduced survival was observed in 0.1 µg/L mirex [R21].

IARC SUMMARY No data are available for humans. Sufficient evidence of carcinogenicity in animals. Overall evaluation: Group 2B: The agent is possibly carcinogenic to humans [R10].

BIOLOGICAL HALF-LIFE In rats, half-lives for biphasic excretion of mirex were 38 hr and 100 days, respectively [R22].

BIODEGRADATION Mirex did not degrade when 1×10^6 cpm of ^{14}C mirex was added to 100 ml of medium containing 0.5 g yeast extract/L with 13 bacterial isolates obtained from mirex bait incubated for 2 mo on moist soil (1). Generally mirex is resistant to attack by bacteria and fungi, and can inhibit the growth of Actinomycetes (1). Although mirex is taken up by microorganisms (1), plants (2,4), and higher animals including fish (5) and rats (2), it is not metabolized (2). Yet analysis of soils from spill sites, 5 and 12 years after the accidents, suggests that dechlorination takes place very slowly, and kepone is a biotransformation product of mirex (3). Kepone was also identified as a transformation product of mirex in estuaries (4). Neither ^{14}C–labeled mirex or kepone degraded in aerobic or anaerobic hydrosoils from the Little Dixie Reservoir and James River tributary, Richmond, VA (5). In addition mirex did not degrade at a concentration of 0.5 g/100 g dry weight in four anaerobic lake sediments after 6 mo incubation (1). However, a loss of mirex was attributed to sludge worms under anaerobic conditions (6) [R23].

BIOCONCENTRATION There is evidence for degradation of mirex to chlordecone (kepone) in the environment. Both mirex and kepone are highly persistent and have high lipid:water partition coefficient and have been shown to bioconcentrate several thousandfold in food chains. [R11, 551].

FIFRA REQUIREMENTS All registered products containing mirex were effectively cancelled on December 1, 1977. (A technical mirex product made by Hooker Chemical Company is unaffected by this Settlement Agreement. However, because mirex produced under this registration may be used only in the formulation of other pesticide products, the registration was useless after December 1, 1977). All existing stocks of mirex within the continental United States were not to be sold, distributed, or used after June 30, 1978. [R24, 16].

REFERENCES
1. U.S. Coast Guard, Department of Transportation. *CHRIS—Hazardous Chemical Data*. Volume II. Washington, DC: U.S. Government Printing Office, 1984–5.

2. *The Merck Index*. 10th ed. Rahway, NJ: Merck Co., Inc., 1983. 889.

3. *Farm Chemicals Handbook* 1989. Willoughby, OH: Meister Publishing Co., 1989, p. C–198.

4. SRI.

5. Martin, H., and C. R. Worthing (eds.). *Pesticide Manual*. 4th ed. Worcestershire, England: British Crop Protection Council, 1974. 360.

6. Verschueren, K. *Handbook of Environmental Data of Organic Chemicals*. 2nd ed. New York, NY: Van Nostrand Reinhold Co., 1983.

7. IARC. *Monographs on the Evaluation of the Carcinogenic Risk of Chemicals to Man*. Geneva: World Health Organization, International Agency for Research on Cancer, 1972–present (multivolume work), p. V20 284 (1979).

8. Ruth, J.H. *Am Ind Hyg Assoc J* 47: A–142–51 (1986).

9. Montesano, R., H. Bartsch, E. Boyland, G. Della Porta, L. Fishbein, R. A. Griesemer, A. B. Swan, L. Tomatis, and W. Davis (eds.). *Handling Chemical Carcinogens in the Laboratory: Problems of Safety*. IARC Scientific Publications No. 33. Lyon, France: International Agency for Research on Cancer, 1979.

10. IARC. *Monographs on the Evaluation of the Carcinogenic Risk of Chemicals to Man*. Geneva: World Health Organization, International Agency for Research on Cancer, 1972–present (multivolume work), p. S7 66 (1987).

11. Doull, J., C. D. Klassen, and M. D. Amdur (eds.). *Casarett and Doull's Toxicology*. 3rd ed., New York: Macmillan Co., Inc., 1986.

12. U.S. Department of the Interior, Fish and Wildlife Service. *Handbook of Toxicity of Pesticides to Wildlife*. Resource Publication 153. Washington, DC: U.S. Government Printing Office, 1984. 55.

13. WHO. *Environ Health Criteria: Mirex*, p. 26 (1984).

14. Gaines, T. B. *Toxicol Appl Pharmacol* 14: 5–534 (1969) as cited in WHO, *Environ Health Criteria: Mirex* p. 26 (1984).

15. Larson, P. S., et al. *Toxicol Appl Pharmacol* 49: 271–7 (1979) as cited in WHO, *Environ Health Criteria: Mirex*, p. 26 (1984).

16. Cabral, J.R.P., et al. *Toxicol Appl Pharmacol* 48: A192 (1979) as cited in WHO, *Environ Health Criteria: Mirex*, p. 26 (1984).

17. Kendall, M. W., *Bull Environ Contam Toxicol* 12: 617–21 (1974) as cited in WHO, *Environ Health Criteria: Mirex*, p. 26 (1984).

18. U.S. Department of the Interior, Fish and Wildlife Service, Bureau of Sports Fisheries and Wildlife. *Lethal Dietary Toxicities of Environmental Pollutants to Birds*. Special Scientific Report—*Wildlife* No. 191. Washington, DC: U.S. Government Printing Office, 1975. 27.

19. U.S. Department of Interior, Fish and Wildlife Service. *Handbook of Acute Toxicity of Chemicals to Fish and Aquatic Invertebrates*. Resource Publication No. 137. Washington, DC: U.S. Government Printing Office, 1980. 53.

20. Brookhout, C. G., et al. *Water, Air, and Soil Pollut* 1: 165 (1972) as cited in USEPA/OWRS, *Quality Criteria for Water 1986 Mirex* (1986) EPA 440/5–86–001].

21. Brookhout, C. G., et al. *Water, Air, and Soil Pollut* 1: 165 (1972) as cited in USEPA/OWRS, *Quality Criteria for Water 1986 Mirex* (1986) EPA 440/5–86–001.

22. *Foreign Compound Metabolism in Mammals* Volume 3. London: The Chemical Society, 1975. 409.

23. (1) Jones, A. S., C. S. Hodges. *J Agr Food Chem* 22: 435–9 (1974); (2) Mehendale, H. M., et al. *Bull Environ Contam Toxicol* 8: 200 (1972); (3) Carlson, D. A., et al. *Science* 194: 939–41 (1976); (4) Brown, L. R. et al. *Effect of Mirex and Carbofuran on Estuarine Microorganisms* USEPA–600/3–75–024 (NTIS PB–247147) p. 57 (1975); (5) Huckins, J. N., et al. *J Agr Food Chem* 30: 1020–7 (1982); (6) Andrade, P. S. L., W. B. Wheeler. *Bull Environ Contam Toxicol* 11: 415–6 (1974).

24. Environmental Protection Agency/ OPTS. *Suspended, Cancelled, and Restricted Pesticides*. 3rd Revision. Washington, DC: Environmental Protection Agency, January 1985.

■ 1,3,4–METHENO–2H– CYCLOBUTA(cd) PENTALEN– 2–ONE,1,1a,3,3a,4,5,5,5a, 5b, 6–DECACHLORO– OCTAHYDRO

CAS # 143–50–0

DOT # NA2761

SIC CODES 2879.

SYNONYMS 1,3,4-metheno-2H-cyclobuta(cd) pentalen-2-one, 1,1a,3,3a,4,5,5,5a,5b, 6-decachlorooctahydro- • clordecone • decachloro-octahydro-1,3,4-metheno-2H-cyclobuta(cd) pentalen-2-one • decachloroketone • decachlorooctahydro-1,3,4-metheno-2h-cyclobuta(cd) pentalen-2-one • decachloropentacyclo (5.2.1.0 (2,6).0 (3,9).0 (5),(8))decan-4-one • decachlorotetracyclodecanone • decachlorotetrahydro-4,7-methanoindenone • ent-16,391 • ent-16391 • kepone • kepone-2-one, decachlorooctahydro- • NCI-C00191 • compound-1189 • gc-1189 • General chemicals 1189 • Ciba-8514 • merex • perchloropentacyclo[5.3.0.0 (2,6).0 (3,9).0 (4,8) decan-5-one • 1,2,3,4,5,5,6,7,8,9,10,10-dodecachlorooctahydro-1,3,4-metheno-2-cyclobuta-(c, d)pentalone • 1,1a,3,3a,

4,5,5,5a,5b,6-decachloro-octahydro-1,3,4-metheno-2h-cyclobuta[cd] pentalen-2-one • 2,3,3a, 4,5,6,7,7a, 8,8a-decachloro-3a, 4,7,7a-tetrahydro-4,7-methanoinden-1-one

MF $C_{10}Cl_{10}O$

MW 490.60

COLOR AND FORM Crystals; tan-to-white solid.

ODOR Odorless.

MELTING POINT 350°C (decomposes).

FIREFIGHTING PROCEDURES Extinguish fire using agent suitable for type of surrounding fire. [R12, 403].

SOLUBILITY Soluble in strongly alkaline aqueous solution; readily soluble in acetone, less soluble in benzene and light petroleum; soluble in hydrocarbon solvents; soluble in alcohols, ketones, acetic acid; soluble in hexane; 0.4% in water at 100°C; 0.470 lb/100 lb water at 212°F; soluble in organic solvents such as benzene and hexane; soluble in light petroleum and may be recrystallized from 85–90% aqueous ethanol.

VAPOR PRESSURE Less than 3×10^{-7} mm Hg at 25°C.

CORROSIVITY Noncorrosive.

STABILITY AND SHELF LIFE Stable to about 350°C; readily hydrates on exposure to room temperature and humidity. [R3].

OTHER CHEMICAL/PHYSICAL PROPERTIES The hydrate (of chlordecone) is less soluble in polar solvents. Oxygenated solvents such as ketones and alcohols are recommended for the hydrate.

PROTECTIVE EQUIPMENT Employees working in areas where contact of skin or eyes with kepone (wet or dry) may occur should wear impervious full-body protective clothing, protecting neck, head, hands, and feet. Where personal respiratory protective equipment is needed to reduce employee exposure, equipment suitable for the use and exposure to be encountered should be provided. [R11, 1170].

PREVENTATIVE MEASURES Emissions of airborne particulates (dust, mist, spray, etc.) of kepone should be controlled at the sources of dispersion by means of effective and properly maintained methods, such as fully enclosed operations and local exhaust ventilation. Ensure that employees do not eat or smoke in areas where kepone is handled, processed, or stored, so as to prevent ingestion of kepone. [R11, 1170].

CLEANUP In case of damage to or leaking from containers of this material, contact the pesticide safety team network. Telephone: 800–424–9300. [R12, 403].

COMMON USES Insecticide for bananas, non-bearing citrus trees (former use), tobacco (former use), ornamental shrubs, insects in buildings (former use), lawns, turf, and flowers (former use) [R6]; base material for the manufacture of kelevan [R7]; normally used as mono– to trihydrate. It is effective against leaf-eating insects, less effective against sucking insects, and useful as larvicide against flies (former use) [R2]; acaricide (former use) [R8]; as a fungicide against apple scab and powdery mildew (former use) and to control the Colorado potato beetle, rust mite on non-bearing citrus, and potato and tobacco wireworm on gladioli and other plants [R4]; kepone was (formerly) registered for the control of rootborers on bananas with a residue tolerance of 0.01 ppm. This constituted the only food or feed use of kepone. Nonfood uses included wireworm control in tobacco fields and bait to control ants and other insects in indoor and outdoor areas [R5, 205]; chlordecone is reported to have fungicidal activity against apple scab and powdery mildew (former use) [R9]; kepone is used as stomach poison in the form of bait, and it controls slugs, snails, and fire ants. [R10, 62].

STANDARD CODES EPA 311; TSCA; not in IATA; Coast Guard classification poison B, poison label; CFR 49 poison class B nos poison; label, 50-lb passenger, 200-lb cargo, storage code 1, 2; CFR 14 CAB code 8; ICC STCC tariff no. 1–C 28 799 31; Superfund designated (hazardous substances) list.

PERSISTENCE Persistent for long periods in soil and water. Kepone is not accumulated in plant life, accumulation is found only in fish, birds, mammals, etc. No significant degradation, either biochemical or photochemical, occurs over the short term.

GENERAL SENSATION Can cause nausea, motor difficulties, slurred speech, tremors, complete prostration, and death. Sensory stimulation triggered or intensified tremors, even as long as 4 weeks after exposure.

PERSONAL SAFETY Protect against both inhalation and absorption through the skin. Must wear protective clothing, including NIOSH–approved rubber gloves and boots, safety goggles or face mask, hooded suit, and either a respirator whose canister is specifically approved for this material, or a self-contained breathing apparatus. Decontaminate fully or dispose of all equipment after use.

ACUTE HAZARD LEVEL This very toxic substance has a oral LD_{50} values of 96 and 65 mg/kg for rats and rabbits, respectively. The lowest published toxic oral dose for rats is 672 mg/kg with a carcinogenic effect noted. The 96-hour LC_{50} for sheepshead minnows is 69.5 ppb and for spot it is 6.6 ppb. In salt water, kepone has a 96-hour LC_{50} of 121 ppb for grass shrimp. The probable human toxicity lethal dose is between 1 teaspoonful and 1 ounce.

CHRONIC HAZARD LEVEL Kepone has been shown to be carcinogenic to rats and mice. EPA has designated this material a category x, with an aquatic LC_{50} less than 0.1 mg/L. In mice kepone induces a dose-dependent reproductive inhibition at all dose levels tested.

DEGREE OF HAZARD TO PUBLIC HEALTH A suspected carcinogen. Material is extremely toxic when ingested or inhaled.

OCCUPATIONAL RECOMMENDATIONS
NIOSH Carcinogen; 10 hr TWA 1 μg/m³; liver function testing required [R22].

ACTION LEVEL Avoid contact and inhalation of the spilled cargo. Stay upwind. Notify local fire, air, and water authorities

of the accident. Evacuate all people to a distance of at least 200 feet upwind and 1,000 feet downwind of the spill.

ON-SITE RESTORATION Dam stream to prevent flow and to retard dissipation by water movement. Apply carbon to adsorb dissolved material. Bottom pumps, dredging, or underwater vacuum systems may be employed in small bodies of water. Dredging may be effective in larger bodies to remove undissolved material and adsorbent from the bottom sediments. An alternative method, under controlled conditions, is to pump the water into a suitable container and then pass it through a filtration system with an activated-carbon filter. For more details see ENVIREX Manual EPA 600/2–77–277. Seek professional environmental engineering assistance through EPA's Environmental Response Team (ERT), Edison, NJ, 24-hour no. 908–548–8730.

BEACH OR SHORE RESTORATION Close beach and shore to public until material has been removed. If tidal, scrape affected area at low tide with mechanical scraper, avoid human contact.

AVAILABILITY OF COUNTERMEASURE MATERIAL Bottom pumps—available through fire departments, EPA regional offices, U.S. Coast Guard, or Army Corps of Engineers; dredging—Army Corps of Engineers; underwater vacuum systems—swimming-pool operators; carbon—water treatment plants and chemical plants.

DISPOSAL After the material has been contained, remove with contaminated soil and place in impervious containers. The material may be incinerated in a pesticide incinerator at the specified temperature/dwell-time combination. Any liquids, sludges, or solid residues generated should be disposed of in accordance with all applicable federal, state, and local pollution control requirements. If appropriate incineration facilities are not available, material may be buried in a chemical waste landfill. Not acceptable at a municipal sewage treatment plant. SRP: At the time of review, criteria for land treatment or burial (sanitary landfill) disposal practices are subject to significant revision.

Prior to implementing land disposal of waste residue (including waste sludge), consult with environmental regulatory agencies for guidance on acceptable disposal practices.

DISPOSAL NOTIFICATION Notify local and state health authorities, local solid waste disposal authorities, supplier, and shipper of material.

INDUSTRIAL FOULING POTENTIAL Not significant.

MAJOR WATER USE THREATENED All downstream users of water, monitor for water concentrations.

PROBABLE LOCATION AND STATE White crystals, will sink to the bottom. Quick action could recover most of the material from the bottom before it dissolves.

WATER CHEMISTRY The anhydrous compound readily takes up water. Volatilization does not take place, due to the strong sorption onto particulate matter.

FOOD-CHAIN CONCENTRATION POTENTIAL Experiments show bioaccumulation of kepone in the food chain.

NON-HUMAN TOXICITY LD_{50} rat oral 95 mg/kg [R1, 172]; LD_{50} dog, oral 250 mg/kg [R8]; LD_{50} rat subcutaneous >2,000 mg/kg [R8]; LD_{50} rabbit subcutaneous 345–475 mg/kg [R8]; LD_{50} rabbit oral 65 mg/kg [R14]; LD_{50} chicken oral 480 mg/kg [R14]; LD_{50} pig oral 2,550 mg/kg (approximated) [R14].

ECOTOXICITY EC_{50} *Chironomus* 320 µg/L/48 hr at 22°C (95% confidence limit 220–450 µg/L), fourth instar, static bioassay (technical material, 90.7%) [R15]; EC_{50} *Daphnia magna* 260 µg/L/48 hr at 17°C (95% confidence limit 200–345 µg/L), first instar, static bioassay (technical material, 90.7%) [R15]; LC_{50} *Gammarus pseudolimnaeus* 180 µg/L/96 hr at 17°C (95% confidence limit 110–290 µg/L), mature, static bioassay (technical material, 90.7%) [R15]; LC_{50} *Salmo gairdneri* (rainbow trout) 30 µg/L/96 hr at 12°C (95% confidence limit 24–38 µg/L), wt 1.1 g, static bioassay (technical material, 90.7%) [R15]; LC_{50} *Ictalurus punctatus* (channel catfish) 225 µg/L/96 hr at 18°C, wt 1.6 g,

static bioassay (technical material, 90.7%) [R15]; LC_{50} *Lepomis macrochirus* (bluegill sunfish) 72 µg/L/96 hr at 24°C, wt 2–5 g, static bioassay (technical material, 90.7%) [R15]; LC_{50} redear sunfish 44 µg/L/96 hr at 24°C (95% confidence limit 41–47 µg/L), wt 1.0 g, static bioassay (technical material, 90.7%) [R15]; LC_{50} *Cyprinodon variegatus* (sheepshead minnow) 69.5 µg/L/96 hr in a flow-through bioassay [R16]; LC_{50} trout 0.066 ppm/24 hr; 0.038 ppm/48 hr; 0.02 ppm/96 hr (conditions of bioassay not specified) [R16]; LC_{50} *Palaemonetes pugio* (decapod) 120.9 µg/L/96 hr in a flow-through bioassay [R16]; LC_{50} *Callinectes sapidus* (decapod) <210 µg/L/96 hr in a flow-through bioassay [R16]; LC_{50} fiddler crab >1.6 mg/L/24 hr; 1.47 mg/L/96 hr (conditions of bioassay not specified) [R16]; EC_{50} *Chlorococcum* species (algae) 0.35 µg/L/7 days at a temperature of 20–0.5°C in a static system experienced growth retardation [R17]; EC_{50} *Dunaliella tertiolecta* (algae) 0.58 µg/L/7 days at a temperature of 20–0.5°C in a static system experienced growth retardation [R17]; EC_{50} *Nitzschia* species (algae) 0.60 µg/L/7 days at a temperature of 20–0.5°C in a static system experienced growth retardation [R17]; EC_{50} *Thalassiosira pseudonana* (algae) 0.60 µg/L/7 days at a temperature of 20–0.5°C in a static system experienced growth retardation [R17]; LC_{50} *Brevoortia tyrannus* (Atlantic menhaden) 17.4 µg/L/96 hr (conditions of bioassay not specified) [R18]; LC_{50} *Menidia menidia* (Atlantic silverside) 28.8 µg/L/96 hr (conditions of bioassay not specified) [R18].

IARC SUMMARY No data are available for humans. Sufficient evidence of carcinogenicity in animals. Overall evaluation: Group 2B: The agent is possibly carcinogenic to humans [R13].

BIOLOGICAL HALF-LIFE Untreated patients showed blood clearance of kepone half-life: 165–27 days; fat clearance half-life: 125 days; with cholestyramine treatment, blood half-life: 80–4 days, fat half-life: 64 days [R19].

BIODEGRADATION Amount recovered after exposure to marine sediments in seawater under anaerobic and aerobic conditions for 1 yr: 95% (1). No evidence of any degradation was detected for chlordecone exposed to hydrosoils from a reservoir (not previously exposed to chlordecone) and a creek (contaminated with chlordecone) under anaerobic or aerobic conditions for 56 days (2). No degradation of chlordecone exposed to sewage sludge observed under anaerobic conditions for 120 hr (3). No degradation reported of chlordecone exposed to contaminated James River sediments with added autoclaved silty clay loam soil for 52 days at pH 7 (4) [R20].

BIOCONCENTRATION BCF of chlordecone: fathead minnow, 1,100–2,200 (1); *Cyprinodon variegatus*, 1548, *Leiostomus xanthrus*, 1,221; *Palaemonetes pugio*, 698; *Callinectes sapidus*, 8.1 (3); no species reported, 8,400 (2); *Brevoortia tyrannus* (Atlantic menhaden), 2,300–9,750; *Menidia menidia* (Atlantic silversides), 21,700–60,200 (4) [R21].

RCRA REQUIREMENTS As stipulated in 40 CFR 261.33, when kepone, as a commercial chemical product or manufacturing chemical intermediate or an off-specification commercial chemical product or a manufacturing chemical intermediate, becomes a waste, it must be managed according to federal or state hazardous waste regulations. Also defined as a hazardous waste is any residue, contaminated soil, water, or other debris resulting from the cleanup of a spill, into water or on dry land, of this waste. Generators of small quantities of this waste may qualify for partial exclusion from hazardous waste regulations (see 40 CFR 261.5) [R23].

FIFRA REQUIREMENTS Under section 3 (c) (2) (b) of FIFRA, the Data Call–in Program, existing registrants are required to provide EPA with needed studies. For chlordecone, an intent to cancel registration of pesticide products containing chlordecone (kepone) was effective as of May 11, 1977 [R24].

GENERAL RESPONSE Avoid contact with solid or dust. Keep people away. Wear goggles, self-contained breathing apparatus, rubber overclothing (including gloves). Stop discharge if possible. Isolate, and remove discharged material. Notify local health and pollution control agencies.

FIRE RESPONSE Fire data not available.

EXPOSURE RESPONSE Call for medical aid. Solid or dust is poisonous if swallowed or inhaled, or if skin is exposed. Flush affected areas with plenty of water. If in eyes, hold eyelids open, and flush with plenty of water. If swallowed and victim is conscious, have victim drink water or milk, and have victim induce vomiting. If swallowed and victim is unconscious or having convulsions, do nothing except keep victim warm.

RESPONSE TO DISCHARGE Issue warning—poison, water contaminant. Restrict access. Should be removed. Provide chemical and physical treatment.

WATER RESPONSE Harmful to aquatic life in very low concentrations. Dangerous if it enters water intakes. Notify local health and wildlife officials. Notify operators of nearby water intakes.

EXPOSURE SYMPTOMS Inhalation and ingestion: these symptoms present in all affected patients—neurologic impairment, anxiety, irritability, memory disturbance, headache, tremors, opsiclonus, stuttering, slurred speech, and abnormal tandem gait.

EXPOSURE TREATMENT Call a physician. Inhalation: remove from exposure. Eyes: flush with copious amounts of water. Skin: wash thoroughly with soap and water. Ingestion: induce emesis or perform gastric lavage. Give saline cathartic. Barbiturates to control tremors or convulsions.

REFERENCES

1. Sittig, M. (ed.) *Pesticide Manufacturing and Toxic Materials Control Encyclopedia*. Park Ridge, NJ: Noyes Data Corporation. 1980.

2. Martin, H., and C. R. Worthing (eds.). Pesticide Manual. 4th ed. Worcestershire, England: British Crop Protection Council, 1974. 96.

3. IARC. *Monographs on the Evaluation of the Carcinogenic Risk of Chemicals to Man.* Geneva: World Health Organization, International Agency for Research on Cancer, 1972–present (multivolume work), p. V20 68 (1979).

4. WHO. *Environ Health Criteria: Chlordecone,* p. 15 (1984).

5. Sittig, M. *Handbook of Toxic and Hazardous Chemicals and Carcinogens,* 1985. 2nd ed. Park Ridge, NJ: Noyes Data Corporation, 1985.

6. SRI.

7. WHO. *Environ Health Criteria: Chlordecone,* p. 9 (1984).

8. Larson, L. L., E. E. Kenaga, R. W. Morgan. *Commercial and Experimental Organic Insecticides.* 1985 Revision. College Park, MD: Entomological Society of America, 1985. 22.

9. Spencer, E. Y. *Guide to the Chemicals Used in Crop Protection.* 7th ed. Publication 1093. Research Institute, Agriculture Canada, Ottawa, Canada: Information Canada, 1982. 102.

10. Matsumura, F. *Toxicology of Insecticides.* 2nd ed. New York, NY: Plenum Press, 1985.

11. International Labour Office. *Encyclopedia of Occupational Health and Safety.* Vols. I and II. Geneva, Switzerland: International Labour Office, 1983.

12. Association of American Railroads. *Emergency Handling of Hazardous Materials in Surface Transportation.* Washington, DC: Assoc. of American Railroads, Hazardous Materials Systems (BOE), 1987.

13. IARC. *Monographs on the Evaluation of the Carcinogenic Risk of Chemicals to Man.* Geneva: World Health Organization, International Agency for Research on Cancer, 1972–present (multivolume work), p. S7 59 (1987).

14. WHO. *Environ Health Criteria: Chlordecone,* p. 24 (1984).

15. U.S. Department of Interior, Fish and Wildlife Service. *Handbook of Acute Toxicity of Chemicals to Fish and Aquatic Invertebrates.* Resource Publication No. 137. Washington, DC: U.S. Government Printing Office, 1980. 45.

16. Verschueren, K. *Handbook of Environmental Data of Organic Chemicals.* 2nd ed. New York, NY: Van Nostrand Reinhold Co., 1983. 788.

17. Walsh, G. E., et al. *Chesapeake Sci* 18: 222–223 (1977) as cited in WHO, *Environ Health Criteria: Chlordecone,* p. 32 (1984).

18. Roberts, M. H., D. J. Fisher. *Arch Environ Contam Toxicol* 14 (1): 1–6 (1985).

19. Cohn, W. J., et al. in *Engl J Med* 298 (5): 243–8 (1978).

20. (1)) Vind, H. P., et al. *Biodeterioration of Navy Insecticides in the Ocean* NTIS AD–77310 (1973); (2) Huckins, J. N., et al. *J Agric Food Chem* 30: 1020–7 (1982); (3) Geer, R. D. *Predicting the Anaerobic Degradation of Organic Chemical Pollutants in Waste Water Treatment Plants from Their Electrochemical Reduction Behavior* NTIS PB–289 224 (1978); (4) Gambrell, R. P., et al. *J Water Pollut Control Fed* 56: 174–82 (1984).

21. (1) Spehar, R. L. *J Water Pollut Control Fed* 54: 877–922 (1982); (2) Kenaga, E. E., *Ecotox Environ Safety* 4: 26–38 (1980); (3) Reisch, D. J., et al. *Marine and Estuarine Pollut* 50: 1424–69 (1978); (4) Roberts, M. H., D. J. Fisher *Arch Environ Contam Toxicol* 14: 1–6 (1985).

22. NIOSH/CDC. *NIOSH Recommendations for Occupational Safety and Health Standards 1988,* Aug. 1988 (Suppl. to *Morbidity, and Mortality Wkly.* Vol. 37 No. 5–7, Aug. 26, 1988). Atlanta, GA: National Institute for Occupational Safety, and Health, CDC, 1988. 17.

23. 40 CFR 261.33 (7/1/88).

24. USEPA/OPP. *The Status of Chemicals in Special Review Program, Registration Standards Program, and Data Call–In Program* p. 9 (1987).

■ (L)-METHIONINE

CAS # 63–68–3

SYNONYMS acimetion • alpha-amino-gamma-methylmercaptobutyric acid • 1-alpha-amino-gamma-methylmercapto-butyric acid • 2-amino-4-methylthiobutanoic acid • 1 (-) -amino-gamma-methylthiobutyric acid • 2-amino-4-(methylthio) butyric acid • banthionine • butanoic acid, 2-amino-4-(methylthio)-, (s)- • butyric acid, 2-amino-4-(methylthio)- • cymethion • cynaron • dyprin • lobamine • meonine • mertionin • met • methilanin • methionine • methionine, 1 • s-methionine • gamma-methylthio-alpha-aminobutyric acid • 1-gamma-methylthio-alpha-aminobutyric acid • metione • neo-methidin • neston • thiomedon

MF $C_5H_{11}NO_2S$

MW 149.23

pH (1% aqueous solution) = 5.6–6.1.

COLOR AND FORM Minute hexagonal plates from dilute alcohol.

ODOR Faint.

TASTE Sulfurous.

DISSOCIATION CONSTANTS pK_a = 2.28, pK_2 = 9.21.

BOILING POINT 186°C, sublimes.

MELTING POINT 280–282°C (decomp, sealed capillary).

SOLUBILITY Soluble in water; crystals are somewhat water-repellant at first; soluble in warm dilute alcohol. Insoluble in abs alcohol and ether. Insoluble in petroleum ether, benzene, acetone. Slightly soluble in acetic acid. Water solubility = 56.6 g/L at 25°C.

OCTANOL/WATER PARTITION COEFFICIENT log P = –1.87.

OTHER CHEMICAL/PHYSICAL PROPERTIES IR: 2549 (Coblentz Society Spectral Collection) (Methionine(DL)).

DISPOSAL SRP: At the time of review, criteria for land treatment or burial (sanitary landfill) disposal practices are subject to significant revision. Prior to implementing land disposal of waste residue (including waste sludge), consult with environmental regulatory agencies for guidance on acceptable disposal practices.

COMMON USES Medication: lipotropic agent; (vet) essential nutrient, lipotropic agent, has been used to regulate urinary pH [R1].

OCCUPATIONAL RECOMMENDATIONS None listed.

BIODEGRADATION The biotransformation of methionine in anoxic sediment slurries was found to produce 3–mercapto-propionate (1). In salt marsh sediment slurries, microbial metabolism of methionine yielded methanethiol as the major volatile organosulfur product, with the formation of lesser amounts of dimethyl-sulfide (2); the decomposition of methionine occurred rapidly in anoxic salt marsh sediments (2). In a laboratory activated sludge system, methionine had an 80% BOD reduction after 16 days of incubation (3). Methionine was highly bioconvertible (68–90% CO_2 evolution) in 35–78-day anaerobic degradation studies using waste activated sludge from the San Jose–Santa Clara Water Pollution Control Plant (4). In a Warburg respirometer study using activated sludge, methionine (at a concentration of 500 mg/L) had a theoretical BOD of 2.6% over a 24-hr incubation period (5). In an activated sludge system that had been acclimated to phenol, methionine had a theoretical oxidation of 16% after 12 hr of aeration (6) [R2].

BIOCONCENTRATION Based upon a water solubility of 56,600 mg/L at 25°C (1) and an experimental $logK_{ow}$ of –1.87 (2), respective BCFs of 1.3 and 0.02 can be estimated for L–methionine from recommended regression-derived equations (2,SRC); these estimated BCF values suggest that bioconcentration in aquatic organisms is not an important environmental fate process (SRC) [R3].

FDA Manufacturers, packers, and distributors of drug and drug products for human use are responsible for complying with the labeling, certification, and usage requirements as prescribed by the Federal Food, Drug, and Cosmetic Act, as amended (secs 201 – 902, 52 Stat. 1,040 et seq., as amended; 21 U.S.C. 321–392). [R4].

REFERENCES

1. Budavari, S (ed.). *The Merck Index— Encyclopedia of Chemicals, Drugs, and Biologicals.* Rahway, NJ: Merck and Co., Inc., 1989. 943.

2. (1) Kiene, R. P., B. F. Taylor. *Nature* 332: 148–50 (1988). (2) Kiene, R. P., P. T. Visscher. *Appl Environ Microbiol* 53: 2426–34 (1987). (3) Engelbrecht, R. S., R. E. McKinney. *Sew Indust Wastes* 29: 1350–62 (1957). (4) Stuckey D,C., P. L. McCarty. *Water Res* 18: 1343–53 (1984). (5) Malaney, G.W., R. M. Gerhold. *J Water Pollut Control Fed* 41: R18–R33 (1969). (6) McKinney, R. E., et al. *Sew Indust Wastes* 28: 547–57 (1956).

3. (1) Yalkowsky, S. H., R. M. Dannenfelser. *Arizona Database of Aqueous Solubilities.* Univ of AZ, College of Pharmacy (1992). (2) Hansch, C., and A. J. Leo. *Medchem Project Issue No 26.* Claremont, CA: Pomona College (1985). (3) Lyman, W. J., et al. *Handbook of Chemical Property Estimation Methods,* Washington, DC: Amer Chemical Soc p. 5–4, 5–10 (1990).

4. Kleemann, A., et al. pp. 76 in *Ullmann's Encycl of Indust Chem,* 5th ed. VCH Publishers, NY. VolA2 (1985).

■ METHOXYETHYLMERCURIC ACETATE

CAS # 151–38–2

SYNONYMS methoxyethylmercury acetate • mea • mema • 2-methoxyethylmercury acetate • aceto (2-methoxyethyl) mercury • panogen-metox • panogen-m • cekusil-universal-a

MF $C_5H_{10}HgO_3$

MW 318.7

COLOR AND FORM White crystals.

MELTING POINT 40–42°C.

SOLUBILITY Soluble in water; readily soluble in polar solvents, including methanol and ethylene glycol.

CORROSIVITY Corrosive to iron and other metals.

STABILITY AND SHELF LIFE Stable at pH 9.5 to 11.5 [R1].

OTHER CHEMICAL/PHYSICAL PROPERTIES Methoxyethylmercury radicals have been described as resembling sodium ion in that they are strongly alkaline and form highly ionized salts that are generally water-soluble and appreciably volatile. These compounds are quantitatively decomposed by strong acids. (methoxyethylmercury radicals).

PROTECTIVE EQUIPMENT Wear rubber gloves, respirators, filter, and goggles [R1].

DISPOSAL SRP: At the time of review, criteria for land treatment or burial (sanitary landfill) disposal practices are subject to significant revision. Prior to implementing land disposal of waste residue (including waste sludge), consult with environmental regulatory agencies for guidance on acceptable disposal practices.

COMMON USES Seed treatment for cotton and small grains [R2]; fungicidal seed dressing for controlling seedborne diseases of wheat, barley, oats, and rye [R1].

OCCUPATIONAL RECOMMENDATIONS None listed.

THRESHOLD LIMIT 8-hr time-weighted average (TWA) 0.1 mg/m³, skin (1982) (mercury aryl compounds, as Hg) [R4, 24]. BEI (Biological Exposure Index): Total inorganic mercury in urine (preshift): 35 µg/g creatinine (mercury) [R4, 60]. BEI (Biological Exposure Index): Total inorganic mercury in blood (end of shift at end of workweek): 15 µg/L. (mercury) [R4, 60].

NON-HUMAN TOXICITY LD₅₀ rat, oral 25 mg/kg [R1].

IARC SUMMARY Evaluation: There is inadequate evidence in humans for the carcinogenicity of mercury and mercury compounds. Overall evaluation: Metallic mercury and inorganic mercury compounds are not classifiable as to their carcinogenicity to humans (Group 3) [R3].

BIOLOGICAL HALF-LIFE Biological half-life in rats is 4–10 days [R1].

BIODEGRADATION In soil all the organomercury compounds are decomposed to mercury salts or to metallic mercury, which are the active fungicides It has been suggested that this decomposition takes place through base exchange to form organomercury clays, which subsequently form mercury salts by further base exchange. These mercuric salts are then reduced to mercurous salts and to mercury. The metallic mercury liberated in the soil is ultimately converted to mercury sulfide by reaction with hydrogen sulfide liberated by soil microorganisms. (mercury compounds) [R5, 22].

BIOCONCENTRATION Mercury compounds pose a severe environmental hazard through concentration and accumulation in food chains as biologically formed dimethyl mercury. (mercury compounds) [R5, 24].

FIFRA REQUIREMENTS All uses of mercury are cancelled except the following: (1) as a fungicide in the treatment of textiles and fabrics intended for continuous outdoor use; (2) as a fungicide to control brown mold on freshly sawn lumber; (3) as a fungicide treatment to control Dutch elm disease; (4) as an in-can preservative in water-based paints and coatings; (5) as a fungicide in water-based paints and coatings used for exterior application; (6) as a fungicide to control "winter turf diseases" such as *Sclerotinia boreales*, and gray and pink snow mold, subject to the following: a. the use of these products shall be prohibited within 25 feet of any water body where fish are taken for human consumption; b. these products can be applied only by or under the direct supervision of golf course superintendents; c. the products will be classified as restricted-use pesticides when they are reregistered and classified in accordance with section 4(c) of FEPCA. (mercury) [R6].

REFERENCES

1. Hartley, D., and H. Kidd (eds.). *The Agrochemicals Handbook*. 2nd ed. Lechworth, Herts, England: The Royal Society of Chemistry, 1987, p. A461/Aug 87.

2. *Farm Chemicals Handbook* 1989. Willoughby, OH: Meister Publishing Co., 1989, p. C–192.

3. IARC. *Monographs on the Evaluation of the Carcinogenic Risk of Chemicals to Man*. Geneva: World Health Organization, International Agency for Research on Cancer, 1972–present (multivolume work), p. 58 324 (1993).

4. American Conference of Governmental Industrial Hygienists. *Threshold Limit Values for Chemical Substances and Physical Agents and Biological Exposure Indices for 1994–1995*. Cincinnati, OH: ACGIH, 1994.

5. White-Stevens, R. (ed.). *Pesticides in the Environment*. Volume 1, Part 1, Part 2. New York: Marcel Dekker, Inc., 1971.

6. Environmental Protection Agency/OPTS. *Suspended, Cancelled and Restricted Pesticides*. 3rd Revision. Washington, DC: Environmental Protection Agency, January 1985. 16.

■ METHYL ACETOACETATE

CAS # 105–45–3

SYNONYMS 1-methoxybutane-1,3-dione • 3-oxobutanoic acid methyl ester • acetoacetic acid, methyl ester • acetoacetic methyl ester • butanoic acid, 3-oxo-, methyl ester • methyl 3-oxobutyrate • methyl acetylacetate • methyl acetylacetonate • methylacetoacetate

MF $C_5H_8O_3$

MW 116.13

COLOR AND FORM Colorless liquid.

DENSITY 1.0762 at 20°C.

VISCOSITY 1.704 cP at 20°C.

BOILING POINT 171.7°C.

MELTING POINT −80°C.

AUTOIGNITION TEMPERATURE 536°F (280°C) [R4].

FIREFIGHTING PROCEDURES Use foam, carbon dioxide, dry chemical [R3]. If material is on fire or involved in a fire do not extinguish fire unless flow can be stopped. Use water in flooding quantities as fog. Solid streams of water may be ineffective. Cool all affected containers with flooding quantities of water. Apply water from as far a distance as possible. Use "alcohol" foam, dry chemical, or carbon dioxide [R5].

SOLUBILITY Soluble in 2 parts water; miscible with alcohol and ether; solubility in water: 38 g/100 ml.

VAPOR PRESSURE 0.7 mm Hg at 20°C.

OCTANOL/WATER PARTITION COEFFICIENT Log K_{ow} = -0.264.

OTHER CHEMICAL/PHYSICAL PROPERTIES Solidifies, melting point at -80°C; slightly decomposition between 169 and 171°C; gives a deep-red color with ferric chloride; conversion factors (wt/vol): 4.74 mg/m³ is equivalent to 1 ppm; bulk density: 9 lb/gal at 20°C; vapor pressure = 0.892 mm Hg at 25°C (est); Henry's Law constant = 1.18×10^{-7} atm-m³/mole at 25°C (est).

PROTECTIVE EQUIPMENT Wear appropriate chemical protective gloves, boots, and goggles [R5].

PREVENTATIVE MEASURES If material is not on fire and not involved in a fire keep sparks, flames, and other sources of ignition away. Keep material out of water sources and sewers. Build dikes to contain flow as necessary. Use water spray to knock down vapors [R5]. Avoid breathing vapors. Keep upwind. Do not handle broken packages unless wearing appropriate personal protective equipment [R5]. SRP: The scientific literature supports the wearing of contact lenses in industrial environments, as part of a program to protect the eye against chemical compounds and minerals causing eye irritation. However, there may be individual substances whose irritating or corrosive properties are such that the wearing of contact lenses would be harmful to the eye. In those specific cases contact lenses should not be worn.

COMMON USES solvent for cellulose ethers; component of solvent mixtures for cellulose esters; chemical intermediate for organic compounds [R1]. Used in synthesis of nifedipine [R2].

OCCUPATIONAL RECOMMENDATIONS None listed.

NON-HUMAN TOXICITY LD_{50} rat oral 3.0 g/kg [R6]; LD_{50} rabbit dermal >10 ml/kg. [R10, 2282].

BIODEGRADATION Methyl acetoacetate was degraded more than 90% in 2 days at 22°C using activated sludge inocula in a Zahn–Wellens screening test (1) [R7].

BIOCONCENTRATION Based on an estimated log K_{ow} of -0.264 (3) the BCF for methyl acetoacetate can be estimated to be 0.37 using a recommended regression-derived equation (1). Methyl acetoacetate has been reported to be soluble in 2 parts water (no temperature reported) (2) (33% weight; 333,000 mg/L). Based on this water solubility value, the BCF for methyl acetoacetate can be estimated to be 0.48 using a recommended regression-derived equation (1). These BCF values suggest that methyl acetoacetate would not bioconcentrate in aquatic organisms (SRC) [R8].

ATMOSPHERIC STANDARDS This action promulgates standards of performance for equipment leaks of volatile organic compounds (VOC) in the synthetic organic chemical manufacturing industry (SOCMI). The intended effect of these standards is to require all newly constructed, modified, and reconstructed SOCMI process units to use the best demonstrated system of continuous emission reduction for equipment leaks of VOC, considering costs, non-air-quality health, and environmental impact and energy requirements. Methyl acetoacetate is produced, as an intermediate or a final product, by process units covered under this subpart [R9]. Fire potential fire hazard: moderate, when exposed to heat, flame, or oxidizers [R3].

GENERAL RESPONSE Stop discharge if possible. Keep people away. Call fire department. Shut off all sources of ignition.

Isolate and remove discharged material. Notify local health and pollution agencies.

FIRE RESPONSE Combustible. Wear self-contained breathing apparatus and protective clothing. Extinguish with dry chemical, alcohol foam, or CO_2. Use water sprays to cool fire-exposed containers.

EXPOSURE RESPONSE Call for medical aid. Liquid irritating to eyes. Harmful if swallowed. Remove contaminated clothing and shoes. Flush affected areas with plenty of water. If in eyes, hold eyelids open and flush with plenty of water.

WATER RESPONSE Effect of low concentrations on aquatic life is unknown. May be dangerous if it enters water intakes. Notify local health and wildlife officials. Notify operators of nearby water intakes.

EXPOSURE SYMPTOMS May be harmful by inhalation, ingestion, or skin absorption. Causes eye irritation. May cause skin irritation.

EXPOSURE TREATMENT Inhalation: call for medical aid. Remove to fresh air. If not breathing, give aritificial respiration. If breathing is difficult, give oxygen. Eyes: flush eyes with copious amounts of water for at least 15 minutes. Skin: wash with soap and copious amounts of water.

REFERENCES

1. SRI.

2. *Kirk-Othmer Encyclopedia of Chemical Technology.* 3rd ed., Volumes 1–26. New York, NY: John Wiley and Sons, 1978–1984, p. V4 919 (1978).

3. Sax, N. I. *Dangerous Properties of Industrial Materials,.* 5th ed. New York: Van Nostrand Reinhold, 1979. 805.

4. National Fire Protection Guide. *Fire Protection Guide on Hazardous Materials.* 10th ed. Quincy, MA: National Fire Protection Association, 1991, p. 325M-65.

5. Association of American Railroads. *Emergency Handling of Hazardous Materials in Surface Transportation.* Washington, DC: Assoc. of American Railroads, Hazardous Materials Systems (BOE), 1987. 446.

6. Budavari, S. (ed.). *The Merck Index— Encyclopedia of Chemicals, Drugs, and Biologicals.* Rahway, NJ: Merck and Co., Inc., 1989. 948.

7. Zahn, R., H. Wellens, Z. Wasser. *Abwasser Forsch* 13: 1–7 (1980).

8. (1) Lyman, W. J., et al., *Handbook of Chemical Property Estimation Methods.* Washington DC: Amer Chemical Soc pp. 5–4, 5–10 (1990). (2) Windholz, M., et al., *The Merck Index* 10th ed. Rahway, NJ: Merck and Co., Inc., p. 863 (1983). (3) GEMS. *Graphical Exposure Modeling System.* PCGEMS (1987).

9. 40 CFR 60.489 (7/1/90).

10. Clayton, G. D., and F. E. Clayton (eds.). *Patty's Industrial Hygiene and Toxicology.* Volume 2A, 2B, 2C: *Toxicology.* 3rd ed. New York: John Wiley and Sons, 1981–1982.

■ METHYL CHLOROFORMATE

CAS # 79–22–1

SYNONYMS carbonochloridic acid, methyl ester • Chlorameisensaeure-methylester (German) • chlorocarbonate de methyle (French) • chlorocarbonic acid methyl ester • chloroformiate de methyle (French) • chloroformic acid methyl ester • formic acid, chloro-, methyl ester • k-stoff • methoxycarbonyl chloride • methyl chlorocarbonate • methylchloorformiaat (Dutch) • metilcloroformiato (Italian) • tl-438 • mcf

MF $C_2H_3ClO_2$

MW 94.50

COLOR AND FORM Clear liquid.

ODOR Unpleasant, acrid.

DENSITY 1.223 at 20°C.

SURFACE TENSION 0.044116 newtons/meter at 192 K.

BOILING POINT 71.0°C.

MELTING POINT Less than −114°F = less than −81°C = less than 192 K.

HEAT OF COMBUSTION -6.89×10^8 J/kmol.

HEAT OF VAPORIZATION 4.2599×10^7 J/kmol at 192 K.

AUTOIGNITION TEMPERATURE 940°F [R1].

EXPLOSIVE LIMITS AND POTENTIAL Containers may explode in fire. Vapor may explode if ignited in an enclosed area [R1].

FLAMMABILITY LIMITS Lower: 6.7% [R1].

FIREFIGHTING PROCEDURES Water, dry chem, foam, carbon dioxide. Cool exposed containers with water [R1]. Do not extinguish fire unless flow can be stopped. Use alcohol foam, dry chemical, or carbon dioxide. Cool all affected containers with flooding quantities of water. Apply water from as far a distance as possible. Do not use water on material itself. If large quantities of combustibles are involved, use water in flooding quantities as spray and fog. Use water spray to knock down vapors [R4].

TOXIC COMBUSTION PRODUCTS Irritating and toxic hydrogen chloride and phosgene may be formed [R1].

SOLUBILITY Soluble in chloroform, benzene, in all proportions in alcohol, ether.

VAPOR PRESSURE 108.5 mm Hg at 25°C.

CORROSIVITY Corrodes rubber.

OTHER CHEMICAL/PHYSICAL PROPERTIES Ratio of specific heats of vapor (gas): 1.1544. Decomposed by water. Liquid molar volume = 0.077878 m^3/kmol; heat of formation = -4.24×10^8 J/kmol; autoignition temperature = 777.15 K; Flammability limits = 6.7 to 15.6 volume percent. Aqueous hydrolysis rate constant in distilled water is 0.000564/sec at 25°C, which corresponds to a hydrolysis half-life of about 20.5 min. Chloroformates are reactive intermediates that combine acid chloride and ester functions. They undergo many reactions of acid chlorides; however, the rates are usually lower. Reactions of chloroformates, like other acid chlorides, proceed faster with better yields when alkali hydroxides or tertiary amines are present to react with the HCl as it forms (chloroformates).

PROTECTIVE EQUIPMENT Acid or organic-canister mask or self-contained breathing apparatus; goggles or face shield; plastic gloves [R1].

PREVENTATIVE MEASURES If material not on fire and not involved in Fire: Keep sparks, flames, and other sources of ignition away. Keep material out of water sources and sewers. Build dikes to contain flow as necessary. Attempt to stop leak if without undue personnel hazard. Use water spray to knock down vapors. Do not use water on material itself. Neutralize spilled material with crushed limestone, soda ash, or lime [R4].

DISPOSAL Generators of waste (equal to or greater than 100 kg/mo) containing this contaminant, EPA hazardous waste number U156, must conform with USEPA regulations in storage, transportation, treatment, and disposal of waste [R5].

COMMON USES Organic synthesis, insecticides [R2]; as a warfare agent during World War I (former use) [R10, 2387]; reactive intermediate to pesticides, pharmaceuticals, and other chemicals; pharmaceutical synthesis, agricultural products (herbicides, insecticides) [R3]; used as solvent in the photographic industry employed in the production of carbamates that are used to synthesize dyes, drugs, veterinary medicines, herbicides, and insecticides. Used as starting material for the synthesis of the preservative velcorin. [R11, 562].

OCCUPATIONAL RECOMMENDATIONS None listed.

HUMAN TOXICITY A concentration of 190 ppm (1 mg/L) has been lethal in 10 min. [R10, 2387].

NON-HUMAN TOXICITY LD_{50} rat, oral <0.05 g/kg [R10, 2386].

BIODEGRADATION No data were located concerning the biodegradation of methyl chloroformate either in natural systems or in laboratory studies (SRC). Because methyl chloroformate hydrolyzes with a calculated half-life of 34.8–min at 19.6°C (1,SRC), biodegradation probably will not be a significant process in the environment (SRC) [R6].

BIOCONCENTRATION Because methyl chloroformate hydrolyzes in water relatively rapidly (1), bioconcentration in aquatic organisms is not expected to be a significant process (SRC) [R6].

CERCLA REPORTABLE QUANTITIES Persons in charge of vessels or facilities are required to notify the National Response Center (NRC) immediately, when there is a release of this designated hazardous substance, in an amount equal to or greater than its reportable quantity of 1,000 lb or 454 kg. The toll-free number of the NRC is (800) 424–8802; in the Washington DC metropolitan area (202) 426–2675. The rule for determining when notification is required is stated in 40 CFR 302.4 (section IV. D. 3.b) [R7].

RCRA REQUIREMENTS U156; as stipulated in 40 CFR 261.33, when methyl chlorocarbonate, as a commercial chemical product or manufacturing chemical intermediate or an off-specification commercial chemical product or a manufacturing chemical intermediate, becomes a waste, it must be managed according to federal or state hazardous waste regulations. Also defined as a hazardous waste is any residue, contaminated soil, water, or other debris resulting from the cleanup of a spill, into water or on dry land, of this waste. Generators of small quantities of this waste may qualify for partial exclusion from hazardous waste regulations (40 CFR 261.5) [R8].

DOT *Health Hazards:* Poisonous; may be fatal if inhaled, swallowed, or absorbed through skin. Contact may cause burns to skin and eyes. Runoff from fire control or dilution water may cause pollution [R9, p. G–28].

Fire or Explosion: Flammable/combustible material; may be ignited by heat, sparks, or flames. Vapors may travel to a source of ignition, and flash back. Container may explode in heat of fire. Vapor explosion and poison hazard indoors, outdoors, or in sewers. Runoff to sewer may create fire or explosion hazard [R9, p. G–28].

Emergency Action: Keep unnecessary people away; isolate hazard area, and deny entry. Stay upwind; keep out of low areas. Positive-pressure self-contained breathing apparatus (SCBA) and chemical protective clothing that is specifically recommended by the shipper or manufacturer may be worn. It may provide little or no thermal protection. Structural firefighters' protective clothing is not effective for these materials. See the Table of Initial Isolation and Protective Action Distances. If you find the ID Number and the name of the material there, begin protective action. Isolate for 1/2 mile in all directions if tank, rail car, or tank truck is involved in fire. Call emergency response number on shipping paper first. If shipping paper not available or no answer, call CHEMTREC at 1–800–424–9300 [R9, p. G–28].

Fire: Small fires: Dry chemical, CO_2, water spray, or alcohol-resistant foam. Large Fires: Water spray, fog, or alcohol-resistant foam. Move container from fire area if you can do so without risk. Dike fire-control water for later disposal; do not scatter the material. Apply cooling water to sides of containers that are exposed to flames until well after fire is out. Stay away from ends of tanks. Withdraw immediately in case of rising sound from venting safety device or any discoloration of tank due to fire [R9, p. G–28].

Spill or Leak: Shut off ignition sources; no flares, smoking, or flames in hazardarea. Fully encapsulating, vapor-protective clothing should be worn for spills and leaks with no fire. Do not touch or walk through spilled material; stop leak if you can do so without risk. Water spray may reduce vapor; but it may not prevent ignition in closed spaces.

Small Spills: Take up with sand or other noncombustible absorbent material and place into containers for later disposal. Large Spills: Dike far ahead of liquid spill for later disposal [R9, p. G–28].

First Aid: Move victim to fresh air, and call for emergency medical care; if not breathing, give artificial respiration; if breathing is difficult, give oxygen. In case of contact with material, immediately flush skin or eyes with running water for at least 15 minutes. Remove and isolate contaminated clothing and shoes at the site. Keep victim quiet and maintain normal body temperature. Effects may be delayed; keep victim under observation [R9, p. G–28].

Initial Isolation and Protective Action Distances: Small Spills (from a small package, or small leak from a large package): First, isolate in all directions 500 feet; then, protect persons downwind 0.2 mile (day), or 1.4 miles (night). Large spills (from a large package or from many small packages): First, isolate in all directions 500 feet; then, protect persons downwind 0.2 mile (day), or 1.8 miles (night) [R9, p. G–Table].

FIRE POTENTIAL Very dangerous when exposed to heat sources, sparks, flame, or oxidizers. [R10, 1826].

REFERENCES

1. U.S. Coast Guard, Department of Transportation. *CHRIS—Hazardous Chemical Data.* Volume II. Washington, DC: U.S. Government Printing Office, 1984-5.

2. Sax, N. I., and R. J. Lewis, Sr. (eds.). *Hawley's Condensed Chemical Dictionary.* 11th ed. New York: Van Nostrand Reinhold Co., 1987. 764.

3. Kuney, J. H., J. M. Mullican (eds.). *Chemcyclopedia.* Washington, DC: American Chemical Society, 1994. 90.

4. Association of American Railroads. *Emergency Handling of Hazardous Materials in Surface Transportation.* Washington, DC: Association of American Railroads, Bureau of Explosives, 1992. 624.

5. 40 CFR 240–280, 300–306, 702–799 (7/1/90).

6. Queen, A. *Can J Chemical* 45: 1619–29 (1967).

7. 40 CFR 302.4 (7/1/92).

8. 40 CFR 261.33 (7/1/92).

9. U.S. Department of Transportation. *Emergency Response Guidebook 1993.* DOT P 5800.6. Washington, DC: U.S. Government Printing Office, 1993.

10. Clayton, G. D., and F. E. Clayton (eds.). *Patty's Industrial Hygiene and Toxicology.* Volume 2A, 2B, 2C: *Toxicology.* 3rd ed. New York: John Wiley and Sons, 1981–1982.

11. Gerhartz, W. (exec ed.). *Ullmann's Encyclopedia of Industrial Chemistry.* 5th ed. Vol A1: Deerfield Beach, FL: VCH Publishers, 1985 to present, p. VA6.

■ METHYL ISOTHIOCYANATE

CAS # 556–61–6

SYNONYMS methane, isothiocyanato • ai3-28257 • Caswell no 573 • EPA pesticide chemical code 068103 • isothiocyanate de methyle (French) • isothiocyanatomethane • isothiocyanic acid, methyl ester • isotiocianato de metilo (Spanish) • isotiocianato di metile (Italian) • metile isotiocianato (Italian) • methylisothiocyanat (Denmark) • methyl mustard • methyl mustard oil • Methylisothiocyanat (German) • methylisothiocyanaat (Netherlands) • methylisothiokyanat (Czech) • Methylsenfoel (German) • morton ep-161e • trapex • trapexide • vorlex (Nor-Am) • vorlex 201 (Nor-Am) • wn-12 • mit • mitc

MF C_2H_3NS

MW 73.12

COLOR AND FORM Colorless crystals.

ODOR Pungent horseradish-like odor.

DENSITY 1.0691 at 37°C.

BOILING POINT 119°C at 758 mm Hg.

MELTING POINT 35°C.

SOLUBILITY 7.6 g/L in water at 20°C; readily soluble in acetone, benzene, cyclohexanone, dichloromethane, light petroleum, and methanol; readily soluble in common organic solvents, such as chloroform, carbon tetrachloride, xylene, and mineral oils; soluble in alcohol and ether.

VAPOR PRESSURE 19 mm Hg at 20°C.

OCTANOL/WATER PARTITION COEFFICIENT
Log K_{ow} = 0.94.

CORROSIVITY Corrosive to iron, zinc, and other metals.

STABILITY AND SHELF LIFE Unstable and reactive. Sensitive to oxygen and to light [R5].

OTHER CHEMICAL/PHYSICAL PROPERTIES Boiling point: 117–119°C (technical grade, 96% pure); vapor pressure: 30 mm Hg at 30°C; saturated concentration: 75.6 g/m³ at 20°C; 115 g/m³ at 30°C; rapidly hydrolyzed by alkalis, more slowly in acidic and neutral solutions. Boiling point: 88°C; density: 1.520 g/m³; colorless liquid.

PROTECTIVE EQUIPMENT Wear respirator and polyethylene gloves and footwear, because methyl isothiocyanate can penetrate rubber [R5].

PREVENTATIVE MEASURES Do not reuse container. Do not cut or weld container because an explosion may result even if the container is empty. (vorlex formulation) [R3].

DISPOSAL Group I Containers: Combustible containers from organic or metallo-organic pesticides (except organic mercury, lead, cadmium, or arsenic compounds) should be disposed of in pesticide incinerators or in specified landfill sites. (organic or metallo-organic pesticides) [R8].

COMMON USES Agricultural insecticide [R6]. Preplant for weeds, fungi, insects, and nematodes in potatoes, tobacco, vegetables, and, ornamentals (vorlex formulation) [R3]. A possible military poison [R2].

OCCUPATIONAL RECOMMENDATIONS None listed.

NON-HUMAN TOXICITY LD_{50} rat oral 175 mg/kg [R1]; LD_{50} rat percutaneous 961 mg/kg (vorlex formulation) [R3]; LD_{50} rabbit percutaneous 1,243 mg/kg (vorlex formulation) [R3]; LD_{50} mouse male oral 90 mg/kg [R4]; LD_{50} mouse male percutaneous 1,870 mg/kg [R4].

ECOTOXICITY LC_{50} bluegill sunfish 0.13 mg/L/96 hr (conditions of bioassay not specified) [R5]; LC_{50} rainbow trout 0.37 mg/L/96 hr (conditions of bioassay not specified) [R5]; LC_{50} carp 0.37 mg/L/96 hr (conditions of bioassay not specified) [R5]; LD_{50} duck oral 136 mg/kg [R4].

BIODEGRADATION In a soil biodegradation study, the biodegradation of methyl isothiocyanate generally followed first-order kinetics in soils previously untreated with the compound. In soils previously exposed to methyl isothiocyanate, the biodegradation was generally much faster, and the biodegradation did not follow first-order kinetics (1). Even in soils that were previously untreated with methyl isothiocyanate, an accelerated transformation (which did follow first-order kinetics) was observed after an initial period (8–15 days) of first-order transformation (1). Using the first-order kinetics, the biodegradation half-life was estimated to range 0.5 to 50 days (2). Generally, the transformation was appreciably faster in soils that had been previously treated with the chemical frequently (2). At or above concns of 0.8 mg/L, methyl isothiocyanate inhibited nitrification in the activated-sludge process of sewage disposal (2) [R9].

BIOCONCENTRATION Based on two regression equations, one involving water solubility and the other soil adsorption coefficient, the bioconcentration factor of methyl isothiocyanate was estimated to be in the range 0.2 to 4 (1). These values indicate bioconcentration of the compound in aquatic organisms may not be important (SRC) [R10].

DOT *Health Hazards:* Poisonous; may be fatal if inhaled, swallowed, or absorbed through skin. Contact may cause burns to skin and eyes. Runoff from fire control or dilution water may cause pollution [R7].

Fire or Explosion: Flammable/combustible material; may be ignited by heat, sparks, or flames. Vapors may travel to a source of ignition, and flash back. Container may explode in heat of fire. Vapor explosion and poison hazard indoors, outdoors, or in sewers. Runoff to sewer may create fire or explosion hazard [R7].

Emergency Action: Keep unnecessary people away; isolate hazard area and deny entry. Stay upwind; keep out of low areas.

Positive-pressure self-contained breathing apparatus (SCBA) and chemical protective clothing that is specifically recommended by the shipper or manufacturer may be worn. It may provide little or no thermal protection. Structural firefighters' protective clothing is not effective for these materials. Isolate for 1/2 mile in all directions if tank, rail car, or tank truck is involved in fire. Call CHEMTREC at 1–800–424–9300 for emergency assistance [R7].

Fire: Small fires: Dry chemical, CO_2, water spray, or alcohol-resistant foam. Large Fires: Water spray, fog, or alcohol-resistant foam. Move container from fire area if you can do so without risk. Dike fire-control water for later disposal; do not scatter the material. Apply cooling water to sides of containers that are exposed to flames until well after fire is out. Stay away from ends of tanks. Withdraw immediately in case of rising sound from venting safety device or any discoloration of tank due to fire [R7].

Spill or Leak: Shut off ignition sources; no flares, smoking, or flames in hazard area. Fully encapsulating, vapor– protective clothing should be worn for spills and leaks with no fire. Do not touch or walk through spilled material; stop leak if you can do so without risk. Water spray may reduce vapor; but it may not prevent ignition in closed spaces.

Small Spills: Take up with sand or other noncombustible absorbent material and place into containers for later disposal. Large Spills: Dike far ahead of liquid spill for later disposal [R7].

First Aid: Move victim to fresh air and call for emergency medical care; if not breathing, give artificial respiration; if breathing is difficult, give oxygen. In case of contact with material, immediately flush skin or eyes with running water for at least 15 minutes. Remove and isolate contaminated clothing and shoes at the site. Keep victim quiet and maintain normal body temperature. Effects may be delayed; keep victim under observation [R7].

FIRE POTENTIAL Flammable [R6].

REFERENCES

1. Sittig, M., (ed.) *Pesticide Manufacturing and Toxic Materials Control Encyclopedia*. Park Ridge, NJ: Noyes Data Corporation. 1980. 523.

2. Sax, N. I., and R. J. Lewis, Sr. (eds.). *Hawley's Condensed Chemical Dictionary*. 11th ed. New York: Van Nostrand Reinhold Co., 1987. 773.

3. *Farm Chemicals Handbook* 1991. Willoughby, OH: Meister, 1991, p. C–323.

4. Worthing, C. R., and S. B. Walker (eds.). *The Pesticide Manual—A World Compendium*. 8th ed. Thornton Heath, UK: The British Crop Protection Council, 1987. 564.

5. Hartley, D., and H. Kidd (eds.). *The Agrochemicals Handbook*. 2nd ed. Lechworth, Herts, England: The Royal Society of Chemistry, 1987, p. A274/Aug 87.

6. ITII. *Toxic and Hazardous Industrial Chemicals Safety Manual*. Tokyo, Japan: The International Technical Information Institute, 1988. 341.

7. U.S. Department of Transportation. *Emergency Response Guidebook 1990*. DOT P 5800.5. Washington, DC: U.S. Government Printing Office, 1990, p. G–28.

8. 40 CFR 165.9 (a) (7/1/90).

9. (1) Tomlinson, T. G., et al., *J Appl Bact* 29: 266–91 (1966). (2) Smelt, J. H., et al. *J Environ Sci Health* B24: 437–55 (1989).

10. Kenega, E. E. *Ecotox Environ Safety* 4: 26–38 (1980).

■ N-METHYLANILINE

CAS # 100–61–8

DOT # 2294

SYNONYMS (methylamino) benzene • aniline, N-methyl • anilinomethane • benzeneamine, N-methyl • methylaniline • methylphenylamine • monomethylaniline • N-methylaminobenzene • N-methylben- zenamine • N-methylphenylamine • N-monomethylaniline • N-phenylmethylamine

MF C₇H₉N

MW 107.17

COLOR AND FORM Colorless to reddish-brown oily liquid.

DENSITY 0.989 at 20°C.

BOILING POINT 194–196°C.

MELTING POINT –57°C.

SOLUBILITY Soluble in alcohol, ether, chloroform; insoluble in water.

VAPOR PRESSURE 1 mm Hg at 36.0°C.

STABILITY AND SHELF LIFE Turns reddish brown on standing [R4].

PREVENTATIVE MEASURES Basic ventilation methods are local exhaust ventilation and dilution or general ventilation [R5].

COMMON USES Acid acceptor [R2]. Solvent for organic reactions and for nitrocellulose [R1].

OCCUPATIONAL RECOMMENDATIONS None listed.

THRESHOLD LIMIT 8-hr time-weighted average (TWA) 0.5 ppm, 2.2 mg/m³, skin (1986) [R7, 25].

BIODEGRADATION Thirty-eight process wastewaters and 37 organic substances identified in the wastewater of the Kashima petrochemical complex were subjected to a biodegradability test using the respiration method. The test used the activated sludge of the Fukashiba industrial wastewater treatment plant, which was acclimatized to the wastewater and organic substances. Using the respiration meter, the oxygen uptake and decr in total organic carbon after 15 days cultivation were measured and the biological degradability was calculated. An initial concentration of 100 mg/L N–methylaniline resulted in 42% total organic carbon [R6].

DOT *Health Hazards:* Poisonous; may be fatal if inhaled, swallowed, or absorbed through skin. Contact may cause burns to skin and eyes. Runoff from fire control or dilution water may cause pollution [R3].

Fire or Explosion: May be ignited by heat, sparks, or flames. Container may explode in heat of fire. Vapor explosion and poison hazard indoors, outdoors, or in sewers [R3].

Emergency Action: Keep unnecessary people away; isolate hazard area and deny entry. Stay upwind, out of low areas, and ventilate closed spaces before entering. Self-contained breathing apparatus and chemical protective clothing that is specifically recommended by the shipper or producer may be worn but they do not provide thermal protection unless that is stated by the clothing manufacturer. Structural firefighter's protective clothing is not effective with these materials. Call CHEMTREC at 1–800–424–9300, as soon as possible, especially if there is no local hazardous materials team available [R3].

Fire: Small fires: Dry chemical, CO₂, halon, water spray, or standard foam. Large Fires: Water spray, fog, or standard foam is recommended. Move container from fire area if you can do so without risk. Cool container with water using unmanned device until well after fire is out. Fight fire from maximum distance. Stay away from ends of tanks. Dike fire control water for later disposal; do not scatter the material [R3].

Spill or Leak: Shut off ignition sources; no flares, smoking, or flames in hazard area. Do not touch spilled material; stop leak if you can do so without risk. Water spray may reduce vapor; but it may not prevent ignition in closed spaces. Small Spills: Take up with sand or other noncombustible absorbent material and place into containers for later disposal. Small Dry Spills: With clean shovel place material into clean, dry container, and cover; move containers from spill area. Large Spills: Dike far ahead of liquid spill for later disposal [R3].

First Aid: Move victim to fresh air and call for emergency medical care; if not breathing, give artificial respiration; if breathing is difficult, give oxygen. In case of contact with material, immediately flush skin or eyes with running water for at least 15 minutes. Speed in removing material from skin is of extreme importance. Remove and isolate contaminated clothing and shoes at the site. Keep victim quiet and maintain normal body temperature. Effects may be delayed; keep victim under observation [R3].

GENERAL RESPONSE Avoid contact with liquid. Keep people away. Stop discharge if possible. Call fire department. Isolate and remove discharged material. Notify local health and pollution control agencies.

FIRE RESPONSE Combustible. Poisonous gases may be produced in fire. Wear goggles and self-contained breathing apparatus. Extinguish with dry chemicals, foam, or carbon dioxide. Water may be ineffective on fire. Cool exposed containers with water.

EXPOSURE RESPONSE Call for medical aid. Liquid irritating to skin and eyes. Harmful if swallowed. Remove contaminated clothing and shoes. Flush affected areas with plenty of water. If swallowed and victim is conscious, have victim drink water or milk. If swallowed, and victim is unconscious or having convulsions, do nothing except keep victim warm.

RESPONSE TO DISCHARGE Issue warning—poison, water contaminant. Restrict access. Mechanical containment. Should be removed. Chemical and physical treatment.

WATER RESPONSE Effect of low concentrations on aquatic life is unknown. Fouling to shoreline. May be dangerous if it enters water intakes. Notify local health and wildlife officials. Notify operators of nearby water intakes.

EXPOSURE SYMPTOMS Inhalation causes dizziness and headache. Ingestion causes bluish discoloration (cyanosis) of lips, ear lobes, and fingernail beds. Liquid irritates eyes. Absorption through skin produces same symptoms as for ingestion.

EXPOSURE TREATMENT Inhalation: remove victim to fresh air and call a physician at once; administer oxygen until phy-

sician arrives. Ingestion: give large amount of water; get medical attention at once. Eyes or skin: flush with plenty of water for at least 15 min; if cyanosis is present, shower with soap and warm water, with special attention to scalp and finger nails; remove any contaminated clothing.

REFERENCES

1. SRI.

2. Hawley, G. G. *The Condensed Chemical Dictionary.* 9th ed. New York: Van Nostrand Reinhold Co., 1977. 560.

3. Department of Transportation. *Emergency Response Guidebook 1987.* DOT P 5800.4. Washington, DC: U.S. Government Printing Office, 1987, p. G–57.

4. American Conference of Governmental Industrial Hygienists, Inc. *Documentation of the Threshold Limit Values,* 4th ed., 1980. Cincinnati, OH: American Conference of Governmmental Industrial Hygienists, Inc., 1980. 265.

5. Sax, N. I. *Dangerous Properties of Industrial Materials,.* 4th ed. New York: Van Nostrand Reinhold, 1975. 910.

6. Matsui, S., et al., *Water Sci Technol* 20 (10): 201–10 (1989).

7. American Conference of Governmental Industrial Hygienists. *Threshold Limit Values for Chemical Substances and Physical Agents and Biological Exposure Indices for 1994–1995.* Cincinnati, OH: ACGIH, 1994.

■ 3–METHYLCHOLANTHRENE

CAS # 56–49–5

SYNONYMS 3-MC • 3-MCh • 3-methyl-i,j-cyclopentabenz (a) anthracene • 3-methylbenz (j) aceanthrene • 3-methylcyclopentabenzophenanthrene • 20-mc • 20-methylcholanthrene • benz (j) aceanthrylene, 1,2-dihydro-3-methyl • 3-meca • 3-Methylcholanthren (German)

COLOR AND FORM Pale-yellow, slender prisms from benzene plus ether.

DENSITY 1.28 at 20°C.

BOILING POINT 280°C at 80 mm Hg.

MELTING POINT 179°C.

SOLUBILITY Soluble in benzene, xylene, toluene; slightly soluble in amyl alcohol; water solubility of $1.06 \times 10^{-8} m^2$ at 25°C; 3–MC solubility in a mixed lipid/sodium taurocholate bile salt system was minimal below the critical miocelle concentration of the bile salt.

VAPOR PRESSURE Estimated at 3.8×10^{-6} mm Hg.

OCTANOL/WATER PARTITION COEFFICIENT Log K_{ow} = 6.42 (reported).

OTHER CHEMICAL/PHYSICAL PROPERTIES Conversion factor (wt/vol): 10.98 mg/m³ is equivalent to 1 ppm.

PROTECTIVE EQUIPMENT For carcinogens: dispensers of liquid detergent safety pipettes for all pipetting. In animal laboratory protective suits (preferably disposable, one-piece, and close-fitting at ankles and wrists), gloves, hair covering, and overshoes. In chemical laboratory, gloves and gowns should always be worn; however, gloves should not be assumed to provide full protection. Carefully fitted masks or respirators when working with particulates or gases, and disposable plastic aprons gowns should be of distinctive color. (chemical carcinogens) [R8, 8].

PREVENTATIVE MEASURES For carcinogens: smoking, drinking, eating, storage of food, or of food and beverage containers, or utensils, and the application of cosmetics should be prohibited in any laboratory. All personnel should remove gloves, if worn, after completion of procedures in which carcinogens have been used. They should wash hands, preferably using dispensers of liquid detergent, and rinse thoroughly. Consideration should be given to appropriate methods for cleaning the skin, depending on nature of the contaminant. No standard procedure can be recommended, but the use of organic solvents should be avoided. Safety pipettes should be used for all pipetting. (chemical carcinogens) [R8, 8].

CLEANUP The particle-bound portion of polycyclic aromatic hydrocarbons (PAH) can be removed by sedimentation, flocculation, and filtration processes. The remaining one-third dissolved PAH usually requires oxidation for partial removal/transformation (polycyclic aromatic hydrocarbons) [R9]. Precautions for carcinogens: A high-efficiency particulate arrestor (HEPA) or charcoal filters can be used to minimize amt of carcinogen in exhausted-air-ventilated safety cabinets, lab hoods, glove boxes, or animal rooms. Filter housing that is designed so that used filters can be transferred into plastic bag without contaminating maintenance staff is available commercially. Filters should be placed in plastic bags immediately after removal. The plastic bag should be sealed immediately. The sealed bag should be labeled properly. Waste liquids should be placed or collected in proper containers for disposal. The lid should be secured and the bottles properly labeled. Once filled, bottles should be placed in plastic bag, so that outer surface is not contaminated. The plastic bag should also be sealed and labeled. Broken glassware should be decontaminated by solvent extraction, by chemical destruction, or in specially designed incinerators. (chemical carcinogens) [R8, 15].

DISPOSAL At the time of review, criteria for land treatment or burial (sanitary landfill) disposal practices are subject to significant revision. Prior to implementing land disposal of waste residue (including waste sludge), consult with environmental regulatory agencies for guidance on acceptable disposal practices [R3].

COMMON USES Research chemical [R1].

OCCUPATIONAL RECOMMENDATIONS.
OSHA Meets criteria for OSHA medical records rule [R6].

BIODEGRADATION Methylcholanthrene (MC) is essentially inert to oxidation by activated sludge (1) [R4].

BIOCONCENTRATION BCF *Daphnia magna*, 775; BCF increased by presence of Aldrich humic acids, although increase is not proportional to concentration of humic acids; e.g. BCF (dissolved, organic carbon, ppm): 1,823 (2.0), 2,179 (6.0), 1,096 (10.0) (1). Using a reported log K_{ow} of 6.42 (2), a BCF of 45,000 was estimated (3,SRC). Based on this predicted BCF, 3–methylcholanthrene will be expected to bioconcentrate in aquatic organisms (SRC) [R5].

RCRA REQUIREMENTS When 3–methylcholanthrene, as a commercial chemical product or manufacturing chemical intermediate or an off-specification commercial chemical product or a manufacturing chemical intermediate, becomes a waste, it must be managed according to federal or state hazardous waste regulations. Also defined as a hazardous waste is any residue, contaminated soil, water, or other debris resulting from the cleanup of a spill, into water or on dry land, of this waste. Generators of small quantities of this waste may qualify for partial exclusion from hazardous waste regulations (see 40 CFR 261.5) [R7].

DOT *Fire or Explosion:* Flammable/combustible material; may be ignited by heat, sparks, or flames. May burn rapidly with flare-burning effect (naphthalene) [R6].

Health Hazards: Fire may produce irritating or poisonous gases. Contact may cause burns to skin and eyes. Runoff from fire control or dilution water may cause pollution (naphthalene) [R6].

Fire: Small fires: Dry chemical, sand, water spray, or foam. Large Fires: Water spray, fog, or foam. Move container from fire area if you can do so without risk. Cool containers that are exposed to flames with water from the side until well after fire is out. For massive fire in cargo area, use unmanned hose holder or monitor nozzles; if this is impossible, withdraw from area, and let fire burn. Magnesium Fires: Use dry sand, Met–L–X powder, or G–1 graphite powder (naphthalene) [R6].

Spill or Leak: Shut off ignition sources; no flares, smoking, or flames in hazard area. Do not touch spilled material. Small Dry Spills: With clean shovel, place material into clean, dry container, and cover; move containers from spill area. Large

Spills: Wet down with water and dike for later disposal (naphthalene) [R6].

First Aid: Move victim to fresh air; call for emergency medical care. In case of contact with material, immediately flush skin or eyes with running water for at least 15 min. Remove and isolate contaminated clothing and shoes at the site (naphthalene) [R6].

Emergency Action: Keep unnecessary people away; isolate hazard area and deny entry. Stay upwind; keep out of low areas. Wear self-contained (positive-pressure if available) breathing apparatus and full protective clothing. If water pollution occurs, notify appropriate authorities. For emergency assistance call CHEMTREC, (800) 424–9300 (naphthalene) [R6].

REFERENCES

1. SRI.

2. USEPA. *Ambient Water Quality Criteria Doc; Polycyclic Aromatic Hydrocarbons (Draft)* p. C–4 (1980).

3. SRP.

4. Lutin, P. A., et al. *Purdue U Eng Bull Ext Ser* 118: 131–45 (1965).

5. (1) Leversee, G. J., et al. *Can J Fish Aquat Sci* 40: 63–9 (1983) (2) Hansch, C., and A. J. Leo. *Medchem Project* Issue no. 26, Claremont, CA: Pomona College (1985) (3) Lyman, W. J., et al., *Handbook of Chemical Property Estimation Methods, Environmental Behavior of Organic Commpounds.* New York: McGraw-Hill, p. 5–5 (1983).

6. 29 CFR 1910.20 (7/1/86).

7. 40 CFR 261.33 (7/1/87).

8. Montesano, R., H. Bartsch, E. Boyland, G. Della Porta, L. Fishbein, R. A. Griesemer, A.B. Swan; L. Tomatis, and W. Davis (eds.). *Handling Chemical Carcinogens in the Laboratory: Problems of Safety.* IARC Scientific Publications No. 33. Lyon, France: International Agency for Research on Cancer, 1979.

9. "Toxic Substances," U.S. Department of Health, Education, and Welfare, Rockville, MD, June 1973.

■ METHYLCYCLOHEXANE

CAS # 108–87–2

SYNONYMS cyclohexane, methyl • cyclohexylmethane • hexahydrotoluene • metylocykloheksan (Polish) • sextone-b • toluene hexahydride • toluene, hexahydro-

MF C_7H_{14}

MW 98.21

COLOR AND FORM Colorless liquid.

ODOR THRESHOLD 630 µg/L [R6].

DENSITY 0.7694 at 20°C.

BOILING POINT 100.9°C at 760 mm Hg.

MELTING POINT –126.6°C.

HEAT OF COMBUSTION–4,565.3 kJ/mol at 25°C (liquid).

HEAT OF VAPORIZATION 31.27 kJ/mol at boiling pt; 35.36 kJ/mol at 25°C.

AUTOIGNITION TEMPERATURE 482°F [R11, 839].

EXPLOSIVE LIMITS AND POTENTIAL Moderate explosion hazard when exposed to heat or flame. Lower explosive limit 1.2%, upper-limit 6.7%. [R11, 838].

FIREFIGHTING PROCEDURES Foam, carbon dioxide, dry chemical [R11, 839]; water may be ineffective [R3]. Do not extinguish fire unless flow can be stopped. Use water in flooding quantities as fog. Solid streams of water may be ineffective. Cool all affected containers with flooding quantities of water. Apply water from as far a distance as possible. Use foam, dry chemical, or carbon dioxide [R5].

SOLUBILITY Soluble in alcohol, ether, acetone, benzene; miscible with petroleum ether, carbon tetrachloride; 14.0 mg/L at 20°C (water).

VAPOR PRESSURE 43 mm Hg at 25°C.

OTHER CHEMICAL/PHYSICAL PROPERTIES Refractivity of 1% vapor in air at 25°C and 760 mm Hg is 17.5×10^{-6}; specific dispersion 97.8; enthalpy of fusion: 16.43 cal/g; ionization potential, 9.85 eV.

PROTECTIVE EQUIPMENT SRP: Local exhaust ventilation should be applied wherever there is an incidence of point source emissions or dispersion of regulated contaminants in the work area. Ventilation control of the contaminant as close as possible to its point of generation is both the most economical and safest method to minimize personnel exposure to airborne contaminants.

PREVENTATIVE MEASURES Contact lenses should not be worn when working with this chemical [R4].

CLEANUP Absorb on paper. Evaporate on a glass or iron dish in hood. Burn the paper [R7]. If material not on fire and not involved in.

Fire: Keep sparks, flames, and other sources of ignition away. Keep material out of water sources and sewers. Build dikes to contain flow as necessary. Attempt to stop leak if without undue personal hazard. Use water spray to knock down vapors [R5].

DISPOSAL Spray into the furnace. Incineration will become easier by mixing with a more flammable solvent [R7].

COMMON USES Used commercially as a solvent for cellulose derivatives, particularly with other solvents, and as an organic intermediate in organic synthesis. It is one of the components found in jet fuel [R1].

OCCUPATIONAL RECOMMENDATIONS

OSHA Table Z–1 8–hour time-weighted average: 500 ppm (2,000 mg/m³) [R9].

NIOSH 10-hr time-weighted average: 400 ppm (1,600 mg/m³) [R4].

THRESHOLD LIMIT 8-hr time-weighted average (TWA) 400 ppm, 1,610 mg/m³ (1987) [R12, 26].

NON-HUMAN TOXICITY LC25 rabbit inhalation 7,300 ppm, 6 hr/day, 5 day/wk, 2 wk [R8].

ECOTOXICITY LC50 golden shiner 72.0 mg/L/96 hr (emulsion) (conditions of bioassay not specified) [R8].

BIODEGRADATION Biodegradation: 75% after 192 hr at 13°C (initial concentration: 0.05 µg/L) [R8].

ATMOSPHERIC STANDARDS This action promulgates standards of performance for equipment leaks of volatile organic compounds (VOC) in the synthetic organic chemical manufacturing industry (SOCMI). The intended effect of these standards is to require all newly constructed, modified, and reconstructed SOCMI process units to use the best demonstrated system of continuous emission reduction for equipment leaks of VOC, considering costs, non-air-quality health, and environmental impact and energy requirements. Methylcyclohexane is produced, as an intermediate or a final product, by process units covered under this subpart [R10].

DOT *Fire or Explosion:* Flammable/combustible material; may be ignited by heat, sparks, or flames. Vapors may travel to a source of ignition, and flash back. Container may explode in heat of fire. Vapor explosion hazard indoors, outdoors, or in sewers. Runoff to sewer may create fire or explosion hazard [R2].

Health Hazards: May be poisonous if inhaled or absorbed through skin. Vapors may cause dizziness or suffocation. Contact may irritate or burn skin and eyes. Fire may produce irritating or poisonous gases. Runoff from fire control or dilution water may cause pollution [R2].

Emergency Action: Keep unnecessary people away; isolate hazard area and deny entry. Stay upwind; keep out of low areas. Positive-pressure self-contained breathing apparatus (SCBA) and structural firefighters' protective clothing will provide limited protection. Isolate for 1/2 mile in all directions if tank, rail car, or tank truck is involved in fire. Call CHEMTREC at 1-800-424-9300 for emergency assistance. If water pollution occurs, notify the appropriate authorities [R2].

Fire: Small fires: Dry chemical, CO_2, water spray, or regular foam. Large Fires: Water spray, fog, or regular foam. Move container from fire area if you can do so without risk. Apply cooling water to sides of containers that are exposed to flames until well after fire is out. Stay away from ends of tanks. For massive fire in cargo area, use unmanned hose holder or monitor nozzles; if this is impossible, withdraw from area, and let fire burn. Withdraw immediately in case of rising sound from venting safety device or any discoloration of tank due to fire [R2].

Spill or Leak: Shut off ignition sources; no flares, smoking, or flames in hazard area. Stop leak if you can do so without risk. Water spray may reduce vapor; but it may not prevent ignition in closed spaces.

Small Spills: Take up with sand or other noncombustible absorbent material and place into containers for later disposal. Large Spills: Dike far ahead of liquid spill for later disposal [R2].

First Aid: Move victim to fresh air and call for emergency medical care; if not breathing, give artificial respiration; if breathing is difficult, give oxygen. In case of contact with material, immediately flush eyes with running water for at least 15 minutes. Wash skin with soap and water. Remove and isolate contaminated clothing and shoes at the site [R2].

FIRE POTENTIAL Dangerous fire hazard when exposed to heat, flame, or oxidizers [R11, 839].

REFERENCES

1. Sittig, M. *Handbook of Toxic and Hazardous Chemicals and Carcinogens,* 1985. 2nd ed. Park Ridge, NJ: Noyes Data Corporation, 1985. 592.

2. U.S. Department of Transportation. *Emergency Response Guidebook 1990.* DOT P 5800.5. Washington, DC: U.S. Government Printing Office, 1990, p. G-27.

3. National Fire Protection Guide. *Fire Protection Guide on Hazardous Materials.* 10th ed. Quincy, MA: National Fire Protection Association, 1991, p. 325M-68.

4. NIOSH. *NIOSH Pocket Guide to Chemical Hazards.* DHHS (NIOSH) Publication No. 90-117. Washington, DC: U.S. Government Printing Office, June 1990, 148.

5. Association of American Railroads. *Emergency Handling of Hazardous Materials in Surface Transportation*. Washington, DC: Association of American Railroads, Bureau of Explosives, 1992. 644.

6. Amoore, J.E., E. Hautala. *J Appl Toxicol* 3 (6): 272–90 (1983).

7. ITII. *Toxic and Hazardous Industrial Chemicals Safety Manual*. Tokyo, Japan: The International Technical Information Institute, 1988. 333.

8. Verschueren, K. *Handbook of Environmental Data of Organic Chemicals*. 2nd ed. New York, NY: Van Nostrand Reinhold Co., 1983. 842.

9. 29 CFR 1910.1000 (58 FR 35338 (6/30/93)).

10. 40 CFR 60.489 (7/1/92).

11. Sax, N.I. *Dangerous Properties of Industrial Materials*. 6th ed. New York, New York: Van Nostrand Reinhold, 1984.

12. American Conference of Governmental Industrial Hygienists. *Threshold Limit Values for Chemical Substances and Physical Agents and Biological Exposure Indices for 1994–1995*. Cincinnati, OH: ACGIH, 1994.

■ METHYLDICHLOROSILANE

CAS # 75-54-7

DOT # 1242

SYNONYMS dichlorohydridomethylsilicon • dichloromethylsilane • methyldichlorsilan (Czech) • monomethyldichlorosilane • silane, dichloromethyl

MF CH_4Cl_2Si

MW 115.04

COLOR AND FORM Colorless liquid.

ODOR Acrid odor.

DENSITY 1.10 at 27°C.

BOILING POINT 41°C.

MELTING POINT –93°C.

HEAT OF COMBUSTION –4,700 Btu/lb.

HEAT OF VAPORIZATION 106 Btu/lb.

AUTOIGNITION TEMPERATURE Greater than 600°F (316°C) [R6, p. 325M–68].

EXPLOSIVE LIMITS AND POTENTIAL Silanes form spontaneously explosive mixtures with air (silanes) [R3].

FLAMMABILITY LIMITS 6.0%–55% [R6, p. 49–62].

FIREFIGHTING PROCEDURES Dry chemical or carbon dioxide used to extinguish small fires. Flooding with water may be necessary to prevent reignition. Water may be used if large amounts of combustible material are involved and firefighters are protected by distance or barrier from violent methyldichlorosilane-water reaction. Water may be used to keep fire-exposed containers cool [R6, p. 49–65]. Water may be ineffective [R6, p. 49–62].

TOXIC COMBUSTION PRODUCTS Toxic hydrogen chloride and phosgene gases may be formed [R1].

SOLUBILITY Soluble in benzene, ether, heptane; soluble in water.

OTHER CHEMICAL/PHYSICAL PROPERTIES Releases hydrogen in an alkaline mixture. Liquid is heavier than water (specific gravity, greater than 1.1). Liquid surface tension (est) 35 dynes/cm = 0.035 N/m at 20°C. Organosilanes containing one or more Si—H bonds have excellent reducing capabilities (organosilanes). The polarity of the Si—H bond is opposite to that of the C—H bond. This difference in polarity imparts hydride character to the Si—H bonds of organosilanes. This difference in electronegativities is not as great as in ionic hydrides, and the Si—H bond is still largely (98%) covalent (organosilanes). Organofunctional silanes are noted for their ability to bond organic polymer systems to inorganic substrates (silane compounds). The reaction of organosilanes with halogens and halogen compounds usually proceeds in good yield through cleavage of the Si—H bond and formation of the silicon-halogen bond. Reaction with fluorine, however, does not proceed satisfactorily because of cleavage of not only the Si—H but also C—Si and C—H bonds. Direct halogenation with chlorination, bromine, and iodine proceeds smoothly, however. (organosilanes).

PROTECTIVE EQUIPMENT Wear full protective clothing [R6, p. 49–63]. Where closed-circuit apparatus is not possible rely on suitable personal protective equipment such as corrosion-resistant and impervious suits or overalls, foot protection, hand and arm protection, head, eye, and face protection; where corrosive gases are expected, appropriate respiratory protective equipment is required, ranging from simple masks or respirators to air or oxygen lines or self-contained breathing apparatus, coupled with gas-tight goggles. No universally suitable material for personal equipment can be indicated but natural rubbers, synthetic rubbers, polyvinyl chloride, polypropylene, or polyethylene either in sheet form or with fabric backing are suitable. Where aprons are used, these should have bibs; sleeves should be worn outside gauntlets or gloves, and trouser legs should cover tops of shoes in order to prevent splashes. (corrosive substances) [R5].

PREVENTATIVE MEASURES Provide adequate protection against generated hydrogen chloride. Do not allow water to get in container, because resulting pressure could cause container to rupture [R6, p. 49–63]. Avoid contact with skin and eyes. Use effective fume removal device (silanes) [R7, 43.231]. Corrosive substances should not be allowed into contact with various materials that they attack. High standard of maintenance and good housekeeping are essential. Adequate ventilation and exhaust arrangements, whether general or local, should be provided whenever corrosive gases or dusts are present. Most satisfactory method of ensuring worker protection is to prevent, from outset, any planned, or accidental contact with corrosive substances by utilizing only closed-circuit apparatus. All workers required to handle corrosive substances or likely to come into contact with them should be fully informed of hazards involved and trained in appropriate safe working practices and first-aid measures (corrosive

substances) [R5]. Caution: silanes are toxic. Avoid contact with skin and eyes. Use effective fume-removal device (silanes) [R7, 43.321]. SRP: Local exhaust ventilation should be applied wherever there is an incidence of point-source emissions or dispersion of regulated contaminants in the work area. Ventilation control of the contaminant as close as possible to its point of generation is both the most economical and safest method to minimize personnel exposure to airborne contaminants. SRP: Contaminated protective clothing should be segregated in such a manner so that there is no direct personal contact by personnel who handle, dispose of, or clean the clothing. Quality assurance to ascertain the completeness of the cleaning procedures should be implemented before the decontaminated protective clothing is returned for reuse by the workers. Contaminated clothing should not be taken home at end of shift, but should remain at employee's place of work for cleaning. SRP: The scientific literature supports the wearing of contact lenses in industrial environments as part of a program to protect the eye against chemical compounds and minerals causing eye irritation. However, there may be individual substances whose irritating or corrosive properties are such that the wearing of contact lenses would be harmful to the eye. In those specific cases contact lenses should not be worn. (chlorosilanes).

CLEANUP Spills can be neutralized by flushing with large quantities of water followed by treatment with sodium bicarbonate. [R6, p. 49–63].

COMMON USES Monomer for silicone fluids [R2].

OCCUPATIONAL RECOMMENDATIONS None listed.

BIOCONCENTRATION Chlorosilanes are not expected to persist long enough to bioconcentrate (chlorosilanes) [R4].

GENERAL RESPONSE Avoid contact with liquid and vapor. Keep people away. Wear goggles and self-contained breathing apparatus. Shut off ignition sources. Call fire department. Stop discharge if possible. Isolate and remove discharged material.

Notify local health and pollution control agencies.

FIRE RESPONSE Flammable. Poisonous gases may be produced in fire. Containers may explode in fire. Flashback along vapor trail may occur. Vapor may explode if ignited in an enclosed area. Extinguish with dry chemicals or carbon dioxide. Do not use water or foam on fire. Do not use water or foam on adjacent fires.

EXPOSURE RESPONSE Call for medical aid. Vapor irritating to eyes, nose, and throat. If inhaled will cause difficult breathing. Move victim to fresh air. If breathing has stopped, give artificial respiration. If breathing is difficult, give oxygen. Liquid will burn skin, and eyes. Harmful if swallowed. Remove contaminated clothing and shoes. Flush affected areas with plenty of water. If swallowed and victim is conscious, have victim drink water or milk. Do not induce vomiting.

RESPONSE TO DISCHARGE Issue warning—high flammability. Restrict access. Evacuate area. Disperse and flush with care.

WATER RESPONSE Effect of low concentrations on aquatic life is unknown. May be dangerous if it enters water intakes. Notify local health and wildlife officials. Notify operators of nearby water intakes.

EXPOSURE SYMPTOMS Inhalation causes irritation of respiratory tract; heavy exposure can cause pulmonary edema. Contact of liquid with skin or eyes causes severe burns. Ingestion causes burns of mouth and stomach.

EXPOSURE TREATMENT Get medical attention following all exposures to this compound. Inhalation: remove victim from exposure; if breathing has stopped, begin artificial respiration. Eyes: flush with water for 15 min. Skin: flush with water. Ingestion: do not induce vomiting; give large amounts of water.

REFERENCES

1. U.S. Coast Guard, Department of Transportation. *CHRIS—Hazardous Chemical Data*. Volume II. Washington, DC: U.S. Government Printing Office, 1984–5.

2. SRI.

3. Clayton, G.D., F.E. Clayton (eds.). *Patty's Industrial Hygiene and Toxicology*: Volume 2A, 2B, 2C: Toxicology. 3rd ed. New York: John Wiley and Sons, 1981–1982. 3035.

4. ITC/USEPA. *Information Review #223 (Draft) Chlorosilanes*, p. 10 (1981).

5. International Labour Office. *Encyclopedia of Occupational Health and Safety*. Vols. I and II. Geneva, Switzerland: International Labour Office, 1983. 554.

6. National Fire Protection Association. *Fire Protection Guide on Hazardous Materials*. 9th ed. Boston, MA: National Fire Protectioon Association, 1986.

7. Association of Official Analytical Chemists. *Official Methods of Analysis*. 10th ed. and supplements. Washington, DC: Association of Official Analytical Chemists, 1965. New editions through 13th ed. plus supplements, 1982, p. 13/777.

■ 2–METHYLNAPHTHALENE

CAS # 91–57–6

SYNONYMS beta-methylnaphthalene • naphthalene, 2-methyl • naphthalene, beta-methyl

MF $C_{11}H_{10}$

MW 142.2

COLOR AND FORM Solid.

ODOR THRESHOLD 1.00×10^{-2} ppm at room temperature (detection in air, chemically pure) [R4]. 2.00×10^{-2} ppm at room temp and 60 °C (detection in air, chemically pure) [R7]. 5.00×10^{-2} ppm at 60 °C (detection in air, chemically pure) [R7] .

DENSITY 1.0058 at 20°C.

BOILING POINT 241°C.

MELTING POINT 34.6°C.

SOLUBILITY Very soluble in alcohol and ether; soluble in benzene.

VAPOR PRESSURE Vapor pressure = 6.81×10^{-2} mm Hg at 25°C.

OCTANOL/WATER PARTITION COEFFICIENT
Log K_{ow} = 3.86.

OTHER CHEMICAL/PHYSICAL PROPERTIES
Heat of fusion: 20.11 cal/g = 89.14 J/g = 11,965 J/mol; enthalpy of formation: 10.72 kcal/mole; Gibbs (free) energy of formation: 46.03 kcal/mole; entropy: 52.58 cal/°C–mole; heat capacity: 46.84 cal/°C–mole; critical volume: 462 m³/mole; monoclinic crystals from alcohol.

DISPOSAL
SRP: At the time of review, criteria for land treatment or burial (sanitary landfill) disposal practices are subject to significant revision. Prior to implementing land disposal of waste residue (including waste sludge), consult with environmental regulatory agencies for guidance on acceptable disposal practices. The following wastewater treatment technology has been investigated for naphthalene: Biological treatment. (naphthalene) [R10].

COMMON USES
Organic synthesis; insecticides [R1]; (SRP): pesticide adjuvant; used as dye carrier [R2]; Pure 2–methylnaphthalene is primarily used in vitamin K production and as a chemical intermediate [R3].

OCCUPATIONAL RECOMMENDATIONS
None listed.

ECOTOXICITY
LC$_{50}$ Cancer magister (dungeness crab) larvae 5.0 mg/L/48 hr [R5]; LC$_{50}$ Cancer magister (dungeness crab) larvae 1.3 mg/L/96 hr [R5].

BIOLOGICAL HALF-LIFE
The uptake, disposition, biotransformation, and elimination of ^{14}C–labeled 2–methylnaphthalene was studied in several species of fish. Half-lives of elimination of ^{14}C–2– methylnaphthalene from tissues, following exposure of rainbow trout fingerlings for 8 hr, were less than 24 hr. After exposure for 4 wk, half-life of elimination of ^{14}C was biphasic: half-life of rapid phase less than 24 hr; slow phase, hundreds of hr [R6].

BIODEGRADATION
Samples from different marine environments challenged with Prudhoe Bay crude oil. Studies using nutrient supplemented samples showed that simple aromatics (e.g., 2–methylnaphthalene) were more readily degraded by microorganisms than were n-alkanes [R7].

Microbial degradation to carbon dioxide in seawater at 12° C in the dark after 24-hr incubation at 50 μg/L: 0.10 μg/L/day, turnover time: 500 days; after addition of aq extract of fuel oil 2: degradation rate: 0.26 μg/L/day turnover time: 200 days [R11, 864]. Degradation in seawater by oil oxidizing microorganisms: 17.1% breakdown after 21 days at 22° C in stoppered bottles containing a 100 ppm mixture of alkanes, cycloalkanes, and aromatics [R11, 864]. Aerobic aqueous screening test data showed an 84 and 95% loss of 0.1 ppm methylnaphthalene in 1 and 5.6 days, respectively, for acclimated sewage inoculum, and there was no degradation with unacclimated sewage(1). When marine water was used to inoculate, 2-methylnaphthalene at a concn of 0.067 ppm disappeared within 10 days under aerobic conditions at 25°C (2). Less than 5% degradation of 2-methylnaphthalene occurred in 28 days using the Japanese MITI I procedure; however, for the MITI II test with a freshwater inoculum, 72% was lost in 28 days under aerobic conditions at 25°C (3). 2-Naphthoic acid was identified as a microbial co-oxidation product of 2-methylnaphthalene by mixed cultures of Nocardia sp. isolated from soil (4) [R12]. 2-Methylnaphthalene at a concn of 0.5 ppm was completely removed within 14 days from acclimated fresh-wellwater grab samples from Tuffenwies and Zurich, Switzerland, with a pH of 8.0, at 10 and 25°C and microbial populations of 300–400 cells/mL (1). Grab samples of groundwater aquifer soil that had acclimated to creosote wastes containing 2-methylnaphthalene were able to degrade 2-methylnaphthalene at concn between 0.02 and 0.12 ppm under aerobic conditions at 25° C for a 56-day period (2). An average loss of 6.5% per week was observed for autoclaved controls (2). Unacclimated material from the same aquifer degraded 2- methylnaphthalene at an average rate of 3.5% per week; however, autoclaved controls lost 2-methylnaphthalene at an overall rate of 11.1% per week (2). A marine water die-away study with sediment inoculum from Dunstaffnage Bay, Oban, Scotland, showed a 88.5% loss of 2-methylnaphthalene contained in crude oil after 7 days at 20°C (3). After standardization to controls, the measure of radiolabelled CO_2 evolution from radiolabeled methylnapthalene contained in crude oil was 0.8, 2.2, and 0.8% for marine water grab samples from Saanich Inlet, Canada, incubated at 12 °C for 1, 2, and 3 days, respectively (4). [R13].

BIOCONCENTRATION Uptake of naphthalenes from sand and detritus contaminated with Prudhoe Bay crude oil was examined in a detritivorous clam. Concentration of naphthalenes were determined by UV spectrophometry and the use of radiolabeled ^{14}C-2-methylnaphthalene. Naphthalenes released from sediment to water were available for uptake by clams [R8].

ATMOSPHERIC STANDARDS This action promulgates standards of performance for equipment leaks of volatile organic compounds (VOC) in the synthetic organic manufacturing industry (SOCMI). The intended effect of these standards is to require all newly constructed, modified, and reconstructed SOCMI process units to use the best demonstrated system of continuous emission reduction for equipment leaks of VOC, considering costs, non-air-quality health, and environmental impact and energy requirements. Alkyl naphthalenes are produced, as intermediates or final products, by process units covered under this subpart. (alkyl naphthalenes) [R9].

REFERENCES

1. Sax, N. I., and R. J. Lewis, Sr. (eds.). *Hawley's Condensed Chemical Dictionary*. 11th ed. New York: Van Nostrand Reinhold Co., 1987. 775.

2. *Kirk–Othmer Encyclopedia of Chemical Technology*. 3rd ed., Volumes 1–26. New York, NY: John Wiley and Sons, 1978–1984, p. 8 (79) 153.

3. Gaydos, R. M., *Kirk–Othmer Encycl Chemical Tech 3rd* New York, NY: Wiley 15: 698–719 (1981).

4. Fazzalari, F. A. (ed.). *Compilation of Odor and Taste Threshold Values Data*. ASTM Data Series DS 48A (Committee E–18). Philadelphia, PA: American Society for Testing and Materials, 1978. 103.

5. Caldwell, et al. *Fate Eff Pet Hydrocarbons Mar Ecosyst Org, Proc Symp*, 1977, 210–20.

6. Melancon, M. J., Jr., J. J. Lech, *Astm Tech Publ; Iss Stp 667, Aquat Toxicol* 5–22 (1979).

7. Fedorak, P. M., D.W.S. Westlake, *Can J Microbiol 27* (4): 432–43 (1981).

8. Roesijadi, G., et al. *Environ Pollut 15* (3): 223–30 (1978).

9. 40 CFR 60.489 (7/1/89).

10. USEPA. *Management of Hazardous Waste Leachate*, EPA Contract No.68-03-2766 p. E-3-E-22 (1982).

11. Verschueren, K. *Handbook of Environmental Data of Organic Chemicals*. 2nd ed. New York, NY: Van Nostrand Reinhold Co., 1983.

12. (1) Gaffney, P. E. *J Water Pollut Control Fed* 48: 2731-7 (1976). (2) Vanderlinden, A. C. *Dev Biodegrad Hydrocarbons* 1: 165–200 (1978). (3) Yoshida, K., et al. *Aromatikkusu* 35: 287–92 (1983). (4) Raymond, R. L., et al. *Applied Microbiol* 15: 857–65 (1967).

13. (1) Kappeler, T., K. Wuhrmann. *Water Res* 12: 327–33 (1978). (2) Wilson, J. T., et al. *Environ Toxicol Chem* 4: 721–6 (1985). (3) Rowaland, S. J., et al. *Org Geochem* 9: 153–61 (1986). (4) Lee, R. F., J. W. Anderson. *Bull Mar Sci* 27: 127–34 (1977).

■ 2-METHYLPENTANE

CAS # 107–83–5

DOT # 1208

SYNONYMS dimethylpropylmethane • isohexane • methylpentane • pentane, 2-methyl-

MF C_6H_{14}

MW 86.20

COLOR AND FORM Colorless liquid.

ODOR THRESHOLD Odor low: 0.2886 mg/m^3; odor high: 0.2886 mg/m^3 [R4].

DENSITY 0.6532 at 20°C.

BOILING POINT 60.3°C.

MELTING POINT −153.7°C.

HEAT OF VAPORIZATION 27.79 kJ/mole.

AUTOIGNITION TEMPERATURE 583°F (306°C) [R2].

EXPLOSIVE LIMITS AND POTENTIAL Upper explosive limit: 7.0% by volume; lower explosive limit: 1.2% by volume [R3].

FLAMMABILITY LIMITS Lower flammable limit: 1.2% by volume; upper flammable limit: 7.0% by volume [R2].

FIREFIGHTING PROCEDURES Water may be ineffective [R2].

SOLUBILITY Soluble in heptane; soluble in alcohol, ether, in all proportions in acetone, benzene, chloroform; water solubility = 14 mg/L.

VAPOR PRESSURE 400 mm Hg at 41.6°C.

OCTANOL/WATER PARTITION COEFFICIENT Log K_{ow} = 3.74 (est).

OTHER CHEMICAL/PHYSICAL PROPERTIES Henry's Law constant = 1.71 atm–m³/ mole (calc).

PREVENTATIVE MEASURES SRP: Local exhaust ventilation should be applied wherever there is an incidence of point-source emissions or dispersion of regulated contaminants in the work area. Ventilation control of the contaminant as close as possible to its point of generation is both the most economical and safest method to minimize personnel exposure to airborne contaminants.

COMMON USES Organic synthesis, solvent [R7, 778].

OCCUPATIONAL RECOMMENDATIONS None listed.

BIODEGRADATION A quarter of the microorganisms, most notably of the *Nocardia* sp., isolated from gasoline-contaminated groundwater supported growth of 2–methylpentane (1). Generally iso alkanes are significantly more resistant to microbial attack than n-alkanes (2). However, a soil microorganism, *Corynebacterium* sp., oxidized 2–methylpentane at about the same rate as n-pentane, although at a significantly lower rate than n-hexane (2) [R5].

BIOCONCENTRATION Using the estimated log octanol/water partition coefficient for 2–methylpentane of 3.74 (1) and a water solubility of 14 mg/L (2), one can estimate a BCF ranging from 100 to 408, using six different regression equations (3). Therefore 2–methylpentane has a low to moderate bioconcentration potential (SRC) [R6].

DOT *Fire or Explosion:* Flammable/combustible material; may be ignited by heat, sparks, or flames. Vapors may travel to a source of ignition, and flash back. Container may explode in heat of fire. Vapor explosion hazard indoors, outdoors, or in sewers. Runoff to sewer may create fire or explosion hazard (methylpentane) [R1].

Health Hazards: May be poisonous if inhaled or absorbed through skin. Vapors may cause dizziness or suffocation. Contact may irritate or burn skin and eyes. Fire may produce irritating or poisonous gases. Runoff from fire control or dilution water may cause pollution (methylpentane) [R1].

Emergency Action: Keep unnecessary people away; isolate hazard area and deny entry. Stay upwind; keep out of low areas. Positive-pressure self-contained breathing apparatus (SCBA) and structural firefighters' protective clothing will provide limited protection. Isolate for 1/2 mile in all directions if tank, rail car, or tank truck is involved in fire. Call CHEMTREC at 1–800–424–9300 for emergency assistance. If water pollution occurs, notify the appropriate authorities (methylpentane) [R1].

Fire: Small fires: Dry chemical, CO_2, water spray, or alcohol-resistant foam. Large Fires: Water spray, fog, or alcohol-resistant foam. Do not use dry chemical extinguishers to control fires involving nitromethane or nitroethane. Move container from fire area if you can do so without risk. Apply cooling water to sides of containers that are exposed to flames until well after fire is out. Stay away from ends of tanks. For massive fire in cargo area, use unmanned hose holder or monitor nozzles; if this is impossible, withdraw from area, and let fire burn. Withdraw immediately in case of rising sound from venting safety device or any discoloration of tank due to fire (methylpentane) [R1].

Spill or Leak: Shut off ignition sources; no flares, smoking, or flames in hazard area. Stop leak if you can do so without risk. Water spray may reduce vapor; but it may not prevent ignition in closed spaces.

Small Spills: Take up with sand or other noncombustible absorbent material and place into containers for later disposal. Large Spills: Dike far ahead of liquid spill for later disposal (methylpentane) [R1].

First Aid: Move victim to fresh air and call for emergency medical care; if not breathing, give artificial respiration; if breathing is difficult, give oxygen. In case of contact with material, immediately flush eyes with running water for at least 15 minutes. Wash skin with soap and water. Remove and isolate contaminated clothing and shoes at the site (methylpentane) [R1].

FIRE POTENTIAL Flammable, dangerous fire risk [R7, 778].

GENERAL RESPONSE Stop discharge if possible. Keep people away. Shut off ignition sources and call fire department. Stay upwind and use water spray to "knock down" vapor. Isolate and remove discharged material. Notify local health and pollution control agencies.

FIRE RESPONSE Flammable. Flashback along vapor trail may occur. Vapor may explode if ignited in an enclosed area. Extinguish with dry chemical, foam, or carbon dioxide. Water may be ineffective on fire. Cool exposed containers with water.

EXPOSURE RESPONSE Call for medical aid. Vapor irritating to eyes, nose, and throat. If inhaled, will cause dizziness, headache, difficult breathing, or loss of consciousness. Move to fresh air. If breathing has stopped, give artificial respiration. If breathing is difficult, give oxygen. Liquid irritating to skin and eyes. If swallowed, will cause nausea or vomiting. Remove

contaminated clothing, and shoes. Flush affected areas with plenty of water. If swallowed and victim is conscious, have victim drink water or milk. Do not induce vomiting.

RESPONSE TO DISCHARGE Issue warning—high flammability. Evacuate area. Disperse and flush.

WATER RESPONSE Effect of low concentrations on aquatic life is unknown. May be dangerous if it enters water intakes. Notify local health and wildlife officials. Notify operators of nearby water intakes.

EXPOSURE SYMPTOMS Inhalation causes irritation of respiratory tract, cough, mild depression, cardiac arrhythmias. Aspiration causes severe lung irritation, coughing, pulmonary edema; excitement followed by depression. Ingestion causes nausea, vomiting, swelling of abdomen, headache, depression.

EXPOSURE TREATMENT Inhalation: maintain respiration, give oxygen if needed. Aspiration: enforce bed rest; give oxygen. Ingestion: do not induce vomiting; call a doctor. Eyes: wash with copious amount of water. Skin: wipe off, wash with soap and water.

REFERENCES

1. U.S. Department of Transportation. *Emergency Response Guidebook.* 1990. DOT P 5800.5. Washington, DC: U.S. Government Printing Office, 1990, p. G–26.

2. National Fire Protection Guide. *Fire Protection Guide on Hazardous Materials.* 10th ed. Quincy, MA: National Fire Protection Association, 1991, p. 325M–71.

3. Sax, N. I. *Dangerous Properties of Industrial Materials,.* 6th ed. New York, NY: Van Nostrand Reinhold, 1984. 1901.

4. Ruth, J.H., *Am Ind Hyg Assoc J* 47: A–142–51 (1986).

5. (1) Jamison, V. W., et al., pp. 187–96 in *Proc intermediate Biodeg Symp* 3rd, Sharpley, J. M., A. M. Kapalan. (eds.), Essex, England (1976). (2) Buswell, J. A.,

P. Jurtshuk. *Arch Microbiol* 64: 215–22 (1969).

6. (1) USEPA. *CLOGP* (1986). (2) Riddick, J. A., et al., *Organic Solvents* 4th ed. New York: Wiley (1986). (3) Lyman, W. J., et al., *Handbook of Chemical Property Estimation Methods*. New York: McGraw-Hill, Chapt 5 (1982).

7. Sax, N.I., and R.J. Lewis, Sr. (eds.). *Hawley's Condensed Chemical Dictionary*. 11th ed. New York: Van Nostrand Reinhold Co., 1987.

■ 2-METHYLPROPYL METHACRYLATE

CAS # 97–86–9

SYNONYMS 2-propenoic acid, 2-methyl-, 2-methylpropyl ester • isobutyl 2-methyl-2-propenoate • isobutyl alpha-methacrylate • isobutyl alpha-methylacrylate • isobutyl methacrylate • methacrylic acid, isobutyl ester • propenoic acid, 2-methyl, isobutyl ester

MF $C_8H_{14}O_2$

MW 142.22

COLOR AND FORM Liquid.

DENSITY 0.8858 at 20°C.

BOILING POINT 155°C at 760 mm Hg.

FIREFIGHTING PROCEDURES Use water in flooding quantities as fog. Do not extinguish fire unless flow can be stopped. Cool all affected containers with flooding quantities of water. Apply water from as far a distances as possible. Use foam, dry chemical, or carbon dioxide [R2]. Wear positive-pressure self-contained breathing apparatus when fighting fires involving this material [R2].

TOXIC COMBUSTION PRODUCTS When heated to high-temperature it may release acrid smoke and fumes [R2].

SOLUBILITY >10% in alcohol; >10% in ether; water solubility of 594 mg/L at 25°C.

VAPOR PRESSURE 1.8 mm Hg at 20°C.

OCTANOL/WATER PARTITION COEFFICIENT Log K_{ow} = 2.66.

OTHER CHEMICAL/PHYSICAL PROPERTIES Conversion factors: 5.80 mg/m³ = 1 ppm; bulk density: 0.882 g/ml at 25°C.

PROTECTIVE EQUIPMENT Suitable protective clothing and self-contained resp protective apparatus should be available for use by those who may have to rescue persons overcome by fumes. (acrylic acid and derivatives) [R6].

PREVENTATIVE MEASURES Hazard is the generation of considerable exothermic heat in some of the reactions, so that high pressures and temperature may develop. This danger should be borne in mind when designing plant. Awareness of the dangers and of good engineering design are essential to safety. Employees should be instructed about the necessity of cleansing the skin if it is contaminated by materials that are irritants or skin-absorbed. With careful design, however, and complete enclosure of those processes where toxic chemicals or intermediates occur, dangerous exposures can be avoided. (acrylic acid and derivatives) [R6].

COMMON USES Used as monomer for acrylic resins [R2]. In hydrogel lenses [R3]. Concrete can be made water-repellent by the polymerization of vinyl monomers on the surface. A treatment that may be practical for highway bridge decks is the application of methyl methacrylate, isodecyl methacrylate, or isobutyl methacrylate [R4].

OCCUPATIONAL RECOMMENDATIONS None listed.

BIODEGRADATION 2–Methylpropyl methacrylate may significantly biodegrade based upon the biodegradability of methyl methacrylate, which was confirmed to be significantly degradable according to the MITI test, the biodegradability screening test of the Japanese Ministry of International Trade and Industry (1) [R7].

BIOCONCENTRATION Using a reported log K_{ow} of 2.66 (1), a BCF of 62 has been calculated using a recommended regression-derived equation (2,SRC). This estimated BCF indicates that 2–methylpropyl methacrylate will not be expected to bio-concentrate in aquatic organisms (SRC) [R8].

TSCA REQUIREMENTS Section 8 (a) of TSCA requires manufacturers of this chemical substance to report preliminary assessment information concerned with production, use, and exposure to EPA as cited in the preamble of the 51 FR 41329 [R9].

DOT *Fire or Explosion:* Flammable/combustible material; may be ignited by heat, sparks, or flames. Vapors may travel to a source of ignition, and flash back. Container may explode in heat of fire. Vapor explosion hazard indoors, outdoors, or in sewers. Runoff to sewer may create fire or explosion hazard [R5].

Health Hazards: May be poisonous if inhaled or absorbed through skin. Vapors may cause dizziness or suffocation. Contact may irritate or burn skin and eyes. Fire may produce irritating or poisonous gases. Runoff from fire control or dilution water may cause pollution [R5].

Emergency Action: Keep unnecessary people away; isolate hazard area and deny entry. Stay upwind; keep out of low areas. Self-contained breathing apparatus (SCBA) and structural firefighter's protective clothing will provide limited protection. Isolate for 1/2 mile in all directions if tank car or truck is involved in fire. Call CHEMTREC at 1–800–424–9300 for emergency assistance. If water pollution occurs, notify the appropriate authorities [R5].

Fire: Small fires: Dry chemical, CO_2, halon, water spray, or standard foam. Large Fires: Water spray, fog, or standard foam is recommended. Move container from fire area if you can do so without risk. Cool containers that are exposed to flames with water from the side until well after fire is out. Stay away from ends of tanks. For massive fire in cargo area, use unmanned hose holder or monitor nozzles; if this is impossible, withdraw from area, and let fire burn. Withdraw immediately in case of rising sound from venting

safety device or any discoloration of tank due to fire [R5].

Spill or Leak: Shut off ignition sources; no flares, smoking, or flames in hazard area. Stop leak if you can do so without risk. Water spray may reduce vapor; but it may not prevent ignition in closed spaces.

Small Spills: Take up with sand or other noncombustible absorbent material and place into containers for later disposal. Large Spills: Dike far ahead of liquid spill for later disposal [R5].

First Aid: Move victim to fresh air and call emergency medical care; if not breathing, give artificial respiration; if breathing is difficult, give oxygen. In case of contact with material, immediately flush eyes with running water for at least 15 minutes. Wash skin with soap and water. Remove and isolate contaminated clothing and shoes at the site [R5].

FIRE POTENTIAL Moderate fire risk [R1].

REFERENCES

1. Sax, N. I., and R. J. Lewis, Sr. (eds.). *Hawley's Condensed Chemical Dictionary.* 11th ed. New York: Van Nostrand Reinhold Co., 1987. 654.

2. Association of American Railroads. *Emergency Handling of Hazardous Materials in Surface Transportation.* Washington, DC: Assoc. of American Railroads, Hazardous Materials Systems (BOE), 1987. 393.

3. *Kirk–Othmer Encyclopedia of Chemical Technology.* 3rd ed., Volumes 1–26. New York, NY: John Wiley and Sons, 1978–1984, p. 6 (79) 7330.

4. *Kirk–Othmer Encyclopedia of Chemical Technology.* 3rd ed., Volumes 1–26. New York, NY: John Wiley and Sons, 1978–1984, p. 24 (84) 461.

5. Department of Transportation. *Emergency Response Guidebook 1987.* DOT P 5800.4. Washington, DC: U.S. Government Printing Office, 1987, p. G–27.

6. International Labour Office. *Encyclopedia of Occupational Health and Safety.* Vols. I and II. Geneva, Switzerland: International Labour Office, 1983. 53.

7. Sasaki, S. pp. 283–98 in *Aquat Pollutants: Transform, and Biolog Effects.* Hutzinger, O., et al. (ed.), Oxford: Pergamon Press (1978).

8. (1) Hansch, C., and A. J. Leo. *Medchem Project* Issue No 26. Claremont, CA: Pomona College (1985). (2) Lyman, W. J., et al., *Handbook of Chemical Property Estimation Methods.* New York: McGraw-Hill, p. 5–5 (1982).

9. 40 CFR 712.30 (7/1/88).

■ METHYLTRICHLOROSILANE

CAS # 75–79–6

DOT # 1250

SYNONYMS Methyl-trichlorsilan (Czech) • methylsilyl trichloride • silane, methyltrichloro- • silane, trichloromethyl- • trichloromethylsilane • trichloromethylsilicon

MF CH_3Cl_3Si

MW 149.48

COLOR AND FORM Colorless liquid.

ODOR Acrid odor, sharp like hydrochloric acid.

DENSITY 1.27 at 25°C.

SURFACE TENSION 20.3 dynes/cm = 0.0203 N/m at 20°C.

BOILING POINT 66.4°C.

MELTING POINT −90°C.

HEAT OF COMBUSTION −3,000 Btu/lb (est).

HEAT OF VAPORIZATION 89.3 Btu/lb.

AUTOIGNITION TEMPERATURE Greater than 760°F [R2].

EXPLOSIVE LIMITS AND POTENTIAL May form explosive mixture with air. [R4, 692].

FLAMMABILITY LIMITS 7.6%, and greater than 20% [R2].

FIREFIGHTING PROCEDURES Use dry chemical or carbon dioxide to extinguish small fires. Flooding with water may be necessary to prevent reignition. Water may be used if large amount of combustible materials are involved and if firefighters can protect themselves by distance or barriers from violent methyltrichlorosilane-water reaction. Water may be used to keep fire-exposed containers cool [R2].

TOXIC COMBUSTION PRODUCTS Toxic hydrogen chloride and phosgene gases are formed in fires [R1].

VAPOR PRESSURE 167 mm Hg at 25°C.

STABILITY AND SHELF LIFE Decomposition in moist air, creating hydrochloric acid with odor threshold of 1 ppm [R1].

OTHER CHEMICAL/PHYSICAL PROPERTIES Organo-functional silanes are noted for their ability to bond organic polymer systems to inorganic substrates. (silane compounds).

PROTECTIVE EQUIPMENT Protective equipment should incl corrosion-resistant and impervious suits, foot, hand and arm, head, and eye and face protection; where corrosive gases may be expected resp protective equipment is required; natural rubbers, synthetic rubbers, polyvinyl chloride, polypropylene, or polyethylene either in sheet form or with fabric backing are suitable for personal protective equipment. (corrosive substances) [R5, 554].

PREVENTATIVE MEASURES Preventive measures should be directed primarily at preventing or minimizing contact between corrosive substances and skin, mucous membranes, and eyes. Adequate ventilation and exhaust arrangements, whether general or local, should be provided whenever corrosive gases or dusts are present. (corrosive substances) [R5, 553].

CLEANUP Spills can be neutralized by flushing with large quantities of water followed by treatment with sodium bicarbonate. Provide adequate protection against generated hydrogen chloride. Do not allow water to get into container, because resulting pressure could cause container to rupture [R2].

COMMON USES Intermediate for silicones [R4, 783].

OCCUPATIONAL RECOMMENDATIONS None listed.

NON-HUMAN TOXICITY LD$_{50}$ rat oral 0.8 g/kg (from table) [R6, 2398].

BIODEGRADATION Atmospheric fate: If released to the atmosphere, methyltrichlorosilane is expected to undergo hydrolysis with the water vapor in the air (1,2,SRC) [R3].

BIOCONCENTRATION Methyltrichlorosilane will decompose upon contact with the water in biological tissues; thus, consideration of the potential for bioaccumulation is not warranted (SRC).

GENERAL RESPONSE Avoid contact with liquid and vapor. Keep people away. Wear goggles, self-contained breathing apparatus, and rubber overclothing (including gloves). Shut off ignition sources. Call fire department. Stop discharge if possible. Isolate and remove discharged material. Notify local health and pollution control agencies.

FIRE RESPONSE Flammable. Poisonous gases may be produced in fire. Containers may explode in fire. Flashback along vapor trail may occur. Vapor may explode if ignited in an enclosed area. Wear goggles and self-contained breathing apparatus. Extinguish with dry chemicals or carbon dioxide. Do not use water or foam on fire. Do not use water or foam on adjacent fires.

EXPOSURE RESPONSE Call for medical aid. Vapor irritating to eyes, nose, and throat. If inhaled will cause difficult breathing. Move victim to fresh air. If breathing has stopped, give artificial respiration. If breathing is difficult, give oxygen. Liquid will burn skin and eyes. Harmful if swallowed. Remove contaminated clothing and shoes. Flush affected areas with plenty of water. If swallowed and victim is conscious, have victim drink water or milk. Do not induce vomiting.

RESPONSE TO DISCHARGE Issue warning—high flammability, air contaminant. Restrict access. Evacuate area. Disperse and flush with care.

WATER RESPONSE Effect of low concentrations on aquatic life is unknown. May be dangerous if it enters water intakes. Notify local health and wildlife officials. Notify operators of nearby water intakes.

EXPOSURE SYMPTOMS Inhalation causes irritation of mucous membranes. Contact with liquid causes severe burns of eyes and skin. Ingestion causes severe burns of mouth and stomach.

EXPOSURE TREATMENT Get medical attention at once following all exposures to this compound. Inhalation: remove victim from exposure; give artificial respiration if breathing has ceased. Eyes: flush with water for 15 min. Skin: flush with water. Ingestion: do not induce vomiting; give large amounts of water.

REFERENCES

1. U.S. Coast Guard, Department of Transportation. *CHRIS—Hazardous Chemical Data. Volume II.* Washington, DC: U.S. Government Printing Office, 1984–5.

2. National Fire Protection Association. *Fire Protection Guide on Hazardous Materials.* 9th ed. Boston, MA: National Fire Protection Association, 1986, p. 49–64.

3. (1) Sax, N. I., R.J. Lewis, Sr. *Hawley's Condensed Chemical Dictionary,* New York: Van Nostrand Reinhold Co. 11: 783 (1987). (2) Collin, W. "Silicon Compounds (Silicon Halides) in: Kirk–Orthmer Encycl Chemical Tech, 3rd Ed. New York: John Wiley, 20: 881–7 (1982).

4. Sax, N.I., and R.J. Lewis, Sr. (eds.). *Hawley's Condensed Chemical Dictionary.* 11th ed. New York: Van Nostrand Reinhold Co., 1987.

5. International Labour Office. *Encyclopedia of Occupational Health and Safety.* Vols. I and II. Geneva, Switzerland: International Labour Office, 1983.

6. Clayton, G. D., and F. E. Clayton (eds.). *Patty's Industrial Hygiene and Toxicology.* Volume 2A, 2B, 2C: *Toxicology.* 3rd ed. New York: John Wiley and Sons, 1981–1982.

■ 2-METHYL-2,4-PENTANE-DIOL

CAS # 107–41–5

SIC CODES 2844; 2992.

SYNONYMS 1,1,3-trimethyltrimethylenediol • 2,4-dihydroxy-2-methylpentane • 2,4-pentanediol, 2-methyl • 2-methyl pentane-2,4-diol • 2-methyl-2,4-pentandiol • 4-methyl-2,4-pentanediol • alpha, alpha, alpha′-trimethyltrimethylene glycol • diolane • hexylene glycol • isol • pinakon

MF $C_6H_{14}O_2$

MW 118.20

COLOR AND FORM Colorless liquid.

ODOR Mild, sweetish; atmospheres essentially saturated at room temperature are detectable by odor.

DENSITY 0.9254 at 17°C.

SURFACE TENSION 33.1 dyne/cm at 20°C.

VISCOSITY 34 cP at 20°C.

BOILING POINT 198°C at 760 mm Hg.

MELTING POINT Freezing point = –50°C (glass or vitreous condition).

HEAT OF COMBUSTION Standard net heat of combustion = -3.4356×10^9 J/kmol.

HEAT OF VAPORIZATION 13.7 kcal/mol at the boiling point.

AUTOIGNITION TEMPERATURE Autoignition temperature = 579 K [R7].

FLAMMABILITY LIMITS Flammability limits = 1.3–9 vol% [R7].

SOLUBILITY Soluble in alcohol, ether; soluble in lower aliphatic hydrocarbons; soluble in a variety of organic solvents; miscible with fatty acids; infinite water solubility.

VAPOR PRESSURE 0.05 mm Hg at 20°C.

OTHER CHEMICAL/PHYSICAL PROPERTIES Percent in saturated air: 0.0066 (20°C); wt/gal 7.69 lb; dipole moment: 2.8; vapor pressure = 0.013 mm Hg at 25°C; density = 0.92109 at 20°C; 0.9181 at 25°C; heat of formation = -5.3476×10^8 J/kmol; heat of fusion at melting point = 1.48×10^7 J/kmol; triple point temperature and pressure = 223.15°C and 9.5609×10^{-6} Pa.

PREVENTATIVE MEASURES SRP: The scientific literature for the use of contact lenses in industry is conflicting. The benefit or detrimental effects of wearing contact lenses depend not only upon the substance, but also on factors including the form of the substance, characteristics and duration of the exposure, the uses of other eye protection equipment, and the hygiene of the lenses. However, there may be individual substances whose irritating or corrosive properties are such that the wearing of contact lenses would be harmful to the eye. In those specific cases, contact lenses should not be worn. In any event, the usual eye protection equipment should be worn even when contact lenses are in place.

DISPOSAL Spray into incinerator. Flammable solvent may be added. SRP: At the time of review, criteria for land treatment or burial (sanitary landfill) disposal practices are subject to significant revision. Prior to implementing land disposal of waste residue (including waste sludge), consult with environmental regulatory agencies for guidance on acceptable disposal practices.

COMMON USES In cosmetics [R2]; used in textile dye vehicles; recommended as solvent in petroleum refining [R3]; as cement additive; as chemical intermediate [R4]; fuel and lubricant additive [R5]; coupling agent in hydraulic brake fluids and printing inks; gasoline anti-icer additive [R1].

STANDARD CODES ICC—no; USCG—grade E combustible liquid. NFPA—1, 1, 0.

PERSISTENCE Glycols biodegrade quite rapidly.

DIRECT CONTACT Skin, eyes, mucous membranes.

GENERAL SENSATION Mild sweetish odor. Large oral doses can produce narcosis. Eye irritation grade 4—small burns diluted.[R12]; Symptoms may include stimulation of central nervous system followed by depression, vomiting, drowsiness, coma, respiratory failure, convulsions, renal damage, and death.

PERSONAL SAFETY Wear protective clothing and canister-type mask.

ACUTE HAZARD LEVEL Irritant. Slightly toxic via ingestion and inhalation. May produce BOD problem.

CHRONIC HAZARD LEVEL Mild chronic irritant.

DEGREE OF HAZARD TO PUBLIC HEALTH Irritant. Slight ingestive and inhalative toxicity. Mild chronic irritant.

OCCUPATIONAL RECOMMENDATIONS None listed.

THRESHOLD LIMIT Ceiling limit 25 ppm, 121 mg/m^3 (1977) [R11].

ACTION LEVEL Notify fire authority. Isolate from oxidizing material. Remove ignition sources.

ON-SITE RESTORATION Carbon may be ineffective. seek professional environmental engineering assistance through EPA's environmental response team (ERT), Edison, NJ, 24–hour no., 908–548–8730.

BEACH OR SHORE RESTORATION Burn off.

AVAILABILITY OF COUNTERMEASURE MATERIAL Carbon—water treatment plants, sugar refineries.

MAJOR WATER USE THREATENED Fisheries, recreation.

PROBABLE LOCATION AND STATE Liquid. Will dissolve.

COLOR IN WATER Colorless

NON-HUMAN TOXICITY LD$_{50}$ rat oral 4.79 g/kg [R4]; LD$_{50}$ mouse oral 3.5 g/kg [R4].

ECOTOXICITY LC$_{50}$ *Pimephales promelas* (fathead minnow) 10,700 mg/L 96 hr flow-through bioassay, wt 0.12 g, water hardness 45.5 mg/L CaCO$_3$, temp: 25 ± 1°C, pH 7.5, dissolved oxygen greater than 60% of saturation [R8].

BIODEGRADATION Using the Japanese MITI protocol (2–week incubation period with activated sludge inoculum), 2–methyl–2,4–pentanediol was found to be "well biodegradable," as theoretical BODs of 35–76% have been observed (1–3). Using a standard BOD dilution test and an inoculum from a sanitary waste treatment facility, a theoretical BOD of 95% was observed over a 5-day period (4). A 5-day theoretical BOD of 56% was observed using acclimated mixed microbial cultures as inoculum (5) [R9].

BIOCONCENTRATION
2–Methyl–2,4–pentanediol is miscible in water (1); therefore, bioconcentration in aquatic organisms in not expected to be an important fate process (SRC) [R10].

FIRE POTENTIAL Fire hazard: low, when exposed to heat or flame; can react with oxidizing materials [R6].

REFERENCES

1. SRI.

2. Budavari, S.(ed.). *The Merck Index— Encyclopedia of Chemicals, Drugs, and Biologicals*. Rahway, NJ: Merck and Co., Inc., 1989. 744.

3. Gosselin, R. E., H. C. Hodge, R. P. Smith, and M. N. Gleason. *Clinical Toxicology of Commercial Products*. 4th ed. Baltimore: Williams and Wilkins, 1976, p. II–120.

4. American Conference of Governmental Industrial Hygienists, Inc. *Documentation of the Threshold Limit Values, and Biological Exposure Indices*. 6th ed. Volumes I, II, III. Cincinnati, OH: ACGIH, 1991. 759.

5. Sax, N. I., and R. J. Lewis, Sr. (eds.). *Hawley's Condensed Chemical Dictionary*. 11th ed. New York: Van Nostrand Reinhold Co., 1987. 602.

6. Sax, N. I. *Dangerous Properties of Industrial Materials,*. 6th ed. New York, NY: Van Nostrand Reinhold, 1984. 1075.

7. Daubert, T. E., R. P. Danner. *Physical and Thermodynamic Properties of Pure Chemicals Data Compilation*. Washington, DC: Taylor, and Francis, 1989.

8. Vieth, G.D., et al., *Canadian J Fisheries Aquat Sci*. 40 (6): 743–8 (1983).

9. (1) Chemicals Inspection, and Testing Institute. *Biodegradation and Bioaccumulation Data of Existing Chemicals Based on the CSCL Japan*. Japan Chemical Industry Ecology—Toxicology and Information Center. ISBN 4–89074–101–1, pp. 2–41 (1992). (2) Sasaki, S. pp. 283–98 in *Aquatic Pollutants: Transformation and Biodegradation Effects*. Hutzinger, O., et al. (eds.), Oxford: Pergamon Press (1978). (3) Kawasaki, M., *Ecotox Environ Safety* 4: 444–54 (1980). (4) Bridie, A. L., et al., *Water Res* 13: 627–30 (1979). (5) Babeu, L., D. D. Vaishnav. *J Indust Microbiol* 2: 107–15 (1987).

10. Riddick, J.A., et al., *Organic Solvents Techniques of chemistry*, 4th ed. New York: Wiley–Interscience pp. 275–6 (1986).

11. American Conference of Governmental Industrial Hygienists. *Threshold Limit Values for Chemical Substances and Physical Agents and Biological Exposure Indices for 1994–1995*. Cincinnati, OH: ACGIH, 1994. 23.

12. Smyth, H. F., Jr., C. P. Carpenter, C. S. Weil, U. C. Pozzani, J. A. Striegel, and J. S. Nycum, "Range–Finding Toxicity Data: List VII," *American Industrial Hygiene Association Journal*, 30: 470–476. 1969. Smyth, H. F., C. P. Carpenter, and C. S. Weil, "Range-finding Toxicity Data: List IV," *AMA Archives of Industrial Hygiene, and Occupational Medicine*, 4: 119–122, 1951. Smyth, H. F., C. P. Carpenter, C. S. Weil, U. C. Pozzani, and J. A. Striegel, "Range–Finding Toxicity Data: List VI," *American Industrial Hygiene Association Journal*, 23: 95–107, 1962. Smyth, H. F., C. P. Carpenter, C. S. Weil, and U. C. Pozzani, "Range–Finding Toxicity Data: List V," *AMA Archives of Industrial Hygiene, and Occupational Medicine*, 10: 61–68, 1954. Smyth, H. F., C. P. Carpenter, and C. S. Weil, "Range–Finding Toxicity Data: List 111," *Journal of Industrial Hygiene and Toxicology*, 31: 60–62, 1949. Smyth, H. F., 1. Seaton, and L. Fischer, "The Single Dose Toxicity of Some Glycols, and Derivatives," *Journal of Industrial Hygiene and Toxicology*, 23 (6): 259–268, 1941. Smyth, H. F., and C.

P. Carpenter, "Further Experience with the Range–Finding Test in the Industrial Toxicology Laboratory," *Journal of Industrial Hygiene and Toxicology*, 30: 63–68, 1948. Smyth, H. F., and C. P. Carpenter, "The Place of the Range–Finding Test in the Industrial Toxicology Laboratory," *Journal of Industrial Hygiene and Toxicology*, 26: 269–273, 1 944.

■ METHYL SALICYLATE

CAS # 119–36–8

SYNONYMS oil of wintergreen • 2-(methoxycarbonyl) phenol • 2-hydroxybenzoic acid methyl ester • analgit • benzoic acid, 2-hydroxy-, methyl ester • betula • betula oil • exagien • flucarmit • gaultheria oil • gaultheria oil, artificial • methyl 2-hydroxybenzoate • methyl o-hydroxybenzoate • metylester-kyselinysalicylove (Czech) • natural wintergreen oil • o-hydroxybenzoic acid methyl ester • o-hydroxybenzoic acid, methyl ester • salicylic acid, methyl ester • sweet birch oil • synthetic wintergreen oil • teaberry oil • wintergreen oil • wintergreen oil, synthetic • fema number 2745

MF $C_8H_8O_3$

MW 152.16

COLOR AND FORM Colorless, yellowish, or reddish oily liquid.

ODOR Liquid having the characteristic odor of wintergreen.

TASTE Liquid having the characteristic taste of wintergreen.

DENSITY 1.184 at 25°C.

SURFACE TENSION 44.2 dynes/cm at −19.8°C; 19.8 dynes/cm at 212.2°C.

DISSOCIATION CONSTANT $pK_a = 9.8$.

BOILING POINT 220–224°C.

MELTING POINT −8.6°C.

HEAT OF COMBUSTION −902.2 kcal/mole at 25°C.

HEAT OF VAPORIZATION 11.155 kcal/mol at the boiling point.

AUTOIGNITION TEMPERATURE 850°F [R8].

FIREFIGHTING PROCEDURES Water, foam, carbon dioxide, dry chemical [R7].

SOLUBILITY Soluble in chloroform, ether; miscible with alcohol, glacial acetic acid; soluble in diethyl ether; soluble in most common organic solvents; soluble in water: 0.74%w at 30°C.

VAPOR PRESSURE 0.0343 mm Hg at 25°C.

OCTANOL/WATER PARTITION COEFFICIENT Log K_{ow} = 2.55.

STABILITY AND SHELF LIFE Sensitive to light and heat [R15, 77].

OTHER CHEMICAL/PHYSICAL PROPERTIES Density of natural ester is about 1.180; wintergreen essential oil: ester value: 354–356; optical rotation: –0°25' to –1°30'; 1 mm Hg at 54.0°C; heat capacity = 59.46 cal/deg K–mol at 15–30°C.

DISPOSAL SRP: At the time of review, criteria for land treatment or burial (sanitary landfill) disposal practices are subject to significant revision. Prior to implementing land disposal of waste residue (including waste sludge), consult with environmental regulatory agencies for guidance on acceptable disposal practices.

COMMON USES Oil: in fern and cypress type perfumes and in toothpaste [R11, 490]; for flavoring candies, etc.; counterirritant; medication (vet): counterirritant [R12, 962]; medication (vet): carminative, odorant [R1]; UV–absorber in sunburn lotions; flavor in foods, beverages, pharmaceuticals; odorant [R3]; chewing gum; fragrance in detergents; local analgesic for human toxicity and veterinary medicine; solvent for insecticides, polishes, and inks [R4]; reported uses: non-alcoholic beverages 10 ppm; candy 900–5,000 ppm [R11, 490]; in perfumery as a modifier in blossom fragrances, and as a mild antiseptic in oral hygiene products [R5]; medication: as a pharmaceutical, it is used in liniments and ointments for the relief of pain in the lumbar and sciatic regions, and for rheumatic conditions. Other miscella-

neous applications for methyl salicylate are as a dye carrier, UV light stabilizer in acrylic resins, and chemical intermediate [R6]; as pharmaceutical necessity, it is used to flavor official aromatic cascara sagrada fluid extract [R2].

OCCUPATIONAL RECOMMENDATIONS None listed.

HUMAN TOXICITY LD_{lo} child oral 0.17 g/kg; LD_{50} adult oral 0.5 g/kg [R14, 2310].

NON-HUMAN TOXICITY LD_{50} rat, oral 0.887 g/kg [R14, 2310]; LD_{50} rabbit, oral 2.8 g/kg [R14, 2310]; LD_{50} guinea pig, oral 1.060 g/kg [R14, 2310]; LD_{50} dog, oral 2.1 g/kg [R14, 2310]; LD_{50} guinea pig dermal 0.70 ml/kg [R14, 2310].

BIOLOGICAL HALF-LIFE The plasma half-life for salicylate is 2 to 3 hr in low doses and about 12 hr at usual antiinflammatory doses. The half-life of salicylate may be as long as 15 to 30 hr at high therapeutic doses or when there is intoxication. (salicylates) [R13, 650].

BIODEGRADATION Methyl salicylate in a five-day BOD test exhibited a value of 55–57% of the theoretical BOD (1,2). Another 5-day BOD determination yielded 65% of the theoretical BOD (3). Methyl salicylate was completely degraded by a microbial mixture when incubated for 7 days at 30°C (4). Significant biodegradation of methyl salicylate in the environment would be expected from this result; however, no data concerning biodegradation in natural waters or soil could be located [R9].

BIOCONCENTRATION The log octanol/water partition coefficient for methyl salicylate is 2.55 (1). The BCF estimated from this log K_{ow} using a regression equation is 4, which indicates that the ester will not bioconcentrate in fish (SRC) [R10].

FIRE POTENTIAL Slight, when exposed to heat or flame; can react with oxidizing materials [R7].

REFERENCES

1. Rossoff, I. S. *Handbook of Veterinary Drugs.* New York: Springer Publishing Company, 1974. 360.

2. Osol, A., and J. E. Hoover, et al. (eds.). *Remington's Pharmaceutical Sciences.* 15th ed. Easton, PA: Mack Publishing Co., 1975. 1232.

3. Sax, N. I., and R. J. Lewis, Sr. (eds.). *Hawley's Condensed Chemical Dictionary.* 11th ed. New York: Van Nostrand Reinhold Co., 1987. 781.

4. SRI.

5. Gerhartz, W. (exec ed.). *Ullmann's Encyclopedia of Industrial Chemistry.* 5th ed. Vol A1: Deerfield Beach, FL: VCH Publishers, 1985 to Present., p. 202.

6. *Kirk–Othmer Encyclopedia of Chemical Technology.* 3rd ed., Volumes 1–26. New York, NY: John Wiley and Sons, 1978–1984, p. V20 513.

7. Sax, N. I. *Dangerous Properties of Industrial Materials.* 5th ed. New York: Van Nostrand Reinhold, 1979. 832.

8. National Fire Protection Guide. *Fire Protection Guide on Hazardous Materials.* 10th ed. Quincy, MA: National Fire Protection Association, 1991, p. 325M-72.

9. (1) Maggio, P., et al., *Ind Carta* 14: 105–11 (1976). (2) Maggio, P., et al., *Tinctoria* 73: 15–20 (1976). (3) Crespi–Rosell, M., J. Cegarra–Sanchez. *Bol Inst Invest Text Coop Ind* 77: 41–57 (1980). (4) Goulding, C., et al., *J Appl Bacteriol* 65: 1–5 (1988).

10. (1) Hansch, C., and A. J. Leo. *Medchem Project* Issue No 26, Claremont, CA: Pomona College (1985) (2) Lyman W.J., et al. (eds.); *Handbook of Chemical Property Estimation Methods.* New York: McGraw-Hill, Chapt 5 (1982).

11. *Fenaroli's Handbook of Flavor Ingredients.* Volume 1. Edited, translated, and revised by T.E. Furia and N. Bellanca. 2nd ed. Cleveland: The Chemical Rubber Co., 1975.

12. Budavari, S. (ed.). *The Merck Index— Encyclopedia of Chemicals, Drugs and Biologicals.* Rahway, NJ: Merck and Co., Inc., 1989.

13. Gilman, A.G., T.W. Rall, A.S. Nies, and P. Taylor (eds.). *Goodman and Gilman's The Pharmacological Basis of Therapeutics.* 8th ed. New York, NY. Pergamon Press, 1990.

14. Clayton, G. D., and F. E. Clayton (eds.). *Patty's Industrial Hygiene and Toxicology.* Volume 2A, 2B, 2C: Toxicology. 3rd ed. New York: John Wiley and Sons, 1981–1982.

15. Sunshine, I. (ed.). *CRC Handbook of Analytical Toxicology.* Cleveland: The Chemical Rubber Co., 1969.

■ METHYL VINYL KETONE

CAS # 78-94-4

DOT # 1251

SYNONYMS 1-buten-3-one • 2-butenone • 3-buten-2-one • 3-butene-2-one • acetone, methylene • acetyl-ethylene • butenone • delta (3)-2-butenone • gamma-oxo-alpha-butylene • ketone, methyl vinyl • methyl vinyl cetone (French) • methylene acetone • methylvinyl ketone • Methylvinylketon (German) • vinyl methyl ketone

MF C_4H_6O

MW 70.10

COLOR AND FORM Colorless liquid.

ODOR Pungent odor; odor concentration (low) = 0.5720 mg/m³.

ODOR THRESHOLD 0.5 mg/m³ [R4].

DENSITY 0.8636 at 20°C; 0.8407 at 25°C.

SURFACE TENSION 24 dynes/cm = 0.024 N/m at 20°C (est).

VISCOSITY 0.807 cP at 70°F.

BOILING POINT 81.4°C.

MELTING POINT FP: 20°F = −7°C = 266 K.

HEAT OF COMBUSTION −14,600 Btu/lb = −8,100 cal/g = −340×10⁵ J/kg (est).

HEAT OF VAPORIZATION 203 Btu/lb = 113 cal/g = 4.73×10⁵ J/kg.

AUTOIGNITION TEMPERATURE 491°C [R7, p. 325M-73].

FLAMMABILITY LIMITS Lower: 2.1%; upper 15.6% [R7, p. 49–64].

FIREFIGHTING PROCEDURES In advanced or massive fires, firefighting should be done from a protected location. Use dry chemical, alcohol foam, or carbon dioxide. Water spray may be ineffective as an extinguishing agent but water should be used to keep fire-exposed containers cool. Direct hose streams from a protected location. If a leak or spill has not ignited, use water spray to disperse the vapors. If it is necessary to stop a leak, use water spray to protect men attempting to do so. Water spray may be used to flush spills away from exposures and to dilute spills to nonflammable mixtures [R7, p. 49–65]. Do not extinquish fire unless flow can be stopped. Use water in flooding quantities as fog. Solid streams of water may be ineffective. Apply water from as far a distance as possible [R3]. If fire becomes uncontrollable or container is exposed to direct flame consider evacuation of one-half (1/2) mile radius [R3].

TOXIC COMBUSTION PRODUCTS Dangerous, upon exposure to heat or flame; emits toxic and irritating fumes [R2].

SOLUBILITY Soluble in methanol, glacial acetic acid; slightly soluble in hydrocarbons; soluble in benzene; >10% in water; >10% in acetone; >10% in ether; >10% in ethanol.

VAPOR PRESSURE 83.9 mm Hg at 25°C (SRC).

OCTANOL/WATER PARTITION COEFFICIENT Estimated log K_{ow} = 0.117.

STABILITY AND SHELF LIFE Polymerizes on standing [R1].

OTHER CHEMICAL/PHYSICAL PROPERTIES Forms binary azeotrope with water; BP: 75°C at 760 mm Hg (12% water).

PROTECTIVE EQUIPMENT Wear full protective clothing. [R7, p. 49–65].

PREVENTATIVE MEASURES If material leaking (not on fire) consider evacuation from downwind area based on amount of material spilled, location, and weather conditions [R3].

DISPOSAL SRP: At the time of review, criteria for land treatment or burial (sanitary landfill) disposal practices are subject to significant revision. Prior to implementing land disposal of waste residue (including waste sludge), consult with environmental regulatory agencies for guidance on acceptable disposal practices.

COMMON USES Alkylating agent; commercial starting material for plastics; as intermediate in synthesis of steroids and vitamin A [R1].

OCCUPATIONAL RECOMMENDATIONS None listed.

BIODEGRADATION A theoretical BOD of 10% was determined for methyl vinyl ketone over a 5-day incubation period using the French AFNOR test (1). Using a standard BOD technique with acclimated sewage inoculum, a theoretical BOD of 0% was measured for methyl vinyl ketone over a 5-day incubation period (2) [R5].

BIOCONCENTRATION Based on an estimated log K_{ow} of 0.117 (2), the BCF for methyl vinyl ketone can be estimated to be 0.72 using a recommended regression derived equation (1,SRC). This BCF value suggests that bioconcentration in aquatic organisms is not significant (SRC) [R6].

GENERAL RESPONSE Avoid contact with liquid. Keep people away. Wear rubber overclothing (including gloves). Shut off ignition sources. Call fire department. Stop discharge if possible. Stay upwind. Use water spray to "knock down" vapor. Isolate and remove discharged material. Notify local health and pollution control agencies.

FIRE RESPONSE Flammable. Containers may explode in fire. Flashback along vapor trail may occur. Vapor may explode if ignited in an enclosed area. Extinguish with dry chemicals, alcohol foam, or carbon dioxide. Water may be ineffective on fire. Cool exposed containers with water.

EXPOSURE RESPONSE Call for medical aid. Vapor irritating to eyes, nose, and throat. If inhaled will cause coughing or difficult breathing. Move victim to fresh air. If breathing has stopped, give artificial respiration. If breathing is difficult, give oxygen. Liquid poisonous if swallowed. Will burn skin and eyes. Remove contaminated clothing and shoes. Flush affected areas with plenty of water. If swallowed and victim is conscious, have victim drink water or milk. Do not induce vomiting.

RESPONSE TO DISCHARGE Issue warning—high flammability, water contaminant. Restrict access. Evacuate area. Disperse and flush.

WATER RESPONSE Effect of low concentrations on aquatic life is unknown. May be dangerous if it enters water intakes. Notify local health and wildlife officials. Notify operators of nearby water intakes.

EXPOSURE SYMPTOMS Inhalation causes irritation of nose and throat. Vapor causes tears; contact with liquid can burn eyes. Liquid irritates skin and will cause burn if not removed at once. Ingestion causes irritation of mouth and stomach.

EXPOSURE TREATMENT Get medical attention for all exposures to this compound. Inhalation: move victim to fresh air; administer artificial respiration if necessary. Eyes or skin: flush with copious quantity of water for 15 min. Ingestion: do not induce vomiting.

REFERENCES

1. *The Merck Index.* 10th ed. Rahway, NJ: Merck Co., Inc., 1983. 878.

2. Sax, N. I. *Dangerous Properties of Industrial Materials.* 6th ed. New York, NY: Van Nostrand Reinhold, 1984. 552.

3. Association of American Railroads. *Emergency Handling of Hazardous Materials in Surface Transportation.* Washington, DC: Assoc. of American Railroads, Hazardous Materials Systems (BOE), 1987. 461.

4. Verschueren, K. *Handbook of Environmental Data of Organic Chemicals.* 2nd ed. New York, NY: Van Nostrand Reinhold Co., 1983. 876.

5. (1) Dore, M., et al., *Trib Cebedeau 28*: 3–11 (1975). (2) Niemi, G. J., et al., *Environ Toxicol Chem 6*: 515–27 (1987).

6. (1) Lyman, W. J., et al., *Handbook of Chemical Estimation Methods.* New York: McGraw-Hill, p. 5–4 (1982) (2) GEMS; Graphical Exposure Modeling System. PCGEMS (1987).

7. National Fire Protection Association. *Fire Protection Guide on Hazardous Materials.* 9th ed. Boston, MA: National Fire Protection Association, 1986.

■ METRIBUZIN

CAS # 21087–64–9

SYNONYMS sencoral • sengoral • 4-amino-6-tert-butyl-3-(methylthio)as-triazin-S (4H)-one • Bayer-6159h • Bayer-6443h • dic-1468 • sencorex • metribuzine • 4-amino-6-tert-butyl-4,5-dihydro-3-methylthio-1,2,4-triazi n-5-one • 4-amino-6-tert-butyl-3-(methylthio)-as-triazin-5 (4h)-one • lexone-df • sencor-4 • sencor-df • 4-amino-6-(1,1-dimethylethyl)-3-methylthio-1,2,4-triazin-5 (4h) one • 4-amino-6-(1,1-dimethylethyl)-3-(methylthio)-1,2,4-triazin-5 (4h)-one • lexone • sencor • bay-dic-1468 • lexone-4l • Bay-94337

MF $C_8H_{14}N_4OS$

MW 214.3

COLOR AND FORM Colorless crytals; white crystalline solid.

ODOR Mild chemical odor.

DENSITY 1.28 at 20°C.

MELTING POINT 125–126.5°C.

SOLUBILITY Soluble in methanol and ethanol; very slightly soluble in water (1,200 ppm); soluble in glycol ether acetate; in water at 20°C, 1.2 g/L; in dimethylformamide 1,780, cyclohexanone 1,000, chloroform 850, acetone 820, methanol 450, dichloromethane 333, benzene 220, n-butanol 150, ethanol 190, toluene 120, xylene 90, n-hexane 2 (all in g/kg at 20°C).

VAPOR PRESSURE $<10^{-5}$ mm Hg at 20°C.

CORROSIVITY Noncorrosive.

OTHER CHEMICAL/PHYSICAL PROPERTIES Slight sulfurous odor (technical metribuzin); Henry's Law constant = $<2.31\times10^{-9}$ atm-m³/mole at 20°C (est).

PREVENTATIVE MEASURES Avoid contact with skin, eyes, and clothing. Remove contaminated clothing and wash with soap and hot water before reuse [R1].

DISPOSAL Bury empty container or product that cannot be used in a safe place away from water supplies, or dispose of by alternative procedures recommended by federal, state, or local authorities. Open dumping is prohibited [R1].

COMMON USES Controls a large number of grass and broadleaf weeds infesting agricultural crops [R7]. Controls a large number of grass and broadleaf weeds infesting agricultural crops [R8]; control of many annual broad leaved and grass weeds in asparagus, potatoes, tomatoes, lucerne, sainfoin, peas, lentils, soya beans, sugar cane, pineapples, and cereals [R9]; pre- and post-emergence triazone herbicide [R10].

OCCUPATIONAL RECOMMENDATIONS .
OSHA 8-hr time-weighted average: 5 mg/m³. Final rule limits were achieved by any combination of engineering controls, work practices, and personal protective equipment during the phase-in period, Sept 1, 1989 through Dec 30, 1993. Final rule limits became effective Dec 31, 1993 [R4].

THRESHOLD LIMIT 8-hr time-weighted average (TWA) 5 mg/m³ (1984) [R5].

NON-HUMAN TOXICITY LD_{50} rat oral 1,100 mg/kg [R2]; LD_{50} rat dermal >2,000 mg/kg [R11]; LD_{50} rabbit dermal >2,000 mg/kg [R11]; LD_{50} mouse oral 698–711 mg/kg [R9]; LD_{50} guinea pig oral 250 mg/kg [R9]; LD_{50} cat oral >500 mg/kg [R9]; LD_{50} rat percutaneous >20,000 mg/kg [R9].

BIODEGRADATION In mineral and muck soils, metribuzin metabolism via deamination and thiodealkylation produced: 6-(1,1-dimethylethyl)-3-methylthio-1,2,

4-triazin-5-(4H)-one metribuzin; 4-amino-6-(1,1-dimethylethyl)-1,2,4-triazin-3,5-(2H, 4H)-dione 3,5-diketo; and 6-(1,1-dimethylethyl)-1,2,4-triazin-3,5-(2H, 4H)-dione deaminated diketo. Over a pH range of 4.5 to 6.9 in sandy clay loam, microbial degradation gave $^{14}CO_2$ from ^{14}C ring-labeled metribuzin. It was also observed that metribuzin degradation by soil microorganisms decreased with increasing soil pH. [R12, 536].

BIOCONCENTRATION The bioconcentration factor (BCF) of metribuzin in the golden ide fish (*Leuciscus idus melanotus*) was experimentally determined to be 10 in a 3-day static test (1). Based on this BCF value, metribuzin is not expected to bioconcentrate in fishes (SRC) [R3].

FIFRA REQUIREMENTS In 1988, Congress amended FIFRA to strengthen and accelerate EPA's reregistration program. The nine-year reregistration scheme mandated by "FIFRA 88" applies to each registered pesticide product containing an active ingredient initially registered before November 1, 1984. List A consists of the 194 chemical cases (or 350 individual active ingredients) for which EPA had issued Registration Standards prior to the effective date of FIFRA '88. List: A; Case: Metribuzin; Case No. 0181; Pesticide type: Herbicide; Registration Standard Date: 06/30/85; Case Status: Awaiting Data/Data in Review: OPP awaits data from the pesticide's producer(s) regarding its human health or environmental effects, or OPP has received and is reviewing such data, in order to reach a decision about the pesticide's eligibility for reregistration. Active Ingredient (AI): Metribuzin; Data Call-in (DCI) Date: 07/31/91; AI Status: The producer(s) of the pesticide has made commitments to conduct the studies and pay the fees required for reregistration, and is meeting those commitments in a timely manner [R6].

REFERENCES

1. *Farm Chemicals Handbook* 1992. Willoughby, OH: Meister Publishing Co., 1992, p. C–224.

2. Morgan, D. P. *Recognition and Management of Pesticide Poisonings.* EPA

540/9–80–005. Washington, DC: U.S. Government Printing Office, Jan. 1982. 84.

3. Freitag, D., et al., *Chemosphere* 14: 1589–616 (1985).

4. 29 CFR 1910.1000 (7/1/91).

5. American Conference of Governmental Industrial Hygienists. *Threshold Limit Values for Chemical Substances and Physical Agents and Biological Exposure Indices for 1994–1995.* Cincinnati, OH: ACGIH, 1994. 27.

6. USEPA/OPP; *Status of Pesticides in Reregistration, and Special Review* p. 100 (Mar, 1992) EPA 700–R–92–004.

7. Worthing, C. R., and S. B. Walker (eds.). *The Pesticide Manual—A World Compendium.* 8th ed. Thornton Heath, UK: The British Crop Protection Council, 1987.

8. *Farm Chemicals Handbook 1992.* Willoughby, OH: Meister Publishing Co., 1992, p. C–224.

9. Hartley, D., and H. Kidd (eds.). *The Agrochemicals Handbook.* 2nd ed. Lechworth, Herts, England: The Royal Society of Chemistry, 1987, p. A280/Aug 87.

10. Budavari, S. (ed.). *The Merck Index— Encyclopedia of Chemicals, Drugs, and Biologicals.* Rahway, NJ: Merck and Co., Inc., 1989. 968.

11. American Conference of Governmental Industrial Hygienists. *Documentation of the Threshold Limit Values and Biological Exposure Indices.* 5th ed. Cincinnati, OH: American Conference of Governmental Industrial Hygienists, 1986. 411.

12. Menzie, C.M. *Metabolism of Pesticides—Update III.* Special Scientific Report—*Wildlife* No. 232. Washington, DC: U.S. Department of the Interior, Fish and Wildlife Service, 1980.

■ METRONIDAZOLE

CAS # 443–48–1

SYNONYMS 1-(2-hydroxy-1-ethyl) -2-methyl-5-nitroimidazole • 1-(2-hydroxyethyl)-2-methyl-5-nitroimidazole • 1-(beta-ethylol)-2-methyl-5-nitro-3-azapyrrole • 1-(beta-hydroxyethyl)-2-methyl-5-nitroimidazole • 1-hydroxyethyl-2-methyl-5-nitroimidazole • 2-methyl-1-(2-hydroxyethyl)-5-nitroimidazole • 2-methyl-3-(2-hydroxyethyl)-4-nitroimidazole • 2-methyl-5-nitro-1-imidazoleethanol • 2-methyl-5-nitroimidazole-1-ethanol • acromona • atrivyl • bexon • cont • danizol • deflamon-wirkstoff • efloran • elyzol • entizol • eumin • flagemona • flagyl • fossyol • giatricol • klion • meronidal • metronidaz • metronidazol • mexibol 'silanes' • monagyl • nalox • neo-tric • nida • novonidazol • sanatrichom • sc-10295 • trichazol • trichex • trichocide • trichomonacid-'pharmachim' • trichopol • tricocet • tricom • tricowas-b • trikacide • trikamon • trikojol • trikozol • trimeks • vagilen • vertisal

MF $C_6H_9N_3O_3$

MW 171.16

pH pH of saturated aqueous solution is 5.8.

COLOR AND FORM Cream-colored crystals; white to pale-yellow crystals or crystalline powder.

ODOR Odorless.

DISSOCIATION CONSTANT 2.38.

MELTING POINT 158–160°C.

SOLUBILITY G/100 ml at 20°C: 1.0 in water, 0.5 in ethanol, less than 0.05 in ether, chloroform; soluble in dilute acids; sparingly soluble in dimethylformamide; 11,000 mg/L in water.

OCTANOL/WATER PARTITION COEFFICIENT Log K_{ow} = −0.02 at 25°C.

STABILITY AND SHELF LIFE Stable in air but darkens on exposure to light [R2].

PROTECTIVE EQUIPMENT Dispensers of liquid detergent should be available. Safety pipettes should be used for all pipetting. In animal laboratory, personnel should wear protective suits (preferably disposable, one-piece, and close-fitting at ankles and wrists), gloves, hair covering, and overshoes. In chemical laboratory, gloves and gowns should always be worn; however, gloves should not be assumed to provide full protection. Carefully fitted masks or respirators may be necessary when working with particulates or gases, and disposable plastic aprons might provide addnl protection. Gowns should be of distinctive color, as a reminder that they are not to be worn outside the laboratory. (chemical carcinogens) [R4, 8].

PREVENTATIVE MEASURES Smoking, drinking, eating, storage of food, or of food and beverage containers, or utensils, and the application of cosmetics should be prohibited in any laboratory. All personnel should remove gloves, if worn, after completion of procedures in which carcinogens have been used. They should wash hands, preferably using dispensers of liquid detergent, and rinse thoroughly. Consideration should be given to appropriate methods for cleaning the skin, depending on nature of the contaminant. No standard procedure can be recommended, but the use of organic solvents should be avoided. Safety pipettes should be used for all pipetting (chemical carcinogens) [R4, 8]. Precautions for "carcinogens": In animal laboratory, personnel should remove their outdoor clothes and wear protective suits (preferably disposable, one-piece and close-fitting at ankles and wrists), gloves, hair covering, and overshoes. Clothing should be changed daily but discarded immediately if obvious contamination occurs. Also, workers should shower immediately. In chemical laboratory, gloves and gowns should always be worn however, gloves should not be assumed to provide full protection. Carefully fitted masks or respirators may be necessary when working with particulates or gases, and disposable plastic aprons might provide addnl protection. Gowns should be of distinctive color, as a reminder that they should not be worn outside of lab (chemical carcinogens) [R4, 8]. Precautions for "carcinogens": Operations connected with synth and purification should be carried out under well-ventilated hood. Analytical

procedures should be carried out with care and vapors evolved during procedures should be removed. Expert advice should be obtained before existing fume cupboards are used and when new fume cupboards are installed. It is desirable that there be means for decreasing the rate of air extraction, so that carcinogenic powders can be handled without powder being blown around the hood. Glove boxes should be kept under negative air pressure. Air changes should be adequate, so that concn of vapors of volatile carcinogens will not occur (chemical carcinogens) [R4, 8]. Precautions for carcinogens: Vertical laminar-flow biological safety cabinets may be used for containment of in-vitro procedures provided that the exhaust air flow is sufficient to provide an inward air flow at the face opening of the cabinet, and contaminated air plenums that are under positive pressure are leak-tight. Horizontal laminar-flow hoods or safety cabinets, where filtered air is blown across the working area towards the operator, should never be used. Each cabinet or fume cupboard to be used should be tested before work is begun (e.g., with fume bomb) and label fixed to it, giving date of test and avg airflow measured. This test should be repeated periodically and after any structural changes (chemical carcinogens) [R4, 9]. Precautions for carcinogens: Principles that apply to chem or biochem lab also apply to microbiological and cell-culture labs. Special consideration should be given to route of admin. Safest method of administering volatile carcinogen is by injection of a solution. Admin by topical application, gavage, or intratracheal instillation should be performed under hood. If chem will be exhaled, animals should be kept under hood during this period. Inhalation exposure requires special equipment. Unless specifically required, routes of admin other than in the diet should be used. Mixing of carcinogen in diet should be carried out in sealed mixers under fume hood, from which the exhaust is fitted with an efficient particulate filter. Techniques for cleaning mixer and hood should be devised before expt begun. When mixing diets, special protective clothing and, possibly, respirators may be required (chemical carcinogens) [R4, 9]. Precautions for carcinogens: When admin in diet or applied to skin, animals should be kept in cages with solid bottoms and sides and fitted with a filter top. When volatile carcinogens are given, filter tops should not be used. Cages which have been used to house animals that received carcinogens should be decontaminated. Cage-cleaning facilities should be installed in area in which carcinogens are being used, to avoid moving of contaminated cages. It is difficult to ensure that cages are decontaminated, and monitoring methods are necessary. Situations may exist in which the use of disposable cages should be recommended, depending on type and amount of carcinogen and efficiency with which it can be removed (chemical carcinogens) [R4, 10]. Precautions for "carcinogens": To eliminate risk that contamination in lab could build up during conduct of expriment, periodic checks should be carried out on lab atmospheres, surfaces, such as walls, floors and benches, and interior of fume hoods and airducts. As well as regular monitoring, check must be carried out after cleanup of spillage. Sensitive methods are required when testing lab atmospheres. Methods should, where possible, be simple and sensitive (chemical carcinogens) [R4, 10]. Precautions for carcinogens: Rooms in which obvious contamination has occurred, such as spillage, should be decontaminated by lab personnel engaged in experiment. Design of experiment should avoid contamination of permanent equipment. Procedures should ensure that maintenance workers are not exposed to carcinogens. Particular care should be taken to avoid contamination of drains or ventilation ducts. In cleaning labs, procedures should be used that do not produce aerosols or dispersal of dust, i.e., wet mop or vacuum cleaner equipped with high-efficiency particulate filter on exhaust, which are avail commercially, should be used. Sweeping, brushing and use of dry dusters or mops should be prohibited. Grossly contaminated cleaning materials should not be re-used. If gowns or towels are contaminated, they should not be sent to laundry, but decon-

taminated or burnt, to avoid any hazard to laundry personnel (chemical carcinogens) [R4, 10]. Precautions for carcinogens: Doors leading into areas where carcinogens are used should be marked distinctively with appropriate labels. Access limited to persons involved in experiment. A prominently displayed notice should give the name of the scientific investigator or other person who can advise in an emergency and who can inform others (such as firemen) on the handling of carcinogenic substances. (chemical carcinogens) [R4, 11].

CLEANUP A high-efficiency particulate arrestor (HEPA) or charcoal filters can be used to minimize amount of carcinogen in exhausted-air-ventilated safety cabinets, lab hoods, glove boxes, or animal rooms. Filter housing that is designed so that used filters can be transferred into plastic bag without contaminating maintenance staff is avail commercially. Filters should be placed in plastic bags immediately after removal. The plastic bag should be sealed immediately. The sealed bag should be labelled properly. Waste liquids should be placed or collected in proper containers for disposal. The lid should be secured and the bottles properly labelled. Once filled, bottles should be placed in plastic bag, so that outer surface is not contaminated. The plastic bag should also be sealed and labelled. Broken glassware should be decontaminated by solvent extraction, by chemical destruction, or in specially designed incinerators. (chemical carcinogens) [R4, 15].

DISPOSAL SRP: At the time of review, criteria for land treatment or burial (sanitary landfill) disposal practices are subject to significant revision. Prior to implementing land disposal of waste residue (including waste sludge), consult with environmental regulatory agencies for guidance on acceptable disposal practices.

COMMON USES Vet: antiprotozoal; treponemicide [R1]; medication: antiprotozoal (*Trichomonas*) [R1].

OCCUPATIONAL RECOMMENDATIONS None listed.

IARC SUMMARY Inadequate evidence of

carcinogenicity in humans. Sufficient evidence of carcinogenicity in animals. Overall evaluation: Group 2B: The agent is possibly carcinogenic to humans [R3].

BIOLOGICAL HALF-LIFE The normal half-life of metronidazole is approximately eight hours. [R5, 1546].

FIFRA REQUIREMENTS In 1988, Congress amended FIFRA to strengthen and accelerate EPA's reregistration program. The nine-year reregistration scheme mandated by "FIFRA 88" applies to each registered pesticide product containing an active ingredient initially registered before November 1, 1984. Pesticides for which EPA had not issued Registration Standards prior to the effective date of FIFRA '88 were divided into three lists based upon their potential for exposure and other factors, with List B being of highest concern and D of least. List: C; Case: Metronidazole; Case No.: 3096; Pesticide type: antimicrobial; Case Status: Awaiting Data/Data in Review: OPP awaits data from the pesticide's producer(s) regarding its human health and/or environmental effects, or OPP has received and is reviewing such data, in order to reach a decision about the pesticide's eligibility for reregistration. Active Ingredient (AI): Metronidazole; AI Status: The producer(s) of the pesticide has made commitments to conduct the studies and pay the fees required for reregistration, and is meeting those commitments in a timely manner. [R6].

REFERENCES

1. Budavari, S. (ed.). *The Merck Index—Encyclopedia of Chemicals, Drugs, and Biologicals.* Rahway, NJ: Merck and Co., Inc., 1989. 968.

2. Osol, A., and J. E. Hoover, et al. (eds.). *Remington's Pharmaceutical Sciences.* 15th ed. Easton, PA: Mack Publishing Co., 1975. 1161.

3. IARC. *Monographs on the Evaluation of the Carcinogenic Risk of Chemicals to Man.* Geneva: World Health Organization, International Agency for Research on Cancer, 1972–present (multivolume work), p. S7 66 (1987).

4. Montesano, R., H. Bartsch, E. Boyland, G. Della Porta, L. Fishbein, R. A. Griesemer, A.B. Swan, L. Tomatis, and W. Davis (eds.). *Handling Chemical Carcinogens in the Laboratory: Problems of Safety.* IARC Scientific Publications No. 33. Lyon, France: International Agency for Research on Cancer, 1979.

5. American Medical Association, Council on Drugs. *AMA Drug Evaluations Annual 1994.* Chicago, IL: American Medical Association, 1994.

6. USEPA/OPP. *Status of Pesticides in Reregistration and Special Review.* p. 202 (Mar. 1992) EPA 700-R-92-004.

■ MGK-264

CAS # 113-48-4

SYNONYMS 2-(2-ethylhexyl)-3a, 4,7,7a-tetrahydro-4,7-methano-1h-isoindole-1,3 (2h)-dione • 4,7-methano-1h-isoindole-1,3 (2h)-dione, 2-(2-ethylhexyl)-3a, 4,7,7a-tetrahydro • 5-norbornene-2,3-dicarboximide, N-(2-ethylhexyl)- • bicyclo (2.2.1) heptene-2-dicarboxylic acid, 2-ethylhexylimide • endomethylenetetrahydrophthalic acid, n-2-ethylhexyl imide • ent-8,184 • n-(2-ethylhexyl)-5-norbornene-2,3-dicarboximide • n-(2-ethylhexyl) bicyclo-(2,2,1)-hept-5-ene-2,3-dicarboximide • n-2-ethylhexylimide endomethylenetetrahydrophthalic acid • n-octyl bicycloheptene dicarboximide • n-octylbicyclo-(2.2.1)-5-heptene-2,3-dicarboximide • octacide-264 • pyrodone • sinepyrin-222 • synergist-264 • Van Dyk 264 • mgk-264 • n-(2-ethylhexyl) -8,9,10-trinorborn-5-ene-2,3-dicarboximide

MF $C_{17}H_{25}NO_2$

MW 275.4

COLOR AND FORM Very-light-yellow liquid.

TASTE Bitter taste.

BOILING POINT 158.2°C at 2 mm Hg.

MELTING POINT FP: below −20°C.

SOLUBILITY Practically insoluble in water; miscible with most organic solvents, including petroleum oils and fluorinated hydrocarbons.

CORROSIVITY Noncorrosive.

STABILITY AND SHELF LIFE Stable in range pH 6–8 to light and heat [R2].

OTHER CHEMICAL/PHYSICAL PROPERTIES Density: 1.040–1.060 at 20°C (technical grade); stable in range pH 6–8 to light and heat.

DISPOSAL SRP: At the time of review, criteria for land treatment or burial (sanitary landfill) disposal practices are subject to significant revision. Prior to implementing land disposal of waste residue (including waste sludge), consult with environmental regulatory agencies for guidance on acceptable disposal practices.

COMMON USES Synergist for pyrethroids in aerosol sprays for household and veterinary use [R1]. Synergist for pyrethrins, allethrin, pyethroids, and rotenone. Often used in combination with piperonyl butoxide in aerosols, household, and industrial sprays. Stabilizes and prolongs the active life of pyrethrins, allethrin, and MGK repellent 874. [R4, p. C–239].

OCCUPATIONAL RECOMMENDATIONS None listed.

BIOCONCENTRATION Using an estimated log K_{ow} of 3.62 (1), one would estimate a BCF of 330 for MGK 264 using a recommended regression equation (2). This would indicate that MGK 264 would bioconcentrate moderately in aquatic organisms (SRC) [R3].

FIFRA REQUIREMENTS In 1988, Congress amended FIFRA to strengthen and accelerate EPA's reregistration program. The nine-year reregistration scheme mandated by "FIFRA 88" applies to each registered pesticide product containing an active ingredient initially registered before November 1, 1984. Pesticides for which EPA had not issued Registration Standards prior to the effective date of FIFRA '88 were divided into three lists based upon their potential for exposure and other factors, with List B being of highest concern and D of least. List: B; Case: MGK–264; Case No.: 2,430; Pesticide type: Insecticide, fungicide, antimicrobial; Case Status: Awaiting Data/Data in Review: OPP awaits data from the pesticide's producer(s) regarding its human health and/or environmental effects, or OPP has received and is reviewing such data, in order to reach a decision about the pesticide's eligibility for reregistration. Active Ingredient (AI): n-Octyl bicycloheptenedicarboximide; Data Call-in (DCI) Date: 06/03/91; AI Status: The producer(s) of the pesticide has made commitments to conduct the studies and pay the fees required for reregistration, and is meeting those commitments in a timely manner [R5]. Tolerances are established for residues of the insecticide n-octyl bicycloheptene-dicarboximide, resulting from dermal application, in raw agricultural commodities as follows: cattle (fat); goats (fat); hogs (fat); horses (fat); milk (fat); sheep (fat). [R6].

REFERENCES

1. Worthing, C. R., and S. B. Walker (eds.). *The Pesticide Manual—A World Compendium.* 8th ed. Thornton Heath, UK: The British Crop Protection Council, 1987. 359.

2. Worthing, C. R., S. B. Walker (eds.). *The Pesticide Manual—A World Compendium.* 7th ed. Lavenham, Suffolk, Great Britain: The Lavenham Press Limited, 1983. 250.

3. (1) Meylan, W. M., P. H. Howard. *J Pharm Sci* 84: 83–92 (1995). (2) Lyman, W. J., et al., *Handbook of Chemical Property Estimation Methods.* New York: McGraw-Hill, Chapt 5, Eqn 5–2 (1982).

4. *Farm Chemicals Handbook 1994.* Willoughby, OH: Meister, 1994.

5. USEPA/OPP. *Status of Pesticides in Reregistration and Special Review.* p.141 (Mar. 1992) EPA 700-R-92-004.

6. 40 CFR 180.367 (7/1/94).

■ MITOMYCIN-C

CAS # 50-07-7

SYNONYMS ametycine • mutamycin • nsc-26980 • azirino (2′,3′:3,4) pyrrolo (1,2-a) indole-4,7-dione, 6-amino-8-(((aminocarbonyl)oxy)methyl)-1,1a, 2,8,8a,8b-hexahydro-8-a-methoxy-5-methyl-, (1aS-(1a-alpha, 8beta, 8a-alpha, 8balpha))-(9CI) • Mitomycyna-C (Polish) • NCI-C04706

MF $C_{15}H_{18}N_4O_5$

MW 334.37

COLOR AND FORM Blue-violet crystals.

MELTING POINT Above 360°C.

SOLUBILITY Soluble in water, methanol, butyl acetate, acetone, and cyclohexanone; slightly soluble in benzene, ether, and carbon tetrachloride; practically insoluble in petroleum ether.

STABILITY AND SHELF LIFE Commercially available mitomycin powder is stable for at least 4 years at room temperature [R3].

PROTECTIVE EQUIPMENT For carcinogens: dispensers of liquid detergent should be available. Safety pipettes should be used for all pipetting. In animal laboratory, personnel should wear protective suits (preferably disposable, one-piece, and close-fitting at ankles and wrists), gloves, hair covering, and overshoes. In chemical laboratory, gloves and gowns should always be worn; however, gloves should not be assumed to provide full protection. Carefully fitted masks or respirators may be necessary when working with particulates or gases, and disposable plastic aprons might provide addnl protection. Gowns should be of distinctive color, as a reminder that they are not to be worn outside the laboratory. (chemical carcinogens) [R8, 8].

PREVENTATIVE MEASURES For carcinogens: smoking, drinking, eating, storage of food, or of food and beverage containers, or utensils, and the application of cosmetics should be prohibited in any laboratory. All personnel should remove gloves, if worn, after completion of procedures in which carcinogens have been used. They should wash hands, preferably using dispensers of liquid detergent, and rinse thoroughly. Consideration should be given to appropriate methods for cleaning the skin, depending on nature of the contaminant. No standard procedure can be recommended, but the use of organic solvents should be avoided. Safety pipettes should be used for all pipetting. (chemical carcinogens) [R8, 8].

CLEANUP For carcinogens: a high-efficiency particulate arrestor (HEPA) or charcoal filters can be used to minimize amount of carcinogen in exhausted-air-ventilated safety cabinets, lab hoods, glove boxes, or animal rooms. Filter housing that is designed so that used filters can be transferred into plastic bag without contaminating maintenance staff is avail commercially. Filters should be placed in plastic bags immediately after removal. The plastic bag should be sealed immediately. The sealed bag should be labelled properly. Waste liquids should be placed or collected in proper containers for disposal. The lid should be secured and the bottles properly labelled. Once filled, bottles should be placed in plastic bag, so that outer surface is not contaminated. The plastic bag should also be sealed and labelled. Broken glassware should be decontaminated by solvent extraction, by chemical destruction, or in specially designed incinerators. (chemical carcinogens) [R8, 15].

DISPOSAL Generators of waste (equal to, or greater than 100 kg/mo) containing this contaminant, EPA hazardous waste number U010, must conform with USEPA regulations in storage, transportation, treatment, and disposal of waste [R4].

COMMON USES Medication [R2].

OCCUPATIONAL RECOMMENDATIONS None listed.

NON-HUMAN TOXICITY LD_{50} mouse iv 5 mg/kg [R1].

IARC SUMMARY No data are available for humans. Sufficient evidence of carcinogenicity in animals. Overall evaluation:

Group 2B: The agent is possibly carcinogenic to humans [R5].

BIOLOGICAL HALF-LIFE After doses of 20 mg/m² mitomycin is cleared from plasma with a half-time of approximately 1 hour [R2].

CERCLA REPORTABLE QUANTITIES Persons in charge of vessels or facilities are required to notify the National Response Center (NRC) immediately, when there is a release of this designated hazardous substance, in an amount equal to or greater than its reportable quantity of 10 lb, or 4.54 kg. The toll-free number of the NRC is (800) 424–8802; in the Washington DC metropolitan area (202) 426–2675. The rule for determining when notification is required is stated in 40 CFR 302.4 (section IV. D. 3.b) [R6].

RCRA REQUIREMENTS U010; as stipulated in 40 CFR 261.33, when mitomycin C, as a commercial chemical product or manufacturing chemical intermediate or an off-specification commercial chemical product or a manufacturing chemical intermediate, becomes a waste, it must be managed according to federal or state hazardous waste regulations. Also defined as a hazardous waste is any residue, contaminated soil, water, or other debris resulting from the cleanup of a spill, into water, or on dry land, of this waste. Generators of small quantities of this waste may qualify for partial exclusion from hazardous waste regulations (40 CFR 261.5) [R7].

REFERENCES

1. Budavari, S. (ed.). *The Merck Index— Encyclopedia of Chemicals, Drugs, and Biologicals*. Rahway, NJ: Merck and Co., Inc., 1989. 979.

2. Gilman, A. G., T. W. Rall, A. S. Nies, and P. Taylor (eds.). *Goodman and Gilman's The Pharmacological Basis of Therapeutics*. 8th ed. New York, NY. Pergamon Press, 1990. 1247.

3. McEvoy, G. K. (ed.). *AHFS Drug Information 90*. Bethesda, MD: American Society of Hospital Pharmacists, Inc., 1990 (Plus Supplements 1990). 537.

4. 40 CFR 240–280, 300–306, 702–799 (7/1/89).

5. IARC. *Monographs on the Evaluation of the Carcinogenic Risk of Chemicals to Man*. Geneva: World Health Organization, International Agency for Research on Cancer, 1972–present (multivolume work), p. S7 67 (1987).

6. 54 FR 33419 (8/14/89).

7. 40 CFR 261.33 (7/1/90).

8. Montesano, R., H. Bartsch, E. Boyland, G. Della Porta, L. Fishbein, R. A. Griesemer, A.B. Swan, L. Tomatis, and W. Davis (eds.). *Handling Chemical Carcinogens in the Laboratory: Problems of Safety*. IARC Scientific Publications No. 33. Lyon, France: International Agency for Research on Cancer, 1979.

■ MORPHOLINE, N–NITROSO

CAS # 59–89–2

SYNONYMS 4-nitrosomorpholine • morpholine, 4-nitroso • N-Nitrosomorpholin (German) • nci-c02164 • nitrosomorpholine • nmor

MF C₄H₈N₂O₂

MF $C_4H_8N_2O_2$

MW 116.14

COLOR AND FORM Yellow crystals.

BOILING POINT 224–225°C at 747 mm Hg.

MELTING POINT 29°C.

SOLUBILITY Miscible in water in all prop; soluble in organic solvents; soluble in water.

OCTANOL/WATER PARTITION COEFFICIENT Log K_{ow} = –0.44.

STABILITY AND SHELF LIFE Stable at room temperature for more than 14 days in neutral and alk aqueous solution in dark; slightly less stable in acidic solution [R2].

OTHER CHEMICAL/PHYSICAL PROPERTIES

Strong oxidants (peracids) oxidize it to corresponding nitramine; can be reduced to corresponding hydrazine or amine.

PROTECTIVE EQUIPMENT Precautions for carcinogens: Dispensers of liquid detergent should be available. Safety pipettes should be used for all pipetting. In animal laboratory, personnel should wear protective suits (preferably disposable, one-piece, and close-fitting at ankles and wrists), gloves, hair covering, and overshoes. In chemical laboratory, gloves and gowns should always be worn; however, gloves should not be assumed to provide full protection. Carefully fitted masks or respirators may be necessary when working with particulates or gases, and disposable plastic aprons might provide addnl protection. Gowns should be of distinctive color, as a reminder that they are not to be worn outside the laboratory (chemical carcinogens) [R6, 8].

PREVENTATIVE MEASURES Precautions for carcinogens: Smoking, drinking, eating, storage of food, or of food and beverage containers, or utensils, and the application of cosmetics should be prohibited in any laboratory. All personnel should remove gloves, if worn, after completion of procedures in which carcinogens have been used. They should wash hands, preferably using dispensers of liquid detergent, and rinse thoroughly. Consideration should be given to appropriate methods for cleaning the skin, depending on nature of the contaminant. No standard procedure can be recommended, but the use of organic solvents should be avoided. Safety pipettes should be used for all pipetting (chemical carcinogens) [R6, 8].

CLEANUP Precautions for carcinogens: A high-efficiency particulate arrestor (HEPA) or charcoal filters can be used to minimize amount of carcinogen in exhausted-air-ventilated safety cabinets, lab hoods, glove boxes, or animal rooms. Filter housing that is designed so that used filters can be transferred into plastic bag without contaminating maintenance staff is avail commercially. Filters should be placed in plastic bags immediately after removal. The plastic bag should be sealed immediately. The sealed bag should be labelled properly. Waste liquids should be placed or collected in proper containers for disposal. The lid should be secured and the bottles properly labelled. Once filled, bottles should be placed in plastic bag, so that outer surface is not contaminated. The plastic bag should also be sealed and labelled. Broken glassware should be decontaminated by solvent extraction, by chemical destruction, or in specially designed incinerators (chemical carcinogens) [R6, 15].

DISPOSAL Precautions for carcinogens: There is no universal method of disposal that has been proved satisfactory for all carcinogenic compounds, and specific methods of chemical destruction published have not been tested on all kinds of carcinogen-containing waste. Summary of avail methods and recommendations given must be treated as guide only (chemical carcinogens) [R6, 14]. Precautions for "carcinogens": Incineration may be only feasible method for disposal of contaminated laboratory waste from biological expt. However, not all incinerators are suitable for this purpose. The most efficient type is probably the gas-fired type, in which a first-stage combustion with a less-than-stoichiometric air:fuel ratio is followed by a second stage with excess air. Some are designed to accept aqueous and organic-solvent solutions; otherwise, it is necessary to absorb soln onto suitable combustible material, such as sawdust. Alternatively, chem destruction may be used, esp when small quantities are to be destroyed in laboratory (chemical carcinogens) [R6, 15]. Precautions for carcinogens: HEPA (high-efficiency particulate arrestor) filters can be disposed of by incineration. For spent charcoal filters, the adsorbed material can be stripped off at high temp and carcinogenic wastes generated by this treatment conducted to and burned in an incinerator. Liquid waste: Disposal should be carried out by incineration at temp that ensure complete combustion. Solid waste: Carcasses of lab animals, cage litter, and misc solid wastes should be disposed of by incineration at temp high enough to ensure destruction of chem carcinogens or their metabolites (chemical carcinogens) [R6, 15]. Precautions for carcinogens: Small quantities of some carcinogens can be destroyed using chem reactions, but no general rules can be given. As a general technique treatment with sodium dichromate in strong sulfuric acid can be used. The time necessary for destruction is seldom known but 1–2 days is generally considered sufficient when freshly prepd reagent is used. Carcinogens that are easily oxidizable can be destroyed with milder oxidative agents, such as saturated soln of potassium permanganate in acetone, which appears to be a suitable agent for destruction of hydrazines or of compounds containing isolated carbon-carbon double bonds. Concn or 50% aqueous sodium hypochlorite can also be used as an oxidizing agent (chemical carcinogens) [R6, 16]. Precautions for carcinogens: Carcinogens that are alkylating, arylating or acylating agents per se can be destroyed by reaction with appropriate nucleophiles, such as water, hydroxyl ions, ammonia, thiols, and thiosulfate. The reactivity of various alkylating agents varies greatly and is also influenced by sol of agent in the reaction medium. To facilitate the complete reaction, it is suggested that the agents be dissolved in ethanol or similar solvents. No method should be applied until it has been thoroughly tested for its effectiveness and safety on material to be inactivated. For example, in case of destruction of alkylating agents, it is possible to detect residual compounds by reaction with 4(4-nitrobenzyl) pyridine (chemical carcinogens) [R6, 17].

COMMON USES Solvent for polyacrylonitrile and chemical intermediate in synthesis of N–aminomorpholine [R1].

OCCUPATIONAL RECOMMENDATIONS None listed.

NON-HUMAN TOXICITY LD$_{50}$ Rat (route not specified) 320 mg/kg [R4].

IARC SUMMARY No data are available for humans. Sufficient evidence of carcinogenicity in animals. Overall evaluation:

Group 2B: The agent is possibly carcinogenic to humans [R3].

BIOCONCENTRATION A bioconcentration factor (BCF) of <1 was estimated for N–nitrosomorpholine based on a measured log K_{ow} of −0.44 (1,2,SRC). This BCF value and the complete water solubility of N–nitrosomorpholine suggest that this compound will not bioaccumulate appreciably in aquatic organisms (3,SRC) [R5].

REFERENCES

1. IARC. *Monographs on the Evaluation of the Carcinogenic Risk of Chemicals to Man*. Geneva: World Health Organization, International Agency for Research on Cancer, 1972–present (multivolume work), p. V17 264 (1978).

2. IARC. *Monographs on the Evaluation of the Carcinogenic Risk of Chemicals to Man*. Geneva: World Health Organization, International Agency for Research on Cancer, 1972–present (multivolume work), p. V17 263.

3. IARC. *Monographs on the Evaluation of the Carcinogenic Risk of Chemicals to Man*. Geneva: World Health Organization, International Agency for Research on Cancer, 1972–present (multivolume work), p. S7 68 (1987).

4. *Kirk–Othmer Encyclopedia of Chemical Technology*. 3rd ed., Volumes 1–26. New York, NY: John Wiley and Sons, 1978–1984, p. V15 992 (1981).

5. (1) Hansch, C., and A. J. Leo. *Medchem Project* Issue No. 26, Claremont, CA: Pomona College (1985). (2) Lyman, W. J., et al., *Handbook of Chemical Property Estimation Methods*. New York: McGraw-Hill, p. 5–5 (1982). (3) IARC; N–Nitrosomorpholine 17: 263–5 (1978)

6. Montesano, R., H. Bartsch, E. Boyland, G. Della Porta, L. Fishbein, R. A. Griesemer, A.B. Swan, L. Tomatis, and W. Davis (eds.). *Handling Chemical Carcinogens in the Laboratory: Problems of Safety*. IARC Scientific Publications No. 33. Lyon, France: International Agency for Research on Cancer, 1979.

■ 1–NAPHTHYLAMINE

CAS # 134–32–7

DOT # 2077

SYNONYMS 1-aminonaftalen (Czech) • 1-aminonaphthalene • 1-naftilamina (Spanish) • 1-naftylamine (Dutch) • 1-naphthalamine • 1-naphthalenamine • 1-Naphthylamin (German) • alfa-naftyloamina (Polish) • alfanaftilamina (Italian) • alpha-aminonaphthalene • alpha-naftylamin (Czech) • alpha-naphthylamine • ci-37265 • ci-azoic-diazo-component-114 • fast-garnet-b-base • fast-garnet-base-b • naphthalidam • naphthalidine

MF $C_{10}H_9N$

MW 143.20

pH Weak base.

COLOR AND FORM Needles from dilute ethanol and ether; yellow rhombic needles; white crystals; needles, become red on exposure to air, or a reddish, crystalline mass.

ODOR Weak ammonia-like odor.

DENSITY Sp gr: 1.1229 at 25°C.

DISSOCIATION CONSTANT pK_a of 3.92 at 25°C.

BOILING POINT 300.8°C at 760 mm Hg.

MELTING POINT 50°C.

HEAT OF COMBUSTION −15,290 Btu/lb = −8,495 cal/g = −355.4×10⁵ J/kg.

FIREFIGHTING PROCEDURES Dry chemical, carbon dioxide, mist, spray [R4].

TOXIC COMBUSTION PRODUCTS Toxic nitrogen oxides are produced in fire [R5].

SOLUBILITY Soluble in 590 parts water; freely soluble in alcohol, ether; water solubility of 1,700 ppm at 20°C.

VAPOR PRESSURE 1 mm Hg at 104.3°C.

OCTANOL/WATER PARTITION COEFFICIENT Log K_{ow} = 168.49.

STABILITY AND SHELF LIFE Oxidizes in air [R7].

OTHER CHEMICAL/PHYSICAL PROPERTIES Sublimes; reduces warm ammoniacal silver nitrate. Has general characteristics of primary aromatic amines.

PREVENTATIVE MEASURES Contact lenses should not be worn when working with this chemical [R6].

DISPOSAL Generators of waste (equal to or greater than 100 kg/mo) containing this contaminant, EPA hazardous waste number U167, must conform with USEPA regulations in storage, transportation, treatment, and disposal of waste [R8].

COMMON USES Toning prints made with cerium salts [R2]. Chemical intermediate for dyes, eg, azoic diazo component; chemical intermediate for azo dye coupling agents, for N–1–naphthylphthalmic acid herbicide, for N–phenyl-1–naphthylamine rubber antioxidant, for an imidazoline adrenergic agent; chemical intermediate for 1–naphthylthiourea rodenticide and for a fluoroacetamide miticide (former uses) [R1].

OCCUPATIONAL RECOMMENDATIONS None listed.

OSHA Workers' exposure to alpha-naphthylamine is to be controlled through the required use of engineering controls, work practices, and personal protective equipment, including respirators. No PELs are listed for this substance [R13].

NIOSH NIOSH recommends that 1–naphthylamine be regulated as a potential human carcinogen [R6].

ECOTOXICITY TLM *Oryzias latipes* 15 mg/L/24 hr [R10]; TLM *Oryzias latipes* 7.0 mg/L/48 hr [R10].

IARC SUMMARY Inadequate evidence of carcinogenicity in humans. Inadequate evidence of carcinogenicity in animals. Overall evaluation: Group 3: The agent is not classifiable as to its carcinogenicity to humans [R9].

BIODEGRADATION Using activated sludge

from both domestic and industrial sources and the Warburg technique, 1–naph- thy- lamine depletion after 6 hrs at 25°C was measured to be 80–84% from an initial concentration of 20 ppm (1). 1–Naph- thylamine, at a concentration of 500 ppm, was extensively oxidized by an aniline- acclimated activated sludge after 19 hr in a Warburg respirometer (2). Over a 20-day period, no biodegradation of 200 ppm 1–naphthylamine as a sole carbon source was observed in a batch system contain- ing an adapted activated sludge (3). 1–Naphthylamine at 500 ppm was found to inhibit oxygen uptake in a Warburg respirometer containing a municipal activated sludge (4). At 1,000 ppm, 1–naphthylamine was toxic to an activat- ed sludge (5). 1–Naphthylamine degrada- tion in six different soils after 308 days of incubation at 23°C varied from 16.6– 30.7% as measured by radio-labelled car- bon dioxide evolution (6); comparison of carbon dioxide evolution in sterilized (via gamma radiation) versus unsterilized soil suggested that the degradation was pre- dominantly microbial in nature (6); ap- proximately 16, 28, and 33% of added 1–naphthylamine degraded in Russell soil at temperatures of 12, 23, and 30°C, respectively, indicating the effect of tem- perature on microbial degradation (6) [R11].

BIOCONCENTRATION Based on a log K_{ow} of 2.25 (1) and a water solubility of 1,700 ppm at 20°C (2), BCF values of 30 and 9 are estimated, respectively, from recom- mended equations (3,SRC); these BCF val- ues indicate that 1-naphthylamine is not expected to bioconcentrate significantly in aquatic organisms (SRC) [R12].

CERCLA REPORTABLE QUANTITIES U167; persons in charge of vessels or facilities are required to notify the National Re- sponse Center (NRC) immediately, when there is a release of this designated haz- ardous substance, in an amount equal to or greater than its reportable quantity of 100 lb or 45.4 kg. The toll-free number of the NRC is (800) 424–8802; in the Washington DC metropolitan area (202) 426–2675. The rule for determining when

notification is required is stated in 40 CFR 302.4 (section IV. D. 3.b) [R14].

RCRA REQUIREMENTS U167; as stipulated in 40 CFR 261.33, when 1–naphthalena- mine, as a commercial chemical product or manufacturing chemical intermediate or an off-specification commercial chemi- cal product or a manufacturing chemical intermediate, becomes a waste, it must be managed according to federal or state hazardous waste regulations. Also defined as a hazardous waste is any residue, contaminated soil, water, or other debris resulting from the cleanup of a spill, into water or on dry land, of this waste. Gener- ators of small quantities of this waste may qualify for partial exclusion from hazard- ous waste regulations (40 CFR 261.5) [R15].

DOT *Health Hazards:* Poisonous; may be fatal if inhaled, swallowed, or absorbed through skin. Contact may cause burns to skin and eyes. Runoff from fire control or dilution water may give off poisonous gases and cause water pollution. Fire may produce irritating or poisonous gases [R3].

Fire or Explosion: Some of these materi- als may burn, but none of them ignites readily. Container may explode violently in heat of fire [R3].

Emergency Action: Keep unnecessary people away; isolate hazard area and deny entry. Stay upwind, out of low areas, and ventilate closed spaces before entering. Positive-pressure self-contained breathing apparatus (SCBA) and chemical protective clothing that is specifically recommended by the shipper or manufacturer may be worn. It may provide little or no thermal protection. Structural firefighters' protec- tive clothing is not effective for these materials. Remove and isolate contami- nated clothing at the site. Call CHEM- TREC at 1–800–424–9300 as soon as possible, especially if there is no local hazardous materials team available [R3].

Fire: Small fires: Dry chemical, CO_2, water spray, or regular foam. Large Fires: Water spray, fog, or regular foam. Move container from fire area if you can do so without risk. Fight fire from maximum distance. Stay away from ends of tanks. Dike fire-control water for later disposal; do not scatter the material [R3].

Spill or Leak: Do not touch spilled material; stop leak if you can do so without risk. Fully encapsulating, vapor-protective clothing should be worn for spills and leaks with no fire. Use water spray to reduce vapors. Small Spills: Take up with sand or other noncombustible absorbent material and place into containers for later disposal. Small Dry Spills: With clean shovel place material into clean, dry container, and cover loosely; move containers from spill area. Large Spills: Dike far ahead of liquid spill for later disposal [R3].

First Aid: Move victim to fresh air and call emergency medical care; if not breathing, give artificial respiration; if breathing is difficult, give oxygen. In case of contact with material, immediately flush skin or eyes with running water for at least 15 minutes. Speed in removing material from skin is of extreme importance. Remove and isolate contaminated clothing and shoes at the site. Keep victim quiet and maintain normal body temperature. Effects may be delayed; keep victim under observation [R3].

FDA Alpha-Naphthylamine is an indirect food additive for use only as a component of adhesives [R16].

GENERAL RESPONSE Avoid contact with solid and dust. Keep people away. Wear dust respirator and rubber overclothing (including gloves). Isolate and remove discharged material. Notify local health and pollution control agencies.

FIRE RESPONSE Combustible. Poisonous gases are produced in fire. Irritating gases are produced when heated. Extinguish with water, dry chemicals, foam, or carbon dioxide.

EXPOSURE RESPONSE Call for medical aid. Dust poisonous if inhaled or if skin is exposed. Irritating to eyes. Move victim to fresh air. If breathing is difficult, give oxygen. Solid poisonous if swallowed or if skin is exposed. Remove contaminated clothing and shoes. Flush affected areas

with plenty of water. If swallowed and victim is conscious, have victim drink water or milk, and have victim induce vomiting. If swallowed and victim is unconscious or having convulsions, do nothing except keep victim warm.

RESPONSE TO DISCHARGE Issue warning—poison, water contaminant. Restrict access. Should be removed. Provide chemical and physical treatment.

WATER RESPONSE Effect of low concentrations on aquatic life is unknown. May be dangerous if it enters water intakes. Notify local health and wildlife officials. Notify operators of nearby water intakes.

EXPOSURE SYMPTOMS Inhalation may cause cyanosis (blue color in lips and under fingernails). Contact with liquid causes local irritation of eyes. Neither ingestion nor contact with skin produces any recognized immediate effects.

EXPOSURE TREATMENT Persons undergoing severe exposure to this compound should have continuing medical attention for possible development of cancer. Inhalation: obtain medical attention for cyanosis. Eyes: flush with water for at least 15 min. Skin: wash carefully with soap and water. Ingestion: get medical attention.

REFERENCES

1. SRI.

2. Budavari, S. (ed.). *The Merck Index—Encyclopedia of Chemicals, Drugs, and Biologicals.* Rahway, NJ: Merck and Co., Inc., 1989. 1012.

3. U.S. Department of Transportation. *Emergency Response Guidebook 1990.* DOT P 5800.5. Washington, DC: U.S. Government Printing Office, 1990, p. G–55.

4. Sax, N. I. *Dangerous Properties of Industrial Materials,.* 6th ed. New York, NY: Van Nostrand Reinhold, 1984. 845.

5. U.S. Coast Guard, Department of Transportation. *CHRIS—Hazardous Chemical Data.* Volume II. Washington, DC: U.S. Government Printing Office, 1984–5.

6. NIOSH. *Pocket Guide to Chemical Hazards.* 2nd Printing. DHHS (NIOSH) Publ.

No. 85–114. Washington, DC: U.S. Dept. of Health and Human Services, NIOSH/ Supt. of Documents, GPO, February 1987. 170.

7. IARC. *Monographs on the Evaluation of the Carcinogenic Risk of Chemicals to Man.* Geneva: World Health Organization, International Agency for Research on Cancer, 1972–present (multivolume work), p. V4 87 (1974).

8. 40 CFR 240–280, 300–306, 702–799 (7/1/89).

9. IARC. *Monographs on the Evaluation of the Carcinogenic Risk of Chemicals to Man.* Geneva: World Health Organization, International Agency for Research on Cancer, 1972–present (multivolume work), p. S7 67 (1987).

10. Tonogai, Y., et al., *J Toxicol Sci* 7 (3): 193–203 (1982).

11. (1) Baird, R., et al., *J Water Pollut Control Fed* 49: 1609 (1977), (2) Malaney, G. W. *J Water Pollut Control Fed* 32: 1300 (1960). (3) Pitter, P. *Water Res* 10: 231 (1976). (4) Lutin, P. A., et al., Purdue Univ Eng Bull, Ext Series 118: 131 (1965). (5) Bosch, F. M., E. Van Vaerenbergh. *Doc Eur Sewage Refuse Symp Eas* 4th, pp. 272 (1978). (6) Graveel, J. G., *J Environ Qual* 15: 53 (1986).

12. (1) Hansch, C., and A. J. Leo. *Medchem Project* 26 Claremont, CA: Pomona College (1985). (2) Dragun, J., C. S. Helling. *Soil Sci* 139: 100 (1985). (3) Lyman, W. J., et al., *Handbook of Chemical Property Estimation Methods Environ Behavior of Organic Compounds*, New York: McGraw-Hill, pp. 5–4,5–10 (1982).

13. 29 CFR 1910.1004 (7/1/88).

14. 54 FR 33419 (8/14/89).

15. 40 CFR 261.33 (7/1/88).

16. 21 CFR 175.105 (4/1/90).

■ 2-NAPHTHYLAMINE, N-PHENYL

CAS # 135-88-6

SYNONYMS 2-anilinonaphthalene • 2-naphthalenamine, N-phenyl • 2-naphthylphenylamine • 2-phenylaminonaphthalene • aceto-pbn • agerite powder • antioxidant-116 • antioxidant-pbn • beta-naphthyl-phenylamine • N-(2-naphthyl) aniline • N-phenyl-beta-naphthylamine • neosone-d • neozon-d • neozone • neozone-d • nilox-pbna • nonox-d • nonox-dn • pbna • phenyl-2-naphthylamine • phenyl-beta-naphthylamine • stabilizator-ar • vulkanox-pbn

MF $C_{16}H_{13}N$

MW 219.29

COLOR AND FORM Needles from methanol; white to yellowish crystals; gray to tan flakes or powder.

DENSITY 1.24.

BOILING POINT 395.5°C.

MELTING POINT 108°C.

SOLUBILITY Solubility in ethanol 50 g/L, in benzene 27 g/L, in acetone 640 g/L; insoluble in water; soluble in alcohol, ether, acetic acid (bluish fluorescence).

OCTANOL/WATER PARTITION COEFFICIENT Log K_{ow} = 4.38.

OTHER CHEMICAL/PHYSICAL PROPERTIES May be oxidized to 7–phenyl–7h–dibenzo[c,g]carbazole.

DISPOSAL SRP: At the time of review, criteria for land treatment or burial (sanitary landfill) disposal practices are subject to significant revision. Prior to implementing land disposal of waste residue (including waste sludge), consult with environmental regulatory agencies for guidance on acceptable disposal practices.

COMMON USES Rubber antioxidant; lubricant; inhibitor (butadiene) [R2]; stabilizer in electrical-insulating silicone enamels; heat and light stabilizer; vulcanization accelerator; component of rocket fuels; in surgical plasters; in tin-electroplating baths; chemical intermediate; in dyes; catalyst and polymerization inhibitor [R1].

OCCUPATIONAL RECOMMENDATIONS None listed.

THRESHOLD LIMIT A2; A2 = suspected human carcinogen (1979) [R5].

NON-HUMAN TOXICITY LD$_{50}$ rat oral 8,730 mg/kg [R8, 1211]; LD$_{50}$ mouse, oral 1,450 mg/kg [R8, 1211].

IARC SUMMARY Inadequate evidence of carcinogenicity in humans. Limited evidence of carcinogenicity in animals. Overall evaluation: Group 3: The agent is not classifiable as to its carcinogenicity to humans [R4].

BIODEGRADATION Using an initial concentration of 100 mg/L N–phenyl–2–naphthylamine, 0% BOD was observed after a 2–week period in a biodegradation screening test using 30 mg/L sludge (1). N–Phenyl–2–naphthylamine, initial concentration of 100 ppm, exhibited a 0–29% theoretical BOD in a Japanese MITI test after 14 days of incubation at 25°C (2). N–phenyl–2–naphthylamine was shown to degrade to beta-naphthylamine in a screening study using activated-sludge inoculum at 21°C (3). In this study, 44.2 and 9.3% CO_2 production was observed in 1 day at initial concns of 2 and 5 ppm, respectively (3). At an initial concentration of 8 ppm, N–phenyl–2–naphthylamine was observed to be inhibitory to microorganisms under the same test conditions (3). Furthermore, adsorption to solids was found to occur and, hence, reduce biodegradation (3) [R6].

BIOCONCENTRATION The BCF value for N–phenyl–2–naphthylamine was determined to be 147 for the fathead minnow (*Pimephales promelas*) in a 32-day exposure flow-through test of Lake Superior water at 25°C (1) [R7].

FIRE POTENTIAL Fire hazard: moderate, when exposed to heat or flame; can react with oxidizing materials [R3].

REFERENCES

1. IARC. *Monographs on the Evaluation of the Carcinogenic Risk of Chemicals to Man.* Geneva: World Health Organization, International Agency for Research on Cancer, 1972–present (multivolume work), p. V16 327 (1978).

2. Lewis, R. J., Sr. (ed.). *Hawley's Con-* densed Chemical Dictionary. 12th ed. New York, NY: Van Nostrand Reinhold Co., 1993 903.

3. Sax, N. I. *Dangerous Properties of Industrial Materials,.* 4th ed. New York: Van Nostrand Reinhold, 1975. 1018.

4. IARC. *Monographs on the Evaluation of the Carcinogenic Risk of Chemicals to Man.* Geneva: World Health Organization, International Agency for Research on Cancer, 1972–present (multivolume work), p. S7 70 (1987).

5. American Conference of Governmental Industrial Hygienists. *Threshold Limit Values for Chemical Substances and Physical Agents and Biological Exposure Indices for 1994–1995.* Cincinnati, OH: ACGIH, 1994. 29.

6. (1) Chemicals Inspection and Testing Institute. *Japan Chemical Industry Ecology—Toxicology and Information Center* ISBN 4–89074–101–1 (1992). (2) Sasaki, S. pp. 283–98 in *Aquatic Pollutants.* Hutzinger, O., et al. (eds.). Oxford: Pergamon Press (1978). (3) Ku, Y., G. H. Alverez. *Chemosphere* 11: 41–6 (1982).

7. Vieth, G.D., et al., *J Fish Res Board Can* 36: 1040–8 (1979).

8. American Conference of Governmental Industrial Hygienists, Inc. *Documentation of the Threshold Limit Values and Biological Exposure Indices.* 6th ed. Volumes I, II, III. Cincinnati, OH: ACGIH, 1991.

■ NICKEL NITRATE

CAS # 13138–45–9

SYNONYMS nickel dinitrate • nickel nitrate [Ni(NO$_3$)$_2$]; • nickel(2) nitrate • nickel(II) nitrate (1: 2) • nickelous nitrate • nitric acid, nickel(2) salt • nitric acid, nickel(II) salt

MF N$_2$O$_6$•Ni

MW 182.73

COLOR AND FORM Green crystals.

EXPLOSIVE LIMITS AND POTENTIAL Prolonged exposure to fire or heat may result in an explosion [R11].

FIREFIGHTING PROCEDURES Use flooding amount of water in early stages of fire. When large quantities are involved in fire, nitrate may fuse or melt, in which condition application of water may result in extensive scattering of molten material [R10]. If material is on fire or is involved in Fire: flood with water, cool all affected containers with flooding quantities of water, and apply water from as far a distance as possible [R11]. If fire becomes uncontrollable, evacuate for a radius of 2,500 feet [R11]. Personnel protection in firefighting: Wear appropriate protective gloves, boots, and goggles. Do not handle broken packages unless wearing appropriate personal protective equipment; wash away any material that may have contacted the body with copious amount of water or soap and water; wear positive-pressure self-contained breathing apparatus when fighting fires involving this material. Evacuation: If fire becomes uncontrollable, consider evacuation of one-half-mile radius [R11].

TOXIC COMBUSTION PRODUCTS Toxic oxides of nitrogen are produced in fires involving nickel nitrate [R11].

BOILING POINT 137°C (hexahydrate).

MELTING POINT 56.7°C (hexahydrate).

DENSITY 2.05 (hexahydrate).

pH Aqueous solution is acid, pH about 4 (hexahydrate).

SOLUBILITY 238.5 g/dl water at 0°C (hexahydrate); soluble in alcohol; ammonium hydroxide (hexahydrate); very soluble in hot water (hexahydrate); 95.030 lb/100 lb water at 70°F (hexahydrate); 3 g/100 ml hydrazine at 20°C; 7.5 wt% in ethylene glycol at 20°C; 48.5 wt% in water at 20°C.

OTHER CHEMICAL/PHYSICAL PROPERTIES Strong oxidizing agent (hexahydrate); deliquescent (hexahydrate); heat of solution: 1.1×10^5 J/kg = 26 cal/g = 47 Btu/lb (hexahydrate). Nickel nitrate hexahydrate loses water on heating and eventually decomposes, forming nickel oxide. The loss of the individual waters of hydration upon heating the hexahydrate can be studied and the existence of the anhydrous covalent compound $Ni(NO_3)_2$ can be observed before it decomposes, using differential thermal analysis and thermogravimetric analysis techniques. Green monoclinic crystals (hexahydrate).

PROTECTIVE EQUIPMENT Respirator selection: Upper-limit devices recommended by NIOSH: at any detectable concentration: any self-contained breathing apparatus with a full facepiece and operated in a pressure-demand or other positive-pressure mode or any supplied-air respirator with a full facepiece and operated in a pressure-demand or other positive-pressure mode in combination with an auxiliary self-contained breathing apparatus operated in pressure-demand or other positive-pressure mode; escape: any air-purifying full facepiece respirator with a high-efficiency particulate filter; any appropriate escape-type self-contained breathing apparatus (nickel, metal, and soluble compounds (as Ni)) [R21, 173].

PREVENTATIVE MEASURES Stringent requirements for warning labels and signs, employee instruction, work practices, sanitation, environmental monitoring, and recordkeeping are recommended where employees are exposed to nickel compounds. (nickel and compounds) [R22, 1439].

CLEANUP For land spill, dig a pit, pond, lagoon, or holding area to contain liquid or solid material. (SRP: If time permits, pits, ponds, lagoons, soak holes, or holding areas should be sealed with an impermeable flexible-membrane liner.) Cover solids with a plastic sheet to prevent dissolving in rain or firefighting water [R11]. Add soda ash. Use mechanical dredges or lifts to remove immobilized masses of pollutants and precipitates [R11]. Indoors: 1. Ventilate area of spill. 2. Collect spilled material in the most convenient and safe manner for reclamation. Liquid containing nickel should be absorbed in vermiculite, dry sand, earth, or a similar material (nickel, metal, and soluble nickel compounds) [R20, 4].

DISPOSAL SRP: At the time of review, criteria for land treatment or burial (sanitary landfill) disposal practices are subject to significant revision. Prior to implementing land disposal of waste residue (including waste sludge), consult with environmental regulatory agencies for guidance on acceptable disposal practices.

COMMON USES Nickel plating; preparation of nickel catalysts; manufacture of of brown ceramic colors (hexahydrate) [R2, 820]; in batteries [R3]; used as catalyst to remove waste gas containing ozone from waste water [R4]; used as marker to distinguish between treated and untreated horn flies when placed in larval rearing medium at levels that might be detected in adults by absorption spectroscopy [R5]; has been tested for control of *Melamspora lini* on flax [R6]; colorant for glass [R7]; nickel nitrate was most effective (in the treatment of brown spot disease on rice seedlings. [R8].

OCCUPATIONAL RECOMMENDATIONS

OSHA Meets criteria for OSHA medical records rule. (nickel, metal, and soluble compounds, as Ni) [R16].

NIOSH NIOSH recommends that the substance be treated as a potential human carcinogen (nickel, inorganic compounds) [R17]; threshold-limit 8–hr time-weighted average (TWA) 0.1 mg/m³ (nickel, soluble compounds, as Ni) [R23, 27].

INTERNATIONAL EXPOSURE LIMITS Max allowable concentration (max), former USSR 0.005 mg/m³ as Ni (nickel salts and aerosols) [R22, 1438].

ECOTOXICITY TLm stickleback 0.8 ppm/ 10 days/fresh water (conditions of bioassay not specified) (hexahydrate) [R1]; LC_{50} *Channa punctatus* 306.9 mg/L/96 hr (conditions of bioassay not specified) (nickel ion) [R13]; LC_{50} *Daphnia magna* 0.13 mg/L/3 weeks (conditions of bioassay not specified) (nickel ion) [R14]; LC_{50} *Acroneuria lycoria* 4 mg/L/96 hr (conditions of bioassay not specified) (nickel ion) [R14]; LC_{50} *Artemia salina* 163.0 mg/L/48 hr (conditions of bioassay not specified) (nickel ion) [R15].

IARC SUMMARY Evaluation: There is suffi-cient evidence in humans for the carcinogenicity of nickel sulfate and of the combinations of nickel sulfides and oxides encountered in the nickel refining industry. There is inadequate evidence in humans for the carcinogenicity of metallic nickel and nickel alloys. There is sufficient evidence in experimental animals for the carcinogenicity of metallic nickel, nickel monoxides, nickel hydroxides, and crystalline nickel sulfides. There is limited evidence in experimental animals for the carcinogenicity of nickel alloys, nickelocene, nickel carbonyl, nickel salts, nickel arsenides, nickel antimonide, nickel selenides, and nickel telluride. There is inadequate evidence in experimental animals for the carcinogenicity of nickel trioxide, amorphous nickel sulfide, and nickel titanate. The Working Group made the overall evaluation on nickel compounds as a group on the basis of the combined results of epidemiological studies, carcinogenicity studies in experimental animals, and several types of other relevant data, supported by the underlying concept that nickel compounds can generate nickel ions at critical sites in their target cells. Overall evaluation: Nickel compounds are carcinogenic to humans (Group 1). Metallic nickel is possibly carcinogenic to humans (Group 2B) [R12].

BIOLOGICAL HALF-LIFE For the first 50 hr after exposure, the biological half-life of nickel in rat plasma was 6.3 hr, and in rabbit plasma it was 7.5 hr. (nickel salts) [R24, 469].

BIOCONCENTRATION Although nickel is bioaccumulated, the concentration factors are such as to suggest that partitioning into the biota is not a dominant fate process. (nickel) [R25, p. 15–6].

RCRA REQUIREMENTS The Environmental Protection Agency has promulgated regulations concerning ground-water monitoring with regard to screening suspected contamination at land-based hazardous waste treatment, storage, and disposal facilities. There are new requirements to analyze for a specified core list of chemicals plus those chemicals specified by the

Regional Administrator on a site-specific basis (total nickel, all species) [R18].

DOT *Fire or Explosion:* May ignite other combustible materials (wood, paper, oil, etc). Reaction with fuels may be violent. Runoff to sewer may create fire or explosion hazard [R9]. Health Hazards: Contact may cause burns to skin and eyes. Vapors or dust may be irritating. Fire may produce irritating or poisonous gases. Runoff from fire control or dilution water may cause pollution [R9]. Emergency Action: Keep unnecessary people away; isolate hazard area and deny entry. Self-contained breathing apparatus (SCBA) and structural firefighter's protective clothing will provide limited protection. Call CHEMTREC at 1–800–424–9300 For emergency assistance. If water pollution occurs, notify the appropriate authorities [R9].

Fire: Small fires: Dry chemical, CO_2, halon, or water spray. Large Fires: Water spray or fog. Move container from fire area if you can do so without risk. Cool containers that are exposed to flames with water from the side until well after fire is out. Stay away from ends of tanks. For massive fire in cargo area, use unmanned hose holder or monitor nozzles; if this is impossible, withdraw from area, and let fire burn [R9].

Spill or Leak: Do not touch spilled material. Keep combustibles (wood, paper, oil, etc.) away from spilled material. Small Dry Spills: With clean shovel place material into clean, dry container, and cover; move containers from spill area. Small Liquid Spills: Take up with sand, earth, or other noncombustible absorbent material. Large Spills: Dike far ahead of liquid spill for later disposal [R9].

First Aid: Move victim to fresh air; call emergency medical care. Remove and isolate contaminated clothing and shoes at the site. In case of contact with material, immediately flush skin or eyes with running water for at least 15 minutes [R9].

FIRE POTENTIAL In contact with easily oxidizable substances it may react rapidly enough to cause ignition; violent combustion increases flammability of any combustible substance [R10]. Contact of solid with wood or paper may cause fires (hexahydrate) [R1]. Dangerous fire risk. (hexahydrate) [R2, 820].

FDA Bottled water shall, when a composite of analytical units of equal volume from a sample is examined by the methods described in paragraph (d) (1) (II) of this section, meet the standards of chemical quality, and shall not contain nitrate in excess of 10.0 mg/L. (nitrate as N) [R19].

REFERENCES

1. U.S. Coast Guard, Department of Transportation. *CHRIS—Hazardous Chemical Data.* Volume II. Washington, DC: U.S. Government Printing Office, 1984–5.

2. Sax, N.I., and R.J. Lewis, Sr. (eds.). *Hawley's Condensed Chemical Dictionary.* 11th ed. New York: Van Nostrand Reinhold Co., 1987.

3. IARC. *Monographs on the Evaluation of the Carcinogenic Risk of Chemicals to Man.* Geneva: World Health Organization, International Agency for Research on Cancer, 1972–present (multivolume work), p. V11 84 (1976).

4. Aoki, M., et al., *Japan Kokai Patent* 78 26291 03/10/78 Mitsubishi Chemical Industries Co., Ltd.

5. Chamberlain, W.F., et al., *Southwest Entomol* 2 (2): 73 (1977).

6. Krzysztalowska, H. *PR Inst Przem Wlok Lykowych* 18: 29 (1971).

7. SRI.

8. Giri, D.N., A.K. Sinha. *Ann Appl Biol* 103 (2): 229–35 (1983).

9. Department of Transportation. *Emergency Response Guidebook 1987.* DOT P 5800.4. Washington, DC: U.S. Government Printing Office, 1987, p. G–35.

10. National Fire Protection Association. *Fire Protection Guide on Hazardous Materials.* 9th ed. Boston, MA: National Fire Protection Association, 1986, p. 49–67.

11. Association of American Railroads. *Emergency Handling of Hazardous Materials in Surface Transportation.* Wash-

ington, DC: Assoc. of American Railroads, Hazardous Materials Systems (BOE), 1987. 483.

12. IARC. *Monographs on the Evaluation of the Carcinogenic Risk of Chemicals to Man.* Geneva: World Health Organization, International Agency for Research on Cancer, 1972–present (multivolume work), p. 49 410 (1990).

13. Saxena, O.P.; *J Environ Biol* 4 (2): 91 (1983).

14. Nat'l Research Council Canada. *Effects of Nickel in the Canadian Environment* p. 28 (1981) NRCC 18568.

15. Kissa, E., et al., *Arch Hydrobiol* 102 (2): 255–64 (1984).

16. 29 CFR 1910.20 (7/1/87).

17. NIOSH/CDC. *NIOSH Recommendations for Occupational Safety and Health Standards Sept. 1986* (supplement to *Morbidity and Mortality Weekly Report* 35 No. 15, Sept. 26, 1986), p. 24S.

18. 52 FR 25942 (7/9/87).

19. 21 CFR 103.35 (4/1/86).

20. Mackison, F. W., R. S. Stricoff, and L. J. Partridge, Jr. (eds.). *NIOSH/OSHA—Occupational Health Guidelines for Chemical Hazards.* DHHS (NIOSH) Publication No. 81–123 (3 vols). Washington, DC: U.S. Government Printing Office, Jan. 1981.

21. NIOSH. *Pocket Guide to Chemical Hazards.* 5th Printing/Revision. DHHS (NIOSH) Publ. No. 85–114. Washington, DC: U.S. Dept. of Health and Human Services, NIOSH/Supt. of Documents, GPO, Sept. 1985.

22. International Labour Office. *Encyclopedia of Occupational Health and Safety.* Vols. I and II. Geneva, Switzerland: International Labour Office, 1983.

23. American Conference of Governmental Industrial Hygienists. *Threshold Limit Values for Chemical Substances and Physical Agents and Biological Exposure Indices for 1994–1995.* Cincinnati, OH: ACGIH, 1994.

24. Friberg, L., G.F. Nordberg, E. Kessler and V.B. Vouk (eds.). *Handbook of the Toxicology of Metals.* 2nd ed. Vols I, II.: Amsterdam: Elsevier Science Publishers B.V., 1986, p. V2.

25. Callahan, M.A., M.W. Slimak, N.W. Gabel, et al. *Water-Related Environmental Fate of 129 Priority Pollutants.* Volume I. EPA-440/4 79-029a. Washington, DC: U.S. Environmental Protection Agency, December 1979.

◼ NICKEL OXIDE

CAS # 1313–99–1

SYNONYMS bunsenite • ci-77777 • green nickel oxide • mononickel oxide • nickel protoxide • nickel(2) oxide • nickel(II) oxide (1:1) • nickel(II) oxide • nickelous oxide • nickel oxide (NiO)

MF NiO

MW 74.71

COLOR AND FORM Green powder; greenish-black cubic crystals; yellow when hot.

DENSITY 6.67.

MELTING POINT 1,984°C.

SOLUBILITY 0.11 mg/100 ml water at 20°C; soluble in acids, ammonium hydroxide; insoluble in caustic solutions; soluble in potassium cyanide.

OTHER CHEMICAL/PHYSICAL PROPERTIES Absorbs oxygen at 400°C, forming nickelic oxide, which is reduced to nickel oxide at 600°C. Reacts with acids to form nickel salts and soaps.

PROTECTIVE EQUIPMENT Dispensers of liquid detergent should be available. Safety pipettes should be used for all pipetting. In animal laboratory, personnel should wear protective suits (preferably disposable, one-piece, and close-fitting at ankles and wrists), gloves, hair covering, and overshoes. In chemical laboratory, gloves and gowns should always be worn; however, gloves should not be assumed to provide full protection. Carefully fitted masks or respirators may be necessary when working with particulates or gases, and disposable plastic aprons might provide addnl protection. Gowns should be of distinctive color, as a reminder that they are not to be worn outside the laboratory. (chemical carcinogens) [R12, 8].

PREVENTATIVE MEASURES Smoking, drinking, eating, storage of food, or of food and beverage containers, or utensils, and the application of cosmetics should be prohibited in any laboratory. All personnel should remove gloves, if worn, after completion of procedures in which carcinogens have been used. They should wash hands, preferably using dispensers of liquid detergent, and rinse thoroughly. Consideration should be given to appropriate methods for cleaning the skin, depending on nature of the contaminant. No standard procedure can be recommended, but the use of organic solvents should be avoided. Safety pipettes should be used for all pipetting. (chemical carcinogens) [R12, 8].

CLEANUP A high-efficiency particulate arrestor (HEPA) or charcoal filters can be used to minimize amount of carcinogen in exhausted-air-ventilated safety cabinets, lab hoods, glove boxes, or animal rooms Filter housing that is designed so that used filters can be transferred into plastic bag without contaminating maintenance staff is avail commercially. Filters should be placed in plastic bags immediately after removal. The plastic bag should be sealed immediately. The sealed bag should be labelled properly. Waste liquids should be placed or collected in proper containers for disposal. The lid should be secured and the bottles properly labelled. Once filled, bottles should be placed in plastic bag, so that outer surface is not contaminated. The plastic bag should also be sealed and labelled. Broken glassware should be decontaminated by solvent extraction, by chemical destruction, or in specially designed incinerators. (chemical carcinogens) [R12, 15].

DISPOSAL SRP: At the time of review, criteria for land treatment or burial (sanitary landfill) disposal practices are subject to significant revision. Prior to implementing land disposal of waste residue (includ-

ing waste sludge), consult with environmental regulatory agencies for guidance on acceptable disposal practices.

COMMON USES Painting on porcelain [R2]; in fuel cell electrodes [R1]. Miscellaneous applications of nickel oxide include its use in: (a) the. manufacture of ferrites (e.g., $NiOFe_2O_3$), which find use in the electronics field because of their magnetic properties; (b) the manufacture of nickel salts (e.g., chloride, nitrate, and sulfate), which can be used to make refined nickel oxide; (c) the production of active nickel catalysts; (d) electroplating; and coloring and decolorizing glass [R3]. Secondary (rechargeable) cells with zinc anodes contain alkaline zinc nickel oxide and zinc chloride [R4]. Nonmetallic resistance thermometers or thermistors are generally temperature-sensitive semiconducting ceramics. The thermister material is usually a metal oxide, e.g., nickel oxide [R5]. Used in ceramic matrices [R6]. Brown ceramic colorant [R13, 555]; gray ceramic colorant [R13, 556].

OCCUPATIONAL RECOMMENDATIONS

OSHA Meets criteria for OSHA medical records rule. (nickel, metal, and soluble compounds, as Ni) [R9].

NIOSH NIOSH recommends that the substance be treated as a potential human carcinogen. (nickel, inorganic compounds) [R10].

THRESHOLD LIMIT 8–hr time-weighted average (TWA) 1 mg/m³ (nickel, insoluble compounds, as Ni) [R15, 27].

INTERNATIONAL EXPOSURE LIMITS Max allowable concentration (MAC), former USSR 0.5 mg/m³ (nickel, nickel oxide, and nickel sulfides as dust) [R14, 1438].

IARC SUMMARY Evaluation: There is sufficient evidence in humans for the carcinogenicity of nickel sulfate, and of the combinations of nickel sulfides and oxides encountered in the nickel refining industry. There is inadequate evidence in humans for the carcinogenicity of metallic nickel and nickel alloys. There is sufficient evidence in experimental animals for the carcinogenicity of metallic nickel, nickel monoxides, nickel hydroxides, and crystalline nickel sulfides. There is limited evidence in experimental animals for the carcinogenicity of nickel alloys, nickelocene, nickel carbonyl, nickel salts, nickel arsenides, nickel antimonide, nickel selenides, and nickel telluride. There is inadequate evidence in experimental animals for the carcinogenicity of nickel trioxide, amorphous nickel sulfide, and nickel titanate. The Working Group made the overall evaluation on nickel compounds as a group on the basis of the combined results of epidemiological studies, carcinogenicity studies in experimental animals, and several types of other relevant data, supported by the underlying concept that nickel compounds can generate nickel ions at critical sites in their target cells. Overall evaluation: Nickel compounds are carcinogenic to humans (Group 1). Metallic nickel is possibly carcinogenic to humans (Group 2B) [R7].

BIOLOGICAL HALF-LIFE Male rats were exposed to nickel oxide (NiO) aerosols (mass median aerodynamic diameter, 1.2 and 4.0 mm). The average exposure concentration was controlled from a low level of 0.6 mg/m^3 to a high level of 70 mg/m^3 and total exposure time was 140 hr. Some rats were sacrificed just after the exposure, whereas others were exposed for 1 mo, and kept for 12- and 20-mo clearance periods before sacrifice. There were no differences in body wt gain between NiO exposure groups and controls. Nickel concentrations in lung of exposure groups were much higher than those of controls and decreased with increased clearance time. No apparent deposition of nickel was observed in the liver, kidney, spleen, and blood immediately after the exposure, but, in the high-exposure groups, nickel concentration in the liver, spleen, and blood increased slightly with increasing clearance time. The biological half-time of NiO deposited in the lung, assuming that the amount of the clearance is proportional to the amount of the NiO deposited, was 11.5 and 21 mo for 1.2 and 4.0 mm, respectively [R8].

RCRA REQUIREMENTS The Environmental Protection Agency has promulgated regulations concerning groundwater monitoring with regard to screening suspected contamination at land-based hazardous waste treatment, storage, and disposal facilities. There are new requirements to analyze for a specified core list of chemicals plus those chemicals specified by the Regional Administrator on a site-specific basis. Total nickel (all species) is included on this list. (total nickel (all species)) [R11].

REFERENCES

1. Sax, N. I., and R. J. Lewis, Sr. (eds.). *Hawley's Condensed Chemical Dictionary.* 11th ed. New York: Van Nostrand Reinhold Co., 1987. 821.

2. *The Merck Index.* 10th ed. Rahway, NJ: Merck Co., Inc., 1983. 933.

3. NTP. *Chemical Selection Working Group Profile: Nickel Oxide* p. 2 (1979).

4. *Kirk–Othmer Encyclopedia of Chemical Technology.* 3rd ed., Volumes 1–26. New York, NY: John Wiley and Sons, 1978–1984, p. 24 (84) 807–51.

5. *Kirk–Othmer Encyclopedia of Chemical Technology.* 3rd ed., Volumes 1–26. New York, NY: John Wiley and Sons, 1978–1984, p. 22 (83) 679–709.

6. *Kirk–Othmer Encyclopedia of Chemical Technology.* 3rd ed., Volumes 1–26. New York, NY: John Wiley and Sons, 1978–1984, p. 12 (80) 476.

7. IARC. *Monographs on the Evaluation of the Carcinogenic Risk of Chemicals to Man.* Geneva: World Health Organization, International Agency for Research on Cancer, 1972–present (multivolume work), p. 49 410 (1990).

8. Tanaka, I., et al., *Biol Trace Elem Res 8* (3): 203–10 (1985).

9. 29 CFR 1910.20 (7/1/87).

10. NIOSH/CDC. *NIOSH Recommendations for Occupational Safety and Health Standards.* Sept. 1986 (Supplement to *Morbidity and Mortality Weekly Report* 35 No. 15, Sept. 26, 1986), p. 24S.

11. 52 FR 25942 (7/9/87).

12. Montesano, R., H. Bartsch, E. Boyland, G. Della Porta, L. Fishbein, R. A.

Griesemer, A.B. Swan, L. Tomatis, and W. Davis (eds.). *Handling Chemical Carcinogens in the Laboratory: Problems of Safety.* IARC Scientific Publications No. 33. Lyon, France: International Agency for Research on Cancer, 1979.

13. *Kirk-Othmer Encyclopedia of Chemical Technology.* 3rd ed., Volumes 1–26. New York, NY: John Wiley and Sons, 1978–1984, p. 6(79).

14. International Labour Office. *Encyclopedia of Occupational Health and Safety.* Vols. I and II. Geneva, Switzerland: International Labour Office, 1983.

15. American Conference of Governmental Industrial Hygienists. *Threshold Limit Values for Chemical Substances and Physical Agents and Biological Exposure Indices for 1994–1995.* Cincinnati, OH: ACGIH, 1994.

■ NITRILOTRIACETIC–ACID

CAS # 139–13–9

SIC CODES 2842; 2879; 2844.

SYNONYMS acetic acid, nitrilotri- • ai3-52483 • alpha,alpha',alpha''-trimethyl-aminetricarboxylic acid • aminotriacetic acid • chel-300 • complexon-i • glycine, N,N-bis (carboxymethyl)- • hampshire-nta-acid • kyselina-nitrilotrioctova (Czech) • N,N-bis(carboxymethyl) glycine • nci-c02766 • nitrilo-2,2',2''-triacetic acid • nta • titriplex-i • tri (carboxymethyl) amine • triglycine • triglycollamic acid • trilon-a • versene-nta-acid

MF $C_6H_9NO_6$

MW 191.16

pH Saturated aqueous solution, 2.3.

COLOR AND FORM Prismatic crystals from hot water; white crystalline powder.

DENSITY Greater than 1 at 20°C (solid).

DISSOCIATION CONSTANT pK_{a1} 3.03.

MELTING POINT 242°C with decomposition.

SOLUBILITY Soluble in hot alcohol; 1.28 g dissolves in 1 L water at 22.5 °C; soluble in most organic solvents; relatively soluble in water (59,000 mg/L at 25°C).

OTHER CHEMICAL/PHYSICAL PROPERTIES Forms mono–, di–, and tribasic water-soluble salts; vapor pressure = 3×10^{-5} mm Hg at 25°C (est).

PROTECTIVE EQUIPMENT (NIOSH certified respirator) rubber gloves; chemical safety goggles (nitrilotriacetic acid and salts) [R5].

CLEANUP This study used acidity, ethylenediaminetetraacetic acid, and nitrilotriacetic acid treatment processes to explore the removal efficiencies of heavy metals from urban and industrial sludges. The results indicate that the optimum treatment efficiencies of heavy metals extraction from sludge are related to the species of heavy metals in sludge, dosage of extractants, and the reaction time. The removal efficiency of a three stage counter-current process was higher than those of single-stage processes. The reaction kinetics can be expressed by the equation $dC/dt = -KC$. The cost of the acid treatment process per unit weight of heavy metal extracted was lowest in conditions of high heavy-metal concentrations, but the ethylenedinitrilotetraacetic process was the cheapest with low heavy-metal concentrations [R6].

COMMON USES Sequestering agent; (SRP: proposed) builder in synthetic detergents [R4]; eluting agent in purification of rare-earth elements; synthesis [R2]; chelating agent in cleaning and separation of metals, textile processing, soaps, rubber, and emulsion polymerization [R1]; boiler feed-water additive; water treatment; pulp and paper processing [R3].

STANDARD CODES No NFPA code.

PERSISTENCE Can be degraded by pseudomonads [R13]. Biodegradation requires acclimation periods of 2–3 weeks, but can be achieved in river water . Forms iron complex immediately [R11]. Direct sunlight can cause photo-oxidation of iron complexes within 14 days.

DIRECT CONTACT Slight to eyes.

GENERAL SENSATION Pollard found 25% aqueous solution nonirritating to rabbit skin [R10].

ACUTE HAZARD LEVEL Lethality at 100 ppm has been reported due to pH drop [R12]. Produced emesis in dogs at 1 ppm level [R10]; toxic to fish in 100–ppm range.

CHRONIC HAZARD LEVEL Can act as nitrogen source for some algae [R12]. No toxic effect on rats after 14 weeks at 5% [R10]. Rats fed 2 g/kg/day had enlarged kidneys.

OCCUPATIONAL RECOMMENDATIONS None listed.

ON-SITE RESTORATION Carbon may prove effective. Seek professional environmental engineering assistance through EPA's Environmental Response Team (ERT), Edison, NJ, 24–hour no.: 908–548–8730.

AVAILABILITY OF COUNTERMEASURE MATERIAL Carbon—water treatment plants, sugar refineries.

DISPOSAL METHOD Spray into incinerator.

MAJOR WATER USE THREATENED Fisheries, irrigation.

PROBABLE LOCATION AND STATE Rapidly forms iron complexes in natural waters [R11].

WATER CHEMISTRY Forms acid solution. Iron complexes form immediately. Subject to photo-oxidation. Can also be attacked biologically.

IARC SUMMARY Evaluation: There is sufficient evidence for the carcinogenicity of nitrilotriacetic acid and its sodium salts in experimental animals. No data were available from studies in humans on the carcinogenicity of nitrilotriacetic acid and its salts. In formulating the overall evaluation, the Woring Group took note of the fact that nitrilotriacetic acid is liberated to some extent from nitrilotriacetate salts in solution. Overall evaluation: Nitrilotriacetic acid and its salts are possibly carcinogenic in humans (Group 2B) [R7].

BIODEGRADATION NTA is readily decomposed by soil microorganisms under an-aerobic or aerobic conditions. NTA nitrogens converted to nitrate under aerobic conditions, indicating use of detergents containing NTA leading to nitrate enrichment of water resources [R8].

BIOCONCENTRATION Based upon a measured water solubility of 59,000 mg/L at 25°C (1), the BCF for nitrilotriacetic acid can be estimated to be about 10 from a recommended regression-derived equation (2,SRC). This BCF suggests that bioconcentration in aquatic organisms is not important environmentally (SRC) [R9].

REFERENCES

1. SRI.

2. Hawley, G. G. *The Condensed Chemical Dictionary.* 9th ed. New York: Van Nostrand Reinhold Co., 1977. 613.

3. Sittig, M. *Hdbk Tox and Hazard Chemical and Carcinogens.* 2nd ed. 1985 p. 649.

4. *The Merck Index.* 9th ed. Rahway, NJ: Merck and Co., Inc., 1976. 854.

5. U.S. Coast Guard, Department of Transportation. *CHRIS—Hazardous Chemical Data.* Manual Two. Washington, DC: U.S. Government Printing Office, Oct., 1978.

6. Loo, K.S.L., Y. H. Chen. *Sci Total Environ* 90: 99–116 (1990).

7. IARC. *Monographs on the Evaluation of the Carcinogenic Risk of Chemicals to Man.* Geneva: World Health Organization, International Agency for Research on Cancer, 1972–present (multivolume work), p. 48 204 (1990).

8. Tabatabai, M. A., J. M. Bremner. "Decomposition of Nitrilotriacetate (NTA) in Soils"; *Soil Biol Biochemical* 7 (2) 103 (1975).

9. (1) Yalkowsky, S. H., *Arizona Database of Aqueous Solubilities.* Univ of AZ, College of Pharmacy (1989). (2) Lyman, W. J., et al., *Handbook of Chemical Property Estimation Methods,* Washington, DC: Amer Chemical Soc p. 5–10 (1990).

10. Nixon, G. A., "Toxicity Evaluation of Trisodium Nitrilotriacetate," *Toxicology*

and Applied Pharmacology, Vol. 18, 398–4066, 1971.

11. Tratt, T., R. W. Henwood, and C. M. Langford, "Sunlight Photochemistry of Ferric Nitrilotriacetate Complexes," *Environmental Science and Technology*, Vol. 6, No. 4, April 1972.

12. Thom, W. S., *Nitrilotriacetic Acid: A Literature Survey, Wnter Research*, 1971, Pergamon Press, Vol. 5, pp. 391–399.

13. Thom, W. S., *Nitrolotriacetic Acid: A Literature Survey, Wnter Research*, 1971, Pergamon Press, Vol. 5, pp. 391–399.

■ p–NITROBIPHENYL

CAS # 92–93–3

SYNONYMS 1,1'-biphenyl, 4-nitro- • 4-nitrobiphenyl • 4-nitrodiphenyl • 4-phenyl-nitrobenzene • biphenyl, 4-nitro- • p-nitrodiphenyl • p-phenyl-nitrobenzene • pnb

MF $C_{12}H_9NO_2$

MW 199.22

COLOR AND FORM Yellow needles from alcohol; white needles.

ODOR Sweetish odor.

BOILING POINT 340°C at 760 mm Hg.

MELTING POINT 114°C.

SOLUBILITY Soluble in ether, benzene, chloroform, acetic acid; slightly soluble in cold alcohol.

OCTANOL/WATER PARTITION COEFFICIENT Log K_{ow} = 3.77 (est).

OTHER CHEMICAL/PHYSICAL PROPERTIES IR: 6261 (Coblentz Society spectral collection) (biphenyl, 2–nitro); UV: 373 (Sadtler research laboratories spectral collection) (biphenyl, 2–nitro); NMR: 86 (Sadtler research laboratories spectral collection) (biphenyl, 2–nitro); IR: 20340 (Sadtler research laboratories prism collection) (biphenyl, 3–nitro); UV: 6834 (Sadtler research laboratories spectral collection) (biphenyl, 3–nitro); Henry's Law Constant = 3.54×10^{-6} atm–m³/mole at 25°C (est).

PROTECTIVE EQUIPMENT The following types of respirators should be selected under the prescribed concentration: (1) At any detectable concentration: any self-contained breathing apparatus that has a full facepiece and is operated in a pressure-demand or other positive-pressure mode or any supplied-air respirator that has a full facepiece and is operated in a pressure-demand or other positive-pressure mode in combination with an auxiliary self-contained breathing apparatus operated in a pressure-demand or other positive-pressure mode. (2) Escape from suddenly occurring respiratory hazards: any air-purifying, full facepiece respirator with a high-efficiency particulate filter, or any appropriate escape-type, self-contained breathing apparatus. [R9, 163].

PREVENTATIVE MEASURES Contact lenses should not be worn when working with this chemical [R9, 163]. SRP: The scientific literature supports the wearing of contact lenses in industrial environments, as part of a program to protect the eye against chemical compounds and minerals causing eye irritation. However, there may be individual substances whose irritating or corrosive properties are such that the wearing of contact lenses would be harmful to the eye. In those specific cases contact lenses should not be worn.

CLEANUP Research program was conducted to determine the capability of biol treatment and activated-carbon adsorption to remove chemical carcinogens from water and wastewater. Compounds studied included 4–nitrobiphenyl. All compounds tested exhibited some degree of biol degradation. Carbon adsorption was also effective in removing the compound from aqueous solution [R3].

COMMON USES Formerly in preparation of p-biphenylamine [R2].

OCCUPATIONAL RECOMMENDATIONS

OSHA Workers' exposure to 4–nitro-diphenyl is to be controlled through the required use of engineering controls, work practices, and personal protective equipment, including respirators. Identified as an occupational carcinogen without establishing a PEL [R8].

THRESHOLD LIMIT A1; A1 = Confirmed human carcinogen (1976) [R4].

NON-HUMAN TOXICITY LD_{50} rabbit oral 1.97 g/kg [R1]; LD_{50} rat oral 2.23 g/kg [R1].

IARC SUMMARY No data are available for humans. Inadequate evidence of carcinogenicity in animals. Overall evaluation: Group 3: The agent is not classifiable as to its carcinogenicity to humans [R5].

BIODEGRADATION p–Nitrobiphenyl exhibited fast biodegradation with rapid adaptation using a static-culture screening procedure with a settled domestic wastewater microbial inocula (1); degradation was determined to be 97.5% after 7 days of incubation (1) [R6].

BIOCONCENTRATION Based upon an estimated log K_{ow} of 3.77 (1), the BCF for p-nitrobiphenyl can be estimated to be 432 from a recommended regression equation (2,SRC). This estimated BCF value suggests that moderate bioconcentration may occur (SRC) [R7].

REFERENCES

1. American Conference of Governmental Industrial Hygienists. *Documentation of the Threshold Limit Values and Biological Exposure Indices.* 5th ed. Cincinnati, OH: American Conference of Governmental Industrial Hygienists, 1986. 433.

2. Budavari, S. (ed.). *The Merck Index— Encyclopedia of Chemicals, Drugs, and Biologicals.* Rahway, NJ: Merck and Co., Inc., 1989. 1043.

3. Fochtman, E. G., *Biodegradation and Carbon Adsorption of Carcinogenic and Hazardous Organic Compounds*; Report; ISS EPA–600/2–81–032; Order no. PB81–171852, 45 pages (1981).

4. American Conference of Governmental Industrial Hygienists. *Threshold Limit Values for Chemical Substances and Physical Agents and Biological Exposure Indices for 1994–1995.* Cincinnati, OH: ACGIH, 1994. 28.

5. IARC. *Monographs on the Evaluation of the Carcinogenic Risk of Chemicals to Man.* Geneva: World Health Organization, International Agency for Research on Cancer, 1972–present (multivolume work), p. S7 67 (1987).

6. Tabak, H. H., et al., pp. 267–328 in *Test Protocols for Environmental Fate and Movement of Toxicants.* Proc of a Symp Assoc of Official Anal Chemical 94th Ann Mtg, Washington, DC (1981).

7. (1) GEMS. *Graphical Exposure Modeling System.* CLOGP. USEPA (1987). (2) Lyman, W. J., et al., *Handbook of Chemical Property Estimation Methods.* New York: McGraw-Hill, p. 5–4 (1982).

8. 29 CFR 1910.1003 (7/1/91).

9. NIOSH. *NIOSH Pocket Guide to Chemical Hazards.* DHHS (NIOSH) Publication No. 90–117. Washington, DC: U.S. Government Printing Office, June 1990.

■ OCTACHLORODIBENZO–p–DIOXIN

CAS # 3268–87–9

SYNONYMS 1,2,3,4,6,7,8,9-octachlorodibenzo (b,e) (1,4) dioxin • 1,2,3,4,6,7,8,9-octachlorodibenzodioxin • octachloro-para-dibenzodioxin • dibenzo-p-dioxin, 1,2,3,4,6,7,8,9-octachloro • ocdd

MF $C_{12}Cl_8O_2$

MW 459.72

MELTING POINT 330°C.

SOLUBILITY Solubility (g/L): acetic acid 0.048; anisole 1.730; chloroform 0.562; ortho-dichlorobenzene 1.832; dioxane 0.384; diphenyl oxide 0.841; pyridine 0.400; xylene 3.575. Solubility in water: (0.4 or $\pm$ –0.1) $\times 10^{-9}$ g/L at 20°C.

VAPOR PRESSURE 8.25×10^{-13} mm Hg at 25°C (experimental).

OCTANOL/WATER PARTITION COEFFICIENT Log K_{ow} = 8.78–13.37.

STABILITY AND SHELF LIFE Extremely stable, even on heating to 700°C [R1].

OTHER CHEMICAL/PHYSICAL PROPERTIES Henry's law constant = 6.74×10^{-6} atm–m³/mole (est).

CLEANUP Irradiation experiments (254 nm) were performed to decrease high polychlorinated dibenzo-p-dioxins/furans surface contaminations after a malfunction of a capacitor filled with a technique PCB formulation. Destruction efficiencies >89% were achieved [R2].

OCCUPATIONAL RECOMMENDATIONS None listed.

IARC SUMMARY Classification of carcinogenicity: (1) evidence in humans: no data; (2) evidence in animals: insufficient. Overall summary evaluation of carcinogenic risk to humans is Group 3: The agent is not classifiable as to its carcinogenicity to humans. (chlorinated dibenzodioxins (other than TCDD); from table).

BIOLOGICAL HALF-LIFE The half-lives of 1,2,3,6,7,8-hexachlorodibenzo-p–dioxin, 1,2,3,4,6,7,8–heptachlorodibenzo-p–dioxin, and octachlorodibenzo-p–dioxin (were calculated) to be about 3.5, 3.6, and 2 years, respectively. The estimation was based on the analysis of fat-tissue biopsies collected with an interval of 28 months from one 14–year-old girl who for a period of about 2–3 years had been exposed to technical pentachlorophenol [R3].

BIOCONCENTRATION Using a continuous-flow-through system, octachlorodibenzo-p–dioxin BCFs of 34–136 were measured in rainbow trout, and 2,226 in fathead minnows (1–2); however, because all of the octachlorodibenzo-p–dioxin present in the system was not in the dissolved phase ("true solution"), recalculation to "true solution" yields BCFs of 8,500 in rainbow trout, and 22,300 in fathead minnow (1). Over a 64-day exposure period in an aquarium, the BCF measured for octachlorodibenzo-p–dioxin was 45.5 (2); bioaccumulation of the tetrachloro-isomers was an order a magnitude higher than for octachlorodibenzo-p–dioxin and the BCF was observed to decrease with an increase of chlorination (2). The BCFs observed for guppies (*Poecilia reticulata*) were 708 (wet basis) and 7,762 (lipid basis) (4); these relatively low BCFs (when compared to the log K_{ow} of 8.2 and measured BCFs for PCBs) were found to result from rapid depuration rates, which were attributed to the ability of the guppies to metabolize octachlorodibenzo-p–dioxin (4); the metabolism appeared to involve hydroxylation that was mediated by a mixed-function oxidase system (4) [R4].

REFERENCES

1. IARC. *Monographs on the Evaluation of the Carcinogenic Risk of Chemicals to Man.* Geneva: World Health Organization, International Agency for Research on Cancer, 1972–present (multivolume work), p. V15 48 (1977).

2. Borwitzky, H., K. W. Schramm. *Chemosphere* 22 (5–6): 485–94 (1991).

3. WHO. *Environmental Health Criteria 88: Polychlorinated Dibenzo-para–Dioxins and Dibenzofurans* p. 262 (1989).

4. (1) Muir, D.C.G., et al. *Chemosphere* 14: 829–33 (1985). (2) Muir, D.C.G., et al. *Environ Toxicol Chem* 261–72 (1986). (3) Miyata, H., et al. *Chemosphere* 19: 517–20 (1989). (4) Gobas, F.A.P.C., S. M. Schrap. *Chemosphere* 20: 495–512 (1990).

■ OCTACHLORONAPHTHALENE

CAS # 2234-13-1

SYNONYMS 1,2,3,4,5,6,7,8-octachloronaphthalene • naphthalene, octachloro- • perchloronaphthalene • perna

MF $C_{10}Cl_8$

MW 403.70

COLOR AND FORM Pale-yellow needles from benzene and carbon tetrachloride; waxy yellow solid.

ODOR Aromatic odor.

DENSITY 2.00.

BOILING POINT 440°C.

MELTING POINT 192°C.

TOXIC COMBUSTION PRODUCTS Toxic gases and vapors (such as carbon monoxide and toxic chloride fumes) may be released in a fire involving octachloronaphthalene [R2].

SOLUBILITY Slightly soluble in alcohol; soluble in benzene and chloroform; very soluble in petroleum ether; water sol: 0.08 μg/L at 22°C.

VAPOR PRESSURE 1×10^{-8} mm Hg at 25°C.

OCTANOL/WATER PARTITION COEFFICIENT Log K_{ow} = 6.42 (est).

OTHER CHEMICAL/PHYSICAL PROPERTIES The Henry's Law constant for octachloronaphthalene can be estimated to be 0.0000478 atm-m³/mole. Vapor pressure: >1 mm Hg at 20°C (68°F).

PROTECTIVE EQUIPMENT Employees should be provided with and required to use impervious clothing, gloves, face shields (eight-inch minimmum), and other appropriate protective clothing necessary to prevent any possibility of skin contact with molten octachloronaphthalene, repeated or prolonged skin contact with solid octachloronaphthalene or liquids containing octachloronaphthalene, or skin contact with octachloronaphthalene vapors from heated material [R2]. Wear appropriate clothing to prevent any possibility of skin contact with molten material or solutions. Wear eye protection and wash promptly when skin is contaminated (chloronaphthalenes) [R8].

PREVENTATIVE MEASURES Contact lenses should not be worn when working with this chemical [R9, 169]. SRP: The scientific literature supports the wearing of contact lenses in industrial environments, as part of a program to protect the eye against chemical compounds and minerals causing eye irritation. However, there may be individual substances whose irritating or corrosive properties are such that the wearing of contact lenses would be harmful to the eye. In those specific cases contact lenses should not be worn. Non-impervious clothing that becomes contaminated with molten or solid octachloronaphthalene or liquids containing octachloronaphthalene should be removed promptly and not reworn until the octachloronaphthalene is removed from the clothing [R2]. If employess' clothing may have become contaminated with solid octachloronaphthalene, employees should change into uncontaminated clothing before leaving the work premises [R2]. Clothing contaminated with octachloronaphthalene should be placed in closed containers for storage until it can be discarded or until provision is made for the removal of octachloronaphthalene from the clothing. If the clothing is to be laundered or otherwise cleaned to remove the octachloronaphthalene, the person performing the operation should be informed of octachloronaphthalene's hazardous properties [R2]. Condenser impregnation and other operations involving melting of chloronaphthalene should be enclosed or provided with effective local exhaust ventilation (chloronaphthalenes) [R10]. SRP: Local exhaust ventilation should be applied wherever there is an incidence of point-source emissions or dispersion of regulated contaminants in the work area. Ventilation control of the contaminant as close as possible to its point of generation is both the most economical and safest method to minimize personnel exposure to airborne contaminants.

CLEANUP 1. Ventilate area of spill. 2. Collect spilled material in most convenient and safe manner for reclamation or for disposal in secured sanitary landfill. Liquid should be absorbed in vermiculite, dry sand, earth, or a similar material [R2].

DISPOSAL Octachloronaphthalene may be disposed of in a secured sanitary landfill [R2]. Incineration, preferably after mixing with another combustible fuel. Care must be exercised to assure complete combustion to prevent the formation of phosgene. An acid scrubber is necessary to remove the halo acids produced. Recommendable method: Incineration. peer-review: Ensure plentiful supply of hydrocarbon fuel. (Peer-review conclusions of an IRPTC expert consultation (May 1985)) [R11].

COMMON USES Has been utilized as fireproof, waterproof additive in cable insulation, and in other protective coating materials and as additive to lubricants [R1].

OCCUPATIONAL RECOMMENDATIONS

OSHA 15–min short-term exposure limit: 0.3 mg/m³. Final rule limits were achieved by any combination of engineering controls, work practices, and personal protective equipment—effective Sept 1, 1989 through Dec 30, 1992. Final rule limits became effective Dec 31, 1992. Skin absorption designation in effect as of Sept 1, 1989 [R5]. 8–hr time-weighted avg: 0.1 mg/m³. Skin absorption designation. [R5].

THRESHOLD LIMIT 8–hr time-weighted average (TWA) 0.1 mg/m³, skin short-term exposure limit (STEL) 0.3 mg/m³, skin (1976) [R6].

BIODEGRADATION A predictive method based upon evaluated biodegradation data and the fact that octachloronaphthalene contains 8 aromatic chlorine substructures predicts that octachloronaphthalene has a low probability of biodegrading fast in the environment (1). Octachloronaphthalene is reported to have poor biodegradability (2) [R3].

BIOCONCENTRATION In a continuous-flow-through system, an octachloronaphthalene BCF of 0 was measured in guppies (*Poecillia reticula*) over a 7-day exposure period (1–2); it was proposed that the lack of uptake was due to the size of the octachloronaphthalene molecule and its inability to permeate the membrane (1–2). In a static system, a mean BCF of 340 was measured in rainbow trout over a 96-day exposure period (3) [R4]. Uptake rate constants by fish of di-, tetra-, hexa-, octa-, and decachlorobiphenyls are independent of the solute's hydrophobicity. By combining a 2-compartment bioconcentration model with a modified membrane permeation model, simple relationships between uptake rate constants for fish and the solutes' hydrophobicities and molecular configurations are obtained. The observed lack of uptake by fish of hexabromobenzene, octachloronaphthalene, and octachlorodibenzo-p-dioxin is not due to insufficient exposure concentration, because the exposure concns of these cmpd, in the continuous-flow water satn system were significantly higher than those of some PCB congeners. The lack of uptake of these chemicals can be explained by proposing influence of membrane permeation on the mechanism of bioconcentration. For these cmpds, the size, rather than octan-1-ol/water partition coefficients or aq solubility, causes a lack of uptake by fish. [R12].

TSCA REQUIREMENTS Pursuant to section 8 (d) of TSCA, EPA promulgated a model health and safety data reporting rule. The section 8 (d) model rule requires manufacturers, importers, and processors of listed chemical substances and mixtures to submit to EPA copies and lists of unpublished health and safety studies. Octachloronaphthalene is included on this list [R7].

FIRE POTENTIAL Not combustible [R2].

REFERENCES

1. American Conference of Governmental Industrial Hygienists. *Documentation of*

the *Threshold Limit Values and Biological Exposure Indices.* 5th ed. Cincinnati, OH: American Conference of Governmental Industrial Hygienists, 1986. 447.

2. Mackison, F. W., R. S. Stricoff, and L. J. Partridge, Jr. (eds.). *NIOSH/OSHA—Occupational Health Guidelines for Chemical Hazards.* DHHS (NIOSH) Publication No. 81–123 (3 vols). Washington, DC: U.S. Government Printing Office, Jan. 1981.

3. (1) Howard, P. H., et al. *Environ Toxicol Chem* 11: 593–603 (1992). (2) Koda, Y., et al., *Nagoya Kogyo Gijutsu Shikensho Hokoku* 31: 299–305 (1982).

4. (1) Opperhuizen, A. *Chemosphere* 14: 1871–96 (1985). (2) Opperhuizen, A. *ASTM Spec Tech Publ 921 (Aquat Toxicol Environ Fate)* 9: 304–15 (1986). (3) Oliver, B.G., and A.J. Niimi. *Environ Sci Technol* 19: 842–9 (1985).

5. 29 CFR 1910.1000 (7/1/91).

6. American Conference of Governmental Industrial Hygienists. *Threshold Limit Values for Chemical Substances and Physical Agents and Biological Exposure Indices for 1994–1995.* Cincinnati, OH: ACGIH, 1994. 28.

7. 40 CFR 716.120 (7/1/91).

8. Sittig, M. *Handbook of Toxic and Hazardous Chemicals.* Park Ridge, NJ: Noyes Data Corporation, 1981. 153.

9. NIOSH. *NIOSH Pocket Guide to Chemical Hazards.* DHHS (NIOSH) Publication No. 90–117. Washington, DC: U.S. Government Printing Office, June 1990.

10. International Labour Office. *Encyclopedia of Occupational Health and Safety.* Vols. I and II. Geneva, Switzerland: International Labour Office, 1983. 466.

11. United Nations. *Treatment and Disposal Methods for Waste Chemicals (IRPTC File).* Data Profile Series No. 5. Geneva, Switzerland: United Nations Environmental Programme, Dec. 1985. 213.

12. Opperhuizen, A. *ASTM Spec Tech Publ 921 (Aquat Toxicol Environ Fate).* 9: 304–15 (1986).

■ 2–OCTANOL

CAS # 123–96–6

SIC CODES 2844.

SYNONYMS 1-methyl-1-heptanol • 1-methylheptyl alcohol • 2-hydroxy-n-octane • 2-hydroxyoctane • 2-octyl alcohol • beta-octyl alcohol • capryl-alcohol • hexylmethylcarbinol • methyl-hexyl-carbinol • methylhexylcarbinol • n-octan-2-ol • s-octyl alcohol • sec-capryl alcohol • sec-caprylic alcohol • sec-n-octyl alcohol • secondary caprylic alcohol • fema-number-2801

MF $C_8H_{18}O$

MW 130.26

COLOR AND FORM Colorless.

SOLUBILITY Miscible with aromatic and aliphatic hydrocarbons; soluble in most common organic solvents; water solubility: 1,120 mg/L at 25°C.

VAPOR PRESSURE 2.42×10^{-1} mm Hg at 25°C.

OCTANOL/WATER PARTITION COEFFICIENT log K_{ow} = 2.72.

OTHER CHEMICAL/PHYSICAL PROPERTIES Oily, refractive liquid; solubility: 0.096 ml/100 ml water; aromatic, yet somewhat unpleasant odor, particularly on heating; density 0.8193 at 20°C; BP: 178.5°C at 760 mm Hg; index of refraction: 1.42025 at 20°C (dl-form); BP: 86°C at 20 mm Hg; specific optical rotation: 9.9°C at 17°C; index of refraction: 0.8216 at 20°C (D-form); BP: 86°C at 20 mm Hg; specific optical rotation: −9.9°C at 17°C (L-form); soluble in alcohol, ether, acetone; max absorption (vapor): 196 nm; max absorption (hexane): 175 nm (log E = 2.5); MP: −31.6°C (DL-form); density: 0.8201 at 20°C; miscible with water; soluble in alcohol, ether, acetone; index of refraction: 1.4264 at 20°C (L-form); 0.132% in saturated air at 32.8°C; density of saturated air (air = 1): 1.001; 1 mg/L = 187.8 ppm at 25°C, 760 mm Hg; 1 ppm = 5.32 mg/m³ at 25°C, 760 mm Hg; miscible with water; soluble in alcohol, ether, acetone; index of refraction: 1.4264 at 20°C (D–form); vapor pressure of 1 mm Hg at 32.8°C.

COMMON USES In manufacture of perfumes; in disinfectant soaps; to prevent foaming; solvent for fats and waxes [R2]. In lacquers, enamels; as chemical intermediate in manufacture of plasticizers [R9, 4618]. Chemical intermediate for alcohol-derived surfactants [R3]. Solvent, manufacture of plasticizers, wetting agents, foam control agents, hydraulic oils, petroleum additives, perfume intermediates, masking of industrial odors [R1].

STANDARD CODES NFPA—1, 2, 0.

PERSISTENCE Biodegrades at moderate rate.

MAJOR SPECIES THREATENED Waterfowl.

DIRECT CONTACT Contact may dry skin by extracting oils.

GENERAL SENSATION Aromatic, unpleasant odor.

PERSONAL SAFETY Wear skin protection and canister-type mask.

ACUTE HAZARD LEVEL Details unknown. Considered slightly toxic. Can produce tastes in water at low concentrations. Will produce BOD.

CHRONIC HAZARD LEVEL Unknown.

DEGREE OF HAZARD TO PUBLIC HEALTH Details unknown. Considered slightly toxic.

OCCUPATIONAL RECOMMENDATIONS None listed.

ACTION LEVEL Notify fire authority. Remove ignition source. Attempt to contain slick.

ON-SITE RESTORATION Oil-skimming equipment and sorbent foams can be applied to slicks. Carbon or peat can be used on dissolved portion. Seek professional environmental engineering assistance through EPA's Environmental Response Team (ERT), Edison, NJ, 24–hour no.: 908–548–8730.

BEACH OR SHORE RESTORATION Burn off.

AVAILABILITY OF COUNTERMEASURE MATERIAL Oil-skimming equipment—stored at major ports; sorbent foams (polyurethane)—upholstery shops; carbon—water treatment plants, sugar refineries; peat—nurseries, floral shops.

DISPOSAL Spray into incinerator or burn in paper packaging. Additional flammable solvent may be added.

INDUSTRIAL FOULING POTENTIAL Slicks may reduce heat transfer or cause hot spots and scaling. May pose rupture hazard if confined in boiler feed water.

MAJOR WATER USE THREATENED Recreation, potable supply, industrial, fisheries.

PROBABLE LOCATION AND STATE Will float on surface of water as a slick slowly dissolving. Colorless oily liquid.

COLOR IN WATER Colorless.

NON-HUMAN TOXICITY LD$_{50}$ rat oral >3.2 g/kg [R4].

ECOTOXICITY Capryl alcohol was lethal to rainbow trout at less than 100 ppm (mg/L or µl/L) [R5].

BIODEGRADATION River die-away tests and grab sample data pertaining to the biodegradation of 2-octanol in natural waters and soil were not located in the available literature. Yet, a number of aerobic and anaerobic biological screening studies, which utilized settled waste water, sewage, or activated sludge for inocula, have demonstrated that 2-octanol is biodegradable (1–5). These studies indicate 2-octanol should degrade rapidly when acclimation has occurred (SRC) [R6].

BIOCONCENTRATION Based upon a water solubility of 1,120 mg/L at 25°C (1) and log K$_{ow}$ of 2.72 (2), the bioconcentration factor (log BCF) for 2-octanol has been calculated to be 1.07 and 1.84, respectively, from recommended regression-derived equations (3). These BCF values indicate the potential for 2-octanol to bioconcentrate in aquatic organisms is low (SRC) [R7].

FDA 2-Octanol is a food additive permitted for direct addition to food for human consumption, as long as (1) the quantity added to food does not exceed the amount reasonably required to accomplish its intended physical, nutritive, or other technical effect in food, and (2) when intended for use in or on food it is of appropriate food grade and is prepared and handled as a food ingredient. Used as a synthetic flavoring substance or adjuvant [R8].

REFERENCES

1. Sax, N. I., and R. J. Lewis, Sr. (eds.). *Hawley's Condensed Chemical Dictionary.* 11th ed. New York: Van Nostrand Reinhold Co., 1987. 850.

2. Budavari, S. (ed.). *The Merck Index—Encyclopedia of Chemicals, Drugs, and Biologicals.* Rahway, NJ: Merck and Co., Inc., 1989. 1070.

3. SRI.

4. Verschueren, K. *Handbook of Environmental Data of Organic Chemicals.* 2nd ed. New York, NY: Van Nostrand Reinhold Co., 1983. 934.

5. Spague, J. B., W. J. Logan. *J Mar Res* 19 (4): 269 (1979).

6. (1) Babeu, L., D. Vaishnav. *J Indust Microb* 2: 107–15 (1987). (2) Gerhold, R.M., G. W. Malaney. *J Water Pollut Contr Fed* 38: 562–79 (1966). (3) Shelton, D. R., J. M. Tiedje. *Development of Tests for Determining Anaerobic Biodegradation Potential.* USEPA–560/5–81–013 (NTIS PB84–166495) pp. 92 (1981) (4) Shelton, D.R., J.M. Tiedje. *Appl Environ Microbiol* 47: 850–7 (1981). (5) Yonezawa, Y., Y. Urushigawa. *Chemosphere* 8: 139–42 (1979).

7. (1) Amidon, G. L., et al., *J Pharm Sci* 63: 1858–66 (1974). (2) CLOGP. *PCGEMS Graphical Exposure Modeling System.* USEPA (1986). (3) Lyman, W. J., et al., *Handbook of Chemical Property Estimation Methods.* New York: McGraw-Hill, pp. 5–4, 5–10 (1982).

8. 21 CFR 172.515 (4/1/90).

9. Clayton, G. D., and F. E. Clayton (eds.). *Patty's Industrial Hygiene and Toxicology:* Volume 2A, 2B, 2C: *Toxicology.* 3rd ed. New York: John Wiley and Sons, 1981–1982.

■ OMETHOATE

CAS # 1113–02–6

SYNONYMS o,o-dimethyl s-methylcarbamoylmethyl phosphorothioate • 2-dimethoxyphosphinoylthio-n-methylacetamide • o,o-dimethyl s-(2-(methylamino)-2-oxoethyl)phosphorothioate • o,o-dimethyl phosphorothioate s-ester with 2-mercapto-n-methylacetamide • dimethoate-met • folimat • bay-45432 • s-6876 • bayer-45,432 • bayer S-6876

MF $C_5H_{12}NO_4PS$

MW 213.19

COLOR AND FORM Colorless to yellow oil.

ODOR Leek-like odor.

DENSITY 1.32 at 20°C.

SOLUBILITY Readily soluble in water, alcohols, acetone, and many hydrocarbons. Slightly soluble in diethyl ether. Almost insoluble in petroleum ether. Soluble in toluene. Miscible in dichloromethane, 2–propanol, water. Nearly insoluble in n-hexane.

VAPOR PRESSURE 3.2 mPa at 20°C.

CORROSIVITY Noncorrosive.

STABILITY AND SHELF LIFE Hydrolyzed in alkaline media. Stable under neutral aqueous conditions [R1].

OTHER CHEMICAL/PHYSICAL PROPERTIES Compatible with most other pesticides, but incompatible with alkaline materials. On hydrolysis at 24°C, 50% decomposition occurs in 2.5 days at pH 7. Decomposes at ca 135°C.

PREVENTATIVE MEASURES SRP: The scientific literature supports the wearing of contact lenses in industrial environments, as part of a program to protect the eye against chemical compounds and minerals causing eye irritation. However, there may be individual substances whose irritating or corrosive properties are such that the wearing of contact lenses would be harmful to the eye. In those specific cases contact lenses should not be worn.

COMMON USES Systemic acaricide and insecticide [R2]. Control of spider mites, aphids (including woolly aphids), beetles, caterpillars, scale insects, thrips, psyllids, fruit flies, etc., on pome fruit, stone fruit, citrus fruit, vines, hops, beet, cotton, coffee, sugarcane, cereals, rice, potatoes, ornamentals, and other crops, and in forestry [R1].

OCCUPATIONAL RECOMMENDATIONS None listed.

NON-HUMAN TOXICITY LD_{50} rat oral 50 mg/kg [R1]; LD_{50} rabbit oral 50 mg/kg [R1]; LD_{50} cat oral 50 mg/kg [R1]; LD_{50} guinea pig oral 100 mg/kg [R1]; LD_{50} rat percutaneous 700 mg/kg/7 days [R1]; LC_{50} rat male inhalation >1.5 mg/L/1 hr [R1]; LD_{50} hen oral 125 mg/kg [R1]; in 90-day feeding trials, no-effect level for rats was 1 mg/kg diet [R1].

ECOTOXICITY LC_{50} goldfish 10–100 mg/L/ 96 hr (conditions of bioassay not specified) [R1].

BIODEGRADATION A predictive method based upon evaluated biodegradation data and the fact that omethoate contains a phosphate ester substructure predicts that omethoate has a high probability of biodegrading in the environment (1) [R3].

BIOCONCENTRATION Based on an estimated log K_{ow} of −0.377 (1), the BCF for omethoate can be estimated to be about 0.3 from a regression-derived equation (2,SRC). This value of BCF indicates that the bioconcentration of omethoate in aquatic organisms is not expected to be an important fate process (SRC) [R4].

DOT *Health Hazards:* Poisonous; may be fatal if inhaled, swallowed, or absorbed through skin. Contact may cause burns to skin and eyes. Runoff from fire control or dilution water may cause pollution (organophosphorus pesticide, liquid, flammable, poisonous, not otherwise specified) [R5, p. G–28].

Fire or Explosion: Flammable/combustible material; may be ignited by heat, sparks, or flames. Vapors may travel to a source of ignition, and flash back. Container may explode in heat of fire. Vapor explosion and poison hazard indoors, outdoors, or in sewers. Runoff to sewer may create fire or explosion hazard (organophosphorus pesticide, liquid, flammable, poisonous, not otherwise specified) [R5, p. G–28].

Emergency Action: Keep unnecessary people away; isolate hazard area, and deny entry. Stay upwind; keep out of low areas. Positive-pressure self-contained breathing apparatus (SCBA) and chemical protective clothing that is specifically recommended by the shipper or manufacturer may be worn. It may provide little or no thermal protection. Structural firefighters' protective clothing is not effective for these materials. Isolate for 1/2 mile in all directions if tank, rail car, or tank truck is involved in fire. Call CHEMTREC at 1–800–424–9300 for emergency assistance (organophosphorus pesticide, liquid, flammable, poisonous, not otherwise specified) [R5, p. G–28].

Fire: Small fires: Dry chemical, CO_2, water spray, or alcohol-resistant foam. Large Fires: Water spray, fog, or alcohol-resistant foam. Move container from fire area if you can do so without risk. Dike fire-control water for later disposal; do not scatter the material. Apply cooling water to sides of containers that are exposed to flames until well after fire is out. Stay away from ends of tanks. Withdraw immediately in case of rising sound from venting safety device or any discoloration of tank due to fire (organophosphorus pesticide, liquid, flammable, poisonous, not otherwise specified) [R5, p. G–28].

Spill or Leak: Shut off ignition sources; no flares, smoking, or flames in hazard area. Fully encapsulating, vapor-protective clothing should be worn for spills and leaks with no fire. Do not touch or walk through spilled material; stop leak if you can do so without risk. Water spray may reduce vapor; but it may not prevent ignition in closed spaces. Small Spills: Take up with sand or other noncombustible absorbent material and place into containers for later disposal. Large Spills:

Dike far ahead of liquid spill for later disposal (organophosphorus pesticide, liquid, flammable, poisonous, not otherwise specified) [R5, p. G–28].

First Aid: Move victim to fresh air, and call for emergency medical care; if not breathing, give artificial respiration; if breathing is difficult, give oxygen. In case of contact with material, immediately flush skin or eyes with running water for at least 15 minutes. Remove and isolate contaminated clothing and shoes at the site. Keep victim quiet and maintain normal body temperature. Effects may be delayed; keep victim under observation. (organophosphorus pesticide, liquid, flammable, poisonous, not otherwise specified) [R5. p. G–28].

REFERENCES

1. Hartley, D., and H. Kidd (eds.). *The Agrochemicals Handbook.* 2nd ed. Lechworth, Herts, England: The Royal Society of Chemistry, 1987, p. A304 (1987).

2. Worthing, C. R., and S. B. Walker (eds.). *The Pesticide Manual—A World Compendium.* 8th ed. Thornton Heath, UK: The British Crop Protection Council, 1987. 614.

3. Howard, P. H., et al., *Environ Toxicol Chem* 11: 593–603 (1992).

4. (1) CLOGP. *GEMS–Graphic Exposure Modelling System.* CLOGP USEPA (1986). (2) Lyman, W. J., et al., *Handbook of Chemical Property Estimation Methods.* Washington DC: Amer Chemical Soc, pp. 4–9 (1990).

5. U.S. Department of Transportation. *Emergency Response Guidebook 1990.* DOT P 5800.5. Washington, DC: U.S. Government Printing Office, 1990.

■ OXYCARBOXIN

CAS # 5259–88–1

SYNONYMS 1,4-oxathiin, 2,3-dihydro-5-carboxanilido-6methyl-, 4,4-dioxide • 1,4-oxathiin-3-carboxamide, 5,6-dihydro-2-methyl-N-phenyl-, 4,4-dioxide •

1,4-oxathiin-3-carboxanilide, 5,6-dihydro-2-methyl-, 4,4-dioxide • 2,3-dihydro-5-carboxanilido-6-methyl-1,4-oxathiin, 4,4-dioxide • 5,6-dihydro-2-methyl-1,4-oxathiin-3-carboxanilide 4,4-dioxide • carboxin-sulfone • dcmod • dioxide of vitavax • f-461 • f-461 [pesticide] • f461 • oxicarboxin • oxycarboxine • plant-vax • plantvax • plantvax-20 • vitavaxsulfone • vitavex • 2,3-dihydro-6-methyl-5-phenylcarbamoyl-1,4-oxathiin 4,4-dioxide • 5,6-dihydro-2-methyl-1,4-oxathi-ine-3-carboxanilide 4,4-dioxide

MF $C_{12}H_{13}NO_4S$

MW 267.32

COLOR AND FORM Off-white crystals; colorless crystals; white solid.

MELTING POINT 127.5–130°C.

SOLUBILITY 0.1 g in 100 g water; 7 g in 100 g methanol; 3 g in 100 g ethanol; 36 g in 100 g acetone; 3.4 g in 100 g benzene; 223 g in 100 g dimethyl sulfoxide (all at 25°C).

VAPOR PRESSURE Less than 1 mPa at 20°C.

CORROSIVITY Corrosive.

STABILITY AND SHELF LIFE Stable, except under highly acidic or highly alkaline conditions [R7].

OTHER CHEMICAL/PHYSICAL PROPERTIES Henry's Law constant is 3.5×10^{-9} atm–m³/mole at 25°C (est).

PREVENTATIVE MEASURES Do not inhale dust or spray mist; avoid contact with skin and mucous membranes [R6].

COMMON USES Foliar application for rust on carnations and geraniums (greenhouse only) [R1]. Plantvax was effective in controlling udbatta disease of rice caused by *Ephelis oryzae* [R2]. Oxycarboxin was excellent against stripe rust (*Puccinia striiformis*), good against stripe smut (*Ustilago striiformis*), and completely eradicated flag smut (*Urocystis agropyri*) after soil application [R3]. 500 ppm oxycarboxin protected soybean against rust and incr seed wt [R4]. Soil drenches with oxycarboxin prevented mal secco disease development in lemon plant stems inoculated 1 month later with *Deuterophoma trachelphila* [R5].

OCCUPATIONAL RECOMMENDATIONS None listed.

NON-HUMAN TOXICITY LD_{50} rabbit percutaneous >16,000 mg/kg [R8]; LC_{50} rat oral 2,000 mg/kg [R7].

ECOTOXICITY LC_{50} bobwhite quail oral >10,000 mg/kg in 8-day diet [R8]; LC_{50} mallard ducks oral >4,640 mg/kg in 8-day diet [R8]; LC_{50} bluegill 28.1 mg/L/96 hr (conditions of bioassay not specified) [R8]; LC_{50} rainbow trout 19.9 mg/L/96 hr (conditions of bioassay not specified) [R8]; LC_{50} water flea 69.1 mg/L/96 hr (conditions of bioassay not specified) [R8].

BIODEGRADATION Microorganisms found in wet soils were examined for their ability to degrade fungicides. The blue-green algae *Anabaena, Nostoc,* and *Tolypothrix* extensively degraded oxycarboxin and CARBOXIN. The photosynthetic bacterium *Rhodospirillum* degraded oxycarboxin to a greater extent than carboxin. The protozoa *Colpoda* brought about an extensive degradation of carboxin, but not of oxycarboxin [R9]. When incubated with a *Nocardia* species isolated from soil, oxycarboxin was found to undergo amide hydrolysis (1). Blue-green algae (*Anabaena, Nostoc, Tolypothrix,* and *Chlorella vulgaris*) isolated from wet soil were found to degrade oxycarboxin (2). Five different soil types including 3 monoionic clays (red sandy loam, black clayey, laterite, saline alkali, and coffee plantation), were found to degrade oxycarboxin in 10 days under aerobic conditions (3). [R11].

BIOCONCENTRATION Based upon its water solubility, 1,000 mg/L at 25°C (1), and the octanol/water partition coefficient, 0.9 (2), the bioconcentration factor for oxycarboxin can be estimated to range from 3 to 12 using regression-derived equations (3,SRC). The magnitude of these values indicates that bioconcentration in fish and aquatic organisms will not be an important fate process for oxycarboxin (SRC) [R10].

FIFRA REQUIREMENTS In 1988, Congress amended FIFRA to strengthen and accelerate EPA's reregistration program. The nine-year reregistration scheme mandated by "FIFRA 88" applies to each registered pesticide product containing an active ingredient initially registered before November 1, 1984. List A consists of the 194 chemical cases (or 350 individual active ingredients) for which EPA had issued Registration Standards prior to the effective date of FIFRA '88. List: A; Case: Carboxin; Case No. 0012; Pesticide type: Fungicide; Registration Standard Date: 08/08/81; Case Status: Awaiting Data/ Data in Review: OPP awaits data from the pesticide's producer(s) regarding its human health or environmental effects, or OPP has received and is reviewing such data, in order to reach a decision about the pesticide's eligibility for reregistration. Active Ingredient (AI): Oxycarboxin; AI Status: The producer(s) of the pesticide has made commitments to conduct the studies and pay the fees required for reregistration, and is meeting those commitments in a timely manner. [R33].

REFERENCES

1. *Farm Chemicals Handbook* 1993. Willoughby, OH: Meister Publishing Co., 1993, p. C–254.

2. Mohanty, N. N., *Proc Indian Nat Sci Acad.* Part B; 37 (6): 433–40 (1971).

3. Hardison, J. R. *Phytopathology.* 61 (6): 731–5 (1971).

4. Jan, C., L. Wu. L. I. Kuo. *Tai–Wan Ta Hsueh Nung Hsueh Yuan Yen Chiu Pao Kao* 12 (1): 173–90 (1971).

5. Solel, Z., et al. *Phytopathology* 62 (9): 1007–13 (1972).

6. Hartley, D., and H. Kidd (eds.). *The Agrochemicals Handbook.* Old Woking, Surrey, United Kingdom: Royal Society of Chemistry/Unwin Brothers Ltd., 1983, p. A308/Oct 83.

7. Hartley, D., and H. Kidd (eds.). *The Agrochemicals Handbook.* 2nd ed. Lechworth, Herts, England: The Royal Society of Chemistry, 1987, p. A308/Aug 87.

8. Worthing, C. R., and S. B. Walker (eds.). *The Pesticide Manual—A World Com-*

pendium. 8th ed. Thornton Heath, UK: The British Crop Protection Council, 1987. 624.

9. Balasubramanya, R. H., R. B. Patil. *Plant Soil* 57 (2–3): 457–61 (1980).

10. (1) Briggs, G. G. *J Agric Food Chem* 29: 1050–9 (1981). (2) Worthing, C. R., S. B. Walker. *The Pesticide Manual* 8th ed. Croydon, England: The British Crop Protection Council (1987). (3) Lyman, W. J., et al., *Handbook of Chemical Property Estimation Methods.* New York: McGraw-Hill, Chapt 5 (1982).

11. (1) Menzie, C. M., *Metabolism of Pesticides.* Update III. Fish and Wildlife Service: Washington, DC, NTIS PB83–165498 (1980). (2) Balasubramanya, R. H., R. B. Patil. *Plant Soil* 57: 457–61 (1980). (3) Balasubramanya, R. H., R. B. Patil. *Plant Soil* 57: 195–201 (1980).

12. USEPA/OPP. *Status of Pesticides in Reregistration and Special Review.* p. 71 (Mar, 1992) EPA 700–R–92–004.

■ OXYDEMETON–METHYL

CAS # 301–12–2

SYNONYMS bay-21097 • bayer-21097 • demeton-methyl sulphoxide • demeton-o-methyl sulfoxide • demeton-s methyl sulfoxide • Demeton-s-methyl sulfoxid [German] • ent-24,964 • ethanethiol, 2-(ethylsulfinyl)-, s-ester with o,o-dimethyl phosphorothioate • isomethylsystox sulfoxide • metaisosystox sulfoxide • metasystemox • metasystox-r • methyl-oxydemeton-s • metilmercaptofosoksid (former USSR) • o,o-dimethyl s-(2-(ethylsufinyl) ethyl) thiophosphate • o,o-dimethyl s-(2-(ethylsulfinyl) ethyl) monothiophosphate • o,o-dimethyl s-(2-(ethylsulfinyl) ethyl) phosphorothioate • o,o-dimethyl-s-(2-ethylsulfinyl-ethyl)-monothiofosfaat [Dutch] • o,o-Dimethyl-s-(3-oxo-3-thia-pentyl) -monothiophosphat [German] • o-o-dimetil-s-(2-etil-solfinil-etil)-monotiofosfato [Italian] • oxydemeton- metile [Italian] • phosphorothioic acid, s-(2-(ethylsulfinyl) ethyl) o,o-dimethyl ester • r-2170 • s-(2-(ethylsulfinyl) ethyl) o,o-dimethl phosphorothioate • thiophosphate de o,o-dimethyle et de s-2-ethylsulfinylethyle [French]

MF $C_6H_{15}O_4PS_2$

MW 246.31

COLOR AND FORM Clear amber liquid.

DENSITY 1.289 at 20°C.

BOILING POINT 106°C at 0.01 mm Hg.

MELTING POINT Below –10°C.

SOLUBILITY Miscible with water; soluble at 20°C: 100–1,000 g/L dichloromethane, propan–2–ol; sparingly soluble in light petroleum.

VAPOR PRESSURE 3.8 mPa (2.85×10^{-5} mm Hg) at 20°C.

STABILITY AND SHELF LIFE Subject to hydrolysis and the rate is fairly rapidly in alkaline media, oxidizes to sulfone, decomposes at elevated temperatures. [R6, 433].

OTHER CHEMICAL/PHYSICAL PROPERTIES Hydrolyzed in alkaline media.

DISPOSAL SRP: At the time of review, criteria for land treatment or burial (sanitary landfill) disposal practices are subject to significant revision. Prior to implementing land disposal of waste residue (including waste sludge), consult with environmental regulatory agencies for guidance on acceptable disposal practices.

COMMON USES Systemic insecticide, acaricide, effective control by contact and systemic action of many destructive pests that attack certain vegetable, fruit, and field crops. Primarily for aphids, mites, thrips, leafhoppers, and other sucking pests [R1].

OCCUPATIONAL RECOMMENDATIONS None listed.

ECOTOXICITY LD$_{50}$ *Anas platyrhynchos* (mallard), males 4 mo old, oral 53.9 mg/kg (95% confidence limit: 38.9–74.8 mg/kg) [R5, 60]. LD$_{50}$ *Phasianus colchicus* (pheasant), males 3–4 mo old, oral 42.4

mg/kg (95% confidence limit: 30.6–58.8 mg/kg) [R5, 60]. LD_{50} *Alectoris chukar* males and females, 3–4 mo old, oral 84.7–157 mg/kg [R5, 60]. LD_{50} *Passer domesticus*,(house sparrow) males, oral 70.8 mg/kg (95% confidence limit: 43.4–116 mg/kg) [R5, 61]. LD_{50} *Callipepla californica*, males, oral 47.6 mg/kg (95% confidence limit: 34.3–66.0 mg/kg) [R5, 61]. LD_{50} *Columba livia*, males and females, oral 14.0 mg/kg (95% confidence limit: 8.84–22.3 mg/kg [R5, 61]. LC_{50} *Asellus brevicaudus*, mature 1.4 mg/L/96 hr at 15°C. Static bioassay without aeration, pH 7.2–7.5, water hardness 40–50 mg/L as calcium carbonate, and alkalinity of 30–35 mg/L [R2]. LC_{50} *Grammarus fasciatus*, mature 1.0 mg/L/96 hr at 15°C (95% confidence limit: 0.06–1.7 mg/L). Static bioassay without aeration, pH 7.2–7.5, water hardness 40–50 mg/L as calcium carbonate, and alkalinity of 30–35 mg/L [R2]. LC_{50} *Salmo gairdneri* (rainbow trout), wt 1.1 g, 6.4 mg/L/96 hr at 15°C (95% confidence limit: 4.4–9.2 mg/L). Static bioassay without aeration, pH 7.2–7.5, water hardness 40–50 mg/L as calcium carbonate, and alkalinity of 30–35 mg/L [R2]. LC_{50} *Ictalurus punctatus* (channel catfish), wt 1.4 g, less than 18 mg/L/96–hr at 18°C. Static bioassay without aeration, pH 7.2–7.5, water hardness 40–50 mg/L as calcium carbonate, and alkalinity of 30–35 mg/L [R2]. LC_{50} *Lepomis macrochirus* (bluegill), wt 1.0 g, 13.0 mg/L/96 hr at 18°C in hard water, 272 ppm $CaCO_3$ (95% confidence limit: 10.6–16.7 mg/L). Static bioassay without aeration, pH 7.2–7.5, water hardness 40–50 mg/L as calcium carbonate, and alkalinity of 30–35 mg/L [R2]. LC_{50} *Micropterus salmoides* (largemouth bass), wt 0.7 g, 31.15 mg/L/96 hr at 18°C in hard water, 272 ppm $CaCO_3$ (95% confidence limit: 27.4–36.2 mg/L). Static bioassay without aeration, pH 7.2–7.5, water hardness 40–50 mg/L as calcium carbonate, and alkalinity of 30–35 mg/L [R2]. LC_{50} *Stizostedion vitreum vitreum* (walleye), wt 1.4 g, 18.0 mg/L/96 hr at 18°C (95% confidence limit: 15.6–20.8). Static bioassay without aeration, pH 7.2–7.5, water hardness 40–50 mg/L as calcium

carbonate, and alkalinity of 30–35 mg/L [R2].

BIODEGRADATION Although no biodegradation rates were located for screening tests or natural systems, oxydemeton-methyl has been shown to be degraded by many soil microorganisms (1). One microorganism, a *Pseudomonas putida*, cleaved the thioester bond to form 2-(ethylsulfinyl) ethanethiol, which was subsequently oxidized to bis[2-(ethylsulfinyl) ethyl]disulfide, and 2-(ethylsulfonyl) ethanethiol (1). Another microorganism, a *Nocardia* sp., produced bis[2-(ethylsulfonyl) ethyl]disulfide, bis[2-(ethylsulfinyl)ethyl]sulfide, and bis[2-(ethylsulfonyl)ethyl]sulfide via oxidation, beta-elimination, and hydrolysis (1) [R3].

BIOCONCENTRATION Using an estimated log K_{ow} of −1.03 (1), one would estimate a BCF of 9.7 for oxydemeton-methyl using a recommended regression equation (2). This would indicate that oxydemeton-methyl would not bioconcentrate in aquatic organisms (SRC) [R4].

REFERENCES

1. *Farm Chemicals Handbook 1994*. Willoughby, OH: Meister, 1994, p. C–266.

2. U.S. Department of Interior, Fish and Wildlife Service. *Handbook of Acute Toxicity of Chemicals to Fish and Aquatic Invertebrates*. Resource Publication No. 137. Washington, DC: U.S. Government Printing Office, 1980. 55.

3. Ziegler, W., et al. *J Agri Food Chem* 28: 1102–6 (1980).

4. Meylan, W. M., P. H. Howard. *J Pharm Sci* 84: 83–92 (1995). (2) Lyman, W. J., et al., *Handbook of Chemical Property Estimation Methods*, New York: McGraw-Hill, Chapt 5, Eqn 5–2 (1982).

5. U.S. Department of the Interior, Fish and Wildlife Service. *Handbook of Toxicity of Pesticides to Wildlife*. Resource Publication 153. Washington, DC: U.S. Government Printing Office, 1984.

6. Spencer, E. Y. *Guide to the Chemicals Used in Crop Protection*. 7th ed. Publication 1093. Research Institute, Agriculture

Canada, Ottawa, Canada: Information Canada, 1982.

■ OXYTHIOQUINOX

CAS # 2439-01-2

SYNONYMS bay-36205 • bayer-36205 • bayer-4964 • bayer-ss2074 • carbonic acid, dithio-, cyclic s, s-(6-methyl-2,3-quinoxalinediyl) ester • chinomethionat • chinomethionate • cyclic s, s-(6-methyl-2,3-quinoxalinediyl) dithiocarbonate • 1,3-dithiolo (4,5-b)quinoxalin-2-one, 6-methyl • dithiolo (4,5-b)quinoxalin-2-one, 6-methyl • dithioquinox • ent-25,606 • erade • erazidon • forstan • 6-methyl-1,3-dithiolo (4,5-b)quinoxalin-2-one • 6-methyl-2-oxo-1,3-dithiolo (4,5-b) quinoxaline • 6-methyl-2-oxo-1,3-dithio[4,5-b]quinoxaline • 6-methyl-2,3-quinoxaline dithiocarbonate • 6-methyl-2,3-quinoxalinedithiol cyclic carbonate • 6-methyl-2,3-quinoxalinedithiol cyclic s,s-dithiocarbonate • 6-methyl-2,3-quinoxalinedithiol cyclic dithiocarbonate • 6-methyl-quinoxaline-2,3-dithiolcyclocarbonate • morestan • morestan-2 • morestane • mqd • quinomethionate • 2,3-quinoxalinedithiol, 6-methyl-, cyclic carbonate • 2,3-quinoxalinedithiol, 6-methyl-, cyclic dithiocarbonate (ester) • ss-2074

MF $C_{10}H_6N_2OS_2$

MW 234.29

COLOR AND FORM Yellow crystals from benzene.

ODOR Odorless.

MELTING POINT 172°C.

SOLUBILITY Practically insoluble in water; freely soluble in DMF; soluble in hot benzene, toluene, dioxane; slightly soluble in methanol, ethanol, acetone; wt/wt in dimethylformamide 1.8%, in cyclohexanone 1.8%, in petroleum oils 0.4%; 18 g/kg cyclohexanone; 10 g/kg dimethylformamide; 4 g/kg petroleum oils.; 1 mg/L in water, temperature 20–25°C.

VAPOR PRESSURE 2×10^{-7} mm Hg at 20°C.

OTHER CHEMICAL/PHYSICAL PROPERTIES Closely related to thioquinox but more stable to oxidn; subject to hydrolysis under alkaline conditions; sublimes when heated. Relatively stable under normal conditions. Hydrolyzed in alkaline media.

DISPOSAL SRP: At the time of review, criteria for land treatment or burial (sanitary landfill) disposal practices are subject to significant revision. Prior to implementing land disposal of waste residue (including waste sludge), consult with environmental regulatory agencies for guidance on acceptable disposal practices.

COMMON USES Non-systemic fungicidal/acaricidal control of mites, mite eggs, pear psylla, and powdery mildew; red spider mite in gooseberry, strawberry, marrow, apple, and blackcurrant; powdery mildews in blackcurrant, marrow, and gooseberry; and leaf spot in blackcurrant [R1]. Insecticide [R2]. Awaiting data; data in review [R3]. Selective non-systemic contact fungicide with protective and eradicant action. Also has some acaricidal activity. Control of powdery mildews and spider mites on fruit (including citrus), ornamentals, cucurbits, cotton, coffee, tea, tobacco, walnuts, vegetables, and glasshouse crops; American gooseberry mildew on gooseberries and currants [R4].

OCCUPATIONAL RECOMMENDATIONS None listed.

BIODEGRADATION The nitrogen-containing ring structure of oxythioquinox is apparently not cleaved by anaerobic ecosystems (1) [R5].

BIOCONCENTRATION Based on a water solubility of 1 mg/L (1) and an estimated log K_{ow} of 2.87 (2), respective BCFs of 89 and 618 can be estimated for oxythioquinox using a recommended regression-derived equation (3,SRC). These BCF values suggest that bioconcentration in aquatic organisms may not be an important fate process (SRC) [R6].

FIFRA REQUIREMENTS In 1988, Congress amended FIFRA to strengthen and accelerate EPA's reregistration program. The nine-year reregistration scheme mandated by "FIFRA 88" applies to each registered pesticide product containing an active ingredient initially registered before November 1, 1984. Pesticides for which EPA had not issued Registration Standards prior to the effective date of FIFRA '88 were divided into three lists based upon their potential for exposure and other factors, with List B being of highest concern and D of least. List: B; Case: Oxythioquinox; Case No.: 2495; Pesticide type: Insecticide, fungicide; Case Status: Awaiting data/Data in Review; Active Ingredient (AI): 6-Methyl-2,3-quinoxalinedithiol cyclic s,s-dithiocarbonate; Data Call-in (DCI) Date: 2/21/91; AI Status: Supported. [R7].

REFERENCES

1. Hartley, D., and H. Kidd (eds.). *The Agrochemicals Handbook*. 2nd ed. Lechworth, Herts, England: The Royal Society of Chemistry, 1987.

2. *Farm Chemicals Handbook 1981*. Willoughby, OH: Meister, 1981, p. C–228.

3. USEPA. *Status of Pesticides in Reregistration and Special Review*. Washington, DC: USEPA–738–R–94–008 (1994).

4. Tomlin, C. *The Pesticide Manual*. 10th ed. British Crop Protection Council: The Royal Society of Chemistry p. 163 (1994).

5. Williams, P. P. *Res Rev* 66: 63–135 (1977).

6. (1) Wauchope, R. D., et al. *Rev Environ Contam Toxicol* 123: 1–36 (1991). (2) Meylan, W. M., P. H. Howard. *J Pharm Sci* 84: 83–92 (1995). (3) Lyman, W. J., et al., *Handbook of Chemical Property Estimation Methods*. Washington, DC: Amer Chemical Soc, p. 5–4 (1990).

7. SRI. *1994 Directory of Chemical Producers—United States of America*. Menlo Park, CA: SRI International, 1994. 801.

■ PARALDEHYDE

CAS # 123–63–7

DOT # 1264

SYNONYMS 1,3,5-trimethyl-2,4,6-trioxane • 1,3,5-trioxane, 2,4,6-trimethyl • 2,4,6-trimethyl-1,3,5-trioxaan (Dutch) • 2,4,6-trimethyl-1,3,5-trioxane • 2,4,6-trimethyl-s-trioxane • 2,4,6-trimetil-1,3,5-triossano (Italian) • acetaldehyde, trimer • elaldehyde • p-acetaldehyde • paraacetaldehyde • paracetaldehyde • paral • paraldehyde-draught (BPC1973) • paraldehyde-enema (BPC1973)

MF $C_6H_{12}O_3$

MW 132.18

COLOR AND FORM Colorless, transparent liquid.

ODOR Characteristic aromatic odor; agreeable odor.

ODOR THRESHOLD 0.02–0.025 mg/m³ (recognition) [R8].

TASTE Disagreeable taste; burning taste.

DENSITY 0.9923 at 20°C.

SURFACE TENSION 27.82 dynes/cm.

VISCOSITY 1.128 cP at 70°F.

BOILING POINT 124.5°C.

MELTING POINT 12.6°C.

HEAT OF COMBUSTION −5,652 cal/g (236×10^5 J/kg) at 25°C.

HEAT OF VAPORIZATION 75 cal/g (latent heat of vaporization).

AUTOIGNITION TEMPERATURE 460°F [R12, p. 49–132].

EXPLOSIVE LIMITS AND POTENTIAL Explosion hazard is slight when exposed to heat or flame [R6].

FLAMMABILITY LIMITS Lower limit, 1.3% [R12, p. 49–132].

FIREFIGHTING PROCEDURES Use water spray, dry chemical, "alcohol resistant" foam, or carbon dioxide. Water may be ineffective. Use water spray to keep fire-exposed containers cool. Approach fire from upwind to avoid hazardous vapors and toxic decomposition products [R12, p. 49–132]. If material is on fire or involved in a Fire: Do not extinguish fire unless flow can be stopped. Use water in flooding quantities as fog. Solid streams of water may spread fire. Cool all affected containers with flooding quantities of water. Apply water from as far a distance as possible. Use "alcohol" foam, dry chemical, or carbon dioxide [R7].

TOXIC COMBUSTION PRODUCTS Emits toxic fumes on heating [R6].

SOLUBILITY Soluble in 8 parts water at 25°C; 17 parts boiling water; soluble in water: 1 in 9 parts at 20°C, 1 in 10 parts at 30°C, and 1 in 11.5 parts at 37°C; soluble in sodium chloride solution: 1 in 9.5 parts at 20°C, 1 in 11 parts at 30°C, and 1 in 12.5 parts at 37°C; miscible with alcohol, chloroform, ether, oils; water solubility = 12g/100 g water at 13°C.

VAPOR PRESSURE 25.3 mm Hg at 20°C.

OCTANOL/WATER PARTITION COEFFICIENT Log K_{ow} = 0.67.

CORROSIVITY Has solvent like action upon rubber, polystyrene, and styrene-acrylonitrile copolymer.

STABILITY AND SHELF LIFE On exposure to light and air, it decomposes to acetaldehyde, and is oxidized to acetic acid [R17, 131].

OTHER CHEMICAL/PHYSICAL PROPERTIES Gives acetaldehyde on heating with dilute HCl or on warming with several drops concentrated H_2SO_4; congeals not below 11°C and distills between 120°C and 126°C; specific heat 0.434; bulk density 8.27 lb/gal (20°C); vapor pressure = 11.03 mm Hg at 25 C°; heat capacity = 1.947 J/g-degK at 25 °C; heat of formation from acetaldehyde = −113.0 kJ/mole.

PROTECTIVE EQUIPMENT Personal protective equipment: wear rubber gloves, self-contained breathing apparatus [R9].

PREVENTATIVE MEASURES If material is not on fire and not involved in a Fire: Keep sparks, flames, and other sources of ignition away. Keep material out of water sources and sewers. Build dikes to contain flow as necessary. Attempt to stop leak if without undue personnel hazard. Use water spray to disperse vapors and dilute standing pools of liquid [R7].

DISPOSAL SRP: At the time of review, criteria for land treatment or burial (sanitary landfill) disposal practices are subject to significant revision. Prior to implementing land disposal of waste residue (including waste sludge), consult with environmental regulatory agencies for guidance on acceptable disposal practices.

COMMON USES Substitute for acetaldehyde; manufacture of dyestuff intermediates; solvent for fats, oils, waxes, gums, resins; leather, mixtures for cellulose derivatives [R18, 874]; manufacture of organic compounds; sedative, hypnotic agent for therapeutic use [R1]; industrially it has been used as solvent and also as rubber activator and antioxidant [R2]; Used in chemical synthesis as a source of acetaldehyde whereby resin formation and other secondary reactions are largely eliminated—for instance, production of pyridines and chlorination of chloral [R3]. Used rectally in status epilepticus, primarily in children, when other agents are not effective [R4]. Effective in preventing the development of alcohol withdrawal symptoms or suppressing the syndrome once it develops. Reduces anxiety associated with withdrawal of other drugs such as opiates or barbiturates [R20, 1532]. Used to control seizures arising from tetanus or eclampsia, or induced by poisons, as well as in the treatment of status epilepticus [R20, 1532]. Used to relieve intractable pain not responding to opiates, but it does not produce effective analgesia in subanesthetic doses, and may produce excitement or delirium in the presence of pain. [R20, 1532].

OCCUPATIONAL RECOMMENDATIONS None listed.

NON-HUMAN TOXICITY LD$_{50}$ dog oral (acute) 3.0 to 4.0 g/kg [R8].

BIOLOGICAL HALF-LIFE The biological half-life of paraldehyde (in humans) has been reported to be 3.5 to 9.5 hours with an average of about 7.5 hours. [R22, 1452].

BIODEGRADATION Japanese MITI, initial concentration 100 ppm, 14 days <30% BODT, activated-sludge inoculum (1). A bacterium isolated from sewage, C. paraldehydium KY 4359, was found to degrade paraldehyde to acetaldehyde and acetic acid (2) [R10].

BIOCONCENTRATION Based on a water solubility of 125,000 mg/L at 25°C and a log K_{ow} of 0.67, bioconcentration factors (BCF) of 0.8–2 were estimated for paraldehyde (1,2,3,SRC). These BCF values suggest that paraldehyde would not bioaccumulate significantly in aquatic organisms (SRC) [R11].

ATMOSPHERIC STANDARDS This action promulgates standards of performance for equipment leaks of volatile organic compounds (VOC) in the synthetic organic chemical manufacturing industry (SOCMI). These standards implement Section 111 of the Clean Air Act and are based on the Administrator's determination that emissions from the SOCMI cause, or contribute significantly to, air pollution that may reasonably be anticipated to endanger public health or welfare. The intended effect of these standards is to require all newly constructed, modified, and reconstructed SOCMI process units to use the best demonstrated system of continuous emission reduction for equipment leaks of VOC, considering costs, non-air-quality health, and environmental impact and energy requirements. Paraldehyde is produced, as an intermediate or final product, by process units covered under this subpart. These standards of performance become effective upon promulgation but apply to affected facilities for which construction or modification commenced after January 5, 1981 [R12].

CERCLA REPORTABLE QUANTITIES Persons in charge of vessels or facilities are required to notify the National Response Center (NRC) immediately, when there is a release of this designated hazardous substance, in an amount equal to or greater than its reportable quantity of 1,000 lb, or 454 kg. The toll-free number of the NRC is (800) 424–8802; in the Washington DC metropolitan area (202) 426–2675. The rule for determining when notification is required is stated in 40 CFR 302.4 (section IV. D. 3.b) [R13].

RCRA REQUIREMENTS U182; as stipulated in 40 CFR 261.33, when paraldehyde, as a commercial chemical product or manufacturing chemical intermediate or an off-specification commercial chemical product or a manufacturing chemical intermediate, becomes a waste, it must be managed according to federal or state hazardous waste regulations. Also defined as a hazardous waste is any residue, contaminated soil, water, or other debris resulting from the cleanup of a spill, into water or on dry land, of this waste. Generators of small quantities of this waste may qualify for partial exclusion from hazardous waste regulations (40 CFR 261.5) [R14].

DOT *Fire or Explosion:* Flammable/combustible material; may be ignited by heat, sparks, or flames. Vapors may travel to a source of ignition, and flash back. Container may explode in heat of fire. Vapor explosion hazard indoors, outdoors, or in sewers. Runoff to sewer may create fire or explosion hazard [R5].

Health Hazards: May be poisonous if inhaled or absorbed through skin. Vapors may cause dizziness or suffocation. Contact may irritate or burn skin and eyes. Fire may produce irritating or poisonous gases. Runoff from fire control or dilution water may give off poisonous gases and cause water pollution [R5].

Emergency Action: Keep unnecessary people away; isolate hazard area and deny entry. Stay upwind; keep out of low areas. Positive-pressure self-contained breathing apparatus (SCBA) and structural firefighters' protective clothing will provide limited protection. Isolate for 1/2 mile in all directions if tank, rail car, or tank truck is involved in fire. Call emergency response telephone number on shipping paper first. If shipping paper not available or no answer, call CHEMTREC at 1–800–424–

9300. If water pollution occurs, notify the appropriate authorities [R5].

Fire: Small fires: Dry chemical, CO_2, water spray, or alcohol-resistant foam. Large Fires: Water spray, fog, or alcohol-resistant foam. Move container from fire area if you can do so without risk. Apply cooling water to sides of containers that are exposed to flames until well after fire is out. Stay away from ends of tanks. For massive fire in cargo area, use unmanned hose holder or monitor nozzles; if this is impossible, withdraw from area, and let fire burn. Withdraw immediately in case of rising sound from venting safety device or any discoloration of tank due to fire [R5].

Spill or Leak: Shut off ignition sources; no flares, smoking, or flames in hazard area. Stop leak if you can do so without risk. Water spray may reduce vapor; but it may not prevent ignition in closed spaces. Small Spills: Take up with sand or other noncombustible absorbent material and place into containers for later disposal. Large Spills: Dike far ahead of liquid spill for later disposal [R5].

First Aid: Move victim to fresh air and call for emergency medical care; if not breathing, give artificial respiration; if breathing is difficult, give oxygen. In case of contact with material, immediately flush eyes with running water for at least 15 minutes. Wash skin with soap and water. Remove and isolate contaminated clothing and shoes at the site [R5].

FIRE POTENTIAL Dangerous, when exposed to heat, flame, or oxidizers [R6]. Flammable liquid. [R12, p. 49–132].

FDA Paraldehyde is subject to control under the Federal Control Substance Act of 1970 as a schedule IV (C–IV) drug [R19, 1172]. Paraldehyde is an indirect food additive for use as a component of adhesives [R15]. Derivatives designated as a habit forming drug [R16].

GENERAL RESPONSE Stop discharge if possible. Keep people away. Shut off ignition sources and call fire department. Avoid contact with liquid. Wear goggles and self-contained breathing apparatus.

Isolate and remove discharged material. Notify local health and pollution control agencies.

FIRE RESPONSE Flammable flashback along vapor trail may occur. Vapor may explode if ignited in an enclosed area. Poisonous gases are produced in fire. Wear goggles and self-contained breathing apparatus. Extinguish with alcohol foam, carbon dioxide, or dry chemical. Water may be ineffective on fire.

EXPOSURE RESPONSE Call for medical aid. Vapor harmful if inhaled. Move to fresh air. If breathing has stopped, give artificial respiration. If breathing is difficult, give oxygen. Liquid, if swallowed, will cause headache, incoordination, drowsiness, or coma. Irritating to eyes and skin. Remove contaminated clothing and shoes. Flush affected areas with plenty of water. If swallowed and victim is conscious, have victim drink water or milk.

RESPONSE TO DISCHARGE Issue warning—high flammability. Evacuate area. Provide chemical and physical treatment. Disperse and flush.

WATER RESPONSE Effects of low concentrations on aquatic life are unknown. May be dangerous if it enters water intakes. Notify local health and wildlife officials. Notify operators of nearby water intakes.

EXPOSURE SYMPTOMS Inhalation and ingestion: irritation, headache, bronchitis, pulmonary edema. Irritating to digestive tract. Hypnotic, and analgesic properties. Incoordination and drowsiness, followed by sleep. Larger doses—coma, weak pulse, and shallow respiration, cyanosis, death from respiratory paralysis. Eyes: irritation, can cause serious injury. Skin: dermatitis (skin inflammation).

EXPOSURE TREATMENT Call a doctor. Inhalation: remove from exposure, give artificial respiration, or oxygen if indicated. Eyes: irrigate with water for 15 minutes. Skin: wash contaminated area with soap and water. Ingestion: gastric lavage, saline catharsis.

REFERENCES

1. Budavari, S. (ed.). *The Merck Index—Encyclopedia of Chemicals, Drugs, and Biologicals.* Rahway, NJ: Merck and Co., Inc., 1989. 1112.

2. International Labour Office. *Encyclopedia of Occupational Health and Safety.* Vols. I and II. Geneva, Switzerland: International Labour Office, 1983. 36.

3. Gerhartz, W. (exec ed.). *Ullmann's Encyclopedia of Industrial Chemistry.* 5th ed. Vol A1: Deerfield Beach, FL: VCH Publishers, 1985 to present, p. VA1 41.

4. American Medical Association, Council on Drugs. *AMA Drug Evaluations Annual 1994.* Chicago, IL: American Medical Association, 1994. 377.

5. U.S. Department of Transportation. *Emergency Response Guidebook 1993.* DOT P 5800.6. Washington, DC: U.S. Government Printing Office, 1993, p. G–26.

6. Sax, N. I. *Dangerous Properties of Industrial Materials,.* 6th ed. New York, NY: Van Nostrand Reinhold, 1984. 2117.

7. Association of American Railroads. *Emergency Handling of Hazardous Materials in Surface Transportation.* Washington, DC: Association of American Railroads, Bureau of Explosives, 1992. 726.

8. Verschueren, K. *Handbook of Environmental Data of Organic Chemicals.* 2nd ed. New York, NY: Van Nostrand Reinhold Co., 1983. 942.

9. U.S. Coast Guard, Department of Transportation. *CHRIS—Hazardous Chemical Data.* Volume II. Washington, DC: U.S. Government Printing Office, 1984–5.

10. (1) Kawasaki, M., *Ecotox Environ Safety* 4: 444–54 (1980). (2) Takayama, K., et al. *Hakko Kogaku Zasshi* 48: 669–75 (1970).

11. (1) Windholz, M. *The Merck Index* 10th ed Rahway, NJ: Merck and Co. p. 1008 (1983). (2) Hansch, C., and A. J. Leo. *Medchem Project* Issue No. 26, Claremont, CA: Pomona College (1985). (3) Lyman, W. J., et al., *Handbook of Chemical Property Estimation Methods.* New York: McGraw-Hill, p. 5–5 (1982).

12. 40 CFR 60.489 (7/1/92).

13. 40 CFR 302.4 (7/1/92).

14. 40 CFR 261.33 (7/1/92).

15. 21 CFR 175.105 (4/1/93).

16. 21 CFR 329.1 (4/1/86).

17. Goodman, L.S., and A. Gilman. (eds.) *The Pharmacological Basis of Therapeutics.* 5th ed. New York: Macmillan Publishing Co., Inc., 1975.

18. Sax, N.I., and R.J. Lewis, Sr. (eds.). *Hawley's Condensed Chemical Dictionary.* 11th ed. New York: Van Nostrand Reinhold Co., 1987.

19. ASHP. *American Hospital Formulary Service—Drug Information 87.* Bethesda, MD: American Society of Hospital Pharmacists, 1987 (plus supplements, 1987).

20. McEvoy, G.K. (ed.). *American Hospital Formulary Service–Drug Information 94.* Bethesda, MD: American Society of Hospital Pharmacists, Inc. 1994 (plus supplements).

21. National Fire Protection Guide. *Fire Protection Guide on Hazardous Materials.* 10th ed. Quincy, MA: National Fire Protection Association, 1991.

22. McEvoy, G.K. (ed.). *American Hospital Formulary Service—Drug Information 93.* Bethesda, MD: American Society of Hospital Pharmacists, Inc., 1993 (plus supplements, 1993).

■ PARAOXON

CAS # 311–45–5

SYNONYMS diethyl 4-nitrophenyl phosphate • diethyl p-nitrophenyl phosphate • diethyl paraoxon • diethyl-p-nitrofenyl ester kyseliny fosforecne (Czech) • ent-16,087 • ethyl-paraoxon • o, o'-Diethyl-p-nitrophenylphosphat (German) • o,o-diethyl o-p-nitrophenyl phosphate • o,o-diethyl phosphoric acid o-p-nitrophenyl ester • o,o-dietyl-o-p-nitrofenylfosfat (Czech) • oxyparathion • p-nitrophenyl diethyl phosphate • p-nitrophenyl diethylphosphate • phenol, p-nitro-, ester with diethyl phosphate • phosphoric acid, diethyl 4-nitrophenyl ester • phosphoric acid, diethyl p-nitrophenyl ester • para-oxon • mintacol • chinorto • ester-25 • eticol • fosfakol • hc-2072 • mintaco • mintisal • miotisal-a • paraoxone • paroxan • pestox-101 • phosphachole • phosphacol • phosphakol • soluglaucit • ts-219 • e-600 (pesticide) • e-600

MF $C_{10}H_{14}NO_6P$

MW 275.22

COLOR AND FORM Oily liquid; reddish yellow oil.

ODOR Slight odor.

DENSITY 1.2683 at 20°C.

BOILING POINT 169–170°C at 1.0 mm Hg.

FIREFIGHTING PROCEDURES Use water spray, dry chemical, foam, or carbon dioxide. Use water to keep fire-exposed containers cool. If a leak or spill has not ignited, use water spray to disperse vapors, and to provide protection for firefighters. Water spray may be used to flush spills away from exposures. In advanced or massive fires, firefighting should be done from safe distance or from protected location [R2].

SOLUBILITY 2,400 μg/ml in water at 25°C; freely soluble in ether and other organic solvents.

OCTANOL/WATER PARTITION COEFFICIENT Log K_{ow} = 1.59.

OTHER CHEMICAL/PHYSICAL PROPERTIES Aqueous solution stable up to pH 7; roughly 300 times more stable to hydrolysis than tetraethyl pyrophosphate. Decomposes rapidly in alkaline solutions.

DISPOSAL Generators of waste (equal to or greater than 100 kg/mo) containing this contaminant, EPA hazardous waste number P041, must conform with USEPA regulations in storage, transportation, treatment, and disposal of waste [R3].

COMMON USES Medication: miotics (SRP: formerly) used for treating glaucoma include paraoxon [R1].

OCCUPATIONAL RECOMMENDATIONS None listed.

BIOLOGICAL HALF-LIFE Hydrolysis and co-valent binding to nonessential esterases are two biochemical processes that can prevent paraoxon from reacting with the essential enzyme acetylcholinesterase. Both processes have been proposed as the primary route of paraoxon detoxification in vivo. These experiments were designed to assess the relative contribution of each pathway to the disappearance of paraoxon in the rabbit. In vitro, paraoxon disappeared from whole rabbit blood with a half-life of 17.7 sec. Hydrolysis by paraoxonase accounted entirely for this disappearance, and covalent binding contributed essentially nothing. In vivo, following an iv injection of 0.15 mg/kg paraoxon, serum paraoxonase hydrolyzed as much as 41% of the injected dose within the first 30 sec. Pretreatment of rabbits with an ip injection of tri-o-tolyl phosphate eliminated more than 95% of the paraoxon binding sites. However, pretreatment with tri-o-tolyl phosphate had no significant effect on the half-life or volume of distribution of paraoxon, indicating that covalent binding sites did not contribute significantly to the clearance of paraoxon from whole rabbits under these conditions. Hydrolysis of paraoxon by tissue paraoxonases, in addition to that catalyzed by paraoxonase in the blood, could account for its rapid metabolism. These findings demonstrate that paraoxonase has a major role in the disappearance of paraoxon in the rabbit. This suggests that susceptibility of people to chronic paraoxon poisoning may vary according to their inherited level and type of serum paraoxonase [R4].

BIODEGRADATION In a study of 32 California orange groves chosen at random, it was found that paraoxon residues disappeared with a half-life of 5.55 days. [R7, 383].

CERCLA REPORTABLE QUANTITIES Persons in charge of vessels or facilities are required to notify the National Response Center (NRC) immediately, when there is a release of this designated hazardous substance, in an amount equal to or greater than its reportable quantity of 100 lb or 45.4 kg. The toll-free number of the NRC is (800) 424–8802; in the Washington DC metropolitan area (202) 426–2675. The rule for determining when notification is required is stated in 40 CFR 302.4 (section IV. D. 3.b) [R5].

RCRA REQUIREMENTS As stipulated in 40 CFR 261.33, when paraoxon, as a commercial chemical product or manufacturing chemical intermediate or an off-specification commercial chemical product or a manufacturing chemical intermediate, becomes a waste, it must be managed according to federal or state hazardous waste regulations. Also defined as a hazardous waste is any container or inner liner used to hold this waste or any residue, contaminated soil, water, or other debris resulting from the cleanup of a spill, into water or on dry land, of this waste. Generators of small quantities of this waste may qualify for partial exclusion from hazardous waste regulations (40 CFR 261.5 (e)) [R6].

REFERENCES

1. Grant, W. M. *Toxicology of the Eye.* 3rd ed. Springfield, IL: Charles C. Thomas Publisher, 1986. 678.

2. National Fire Protection Association. *Fire Protection Guide on Hazardous Materials.* 9th ed. Boston, MA: National Fire Protection Association, 1986, p. 49–71.

3. 40 CFR 240–280, 300–306, 702–799 (7/1/89).

4. Butler, E. G., et al., *Drug Metab Dispos* 13 (6): 640–4 (1985).

5. 40 CFR 302.4 (7/1/88).

6. 40 CFR 261.33 (7/1/88).

7. Hayes, Wayland J., Jr. *Pesticides Studied in Man.* Baltimore/London: Williams and Wilkins, 1982.

■ PARAQUAT

CAS # 4685–14–7

SYNONYMS 1,1'-dimethyl-4,4'-

bipyridinium • 1,1'-dimethyl-4,4'-bipyridinium cation • 4,4'-bipyridinium, 1,1'-dimethyl • N,N'-dimethyl-4,4'-bipyridinium • N,N'-dimethyl-4,4'-bipyridinium dication • paraquat-dication • paraquat-ion • dextrone • gramoxone • weedol • esgram • 1,1'-dimethyl-4,4'-bipyridyldiylium • dextrone-x • N,N'-dimethyl-gamma, gamma'-dipyridylium • methyl-viologen (2) • 1,1'-dimethyl-4,4'-bipyridinium salt

MF $C_{12}H_{14}N_2$

MW 186

COLOR AND FORM Colorless solid; yellow solid.

ODOR Odorless.

DENSITY 1.24 at 20°C.

BOILING POINT (760 mm Hg): Decomposes at 175–180°C (347–356°F).

EXPLOSIVE LIMITS AND POTENTIAL Not combustible [R22, 146].

FIREFIGHTING PROCEDURES Respirator selection for firefighting: self-contained breathing apparatus with a full facepiece operated in pressure-demand or other positive-pressure mode [R4].

SOLUBILITY Practically insoluble in organic solvents, soluble in water.

VAPOR PRESSURE Approximately 0 mm at 20°C.

OCTANOL/WATER PARTITION COEFFICIENT Log K_{ow} = −4.22 at pH 7.4.

CORROSIVITY Paraquat is corrosive to metals.

STABILITY AND SHELF LIFE Decomposition in presence of UV light [R18, 2753].

OTHER CHEMICAL/PHYSICAL PROPERTIES Conversion factors: 1 mg/kg = 131 ppm; 1 ppm = 7.61 mg/m³. Conversion factors: 1 mg/kg = 95 ppm; 1 ppm = 10.51 mg/m³. MP: 75–180°C (paraquat dichloride). Faint ammoniacal odor; hygroscopic; BP: decomposition at high temp; specific gravity: 1.24–1.26 at 20°C (paraquat dichloride). Vapor pressure of salts is very low, below 1×10^{-7} mm Hg (paraquat salts). White crystalline solid (paraquat dichloride and di(methyl sulfate) salts). Dark-red solution (technical paraquat). Yellow solid. Freely soluble in water; insoluble in hydrocarbons; sparingly soluble in lower alcohols (paraquat dichloride). Inactivated by inert clays and anionic surfactants; corrosive to metal; nonvolatile. Colorless crystals; mp: 300°C (decomposes); hydrolyzed by alkali (paraquat dichloride). Henry's Law constant = $<1 \times 10^{-9}$ atm–m³/mole.

PROTECTIVE EQUIPMENT Employees should be provided with and required to use impervious clothing, gloves, face shields (eight-inch minimum), and other appropriate protective clothing necessary to prevent skin contact with paraquat or solutions containing paraquat. Employees should be provided with and required to use dust– and splash-proof safety goggles where paraquat or solutions containing paraquat may contact the eyes [R4, 3]. Wear appropriate equipment to prevent reasonable probability of skin contact (paraquat chloride) [R5]. Wear eye protection to prevent reasonable probability of eye contact (paraquat chloride) [R5]. The following equipment should be available: Quick drench (paraquat chloride) [R5]. Recommendations for respirator selection. Max concentration for use: 1 mg/m³. Respirator Classes: Any chemical-cartridge respirator with organic vapor cartridge(s) in combination with a dust, mist, and fume filter. Any supplied-air respirator. Any self-contained breathing apparatus. May require eye protection (paraquat chloride) [R5]. Recommendations for respirator selection. Max concentration for use: 1.5 mg/m³. Respirator Classes: Any chemical-cartridge respirator with a full facepiece and organic vapor cartridge(s) in combination with a dust, mist, and fume filter. Any supplied-air respirator operated in a continuous-flow mode. Any self-contained breathing apparatus with a full facepiece. Any supplied-air respirator with a full facepiece. Any powered, air-purifying respirator with organic vapor cartridge(s) in combination with a dust, mist, and fume filter. May require eye protection (paraquat chloride) [R5]. Recommendations for respirator selection. Condition: Emergency or planned entry into unknown concentration or idlh conditions: Respirator Classes: Any self-contained breathing apparatus that has a full facepiece and is operated in a pressure-demand or other positive-pressure mode. Any supplied-air respirator with a full facepiece and operated in pressure-demand or other positive-pressure mode in combination with an auxiliary self-contained breathing apparatus operated in pressure-demand or other positive-pressure mode (paraquat chloride) [R5]. Recommendations for respirator selection. Condition: Escape from suddenly occurring respiratory hazards: Respirator Classes: Any air-purifying, full facepiece respirator (gas mask) with a chin-style, front- or back-mounted organic vapor canister having a high-efficiency particulate filter. Any appropriate escape-type, self-contained breathing apparatus (paraquat chloride) [R5].

PREVENTATIVE MEASURES Clothing contaminated with paraquat should be placed in closed containers for storage until it can be discarded or until provision is made for the removal of paraquat from the clothing. Eating and smoking should not be permitted in area where paraquat or solutions containing paraquat are handled, processed, or stored. Employees who handle paraquat or solutions containing paraquat should wash their hands thoroughly before eating, smoking, or using toilet facilities [R4, 3]. Contact lenses should not be worn when working with this chemical [R16]. Workers should wash immediately when skin becomes contaminated (paraquat chloride) [R5]. Remove clothing immediately if it is non-impervious clothing that becomes contaminated (paraquat chloride) [R5]. SRP: The scientific literature supports the wearing of contact lenses in industrial environments, as part of a program to protect the eye against chemical compounds and minerals causing eye irritation. However, there may be individual substances whose irritating or corrosive properties are such that the wearing of contact lenses would be harmful to the eye. In those specific cases contact lenses should not be worn (paraquat chloride). Contact lenses should not

be worn when working with this chemical (paraquat chloride) [R5].

CLEANUP If paraquat is spilled, ventilate area of spill. Collect spilled material inthe most convenient and ,safe manner and deposit in sealed containers for reclamation. Liquid containing paraquat should be absorbed in vermiculite, dry sand, earth, or a similar material. [R4, 4].

DISPOSAL If incineration facilities are unavailable for a particular organic or metallo-organic pesticide (except organic mercury, lead, cadmium, and arsenic compounds), additional disposal methods include soil injection, chemical degradation, burial (in a designated landfill), or well injection. However, persons considering the chemical degradation method should contact EPA's Regional Administrator prior to attempting disposal, while the well-injection method should be considered only after all reasonable alternative measures have been explored and found to be less satisfactory in terms of environmental protection. If the above approved disposal methods are unavailable, temporary storage of organic, and metallo-organic pesticides (except, organic mercury, lead, cadmium, and arsenic compounds) may be undertaken (organic, or metallo-organic pesticides) [R6]. Group I Containers: Combustible containers from organic or metallo-organic pesticides (except organic mercury, lead, cadmium, or arsenic compounds) should be disposed of in pesticide incinerators or in specified landfill sites (organic or metallo-organic pesticides) [R17]. Group II Containers: Non-combustible containers from organic or metallo-organic pesticides (except organic mercury, lead, cadmium, or arsenic compounds) must first be triple-rinsed. Containers that are in good condition may be returned to the manufacturer or formulator of the pesticide product, or to a drum reconditioner for reuse with the same type of pesticide product, if such reuse is legal under Department of Transportation regulations (e.g., 49 CFR 173.28). Containers that are not to be reused should be punctured and transported to a scrap metal facility for recycling, disposal, or burial in a designated landfill (organic or metallo-organic pesticides) [R6].

COMMON USES For stubble cleaning (140–840 g ai/hectare); pasture renovation (140–2,210 g/hectare), the killing of unproductive grass in such a way that the ground can be resown without ploughing [R23, 630]. Contact herbicide for weed control in grass seed crops, orchard; no-tillage corn use [R1]. Crop desiccant and defoliant on cotton and potato vines; harvest aid for soybeans [R1]. Paraquat is a total contact herbicide used to control broad-leaved and grassy weeds [R2]. Paraquat is a quick-acting herbicide that destroys green plant tissue by contact action and some translocation. It is used as a plant desiccant for preharvest of cotton and potatoes [R24, 1357].

OCCUPATIONAL RECOMMENDATIONS

OSHA 8-hr time-weighted average: 0.5 mg/m^3. Transitional limits must continue to be achieved by any combination of engineering controls, work practices, and personal protective equipment during the phase-in period, Sept 1, 1989 through Dec 30, 1993. Final rule limits became effective Dec 31, 1993 (paraquat, respirable dust) [R15].

NIOSH 10–hr time-weighted average: 1.5 mg/m^3 (paraquat chloride) [R5].

THRESHOLD LIMIT 8–hr time-weighted average (TWA) 0.1 mg/m^3 (1978) (paraquat, respirable fraction) [R18, 28]. 8–hr time weighted average (TWA) 0.5 mg/m^3 (1978) (paraquat, total dust) [R18, 28]. Excursion Limit Recommendation: Excursions in worker exposure levels may exceed three times the TLV–TWA for no more than a total of 30 min during a work day, and under no circumstances should they exceed five times the TLV–TWA. (paraquat, respirable fraction; total dust) [R18, 5].

NON-HUMAN TOXICITY LD_{50} turkey, oral 250–280 mg/kg [R8]; LD_{50} cow oral 50–75 mg/kg [R9]; LD_{50} sheep oral 50–75 mg/kg [R9]; LD_{50} rat male dermal 80 mg/kg [R24, 1358]; LD_{50} rat female dermal 90 mg/kg [R24, 1358]; LD_{50} rat female ip 19 mg/kg [R24, 1358]; LD_{50} mouse, oral 98 mg/kg [R24, 1358]; LD_{50} guinea pig, oral 22 mg/kg [R24, 1358]; LD_{50} guinea pig female ip 3 mg/kg [R24, 1358]; LD_{50} rabbit dermal 236 mg/kg [R24, 1358]; LD_{50} cat female, oral 35 mg/kg [R24, 1358]; LD_{50} monkey, oral 50 mg/kg [R24, 1358]; LD_{50} chicken female, oral 262 mg/kg [R24, 1358]; LD_{50} rat, oral 155–203 mg/kg (paraquat dichloride) [R10]; LD_{50} rabbit dermal >663 mg/kg (paraquat dichloride) [R10].

ECOTOXICITY LC_{50} *Selenastrum capricornutum* 1.8 mg/L (free cell culture growth inhibition) [R7]; LC_{50} *Selenastrum capricornutum* 7.8 mg/L (immobilized cell culture growth inhibition) [R11]; LC_{50} *Ictalurus punctatus* (channel catfish) >100 mg/L/96hr at 18°C, wt 1.4 g in a static bioassay. Static bioassay without aeration, pH 7.2–7.5, water hardness 40–50 mg/L as calcium carbonate, and alkalinity of 30–35 mg/L (technical material 42% (cation)) [R12]. EC_{50} *Simocephalus serrulatus* 3.7 mg/L/48 hr 1st instar at 16°C (95% confidence interval 2.8–4.8 mg/L). Static bioassay without aeration, pH 7.2–7.5, water hardness 40–50 mg/L as calcium carbonate, and alkalinity of 30–35 mg/L. Soluble concentrate, 2 lb (cation)/gal (paraquat dichloride) [R12]. EC_{50} *Daphnia pulex* 4.0 mg/L/48 hr 1st instar at 16°C (95% confidence interval 2.7–6.0 mg/L). Static bioassay without aeration, pH 7.2–7.5, water hardness 40–50 mg/L as calcium carbonate, and alkalinity of 30–35 mg/L. Soluble concentrate, 2 lb (cation)/gal (paraquat dichloride) [R12]. LC_{50} *Gammarus fasciatus* 11 mg/L/96 hr mature at 21°C (95% confidence interval 8.1–15 mg/L). Static bioassay without aeration, pH 7.2–7.5, water hardness 40–50 mg/L as calcium carbonate, and alkalinity of 30–35 mg/L. Soluble concentrate, 2 lb (cation)/(gal (paraquat dichloride) [R12]. LC_{50} *Pteronarcys* >100 mg) L/96 hr second year class at 16°C. Static bioassay without aeration, pH 7.2–7.5, water hardness 40–50 mg/L as calcium carbonate, and alkalinity of 30–35 mg/L. Soluble concentrate, 2 lb (cation)/gal (paraquat dichloride) [R12]. LC_{50} *Salmo gairdneri* (Rainbow trout) 15 mg/L/96 hr at 13°C (95% confidence interval 11–19 mg/L), wt 0.5 g. Static bioassay without aeration, pH 7.2–7.5, water hardness 40–50 mg/L as calcium carbonate, and alkalinity of 30–35 mg/L. Soluble concentrate, 2 lb (cation)/gal (paraquat dichloride) [R12]. LC_{50} *Lepomis macrochirus* (bluegill) 13.0 mg/L/96 hr at 24°C (95% confidence interval 8.5–19.0 mg/L), wt 0.9 g. Static bioassay without aeration, pH 7.2–7.5, water hardness 40–50 mg/L as calcium carbonate, and alkalinity of 30–35 mg/L. Soluble concentrate, 2 lb (cation)/gal (paraquat dichloride) [R12]. LC_{50} *Poecilia mexicana* 12.53 mg/L/24 hr, static bioassay [R10]. LD_{50} *Anas platyrhynchos* (mallard) oral 199 mg/kg (95% confidence limit 144–276 mg/kg) [R25, 61]. LD_{50} *Anas platyrhynchos* (mallard) percutaneous 600 mg/kg (95% confidence limit 424–848 mg/kg) [R25, 61].

BIOLOGICAL HALF-LIFE Half-life for absorption of radioactivity from lung of hamster admin dichloride salt intratracheally was 21 hr. [R26, (1978)].

BIODEGRADATION Degradation of dichloride salt in soil by reduction ($NaBH_4$ or powd Zn–Ac OH treatments) or alkaline hydrolysis (10% NaOH) studied. All treatments were effective in decr paraquat content of simulated spills. Sodium borohydride was particularly effective [R13].

BIOCONCENTRATION Bioconcentration factor predicted from water solubility = 0.3 (calculated); predicted from soil adsorption coefficient = 1,600 (calculated) (from table) [R14].

FIFRA REQUIREMENTS Classified for restricted use, limited to use by or under the direct supervision of a certified applicator. Formulation: All formulations and concentrations except those listed below. Use pattern: All uses. Classification: Restricted. Criteria influencing restriction: Other hazards; use and accident history, human toxicological data (paraquat (dichloride), and paraquat bis(methyl sulfate)) [R19]. Classified for restricted use, limited to use by or under the direct supervision of a certified applicator. Formulation: Pressurized spray formulations containing 0.44% paraquat bis(methyl sulfate) and 15% petroleum distillates as active ingredients. Use pattern: Spot weed and grass control. Classification: Restricted (paraquat (di-

chloride) and paraquat bis(methyl sulfate)) [R19]. Classified for restricted use, limited to use by or under the direct supervision of a certified applicator. Formulation: Liquid fertilizers containing concentrations of 0.025% paraquat dichloride and 0.03% atrazine; 0.03% paraquat dichloride and 37% atrazine, 0.04% paraquat dichloride, and 0.49% atrazine. Use pattern: All uses. Classification: Unclassified paraquat (dichloride) and paraquat bis(methyl sulfate)) [R19]. Tolerances are established for residues of the desiccant, defoliant, and herbicide paraquat (1,1'-dimethyl–4,4'-bipyridinium ion) derived from the application of either the bis(methyl sulfate) or the dichloride salt (both calculated as the cation) in or on the following raw agricultural commodities: acerola; alfalfa; almond hulls; apples; apricots; asparagus; avocados; bananas; barley grain; bean (straw, dry, forage, hay, lima, and snap; beets, sugar, and sugar (tops); birdsfoot trefoil; broccoli; cabbage; carrots; cattle (fat, meat, kidney, and mpyp); cauliflower; cherries; chinese cabbage; citrus fruits; clover; coffee beans; collards; corn (fresh, fodder, forage, and grain); cottonseed; cucurbits; eggs; figs; goats (fat, meat, kidney, and mbyp); grass, (pasture and range); guar beans; guava; hogs (fat, meat, kidney, and mbyp); hops (fresh); hop vines; horses (fat, meat, kidney, and mbyp); kiwifruit; lettuce; milk; mint (hay); nectarines; nuts; oat grain; olives; onions (dry bulb and green); papayas; passion fruit; peaches; peanuts; peanut (hay, hulls, and vines); pears; peas (succulent, forage, and hay; pineapples; pistachio nuts; plums (fresh prunes); potatoes; poultry (fat, meat, and mbyp); rhubarb; rye grain; safflower seed; sheep (fat, meat, kidney, and mbyp); small fruit; sorghum forage; sorghum grain; soybeans; soybean forage; strawberries; sugarcane; sunflower seeds; turnips (roots and tops); vegetables (fruiting); and wheat grain [R20]. Tolerances with regional registration as defined in 180 (n), are established for residues of the pesticide paraquat (1,1'-dimethyl–4,4'-bipyridinium ion) derived from the application of either the bis(methyl sulfate) or the dichloride salt (both calculated as the cation) in or on the following raw agricultural commodities:

cassava, pigeon peas, taniers, tyfon, yams, and taro (corms). [R20]. In 1988, Congress amended FIFRA to strengthen and accelerate EPA's reregistration program. The nine-year reregistration scheme mandated by "FIFRA 88" applies to each registered pesticide product containing an active ingredient initially registered before November 1, 1984. List A consists of the 194 chemical cases (or 350 individual active ingredients) for which EPA had issued Registration Standards prior to the effective date of FIFRA '88. List: A; Case: Paraquat Dichloride; Case No. 0262; Pesticide type: Herbicide; Registration Standard Date: 05/01/88; Case Status: Awaiting Data/Data in Review: OPP awaits data from the pesticide's producer(s) regarding its human health or environmental effects, or OPP has received and is reviewing such data, in order to reach a decision about the pesticide's eligibility for reregistration. Active Ingredient (AI): Paraquat bis(methysulfate); AI Status: The active ingredient is no longer contained in any registered products. Therefore, EPA is characterizing it as "cancelled." Under FIFRA, pesticide producers may voluntarily cancel their registered products. EPA also may cancel pesticide registrations if registrants fail to pay required fees, to make or meet certain reregistration commitments, or when the Agency reaches findings of unreasonable adverse effects (paraquat bis(methylsulfate)) [R21]. In 1988, Congress amended FIFRA to strengthen and accelerate EPA's reregistration program. The nine-year reregistration scheme mandated by "FIFRA 88" applies to each registered pesticide product containing an active ingredient initially registered before November 1, 1984. List A consists of the 194 chemical cases (or 350 individual active ingredients) for which EPA had issued Registration Standards prior to the effective date of FIFRA '88. List: A; Case: Paraquat Dichloride; Case No. 0262; Pesticide type: Herbicide; Registration Standard Date: 05/01/88; Case Status: Awaiting Data/Data in Review: OPP awaits data from the pesticide's producer(s) regarding its human health or environmental effects, or OPP has received and is reviewing

such data, in order to reach a decision about the pesticide's eligibility for reregistration. Active Ingredient (AI): Paraquat dichloride; Data Call-in (DCI) Date: 12/31/91; AI Status: The producer(s) of the pesticide has made commitments to conduct the studies and pay the fees required for reregistration, and is meeting those commitments in a timely manner. (paraquat dichloride) [R21].

FIRE POTENTIAL Nonflammable in aqueous formulations [R3].

REFERENCES

1. SRI.

2. WHO. *Environ Health Criteria: Paraquat and Diquat.* p. 25 (1985).

3. WHO. *Environ Health Criteria: Paraquat and Diquat.* p. 19 (1984).

4. Mackison, F. W., R. S. Stricoff, and L. J. Partridge, Jr. (eds.). *NIOSH/OSHA—Occupational Health Guidelines for Chemical Hazards.* DHHS (NIOSH) Publication No. 81–123 (3 vols). Washington, DC: U.S. Government Printing Office, Jan. 1981, p. 5.

5. NIOSH. *NIOSH Pocket Guide to Chemical Hazards.* DHHS (NIOSH) Publication No. 90–117. Washington, DC: U.S. Government Printing Office, June 1990, 172.

6. 40 CFR 165 (7/1/85).

7. Bozeman, J., et al. *Aquat Toxicol* 14 (4): 345–52 (1989).

8. Smalley, H. E. *Poultry Sci* 52: 1625–29 (1973) as cited in WHO, *Environ Health Criteria: Paraquat and Diquat.* p. 70 (1984).

9. WHO. *Environ Health Criteria: Paraquat and Diquat* p. 71 (1984).

10. Verschueren, K. *Handbook of Environmental Data of Organic Chemicals.* 2nd ed. New York, NY: Van Nostrand Reinhold Co., 1983. 943.

11. Bozeman, J., et al. *Aquat Toxicol* 14 (4): 345–52 (1989).

12. U.S. Department of Interior, Fish and Wildlife Service. *Handbook of Acute Toxicity of Chemicals to Fish and Aquatic*

Invertebrates. Resource Publication No. 137. Washington, DC: U.S. Government Printing Office, 1980. 56.

13. Staiff, et al.; *Bull Environ Contam Toxicol* 26 (1): 18 (1981).

14. Kenaga, E. E., *Ecotoxicology and Environmental Safety* 4: 30 (1980).

15. 29 CFR 1910.1000 (7/1/91).

16. NIOSH. *Pocket Guide to Chemical Hazards.* 2nd Printing. DHHS (NIOSH) Publ. No. 85–114. Washington, DC: U.S. Dept. of Health and Human Services, NIOSH/Supt. of Documents, GPO, February 1987. 185 17. 40 CFR 165 (165 (7/1/85).

18. American Conference of Governmental Industrial Hygienists. *Threshold Limit Values for Chemical Substances and Physical Agents and Biological Exposure Indices for 1994–1995.* Cincinnati, OH: ACGIH, 1994.

19. 40 CFR 152.175 (7/1/91).

20. 40 CFR 180.205 (a) (7/1/91).

21. USEPA/OPP. *Status of Pesticides in Reregistration and Special Review.* p. 106 (Mar. 1992) EPA 700–R–92–004.

22. Mackison, F. W., R. S. Stricoff, L. J. Partridge, Jr. (eds.). *NIOSH/OSHA Pocket Guide to Chemical Hazards.* DHEW (NIOSH) . Publication No. 78–210. Washington, DC: U.S. Government Printing Office, 1980.

23. Worthing, C. R., and S. B. Walker (eds.). *The Pesticide Manual—A World Compendium.* 8th ed. Thornton Heath, UK: The British Crop Protection Council, 1987.

24. Hayes, W. J., Jr., E. R. Laws, Jr. (eds.). *Handbook of Pesticide Toxicology.* Volume 3. *Classes of Pesticides.* New York, NY: Academic Press, Inc., 1991.

25. U.S. Department of the Interior, Fish and Wildlife Service. *Handbook of Toxicity of Pesticides to Wildlife.* Resource Publication 153. Washington, DC: U.S. Government Printing Office, 1984.

26. Abou-donia, M. B., A. A. Komeil. *Toxicol Appl Pharmacol* 45 (1): 280.

■ PENTACHLOROPHENOL, SODIUM SALT

CAS # 131–52–2

SYNONYMS pcp sodium salt • pcp sodium • pentachlorophenate sodium • pentachlorophenol sodium salt • pentachlorophenoxy-sodium • pentaphenate • phenol, pentachloro-, sodium salt • sodium pcp • sodium pentachlorophenol • sodium pentachlorophenolate • sodium pentachlorophenoxide • sodium pentachlorphenate • sodium, (pentachlorophenoxy)- • sodium pentachlorophenate • ai3-16418 • Caswell No. 784 • EPA pesticide chemical code 063003 • PKhFN • dow dormant fungicide • dowicide-g • dowicide g-st • mystox-d • napclor-g • santobrite • sapco-25 • weedbeads • gr 48-11ps • gr 48-32s

MF C_6Cl_5O • Na

MW 288.30

pH Aqueous solution has alkaline reaction.

COLOR AND FORM Buff-colored flakes; white or tan powder.

FIREFIGHTING PROCEDURES Extinguish fire using agent suitable for type of surrounding fire. Material itself does not burn or burns with difficulty. Use "alcohol" foam, carbon dioxide, or dry chemical. Use water spray to absorb vapors. Cool all affected containers with flooding quantities of water. Apply water from as far a distance as possible. Keep run-off water out of sewers and water sources [R6].

SOLUBILITY 21.0% (wt/vol) in water at 5°C; 29.0% (wt/vol) in water at 40°C; 33% in water at 25°C; soluble in ethanol, acetone; insoluble in benzene; insoluble in petroleum oils.

PROTECTIVE EQUIPMENT Following challenge with a 4.2% sodium pentachlorophenate solution, only the Best (natural rubber) glove allowed breakthrough; this only 30 sec after exposure. Neither the Dayton (5 hr), Playtex (7.5 hr), Edmont

(15.5 hr), nor Granet (15.5 hr) gloves had been permeated following completion of testing after the listed duration [R8].

PREVENTATIVE MEASURES Clothing should be removed immediately if a solution has been spilled on it. All clothing worn during one spraying operation should be laundered before reuse. Routine precautions include washing hands, arms, and face with soap and water before eating, drinking, or smoking. At end of day, workman should shower and change into clean clothing [R7].

CLEANUP 1. Ventilate area of spill; 2. Collect spilled material in most convenient and safe manner, deposit in sealed containers for reclamation. Liquid containing pentachlorophenol should be absorbed in vermiculite, dry sand, earth, or similar material (pentachlorophenol) [R9]. Biological treatment is principal secondary treatment method but other methods employed at some wood preservative plants are carbon absorption, membrane filtration, and oxidation by chlorine, hydrogen peroxide, and ozone. The reduction in concentration of pentachlorophenol (PCP) (in biological treatment) is thought to occur by adsorption upon biomass rather than by degradation. (pentachlorophenol) [R22, 400].

DISPOSAL The following wastewater treatment technologies have been investigated for sodium pentachlorophenol: Concentration process: Biological treatment [R10].

COMMON USES Contact and preemergence herbicide [R2]; General disinfectant, e.g., for trays in mushroom houses [R1]; in cooling towers to control growth of microorganisms, particularly algae [R3]; the sodium salt of pentachlorophenol is used as an antifungal and antibacterial, and has applications in the following areas: adhesives-antimicrobial protection of adhesives based on starch, vegetable, and animal proteins during manufacture, storage, and service life; construction materials—control of mold growth on inert surfaces as with pentachlorophenol; leather—prevention of hide deterioration and in the treatment of solutions during tanning; paint—aid in the self preserva-

tion of protein-based latex paints; petroleum—to prevent the growth of bacteria in drilling muds; photographic solutions—control of fungus and slime; pulp and paper—protection and preservation of processing materials, stored pulp, and fiberboard against mildew rot and termites; textiles—protection of finished yarns and cloth against molding during storage; water treatment—control of algal, fungal, and bacterial induced slimes in industrial recirculating water. The sodium salt of pentachlorophenol is utilized primarily in an aqueous solution [R4].

OCCUPATIONAL RECOMMENDATIONS None listed.

NON-HUMAN TOXICITY LD_{50} rat oral 210 mg/kg [R14].

ECOTOXICITY LC_{50} (*Oncorhynchus tshawytscha*) Chinook salmon 68 µg/L/96 hr at 10°C (95% confidence limit 48–95 µg/L), wt 1.0 g. Static bioassay without aeration, pH 7.2–7.5, water hardness 40–50 mg/L as $CaCO_3$, and alkalinity of 30–35 mg/L [R15]. LC_{50} (*Salmo gairdneri*) rainbow trout 55 µg/L/96 hr at 12°C (95% confidence limit 47–64 µg/L), wt 1.0 g. Static bioassay without aeration, pH 7.2–7.5, water hardness 40–50 mg/L as $CaCO_3$, and alkalinity of 30–35 mg/L [R15]. LC_{50} (*Ictalurus punctatus*) Channel catfish 77 µg/L/96 hr at 20°C (95% confidence limit 61–98 µg/l), wt 0.8 g. Static bioassay without aeration, pH 7.2–7.5, water hardness 40–50 mg/L as $CaCO_3$, and alkalinity of 30–35 mg/L [R15]. LC_{50} (*Lepomis microchirus*) bluegill 44 µg/L/96 hr at 15°C (95% confidence limit 25–78 µg/L), wt 0.4 g. Static bioassay without aeration, pH 7.2–7.5, water hardness 40–50 mg/L as $CaCO_3$, and alkalinity of 30–35 mg/L [R15]. LC_{50} (*Palaemonetes pugio*) grass shrimp greater than 515 µg/L/96 hr, flow-through bioassay [R16]. LC_{50} (*Penaeus aztecus*) brown shrimp greater than 195 µg/L/96 hr, flow-through bioassay [R16]; LC_{50} (*Fundulus similis*) killifish greater than 306 µg/L/96 hr, flow-through bioassay [R16]. LC_{50} (*Lagodon rhomboides*) pinfish 53.2 µg/L/96 hr, flow-through bioassay [R16]. LC_{50} (*Mugil cephalus*) stripped mullet 112 µg/L/96 hr, flow-through bioassay [R16]. EC$_{50}$ (shell deposition) (*Crassostrea virginica*) Eastern oyster 76.5 µg/L/192 hr, flow-through bioassay [R16]. EC$_{50}$ Eastern oyster embryos 40 µg/L/48–hr toxic effect: (abnormal development) in a flow-through bioassay [R12]. LC$_{50}$ (*Lymnaea acuminata*) freshwater pulmonate snails 0.19 mg/L/ 96 hr (95% confidence limit 0.161–0.224 mg/L), static bioassay [R17]. LC$_{50}$ *Salmo gairdneri* 10 wk old fry 66 µg/L/96 hr in a flow-through bioassay [R13]. LC$_{50}$ *Cyprinus carpio* (carp) fry 0.028 ppm/24 hr pH 5.5; 0.044 ppm/24 hr at pH 6.5; 0.208 ppm/24 hr at pH 7.5; 0.40 ppm/24 hr at pH 8.5; 3.126 ppm/24 hr at pH 9.5; all values derived from static bioassay conditions [R18].

IARC SUMMARY Classification of carcinogenicity: (1) evidence in humans: limited. Overall summary evaluation of carcinogenic risk to humans is Group 2B: The agent is possibly carcinogenic to humans. (from table) (chlorophenols) [R11].

BIODEGRADATION Sodium pentachlorophenate is not persistent in water, sewage, or soil because of bacterial decomposition. [R23, 963].

BIOCONCENTRATION Rainbow trout exposed to waterborne sodium pentachlorophenate averaging 35 and 660 ng/L for about 115 days accumulated levels of pentachlorophenate that were related to the concentration and duration of exposure. Pentachlorophenate levels in organs (liver and gall bladder), the remaining fish, and the percent of total pentachlorophenate in the organs remained relatively uniform among the control fish over the 115-day period. Fish exposed to 35 ng/L contained slightly higher pentachlorophenate levels with a higher percentage found in the organs. The highest levels of pentachlorophenate occurred in fish exposed to 660 ng/L, where concentration in the organs averaged 2,200 µg/kg after 115 days. Thus, uptake from water is an important pathway for the accumulation of pentachlorophenate by rainbow trout [R19].

DOT *Health Hazards:* Poisonous if swallowed. Inhalation of dust poisonous. Fire may produce irritating or poisonous gases. Runoff from fire control or dilution water may cause pollution [R5].

Fire or Explosion: Some of these materials may burn, but none of them ignites readily [R5].

Emergency Action: Keep unnecessary people away; isolate hazard area and deny entry. Stay upwind; keep out of low areas. Positive-pressure self-contained breathing apparatus (SCBA) and structural firefighters' protective clothing will provide limited protection. Call CHEMTREC at 1–800–424–9300 for emergency assistance. If water pollution occurs, notify the appropriate authorities [R5].

Fire: Small fires: Dry chemical, CO$_2$, water spray, or regular foam. Large Fires: Water spray, fog, or regular foam. Move container from fire area if you can do so without risk [R5].

Spill or Leak: Do not touch or walk through spilled material; stop leak if you can do so without risk. Small Spills: Take up with sand or other noncombustible absorbent material and place into containers for later disposal. Small Dry Spills: With clean shovel place material into clean, dry container, and cover loosely; move containers from spill area. Large Spills: Dike far ahead of liquid spill for later disposal [R5].

First Aid: Move victim to fresh air; call emergency medical care. In case of contact with material, immediately flush skin or eyes with running water for at least 15 minutes. Remove and isolate contaminated clothing and shoes at the site [R5].

FDA Sodium pentachlorophenate for use as a preservative only is an indirect food additive for use only as a component of adhesives [R20]. Substances used in the manufacture of paper and paperboard products used in food packaging shall incl: sodium pentachlorophenate as a slime control agent. Footnote: Under the conditions of normal use these substances would not reasonably be expected to migrate to food, based on available scientific information and data [R21].

REFERENCES

1. Worthing, C. R., and S. B. Walker (eds.). *The Pesticide Manual—A World Compendium*. 8th ed. Thornton Heath, UK: The British Crop Protection Council, 1987. 641.

2. *Farm Chemicals Handbook* 1991. Willoughby, OH: Meister, 1991, p. C–279.

3. Kavaler, A. R. (ed.). *Chemical Marketing Reporter*. New York, NY: Schnell Publishing Co., Inc., 1984, p. V231 (17) 33.

4. *Kirk-Othmer Encyclopedia of Chemical Technology*. 3rd ed., Volumes 1–26. New York, NY: John Wiley and Sons, 1978–1984, p. V5 867 (1979).

5. U.S. Department of Transportation. *Emergency Response Guidebook 1990*. DOT P 5800.5. Washington, DC: U.S. Government Printing Office, 1990, p. G–53.

6. Association of American Railroads. *Emergency Handling of Hazardous Materials in Surface Transportation*. Washington, DC: Assoc. of American Railroads, Hazardous Materials Systems (BOE), 1987. 637.

7. International Labour Office. *Encyclopedia of Occupational Health and Safety*. Vols. I and II. Geneva, Switzerland: International Labour Office, 1983. 1672.

8. Silkowski, J. B., et al., *Am Ind Hyg Assoc J* 45 (8): 501–4 (1984).

9. Mackison, F. W., R. S. Stricoff, and L. J. Partridge, Jr. (eds.). *NIOSH/OSHA—Occupational Health Guidelines for Chemical Hazards*. DHHS (NIOSH) Publication No. 81–123 (3 vols). Washington, DC: U.S. Government Printing Office, Jan. 1981. 4.

10. USEPA. *Management of Hazardous Waste Leachate*. EPA Contract No. 68–03–2766 p. E–61 (1982).

11. IARC. *Monographs on the Evaluation of the Carcinogenic Risk of Chemicals to Man*. Geneva: World Health Organization, International Agency for Research on Cancer, 1972–present (multivolume work), p. S7 60 (1987).

12. Borthwick, P. W., S. C. Schimmel, in: *Pentachlorophenol*. Rao, K.R., (ed.) New York: Plenum Press, 141–6 (1977).

13. Dominguez, S. E., G. A. Chapman. *Arch Environ Contam Toxicol* 13 (6): 739–44 (1984).

14. *Farm Chemicals Handbook 87*. Willoughby, OH: Meister Publishing Co., 1987, p. C–232.

15. U.S. Department of Interior, Fish and Wildlife Service. *Handbook of Acute Toxicity of Chemicals to Fish and Aquatic Invertebrates*. Resource Publication No. 137. Washington, DC: U.S. Government Printing Office, 1980. 58.

16. Schimmel, S. C., et al. *Environ Sci Res 12 (Pentachlorophenol: Chem Pharmacol Environ Toxicol)*: 147–55 (1978).

17. Gupta, P. K., P. S. Rao. *Arch Hydrobiol* 94 (2): 210–217 (1982).

18. Khangarot, B. S., et al. *Arch Hydrobiol* 103 (3): 375–79 (1985).

19. Niimi, A. J., C. A. McFadden. *Bull Environ Contam Toxicol* 28 (1): 11–9 (1982).

20. 21 CFR 175.105 (4/1/91).

21. 21 CFRR 181.30 (4/1/91).

22. Parr, J.F., P.B. Marsh, and J.M. Kla (eds.). *Land Treatment of Hazardous Wastes*. Park Ridge, New Jersey: Noyes Data Corporation, 1983.

23. Booth, N.H., L.E. McDonald (eds.). *Veterinary Pharmacology and Therapeutics*. 5th ed. Ames, Iowa: Iowa State University Press, 1982.

■ 3-PENTANONE

CAS # 96–22–0

SYNONYMS dek • diethyl ketone • diethylcetone (French) • dimethylacetone • ethyl ketone • metacetone • methacetone • pentanone-3 • propione • ethyl-propionyl

MF $C_5H_{10}O$

MW 86.15

COLOR AND FORM Colorless, mobile liquid.

ODOR Acetone odor.

DENSITY 0.8138 at 20°C.

VISCOSITY 0.493 cP at 15°C.

BOILING POINT 101.7°C.

MELTING POINT −39.8°C.

HEAT OF COMBUSTION −735.6 kcal/g mol wt at 20°C.

HEAT OF VAPORIZATION 11,183.0 cal/mole.

AUTOIGNITION TEMPERATURE 452°C [R1].

EXPLOSIVE LIMITS AND POTENTIAL Lower: 1.6% [R3].

FLAMMABILITY LIMITS Lower limit 1.6% by volume [R4].

FIREFIGHTING PROCEDURES Water may be ineffective fighting fires with low flash point [R4]. Do not extinquish fire unless flow can be stopped. Use water in flooding quantities as fog. Solid streams of water may spread fire. Cool all affected containers with flooding quantities of water. Apply water from as far a distance as possible. Use "alcohol" foam, dry chemical, or carbon dioxide [R6].

SOLUBILITY Soluble in all proportions in ether; >10% in acetone; >10% in ethanol; 47,000 mg/L water at 20°C; 38,000 mg/L water at 100°C.

VAPOR PRESSURE 35.43 mm Hg at 25°C.

OCTANOL/WATER PARTITION COEFFICIENT Log K_{ow} = 0.99 (est).

OTHER CHEMICAL/PHYSICAL PROPERTIES Heat capacity: 190.1 J/mol-dK; mobile; enthalpy of formation: −70.87 kcal/mole; 2.6 wt% water in 3-pentanone.

PREVENTATIVE MEASURES If material not on fire and not involved in.

Fire: Keep sparks, flames, and other sources of ignition away. Keep material out of water sources and sewers. Build dikes to contain flow as necessary. Attempt to stop leak if without undue personnel hazard. Use water spray to disperse vapors and dilute standing pools of liquid [R6].

DISPOSAL SRP: At the time of review, criteria for land treatment or burial (sanitary landfill) disposal practices are subject to significant revision. Prior to implementing land disposal of waste residue (including waste sludge), consult with environmental regulatory agencies for guidance on acceptable disposal practices.

COMMON USES Medicine; organic synthesis [R1].

OCCUPATIONAL RECOMMENDATIONS
OSHA 8-hr time-weighted average: 200 ppm (705 mg/m³). Final rule limits were achieved by any combination of engineering controls, work practices, and personal protective equipment during the phase-in period, Sept 1, 1989 through Dec 30, 1992. Final rule limits became effective Dec 31, 1992 [R10].

THRESHOLD LIMIT 8-hr time-weighted average (TWA) 200 ppm, 705 mg/m³ (1981) [R11, 19].

NON-HUMAN .

TOXICITY LD_{50} rat oral 2.14 g/kg [R5].

ECOTOXICITY LC_{50} *Pimephales promelas* (fathead minnow) 27–28 days old 1,540 mg/L/96 hr (confidence limit: 1,470–1,600 mg/L) at 24.2°C (hardness 46.2 mg/L calcium carbonate, pH 7.88) (purity 98%; conditions of bioassay not specified) [R7].

BIODEGRADATION Several investigators have shown that 3-pentanone readily biodegrades in screening tests using a sewage seed (1–4). They report: 66.4% of theoretical BOD after 5 days incubation with acclimated cultures (1), 89% of theoretical BOD after 5 days (2), and 50.8% and 38% of theoretical BOD after 10 days (3,4). In a semi-continuous activated sludge (SCAS) biological treatment simulation test, 38% of the theoretical BOD of 3-pentanone was lost during the 24-hr treatment (5) [R8].

BIOCONCENTRATION The bioconcentration factor (BCF) estimated from its octanol/water partition coefficient of 0.99 (2) is 3.3 (1,SRC). This indicates that 3-pentanone will have a negligible tendency for bioconcentrating in fish (SRC) [R9].

DOT *Fire or Explosion:* Flammable/combustible material; may be ignited by heat, sparks, or flames. Vapors may travel to a source of ignition, and flash back. Container may explode in heat of fire. Vapor explosion hazard indoors, outdoors, or in sewers. Runoff to sewer may create fire or explosion hazard [R2].

Health Hazards: May be poisonous if inhaled or absorbed through skin. Vapors may cause dizziness or suffocation. Contact may irritate or burn skin and eyes. Fire may produce irritating or poisonous gases. Runoff from fire control or dilution water may cause pollution [R2].

Emergency Action: Keep unnecessary people away; isolate hazard area and deny entry. Stay upwind; keep out of low areas. Self-contained breathing apparatus (SCBA) and structural firefighter's protective clothing will provide limited protection. Isolate for 1/2 mile in all directions if tank car or truck is involved in fire. Call CHEMTREC at 1-800-424-9300 for emergency assistance. If water pollution occurs, notify the appropriate authorities [R2].

Fire: Small fires: Dry chemical, CO_2, halon, water spray, or alcohol foam. Large Fires: Water spray, fog, or alcohol foam is recommended. Move container from fire area if you can do so without risk. Cool containers that are exposed to flames with water from the side until well after fire is out. Stay away from ends of tanks. For massive fire in cargo area, use unmanned hose holder, or monitor nozzles; if this is impossible, withdraw from area, and let fire burn. Withdraw immediately in case of rising sound from venting safety device or any discoloration of tank due to fire [R2].

Spill or Leak: Shut off ignition sources; no flares, smoking, or flames in hazard area. Stop leak if you can do so without risk. Water spray may reduce vapor; but it may not prevent ignition in closed spaces. Small Spills: Take up with sand or other noncombustible absorbent material and

place into containers for later disposal. Large Spills: Dike far ahead of liquid spill for later disposal [R2].

First Aid: Move victim to fresh air and call emergency medical care; if not breathing, give artificial respiration; if breathing is difficult, give oxygen. In case of contact with material, immediately flush eyes with running water for at least 15 minutes. Wash skin with soap and water. Remove and isolate contaminated clothing and shoes at the site [R2].

FIRE POTENTIAL Dangerous, when exposed to heat or flame [R3].

REFERENCES

1. Sax, N. I., and R. J. Lewis, Sr. (eds.). *Hawley's Condensed Chemical Dictionary.* 11th ed. New York: Van Nostrand Reinhold Co., 1987. 394.

2. Department of Transportation. *Emergency Response Guidebook 1987.* DOT P 5800.4. Washington, DC: U.S. Government Printing Office, 1987, p. G-26.

3. Sax, N. I. *Dangerous Properties of Industrial Materials.* 6th ed. New York, NY: Van Nostrand Reinhold, 1984. 2135.

4. National Fire Protection Association. *Fire Protection Guide on Hazardous Materials.* 9th ed. Boston, MA: National Fire Protection Association, 1986, p. 325M-38.

5. American Conference of Governmental Industrial Hygienists. *Documentation of the Threshold Limit Values and Biological Exposure Indices.* 5th ed. Cincinnati, OH: American Conference of Governmental Industrial Hygienists, 1986. 199.

6. Association of American Railroads. *Emergency Handling of Hazardous Materials in Surface Transportation.* Washington, DC: Assoc. of American Railroads, Hazardous Materials Systems (BOE), 1987. 253.

7. Brooke, L. T., D. J. Call, D. T. Geiger, and C. E. Northcott (eds.). *Acute Toxicities of Organic Chemicals to Fathead Minnows (Pimephales promelas).* Superior, WI: Center for Lake Superior Environ-

mental Studies, Univ. of Wisconsin–Superior, 1984. 123.

8. (1) Babeu, L., D. D. Vaishnav. *J Indust Microbiol* 2: 107–15 (1987). (2) Bridie, A. L., et al. *Wat Res* 13: 627–30 (1979). (3) Ettinger, M. B. *Ind Eng Chemical* 48: 256–9 (1956). (4) Heukelekian, H., M. C. Rand. *J Wat Pollut Control Assoc* 29: 1040–53 (1955). (5) Mills, E. J., Jr., V. T. Stack, Jr. *Proc 8th Indust Waste Conf. Eng Bulletin* Purdue Univ, Eng. Ext. Ser. pp. 492–517 (1954).

9. (1) Lyman, W. J., et al., *Handbook of Chemical Property Estimation Methods.* New York: McGraw-Hill, (1982). (2) Tewari, Y. B., et al. *J Chemical Eng Data* 27: 451–4 (1982).

10. 54 FR 2920 (1/19/89).

11. American Conference of Governmental Industrial Hygienists. *Threshold Limit Values for Chemical Substances and Physical Agents and Biological Exposure Indices for 1994–1995.* Cincinnati, OH: ACGIH, 1994.

■PERACETIC ACID

CAS # 79-21-0

DOT # 2131

SYNONYMS acetic peroxide • acetyl hydroperoxide • acide peracetique (French) • desoxon-1 • ethaneperoxoic acid • hydroperoxide, acetyl • osbon-ac • paa • peroxoacetic acid • peroxyacetic acid • acide peroxyacetique (French) • acido peroxiacetico (Spanish) • Caswell No. 644 • EPA pesticide chemical code 063201 • estosteril • Kyselina-peroxyoctova (Czech) • monoperacetic acid • proxitane-4002

MF $C_2H_4O_3$

MW 76.06

COLOR AND FORM Colorless liquid (less than 40%).

ODOR Acrid.

DENSITY 1.226 at 15°C.

VISCOSITY 3.280 cP at 78°F.

DISSOCIATION CONSTANTS $pK_a = 8.20$ at 25°C.

BOILING POINT 105°C.

MELTING POINT 0.1°C.

AUTOIGNITION TEMPERATURE 392°F (200°C) (peracetic acid (less than 40%)) [R8, p. 49–134].

EXPLOSIVE LIMITS AND POTENTIAL Explodes at 110°C; explosion hazard: severe when exposed to heat or by spontaneous chemical reaction; a powerful oxidizing agent [R4].

FIREFIGHTING PROCEDURES Use flooding quantities of water. Use water spray to keep fire-exposed containers cool. Fight fire from protected location or maximum possible distance. Approach fire from upwind to avoid hazardous vapors and toxic decomposition products. (peracetic acid (less than 40%)) [R8, p. 49–133].

SOLUBILITY Very soluble in water, ether, sulfuric acid; soluble in ethanol.

VAPOR PRESSURE 14.5 mm Hg at 25°C.

OCTANOL/WATER .

PARTITION COEFFICIENT $\log K_{ow} = -0.924$ (est).

CORROSIVITY Corrosive to most metals, including aluminum.

STABILITY AND SHELF LIFE Thermally unstable [R8, p. 49–133].

OTHER CHEMICAL/PHYSICAL PROPERTIES Oxidizing material dangerous in contact with organic materials, explodes at 110°C.

PROTECTIVE EQUIPMENT Self-contained breathing apparatus; full protective clothing (goggles, rubber gloves, etc.) [R1].

PREVENTATIVE MEASURES Methods are adopted for handling peracetic acid in air-conditioned lab for germ-free animals to minimize amount of free acid vapor circulating. Techniques helped to confine and neutralize peracetic acid vapor [R5].

CLEANUP Cover with weak reducing agents such as hypo, bisulfites, or ferrous salts. Bisulfites or ferrous salts need additional promoter of some 3M sulfuric acid for rapid reaction. Transfer the slurry (or sludge) into a large container of water and neutralize with soda ash [R3].

COMMON USES Bactericide and fungicide, especially in food-processing; reagent in making caprolactam; synthetic glycerol [R2].

OCCUPATIONAL RECOMMENDATIONS None listed.

NON-HUMAN TOXICITY LD_{50} rat oral 1,540 mg/kg [R3].

BIOCONCENTRATION Based upon an estimated log K_{ow} of −0.924 (1), the BCF for peracetic acid can be estimated to be 0.12 (SRC) from a recommended regression-derived equation (2). This indicates that bioconcentration in aquatic organisms is not important and, in any case, peracetic acid would rapidly react with organic material (SRC) [R6].

FIFRA REQUIREMENTS In 1988, Congress amended FIFRA to strengthen and accelerate EPA's reregistration program. The nine-year reregistration scheme mandated by "FIFRA 88" applies to each registered pesticide product containing an active ingredient initially registered before November 1, 1984. Pesticides for which EPA had not issued Registration Standards prior to the effective date of FIFRA '88 were divided into three lists based upon their potential for exposure and other factors, with List B being of highest concern, and D of least. List: D; Case: Peroxy compounds; Case No. 4072; Pesticide type: Fungicide, Herbicide, Rodenticide, Antimicrobial; Active Ingredient (AI): Peroxyacetic acid; Case Status: Awaiting Data/Data in Review: OPP awaits data from the pesticide's producer(s) regarding its human health or environmental effects, or OPP has received and is reviewing such data, in order to reach a decision about the pesticide's eligibility for reregistration [R7].

GENERAL RESPONSE Stop discharge if possible. Keep people away. Call fire department. Isolate and remove discharged material. Notify local health and pollution control agencies.

FIRE RESPONSE Combustible. May cause fire on contact with combustibles. Containers may explode in fire. Flood discharge area with water. Cool exposed containers with water.

EXPOSURE RESPONSE Call for medical aid. Vapor irritating to eyes, nose, and throat. Move victim to fresh air. If breathing has stopped, give artificial respiration. If breathing is difficult, give oxygen. Liquid irritating to skin and eyes. Harmful if swallowed. Remove contaminated clothing and shoes. Flush affected areas with plenty of water. If swallowed and victim is conscious, have victim drink water or milk.

RESPONSE TO DISCHARGE Issue warning—oxidizing material, water contaminant. Restrict access. Disperse and flush.

WATER RESPONSE Effect of low concentrations on aquatic life is unknown. May be dangerous if it enters water intakes. Notify local health and wildlife officials. Notify operators of nearby water intakes.

EXPOSURE SYMPTOMS Inhalation causes severe irritation of mucous membrane. Contact with liquid causes severe irritation of eyes and skin. Ingestion causes severe distress, including burns of mouth and stomach.

EXPOSURE TREATMENT Inhalation: remove victim to fresh air; if he is not breathing, apply artificial respiration and oxygen; call a doctor. Eyes: flush with water for at least 15 min.; call a doctor. Skin: flush with water and treat burns. Ingestion: give plenty of warm water; call a doctor.

REFERENCES

1. U.S. Coast Guard, Department of Transportation. *CHRIS—Hazardous Chemical Data*. Volume II. Washington, DC: U.S. Government Printing Office, 1984–5.

2. Sax, N. I., R. J. Lewis, Sr (eds.). *Hawley's Condensed Chemical Dictionary*. 11th ed. New York: Van Nostrand Reinhold Co., 1987. 886.

3. ITII. *Toxic and Hazardous Industrial Chemicals Safety Manual*. Tokyo, Japan:

The International Technical Information Institute, 1988. 400.

4. Sax, N. I. *Dangerous Properties of Industrial Materials*. 6th ed. New York, NY: Van Nostrand Reinhold, 1984. 2148.

5. Fordham J.P. *Lab Anim* 12 (4): 247–8 (1978)

6. (1) USEPA, *Graphical Exposure Modeling System (GEMS)*. CLOGP (1987). (2) Lyman, W. J., et al., *Handbook of Chemical Property Estimation Methods*. Washington, DC: Amer Chemical Soc, p. 5–4 (1990).

7. USEPA/OPP. *Status of Pesticides in Reregistration and Special Review*. p. 252 (Mar. 1992) EPA 700–R–92–004.

8. National Fire Protection Guide. *Fire Protection Guide on Hazardous Materials*. 10th ed. Quincy, MA: National Fire Protection Association, 1991.

■ PERCHLORIC ACID, AMMONIUM SALT

CAS # 7790–98–9

DOT # 1442

SIC CODES 2892.

SYNONYMS ammonium perchlorate [NH$_4$ClO$_4$] • perchloric acid, ammonium salt • pkha

MF ClO$_4$•H$_4$N

MW 117.50

COLOR AND FORM Colorless rhombic crystals; orthorhombic crystals; white crystals.

DENSITY 1.95.

EXPLOSIVE LIMITS AND POTENTIAL Shock-sensitive; may explode when exposed to heat or by spontaneous chemical reaction. Sensitive high explosive when contaminated with reducing materials [R2].

SOLUBILITY In water: 10.74 g/100 cc at 0°C; 42.45 g/100 cc at 85°C; soluble in acetone; slightly soluble in ethanol; soluble in methanol; almost insoluble in ether, ethyl acetate.

OTHER CHEMICAL/PHYSICAL PROPERTIES Melting point: decomposition.

COMMON USES Analytical chemistry; etching agent [R2]; animal fattening agent; oxidizing agent in solid rocket propellants; component in explosive mixtures and pyrotechnics; chemical intermediate for alkali and alkaline metal perchlorates; engraving agent [R1].

BINARY REACTANTS Metals, organic material, sulfur.

STANDARD CODES NFPA 2, 1, 4; ICC—oxidizing material, yellow label, 100 lb in an outside container; USCG—oxidizing material, yellow label; IATA—oxidizing material, yellow label, 12-kg passenger, 45-kg cargo.

PERSISTENCE Not considered persistent. Perchlorate will break down slowly to chlorides. Ammonium is subject to biodegradation.

DIRECT CONTACT Irritating to skin and mucous membranes.

PERSONAL SAFETY Wear eye goggles, protective clothing, and self-contained breathing apparatus. Equipment should be resistant to oxidation.

ACUTE HAZARD LEVEL Moderate irritation, inhalation, and ingestion hazard. Toxicity is directly related to the level of un-ionized ammonia. Threshold concentration for fish 0.5 ppm free NH$_3$ [R3].

CHRONIC HAZARD LEVEL Can cause chronic ingestion problems.

DEGREE OF HAZARD TO PUBLIC HEALTH Explosion hazard is great. Irritating material is toxic by ingestion and inhalation.

ACTION LEVEL Notify fire and air authority. Evacuate area. Enter from upwind side. Warn civil defense of potential explosion.

ON-SITE RESTORATION Ammonium ions can be exchanged on natural zeolites (clinoptilolite). Seek professional environmental engineering assistance through EPA's Environmental Response Team

(ERT), Edison, NJ, 24–hour no. 908–548–8730.

AVAILABILITY OF COUNTERMEASURE MATERIAL Zeolites—local quarries, suppliers.

DISPOSAL Add to large volume of concentrated reducer solution (hypo, a bisulfite, or a ferrous salt, and acidify with 3 M H_2SO_4). When reduction is complete, add soda ash or dilute HCl to neutralize. Route to sewage plant.

DISPOSAL NOTIFICATION Contact local sewage authority.

INDUSTRIAL FOULING POTENTIAL <1 ppm ammonium nitrogen recommended industrial use level.

MAJOR WATER USE THREATENED Potable supply.

PROBABLE LOCATION AND STATE Colorless granules. Will dissolve.

WATER CHEMISTRY See a text on ammonia for solution chemistry. Chlorates break down to chloride.

FOOD-CHAIN CONCENTRATION POTENTIAL Negative.

GENERAL RESPONSE Wear goggles and self-contained breathing apparatus. Isolate and remove discharged material. Notify local health and pollution control agencies.

FIRE RESPONSE May cause fire and explode on contact with combustibles. Containers may explode in fire. Poisonous gases may be produced in fire. Wear goggles and self-contained breathing apparatus. Evacuate surrounding area. Combat large fires from protected location with unmanned hose holder or monitor nozzle.

EXPOSURE RESPONSE Call for medical aid. Solid is irritating to eyes and skin. Harmful if swallowed. Flush affected areas with plenty of water. If in eyes, hold eyelids open, and flush with plenty of water. If swallowed and victim is conscious, have victim drink water or milk.

RESPONSE TO DISCHARGE Issue warning—oxidizing material. Disperse and flush.

WATER RESPONSE Effect of low concentrations on aquatic life is unknown. May be dangerous if it enters water intakes. Notify local health and wildlife officials. Notify operators of nearby water intakes.

EXPOSURE SYMPTOMS Irritating to skin and mucous membranes.

REFERENCES

1. SRI.

2. Hawley, G. G. *The Condensed Chemical Dictionary.* 9th ed. New York: Van Nostrand Reinhold Co., 1977. 52.

3. Todd, D. K. *The Water Encyclopedia,* Maple Press, 1970.

■ BIS (4–CHLOROBENZOYL) PEROXIDE

CAS # 94–17–7

SYNONYMS 4-chlorobenzoyl peroxide • bis (p-chlorobenzoyl) peroxide • cadox-ps • luperco-bdb • p, p′-dichlorobenzoyl peroxide • p,p′-dichlorodibenzoyl peroxide • p-chlorobenzoyl peroxide • peroxide, bis (p-chlorobenzoyl)- • di-(p-chlorobenzoyl) peroxide

MF $C14H8Cl2O4$

MW 311.12

COLOR AND FORM White, granular; powder.

ODOR Odorless.

DENSITY Greater than 1.1 at 20°C (solid).

HEAT OF COMBUSTION (est) −9,000 Btu/lb = −5,000 cal/g = −210×10⁵ J/kg.

EXPLOSIVE LIMITS AND POTENTIAL This material may be caused to explode by heat (over 38°C) or contamination. Any contaminant that acts as an accelerator to polymerization or on decomposition of this material can cause explosion. Heat or contact with certain fumes or mists can cause it to explode [R4].

FIREFIGHTING PROCEDURES Flood with water, or use dry chemical, foam, carbon dioxide [R1]. Organic peroxides threatened by fire should be wetted from a safe distance for cooling (organic peroxides) [R6, 1613]. Solid streams of water may be ineffective. Dangerously explosive (p-chlorobenzoyl peroxide (not more than 52% as a paste); p-chlorobenzoyl peroxide (not more than 75% with water)) [R5]. Use water in flooding quantities as fog. Cool all affected containers with flooding quantities of water. Apply water from as far a distance as possible. Use "alcohol" foam, dry chemical, or carbon dioxide (p-chlorobenzoyl peroxide; p-chlorobenzoyl peroxide—not more than 52% as a paste); p-chlorobenzoyl peroxide (not more than 75% with water)) [R5]. If the material is on fire or involved in fire consider evacuation of one-half mile radius. (p-chlorobenzoyl peroxide; p-chlorobenzoyl peroxide (not more than 52% as a paste); p-chlorobenzoyl peroxide (not more than 75% with water) [R5].

TOXIC COMBUSTION PRODUCTS Toxic chlorinated biphenyls are formed in fires [R1].

SOLUBILITY Insoluble in water; soluble in organic solvents.

STABILITY AND SHELF LIFE Stable if below 80°F [R1].

OTHER CHEMICAL/PHYSICAL PROPERTIES Strong oxidizer.

PROTECTIVE EQUIPMENT Wear appropriate chemical protective gloves, boots, and goggles. Wear positive-pressure self-contained breathing apparatus when fighting fires involving this material. (p-chlorobenzoyl peroxide; p-chlorobenzoyl peroxide (not more than 52% as a paste); p-chlorobenzoyl peroxide (not more than 75% with water)) [R5].

PREVENTATIVE MEASURES The safety of many organic peroxides is greatly improved by dispersing them in solvent or non-solvent diluents that absorb the heat of decomposition (e.g., water or plasticizer) or reduce shock sensitivity. Process equipment should be electrically bonded together and grounded. Non-sparking equipment should be used for handling. Peroxides must never be mixed directly with accelerators or other reactive materials, added to

hot materials, or placed in heated reaction vessels. Contamination should be carefully avoided by keeping the work area and equipment clean, handling peroxides in vessels of glass, polyethylene, teflon, or stainless steel (304 or 316) to prevent contamination by corrosion, keeping containers closed, and disposing of excess peroxide promptly, never returning it to the original container. Smoking should be prohibited. Peroxides should not be stored in refrigerators containing food or drink. Laboratory reactions should be carried out behind a safety shield. (organic peroxides) [R6, 1613].

CLEANUP Spills should be cleaned up promptly using non-sparking tools and an inert moist diluent such as vermiculite or sand. Sweepings may be placed in open containers or polyethylene bags and the area washed with water and detergent. Spilled, contaminated, waste, or questionable peroxides should be destroyed. Incineration or burning is generally preferred. Burning may be done by spreading waste in a trench and igniting it from a distance. (peroxides, organic) [R6, 1613].

DISPOSAL Incineration or burning is generally preferred method of destroying peroxides. Burning may be done by spreading waste in a trench and igniting it from a distance. (SRP: Although burning of chemical wastes is frequently recommended, the SRP is of the opinion that incineration technology has not uniformly been sufficiently perfected to ensure that hazardous products are not released into the environment during the incineration process. Incineration can be an appropriate method of waste disposal if it can be demonstrated that the specific method and conditions in use result in no further release of potentially dangerous emissions.) Most organic peroxides can also be hydrolyzed by adding them slowly with stirring to about ten times their weight of cold 10% sodium hydroxide solution. The reaction may require several hours. (peroxides, organic) [R6, 1613].

COMMON USES Bleaching agent; polymerization catalyst [R2].

OCCUPATIONAL RECOMMENDATIONS None listed.

DOT *Fire or Explosion:* May be ignited by heat, sparks, or flames. Container may explode in heat of fire. May explode from heat or contamination. Runoff to sewer may create fire or explosion hazard [R3].

Health Hazards: Contact may cause burns to skin and eyes. Fire may produce irritating or poisonous gases. Runoff from fire control or dilution water may cause pollution [R3].

Emergency Action: Keep unnecessary people away; isolate hazard area and deny entry. Stay upwind; keep out of low areas. Positive-pressure self-contained breathing apparatus (SCBA) and structural firefighters' protective clothing will provide limited protection. Call CHEMTREC at 1–800–424–9300 for emergency assistance. If water pollution occurs, notify the appropriate authorities [R3].

Fire: Small fires: Dry chemical, CO_2, water spray, or regular foam. Large Fires: Flood fire area with water. Apply cooling water to sides of containers that are exposed to flames until well after fire is out. Stay away from ends of tanks. For massive fire in cargo area, use unmanned hose holder, or monitor nozzles; if this is impossible, withdraw from area, and let fire burn [R3].

Spill or Leak: Shut off ignition sources; no flares, smoking, or flames in hazard area. Do not touch or walk through spilled material; stop leak if you can do so without risk. Small Spills: Take up with inert, damp noncombustible material; move containers from spill area. Large Spills: Wet down with water and dike for later disposal [R3].

First Aid: Move victim to fresh air; call for emergency medical care. In case of contact with material, immediately flush eyes with running water for at least 15 minutes. Wash skin with soap and water. Remove and isolate contaminated clothing and shoes at the site. Keep victim quiet and maintain normal body temperature [R3].

FIRE POTENTIAL Powerful oxidizer. Dangerous when involved in fire [R4]. Dangerous fire risk [R2]. Burns very rapidly when ignited. Smoke is unusually heavy when paste form is involved [R1].

REFERENCES

1. U.S. Coast Guard, Department of Transportation. *CHRIS—Hazardous Chemical Data.* Volume II. Washington, DC: U.S. Government Printing Office, 1984–5.

2. Sax, N. I., and R. J. Lewis, Sr. (eds.). *Hawley's Condensed Chemical Dictionary.* 11th ed. New York: Van Nostrand Reinhold Co., 1987. 264.

3. U.S. Department of Transportation. *Emergency Response Guidebook 1990.* DOT P 5800.5. Washington, DC: U.S. Government Printing Office, 1990, p. G–48.

4. Sax, N. I. *Dangerous Properties of Industrial Materials.* 5th ed. New York: Van Nostrand Reinhold, 1979. 489.

5. Association of American Railroads. *Emergency Handling of Hazardous Materials in Surface Transportation.* Washington, DC: Assoc. of American Railroads, Hazardous Materials Systems (BOE), 1987, p. 160–1.

6. International Labour Office. *Encyclopedia of Occupational Health and Safety.* Vols. I and II. Geneva, Switzerland: International Labour Office, 1983.

■ PERTHANE

CAS # 72–56–0

SIC CODES 2879.

SYNONYMS 1,1-bis (p-ethylphenyl)-2,2-dichloroethane • 1,1-dichloro-2,2-bis (p-ethylphenyl) ethane • 2,2-bis (p-ethylphenyl)-1,1-dichloroethane • 2,2-dichloro-1,1-bis (p-ethylphenyl) ethane • a, a-dichloro-2,2-bis (p-ethylphenyl) ethane • benzene, 1,1'-(2,2-dichloroethylidene) bis (4-ethyl) • di (p-ethylphenyl) dichloroethane • diethyldiphenyl dichloroethane • ethane, 1,1-dichloro-2,2-bis (p-ethylphenyl)- • ethane, 2,2-bis (p-ethylphenyl)-1,1-

dichloro • ethylan • nci-c02868 • p,p'-ethyl-ddd • p,p-ethyl ddd • q-137

MF $C_{18}H_{20}Cl_2$

MW 307.28

COLOR AND FORM Crystals from ethanol; when pure, a crystalline solid; the technical product is a wax.

BOILING POINT Decomposes on distillation.

MELTING POINT 56°C.

SOLUBILITY Practically insoluble in water. Soluble in acetone, kerosene, diesel fuel. Readily soluble in common organic solvents, particularly chlorinated hydrocarbons and aromatics. Water solubility = 100 µg/L at 24°C.

CORROSIVITY Slightly corrosive to iron, zinc, and aluminum.

OTHER CHEMICAL/PHYSICAL PROPERTIES Decomposition above 52°C.

DISPOSAL Decomposes above 52°C. SRP: At the time of review, criteria for land treatment or burial (sanitary landfill) disposal practices are subject to significant revision. Prior to implementing land disposal of waste residue (including waste sludge), consult with environmental regulatory agencies for guidance on acceptable disposal practices. Ethylan is readily dehydrochlorinated, like DDT. It undergoes some thermal decomposition above 50°C and the ethyl groups are readily oxidized to carboxylic acid groups. Recommendable methods: Incineration and landfill. Peer-review: Waste material should be burned in an incinerator equipped with scrubbers to absorb hydrogen chloride. Incinerate at temp above 1,000°C for 1–2 sec (peer-review conclusions of an IRPTC expert consultation (May 1985)) [R7].

COMMON USES Insecticidal control of pear psylla and of leaf hoppers and larvae on vegetables; used to control moths and carpet beetles in the textile and dry-cleaning industries (former use) [R1].

PERSONAL SAFETY Self-contained breathing apparatus must be worn. Rubber gloves, hats, suits, and boots must be worn.

OCCUPATIONAL RECOMMENDATIONS None listed.

ACTION LEVEL Evacuate area. Enter from upwind side. Notify local air authority and National Agricultural Chemicals Association.

ON-SITE RESTORATION Carbon or peat may be used as sorbents. Seek professional environmental engineering assistance through EPA' S Environmental Response Team (ERT), Edison, NJ, 24-hour No. 908–548–8730.

AVAILABILITY OF COUNTERMEASURE MATERIAL Carbon—water treatment plants, sugar refineries; peat—nurseries, floral shops.

PROBABLE LOCATION AND STATE Cream to tan semi-solid. formulated as emulsifiable concentrate, wettable powder, or dust. Pure material will sink. Wettable forms will dissolve.

NON-HUMAN TOXICITY LD$_{50}$ mouse oral 9,340 mg/kg [R5, 208]; LD$_{50}$ mouse, oral 9,340 mg/kg [R5, 208]; LD$_{50}$ rat iv 73 mg/kg [R5, 208]; LD$_{50}$ mouse iv 173 mg/kg [R5, 208]; LD$_{50}$ rat, oral 8,170 mg/kg [R5, 208].

ECOTOXICITY LC$_{50}$ bobwhite quail oral greater than 5,000 ppm, 10% mortality at 5,000 ppm [R2]; LC$_{50}$ Japanese quail, oral greater than 5,000 ppm [R2]; LC$_{50}$ pheasants oral greater than 5,000 ppm [R2]; LC$_{50}$ mallards oral greater than 5,000 ppm [R2]; LC$_{50}$ rainbow trout 0.004 mg/L/96 hr, wt 0.7 g at 12°C (95% confidence limit 0.003–0.006 mg/L) [R6]; LC$_{50}$ bluegill 0.020 mg/L/96 hr, wt 0.9 g at 24°C (95% confidence limit 0.016–0.025 mg/L) [R6].

BIOCONCENTRATION Based upon an estimated log K_{ow} of 6.66 (1), the BCF of perthane can be estimated to be approximately 7×10^4 from a regression-derived equation (2,SRC). This estimated BCF value suggests that bioconcentration in aquatic organisms may be a very important fate process (SRC) [R3].

FIFRA REQUIREMENTS This rule revokes tolerances for residues of the pesticide 1,1–dichloro–2,2–bis (p-ethylphenyl) ethane (also known as perthane, ethylan) in or on agricultural commodities because all registrations have been cancelled [R4].

REFERENCES

1. Hartley, D., and H. Kidd (eds.). *The Agrochemicals Handbook.* 2nd ed. Lechworth, Herts, England: The Royal Society of Chemistry, 1987.

2. U.S. Department of the Interior, Fish and Wildlife Service, Bureau of Sports Fisheries and Wildlife. *Comparative Dietary Toxicities of Pesticides to Birds.* Special Scientific Report—*Wildlife* No. 152. Washington, DC: U.S. Government Printing Office, 1972. 41.

3. (1) Meylan, W. M., P. H. Howard. *J Pharm Sci* 84: 83–92 (1995) (2) Lyman, W. J., et al., *Handbook of Chemical Property Estimation Methods* Washington, DC: Amer Chemical Soc, p. 5–4 (1990).

4. 59 FR 44930 (8/31/94).

5. Hayes, Wayland J., Jr. *Pesticides Studied in Man.* Baltimore/London: Williams and Wilkins, 1982.

6. U.S. Department of Interior, Fish and Wildlife Service. *Handbook of Acute Toxicity of Chemicals to Fish and Aquatic Invertebrates.* Resource Publication No. 137. Washington, DC: U.S. Government Printing Office, 1980. 85.

7. United Nations. *Treatment and Disposal Methods for Waste Chemicals* (IRPTC File). Data Profile Series No. 5. Geneva, Switzerland: United Nations Environmental Programme, Dec. 1985. 167.

■ PHENOL, m–CHLORO–

CAS # 108–43–0

SIC CODES 2869.

SYNONYMS 3-chloro-1-hydroxybenzene • 3-hydroxychlorobenzene • m-chlorophenol • phenol, 3-chloro- • phenol, m-chloro-

MF C_6H_5ClO

MW 128.56

COLOR AND FORM Needles; white crystals.

ODOR THRESHOLD Odor detection in air: 1.8×10^{-4} mg/L (gas), in water 1.24 ppm, chemically pure (chlorophenol) [R3].

DENSITY 1.268 at 25°C.

VISCOSITY 11.55 cP at 25°C.

DISSOCIATION CONSTANT 1.4×10^{-9} at 25°C.

BOILING POINT 214°C at 760 mm Hg.

MELTING POINT 33°C.

HEAT OF VAPORIZATION 11,979.7 g cal/g mole.

TOXIC COMBUSTION PRODUCTS When heated to decomposition it emits toxic fumes of chlorine. [R2].

SOLUBILITY Soluble in alcohol, ether, benzene; 2.6 parts soluble in 100 parts water at 20°C; soluble in caustic alkali solution.

VAPOR PRESSURE 0.23 mm Hg at 20°C (calculated).

OCTANOL/WATER PARTITION COEFFICIENT Log K_{ow} = 2.47–2.52.

STABILITY AND SHELF LIFE Discolors on exposure to air [R1].

OTHER CHEMICAL/PHYSICAL PROPERTIES Dipole moment in benzene: 2.10 d.

PREVENTATIVE MEASURES Immediately wash contaminated areas of skin with concentrated soap solution. Contaminated gloves, clothing, shoes should be removed without delay, and disposed of by incineration [R4].

CLEANUP Phenolic compounds in wastewater are oxidized with hydrogen peroxide catalyzed by iron(3)–iron(2). When the wt ratio of phenol:drogen peroxide is 1:3 and iron 5–100 ppm, more than 95% of the phenols are removed in 30 min from a 500–ppm phenol solution at pH 5–6 and 25–50°C. (phenolic compounds) [R5].

DISPOSAL Product residues and sorbent media may be packaged in 17h epoxy-lined drums and disposed of at an EPA–approved disposal site. Other disposal alternatives include deep-well injection and high-temperature destruction using incinerators with scrubbing equipment. Confirm disposal procedures with responsible environmental engineer and regulatory officials. SRP: At the time of review, criteria for land treatment or burial (sanitary landfill) disposal practices are subject to significant revision. Prior to implementing land disposal of waste residue (including waste sludge), consult with environmental regulatory agencies for guidance on acceptable disposal practices.

COMMON USES Intermediate in organic synthesis [R1].

PERSISTENCE 0% biological degradation in 7-day tests for original, 1st, 2nd, and 3rd subculture: >72 days for complete decomposition in soil suspension and >64-day decomposition period by a soil microflora.

PERSONAL SAFETY Wear butyl rubber gloves, protective clothing, self-contained breathing apparatus, and safety shoes.

OCCUPATIONAL RECOMMENDATIONS None listed.

INTERNATIONAL EXPOSURE LIMITS No standards for permissible exposure have been set. [R6, 2615].

ON-SITE RESTORATION Seek professional environmental engineering assistance through EPA's Environmental Response Team (ERT), Edison, NJ, 24–hour no. 908–548–8730. Contain and isolate spill to limit spread. Construct clay/bentonite swale to divert uncontaminated portion of watershed around contaminated portion. Isolation procedures include construction of bentonite or epoxy-lined dams, interceptor trenches, or impounds. Seek professional help to evaluate problem and implement containment procedures. Conduct bench-scale and pilot-scale tests prior to decontamination program implementation. Treatment alternatives for activated carbon sorption, granular activated carbon filtration, biodegradation, bentonite, or peat moss sorption. Treatment alternatives for contaminated soils include well-point collection and treatment of leachates as for contaminated waters, bentonite/cement ground injection to immobilize spill, physical removal of immobilized residues, and packaging for disposal. Confirm treatment procedures with responsible environmental engineer and regulatory officials.

MAJOR WATER USE THREATENED Odor threshold in water 0.100 to 0.200 ppm. Taste threshold in water 0.900 to 1.0 ppm.

PROBABLE LOCATION AND STATE Colorless crystalline solid in pure form. Technical grades may be light tan or slightly pink due to impurities. Will sink and slowly dissolve.

WATER CHEMISTRY Decomposition rate for complete disappearance in soil suspensions is >72 days.

NON-HUMAN TOXICITY LD_{50} rat oral 0.56 ml/kg [R7].

ECOTOXICITY LC_{50} *Salmo gairdneri* (rainbow trout) 10,000 μg/L/48 hr (conditions of bioassay not specified) [R8].

BIODEGRADATION Biological degradation of chlorophenols in activated-sludge 3–chlorophenol was found at a level of 100 mg/L, and the compound was completely degraded in three days with 100% ring degradation [R9].

BIOCONCENTRATION Using a reported solubility of 26,000 ppm at 20°C (1) and a reported log octanol/water partition coefficient of 2.50 (2), BCFs of 15 and 352 were estimated, respectively (3,SRC). Based on these estimated BCF, 3–chlorophenol will not be expected to bioconcentrate in aquatic organisms (SRC) [R10].

ATMOSPHERIC STANDARDS This action promulgates standards of performance for equipment leaks of volatile organic compounds (VOC) in the synthetic organic chemical manufacturing industry (SOCMI). The intended effect of these standards is to require all newly constructed, modified, and reconstructed SOCMI process units to use the best demonstrated system of continuous emission reduction for equipment leaks of VOC, considering costs, non-air-quality health, and environmental impact and energy requirements. Chlorophenols are produced, as an intermediate or final product, by process units covered under this subpart. (chlorophenols) [R11].

REFERENCES

1. Sax, N. I., and R. J. Lewis, Sr. (eds.). *Hawley's Condensed Chemical Dictionary*. 11th ed. New York: Van Nostrand Reinhold Co., 1987. 270.

2. Sax, N. I. *Dangerous Properties of Industrial Materials*. 6th ed. New York, NY: Van Nostrand Reinhold, 1984. 750.

3. Fazzalari, F. A. (ed.). *Compilation of Odor and Taste Threshold Values Data*. ASTM Data Series DS 48A (Committee E–18). Philadelphia, PA: American Society for Testing and Materials, 1978. 36.

4. ITII. *Toxic and Hazarous Industrial Chemicals Safety Manual*. Tokyo, Japan: The International Technical Information Institute, 1982. 126.

5. Greenberg, E. S. *Pap Synthesis Conf (Proc)*, 293–303 (1979).

6. Clayton, G.D., and F.E. Clayton (eds.). *Patty's Industrial Hygiene and Toxicology*: Volume 2A, 2B, 2C: *Toxicology*. 3rd ed. New York: John Wiley and Sons, 1981–1982.

7. Deichmann, W. B., E. G. Mergard. *J Ind Hyg Toxicol* 20: 373 (1948) as cited in USEPA, *Ambient Water Quality Criteria Doc: Chlorinated Phenols* p. C–11 (1980) EPA 440/5–80–032.

8. Shumay, D. L., J. R. Palensky. *Impairment of the Flavor of Fish by Pollutants* EPA R3– 73–010 (1973) as cited in USEPA. *Ambient Water Quality Criteria Doc; Chlorinated Phenols* p. B–11 (1980) EPA 440/5–80–032.

9. Ingols, R. S., et al. *J Water Pollut* 3 38: 629 (1966) as cited in USEPA, *Ambient Water Quality Criteria Doc; Chlorinated Phenols* p. C–4 (1980) EPA 440/5–80–032.

10. (1) Verschueren, K. *Handbook of Environmental Data on Organic Chemicals*. 2nd ed. New York: Van Nostrand Reinhold, pp 377–8 (1983); (2) Hansch, C., and A. J. Leo. *Medchem Project*, Claremont, CA: Pomona College, Issue No. 26 (1985); (3) Lyman, W. J., et al., *Handbook of Chemical Property Estimation Methods. Environmental Behavior of Organic Compounds*. New York: McGraw-Hill, p 5–5 (1982).

11. 40 CFR 60.489 (7/1/87).

■ PHENOL, 3,4–DICHLORO–

CAS # 95–77–2

SYNONYMS phenol, 3,4-dichloro-

MF $C_6H_4Cl_2O$

MW 163.00

COLOR AND FORM Needles from benzene-petroleum ether.

ODOR THRESHOLD In water: 100 µg/L [R4].

TASTE Taste threshold in water: 0.3 µg/L.

DISSOCIATION CONSTANT $pK_a = 8.59$.

BOILING POINT 253.5°C at 767 mm Hg.

MELTING POINT 68°C.

EXPLOSIVE LIMITS AND POTENTIAL During vacuum fractionation of the mixed dichlorophenols produced by partial hydrolysis of trichlorobenzene, rapid admission of air to the receiver caused the column contents to be forced down into the boiler at 210°C, and a violent explosion ensued. (dichlorophenol mixed isomers) [R3].

FIREFIGHTING PROCEDURES Extinguish fire using agent suitable for type of surrounding fire. Material itself does not burn or burns with difficulty. (trichlorophenol) [R2].

SOLUBILITY Soluble in alcohol, ether, and benzene; slightly soluble in water.

VAPOR PRESSURE 0.00217 mm Hg at 25°C (calculated).

OCTANOL/WATER PARTITION COEFFICIENT Log $K_{ow} = 3.33$ (measured).

OTHER CHEMICAL/PHYSICAL PROPERTIES Aqueous solutions of Fenton's reagent iron(2) hydrogen peroxide have been used to effect the total decomposition of the chlorophenols: 2–chlorophenol, 3–chlorophenol, 4–chlorophenol, 3,4–dichlorophenol, and 2,4,5–trichorophenol. The mineralization of these chlorinated aromatic substrates to carbon dioxide and free chloride has been studied as a function of iron(2) hydrogen peroxide and perchloric acid. Increasing the concentration of iron(2) hydrogen peroxide enhances the decomposition process, whereas an increase in the concentration of perchloric acid inhibits the reaction. The presence of iron(3) alone without any iron(2) with hydrogen peroxide has no effect on the degradation of the chlorophenols. In all cases, the stoichiometric quantity of free chloride was obtained at the completion of the decomposition reaction, but the rates of disappearance of the chlorophenol and of the formation of the chloride are not similar.

PROTECTIVE EQUIPMENT Wear boots, protective gloves, and goggles. (trichlorophenol) [R2].

PREVENTATIVE MEASURES SRP: contaminated protective clothing should be segregated in such a manner that there is no direct personal contact by personnel who handle, dispose of, or clean the clothing. Quality assurance to ascertain the completeness of the cleaning procedures should be implemented before the decontaminated protective clothing is returned for reuse by the workers.

CLEANUP Land spill: Dig a pit, pond, lagoon, or holding area. (SRP: If time permits, pits, ponds, lagoons, soak holes, or holding areas should be sealed with an impermeable flexible-membrane liner to contain liquid or solid material. Cover solids with plastic sheet to prevent dissolving in rain or firefighting water.) (trichlorophenol) [R2].

DISPOSAL SRP: At the time of review, criteria for land treatment or burial (sanitary landfill) disposal practices are subject to significant revision. Prior to implementing land disposal of waste residue (including waste sludge), consult with environmental regulatory agencies for guidance on acceptable disposal practices.

COMMON USES Chemical intermediate for

2–chloro–1,4–dihydroxyanthraquinone and 2,3,4–trichlorophenol [R1].

OCCUPATIONAL RECOMMENDATIONS None listed.

NON-HUMAN TOXICITY LD_{50} mouse (male CD–1 ICR) oral 1,685 mg/kg [R5]; LD_{50} mouse (female D–1 ICR) oral 2,046 mg/kg [R5].

BIODEGRADATION The anaerobic degradation of mono– and dichlorophenol isomers by fresh (unacclimated) sludge and by sludge acclimated to either 2–chlorophenol, 3–chlorophenol, or 4–chlorophenol was investigated. Biodegradation was evaluated by monitoring substrate disappearance and, in selected cases, production of $^{14}CH_4$ from labeled substrates. For the dichlorophenols in unacclimated sludge, reductive dechlorination of the Cl group ortho to phenolic OH was observed, and the monochlorophenol compounds released were subsequently degraded. 3,4–Dichlorophenol and 3,5–dichlorophenol were persistent. Sludge acclimated to 3–chlorophenol cross-acclimated to 4–chlorophenol. This sludge degraded 3,4– and 3,5–dichlorophenol. The specific cross-acclimation patterns observed for monochlorophenol degradation demonstrated the existence of two unique microbial activities that were in turn different from fresh sludge [R6].

BIOCONCENTRATION Based on a measured log K_{ow} of 3.33 (1), the BCF of 3,4–dichlorophenol can be estimated to be 200 from a recommended regression-derived equation (2,SRC). This estimated BCF range suggests only a moderate potential for significant bioaccumulation in aquatic organisms (SRC) [R7].

ATMOSPHERIC STANDARDS This action promulgates standards of performance for equipment leaks of volatile organic compounds (VOC) in the synthetic organic chemical manufacturing industry (SOCMI). The intended effect of these standards is to require all newly constructed, modified, and reconstructed SOCMI process units to use the best demonstrated system of continuous emission reduction for equipment leaks of VOC, considering costs, non-air-quality health, and environmental impact and energy requirements. Chlorophenol is produced, as an intermediate or final product, by process units covered under this subpart. (chlorophenols) [R8].

REFERENCES

1. SRI.

2. Association of American Railroads. *Emergency Handling of Hazardous Materials in Surface Transportation.* Washington, DC: Assoc. of American Railroads, Hazardous Materials Systems (BOE), 1987. 694.

3. Bretherick, L. *Handbook of Reactive Chemical Hazards.* 3rd ed. Boston, MA: Butterworths, 1985. 561.

4. Deitz, F., J. Travd. GWf–Wasser/ Abwasser 199: 318 (1978) as cited in USEPA, *Ambient Water Quality Criteria Doc; Chlorinated Phenols,* p. C–31 (1980) EPA 440/5–80–032.

5. Borzelleca, J. F., et al. *Toxicol Lett* 29: 39–42 (1985).

6. Boyd, S. A., D. R. Shelton. *Appl Environ Microbiol* 47 (2): 272–7 (1984).

7. (1) Hansch, C., and A. J. Leo. *Medchem Project* Issue No 26. Claremont, CA: Pomona College (1985); (2) Lyman, W. J., et al. *Handbook of Chemical Property Estimation Methods,* New York: McGraw-Hill p. 5–4 (1982).

8. 40 CFR 60.489 (7/1/87).

■ PHENOL, m–NITRO–

CAS # 554–84–7

DOT # 1663

SIC CODES 2815.

SYNONYMS 3-hydroxynitrobenzene • m-hydroxynitrobenzene • phenol, 3-nitro- • phenol, m-nitro- • m-nitrophenol • Crump leather-lasting dressing

MF $C_6H_5NO_3$

MW 139.12

COLOR AND FORM Monoclinic prisms from ether and dilute hydrochloric acid; colorless to yellow monoclinic form.

ODOR THRESHOLD 3.0 mg/m³ in air [R3].

DENSITY 1.485 at 20°C; 1.2797 at 100°C.

DISSOCIATION CONSTANT K at 18°C = 4.6×10^{-9}.

BOILING POINT 194°C at 70 mm Hg.

MELTING POINT 97°C.

HEAT OF COMBUSTION –684.4 cal/mol wt at 20°C (solid).

FIREFIGHTING PROCEDURES Do not extinguish fire unless flow can be stopped. Use water in flooding quantities. Apply water from as great a distance as possible; solid streams of water may be ineffective. Use "alcohol" foam, carbon dioxide, or dry chemical. Wear self-contained breathing apparatus when fighting fires involving this material. (nitrophenol) [R2].

TOXIC COMBUSTION PRODUCTS Toxic oxides of nitrogen are produced during combustion. (nitrophenol) [R2].

SOLUBILITY Soluble in caustic solution; insoluble in petroleum ether; 169.35 g/100 g in acetone at 0.2°C; 1,305.9 g/100 g in alcohol at 84°C; 116.9 g/100 g in alcohol at 1°C; 1,105.25 g/100 g in alcohol at 85°C; 105.9 g/100 g in ether at 0.2°C; 1,065.8 g/100 g in acetone at 83°C; very soluble in hot water, soluble in hot chloroform; soluble in benzene; 13,500 mg/L in water at 25°C; 133,000 mg/L in water at 90°C.

VAPOR PRESSURE 0.75 torr at 20°C.

OCTANOL/WATER PARTITION COEFFICIENT Log K_{ow} = 2.00.

OTHER CHEMICAL/PHYSICAL PROPERTIES Heat of fusion: 32.98 cal/g = 137.99 J/g = 191,963 J/mol.

PROTECTIVE EQUIPMENT Self-contained breathing apparatus for fumes; rubber gloves; goggles [R4].

PREVENTATIVE MEASURES Keep sparks, flames, and other sources of ignition away; keep material out of water sources and sewers; build dikes to contain flow as necessary. (nitrophenols) [R2].

CLEANUP Land spill: Dig a pit, pond, lagoon, or holding area to contain liquid or solid material. (SRP: If time permits, pits, ponds, lagoons, soak holes, or holding areas should be sealed with an impermeable flexible-membrane liner.) Cover solids with a plastic sheet to prevent dissolving in rain or firefighting water. Neutralize with agricultural lime (slaked lime), crushed limestone, or sodium bicarbonate. (nitrophenols) [R2].

DISPOSAL (1) Pour onto sodium bicarbonate or a sand-soda-ash mixture (90/10). Mix in heavy paper cartons and burn in incinerator. May augment fire with wood or paper. (2) Burn packages of no. 1 in incinerator with afterburner and alkaline scrubber. (3) Dissolve in flammable solvent and burn in incinerator. SRP: At the time of review, criteria for land treatment or burial (sanitary landfill) disposal practices are subject to significant revision. Prior to implementing land disposal of waste residue (including waste sludge), consult with environmental regulatory agencies for guidance on acceptable disposal practices.

COMMON USES As indicator in 0.3% solution in 50% alcohol pH: 6.8 colorless, 8.6 yellow [R1].

STANDARD CODES EPA 311; NFPA—3, –, –; no ICC; no USCG; Superfund designated (hazardous substances) list.

PERSISTENCE Biodegrades at low rate.

MAJOR SPECIES THREATENED Aquatic and terrestrial life.

GENERAL SENSATION Para form is most toxic of isomers. Causes central nervous system depression, methemoglobinemia, and hyperthermia. Has produced kidney damage.

PERSONAL SAFETY Wear self-contained breathing apparatus and skin protection.

ACUTE HAZARD LEVEL THRESHOLD Concentration for fish, as phenol, freshwater—1 ppm, salt—5 ppm [R7]. Phenols are toxic to most forms of animal life at relatively low concentrations. Highly toxic to man when ingested or inhaled. Emits toxic vapors when burned. Phenols will produce tastes in water and can be irritating at low levels.

CHRONIC HAZARD LEVEL For fishing and boating, concentration should be <10 ppm [R7]. Highly toxic when ingested or inhaled over prolonged periods; undissolved solid can remain on bottom as continuing source of phenols.

DEGREE OF HAZARD TO PUBLIC HEALTH Highly toxic by ingestion or inhalation at acute or chronic exposure levels. Emits toxic vapors when burned.

OCCUPATIONAL RECOMMENDATIONS Not listed.

ACTION LEVEL Notify air authority. Enter from upwind. Isolate solid from heat source. Remove ignition sources.

ON-SITE RESTORATION Dredge solids from bottom. Use carbon or peat on dissolved portion. Seek professional environmental engineering assistance through EPA's Environmental Response Team (ERT), Edison, NJ, 24–hour no. 908–548–8730.

BEACH OR SHORE RESTORATION Do not burn.

AVAILABILITY OF COUNTERMEASURE MATERIAL Carbon—water treatment plants, sugar refineries; peat—nurseries, floral shops.

DISPOSAL NOTIFICATION Contact local air authority.

MAJOR WATER USE THREATENED Potable supply, fisheries, recreation.

PROBABLE LOCATION AND STATE Yellow crystals; will dissolve at moderate rate.

WATER CHEMISTRY Susceptible to addition of chlorine. Products of chlorination are more toxic.

NON-HUMAN TOXICITY LD_{50} rat oral 930 (640–1,350) mg/kg [R5].

ECOTOXICITY TLm Vairon 9–10 mg/L (6 hr distilled water); 20–22 mg/L (hard water) (conditions of bioassay not specified) [R3].

BIODEGRADATION Decomposition by a soil microflora in 4 days; adapted activated sludge (bench scaled activated sludge, fill, and draw operations) at 20°C product is sole carbon source: 95.0% COD (chemical oxygen demand) removal at 17.5 mg COD/g dry inoculum/hr; lag period for degradation of 16 mg/L by wastewater or by soil at pH 7.3 and 30°C: 3–5 days; impact on biodegradation process: inhibition of degradation of glucose by *Pseudomonas fluorescens* at 20 mg/L; by *Escherichia coli*: 300 mg/L [R3].

BIOCONCENTRATION The log octanol/water partition coefficient for 3–nitrophenol is 2.00 (1), from which one can estimate a BCF of 19 (2,SRC), indicating little tendency for bioconcentration in fish [R6].

GENERAL RESPONSE Avoid contact with solid or dust. Keep people away. Wear goggles, self-contained breathing apparatus, and rubber overclothing (including gloves). Stop discharge if possible. Isolate and recover discharged material. Notify local health and pollution control agencies.

FIRE RESPONSE Fire data not available. Poisonous gases may be produced when heated. Wear goggles and self-contained breathing apparatus.

EXPOSURE RESPONSE Call for medical aid. Solid or dust, if inhaled or swallowed, or if skin is exposed, may cause headache, lethargy, nausea, and cyanosis. Irritating to eyes. Move to fresh air. Remove contaminated clothing and shoes. Flush affected areas with plenty of water. If in eyes, hold eyelids open, and flush with plenty of water. If swallowed and victim is conscious, have victim drink water or milk, and have victim induce vomiting. If breathing has stopped, give artificial respiration. If breathing is difficult, give oxygen.

RESPONSE TO DISCHARGE Issue warning—water contaminant. Should be removed. Provide chemical and physical treatment.

WATER RESPONSE Harmful to aquatic life in very low concentrations. May be dangerous if it enters water intakes. Notify local health and wildlife officials. Notify operators of nearby water intakes.

EXPOSURE SYMPTOMS Inhalation: inhalation or ingestion causes headaches, drowsiness, nausea, and blue color in lips, ears, and fingernails (cyanosis). Eyes: contact with eyes causes irritation. Skin: can be absorbed through intact skin to give same symptoms as for inhalation.

EXPOSURE TREATMENT Call a physician. Inhalation: remove victim to fresh air; give artificial respiration if needed. Eyes: flush with water for 15 minutes. Skin: wash contaminated areas with soap and water. Ingestion: remove by gastric lavage or emesis and catharsis.

REFERENCES

1. *The Merck Index.* 10th ed. Rahway, NJ: Merck Co., Inc., 1983. 950.

2. Association of American Railroads. *Emergency Handling of Hazardous Materials in Surface Transportation.* Washington, DC: Assoc. of American Railroads, Hazardous Materials Systems (BOE), 1987. 499.

3. Verschueren, K. *Handbook of Environmental Data of Organic Chemicals.* 2nd ed. New York, NY: Van Nostrand Reinhold Co., 1983. 919.

4. U.S. Coast Guard, Department of Transportation. *CHRIS—Hazardous Chemical Data.* Volume II. Washington, DC: U.S. Government Printing Office, 1984–5.

5. Vernot, E. H., et al. *Toxicol, and Appl Pharm* 42: 417–23 (1977).

6. (1) Hansch, C., and A. J. Leo. *Medchem Project* Issue No. 19, Claremont, CA: Pomona College (1981); (2) Lyman, W. J., et al. *Handbook of Chemical Property Estimation Methods.* New York: McGraw-Hill, pp. 5.1–5.30 (1982).

7. Todd, D. K. *The Water Encyclopedia,* Maple Press, 1970.

■ 2-PHENOXYETHANOL

CAS # 122–99–6

SYNONYMS 1-hydroxy-2-phenoxyethane • 2-fenoxyethanol (Czech) • 2-

hydroxyethyl phenyl ether • 2-phenoxyethyl alcohol • arosol • beta-hydroxyethyl phenyl ether • beta-phenoxyethanol • beta-phenoxyethyl alcohol • dowanol-ep • ethanol, 2-phenoxy • ethylene glycol monophenyl ether • ethylene glycol phenyl ether • Fenyl-cellosolve (Czech) • glycol monophenyl ether • phenoxethol • phenoxetol • phenoxyethanol • phenoxyethyl-alcohol • phenoxytol • phenyl cellosolve • phenylmonoglycol ether • Plastiazan-41 (Russian)

MF $C_8H_{10}O_2$

MW 138.18

COLOR AND FORM Oily liquid; colorless liquid.

ODOR Faint aromatic odor.

TASTE Burning taste.

DENSITY 1.1094 at 20°C.

BOILING POINT 245.2°C at 760 mm Hg.

MELTING POINT 14°C.

SOLUBILITY 2.67 g/100 ml water; freely soluble in alcohol, ether, and sodium hydroxide; soluble in alkali.

VAPOR PRESSURE 0.03 mm Hg at 25°C.

OCTANOL/WATER PARTITION COEFFICIENT Log K_{ow} = 1.16 (measured).

STABILITY AND SHELF LIFE Stable in presence of acids and alkalies [R11, 490].

OTHER CHEMICAL/PHYSICAL PROPERTIES Bulk density 9.2 lb/gal; percent in saturated air 0.00096 at 25°C.

PREVENTATIVE MEASURES Reasonable handling precautions, plus particular care to prevent contact with the eyes, should prevent any serious toxic effects [R3].

COMMON USES Fixative for perfumes, in organic synthesis; as bactericide in conjuction with quaternary ammonium compounds; as insect repellent [R1]. Medication: topical antiseptic [R1]; solvent for stamp-pad, ball-point, and specialty inks [R2]. Fixative for cosmetics and soaps [R2]; textile dye carrier [R2]; solvent for cleaners [R2]; chemical intermediate for carboxylic acid esters—e.g., acrylate, ma-leate [R2]. Chemical intermediate for 2–phenoxyethanol phosphate and salts [R2]; chemical intermediate for polymers (e.g., with formaldehyde and melamine) [R2]. Solvent for cellulose acetate, dyes; inks, resins, organic synthesis of plasticizers, germicides, pharmaceuticals [R11, 489]. Long-term preservation of human anatomical specimens for dissection and demonstration purposes. [R11, 490].

OCCUPATIONAL RECOMMENDATIONS None listed.

NON-HUMAN TOXICITY LD$_{50}$ rat oral 1.26 g/kg [R1]; LD$_{50}$ rat oral 13.7 g/kg plastiazan–41 (ethylene glycol phenyl ether) [R5]. LD$_{50}$ mouse oral 16.5 g/kg plastiazan–41 (ethylene glycol phenyl ether) [R5].

ECOTOXICITY EC$_{50}$ *Pimephales promelas* (fathead minnow) 344 mg/L/96 hr (confidence limit 337–352 mg/L), flow-through bioassay with measured concentrations, 26.6°C, dissolved oxygen 6.0 mg/L, hardness 45.0 mg/L calcium carbonate, alkalinity 42.0 mg/L calcium carbonate, and pH 7.62. Effect: loss of equilibrium [R4]. LC$_{50}$ *Pimephales promelas* (fathead minnow) 344 mg/L/96 hr (confidence limit 337–352 mg/L), flow-through bioassay with measured concentrations, 26.6°C, dissolved oxygen 6.0 mg/L, hardness 45.0 mg/L calcium carbonate, alkalinity 42. mg/L calcium carbonate, and pH 7.62 [R4].

BIODEGRADATION Theoretical BODs of 21% (5-day), 66% (10-day), and 75% (20-day) have been measured (1); these BODs indicate that 2–phenoxyethanol will largely be removed during biological waste treatment (1) [R6].

BIOCONCENTRATION Based upon a water solubility of 26,940 mg/L at 25°C (1), the BCF for 2–phenoxyethanol can be estimated to be 2 from a regression-derived equation (2,SRC). Based upon a measured log K_{ow} of 1.16 (3), the BCF for 2–phenoxyethanol can be estimated to be 4.5 from a regression-derived equation (2,SRC). These BCF values suggest that 2–phenoxyethanol will not bioconcentrate significantly in aquatic organisms (SRC) [R7].

ATMOSPHERIC STANDARDS This action promulgates standards of performance for equipment leaks of volatile organic compounds (VOC) in the synthetic organic chemical manufacturing industry (SOCMI). The intended effect of these standards is to require all newly constructed, modified, and reconstructed SOCMI process units to use the best demonstrated system of continuous emission reduction for equipment leaks of VOC, considering costs, non-air-quality health, and environmental impact and energy requirements. Ethylene glycol monophenyl ether is produced, as an intermediate or final product, by process units covered under this subpart [R8].

TSCA REQUIREMENTS Pursuant to section 8 (d) of TSCA, EPA promulgated a model health and safety data reporting rule. The section 8 (d) model rule requires manufacturers, importers, and processors of listed chemical substances and mixtures to submit to EPA copies and lists of unpublished health and safety studies. 2–Phenoxyethanol is included on this list [R9].

FDA Ethylene glycol monophenyl ether is an indirect food additive for use only as a component of adhesives [R10].

REFERENCES

1. Budavari, S. (ed.). *The Merck Index— Encyclopedia of Chemicals, Drugs, and Biologicals*. Rahway, NJ: Merck and Co., Inc., 1989. 1153.

2. SRI.

3. Clayton, G.D., and F.E. Clayton (eds.). *Patty's Industrial Hygiene and Toxicology*: Volume 2A, 2B, 2C: *Toxicology*. 3rd ed. New York: John Wiley and Sons, 1981–1982. 3944.

4. Brooke, L. T., D. J. Call, D. T. Geiger, and C. E. Northcott (eds.). *Acute Toxicities of Organic Chemicals to Fathead Minnows (Pimephales promelas)*. Superior, WI: Center for Lake Superior Environmental Studies, Univ. of Wisconsin–Superior, 1984. 295.

5. Loseva, I. E. *TR Azerb Nauchno–Issled Inst Gig Tr Prof Zabol* 9: 28 (1974)

6. Dow Chemical Co. *The Glycol Ethers Handbook*. Midland, MI: Dow Chemical Co. (1981).

7. (1) Valvani, S. C., et al., *J Pharm Sci* 70: 502–7 (1981). (2) Lyman, W. J., et al., *Handbook of Chemical Property Estimation Methods* New York: McGraw-Hill pp. 5–4, 5–10 (1982) (3) Hansch, C., and A. J. Leo. *Medchem Project* Issue No 26. Claremont, CA: Pomona College (1985).

8. 40 CFR 60.489 (7/1/90).

9. 40 CFR 716.120 (7/1/90).

10. 21 CFR 175.105 (4/1/90).

11. Sax, N.I., and R.J. Lewis, Sr. (eds.). *Hawley's Condensed Chemical Dictionary*. 11th ed. New York: Van Nostrand Reinhold Co., 1987.

■ O–PHENYLENEDIAMINE

CAS # 95–54–5

SYNONYMS 1,2-benzenediamine • 1,2-diaminobenzene • 1,2-phenylenediamine • 2-aminoaniline • ci-76010 • ci oxidation base-16 • o-benzenediamine • o-diaminobenzene • orthamine • EK-1700 • NSC-5354 • SQ-15500

MF $C_6H_8N_2$

MW 108.14

COLOR AND FORM Brownish-yellow leaf from water; plates from chloroform; white solid.

DISSOCIATION CONSTANTS $pK_1 < 2$, $pK_2 = 4.47$.

BOILING POINT 256–258°C.

MELTING POINT 103–104°C.

FLAMMABILITY LIMITS Lower 1.5% [R6].

SOLUBILITY Slightly soluble in water; freely soluble in alcohol, chloroform, ether; soluble in benzene.

OCTANOL/WATER PARTITION COEFFICIENT 0.15.

STABILITY AND SHELF LIFE Darkens in air [R8].

DISPOSAL SRP: At the time of review, criteria for land treatment or burial (sanitary landfill) disposal practices are subject to significant revision. Prior to implementing land disposal of waste residue (including waste sludge), consult with environmental regulatory agencies for guidance on acceptable disposal practices.

COMMON USES Detection of nitrite; textile developing agent; laboratory reagent; vulcanizing agent; ion-exchange resins; block polymers; corrosion inhibitors; photography [R2]; hair dye constituent [R3]; chemical intermediate for dyes: vat red 14 and 15, vat orange 7; chemical intermediate for pigments: red 194, and orange 43; chemical intermediate for fungicides: benomyl and thiabendazole; chemical intermediate for 1,10–phenanthroline, a paint drier; oxidation base dye for furs and hair dye; identification agent for 1,2–diketones and carboxylic acids; identification agent for aromatic and aliphatic aldehydes [R1].

OCCUPATIONAL RECOMMENDATIONS None listed.

THRESHOLD LIMIT 8–hr time-weighted average (TWA) 0.1 mg/m^3 (1991) [R9].

NON-HUMAN TOXICITY LD_{50} rat oral 660–1,284 mg/kg, females more sensitive than males [R13, 1214]; LD_{50} mouse, oral 331–450 mg/kg [R13, 1214]; LD_{50} guinea pig, oral 360 mg/kg [R13, 1214]; LD_{50} rat ip 516 mg/kg [R13, 1214]; LD_{50} rat percutaneous (24 hours) >5,000 mg/kg [R13, 1214]; LD_{LO} mouse subcutaneous 600 mg/kg [R7].

BIODEGRADATION In one aerobic screening study using activated-sludge inoculum, o-phenylenediamine (initial concentration of 25–30 ppm) exhibited 33% removal after a 5-day incubation period when acclimated for 20 days at 20°C and pH 7.2 (1). In a screening test using soil microflora and an initial concentration of 10 µg/ml o-phenylenediamine, the decomposition period was observed to be >64 days for total loss of ultraviolet absor-

bency at a wavelength of 294 nm (2). o–Phenylenediamine (initial concentration of 500 ppm) exhibited a 44.5% theoretical BOD after 8 days in a Warburg respirometer using aniline-acclimated sludge at 20°C (3) [R10].

BIOCONCENTRATION Based on an experimental log octanol-water partition coefficient of 0.15 (2) and a recommended regression-derived equation (1), the BCF for o-phenylenediamine can be estimated to be 0.77 (SRC). This BCF value indicates that o-phenylenediamine will not bioconcentrate in aquatic organisms (SRC) [R11].

TSCA REQUIREMENTS Section 8 (a) of TSCA requires manufacturers of this chemical substance to report preliminary assessment information concerned with production, use, and exposure to EPA as cited in the preamble in 51 FR 41329 [R12].

DOT *Health Hazards:* Poisonous if swallowed. Inhalation of dust poisonous. Fire may produce irritating or poisonous gases. Runoff from fire control or dilution water may cause pollution (phenylenediamines) [R4].

Fire or Explosion: Some of these materials may burn, but none of them ignites readily (phenylenediamines) [R4].

Emergency Action: Keep unnecessary people away; isolate hazard area and deny entry. Stay upwind; keep out of low areas. Positive-pressure self-contained breathing apparatus (SCBA) and structural firefighters' protective clothing will provide limited protection. Call emergency responce telephone number on shipping paper first. If shipping paper not available or no answer, call CHEMTREC at 1–800–424–9300. If water pollution occurs, notify the appropriate authorities (phenylenediamines) [R4].

Fire: Small fires: Dry chemical, CO_2, water spray, or regular foam. Large Fires: Water spray, fog, or regular foam. Move container from fire area if you can do so without risk (phenylenediamines) [R4].

Spill or Leak: Do not touch or walk through spilled material; stop leak if you can do so without risk. Small Spills: Take up with sand or other noncombustible absorbent material and place into containers for later disposal. Small Dry Spills: With clean shovel place material into clean, dry container, and cover loosely; move containers from spill area. Large Spills: Dike far ahead of liquid spill for later disposal (phenylenediamines) [R4].

First Aid: Move victim to fresh air; call emergency medical care. In case of contact with material, immediately flush skin or eyes with running water for at least 15 minutes. Remove and isolate contaminated clothing and shoes at the site (phenylenediamines) [R4].

FIRE POTENTIAL Slight [R5].

REFERENCES

1. SRI.

2. Sax, N. I., and R. J. Lewis, Sr. (eds.). *Hawley's Condensed Chemical Dictionary.* 11th ed. New York: Van Nostrand Reinhold Co., 1987. 901.

3. Burnett, C., et al. *J Toxicol Environ Health* 2 (3) 657–662 (1977).

4. U.S. Department of Transportation. *Emergency Response Guidebook 1993.* DOT P 5800.6. Washington, DC: U.S. Government Printing Office, 1993, p. G–53.

5. Sax, N. I. *Dangerous Properties of Industrial Materials,.* 5th ed. New York: Van Nostrand Reinhold, 1979. 902.

6. National Fire Protection Guide. *Fire Protection Guide on Hazardous Materials.* 10th ed. Quincy, MA: National Fire Protection Association, 1991, p. 325M–79.

7. Sax, N. I. *Dangerous Properties of Industrial Materials,.* 6th ed. New York, NY: Van Nostrand Reinhold, 1984. 2184.

8. Hawley, G. G. *The Condensed Chemical Dictionary.* 10th ed. New York: Van Nostrand Reinhold Co., 1981. 800.

9. American Conference of Governmental Industrial Hygienists. *Threshold Limit Values for Chemical Substances and Physical Agents and Biological Exposure Indices for 1994–1995.* Cincinnati, OH: ACGIH, 1994. 29.

10. (1) Pitter, P. *Water Res* 10: 231–35 (1976). (2) Alexander, M., B. K. Lustigman, *J Agric Food Chem* 14: 410–3 (1966). (3) Malaney, G. W. *J Water Pollut Control Fed* 32: 1300–11 (1960).

11. (1) Lyman, W. J., et al., *Handbook of Chemical Property Estimation Methods.* Washington DC: Amer Chemical Soc p. 5–4 (1990). (2) Hansch, C., and A. J. Leo. *Medchem Project* Issue No 26 Claremont, CA: Pomona College (1985).

12. 40 CFR 712.30 (7/1/94).

13. American Conference of Governmental Industrial Hygienists, Inc. *Documentation of the Threshold Limit Values and Biological Exposure Indices.* 6th ed. Volumes I, II, III. Cincinnati, OH: ACGIH, 1991.

■ PHENYLMERCURIC CHLORIDE

CAS # 100–56–1

SYNONYMS (chloromercuric) benzene • benzene, (chloromercuric) - • benzene, (chloromercuric) - • chlorid-fenylrtutnaty (Czech) • fenylmercurichlorid (Czech) • mercuriphenyl chloride • phenyl chloromercury • phenylmercury chloride • Phenylquecksilberchlorid (German) • pmc • phenyl mercuric chloride • stopspot • mersolite-2 • merfazin • hexason • agrenal • agronal • chlorophenylmercury

MF C_6H_5ClHg

MW 313.18

COLOR AND FORM White satiny leaflets.

MELTING POINT 250–252°C.

SOLUBILITY Soluble in about 20,000 parts cold water; slightly soluble in hot alcohol; soluble in benzene, ether, pyridine.

PREVENTATIVE MEASURES Use disposable uniforms, so that a contaminated uniform is not a source of absorption through the skin. Preventive measures: adequate ventilation; careful attention to good housekeeping, e.g., avoidance of spills, and prompt and proper cleaning if a spill occurs; all containers of mercury and its compounds should be kept tightly closed; floors should be washed on a regular basis with dilute calcium sulfide solution or other suitable reactant; floors should be nonporous; all workers directly involved in the plant operation should shower thoroughly each day before leaving. (mercury compounds) [R6].

CLEANUP Mercury removal from waste water can be accomplished by the BMS process. Chlorine is added to the waste water, oxidizing any mercury present to the ionic state. The BMS adsorbent (an activated carbon concentrate with sulfur compound on its surface) is used to collect ionic mercury. The spent adsorbent is then distilled to recover the mercury, leaving a carbon residue for reuse or disposal. The TMR IMAC Process can also be used; Waste water is fed into a reactor, whereby a slight excess of chlorine is maintained, oxidizing any mercury present to ionic mercury. The liquid is then passed through the TMR IMAC ion-exchange resin, where mercury ions are adsorbed. The mercury is then stripped from the spent resin with hydrochloric acid solution (mercury compounds) [R7]. Spilled mercury compounds or solution can be cleaned up by any method that does not cause excessive airborne contamination or skin contact. (mercury compounds) [R8].

DISPOSAL SRP: At the time of review, criteria for land treatment or burial (sanitary landfill) disposal practices are subject to significant revision. Prior to implementing land disposal of waste residue (including waste sludge), consult with environmental regulatory agencies for guidance on acceptable disposal practices.

COMMON USES Pharmaceutic aid (antimicrobial agent) [R2]; antiseptic agent [R1]; when applied to skin and mucous membranes, phenylmercuric salts inhibit growth of Gram-positive and Gram-negative bacteria, trichomonas, candida, and dermatophytic fungi (phenylmercuric

salts) [R3]. The aryl mercury compounds are used as denaturants in ethyl alcohol (aryl mercury compounds) [R4].

OCCUPATIONAL RECOMMENDATIONS
OSHA Meets criteria for OSHA medical records rule [R12].

THRESHOLD LIMIT Time-weighted average (TWA) 0.1 mg/m^3, skin (1982) (mercury, aryl, and inorganic compounds, as Hg) [R20, 25].

HUMAN TOXICITY Lethal blood level: The concentration of organic mercury present in the blood (serum or plasma) that has been reported to cause death in humans is: >0.06% (i.e., >0.6 µg/ml). (organic mercury) [R9].

ECOTOXICITY LC$_{50}$ *Plumaria elegans* (red alga, sporing) 54 µg/L/18 hr (after 7 days) (conditions of bioassay not specified) [R10].

BIOLOGICAL HALF-LIFE In general, body burden of mercury in man has a half-life of about 60 days. (mercury) [R19, 1612].

BIODEGRADATION All forms of mercury (metal, vapor, inorganic, or organic) are converted to methylmercury. Inorganic forms are converted by microbial action in the atmosphere to methylmercury. (mercurial compounds) [R11].

ATMOSPHERIC STANDARDS Emissions to the atmosphere from sludge incineration plants, sludge drying plants, or a combination of these sludge wastewater treatment plant processes shall not exceed 3,200 grams of mercury per 24–hour period. (total mercury) [R13].

RCRA REQUIREMENTS A solid waste containing chlorophenylmercury may become characterized as a hazardous waste when subjected to the toxicant extraction procedure listed in 40 CFR 261.24, and, if so characterized, must be managed as a hazardous waste [R14].

DOT *Health Hazards:* Poisonous if swallowed. Inhalation of dust poisonous. Fire may produce irritating or poisonous gases. Runoff from fire control or dilution water may cause pollution (phenylmercuric compounds, solid, not otherwise specified) [R5].

Fire or Explosion: Some of these materials may burn, but none of them ignites readily (phenylmercuric compounds, solid, not otherwise specified) [R5].

Emergency Action: Keep unnecessary people away; isolate hazard area and deny entry. Stay upwind; keep out of low areas. Self-contained breathing apparatus (SCBA) and structural firefighter's protective clothing will provide limited protection. Call CHEMTREC at 1–800–424–9300 for emergency assistance. If water pollution occurs, notify the appropriate authorities (phenylmercuric compounds, solid, not otherwise specified) [R5].

Fire: Small fires: Dry chemical, CO_2, halon, water spray, or standard foam. Large Fires: Water spray, fog, or standard foam is recommended. Move container from fire area if you can do so without risk (phenylmercuric compounds, solid, not otherwise specified) [R5].

Spill or Leak: Do not touch spilled material; stop leak if you can do so without risk. Small Spills: Take up with sand or other noncombustible absorbent material and place into containers for later disposal. Small Dry Spills: With clean shovel place material into clean, dry container, and cover; move container from spill area. Large Spills: Dike far ahead of liquid spill for later disposal (phenylmercuric compounds, solid, not otherwise specified) [R5].

First Aid: Move victim to fresh air; call emergency medical care. Remove and isolate contaminated clothing and shoes at the site. In case of contact with material, immediately flush skin or eyes with running water for at least 15 minutes (phenylmercuric compounds, solid, not otherwise specified) [R5].

FDA Bottled water shall, when a composite of analytical units of equal volume from a sample is examined by the methods described in paragraph (d) (1) (II) of this section, meet the standards of chemical quality, and shall not contain mercury in excess of 0.002 mg/L (total mercury)

[R15]. The color additive FD and C Blue Number 2 shall conform to the specifications in the CFR 74.102 and were free from impurities other than those named, including mercury (as Hg) in not more than 1 part per million, to the extent that such other impurities may be avoided by current good manufacturing practice (total mercury) [R16]. The color additive FD and C Green Number 3 shall conform to the specifications in the CFR 74.203 and were free from impurities other than those named, including mercury (as Hg) in not more than 1 part per million, to the extent that such other impurities may be avoided by current good manufacturing practice (total mercury) [R17]. The color additive FD and C Yellow Number 5 shall conform to the specifications in the CFR 74.705 and were free from impurities other than those named, including mercury (as Hg) in not more than 1 part per million, to the extent that such other impurities may be avoided by current good manufacturing practice. (total mercury) [R18].

REFERENCES

1. Sax, N. I., and R. J. Lewis, Sr. (eds.). *Hawley's Condensed Chemical Dictionary.* 11th ed. New York: Van Nostrand Reinhold Co., 1987. 904.

2. *The Merck Index.* 10th ed. Rahway, NJ: Merck Co., Inc., 1983. 1052.

3. American Medical Association, AMA Department of Drugs, *AMA Drug Evaluations.* 3rd ed. Littleton, Massachusetts: PSG Publishing Co., Inc., 1977. 888.

4. Sittig, M. *Handbook of Toxic and Hazardous Chemicals,* Park Ridge, NJ: Noyes Data Corp. p. 421 (1981).

5. Department of Transportation. *Emergency Response Guidebook 1987.* DOT P 5800.4. Washington, DC: U.S. Government Printing Office, 1987, p. G–53.

6. *Kirk–Othmer Encyclopedia of Chemical Technology.* 3rd ed., Volumes 1–26. New York, NY: John Wiley and Sons, 1978–1984, p. 15 (81) 167.

7. Environment Canada. *Tech Info for Problem Spills:* Mercury (Draft), p. 59 (1982).

8. National Research Council. *Prudent Practices for Handling Hazardous Chemicals in Laboratories.* Washington, DC: National Academy Press, 1981. 53.

9. Winek, C. L. *Drug and Chemical Blood-Level Data 1985.* Pittsburgh, PA: Allied Fischer Scientific, 1985.

10. Boney, A. D., et al., *Biochemical Pharmacol* 2: 37 (1959) as cited in USE-PA, Ambient Water Quality Criteria Doc; Mercury p. 72 (1984) EPA 440/5–84–026.

11. Environment Canada. *Tech Info for Problem Spills* p. 41 (1982).

12. 29 CFR 1910.20 (7/1/87).

13. 40 CFR 61.52 (b) (7/1/87).

14. 40 CFR 261.24 (7/1/87).

15. 21 CFR 103.35 (4/1/88).

16. 21 CFR 74.102 (4/1/88).

17. 21 CFR 74.203 (4/1/88).

18. 21 CFR 74.705 (4/1/88).

19. Gilman, A.G., L.S. Goodman, and A. Gilman. (eds.). *Goodman and Gilman's The Pharmacological Basis of Therapeutics.* 7th ed. New York: Macmillan Publishing Co., Inc., 1985.

20. American Conference of Governmental Industrial Hygienists. *Threshold Limit Values for Chemical Substances and Physical Agents and Biological Exposure Indices for 1994–1995.* Cincinnati, OH: ACGIH, 1994.

■ PHENYLMERCURIC NITRATE

CAS # 55–68–5

SYNONYMS Mercuriphenyl nitrate • mercury, (nitrato-o) phenyl • mercury, nitratophenyl • nitric acid, phenylmercury salt • phenylmercuric nitrate, basic • phenylmercury nitrate • phenmerzyl nitrate • phenylmercury(II) nitrate • merphenyl nitrate • merpectogel • mersolite-7 • phe-mer-nite • phenalco • phermernite • phenitol

MF $C_6H_5HgNO_3$

MW 339.71

COLOR AND FORM Crystals.

MELTING POINT 176–186°C.

TOXIC COMBUSTION PRODUCTS When heated to decomposition it emits very toxic fumes of mercury and NO_x [R1].

SOLUBILITY 1g soluble in 600 mL water; slightly soluble in alcohol and glycerin; more soluble in the presence of either nitric acid or alkali hydroxides.

STABILITY AND SHELF LIFE Unstable, decomposition into basic compound on contact with water [R10, 1106].

OTHER CHEMICAL/PHYSICAL PROPERTIES Saturated aqueous solution is acid to litmus; melts between 175°C and 185 °C.

PREVENTATIVE MEASURES Use disposable uniforms, so that a contaminated uniform is not a source of absorption through the skin. Preventative measure: adequate ventilation; careful attention to good housekeeping, e.g., avoidance of spills, and prompt and proper cleaning if a spill occurs; all containers of mercury and its compounds should be kept tightly closed; floors should be washed on a regular basis with dilute calcium sulfide solution or other suitable reactant; floors should be nonporous; all workers directly involved in the plant operation should shower thoroughly each day before leaving. (mercury compounds) [R2].

CLEANUP Mercury removal from waste water can be accomplished by the BMS process. Chlorine is added to the waste water, oxidizing any mercury present to the ionic state. The BMS adsorbent (an activated carbon concentrate of sulfur compound on its surface) is used to collect ionic mercury. The spent adsorbent is then distilled to recover the mercury, leaving a carbon residue for reuse or disposal. The TMR IMAC process can also be used: Waste water is fed into a reactor, whereby a slight excess of chlorine is maintained, oxidizing any mercury present to ionic mercury. The liquid is then passed through the TMR IMAC ion-exchange resin, where mercury ions are adsorbed. The mercury is then stripped from the spent resin with hydrochloric acid solution. (mercury compounds) [R3].

DISPOSAL SRP: At the time of review, criteria for land treatment or burial (sanitary landfill) disposal practices are subject to significant revision. Prior to implementing land disposal of waste residue (including waste sludge), consult with environmental regulatory agencies for guidance on acceptable disposal practices.

COMMON USES In vaginal preparation [R10, 1106].

OCCUPATIONAL RECOMMENDATIONS .

OSHA During an 8-hr work shift, an employee may not be exposed to a concentration of mercury above 1 mg/m³ (mercury) [R6]; threshold-limit 8–hr time-weighted average (TWA) 0.1 mg/m³, skin (1982) (mercury, aryl compounds, as Hg) [R13, 24].

HUMAN TOXICITY Lethal blood level: The concentration of organic mercury present in the blood (serum or plasma) that has been reported to cause death in humans is: >0.06%; (i.e., >0.6 μg/ml). (organic mercury) [R5].

IARC SUMMARY Evaluation: There is inadequate evidence in humans for the carcinogenicity of mercury and mercury compounds. Overall evaluation: metallic mercury and inorganic mercury compounds are not classifiable as to their carcinogenicity to humans (Group 3) [R4].

BIOLOGICAL HALF-LIFE In general, body burden of mercury in man has a half-life of about 60 days. (mercury) [R11, 1612].

BIODEGRADATION Upon entering an aqueous system, virtually any mercurial compound may be microbially converted to methylmercury. (mercury compounds) [R12, p. 14–9].

ATMOSPHERIC STANDARDS Emissions to the atmosphere from sludge incineration plants, sludge drying plants, or a combination of these sludge wastewater treatment plant processes shall not exceed 3,200 grams of mercury per 24–hour period. (total mercury) [R7].

RCRA REQUIREMENTS A solid waste containing phenylmercuric nitrate may become characterized as a hazardous waste when subjected to the toxicant extraction procedure listed in 40 CFR 261.24, and, if so characterized, must be managed as a hazardous waste [R8].

FIFRA REQUIREMENTS All uses of mercury are cancelled except the following: (1) as a fungicide in the treatment of textiles and fabrics intended for continuous outdoor use; (2) as a fungicide to control brown mold on freshly sawn lumber; (3) as a fungicide treatment to control Dutch elm disease; (4) as an in-can preservative in water-based paints and coatings; (5) as a fungicide in water-based paints and coatings used for exterior application; (6) as a fungicide to control "winter turf diseases" such as *Sclerotinia boreales*, and gray and pink snow mold, subject to the following: a. the use of these products were prohibited within 25 feet of any water body where fish are taken for human consumption; b. These products can be applied only by or under the direct supervision of golf course superintendents; c. the products will be classified as restricted-use pesticides when they are reregistered and classified in accordance with section 4 (c) of FEPCA. (mercury) [R9].

REFERENCES

1. Sax, N. I. *Dangerous Properties of Industrial Materials.* 6th ed. New York, NY: Van Nostrand Reinhold, 1984. 1748.

2. *Kirk–Othmer Encyclopedia of Chemical Technology.* 3rd ed., Volumes 1–26. New York, NY: John Wiley and Sons, 1978–1984, p. 15 (81) 167.

3. Environment Canada. *Tech Info for Problem Spills: Mercury (Draft),* p. 59 (1982).

4. IARC. *Monographs on the Evaluation of the Carcinogenic Risk of Chemicals to Man.* Geneva: World Health Organization, International Agency for Research on Cancer, 1972–present (multivolume work), p. 58 324 (1993).

5. Winek, C. L. *Drug and Chemical Blood–Level Data 1985.* Pittsburgh, PA: Allied Fischer Scientific, 1985.

6. 29 CFR 1910.1000 (7/1/87).

7. 40 CFR 61.52 (b) (7/1/87).

8. 40 CFR 261.24 (7/1/87).

9. Environmental Protection Agency/OPTS. *Suspended, Cancelled, and Restricted Pesticides.* 3rd Revision. Washington, DC: Environmental Protection Agency, January 1985. 16.

10. Osol, A. (ed.). *Remington's Pharmaceutical Sciences.* 16th ed. Easton, Pennsylvania: Mack Publishing Co., 1980.

11. Gilman, A.G., L.S. Goodman, and A. Gilman (eds.). *Goodman and Gilman's The Pharmacological Basis of Therapeutics.* 7th ed. New York: Macmillan Publishing Co., Inc., 1985.

12. Callahan, M.A., M.W. Slimak, N.W. Gabel, et al. *Water-Related Environmental Fate of 129 Priority Pollutants.* Volume I. EPA-440/4 79-029a. Washington, DC: U.S. Environmental Protection Agency, December 1979.

13. American Conference of Governmental Industrial Hygienists. *Threshold Limit Values for Chemical Substances and Physical Agents and Biological Exposure Indices for 1994–1995.* Cincinnati, OH: ACGIH, 1994.

■ PHOSPHAMIDON

CAS # 13171–21–6

SYNONYMS (2-chloor-3-diethylamino-1-methyl-3-oxo-prop-1-en-yl)-dimethyl-fosfaat [Dutch] • (2-Chlor-3-diaethylamino-1-methyl-3-ossido-prop-1-en-yl)-dimethyl-phosphat [German] • (2-cloro-3-dietilamino-1-metil-3-oxo-prop-1-en-il)-dimetil-fosfato [Italian] • (o, o-dimethyl-o-(1-methyl-2-chloro-2-diethylcarbamoyl-vinyl) phosphate • 1-chloro-diethylcarbamoyl-1-propen-2-yl dimethyl phosphate • 2-chloro-2-diethylcarbamoyl-1-methylvinyl dimethyl phosphate • 2-chloro-2-diethylcarbamyl-1-methylvinyl-dimethyl phosphate • 2-chloro-3-(diethylamino) -1-methyl-3-oxo-1-propenyl dimethyl phosphate • 2-chloro-N,N-diethyl-3-hydroxycrotonamide dimethyl phosphate • apamidon • c-570 • ciba-570 • crotonamide, 2-chloro-N, N-diethyl-3-hydroxy-, dimethyl phosphate • dimecron • dimethyl 2-chloro-2-diethylcarbamoyl-1-methylvinyl phosphate • dimethyl diethylamido-1-chlorocrotonyl (2) phosphate • dimethyl phosphate of 2-chloro-N, N-diethyl-3-hydroxycrotonamide • Dixon • ent-25515 • famfos • fosfamidon [Dutch] • fosfamidone [Italian] • foszfamidon [Hungarian] • merkon • ml-97 • N, N-diethyl 2-chloro-3-dimethylphosphate crotonamide • o,o-dimethyl o-(2-chloro-2-(N, N-diethylcarbamoyl) -1-methylvinyl) phosphate • phosphamidone • phosphate de dimethyle et de (2-chloro-2-diethylcarbamoyl-1-methyl-vinyle) [French] • phosphoric acid, 2-chloro-3-(diethylamino)-1-methyl-3-oxo-1-propenyl dimethyl ester • phosphoric acid, dimethyl ester, ester with 2-chloro-N,N-diethyl-3-hydroxycrotonamide • sundaram-1975

MF $C_{10}H_{19}ClNO_5P$

MW 299.69

COLOR AND FORM Yellow liquid (pure); colorless liquid; oil.

ODOR Faint odor.

DENSITY 1.2132 at 25.

VISCOSITY 1.22 at 20°C; viscosity: 70 cP at 25°C (technical product).

BOILING POINT 120°C at 0.001 mm Hg; 162°C 1.5 mm Hg.

MELTING POINT −45°C.

SOLUBILITY Miscible with water and most organic solvents except saturated hydrocarbons; 1 g dissolves in about 30 g hexane.

VAPOR PRESSURE 2.5×10^{-5} mm Hg at 20°C.

CORROSIVITY Corrosive to iron, tin, aluminum; noncorrosive to polyethylene.

STABILITY AND SHELF LIFE Isomeric mixture stable in neutral and acid media; half-life at 23°C = 13.8 days at pH 7, 2.2 days at pH 10 [R4].

OTHER CHEMICAL/PHYSICAL PROPERTIES Stable in neutral or acid media; hydrolyzed by alkali. At 20°C 50% hydrolysis (calculated) occurs in 60 d at pH 5, 54 d at pH 7, and 12 d at pH 9.

PROTECTIVE EQUIPMENT Wear a pesticide respirator jointly certified by the Mining Enforcement and Safety Administration and by the National Institute for Occupational Safety and Health. Wear natural rubber gloves, protective clothing, and goggles [R3].

PREVENTATIVE MEASURES Poisonous if swallowed, inhaled, or absorbed through skin. Do not get in eyes, on skin, or on clothing. Do not breathe mist. Wash thoroughly after handling. Do not drink any alcoholic beverage before or during spraying, because alcohol promotes absorption of organic phosphates [R3].

DISPOSAL SRP: At the time of review, criteria for land treatment or burial (sanitary landfill) disposal practices are subject to significant revision. Prior to implementing land disposal of waste residue (including waste sludge), consult with environmental regulatory agencies for guidance on acceptable disposal practices.

COMMON USES Insecticide for citrus and cotton crops, deciduous fruit, and nut crops [R2]. Systemic insecticide with strong stomach action. Used for sucking insects, stem borers in rice, aphids in various crops [R1]. It is effective against sap-feeding insects at rates of 300–600 g/ha and other pests including sugarcane stem borers and rice leaf beetles at rates of 500–1,000 g/ha. [R12, 668].

OCCUPATIONAL RECOMMENDATIONS None listed.

NON-HUMAN TOXICITY LD_{50} rat oral 17.4 mg/kg [R12, 668]; LD_{50} rat percutaneous 374 mg/kg [R12, 668]; LD_{50} mouse sc 26 mg/kg [R11, 373]; LD_{50} mouse ip 5.8 mg/kg [R11, 373]; LD_{50} mouse iv 6 mg/kg [R11, 373]; LD_{50} mouse, oral 13 mg/kg [R11, 373].

ECOTOXICITY LD_{50} *Anas platyrhynchos* (mallard), oral 3.05 mg/kg, 3-mo-old females (95% confidence limit 2.91–5.00) (sample purity 80%) [R5]; LD_{50} *Perdix perdix* (partridge) oral 9.7 mg/kg, 3-mo-old females, and males (95% confidence limit 8.3–11.3) [R6]; LD_{50} *Zenaida macroura* (mourning dove) oral males and females 2.0–4.0 mg/kg [R6]; LD_{50} *Zenaida asiatica* (white-winged dove) oral, adult male, and female, 2.93 mg/kg (95% confidence limit 2.44–3.66 mg/kg) (sample purity 80%) [R5]; EC_{50} *Simocephalus serrulatus* 0.012 mg/L/48 hr at 15°C, first instar (95% confidence limit 0.0079–0.018 mg/L); static bioassay without aeration, pH 7.2–7.5, water hardness 40–50 mg/L as calcium carbonate, and alkalinity of 30–35 mg/L (technical material, 80%) [R7]; EC_{50} *Daphnia pulex* 0.010 mg/L/48 hr at 15°C, first instar (95% confidence limit 0.0067–0.015 mg/L); static bioassay without aeration, pH 7.2–7.5, water hardness 40–50 mg/L as calcium carbonate, and alkalinity of 30–35 mg/L (technical material, 80%) [R7]; LC_{50} *Gammarus fasciatus* 0.013 mg/L/96 hr at 15°C, mature (95% confidence limit 0.006–0.028 mg/L); static bioassay without aeration, pH 7.2–7.5, water hardness 40–50 mg/L as calcium carbonate, and alkalinity of 30–35 mg/L (technical material, 80%) [R7]; LC_{50} *Pteronarcys californica* 1.5 mg/L/96 hr at 15°C, second year class (95% confidence limit 0.77–2.92 mg/L); static bioassay without aeration, pH 7.2–7.5, water hardness 40–50 mg/L as calcium carbonate, and alkalinity of 30–35 mg/L (technical material, 80%) [R7]; LC_{50} *Salmo gairdneri* (rainbow trout) 7.8 mg/L/96 hr at 15°C, wt 0.8 g (95% confidence limit 6.2–9.8 mg/L); static bioassay without aeration, pH 7.2–7.5, water hardness 40–50 mg/L as calcium carbonate, and alkalinity of 30–35 mg/L (technical material, 80%) [R7]; LC_{50} *Pimephales promelas* (fathead minnow) 100 mg/L/96 hr at 18°C, wt 1.0 g (95% confidence limit 91–110 mg/L); static bioassay without aeration, pH 7.2–7.5, water hardness 40–50 mg/L as calcium carbonate, and alkalinity of 30–35 mg/L (technical material, 80%) [R7]; LC_{50} *Ictalurus punctatus* (channel catfish) 70 mg/L/96 hr at 18°C, wt 0.8 g (95% confidence limit 67–74 mg/L); static bioassay without aeration, pH 7.2–7.5, water hardness 40–50 mg/L as calcium carbonate, and alkalinity of 30–35 mg/L (technical material, 80%) [R7]; LC_{50} *Lepomis macrochirus* (bluegill) 3.4 mg/L/96 hr at 24°C, wt 0.5 g (95% confidence limit 2.4–4.9 mg/L); static bioassay without aeration, pH 7.2–7.5, water hardness 40–50 mg/L as calcium carbonate, and alkalinity of 30–35 mg/L (technical material, 80%) [R7]; LD_{50} *Coturnix japonica* (Japanese quail), oral 3.60 mg/kg, adult females (95% confidence limit 1.80–7.20 mg/kg) (sample purity 85%) [R5]; LD_{50} *Phasianus colchicus* (pheasants) oral 4.24 mg/kg, 4–month-old female (95% confidence limit 3.37–5.34 mg/kg) (sample purity 85%) [R5]; LD_{50} *Alectoris chukar* oral 11.8 mg/kg, 3–to–5–month-old males and females (95% confidence limit 10.1–13.8 mg/kg) (sample purity 80%) [R5]; LC_{50} *Coturnix japonica* (Japanese quail) oral 90 ppm [R8].

BIODEGRADATION When applied at 5 ppm, the half-lives of phosphamidon in loam, loamy sand, and sand was approximately 6, 3, and <3 days, respectively (1,2). At an application rate of 1 ppm, the half-life in loam and silt was three to four wks (1,2). No data on sterile controls were presented, and the pH of the soils studied were not reported [R9].

BIOCONCENTRATION Using an estimated log K_{ow} of 0.38 (1), one would estimate a BCF of 1.1 for phosphamidon using a recommended regression equation (2). This would indicate that phosphamidon would not bioconcentrate in aquatic organisms. Bioconcentration of phosphamidon is also unlikely because it is rapidly metabolized in animals (3) [R10].

REFERENCES

1. *Farm Chemicals Handbook 1994.* Willoughby, OH: Meister, 1994, p. C–279.

2. SRI.

3. *Farm Chemicals Handbook 1995.* Willoughby, OH: Meister, 1995, p. C–291.

4. Martin, H., and C. R. Worthing (eds.). *Pesticide Manual.* 4th ed. Worcestershire, England: British Crop Protection Council, 1974. 409.

5. U.S. Department of the Interior, Fish and Wildlife Service. *Handbook of Toxicity of Pesticides to Wildlife.* Resource Publication 153. Washington, DC: U.S. Government Printing Office, 1984. 64.

6. U.S. Department of the Interior, Fish and Wildlife Service, Bureau of Sport Fisheries and Wildlife. *Handbook of Toxicity of Pesticides to Wildlife.* Washington, DC: U.S. Government Printing Office, 1970. 94.

7. U.S. Department of Interior, Fish and Wildlife Service. *Handbook of Acute Toxicity of Chemicals to Fish and Aquatic Invertebrates.* Resource Publication No. 137. Washington, DC: U.S. Government Printing Office, 1980. 64.

8. Hill, E. F., and Camardese, M. B. *Lethal Dietary Toxicities of Environmental Contaminants, and Pesticides to Coturnix.* Fish and Wildlife Technical Report 2. Washington, DC: United States Department of Interior, Fish and Wildlife Service, 1986. 116.

9. (1) Voss, G. H. Geissbuhler. *Res Rev* 37: 133–52 (1971). (2) Benyon, K.I., et al., *Res Rev* 47: 55–142 (1973).

10. (1) Meylan, W.M., P.H. Howard. *J Pharm Sci* 84: 83–92 (1995). (2) Lyman, W. J., et al., *Handbook of Chemical Property Estimation Methods,* New York: McGraw-Hill, Chapt 5, Eqn 5–2 (1982). (3) Geissbuhler, H., et al., *Res Rev* 37: 39–60 (1971).

11. Hayes, Wayland J., Jr. *Pesticides Studied in Man.* Baltimore/London: Williams and Wilkins, 1982.

12. Worthing, C.R., and S.B. Walker (eds.). *The Pesticide Manual—A World Compendium.* 8th ed. Thornton Heath, UK: The British Crop Protection Council, 1987.

■ PHOSPHONIC ACID, (2,2,2–TRICHLORO–1–HYDROXYETHYL) –, DIMETHYL ESTER

$$CH_3O \overset{\overset{O}{\|}}{\underset{\underset{CH_3O}{}}{P}} - \overset{OH}{\underset{}{C}}HCCl_3$$

CAS # 52–68–6

DOT # 2783

SIC CODES 2879; 2842.

SYNONYMS ((2,2,2-trichloro-1-hydroxyethyl) dimethylphosphonate) • 1-hydroxy-2,2,2-trichloroethylphosphonic acid, dimethyl ester • dimethoxy-2,2,2-trichloro-1-hydroxy-ethyl phosphine oxide • dimethyltrichlorohydroxyethyl phosphonate • ent—19,763 • nci—c54831 • o,o-dimethyl (1-hydroxy-2,2,2-trichloroethyl) phosphonate • o,o-dimethyl 2,2,2-trichloro-1-hydroxyethylphosphonate • o,o-dimethyl-(1-hydroxy-2,2,2-trichloro) ethyl phosphate • o,o-dimethyl-1-hydroxy-2,2,2-trichloroethylphosphonate • o,o-Dimethyl-(1-hydroxy-2,2,2-trichloraethyl)-phosphat (German) • o,o-dimethyl-(2,2,2-trichloor-1-hydroxy-ethyl)-fosfonata (Dutch) • o, o-dimethyl-(2,2,2-trichloro-1-idrossi-etil)-fosfonato (Italian) • Bayer-l-13/59 • trichlorofon • trichlorophon • trichlorphon • Bayer-l-1359 • trichloorfon (Dutch) • anthon • dipterex • dylox • dyrex • foschlor • neguvon • masoten • proxol • tugon • Bayer-15, 922 • ditrifon • dep • metriphonate • dimethyl 2,2,2-trichloro-1-hydroxyethylphosphonate • dimethyl (2,2,2-trichloro-1-hydroxyethyl) phosphonate • oms-800 • ent-19-763 • Bayer-15-922 • trichlorphon • nevugon • cekufon • briten • denkaphon • o, o-dimethyl (2,2,2-tri-chloro-1-hydroxyethyl) phosphonate • equino-aid • leivasom • trinex • bilarcil • (2,2,2-trichloro-1-hydroxyethyl)-phosphonic acid dimethyl ester • chlorofos • metrifonate • trichlorphene • vermicide-Bayer-2349 • combot-equine • danex •
ditriphon-50 • agroforotox • chlorophos • dimetox • dyvon • forotox • foschlor-r • loisol • methyl-chlorophos • metriphonate • phoschlor • vermicide-Bayer-2349 • wotexit • aerol-1 • chlorak • chlorfos • chloroftalm • chlorophthalm • bovinox • combot-equine • dylox • dyrex • equino-aid • Foschlorem (Polish) • detf • dicontal-fort

MF $C_4H_8Cl_3O_4P$

MW 257.44

COLOR AND FORM White crystals; pale-yellow crystals; colorless crystals.

ODOR Ethyl ether-like.

DENSITY 1.73 at 20°C.

BOILING POINT 100°C at 1 mm Hg.

MELTING POINT 83–84°C.

FIREFIGHTING PROCEDURES Extinguish fire using agent suitable for type of surrounding fire. Material itself does not burn or burns with difficulty. (trichlorfon, agricultural insecticides, not elsewhere classified, other than liquid) [R12].

SOLUBILITY At 25°C: water 15.4 g/100 ml, ether 17 g/100 ml, chloroform 75 g/100 ml, benzene 15.2 g/100 ml; very slightly soluble in pentane and hexane; soluble in alcohols, toluene; soluble in most chlorinated hydrocarbons; insoluble in petroleum oils; poorly soluble in carbon tetrachloride; soluble in ketones; slightly soluble in aromatic solvents; readily soluble in dichloromethane, 2–propanol; nearly insoluble in n-hexane.

VAPOR PRESSURE 7.8×10^{-6} mm Hg at 20°C.

OCTANOL/WATER PARTITION COEFFICIENT Log K_{ow} = 5.75.

CORROSIVITY Corrosive to metals.

STABILITY AND SHELF LIFE Stable at room temperature [R16].

OTHER CHEMICAL/PHYSICAL PROPERTIES Decomposed by alkali.

PROTECTIVE EQUIPMENT Wear goggles, self-contained breathing apparatus, and rubber overclothing (including gloves) [R13].

PREVENTATIVE MEASURES No food or smoking in the working area. Administer with good ventilation. After the day's work enact deep respirations [R15].

CLEANUP Spillages of pesticides at any stage of their storage or handling should be treated with great care. Liquid formulations may be reduced to solid phase by evaporation. Dry sweeping of solids is always hazardous: These should be removed by vacuum cleaning or by dissolving them in water or other solvent in the factory environment. (pesticides) [R14, 1619].

DISPOSAL Group I containers: combustible containers from organic or metallo-organic pesticides (except organic mercury, lead, cadmium, or arsenic compounds) should be disposed of in pesticide incinerators or in specified landfill sites. (organic or metallo-organic pesticides) [R17]. After the material has been contained, remove with contaminated soil and place in plastic-lined impervious containers. The material may be incinerated in a pesticide incinerator at the specified temperature/dwell-time combination. Any sludges or solid residues generated should be disposed of in accordance with all applicable federal, state and local pollution control requirements. If appropriate incineration facilities are not available, material may be buried in a chemical waste landfill. Material is not amenable to treatment at a municipal sewage treatment plant.

COMMON USES Insecticide for non-agricultural uses—e.g., forests; insecticide for agricultural uses—e.g., vegetables, cotton, alfalfa, corn, deciduous fruits, and nuts; insecticide for livestock—e.g., beef cattle, horses, poultry, and for animal buildings; insecticide for outdoor and aquatic areas and ornamentals, non-food fishery uses, domestic dwellings; anthelmintic agent for horses [R6]; do not apply to lime-treated surfaces [R7]; "pour-on" against grubs in cattle. ointment against mange mites, ringworm, and *Stephanofilaria*, topical application daily for 7 days to cure ringworm on calves; and orally against *Schistosoma bovis, Ostertagia ostertagi, Mecistocirrus* sp, *Hemonchus*

contortus, Cooperia, sp, and *Oesophagostum radiatum*; simultaneous use benefits of atropine are noted particularly in horse and dog trials without any apparent reduction in drug efficacy [R8, 614]. Trichlorfon has been used to treat helminthiasis, including ankylostomiasis, ascariasis, trichuriasis, and creeping eruption in man. Expected pharmacological effects did appear as side effects but no more severe or frequent than for other anthelminthics. Was also tried as treatment for schistosomiasis but was essentially ineffective [R9]. Trichlorfon is a contact and ingested insecticide with penetrant action recommended for agricultural use against fruit flies and lepidopterous larvae at 75–120 g AI/100 L, against coleopterous larvae in lawns, and for controlling household pests, particularly flies, and for the control of ectoparasites of domestic animals [R2, 820]; principal uses are in vegetables, maize, rice, sugarcane, deciduous fruit, grapes, and forestry, being particularly effective against Lepidoptera, Diptera, and Heteroptera; registered for control of cockroaches, crickets, silverfish, bedbugs, and fleas (SRP: former use) [R10]; a paste containing 100 mg/g mebendazole and 454 mg/g of trichlorfon may be used as an oral anthelmintic and boticide in horses, under the federal Food, Drug, and Cosmetic Act [R11]; used to control flies and roaches; it is also an anthelmintic [R3, 1001]. Insecticidal control in agriculture, horticulture, forestry, food storage, gardening, households, and animal husbandry. In particular, control of Diptera, Lepidoptera, Hymenoptera, Hemiptera, and Coleoptera that attack fruit (including citrus), vines, vegetables, cereals, ornamentals, grassland, trees, cotton, safflowers, sunflowers, lucerne, soya beans, sorghum, maize, rice, sugar beet, sugar cane, oil palms, tobacco, olives. Also used to control household pests such as flies, cockroaches, fleas, bed bugs, silverfish, ants; as a fly bait in farm buildings and animal houses; and for control of ectoparasites on domestic animals [R4]; insecticide for many different species in banana, cereals, chickpea, citrus, coffee, cotton, fruit trees, grapes, oleiferous palm, olive groves, pastures, red currant, rice, sugar beet and cane, sunflower, vegetables, tea, tobacco, flowers [R5]; medication (vet): anthelmintic for grubs, bots, screwworms [R1].

BINARY REACTANTS Decomposed by alkali to dichlorvos.

STANDARD CODES EPA 311; TSCA; UN class 6; not listed—IATA; CFR—49 poison class B, poison label, 50 lb passenger, 200 lb cargo; Superfund designated (hazardous substances) list. Storage code 1, 2; CFR—14 CAB code 8; not listed—NFPA; not listed—AAR; ICC—STCC Tariff no. 1–C 28 799 31.

PERSISTENCE Hydrolysis half-life (pH 6, 70°C, ethanol) 3.2 hour; alkali converts to dichlorvos, acid demethylates [R26].

MAJOR SPECIES THREATENED Aquatic microlife and larvae.

DIRECT CONTACT Readily absorbed through the skin to give central nervous system disorders.

GENERAL SENSATION Acute poisoning from inhalation or skin absorption produces headache, weakness, dizziness, anxiety, tremors of the tongue and eyelids, and impairment of visual acuity. Prolonged contact may result in salivation, tearing, abdominal cramps, vomiting, sweating, and muscular fasciculations. Death can occur from respiratory difficulty, cyanosis, and convulsions.

PERSONAL SAFETY Protect against both inhalation and absorption through the skin. Wear full protective clothing including rubber boots and gloves, safety goggles, hooded suit, and a NIOSH–approved self-contained breathing apparatus. Decontaminate fully or dispose of all equipment and clothing after use.

ACUTE HAZARD LEVEL Trichlorfon is very toxic to humans, toxicity by ingestion; between 1 teaspoonful and 1 ounce may be fatal. It is a potent cholinesterase inhibitor *in vitro*. Rapid and complete recovery from sublethal doses has been observed in laboratory animals. The lethal dose in man is approximately 360 mg/kg.

CHRONIC HAZARD LEVEL There are no

U.S. standards for trichlorfon in the workspace air. In Bulgaria the TLV is 1 mg/cu m and in Russia the TLV is 0.5 mg/cu m, and absorption through the skin is recognized. Trichlorfon is carcinogenic, hepatotoxic, and hematotoxic when rats are given intramuscular injections or percutaneously to mice twice weekly at a rate of 15 mg/kg. Alkylating the substance to reduce its chemical toxicity is a possible prophylactic measure to prevent poisoning in man as a result of industrial or agricultural exposure.

DEGREE OF HAZARD TO PUBLIC HEALTH Trichlorfon is a relatively weak cholinesterase inhibitor but is readily converted into the strong inhibitor dichlorvos at physiological pH conditions. Trichlorfon is therefore toxic by all routes of contact. It is also a suspected carcinogen.

OCCUPATIONAL RECOMMENDATIONS None listed.

ACTION LEVEL Avoid contact with the spilled cargo. Stay upwind. Notify local air, fire, and water authorities of the accident. Evacuate all people to a distance of at least 200 feet upwind and 1,000 feet downwind of the spill.

ON-SITE RESTORATION Dam the stream to reduce the flow and to prevent further dissipation of the material by water movement. Activated carbon can be applied to the water to adsorb any dissolved material. Dredges, pumps, or underwater vacuum systems may be employed to remove the undissolved material and sorbent from the bottom. An alternative method, done under controlled conditions, is to pump the water into a suitable container, then pass it through a filtration system that contains an activated-carbon filter. For more details, see ENVIREX Manual, EPA 600/2–77–227. Seek professional environmental engineering assistance through EPA's Environmental Response Team (ERT), Edison, NJ, 24–hour no. 908–548–8730.

BEACH OR SHORE RESTORATION Close beach and shore to the public until material has been removed. If tidal, scrape soil at low tide with a mechanical scraper and place in impervious containers. Avoid human contact.

AVAILABILITY OF COUNTERMEASURE MATERIAL Dredges, pumps—Army Corps of Engineers, fire department; underwater vacuum system—swimming pool operators; activated carbon—water treatment plants, chemical companies.

DISPOSAL NOTIFICATION Notify local and state health authorities, local solid waste disposal authorities, supplier and shipper of the material.

INDUSTRIAL FOULING POTENTIAL Corrosive to iron and steel; caution must be used when contaminated water is used in boiler feed and cooling waters.

MAJOR WATER USE THREATENED All downstream users of water, monitor water for concentration.

PROBABLE LOCATION AND STATE White crystals; will dissolve in the water before cleanup personnel can arrive.

WATER CHEMISTRY Hydrolyzes rapidly. This is accelerated by alkali. Hydrolysis products may still be toxic.

FOOD-CHAIN CONCENTRATION POTENTIAL Unlikely to accumulate in the food chain.

NON-HUMAN TOXICITY LD_{50} rat oral 438 mg/kg [R3, 1001]; LD_{50} rat weanling ip 190 mg/kg [R3, 1001]; LD_{50} rat adult ip 250 mg/kg [R3, 1001]; LD_{50} mouse, oral 579 mg/kg [R3, 1001]; LD_{50} mouse ip 500 mg/kg [R3, 1001]; LD_{50} guinea pig ip 300 mg/kg [R3, 1001]; LD_{50} calf, oral 600 mg/kg [R3, 1001]; LD_{50} chicken, oral 125 mg/kg [R3, 1001]; LD_{50} rat male, oral 560 mg/kg [R4]; LD_{50} rat percutaneous >2,000 mg/kg [R4].

ECOTOXICITY LC_{50} *Gammarus fasciatus* 40 µg/L/96 hr at 21°C (95% confidence limit 26–60 µg/L), mature; static bioassay without aeration, pH 7.2–7.5, water hardness 40–50 mg/L as calcium carbonate and alkalinity of 30–35 mg/L (technical material 98%) [R19, 78]; LC_{50} *Procambarus* (crayfish) 7,800 µg/L/96 hr at 12°C (95% confidence limit 6,520–9,330 µg/L), mature; static bioassay without aeration, pH 7.2–7.5, water hardness 40–50 mg/L

as calcium carbonate and alkalinity of 30–35 mg/L (technical material 98%) [R19, 78]; LC_{50} *Pteronarcella badia* 11 µg/L/96 hr at 16°C (95% confidence limit 7.6–16 µg/L), naiad; static bioassay without aeration, pH 7.2–7.5, water hardness 40–50 mg/L as calcium carbonate and alkalinity of 30–35 mg/L (technical material 98%) [R19, 78]; LC_{50} *Pteronarcys californica* 35 µg/L/96 hr at 16°C (95% confidence limit 22–55 µg/L), second year class; static bioassay without aeration, pH 7.2–7.5, water hardness 40–50 mg/L as calcium carbonate and alkalinity of 30–35 mg/L (technical material 98%) [R19, 78]; LC_{50} *Claassenia sabulosa* 22 µg/L/96 hr at 16°C (95% confidence limit 16–29 µg/L), first year class, static bioassay without aeration, pH 7.2–7.5, water hardness 40–50 mg/L as calcium carbonate and alkalinity of 30–35 mg/L (technical material 98%) [R19, 78]; LC_{50} *Skwala* 24 µg/L/96 hr at 7°C (95% confidence limit 17–32 µg/L), naiad; static bioassay without aeration, pH 7.2–7.5, water hardness 40–50 mg/L as calcium carbonate and alkalinity of 30–35 mg/L (technical material 98%) [R19, 78]; LC_{50} *Salmo clarki* (cutthroat trout) 2,700 µg/L/96 hr at 12°C (95% confidence limit 1,920–3,800 µg/L), wt 0.6 g; static bioassay without aeration, pH 7.2–7.5, water hardness 40–50 mg/L as calcium carbonate and alkalinity of 30–35 mg/L (technical material 98%) [R19, 78]; LC_{50} *Salmo gairdneri* (rainbow trout) 1,750 µg/L/96 hr at 12°C (95% confidence limit 1,250–2,460 µg/L), wt 0.5 g; static bioassay without aeration, pH 7.2–7.5, water hardness 40–50 mg/L as calcium carbonate and alkalinity of 30–35 mg/L (technical material 98%) [R19, 78]; LC_{50} *Salmo saler* (atlantic salmon) 1,400 µg/L/96 hr at 12°C (95% confidence limit 1,130–1,730 µg/L), wt 0.5 g; static bioassay without aeration, pH 7.2–7.5, water hardness 40–50 mg/L as calcium carbonate and alkalinity of 30–35 mg/L (technical material 98%) [R19, 78]; LC_{50} *Salmo trutta* (brown trout) 3,500 µg/L/96 hr at 12°C (95% confidence limit 2,490–4,910 µg/L), wt 4.6 g; static bioassay without aeration, pH 7.2–7.5, water hardness 40–50 mg/L as calcium carbonate and alkalinity of 30–35 mg/L (technical material 98%) [R19, 78]; LC$_{50}$ *Salvelinus fontinalis* (brook trout) 2,500 µg/L/96 hr at 12°C (95% confidence limit 2,180–2, 860 µg/L), wt 0.8 g; static bioassay without aeration, pH 7.2–7.5, water hardness 40–50 mg/L as calcium carbonate and alkalinity of 30–35 mg/L (technical material 98%) [R19, 78]; LC$_{50}$ *Salvelinus namaycush* (lake trout) 550 µg/L/96 hr at 12°C (95% confidencelimit 354–854 µg/L), wt 2.3 g static bioassay without aeration, pH 7.2–7.5, water hardness 40–50 mg/L as calcium carbonate and alkalinity of 30–35 mg/L (technical material 98%) [R19, 78]; LC$_{50}$ *Pimephales promelas* (fathead minnow) 7,900 µg/L/96 hr at 18°C (95% confidence limit 6,740–9,260 µg/L), wt 0.9 g static bioassay without aeration, pH 7.2–7.5, water hardness 40–50 mg/L as calcium carbonate and alkalinity of 30–35 mg/L (technical material 98%) [R19, 78]; LC$_{50}$ *Ictalurus punctatus* (channel catfish) 880 µg/L/96 hr at 18°C (95% confidence limit 766–1,010 µg/L), wt 1.6 g static bioassay without aeration, pH 7.2–7.5, water hardness 40–50 mg/L as calcium carbonate and alkalinity of 30–35 mg/L (technical material 98%) [R19, 78]; LC$_{50}$ *Lepomis macrochirus* (bluegill) 3,170 µg/L/96 hr at 18°C (95% confidence limit 2,680–3,750 µg/L), wt 1.0 g; static bioassay without aeration, pH 7.2–7.5, water hardness 40–50 mg/L as calcium carbonate and alkalinity of 30–35 mg/L (technical material 98%) [R19, 78]; LC$_{50}$ *Micropterus salmoides* (largemouth bass) 3,450 µg/L/96 hr at 18°C (95% confidence limit 3,030–3,930 µg/L), wt 0.8 g; static bioassay without aeration, pH 7.2–7.5, water hardness 40–50 mg/L as calcium carbonate and alkalinity of 30–35 mg/L (tested in hard water, 272 ppm calcium carbonate, technical material 98%) [R19, 78]; LC$_{50}$ *Skwala* 12 µg/L/96 hr at 7°C (95% confidence limit 8.9–15 µg/L), naiad; static bioassay without aeration, pH 7.2–7.5, water hardness 40–50 mg/L as calcium carbonate and alkalinity of 30–35 mg/L (soluble powder 80%) [R19, 78]; LC$_{50}$ *Salmo clarki* (cutthroat trout) 3,250 µg/L/96 hr at 12°C (95% confidence limit 2,740–3,860 µg/L), wt 0.9 g; static bioassay without aeration, pH 7.2–7.5, water hardness 40–50 mg/L as calcium carbon-

ate and alkalinity of 30–35 mg/L (soluble powder 80%) [R19, 79]; LC$_{50}$ *Salmo gairdneri* (rainbow trout) 700 µg/L/96 hr at 12°C (95% confidence limit 500–969 µg/L), wt 1.0 g; static bioassay without aeration, pH 7.2–7.5, water hardness 40–50 mg/L as calcium carbonate and alkalinity of 30–35 mg/L (soluble powder 80%) [R19, 79]; LC$_{50}$ *Salmo trutta* (brook trout) 9,200 µg/L/96 hr at 12°C (95% confidence limit 6,740–12,500 µg/L), wt 0.7 g; static bioassay without aeration, pH 7.2–7.5, water hardness 40–50 mg/L as calcium carbonate and alkalinity of 30–35 mg/L (soluble powder 80%) [R19, 79]; LC$_{50}$ *Lepomis macrochirus* (bluegill) 940 µg/L/96 hr at 18°C (95% confidence limit 645–1,360 µg/L), wt 0.8 g static bioassay without aeration, pH 7.2–7.5, water hardness 40–50 mg/L as calcium carbonate and alkalinity of 30–35 mg/L (soluble powder 80%) [R19, 79]; LC$_{50}$ *Salmo gairdneri* (rainbow trout) 1,400 µg/L/96 hr at 12°C (95% confidence limit 1,050–1,850 µg/L), wt 0.6 g; static bioassay without aeration, pH 7.2–7.5, water hardness 40–50 mg/L as calcium carbonate and alkalinity of 30–35 mg/L (liquid 40%) [R19, 79]; LC$_{50}$ bobwhite quail oral 720 ppm (95% confidence limit 591–871 ppm), age 10 days (98% pure) [R20]; LC$_{50}$ Japanese quail 1,901 ppm (95% confidence limit 1,601–2,255 ppm), age 12 days (98% pure) [R20]; LC$_{50}$ ring-necked pheasant 3,401 ppm (95% confidence limit 2,927–3,957 ppm), age 10 days (98% pure) [R20]; LC$_{50}$ mallard duck greater than 5,000 ppm (no mortality to 1,581 ppm, 30% at 5,000 ppm), age 10 days (98% pure) [R20]; LC$_{50}$ *Pteronarcys californica* 9.8 µg/L/30 days [R21]; LC$_{50}$ *Coturnix* 1,899 ppm for 5 day in diet (95% confidence interval 1,510–2,388 ppm), slope: 4: 92, standard error: 1.07 [R22];.

IARC SUMMARY No data are available for humans. Inadequate evidence of carcinogenicity in animals. Overall evaluation: Group 3: The agent is not classifiable as to its carcinogenicity to humans [R23].

BIOLOGICAL HALF-LIFE The biological half-life of trichlorfon was 80 min in mice [R18].

BIODEGRADATION Ammonifying soil microorganisms decompose trichlorfon [R16].

BIOCONCENTRATION Bioconcentration predicted from water solubility = 0.7 (calculated from water solubility by regression equations from table) [R24];.

FIFRA REQUIREMENTS In 1988, Congress amended FIFRA to strengthen and accelerate EPA's reregistration program. The nine-year reregistration scheme mandated by "FIFRA 88" applies to each registered pesticide product containing an active ingredient initially registered before November 1, 1984. "List A" consists of the 194 chemical cases (or 350 individual active ingredients) for which EPA had issued registration standards prior to the effective date of FIFRA '88. List: A; case: trichlorfon; Case No. 0104; pesticide type: insecticide; registration standard date: 06/30/84; case status: awaiting data/data in review: OPP awaits data from the pesticide's producer(s) regarding its human-health or environmental effects, or OPP has received and is reviewing such data, in order to reach a decision about the pesticide's eligibility for reregistration. Active ingredient (AI): trichlorfon; data call-in (DCI) date: 09/10/91; AI Status: the producer(s) of the pesticide has made commitments to conduct the studies and pay the fees required for reregistration, and is meeting those commitments in a timely manner [R25].

GENERAL RESPONSE Avoid contact with solid. Keep people away. Wear goggles, self-contained breathing apparatus, and rubber overclothing (including gloves). Stop discharge if possible. Isolate and remove discharged material. Notify local health and pollution control agencies.

FIRE RESPONSE Fire data not available.

EXPOSURE RESPONSE Call for medical aid. Solid is poisonous if swallowed or if skin is exposed. Irritating to eyes. Remove contaminated clothing and shoes. Flush affected areas with plenty of water. If in eyes, hold eyelids open, and flush with plenty of water. If swallowed and victim is conscious, have victim drink water or milk, and have victim induce vomiting. If swallowed and victim is unconscious or having convulsions, do nothing except keep victim warm.

RESPONSE TO DISCHARGE Issue warning—poison, water contaminant. Restrict access. Should be removed. Provide chemical and physical treatment.

WATER RESPONSE Harmful to aquatic life in very low concentrations. May be dangerous if it enters water intakes. Notify local health and wildlife officials. Notify operators of nearby water intakes.

EXPOSURE SYMPTOMS Inhalation, ingestion, and skin absorption. Inhibits cholinesterase. Headache, depressed appetite, nausea, miosis are symptoms of light exposures. Moderate effects are peritoneal paralysis, diarrhea, salivation, lacrimation, sweating, dyspnea, substernal tightness, slow pulse, tremors, muscular cramps, and ataxia. Severe symptoms are: pyrexia, cyanosis, pulmonary edema, areflexia, loss of sphincter control, paralysis, coma, heart block, shock, and respiratory failure. Eyes: increases permeability of blood vessels in anterior eye. Reduces corneal sensitivity with glaucoma, abnormalities in intraocular tension, or decreased visual acuity.

EXPOSURE TREATMENT Call a physician. Inhalation: artificial respiration and oxygen if required. In the presence of symptoms give atropine sulfate, 2 to 4 mg at 5– to 10–minute intervals until signs of atropinization appear. Eyes: flush with copious quantities of water. Skin: wash with copious amounts of water and soap. Ingestion: gastric lavage with tap water or emesis induced by syrup of ipecac, followed by saline catharsis. If symptoms occur use treatment as indicated above.

REFERENCES

1. Budavari, S. (ed.). *The Merck Index— Encyclopedia of Chemicals, Drugs, and Biologicals.* Rahway, NJ: Merck and Co., Inc., 1989. 1514;.

2. Worthing, C. R., and S. B. Walker (eds.). *The Pesticide Manual—A World Compendium.* 8th ed. Thornton Heath, UK:

The British Crop Protection Council, 1987.

3. Hayes, W. J., Jr., E. R. Laws, Jr. (eds.). *Handbook of Pesticide Toxicology.* Volume 2. *Classes of Pesticides.* New York, NY: Academic Press, Inc., 1991.

4. Hartley, D., and H. Kidd (eds.). *The Agrochemicals Handbook.* 2nd ed. Lechworth, Herts, England: The Royal Society of Chemistry, 1987, p. A408/Aug 87.

5. *Farm Chemicals Handbook* 1992. Willoughby, OH: Meister Publishing Co., 1992, p. C–344.

6. SRI.

7. Spencer, E. Y. *Guide to the Chemicals Used in Crop Protection.* 7th ed. Publication 1093. Research Institute, Agriculture Canada, Ottawa, Canada: Information Canada, 1982. 572.

8. Rossoff, I. S. *Handbook of Veterinary Drugs.* New York: Springer Publishing Company, 1974.

9. Hayes, W. J., Jr. *Toxicology of Pesticides* Baltimore: Williams and Wilkins, 1975. 76.

10. *Farm Chemicals Handbook* 87. Willoughby, OH: Meister Publishing Co., 1987, p. C–257.

11. 51 FR 13212 (4/18/86).

12. Bureau of Explosives. *Emergency Handling of Haz Matl in Surface Trans,* p. 507 (1981).

13. U.S. Coast Guard, Department of Transportation. *CHRIS—Hazardous Chemical Data.* Volume II. Washington, DC: U.S. Government Printing Office, 1984–5.

14. International Labour Office. *Encyclopedia of Occupational Health and Safety.* Vols. I and II. Geneva, Switzerland: International Labour Office, 1983.

15. ITII. *Toxic and Hazarous Industrial Chemicals Safety Manual.* Tokyo, Japan: The International Technical Information Institute, 1982. 204.

16. IARC. *Monographs on the Evaluation of the Carcinogenic Risk of Chemicals*

to Man. Geneva: World Health Organization, International Agency for Research on Cancer, 1972–present (multivolume work), p. V30 208 (1983).

17. 40 CFR 165 (7/1/86).

18. IARC. *Monographs on the Evaluation of the Carcinogenic Risk of Chemicals to Man.* Geneva: World Health Organization, International Agency for Research on Cancer, 1972–present (multivolume work), p. V30 216 (1983).

19. U.S. Department of Interior, Fish and Wildlife Service. *Handbook of Acute Toxicity of Chemicals to Fish and Aquatic Invertebrates.* Resource Publication No. 137. Washington, DC: U.S. Government Printing Office, 1980.

20. U.S. Department of the Interior, Fish and Wildlife Service, Bureau of Sports Fisheries and Wildlife. *Lethal Dietary Toxicities of Environmental Pollutants to Birds.* Special Scientific Report—*Wildlife* No. 191. Washington, DC: U.S. Government Printing Office, 1975. 35.

21. Verschueren, K. *Handbook of Environmental Data of Organic Chemicals.* 2nd ed. New York, NY: Van Nostrand Reinhold Co., 1983. 567.

22. Hill, E. F., and Camardese, M. B. *Lethal Dietary Toxicities of Environmental Contaminants and Pesticides to Coturnix.* Fish and Wildlife Technical Report 2. Washington, DC: United States Department of Interior, Fish and Wildlife Service, 1986. 136.

23. IARC. *Monographs on the Evaluation of the Carcinogenic Risk of Chemicals to Man.* Geneva: World Health Organization, International Agency for Research on Cancer, 1972–present (multivolume work), p. S7 73 (1987).

24. Kenaga, E. E. *Ecotoxicology and Environmental Safety* 4: 35 (1980).

25. USEPA/OPP. *Status of Pesticides in Reregistration and Special Review,* p. 117 (Mar, 1992) EPA 700–R–92–004.

26. *Isophorone—Technical Bulletin,* Exxon Chemical Company, 1974.

■ PHOSPHONODITHIOIC ACID, ETHYL-, -ETHYL S-PHENYL ESTER

$$CH_3CH_2-\overset{\overset{\displaystyle S}{\|}}{P}\overset{\displaystyle O-C_2H_5}{\underset{\displaystyle S-C_6H_5}{}}$$

CAS # 944-22-9

SYNONYMS difonate • difonatul • dyfonat • dyfonate • ent-25,796 • ent-25,796 • ethylphosphonodithioic acid o-ethyl S-phenyl ester • fonofos • n-2790 • n-2790 • o-Aethyl-S-phenyl-aethyl-dithiophosphonat (German) • o-ethyl S-phenyl ethyldithiophosphonate • o-ethyl S-phenyl(rs)-ethylphosphonodithioate • o-ethyl s-phenyl ethylphosphono-dithioate • (−)-o-ethyl S-phenyl ethylphosphonodithioate • o-ethyl S-phenyl ethylphosphonothiolothionate • phosphonodithioic acid, ethyl-, o-ethyl S-phenyl ester • Stauffer-N-2790 • dyfonate tillam 1-4e • doubledown

MF C_{10}H_{15}OPS_2

MW 246.34

COLOR AND FORM Light-yellow liquid; colorless.

ODOR Aromatic; pungent mercaptan.

DENSITY 1.16 at 25°C.

BOILING POINT 130°C at 0.1 mm Hg.

FIREFIGHTING PROCEDURES Dry chemical, carbon dioxide, foam, and water spray to cool [R6].

TOXIC COMBUSTION PRODUCTS Emits oxides of phosphorus fumes when burned [R6].

SOLUBILITY Miscible with acetone, ethanol, kerosene, 4−methylpentan−2−one, xylene; solubility in water: 13 ppm at 22°C.

VAPOR PRESSURE 28 mPa at 25°C.

OCTANOL/WATER PARTITION COEFFICIENT Log K_{ow}= 3.94.

CORROSIVITY Corrosive to steel.

STABILITY AND SHELF LIFE Stable <100°C [R2, 429].

OTHER CHEMICAL/PHYSICAL PROPERTIES Henry's law constant = 5.4×10^{-6} atm-cu m/mole (est).

PROTECTIVE EQUIPMENT Protective clothing should include goggles, because great care should be taken to avoid eye contact [R7].

PREVENTATIVE MEASURES SRP: Contaminated protective clothing should be segregated in such a manner so that there is no direct personal contact by personnel who handle, dispose of, or clean the clothing. Quality assurance to ascertain the completeness of the cleaning procedures should be implemented before the decontaminated protective clothing is returned for reuse by the workers. Contaminated clothing should not be taken home at end of shift, but should remain at employee's place of work for cleaning.

DISPOSAL Hydrolysis and landfill (of treated residues): alkaline hydrolysis produces methylphosphoramidoate and methyl mercaptan. Those products are less hazardous. Acid hydrolysis yields dimethylphosphorathioate and ammonia. Both products are nontoxic. Containers should be thoroughly drained and rinsed with aqueous alkali. Recommended method: Incineration. Peer-review: large amounts, incinerate in a unit with effluent gas scrubbing (peer-review conclusions of an IRPTC expert consultation (May 1985)) (monitor) [R8].

COMMON USES Dyfonate is soil insecticide useful in controlling soil insects such as corn rootworms (*Diabrotica* species), wireworms (Elateridae), garden symphylan (*Scutigerella immaculata*), root maggots (*Hylemya* species), crickets (Gryllidae), and others [R3]. Control of soil insects in cereals, maize, sorghum, vegetables, ornamentals, fruit (including citrus and bananas), vines, olives, potatoes, sugar beet, sugar cane, groundnuts, tobacco, and turf [R4]; soil fumigant [R5]; controls corn borers and rootworms, cutworms, symphylans (garden centipedes), wireworms, other soil foliar pests [R1].

OCCUPATIONAL RECOMMENDATIONS .

OSHA 8–hr time-weighted average: 0.1 mg/m³. Final rule limits were achieved by any combination of engineering controls, work practices, and personal protective equipment during the phase-in period, Sept 1, 1989 through Dec 30, 1993. Final rule limits became effective Dec 31, 1993. Skin absorption designation in effect as of Sept 1, 1989 [R13]. Threshold limit 8-hr time-weighted average (TWA) 0.1 mg/m³, skin (1977) [R14, 22].

NON-HUMAN TOXICITY LD_{50} rat male, oral 24.5 mg/kg (− dyphonate) [R2, 429]; LD_{50} rat female, oral 10.8 mg/kg (− dyphonate) [R2, 429]; LD_{50} guinea pig percutaneous 278 mg/kg [R2, 429]; LD_{50} rabbit percutaneous 159 mg/kg [R2, 429]; LD_{50} rat, oral 8–17.5 mg/kg [R4]; LC_{50} rat inhalation 1.9 mg/L air/1 hr [R4].

ECOTOXICITY LD_{50} *Salmo gairdneri*(rainbow trout) 0.020 mg/L/96 hr at 13°C (95% confidence limit 0.016–0.025 mg/L), wt 1.7 g; static bioassay without aeration, pH 7.2–7.5, water hardness 40–50 mg/L as calcium carbonate, and alkalinity of 30–35 mg/L [R10]; LD_{50} *Lepomis macrolophus* (bluegill) 0.007 mg/L/96 hr at 24°C (95% confidence limit 0.005–0.009 mg/L), wt 1.0 g; static bioassay without aeration, pH 7.2–7.5, water hardness 40–50 mg/L as calcium carbonate, and alkalinity of 30–35 mg/L [R10]; LD_{50} *Anas platyrhynchos* (mallard) oral 16.9 mg/kg (95% confidence limit 13.4–21.3 mg/L), males 3–4 mo of age [R9].

BIODEGRADATION Two fungal species isolated from soil readily degraded dyphonate (1). Metabolites isolated were dyphonate-oxon, ethylethoxyphosphonothioic acid, ethylethoxyphosphonic acid, methyl phenyl sulfoxide, and methyl phenyl sulfone (1). Biodegradation was the major process for the loss of dyphonate in soil (2). Biodegradation of dyphonate is faster in dyphonate-treated soil than in untreated soil (3–4). The major metabolite of biodegradation of dyphonate in dyphonate-treated soil is carbon dioxide with dyphonate-oxon, methyl phenyl sulfone, and other unidentified polar products as minor metabolites (5) [R11].

BIOCONCENTRATION Based on regression equations (1), log BCF values of 2.11 and 2.76 are estimated assuming a water solubility of 16 mg/L at 25°C (2) and a log K_{ow} value of 3.94 (3), respectively (SRC). In a model ecosystem study, log BCF for dyphonate in mosquito fish (*Gambusia affinis*) was less than 2 (4). The difference between experimentally measured and estimated bioconcentration of dyphonate in aquatic organisms may be due to metabolism of dyphonate (SRC) and it has been concluded (5) that bioconcentration is not important [R12].

FIFRA REQUIREMENTS Classified for restricted use limited to use by or under the direct supervision of a certified applicator. Formulation: emulsifiable concentrates 44% and greater. Use pattern: All uses. Classification: restricted. Criterion influencing restriction: acute dermal toxicity [R15].

REFERENCES

1. *Farm Chemicals Handbook* 1992. Willoughby, OH: Meister Publishing Co., 1992, p. C–132.

2. Worthing, C. R., and S. B. Walker (eds.). *The Pesticide Manual—A World Compendium*. 8th ed. Thornton Heath, UK: The British Crop Protection Council, 1987.

3. Spencer, E. Y. *Guide to the Chemicals Used in Crop Protection*. 7th ed. Publication 1093. Research Institute, Agriculture Canada, Ottawa, Canada: Information Canada, 1982. 308.

4. Hartley, D., and H. Kidd (eds.). *The Agrochemicals Handbook*. 2nd ed. Lechworth, Herts, England: The Royal Society of Chemistry, 1987, p. A214/Aug 87.

5. Lewis, R. J., Sr. (ed.). *Hawley's Condensed Chemical Dictionary*. 12th ed. New York, NY: Van Nostrand Reinhold Co., 1993. 535.

6. Sax, N. I. *Dangerous Properties of Industrial Materials*, Reports. New York: Van Nostrand Reinhold, 1987, p. 10: 6/1990.

7. American Conference of Governmental Industrial Hygienists. *Documentation of*

the *Threshold Limit Values and Biological Exposure Indices.* 5th ed. Cincinnati, OH: American Conference of Governmental Industrial Hygienists, 1986. 275.

8. United Nations. *Treatment and Disposal Methods for Waste Chemicals* (IRPTC File). Data Profile Series No. 5. Geneva, Switzerland: United Nations Environmental Programme, Dec. 1985. 233.

9. U.S. Department of the Interior, Fish and Wildlife Service. *Handbook of Toxicity of Pesticides to Wildlife.* Resource Publication 153. Washington, DC: U.S. Government Printing Office, 1984. 45.

10. U.S. Department of Interior, Fish and Wildlife Service. *Handbook of Acute Toxicity of Chemicals to Fish and Aquatic Invertebrates.* Resource Publication No. 137. Washington, DC: U.S. Government Printing Office, 1980. 83.

11. (1) Tu, C. M., J. R. W. Miles. *Res Rev* 64: 17–65 (1976); (2) Miles, J.R.W., et al. *Bull Environ Contam Toxicol* 22: 312–8 (1979); (3) Ahmad, N., et al. *Bull Environ Contam Toxicol* 23: 423–9 (1979); (4) Wilde, G. *J Econ Entomol* 83: 1250–3 (1990); (5) Racke, K. D., J. R. Coats. *J Agric Food Chem* 36: 193–9 (1988).

12. (1) Lyman, W. J., et al. *Handbook of Chemical Property Estimation Methods.* Amer Chem Soc, Washington, DC, p. 5–5 (1990); (2) Racke, K. D. pp. 47–78 in *Organophosphates: Chemistry, Fate, and Effects.* Chambers, J. E., P. E. Levi (eds.), San Diego, CA: Academic Press, Inc. (1992); (3) Hansch, C., and A. J. Leo. *Medchem Project.* Issue No. 26 Claremont, CA: Pomona College (1985); (4) Metcalf, R. L. *Adv Environ Sci Technol* 8: 195–221 (1977); (5) Kenaga, E. E., *Ecotox Environ Safety* 4: 26–38 (1980).

13. 29 CFR 1910.1000 (7/1/91).

14. American Conference of Governmental Industrial Hygienists. *Threshold Limit Values for Chemical Substances and Physical Agents and Biological Exposure Indices for 1993–1994.* Cincinnati, OH: ACGIH, 1993.

15. 40 CFR 152.175 (7/1/91).

■ PHOSPHORIC ACID, 2-CHLORO-1-(2,4-DI-CHLOROPHENYL) VINYL DIETHYL ESTER

CAS # 470–90–6

SIC CODES 2879.

SYNONYMS 2,4-dichloro-alpha-(chloromethylene) benzyldiethyl phosphate • 2-chloro-1-(2,4-dichlorophenyl) vinyl diethyl phosphate • benzyl alcohol, 2,4-dichloro-alpha-(chloromethylene)-, diethyl phosphate • beta-2-chloro-1-(2',4'-dichlorophenyl) vinyl diethylphosphate • birlan • birlane • c8949 • chlofenvinphos • chlorfenvinfos • chlorphenvinfos • chlorphenvinphos • compound-4072 • cvp • cvp [pesticide] • dermaton • diethyl 1-(2,4-dichlorophenyl)-2-chlorovinyl phosphate • diethyl-2-chloro-1-(2,4-dichlorophenyl) vinyl phosphate • ent-24969 • gc-4072 • o-2-chloor-1-(2,4-dichloor-fenyl)-vinyl-o,o-diethylfosfaat (Dutch) • o-2-Chlor-1-(2,4-dichlor-phenyl)-vinyl-o,o-diaethylphosphat (German) • o-2-cloro-1-(2,4-dicloro-fenil)-vinil-o,o-dietilfosfato (Italian) • oms-1328 • phosphoric acid, 2-chloro-1-(2,4-dichlorophenyl) ethenyl diethyl ester • phosphoric acid, 2-chloro-1-(2,4-dichlorophenyl) vinyl diethyl ester • sapecron • sd-4072 • shell-4072 • supone • vinyphate • chlorfenvinphos • clofenvinfos • enolofos • phosphate de o,o-diethyle et de o-2-chloro-1-(2,4-dichlorophenyl) vinyle (French) • sd-7859 • tarene • o,o-diethyl o-(2-chloro-1-(2',4'-dichlorophenyl) vinyl) phosphate • 2,4-dichloro-alpha-(chloromethylene) benzyl alcohol diethyl phosphate • phosphoric acid, 2-chloro-1-(2,4-dichlorophenyl) ethenyl diethyl ester • o,o-diethyl o-(2-chloro-1-(2,4-dichlorophenyl) vinyl) phosphate • steladone • 2-chloro-1-(2,4-dichlorophenyl)-ethenyl diethyl phosphate • sd-7859 • cfv • vinylphate • cga-26351 • haptarax • haptasol • c-8949 • apachlor • 2-chloro-1-(2,4-dichlorophenyl) vinyldiethylphosphate •

o,o-diethyl-o-1-(2',4'-dichlorophenyl)-2-chlorovinylphosphate • supona

MF $C_{12}H_{14}Cl_3O_4P$

MW 359.58

COLOR AND FORM Colorless liquid.

ODOR Mild.

DENSITY 1.36 g/cu cm at 20°C.

BOILING POINT 167–170°C at 0.5 mm Hg.

MELTING POINT –19 to –23°C.

FIREFIGHTING PROCEDURES Fire-extinguishing media: alcohol-resistant foam, dry powder, carbon dioxide [R8, p. C–76]. Do not extinguish fire unless flow can be stopped. Solid streams of water may be ineffective. Cool all affected containers with flooding quantities of water. Apply water from as great a distance as possible. Use "alcohol" foam, carbon dioxide, or dry chemical. (organophosphorus pesticides, liquid, NOS) [R9].

SOLUBILITY 145 mg/L water at 23°C; miscible with acetone, ethanol, kerosene, xylene, propylene glycol, dichloromethane, hexane.

VAPOR PRESSURE 0.53 mPa at (extrapolated) 20°C.

OCTANOL/WATER PARTITION COEFFICIENT Log K_{ow}= 3.82 (measured).

CORROSIVITY May corrode brass, iron, and steel on prolonged contact; emulsifiable concentrate formulations are corrosive to tin plate.

STABILITY AND SHELF LIFE Slowly hydrolyzed by water or acid, 50% decomposition occurs at 38°C in greater than 700 hr at pH 1.1, greater than 400 hr at pH 9.1, but unstable in alkali—at 20°C 50% loss occurs in 1.28 hr at pH 13. [R1, 152].

OTHER CHEMICAL/PHYSICAL PROPERTIES It is a mixture of the cis– and trans– isomers of the phosphate.

PROTECTIVE EQUIPMENT Wear face mask and neoprene apron when unloading or handling containers containing powder, dust, or granular formulations. [R8, p. C–75].

PREVENTATIVE MEASURES Do not drink any alcoholic beverages before or during spraying because alcohol promotes absorption of organic phosphates. [R8, p. C–75].

DISPOSAL Hydrolysis and landfill: Chlorfenvinphos should be treated by alkali and then mixed with a portion of soil that is rich in organic matter (peer-review conclusions of an IRPTC expert consultation (May 1985)) [R10]. Product residues and sorbent media may be packaged in 17h epoxy-lined drums and disposed of at approved EPA disposal site. Soda ash or caustic destruction at pH's over 7, incineration with hydrochloric acid scrubber, or microwave plama detoxification. Encapsulation by organic polyester resin or silicate fixation. Confirm disposal procedures with responsible environmental engineer and regulatory officials.

COMMON USES Insecticide; acaricide [R4]; it is used for control of ticks, flies, lice, and mites on cattle, and blowfly, lice, ked, and itchmite on sheep [R5]; it is insecticide for control of root flies, rootworms, and cutworms at 2–4 kg/hectare. It is foliage insecticide for control of Colorado beetle on potatoes and leafhoppers on rice at 0.2–0.5 kg/hectare and stem borers in maize, sugar cane, and rice at 0.5–2.0 kg/hectare [R2]; it is used against ticks and flies on horses and goats, as well as fleas on dogs, and on organic wastes and breeding places of fly larvae including dairy barns. Do not use on cats [R6]; nematocide, parasiticide [R7]; soil application or seed treatment for control of fruit flies in maize; wheatbulb flies in wheat; bean seed flies; and phorid and sciarid flies in mushrooms [R3]. Chlorfenvinphos may be used either as a soil insecticide for the control of cutworms, root flies, and root worms at 2–4 kg AI/hectare, or as a foliar insecticide to control *Leptinotarsa decemlineata* on potato, scale insects on citrus at 200–400 g/ha (where it also exhibits ovicidal activity against mite eggs), and of stem borers on maize, rice, and sugarcane at 550–2,200 g/hectare. It controls whiteflies (*Bemisia* species) on cotton at 400–750 g/hectare, but whitefly parasites are not affected. It also controls ectoparasites (*Dalmania bovis*, *Bovicola bovis*, and *Haematopinus quadripertusus*) of cattle at 0.3–0.7 g/L, *Lucilia sericata* and *Ixodes ricinus* of sheep at 0.5 g/L, and *D. ovis*, *Melphagus ovinus*, and *Linognathus ovillus* at 0.1 g/L. Chlorfenvinphos may also be used in public health programs, especially against mosquito larvae. [R1, 152].

STANDARD CODES UN 2783; DOT–UN class 6 poison, poison B, requires poison label.

PERSISTENCE Slowly hydrolyzes in water; half-life in soil averages 12 weeks with a range of 2–23 weeks. Large doses will break down proportionately more slowly than small doses. Decomposes logarithmically (soil). Chlorfenvinphos is unlikely to persist more than one season if applied at recommended dosage rates. (in highly organic soils, 2, 4–dichlorophenyl chloride accumulated to a maximum of 0.11 ppm 15 weeks after application of 9 kg /ha. It persists >6 months in Coachella fine sand as bioassayed by *Hippelates collusor* larvae. An average 20% of that applied on a soil remains after 4 mo. Chlorfenvinphos is considered quite persistent in soil.

MAJOR SPECIES THREATENED Mammals, fish.

ACUTE HAZARD LEVEL High toxicity by all routes of exposure.

OCCUPATIONAL RECOMMENDATIONS None listed.

ON-SITE RESTORATION Seek professional environmental engineering assistance through EPA's Environmental Response Team (ERT), Edison, NJ, 24–hour no., 908–548–8730. Contain and isolate spill by using clay/bentonite dams, interceptor trenches, or impoundments. Construct swale to divert uncontaminated portion of watershed around contaminated portion. Seek professional help to evaluate problem. Implement containment measures and conduct bench-scale and pilot-scale tests prior to full-scale decontamination program implementation. Density stratification and impoundments—remove product from bottom layer by pumping through manifold, polyethylene rope mop collec-

tion, or remove clarified upper portion by skimmers or siphoning. Treatment is required for both clarified and concentrated product fractions. Treatment alternatives include decomposition with soda ash or caustic with pH adjusted above 7, powdered activated carbon, granular activated carbon, biodegradation, bentonite sorption, evaporation, aeration, and chemical oxidation. Treatment alternatives for contaminated soils include spill immobilization with bentonite/clay and collection and treatment of leachate as per contaminated water and physical removal of the immobilized residues with pH adjustment to 8 and land application on an approved sanitary landfill. Residues may be packaged for disposal.

PROBABLE LOCATION AND STATE An oily amber liquid with a mild odor. Spills would tend to settle out in the bottom mud.

WATER CHEMISTRY Slowly hydrolyzes in water. The half-life at 38°C and pH 9.1 is >400 hours; at pH 1.1 >700 hours.

NON-HUMAN TOXICITY LD_{50} mouse oral 117–200 mg/kg [R3]; LD_{50} pigeons oral 16 mg/kg [R1, 153]; LC_{50} rat inhalation 0.05 mg/L air/4 hr [R1, 153]; LD_{50} rabbit, oral 280–400 mg/kg [R13]; LD_{50} rabbit dermal 3,200–4,700 mg/kg [R13]; LD_{50} rat dermal 31–108 mg/kg [R13]; LD_{50} rat oral 9.66 mg/kg [R14]; LD_{50} rat ip 8.5 mg/kg [R14]; LD_{50} rat iv 6.6 mg/kg [R14]; LD_{50} mouse ip 37 mg/kg [R14]; LD_{50} guinea pig oral 123–250 mg/kg [R14]; LD_{50} dog oral >5,000 mg/kg [R12, 1062]; LD_{50} dog iv 50.4 mg/kg [R14].

ECOTOXICITY LD_{50} *Anas platyrhynchos* (mallard) oral 85.5 mg/kg (95% confidence limit 44.5–164 mg/kg), 3–4–mo-old females (91% beta, 8% alpha isomers) [R11]; LD_{50} *Colinus virginianus* (bobwhite) oral 80.0–160 mg/kg, 12–mo-old males (91% beta, 8% alpha isomers) [R11]; LD_{50} (pheasant), oral 63.5 mg/kg (95% confidence limit 45.8–88.1 mg/kg), 3–4–mo-old females (81% beta, 8% alpha isomers) [R11]; LC_{50} *Rasbora heteromorpha* (harlequin fish) 0.27 mg/L/48 hr (conditions of bioassay not specified) [R1, 153]; LC_{50} *Libistes reticulatus* (guppy) 0.53 mg/L/48 hr (conditions of bioassay not specified) [R1, 153].

BIODEGRADATION In one test using sterilized soil and sand, 95% of initial chlorfenvinhos remained after 45 days of incubation, while in nonsterile soil, 70–98% was degraded, suggesting a biological route of disappearance (1). In laboratory studies, the disappearance rate of chlorfenvinhos was more than 15 times faster in nonsterile soil as compared to sterilized soil (2). Using a sandy loam soil and an organic soil, more than 24 weeks of incubation were required to reach 50% degradation of chlorfenviphos when the soils were sterilized (3); in nonsterilized soils, only one week of incubation or less was needed to reach 50% degradation (3). Persistence of chlorfenvinphos was measured in sandy loam and muck soils (both sterile and nonsterile soils) at temperatures of 3, 15, and 28°C (4); degradation was much faster in the nonsterile soils (half-lives less than 10 weeks at 15 and 28°C) as compared to sterile soil (all half-lives much greater than 24 weeks) (4); degradation was slower with decreasing temperatures (4); at 28°C, degradation in the nonsterile soils was found to increase with an increase in soil moisture (0–60% capacity) (5) [R15].

BIOCONCENTRATION Based upon a measured log K_{ow} of 3.82 (1) and a water solubility of 124 mg/L at 20°C (2), the BCF of chlorfenviphos can be estimated to be 470 and 40, respectively, from regression-derived equations (3,SRC). These estimated BCF values suggest that bioconcentration in aquatic organisms may have some environmental importance (SRC). An experimental BCF of 200 has been reported for earthworms (4–5) [R16].

FIFRA REQUIREMENTS In 1988, Congress amended FIFRA to strengthen and accelerate EPA's reregistration program. The nine-year reregistration scheme mandated by "FIFRA 88" applies to each registered pesticide product containing an active ingredient initially registered before November 1, 1984. Pesticides for which EPA had not issued Registration Standards prior to the effective date of FIFRA '88 were divided into three lists based

upon their potential for exposure and other factors, with List B being of highest concern, and D of least. List: B; Case: Chlorfenvinphos; case no. 2090; pesticide type: Insecticide; case status: the pesticide is no longer an active ingredient in any registered pesticide products. Therefore, EPA is characterizing it as "cancelled." Under FIFRA, pesticide producers may voluntarily cancel their registered products. EPA also may cancel pesticide registrations if registrants fail to pay required fees, to make or meet certain reregistration commitments, or when the Agency reaches findings of unreasonable adverse effects. Active Ingredient (AI): 2–chloro–1–(2,4–dichlorophenyl) vinyl diethyl phosphate; AI status: the active ingredient is no longer contained in any registered products. [R17].

REFERENCES

1. Worthing, C. R., and S. B. Walker (eds.). *The Pesticide Manual—A World Compendium*. 8th ed. Thornton Heath, UK: The British Crop Protection Council, 1987.

2. Spencer, E. Y. *Guide to the Chemicals Used in Crop Protection*. 7th ed. Publication 1093. Research Institute, Agriculture Canada, Ottawa, Canada: Information Canada, 1982. 106.

3. Hartley, D., and H. Kidd (eds.). *The Agrochemicals Handbook*. 2nd ed. Lechworth, Herts, England: The Royal Society of Chemistry, 1987, p. A078/Aug 87.

4. Budavari, S. (ed.). *The Merck Index—Encyclopedia of Chemicals, Drugs, and Biologicals*. Rahway, NJ: Merck and Co., Inc., 1989. 322.

5. *Farm Chemicals Handbook 1984*. Willoughby, OH: Meister Publishing Co., 1984, p. C–51.

6. Rossoff, I. S. *Handbook of Veterinary Drugs*. New York: Springer Publishing Company, 1974. 101.

7. Lewis, R. J., Sr. (ed.). *Hawley's Condensed Chemical Dictionary*. 12th ed. New York, NY: Van Nostrand Reinhold Co., 1993 259.

8. *Farm Chemicals Handbook* 1993. Wil-

loughby, OH: Meister Publishing Co., 1993.

9. Association of American Railroads. *Emergency Handling of Hazardous Materials in Surface Transportation.* Washington, DC: Association of American Railroads, Bureau of Explosives, 1992. 713.

10. United Nations. *Treatment and Disposal Methods for Waste Chemicals* (IRPTC File). Data Profile Series No. 5. Geneva, Switzerland: United Nations Environmental Programme, Dec. 1985. 234.

11. U.S. Department of the Interior, Fish and Wildlife Service. *Handbook of Toxicity of Pesticides to Wildlife.* Resource Publication 153. Washington, DC: U.S. Government Printing Office, 1984. 22.

12. Hayes, W. J., Jr., E. R. Laws, Jr. (eds.). *Handbook of Pesticide Toxicology. Volume 2. Classes of Pesticides.* New York, NY: Academic Press, Inc., 1991.

13. Verschueren, K. *Handbook of Environmental Data of Organic Chemicals.* 2nd ed. New York, NY: Van Nostrand Reinhold Co., 1983. 354.

14. Hayes, Wayland J., Jr. *Pesticides Studied in Man.* Baltimore/London: Williams and Wilkins, 1982. 396.

15. (1) Rouchard, J., et al. *Bull Environ Contam Toxicol* 40: 47–53 (1988); (2) Rouchard, J., et al. *Bull Environ Contam Toxicol* 42: 409–16 (1989); (3) Miles, J.R.W., et al. *Bull Environ Contam Toxicol* 22: 312–8 (1979); (4) Miles, J.R.W., et al. *J Environ Sci Health* B18: 705–12 (1983); (5) Miles, J.R.W., et al. *J Environ Sci Health* B19: 237–43 (1984).

16. (1) Hansch, C., and A. J. Leo. *Medchem Project* Issue No. 26, Claremont, CA: Pomona College (1985); (2) Bowman, B. T., W. W. Sans. *J Environ Sci Health* B18: 221–7 (1983); (3) Lyman, W. J., et al. *Handbook of Chemical Property Estimation Methods.* Washington, DC: Amer Chemical Soc, pp. 5–4, 5–10 (1990); (4) Lord, K. A., et al. *Pestic Sci* 11: 401–8 (1980); (5) Connell, D. W., R. D. Markwell. *Chemosphere* 20: 91–100 (1990).

17. USEPA/OPP. *Status of Pesticides in Reregistration and Special Review,* p. 125 (Mar, 1992) EPA 700–R–92–004.

■ PHOSPHORIC ACID, DIMETHYL ESTER, ESTER with (E) –3–HYDROXY–N,N–DIMETHYLCROTONAMIDE

CAS # 141–66–2

SIC CODES 2879.

SYNONYMS 3-(dimethoxyphosphinyloxy)-N, N-dimethyl-cis-crotonamide • 3-(dimethylamino)-1-methyl-3-oxo-1-propenyl dimethyl phosphate (e)-isomer • 3-dimethoxyphosphinyloxy-N,N-dimethylisocrotonamide • 3-hydroxy-N,N-dimethyl-cis-crotonamide dimethyl phosphate • 3-hydroxydimethyl crotonamide dimethyl phosphate • bidrin • c-709 • carbicron • carbomicron • Ciba-709 • cis-2-dimethylcarbamoyl-1-methylvinyl dimethylphosphate • crotonamide, 3-hydroxy-N,N-dimethyl-, cis-, dimethyl phosphate • dicrotofos [Dutch] • dimethyl (e) -2-dimethyl-carbamoyl-1-methylvinyl phosphate • dimethyl 1-dimethylcarbamoyl-1-propen-2-yl phosphate • dimethyl 2-dimethylcarbamoyl-1-methyl-vinyl phosphate • dimethyl ester with (e) -3-hydroxy-N,N-dimethylcrotonamide phosphoric acid • dimethyl phosphate ester with 3-hydroxy-N, N-dimethylcrotonamide • dimethyl phosphate of 3-hydroxy-N,N-dimethyl-cis-crotonamide • ektafos • ent-24,482 • karbicron • o,o-dimethyl o-(N,N-dimethylcarbamoyl-1-methylvinyl) phosphate • o,o-dimethyl-o-(1,4-dimethyl-3-oxo-4-aza-pent-1-enyl) phosphate • o,o-Dimethyl-o-(2-dimethyl-carbamoyl-1-methyl-vinyl) phosphat [German] • o, o-dimetil-o-(1,4-dimetil-3-

oxo-4-aza-pent-1-enil)-fosfato [Italian] • oleobidrin • phosphate de dimethyle et de 2-dimethyl carbamoyl 1-methyl vinyle [French] • phosphoric acid, dimethyl ester, ester with (e)-3-hydroxy-N,N-dimethylcrotonamide • sd-3562 • Shell sd-3562 • o,o-dimethyl-o-(1,4-dimethyl-3-oxo-4-aza-pent-1-enyl) fosfaat [Dutch] • phosphoric acid, 3-(dimethylamino)-1-methyl-3-oxo-1-propenyl dimethyl ester, (e) - • crotonamide, 3-hydroxy-N-N-dimethyl-, dimethyl phosphate, cis • crotonamide, 3-hydroxy-N-N-dimethyl-, dimethyl phosphate, (e) • (e)-phosphoric acid, 3-(dimethylamino) -1-methyl-3-oxo-1-propenyl dimethyl ester

MF $C_8H_{16}NO_5P$

MW 237.22

COLOR AND FORM Yellowish liquid; yellow-brown liquid.

DENSITY 1.216 at 15°C.

BOILING POINT 400°C at 760 mm Hg.

SOLUBILITY Miscible with isobutanol, hexylene glycol, acetone; miscible with methylene chloride, chloroform, acetonitrile, water, xylene, alcohol; barely soluble in mineral oils; miscible with 4–hydroxy–4–methylpentan–2–one; less than 10 g/kg diesel oil, kerosene; miscible with isopropyl oxitol, phentoxone, and ethyl, isopropyl, and diacetone alcohols.

VAPOR PRESSURE 0.0115 Pa at 20°C.

CORROSIVITY Relatively noncorrosive to monel, copper, nickel, and aluminum, but somewhat corrosive to cast iron, mild steel, brass, and stainless steel 304; does not attack glass, polythene, stainless steel 316.

STABILITY AND SHELF LIFE It is stable when stored in glass or polythene containers up to 40°C, but is decomposed after prolonged storage at 55°C; on hydrolysis, at 20°C, 50% loss (calculated) occurs in 88 days at pH 5, 23 days at pH 9 [R2].

OTHER CHEMICAL/PHYSICAL PROPERTIES Technical grade containing about 85% (e)–isomer is an amber liquid: bp 400°C at

760 mm Hg; vapor pressure 13 mPa at 20°C; density 1.21 at 20°C.

SOLUBILITY: Totally miscible with water and organic solvents such as acetone, ethanol, 4–hydroxy–4–methylpentan–2–one, propan–2–ol; less than 10 g/kg diesel oil, kerosene (technical grade).

PREVENTATIVE MEASURES Do not get on clothing. Do not breathe vapors. [R3, p. C–77].

DISPOSAL Product residues and sorbent media may be packaged in 17h epoxy-lined drums and disposed of at an approved EPA disposal site. Destruction by high-temperature incineration or microwave plasma detoxification, if available. Encapsulation by organic polyester resin or silicate fixation. Confirm disposal procedures with responsible environmental engineer and regulatory officials. SRP: At the time of review, criteria for land treatment or burial (sanitary landfill) disposal practices are subject to significant revision. Prior to implementing land disposal of waste residue (including waste sludge), consult with environmental regulatory agencies for guidance on acceptable disposal practices.

COMMON USES Systemic insecticide and acaricide for cotton, apples, and other crops [R1].

STANDARD CODES NA2783; IMCO—pesticide, low hazard, (B) liquid; IATA—poisonous articles, class B, organic phosphates, poison label required; CFR—organic phosphate, poison B, poison label required; NFPA–none.

PERSISTENCE Moderate.

MAJOR SPECIES THREATENED No adverse effects on wildlife from spraying of forest with dicrotophos. Dicrotophos has been classified as "highly toxic to honey bees" and reduces the number of beneficial insects on cotton and on alfalfa.

DIRECT CONTACT In case of accidental contact, remove all contaminated clothing and wash skin with soap and water. If in eyes, immediately flush with water for at least 10 min and get medical attention. Repeated inhalation or skin contact may, with symptoms, progressively increase susceptibility to dicrotophos poisoning.

GENERAL SENSATION Acute intoxication produces general weakness, headache, tightness in chest, blurred vision, non-reactive pupils, salivation, sweating, nausea, vomiting, diarrhea, abdominal cramps, or convulsions.

PERSONAL SAFETY Do not breathe vapors, swallow, or get in eyes, on skin, or on clothing. Wear clean natural-rubber gloves, clean protective clothing, and goggles; during prolonged exposure or commercial application, use an approved respirator.

OCCUPATIONAL RECOMMENDATIONS None listed.

THRESHOLD LIMIT 8–hr time-weighted average (TWA) 0.25 mg/m^3, skin (1977) [R6, 19].

ON-SITE RESTORATION Seek professional environmental engineering assistance through EPA's Environmental Response Team (ERT), Edison, NJ, 24–hour no. 908–548–8730. Contain and isolate spill by using clay/bentonite dams, interceptor trenches, or impoundments. Construct swale to divert uncontaminated portion of watershed around contaminated portion. Seek professional help to evaluate problem, implement containment measures, and conduct bench-scale and pilot-scale tests prior to full-scale decontamination program implementation. Treatment alternatives for contaminated water include powdered activated carbon, granular activated carbon, aeration, biodegradation, and extended holding periods for alkaline hydrolysis. Treatment alternatives for soils include well-point collection and treatment of leachates as for contaminated water, or bentonite/cement ground injection to immobilize spill. Contaminated soils may be packaged for disposal.

PROBABLE LOCATION AND STATE Amber liquid with mild ester odor. Will dissolve in water.

WATER CHEMISTRY Neutral or acid solutions are relatively stable; the half-life of aqueous solutions at pH 9.1 and 38°C is 1,200 hours; and at pH 1.1 and 38°C it is 2,400 hours.

NON-HUMAN TOXICITY LD$_{50}$ rats percutaneous 111–136 mg/kg to 148–181 mg/kg, acute (form not identified) [R2].

ECOTOXICITY LD$_{50}$ mallard oral 4.24 mg/kg (95% confidence limit 3.06–5.88 mg/kg), 3–mo-old-males (sample purity 98% alpha isomer) [R4, 30].

BIODEGRADATION Bidrin is rapidly degraded in soil, and the rate of degradation is directly related to microbial content. Sterilization of soil by heat or by action of selective biocides drastically lowered the rate of bidrin degradation, providing evidence that microorganisms are responsible for decomposition of bidrin in soil. [R5, 183].

FIFRA REQUIREMENTS The farm worker reentry time following application of bidrin is 48 hr [R7].

REFERENCES

1. SRI.

2. Worthing, C. R., S. B. Walker (eds.). *The Pesticide Manual—A World Compendium.* 7th ed. Lavenham, Suffolk, Great Britain: The Lavenham Press Limited, 1983. 190.

3. *Farm Chemicals Handbook* 1984. Willoughby, OH: Meister Publishing Co., 1984.

4. U.S. Department of the Interior, Fish and Wildlife Service. *Handbook of Toxicity of Pesticides to Wildlife.* Resource Publication 153. Washington, DC: U.S. Government Printing Office, 1984.

5. White–Stevens, R. (ed.). *Pesticides in the Environment:* Volume 1, Part 1, Part 2. New York: Marcel Dekker, Inc., 1971.

6. American Conference of Governmental Industrial Hygienists. *Threshold Limit Values for Chemical Substances and Physical Agents and Biological Exposure Indices for 1993–1994.* Cincinnati, OH: ACGIH, 1993.

7. *Worker Protection Standards for Agricultural Pesticides,* U.S. Code of Federal

Regulations, Title 40, Part 170, p. 131 (1984).

■ PHOSPHORIC ACID, DIPHENYL TOLYL ESTER

CAS # 26444–49–5

SIC CODES 2869.

SYNONYMS cresol diphenyl phosphate • cresyl diphenyl phosphate • cresyldiphenyl phosphate • diphenyl cresol phosphate • diphenyl cresyl phosphate • diphenyl p-tolylphosphate • cdp • diphenyl tolyl phosphate • disflamoll-dpk • kronitex-cdp • methylphenyl diphenyl phosphate • monocresyl diphenyl phosphate • phosflex-112 • phosphoric acid, cresyl diphenyl ester • phosphoric acid, diphenyl tolyl ester • phosphoric acid, methylphenyl diphenyl ester • santicizer-140 • tolyl diphenyl phosphate

MF $C_{19}H_{17}O_4P$

MW 340.33

COLOR AND FORM Colorless transparent liquid.

ODOR Very slight odor.

DENSITY 1.208 at 25°C.

BOILING POINT 390°C at 760 mm Hg.

MELTING POINT (freezing point) –38°C.

FIREFIGHTING PROCEDURES Water or foam may cause frothing [R2].

SOLUBILITY Insoluble in water, glycerol; soluble in most organic solvents except glycerol.

OTHER CHEMICAL/PHYSICAL PROPERTIES 13.89 mg/m³ is equivalent to 1 ppm.

PROTECTIVE EQUIPMENT Respiratory protection (supplied-air respirator with full facepiece or self-contained breathing apparatus) should be available where these compounds are manufactured or used and should be worn in case of emergency and overexposure. (phosphorus compounds) [R3].

COMMON USES Plasticizer; extreme-pressure lubricant; hydraulic fluids; gasoline additive; food packaging [R1]; plasticizer and flame-retardant, e.g., for polyvinyl plastics [R6]; plasticizer, e.g., for cellulosic plastics, and polystyrene [R6]; plasticizer—e.g., for polycarbonates and butadiene rubbers [R6]; additive for functional fluids and gasoline [R6].

DIRECT CONTACT Respiratory tract irritant. May have irritating vapors at high-temperatures. CNS moderate to severe toxicity. Odorless. Impossible to have overexposure from vapor at normal temperatures.

ACUTE HAZARD LEVEL Low.

DEGREE OF HAZARD TO PUBLIC HEALTH Low level of hazard. Vapors may be irritating if fire or intense heat accompanies spill.

OCCUPATIONAL RECOMMENDATIONS None listed.

ACTION LEVEL Isolate from water. Locate submerged pool and keep from moving downstream.

NON-HUMAN TOXICITY Orl—rat LD_{50} 6,400 mg/kg, *Raw Material Data Handbook*, Vol. 2,14,75 (NPIRI); skn—rbt LDLo: 3,160 mg/kg, *Office of Toxic Substances Report.* OTS 206227 (TSCAT);, orl—ckn LD_{50} >10 g/kg, *Toxicology and Applied Pharmacology.* 41,291,77 (TXAPA9).

ON-SITE RESTORATION Adsorb on activated carbon or peat moss. Seek professional environmental engineering assistance through EPA's Environmental Response Team (ERT), Edison, NJ, 24–hour no. 908–548–8730.

BEACH OR SHORE RESTORATION Burn off.

AVAILABILITY OF COUNTERMEASURE MATERIAL Activated carbon—water treatment plants, sugar refineries; peat moss —nurseries.

DISPOSAL Incinerate.

DISPOSAL NOTIFICATION Local air authority.

PROBABLE LOCATION AND STATE Colorless liquid. Fair flame retardant. Good low-temperature flexibility. Will sink and remain associated with sediments.

TSCA REQUIREMENTS Pursuant to section 8 (d) of TSCA, EPA promulgated a model health and safety data reporting rule. The section 8 (d) model rule requires manufacturers, importers, and processors of listed chemical substances and mixtures to submit to EPA copies and lists of unpublished health and safety studies. Phosphoric acid, methylphenyl diphenyl ester is included on this list [R4].

RCRA REQUIREMENTS Section 8 (a) of TSCA requires manufacturers of this chemical substance to report preliminary assessment information concerned with production, use, and exposure to EPA as cited in the preamble of the 51 FR 41329 [R5].

REFERENCES

1. Sax, N. I., and R. J. Lewis, Sr. (eds.). *Hawley's Condensed Chemical Dictionary.* 11th ed. New York: Van Nostrand Reinhold Co., 1987. 322.

2. National Fire Protection Guide. *Fire Protection Guide on Hazardous Materials.* 10th ed. Quincy, MA: National Fire Protection Association, 1991, p. 325M–29.

3. International Labour Office. *Encyclopedia of Occupational Health and Safety.* Vols. I and II. Geneva, Switzerland: International Labour Office, 1983. 1684.

4. 40 CFR 716.120 (7/1/90).

5. 40 CFR 712.30 (7/1/90).

6. R1: SRI.

■ PHOSPHORIC ACID, TRI–o–TOLYL ESTER

CAS # 78–30–8

SYNONYMS o-cresyl phosphate • o-tolyl phosphate • o-Trikresylphosphate (German) • phosflex-179c • phosphoric acid, tri (2-tolyl) ester • phosphoric acid, tri-o-cresyl ester • phosphoric acid, tri-o-tolyl ester • phosphoric acid, tris (2-methylphenyl) ester • tocp • totp • tri 2-methylphenyl phosphate • tri-2-tolyl phosphate • tri-o-tolyl phosphate • triorthocresyl-phosphate • tris (o-tolyl) phosphate • Trojkrezylu-fosforan (Polish)

MF $C_{21}H_{21}O_4P$

MW 368.39

COLOR AND FORM Colorless or pale-yellow liquid.

ODOR Practically odorless.

DENSITY 1.1955 at 20°C.

BOILING POINT 410°C.

MELTING POINT 11°C.

AUTOIGNITION TEMPERATURE 725°F (385°C) (from table) [R4].

FIREFIGHTING PROCEDURES CO_2, dry chemical [R3]. Extinguish with dry chemical, foam, or carbon dioxide. Water may be ineffective on fire. Cool exposed containers with water (tricresyl phosphate) [R5].

SOLUBILITY Sparingly soluble in water; freely soluble in alcohol, benzene, ether; very soluble in carbon tetrachloride, toluene; soluble in acetic acid.

VAPOR PRESSURE 10 mm Hg at 265°C.

STABILITY AND SHELF LIFE Stable, nonvolatile (tricresyl phosphate) [R7].

OTHER CHEMICAL/PHYSICAL PROPERTIES BP: about 410°C with slight decomposition. Can react with oxidizing materials.

PROTECTIVE EQUIPMENT Wear appropriate equipment to prevent repeated or prolonged skin contact [R15, 220]. Recommendations for respirator selection. Max concn for use: 0.5 mg/m³: Respirator Class: Any dust and mist respirator [R15, 220]. Recommendations for respirator selection. Max concn for use: 1 mg/m³: Respirator Classes: Any dust and mist respirator except single-use and quarter-mask respirators. Any supplied-air respirator. Any self-contained breathing apparatus [R15, 220]. Recommendations for respirator selection. Max concn for use: 2.5 mg/m³: Respirator Classes: Any powered air-purifying respirator with a dust and mist filter. Any supplied-air respirator operated in a continuous-flow mode [R15, 220]. Recommendations for respirator selection. Max concn for use: 5 mg/m³: Respirator Classes: Any air-purifying, full facepiece respirator with a high-efficiency particulate filter. Any powered, air-purifying respirator with a tight-fitting facepiece, and a high-efficiency particulate filter. Any self-contained breathing apparatus with a full facepiece. Any supplied-air respirator with a full facepiece. Any supplied-air respirator that has a tight-fitting facepiece and is operated in a continuous-flow mode [R15, 220]. Recommendations for respirator selection. Max concn for use: 40 mg/m³: Respirator Class: Any supplied-air respirator operated in a pressure-demand or other positive-pressure mode [R15, 220]. Recommendations for respirator selection. Condition: Emergency or planned entry into unknown concn or IDLH conditions: Respirator Classes: Any self-contained breathing apparatus that has a full facepiece and is operated in a pressure-demand or other positive-pressure mode. Any supplied-air respirator with a full facepiece and operated in pressure-demand or other positive-pressure mode in combination with an auxiliary self-contained breathing apparatus operated in pressure-demand or other positive-pressure mode [R15, 220]. Recommendations for respirator selection. Condition: Escape from suddenly occuring respiratory hazards: Respirator Classes: Any air-purifying, full facepiece respirator with a high-efficiency particulate filter. Any appropriate escape-type, self-contained breathing apparatus [R15, 220]. Wear safety glasses, gas mask, long-sleeve coveralls with tight collars, and cuffs, gloves, and boots (tricresyl phosphate) [R6]. Respiratory protection (supplied-air respirator with full facepiece or self-contained breathing apparatus) should be available where these compounds are manufactured or used and should be worn in case of emergency and overexposure. (phosphorus compounds) [R16,1684].

PREVENTATIVE MEASURES Contact lenses should not be worn when working with this chemical. [R15, 220]. SRP: The scientific literature supports the wearing of contact lenses in industrial environments, as part of a program to protect the eye against chemical compounds and minerals causing eye irritation. However, there may be individual substances whose irritating or corrosive properties are such that the wearing of contact lenses would be harmful to the eye. In those specific cases contact lenses should not be worn. Workers should wash promptly when skin becomes contaminated [R15, 220]. Remove clothing promptly if it is nonimpervious clothing that becomes contaminated [R15, 220]. Prolonged skin or mucous membrane contact should be avoided by use of mechanized processes, protective clothing (including gloves and goggles) and barrier creams; in addition, scrupulous personal hygiene should be encouraged (tricresyl phosphates) [R16, 2217]. Where a process entails the heating of tricresyl phosphate, exhaust ventilation should be installed to collect vapors at source or, under special circumstances, workers may be required to wear respiratory protective equipment; a high standard of general ventilation for the workplace is also necessary, and atmospheric determinations should be carried out at regular intervals to ensure that the concentration of tri-o–cresyl phosphate in air is maintained below recommended permissible levels [R16, 2217]. Workers should not be allowed to eat, drink, or smoke at the workplace. (tricresyl phosphates) [R16, 2217].

DISPOSAL Absorb with paper towels. Place in a plastic bag. Burn in an open pan with help of flammable solvent or in a furnace (tricresyl phosphate) [R6].

COMMON USES Plasticizer in lacquers and varnishes [R1]. Flame retardent and plasticizer, in hot extrusion molding, bulk forming of plasticized polyvinyl chloride, in coatings and adhesives, as a gasoline additive to control pre-ignition, in hydrolic fluids, as a heat exchange medium, as a waterproofing agent, as a component of adhesives for air filter media, in a solvent mixture for various resins, as an extraction solvent, and as an intermediate in the synthesis of pharmaceuticals [R11]. As additive to extreme-pressure lubricants; as a non-flammable fluid in hydraulic systems; as plasticizer in lacquers and varnishes; as plasticizer in vinyl plastics manufacture; as flame retardant; as lead scavenger in gasoline; and as solvent for nitrocellulose, in cellulosic molding compositions [R12, 462]. Aryl phosphates such as tricresyl phosphate or cresyl diphenyl phosphates, when added in small percentages to the gasoline, combat misfires. (aryl phosphates) [R13].

OCCUPATIONAL RECOMMENDATIONS

OSHA 8-hr time-weighted average: 0.1 mg/m³. Skin absorption designation [R9].

NIOSH 10-hr time-weighted average: 0.1 mg/m³, skin [R15, 220].

THRESHOLD LIMIT 8–hr time-weighted average (TWA) 0.1 mg/m³, skin (1986) [R17, 35]. Excursion Limit Recommendation: Excursions in worker exposure levels may exceed three times the TLV–TWA for no more than a total of 30 min during a work day, and under no circumstances should they exceed five times the TLV–TWA. [R17, 5].

INTERNATIONAL EXPOSURE LIMITS East Germany and the former USSR: 0.1 mg/m³ [R2].

HUMAN TOXICITY Toxicity LD human toxicity oral 1.0 g/kg [R2].

NON-HUMAN TOXICITY LD$_{50}$ long-evans hooded, male rat, oral 1,160 mg/kg [R12, 465].

ECOTOXICITY LC$_{50}$ *Lepomis macrochirus* 7,000 ppm/96 hr, static bioassay in fresh water at 23°C, mild aeration applied 24 hr [R8].

TSCA REQUIREMENTS Pursuant to section 8 (d) of TSCA, EPA promulgated a model health and safety data reporting rule. The section 8 (d) model rule requires manufacturers, importers, and processors of listed chemical substances and mixtures to submit to EPA copies and lists of unpublished health and safety studies. Phosphoric acid, tris (2–methylphenyl) ester is included on this list [R10]. Section 8(a) of TSCA requires manufacturers of this chemical substance to report preliminary assessment information concerned with production, use, and exposure to EPA as cited in the preamble of the 51 FR 41329. [R18].

DOT *Health Hazards:* Poisonous; may be fatal if inhaled, swallowed, or absorbed through skin. Contact may cause burns to skin and eyes. Runoff from fire control or dilution water may give off poisonous gases and cause water pollution. Fire may produce irritating or poisonous gases (tricresyl phosphate) [R14].

Fire or Explosion: Some of these materials may burn, but none of them ignites readily. Container may explode violently in heat of fire (tricresyl phosphate) [R14].

Emergency Action: Keep unnecessary people away; isolate hazard area and deny entry. Stay upwind, out of low areas, and ventilate closed spaces before entering. Positive-pressure self-contained breathing apparatus (SCBA) and chemical protective clothing that is specifically recommended by the shipper or manufacturer may be worn. It may provide little or no thermal protection. Structural firefighters' protective clothing is not effective for these materials. Remove and isolate contaminated clothing at the site. Call CHEMTREC at 1–800–424–9300 as soon as possible, especially if there is no local hazardous materials team available (tricresyl phosphate) [R14].

Fire: Small fires: Dry chemical, CO$_2$, water spray, or regular foam. Large Fires: Water spray, fog, or regular foam. Move container from fire area if you can do so without risk. Fight fire from maximum distance. Stay away from ends of tanks. Dike fire-control water for later disposal; do not scatter the material (tricresyl phosphate) [R14].

Spill or Leak: Do not touch or walk through spilled material; stop leak if you can do so without risk. Fully encapsulating, vapor-protective clothing should be worn for spills and leaks with no fire. Use water spray to reduce vapors. Small Spills: Take up with sand or other noncombustible absorbent material and place into containers for later disposal. Small Dry

Spills: With clean shovel place material into clean, dry container, and cover loosely; move containers from spill area. Large Spills: Dike far ahead of liquid spill for later disposal (tricresyl phosphate) [R14].

First Aid: Move victim to fresh air and call emergency medical care; if not breathing, give artificial respiration; if breathing is difficult, give oxygen. In case of contact with material, immediately flush skin or eyes with running water for at least 15 minutes. Speed in removing material from skin is of extreme importance. Remove and isolate contaminated clothing and shoes at the site. Keep victim quiet and maintain normal body temperature. Effects may be delayed; keep victim under observation (tricresyl phosphate) [R7].

GENERAL RESPONSE Avoid contact with liquid or vapor. Keep people away. Wear positive-pressure breathing apparatus and special chemical protective clothing. Stop discharge if possible. Call fire department. Isolate and remove discharged material. Notify local health and pollution control agencies.

FIRE RESPONSE Combustible poisonous gases are produced in fire. Wear positive-pressure breathing apparatus and special chemical protective clothing. Extinguish with dry chemical, carbon dioxide, water spray, fog, or foam. (Water or foam may cause frothing.) Cool exposed containers with water.

EXPOSURE RESPONSE Call for medical aid. Vapor or liquid poisonous. May be fatal if inhaled, swallowed, or absorbed through skin. Exposure causes nausea, vomiting, diarrhea, and abdominal pain. Delayed effects begin in 1–3 weeks after initial effects. Vapor move to fresh air. If in eyes, hold eyelids open, and flush with running water for 15 minutes. If breathing has stopped, give artificial respiration. If breathing is difficult, give oxygen. Liquid remove contaminated clothing, and shoes and isolate. If in eyes, flush with running water for at least 15 minutes; lift upper and lower eyelids occasionally. If on skin, wash with soap and mild detergent and flush with running water for at least 15 minutes. If swallowed and victim is con-

scious, have victim drink water, and induce vomiting by touching back of throat with finger. If swallowed and victim is unconscious or having convulsions, do nothing except keep victim quiet, and maintain body temperature.

RESPONSE TO DISCHARGE Should be removed. Provide chemical and physical treatment.

WATER RESPONSES: Effect of low concentrations on aquatic life is unknown. May be dangerous if it enters water intakes. Notify local health and wildlife officials. Notify operators of nearby water intakes.

EXPOSURE SYMPTOMS Transient gastrointestinal upset accompanied by nausea, vomiting, diarrhea, and abdominal pain, in 1–3 weeks after initial symptoms, soreness of the lower leg muscles, and numbness of toes and fingers occur. Later, weakness of the toes and bilateral wrist drop develop. Eyes: contact may cause burns.

EXPOSURE TREATMENT Inhalation: get medical attention immediately. Move to fresh air. If not breathing, give artificial respiration. If breathing is difficult, give oxygen. Eyes: in case of contact with material, immediately flush eyes with running water for at least 15 minutes, lifting the upper and lower eyelids occasionally. Skin: immediately wash contaminated skin with soap or mild detergent; continue to flush with running water for at least 15 minutes. Speed in removing material from the skin is of extreme importance. Remove and isolate contaminated clothing and shoes at the site. Keep victim quiet and maintain normal body temperature. Effects may be delayed; keep victim under observation. Ingestion: if conscious, give large quantities of water, and induce vomiting by having the victim touch the back of his throat. If unconscious or having convulsions, do nothing except keep the victim quiet, and maintain normal body temperature.

REFERENCES

1. Budavari, S. (ed.). *The Merck Index— Encyclopedia of Chemicals, Drugs, and*

Biologicals. Rahway, NJ: Merck and Co., Inc., 1989. 1535.

2. American Conference of Governmental Industrial Hygienists. *Documentation of the Threshold Limit Values and Biological Exposure Indices.* 5th ed. Cincinnati, OH: American Conference of Governmental Industrial Hygienists, 1986. 611.

3. Sax, N. I. *Dangerous Properties of Industrial Materials.* 6th ed. New York, NY: Van Nostrand Reinhold, 1984. 2698.

4. National Fire Protection Guide. *Fire Protection Guide on Hazardous Materials.* 10th ed. Quincy, MA: National Fire Protection Association, 1991, p. 325M–88.

5. U.S. Coast Guard, Department of Transportation. *CHRIS—Hazardous Chemical Data. Volume II.* Washington, DC: U.S. Government Printing Office, 1984–5.

6. ITII. *Toxic and Hazardous Industrial Chemicals Safety Manual.* Tokyyo, Japan: The International Technical Information Institute, 1988. 540.

7. Sax, N. I., R. J. Lewis, Sr. (eds.). *Hawley's Condensed Chemical Dictionary.* 11th ed. New York: Van Nostrand Reinhold Co., 1987. 1178.

8. Verschueren, K. *Handbook of Environmental Data of Organic Chemicals.* 2nd ed. New York, NY: Van Nostrand Reinhold Co., 1983. 1149.

9. 29 CFR 1910.1000 (7/1/90).

10. 40 CFR 716.120 (7/1/90).

11. American Conference of Governmental Industrial Hygienists. *Documentation of the Threshold Limit Values and Biological Exposure Indices.* 5th ed. Cincinnati, OH: American Conference of Governmental Industrial Hygienists, 1986. 611.

12. Snyder, R. (ed.). *Ethyl Browning's Toxicity and Metabolism of Industrial Solvents.* 2nd ed. Volume II: *Nitrogen and Phosphorus Solvents.* Amsterdam–New York–Oxford: Elsevier, 1990.

13. Toy, A.D.F., E. N. Walsh. *Phosphorus Chemistry in Everyday Living.* 2nd ed. p. 230 (1987).

14. U.S. Department of Transportation. *Emergency Response Guidebook 1990.* DOT P 5800.5. Washington, DC: U.S. Government Printing Office, 1990, p. G–55.

15. NIOSH. *NIOSH Pocket Guide to Chemical Hazards.* DHHS (NIOSH) Publication No. 90–117. Washington, DC: U.S. Government Printing Office, June 1990.

16. International Labour Office. *Encyclopedia of Occupational Health and Safety.* Vols. I and II. Geneva, Switzerland: International Labour Office, 1983.

17. American Conference of Governmental Industrial Hygienists. *Threshold Limit Values for Chemical Substances and Physical Agents and Biological Exposure Indices for 1994–1995.* Cincinnati, OH: ACGIH, 1994.

18. 40 CFR 712.30 (7/1/90).

■ PHOSPHORODITHIOIC ACID, S–(2–CHLORO–1–(1,3–DIHYDRO–1,3–DIOXO–2H–ISOINDOL–2–YL) ETHYL) O,O DIETHYL ESTER

CAS # 10311–84–9

SIC CODES 2879.

SYNONYMS dialifos • dialiphor • ent-27320 • Hercules-14503- • o,o-diethyl S-(2-chloro-1-phthalimidoethyl) phosphorodithioate • phosphorodithioic acid, o,o-diethyl ester, S-ester with N-(2-chloro-1-mercaptoethyl)phthalimide • phosphorodithioic acid, S-(2-chloro-1-phthalimidoethyl) o,o-diethyl ester • phosphorodithioic acid, S-(2-chloro-1-(1,3-dihydro-1,3-dioxo-2H-isoindol-2-yl)ethyl) o,o-diethyl ester • S-(2-chloro-1-(1,3-dihydro-1,3-dioxo-2h-isoindol-2-yl)ethyl)o,o-diethyl phosphorodithioate • S-(2-chloro-1-phthalimidoethyl) o,o-diethyl phosphorodithioate • torak- • dialiphos- • S-(2-chloro-1-phthalmido-ethyl)-o,o-diethylphosphoro-thionate • S-(2-chloro-1-(1,3-dihydro-1,3-dioxy-2H-isoindol-2-yl)ethyl)o,o-diethyl phosphorodithioate • phosphorodithioic acid, S-(2-chloro-1-phthalimidoethyl)-o,o-diethyl ester • phosphorodithioic acid 5-(2-chloro-1-(1,3-dihydro-1,3-dioxo-2H-isoindol-2-yl) ethyl) o,o-diethyl ester • phosphorodithioic acid S-(2-chloro-1-(1,3-dihydro-1,3-dioxo-2H-isoindol-2-yl) ethyl) o,o-diethyl ester • phosphoro-dithioic acid o,o-diethyl ester S-ester with N-(2-chloro-1-mercaptoethyl)phthalimide • S-2-chloro-1-phthalimidoethyl o,o-diethyl phosphorodithioate • N-(2-chloro-1-(diethoxyphosphinothioylthio)ethyl)-phthalimide • o,o-diethyl phosphorodithioate S-ester with N-(2-chloro-1-mercaptoethyl) phthalimide • Hercules-14-503

MF $C_{14}H_{17}ClNO_4PS_2$

MW 393.86

COLOR AND FORM White crystalline solid; also reported as oil; colorless.

MELTING POINT 67–69°C (solid); 62–64°C when recrystallized from toluene and hexane.

FIREFIGHTING PROCEDURES Do not extinguish fire unless flow can be stopped. Solid streams of water may be ineffective. Cool all affected containers with flooding quantities of water. Apply water from as great a distance as possible. Use "alcohol" foam, carbon dioxide, or dry chemical (organophosphorus pesticides, liquid, NOS) [R6]. Extinguish fire using agent suitable for type of surrounding fire (material itself does not burn or burns with difficulty). Use water in flooding quantities as fog. (organophosphorus pesticides, solid, NOS) [R6].

SOLUBILITY Insoluble in water; soluble in aromatic hydrocarbons, ethers, esters, ketones; very soluble in cyclohexanone; slightly soluble in aliphatic hydrocarbons, alcohols; in acetone 760, isophorone 400, chloroform 620, xylene 570, diethyl ether

500, ethanol <10, hexane <10 (all in g/kg at 20°C).

VAPOR PRESSURE 6.2×10^{-8} mm Hg at 20–25°C.

OCTANOL/WATER PARTITION COEFFICIENT Log K_{ow}= 4.69 (measured).

CORROSIVITY Noncorrosive.

STABILITY AND SHELF LIFE The technical product and its formulations are stable for over 2 yr under normal conditions but are hydrolyzed readily by strong alkali [R2].

OTHER CHEMICAL/PHYSICAL PROPERTIES Hydrolyzed by alkali, at room temperature 50% loss at pH 8 in 2.5 hours (technical dialifos).

PROTECTIVE EQUIPMENT Protective clothing, including gloves, should be worn [R7].

PREVENTATIVE MEASURES If material not on fire and not involved in Fire: Keep sparks, flames, and other sources of ignition away. Keep material out of water sources and sewers. Build dikes to contain flow as necessary. Use water spray to knock down vapors. (organophosphorus pesticides, liquid, NOS) [R6].

COMMON USES Dialifos is a non-systemic insecticide and acaricide effective in controlling many insects and mites on apples, citrus, grapes, nut trees, oilseed rape, potatoes, and vegetables [R1]; pesticide against codling moth, red spider mite of deciduous fruit (former use) [R3]; control of chewing and sucking insects and spider mites on potatoes, vegetables, pome fruit, stone fruit, citrus fruit, vines, nut trees, cotton, beet, oilseed rape, and roses (former use) [R4]; (former use) product registered for use on pecans, citrus (Florida, and Texas), apples, and grapes [R5].

OCCUPATIONAL RECOMMENDATIONS None listed.

ON-SITE RESTORATION Seek professional environmental engineering assistance through EPA's Environmental Response Team (ERT), Edison, NJ, 24–hour no., 908–548–8730. Contain and isolate spill to limit spread. Construct clay/bentonite swale to divert uncontaminated portion of watershed around contaminated portion.

Isolation procedures include construction of bentonite, polyethylene-lined dams, interceptor trenches, or impoundments. Seek professional help to evaluate problem, and implement containment procedures. Conduct bench-scale and pilot-scale tests prior to decontamination program implementation. Solids may be removed in settling basins. Treatment alternatives for diluted contaminated waters include powdered activated carbon sorption, granular activated carbon filtration, alkaline hydrolysis, ozone/ultraviolet irradiation, and impoundment in a lined pit with leachate collection system and domed cover. Treatment alternatives for contaminated soils include well-point collection and treatment of leachates as for contaminated waters; bentonite/cement ground injection to immobilize spill, physical removal of immobilized residues, and placement in a lined pit with leachate collection system and domed cover. Contaminated soil residues may be packaged for disposal. Confirm all treatment procedures with responsible environmental engineer and regulatory officials.

DISPOSAL Alkaline hydrolysis or incineration [R8]. Product residues and sorbent media may be packaged in 17h epoxy-lined drums and disposed of at an EPA–approved disposal site. Other disposal alternatives include deep well injection and destruction by high-temperature incineration with scrubbing equipment.

PROBABLE LOCATION AND STATE White or colorless crystalline solid, insoluble in water.

WATER CHEMISTRY Readily hydrolyzed by strong alkali.

NON-HUMAN TOXICITY LD_{50} rat oral 5–71 mg/kg [R10]; LD_{50} rabbit percutaneous 145 mg/kg [R1]; LD_{50} mouse oral 39 mg/kg, male [R4]; LD_{50} rabbit, oral 58–71 mg/kg [R4]; LD_{50} rabbit dermal 145 mg/kg [R9].

ECOTOXICITY LC_{50} *Salmo gairdneri* (rainbow trout) 0.55–1.08 mg/L/24 hr (conditions of bioassay not specified) [R1]; LD_{50} (*Apis mellifera*) honeybee 0.034–0.038 mg/bee [R1]; LD_{50} mallard duck oral 940 mg/kg [R4]; LC_{50} goldfish 1.80–8.30 mg/L/96 hr (conditions of bioassay not specified) [R4].

BIOCONCENTRATION Based upon a measured log K_{ow} of 4.69 (1) and a water solubility of 0.18 mg/L at 20–25°C (2), the BCF of dialifor can be estimated to be 2,000 and 1,600, respectively, from regression-derived equations (3,SRC). These values suggest that bioconcentration in aquatic organisms may be important, but dialifor's rapid hydrolysis probably prevents bioconcentration, especially under basic conditions [R11].

FIFRA REQUIREMENTS In 1988, Congress amended FIFRA to strengthen and accelerate EPA's reregistration program. The nine-year reregistration scheme mandated by "FIFRA 88" applies to each registered pesticide product containing an active ingredient initially registered before November 1, 1984. List A consists of the 194 chemical cases (or 350 individual active ingredients) for which EPA had issued Registration Standards prior to the effective date of FIFRA '88. List: A; Case: Dialifor; Case No. 0010; Pesticide type: Insecticide (acaricide); Registration Standard Date: 08/01/81; Case Status: The pesticide is no longer an active ingredient in any registered pesticide products. Therefore, EPA is characterizing it as "cancelled." Under FIFRA, pesticide producers may voluntarily cancel their registered products. EPA also may cancel pesticide registrations if registrants fail to pay required fees, to make or meet certain reregistration commitments, or when the Agency reaches findings of unreasonable adverse effects. Active Ingredient (AI): Dialifor; AI Status: The active ingredient is no longer contained in any registered products. Therefore, EPA is characterizing it as "cancelled." Under FIFRA, pesticide producers may voluntarily cancel their registered products. EPA also may cancel pesticide registrations if registrants fail to pay required fees, to make or meet certain reregistration commitments, or when the Agency reaches findings of unreasonable adverse effects [R12].

REFERENCES

1. Worthing, C. R., and S. B. Walker (eds.). *The Pesticide Manual—A World Compendium.* 8th ed. Thornton Heath, UK: The British Crop Protection Council, 1987. 245.

2. Hayes, W. J., Jr., E. R. Laws, Jr. (eds.). *Handbook of Pesticide Toxicology.* Volume 2. Classes of Pesticides. New York, NY: Academic Press, Inc., 1991. 1068.

3. Lewis, R. J., Sr. (ed.). *Hawley's Condensed Chemical Dictionary.* 12th ed. New York, NY: Van Nostrand Reinhold Co., 1993 360.

4. Hartley, D., and H. Kidd (eds.). *The Agrochemicals Handbook.* 2nd ed. Lechworth, Herts, England: The Royal Society of Chemistry, 1987, p. A0129/Jun 89.

5. *Farm Chemicals Handbook* 1981. Willoughby, OH: Meister, 1981, p. C–339.

6. Association of American Railroads. *Emergency Handling of Hazardous Materials in Surface Transportation.* Washington, DC: Association of American Railroads, Bureau of Explosives, 1992. 713.

7. Hartley, D., and H. Kidd (eds.). *The Agrochemicals Handbook.* 2nd ed. Lechworth, Herts, England: The Royal Society of Chemistry, 1987, p. A129/Jun 89.

8. Sittig, M. *Handbook of Toxic and Hazardous Chemicals and Carcinogens,* 1985. 2nd ed. Park Ridge, NJ: Noyes Data Corporation, 1985. 298.

9. Hayes, Wayland J., Jr. *Pesticides Studied in Man.* Baltimore/London: Williams and Wilkins, 1982. 399.

10. *Kirk–Othmer Encyclopedia of Chemical Technology.* 3rd ed., Volumes 1–26. New York, NY: John Wiley and Sons, 1978–1984, p. V13 442 (1981).

11. (1) Hansch, C., and A. J. Leo. *Medchem Project* Issue No. 26, Claremont, CA: Pomona College (1985); (2) Freed, V. H., et al. *Pestic Biochemical Physiol* 10: 203–11 (1979); (3) Lyman, W. J., et al. *Handbook of Chemical Property Estimation Methods.* Washington, DC: Amer Chemical Soc. pp. 5–4,5–10 (1990).

12. USEPA/OPP. *Status of Pesticides in Reregistration and Special Review,* p. 82 (Mar, 1992) EPA 700–R–92–004.

■ PHOSPHORODITHIOIC ACID, o, o–DIISOPROPYL ESTER, S–ESTER with N–(2–MERCAPTOETHYL) BENZENESULFONAMIDE

CAS # 741–58–2

SIC CODES 2879.

SYNONYMS benzenesulfonamide, N-(2-mercaptoethyl) -, S-ester with o,o-diisopropylphosphorodithioate • benzulfide • betasan • disan • disan (pesticide) • exporsan • N-(2-(o,o-diisopropyldithiophosphoryl) ethyl) benzenesulfonamide • N-(2-mercaptoethylbenzene) sulfonamide S-(o,o-diisopropylphosphorodithioate) m. N-(beta-o,o-diisopropyldithio-phosphorylethyl) benzenesulfonamide • o,o-bis (1-methylethyl) S-(2-((phenylsulfonyl) amino) ethyl) phosphorodithioate • o,o-diisopropyl 2-(benzenesulfonamido) ethyl dithiophosphate • o,o-diisopropyl phosphorodithioate S-ester with N-(2-mercaptoethyl) benzenesulfonamide • kayaphenone • o,o-diisopropyl S-(2-benzenesulfonylaminoethyl) phosphorodithioate • o,o-di-isopropyl S-2-phenylsulphonylaminoethyl phosphorodithioate • phosphorodithioic acid, o,o-bis (1-methylethyl) S-(2-((phenylsulfonyl) amino) ethyl) ester • phosphorodithioic acid, o,o-diisopropyl ester, S-ester with N-(2-mercaptoethyl) benzenesulfonamide • pre-san • prefar • r-4461 • r-4461 • S-(o,o-diisopropyl phosphorodithioate) ester of N-(2-mercaptoethyl) benzenesulfonamide • S-2-benzenesulphonamidoethyl o,o-di-isopropyl phosphorodithioate • S-beta-(benzenesulfonamido) ethyl o,o-diisopropyl dithiophosphate • sap • sap (herbicide) • o,o-diisopropyl dithiophosphate S-ester with N-(2-

mercaptoethyl) benzenesulphonamide • s-2-benzenesulphonamidoethyl o,o-diisopropyl phosphorodithioate • S-2-benzenesulfonamidoethyl o,o-di-isopropyl phosphorodithioate • betasan-e • prefar-e • betasan-g • (N-(2-mercaptoethyl) benzenesulfonamide) • o,o-bis (1-methylethyl)-S-[2[(phenylsulfonyl) amino]ethyl] phosphorodithioate • (N-(2-ethylthio) benzene sulphonamide-S, o,o-diisopropylphosphorodithioate • S-(o, o-diisopropylphosphorodithioate) of N-(2-mercaptoethyl) benzenesulfonamide

MF $C_{14}H_{24}NO_4PS_3$

MW 397.54

COLOR AND FORM Viscous amber liquid above 34.4°C, solid below; colorless liquid or white crystalline solid.

DENSITY 1.23 at 20°C.

MELTING POINT 34.4°C.

SOLUBILITY Solubility at 20°C: 25 mg/L water; 300 mg/L kerosene; miscible with acetone, ethanol, 4–methylpentan–2–one, xylene; miscible with methyl isobutyl ketone.

VAPOR PRESSURE Less than 0.133 mPa at 20°C.

CORROSIVITY Corrosive to copper.

STABILITY AND SHELF LIFE Stable at 80°C for 50 hr but decomposes at 200°C in 18–40 hr [R6].

OTHER CHEMICAL/PHYSICAL PROPERTIES Amber solid or supercooled liquid (technical bensulide).

PREVENTATIVE MEASURES Avoid contact with eyes and skin and inhalation of spray mists. Guard against drift [R4].

COMMON USES Selective preemergence herbicide [R1]; for the control of undesirable grasses in lawns and broadleaf weeds [R5]; seasonal control of carbgrass, annual bluegrass in grass dichondra lawns; prefar is for crop use in carrots, cole crops [R1]; betasan is registered for control of redroot, pigweed, watergrass, lambsquarters, shepherdspurse, goosegrass, and deadnettle in grass and dichondra lawns

[R3, 48]; pre-emergence control of annual grasses and broad-leaved weeds in established turf, dichonda, brassicas, cucurbits, watermelons, cotton, lettuce, tomatoes, carrots, capsicums, rice, and herbage seed crops [R4].

PERSISTENCE Bensulide, lasting 4 to 12 months in soils, is the most persistent of the organophosphates. It degrades more quickly in conditions of high soil moisture, temperature, and acidity.

GENERAL SENSATION Typical symptoms of organophosphate poisoning are anorexia, nausea, vomiting, diarrhea, excessive salivation, pupillary constriction, bronchoconstriction, muscle twitching, convulsions, coma, and death by respiratory failure. Effects are cumulative.

OCCUPATIONAL RECOMMENDATIONS None listed.

ON-SITE RESTORATION Seek professional environmental engineering assistance through EPA's Environmental Response Team (ERT), Edison, NJ, 24–hour no., 908–548–8730. Contain and isolate spill by using clay/bentonite dams, interceptor trenches or impounds. Construct swale to divert uncontaminated portion of watershed around contaminated portion. Seek professional help to evaluate problem, implement containment measures, and conduct bench scale and pilot scale tests prior to full-scale decontamination program implementation. Density stratification and impoundment—remove product from bottom layer by pumping through manifold, polyethylene rope mop collection, or remove clarified upper portion by skimmers or siphoning. Treatment is required for both clarified and concentrated product fractions. Treatment alternatives include acid hydrolysis, powdered activated carbon sorption, granular activated carbon filtration, and biodegradation. Soil treatment alternatives for contaminated soils include well-point collection and treatment of leachates as for contaminated water; bentonite/cement ground injection to immobilize spill. Contaminated soil residues may be packaged for disposal.

DISPOSAL Product residues and sorbent media may be packaged in 17h epoxy-lined drums and disposed of at an approved EPA disposal site. Destruction by high-temperature incineration with scrubbing equipment, microwave plasma detoxification, if available, or by acidification. Encapsulation by organic polyester resin or silicate fixation. Confirm disposal procedures with responsible environmental engineer and regulatory officials.

PROBABLE LOCATION AND STATE Amber-colored liquid forms a solid below 34.4°C. Insoluble and heavier than water, it will sink.

NON-HUMAN TOXICITY LD_{50} rat (male, albino), oral 360 mg/kg [R2, 60]; LD_{50} rat (female, albino), oral 270 mg/kg [R2, 60]; LD_{50} rat percutaneous 3,950 mg/kg [R2, 60]; LD_{50} rat, oral 770 mg/kg [R3, 49]; LC_{50} rat inhalation >8.0 mg/L (1 hr) [R3, 50]; LD_{50} rabbit percutaneous 3,950 mg/kg [R4].

ECOTOXICITY LC_{50} *Gammarus fasciatus* 1.4 mg/L/96 hr at 15°C, mature (95% confidence limit 0.4–5.1 mg/L); static bioassay without aeration, pH 7.2–7.5, water hardness 40–50 mg/L as calcium carbonate, and alkalinity of 30–35 mg/L [R7]; LC_{50} *Salmo gairdneri* (rainbow trout) 0.7 mg/L/96 hr at 13°C, wt 1.6 g; static bioassay without aeration, pH 7.2–7.5, water hardness 40–50 mg/L as calcium carbonate, and alkalinity of 30–35 mg/L [R7]; LC_{50} *Penaeus azetucus* (brown Gulf shrimp), loss of equilibrium or death >1 ppm/96 hr (conditions of bioassay not specified) [R3, 49]; LC_{50} common goldfish 1–2 ppm/96 hr (conditions of bioassay not specified) [R3, 49]; LC_{50} *Lepomis macrochirus* (bluegill) 0.8 mg/L/96 hr at 24°C, wt 0.2 g specified); static bioassay without aeration, pH 7.2–7.5, water hardness 40–50 mg/L as calcium carbonate, and alkalinity of 30–35 mg/L [R7]; LC_{50} channel catfish 379 µg/L (96 hr) [R8].

BIODEGRADATION Bensulide is degraded slowly in the soil by soil microorganisms. [R3, 49].

FIFRA REQUIREMENTS In 1988, Congress amended FIFRA to strengthen and accelerate EPA's reregistration program. The nine-year reregistration scheme mandated by "FIFRA 88" applies to each registered pesticide product containing an active ingredient initially registered before November 1, 1984. Pesticides for which EPA had not issued Registration Standards prior to the effective date of FIFRA '88 were divided into three lists based upon their potential for exposure and other factors, with List B being of highest concern, and D of least. List: B; Case: Bensulide; Case No. 2035; pesticide type: herbicide; Case Status: Awaiting Data/Data in Review: OPP awaits data from the pesticide's producer(s) regarding its human health or environmental effects, or OPP has received and is reviewing such data, in order to reach a decision about the pesticide's eligibility for reregistration. Active Ingredient (AI): S–(o,o-diisopropyl-phosphorodithioate ester of N–(2–mer-captoethyl) benzenesulfonamide; Data Call-in (DCI) Date: 06/11/91; AI Status: The producer(s) of the pesticide has made commitments to conduct the studies and pay the fees required for reregistration, and is meeting those commitments in a timely manner [R9].

REFERENCES

1. *Farm Chemicals Handbook* 1993. Willoughby, OH: Meister Publishing Co., 1993, p. C–45.

2. Worthing, C. R., and S. B. Walker (eds.). *The Pesticide Manual—A World Compendium.* 8th ed. Thornton Heath, UK: The British Crop Protection Council, 1987.

3. Weed Science Society of America. *Herbicide Handbook.* 5th ed. Champaign, Illinois: Weed Science Society of America, 1983.

4. Hartley, D., and H. Kidd (eds.). *The Agrochemicals Handbook.* 2nd ed. Lechworth, Herts, England: The Royal Society of Chemistry, 1987, p. A033/Aug 87.

5. *Kirk-Othmer Encyclopedia of Chemical Technology.* 3rd ed., Volumes 1–26. New York, NY: John Wiley and Sons, 1978–1984, p. V12 344 (1980).

6. Worthing, C. R. (ed.). *Pesticide Manual.* 6th ed. Worcestershire, England: British Crop Protection Council, 1979. 33.

7. U.S. Department of Interior, Fish and Wildlife Service. *Handbook of Acute Toxicity of Chemicals to Fish and Aquatic Invertebrates.* Resource Publication No. 137. Washington, DC: U.S. Government Printing Office, 1980. 81.

8. Verschueren, K. *Handbook of Environmental Data of Organic Chemicals.* 2nd ed. New York, NY: Van Nostrand Reinhold Co., 1983. 234.

9. USEPA/OPP. *Status of Pesticides in Reregistration and Special Review* p. 120 (Mar, 1992) EPA 700–R–92–004.

■ PHOSPHORODITHIOIC ACID, O,O–DIMETHYL ESTER, S–ESTER with 4-(MERCAPTOMETHYL) –2–METHOXY delta (sup 2)–1,3,4–THIADIAZOLIN–5–ONE

CAS # 950–37–8

SYNONYMS (O,O-dimethyl)-S-(-2-methoxy-delta (sup (2)-1,3,4-thiadiazolin-5-on-4-ylmethyl) dithiophosphate • dmtp • ent-27193 • Geigy-13005 • Geigy gs-13005 • gs-13005 • methidathion • O,O-dimethyl s-(5-methoxy-1,3,4-thiadiazolinyl-3-methyl) dithiophosphate • O,O-dimethyl-S-((2-methoxy-1,3,4 (4h) thiodiazol-5-on-4-yl)-methyl)-dithiofosfaat (Dutch) • O,O-dimethyl-S-(2-methoxy-1,3,4-thiadiazol-5 (4h) -onyl-(4)-methyl) phosphorodithioate • O,O-Dimethyl-S-(2-methoxy-1,3,4-thiadiazol-5-(4h)-onyl-(4)-methyl)-dithiophosphat (German) • O,O-dimetil-S-((2-metossi-1,3,4-(4h)-tiadiazol-5-on-4-il)-metil)-ditiofosfato (Italian) • phosphorodithioic acid, O,O-dimethyl ester, S-ester with 4-(mercaptomethyl)-2-methoxy-delta (2)-1,3,4-thiadiazolin-5-one • phosphorodithioic acid, O,O-dimethyl ester, S-ester with 4-(mercaptomethyl)-2-methoxy-delta (sup (2) -1,3,4-thiadiazolin-5-one • phosphorodithioic acid, S-(5-methoxy-2-oxo-1,3,4-thiadiazol-3 (2h)-yl) methyl) O,O-dimethyl ester • S-((5-methoxy-2-oxo-1,3,4-thiadiazol-3 (2h) -yl) methyl) O,O-dimethyl phosphorodithioate • S-2,3-dihydro-5-methoxy-2-oxo-1,3,4-thiadiazol-3-ylmethyl O,O-dimethyl phosphorodithioate • somonil • supracid • ultracid • ultracid-40 • ultracide • ultracide Ciba-Geigy • Ciba-Geigy gs 13005 • fisons-nc-2964 • dithiophosphoric acid O,O'-dimethyl-S-((2-methoxy-1,3,4-thiadiazol-5 (4h)-on-4-yl) methyl) ester • dithiophosphoric acid O,O'-dimethyl-S-((5-methoxy-1,3,4-thiadiazol-2 (3h)-one-3-yl) methyl) ester • O,O'-dimethyl-S-((2-methoxy-1,3,4-thiadiazole-5 (4h)-one-4-yl) methyl) dithiophosphate • phosphorodithioic acid S-[(5-methoxy-2-oxo-1,3,4-thiadiazol-3 (2h) -yl) methyl] O,O-dimethyl ester • 3-dimethoxyphosphinothioylthiomethyl-5-methoxy-1,3,4-thiadiazol-2 (3h)-one • oms-844 • ent-27-193 • gs-13-005 • suprathion • S[[5-methoxy-2-oxo-1,3,4-thiadiazol-3 (2h) -yl]methyl]O,O-dimethyl phosphorodithioate • s-(2,3-dihydro-5-methoxy-2-oxo-1,3,4-thiadiazol-3-methyl) dimethyl phosphorothiolothionate • gs-13005 • ultracide Ciba-Geigy • supracide-ulvair • ultracide-ulvair-250

MF C$_6$H$_{11}$N$_2$O$_4$PS$_3$

MW 302.34

COLOR AND FORM Colorless crystals; crystals from methanol.

DENSITY 1.495 g/cm^3 (20°C).

MELTING POINT 39–40°C.

FIREFIGHTING PROCEDURES Use water spray, dry chemical, foam, or carbon dioxide. Use water to keep fire-exposed containers cool. If a leak or spill has not ignited, use water spray to disperse vapors and to provide protection for firefighters. Water spray may be used to flush spills away from exposures. In advanced or massive fires, firefighting is done from safe distance or from protected location (parathion) [R7]. Do not extinguish fire unless flow can be stopped. Use water in flooding quantities as fog. Solid streams of water may be ineffective (parathion) [R8, 526]. Self-contained breathing apparatus with a full facepiece operated in pressure-demand or other positive-pressure mode should be used in firefighting. (parathion) [R9, 5].

SOLUBILITY At 20°C: 250 mg/L water; 690 g/kg acetone; 850 g/kg cyclohexanone; 260 g/kg ethanol; 53 g/kg octan-1-ol; 600 g/kg xylene; solubility in water <1%; readily soluble in benzene, acetone, methanol, xylene, and other organic solvents; water solubility = 187 mg/L at 20°C.

VAPOR PRESSURE 0.186 mPa at 20°C.

CORROSIVITY Noncorrosive.

STABILITY AND SHELF LIFE Relatively stable to hydrolysis in neutral or slightly acidic media, less stable in more acidic (pH 1) or in alkaline media (ph 13, 50% loss in 30 min at 25°C) [R2, 546].

OTHER CHEMICAL/PHYSICAL PROPERTIES Volatility slight.

PROTECTIVE EQUIPMENT Neoprene-coated gloves; rubber workshoes or overshoes; latex rubber apron; goggles; respirator or mask approved for toxic dusts and organic vapors (parathion) [R10].

CLEANUP 1. Ventilate area of spill or leak. 2. Collect for reclamation or absorb in vermiculite, dry sand, earth, or a similar material. (parathion) [R9, 4].

DISPOSAL Potential candidate for rotary kiln incineration, with a temperature range of 820 to 1,600°C, and a residence time of seconds. Also, a potential candidate for fluidized bed incineration, with a temperature range of 450 to 980°C, and a residence time of seconds. Also, a potential candidate for liquid injection incineration with a temperature range of 650 to 1,600°C, and a residence time of 0.1 to 2 seconds. (parathion) [R11].

COMMON USES Methidathion is used in the control of scale insects, leaf eaters, and some mites [R1, 1028]; controls alfalfa weevil and certain other insects in alfalfa, scales in citrus, spider mites, boll-

worm, budworm, lygus bug, pink bollworm, and whitefly in cotton [R4]. It is a non-systemic insecticide controlling a wide range of sucking and leaf-eating insects with a specific use against scale insects. Foliar penetration enables it to be used against leafrollers [R2, 546]. For use on sunflower, artichokes, apples, almonds, cherries, apricots, pears, nectarines, plums, prunes, walnuts, peaches, and pecans [R6]. Control of a wide range of sucking and chewing insects (especially scale insects) and spider mites in many crops, e.g., pome fruit, stone fruit, citrus fruit, vines, olives, hops, cotton, potatoes, beet, lucerne, oilseed rape, sunflowers, safflowers, tobacco, hazels, and some vegetables [R3].

OCCUPATIONAL RECOMMENDATIONS None listed.

NON-HUMAN TOXICITY LD_{50} mouse female 18 mg/kg [R5, 371]; LD_{50} guinea pig female 25 mg/kg [R5, 371]; LD_{50} rabbit, oral 80 mg/kg [R5, 371]; LD_{50} dog, oral 200 mg/kg [R5, 371]; LD_{50} rat (adult male), oral 31 mg/kg [R12]; LD_{50} rat (adult female) oral 32 mg/kg [R12]; LD_{50} rat oral 25–54 mg technical/kg [R2, 546]; LD_{50} rat percutaneous 1,546 mg/kg [R2, 546]; LD_{50} hen, oral 80 mg/kg [R3].

ECOTOXICITY LD_{50} *Branta canadensis* (Canada goose) oral 8.41 mg/kg (95% confidence limit 4.20–16.8 mg/kg), adult males, and females [R13]; LD_{50} *Anas platyrhynchos* (mallard), oral 23.6 mg/kg (95% confidence limit 16.5–33.8 mg/kg), 3–4–mo–old females [R13]; LD_{50} pheasant, oral 33.2 mg/kg (95% confidence limit 17.3–63.5 mg/kg), 4–mo–old females [R13]; LD_{50} *Alectoris chukar* (chukar), oral 225 mg/kg (95% confidence limit 178–283 mg/kg), 12–24–mo–old male and females [R13]; LC_{50} *Salmo gairdneri* (rainbow trout) 0.014 mg/L/96 hr at 12°C (95% confidence limit 0.009–0.022 mg/L), wt 0.8 g; static bioassay without aeration, pH 7.2–7.5, water hardness 40–50 mg/L as calcium carbonate, and alkalinity of 30–35 mg/L [R14]; LC_{50} *Lepomis macrochirus* (bluegill) 0.009 mg/L/96 hr at 24°C (95% confidence limit 0.006–0.013 mg/L), wt 0.7 g; static bioassay without aeration, pH 7.2–7.5, water hardness 40–50 mg/L as calcium carbonate, and alkalinity of 30–35 mg/L [R14]; LC$_{50}$ *Coturnix japonica* (Japanese quail), 14 days old, oral (5-day ad libitum in diet) 980 ppm (95% confidence intervals 793–1,193 ppm) [R15].

BIOCONCENTRATION Based on an estimated bioconcentration factor of 28 (1), methidathion is not expected to bioconcentrate in fish (SRC) [R16].

FIFRA REQUIREMENTS In 1988, Congress amended FIFRA to strengthen and accelerate EPA's reregistration program. The nine-year reregistration scheme mandated by "FIFRA 88" applies to each registered pesticide product containing an active ingredient initially registered before November 1, 1984. List A consists of the 194 chemical cases (or 350 individual active ingredients) for which EPA had issued Registration Standards prior to the effective date of FIFRA '88. List: A; Case: methidathion; Case No. 0034; pesticide type: Insecticide (acaricide); Registration Standard Date: 01/31/81; Case Status: awaiting Data/Data in Review: OPP awaits data from the pesticide's producers regarding its human health or environmental effects, or OPP has received and is reviewing such data to reach a decision about the pesticide's eligibility for reregistration. Active Ingredient (AI): methidathion; data call-in (DCI) Date: 09/28/90; AI Status: the producers of the pesticide have made commitments to conduct the studies and pay the fees required for reregistration, and is meeting those commitments in a timely manner [R17].

REFERENCES

1. Hayes, W. J., Jr., E. R. Laws, Jr. (eds.). *Handbook of Pesticide Toxicology.* Volume 2. *Classes of Pesticides.* New York, NY: Academic Press, Inc., 1991.

2. Worthing, C. R., and S. B. Walker (eds.). *The Pesticide Manual—A World Compendium.* 8th ed. Thornton Heath, UK: The British Crop Protection Council, 1987.

3. Hartley, D., and H. Kidd (eds.). *The Agrochemicals Handbook.* 2nd ed. Lech-

worth, Herts, England: The Royal Society of Chemistry, 1987, p. A268/Aug 87.

4. *Farm Chemicals Handbook* 1993. Willoughby, OH: Meister Publishing Co., 1993, p. C–222.

5. Hayes, Wayland J., Jr. *Pesticides Studied in Man.* Baltimore/London: Williams and Wilkins, 1982.

6. *Farm Chemicals Handbook* 1993. Willoughby, OH: Meister Publishing Co., 1993, p. C 222.

7. National Fire Protection Association. *Fire Protection Guide on Hazardous Materials.* 9th ed. Boston, MA: National Fire Protection Association, 1986, p. 49–71.

8. Association of American Railroads. *Emergency Handling of Hazardous Materials in Surface Transportation.* Washington, DC: Assoc. of American Railroads, Hazardous Materials Systems (BOE), 1987.

9. Mackison, F. W., R. S. Stricoff, and L. J. Partridge, Jr. (eds.). *NIOSH/OSHA—Occupational Health Guidelines for Chemical Hazards.* DHHS (NIOSH) Publication No. 81–123 (3 vols). Washington, DC: U.S. Government Printing Office, Jan. 1981.

10. U.S. Coast Guard, Department of Transportation. *CHRIS—Hazardous Chemical Data.* Volume II. Washington, DC: U.S. Government Printing Office, 1984–5.

11. USEPA. *Engineering Handbook for Hazardous Waste Incineration,* p. 3–10 (1981) EPA 68–03–3025.

12. Budavari, S. (ed.). *The Merck Index— Encyclopedia of Chemicals, Drugs, and Biologicals.* Rahway, NJ: Merck and Co., Inc., 1989. 942.

13. U.S. Department of the Interior, Fish and Wildlife Service. *Handbook of Toxicity of Pesticides to Wildlife.* Resource Publication 153. Washington, DC: U.S. Government Printing Office, 1984. 51.

14. U.S. Department of Interior, Fish and Wildlife Service. *Handbook of Acute Toxicity of Chemicals to Fish and Aquatic Invertebrates.* Resource Publication No.

137. Washinngton, DC: U.S. Government Printing Office, 1980. 84.

15. Hill, E. F., and M. B. Camardese. *Lethal Dietary Toxicities of Environmental Contaminants and Pesticides to Coturnix.* Fish and Wildlife Technical Report 2. Washington, DC: United States Department of Interior, Fish and Wildlife Service, 1986. 96.

16. Kenaga, E. E. *Ecotox Environ Safety* 4: 26–38 (1980).

17. USEPA/OPP. *Status of Pesticides in Reregistration and Special Review*, p. 98 (Mar, 1992) EPA 700–R–92–004.

■ PHOSPHORODITHIOIC ACID, O,O–DIMETHYL ESTER, S–ESTER with N–(MERCAPTOMETHYL) PHTHALIMIDE

CAS # 732–11–6

SYNONYMS (O,O-dimethyl-phthalimidiomethyl-dithiophosphate) • decemthion • decemthion p-6 • ent-25,705 • ftalophos • imidathion • N-(mercaptomethyl) phthalimide S-(o,o-dimethyl phosphorodithioate) • O,O-dimethyl S-(phthalimidomethyl) dithiophosphate • O,O-dimethyl S-phthalimidomethyl phosphorodithioate • O,O-dimethyl S-phthalimidomethyl phosphorothionate • percolate • imidan • phosphorodithioic acid S-((1,3-dihydro-1,3-dioxo-2h-isoindol-2-yl) methyl) O,O-dimethyl ester • phosphorodithioic acid, O,O-dimethyl ester, S-ester with N-(mercaptomethyl) phthalimide • phosphorodithioic acid, s-((1,3-dihydro-1,3-dioxo-2h-isoindol-2-yl) methyl) O,O-dimethyl ester • phosphorodithioic acid, S-((1,3-dihydro-1,3-dioxo-isoindol-2-yl) methyl) O,O-dimethyl ester • phthalimide, N-(mercaptomethyl)-, S-ester with O,O-dimethyl phosphorodithioate • phthalimido O,O-dimethyl phosphorodithioate • phthalimidomethyl O,O-dimethyl phosphorodithioate • phthalophos • pmp

(pesticide) • prolate • r-1504 • safidon • smidan • stauffer-r-1504 • simidan • pmp • appa • O,O-dimethyl S-(N-phthalimidomethyl) dithiophosphate • kemolate • S-((1,3-dihydro-1,3-dioxo-2h-isoindol-2-yl) methyl) O,O-dimethy phosphorodithioate • r-1504 • O,O-dimethyl S-(N-phthalimidomethyl) dithiophosphate • S-((1,3-dioxo-2h-isoindol-2-yl) methyl) O,O-dimethyl phosphorodithioate • phtalofos • N-(dimethoxyphosphinothioylthiomethyl) phthalimide • S-((1,3-dihydro-1,3-dioxo-2h-isoindol-2-yl) methyl) O,O-dimethyl phosphorodithioate • O,O-dimethyl phosphorodithioate S-ester with N-(mercaptomethyl) phthalimide • prolate-e • prolate 8-os • imidan-wp • imidan-5-dust • prolate-5-dust • n-(mercaptomethyl) phthalimide-S-(O,O-dimethylphosphorodithioate) • phosphorodithioic acid O,O-dimethyl ester S-ester with N-(mercaptomethyl) phthalimide • phthalophos • prolate • ent-25705 • decemtion p-6 • N-(mercaptomethyl) phthalamide S-(O,O-dimethyl phosphorodithioate) • N-(mercaptomethyl) phthalimide S-(O,O-dimethyl) phosphorodithioate • O,O-dimethyl S-(phthalimidomethyl) phosphorodithioate • phthalimidomethyl O,O-dimethyl phosphorodithioate

MF $C_{11}H_{12}NO_4PS_2$

MW 317.33

COLOR AND FORM Off-white crystalline solid; colorless crystals.

ODOR Offensive.

DENSITY 1.03 at 20°C.

MELTING POINT 71.9°C.

EXPLOSIVE LIMITS AND POTENTIAL Containers may explode in the heat of fire [R6].

FIREFIGHTING PROCEDURES For small fires, use dry chemical, carbon dioxide, water spray, or foam. For large fires, use water spray, fog, or foam. (organophosphorus pesticide NOS) [R6].

SOLUBILITY 100 g/kg in dichloromethane, 4–methylpent–3–en–2–one, butanone;

soluble in most organic solvents; methyl isobutyl ketone 300 g/L at 25°C; at 25°C: 22 mg/L water; 650 g/L acetone; 600 g/L benzene; 5 g/L kerosene; 50 g/L methanol; 300 g/L toluene; 250 g/L xylene; water solubility = 24.4 mg/L at 20°C.

VAPOR PRESSURE 4.9×10^{-7} mm Hg at 20–25°C.

OCTANOL/WATER PARTITION COEFFICIENT Log K_{ow}= 2.83.

CORROSIVITY Slightly corrosive to metals.

STABILITY AND SHELF LIFE Stable under normal storage conditions [R7, 4822].

OTHER CHEMICAL/PHYSICAL PROPERTIES Mp of technical grade (95–98% pure) 66.5–69.5°C (technical imidan).

PROTECTIVE EQUIPMENT Hat, long-sleeved shirt, long-legged trousers or coveralls; chemical-resistant gloves for mixing [R3].

PREVENTATIVE MEASURES Stay upwind; keep out of low areas. Ventilate closed spaces before entering them. Remove and isolate contaminated clothing at the site. Do not touch spilled material. Use water spray to reduce vapors. Take up small spills with sand or other noncombustible absorbent material and place in containers for later disposal. Take up small, dry spills with clean shovel, and place in clean, dry container. (organophosphorus pesticides) [R6].

COMMON USES Insecticide for hornflies on beef cattle and for cattle grubs; for weevils on sweet potatoes in storage and on alfalfa [R2]; used on a variety of crops including alfalfa, almonds, apples, apricots, cherries (tart), citrus, corn, cotton, cranberries, pecans, blueberries, grapes, nectarines, peaches, pears, peas (Pacific Northwest), potatoes, plums/prunes, and certain deciduous shade and ornamental trees and woody evergreens; active against a wide range of insects such as alfalfa weevil, boll weevil, codling moth, leafrollers, plum curculio, grape berrymoth, oriental fruit moth, and many others [R3]; also registered for use on animals for grubs, lice, ticks, scabies, mites [R4]; control of lepidopterous larvae, aphids, psyllids, fruit flies, and spider mites on pome fruit, stone fruit, citrus fruit, and vines; Colorado beetles on potatoes; boll weevils on cotton; olive moths and olive thrips on olives; blossom beetles on oilseed rape; leaf beetles and weevils on lucerne; European corn borers on maize and sorghum; sweet potato weevils on sweet potatoes in storage [R5]; non-systemic acaricide and insecticide, used on top fruit citrus, grapes, potatoes, and in forestry at rates (0.5–1.0 kg AI/hectare) such that it is safe for a range of predators of mites and therefore useful in integrated control programs. It is also used to control mites and warble fly of cattle [R1].

OCCUPATIONAL RECOMMENDATIONS None listed.

NON-HUMAN TOXICITY LD_{50} rats oral 113 mg/kg [R1]; LD_{50} rabbits percutaneous >5,000 mg/kg, albino [R1]; LC_{50} rat inhalation 2.76 mg/L air/1 hr [R5];.

ECOTOXICITY LD_{50} pheasant female oral 237 mg/kg, 3–4 mo old (95% confidence limit 171–329 mg/kg) [R8]; LD_{50} pheasant male oral greater than 250 mg/kg, 3 mo old [R8]; LD_{50} *Anas platyrhynchos*, mallard duck, oral 1,830 mg/kg, 3–4–mo-old males (95% confidence limit 1,270–2,630 mg/kg) [R8]; LC_{50} chinook salmon 180 µg/L/24 hr, flow-through bioassay (technical grade) [R9]; LC_{50} channel catfish 13,000 µg/L/24 hr, flow-through bioassay (technical grade) [R9]; LC_{50} bluegill 70 µg/L/96 hr, flow-through bioassay (technical grade) [R9]; LC_{50} channel catfish 11,000 µg/L/96 hr, flow-through bioassay (technical grade) [R9]; LC_{50} rainbow trout yolk sac fry more than 10,000 µg/L/24 hr, flow-through bioassay (technical grade) [R9]; LC_{50} rainbow trout eyed egg more than 10,000 µg/L/96 hr, flow-through bioassay (technical grade) [R9]; LC_{50} rainbow trout yolk sac fry more than 10,000 µg/L/24 hr, flow-through bioassay (50% wettable powder) [R9]; LC_{50} rainbow trout fingerling 760 µg/L/24 hr, flow-through bioassay (technical grade) [R9]; LC_{50} rainbow trout fingerling 560 µg/L/96 hr. flow-through bioassay (technical grade) [R9]; LC_{50} bluegill 1,300 µg/L/24 hr at 10°C flow-through bioassay (technical grade) [R9]; LC_{50} bluegill 86 µg/L/24 hr at 25°C, flow-through bioassay (technical grade) [R9]; LC$_{50}$ bluegill 640 µg/L/96 hr at pH 8.5, flow-through bioassay (technical grade) [R9]; LC$_{50}$ bluegill 22 µg/L/96 hr at pH 6.5, flow-through bioassay (technical grade) [R9]; LC$_{50}$ bluegill 800 µg/L/24 hr in water containing 58 mg/L CaCO$_3$, flow-through bioassay (tecnnical grade) [R9]; LC$_{50}$ bluegill 1,500 µg/L/24 hr in water containing 160 mg/L CaCO$_3$, flow-through bioassay (technical grade) [R9]; LC$_{50}$ bluegill 800 µg/L/96 hr in water containing 58 mg/L CaCO$_3$, flow-through bioassay (technical grade) [R9]; LC$_{50}$ bluegill 1,400 µg/L/96 hr in water containing 160 mg/L CaCO$_3$, flow-through bioassay (technical grade) [R9]; LC$_{50}$ bluegill 420 µg/L/96 hr, flow-through bioassay (technical grade; fresh solution) [R9]; LC$_{50}$ bluegill more than 10,000 µg/L/96 hr, flow-through bioassay (technical grade; solution aged 4 days) [R9].

BIODEGRADATION Phthalamic and phthalic acids were found as hydrolysis end products in water. Hydrolysis of imidan occurred in soils; tests indicated that this was not dependent on moisture alone but was due in some degree to microbial action. In dry sandy loam soil, time for 50% degradation was 19 days, compared to 3 days in moist soil [R10].

BIOCONCENTRATION Accumulation factors for technical imidan after 24 hr ranged from 10 for fathead minnow, channel catfish, and bluegill to 1 for damselfly. Accumulation factors after 46 hr ranged from 11 for channel catfish to 2 for clam shrimp and damselfly [R9].

FIFRA REQUIREMENTS In 1988, Congress amended FIFRA to strengthen and accelerate EPA's reregistration program. The nine-year reregistration scheme mandated by "FIFRA 88" applies to each registered pesticide product containing an active ingredient initially registered before November 1, 1984. List A consists of the 194 chemical cases (or 350 individual active ingredients) for which EPA had issued Registration Standards prior to the effective date of FIFRA '88. List: A; Case: Phosmet; Case No. 0242; Pesticide type: Insecticide; Registration Standard Date: 09/26/86; Case Status: Awaiting Data/ Data in Review: OPP awaits data from the pesticide's producer(s) regarding its human health or environmental effects, or OPP has received and is reviewing such data, in order to reach a decision about the pesticide's eligibility for reregistration. Active Ingredient (AI): Phosmet; Data Call-in (DCI) Date: 08/30/91; AI Status: The producer of the pesticide has made commitments to conduct the studies and pay the fees required for reregistration, and is meeting those commitments in a timely manner [R11].

REFERENCES

1. Worthing, C. R., and S. B. Walker (eds.). *The Pesticide Manual—A World Compendium.* 8th ed. Thornton Heath, UK: The British Crop Protection Council, 1987. 666.

2. SRI.

3. *Farm Chemicals Handbook* 1993. Willoughby, OH: Meister Publishing Co., 1993, p. C–266.

4. *Farm Chemicals Handbook* 1984. Willoughby, OH: Meister Publishing Co., 1984, p. C–179.

5. Hartley, D., and H. Kidd (eds.). *The Agrochemicals Handbook.* 2nd ed. Lechworth, Herts, England: The Royal Society of Chemistry, 1987, p. A325/Aug 87.

6. Sax, N. I. *Dangerous Properties of Industrial Materials,* Reports. New York: Van Nostrand Reinhold, 1987, p. 11: 3 (1991).

7. Clayton, G.D., and F.E. Clayton (eds.). *Patty's Industrial Hygiene and Toxicology:* Volume 2A, 2B, 2C: *Toxicology.* 3rd ed. New York: John Wiley and Sons, 1981–1982.

8. U.S. Department of the Interior, Fish and Wildlife Service. *Handbook of Toxicity of Pesticides to Wildlife.* Resource Publication 153. Washington, DC: U.S. Government Printing Office, 1984. 47.

9. Julin, A. M., H. O. Sanders. *Trans Am Fish Soc* 106 (4): 386–92 (1977).

10. Menzie, C. M. *Metabolism of Pesticides.* U.S. Department of the Interior,

Bureau of Sport Fisheries and Wildlife, Publication 127. Washington, DC: U.S. Government Printing Office, 1969. 225.

11. USEPA/OPP. *Status of Pesticides in Reregistration and Special Review*, p. 107 (Mar, 1992) EPA 700–R–92–004.

■ PHOSPHORODITHIOIC ACID, O,O–DIMETHYL ESTER, S–ESTER with 2–MERCAPTO–N–METHYL–ACETAMIDE

CAS # 60–51–5

SYNONYMS (O,O-Dimethyl-S-(N-methyl-carbamoyl-methyl)-dithiophosphat) (German) • acetic acid, O,O-dimethyldithiophosphoryl-, N-monomethylamide salt • ent-24650 • nci-c00135 • O,O-dimethyl S-(N-methylcarbamoylmethyl) dithiophosphate • O,O-dimethyl S-(N-methylcarbamoylmethyl) phosphorodithioate • O,O-dimethyl S-methylcarbamoylmethyl phosphorodithioate • O,O-dimethyl-S-(N-methyl-carbamoyl)-methyl-dithiofosfaat (Dutch) • O,O-dimethyldithio-phosphorylacetic acid, N-monomethylamide salt • phosphorodithioic acid O,O-dimethyl ester, ester with 2-mercapto-N-methylacetamide • phosphorodithioic acid, O,O-dimethyl ester, S-ester with 2-mercapto-N-methylacetamide • phosphorodithioic acid, O,O-dimethyl s-(2-(methylamino)-2-oxoethyl) ester • S-methylcarbamoylmethyl O,O-dimethyl phosphorodithioate • dithiophosphate de O,O-dimethyle et de S(-N-methylcarbamoyl-methyle) (French) • O,O-Dimethyl-dithiophosphorylessigsaeure monomethylamid (German) • O,O-Dimethyl-S-(2-oxo-3-aza-butyl)-dithiophosphat (German) • O,O-dimetil-S-(N-metil-carbamoil-metil)-ditiofosfato (Italian) • N-monomethylamide of O,O-dimethyldithiophosphorylacetic acid • American Cyanamid 12,880 • ac-12880

• ac-18682 • l-395 • cl-12880 • cygon-4e • cygon • perfekthion • Rogor-201 • Rogor-40 • Rogor-l • Rogor-p • roxion • fortion-nm • 8014-bis-hc • bi-58 • cygon-insecticide • daphene • de-fend • demos-140 • dimetate • dimethoaat (Dutch) • Dimethoat (German) • dimethogen • dimeton • dimevur • experimental insecticide 12,880 • fip • fosfotox • fosfotox-r • fosfotox-r-35 • fostion-mm • lurgo • pei-75 • perfecthion • phosphamid • phosphamide • racusan • sinoratox • solut • systoate • bi-58 • fosfatox-r • systemin • rebelate • roxion-ua • trimetion • Rogor • fosfamid (former USSR) • devigon • cekuthoate • dimet • dimethoat-tech-95% • de-fend • end-24650 • fostion • Caswell number 358 • EPA pesticide code 035001

MF $C_5H_{12}NO_3PS_2$

MW 229.27

COLOR AND FORM Colorless crystals.

ODOR Camphor-like odor; mercaptan odor.

DENSITY 1.277 at 65°C.

BOILING POINT 107°C at 0.05 mm Hg.

MELTING POINT 51–52°C.

FIREFIGHTING PROCEDURES Dry chemicals, carbon dioxide for small fires. Water spray or foam for larger fires [R4].

SOLUBILITY Soluble in most organic solvents, except saturated hydrocarbons; very soluble in ethanol, chloroform, acetone; slightly soluble in diethyl ether; insoluble in petroleum ether; slightly soluble in aromatic hydrocarbons; more than 5,000 mg/L water at 20 −1.5°C; 25 g/L water at 21°C; soluble in cyclohexanone; low solubility in xylene, hexane; >300 g/kg alcohol at 20°C; >300 g/kg benzene at 20°C; >300 g/kg chloroform at 20°C; >300 g/kg dichloromethane at 20°C; >300 g/kg ketones at 20°C; >300 g/kg toluene at 20°C; >50 g/kg carbon tetrachloride; >50 g/kg saturated hydrocarbons; >50 g/kg octan–1–ol.

VAPOR PRESSURE 1.1 mPa at 25°C.

OCTANOL/WATER PARTITION COEFFICIENT
Log K_{ow}= 0.50 and 0.78.

CORROSIVITY Slightly corrosive to iron.

STABILITY AND SHELF LIFE The biological activity remains practically unvaried for 2 yr under environmental conditions, provided stored in unopened, and undamaged, original containers, in shaded, cool, well-aired places. Crystals may form in formulations stored at <32°F/0°C. Stable a minimum of 1 yr at <25–30°C (77–86°F) [R4].

OTHER CHEMICAL/PHYSICAL PROPERTIES It decomposes on heating, initially forming the o, S–dimethyl analogue.

PROTECTIVE EQUIPMENT Wear impervious gloves, boots, body covering [R4].

PREVENTATIVE MEASURES Avoid eye, skin, clothing contact. Do not breathe dust. Use with adequate ventilation. Wash thoroughly after handling [R4].

CLEANUP Use of granular activated carbon in the adsorption of pesticides from wastewater is reported [R7]. Absorb with paper towels. Place in a plastic bag. Burn in an open pan with the help of flammable solvents or in a furnace [R26].

DISPOSAL Generators of waste (equal to or greater than 100 kg/mo) containing this contaminant, EPA hazardous waste number P044, must conform with USEPA regulations in storage, transportation, treatment, and disposal of waste [R8].

COMMON USES Systemic insecticide-acaricide used for a wide range of insects such as aphids, thrips, planthoppers, white flies, mites on ornamental plants, alfalfa, apples, corn, cotton, grapefruit, grapes, lemons, melons, oranges, pears, pecans, safflower, sorghum, soybeans, tangerines, tobacco, tomatoes, watermelons, wheat, other vegetables; residual wall spray in farm buildings for houseflies [R4]; insecticide for deciduous fruits and nuts and commercial/industrial uses [R1]. It is effective against Diptera of medical importance [R3, 298]; for fruit fly larvae [R5, 100].

OCCUPATIONAL RECOMMENDATIONS None listed.

NON-HUMAN TOXICITY LD_{50} rat female oral 240–336 mg/kg technical material (from table) [R10]; LD_{50} mouse subcutaneous 60 mg/kg technical material (from table) [R10]; LD_{50} hamster male sc 60 mg/kg technical material (from table) [R10]; LD_{50} rabbit oral 300 mg/kg technical material (from table) [R10]; LD_{50} guinea pig oral 350–400 mg/kg technical material (from table) [R10]; LD_{50} mice female oral 60 mg/kg technical material (from table) [R10]; LD_{50} rat oral 500–680 mg/kg [R2]; LD_{50} guinea pig oral 600 mg/kg [R2]; LD_{50} rabbit oral 400–500 mg/kg [R2]; LD_{50} rat percutaneous >800 mg/kg [R2]; LD_{50} guinea pig percutaneous >1,000 mg/kg [R2].

ECOTOXICITY LC_{50} *Gammarus lacustris* (scuds) 0.20 mg/L/96 hr (95% confidence limit 0.15–0.27 mg/L), mature, static bioassay, temperature 21°C (technical, 97.4%) [R11]; LC_{50} *Pteronarcys* 0.043 mg/L/96 hr (95% confidence limit 0.036–0.051 mg/L), second year class, static bioassay, temperature 21°C (technical, 97.4%) [R11]; LC_{50} *Salmo gairdneri* (rainbow trout) 6.2 mg/L/96 hr (95% confidence limit 4.1–9.3 mg/L), wt 1.5 g, static bioassay, temperature 21°C (technical, 97.4%) [R11]; LC_{50} *Salmo gairdneri* (rainbow trout) 20.0 ppm/24 hr (conditions of bioassay not given) [R12]; LC_{50} *Chingatta* 4.48 mg/L/96 hr (30% emulifiable concentrate; conditions of bioassay not given) [R12]; LC_{50} *Lapomis macrochirus* (bluegill) 6.0 mg/L/96 hr, wt 0.3 g, static bioassay, temperature 24°C (technical, 97.4%) [R11]; LC_{50} *Lapomis macrochirus* (bluegill) 28.0 ppm/24 hr (conditions of bioassay not given) [R12]; LC_{50} *Coturnix japonica* (Japanese quail) 346 mg/L in 5-day diet (95% confidence limit 303–394 mg/L) age 14 days (technical, 99%) [R13]; LC_{50} ring-necked pheasants 332 mg/L in 5-day diet (95% confidence limit 293–376 mg/L), age 10 days (technical, 99%) [R13]; LC_{50} mallards 1,011 mg/L in 5-day diet (95% confidence limit 707–1,372 mg/L), age 10 days (technical, 99%) [R13]; LD_{50} redwinged blackbird, oral 6.60–17.8 mg/kg [R14]; LD_{50} starlings oral 31.6 mg/kg [R14]; LC_{50} mosquito fish 40–60 mg/L/96 hr (conditions of bioassay not specified) [R15]; LD50 honey bees 0.9 mg/bee [R15]; LC50 *Salmo gairdneri* (rainbow trout) 58.0 mg/L/24 hr; 27.0 mg/L/48 hr (Rogor 40, 32% dimethoate, from table; conditions of bioassay not specified) [R16]; LC50 *Cyprinus carpio* (carp) 22.39 mg/L/168 hr (from table; conditions of bioassay not specified) [R17]; LD50 mallards (male) oral 41.7 mg/kg (95% confidence limit 30.1–57.8 mg/kg), age 3–4 mo (sample purity 97%) [R9]; LD50 mallards (female), oral 63.5 mg/kg (95% confidence limit 45.8–88.1 mg/kg), age 3–4 mo (sample purity 99.8%) [R9]; LD50 pheasant (male), oral 20.0 mg/kg (95% confidence limit 15.9–25.2 mg/kg), age 3–4 mo (sample purity 97%) [R9]; LC50 *Coturnix japonica* (Japanese quail) oral 496 ppm in 5-day diet (95% confidence limit 373–659 ppm), age 14-day (dimethoate Cygon 2E) [R18]; LC50 *Daphnia magna* 2.50 mg/L/48 hr (from table; conditions of bioassay not specified); LC50 *Pteronarcys californica* (stonefly) 0.043 mg/L/96 hr (conditions of bioassay not specified) [R19]; EC50 *Skeletonema costatum* (marine alga) decreased dry weight is 9.5 mg/L/96 hr (conditions of bioassay not specified) [R20]; EC50 *Skeltonema costatum* (marine alga) protein content 10.65 mg/L/96 hr (conditions of bioassay not specified) [R21, EACO–CIN–PO81]; EC50 *Skeletonema costatum* (marine alga) cell count 11.2 mg/L/96 hr (conditions of bioassay not specified) [R22]; EC50 *Skeletonema costatum* (marine alga) carbohydrate and chlorophyll a content 10.8 mg/L/96 hr (conditions of bioassay not specified) [R21, ECAO–CIN–PO81].

BIODEGRADATION Concentration of dimethoate left (initial concentration 10 ppb) after various times in raw water from Little Miami River at pH 7.3–8.0: 1 hr, 10 ppb, 1 wk, 10 ppb; 2 wk, 8.5 ppb; 4 wk, 7.5 ppb; 8 wk; 5.0 ppb (1). Percent degradation in chehalis clay loam soil in 2 wk, non-sterile, 77%; autoclaved, 18%; irradiated, 20%(2). Half-lives in soil in June–July averaged 11 days, and less than 2% of applied dimethoate residue detected after 10 months (2). In laboratory experiments at 20–30°C half-lives for degradation were 28.9 and 36.7 days (2). In moist soils, dimethoate is readily oxidized to dimethoxon, but the role of microbial degradation on the removal of dimethoate from the environment is uncertain (3). Recovery of dimethoate incubated with culture from enrichment studies using raw sewage: 0 days, 54 ppm; 0.5 days, 54 ppm; 1 day, 52.5 ppm; 6 days, 22.4 ppm; 9 days, 13.5 ppm; 12 days, not detected (4) [R23].

BIOCONCENTRATION An estimated BCF of 2.0 has been calculated using a water solubility of 25,000 ppm (1). Based on this estimated BCF, dimethoate should not bioconcentrate in aquatic organisms [R24].

RCRA REQUIREMENTS As stipulated in 40 CFR 261.33, when dimethoate, as a commercial chemical product or manufacturing chemical intermediate or an off-specification commercial chemical product or a manufacturing chemical intermediate, becomes a waste, it must be managed according to federal or state hazardous waste regulations. Also defined as a hazardous waste is any container or inner liner used to hold this waste or any residue, contaminated soil, water, or other debris resulting from the cleanup of a spill, into water or on dry land, of this waste. Generators of small quantities of this waste may qualify for partial exclusion from hazardous waste regulations (40 CFR 261.5 (e)) [R25].

FIFRA REQUIREMENTS Unconditional denial of all applications for registration of dimethoate products for use in dust formulations. [R6, 4].

REFERENCES
1. SRI.

2. Hartley, D., and H. Kidd (eds.). *The Agrochemicals Handbook.* 2nd ed. Lechworth, Herts, England: The Royal Society of Chemistry, 1987, p. A153/Aug 87.

3. Worthing, C. R., and S. B. Walker (eds.). *The Pesticide Manual—A World Compendium.* 8th ed. Thornton Heath, UK: The British Crop Protection Council, 1987.

4. *Farm Chemicals Handbook 1989*. Willoughby, OH: Meister Publishing Co., 1989, p. C–104.

5. White–Stevens, R. (ed.). *Pesticides in the Environment*: Volume 1, Part 1, Part 2. New York: Marcel Dekker, Inc., 1971.

6. Environmental Protection Agency/OPTS. *Suspended, Cancelled, and Restricted Pesticides*. 3rd Revision. Washington, DC: Environmental Protection Agency, January 1985.

7. Dennis, W. H., Jr., et al. *J Environ Sci Health* B18 (3): 317–31 (1983).

8. 40 CFR 240–280, 300–306, 702–799 (7/1/89).

9. U.S. Department of the Interior, Fish and Wildlife Service. *Handbook of Toxicity of Pesticides to Wildlife*. Resource Publication 153. Washington, DC: U.S. Government Printing Office, 1984. 32.

10. Sanderson, D. M., E. P. Edson. *Br J Ind Med* 21: 52–64 (1964) as cited in USEPA/ECAO, *Health Effects Profile for Dimethoate (Final Draft)*, p. 25 (1984) ECAO–CIN–PO81.

11. U.S. Department of Interior, Fish and Wildlife Service. *Handbook of Acute Toxicity of Chemicals to Fish and Aquatic Invertebrates*. Resource Publication No. 137. Washington, DC: U.S. Government Printing Office, 1980. 31.

12. Verschueren, K. *Handbook of Environmental Data of Organic Chemicals*. 2nd ed. New York, NY: Van Nostrand Reinhold Co., 1983. 541.

13. U.S. Department of the Interior, Fish and Wildlife Service, Bureau of Sports Fisheries, and Wildlife. *Lethal Dietary Toxicities of Environmental Pollutants to Birds*. Special Scientific Report—*Wildlife* No. 191. Washington, DC: U.S. Government Printing Office, 1975. 19.

14. Schafer, E. W., Jr., et al. *Arch Environ Contam Toxicol* 12 (3): 355–82 (1983).

15. Hussar, D. A. (ed.). *Modell's Drugs in Current Use, and New Drugs*. 34th ed. New York, NY: Springer Verlag Publishing Co., 1988. 298.

16. Alabaster, J. S. *International Pest Control* 11 (2): 29–35 (1969) as cited in USEPA/ECAO, *Health Effects Profile for Dimethoate (Final Draft)*, p. 28 (1984) ECAO–CIN–PO81.

17. Basak, P. K., S. K. Konar. *Ind J Fish* 25 (1–2): 141–55 (1978) as cited in USEPA/ECAO, *Health Effects Profile for Dimethoate (Final Draft)*, p. 28 (1984) ECAO–CIN–PO81.

18. Hill, E. F., and M. B. Camardese. *Lethal Dietary Toxicities of Environmental Contaminants and Pesticides to Coturnix*. Fish and Wildlife Technical Report 2. Washington, DC: United States Department of Interior, Fish and Wildlife Service, 1986. 61.

19. Sanders, H. O., O. B. Cope. *Limnol Oceanogr* 13 (1): 112–17 as cited in USEPA/ECAO, *Health Effects Profile for Dimethoate (Final Draft)* p. 27 (1984) ECAO–CIN–PO81.

20. Ibramhim, E. A. *Aquat Toxicol* 3 (1): 1–14 (1983) as cited in USEPA/ECAO, *Health Effects Profile for Dimethoate (Final Draft)*, p. 30 (1984) ECAO–CIN–PO81.

21. Ibrahim, E. A. *Aquat Toxicol* 3 (1): 1–14 (1983) as cited in USEPA/ECAO, *Health Effects Profile for Dimethoate (Final Draft)* p. 30 (1984).

22. Ibrahim, E. A. *Aquat Toxicol* 3 (1): 1–14 (83) as cited in USEPA/ECAO, *Health Effects Profile for Dimethoate (Final Draft)*, p. 30 (1984) ECAO–CIN–PO81.

23. (1) Eichelberger, J. W., J. J. Litchtenberg. *Environ Sci Technol* 5: 541–4 (1971); (2) Getzin, L. W., I. Rosefield. *J Agric Food Chem* 16: 598–601 (1968); (3) USEPA. *Health and Environ Effects Profile for Dimethoate (1985)* EPA ECAO–CIN–PO81 246 pp.; (4) Barik, S., et al. *Agric Wastes* 10: 81–94 (1984).

24. Kenaga, E. E. *Ecotox Environ Safety* 24: 26–38 (1980).

25. 40 CFR 261.33 (7/1/88).

26. ITTI. *Toxic and Hazardous Industrial Chemicals Safety Manual*. Tokyo Japan: The International Technical Information Institute, 1988, p. 187.

PHOSPHORODITHIOIC ACID, O,O-DIMETHYL ESTER, S-ESTER with 3-(MERCAPTOMETHYL)-1,2,3-BENZOTRIAZIN-4 (3H)-ONE

CAS # 86-50-0

DOT # 2783

SIC CODES 2879; 2842.

SYNONYMS 1,2,3-benzotriazin-4 (3H)-one, 3-(mercaptomethyl)-, O,O-dimethyl phosphorodithioate • 3-(mercaptomethyl)-1,2,3-benzotriazin-4 (3H)-one O,O-dimethyl phosphorodithioate S-ester • benzotriazine derivative of a methyl dithiophosphate • benzotriazinedithiophosphoric acid, dimethoxy ester • dimethyldithiophosphoric acid, N-methylbenzazimide ester • ent-23,233 • N-methylbenzazimide, dimethyldithiophosphoric acid ester • NCI-C00066 • O,O-dimethyl S-(3,4-dihydro-4-keto-1,2,3-benzotriazinyl-3-methyl) dithiophosphate • O,O-dimethyl S-(4-oxo-1,2,3-benzotriazino (3)-methyl) thiothionophosphate • O,O-dimethyl S-(4-oxo-3H-1,2,3-benzotriazine-3-methyl) phosphorodithioate • O,O-dimethyl s-(4-oxobenzotriazino-3-methyl) phosphorodithioate • O,O-dimethyl S-4-oxo-1,2,3-benzotriazin-3 (4H) -ylmethyl phosphorodithioate • O,O-dimethyl-S-((4-oxo-3h-1,2,3-benzotriazin-3-yl)-methyl)-dithiofosfaat (Dutch) • O,O-Dimethyl-S-((4-oxo-3h-1,2,3-benzotriazin-3-yl)-methyl)-dithiophosphat (German) • O,O-dimethyl-S-(1,2,3-benzotriazinyl-4-keto) methyl phosphorodithioate • O,O-dimethyl-S-(benzaziminomethyl) dithiophosphate • O,O-dimetil-S-((4-oxo-3h-1,2,3-benzotriazin-3-il) -metil)-ditiofosfato (Italian) • phosphorodithioic acid, O,O-dimethyl ester, S-ester with 3-(mercaptomethyl)-1,2,3-benzotriazin-4 (3H)-one • phosphorodithioic acid, O,O-dimethyl S-((4-oxo-1,2,3-benzotriazin-3 (4h) -yl) methyl) ester • S-(3,4-dihydro-4-oxo-1,2,3-benzotriazin-3-ylmethyl) o,o-dimethyl phosphorodithioate • S-(3,4-dihydro-4-oxo-benzo (alpha) (1,2,3) triazin-3-ylmethyl) O,O-dimethyl phosphorodithioate • O,O-Dimethyl-S-(4-oxobenzotriazin-3-methyl)-dithiophosphat (German) • 3-(mercaptomethyl)-1,2,3-benzotriazin-4 (3H)-one O,O-dimethyl phosphorodithioate S-ester • guthion • azinfos-methyl (Dutch) • azinophos-methyl • azinphos-methyl • azinphos-metile (Italian) • bay-17147 • bay-9027 • Bayer-17147 • Bayer-9027 • azinphos-methyl • carfene • cotneon • cotnion • cotnion-methyl • crysthion-21 • crysthyon • dbd • gothnion • gusathion • gusathion-25 • gusathion-k • gusathion-m • gusathion-methyl • gusathion-20 • methylazinphos • methylgusathion • metiltriazotion • EPA-Shaughnessy-#058001

MF $C_{10}H_{12}N_3O_3PS_2$

MW 317.34

COLOR AND FORM Colorless crystals; brown waxy solid.

ODOR THRESHOLD Detection: 0.0002 mg/kg water [R3, 226].

DENSITY 1.44 at 20°C.

MELTING POINT 73-74°C.

HEAT OF COMBUSTION -8,600 Btu/lb = -4,800 cal/g = -200×10⁵ J/kg (est).

EXPLOSIVE LIMITS AND POTENTIAL When exposed to high temperatures or flame, the containers may explode. [R5, 71].

FIREFIGHTING PROCEDURES Self-contained breathing apparatus with a full facepiece operated in pressure-demand or other positive-pressure mode (when fighting fire) [R4, 5]. Use water in flooding quantities as fog. Solid streams of water may be ineffective. Cool all affected containers with flooding quantities of water. Apply water from as great a distance as possible. Use foam, dry chemical, or carbon dioxide. Keep runoff water out of sewers and water sources. (azinphos methyl, or guthion (agricultural insecticides, not elsewhere classified, other than liquid)) [R5, 71].

TOXIC COMBUSTION PRODUCTS Hazardous decomposition products: toxic gases and vapors (such as sulfur dioxide, oxides of nitrogen, phosphoric acid mist, and carbon monoxide) may be released from a fire involving azinphos methyl. [R4, 2].

SOLUBILITY Soluble in methanol, ethanol, propylene glycol, xylene, other organic solvents; 33 ppm in water at 20°C; >1 kg/L dichloromethane; >1 kg/L toluene.

VAPOR PRESSURE Less than 1 mPa at 20°C.

OCTANOL/WATER PARTITION COEFFICIENT Log K_{ow} = 2.75 (measured).

CORROSIVITY Noncorrosive.

STABILITY AND SHELF LIFE It is unstable at temperatures above 200°C. [R1, 41].

OTHER CHEMICAL/PHYSICAL PROPERTIES Recrystallizable from methanol and isopropanol.

PROTECTIVE EQUIPMENT Upper-limit devices recommended by OSHA: Respirator selection: 2 mg/m³: any supplied-air respirator or any self-contained breathing apparatus or any chemical-cartridge respirator with a full facepiece and organic vapor cartridge(s) in combination with a dust, mist, and fume filter; 5 mg/m³: any powered air-purifying respirator with organic vapor cartridge(s) in combination with a dust, mist, and fume filter, or any supplied-air respirator operated in a continuous-flow mode; or any chemical-cartridge respirator with a full facepiece and organic vapor cartridges in combination with a high-efficiency particulate filter; or any supplied-air respirator with a tight-fitting facepiece operated in a continuous-flow mode; or any self-contained breathing apparatus with a full facepiece; or any supplied-air respirator with a full facepiece; or any air-purifying full facepiece respirator (gas mask) with a chin-style or front– or back-mounted organic vapor canister; or

any air-purifying respirator with a high-efficiency particulate filter; emergency or planned entry in unknown concentration or IDLH conditions: any self-contained breathing apparatus with a full facepiece and operated in a pressure-demand or other positive-pressure mode; or any supplied-air respirator with a full facepiece and operated in a pressure-demand or other positive-pressure mode in combination with an auxiliary self-contained breathing apparatus operated in pressure-demand or other positive-pressure mode; escape: any air-purifying full facepiece respirator (gas mask) with a chin-style or front– or back– mounted organic vapor canister having a high-efficiency particulate filter; or any appropriate escape-type self-contained breathing apparatus [R6].

PREVENTATIVE MEASURES Contact lenses should not be worn when working with this chemical [R6].

CLEANUP Environmental considerations: land spill: dig a pit, pond, lagoon, or holding area to contain liquid or solid material. (SRP: If time permits, pits, ponds, lagoons, soak holes, or holding areas should be sealed with an impermeable flexible-membrane liner.) Cover solids with a plastic sheet to prevent dissolving in rain or firefighting water. (azinphos methyl or guthion (agricultural insecticides, not elsewhere classified, other than liquid)) [R5, 71].

DISPOSAL 1. By making packages of azinphos-methyl in paper or other flammable material and burning in a suitable combustion chamber equipped with an appropriate effluent-gas-cleaning device. 2. By dissolving azinphos-methyl in a flammable solvent (such as alcohol) and atomizing in a suitable combustion chamber equipped with an appropriate effluent gas-cleaning-device [R4, 4]. Decomposes with heat. Subject to oxidation. Hydrolyzes in acid, rapidly in cold alkali.

COMMON USES Nonsystemic insecticide and acaricide [R3, 226].

STANDARD CODES EPA 311; class B poison; Superfund designated (hazardous substances) list.

PERSISTENCE The half-life of spray and dust has been reported as 2–4 days on cotton leaves. Hydrolysis half-life (pH 6, 70°C, ethanol), in 10.4 hours, changes by factor of 10 [R13]; for each pH unit under alkaline solutions (pH > 8) [R12], loses insecticidal activity in 2–4 weeks. Also hydrolyzes in acid.

MAJOR SPECIES THREATENED Fish and animal life.

DIRECT CONTACT Skin contact limit 0.2 mg/m^3 [R11].

GENERAL SENSATION Can be absorbed through skin.

PERSONAL SAFETY Guthion is absorbed through the skin. If high concentrations abound, impervious clothing and breathing apparatus are required.

ACUTE HAZARD LEVEL Lethal dose for man is estimated to be 3 mg/kg.

CHRONIC HAZARD LEVEL Goldfish are affected by 0.01 mg/L.

DEGREE OF HAZARD TO PUBLIC HEALTH Highly toxic via all routes.

OCCUPATIONAL RECOMMENDATIONS .
OSHA Meets criteria for OSHA medical records rule [R9]. Threshold-limit 8–hr time-weighted average (TWA) 0.2 mg/m^3, skin (1986) [R10, 13].

ACTION LEVEL Evacuate area. Enter from upwind side. Notify local air authority and National Agricultural Chemicals Association.

ON-SITE RESTORATION Carbon or peat may be used to adsorb spills. Seek professional environmental engineering assistance through EPA's Environmental Response Team (ERT), Edison, NJ, 24–hour no. 908–548–8730.

BEACH OR SHORE RESTORATION May be burned or heated off sand.

AVAILABILITY OF COUNTERMEASURE MATERIAL Carbon—water treatment plants, sugar refineries; peat—nurseries, floral shops.

MAJOR WATER USE THREATENED Potable supply, fisheries, recreation.

PROBABLE LOCATION AND STATE Will sink and dissolve very very slowly. Brown waxy solid or white powder, formulated as emulsifiable concentrate, wettable powder, or liquid. Wettable forms will dissolve.

WATER CHEMISTRY Hydrolyzes in water; process accelerates under alkaline conditions.

NON-HUMAN TOXICITY LD$_{50}$ guinea pig male oral 80 mg/kg [R3, 227].

ECOTOXICITY TLm *Crassostrea virginica* (American oyster) eggs 620 ppb/48 hr in a static lab bioassay [R3, 226]; TLm *Mercenaria mercenaria* (hard clam) eggs 860 ppb/48 hr in a static lab bioassay [R3, 226]; TLm *Mercenaria mercenaria* (hard clam) larvae 860 ppb/12 hr in a static lab bioassay [R3, 226]; LC$_{50}$ *Gammarus lacustris* 0.15 μg/L/96 hr (conditions of bioassay not specified) [R3, 226]; LC$_{50}$ *Gammarus fasciatus* 0.10 μg/L/96 hr (conditions of bioassay not specified) [R3, 226].

BIODEGRADATION After 44 and 197 days incubation in a soil, about 50 and 93%, respectively, of the radiolabeled azinphosmethyl was degraded and after 222 days incubation, 18.6% of the radiolabel was recovered as $^{14}CO_2$ (1). Seventeen metabolites were identified (1). Azinphosmethyl was degraded in batch and continuous culture by mixed cultures of microorganisms that were collected from soil, raw sewage, a trickling filter, activated sludge, and settled sludge (2). Azinphosmethyl concentration decreased from 99 mg/L to 49 mg/L after 4 days incubation in a stirred flask containing azinphosmethyl as the sole carbon source and a mixed culture (2). The main degradation products of azinphosmethyl in soil and by selected soil microorganisms are benzazimide, thiomethylbenzazimide, bis–(benzazimidyl-methyl) disulfide, and anthranilic acid (3) [R7].

BIOCONCENTRATION Using a measured log octanol/water partition coefficient (log K$_{ow}$) of 2.75 (1) and the recommended regression equation (2), an estimated bioconcentration factor (BCF) for azinphosmethyl is 72 (2). This indicates a low potential for bioconcentration (2). No azinphosmethyl was detected in fish samples taken from a lake after insecticide applications in 1981–83 (3) [R8].

FIFRA REQUIREMENTS All liquid formulations with concentrations of 13.5% or greater are currently classified as restricted-use chemicals. [R2, (1987)].

GENERAL RESPONSE Avoid contact with solid and dust. Keep people away. Wear goggles and self-contained breathing apparatus. Stay upwind. Use water spray to "knock down" dust. Isolate and remove discharged material. Notify local health and pollution control agencies.

FIRE RESPONSE Not flammable. Poisonous gases may be produced when heated.

EXPOSURE RESPONSE Call for medical aid. Dust is poisonous if inhaled. Move victim to fresh air. If breathing is difficult, give oxygen. Solid is poisonous if swallowed. If swallowed, and victim is conscious, have victim drink water or milk, and have victim induce vomiting. If swallowed and victim is unconscious or having convulsions, do nothing except keep victim warm.

RESPONSE TO DISCHARGE Issue warning—water contaminant. Should be removed. Provide chemical and physical treatment.

WATER RESPONSE Harmful to aquatic life in very low concentrations. May be dangerous if it enters water intakes. Notify local health and wildlife officials. Notify operators of nearby water intakes.

EXPOSURE SYMPTOMS Dust irritates eyes. Inhalation or ingestion causes sweating, constriction of pupils of eyes, asthmatic symptoms, cramps, weakness, convulsions, collapse.

EXPOSURE TREATMENT Inhalation: remove to fresh air; keep warm; call doctor. Eyes: flush with water for at least 15 min. Skin: flush with water, wash with soap and water. Ingestion: get medical attention at once; give water slurry of charcoal; do not give milk or alcohol.

REFERENCES

1. Worthing, C. R., and S. B. Walker (eds.). *The Pesticide Manual—A World Compendium.* 8th ed. Thornton Heath, UK: The British Crop Protection Council, 1987.

2. Purdue University, *National Pesticide Information Retrieval System.*

3. Verschueren, K. *Handbook of Environmental Data of Organic Chemicals.* 2nd ed. New York, NY: Van Nostrand Reinhold Co., 1983.

4. Mackison, F. W., R. S. Stricoff, and L. J. Partridge, Jr. (eds.). *NIOSH/OSHA—Occupational Health Guidelines for Chemical Hazards.* DHHS (NIOSH) Publication No. 81–123 (3 vols). Washington, DC: U.S. Government Printing Office, Jan. 1981.

5. Association of American Railroads. *Emergency Handling of Hazardous Materials in Surface Transportation.* Washington, DC: Assoc. of American Railroads, Hazardous Materials Systems (BOE), 1987.

6. NIOSH. *Pocket Guide to Chemical Hazards.* 2nd Printing. DHHS (NIOSH) Publ. No. 85–114. Washington, DC: U.S. Dept. of Health and Human Services, NIOSH/Supt. of Documents, GPO, February 1987. 55.

7. (1) Engelhardt, G., et al. *J Agric Food Chem* 32: 102–8 (1984); (2) Barik, S., et al. *Agric Wastes* 10: 81–94 (1984); (3) Engelhardt, G, P. R. Wallnoefer. *Chemosphere* 12: 955–60 (1983).

8. (1) Hansch, C., and A. J. Leo. *Medchem Project,* Claremont, CA: Pomona College, Issue No. 26 (1985); (2) Lyman, W. J., et al., *Handbook of Chemical Property Estimation Methods. Environmental Behavior of Organic Compounds.* New York: McGraw-Hill (1982); (3) Bush, P. B., et al. *Water Res Bull* 22: 817–29 (1986).

9. 29 CFR 1910.20 (7/1/88).

10. American Conference of Governmental Industrial Hygienists. *Threshold Limit Values for Chemical Substances and Physical Agents and Biological Exposure*

Indices for 1993–1994. Cincinnati, OH: ACGIH, 1993.

11. *Threshold Limit Values for Airborne Contaminants for 1968*, American Conference of Governmental Industrial Hygienists, 1968, Cincinnati, May 13, 1968 (1970 edition also used but not referenced).

12. Elo, M. *Organophosphorus Pesticides: Organic and Biological Chemistry*, Cleveland, OH: CRC Press Inc., 1974.

13. McKee, J. E., H. W. Wolf. *Water Quality Criteria*, Caifornia State Water Quality Control Board, 1963. 2nd ed.

■ PHOSPHOROTHIOIC ACID, O–(2–CHLORO–4–NITRO-PHENYL) O,O–DIMETHYL ESTER

CAS # 2463–84–5

SIC CODES 2879; 2842.

SYNONYMS AC-4124 • American Cyanamid 4,124 • bay-14981 • Bayer-22/190 • captec • chlorthion • dicaptan • dicapthion • dicapton • dimethyl 2-chloronitrophenyl thiophosphate • ent-17,035 • experimental insecticide 4124 • insecticide-acc-4124 • isochloorthion (Dutch) • isochlorothion • isochlorthion • isomeric-chlorthion • O,O-dimethyl O-2-chloro-4-nitrophenyl phosphorothioate • O,O-dimethyl-O-(2-chloro-4-nitrophenyl) thionophosphate • O-(2-chloro-4-nitrophenyl) O,O-dimethyl phosphorothioate • O-(4-chloor-3-nitro-fenyl)-O, O-dimethylmonothiofosfaat (Dutch) • O-(4-Chlor-3-nitro-phenyl) - O,O-dimethyl-monothiophosphat (German) • O-(4-cloro-3-nitro-fenil) -O, O-dimetil-monotiofosfato (Italian) • oms-214 • p-nitro-O-chlorophenyl dimethyl thionophosphate • phenol, 2-chloro-4-nitro-, O-ester with O,O-dimethyl phosphorothioate • phosphorothioic acid, O-(2-chloro-4-nitrophenyl) O, O-dimethyl ester • thiophosphate de O,O-dimethyle et de O-4-chloro-3-nitrophenyle (French) • O-(2-chloro-4-nitrophenyl) O,O-dimethyl ester • acc-4124 • isomeric-clorthio • di-captan • American Cyanamid 4124 • O-(2-chloro-4-nitrophenyl) O,O-dimethyl phosphorothioate • phosphorothioic acid O-(2-chloro-4-nitrophenyl) O,O-dimethyl ester • O,O-dimethyl O-(2-chloro-4-nitrophenyl) phosphorothioate • O-2-chloro-4-nitrophenyl O,O-dimethyl phosphorothioate • oms-214 • ent-17-035 • O-(2-chloro-4-nitrophenyl) O,O-dimethylphosphorothioate

MF $C_8H_9ClNO_5PS$

MW 297.66

COLOR AND FORM Crystals from methanol; white solid from methanol.

MELTING POINT 53°C.

SOLUBILITY Soluble in acetone, cyclohexanone, ethyl acetate, toluene, xylene, ethylene glycol, propylene glycol, some oils; practically insoluble in water; soluble in cyclohexane; water solubility = 14.7 mg/L at 20°C.

VAPOR PRESSURE 3.6×10^{-6} mm Hg.

OCTANOL/WATER PARTITION COEFFICIENT Log K_{ow}= 3.58 at 20°C.

OTHER CHEMICAL/PHYSICAL PROPERTIES Henry's law constant = 9.59×10^{-8} atm-cu m/mole (est).

PROTECTIVE EQUIPMENT Use of rubber gloves, goggles, respirator, and other protective clothing is advisable. (organophosphate insecticides) [R2, 110].

PREVENTATIVE MEASURES Great care should be exercised in handling. When spraying and dusting contaminated clothing should be changed frequently. (organophosphate insecticides) [R2, 110].

COMMON USES Insecticide [R1]; aphicide [R3]; acaracide [R4].

MAJOR SPECIES THREATENED Man, animals, plants, and aquatic life.

PERSONAL SAFETY Absorbed through skin. Protective clothing and filter mask are recommended.

ACUTE HAZARD LEVEL Acute oral toxicity

to rats is near 400 mg/kg [R8]. Highly toxic material.

CHRONIC HAZARD LEVEL No retarded growth with rats fed 25 ppm in diet for 1 year [R9].

DEGREE OF HAZARD TO PUBLIC HEALTH Highly toxic. Potential mutagenic effects.

OCCUPATIONAL RECOMMENDATIONS None listed.

ACTION LEVEL Evacuate area. Enter from upwind side. Notify local air authority and National Agricultural Chemicals Association.

ON-SITE RESTORATION Subject to adsorption on carbon or peat. Seek professional environmental engineering assistance through EPA's Environmental Response Team (ERT), Edison, NJ, 24–hour no., 908–548–8730.

AVAILABILITY OF COUNTERMEASURE MATERIAL Carbon—water treatment plants, sugar refineries; peat—nurseries, floral shops.

MAJOR WATER USE THREATENED Potable supplies, fisheries.

PROBABLE LOCATION AND STATE Will not dissolve readily unless accompanied by wetting agents. Will sink. Crystalline powder.

FOOD CHAIN CONCENTRATION POTENTIAL Negative.

NON-HUMAN TOXICITY LD_{50} rat oral 500–1,000 mg/kg [R5]; LD_{50} mouse oral 475 mg/kg [R4]; LD_{50} rat female oral 330 mg/kg [R3]; LD_{50} rat male dermal 790 mg/kg [R3]; LD_{50} rat female dermal 1,250 mg/kg [R3].

ECOTOXICITY TLm *Mercenaria mercenaria* (hard clam) 3.34 mg/L/48 hr, eggs (conditions of bioassay not specified) [R4].

BIOCONCENTRATION Using a flow-through system and a 3–11-day exposure period, a dicapthon BCF (extractable lipid basis) of 891 was measured in guppy fish (*Poecilia reticulata*) (1). Based upon a measured log K_{ow} of 3.58 (2) and a water solubility of 14.7 mg/L at 20°C (3), the BCF of dicapthon can be estimated to be 310 and 136,

respectively, from regression-derived equations (4,SRC). These BCF values suggest that some bioconcentration may occur in aquatic organisms (SRC) [R6].

FIFRA REQUIREMENTS In 1988, Congress amended FIFRA to strengthen and accelerate EPA's reregistration program. The nine-year reregistration scheme mandated by "FIFRA 88" applies to each registered pesticide product containing an active ingredient initially registered before November 1, 1984. Pesticides for which EPA had not issued Registration Standards prior to the effective date of FIFRA '88 were divided into three lists based upon their potential for exposure and other factors, with List B being of highest concern, and D of least. List: B; Case: dicapthon; case no. 2785; pesticide type: insecticide; case status: the pesticide is no longer an active ingredient in any registered pesticide products. Therefore, EPA is characterizing it as "cancelled." Under FIFRA, pesticide producers may voluntarily cancel their registered products. EPA also may cancel pesticide registrations if registrants fail to pay required fees, to make or meet certain reregistration commitments, or when the Agency reaches findings of unreasonable adverse effects. Active Ingredient (AI): O–(2–chloro–4–nitrophenyl) O, O–dimethyl phosphorothioate; AI status: the active ingredient is no longer contained in any registered products. Therefore, EPA is characterizing it as "cancelled." Under FIFRA, pesticide producers may voluntarily cancel their registered products. EPA also may cancel pesticide registrations if registrants fail to pay required fees, to make or meet certain reregistration commitments, or when the Agency reaches findings of unreasonable adverse effects [R7].

REFERENCES

1. *Farm Chemicals Handbook* 1993. Willoughby, OH: Meister Publishing Co., 1993, p. C–112.

2. White–Stevens, R. (ed.). *Pesticides in the Environment*: Volume 1, Part 1, Part 2. New York: Marcel Dekker, Inc., 1971.

3. Hayes, Wayland J., Jr. *Pesticides Stud-*

ied in Man. Baltimore/London: Williams and Wilkins, 1982. 360.

4. Verschueren, K. *Handbook of Environmental Data of Organic Chemicals.* 2nd ed. New York, NY: Van Nostrand Reinhold Co., 1983. 471.

5. *Kirk-Othmer Encyclopedia of Chemical Technology.* 3rd ed., Volumes 1–26. New York, NY: John Wiley and Sons, 1978–1984, p. V13 440 (1981).

6. (1) De Bruijn, J., J. Hermens. *Environ Toxicol Chem* 10: 791–804 (1991); (2) Hansch, C., and A. J. Leo. *Medchem Project* Issue No. 26, Claremont, CA: Pomona College (1985); (3) Bowman, B. T., W. W. Sans. *J Environ Sci Health* B18: 221–7 (1983); (4) Lyman, W. J., et al. *Handbook of Chemical Property Estimation Methods*, Washington, DC: Amer Chemical Soc, pp. 5–4,5–10 (1990).

7. USEPA/OPP. *Status of Pesticides in Reregistration and Special Review*, p. 129 (Mar, 1992) EPA 700–R–92–004.

8. McKee, J. E., H. W. Wolf. *Water Quality Criteria*, Caifornia State Water Quality Control Board, 1963. 2nd Ed.

9. Elo, M. *Organophosphorus Pesticides: Organic and Biological Chemistry*, Cleveland: OH: CRC Press Inc.,1974.

■ PHOSPHOROTHIOIC ACID, O, O–DIETHYL O–(p–NITROPHENYL) ESTER

$$C_2H_5O\underset{C_2H_5O}{\overset{\overset{\overset{S}{\|}}{P}}{}}-O-\langle\text{ring}\rangle-NO_2$$

CAS # 56–38–2

DOT # 2784

SIC CODES 2879; 2842.

SYNONYMS Aatp • aralo • bladan-f • rhodiatox • snp • stathion • sulphos • panthion • paramar • paraphos • parawet • fostox • genithion • niran • folidol • alkron • alleron • aphamite •

bladan • etilon • fosferno • phosphenol • thiophos • phosphemol • stabilized ethyl-parathion • diethyl para-nitrophenol thiophosphate • O,O-diethyl-O-(4-nitrophenyl) phosphorothioate • diethyl p-nitrophenyl thiophosphate • O,O-diethyl O-p-nitrophenyl thiophosphate • dpp • phenol, p-nitro-, O-ester with O,O-diethylphosphoro-thioate • phosphorothioic acid, O,O-diethyl O-(4-nitrophenyl) ester • ent-15,108 • Nitrostigmin (German) • penncap-e • American Cyanamid-3422 • Bayer e-605 • e-605 • ac-3422 • acc-3422 • Bay e-605 • e-605-forte • folicol-e605 • niran e-4 • t-47 • tox-47 • strathion • tiofos • rhodiatrox • sixty-three special ec insecticide • soprathion • fosfermo • fosfex • fosfive • fosova • kolphos • kypthion • lethalaire g-54 • murfos • oleoparaphene • orthophos • pac • paradust • pestox-plus • pethion • drexel-parathion-8e • ecatox • ekatin-wf-and-wf-ulv • ethlon • folidol-e-and-e-605 • super-rodiatox • thiomex • thiophos-3422 • vitrex • rb • rhodiasol • selephos • oleofos-20 • pacol • paramar-50 • parathene • folidol-e • gearphos • lirothion • niuif-100 • nourithion • ekatox • vapophos • aat • parathion-ethyl • corothion • niran • phoskil • soprothion • EPA pesticide code-057501 • EPA-Shaughnessy code 057701

MF $C_{10}H_{14}NO_5PS$

MW 291.28

COLOR AND FORM The pure material is a yellowish liquid at temperatures above 6°C.

ODOR Usually has a faint odor; garlic like.

ODOR THRESHOLD 4.00×10^{-2} ppm (detection in water; purity not specified) [R8].

DENSITY 1.26 at 25°C.

SURFACE TENSION 39.2 dynes/cm at 25°C.

VISCOSITY 15.30 cP at 25°C.

BOILING POINT 375°C at 760 mm Hg.

MELTING POINT 6°C.

HEAT OF COMBUSTION –9,240 Btu/lb = –5,140 cal/g = -215×10^5 J/kg.

EXPLOSIVE LIMITS AND POTENTIAL Containers may explode when heated [R7].

FIREFIGHTING PROCEDURES Use water spray, dry chemical, foam, or carbon dioxide. Use water to keep fire-exposed containers cool. If a leak or spill has not ignited, use water spray to disperse vapors, and to provide protection for firefighters. Water spray may be used to flush spills away from exposures. In advanced or massive fires, firefighting should be done from safe distance or from a protected location [R5]. Do not extinguish fire unless flow can be stopped. Use water in flooding quantities as fog. Solid streams of water may be ineffective [R4, 526]. Self-contained breathing apparatus with a full facepiece operated in pressure-demand or other positive-pressure mode should be used in firefighting [R6, 5]. Extinguish fire using agent suitable for type of surrounding fire.

TOXIC COMBUSTION PRODUCTS When heated to decomposition, can emit additional toxic fumes of oxides of nitrogen, phosphorus, and sulfur [R5].

SOLUBILITY 12.4 ± 1.0 μg/ml water at 20.0°C; 24 mg/L water at 25°C; practically insoluble in petroleum ether, kerosene, and usual spray oils; completely soluble in alcohols, esters, ethers, ketones, aromatic hydrocarbons, animal, and vegetable oils; soluble in chloroform.

VAPOR PRESSURE 5.0 mPa at 20°C.

OCTANOL/WATER PARTITION COEFFICIENT Log K_{ow} = 3.93, 3.40, and 3.47 (measured).

STABILITY AND SHELF LIFE Stable in distilled water in acid solution; hydrolyzed in presence of alkaline material and slowly decomposes in air [R9].

OTHER CHEMICAL/PHYSICAL PROPERTIES Technical product is a brown liquid with a garlic-like odor (technical product).

PROTECTIVE EQUIPMENT Neoprene-coated gloves; rubber workshoes or overshoes; latex rubber apron; goggles; respirator or mask approved for toxic dusts and organic vapors [R7].

PREVENTATIVE MEASURES At the end of the work day the protective clothes should be removed and a shower taken before returning to street clothes. If clothing becomes contaminated with parathion, it should be immediately removed, and a shower taken without delay. Smoking or eating should not be permitted when handling parathion until all outer working clothing is removed and the hands and face are washed. [R3, 1593].

CLEANUP 1. Ventilate area of spill or leak. 2. Collect for reclamation, or absorb in vermiculite, dry sand, earth, or a similar material. [R6, 4].

DISPOSAL Decomposes slightly under ultraviolet light. Hydrolyzes in alkali. (1) Mix with equal parts sand and limestone in paper boxes. Place in open incinerator and wet down with flammable solvent. Ignite with excelsior train. Stay upwind. (2) Shovel mixture of #1 into incinerator with afterburner and alkaline scrubber. SRP: At the time of review, criteria for land treatment or burial (sanitary landfill) disposal practices are subject to significant revision. Prior to implementing land disposal of waste residue (including waste sludge), consult with environmental regulatory agencies for guidance on acceptable disposal practices.

COMMON USES Insecticide, e.g., for wheat and nuts; acaricide [R2].

STANDARD CODES EPA 311; NFPA—4, 1, 0; class B poison; Superfund designated (hazardous substances) list.

PERSISTENCE Parathion when mixed with river water was not persistent past 4 weeks [R20]. Can hydrolyze or go to aminoparathion through action of yeasts. Sulfur atom subject to attack.

MAJOR SPECIES THREATENED Aquatic beetles and mosquito larvae are most adversely affected at 0.05 lb/acre doses.

INHALATION LIMIT 0.1 mg/m³.

DIRECT CONTACT Absorbed through skin.

PERSONAL SAFETY Parathion is absorbed through the skin. Wear protective clothing. Wash contacted areas well.

ACUTE HAZARD LEVEL Acute lethal oral

dose in man is 10 to 20 mg [R19]. Lethal threshold values minimum LC values, trout—19.2, *Cambarus*—1.2, carp—144.0.

CHRONIC HAZARD LEVEL Absorption of 5 mg will produce symptoms in man [R18].

DEGREE OF HAZARD TO PUBLIC HEALTH Dermal application dangerous at 0.6 mg for single dose and. 3 mg for repeated dose.

OCCUPATIONAL RECOMMENDATIONS

OSHA Meets criteria for OSHA medical records rule [R13].

NIOSH 0.05 mg/m³, 10–hr TWA; prevent skin contact; blood monitoring is required [R14]. Threshold-limit 8–hr time-weighted average (TWA) 0.1 mg/m³, skin (1986) [R15, 28].

INTERNATIONAL EXPOSURE LIMITS Permissible levels of parathion in working environment in 14 countries by regulation or recommended guidelines: Australia 1978 ceiling 0.1 mg/m³ guideline; Belgium 1978 ceiling 0.1 mg/m³ regulation; Bulgaria 1971 maximum 0.05 mg/m³ regulation; Finland 1975 ceiling 0.1 mg/m³ regulation; Federal Republic of Germany 1979 TWA 0.1 mg/m³ guideline; Hungary 1974 TWA 0.05 mg/m³ regulation; Italy 1978 TWA 0.1 mg/m³ guideline; Japan 1978 ceiling 0.1 mg/m³ guideline; The Netherlands 1978 ceiling 0.1 mg/m³ guideline; Romania 1975 twa 0.05 mg/m³ regulation; Switzerland 1978 TWA 0.1 mg/m³ regulation; former USSR 1977 maximum 0.05 mg/m³ regulation; Yugoslavia 1971 ceiling 0.1 mg/m³ regulation; ACGIH 1981 TWA 0.1 mg/m³ guideline (from table) [R16].

ACTION LEVEL Evacuate area. Enter from upwind side. Notify local air authority and National Agricultural Chemicals Association.

ON-SITE RESTORATION Carbon or peat may be used as sorbents. Seek professional environmental engineering assistance through EPA's Environmental Response Team (ERT), Edison, NJ, 24–hour no. 908–548–8730.

AVAILABILITY OF COUNTERMEASURE MATERIAL Carbon—water treatment plants, sugar refineries; peat—nurseries, floral shops.

DISPOSAL NOTIFICATION Contact local air authority.

PROBABLE LOCATION AND STATE Yellow liquid, will sink to bottom and dissolve very, very slowly.

NON-HUMAN TOXICITY LD₅₀ *Capra hircus* (domestic goat) oral 28–56 mg/kg, 12–72–mo-old females (sample purity, 98.76%) [R11];.

ECOTOXICITY LD₅₀ *Dendrocygna bicolor* (fulvous whistling-duck) oral 0.125–0.250 mg/kg (sample purity, 98.76%) [R11].

IARC SUMMARY No data are available for humans. Inadequate evidence of carcinogenicity in animals. Overall evaluation: Group 3: The agent is not classifiable as to its carcinogenicity to humans [R10].

BIODEGRADATION Parathion biodegrades in acclimated natural waters within several weeks (1,2). 5 ppm of parathion completely degraded within 2 weeks in the well-acclimated water from Holland Marsh, a vegetable growing area in Ontario, being almost quantitatively converted to aminoparathion, whereas only 10% degradation occurred in 16 weeks when the water was sterilized (1). In the waters of the Little Miami River, a small stream that receives domestic and industrial waste as well as farm runoff, 50% of parathion (10 ppb) degraded in 1 week, and none could be detected after 4 weeks (2). Surface water of varying salinity (0–28 ppt), collected from the Mississippi Sound estuary system degraded parathion with a 45-day half-life at 30°C, which was independent of salinity (3). Both chemical action and biological action of marine plankton were responsible (4). When the water was filtered to remove the plankton, the half-life at 27°C increased from 41 to 56 days (4). Parathion (1.4–28 ppm) degradation in soil with 3 different moisture contents increased with parathion concentration and moisture content (5). The percentage parathion degraded after 11 days ranged from 96% at high concentration and moisture levels, to 20% at low concentrations and moisture levels (5).

Further degradation was very slow and the degradation was never complete (5). One reference claims that residues of parathion persisted for >16 yr, but the level of application was extremely high (6). After 8 weeks of incubation in an organic and a mineral soil, <2, and 6%, respectively, of the 1 ppm parathion applied remained, whereas in sterilized controls 80 and 95% remained (7). The half-life of parathion (10 ppm) in a sandy loam and organic soil was <1 and 1 1/2 week with only 5% remaining after 3 and 10 weeks, respectively (9). A lag of approximately 2 weeks occurred when 0.1 and 1 ppm parathion was incubated in Willamette clay loam (Oregon) at moisture levels of 50% field capacity; half-lives were 16 and 26 weeks, respectively (11). Metabolic pathways involve both oxidative and reductive reactions (9). The primary oxidative pathway involves an initial hydrolysis to p-nitrophenol and diethylthiophosphoric acid (9). A second oxidative pathway involves oxidation to paraoxon (9). Under low oxygen conditions reduction to aminoparathion occurs (9). When parathion (500 ppm) is incubated in flooded alluvial soil, 43% remained after 6 days, and 0.09% after 12 days (12). The parathion is reduced to aminoparathion under these anaerobic conditions. Amendation of the soil with rice straw increases the rate of degradation (12). In parallel experiments in which parathion was incubated for 30 min in soil suspensions of 5 30 day flooded (anaerobic) soils and aerobic soils, no degradation occurred in the aerobic soils, whereas 35–68% degradation occurred in the anaerobic soils. The most-reduced soils effected the most rapid degradation (13). After repeated application of parathion to flooded soils the degradation pathway shifts from reduction to hydrolysis (14). Parathion is more persistent in flooded saline soils than in nonsaline soils. Whereas degradation was completed in 20 days in a nonsaline soil, the degradation rate decreased with salinity in 5 soils, ranging from 10 to 50% in 20 days (15). Parathion is degraded in activated sludge treatment plants. With adequate aeration high levels of parathion wastes were destroyed within 7–10 days in a treatment plant (8). Parathion is destroyed during composting of agricultural wastes (10) [R12].

BIOCONCENTRATION Although no specific data are available on possible bioaccumulation or biomagnification of parathion, physical, chemical, and biological properties make it unlikely that these phenomena will occur in food chains or food webs. [R1, 627].

RCRA REQUIREMENTS As stipulated in 40 CFR 261.33, when parathion, as a commercial chemical product or manufacturing chemical intermediate or an off-specification commercial chemical product or a manufacturing chemical intermediate, becomes a waste, it must be managed according to federal or state hazardous waste regulations. Also defined as a hazardous waste is any container or inner liner used to hold this waste or any residue, contaminated soil, water, or other debris resulting from the cleanup of a spill, into water or on dry land, of this waste. Generators of small quantities of this waste may qualify for partial exclusion from hazardous waste regulations (40 CFR 261.5 (e)) [R17].

FIFRA REQUIREMENTS In USA no worker is allowed to enter a field treated with parathion until 48 hr after treatment (USEPA, 1980F) [R16].

GENERAL RESPONSE Avoid contact with liquid and vapor. Keep people away. Wear goggles, self-contained breathing apparatus, and rubber overclothing (including gloves). Stop discharge if possible. Isolate and remove discharged material. Notify local health and pollution control agencies.

FIRE RESPONSE Not flammable. Poisonous gases are produced when heated.

EXPOSURE RESPONSE Call for medical aid. Liquid is poisonous if swallowed or if skin is exposed. Remove contaminated clothing and shoes. Flush affected areas with plenty of water. If in eyes, hold eyelids open, and flush with plenty of water. If swallowed and victim is conscious, have victim drink water or milk, and have victim induce vomiting. If swallowed and victim is unconscious or having convulsions, do nothing except keep victim warm.

RESPONSE TO DISCHARGE Issue warning—poison, water contaminant. Restrict access. Should be removed. Provide chemical and physical treatment.

WATER RESPONSE Harmful to aquatic life in very low concentrations. May be dangerous if it enters water intakes. Notify local health and wildlife officials. Notify operators of nearby water intakes.

EXPOSURE SYMPTOMS Inhalation of mist, dust, or vapor (or ingestion or absorption through the skin) causes dizziness, usually accompanied by constriction of the pupils, headache, and tightness of the chest. Nausea, vomiting, abdominal cramps, diarrhea, muscular twitchings, convulsions, and possibly death may follow. An increase in salivary and bronchial secretions may result that simulates severe pulmonary edema. Contact with eyes causes irritation.

EXPOSURE TREATMENT Call a doctor for all exposures to this compound. Inhalation: remove victim from exposure immediately; have physician treat with atropine injections until full atropinization; 2–PAM may also be administered by physician. Eyes: flush with water immediately after contact for at least 15 min. Skin: remove all clothing and shoes immediately; quickly wipe off the affected area with a clean cloth; follow immediately with a shower, using plenty of soap. If a complete shower is impossible, wash the affected skin repeatedly with soap and water. Ingestion: if victim is conscious, induce vomiting, and repeat until vomit fluid is clear; make victim drink plenty of milk or water; have victim lie down and keep warm.

REFERENCES

1. National Research Council. *Drinking Water and Health* Volume 1. Washington, DC: National Academy Press, 1977.

2. SRI.

3. International Labour Office. *Encyclopedia of Occupational Health and Safety.* Vols. I and II. Geneva, Switzerland: International Labour Office, 1983.

4. Association of American Railroads. *Emergency Handling of Hazardous Materials in Surface Transportation.* Washington, DC: Assoc. of American Railroads, Hazardous Materials Systems (BOE), 1987.

5. National Fire Protection Association. *Fire Protection Guide on Hazardous Materials.* 9th ed. Boston, MA: National Fire Protection Association, 1986, p. 49–71.

6. Mackison, F. W., R. S. Stricoff, and L. J. Partridge, Jr. (eds.). *NIOSH/OSHA—Occupational Health Guidelines for Chemical Hazards.* DHHS (NIOSH) Publication No. 81–123 (3 vols). Washington, DC: U.S. Government Printing Office, Jan. 1981.

7. U.S. Coast Guard, Department of Transportation. *CHRIS—Hazardous Chemical Data.* Volume II. Washington, DC: U.S. Government Printing Office, 1984–5.

8. Fazzalari, F. A. (ed.). *Compilation of Odor and Taste Threshold Values Data.* ASTM Data Series DS 48A (Committee E–18). Philadelphia, PA: American Society for Testing and Materials, 1978. 125.

9. Sax, N. I., and R. J. Lewis, Sr. (eds.). *Hawley's Condensed Chemical Dictionary.* 11th ed. New York: Van Nostrand Reinhold Co., 1987. 874.

10. IARC. *Monographs on the Evaluation of the Carcinogenic Risk of Chemicals to Man.* Geneva: World Health Organization, International Agency for Research on Cancer, 1972–present (multivolume work), p. S7 69 (1987).

11. U.S. Department of the Interior, Fish and Wildlife Service. *Handbook of Toxicity of Pesticides to Wildlife.* Resource Publication 153. Washington, DC: U.S. Government Printing Office, 1984. 62.

12. (1) Sharom, M. S., et al. *Water Res* 14: 1089–93 (1980); (2) Eichelberger, J. W., J. J. Lichtenberg. *Environ Sci Technol* 5: 541–4 (1971); (3) Walker, W. W. *Insecticide Persistence in Natural Seawater As Affected by Salinity, Temperature, and Sterility* p 25 USEPA–600/3–78–044 (1978); (4) Wade, M. J. *Diss Abst Int* 40: 4704 (1979); (5) Gerstl, Z., et al., *Soil Sci Soc Am J* 43: 843–8 (1979); (6) M. Alexan-

der, *Biotech Bioeng* 15: 611–47 (1973); (7) Chapman, R. A., et al. *Bull Environ Contam Toxicol* 26: 513–9 (1981); (8) Sethunathan, N., et al. *Res Rev* 68: 91–122 (1977); (9) Miles, J.R.W., et al. *Bull Environ Contam Toxicol* 22: 312–8 (1979); (10) Vogtmann, H., et al. pp. 357–78 in Proc 2nd Internl Symp Bet Degan, Isreal (1983); (11) Freed, V. H., et al. *J Agric Food Chem* 27: 706–8 (1979); (12) Adhya, T. K., et al. *J Agric Food Chem* 29: 90 (1981); (13) Adhya, T. K., et al. *Pest Biochemical Phys* 16: 14 (1981); (14) *Food Chem* 27: 1391–2 (1979); (15) Reddy, B. R., N. Sethunathan. *Soil Biol Biochemical* 17: 235 (1985).

13. 29 CFR 1910.20 (7/1/88).

14. NIOSH/CDC. *NIOSH Recommendations for Occupational Safety and Health Standards* 1988, Aug. 1988 (Suppl. to *Morbidity, and Mortality Wkly.* Vol. 37 No. 5–7, Aug. 26, 1988). Atlanta, GA: National Institute for Occupational Safety and Health, CDC, 1988. 23.

15. American Conference of Governmental Industrial Hygienists. *Threshold Limit Values for Chemical Substances and Physical Agents and Biological Exposure Indices for 1993–1994.* Cincinnati, OH: ACGIH, 1993.

16. IARC. *Monographs on the Evaluation of the Carcinogenic Risk of Chemicals to Man.* Geneva: World Health Organization, International Agency for Research on Cancer, 1972–present (multivolume work), p. V30 156 (1983).

17. 40 CFR 261.33 (7/1/88).

18. McKee, J. E., H. W. Wolf. *Water Quality Criteria,* California State Water Quality Control Board, 1963. 2nd Ed.

19. Sax, N. I., *Dangerous Properties of Industrial Materials,* New York: Van Nostrand Reinhold Company, 1969.

20. Eichelberger, J. W., J. J. Lichtenberg, "Persistence of Pesticides in River Water," *Environmental Science, and Technology,* 1971, Vol. 5, No. 6. June.

■ PHOSPHOROTHIOIC ACID, O,O–DIETHYL O-(2-(ETHYLTHIO) ETHYL) ESTER, mixed with O,O DIETHYL S-(2-(ETHYLTHIO) ETHYL) PHOSPHOROTHIOATE (7: 3)

CAS # 8065–48–3

SIC CODES 2879; 2842.

SYNONYMS Bay-10756 • Bayer-8169 • demeton • demeton-o and demeton-s • demox • diethoxy thiophosphoric acid ester of 2-ethylmercaptoethanol • e-1059 • ent-17,295 • ethyl-systox • mercaptofos • mercaptophos • O,O-diethyl 2-ethylmercaptoethyl thiophosphate • O,O-diethyl O (and (s)-2-(ethylthio)) ethyl phosphorothioate mixture • phosphorothioic acid, O,O-diethyl O-(2-(ethylthio) ethyl) ester, mixed with O,O-diethyl S-(2-(ethylthio) ethyl) ester (7:3) • phosphorothioic acid, O,O-diethyl O-(2-(ethylthio) ethyl) ester, mixture with O,O-diethyl S-(2-(ethylthio) ethyl) phosphorothioate • septox • systemox • ulv • O,O-diethyl O-[2-(ethylmercapto) ethyl] thionophosphate • dematon • beta-ethylmercaptoethyl diethyl thionophosphate • diethoxythiophosphoric acid ester of 2-ethylmercaptoethanol

MF $C_8H_{19}O_3PS_2 \cdot C_8H_{19}O_3S_2$.

MW 516.72

COLOR AND FORM Oily liquid; amber color.

ODOR Odor of sulfur.

DENSITY 1.1183 at 20°C.

BOILING POINT 134°C at 2 mm Hg.

FLAMMABILITY LIMITS 1.0%–5.3% [R4].

FIREFIGHTING PROCEDURES Dry chemical, foam, carbon dioxide. Water may be ineffective on fire [R4].

TOXIC COMBUSTION PRODUCTS Irritating fumes of sulfur dioxide and phosphoric acid may form in fire [R4].

SOLUBILITY Soluble in propylene glycol,

ethanol, toluene, and similar hydrocarbons; 0.01% in water.

VAPOR PRESSURE 3.4×10⁻⁴ mbar at 20°C.

CORROSIVITY Noncorrosive.

STABILITY AND SHELF LIFE Thin films of demeton-s undergo rapid conversion to more-hydrophilic compound when exposed to air and light. (demeton-s) [R6, 509].

OTHER CHEMICAL/PHYSICAL PROPERTIES Dense, oily, brown liquid with pungent, unpleasant smell (commercial grade).

PROTECTIVE EQUIPMENT Take precautions with respect to its easy absorption (protective clothing) and high volatility (respiratory protection); in other respects, as for other cholinesterase-inhibiting organophosphates. [R2].

PREVENTATIVE MEASURES Contact lenses should not be worn when working with this chemical [R5].

CLEANUP 1. Ventilate area of spill or leak. 2. For small quantities, absorb on paper towels. Evaporate in a safe place (such as a fume hood). Allow sufficient time for evaporating vapors to completely clear the hood ductwork. Burn the paper in a suitable location away from combustible materials. Large quantities can be reclaimed or collected and atomized in a suitable combustion chamber equipped with an appropriate effluent-gas-cleaning device [R7].

DISPOSAL Decomposes at elevated temperatures. Hydrolyzes in alkali, acid, boiling water. Mix with equal parts sand and crushed limestone and (1) wet with flammable solvent in open incinerator, igniting with excelsior train. Stay upwind or (2) burn in incinerator equipped with afterburner and alkaline scrubber. SRP: At the time of review, criteria for land treatment or burial (sanitary landfill) disposal practices are subject to significant revision. Prior to implementing land disposal of waste residue (including waste sludge), consult with environmental regulatory agencies for guidance on acceptable disposal practices.

COMMON USES Systemic nematocide [R3]; systemic insecticide and acaricide for vegetables, field crops, orchard crops, ornamentals [R1].

STANDARD CODES NFPA 3, –, –.

PERSISTENCE Hydrolysis half-life (pH 6, 70°C, ethanol) 18 hours. Changes by factor of 10 for each pH unit above pH 8. Chemical, not biochemical, hydrolysis initiates degradation [R14]. Persisted in soil for 23 days [R13].

INHALATION LIMIT 0.1 mg/m³.

PERSONAL SAFETY As for parathion. Self-contained breathing apparatus must be worn. Rubber gloves, hats, suits, and boots must be worn.

ACUTE HAZARD LEVEL Estimated lethal oral dose for man is 1.4 mg/kg. Lethal threshold values minimum LC values trout—2, 4, *Cambarus*—120, carp—360.0.

CHRONIC HAZARD LEVEL As little as 2 mg/kg causes adverse effects in dogs [R12].

OCCUPATIONAL RECOMMENDATIONS None listed.

THRESHOLD LIMIT 8–hr time-weighted average (TWA) 0.01 ppm, 0.11 mg/m³, skin (1986) [R10, 17].

ACTION LEVEL Evacuate area. Enter from upwind side. Notify local air authority and National Agricultural Chemicals Association.

ON-SITE RESTORATION Carbon or peat may be used as sorbents. Seek professional environmental assistance through EPA's Environmental Response Team (ERT), Edison, NJ, 24–hour no. 908–548–8730.

AVAILABILITY OF COUNTERMEASURE MATERIAL Carbon—water treatment plants, sugar refineries; peat—nurseries, floral shops.

DISPOSAL NOTIFICATION Contact local air authority.

PROBABLE LOCATION AND STATE Will sink and dissolve at extremely slow rate. Light-brown or colorless oil or liquid. Formulated as liquid or emulsifiable concentrate.

NON-HUMAN TOXICITY LD_{50} domestic goats oral 8.0–18.0 mg/kg, males (92% pure) [R8, 28].

ECOTOXICITY LC_{50} bobwhite quail 596 ppm in 5-day diets (95% confidence limit 472–789 ppm), age 14 days (purity 96%) [R9].

FIFRA REQUIREMENTS EPA farm worker field reentry interval: 48 hr [R11].

GENERAL RESPONSE Avoid contact with liquid and vapor. Keep people away. Wear goggles and self-contained breathing apparatus. Shut off ignition sources. Call fire department. Stop discharge if possible. Isolate and remove discharged material. Notify local health and pollution control agencies.

FIRE RESPONSE Solution in a combustible solvent. Poisonous gases may be produced in fire. Flashback along vapor trail may occur. Vapor may explode if ignited in an enclosed area. Wear goggles and self-contained breathing apparatus. Extinguish with dry chemicals, foam, or carbon dioxide. If breathing is difficult, give oxygen. Liquid is poisonous if swallowed. Irritating to skin and eyes. If swallowed will cause nausea, vomiting, or loss of consciousness. Remove contaminated clothing and shoes. Flush affected areas with plenty of water. If in eyes, hold eyelids open, and flush with plenty of water. If swallowed and victim is conscious, have victim drink water or milk, and have victim induce vomiting. If swallowed and victim is unconscious or having convulsions, do nothing except keep victim warm.

RESPONSE TO DISCHARGE Issue warning—poison, water contaminant, high flammability. Restrict access. Mechanical containment. Should be removed. Provide chemical and physical treatment.

WATER RESPONSE Effect of low concentrations on aquatic life is unknown. May be dangerous if it enters water intakes. Notify local health and wildlife officials. Notify operators of nearby water intakes.

EXPOSURE SYMPTOMS Inhalation causes headache, vertigo, blurred vision, lachrymation, salivation, sweating, muscular weakness, ataxia, dyspnea, diarrhea, abdominal cramps, vomiting, coma, pulmonary edema, and death. Ingestion causes nausea, vomiting, muscle twitching, coma. Contact with eyes or skin causes irritation.

EXPOSURE TREATMENT Speed is essential. Call a physician after all overexposure to demeton. Inhalation: move to fresh air; if needed, begin artificial respiration. Ingestion: administer milk, water, or salt water, and induce vomiting repeatedly. Eyes: flush with water for at least 15 min. Skin: flood and wash exposed skin areas thoroughly with water; remove contaminated clothing under a shower; wash with soap and water.

REFERENCES

1. SRI.

2. Hartley, D., and H. Kidd (eds.). *The Agrochemicals Handbook.* Old Woking, Surrey, United Kingdom: Royal Society of Chemistry/Unwin Brothers Ltd., 1983, p. A123/OCT 83.

3. White–Stevens, R. (ed.). *Pesticides in the Environment:* Volume.

2. New York: Marcel Dekker, Inc., 1976. 235.

4. U.S. Coast Guard, Department of Transportation. *CHRIS—Hazardous Chemical Data.* Manual Two. Washington, DC: U.S. Government Printing Office, Oct., 1978.

5. NIOSH. *Pocket Guide to Chemical Hazards.* 2nd Printing. DHHS (NIOSH) Publication No. 127. Washington, DC: U.S. Government Printing Office, 1969.

6. Sunshine, I. (Ed.). *CRC Handbook of Analytical Toxicology,* Cleveland, OH: Chemical Rubber Co., 1969, p. 509.

7. Mackison, F. W., R. S. Stricoff, and L. J. Partridge, Jr. (eds.). *NIOSH/OSHA—Occupational Health Guidelines for Chemical Hazards.* DHHS (NIOSH) Publication No. 81–123 (3 vols). Washington, DC: U.S. Government Printing Office, Jan. 1981.

8. U.S. Department of the Interior, Fish and Wildlife Service. *Handbook of Toxici-

ty of Pesticides to Wildlife. Resource Publication 153. Washington, DC: U.S. Government Printing Office, 1984.

9. U.S. Department of the Interior, Fish and Wildlife Service, Bureau of Sports Fisheries, and Wildlife. *Lethal Dietary Toxicities of Environmental Pollutants to Birds. Special Scientific Report—Wildlife* No. 191. Washington, DC: U.S. Government Printing Office, 1975. 17.

10. American Conference of Governmental Industrial Hygienists. *Threshold Limit Values for Chemical Substances and Physical Agents and Biological Exposure Indices for 1993–1994*. Cincinnati, OH: ACGIH, 1993.

11. US Code of Federal Regulations, Title 40, part 162.42 (1984).

12. A. D. Little Inc., "Relationship between Organic Chemical Pollution of Fresh Water and Health," FWQA, 1970, 71632, Dec.

13. Pimental, David. *Ecological Effects of Pesticides on Non–Target Species*, Presidential Report, Office of Science and Technology, 1971, June.

14. Elo, M. *Organophosphorus Pesticides: Organic and Biological Chemistry*, Cleveland, OH: CRC Press Inc., 1974.

■ PHOSPHOROTHIOIC ACID, O,O–DIETHYL O–(2-ISOPROPYL-6–METHYL–4–PYRIMIDINYL) ESTER

CAS # 333–41–5

DOT # 1615

SIC CODES 2879; 2842.

SYNONYMS alfa-tox • antigal • basudin • basudin-10-g • bazuden • ciazinon • dacutox • dassitox • dazzel • diazajet • diazide • diazinon-ag-500 • diazinone • diazitol • diazol • dicid • diethyl 2-isopropyl-4-methyl-6-pyrimidinyl phosphorothionate • diethyl 2-isopropyl-4-methyl-6-pyrimidyl thionophosphate • diethyl 4-(2-isopropyl-6-methylpyrimidinyl) phosphorothionate • dimpylat • dimpylate • dipofene • dizinon • dyzol • ent-19,507 • exodin • flytrol • g-301 • g-24480 • galesan • garden-tox • gardentox • geigy-24480 • isopropylmethylpyrimidyl-diethyl-thiophosphate • kayazinon • kayazol • knox-out • NCI-co8673 • nedcidol • neocidol • neocidol (oil) • nipsan • nucidol • O,O-Diaethyl-o-(2-isopropyl-4-methyl-pyrimidin-6-yl)-monot hiophosphat (German) • O,O-diethyl 2-isopropyl-4-methylpyrimidyl-6-thiophosphate • O,O-diethyl O-(2-isopropyl-4-methyl-6-pyrimidyl) thionophosphate • O,O-diethyl o-(2-isopropyl-6-methyl-4-pyrimidinyl) phosphorothioate • O,O-diethyl o-6-methyl-2-isopropyl-4-pyrimidinyl phosphorothioate • O,O-diethyl-o-(2-isopropyl-4-methyl-6-pyrimidinyl)-phosphorothioate • O,O-diethyl-o-(2-isopropyl-4-methyl-6-pyrimidyl) phosphorothioate • O,O-diethyl-o-(2-isopropyl-4-methyl-pyrimidin-6-yl)-monoth iofosfaat (Dutch) • O,O-dietil-o-(2-isopropil-4-metil-pirimidin-6-il)-monotiof osfato (Italian) • O-2-isopropyl-4-methylpyrimidyl-O,O-diethyl phosphorothioate • oleodiazinon • phosphorothioate, O,O-diethyl O-6-(2-isopropyl-4-methylpyrimidyl) • phosphorothioic acid, O,O-diethyl O-(2-isopropyl-6-methyl-4-pyrimidinyl) ester • phosphorothioic acid, O,O-diethyl O-(6-methyl-2-(1-methylethyl)-4-pyrimidinyl) ester • phosphorothioic acid, O,O-diethyl O-(isopropylmethylpyrimidinyl) ester • sarolex • spectracide • spectracide-25ec • thiophosphoric acid 2-isopropyl-4-methyl-6-pyrimidyl diethyl ester

MF $C_{12}H_{21}N_2O_3PS$

MW 304.38

COLOR AND FORM Colorless liquid.

ODOR Faint ester-like odor.

DENSITY 1.116–1.118 at 20°C.

BOILING POINT 83–84°C at 2×10^{-3} mm Hg.

SOLUBILITY 0.004% at 20°C; miscible with

petroleum ether, alcohol, ether, cyclohexane, benzene, and similar hydrocarbons; soluble in organic solvents; freely soluble in ketones.

VAPOR PRESSURE 1.4×10^{-4} mm Hg at 20°C.

STABILITY AND SHELF LIFE More stable in alkaline formulations than when at neutral or acid pH [R7].

OTHER CHEMICAL/PHYSICAL PROPERTIES Hydrolyzes slowly in water and dilute acid.

DISPOSAL Hydrolysis: diazinon is hydrolyzed in acid media. In excess water this compound yields diethylthiophosphoric acid and 2–isopropyl–4–methyl–6–hydroxypyrimidine. With insufficient water, highly toxic tetraethyl monothiopyrophosphate is formed [R8, 250]. Sensitive to oxidation above 100°C; degradation above 120°C. Hydrolyzes slowly in water and dilute acid; reacts with strong acid and alkali.

COMMON USES Insecticide for non-agricultural use (e.g., home and garden) and for agricultural use (e.g., alfalfa and corn) [R1]. Acaricide [R4]. Insecticide (use against fire ants permitted by EPA) [R5]. A non-systemic insecticide, its main applications are in fruit trees, horticultural crops, maize, potatoes, rice, sugarcane, tobacco, and vineyards for a wide range of sucking and leaf-eating insects. Also, used against flies in glasshouses, mushroom houses [R6]. To control ticks and insects on animals and premises, controlling face fly larvae in manure [R3]. Vet: used against flies and ticks in veterinary practice [R6].

STANDARD CODES EPA 311; Superfund designated (hazardous substances) list.

PERSISTENCE Hydrolysis half-life (pH 6, 70°C ethanol) 37 hours. pH 7.4 at 20°C 155 days, pH 3.1 0.5 day, at pH 10.1 6 days. Decomposition is catalyzed by the presence of silty clay [R24]. With copper montemorillonite at 20°C half-life was 4 hours. Biochemical aspects are largely overshadowed by hydrolysis. Bacteria then degrades hydrolysis by-products.

Bacteria are more important in aqueous media [R102]. Persistent in soil for 12 weeks [R25]. Some volatilization can be expected [R23].

MAJOR SPECIES THREATENED Aquatic life, particularly marine species.

INHALATION LIMIT 0.1 mg/m³.

PERSONAL SAFETY Absorbed through skin. Protective clothing and filter mask recommended.

ACUTE HAZARD LEVEL Fatal dose in man estimated to be 360 mg/kg.

CHRONIC HAZARD LEVEL Goldfish are adversely affected at 0.1 mg/L. Surface waters should never exceed 0.009 ppb [R26]. No gross effects in rats fed 100 ppm in diet for 2 years [R23].

DEGREE OF HAZARD TO PUBLIC HEALTH High under disaster circumstances. Potential mutagenic and teratogenic effects.

OCCUPATIONAL RECOMMENDATIONS
OSHA 8–hr time-weighted average: 0.1 mg/m³. Skin absorption designation. Final rule limits were met by any combination of engineering controls, work practices, and personal protective equipment—effective Sept 1, 1989 [R21]. Threshold-limit 8–hr time-weighted average (TWA) 0.1 mg/m³. Skin designation (1986). [R22, 18].

ACTION LEVEL Evacuate area. Enter from upwind side. Notify local air authority and National Agricultural Chemicals Association.

ON-SITE RESTORATION Can be adsorbed with carbon or peat. Seek professional environmental engineering assistance through EPA's Environmental Response Team (ERT), Edison, NJ, 24–hour no. 908–548–8730.

BEACH OR SHORE RESTORATION Heating contaminated stretches will decompose. Fumes are toxic, however, and must not be inhaled.

AVAILABILITY OF COUNTERMEASURE MATERIAL Carbon—water treatment plants, sugar refineries; peat—nurseries, floral shops.

MAJOR WATER USE THREATENED Recreation waters, fisheries, potable supplies.

PROBABLE LOCATION AND STATE Will sink and dissolve very slowly unless accompanied by wetting agents. Typical product forms—dust, emulsifiable concentrate, granular, solution, or wettable powder. Material is light-yellow to dark-brown liquid when it has not been purified to its natural colorless state.

WATER CHEMISTRY Weak base. Hydrolyzes fairly rapidly. Accerated degradation in acid or alkali. By-products are biodegradable.

FOOD-CHAIN CONCENTRATION POTENTIAL Negative.

HUMAN TOXICITY 0.02 mg/kg/day is a no-effect level in man [R9, 387]. Estimated adult oral fatal dose is approximately 25 g. [R11].

NON-HUMAN TOXICITY LD_{50} rat male oral 250 mg/kg [R7]; LD_{50} rat female oral 285 mg/kg [R7]; LD_{50} rat male oral 108 mg/kg [R10]; LD_{50} rat female oral 76 mg/kg [R10]; LD_{50} guinea pig oral 240–320 mg/kg [R10]; LD_{50} rabbit acute 130 mg/kg [R10]; LD_{50} rat oral 300–400 mg/kg (technical grade) [R14, 459]; LD_{50} rat male dermal 900 mg/kg [R14, 459]; LD_{50} rat female dermal 445 mg/kg [R14, 459].

ECOTOXICITY LD_{50} mallard, males 3–4 mo oral 3.54 mg/kg (95% confidence limits 2.37–5.27) [R12]; LD_{50} pheasant, males 3–4 mo old oral 4.33 mg/kg (95% confidence limits 3.02–6.22) [R12]; LD_{50} bullfrog oral greater than 2,000 mg/kg, female [R12]; LC_{50} *Gillia altilis* (snail) 40 μm (11 ppm) /96 hr (static renewal bioassay) [R13]; LC_{50} rainbow trout 90 μg/L/96 hr (conditions of bioassay not specified) [R15]; LC_{50} *Pteronarcys* (stonefly) 25 μg/L/96 hr (conditions of bioassay not specified) [R15]; LC_{50} bluegill 168 μg/L/96 hr (conditions of bioassay not specified) [R15]; LC_{50} *Pimephales promelas* (fathead minnow) 31-day-old 9.35 mg/L/96 hr. Affected fish lost schooling behavior, were hyperactive, and swam in a corkscrew/spiral pattern. They were also overreactive to external stimuli, had increased respiration, convulsions, rigid muscula-

ture, and hemorrhaging. In addition, they had spinal deformities, and lost equilibrium prior to death (conditions of bioassay not specified) [R16]; LC_{50} *Coturnix* oral 167 ppm (99% active ingredient) [R17]; LC_{50} *Coturnix* oral 101 ppm (48% active ingredient) [R17]; LC_{50} *Acartia tonsa* (marine calanoid copepod) 2.57 μg/L/96 hr (conditions of bioassay not specified) [R14, 458]; LC_{50} *Gammarus lacustris* (freshwater scud) 200 μg/L/96 hr (conditions of bioassay not specified) [R14, 458]; LC_{50} *Simocephalus serrulatus* (water flea) 1.4 μg/L/48 hr (conditions of bioassay not specified) [R14, 458]; LC_{50} *Daphnia pulex* 0.90 μg/L/96 hr (conditions of bioassay not specified) [R14, 458]; LC_{50} *Pteronarcys californica* (stonefly) 25 μg/L/96 hr (conditions of bioassay not specified) [R14, 458]; LC_{50} bluegill 0.052 ppm/24 hr (conditions of bioassay not specified) [R14, 459]; LC_{50} rainbow trout 0.380 ppm/24 hr (conditions of bioassay not specified) [R14, 459]; LC_{50} carp 3.18 mg/L/24 hr (conditions of bioassay not specified) [R14, 459].

BIOLOGICAL HALF-LIFE Oral dose of ectoparasiticide, ^{14}C diazinon, was rapidly eliminated from rat (biological $t_{1/2}$ was 12 hr). 80% of ^{14}C was excreted in urine and 18% in feces. Comparable excretion pattern and lower biological $t_{1/2}$ of 9 hr was obtained after iv admin of 3 ^{14}C metabolites of diazinon. [R18, 148].

BIODEGRADATION Degraded in soil to 10% in 3 wk [R2].

BIOCONCENTRATION Average bioconcentration ratios were: in fish, from 152 for topmouth gudgeon to 17.5 for guppies; in snails, 17.0 for red snail to 5.9 for pond snails; and in crayfish, 4.9. [R19].

FIFRA REQUIREMENTS Tolerances are established for residues of the insecticide diazinon (o,o-diethyl o–(6–methyl–2–(1–methylethyl)–4–pyrimidinyl) phosphorothioate) in or on the following raw agricultural commodities: alfalfa, fresh; alfalfa, hay; almonds; almonds, hull; apples; apricots; bananas (NMT 0.1 ppm were present in the pulp after peel is removed); beans, forage; beans, hay; beans, guar; beans, guar, forage; beans, lima; beans, snap; beets, roots; beets, sugar, roots;

beets, sugar, tops; beets, tops; birdsfoot trefoil; birdsfoot trefoil, hay; blackberries; blueberries; boysenberries; carrots; cattle, fat (pre-slaughter application); cattle, meat (fat basis) (pre-s appli); cattle mbyp (fat basis) (pre-s appli); celery; cherries; chicory, red (tops) (also known as radicchio); citrus; clover (fresh); clover, hay; coffee beans; corn, forage; corn (inc sweet k + CWHR); cottonseed; cowpeas; cowpeas, forage; cranberries; cucumbers; dandelions; dewberries; endive (escarole); figs; filberts; ginseng; grapes; grass (NMT 40 ppm shall remain 24 hours after appli); grass, hay; hops; kiwi fruit; lespedeza; lettuce; loganberries; melons; mushrooms; nectarines; olives; onions; parsley; parsnips; peaches; peanuts; peanuts, forage; peanuts, hay; peanuts, hulls; pears; peavine hay; peavines; peas with pods (determined on peas afer removing any shell present when marketed); pecans; peppers; pineapples; pineapples, forage; plums (fresh prunes); potatoes; potatoes, sweet; radishes; raspberries; rutabagas; sheep, fat (pre-s appli); sheep, meat (fat basis) (pre-s appli); sheep, mbyp (fat basis) (pre-s appli); sorghum, forage; sorghum, grain; soybeans; soybeans, forage; spinach; squash, summer; squash, winter; strawberries; sugarcane; Swiss chard; tomatoes; turnips, roots; turnips, tops; vegetables, leafy, brassica (cole); walnuts; and watercress. [R20].

GENERAL RESPONSE Stop discharge if possible. Keep people away. Isolate and remove discharged material. Notify local health and pollution control agencies.

FIRE RESPONSE Not flammable. Poisonous gases are produced when heated.

EXPOSURE RESPONSE Call for medical aid. Liquid is poisonous if swallowed. Irritating to skin and eyes. Remove contaminated clothing and shoes. Flush affected areas with plenty of water. If in eyes, hold eyelids open, and flush with plenty of water. If swallowed and victim is conscious, have victim drink water or milk.

RESPONSE TO DISCHARGE Issue warning—poison, water contaminant, high flammability (if solution). Restrict access.

Should be removed. Provide chemical and physical treatment.

WATER RESPONSE Harmful to aquatic life in very low concentrations. May be dangerous if it enters water intakes. Notify local health and wildlife officials. Notify operators of nearby water intakes.

EXPOSURE SYMPTOMS Ingestion or prolonged inhalation of mist causes headache, giddiness, blurred vision, nervousness, weakness, cramps, diarrhea, discomfort in the chest, sweating, miosis, tearing, salivation, and other excessive respiratory tract secretion, vomiting, cyanosis, papilledema, uncontrollable muscle twitches, convulsions, coma, loss of reflexes, and loss of sphincter control. Liquid irritates eyes and skin.

EXPOSURE TREATMENT Inhalation: remove to fresh air; keep warm; get medical attention at once. Eyes: flush with plenty of water for at least 15 min. and get medical attention. Skin: wash contaminated area with soap and water. Ingestion: get medical attention at once; give water slurry of charcoal; do not give milk or alcohol.

REFERENCES

1. SRI.

2. Spencer, E. Y. *Guide to the Chemicals Used in Crop Protection*. 7th ed. Publication 1093. Research Institute, Agriculture Canada, Ottawa, Canada: Information Canada, 1982. 178.

3. Rossoff, I. S. *Handbook of Veterinary Drugs*. New York: Springer Publishing Company, 1974. 162.

4. Clayton, G.D., and F.E. Clayton (eds.). *Patty's Industrial Hygiene and Toxicology*: Volume 2A, 2B, 2C: *Toxicology*. 3rd ed. New York: John Wiley and Sons, 1981–1982. 4815.

5. Sax, N. I., and R. J. Lewis, Sr. (eds.). *Hawley's Condensed Chemical Dictionary*. 11th ed. New York: Van Nostrand Reinhold Co., 1987. 364.

6. Worthing, C. R., and S. B. Walker (eds.). *The Pesticide Manual—A World Compendium*. 8th ed. Thornton Heath, UK:

The British Crop Protection Council, 1987. 248.

7. Budavari, S. (ed.). *The Merck Index— Encyclopedia of Chemicals, Drugs, and Biologicals.* Rahway, NJ: Merck and Co., Inc., 1989. 472.

8. United Nations. *Treatment and Disposal Methods for Waste Chemicals* (IRPTC File). Data Profile Series No. 5. Geneva, Switzerland: United Nations Environmental Programme, Dec. 1985.

9. Hayes, Wayland J., Jr. *Pesticides Studied in Man.* Baltimore/London: Williams and Wilkins, 1982.

10. American Conference of Governmental Industrial Hygienists. *Documentation of the Threshold Limit Values and Biological Exposure Indices.* 5th ed. Cincinnati, OH: American Conference of Governmental Industrial Hygienists, 1986. 172.

11. Ellenhorn, M. J., and D. G. Barceloux. *Medical Toxicology—Diagnosis and Treatment of Human Poisoning.* New York, NY: Elsevier Science Publishing Co., Inc. 1988. 1071.

12. U.S. Department of the Interior, Fish and Wildlife Service, Bureau of Sport Fisheries and Wildlife. *Handbook of Toxicity of Pesticides to Wildlife.* Washington, DC. U.S. Government Printing Office, 1970. 44.

13. Robertson, J. B., C. Mazzella. *Bull Environ Contam Toxicol* 42 (3): 320–4 (1989).

14. Verschueren, K. *Handbook of Environmental Data of Organic Chemicals.* 2nd ed. New York, NY: Van Nostrand Reinhold Co., 1983.

15. U.S. Department of Interior, Fish and Wildlife Service. *Handbook of Acute Toxicity of Chemicals to Fish and Aquatic Invertebrates.* Resource Publication No. 137. Washington, DC: U.S. Government Printing Office, 1980. 26.

16. Geiger, D. L., D. J. Call, L. T. Brooke (eds.). *Acute Toxicities of Organic Chemicals to Fathead Minnows (Pimephales promelas).* Vol. IV. Superior, Wisconsin: University of Wisconsin–Superior, 1988. 279.

17. Hill, E. F., and M. B. Camardese. *Lethal Dietary Toxicities of Environmental Contaminants and Pesticides to Coturnix.* Fish and Wildlife Technical Report 2. Washington, DC: United States Department of Interior, Fish and Wildlife Service, 1986. 54.

18. The Chemical Society. *Foreign Compound Metabolism in Mammals.* Volume 2: *A Review of the Literature Published Between 1970 and 1971.* London: The Chemical Society, 1972.

19. Kanazawa, J. *Bull Environ Contam Toxicol* 20 (5): 613 (1978).

20. 40 CFR 180.153 (7/1/88).

21. 54 FR 2920 (1/19/89).

22. American Conference of Governmental Industrial Hygienists. *Threshold Limit Values for Chemical Substances and Physical Agents and Biological Exposure Indices for 1993–1994.* Cincinnati, OH: ACGIH, 1993.

23. Midwest Research Institute and RVR Consultants. *Production, Distribution, Use, and Environmental Impact Potential of Selected Pesticides,* NTIS PB–238 795, March 15, 1974.

24. Elo, M. *Organophosphoros Pesticides: Organic and Biological Chemistry,* Cleveland: Ohio: CRC Press Inc., 1974.

25. Dinauer, R. C., et. al., *Pesticides in Soil and Water,* Madison, Wisconsin: Soil Science Society of America, Inc., 1974.

26. *Proposed Criteria For Water Quality,* Volume I, U.S. Environmental Protection Agency. October 1973.

■ PHOSPHOROTHIOIC ACID, O,O–DIETHYL O–(3,5,6–TRICHLORO–2– PYRIDYL) ESTER

CAS # 2921–88–2

SIC CODES 2879.

SYNONYMS 2-pyridinol, 3,5,6-trichloro-, o-ester with O,O-diethyl phosphorothioate • chloropyrifos • chloropyriphos • chlorpyrifos-ethyl • chlorpyriphos • ent-27,311 • ent-27311 • O,O-Diethyl-o-3,5,6-trichlor-2-pyridylmonothiophosphat (German) • O,O-diethyl O-3,5,6-trichloro-2-pyridyl phosphorothioate • phosphorothioic acid, O,O-diethyl O-(3,5,6-trichloro-2-pyridinyl) ester • phosphorothioic acid, O,O-diethyl O-(3,5,6-trichloro-2-pyridyl) ester • O,O-diethyl O-(3,5,6-trichloro-2-pyridinyl) phosphorothioate • dowco-179 • oms-0971 • brodan • detmol-ua • dursban • dursban-4e • dursban-f • eradex • killmaster • pyrinex • suscon • lorsban • EPA pesticide code-059101

MF $C_9H_{11}Cl_3NO_3PS$

MW 350.59

COLOR AND FORM White granular crystals; colorless crystals.

ODOR Mild mercaptan odor.

DENSITY 1.398 at 43.5°C (liquid).

MELTING POINT 41–42°C.

FIREFIGHTING PROCEDURES Extinguish fire using an agent suitable for the type of surrounding fire. Material itself does not burn, or burns with difficulty [R5].

SOLUBILITY 0.7 ppm in water at 20°C; solubility at 25°C: 2 mg/L of water; solubility at 25°C: 6.5 kg/kg acetone; solubility at 25°C: 7.9 kg/kg benzene; solubility at 25°C: 6.3 kg/kg chloroform; solubility at 25°C: 450 g/kg methanol; solubility at 25°C: in isooctane 79% wt/wt; soluble in most organic solvents.

VAPOR PRESSURE 1.87×10^{-5} mm Hg at 25°C.

OCTANOL/WATER PARTITION COEFFICIENT Log K_{ow} = 4.96.

CORROSIVITY Corrosive to copper and brass.

STABILITY AND SHELF LIFE Very stable under neutral or slightly acid conditions at room temperature [R7].

OTHER CHEMICAL/PHYSICAL PROPERTIES Heat of sublimation 26,800 cal/mol (from table).

PROTECTIVE EQUIPMENT Wear appropriate chemical protective gloves, boots, and goggles [R5].

PREVENTATIVE MEASURES (SRP) Contact lenses should not be worn when working with this chemical.

CLEANUP Environmental considerations: Land spill: Dig a pit, pond, lagoon, or holding area to contain liquid or solid material. (SRP: If time permits, pits, ponds, lagoons, soak holes, or holding areas should be sealed with an impermeable flexible-membrane liner.) Dike material surface flow using soil, sandbags, foamed polyurethane, or foamed concrete; absorb bulk liquid with fly ash or cement powder. (chlorpyrifos, agricultural insecticides, not elsewhere classified, liquid) [R5].

DISPOSAL This compound should be susceptible to removal from wastewater by air stripping [R8]. After the material has been contained, apply a sorbent material (peat, straw, sand, sawdust, etc.) to the contaminated area. Remove all contaminated sorbent and soil and place in impervious containers. The material may be incinerated in a pesticide incinerator at the specified temperature/dwell-time combination. Any liquids, sludges, or solid residues generated should be disposed of in accordance with all applicable federal, state, and local pollution control requirements. If appropriate incineration facilities are not available, material may be buried in an approved chemical waste landfill away from domestic water supplies. Not acceptable at a municipal sewage treatment plant.

COMMON USES Acaricide [R3]. It has a broad range of insecticidal activity and is effective by contact, ingestion, and vapor action, but is not systemic. Used for control of flies, household pests, mosquitoes (larvae and adults) and of various crop pests in soil and on foliage; also used for control of ectoparasites on cattle and sheep [R2]. Effective against cattle and sheep ticks and psoroptic mange in sheep by dipping (0.1%) and turkey chiggers by soil treatment (4 lb/acre or 448 mg/m²) [R4].

STANDARD CODES EPA 311; TSCA; Coast Guard Classification poison B, poison label; not listed IATA; CFR 49 poison class B, poison label, 50 lb. passenger, 200 lb. cargo, storage code 1, 2; CFR 14 CAB code 8; ICC STCC tariff no. 1–C 28 799 31.

PERSISTENCE Nonpersistent. Most organophosphates are quite readily hydrolyzed, therefore are not persistent in soils and water. Those not readily hydrolyzed might be expected to be taken up by a wide range of aquatic organisms; however, organophosphates are relatively short-lived in biological systems.

MAJOR SPECIES THREATENED Highly toxic to bees and fishes.

INHALATION LIMIT 0.2 mg/m³.

DIRECT CONTACT Readily absorbed through the skin to give toxic effects. Prolonged, confined contact may product appreciable irritation and slight burn.

GENERAL SENSATION After inhalation or skin absorption signs and symptoms may begin within 30 to 60 minutes and are at a maximum in 2–8 hours. Headache, dizziness, weakness, anxiety, tremors of the tongue and eyelids, and impairment of visual acuity. At higher levels of exposure nausea, vomiting, salivation, tearing, abdominal cramps, muscular fasciculations, respiratory difficulty, pulmonary edema, cyanosis, convulsions, and coma. At fire temperatures may be irritating to eyes, mucous membranes, and lungs.

PERSONAL SAFETY Protect against both inhalation and absorption through the skin. Must wear protective clothing, including NIOSH–approved rubber gloves and boots, safety goggles or face mask, hooded suit, and either a respirator whose canister is specifically approved for this material, or a self-contained breathing apparatus. Decontaminate fully or dispose of all equipment after use.

ACUTE HAZARD LEVEL This material is very toxic to humans; between 1 teaspoonful and 1 ounce may be fatal. The oral LD_{50} for rats is 145 mg/kg and for guinea pigs the oral LD_{50} is 500 mg/kg. The oral LD_{50} for wild birds is 8 mg/kg. The skin LD_{50} for rats is 202 mg/kg, and for rabbits is 2,000 mg/kg. In freshwater, the 48-hour LC_{50} for rainbow trout is 0.02 mg/L. There will be slight conjunctival irritation upon eye contact, but no corneal injury.

CHRONIC HAZARD LEVEL The threshold-limit-value-time-weighted average for a 40–hour workweek is 0.2 mg/m³, and the short-term exposure limit for 15 minutes is 0.6 mg/m³. EPA has designated this material a category X, with an aquatic LC_{50} value less than 0.1 mg/L. Prolonged skin contact may produce appreciable irritation and slight burn.

DEGREE OF HAZARD TO PUBLIC HEALTH A cholinesterase inhibitor. Highly toxic through inhalation and skin absorption. Small doses at frequent intervals are largely additive.

OCCUPATIONAL RECOMMENDATIONS None listed.

THRESHOLD LIMIT 8-hr time-weighted average (TWA) 0.2 mg/m³, skin (1990) [R27, 16].

ACTION LEVEL Avoid contact with and inhalation of the spilled cargo. Stay upwind. Notify local fire, air, and water authorities of the accident. Evacuate all people to a distance of at least 200 feet upwind and 1,000 feet downwind of the spill.

ON-SITE RESTORATION Dam stream to prevent flow and to retard dissipation by water movement. Apply carbon to adsorb the dissolved material. Bottom pumps, dredging, or underwater vacuum systems may be employed in small bodies of water; may be effective in larger bodies to remove undissolved material and adsorbent. Under controlled conditions, an alternative method is to pump the water into a suitable container, then pass it through a filtration system with an activated-carbon filter. For more details see ENVIREX Manual EPA 600/2–77–227. Seek professional environmental engineering assistance through EPA's Environmental Response

Team (ERT), Edison, NJ, 24–hour no. 908–548–8730.

BEACH OR SHORE RESTORATION Close beach and shore to public until material has been removed. If tidal, scrape affected area at low tide with mechanical scraper; avoid human contact.

AVAILABILITY OF COUNTERMEASURE MATERIAL Bottom pumps—available through fire departments, EPA Regional Offices, U.S. Coast Guard, or Army Corps of Engineers; dredging—Army Corps of Engineers; underwater vacuum systems—swimming-pool operators; carbon—water treatment plants and chemical plants.

DISPOSAL NOTIFICATION Notify local and state health authorities, local solid waste disposal authorities, supplier, and shipper of material.

INDUSTRIAL FOULING POTENTIAL Not considered significant.

MAJOR WATER USE THREATENED All downstream users of water, monitor for water concentrations.

PROBABLE LOCATION AND STATE White granular crystals; will sink and are almost completely recoverable by physical means. Would appear as a layer on the bottom.

WATER CHEMISTRY The material is stable in acidic water but will hydrolyze in alkaline media, and hydrolysis also increases with a temperature rise.

FOOD-CHAIN CONCENTRATION POTENTIAL Negative. Unlikely to accumulate in the food chain.

NON-HUMAN TOXICITY LD_{50} albino rats males oral 151 mg/kg (95% confidence limit 179–252 mg/kg) (purity 99%) [R9]; LD_{50} rock doves (domestic pigeons) oral 26.9 mg/kg (95% confidence limit 19.0–38 mg/kg) (purity 94.5%) [R9]; LD_{50} domestic goats females oral 500–1,000 mg/kg (purity 94.5%) [R9].

ECOTOXICITY LC_{50} Coturnix (Japanese quail) oral 293 ppm (95% confidence limit 112–767 ppm) (technical material, 97% active ingredient) [R10]; LD_{50} mallards female oral 75.6 mg/kg (95% confidence

limit 35.4–161 mg/kg) (purity 99%) [R9]; LD_{50} pheasant 3–5–month-old males, oral 8.41 mg/kg (95% confidence limit 2.77–25.5 mg/kg) (purity 99%) [R9]; LD_{50} chukar 3–5–month-old males, oral 60.7 mg/kg (95% confidence limit 43.8–84.1 mg/kg) (purity 99%) [R9]; LD_{50} Coturnix coturnix (Japanese quail) 2.5–month-old males, oral 15.9 mg/kg (95% confidence limit 10.5–24.0 mg/kg) (purity 94.5%) [R9]; LD_{50} house sparrows males, oral 21.0 mg/kg (95% confidence limit 5.59–79.1 mg/kg) (purity 94.5%) [R9]; LD_{50} Branta canadensis (Canadian geese) males and females oral more than 80 mg/kg (purity 99%) [R9]; LD_{50} lesser sandhill crane males oral 25–50 mg/kg (purity 94.5% and 99%) [R9]; LD_{50} Anas platyrhynchos (mallard ducklings) 15–19-day-old males and females oral 167 mg/kg (95% confidence limit 11.5–1,089 mg/kg) (purity 99%) [R9]; LD_{50} pheasant (Pheasant sp) 3–5–month old females oral 17.7 mg/kg (95% confidence limit 12.5–25.0 mg/kg) (purity 99%) [R9]; LD_{50} chukar 3–5–mo-old females oral 61.6 mg/kg (95% confidence limit 47.5–78.6 mg/kg) (purity 99%) [R9]; LD_{50} Coturnix coturnix (Japanese quail) 2–mo-old males oral 17.8 mg/kg (95% confidence limit 15.0–21.2 mg/kg) (purity 94.5%) [R9]; LD_{50} Rana catesbiana (bullfrogs) males oral more than 400 mg/kg (purity 94.5%) (conditions of bioassay not specified) [R9]; LD_{50} Quiscalus quiscula (common grackle) oral 13 mg/kg adult [R11]; LD_{50} Corvus brachyrhynchos (crow) oral >32 mg/kg adult [R11]; LC_{50} Cyprinodon variegatus (sheepshead minnow) juvenile >1,000 µg/L/24 hr at a salinity of 24 g/kg (99% purity) (conditions of bioassay not specified) [R12]; LC_{50} Leiostomus xanthurus (spot) juvenile 7 µg/L/48 hr at a salinity of 26 g/kg (99% purity (conditions of bioassay not specified) [R12]; LC_{50} Fundulus similis (longnose killifish) 3.2 µg/L/48 hr at a salinity of 24 g/kg (99% purity) (conditions of bioassay not specified) [R12]; LC_{50} Poecilia reticulata (guppy) 220 µg/L/48 hr (technical; conditions of bioassay not specified) [R13]; LC_{50} Anopheles freeborni (mosquito) 4th instar 0.9–7.0 µg/L/24 hr [R14]; LC_{90-95} Hydropsyche pellucidula (Trichloptera) >0.5 ppm/1 hr (conditions of bioassay not specified) [R6,

392]; LC$_{90-95}$ Simulium, ornatum (Diptera) 0.05–0.1 ppm/1 hr (conditions of bioassay not specified) [R6, 392]; TL$_{50}$ Palaemon macrodactylus (Korean shrimp) 0.25 (0.10–0.63) ppm/96 hr static lab bioassay [R6, 391]; TL$_{50}$ Palaemon macrodactylus (Korean shrimp) 0.01 (0.002–0.046) ppm/96 hr intermittent flow lab bioassay [R6, 391]; LC$_{90-95}$ Bactis rhodani (Ephimetoptera) 0.01–0.02 ppm/1 hr (conditions of bioassay not specified) [R6, 392]; LC$_{90-95}$ Brachycentrus subnubilis (Trichoptera) 0.2–0.5 ppm/1 hr (conditions of bioassay not specified) [R6, 392]; LC$_{50}$ Gammarus lacustris (crustacean) 0.11 µg/L/96 hr (conditions of bioassay not specified) [R6, 391]; LC$_{50}$ Gammarus fasciatus (crustacean) 0.32 µg/L/96 hr (conditions of bioassay not specified) [R6, 391]; TLm Cymatogaster aggregata (shiner perch) 3.5 ppb/96 hr static lab bioassay [R6, 391]; TLm Cymatogaster aggregata (shiner perch) 3.7 ppb/96 hr flowing water bioassay [R6, 391]; LC$_{50}$ Salmo gairdneri (rainbow trout) 15 µg/L/96 hr at 7.2°C; 51 µg/L/96 hr at 1.6°C (technical material) [R15]; LD$_{50}$ Periplaneta americana (American cockroach) 5.7 µg/g/24 hr nymph; 0.67 µg/insect/24 hr adult (topical application) [R16]; LD$_{50}$ Blatella germanica (German cockroach) adult male 1.92 µg/g (0.092 µg/insect)/24 hr (topical application) [R17]; LD$_{50}$ Leptocoris trivittatus (boxelder bug) >4.9 µg/g/24 hr, >0.2 µg/insect nymph (topical application) [R17]; LD$_{50}$ Schizaphis graminum (green bug) <41.6 µg/g/24 hr, <0.5 µg/insect adult (topical application) [R17]; LD$_{50}$ Sinea diadema (assassin bug) <0.5 µg/insect/24 hr (topical application) [R17]; LD$_{50}$ Leptinotarsa decemlineata (Colorado potato beetle) >2.0 µg/insect/24 hr larva (topical application) [R17]; LD$_{50}$ Apis mellifera (honeybee) approximately 1.14 µg/bee (as dust) adult worker (topical application) [R17]; LD$_{50}$ Musca domestica (housefly) 2.2 µg/g/24 hr, 0.075 µg/fly adult female (topical application) [R17]; LD$_{50}$ Stomoxys calcitrans (stable fly) 1.5 µg/g/24 hr, 0.024 µg/fly adult female; 1.13 µg/g/24 hr, 0.093 µg/fly adult male (topical application) [R17]; LC$_{50}$ Culex pipens (mosquito) 4th instar 1.2 µg/L/24 hr (technical material) [R18]; LC$_{50}$ Aedes species (mosquito) 4th instar 0.5–3.5 µg/L/24 hr (technical material) [R19]; LC$_{50}$ Aedes aegypti (mosquito) 2nd instar 0.0011 µg/L/24 hr; 4th instar 0.0014 µg/L/24 hr (technical material, 96%) [R20]; LC$_{50}$ Aedes aegypti (mosquito) 3rd, 4th instar, 10 µg/L/18 hr (technical material, 96%) [R21]; LC$_{50}$ Belostoma sp (giant water bug) adult, 15 µg/L/24 hr (technical material) [R22]; LC$_{50}$ Hygrotus sp (predaceous diving beetle) adult, 40 µg/L/24 hr (technical material) [R22]; LC$_{50}$ Laccophilus decipiens (predaceous diving beetle) adult, 4.6 µg/L/24 hr (technical material) [R22]; LC$_{50}$ Thermonectus basillaris (predaceous diving beetle) 6 µg/L/24 hr (technical material) [R22]; LC$_{50}$ Berosus styliferus (water scavenger beetle) adult 9 µg/L/24 hr (technical material) [R22]; LC$_{50}$ Hydrophilus triangularis (water scavenger beetle) larva, 20 µg/L/24 hr; 30 µg/L/24 hr, adult (technical material) [R22]; LC$_{50}$ Tropisternus lateralis (water scavenger beetle) larva, 52 µg/L/24 hr; adult, 8 µg/L/24 hr (technical material) [R22]; LC$_{50}$ Daphnia sp (cladoceran) 0.88 µg/L/4 hr (encapsulated formulation, conditions of bioassay not specified) [R23]; LC$_{50}$ Hyalella azteca (amphipod) 1.28 µg/L/24 hr (encapsulated formulation) [R24]; LC$_{50}$ Ephemerella sp (mayfly) 0.33 µg/L/72 hr (Encapsulated formulation) [R23]; LC$_{50}$ Neoplea striola (pygmy backswimmer) 0.97 µg/L/144 hr (encapsulated formulation) [R23].

BIODEGRADATION Measured half-life of 4 weeks (clay loam) and 12 weeks (silt loam) in non-sterile soils versus 24 weeks in both soils sterilized by autoclaving was indicative of significant biodegradation (1). Half-lives of one week (sandy loam) and 2.5 weeks (organic) in non-sterile soils versus half-life of 17 and 40 weeks, respectively, in the sterilized soils (2). After 4 weeks of incubation, 33–38% of applied chlorpyrifos was degraded in a clay loam sterilized by autoclaving or gamma irradiation while 62% was degraded in the non-sterile soil (3). No significant difference in the degradation rate was observed in a natural water versus the natural water that had been sterilized (4). The degradation rate in non-sterile sandy loam and muck soils was found to be significantly faster than in the sterilized soils with the degradation rate in non-sterile soil decreasing with a decrease in temperature (3 to 28°C) and variable with moisture content (5,6). The half-life of chlorpyrifos in a seawater-sediment system was 24 days but was well in excess of the 28-day experimental period when the system was sterilized with formalin (7). In a shake-flask screening test similar to a river die-away test, chlorpyrifos degraded about 40% faster in active (natural) water as compared to the same water that had been sterilized with formalin (8) [R25].

BIOCONCENTRATION Experimental log BCF value of 2.67 determined from a 35-day flowing-water study using mosquito fish (1). Experimental log BCF value of 2.50 determined from a static ecosystem study using mosquito fish (2). Based on an experimentally measured log K_{ow} of 4.96 (3), the log BCF is estimated to be 3.54 from a recommended regression equation (4,SRC). Based on an experimentally measured water solubility of 0.7 ppm at 20°C (5), the log BCF is estimated to be 2.88 from a recommended regression equation (4,SRC) [R26].

FIFRA REQUIREMENTS Additional residue data on various processed commodities are required. Also, additional chronic toxicity, oncogenicity, and mutagenicity testing is needed to better determine the long-term effects of this chemical. Plant, animal, and exposure data are required to better qualify and quantify human exposure to residues from dietary and non-dietary sources. Other requirements include studies on: acute inhalation, general metabolism, hydrolysis, photodegradation, soil metabolism, mobility, dissipation, accumulation, fish embryo-larvae, large-scale field testing, monitoring for crop runoff, phytotoxic effects on algae and other aquatic plants; and indoor monitoring. [R1, (1988)].

REFERENCES

1. Purdue University, *National Pesticide Information Retrieval System.*

2. Worthing, C. R., and S. B. Walker (eds.). *The Pesticide Manual—A World Compendium.* 8th ed. Thornton Heath, UK: The British Crop Protection Council, 1987. 179.

3. Carmichael, W. W. (ed.). *Environmental Science Research.* Volume 20. *The Water Environment–Algal Toxins and Health.* New York/London: Plenum Press, 1981. 310.

4. Rossoff, I. S. *Handbook of Veterinary Drugs.* New York: Springer Publishing Company, 1974. 171.

5. Association of American Railroads. *Emergency Handling of Hazardous Materials in Surface Transportation.* Washington, DC: Assoc. of American Railroads, Hazardous Materials Systems (BOE), 1987. 167.

6. Verschueren, K. *Handbook of Environmental Data of Organic Chemicals.* 2nd ed. New York, NY: Van Nostrand Reinhold Co., 1983.

7. Menzie, C. M. *Metabolism of Pesticides.* U.S. Department of the Interior, Bureau of Sport Fisheries, and Wildlife, Publication 127. Washington, DC: U.S. Government Printing Office, 1969. 194.

8. USEPA/ORD. *Innovative and Alternative Technology Assessment Manual* pp. 3–5, 3–11–2 (1980) EPA 430/9–78–009.

9. U.S. Department of the Interior, Fish and Wildlife Service. *Handbook of Toxicity of Pesticides to Wildlife.* Resource Publication 153. Washington, DC: U.S. Government Printing Office, 1984. 23.

10. Hill, E. F., and M. B. Camardese. *Lethal Dietary Toxicities of Environmental Contaminants and Pesticides to Coturnix.* Fish and Wildlife Technical Report 2.Washington, DC: United States Department of Interior, Fish and Wildlife Service, 1986. 44.

11. Schafer, E. W. *Toxicol Appl Pharmacol* 21 (3): 315–30 (1972) as cited in Nat'l Research Council Canada, *Ecotoxicology of Chlorpyrifos,* p. 144 (1978) NRCC No. 16079.

12. USEPA. *Ambient Water Quality Criteria Doc; Chlorpyrifos,* p. 40 (1986) EPA 440/5–86–005.

13. Rongsriyam, Y. S., et al. *Bull Wrld*

Health Org 39: 977–80 (1968) as cited in USEPA, *Ambient Water Quality Criteria Doc; Chlorpyrifos* p. 38 (1986) EPA 440/5–86–005.

14. Womeldorf, D. J., et al. *Mosq News* 30: 375–82 (1970) as cited in USEPA, *Ambient Water Quality Criteria Doc; Chlorpyrifos*, p. 35 (1986) EPA 440/5–86–005.

15. Macek, K. J., et al. *Bull Environ Contam Toxicol* 4: 174–83 (1969) as cited in USEPA, *Ambient Water Quality Criteria Doc; Chlorpyrifos*, p. 36 (1986) EPA 440/5–86–005.

16. Kenaga, E. E. *Bull World Health Org* 44 p. 163 (1978) NRCC No. 16079.

17. Kenaga, E. E., et al. *J Econ Entomol* 58 (6): 1043–50 (1965) as cited in Nat'l Research Council Canada, *Ecotoxicology of Chlorpyrifos*, p. 163 (1978) NRCC No. 16079.

18. Rettich, F. *Mosq News* 37: 252–7 (1977) as cited in USEPA, *Ambient Water Quality Criteria Doc; Chlorpyrifos* p. 35 (1986) EPA 440/5–86–005.

19. Rettich, F. *Mosq News* 37: 252–7 (1977) as cited in USEPA, *Ambient Water Quality Criteria Doc; Chlorpyrifos* p. 34 (1986) EPA 440/5–86–005.

20. Saleh, M. S., et al. *J Agric Sci* 97: 87–96 (1981) as cited in USEPA, *Ambient Water Quality Criteria Doc; Chlorpyrifos* p. 34 (1986) EPA 440/5–86–005.

21. Verma, K. V., S. J. Rahman. *J Commun Dis* 16: 162–4 (1984) as cited in USEPA, *Ambient Water Quality Criteria Doc; Chlorpyrifos*, p. 34 (1986) EPA 440/5–86–005.

22. Ahmed, W. *The Effectiveness of Predators of Rice Field Mosquitoes in Relation to Pesticide Use in Rice Cultures.* Ph. D. Thesis University of California–Davis (1977) as cited in USEPA, *Ambient Water Quality Criteria Doc; Chlorpyrifos*, p. 33 (1986) EPA 440/5–86–005.

23. USEPA. *Ambient Water Quality Criteria Doc; Chlorpyrifos*, p. 33 (1986) EPA 440/5–86–005.

24. USEPA. *Ambient Water Quality Criteria Doc*, p. 33 (1986) EPA 440/5–86–005.

25. (1) Getzin, L. W. *J Econ Entomol* 74: 158 (1981); (2) Miles, J.R.W. *Bull Environ Contam Toxicol* 22: 312 (1979); (3) Getzin, L. W., I. Rosefield. *J Agric Food Chem* 16: 598 (1968); (4) Sharom, M. S., et al. *Water Res* 14: 1089 (1980); (5) Miles, J.R.W., et al. *J Environ Sci Health* B18: 705 (1983); (6) Miles, J.R.W., et al. *Journal Environ Sci Health* B19: 237 (1984); (7) Schimmel, S. C., et al. *J Agric Food Chem* 31: 104 (1983); (8) Walker, W. W. *Development of a Fate/Toxicity Screening Test* USEPA–600/S4–84–074 (1984).

26. (1) Vieth, G.D., et al. *J Fish Res Board Can* 36: 1040 (1979); (2) Kenaga, E. E. *Environ Sci Technol* 14: 553 (1980); (3) GEMS. *Graphical Exposure Modeling System Octanol Water Partition Coefficient (CLOGP) Data Base* Office of Toxic Substances USEPA (1986); (4) Lyman, W. J., et al. *Handbook of Chemical Property Estimation Methods. Environmental Behavior of Organic Coompounds.* New York: McGraw-Hill, p 5–4, 5–10; (5) Bowman, B. T., W. W. Sans. *J Environ Sci Health* B18: 221 (1983).

27. American Conference of Governmental Industrial Hygienists. *Threshold Limit Values for Chemical Substances and Physical Agents and Biological Exposure Indices for 1993–1994.* Cincinnati, OH: ACGIH, 1993.

■ PHOSPHOROTRITHIOIC ACID, S,S,S–TRIBUTYL ESTER

CAS # 78–48–8

SYNONYMS b-1,776 • butifos • butiphos • butyl-phosphorotrithioate • butyl-phosphorotrithioate ((BuS)$_3$ PO) • chemagro-1,776 • chemagro b-1776 • de-green • def • def-defoliant • e-z-off d • fos-fall "a" • fossfall • ortho-phosphate-defoliant • phosphorotrithioic acid, S,S,S-tributyl ester • S,S,S-tributyl trithiophosphate • tbtp

MF $C_{12}H_{27}OPS_3$

MW 314.54

COLOR AND FORM Colorless to pale-yellow liquid.

ODOR Mercaptan-like odor.

DENSITY 1.06 at 20°C.

BOILING POINT 150°C at 0.3 mm Hg.

MELTING POINT Below −25°C.

SOLUBILITY Soluble in chloroform, benzene, petroleum ether; soluble in most organic solvents.

VAPOR PRESSURE 1.38×10^{-5} mm Hg at room temperature.

OCTANOL/WATER PARTITION COEFFICIENT log K_{ow} = 5.7 (measured).

STABILITY AND SHELF LIFE Relatively stable to heat and to acids but is slowly hydrolyzed under alkaline conditions [R2].

COMMON USES Defoliant on cotton [R1].

OCCUPATIONAL RECOMMENDATIONS None listed.

NON-HUMAN TOXICITY LD_{50} rat oral 200 mg/kg [R3]; LD_{50} rat dermal > 1000 mg/kg [R8]; LD_{50} rat female oral 325 mg/kg [R9, p. II-301]; LD_{50} guinea pig male oral 260 mg/kg [R9, p. II-301].

ECOTOXICITY LC_{50} *Grammarus fasciatus* 100 µg/L/96 hr at 21°C (95% confidence interval, 68 to 150 µg/L/96 hr). Static conditions without aeration [R4]. LC_{50} *Pteronarcys* 2,100 µg/L/96 hr at 15°C (95% confidence interval, 1,500 to 2,900 µg/L/96 hr). Static conditions without aeration [R4]. LC_{50} rainbow trout 660 µg/L/96 hr at 13 °C (95% confidence interval, 560 to 750 µg/L/96 hr). Static conditions without aeration [R4]. LC_{50} bluegill 620 µg/L/96 hr at 18°C (95% confidence interval, 390 to 975 µg/L/96 hr). Static conditions without aeration [R4]. LC_{50} *Grammarus lacustris* 100 µg/L/96 hr (conditions of bioassay not specified) [R8]. LC_{50} *Pteronarcys californica* 2,100 µg/L/96 hr (conditions of bioassay not specified) [R8].

BIODEGRADATION In experiments studying the persistence of pesticide merphos conducted in water from the Little Miami River in sealed glass jars exposed to sunlight and artificial fluorescent light, the amount of S,S,S–tributyl phosphorotrithioate (formed in an apparently quantitative yield from merphos (S, S,S–tributyl phosphorotrithioite) within 1 hr) recovered was 50%, 30%, 10%, and <5% after 1, 2, 4, and 8 weeks, respectively (1). The study did not determine whether the observed degradation of S,S,S–tributyl phosphorotrithioate was due to abiotic processes (such as photolysis or hydrolysis) or biological processes (1). Residues of S,S,S–tributyl phosphorotrithioate contained in cotton gin wastes were stable to composting under anaerobic and aerobic conditions (2). The amount of S,S,S–tributyl phosphorotrithioate recovered from an untreated seawater-sediment mixture after 5 days was 20% versus a 77% recovery after 7 days from mixtures that were sterilized with formalin; no volatilization losses were observed in experiments with sediment present (3). A 50% loss was observed in untreated seawater without sediment after accounting for volatilization losses, but no sterilized control experiments were run in seawater alone (3). Some of the losses observed in these experiments may have been due to abiotic hydrolysis (3,SRC) [R5].

BIOCONCENTRATION A BCF of 350 was determined for pinfish (*Lagodon rhomboides*) in a 96–hr experiment, but the investigators concluded that the value was a low estimate of the true BCF because the experiment was not run for a long enough time for equilibrium to be reached (1). Based upon a reported log K_{ow} of 5.7 (1), a BCF of 12,600 has been estimated using a recommended regression equation (2) [R6].

FIFRA REQUIREMENTS Tolerances are established for residues of the defoliant S, S,S–tributylphosphorotrithioate in or on the following raw agricultural commodities: cottonseed; cattle, goats, and sheep (meat, fat, and meat by-products); and milk [R7]. Under section 3(c)(2)(b) of FIFRA, the Data Call-in Program, existing registrants are required to provide EPA with needed studies. For S,S,S-tributylphosphorotrithioate, responses to the

Data Call-in are under review [R10]. Criteria of Concern: neurotoxicity. Returned to the registration process based on a finding of no unreasonable adverse effects. Registrants agreed to protective clothing requirement. [R11].

REFERENCES

1. Worthing, C. R., S. B. Walker (eds.). *The Pesticide Manual—A World Compendium*. 8th ed. Thornton Heath, UK: The British Crop Protection Council, 1987. 818.

2. Worthing, C. R. (ed.). *Pesticide Manual*. 6th ed. Worcestershire, England: British Crop Protection Council, 1979. 527.

3. Verschueren, K. *Handbook of Environmental Data of Organic Chemicals*. 2nd ed. New York, NY: Van Nostrand Reinhold Co., 1983. 1120.

4. U.S. Department of Interior, Fish and Wildlife Service. *Handbook of Acute Toxicity of Chemicals to Fish and Aquatic Invertebrates*. Resource Publication No. 137. Washington, DC: U.S. Government Printing Office, 1980. 26.

5. (1) Eichelberger, J. W., J. J. Lichtenberg. *Environ Sci Technol* 5: 541–4 (1971). (2) Winterlin, W. L., et al., *J Environ Sci Health* B21: 507–28 (1986). (3) Environ Res Lab. *Acephate, Alicarb, Carbophenothion, DEF, EPN, Ethoprop, Methyl Parathion, and Phorate* USEPA–600/4–81–041 (NTIS PB81–244477) Washington, DC: pp. 19–35 (1981).

6. (1) Environ Res Lab. *Acephate, Aldicarb, Carbophenothion, DEF, EPN, Ethoprop, Methyl Parathion, and Phorate* USEPA–600/4–81–041 (NTIS PB81–244477). Washington, DC: pp. 19–35 (1981). (2) Lyman, W. J., et al., *Handbook of Chemical Property Estimation Methods*, New York: McGraw-Hill, p. 5–5 (1982).

7. 40 CFR 180.272 (7/1/90).

8. Verschueren, K. *Handbook of Environmental Data of Organic Chemicals*. 2nd ed. New York, NY: Van Nostrand Reinhold Co., 1983. 1120.

9. Gosselin, R.E., R.P. Smith, H.C. Hodge. *Clinical Toxicology of Commercial Products*. 5th ed. Baltimore: Williams and Wilkins, 1984.

10. USEPA/OPP. *Report on the Status of Chemicals in the Special Review Program*. Registration Standards Program, Data Call-in Program, and Other Registration Activities p. 67 (1988) EPA 540/09-89-037.

11. USEPA. *Report on the Status of Chemicals in the Special Review Program and Registration Standards in the Reregisteration Program*. p. 1–26 (1989).

■ 2-PICOLINE, 5-ETHYL

CAS # 104–90–5

DOT # 2300

SYNONYMS 5–ethyl–alpha–picoline • aldehyde–collidine • aldehydine • 6–methyl–3–ethylpyridine • 5–ethyl–4–picoline • mep • 2–methyl–5–ethylpyridine • 5–ethyl–2–methyl pyridine • 5–ethyl–2–picoline

MF $C_8H_{11}N$

MW 121.20

COLOR AND FORM Liquid; colorless.

ODOR THRESHOLD 0.03 mg/m³ (odor low); 94.1 mg/m³ (odor high) [R8].

DENSITY Specific gravity: 0.9184 at 23°C.

BOILING POINT 177.8°C at 747 mm Hg.

MELTING POINT –70.3°C.

AUTOIGNITION TEMPERATURE 939°F [R5].

EXPLOSIVE LIMITS AND POTENTIAL These materials (nitric acid and 5-ethyl-2-methyl pyridine) were placed in small autoclave and heated and stirred for 40 min. Emergency vent was opened due to sudden pressure rise. A violent explosion occurred 90 seconds later (nitric acid) [R6, p. 291M-

90 seconds later (nitric acid) [R6, p. 291M-137]. Vapor forms explosive mixtures with air. [R6, p. 49–50].

FIREFIGHTING PROCEDURES Use water spray, dry chemical, "alcohol" foam or carbon dioxide. Use water spray to disperse vapors and to protect men attempting to stop a leak. Water spray may be used to flush spills from exposures. [R6, p. 49–50].

TOXIC COMBUSTION PRODUCTS Irritating vapors are generated when heated [R7].

STANDARD CODES NFPA, ICC, USCG—no.

VAPOR PRESSURE 1.4 mm Hg at 25 °C.

SURFACE TENSION 36 dynes/cm = 0.036 N/m 20°C.

OTHER CHEMICAL/PHYSICAL PROPERTIES Water is soluble in liquid.

PROTECTIVE EQUIPMENT AND CLOTHING Wear self-contained breathing apparatus; wear goggles if eye protection not provided [R6, p. 49–50]. Depending on the extent of possible contact, workers should be provided with personal protective equipment. A charcoal gas mask canister respirator has been found to be effective against a 2% pyridine concentration at 30 L/min for 1 hr. Rubber and plastic gloves should not be relied upon to prevent skin contact because pyridine and many of its derivatives penetrate these materials. (pyridine, homologs and derivatives) [R9].

PREVENTIVE MEASURES Contact lenses should not be worn when working with this chemical (pyridine) [R10]. Contact lens use in industry is controversial. A survey of 100 corporations resulted in the recommendation that each company establish its own contact lens use policy. One presumed hazard of contact lens use is possible chemical entrapment. It was found that contact lenses minimized injury or protected the eye. The eye was afforded more protection from liquid irritants. Soft contact lenses did not worsen corneal damage from strong chemicals and in some cases could actually protect the eye. Overall, the literature supports the wearing of contact lenses in industrial environments as part of the standard eye

protection, e.g., face shields; however, more data are needed to establish the value of contact lenses. [R11].

COMMON USES Mfr of nicotinic acid and nicotinamide; vinyl pyridines for copolymers; intermediate for germicides and textile finishes; corrosion inhibitor for chlorinated solvents [R2]. Intermediate for insecticides [R3, 2734].

MAJOR SPECIES THREATENED Waterfowl.

DIRECT CONTACT Possibly a mild irritant—skin.

GENERAL SENSATION Skin irritation grade 6—necrosis undiluted; eye irritation grade 9—severe burns [R1].

PERSONAL SAFETY Wear skin protection and canister type mask.

ACUTE HAZARD LEVEL Details unknown —felt to be moderately toxic. Low threshold in water. Emits toxic vapors when burned.

CHRONIC HAZARD LEVEL Unknown.

DEGREE OF HAZARD TO PUBLIC HEALTH Moderately toxic. Emits toxic vapors when burned.

OCCUPATIONAL RECOMMENDATIONS None listed.

ACTION LEVEL Notify fire authority. Remove ignition source. Attempt to contain slick.

ON-SITE RESTORATION Oil-skimming equipment and sorbent foams can be applied to slick. Seek professional environmental engineering assistance through EPA's environmental response team (ERT), Edison, NJ, 24–hour no., 908-548-8730.

BEACH OR SHORE RESTORATION Do not burn.

AVAILABILITY OF COUNTERMEASURE MATERIAL Oil-skimming equipment—stored at major ports; sorbent foams–upholstery shops (polyurethane).

DISPOSAL Generators of waste (equal to or greater than 100 kg/mo) containing this contaminant, EPA hazardous waste number U191, must conform with USEPA regulations in storage, transportation, treatment, and disposal of waste [R12].The following wastewater treatment technologies have been investigated for 2-methyl-5-ethylpyridine: Concentration process: Activated carbon. [R13].

DISPOSAL NOTIFICATION Contact local air authority.

NON-HUMAN TOXICITY LD_{50} rat oral 0.8–1.8 g/kg (from table) [R3, 2723]; LD_{50} rat ip 0.2–0.4 g/kg (from table) [R3, 2723]; LD_{50} mouse oral 0.8–1.6 g/kg (from table) [R3, 2723]; LD_{50} mouse ip 0.1–0.2 g/kg (from table) [R3, 2723]; LD_{50} guinea pig percutaneous 2.5–5.0 ml/kg (from table) [R3, 2723].

ECOTOXICITY LC_{50} *Pimephales promelas* (fathead minnow) 81.1 mg/L/96 hr (confidence limit 77.6–84.8 mg/L), flow-through bioassay with measured concentrations, 26.2°C, dissolved oxygen 5.9 mg/L, hardness 45.5 mg/L $CaCO_3$, alkalinity 42.0 mg/L $CaCO_3$, and pH 7.49 [R14]. EC_{50} *Pimephales promelas* (fathead minnow) 69.8 mg/L/96 hr (confidence limit 61.2–79.6 mg/l), flow-through bioassay with measured concentrations, 26.2°C, dissolved oxygen 5.9 mg/L, hardness 45.5 mg/L $CaCO_3$, alkalinity 42.0 mg/L $CaCO_3$, and pH 7.49. Effect: loss of equilibrium. [R14].

BIODEGRADATION BOD: (theoretical) 4.4%, 5 days; 56.6%, 20 days; 0.12–2.14 lb/lb, 5 days [R7].

BIOCONCENTRATION Based on a water solubility of 12,000 mg/L at 25°C (1) and an estimated log K_{ow} of 2.49 (2), respective bioconcentration factors (log BCF) of 0.49 and 1.66 for 2-methyl-5-ethylpyridine have been calculated using recommended regression-derived equations (3,SRC). These BCF values indicate that 2-methyl-5-ethylpyridine should not bioconcentrate among aquatic organisms (SRC). [R14].

INDUSTRIAL FOULING POTENTIAL Slick may reduce heat transfer or cause hot spots and scaling.

MAJOR WATER USE THREATENED Potable supply, industrial.

PROBABLE LOCATION AND STATE Liquid. Will float as a slick on surface of water.

COLOR IN WATER Colorless.

DOT *Health Hazards:* Contact causes burns to skin and eyes. If inhaled, may be harmful. Fire may produce irritating or poisonous gases. Runoff from fire control or dilution water may cause pollution [R4].

Fire or Explosion: Some of these materials may burn, but none of them ignites readily. Flammable/poisonous gases may accumulate in tanks and hopper cars. Some of these materials may ignite combustibles (wood, paper, oil, etc.) [R4].

Emergency Action: Keep unnecessary people away; isolate hazard area and deny entry. Stay upwind; keep out of low areas. Self-contained breathing apparatus (SCBA) and structural firefighter's protective clothing will provide limited protection. Call CHEMTREC at 1–800–424–9300 for emergency assistance. If water pollution occurs, notify the appropriate authorities [R4].

Fire: Some of these materials may react violently with water. Small Fires: Dry chemical, CO_2, halon, water spray or standard foam. Large Fires: Water spray, fog or standard foam is recommended. Move container from fire area if you can do it without risk. Cool containers that are exposed to flames with water from the side until well after fire is out. Stay away from ends of tanks [R4].

Spill or Leak: Do not touch spilled material; stop leak if you can do it without risk. Small Spills: Take up with sand or other noncombustible absorbent material and place into containers for later disposal. Small Dry Spills: With clean shovel place material into clean, dry container, and cover; move containers from spill area. Large Spills: Dike far ahead of liquid spill for later disposal [R4].

First Aid: Move victim to fresh air; call emergency medical care. Remove and isolate contaminated clothing and shoes at the site. In case of contact with material, immediately flush skin or eyes with running water for at least 15 minutes. Keep victim quiet and maintain normal body temperature. [R4].

GENERAL RESPONSE Avoid contact with liquid. Keep people away. Wear goggles, self-contained breathing apparatus, and rubber overclothing (including gloves). Stop discharge if possible. Call fire department. Isolate and remove discharged material. Notify local health and pollution control agencies.

FIRE RESPONSE Combustible. Wear goggles, self-contained breathing apparatus, and rubber overclothing (including gloves). Extinguish with water, dry chemical, alcohol foam, or carbon dioxide. Cool exposed containers with water.

EXPOSURE RESPONSE Call for medical aid. Liquid will burn skin and eyes. Harmful if swallowed. Remove contaminated clothing and shoes. Flush affected areas with plenty of water. If swallowed and victim is conscious, have victim drink water or milk.

RESPONSE TO DISCHARGE Mechanical containment. Should be removed. Chemical and physical treatment.

WATER RESPONSE Effect of low concentrations on aquatic life is unknown. Fouling to shoreline. May be dangerous if it enters water intakes. Notify local health and wildlife officials. Notify operators of nearby water intakes.

EXPOSURE SYMPTOMS Breathing of vapors will cause vomiting and chest discomfort. Contact with liquid causes skin and eye burns.

EXPOSURE TREATMENT Inhalation: remove victim to fresh air; give oxygen if breathing is difficult; call a physician. Skin or eyes: immediately flush with plenty of water for at least 15 min; get medical care for eyes.

REFERENCES

1. Smyth, H. F., Jr., C. P. Carpenter, C. S. Weil, U. C. Pozzani, J. A. Striegel, and J. S. Nycum, "Range–Finding Toxicity Data: List VII," *American Industrial Hygiene Association Journal*, 30: 470–476. 1969. Smyth, H. F., C. P. Carpenter, and C. S. Weil, "Range–Finding Toxicity Data: List IV," *AMA Archives of Industrial Hygiene and Occupational Medicine*, 4: 119–122, 1951. Smyth, H. F., C. P. Carpenter, C. S. Weil, U. C. Pozzani, and J. A. Striegel, "Range–Finding Toxicity Data: List VI," *American Industrial Hygiene Association Journal*, 23: 95–107, 1962. Smyth, H. F., C. P. Carpenter, C. S. Weil, and U. C. Pozzani, "Range–Finding Toxicity Data: List V," *AMA Archives of Industrial Hygiene and Occupational Medicine*, 10: 61–68, 1954. Smyth, H. F., C. P. Carpenter, and C. S. Weil, "Range–Finding Toxicity Data: List III," *Journal of Industrial Hygiene and Toxicology*, 31: 60–62, 1949. Smyth, H. F., L. Seaton, and L. Fischer, "The Single Dose Toxicity of Some Glycols and Derivatives," *Journal of Industrial Hygiene and Toxicology*, 23 (6): 259–268, 1941. Smyth, H. F., and C. P. Carpenter, "Further Experience with the Range–Finding Test in the Industrial Toxicology Laboratory," *Journal of Industrial Hygiene and Toxicology*, 30: 63–68, 1948. Smyth, H. F., and C. P. Carpenter, "The Place of the Range–Finding Test in the Industrial Toxicology Laboratory," *Journal of Industrial Hygiene and Toxicology*, 26: 269–273, 1 944.

2. Sax, N.I., and R.J. Lewis, Sr. (eds.). *Hawley's Condensed Chemical Dictionary*. 11th ed. New York: Van Nostrand Reinhold Co., 1987. 769.

3. Clayton, G. D., and F. E. Clayton (eds.). *Patty's Industrial Hygiene and Toxicology*. Volume 2A, 2B, 2C: *Toxicology*. 3rd ed. New York: John Wiley and Sons, 1981–1982.

4. Department of Transportation. *Emergency Response Guidebook 1987*. DOT P 5800.4. Washington, DC: U.S. Government Printing Office, 1987, p. G-60.

5. U.S. Coast Guard, Department of Transportation. *CHRIS—Hazardous Chemical Data*. Manual Two. Washington, DC: U.S. Government Printing Office, Oct. 1978.

6. National Fire Protection Association. *Fire Protection Guide on Hazardous Materials*. 9th ed. Boston, MA: National Fire Protection Association, 1986.

7. U.S. Coast Guard, Department of Trans-

portation. *CHRIS—Hazardous Chemical Data.* Volume II. Washington, DC: U.S. Government Printing Office, 1984–5.

8. Ruth, J. H. *Am Ind Hyg Assoc J.* 47: A-142–51 (1986).

9. International Labour Office. *Encyclopedia of Occupational Health and Safety.* Vols. I and II. Geneva, Switzerland: International Labour Office, 1983. 1810.

10. NIOSH. *Pocket Guide to Chemical Hazards.* 2nd Printing. DHHS (NIOSH) Publ. No. 85–114. Washington, DC: U.S. Dept. of Health and Human Services, NIOSH/Supt. of Documents, GPO, February 1987. 202.

11. Randolph, S. A., M. R. Zavon. *J Occup Med* 29: 237–42 (1987).

12. 40 CFR 240–280, 300–306, 702–799 (7/1/89).

13. USEPA. *Management of Hazardous Waste Leachate.* EPA Contract No. 68–03–2766 p. E-139 (1982).

14. Brooke, L.T., D.J. Call, D.T. Geiger, and C.E. Northcott (eds.). *Acute Toxicities of Organic Chemicals to Fathead Minnows (Pimephales promelas).* Superior, WI: Center for Lake Superior Environmental Studies, Univ. of Wisconsin–Superior, 1984. 301.

■ PINDONE

CAS # 83–26–1

SYNONYMS 1,3-indandione, 2-pivaloyl • 1h-indene-1,3 (2h)-dione, 2-(2,2-dimethyl-1-oxopropyl)- • 2-(2,2-dimethyl-1-oxopropyl)-1h-indene-1,3 (2h)-dione • 2-(trimethylacetyl)-1,3-indandione • 2-(trimetil-acetil)-indan-1,3-dione [Italian] • 2-pivaloyl-1,3-indandione • 2-pivaloyl-indaan-1,3-dion [Dutch] • 2-Pivaloyl-indan-1,3-dion [German] • 2-pivaloylindane-1,3-dione • 2-pivalyl-1,3-indandione • chemrat • pindon [Dutch] • pivacin • pival • pivaldion [Italian] • pivaldione [French] • pivalyl • pivalyl indan-1,3-dione • pivalyl-indandione •

pivalyl-valone • pivalyn • tri-ban • 2-pivaloylindan-1,3-dione

MF $C_{14}H_{14}O_3$

MW 230.25

COLOR AND FORM Bright-yellow crystals from ethanol; bright-yellow powder; yellow-brown cystalline solid.

ODOR Odorless; almost no odor.

TASTE Tasteless.

DENSITY 1.06.

BOILING POINT 180°C at 1 mm Hg.

MELTING POINT 108–110°C.

FIREFIGHTING PROCEDURES Extinguish fire using agent suitable for type of surrounding fire. (Material itself does not burn or burns with difficulty.) Use water in flooding quantities as fog. Use foam, dry chemical, or carbon dioxide [R4].

SOLUBILITY Soluble in alcohol, ether, acetone, aqueous alkali solution; 18 ppm in water at 25°C; soluble in most organic solvents; soluble in aqueous alkali or ammonia to give bright-yellow salts.

VAPOR PRESSURE 1 mm Hg at 180°C.

OTHER CHEMICAL/PHYSICAL PROPERTIES Bright-yellow crystals; soluble in water; MP: 205–210°C (sodium salt).

PROTECTIVE EQUIPMENT Recommendations for respirator selection. Max concentration for use: 0.5 mg/m³. Respirator Class(es): Any dust and mist respirator [R5]. Recommendations for respirator selection. Max concn for use: 1 mg/cu m. Respirator Class(es): Any dust and mist respirator except single-use and quarter-mask respirators. Any supplied-air respirator. Any self-contained breathing apparatus [R5]. Recommendations for respirator selection. Max concn for use: 2.5 mg/m³. Respirator Class(es): Any powered, air-purifying respirator with a dust and mist filter. Any supplied-air respirator operated in a continuous-flow mode [R5]. Recommendations for respirator selection. Max concn for use: 5 mg/m³. Respirator Class(es): Any powered, air-purifying respirator with a tight-fitting facepiece and a high-efficiency particulate filter. Any self-contained breathing apparatus with a full facepiece. Any supplied-air respirator with a full facepiece. Any air-purifying, full facepiece respirator with a high-efficiency particulate filter. Any supplied-air respirator that has a tight-fitting facepiece and is operated in a continuous-flow mode [R5]. Recommendations for respirator selection. Max concn for use: 200 mg/m³. Respirator Class(es): Any supplied-air respirator with a full facepiece and operated in a pressure-demand or other positive-pressure mode. [R5]. Recommendations for respirator selection. Condition: Emergency or planned entry into unknown concentration or IDLH conditions: Respirator Class(es): Any self-contained breathing apparatus that has a full facepiece and is operated in a pressure-demand or other positive-pressure mode. Any supplied-air respirator with a full facepiece and operated in pressure-demand or other positive-pressure mode in combination with an auxiliary self-contained breathing apparatus operated in pressure-demand or other positive-pressure mode. [R5]. Recommendations for respirator selection. Condition: Escape from suddenly occurring respiratory hazards: Respirator Class(es): Any air-purifying, full facepiece respirator with a high-efficiency particulate filter. Any appropriate escape-type, self-contained breathing apparatus. [R5].

PREVENTATIVE MEASURES Contact lenses should not be worn when working with this chemical [R5]. SRP: The scientific literature for the use of contact lenses in industry is conflicting. The benefit or detrimental effects of wearing contact lenses depend not only upon the substance, but also on factors including the form of the substance, characteristics and duration of the exposure, the uses of other eye protection equipment, and the hygiene of the lenses. However, there may be individual substances whose irritating or corrosive properties are such that the wearing of contact lenses would be harmful to the eye. In those specific cases, contact lenses should not be worn. In any event, the usual eye protection equipment should be worn even when contact lenses are in place. In emergencies during mfr or

use when there may be danger of heavy exposure, suitable resp protective equipment should be worn to prevent dust from entering mouth. Personal hygiene should be practiced to ensure that food eaten by people does not become contaminated (coumarin and indandione derivative) [R8, 561]. If material not on fire and not involved in Fire: Keep sparks, flames, and other sources of ignition away. Keep material out of water sources and sewers [R9]. Personnel protection: Avoid breathing dusts, and fumes from burning material. Keep upwind. Avoid bodily contact with the material. Do not handle broken packages unless wearing appropriate personal protective equipment [R9]. Work clothing should be changed daily if it is reasonably probable that the clothing may be contaminated. [R5].

CLEANUP 1. Ventilate area of spill. 2. For small quantites, sweep onto paper or other suitable material, place in appropriate container, and burn in safe place (such as fume hood). 2. Large quantities may be reclaimed; however, if this is not practical, dissolve in flammable solvent (such as alcohol), and atomize in suitable combustion chamber equipped with appropriate effluent-gas-cleaning device [R6].

DISPOSAL SRP: At the time of review, criteria for land treatment or burial (sanitary landfill) disposal practices are subject to significant revision. Prior to implementing land disposal of waste residue (including waste sludge), consult with environmental regulatory agencies for guidance on acceptable disposal practices.

COMMON USES Pharmaceutical intermediate; insecticide; rodenticide [R3]; was originally suggested as a substitute for the greater part of the pyrethrins in pyrethrum sprays. It is used as an anticoagulant rodenticide in baits containing 250 mg/kg [R1].

OCCUPATIONAL RECOMMENDATIONS
OSHA Table Z–1 8-hr time-weighted average: 0.1 mg/m³ [R7].

NIOSH 10–hr time-weighted average: 0.1 mg/m³ [R5].

THRESHOLD LIMIT 8-hr time-weighted average (TWA) 0.1 mg/m³ (1987) [R34, 30].

NON-HUMAN TOXICITY LD₅₀ rat oral 10.3 mg/kg [R2]; LD₅₀ dog oral is 75–100 mg/kg [R23, 175].

REFERENCES

1. Worthing, C. R., and S. B. Walker (eds.). *The Pesticide Manual—A World Compendium.* 8th ed. Thornton Heath, UK: The British Crop Protection Council, 1987. 676.

2. American Conference of Governmental Industrial Hygienists, Inc. *Documentation of the Threshold Limit Values and Biological Exposure Indices.* 6th ed. Volumes I, II, III. Cincinnati, OH: ACGIH, 1991. 1274.

3. Lewis, R. J., Sr. (ed.). *Hawley's Condensed Chemical Dictionary.* 12th ed. New York, NY: Van Nostrand Reinhold Co., 1993, 921.

4. Association of American Railroads. *Emergency Handling of Hazardous Materials in Surface Transportation.* Washington, DC: Association of American Railroads, Bureau of Explosives, 1992. 764.

5. NIOSH. *NIOSH Pocket Guide to Chemical Hazards.* DHHS (NIOSH) Publication No. 90–117. Washington, DC: U.S. Government Printing Office, June 1990, 184.

6. Mackison, F. W., R. S. Stricoff, and L. J. Partridge, Jr. (eds.). *NIOSH/OSHA—Occupational Health Guidelines for Chemical Hazards.* DHHS (NIOSH) Publication No. 81–123 (3 vols). Washington, DC: U.S. Government Printing Office, Jan. 1981.

7. 29 CFR 1910.1000 (58 FR 35338 (6/30/93)).

8. International Labour Office. *Encyclopedia of Occupational Health and Safety.* Vols. I and II. Geneva, Switzerland: International Labour Office, 1983.

9. Association of American Railroads. *Emergency Handling of Hazardous Materials in Surface Transportation.* Washington, DC: Association of American Railroads, Bureau of Explosives, 1992. 764.

■ PIPERAZINE

CAS # 110–85–0

DOT # 2579

SYNONYMS 1,4-diazacyclohexane • 1,4-diethylenediamine • 1,4-piperazine • antiren • diethylenediamine • diethyleneimine • dispermine • eraverm • hexahydro-1,4-diazine • hexahydropyrazine • lumbrical • piperazidine • Piperazin (German) • piperazine, anhydrous • pipersol • pyrazine-hexahydride • pyrazine, hexahydro • uvilon • vermex • Worm-a-ton • wurmirazin

MF C₄H₁₀N₂

MW 86.16

pH 10.8–11.8 (10% aqueous solution).

COLOR AND FORM Plates or leaflets from ethanol; white to slightly off-white lumps or flakes; colorless, transparent, needle-like crystals.

ODOR Characteristic amine odor; ammoniacal odor.

TASTE Salty taste.

DENSITY 1.1.

DISSOCIATION CONSTANT Strong base: pKₐ = 4.19.

BOILING POINT 146°C at 760 mm Hg.

MELTING POINT 106°C.

HEAT OF COMBUSTION −2.7380×10⁹ J/kmol.

HEAT OF VAPORIZATION 47.3 J/g.

AUTOIGNITION TEMPERATURE 728.12°K [R8].

FLAMMABILITY LIMITS Flammability limits = 1.6 to 12.5 vol% [R8].

FIREFIGHTING PROCEDURES Extinguish fire using agent suitable for type of surrounding fire. (Material itself does not burn or burns with difficulty.) Use water in flooding quantities as fog. Apply water from as far a distance as possible. Use water spray to knock down vapors [R9].

SOLUBILITY Insoluble in ether; freely soluble in water, glycerol, and glycols; 1 g dissolves in 2 ml of 95% alcohol; soluble in water; insoluble in alcohol and ether (citrate); solubility in water: 35% at 0°C, 41% at 20°C, 48% at 50°C; insoluble in organic solvents (piperazine hydrochloride).

VAPOR PRESSURE 0.16 mm Hg at 20°C.

OCTANOL/WATER PARTITION COEFFICIENT Log K_{ow} = −1.17.

STABILITY AND SHELF LIFE Stable at temperature to 270°C and in neutral or acid media (piperazine hydrochloride) [R4].

OTHER CHEMICAL/PHYSICAL PROPERTIES Absorbs water and CO_2 from air; forms soluble compd with theophylline; pH of 10% aqueous solution is 10.8–11.8; BP: 125 to 130°C (hexahydrate); hygroscopic; crystals from water (hexahydrate form); freely soluble in water; soluble in alcohol (about 1:2); practically insoluble in ether (hexahydrate); heat of fusion = 298.9 J/g; heat of formation = 2.230×10^7 J/kmol; liquid molar volume = 0.129371 m³/kmol; vapor pressure = 2.6372×10^4 Pa at 379.15 K; slightly hygroscopic (piperazine hydrochloride); 82–83°C (piperazine hydrochloride monohydrate); cream-colored crystalline powder (piperazine hydrochloride).

PREVENTATIVE MEASURES If material not on fire and not involved in Fire: Keep sparks, flames, and other sources of ignition away. Keep material out of water sources and sewers. Build dikes to contain flow as necessary [R9].

DISPOSAL SRP: At the time of review, criteria for land treatment or burial (sanitary landfill) disposal practices are subject to significant revision. Prior to implementing land disposal of waste residue (including waste sludge), consult with environmental regulatory agencies for guidance on acceptable disposal practices.

COMMON USES Corrosion inhibitor; insecticides; accelerator for curing polychloroprene; antihelmintic [R2]. Anthelmintic for pinworm and roundworm infections in human toxicity toxicitys, swine, poultry, other animals; chemical intermediate for piperazine salts [R1]. Medication: treatment of oxyuriasis (pinworms) in man and animals [R5]. Piperazine and various salts of piperazine are widely used for the treatment on intestinal worms [R14, 295]. In manufacture of anthelmintics, antifilarials, antihistamines, tranquilizers, etc., corrision inhibitors, surface-active agents [R3]. Used in making insecticides and fibers (piperazine hydrochloride) [R6].

OCCUPATIONAL RECOMMENDATIONS None listed.

THRESHOLD LIMIT 8–hr time-weighted average (TWA) 5 mg/m³ (1982) (piperazine dihydrochloride) [R13, 30].

NON-HUMAN TOXICITY LD_{50} rat, oral 4,900 mg/kg (piperazine hydrochloride) [R12, 2235].

DOT *Health Hazards:* Contact causes burns to skin and eyes. If inhaled, may be harmful. Fire may produce irritating or poisonous gases. Runoff from fire control or dilution water may cause pollution [R7].

Fire or Explosion: Some of these materials may burn, but none of them ignites readily. Flammable/poisonous gases may accumulate in tanks and hopper cars. Some of these materials may ignite combustibles (wood, paper, oil, etc) [R7].

Emergency Action: Keep unnecessary people away; isolate hazard area and deny entry. Stay upwind; keep out of low areas. Positive-pressure self-contained breathing apparatus (SCBA) and structural firefighters' protective clothing will provide limited protection. Call emergency response telephone number on shipping paper first. If shipping paper not available or no answer, call CHEMTREC at 1–800–424–9300. If water pollution occurs, notify the appropriate authorities [R7].

Fire: Some of these materials may react violently with water. Small Fires: Dry chemical, CO_2, water spray, or regular foam. Large Fires: Water spray, fog, or regular foam. Move container from fire area if you can do so without risk. Apply cooling water to sides of containers that are exposed to flames until well after fire is out. Stay away from ends of tanks [R7].

Spill or Leak: Do not touch or walk through spilled material; stop leak if you can do so without risk. Small Spills: Take up with sand or other noncombustible absorbent material and place into containers for later disposal. Small Dry Spills: With clean shovel place material into clean, dry container, and cover loosely; move containers from spill area. Large Spills: Dike far ahead of liquid spill for later disposal [R7].

First Aid: Move victim to fresh air; call emergency medical care. In case of contact with material, immediately flush skin or eyes with running water for at least 15 minutes. Remove and isolate contaminated clothing and shoes at the site. Keep victim quiet and maintain normal body temperature [R7].

FIRE POTENTIAL Moderate via heat, flames, and oxidizers [R12, 2234].

FDA Piperazine is an indirect food additive for use as a component of adhesives [R10]. Manufacturers, packers, and distributors of drug and drug products for human use are responsible for complying with the labeling, certification, and usage requirements as prescribed by the Federal Food, Drug, and Cosmetic Act, as amended (secs 201–902, 52 Stat. 1040 et seq., as amended; 21 U.S.C. 321–392) [R11].

GENERAL RESPONSE Avoid contact with solid and dust. Keep people away. Stop discharge if possible. Isolate and remove discharged material. Notify local health and pollution control agencies.

FIRE RESPONSE Combustible. Poisonous gases may be produced in fire. Wear goggles and self-contained breathing apparatus. Extinguish with dry chemicals, alcohol foam, or carbon dioxide. Water may be ineffective on fire. Cool exposed containers with water.

EXPOSURE RESPONSE Call for medical aid. Dust irritating to eyes, nose, and throat. If inhaled will cause coughing or difficult breathing. If breathing has stopped, give artificial respiration. If breathing is diffi-

cult, give oxygen. Solid will burn eyes. Irritating to eyes. If swallowed will cause nausea and vomiting. Remove contaminated clothing and shoes. Flush affected areas with plenty of water. If swallowed and victim is conscious, have victim drink water or milk, and have victim induce vomiting. If swallowed and victim is unconscious, or having convulsions, do nothing except keep victim warm.

RESPONSE TO DISCHARGE Issue warning—water contaminant. Disperse and flush.

WATER RESPONSE Effect of low concentrations on aquatic life is unknown. May be dangerous if it enters water intakes. Notify local health and wildlife officials. Notify operators of nearby water intakes.

EXPOSURE SYMPTOMS Inhalation of dust irritates nose and throat. Ingestion causes irritation of mouth and stomach; has been known to cause severe allergic reaction. Contact with eyes causes burns. Repeated contact with skin may cause irritation and sensitization.

EXPOSURE TREATMENT Inhalation: move to fresh air. Ingestion: give large amount of water; induce vomiting; get medical attention. Eyes: flush with plenty of water for at least 15 min; get medical attention. Skin: wash with soap and water.

REFERENCES

1. SRI.

2. Sax, N. I., and R. J. Lewis, Sr. (eds.). *Hawley's Condensed Chemical Dictionary.* 11th ed. New York: Van Nostrand Reinhold Co., 1987. 921.

3. Kuney, J. H., J. M. Mullican (eds.). *Chemcyclopedia.* Washington, DC: American Chemical Society, 1994. 102.

4. Spencer, E. Y. *Guide to the Chemicals Used in Crop Protection.* 6th ed. Publication 1093, Research Institute, Agriculture Canada, Ottawa, Canada: Information Canada, 1973. 416.

5. Gosselin. *CTCP* 5th ed 1984 p. II–390.

6. American Conference of Governmental Industrial Hygienists, Inc. *Documentation of the Threshold Limit Values,* 4th ed.,

1980. Cincinnati, OH: American Conference of Governmmental Induustrial Hygienists, Inc., 1980. 342.

7. U.S. Department of Transportation. *Emergency Response Guidebook 1993.* DOT P 5800.6. Washington, DC: U.S. Government Printing Office, 1993, p. G–60.

8. Daubert, T. E., R. P. Danner. *Physical and Thermodynamic Properties of Pure Chemicals Data Compilation.* Washington, DC: Taylor and Francis, 1989.

9. Association of American Railroads. *Emergency Handling of Hazardous Materials in Surface Transportation.* Washington, DC: Association of American Railroads, Bureau of Explosives, 1992. 766.

10. 21 CFR 175.105 (4/1/93).

11. 21 CFR 200–299, 300–499, 820, and 860 (4/1/93).

12. Sax, N.I. *Dangerous Properties of Industrial Materials.* 6th ed. New York, NY: Van Nostrand Reinhold, 1984.

13. American Conference of Governmental Industrial Hygienists. *Threshold Limit Values for Chemical Substances and Physical Agents and Biological Exposure Indices for 1994–1995.* Cincinnati, OH: ACGIH, 1994.

14. *Kirk-Othmer Encyclopedia of Chemical Technology.* 3rd ed., Volumes 1–26. New York, NY: John Wiley and Sons, 1978–1984, p. V2.

■ PIPERONYL BUTOXIDE

CAS # 51–03–6

SYNONYMS 1,3-benzodioxole, 5-((2-(2-butoxyethoxy) ethoxy) methyl)-6-propyl; • butacide; • butocide; • butoxide; alpha-(2-(2-n-butoxyethoxy)-ethoxy) -4,5-methylenedioxy-2-propyltoluene • 5-((2-(2-butoxyethoxy) ethoxy) methyl-6-propyl-1,3-benzodioxole • butyl carbitol 6-propylpiperonyl ether • butyl-carbityl (6-propylpiperonyl) ether • ENT-14,250; FMC-5273 • 3,4-Methylenedioxy-6-propylbenzyl-n-butyl-diaethyl-englykolaether (German) • (3,4-methylenedioxy-6-propylbenzyl) (butyl) diethylene glicol ether • 3,4-methylenedioxy-6-propylbenzyl n-butyl diethyleneglycol ether • NCI-C02813 • NIA-5273 • PB • 6-(propylpiperonyl)-butyl carbityl ether • 6-propylpiperonyl butyl diethylene glycol ether • 5-Propyl-4-(2,5,8-trioxa-dodecyl)-1,3-benzodioxol (German) • pyrenone-606 • toluene, alpha-(2-(2-butoxyethoxy) ethoxy)-4,5-(methylenedioxy)-2-propyl

COLOR AND FORM Light-brown liquid; pale-yellowish liquid.

ODOR Odorless; mild.

MW 338.49

MF $C_{19}H_{30}O_5$

TASTE Faint bitter taste.

DENSITY 1.05–1.07 at 25°C.

BOILING POINT 180°C at 1 mm Hg.

FIREFIGHTING PROCEDURES Foam, carbon dioxide, dry chemical [R3].

SOLUBILITY Miscible with methanol, ethanol, benzene, freons, geons, petroleum oils, and other organic solvents; very slightly soluble in water.

STABILITY AND SHELF LIFE Stable to light [R2].

OTHER CHEMICAL/PHYSICAL PROPERTIES Resistant to hydrolysis and noncorrosive.

DISPOSAL SRP: At the time of review, criteria for land treatment or burial (sanitary landfill) disposal practices are subject to significant revision. Prior to implementing land disposal of waste residue (including waste sludge), consult with environmental regulatory agencies for guidance on acceptable disposal practices.

COMMON USES Insecticide synergist, esp for pyrethrum and rotenone [R1].

OCCUPATIONAL RECOMMENDATIONS None listed.

NON-HUMAN TOXICITY LD$_{50}$ rat oral 11.5 g/kg [R5].

ECOTOXICITY LC$_{50}$ Japanese quail (*Coturnix japonica*), 14 days old, oral (5-day ad libitum in diet) >5,000 ppm (technical grade, 100% active ingedient) [R6].

IARC SUMMARY No data are available for humans. Inadequate evidence of carcinogenicity in animals. Overall evaluation: Group 3: The agent is not classifiable as to its carcinogenicity to humans [R4].

BIODEGRADATION No information was located concerning the biodegradation of piperonyl butoxide. However, there are data to indicate that the side chains on the benzodioxole moiety, as well as the benzodioxole itself, biodegrade. It is known that ethoxylate chains degrade, undergoing rapid stepwise removal of the ethoxy groups, during wastewater treatment using activated sludge (1,2,3), and they also degrade under anaerobic conditions using enriched cultures from digester sludge (4). Linear side chains on benzene rings also biodegrade (5,6,7). Biodegradability has also been shown for two compounds, piperonal and piperonylic acid, that are related to piperonyl butoxide (without side chains). Piperonal readily biodegraded in a screening study using an activated sludge seed (8) and piperonylic acid is oxidized by soil bacteria (9). These data suggest that piperonyl butoxide may biodegrade in the environment (SRC) [R7].

BIOCONCENTRATION Using the estimated log K$_{ow}$ of 4.29 (1) for piperonyl butoxide, one would estimate a BCF of 1,100 using a recommended regression equation (2). This would indicate that piperonyl butoxide would bioconcentrate in aquatic organisms (SRC) [R8].

FIFRA REQUIREMENTS In 1988, Congress amended FIFRA to strengthen and accelerate EPA's reregistration program. The nine-year reregistration scheme mandated by "FIFRA 88" applies to each registered pesticide product containing an active ingredient initially registered before November 1, 1984. Pesticides for which EPA had not issued Registration Standards prior to the effective date of FIFRA '88 were divided into three lists based upon their potential for exposure and other factors, with List B being of highest concern, and D of least. List: B; Case: Piperonyl butoxide; Case No. 2525; Pesticide type: insecticide, fungicide, rodenticide, antimicrobial; case status: awaiting data/data in review: OPP awaits data from the pesticide's producer(s) regarding its human health or environmental effects, or OPP has received and is reviewing such data, in order to reach a decision about the pesticide's eligibility for reregistration. Active Ingredient (AI): butylcarbitol (6-propylpiperonyl) ether; Data Call-in (DCI) Date: 05/13/91; AI Status: The producer(s) of the pesticide has made commitments to conduct the studies and pay the fees required for reregistration, and is meeting those commitments in a timely manner [R9].

REFERENCES

1. *The Merck Index.* 9th ed. Rahway, NJ: Merck and Co., Inc., 1976. 973.

2. Worthing, C. R., and S. B. Walker (eds.). *The Pesticide Manual—A World Compendium.* 8th ed. Thornton Heath, UK: The British Crop Protection Council, 1987. 677.

3. Sax, N. I. *Dangerous Properties of Industrial Materials,.* 6th ed. New York, NY: Van Nostrand Reinhold, 1984. 2242.

4. IARC. *Monographs on the Evaluation of the Carcinogenic Risk of Chemicals to Man.* Geneva: World Health Organization, International Agency for Research on Cancer, 1972–present (multivolume work), p. S7 70 (1987).

5. Gosselin, R. E., R. P. Smith, H. C. Hodge. *Clinical Toxicology of Commercial Products.* 5th ed. Baltimore: Williams and Wilkins, 1984, p. II–3310.

6. Hill, E. F., and M. B. Camardese *Lethal Dietary Toxicities of Environmeental Contaminants and Pesticides to Coturnix.* Fish and Wildlife Technical Report 2. Washington, DC: United States Department of Interior, Fish and Wildlife Service, 1986. 118.

7. (1) Giger, W., et al. *Science* 225: 623–5 (1984). (2) Kravetz, L., et al. *Soap Cosmet Chemical Spec* 58: 34–8, 40–2, 102B (1982). (3) Brueschweiler, H., et al. *Tenside Deterg* 20: 317–24 (1983). (4) Dwyer, D. F., J. M. Tiedje. *Appl Environ Microbiol* 46: 185–90 (1983). (5) Babeu, L., D. D. Vaishnav. *J Ind Microbiol* 2: 107–15 (1987). (6) Van Der Linden, A. C. *Dev Biodegrad Hydrocarbons* 1: 165–200 (1978). (7) Marion, C. V., G. W. Malaney, *Proc Ind Waste Cong* 18: 297–308 (1964). (8) Chemicals Inspection and Testing Institute. *Biodegradation and Bioaccumulation Data of Existing Chemicals Based on the CSCL Japan.* Japan Chemical Industry Ecology–Toxicology and Information Center. ISBN 4–89074–101–1 (1992). (9) Vasavada, P. C. *J Environ Sci Health* A11: 213–23 (1976).

8. (1) Meylan, W. M., P. H. Howard. *J Pharm Sci* 84: 83–92 (1995). (2) Lyman, W. J., et al. (eds). *Handbook of Chemical Property Estimation Methods*, New York: McGraw-Hill, Chapt 5, Eqn 5–2 (1982).

9. USEPA/OPP. *Status of Pesticides in Reregistration and Special Review.* p. 153 (Mar, 1992) EPA 700–R–92–004.

■ PIVALIC ACID

CAS # 75–98–9

SYNONYMS 2,2-dimethylpropionic acid • alpha, alpha-dimethylpropionic acid • neopentanoic acid • pivalic acid • propanoic acid, 2,2-dimethyl • tert-pentanoic acid • trimethylacetic acid • Versatic-5

MF $C_5H_{10}O_2$

MW 102.15

COLOR AND FORM Colored crystals.

DENSITY 0.905 at 50°C.

DISSOCIATION CONSTANT K = $9.76×10^{-6}$ at 25°C.

BOILING POINT 163.8°C.

MELTING POINT 35.5°C.

SOLUBILITY 1 g dissolves in 40 ml water; freely soluble in alcohol, ether; water solubility = 21,700 mg/L at 20°C.

OTHER CHEMICAL/PHYSICAL PROPERTIES Henry's Law constant = $1.28×10^{-6}$ atm-m³/mole at 25°C.

COMMON USES Intermediate, as a replacement for some natural materials [R1].

OCCUPATIONAL RECOMMENDATIONS None listed.

NON-HUMAN TOXICITY LD$_{50}$ rat oral between 500 and 5,000 mg/kg [R2].

ECOTOXICITY LD$_{50}$ goldfish 400 to 375 mg/L/(24 to 96 hr) at pH 5 (conditions of bioassay not specified) [R2]; LD$_{50}$ goldfish 4,500 mg/L/24 hr at pH 7 (conditions of bioassay not specified) [R2].

BIODEGRADATION 2,2– Dimethylpropanoic acid at a high initial concentration of 100 ppm exhibited a 24% BODT over an incubation period of 5 days in an aerobic screening study at 21°C using sewage inoculum (1). An aerobic screening study using activated-sludge inoculum at pH 7 showed only a small removal of 2,2–dimethylpropanoic acid (high initial concentration of 500 ppm) in a Warburg respirometer at pH 7 (5). 2,2–Dimethylpropanoic acid exhibited a 10% BODT over an incubation period of 5 days in an aerobic screening study at 20°C using sewage inoculum (2). 2,2–Dimethylpropanoic acid at an initial concentration of 3.4 ppm exhibited a 52% BODT over an incubation period of 10 days in an aerobic screening study at 25°C and pH 6.5–7.5 using sewage inoculum (3). 2,2–Dimethylpropanoic acid at an initial concentration of 4.1 ppm exhibited 1.2, 0, 14, and 86% BODT over respective incubation periods of 2, 5, 10, and 20 days in an aerobic screening study at 25°C using soil inoculum (4,SRC) [R3].

BIOCONCENTRATION Based on a reported water solubility of 21,700 mg/L at 20°C (2) and a regression derived equation (1), a BCF of about 111 can be estimated for 2,2–dimethylpropanoic acid (SRC), which suggests that bioconcentration in aquatic organisms will not be important [R4].

ATMOSPHERIC STANDARDS This action promulgates standards of performance for equipment leaks of volatile organic compounds (VOC) in the synthetic organic chemical manufacturing industry (SOCMI). The intended effect of these standards is to require all newly constructed, modified, and reconstructed SOCMI process units to use the best demonstrated system of continuous emission reduction for equipment leaks of VOC, considering costs, non-air-quality health, and environmental impact and energy requirements. Neopentanoic acid is produced, as an intermediate or a final product, by process units covered under this subpart [R5].

GENERAL RESPONSE Avoid contact with solid and dust. Keep people away. Wear self-contained positive-pressure breathing apparatus and full protective clothing. Stop discharge if possible. Shut off ignition sources. Call fire department. Isolate and remove discharged material. Notify local health and pollution control agencies.

FIRE RESPONSE Combustible. Produces vapors irritating to eyes and skin. Extinguish *Small fires:* dry chemical, CO_2, water spray or foam; large fires: water spray, fog, or foam.

EXPOSURE RESPONSE Call for medical aid. Solid irritating to eyes and skin. Harmful if swallowed. If in eyes or on skin, flush with running water for at least 15 minutes; hold eyelids open if necessary. Wash skin with soap and water. Remove and isolate contaminated clothing and shoes at the site. If swallowed and victim is unconscious or having convulsions, do nothing except keep victim warm.

RESPONSE TO DISCHARGE Issue warning—water contaminant. Should be removed. Chemical and physical treatment.

WATER RESPONSES Dangerous to aquatic life in high concentrations. May be dangerous if it enters water intakes. Notify local health and wildlife officials. Notify operators of nearby water intakes.

EXPOSURE SYMPTOMS Because of low volatility, it is relatively harmless when inhaled at normal ambient temperature (around 20°C). It is slightly toxic by ingestion or skin absorption. The vapor is irritating at elevated temperatures. Can cause considerable discomfort by oral routes; may cause reversible or irreversible changes to exposed tissue, not permanent injury or death.

EXPOSURE TREATMENT Inhalation: remove victim to fresh air, get medical attention if irritation persists. Eyes: hold eyelids open and flush with plenty of water for at least 15 minutes and get medical attention. Skin: contaminated skin should be washed with soap and water.

REFERENCES

1. Sax, N. I., R. J. Lewis, Sr. (eds.). *Hawley's Condensed Chemical Dictionary.* 11th ed. New York: Van Nostrand Reinhold Co., 1987. 1186.

2. Verschueren, K. *Handbook of Environmental Data of Organic Chemicals.* 2nd ed. New York, NY: Van Nostrand Reinhold Co., 1983. 1008.

3. (1) Babeu, L., D. D. Vaishnav. *J Ind Microbiol 2:* 107–15 (1987). (2) Bridie, A. L., et al., *Water Res 13:* 627–30 (1979). (3) Dias, F. F., M. Alexander. *Appl Microbiol 22:* 1114–8 (1971). (4) Hammond, M. W., M. Alexander, *Environ Sci Technol 6:* 732–5 (1972). (5) Mohanrao, G. J., R. E. McKinney. *Intermediate J Air Water Pollut 6:* 153 (1962).

4. (1) Lyman, W. J., et al., *Handbook of Chemical Property Estimation Methods.* Washington DC: Amer Chemical Soc, pp. 4–9, 5–4, 5–10, 7–4, 7–5, 15–15 to 15–32 (1990). (2) Yalkowsky, S. H. *Arizona Data Base of Water Solubility* (1989).

5. 40 CFR 60.489 (7/1/90).

■ POTASSIUM HYDROXIDE

CAS # 1310–58–3

DOT # 1814

SIC CODES 2841; 2268; 2851.

SYNONYMS caustic potash • hydroxyde

de potassium (French) • Kaliumhydroxid (German) • kaliumhydroxyde (Dutch) • lye • potassa • potasse caustique (French) • potassio (idrossido di)- (Italian) • potassium, (hydroxyde de)- (French) • potassium hydrate • potassium hydroxide (K(OH))

MF HKO

MW 56.11

pH 13.5 (0.1–molar aqueous solution).

COLOR AND FORM White or slightly yellow lumps, rods, pellets; sticks, flakes, or fused masses; white rhombic crystals; colorless watery liquid.

DENSITY 2.044 mg/ml.

BOILING POINT 1,324°C.

MELTING POINT 380°C.

EXPLOSIVE LIMITS AND POTENTIAL A piece of potassium hydroxide causes liquid chlorine dioxide to explode. A reaction between N–nitrosomethylurea and potassium hydroxide in n-butyl ether resulted in an explosion due to formation of diazomethane. [R9, p. 491M–173].

FIREFIGHTING PROCEDURES Flood with water, using care not to spatter or splash this material. [R9, p. 49–76].

SOLUBILITY 1 g soluble in 2.5 ml glycerin at 25°C; insoluble in ether, ammonia; 100 g/90 ml water at 25°C; 100 g/375 ml ethanol at 25°C.

VAPOR PRESSURE 1 mm Hg at 714°C.

OTHER CHEMICAL/PHYSICAL PROPERTIES When dissolved in water or alcohol or solution treated with acid, much heat is generated.

PROTECTIVE EQUIPMENT Water bubbler, eye fountains, and showers must be available where skin or eye contact with alkalies is possible. Tight-fitting goggles, rubber aprons, and rubber gloves must be worn when handling alkalies in concentrated solution. Employees must be trained to use safety equipment constantly. [R11, 202].

PREVENTATIVE MEASURES Exercise great care in handling potassium hydroxide, as it rapidly destroys tissue. Do not handle with bare hands [R3].

CLEANUP Spilled material should be flushed away quickly and never be left unattended. If material is solid, it can be shoveled away, and any remaining traces neutralized with dilute acetic acid [R12].

DISPOSAL Neutralization and discharge to sewer: Carefully dissolve in water and neutralize with dilute acetic acid. Flush to sewer with lots of water, regulations permitting, or dispose of through a licensed contractor. Consider use of waste caustic for neutralizing plant acid wastes [R13]. Dilute and neutralize with acid. Caution against acid spitting or extreme heat buildup. Route to sewage plant.

COMMON USES Medication (vet): caustic; in disbudding calves' horns; in aqueous solution to dissolve scales and hair in skin scrapings [R1]; miscellaneous or general-purpose food additive [R4]; electroplating; photoengraving, and lithography; printing inks; in analytical chemistry and in organic synthesis; manufacture of liquid soap; pharmaceutical aid (as alkalizing agent); mordant for woods; absorbing carbon dioxide; mercerizing cotton; paint and varnish removers [R1]; chemical intermediate for potassium carbonate, tetrapotassium pyrophosphate, soaps, and liquid fertilizers; agent in manufacture of dyestuffs and herbicides [R5]; electrolyte in alkaline storage batteries and some fuel cells; absorbent for hydrogen sulfide [R2]; scrubbing and cleaning operations, e.g., industrial gases [R6]; used in petroleum refining [R7]; medication: potassium hydroxide is a powerful caustic that has been used to remove warts. A 2.5% solution in glycerol may be used as a cuticle solvent [R8].

BINARY REACTANTS Acetic acid, acrolein, acrylonitrile, chlorine dioxide, chloroform and methyl alcohol, 1, 2–dichloroethylene, maleic anhydride, nitroethane, nitrogen trichloride, nitromethane, trichloroethylene, water.

STANDARD CODES EPA 311; NFPA—3, 0, 1; ICC—(solution) corrosive liquid, white label, 10 gallon in an outside container;

USCG—hazardous article; IATA—(solid) other restricted articles, class B, no label required, 12-kg passenger, 45-kg cargo; Superfund designated (hazardous substances) list.

PERSISTENCE Natural carbon dioxide will neutralize very slowly.

MAJOR SPECIES THREATENED All species.

INHALATION LIMIT 0.002 mg/m³.

DIRECT CONTACT Severe burns to eyes and skin.

GENERAL SENSATION Warning properties are not good. Forms caustic solution. Inhalation of dust or concentrated mist can damage respiratory tract extensively. Ingestion causes violent pain in throat and epigastrium, hermatemesis, collapse. If not immediately fatal, stricture of esophagus may develop.

PERSONAL SAFETY Do not handle with bare hands. PVA not recommended for gloves [R17]; dilute solutions can cause dermatitis. Wear caustic-resistant outerwear, e.g., neoprene, rubber. Eye protection is also recommended. In closed quarters or high concentrations of dust, breathing apparatus should be worn. In most cases filter masks will suffice.

ACUTE HAZARD LEVEL Strong irritant. Highly toxic when ingested or inhaled. Can be damaging to livestock and aquatic life in 30–200 ppm range. Produces tastes in water at low concentrations.

CHRONIC HAZARD LEVEL Repeated exposures may worsen effects. Strong chronic irritant. Systemic effects unknown.

DEGREE OF HAZARD TO PUBLIC HEALTH Strongly corrosive. Highly toxic with ingestion or inhalation. Strong chronic irritant.

OCCUPATIONAL RECOMMENDATIONS
OSHA Ceiling value of 2 mg/m³. Final rule limits were achieved by any combination of engineering conrols, work practices, and personal protective equipment during the phase-in period, Sept 1, 1989 through Dec 30, 1992. Final rule limits became effective Dec 31, 1992 [R15]. Threshold-limit ceiling limit 2 mg/m³ (1977). [R16].

ACTION LEVEL Notify air authority. Suppress suspension of dust. Restrict access to affected waters.

ON-SITE RESTORATION Neutralize with cation exchanger or acetic acid. Seek professional environmental engineering assistance through EPA's Environmental Response Team (ERT), Edison, NJ, 24-hour no. 908–548–8730.

BEACH OR SHORE RESTORATION Wash with mild acid.

AVAILABILITY OF COUNTERMEASURE MATERIAL Cation exchangers—water softener suppliers; acetic acid—plastic industries, electronic industries.

DISPOSAL NOTIFICATION Contact local sewage authority.

MAJOR WATER USE THREATENED All uses.

PROBABLE LOCATION AND STATE Dissolved in water. White to gray deliquescent crystals. Will dissolve. May be shipped as solution.

WATER CHEMISTRY Forms caustic solution.

FOOD-CHAIN CONCENTRATION POTENTIAL Radioactive potassium (^{42}K) has been concentrated 4–10 times by crabs and aquatic invertebrates. Negative.

NON-HUMAN TOXICITY LD_{50} rat oral 1.23 g/kg [R1].

ECOTOXICITY TLm mosquito fish 80 ppm/24 hr fresh water (conditions of bioassay not specified) [R10].

FIFRA REQUIREMENTS Residues of potassium hydroxide are exempted from the requirement of a tolerance when used as a neutralizer in accordance with good agricultural practices as inert (or occasionally active) ingredients in pesticide formulations applied to growing crops or to raw agricultural commodities after harvest [R14].

GENERAL RESPONSE Avoid contact with liquid. Keep people away. Wear rubber overclothing (including gloves). Stop discharge if possible. Isolate and remove discharged material. Notify local health and pollution control agencies.

FIRE RESPONSE Not flammable.

EXPOSURE RESPONSE Call for medical aid. Liquid will burn skin and eyes. Harmful if swallowed. Remove contaminated clothing and shoes. Flush affected areas with plenty of water. If in eyes, hold eyelids open, and flush with plenty of water. If swallowed and victim is conscious, have victim drink water or milk. Do not induce vomiting.

RESPONSE TO DISCHARGE Issue warning—corrosive. Restrict access. Disperse and flush.

WATER RESPONSE Harmful to aquatic life in very low concentrations. May be dangerous if it enters water intakes. Notify local health and wildlife officials. Notify operators of nearby water intakes.

EXPOSURE SYMPTOMS Causes severe burns of eyes, skin, and mucous membranes.

EXPOSURE TREATMENT Act quickly. Eyes: flush with water for at least 15 min. Skin: flush with water, then rinse with dilute vinegar (acetic acid). Ingestion: give water and milk. Do not induce vomiting. Call physician at once, even when injury seems to be slight.

REFERENCES

1. Budavari, S. (ed.). *The Merck Index—Encyclopedia of Chemicals, Drugs, and Biologicals.* Rahway, NJ: Merck and Co., Inc., 1989. 1215.

2. Sax, N. I., and R. J. Lewis, Sr. (eds.). *Hawley's Condensed Chemical Dictionary.* 11th ed. New York: Van Nostrand Reinhold Co., 1987. 956.

3. Osol, A., J. E. Hoover, et al. (eds.). *Remington's Pharmaceutical Sciences.* 15th ed. Easton, PA: Mack Publishing Co., 1975. 723.

4. Furia, T. E. (ed.). *CRC Handbook of Food Additives.* 2nd ed. Cleveland: The Chemical Rubber Co., 1972. 925.

5. SRI.

6. *Kirk–Othmer Encyclopedia of Chemical Technology.* 3rd ed., Volumes 1–26. New York, NY: John Wiley and Sons, 1978–1984, p. V18 939.

7. *Chemcyclopedia* 1986 p. 169.

8. Reynolds, J.E.F., A. B. Prasad (eds.) *Martindale—The Extra Pharmacopoeia.* 28th ed. London: The Pharmaceutical Press, 1982. 44.

9. National Fire Protection Association. *Fire Protection Guide on Hazardous Materials.* 9th ed. Boston, MA: National Fire Protection Association, 1986.

10. U.S. Coast Guard, Department of Transportation. *CHRIS—Hazardous Chemical Data.* Volume II. Washington, DC: U.S. Government Printing Office, 1984–5.

11. International Labour Office. *Encyclopedia of Occupational Health and Safety.* Vols. I and II. Geneva, Switzerland: International Labour Office, 1983.

12. International Labour Office. *Encyclopedia of Occupational Health and Safety.* Volumes I and II. New York: McGraw-Hill Book Co., 1971. 78.

13. United Nations. *Treatment and Disposal Methods for Waste Chemicals* (IRPTC File). Data Profile Series No. 5. Geneva, Switzerland: United Nations Environmental Programme, Dec. 1985. 266.

14. 40 CFR 180.1001 (c) (7/1/88).

15. 54 FR 2920 (1/19/89).

16. American Conference of Governmental Industrial Hygienists. *Threshold Limit Values for Chemical Substances and Physical Agents and Biological Exposure Indices for 1993–1994.* Cincinnati, OH: ACGIH, 1993. 30.

17. Gauerke, J. R., "Work Gloves to Meet OSHA Rules," *Chemical Engineering,* April 3, 1972.

■ POTASSIUM IODIDE

CAS # 7681–11–0

SIC CODES 2042; 2833.

SYNONYMS Asmofug-e • dipotassium

diiodide • iodure de potassium • joptone • k1-n • kali-iodide • kalii-iodidum • knollide • potassium iodide [$K_2 I_2$] • potassium iodide [$K_3 I_3$] • potassium monoiodide • tripotassium triiodide

MF IK

MW 166.01

pH Aqueous solution is neutral or usually slightly alkaline, pH 7–9.

COLOR AND FORM Colorless or white, cubical crystals, white granules, or powder; hexahedral crystals, either transparent or somewhat opaque.

TASTE Strong, bitter, saline.

DENSITY 3.13.

BOILING POINT 1,330°C.

MELTING POINT 680°C.

EXPLOSIVE LIMITS AND POTENTIAL A sample of fluorine perchlorate exploded on contact with a potassium iodide solution [R7].

SOLUBILITY 1 g soluble in 0.7 ml water; 0.5 ml boiling water; 1 g soluble in 51 ml absolute alcohol; 8 ml methanol; 1 g soluble in 75 ml acetone; 2 ml glycerol; 1 g soluble in about 2.5 ml glycol; 127.5 g soluble in 100 cc water at 0°C; 1.88 g soluble in 100 cc alcohol at 25°C; 1.31 g soluble in 100 cc acetone at 25°C; soluble in ether, ammonia.

STABILITY AND SHELF LIFE Stable in dry air [R3, 801].

OTHER CHEMICAL/PHYSICAL PROPERTIES Slightly deliquescent in moist air; on long exposures to air becomes yellow due to liberation of iodine, and small quantities of iodate may be formed.

COMMON USES Manufacture of photographic emulsions; in animal and poultry feeds to the extent of 10–30 ppm; in table salts and in some drinking water [R2]; vet: in actinobacillosis, actinomycosis; for simple goiter; in iodine deficiency, and in chronic poisoning with lead or mercury; orally only, not by injection; externally for treatment of bursal enlargements [R2]; photography (precipitating silver); spectroscopy; infrared transmission; scintilla-tion [R5]; in treatment of hyperthyroidism (Lugol's solution—also used as a fixative for histology) [R4]; topical deodorizing agent for livestock manure [R6]; expectorant for treatment of chronic respiratory diseases; antifungal agent (human toxicity and vet use); iodine source in treatment of thyroid disorders; manufacture of photographic emulsions (used with silver nitrate); ingredient in personal hygiene products; lab reagent for analytical chemistry [R1].

STANDARD CODES No NFPA code.

PERSISTENCE Oxidizes to iodine with exposure to light, moisture, and air.

GENERAL SENSATION Prolonged absorption of iodides may produce iodism, which is manifested by a skin rash, running nose, headache, and irritation of mucous membranes; with more severe exposure, pimples, boils, redness, black-and-blue spots, hives and blisters. Weakness, anemia, loss of weight, and general depression also may occur.

PERSONAL SAFETY Wear filter-type mask.

ACUTE HAZARD LEVEL Moderately toxic when ingested or inhaled. Toxic to aquatic invertebrates. Very low taste threshold.

CHRONIC HAZARD LEVEL Moderately toxic with ingestion or inhalation.

DEGREE OF HAZARD TO PUBLIC HEALTH Moderate ingestive and inhalative threat. Emits toxic iodides when heated to decomposition. Moderate chronic exposure hazard with ingestion or inhalation.

OCCUPATIONAL RECOMMENDATIONS None listed.

ACTION LEVEL Suppress suspension of dusts.

ON-SITE RESTORATION Use anion exchanger. Aerate to increase oxidation to iodine. Seek professional environmental engineering assistance through EPA's Environmental Response Team (ERT), Edison, NJ, 24–hour no. 908–548–8730.

AVAILABILITY OF COUNTERMEASURE MATERIAL Anion exchanger—water softener suppliers.

DISPOSAL Landfill or return to supplier for recovery.

MAJOR WATER USE THREATENED Potable supply, fisheries.

PROBABLE LOCATION AND STATE Colorless crystal or white granule. Will dissolve. White at first. Yellows with time.

FOOD-CHAIN CONCENTRATION POTENTIAL Both potassium and iodine can be concentrated by plants. Positive; iodine has been concentrated by factors of 10,000 in marine algae. Crabs and invertebrates have concentrated potassium 4–10 times.

GENERAL RESPONSE Stop discharge if possible. Keep people away. Isolate and remove discharged material. Notify local health and pollution control agencies.

FIRE RESPONSE Not flammable.

EXPOSURE RESPONSE Call for medical aid. Solid is harmful if swallowed. Flush affected areas with plenty of water. If swallowed and victim is conscious, have victim drink water or milk.

RESPONSE TO DISCHARGE Issue warning—water contaminant. Disperse and flush.

WATER RESPONSE Effect of low concentrations on aquatic life is unknown. May be dangerous if it enters water intakes. Notify local health and wildlife officials. Notify operators of nearby water intakes.

EXPOSURE SYMPTOMS May irritate eyes or open cuts.

EXPOSURE TREATMENT Flush all affected areas with water.

REFERENCES

1. SRI.

2. *The Merck Index.* 9th ed. Rahway, NJ: Merck and Co., Inc., 1976. 992.

3. Osol, A., and J. E. Hoover, et al. (eds.). *Remington's Pharmaceutical Sciences.* 15th ed. Easton, PA: Mack Publishing Co., 1975.

4. *American Hospital Formulary Service.* Volumes I and II. Washington, DC: American Society of Hospital Pharmacists, to 1984, p. 48: 00.

5. Hawley, G. G. *The Condensed Chemical Dictionary.* 9th ed. New York: Van Nostrand Reinhold Co., 1977. 714.

6. Riley, H.W.; U.S. Patent Number 4155975 05/22/79.

7. National Fire Protection Association. *Fire Protection Guide on Hazardous Materials.* 7th ed. Boston, MA: National Fire Protection Association, 1978, p. 491M–189.

8. Shelford, V. E., "An Experimental Study of the Effects of Gas Waste upon Fishes with Special Reference to Stream Pollution," *Bull. 111. State Lab. Nat. Hist.,* 1917, 11: 381–412.

■ PROMETONE

CAS # 1610–18–0

SYNONYMS 1,3,5-triazine-2,4-diamine, 6-methoxy-N,N′-bis(1-methylethyl)- • 2,4-bis(isopropylamino)-6-methoxy-s-triazine • 2,6-diisopropylamino-4-methoxytriazine • 2-methoxy-4,6-bis (isopropylamino)-1,3,5-triazine • 2-methoxy-4,6-bis (isopropylamino)-s-triazine • g-31435 • g-31435 • gesafram • gesafram-50 • methoxypropazine • ontracic-800 • ontrack • ontrack-we-2 • primatol • pramitol-5p • prometon • prometrone • s-triazine, 2,4-bis(isopropylamino)-6-methoxy • N,N′-diisopropyl-6-methoxy-1,3,5-triazine-2,4-diamine • 2,4-bis(isopropylamino)-6-methoxy-1,3,5-triazine • 6-methoxy-N,N′-bis(1-methylethyl)-1,3,5-triazine-2,4-diamine

MF $C_{10}H_{19}N_5O$

MW 225.34

COLOR AND FORM Colorless powder; white, crystalline.

DENSITY 1.088 g/cm³ at 20°C.

DISSOCIATION CONSTANT pK$_a$: 4.3 at 21°C.

MELTING POINT 91–92°C.

FIREFIGHTING PROCEDURES Extinguish fire using agent suitable for type of surrounding fire. Use water in flooding quantities as fog. Use "alcohol foam" Wear positive-pressure self-contained breathing apparatus when fighting fires involving this material—triazine pesticide (compounds and preparations), solid (insecticides, agricultural, not elsewhere classified, other than liquid)) [R3]. Avoid breathing dusts and fumes from burning material. Keep upwind (triazine pesticide, solid, not otherwise specified (compounds, and preparations) (agricultural insecticides, not elsewhere classified, other than liquid) [R3].

SOLUBILITY 750 mg/L water at 20°C; >500 g/L acetone at 20°C; readily soluble in chloroform; freely soluble in alcohol; 150 g/L octan–1–ol; >250 g/L benzene at 20°C; 250 g/L toluene at 20°C; >500 g/L methanol at 20°C.

VAPOR PRESSURE 2.3×10^{-6} mm Hg at 20°C.

CORROSIVITY Noncorrosive under common application conditions.

STABILITY AND SHELF LIFE Stable to hydrolysis at 20°C in neutral, alkaline, or slightly acidic media [R6, 698].

OTHER CHEMICAL/PHYSICAL PROPERTIES Heating prometone at 260°C produces isomerization and dismutation products.

PROTECTIVE EQUIPMENT The use of personal protective equipment such as glasses, synthetic gloves, and NIOSH–approved breathing apparatus is important. (herbicides) [R4].

PREVENTATIVE MEASURES Avoid inhalation of dust. Avoid contact with skin. [R7, 392].

DISPOSAL SRP: At the time of review, criteria for land treatment or burial (sanitary landfill) disposal practices are subject to significant revision. Prior to implementing land disposal of waste residue (including waste sludge), consult with environmental regulatory agencies for guidance on acceptable disposal practices.

COMMON USES Nonselective preemergence and postemergence herbicide that controls most annual and perennial broadleaf and grassy weeds [R7, 391]. A non-selective herbicide used for control of most annual and perennial broad-leaved, grass, and brush weeds on non-crop areas at 10–20 kg AI/ha and may be applied to the ground before laying asphalt [R6, 698]. Prometone has been used for control of deeprooted perennial grasses such as bermuda [R2].

OCCUPATIONAL RECOMMENDATIONS None listed.

NON-HUMAN TOXICITY LD$_{50}$ rat oral 2,980 mg/kg (technical) [R1]; LD$_{50}$ rabbit percutaneous >2,000 mg/kg (technical) [R1]; LC$_{50}$ rat inhalation >3.26 mg/L/4 hr (technical) [R1]; LD$_{50}$ rat oral 2,276 mg/kg (pramitol 25E) [R1]; LD$_{50}$ rabbit percutaneous 2,200 mg/kg (pramitol 25E) [R1]; LC$_{50}$ rat inhalation 36.0 mg/L/4 hr (pramitol 25E) [R1]; LD$_{50}$ rabbit percutaneous 1,500–2,000 mg/kg, abraded skin [R7, 393].

ECOTOXICITY LC$_{50}$ *Salmo gairdneri* (Rainbow trout) 20 ppm/96 hr (technical; conditions of bioassay not specified) [R7, 393]; LC$_{50}$ Crucian carp 70 mg/L/96 hr (conditions of assay not given) [R6, 698]; LC$_{50}$ *Lepomis macrochirus* (bluegill sunfish) >32 ppm/96 hr (technical; conditions of assay not given) [R7, 393]; LC$_{50}$ *Carassius auratus* (goldfish) 8.6 mg/L/96 hr (conditions of bioassay not specified) [R1]; LC$_{50}$ *Colinus virginianus* (bobwhite quail) chicks oral >5,080 ppm (8-day diet) (pramitol 25E) [R7, 393]; LC$_{50}$ *Anas platyrhynchos* (mallard duck), oral 4,572 ppm (8-day diet) (pramitol 25E) [R7, 393].

BIODEGRADATION Microbial action probably accounts for the major breakdown of prometone in soil. A range of soil microorganisms can utilize it as source of energy and nitrogen. [R7, 392].

BIOCONCENTRATION The BCF of prometone estimated from its water solubility, using a recommended regression equation, is 15 (1). Therefore prometone would not be expected to bioconcentrate in aquatic organisms (SRC) [R5].

DOT *Health Hazards:* Poisonous if swallowed. Inhalation of dust poisonous. Fire may produce irritating or poisonous gases. Runoff from fire control or dilution water may cause pollution (triazine pesticide, solid, not otherwise specified).

Fire or Explosion: Some of these materials may burn, but none of them ignites readily (triazine pesticide, solid, not otherwise specified).

Emergency Action: Keep unnecessary people away; isolate hazard area and deny entry. Stay upwind; keep out of low areas. Self-contained breathing apparatus (SCBA) and structural firefighter's protective clothing will provide limited protection. Call CHEMTREC at 1–800–424–9300 for emergency assistance. If water pollution occurs, notify the appropriate authorities (triazine pesticide, solid, not otherwise specified).

Fire: Small fires: Dry chemical, CO_2, halon, water spray, or standard foam. Large Fires: Water spray, fog, or standard foam is recommended. Move container from fire area if you can do so without risk (triazine pesticide, solid, not otherwise specified).

Spill or Leak: Do not touch spilled material; stop leak if you can do so without risk. Small Spills: Take up with sand or other noncombustible absorbent material and place into containers for later disposal. Small Dry Spills: With clean shovel place material into clean, dry container, and cover; move containers from spill area. Large Spills: Dike far ahead of liquid spill for later disposal (triazine pesticide, solid, not otherwise specified).

First Aid: Move victim to fresh air; call emergency medical care. Remove and isolate contaminated clothing and shoes at the site. In case of contact with material, immediately flush skin or eyes with running water for at least 15 minutes (triazine pesticide, solid, not otherwise specified).

FIRE POTENTIAL Nonflammable [R7, 392].

REFERENCES
1. *Farm Chemicals Handbook* 1989. Willoughby, OH: Meister Publishing Co., 1989, p. C–235.

2. White–Stevens, R. (ed.). *Pesticides in the Environment,* Volume 1, Part 1, Part 2. New York: Marcel Dekker, Inc., 1971. 60.

3. Association of American Railroads. *Emergency Handling of Hazardous Materials in Surface Transportation.* Washington, DC: Assoc. of American Railroads, Hazardous Materials Systems (BOE), 1987. 692.

4. International Labour Office. *Encyclopedia of Occupational Health and Safety.* Vols. I and II. Geneva, Switzerland: International Labour Office, 1983. 1039.

5. Kenega, E. E. *Ecotoxicol Environ Safety* 4: 26–38 (1980).

6. Worthing, C.R., and S.B. Walker (eds.). *The Pesticide Manual—A World Compendium.* 8th ed. Thornton Heath, UK: The British Crop Protection Council, 1987.

7. Weed Science Society of America. *Herbicide Handbook.* 5th ed. Champaign, Illinois: Weed Science Society of America, 1983.

■ PROPACHLOR

CAS # 1918–16–7

SYNONYMS 2-chloro-N-isopropyl-N-phenylacetamide • 2-chloro-N-isopropylacetanilide • acetamide, 2-chloro-N-(1-methylethyl)-N-phenyl • acetanilide, 2-chloro-N-isopropyl • acilid • alpha-chloro-N-isopropylacetanilide • bexton-41 • cp-31393 • kartex-a • N-isopropyl-2-chloroacetanilide • N-isopropyl-alpha-chloroacetanilide • niticid • propachlore • ramrod • ramrod-65 • satecid • 2-chloro-N-(1-

methylethyl)-N-phenylacetamide • prolex • propachlore • 2-chloro-N-isopropyl acetanilide

MF $C_{11}H_{14}ClNO$

MW 211.71

COLOR AND FORM Light-tan solid; white, crystalline solid.

DENSITY 1.242 g/ml at 25°C.

BOILING POINT 110°C at 0.03 mm Hg.

MELTING POINT 67–76°C.

SOLUBILITY Soluble in common organic solvents except aliphatic hydrocarbons; at 25°C: 44.8% in acetone, 73.7% in benzene, 17.4% in carbon tetrachloride, 40.8% in ethyl alcohol, 23.9% in xylene; 60.2% in chloroform, 21.9% in ether, 34.2% in toluene; 613 ppm by wt in water; slightly soluble in diethyl ether.

VAPOR PRESSURE 2.3×10^{-4} mm Hg at 25°C.

CORROSIVITY No corrosion to number 316 and 304 stainless steel, aluminum, and heresite; corrosive to ordinary steel.

STABILITY AND SHELF LIFE Stable for at least 4 years, not sensitive to light [R8, 403]. Stable to UV light. [R3].

OTHER CHEMICAL/PHYSICAL PROPERTIES Decomposition temp: 170°C; vapor pressure at 110°C: 0.03 mm Hg; mp: 77°C (pure propachlor); decomposed in alkaline and strongly acidic media; log P = 2.80 (calc); Henry's Law constant = 1.09×10^{-7} atm m³/mole at 20°C.

PROTECTIVE EQUIPMENT Wear rubber gloves and wash thoroughly after handling. Contaminated clothing should be laundered before reuse. [R8, 403].

DISPOSAL Propachlor should be incinerated in a unit operating at 850°C equipped with off- gas-scrubbing equipment. Recommendable method: incineration [R5].

COMMON USES Herbicide for grasses and broadleaf weeds on corn, sorghum, soybeans, sugar beets, and other field crops, forage crops, pasture land, and range land, summer fallow, and on vegetables [R1]. Herbicide effective against annual grasses and some broad-leaved weeds in beans, brassicas, cotton, groundnuts, leeks, maize, onions, peas, roses, ornamental trees, and shrubs, and sugarcane [R4]. Selective herbicide for control of annual grasses, and some annual broad-leaved weeds in garlic, fennel, flower crops, flax, oilseed rape, strawberries, pumpkins [R3].

OCCUPATIONAL RECOMMENDATIONS None listed.

NON-HUMAN TOXICITY LD_{50} rabbit percutaneous >20,000 mg/kg [R4]; LD_{50} rat oral 1,800 mg/kg (technical propachlor) [R4].

ECOTOXICITY LD_{50} pheasant oral 735 mg/kg (technical material) [R6]; LC_{50} mallard duck dietary greater than 5,000 ppm/8 day (technical propachlor) [R8, 403]; LC_{50} bobwhite quail dietary greater than 5,000 ppm/8 day (technical propachlor) [R8, 403]; TLM fathead minnows 0.49 mg/L/96 hr (technical material) (conditions of bioassay not specified) [R6]; TLM bluegill fingerling 1.30 mg/L/96 hr (technical material) (conditions of bioassay not specified) [R6]; LC_{50} rainbow trout 0.17 mg/L/96 hr (conditions of bioassay not specified) [R3]; LC_{50} bluegill sunfish >1.4 mg/L/96 hr (conditions of bioassay not specified) [R3].

BIODEGRADATION Rapidly degradable; AI disappears in 4–6 weeks in light soil [R2]. A pure culture of *Fusarium oxyporium* degrades propachlor (1–2). Dehalogenation was the major degradation mechanism and 2-hydroxy-N-isopropylacetanilide and bis(N-isopropylanilido-N-carboxymethylene) oxide were identified as metabolites(2). Low concentration range of propachlor (µg/L range) was not mineralized in a sewage and a lake water, but was extensively metabolized to a number of unidentified polar products (3). In a lake water, 59% of propachlor metabolized in 6 weeks (3). At higher concentration (mg/L range), the second-order biodegradation rate constant for propachlor was estimated to be 1.1×10^{-9} L/organism-hr (4) [R9]. Propachlor degraded about 33 times faster in a non-sterilized silt soil than the sterilized soil, leading to the conclusion that microbial transformation is the major deg-gradation pathway for propachlor (1). The degradation proceeds via microbial hydrolysis with the formation of hydroxypropachlor or N-isopropylaniline (2,7). The hydrolyzed products may form conjugated compounds with active soil sites and may remain strongly sorbed to these active sites (2,6). Identified propachlor metabolites in soil are N-isopropylaniline, N-isopropylanilide, N-(1-hydroxy-isopropyl) acetanilide and N-isopropyl-2-acetoxy-acetanilide (3). In addition to degradation via cometabolism, propachlor also mineralizes in soil (4). Soil suspensions mineralized 0.6–63% of ring-labeled propachlor to carbon dioxide in 30 days (4). It was suggested that mineralization is the major means for destruction of propachlor in soil (4). Increase in moisture content and temperature of soil enhanced the biodegradation of propachlor (5). At high concentration (500 mg/kg), the biodegradation of propachlor was relatively slower compared to lower concn (6). [R10].

BIOCONCENTRATION In a model ecosystem study, the bioconcentration factor (BCF) for propachlor in mosquito fish (*Gambusia affinis*) was detected to be approximately 1 (1–2). Based on water solubility and K_{oc}, and regression equations, the BCF for propachlor was estimated to be 14 to 17 (3). Therefore, bioconcentration of propachlor in aquatic-ecosystem organisms will not be important (3,SRC) [R7].

FIFRA REQUIREMENTS Tolerances are established for residues of the herbicide 2-chloro-N-isopropylacetanilide and its metabolites (calculated as 2-chloro-N-isopropylacetanilide) in or on the following raw agricultural commodities: beets (sugar, roots); beets (sugar, tops); cattle (fat, mbyp, and meat); corn (forage); corn (grain and sweet); cottonseed; eggs; flax (seed); flax (straw); goats (fat, mbyp, and meat); hogs (fat, mbyp, and meat); horses (fat, mbyp, and meat); milk; peas (with pods, determined on peas after removing any pod present when marketed); peas (forage); poultry (fat, mbyp, and meat); pumpkins; sheep (fat, mbyp, meat); sorghum (fodder and forage); and sorghum (grain) [R11]. In 1988, Congress amended FIFRA to strengthen and accelerate EPA's rere-

gistration program. The nine-year reregistration scheme mandated by "FIFRA 88" applies to each registered pesticide product containing an active ingredient initially registered before November 1, 1984. List A consists of the 194 chemical cases (or 350 individual active ingredients) for which EPA had issued Registration Standards prior to the effective date of FIFRA '88. List: A; Case: Propachlor; Case No.: 0177; Pesticide type: Herbicide; Registration Standard Date: 12/01/84; Case Status: Awaiting Data/Data in Review: OPP awaits data from the pesticide's producer(s) regarding its human health and/or environmental effects, or OPP has received and is reviewing such data, in order to reach a decision about the pesticide's eligibility for reregistration. Active Ingredient (AI): Propachlor; Data Call-in (DCI) Date: 09/23/91; AI Status: The producer(s) of the pesticide has made commitments to conduct the studies and pay the fees required for reregistration, and is meeting those commitments in a timely manner. [R12].

FIRE POTENTIAL Fire point, 316°C [R8, 403], nonflammable [R4].

REFERENCES

1. SRI.

2. *Farm Chemicals Handbook 1992.* Willoughby, OH: Meister Publishing Co., 1992, p. C–276.

3. Hartley, D., and H. Kidd (eds.). *The Agrochemicals Handbook.* 2nd ed. Lechworth, Herts, England: The Royal Society of Chemistry, 1987, p. A341/Aug 87.

4. Worthing, C. R., and S. B. Walker (eds.). *The Pesticide Manual—A World Compendium.* 8th ed. Thornton Heath, UK: The British Crop Protection Council, 1987. 702.

5. United Nations. *Treatment and Disposal Methods for Waste Chemicals (IRPTC File).* Data Profile Series No. 5. Geneva, Switzerland: United Nations Environmental Programme, Dec. 1985. 66.

6. Weed Science Society of America. *Herbicide Handbook.* 4th ed. Champaign, IL:

Weed Science Society of America, 1979. 374.

7. (1) Yu, C–C., et al. *J Agric Food Chem* 23: 877–79 (1975). (2) Metcalf, R. L. *Adv Environ Sci Technol* 8: 195–221 (1977). (3) Kenaga, E. E., *Ecotox Environ Safety* 4: 26–38 (1980).

8. Weed Science Society of America. *Herbicide Handbook.* 5th ed. Champaign, Illinois: Weed Science Society of America, 1983.

9. (1) Munnecke, D. M., *FEMS Symp 12 (Microb Degrad Xenobiotics Recalcitrant Compounds).* 251–69 (1981). (2) Jaworski, E. G., pp. 359–76 in *Herbicides.* Vol 1. Kearney, P. C., D. D. Kaufman (eds.). New York: Marcel Dekker, Inc (1975). (3) Novick, N. J., M. Alexander. *Appl Environ Microbiol* 49: 737–43 (1985). (4) Steen, W. C., T. W, Collette. *Appl Environ Microbiol* 55: 2545–49 (1989).

10. (1) Beestman, G. B., J. M. Deming. *Agron J* 66: 308–11 (1974). (2) Frank, R., et al. *Can J Plant Sci* 57: 473–77 (1977). (3) Lee, J. K., et al. *J Korean Agric Chem Soc* 25: 44–54 (1982). (4) Novick, N. J., et al. *J Agric Food Chem* 34: 721–5 (1986). (5) Zimdahl, R. L., S. K. Clark. *Weed Sci* 30: 545–8 (1982). (6) USEPA. *Drinking Water Health Advisory: Pesticides,* Chelsea, MI: Lewis Publ., Inc. p. 665–76 (1989). (7) Clegg, B. S., et al. *Bull Environ Contam Toxicol* 47: 104–11 (1991).

11. 40 CFR 180.211 (7/1/91).

12. USEPA/OPP. *Status of Pesticides in Reregistration and Special Review.* p. 110 (Mar, 1992) EPA 700-R-92-004.

■ PROPANE, DICHLORO– mixed with PROPENE, DICHLORO–

CAS # 8003–19–8

DOT # 2047

SIC CODES 2879.

SYNONYMS 1-propene, 1,3-dichloro-, mixture with 1,2-dichloropropane • ent-8,420 • 1,3-dichloropropene and 1,2-dichloropropane mixture • Dichlorpropan-dichlorpropengemisch (German) • propane, dichloro- mixed with propene, dichloro- • mixture of 1,3-dichloropropane, 1,3-dichloropropene, and related C3 compounds • dowfume-n • d-d

MF $C_3H_6Cl_2 \cdot C_3H_4Cl_2$

MW 223.96

COLOR AND FORM Clear amber liquid.

ODOR Pungent.

DENSITY About 1.4 at 4°C.

FIREFIGHTING PROCEDURES Do not extinguish fire unless flow can be stopped. Use water in flooding quantities as fog. Solid streams of water may spread fire. Cool all affected containers with flooding quantities of water. Apply water from as great a distance as possible. Use foam, dry chemical, or carbon dioxide. (dichloropropene and propylene dichloride mixture) [R5].

TOXIC COMBUSTION PRODUCTS Toxic and irritating gases (of phosgene and HCl) may be generated. (dichloropropene) [R6].

SOLUBILITY About 2 g/kg water at room temp; fully miscible with esters, halogenated solvents, hydrocarbons, ketones.

VAPOR PRESSURE 4.6 kPa at 20°C.

CORROSIVITY Corrosive to iron, aluminum, magnesium, and other metals and alloys.

STABILITY AND SHELF LIFE Stable in neutral and dilute acidic media; decomposition by alkalis, concentrated acids, halogens, and some metal salts [R3].

OTHER CHEMICAL/PHYSICAL PROPERTIES It flash-distills over range of 59–115°C.

PROTECTIVE EQUIPMENT Wear appropriate chemical protective gloves, boots, and goggles. Wear positive-pressure self-contained breathing apparatus when fighting fires involving this material. (dichloropropene and propylene dichloride mixture) [R5].

PREVENTATIVE MEASURES Do not cut or weld container [R4].

CLEANUP Land spill: Dig a pit, pond, lagoon, or holding area to contain liquid or solid material. (SRP: If time permits, pits, ponds, lagoons, soak holes, or holding areas should be sealed with an impermeable flexible-membrane liner.) Dike surface flow using soil, sandbags, foamed polyurethane, or foamed concrete. Absorb bulk liquid with fly ash, cement powder, sawdust, or commercial sorbents. Apply appropriate foam to diminish vapor and fire hazard. (dichloropropene and propylene dichloride mixture) [R5].

DISPOSAL Group I containers: combustible containers from organic or metallo-organic pesticides (except organic mercury, lead, cadmium, or arsenic compounds) should be disposed of in pesticide incinerators or in specified landfill sites (organic or metallo-organic pesticides) [R7]. Use an absorbent on the spill. Collect material and contaminated soil, place in lined metal drums and ship back to the supplier. Material may also be incinerated in a chemical incinerator. If incineration is not available or practical, material may be buried in a specially designated chemical landfill.

COMMON USES The mixture is a pre-plant nematicide effective against nematodes including root knot, meadow, sting, dagger, spiral, and sugar beet nematodes [R2, 263]; dorlone (Dow Chemical) is designed for use against mixed-nematode infestation, such as stylet and lesion nematodes of tobacco, by row treatment with at least 6 U.S. gal/acre. [R1, 197].

BINARY REACTANTS Dilute inorganic bases; concentrated acids; metal salts; halogens; active metals.

STANDARD CODES EPA 311; TSCA; IATA corrosive material, corrosive label, 1 L passenger, 40 L cargo; DOT flammable liquid, flammable liquid label; AAR, Bureau of Explosives corrosive material, flammable, STCC 4936329; CFR 14 CAB code 8; Superfund designated (hazardous substances) list.

PERSISTENCE Nonpersistent. The dichloropropene component of the mixture is biodegraded in the soil. The degradation products are strongly bound to the soil particles and are unlikely to be mobile in the soil water and should not leach into ground waters. There is little evidence of biodegradation of the dichloropropane component, but it is volatile.

INHALATION LIMIT 350 mg/m^3.

DIRECT CONTACT Severe skin irritation, will be absorbed by skin. Vapors have very strong odor that warns of danger and reduces the exposure hazard.

GENERAL SENSATION Ingestion will cause acute gastrointestinal distress, with congestion and edema of lungs. Inhalation will cause gasping, refusal to breathe, respiratory distress. Eye and upper respiratory irritation appears soon after exposure to vapors. Repeated exposure may give rise to central nervous system depression. Dermal exposure will cause severe skin irritation with marked inflammation of skin and cutaneous tissue.

PERSONAL SAFETY Protect against both inhalation and absorption through the skin. Must wear protective clothing, self-contained breathing apparatus, or full facemask equipped with a canister approved by NIOSH for organic vapors. Use canister for less than 1 hour or until odor comes through, whichever occurs first. Material will penetrate ordinary rubber protective equipment such as boots and gloves, so cover shoes and hands with polyethylene bags. Wear spark-proof boots.

ACUTE HAZARD LEVEL D–d mixture is very toxic to mammals by ingestion and inhalation, and moderately toxic by skin absorption. Between 1 teaspoonful and 1 ounce may be fatal to humans. Its odor and intense irritation of eyes, skin, and respiratory mucosa warn of danger and reduce the exposure hazard. The mean lethal dose in rats is about 140 mg/kg by ingestion. Direct skin contact will lead to severe irritation. The 96-hour LC$_{50}$ to tidewater silversides is 240 ppm, based on 1,2–dichloropropane.

CHRONIC HAZARD LEVEL Chronic exposure can lead to central nervous system depression, and liver and kidney damage.

Inhalation is the chief hazard, and symptoms abate promptly after exposure ceases.

DEGREE OF HAZARD TO PUBLIC HEALTH
Vapors irritating to respiratory tract, eyes, skin, digestive tract. Contact with liquid will cause burns to exposed surfaces. May cause liver and kidney damage.

OCCUPATIONAL RECOMMENDATIONS None listed.

ACTION LEVEL Avoid contact with the spilled material. Eliminate all ignition sources. Stay upwind. Notify local air, water, and fire authorities. Evacuate all people to a distance of at least 200 feet upwind and 1,000 feet downwind of the spill.

ON-SITE RESTORATION Dam stream to reduce flow and prevent further dissipation by water movement. Apply activated carbon to adsorb dissolved material. Bottom pumps or underwater vacuum systems may be employed to remove any undissolved liquid and adsorbent from the bottom. Seek professional environmental engineering assistance through EPA's Environmental Response Team (ERT), Edison, NJ, 24–hour no. 908–548–8730.

BEACH OR SHORE RESTORATION Close beach and shore to the public until material has been removed. If tidal, scrape affected area at low tide with a mechanical scraper and avoid human contact. Do not burn material.

AVAILABILITY OF COUNTERMEASURE MATERIAL Carbon—water treatment plants, chemical companies; vacuum systems—swimming-pool operators; bottom pumps—fire departments, Army Corps of Engineers.

DISPOSAL NOTIFICATION Notify local and state health authorities, local solid waste disposal authorities, supplier, and shipper of material.

PROBABLE LOCATION AND STATE Straw- to amber-colored liquid; will sink; may form a layer above the bottom sediments and slowly dissolve. Color in water, straw to amber.

NON-HUMAN TOXICITY LD$_{50}$ rat percuta-neous 779 mg/kg [R8]; LD$_{50}$ mouse oral 314 mg/kg [R2, 264]; LD$_{50}$ rat skin 2,100 mg/kg [R9]; LD$_{50}$ rabbit skin 2,100 mg/kg [R9]; LD$_{50}$ rat, oral 140 mg/kg [R9].

ECOTOXICITY LC$_{50}$ *Salmo clarki* (cutthroat trout) 1.0–10.0 mg/L/96 hr at 12°C, wt 1.0 g (technical material, 100%). Static bioassay without aeration, pH 7.2–7.5, water hardness 40–50 mg/L as calcium carbonate, and alkalinity of 30–35 mg/L [R10]. LC$_{50}$ *Salmo gairdneri* (rainbow trout) 5.5 mg/L/96 hr at 12°C, wt 1.1 g (95% confidence limit 3.6–8.4 mg/L) (technical material, 100%). Static bioassay without aeration, pH 7.2–7.5, water hardness 40–50 mg/L as calcium carbonate, and alkalinity of 30–35 mg/L [R10]. LC$_{50}$ *Ictalurus punctatus* (channel catfish) 414 mg/L/96 hr at 18°C, wt 1.1 g (technical material, 100%). Static bioassay without aeration, pH 7.2–7.5, water hardness 40–50 mg/L as calcium carbonate, and alkalinity of 30–35 mg/L [R10]. LC$_{50}$ *Micropterus salmoides* (largemouth bass) 3.4 mg/L/96 hr at 18°C, wt 0.9 g (technical material, 100%). Static bioassay without aeration, pH 7.2–7.5, water hardness 40–50 mg/L as calcium carbonate, and alkalinity of 30–35 mg/L [R10]. LC$_{50}$ *Lepomis macrochirus* (bluegill) 3.9 mg/L/96 hr at 18°C, wt 1.4 g (technical material, 100%). Static bioassay without aeration, pH 7.2–7.5, water hardness 40–50 mg/L as calcium carbonate, and alkalinity of 30–35 mg/L [R10]. LC$_{50}$ *Stizostedion vitreum vitreum* (walleye) 1.0 mg/L/96 hr at 18°C, wt 1.3 g (technical material, 100%). Static bioassay without aeration, pH 7.2–7.5, water hardness 40–50 mg/L as calcium carbonate, and alkalinity of 30–35 mg/L [R10]. LC$_{50}$ harlequin fish 4–5 mg/L/96 hr [R8]; LD$_{50}$ honeybee above 60 µg/bee [R8].

BIODEGRADATION Following fumigation of Yolo loam with D–D, soluble chloride is released after a 10-day lag; the rate of release then decreasing slowly to zero in approximately 300 days (1). In laboratory experiments there was little evidence of 1,2–dichloropropane biodegradation in sandy loam because after 12 weeks of incubation with radiolabeled chemical, 98% was recovered as the parent compound (1). The situation was similar with medium loam, although up to 5% of the radioactivity was unextractable from soil after 20 wk (1). With both soils, no volatile degradation products were detected (1). When incubated with soil, 1,3–dichloropropene is converted to the respective 3–chloroallyl alcohol and 3–chloroacrylic acid (1). Twelve weeks after labeled *cis–* or *trans*–1,3–dichloropropene was added to soil and stored in sealed containers, 19% of the cis–, and 18% of the trans–isomer remained in sandy loam, and 10% of the cis– and 22% of the trans–isomer remained in medium loam (1). After 20 wk, 5 and 4% of the cis– and trans–isomers, respectively, remained in the sandy loam, and 3 and 14%, respectively, remained in the medium loam (1). The half-lives of the applied dichloropropenes were 3–4 wk (1). It is possible that some of the parent compound was lost in these experiments by volatilization (1). In a separate experiment, 0.3 and 0.1% of radioactivity was recovered as CO_2 when *cis–* and *trans*–1,3–dichloropropene were incubated in sandy soil for 3 days (1). In screening tests using activated-sludge inocula, 42 and 55% degradation was reported in 7 days for 1,2–dichloropropane and 1,3–dichloropropene, respectively (3). However, no degradation of 1,2–dichloropropane occurred in 2 weeks in another screening test with a mixed inoculum of soil, surface water, and sludge (4). Although >99% removal occurred for both chemicals in activated sludge treatment plants (5), loss was probably due entirely to volatilization for 1,2–dichloropropane (2) [R11].

BIOCONCENTRATION The BCF for 1,3–dichloropropene estimated from its water solubility is 13 (2), and for 1,2–dichloropropane, as estimated from its calculated log octanol/water coefficient, 1.99 (1), is 18 (3), indicating that bioconcentration in fish is not significant for either of these chemicals [R12].

GENERAL RESPONSE Avoid contact with liquid and vapor. Keep people away. Wear self-contained positive-pressure breathing apparatus and chemical protective suit. Shut off ignition sources. Call fire department. Stop discharge if possible. Evacuate area. Stay upwind and use water spray to "knock down" vapor. Isolate and remove discharged material. Notify local health and pollution control agencies.

FIRE RESPONSE Flammable. Poisonous gases are produced in fire. Containers may explode in fire. Vapor may explode indoors, outdoors, or in sewers. Flashback along vapor trail may occur. Extinguish.

Small fires: dry chemical, CO_2, water spray, or foam; large fires: water spray, fog, or foam. Cool exposed containeers with water. Combat fires from safe distances or protected location.

EXPOSURE RESPONSE Call for medical aid. Vapor is irritating to eyes, respiratory tract, skin, and digestive tract. Inhalation will cause gasping, refusal to breathe, and respiratory distress; may be fatal. Move victim to fresh air. If breathing has stopped, give artificial respiration. If breathing is difficult, give oxygen. Liquid may be fatal if swallowed or absorbed through skin. Will burn exposed tissues. If in eyes, hold eyelids open, and flush with water for 15 minutes. If on skin, flush with water for 15 minutes; wash with soap and water. Remove and isolate contaminated clothing and shoes at the site. If swallowed and victim is conscious, have victim drink water, and induce vomiting. If swallowed and victim is unconscious or having convulsions, do nothing except keep victim warm.

RESPONSE TO DISCHARGE Issue warning—high flammability, water contaminant. Restrict access. Evacuate area. Should be removed. Provide chemical and physical treatment.

WATER RESPONSE Dangerous to aquatic life in high concentrations. May be dangerous if it enters water intakes. Notify local health and wildlife officials. Notify operators of nearby water intakes.

EXPOSURE SYMPTOMS Toxic; may be fatal if inhaled, swallowed, or absorbed through the skin. Inhalation causes gasping, refusal to breathe, and respiratory distress. Contact may cause burns to skin and eyes. Ingestion may cause acute gastrointestinal distress with congestion and edema of the lungs.

EXPOSURE TREATMENT Inhalation: move victim to fresh air. If breathing has stopped, give artificial respiration. If breathing is difficult, give oxygen. Get medical attention. Eyes: immediately flush eyes with running water for at least 15 min; hold eyelids open if necessary. Continue to flush eyes with water during transport to treatment facilities. Skin: immediately deluge exposed area with water for at least 15 min; remove and isolate contaminated clothing and shoes at the site. Wash contaminated area with soap and water. Ingestion: if victim is conscious, give no more than 2 glasses of water. Induce vomiting either by giving 30 cc (2 tablespoons) syrup of ipecac or by touching the back of the throat. If victim is unconscious or having convulsions, do nothing except keep victim quiet, and maintain his or her normal body temperature.

REFERENCES

1. Spencer, E. Y. *Guide to the Chemicals Used in Crop Protection.* 7th ed. Publication 1093. Research Institute, Agriculture Canada, Ottawa, Canada: Information Canada, 1982.

2. Worthing, C. R., and S. B. Walker (eds.). *The Pesticide Manual—A World Compendium.* 8th ed. Thornton Heath, UK: The British Crop Protection Council, 1987.

3. Hartley, D., and H. Kidd (eds.). *The Agrochemicals Handbook.* 2nd ed. Lechworth, Herts, England: The Royal Society of Chemistry, 1987, p. A117/Aug 87.

4. *Farm Chemicals Handbook* 1986. Willoughby, OH: Meister Publishing Co., 1986, p. C-74.

5. Association of American Railroads. *Emergency Handling of Hazardous Materials in Surface Transportation.* Washington, DC: Assoc. of American Railroads, Hazardous Materials Systems (BOE), 1987. 246.

6. U.S. Coast Guard, Department of Transportation. *CHRIS—Hazardous Chemical Data.* Volume II. Washington, DC: U.S. Government Printing Office, 1984-5.

7. 40 CFR 165.9 (a) (7/1/91).

8. Worthing, C. R., S. B. Walker (eds.). *The Pesticide Manual—A World Compendium.* 7th ed. Lavenham, Suffolk, Great Britain: The Lavenham Press Limited, 1983. 182.

9. Sax, N. I. *Dangerous Properties of Industrial Materials.* 6th ed. New York, NY: Van Nostrand Reinhold, 1984. 963.

10. U.S. Department of Interior, Fish and Wildlife Service. *Handbook of Acute Toxicity of Chemicals to Fish and Aquatic Invertebrates.* Resource Publication No. 137. Washington, DC: U.S. Government Printing Office, 1980. 24.

11. (1) Martin, J. P., *Pestic Eff Soils Water,* pp. 95-108 Symp Pap, Columbus, OH (1965); (2) Roberts, T. R., G. Stoydm. *Pestic Sci* 7: 325-35 (1976); (3) Tabak, H. H., et al., *J Water Pollut Control Fed* 53: 1503-18 (1981); (4) USEPA. *Treatability Manual* Volume I, pp. I. 12.13-1 to I. 12.14-5 USEPA-600/2-82-001a (1981); (5) Kincannon, D. F., et al. pp. 641-50 in *Proc 37th Industrial Waste Conf,* Bell, J.M. (ed.), Ann Arbor Sci Pub (1983) 12 (1) GEMS; *Graphical Exposure Modeling System. CLOGP3* (1986); (2) Kenaga, E. E., *Ecotox Env Safety* 4: 26-38 (1980); (3) Lyman, W. J., et al. *Handbook of Chemical Property Estimation Methods. Environmental Behavior of Organic Compounds.* New York: McGraw-Hill, pp. 5.1-5.30 (1982).

■ PROPANE, 1,1-DICHLORO

CAS # 78-99-9

DOT # 1279

SYNONYMS propane, 1,1-dichloro • propylidene chloride • alpha, alpha-dichloropropane • alpha, alpha-propylene dichloride

MF $C_3H_6Cl_2$

MW 112.99

COLOR AND FORM Liquid.

ODOR Sweet.

DENSITY 1.1321 at 20°C.

SURFACE TENSION 29 dynes/cm = 0.0261 N/m (est) at 20°C.

BOILING POINT 88.1°C at 760 mm Hg.

HEAT OF COMBUSTION −6,667 Btu/lb = −3,704 cal/g.

AUTOIGNITION TEMPERATURE 1035°F (est) [R3].

EXPLOSIVE LIMITS AND POTENTIAL Vapor may explode if ignited in enclosed area [R3].

FLAMMABILITY LIMITS Flammable limits in air: 3.4%–14.5% (est) [R3].

TOXIC COMBUSTION PRODUCTS Emits fumes of phosgene [R3].

SOLUBILITY Soluble in alcohol, ether, benzene, chloroform; very slightly soluble in water; soluble in many organic solvents; water solubility is 0.63 g/L (calculated).

VAPOR PRESSURE Vapor pressure of 65.9 torr at 25°C.

OCTANOL/WATER PARTITION COEFFICIENT log K_{ow} = 2.34 (calc).

OTHER CHEMICAL/PHYSICAL PROPERTIES Gas phase pyrolysis of 1,1-dichloropropane was reinvestigated at 9–185 torr between 389.2 and 456.5°C. The observed cis:trans ratio of the 1-chloropropene produced was 1.20 under these conditions. The comparable equilibrium cis:trans ratio at these temperatures was 1.90. The pressure change obeys first-order kinetics.

PROTECTIVE EQUIPMENT Breakthrough times for dichloropropane on chlorinated polyethylene are less (usually significantly) than one hour, as reported by two or more testers [R4].

PREVENTATIVE MEASURES SRP: The scientific literature supports the wearing of contact lenses in industrial environments, as part of a program to protect the eye against chemical compounds and minerals causing eye irritation. However, there may be individual substances whose irritating or corrosive properties are such that the wearing of contact lenses would be harmful to the eye. In those specific cases contact lenses should not be worn.

COMMON USES Solvent for pesticide formulations [R2].

OCCUPATIONAL RECOMMENDATIONS None listed.

NON-HUMAN TOXICITY LD_{50} rat oral 6.5 g/kg [R1].

ECOTOXICITY LC_{50} *Lepomis macrochirus* (bluegill) 97,900 µg/L/96 hr [R5].

BIOCONCENTRATION Steady state BCF = 19.4 (est) weighted average BCF = 7.66 [R6].

CERCLA REPORTABLE QUANTITIES Persons in charge of vessels or facilities are required to notify the National Response Center (NRC) immediately, when there is a release of this designated hazardous substance, in an amount equal to or greater than its reportable quantity of 1,000 lb or 454 kg. The toll-free number of the NRC is (800) 424–8802; in the Washington DC metropolitan area (202) 426–2675. The rule for determining when notification is required is stated in 40 CFR 302.4 (section IV. D. 3.b) [R7].

TSCA REQUIREMENTS Pursuant to section 8 (d) of TSCA, EPA promulgated a model health and safety data reporting rule. The section 8 (d) model rule requires manufacturers, importers, and processors of listed chemical substances and mixtures to submit to EPA copies and lists of unpublished health and safety studies. Propane, 1,1-dichloro is included on this list [R8].

GENERAL RESPONSE Stop discharge if possible. Keep people away. Shut off ignition sources and call fire department. Stay upwind and use water spray to "knock down" vapor. Avoid contact with liquid and vapor. Isolate and remove discharged material. Notify local health and pollution control agencies.

FIRE RESPONSE Flammable. poisonous gases are produced in fire. Flashback along vapor trail may occur. Vapor may explode if ignited in an enclosed area. Wear goggles and self-contained breathing apparatus. Extinguish with foam, dry chemical, or carbon dioxide. Cool exposed containers with water.

EXPOSURE RESPONSE Call for medical aid. Vapor irritating to eyes, nose, and throat. Move to fresh air. If breathing has stopped, give artificial respiration. If breathing is difficult, give oxygen. Liquid irritating to skin and eyes. Harmful if swallowed. Remove contaminated clothing and shoes. Flush affected areas with plenty of water. If swallowed and victim is conscious, have victim drink water or milk.

RESPONSE TO DISCHARGE Issue warning—high flammability. Evacuate area.

WATER RESPONSE Effect of low concentrations on aquatic life is unknown. May be dangerous if it enters water intakes. Notify local health and wildlife officials. Notify operators of nearby water intakes.

EXPOSURE SYMPTOMS Inhalation: may cause some central nervous system depression. Eyes: may cause some pain and irritation. Skin: mild irritation.

EXPOSURE TREATMENT Call a doctor. Inhalation: remove to fresh air. If breathing has stopped, give artificial respiration. Eyes: flush with running water for 15 minutes. Skin: wash thoroughly with soap and water. Ingestion: gastric lavage or emesis and catharsis.

REFERENCES

1. Budavari, S. (ed.). *The Merck Index— Encyclopedia of Chemicals, Drugs, and Biologicals*. Rahway, NJ: Merck and Co., Inc., 1989. 1248.

2. 40 CFR 180.1001 (d) (7/1/91).

3. U.S. Coast Guard, Department of Transportation. *CHRIS—Hazardous Chemical Data*. Volume II. Washington, DC: U.S. Government Printing Office, 1984–5.

4. ACGIH. *Guidelines Select of Chemical Protect Clothing, Volume 1, Field Guide* p. 48 (1983).

5. USEPA. *Ambient Water Quality Criteria Doc; Dichloropropanes and Dichloropropenes* p. B–6 (1980).

6. USEPA. *Ambient Water Quality Crite-*

ria Doc; Dichloropropanes and Dichloropropenes p. C–5 (1980).

7. 40 CFR 302.4 (7/1/91).

8. 40 CFR 716.120 (7/1/91).

■ PROPANENITRILE, 2-METHYL

CAS # 78–82–0

DOT # 22.84

SYNONYMS 1-cyano-1-methylethane • 2-cyanopropane • 2-methylpropane nitrile • 2-methylpropionitrile • dimethylacetonitrile • isobutyronitrile • isopropyl cyanide • isopropyl nitrile • propanenitrile, 2-methyl • propanoic acid, 2-methyl-, nitrile

MF C_4H_7N

MW 69.12

COLOR AND FORM Colorless liquid.

DENSITY 0.7608 at 30°C.

BOILING POINT 103.8°C at 760 mm Hg.

MELTING POINT −71.5°C.

AUTOIGNITION TEMPERATURE 900°F (482°C) [R3].

FIREFIGHTING PROCEDURES "Alcohol" foam. Water may be ineffective [R2].

SOLUBILITY Slightly soluble in water; soluble in acetone; very soluble in alcohol and ether.

VAPOR PRESSURE 32.7 mm Hg at 25°C.

OCTANOL/WATER PARTITION COEFFICIENT Log K_{ow} = 0.44 (est).

OTHER CHEMICAL/PHYSICAL PROPERTIES Specific gravity: 0.773 at 20°C.

PROTECTIVE EQUIPMENT Wear appropriate equipment to prevent any possibility of skin contact. (cyanides) [R4].

PREVENTATIVE MEASURES Workers should wash immediately when skin becomes contaminated. (cyanides) [R4].

DISPOSAL Generators of waste (equal to or greater than 100 kg/mo) containing this contaminant, EPA hazardous waste numbers D003 and P030, must conform with USEPA regulations in storage, transportation, treatment, and disposal of waste. (cyanide compounds) [R5].

COMMON USES Intermediate for insecticides [R1].

OCCUPATIONAL RECOMMENDATIONS

NIOSH 10–hr time-weighted average: 8 ppm (22 mg/m³)[R8].

NON-HUMAN TOXICITY LD_{50} rat oral 100 mg/kg [R10, 346]; LD_{50} rabbit dermal 310 mg/kg [R10, 346].

BIODEGRADATION 2–Methylpropanenitrile has been shown to biodegrade readily using the Japanese MITI screening test (2–wk incubation, 100–ppm concentration, activated-sludge seed) with theoretical BODs of 53.9–66.3% (1–2). A predictive method based upon evaluated biodegradation data and the fact that 2–methylpropanenitrile contains a nitrile substructure predicts that 2–methylpropanenitrile has a high probability of biodegrading fast in the environment (3) [R6].

BIOCONCENTRATION Based upon an estimated log K_{ow} of 0.44 (1), the BCF for 2–methylpropanenitrile can be estimated to be about 1 from a regression-derived equation (2,SRC). This BFC value suggests that bioconcentration in aquatic organisms is not environmentally important (SRC) [R7].

RCRA REQUIREMENTS D003; a solid waste containing a cyanide compound may become characterized as a hazardous waste when subjected to testing for reactivity as stipulated in 40 CFR 261.23, and, if so characterized, must be managed as a hazardous waste. (cyanide compounds) [R9].

GENERAL RESPONSE Avoid contact with liquid and vapor. Keep people away. Wear goggles and self-contained breathing apparatus. Shut off ignition sources and call fire department. Stay upwind and use water spray to "knock down" vapor. Stop discharge if possible. Isolate and remove discharged material. Notify local health and pollution control agencies.

FIRE RESPONSE Flammable. Flashback along vapor trail may occur. Vapor may explode if ignited in an enclosed area. Wear goggles and self-contained breathing apparatus. Combat fires from safe distance or protected location. Extinguish with dry chemical, alcohol foam, or carbon dioxide. Water may be ineffective on fire. Cool exposed containers with water.

EXPOSURE RESPONSE Call for medical aid. Vapor poisonous if inhaled. Irritating to eyes, nose, and throat. Move to fresh air. If breathing has stopped, give artificial respiration. If breathing is difficult, give oxygen. Liquid poisonous if swallowed or if skin is exposed. Irritating to skin and eyes. Remove contaminated clothing and shoes. Flush affected areas with plenty of water. If swallowed and victim is conscious, have victim drink water or milk.

RESPONSE TO DISCHARGE Issue warning—high flammability, poison; restrict access; evacuate area; mechanical containment; should be removed; chemical and physical treatment.

WATER RESPONSE Effect of low concentrations on aquatic life is unknown. May be dangerous if it enters water intakes. Notify local health and wildlife officials. Notify operators of nearby water intakes.

EXPOSURE SYMPTOMS Inhalation, ingestion, or skin contact causes weakness, headache, confusion, nausea, vomiting; acute cyanide poisoning may result. Contact with eyes causes irritation.

EXPOSURE TREATMENT Get medical attention following all overexposures to this chemical. Watch for symptoms of cyanide poisoning. Inhalation: move patient to fresh air; apply artificial respiration if breathing stops. Ingestion: break an amyl nitrite pearl in a cloth and hold lightly under patient's nose for 15 sec; if conscious, induce vomiting, and repeat until vomit is clear; repeat inhalation of amyl nitrite 5 times at 15–sec intervals. Eyes: flush with water for at least 15 min. Skin: flush with water; remove contaminated clothing; destroy contaminated shoes.

REFERENCES

1. Sax, N. I., R. J. Lewis, Sr. (eds.). *Haw-*

ley's Condensed Chemical Dictionary. 11th ed. New York: Van Nostrand Reinhold Co., 1987. 655.

2. National Fire Protection Guide. *Fire Protection Guide on Hazardous Materials*. 10th ed. Quincy, MA: National Fire Protection Association, 1991, p. 325M–61.

3. Hawley, G. G. *The Condensed Chemical Dictionary*. 9th ed. New York: Van Nostrand Reinhold Co., 1977. 479.

4. NIOSH. *NIOSH Pocket Guide to Chemical Hazards*. DHHS (NIOSH) Publication No. 90–117. Washington, DC: U.S. Government Printing Office, June 1990, 76.

5. 40 CFR 240–280, 300–306, 702–799 (7/1/91).

6. (1) Kitano, M. *Biodegradation and Bioaccumulation Test on Chemical Substances*. OECD Tokyo Meeting. Reference Book TSU–No. 3 (1978). (2) Sasaki, S. pp. 283–98 in *Aquatic Pollutants: Transformation and Biodegradation Effects*. Hutzinger, O., et al. (eds.), Oxford: Pergamon Press (1978). (3) Howard, P. H. et al., *Environ Toxicol Chem* 11: 593–603 (1992).

7. (1) USEPA. *Graphical Exposure Modeling System (GEMS)*. CLOGP (1987). (2) Lyman, W. J., et al. *Handbook of Chemical Property Estimation Methods*. Washington, DC: Amer Chemical Soc, p. 5–4 (1990).

8. NIOSH/CDC. NIOSH *Recommendations for Occupational Safety and Health Standards 1988*, Aug. 1988 (Suppl. to *Morbidity and Mortality Wkly*. Vol. 37, No. 5–7, Aug. 26, 1988). Atlanta, GA: National Institute for Occupational Safety and Health, CDC, 1988, p. V37 (S7) 21.

9. 40 CFR 261.23 (7/1/91).

10. Snyder, R. (ed.). *Ethyl Browning's Toxicity and Metabolism of Industrial Solvents*. 2nd ed. Volume II: *Nitrogen and Phosphorus Solvents*. Amsterdam–New York–Oxford: Elsevier, 1990.

■ 2-PROPANONE, BROMO-

CAS # 598–31–2

DOT # 1569

SIC CODES 2818.

SYNONYMS 1-bromo-2-propanone • 2-propanone, 1-bromo- • 2-propanone, bromo- • acetonyl bromide • acetylmethyl bromide • bromo-2-propanone • bromomethyl methyl ketone • monobromoacetone

MF C_3H_5BrO

MW 136.99

COLOR AND FORM Colorless liquid; rapidly becomes violet even in absence of air.

ODOR Pungent odor.

DENSITY 1.634 at 23°C.

BOILING POINT 137°C; 63.5–64°C at 50 mm Hg.

MELTING POINT –36.5°C.

FIREFIGHTING PROCEDURES If material on fire or involved in Fire: Do not extinguish fire unless flow can be stopped. Use water in flooding quantities as fog. Solid streams of water may be ineffective. Cool all affected containers with flooding quantities of water. Apply water from as great a distance as possible. Use "alcohol" foam, carbon dioxide, or dry chemical [R3].

SOLUBILITY Soluble in alcohol, acetone; soluble in ether, benzene; water solubility = 130,000 ppm.

VAPOR PRESSURE 9 mm Hg at 20°C.

OCTANOL/WATER PARTITION COEFFICIENT Log K_{ow} = 0.485.

STABILITY AND SHELF LIFE Rapidly turns violet even in absence of air [R4, 390].

OTHER CHEMICAL/PHYSICAL PROPERTIES Partial decomposition at boiling point: 136°C.

PROTECTIVE EQUIPMENT Personnel exposed to its vapors should wear gastight chemical safety goggles and respiratory protective equipment. [R4, 39].

PREVENTATIVE MEASURES Whenever possible, this material should be used in enclosed systems. [R4, 39].

DISPOSAL At the time of review, criteria for land treatment or burial (sanitary landfill) disposal practices are subject to significant revision. Prior to implementing land disposal of waste residue (including waste sludge), consult with environmental regulatory agencies for guidance on acceptable disposal practices [R5]. Send to Army Chemical Corps.

COMMON USES Chemical war gas (former use) [R1]; organic synthesis; tear gas [R2].

STANDARD CODES NFPA, USCG—no.; ICC—class A poison, poison A label, not accepted in an outside container.

MAJOR SPECIES THREATENED All species; may smother benthic life.

DIRECT CONTACT Extremely irritating and hazardous to the eyes.

GENERAL SENSATION Violent lacrimator. Recognition odor in air, 0.5 mg/m³ [R9].

PERSONAL SAFETY Gastight eye and skin protection. Self-contained breathing apparatus approved for warfare agents.

ACUTE HAZARD LEVEL Extreme hazard.

CHRONIC HAZARD LEVEL Undissolved portion will remain on bottom and act as source for equilibrium quantities to dissolve over a long period of time. Must be removed before use of water can be permitted.

DEGREE OF HAZARD TO PUBLIC HEALTH Violent lacrimator; extremely toxic.

OCCUPATIONAL RECOMMENDATIONS

OSHA Meets criteria for OSHA medical records rule [R7].

ACTION LEVEL Notify fire, air, and civil defense authority. Call department of defense immediately. Evacuate area. Do not enter.

ON-SITE RESTORATION Carbon will adsorb soluble fraction. Heavy undissolved layer should be pumped or vacuumed up and sealed in containers. Seek professional environmental engineering assistance through EPA's Environmental Response Team (ERT), Edison, NJ, 24-hour no. 908-548-8730.

BEACH OR SHORE RESTORATION Do not burn.

AVAILABILITY OF COUNTERMEASURE MATERIAL Carbon—water treatment plants, sugar refineries; pumps—fire department; vacuum—swimming-pool suppliers.

DISPOSAL NOTIFICATION Department of Defense—Army.

INDUSTRIAL FOULING POTENTIAL Extreme danger to food processing waters; may pose explosive hazard in boiler water or cooling systems.

MAJOR WATER USE THREATENED All uses.

PROBABLE LOCATION AND STATE Turns violet rapidly. Will sink to bottom of water course and dissolve very slowly.

BIOCONCENTRATION The relatively low log K_{ow} (0.485) for bromoacetone and the relatively high water solubility (130,000 ppm) indicate that bioconcentration in aquatic organisms is not likely to be significant (1) [R6].

RCRA REQUIREMENTS As stipulated in 40 CFR 261.33, when bromoacetone, as a commercial chemical product or manufacturing chemical intermediate or an off-specification commercial chemical product or a manufacturing chemical intermediate, becomes a waste, it must be managed according to federal or state hazardous waste regulations. Also defined as a hazardous waste is any container or inner liner used to hold this waste or any residue, contaminated soil, water, or other debris resulting from the cleanup of a spill, into water or on dry land, of this waste. Generators of small quantities of this waste may qualify for partial exclusion from hazardous waste regulations (40 CFR 261.5 (e)) [R8].

GENERAL RESPONSE Avoid contact with liquid and vapor. Keep people away. Wear self-contained breathing apparatus and full protective clothing. Shut off ignition sources. Call fire department. Stop discharge if possible. Isolate and remove discharged material. Notify local health and pollution control agencies.

FIRE RESPONSE Flammable: Emits toxic fumes in fire. Flashback may occur along vapor trail. Containers may explode in fire. Wear self-contained breathing apparatus and full protective clothing. Extinguish with CO_2, dry chemical, or foam.

EXPOSURE RESPONSE Call for medical aid. Vapor is extremely irritating to the eyes, nose, throat, and upper respiratory system. May be harmful if inhaled or absorbed through the skin. Move victim to fresh air. If breathing has stopped, give artificial respiration. If breathing is difficult, give oxygen. Liquid is corrosive to eyes, skin, and upper respiratory tract. Harmful if swallowed or absorbed through the skin. If in eyes: hold eyelids open and flush with water for at least 15 minutes. Remove contaminated clothing and shoes, flush affected areas with plenty of water for at least 15 minutes. If swallowed: do nothing except keep victim warm. Do not induce vomiting.

RESPONSE TO DISCHARGE Evacuate area. Issue warning—high flammability, poison, corrosive. Should be removed. Provide chemical and physical treatment.

WATER RESPONSE Toxic to aquatic life in low concentrations. May be dangerous if it enters water intakes. Notify local health and wildlife officials. Notify operators of local water intakes.

EXPOSURE SYMPTOMS Very powerful lacrimator and upper-respiratory-tract irritant. Intensely irritating to the eyes, nose, throat, and lungs. Corrosive to the skin.

EXPOSURE TREATMENT Call a physician. Eyes: hold eyelids open, flush with running water for at least 15 minutes. Skin: remove contaminated clothing and shoes, flush affected areas with plenty of water for at least 15 minutes. Inhalation: move victim to fresh air. If breathing has stopped, give artificial respiration. If breathing is difficult, give oxygen. Ingestion: do nothing except keep victim warm.

REFERENCES

1. *The Merck Index.* 10th ed. Rahway, NJ: Merck Co., Inc., 1983. 193.

2. Sax, N. I., and R. J. Lewis, Sr. (eds.). *Hawley's Condensed Chemical Dictionary.* 11th ed. New York: Van Nostrand Reinhold Co., 1987. 170.

3. Association of American Railroads. *Emergency Handling of Hazardous Materials in Surface Transportation.* Washington, DC: Assoc. of American Railroads, Hazardous Materials Systems (BOE), 1987. 102.

4. International Labour Office. *Encyclopedia of Occupational Health and Safety.* Vols. I and II. Geneva, Switzerland: International Labour Office, 1983.

5. SRP.

6. USEPA. *Health and Environmental Effects Profile for Bromoacetone.* ECAO CIN–P167 (Final Draft) p. 5 (1986).

7. 29 CFR 1910.20 (7/1/87).

8. 53 FR 13382 (4/22/88).

9. Sullivan, R. J. *Air Pollution Aspects of Odorous Compounds,* NTIS PB 188 089, September 1969.

■ PROPAZINE

$(CH_3)_2CHNH$ ... Cl ... $NHCH(CH_3)_2$

CAS # 139–40–2

SYNONYMS 1,3,5-triazine-2,4-diamine, 6-chloro-N,N′-bis(1-methylethyl)- • 2,4-bis (isopropylamino) -6-chloro-s-triazine • 2-chloro-4,6-bis (isopropylamino) -s-triazine • 2-chloro-4,6-di(isopropylamino) -1,3,5-triazine • 6-chloro-N,N′-bis(1-methylethyl)-1,3,5-triazine-2,4-diamine • g-30028 • geigy-30,028 • gesamil • milogard • plantulin • primatol-p • propasin • propazin • propazine (herbicide) • prozinex • s-triazine, 2-chloro-4,6-bis(isopropylamino)- • Maxx-90- •

Caswell no. 184 • EPA pesticide code 080808.

MF $C_9H_{16}ClN_5$

MW 229.75

COLOR AND FORM Colorless powder.

DENSITY 1.162 g/cm³ at 20°C.

MELTING POINT 212–214°C.

FIREFIGHTING PROCEDURES Extinguish fire using agent suitable for type of surrounding fire. Use water in flooding quantities as fog. Use "alcohol foam" Wear positive-pressure self-contained breathing apparatus when fighting fires involving this material (triazine pesticide (compounds and preparations), solid (insecticides, agricultural, not elsewhere classified, other than liquid)) [R4]. Avoid breathing dusts and fumes from burning material. Keep upwind. (triazine pesticide, solid, not otherwise specified (compounds and preparations) (agricultural insecticides, not elsewhere classified, other than liquid)) [R4].

SOLUBILITY 5.0 mg/L water at 20°C; 6.2 g/kg benzene at 20°C; 6.2 g/kg toluene at 20°C; 2.5 g/kg carbon tetrachloride at 20°C; 8.6 ppm in water at 25°C.

VAPOR PRESSURE 0.0039 mPa at 20°C.

OCTANOL/WATER PARTITION COEFFICIENT Log K_{ow} = 2.93.

CORROSIVITY Noncorrosive under normal use conditions.

STABILITY AND SHELF LIFE Very stable over several yr of shelf life, with only slight sensitivity to light and temperature [R8, 411]. Stable in neutral, slightly acid, or slightly alkaline media [R10].

OTHER CHEMICAL/PHYSICAL PROPERTIES Stable to hydrolysis in neutral, slightly acidic, or slightly alkaline media.

PROTECTIVE EQUIPMENT The use of personal protective equipment such as glasses, synthetic gloves, and NIOSH–approved breathing apparatus is important. (herbicides) [R5].

PREVENTATIVE MEASURES Avoid contact with eyes, contact with skin, and inhalation of dust. [R8, 411].

DISPOSAL SRP: At the time of review, criteria for land treatment or burial (sanitary landfill) disposal practices are subject to significant revision. Prior to implementing land disposal of waste residue (including waste sludge), consult with environmental regulatory agencies for guidance on acceptable disposal practices.

COMMON USES Selective post-emergence herbicide for carrots, celery, and fennel [R3]; it is herbicide recommended for control of broad-leaved and grass weeds in sorghum and umbelliferous crops at 0.5 to 3 kg AI/ha. [R7, 709].

OCCUPATIONAL RECOMMENDATIONS None listed.

NON-HUMAN TOXICITY LD_{50} rat oral >7,700 mg/kg (techical grade) [R7, 709]; LD_{50} rat percutaneous >3,100 mg/kg [R7, 709]; LD_{50} mouse, oral >5,000 mg/kg [R9, 535]; LD_{50} rat, oral >5,000 mg/kg (Milogard 80W) [R2]; LD_{50} rabbit percutaneous >10,200 mg/kg (Milogard 80W) [R2]; LC_{50} rabbit inhalation 2.04 mg/L/4 hr (Milogard 80W) [R2]; LD_{50} rat oral >15,380 mg/kg (Milogard 4L) [R2]; LC_{50} rabbit inhalation >14.2 mg/L/4 hr (Milogard 4L) [R2]; LD_{50} rabbit percutaneous >3,000 mg/kg (Milogard 4L) [R2]; LD_{50} rat oral >5,050 mg/kg (Milogard; Maxx 90) [R2]; LD_{50} rabbit percutaneous >2,000 mg/kg (Milogard; Maxx 90) [R2]; LC_{50} rabbit inhalation 105.6 mg/L/4 hr (Milogard; Maxx 90) [R2].

ECOTOXICITY LC_{50} *Salmo gairdneri* (rainbow trout) 18 mg/L/96 hr (technical; conditions of bioassay not specified) [R8, 413]; LC_{50} *Lepomis macrochirus* (bluegill sunfish) >100 mg/L/96 hr (technical; conditions of bioassay not specified) [R8, 413]; LC_{50} *Carassius auratus* (goldfish) >32.0 mg/L/96 hr (conditions of bioassay not specified) [R1]; LC_{50} bobwhite quail oral >10,000 ppm (8-day diet) (technical) [R8, 413]; LC_{50} mallard ducks, oral >10,000 ppm (8-day diet) (technical) [R8, 413].

BIODEGRADATION Microbial activity possibly accounts for decomposition of significant portion of propazine in soil. Several soil microorganisms have been reported to utilize propazine as a source of energy and nitrogen. Microorganisms either indirectly or directly affect the fate of propazine degradation via influencing hydroxylation and dealkylation. [R8, 412].

BIOCONCENTRATION Bioconcentration factor predicted from soil adsorption coefficient = 8 (calculated) (from table) [R6].

DOT *Health Hazards:* Poisonous if swallowed. Inhalation of dust poisonous. Fire may produce irritating or poisonous gases. Runoff from fire control or dilution water may cause pollution (triazine pesticide, solid, not otherwise specified).

Fire or Explosion: Some of these materials may burn, but none of them ignites readily (triazine pesticide, solid, not otherwise specified).

Emergency Action: Keep unnecessary people away; isolate hazard area and deny entry. Stay upwind; keep out of low areas. Self-contained breathing apparatus (SCBA) and structural firefighter's protective clothing will provide limited protection. Call CHEMTREC at 1–800–424–9300 for emergency assistance. If water pollution occurs, notify the appropriate authorities (triazine pesticide, solid, not otherwise specified).

Fire: Small fires: Dry chemical, CO_2, halon, water spray, or standard foam. Large Fires: Water spray, fog, or standard foam is recommended. Move container from fire area if you can do so without risk (triazine pesticide, solid, not otherwise specified).

Spill or Leak: Do not touch spilled material; stop leak if you can do so without risk. Small Spills: Take up with sand or other noncombustible absorbent material and place into containers for later disposal. Small Dry Spills: With clean shovel place material into clean, dry container, and cover; move containers from spill area. Large Spills: Dike far ahead of liquid spill for later disposal (triazine pesticide, solid, not otherwise specified).

First Aid: Move victim to fresh air; call for emergency medical care. Remove and isolate contaminated clothing and shoes at the site. In case of contact with material, immediately flush skin or eyes with running water for at least 15 minutes (triazine pesticide, solid, not otherwise specified).

FIRE POTENTIAL Nonflammable [R8, 411].

REFERENCES

1. Hartley, D., and H. Kidd (eds.). *The Agrochemicals Handbook.* 2nd ed. Lechworth, Herts, England: The Royal Society of Chemistry, 1987, p. A344/Aug 87.

2. *Farm Chemicals Handbook* 1989. Willoughby, OH: Meister Publishing Co., 1989, p. C–240.

3. White–Stevens, R. (ed.). *Pesticides in the Environment:* Volume 1, Part 1, Part 2. New York: Marcel Dekker, Inc., 1971. 60.

4. Association of American Railroads. *Emergency Handling of Hazardous Materials in Surface Transportation.* Washington, DC: Assoc. of American Railroads, Hazardous Materials Systems (BOE), 1987. 692.

5. International Labour Office. *Encyclopedia of Occupational Health and Safety.* Vols. I and II. Geneva, Switzerland: International Labour Office, 1983. 1039.

6. Kenaga, E. E. *Ecotoxicol Environ Safety* 4: p. 31 (1980).

7. Worthing, C.R., and S.B. Walker (eds.). *The Pesticide Manual—A World Compendium.* 8th ed. Thornton Heath, UK: The British Crop Protection Council, 1987.

8. Weed Science Society of America. *Herbicide Handbook.* 5th ed. Champaign, Illinois: Weed Science Society of America, 1983.

9. National Research Council. *Drinking Water and Health* Volume 1. Washington, DC: National Academy Press, 1977.

10. USEPA. *Health and Environmental Effects Profile for Propazine.* 1994. Publication umber ECAO-CIN-P068.

■ PROPENE, 2,3–DICHLORO–

CAS # 78–88–6

DOT # 2047

SYNONYMS 1-propene, 2,3-dichloro- • 2,3-dichloropropene • 2,3-dichloropropylene • 2-chloroallyl chloride • propene, 2,3-dichloro • propylene, 2,3-dichloro-

MF $C_3H_4Cl_2$

MW 110.97

COLOR AND FORM Straw-colored liquid.

ODOR Pungent.

DENSITY 1.211 at 20°C.

BOILING POINT 94°C at 760 mm Hg.

FLAMMABILITY LIMITS Lower flammable limit: 5.3% by volume; upper flammable limit: 14.5% by volume [R4].

FIREFIGHTING PROCEDURES Fire extinguishing agents: small fires: dry chemical or carbon dioxide. Large fires: water fog, spray, or foam [R3].

TOXIC COMBUSTION PRODUCTS When heated to decomposition it emits toxic fumes of hydrochloric acid and phosgene [R5].

SOLUBILITY Water solubility: 2,150 mg/L; soluble in alcohol, ether, benzene, chloroform; water solubility: 0.0194 g/L.

VAPOR PRESSURE 53 mm Hg at 25°C.

OCTANOL/WATER PARTITION COEFFICIENT Log K_{ow} = 1.88 (calc).

OTHER CHEMICAL/PHYSICAL PROPERTIES Henry's Law constant: 0.0036 atm-cu m/mol (calc).

PROTECTIVE EQUIPMENT Rubber gloves, self-contained breathing apparatus, protective clothing [R3].

PREVENTATIVE MEASURES SRP: The scientific literature supports the wearing of contact lenses in industrial environments, as part of a program to protect the eye against chemical compounds and minerals causing eye irritation. However, there may be individual substances whose irri-

tating or corrosive properties are such that the wearing of contact lenses would be harmful to the eye. In those specific cases contact lenses should not be worn.

DISPOSAL Group I containers: Combustible containers from organic or metalloorganic pesticides (except organic mercury, lead, cadmium, or arsenic compounds) should be disposed of in pesticide incinerators or in specified landfill sites. (organic or metallo-organic pesticides) [R6].

COMMON USES Component of dichloropropene-dichloropropane mixture (fumigant) (minor use) [R1]; chemical intermediate for sulfallate (CDEC) herbicide (former use) [R1]; used as a copolymer and chemical intermediate in agricultural and pharmaceutical products [R2].

NON-HUMAN TOXICITY LD_{50} rat oral 320 mg/kg [R5]; LD_{50} rabbit skin 1,580 mg/kg [R5].

BIODEGRADATION The total degradation of 1,2–dichloropropane and 1,3– and 2,3–dichloropropenes in soil at normal field rates was extremely slow. The half-life of 2,3–dichloropropene was 4 times as long as those of the other compounds tested [R7].

BIOCONCENTRATION Based on the calculated log octanol/water partition coefficient (1.88) (1,SRC), the estimated log BCF is 1.20 (2,SRC). Therefore bioconcentration in fish will be insignificant [R8].

TSCA REQUIREMENTS Pursuant to section 8 (d) of TSCA, EPA promulgated a model health and safety data reporting rule. The section 8 (d) model rule requires manufacturers, importers, and processors of listed chemical substances and mixtures to submit to EPA copies and lists of unpublished health and safety studies. 1–Propene, 2,3–dichloro– is included on this list [R9].

GENERAL RESPONSE Avoid contact with liquid and vapor. Keep people away. Wear goggles, self-contained breathing apparatus, and rubber overclothing (including gloves). Shut off ignition sources and call fire department. Stop discharge if possible. Isolate and remove discharged material. Notify local health and pollution control agencies.

FIRE RESPONSE Flammable. Poisonous gases are produced in fire. Wear goggles and self-contained breathing apparatus. Extinguish with water, dry chemical, foam, or carbon dioxide. Cool exposed containers with water.

EXPOSURE RESPONSE Call for medical aid. Vapor irritating to eyes, nose, and throat. Move to fresh air. If breathing has stopped, give artificial respiration. If breathing is difficult, give oxygen. Liquid will burn skin and eyes. Harmful if swallowed. Remove contaminated clothing and shoes. Flush affected areas with plenty of water. If in eyes, hold eyelids open, and flush with plenty of water. If swallowed and victim is conscious, have victim drink water or milk, and have victim induce vomiting. If swallowed and victim is unconscious or having convulsions, do nothing except keep victim warm.

RESPONSE TO DISCHARGE Issue warning—high flammability, water contaminant. Restrict access. Should be removed. Provide chemical and physical treatment.

WATER RESPONSE Harmful to aquatic life in very low concentrations. May be dangerous if it enters water intakes. Notify local health and wildlife officials. Notify operators of nearby water intakes.

EXPOSURE SYMPTOMS Inhalation: vapors are poisonous, painful, and irritating. Headache and dizziness may occur. Overexposure may cause liver and kidney damage and even death. Eyes: irritation and lacrimation. May cause transient corneal injury. Skin: slight irritation, readily absorbed in toxic amounts, causing headache, dizziness, and other systematic symptoms. Ingestion: acute gastrointestinal distress with pulmonary congestion and edema, central nervous system depression.

EXPOSURE TREATMENT Inhalation: Get medical aid. Remove from exposure. Administer oxygen to relieve cyanosis and pulmonary edema. If respiration stops give mouth to-mouth resuscitation. Eyes: flush with water for 15 min. Skin: remove and discard contaminated clothing. Wash contaminated skin with soap and water. Ingestion: remove gastric aspiration and lavage. Use water as lavage fluid. Use demulcents such as alumina gels.

REFERENCES

1. SRI.

2. *Chemocyclopedia*; American Chemical Society (1986).

3. U.S. Coast Guard, Department of Transportation. *CHRIS—Hazardous Chemical Data.* Volume II. Washington, DC: U.S. Government Printing Office, 1984–5.

4. National Fire Protection Guide. *Fire Protection Guide on Hazardous Materials.* 10th ed. Quincy, MA: National Fire Protection Association, 1991, p. 325M–35.

5. Sax, N. I. *Dangerous Properties of Industrial Materials,.* 6th ed. New York, NY: Van Nostrand Reinhold, 1984. 964.

6. 40 CFR 165.9 (a) (7/1/91).

7. Van Dijk, H. *Pestic Sci* 11 (6): 625–32 (1980).

8. (1) *GEMS; Graphical Exposure Modeling System.* CLOG3 (1986); (2) Lyman, W. J., et al. *Handbook of Chemical Property Estimation Methods* New York: McGraw-Hill pp. 5.1–5.30 (1982).

9. 40 CFR 716.120 (7/1/91).

■ 2-(3-CHLOROPHENOXY) PROPIONIC ACID

CAS # 101–10–0

SYNONYMS 2-(m-chlorophenoxy) propionic acid • cloprop • amchemical 3-CP • Caswell No. 206 • EPA pesticide chemical code 021201; metachlorphenprop; 2-(3-chlorophenoxy) propanoic acid; propanoic acid, 2-(3-chlorophenoxy)

MF $C_9H_9ClO_3$

CORROSIVITY Noncorrosive.

STABILITY AND SHELF LIFE Highly stable [R4].

PROTECTIVE EQUIPMENT Dispensers of liquid detergent should be available. Safety pipettes should be used for all pipetting. In animal laboratory, personnel should wear protective suits (preferably disposable, one-piece, and close-fitting at ankles and wrists), gloves, hair covering, and overshoes. In chemical laboratory, gloves and gowns should always be worn; however, gloves should not be assumed to provide full protection. Carefully fitted masks or respirators may be necessary when working with particulates or gases, and disposable plastic aprons might provide addnl protection. Gowns should be of distinctive color, as a reminder that they are not to be worn outside the laboratory. (chemical carcinogens) [R8, 8].

PREVENTATIVE MEASURES SRP: The scientific literature supports the wearing of contact lenses in industrial environments, as part of a program to protect the eye against chemical compounds and minerals causing eye irritation. However, there may be individual substances whose irritating or corrosive properties are such that the wearing of contact lenses would be harmful to the eye. In those specific cases, contact lenses should not be worn.

CLEANUP A high-efficiency particulate arrestor (HEPA) or charcoal filters can be used to minimize amount of carcinogen in exhausted-air-ventilated safety cabinets, lab hoods, glove boxes, or animal rooms. Filter housing that is designed so that used filters can be transferred into plastic bag without contaminating maintenance staff is avail commercially. Filters should be placed in plastic bags immediately after removal. The plastic bag should be sealed immediately. The sealed bag should be labelled properly Waste liquids should be placed or collected in proper containers for disposal. The lid should be secured and the bottles properly labelled. Once filled, bottles should be placed in plastic bag, so that outer surface is not contaminated. The plastic bag should also be sealed and labelled. Broken glassware should be decontaminated by solvent extraction, by

chemical destruction, or in specially designed incinerators. (chemical carcinogens) [R8, 15].

DISPOSAL Group I containers: combustible containers from organic or metallo-organic pesticides (except organic mercury, lead, cadmium, or arsenic compounds) should be disposed of in pesticide incinerators or in specified landfill sites. (organic or metallo-organic pesticides) [R5].

COMMON USES Fruit thinner, plums, prunes [R1]; size increaser for pineapple [R2]. Control of broad-leafed plants in row crops; defoliants, general brush control (chlorophenoxy compounds; from table) [R3].

OCCUPATIONAL RECOMMENDATIONS None listed.

NON-HUMAN TOXICITY LD_{50} rat oral >750 mg/kg [R4].

IARC SUMMARY Classification of carcinogenicity: (1) evidence in humans: limited; overall summary evaluation of carcinogenic risk to humans is Group 2B: The agent is possibly carcinogenic to humans. (chlorophenoxy herbicides; from table) [R6].

DOT *Health Hazards:* Poisonous; may be fatal if inhaled, swallowed, or absorbed through skin. Contact may cause burns to skin and eyes. Runoff from fire control or dilution water may cause pollution (phenoxy pesticide, liquid, flammable, poisonous, not otherwise specified) [R7, p. G–28].

Fire or Explosion: Flammable/combustible material; may be ignited by heat, sparks, or flames. Vapors may travel to a source of ignition, and flash back. Container may explode in heat of fire. Vapor explosion and poison hazard indoors, outdoors, or in sewers. Runoff to sewer may create fire or explosion hazard (phenoxy pesticide, liquid, flammable, poisonous, not otherwise specified) [R7, p. G–28].

Emergency Action: Keep unnecessary people away; isolate hazard area, and deny entry. Stay upwind; keep out of low areas. Positive-pressure self-contained breathing apparatus (SCBA) and chemical protective clothing that is specifically recommended by the shipper or manufacturer may be worn. It may provide little or no thermal protection. Structural firefighters' protective clothing is not effective for these materials. Isolate for 1/2 mile in all directions if tank, rail car, or tank truck is involved in fire. Call CHEMTREC at 1–800–424–9300 for emergency assistance (phenoxy pesticide, liquid, flammable, poisonous, not otherwise specified) [R7, p. G–28].

Fire: Small fires: Dry chemical, CO_2, water spray, or alcohol-resistant foam. Large Fires: Water spray, fog, or alcohol-resistant foam. Move container from fire area if you can do so without risk. Dike fire-control water for later disposal; do not scatter the material. Apply cooling water to sides of containers that are exposed to flames until well after fire is out. Stay away from ends of tanks. Withdraw immediately in case of rising sound from venting safety device or any discoloration of tank due to fire (phenoxy pesticide, liquid, flammable, poisonous, not otherwise specified) [R7, p. G–28].

Spill or Leak: Shut off ignition sources; no flares, smoking, or flames in hazard area. Fully encapsulating, vapor-protective clothing should be worn for spills and leaks with no fire. Do not touch or walk through spilled material; stop leak if you can do so without risk. Water spray may reduce vapor; but it may not prevent ignition in closed spaces. Small Spills: Take up with sand or other noncombustible absorbent material and place into containers for later disposal. Large Spills: Dike far ahead of liquid spill for later disposal (phenoxy pesticide, liquid, flammable, poisonous, not otherwise specified) [R7, p. G–28].

First Aid: Move victim to fresh air, and call for emergency medical care; if not breathing, give artificial respiration; if breathing is difficult, give oxygen. In case of contact with material, immediately flush skin or eyes with running water for at least 15 minutes. Remove and isolate contaminated clothing and shoes at the site. Keep victim quiet and maintain normal body temperature. Effects may be delayed; keep victim under observation (phenoxy pesticide, liquid, flammable, poisonous, not otherwise specified) [R7, p. G–28].

REFERENCES

1. *Farm Chemicals Handbook* 1991. Willoughby, OH: Meister, 1991, p. C72.

2. *Kirk–Othmer Encyclopedia of Chemical Technology.* 3rd ed., Volumes 1–26. New York, NY: John Wiley and Sons, 1978–1984, p. V18 3 (1982).

3. Zenz, C. *Occupational Medicine—Principles and Practical Applications.* 2nd ed. St. Louis, MO: Mosby–Yearbook, Inc, 1988. 954.

4. Hartley, D., and H. Kidd (eds.). *The Agrochemicals Handbook.* 2nd ed. Lechworth, Herts, England: The Royal Society of Chemistry, 1987, p. A0931.

5. 40 CFR 165.9 (a) (7/1/90).

6. IARC. *Monographs on the Evaluation of the Carcinogenic Risk of Chemicals to Man.* Geneva: World Health Organization, International Agency for Research on Cancer, 1972–present (multivolume work), p. S7 60 (1987).

7. U.S. Department of Transportation. *Emergency Response Guidebook 1990.* DOT P 5800.5. Washington, DC: U.S. Government Printing Office, 1990.

8. Montesano, R., H. Bartsch, E. Boyland, G. Della Porta, L. Fishbein, R. A. Griesemer, A.B. Swan, L. Tomatis, and W. Davis (eds.). *Handling Chemical Carcinogens in the Laboratory:Problems of Safety.* IARC Scientific Publications No. 33. Lyon, France: International Agency for Research on Cancer, 1979.

■ PROPYL MERCAPTAN

CAS # 107–03–9

DOT # 2704

SYNONYMS 1-propanethiol • 3-mercaptopropanol • n-propyl mercaptan • n-thiopropyl alcohol • propane-1-thiol • propanethiol • 1-propylmercaptan

MF C_3H_8S

MW 76.17

COLOR AND FORM Colorless, mobile liquid.

ODOR Offensive-smelling liquid; characteristic odor of cabbage.

ODOR THRESHOLD 0.00075 ppm [R6].

TASTE Below 2–3 ppm, sweet onion and cabbage-like flavor.

DENSITY 0.8411 g/mL at 20°C.

SURFACE TENSION 24.7 dynes/cm = 0.0247 N/m at 20°C.

BOILING POINT 67–68°C.

MELTING POINT −113.3°C.

HEAT OF COMBUSTION −15,990 Btu/lb = −8,890 cal/g = −372×10⁵ J/kg.

HEAT OF VAPORIZATION 179 Btu/lb = 99 cal/g = 4.16×10⁵ J/kg.

EXPLOSIVE LIMITS AND POTENTIAL Containers may explode in fire. Vapor may explode if ignited in an enclosed area [R6].

FIREFIGHTING PROCEDURES Dry chemical, foam, carbon dioxide. Water may be ineffective [R6].

TOXIC COMBUSTION PRODUCTS In fires toxic sulfur dioxide is generated [R6].

SOLUBILITY Soluble in ether, acetone, benzene, and alcohol; soluble in propylene glycol; slightly soluble in water.

VAPOR PRESSURE 154.2 mm Hg at 25°C.

OCTANOL/WATER PARTITION COEFFICIENT Log K_{ow} of 1.70 (est).

OTHER CHEMICAL/PHYSICAL PROPERTIES Forms binary azeotropic mixture with propyl alcohol (91.35%) and ternary azeotrope with water and propyl alcohol; Henry's Law constant: 0.00461 atm–m³/mole (est).

PROTECTIVE EQUIPMENT Goggles or face shield; rubber gloves; self-contained breathing apparatus or organic canister mask [R2].

PREVENTATIVE MEASURES Workers should wash promptly when skin becomes contaminated. (ethyl mercaptan) [R7].

DISPOSAL Ethyl mercaptan is a waste chemical stream constituent that may be subjected to ultimate disposal by controlled incineration (2,000°F) followed by scrubbing with a caustic solution. (ethyl mercaptan) [R8].

COMMON USES Synthetic flavoring agent [R1]; chemical intermediate; herbicide [R3]; synthetic chemical reactions; as an odorant of odorless, toxic gases; as a fungicide or as fungicidal component [R14, 2074].

OCCUPATIONAL RECOMMENDATIONS
NIOSH 15–min ceiling value: 0.5 ppm (1.6 mg/m³) [R12].

NON-HUMAN TOXICITY LC_{50} rat inhalation 7,300 ppm/4 h [R9]; LD_{50} rat oral 1,730 mg/kg [R9]; LD_{50} rat ip 515 mg/kg [R9].

BIODEGRADATION Using a Warburg respirometer, propyl mercaptan was slowly bio-oxidized by a pure culture seed of *Alcaligenes faecalis* bacteria that was isolated from an activated sludge (1) [R10].

BIOCONCENTRATION Based upon an estimated log K_{ow} of 1.70 (1), the BCF for propyl mercaptan can be estimated to be about 12 from a regression-derived equation (2,SRC). This BCF value suggests that bioconcentration in aquatic organisms is not environmentally important (SRC) [R11].

DOT *Fire or Explosion:* Flammable/combustible material; may be ignited by heat, sparks, or flames. Vapors may travel to a source of ignition, and flash back. Container may explode in heat of fire. Vapor explosion hazard indoors, outdoors, or in sewers. Runoff to sewer may create fire or explosion hazard [R4].

Health Hazards: May be poisonous if inhaled or absorbed through skin. Vapors may cause dizziness or suffocation. Contact may irritate or burn skin and eyes. Fire may produce irritating or poisonous gases. Runoff from fire control or dilution water may cause pollution [R4].

Emergency Action: Keep unnecessary people away; isolate hazard area and deny entry. Stay upwind; keep out of low areas. Positive-pressure self-contained breathing apparatus (SCBA) and structural firefighters' protective clothing will provide limited protection. Isolate for 1/2 mile in all directions if tank, rail car, or tank truck is involved in fire. Call CHEMTREC at 1-800-424-9300 for emergency assistance. If water pollution occurs, notify the appropriate authorities [R4].

Fire: Small fires: Dry chemical, CO₂, water spray, or regular foam. Large Fires: Water spray, fog, or regular foam. Move container from fire area if you can do so without risk. Apply cooling water to sides of containers that are exposed to flames until well after fire is out. Stay away from ends of tanks. For massive fire in cargo area, use unmanned hose holder, or monitor nozzles; if this is impossible, withdraw from area, and let fire burn. Withdraw immediately in case of rising sound from venting safety device or any discoloration of tank due to fire [R4].

Spill or Leak: Shut off ignition sources; no flares, smoking, or flames in hazard area. Stop leak if you can do so without risk. Water spray may reduce vapor; but it may not prevent ignition in closed spaces. Small Spills: Take up with sand or other noncombustible absorbent material and place into containers for later disposal. Large Spills: Dike far ahead of liquid spill for later disposal [R4].

First Aid: Move victim to fresh air and call emergency medical care; if not breathing, give artificial respiration; if breathing is difficult, give oxygen. In case of contact with material, immediately flush eyes with running water for at least 15 minutes. Wash skin with soap and water. Remove and isolate contaminated clothing and shoes at the site [R4].

FIRE POTENTIAL Very dangerous fire hazard [R5].

FDA Propyl mercaptan is a food additive permitted for direct addition to food for human consumption, as long as (1) the quantity added to food does not exceed the amount reasonably required to accomplish its intended physical, nutritive, or other technical effect in food, and (2) when intended for use in or on food it is of appropriate food grade and is prepared and handled as a food ingredient. Synthetic flavoring substances and adjuvants may be safely used alone or in combination with flavoring substances and adjuvants generally recognized as safe in food, prior-sanctioned for such use, or regulated [R13].

GENERAL RESPONSE Stop discharge if possible. Keep people away. Shut off ignition sources. Call fire department. Isolate and remove discharged material. Notify local health and pollution control agencies.

FIRE RESPONSE Flammable. Poisonous gases are produced in fire. Containers may explode in fire. Flashback along vapor trail may occur. Vapor may explode if ignited in an enclosed area. Wear goggles and self-contained breathing apparatus. Extinguish with dry chemicals, foam, or carbon dioxide. Water may be ineffective on fire. Cool exposed containers with water.

EXPOSURE RESPONSE Call for medical aid. Vapor if inhaled will cause difficult breathing. Move victim to fresh air. If breathing has stopped, give artificial respiration. If breathing is difficult, give oxygen. Liquid irritating to skin and eyes. Harmful if swallowed. Remove contaminated clothing and shoes. Flush affected area with plenty of water. If swallowed and victim is conscious, have victim drink water or milk, and have victim induce vomiting. If swallowed and victim is unconscious or having convulsions, do nothing except keep victim warm.

RESPONSE TO DISCHARGE Issue warning—high flammability, water contaminant, air contaminant; restrict access; evacuate area; mechanical containment; should be removed; chemical and physical treatment.

WATER RESPONSE Effect of low concentrations on aquatic life is unknown. Fouling to shoreline. May be dangerous if it enters water intakes. Notify local health and wildlife officials. Notify operators of nearby water intakes.

EXPOSURE SYMPTOMS Inhalation causes

muscular weakness, convulsions, and respiratory paralysis; high concentrations may cause pulmonary irritation. Contact with liquid causes irritation of eyes and skin. Ingestion causes irritation of mouth and stomach.

EXPOSURE TREATMENT Inhalation: remove victim from contaminated atmosphere; give artificial respiration and oxygen if needed; observe for premonitory signs of pulmonary edema. Eyes: flush with water for 15 min; if irritation persists, see a physician. Skin: flush with water; wash with soap, and water. Ingestion: induce vomiting and follow with gastric lavage.

REFERENCES

1. *Fenaroli's Handbook of Flavor Ingredients*. Volume 2. Edited, translated, and revised by T. E. Furia and N. Bellanca. 2nd ed. Cleveland: The Chemical Rubber Co., 1975. 500.

2. U.S. Coast Guard, Department of Transportation. *CHRIS—Hazardous Chemical Data*. Manual Two. Washington, DC: U.S. Government Printing Office, Oct., 1978.

3. Sax, N. I., and R. J. Lewis, Sr. (eds.). *Hawley's Condensed Chemical Dictionary*. 11th ed. New York: Van Nostrand Reinhold Co., 1987. 969.

4. U.S. Department of Transportation. Emergency Response Guidebook 1990. DOT P 5800.5. Washington, DC: U.S. Government Printing Office, 1990, p. G-27.

5. Sax, N. I. *Dangerous Properties of Industrial Materials*. 6th ed. New York, NY: Van Nostrand Reinhold, 1984. 2294.

6. U.S. Coast Guard, Department of Transportation. *CHRIS—Hazardous Chemical Data*. Volume II. Washington, DC: U.S. Government Printing Office, 1984–5.

7. NIOSH. *NIOSH Pocket Guide to Chemical Hazards*. DHHS (NIOSH) Publication No. 90–117. Washington, DC: U.S. Government Printing Office, June 1990, 112.

8. USEPA. *Engineering Handbook for Hazardous Waste Incineration* p. 2–7 (1981) EPA 68-03-3025.

9. *Kirk-Othmer Encyclopedia of Chemical Technology*. 3rd ed., Volumes 1–26. New York, NY: John Wiley and Sons, 1978–1984, p. V22 959 (1983).

10. Marion, C. V., G. W. Malaney. *J Water Pollut Control Fed* 35: 1269–84 (1963)

11. (1) USEPA. *Graphical Exposure Modeling System* (GEMS). CLOGP (1987). (2) Lyman, W. J., et al., *Handbook of Chemical Property Estimation Methods*. Washington, DC: Amer Chemical Soc, p. 5–4 (1990).

12. NIOSH/CDC. *NIOSH Recommendations for Occupational Safety and Health Standards 1988*, Aug. 1988 (Suppl. to *Morbidity and Mortality Wkly*. Vol. 37 No. 5–7, Aug. 26, 1988). Atlanta, GA: National Institute for Occupational Safety and Health, CDC, 1988, p. V37 (S7) 26.

13. 21 CFR 172.515 (4/1/91).

14. Clayton, G. D., and F. E. Clayton (eds.). *Patty's Industrial Hygiene and Toxicology*. Volume 2A, 2B, 2C: *Toxicology*. 3rd ed. New York: John Wiley and Sons, 1981–1982.

■ PYRENE

CAS # 129–00–0

SYNONYMS benzo-(d,e,f)-phenanthrene • benzo (def) phenanthrene • beta-pyrene

MF $C_{16}H_{10}$

MW 202.26

COLOR AND FORM Monoclinic prismatic tablets from alcohol or by sublimation; pure pyrene is colorless; pale-yellow plates (when recrystallized from toluene); colorless solid (tetracene impurities give yellow color).

DENSITY 1.271 at 23°C.

BOILING POINT 393°C at 760 mm Hg; 260°C at 60 mm Hg.

MELTING POINT 156°C.

FIREFIGHTING PROCEDURES Self-contained breathing apparatus with a full facepiece operated in pressure-demand or other positive-pressure mode (coal tar pitch volatiles) [R12, 6]. Extinguishant: Foam, dry chemical, and carbon dioxide. (coal tar pitch volatiles) [R12, 3].

SOLUBILITY 0.135 mg/L ($\pm$ —0.005 mg/L) in water at 25°C; soluble in carbon disulfide and toluene; soluble in alcohol, ether, benzene, and petroleum ether.

VAPOR PRESSURE 6.85×10^{-7} mm Hg at 20°C.

OCTANOL/WATER PARTITION COEFFICIENT Log K_{ow} = 4.88.

OTHER CHEMICAL/PHYSICAL PROPERTIES Absorption coefficient for thermal electrons: 6.0; ionization potential: 7.58; coefficient of highest filled or lowest empty molecular orbital: 0.445; K–region: bond = 4,5, bond order = 1.777, (CLE) min carbon localization energy plus (BLE) bond localization energy = 3.33; solid and solution have slight blue fluorescence.

PROTECTIVE EQUIPMENT Employees should be provided with and required to use impervious clothing, gloves, face shields (eight-inch minimum), and other appropriate protective clothing necessary to prevent any possibility of skin contact with coal tar pitch volatiles. Employees should be provided with and required to use splash-proof goggles where there is any possibility of liquid coal tar volatiles contacting the eyes. (coal tar pitch volatiles) [R12, 3].

PREVENTATIVE MEASURES Areas in which exposure to coal tar pitch volatiles may occur should be identified by signs or other appropriate means, and access to these areas should be limited to authorized persons. (coal tar pitch volatiles) [R12, 4].

CLEANUP In surface waters, one-third of the total polycyclic aromatic hydrocarbons (PAH) is bound to larger suspended particles, one-third is bound to finely dispersed particles, and the last third is present in dissolved form. The particle-bound portion of PAH can be removed by sedimentation, flocculation, and filtration processes. The remaining one-third dissolved PAH usually requires oxidation for partial removal/transformation. (polynuclear aromatic hydrocarbons) [R4].

DISPOSAL SRP: At the time of review, criteria for land treatment or burial (sanitary landfill) disposal practices are subject to significant revision. Prior to implementing land disposal of waste residue (including waste sludge), consult with environmental regulatory agencies for guidance on acceptable disposal practices.

COMMON USES Biochemical research [R1]; pyrene from coal-tar has been used as a starting material for the synthesis of benzo (a) pyrene [R2].

OCCUPATIONAL RECOMMENDATIONS
OSHA Meets criteria for OSHA medical records rule [R9].

NIOSH NIOSH considers coal tar products carcinogenic and conditions should be made to keep exposures as low as possible. Current NIOSH research indicates that asphalt products are carcinogenic to laboratory animals and therefore may be more toxic to humans than previously believed. (coal tar products) [R10].

ECOTOXICITY TLm (median threshold-limit) mosquito fish 0.0026 mg/L/96 hr at 24–27°C in a static bioassay [R5].

IARC SUMMARY No data are available for humans. Inadequate evidence of carcinogenicity in animals.

OVERALL EVALUATION Group 3: The agent is not classifiable as to its carcinogenicity to humans [R6].

BIODEGRADATION Bacteria isolated from Colgate Creek, Chesapeake Bay, and cultured in bay water degraded 19.6–22.4% pyrene present, unspecfied time; Eastern Bay bacteria similiarly cultured effected 2.0–8.2% degradation (1). Pyrene at 5 ppm was 71% degraded with rapid adaptation after 7 days by microbes in settled domestic wastewater, 100% degraded 7 days after addition of the second subculture. At 10 ppm only 13% degradation was observed after 28 total days and the addition of three subcultures (2). Degradation in Center Hill Reservoir water, 4 weeks, 0–57.1%, average 16.7% (3). In natural water from polluted steam, Urbana, IL, pyrene was cometabolized 63.3% (naphthalene as growth substrate), and 53.8% (phenanthrene growth substrate) in 4 weeks (4). Pyrene levels from sewage treatment plant: raw sewage, <5 ppb, final effluent, <5 ppb, waste activated sludge, <5–423 ppb (wet weight), digester sludge, 50.5–563 ppb (wet weight) (5). Pyrene in contact with organics that are undergoing a high rate of composting was degraded 0 and 31% in 30-day duplicate tests; rate of degradation was higher over the last 23 days than it was the first 7 days (6). Pyrene has been shown to be bioconverted via oxidation by isolated soil bacteria without any growth substrate present (7) [R7].

BIOCONCENTRATION Reported BCF: rainbow trout, 72 (1); goldfish, 457 (2); fathead minnow, 600–970 (3). Based on these values, minimal to moderate bioconcentration of pyrene in aquatic organisms would be expected (SRC) [R8].

CERCLA REPORTABLE QUANTITIES Persons in charge of vessels or facilities are required to notify the National Response Center (NRC) immediately, when there is a release of this designated hazardous substance, in an amount equal to or greater than its reportable quantity of 5,000 lb or 2,270 kg. The toll-free telephone number of the NRC is (800) 424–8802; in the Washington metropolitan area (202) 426–2675. The rule for determining when notification is required is stated in 40 CFR 302.6 (section IV. D. 3.b) [R11].

DOT *Fire or Explosion:* Flammable/combustible material; may be ignited by heat, sparks, or flames. May burn rapidly with flare-burning effect (naphthalene) [R3].

Health Hazards: Fire may produce irritating or poisonous gases. Contact may cause burns to skin and eyes. Runoff from fire control or dilution water may cause pollution (naphthalene) [R3].

Emergency Action: Keep unnecessary people away; isolate hazard area and deny entry. Stay upwind; keep out of low areas. Wear self-contained (positive-pressure if available) breathing apparatus and full protective clothing. If water pollution occurs, notify appropriate authorities. For emergency assistance call CHEMTREC, (800) 424–9300 (naphthalene) [R3].

Spill or Leak: Shut off ignition sources; No flares, smoking, or flames in hazard area. Do not touch spilled material. Small Dry Spills: With clean shovel, place material into clean, dry container, and cover; move containers from spill area. Large Spills: Wet down with water and dike for later disposal (naphthalene) [R3].

First Aid: Move victim to fresh air; call for emergency medical care. In case of contact with material, immediately flush skin or eyes with running water for at least 15 min. Remove and isolate contaminated clothing and shoes at the site (naphthalene) [R3].

REFERENCES

1. Sax, N. I., and R. J. Lewis, Sr. (eds.). *Hawley's Condensed Chemical Dictionary.* 11th ed. New York: Van Nostrand Reinhold Co., 1987. 981.

2. *IARC Monographs* 1972–present, V32, p. 432.

3. Department of Transportation. *Emergency Response Guidebook 1984* DOT P 5800.3 Washington, DC: U.S. Government Printing Office, 1984, p. G–32.

4. USEPA. *Ambient Water Quality Criteria Doc; Polynuclear Aromatic Hydrocarbons (Draft)* p. C–4 (1980).

5. Verschueren, K. *Handbook of Environmental Data of Organic Chemicals.* 2nd ed. New York, NY: Van Nostrand Reinhold Co., 1983. 1034.

6. IARC. *Monographs on the Evaluation of the Carcinogenic Risk of Chemicals to Man.* Geneva: World Health Organization, International Agency for Research on Cancer, 1972–present (multivolume work), p. S7 71 (1987).

7. (1) Walker, J. D., R. R. Colwell. p 783–90 in *International Conference on Water Pollution Research.* Volume 7 (1974). (2) Tabak, H. H., et al., *J Water Pollut Control Fed* 53: 1503–18 (1981). (3) Sayler, G. S., T. W. Serrill. "Bacterial Degradation of Coal Conversion By-prod-

ucts (Polycyclic Aromatic Hydrocarbons)," in *Aquatic Environments Office of Water Res and Technol* NTIS PB 83–187161 (1981). (4) McKenna, E. J., *Biodegradation of Polynuclear Aromatic Hydrocarbon Pollutants by Soil and Water Microorganisms Water Resources Center* 113: 1–25 (1976). (5) Lue–Hing, C., et al., *AIChE Symp Ser* 77: 144–50 (1981). (6) Snell Environmental Group Inc. *Rate of Biodegradation of Toxic Compounds While in Contact with Organics Which are Actively Composting.* Lansing, MI NTIS PB 84–193150 (1982). (7) Sims, R. C., M. R. Overcash. *Res Rev* 88: 1–68 (1983).

8. (1) Spehar, R. L., et al., *J Water Pollut Control Fed* 52: 1703–74 (1980). (2) Ogata, M., et al., *Bull Environ Contam Toxicol* 33: 561–7 (1984). (3) Carlson, R. M., et al., *Implications to the Aquatic Environment of Polynuclear Aromatic Hydrocarbons Liberated from Northern Great Plains Coal* USEPA–600/2–79–093 (1979).

9. 29 CFR 1910.20 (7/1/87).

10. NIOSH. *Health Hazard Evaluation.* Report No. HETA 82–067–1253 13pp. (1984).

11. 51 FR 34534 (9/29/86).

12. Mackison, F. W., R. S. Stricoff, and L. J. Partridge, Jr. (eds.). *NIOSH/OSHA— Occupational Health Guidelines for Chemical Hazards.* DHHS(NIOSH) Publication No. 81–123 (3 VOLS). Washington, DC: U.S. Government Printing Office, Jan. 1981.

■ PYRETHRIN–I

CAS # 121–21–1

SIC CODES 2879.

SYNONYMS cyclopropanecarboxylic acid, 2,2-dimethyl-3-(2-methyl-1-propenyl)-, 2-methyl-4-oxo-3-(2,4-pentadienyl)-2-cyclopenten-1-yl ester, (1R-(1alpha (S (Z)),3beta)- • (+)-pyrethronyl (+)-trans-chrysanthemate • cyclopropaneacrylic acid, 2,2-dimethyl-3-(2-methylpropenyl)-,

ester with 4-hydroxy-3-methyl-2-(2,4-pentadienyl)-2-cyclopenten-1-one • piretrina-1 (Portuguese) • 2-methyl-4-oxo-3-(2,4-pentadienyl)-2-cyclopenten-2,2-dimethyl-3-(2-methyl 1-propenyl) cyclopropane carboxylate • 2,2-dimethyl-3-(2-methyly-1-propenyl) cyclopropanecarboxylic acid 2-methyl-4-oxo 3-(2,4-pentadienyl)-2-cycloenten-1-yl ester

MF $C_{21}H_{28}O_3$

MW 328.49

COLOR AND FORM Viscous liquid.

BOILING POINT 146–150°C at 0.0005 mm Hg.

SOLUBILITY Practically insoluble in water; soluble in alcohol, petr ether, kerosene, carbon tetrachloride, ethylene dichloride, nitromethane; soluble in diethyl ether.

STABILITY AND SHELF LIFE Oxidizes readily and becomes inactive in air. [R6, 1266].

OTHER CHEMICAL/PHYSICAL PROPERTIES Hydrolyzed in water, and the process is speeded by acid or alkali; BP: 170°C at 0.1 mm Hg (with decomposition).

PROTECTIVE EQUIPMENT Wear appropriate equipment to prevent repeated or prolonged skin contact. (pyrethrum and pyrethrins) [R2].

PREVENTATIVE MEASURES SRP: The scientific literature supports the wearing of contact lenses in industrial environments, as part of a program to protect the eye against chemical compounds and minerals causing eye irritation. However, there may be individual substances whose irritating or corrosive properties are such that the wearing of contact lenses would be harmful to the eye. In those specific cases contact lenses should not be worn.

CLEANUP Environmental consideration— land spill: Dig a pit, pond, lagoon, or holding area to contain liquid or solid material. (SRP: If time permits, pits, ponds, lagoons, soak holes, or holding areas should be sealed with an impermeable flexible-membrane liner.) Dike surface flow using soil, sandbags, foamed polyurethane, or foamed concrete. Absorb bulk liquid with fly ash or cement powder (pyrethrins) [R1]. Environmental consideration—water spill: If pyrethrins are dissolved, apply activated carbon at ten times the spilled amount in the region of 10 ppm or greater concentration. Use mechanical dredges or lifts to remove immobilized masses of pollutants and precipitates. (pyrethrins) [R1].

DISPOSAL After material has been contained, remove with contaminated soil, and place in impervious containers. The material may be incinerated in a pesticide incinerator at the specified temperature/dwell-time combination. Any liquids, sludges, or solid residues generated should be disposed of in accordance with all applicable federal, state, and local pollution control requirements. If appropriate incineration facilities are not available, material may be buried in a chemical waste landfill. Not amenable to biological treatment at sewage treatment plant. SRP: At the time of review, criteria for land treatment or burial (sanitary landfill) disposal practices are subject to significant revision. Prior to implementing land disposal of waste residue (including waste sludge), consult with environmental regulatory agencies for guidance on acceptable disposal practices.

STANDARD CODES EPA 311; TSCA; IATA not restricted; NFPA 704M system 2, 1, –; not listed CFR 49; CFR 14 cab code 8; ICC STCC tariff 1–C 28 799 31; not listed AAR; Superfund designated (hazardous substances) list.

PERSISTENCE Persistent.

MAJOR SPECIES THREATENED Fish are especially sensitive to pyrethrins.

DIRECT CONTACT Pyrethrins may cause an allergic reaction to the skin of people who are sensitive to ragweed.

GENERAL SENSATION Contact with skin may cause dermatitis in sensitive persons and in about 50% of those who are sensitive to ragweed. Dermatitis may be severe and associated with rhinitis and asthma. Inhalation or absorption through the skin may produce nausea, vomiting, gastroenteritis with diarrhea. High concentrations may produce hyperexcitability, incoordination, tremors, muscular paralysis, and death due to respiratory failure. Eye contact may produce severe eye irritation. Solvents used in formulations may produce pulmonary edema.

PERSONAL SAFETY Protect against inhalation and absorption through the skin. Must wear protective clothing including NIOSH–approved rubber gloves and boots, safety goggles, and respirator whose canister is specifically approved for this material. Care must be exercised to decontaminate fully all equipment and clothing after use.

ACUTE HAZARD LEVEL Pyrethrin I is only moderately toxic to warm blooded animals by oral administration but highly toxic by parenteral routes. It is readily detoxified by hydrolysis; thus, in an experiment, rats can ingest over a 24–hour period a dose that is lethal if taken at one time, and can maintain this intake every day of their lives without apparent injury. The fatal oral dose in man, as inferred from data on rats and guinea pigs, is 1 to 2 g/kg, expressed as pyrethrin. Because the pyrethrin concentration in the various formulations is quite low, serious, poisonings are highly improbable. Kerosene, and naphtha, the common solvents in pyrethrum sprays, are generally more hazardous than the active ingredient. The synergists commonly found in formulations possess even lower acute toxicities in lab mammals than do pyrethrins. Although they enhance insecticidal potency, there is no evidence that they increase the mammalian toxicity of fresh pyrethrin.

CHRONIC HAZARD LEVEL The threshold-limit-value-time-weighted average for a 40–hour workweek is 5 mg/m^3. There is a possibility of allergic reactions in sensitive people, especially those who are sensitive to ragweed pollen. Severe dermatitis has also been reported.

DEGREE OF HAZARD TO PUBLIC HEALTH Pyrethrin I is not very toxic by oral ingestion, but irritating to eyes, skin, and respiratory tract. It is an allergen to some people, especially to those sensitive to ragweed pollen.

OCCUPATIONAL RECOMMENDATIONS

NIOSH 10-hr time-weighted average: 5 mg/m^3 (pyrethrum) [R2].

ACTION LEVEL Avoid contact with the spilled cargo. Stay upwind. Notify local air, water, and fire authorities of the accident. Evacuate area when oil-based formulation is present, due to its high flammability.

ON-SITE RESTORATION Dam stream if possible to reduce flow and prevent further dispersion of the material. Dust formulations can be dredged from the bottom. Oil formulations may float, so booms and surface skimmers may be employed. Apply activated carbon to the water to adsorb the material. Bottom pumps, vacuum systems, or dredges may be employed to remove undissolved material and sorbent from the bottom. An alternative method, under controlled conditions, is to pump the water into a suitable container, use gravity separation tanks, dual filtration systems, and an activated-carbon filter. For more details, see Envirex manual, EPA 600/2-77-227. Seek professional environmental engineering assistance through EPA's Environmental Response Team (ERT), Edison, NJ, 24-hour no., 908-548-8730.

BEACH OR SHORE RESTORATION Close beach to the public until material has been removed. Scrape effected area with mechanical scraper and avoid human toxicity contact. Do not burn material.

AVAILABILITY OF COUNTERMEASURE MATERIAL Booms and surface skimmers—Army Corp of Engineers; carbon—water treatment plants, chemical companies; pumps—fire stations; dredges—Army Corps of Engineers; vacuum systems—swimming-pool operators.

DISPOSAL NOTIFICATION Notify local and state health authorities, local solid waste disposal authorities, supplier, and shipper of material.

INDUSTRIAL FOULING POTENTIAL Industrial fouling potential believed low.

PROBABLE LOCATION AND STATE Solid formulations will sink to the bottom and become incorporated with the sediments without dissolving. Liquid formulations will likely float on the surface, due to the hydrocarbon solvents.

WATER CHEMISTRY No reaction.

FOOD-CHAIN CONCENTRATION POTENTIAL Moderate; moderately persistent in the food chain.

ECOTOXICITY LD$_{50}$ frog injection 0.80 mg/kg [R7, 76]; LC$_{50}$ Channel catfish 114 µg/L/96 hr (static test); 132 µg/L/96 hr (flow-through test) (pyrethrins) [R3]; LC$_{50}$ *Gammarus lacustris* 12 µg/L/96 hr (conditions of bioassay not specified; pyrethrins) [R3]; LC$_{50}$ *Simocephalus serrulatus* 42 µg/L/48 hr (conditions of bioassay not specified; pyrethrins) [R3]; LC$_{50}$ *Daphnia pulex* 25 µg/L/48 hr (conditions of bioassay not specified; pyrethrins) [R3]; LC$_{50}$ *Pteronarcys californica* 1.0 µg/L/96 hr (conditions of bioassay not specified; pyrethrins) [R3]; LC$_{50}$ rainbow trout 56 ppb/24 hr (conditions of bioassay not specified; pyrethrins) [R3]; LD$_{50}$ mallards (females 1–4 mo) >10,000 mg/kg (pyrethrum) [R4].

BIOCONCENTRATION Based upon an estimated log K_{oc} of 3.599 (1), the BCF for pyrethrin I can be estimated to be 2,908 from a recommended regression-derived equation (2,SRC). Although this BCF value suggests that bioconcentration may be important in aquatic organisms, such bioconcentration is unlikely because of rapid degradation in water and metabolism in fish (SRC) [R5].

FIRE POTENTIAL Pyrethrins burn with difficulty. (pyrethrins) [R1].

REFERENCES

1. Bureau of Explosives. *Emergency Handling of Haz Matl in Surface Trans* p. 434 (1981).

2. NIOSH. *NIOSH Pocket Guide to Chemical Hazards.* DHHS (NIOSH) Publication No. 90-117. Washington, DC: U.S. Government Printing Office, June 1990, 190.

3. Verschueren, K. *Handbook of Environmental Data of Organic Chemicals.* 2nd ed. New York, NY: Van Nostrand Reinhold Co., 1983. 1035.

4. U.S. Department of the Interior, Fish and Wildlife Service, Bureau of Sport Fisheries and Wildlife. *Handbook of Toxicity of Pesticides to Wildlife.* Washington, DC, U.S. Government Printing Office, 1970. 96.

5. (1) USEPA. *Graphical Exposure Modeling System* (GEMS). CLOGP (1987). (2) Lyman, W. J., et al., *Handbook of Chemical Property Estimation Methods.* Washington, DC: Amer Chemical Soc, p. 5–5 (1990).

6. Budavari, S. (ed.). *The Merck Index— Encyclopedia of Chemicals, Drugs and Biologicals.* Rahway, NJ: Merck and Co., Inc., 1989.

7. Hayes, Wayland J., Jr. *Pesticides Studied in Man.* Baltimore/London: Williams and Wilkins, 1982.

■ PYRETHRIN-II

CAS # 121–29–9

SIC CODES 2879.

SYNONYMS cyclopropanecarboxylic acid, 3-(3-methoxy-2-methyl-3-oxo-1-propenyl)-2,2-dimethyl-, 2-methyl-4-oxo-3-(2,4-pentadienyl)-2-cyclopenten-1-yl ester, (1R-(1alpha (S (Z)),3beta (E)) - • (+)-pyrethronyl • (+)-pyrethrate • cyclopropaneacrylic acid, 3-carboxy-alpha, 2,2-trimethyl-, 1-methyl ester, ester with 4-hydroxy-3-methyl-2-(2,4-pentadienyl)-2-cyclopenten-1-one • ent-7,543- • 3-(3-methoxy-2-methyl-3-oxo-1-propenyl)-2,2-dimethyl cyclopropanecarboxylic acid 2-methyl-4-oxo-3-(2,4-pentadienyl)-2-cyclopenten-1-yl ester • 2-methyl-4-oxo-3-(2,4-pentadienyl) -2-cyclopenten-1-yl 3-(3-methoxy-2-methyl-3-oxo-1-propenyl)-2,2-dimethyl cyclopropane carboxylate • chrysanthemum-dicarboxylic acid monomethyl ester pyrethrolone ester

MF $C_{22}H_{28}O_5$

MW 372.50

COLOR AND FORM Viscous liquid.

BOILING POINT 192–193°C at 0.007 mm Hg.

SOLUBILITY Soluble in alcohol, ether, petroleum ether. Practically insoluble in water; soluble in petroleum ether (less soluble than pyrethrin (I), kerosene, carbon tetrachloride, ethylene dichloride, nitromethane.

OCTANOL/WATER PARTITION COEFFICIENT Log K_{ow} = 3.599 (estimated by fragment method).

CORROSIVITY Noncorrosive.

STABILITY AND SHELF LIFE Oxidizes rapidly and becomes inactive in air. [R11, 1267].

OTHER CHEMICAL/PHYSICAL PROPERTIES Hydrolyzed in water, and the process is speeded by acid or alkali. BP: 200°C at 1 mm Hg (with decomposition).

PROTECTIVE EQUIPMENT Wear appropriate equipment to prevent repeated or prolonged skin contact. (pyrethrum and pyrethrins) [R3].

PREVENTATIVE MEASURES SRP: The scientific literature supports the wearing of contact lenses in industrial environments, as part of a program to protect the eye against chemical compounds and minerals causing eye irritation. However, there may be individual substances whose irritating or corrosive properties are such that the wearing of contact lenses would be harmful to the eye. In those specific cases contact lenses should not be worn.

CLEANUP Environmental consideration— land spill: Dig a pit, pond, lagoon, or holding area to contain liquid or solid material. (SRP: If time permits, pits, ponds, lagoons, soak holes, or holding areas should be sealed with an impermeable flexible-membrane liner.) Dike surface flow using soil, sandbags, foamed polyurethane, or foamed concrete. Absorb bulk liquid with fly ash or cement powder (pyrethrins) [R2]. Environmental consideration—water spill: If pyrethrins are dissolved, apply activated carbon at ten times the spilled amount in the region of 10 ppm or greater concentration. Use mechanical dredges or lifts to remove immobilized masses of pollutants and precipitates. (pyrethrins) [R2].

DISPOSAL Pyrethrin is highly unstable in the presence of light, moisture, and air. It is rapidly oxidized and inactivated by air. Most of the insecticidal activity of the product is destroyed by minor changes in the molecule. It could be dumped into a landfill or buried in noncrop land away from water. In each of these cases it would be better to mix the product with lime. Incineration would be an effective disposal procedure where permitted. If an efficient incinerator is not available, the product should be mixed with large amount of combustible material. Recommendable methods: Landfill, incineration, and open burning. Not recommendable method: Discharge to sewer. Peer-review: Mix with sawdust and burn at a remote place (peer-review conclusions of an IRPTC expert consultation (May 1985)) [R4]. After material has been contained, remove with contaminated soil, and place in impervious containers. The material may be incinerated in a pesticide incinerator at the specified temperature/dwell-time combination. Any liquids, sludges, or solid residues generated should be disposed of in accordance with all applicable federal, state, and local pollution control requirements. If appropriate incineration facilities are not available, material may be buried in a chemical waste landfill. Not amenable to biological treatment at sewage treatment plant.

COMMON USES A botanical insecticide, the active principles of which are pyrethrins I and II (esters of pyrethrolone and chrysanthemic acid and pyrethroic acid), cinerins I and II (esters of cinerolone and chrysanthemic and pyrethroic acids), and jasmolin I and II (jasmoline and chrysanthemic and pyrethroic acids), collectively known as the "pyrethrins" [R1]. Insecticide (pyrethrins) [R11, 1267].

STANDARD CODES EPA 311; TSCA; IATA not restricted; NFPA 704M system 2, 1, −; not listed CFR 49; CFR cab code 8; ICC STCC tariff 1−C 28 799 31; not listed AAR; Superfund designated (hazardous substances) list.

PERSISTENCE Persistent.

MAJOR SPECIES THREATENED Fish are especially sensitive to pyrethrins.

DIRECT CONTACT Pyrethrins may cause an allergic reaction to the skin of people who are sensitive to ragweed.

GENERAL SENSATION Contact with skin may cause dermatitis in sensitive persons and in about 50% of those who are sensitive to ragweed. Dermatitis may be severe and associated with rhinitis and asthma. Inhalation or absorption through the skin may produce nausea, vomiting, gastroenteritis with diarrhea. High concentrations may produce hyperexcitability, incoordination, tremors, muscular paralysis, and death due to respiratory failure. Eye contact may produce severe eye irritation. Solvents used in formulations may produce pulmonary edema.

PERSONAL SAFETY Protect against inhalation and absorption through the skin. Must wear protective clothing including NIOSH−approved rubber gloves and boots, safety goggles, and respirator whose canister is specifically approved for this material. Care must be exercised to decontaminate fully all equipment and clothing after use.

ACUTE HAZARD LEVEL Pyrethrin II is only moderately toxic to warmblooded animals by oral administration but highly toxic by parenteral routes. It is readily detoxified by hydrolysis; thus, in an experiment, rats can ingest over a 24−hour period a dose that is lethal if taken at one time; and can maintain this intake every day of their lives without apparent injury. The fatal oral dose in man, as inferred from data on rats and guinea pigs, is 1 to 2 g/kg, expressed as pyrethrin. Because the pyrethrin concentration in the various formulations is quite low, serious poisonings are highly improbable. Kerosene and naphtha, the common solvents in pyrethrum sprays, are generally more hazardous than the active ingredient. The synergists commonly found in formulations possess even lower acute toxicities in lab mammals than do pyrethrins. Although they enhance insecticidal potency, there is no evidence that they increase the mammalian toxicity of fresh pyrethrin.

CHRONIC HAZARD LEVEL The threshold-limit-value-time-weighted average for a 40–hour workweek is 5 mg/m³. There is a possibility of allergic reactions in sensitive people, especially those who are sensitive to ragweed pollen. Severe dermatitis has also been reported.

DEGREE OF HAZARD TO PUBLIC HEALTH Pyrethrin II is not very toxic by oral ingestion, but irritating to eyes, skin, and respiratory tract. It is an allergen to some people, especially to those sensitive to ragweed pollen. It is also less toxic than pyrethrin I.

OCCUPATIONAL RECOMMENDATIONS

NIOSH 10–hr time-weighted average: 5 mg/m³ (pyrethrum) [R3].

ACTION LEVEL Avoid contact with the spilled cargo. Stay upwind. Notify local air, water, and fire authorities of the accident. Evacuate area when oil-based formulation is present, due to its high flammability.

ON-SITE RESTORATION Dam stream if possible to reduce flow and prevent further dispersion of the material. Dust formulations can be dredged from the bottom. Oil formulations may float, so booms and surface skimmers may be employed. Apply activated carbon to the water to adsorb the material. Bottom pumps, vacuum systems, or dredges may be employed to remove undissolved material and sorbent from the bottom. An alternative method, under controlled conditions, is to pump the water into a suitable container, use gravity separation tanks, dual filtration systems, and an activated carbon filter. For more details, see Envirex manual, EPA 600/2–77–227. Seek professional environmental engineering assistance through EPA's Environmental Response Team (ERT), Edison, NJ, 24–hour no., 908–548–8730.

BEACH OR SHORE RESTORATION Close beach to the public until material has been removed. Scrape effected area with mechanical scraper and avoid human toxicity contact. Do not burn material.

AVAILABILITY OF COUNTERMEASURE MATERIAL Booms and surface skimmers—Army Corp of Engineers; carbon—water treatment plants, chemical companies; pumps—fire stations; dredges—Army Corps of Engineers; vacuum systems—swimming-pool operators.

DISPOSAL NOTIFICATION Notify local and state health authorities, local solid waste disposal authorities, supplier, and shipper of material.

INDUSTRIAL FOULING POTENTIAL Industrial fouling potential believed low.

PROBABLE LOCATION AND STATE Solid formulations will sink to the bottom and become incorporated with the sediments without dissolving. Liquid formulations will likely float on the surface, due to the hydrocarbon solvents.

WATER CHEMISTRY No reaction.

FOOD-CHAIN CONCENTRATION POTENTIAL Moderate; moderately persistent in the food chain.

NON-HUMAN TOXICITY LD_{50} rat male oral greater than 600 mg/kg [R10, 78]; LD_{50} mouse intraperitoneal less than 240 mg/kg [R10, 78]; LD_{50} cat female intravenous 1 mg/kg [R10, 78]; LD_{50} rat, oral 1.2 g/kg [R11, 1267];.

ECOTOXICITY LD_{50} frog injection 0.75 mg/kg [R10, 76]; LC_{50} *Coturnix* 75,000 ppm. No overt signs of toxicity to 5,000 ppm [R5]. LC_{50} channel catfish 114 µg/L/96 hr (static test); 132 µg/L/96 hr (flow-through test) (pyrethrins) [R6]; LC_{50} *Gammarus lacustris* 12 µg/L/96 hr (conditions of bioassay not specified; pyrethrins) [R6]; LC_{50} *Simocephalus serrulatus* 42 µg/L/48 hr (conditions of bioassay not specified; pyrethrins) [R6]; LC_{50} *Daphnia pulex* 25 µg/L/48 hr (conditions of bioassay not specified; pyrethrins) [R6]; LC_{50} *Pteronarcys californica* 1.0 µg/L/96 hr (conditions of bioassay not specified; pyrethrins) [R6]; LC_{50} rainbow trout 56 ppb/24 hr (conditions of bioassay not specified; pyrethrins) [R6]; LD_{50} mallards (females 1–4 mo) >10,000 mg/kg (pyrethrum) [R7].

BIODEGRADATION A few pyrethroid insecticides, including Permethrin, are theoretically susceptible to biological degradation (1). Permethrin, having similar cyclopropane ring structure as pyrethrin II, was actually found to undergo significant biological degradation in organic and mineral soil (1). Therefore, biodegradation is expected to play an important role for pyrethrin II loss in soil (SRC) [R8].

BIOCONCENTRATION Based upon an estimated log K_{ow} of 3.599 (1), the BCF for pyrethrin II can be estimated to be 320 from a recommended regression-derived equation (2,SRC). Although this BCF value suggests that bioconcentration may be important in aquatic organisms, such bioconcentration is unlikely because of rapid degradation in water and metabolism in fish (SRC) [R9].

FIRE POTENTIAL Pyrethrins burn with difficulty. (pyrethrins) [R2].

REFERENCES

1. *Farm Chemicals Handbook* 1991. Willoughby, OH: Meister, 1991, p. C–257.

2. Bureau of Explosives. *Emergency Handling of Haz Matl in Surface Trans* p. 434 (1981).

3. NIOSH. *NIOSH Pocket Guide to Chemical Hazards.* DHHS (NIOSH) Publication No. 90–117. Washington, DC: U.S. Government Printing Office, June 1990, 190.

4. United Nations. *Treatment and Disposal Methods for Waste Chemicals* (IRPTC File). Data Profile Series No. 5. Geneva, Switzerland: United Nations Environmental Programme, Dec. 1985. 156.

5. Hill, E. F., and M. B. Camardese. *Lethal Dietary Toxicities of Environmental Contaminants and Pesticides* to Coturnix. Fish and Wildlife Technical Report 2.Washington, DC: United States Department of Interior, Fish and Wildlife Service, 1986. 121.

6. Verschueren, K. *Handbook of Environmental Data of Organic Chemicals.* 2nd ed. New York, NY: Van Nostrand Reinhold Co., 1983. 1035.

7. U.S. Department of the Interior, Fish and Wildlife Service, Bureau of Sport Fisheries and Wildlife. *Handbook of Toxicity of Pesticides to Wildlife.* Washington, DC U.S. Government Printing Office, 1970. 96.

8. (1) Chapman, R. A., et al. *Bull Environ Contam Toxicol* 26: 513–9 (1981). (2) Worthing, C. R., S. B. Walker. *The Pesticide Manual* 8th ed. Suffolk, UK: The Lavenham Press Ltd. pp. 647, 726–30 (1987).

9. (1) USEPA. *Graphical Exposure Modeling System* (GEMS). CLOGP (1987). (2) Lyman, W. J., et al., *Handbook of Chemical Property Estimation Methods.* New York: McGraw-Hill, pp. 5–5 (1982).

10. Hayes, Wayland J., Jr. *Pesticides Studied in Man.* Baltimore/London: Williams and Wilkins, 1982.

11. Budavari, S. (ed.). *The Merck Index—Encyclopedia of Chemicals, Drugs and Biologicals.* Rahway, NJ: Merck and Co., Inc., 1989.

■ QUININE

CAS # 130–95–0

SIC CODES 2830; 8000; 722.

SYNONYMS (-)-quinine • 2-quinuclidine-methanol, alpha-(6-methoxy-4-quinolyl)-5-vinyl • 6′-methoxy-cinchonan-9-ol • 6-methoxy-alpha-(5-vinyl-2-quin-uclidinyl)-4-quinolinemethanol • 6-methoxycinchonine • alpha-(6-methoxy-4-quinolyl)-5-vinyl-2-quinuclidinemethanol • alpha-(6-methoxy-4-quinolyl)-5-vinyl-2-quinclidinemethanol • chinine • cinchonan-9-ol, 6′-methoxy-, (8alpha, 9r)-• Chinin (German)

MF $C_{20}H_{24}N_2O_2$

MW 324.46

pH saturated aqueous solution, 8.8.

COLOR AND FORM Triboluminescent, orthorhombic needles from absolute alcohol; bulky, white amorphous powder or crystals; crystals turn brown on exposure to air.

ODOR Odorless.

TASTE Bitter taste.

DISSOCIATION CONSTANTS $K_1 = 1.08 \times 10^{-6}$; $K_2 = 1.5 \times 10^{-10}$; pK_1 at 18°C = 5.07; $pK_2 = 9.7$

MELTING POINT Melts with some decomposition at 177°C.

SOLUBILITY 1 g dissolves in: 1,900 mL water, 760 ml boiling water, 80 ml benzene (18 ml at 50°C), 1.2 ml chloroform, 250 ml dry ether, 20 ml glycerol, 0.8 ml alcohol, 1,900 ml of 10% ammonia water; almost insoluble in petroleum ether; soluble in carbon disulfide, alkalies, and acids (with formation of salts); soluble in pyrimidine.

STABILITY AND SHELF LIFE Quinine sulfate darkens on exposure to light. Quinine sulfate capsules should be stored in tight, light-resistant containers at a temperature less than 40°C, preferably between 15–30°C. Quinine sulfate tablets should be stored in well-closed containers at a temperature less than 40°C, preferably between 15–30°C. (quinine sulfate) [R5].

OTHER CHEMICAL/PHYSICAL PROPERTIES Sublimes in high vacuum at 170–180°C; blue fluorescence is especially strong in dilute sulfuric acid; max absorption (water): 347.5 nm (log e = 3.74) (hydrate); quinine sulfate occurs as fine, needle-like, white crystals, which are usually lusterless and make a light and readily compressible mass; the drug has a persistent, very bitter taste. Quinine sulfate has solubilities of approximately 2 mg/ml in water and 8.3 mg/ml in alcohol at 25°C (quinine sulfate); microcrystalline powder; mp: 57°C; efflorescent; loses one water molecule in air, two water molecules over sulfuric acid; anhydrous at 125°C (quinine trihydrate). Crystals from ethanol; mp: 211–212.5°C; specific optical rotation: −156.4°C at 25°C/D (concentration by volume = 0.97 g in 100 ml in methanol). (quinine tartrate). Levorotatory.

COMMON USES Medication (vet): bitter stomachic, analgesic, antipyretic [R3]; medication: antimalarial agent; analgesic, and antipyretic for headache and neuralgia; medicinal to initiate labor; agent for treatment of muscle diseases; agent for treatment of varicose veins; bitter agent in carbonated beverages (trace amounts) [R1]; medicine (antimalarial) as the alkaloid or as numerous salts and derivatives; flavoring in carbonated beverages [R2].

DIRECT CONTACT Irritant to mucous membranes.

GENERAL SENSATION Upon contact with this material, the eyes become swollen, watery, and exude a sticky, viscous liquid that forms yellowish crusts. Upon ingestion, it causes dilation of the pupils. The optic nerve becomes pale and atrophic, and retina shows thready arteries; ptosis and clonic spasms of the lids result. Vision returns in from 24 to 28 hours and gradually improves. Quinine dermatitis is an occupational hazard to barbers, particularly and generally to people who work with quinine tonics, medicaments, or cosmetics. Quinine has no influence upon sound skin, but it is distinctly irritant to mucous membranes and raw surfaces. Internally it can cause a sense of fullness in the head, tinnitis aureum, slight deafness, disorders of vision, and sometimes blindness. Its physiological effects vary with the individual. Occasionally, it can cause cutaneous eruptions, such as erythemia, urticaria, herpes, purpuria, and even gangrenous affections.

PERSONAL SAFETY Wear skin protection. If intense heat or flame prevails, self-contained breathing apparatus may be required.

ACUTE HAZARD LEVEL Irritant and allergen; moderately toxic when ingested. Produces tastes in water. Toxic to fish.

CHRONIC HAZARD LEVEL Chronic allergen; moderately toxic with repeated ingestion

DEGREE OF HAZARD TO PUBLIC HEALTH Irritant and allergen; moderately toxic with acute or chronic ingestion. Emits toxic vapors when heated to decomposition.

OCCUPATIONAL RECOMMENDATIONS None listed.

ACTION LEVEL Suppress suspension of dusts. Isolate from heat.

ON-SITE RESTORATION Dredge undissolved solids from bottom. Apply carbon or peat to spill. Seek professional environmental engineering assistance through EPA's Environmental Response Team (ERT), Edison, NJ, 24-hour no., 908-548-8730.

AVAILABILITY OF COUNTERMEASURE MATERIAL Carbon—water treatment plants, sugar refineries; peat—nurseries, floral shops

DISPOSAL 1. A gas—pipe the gas into the incinerator; or lower into a pit and allow it to burn away. 2. A liquid—atomize into an incinerator. combustion may be improved by mixing with a more flammable solvent. 3. A solid—make up packages in paper or other flammable material. Burn in the incinerator; the solid may be dissolved in a flammable solvent and sprayed into the fire chamber.

DISPOSAL NOTIFICATION Notify local air authority.

INDUSTRIAL FOULING POTENTIAL Pollution of water by the effluent from a quinine factory had very little injurious effect on fish.

MAJOR WATER USE THREATENED Potable supply, recreation, fisheries.

PROBABLE LOCATION AND STATE Triboluminescent, orthorhombic needles, or bulky, white, amorphous powder. Will sink, and very little will dissolve.

WATER CHEMISTRY Saturated solution has pH of 8.8; presence of mineral acids will sponsor production of associated salts.

COLOR IN WATER Colorless.

HUMAN TOXICITY Fatal dose (adults) oral approximately 2–8 g [R8, 992].

BIOLOGICAL HALF-LIFE The plasma elimination half-life of quinine reportedly averages 8–21-hr in adults with malaria and 7–13-hr in healthy or convalescing adults. In children 1–12 years of age, the plasma elimination half-life of quinine reportedly averages 11–12 hr in those with malaria and 6 hr in those convalescing from the disease [R5].

FIRE POTENTIAL Slight [R4].

FDA Manufacturers, packers, and distributors of drug and drug products for human use are responsible for complying with the labeling, certification, and usage requirements as prescribed by the Federal Food, Drug, and Cosmetic Act, as amended (secs 201–902, 52 Stat. 1,040 et seq., as amended; 21 U.S. C. 321–392) [R6]. Quinine is a food additive permitted for direct addition to food for human consumption, as long as (1) the quantity added to food does not exceed the amount reasonably required to accomplish its intended physical, nutritive, or other technical effect in food, and (2) when intended for use in or on food, it is of appropriate food grade, and is prepared and handled as a food ingredient. Quinine may be used in food in accordance with specified conditions [R7].

REFERENCES

1. SRI.

2. Lewis, R. J., Sr. (ed.). *Hawley's Condensed Chemical Dictionary*. 12th ed. New York, NY: Van Nostrand Reinhold Co., 1993, 987.

3. Budavari, S. (ed.). *The Merck Index— Encyclopedia of Chemicals, Drugs, and Biologicals*. Rahway, NJ: Merck and Co., Inc., 1989. 1283.

4. Sax, N. I. *Dangerous Properties of Industrial Materials*. 6th ed. New York, NY: Van Nostrand Reinhold, 1984. 2349.

5. McEvoy, G. K. (ed.). *American Hospital Formulary Service—Drug Information 93*. Bethesda, MD: American Society of Hospital Pharmacists, Inc., 1993 (plus supplements, 1993). 437.

6. 21 CFR 200–299, 300–499, 820, and 860 (4/1/91).

7. 21 CFR 172.575 (4/1/91).

8. Gilman, A.G., T.W. Rall, A.S. Nies, and P. Taylor (eds.). *Goodman and Gilman's The Pharmacological Basis of Therapeutics*. 8th ed. New York, NY. Pergamon Press, 1990.

■ SACCHARIN

CAS # 81–07–2

SYNONYMS 1,2-benzisothiazol-3 (2h)-one, 1,1-dioxide • 1,2-benzisothiazolin-3-one, 1,1-dioxide • 1,2-dihydro-2-ketobenzisosulfonazole • 2,3-dihydro-3-oxobenzisosulfonazole • 2-sulphobenzoic imide • 3-benzisothiazolinone 1,1-dioxide • 3-hydroxybenzisothiazole-s,s-dioxide • anhydro-o-sulfaminebenzoic acid • benzo-2-sulphimide • benzo-sulphinide • benzoic-sulfimide • benzosulfinide • benzosulphimide • garantose • glucid • gluside • hermesetas • insoluble saccharin • kandiset • natreen • o-benzoic sulfimide • o-benzosulfimide • o-benzosulphimide • o-benzoyl sulfimide • o-benzoyl sulphimide • o-sulfobenzimide • o-sulfobenzoic acid imide • sacarina • saccharimide • saccharin acid • saccharin insoluble • saccharina • saccharine • saccharinol • saccharinose • saccharol • saxin • sucre edulcor • sucrette • sykose • zaharina • benzoylsulfonic imide • 1,2-benzisothiazolinone, 1,1-dioxide • 2-sulphobenzoic imide

MF $C_7H_5NO_3S$

MW 183.19

pH acid to litmus; pH of 0.35% aqueous solution 2.0.

COLOR AND FORM Monoclinic crystals; prisms from alcohol, leaves from water; white crystals or white crystalline powder.

ODOR Odorless or has faint aromatic odor.

TASTE In dilute aqueous solution it is 500 times as sweet as sugar; sweet taste detectable in 1:100,000 diln.

DENSITY 0.828.

MELTING POINT 228.8–229.7°C.

HEAT OF COMBUSTION At constant volume: 4,753.1 cal/g.

SOLUBILITY 1 g dissolves in 290 ml water, 25 ml boiling water, 31 ml alcohol, 12 ml acetone, and in about 50 ml glycerol; freely soluble in solution of alkali carbonates; slightly soluble in chloroform, ether; slightly soluble in benzene; dissolved by dilute solution of ammonia, solution of alkali hydroxides; water solubility of 4,000 mg/L at 25°C.

VAPOR PRESSURE 9.11×10^{-7} mm Hg at 25°C.

OCTANOL/WATER PARTITION COEFFICIENT Log K_{ow} = 0.91.

OTHER CHEMICAL/PHYSICAL PROPERTIES Perfect 100 cleavage; acicular crystals by vacuum sublimation. 1 g soluble in 2.6 ml water, 4.7 ml alcohol (calcium salt). Sol: 30.5 g/100 g 92.5% ethanol-water mixture at 25°C; 35.1 g/100 g propylene at 25°C; 36.3 g/100 ml glycol at 25°C; 13.3 g/100 ml glycerin at 25°C (calcium salt). Freely soluble in water (ammonium salt). 1 g dissolves in 1.2 ml water, about 50 ml alcohol (sodium salt). Sol: 2.6 g/100 g 92.5% ethanol-water at 25°C; 44.7 g/100 g propylene at 25°C; 46.2 g/100 ml glycol at 25°C; 55.8 g/100 g glycerin at 25°C (sodium salt). Index of refraction: 1.480 (alpha), greater than 1.523 (beta), 1.692 (gamma) (calcium salt). Index of refraction: 1.560 (alpha); 1.642 (beta); 1.733 (gamma) (sodium salt). Aqueous solutions are neutral or alkaline to litmus, but not to phenolphthalein; effloresces in dry air (sodium salt). Sublimes in vacuum.

PROTECTIVE EQUIPMENT Precautions for carcinogens: Dispensers of liquid detergent should be available. Safety pipettes should be used for all pipetting. In animal laboratory, personnel should wear protective suits (preferably disposable, one-piece, and close-fitting at ankles and wrists), gloves, hair covering, and overshoes. In chemical laboratory, gloves and gowns should always be worn; however, gloves should not be assumed to provide full protection. Carefully fitted masks or respirators may be necessary when working with particulates or gases, and disposable plastic aprons might provide addnl protection. Gowns should be of distinctive color, as a reminder that they are not to be worn outside the laboratory. (chemical carcinogens) [R9, 8].

PREVENTATIVE MEASURES For carcinogens: Smoking, drinking, eating, storage of food, or of food and beverage containers, or utensils, and the application of cosmetics should be prohibited in any laboratory. All personnel should remove gloves, if worn, after completion of procedures in which carcinogens have been used. They should wash hands, preferably using dispensers of liquid detergent, and rinse thoroughly. Consideration should be given to appropriate methods for cleaning the skin, depending on nature of the contaminant. No standard procedure can be recommended, but the use of organic solvents should be avoided. Safety pipettes should be used for all pipetting (chemical carcinogens) [R9, 8].

CLEANUP For carcinogens: A high-efficiency particulate arrestor (HEPA) or charcoal filters can be used to minimize amount of carcinogen in exhausted-air-ventilated safety cabinets, lab hoods, glove boxes, or animal rooms. Filter housing that is designed so that used filters can be transferred into plastic bag without contaminating maintenance staff is avail commercially. Filters should be placed in plastic bags immediately after removal. The plastic bag should be sealed immediately. The sealed bag should be labelled properly. Waste liquids should be placed or collected in proper containers for disposal. The lid should be secured and the bottles properly labelled. Once filled, bottles should be placed in plastic bag, so that outer surface is not contaminated. The plastic bag should also be sealed and labelled. Broken glassware should be decontaminated by solvent extraction, by chemical destruction, or in specially designed incinerators (chemical carcinogens) [R9, 15].

DISPOSAL Generators of waste (equal to, or greater than 100 kg/mo) containing this contaminant, EPA hazardous waste number U202, must conform with USEPA regulations in storage, transportation, treatment, and disposal of waste [R3].

COMMON USES Non-caloric synthetic sweetener in tablets, liquid products, soft drinks, dietetic foods, miscellaneous pharmaceuticals, chewing gum, toothpaste,

smokeless tobacco products; electroplating bath additive; cattle feed additive; antiseptic (former use); preservative to retard fermentation of food (former use) [R1]. Saccharin has been used in the sweetening of pharmaceutical tablets and in processing of tobacco [R2].

OCCUPATIONAL RECOMMENDATIONS None listed.

ECOTOXICITY LC$_{50}$ *Pimephales promelas* (fathead minnow) 30-dayold 18.3 g/L/96 hr (confidence limit 16.4–20.4 g/L). Dilution factor: 0%, 20%, 40%, 60%, 80%, and 100% of the stock solution. Behavioral observations were not recorded. Increased alkalinity values were due to a reaction between the titrant and toxicant. (saccharin, sodium salt hydrate) [R5].

IARC SUMMARY Inadequate evidence of carcinogenicity in humans. Sufficient evidence of carcinogenicity in animals. Overall evaluation: Group 2B: The agent is possibly carcinogenic to humans [R4].

BIOCONCENTRATION Based on a water solubility of 4,000 mg/L at 25°C (1) and a log K$_{ow}$ of 0.91 (2), respective bioconcentration factors (log BCF) of 0.46, and 0.75 for saccharin have been calculated using recommended regression-derived equations (3,SRC). These BCF values indicate that saccharin should not bioconcentrate among aquatic organisms (SRC) [R6].

CERCLA REPORTABLE QUANTITIES Persons in charge of vessels or facilities are required to notify the National Response Center (NRC) immediately, when there is a release of this designated hazardous substance, in an amount equal to or greater than its reportable quantity of 100 lb or 45.4 kg. The toll free number of the NRC is (800) 424–8802; in the Washington DC metropolitan area (202) 426–2675. The rule for determining when notification is required is stated in 40 CFR 302.4 (section IV. D. 3.b). (saccharin and salts) [R7].

RCRA REQUIREMENTS As stipulated in 40 CFR 261.33, when saccharin, as a commercial chemical product or manufacturing chemical intermediate or an off-specification commercial chemical product or a manufacturing chemical intermediate, becomes a waste, it must be managed according to federal or state hazardous waste regulations. Also defined as a hazardous waste is any residue, contaminated soil, water, or other debris resulting from the cleanup of a spill, into water or on dry land, of this waste. Generators of small quantities of this waste may qualify for partial exclusion from hazardous waste regulations (40 CFR 261.5). (saccharin and salts) [R8].

REFERENCES

1. SRI.

2. IARC. *Monographs on the Evaluation of the Carcinogenic Risk of Chemicals to Man.* Geneva: World Health Organization, International Agency for Research on Cancer, 1972–present (multivolume work), p. V22 125 (1980).

3. 40 CFR 240–280, 300–306, 702–799 (7/1/89).

4. IARC. *Monographs on the Evaluation of the Carcinogenic Risk of Chemicals to Man.* Geneva: World Health Organization, International Agency for Research on Cancer, 1972–present (multivolume work), p. S7 71 (1987).

5. Geiger D. L., D. J. Call, L. T. Brooke (eds.). *Acute Toxicities of Organic Chemicals to fathead Minnows (Pimephales promelas).* Vol. IV. Superior Wisconsin: University of Wisconsin–Superior, 1988. 141.

6. (1) Seidell, A. *Solubilities of Organic Compounds.* New York: Van Nostrand Co Inc (1941) (2) Hansch, C., and A. J. Leo. *Medchem Project* Issue No 26. Claremont CA: Pomona College (1985) (3) Lyman, W. J., et al., *Handbook of Chemical Property Estimation Methods.* New York McGraw-Hill, pp. 5–4, 5–10 (1982).

7. 54 FR 33419 (8/14/89).

8. 40 CFR 261.33 (7/1/88).

9. Montesano, R., H. Bartsch, E. Boyland, G. Della Porta, L. Fishbein, R. A. Griesemer, A.B. Swan, L. Tomatis, and W. Davis (eds.). *Handling Chemical Carcinogens in the Laboratory: Problems of Safety.* IARC Scientific Publications No. 33. Lyon, France: International Agency for Research on Cancer, 1979.

■ SARIN

$$(CH_3)_2CHO \diagdown \overset{\overset{\textstyle O}{\|}}{P}-F$$
$$H_3C \diagup$$

CAS # 107–44–8

SYNONYMS methylphosphonofluoridic acid 1-methyl-ethyl ester • Isopropyl-methylphosphonofluoridate • GB • isopropoxymethylphoshoryl fluoride • methlyfluorophosphonic acid isopropyl ester • phosphonofluoridic acid, methyl-, 1-methylethyl ester • isopropylester-kyseliny-methylfluorfosfonove (Czech) • methylphosphonofluoridic acid isopropyl ester • methylphosphonofluoride acid, isopropyl ester • Methylfluorphosphorsaeureisopropylester (German) • o-isopropyl methylphosphonofluoridate • phosphonofluoridic acid, methyl-, isopropyl ester

MF $C_4H_{10}FO_2P$

MW 140.09

COLOR AND FORM Liquid.

ODOR Almost odorless in pure state.

DENSITY 1.10 at 20°C.

BOILING POINT 147°C at 760 mm Hg; 56°C at 16 mm Hg.

MELTING POINT –57°C.

SOLUBILITY Miscible with water.

VAPOR PRESSURE 2.9 mm Hg at 25°C.

CORROSIVITY Slightly corrosive to steel.

STABILITY AND SHELF LIFE Unstable in the presence of water [R1].

OTHER CHEMICAL/PHYSICAL PROPERTIES Hydrolyzed by water; freezing point = –56°C; volatility: 21,900 mg/m³ at 25°C; G–agents are miscible in both polar and nonpolar solvents. They hydrolyze slowly in water at neutral or slightly acidic pH and more rapidly under strong acid or alkaline conditions. The hydrolysis products are considerablly less toxic than the original agent (G–agents); Although the G–agents are liquids under ordinary atmospheric conditions, their relatively high volatility permits them to be disseminated in vapor form (G–agents); colorless, odorless (G–agents); rapidly hydrolyzed by dilute aqueous sodium hydroxide, or sodium carbonate forming relatively non-toxic products. In the nerve gas known as sarin, one of the isopropoxy group of diisopropylphosphorofluoridate (DFP) is replaced by a methyl group.

PROTECTIVE EQUIPMENT The primary item of individual protection against toxic chemicals is the protective (gas) mask. The current US Army standard mask is the M17A1, which has as its basic component a molded rubber facepiece with large cheek pouches that hold filter pads. The filter pads consist of 6 sheets of core layer laminated between two sheets of backing layer. A speech diaphragm, incorporating an outlet valve, is attached to the faceblank at the mouth position. Large eyepieces in the faceblank provide a wide field of vision. The M17A1 mask provides complete respiratory protection against all known military toxic chemical agents, but it does not afford protection against industrial toxics such as ammonia and carbon monoxide (chemicals in war) [R3]. Airtight, impermeable clothing was developed for personnel who must enter heavily contaminated areas. This clothing is made of butyl rubber or a coated fabric and provides complete protection against liquid agents (chemicals in war) [R3]. Respiratory protection (supplied-air respirator with full facepiece or self-contained breathing apparatus) should be available where these compounds are manufactured or used and should be worn in case of emergency and overexposure. (phosphorus compounds) [R9].

PREVENTATIVE MEASURES Although resistant to liquid chemical agents, impermeable protective clothing may be penetrated after a few hr of exposure to heavy concentration of agent. Consequently, liquid contamination on the clothing must be neutralized or removed as soon as possible (chemicals in war) [R4]. Collective protection involves primarily the use of shelters where personnel can work or rest. Such shelters must be airtight to prevent the inward seepage of chemical agents and, thus, require a means of providing uncontaminated air. Such an air supply can be obtained by two methods. In small shelters a filter material called diffusion board can be incorporated into the construction or used as a liner in existing structures. For large shelters, a mechanical collective protector can be used (chemicals in war) [R10]. The U.S. Army's Decontamination and Reimpregnating Kit is used by the individual soldier for decontamination of personal equipment and clothing, and for reimpregnating the Clothing Outfit Chemical Protective Liner System. The M13 kit contains a pad of Fuller's earth powders for decontaminating the inside of the protective mask and, in the absence of the M258 Skin Decontaminating Kit, for exposed skin; two bags of chloramide powder, each containing a dye capsule, for decontaminating clothing, and reimpregnating the Clothing Outfit Chemical Protective and a cutter for cutting away heavily contaminated areas of clothing. (chemicals in war) [R11].

CLEANUP If decontamination cannot be left to natural processes, chemical neutralizers or removal must be used. Decontaminating agent DS2 is a general-purpose decontaminant. It consists of 70% diethylenetriamine, 28% ethylene glycol monomethyl ether, and 2% sodium hydroxide. DS2 reacts with both the nerve agents and blister agents to effectively reduce their hazards. Important limitations in the use of DS2 are: (a) personnel must remain masked because of the vapor; (b) rubber gloves must be worn to protect the hands; (c) it is a combustible liquid, therefore it must not be allowed to get on hot metal surfaces such as running engines or exhaust pipes. (chemicals in war) [R5].

COMMON USES Chemical warfare agent [R2].

OCCUPATIONAL RECOMMENDATIONS None listed.

HUMAN TOXICITY The lethal dose for man is estimated to be 0.01 mg/kg [R6].

NON-HUMAN TOXICITY LD_{50} rat oral 550 µg/kg [R7].

BIOCONCENTRATION Sarin is miscible in water (1); therefore, bioconcentration in aquatic organisms is not expected to be an important fate process (SRC) [R8].

REFERENCES

1. *Kirk–Othmer Encyclopedia of Chemical Technology.* 3rd ed., Volumes 1–26. New York, NY: John Wiley and Sons, 1978–1984, p. V5 398 (1979).

2. Budavari, S. (ed.). *The Merck Index— Encyclopedia of Chemicals, Drugs, and Biologicals.* Rahway, NJ: Merck and Co., Inc., 1989. 1328.

3. *Kirk–Othmer Encyclopedia of Chemical Technology.* 3rd ed., Volumes 1–26. New York, NY: John Wiley and Sons, 1978–1984, p. V5 409 (11979).

4. *Kirk–Othmer Encyclopedia of Chemical Technology.* 3rd ed., Volumes 1–26. New York, NY: John Wiley and Sons, 1978–1984, p. V5 410 (1979).

5. *Kirk–Othmer Encyclopedia of Chemical Technology.* 3rd ed., Volumes 1–26. New York, NY: John Wiley and Sons, 1978–1984, p. V5 412 (1979).

6. Zenz, C. *Occupational Medicine— Principles and Practical Applications.* 2nd ed. St. Louis, MO: Mosby–Yearbook, Inc, 1988. 676.

7. *Kirk–Othmer Encyclopedia of Chemical Technology.* 3rd ed., Volumes 1–26. New York, NY: John Wiley and Sons, 1978–1984, p. V10 786 (1980).

8. (1) Britton, K. B., *Low Temperature Effects on Sorption, Hydrolysis, and Photolysis of Organophosphates: A Literature Review.* Washington DC: U.S. Army Corps of Engineers. Special Report 86–38. NTIS AD–A178349/7/GAR p. 23 (1986).

9. International Labour Office. *Encyclopedia of Occupational Health and Safety.* Vols. I & II. Geneva, Switzerland: International Labour Office, 1983. 1684.

10. *Kirk–Othmer Encyclopedia of Chemical Technology.* 3rd ed., Volumes 1–26. New York, NY: John Wiley and Sons, 1978–1984, p. V5 411 (1979).

11. *Kirk–Othmer Encyclopedia of Chemical Technology.* 3rd ed., Volumes 1–26. New York, NY: John Wiley and Sons, 1978–1984, p. V5 414 (1979).

■ SELENIOUS ACID, DISODIUM SALT

CAS # 10102–18–8

DOT # 2630

SIC CODES 3210.

SYNONYMS disodium selenite • disodium selenium trioxide • Natriumselenit (German) • selenious acid (H_2SeO_3), disodium salt • selenious acid, disodium salt

MF $O_3Se•2Na$.

MW 172.94

COLOR AND FORM Tetragonal prisms; white crystals.

DENSITY 3.1 (water = 1).

BOILING POINT Decomposes.

MELTING POINT 710°C (1,310°F) (decomposes).

FIREFIGHTING PROCEDURES Extinguish fire using agent suitable for type of surrounding fire. Material itself does not burn or burns with difficulty. Use water in flooding quantities as fog. Use foam, dry chemical, or carbon dioxide. Wear positive-pressure self-contained breathing apparatus when fighting fires involving this material [R5].

TOXIC COMBUSTION PRODUCTS Toxic gases and vapors may be released in a fire involving sodium selenite. [R6, 3].

SOLUBILITY Freely soluble in water, insoluble in alcohol; 85 g/100 g water at 20°C.

STABILITY AND SHELF LIFE Stable in air [R1].

OTHER CHEMICAL/PHYSICAL PROPERTIES Freely soluble in water to form a slightly alkaline solution (pentahydrate).

PROTECTIVE EQUIPMENT Respirator selection, recommended by OSHA: 2 mg/m³: any dust and mist respirator with a full facepiece, if not present as fume, or any supplied-air respirator or any appropriate escape-type self-contained breathing apparatus; may require eye protection; 75 mg/m³: any powered air-purifying respirator with a dust and mist filter, if not present as fume, or any supplied-air respirator operated in a continuous-flow mode, may require eye protection; 10 mg/m³: any air-purifying full facepiece respirator with a high-efficiency particulate filter or any self-contained breathing apparatus with a full facepiece or any supplied-air respirator with full facepiece; 100 mg/m³: any supplied air respirator with a full facepiece and operated in a pressure-demand or other positive-pressure mode. Emergency or planned entry in unknown concentration or IDLH conditions: any self-contained breathing apparatus with a full facepiece and operated in a pressure-demand or other positive-pressure mode or any supplied-air respirator with a full facepiece and operated in pressure-demand or other positive-pressure mode in combination with an auxiliary self-contained breathing apparatus operated in pressure-demand or other positive-pressure mode. Escape: any air-purifying full facepiece respirator with a high-efficiency particulate filter or any appropriate escape-type self-contained breathing apparatus (selenium and selenium compounds (as Se)) [R7].

PREVENTATIVE MEASURES If the concentration of selenium in the urine of workers on a particular process is above 0.1 mg/L then action should be taken to improve the industrial hygiene of that process (selenium compounds) [R8, 2019].

CLEANUP Land spill: dig a pit, pond, lagoon, or holding area to contain liquid or solid material (SRP: if time permits, pits, ponds, lagoons, soak holes, or holding areas should be contained with a flexible impermeable membrane liner.) Cover solids with a plastic sheet to prevent dissolving in rain or firefighting water [R5].

DISPOSAL Liquid or solid—make a solution strongly acidic with hydrochloric acid. Slowly add sodium sulfite to the cold solution with stirring, thus producing sulfur dioxide, the reducer. Upon heating, dark-grey selenium and black tellurium form. Let stand overnight. Filter and dry. Ship to supplier.

SRP: at the time of review, criteria for land treatment or burial (sanitary landfill) disposal practices are subject to significant revision. Prior to implementing land disposal of waste residue (including waste sludge), consult with environmental regulatory agencies for guidance on acceptable disposal practices.

COMMON USES Removing green color from glass during its mfr; alkaloidal reagent [R1]; reagent in bacteriology; testing germination of seeds; decorating porcelain [R2]; veterinary: feed additive; mineral supplement and nutritional source of selenium for domestic animals, especially chickens and swine [R3]; in Douglas fir seedlings no browsing damage from deer was found over a 4-wk period when the soil was treated with sodium selenite (Na_2SeO_3). Similarly, rabbits preferred untreated Douglas fir foliage (10–20 µg Se/g foliage) [R4].

STANDARD CODES EPA 311; Superfund designated (hazardous substances) list.

PERSISTENCE Can persist indefinitely.

MAJOR SPECIES THREATENED >6.4 ppm in diet of rats causes growth retardation, 8 ppm in diet lethal to rats.

INHALATION LIMIT 0.2 mg/m³.

DIRECT CONTACT Irritant—dust affects respiratory tract.

GENERAL SENSATION Symptoms include garlic odor of breath. Pallor, nervousness, depression, digestive disturbances have been reported with chronic exposure.

PERSONAL SAFETY Good housekeeping, adequate ventilation, personal cleanliness, frequent change of clothing. Wear gloves and either safety glasses or chemical goggles. Wash exposed skin with water and treat with sodium thiosulfate solution. Industrial filter mask may be required.

ACUTE HAZARD LEVEL Irritant. Moderately toxic when ingested or inhaled. Toxic to fish and livestock.

CHRONIC HAZARD LEVEL Chronic irritant. Marine waters should not exceed 1/100 of 96–hour LC$_{50}$ [R21]; moderately toxic by all routes.

DEGREE OF HAZARD TO PUBLIC HEALTH Irritant. Moderately toxic by all routes. Can be absorbed through skin at chronically toxic levels. Emits toxic NaO and Se dusts when heated to decomposition.

OCCUPATIONAL RECOMMENDATIONS

OSHA 8–hr time-weighted average: 0.2 mg/m^3 (selenium and selenium compounds (as Se)) [R17]; threshold-limit 8–hr time-weighted average (TWA) 0.2 mg/m^3 (1977) (selenium and compounds, as Se) [R18, 31].

ACTION LEVEL Suppress suspension of dusts.

ON-SITE RESTORATION Use anion exchanger or scavenge on alum floc. Seek professional environmental engineering assistance through EPA's Environmental Response Team (ERT), Edison, NJ, 24–hour no. 908–548–8730.

AVAILABILITY OF COUNTERMEASURE MATERIAL Anion exchanger—water softener suppliers; alum—water treatment plants.

MAJOR WATER USE THREATENED Potable supply, recreation, fisheries.

PROBABLE LOCATION AND STATE Tetragonal prisms. Will dissolve.

FOOD-CHAIN CONCENTRATION POTENTIAL Instances reported of poisoning among human beings who consumed products of animals that had ingested seleniferous plants [R20]. Positive.

NON-HUMAN TOXICITY LD$_{50}$ lamb oral 119 mg/kg [R10, 71].

ECOTOXICITY LC$_{50}$ *Apeltes quadracus* 17,350 ppb/96 hr (conditions of bioassay not specified) (selenium salts) [R11]; LC$_{50}$ *Cyprinodon variegatus* (adult) 7,400–67,100 ppb/96 hr (conditions of bioassay not specified) (selenium salts) [R11]; LC$_{50}$ *Lagodon rhomboides* 4,400 ppb/96 hr (conditions of bioassay not specified) (selenium salts) [R11]; LC$_{50}$ *Melanogrammus aeglifinus* (larvae) 600 ppb/96 hr (conditions of bioassay not specified) (selenium salts) [R11]; LC$_{50}$ *Menidia menidia* 9,725 ppb/96 hr (conditions of bioassay not specified) (selenium salts) [R11]; LC$_{50}$ *Parlichthys detatus* (larvae) 3,500 ppb/96 hr (conditions of bioassay not specified) (selenium salts) [R11]; LC$_{50}$ *Pseudopleuronectes americanus* (larvae) 14,250–15,000 ppb/96 hr (conditions of bioassay not specified) (selenium salts) [R11]; LC$_{50}$ *Acartia clausi* 1,740 ppb/96 hr (conditions of bioassay not specified) (selenium salts) [R12]; LC$_{50}$ *Acartia tonsa* 800 ppb/96 hr (conditions of bioassay not specified) (selenium salts) [R12]; LC$_{50}$ *Callinectes sapidus* 4,600 ppb/96 hr (conditions of bioassay not specified) (selenium salts) [R12]; LC$_{50}$ *Cancer magister* (larvae) 1,040 ppb/96 hr (conditions of bioassay not specified) (selenium salts) [R12]; LC$_{50}$ *Mysidopsis bahia* (adult) 1,500 ppb/96 hr; (juvenile) 600 ppb/96 hr (conditions of bioassay not specified) (selenium salts) [R12]; LC$_{50}$ *Penaeus aztecus* 1,200 ppb/96 hr (conditions of bioassay not specified) (selenium salts) [R12]; LC$_{50}$ *Lepomis macrochirus* 28,500 ppb/96 hr (conditions of bioassay not specified) (selenium salts) [R13]; LC$_{50}$ *Pimephales promelas* (fry) 2,100 ppb/96 hr (conditions of bioassay not specified) (selenium salts) [R13]; LC$_{50}$ *Pimephales promelas* (juvenile) 5,200 ppb/96 hr (conditions of bioassay not specified) (selenium salts) [R13]; LC$_{50}$ *Pimephales promelas* (adult) 620–970 ppb/96 hr (conditions of bioassay not specified) (selenium salts) [R12]; LC$_{50}$ *Salmo gairdneri* 8,100 ppb/96 hr (conditions of bioassay not specified) (selenium salts) [R12]; LC$_{50}$ *Salvelinus* fontinalis 10,200 ppb/96 hr (conditions of bioassay not specified) (selenium salts) [R12]; LC$_{50}$ *Carassius auratus* 26,100 ppb/96 hr (conditions of bioassay not specified) (selenium salts) [R13]; LC$_{50}$ *Catostomus com-*

mersoni 31,400 ppb/96 hr (conditions of bioassay not specified) (selenium salts) [R13]; LC$_{50}$ Cyprinus carpio 35,000 ppb/96 hr (conditions of bioassay not specified) (selenium salts) [R13]; LC$_{50}$ Gambusia affinis 12,600 ppb/96 hr (conditions of bioassay not specified) (selenium salts) [R13]; LC$_{50}$ Ictalurus punctatus 13,600 ppb/96 hr (conditions of bioassay not specified) (selenium salts) [R13]; LC$_{50}$ Jordanella floridae 6,500 ppb/96 hr (conditions of bioassay not specified) (selenium salts) [R13]; LC$_{50}$ Daphnia magna 710 ppb/96 hr (conditions of bioassay not specified) (selenium salts) [R14]; LC$_{50}$ Hyallela azteca 340 ppb/96 hr (conditions of bioassay not specified) (selenium salts) [R14]; LC$_{50}$ Daphnia pulex 600–800 ppb/96 hr (conditions of bioassay not specified) (selenium salts) [R13]; LC$_{50}$ Xenopus laevis, embryo 4,000 ppb/96 hr (conditions of bioassay not specified) (selenium salts) [R13]; LC$_{50}$ Xenopus laevis, tadpole 1,500 ppb/7-day (conditions of bioassay not specified) (selenium salts) [R13]; LC$_{50}$ Anabaena variabilis 15,000–17,000 ppb/96 hr (conditions of bioassay not specified) (selenium salts) [R14]; LC$_{50}$ Anacystis ridulans 30,000–4,000 ppb/96 hr (conditions of bioassay not specified) (selenium salts) [R14]; LC$_{50}$ Snail, Physa species 24,000 ppb/96 hr (conditions of bioassay not specified) (selenium salts) [R14]; LC$_{50}$ Tanytarsus dissimilis 42,400 ppb/96 hr (conditions of bioassay not specified) (selenium salts) [R14].

BIOLOGICAL HALF-LIFE In studies on humans, three phases of elimination can usually be identified after selenite administration. The half-time of the first phase is about 1 day, the second 8–20 days, and the third 65–116 days. In some studies, the half-time of the terminal phase for selenium elimination was longer when selenomethionine was used instead of selenite. (selenite) [R9, 496].

BIODEGRADATION One suggested mechanism for selenite reduction in certain microorganisms involves attachment to a carrier protein and transformation from selenite to elemental selenium, which in turn may be oxidized to selenite by the action of Bacillus sp, as one example. (selenite) [R15].

BIOCONCENTRATION It is known that selenium accumulates in living tissues. For example, the selenium content of human blood is about 0.2 ppm. This value is about 1,000 times greater than the selenium found in surface waters on the planet earth. It is clear that the human body does accumulate or concentrate selenium with respect to the environmental levels of selenium. Selenium has been found in marine fish meal at levels of about 2 ppm. This amount is around 50,000 times greater than the selenium found in seawater. It seems obvious that marine fish are efficient concentrators of selenium. (selenium compounds) [R16].

RCRA REQUIREMENTS A solid waste containing selenium (such as sodium selenite) may become characterized as a hazardous waste when subjected to the toxicant extraction procedure listed in 40 CFR 261.24, and, if so characterized, must be managed as a hazardous waste [R19].

GENERAL RESPONSE Avoid contact with solid and dust. Keep people away. Wear goggles, self-contained breathing apparatus, and rubber overclothing (including gloves). Stop discharge if possible. Isolate and remove discharged material. Notify local health and pollution control agencies.

FIRE RESPONSE Not flammable. Poisonous gases may be produced in fire. Wear goggles, self-contained breathing apparatus, and rubber overclothing (including gloves).

EXPOSURE RESPONSE Call for medical aid. Dust or solid poisonous if inhaled or swallowed. Irritating to skin and eyes. Move to fresh air. If breathing has stopped, give artificial respiration. If breathing is difficult, give oxygen. Remove contaminated clothing, and shoes. Flush affected areas with plenty of water. If swallowed and victim is conscious, have victim drink water or milk, and have victim induce vomiting. If swallowed and victim is unconscious or having convulsions, do nothing except keep victim warm.

RESPONSE TO DISCHARGE Issue warning—water contamination. Should be removed. Provide chemical and physical treatment.

WATER RESPONSE Harmful to aquatic life in very low concentrations. May be dangerous if it enters water intakes. Notify local health and wildlife officials. Notify operators of nearby water intakes.

EXPOSURE SYMPTOMS Inhalation or ingestion: nervousness and fear followed by vomiting, then quietness and somnolence. Respiration becomes difficult, dyspnea develops, followed by opisthotonos, tetanic spasm, colonic spasm, falling blood pressure, and respiratory failure. Eyes: can cause severe damage. Skin: severe irritation may occur.

EXPOSURE TREATMENT Call a physician. Inhalation: follow general procedures for poisons. Eyes: wash with water. Skin: wash with copious amounts of water. Ingestion: induce vomiting until vomit is clear, follow with gastric lavage and saline cathartics.

REFERENCES

1. *The Merck Index*. 10th ed. Rahway, NJ: Merck Co., Inc., 1983. 1241.

2. Sax, N. I., and R. J. Lewis, Sr. (eds.). *Hawley's Condensed Chemical Dictionary*. 11th ed. New York: Van Nostrand Reinhold Co., 1987. 1072.

3. Kuney, J. H., and J. N. Nullican (eds.) *Chemcyclopedia*. Washington, DC: American Chemical Society, 1988. 212.

4. Allan, G. G. U.S. Patent No. 4388303 06/14/83 (University of Washington).

5. Association of American Railroads. *Emergency Handling of Hazardous Materials in Surface Transportation*. Washington, DC: Assoc. of American Railroads, Hazardous Materials Systems (BOE), 1987. 640.

6. Mackison, F. W., R. S. Stricoff, and L. J. Partridge, Jr. (eds.). *NIOSH/OSHA—Occupational Health Guidelines for Chemical Hazards*. DHHS (NIOSH) Publication No. 81-123 (3 vols). Washington, DC: U.S. Government Printing Office, Jan. 1981.

7. NIOSH. *Pocket Guide to Chemical Hazards*. 5th Printing/Revision. DHHS (NIOSH) Publ. No. 85-114. Washington, DC: U.S. Dept. of Health and Human Services, NIOSH/Supt. of Documents, GPO, Sept. 1985. 207.

8. International Labour Office. *Encyclopedia of Occupational Health and Safety*. Vols. I and II. Geneva, Switzerland: International Labour Office, 1983.

9. Friberg, L., G. F. Nordberg, E. Kessler, and V. B. Vouk (eds.). *Handbook of the Toxicology of Metals*. 2nd ed. Vols I, II. Amsterdam: Elsevier Science Publishers B. V., 1986, p. V2.

10. Clarke, M. L., D. G. Harvey, and D. J. Humphreys. *Veterinary Toxicology*. 2nd ed. London: Bailliere Tindall, 1981.

11. U.S. Dept of Interior/Fish and Wildlife Service Contaminant Reviews. *Selenium Hazards to Fish, Wildlife, and Invertebrates: A Synoptic Review*, Biol Rept No. (85) 1.5 p. 34 (1985).

12. U.S. Dept of Interior/Fish and Wildlife Service Contaminant Reviews. *Selenium Hazards to Fish, Wildlife, and Invertebrates: A Synoptic Review*, Biol Rept No. (85) 1.5 p. 33 (1985).

13. U.S. Dept of Interior/Fish and Wildlife Service Contaminant Reviews. *Selenium Hazards to Fish, Wildlife, and Invertebrates: A Synoptic Review*, Biol Rept No. (85) 1.5 p. 32 (1985).

14. U.S. Dept of Interior/Fish and Wildlife Service Contaminant Reviews. *Selenium Hazards to Fish, Wildlife, and Invertebrates: A Synoptic Review*, Biol Rept No. (85) 1.5 p. 31 (1985).

15. U.S. Dept of Interior/Fish and Wildlife Service Contaminant Reviews. *Selenium Hazards to Fish, Wildlife, and Invertebrates: A Synoptic Review*, Biol Rept No. (85) 1.5 p. 4 (1985).

16. Wilber, C. G. *Clin Toxicol* 17 (2): 177 (1980).

17. 29 CFR 1910.1000 (7/1/87).

18. American Conference of Governmental Industrial Hygienists. *Threshold Limit Values for Chemical Substances and*

Physical Agents and Biological Exposure Indices for 1993–1994. Cincinnati, OH: ACGIH, 1993.

19. 40 CFR 261.24 (7/1/87).

20. Jones, H. R. *Environmental Control in the Organic and Petrochemical Industries,* Noyes Data Corporation, Park Ridge, NJ. DD. 8–25, 1971.

21. *Proposed Criteria For Water Quality,* Volume I, U.S. Environmental Protection Agency. October 1973.

■ SELENIUM(IV) DISULFIDE (1:2)

CAS # 7488–56–4

SYNONYMS selenium sulfide (SeS$_2$) • Caswell No. 732A • exsel • seleen • selenium disulfide • selsun • selsun blue • sulfur selenide • selenex • sel-o-rinse • EPA pesticide chemical code 072003-

MF S$_2$Se

MW 143.08

pH 2–6.

COLOR AND FORM Brown to red-yellow; bright-orange powder.

ODOR Faint odor.

MELTING POINT <100°C.

SOLUBILITY Soluble in ammonium monosulfide; decomposes in aqueous nitric acid. Practically insoluble in water and organic solvents.

OTHER CHEMICAL/PHYSICAL PROPERTIES Decomposes in aqua regia and in nitric acid.

PROTECTIVE EQUIPMENT Workers should be provided with hand protection, overalls, eye and face protection and masks, although supplied-air respiratory protective equipment is necessary in cases where good extraction is not possible, such as in the cleaning of ventilation ducts (selenium) [R6]. Recommendations for respirator selection. Max concentra-tion for use: 2 mg/m³: Respirator Classes: Any dust and mist respirator with a full facepiece (if not present as a fume). Any supplied-air respirator; may require eye protection. Any self-contained breathing apparatus; may require eye protection (selenium compounds, as Se) [R6]. Recommendations for respirator selection. Max concentration for use: 5 mg/m³: Respirator Classes: Any powered, air-purifying respirator with a dust and mist filter (if not present as a fume); may require eye protection. Any supplied-air respirator operated in a continuous-flow mode; may require eye protection (selenium compounds, as Se) [R6]. Recommendations for respirator selection. Max concentration for use: 10 mg/m³: Respirator Classes: Any air-purifying, full facepiece respirator with a high-efficiency particulate filter. Any self-contained breathing apparatus with a full facepiece. Any supplied-air respirator with a full facepiece (selenium compounds, as Se) [R6]. Recommendations for respirator selection. Max concentration for use: 100 mg/m³: Respirator Class: Any supplied-air respirator with a full facepiece and operated in a pressure-demand or other positive-pressure mode (selenium compounds, as Se) [R6]. Recommendations for respirator selection. Condition: Emergency or planned entry into unknown concentration or IDLH conditions: Respirator Classes: Any self-contained breathing apparatus that has a full facepiece and is operated in a pressure-demand or other positive-pressure mode. Any supplied-air respirator with a full facepiece and operated in pressure-demand or other positive-pressure mode in combination with an auxiliary self-contained breathing apparatus operated in pressure-demand or other positive-pressure mode (selenium compounds, as Se) [R6]. Recommendations for respirator selection. Condition: Escape from suddenly occurring respiratory hazards: Respirator Classes: Any air-purifying, full facepiece respirator with a high-efficiency particulate filter. Any appropriate escape-type, self-contained breathing apparatus (selenium compounds, as Se) [R6]. Dispensers of liquid detergent should be available. Safety pipettes should be used for all pipetting. In animal laboratory, personnel should wear protective suits (preferably disposable, one-piece, and close-fitting at ankles and wrists), gloves, hair covering, and overshoes. In chemical laboratory, gloves and gowns should always be worn; however, gloves should not be assumed to provide full protection. Carefully fitted masks or respirators may be necessary when working with particulates or gases, and disposable plastic aprons might provide addnl protection. Gowns should be of distinctive color, as a reminder that they are not to be worn outside the laboratory (chemical carcinogens) [R16, 8].

PREVENTATIVE MEASURES Smoking, eating, and drinking at workplace prohibited, and messrooms and sanitary facilities including showers and locker rooms should be provided distant from exposure areas (selenium compounds) [R6]. SRP: The scientific literature supports the wearing of contact lenses in industrial environments, as part of a program to protect the eye against chemical compounds and minerals causing eye irritation. However, there may be individual substances whose irritating or corrosive properties are such that the wearing of contact lenses would be harmful to the eye. In those specific cases contact lenses should not be worn. Smoking, drinking, eating, storage of food, or of food and beverage containers, or utensils, and the application of cosmetics should be prohibited in any laboratory. All personnel should remove gloves, if worn, after completion of procedures in which carcinogens have been used. They should wash hands, preferably using dispensers of liquid detergent, and rinse thoroughly. Consideration should be given to appropriate methods for cleaning the skin, depending on nature of the contaminant. No standard procedure can be recommended, but the use of organic solvents should be avoided. Safety pipettes should be used for all pipetting (chemical carcinogens) [R16, 8]. In animal laboratory, personnel should remove their outdoor clothes, and wear protective suits (preferably disposable, one-piece, and close-fitting at ankles and wrists), gloves, hair covering, and overshoes. Clothing should be changed daily but discarded immediately if obvious contamination occurs; also workers should shower immediately. In chemical laboratory, gloves, and gowns should always be worn; however, gloves should not be assumed to provide full protection. Carefully fitted masks or respirators may be necessary when working with particulates or gases, and disposable plastic aprons might provide addnl protection. If gowns are of distinctive color, this will serve as a reminder that they should not be worn outside of lab (chemical carcinogens) [R16, 8]. Operations connected with synthesis and purification should be carried out under well-ventilated hood. Analytical procedures should be carried out with care and vapors evolved during procedures should be removed. Expert advice should be obtained before existing fume cupboards are used and when new fume cupboards are installed. It is desirable that there be means for decreasing the rate of air extraction, so that carcinogenic powders can be handled without powder being blown around the hood. Glove boxes should be kept under negative air pressure. Air changes should be adequate, so that concentration of vapors of volatile carcinogens will not occur (chemical carcinogens) [R16, 8]. Vertical laminar-flow biological safety cabinets may be used for containment of *in vitro* procedures, provided that the exhaust-airflow is sufficient to provide an inward air flow at the face opening of the cabinet, and contaminated air plenums that are under positive pressure are leak-tight. Horizontal laminar-flow hoods or safety cabinets, where filtered air is blown across the working area toward the operator, should never be used. Each cabinet or fume cupboard to be used should be tested before work is begun (e.g., with fume bomb), and a label fixed to it, giving date of test and average air flow measured. This test should be repeated periodically and after any structural changes (chemical carcinogens) [R16, 9]. Principles that apply to chemical or biochemical lab also apply to microbiological and cell-culture labs Special consideration should be given to route of admin. Safest method of administering volatile carcinogen is by injection of a solution. Admin by topical application, gavage, or intratracheal instillation should be performed under hood. If chemical will be exhaled, animals should be kept under hood during this period. Inhalation exposure requires special equipment. Unless specifically required, routes of admin other than in the diet should be used. Mixing of carcinogen in diet should be carried out in sealed mixers under fume hood, from which the exhaust is fitted with an efficient particulate filter. Techniques for cleaning mixer and hood should be devised before expt is begun. When mixing diets, special protective clothing and, possibly, respirators may be required (chemical carcinogens) [R16, 9]. When admin in diet or applied to skin, animals should be kept in cages with solid bottoms and sides and fitted with a filter top. When volatile carcinogens are given, filter tops should not be used. Cages that have been used to house animals that received carcinogens should be decontaminated. Cage-cleaning facilities should be installed in area in which carcinogens are being used, to avoid moving of contaminated cages. It is difficult to ensure that cages are decontaminated, and monitoring methods are necessary. Situations may exist in which the use of disposable cages should be recommended, depending on type and amount of carcinogen and efficiency with which it can be removed (chemical carcinogens) [R16, 10]. To eliminate risk that contamination in lab could build up during conduct of expt, periodic checks should be carried out on lab atmospheres, surfaces, such as walls, floors, and benches, and interior of fume hoods and airducts. As well as regular monitoring, check must be carried out after cleaning-up of spillage. Sensitive methods are required when testing lab atmospheres. Methods should, where possible, be simple and sensitive (chemical carcinogens) [R16, 10]. Rooms in which obvious contamination has occurred, such as spillage, should be decontaminated by lab personnel engaged in expt. Design of expt should avoid contamination of permanent equipment. Procedures should ensure that maintenance workers are not exposed to carcinogens. Particular care should be taken to avoid contamination of drains or ventilation ducts. In cleaning labs, procedures should be used that do not produce aerosols or dispersal of dust, i.e., wet mop or vacuum cleaner equipped with high-efficiency particulate filter on exhaust, which are avail commercially, should be used. Sweeping, brushing, and use of dry dusters or mops should be prohibited. Grossly contaminated cleaning materials should not be re-used If gowns or towels are contaminated, they should not be sent to laundry, but decontaminated or burned, to avoid any hazard to laundry personnel (chemical carcinogens) [R16, 10]. Doors leading into areas where carcinogens are used should be marked distinctively with appropriate labels. Access limited to persons involved in expt. A prominently displayed notice should give the name of the scientific investigator or other person who can advise in an emergency and who can inform others (such as firemen) on the handling of carcinogenic substances (chemical carcinogens) [R16, 11].

CLEANUP Persons not wearing protective equipment and clothing should be restricted from areas of spills until cleanup has been completed (selenium and its inorganic compounds) [R7]. If selenium or its inorganic compounds are spilled, the following steps should be taken: 1. Ventilate area of spill. 2. Collect spilled material in the most convenient and safe manner and deposit in sealed containers for reclamation or for disposal in a secured sanitary landfill. Liquid containing selenium and its inorganic compounds should be absorbed in vermiculite, dry sand, earth, or a similar material. Waste disposal method: Selenium and its inorganic compounds may be disposed of in sealed containers in a secured sanitary landfill (selenium and its inorganic compounds) [R7]. A high-efficiency particulate arrestor (HEPA) or charcoal filters can be used to minimize amount of carcinogen in exhausted-air-ventilated safety cabinets, lab hoods, glove boxes, or animal rooms. Filter housing that is designed so that used filters can be transferred into plastic bag without contaminating maintenance staff is avail commercially. Filters should be placed in plastic bags immediately after removal.

The plastic bag should be sealed immediately. The sealed bag should be labeled properly. Waste liquids should be placed or collected in proper containers for disposal. The lid should be secured and the bottles properly labeled. Once filled, bottles should be placed in plastic bag, so that outer surface is not contaminated. The plastic bag should also be sealed and labeled. Broken glassware should be decontaminated by solvent extraction, by chemical destruction, or in specially designed incinerators. (chemical carcinogens) [R16, 15].

DISPOSAL Generators of waste (equal to, or greater than 100 kg/mo) containing this contaminant, EPA hazardous waste number U205, must conform with USEPA regulations in storage, transportation, treatment, and disposal of waste [R8]. A poor candidate for incineration [R17]. The following wastewater treatment technologies have been investigated for selenium: Concentration process: Chemical precipitation (selenium) [R21]. The following wastewater treatment technologies have been investigated for selenium: Concentration process: Activated carbon (selenium) [R18]. Recommendable: Chemical reduction, solidification landfill. Not recommendable: Thermal destruction, discharge to sewer. Because the price of selenium is very high, recovery of it by processing selenium-containing waste is economical. For example, selenium refineries are equipped with wet scrubbers that employ an approximately 50% solution of hydrobromic acid containing free bromine. The acid fumes are adsorbed in a lime and soda bed, and the selenium separated from the hydrobromic acid solution by distillation is recycled. If possible, convert selenium compounds to an insoluble form with sulfur dioxide before landfill or solidification. Do not use metal and acid to reduce selenium compounds. This will produce toxic gaseous hydrogen selenide (selenium compounds) [R19]. There is no universal method of disposal that has been proved satisfactory for all carcinogenic compounds, and specific methods of chemical destruction published have not been tested on all kinds of carcinogen-

containing waste. Summary of avail methods and recommendations given must be treated as guide only (chemical carcinogens) [R16, 14]. Incineration may be only feasible method for disposal of contaminated laboratory waste from biological expt. However, not all incinerators are suitable for this purpose. The most efficient type is probably the gas-fired type, in which a first-stage combustion with a less-than-stoichiometric air:fuel ratio is followed by a second stage with excess air. Some are designed to accept aqueous and organic-solvent solutions; otherwise, it is necessary to absorb solution onto suitable combustible material, such as sawdust. Alternatively, chemical destruction may be used, esp when small quantities are to be destroyed in laboratory (chemical carcinogens) [R16, 15]. HEPA (high-efficiency particulate arrestor) filters can be disposed of by incineration. For spent charcoal filters, the adsorbed material can be stripped off at high temperature, and carcinogenic wastes generated by this treatment conducted to and burned in an incinerator. Liquid waste: Disposal should be carried out by incineration at temperature that ensure complete combustion. Solid waste: Carcasses of lab animals, cage litter, and miscellaneous solid wastes should be disposed of by incineration at temperature high enough to ensure destruction of chemical carcinogens or their metabolites (chemical carcinogens) [R16, 15]. Small quantities of some carcinogens can be destroyed using chemical reactions but no general rules can be given. As a general technique, treatment with sodium dichromate in strong sulfuric acid can be used. The time necessary for destruction is seldom known but 1–2 days is generally considered sufficient when freshly prepd reagent is used. Carcinogens that are easily oxidizable can be destroyed with milder oxidative agents, such as saturated solution of potassium permanganate in acetone, which appears to be a suitable agent for destruction of hydrazines or of compounds containing isolated carbon-carbon double bonds. Concentrated or 50% aqueous sodium hypochlorite can also be used as an oxidizing agent (chemical carcinogens) [R16, 16]. Carcinogens that are alkylating, arylating, or acylating agents *per se* can be destroyed by reaction with appropriate nucleophiles, such as water, hydroxyl ions, ammonia, thiols, and thiosulfate. The reactivity of various alkylating agents varies greatly and is also influenced by solubility of agent in the reaction medium. To facilitate the complete reaction, it is suggested that the agents be dissolved in ethanol, or similar solvents. No method should be applied until it has been thoroughly tested for its effectiveness and safety on material to be inactivated. For example, in case of destruction of alkylating agents, it is possible to detect residual compounds by reaction with 4 (4–nitrobenzyl)–pyridine (chemical carcinogens) [R16, 17].

COMMON USES Medication [R15, 1031]; medication (vet) (selenium salts) [R1].

OCCUPATIONAL RECOMMENDATIONS

OSHA 8-hr time-weighted average: 0.2 mg/m^3 (selenium compounds, as Se) [R10].

NIOSH 10-hr time-weighted average: 0.2 mg/m^3 (selenium compounds, as Se) [R3].

THRESHOLD LIMIT 8–hr time-weighted average (TWA) 0.2 mg/m^3 (1977) (selenium and compounds, as Se) [R20, 31]. Excursion Limit Recommendation: Excursions in worker exposure levels may exceed three times the TLV–TWA for no more than a total of 30 min during a work day, and under no circumstances should they exceed five times the TLV–TWA (selenium and compounds, as Se) [R20, 5].

INTERNATIONAL EXPOSURE LIMITS Germany, Sweden, former USSR: 0.1 mg/m^3 (selenium compounds, as Se) [R5].

NON-HUMAN TOXICITY LD$_{50}$ rat oral 138 mg/kg [R4]; LD$_{50}$ mouse oral 370 mg/kg [R5].

IARC SUMMARY Classification of carcinogenicity: (1) evidence in humans: inadequate; (2) evidence in animals: inadequate. Summary evaluation of carcinogenic risk to humans 3: The agent is not classifiable as to its carcinogenicity to humans (from table, selenium and selenium compounds) [R9].

CERCLA REPORTABLE QUANTITIES Per-

sons in charge of vessels or facilities are required to notify the National Response Center (NRC) immediately, when there is a release of this designated hazardous substance, in an amount equal to or greater than its reportable quantity of 10 lb or 4.54 kg. The toll-free number of the NRC is (800) 424–8802; in the Washington DC metropolitan area (202) 426–2675. The rule for determining when notification is required is stated in 40 CFR 302.4 (section IV. D. 3.b) [R11].

RCRA REQUIREMENTS U205; as stipulated in 40 CFR 261.33, when selenium sulfide, as a commercial chemical product or manufacturing chemical intermediate or an off-specification commercial chemical product or a manufacturing chemical intermediate, becomes a waste, it must be managed according to federal or state hazardous waste regulations. Also defined as a hazardous waste is any residue, contaminated soil, water, or other debris resulting from the cleanup of a spill, into water or on dry land, of this waste. Generators of small quantities of this waste may qualify for partial exclusion from hazardous waste regulations (40 CFR 261.5) [R12].

DOT *Health Hazards:* Poisonous; may be fatal if inhaled, swallowed, or absorbed through skin. Contact may cause burns to skin and eyes. Runoff from fire control or dilution water may give off poisonous gases, and cause water pollution. Fire may produce irritating or poisonous gases [R2].

Fire or Explosion: Some of these materials may burn, but none of them ignites readily. Container may explode violently in heat of fire [R2].

Emergency Action: Keep unnecessary people away; isolate hazard area and deny entry. Stay upwind, out of low areas, and ventilate closed spaces before entering. Positive-pressure self-contained breathing apparatus (SCBA) and chemical protective clothing that is specifically recommended by the shipper or manufacturer may be worn. It may provide little or no thermal protection. Structural firefighters' protective clothing is not effective for these materials. Remove and isolate contaminated clothing at the site. Call CHEM-TREC at 1–800–424–9300 as soon as possible, especially if there is no local hazardous materials team available [R2].

Fire: Small fires: Dry chemical, CO_2, water spray, or regular foam. Large Fires: Water spray, fog, or regular foam. Move container from fire area if you can do so without risk. Fight fire from maximum distance. Stay away from ends of tanks. Dike fire-control water for later disposal; do not scatter the material [R2].

Spill or Leak: Do not touch or walk through spilled material; stop leak if you can do so without risk. Fully encapsulating, vapor-protective clothing should be worn for spills and leaks with no fire. Use water spray to reduce vapors. Small Spills: Take up with sand or other noncombustible absorbent material and place into containers for later disposal. Small Dry Spills: With clean shovel place material into clean, dry container, and cover loosely; move containers from spill area. Large Spills: Dike far ahead of liquid spill for later disposal [R2].

First Aid: Move victim to fresh air and call for emergency medical care; if not breathing, give artificial respiration; if breathing is difficult, give oxygen. In case of contact with material, immediately flush skin or eyes with running water for at least 15 minutes. Speed in removing material from skin is of extreme importance. Remove and isolate contaminated clothing and shoes at the site. Keep victim quiet and maintain normal body temperature. Effects may be delayed; keep victim under observation [R2].

FDA Bottled water shall, when a composite of analytical units of equal volume from a sample is examined by the methods described in paragraph (d) (1) (II) of this section, meet the standards of chemical quality, and shall not contain selenium in excess of 0.01 mg/L (selenium) [R13]. (a) Specifications: The product contains 0.9 percent weight in weight (w/w) selenium disulfide (1 percent weight in volume (w/v)). (c)(1) Indications for use: For use on dogs as a cleansing shampoo and as an agent for removing skin debris associated with dry eczema, seborrhea, and nonspecific dermatoses. (c)(2) Amount: One to 2 ounces per application. (c)(3) Limitations: Use carefully around scrotum and eyes, covering scrotum with petrolatum. Allow the shampoo to remain for 5 to 15 minutes before thorough rinsing. Repeat treatment once or twice a week. If conditions persist or if rash or irritation develops, discontinue use, and consult a veterinarian. (selenium disulfide suspension) [R14].

REFERENCES

1. IARC. *Monographs on the Evaluation of the Carcinogenic Risk of Chemicals to Man.* Geneva: World Health Organization, International Agency for Research on Cancer, 1972–present (multivolume work), p. V9 248 (1975).

2. U.S. Department of Transportation. *Emergency Response Guidebook 1990.* DOT P 5800.5. Washington, DC: U.S. Government Printing Office, 1990, p. G–55.

3. NIOSH. *NIOSH Pocket Guide to Chemical Hazards.* DHHS (NIOSH) Publication No. 90–117. Washington, DC: U.S. Government Printing Office, June, 1990 194.

4. Sax, N. I. *Dangerous Properties of Industrial Materials.* 6th ed. New York, NY: Van Nostrand Reinhold, 1984. 2391.

5. American Conference of Governmental Industrial Hygienists. *Documentation of the Threshold Limit Values and Biological Exposure Indices.* 5th ed. Cincinnati, OH: American Conference of Governmental Industrial Hygienists, 1986. 517.

6. International Labour Office. *Encyclopedia of Occupational Health and Safety.* Vols. I and II. Geneva, Switzerland: International Labour Office, 1983. 2018.

7. Mackison, F. W., R. S. Stricoff, and L. J. Partridge, Jr. (eds.). *NIOSH/OSHA—Occupational Health Guidelines for Chemical Hazards.* DHHS (NIOSH) Publication No. 81–123 (3 vols). Washington, DC: U.S. Government Printing Office, Jan. 1981.

8. 40 CFR 240–280, 300–306, 702–799 (7/1/90).

9. IARC. *Monographs on the Evaluation*

of the Carcinogenic Risk of Chemicals to Man. Geneva: World Health Organization, International Agency for Research on Cancer, 1972–present (multivolume work), p. S7 71 (1987).

10. 29 CFR 1910.1000 (7/1/90).

11. 40 CFR 302.4 (7/1/90).

12. 40 CFR 261.33 (7/1/90).

13. 21 CFR 103.35 (4/1/91).

14. 21 CFR 524.2101 (4/1/91).

15. American Medical Association. *AMA Drug Evaluations Annual* 1991. Chicago, IL: American Medical Association, 1991.

16. Montesano, R., H. Bartsch, E. Boyland, G. Della Porta, L. Fishbein, R. A. Griesemer, A. B. Swan, L. Tomatis, and W. Davis (eds.). *Handling Chemical Carcinogens in the Laboratory: Problems of Safety.* IARC Scientific Publications No. 33. Lyon, France: International Agency for Research on Cancer, 1979.

17. USEPA. *Management of Hazardous Waste Leachate*, EPA Contract No. 68–03–2766 p. E–73 (1982).

18. USEPA. *Management of Hazardous Waste Leachate*, EPA Contract No. 68–03–2766 p. E–165 (1982).

19. United Nations. *Treatment and Disposal Methods for Waste Chemicals* (IRPTC File). Data Profile Series No. 5. Geneva, Switzerland: United Nations Environmental Programme, Dec. 1985. 278.

20. American Conference of Governmental Industrial Hygienists. *Threshold Limit Values for Chemical Substances and Physical Agents and Biological Exposure Indices for 1994–1995.* Cincinnati, OH: ACGIH, 1994.

■ SELENIUM HEXAFLUORIDE

CAS # 7783–79–1

SYNONYMS selenium fluoride • selenium fluoride (SeF_6) • selenium fluoride (SeF_6), (OC-6-11) • UN2194

MF F_6Se

MW 192.95

COLOR AND FORM Colorless gas.

DENSITY 3.25 g/L at −28°C.

BOILING POINT −34.5°C.

MELTING POINT −50.8°C.

EXPLOSIVE LIMITS AND POTENTIAL If involved in a fire the containers may violently rupture or rocket [R2].

FIREFIGHTING PROCEDURES Extinguish fire using agent suitable for type of surrounding fire. Material itself does not burn or burns with difficulty. Cool all affected containers with flooding quantities of water. Apply water from as far a distance as possible. Use foam, dry chemical, or carbon dioxide. Do not apply water to point of leak in tank car or container. Wear positive-pressure self-contained breathing apparatus when fighting fires involving this material. If fire becomes uncontrollable or container is exposed to direct flame, consider evacuation of one-third (1/3) mile radius [R2].

TOXIC COMBUSTION PRODUCTS When heated to high temperatures it may decompose to emit toxic fluoride and selenium fumes [R2].

SOLUBILITY Insoluble in water.

VAPOR PRESSURE 651.2 mm Hg at −48.7°C.

OTHER CHEMICAL/PHYSICAL PROPERTIES Covalently saturated; does not attack glass; sublimes at −63.8°C; heat of formation at 25°C: −1,029.3 kJ/mol (endothermic).

PROTECTIVE EQUIPMENT Respirator Selection, recommended by OSHA: 0.5 ppm: Any self-contained breathing apparatus. 1.25 ppm: Any supplied-air respirator operated in a continuous-flow mode. 2.5 ppm: Any self-contained breathing apparatus with a full facepiece or any supplied-air respirator with a full facepiece or any supplied-air respirator with a tight-fitting facepiece operated in a continuous-flow mode. 5 ppm: any supplied-air respirator with a full facepiece and operated in a pressure-demand or other positive-pressure mode. Emergency or planned entry in unknown concentration or IDLH conditions: Any supplied-air respirator with a full facepiece and operated in pressure-demand or other positive-pressure mode in combination with an auxiliary self-contained breathing apparatus operated in pressure-demand or other positive-pressure mode. Escape: Any air-purifying full facepiece respirator (gas mask) with a chin-style or front– or back-mounted canister providing protection against the compound of concern or any appropriate escape-type self-contained breathing apparatus. (selenium hexafluoride (as Se)) [R3].

PREVENTATIVE MEASURES All sources of selenium fumes should be fitted with exhaust ventilation systems with an air speed of at least 30 m/min (Selenium fumes) [R7, 2018].

CLEANUP If selenium hexafluoride is leaked, the following steps should be taken: 1. Ventilate area of leak. 2. Stop flow of gas. If source of leak is a cylinder and the leak cannot be stopped in place, remove the leaking cylinder to a safe place in the open air, and repair the leak or allow the cylinder to empty [R8, 2]. Because the price of selenium is very high; recovery of it by processing selenium containing waste is economical. For example, selenium refineries are equipped with wet scrubbers that employ an approximately 50% solution of hydrobromic acid containing free bromine. The acid fumes are adsorbed in a lime and soda bed, and the selenium separated from the hydrobromic acid solution by distillation is recycled. If possible, convert selenium compounds to an insoluble form with SO_2 before landfill or solidification. Do not use metal and acid to reduce selenium compounds. This will produce toxic gaseous hydrogen selenide. (selenium compounds) [R4].

DISPOSAL SRP: At the time of review, criteria for land treatment or burial (sanitary landfill) disposal practices are subject to significant revision. Prior to implementing land disposal of waste residue (including waste sludge), consult with environmental regulatory agencies for guidance on acceptable disposal practices.

COMMON USES Gaseous electric insulator [R1].

OCCUPATIONAL RECOMMENDATIONS

OSHA 8–hr time-weighted average: 0.05 ppm (0.4 mg/m³) [R5]; threshold-limit 8–hr time-weighted average (TWA) 0.05 ppm, 0.16 mg/m³, as Se (1979) [R9, 31].

RCRA REQUIREMENTS A solid waste containing selenium (such as selenium hexafluoride) may become characterized as a hazardous waste when subjected to the toxicant extraction procedure listed in 40 CFR 261.24, and, if so characterized, must be managed as a hazardous waste [R6].

DOT *Health Hazards:* Poisonous; may be fatal if inhaled or absorbed through skin. Contact may cause burns to skin and eyes. Contact with liquid may cause frostbite. Runoff from fire control or dilution water may cause pollution [R2, p. G–15].

Fire or Explosion: Some of these materials may burn, but none of them ignites readily. Cylinder may explode in heat of fire [R2, p. G–15].

Emergency Action: Keep unnecessary people away; isolate hazard area, and deny entry. Stay upwind, out of low areas, and ventilate closed spaces before entering. Self-contained breathing apparatus (SCBA) and structural firefighter's protective clothing will provide limited protection for short-term exposure to these materials. Fully encapsulated protective clothing should be worn for spills and leaks with no fire. Evacuate the leak or spill area immediately for at least 50 feet in all directions. Call CHEMTREC at 1–800–424–9300 as soon as possible, especially if there is no local hazardous materials team available [R2, p. G–15].

Fire:

Small fires: Dry chemical, CO_2, or halon. Large Fires: Water spray, fog, or standard foam is recommended. Do not get water inside container. Move container from fire area if you can do so without risk. Cool containers that are exposed to flames with water from the side until well after fire is out. Stay away from ends of tanks. Isolate area until gas has dispersed [R2, p. G–15].

Spill or Leak: Stop leak if you can do so without risk. Use water spray to reduce vapor; do not put water directly on leak or spill area. Small Spills: Flush area with flooding amounts of water. Large Spills: Dike far ahead of liquid spill for later disposal. Do not get water inside container. Isolate area until gas has dispersed [R2, p. G–15].

First Aid: Move victim to fresh air, and call for emergency medical care; if not breathing, give artificial respiration; if breathing is difficult, give oxygen. Remove and isolate contaminated clothing and shoes at the site. In case of contact with material, immediately flush skin or eyes with running water for at least 15 minutes. Keep victim quiet, and maintain normal body temperature. Effects may be delayed; keep victim under observation [R2, p. G–15]

Initial Isolation and Evacuation Distances: For a spill or leak from a drum or small container (or small leak from tank): Isolate in all directions 150 feet. [R2, Table].

REFERENCES

1. *The Merck Index.* 10th ed. Rahway, NJ: Merck Co., Inc., 1983. 1213.

2. Association of American Railroads. *Emergency Handling of Hazardous Materials in Surface Transportation.* Washington, DC: Assoc. of American Railroads, Hazardous Materials Systems (BOE), 1987. 614.

3. NIOSH. *Pocket Guide to Chemical Hazards.* 5th Printing/Revision. DHHS (NIOSH) Publ. No. 85–114. Washington, DC: U.S. Dept. of Health, and Human Services, NIOSH/Supt. of Documents, GPO, Sept. 1985. 207.

4. United Nations. *Treatment, and Disposal Methods for Waste Chemicals* (IRPTC File). Data Profile Series No. 5. Geneva, Switzerland: United Nations Environmental Programme, Dec. 1985. 278.

5. 29 CFR 1910.1000 (7/1/87).

6. 40 CFR 261.24 (7/1/87).

7. International Labour Office. *Encyclopedia of Occupational Health and Safety.* Vols. I and II. Geneva, Switzerland: International Labour Office, 1983.

8. Mackison, F. W., R. S. Stricoff, and L. J. Partridge, Jr. (eds.). *NIOSH/OSHA—Occupational Health Guidelines for Chemical Hazards.* DHHS(NIOSH) Publication No. 81–123 (3 vols). Washington, DC: U.S. Government Printing Office, Jan. 1981.

9. American Conference of Governmental Industrial Hygienists. *Threshold Limit Values for Chemical Substances and Physical Agents and Biological Exposure Indices for 1994–1995.* Cincinnati, OH: ACGIH, 1994.

■ SELENIUM SULFIDE

CAS # 7446–34–6

SYNONYMS NCI-c50033 • selenium monosulfide • selenium sulfide (SeS) • selenium sulphide • sulfur selenide (SSe) • Selensulfid (German)

MF SSe

MW 111.02

COLOR AND FORM Orange-yellow tablets or powder; bright-orange powder.

DENSITY 3.056 at 0°C.

SOLUBILITY soluble in carbon disulfide; insoluble in water, ether.

OTHER CHEMICAL/PHYSICAL PROPERTIES Paper coated with selenium sulfide blackens in mercury vapor because of formation of hydrogen sulfide. MP: decomposition at 118–119°C; decomposition in alcohol (crystalline form). Not decomposed by water or dilute acid to selenic and selenious acids.

PROTECTIVE EQUIPMENT Respirator Selection, recommended by OSHA: 2 mg/m³: any dust and mist respirator with a full facepiece, if not present as fume, or any supplied-air respirator or any appropriate escape-type self-contai... breathing apparatus may require eye protection; 75 mg/m^3: any powered air-purifying respirator with a dust and mist filter, if not present as fume, or any supplied-air respirator operated in a continuous-flow mode; may require eye protection; 10 mg/m^3: any air-purifying full facepiece respirator with a high-efficiency particulate filter or any self-contained breathing apparatus with a full facepiece or any supplied-air respirator with full facepiece; 100 mg/m^3: any supplied-air respirator with a full facepiece and operated in a pressure-demand or other positive-pressure mode. Emergency or planned entry in unknown concentration or IDLH conditions: any self-contained breathing apparatus with a full facepiece and operated in a pressure-demand or other positive-pressure mode or any supplied-air respirator with a full facepiece and operated in pressure-demand or other positive-pressure mode in combination with an auxiliary self-contained breathing apparatus operated in pressure-demand or other positive-pressure mode. Escape: any air-purifying full facepiece respirator with a high-efficiency particulate filter or any appropriate escape-type self-contained breathing apparatus. (selenium and selenium compounds (as Se)) [R3].

PREVENTATIVE MEASURES If the concentration of selenium in the urine of workers on a particular process is above 0.1 mg/L then action should be taken to improve the industrial hygiene of that process (selenium compounds) [R10, 2019]. Smoking, eating, and drinking at the workplace should be prohibited, and messrooms and sanitary facilities including showers and locker rooms should be provided at a point distant from exposure areas. (selenium compounds) [R10, 2018].

CLEANUP Because the price of selenium is very high, recovery of it by processing selenium-containing waste is economical. For example, selenium refineries are equipped with wet scrubbers that employ an approximately 50% solution of hydrobromic acid containing free bromine. The acid fumes are adsorbed in a lime and soda bed, and the selenium separated from the hydrobromic acid solution by

distillation is recycled. If possible, convert selenium compounds to an insoluble form with SO$_2$ before landfill or solidification. Do not use metal and acid to reduce selenium compounds. This will produce toxic gaseous hydrogen selenide (selenium compounds) [R4].

DISPOSAL SRP: At the time of review, criteria for land treatment or burial (sanitary landfill) disposal practices are subject to significant revision. Prior to implementing land disposal of waste residue (including waste sludge), consult with environmental regulatory agencies for guidance on acceptable disposal practices. Poor candidate for incineration [R11]. Depending upon the industrial source, selenium occurs in either the particulate or the soluble form. For the former, physical treatment processes such as sedimentation or filtration are quite effective. Lime precipitation of soluble selenium has been shown to be ineffective. Alum coagulation/coprecipitation is somewhat effective, with removal efficiencies approaching 50% achievable. Iron coagulation/coprecipitation is capable of above 80% selenium removal, with efficiency increasing with decreasing pH. Activated-carbon adsorption is not effective, except in one reported instance involving high-pH pretreatment. The most efficient process for soluble selenium appears to be anion exchange. Preoxidation of the selenite to selenate is reported to enhance exchange capacity (selenium compounds) [R12]. Chemical Treatability of Selenium; Concentration Process: Chemical Precipitation; Chemical Classification: Metals; Scale of Study: Full scale/continous flow; Type of Wastewater Used: Domestic; Influent concentration: <2.5 ppb/ 6.5 ppb. Results of Study: 0% reduction with lime; (lime dose of 350–400 ppm as calcium oxide at pH = 11.3) (selenium compounds) [R13]. Chemical Treatability of Selenium; Concentration Process: Chemical Precipitation; Chemical Classification: Metal; Scale of Study: Pilot Scale; Type of Wastewater Used: Synthetic; Influent concentration: 100 ppb Results of Study: 75% reduction with ferric chloride; (3 coagulants used: 220 ppm of alum at pH = 6.4;

40 ppm of ferric chloride at pH = 6.2; 415 ppm of lime at pH = 11.5; Chemical coagulation was followed by dual-media filtration (selenium compounds) [R14].

COMMON USES Vet: topically in eczemas and dermatomycoses [R1]. Manufacture of glass, photoelectric cells [R2]. Detergent preparation as topical antiseborrheic [R1]. Useful in treatment of *Tinea versicolor* (pityriasis versicolor), a chronic fungal infection of the skin caused by *Malassezia furfur* [R9, 1510]. As an ointment for application to the lid margins in treatment of seborrheic blepharitis. [R16, 808].

OCCUPATIONAL RECOMMENDATIONS

OSHA 8–hr time-weighted average: 0.2 mg/m^3 (selenium and selenium compounds (as Se)) [R6].

THRESHOLD LIMIT 8–hr time-weighted average (TWA) 0.2 mg/m^3 (1977) (selenium and compounds, as Se) [R15, 31]. Excursion Limit Recommendation: Excursions in worker exposure levels may exceed three times the TLV–TWA for no more than a total of 30 min during a work day, and under no circumstances should they exceed five times the TLV–TWA, provided that the TLV–TWA is not exceeded (selenium and compounds, as Se) [R15, 5].

NON-HUMAN TOXICITY LD$_{50}$ rat, oral 138 mg/kg [R5].

RCRA REQUIREMENTS As stipulated in 40 CFR 261.33, when selenium sulfide, as a commercial chemical product or manufacturing chemical intermediate or an off-specification commercial chemical product or a manufacturing chemical intermediate, becomes a waste, it must be managed according to federal or state hazardous waste regulations. Also defined as a hazardous waste is any residue, contaminated soil, water, or other debris resulting from the cleanup of a spill, into water or on dry land, of this waste. Generators of small quantities of this waste may qualify for partial exclusion from hazardous waste regulations (40 CFR 261.5) [R7].

FDA Bottled water shall, when a composite of analytical units of equal volume from a sample is examined by the methods described in paragraph (d) (1) (II) of this section, meet the standards of chemical quality, and shall not contain selenium in excess of 0.01 mg/L (selenium compounds) [R8].

REFERENCES

1. *The Merck Index*. 10th ed. Rahway, NJ: Merck Co., Inc., 1983. 1213.

2. Arena, J. M., and R. H. Drew (eds.). *Poisoning—Toxicology, Symptoms, Treatments*. 5th ed. Springfield, IL: Charles C. Thomas Publisher, 1986. 664.

3. NIOSH. *Pocket Guide to Chemical Hazards*. 5th Printing/Revision. DHHS (NIOSH) Publ. No. 85–114. Washington, DC: U.S. Dept. of Health and Human Services, NIOSH/Supt. of Documents, GPO, Sept. 1985. 207.

4. United Nations. *Treatment and Disposal Methods for Waste Chemicals* (IRPTC File). Data Profile Series No. 5. Geneva, Switzerland: United Nations Environmental Programme, Dec. 1985. 278.

5. *Kirk–Othmer Encyclopedia of Chemical Technology*. 3rd ed., Volumes 1–26. New York, NY: John Wiley and Sons, 1978–1984, p. 5 (79) 500.

6. 29 CFR 1910.1000 (7/1/87).

7. 53 FR 13382 (4/22/88).

8. 21 CFR 103.35 (4/1/86).

9. American Medical Association, Department of Drugs. *Drug Evaluations*. 6th ed. Chicago, IL: American Medical Association, 1986.

10. International Labour Office. *Encyclopedia of Occupational Health and Safety*. Vols. I and II. Geneva, Switzerland: International Labour Office, 1983.

11. USEPA. *Engineering Handbook for Hazardous Waste Incineration* p. 3–15 (1981) EPA 68–03–3025.

12. Patterson, J. W. *Industrial Wastewater Treatment Technology*. 2nd Edition, p. 402 (1985).

13. McCarty, P. L., et al., *Water Factory. 21:* Reclaimed Water, Volatile Organics, Virus, and Treatment Performance (1978) EPA–600/2–78–076 as cited in USEPA,

Management of Hazardous Waste Leachate, EPA Contract No. 68–03–2766 p. E–73 (1982).

14. Hannah, S.A., et al., *J Water Pollut Cont Fed* 49 (11): 2297–2309 (1977) as cited in USEPA, *Management of Hazardous Waste Leachate*, EPA Contract No. 68–03–2766 p. E–73 (1982).

15. American Conference of Governmental Industrial Hygienists. *Threshold Limit Values for Chemical Substances and Physical Agents and Biological Exposure Indices for 1994–1995*. Cincinnati, OH: ACGIH, 1994.

16. Grant, W. M. *Toxicology of the Eye*. 3rd ed. Springfield, IL: Charles C. Thomas Publisher, 1986.

■ SELENOUREA

CAS # 630–10–4

SYNONYMS 2-selenourea • carbamimidoselenoic acid • selenouronium • urea, 2-seleno • urea, seleno

MF CH_4N_2Se

MW 123.03

COLOR AND FORM Prisms or needles from water.

MELTING POINT 200°C (decomp).

SOLUBILITY Soluble in water.

PROTECTIVE EQUIPMENT Recommendations for respirator selection. Max concentration for use: 2 mg/m³: Respirator Classes: Any dust and mist respirator with a full facepiece (if not present as a fume). Any supplied-air respirator; may require eye protection. Any self-contained breathing apparatus; may require eye protection (selenium compounds, as Se) [R2]. Recommendations for respirator selection. Max concn for use: 5 mg/m³: Respirator Classes: Any powered, air-purifying respirator with a dust and mist filter (if not present as a fume); may require eye protection. Any supplied-air respirator operated in a continuous-flow mode; may require eye protection (selenium compounds, as Se) [R2]. Recommendations for respirator selection. Max concn for use: 10 mg/m³: Respirator Classes: Any air-purifying, full facepiece respirator with a high-efficiency particulate filter. Any self-contained breathing apparatus with a full facepiece. Any supplied-air respirator with a full facepiece (selenium cmpds, as Se) [R2]. Recommendations for respirator selection. Max concn for use: 100 mg/m³: Respirator Class: Any supplied-air respirator with a full facepiece and operated in a pressure-demand or other positive-pressure mode (selenium compounds, as Se) [R2]. Recommendations for respirator selection. Condition: Emergency or planned entry into unknown concn or IDLH conditions: Respirator Classes: Any self-contained breathing apparatus that has a full facepiece and is operated in a pressure-demand or other positive-pressure mode. Any supplied-air respirator with a full facepiece and operated in pressure-demand or other positive-pressure mode in combination with an auxiliary self-contained breathing apparatus operated in pressure-demand or other positive-pressure mode (selenium compounds, as Se) [R2]. Recommendations for respirator selection. Condition: Escape from suddenly occurring respiratory hazards: Respirator Classes: Any air-purifying, full facepiece respirator with a high- efficiency particulate filter. Any appropriate escape-type, self-contained breathing apparatus (selenium compounds, as Se) [R2].

PREVENTATIVE MEASURES SRP: The scientific literature supports the wearing of contact lenses in industrial environments, as part of a program to protect the eye against chemical compounds and minerals causing eye irritation. However, there may be individual substances whose irritating or corrosive properties are such that the wearing of contact lenses would be harmful to the eye. In those specific cases contact lenses should not be worn.

DISPOSAL Generators of waste (equal to or greater than 100 kg/mo) containing this contaminant, EPA hazardous waste number P103, must conform with USEPA regulations in storage, transportation, treatment, and disposal of waste [R5]. A poor candidate for incineration. [R11].

COMMON USES Experimental radioprotective agent [R1].

OCCUPATIONAL RECOMMENDATIONS

OSHA 8-hr time-weighted average: 0.2 ppm (0.2 mg/m³) (selenium compounds, as Se) [R7].

NIOSH 10-hr time-weighted average: 0.2 mg/m³ (selenium compounds, as Se) [R2].

THRESHOLD LIMIT 8–hr time-weighted average (TWA) 0.2 mg/m³ (1977) (selenium and compounds, as Se) [R12, 31]. Excursion Limit Recommendation: Excursions in worker exposure levels may exceed three times the TLV–TWA for no more than a total of 30 min during a work day, and under no circumstances should they exceed five times the TLV–TWA. (selenium and compounds, as Se) [R12, 5].

INTERNATIONAL EXPOSURE LIMITS Germany, Sweden, former USSR: 0.1 mg/m³ (selenium compounds, as Se) [R4].

NON-HUMAN TOXICITY LD$_{50}$ rat oral 50 mg/kg [R3]; LD$_{50}$ mouse iv 56 mg/kg [R3].

IARC SUMMARY Classification of carcinogenicity: (1) evidence in humans: inadequate; (2) evidence in animals: inadequate. Summary evaluation of carcinogenic risk to humans is level 3: The agent is not classifiable as to its carcinogenicity to humans. (from table, selenium and selenium compounds) [R6].

CERCLA REPORTABLE QUANTITIES Persons in charge of vessels or facilities are required to notify the National Response Center (NRC) immediately, when there is a release of this designated hazardous substance, in an amount equal to or greater than its reportable quantity of 1,000 lb or 454 kg. The toll-free number of the NRC is (800) 424–8802; in the Washington DC metropolitan area (202) 426–2675. The rule for determining when notification is required is stated in 40 CFR 302.4 (section IV. D. 3.b) [R8].

RCRA REQUIREMENTS P103; As stipulated in 40 CFR 261.33, when selenourea, as a

commercial chemical product or manufacturing chemical intermediate or an off-specification commercial chemical product or a manufacturing chemical intermediate, becomes a waste, it must be managed according to federal or state hazardous waste regulations. Also defined as a hazardous waste is any container or inner liner used to hold this waste or any residue, contaminated soil, water, or other debris resulting from the cleanup of a spill, into water or on dry land, of this waste. Generators of small quantities of this waste may qualify for partial exclusion from hazardous waste regulations (40 CFR 261.5 (e)) [R9].

FDA Bottled water shall, when a composite of analytical units of equal volume from a sample is examined by the methods described in paragraph (d) (1) (II) of this section, meet the standards of chemical quality, and shall not contain selenium in excess of 0.01 mg/L. (selenium) [R10].

REFERENCES

1. *Kirk–Othmer Encyclopedia of Chemical Technology*. 3rd ed., Volumes 1–26. New York, NY: John Wiley and Sons, 1978–1984, p. V17 821 (1982).

2. NIOSH. *NIOSH Pocket Guide to Chemical Hazards*. DHHS (NIOSH) Publication No. 90–117. Washington, DC: U.S. Government Printing Office, June 1990, 194.

3. Sax, N. I. *Dangerous Properties of Industrial Materials*. 6th ed. New York, NY: Van Nostrand Reinhold, 1984. 2392.

4. American Conference of Governmental Industrial Hygienists. *Documentation of the Threshold Limit Values and Biological Exposure Indices*. 5th ed. Cincinnati, OH: American Conference of Governmental Industrial Hygienists, 1986. 517.

5. 40 CFR 240–280, 300–306, 702–799 (7/1/90).

6. IARC. *Monographs on the Evaluation of the Carcinogenic Risk of Chemicals to Man*. Geneva: World Health Organization, International Agency for Research on Cancer, 1972–present (multivolume work), p. S7 71 (1987).

7. 29 CFR 1910.1000 (7/1/90).

8. 40 CFR 302.4 (7/1/90).

9. 40 CFR 261.33 (7/1/90).

10. 21 CFR 103.35 (4/1/91).

11. USEPA. *Engineering Handbook for Hazardous Waste Incineration.* p. 3–10 (1981) EPA 68–03–3025.

12. American Conference of Governmental Industrial Hygienists. *Threshold Limit Values for Chemical Substances and Physical Agents and Biological Exposure Indices for 1994–1995.* Cincinnati, OH: ACGIH, 1994.

■ SILICON DIOXIDE

CAS # 7631–86–9

SYNONYMS acticel • aerogel-200 • aerosil • aerosil-300 • aerosil-380 • aerosil-a-300 • aerosil bs-50 • aerosil-e-300 • aerosil-k-7 • aerosil m-300 • aerosil-degussa • amorphous-silica-dust • cab-o-sil m-5 • cabosil-N-5 • cabosil st-1 • carplex • carplex-30 • carplex-80 • celite-superfloss • colloidal-silicon-dioxide • corasil-ii • diatomaceous-earth • ent-25,550 • extrusil • fossil-flour • fused-silica • hk-400 • ludox-hs-40 • manosil-vn-3 • n1030 • nalco-1050 • nalfloc-N-1050 • neosyl • porasil • positive-sol-130m • positive-sol-232 • quso-51 • quso-g-30 • sg-67 • silanox-101 • silica (sio2) • silica,-amorphous-fused • silicic-anhydride • sillikolloid • siloxid • snowtex-30 • snowtex-o • superfloss • syton-2× • tokusil-tplm • u-333 • ultrasil-vh-3 • ultrasil-vn-3 • vitasil-220 • vulkasil-s • white-carbon • zeofree-80 • zipax • zorbax-sil • pigment-white-27 • ci-7811 • silica • wessalon • sipernat • opal • chalcedony • tridymite • cristobalite • quartz

MF O_2Si

MW 60.09

COLOR AND FORM Transparent crystals or amorphous powder.

TASTE Tasteless.

DENSITY 2.20 g/cm (corning 7,940 fused silica).

SURFACE TENSION 5.200 at 298 K (calculated) (Corning 7,940 fused silica).

BOILING POINT 2,503.20 K.

MELTING POINT 1,610°C.

HEAT OF COMBUSTION Reported as zero.

EXPLOSIVE LIMITS AND POTENTIAL Heating a mixture of powdered magnesium and silica (later found not to be absolutely dry) caused a violent explosion rather than the vigorous reaction anticipated [R12, 1271]. Mixtures of silica gel, and the liquid difluoride sealed in tubes at 334 mbar exploded above −196°C, presence of moisture rendering the mixture shock-sensitive at this temperature [R12, 1062]. Interaction of the yellow hexafluoride with silica to give xenon tetrafluoride oxide must be interrupted before completion to avoid the possibility of formation and detonation of xenon trioxide [R12, 1079]. Silica gel at −78°C adsorbs 4.5 wt % of ozone, and below this temperature the concentration increases rapidly. At below −112°C ozone liquefies and there is a potential explosion hazard at temperatures below −100°C if organic material is present. [R12, 1363].

SOLUBILITY Practically insoluble in water or acids, except hydrofluoric acid; very slightly soluble in alkali. Soluble in hot potassium hydroxide and hot sodium hydroxide solutions. Insoluble in ethanol.

VAPOR PRESSURE 10 mm Hg at 1,732°C.

CORROSIVITY Noncorrosive.

OTHER CHEMICAL/PHYSICAL PROPERTIES Silica occurs naturally in crystalline (quartz, cristobalite, and tridymite), cryptocrystalline (e.g., chalcedony), and amorphous (e.g., opal) forms, and the specific gravity and melting point depend on the crystalline form. Density: 2.2 (amorphous); 2.65 at 0°C (quartz); melts to a glass; lowest coefficient of expansion by heat of any known substance; dissolves (in hydrofluoric acid), forming the gas silicon tetrafluoride; slowly attacked by heating with concentration phosphoric acid; crystallized forms of silica are scarcely attacked by alkalies, whereas amorphous is sol, esp when finely divided. White to light-gray to pale-buff powder (fossil flour). Combines with alkalies under suitable conditions to form silcates. In crystalline silica, the silicon and oxygen atoms are arranged in a definite regular pattern throughout the crystal. The characteristic crystal faces of a crystalline form of silica are the outward expression of this regular arrangement of the atoms. In amorphous silica the different molecules are in a dissimilar spatial relationship one to another, with the result that there is no definite regular pattern between molecules some distance apart. Thermal conductivity = 1.37 W/m–K at 298 K (corning 7,940 fused silica). Heat of Formation = –903.2 kJ/mol at 298 K (Corning 7,940 fused silica). Heat Capacity = 37.94 J/mol–K at 298 K (Corning 7,940 fused silica). Dielectric Constant = 3.8 (Corning 7,940 fused silica).

PROTECTIVE EQUIPMENT Frequent quant dust counts and analyses must be made. Particle counts must be kept within safe limits. Workers exposed should have yearly chest exam. Airline face masks and Protective suits must be worn in situations where dust cannot be controlled [R14, 257]. Recommendations for respirator selection. Max concentration for use: 30 mg/m^3. Respirator Class(es): Any dust and mist respirator [R13]. Recommendations for respirator selection. Max concentration for use: 60 mg/m^3. Respirator Class(es): Any dust and mist respirator except single-use and quarter-mask respirators. Any supplied-air respirator. Any self-contained breathing apparatus [R15]. Recommendations for respirator selection. Max concentration for use: 150 mg/m^3. Respirator Class(es): Any powered, air-purifying respirator with a dust and mist filter. Any supplied-air respirator operated in a continuous-flow mode [R13]. Recommendations for respirator selection. Max concentration for use: 300 mg/m^3. Respirator Class(es): Any air-purifying, full facepiece respirator with a high-efficiency particulate filter. Any powered, air-purifying respirator with a tight-fitting facepiece

and a high-efficiency particulate filter. Any self-contained breathing apparatus with a full facepiece. Any supplied-air respirator with a full facepiece. Any supplied-air respirator that has a tight-fitting facepiece and is operated in a continuous-flow mode [R13]. Recommendations for respirator selection. Max concentration for use: 3,000 mg/m^3. Respirator Class(es): Any supplied-air respirator operated in a pressure-demand or other positive-pressure mode [R13]. Recommendations for respirator selection. Condition: Emergency or planned entry into unknown concentration or IDLH conditions: Respirator Class(es): Any self-contained breathing apparatus that has a full facepiece and is operated in a pressure-demand or other positive-pressure mode. Any supplied-air respirator with a full facepiece and operated in pressure-demand or other positive-pressure mode in combination with an auxiliary self-contained breathing apparatus operated in pressure-demand or other positive-pressure mode [R15]. Recommendations for respirator selection. Condition: Escape from suddenly occurring respiratory hazards: Respirator Class(es): Any air-purifying, full facepiece respirator with a high-efficiency particulate filter. Any appropriate escape-type, self-contained breathing apparatus [R15].

PREVENTATIVE MEASURES Wetting processes to control dusts must be used wherever feasible. Dust-producing operations should be segregated [R14, 257]. Contact lenses should not be worn when working with this chemical [R13]. SRP: The scientific literature for the use of contact lenses in industry is conflicting. The benefit or detrimental effects of wearing contact lenses depend not only upon the substance, but also on factors including the form of the substance, characteristics, and duration of the exposure, the uses of other eye protection equipment, and the hygiene of the lenses. However, there may be individual substances whose irritating or corrosive properties are such that the wearing of contact lenses would be harmful to the eye. In those specific cases, contact lenses should not be worn. In any event, the usual eye protection equipment should be worn even when contact lenses are in place.

DISPOSAL SRP: At the time of review, criteria for land treatment or burial (sanitary landfill) disposal practices are subject to significant revision. Prior to implementing land disposal of waste residue (including waste sludge), consult with environmental regulatory agencies for guidance on acceptable disposal practices.

COMMON USES Manufacture of glass, water glass, refractories, abrasives, ceramics, enamels; in scouring-and grinding compd, ferrosilicon; molds for castings; decolorizing and purifying oils, petroleum products, etc [R10, 1346]. Clarifying agent; filtering liquids; manufacture of heat insulators, fire brick, and fire– and acid-proof packing materials; paints; filler for paper, paints; adsorbent dynamite; in metal polishes, dentifrices; in nail polishes; in chromatography (fossil flour) [R10, 787]. food additive (diatomite) [R4]. Filtration agent; abrasive; industrial filler; in lightweight aggregate; in pozzolans; inert carrier; coating agent; insulating agent [R1]. Synthetic quartz is used for frequency control in electrical oscillators and filters and electromechanical transducers [R5]. Fossil flour is largely used as absorbent for liquid and for dispensing fluid extracts in powder form; also in cataplasms and as constituent of and excipient for pill masses [R10, 787]. Post harvest additive to control insects in stored grain, beans, peas, and soybeans. Not for use on food grains [R6]. Abrasive uses: diatomaceous earth for scouring and polishing soaps and powders; all forms for wood polishing and finishing. chemical industries: diatomaceous earth as filtering medium [R14, 430]. Pharmaceutically (as anticaking, thixotropic, thickening, gelling, tabletting agent) [R3]. Used to reduce russetting on apples [R2]. Wessalon is a flow conditioner. A carrier for liquid active substances of all types. Anticaking agent in high-concentration dry pesticide formulations. Grinding aid with waxy pesticides. Good compatibility with most reactive pesticides [R11, p. C–367].

STANDARD CODES Truck (LTL) class 150

TL class 100 rail class 70 (12, 000 lb) rail class 50 (20, 000 lb).

OCCUPATIONAL RECOMMENDATIONS None listed.

ON-SITE RESTORATION Seek professional environmental engineering assistance through EPA's Environmental Response Team (ERT), Edison, NJ, 24–hour no., 908–548–8730.

NON-HUMAN TOXICITY LD_{50} rat oral >22,500 mg/kg [R2]; LD_{50} mouse oral >15,000 mg/kg [R2].

ECOTOXICITY LC_{50} carp >10,000 mg/L/72 hr (conditions of bioassay not specified) [R2].

FIFRA REQUIREMENTS: In 1988, Congress amended FIFRA to strengthen and accelerate EPA's reregistration program. The nine-year reregistration scheme mandated by "FIFRA 88" applies to each registered pesticide product containing an active ingredient initially registered before November 1, 1984. Pesticides for which EPA had not issued Registration Standards prior to the effective date of FIFRA '88 were divided into three lists based upon their potential for exposure and other factors, with List B being of highest concern, and D of least. List: D; Case: Silica and silicates; Case No. 4081; Pesticide type: Insecticide, Fungicide, and Antimicrobiol; Case Status: RED Approved 09/91; OPP has reached a decision that some or all uses of the pesticide are eligible for reregistration, and issued a Reregistration Eligibility Document (RED). Active Ingredient (AI): silicon dioxide; AI Status: OPP has completed a Reregistration Eligibility Document (RED) for the active ingredient/case [R18]. Residues of silicon dioxide, fumed, amorphous, are exempted from the requirement of a tolerance when used as a flow control, anticaking, and carrier agent in accordance with good agricultural practices as inert (or occasionally active) ingredients in pesticide formulations applied to growing crops or to raw agricultural commodities after harvest [R16]. Silicon dioxide, fumed, amorphous exempted from the requirement of a tolerance when used as a flow control, anticaking, and carrier agent in accordance with good agricultural practice as inert (or occasionally active) ingredients in pesticide formulations applied to growing crops only [R17].

FDA The food additive silicon dioxide may be safely used in food in accordance with specified conditions [R7]. Silicon dioxide may be safely used as a component of articles intended for use in packaging, transporting, or holding food in accordance with prescribed conditions [R8]. The food additive silicon dioxide may be safely used in animal feed in accordance with (specified) conditions [R9].

REFERENCES

1. SRI.

2. Hartley, D., and H. Kidd (eds.). *The Agrochemicals Handbook*. 2nd ed. Lechworth, Herts, England: The Royal Society of Chemistry, 1987.

3. Roossoff, I. S. *Handbook of Veterinary Drugs*. New York: Springer Publishing Company, 1974. 526.

4. Furia, T. E. (ed.). *CRC Handbook of Food Additives*. 2nd ed. Cleveland: The Chemical Rubber Co., 1972. 937.

5. *Kirk-Othmer Encyclopedia of Chemical Technology*. 3rd ed., Volumes 1–26. New York, NY: John Wiley and Sons, 1978–1984, p. V20 763 (1982).

6. Spencer, E. Y. *Guide to the Chemicals Used in Crop Protection*. 6th ed. Publication 1093, Research Institute, Agriculture Canada, Ottawa, Canada: Information Canada, 1973. 457.

7. 21 CFR 172.480 (4/1/93).

8. 21 CFR 175.105 (4/1/93).

9. 21 CFR 573.940 (4/1/93).

10. Budavari, S. (ed.). *The Merck Index—Encyclopedia of Chemicals, Drugs, and Biologicals*. Rahway, NJ: Merck and Co., Inc., 1989.

11. *Farm Chemicals Handbook 1993*. Willoughby, OH: Meister Publishing Co., 1993.

12. Bretherick, L. *Handbook of Reactive Chemical Hazards*. 3rd ed. Boston, MA: Butterworths, 1985.

13. NIOSH. *NIOSH Pocket Guide to Chemical Hazards*. DHHS (NIOSH) Publication No. 90–117. Washington, DC: U.S. Government Printing Office, June 1990, 194.

14. Dreisbach, R. H. *Handbook of Poisoning*. 9th ed. Los Altos, California: Lange Medical Publications, 1977.

15. NIOSH. *NIOSH Pocket Guide to Chemical Hazards*. DHHS (NIOSH) Publication No. 90–117. Washington, DC: U.S. Government Printing Office, June 1990, p. 194.

16. 40 CFR 180.1001 (c) (7/1/92).

17. 40 CFR 180.1001 (d) (7/1/92).

18. USEPA/OPP. *Status of Pesticides in Reregistration and Special Review*. p. 255 (Mar, 1992) EPA 700–R–92–004.

■ SODIUM ACETATE

CAS # 127–09–3

SIC CODES 2833.

SYNONYMS acetic acid, sodium salt • anhydrous sodium acetate • natrium aceticum • Natriumacetat (German) • sodii acetas • fema number 3024

MF $C_2H_3O_2$ • Na

MW 82.04

COLOR AND FORM White granular powder or monoclinic crystals; colorless.

ODOR Odorless.

DENSITY 1.528 g/ml.

MELTING POINT 324°C.

AUTOIGNITION TEMPERATURE 1,125°F (sodium acetate trihydrate) [R2].

EXPLOSIVE LIMITS AND POTENTIAL Combustible [R2].

SOLUBILITY Water: 119 g/100 mL at 0°C, 170.15 g/100 mL at 100°C; slightly soluble in alcohol.

STABILITY AND SHELF LIFE Solutions are stable; effloresces in warm, dry air (sodium acetate trihydrate, USP) [R1].

OTHER CHEMICAL/PHYSICAL PROPERTIES Hygroscopic; becomes anhydrous at 120°C, decomposition at higher temp; pH of 0.1–m aqueous solution at 25°C is 8.9; efflorescent in warm air (sodium acetate trihydrate); colorless, transparent crystals or granular, crystalline powder; odorless or has faint acetous odor (sodium acetate trihydrate, USP); sol: 76.2 g/100 ml water at 0°C; 138.8 g/100 ml water at 50°C; 2.1 g/100 ml alcohol at 18°C; soluble in ether; index of refraction: 1.464 (beta); density 1.45 g/ml; MP 58°C; colorless monoclinic prisms (sodium acetate trihydrate).

COMMON USES Auxiliary in acetylations (anhydrous sodium acetate) [R4]; in photography; as reagent to eliminate effect of strong acids (buffer); for foot warmers and milk bottle warmers (sodium acetate trihydrate) [R4]. In manufacture of pharmaceuticals; soaps; purification of glucose; meat preservation; electroplating; tanning; dehydrating agent [R2]; buffer [R2]; buffer in manufacture of cellulose acetate from the triacetate; catalyst for p–(2,4–dinitroanilino) phenol for dyes; catalyst in manufacture of polyester resins; mordant in dye mfr; chemical intermediate for acetyl chloride; systemic and urinary alkalizer in human toxicity medicine; agent to treat bovine ketosis in vet medicine [R5]. Preservative in foods (as hydrate) [R5]; It is included in USP as pharmaceutic aid used in solution for hemodialysis and peritoneal dialysis (sodium acetate trihydrate, USP) [R1]. Sodium acetate is often used in total parenteral nutrition as bicarbonate precursor [R3]. Synthesis of cinnamic acid and related compounds [R6]. Used in textile dyeing, as a buffer in photography [R6] (vet): as an electrolyte source in numerous parenteral and oral mixtures. Its dietary use has been associated with incr milk production and an incr in fat content of milk [R7].

BINARY REACTANTS Potassium nitrate.

STANDARD CODES No NFPA

PERSISTENCE Acetate biodegrades at moderate rate.

DIRECT CONTACT Mild irritant, skin.

PERSONAL SAFETY Wear skin protection. If intense heat or flame are present, canister mask may be required to guard against sodium oxide dust.

ACUTE HAZARD LEVEL 32, 600 ppm was required to stimulate half of the water beetles tested. Slight ingestive hazard, but generally nontoxic. Can produce some BOD.

CHRONIC HAZARD LEVEL Unknown systemic effects.

DEGREE OF HAZARD TO PUBLIC HEALTH Slight ingestive hazard.

OCCUPATIONAL RECOMMENDATIONS None listed.

ACTION LEVEL If intense heat or flame is present, notify air authority of potential oxide dust emission.

ON-SITE RESTORATION Seek professional environmental engineering assistance through EPA's Environmental Response Team (ERT), Edison, NJ, 24–hour no., 908–548–8730. Use mixed ion-exchange resins.

AVAILABILITY OF COUNTERMEASURE MATERIAL Ion exchange resins—water softener suppliers

DISPOSAL Add to large volume water. Stir in light excess soda ash. (Add slaked lime in presence of fluoride.) Decant and neutralize in second container with 6M HCl. Route to sewage plant. Landfill sludge.

DISPOSAL NOTIFICATION Contact local sewage and solid waste authority.

INDUSTRIAL FOULING POTENTIAL >50 ppm sodium and potassium can cause foaming in boiler-feed water.

MAJOR WATER USE THREATENED Fisheries, potable supply.

PROBABLE LOCATION AND STATE Colorless crystals or white granules. Will dissolve.

COLOR IN WATER Colorless

NON-HUMAN TOXICITY LD$_{50}$ rat oral 3,500

mg/kg [R8]; LD$_{50}$ mouse oral 4,960 mg/kg [R8].

TSCA REQUIREMENTS This action promulgates standards of performance for equipment leaks of volatile organic compounds (VOC) in the synthetic organic chemical manufacturing industry (SOCMI). The intended effect of these standards is to require all newly constructed, modified, and reconstructed SOCMI process units to use the best demonstrated system of continuous emission reduction for equipment leaks of VOC, considering costs, non-air-quality health, and environmental impact and energy requirements. Sodium acetate is produced, as an intermediate or a final product, by process units covered under this subpart [R9].

FDA Manufacturers, packers, and distributors of drug and drug products for human use are responsible for complying with the labeling, certification, and usage requirements as prescribed by the Federal Food, Drug, and Cosmetic Act, as amended (secs 201–902, 52 Stat. 1040 et seq., as amended; 21 U.S. C. 321–392) [R10]. The Approved Drug Products with Therapeutic Equivalence Evaluations List identifies currently marketed prescription drug products, including sodium acetate, approved on the basis of safety and effectiveness by FDA under sections 505 and 507 of the Federal Food, Drug, and Cosmetic Act [R11]. Substances migrating to food from cotton and cotton fabrics used in dry food packaging that are generally recognized as safe for their intended use include sodium acetate [R12]. Substance added directly to human food affirmed as generally recognized as safe (GRAS) when in compliance with specified requirements [R13]. Sodium acetate used as a general-purpose food additive in animal drugs, feeds, and related products is generally recognized as safe when used in accordance with good manufacturing or feeding practice [R14].

REFERENCES

1. Osol, A., and J. E. Hoover, et al (eds.). *Remington's Pharmaceutical Sciences.* 15th ed. Easton, PA: Mack Publishing Co., 1975. 772.

2. Lewis, R. J., Sr. (ed.). *Hawley's Condensed Chemical Dictionary.* 12th ed. New York, NY: Van Nostrand Reinhold Co., 1993 1047.

3. American Medical Association. *AMA Drug Evaluations Annual 1991.* Chicago, IL: American Medical Association, 1991. 723.

4. Budavari, S. (ed.). *The Merck Index—Encyclopedia of Chemicals, Drugs, and Biologicals.* Rahway, NJ: Merck and Co., Inc., 1989. 1355.

5. SRI.

6. *Kirk-Othmer Encyclopedia of Chemical Technology.* 3rd ed., Volumes 1–26. New York, NY: John Wiley and Sons, 1978–1984, p. V1 142 (1978).

7. Rossoff, I. S. *Handbook of Veterinary Drugs.* New York: Springer Publishing Company, 1974. 529.

8. *Kirk-Othmer Encyclopedia of Chemical Technology.* 3rd ed., Volumes 1–26. New York, NY: John Wiley and Sons, 1978–1984, p. V15 581 (1981).

9. 40 CFR 60.489 (7/1/91).

10. 21 CFR 200–299, 300–499, 820, and 860 (4/1/91).

11. DHHS/ FDA. *Approved Drug Products with Therapeutic Equivalence Evaluations* 12th edition p. 3–247 (1992).

12. 21 CFR 182.70 (4/1/91).

13. 21 CFR 184.1721 (4/1/91).

14. 21 CFR 582.1721 (4/1/91).

■ SODIUM CHLORIDE

CAS # 7647–14–5

SYNONYMS Ayr • Caswell no. 754 • common salt • EPA pesticide chemical code 013905 • hypersal • Natriumchlorid (German) • rock salt • saline • salt • sea salt • ss salt • table salt • halite

MF NaCl

MW 58.44

pH 6.7–7.3; its aqueous solution is neutral.

COLOR AND FORM Colorless, transparent crystals, or white, crystalline powder.

TASTE SRP: salty.

DENSITY 2.165 at 25°C.

BOILING POINT 1,413°C.

MELTING POINT 801°C.

EXPLOSIVE LIMITS AND POTENTIAL Noncombustible [R1].

SOLUBILITY Soluble in glycerol, very slightly soluble in alcohol; 35.7 g/100 ml of water at 0°C; 39.12 g/100 ml of water at 100°C; insoluble in hydrochloric acid; soluble in ammonia.

VAPOR PRESSURE 1 mm Hg at 865°C.

STABILITY AND SHELF LIFE Store below 40°C (104°F), preferably between 15 and 30°C (59 and 86°F), unless otherwise specified by manufacturer. Protect from freezing. (sodium chloride 20% injection USP) [R4].

OTHER CHEMICAL/PHYSICAL PROPERTIES Density of saturated aqueous solution at 25°C is 1.202. Somewhat hygroscopic. A 23% aqueous solution of sodium chloride freezes at −20.5°C.

PREVENTATIVE MEASURES SRP: The scientific literature for the use of contact lenses in industry is conflicting. The benefit or detrimental effects of wearing contact lenses depend not only upon the substance, but also on factors including the form of the substance, characteristics and duration of the exposure, the uses of other eye protection equipment, and the hygiene of the lenses. However, there may be individual substances whose irritating or corrosive properties are such that the wearing of contact lenses would be harmful to the eye. In those specific cases, contact lenses should not be worn. In any event, the usual eye protection equipment should be worn even when contact lenses are in place.

CLEANUP Solids: Collect and remove with a broom in a large bucket. Dilute with water. Drain into the sewer with sufficient water. For solution: Wipe with mop or use water aspiration. Drain into the sewer with sufficient water [R5].

DISPOSAL Put into large vessel containing water. Discharge into the sewer with sufficient water [R5].

COMMON USES Essential in diet to maintain chloride balance in body. In the production of chemicals (sodium hydroxide, soda ash, hydrogen chloride, chlorine, metallic sodium), ceramic glazes, metallurgy, curing hides, food preservative, mineral waters, soap manufacture (salting out), home water softeners, highway deicing, regeneration of ion-exchange resins, photography, food seasoning, herbicide, fire extinguishing, nuclear reactors, mouthwash, medicine (heat exhaustion), salting out dyestuffs, supercooled solutions. Single crystals are used for spectroscopy, UV, and infrared transmissions [R1]. Medication: Electrolyte replenisher, emetic; topical anti-inflammatory [R2]. Medication (vet): Essential nutrient factor. May be given orally as emetic, stomachic, laxative, or to stimulate thirst (prevention of calculi). Intravenously as isotonic solution to raise blood volume, to combat dehydration. Locally as wound irrigant, rectal douche [R2]. Used in postemergence weed control in table beets (humid regions) [R3].

OCCUPATIONAL RECOMMENDATIONS None listed.

HUMAN TOXICITY An ingestion of 0.5–1 g/kg can be toxic to most patients. [R13, 545].

NON-HUMAN TOXICITY LD_{50} rat, oral 3,000 mg/kg [R5]; LD_{50} mouse ip 2,602 mg/kg [R5]; LD_{50} mouse oral 4,000 mg/kg [R6].

FIFRA REQUIREMENTS: Residues of sodium chloride are exempted from the requirement of a tolerance when used as a solid diluent carrier in accordance with good agricultural practices as inert (or occasionally active) ingredients in pesticide formulations applied to growing crops or to raw agricultural commodities after harvest [R11]. In 1988, Congress amended FIFRA to strengthen and accelerate EPA's reregistration program. The nine-year reregistration scheme mandated by "FIFRA

88" applies to each registered pesticide product containing an active ingredient initially registered before November 1, 1984. Pesticides for which EPA had not issued Registration Standards prior to the effective date of FIFRA '88 were divided into three lists based upon their potential for exposure and other factors, with List B being of highest concern, and D of least. List: D; Case: Inorg. halides; Case No. 4051; Pesticide type: Insecticide (molluscicide), fungicide, herbicide, and antimicrobiol; Case Status: Awaiting Data/Data in Review: OPP awaits data from the pesticide's producer(s) regarding its human health or environmental effects, or OPP has received and is reviewing such data, in, order to reach a decision about the pesticide's eligibility for reregistration; Active Ingredient (AI): Sodium chloride; AI Status: The producer(s) of the pesticide has made commitments to conduct the studies and pay the fees required for reregistration, and is meeting those commitments in a timely manner [R12].

FDA Manufacturers, packers, and distributors of drug and drug products for human use are responsible for complying with the labeling, certification, and usage requirements as prescribed by the Federal Food, Drug, and Cosmetic Act, as amended (secs 201–902, 52 Stat. 1040 et seq., as amended; 21 U.S. C. 321–392) [R7]. The Approved Drug Products with Therapeutic Equivalence Evaluations List identifies currently marketed prescription drug products, including sodium chloride, approved on the basis of safety and effectiveness by FDA under sections 505 and 507 of the Federal Food, Drug, and Cosmetic Act [R8]. Substances migrating to food from cotton and cotton fabrics used in dry food packaging that are generally recognized as safe for their intended use include sodium chloride [R9]. Substances migrating to food from paper and paperboard products used in food packaging that are generally recognized as safe for their intended use include sodium chloride [R10].

REFERENCES

1. Lewis, R. J., Sr. (ed.). *Hawley's Condensed Chemical Dictionary.* 12th ed. New York, NY: Van Nostrand Reinhold Co., 1993, 1052.

2. Budavari, S. (ed.). *The Merck Index— Encyclopedia of Chemicals, Drugs, and Biologicals.* Rahway, NJ: Merck and Co., Inc., 1989. 1359.

3. *Farm Chemicals Handbook 1992.* Willoughby, OH: Meister Publishing Co., 1992, p. C–309.

4. U.S. Pharmacopeial Convention. *U.S. Pharmacopeia Dispensing Information* (USP DI); Drug Information for the Health Care Professional, 12th ed., V. I p. 2482 (1992).

5. ITII. *Toxic and Hazarous Industrial Chemicals Safety Manual.* Tokyo, Japan: The International Technical Information Institute, 1982. 472.

6. *Kirk–Othmer Encyclopedia of Chemical Technology.* 3rd ed., Volumes 1–26. New York, NY: John Wiley and Sons, 1978–1984, p. V15 581 (1981).

7. 21 CFR 200–299, 300–499, 820, and 860 (4/1/91).

8. DHHS/FDA. *Approved Drug Products with Therapeutic Equivalence Evaluations.* 12th edition p. 3–248 (1992).

9. 21 CFR 182.70 (4/1/91).

10. 21 CFR 182.90 (4/1/91).

11. 40 CFR 180.1001 (c) (7/1/91).

12. USEPA/OPP. *Status of Pesticides in Reregistration and Special Review.* p. 242 (Mar, 1992) EPA 700–R–92–004.

13. Ellenhorn, M. J., and D. G. Barceloux. *Medical Toxicology—Diagnosis and Treatment of Human Poisoning.* New York, NY: Elsevier Science Publishing Co., Inc. 1988.

■ SODIUM FERROCYANIDE

CAS # 13601–19–9

SIC CODES 3861; 3356; 2818.

SYNONYMS sodium hexacyanoferrate(II) •

ferrate(4-), hexacyano-, tetrasodium • ferrate(4-), hexakis (cyano-c)-, tetrasodium, (OC-6-11) • sodium ferrocyanide ($Na_4(Fe(CN)_6)$) • sodium hexacyanoferrate(II) • sodium hexacyanoferrate(II) • sodium prussiate yellow • tetrasodium ferrocyanide • tetrasodium hexacyanoferrate • tetrasodium hexacyanoferrate(4-) • yellow prussiate of soda

MF $C_6FeN_6 \cdot 4Na$

MW 267.06

COLOR AND FORM Yellow solid.

ODOR Odorless.

DENSITY 1.458.

STABILITY AND SHELF LIFE Slightly efflorescent; steady dehydration occurs above 50°C (decahydrate) [R7, 1113].

OTHER CHEMICAL/PHYSICAL PROPERTIES Solubility (% in water): 10.2% at 1°C; 14.7% at 17°C; 17.6% at 25°C; 28.1% at 53°C; 39% at 85°C; 39.7% at 96.6°C (calculated as anhydrous); solubility (g/100 cc water): 31.85 g at 20°C, 156.5 g at 98°C; index of refraction: 1.519, 1.530, 1.544; density 1.458 (decahydrate) pale-yellow, monoclinic crystals; practically insoluble in most organic solvents; becomes anhydrous at 81.5°C; decomposition 435°C, forming sodium cyanide, iron, carbon, and nitrogen (decahydrate).

PROTECTIVE EQUIPMENT U.S. Bureau of Mines respirator for excessive dustiness; safety goggles [R3].

CLEANUP Heated waste water containing iron cyanide complex is treated with sodium hypochlorite [R4].

COMMON USES In ore flotation; in photography for bleaching, toning, and fixing; emulsion polymerization catalyst; to prevent caking of rock salt [R7, 1113]; additive to pickling baths; peptizing agent in rubber; arc stabilizer in welding rods coatings [R7, 1114]; manufacture of sodium ferricyanide; blue pigments; blue print paper; metallurgy; tanning; dyes; anticaking agent for sodium chloride (decahydrate) [R1]; as food additive permitted in feed and drinking water of animals or for treatment of food producing animals, also permitted in food for human consumption [R2].

BINARY REACTANTS Acids.

STANDARD CODES No NFPA.

PERSISTENCE Sunlight will degrade to much more toxic CN^-.

MAJOR SPECIES THREATENED Aquatic life.

DIRECT CONTACT Mild skin irritant.

PERSONAL SAFETY Wear skin protection. Intense heat or acid may necessitate use of self-contained breathing apparatus or HCN–approved mask.

ACUTE HAZARD LEVEL Mild irritant. Sunlight can release levels of CN toxic to fish from 1.45 ppm ferrocyanide [R6]; Slight ingestive hazard. Sunlight may force production of toxic CN^-. Emits poisonous HCN when heated to decomposition or contacted with acid.

CHRONIC HAZARD LEVEL Slight chronic ingestion and inhalation threat. Exposure to sunlight converts to CN and hence increases hazard many fold.

DEGREE OF HAZARD TO PUBLIC HEALTH Mild irritant. Slight acute and chronic ingestion hazard. Slight hazard from chronic inhalation. Emits highly poisonous HCN when contacted with acid or heated to decomposition. Major hazard is water exposed to sunlight for some time, allowing CN to form.

OCCUPATIONAL RECOMMENDATIONS None listed.

ACTION LEVEL If intense heat is present notify air authority and evacuate area. Prevent suspension of dusts.

ON–SITE ACTION Seek professional environmental engineering assistance through EPA's Environmental Response Team (ERT), Edison, NJ, 24–hour no. 908–548–8730 Use anion exchanger or precipitate with ferric salt.

AVAILABILITY OF COUNTERMEASURE MATERIAL Anion exchanger—water softener supplier; $FeCl_3$—water treatment plants, sugar refineries.

DISPOSAL Landfill, or return to supplier.

DISPOSAL NOTIFICATION Contact local landfill authority.

MAJOR WATER USE THREATENED Fisheries, potable supply, recreation.

PROBABLE LOCATION AND STATE Yellow crystal. Will dissolve.

WATER CHEMISTRY Sodium ferrocyanide oxidizes to ferricyanide in air and light. Decomposes with heating. Heat and intense sunlight ultimately will cause release of CN at a yield of 6.5 ppm CN from 100 ppm, 3 ppm CN from 20 ppm, 1.5 ppm CN from 10 ppm, and 0.5 ppm CN from 5 ppm [R6]. Can form insoluble salts with heavy metals such as copper.

COLOR IN WATER Yellow

FDA It is used as anticaking agent in salt for oral use at level not to exceed 13 ppm or 0.00013% by FDA regulations [R5].

REFERENCES

1. Hawley, G. G. *The Condensed Chemical Dictionary*. 9th ed. New York: Van Nostrand Reinhold Co., 1977. 791.

2. Sax, N. I. *Dangerous Properties of Industrial Materials*,,. 5th ed. New York: Van Nostrand Reinhold, 1979. 981.

3. U.S. Coast Guard, Department of Transportation. *CHRIS—Hazardous Chemical Data. Manual* Two. Washington, DC: U.S. Government Printing Office, Oct., 1978.

4. Shibuya, et al., *Treatment of Waste Water Containing an Iron Cyanide Complex by Heating to above 130°C*; Japan patent number 74 32269 8/29/74 (Dainichi–Nippon Cables, Ltd).

5. Rossoff, I. S. *Handbook of Veterinary Drugs*. New York: Springer Publishing Company, 1974. 539.

6. Burdick, G. E., and M. Lipshuitz, "Toxicity of Ferro and Ferricyanide Solutions to Fish and Determination of the Cause of Mortality," *Trans American Fisheries Society*, Vol. 78, 1 94%.

7. *The Merck Index*. 9th ed. Rahway, New Jersey: Merck and Co., Inc., 1976.

■ SODIUM 2,4–DICHLOROPHENOXYETHYL SULFATE

CAS # 136–78–7

SIC CODES 2879.

SYNONYMS 2,4-des sodium • 2,4-des-Na • 2,4-Desnatrium (German) • 2,4-dichlorophenoxyethyl hydrogen sulfate sodium salt • 2,4-dichlorophenoxyethyl sulfate, sodium salt • 2-(2,4-dichlorophenoxy) ethanol hydrogen sulfate sodium salt • crag herbicide • crag herbicide-1- • crag sesone • disul • disul-Na • disul-sodium • ethanol 2-(2,4-dichlorophenoxy)-, hydrogen sulfate sodium salt • ethanol, 2-(2,4-dichlorophenoxy)-, hydrogen sulfate, sodium salt • experimental herbicide-1 • Natrium-2,4-dichlorphenoxyaethylsulfat (German) • ses • ses-t • seson • sesone • sodium 2,4-dichlorophenoxyethyl sulfate • sodium 2,4-dichlorophenyl cellosolve sulfate • sodium 2-(2,4-dichlorophenoxy)ethyl sulfate • crag-1 • 2, 4-des-na • ses • disul-na • crag-ses • 2, 4-diehlorophenoxyethyl hydrogen sulfate, sodium salt

MF $C_8H_7Cl_2O_5S \cdot Na$

MW 309.10

COLOR AND FORM White crystalline solid; colorless solid.

ODOR Odorless.

DENSITY 1.70 at 20°C.

MELTING POINT 245°C decomposition.

SOLUBILITY Solubility in hard water containing 260 ppm calcium carbonate: 4%; solubility in water: 25.5% at 25°C; insoluble in most organic solvents except methanol; very soluble in benzene; slightly soluble in acetone; 90% water-soluble.

VAPOR PRESSURE 0.1 mm Hg at 20°C.

STABILITY AND SHELF LIFE Non-volatile [R1].

OTHER CHEMICAL/PHYSICAL PROPERTIES Precipitated by soluble calcium salts; readily hydrolyzed at pH values below 5.5.

PROTECTIVE EQUIPMENT Wear appropriate equipment to prevent repeated or prolonged skin contact [R3]. Wear eye protection to prevent reasonable probability of eye contact [R3]. Recommendations for respirator selection. Max concn for use: 50 mg/m³: Respirator Class(es): Any dust and mist respirator [R3]. Recommendations for respirator selection. Max concentration for use: 100 mg/m³: Respirator Class(es): Any dust and mist respirator except single-use and quarter-mask respirators. Any supplied-air respirator. Any self-contained breathing apparatus [R3]. Recommendations for respirator selection. Max concn for use: 250 mg/m³: Respirator Class(es): Any powered, air-purifying respirator with a dust and mist filter. Any supplied-air respirator operated in a continuous-flow mode [R3]. Recommendations for respirator selection. Max concn for use: 500 mg/m³: Respirator Class(es): Any air-purifying, full facepiece respirator with a high-efficiency particulate filter. Any self-contained breathing apparatus with a full facepiece. Any supplied-air respirator with a full facepiece. Any powered, air-purifying respirator with a tight-fitting facepiece, and a high-efficiency particulate filter. Any supplied-air respirator that has a tight-fitting facepiece and is operated in a continuous-flow mode. May require eye protection [R3]. Recommendations for respirator selection. Max concn for use: 5,000 mg/m³: Respirator Class(es): Any supplied-air respirator with a full facepiece and operated in a pressure-demand or other positive-pressure mode [R3]. Recommendations for respirator selection. Condition: Emergency or planned entry into unknown concn or IDLH conditions: Respirator Class(es): Any self-contained breathing apparatus that has a full facepiece and is operated in a pressure-demand or other positive-pressure mode. Any supplied-air respirator with a full facepiece and operated in pressure-demand or other positive-pressure mode in combination with an auxiliary self-contained breathing apparatus operated in pressure-demand or other positive-pressure mode [R3]. Recommendations for respirator selection. Condition: Escape from suddenly occurring respiratory hazards: Respirator Class(es): Any air-purifying, full facepiece respirator with a high-efficiency particulate filter. Any appropriate escape-type, self-contained breathing aparatus. [R3].

PREVENTATIVE MEASURES Workers should wash promptly when skin becomes contaminated [R3]. Work clothing should be changed daily if it is reasonably probable that the clothing may be contaminated [R3]. Remove clothing promptly if it is non-impervious clothing and becomes contaminated [R8]. SRP: The scientific literature supports the wearing of contact lenses in industrial environments, as part of a program to protect the eye against chemical compounds and minerals causing eye irritation. However, there may be individual substances whose irritating or corrosive properties are such that the wearing of contact lenses would be harmful to the eye. In those specific cases contact lenses should not be worn. Contact lenses should not be worn when working with this chemical. [R3].

CLEANUP 1. Ventilate area of spill. 2. Collect material in most convenient and safe manner and deposit in sealed containers for reclamation or for disposal in secured sanitary landfill. 2. Liquid containing crag herbicide should be absorbed in vermiculite, dry sand, earth, or similar material. [R10, 324].

COMMON USES Used as an herbicide [R10, p. II–224].

PERSISTENCE Converted in soil to ethanol form, which oxidizes to 2,4–d. applied at 2.1 lb/acre persisted in soil for 6 weeks.

DIRECT CONTACT Strong solutions may be irritating to the skin. Ingestion.

GENERAL SENSATION May cause liver, kidney damage [R7].

PERSONAL SAFETY Self-contained breathing apparatus must be worn. Rubber gloves, hats, suits, and boots must be worn.

CHRONIC HAZARD LEVEL Dogs fed 360 mg/kg/day in diet for one year had no adverse effects.

OCCUPATIONAL RECOMMENDATIONS

OSHA 8-hr time-weighted average: 15 mg/m³ (crag herbicide (sesone), total dust) [R6]. 8-hr time-weighted avg: 5 mg/m³ (crag herbicide (sesone), respirable fraction) [R6].

NIOSH 10-hr time-weighted average: 10 mg/m³ (total) [R3]. 10-hr time-weighted avg: 5 mg/m³ (resp) [R3].

THRESHOLD LIMIT 8-hr time-weighted average (TWA) 10 mg/m³ (1986) [R9, 31]. Excursion Limit Recommendation: Excursions in worker exposure levels may exceed three times the TLV–TWA for no more than a total of 30 min during a work day, and under no circumstances should they exceed five times the TLV–TWA. [R9, 5].

ACTION LEVEL Evacuate area. Enter from upwind side. Notify local air authority and National Agricultural Chemicals Association.

ON-SITE RESTORATION Carbon or peat may be used as sorbents. Seek professional environmental engineering assistance through EPA's Environmental Response Team (ERT), Edison, NJ, 24–hour no., 908–548–8730.

AVAILABILITY OF COUNTERMEASURE MATERIAL Carbon—water treatment plants, sugar refineries; peat—nurseries, floral shops.

PROBABLE LOCATION AND STATE Will dissolve. Microbiologically converted to 2, 4–d [R8].

NON-HUMAN TOXICITY LD$_{50}$ rat oral 1,500 mg/kg [R5]; LD$_{50}$ mouse oral 730 mg/kg [R5].

BIODEGRADATION Decomposition in soil is aided by microorganisms such as *Bacillus cereus var mycoides* and takes place within a few years [R2].

FIFRA REQUIREMENTS The following toler-

ances are established for residues of the herbicide sesone (sodium 2,4–dichlorophenoxyethyl sulfate) in or on the raw agricultural commodities indicated: potatoes, peanuts, peanut hulls, peanut hay, asparagus, and strawberries. [R4].

REFERENCES

1. Spencer, E. Y. *Guide to the Chemicals Used in Crop Protection.* 7th ed. Publication 1093. Research Institute, Agriculture Canada, Ottawa, Canada: Information Canada, 1982. 244.

2. White–Stevens, R. (ed.). *Pesticides in the Environment*: Volume 1, Part 1, Part 2. New York: Marcel Dekker, Inc., 1971. 38.

3. NIOSH. *NIOSH Pocket Guide to Chemical Hazards.* DHHS (NIOSH) Publication No. 90–117. Washington, DC: U.S. Government Printing Office, June 1990, 74.

4. 40 CFR 180.102 (7/1/91).

5. American Conference of Governmental Industrial Hygienists. *Documentation of the Threshold Limit Values and Biological Exposure Indices.* 5th ed. Cincinnati, OH: American Conference of Governmental Industrial Hygienists, 1986. 519.

6. 29 CFR 1910.1000 (58 FR 35338 (6/30/93).

7. Stecher, P. G. (ed.). *Merck Index,* 1968, 8th ed., Rahway, NJ.: Merck and Company,

8. *Oregon Weed Control Handbook,* Extension Services, Oregon State University, Corvallis, Oregon, 1975.

9. American Conference of Governmental Industrial Hygienists. *Threshold Limit Values for Chemical Substances and Physical Agents and Biological Exposure Indices for 1994–1995.* Cincinnati, OH: ACGIH, 1994.

10. Gosselin, R. E., R. P. Smith, H. C. Hodge. *Clinical Toxicology of Commercial Products.* 5th ed. Baltimore: Williams and Wilkins, 1984.

■ SODIUM HYDROSULFITE

CAS # 7775–14–6

SIC CODES 2819.

SYNONYMS Blankit • burmol • disodium-dithionite • disodium hydrosulfite • dithionous acid, disodium salt • hydros • sodium dithionite • sodium dithionite [Na₂(S₂O₄)] • sodium hydrosulfite (Na₂S₂O₄) • sodium hydrosulphite • sodium hyposulfite • sodium sulfoxylate • v-brite b • vatrolite

MF O₄S₂•2 Na

MW 174.10

COLOR AND FORM White or grayish-white crystalline powder.

ODOR Faint sulfurous.

FIREFIGHTING PROCEDURES Avoid use of water unless flooding amounts are available for application and flushing; carbon dioxide, dry chemical, and dry sand are best [R5, p. 49–267]. Personal protection: in fire conditions wear self-contained breathing apparatus. [R5, p. 40–266].

SOLUBILITY Slightly soluble in alcohol.

STABILITY AND SHELF LIFE Oxidizes in air, especially in presence of moisture or in solution, to bisulfite and bisulfate and acquires acid reaction [R3].

OTHER CHEMICAL/PHYSICAL PROPERTIES MP: 52°C, Decomposition; decomposition in hot water and acid; colorless monoclinic crystals, or yellow-white powder; solubility in water: 25.4 g/100 cc at 20°C; soluble in alkali, insoluble in alcohol (dihydrate). Light lemon-colored solid in powder or flake form; also available in liquid form (dihydrate). Density: 2.189; loses all of its water of crystallization at 110°C; when heated to 267°C it is dissociated into sodium sulfate and sulfur dioxide (dihydrate).

COMMON USES Reducing agent to strip certain dyes from fabrics and in vat dyeing; bleaching agent for ground wood pulp, soap, straws, sugar [R1]. Oxygen scavenger for synthetic rubbers [R2].

STANDARD CODES NFPA–No; IC—flammable solid, yellow label, 100 lb in an outside container; USCG—nonflammable solid, yellow label; IATA—flammable solid, yellow label, 12 kg passenger, 45 kg cargo.

PERSISTENCE Oxidizes to bisulfite then bisulfate.

GENERAL SENSATION Forms acidic solution. Slight characteristic odor. Large doses can be tolerated, though stomach irritation may result from formation of sulfurous acid. Effects include retarded growth, nerve irritation, atrophy of bone marrow, depression, and paralysis.

PERSONAL SAFETY Wear self-contained breathing apparatus and acid-resistant skin protection.

ACUTE HAZARD LEVEL Moderately toxic with ingestion or inhalation. Forms acid bisulfate solution, which is corrosive to skin. Emits toxic vapors when heated to decomposition.

CHRONIC HAZARD LEVEL Slight ingestive and inhalative hazard.

DEGREE OF HAZARD TO PUBLIC HEALTH Sulfites lower the pH and increase corrosivity. Moderately toxic with ingestion and inhalation. Chronic hazard is slight. Emits toxic vapors when heated to decomposition.

OCCUPATIONAL RECOMMENDATIONS None listed.

ACTION LEVEL Notify fire and air authority. Restrict access to affected water. If intense heats prevail, evacuate area. Enter from upwind. Remove ignition source. Isolate from water.

ON-SITE RESTORATION Seek professional environmental engineering assistance through EPA's Environmental Response Team (ERT), Edison, NJ, 24–hour no., 908–548–8730. Neutralize with NaHCO₃.

BEACH OR SHORE RESTORATION Wash with mild caustic solution.

AVAILABILITY OF COUNTERMEASURE MATERIAL NaHCO₃—grocery distributors, bakeries.

DISPOSAL Landfill, or neutralize and route to sewer plant.

DISPOSAL NOTIFICATION Contact local landfill or sewage authority.

INDUSTRIAL FOULING POTENTIAL Can be corrosive to equipment.

MAJOR WATER USE THREATENED Recreation, potable supply.

PROBABLE LOCATION AND STATE Colorless crystal or yellow-white powder. Will dissolve quickly. Will be dissolved in water. Will soon change to bisulfate form.

COLOR IN WATER Colorless.

DOT *Fire or Explosion:* Flammable/combustible material. May ignite itself if exposed to air. May re-ignite after fire is extinguished. May burn rapidly with flare-burning effect. Runoff to sewer may create fire or explosion hazard [R4].

Health Hazards: If inhaled, may be harmful. Contact may cause burns to skin and eyes. Fire may produce irritating or poisonous gases. Runoff from fire control or dilution water may cause pollution [R4].

Emergency Action: Keep unnecessary people away; isolate hazard area and deny entry. Stay upwind; keep out of low areas. Self-contained breathing apparatus (SCBA) and structural firefighter's protective clothing will provide limited protection. Call CHEMTREC at 1–800–424–9300 for emergency assistance. If water pollution occurs, notify the appropriate authorities [R4].

Fire: Some of these materials may react violently with water. Small Fires: Dry chemical, soda ash, lime, or sand. Large Fires: Flood fire area with water from a distance. Do not get water inside container. Move container from fire area if you can do so without risk. Cool containers that are exposed to flames with water from the side until well after fire is out. Stay away from ends of tanks. For massive fire in cargo area, use unmanned hose holder, or monitor nozzles; if this is impossible, withdraw from area, and let fire burn [R4].

Spill or Leak: Do not touch spilled material; stop leak if you can do so without risk. Do not get water inside container. Small Spills: Flush area with flooding amounts of water. Large Spills: Dike liquid spill for later disposal [R4].

First Aid: Move victim to fresh air; call emergency medical care. Remove and isolate contaminated clothing and shoes at the site. In case of contact with material, immediately flush skin or eyes with running water for at least 15 minutes [R4].

FIRE POTENTIAL Burns slowly, about like sulfur. Heats spontaneously in contact with moisture and air, and may ignite nearby combustible materials. [R5, p. 49–266].

REFERENCES

1. SRI.

2. Hawley, G. G. *The Condensed Chemical Dictionary.* 9th ed. New York: Van Nostrand Reinhold Co., 1977. 790.

3. *The Merck Index.* 9th ed. Rahway, NJ: Merck and Co., Inc., 1976. 1115.

4. Department of Transportation. *Emergency Response Guidebook 1987.* DOT P 5800.4. Washington, DC: U.S. Government Printing Office, 1987, p. G–37.

5. National Fire Protection Association. *Fire Protection Guide on Hazardous Materials.* 7th ed. Boston, MA: National Fire Protection Association, 1978.

■ BIS (2-ETHYLHEXYL) SODIUM SULFOSUCCINATE

CAS # 577–11–7

SYNONYMS 1,4-bis(2-ethylhexyl) sodium sulfosuccinate • aerosol-ot • alphasol-ot • bis(2-ethylhexyl) s-sodium sulfosuccinate • bis(2-ethylhexyl) sodium sulfosuccinate • bis(2-ethylhexyl) sulfosuccinate sodium salt • butanedioic acid, sulfo-, 1,4-bis(2-ethylhexyl) ester, sodium salt • colace • complemix • constonate • coprol • dioctlyn • dioctyl sodium sulfosuccinate • dioctyl-medo forte • dioctylal • diomedicone •

diosuccin • diotilan • diovac • diox • disonate • doxinate • doxol • dss • dulsivac • duosol • konlax • kosate • laxinate • manoxol-ot • molatoc • molcer • molofac • monawet-md-70e • nevax • norval • regutol • revac • sobital • sodium bis(2-ethylhexyl) sulfosuccinate • softil • succinic acid, sulfo-, 1,4-bis(2-ethylhexyl) ester, sodium salt • sulfosuccinic acid bis(2-ethylhexyl) ester sodium salt • sv-102 • velmol

MF $C_{20}H_{38}O_7S$ • Na

MW 445.63

COLOR AND FORM White, wax-like solid.

ODOR Characteristic odor.

SURFACE TENSION Dynes/cm in water at 25°C: 62.8% (0.001%); 28.7 (0.1%); 26.0 (1.0%).

SOLUBILITY Solubility in water (g/L): 15 (25°C), 23 (40°C), 30 (50°C), 55 (70°C); soluble in carbon tetrachloride, petroleum ether, naphtha, xylene, dibutyl phthalate, liquid petrolatum, acetone, alcohol, vegetable oils; very soluble in water-miscible, organic solvents; freely soluble in glycerin; water solubility = 71,000 mg/L at 25°C.

STABILITY AND SHELF LIFE Stable in acid and neutral solution; hydrolyzes in alkaline solution [R6].

DISPOSAL SRP: At the time of review, criteria for land treatment or burial (sanitary landfill) disposal practices are subject to significant revision. Prior to implementing land disposal of waste residue (including waste sludge), consult with environmental regulatory agencies for guidance on acceptable disposal practices.

COMMON USES Wetting agent in industrial, pharmaceutical, cosmetic, and food applications; dispersing and solubilizing agent in foods; adjuvant in tablet formation [R1]. Food additive (processing aid in sugar industry, stabilizer for hydrophilic colloids); dispersant; emulsifier [R2]. Employed as dispersing and emulsifying agent in various dermatological preparation [R3]. Used in powderless etching systems [R4]. Wetting agent which has been used primarily as ingredient in certain detergents. It has been found to have potentiating powers, particularly in adding previously unachievable flavor, or freshness to canned milks [R5].

OCCUPATIONAL RECOMMENDATIONS None listed.

NON-HUMAN TOXICITY LD_{50} mouse oral 2.64 g/kg [R7].

BIODEGRADATION In a river die-away screen test of river water, bis(2–ethylhexyl) sodium sulfosuccinate biodegraded 95% (12 days), 91% (12 days), 91% (17 days), 97.3% (6 days), and 97.7% (3 days), at concentrations of 12.9, 4.5, 3.3, 11.3, and 12.9 ppm, respectively, with a lag period of 6 days (1). This study also conducted a sterile control in which there was 9% loss of bis(2–ethylhexyl) sodium sulfosuccinate (1). A BOD test of aerobic activated sludge biodegraded bis(2–ethylhexyl) sodium sulfosuccinate 80–95% after 8 hours from initial concentrations of 2–13 ppm with a 5–7-week lag (2). This same study tested sewage in the same manner and obtained 60–80% biodegradation of bis(2–ethylhexyl)sodium sulfosuccinate after a 3–9-week lag (2). A study using DOC found that bis(2–ethylhexyl) sodium sulfosuccinate (40 ppm) biodegraded 83% after 20 days in aerobic sewage (3). In an aerobic closed-bottle screening study using activated sludge and soil inoculum, 100 mg/L bis(2–ethylhexyl)sodium sulfosuccinate had a 4–week theoretical BOD of 0–9% (4). With 1 mg added to 10 ml sediment, bis (2–ethylhexyl) sodium sulfosuccinate biodegraded 55–94% in river sediments, 8% in sand, and 13% in clay after 3 days (5) [R8].

BIOCONCENTRATION Based upon an experimental water solubility of 71,000 mg/L (1), the BCF of bis(2–ethylhexyl)sodium sulfosuccinate can be estimated to be approximately 1.13 from a regression-derived equation (2). The BCF for bis(2–ethylhexyl)sodium sulfosuccinate has also been experimentally determined to be <0.9 at 0.5 mg/L and <9.3 at 0.05 mg/L for a 6–week duration (1). Based on these BCF values, bioconcentration is not expected to be an important fate process (SRC) [R9].

REFERENCES

1. Budavari, S. (ed.). *The Merck Index— Encyclopedia of Chemicals, Drugs, and Biologicals*. Rahway, NJ: Merck and Co., Inc., 1989. 535.

2. Lewis, R. J., Sr. (ed.). *Hawley's Condensed Chemical Dictionary*. 12th ed. New York, NY: Van Nostrand Reinhold Co., 1993 425.

3. Gosselin, R. E., H. C. Hodge, R. P. Smith, and M. N. Gleason. *Clinical Toxicology of Commercial Products*. 4th ed. Baltimore: Williams and Wilkins, 1976, p. II–178.

4. *Kirk-Othmer Encyclopedia of Chemical Technology*. 3rd ed., Volumes 1–26. New York, NY: John Wiley and Sons, 1978–1984, p. V19 137.

5. Furia, T. E. (ed.). *CRC Handbook of Food Additives*. 2nd ed. Cleveland: The Chemical Rubber Co., 1972. 519.

6. *The Merck Index*. 9th ed. Rahway, NJ: Merck and Co., Inc., 1976. 438.

7. Case, M. T., et al. *Drug Chemical Toxicol* 1 (1) 89–101 (1978)

8. (1) Hammerton, C. *J Appl chemical* 5: 517–24 (1955) (2) Jenkins, S. H., et al., *Water Research* 1: 31–53 (1967) (3) Cordon, T. C., et al., *J Amer Oil Chemical Soc* 47: 203–6 (1970) (4) Chemical Inspection and Testing Institute. *Biodegradation and Bioaccumulation Data of Existing Chemicals Based on the CSCL Japan* p. 2–100 (1992) (5) Lenhard, G., A. Du Plooy. *Hydrobiol* 22: 317–23 (1963)

9. (1) Chemicals Inspection and Testing Institute. *Biodegradation and Bioaccumulation Data of Existing Chemicals Based on the CSCL Japan* p 2–100 (1992) (2) Lyman, W. J., et al., *Handbook of Chemical Property Estimation Methods*. Washington DC: Amer Chemical Soc pp. 5– 10 (1990).

■ SODIUM METHYLDITHIOCARBAMATE

CAS # 137–42–8

SIC CODES 2879.

SYNONYMS sodium-metam • methyldithiocarbamic acid, sodium salt • sodium methyldithiocarbamate • N-methylaminodithioformicacid, sodium salt • sodium-N-methyl-aminodithioformate • N-methyl-aminomethanethionothiolicacid, sodium salt • smdc • sodium-N-methylaminomethanethionothiolate • metham-sodium • sodium-metham • metam-sodium • sistan • trimatron • vpn • maposol • a7-vapam • basamid-fluid • carbam • carbamic acid, methyldithio-, monosodium salt • carbamic acid, N-methyldithio, sodium salt • carbamodithioic acid, methyl-, monosodium salt • carbathion • carbathione • carbation • karbation • masposol • metam-sodium • metam-fluid basf • metam-sodium (Dutch, French, German, Italian) • Metham (German) • metham-sodium • methylcarba-modithioic acid sodium salt • methyl-dithiocarbamic acid, sodium salt • N-methylaminodithioformic acid sodium salt • N-methylamino-methanethionothiolic acid sodium salt • N-methyldithiocarbamate de sodium (French) • N-metil-ditiocarbammato di sodio (Italian) • natrium-N-methyl-dithiocarbamaat (Dutch) • Natrium-N-methyl-dithiocarbamat (German) • sistan • smdc • sodium-metham • sodium methylcarbamodithioate • sodium monomethyldithiocarbamate • sodium N-methylaminodithioformate • sodium N-methylaminomethanethionothiolate • sodium N-methyldithiocarbamate • solesan-500 • sometam • trapex • trimaton • vdm • vpm • vpm (fungicide) • vapam • mapasol • metam • nematin

MF $C_2H_4NS_2$ • Na

MW 129.18

COLOR AND FORM White crystalline solid.

FLAMMABILITY LIMITS Nonflammable (sodium methyldithiocarbamate, dihydrate) [R8, 937].

SOLUBILITY Slightly soluble in acetone, xylene, and kerosene at 20°C; in water: 722 g/L at 20°C; insoluble in most organic solvents.

CORROSIVITY The aqueous solution is corrosive to brass, copper, and zinc.

STABILITY AND SHELF LIFE Stable with only slight decomposition on storage for several yr in glass containers as 32.7% solution in alkaline water; pure solid decomposes in several weeks. [R10, 304].

OTHER CHEMICAL/PHYSICAL PROPERTIES Crystals; anhydrous at 130°C; moderately soluble in alcohol; 72.2 g/100 ml water at 20°C; moderately soluble in alcohol; sparingly soluble in other solvents; unpleasant odor, similar to that of disulfide (sodium methyldithiocarbamate dihydrate). In dilute aqueous solution, at pH 9.5, vapam decomposes to methyl isothiocyanate and sulfur; under acid conditions, carbon disulfide, hydrogen sulfide, N,N'-dimethylthiuram disulfide, methylamine, methyl isothiocyanate are formed. A soil fungicide, nematocide, and herbicide reputed to act by decomposition in moist soil to methyl isothiocyanate. If true, this may be a unique property of metham, as opposed to many closely related chemicals used for much of the same purposes, e.g., ferbam, nabam, ziram, zineb, and maneb. BP: 110°C (commercial product); nonvolatile.

PROTECTIVE EQUIPMENT Clothing and equipment should be consistent with good pesticide handling [R1].

PREVENTATIVE MEASURES Avoid contact with skin and eyes. Avoid breathing vapors after application to soil. If skin or eyes are contacted, flush immediately with large quantities of running water. After treating soil, leave the immediate area. [R10, 304].

DISPOSAL Landfill: Metham sodium should be treated by acid before burial. The reactions usually proceed more rapidly in solution (such as by stirring in a plastic container with excess hydrochloric acid), but this method is somewhat hazardous. The preferred method is to mix the pesticide with sand (or other adsorbent) in a pit or trench at least 0.5 m deep in a clay soil and then add the acid. Recommendable method: Incineration. Not recommendable methods: Wet treatment, evaporation, and discharge to sewer. Peer-review: Large amount should be incinerated in a unit with effluent gas scrubbing. Metham sodium liberates methyl isocyanate on reaction with water (peer-review conclusions of an IRPTC expert consultation (May 1985)) [R3]. Unstable as a solid. decomposes slowly in dilute aqueous solution. Unstable with acid.

COMMON USES Soil fumigant for fungi, bacteria, nematodes, insects [R7, 52]. Metham is soil fumigant used to kill the germinating seeds, rhizomes, tubers, roots, and stems of weeds in the soil. Weeds controlled include crabgrass, foxtail, annual bluegrass, barnyardgrass, pigweed, purslane, chickweed, dandelion, ragweed, henbit, lambsquarters, field bindweed, bermudagrass, johnsongrass, nutgrass, cudweed, and many other troublesome weeds. It is often used as a pre-planting treatment in areas that are to be planted with lawngrasses, seed beds, flowers, ornamentals, nurseries, grapevines, fruit trees, and row crop. It is also used to kill roots of trees growing in sewer pipes and roots forming grafts between trees which may transmit Dutch elm disease and oak wilt [R10, 303]. Chemical control of forest diseases caused by certain organisms (pythium, rhizoctonia, fusarium) [R2].

PERSISTENCE In soil, 2 weeks. low pH, <7, and microbial life degrade vapam quickly.

PERSONAL SAFETY Self-contained breathing apparatus must be worn; rubber gloves, hats, suits, and boots must be worn.

OCCUPATIONAL RECOMMENDATIONS None listed.

ACTION LEVEL Evacuate area. Enter from upwind side. Notify local air authority and National Agricultural Chemicals Association.

ON-SITE RESTORATION Carbon or peat may be used as sorbents. Seek professional environmental assistance through EPA's Environmental Response Team (ERT), Edison, NJ, 24–hour no., 908–548–8730.

AVAILABILITY OF COUNTERMEASURE MATERIAL Carbon—water treatment plants, sugar refineries; peat—nurseries floral shops.

PROBABLE LOCATION AND STATE Colorless crystalline solid formulated in solution.

NON-HUMAN TOXICITY LD_{50} rat oral 820 mg/kg (sodium methyldithiocarbamate, dihydrate) [R8, 937]; LD_{50} mouse, oral 285 mg/kg (sodium methyldithiocarbamate, dihydrate) [R8, 937]; LD_{50} rabbit percutaneous 2,000 mg/kg [R10, 305].

ECOTOXICITY LC_{50} *Colinus virginianus* (bobwhite quail) oral >5,000 ppm in 5-day diet ad libitum, age 14 days (vapam; purity: technical) [R4]; LC_{50} *Coturnix japonica* (Japanese quail) oral >5,000 ppm in 5-day diet ad libitum age 14 days (vapam; purity: technical) [R4]; LC_{50} *Phasianus colchicus* (ring-necked pheasant) >5,000 ppm in 5-day diet ad libitum, age 14 days (vapam; purity: technical) [R4]; LC_{50} *Anas platyrhynchos* (mallard) oral >5,000 ppm in 5-day diet ad libitum, age 14 days (vapam; purity: technical) [R4].

BIODEGRADATION Sodium methyldithiocarbamate in soil breaks down by chemical means rather than by soil microorganisms (1). No data on the biodegradation of sodium methyldithiocarbamate in aquatic systems could be located; however, microbial degradation of this compound is expected to be insignificant in light of its rapid rate of chemical breakdown (SRC) [R5].

BIOCONCENTRATION A calculated bioconcentration factor of 0.3 (1) implies that sodium methyldithiocarbamate will not significantly bioconcentrate in fish and aquatic organisms (SRC) [R6].

DOT *Health Hazards:* Poisonous; may be fatal if inhaled, swallowed, or absorbed through skin. Contact may cause burns to skin and eyes. Runoff from fire control or dilution water may cause pollution (dithiocarbamate pesticide, flammable liquid, NOS) [R9, p. G–28].

Fire or Explosion: Flammable/combustible material; may be ignited by heat, sparks, or flames. Vapors may travel to a source of ignition, and flash back. Container may explode in heat of fire. Vapor explosion and poison hazard indoors, outdoors, or in sewers. Runoff to sewer may create fire or explosion hazard (dithiocarbamate pesticide, flammable liquid, NOS) [R9, p. G–28].

Emergency Action: Keep unnecessary people away; isolate hazard area, and deny entry. Stay upwind; keep out of low areas. Positive-pressure self-contained breathing apparatus (SCBA) and chemical protective clothing that is specifically recommended by the shipper or manufacturer may be worn. It may provide little or no thermal protection. Structural firefighters' protective clothing is not effective for these materials. Isolate for 1/2 mile in all directions if tank, rail car, or tank truck is involved in fire. Call CHEMTREC at 1–800–424–9300 for emergency assistance (dithiocarbamate pesticide, flammable liquid, NOS) [R9, p. G–28].

Fire: Small fires: Dry chemical, CO_2, water spray, or alcohol-resistant foam. Large Fires: Water spray, fog, or alcohol-resistant foam. Move container from fire area if you can do so without risk. Dike fire-control water for later disposal; do not scatter the material. Apply cooling water to sides of containers that are exposed to flames until well after fire is out. Stay away from ends of tanks. Withdraw immediately in case of rising sound from venting safety device or any discoloration of tank due to fire (dithiocarbamate pesticide, flammable liquid, NOS) [R9, p. G–28].

Spill or Leak: Shut off ignition sources; no flares, smoking, or flames in hazard area. Fully-encapsulating, vapor-protective clothing should be worn for spills and leaks with no fire. Do not touch or walk through spilled material; stop leak if you can do so without risk. Water spray may reduce vapor, but it may not prevent ignition in closed spaces. Small Spills: Take up with sand or other noncombustible absorbent material and place into containers for later disposal. Large Spills: Dike far ahead of liquid spill for later disposal (dithiocarbamate pesticide, flammable liquid, NOS) [R9, p. G–28].

First Aid: Move victim to fresh air, and call for emergency medical care; if not breathing, give artificial respiration; if breathing is difficult, give oxygen. In case of contact with material, immediately flush skin or eyes with running water for at least 15 minutes. Remove and isolate contaminated clothing and shoes at the site. Keep victim quiet and maintain normal body temperature. Effects may be delayed; keep victim under observation (dithiocarbamate pesticide, flammable liquid, NOS) [R9, p. G–28].

REFERENCES

1. *Farm Chemicals Handbook* 1991. Willoughby, OH: Meister, 1991, p. C–197.

2. White–Stevens, R. (ed.). *Pesticides in the Environment*: Volume.

3. New York: Marcel Dekker, Inc., 1977. 206.

3. United Nations. *Treatment, and Disposal Methods for Waste Chemicals* (IRPTC File). Data Profile Series No. 5. Geneva, Switzerland: United Nations Environmental Programme, Dec. 1985. 128.

4. U.S. Department of the Interior, Fish and Wildlife Service, Bureau of Sports Fisheries, and Wildlife. *Lethal Dietary Toxicities of Environmental Pollutants to Birds.* Special Scientific Report—*Wildlife* No. 191. Washington, DC: U.S. Government Printing Office, 1975. 35.

5. Weed Science Society of America. *Herbicide Handbook* 6th ed: 176–7 (1989).

6. Kenaga, E. E., *Ecotox Env Safety* 4: 26–38 (1980).

7. White-Stevens, R. (ed.). *Pesticides in the Environment.* Volume 1, Part 1, Part 2. New York: Marcel Dekker, Inc., 1971.

8. Budavari, S. (ed.). *The Merck Index—Encyclopedia of Chemicals, Drugs and Biologicals.* Rahway, NJ: Merck and Co., Inc., 1989.

9. U.S. Department of Transportation. *Emergency Response Guidebook 1990.* DOT P 5800.5. Washington, DC: U.S. Government Printing Office, 1990.

10. Weed Science Society of America. *Herbicide Handbook.* 5th ed. Champaign, Illinois: Weed Science Society of America, 1983.

■ SODIUM NITRATE

CAS # 7631–99–4

SYNONYMS Chile saltpeter • cubic niter • niter • nitrate de sodium (French) • nitrate of soda • nitratine • nitric acid sodium salt • nitric acid, sodium salt • saltpeter [chile] • soda niter • sodium nitrate (dot) • sodium(i) nitrate (1:1)

MF $NO_3 \cdot Na$

MW 85.00

pH Aqueous solution is neutral.

COLOR AND FORM Colorless, trigonal, or rhombohedron crystals; white granules or powder.

TASTE Saline, slightly bitter taste.

DENSITY 2.26.

BOILING POINT 380°C with decomposition.

MELTING POINT 308°C.

EXPLOSIVE LIMITS AND POTENTIAL Explodes when heated to over 1,000°C [R7].

FIREFIGHTING PROCEDURES Use abundant amount of water in early stages of fire. When large quantities are involved in fire, nitrate may fuse, or melt; in such conditions, application of water may result in extensive scattering of molten material [R5]. Flood with water. Cool all affected containers with flooding quantities of water. Apply water from as far a distance as possible [R6]. Evacuation: If fire becomes uncontrollable consider evacuation of one-half (1/2) mile radius [R6].

SOLUBILITY 92.1 g/100 cc water at 25°C, 180 g/100 cc water at 100°C; very soluble in ammonia; very slightly soluble in acetone, glycerol; 1 g dissolves in 125 ml alcohol, 52 ml boiling alcohol, 3,470 ml absolute alcohol, 300 ml methanol; water solubility = 730,000 mg/L at 0°C; is soluble in liquid ammonia and forms Na $NO_3 \cdot 4NH_3$ below −42°C. The solubility in anhydrous methanol is 2.8 wt% at 25°C.

OTHER CHEMICAL/PHYSICAL PROPERTIES When dissolved in water, temperature of solution is lowered; deliquesces in moist air. Liquid molar volume = 0.044616 m^3/kmol. Heat of formation = −4.6785 × 10^8 J/kmol. Heat of fusion at 580.15 K = 1.4602×10^7 J/kmol. Enthalpy of formation = −466.8 kJ/mol. Enthalpy of fusion = 15.7 kJ/mol. Specific heat capacity = 93.1 J/mol–K at 298 K.

PROTECTIVE EQUIPMENT Rubber gloves, goggles, laboratory coat [R7].

PREVENTATIVE MEASURES If material not on fire and not involved in Fire: Keep sparks, flames, and other sources of ignition away. Keep material out of water sources and sewers [R6]. Personnel protection: Do not handle broken packages unless wearing appropriate personal protective equipment. Wash away any material that may have contacted the body with copious amounts of water or soap and water. Approach fire with caution. [R6].

CLEANUP SRP: (Laboratory quantities) For solid: Sweep into a beaker. Dilute with sufficient water. Ad soda ash. Mix and neutralize with 6M HCl. Drain into the sewer with abundant water. For solution: Cover with soda ash. After mixing, transfer into a beaker containing water. Neutralize with 6M HCl. Drain into the sewer with abundant water [R5].

DISPOSAL SRP: At the time of review, criteria for land treatment or burial (sanitary landfill) disposal practices are subject to significant revision. Prior to implementing land disposal of waste residue (including waste sludge), consult with environmental regulatory agencies for guidance on acceptable disposal practices.

COMMON USES Manufacture of enamels for pottery, nitric acid, sodium nitrite; in matches; catalyst in manufacture of sulfuric acid; pickling meats [R2]. Fertilizer for cotton, tobacco, and vegetable crops; oxidizing component of explosives and blasting agents; oxidizer and fluxing agent in the manufacture of glass and enamels; component of charcoal briquettes, heat-transfer salt; curing agent and preservative in meats; chemical for recovery of tin from scrap; oxidizing agent (e.g., in metal-coloring solns); chemical intermediate (e.g., for potassium nitrate) [R1]. Oxidizing agent in manufacture of pharmaceuticals, anaphrodisiac, modifying burning properties of tobacco [R3]. More than half of the sodium nitrate produced worldwide is used as a fertilizer for crops such as cotton, tobacco, and vegetables; in the U.S. it is of minor importance compared to other fertilizers. The major industrial use is in the explosives industry. Also, used in exhaust-gas combustion devices to promote combustion, and in the production of cleaning agents for plugged drain pipes [R12, 270].

OCCUPATIONAL RECOMMENDATIONS None listed.

NON-HUMAN TOXICITY The minimum lethal dose of sodium nitrate in cattle has been estimated to be 0.65–0.75 g/kg. [R16, 65].

BIODEGRADATION In two simulated biological treatment plant using activated sludge, 45% removal of 2–chloroethanol in 24 hr was obtained with about 5% due to evaporation (1) and 45% of theoretical BOD was removed in 10 days (2). Conflicting results were obtained in a biodegradability test performed using process wastewater and activated sludge that resulted in 20% COD removal and 1% total organic carbon (TOC) removal in 24 hours (4). 2–Chloroethanol's degradation pathway is via 2–chloroacetaldehyde and monochloroacetate to glycolic acid (3) [R8]. 2– Chloroethanol is readily biodegradable in screening tests and biological treatment simulations using sewage and activated-sludge inocula (1–6). Various investigators have obtained the following results in percent of theoretical BOD in screening tests using sewage inocula: 57% in 20 days (1); 50% in 10 days (2); and 87% in 10 days (3). The results of these screening tests indicate that acclimation is important in the biodegradation process [R13].

BIOCONCENTRATION Using the log octanol/water partition coefficient for 2–chloroethanol, 0.03 (1), one can estimate a BCF of 0.62 using a recommended regression equation (2,SRC). Therefore, 2–chloroethanol would not be expected to bioconcentrate in aquatic organisms [R9].

DOT *Fire or Explosion:* May ignite other combustible materials (wood, paper, oil, etc.). These materials will accelerate burning when they are involved in a fire; some will react violently with fuels. Runoff to sewer may create fire or explosion hazard [R4].

Health Hazards: Contact may cause burns to skin and eyes. Vapors or dust may be irritating. Fire may produce irritating or poisonous gases. Runoff from fire control or dilution water may cause pollution [R4].

Emergency Action: Keep unnecessary people away; isolate hazard area and deny entry. Positive pressure self-contained breathing apparatus (SCBA) and structural firefighters' protective clothing will provide limited protection. Call emergency response telephone number on shipping paper first. If shipping paper not available or no answer, call CHEMTREC at 1–800–424–9300. If water pollution occurs, notify the appropriate authorities [R4].

Fire: Small fires: Water only; no dry chemical, CO_2, or halon. Large Fires: Flood fire area with water from a distance. Move container from fire area if you can do so without risk. Apply cooling water to sides of containers that are exposed to flames until well after fire is out. For massive fire in cargo area, use unmanned hose holder, or monitor nozzles; if this is impossible, withdraw from area, and let fire burn [R4].

Spill or Leak: Do not touch or walk through spilled material. Keep combusti-

bles (wood, paper, oil, etc.) away from spilled material. Small Dry Spills: With clean shovel place material into clean, dry container, and cover loosely; move containers from spill area. Small Liquid Spills: Take up with sand, earth, or other noncombustible absorbent material. Large Spills: Dike far ahead of liquid spill for later disposal [R4].

First Aid: Move victim to fresh air; call for emergency medical care. In case of contact with material, immediately flush skin or eyes with running water for at least 15 minutes. Remove and isolate contaminated clothing and shoes at the site [R4].

FIFRA REQUIREMENTS In 1988, Congress amended FIFRA to strengthen and accelerate EPA's reregistration program. The nine-year reregistration scheme mandated by "FIFRA 88" applies to each registered pesticide product containing an active ingredient initially registered before November 1, 1984. Pesticides for which EPA had not issued Registration Standards prior to the effective date of FIFRA '88 were divided into three lists based upon their potential for exposure and other factors, with List B being of highest concern, and D of least. List: D; Case: Inorg. nitrate/nitrite; Case No. 4052; Pesticide type: Rodenticide, Antimicrobial; Case Status: RED Approved 09/91; OPP has reached a decision that some or all uses of the pesticide are eligible for reregistration, and issued a Reregistration Eligibility Document (RED). Active Ingredient (AI): Sodium nitrate; AI Status: OPP has completed a Reregistration Eligibility Document (RED) for the active ingredient/case [R15]. Sodium nitrate is exempted from the requirement of a tolerance when used as a solid diluent in accordance with good agricultural practice as inert (or occasionally active) ingredients in pesticide formulations applied to growing crops only [R14].

GENERAL RESPONSE Call fire department. Keep people away. Stop discharge if possible. Isolate and remove discharged material. Notify local health and pollution control agencies.

FIRE RESPONSE Not flammable. May cause fire and explode on contact with combustibles. Poisonous gases may be produced in fire. Wear goggles and self-contained breathing apparatus. Flood discharged area with water.

EXPOSURE RESPONSE Call for medical aid. Solid, if swallowed, may cause dizziness, abdominal cramps, vomiting, convulsions, and collapse. Flush exposed areas with plenty of water. If swallowed and victim is conscious, have victim drink water or milk, and have victim induce vomiting. If swallowed and victim is unconscious or having convulsions, do nothing except keep victim warm.

RESPONSE TO DISCHARGE Issue warning—oxidizing material. Provide chemical and physical treatment. Disperse and flush.

WATER RESPONSE: Dangerous to aquatic life in very high concentrations. May be dangerous if it enters water intakes. Notify local health and wildlife officials. Notify operators of nearby water intakes.

EXPOSURE SYMPTOMS Ingestion: dizziness, abdominal cramps, vomiting, bloody diarrhea, weakness, convulsions, and collapse. Small repeated doses may cause headache and mental impairment.

EXPOSURE TREATMENT: See a physician. Eyes: rinse with water. Skin: wash with water for 15 minutes. Ingestion: drink water, milk, or activated charcoal; then induce vomiting or gastric lavage followed by catharsis.

FIRE POTENTIAL Flames up when heated to 540°C [R5].

FDA Sodium nitrate is a food additive permitted for direct addition to food for human consumption, as long as (1) the quantity added to food does not exceed the amount reasonably required to accomplish its intended physical, nutritive, or other technical effect in food, and (2) when intended for use in or on food it is of appropriate food grade, and is prepared and handled as a food ingredient [R10]. Sodium nitrate is an indirect food additive for use as a component of adhesives [R11].

REFERENCES

1. SRI.

2. *The Merck Index.* 10th ed. Rahway, NJ: Merck Co., Inc., 1983. 1238.

3. Sax. *Hawley's Condensed Chemical Dictnry.* 11th ed. 1987, p. 1066.

4. U.S. Department of Transportation. *Emergency Response Guidebook 1993.* DOT P 5800.6. Washington, DC: U.S. Government Printing Office, 1993, p. G–35.

5. ITII. *Toxic and Hazardous Industrial Chemicals Safety Manual.* Tokyo, Japan: The International Technical Information Institute, 1988. 484.

6. Association of American Railroads. *Emergency Handling of Hazardous Materials in Surface Transportation.* Washington, DC: Association of American Railroads, Bureau of Explosives, 1992. 845.

7. U.S. Coast Guard, Department of Transportation. *CHRIS—Hazardous Chemical Data.* Volume II. Washington, DC: U.S. Government Printing Office, 1984–5.

8. (1) Matsui, S., et al. *Prog Water Technol.* 7: 645–59 (1975) (2) Mills, E. J. Jr., V. T. Stack, Jr. pp. 492–517 in *Proc 8th Indust Waste Conf Eng Bull* Purdue Univ Eng Ext Ser (1954) (3) Stucki, G., et al. *FEMS Symp 12 Microb Degrad Zenobiotics Recalitrant Compounds*: 131–7 (1981) (4) Klockner, D., *Chemical Ind* 39: 92–4 (1987).

9. (1) Hansch, C., and A. J. Leo. *Medchem Project* Issue No 26, Claremont, CA: Pomona College (1985) (2) Lyman, W. J., et al., *Handbook of Chemical Property Estimation Methods*, New York: McGraw-Hill, Chapt 5 (1982).

10. 21 CFR 172.170 (4/1/93).

11. 21 CFR 175.105 (4/1/93).

12. Gerhartz, W. (exec ed.). *Ullmann's Encyclopedia of Industrial Chemistry.* 5th ed. Vol A1: Deerfield Beach, FL: VCH Publishers, 1985 to present, p. VA17

13. (1) Conway, R. A., et al. *Environ Sci Technol* 17: 107–12 (1983) (2) Heukelekian, H., M. C. Rand. *J Water Pollut Contr*

Assoc 29: 1040–53 (1955) (3) Lamb, C. B., G. F. Jenkins. pp. 326–9 in *Proc 8th Industrial Waste Conf.* Purdue Univ (1952) (4) Matsui, S., et al. *Prog Water Technol.* 7: 645–59 (1975) (5) Mills, E. J., Jr., V. T. Stack, Jr. pp. 492–517 in *Proc 8th Indust Waste Conf Eng Bull.* Purdue Univ Eng Ext Ser (1954) (6) Klockner, D., *Chemical Ind* 39: 92–4 (1987).

14. 40 CFR 180.1001 (d) (7/1/92).

15. USEPA/OPP. *Status of Pesticides in Reregistration and Special Review.* p. 242 (Mar, 1992) EPA 700–R–92–004.

16. Humphreys, D. J. *Veterinary Toxicology.* 3rd ed. London, England: Bailliere Tindell, 1988.

■ SODIUM SILICATE

CAS # 1344–09–8

SIC CODES 2841.

SYNONYMS s ilicic acid, sodium salt • sodium beta-silicate • waterglass • 49fg • as-bond-1001 • carsil • dp-222 • dryseq • dupont-26 • l-96 (salt) • metso-99 • n-38 • pyramid-1 • pyramid-8 • q-70 • sikalon • silica-e • silica-k • silica-n • silica-r • sodium sesquisilicate • sodium siliconate • water-glass • sodium water-glass • soluble glass • agrosil-lr • agrosil-s • britesil • britesil-h-20 • britesil-H-24 • carsil (silicate) • Caswell no. 792 • EPA pesticide chemical code 072603 • hk-30 (van) • portil-a • star • silican • sodium silicate glass • sodium silicate solution

MF Variable

MW Variable

pH Aqueous solutions are strongly alkaline.

COLOR AND FORM Colorless to white or grayish-white, crystal-like pieces or lumps; lumps of greenish glass; white powders; liquids cloudy or clear.

EXPLOSIVE LIMITS AND POTENTIAL Noncombustible [R23, 1072].

SOLUBILITY Very slightly soluble or almost insoluble in cold water; less readily soluble in large amount of water than in small amt; anhydrous dissolves with more difficulty than hydrate in water; glass form soluble in steam under pressure; partially miscible with primary alcohol and ketones; miscible with some polyhydric alcohol; insoluble in alcohol, potassium, and sodium salts; brought into solution by heating with water under pressure; silicates containing more sodium dissolve more readily.

OTHER CHEMICAL/PHYSICAL PROPERTIES Coagulated by brine; precipitated by alkaline earth and heavy-metal ions. In commerce mole ratios of SiO_2–Na_2O vary from 1.5–3.5; the lower the ratio, the lower the solubility and alkalinity. Forms gelatinous mixtures with water. Deliquescent. Gels form with acids between pH 3 and 9. Sodium by electrodialysis or with an ion-exchange resin that forms active silica. Liquids varying from highly fluid to extreme viscosity; viscosity range from 0.4 to 600,000 P. Freezing point slightly below that of water.

PROTECTIVE EQUIPMENT Rubber or polyethylene gloves; safety glasses or other form of eye protection; self-contained breathing apparatus or one that absorbs organic vapors. (ethyl silicate) [R18].

PREVENTATIVE MEASURES SRP: The scientific literature supports the wearing of contact lenses in industrial environments, as part of a program to protect the eye against chemical compounds and minerals causing eye irritation. However, there may be individual substances whose irritating or corrosive properties are such that the wearing of contact lenses would be harmful to the eye. In those specific cases contact lenses should not be worn.

COMMON USES Lining bessemer converters, acid concentraters; manufacture of grindstones, abrasive wheels (as binder only) [R2]. Catalysts and silica gels; soaps and detergents; adhesives, especially sealing and laminating paper-board containers [R23, 1073]. Water treatment; bleaching, and sizing of textiles and paper pulp; soil solidification; glass foam [R23, 1073].

Pigments; drilling fluids; foundry cores and molds; waterproofing mortars, and cements; enhanced oil recovery [R23, 1073]. Used for preserving eggs, in cements, in water softeners, as detergents, in fireproofing fabrics [R5]. Chemical intermediate for silica catalysts, silica pigments, and a component for cleaners and detergents (as sodium silicates) [R1]. It is employed in soaps, in fireproofing mixtures, and in boiler compound [R21, 936]. Flotation method (to detect-insect infested rice or maize kernels) employs a mixture of water glass and water at specific gravity of 1.16–1.19 in which infested kernels promptly float to surface [R6]. Sodium silicate is used in zinc electrodeposition [R7]. Sodium silicate is used in wool bleaching [R8]. Sodium silicate is used as an agglomeration liquid [R9]. Greases have been made from copper phthalocyanine with silicones or with sodium silicate [R10]. Sodium silicate is used as a metal cleaner [R11]. Sodium bicarbonate can be used with sodium silicate grouts where low-strength, semipermanent grouts are required, e.g., in tunnel face stabilizations that need last only until the tunnel rings are in place [R4]. Sodium silicate is used as a hydrogen peroxide stabilizer [R12]. Corrugated and paperboard products have used sodium silicate based adhesives for many years [R13]. In ore flotation as dispersant for siliceous gangue, slime dispersant, and surface conditioner in a variety of ores [R22, p. 1029]. Flame retardant; chemical equipment lining [R14]. Impregnating wood [R15].

STANDARD CODES NFPA—1, –,—; IATA—other restricted articles, Class B, no label required, no limit.

PERSISTENCE Can persist indefinitely.

DIRECT CONTACT Skin, mucous membranes.

GENERAL SENSATION Forms alkaline solution. Swallowing results in vomiting and diarrhea.

PERSONAL SAFETY Wear skin protection. Intense heat may necessitate use of filter mask.

ACUTE HAZARD LEVEL Slight ingestive hazard. Mild irritant. Moderately toxic to aquatic life. Emits toxic dusts when heated to decomposition. Silicates at concentration >50 ppm can cause turbidity problems.

CHRONIC HAZARD LEVEL Mild chronic irritant. An abundance of silica in water along with other nutrients can stimulate growth of diatoms.

DEGREE OF HAZARD TO PUBLIC HEALTH Mild irritant. Slight ingestive hazard. Emits toxic dusts when heated to decomposition.

OCCUPATIONAL RECOMMENDATIONS None listed.

ACTION LEVEL Prevent suspension of dust. Restrict access to affected waters in immediate area.

ON-SITE RESTORATION Seek professional environmental engineering assistance through EPA's Environmental Response Team (ERT), Edison, NJ, 24–hour no., 908–548–8730. Acidify and flocculate with diluted acetic acid.

AVAILABILITY OF COUNTERMEASURE MATERIAL Acetic acid—plastic and electronic industries.

DISPOSAL Add to large volume water. Stir in light excess soda ash. (Add slaked lime in presence of fluoride.) Decant and neutralize in second container with 6M HCl. Route to sewage plant. Landfill sludge.

DISPOSAL NOTIFICATION Contact local sewage and solid waste authority.

MAJOR WATER USE THREATENED Fisheries, potable supply.

PROBABLE LOCATION AND STATE Colorless deliquescent amorphous solid. Will be dissolved in water. May form dispersion at first that will dissolve slowly.

WATER CHEMISTRY Forms alkaline solution. As little as 247 ppm raises pH to 9.1. Calcium salt is less soluble. Silicate >50 ppm is associated with turbidity.

COLOR IN WATER Colorless.

NON-HUMAN TOXICITY LD_{50} rat oral 1.1–1.6 g/kg [R3]; LD_{50} mouse oral 1,100 mg/kg [R19].

BIOLOGICAL HALF-LIFE Sodium silicate half-life 24 hr. First-order excretion kinetics were followed in determining urinary silicon exretion by rats [R20].

DOT *Fire or Explosion:* Flammable/combustible material; may be ignited by heat, sparks, or flames. May burn rapidly with flare-burning effect (silicon powder, amorphous) [R16].

Health Hazards: Fire may produce irritating or poisonous gases. Contact may cause burns to skin and eyes. Runoff from fire control or dilution water may cause pollution (silicon powder, amorphous) [R16].

Emergency Action: Keep unnecessary people away; isolate hazard area and deny entry. Stay upwind; keep out of low areas. Positive-pressure self-contained breathing apparatus (SCBA) and structural firefighters' protective clothing will provide limited protection. Call CHEMTREC at 1–800–424–9300 for emergency assistance. If water pollution occurs, notify the appropriate authorities (silicon powder, amorphous) [R16].

Fire: Small fires: Dry chemical, sand, earth, water spray, or regular foam. Large Fires: Water spray, fog, or regular foam. Move container from fire area if you can do so without risk. Apply cooling water to sides of containers that are exposed to flames until well after fire is out. Stay away from ends of tanks. For massive fire in cargo area, use unmanned hose holder, or monitor nozzles; if this is impossible, withdraw from area, and let fire burn. Magnesium Fires: Use dry sand, Met–L–X powder, or G–1 graphite powder (silicon powder, amorphous) [R16].

Spill or Leak: Shut off ignition sources; no flares, smoking, or flames in hazard area. Do not touch or walk through spilled material. Small Dry Spills: With clean shovel place material into clean, dry container, and cover loosely; move containers from spill area. Large Spills: Wet down with water and dike for later disposal (silicon powder, amorphous) [R16].

First Aid: Move victim to fresh air; call for emergency medical care. In case of contact with material, immediately flush skin or eyes with running water for at least 15 minutes. Removal of solidified molten material from skin requires medical assistance. Remove and isolate contaminated clothing and shoes at the site (silicon powder, amorphous) [R16].

FIRE POTENTIAL Flammable solid (silicon powder, amorphous) [R17].

REFERENCES

1. SRI.

2. Budavari, S. (ed.). *The Merck Index— Encyclopedia of Chemicals, Drugs, and Biologicals.* Rahway, NJ: Merck and Co., Inc., 1989. 1368.

3. Gosselin, R. E., R. P. Smith, H. C. Hodge. *Clinical Toxicology of Commercial Products.* 5th ed. Baltimore: Williams and Wilkins, 1984, p. II–105.

4. *Kirk–Othmer Encyclopedia of Chemical Technology.* 3rd ed., Volumes 1–26. New York, NY: John Wiley and Sons, 1978–1984, p. V5 368–74 (79).

5. Patty, F. (ed.). *Industrial Hygiene, and Toxicology: Volume II: Toxicology.* 2nd ed. New York: Interscience Publishers, 1963. 869.

6. White–Stevens, R. (ed.). *Pesticides in the Environment:* Volume 3. New York: Marcel Dekker, Inc., 1977. 269.

7. *Kirk–Othmer Encyclopedia of Chemical Technology.* 3rd ed., Volumes 1–26. New York, NY: John Wiley and Sons, 1978–1984, p. 24 807–8551 (84).

8. *Kirk–Othmer Encyclopedia of Chemical Technology.* 3rd ed., Volumes 1–26. New York, NY: John Wiley and Sons, 1978–1984, p. 24 612–644 (84).

9. *Kirk–Othmer Encyclopedia of Chemical Technology.* 3rd ed., Volumes 1–26. New York, NY: John Wiley and Sons, 1978–1984, p. 24 332–432 (84).

10. *Kirk–Othmer Encyclopedia of Chemical Technology.* 3rd ed., Volumes 1–26. New York, NY: John Wiley and Sons, 1978–1984, p. V17 777–87 (82).

11. *Kirk–Othmer Encyclopedia of Chemical Technology.* 3rd ed., Volumes 1–26.

New York, NY: John Wiley and Sons, 1978–1984, p. V8 826–69 (79).

12. *Kirk–Othmer Encyclopedia of Chemical Technology*. 3rd ed., Volumes 1–26. New York, NY: John Wiley and Sons, 1978–1984, p. V3 938–58 (78).

13. *Kirk–Othmer Encyclopedia of Chemical Technology*. 3rd ed., Volumes 1–26. New York, NY: John Wiley and Sons, 1978–1984, p. V1 488–510 (78).

14. Sax. *Hawley's Condensed Chemical Dictionary*. 11th ed. 1987, p. 952.

15. Hawley, G. G. *The Condensed Chemical Dictionary*. 9th ed. New York: Van Nostrand Reinhold Co., 1977. 800.

16. U.S. Department of Transportation. *Emergency Response Guidebook 1990*. DOT P 5800.5. Washington, DC: U.S. Government Printing Office, 1990, p. G–32.

17. Bureau of Explosives. *Hazardous Materials* Regs of DOT. p. 140 (1985) ICC No. BOE–6000E.

18. U.S. Coast Guard, Department of Transportation. *CHRIS—Hazardous Chemical Data*. Manual Two. Washington, DC: U.S. Government Printing Office, Oct., 1978.

19. IITII. *Toxic and Hazardous Industrial Chemicals Safety Manual*. Tokyo, Japan: The International Technical Information Institute, 1988. 487.

20. Benke, G. M., T. W. Osborn. *Food Cosmet Toxicol* 17 (2): 123 (1979).

21. Grant, W.M. *Toxicology of the Eye*. 3rd ed. Springfield, IL: Charles C. Thomas Publisher, 1986.

22. Considine. *Chemical and Process Technol Encyc*. 1974.

23. Sax, N.I., and R.J. Lewis, Sr. (eds.). *Hawley's Condensed Chemical Dictionary*. 11th ed. New York: Van Nostrand Reinhold Co., 1987.

■ SODIUM SULFATE

CAS # 7757–82–6

SIC CODES 2815; 2891.

SYNONYMS bisodium sulfate • disodium monosulfate • disodium sulfate • disodium sulphate • Natriumsulfat (German) • sodium sulfate (2:1) • sodium sulfate anhydrous • sodium sulfate (Na_2SO_4) • sodium sulphate • sulfuric acid disodium salt • sulfuric acid, disodium salt • thenardite

MF O_4S • 2Na

MW 142.04

COLOR AND FORM White powder or orthorhombic bipyramidal crystals; monoclinic crystals from 160–185°C; transition point to hexagonal crystals is about 241°C.

ODOR Odorless.

TASTE Bitter saline taste.

DENSITY 2.671.

VISCOSITY 2.48 (22% solution at 20°C).

MELTING POINT 888°C.

FIREFIGHTING PROCEDURES Most fire extinguishing agents may be used in fires involving sodium sulfate [R9].

TOXIC COMBUSTION PRODUCTS When heated in fire, sodium sulfate may emit toxic fumes of sulfur oxides [R9].

SOLUBILITY Soluble in about 3.6 parts water; 1 in 2 parts water is max at 33°C; at 100°C it requires 2.4 parts water; insoluble in alcohol; soluble in hydrogen iodide.

CORROSIVITY The rates of corrosion of iron and steel in water are a function of the specific mineral quality as well as the alkalinity and pH values. Sodium sulfate is a strong contributor to the rate of corrosion. For example, in water with 400 mg/L of alkalinity (as $CaCO_3$) at pH 7, the corrosion rate will be zero at 200 mg/L of Na_2SO_4, but when the concentration of sodium sulfate is 400 mg/L, the corrosion rate will be about 100 mg per square cm per day.

STABILITY AND SHELF LIFE Effloresces rapidly in air (decahydrate) [R17, 952].

OTHER CHEMICAL/PHYSICAL PROPERTIES Transition point to anhydrous form is 24.4°C; white, rhombic, or tetragonal; 19.5 g/100 cc water at 0°C, 44 g/100 CC water at 20°C (heptahydrate); density: 1.46; loses all its water at 100°C; objectionable taste; efflorescent crystals or granules; odorless; aqueous solution is neutral; pH 6 to 7.5; soluble in 1.5 parts water at 25°C, in 3.3 parts water at 15°C (solubility in water decr by sodium chloride); soluble in glycerol; insoluble in alcohol (decahydrate); transparent, monoclinic crystals; MP: 32.38°C; index of refraction: 1.394, 1.396, 1.398 ((decahydrate) decahydrate); 4.76 g/100 cc water at 0°C, 42.7 g/100 cc water at 100°C; soluble in glycerol; insoluble in alcohol; index of refraction: 1.484, 1.477, 1.471 /natural thenardite; latent heat of fusion: 24.4 kJ at melting point; heat of formation: −1,387 kJ/mol; entropy: 149.6 J/(mole x K); heat of solution: 1.17 kJ/mole at 18°C.

PROTECTIVE EQUIPMENT If the spilled material is known to be sodium sulfate: Tight-fitting safety goggles, rubber gloves, and protective clothing should be worn [R9].

CLEANUP Spills on land: Contain if possible by forming mechanical barriers to prevent spreading. Shovel material into containers for recovery or disposal [R9].

ON-SITE RESTORATION Apply anion exchangers. Seek professional environmental engineering assistance through EPA's Enviromental Response Team (ERT), Edison, NJ, 24–hour no (201) 321–6660 [R2].

DISPOSAL Waste sodium sulfate must never be discharged directly into sewers or surface waters. Recovered sodium sulfate may be disposed of by burial in a landfill [R10].

COMMON USES Anhydrous form for drying organic liquid; Kjeldahl nitrogen determination; in manufacture of glass, ultramarine [R3]. Freezing mixtures; manufacture of sodium salts, and ceramic glazes [R17, 952]. Pulping liquor component in kraft wood pulping process; filler, diluent, and processing aid in dry detergents; melting aid in float glass production; dyebath additive to standardize and level color; component of spinning bath in manufacture of viscose rayon; sulfur source in animal feeds; coal treatment to improve furnace performance [R1]. Tanning; pharmaceuticals; freezing mixtures; laboratory reagent; food additiv [R4]. Therapeutic category: cathartic; (veterinary) purgativ [R5]. In moderation is used medically as a diuretic. It is also a component of laxatives and antacids [R6]. Molybdoate orange is prepared by the addition of $NaCrO_4$, Na_2MoO_4, and Na_2SO_4 to $Pb(NO_3)$ [R7].

BINARY REACTANTS Aluminum.

STANDARD CODES No NFPA.

PERSISTENCE Can persist indefinitely.

MAJOR SPECIES THREATENED Water containing 4,546 to 7,369 mg/L of salts, mostly sodium sulfate, was not harmful to two cows over a two-year period. However, a sudden change from normal water to this water caused diarrhea. At a level of 7,000 mg/L there was some evidence of weight reduction in cattle and at 10, 000 mg/L weight losses were severe. Rats were not harmed by water containing 15,000 mg/L of sodium sulfate. Poultry, however, experienced a 33 percent mortality after drinking, for 15 days, water containing 7,500 mg/L of sodium sulfate. Ballantyne sets the threshold limit of Na_2SO_4 for all livestock at 2,050 mg/L.

GENERAL SENSATION Odorless.

PERSONAL SAFETY Intense heat may necessitate use of self-contained breathing apparatus.

ACUTE HAZARD LEVEL Low. Emits toxic vapors when heated to decomposition.

CHRONIC HAZARD LEVEL Unknown.

DEGREE OF HAZARD TO PUBLIC HEALTH Water containing 500 mg/L of sodium sulfate and 1,200 mg/L of magnesium sulfate occasionally has caused diarrhea among human beings. Under normal conditions such water would be regarded as unsuitable for domestic use. Emits toxic vapors when heated to decomposition.

OCCUPATIONAL RECOMMENDATIONS None listed.

ACTION LEVEL If intense heat prevails, notify air authority. Suppress suspension of dusts.

AVAILABILITY OF COUNTERMEASURE MATERIAL Anion exchangers—water softener suppliers.

DISPOSAL NOTIFICATION Local landfill authority.

INDUSTRIAL FOULING POTENTIAL For boiler waters, the ratio between the concentrations of sodium sulfate and sodium sulfite is significant. The following ratios are specified as limits: boiler pressure in psi ratio—Na_2SO_4:Na_2SO_2, not contain more than about 100 mg/L of sodium sulfate, and for ice making, not more than 300 mg/L.

MAJOR WATER USE THREATENED None.

PROBABLE LOCATION AND STATE Colorless crystals. Will dissolve.

COLOR IN WATER Colorless.

NON-HUMAN TOXICITY LD_{50} rabbit percutaneous >4.0 g/kg [R12].

ECOTOXICITY LC_{50} *Aspergillus* 80 ppm/40 hr at 32°C [R2]; LC_{50} *Nitzschia* 1,900 ppm/120-hr static bioassay [R2]; TLm bluegill 12,750 ppm/96 hr [R2]; LC_{50} mosquito fish 17,500 mg/L/48 hrs in turbid wate [R13]; LC_{50} fathead minnow 13,500 to 14,000 mg/L/24 to 96 hr in soft wate [R13]; LC_{50} *Daphnia magna* 4,547 mg/L/96 hr [R14]; LC_{50} caddis fly (*Stenonema ares*) 320 mg/L/96 hr in soft wate [R15].

BIOCONCENTRATION Sodium sulfate may persist indefinitely in the environment. It does not show bioaccumulation or food-chain contamination effects [R11].

DOT

Spill or Leak: Shut off ignition sources; no flares, smoking, or flames in hazard area. Do not touch spilled material. Small Dry Spills: Clean shovel to place material into clean, dry container, and cover; move containers from spill area. Large Spills: Wet down with water and dike for later disposal (sodium sulfide, anhydrous) [R8].

Fire or Explosion: Flammable, combustible material; may be ignited by heat, sparks, or flames. May burn rapidly with flare-burning effect (sodium sulfide, anhydrous) [R8].

First Aid: Move victim to fresh air; call for emergency medical care. In case of contact with material, immediately flush skin or eyes with running water for at least 15 min. Remove and isolate contaminated clothing and shoes at the site (sodium sulfide, anhydrous) [R8].

Health Hazard: Poisonous if swallowed. Skin contact poisonous. Contact may cause burns to skin and eyes. Fire may produce irritating or poisonous gases. Runoff from fire control or dilution water may cause pollution (sodium sulfide, anhydrous) [R8].

Emergency Action: Keep unnecessary people away; isolate hazard area and deny entry. Stay upwind; keep out of low areas. Wear self-contained (positive-pressure, if available) breathing apparatus and full protective clothing. If water pollution occurs, notify appropriate authorities. For emergency assistance call CHEMTREC (800) 424–9300 (sodium sulfide, anhydrous) [R8].

Fire: Small fires: Dry chemical, water spray, or foam. Large fires: Water spray, fog, or foam. Move container from fire area if you can do so without risk. Cool containers that are exposed to flames with water from the side until well after fire is out. For massive fire in cargo area, use unmanned hose holder, or monitor nozzles; if this is impossible, withdraw from area, and let fire burn (sodium sulfide, anhydrous) [R8].

FIRE POTENTIAL Noncombustible solid [R9]. Nonflammable [R2].

FDA Direct food additives FDA 121.1059, limitations: miscellaneous food additive in chewing-gum base; FDA 8.303, limitations: in production of caramel [R16].

REFERENCES

1. SRI.

2. NIH/EPA. *OHM/TADS* (1986).

3. *The Merck Index.* 10th ed. Rahway, NJ: Merck Co., Inc., 1983. 1242.

4. SAX. *Hawley's Condensed Chemical Dictionary.* 11th ed. 1987, p. 952.

5. *Merck index.* 10th ed. 1983, p. 1242.

6. IARC *Monographs* 1972–present, V21 p. 254.

7. *Kirk–Othmer Encyc Chemical Tech.* 3rd ed. 1978–present, V17 p. 820.

8. Department of Transportation. *Emergency Response Guidebook 1984.* DOT P 5800.3 Washington, DC: U.S. Government Printing Office, 1984, p. G–34.

9. Environment Canada. *Tech Info for Problem Spills: Sodium Sulfate.* p. 47 (1985).

10. Environment Canada. *Tech Info for Problem Spills: Sodium Sulfate.* p. 47 (1985).

11. Environment Canada. *Tech Info for Problem Spills: Sodium Sulfate.* p. 41 (1985).

12. Environment Canada. *Tech Info for Problem Spills: Sodium Sulfate* (Draft), p. 44 (1985).

13. Environment Canada. *Tech Info for Problem Spills: Sodium Sulfate* p. 39 (1985).

14. Environment Canada. *Tech Info for Problem Spills: Sodium Sulfate.* p. 40 (1985).

15. Environment Canada. *Tech Info for Problem Spills: Sodium Sulfate.* p. 40 (1985).

16. Furia, T. E. (ed.). *CRC Handbook of Food Additives.* 2nd ed. Cleveland: The Chemical Rubber Co., 1972. 945.

17. Hawley, G. G. *The Condensed Chemical Dictionary.* 10th ed. New York: Van Nostrand Reinhold Co., 1981.

■ SODIUM TETRABORATE

CAS # 1330–43–4

SYNONYMS anhydrous borax • borax glass • borax, dehydrated • boric acid ($H_2B_4O_7$) disodium salt • boric acid ($H_2B_4O_7$) sodium salt • boron sodium oxide ($B_4Na_2O_7$) • disodium tetraborate • fr-28 • fused borax • fused sodium borate • rasorite-65 • sodium borate • sodium borate, anhydrous

MF $B_4H_2O_7$ • 2Na

MW 201.3

COLOR AND FORM Powder or glass-like plates; white, free-flowing crystals; light-grey solid.

ODOR Odorless.

DENSITY 2.367.

BOILING POINT Decomposes at boiling point (1,575°C).

MELTING POINT 741°C.

SOLUBILITY 1.06 g/100 cc water at 0°C; 8.79 g/100 cc water at 40°C; insoluble in alcohol; 2.56 g/100 g water at 20°C; 30 wt% in ethylene glycol.

CORROSIVITY Solutions are not a corrosion hazard to ferrous metals.

STABILITY AND SHELF LIFE Becomes opaque on exposure to air [R2].

OTHER CHEMICAL/PHYSICAL PROPERTIES Hygroscopic; heat of formation (glass): −3.2566 MJ/mol (−778.4 kcal/mol), heat of formation (alpha crystalline form): −3.2767 MJ/mol (−783.2 kcal/mol), heat of solution: −213.8 kJ/kg (−92 Btu/lb); heat of fusion: 81.2 kJ/mol (19.4 kcal/mol).

PROTECTIVE EQUIPMENT Wear goggles or face shield when handling [R12, 64].

PREVENTATIVE MEASURES Contact lens use in industry is controversial. A survey of 100 corporations resulted in the recommendation that each company establish their own contact lens use policy. One presumed hazard of contact lens use is possible chemical entrapment. It was found that contact lenses minimized injury or protected the eye. The eye was afforded more protection from liquid irritants. Soft contact lenses do not worsen corneal damage from strong chemicals and in some cases could actually protect the eye. Overall, the literature supports the wearing of contact lenses in industrial environments as part of the standard eye protection, e.g., face shields; however, more data are needed to establish the value of contact lenses [R9].

DISPOSAL SRP: At the time of review, criteria for land treatment, or burial (sanitary landfill) disposal practices are subject to significant revision. Prior to implementing land disposal of waste residue (including waste sludge), consult with environmental regulatory agencies for guidance on acceptable disposal practices.

COMMON USES Manufacture of glass, enamels, and other ceramic products [R1]. Soldering metals; manufacture of glazes; tanning; in cleaning compound; artificially aging wood; curing and preserving skins [R2]. As preservative either alone or with other antiseptics against wood fungus; fireproofing fabrics and wood [R2]. Medication (vet) [R2]; Sodium borate is included as buffer in many products [R3]. Fluxing agent for porcelain enamels and ceramic glazes); non-selective herbicide; alkalizing agent in pharmaceuticals; emulsifying agent in cosmetics [R4]. In the glass industry, for fiberglass insulation, borosilicate glass, and enamels. Also used as a fire retardant, an antifreeze additive, and as an algicide in industrial water. Used in fertilizers [R11, 98]. Corrosion inhibitor [R5]. Has been used as an exptl growth promoter in poultry feeds [R6]. Insecticidal, fungicidal, and nematocidal at high concentrations. Is useful in fertilizers to supply trace amounts of boron essential to plant growth [R12, 64]. Nonselective vegetation control adapted to preventing emergence under asphalt. Ground-dry application by hand or mechanical spreaders. Rates: 2.5 to 3.7 kg/10 m³. Usual carrier: none, or slurry with water [R12, 64]. Boric acid, borates, and perborates have been used as mild antiseptics or bacteriostats in eyewashes, mouthwashes, burn dressings, and diaper-rash powders; however, the effectiveness of boric acid has largely been discredited. [R13, 131].

OCCUPATIONAL RECOMMENDATIONS

OSHA 8-hr time-weighted average: 10 mg/m³. Final rule limits were achieved by any combination of engineering controls, work practices, and personal protective equipment during the phase-in period, Sept 1, 1989 through Dec 30, 1992. Final rule limits became effective Dec 31, 1992. (borates, tetra, sodium salts, anhydrous) [R10].

THRESHOLD LIMIT 8–hr time-weighted average (TLV) 1 mg/m³ (1989–90) (borates, tetra, sodium salts, anhydrous) [R14, 14].

INTERNATIONAL EXPOSURE LIMITS Belgium (1974) and the Netherlands (1976) have adopted the value 1 mg/m³ [R8].

BIOCONCENTRATION Accumulates in the plant [R12, 64].

FIRE POTENTIAL Not flammable [R7].

REFERENCES

1. Sax, N. I., and R. J. Lewis, Sr. (eds.). *Hawley's Condensed Chemical Dictionary.* 11th ed. New York: Van Nostrand Reinhold Co., 1987. 162.

2. *The Merck Index.* 10th ed. Rahway, NJ: Merck Co., Inc., 1983. 1231.

3. Gilman, A. G., L. S. Goodman, and A. Gilman (eds.). *Goodman and Gilman's The Pharmacological Basis of Therapeutics.* 6th ed. New York: Macmillan Publishing Co., Inc. 1980. 972.

4. SRI.

5. 29 CFR 180.1001 (7/1/88).

6. Rossoff, I. S. *Handbook of Veterinary Drugs.* New York: Springer Publishing Company, 1974. 50.

7. U.S. Coast Guard, Department of Transportation. *CHRIS—Hazardous Chemical Data.* Volume II. Washington, DC: U.S. Government Printing Office, 1984–5.

8. American Conference of Governmental Industrial Hygienists. *Documentation of the Threshold Limit Values and Biological Exposure Indices.* 5th ed. Cincinnati, OH: American Conference of Governmental Industrial Hygienists, 1986. 60.

9. Randolph, S. A., M. R. Zavon. *J Occup Med* 29: 237–42 (1987).

10. 54 FR 2920 (1/19/89).

11. *Kirk-Othmer Encyclopedia of Chemical Technology*. 3rd ed., Volumes 1–26. New York, NY: John Wiley and Sons, 1978–1984., p. 4(78).

12. Weed Science Society of America. *Herbicide Handbook*. 5th ed. Champaign, Illinois: Weed Science Society of America, 1983.

13. Seiler, H.G., H. Sigel and A. Sigel (eds.). *Handbook on the Toxicity of Inorganic Compounds*. New York, NY: Marcel Dekker, Inc. 1988.

14. American Conference of Governmental Industrial Hygienists. *Threshold Limit Values for Chemical Substances and Physical Agents and Biological Exposure Indices for 1994–1995*. Cincinnati, OH: ACGIH, 1994.

■ STANNIC CHLORIDE

CAS # 7646-78-8

SYNONYMS etain (tetrachlorure d') (French) • libavius fuming spirit • stagno (tetracloruro di) (Italian) • stannane, tetrachloro • stannic chloride, anhydrous • stannic tetrachloride • tetrachlorostannane • tetrachlorotin • tin chloride ($SnCl_4$) • tin perchloride • tin tetrachloride • tin tetrachloride, anhydrous • tin(IV) chloride • tin(IV) tetrachloride • tintetrachloride (Dutch) • Zinntetrachlorid (German)

MF Cl_4Sn

MW 260.53

COLOR AND FORM Colorless liquid, cubic solid; colorless to yellow liquid.

ODOR Acrid.

DENSITY 2.226 (liquid).

BOILING POINT 115°C.

MELTING POINT −33°C.

FIREFIGHTING PROCEDURES Do not use hand-held water-base extinguishers on fires in vicinity of tin tetrachloride. Use water spray if necessary to keep containers cool [R14, p. 49–84]. In fighting fires wear full protective clothing [R14, p. 49–84]. Use dry chemical, dry sand, or carbon dioxide. Do not use water on material itself. If large quantities of combustibles are involved, use water in flooding quantities as spray and fog. Use water to knock down vapors. Apply water from as far a distance as possible [R4]. Wear a self-contained breathing apparatus with full facepiece operated in pressure-demand or other positive-pressure mode. (inorganic tin compounds (as tin)) [R13, 6].

TOXIC COMBUSTION PRODUCTS Toxic gases and vapors may be released in a fire involving the tin chlorides and stannous sulfate [R13, 2]. When heated to decomposition, it emits toxic fumes of chlorine. [R15].

SOLUBILITY Soluble in cold water, ether; soluble in acetone, ethanol, gasoline, benzene, carbon tetrachloride, toluene, kerosene, methanol; soluble in carbon disulfide.

VAPOR PRESSURE 18 mm Hg at 20°C.

CORROSIVITY Corrosive to metals.

OTHER CHEMICAL/PHYSICAL PROPERTIES Fumes in moist air. White or slightly yellow crystals or fused small lumps; slight hydrochloric acid odor; very soluble in water; soluble in alcohol (stannic chloride pentahydrate). MP: approximately 56°C (decomposes); density: 2.04 g/cu cm at 25°C (stannic chloride, pentahydrate). Monoclinic crystals (stannic chloride, pentahydrate) (deliquescent stannic chloride, pentachloride). Hydrolyzes in hot water. Opaque; bp: stable 56 to 83°C; soluble in cold water (stannic chloride, tetrahydrate). Colorless monoclinic crystals; mp: 80°C; bp: stable 64 to 83°C; soluble in cold water (stannic chloride, trihydrate). Heat of fusion: 2,190 cal/g mole. Dielectric constant: 2.97 at 20°C. Tin forms two series of compound, the stannous compound, tin(2), of bivalent tin, and the stannic compound, tin(4), of quadrivalent tin. At pH above 6, stannous compounds are easily oxidized, and are strong reducing agents. Bivalent tin can be in the ionic form, but stannic compounds are covalent, and the ionic form of tin(4) does not exist. (inorganic tin).

PROTECTIVE EQUIPMENT Employees should be provided with and required to use impervious clothing, gloves, face shields (eight-inch minimum), and other appropriate protective clothing necessary to prevent any possibility of skin contact with liquid stannic chloride [R13, 3]. Employees should be provided with and required to use dust– and splash-proof goggles where there is any possibility of liquid stannic chloride contacting the eyes [R13, 3]. Wear appropriate chemical protective boots [R4]. Skin contact should be prevented by protective clothing (tin and inorganic tin compounds) [R17, 863]. Respirator selection: Upper-limit devices recommended by OSHA: 10 mg/m³: any dust and mist respirator except single-use respirators; 20 mg/m³: any dust and mist respirator except single-use and quarter-mask respirators or any supplied-air respirator or any self-contained breathing apparatus; 50 mg/m³: any powered air-purifying respirator with a dust and mist filter or any supplied-air respirator operated in a continuous-flow mode; 100 mg/m³: any air-purifying full facepiece respirator with a high-efficiency particulate filter or any self-contained breathing apparatus or any supplied-air respirator with a full facepiece; 400 mg/m³: any supplied-air respirator with a full facepiece and operated in a pressure-demand or other positive-pressure mode; emergency or planned entry into unknown concentration or IDLH conditions: any self-contained breathing apparatus with full facepiece and operated in a pressure-demand or other positive-pressure mode or any supplied air respirator with a full facepiece and operated in pressure-demand or other positive-pressure mode in combination with an auxiliary self-contained breathing apparatus operated in pressure-demand or other positive-pressure mode; escape: any air-purifying full facepiece respirator with high-efficiency particulate filter or any appropriate es-

cape-type self-contained breathing apparatus. (tin, inorganic compounds except oxides (as Sn)) [R16, 225].

PREVENTATIVE MEASURES Contact lens use in industry is controversial. A survey of 100 corporations resulted in the recommendation that each company establish their own contact lens use policy. One presumed hazard of contact lens use is possible chemical entrapment. It was found that contact lens minimized injury or protected the eye. The eye was afforded more protection from liquid irritants. Soft contact lenses do not worsen corneal damage from strong chemicals and in some cases could actually protect the eye. Overall, the literature supports the wearing of contact lenses in industrial environments as part of the standard eye protection, e.g., face shields; however, more data are needed to establish the value of contact lenses [R5]. Nonimpervious clothing that becomes contaminated with stannic chloride should be removed immediately and not reworn until the contaminant is removed from the clothing [R13, 3]. Tin tetrachloride can be handled in iron or steel containers when dry, but not so in a moist vapor-air mixture: any fans, ducts, or enclosures for control of the vapors must either be replaced frequently or be constructed of material resistant to hydrochloric acid [R18]. Clothing contaminated with stannous chloride or stannic chloride should be placed in closed containers for storage until it can be discarded or until provision is made for the removal of contaminant from the clothing. If the clothing is to be laundered or otherwise cleaned to remove the contaminant, the person performing the operation should be informed of contaminant's hazardous properties [R13, 3]. Where there is a possibility of exposure liquid stannic chloride, facilities for quick drenching of the body should be provided for emergency use. Where there is any possibility that employees' eyes may be exposed to liquid stannic chloride, an eye wash fountain should be provided within the immediate work area for emergency use [R13, 3]. Skin that becomes contaminated with stannic chloride should be immediately washed or showered to remove any contaminant. Employees who handle stannic chloride should wash their hands thoroughly before eating, smoking, or using toilet facilities [R13, 4]. If material not involved in fire: Keep material out of water sources and sewers. Build dikes to contain flow as necessary. Use water spray to knock down vapors. Do not use water on material itself [R15]. Avoid breathing vapors. Keep upwind. Avoid bodily contact with material. Do not handle broken packages unless wearing appropriate personal protective equipment [R4]. Keep away from residential areas [R11, 491]. SRP: Contaminated protective clothing should be segregated in such a manner so that there is no direct personal contact by personnel who handle, dispose of, or clean the clothing. Quality assurance to ascertain the completeness of the cleaning procedures should be implemented before the decontaminated protective clothing is returned for reuse by the workers. Contaminated clothing should not be taken home at end of shift, but should remain at employee's place of work for cleaning. Contact lenses should not be worn when working with this chemical (tin, inorganic compounds except oxides (as Sn)) [R16, 225]. If inorganic tin compounds get on the skin, immediately wash the skin using soap or mild detergent, and water. If inorganic tin compounds penetrate through the clothing, remove clothing immediately, and wash the skin using soap or mild detergent, and water. If irritation persists after washing, get medical attention (inorganic tin compounds (as tin)) [R13, 5]. Persons not wearing protective equipment and clothing should be restricted from areas of spills or leaks until cleanup has been completed. (inorganic tin compound (as tin)) [R13, 5].

CLEANUP Cover spills with sufficient amount of sodium bicarbonate. Remove the mixture in a container such as a fiber drum, plastic bag, or carton box for easy disposal [R11, 492]. Neutralize spilled material with crushed limestone, soda ash, or lime [R4]. If tin, stannous chloride, stannic chloride, stannous sulfate, or potassium stannate is spilled or leaked, the following

steps should be taken: 1. Ventilate area of spill or leak. 2. Collect spilled or leaked material in the most convenient and safe manner for reclamation or for disposal. Liquid containing inorganic tin compounds should be absorbed in vermiculite, dry sand, earth, or a similar material. [R13, 5].

DISPOSAL SRP: At the time of review, criteria for land treatment or burial (sanitary landfill) disposal practices are subject to significant revision. Prior to implementing land disposal of waste residue (including waste sludge), consult with environmental regulatory agencies for guidance on acceptable disposal practices. Chemical treatability of tin. Concentration Process: Chemical precipitation; Chemical Classification: Metals; Scale of Study: Pilot scale; Type of Wastewater Used: Synthetic wastewater; Results of Study: At 600 ppm, 95.3% reduction with alum. At 500 ppm, 98% reduction with ferric chloride, 92% reduction with lime; three coagulants used: 200 mg of alum at pH = 6.4, 40 ppm of ferric chloride at pH = 6.2, 41 ppm of lime at pH = 11.5. Chemical coagulation was followed by dual media filtration. (total tin) [R19].

COMMON USES As mordant; reviving colors; dehydrating agent in organic syntheses; stabilizer for colors and perfumes in soap; in dyeing of fabrics, weighting silk, tinning vessels, in ceramics to produce abrasion-resistant or light-reflecting coatings [R2]. Principally as a chemical intermediate for other tin compounds especially organotins; surface treatment of glass and other nonconductive materials; used to strengthen glassware for returnable and nonreturnable foodstuff bottles and jars for restaurant and catering glasses; as a catalyst in Friedel–Crafts acylation, alkylation, and cyclization reactions, esterifications, halogenations, and curing [R10, 47]. In electroconductive and electroluminescent coatings; manufacture of fuchsin; bacteria and fungi control in soaps [R1]. Stabilizers for plastics [R11, 491]. Manufacture of blueprint and other sensitized paper and antistatic agent for synthetic fibers [R10, 47]. Bleaching agent for sugar

[R12, 1942]. In preparation of lubricating-oil additives. [R13, 4].

OCCUPATIONAL RECOMMENDATIONS

OSHA 8-hr time-weighted average: 2 mg/m³ (tin inorganic compounds, except oxides) [R7]. Meets criteria for OSHA medical records rule. [R21].

THRESHOLD LIMIT 8-hr time-weighted average (TWA) 2 mg/m³ (1986) (tin oxide and inorganic compounds, except SnH_4, as Sn) [R22, 34]. Excursion Limit Recommendation: Excursions in worker exposure levels may exceed three times the TLV–TWA for no more than a total of 30 min during a work day, and under no circumstances should they exceed five times the TLV–TWA (tin oxide, and inorganic compounds, except SnH_4, as Sn) [R22, 5].

INTERNATIONAL EXPOSURE LIMITS In the Federal Republic of Germany the maximum concentration in the air at the workplace for inorganic compounds, calculated as tin, is 2 mg/m³. (inorganic tin compounds (as tin)) [R8].

NON-HUMAN TOXICITY LD_{50} mouse intraperitoneal 46 mg/kg [R23, 183].

BIOLOGICAL HALF-LIFE From bone the biological half-time is about 400 days (tin) [R6]. In humans, 20% of absorbed tin cleared with half-time of 4 days, 20% with 25 days, and 60% with 400 days (soluble tin compounds) [R20, 575].

BIOCONCENTRATION The insolubility of tin at a neutral to alkaline pH range prevents plant uptake and subsequent food-chain contamination. (tin compounds) [R24, 294].

DOT *Health Hazards:* Contact causes burns to skin and eyes. If inhaled, may be harmful. Fire may produce irritating or poisonous gases. Runoff from fire-control or dilution water may cause pollution (stannic chloride, hydrated) [R3].

Fire or Explosion: Some of these materials may burn, but none of them ignites readily. Flammable/poisonous gases may accumulate in tanks and hopper cars. Some of these materials may ignite combustibles (wood, paper, oil, etc.) (stannic chloride, hydrated) [R3].

Emergency Action: Keep unnecessary people away; isolate hazard area and deny entry. Stay upwind; keep out of low areas. Self-contained breathing apparatus (SCBA) and structural firefighter's protective clothing will provide limited protection. Call CHEMTREC at 1–800–424–9300 for emergency assistance. If water pollution occurs, notify the appropriate authorities (stannic chloride, hydrated) [R3].

Fire: Some of these materials may react violently with water. Small Fires: Dry chemical, CO_2, halon, water spray, or standard foam. Large Fires: Water spray, fog, or standard foam is recommended. Move container from fire area if you can do so without risk. Cool containers that are exposed to flames with water from the side until well after fire is out. Stay away from ends of tanks (stannic chloride, hydrated) [R3].

Spill or Leak: Do not touch spilled material; stop leak if you can do so without risk. Small Spills: Take up with sand or other noncombustible absorbent material and place into containers for later disposal. Small Dry Spills: With clean shovel place material into clean, dry container, and cover; move containers from spill area. Large Spills: Dike far ahead of liquid spill for later disposal (stannic chloride, hydrated) [R3].

First Aid: Move victim to fresh air; call emergency medical care. Remove and isolate contaminated clothing and shoes at the site. In case of contact with material, immediately flush skin or eyes with running water for at least 15 minutes. Keep victim quiet, and maintain normal body temperature (stannic chloride, hydrated) [R3].

FIRE POTENTIAL Nonflammable [R11, 491].

FDA Bottled water shall, when a composite of analytical units of equal volume from a sample is examined by the methods described in paragraph (d) (1) (II) of this section, meet the standards of chemical quality and shall not contain chloride in excess of 250.0 mg/L. (chloride) [R9].

REFERENCES

1. Sax, N. I., and R. J. Lewis, Sr. (eds.). *Hawley's Condensed Chemical Dictionary.* 11th ed. New York: Van Nostrand Reinhold Co., 1987. 1087.

2. *The Merck Index.* 10th ed. Rahway, NJ: Merck Co., Inc., 1983. 1256.

3. Department of Transportation. *Emergency Response Guidebook 1987.* DOT P 5800.4. Washington, DC: U.S. Government Printing Office, 1987, p. G-60.

4. Association of American Railroads. *Emergency Handling of Hazardous Materials in Surface Transportation.* Washington, DC: Assoc. of American Railroads, Hazardous Materials Systems (BOE), 1987. 683.

5. Randolph, S.A., M.R. Zavon. *J Occup Med* 29: 237–42 (1987).

6. Friberg, L., G. F. Nordberg, E. Kessler, and V. B. Vouk (eds.). *Handbook of the Toxicology of Metals.* 2nd ed. Vols I, II. Amsterdam: Elsevier Science Publishers B. V., 1986, p. V2 568.

7. 29 CFR 1910.1000 (7/1/88).

8. Seiler, H. G., H. Sigel, and A. Sigel (eds.). *Handbook on the Toxicity of Inorganic Compounds.* New York, NY: Marcel Dekker, Inc. 1988. 702.

9. 21 CFR 103.35 (4/1/88).

10. *Kirk-Othmer Encyclopedia of Chemical Technology.* 3rd ed., Volumes 1–26. New York, NY: John Wiley and Sons, 1978–1984, p. 23 (83).

11. ITII. *Toxic and Hazarous Industrial Chemicals Safety* Manual. Tokyo, Japan: The International Technical Information Institute, 1982.

12. Clayton, G.D., and F.E. Clayton (eds.). *Patty's Industrial Hygiene and Toxicology:* Volume 2A, 2B, 2C: *Toxicology.* 3rd ed. New York: John Wiley and Sons, 1981–1982.

13. Mackison, F. W., R. S. Stricoff, and L. J. Partridge, Jr. (eds.). *NIOSH/OSHA— Occupational Health Guidelines for Chemical Hazards.* DHHS (NIOSH) Publication No. 81-123 (3 vols). Washington, DC: U.S. Government Printing Office, Jan. 1981.

14. National Fire Protection Association. *Fire Protection Guide on Hazardous Materials.* 9th ed. Boston, MA: National Fire Protection Association, 1986.

15. Sax, N. I. *Dangerous Properties of Industrial Materials.* 6th ed. New York, NY: Van Nostrand Reinhold, 1984. 2582.

16. NIOSH. *Pocket Guide to Chemical Hazards.* 2nd Printing. DHHS (NIOSH) Publ. No. 85–114. Washington, DC: U.S. Dept. of Health and Human Services, NIOSH/Supt. of Documents, GPO, February 1987.

17. Sittig, M. *Handbook of Toxic and Hazardous Chemicals and Carcinogens,* 1985. 2nd ed. Park Ridge, NJ: Noyes Data Corporation, 1985.

18. Patty, F. (ed.). *Industrial Hygiene and Toxicology:* Volume II: *Toxicology.* 2nd ed. New York: Interscience Publishers, 1963. 2255.

19. Hannah, S. A., et al. *J Water Pollut Control Fed* 49 (11): 2297–309 (1977) as cited in USEPA, *Management of Hazardous Waste Leachate,* EPA Contract No. 68–03–2766 p. E–74 (1982).

20. Friberg, L., G. F. Nordberg, E. Kessler, and V. B. Vouk (eds.). *Handbook of the Toxicology of Metals.* 2nd ed. Vols I, II. Amsterdam: Elsevier Science Publishers B.V., 1986, p. V2.

21. 29 CFR 1910.20 (7/1/88)

22. American Conference of Governmental Industrial Hygienists. *Threshold Limit Values for Chemical Substances and Physical Agents and Biological Exposure Indices for 1994–1995.* Cincinnati, OH: ACGIH, 1994.

23. Venugopal, B., and T.D. Luckey. *Metal Toxicity in Mammals.* 2. New York: Plenum Press, 1978.

24. Brown, K.W., G. B. Evans, Jr., B.D. Frentrup (eds.). *Hazardous Waste Land Treatment.* Boston, MA: Butterworth Publishers, 1983.

■ STANNOUS FLUORIDE

CAS # 7783–47–3

SYNONYMS fluoristan • stannous fluoride (SnF_2) • tin bifluoride • tin difluoride • tin fluoride (SnF_2) • tin fluoride • easygel • gel-kam • gel-tin • Omnii-gel • Stop-home-treatment • Oral-B rinsing solution, concentrate • AIM • Cap-Tin Mouthrinse • Crest • Iradicar-SnF_2• Iradicar Stannous Fluoride • King's Gel-Tin • Stancare • Stanide • Iradicav-SnF_2• Iradicav

MF F_2Sn

MW 156.70

pH 2.8–3.5 (0.4% aqueous solution).

COLOR AND FORM Monoclinic, lamellar plates; white, crystalline powder.

TASTE Bitter, salty taste.

DENSITY 4.57 at 25°C.

BOILING POINT 850°C.

MELTING POINT 213°C.

FIREFIGHTING PROCEDURES Wear a self-contained breathing apparatus with full facepiece operated in pressure-demand or other positive-pressure mode. (inorganic tin compounds (as tin)) [R13, 6].

SOLUBILITY 30–39% in water at 20°C; practically insoluble in alcohol; practically insoluble in ether and chloroform; 72–82% in anhydrous hydrogen fluoride at 20°C; 42 lb/100 lb water at 64.4°F.

STABILITY AND SHELF LIFE Forms oxyfluoride, stannous oxyfluoride, on exposure to air [R1].

OTHER CHEMICAL/PHYSICAL PROPERTIES Tin forms two series of compounds, the stannous compounds, tin(2), of bivalent tin, and the stannic compounds, tin(4), of quadrivalent tin. At pH above 6, stannous compounds, are easily oxidized, and are strong reducing agents. Bivalent tin can be in the ionic form, but stannic compounds are covalent, and the ionic form of tin(4) does not exist. (inorganic tin).

PROTECTIVE EQUIPMENT Wear goggles and self-contained respirator (NIOSH–approved respirator) [R2].

PREVENTATIVE MEASURES Contact lenses should not be worn when working with this chemical. (tin, inorganic compounds except oxides (as Sn)) [R12, 225].

DISPOSAL SRP: At the time of review, criteria for land treatment or burial (sanitary landfill) disposal practices are subject to significant revision. Prior to implementing land disposal of waste residue (including waste sludge), consult with environmental regulatory agencies for guidance on acceptable disposal practices.

COMMON USES Dental caries prophylatic [R1]. Ingredient of caries-preventing toothpastes [R1]; in mouthwash [R3].

OCCUPATIONAL RECOMMENDATIONS
OSHA Meets criteria for OSHA medical records rule (fluoride as (F)) [R7]; threshold-limit 8–hr time weighted average (TWA) 2 mg/m³ (1986) (tin oxide and inorganic compounds, except SnH_4, as Sn) [R11, 34].

INTERNATIONAL EXPOSURE LIMITS In the Federal Republic of Germany the maximum concentration in the air at the workplace for inorganic compounds, calculated as tin, is 2 mg/m³. (inorganic tin compounds (as tin)) [R8].

NON-HUMAN TOXICITY LD_{50} mouse oral 112 mg/kg [R4]; LD_{50} rat oral 200–300 mg/kg [R4]; LD_{50} Swiss white mouse oral (admin by stomach tube) 31.2 mg/kg body wt [R5]; LD_{50} Rochester rat oral (admin by stomach tube) 45.7 mg/kg body wt [R5]; LD_{50} Swiss white mouse oral (admin by stomach tube) 25.5 mg/kg body wt [R5].

BIOLOGICAL HALF-LIFE The fate of (113)tin has been evaluated in the rat following the administration of (113)tin(II) and (113)tin(IV). The biological half-life of the tin in bones was calculated to be 20 to 40 days [R6].

BIOCONCENTRATION The insolubility of tin at a neutral to alkaline pH range prevents plant uptake and subsequent food-chain contamination. (tin compounds) [R10, 294].

FIRE POTENTIAL Not flammable [R2].

FDA Bottled water packaged in the USA to which no fluoride is added shall not contain fluoride in excess of 1.8 mg/L at 63.9–70.6°F. Bottled water packaged in the USA to which fluoride is added shall not contain fluoride in excess of 1.2 mg/L at 63.9–70.6°F. Imported bottled water to which no fluoride is added and imported bottled water to which fluoride is added shall not contain fluoride in excess of 1.4 mg/L and 0.8 mg/L, respectively. (fluoride) [R9].

REFERENCES

1. *The Merck Index*. 10th ed. Rahway, NJ: Merck Co., Inc., 1983. 1257.

2. U.S. Coast Guard, Department of Transportation. *CHRIS—Hazardous Chemical Data*. Volume II. Washington, DC: U.S. Government Printing Office, 1984–5.

3. Plies, S. M. *Tex Rep Biol Med*. 30 (2): 163–80 (1972).

4. Gosselin, R. E., R. P. Smith, H. C. Hodge. *Clinical Toxicology of Commercial Products*. 5th ed. Baltimore: Williams and Wilkins, 1984, p. II–146.

5. IARC. *Monographs on the Evaluation of the Carcinogenic Risk of Chemicals to Man*. Geneva: World Health Organization, International Agency for Research on Cancer, 1972–present (multivolume work), p. V27 273 (1982).

6. Hiles, R. A. *Toxicol Appl Pharmacol*. 27 (2): 366 (1974).

7. 29 CFR 1910.20 (7/1/88).

8. Seiler, H. G., H. Sigel, and A. Sigel (eds.). *Handbook on the Toxicity of Inorganic Compounds*. New York, NY: Marcel Dekker, Inc. 1988. 702.

9. 21 CFR 103.35 (4/1/88).

10. Brown, K.W., G. B. Evans, Jr., B.D. Frentrup (eds.). *Hazardous Waste Land Treatment*. Boston, MA: Butterworth Publishers, 1983.

11. American Conference of Governmental Industrial Hygienists. *Threshold Limit Values for Chemical Substances and Physical Agents and Biological Exposure Indices for 1994–1995*. Cincinnati, OH: ACGIH, 1994.

12. NIOSH. *Pocket Guide to Chemical Hazards*. 2nd Printing. DHHS (NIOSH) Publ. No. 85–114. Washington, DC: U.S. Dept. of Health and Human Services, NIOSH/Supt.of Documents, GPO, February 1987.

13. Mackison, F. W., R. S. Stricoff, and L. J. Partridge, Jr. (eds.). *NIOSH/OSHA— Occupational Health Guidelines for Chemical Hazards*. DHHS (NIOSH) Publication No. 81–123 (3 vols). Washington, DC: U.S. Government Printing Office, Jan. 1981.

■ STEARIC ACID, LEAD SALT

CAS # 7428–48–0

DOT # 2811

SIC CODES 2851; 2992.

SYNONYMS stearic acid, lead salt • octadecanoic acid, lead salt • neutral lead stearate • austrostab 110 e • listab 28 • lead octadecanoate

MF $C_{18}H_{36}O_2$ • x Pb

MW 1734.87 (774.17 as solid—CHRIS)

STANDARD CODES EPA 311; TSCA; IATA poison class B NOS, poison label, 25 kg passenger, 95 kg cargo; CFR 49 poison class B, poison label, 50 lb passenger, 200 lb cargo, storage code 1, 2; CFR 14 CAB code 8; not listed NFPA; not listed AAR; ICC STCC Tariff no. 1–c 28 162 30; Superfund designated (hazardous substances) list.

PERSISTENCE Persistent.

MAJOR SPECIES THREATENED Fish species such as salmon, trout, minnows, and catfish are especially sensitive to lead. Toxicity intensified in soft water.

INHALATION LIMIT 0.15 mg/m³.

DIRECT CONTACT Can be absorbed through the skin.

GENERAL SENSATION Inhalation or skin absorption may produce abdominal pain, loss of appetite, weight loss, constipation, apathy or irritability, vomiting, fatigue, headache, weakness, metallic taste, and muscle incoordination. At fire temperatures causes coughing, choking, headache, dizziness, weakness, difficult breathing, and cyanosis. Bloody sputum may be noted. Skin contact produces pain and severe burns. Eye contact produces corneal destruction.

PERSONAL SAFETY Protect against inhalation and absorption through the skin. Must wear protective clothing including NIOSH-approved rubber gloves and boots, safety goggles or face mask, and a respirator whose canister is specifically approved for this material. Decontaminate fully or dispose of all equipment and clothing after use.

ACUTE HAZARD LEVEL Lead stearate may be moderately to very toxic to human beings by ingestion; between 1 ounce and 1 pound may be fatal. When lead compounds are ingested, much passes through the body unabsorbed. Long periods of exposure are necessary to produce symptoms. Lead compounds are easily absorbed in the respiratory tract when inhaled. Smoke resulting from a fire containing this material will be quite irritating to the skin, eyes, and lungs. Ingestion of large amounts of lead stearate will lead to gastrointestinal distress. The 96-hour TLM aquatic toxicity to bluegill sunfish is approximately 265 ppm, based on 69 ppm of the lead ion.

CHRONIC HAZARD LEVEL Lead is a cumulative poison. Repeated exposure through inhalation or ingestion will lead to damage to the liver, kidney, blood, and nervous system. Lead is a suspected carcinogen of the lungs and kidney. The threshold-limit value, time-weighted average for a 40-hour workweek is 0.15 mg/m³, and the short-term exposure limit (1.5 minutes) is 0.45 mg/m³. The OSHA standard for an 8-hour workday is 0.2 mg/m³ for inorganic lead compounds.

DEGREE OF HAZARD TO PUBLIC HEALTH Can be absorbed through the lungs or through skin contact. A suspected carcinogen of the lungs and kidney. Lead is

accumulated in the body, and chronic poisoning is a serious hazard.

OCCUPATIONAL RECOMMENDATIONS None listed.

ACTION LEVEL Avoid contact with the spilled cargo. Stay upwind. Notify local air, water, and fire authorities of the accident. Evacuate all people to a distance of at least 200 feet upwind and 1,000 feet downwind of the spill.

ON-SITE RESTORATION Dam stream if possible to reduce the flow and prevent further dispersion of the material by water movement. Bottom pumps or underwater vacuum systems may be employed in small bodies of water; dredging may be effective in larger bodies, to remove any undissolved material from the bottom. If treatment is to be undertaken, pump water into a suitable container. Under controlled conditions add calcium hydroxide (Ca(OH₂) to a pH of 8.5. Precipitate lead and filter discharge. Use carbon as a polishing step. For more details see Envirex Manual EPA 600/2-77-227. Seek professional environmental engineering assistance through EPA's Environmental Response Team (ERT), Edison, NJ, 24-hour no., 908-548-8730.

BEACH OR SHORE RESTORATION Close beach and shore to the public until material has been removed. If tidal, scrape affected area at low tide with a mechanical scraper, and avoid human contact. Do not burn material.

AVAILABILITY OF COUNTERMEASURE MATERIAL Bottom pumps, dredges—fire departments, U.S. Coast Guard, Army Corps of Engineers; vacuum systems—swimming-pool operators; activated carbon—water treatment plants, chemical companies.

DISPOSAL After the material has been contained, remove it and the contaminated soil and place in impervious containers. If practical, transport material back to the supplier for recovery. If this is not practical, the material should be buried in a specially designated chemical landfill. Not amenable to treatment at a municipal sewage treatment plant.

DISPOSAL NOTIFICATION Notify local and state health authorities, local solid waste disposal authorities, supplier, and shipper.

INDUSTRIAL FOULING POTENTIAL Very small fouling potential.

PROBABLE LOCATION AND STATE White powder, will sink.

COLOR IN WATER Colorless.

FOOD-CHAIN CONCENTRATION POTENTIAL Positive. Lead compounds are concentrated in the food chain.

GENERAL RESPONSE Shut off ignition sources and call fire department. Wear goggles, self-contained breathing apparatus, and protective clothing (including gloves). Stop discharge if possible. Keep people away. Isolate and remove discharged material. Notify local health and pollution control agencies.

FIRE RESPONSE Not flammable. Poisonous fumes produced at high temperatures. Dusy may explode at high temperatures or source of ignition. Extinguish with CO₂ or foam.

EXPOSURE RESPONSE Call for medical aid. Dust poisonous if inhaled. Move to fresh air. Keep victim quiet and warm. Solid if swallowed will cause headache, abdominal pain, nausea, and vomiting. Flush affected areas with water. If swallowed and victim is conscious, have victim drink milk or water, and have victim induce vomiting.

RESPONSE TO DISCHARGE Issue warning—water contaminant. Restrict access. Should be removed. Provide chemical and physical treatment.

WATER RESPONSE Harmful to aquatic life in very low concentrations. May be dangerous if it enters water intakes. Notify local health and wildlife officials. Notify operators of nearby water intakes.

EXPOSURE SYMPTOMS Inhalation: joint and muscle pains, headache, dizziness, and insomnia. Weakness, frequently of extensor muscles of hand and wrist (unilateral or bilateral). Heavy contamination—brain damage. Stupor progressing to coma, with or without convulsion, often death. Excitation, confusion, and mania less common. Cerebrospinal pressure may be increased. Ingestion: abdominal pain, diarrhea, constipation, loss of appetite, muscular weakness, headache, blue line on gums, metallic taste, nausea, and vomiting.

EXPOSURE TREATMENT Call a physician. Inhalation: remove from the source of exposure. Keep victim quiet and warm. Eyes: wash with water for 15 to 20 minutes. Skin: wash with soap and water. Ingestion: call a physician at once. Induce vomiting, give milk and magnesium sulfate (epsom salts).

REFERENCES

1. RTECS Database (NIOSH).

2. CHRIS Database (DOT).

■ SUCCINIC ACID, 2,2–DIMETHYLHYDRAZIDE

CAS # 1596–84–5

SYNONYMS alar • alar-85 • alar-85 • aminozid • aminozide • b-995 • b-9 • b-nine • Bernsteinsaeure-2,2-dimethylhydrazid (German) • butanedioic acid, mono (2,2-dimethylhydrazide) • daminozide • dimas • dimethylaminosuccinamic acid • dmasa • dmsa • dyak • kylar • n-(dimethylamino) succinamic acid • N-dimethyl amino-beta-carbamyl propionic acid • N-Dimethylamino, succinamidsaeure [German] • N-Dimethylamino-succinamidsaeure (German) • nci-c03827 • sadh • succinic 1,1-dimethyl hydrazide • succinic acid 2,2-dimethylhydrazide • succinic acid N,N-dimethylhydrazide • succinic acid, mono (2,2-dimethylhydrazide) • succinic N′,N′-dimethylhydrazide

MF $C_6H_{12}N_2O_3$

MW 160.20

pH 3.8 (5% aqueous solution).

COLOR AND FORM White crystalline solid.

MELTING POINT 154–155°C.

SOLUBILITY Solubility at 25°C: 100 g/kg water; 25 g/kg acetone; 50 g/kg methanol; it is insoluble in simple hydrocarbons.

VAPOR PRESSURE Less than 0.01 mP$_a$ at 20°C.

STABILITY AND SHELF LIFE Hydrolyzed by acids and alkalis on heating [R1].

OTHER CHEMICAL/PHYSICAL PROPERTIES Henry's Law constant = 2.22×10^{-15} atm-m^3/mole (est).

COMMON USES (SRP: former use) A plant growth regulating agent used in orchard crops and ornamentals [R2]. (SRP: former use) To induce flowering in fruit production and prevent or delay flowering where this is desirable [R3].

OCCUPATIONAL RECOMMENDATIONS None listed.

NON-HUMAN TOXICITY LD$_{50}$ rat oral 8,400 mg/kg [R4]; LD$_{50}$ rabbit percutaneous >5,000 mg/kg [R1].

ECOTOXICITY LC$_{50}$ rainbow trout 360 mg/L/96 hr (conditions of bioassay not specified) [R5]; LC$_{50}$ bluegill 650 mg/L/96 hr (conditions of bioassay not specified) [R5].

BIODEGRADATION N–Methyl–^{14}C–labeled alar applied to 4 soils under greenhouse conditions. Microbial degradation major route of dissipation from soil. Half-life of alar was 3 to 4 days on all soils and major degradation product was ^{14}CO$_2$. In 14 days, about 84% of label recovered as ^{14}CO$_2$ remainder assoc with soil organic matter [R6]. In soil incubation studies using a sterile and non-sterile sandy loam soil, ^{14}C–labelled succinic acid, 2,2–dimethylhydrazide degraded much more rapidly in the non-sterile soil, indicating that microbial degradation was the major route of dissipation in the soil (1); in the sterile soil, only 0.2% of applied succinic acid, 2,2–dimethylhydrazide was recovered as ^{14}CO$_2$ after 10 days of incubation, whereas in the non-sterile soil, 75.4% was recovered as ^{14}CO$_2$ after 14 days (1). [R8].

BIOCONCENTRATION Based upon a water solubility of 100,000 ppm, the BCF of succinic acid, 2,2–dimethylhydrazide has been estimated to be 0.9 (1); this BCF value suggests that bioconcentration in aquatic organisms is not an important fate process (SRC). Succinic acid, 2,2–dimethylhydrazide does not bioconcentrate in fish, nor does it accumulate in rotational crops (2) [R7].

FIFRA REQUIREMENTS Criteria of Concern: Oncogenicity. Action/Use Affected: Cancelled, all daminozide products, for food uses. Remaining registrations for nonfood uses include cut chrysanthemums and bedding plants. Special review of nonfood uses continued, pending an evaluation of the cancer studies in rats, and mice with unsymmetrical dimethyl hydrazine, a degradate, and metabolite of daminozide [R9]. Existing Stocks Provisions: The sale, distribution, shipment, and use of existing stocks of daminozide for food uses is prohibited in the USA [R9].

REFERENCES

1. Hartley, D., and H. Kidd (eds.). *The Agrochemicals Handbook*. 2nd ed. Lechworth, Herts, England: The Royal Society of Chemistry, 1987, p. A113/Aug 87.

2. Gosselin, R. E., R. P. Smith, H. C. Hodge. *Clinical Toxicology of Commercial Products*. 5th ed. Baltimore: Williams and Wilkins, 1984, p. II–351.

3. *Kirk–Othmer Encyclopedia of Chemical Technology*. 3rd ed., Volumes 1–26. New York, NY: John Wiley and Sons, 1978–1984, p. V12 345 ((1980).

4. *Kirk–Othmer Encyclopedia of Chemical Technology*. 3rd ed., Volumes 1–26. New York, NY: John Wiley and Sons, 1978–1984, p. V18 7 (1982).

5. *Farm Chemicals Handbook* 1993. Willoughby, OH: Meister Publishing Co., 1993, p. C–104.

6. Menzie, C. M. *Metabolism of Pesticides*, Update II. U.S. Department of the Interior, Fish Wildlife Service, Special Scientific Report—*Wildlife* No. 212. Washington, DC: U.S. Government Printing Office, 1978. 1.

7. (1) Kenaga, E. E., *Ecotoxicol Environ Safety* 4: 26–38 (1980) (2) USEPA. *Pesti-*

cide Fact Book. Park Ridge, NJ: Noyes Data Corp p. 232–4 (1988).

8. Dannals, L. E., et al. *Arch Environ Contam Toxicol* 2: 213–21 (1974).

9. Environmental Protection Agency/OPTS. *Suspended, Cancelled, and Restricted Pesticides.* 5th ed. Washington, DC: Environmental Protection Agency, February 1990.

■ SULFAMIC ACID

CAS # 5329-14-6

SYNONYMS amidosulfonic acid • amidosulfuric acid • aminosulfonic acid • aminosulfuric acid • jumbo • kyselina-amidosulfonova (Czech) • kyselina-sulfaminova (Czech) • sulfamidic acid • sulphamic acid

MF H_3NO_3S

MW 97.10

pH 1N, pH = 0.41; 0.75N, pH = 0.5; 0.5N, pH = 0.63; 0.25N, pH = 0.87; 0.1N, pH = 1.18; 0.05N, pH =1.41; 0.01N, pH =2.02.

COLOR AND FORM Orthorhombic crystals; white crystalline solid; granular grade is off-white in color.

ODOR Odorless.

DENSITY 2.15.

DISSOCIATION CONSTANT Dissociation constant at 25°C = 0.101.

MELTING POINT 205°C.

HEAT OF COMBUSTION 1.53×10^7 J/kmol.

FLAMMABILITY LIMITS 9.3 vol% [R5].

FIREFIGHTING PROCEDURES Extinguish fire using agent suitable for type of surrounding fire. (Material itself does not burn or burns with difficulty.) Use dry chemical, dry sand, or carbon dioxide. Do not use water on material itself. If large quantities of combustibles are involved, use water in flooding quantities as spray and fog [R6].

SOLUBILITY Sparingly soluble in alcohol, methanol; slightly soluble in acetone; in-soluble in ether; freely soluble in nitrogenous bases and in nitrogen-containing organic solvents; insoluble in carbon disulfide and carbon tetrachloride; 12.8 wt% in water at 0°C; 17.57 wt% in water at 20°C; 22.77 wt% in water at 40°C; 0.1667 wt% in formamide at 25°C; 0.0412 wt% in methanol at 25°C; 0.0167 wt% in ethanol (2% benzene) at 25°C; 0.0040 wt% in acetone at 25°C; 0.0001 wt% in ether at 25°C.

CORROSIVITY H_2SO_4 (3–wt%) and HCl (3–wt%) are 2.6 and 4.2 times more corrosive, respectively, than a 3–wt% solution of sulfamic acid on 1,010 steel. On the same scale, H_2SO_4 and HCl (3–wt% solutions) are 1.5 and 6.7, respectively, times more corrosive on copper; 1.5 and 2.8 more corrosive on brass; 4.0 and 7.0 more corrosive on bronze; 0.6 and 5.3 more corrosive on aluminum.

STABILITY AND SHELF LIFE Stable when dry but slowly hydrolyzes in solution, forming ammonium bisulfate [R8].

OTHER CHEMICAL/PHYSICAL PROPERTIES Non-volatile; nonhygroscopic; aqueous solutions are highly ionized, giving pH values lower than solutions of formic, phosphoric, and oxalic acids. Decomposition at boiling point. Slowly hydrolyzes in solution to form ammonium bisulfate. Decomposition temperature = 209°C. Heat of formation = -6.749×10^8 J/kmol. Heat of solution = 19.0 kJ/mol. Strong acid; pH of 1% solution at 25°C 1.18. Titrated with bases by means of indicators showing color change between pH 4.5–9.

PREVENTATIVE MEASURES SRP: The scientific literature for the use of contact lenses in industry is conflicting. The benefit or detrimental effects of wearing contact lenses depend not only upon the substance, but also on factors including the form of the substance, characteristics, and duration of the exposure, the uses of other eye protection equipment, and the hygiene of the lenses. However, there may be individual substances whose irritating or corrosive properties are such that the wearing of contact lenses would be harmful to the eye. In those specific cases, contact lenses should not be worn. In any event, the usual eye protection equipment should be worn even when contact lenses are in place. If material not on fire and not involved in fire: Keep sparks, flames, and other sources of ignition away. Keep material out of water sources and sewers [R9]. Personnel protection: Avoid breathing dusts and fumes from burning material. Avoid bodily contact with the material. If contact with the material is anticipated, wear appropriate chemical protective clothing. [R9].

CLEANUP Wear eye protection, laboratory coat, and nitrile rubber gloves. Clean up with dust pan and brush. May be flushed away to waste with at least 50 volume of water or mixed with sand and disposed of as normal refuse [R7].

DISPOSAL SRP: At the time of review, criteria for land treatment or burial (sanitary landfill) disposal practices are subject to significant revision. Prior to implementing land disposal of waste residue (including waste sludge), consult with environmental regulatory agencies for guidance on acceptable disposal practices. Package Lots: Place in a separate labeled container for recycling or disposal by burning in a furnace equipped with an afterburner and scrubber. Small Quantities: Wear eye protection, laboratory coat, and nitrile rubber gloves. Work in the fume hood. Sift slowly into a large container of cold water, with agitation. When dissolved, neutralize with sodium carbonate, and pour into drain with at least 50 times its volume of water. [R10].

COMMON USES As standard in alkalimetry; in acid cleaning; in nitrite removal; in chlorine stabilization for use in swimming pools, cooling towers, paper mills [R2]. Gas-liberating compositions;, organic synthesis; catalyst for urea-formaldehyde resins [R3]. Chemical intermediate for its salts; metal cleaning component; wood pulp bleaching agent; analytical reagent; agent in dying of wool with acid dyes; bactericide in pools and paper mills [R1].

OCCUPATIONAL RECOMMENDATIONS None listed.

FIFRA REQUIREMENTS In 1988, Congress amended FIFRA to strengthen and accel-

erate EPA's reregistration program. The nine-year reregistration scheme mandated by "FIFRA 88" applies to each registered pesticide product containing an active ingredient initially registered before November 1, 1984. Pesticides for which EPA had not issued Registration Standards prior to the effective date of FIFRA '88 were divided into three lists based upon their potential for exposure and other factors, with List B being of highest concern, and D of least. List: D; Case: Sulfamic acid; Case No. 4085; Pesticide type: Antimicrobial; Case Status: Awaiting Data/Data in Review: OPP awaits data from the pesticide's producer(s) regarding its human health or environmental effects, or OPP has received and is reviewing such data, in order to reach a decision about the pesticide's eligibility for reregistration. Active Ingredient (AI): Sulfamic acid; AI Status: The producer(s) of the pesticide has made commitments to conduct the studies and pay the fees required for reregistration, and is meeting those commitments in a timely manner [R12]. A tolerance is established for residues of sulfamate ion, expressed as sulfamic acid, from the postharvest application of the fungicide chlorosulfamic acid in or on asparagus, carrots, cauliflower, celery, potatoes, and radishes. [R11].

DOT *Health Hazards:* Contact causes burns to skin and eyes. If inhaled, may be harmful. Fire may produce irritating or poisonous gases. Runoff from fire control or dilution water may cause pollution [R4].

Fire or Explosion: Some of these materials may burn, but none of them ignites readily. Flammable/poisonous gases may accumulate in tanks and hopper cars. Some of these materials may ignite combustibles (wood, paper, oil, etc.) [R4].

Emergency Action: Keep unnecessary people away; isolate hazard area and deny entry. Stay upwind; keep out of low areas. Positive-pressure self-contained breathing apparatus (SCBA) and structural firefighters' protective clothing will provide limited protection. Call emergency response telephone number on shipping paper first. If shipping paper not available or no answer, call CHEMTREC at 1–800–424–9300. If water pollution occurs, notify the appropriate authorities [R4].

Fire: Some of these materials may react violently with water. Small Fires: Dry chemical, CO_2, water spray, or regular foam. Large Fires: Water spray, fog, or regular foam. Move container from fire area if you can do so without risk. Apply cooling water to sides of containers that are exposed to flames until well after fire is out. Stay away from ends of tanks [R4].

Spill or Leak: Do not touch or walk through spilled material; stop leak if you can do so without risk. Small Spills: Take up with sand or other noncombustible absorbent material and place into containers for later disposal. Small Dry Spills: With clean shovel place material into clean, dry container, and cover loosely; move containers from spill area. Large Spills: Dike far ahead of liquid spill for later disposal [R4].

First Aid: Move victim to fresh air; call emergency medical care. In case of contact with material, immediately flush skin or eyes with running water for at least 15 minutes. Remove and isolate contaminated clothing and shoes at the site. Keep victim quiet and maintain normal body temperature [R4].

REFERENCES

1. SRI.

2. Budavari, S. (ed.). *The Merck Index—Encyclopedia of Chemicals, Drugs, and Biologicals*. Rahway, NJ: Merck and Co., Inc., 1989. 1408.

3. Sax, N. I., and R. J. Lewis, Sr. (eds.). *Hawley's Condensed Chemical Dictionary*. 11th ed. New York: Van Nostrand Reinhold Co., 1987. 1104.

4. U.S. Department of Transportation. *Emergency Response Guidebook 1993*. DOT P 5800.6. Washington, DC: U.S. Government Printing Office, 1993, p. G–60.

5. Daubert, T. E., R. P. Danner. *Physical and Thermodynamic Properties of Pure Chemicals Data Compilation*. Washington, DC: Taylor and Francis, 1989.

6. Association of American Railroads. *Emergency Handling of Hazardous Materials in Surface Transportation*. Washington, DC: Association of American Railroads, Bureau of Explosives, 1992. 869.

7. Armour, M. A. *Hazardous Laboratory Chemicals Disposal Guide*. Boca Raton, FL: CRC Press Inc., 1991. 413.

8. *The Merck Index*. 9th ed. Rahway, NJ: Merck and Co., Inc., 1976. 1154.

9. Association of American Railroads. *Emergency Handling of Hazardous Materials in Surface Transportation*. Washington, DC: Association of American Railroads, Bureau of Explosives, 1992. 869.

10. Armour, M. A. *Hazardous Laboratory Chemicals Disposal Guide*. Boca Raton, FL: CRC Press Inc., 1991. 413.

11. 40 CFR 180.201 (7/1/92) 12 USEPA/OPP. *Status of Pesticides in Reregistration and Special Review*. p. 258 (Mar, 1992) EPA 700–R–92–004.

■ SULFUR MONOCHLORIDE

DOT # 1828

SIC CODES 2810; 2879; 2822; 2815.

SYNONYMS sulfur chloride • sulfur subchloride • disulfur dichloride

MW 135.03

MF Cl_2S_2

MELTING POINT –77.

CORROSIVITY Sulfur chlorides are noncorrosive to carbon steel and iron when dry. When wet they attack steel, cast iron, aluminum, stainless steel, copper and copper alloys, and many nickel-based materials (sulfur chlorides). Sulfur monochloride will attack some forms of plastics, rubber, and coatings.

DENSITY 1.6885 at 15.5°C.

HEAT OF VAPORIZATION 115 Btu/lb = 63.8 cal/g = 2.67×10^5 J/kg.

SOLUBILITY Soluble in alc, benzene, ether, carbon disulfide, carbon tetrachloride, oils; soluble in amyl acetate; insoluble in water.

BOILING POINT 138 °C.

VAPOR DENSITY 4.66 (air = 1) .

VAPOR PRESSURE 6.8 torr at 20°C.

BINARY REACTANTS Chromyl chloride, phosphorus trioxide, water, organic matter.

STANDARD CODES EPA 311; ICC corrosive liquid, white label, 1 gal in an outside container; USCG corrosive liquid, white label; IATA corrosive liquid, white label, not acceptable passenger, 5 L cargo; NFPA—2, 1, 1; Superfund designated (hazardous substances) list.

PERSISTENCE Decomposes on contact with water.

DIRECT CONTACT Vapors irritate, cause tears in eyes.

GENERAL SENSATION Affects breathing. Recognition odor 0.001 ppm in air [R2]; penetrating odor. Corrosive vapors affect eyes, lungs, and mucous membranes. Vapors decompose in lung moisture to corrosive products.

PERSONAL SAFETY Avoid contact of liquid and vapors with skin, eyes, or clothing; avoid breathing vapors. Wear self-contained breathing apparatus.

ACUTE HAZARD LEVEL Strong irritant. Highly toxic when ingested or inhaled. A concentration of 150 ppm in air has been fatal to mice in 1 min [R1].

CHRONIC HAZARD LEVEL Moderately toxic with ingestion or inhalation.

DEGREE OF HAZARD TO PUBLIC HEALTH Strong irritant. May react violently with water; Highly toxic when ingested or inhaled. Emits toxic vapors when contacted with acid or water, or when heated to decomposition.

OCCUPATIONAL RECOMMENDATIONS None listed.

ACTION LEVEL Notify air authority. Isolate from water. Remove ignition sources. If massive spill occurs in water, evacuate

area to avoid H_2S exposure. Restrict access to affected waters.

ON-SITE RESTORATION Neutralize with anion exchanger, lime, or $NaHCO_3$.Seek professional environmental engineering assistance through EPA's Environmental Response Team (ERT), Edison, NJ, 24–hour no. 908–548–8730.

AVAILABILITY OF COUNTERMEASURE MATERIAL Anion exchanger—water softener suppliers; lime—cement plants; $NaHCO_3$—grocery distributors, bakeries.

DISPOSAL Sprinkle or sift onto a thick layer of mixed dry soda ash and slaked lime (50–50) from behind a body shield. Mix and spray water cautiously with an atomizer. Scoop up and sift cautiously into a large volume of water. Neutralize and wash down the drain with large excess of water.

INDUSTRIAL FOULING POTENTIAL Cl tolerance limits: pulp and paper—75 ppm, textile, and brewing—100 ppm, carbonated drinks—250 ppm. Decomposition products are highly corrosive.

MAJOR WATER USE THREATENED Potable supply, recreation, fisheries, industrial.

PROBABLE LOCATION AND STATE Light-amber to yellowish-red, fuming, oily liquid. Will evolve gas; will be present as H_2S, sulfur, sulfite, thiosulfuric acid, and HCl.

WATER CHEMISTRY Decomposes upon contact with water to form H_2S, sulfur, sulfites, thiosulfuric acid, and HCl.

COLOR IN WATER Colorless.

GENERAL RESPONSE Avoid contact with liquid and vapor. Keep people away. Wear goggles, self-contained breathing apparatus, and rubber overclothing (including gloves). Stop discharge if possible. Call fire department. Isolate and remove discharged material. Notify local health and pollution control agencies.

FIRE RESPONSE Combustible. Wear goggles, self-contained breathing apparatus, and rubber overclothing (including gloves). Extinguish with dry chemical or carbon dioxide. Cool exposed containers with water. Water reacts violently with compound.

EXPOSURE RESPONSE Call for medical aid. Vapor is irritating to eyes. Poisonous if inhaled. Move to fresh air. If breathing has stopped, give artificial respiration. If breathing is difficult, give oxygen. liquid will burn skin and eyes. Poisonous if swallowed. Remove contaminated clothing and shoes. Flush affected areas with plenty of water. If swallowed and victim is conscious, have victim drink water or milk. Do not induce vomiting.

RESPONSE TO DISCHARGE Issue warning—corrosive, air contaminant, water contaminant. Restrict access. Provide chemical and physical treatment.

WATER RESPONSE Dangerous to aquatic life in high concentrations. May be dangerous if it enters water intakes. Notify local health and wildlife officials. Notify operators of nearby water intakes.

EXPOSURE SYMPTOMS Vapors irritate eyes and respiratory system; pulmonary edema may result. Liquid burns and damages eyes. Unless removed at once, it burns the skin. Ingestion causes severe damage to mouth and stomach.

EXPOSURE TREATMENT Inhalation: remove to fresh air; use artificial respiration and oxygen if required; call a doctor. Ingestion: give water; do not induce vomiting; call a doctor. Eyes: flush with water for at least 15 min.; obtain medical attention at once. Skin: flush with water; remove contaminated clothing under shower.

REFERENCES

1. Sax, N. I., *Dangerous Properties of Industrial Materials*. New York: Van Nostrand Reinhold Company, 1969.

2. Sullivan, R. J., *Air Pollution Aspects of Odorous Compounds*, NTIS PB 188 089, September 1969.

■ TABUN

$$(CH_3)_2N \diagdown \underset{\underset{C_2H_5O}{\diagup}}{\overset{\overset{O}{\parallel}}{P}} - CN$$

CAS # 77–81–6

SYNONYMS ethyl N-dimethylphosphoramidocyanidate • ethyl 'm; N,N-dimethylphosphoramidocyanidate • dimethylphosphoramidocyanidic acid, ethyl ester • dimethylamidoethoxyphosphoryl cyanide • GA • ethyl phosphorodimethylamidocyanidate • O-ethyl N,N-dimethylphosphoramidocyanidate

MF $C_5H_{11}N_2O_2P$

MW 162.15

COLOR AND FORM Liquid.

ODOR Fruity odor reminiscent of bitter almonds.

DENSITY 1.077.

BOILING POINT 240°C at 760 mm Hg.

MELTING POINT Freezing point: –50°C.

SOLUBILITY Readily soluble in organic solvents. Miscible with water.

VAPOR PRESSURE 0.07 mm Hg at 25°C.

OTHER CHEMICAL/PHYSICAL PROPERTIES Quickly hydrolyzed (by water). Volatility at 25°C: 610 mg/m³. G-agents are miscible in both polar and nonpolar solvents. They hydrolyze slowly in water at neutral or slightly acidic pH and more rapidly under strong acid or alkaline conditions. The hydrolysis products are considerably less toxic than the original agent (G-agents). Although the G-agents are liquids under ordinary atmospheric conditions, their relatively high volatility permits them to be disseminated in vapor form. (G-agents, colorless, odorless).

PROTECTIVE EQUIPMENT The primary item of individual protection against toxic chemicals is the protective ("gas") mask. The current U.S. Army standard mask is the M17A1, which has as its basic component a molded rubber facepiece with large cheek pouches that hold filter pads. The filter pads consist of 6 sheets of core layer laminated between two sheets of backing layer. A speech diaphragm, incorporating an outlet valve, is attached to the faceblank at the mouth position. Large eyepieces in the faceblank provide a wide field of vision. The M17A1 mask provides complete respiratory protection against all known military toxic chemical agents, but it does not afford protection against industrial toxics such as ammonia and carbon monoxide (chemicals in war) [R2]. Airtight, impermeable clothing was developed for personnel who must enter heavily contaminated areas. This clothing is made of butyl rubber or a coated fabric and provides complete protection against liq agents (chemicals in war) [R2]. Respiratory protection (supplied-air respirator with full facepiece or self-contained breathing apparatus) should be available where these compounds are manufactured or used and should be worn in case of emergency and overexposure. (phosphorus compounds) [R6].

PREVENTATIVE MEASURES Although resistant to liquid chemical agents, impermeable protective clothing may be penetrated after a few hr of exposure to heavy concentration of agent. Consequently, liquid contamination on the clothing must be neutralized or removed as soon as possible (chemicals in war) [R3]. Collective protection involves primarily the use of shelters where personnel can work or rest. Such shelters must be airtight to prevent the inward seepage of chemical agents and, thus, require a means of providing uncontaminated air. Such an air supply can be obtained by two methods. In small shelters, a filter material called diffusion board can be incorporated into the construction or used as a liner in existing structures. For large shelters, a mechanical collective protector can be used (chemicals in war) [R7]. The U.S. Army's Decontamination and Reimpregnating Kit is used by the individual soldier for decontamination of personal equipment and clothing, and for reimpregnating the clothing outfit chemical protective (liner system). The M13 kit contains a pad of

Fuller's earth powders for decontaminating the inside of the protective mask and, in the absence of the M258 Skin Decontaminating Kit, for exposed skin; two bags of chloramide powder, each containing a dye capsule, for decontaminating clothing and reimpregnating the clothing outfit chemical protective (liner system); and a cutter for cutting away heavily contaminated areas of clothing. (chemicals in war) [R8].

CLEANUP If decontamination cannot be left to natural processes, chemical neutralizers or removal must be used. Decontaminating agent DS2 is a general-purpose decontaminant. It consists of 70% diethylenetriamine, 28% ethylene glycol monomethyl ether, and 2% sodium hydroxide. DS2 reacts with both the nerve agents and blister agents to effectively reduce their hazards. Important limitations in the use of DS2 are: (a) personnel must remain masked because of the vapor; (b) rubber gloves must be worn to protect the hands; (c) it is a combustible liquid, therefore it must not be allowed to get on hot metal surfaces such as running engines or exhaust pipes. (chemicals in war) [R4].

COMMON USES Chemical warfare agent; military nerve gas [R1].

OCCUPATIONAL RECOMMENDATIONS None listed.

HUMAN TOXICITY The lethal dose for man may be as low as 0.01 mg/kg [R1].

NON-HUMAN TOXICITY LD_{50} mouse ip 0.6 mg/kg [R1]; LD_{50} mouse iv 0.287 mg/kg [R9].

BIOCONCENTRATION Tabun hydrolyzes readily in water (1); therefore, bioconcentration in aquatic organisms is not expected to be an important environmental fate process (SRC) [R5].

REFERENCES

1. Budavari, S. (ed.). *The Merck Index—Encyclopedia of Chemicals, Drugs, and Biologicals.* Rahway, NJ: Merck and Co., Inc., 1989. 1427.

2. *Kirk–Othmer Encyclopedia of Chemical Technology.* 3rd ed., Volumes 1–26. New York, NY: John Wiley and Sons, 1978–1984, p. V5 409 (11979).

3. *Kirk–Othmer Encyclopedia of Chemical Technology.* 3rd ed., Volumes 1–26. New York, NY: John Wiley and Sons, 1978–1984, p. V5 410 (1979).

4. *Kirk–Othmer Encyclopedia of Chemical Technology.* 3rd ed., Volumes 1–26. New York, NY: John Wiley and Sons, 1978–1984, p. V5 412 (1979).

5. Epstein, J., et al., *Summary Report on a Data Base for Predicting Consequences of Chemical Disposal Operations.* EASP–1200–12. Edgewood Arsenal, MD: Dept of the Army NTIS–ADB 955399 002 (1973).

6. International Labour Office. *Encyclopedia of Occupational Health and Safety.* Vols. I and II. Geneva, Switzerland: International Labour Office, 1983. 1684.

7. *Kirk-Othmer Encyclopedia of Chemical Technology.* 3rd ed., Volumes 1–26. New York, NY: John Wiley and Sons, 1978–1984, p. V5 411 (1979).

8. *Kirk-Othmer Encyclopedia of Chemical Technology.* 3rd ed., Volumes 1–26. New York, NY: John Wiley and Sons, 1978–1984, p. V5 414 (1979).

9. Tripathi, H. L., W. L. Dewey. *J Toxicol Environ Health* 26 (4): 437–46 1989.

■ TEBUTHIURON

CAS # 34014–18–1

SYNONYMS urea, 1-(5-(t-butyl) -1,3,4-thiadiazol-2-yl)-1,3-dimethyl • N-[5-(1,1-dimethylethyl) -1,3,4-thiadiazol-2-yl]N,N'-dimethylurea • el-103 • sha-105501 • brush-bullet • graslan • perflan • Caswell no. 366aa

MF $C_9H_{16}N_4OS$

MW 228.35

COLOR AND FORM Gray to dark-brown pellet; colourless solid (technical grade).

ODOR Faint musty odor.

MELTING POINT 161.5–164°C.

SOLUBILITY Dry flowable and wettable powder disperse in water. Pellets are not soluble, but disintegrate in water. Granules not soluble; powder, 2,500 mg/L water at 25°C; solubility at 25°C: 70 g/L acetone; 60 g/L acetonitrile; 6.1 g/L hexane; 170 g/L methanol; 60 g/L 2-methoxyethanol.

VAPOR PRESSURE 0.27 mPa at 25°C.

OCTANOL/WATER PARTITION COEFFICIENT Log K_{ow} = 1.79.

CORROSIVITY Noncorrosive.

STABILITY AND SHELF LIFE Stable to light [R1].

OTHER CHEMICAL/PHYSICAL PROPERTIES Stable in aqueous media between pH 5 and 9. Hydrolyzed at higher temperatures by strong alkalis and strong acids.

PROTECTIVE EQUIPMENT Safety glasses or goggles, impermeable gloves, waterproof boots, long-sleeved shirt, and long pants [R2].

PREVENTATIVE MEASURES Keep away from children. Do not reuse empty container [R2]. Harmful if swallowed. Avoid contact with skin, eyes, or clothing. In case of contact, flush with water. Avoid breathing mist or dust. Do not apply, drain, or flush equipment on or near desirable trees or other plants, or on areas to which their roots extend. Do not contaminate any body of water [R6, 452].

COMMON USES Herbicide for non-cropland areas, rangelands, rights of way, and industrial sites [R2]. It is a broad-spectrum herbicide for control of herbaceous and woody plants. Uses include: total vegetation control in non-crop areas; control of undesireable woody plants in pastures and rangeland; control of broad-leaved weeds in sugarcane [R1].

OCCUPATIONAL RECOMMENDATIONS None listed.

NON-HUMAN TOXICITY LD_{50} rat oral 644 mg/kg [R2]; LD_{50} mouse oral 579 mg/kg [R7]; LD_{50} rabbit oral 286 mg/kg [R7]; LD0 dog oral >500 mg/kg [R6, 454]; TL_{50} bluegill, oral 112 ppm [R6, 454]; LD_{50} rat, oral 640 mg/kg [R8].

BIODEGRADATION Thiadiazole ring-labeled ^{14}C–tebuthiuron in loam soil degraded from 8 ppm immediately post treatment to 5.7 ppm at 273 days post treatment, indicating a half-life of greater than 273 days [R3].

BIOCONCENTRATION Using the estimated log octanol/water partition coefficient, 1.78 (1), one can estimate a BCF of 13 using a recommended regression equation (2). Therefore tebuthiuron would not be expected to bioconcentrate in aquatic organisms (SRC) [R4].

FIFRA REQUIREMENTS In 1988, Congress amended FIFRA to strengthen and accelerate EPA's reregistration program. The nine-year reregistration scheme mandated by "FIFRA 88" applies to each registered pesticide product containing an active ingredient initially registered before November 1, 1984. List A consists of the 194 chemical cases (or 350 individual active ingredients) for which EPA had issued Registration Standards prior to the effective date of FIFRA '88. List: A; Case: Tebuthiuron; Case No. 0054; Pesticide type: Herbicide; Registration Standard Date: 07/30/87; Case Status: Awaiting Data/Data in Review: OPP awaits data from the pesticide's producer(s) regarding its human health or environmental effects, or OPP has received and is reviewing such data, in order to reach a decision about the pesticide's eligibility for reregistration. Active Ingredient (AI): Tebuthiuron; Data Call-in (DCI) Date: 07/31/91; AI Status: The producer(s) of the pesticide has made commitments to conduct the studies and pay the fees required for reregistration, and is meeting those commitments in a timely manner [R5]. Tolerances are established for the residues of the herbicide tebuthiuron and its metabolites containing the dimethylethyl thiadiazole moiety in or on the following raw agricultural commodities: cattle, fat; cattle, meat by-products; cattle, meat; goats, fat; goats, meat by-products; goats, meat; grass, hay; grass, rangeland forage; horses, fat; horses meat by-products; horses, meat; milk; sheep, fat; sheep, meat by-products; sheep, meat [R9].

REFERENCES

1. Worthing, C. R., and S. B. Walker (eds.). *The Pesticide Manual—A World Compendium.* 8th ed. Thornton Heath, UK: The British Crop Protection Council, 1987. 768.

2. *Farm Chemicals Handbook* 1994. Willoughby, OH: Meister, 1994, p. C 348.

3. USEPA/ODW; *Health Advisories for 50 Pesticides* (1988).

4. (1) Meylan, W. M., P. H. Howard. *Group Contribution Method for Estimating Octanol–Water Partition Coefficients.* SETAC Meeting Cincinnati, OH. Nov 8–12, (1992) (2) Lyman, W. J., et al., *Handbook of Chemical Property Estimation Methods*, New York: McGraw-Hill, Chapt 5 (1982).

5. USEPA/OPP; *Status of Pesticides in Reregistration and Special Review* p. 114 (Mar, 1992) EPA 700–R–92–004.

6. Weed Science Society of America. *Herbicide Handbook.* 5th ed. Champaign, Illinois: Weed Science Society of America, 1983.

7. Todd, G.C., et al., *Food Cosmet Toxicol* 12: 461–70 (1974) as cited in USEPA/ODW, *Health Advisories for 50 Pesticides* (1988).

8. Gosselin, R. E., R. P. Smith, H. C. Hodge. *Clinical Toxicology of Commercial Products.* 5th ed. Baltimore: Williams and Wilkins, 1984, p. II–333.

9. 40 CFR 180.390 (7/1/92).

■ TERBACIL

CAS # 5902–51–2

SYNONYMS 2,4 (1H, 3H) - pyrimidinedione, 5-chloro-3-(1,1-dimethylethyl)-6-methyl • 3-t-butyl-5-chloro-6-methyluracil • 3-tert-butyl-5-chloro-6-methyluracil • 5-chloro-3-tert-butyl-6-methyluracil • compound-732 • Du Pont-732 • experimental herbicide 732 • sinbar • turbacil • uracil, 3-tert-butyl-5-chloro-6-methyl • 5-chloro-3-(1,1-dimethylethyl)-6-methyl-2,4 (1H, 3H)-pyrimidinedione • Geonter • DuPont Herbicide-732

MF $C_9H_{13}ClN_2O_2$

MW 216.69

COLOR AND FORM White crystalline powder; colorless crystals.

ODOR Odorless.

DENSITY 1.34 at 25°C.

MELTING POINT 175–177°C.

SOLUBILITY Solubility in water at 25°C: 710 ppm; soluble in dimethylacetamide; moderately soluble in methyl isobutyl ketone, butyl acetate; 33.7 g soluble in 100 g dimethyl formamide; 22 g soluble in 100 g cyclohexanone; 6.5 g soluble in 100 g xylene; barely soluble in mineral oils and aliphatic hydrocarbons. Readily soluble in strong aqueous alkalis.

VAPOR PRESSURE 4.7×10^{-7} mmHg at 29.5°C; 2.9×10^{-7} mmHg at 25°C (estimated SRC).

CORROSIVITY Noncorrosive.

STABILITY AND SHELF LIFE Active ingredient is chemically stable under normal storage conditions [R7, 456].

OTHER CHEMICAL/PHYSICAL PROPERTIES Sublimation occurs. Henry's Law constant = 1.2×10^{-10} atm–m³/mole at 25°C (est).

PROTECTIVE EQUIPMENT Protective clothing: use good sanitary practices [R2]. Preventative measures avoid breathing dust or spray mist. Avoid contact with skin, eyes, and clothing [R7, 456]. SRP: The scientific literature supports the wearing of contact lenses in industrial environments, as part of a program to protect the eye against chemical compounds and minerals causing eye irritation. However, there may be individual substances whose irritating or corrosive properties are such that the wearing of contact lenses would be harmful to the eye. In those specific cases contact lenses should not be worn.

DISPOSAL Do not contaminate water, food, or feed by disposal. Dispose of or bury empty container away from water supplies in accordance with federal, state, and local regulations. Open dumping is prohibited. [R7, 456].

COMMON USES Used for selective control of many annual and some perennial weeds in apples, blueberries, peaches, citrus, alfalfa, mint, and sugarcane [R7, 455]. Selective pre-emergence herbicide for partial control of some grasses [R1]. Control of most annual grasses and broad-leaved weeds, and some perennial weeds in established asparagus lucerne pecans, strawberries [R3].

OCCUPATIONAL RECOMMENDATIONS None listed.

NON-HUMAN TOXICITY LD$_{50}$ rat oral >5,000 mg/kg [R3].

ECOTOXICITY LC$_{50}$ Peking duckling >56,000 mg/kg diet/8 day [R3]; LC$_{50}$ pheasant chicks >31,450 mg/kg diet/8 day [R3]; LC$_{50}$ pumpkinseed sunfish 86 mg/L/48 hr (conditions of bioassay not specified) [R3]; LC$_{50}$ fiddler crab >1,000 mg/L/48 hr (conditions of bioassay not specified) [R4].

BIODEGRADATION One study reviewed several biodegradation tests under aerobic and anaerobic conditions (1). Terbacil, 100 ppm initial concentration, biodegraded 20 percent after 32 weeks in an aerobic sandy loam soil at 23°C (1). In an aerobic loam soil at 32°C, 8 ppm terbacil exhibited a half-life of about 5 months (1). Half-lives of 2–3 months were observed for 2 ppm terbacil in aerobic silt loam and sandy loam soils (1). At an initial concentration of 2.88 ppm, 28 percent biodegradation was observed for radio-labelled terbacil (unspecified placement of the label) after 600 days in an aerobic sandy loam soil (2); in the same study, a trace amount of 2,88 ppm terbacil (unspecified placement of the label) biodegraded in an anaerobic sandy loam after 145 days (2). Furthermore, 2.1 ppm terbacil degraded less than 5 percent in anaerobic silt loam and sandy soil after 60 days at 20°C in the dark (1). After 90 days of incubation in sterile and nonsterile soils, at least 90 percent of 2 ppm terbacil remained; 0.8–1.5 percent of applied ^{14}C evolved as CO$_2$ from nonsterile soil, whereas 0.01 percent was evolved

from sterile soil (1). Therefore, terbacil is only slowly metabolized by microbes in soil [R5].

BIOCONCENTRATION Based on estimated bioconcentration factors of 2 and 15 (1), terbacil is not expected to bioconcentrate in fish (SRC) [R6].

FIFRA REQUIREMENTS In 1988, Congress amended FIFRA to strengthen and accelerate EPA's reregistration program. The nine-year reregistration scheme mandated by "FIFRA 88" applies to each registered pesticide product containing an active ingredient initially registered before November 1, 1984. List A consists of the 194 chemical cases (or 350 individual active ingredients) for which EPA had issued Registration Standards prior to the effective date of FIFRA '88. List: A; Case: Terbacil; Case No. 0039; Pesticide type: Herbicide; Registration Standard Date: 09/12/89; Case Status: Awaiting Data/ Data in Review: OPP awaits data from the pesticide's producer(s) regarding its human health or environmental effects, or OPP has received and is reviewing such data, in order to reach a decision about the pesticide's eligibility for reregistration. Active Ingredient (AI): Terbacil; AI Status: The producer(s) of the pesticide has made commitments to conduct the studies and pay the fees required for reregistration, and is meeting those commitments in a timely manner [R10]. Tolerances are established for residues of the herbicide terbacil (3–tert–buty–5–chloro–6–methyluracil) in or on the following raw agricultural commodities: apples; citrus fruits; peaches; pears; and sugarcane [R8]. Tolerances are established for combined residues of the herbicide terbacil (3–tert-buty–5–chloro–6–methyluracil) and its metabolites 3–tert-butyl–5–chloro–6–hydroxymethyluracil, 6–chloro–2, 3–dihydro 7–hydroxymethyl–3,3–dimethyl–5H–oxazolo (3,2–a) pyrimidin–5–one, and 6–chloro–2,3–dihydro–3,3,7–trimethyl–5H–oxazolo (3,2–a) pyrimidine–5–one (calculated as terbacil) in or on the following raw agricultural commodities: alfalfa (forage and hay); asparagus; blueberries; cranberries (blackberries, boysenberries, dewberries, loganberries, raspberries, and youngberries); cattle (fat, mbyp, and meat); goats (fat, mbyp, and meat); hogs (fat, mbyp, and meat); horses (fat, mbyp, and meat); milk (fat); mint hay (peppermint and spearmint); pecans; sainfoin (forage and hay); sheep (fat, mbyp, and meat); and strawberries. [R9].

FIRE POTENTIAL Nonflammable [R7, 456].

REFERENCES

1. Spencer, E. Y. *Guide to the Chemicals Used in Crop Protection.* 7th ed. Publication 1093. Research Institute, Agriculture Canada, Ottawa, Canada: Information Canada, 1982. 541.

2. *Farm Chemicals Handbook 1993.* Willoughby, OH: Meister Publishing Co., 1993, p. C–335.

3. Hartley, D., and H. Kidd (eds.). *The Agrochemicals Handbook.* 2nd ed. Lechworth, Herts, England: The Royal Society of Chemistry, 1987, p. A381/Aug 87.

4. Worthing, C. R., and S. B. Walker (eds.). *The Pesticide Manual—A World Compendium.* 8th ed. Thornton Heath, UK: The British Crop Protection Council, 1987. 775.

5. (1) USEPA. *Pesticides Pollution, and Control Terbacil–Registration Standard.* USEPA/540/RS–82/013 Washington, DC: Off Pest Programs (1982). (2) Wolf, D. C., J. P. Martin. *Soil Sci Amer Proc* 38: 921–5 (1974).

6. Kenaga, E. E., *Ecotox Environ Safety* 4: 26–38 (1980).

7. Weed Science Society of America. *Herbicide Handbook.* 5th ed. Champaign, IL: Weed Science Society of America, 1983.

8. 40 CFR 180.209 (a) (7/1/91).

9. 40 CFR 180.209 (b) (7/1/91).

10. USEPA/OPP. *Status of Pesticides in Reregistration and Special Review.* p. 115 (Mar, 1992) EPA 700–R–92–004.

■ TERT-BUTYL HYDROPEROXIDE

$$CH_3-\underset{\underset{CH_3}{|}}{\overset{\overset{CH_3}{|}}{C}}-O-O-H$$

CAS # 75–91–2

DOT # 1949

SYNONYMS 2-hydroperoxy-2-methylpropane • cadox-tbh • hydroperoxide, 1,1-dimethylethyl • hydroperoxide, tert-butyl • hydroperoxyde de butyle tertiaire (French) • perbutyl-h • terc-butylhydroperoxid (Czech) • trigonox a-75 (Czech)

MF $C_4H_{10}O_2$

MW 90.14

COLOR AND FORM Water-white liquid.

DENSITY 0.8960 at 20°C.

BOILING POINT 35°C at 20 mm Hg.

MELTING POINT –8°C.

EXPLOSIVE LIMITS AND POTENTIAL Vapors form explosive mixtures with air. It is shock sensitive. Spontaneous explosion may occur if mixed with readily oxidizable, organic, or flammable materials. [R3, p. 49–73].

FIREFIGHTING PROCEDURES Alcohol foam, carbon dioxide, dry chemical [R2]. Fight fires with water from an explosion-resistant location. In advanced or massive fires, area should be evacuated. If fire occurs in vicinity of this material water should be used to keep containers cool. Clean-up and salvage operations should not be attempted until all of peroxide has cooled completely. [R3, p. 49–73].

SOLUBILITY Soluble in water, ethanol, ether, and chloroform; very soluble in alkali metal hydroxide solution.

STABILITY AND SHELF LIFE Stable to 75°C; slow 1st-order decomposition can be accelerated by presence of 1 mole-% of copper, cobalt, and manganese salts [R1].

OTHER CHEMICAL/PHYSICAL PROPERTIES

A highly reactive peroxy compound. Self-accelerating decomposition temperature 190–200°F. Decomposition at 89°C.

CLEANUP In event of spillage as result of fire or use, the spilled material should be absorbed with noncombustible absorbent, such as vermiculite. Sweep up, and place in plastic container for immediate disposal. Do not use spark-generating metals or cellulosic materials. Dispose of absorbed peroxide by placing it on ground in remote outdoor area and igniting with a long torch. Empty peroxide containers should be washed with 10% sodium hydroxide solution. [R3, p. 49–73].

COMMON USES To introduce peroxy group into organic mol; in radical substitution reactions [R1]. Oxidation and sulfonation catalyst; bleaching; deodorizing [R4]. Principally chemical intermediate for propylene oxide and t-butyl alcohol; small amount as curing agent for thermoset polyesters; initiator for vinyl chloride polymerization and copolymerization; small amount as chemical intermediate for adhesives, plastics, rubber and elastomers [R5]. t-Butyl hydroperoxide/hydroxylamine hydrochloride is one of the most widely used cationic emulsion polymerization initiator systems [R6]. Used to prepare epichlorohydrin by an epoxidation reaction with allyl chloride over vanadium, tungsten, or molybdenum complexes or by oxidation with air over cobalt [R7].

OCCUPATIONAL RECOMMENDATIONS None listed.

GENERAL RESPONSE DATA Stop discharge if possible. Keep people away. Evacuate area in case of large discharge. Shut off ignition sources and call fire department. Stay upwind and flood spill area with water. Avoid contact with liquid. Notify local health and pollution control agencies.

FIRE RESPONSE Flammable. May explode if subjected to heat, flame, or shock. May cause fire, and explode on contact with combustibles. Vapor may explode if ignited in an enclosed area. Evacuate surrounding area. Combat fires from safe distance or protected location. Flood discharge area with water. Extinquish with dry chemical, foam, or ion. If breathing is difficult, give oxygen. Liquid irritating to skin and eyes. Harmful if swallowed. Remove contaminated clothing and shoes. Flush affected areas with plenty of water. If swallowed and victim is conscious, have victim drink water or milk, and have victim induce vomiting. If swallowed and victim is unconscious or having convulsions, do nothing except keep victim warm.

RESPONSE TO DISCHARGE Issue warning—high flammability. Restrict access. Mechanical containment. Chemical and physical treatment.

WATER RESPONSE Effect of low concentrations on aquatic life is unknown. May be dangerous if it enters water intakes. Notify local health and pollution control officials. Notify operators of nearby water intakes.

EXPOSURE SYMPTOMS Liquid causes severe burns of skin and eyes.

EXPOSURE TREATMENT Ingestion: induce vomiting and follow with gastric lavage. Inhalation: remove individual from contaminated atmosphere; give artificial respiration and oxygen if needed. skin, eye, and mucous membrane contact: flood affected tissues with water.

REFERENCES

1. *The Merck Index.* 9th ed. Rahway, NJ: Merck and Co., Inc., 1976. 200.

2. Sax, N. I. *Dangerous Properties of Industrial Materials,* 5th ed. New York: Van Nostrand Reinhold, 1979. 447.

3. National Fire Protection Association. *Fire Protection Guide on Hazardous Materials.* 7th ed. Boston, MA: National Fire Protection Association, 1978.

4. Hawley, G.G. *The Condensed Chemical Dictionary.* 9th ed. New York: Van Nostrand Reinhold Co., 1977. 138.

5. SRI.

6. *Kirk-Othmer Encyc Chem Tech.* 3rd ed. 1978-present, V14 p. 93.

7. *Kirk-Othmer Encyc Chem Tech.* 3rd ed. 1978-present, V5 p. 861.

■ TERRAZOLE

CAS # 2593-15-9

SYNONYMS 1,2,4-thiadiazole, 5-ethoxy-3-(trichloromethyl)- • 3-(Trichloromethyl)-5-ethoxy-1,2,4-thiadiazole • 5-aethoxy-3-trichlormethyl-1,2,4-thiadiazol [German] • 5-ethoxy-3-trichloromethyl-1,2,4-thiadiazole • aaterra • dwell • echlomezol • echlomezole • etcmtb • ethazol • ethazole [fungicide] • etmt • etridiazol • etridiazole • koban • mF-344 • Olin-Mathieson-2,424 • om-2424 • pansoil • terrachlor-super x • terracoat-121 • terraflo • truban

MF C_5H_5CL_3N_2OS

MW 247.53

COLOR AND FORM Pale-yellow liquid when pure.

ODOR Mild persistent odor.

DENSITY 1.503 at 25°C.

BOILING POINT 95°C at 1 mm Hg.

MELTING POINT FP: 20°C.

SOLUBILITY Soluble in acetone, ether, ethanol, xylene, carbon tetrachloride; solubility in water: 50 mg/L at 25°C; It is practically insoluble in water but soluble in many organic solvents. In water at 20°C, less than 200 mg/L. Very readily soluble in common, organic solvents, e.g., acetone, cyclohexanone, chloroform, carbon tetrachloride, benzene, xylene, ethanol.

VAPOR PRESSURE 1.3×10^{-4} mbar at 20°C.

STABILITY AND SHELF LIFE Stable and suffered no loss of biological activity during 3 yr storage under normal conditions [R5]. Thermally stable to 165°C, not degraded by UV light or oxygen [R8]. Hydrolyzed on contact with alkali media [R8].

OTHER CHEMICAL/PHYSICAL PROPERTIES Reddish-brown liquid (technical product). Stable to elevated temperatures (up to 165°C) and to UV radiation. Hydrolyzed by strong acids. Vapor pressure: 9.75×10^{-5} mm Hg at 25°C.

DISPOSAL SRP: At the time of review, criteria for land treatment or burial (sanitary landfill) disposal practices are subject to significant revision. Prior to implementing land disposal of waste residue (including waste sludge), consult with environmental regulatory agencies for guidance on acceptable disposal practices.

COMMON USES Fungicide for seed treatment of field crops [R1]. A soil fungicide with preventative and curative properties. It is effective against *Phytophthora* and *Pythium* spp and is used on turf, ornamentals, vegetables, avocados, and other crops [R2]. Fungicidal control of *Phytophthora* and *Pythium* species before sowing or planting bedding plants, pot plants, and tulips, and in the setting of seedlings of tomatoes, cucumbers, cauliflowers, and celery. Control of red core in spring-planted strawberries [R4]. Effective against the seedling disease complex (*Fusarium, Rhizoctonia, Pythium*) of beans, corn, cotton, peanuts, peas, sorghum, soybeans, safflower, sugar beets, and wheat; as a seed treatment, or as in-furrow soil applications, and root rot (*Phytophthora cinnamomi*) of avocado: *Pythium* on strawberries; used on ornamentals, turf, vegetables [R3]. Received Experimental Use Permit as a nitrification inhibitor; full label received on cotton; expect registration on corn and wheat [R3].

OCCUPATIONAL RECOMMENDATIONS None listed.

ECOTOXICITY LD_{50} mallard duck oral >2,000 mg/kg, males 4–5 mo of age [R6].

BIOCONCENTRATION Based on an experimental water solubility of 50 mg/L at room temperature (1), the bioconcentration factor for terrazole can be estimated to be about 70 from a regression-derived equation (2,SRC). This estimated bioconcentration factor indicates that terrazole is not expected to bioconcentrate in fish and aquatic organisms (SRC) [R7].

FIFRA REQUIREMENTS In 1988, Congress amended FIFRA to strengthen and accelerate EPA's reregistration program. The nine-year reregistration scheme mandated by "FIFRA 88" applies to each registered pesticide product containing an active ingredient initially registered before November 1, 1984. List A consists of the 194 chemical cases (or 350 individual active ingredients) for which EPA had issued Registration Standards prior to the effective date of FIFRA '88. List: A; Case: Terrazole; Case No.: 0009; Pesticide type: Fungicide; Registration Standard Date: 09/01/80; Case Status: Awaiting Data/Data in Review: OPP awaits data from the pesticide's producer(s) regarding its human health and/or environmental effects, or OPP has received and is reviewing such data, in order to reach a decision about the pesticide's eligibility for reregistration. Active Ingredient (AI): Etridiazole; AI Status: The producer(s) of the pesticide has made commitments to conduct the studies and pay the fees required for reregistration, and is meeting those commitments in a timely manner. [R9].

REFERENCES

1. SRI.

2. Gerhartz, W. (exec ed.). *Ullmann's Encyclopedia of Industrial Chemistry.* 5th ed. Vol A1: Deerfield Beach, FL: VCH Publishers, 1985 to present., p. A12 95.

3. *Farm Chemicals Handbook 1984.* Willoughby, OH: Meister Publishing Co., 1984, p. C–222.

4. Hartley, D., and H. Kidd (eds.). *The Agrochemicals Handbook.* 2nd ed. Lechworth, Herts, England: The Royal Society of Chemistry, 1987, p. A188/OCT 83.

5. Worthing, C. R., S. B. Walker (eds.). *The Pesticide Manual—A World Compendium.* 7th ed. Lavenham, Suffolk, Great Britain: The Lavenham Press Limited, 1983. 252.

6. U.S. Department of the Interior, Fish and Wildlife Service. *Handbook of Toxicity of Pesticides to Wildlife.* Resource Publication 153. Washington, DC: U.S. Government Printing Office, 1984. 79.

7. Worthing, C. R., S. B. Walker, eds. *The Pesticide Manual.* 8th ed Croydon, England: The British Crop Protection Council (1987).

8. Spencer, E. Y. *Guide to the Chemicals Used in Crop Protection.* 7th ed. Publication 1093. Research Institute, Agriculture Canada, Ottawa, Canada: Information Canada, 1982. 279.

9. USEPA/OPP. *Status of Pesticides in Reregistration and Special Review* p. 116 (Mar, 1992) EPA 700-R-92-004.

■ 1,2,3,4–TETRACHLORO-BENZENE

CAS # 634–66–2

SYNONYMS benzene, 1,2,3,4-tetrachloro

MF $C_6H_2Cl_4$

MW 215.88

COLOR AND FORM Needles from alcohol; white crystals.

BOILING POINT 254°C at 760 mm Hg.

MELTING POINT 47.5°C.

SOLUBILITY Insoluble in water; soluble in hot alcohol; slightly soluble in alcohol; very soluble in ether, carbon disulfide, acetic acid, petroleum ether; 3.5 ppm (22°C).

OTHER CHEMICAL/PHYSICAL PROPERTIES The solubilities of several chlorobenzenes and other mixtures in water were determined. The results varied with the phase of the solute mixture and the hydrophilicity of the components and were interpreted through activity coefficients calculated by the UNIFAC equation. Mixtures of structurally related hydrophobic liquids were near ideal in the organic phase, and, in the aqueous phase, the activity coefficient of a component was unaffected by the presence of cosolutes. Increasing hydrophilicity of the solutes led to deviations from ideality in the organic phase, but these could be accounted for by the UNIFAC equation. For mixtures of solids, that did not interact, the components tended to behave independent of one another, and their solubilities were approximately additive. The behavior of mixtures of liquids and solids was intermediate between that of liquid mixtures and that of mixtures of solids. The application of these results to the toxicity of organic mixtures in water was also discussed. (chlorobenzenes).

CLEANUP Contain and isolate spill to limit spread. Construct clay bentonite swale to divert uncontaminated portion of watershed around contaminated portion. Isolation procedures include construction of bentonite-lined dams, interceptor trenches, or impoundments. Seek professional help to evaluate problem and implement containment procedures. Conduct bench-scale and pilot-scale tests prior to implementation of full-scale decontamination program. For density stratification and impoundment, remove product from bottom layer by pumping through manifold or by polyethylene rope collection, or remove clarified upper portion by skimmers or siphoning. Treatment alternatives for contaminated water include sorption with powdered activated charcoal; filtration through a granular activated carbon bed; chemical oxidation with potassium permanganate and biodegradation. Contaminated water may be impounded in a lined pit with leachate collection systems and domed cover. Treatment alternatives for contaminated soils include well-point collection with treatment of leachates as for contaminated waters and bentonite/cement ground injection to immobilize spill. Place immobilized residues in a lined pit with leachate collection system and domed cover. Contaminated soil or immobilized residues may be packaged for disposal. Confirm all treatment procedures with responsible environmental engineer and regulatory officials [R3].

DISPOSAL Halogenated compounds may be disposed of by incineration provided they are blended with other compatible wastes or fuels so that the composite contains less than 30% halogens and the heating value is from 7,000 to 9,000 Btu/lb. Liquid injection, rotary kiln, and fluidized bed incinerators are typically used to destroy liquid halogenated wastes. Temperatures of at least 2,000–2,200°F and residence times of more than 2 sec are required for the destruction of halogenated aromatic hydrocarbons (halogenated aromatic hydrocarbons) [R4]. Product residues and sorbent media may be packaged in 17H epoxy-lined drums and disposed of at an EPA–approved site. Destroy by high-temperature incineration with hydrochloric acid scrubber or microwave plasma treatment, if available. [R7].

COMMON USES Component of dielectric fluids; synthesis [R2]; chemical intermediate (mixed with 1,2,4,5–tetrachlorobenzene) for pentachloronitrobenzene [R1].

OCCUPATIONAL RECOMMENDATIONS None listed.

ECOTOXICITY LC_{50} *Pimephales promelas* (fathead minnow) 1.1 mg/L 96 hr flow-through bioassay, wt 0.12 g, water hardness 45.5 mg/L $CaCO_3$, temp: $25 \pm 1°C$, pH 7.5, dissolved oxygen greater than 60% of saturation [R5].

TSCA REQUIREMENTS Chemical manufacturers including certain producers and importers are required to submit information that includes data on the amount of chemicals produced, amounts directed to certain classes of uses, and potential exposures and environmental releases associated with the manufacturer's own and his or her immediate customer's processing of the chemicals [R6]. Section 8(a) of TSCA requires manufacturers of this chemical substance to report preliminary assessment information concerned with production, use, and exposure to EPA. [R8].

REFERENCES

1. SRI.

2. Hawley, G. G. *The Condensed Chemical Dictionary*. 9th ed. New York: Van Nostrand Reinhold Co., 1977. 845.

3. NIH/EPA. OHMTADS (1985).

4. 40 CFR 260.340–260.351 (1985).

5. Vieth, G.D., et al., *Canadian J Fisheries Aquat Sci* 40 (6): 743–8 (1983).

6. 47 FR 26992 (6/22/82).

7. NIH/EPA. OHM/TADS (1985).

8. 40 CFR 712.30 (7/1/82).

■ 1,2,3,5–TETRACHLORO-
BENZENE

CAS # 634–90–2

SYNONYMS benzene, 1,2,3,5-tetrachloro

MF $C_6H_2Cl_4$

MW 215.88

COLOR AND FORM Needles from alcohol; white flakes.

BOILING POINT 246°C.

MELTING POINT 54.5°C.

SOLUBILITY Soluble in hot water, ether, benzene; slightly soluble in alcohol; very soluble in carbon disulfide, petroleum ether; 2.4 ppm in water (22°C).

OTHER CHEMICAL/PHYSICAL PROPERTIES Distillation range 240–246°C. The solubilities of several chlorobenzenes and other mixtures in water were determined. The results varied with the phase of the solute mixture and the hydrophilicity of the components and were interpreted through activity coefficients calculated by the UNIFAC equation. Mixtures of structurally related hydrophobic liquids were near ideal in the organic phase and, in the aqueous phase, the activity coefficient of a component was unaffected by the presence of cosolutes. Increasing hydrophilicity of the solutes led to deviations from ideality in the organic phase but these could be accounted for by the UNIFAC equation. For mixtures of solids, that did not interact, the components tended to behave independently of one another, and their solubilities were approximately additive. The behavior of mixtures of liquids and solids was intermediate between that of liquid mixtures and that of mixtures of solids. The application of these results to the toxicity of organic mixtures in water was also discussed. (chlorobenzenes).

CLEANUP Contain and isolate spill to limit spread. Construct clay bentonite swale to divert uncontaminated portion of water-shed around contaminated portion. Isolation procedures include construction of bentonite-lined dams, interceptor trenches, or impoundments. Seek professional help to evaluate problem and implement containment procedures. Conduct bench-scale and pilot-scale tests prior to implementation of full-scale decontamination program. For density stratification and impoundment, remove product from bottom layer by pumping through manifold or by polyethylene rope collection, or remove clarified upper portion by skimmers or siphoning. Treatment alternatives for contaminated water include sorption with powdered activated charcoal; filtration through a granular activated carbon bed; chemical oxidation with potassium permanganate and biodegradation. Contaminated water may be impounded in a lined pit with leachate collection systems and domed cover. Treatment alternatives for contaminated soils include well-point collection with treatment of leachates as for contaminated waters and bentonite/cement ground injection to immobilize spill. Place immobilized residues in a lined pit with leachate collection system and domed cover. Contaminated soil or immobilized residues may be packaged for disposal. Confirm all treatment procedures with responsible environmental engineer and regulatory officials [R3].

DISPOSAL Halogenated compounds may be disposed of by incineration provided they are blended with other compatible wastes or fuels so that the composite contains less than 30% halogens and the heating value is from 7,000 to 9,000 Btu/lb. Liquid injection, rotary kiln, and fluidized bed incinerators are typically used to destroy liquid halogenated wastes. Temperatures of at least 2,000–2,200°F and residence times (of more than 2 secs) are required for the destruction of halogenated aromatic hydrocarbons (halogenated aromatic hydrocarbons) [R4]. Product residues and sorbent media may be packaged in 17H epoxy-lined drums and disposed of at an EPA–approved site. Destroy by high temperature incineration with hydrochloric acid scrubber or microwave plasma treatment, if available. [R11].

COMMON USES Research chemical [R1]; intermediate for herbicides and defoliants; insecticide; impregnant for moisture resistance; electrical insulation [R2]; temporary protection in packing [R2].

OCCUPATIONAL RECOMMENDATIONS None listed.

NON-HUMAN TOXICITY LD_{50} rats oral 1,727 mg/kg [R6]; TDLo rats oral 2mg/kg (6–15 days preg), toxic effects: reproductive fertility [R6].

ECOTOXICITY LC_{50} bluegill sunfish 57.8 mg/L/24 hr; 11.5 mg/L/48 hr; 6.42 mg/L/96 hr [R7]. LC_{50} sheepshead minnow >7.5 mg/L/24 hr; 5.59 mg/L/48hr; 3.67 mg/L/96 hr [R8]. LC_{50} mysid shrimp 0.34 ppm/96 hr [R5].

BIOLOGICAL HALF-LIFE Bluegills were exposed for 28 days. Half-life was 2 to 4 days [R3].

BIOCONCENTRATION A measured steady-state bioconcentration factor of 1,800 was obtained for 1,2,3,5–tetrachlorobenzene using bluegill [R9].

TSCA REQUIREMENTS Chemical manufacturers including certain producers and importers are required to submit information that includes data on the amount of chemicals produced, amounts directed to certain classes of uses, and potential exposures and environmental releases associated with the manufacturer's own and his or her immediate customer's processing of the chemicals [R10]. Section 8(a) of TSCA requires manufacturers of this chemical substance to report preliminary assessment information concerned with production, use, and exposure to EPA. [R12].

REFERENCES

1. SRI.

2. Hawley, G. G. *The Condensed Chemical Dictionary*. 9th ed. New York: Van Nostrand Reinhold Co., 1977. 845.

3. NIH/EPA. OHMTADS (1985).

4. 40 CFR 260.340–260.351 (1985).

5. NIH/EPA. OHM/TADS (1985).

6. NIOSH. *Current Awareness Listing* (1985).

7. USEPA. *Health Assessment Document* p. 6–6 (1985) EPA 600/8–84–015.

8. USEPA. *Health Assessment Document* p. 6–6 (1985) EPA–660/8–84–015.

9. U.S. Environ Prot Agency. No. 68–01–4646 as cited in USEPA. *Ambient Water Quality Criteria Doc; Chlorinated Benzenes* p. C. 53 (1980) EPA 440/5–80–028.

10. 47 FR 26992 (6/22/82).

11. NIH/EPA; OHM/TADS (1985).

12. 40 CFR 712.30 (7/1/82).

■ TETRACHLORONAPHTHALENE

CAS # 1335–88–2

SYNONYMS halowax • tetrachloro-naphthalene (ACGIH: OSHA) • naphthalene, tetrachloro

MF $C_{10}H_4Cl_4$

MW 265.94

COLOR AND FORM Crystals; pale-yellow solid; colorless to pale-yellow solid.

ODOR Aromatic.

DENSITY 1.59 to 1.65 (water = 1).

BOILING POINT 311.5 to 360°C.

MELTING POINT 182°C.

AUTOIGNITION TEMPERATURE None to boiling point [R2, 2]

FIREFIGHTING PROCEDURES Extinguishant: foam, carbon dioxide, dry chemical. Also wear a self-contained breathing apparatus (SCBA) with a full facepiece operated in the pressure-demand or other positive-pressure mode. [R2, 2].

TOXIC COMBUSTION PRODUCTS Toxic gases and vapors (such as hydrogen chloride, phosgene, and carbon monoxide) may be released in a fire involving tetrachloronaphthalene [R2].

SOLUBILITY Insoluble in water.

VAPOR PRESSURE 1×10^{-6} mm Hg at 25°C (est).

OTHER CHEMICAL/PHYSICAL PROPERTIES MP: 115°C; Henry's Law constant = 0.000159 atm−m³/mole (est).

PROTECTIVE EQUIPMENT Wear appropriate clothing to prevent any possibility of skin contact with molten material or solutions. Wear eye protection and wash promptly when skin is contaminated (chloronaphthalenes) [R3]. Persons should be provided with and required to use impervious clothing, gloves, face shields (eight-inch minimum) and other appropriate clothing to prevent skin contact with liquid or molten tetrachloronaphthalene or its fumes. [R2, 4].

PREVENTATIVE MEASURES Contact lenses should not be worn when working with this chemical [R9, 209]. SRP: The scientific literature supports the wearing of contact lenses in industrial environments, as part of a program to protect the eye against chemical compounds and minerals causing eye irritation. However, there may be individual substances whose irritating or corrosive properties are such that the wearing of contact lenses would be harmful to the eye. In those specific cases contact lenses should not be worn. If employees' clothing may have become contaminated with solid tetrachloronaphthalene, employees should change into uncontaminated clothing before leaving the work premises [R2]. Clothing contaminated with tetrachloronaphthalene should be placed in closed containers for storage until it can be discarded or until provision is made for the removal of tetrachloronaphthalene from the clothing. If the clothing is to be laundered or otherwise cleaned to remove the tetrachloronaphthalene, the person performing the operation should be informed of tetrachloronaphthalene's hazardous properties [R2]. Non-impervious clothing that becomes contaminated with molten tetrachloronaphthalene should be removed immediately and not reworn until the tetrachloronaphthalene is removed from the clothing [R2]. Employees should be provided with and required to use impervious clothing, gloves, face shields (eight-inch minimum), and other appropriate protective clothing necessary to: prevent any possibility of skin contact with molten tetrachloronaphthalene; prevent repeated or prolonged skin contact with solid tetrachloronaphthalene or liquids containing tetrachloronaphthalene; or prevent skin contact with tetrachloronaphthalene fumes from heated material [R2]. Condenser impregnation and other operations involving melting of chloronaphthalene should be enclosed or provided with effective local exhaust ventilation (chloronaphthalenes) [R10]. SRP: Local exhaust ventilation should be applied wherever there is an incidence of point-source emissions or dispersion of regulated contaminants in the work area. Ventilation control of the contaminant as close as possible to its point of generation is both the most economical and safest method to minimize personnel exposure to airborne contaminants.

DISPOSAL Incineration, preferably after mixing with another combustible fuel. Assure complete combustion to prevent the formation of phosgene. An acid scrubber is necessary to remove the halo acids produced. Recommendable method: Incineration. Peer-review: Ensure plentiful supply of hydrocarbon fuel (peer-review conclusions of an IRPTC expert consultation (May 1985)) [R4].

COMMON USES Pesticide (SRP: former use) [R2, 94]. Tetrachloronaphthalene is used in electrical insulating materials, as a component of resins or polymers for coating or impregnating textiles, wood, and paper, and as an additive in cutting oils and lubricants [R1].

OCCUPATIONAL RECOMMENDATIONS
OSHA 8−hr time-weighted average: (2 mg/m³). Skin absorption designation [R7].

THRESHOLD LIMIT 8−hr-time weighted average (TWA) 2 mg/m³ (1986) [R11, 34]. Excursion Limit Recommendation: Excursions in worker exposure levels may exceed three times the TLV−TWA for no more than a total of 30 min during a work day, and under no circumstances should they exceed five times the TLV−TWA. [R11, 5].

BIODEGRADATION Tetrachloronaphthalene is reported to have "poor" biodegradability (1) [R5].

BIOCONCENTRATION In a continuous-flow-through system, tetrachloronaphthalene BCFs of 25,000–33,000 were measured in guppies (*Poecillia reticula*) over a 7-day exposure period with three different tetrachloronaphthalene isomers (1). In a static system, a mean BCF of 5,100 was measured in rainbow trout (for the 1,2,3,4–tetrachloronaphthalene isomer) over a 96-day exposure period (2). In a continuous-flow-through system, a mean BCF of 21,000 was measured in oligochaete worms (for the 1,2,3,4–tetrachloronaphthalene isomer) over a 79-day exposure period (3). Fish BCFs of 5,000–20,000 have been reported for the Japanese MITI bioaccumulation test (4) [R6].

TSCA REQUIREMENTS Pursuant to section 8 (d) of TSCA, EPA promulgated a model health and safety data reporting rule. The section 8 (d) model rule requires manufacturers, importers, and processors of listed chemical substances and mixtures to submit to EPA copies and lists of unpublished health and safety studies. Tetrachloronaphthalene is included on this list [R8].

REFERENCES

1. American Conference of Governmental Industrial Hygienists. *Documentation of the Threshold Limit Values and Biological Exposure Indices.* 5th ed. Cincinnati, OH: American Conference of Governmental Industrial Hygienists, 1986. 562.

2. Mackison, F. W., R. S. Stricoff, and L. J. Partridge, Jr. (eds.). *NIOSH/OSHA—Occupational Health Guidelines for Chemical Hazards.* DHHS (NIOSH) Publication No. 81–123 (3 vols). Washington, DC: U.S. Government Printing Office, Jan. 1981.

3. Sittig, M. *Handbook of Toxic and Hazardous Chemicals.* Park Ridge, NJ: Noyes Data Corporation, 1981. 153.

4. United Nations. *Treatment and Disposal Methods for Waste Chemicals* (IRPTC File). Data Profile Series No. 5. Geneva, Switzerland: United Nations Environmental Programme, Dec. 1985. 212.

5. Koda, Y., et al., *Nagoya Kogyo Gijutsu Shikensho Hokoku* 31: 299–305 (1982).

6. (1) Opperhuizen, A. *Chemosphere* 14: 1871–96 (1985). (2) Oliver, B.G., and A.J. Niimi, *Environ Sci Technol* 19: 842–9 (1985). (3) Oliver, B. G. *Environ Sci Technol* 21: 785–90 (1987). (4) Kawasaki, M., *Ecotox Environ Safety* 4: 444–54 (1980).

7. 29 CFR 1910.1000 (7/1/91).

8. 40 CFR 716.120 (7/1/91).

9. NIOSH. *NIOSH Pocket Guide to Chemical Hazards.* DHHS (NIOSH) Publication No. 90–117. Washington, DC: U.S. Government Printing Office, June 1990.

10. International Labour Office. *Encyclopedia of Occupational Health and Safety.* Vols. I and II. Geneva, Switzerland: International Labour Office, 1983. 466.

11. American Conference of Governmental Industrial Hygienists. *Threshold Limit Values for Chemical Substances and Physical Agents and Biological Exposure Indices for 1994–1995.* Cincinnati, OH: ACGIH, 1994.

■ 1–TETRADECENE

CAS # 1120–36–1

SYNONYMS 1-tetradecylene • alpha-tetradecene • n-tetradec-1-ene • ai3-10509

MF $C_{14}H_{28}$

MW 196.38

COLOR AND FORM Colorless liquid.

ODOR Mild, pleasant.

DENSITY 0.7745 at 20°C.

SURFACE TENSION 25.0 dynes/cm at 20°C.

VISCOSITY 2.61 m^3m/s (kinematic).

BOILING POINT 232–234°C at 760 mm Hg.

MELTING POINT −12°C.

HEAT OF COMBUSTION −17,600 Btu/lb = −9,779 cal/g.

HEAT OF VAPORIZATION 103 Btu/lb = 57.1 cal/g.

AUTOIGNITION TEMPERATURE 455°F (235°C) [R4].

FIREFIGHTING PROCEDURES Carbon dioxide, dry chemical [R3]. Water or foam may cause frothing [R4]. Extinguish with dry chemical, foam, or carbon dioxide. Water may be ineffective on fire [R4].

SOLUBILITY Soluble in ethanol, ether, benzene; water solubility = 4.0×10^{-4} mg/L at 25°C (est).

VAPOR PRESSURE 1.5×10^{-2} mm Hg at 25°C.

OCTANOL/WATER PARTITION COEFFICIENT Log K_{ow} = 7.3 (est).

OTHER CHEMICAL/PHYSICAL PROPERTIES Free energy of formation at 25°C 146.4 kJ/mol; liquid water interfacial tension = 32.8 dynes/m at 22.7°C; ratio of specific heats of vapor (gas): 1.027; Henry's Law constant = 8.48 atm–m³/mole at 25°C (est).

PROTECTIVE EQUIPMENT Personal protective equipment. Goggles or face shield [R5].

PREVENTATIVE MEASURES SRP: The scientific literature supports the wearing of contact lenses in industrial environments, as part of a program to protect the eye against chemical compounds and minerals causing eye irritation. However, there may be individual substances whose irritating or corrosive properties are such that the wearing of contact lenses would be harmful to the eye. In those specific cases contact lenses should not be worn.

COMMON USES Chemical intermediate (as part of alkene mixt) [R1]; in linear alpha-olefin mixture for production of sulfonates; chemical intermediate for n-tetradecyl dimethyl amine [R1]; solvent in perfumes, flavors, medicines, dyes, oils, resins [R2].

STANDARD CODES NFPA, ICC, USCG—no.

PERSISTENCE Unsaturated bond subject to photochemical attack.

MAJOR SPECIES THREATENED Waterfowl.

DIRECT CONTACT Potentially skin.

GENERAL SENSATION May be narcotic at high concentrations.

PERSONAL SAFETY Wear skin protection and canister mask.

ACUTE HAZARD LEVEL Irritant, possibly narcotic at high concentrations.

CHRONIC HAZARD LEVEL Unknown.

DEGREE OF HAZARD TO PUBLIC HEALTH Thought to be irritant and narcotic at high concentrations.

OCCUPATIONAL RECOMMENDATIONS None listed.

ACTION LEVEL Notify fire and air authority. Enter from upwind and remove ignition source. Attempt to contain slick.

ON-SITE RESTORATION Oil-skimming equipment and sorbent foams can be applied to slick. Seek professional environmental engineering assistance through EPA's Environmental Response Team (ERT), Edison, NJ, 24–hour no., 908–548–8730.

BEACH OR SHORE RESTORATION Burn off.

AVAILABILITY OF COUNTERMEASURE MATERIAL Oil-skimming equipment—stored at major ports; sorbent foams (polyurethane)—upholstery shops.

DISPOSAL Spray into incinerator or burn in paper packaging. Additional flammable solvent may be added.

INDUSTRIAL FOULING POTENTIAL Slick may reduce heat transfer or cause hot spots or scaling.

MAJOR WATER USE THREATENED Recreation.

PROBABLE LOCATION AND STATE Colorless liquid will form slick on top of water. Will float.

COLOR IN WATER Colorless.

BIODEGRADATION Incubation of 1–tetradecene with fungi resulted in the formation of tetradecanoic acid, tetradecen–4–ol, 13–tetradecen–4–ol, tetradecen–3–ol, 13–tetradecene–4–on, tetradecan–2–one, and tetradecen–3–one (1–2). Pure cultures of *Corynebacterium sp.*(3),

Pseudomonas aerginosa (4), *Micrococcus cerficans* (5), and members of the genus *Candida* (6) were found to oxidize 1–tetradecene under aerobic conditions [R6].

BIOCONCENTRATION Estimated bioconcentration factors ranging from 17,500 to 51,000 (1,SRC) can be calculated, for 1–tetradecene using appropriate regression equations (1), and its estimated water solubility, 4.0×10^{-4} mg/L at 25°C (1,SRC) obtained from a vapor pressure of 1.5×10^{-2} mm Hg at 25°C (2), and estimated Henry's Law constant of 8.48 atm-m³/mole at 25°C (3,SRC), and its estimated octanol/water partition coefficient, 7.3 (1,SRC), obtained from its estimated water solubility. These values indicate that 1–tetradecene will bioconcentrate in fish and aquatic organisms (SRC) [R7].

FIRE POTENTIAL Fire hazard: low; can react with oxidizing materials [R3].

REFERENCES

1. SRI.

2. Sax, N. I., and R. J. Lewis, Sr. (eds.). *Hawley's Condensed Chemical Dictionary.* 11th ed. New York: Van Nostrand Reinhold Co., 1987. 1133.

3. Sax, N. I. *Dangerous Properties of Industrial Materials,*. 5th ed. New York: Van Nostrand Reinhold, 1979. 1014.

4. National Fire Protection Guide. *Fire Protection Guide on Hazardous Materials.* 10th ed. Quincy, MA: National Fire Protection Association, 1991, p. 325M–85.

5. U.S. Coast Guard, Department of Transportation. *CHRIS—Hazardous Chemical Data.* Volume II. Washington, DC: U.S. Government Printing Office, 1984–5.

6. (1) Allen, J. E., A. J. Markovetz. *J Bacteriol* 103: 426–34 (1970). (2) Allen, J. E., et al., *Lipids* 6: 448–52 (1971). (3) Buswell, J. A., P. Jurtshuk. *Arch Mikrobiol* 64: 215–22 (1969). (4) Markovetz, A. J., et al., *J Bacteriol* 93: 1289–92 (1967). (5) *J Bacteriology* 95: 2108–11 (1968). (6) Klug, M. J., A. J. Markovetz. *App Microbiol* 15: 690–3 (1967).

7. (1) Lyman, W. J., et al., *Handbook of Chemical Property Estimation Methods.* New York: McGraw-Hill, Chapt 2, 5 and 15 (1982). (2) Daubert, T. E., R. P. Danner. *Physical and Thermodynamic Properties of Pure Chemicals* New York, NY: Hemisphere Pub Corp (1989). (3) Meylan, W. M., P. H. Howard. *Environ Toxicol Chem* 10: 1283–93 (1991).

■ TETRADIFON

CAS # 116–29–0

SYNONYMS 1,2,4-trichloro-5-((4-chlorophenyl)-sulfonyl) benzene • 2,4,4′,5-tetrachloor-difenyl-sulfon (Dutch) • 2,4,4′,5-Tetrachlor-diphenyl-sulfon (German) • 2,4,4′,5-tetrachlorodiphenyl sulfone • 2,4,4′,5-tetrachlorodiphenyl sulphone (British) • 2,4,4′,5-tetracloro-difenil-solfone (Italian) • 2,4,5,4′-tetrachlorodiphenyl sulfone • 4-chlorophenyl, 2,4,5-trichlorophenyl sulfone • benzene, 1,2,4-trichloro-5-((4-chlorophenyl) sulfonyl)- • p-chlorophenyl 2,4,5-trichlorophenyl sulfone • sulfone, p-chlorophenyl 2,4,5-trichlorophenyl • Duphar • akaritox • aredion • fmc-5488 • mition • nia-5488 • roztoczol • polacaritox • roztoczol-extra • roztozol • tedion • tetradichlone • tedion V-18 • childion • tedane • dorvert • tedane-combi • turbair-acaricide • tedion-V18 • aracnol-K

MF $C_{12}H_6Cl_4O_2S$

MW 356.04

COLOR AND FORM Colorless crystalline solid; crystals from benzene; white, crystalline powder.

ODOR Odorless.

DENSITY 1.515 at 20°C.

MELTING POINT 146.5–147.5°C.

SOLUBILITY Solubility at 10°C: 50 µg/l water; at 20°C: 82 g/L acetone; 148 g/L benzene; 255 g/L chloroform; 200 g/L cyclohexanone; 10 g/L kerosene; 10 g/L methanol; 135 g/L toluene; 115 g/L xylene; soluble in aromatic hydrocarbon solvents, dioxan; at 18°C: 0.4 g/100 g petroleum ether; 7.1 g/100 g ethyl acetate; 10.5 g/100 g methyl ethyl ketone; and 1.6 g/100 g carbon tetrachloride; solubility in water at 20°C, 0.08 mg/L.

VAPOR PRESSURE 0.032 mPa at 20°C.

OCTANOL/WATER PARTITION COEFFICIENT Log K_{ow} = 4.72 (est).

CORROSIVITY Noncorrosive.

STABILITY AND SHELF LIFE Stable to concentrated and dilute alkalis, mineral acids, high temperature, and UV light [R1].

OTHER CHEMICAL/PHYSICAL PROPERTIES Melting point: 144°C (technical product); off-white to light-yellowish substance (technical product); 100°C, sublimes; it is resistant to acid or alkali hydrolysis. Henry's Law constant = 7.5×10^{-8} atm−m³/mole at 25°C (est). Tetradifon is resistant to strong oxidizing agents and stable to sunlight. Compatible with other pesticides. Henry's Law constant = 7.5×10^{-8} atm−m³/mole at 25°C (low est).

DISPOSAL Tetradifon should be incinerated in a unit operating at 1,000°C for 1−2 seconds and equipped with afterburner. Recommendable method: Incineration [R5].

COMMON USES Acaricide; ovicide on deciduous fruits, citrus, cotton, and other crops [R1]. Insecticide [R3]; smoke generators [R12, p. C−228]; control of eggs, and young active stages of phytophagous mites on fruit trees (including citrus and nuts), vines, vegetables, ornamentals, cotton, hops, coffee, tea, and rice [R2]. A food additive permitted in food for human consumption [R4].

OCCUPATIONAL RECOMMENDATIONS None listed.

NON-HUMAN TOXICITY LD_{50} rat oral >14,700 mg/kg [R7]; LD_{50} rat oral 5,000−14,700 mg/kg (technical tetradifon) [R8]; LD_{50} dog oral 2,000 mg/kg [R9]; mice ip 75 mg/kg [R9].

ECOTOXICITY LD_{50} mallard oral >2,000 mg/kg, male, 4 mo old [R6]; LC_{50} carp >10 mg/L/3 hr (conditions of bioassay not specified) [R2]; LC_{50} bobwhite quail dietary >5,000 mg/kg/8 day [R2].

BIODEGRADATION Incubation of tetradifon with inocula of rumina bacteria or ciliated protozoa under anaerobic conditions suggests that tetradifon is not nutritionally used for growth of the organism and that this compound does not stimulate endogenous gas production (1) [R10].

BIOCONCENTRATION Based on an estimated log K_{ow} of 4.72 (2) and a regression-derived equation (1), the BCF for tetradifon can be estimated to be about 2,000 (SRC) [R11].

DOT *Health Hazards:* Poisonous; may be fatal if inhaled, swallowed, or absorbed through skin. Contact may cause burns to skin and eyes. Runoff from fire control or dilution water may cause pollution (organochlorine pesticides, liquid, flammable, poisonous, not otherwise specified) [R13, p. G−28].

Fire or Explosion: Flammable/combustible material; may be ignited by heat, sparks, or flames. Vapors may travel to a source of ignition, and flash back. Container may explode in heat of fire. Vapor explosion and poison hazard indoors, outdoors, or in sewers. Runoff to sewer may create fire or explosion hazard (organochlorine pesticides, liquid, flammable, poisonous, not otherwise specified) [R13, p. G−28].

Emergency Action: Keep unnecessary people away; isolate hazard area, and deny entry. Stay upwind; keep out of low areas. Positive-pressure self-contained breathing apparatus (SCBA) and chemical protective clothing that is specifically recommended by the shipper or manufacturer may be worn. It may provide little or no thermal protection. Structural firefighters' protective clothing is not effective for these materials. Isolate for 1/2 mile in all directions if tank, rail car, or tank truck is involved in fire. Call CHEMTREC at 1−800−424−9300 for emergency assistance (organochlorine pesticides, liquid, flammable, poisonous, not otherwise specified) [R13, p. G−28].

Fire: Small fires: Dry chemical, CO_2,

water spray, or alcohol-resistant foam. Large Fires: Water spray, fog, or alcohol-resistant foam. Move container from fire area if you can do so without risk. Dike fire-control water for later disposal; do not scatter the material. Apply cooling water to sides of containers that are exposed to flames until well after fire is out. Stay away from ends of tanks. Withdraw immediately in case of rising sound from venting safety device or any discoloration of tank due to fire (organochlorine pesticides, liquid, flammable, poisonous, not otherwise specified) [R13, p. G–28].

Spill or Leak: Shut off ignition sources; no flares, smoking, or flames in hazard area. Fully encapsulating, vapor-protective clothing should be worn for spills and leaks with no fire. Do not touch or walk through spilled material; stop leak if you can do so without risk. Water spray may reduce vapor; but it may not prevent ignition in closed spaces. Small Spills: Take up with sand or other noncombustible absorbent material and place into containers for later disposal. Large Spills: Dike far ahead of liquid spill for later disposal (organochlorine pesticides, liquid, flammable, poisonous, not otherwise specified) [R13, p. G–28].

First Aid: Move victim to fresh air, and call for emergency medical care; if not breathing, give artificial respiration; if breathing is difficult, give oxygen. In case of contact with material, immediately flush skin or eyes with running water for at least 15 minutes. Remove and isolate contaminated clothing and shoes at the site. Keep victim quiet and maintain normal body temperature. Effects may be delayed; keep victim under observation (organochlorine pesticides, liquid, flammable, poisonous, not otherwise specified) [R13, p. G–28].

REFERENCES

1. Budavari, S. (ed.). *The Merck Index— Encyclopedia of Chemicals, Drugs, and Biologicals.* Rahway, NJ: Merck and Co., Inc., 1989. 1,450.

2. Hartley, D., and H. Kidd (eds.). *The Agrochemicals Handbook.* 2nd ed. Lech-worth, Herts, England: The Royal Society of Chemistry, 1987, p. A387/Aug 87.

3. Sunshine, I. (ed.). *CRC Handbook of Analytical Toxicology.* Cleveland: The Chemical Rubber Co., 1969. 534.

4. Sax, N. I. *Dangerous Properties of Industrial Materials,.* 6th ed. New York, NY: Van Nostrand Reinhold, 1984. 765.

5. United Nations. *Treatment and Disposal Methods for Waste Chemicals* (IRPTC File). Data Profile Series No. 5. Geneva, Switzerland: United Nations Environmental Programme, Dec. 1985. 286.

6. U.S. Department of the Interior, Fish and Wildlife Service. *Handbook of Toxicity of Pesticides to Wildlife.* Resource Publication 153. Washington, DC: U.S. Government Printing Office, 1984. 80.

7. *Kirk–Othmer Encyclopedia of Chemical Technology.* 3rd ed., Volumes 1–26. New York, NY: John Wiley and Sons, 1978–1984, p. V13 461 (1981).

8. WHO. *Environ Health Criteria* 67: Tetradifon p. 9 (1986).

9. WHO. *Environ Health Criteria* 67: Tetradifon p. 23 (1986).

10. Williams, P. P. *Res Rev* 66: 102 (1977).

11. (1) Lyman, W. J., et al., *Handbook of Chemical Property Estimation Methods.* Washington DC: Amer Chemical Soc, pp. 4–9, 5–4, 5–10, 7–4, 7–5, 15–15 to 15–32 (1990). (2) CLOGP. *GEMS—Graphic Exposure Modeling System* CLOGP USEPA (1986).

12. *Farm Chemicals Handbook 1984.* Willoughby, Ohio: Meister Publishing Co., 1984.

13. U.S. Department of Transportation. *Emergency Response Guidebook 1990.* DOT P 5800.5. Washington, DC: U.S. Government Printing Office, 1990.

■ 2,3: 4,5–DI (2–BUTENYL) TETRAHYDROFURFURAL

CAS # 126–15–8

SYNONYMS 1,5a,6,9,9a,9b-hexahydro-4a (4h)-dibenzofurancarboxaldehyde • 2,3,4,5-bis(2-butenylene)tetrahydrofurfural • 2,3,4,5-bis (2-butylene tetrahydrofurfural) • 2,3,4,5-bis(delta (2)-butenylene)tetrahydrofurfural • 2,3,4,5-bis(delta (2)-butylene)tetrahydrofurfural • 2,3,4,5-bis(delta (sup (2)-butenylene)tetrahydrofurfural • 2,3,4,5-bis(delta (sup (2)-butylene)tetrahydrofurfural • 2,3:4,5-bis(2-butene-1,4-diyl)tetrahydrofurfural • 2-furaldehyde, 2,3:4,5-bis(2-butenylene)tetrahydro • 2-furancarboxaldehyde, 2,3:4,5-bis(2-butene-1,4-diyl)tetrahydro • 4a (4h)-dibenzofurancarboxaldehyde,1,5a,6,9, 9a,9b-hexahydro • 4a-formyl-1,4,4a,5a,6,9,9a,9b-octahydro-dibenzofuran • ac-r-11 • bis-delta(sup (2)-butylenetetrahydrofurfural • bisbutenylenetetrahydrofurfural • butadien-furfural copolymer • dibutylene-tetrafurfural • ent-17,596 • mgk-11 • mgk-repellent-11- • mgk-repellent-II • phillips r-11 • r-11 [insect repellent] • r-11

MF $C_{13}H_{16}O_2$

MW 204.29

COLOR AND FORM Pale-yellow liquid.

ODOR Fruity odor.

DENSITY 1.10 at 20°C.

BOILING POINT 307°C.

MELTING POINT –80°C.

SOLUBILITY Practically insoluble in water and in dilute alkali; miscible with petroleum oils, toluene, xylene, ethanol.

STABILITY AND SHELF LIFE Stable for long periods in drums [R2].

OTHER CHEMICAL/PHYSICAL PROPERTIES Density: 1.120 at 20°C (technical grade).

DISPOSAL SRP: At the time of review, criteria for land treatment or burial (sanitary landfill) disposal practices are subject to significant revision. Prior to implementing land disposal of waste residue (including waste sludge), consult with environmental regulatory agencies for guidance on acceptable disposal practices.

COMMON USES Insect repellent [R1].

OCCUPATIONAL RECOMMENDATIONS None listed.

NON-HUMAN TOXICITY LD_{50} rat oral 2 g/kg [R3].

TSCA REQUIREMENTS Section 8 (a) of TSCA requires manufacturers of this chemical substance to report preliminary assessment information concerned with production, use, and exposure to EPA as cited in the preamble in 51 FR 41329 [R4]. Pursuant to section 8(d) of TSCA, EPA promulgated a model Health and Safety Data Reporting Rule. The section 8(d) model rule requires manufacturers, importers, and processors of listed chemical substances and mixtures to submit to EPA copies and lists of unpublished health and safety studies. (4a(4H)-Dibenzofurancarboxaldehyde,1,5a,6,9,9a,9b-hexahydro) is included on this list. [R12].

FIFRA REQUIREMENTS In 1988, Congress amended FIFRA to strengthen and accelerate EPA's reregistration program. The nine-year reregistration scheme mandated by "FIFRA 88" applies to each registered pesticide product containing an active ingredient initially registered before November 1, 1984. Pesticides for which EPA had not issued Registration Standards prior to the effective date of FIFRA '88 were divided into three lists based upon their potential for exposure and other factors, with List B being of highest concern and D of least. List: C; Case: Repellant R-11; Case No.: 3127; Pesticide type: Insecticide; Case Status: The pesticide is no longer an active ingredient in any registered pesticide products. Therefore, EPA is characterizing it as "cancelled." Under FIFRA, pesticide producers may voluntarily cancel their registered products. EPA also may cancel pesticide registrations if registrants fail to pay required fees, to make or meet certain reregistration commitments, or when the Agency reaches findings of unreasonable adverse effects. Active Ingredient (AI): Bis(2-butylene)tetrahydro-2-furaldehyde; AI Status: The active ingredient is no longer contained in any registered products. Therefore, EPA is characterizing it as "cancelled." Under FIFRA, pesticide producers may voluntarily cancel their registered products. EPA also may cancel pesticide registrations if registrants fail to pay required fees, to make or meet certain reregistration commitments, or when the Agency reaches findings of unreasonable adverse effects. [R6].

REFERENCES

1. Budavari, S. (ed.). *The Merck Index— Encyclopedia of Chemicals, Drugs, and Biologicals*. Rahway, NJ: Merck and Co., Inc., 1989. 1289.

2. Worthing, C. R. (ed.). *Pesticide Manual.* 6th ed. Worcestershire, England: British Crop Protection Council, 1979. 391.

3. Gosselin, R. E., R. P. Smith, H. C. Hodge. *Clinical Toxicology of Commercial Products*. 5th ed. Baltimore: Williams and Wilkins, 1984, p. II–345.

4. 40 CFR 712.30 (7/1/94).

5. 40 CFR 716.120 (7/1/94).

6. USEPA/OPP. *Status of Pesticides in Reregistration and Special Review.* p.211 (Mar, 1992) EPA 700-R-92-004.

■ TETRAHYDRONAPHTHALENE

CAS # 119–64–2

SIC CODES 2851; 2842; 2842.

SYNONYMS tetralin • tetranap • napthalene-1, 2, 3, 4-tetrahydride • 1, 2, 3, 4-tetrahydronaphthalene

MF $C_{10}H_{12}$

MW 132.22

COLOR AND FORM Colorless liquid.

ODOR THRESHOLD Detection threshold 18 ppm in water, purity not specified [R5].

DENSITY 0.9702 g/ml at 20 °C/4°C.

VISCOSITY 2.012 cP at 25°C.

BOILING POINT 207.6°C at 760 mm.

MELTING POINT–35.8°C.

BINARY REACTANTS Volatile with steam.

EXPLOSIVE LIMITS AND POTENTIAL Oxidizes on exposure to air, leaving explosive resinous residues. [R3, 119].

FIREFIGHTING PROCEDURES Water may be ineffective on fire. Cool exposed containers with water. [R6].

SOLUBILITY Miscible with ethanol, butanol, acetone, benzene, petroleum ether, chloroform, and petroleum ether, decalin; soluble in methanol: 50.6% wt/wt; insoluble in water. Soluble in aniline; soluble in ether.

VAPOR PRESSURE 0.368 mm Hg at 25°C (experimental).

OTHER CHEMICAL/PHYSICAL PROPERTIES Moisture content, none; residue on evaporation, none; bulk density 8 lb/gal. Heat capacity at 25°C: 217.5 J/mol K. Henry's Law constant = 1.7×10^{-3} atm-m^3/mol at 25°C.

PREVENTIVE MEASURES SRP: The scientific literature for the use of contact lenses in industry is conflicting. The benefit or detrimental effects of wearing contact lenses depend not only upon the substance, but also on factors including the form of the substance, characteristics and duration of the exposure, the uses of other eye protection equipment, and the hygiene of the lenses. However, there may be individual substances whose irritating or corrosive properties are such that the wearing of contact lenses would be harmful to the eye. In those specific cases, contact lenses should not be worn. In any event, the usual eye protection equipment should be worn even when contact lenses are in place.

COMMON USES Solvent for camphor, sulfur, and iodine; in paint thinners; as paint remover when mixed with decalin or white spirit; as insecticide for clothes moth; as substitute for turpentine and petrol in Germany [R3, 119]. Solvent for naphthalene, fats, resins, oils, waxes, used instead of turpentine in lacquers, shoe polishes, floor waxes; degreasing agent [R1]. In plant pathology, tetralin has been successful in total destruction of crown gall and olive knot neoplasms [R2, 3242]. A combination of 31% tetrahydronaphthalene and 0.03% cupric oleate (cuprex) is promoted as a pediculicide and miticide, but its true efficacy remains to be determined. [R4].

STANDARD CODES NFPA—1, 2, 0; ICC—no; USCG - Grade E combustible.

PERSISTENCE Does not biodegrade well. Prolonged exposure to air can lead to oxidation and formation of unstable tetralin peroxide.

MAJOR SPECIES THREATENED Waterfowl.

DIRECT CONTACT Skin irritant.

GENERAL SENSATION Odor resembles mixture of benzene and menthol. Narcotic in high concentrations. Can cause cataracts and kidney injury.

PERSONAL SAFETY Wear skin protection and canister-type mask.

ACUTE HAZARD LEVEL Irritant. Moderately toxic when ingested or inhaled. Solubility is too low to pose real threat in drinking water.

CHRONIC HAZARD LEVEL Moderately hazardous with repeated inhalation or ingestion.

DEGREE OF HAZARD TO PUBLIC HEALTH Moderately hazardous when ingested or inhaled under acute or chronic conditions. Irritant.

OCCUPATIONAL RECOMMENDATIONS None listed.

ACTION LEVEL Notify fire authority. Remove ignition source. Attempt to contain slick.

ON-SITE RESTORATION Oil-skimming equipment and sorbent foams can be applied to slick. Seek professional environmental engineering assistance through EPA's Environmental Response Team (ERT), Edison, NJ, 24-hour no. 908-548-8730.

BEACH OR SHORE RESTORATION Burn off.

AVAILABILITY OF COUNTERMEASURE MATERIAL Oil-skimming equipment—stored at major ports; sorbent foams (polyurethane)—upholstery shops.

DISPOSAL Spray into incinerator or burn in paper packaging. Additional flammable solvent may be added.

INDUSTRIAL FOULING POTENTIAL Slick may cause heat transfer or cause hot spots or scaling. May pose rupture hazard when confined in boiler-feed water.

MAJOR WATER USE THREATENED Potable supply, recreation, industrial.

PROBABLE LOCATION AND STATE Colorless liquid. Will float on surface of water.

COLOR IN WATER Colorless.

NON-HUMAN TOXICITY LD$_{50}$ rat oral 2,860 mg/kg [R7, 122]; LD$_{50}$ rabbit dermal 17,300 mg/kg [R7, 122].

BIODEGRADATION Degradation in sea water by oil-oxidizing microorganisms: 31% breakdown after 21 days at 22 °C in stoppered bottles containing a 1,000–ppm mixture of alkanes, cycloalkanes, and aromatics [R8]. One screening test using sewage seed suggests that tetralin (initial concn of 3–10 ppm) biodegrades slowly in synthetic seawater; 3, 3, 2, and 3 percent theoretical BOD in 5, 10, 15, and 20 days, respectively (1). However, 0.0124 ppm tetralin was not detected after 10 days of incubation in artificial seawater at 25°C indicating that tetralin is moderately oxidized (2). An initial concn of 0.0112 ppm tetralin was observed to degrade 100% in 8 days in a groundwater grab sample at 10°C and pH 8 (3). Tetralin (20.1 ppm initial concn) exhibited 18, 40, 58, 65, 70, and 75 percent theoretical CO$_2$ production after 5, 10, 15, 20, 25, and 35 days, respectively, in water from the Ohio River at 22–25°C and pH 7.2; the half-life in this study was 13 days and, after redosing, a half-life of 9 days was observed under the same conditions (4). [R9].

BIOCONCENTRATION A measured BCF in fish was reported to be about 200 (1); this experimental BCF suggests that bioconcentration in aquatic organisms will be important environmentally (SRC). [R10].

GENERAL RESPONSE Stop discharge if possible. Call fire department. Avoid contact with liquid. Isolate and remove discharged material. Notify local health and pollution control agencies.

FIRE RESPONSE Combustible. Extinguish with foam, dry chemical, or carbon dioxide. Water may be ineffective on fire. Cool exposed containers with water.

EXPOSURE RESPONSE Call for medical aid. Liquid irritating to skin and eyes. Harmful if swallowed. Remove contaminated clothing and shoes. Flush affected areas with plenty of water. If in eyes, hold eyelids open and flush with plenty of water. If swallowed and victim is conscious, have victim drink water or milk.

RESPONSE TO DISCHARGE Mechanical containment; should be removed; chemical and physical treatment.

WATER RESPONSE Effect of low concentrations on aquatic life is unknown. Fouling to shoreline. May be dangerous if it enters water intakes. Notify local health and wildlife officials. Notify operators of nearby water intakes.

EXPOSURE SYMPTOMS Liquid may cause nervous disturbance, green coloration of urine, and skin and eye irritation.

EXPOSURE TREATMENT Ingestion: induce vomiting; call a doctor; medical treatment should be aimed at conservation of liver and kidney function. Eyes: flush with water for at least 15 min; call a doctor. Skin: wipe off, wash with soap and water.

REFERENCES

1. Budavari, S. (ed.). The Merck Index— Encyclopedia of Chemicals, Drugs and Biologicals. Rahway, NJ: Merck and Co., Inc., 1989. 1453.

2. Clayton, G. D., and F. E. Clayton (eds.). Patty's Industrial Hygiene and Toxicology: Volume 2A, 2B, 2C: Toxicology. 3rd ed. New York: John Wiley and Sons, 1981–1982.

3. Browning, E. Toxicity and Metabolism of Industrial Solvents. New York: American Elsevier, 1965.

4. Gilman, A. G., L. S. Goodman, and A. Gilman. (eds.). Goodman and Gilman's The Pharmacological Basis of Therapeutics. 6th ed. New York: Macmillan Publishing Co., Inc. 1980. 985.

5. Fazzalari, F.A. (ed.). Compilation of Odor and Taste Threshold Values Data. ASTM Data Series DS 48A (Committee E-18). Philadelphia, PA: American Society for Testing and Materials, 1978. 156.

6. U.S. Coast Guard, Department of Transportation. CHRIS—Hazardous Chemical Data. Volume II. Washington, DC: U.S. Government Printing Office, 1984—5.

7. Hayes, Wayland J., Jr. Pesticides Studied in Man. Baltimore/London: Williams and Wilkins, 1982.

8. Verschueren, K. Handbook of Environmental Data of Organic Chemicals. 2nd ed. New York, NY: Van Nostrand Reinhold Co., 1983. 1089.

9. (1) Price, K. S., et al. J Water Pollut Contr Fed. 46: 63–77 (1974). (2) Vanderlinden, A. C. Dev Biodeg Hydrocarbons 1: 165–200 (1978). (3) Kappeler, T., K. Wuhrmann. Water Res 12: 327–33 (1978). (4) Ludzack, F. J., M. B. Ettinger. Eng Bull Ext Ser no. 115, pp. 278–82 (1963).

10. Sabljic, A. Z Gesamte Hyg Ihre Grenzgeb 33: 493–6 (1987).

■ TETRAMETHRIN

CAS # 7696–12–0

SYNONYMS 1-cyclohexene-1,2-dicarboximidomethyl-2,2-dimethyl-3-(2-met hylpropenyl) cyclopropane carboxylate • fmc-9260 • phthalthrin • neopynamin • 2,2-dimethyl-3-(2-methyl-1-propenyl) cyclopropanecarboxylic acid

(1,3,4,5,6,7-hexahydro-1,3-dioxo-2H-isoindol-2-yl) methyl ester • 2,2-dimethyl-3-(2-methylpropenyl) cyclopropanecarboxylic acid ester with N-(hydroxymethyl) -1-cyclohexene-1,2-dicarboximide • N-(3,4,5,6-tetrahydrophthalimide) methyl-cis, trans-chrysanthemate • N-(chrysanthemoxymethyl)-1-cyclohexene-1,2-dicarboximide • SP-1103 • Weo-Pynamin • (1,3,4,5,6,7-hexahydro-1,3-dioxo-2H-isoindol-2-yl) methyl 2,2-dimethyl-3-(2-methyl-1-propenyl) cyclopropanecarboxylate. • 3,4,5,6-tetrahydrophthalimidomethyl(-)-cis, trans-chrysanthemate • cyclohex-1-ene-1,2-dicarboximidomethyl (1rs)-cis, trans-2,2-dimethyl-3-(2-methyl-prop-1-enyl) cyclopropane carboxylate • tetramethrine • py-kill

MF $C_{19}H_{25}NO_4$

MW 331.42

COLOR AND FORM White crystalline solid; colorless crystals.

ODOR Slight pyrethrum-like odor.

DENSITY 1.108 at 20°C.

BOILING POINT 180–190°C at 0.1 mm Hg.

MELTING POINT 65–80°C.

FIREFIGHTING PROCEDURES Use carbon dioxide, foam, or dry chemical (on fires involving pyrethroids) (pyrethrum) [R10, 2]. Fire-fighting: Self-contained breathing apparatus with a full facepiece operated in pressure-demand or other positive-pressure mode (pyrethrum) [R10, 5]. Extinguish fire using agent suitable for type of surrounding fire (pyrethrins) [R3].

SOLUBILITY Water solubility: 4.6 mg/L at 30°C.; Methanol (53 g/kg), hexane (20 g/kg), xylene (1 g/kg), acetone, toluene. In water at 30°C, 4.6 mg/L. In acetone 400, benzene 500, toluene 400, xylene 500, methyl isobutyl ketone 250, cyclohexanone 350, ethyl acetate 350, methylnaphthalene 350, trichloroethane 400, hexane 20, kerosene 30, Freon 11 (trichlorofluoromethane) 230, freon 12 (dichlorodifluoromethane) 13, piperonyl butoxide 200 (all in g/kg at 25°C).

VAPOR PRESSURE 0.944 mPa at 30°C.

CORROSIVITY Noncorrosive.

STABILITY AND SHELF LIFE Stable under normal storage and use [R1].

OTHER CHEMICAL/PHYSICAL PROPERTIES MP: 60–80°C (technical tetramethrin, ca 90%).

PROTECTIVE EQUIPMENT Employees should be provided with and required to use dust– and splash-proof safety goggles where (pyrethroids) may contact the eyes pyrethroids [R10, 3]. Employees should be provided with and be required to use impervious clothing, gloves, and face shields (eight-inch minimum) (pyrethroids) [R10, 2]. Wear appropriate equipment to prevent repeated or prolonged skin contact (pyrethrum and pyrethrins) [R12]. Wear eye protection to prevent reasonable probability of eye contact (pyrethrins) [R12]. Recommendations for respirator selection. Max concentration for use: 50 mg/m³: Respirator Classes: Any chemical cartridge respirator with organic vapor cartridge(s) in combination with a dust, mist, and fume filter. May require eye protection. Any supplied-air respirator. May require eye protection. Any self-contained breathing apparatus. May require eye protection (pyrethrins) [R12]. Recommendations for respirator selection. Max concentration for use: 125 mg/m³: Respirator Classes: Any supplied-air respirator operated in a continuous-flow mode. May require eye protection. Any powered, air-purifying respirator with organic vapor cartridge(s) in combination with a dust, mist, and fume filter. May require eye protection (pyrethrins) [R12]. Recommendations for respirator selection. Max concentration for use: 250 mg/m³: Respirator Classes: Any chemical cartridge respirator with a full facepiece and organic vapor cartridge(s) in combination with a high-efficiency particulate filter. Any self-contained breathing apparatus with a full facepiece. Any supplied-air respirator with a full facepiece. Any powered, air-purifying respirator with a tight-fitting facepiece, and organic vapor cartridge(s) in combination with a high-efficiency particulate filter. May require eye protection (pyrethrins) [R12]. Recommendations for respirator selection. Max concentration for use: 5,000 mg/m³: Respirator Class: Any supplied-air respirator with a full facepiece and operated in a pressure-demand or other positive-pressure mode (pyrethrins) [R12]. Recommendations for respirator selection. Condition: Emergency or planned entry into unknown concentration or IDLH conditions: Respirator Classes: Any self-contained breathing apparatus that has a full facepiece and is operated in a pressure-demand or other positive-pressure mode. Any supplied-air respirator with a full facepiece and operated in pressure-demand or other positive-pressure mode in combination with an auxiliary self-contained breathing apparatus operated in pressure-demand or other positive-pressure mode (pyrethrins) [R12]. Recommendations for respirator selection. Condition: Escape from suddenly occurring respiratory hazards: Respirator Classes: Any air-purifying, full facepiece respirator (gas mask) with a chin-style, front– or back-mounted organic vapor canister having a high-efficiency particulate filter. Any appropriate escape-type, self-contained breathing apparatus (pyrethrins) [R12].

PREVENTATIVE MEASURES Skin that becomes contaminated with pyrethrum should be promptly washed or showered with soap or mild detergent and water (pyrethrum) [R10, 3]. Clothing contaminated with pyrethrum should be placed in closed containers for storage until provision is made for the removal of pyrethrum from the clothing (pyrethrum) [R10, 2]. Respirators may be used when engineering and work practice controls are not technically feasible, when such controls are in the process of being installed, or when they fail or need to be supplemented. Respirators may also be used for operations that require entry into tanks or closed vessels, and in emergency situations (pyrethrum) [R11, 2]. Employees who handle pyrethrum should wash their hands thoroughly with soap or mild detergent and water before eating, smoking, or using toilet facilities (pyrethrum) [R11, 3]. Avoid contact with skin. Keep out of any

body of water. Do not contaminate water by cleaning of equipment or disposal of waste. Do not reuse empty container. Destroy it by perforating or crushing (pyrethrum) [R13]. SRP: The scientific literature supports the wearing of contact lenses in industrial environments, as part of a program to protect the eye against chemical compounds and minerals causing eye irritation. However, there may be individual substances whose irritating or corrosive properties are such that the wearing of contact lenses would be harmful to the eye. In those specific cases contact lenses should not be worn. Contact lenses should not be worn when working with this chemical (pyrethrins) [R12]. Workers should wash promptly when skin becomes contaminated (pyrethrins) [R12]. Work clothing should be changed daily if it is reasonably probable that the clothing may be contaminated (pyrethrins) [R12]. Remove clothing promptly if it is non-impervious clothing that becomes contaminated (pyrethrins) [R12].

CLEANUP Spillages of pesticides at any stage of their storage or handling should be treated with great care. Liquid formulations may be reduced to solid phase by evaporation. Dry sweeping of solids is always hazardous: these should be removed by vacuum cleaning or by dissolving them in water or other solvent in the factory environment (pesticides) [R4]. Environmental consideration—Land spill: Dig a pit, pond, lagoon, or holding area to contain liquid or solid material. SRP: If time permits, pits, ponds, lagoons, soak holes, or holding areas should be sealed with an impermeable flexible-membrane liner. Dike surface flow using soil, sandbags, foamed polyurethane, or foamed concrete. Absorb bulk liquid with fly ash or cement powder (pyrethrins) [R3]. Environmental consideration—Water spill: If pyrethrins are dissolved, apply activated carbon at ten times the spilled amount in the region of 10 ppm, or greater concentration. Use mechanical dredges or lifts to remove immobilized masses of pollutants and precipitates (pyrethrins) [R3].

DISPOSAL Treatment and disposal methods: Open burning. Burn tetramethrin in a shallow depression well away from any buildings, animals, or susceptible vegetation. Recommendable methods: Incineration, discharge to sewer, landfill. Peer review: Small amounts: Only well-diluted discharge to sewer. Large amounts: Incinerate in a unit with effluent gas scrubbing (peer review conclusions of an IRPTC expert consultation, May 1985) [R5]. Incineration would be an effective disposal procedure where permitted. If an efficient incinerator is not available, the product should be mixed with large amounts of combustible material, and contact with the smoke should be avoided (pyrethrin products) [R14]. Group I Containers: Combustible containers from organic or metallo-organic pesticides (except organic mercury, lead, cadmium, or arsenic compounds) should be disposed of in pesticide incinerators or in specified landfill sites (organic or metallo-organic pesticides) [R15]. Group II Containers: Non-combustible containers from organic or metallo-organic pesticides (except organic mercury, lead, cadmium, or arsenic compounds) must first be triple-rinsed. Containers that are in good condition may be returned to the manufacturer or formulator of the pesticide product, or to a drum reconditioner for reuse with the same type of pesticide product, if such reuse is legal under Department of Transportation regulations (e.g., 49 CFR 173.28). Containers that are not to be reused should be punctured and transported to a scrap metal facility for recycling, disposal, or burial in a designated landfill (organic or metallo-organic pesticides) [R16].

COMMON USES A semi-synthetic pyrethroid that is commonly supplied in combination with piperonyl butoxide or resmethrin in oil- or water-based sprays designed to control household insects [R11, p. II–262]. Insecticide [R11, 1454]. Normally used in combination with synergists (e.g., piperonyl butoxide), and other insecticides for control of flies, cockroaches, mosquitoes, wasps, and other insect pests in public health and home and garden use [R2].

OCCUPATIONAL RECOMMENDATIONS None listed.

NON-HUMAN TOXICITY LD_{50} mouse oral >20,000 mg/kg [R1]; LD_{50} mouse dermal >15,000 mg/kg [R1]; LD_{50} rat oral >20 g/kg [R11, p. II–263]; LD_{50} mouse (male) sc 2,020 mg/kg (1R, cis/trans–) [R6]; LD_{50} mouse (female) sc 1,950 mg/kg ((1R, cis/trans–) [R6]; LD_{50} mouse (male) ip 631 mg/kg ((1R, cis/trans–) [R6]; LD_{50} mouse (female) ip 527 mg/kg ((1R, cis/trans–) [R6]; LD_{50} rat, oral 4,600 mg/kg (racemic) [R6]; LD_{50} rat dermal >10,000 mg/kg (racemic) [R6]; LD_{50} mouse (male) oral (animals not fasted, corn-oil vehicle) 1,920 mg/kg (racemic) [R6]; LD_{50} mouse (female), oral (animals not fasted, corn oil vehicle) 2,000 mg/kg (racemic) [R6]; LD_{50} rat sc >5,000 mg/kg ((1R, cis/trans–) [R6]; LD_{50} rat (male) ip 770 mg/kg ((1R, cis/trans–) [R6]; LD_{50} rat (female) ip 548 mg/kg ((1R, cis/trans–) [R6]; LD_{50} mouse (male), oral (animals not fasted) 1,060 mg/kg ((1R, cis/trans–) [R6]; LD_{50} mouse (female), oral (animals not fasted) 1,040 mg/kg ((1R, cis/trans–) [R6]; LD_{50} rat, oral >20 m/kg [R11, p. II–263]; LD_{50} mouse, oral 5,200 mg/kg [R2]; LD_{50} albino rat male oral >4,640 mg/kg [R2]; LD_{50} rat percutaneous >5,000 mg/kg [R2]; LC_{50} rat inhalation >2.5 mg/L air/3 hr [R2].

ECOTOXICITY LD_{50} bobwhite quail oral >1,000 mg/kg [R2]; LD_{50} mallard duck oral >1,000 mg/kg [R2]; LC_{50} bluegill sunfish 0.069 mg/L/96 hr (conditions of bioassay not specified) [R2]; LC_{50} *Oryzias latipes* (killifish) adult (static system at 25°C) 0.2 mg/L/48 hr (technical) [R7]; LC_{50} *Oryzias latipes* (killifish) adult 0.2 mg/L/48 hr at 25°C static bioassay ((+)–trans) [R7]. LC_{50} *Oryzias latipes* (killifish) adult 0.5 mg/L/48 hr at 25°C static bioassay ((+)–cis) [R7]. LC_{50} *Lepomis macrochirur* (bluegill sunfish) 0.019 mg/L/96 hr (conditions of bioassay not specified) [R7]. LC_{50} *Salmo gairdneri* (Rainbout trout) 0.021 mg/L/96 hr (conditions of bioassay not specified) [R7]. LC_{50} *Dapnia pulex* (static system at 25°C) >50 mg/L/3 hr (technical) [R7]. LC_{50} *Dapnia pulex* (static system at 25°C) >50 mg/L/3 hr ((+)–trans) [R7]. LC_{50} *Dapnia pulex* (static system at 25°C) >50 mg/L/3 hr ((+) –cis) [R7].

DOT *Health Hazards:* Poisonous; may be fatal if inhaled, swallowed, or absorbed through skin. Contact may cause burns to skin and eyes. Runoff from fire control or dilution water may give off poisonous gases and cause water pollution. Fire may produce irritating or poisonous gases (pesticide, liquid, or solid, poisonous, not otherwise specified) [R11, p. G–55].

Fire or Explosion: Some of these materials may burn, but none of them ignites readily. Container may explode violently in heat of fire (pesticide, liquid, or solid, poisonous, not otherwise specified) [R11, p. G–55].

Emergency Action: Keep unnecessary people away; isolate hazard area, and deny entry. Stay upwind, out of low areas, and ventilate closed spaces before entering. Positive-pressure self-contained breathing apparatus (SCBA) and chemical protective clothing that is specifically recommended by the shipper or manufacturer may be worn. It may provide little or no thermal protection. Structural firefighters' protective clothing is not effective for these materials. Remove and isolate contaminated clothing at the site. Call CHEMTREC at 1–800–424–9300 as soon as possible, especially if there is no local hazardous materials team available (pesticide, liquid, or solid, poisonous, not otherwise specified) [R11, p. G–55].

Fire: Small fires: Dry chemical, CO_2, water spray, or regular foam. Large Fires: Water spray, fog, or regular foam. Move container from fire area if you can do so without risk. Fight fire from maximum distance. Stay away from ends of tanks. Dike fire-control water for later disposal; do not scatter the material (pesticide, liquid, or solid, poisonous, not otherwise specified) [R11, p. G–55].

Spill or Leak: Do not touch or walk through spilled material; stop leak if you can do so without risk. Fully encapsulating, vapor-protective clothing should be worn for spills and leaks with no fire. Use water spray to reduce vapors. Small Spills: Take up with sand or other noncombustible absorbent material and place into containers for later disposal. Small Dry

Spills: With clean shovel place material into clean, dry container, and cover loosely; move containers from spill area. Large Spills: Dike far ahead of liquid spill for later disposal (pesticide, liquid, or solid, poisonous, not otherwise specified) [R11, p. G–55].

First Aid: Move victim to fresh air, and call for emergency medical care; if not breathing, give artificial respiration; if breathing is difficult, give oxygen. In case of contact with material, immediately flush skin or eyes with running water for at least 15 minutes. Speed in removing material from skin is of extreme importance. Remove and isolate contaminated clothing and shoes at the site. Keep victim quiet and maintain normal body temperature. Effects may be delayed; keep victim under observation (pesticide, liquid, or solid, poisonous, not otherwise specified) [R11, p. G–55].

FIFRA REQUIREMENTS In 1988, Congress amended FIFRA to strengthen and accelerate EPA's reregistration program. The nine-year reregistration scheme mandated by "FIFRA 88" applies to each registered pesticide product containing an active ingredient initially registered before November 1, 1984. Pesticides for which EPA had not issued Registration Standards prior to the effective date of FIFRA '88 were divided into three lists based upon their potential for exposure and other factors, with List B being of highest concern, and D of least. List: B; Case: Tetramethrin; Case No. 2660; Pesticide type: Insecticide; Active Ingredient (AI): (1–Cyclohexene–1,2–dicarboximido)methyl 2,2–dimethyl–3–(2–methylpropenyl) cyclopropanecarboxylate; Data Call-in (DCI) Date: 6/10/91; Case Status: Awaiting Data/Data in Review: OPP awaits data from the pesticide's producer(s) regarding its human health or environmental effects, or OPP has received and is reviewing such data, in order to reach a decision about the pesticide's eligibility for reregistration. [R6].

FIRE POTENTIAL Pyrethrins burn with difficulty. (pyrethrins) [R3].

REFERENCES

1. Spencer, E. Y. *Guide to the Chemicals Used in Crop Protection.* 7th ed. Publication 1093. Research Institute, Agriculture Canada, Ottawa, Canada: Information Canada, 1982. 549.

2. Hartley, D., and H. Kidd (eds.). *The Agrochemicals Handbook.* 2nd ed. Lechworth, Herts, England: The Royal Society of Chemistry, 1987, p. A388/Aug 87.

3. Bureau of Explosives. *Emergency Handling of Haz Matl in Surface Trans* p. 434 (1981).

4. International Labour Office. *Encyclopedia of Occupational Health and Safety.* Vols. I and II. Geneva, Switzerland: International Labour Office, 1983. 1619.

5. United Nations. *Treatment and Disposal Methods for Waste Chemicals* (IRPTC File). Data Profile Series No. 5. Geneva, Switzerland: United Nations Environmental Programme, Dec. 1985. 158.

6. WHO. *Environmental Health Criteria* 98: Tetramethrin p. 34 (1990).

7. WHO. *Environmental Health Criteria* 98: Tetramethrin p. 33 (1990).

8. Budavari, S. (ed.). *The Merck Index—Encyclopedia of Chemicals, Drugs, and Biologicals.* Rahway, NJ: Merck and Co., Inc., 1989.

9. Gosselin, R. E., R. P. Smith, H. C. Hodge. *Clinical Toxicology of Commercial Products.* 5th ed. Baltimore: Williams and Wilkins, 1984.

10. Mackison, F. W., R. S. Stricoff, and L. J. Partridge, Jr. (eds.). *NIOSH/OSHA—Occupational Health Guidelines for Chemical Hazards.* DHHS (NIOSH) Publication No. 81–123 (3 vols). Washington, DC: U.S. Government Printing Office, Jan. 1981.

11. U.S. Department of Transportation. *Emergency Response Guidebook* 1990. DOT P 5800.5. Washington, DC: U.S. Govvernment Printing Office, 1990.

12. NIOSH. *NIOSH Pocket Guide to Chemical Hazards.* DHHS (NIOSH) Publication No. 90–117. Washington, DC: U.S. Government Printing Office, June 1990, 190.

13. *Farm Chemicals Handbook 1986.* Willoughby, OH: Meister Publishing Co., 1986, p. C–198.

14. Sittig, M. *Handbook of Toxic and Hazardous Chemicals and Carcinogens.* 1985. 2nd ed. Park Ridge, NJ: Noyes Data Corporation, 1985. 762.

15. 40 CFR 165.9 (a) (7/1/90).

16. 40 CFR 165.9 (b) (7/1/90).

17. USEPA/OPP. *Status of Pesticides in Reregistration and Special Review.* p. 162 (Mar, 1992) EPA 700–R–92–004.

■ THIOPHENOL

CAS # 108–98–5

SYNONYMS benzene, mercapto • benzenethiol • mercaptobenzene • phenol, thio • phenyl mercaptan • phenylmercaptan • phenylthiol

MF C_6H_6S

MW 110.18

pH Feebly acidic.

COLOR AND FORM Water-white liquid; prism-like crystals from petroleum ether.

ODOR Repulsive, penetrating, garlic-like.

ODOR THRESHOLD Water: 0.00028 mg/L; air: 0.00094 µl/L; odor safety class B; B = 50–90% of distracted persons perceive warning of TLV [R6].

DENSITY Sp GR: 1.0728 at 25°C.

DISSOCIATION CONSTANT pK_a = 6.62.

BOILING POINT 168.3°C at 760 mm Hg.

MELTING POINT −14.8°C.

EXPLOSIVE LIMITS AND POTENTIAL At normal room temperature the lower-molecular-wt thiols may vaporize to form explosive mixtures with air. (thiols) [R5].

SOLUBILITY Very soluble in alcohol; miscible with ether, benzene, carbon disulfide; water solubility of 836 mg/L at 25°C.

VAPOR PRESSURE 2 mm Hg at 25°C.

OCTANOL/WATER PARTITION COEFFICIENT Log K_{ow} = 2.52.

STABILITY AND SHELF LIFE Oxidizes in air, especially when dissolved in alcoholic ammonia, forming diphenyl disulfide [R1].

OTHER CHEMICAL/PHYSICAL PROPERTIES Entropy: 52.6 at 25°C; hydrogen of SH group is easily replaced by metals; conversion factor: 4.74 mg/m³ = 1 ppm; heat of fusion: 24.90 calories/g; specific heat: 0.3829 at 25°C; Henry's Law Constant for thiophenol has been experimentaly determined to be 3.5×10^{-4} atm–m³/mole.

PROTECTIVE EQUIPMENT Routine work practices should emphasize proper protective equipment such as respirators and eye goggles. (thiols) [R5].

PREVENTATIVE MEASURES Use local exhaust ventilation. Safety showers, eyewash fountains, and fire extinguishers should be located in areas where appreciable amount of thiols are used. Handwashing facilities, soap, and water, should be made available to involved employees. (thiols) [R5].

CLEANUP Spills of thiols can be neutralized with a household bleach solution and flushed with an abundant flow of water. (thiols) [R5].

DISPOSAL Generators of waste (equal to or greater than 100 kg/mo) containing this contaminant, EPA hazardous waste number P014, must conform with USEPA regulations in storage, transportation, treatment, and disposal of waste [R7].

COMMON USES Chemical intermediate for carbophenothion insecticide and acaricide, fungicides, e.g., fonofos; chemical intermediate for its sodium salt, alkyl phenyl sulfides, pharmaceuticals, pentachlorothiophenol, and its zinc salt and intermediate for polymers (e.g., with formaldehyde, other phenols) [R2]. Mosquito larvicide [R3].

OCCUPATIONAL RECOMMENDATIONS

OSHA 8–hr time-weighted average: 0.5 ppm (2 mg/m³). Final rule limits were achieved by any combination of engineering controls, work practices, and personal protective equipment during the phase-in period, Sept 1, 1989 through Dec 30, 1992. Final rule limits became effective Dec 31, 1992 [R10].

THRESHOLD LIMIT 8–hr time-weighted average (TWA) 0.5 ppm, 2.3 mg/m³ (1978) [R15, 29].

NON-HUMAN TOXICITY LD_{50} rat, oral 46.2 mg/kg [R3]; LD_{50} rat ip 25.2 mg/kg [R3]; LD_{50} rat dermal 300 mg/kg [R3]; LD_{50} rabbit dermal 134 mg/kg (est) [R3].

BIODEGRADATION Warburg respirometer, activated sludge inocula, 6 days 30–42% theoretical BOD (1). Warburg respirometer, phenol, or resorcinol-adapted *Trichosporon cutaneum* yeast cells isolated from soil, oxidation of thiophenol occurred after a 10–30–min lag (2). Thiophenol was oxidized by isolated cells of the bacterium *Thiobacillus thiooxidans* (3) [R8].

BIOCONCENTRATION Bioconcentration factors (BCF) of 14–48 were estimated for thiophenol using linear regression equations based on a measured log K_{ow} of 2.52 and a measured water solubility of 836 mg/L at 25°C (1,2,3,SRC). These BCF values suggest that this compound will not bioaccumulate significantly in aquatic organisms (SRC) [R9].

CERCLA REPORTABLE QUANTITIES Persons in charge of vessels or facilities are required to notify the National Response Center (NRC) immediately, when there is a release of this designated hazardous substance, in an amount equal to or greater than its reportable quantity of 100 lb or 45.4 kg. The toll-free number of the NRC is (800) 424–8802; in the Washington DC metropolitan area (202) 426–2675. The rule for determining when notification is required is stated in 40 CFR 302.4 (section IV. D. 3.b) [R11].

TSCA REQUIREMENTS Pursuant to section 8 (d) of TSCA, EPA promulgated a model health and safety data reporting rule. The section 8 (d) model rule requires manufacturers, importers, and processors of listed chemical substances and mixtures to submit to EPA copies and lists of unpublished health and safety studies. Benzenethiol is included on this list [R12].

RCRA REQUIREMENTS P014; as stipulated in 40 CFR 261.33, when thiophenol, as a commercial chemical product or manufacturing chemical intermediate or an off-specification commercial chemical product or a manufacturing chemical intermediate, becomes a waste, it must be managed according to federal or state hazardous waste regulations. Also defined as a hazardous waste is any container or inner liner used to hold this waste or any residue, contaminated soil, water, or other debris resulting from the cleanup of a spill, into water or on dry land, of this waste. Generators of small quantities of this waste may qualify for partial exclusion from hazardous waste regulations (40 CFR 261.5 (e)) [R13].

DOT *Health Hazards:* Poisonous; may be fatal if inhaled, swallowed, or absorbed through skin. Contact may cause burns to skin and eyes. Runoff from fire control, or dilution water may cause pollution [R4].

Fire or Explosion: May be ignited by heat, sparks, or flames. Container may explode in heat of fire. Vapor explosion and poison hazard indoors, outdoors, or in sewers [R4].

Emergency Action: Keep unnecessary people away; isolate hazard area and deny entry. Stay upwind, out of low areas, and ventilate closed spaces before entering. Self-contained breathing apparatus and chemical protective clothing that is specifically recommended by the shipper or producer may be worn but they do not provide thermal protection unless that is stated by the clothing manufacturer. Structural firefighter's protective clothing is not effective with these materials. Call CHEMTREC at 1–800–424–9300 as soon as possible, especially if there is no local hazardous materials team available [R4].

Fire: Small fires: Dry chemical, CO₂, water spray, or regular foam. Large Fires: Water spray, fog, or regular foam. Move container from fire area if you can do so without risk. Cool container with water using unmanned device until well after fire is out. Fight fire from maximum distance. Stay away from ends of tanks. Dike fire-control water for later disposal; do not scatter the material [R4].

Spill or Leak: Shut off ignition sources; no flares, smoking, or flames in hazard area. Fully encapsulating, vapor-protective clothing should be worn for spills and leaks with no fire. Do not touch or walk through spilled material; stop leak if you can do so without risk. Water spray may reduce vapor; but it may not prevent ignition in closed spaces. Small Spills: Take up with sand or other noncombustible absorbent material and place into containers for later disposal. Small Dry Spills: With clean shovel place material into clean, dry container, and cover loosely; move containers from spill area. Large Spills: Dike far ahead of liquid spill for later disposal [R4].

First Aid: Move victim to fresh air and call for emergency medical care; if not breathing, give artificial respiration; if breathing is difficult, give oxygen. In case of contact with material, immediately flush skin or eyes with running water for at least 15 minutes. Speed in removing material from skin is of extreme importance. Remove and isolate contaminated clothing and shoes at the site. Keep victim quiet and maintain normal body temperature. Effects may be delayed; keep victim under observation [R4].

GENERAL RESPONSE DATA Avoid contact with liquid and vapor. Keep people away. Wear positive-pressure breathing apparatus and special chemical protective clothing. Shut off ignition sources and call fire department. Stop discharge if possible. Stay upwind and use water spray to knock down vapor. Isolate and remove discharged material. Notify local health and pollution control agencies.

FIRE RESPONSE Combustible poisonous gases are produced in fire or when heated. Container may explode in heat of fire. Vapor explosion and poison hazard indoors, outdoors or in sewers. Wear positive-pressure breathing apparatus and special chemical protective clothing. Extinguish small fire with: dry chemical, carbon dioxide, water spray or foam; Large fires: water spray, fog, or foam. Cool exposed containers with water. Combat fires from safe distance or protected location (behind barriers) with unmanned monitor nozzle.

EXPOSURE RESPONSE Call for medical aid. Vapor poisonous. May be fatal if inhaled or absorbed through skin. Irritating to eyes, skin, and mucous membranes. Over exposure may cause headache, dizziness, coughing, difficulty in breathing, nausea, and vomiting. Symptoms may be delayed. Move to fresh air. If breathing has stopped, give artificial respiration. If breathing is difficult, give oxygen. Liquid poisonous. May be fatal if swallowed or absorbed through skin. May burn skin and eyes. Speed in removing material from skin is extremely important. If in eyes or on skin, immediately flush with running water for at least 15 minutes; hold eyelids open occasionally if appropriate. Remove and isolate contaminated clothing and shoes at the site. Effects may be delayed; keep victim under observation. If swallowed and victim is conscious, have victim drink several glasses of water and induce vomiting by touching back of throat. If swallowed and victim is unconscious or having convulsions, do nothing except keep victim warm.

RESPONSE TO DISCHARGE Issue warning—poison. Restrict access. Should be removed. Chemical and physical treatment.

WATER RESPONSE Effects of low concentration on aquatic life is unknown. May be dangerous if it enters water intakes. Notify local health and wildlife officials. Notify operators of nearby water intakes.

EXPOSURE SYMPTOMS Poisonous; may be fatal if inhaled, swallowed, or absorbed through skin. Causes eye burns and skin irritation. It is capable of producing severe irritation of the eyes, skin, respiratory and digestive tract. Over exposure may cause headache, dizziness, coughing, difficulty in breathing, nausea, and vomiting. Inhalation of high doses may cause lung damage; prolonged or repeated contact with the skin may cause dermatitis. Symptoms may not appear until several hours after exposure and they are made more severe by physical effort.

EXPOSURE TREATMENT Inhalation: Move victim to fresh air; call emergency medical care. If not breathing, give artifical respiration. If breathing is difficult, give oxygen. Eyes or skin: Immediately flush eyes or skin with running water for at least 15 minutes. Hold upper and lower eyelids open occasionally if appropiate. Speed in removing material from skin is extremely important. Remove and isolate contaminated clothing and shoes at the site. Keep victim quiet and maintain normal body temperature. Effects may be delayed; keep victim under observation. Ingestion: if conscious, have victim drink several glasses of water and induce vomiting by touching the back of the throat with a finger; repeat until vomitus is clear. If unconscious or having convulsions, do nothing except keep victim warm.

FDA Thiophenol is a food additive permitted for direct addition to food for human consumption, as long as (1) the quantity added to food does not exceed the amount reasonably required to accomplish its intended physical, nutritive, or other technical effect in food, and (2) when intended for use in or on food it is of appropriate food grade and is prepared and handled as a food ingredient. Synthetic flavoring substances and adjuvants may consist of thiophenol, used alone, or in combination with flavoring substances and adjuvants generally recognized as safe [R14].

REFERENCES

1. Budavari, S. (ed.). *The Merck Index— Encyclopedia of Chemicals, Drugs, and Biologicals*. Rahway, NJ: Merck and Co., Inc., 1989. 1473.

2. SRI.

3. Clayton, G.D., and F.E. Clayton (eds.). *Patty's Industrial Hygiene and Toxicology*: Volume 2A, 2B, 2C: *Toxicology*. 3rd ed. New York: John Wiley and Sons, 1981–1982. 2080.

4. U.S. Department of Transportation. *Emergency Response Guidebook 1990*. DOT P 5800.5. Washington, DC: U.S.

Government Printing Office, 1990, p. G–57.

5. International Labour Office. *Encyclopedia of Occupational Health and Safety*. Vols. I and II. Geneva, Switzerland: International Labour Office, 1983. 2172.

6. Amoore, J. E., E. Hautala. *J Appl Toxicol* 3 (6): 272–90 (1983).

7. 40 CFR 240–280, 300–306, 702–799 (7/1/89).

8. (1) Lutin, P. A., et al., *Purdue Univ Eng Bull Ext Ser* 118: 131–45 (1965). (2) Neujahr, H. Y., J. M. Varga. *Eur J Biochemical* 13: 37–44 (1970). (3) Tano, T., K. Imai. *Hakko Kyokaishi* 26: 322–7 (1968).

9. (1) Hansch, C., and A. J. Leo. *Medchem Project* Issue no. 26 Claremont, CA: Pomona College (1985). (2) Hine J., P. K. Mookerjee. *Organic Chemical* 40: 292–8 (1975) (3) Lyman, W. J., et al., *Handbook of Chemical Property Estimation Methods*. New York: McGraw-Hill, p. 5–5 (1982).

10. 54 FR 2920 (1/19/89).

11. 54 FR 33419 (8/14/89).

12. 40 CFR 716.120 (7/1/88).

13. 40 CFR 261.33 (7/1/88).

14. 21 CFR 172.515 (4/1/90).

15. American Conference of Governmental Industrial Hygienists. *Threshold Limit Values for Chemical Substances and Physical Agents and Biological Exposure Indices for 1994–1995*. Cincinnati, OH: ACGIH, 1994.

■ TITANIUM DIOXIDE

CAS # 13463–67–7

SIC CODES 2816; 2819; 2851.

SYNONYMS NCI-Co4240 • titandioxid (Sweden) • titanic anhydride • titanic oxide • titanium oxide • titanium oxide (TiO_2) • titanium peroxide (TiO_2) • 1700-white • a-fil cream • anatase • austiox r-cr 3 • bayertitan • baytitan • brookite • C-weiss 7 (German) • calcotone-white-t • ci-77891 • ci-pigment-white-6 • cosmetic white c47-5175 • hombitan • kronos-cl-220 • kronos titanium dioxide • levanox-white-rkb • r-680 • rayox • ro-2 • runa-rh20 • rutile • rutiox-cr • Ti-pure r 900 • tiofine • tiona-td • tioxide • tioxide ad-m • tioxide r-cr • tioxide-rhd • tioxide-rsm • tipaque • titafrance • titanox-2010 • titanox-ranc • unitane • unitane 0-110 • unitane 0-220 • unitane or-150 • unitane or-340 • unitane or-342 • unitane or-350 • unitane or-540 • unitane or-640 • titania • titanox • unitane • uniwhite-or-450 • uniwhite-or-650 • uniwhite-ao • uniwhite-ko • zopaque • trioxide (s) • tronox • tipaque-r-820 • tioxide-r-xl • ti-pure • runa-arh-200 • p-25- (oxide) • orgasol-1002d-white-10-extra-cos • p-25 • kronos • kronos-rn-40p • kronos-rn-56 • kh-360 • horse head a-420 • cosmetic-hydrophobic-tio2-9428 • cosmetic-micro-blend-tio2-9228 • blend-white-9202 • atlas white titanium dioxide • bayeritian • bayertitan-a • bayertitan r-u-f

MF O_2Ti

MW 79.90

pH Suspension in water (1 in 10) is neutral to litmus.

COLOR AND FORM Colorless, tetragonal crystals; amorphous, infusible powder; white powder; rutile: tetragonal crystals.

ODOR Odorless.

TASTE Tasteless.

DENSITY 4.26.

BOILING POINT 2,500–3,000°C.

MELTING POINT 1,830–1,850°C.

FIREFIGHTING PROCEDURES Self-contained breathing apparatus with a full facepiece operated in pressure-demand or other positive-pressure mode. Only NIOSH–approved or MSHA–approved equipment should be used. [R22, 4].

SOLUBILITY Soluble in hot concentrated sulfuric acid; in hydrofluoric acid; insoluble in hydrochloric acid, nitric acid, or dilute sulfuric acid; insoluble in organic solvents; soluble in alkali; insoluble in water.

OTHER CHEMICAL/PHYSICAL PROPERTIES
Gravimetric factor: 0.33279. Anatase: index of refraction: 2.554; 2.493; specific gravity: 3.84. Brookite: index of refraction: 2.583; 2.586; 2.741; specific gravity: 4.17. Rutile: specific gravity: 4.26; index of refraction: 2.616; 2.903. Reactivity: Depends on previous heat treatment; prolonged heating produces less-soluble material; made soluble by fusion with potassium bisulfate or alkali hydroxides or carbonates to form alkali titanates; index of refraction: 2.7 (titania crystals). When substantially pure, a massive single crystal of rutile has properties of precious gem with a very light straw color and with reflectance, refraction, and brilliance measuring greater than those of a diamond. High opacity and tinting values are claimed for rutile-like pigments. Titania is large crystals (translucent water-white or with yellowish cast); these crystals have refractive index higher than that of diamonds but lack hardness of diamonds. Possesses perhaps the greatest hiding power of all inorg white pigments. Rutile: tetragonal crystals; reddish brown to red; sometimes yellowish; bluish. Anatase: tetragonal crystals; brown; yellowish brown; reddish brown; blue; black; green; gray. Brookite: brown; yellowish brown; reddish brown; black. Anatase: brown-black and tetragonal crystals. Brookite: white rhombic crystals. Tetravalent titanium compounds are the most stable (of the variable valence compounds of titanium). (titanium compounds).

PROTECTIVE EQUIPMENT Dust or mist concentration: 75 mg/m³ or less: Any dust and mist respirator. 150 mg/m³ or less: any dust and mist respirator, except single-use or quarter-mask respirator. 750 mg/m³ or less: any fume respirator or high-efficiency particulate filter respirator; or any supplied-air respirator; or any self-contained breathing apparatus with a full facepiece. 7,500 mg/m³ or less: A powered air-purifying respirator with a high-efficiency particulate filter; or a Type C supplied-air respirator operated in pressure-demand or other positive-pressure or

continuous-flow mode. Greater than 7,500 mg/m³ or entry and escape from unknown concentrations: Self-contained breathing apparatus with a full facepiece operated in pressure-demand or other positive-pressure mode; or a combination respirator that includes a Type C supplied-air respirator with a full facepiece operated in pressure-demand or other positive-pressure or continuous-flow mode, and an auxiliary self-contained breathing apparatus operated in pressure-demand or other positive-pressure mode. [R22, 4].

PREVENTATIVE MEASURES If allergic reactions to topical titanium dioxide occur, its use should be discontinued [R2].

CLEANUP 1. Ventilate area of spill. 2. Collect spilled material in the most convenient and safe manner for reclamation or for disposal in a secured sanitary landfill. Liquid containing titanium dioxide should be absorbed in vermiculite, dry sand, earth, or a similar material [R9].

DISPOSAL Titanium dioxide may be disposed of in a secured sanitary landfill [R9]. Attempt to recover product for salvage or sale to industrial waste exchange. If disposal is warranted, package in marked containers, and confirm disposal methods with local authorities.

COMMON USES Component of porcelain enamels and glazes—e.g., as opacifier [R4]; pigment in floor coverings and printing inks [R4]; pigment in coated fabrics—e.g., upholstery [R4]; delustering agent for synthetic fiber—e.g., nylon and rayon [R4]; pigment and semiconductor in electronics mfr—e.g., capacitors [R4]; other pigment uses—e.g., in adhesives and cosmetics [R4]; component of coatings for steel arc-welding electrodes [R4]; shrinking agent for glass fibers [R4]; chemical intermediate for titanium metal (unisolated ore component) [R4]; chemical intermediate for titanium alloys (unisolated ore component) [R4]; principal uses are in sub-coating of confectionery panned goods and in icings. Opacifying power has been utilized in panning or tablet coating. Titanium dioxide is also used alone in sugar syrup for use in sub-coating of tabletted products [R5]. Titanium dioxide is used as pigment in ceramics, roofing, and plastics, and is also used as nutritional marker [R18, 196]. In ointments or lotions it reflects a very high proportion of incident sunlight and hence protects the skin from sunburn. It is also used in cosmetics and as a dusting powder [R19, 732]. Radioactive decontamination of skin. Single crystals are high-temperature transducers [R6]. Airfloated ilmenite is used for titanium pigment manufacture. Rutile sand is suitable for welding-rod-coating materials, as ceramic colorant, as source of titanium metal. As color in the food industry. Anatase titanium dioxide is used for welding-rod coatings, acid-resistant vitreous enamels, in specification paints, exterior white house paints, acetate rayon, white interior air-dry and baked enamels and lacquers, inks, and plastics, for paper filling and coating, in water paints, tanners, leather finishes, shoe whiteners, and ceramics. Topical protectant [R1]. White pigment in paints, paper, rubber, plastics; opacifying agent, cosmetics, floor coverings, glassware, enamel fruits, delustering synthetic fibers, printing inks, welding rods [R3]. Serves as a clouding agent for incorporation in dry beverage mixes, and in tobacco wrapping and tobacco substitutes [R7]. chemical intermediate for titanium compounds—e.g., barium titanate [R4]. Used as an implant material in orthopedics, oral surgery, and neurosurgery. (titanium and its compounds) [R20, 602].

BINARY REACTANTS Violent reaction with lithium [R3].

INHALATION LIMIT 15 mg/m^3.

DIRECT CONTACT In cosmetics, pure TiO_2 paste has not shown any adverse effects. No specific effects of eye contact with pure TiO_2, but eye protection is recommended at air levels greater than the TLV.

PERSONAL SAFETY When airborne levels exceed the TLV, a particulate respirator and local mechanical ventilation may be desirable and eye protection should be worn. If TiO_2 contacts the eyes, they should be flushed with water.

ACUTE HAZARD LEVEL Inhalation prob-

lems expected only as a nuisance dust. Intratracheal injection of 20–50 mg TiO_2 in rats and less than 400 mg in rabbits showed only nonspecific reactions to dust particles. The hazard expected to aquatic lifeforms is simply smothering.

CHRONIC HAZARD LEVEL Slight lung fibrosis. In humans, occupational exposure has shown titanium dioxide to be nontoxic and chemically nonirritating. It is classified as a nuisance dust. Rats that inhaled TiO_2 dust 4 times daily, 5 d/week for up to 13 mo did not show any pathological responses 7 mo after exposure ended. Guinea pigs, however, showed fibrotic effects and eosinophilic infiltrations after repeated inhalations for from 5 d to 4 mo.

OCCUPATIONAL RECOMMENDATIONS

OSHA 8–hr time-weighted average: 15 mg/m^3 (titanium dioxide, total dust) [R11]; threshold-limit 8–hr time-weighted average (TWA) 10 mg/m^3 (1986) [R21, 34].

ON-SITE RESTORATION Physical characteristics dictate that a dust respirator be worn when working with this compound. treatment alternatives for contaminated waters include clarification and settling of contaminated water for ultimate recovery of spilled product. Treatment for contaminated soils include physical collection and packaging in marked containers. Seek professional environmental engineering assistance through EPA's Environmental Response Team (ERT), Edison, NJ, 24–hour no. 908–548–8730.

PROBABLE LOCATION AND STATE White to black powder or crystalline form. Insoluble, will sink. Natural forms usually red or brown-red by transmitted light due to content of metal impurities such as iron.

FOOD-CHAIN CONCENTRATION POTENTIAL Considerable information on titanium in human foods is related to the use of TiO_2 as a food additive. Plants contain approximately 1 mg/kg. Mammalian absorption after oral ingestion is low. For example, rats given TiO_2 at the rate of 250 ppm in the diet excreted more than 90% of the daily intake in the feces.

IARC SUMMARY Evaluation: There is inadequate evidence for the carcinogenicity of titanium dioxide in humans. There is limited evidence for the carcinogenicity of titanium dioxide in experimental animals. Overall evaluation: Titanium dioxide is not classifiable as to its carcinogenicity to humans (Group (3) [R10].

BIOLOGICAL HALF-LIFE The studies available on titanium do not provide adequate data for an estimate of the biological half-time in man or animals. A half-life of 320 days has been calculated. However, a half-life in mice of 640 days has been reported, and it was speculated that the half-life may be even longer in man. (titanium) [R20, 600].

FIRE POTENTIAL Not combustible [R8].

FDA Certification of this color additive when used as a food is not necessary for the protection of the public health and therefore batches thereof are exempt from the requirements of section 706 (c) of the Federal Food, Drug, and Cosmetic Act [R12]. Certification of this color additive when used as a drug is not necessary for the protection of the public health and therefore batches thereof are exempt from the requirements of section 706 (c) of the Federal Food, Drug, and Cosmetic Act [R13]. Certification of this color additive when used as a cosmetic is not necessary for the protection of the public health and therefore batches thereof are exempt from the requirements of section 706 (c) of the Federal Food, Drug, and Cosmetic Act [R14]. Certification of this color additive when used as a medical device is not necessary for the protection of the public health and therefore batches thereof are exempt from the requirements of section 706 (c) of the Federal Food, Drug, and Cosmetic Act [R15]. Titanium dioxide is an indirect food additive for use only as a component of adhesives [R16]. Substances used in the manufacture of paper and paperboard products used in food packaging shall include titanium dioxide. Under the conditions of normal use titanium dioxide would not reasonably be expected to migrate to food, based on available scientific information and data [R17].

REFERENCES

1. Budavari, S. (ed.). *The Merck Index— Encyclopedia of Chemicals, Drugs, and Biologicals.* Rahway, NJ: Merck and Co., Inc., 1989. 1492.

2. *American Hospital Formulary Service.* Volumes I, and II. Washington, DC: American Society of Hospital Pharmacists, to 1984, p. 84: 2412.

3. Sax, N. I., and R. J. Lewis, Sr. (eds.). *Hawley's Condensed Chemical Dictionary.* 11th ed. New York: Van Nostrand Reinhold Co., 1987. 1159.

4. SRI.

5. Furia, T. E. (ed.). *CRC Handbook of Food Additives.* 2nd ed. Cleveland: The Chemical Rubber Co., 1972. 603.

6. Sax. *Hawley's Condensed Chemical Dictionary* 11th ed. 1987 p. 1159.

7. Friberg, L., G. F. Nordberg, E. Kessler, and V. B. Vouk (eds.). *Handbook of the Toxicology of Metals.* 2nd ed. Vols I, II. Amsterdam: Elsevier Science Publisshers B. V., 1986, p. V. II 597.

8. ITII. *Toxic and Hazarous Industrial Chemicals* Safety Manual. Tokyo, Japan: The International Technical Information Institute, 1982. 523.

9. Mackison, F. W., R. S. Stricoff, and L. J. Partridge, Jr. (eds.). *NIOSH/OSHA—Occupational Health Guidelines for Chemical Hazards.* DHHS (NIOSH) Publication No. 81–123 (3 vols). Washington, DC: U.S. Government Printing Office, Jan. 1981.

10. IARC. *Monographs on the Evaluation of the Carcinogenic Risk of Chemicals to Man.* Geneva: World Health Organization, International Agency for Research on Cancer, 1972–present (multivolume work), p. 47 322 (1989).

11. 29 CFR 1910.1000 (7/1/90).

12. 21 CFR 73.575 (4/1/90).

13. 21 CFR 73.1575 (4/1/90).

14. 21 CFR 73.2575 (4/1/90).

15. 21 CFR 73.3126 (4/1/90).

16. 21 CFR 175.105 (4/1/90).

17. 21 CFR 181.30 (4/1/90).

18. Venugopal, B. and T.D. Luckey. *Metal*

Toxicity in Mammals, 2. New York: Plenum Press, 1978.

19. Osol, A. (ed.). *Remington's Pharmaceutical Sciences.* 16th ed. Easton, Pennsylvania: Mack Publishing Co., 1980.

20. Friberg, L., G.F. Nordberg, E. Kessler, and V.B. Vouk (eds.). *Handbook of the Toxicology of Metals.* 2nd ed. Vols I, II.: Amsterdam: Elsevier Science Publishers B.V., 1986, p. V2.

21. American Conference of Governmental Industrial Hygienists. *Threshold Limit Values for Chemical Substances and Physical Agents and Biological Exposure Indices for 1994–1995.* Cincinnati, OH: ACGIH, 1994.

22. Mackison, F. W., R. S. Stricoff, and L. J. Partridge, Jr. (eds.). *NIOSH/OSHA— Occupational Health Guidelines for Chemical Hazards.* DHHS (NIOSH) Publication No. 81–123 (3 vols). Washington, DC: U.S. Government Printing Office, Jan. 1981.

■ TOLUENE, p–CHLORO–

CAS # 106–43–4

DOT # 2238

SYNONYMS 4-chloro-1-methylbenzene • p-chlorotoluene • benzene, 1-chloro-4-methyl- • p-tolyl chloride • toluene, p-chloro-

MF C_7H_7Cl

MW 126.59

COLOR AND FORM Colorless liquid.

DENSITY 1.0697 at 20°C.

SURFACE TENSION 34.60 dynes/cm at 25°C.

BOILING POINT 161.99°C.

MELTING POINT 7.5°C.

HEAT OF VAPORIZATION 42.475 kJ/mol.

FIREFIGHTING PROCEDURES Do not extinguish fire unless flow can be stopped. Use water in flooding quantities as fog. Solid streams of water may be ineffective. Cool all affected containers with flooding quantities of water. Apply water from as great a distance as possible. Use "alcohol" foam, carbon dioxide, or dry chemical [R5].

SOLUBILITY Soluble in alcohol, benzene; soluble in acetic acid; >10% in ethyl ether; >10% in ethanol; >10% in chloroform; soluble in acetone; water solubility of 106 mg/L at 20°C.

VAPOR PRESSURE 2.6 mm Hg at 20°C.

OCTANOL/WATER PARTITION COEFFICIENT Log K_{ow} = 3.33.

OTHER CHEMICAL/PHYSICAL PROPERTIES Freezing point 7.20°C at 99.7 mole–% purity.

PREVENTATIVE MEASURES Avoid prolonged breathing of vapor. Use with adequate ventilation [R4].

DISPOSAL At the time of review, criteria for land treatment or burial (sanitary landfill) disposal practices are subject to significant revision. Prior to implementing land disposal of waste residue (including waste sludge), consult with environmental regulatory agencies for guidance on acceptable disposal practices [R6].

COMMON USES Solvent and intermediate for organic chemicals and dyes [R2]; chemical intermediate; specialty solvent [R3]; disinfectant against Coccidia oocysts, with low toxicity to fowl [R1].

OCCUPATIONAL RECOMMENDATIONS None listed.

BIODEGRADATION Japanese MITI, initial concentration 100 ppm, 14 days <30% BODT, activated-sludge inoculum (1,2). An isolated strain of *Pseudomonas putida* 39/D, oxidized p-chlorotoluene to (+) –cis–4–chloro–2,3–dihydroxy–1–methyl-cyclohexa–4,6–diene (3). p–Chlorotoluene is metabolized via cis-dihydrodiol to its respective catechol, which is resistant to further degradation (4) [R7].

BIOCONCENTRATION Based on a water solubility of 106 mg/L at 20°C and a log K_{ow} of 3.33, bioconcentration factors (BCF) of 45–200 were estimated for p-chlorotoluene (1,2,3,SRC). These BCF values suggest that slight bioaccumulation in aquatic organisms may occur (SRC) [R8].

ATMOSPHERIC STANDARDS This action promulgates standards of performance for equipment leaks of volatile organic compounds (VOC) in the synthetic organic chemical manufacturing industry (SOCMI). These standards implement Section 111 of the Clean Air Act and are based on the Administrator's determination that emissions from the SOCMI cause, or contribute significantly to, air pollution that may reasonably be anticipated to endanger public health or welfare. The intended effect of these standards is to require all newly constructed, modified, and reconstructed SOCMI process units to use the best demonstrated system of continuous emission reduction for equipment leaks of VOC, considering costs, non-air-quality health, and environmental impact and energy requirements. p–Chlorotoluene is produced, as an intermediate or final product, by process units covered under this subpart. These standards of performance become effective upon promulgation but apply to affected facilities for which construction or modification commenced after January 5, 1981 [R10].

FIFRA REQUIREMENTS Chlorotoluene, an isomeric mixture predominantly of ortho–, and para-monochlorotoluene with up to 6 percent unreacted toluene and a boiling range of 110°C to 162°C, is exempted from the requirement of a tolerance when used as a solvent or cosolvent in pesticide formulations with the following restrictions: (a) not for use after edible parts of the plant begin to form, (b) do not graze livestock in treated areas within 48 hrs after application [R9].

GENERAL RESPONSE Avoid contact with liquid. Keep people away. Wear goggles and self-contained breathing apparatus. Stop discharge if possible. Call fire department. Isolate and remove discharged material. Notify local health and pollution control agencies.

FIRE RESPONSE Combustible wear goggles and self-contained breathing apparatus. Extinguish with alcohol foam, carbon dioxide, or dry chemical.

EXPOSURE RESPONSE Call for medical aid. Liquid irritating to skin and eyes. Harmful if swallowed. Remove contaminated clothing and shoes. Flush affected areas with plenty of water. If in eyes, hold eyelids open, and flush with plenty of water. If swallowed and victim is conscious, have victim drink water or milk, and induce vomiting.

RESPONSE TO DISCHARGE Restrict access. Provide chemical and physical treatment. Disperse and flush.

WATER RESPONSE Harmful to aquatic life in very low concentrations. May be dangerous if it enters water intakes. Notify local health and wildlife officials. Notify operators of nearby water intakes.

EXPOSURE SYMPTOMS Inhalation: irritation of respiratory system. Eyes and skin: severe irritation. Ingestion: severe internal damage if swallowed.

EXPOSURE TREATMENT Get medical aid. Inhalation: move to fresh air. Remove contaminated clothing. Keep warm and quiet. If breathing has stopped give artificial respiration. Eyes and skin: wash with plenty of water. Ingestion: give one or two glasses of water or milk. Induce vomiting. Give cathartics.

REFERENCES

1. Tsunoda, K., Heisha, T. Japan Patent 74 07209 02/19/74 (Chuo Kagaku and Co., Ltd.).

2. Sax, N. I., and R. J. Lewis, Sr. (eds.). *Hawley's Condensed Chemical Dictionary.* 11th ed. New York: Van Nostrand Reinhold Co., 1987. 274.

3. SRI.

4. Sax, N. I. *Dangerous Properties of Industrial Materials,.* 4th ed. New York: Van Nostrand Reinhold, 1975. 556.

5. Association of American Railroads. *Emergency Handling of Hazardous Materials in Surface Transportation.* Washington, DC: Assoc. of American Railroads, Hazardous Materials Systems (BOE), 1987. 166.

6. SRP.

7. (1) Kawasaki, M. *Ecotox Environ Safety* 4: 444–54 (1980); (2) Sasaki, S. pp. 283–98 in *Aquatic Pollutants: Transformation and Biological Effects*. Hutzinger, O., et al., eds. Oxford: Pergamon Press (1978); (3) Gibson, D. T., et al. *Biochemistry* 7: 3795–802 (1968); (4) Gibson, D. T. pp. 187–204 in *Aquatic Pollutants: Transformation and Biological Effects*; Hutzinger, O., et al., eds. Oxford: Pergamon Press (1978).

8. (1) Yalkowsky, S. H., et al. *Arizona Data Base* 2nd ed. (1987); (2) Hansch, C., and A. J. Leo. *Medchem Project* Issue No. 26, Claremont, CA: Pomona College (1985); (3) Lyman, W. J., et al. *Handbook of Chemical Property Estimation Methods*, New York: McGraw-Hill p. 5–5 (1982).

9. 40 CFR 180.1045 (1987).

10. 40 CFR 60.489 (7/1/87).

■ p-TOLUIDINE

CAS # 106–49–0

SYNONYMS 1-amino-4-methylbenzene • 4-amino-1-methylbenzene • 4-aminotoluen (Czech) • 4-methylaniline • 4-methylbenzenamine • 4-toluidine • aniline, p-methyl • benzenamine, 4-methyl- • ci-37107 • naphtol as-kg • naphtol as-kgll • p-aminotoluene • p-methylaniline • p-methylbenzenamine • p-toluidin (Czech) • p-toluidine • p-tolylamine • tolyamine • ci azoic coupling component 107

MF C_7H_9N

MW 107.17

COLOR AND FORM Lustrous plates or leaflets; white solid; colorless leaflets.

ODOR Aromatic, winelike.

TASTE Burning.

DENSITY 1.046 at 20°C.

DISSOCIATION CONSTANT pK_a: 5.08.

BOILING POINT 200.5°C.

MELTING POINT 44–45°C.

AUTOIGNITION TEMPERATURE 900°F [R4, p. 49–88].

FIREFIGHTING PROCEDURES Water spray may be used to extinguish the fire because material can be cooled below its flash point [R4, p. 49–88]. Use carbon dioxide, dry chemical, foam, or water spray [R4, p. 49–88]. Do not extinguish fire unless flow can be stopped. Use water in flooding quantities as fog. Solid streams of water may be ineffective. Cool all affected containers with flooding quantities of water. Apply water from as great a distance as possible (toluidine mixture) [R5]. Wear self-contained breathing apparatus when fighting fire involving this material. (toluidine mixture) [R5].

TOXIC COMBUSTION PRODUCTS When heated, it emits highly toxic fumes. Emits toxic fumes of nitrous oxide [R3].

SOLUBILITY Soluble in about 135 parts water; freely soluble in alcohol, ether, acetone, methanol, carbon disulfide, oils, dilute acids; soluble in pyrimidine; water solubility = 6.64 g/L.

VAPOR PRESSURE 0.34 torr.

OCTANOL/WATER PARTITION COEFFICIENT Log K_{ow}= 1.39.

OTHER CHEMICAL/PHYSICAL PROPERTIES VP: 1 mm Hg at 42°C.

PROTECTIVE EQUIPMENT Full protective clothing, including self-contained breathing apparatus, coat, pants, gloves, boots, and bands around legs, arms, and waist should be provided. No skin surface should be exposed. [R4, p. 49–88].

PREVENTATIVE MEASURES If material not on fire and not involved in Fire: Keep sparks, flames, and other sources of ignition away. Keep material out of water sources and sewers. Build dikes to contain flow as necessary. Use water spray to knock down vapors. (toluidine mixture) [R5].

DISPOSAL Generators of waste (equal to or greater than 100 kg/mo) containing this contaminant, EPA hazardous waste number U353, must conform with USEPA regulations in storage, transportation, treatment, and disposal of waste [R6].

COMMON USES Chemical intermediate for many dyes [R2]; test reagent for lignin, nitrite, chloroglucinol; agent in preparation of ion exchange resins [R2].

OCCUPATIONAL RECOMMENDATIONS.

OSHA 8-hr time-weighted average: 2 ppm (9 mg/m³). Final rule limits were achieved by any combination of engineering controls, work practices, and personal protective equipment during the phase-in period, Sept 1, 1989 through Dec 30, 1992. Final rule limits became effective Dec 31, 1992. Skin absorption designation in effect as of Sept 1, 1989 [R11]. Threshold-limit 8-hr time-weighted average (TWA) 2 ppm, 8.8 mg/m³, skin (1986) [R12, 36].

INTERNATIONAL EXPOSURE LIMITS MAC former USSR 3 mg/m³ skin (toluidine) [R1].

NON-HUMAN TOXICITY LD_{50} mouse oral 794 mg/kg [R7].

ECOTOXICITY LD_{50} starling oral 42.2 mg/kg [R8].

BIODEGRADATION 4-Aminotoluene was readily degraded in biodegradability screening tests (1-8). Some test results are: 64% of theoretical BOD used in 5 days with a sewage seed (2); 64% of theoretical BOD in 8 days using an activated sludge inoculum acclimated to aniline (3); 97.7% removal in 5 days with activated sludge (5); and significantly degraded in the MITI test, the biodegradability test of the Japanese Ministry of International Trade and Industry (6). When 4-aminotoluene was incubated with sewage, 90 and 100% degradation occurred after 10 and 14 days, respectively (1). Complete degradation was obtained in 4 days with a soil inoculum (7). 500 ppm of 4-aminotoluene was completely degraded after 9 days in a Chernozen soil, leaving degradation products that persisted for over 90 days (4). When ring-labeled 4-aminotoluene was incubated with 3 silt loam soils with organic carbon contents ranging from 1 to 3.5% for 63 days, 13.5-15.9% CO_2 was released (8). Subsequent experiments established that mineralization was a result of microbial action (9). Under anaerobic conditions, no change in UV absorbancy was evident after 53 days when 4-aminotoluene was incubated with sewage (1). However, two degradation products were identified, namely, 2-methylformanilide and 4-methylacetanilide (1) [R9].

BIOCONCENTRATION The log octanol/water partition coefficient for 4-aminotoluene is 1.39 (1), from which one can estimate a BCF of 6.7 using a recommended regression equation (2,SRC). Therefore, 4-aminotoluene would not be expected to bioconcentrate in aquatic organisms [R10].

ATMOSPHERIC STANDARDS This action promulgates standards of performance for equipment leaks of volatile organic compounds (VOC) in the synthetic organic chemical manufacturing industry (SOCMI). The intended effect of these standards is to require all newly constructed, modified, and reconstructed SOCMI process units to use the best demonstrated system of continuous emission reduction for equipment leaks of VOC, considering costs, non-air-quality health, and environmental impact and energy requirements. Toluidines are produced, as intermediates or final products, by process units covered under this subpart. (toluidines) [R13].

TSCA REQUIREMENTS Pursuant to section 8 (d) of TSCA, EPA promulgated a model health and safety data reporting rule. The section 8 (d) model rule requires manufacturers, importers, and processors of listed chemical substances and mixtures to submit to EPA copies and lists of unpublished health and safety studies. Benzemamine, 4-methyl is included on this list [R14].

RCRA REQUIREMENTS U353; as stipulated in 40 CFR 261.33, when p-toluidine, as a commercial chemical product or manufacturing chemical intermediate or an off-specification commercial chemical product or a manufacturing chemical intermediate, becomes a waste, it must be managed according to federal or state hazardous waste regulations. Also defined as a hazardous waste is any residue, contaminated soil, water, or other debris resulting from the cleanup of a spill, into water or on dry land, of this waste. Generators of small quantities of this waste may qualify for partial exclusion from hazardous waste regulations (40 CFR 261.5) [R15].

REFERENCES

1. International Labour Office. *Encyclopedia of Occupational Health and Safety.* Vols. I and II. Geneva, Switzerland: International Labour Office, 1983. 146.

2. SRI.

3. Sax, N. I. *Dangerous Properties of Industrial Materials,.* 6th ed. New York, NY: Van Nostrand Reinhold, 1984. 2594.

4. National Fire Protection Association. *Fire Protection Guide on Hazardous Materials.* 9th ed. Boston, MA: National Fire Protection Association, 1986.

5. Association of American Railroads. *Emergency Handling of Hazardous Materials in Surface Transportation.* Washington, DC: Assoc. of American Railroads, Hazardous Materials Systems (BOE), 1987. 687.

6. 40 CFR 240–280, 300–306, 702–799 (7/1/91).

7. American Conference of Governmental Industrial Hygienists. *Documentation of the Threshold Limit Values and Biological Exposure Indices.* 5th ed. Cincinnati, OH: American Conference of Governmental Industrial Hygienists, 1986. 590.

8. Schafer, E. W., et al. *Arch Environ Contam Toxicology* 12: 355–82 (1983).

9. (1) Hallas, L. E., M. Alexander, *Appl Environ Microbiol* 45: 1234–41 (1983); (2) Heukelekian, H., M. C. Rand. *J Water Pollut Contr Assoc* 29: 1040–53 (1955); (3) Malaney, G. W. *J Water Pollut Control Fed* 32: 1300–11 (1960); (4) Medvedev, V. A., V. D. Davidov. pp. 245–54 in *Decomposition of Toxic, and Nontoxic Organic Compounds in Soil,* Overcash, M. R., ed. Ann Arbor, MI: Ann Arbor Sci Publ (1981); (5) Pitter, P. *Water Res* 10: 231–5 (1976); (6) Sasaki, S. pp. 283–98 in *Aquatic Pollutants: Transformation and Biological Effects,* Hutzinger, O., et al., ed. Oxford: Pergamon Press (1978); (7) Alex-

ander, M., B. K. Lustigman, *J Agric Food Chem* 14: 410–3 (1966); (8) Graveel, J. G., et al. *Environ Toxicol Chem* 4: 607–13 (1985); (9) Graveel, J. G., et al. *J Environ Qual* 15: 53–9 (1986).

10. (1) Hansch, C., and A. J. Leo. *Medchem Project,* Claremont, CA: Pomona College (1985); (2) Lyman, W. J., et al. pp. 5.1–5.30 in *Handbook of Chemical Property Estimation Methods. Environ Behavior of Organic Compounds.* New York: McGraw-Hill (1982).

11. 29 CFR 1910.1000 (7/1/91).

12. American Conference of Governmental Industrial Hygienists. *Threshold Limit Values for Chemical Substances and Physical Agents and Biological Exposure Indices for 1993–1994.* Cincinnati, OH: ACGIH, 1993.

13. 40 CFR 60.489 (7/1/91).

14. 40 CFR 716.120 (7/1/91).

15. 40 CFR 261.33 (7/1/91).

■ TRALOMETHRIN

CAS # 66841–25–6

SYNONYMS Cyclopropanecarboxylic acid, 2,2-dimethyl-3-(1,2,2,2-tetrabromoethyl)-, cyano (3-phenoxy phenyl) methyl ester • (S)-alpha-cyano-3-phenoxybenzyl (1R,3S)-2,2-dimethyl-3-[(RS)-1,2,2,2-tetrabromoethyl]-cyclopropane-carboxylate. • (S)-alpha-cyano-3-phenoxybenzyl (1R) -cis-2,2-dimethyl-3-[(RS)-1,2,2,2-tetrabromoethyl]-cyclopropanecarboxylate • cyano (3-phenoxyphenyl) methyl 2,2-dimethyl-3-(1,2,2,2-tetrabromoethyl) cyclopropanecarboxylate • scout • ru-25474 • hag-107 • scout x-tra

MF $C_{22}H_{19}Br_4NO_3$

MW 665.06

pH 6.6.

COLOR AND FORM Yellow-orange resinoid.

DENSITY 1.70 at 20°C.

BOILING POINT 138–148°C.

FIREFIGHTING PROCEDURES Use carbon dioxide, foam, or dry chemical (on fires involving pyrethroids) (pyrethrum) [R7, 2]. Fire fighting: Self-contained breathing apparatus with a full facepiece operated in pressure-demand or other positive-pressure mode (pyrethrum) [R7, 5]. Extinguish fire using agent suitable for type of surrounding fire. (pyrethrins) [R2].

SOLUBILITY In water 70 mg/kg. In acetone, dichloromethane, toluene, xylene >1,000; dimethylsulfoxide >500; ethanol >180 (all in g/L).

VAPOR PRESSURE 17 pPa.

STABILITY AND SHELF LIFE Stable for 6 months at 50°C. Acidic media reduce hydrolysis and epimerization [R1].

OTHER CHEMICAL/PHYSICAL PROPERTIES Orange-to-yellow resinous solid (technical tralomethrin, >93%).

PROTECTIVE EQUIPMENT Employees should use dust– and splash-proof safety goggles where pyrethroids may contact the eyes. (pyrethroids) [R7, 3].

PREVENTATIVE MEASURES Skin that becomes contaminated with pyrethrum should be promptly washed or showered with soap or mild detergent and water. (pyrethrum) [R7, 3].

CLEANUP Spillages of pesticides at any stage of their storage or handling should be treated with great care. Liquid formulations may be reduced to solid phase by evaporation. Dry sweeping of solids is always hazardous: these should be removed by vacuum cleaning or by dissolving them in water or other solvent in the factory environment. (pesticides) [R3].

DISPOSAL Incineration would be an effective disposal procedure where permitted. If an efficient incinerator is not available, the product should be mixed with large amounts of combustible material, and contact with the smoke should be avoided. (pyrethrin products) [R4].

COMMON USES Insecticide. Control of foliar insect pests in cotton, and coleopterous, lepidopterous, and other insects in potatoes, soya beans, and vegetables [R1].

NON-HUMAN TOXICITY LD_{50} rat male oral 1,250 mg/kg [R1]; LD_{50} rat female oral 1,070 mg/kg [R1]; LD_{50} rabbit percutaneous >2,000 mg/kg [R1]; LC_{50} rat inhalation >0.286 mg/L air [R1].

ECOTOXICITY LD_{50} quail oral >2,510 mg/kg [R1]; LC_{50} mallard duck dietary 7,716 mg/kg diet/8 day [R1]; LC_{50} quail dietary 4,300 mg/kg diet/8 day [R1]; LC_{50} rainbow trout 0.0016 mg/L/96 hr (conditions of bioassay not specified) [R1]; LC_{50} bluegill sunfish 0.0043 mg/l/96 hr (conditions of bioassay not specified) [R1]; LC_{50} Daphnia 38 ng/L/48 hr (conditions of bioassay not specified) [R1].

FIFRA REQUIREMENTS Tolerances are established for the combined residues of the insecticide tralomethrin and its metabolites (S)–alpha-cyano–3–phenoxybenzyl (1R,3R) –3–(2,2–dibromovinyl–2,2–dimethylcyclopropanecarboxylate and (S)–alpha-cyano–3–phenoxybenzyl (1S,3R)–3–(2,2–dibromovinyl) –2,2–dimethylcyclopropanecarboxylate calculated as the parent in or on the raw agricultural commodities cottonseed and soybeans [R5].

DOT *Health Hazards*: poisonous; may be fatal if inhaled, swallowed or absorbed through skin. Contact may cause burns to skin and eyes. Runoff from fire control or dilution water may give off poisonous gases and cause water pollution. Fire may produce irritating or poisonous gases (pesticide, liquid or solid, poisonous, not otherwise specified) [R6, p. G-55].

Fire or Explosion: Some of these materials may burn, but none of them ignites readily. Container may explode violently in heat of fire (pesticide, liquid or solid, poisonous, not otherwise specified) [R6, p. G-55].

Emergency Action: Keep unnecessary people away; isolate hazard area and deny entry. Stay upwind, out of low areas, and ventilate closed spaces before entering. Positive-pressure self-contained breathing apparatus (SCBA) and chemical protective clothing that is specifically recommended by the shipper or manufacturer may be worn. It may provide little or no thermal protection. Structural firefighters' protective clothing is not effective for these materials. Remove and isolate contaminated clothing at the site. Call CHEMTREC at 1–800–424–9300 as soon as possible, especially if there is no local hazardous materials team available (pesticide, liquid or solid, poisonous, not otherwise specified) [R6, p. G-55].

Fire: Small fires: Dry chemical, CO_2, water spray or regular foam. Large Fires: Water spray, fog, or regular foam. Move container from fire area if you can do so without risk. Fight fire from maximum distance. Stay away from ends of tanks. Dike fire-control water for later disposal; do not scatter the material (pesticide, liquid or solid, poisonous, not otherwise specified) [R6, p. G-55].

Spill or Leak: Do not touch or walk through spilled material; stop leak if you can do so without risk. Fully encapsulating, vapor-protective clothing should be worn for spills and leaks with no fire. Use water spray to reduce vapors. Small Spills: Take up with sand or other noncombustible absorbent material and place into containers for later disposal. Small Dry Spills: With clean shovel place material into clean, dry container and cover loosely; move containers from spill area. Large Spills: Dike far ahead of liquid spill for later disposal (pesticide, liquid or solid, poisonous, not otherwise specified) [R6, p. G-55].

First Aid: Move victim to fresh air and call emergency medical care; if not breathing, give artificial respiration; if breathing is difficult, give oxygen. In case of contact with material, immediately flush skin or eyes with running water for at least 15 minutes. Speed in removing material from skin is of extreme importance. Remove and isolate contaminated clothing and shoes at the site. Keep victim quiet and maintain normal body temperature. Effects may be delayed; keep victim under observation (pesticide, liquid, or solid, poisonous, not otherwise specified) [R6, p. G-55].

REFERENCES

1. Hartley, D., and H. Kidd (eds.). *The Agrochemicals Handbook.* 2nd ed. Lech-worth, Herts, England: The Royal Society of Chemistry, 1987, p. A961/Aug 87.

2. Bureau of Explosives. *Emergency Handling of Haz Matl in Surface Trans* p. 434 (1981).

3. International Labour Office. *Encyclopedia of Occupational Health and Safety.* Vols. I and II. Geneva, Switzerland: International Labour Office, 1983. 1619.

4. Sittig, M. *Handbook of Toxic and Hazardous Chemicals and Carcinogens,* 1985. 2nd ed. Park Ridge, NJ: Noyes Data Corporation, 1985. 762.

5. 40 CFR 180.422 (7/1/91).

6. U.S. Department of Transportation. *Emergency Response Guidebook 1990.* DOT P 5800.5. Washington, DC: U.S. Government Printing Office, 1990.

7. Mackison, F. W., R. S. Stricoff, and L. J. Partridge, Jr. (eds.). *NIOSH/OSHA—Occupational Health Guidelines for Chemical Hazards.* DHHS (NIOSH) Publication No. 81–123 (3 vols). Washington, DC: U.S. Government Printing Office, Jan. 1981.

■ TRIALLATE

CAS # 2303–17–5

SYNONYMS 2,3,3-trichloroallyl diisopropylthiocarbamate • 2,3,3-trichloroallyl N,N-diisopropylthiocarbamate • 2-propene-1-thiol, 2,3,3-trichloro-, diisopropylcarbamate • avadex-bw • carbamic acid, diisopropylthio-, s-(2,3,3-trichloroallyl) ester • carbamothioic acid, bis (1-methylethyl) -, S-(2,3,3-trichloro-2-propenyl) ester • cp-23426- • dipthal • far-go • N,N-Diisopropyl-2,3,3-trichlorallyl-thiolcarbamat (German) • N-diisopropylthiocarbamic acid s-2,3,3-trichloro-2-propenyl ester • S-2,3,3-trichloroallyl diisopropylthiocarbamate • S-2,3,3-trichloroallyl N,N-diisopropylthiocarbamate • thiocarbamic acid, N-diisopropyl-, S-2,3,3-trichloroallyl ester • tri-allate • fargo • S-2,3,3-trichloroallyl di-isopropyl (thiocarbamate) • S-2,3,3-trichloroallyl di-isopropylthiocarbamate • S-(2,3,3-trichloroallyl) diisopropylthiocarbamate • bis (1-methylethyl) carbamothioic acid S-(2,3,3-trichloro-2-propenyl) ester • diisopropylthiocarbamic acid S-(2,3,3-trichloroallyl) ester • 2,3,3-trichloro-2-propene-1-thiol diisopropylcarbamate • S-(2,3,3-trichloro-2-propenyl) bis (1-methylethyl) carbamothiote • S-2,3,3-trichloroallyl-N,N-diisopropyl-thiolcarbamate

MF $C_{10}H_{16}Cl_3NOS$

MW 304.68

COLOR AND FORM Colorless crystals.

DENSITY 1.273 at 25 0.6°C.

BOILING POINT 117°C at 0.3 mm Hg.

MELTING POINT 29–30°C.

EXPLOSIVE LIMITS AND POTENTIAL Combustible [R5].

FLAMMABILITY LIMITS Nonflammable [R7, 465].

FIREFIGHTING PROCEDURES Water spray, foam, dry chemical, carbon dioxide, or other class B agent [R3].

SOLUBILITY Soluble in heptane, ethyl alcohol, acetone, ether, benzene, ethyl acetate, water (4 ppm) at 25°C.

VAPOR PRESSURE 1.2×10^{-4} mm Hg at 25°C.

OCTANOL/WATER PARTITION COEFFICIENT Log P = 3.98 (calc).

CORROSIVITY Noncorrosive.

STABILITY AND SHELF LIFE Indefinitely stable, insensitive to light or heat [R7, 465]. Stable under normal storage conditions. Hydrolyzed by strong acids and alkalis. [R4].

OTHER CHEMICAL/PHYSICAL PROPERTIES Decomposition temp: greater than 200°C; highly resistant to decomposition by UV irradiation; oily amber liquid (technical

triallate); Henry's Law constant = 1.93×10^{-5} atm–m³/mole (est).

PROTECTIVE EQUIPMENT Wear rubber gloves and goggles when handling. [R7, 465].

PREVENTATIVE MEASURES Avoid contact with skin and eyes [R6].

COMMON USES Preemergence selective herbicide [R3]. Control of wild oats and some annual grasses in wheat, barley, rye, field beans, peas, lentils, beet, oilseed rape, maize, flax, lucerne, clover, vetches, sainfoin, safflowers, sunflowers, and vegetables [R4]. Herbicide used to control wild oats in barley, lentils, peas, and spring and winter wheat [R2]. For wild oats in barley, green peas, field dried peas, lentils, wheat (durum, spring, winter) [R3]. Use on canarygrass voluntarily cancelled by Monsanto (in USA) [R1].

OCCUPATIONAL RECOMMENDATIONS None listed.

NON-HUMAN TOXICITY LD_{50} rat oral 1,100 mg/kg [R4]; LD_{50} rabbit percutaneous 8,200 mg/kg [R4].

ECOTOXICITY LD_{50} bluegill sunfish 4.9 mg/L/96 hr/ec (4 lb/gal) (conditions of bioassay not specified) [R7, 446]; LD_{50} rainbow trout 9.6 mg/L/96 hr/ec (4 lb/gal) (conditions of bioassay not specified) [R7, 466]; LD_{50} bobwhite quail, oral greater than 2,251 mg/kg [R2]; LC_{50} mallard ducklings diet greater than 5,000 mg/kg/8 days [R2]; LC_{50} Colinus virginianus (bobwhite quail) diet greater than 5,000 mg/kg/8 days [R2]; LC_{50} Lepomis macrochirus (bluegill) 1.3 mg/L/96 hr (conditions of bioassay not specified) [R2]; LC_{50} Salmo gairdneri (rainbow trout) 1.2 mg/L/96 hr (conditions of bioassay not specified) [R2].

BIODEGRADATION Adsorbed by soil colloids. microbial breakdown: responsible for major decomposition in soil [R7, 465]. Biodegradation studies showed losses of 65% and 75% of added triallate after 16 weeks incubation in Regina heavy clay (pH 7.5, 4.0% organic matter) and Weyburn loam (pH 7.0, 6.5% organic matter), respectively, whereas less than 5% of added triallate was lost from sterile clay and loam samples (1). Triallate was reported to be remaining at levels of 51%, 54%, 63%, and 85% after 12 wk incubation in Regina heavy clay at moisture content levels of 40%, 35%, 30%, and 20%, respectively (2). Triallate was reported to be remaining at levels of 43%, 47%, 48%, and 60% after 12 weeks incubation in Weyburn loam at moisture content levels of 40%, 35%, 30%, and 20%, respectively (2). Half-lives of 8–11 wk have been reported for the primary biodegradation of triallate from Regina heavy clay and Weyburn loam (1). These data indicate that microbial degradation is responsible for the loss of triallate in moist soils (1). [R8].

FIFRA REQUIREMENTS Tolerances are established for residues of the herbicide S-2,3,3-trichloroallyl diisopropylthiocarbamate in or on the following raw agricultural commodities: grass (canary, annual, seed and straw), barley (grain and straw), lentils (including forage and hay), peas (including forage and hay), and wheat (grain and straw) [R9]. In 1988, Congress amended FIFRA to strengthen and accelerate EPA's reregistration program. The nine-year reregistration scheme mandated by "FIFRA 88" applies to each registered pesticide product containing an active ingredient initially registered before November 1, 1984. Pesticides for which EPA had not issued Registration Standards prior to the effective date of FIFRA '88 were divided into three lists based upon their potential for exposure and other factors, with List B being of highest concern and D of least. List: B; Case: Triallate; Case No.: 2695; Pesticide type: Herbicide; Case Status: Awaiting Data/Data in Review: OPP awaits data from the pesticide's producer(s) regarding its human health and/or environmental effects, or OPP has received and is reviewing such data, in order to reach a decision about the pesticide's eligibility for reregistration. Active Ingredient (AI): S-(2,3,3-Trichloroallyl) diisopropylthiocarbamate; Data Call-in (DCI) Date: 06/04/91; AI Status: The producer(s) of the pesticide has made commitments to conduct the studies and pay the fees required for reregistration, and is meeting those commitments in a timely manner. [R10].

REFERENCES

1. *Farm Chemicals Handbook 1992*. Willoughby, OH: Meister Publishing Co., 1992, p. C–147.

2. Worthing, C. R., and S. B. Walker (eds.). *The Pesticide Manual—A World Compendium*. 8th ed. Thornton Heath, UK: The British Crop Protection Council, 1987. 816.

3. *Farm Chemicals Handbook 1993*. Willoughby, OH: Meister Publishing Co., 1993, p. C–148.

4. Hartley, D., and H. Kidd (eds.). *The Agrochemicals Handbook*. 2nd ed. Lechworth, Herts, England: The Royal Society of Chemistry, 1987, p. A403/Aug 87.

5. Lewis, R. J., Sr. (ed.). *Hawley's Condensed Chemical Dictionary*. 12th ed. New York, NY: Van Nostrand Reinhold Co., 1993, 1169.

6. Hartley, D., and H. Kidd (eds.). *The Agrochemicals Handbook*. Old Woking, Surrey, United Kingdom: Royal Society of Chemistry/Unwin Brothers Ltd., 1983, p. A403/Oct 83.

7. Weed Science Society of America. *Herbicide Handbook*. 5th ed. Champaign, Illinois: Weed Science Society of America, 1983.

8. (1) Smith, A. E., *Weed Res 9*: 306–13 (1969), (2) Smith, A. E., *Weed Res 10*: 331–39 (1970).

9. 40 CFR 180.314 (7/1/91).

10. USEPA/OPP. *Status of Pesticides in Reregistration and Special Review*. p. 164 (Mar. 1992) EPA 700-R-92-004.

■ TRICHLOROETHYLSILANE

CAS # 115–21–9

DOT # 1196

SYNONYMS ethyl silicon trichloride • ethyltrichlorosilane • silane, ethyl (trichloro)- • silane, trichloroethyl • silicane, trichloroethyl •

trichloroethylsilicane •
trichloroethylsilicon

MF $C_2H_5Cl_3Si$

MW 163.51

COLOR AND FORM Colorless liquid.

DENSITY 1.2381 at 20°C.

BOILING POINT 97.9°C at 760 mm Hg.

MELTING POINT −105.6°C.

HEAT OF VAPORIZATION 9,457.8 cal/mol.

EXPLOSIVE LIMITS AND POTENTIAL May form explosive mixtures with air [R4].

FIREFIGHTING PROCEDURES Use dry chemical, "alcohol resistant" foam, carbon dioxide. Water streams may be ineffective. Use water spray to keep fire-exposed containers cool. Approach fire from upwind to avoid hazardous vapors and toxic decomposition products [R8, p. 49–90]. Apply water from as far a distance as possible. Do not use water on material itself. If large quantities of combustibles are involved, use water in flooding quantities as spray and fog. Use water spray to knock down vapors [R3].

TOXIC COMBUSTION PRODUCTS Combustion by-products include hydrogen chloride, other irritants and toxic gases. [R8, p. 49–90].

SOLUBILITY Soluble in benzene, ether, heptane, perchloroethylene.

VAPOR PRESSURE 47.18 torr at 25°C.

OTHER CHEMICAL/PHYSICAL PROPERTIES Heat of vaporization: 7.7 kcal/mole at boiling point; density: 1.2342 at 24°C; density: 1.2349 at 20°C; readily hydrolyzed with liberation of hydrogen chloride; decomposes in water and alcohol; boiling point: 99.1°C at 1 atm.

PROTECTIVE EQUIPMENT Those who have to handle chlorosilanes should wear goggles and, if necessary, a gas mask. Handling should be done as far as possible under suitably ventilated hoods. (chlorosilanes) [R5].

PREVENTATIVE MEASURES SRP: Contaminated protective clothing should be segregated in such a manner so that there is no direct personal contact by personnel who handle, dispose of, or clean the clothing. Quality assurance to ascertain the completeness of the cleaning procedures should be implemented before the decontaminated protective clothing is returned for reuse by the workers. Contaminated clothing should not be taken home at end of shift, but should remain at employee's place of work for cleaning.

CLEANUP Eliminate all ignition sources. Stop or control the leak, if this can be done without undue risk. Use water spray to cool and disperse vapors and protect personnel. Control runoff and isolate discharged material for proper disposal [R8, p. 49–90]. Absorb the spills with paper towels or like materials. Place in hood to evaporate. Dispose by burning the towel. [R9, 242].

DISPOSAL SRP: At the time of review, criteria for land treatment or burial (sanitary landfill) disposal practices are subject to significant revision. Prior to implementing land disposal of waste residue (including waste sludge), consult with environmental regulatory agencies for guidance on acceptable disposal practices.

COMMON USES Chemical intermediate for silicones [R1].

OCCUPATIONAL RECOMMENDATIONS None listed.

NON-HUMAN TOXICITY LD_{50} rat oral 1,330 mg/kg [R6]; LD_{50} rat ip approximately 1.0 g/kg [R7]; LD_{50} rat oral 0.8 g/kg [R7].

DOT *Fire or Explosion:* Flammable/combustible material; may be ignited by heat, sparks, or flames. Vapors may travel to a source of ignition, and flash back. Container may explode in heat of fire. Vapor explosion hazard indoors, outdoors, or in sewers. Runoff to sewer may create fire or explosion hazard [R2].

Health Hazards: May be poisonous if inhaled. Contact may cause burns to skin and eyes. Fire may produce irritating or poisonous gases. Runoff from fire control or dilution water may cause pollution [R2].

Emergency Action: Keep unnecessary people away; isolate hazard area and deny

entry. Stay upwind; keep out of low areas. Positive-pressure self-contained breathing apparatus (SCBA) and structural firefighters' protective clothing will provide limited protection. See the table of initial isolation and protective action distances. If you find the ID number and the name of the material there, begin protective action. Isolate for 1/2 mile in all directions if tank, rail car, or tank truck is involved in fire. Call emergency response telephone number on shipping paper first. If shipping paper not available or no answer, call CHEMTREC at 1–800–424–9300. If water pollution occurs, notify the appropriate authorities [R2].

Fire: Some of these materials may react violently with water. Small Fires: Dry chemical, CO_2, water spray, or regular foam. Large Fires: Water spray, fog, or regular foam. Move container from fire area if you can do so without risk. Do not get water inside container. Apply cooling water to sides of containers that are exposed to flames until well after fire is out. Stay away from ends of tanks. Withdraw immediately in case of rising sound from venting safety device or any discoloration of tank due to fire [R2].

Spill or Leak: Shut off ignition sources; no flares, smoking, or flames in hazard area. Do not touch or walk through spilled material; stop leak if you can do so without risk. Use water spray to reduce vapor; do not get water inside container. Small Spills: Take up with sand or other noncombustible absorbent material and place into containers for later disposal. Large Spills: Dike far ahead of liquid spill for later disposal [R2].

First Aid: Move victim to fresh air and call for emergency medical care; if not breathing, give artificial respiration; if breathing is difficult, give oxygen. In case of contact with material, immediately flush skin or eyes with running water for at least 15 minutes. Remove and isolate contaminated clothing and shoes at the site. Keep victim quiet and maintain normal body temperature [R2].

FIRE POTENTIAL Flammable liquid [R8, p. 49–90].

GENERAL RESPONSE Avoid contact with liquid and vapor. Keep people away. Wear goggles, self-contained breathing apparatus, and rubber overclothing (including gloves). Shut off ignition sources. Call fire department. Stop discharge if possible. Isolate and remove discharged material. Notify local health and pollution control agencies.

FIRE RESPONSE Flammable. Poisonous gases are produced in fire. Containers may explode in fire. Flashback along vapor trail may occur. Vapor may explode if ignited in an enclosed area. Wear goggles and self-contained breathing apparatus. Extinguish with dry chemicals, alcohol foam, or carbon dioxide. Water may be ineffective on fire. Cool exposed containers with water.

EXPOSURE RESPONSE Call for medical aid. Vapor irritating to eyes, nose, and throat. If inhaled will cause difficult breathing. Move victim to fresh air. If breathing has stopped, give artificial respiration. If breathing is difficult, give oxygen. Liquid will burn skin and eyes. Harmful if swallowed. Remove contaminated clothing and shoes. Flush affected areas with plenty of water. If swallowed and victim is conscious, have victim drink water or milk. Do not induce vomiting.

RESPONSE TO DISCHARGE Issue warning—high flammability, corrosive. Restrict access. Disperse and flush with care.

WATER RESPONSE Effect of low concentrations on aquatic life is unknown. May be dangerous if it enters water intakes. Notify local health and wildlife officials. Notify operators of nearby water intakes.

EXPOSURE SYMPTOMS Inhalation causes irritation of nose and throat. Contact with liquid causes severe burns of eyes and skin. Ingestion causes burns of mouth and stomach.

EXPOSURE TREATMENT Inhalation: remove victim from exposure; administer artificial respiration if breathing has stopped; call physician. Eyes: flush with water for 15 min; obtain medical attention immediately. Skin: flush with water; obtain medical attention immediately if irritation persists. Ingestion: give large amounts of water; get medical attention.

REFERENCES

1. SRI.

2. U.S. Department of Transportation. *Emergency Response Guidebook 1993.* DOT P 5800.6. Washington, DC: U.S. Government Printing Office, 1993, p. G–29.

3. Association of American Railroads. *Emergency Handling of Hazardous Materials in Surface Transportation.* Washington, DC: Association of American Railroads, Bureau of Explosives, 1992. 440.

4. Hawley, G. G. *The Condensed Chemical Dictionary.* 10th ed. New York: Van Nostrand Reinhold Co., 1981. 444.

5. Lefaux, R. *Practical Toxicology of Plastics.* Cleveland: CRC Press Inc., 1968. 109.

6. Sax, N. I. *Dangerous Properties of Industrial Materials,.* 6th ed. New York, NY: Van Nostrand Reinhold, 1984. 1397.

7. Clayton, G.D., and F.E. Clayton (eds.). *Patty's Industrial Hygiene and Toxicology:* Volume 2A, 2B, 2C: *Toxicology.* 3rd ed. New York: John Wiley and Sons, 1981–1982. 2398.

8. National Fire Protection Guide. *Fire Protection Guide on Hazardous Materials.* 10th ed. Quincy, MA: National Fire Protection Association, 1991.

9. ITII. *Toxic and Hazardous Industrial Chemicals Safety Manual.* Tokyo, Japan: The International Technical Information Institute, 1988.

■ TRICHLOROISOCYANURIC ACID

CAS # 87–90–1

SYNONYMS 1,3,5-triazine-2,4,6 (1h, 3h, 5h)-trione, 1,3,5-trichloro • s-triazine-2,4,6 (1H, 3H, 5H) -trione, 1,3,5-trichloro • 1,3,5-trichloro-2,4,6-trioxohexahydro-s-triazine • 1,3,5-trichloro-s-triazine-2,4,6 (1H, 3H, 5H)-trione • 1,3,5-trichloroisocyanuric acid • acl-85 • cdb-90 • chloreal • fi-clor-91 • fichlor-91 • isocyanuric chloride • Kyselina-trichloisokyanurova (Czech) • N,N′,N″-trichloroisocyanuric acid • neochlor-90 • nsc-405124 • s-triazine-2,4,6 (1H, 3H, 5H)-trione, 1,3,5-trichloro • symclosen • symclosene • trichlorinated isocyanuric acid • trichloro-s-triazine-2,4,6 (1H, 3H, 5H)-trione • trichloro-s-triazinetrione • trichlorocyanuric acid • trichloroiminocyanuric acid • trichloroisocyanic acid

MF $C_3Cl_3N_3O_3$

MW 232.42

pH 1% aqueous solution = 2.7–3.3.

COLOR AND FORM Needles from ethylene chloride; white crystalline powder or granules.

ODOR Strong chlorine odor.

DENSITY Greater than 1 at 20°C (solid).

MELTING POINT 246–47°C (decomp).

EXPLOSIVE LIMITS AND POTENTIAL Containers may explode when heated [R5].

FIREFIGHTING PROCEDURES Approach fire from upwind to avoid hazardous vapors and toxic decomposition products. Fight fire from protected location or maximum possible distance. Use flooding quantities of water on fire-involved containers. If necessary use water spray to keep fire-exposed containers cool. Avoid use of water on non-involved material wherever possible [R9, p. 49–172]. If material involved in fire: Cool all affected containers with flooding quantities of water. Use water in flooding quantities as fog. Apply water from as far a distance as possible. Extinguish fire using agent suitable for type of surrounding fire. (Material itself does not burn or burns with difficulty.) [R4].

TOXIC COMBUSTION PRODUCTS Toxic chlorine or nitrogen trichloride may be formed in fires [R5].

SOLUBILITY Soluble in chlorinated and highly polar solvents; solubility in water at 25°C = 1 g/100 g; solubility in acetone at 30°C = 35.0 g/100 g.

STABILITY AND SHELF LIFE Thermally unstable [R9, p. 49-173].

OTHER CHEMICAL/PHYSICAL PROPERTIES Slightly hygroscopic; loose bulk density (approx) powder 31 lb/cu ft, granular 60 lb/cu ft; decomposes at 225°C.

PROTECTIVE EQUIPMENT NIOSH-approved respirator or chlorine canister mask; goggles; rubber gloves [R5].

PREVENTATIVE MEASURES SRP: The scientific literature for the use of contact lenses in industry is conflicting. The benefit or detrimental effects of wearing contact lenses depend not only upon the substance, but also on factors including the form of the substance, characteristics, and duration of the exposure, the use of other eye protection equipment, and the hygiene of the lenses. However, there may be individual substances whose irritating or corrosive properties are such that the wearing of contact lenses would be harmful to the eye. In those specific cases, contact lenses should not be worn. In any event, the usual eye protection equipment should be worn even when contact lenses are in place.

CLEANUP Cover with weak reducing agent such as hypo, bisulfites, or ferrous salts. Bisulfites or ferrous salts need additional promoter of some $3MH_2SO_4$ for rapid reaction. Transfer the slurry (or sludge) into a large container of water and neutralize with soda ash. Drain into the sewer with abundant water. [R8, 539].

DISPOSAL SRP: At the time of review, criteria for land treatment or burial (sanitary landfill) disposal practices are subject to significant revision. Prior to implementing land disposal of waste residue (including waste sludge), consult with environmental regulatory agencies for guidance on acceptable disposal practices.

COMMON USES Swimming pool sanitizer to reduce bacteria count; active ingredient in indust deodorants [R1]. Used as source of available chlorine in "dry type" bleaches, scouring powders, dishwashing compound, and sanitizing compound [R2].

OCCUPATIONAL RECOMMENDATIONS None listed.

NON-HUMAN TOXICITY LD_{50} rat oral 406 mg/kg [R7]; LD_{50} rabbit skin 20,000 mg/kg [R7]; LD_{50} rat oral 750 mg/kg [R6].

DOT *Health Hazards:* Poisonous if swallowed. Inhalation of dust poisonous. Contact may cause burns to skin and eyes. Fire may produce irritating or poisonous gases. Runoff from fire control or dilution water may cause pollution (trichloroisocyanuric acid, dry) [R3].

Fire or Explosion: May burn rapidly. May ignite other combustible materials (wood, paper, oil, etc.). These materials will accelerate burning when they are involved in a fire; some will react violently with fuels (trichloroisocyanuric acid, dry) [R3].

Emergency Action: Keep unnecessary people away; isolate hazard area, and deny entry. Positive-pressure self-contained breathing apparatus (SCBA) and structural firefighters' protective clothing will provide limited protection. call emergency response telephone number on shipping paper first. If shipping paper not available or no answer, Call CHEMTREC at 1-800-424-9300. If water pollution occurs, notify the appropriate authorities (trichloroisocyanuric acid, dry) [R3].

Fire: Small fires: Water only; no dry chemical, CO_2, or halon. Large Fires: Flood fire area with water from a distance. Move container from fire area if you can do so without risk. Apply cooling water to sides of containers that are exposed to flames until well after fire is out. Stay away from ends of tanks. For massive fire in cargo area, use unmanned hose holder, or monitor nozzles (trichloroisocyanuric acid, dry) [R3].

Spill or Leak: Do not touch or walk through spilled material. Keep combustibles (wood, paper, oil, etc.) away from spilled material. Small Dry Spills: With clean shovel place material into clean, dry container, and cover loosely; move containers from spill area. Large Spills: Dike far ahead of liquid spill for later disposal (trichloroisocyanuric acid, dry) [R3].

First Aid: Move victim to fresh air; call for emergency medical care. In case of contact with material, immediately flush skin or eyes with running water for at least 15 minutes. Remove and isolate contaminated clothing and shoes at the site (trichloroisocyanuric acid, dry) [R3].

GENERAL RESPONSE Keep people away. Call fire department. Isolate and remove discharged material. Notify local health and pollution control agencies.

FIRE RESPONSE Not flammable. May cause fire on contact with combustibles. Poisonous gases are produced in fire. Containers may explode in fire. Wear goggles and self-contained breathing apparatus. Flood discharge area with water. Cool exposed containers with water.

EXPOSURE RESPONSE Call for medical aid. Dust irritating to eyes, nose, and throat. If inhaled will cause coughing or difficult breathing. Move victim to fresh air. If in eyes, hold eyelids open and flush with plenty of water. If breathing is difficult, give oxygen. Solid irritating to skin and eyes. Harmful if swallowed. Remove contaminated clothing and shoes. Flush affected areas with plenty of water. If in eyes, hold eyelids open and flush with plenty of water. If swallowed and victim is conscious, have victim drink water or milk and have victim induce vomiting. If swallowed and victim is unconscious or having convulsions, do nothing except keep victim warm.

RESPONSE TO DISCHARGE Issue warning—oxidizing material, water contaminant. Restrict access. Disperse and flush.

WATER POLLUTION RESPONSE Effect of low concentrations on aquatic life is unknown. May be dangerous if it enters water intakes. Notify local health and wildlife officials. Notify operators of nearby water intakes.

EXPOSURE SYMPTOMS Inhalation causes sneezing and coughing. Contact with dust causes moderate irritation of eyes and itching and redness of skin. Ingestion causes burns of mouth and stomach.

EXPOSURE TREATMENT Inhalation: remove victim to fresh air. Eyes: irrigate with running water for 15 min; call physician. Skin: flush with water. Ingestion: induce vomiting and call physician.

FIRE POTENTIAL Noncombustible, but, due to high reactivity (chlorination, oxidation), may cause ignition by contact with many substances, organic substances easily chlorinated or oxidized. [R8, 538].

REFERENCES

1. SRI.

2. Gosselin, R. E., H. C. Hodge, R. P. Smith, and M. N. Gleason. *Clinical Toxicology of Commercial Products.* 4th ed. Baltimore: Williams and Wilkins, 1976, p. II–77.

3. U.S. Department of Transportation. *Emergency Response Guidebook 1993.* DOT P 5800.6. Washington, DC: U.S. Government Printing Office, 1993, p. G–42.

4. Association of American Railroads. *Emergency Handling of Hazardous Materials in Surface Transportation.* Washington, DC: Association of American Railroads, Bureau of Explosives, 1992. 920.

5. U.S. Coast Guard, Department of Transportation. *CHRIS—Hazardous Chemical Data.* Volume II. Washington, DC: U.S. Government Printing Office, 1984–5.

6. Gosselin, R. E., R. P. Smith, H. C. Hodge. *Clinical Toxicology of Commercial Products.* 5th ed. Baltimore: Williams and Wilkins, 1984, p. II–111.

7. Sax, N. I. *Dangerous Properties of Industrial Materials,.* 6th ed. New York, NY: Van Nostrand Reinhold, 1984. 2624.

8. ITII. *Toxic and Hazardous Industrial Chemicals Safety Manual.* Tokyo, Japan: The International Technical Information Institute, 1988.

9. National Fire Protection Guide. *Fire Protection Guide on Hazardous Materi-*

als. 10th ed. Quincy, MA: National Fire Protection Association, 1991.

■ TRICHLORONAPHTHALENE

CAS # 1321–65–9

SYNONYMS naphthalene, trichloro • nibren • seekay-wax • nibren-wax

MF $C_{10}H_5Cl_3$

MW 231.50

COLOR AND FORM Colorless to pale-yellow solid.

ODOR Aromatic odor.

DENSITY 1.58.

BOILING POINT 304.44–354.44°C.

MELTING POINT 92.78°C.

FIREFIGHTING PROCEDURES Extinguishant: Foam, carbon dioxide, dry chemical. Also wear a self-contained breathing apparatus (SCBA) with a full facepiece operated in the pressure-demand or other positive-pressure mode. [R9, 2].

TOXIC COMBUSTION PRODUCTS Toxic gases and vapors (such as hydrogen chloride, phosgene, and carbon monoxide) may be released in a fire involving trichloronaphthalene. [R9, 2].

SOLUBILITY Insoluble in water.

VAPOR PRESSURE Less than 1 torr at 20°C.

OTHER CHEMICAL/PHYSICAL PROPERTIES Water solubility: 0.0017–0.0064 mg/L at 25°C (trichloronaphthalene isomers); log K_{ow} = 5.12–5.35 (measured) (trichloronaphthalene isomers); Henry's Law constant = 2.14×10^{-4} atm–m³/mole (est).

PROTECTIVE EQUIPMENT Persons should be provided with and required to use impervious clothing, gloves, faceshields (eight-inch minimum), and other appropriate clothing to prevent skin contact with liquid or molten trichloronaphthalene or its fumes [R9, 2]. Wear appropriate clothing to prevent any possibility of skin contact with molten material or solutions.

Wear eye protection and wash promptly when skin is contaminated. (chloronaphthalenes) [R11].

PREVENTATIVE MEASURES Clothing contaminated with trichloronaphthalene should be removed immediately and placed in closed containers until it can be discarded or until provision is made for the removal of trichloronaphthalene from the clothing [R9, 3]. Employees who handle trichloronaphthalene should wash their hands throughly with soap or mild detergent and water before eating, smoking, or using toilet facilities [R1]. Contact lenses should not be worn when working with this chemical [R10, 219]. SRP: The scientific literature supports the wearing of contact lenses in industrial environments, as part of a program to protect the eye against chemical compounds and minerals causing eye irritation. However, there may be individual substances whose irritating or corrosive properties are such that the wearing of contact lenses would be harmful to the eye. In those specific cases contact lenses should not be worn. Condenser impregnation and other operations involving melting of chloronaphthalene should be enclosed or provided with effective local exhaust ventilation (chloronaphthalenes) [R2]. SRP: Local exhaust ventilation should be applied wherever there is an incidence of point-source emissions or dispersion of regulated contaminants in the work area. Ventilation control of the contaminant as close as possible to its point of generation is both the most economical and safest method to minimize personnel exposure to airborne contaminants.

CLEANUP 1. Remove all ignition sources. 2. Ventilate area of spill. 3. For small quantities, sweep onto paper or other flammable material, place in an appropriate container, and burn in a safe place, such as a fume hood [R1].

DISPOSAL Incineration, preferably after mixing with another combustible fuel. Assure complete combustion to prevent the formation of phosgene. An acid scrubber is necessary to remove the halo acids produced. Recommendable method: Incineration. Peer-review: Ensure plentiful supply of hydrocarbon fuel (peer-review conclusions of an IRPTC expert consultation (May 1985)) [R3].

COMMON USES Employed in lubricants and as insulation for electrical wire. [R13, 600].

OCCUPATIONAL RECOMMENDATIONS

OSHA 8–hr time-weighted average: (5 mg/m³). Skin absorption designation [R7].

THRESHOLD LIMIT 8–hr time-weighted average (TWA) 5 mg/m³, skin (1986) [R12, 34]. Excursion Limit Recommendation: Excursions in worker exposure levels may exceed three times the TLV-TWA for no more than a total of 30 min during a work day, and under no circumstances should they exceed five times the TLV-TWA. [R12, 5].

INTERNATIONAL EXPOSURE LIMITS Maximum allowable concentration (MAC) former USSR = 1 mg/m³ [R2].

ECOTOXICITY LC_{50} shrimp 69–90 µg/L/4-day [R4]; LC_{50} horseshoe crab 80 µg/L (chloronaphthalenes) [R4].

BIODEGRADATION No biodegradation data are available on trichloronaphthalene. A predictive method based upon evaluated biodegradation data (1) and the fact that trichloronaphthalene contains 3 aromatic chlorine substructures predicts that trichloronaphthalene has a low probability of being an important environmental fate process (1) [R5].

BIOCONCENTRATION Based upon measured log K_{ow} values of 5.12–5.35 (1) and water solubilities of 0.0017–0.0064 mg/L at 25°C for various trichloronaphthalene isomers (1), the BCF for trichloronaphthalene can be estimated to range from about 2,900 to 6,900 using appropriate regression-derived equations (2,SRC). Thus, bioconcentration should be very important in aquatic organisms that cannot metabolize trichloronaphthalenes (SRC) [R6].

TSCA REQUIREMENTS Pursuant to section 8 (d) of TSCA, EPA promulgated a model health and safety data reporting rule. The section 8 (d) model rule requires manufacturers, importers, and processors of listed chemical substances and mixtures to submit to EPA copies and lists of unpublished health and safety studies. Trichloronaphthalene is included on this list [R8].

REFERENCES

1. Mackison, F. W., R. S. Stricoff, and L. J. Partridge, Jr. (eds.). *NIOSH/OSHA—Occupational Health Guidelines for Chemical Hazards.* DHHS (NIOSH) Publication No. 81–123 (3 vols). Washington, DC: U.S. Government Printing Office, Jan. 1981.

2. International Labour Office. *Encyclopedia of Occupational Health and Safety.* Vols. I and II. Geneva, Switzerland: International Labour Office, 1983. 466.

3. United Nations. *Treatment and Disposal Methods for Waste Chemicals* (IRPTC File). Data Profile Series No. 5. Geneva, Switzerland: United Nations Environmental Programme, Dec. 1985. 212.

4. Health and Welfare Canada. *Chloronaphthalenes: An Environmental Health Perspective.* p. 79 (1982) 83–EHD–96.

5. Howard, P. H., et al. *Environ Toxicol Chem* 11: 593–603 (1992).

6. (1) Opperhuizen, A. *Tox Environ Chemical* 15: 249–64 (1987) (2) Lyman, W. J., et al., *Handbook of Chemical Property Estimation Methods.* Washington, DC: Amer Chemical Soc, p. 4–9 (1990).

7. 29 CFR 1910.1000 (7/1/91).

8. 40 CFR 716.120 (7/1/91).

9. Mackison, F. W., R. S. Stricoff, and L. J. Partridge, Jr. (eds.). *NIOSH/OSHA—Occupational Health Guidelines for Chemical Hazards.* DHHS (NIOSH) Publication No. 81–123 (3 vols). Washington, DC: U.S. Government Printing Office, Jan. 1981.

10. NIOSH. *NIOSH Pocket Guide to Chemical Hazards.* DHHS (NIOSH) Publication No. 90–117. Washington, DC: U.S. Government Printing Office, June 1990.

11. Sittig, M. *Handbook of Toxic and Hazardous Chemicals.* Park Ridge, NJ: Noyes Data Corporation, 1981. 153.

12. American Conference of Governmental Industrial Hygienists. *Threshold Limit Values for Chemical Substances and*

Physical Agents and Biological Exposure Indices for 1994–1995. Cincinnati, OH: ACGIH, 1994.

13. American Conference of Governmental Industrial Hygienists. *Documentation of the Threshold Limit Values and Biological Exposure Indices.* 5th ed. Cincinnati, OH: American Conference of Governmental Industrial Hygienists, 1986.

■ TRICHLOROPENTYLSILANE

CAS # 107–72–2

DOT # 1728

SYNONYMS amyltrichlorosilane • pentylsilicon trichloride • pentyltrichlorosilane • silane, trichloropentyl • trichloroamylsilane

MF $C_5H_{11}Cl_3Si$

MW 205.60

COLOR AND FORM Colorless to yellow liquid.

ODOR Sharp, like hydrochloric acid; pungent.

DENSITY 1.1330 at 20°C.

SURFACE TENSION 20 dynes/cm = 0.020 N/m at 20°C (estimated).

BOILING POINT 168°C.

HEAT OF COMBUSTION –6,630 Btu/lb = –3,680 cal/g = -154×10^5 J/kg (estimated).

HEAT OF VAPORIZATION 86.8 Btu/lb = 48.2 cal/g = 2.02×10^5 J/kg (estimated).

EXPLOSIVE LIMITS AND POTENTIAL Extreme caution is necessary when handling silane in systems with halogenated compounds, as a trace of free halogen may cause violent explosion (silane) [R4].

FIREFIGHTING PROCEDURES Dry chemical, carbon dioxide. Do not use water or foam [R2].

TOXIC COMBUSTION PRODUCTS Irritating hydrogen chloride and toxic phosgene may be formed [R2].

CORROSIVITY Corrodes metals.

STABILITY AND SHELF LIFE Readily hydrolyzed by moisture, with liberation of HCl [R1].

OTHER CHEMICAL/PHYSICAL PROPERTIES A mixture of isomers.

PROTECTIVE EQUIPMENT Protective equipment should incl corrosion-resistant and impervious suits, foot, hand, and arm, head, and eye and face protection; where corrosive gases may be expected, respiratory equipment is required. (corrosive substances) [R5, 554].

PREVENTATIVE MEASURES Preventive measures should be directed primarily at preventing or minimizing contact between corrosive substances and skin, mucous membranes, and eyes. Adequate ventilation and exhaust arrangements, whether general or local, should be provided whenever corrosive gases or dusts are present. The most satisfactory method of ensuring worker protection is to prevent contact with corrosive substances by utilizing only closed-circuit apparatus. It is good practice to install emergency showers at all strategic locations; bath tubs filled with clean water can also provide valuable service in emergency. (corrosive substances) [R5, 553].

COMMON USES Intermediate for silicones [R1].

OCCUPATIONAL RECOMMENDATIONS None listed.

DOT *Fire or Explosion:* Flammable/combustible material; may be ignited by heat, sparks, or flames. Vapors may travel to a source of ignition, and flash back. Container may explode in heat of fire. Vapor explosion hazard indoors, outdoors, or in sewers. Runoff to sewer may create fire or explosion hazard [R3].

Health Hazards: If inhaled, may be harmful; contact may cause burns to skin and eyes. Fire may produce irritating or poisonous gases. Runoff from fire control or dilution water may cause pollution [R3].

Emergency Action: Keep unnecessary people away; isolate hazard area and deny entry. Stay upwind; keep out of low areas. Self-contained breathing apparatus

(SCBA) and structural firefighter's protective clothing will provide limited protection. Isolate for 1/2 mile in all directions if tank car or truck is involved in fire. Call CHEMTREC at 1–800–424–9300 for emergency assistance. If water pollution occurs, notify the appropriate authorities [R3].

Fire: Some of these materials may react violently with water. Small Fires: Dry chemical, CO₂, halon, water spray, or standard foam. Large Fires: Water spray, fog, or standard foam is recommended. Move container from fire area if you can do so without risk. Do not get water inside container. Cool containers that are exposed to flames with water from the side until well after fire is out. Stay away from ends of tanks. Withdraw immediately in case of rising sound from venting safety device or any discoloration of tank due to fire [R3].

Spill or Leak: Shut off ignition sources; no flares, smoking, or flames in hazard area. Do not touch spilled material; stop leak if you can do so without risk. Use water spray to reduce vapor; do not get water inside container. Small Spills: Take up with sand or other noncombustible absorbent material and place into containers for later disposal. Large Spills: Dike far ahead of liquid spill for later disposal [R3].

First Aid: Move victim to fresh air and call emergency medical care; if not breathing, give artificial respiration; if breathing is difficult, give oxygen. Remove and isolate contaminated clothing and shoes at the site. In case of contact with material, immediately flush skin or eyes with running water for at least 15 minutes. Keep victim quiet and maintain normal body temperature [R3].

FIRE POTENTIAL Certain strong corrosives may, on contact with organic matter or other chemicals, cause fire. (corrosive substances) [R5, 553].

GENERAL RESPONSE Avoid contact with liquid and vapor. Keep people away. Wear goggles, self-contained breathing apparatus, and rubber overclothing (including

gloves). Stop discharge if possible. Call fire department. Isolate and remove discharged material. Notify local health and pollution control agencies.

FIRE RESPONSE Combustible. Poisonous gases are produced in fire. Wear goggles and self-contained breathing apparatus. Extinguish with dry chemicals or carbon dioxide. Do not use water or foam. If breathing is difficult, give oxygen. Liquid will burn skin and eyes. Harmful if swallowed. Remove contaminated clothing and shoes. Flush affected areas with plenty of water. If swallowed, and victim is conscious, have victim drink water or milk. Do not induce vomiting.

RESPONSE TO DISCHARGE Issue warning—corrosive restrict access disperse and flush.

WATER RESPONSE Effect of low concentrations on aquatic life is unknown. May be dangerous if it enters water intakes. Notify local health and wildlife officials. Notify operators of nearby water intakes.

EXPOSURE SYMPTOMS Inhalation causes irritation of mucous membrane. Contact of liquid with eyes or skin causes severe burns, and ingestion causes severe burns of mouth and stomach.

EXPOSURE TREATMENT Get medical attention immediately after exposure to this compound. Inhalation: remove from exposure; support respiration. Eyes: flush with water for 15 min. Skin: flush with water. Ingestion: give large amounts of water.

REFERENCES

1. Hawley, G. G. *The Condensed Chemical Dictionary.* 10th ed. New York: Van Nostrand Reinhold Co., 1981. 70.

2. U.S. Coast Guard, Department of Transportation. *CHRIS—Hazardous Chemical Data.* Manual Two. Washington, DC: U.S. Government Printing Office, Oct., 1978.

3. Department of Transportation. *Emergency Response Guidebook.* 1987. DOT P 5800.4. Washington, DC: U.S. Government Printing Office, 1987, p. G–29.

4. Bretherick, L. *Handbook of Reactive*

Chemical Hazards. 2nd ed. Boston MA: Butterworths, 1979. 1000.

5. International Labour Office. Encyclopedia of Occupational Health and Safety. Vols. I and II. Geneva, Switzerland: International Labour Office, 1983.

1,2,3–TRICHLOROPROPANE

CAS # 96–18–4

SYNONYMS allyl trichloride • glycerol trichlorohydrin • glyceryl trichlorohydrin • nci-c60220 • propane, 1,2,3-trichloro • trichlorohydrin

MF $C_3H_5Cl_3$

MW 147.43

COLOR AND FORM Colorless to straw-colored liquid.

ODOR Odor described as being quite similar to that of trichloroethylene or chloroform; strong acrid odor.

DENSITY 1.3889 at 20°C.

BOILING POINT 156.8°C at 760 mm Hg.

MELTING POINT –14.7°C.

HEAT OF VAPORIZATION 10,714.3 cal/mol.

AUTOIGNITION TEMPERATURE 304°C [R5].

FLAMMABILITY LIMITS Lower, 3.2%; upper 12.6% [R14, 601].

FIREFIGHTING PROCEDURES Water (as a blanket), spray, mist, dry chemical [R4].

SOLUBILITY Soluble in alcohol, ether, chloroform; water solubility: 1,750 mg/L at 25°C.

VAPOR PRESSURE 3 torr at 25°C.

OTHER CHEMICAL/PHYSICAL PROPERTIES Dissolves oils, fats, waxes, chlorinated rubber, and numerous resins; 1 mg/m³= 0.16 ppm; 1 ppm = 6.13 mg/m³; forms azeotropes with camphene, alpha-pinene, and 2,7–dimethyloctane; critical molar volume = 348 cu cm/mol; enthalpy of vaporization (at bp) = 8.87 kcal/mol; enthalpy of sublimation (at 298 K = 11.22 kcal/mol; mass: 97 (Aldermaston, Eight Peak Index of Mass Spectra, UK) (propane, 1,1,1–trichloro); IR: 10638 (Sadtler Research Laboratories IR Grating Collection) (propane, 1,1,2–trichloro); NMR: 3652 (Sadtler Research Laboratories Prism Collection) (propane, 1,1,2–trichloro); mass: 815 (Atlas of Mass Spectral Data, John Wiley and Sons, New York) (propane, 1,1,2–trichloro); IR: 23715 (Sadtler Research Laboratories Prism Collection) (propane, 1,2,2–trichloro); NMR: 383 (Varian Associates NMR Spectra Catalogue) (propane, 1,2,2–trichloro); mass: 815 (Atlas of Mass Spectral Data, John Wiley and Sons, New York) (propane, 1,2,2–trichloro).

PROTECTIVE EQUIPMENT Rubber gloves, self-contained breathing apparatus, and overalls [R5].

PREVENTATIVE MEASURES Contact lenses should not be worn when working with this chemical [R6].

CLEANUP 1. Remove all ignition sources. 2. Ventilate area of spill or leak. 3. For small quantities, absorb on paper towels. Evaporate in a safe place (such as a fume hood). Allow sufficient time for evaporating vapors to completely clear the hood ductwork. Burn the paper in a suitable location away from combustible materials. Large quantities can be reclaimed or collected and atomized in a suitable combustion chamber equipped with an appropriate effluent gas cleaning device [R7].

DISPOSAL 1. By absorbing it in vermiculite, dry sand, earth, or a similar material, and disposing in a secured sanitary landfill. 2. By atomizing in a suitable combustion chamber equipped with an appropriate effluent gas cleaning device [R7].

COMMON USES Chemical intermediate for polysulfide liquid polymers; paint and varnish remover; solvent [R2]; degreasing agent [R1]; in the synthesis of hexafluoropropylene [R3].

OCCUPATIONAL RECOMMENDATIONS

OSHA 8-hr time-weighted average: 50 ppm (300 mg/m³). Transitional limits must continue to be achieved by any combination of engineering controls, work practices, and personal protective equipment during the phase-in period, Sept 1, 1989 through Dec 30, 1992. Final rule limits became effective Dec 31, 1992 [R11].

THRESHOLD LIMIT 8-hr time-weighted average (TWA) 10 ppm, 60 mg/m^3, skin (1987) [R15, 35].

NON-HUMAN TOXICITY LD_{50} rat oral 450 mg/kg [R14, 601]; LD_{50} rabbit percutaneous 2,500 mg/kg [R14, 601].

ECOTOXICITY LD_{50} *Poecilia reticulata* (guppies) 42 ppm 7-day exposure (conditions of bioassay not specified) [R9].

IARC SUMMARY Evaluation: There is inadequate evidence in humans for the carcinogenicity of 1,2,3–trichloropropane. There is sufficient evidence in experimental animals for the carcinogenicity of 1,2,3–trichloropropane. Overall evaluation: 1,2,3–trichloropropane is probably carcinogenic to humans (Group 2A). In making the overall evaluation, the working group took into account the following evidence: (1) 1,2,3–trichloropropane causes tumors at multiple sites and at high incidence in mice and rats. (2) The metabolism of 1,2,3–trichloropropane is qualitatively similar in human and rodent microsomes. (3) 1,2,3–Trichloropropane is mutagenic to bacteria and to cultured mammalian cells and binds to the DNA of animals treated in vivo [R8].

BIOCONCENTRATION A bioconcentration factor for 1,2,3–trichloropropane can be estimated at 9, suggesting that bioaccumulation in aquatic organisms should not occur (1,SRC) [R10].

ATMOSPHERIC STANDARDS This action promulgates standards of performance for equipment leaks of volatile organic compounds (VOC) in the synthetic organic chemical manufacturing industry (SOCMI). The intended effect of these standards is to require all newly constructed, modified, and reconstructed SOCMI process units to use the best demonstrated system of continuous emission reduction for equipment leaks of VOC, considering costs, non-air-quality health, and environmental impact and energy requirements. 1,2,3,–Trichloropropane is produced, as an intermediate or final product, by pro-

cess units covered under this subpart [R12].

FIRE POTENTIAL Underwriters classification: 20–25 (between kerosene class and paraffin oil class) [R13, 1188]. Moderate, when exposed to heat, flames (sparks), or powerful oxidizers [R4].

REFERENCES

1. Sax, N. I., and R. J. Lewis, Sr. (eds.). *Hawley's Condensed Chemical Dictionary*. 11th ed. New York: Van Nostrand Reinhold Co., 1987. 1178.

2. SRI.

3. Gangal, S.V., *Kirk–Othmer Encycl Chemical Tech* 3rd Ed. New York: John Wiley, 11: 24–35 (1981).

4. Sax, N. I. *Dangerous Properties of Industrial Materials,*. 6th ed. New York, NY: Van Nostrand Reinhold, 1984. 173.

5. ITII. *Toxic and Hazarous Industrial Chemicals Safety Manual*. Tokyo, Japan: The International Technical Information Institute, 1982. 539.

6. NIOSH. *Pocket Guide to Chemical Hazards*. 2nd Printing. DHHS (NIOSH) Publ. No. 85–114. Washington, DC: U.S. Dept. of Health and Human Services, NIOSH/ Supt. of Documents, GPO, February 1987. 230.

7. Mackison, F. W., R. S. Stricoff, and L. J. Partridge, Jr. (eds.). *NIOSH/OSHA—Occupational Health Guidelines for Chemical Hazards*. DHHS (NIOSH) Publication No. 81–123 (3 vols). Washington, DC: U.S. Government Printing Office, Jan. 1981. 3.

8. IARC. *Monographs on the Evaluation of the Carcinogenic Risk of Chemicals to Man*. Geneva: World Health Organization, International Agency for Research on Cancer, 1972–present (multivolume work), p. 63 240 (1995).

9. Verschueren, K. *Handbook of Environmental Data of Organic Chemicals*. 2nd ed. New York, NY: Van Nostrand Reinhold Co., 1983. 1146

10. Lyman, W. J., et al., *Handbook of Chemical Property Estimation Methods*

New York: McGraw-Hill, pp. 5–1 to 5–30 (1982).

11. 54 FR 2920 (1/19/89).

12. 40 CFR 60.489 (7/1/89).

13. Sax, N.I. *Dangerous Properties of Industrial Materials.* 4th ed. New York: Van Nostrand Reinhold, 1975.

14. American Conference of Governmental Industrial Hygienists. *Documentation of the Threshold Limit Values and Biological Exposure Indices.* 5th ed. Cincinnati, OH: American Conference of Governmental Industrial Hygienists, 1986.

15. American Conference of Governmental Industrial Hygienists. *Threshold Limit Values for Chemical Substances and Physical Agents and Biological Exposure Indices for 1994–1995.* Cincinnati, OH: ACGIH, 1994.

■ TRICHLOROVINYLSILANE

CAS # 75–94–5

DOT # 1305

SYNONYMS A-150 • a-150 [silane] • silane, trichloroethenyl • silane, trichlorovinyl • silane, vinyl trichloro a-150 • trichloro (vinyl) silane • trichlorovinyl silicane • trichlorovinylsilicon • union carbide a-150 • vinylsilicon trichloride • vinyltrichlorosilane • vtcs

MF $C_2H_3Cl_3Si$

MW 161.49

COLOR AND FORM Colorless or pale-yellow liquid.

ODOR Sharp, choking; like hydrochloric acid.

DENSITY 1.3.

SURFACE TENSION 28 dynes/cm = 0.028 N/m at 20°C.

BOILING POINT 91°C.

MELTING POINT FP: −139°F.

HEAT OF COMBUSTION −4,300 Btu/lb = −2,400 cal/g = −100×10⁷ J/kg.

HEAT OF VAPORIZATION 88 Btu/lb = 49 cal/g = 2.0×10⁵ J/kg.

AUTOIGNITION TEMPERATURE 505°F [R2].

EXPLOSIVE LIMITS AND POTENTIAL Vinyl trichlorosilane reacts violently with water or moist air. [R5, p. 491M–426]. Vapor may explode if ignited in an enclosed area. [R2].

FLAMMABILITY LIMITS 3% [R2].

FIREFIGHTING PROCEDURES In advanced or massive fires, firefighting should be done from safe distance, or from protected location [R5, p. 325M–189]. Extinguish with dry chemicals or carbon dioxide. Do not use water or foam on fire. Cool exposed containers with waater [R2].

TOXIC COMBUSTION PRODUCTS Toxic chlorine and phosgene gases may be formed in fires [R2].

SOLUBILITY Soluble in most organic solvents.

OTHER CHEMICAL/PHYSICAL PROPERTIES Reacts with alcohol. Polymerizes easily. Readily hydrolyzed, with liberation of hydrochloric acid.

PROTECTIVE EQUIPMENT Acid-vapor-type respiratory protection; rubber gloves; chemical worker's goggles; other protective equipment as necessary to protect skin and eyes [R2].

PREVENTATIVE MEASURES Handling should be done as far as possible under suitably ventilated hoods (chlorosilanes) [R3].

COMMON USES Monomer for copolymers in water repellents, electrical insulating resins, and high-temperature resins for paints [R1]. Intermediate for silicones; coupling agent in adhesives and bonds [R4].

OCCUPATIONAL RECOMMENDATIONS None listed.

GENERAL RESPONSE DATA Shut off ignition sources. Call fire department. Avoid contact with liquid. Keep people away. Stop discharge if possible. Isolate and remove discharged material. Notify local health and pollution control agencies.

FIRE RESPONSE Flammable. Poisonous gases may be produced in fire. Containers may explode in fire. Flashback along vapor trail may occur. Vapor may explode if ignited in an enclosed area. Extinguish with dry chemicals or carbon dioxide. Do not use water or foam on fire. Cool exposed containers with water.

EXPOSURE RESPONSE Call for medical aid. Vapor irritating to eyes, nose, and throat. Harmful if inhaled. Move victim to fresh air. If breathing is difficult, give oxygen. Liquid will burn skin and eyes. Harmful if swallowed. Remove contaminated clothing and shoes. Flush affected areas with plenty of water. If swallowed and victim is conscious, have victim drink water or milk. Do not induce vomiting.

RESPONSE TO DISCHARGE Issue warning—high flammability, corrosive, air contaminant. Restrict access. Disperse and flush with care.

WATER RESPONSE Effect of low concentrations on aquatic life is unknown. May be dangerous if it enters water intakes. Notify local health and wildlife officials. Notify operators of nearby water intakes.

EXPOSURE SYMPTOMS Inhalation causes irritation of mucous membranes. Vapor irritates eyes. Contact with liquid causes severe burns of eyes and skin. Ingestion causes burns of mouth and stomach.

EXPOSURE TREATMENT Get medical attention following all exposures to this compound. Inhalation: remove victim from exposure; give artificial respiration if required. Eyes: flush with water for 15 min. Skin: flush with water. Ingestion: Do not induce vomiting; give large amount of water.

REFERENCES

1. SRI.

2. U.S. Coast Guard, Department of Transportation. *CHRIS—Hazardous Chemical Data.* Manual Two. Washington, DC: U.S. Government Printing Office, Oct., 1978.

3. Lefaux, R. *Practical Toxicology of Plas-tics.* Cleveland: CRC Press Inc., 1968. 109.

4. Hawley, G.G. *The Condensed Chemical Dictionary.* 10th ed. New York: Van Nostrand Reinhold Co., 1981. 1087.

5. National Fire Protection Association. *Fire Protection Guide on Hazardous Materials.* 7th ed. Boston, MA: National Fire Protection Association, 1978.

■ TRIMETHYLCHLOROSILANE

CAS # 75–77–4

DOT # 1298

SYNONYMS chlorotrimethylsilane • monochlorotrimethylsilicon • silane, chlorotrimethyl • silane, trimethylchloro • silicane, chlorotrimethyl • silylium, trimethyl-, chloride • trimethyl chlorosilane • trimethylsilyl chloride • tl-1163

MF C_3H_9ClSi

MW 108.66

COLOR AND FORM Colorless liquid.

ODOR Sharp, hydrochloric acid-like odor; acrid.

DENSITY 0.854 at 25°C.

SURFACE TENSION 17.8 dynes/cm (est).

BOILING POINT 57°C.

MELTING POINT –57.7°C.

HEAT OF COMBUSTION –10,300 Btu/lb (est).

HEAT OF VAPORIZATION 126 Btu/lb.

AUTOIGNITION TEMPERATURE 743°F [R2].

EXPLOSIVE LIMITS AND POTENTIAL Containers may explode in fire. Vapor may explode if ignited in enclosed area [R2].

FLAMMABILITY LIMITS Lower 1.8% [R2].

TOXIC COMBUSTION PRODUCTS Toxic and irritating hydrogen chloride and phosgene may be formed in fires [R2].

SOLUBILITY Soluble in benzene, ether, perchloroethylene.

OTHER CHEMICAL/PHYSICAL PROPERTIES
Ratio of specific heats of vapor (gas):
1.0683 (est). Readily hydrolyzed with liberation of hydrochloric acid. Organo-functional silanes are noted for their ability to bond organic polymer systems to inorganic substrates (silane compounds). The reaction of organosilanes with halogens and halogen compounds usually proceeds in good yield through cleavage of the Si--H bond and formation of the silicon-halogen bond. Reaction with fluorine, however, does not proceed satisfactorily because of cleavage of not only the Si--H but also C--Si and C--H bonds. Direct halogenation with chlorine, bromine, and iodine proceeds smoothly, however. (organosilanes).

PROTECTIVE EQUIPMENT Prevent, from outset, any planned, or accidental contact with corrosive substances by utilizing only closed-circuit apparatus. Otherwise, corrosion-resistant and impervious suits or overalls, foot, hand and arm, head, eye, and face protection should be relied on (corrosive substances) [R4, 554]. Where aprons are used these should have bibs; sleeves should be worn outside gloves, and trouser legs should cover tops of shoes when exposure is not severe. Barrier creams can often be used instead of gloves (corrosive substances) [R4, 554]. Acid-vapor-type respiratory protection; rubber gloves; chemical worker's goggles; other protective equipment as necessary to protect skin and eyes. [R2].

PREVENTATIVE MEASURES It is good practice to install emergency showers at all strategic locations; these should be actuated automatically by pressure on foot plate; bath tubs filled with clean water can also provide valuable service in emergency (corrosive substances) [R4, 548]. SRP: The scientific literature supports the wearing of coontact lenses in industrial environments, as part of a program to proteect the eye against chemical compounds and minerals causing eye irritation. However, there may be individual substances whose irritating or corrosive properties are such that the wearing of contact lenses would be harmful to the eye. In those specific cases contact lenses

should not be worn. Caution: silanes are toxic. Avoid contact with skin and eyes. Use effective fume-removal device (silanes) [R5]. SRP: Contaminated protective clothing should be segregated in such a manner so that there is no direct personal contact by personnel who handle, dispose of, or clean the clothing. Quality assurance to ascertain the completeness of the cleaning procedures should be implemented before the decontaminated protective clothing is returned for reuse by the workers. Contaminated clothing should not be taken home at end of shift, but should remain at employee's place of work for cleaning.

CLEANUP Shut off ignition sources. Call fire department. Avoid contact with liquid. Keep people away. Stop discharge if possible. Isolate and remove discharged material. Notify local health and pollution control agencies [R3].

COMMON USES Intermediate for silicone fluids, as a chain-terminating agent, imparting water repellency [R1].

OCCUPATIONAL RECOMMENDATIONS None listed.

GENERAL RESPONSE DATA Shut off ignition sources. Call fire department. Avoid contact with liquid. Keep people away. Stop discharge if possible. Isolate and remove discharged material. Notify local health and pollution control agencies.

FIRE RESPONSE Flammable. Poisonous gases may be produced in fire. Containers may explode in fire. Flashback along vapor trail may occur. Vapor may explode if ignited in an enclosed area. Extinguish with dry chemicals or carbon dioxide. Do not use water or foam on fire. Cool exposed containers with water.

EXPOSURE RESPONSE Call for medical aid. Vapor irritating to eyes, nose, and throat. Harmful if inhaled. Move victim to fresh air. If breathing has stopped, give artificial respiration. If breathing is difficult, give oxygen. Liquid will burn skin and eyes. Harmful if swallowed. Remove contaminated clothing and shoes. Flush affected areas with plenty of water. If swallowed and victim is conscious, have victim drink water or milk. Do not induce vomiting.

RESPONSE TO DISCHARGE Issue warning—high flammability, air contaminant, corrosive. Restrict access. Evacuate area. Disperse and flush with care.

WATER RESPONSE Effect of low concentrations on aquatic life is unknown. May be dangerous if it enters water intakes. Notify local health and wildlife officials. Notify operators of nearby water intakes.

EXPOSURE SYMPTOMS Inhalation of vapor irritates mucous membranes. Contact of liquid with eyes or skin causes severe burns. Ingestion causes severe burns of mouth and stomach.

EXPOSURE TREATMENT Get medical attention following all exposures to this compound. Inhalation: remove victim from exposure; if breathing is difficult or stopped, give artificial respiration. Eyes: flush with water for 15 min. Skin: flush with water. Ingestion: do not induce vomiting; give large amount of water.

REFERENCES

1. Sax, N. I., and R. J. Lewis, Sr. (eds.). *Hawley's Condensed Chemical Dictionary.* 11th ed. New York: Van Nostrand Reinhold Co., 1987. 1187.

2. U.S. Coast Guard, Department of Transportation. *CHRIS—Hazardous Chemical Data.* Volume II. Washington, DC: U.S. Government Printing Office, 1984–5.

3. U.S. Coast Guard, Department of Transportation. *CHRIS—Hazardous Chemical Data.* Manual Two. Washington, DC: U.S. Government Printing Office, Oct., 1978.

4. International Labour Office. *Encyclopedia of Occupational Health and Safety.* Vols. I and II. Geneva, Switzerland: International Labour Office, 1983.

5. Association of Official Analytical Chemists. *Official Methods of Analysis.* 10th ed. and supplements. Washington, DC: Association of Official Analytical Chemists, 1965. New editions through 13th ed. plus supplements, 1982, p. 13/777 43.321.

■ 1,3,5-TRINITROBENZENE

CAS # 99–35–4

SYNONYMS benzene, 1,3,5-trinitro- • benzite • s-trinitrobenzene • sym-trinitrobenzene • symmetric trinitrobenzene • syn-trinitrobenzene • tnb • trinitrobenzeen (Dutch) • trinitrobenzene • Trinitrobenzol (German)

MF $C_6H_3N_3O_6$

MW 213.12

COLOR AND FORM Slightly yellowish crystals.

DENSITY 1.76 at 20°C.

BOILING POINT 315°C at 760 mm Hg.

MELTING POINT 122.5°C.

HEAT OF COMBUSTION –663.7 kcal/mol wt at 20°C (solid).

EXPLOSIVE LIMITS AND POTENTIAL Classified as high explosive. Highly sensitive to shock or heat [R5].

FIREFIGHTING PROCEDURES Fight fires from an explosion-resistant location. In advanced or massive fires, the area should be evacuated. If fire occurs in the vicinity of this material, water should be used to keep containers cool [R5]. Dangerously explosive. Do not fight fires in a cargo of explosives. Evacuate area. and let burn (trinitrobenzene, dry (high explosive), class A; trinitrobenzene, wet, containing at least 10% water, over 16 ounces in one outside package (high explosive), class A) [R6]. Dangerously explosive. Flood with water. Cool all affected containers with flooding quantities of water. Apply water from as far a distance as possible (trinitrobenzene, wet (containing at least 10% water)) [R6]. Evacuation: If the material is on fire or involved in fire consider evacuation of one (1) mile radius (trinitrobenzene, dry (high explosive), class A; trinitrobenzene, wet, containing at least 10% water, over 16 ounces in one outside package (high explosive), class A) [R6].

SOLUBILITY 0.035 g/100 g water; 6.2 g/100 g benzene; 4.9 g/100 g methanol; 1.9 g/100 g alcohol; 0.25 g/100 g carbon disulfide; 1.5 g/100 g ether; 0.5 g/100 g petroleum ether; freely soluble in dilute sodium sulfite; soluble in chloroform; very soluble in hot toluene; >10% in acetone.

VAPOR PRESSURE 3.2×10^{-6} mm Hg at 20°C (extrapolated).

OCTANOL/WATER.

PARTITION COEFFICIENT Log K_{ow} = 1.10.

OTHER CHEMICAL/PHYSICAL PROPERTIES Can be sublimed by careful heating; it is dimorphous. The other (rare) form melts at 61°C; orthorhombic bipyramidal plates from glacial acetic acid; rhombic plates from benzene; leaflets from water; enthalpies of formation: –10.40 kcal/mole (crystalline solid); half-wave potentials (vs saturated calomel electrode) of 1,3,5–trinitrobenzene at 25°C: –0.20, –0.29, –0.34 (phthalate buffer, pH 4.1), and –0.34, –0.48, –0.65 (borate buffer, pH 9.2); heat of sublimation: 23.8 kcal/mole at 298 K; IR: 3515 (Documentation of Molecular Spectroscopy Collection) (benzene, 1,2,4–trinitro); UV: 4–51 (Organic Electronic Spectral Data, Phillips et al., John Wiley and Sons, New York) (benzene, 1,2,4–trinitro).

PROTECTIVE EQUIPMENT For rescue operations, use complete protective clothing [R5].

PREVENTATIVE MEASURES If material is not on fire and not involved in fire keep sparks, flames, and other sources of ignition away. Keep spilled material wet. Wet spilled material before picking it up. Do not attempt to sweep up dry material. (trinitrobenzene, dry (high explosive), class A; trinitrobenzene, wet, containing at least 10% water, over 16 ounces in one outside package (high explosive), class A) [R6].

DISPOSAL Generators of waste (equal to or greater than 100 kg/mo) containing this contaminant, EPA hazardous waste number U234, must conform with USEPA

regulations in storage, transportation, treatment, and disposal of waste [R7].

COMMON USES Explosive [R1]; used to vulcanize natural rubber [R2]; used as an acid-base indicator in the pH range of 12.0–14.0 [R3].

OCCUPATIONAL RECOMMENDATIONS None listed.

BIODEGRADATION The microbial degradation of 1,3,5–trinitrobenzene was incomplete and unsustained in Tennessee River water. Nitro group reduction occurred in the presence of added nutrients and lab cultures of Tennessee River microorganisms [R8].

BIOCONCENTRATION Bioconcentration factors (BCF) of 5 and 23 have been estimated for 1,3,5– trinitrobenzene using regression equations (1) based on a log octanol/water partition coefficient of 1.18 (2) and a water solubility of 340 mg/L at 20°C (3), respectively (SRC). These BCF values suggest that this compound would not bioaccumulate significantly in aquatic organisms (4,SRC) [R9].

CERCLA REPORTABLE QUANTITIES Persons in charge of vessels or facilities are required to notify the National Response Center (NRC) immediately, when there is a release of this designated hazardous substance, in an amount equal to or greater than its reportable quantity of 10 lb or 4.54 kg. The toll-free number of the NRC is 1–800–424–8802; in the Washington DC metropolitan area 1–202–426–2675. The rule for determining when notification is required is stated in 40 CFR 302.4 (section IV. D. 3.b) [R10].

RCRA REQUIREMENTS As stipulated in 40 CFR 261.33, when 1,3,5–trinitrobenzene, as a commercial chemical product or manufacturing chemical intermediate or an off-specification commercial chemical product or a manufacturing chemical intermediate, becomes a waste, it must be managed according to federal or state hazardous waste regulations. Also defined as a hazardous waste is any residue, contaminated soil, water, or other debris resulting from the cleanup of a spill, into water or on dry land, of this waste. Generators of small quantities of this waste may qualify for partial exclusion from hazardous waste regulations (40 CFR 261.5) [R11].

DOT *Fire or Explosion:* Flammable/combustible material; may be ignited by heat, sparks, or flames. Dried out material may explode if exposed to heat, flame, or shock; keep material wet with water or treat it as an explosive (Guide 46). Runoff to sewer may create fire or explosion hazard (trinitrobenzene, wet) [R4].

Health Hazards: Contact may cause burns to skin and eyes. Fire may produce irritating or poisonous gases. Runoff from fire control or dilution water may cause pollution (trinitrobenzene, wet) [R4].

Emergency Action: Keep unnecessary people away; isolate hazard area, and deny entry. Stay upwind; keep out of low areas. Self-contained breathing apparatus (SCBA) and structural firefighter's protective clothing will provide limited protection. Call CHEMTREC at 1–800–424–9300 for emergency assistance. If water pollution occurs, notify the appropriate authorities (trinitrobenzene, wet) [R4].

Fire: Tire Fires: Flood with water; if no water is available, use dry chemical or dirt. Caution: Tire fires may start again. Do not move cargo or vehicle if cargo has been exposed to heat. For massive fire in cargo area, use unmanned hose holder, or monitor nozzles; if this is impossible, withdraw from area, and let fire burn (trinitrobenzene, wet) [R4].

Spill or Leak: Shut off ignition sources; no flares, smoking, or flames in hazard area. Do not touch spilled material. Small Spills: Flush area with flooding amounts of water. Large Spills: Wet down with water and dike for later disposal (trinitrobenzene, wet) [R4].

First Aid: Move victim to fresh air; call for emergency medical care. Remove and isolate contaminated clothing and shoes at the site. In case of contact with material, immediately flush skin or eyes with running water for at least 15 minutes. (trinitrobenzene, wet) [R4].

REFERENCES

1. *The Merck Index.* 10th ed. Rahway, NJ: Merck Co., Inc., 1983. 1389.

2. *Kirk–Othmer Encyclopedia of Chemical Technology.* 3rd ed., Volumes 1–26. New York, NY: John Wiley and Sons, 1978–1984, p. 20 (82) 33393.

3. *Kirk–Othmer Encyclopedia of Chemical Technology.* 3rd ed., Volumes 1–26. New York, NY: John Wiley and Sons, 1978–1984, p. 13 (81) 10.

4. Department of Transportation. *Emergency Response Guidebook 1987.* DOT P 5800.4. Washington, DC: U.S. Government Printing Office, 1987, p. G–33.

5. National Fire Protection Association. *Fire Protection Guide on Hazardous Materials.* 9th ed. Boston, MA: National Fire Protection Association, 1986, p. 49–91.

6. Association of American Railroads. *Emergency Handling of Hazardous Materials in Surface Transportation.* Washington, DC: Assoc. of American Railroads, Hazardous Materials Systems (BOE), 1987. 700.

7. 40 CFR 240–280, 300–306, 702–799 (7/1/89).

8. Mitchell, W. R., et al. *Microbial Interactions with Several Munitions Compounds: 1,3–dinitrobenzene, 1,3,5–Trinitrobenzene, and 3,5–Dinitroaniline;* Report, ISS USAMBRDL-TR-8201, Order No AD-A116651, 44 pages (1982).

9. (1) Lyman, W. J., et al., *Handbook of Chemical Property Estimation Methods.* New York: McGraw-Hill, p. 5–5 (1982). (2) Hansch, C., and A. J. Leo. *Medchem Project* Issue No. 25, Claremont, CA: Pomona College (1985). (3) Spanggord, R. J., et al., *Environmental Fate Studies on Certain Munition Wastewater Constituents* Final Report, Phase 1—Literature Review. SRI project no. LSU–7934 Menlo Park, CA: SRI International (1980). (4) Kenaga, E. E., *Ecotox Environ Safety* 4: 26–38 (1980).

10. 54 FR 33418 (8/14/89).

11. 40 CFR 261.33 (7/1/88).

■ TRIPHENYL PHOSPHATE

CAS # 115–86–6

SYNONYMS celluflex-tpp • disflamoll-tp • phenyl phosphate ((PHO)3PO) • phosflex-tpp • phosphoric acid, triphenyl ester • tp • tpp • triphenoxyphosphine oxide • triphenylphosphate

MF $C_{18}H_{15}O_4P$

MW 326.30

COLOR AND FORM Crystals from absolute alcohol-ligroin, prisms from alcohol, needles from ether-ligroin; colorless crystalline powder; white platelets.

ODOR Characteristic odor resembling phenol at room temp; odorless.

DENSITY 1.2055 at 50°C.

BOILING POINT 245°C at 11 mm Hg.

MELTING POINT 49–50°C.

FLAMMABILITY LIMITS Noncombustible [R17, 547].

FIREFIGHTING PROCEDURES Carbon dioxide, dry chemical [R5].

TOXIC COMBUSTION PRODUCTS Hazardous decomposition products: Toxic gases and vapors (such as phosphoric acid fume and carbon monoxide) may be released in a fire involving triphenyl phosphate [R7].

SOLUBILITY 0.002% soluble in water at 54°C; practically insoluble in petrol; soluble in benzene, chloroform, ether, acetone; moderately soluble in alcohol; very soluble in carbon tetrachloride; soluble in most lacquers, solvents, thinners, oils.

VAPOR PRESSURE 1 mm Hg at 193.5°C.

OTHER CHEMICAL/PHYSICAL PROPERTIES 13.32 mg/m³ = approximately 1 ppm; wt/gal = 10.5 lb; the Henry's Law constant for triphenyl phosphate can be estimated to be approximately 5.65×10^{-8} atm m³/mole at 25°C (SRC). The Henry's Law constant for triphenyl phosphate can also be estimated to be 3.98×10^{-8} atm–m³/mole at 25°C using a chemical structure estimation method (4,SRC).

PROTECTIVE EQUIPMENT Recommendations for respirator selection. Max concentration for use: 15 mg/m³: Respirator Class: Any dust respirator [R6].

PREVENTATIVE MEASURES Handle with caution, prevent contact with skin, and keep air concentration below recommended values. [R15, 2374].

CLEANUP If triphenyl phosphate is spilled: ventilate areas of spill. For small quantities, sweep onto paper or other suitable material, place in appropriate container, and burn in a safe place (such as a fume hood). Large quantities may be retained; however, if this is not practical, dissolve in a flammable solvent (such as alcohol), and atomize in a suitable combustion chamber equipped with an appropriate effluent-gas-cleaning device [R7].

DISPOSAL Group I containers: combustible containers from organic or metallo-organic pesticides (except organic mercury, lead, cadmium, or arsenic compounds) should be disposed of in pesticide incinerators or in specified landfill sites. (organic or metallo-organic pesticides) [R8].

COMMON USES Noncombustible substitute for camphor in celluloid; impregnating roofing paper; rendering acetylcellulose, nitrocellulose, airplane "dope" stable and fireproof; plasticizer in lacquers and varnishes [R1]; plasticizer for hot-melt adhesives [R2]; impregnating upholstery [R3]; fire-retarding agent, plasticizer for cellulose acetate and nitrocellulose [R4].

OCCUPATIONAL RECOMMENDATIONS
OSHA 8-hr time-weighted average: 3 mg/m³ [R12].

NIOSH 10-hr time-weighted average: 3 mg/m³ [R6].

THRESHOLD LIMIT 8–hr time-weighted average (TWA) 3 mg/m³ (1986) [R19, 35].

NON-HUMAN TOXICITY LD_{50} mouse, oral 1.32 ± 0.28 g/kg [R3]; LD_{50} rat oral 3.8 ± 0.26 g/kg [R3]; LD_{50} monkey sc 0.5 g/kg [R3]; LD_{50} rat Sprague Dawley oral 10.8 g/kg [R16, 489]; LD_{50} rat, oral 3.8 g/kg (in oil) [R16, 489]; LD_{50} mouse, oral 1.3 g/kg (in oil) [R16, 489]; LD_{50} white leghorn chicken, oral >5.0 g/kg [R16, 489]; LD50 rabbit dermal >7.9 g/kg [R16, 490].

ECOTOXICITY LC50 *Pimephales promelas* (fathead minnow) 0.87 mg/L/96 hr (confidence limit 0.81–0.94 mg/L), flow-through bioassay with measured concentrations, 24.5°C, dissolved oxygen 6.4 mg/L, hardness 45.6 mg/L calcium carbonate, alkalinity 43.4 mg/L calcium carbonate, and pH 7.78 [R9]. EC50 *Pimephales promelas* (fathead minnow) 0.51 mg/L/96 hr (confidence limit 0.47–0.56 mg/L), flow-through bioassay with measured concentrations, 24.5°C, dissolved oxygen 6.4 mg/L, hardness 45.6 mg/L calcium carbonate, alkalinity 43.4 mg/L calcium carbonate, and pH 7.78. Effect: loss of equilibrium [R9]. LC50 *Lepomis macrochirus* 290 ppm/96 hr, static bioassay in fresh water at 23°C, mild aeration applied after 24 hr [R18, 1171]. LC50 *Menidia beryllina* 95 ppm/96 hr, static bioassay in synthetic seawater at 23°C, mild aeration applied after 24 hr [R18, 1172]. LC50 rainbow trout 300 μg/L/96 hr (conditions of bioassay not specified) [R18, 1172].

BIODEGRADATION The ultimate biodegradation of triphenyl phosphate (TPP), initial concentration of 22 ppm, was observed to be 82% of theoretical CO_2 evolution in a CO_2-evolution screening study using acclimated seed (14-day acclimation period) over an incubation period of 27 days (2). The ultimate biodegradation of triphenyl phosphate, initial concentration of 18.3 ppm, was observed to be 61.9, and 81.8% of theoretical CO_2 evolution in a CO_2-evolution screening study using acclimated seed (14-day acclimation period) over an incubation period of 7 and 28 days, respectively (1). Triphenyl phosphate (concn of 3 to 13 ppm) degraded 84 to 96% over 24 hrs in semicontinuous activated sludge (SCAS) tests (1,2,4) and a half-life of 2 to 4 days was observed in river die-away tests for triphenyl phosphate at an initial concentration of 0.05 to 1 ppm (1,2). In 3 freshwater grab sample studies (pH 7.8 to 8.2), triphenyl phosphate exhibited 100% biodegradation after a 2-day lag period (7- to 8-day incubation time) (3) [R10].

BIOCONCENTRATION Rates of uptake by rainbow trout (*Salmo gairdneri*) in short-term static exposures were higher for triphenyl phosphate (TPP) in dechlorinated city water than in river water, suggesting that sorption to suspended solids or dissolved organic matter reduces the bioavailability of triphenyl phosphate in natural waters (4–5). Bioconcentration factors (BCFs) of 180 to 280 have been measured in rainbow trout exposed to Pydraul 50E (a hydraulic fluid containing 35% triphenyl phosphate) for 90 days in flowing water (1). A BCF range of 132 to 364 was observed for rainbow trout exposed to triphenyl phosphate in flowing water for 90 days (2). BCFs of 573 and 561 have been measured using static tests for radiolabelled TPP in rainbow trout and fathead minnows (*Pimephales promelas*), respectively (3). A BCF of 2,590 was determined for triphenyl phosphate in rainbow trout based on the bioaccumulation ratio (the rate of the initial uptake rate constant to the initial clearance rate constant); however, this represents the worse-case estimate based on the total ^{14}C content of whole fish after 24–hr exposure (5) [R11].

TSCA REQUIREMENTS Pursuant to section 8 (d) of TSCA, EPA promulgated a model health and safety data reporting rule. The section 8 (d) model rule requires manufacturers, importers, and processors of listed chemical substances and mixtures to submit to EPA copies and lists of unpublished health and safety studies. Phosphoric acid, triphenyl ester is included on this list [R13].

FIRE POTENTIAL Fire hazard: slight, when exposed to heat or flame; spontaneous heating: no [R5].

FDA Triphenylphosphate is an indirect food additive for use only as a component of adhesives [R14].

REFERENCES

1. Budavari, S. (ed.). *The Merck Index— Encyclopedia of Chemicals, Drugs, and Biologicals.* Rahway, NJ: Merck and Co., Inc., 1989. 1533.

2. SRI.

3. American Conference of Governmental

Industrial Hygienists. *Documentation of the Threshold Limit Values and Biological Exposure Indices.* 5th ed. Cincinnati, OH: American Conference of Governmental Industrial Hygienists, 1986. 613.

4. Sax, N. I., and R. J. Lewis, Sr (eds.). *Hawley's Condensed Chemical Dictionary.* 11th ed. New York: Van Nostrand Reinhold Co., 1987. 1192.

5. Sax, N. I. *Dangerous Properties of Industrial Materials.* 6th ed. New York, NY: Van Nostrand Reinhold, 1984. 2684.

6. NIOSH. *NIOSH Pocket Guide to Chemical Hazards.* DHHS (NIOSH) Publication No. 90–117. Washington, DC: U.S. Government Printing Office, June 1990, 220.

7. Mackison, F. W., R. S. Stricoff, and L. J. Partridge, Jr. (eds.). *NIOSH/OSHA—Occupational Health Guidelines for Chemical Hazards.* DHHS (NIOSH) Publication No. 81–123 (3 vols). Washington, DC: U.S. Government Printing Office, Jan. 1981.

8. 40 CFR 165.9 (a) (7/1/91).

9. Geiger, D. L., S. H. Poirier, L.T. Brooke, D. J. Call (eds.). *Acute Toxicities of Organic Chemicals to Fathead Minnows (Pimephales promelas).* Vol. III. Superior, Wisconsin: University of Wisconsin-Superior, 1986. 311.

10. (1) Saeger, V. W., et al., *Environ Sci Technol* 13: 840–4 (1979). (2) Mayer, F. L., et al., pp. 103–23 in *Aquatic Toxicity and Hazard Assessment:* 4th Conf. ASTM STP 737. Branson, D. R., K. L. Dickson (eds.), *Amer Soc Test Mater* (1981). (3) Howard, P.H., P. G. Deo. *Bull Environ Contam Toxicol* 22: 337–44 (1979). (4) Carson, D. B., et al., *Aquat Toxicol Risk Assess* 13: 48–59 ASTM STP 4 (1990).

11. (1) Lombardo P, I. J. Egry. *J Assoc Offic Anal Chemical* 62: 47–51 (1979). (2) Mayer, F. L., et al., pp. 103–23 in *Aquatic Toxicity and Hazard Assessment:* 4th Conf. ASTM STP 737. Branson, D. R., K. L. Dickson (eds.), *Amer Soc Test Mater* (1981). (3) Muir, D.C.G., et al. *Chemosphere* 12: 155–66 (1983). (4) Boethling, R. S., J. C. Cooper. *Res Rev* 94: 49–99 (1985). (5) Muir, D.C.G., et al., *Chemosphere* 9: 525–32 (1980).

12. 29 CFR 1910.1000 (7/1/90).

13. 40 CFR 716.120 (7/1/90).

14. 21 CFR 175.105 (4/1/91).

15. Clayton, G. D., and F. E. Clayton (eds.). *Patty's Industrial Hygiene and Toxicology.* Volume 2A, 2B, 2C: Toxicology. 3rd ed. New York: John Wiley and Sons, 1981–1982.

16. Snyder, R. (ed.). *Ethyl Browning's Toxicity and Metabolism of Industrial Solvents.* 2nd ed. Volume II: *Nitrogen and Phosphorus Solvents.* Amsterdam–New York–Oxford: Elsevier, 1990.

17. ITII. *Toxic and Hazardous Industrial Chemicals Safety Manual.* Tokyo, Japan: The International Technical Information Institute, 1988.

18. Verschueren, K. *Handbook of Environmental Data of Organic Chemicals.* 2nd ed. New York, NY: Van Nostrand Reinhold Co., 1983.

19. American Conference of Governmental Industrial Hygienists. *Threshold Limit Values for Chemical Substances and Physical Agents and Biological Exposure Indices for 1994–1995.* Cincinnati, OH: ACGIH, 1994.

■ URANYL FLUORIDE

CAS # 13536–84–0

SYNONYMS uranium fluoride oxide • uranium fluoride oxide (UO$_2$F$_2$) • uranium oxyfluoride • uranium, difluorodioxo • uranyl-fluoride (UO$_2$F$_2$)

MF F$_2$O$_2$U

MW 308.00

COLOR AND FORM Pale-yellow, rhombohedral, hygroscopic.

DENSITY 6.37.

MELTING POINT Decomposes at 300°C.

TOXIC COMBUSTION PRODUCTS When heated to decomposition it emits toxic fumes of F-. [R8, 2712].

SOLUBILITY Soluble in water.

OTHER CHEMICAL/PHYSICAL PROPERTIES Green-yellow fluorescence (uranyl ion); green tetravalent uranium and yellow uranyl ion (uranium(2) dioxide) are the only species that are stable in solution. (tetravalent uranium and uranyl ion).

PROTECTIVE EQUIPMENT Respirator selection (NIOSH): 12.5 mg/m^3: any dust and mist respirator except single-use respirators. 25 mg/m^3: any dust and mist respirator except single-use and quarter-mask respirator or any supplied-air respirator or any self-contained breathing apparatus. 62.5 mg/m^3: any powered air-purifying respirator with a dust and mist filter (note: may need acid gas sorbent) or any supplied-air respirator operated in a continuous-flow mode. 125 mg/m^3: any self-contained breathing apparatus with a full facepiece or any air-purifying full facepiece respirator with a high-efficiency particulate filter or any supplied-air respirator with a full facepiece. 500 mg/m^3: any supplied-air respirator with a full facepiece and operated in a pressure-demand or other positive-pressure mode. Emergency or planned entry in unknown or IDLH conditions: any self-contained breathing apparatus with full facepiece and operated in a pressure-demand or other positive-pressure mode or any supplied-air respira-

tor with a full facepiece and operated in pressure-demand or other positive-pressure mode in combination with auxiliary self-contained breathing apparatus operated in pressure-demand or other positive-pressure mode. Escape: any air-purifying full facepiece respirator (gas mask) with a chin-style or front- or back-mounted acid gas canister having a high-efficiency particulate filter or any appropriate escape-type self-contained breathing apparatus. (fluorides (as fluoride)) [R9, 127].

PREVENTATIVE MEASURES Promptly remove nonimpervious clothing that becomes contaminated. Wash promptly when skin becomes contaminated. (fluorides (as fluoride)) [R9, 127].

CLEANUP 1. Ventilate area of spill. 2. collect spilled material in the most convenient and safe manner and deposit in sealed containers for reclamation. Liquid containing soluble uranium compounds should be absorbed in vermiculite, dry sand, earth, or similar material (sol uranium compounds, as uranium) [R11, 4]. 3. U.S. Patent 4,234,555; Nov 18, 1980: assigned to the U.S. Dept of Energy, describes effective method for removing uranium from aqueous hydrofluoric acid solutions containing trace quantities of the same. The method comprises contacting the solution with particulate calcium fluoride to form uranium-bearing particulates, permitting the particulates to settle, and separating the solution from the settled particulates. As applied to dilute solution containing 120 ppm uranium, the method removes at least 92% of the uranium. (uranium) [R1].

DISPOSAL Disposal of wastes (containing uranium) should follow guidelines set forth by the Nuclear Regulatory Commission. (uranium and compounds) [R11, 296].

OCCUPATIONAL RECOMMENDATIONS

OSHA 8-hr time-weighted average: 0.05 mg/m^3 (uranium, soluble compounds) [R4].

NIOSH Time-weighted average (TWA) 2.5 mg/m^3/10 hr (fluorides (as fluoride)) [R9, 126]; threshold-limit 8–hr time-weighted average (TWA) 0.2 mg/m^3; short-term exposure limit (STEL) 0.6 mg/m³ (1976) (uranium (natural) soluble and insoluble compounds, as U) [R5].

INTERNATIONAL EXPOSURE LIMITS Other recommendations: Bulgaria (1971), Poland (1976), former USSR (1977) 0.075 mg/m³ insoluble compounds, 0.015 mg/m³ soluble compounds; West Germany (1976), Switzerland (1976), Yugoslavia (1971) each recommend the former TLVs of 0.25 mg/m³, and 0.05 for insoluble and soluble compounds respectively; Belgium (1974), Finland (1975), Netherlands (1973) list 0.2 mg/m³ for soluble compounds only, while Australia (1973) lists the current TLV 0.2 mg/m³ for both the soluble and insoluble compounds of uranium. (uranium natural soluble and insoluble compounds (as uranium)) [R6].

NON-HUMAN TOXICITY LD$_{50}$ Wistar rat: adult male 2.5 mg/kg; adult female 1 mg/kg [R2] LD$_{50}$ rat young male (50–100 g): 78 mg uranium/kg/24 hr; young female (50–100 g): 78 mg uranium/kg/24 hr; adult male (150–300 g): 87 mg uranium/kg/24 hr; adult female (150–300 g): 87 mg uranium/kg/24 hr; old male (300–400 g): 40 mg uranium/kg/24 hr (from table) [R3].

BIOLOGICAL HALF-LIFE Of the absorbed uranium (following ingestion) 10% goes to the kidneys and 10% is deposited in the skeleton. The kidney retention is believed to be brief, with a biological half-life of 1–2 weeks. Approximately 80% of the skeletal mass is assumed to be compact bone in which about 95% of deposited uranium has a short effective half-life ranging from 1 mo to 1 yr, whereas only 2% of the absorbed uranium in remaining 20% of skeleton may have an average half-life of about 10 years. (uranium compounds) [R12, 92].

FDA Bottled water packaged in the USA to which no fluoride is added shall not contain fluoride in excess of 1.8 mg/L at 63.9–70.6°F. Bottled water packaged in the USA to which fluoride is added shall not contain fluoride in excess of 1.2 mg/L at 63.9–70.6°F. Imported bottled water to which no fluoride is added and imported bottled water to which fluoride is added shall not contain fluoride in excess of 1.4 mg/L and 0.8 mg/L, respectively. (fluoride) [R7].

REFERENCES

1. Duffy, J. I. (ed.). *Treatment, Recovery, and Disposal Processes for Radioactive Wastes—Recent Advances*. Park Ridge, NJ: Noyes Data Corporation, 1983. 154.

2. Wrenn, M.E., et al., *The Potential Toxicity of Uranium in Water* p. 182 (1987) EPA–600/J–87/096.

3. Voegtlin, C., H.C. Hodge (eds.); *Pharmacology and Toxicology of Uranium Compounds* p. V–2 (1953).

4. 29 CFR 1910.1000 (7/1/88).

5. American Conference of Governmental Industrial Hygienists. *Threshold Limit Values for Chemical Substances and Physical Agents and Biological Exposure Indices for 1994–1995*. Cincinnati, OH: ACGIH, 1994. 34.

6. American Conference of Governmental Industrial Hygienists. *Documentation of the Threshold Limit Values and Biological Exposure Indices*. 5th ed. Cincinnati, OH: American Conference of Governmental Industrial Hygienists, 1986. 617.

7. 21 CFR 103.35 (4/1/88).

8. Sax, N.I. *Dangerous Properties of Industrial Materials*. 6th ed. New York, NY: Van Nostrand Reinhold, 1984.

9. NIOSH. *Pocket Guide to Chemical Hazards*. 2nd Printing. DHHS (NIOSH) Publ. No. 85–114. Washington, DC: U.S. Dept. of Health and Human Services, NIOSH/ Supt. of Documents, GPO, February 1987.

10. Mackison, F. W., R. S. Stricoff, and L. J. Partridge, Jr. (eds.). *NIOSH/OSHA— Occupational Health Guidelines for Chemical Hazards*. DHHS (NIOSH) Publication No. 81–123 (3 mols). Washington, DC: U.S. Government Printing Office, Jan. 1981.

11. Brown, K.W., G. B. Evans, Jr., B.D. Frentrup (eds.). *Hazardous Waste Land Treatment*. Boston, MA: Butterworth Publishers, 1983.

12. National Research Council. *Drinking*

Water and Health. Volume 5. Washington, DC: National Academy Press, 1983.

■ URANYL SULFATE

CAS # 1314-64-3

SYNONYMS dioxosulfatouranium • uranium oxide sulfate • uranium oxide sulfate ($UO_2(SO_4)$) • uranium oxide sulfate (UO_2SO_4) • uranium oxysulfate • uranium, dioxo (sulfato(^{2}-)-O)- • uranium, dioxosulfato • uranyl sulfate (UO_2 (SO_4)) • uranyl sulfate (UO_2SO_4)

MF O_6SU

MW 366.09

OTHER CHEMICAL/PHYSICAL PROPERTIES Lemon-yellow, crystalline mass; density 3.28; soluble in about 5 parts water, 25 parts alcohol (trihydrate); slightly soluble in water; soluble in sulfurous acid; MP anhydrous at 300°C (heptahydrate); yellow-green crystals; MP: decomposes at 100°C; sol: 24.3 g/100 cc concentration sulfuric acid at 13°C; sol: 30 g/100 cc concentration hydrochloric acid at 13°C 20.5 g/100 cc water at 15.5°C, 22.2 g/100 cc water at 100°C (trihydrate); odorless; specific gravity (solid): 3.28 at 20°C (trihydrate); monohydrate and trihydrate are stable (hydrates). Green-yellow fluorescence (uranyl ion); green tetravalent uranium and yellow uranyl ion (uranium(2) dioxide) are the only species which are stable in solution. (tetravalent uranium and uranyl ion).

PROTECTIVE EQUIPMENT Approved dust respirator; goggles or face shield; protective clothing (trihydrate) [R1].

PREVENTATIVE MEASURES SRP: Contaminated protective clothing should be segregated in such a manner so that there is no direct personal contact by personnel who handle, dispose of, or clean the clothing. Quality assurance to ascertain the completeness of the cleaning procedures should be implemented before the decontaminated protective clothing is returned for reuse by the workers.

CLEANUP 1. Ventilate area of spill. 2. Collect spilled material in the most convenient and safe manner and deposit in sealed containers for reclamation. Liquid containing soluble uranium compounds should be absorbed in vermiculite, dry sand, earth, or similar material (sol uranium compounds, as uranium) [R6, 4]. U.S. Patent 4,234,555; Nov 18, 1980: assigned to the U.S. Dept of Energy, describes effective method for removing uranium from aqueous hydrofluoric acid solutions containing trace quantities of the same. The method comprises contacting the solution with particulate calcium fluoride to form uranium-bearing particulates, permitting the particulates to settle, and separating the solution from the settled particulates. As applied to dilute solution containing 120 ppm uranium, the method removes at least 92% of the uranium. (uranium) [R2].

DISPOSAL Disposal of wastes containing uranium should follow guidelines set forth by the Nuclear Regulatory Commission (uranium and compounds) [R7, 296]. After the material has been contained, remove it and the contaminated soil and place in impervious containers. If practical, transport back to the supplier or chemical company to recover the radioactive material or to deactivate it. If this is not practical, or facilities not available, the material should be encapsulated and buried in a specially designated chemical landfill, as designated by the U.S. Department of Energy. Not acceptable at a municipal sewage treatment plant.

STANDARD CODES TSCA; not listed IATA; not listed CFR 49; CFR 14 CAB code 8; not listed NFPA; not listed AAR.

PERSISTENCE Persistent.

MAJOR SPECIES THREATENED Uranyl salts are quite toxic to fish and fish food organisms.

INHALATION LIMIT 0.2 mg/m³.

DIRECT CONTACT May produce injury by skin and inhalation on prolonged exposure.

GENERAL SENSATION Causes reddening of

eyes and mucous membranes. Produces coughing and choking following inhalation. Produces a decrease in urine output. Breathing becomes difficult. Produces radiation sickness including nausea, vomiting, loss of hair, and anemia.

PERSONAL SAFETY Protect against both inhalation and contact with skin. Must wear protective clothing including NIOSH-approved rubber gloves and boots, safety goggles, or face mask, and a respirator whose canister is specifically approved for this material. A self-contained breathing apparatus may be required in fire conditions. Decontaminate fully or dispose of all equipment and clothing after use.

ACUTE HAZARD LEVEL Acute contact with uranyl sulfate will cause irritation to the eyes, respiratory system, gastro-intestinal tract, and the skin.

CHRONIC HAZARD LEVEL Chronic exposure may result in carcinomas in the kidney, liver, or lungs. Also, exposure can lead to dermatitis. Radiation sickness can be produced. The threshold-limit-value; time-weighted average for a 40–hour workweek is 0.2 mg/m³, as uranium, and the short-term exposure limit (15 minutes) is 0.6 mg/m³, as uranium.

DEGREE OF HAZARD TO PUBLIC HEALTH Radioactive material is a carcinogen. It is an irritant to the skin, causing dermatitis. Damages kidneys and lungs.

OCCUPATIONAL RECOMMENDATIONS

OSHA 8-hr time-weighted average: 0.05 mg/m³ (uranium, soluble compounds) [R3].

THRESHOLD LIMIT 8-hr time-weighted average (TWA) 0.2 mg/m³; short-term exposure limit (STEL) 0.6 mg/m³ (1976) (uranium (natural) soluble and insoluble compounds, as U) [R4].

INTERNATIONAL EXPOSURE LIMITS Other recommendations: Bulgaria (1971), Poland (1976), former USSR (1977) 0.075 mg/m³ insoluble compounds, 0.015 mg/m³ soluble compounds; former West Germany (1976), Switzerland (1976), Yugoslavia (1971) each recommend the former TLVs of 0.25 mg/m³ and 0.05 for insoluble and soluble compounds, respectively; Belgium

(1974), Finland (1975), Netherlands (1973) list 0.2 mg/m³ for soluble compounds only, whereas Australia (1973) lists the current TLV 0.2 mg/m³ for both the soluble and insoluble compounds of uranium. (uranium natural soluble and insoluble compounds (as uranium)) [R5].

ACTION LEVEL Avoid contact with the spilled material. Stay upwind. Notify local air, water, fire authorities, U.S. Department of Energy, Radiological Assistance Plan, and State Radiological teams of the accident. Evacuate all people to a distance of at least 500 feet upwind and 2,000 feet downwind of the spill. All persons should be checked for radioactivity contamination.

ON-SITE RESTORATION Dam stream if possible to reduce the flow and prevent further dissipation due to water movement. Under controlled conditions the water can be pumped into a suitable container, and lime added to raise the pH, and precipitate out the uranyl sulfate. Filter and then neutralize with hydrochloric acid. For more details see Envirex manual, EPA 600/2–77–227. Seek professional environmental engineering assistance through EPA's Environmental Response Team (ERT), Edison, NJ, 24–hour no., 908–548–8730.

BEACH OR SHORE RESTORATION Close beach and shore to the public until material has been removed. Do not burn material.

AVAILABILITY OF COUNTERMEASURE MATERIAL Pumps—fire departments, Army Corps of Engineers.

DISPOSAL NOTIFICATION Notify local and state helth authorities, local solid waste disposal authorities, supplier, shipper, and U.S. Dept. of Energy.

PROBABLE LOCATION AND STATE Yellowish green crystals; will sink and dissolve.

WATER CHEMISTRY Will give acid solution.

COLOR IN WATER Yellowish green.

BIOLOGICAL HALF-LIFE Of the absorbed uranium (following ingestion) 10% goes to the kidneys and 10% is deposited in the skeleton. The kidney retention is believed to be brief, with a biological half-life of 1–2 weeks. Approximately 80% of the skeletal mass is assumed to be compact bone, in which about 95% of deposited uranium has a short effective half-life ranging from 1 mo to 1 yr, whereas only 2% of the absorbed uranium in remaining 20% of skeleton may have an average half-life of about 10 years (uranium compounds) [R8, 92].

REFERENCES

1. U.S. Coast Guard, Department of Transportation. *CHRIS—Hazardous Chemical Data*. Volume II. Washington, DC: U.S. Government Printing Office, 1984–5.

2. Duffy, J. I. (ed.). *Treatment, Recovery, and Disposal Processes for Radioactive Wastes—Recent Advances*. Park Ridge, NJ: Noyes Data Corporation, 1983. 154.

3. 29 CFR 1910.1000 (7/1/88).

4. American Conference of Governmental Industrial Hygienists. *Threshold Limit Values for Chemical Substances and Physical Agents and Biological Exposure Indices for 1994–1995*. Cincinnati, OH: ACGIH, 1994. 35.

5. American Conference of Governmental Industrial Hygienists. *Documentation of the Threshold Limit Values and Biological Exposure Indices*. 5th ed. Cincinnati, OH: American Conference of Governmental Industrial Hygienists, 1986. 617.

6. Mackison, F. W., R. S. Stricoff, and L. J. Partridge, Jr. (eds.). *NIOSH/OSHA—Occupational Health Guidelines for Chemical Hazards*. DHHS (NIOSH) Publication No. 81–123 (3 vols). Washington, DC: U.S. Government Printing Office, Jan. 1981.

7. Brown, K.W., G.B. Evans, Jr., B.D. Frentrup (eds.). *Hazardous Waste Land Treatment*. Boston, MA: Butterworth Publishers, 1983.

8. National Research Council. *Drinking Water and Health*. Volume 5. Washington, DC: National Academy Press, 1983.

■ VANADIUM

CAS # 7440-62-2

MF V

MW 50.9415

COLOR AND FORM Light-gray or white lustrous powder, fused hard lumps, or body-centered cubic crystals; pure vanadium is a bright white metal.

DENSITY 6.11 at 18.7°C.

BOILING POINT 3,380°C.

MELTING POINT 1,917°C.

HEAT OF VAPORIZATION 458.6 kJ/mol.

SOLUBILITY Insoluble in water.

STABILITY AND SHELF LIFE Oxidizes readily above 660°C [R2].

OTHER CHEMICAL/PHYSICAL PROPERTIES Two naturally occurring isotopes: (51)V (99.75%): (50)V (0.25%); the latter is radioactive: $T_{1/2}$ 6×10^{15} years; artificial isotopes: 46-49; 52-54. Precipitates gold, silver, and platinum from their salts; reduces mercuric salts to mercurous, ferric salts to ferrous; not attacked by bromine water or by cold sulfuric acid. Vanadium is resistant to attack by hydrochloric or dilute sulfuric acid and to alkali solutions. It is also quite resistant to corrosion by seawater but is reactive toward nitric, hydrofluoric, or concentrated sulfuric acids. Galvanic corrosion tests run in simulated seawater indicate that vanadium is anodic with respect to stainless steel and copper but cathodic to aluminum and magnesium. Vanadium exhibits corrosion resistance to liquid metals, e.g., bismuth, and low-oxygen sodium. Forms an alloy with Fe (ferrovanadium) in which there is complete liquid solubility; forms a very hard and stable carbide, V_4C_3 in carbon and most alloy steels. Acts as metal or nonmetal; valences 2, 3, 4, 5. When heated in air at different temperatures, it oxidizes to a brownish-black trioxide, a blue-black tetraoxide, or a reddish-orange pentoxide. It reacts with chlorine at fairly low temperatures (180°C), forming vanadium tetrachloride

and with carbon and nitrogen at high-temperatures forming VC and VN, respectively. The pure metal in massive form is relatively inert toward oxygen, nitrogen, and hydrogen at room temperature. Lattice constant, 0.3026 nm; specific heat, 0.50 J/g at 20-100°C; latent heat of fusion, 16.02 kJ/mol; enthalpy, 5.27 kJ/mol at 25°C; entropy, 29.5 kJ/mol °C at 25°C; thermal conductivity, 0.31 W/cm K at 100°C; electrical resistance, 24.8-26.0 microhm cm at 20°C; recrystallization temp, 800-1,000°C; modulus of elasticity, $(1.2-1.3) \times 10^5$ MPa. Pure vanadium is soft and ductile. Vanadium has good structural strength and a low-fission neutron cross-section, making it useful in nuclear applications.

PROTECTIVE EQUIPMENT The employer shall provide chemical safety goggles or face shields (8-inch minimum) with goggles and shall ensure that employees wear the protective equipment during any operation in which vanadium may enter the eyes, shall provide appropriate clothing and equipment, and shall ensure that employees wear these where needed to prevent gross skin and eye contact [R5]. The employer shall use engineering controls if needed to keep the concentration of airborne vanadium at or below the limits specified and shall provide protective clothing and equipment resistant to the penetration of vanadium when necessary to prevent gross skin and eye contact with liquid vanadium solutions. Protective equipment suitable for emergency use were located at clearly identified stations outside the work area (vanadium) [R5]. Respirator selection, upper-limit devices recommended by NIOSH for vanadium pentoxide dust. For concentration up to 0.5 mg/m³, use high-efficiency particulate filter, or any supplied-air respirator, or any self-contained breathing apparatus. For concentration up to 1.25 mg/m³, use any supplied-air respirator operated in a continuous-flow mode, or any powered air-purifying respirator with a high-efficiency particulate filter. For concentration up to 2.5 mg/m³, use any air purifying full facepiece respirator with a high-efficiency particulate filter, or any self-contained breathing apparatus with a full facepiece, or any supplied-air respirator with a full facepiece and powered air-purifying respirator with a tight-fitting facepiece and a high-efficiency particulate filter. For concentration up to 70 mg/m³, use any supplied-air respirator with a full facepiece, and operated in a pressure-demand or other positive-pressure mode. For entry into IDLH or unknown concentration, use any self-contained breathing apparatus with a full facepiece, and operated in a pressure-demand or other positive-pressure mode; or any supplied-air respirator with a full facepiece and operated in a pressure-demand or other positive-pressure mode in combination with an auxiliary self-contained breathing apparatus operated in pressure-demand or other positive-pressure mode. For escape conditions, use any air-purifying full facepiece respirator with a high-efficiency particulate filter, or any appropriate escape-type self-contained breathing apparatus. (vanadium pentoxide) [R13].

PREVENTATIVE MEASURES In processes assoc with manufacture of metallic vanadium, and in sieving of used catalyst during maintenance operations, the escape of dust should be prevented by the enclosure of process, and by provision of exhaust ventilation. In boiler cleaning in power stations and on ships, maintenance workers may have to enter the boilers to remove soot, and to make repairs. These workers should wear adequate resp protective equipment with full face mask and eye protection [R6]. Contact lenses should not be worn when working with this chemical (vanadium pentoxide dust and fume (as (V)) [R13]. If material not involved in Fire: Keep material out of water sources and sewers. Build dikes to contain flow as necessary. Keep upwind. Avoid breathing vapors or dusts. Wash away any material which may have contacted the body with copious amounts of water or soap and water (vanadium pentoxide) [R14]. Non-impervious clothing that becomes contaminated should be removed promptly and not reworn until the vanadium dust is removed from the clothing. Skin that becomes contaminated should be promptly

washed or showered to remove any pentoxide dust (vanadium pentoxide) [R15]. Contact lens use in industry is controversial. A survey of 100 corporations resulted in the recommendation that each company establish their own contact lens use policy. One presumed hazard of contact lens use is possible chemical entrapment. It was found that contact lenses minimized injury or protected the eye. The eye was afforded more protection from liquid irritants. Soft contact lenses do not worsen corneal damage from strong chemicals and in some cases could actually protect the eye. Overall, the literature supports the wearing of contact lenses in industrial environments as part of the standard eye protection, e.g., face shields; however, more data are needed to establish the value of contact lenses [R16]. SRP: Contaminated protective clothing should be segregated in such a manner so that there is no direct personal contact by personnel who handle, dispose of, or clean the clothing. Quality assurance to ascertain the completeness of the cleaning procedures should be implemented before the decontaminated protective clothing is returned for reuse by the workers. Contaminated clothing should not be taken home at end of shift, but should remain at employee's place of work for cleaning. (vanadium pentoxide).

CLEANUP Land spill: Dig a pit, pond, lagoon, holding area to contain liquid or solid material. (SRP: If time permits, pits, ponds, lagoons, soak holes, or holding areas should be sealed with an impermeable flexible-membrane liner.) Cover solids with a plastic sheet to prevent dissolving in rain or firefighting water. (vanadium pentoxide) [R7].

DISPOSAL SRP: At the time of review, criteria for land treatment or burial (sanitary landfill) disposal practices are subject to significant revision. Prior to implementing land disposal of waste residue (including waste sludge), consult with environmental regulatory agencies for guidance on acceptable disposal practices. Waste materials contaminated with vanadium were disposed of in a manner not hazardous to employees. The disposal method must conform with applicable local, state, and federal regulations and must not constitute a hazard to the surrounding population or environment [R22]. The following wastewater treatment technologies have been investigated for vanadium: Concentration process: chemical precipitation. [R17].

COMMON USES The principal use of vanadium is as an alloying addition to iron and steel, particularly in high-strength steels and, to a lesser extent, in tool steels, and castings [R10, 673]. In the production of aerospace titanium alloys [R1]. Target material for x-rays, vanadium compounds, especially catalysts for synthetic rubber [R11, 1214]. In producing rust-resistant, spring, and high-speed tool steels. It is an important carbide stabilizer in making steels. Vanadium foil is used as a bonding agent in cladding titanium to steel [R2]. Component of microstructure and gas content controller in copper alloys; strength enhancer and heat resistor in aluminum alloys; component of other alloys—e.g., permanent magnet alloys; catalyst in certain petrochemical reactions; component of instruments for experimental programs [R3]. Interest in the intermetallic compound V_3Ga for superconductor applications could lead to expanded use of vanadium in the future. During the 1970s, vanadium alloys were considered for use as cladding material for the fuel in liquid-metal-cooled fast reactors. However, most development programs involving vanadium for this purpose have been reduced because of insufficient funding. [R10, 673].

OCCUPATIONAL RECOMMENDATIONS
OSHA An employee's exposure to vanadium respirable dust (as V_2O_5) shall at no time exceed the ceiling value of 0.5 mg/m^3. Transitional limits must continue to be achieved by any combination of engineering controls, work practices, and personal protective equipment during the phase-in period, Sept 1, 1989 through Dec 30, 1992 (vanadium respirable dust (as V_2O_5)) [R8]. 8–hr time-weighted average: 0.05 mg/m^3. Final rule limits were achieved by any combination of engineering controls, work

practices, and personal protective equipment during the phase-in period, Sept 1, 1989 through Dec 30, 1992. Final rule limits became effective Dec 31, 1992 (vanadium respirable dust (as V_2O_5)) [R18]. Meets criteria for OSHA medical records rule (vanadium respirable dust (as V_2O_5)) [R19].

NIOSH Time-weighted average (TWA) 1 mg vanadium/m^3; health effects on eye, skin, and lungs were considered; pulmonary function testing and periodic chest X–rays are required (August 1977) [R9].

THRESHOLD LIMIT Time-weighted average (TWA) 0.05 mg/m^3 (1982) (respirable dust and fume, as V_2O_5) [R12, 35]. Excursion Limit Recommendation: Excursions in worker exposure levels may exceed three times the TLV–TWA for no more than a total of 30 min during a work day and under no circumstances should they exceed five times the TLV–TWA. (respirable dust, and fume, as V_2O_5) [R12, 5].

FIRE POTENTIAL Moderate, in the form of dust, when exposed to heat, flame, or sparks [R4].

REFERENCES
1. Bureau of Mines. *Mineral Commodity Summaries*. p. 176 (1989).

2. Weast, R. C. (ed.) *Handbook of Chemistry and Physics*. 69th ed. Boca Raton, FL: CRC Press Inc., 1988–1989, p. B–40.

3. SRI.

4. Sax, N. I. *Dangerous Properties of Industrial Materials,*. 6th ed. New York, NY: Van Nostrand Reinhold, 1984. 2717.

5. NIOSH. *Criteria Document: Vanadium*. p. 6 (1977) DHEW Pub. NIOSH 77–222.

6. International Labour Office. *Encyclopedia of Occupational Health and Safety*. Vols. I and II. Geneva, Switzerland: International Labour Office, 1983. 2241.

7. Association of American Railroads. *Emergency Handling of Hazardous Materials in Surface Transportation*. Washington, DC: Assoc. of American Railroads, Hazardous Materials Systems (BOE), 1987. 713.

8. 54 FR 2920 (1/19/89).

9. NIOSH/CDC. *NIOSH Recommendations for Occupational Safety and Health Standards 1988*, Aug. 1988 (Suppl. to *Morbidity and Mortality Wkly.* Vol. 37 No. 5–7, Aug. 26, 1988). Atlanta, GA: National Institute for Occupational Safety and Health, CDC, 1988. 28.

10. *Kirk–Othmer Encyclopedia of Chemical Technology*. 3rd ed., Volumes 1–26. New York, NY: John Wiley and Sons, 1978–1984, p. 23 (83).

11. Sax, N. I., and R. J. Lewis, Sr. (eds.). *Hawley's Condensed Chemical Dictionary*. 11th ed. New York: Van Nostrand Reinhold Co., 1987.

12. American Conference of Governmental Industrial Hygienists. *Threshold Limit Values for Chemical Substances and Physical Agents and Biological Exposure Indices for 1994–1995*. Cincinnati, OH: ACGIH, 1994.

13. NIOSH. *Pocket Guide to Chemical Hazards*. 2nd Printing. DHHS (NIOSH) Publ. No. 85–114. Washington, DC: U.S. Dept. of Health and Human Services, NIOSH/Supt. of Documents, GPO, February 1987. 235.

14. Association of American Railroads. *Emergency Handling of Hazardous Materials in Surface Transportation*. Washington, DC: Assoc. of American Railroads, Hazardous Materials Systems (BOE), 1987. 713.

15. Mackison, F. W., R. S. Stricoff, and L. J. Partridge, Jr. (eds.). *NIOSH/OSHA— Occupational Health Guidelines for Chemical Hazards*. DHHS (NIOSH) Publication No. 81–123 (3 vols). Washington, DC: U.S. Government Printing Office, Jan. 1981. 4.

16. Randolph, S. A., M. R. Zavon. *J Occup Med* 29: 237–42 (1987).

17. NIOSH. *Criteria Document: Vanadium*. p. 12 (1977) DHEW Pub. NIOSH 77–222.

18. 54 FR 2920 (1/19/89).

19. 29 CFR 1910.20 (7/1/88).

■ VANADIUM TETRACHLORIDE

CAS # 7632–51–1

SYNONYMS tetrachlorovanadium • vanadium chloride • vanadium chloride (VCl4) • vanadium chloride (VCl4), (t-4) • vanadium(IV) chloride

MF Cl4V

MW 192.75

COLOR AND FORM Red to brown liquid.

ODOR Pungent.

DENSITY 1.816 at 20°C.

BOILING POINT 148.5°C at 755 mm Hg.

MELTING POINT −28 ±2°C.

FIREFIGHTING PROCEDURES Wear full protective clothing [R4]. Dry chemical and carbon dioxide are the preferred extinguishing agents for fires in the vicinity of vanadium tetrachloride. The use of water must be avoided except to cool surrounding containers where there is no danger of water coming into contact with the vanadium tetrachloride [R4]. If material is involved in fire use dry chemical, dry sand, or carbon dioxide. Do not use water on material itself. If large quantities of combustibles are involved, use water in flooding quantities as spray and fog. Use water spray to knock down vapors. Cool all affected containers with flooding quantities of water. Apply water from as far a distance as possible [R5].

SOLUBILITY Soluble in absolute alcohol, ether, chloroform, acetic acid; soluble in water with decomposition.

VAPOR PRESSURE 7.8 mbar at 20°C.

CORROSIVITY It is corrosive to metal.

STABILITY AND SHELF LIFE Fumes in moist air [R4].

OTHER CHEMICAL/PHYSICAL PROPERTIES Decomposes in cold water. The complex biochemistry of vanadium compounds depends on the oxidation states of the metal from −1 to 5. The vanadium compounds can easily change their oxidation states under physiological conditions, so that vanadium is mostly in the 5 oxidation state (except in the presence of reducing agents) (vanadium compounds). Valence states of -1 and 0 may occur in solid compounds, e.g., carbonyl, and certain complexes. In oxidation state 5, vanadium is diamagnetic, and forms colorless or pale-yellow compounds. In lower oxidation states, the presence of one or more 3d electrons, usually unpaired, results in paramagnetic and colored compounds. All compounds of vanadium having unpaired electrons are colored, but because the absorption spectra may be complex, a specific color does not necessarily correspond to a particular oxidation state (vanadium compounds). Coordination compounds of vanadium are mainly based on 6 coordination, in which vanadium has a pseudooctahedral structure. Coordination number 4 is typical of many vanadates. Coordination numbers 5 and 8 also are known for vanadium compounds, but numbers less than 4 have not been reported. (vanadium compounds).

PROTECTIVE EQUIPMENT Wear goggles and protective clothing [R1]. Because vanadium compounds cause irritation of the respiratory tract, it is recommended that protective equipment be worn while processing those compounds and that workers have periodic medical examinations (vanadium compounds) [R10]. The employer shall use engineering controls if needed to keep the concentration of airborne vanadium at or below the limits specified and shall provide protective clothing and equipment resistant to the penetration of vanadium when necessary to prevent gross skin and eye contact with liquid vanadium solutions. Protective equipment suitable for emergency use were located at clearly identified stations outside the work area (vanadium) [R11]. Respirator selection, upper-limit devices recommended by NIOSH for vanadium pentoxide dust. For concentration up to 0.5 mg/m³, use high-efficiency particulate filter, any supplied-air respirator, or any self-contained breathing apparatus. For concentration up to 1.25 mg/m³, use any supplied-air respirator operated in a con-

tinuous-flow mode, or any powered air-purifying respirator with a high-efficiency particulate filter. For concentration up to 2.5 mg/m³, use any air-purifying full facepiece respirator with a high-efficiency particulate filter, or any self-contained breathing apparatus with a full facepiece, or any supplied-air respirator with a full facepiece and powered air-purifying respirator with a tight-fitting facepiece and a high-efficiency particulate filter. For concentration up to 70 mg/m³, use any supplied-air respirator with a full facepiece, and operated in a pressure-demand or other positive-pressure mode. For entry into IDLH or unknown concentration, use any self-contained breathing apparatus with a full facepiece, and operated in a pressure-demand or other positive-pressure mode, or any supplied-air respirator with a full facepiece and operated in a pressure-demand or other positive-pressure mode in combination with an auxiliary self-contained breathing apparatus operated in pressure-demand or other positive-pressure mode. For escape conditions, use any air-purifying full facepiece respirator with a high-efficiency particulate filter, or any appropriate escape-type self-contained breathing apparatus (vanadium pentoxide) [R12, 235].

PREVENTATIVE MEASURES Contact lens use in industry is controversial. A survey of 100 corporations resulted in the recommendation that each company establish their own contact lens use policy. One presumed hazard of contact lens use is possible chemical entrapment. It was found that contact lens minimized injury or protected the eye. The eye was afforded more protection from liquid irritants. Soft contact lenses do not worsen corneal damage from strong chemicals and in some cases could actually protect the eye. Overall, the literature supports the wearing of contact lenses in industrial environments as part of the standard eye protection, e.g., face shields; however, more data are needed to establish the value of contact lenses [R6]. Open containers only in dry, oxygen-free atmosphere or inert gas [R1]. If material is not involved in fire, keep material out of water sources and sewers. Build dikes to contain flow as necessary. Use water spray to knock down vapors. Do not use water on material itself [R5]. Avoid breathing vapors. Keep upwind. Avoid bodily contact with the material. Do not handle broken packages unless wearing appropriate personal protective equipment. Wash away any material that may have contacted the body with copious amounts of water or soap and water. If contact with the material is anticipated, wear appropriate chemical protective clothing [R5]. Contact lenses should not be worn when working with this chemical [R12]. SRP: Contaminated protective clothing should be segregated in such a manner so that there is no direct personal contact by personnel who handle, dispose of, or clean the clothing. Quality assurance to ascertain the completeness of the cleaning procedures should be implemented before the decontaminated protective clothing is returned for reuse by the workers. Contaminated clothing should not be taken home at end of shift, but should remain at employee's place of work for cleaning. Safety showers and eyewash fountains should be located in or near areas where gross exposures to vanadium compounds are likely to occur and should be properly maintained. If vanadium compounds, especially the halide or oxyhalide liquids, come in contact with the skin, the affected area should be flushed promptly with water. The eyes, if splashed or otherwise contaminated with these reactive halides, should be flushed immediately and thoroughly with water at low pressure. The employee should then be taken promptly to the nearest medical facility to determine the need for further treatment (vanadium compounds) [R13].

CLEANUP In the event of minor spills, flush with large quantities of water, then neutralize with sodium carbonate when fumes subside. For major spills, cover with foam or soda ash, then hydrolyze the product in a controlled manner using a water fog [R4]. Neutralize spilled materials with crushed limestone or lime [R5]. Land spill: Dig a pit, pond, lagoon, holding area to contain liquid or solid material. (SRP: If time permits, pits, ponds, lagoons, soak holes, or holding areas should be sealed with an impermeable flexible-membrane liner.) Cover solids with a plastic sheet to prevent dissolving in rain or firefighting water (vanadium pentoxide) [R5].

DISPOSAL SRP: At the time of review, criteria for land treatment or burial (sanitary landfill) disposal practices are subject to significant revision. Prior to implementing land disposal of waste residue (including waste sludge), consult with environmental regulatory agencies for guidance on acceptable disposal practices.

COMMON USES Preparation of vanadium trichloride, vanadium dichloride, and organovanadium compounds [R1]. As mordant in the dyeing industry [R2]. Medication (former use) (vanadium compounds) [R9, 220]. USA consumptions of vanadium oxytrichloride and some vanadium tetrachloride as catalyst for ethylene propylene diene monomer rubber were 60 tons (1972), 83 tons (1974), and 67 tons (1975) (vanadium oxytrichloride and vanadium tetrachloride) [R8, 701].

OCCUPATIONAL RECOMMENDATIONS
NIOSH Ceiling 0.05 mg vanadium/m³ for 15 minutes for vanadium compounds (Aug 1977) (vanadium compounds) [R7]. A limit of 0.1 mg/L has been suggested in the former USSR as a maximum permissible limit for water basins (vanadium) [R14, 298].

INTERNATIONAL EXPOSURE LIMITS An European Community guideline recommends an every-4-month control at the workplace if a 50 µg vanadium/L urine level is reached and an annual control if the 5 µg vanadium/L urine level is reached. In case of levels of >50 µg vanadium/L urine a temporary removal from risk should be done (vanadium) [R15, 754]. A limit of 0.1 mg/L has been suggested in the former USSR as a maximum permissible limit for water basins. (vanadium) [R14, 298].

BIOLOGICAL HALF-LIFE No adequate biokinetic data on the vanadium half-time in man are available. The ICRP (International Commission on Radiological Protection) estimate for the whole-body retention of vanadium in man is 42 days. Other studies on workers exposed to vanadium in workroom air report that blood and urinary values of vanadium drop to half the initial value within few days after cessation of exposure. (vanadium compounds) [R16, 646].

DOT *Health Hazards:* Poisonous if inhaled or swallowed. Contact causes severe burns to skin and eyes. Runoff from fire control or dilution water may cause pollution [R3].

Fire or Explosion: Some of these materials may burn, but none of them ignites readily. May ignite other combustible materials (wood, paper, oil, etc.). Violent reaction with water. Flammable or poisonous gases may accumulate in tanks and hopper cars. Runoff to sewer may create fire or explosion hazard [R3].

Emergency Action: Keep unnecessary people away; isolate hazard area and deny entry. Stay upwind, out of low areas, and ventilate closed spaces before entering. Self-contained breathing apparatus and chemical protective clothing that is specifically recommended by the shipper or producer may be worn but they do not provide thermal protection unless that is stated by the clothing manufacturer. Structural firefighter's protective clothing is not effective with these materials. Evacuate the leak or spill area immediately for at least 50 feet in all directions. Call CHEMTREC at 1–800–424–9300 for emergency assistance. If water pollution occurs, notify the appropriate authorities [R3].

Fire: Do not get water inside container. Small fires: Dry chemical, carbon dioxide, or halon. Large Fires: Flood fire area with water from a distance. Do not get solid stream of water on spilled material. Move container from fire area if you can do so without risk. Cool containers that are exposed to flames with water from the side until well after fire is out. Stay away from ends of tanks [R3].

Spill or Leak: Do not touch spilled material; stop leak if you can do so without risk. Do not get water inside container. Use water spray to reduce vapor; do not put water directly on leak or spill area. Keep combustibles (wood, paper, oil, etc.) away from spilled material. Spills: Dike for later disposal; do not apply water unless directed to do so. Cleanup only under supervision of an expert [R3].

First Aid: Move victim to fresh air and call for emergency medical care; if not breathing, give artificial respiration; if breathing is difficult, give oxygen. Remove and isolate contaminated clothing and shoes at the site. Speed in removing material from skin is of extreme importance. In case of contact with material, immediately flush skin or eyes with running water for at least 15 minutes. Keep victim quiet and maintain normal body temperature [R3].

REFERENCES

1. Sax, N. I., and R. J. Lewis, Sr. (eds.). *Hawley's Condensed Chemical Dictionary.* 11th ed. New York: Van Nostrand Reinhold Co., 1987. 1216.

2. International Labour Office. *Encyclopedia of Occupational Health and Safety.* Vols. I and II. Geneva, Switzerland: International Labour Office, 1983. 2240.

3. Department of Transportation. *Emergency Response Guidebook* 1987. DOT P 5800.4. Washington, DC: U.S. Government Printing Office, 1987, p. G–39.

4. National Fire Protection Association. *Fire Protection Guide on Hazardous Materials.* 9th ed. Boston, MA: National Fire Protection Association, 1986, p. 49–92.

5. Association of American Railroads. *Emergency Handling of Hazardous Materials in Surface Transportation.* Washington, DC: Assoc. of American Railroads, Hazardous Materials Systems (BOE), 1987. 713.

6. Randolph, S. A., M. R. Zavon. *J Occup Med* 29: 237–42 (1987).

7. NIOSH/CDC. *NIOSH Recommendations for Occupational Safety and Health Standards 1988.* Aug. 1988 (Suppl. to *Morbidity and Mortality Wkly.* Vol. 37 No. 5–7, Aug. 26, 1988). Atlanta, GA: National Institute for Occupational Safety and Health, CDC, 1988. 28.

8. *Kirk-Othmer Encyclopedia of Chemical Technology.* 3rd ed., Volumes 1–26. New York, NY: John Wiley and Sons, 1978–1984, p. V23 (83).

9. Venugopal, B., and T. D. Luckey. *Metal Toxicity in Mammals,* 2. New York: Plenum Press, 1978.

10. Gul'ko, A. G. *Gig Saint.* 21: 24–8 (1956) (Rus) as cited in NIOSH, *Criteria Document: Vanadium.* p. 26 (1977) DHEW Pub. NIOSH 77–222.

11. NIOSH. *Criteria Document: Vanadium.* p. 6 (1977) DHEW Pub. NIOSH 77–222.

12. NIOSH. *Pocket Guide to Chemical Hazards.* 2nd Printing. DHHS (NIOSH) Publ. No. 85–114. Washington, DC: U.S. Dept. of Health and Human Services, NIOSH/Supt. of Documents, GPO, February 1987.

13. NIOSH. *Criteria Document: Vanadium.* p. 91 (1977) DHEW Pub. NIOSH 77–222.

14. National Research Council. *Drinking Water and Health.* Volume 1. Washington, DC: National Academy Press, 1977.

15. Seiler, H. G., H. Sigel, and A. Sigel (eds.). *Handbook on the Toxicity of Inorganic Compounds.* New York, NY: Marcel Dekker, Inc. 1988.

16. Friberg, L., G. F. Nordberg, E. Kessler, and V. B. Vouk (eds.). *Handbook of the Toxicology of Metals.* 2nd ed. Vols I, II. Amsterdam: Elsevier Science Publishers B.V., 1986, p. V2.

■ VINYL FLUORIDE

CAS # 75–02–5

SYNONYMS ethene, fluoro • ethylene, fluoro • fluoroethene • fluoroethylene • monofluoroethylene

COLOR AND FORM Colorless gas.

DENSITY 0.636 (liquid) at 21°C.

SURFACE TENSION 3.4137×10^{-2} newtons/meter at melting point.

VISCOSITY 3.2347×10^{-4} pascal/seconds (liquid) at boiling point.

BOILING POINT −72°C.

MELTING POINT −160.5°C.

HEAT OF COMBUSTION -1.01×10^{8} J/kmol.

HEAT OF VAPORIZATION 2.2243×10^{7} J/kmol at melting point.

AUTOIGNITION TEMPERATURE 725°F (vinyl fluoride, inhibited) [R4].

EXPLOSIVE LIMITS AND POTENTIAL Explosive limits: 2.6% (lower); 21.7% (upper) [R2].

FLAMMABILITY LIMITS Lower flammable limit: 2.6% by volume; upper flammable limit: 21.7% by volume [R3].

FIREFIGHTING PROCEDURES Let fire burn; shut off flow of gas; cool adjacent containers with water (vinyl fluoride, inhibited) [R4]. Do not extinguish fire unless flow can be stopped. Use water in flooding quantities as fog. Cool all affected containers with flooding quantities of water. Apply water from as far a distance as possible (vinyl fluoride, inhibited) [R5]. Evacuation: If fire becomes uncontrollable or container is exposed to direct flame— consider evacuation of one-half (1/2) mile radius (vinyl fluoride, inhibited) [R5].

TOXIC COMBUSTION PRODUCTS Toxic hydrogen fluoride gas is generated in a fire. (vinyl fluoride, inhibited) [R4].

SOLUBILITY Insoluble in water; soluble in alcohol and ether; soluble in acetone; water solubility at 80°C (g/100 g water): 0.94 (at 3.4 mPa), 1.54 (at 6.9 mPa).

VAPOR PRESSURE 2.4 mPa at 21°C.

OTHER CHEMICAL/PHYSICAL PROPERTIES 1 mg/L = 532 ppm; 1 ppm = 1.88 mg/cu m at 25°C, 760 mm Hg.

PROTECTIVE EQUIPMENT Protective gloves; safety glasses; self-contained breathing apparatus (vinyl fluoride, inhibited) [R4].

PREVENTATIVE MEASURES If material not on fire and not involved in Fire: Keep sparks, flames, and other sources of ignition away. Keep material out of water sources and sewers. Attempt to stop leak if without undue personnel hazard. Use water spray to knock down vapors (vinyl fluoride, inhibited) [R5].

DISPOSAL SRP: At the time of review, criteria for land treatment or burial (sanitary landfill) disposal practices are subject to significant revision. Prior to implementing land disposal of waste residue (including waste sludge), consult with environmental regulatory agencies for guidance on acceptable disposal practices.

COMMON USES The main use is in the production of poly(vinyl fluoride) [R1].

OCCUPATIONAL RECOMMENDATIONS None listed.

IARC SUMMARY Evaluation: There is inadequate evidence in humans for the carcinogenicity of vinyl fluoride. There is sufficient evidence in experimental animals for the carcinogenicity of vinyl fluoride. Overall evaluation: Vinyl fluoride is probably carcinogenic to humans (Group 2A). In making the overall evaluation, the working group took into account the following evidence: Vinyl fluoride is closely related structurally to the known carcinogen vinyl chloride. The two chemicals cause the same rare tumor (hepatic haemangiosarcoma) in experimental animals, which is also a tumor caused by vinyl chloride in humans [R6].

BIOCONCENTRATION Based upon an estimated log K_{ow} of 1.19 (1), the BCF of vinyl fluoride can be estimated to be approximately 4.7 from a regression-derived equation (2,SRC). This estimated BCF value suggests that bioconcentration in aquatic organisms is not an important fate process (SRC) [R7].

TSCA REQUIREMENTS Section 8 (a) of TSCA requires manufacturers of this chemical substance to report preliminary assessment information concerned with production, use, and exposure to EPA as cited in the preamble of 51 FR 41329 [R8].

REFERENCES

1. Gerhartz, W. (exec ed.). *Ullmann's Encyclopedia of Industrial Chemistry.* 5th ed. Vol A1: Deerfield Beach, FL: VCH Publishers, 1985 to Present., p. VA11, 363.

2. Sax, N. I. *Dangerous Properties of Industrial Materials,.* 5th ed. New York: Van Nostrand Reinhold, 1979. 1088.

3. National Fire Protection Guide. *Fire Protection Guide on Hazardous Materials.* 10th ed. Quincy, MA: National Fire Protection Association, 1991, p. 325M–93.

4. U.S. Coast Guard, Department of Transportation. *CHRIS—Hazardous Chemical Data.* Volume II. Washington, DC: U.S. Government Printing Office, 1984–5.

5. Association of American Railroads. *Emergency Handling of Hazardous Materials in Surface Transportation.* Washington, DC: Association of American Railroads, Bureau of Explosives, 1992. 950.

6. IARC. *Monographs on the Evaluation of the Carcinogenic Risk of Chemicals to Man.* Geneva: World Health Organization, International Agency for Research on Cancer, 1972–present (multivolume work), p. S7 73 (1987).

7. (1) Meylan, W.M., P.H. Howard. *J Pharm Sci* 84: 83–92 (1995) (2) Lyman, W. J., et al., *Handbook of Chemical Property Estimation Methods,* Washington, DC: Amer Chemical Soc. p. 5–4 (1990).

8. 40 CFR 712.30 (7/1/94).

■ VINYLSTYRENE

CAS # 1321–74–0

SYNONYMS benzene, diethenyl • benzene, divinyl • diethenylbenzene • divinylbenzene • dvb • dvb-22 • dvb-55 • dvb-80 • dvb-27 • dvb-100 • vinylstyrene • m-divinylbenzene

MF $C_{10}H_{10}$

MW 130.20

ODOR THRESHOLD As with other styrene monomers, the odor can be detected by humans at levels below dangerous concentration, i.e., 10–60 ppm. [R9, 799].

EXPLOSIVE LIMITS AND POTENTIAL Lower, 1.1%; upper, 6.2% in air (dvb–22) [R9, 798].

FLAMMABILITY LIMITS 1.1% (lower); 6.2% (upper) [R4].

FIREFIGHTING PROCEDURES In advanced or massive fires, firefighting should be done from safe distance or protected location. Use water spray, dry chemical, foam, or carbon dioxide. Use water to keep fire-exposed containers cool. If leak or spill has not ignited, use water to disperse the vapors. If it is necessary to stop a leak, use water spray to protect men attempting to do so. Water spray may be used to flush spills away from exposures [R4]. Wear self-contained breathing apparatus; wear goggles if eye protection not provided [R4]. If material on fire or involved in fire, use water in flooding quantities as fog. Cool all affected containers with flooding quantities of water. Use foam, dry chemical, or carbon dioxide. (divinyl benzene (combustible liquid, flammable liquid not otherwise specified)) [R5].

SOLUBILITY Water soluble at 50 ppm.

VAPOR PRESSURE 6.58×10^{-1} mm Hg.

OCTANOL/WATER PARTITION COEFFICIENT Log K_{ow} = 3.59 (estimated).

OTHER CHEMICAL/PHYSICAL PROPERTIES BP: 76°C at 14 mm Hg; density, 0.9325 at 22°C; Index of refraction, 1.5767 at 20°C; soluble in acetone, benzene (o-divinylbenzene). BP: 121°C at 76 mm Hg; MP: –52.3°C; density, 0.9294 at 20°C; Index of refraction, 1.5760 at 20°C; soluble in acetone, benzene (m-divinylbenzene). BP: 95–6°C at 18 mm Hg; MP: 31°C; density, 0.913 at 40°C; index of refraction, 1.5835 at 25°C; soluble in acetone, benzene (p-divinylbenzene). Water-white liquid; easily polymerized; viscosity, 109 cP at 20°C (pure m-divinylbenzene). Vapor density (air = 1), 4.48; vapor pressure, 1 mm Hg at 32.7°C; conversion factor, 5.3 mg/m³ is equiv to 1 ppm (m-divinylbenzene); MW, 130.08; index of refraction, 1.5326 at 25°C/D; viscosity, 0.883 cP at 25°C; surface tension, 30.55 dynes/cm at 25°C; density, 0.8979 g/cu cm at 20°C; BP, 180°C (calculated); critical pressure, 2.45 MPa, and critical temperature, 348°C (each calculated); latent heat of vaporization, 320.49 J/g at boiling point; solubility, 0.0065% in water at 25°C; soluble in acetone, carbon tetrachloride, benzene, ethanol. MW, 130.18; index of refraction, 1.5585 at 25°C/D; viscosity, 1.007 cP at 25°C; surface tension, 32.10 dynes/cm at 25°C; density, 0.9162 g/cu cm at 20°C; BP, 195°C (calculated); freezing point, –45°C; critical pressure, 2.45 MPa, and critical temperature, 369°C (each calculated); latent heat of vaporization, 350.62 J/g at boiling point; solubility, 0.0052% in water at 25°C; soluble in acetone, carbon tetrachloride, benzene, ethanol. Pale straw-colored liquid (divinylbenzene 55%).

PROTECTIVE EQUIPMENT Wear appropriate chemical protective gloves, boots, and goggles. (divinyl benzene (flammable liquid, combustible liquid, not otherwise specified)) [R5].

PREVENTATIVE MEASURES Eyes and skin should be protected, and respiratory equipment should be used above the irritant level. [R11, 3323].

DISPOSAL Atomize into incinerator. May add flammable solvent. SRP: At the time of review, criteria for land treatment or burial (sanitary landfill) disposal practices are subject to significant revision. Prior to implementing land disposal of waste residue (including waste sludge), consult with environmental regulatory agencies for guidance on acceptable disposal practices.

COMMON USES Polymerization monomer for special synthetic rubbers, drying oils, ion-exchange resins, casting resins, polyesters [R10, 437]. Monomer for styrene-vinylstyrene ion-exchange resins [R1]. Monomer and cross-linking agent for acrylic polymers [R1]. Monomer for cholestyramine resin (cardiovascular agent) [R1]. Its uses as a monomer or polymer are extensive. The monomer is utilized as an insecticide stabilizer, as an ion exchange resin, as a cross-linking agent in water purification and decolonization, as a sustained release agent, and as a dental filling component, applications for which patents are avail. In biology, it has been applied as an experimental clotting agent for sustained-life research [R11, 3323]. Used in the manufacture of soft contact lenses [R2]. Used in the manufacture of cholestyramine resin which is used as a drug to increase lipoprotein catabolism [R3]. The world market for DVB as 100 wt% DVB was 3,000–4,000 metric tons in 1981. By far the largest use was as a copolymer with styrene in ion-exhange resins. These resins are used primarily in home water softeners as well as for municipal and industrial water conditioning [R9, 798]. Tetrafilcon A, which is used in making the Aquaflex and Aosoft contact lenses has, in addition to its main ingredients 2–hydroxyethyl methacrylate (about 82%), and vinylpyrrolidinone (about 15%), small amounts of methyl methacrylate (about 2%), and divinylbenzene (about 0.5%) as the cross-linking agent [R2]. The bile sequestering quaternary ammonium salt anion-exchange resin cholestyramine is a high-mol-wt copolymer of styrene, which bears a quaternary ammonium moiety, with 2% of divinylbenzene [R3].

BINARY REACTANTS Acids, bases, oxidizing materials, metal salts like ferric and aluminum chloride.

STANDARD CODES NFPA–not listed; USCG—not listed; CFR—not listed.

MAJOR SPECIES THREATENED Waterfowl.

DIRECT CONTACT Irritating with prolonged contact to skin and eyes.

GENERAL SENSATION Possible dizziness or drowsiness from vapors [R16]. Mild eye irritation but no corneal damage [R16]. Disagreeable odor.

PERSONAL SAFETY Wear self-contained breathing apparatus and skin protection.

ACUTE HAZARD LEVEL Moderate inhalative toxicant.

CHRONIC HAZARD LEVEL Moderate inhalation hazard when exposed chronically at sublethal concentrations. Prolonged skin contact may cause irritation.

DEGREE OF HAZARD TO PUBLIC HEALTH Moderate inhalative hazard from both acute and chronic exposures. 1.5–2 cupfuls estimated lethal dose to man [R8].

OCCUPATIONAL RECOMMENDATIONS
OSHA 8-hr time-weighted average: 10 ppm (50 mg/m³). Final rule limits were achieved by any combination of engineering controls, work practices, and personal protective equipment during the phase-in period, Sept 1, 1989 through Dec 30, 1992. Final rule limits became effective Dec 31, 1992 (CAS number: 108–57–6) [R7].

THRESHOLD LIMIT 8-hr time-weighted average (TWA) 10 ppm, 53 mg/m³ (1980) [R12, 20].

ACTION LEVEL Notify local fire and air authority. Enter from upwind and remove all ignition sources. Attempt to contain slick.

ON-SITE RESTORATION Oil-skimming equipment and sorbent foams can be used to treat slick. Seek professional environmental engineering assistance through EPA's Environmental Response Team (ERT), Edison, NJ, 24–hour no., 908–548–8730.

AVAILABILITY OF COUNTERMEASURE MATERIAL Oil-skimming equipment—stored at major ports; sorbent foams (polyurethane)—upholstery shops.

INDUSTRIAL FOULING POTENTIAL Slicks may reduce heat transfer or cause hot spots and scaling. Volatility suggests explosion hazard when confined in boiler feed or cooling waters.

MAJOR WATER USE THREATENED Recreation, industrial.

PROBABLE LOCATION AND STATE Water-white liquid; will float in slick on surface.

BIODEGRADATION Data regarding the biodegradation of vinylstyrene in either soil or aquatic systems were not available (SRC).

BIOCONCENTRATION Based on a water solubility of 50 ppm (1) and an estimated log K_{ow} of 3.59 (2), a range for log BCF between 1.83 and 2.50 has been calculated (3), which indicates vinylstyrene is not expected to bioconcentrate in aquatic systems (SRC) [R6].

REFERENCES

1. SRI.

2. *Kirk-Othmer Encyclopedia of Chemical Technology.* 3rd ed., Volumes 1–26. New York, NY: John Wiley and Sons, 1978–1984, p. 6 (79) 7331.

3. *Kirk-Othmer Encyclopedia of Chemical Technology.* 3rd ed., Volumes 1–26. New York, NY: John Wiley and Sons, 1978–1984, p. 4 (78) 913.

4. National Fire Protection Association. *Fire Protection Guide on Hazardous Materials.* 9th ed. Boston, MA: National Fire Protection Association, 1986, p. 49–45.

5. Association of American Railroads. *Emergency Handling of Hazardous Materials in Surface Transportation.* Washington, DC: Assoc. of American Railroads, Hazardous Materials Systems (BOE), 1987. 280.

6. (1) Santodonato, J., et al., *Invest Selected Potential Environ Contam: Styrene, Ethyl Benzene, and Related Compounds.* USEPA–560/11–80/018 p. 57, 69 (1980). (2) CLOGP, PCGEMS. *Graphical Exposure Modeling System.* USEPA (1986). (3) Lyman, W. J., et al., *Handbook of Chemical Property Estimation Methods.* New York: McGraw-Hill, p. 4–9 (1982).

7. 54 FR 2920 (1/19/89).

8. OSHA–20 form for *Divinyl Benzene* provided by Dow Chemical Co., 1974.

9. *Kirk-Othmer Encyclopedia of Chemical Technology.* 3rd ed., Volumes 1–26. New York, NY: John Wiley and Sons, 1978–1984, p. 21(83).

10. Sax, N.I., and R.J. Lewis, Sr. (eds.). *Hawley's Condensed Chemical Dictionary.* 11th ed. New York: Van Nostrand Reinhold Co., 1987.

11. Clayton, G. D., and F. E. Clayton (eds.). *Patty's Industrial Hygiene and Toxicology.* Volume 2A, 2B, 2C: *Toxicology.* 3rd ed. New York: John Wiley and Sons, 1981–1982.

12. American Conference of Governmental Industrial Hygienists. *Threshold Limit Values for Chemical Substances and Physical Agents and Biological Exposure Indices for 1994–1995.* Cincinnati, OH: ACGIH, 1994.

■ 4-VINYLTOLUENE

CAS # 622–97–9

SYNONYMS benzene, 1-ethenyl-4-methyl • 1-p-tolylethene • p-vinyltoluene • p-methyl styrene • 4-methylstyrene

MF C_9H_{10}

MW 118.19

DENSITY 0.9173 at 25°C.

SURFACE TENSION 0.043227 newtons/m at 239.02 K.

VISCOSITY 0.0020986 Pa-s (liquid) at 239.02 K.

BOILING POINT 172.8°C.

MELTING POINT –34.1°C.

HEAT OF COMBUSTION -4.8229×10^9 J/kmol.

HEAT OF VAPORIZATION 5.1703×10^7 J/kmol at 238.02 K.

FIREFIGHTING PROCEDURES Use water spray, dry chemical, foam, or carbon dioxide. Use water spray to keep fire-exposed containers cool. Fight fire from protected location or maximum possible distance (vinyl toluene) [R5]. Evacuation: If fire becomes uncontrollable or container is exposed to direct flame, consider evacuation of one-third (1/3) mile radius (vinyltoluene, inhibited) [R6]. Do not extinguish fire unless flow can be stopped. Use water in flooding quantities as fog. Solid streams of water may spread fire. Cool all affected containers with flooding quantities of water. Apply water from as far a distance as possible. Use foam, dry chemical, or carbon dioxide. (vinyltoluene, inhibited) [R6].

SOLUBILITY Soluble in benzene.

VAPOR PRESSURE 1.81 mm Hg at 25°C.

OCTANOL/WATER PARTITION COEFFICIENT Log K_{ow} = 3.35.

OTHER CHEMICAL/PHYSICAL PROPERTIES

Critical volume, 3.33 ml/g; critical density, 0.30 g/ml; specific heat of vapor, 1.2284 J/g K at 25°C; heat of formation, 115.48 kJ/mol of liquid at 25°C; heat of polymerization, 66.9± 0.2 kJ/mol; shrinkage upon polymerization, 12.6% by vol; cubical coefficient of expansion, 9.361 × 10⁻⁴ at 20°C (vinyl toluene). Liquid molar volume = 0.129067 m³/kmol; heat of formation = 1.1464×10⁸ J/kmol.

PROTECTIVE EQUIPMENT Respirator selection, upper-limit devices recommended by ACGIH. At concentration up to 500 ppm, use any chemical cartridge respirator with organic vapor cartridge(s) or any supplied-air respirator or any self-contained breathing apparatus. For concentration up to 1,000 ppm, use any powered air-purifying respirator with organic vapor cartridge(s) or any chemical cartridge respirator with a full facepiece and organic vapor cartridge(s). For concentration up to 1,250 ppm, use any supplied-air respirator operated in a continuous-flow mode. For concentration up to 2,500 ppm, use any air-purifying full facepiece respirator (gas mask) with a chin-style or front– or back-mounted organic-vapor canister or any self-contained breathing apparatus with a full facepiece or any supplied-air respirator with a full facepiece. For concentration up to 5,000 ppm, use any supplied-air respirator with a full face-piece, and operated in a pressure-demand or other positive-pressure mode. For emergency or planned entry in unknown concentrations or IDLH conditions, use any self-contained breathing apparatus with a full facepiece, and operated in a pressure-demand or other positive-pressure mode or any supplied-air respirator with a full facepiece and operated in a pressure-demand or other positive-pressure mode in combination with an auxiliary self-contained breathing apparatus operated in pressure-demand or other positive-pressure mode. For escape conditions, use any air-purifying full facepiece respirator (gas mask) with a chin-style or front– or back-mounted organic vapor canister or any appropriate escape-type self-contained breathing apparatus. (vinyl toluene) [R10, p. 188–189].

PREVENTATIVE MEASURES Personnel protection: Avoid breathing vapors. Keep upwind. Do not handle broken packages unless wearing appropriate personal protective equipment. Wash away any material that may have contacted the body with copious amounts of water or soap and water. (vinyltoluene, inhibited) [R6].

COMMON USES Use in mixtures with other vinyltoluene isomers (3–vinyltoluene) as monomers for producing poly(vinyltoluene) [R9, 256]. Used as a monomer in the production of polyester resins [R1]. Copolymerized with isobutylene to produce synthetic elastomers [R2]. Reactive monomer; used in the coatings industry as a modifier for drying oils and oil-modified alkyds; used as a replacement for styrene in unsaturated polyester resins; used as a copolymer with styrene to increase the operating temperature range of paints, coatings, and varnishes. (vinyltoluene) [R3].

OCCUPATIONAL RECOMMENDATIONS None listed.

NON-HUMAN TOXICITY LD_{50} rat single oral para-isomer mixture 4 g/kg [R7].

BIOCONCENTRATION A log BCF of 1.50 (1) for p-methylstyrene in goldfish indicates p-methylstyrene is not expected to bioconcentrate in aquatic organisms (SRC) [R8].

DOT *Fire or Explosion:* Flammable/combustible material; may be ignited by heat, sparks, or flames. Vapors may travel to a source of ignition, and flash back. Container may explode in heat of fire. Vapor explosion hazard indoors, outdoors, or in sewers. Runoff to sewer may create fire or explosion hazard. Material may be transported hot (vinyl toluene, inhibited) [R4].

Health Hazards: May be poisonous if inhaled or absorbed through skin. Vapors may cause dizziness or suffocation. Contact may irritate or burn skin and eyes. Fire may produce irritating or poisonous gases. Runoff from fire control or dilution water may cause pollution (vinyl toluene, inhibited) [R4].

Emergency Action: Keep unnecessary people away; isolate hazard area and deny

entry. Stay upwind; keep out of low areas. Positive-pressure self-contained breathing apparatus (SCBA) and structural firefighters' protective clothing will provide limited protection. Isolate for 1/2 mile in all directions if tank, rail car, or tank truck is involved in fire. Call emergency response telephone number on shipping paper first. If shipping paper not available or no answer, call CHEMTREC at 1–800–424–9300. If water pollution occurs, notify the appropriate authorities (vinyl toluene, inhibited) [R4].

Fire: Small fires: Dry chemical, CO_2, water spray, or regular foam. Large Fires: Water spray, fog, or regular foam. Move container from fire area if you can do so without risk. Apply cooling water to sides of containers that are exposed to flames until well after fire is out. Stay away from ends of tanks. For massive fire in cargo area, use unmanned hose holder, or monitor nozzles; if this is impossible, withdraw from area, and let fire burn. Withdraw immediately in case of rising sound from venting safety device or any discoloration of tank due to fire (vinyl toluene, inhibited) [R4].

Spill or Leak: Shut off ignition sources; no flares, smoking, or flames in hazard area. Stop leak if you can do so without risk. Water spray may reduce vapor, but it may not prevent ignition in closed spaces. Small Spills: Take up with sand or other noncombustible absorbent material and place into containers for later disposal. Large Spills: Dike far ahead of liquid spill for later disposal (vinyl toluene, inhibited) [R4].

First Aid: Move victim to fresh air, and call for emergency medical care; if not breathing, give artificial respiration; if breathing is difficult, give oxygen. In case of contact with material, immediately flush eyes with running water for at least 15 minutes. Wash skin with soap and water. Remove and isolate contaminated clothing and shoes at the site (vinyl toluene, inhibited) [R4].

REFERENCES

1. *Kirk-Othmer Encyclopedia of Chemical Technology.* 4th ed. Volumes 1: New York, NY. John Wiley and Sons, 1991–present., p. V7 29.

2. *Kirk-Othmer Encyclopedia of Chemical Technology.* 4th ed. Volumes 1: New York, NY. John Wiley and Sons, 1991–present., p. V8 938.

3. IARC. *Monographs on the Evaluation of the Carcinogenic Risk of Chemicals to Man.* Geneva: World Health Organization, International Agency for Research on Cancer, 1972–present (multivolume work), p. V60 375 (1994)

4. U.S. Department of Transportation. *Emergency Response Guidebook 1993.* DOT P 5800.6. Washington, DC: U.S. Government Printing Office, 1993, p. G–27.

5. National Fire Protection Guide. *Fire Protection Guide on Hazardous Materials.* 10th ed. Quincy, MA: National Fire Protection Association, 1991, p. 49–182.

6. Association of American Railroads. *Emergency Handling of Hazardous Materials in Surface Transportation.* Washington, DC: Association of American Railroads, Bureau of Explosives, 1992. 952.

7. American Conference of Governmental Industrial Hygienists, Inc. *Documentation of the Threshold Limit Values and Biological Exposure Indices.* 6th ed. Volumes I, II, III. Cincinnati, OH: ACGIH, 1991. 1717.

8. Ogata, M., et al., *Bull Environ Contam Toxicol* 33: 561–7 (1984).

9. Gerhartz, W. (exec ed.). *Ullmann's Encyclopedia of Industrial Chemistry.* 5th ed.Vol A1: Deerfield Beach, FL: VCH Publishers, 1985 to present, p. VA13.

10. U.S. Department of Transportation. *Emergency Response Guidebook 1993.* DOT P 5800.6. Washington, DC: U.S. Government Printing Office, 1993, p. G-27.

■ WHITE PHOSPHORUS

CAS # 7723-14-0

SIC CODES 2842.

SYNONYMS yellow phosphorus • Bonide blue death rat killer • Common sense cockroach and rat preparations • fosforo bianco (Italian) • gelber Phosphor (German) • phosphore blanc (French) • phosphorus (white) • phosphorus white, molten (UN2447) (DOT) • phosphorus, white or yellow, dry or under water or in solution (UN1381) (DOT) • phosphorus yellow • phosphorus, yellow (ACGIH: OSHA) • Rat-Nip • tetrafosfor (Dutch) • Tetraphosphor (German) • UN1381 (DOT) • UN2447 (DOT) • weiss Phosphor (German) • white phosphorus

BINARY REACTANTS Air, alkaline hydroxides, ammonium nitrate, antimony pentafluoride, barium bromate, barium chlorates, beryllium, boron trioxide, bromine, bromine trifluoride, bromoazide, calcium bromate, calcium chlorate, calcium iodate, cesium acetylene, carbide, cesium nitride, charcoal, and air, chlorates, chlorine, chlorine dioxide, chlorine monoxide, chlorine trioxide, chlorosulfonic acid, chromic acid, chromic anhydride, chromium trioxide, chromyl chloride, cyanogen iodide, fluorine, iodates, iodine, iodine pentafluoride, lead dioxide, lithium carbide, lithium silicide, magnesium bromate, magnesium chlorate, magnesium iodate, magnesium perchlorate, manganese, mercuric oxide, mercurous nitrate, nitrates, nitric acid, nitrogen bromide, nitrogen dioxide, nitrogen tribromide, nitrogen trichloride, nitrosyl fluoride potassium bromate, potassium chlorate, potassium hydroxide,. potassium iodate, potassium nitride, potassium permanganate, potassium peroxide, rubidium acetylene carbide, selenium oxychloride, silver nitrate, silver oxide, sodium bromate, sodium carbide, sodium chlorate, sodium hydroxide, sodium iodate, sodium peroxide, sulfur, sulfur trioxide, sulfuric acid, thorium, zinc bromate, zinc chlorate, zinc iodate, zirconium.

MW 123.88

MF P_4

STANDARD CODES EPA 311; NFPA–3, 3, 1; ICC—flammable solid, yellow label, not accepted dry, 25 lbs in water; IATA—(dry) flammable solid, not acceptable (in water) yellow label, not acceptable passenger, 12 kg cargo; USCG—nonflammable solid, yellow label; Superfund designated (hazardous substances) list.

PERSISTENCE Burns spontaneously in air.

INHALATION LIMIT 0.1 mg/m³.

DIRECT CONTACT Irritation generally high—severe burns skin.

GENERAL SENSATION Sharp pungent odor. Affects bones; vapors of burning phosphorus irritate nose, throat, and lungs. Extreme local and systemic effects. Ingestion can cause severe g.i. disturbances, bloody diarrhea, liver damage, skin eruptions, oliguria, circulatory collapse, coma, convulsions, and death. Can cause photophobia with myosis, dilation of pupils, retinal hemorrhage, congestion of blood vessels, and, rarely, an optic neuritis.

PERSONAL SAFETY Safety goggles and air line or self-contained breathing apparatus.

ACUTE HAZARD LEVEL Very toxic to fish. Approximate fatal dose 50 mg. Strong irritant; highly toxic with ingestion or inhalation. Emits highly toxic oxide fumes.

CHRONIC HAZARD LEVEL 1 mg/day causes bone and liver damage. Marine waters should not exceed 1/100 of 96-hour LC_{50} [R2]; chronic poisoning is a real possibility, characterized by bony neurosis, spontaneous fractures, anemia, weight loss. highly toxic with repeated ingestion or inhalation.

DEGREE OF HAZARD TO PUBLIC HEALTH Strong irritant. Highly toxic with acute or chronic ingestion or inhalation. Emits highly toxic vapors.

OCCUPATIONAL RECOMMENDATIONS None listed.

ACTION LEVEL Notify fire and air authority. Keep covered with water.

ON-SITE RESTORATION Dredge from bottom. Seek professional environmental engineering assistance through EPA's Environmental Response Team (ERT), Edison, NJ, 24–hour no. 908–548–8730.

DISPOSAL Burning produces phosphorus pentoxide, which is highly reactive to any tissue, etc. "Phossy" water must be chemically treated or flame treated before release.

DISPOSAL NOTIFICATION Contact local fire and air authorities.

INDUSTRIAL FOULING POTENTIAL Industrial use recommended level 4 ppm.

MAJOR WATER USE THREATENED Fisheries, potable supply, recreation.

PROBABLE LOCATION AND STATE Volatile, sublimes in vacuum at ordinary temperatures when exposed to light. When exposed to air in the dark, emits a greenish light, and gives off white fumes. Will sink to bottom of water course. A yellowish waxlike solid.

WATER CHEMISTRY although there are no definitive published reports on kinetics of oxidation of elemental phosphorus in water, it appears that the rate is highly dependent on the degree of dispersion. At concentration ca 10 µg/L, well below the accepted solubility limit of 3 mg/L, with the dissolved oxygen content unspecified, elemental phosphorus disappears by a first-order process with a half-life of 2 hours at ca 10°C; 0.85 hour at 30°C. At concentrations 50–100 mg/L, well above the solubility limit, with a dissolved oxygen content of 6–7 mg/L, the same reaction has a half-life of 80 hours at 30°C, and 240 hours at 0°C. The relatively small temperature effect, combined with the large inverse concentration effect, is consistent with a diffusion controlled process. The oxidation of colloidal phosphorus in seawater is reported to be measurably slower than in freshwater, suggesting that the high salt content brings about agglomeration of the P particles. Thus, rapidly moving freshwater should lose elemental P faster than quiescent seawater.

COLOR IN WATER May form yellowish white colloidal dispersion.

FOOD-CHAIN CONCENTRATION POTENTIAL Radioactive phosphorus [R3] has been concentrated by waterfowl up to 75, 000 times, and by aquatic life by factors up to 850,000. Positive; cod and other marine fish concentrate up to 1,000 times [R1].

REFERENCES

1. Ducre, J. C., and D. H. Rosenblatt, *Mammalian Toxicology, and Toxicity to Aquatic Organisms of Four Important Types of Waterborne Munitions Pollutants*, NTIS, AD–778 725, March 1974.

2. *Proposed Criteria for Water Quality*, Volume I, U.S. Environmental Protection Agency. October 1973.

3. *Genetika*, 1968, 4.

■ m–XYLENE

CAS # 108–38–3

DOT # 1307

SYNONYMS 1, 3-dimethylbenzene

MF C_8H_{10}

MW 106.1

STANDARD CODES EPA 311; NFPA—2, 3, 0; ICC—flammable liquid, red label, 10 gallon in an outside container; USCG—grade C flammable liquid; IATA—flammable liquid, red label 1 liter passenger, 40 liters cargo; Superfund designated (hazardous substances) list.

PERSISTENCE Does not biodegrade well.

MAJOR SPECIES THREATENED Waterfowl threshold conconcentration for harm to common crops—800–2,400 ppm [R3].

INHALATION LIMIT 435.

DIRECT CONTACT Mild irritant to skin.

GENERAL SENSATION Can be absorbed through skin at slightly hazardous chronic levels. Recognition odor 1.1 ppm in air [R4]; may be narcotic at high concentrations. skin irritation grade 5—necrosis from 0.005 ml; eye irritation grade 1—no effect [R2].

PERSONAL SAFETY Wear skin protection and canister type mask. If intense heat prevails, self-contained breathing apparatus may be required.

ACUTE HAZARD LEVEL Lethal concentrated for mice in air, 6,000 ppm [R1]. Mild irritant. Slight inhalation hazard. Toxic to fish. Produces odor in water at low levels.

CHRONIC HAZARD LEVEL Slight inhalation and skin absorption hazard.

DEGREE OF HAZARD TO PUBLIC HEALTH Mildly irritating. Slight inhalation hazard with acute or chronic exposure.

OCCUPATIONAL RECOMMENDATIONS None listed.

ACTION LEVEL Notify fire and air authority. If intense heat prevails, evacuate area, and warn civil defense of possible explosion. Enter from upwind. Remove ignition source. Attempt to contain slick.

ON-SITE RESTORATION Oil-skimming equipment and sorbent foams can be applied to slick. Use carbon on undissolved portion. Seek professional environmental engineering assistance through EPA's Environmental Response Team (ERT), Edison, NJ, 24–hour no., 908–548–8730.

AVAILABILITY OF COUNTERMEASURE MATERIAL Oil-skimming equipment—stored at major ports; sorbent foams (polyurethane)—upholstery shops; carbon—water treatment plants, sugar refineries.

DISPOSAL Spray into incinerator or burn in paper packaging. Additional flammable solvent may be added.

INDUSTRIAL FOULING POTENTIAL Slick may reduce heat transfer or cause hot spots or scaling; may pose rupture hazard if confined in boiler-feed or cooling-system water.

MAJOR WATER USE THREATENED Recreation, fisheries, industrial.

PROBABLE LOCATION AND STATE Colorless liquid. Will float in slick on surface.

WATER CHEMISTRY Xylene is slightly basic; relatively stable to chemical and biochemical attack.

COLOR IN WATER Colorless.

GENERAL RESPONSE Stop discharge if possible. Keep people away. Call fire department. Avoid contact with liquid and vapor. Isolate and remove discharged material. Notify local health and pollution control agencies.

FIRE RESPONSE Flammable. Flashback along vapor trail may occur. Vapor may explode if ignited in an enclosed area. Wear self-contained breathing apparatus. Extinguish with foam, dry chemical, or carbon dioxide. Water may be ineffective on fire. Cool exposed containers with water.

EXPOSURE RESPONSE Call for medical aid. Vapor irritating to eyes, nose, and throat. If inhaled, will cause headache, difficult breathing, or loss of consciousness. Move to fresh air. If breathing has stopped, give artificial respiration. If breathing is difficult, give oxygen. Liquid irritating to skin and eyes. If swallowed, will cause nausea, vomiting, or loss of consciousness. Remove contaminated clothing and shoes. Flush affected areas with plenty of water. If swallowed and victim is conscious, have victim drink water or milk. Do not induce vomiting.

RESPONSE TO DISCHARGE Issue warning—high flammability. Evacuate area. Should be removed. Chemical and physical treatment.

WATER RESPONSE Harmful to aquatic life in very low concentrations. Fouling to shoreline. May be dangerous if it enters water intakes. Notify local health and wildlife officials. Notify operators of nearby water intakes.

EXPOSURE SYMPTOMS Vapors cause headache and dizziness. Liquid irritates eyes and skin. If taken into lungs, causes severe coughing, distress, and rapidly developing pulmonary edema. If ingested, causes nausea, vomiting, cramps, headache, and coma; can be fatal. Kidney and liver damage can occur.

EXPOSURE TREATMENT Inhalation: remove to fresh air; administer artificial respiration and oxygen if required; call a doctor. Ingestion: do not induce vomiting; call a doctor. Eyes: flush with water for at least 15 min. Skin: wipe off, wash with soap and water.

REFERENCES

1. Stecher, P. G. (ed.). *Merck Index*, NJ: Merck and Company, 1968. 8th Ed., Rahway.

2. Smyth, H. F., Jr., C. P. Carpenter, C. S. Weil, U. C. Pozzani, J. A. Striegel, and J. S. Nycum, "Range–Finding Toxicity Data: List VII," *American Industrial Hygiene Association Journal*, 30: 470–476. 1969. Smyth, H. F., C. P. Carpenter, and C. S. Weil, "Range–Finding Toxicity Data: List IV," *AMA Archives of Industrial Hygiene and Occupational Medicine*, 4: 119–122, 1951. Smyth, H. F., C. P. Carpenter, C. S. Weil, U. C. Pozzani, and J. A. Striegel, "Range–Finding Toxicity Data: List VI," *American Industrial Hygiene Associa-tion Journal*, 23: 95–107, 1962. Smyth, H. F., C. P. Carpenter, C. S. Weil, and U. C. Pozzani, "Range–Finding Toxicity Data: List V," *AMA Archives of Industrial Hygiene and Occupational Medicine*, 10: 61–68, 1954. Smyth, H. F., C. P. Carpenter, and C. S. Weil, "Range–Finding Toxicity Data: List 111," *Journal of Industrial Hygiene and Toxicology*, 31: 60–62,1949. Smyth, H. F., L. Seaton, and L. Fischer, "The Single Dose Toxicity of Some Glycols and Derivatives," *Journal of Industrial Hygiene and Toxicology*, 23 (6): 259–268, 1941. Smyth, H. F., and C. P. Carpenter, "Further Experience with the Range–Finding Test in the Industrial Toxicology Laboratory," *Journal of Industrial Hygiene and Toxicology*, 30: 63–68, 1948. Smyth, H. F., and C. P. Carpenter, "The Place of the Range–Finding Test in the Industrial Toxicology Laboratory," *Journal of Industrial Hygiene and Toxicology*, 26: 269–273, 1944.

3. National Academy of Sciences. *Water Quality Criteria, 1972* (advance copy), U.S. Environmental Protection Agency, December 1973.

4. Sullivan, R. J. *Air Pollution Aspects of Odorous Compounds*. NTIS PB 188 089, September 1969.

■ XYLIDINE

CAS # 1300-73-8

SYNONYMS aminodimethylbenzene • benzenamine, ar, ar-dimethyl • xylidinen (Dutch) • xylidine (Italian) • ai3-24178-x (USDA) • dimethylphenylamine • dimethylaniline • xylidine isomers • xylidine mixed ortho-meta-para isomers

MF $C_8H_{11}N$

MW 121.18

COLOR AND FORM Exists in 6 isomeric forms varying from a light-yellow to a brown liquid; all isomers except ortho–4–xylidine are liquids above 27°C.

ODOR Weak aromatic amine odor.

ODOR THRESHOLD Odor threshold in air = 1.00×10^{12} molecules/cc [R4].

DENSITY 0.97–0.99.

BOILING POINT 213–226°C.

MELTING POINT –36°C.

EXPLOSIVE LIMITS AND POTENTIAL Lower explosive limit in air: 1.5% by volume [R9, 238].

FIREFIGHTING PROCEDURES Use water spray, dry chemical, foam, or CO_2. Wear full protective clothing [R3].

SOLUBILITY Slightly soluble in water; soluble in alcohol.

VAPOR PRESSURE Less than 1 mm Hg at 68°F.

CORROSIVITY Liquid xylidine will attack some forms of plastics, rubber, and coatings.

STABILITY AND SHELF LIFE Heat contributes to instability. [R8, 2].

OTHER CHEMICAL/PHYSICAL PROPERTIES Forms more or less soluble salts with the strong mineral acids.

PROTECTIVE EQUIPMENT The following types of respirators should be selected under the prescribed concentrations: 20 ppm: Any chemical cartridge respirator with organic vapor cartridge(s); any supplied-air respirator; any self-contained breathing apparatus. 50 ppm: Any supplied-air respirator operated in a continuous-flow mode; any powered air-purifying respirator with organic vapor cartridge(s). 100 ppm: Any chemical cartridge respirator with a fullfacepiece and organic-vapor cartridge(s); any powered air-purifying respirator with a tight-fitting facepiece and organic vapor cartridge(s); any air-purifying full facepiece respirator (gas mask) with a chin-style or front– or back-mounted organic vapor canister; any supplied-air respirator with a full facepiece; any self-contained breathing apparatus with a full facepiece. 150 ppm: Any supplied-air respirator with a half mask and operated in a pressure-demand or other positive-pressure mode. Emergency or planned entry in unknown concentration or IDLH condition: Any self-contained breathing apparatus with a full facepiece and operated in a pressure-demand or other positive-pressure mode; Any supplied-air respirator with a full facepiece and operated in a pressure-demand or other positive-pressure mode in combination with an auxiliary self-contained breathing apparatus operated in pressure-demand or other positive-pressure mode. Escape: Any air-purifying full facepiece respirator (gas mask) with a chin-style or front- or back-mounted organic vapor canister; any appropriate escape-type self-contained breathing apparatus. [R9, 239].

PREVENTATIVE MEASURES SRP: Contact lenses should not be worn when working with this chemical.

DISPOSAL SRP: At the time of review, criteria for land treatment or burial (sanitary landfill) disposal practices are subject to significant revision. Prior to implementing land disposal of waste residue (including waste sludge), consult with environmental regulatory agencies for guidance on acceptable disposal practices.

COMMON USES Xylidine is a raw material in the manufacture of dyes, pharmaceuticals [R1]. As a gasoline additive [R8, 3]. Used in manufacture of curing agents, antioxidants, and antiozonants [R8, 3]. Use in polymer production [R8, 3]. In organic synthesis in preparation of wood preservatives, wetting agents for textiles, frothing agents for ore dressing, special lacquers, and metal complexes. [R8, 3].

OCCUPATIONAL RECOMMENDATIONS
OSHA 8-hr time-weighted average: 5 ppm (25 mg/m³), skin [R6].

THRESHOLD LIMIT 8-hr time-weighted average (TWA) 0.5 ppm, 2.5 mg/m³, skin (1990) (xylidine (mixed isomers)) [R10, 36].

INTERNATIONAL EXPOSURE LIMITS MAC former USSR 3 mg/m³ (skin) [R5].

NON-HUMAN TOXICITY LC_{50} mice inhalation 149 ppm/7 hr [R1]; MLD rabbit oral 620 mg/kg [R1]; MLD rabbit iv 240 mg/kg [R1]; MLD cat iv 120 mg/kg [R1].

ATMOSPHERIC STANDARDS This action promulgates standards of performance for equipment leaks of volatile organic compounds (VOC) in the synthetic organic chemical manufacturing industry (SOC-MI). The intended effect of these standards is to require all newly constructed, modified, and reconstructed SOCMI process units to use the best demonstrated system of continuous emission reduction for equipment leaks of VOC, considering costs, non-air-quality health, and environmental impact and energy requirements. Xylidine is produced, as an intermediate or final product, by process units covered under this subpart [R7].

DOT *Health Hazards:* Poisonous; may be fatal if inhaled, swallowed, or absorbed through skin. Contact may cause burns to skin and eyes. Runoff from fire control or dilution water may give off poisonous gases and cause water pollution. Fire may produce irritating or poisonous gases [R2].

Fire or Explosion: Some of these materials may burn, but none of them ignites readily. Container may explode violently in heat of fire [R2].

Emergency Action: Keep unnecessary people away; isolate hazard area and deny entry. Stay upwind, out of low areas, and ventilate closed spaces before entering. Self-contained breathing apparatus and chemical protective clothing that is specifically recommended by the shipper or producer may be worn but they do not provide thermal protection unless it is stated by the clothing manufacturer. Structural firefighter's protective clothing is not effective with these materials. Remove and isolate contaminated clothing at the site. Call CHEMTREC at 1–800–424–9300 as soon as possible, especially if there is no local hazardous materials team available [R2].

Fire: Small fires: Dry chemical, CO_2, halon, water spray, or standard foam. Large Fires: Water spray, fog, or standard foam is recommended. Move container from fire area if you can do so without risk. Fight fire from maximum distance. Stay away from ends of tanks. Dike fire control water for later disposal; do not scatter the material [R2].

Spill or Leak: Do not touch spilled material; stop leak if you can do so without risk. Use water spray to reduce vapors. Small Spills: Take up with sand or other noncombustible absorbent material and place into containers for later disposal. Small Dry Spills: With clean shovel place material into clean, dry container, and cover; move containers from spill area. Large Spills: Dike far ahead of liquid spill for later disposal [R2].

First Aid: Move victim to fresh air and call emergency medical care; if not breathing, give artificial respiration; if breathing is difficult, give oxygen. In case of contact with material, immediately flush skin or eyes with running water for at least 15 minutes. Speed in removing material from skin is of extreme importance. Remove and isolate contaminated clothing and shoes at the site. Keep victim quiet and maintain normal body temperature. Effects may be delayed; keep victim under observation [R2].

REFERENCES

1. American Conference of Governmental Industrial Hygienists. *Documentation of the Threshold Limit Values and Biological Exposure Indices.* 5th ed. Cincinnati, OH: American Conference of Governmental Industrial Hygienists, 1986. 639.

2. Department of Transportation. *Emergency Response Guidebook 1987.* DOT P 5800.4. Washington, DC: U.S. Government Printing Office, 1987, p. G–55.

3. National Fire Protection Association. *Fire Protection Guide on Hazardous Materials.* 9th ed. Boston, MA: National Fire Protection Association, 1986, p. 49–94.

4. Fazzalari, F. A. (ed.). *Compilation of Odor and Taste Threshold Values Data.* ASTM Data Series DS 48A (Committee E–18). Philadelphia, PA: American Society for Testing, and Materials, 1978. 165.

5. International Labour Office. *Encyclopedia of Occupational Health and Safety.* Vols. I and II. Geneva, Switzerland: International Labour Office, 1983. 147.

6. 29 CFR 1910.1000 (7/1/87).

7. 40 CFR 60.489 (7/1/87).

8. Mackison, F. W., R. S. Stricoff, and L. J. Partridge, Jr. (eds.). *NIOSH/OSHA—Occupational Health Guidelines for Chemical Hazards*. DHHS (NIOSH) Publication No. 81–123 (3 vols). Washington, DC: U.S. Government Printing Office, Jan. 1981.

9. NIOSH. *Pocket Guide to Chemical Hazards*. 5th Printing/Revision. DHHS (NIOSH) Publ. No. 85–114. Washington, DC: U.S. Dept. of Health and Human Services, NIOSH/Supt. of Documents, GPO, Sept. 1985.

10. American Conference of Governmental Industrial Hygienists. *Threshold Limit Values for Chemical Substances and Physical Agents and Biological Exposure Indices for 1994–1995*. Cincinnati, OH: ACGIH, 1994.

■ ZINC BORATE

CAS # 1332–07–6

SIC CODES 2819; 2834.

SYNONYMS borax-2335 • boric acid, zinc salt • zb-112 • zb-237 • zn-100 • firebrake—zb

MF $2ZnO_3 \cdot 3B_2O_3 \cdot 3 \cdot 5H_2O$

MW 161.44

COLOR AND FORM White, amorphous powder.

ODOR None.

FIREFIGHTING PROCEDURES Extingiush fire using agent suitable for type of surrounding fire. Material itself does not burn or burns with difficulty [R5].

SOLUBILITY Soluble in dilute acids; slightly soluble in water.

OTHER CHEMICAL/PHYSICAL PROPERTIES MP: 980°C; density/specific gravity: 4.22 crystals; 3.64 powder; soluble in cold water; amorphous powd soluble in HCl; crystals insoluble in HCl; white triclinic crystals or amorphous powder ($3ZnO \cdot 2B_2O_3$); odorless; density: 2.7 at 20°C (solid) ($2ZnO \cdot 3B_2O_3 \cdot 3.5H_2O$). Threshold concentration of taste (in water): approximately 15 ppm; 40 ppm (imparts) a metallic taste. (zinc salts).

PROTECTIVE EQUIPMENT NIOSH-approved respirator; goggles or face shield; protective gloves [R2].

PREVENTATIVE MEASURES If material not on fire and not involved in Fire: Keep material out of water sources and sewers. Build dikes to contain flow as necessary. Keep upwind. Avoid breathing vapors or dust. Wash away any material which may have contacted the body with copious amounts of water or soap and water [R5].

CLEANUP Environmental considerations: Land spill: Dig a pit, pond, lagoon, or holding area to contain liquid or solid material. SRP: If time permits, pits, ponds, lagoons, soak holes, or holding areas should be contained with a flexible impermeable membrane liner. Cover solids with a plastic sheet to prevent dissolving in rain or firefighting water [R5]. Environmental considerations: Water spill: Neutralize with agricultural lime (slaked lime), crushed limestone, or sodium bicarbonate. Add soda ash. Adjust pH to neutral (pH 7). Use mechanical dredges or lifts to remove immobilized masses of pollutants and precipitates or greater concentration [R5].

DISPOSAL After the material has been contained, remove it and the contaminated soil and place in impervious containers. If practical, transport material back to the supplier for recovery. If this is not practical, the material should be buried in a specially designated chemical landfill. Not acceptable at a sewage treatment plant. SRP: At the time of review, criteria for land treatment, or burial (sanitary landfill) disposal practices are subject to significant revision. Prior to implementing land disposal of waste residue (including waste sludge), consult with environmental regulatory agencies for guidance on acceptable disposal practices.

COMMON USES In medicine; fireproofing textiles; fungistat and mildew inhibitor; flux in ceramics [R1]; flame retardant for PVC, cellulosics, and unsaturated halogenated polyesters [R4]. Fire retardant, smoke suppressant, afterglow depressant, biostat in polymers, elastomers, paper, paperboard, textile coatings; biostat for high-temp processing; synergist with antimony oxide and alumina trihydrate [R3].

STANDARD CODES EPA 311; TSCA; not listed IATA; not listed CFR 49; CFR 14 · CAB code 8; not listed NFPA; not listed AAR; ICC STCC Tariff no 1–C 28 195 38; Superfund designated (hazardous substances) list.

PERSISTENCE Persistent.

MAJOR SPECIES THREATENED In general, the salmonids are most sensitive to zinc in soft water; the rainbow trout is the most sensitive in hard water.

INHALATION LIMIT 10.

DIRECT CONTACT Eye, skin, respiratory, and mucous membrane irritant.

GENERAL SENSATION Skin contamination may produce irritation, rash, and reddening. Contamination of mucous membranes produces reddening and irritation. Exposure of eye may produce pain, photophobia, and corneal damage. Inhalation may produce coughing and choking.

PERSONAL SAFETY Protect against both inhalation and absorption through the skin. Must wear protective clothing including NIOSH–approved rubber gloves and boots, safety goggles, or face mask, and a respirator whose canister is specifically approved for this material. Decontaminate fully or dispose of all equipment and clothing after use.

ACUTE HAZARD LEVEL Zinc salts are astringent, corrosive to the skin, irritating to the gastrointestinal tract, and may act as emetics. The emetic concentration in water may be between 675 and 2,280 ppm. When large doses are ingested, fatal collapse may occur as a result of serious damage to the buccal and gastroenteric mucous membranes. The zinc ion is too poorly absorbed to induce acute systemic intoxication. When in fires or highly heated, zinc oxides are emitted that may act as a skin irritant, and when inhaled give symptoms of metal-fume fever. The aquatic toxicity of 97 ppm is based on a 96–hour LC_{50} of 33 ppm for fathead minnows for the zinc ion in hard water.

CHRONIC HAZARD LEVEL Generally, zinc compounds have no cumulative effects. The continued administration of zinc salts in small doses has no effect in man except those of disordered digestion and constipation.

DEGREE OF HAZARD TO PUBLIC HEALTH Irritating to skin, eyes, nose, throat, causes gastrointestinal disturbances.

OCCUPATIONAL RECOMMENDATIONS None listed.

ACTION LEVEL Avoid contact with the spilled cargo. Stay upwind. notify local air, water, and fire authorities of the accident. Evacuate all people to a distance of at least 200 feet upwind and 1,000 feet downwind of the spill.

ON-SITE RESTORATION Dam stream if possible to reduce the flow and prevent further dissipation by water movement. Bottom pumps or underwater vacuum systems may be employed in small bodies of water; dredging may be effective in larger bodies, to remove any undissolved material from the bottom. Pump water into a suitable container. Under controlled conditions add a mixture of soda ash and lime in a 50/50 ratio to a pH of 7.5–8.5. Allow to settle. Filter if needed and neutralize to a pH of 7. For more information see Envirex Manual EPA 600/2–77–227. Seek professional environmental engineering assistance through EPA's Environmental Response Team (ERT), Edison, NJ, 24–hour no., 908–548–8730.

BEACH OR SHORE RESTORATION Close beach and shore to the public until material has been removed. If tidal, scrape affected area at low tide with a mechanical scraper, and avoid contact. Do not burn material.

AVAILABILITY OF COUNTERMEASURE MATERIAL Bottom pumps, dredges—fire departments, U.S. Coast Guard, Army Corps of Engineers; vacuum systems—swimming-pool operators.

DISPOSAL NOTIFICATION Notify local and state health authorities, local solid waste disposal authorities, supplier, and shipper.

INDUSTRIAL FOULING POTENTIAL Low industrial fouling potential.

MAJOR WATER USE THREATENED Fisheries.

PROBABLE LOCATION AND STATE White powder; will sink and slowly dissolve.

WATER CHEMISTRY No reaction.

COLOR IN WATER Colorless.

FOOD-CHAIN CONCENTRATION POTENTIAL Moderate. Zinc will accumulate in some organisms, but is not considered to be bioconcentrative.

CERCLA REPORTABLE QUANTITIES Persons in charge of vessels or facilities are required to notify the National Response Center (NRC) immediately, when there is a release of this designated hazardous substance, in an amount equal to or greater than its reportable quantity of 1,000 lb or

454 kg. The toll-free telephone number of the NRC is (800) 424–8802; in the Washington metropolitan area (202) 426–2675. The rule for determining when notification is required is stated in 40 CFR 302.6 (section IV. D. 3.b) [R6].

FIRE POTENTIAL Nonflammable [R1].

FDA Bottled water shall, when a composite of analytical units of equal volume from a sample is examined by the methods described in paragraph (d) (1) (ii) of this section, meet the standards of chemical quality, and shall not contain zinc in excess of 5.0 mg/L. (soluble zinc salts) [R7].

GENERAL RESPONSE Avoid contact with solid and dust. Keep people away. Stop discharge if possible. Isolate and remove discharged material. Notify local health and pollution control agencies.

FIRE RESPONSE Not flammable.

EXPOSURE RESPONSE Call for medical aid. Dust irritating to eyes, nose, and throat. If inhaled will cause coughing or difficult breathing. If breathing has stopped, give artificial respiration. If breathing is difficult, give oxygen. Solid irritating to skin and eyes. If swallowed will cause nausea and vomiting. Remove contaminated clothing and shoes. Flush affected areas with plenty of water. If swallowed and victim is conscious, have victim drink water or milk, and have victim induce vomiting. If swallowed and victim is unconscious or having convulsions, do nothing except keep victim warm.

RESPONSE TO DISCHARGE Should be removed. Provide chemical and physical treatment.

WATER RESPONSE Effect of low concentrations on aquatic life is unknown. May be dangerous if it enters water intakes. Notify local health and wildlife officials. Notify operators of nearby water intakes.

EXPOSURE SYMPTOMS Inhalation of dust may irritate nose and throat. Ingestion can cause gastrointestinal disturbances, convulsions, central nervous depressions, skin eruptions, shock, and death. Contact with eyes or skin causes irritation.

EXPOSURE TREATMENT Inhalation: move to fresh air. Ingestion: administer gastric lavage with warm tap water; saline catharsis; consult physician. Eyes or skin: flush with water.

REFERENCES

1. Sax, N. I., and R. J. Lewis, Sr. (eds.). *Hawley's Condensed Chemical Dictionary*. 11th ed. New York: Van Nostrand Reinhold Co., 1987. 1251.

2. U.S. Coast Guard, Department of Transportation. *CHRIS—Hazardous Chemical Data*. Volume II. Washington, DC: U.S. Government Printing Office, 1984–5.

3. Kuneyy, J. H., and J. N. Nullican (eds.) *Chemcyclopedia*. Washington, DC: American Chemical Society, 1988. 219.

4. *Kirk–Othmer Encyclopedia of Chemical Technology*. 3rd ed., Volumes 1–26. New York, NY: John Wiley and Sons, 1978–1984, p. 10 (78) 360.

5. Association of American Railroads. *Emergency Handling of Hazardous Materials in Surface Transportation*. Washington, DC: Assoc. of American Railroads, Hazardous Materials Systems (BOE), 1987. 731.

6. 51 FR 34534 (9/29/86).

7. 21 CFR 103.35 (4/1/86).

■ ZINC AMMONIUM CHLORIDE

CAS # 52628–25–8

SIC CODE 2819.

SYNONYMS ammonium pentachlorozincate; ammonium zinc chloride; ammonium tetrachlorozincate

MW Variable, depending on form

MF $ZnCl_2 \cdot 2NH_4Cl$, $ZnCl_2 \leq NH_4Cl$.

STANDARD CODES EPA 311; TSCA; not listed IATA; not listed CFR 49; CFR 14 CAB code 8; not listed NFPA; not listed AAR; ICC STCC tariff no 1–C 28 195 37; Superfund designated (hazardous substances) list.

PERSISTENCE Persistent.

MELTING CHARACTERISTICS Sublimes at 340°C without melting if absolutely dry.

SOLUBILITY Very soluble in water.

DENSITY 1.81.

FLAMMABILITY Nonflammable. At fire temperatures it emits highly toxic and irritating chlorides and ammonia. Inhalation of zinc oxides may produce metal-fume fever. When heated, evolves zinc oxide fumes, which can cause a disease known as brass founders ague or brass chills.

FIREFIGHTING Water, foam, carbon dioxide, dry chemicals may be used on containing fires zinc ammonium chloride. Contain runoff.

COMMON USES In manufacture of dry cells; as flux for welding, soldering, galvanizing.

MAJOR SPECIES THREATENED In general, the salmonids are most sensitive to zinc in soft water; the rainbow trout is the most sensitive in hard water.

DIRECT CONTACT Eye, skin, respiratory, and mucous membrane irritant.

GENERAL SENSATION Contamination of skin may produce reddening, rash, and irritation. Contamination of mucous membranes produces reddening and irritation. Exposure of eye may produce pain, photophobia, and corneal damage. At fire temperatures produces coughing, choking, difficult breathing, and eye irritation.

PERSONAL SAFETY Protect against both inhalation and absorption through the skin. Must wear protective clothing including NIOSH-approved rubber gloves and boots, safety goggles or face mask, and a respirator whose canister is specifically approved for this material. Decontaminate fully or dispose of all equipment and clothing after use.

ACUTE HAZARD LEVEL Zinc salts are astringent, corrosive to the skin, irritating to the gastrointestinal tract, and may act as emetics, where the emetic concentration in water may be between 675, and 2,280 ppm. After large doses have been ingested,

fatal collapse may occur as a result of serious damage to the buccal and gastroenteric mucous membranes. The zinc ion is too poorly absorbed to induce acute systemic intoxication. When in fires or highly heated, zinc oxides are emitted, which may act as a skin irritant, and when inhaled give symptoms of metal-fume fever. The aquatic toxicity of 150 ppm is based on a 96-hour LC_{50} of 33 ppm for fathead minnows, for the zinc ion in hard water.

CHRONIC HAZARD LEVEL Generally, zinc compounds have no cumulative effects. The continued administration of zinc salts in small doses has no effect in man except those of disordered digestion and constipation.

DEGREE OF HAZARD TO PUBLIC HEALTH Emetic when swallowed, irritating, and corrosive to skin, eyes, and gastrointestinal tract.

ACTION LEVEL Avoid contact with the spilled cargo. Stay upwind. Notify local air, water, and fire authorities of the accident. Evacuate all people to a distance of at least 200 feet upwind and 1,000 feet downwind of the spill.

ON-SITE RESTORATION Dam stream if possible to reduce the flow and prevent further dissipation by water movement. Dredging may not be effective due to high solubility. If treatment is to be attempted, pump water into a suitable container. settle, filter, and pass through ion exchange. Add lime and soda ash in a 50/50 ratio to a pH of 7.5–8.5. Settle and filter if needed. Neutralize if necessary. For more information see ENVIREX Manual EPA 600/2-77-227. Seek professional environmental engineering assistance through EPA's Environmental Response Team (ERT), Edison, NJ, 24-hour no., 908-548-8730.

BEACH OR SHORE RESTORATION Close beach and shore to the public until material has been removed. Avoid human contact. Do not burn material.

AVAILABILITY OF COUNTERMEASURE MATERIAL Bottom pumps—fire departments,

U.S. Coast Guard, Army Corps of Engineers.

DISPOSAL After the material has been contained, remove it and the contaminated soil and place in impervious containers. If practical, transport material back to the supplier for recovery. If this is not practical, the material should be buried in a specially designated chemical landfill. Not acceptable at a municipal sewage treatment plant.

DISPOSAL NOTIFICATION Notify local and state health authorities, local solid waste disposal authorities, supplier, and shipper.

INDUSTRIAL FOULING POTENTIAL May foul boiler-feed and cooling water.

MAJOR WATER USE THREATENED Possible effect on fisheries; industrial users.

PROBABLE LOCATION AND STATE White powder or crystals; will dissolve.

WATER CHEMISTRY Will give acid solution.

COLOR IN WATER Colorless.

FOOD-CHAIN CONCENTRATION POTENTIAL Moderate. Zinc will accumulate in some organisms, but is not considered to be bioconcentrative.

NON-HUMAN TOXICITY LC$_{50}$ 150 ppm, 96–hr fathead minnow, est, based on zinc. In general, the salmonids are most sensitive to zinc in soft water; the rainbow trout is the most sensitive in hard water.

REFERENCE
1. USEPA OHMTADS Database.

■ ZINC CHROMATE

CAS # 13530–65–9

SIC CODES 2816; 2899.

SYNONYMS basic zinc chromate • chromic acid (H_2CrO_4), zinc salt (1:1) • chromic acid, zinc salt • zinc tetraoxychromate • zinc tetroxychromate • buttercup yellow • ci-77955 • pigment yellow 36 • zinc yellow

MF CrH_2O_4Zn

MW 183.39

COLOR AND FORM Lemon-yellow prisms; yellow, fine powder.

ODOR Odorless.

DENSITY 3.40.

FIREFIGHTING PROCEDURES Respiratory protection from chromic acid and chromates while fighting fires: self-contained breathing apparatus with a full facepiece operated in pressure-demand or other positive-pressure mode. (chromic acid and chromates) [R19, 7].

SOLUBILITY Insoluble in cold water; soluble in acid, liquid ammonia; insoluble in acetone; sparingly soluble in water.

STABILITY AND SHELF LIFE Trivalent chromium is the most stable oxidation state and hexavalent chromium is the second-most-stable oxidation state. (trivalent and hexavalent chromium) [R3].

OTHER CHEMICAL/PHYSICAL PROPERTIES Hexavalent chromium is acidic (hexavalent chromium). Solid yellow, molecular wt: 307.6 (zinc chromate heptahydrate). Corrosive because of its oxidizing potency. (chromate salts).

PROTECTIVE EQUIPMENT Wear appropriate equipment to prevent any possibility of skin contact. Wear eye protection to prevent any possibility of eye contact. (chromic acid and chromates (as CrO_3)) [R20, 83].

PREVENTATIVE MEASURES Showers and wash basins should be located in the locker area to encourage good personal hygiene. (hexavalent chromium) [R2].

CLEANUP Trivalent chromium can be effectively removed from drinking water by conventional coagulation techniques, but these techniques are inadequate when chromium is in the hexavalent form. Reverse osmosis is effective for removal of both forms of chromium (trivalent and hexavalent chromium) [R4]. A high-efficiency particulate arrestor (HEPA) or charcoal filters can be used to minimize amount of carcinogen in exhausted-air-ventilated safety cabinets, lab hoods, glove boxes, or animal rooms. Filter housing that is designed so that used filters can be transferred into plastic bag without contaminating maintenance staff is available commercially. Filters should be placed in plastic bags immediately after removal. The plastic bag should be sealed immediately. The sealed bag should be labelled properly. Waste liquids should be placed or collected in proper containers for disposal. The lid should be secured and the bottles properly labelled. Once filled, bottles should be placed in plastic bag, so that outer surface is not contaminated. The plastic bag should also be sealed and labelled (chemical carcinogens) [R21, 15]. Where possible, wet methods of cleaning should be used; at other sites, the only acceptable alternative is by vacuum cleaning. Spills of liquid or solid must be removed immediately to prevent dispersion as airborne dust. (chromium compounds) [R22, 472].

DISPOSAL Route to metal salvage facility. SRP: At the time of review, criteria for land treatment or burial (sanitary landfill) disposal practices are subject to significant revision. Prior to implementing land disposal of waste residue (including waste sludge), consult with environmental regulatory agencies for guidance on acceptable disposal practices.

COMMON USES Primarily used in priming paints for metals, for which they provide resistance against corrosion (zinc chromates) [R18, p. 644.1 (86)]. Artists' color; varnishes; pigment in automotive paints; to impart corrosion resistance to epoxy laminates [R1].

PERSISTENCE Accumulates in gill tissue and bone. Oysters can concentrate 100,000 times. Concentration factor for zinc—marine plants 1,000; zinc will eventually precipitate as a basic salt, but chromates can remain in solution indefinitely; invertebrates 100,000; fish 2,000; freshwater, plants 4,000; invertebrates 40,000; and fish 1,000 [R12]; half-life in total human body—933 days. Concentration factor for chromium—marine plants 2,000; freshwater and marine inverte-brates 2,000; marine fish 400; freshwater fish 200.

MAJOR SPECIES THREATENED All species.

DIRECT CONTACT Skin and mucous membranes.

GENERAL SENSATION Soluble zinc salts have a harsh metallic taste; small doses cause nausea and vomiting; larger doses cause violent vomiting and purging. Chromic acid salts have a corrosive action on the skin and mucous membranes. Chromate salts have been associated with cancer of the lungs. Lesions and ulcers of exposed skin on hands and forearms may occur. They are slow healing. Odorless. Wear filler mask.

PERSONAL SAFETY Skin protection required. Self-contained breathing apparatus required if flame or intense heat prevails.

ACUTE HAZARD LEVEL Threshold concentration for fish 0.1 ppm Zn, 0.05 ppm Cr(VI) [R10]. Highly hazardous by ingestion or inhalation. Strong irritant. Toxic to aquatic life. Will produce tastes in water, and can upset sewage treatment processes.

CHRONIC HAZARD LEVEL Maximum allowable toxicant concentration (MATC) for fathead minnow 0.032–0.18. Application factor for extrapolating 96hr TL_{50} data 0.003–0.02 [R13]. Daphnids suffered decreased reproduction when exposed to 0.102 ppm Zn for 3 weeks, as did fathead minnows exposed to 0.18 ppm for 10 months [R14]. Freshwater should not exceed 0.003 of the 96–hour LC_{50} and marine waters 0.01 of the 96–hour LC_{50} [R15]. A low but significant mortality has been found among rainbow trout exposed continuously for 4 months to constant concentrations of 0.2 of the 5-day LC_{50} and among rudd exposed for 8.5 months to 0.3 of the 7-day LC_{50}. Avoidance reactions have also been observed at 0.35–0.43 of the 7-day LC_{50} by migrating Atlantic salmon in a river polluted with copper and zinc. Carp and goldfish show avoidance of 0.3–0.45 of lethal concentrations under laboratory conditions [R16]. Chronic ingestive and inhalative toxin, chronic irritant. Chronic rainbow trout studies show impaired reproduction over a 2–year period at 0.4 ppm for hexavalent chromium. Freshwater should not exceed 0.03 ppm Cr, marine waters 0.01 96–hr LC_{50} [R17]. Application to convert 96–hour LC_{50} for Cr^{+6} to chronic safe limit—0.03 for fathead minnow, 0.01 for brook trout, and 0.003 for rainbow trout [R11].

DEGREE OF HAZARD TO PUBLIC HEALTH Strong irritant. Highly hazardous with acute or chronic ingestion of inhalation. Emits toxic vapors when heated to decomposition. Potential carcinogen.

OCCUPATIONAL RECOMMENDATIONS.

OSHA 1 mg/m³ (8–hr TWA) (chromium, metal, and insoluble salts (as Cr)) [R20, 82]; threshold-limit 8–hr time-weighted average (TWA) 0.01 mg/m³ (1988) (zinc chromates, as Cr) [R6].

INTERNATIONAL EXPOSURE LIMITS Acceptable average concentration of occupational exposure based on 8–hr exposures for chromium, metal, and insoluble salts: 1,000 µg/m³ (from table; chromium metal and insoluble salts) [R23, 456].

ACTION LEVEL Isolate from heat and flame. Prevent suspension of dusts.

ON-SITE RESTORATION Add ferric salt and sodium hydrogen carbonate to precipitate ferric chromate, which can be dredged. Add lime to precipitate zinc. Seek professional environmental engineering assistance through EPA's Environmental Response Team (ERT), Edison, NJ, 24–hour no. 908–548–8730.

AVAILABILITY OF COUNTERMEASURE MATERIAL Lime—cement plants; ferric chloride—water treatment plants, photography shops; sodium hydrogen carbonate—grocery stores and bakeries.

MAJOR WATER USE THREATENED Potable supply; fisheries; recreation.

PROBABLE LOCATION AND STATE Lemon-yellow prisms. pale-green-yellow pigment. Will sink and partially dissolve.

COLOR IN WATER Yellow to green.

FOOD-CHAIN CONCENTRATION POTENTIAL Positive—radioactive zinc (^{65}Zn) has been found to concentrate in plants, milk, bone, mollusks, and aquatic life. Zinc can pass from water to forage to cows' milk.

IARC SUMMARY Classification of carcinogenicity: (1) evidence in humans: sufficient; (2) evidence in animals: sufficient; (3) evidence for activity in short-term tests: sufficient for Cr(VI). Summary evaluation of carcinogenic risk to humans, 1: Chromium and certain chromium compounds are carcinogenic to humans. (chromium and chromium compounds) [R5].

BIOCONCENTRATION Bioconcentration factors for chromium(VI) range from 125 to 236 for bivalve molluscs and polychaetes. (hexavalent chromium) [R7].

RCRA REQUIREMENTS A solid waste containing zinc chromate may become characterized as a hazardous waste when subjected to the toxicant extraction procedure listed in 40 CFR 261.24, and, if so characterized, must be managed as a hazardous waste [R8].

FDA Rubber articles intended for repeated use may be safely used in producing, manufacturing, packing, processing, preparing, treating, packaging, transporting, or holding food, subject to the provisions of this section. Zinc chromate is used as a colorant in which the total may not exceed 10% by weight of the rubber product [R9].

REFERENCES

1. Hawley, G. G. *The Condensed Chemical Dictionary*. 9th ed. New York: Van Nostrand Reinhold Co., 1977. 938.

2. NIOSH. *Criteria Document: Chromium(VI)* p. 16 (1976) DHEW Pub. NIOSH 76–129.

3. USEPA. *Health Assessment Document* p. 2–2 (1984) EPA 600/8–83–014F.

4. 48 FR 45512 (10/5/83).

5. IARC. *Monographs on the Evaluation of the Carcinogenic Risk of Chemicals to Man.* Geneva: World Health Organization, International Agency for Research on Cancer, 1972–present (multivolume work), p. S7 60 (1987).

6. American Conference of Governmental

Industrial Hygienists. *Threshold Limit Values for Chemical Substances and Physical Agents and Biological Exposure Indices for 1994–1995.* Cincinnati, OH: ACGIH, 1994. 36.

7. USEPA. *Ambient Water Quality Criteria Doc; Chromium* p. 18 (1984) EPA 440/5-84-029.

8. 40 CFR 261.24 (7/1/87).

9. 21 CFR 177.2600 (4 vi) (4/1/87).

10. Todd, D. K., *The Water Encyclopedia,* Maple Press, 1970.

11. Environmental Protection Agency, *Effects of Pesticides in Water,* NTIS PB-222 320, 1972.

12. Chapman, W. H., M. L. Fisher, and M. W. Pratt, *Concentration Factors of Chemical Elements in Edible Aquatic Organisms,* UCRL-50564, Lawrence Radiation Laboratory, Livermore, California, December 30, 1968.

13. Pickering, Q. H., "Chronic Toxicity of Nickel to the Fathead Minnow," *JWPCF,* Vol. 46, No. 4, April 1974.

14. Environmental Protection Agency, *Report of the Pesticide Technical Committee to the Lake Michigan Enforcement Conference on Selected Trace Metals,* NTIS PB-220 361, September 1972.

15. *Proposed Criteria For Water Quality, Volume I* , U.S. Environmental Protection Agency. October 1973.

16. *Water Quality Criteria for European Freshwater Fish,* Water Research EFAC Technical Paper No. 21, Vol. 8, pp. 683–684, 1974.

17. Bond, C. E., Lewis, R. H., and Fryer, W. L., *Toxicity of Various Herbicidal Materials to Fishes,* C. M. Tarzwell (Comp.), *Biological Problems in Water Pollution,* Trans. 1959 Seminar. Cincinnati, Ohio, Robt. A. Taft San. Eng. Center. Tech. Rept. W60-3, 1960, pp. 96–101.

18. American Conference of Governmental Industrial Hygienists. *Documentation of the Threshold Limit Values and Biological Exposure Indices.* 5th ed. Cincinnati, OH: American Conference of Governmental Industrial Hygienists, 1986.

19. Mackison, F. W., R. S. Stricoff, and L. J. Partridge, Jr. (eds.). *NIOSH/OSHA— Occupational Health Guidelines for Chemical Hazards.* DHHS (NIOSH) Publication No. 81–123 (3 vols). Washington, DC: U.S. Government Printing Office, Jan. 1981.

20. NIOSH. *Pocket Guide to Chemical Hazards.* 5th Printing/Revision. DHHS (NIOSH) Publ. No. 85–114. Washington, DC: U.S. Dept. of Health and Human Services, NIOSH/Supt. of Documents, GPO, Sept. 1985.

21. Montesano, R., H. Bartsch, E. Boyland, G. Della Porta, L. Fishbein, R. A. Griesemer, A.B. Swan, L. Tomatis, and W. Davis (eds.). *Handling Chemical Carcinogens in the Laboratory: Problems of Safety.* IARC Scientific Publications No. 33. Lyon, France: International Agency for Research on Cancer, 1979.

22. International Labour Office. *Encyclopedia of Occupational Health and Safety.* Vols. I and II. Geneva, Switzerland: International Labour Office, 1983.

23. Casarett, L.J., and J. Doull. *Toxicology: The Basic Science of Poisons.* New York: Macmillan Publishing Co., 1975.

■CAS NUMBERS INDEX, VOLUMES 1–3

SYNONYM Index, Volumes I–III

3-(acetonylbenzyl) 4-hydroxycoumarin [81-81-2] **Vol. I:** 525

acetonyl bromide [598-31-2] **Vol. III:** 800

acetonyl chloride [78-95-5] **Vol. II:** 828

acetonyldimethylcarbinol [123-42-2] **Vol. II:** 688

aceto-pbn [135-88-6] **Vol. III:** 640

acetophenon [98-86-2] **Vol. I:** 86

acetophenone, 2-chloro [532-27-4] **Vol. I:** 89

acetoquat-cpc [123-03-5] **Vol. III:** 196

acetosal [79-34-5] **Vol. III:** 414

aceto tetd [137-26-8] **Vol. I:** 632

6-acetoxy-2,4-dimethyl-1,3-dioxane [828-00-2] **Vol. III:** 363

6-acetoxy-2,4-dimethyl-meta-dioxane [828-00-2] **Vol. III:** 363

acetoxyethane [141-78-6] **Vol. II:** 19

1-acetoxy-2-ethoxyethane [111-15-9] **Vol. II:** 365

1-acetoxyethylene [108-05-4] **Vol. I:** 69

(acetoxymercuri) benzene [62-38-4] **Vol. II:** 535

2-acetoxypentane [626-38-0] **Vol. I:** 52

acetoxyphenylmercury [62-38-4] **Vol. II:** 535

1-acetoxypropane [109-60-4] **Vol. I:** 55

2-acetoxypropane [108-21-4] **Vol. I:** 46

alpha-acetoxytoluene [140-11-4] **Vol. III:** 115

acetylaminobenzene [103-84-4] **Vol. III:** 3

acetyl anhydride [108-24-7] **Vol. I:** 73

acetylaniline [103-84-4] **Vol. III:** 3

N-acetylaniline [103-84-4] **Vol. III:** 3

acetylbenzene [98-86-2] **Vol. I:** 86

acetyl chloride, chloro [79-04-9] **Vol. III:** 214

acetyldimethylamine [127-19-5] **Vol. I:** 10

acetylene black [7440-44-0] **Vol. III:** 179

acetylene dichloride [540-59-0] **Vol. II:** 406

acetylene tetrachloride [79-34-5] **Vol. III:** 414

acetylene trichloride [79-01-6] **Vol. II:** 427

acetylenogen [75-20-7] **Vol. I:** 391

acetylenylcarbinol [107-19-7] **Vol. II:** 864

acetyl ether [108-24-7] **Vol. I:** 73

acetyl-ethylene [78-94-4] **Vol. III:** 624

acetyl hydroperoxide [79-21-0] **Vol. III:** 682

acetylmethyl bromide [598-31-2] **Vol. III:** 800

acetyl peroxide [110-22-5] **Vol. III:** 301

acid ammonium carbonate [1066-33-7] **Vol. II:** 62

acid ammonium fluoride [1341-49-7] **Vol. II:** 64

acide acetique (French) [64-19-7] **Vol. I:** 12

acide acrylique (French) [79-10-7] **Vol. I:** 110

acide arsenieux (French) [1327-53-3] **Vol. I:** 203

acide arsenique liquide (French) [7778-39-4] **Vol. II:** 91

acide benzoique (French) [65-85-0] **Vol. I:** 293

acide cacodylique (French) [75-60-5] **Vol. I:** 214

acide carbolique (French) [108-95-2] **Vol. I:** 920

acide chlorhydrique (French) [7647-01-0] **Vol. I:** 738

acide chloroacetique (French) [79-11-8] **Vol. I:** 22

acide chromique (French) [7738-94-5] **Vol. II:** 239

acide cyanacetique (French) [372-09-8] **Vol. I:** 29

acide cyanhydrique (French) [74-90-8] **Vol. I:** 743

acide 2,4-dichloro phenoxyacetique (French) [94-75-7] **Vol. I:** 30

acide dichloro-3,6 picolinique (French) [1702-17-6] **Vol. III:** 239

acide dimethylarsenique (French) [75-60-5] **Vol. I:** 214

acide ethylenediaminetetracetique (French) [60-00-4] **Vol. I:** 36

acide fluorhydrique (French) [7664-39-3] **Vol. I:** 750

acide formique (French) [64-18-6] **Vol. I:** 711

acide monochloracetique (French) [79-11-8] **Vol. I:** 22

acide nitrique (French) [7697-37-2] **Vol. I:** 879

acide oxalique (French) [144-62-7] **Vol. I:** 904

acide peracetique (French) [79-21-0] **Vol. III:** 682

acide peroxyacetique (French) [79-21-0] **Vol. III:** 682

acide phosphorique (French) [7664-38-2] **Vol. I:** 942

acide picrique (French) [88-89-1] Vol. II: 793

acide propionique (French) [79-09-4] Vol. I: 1012

acide sulfhydrique (French) [7783-06-4] Vol. II: 485

acide sulfurique (French) [7664-93-9] Vol. II: 969

acide sulphhydrique [7783-06-4] Vol. II: 485

acide tannique (French) [1401-55-4] Vol. II: 237

acide thioglycolique (French) [68-11-1] Vol. III: 538

acide trichloracetique (French) [76-03-9] Vol. II: 32

acide 2,4,5-trichloro phenoxyacetique (Frenchs) [93-76-5] Vol. I: 59

acide 2-(2,4,5-trichloro-phenoxy) propionique (French) [93-72-1] Vol. II: 845

acido acetico (Italian) [64-19-7] Vol. I: 12

acido acrilio (Spanish) [79-10-7] Vol. I: 110

acido cianidrico (Italian) [74-90-8] Vol. I: 743

acido clorhidrico (Spanish) [7647-01-0] Vol. I: 738

acido cloridrico (Italian) [7647-01-0] Vol. I: 738

acido (2,4-dichloro fenossi) acetico (Italian) [94-75-7] Vol. I: 30

acido (3,6-dicloro-2-metossi)-benzoico (Italian) [1918-00-9] Vol. II: 159

acido fluorhidrico (Spanish) [7664-39-3] Vol. I: 750

acido fluoridrico (Italian) [7664-39-3] Vol. I: 750

acido formico (Italian) [64-18-6] Vol. I: 711

acido fosforico (Italian) [7664-38-2] Vol. I: 942

acidomonocloroacetico (Italian) [79-11-8] Vol. I: 22

acido nitrico (Italian) [7697-37-2] Vol. I: 879

acido nitrico (Spanish) [7697-37-2] Vol. I: 879

acido ossalico (Italian) [144-62-7] Vol. I: 904

acido peroxiacetico (Spanish) [79-21-0] Vol. III: 682

acido picrico (Italian) [88-89-1] Vol. II: 793

acido solforico (Italian) [7664-93-9] Vol. II: 969

acido sulfurico (Spanish) [7664-93-9] Vol. II: 969

acido tricloroacetico (Italian) [76-03-9] Vol. II: 32

acido 2-(2,4,5-tricloro-fenossi)-propionico (Italian) [93-72-1] Vol. II: 845

acid, tannic [1401-55-4] Vol. II: 237

acifloctin [124-04-9] Vol. II: 48

aciletten [77-92-9] Vol. I: 494

acilid [1918-16-7] Vol. III: 789

acimetion [63-68-3] Vol. III: 593

acinetten [124-04-9] Vol. II: 48

acl-85 [87-90-1] Vol. III: 936

acme lv-4- [94-75-7] Vol. I: 30

acme lv-6- [94-75-7] Vol. I: 30

acomethylene [334-88-3] Vol. III: 306

acp-m-728 [133-90-4] Vol. III: 44

acqninite [76-06-2] Vol. II: 606

acquinite [107-02-8] Vol. I: 99

acquinite [76-06-2] Vol. II: 606

ac-r-11 [126-15-8] Vol. III: 909

acraldehyde [107-02-8] Vol. I: 99

acricid [485-31-4] Vol. II: 261

acroleic acid [79-10-7] Vol. I: 110

acroleina (Italian) [107-02-8] Vol. I: 99

acroleine (Dutch, French) [107-02-8] Vol. I: 99

acrolein, 2-ethyl-3-propyl- [645-62-5] Vol. II: 464

acromona [443-48-1] Vol. III: 628

acrylagel [79-06-1] Vol. I: 106

acrylaldehyde [107-02-8] Vol. I: 99

Acrylaldehyd (German) [107-02-8] Vol. I: 99

acrylamide monomer [79-06-1] Vol. I: 106

acrylate de methyle (French) [96-33-3] Vol. I: 119

acrylate d'ethyle (French) [140-88-5] Vol. I: 114

acrylic acid amide [79-06-1] Vol. I: 106

acrylic acid, butyl ester [141-32-2] Vol. II: 842

acrylic acid, ethyl ester [140-88-5] Vol. I: 114

acrylic acid, 2-ethylhexyl ester [103-11-7] Vol. II: 45

acrylic acid, glacial [79-10-7] Vol. I: 110

acrylic acid, 2-hydroxyethyl ester [818-61-1] Vol. III: 484

acrylic acid isobutyl ester [106-63-8] Vol. III: 15

acrylic acid, isobutyl ester [106-63-8] Vol. III: 15

acrylic acid, isodecyl ester [1330-61-6] Vol. III: 495

acrylic acid, 2-methyl-, dodecyl ester [142-90-5] Vol. III: 391

acrylic acid methyl ester [96-33-3] Vol. I: 119

acrylic acid, methyl ester [96-33-3] Vol. I: 119

acrylic acid, 2-methyl-, methyl ester [80-62-6] Vol. II: 567

acrylic acid, n-butyl ester [141-32-2] Vol. II: 842

acrylicaldehyde [107-02-8] Vol. I: 99

acrylic amide [79-06-1] Vol. I: 106

acrylnitril (German,Dutch) [107-13-1] Vol. I: 123

acrylon [107-13-1] Vol. I: 123

acrylonitrile (DOT) [107-13-1] Vol. I: 123

acrylonitrilemonomer [107-13-1] Vol. I: 123

2-(acryloyloxy) ethanol [818-61-1] Vol. III: 484

Acrylsaeureaethylester (German) [140-88-5] Vol. I: 114

Acrylsaeuremethylester (German) [96-33-3] Vol. I: 119

acryptan [133-07-3] Vol. III: 461

acticarbone [7440-44-0] Vol. III: 179

acticel [16871-71-9] Vol. II: 911

acticel [7631-86-9] Vol. III: 846

actinite-pk [1912-24-9] Vol. II: 1050

actinium-X [872-50-4] Vol. II: 893

activated carbon [7440-44-0] Vol. III: 179

adacene-12 [112-41-4] Vol. III: 390

add f [64-18-6] Vol. I: 711

adilactetten [124-04-9] Vol. II: 48

adipate [124-04-9] Vol. II: 48

adipic acid, bis (2-ethylhexyl) ester [103-23-1] Vol. II: 291

adipic acid dinitrile [111-69-3] Vol. I: 128

adipic acid nitrile [111-69-3] Vol. I: 128

adipinic acid [124-04-9] Vol. II: 48

adipodinitrile [111-69-3] Vol. I: 128

adipol-2eh [103-23-1] Vol. II: 291

adipyldinitrile [111-69-3] Vol. I: 128

adol-52 [36653-82-4] Vol. III: 193

adol-54 [36653-82-4] Vol. III: 193

adol-52nf [36653-82-4] Vol. III: 193

adol-52-nf [36653-82-4] Vol. III: 193

adonal [50-06-6] Vol. I: 223

adronal [108-93-0] Vol. I: 580

adronol [108-93-0] Vol. I: 580

adsorbit [7440-44-0] Vol. III: 179

ad-6(suspendingagent) [75-56-9] Vol. I: 991

aer fixus [124-38-9] Vol. III: 182

aero cyanamid [156-62-7] Vol. II: 265

aero cyanamid granular [156-62-7] Vol. II: 265

aero cyanamid special grade [156-62-7] Vol. II: 265

aerogel-200 [16871-71-9] Vol. II: 911

aerogel-200 [7631-86-9] Vol. III: 846

aerol-1 [52-68-6] Vol. III: 707

aero liquid hcn [74-90-8] Vol. I: 743

aerosil [16871-71-9] Vol. II: 911

aerosil-300 [16871-71-9] Vol. II: 911

aerosil-300 [7631-86-9] Vol. III: 846

aerosil-380 [16871-71-9] Vol. II: 911

aerosil-380 [7631-86-9] Vol. III: 846

aerosil [7631-86-9] Vol. III: 846

aerosil-a-300 [16871-71-9] Vol. II: 911

aerosil-a-300 [7631-86-9] Vol. III: 846

aerosil bs-50 [7631-86-9] Vol. III: 846

aerosil-degussa [16871-71-9] Vol. II: 911

aerosil-degussa [7631-86-9] Vol. III: 846

aerosil-e-300 [16871-71-9] Vol. II: 911

aerosil-e-300 [7631-86-9] Vol. III: 846

aerosil-k-7 [16871-71-9] Vol. II: 911

aerosil-k-7 [7631-86-9] Vol. III: 846

aerosil m-300 [16871-71-9] Vol. II: 911

aerosil m-300 [7631-86-9] Vol. III: 846

aerosol-ot [577-11-7] Vol. III: 859

aerothene mm [71-55-6] Vol. I: 684

aerothene tt [71-55-6] Vol. I: 684

Aethanethiol (German) [75-08-1] Vol. III: 419

aethanolamin (German) [141-43-5] Vol. II: 356

aethanol (German) [64-17-5] Vol. II: 394

2-aethoxy-aethylacetat (German) [111-15-9] Vol. II: 365

S-(1,2-bis (aethoxy-carbonyl)-aethyl)-o, o-dimethyl-dithiophosphat (German) [121-75-5] Vol. II: 950

5-aethoxy-3-trichlormethyl-1,2,4-thiadiazol [German] [2593-15-9] Vol. III: 898

aethylacetat (German) [141-78-6] Vol. II: 19

Aethylacrylat (German) [140-88-5] Vol. I: 114

aethylalkohol (German) [64-17-5] Vol. II: 394

2-aethylamino-4-chlor-6-isopropyl amino-1,3,5-triazin (German) [1912-24-9] Vol. II: 1050

4-aethylamino-2-tert-butylamino-6-methyl-thio-s-triazin (German) [886-50-0] Vol. II: 1045

Aethylbenzol (German) [100-41-4] Vol. I: 264

aethylcarbamat (German) [51-79-6] Vol. II: 224

aethylchloride (German) [75-00-3] Vol. II: 321

Aethylenchlorid (German) [107-06-2] Vol. I: 643

Aethylenechlorhydrin (German) [107-07-3] Vol. III: 220

aethylenglykolaetheracetat (German) [111-15-9] Vol. II: 365

aethylenglykol-monomethylaether (German) [109-86-4] Vol. II: 370

Aethylenimin (German) [151-56-4] Vol. I: 671

aethylenoxid (German) [75-21-8] Vol. II: 401

aethylformiat (German) [109-94-4] Vol. II: 438

aethylidenchlorid (German) [75-34-3] Vol. II: 325

aethylis chloridum [75-00-3] Vol. II: 321

Aethylmercaptan (German) [75-08-1] Vol. III: 419

Aethylmethylketon (German) [78-93-3] Vol. I: 347

S-Aethyl-N-hexahydro-1H-azepinthiolcarba-mat (German) [2212-67-1] Vol. III: 77

o-Aethyl-o-(3-methyl-4-methylthio phenyl)-isopropylamido-phosphor saeureester (German) [22224-92-6] Vol. II: 743

o-Aethyl-o-(4-nitro-phenyl)-phenylmono-thiophosphonat (German) [2104-64-5] Vol. II: 738

o-Aethyl-s-phenyl-aethyl-dithiophos phonat (German) [944-22-9] Vol. II: 735

o-Aethyl-S-phenyl-aethyl-dithiophosphonat (German) [944-22-9] Vol. III: 714

Aethylurethan (German) [51-79-6] Vol. II: 224

af-101 [330-54-1] Vol. II: 1078

af-260 [21645-51-2] Vol. II: 57

Af-260 [21645-51-2] Vol. III: 35

afalon [330-55-2] Vol. II: 1084

afalon-inuron [330-55-2] Vol. II: 1084

a-fil cream [13463-67-7] Vol. III: 921

Aflatoxin [1402-68-2] Vol. III: 18

after-damp [124-38-9] Vol. III: 182

ag-3 (adsorbent) [7440-44-0] Vol. III: 179

ag-5 (adsorbent) [7440-44-0] Vol. III: 179

agent blue [75-60-5] Vol. I: 214

agent sa [7784-42-1] Vol. I: 212

agerite powder [135-88-6] Vol. III: 640

agrenal [100-56-1] Vol. III: 699

agricide maggot killer [8001-35-2] Vol. I: 1118

agricorn d [94-75-7] Vol. I: 30

agritox [94-74-6] Vol. I: 25

agroceres [76-44-8] Vol. II: 628

Agro chemical brand torbidan 28 [8001-35-2] Vol. I: 1118

Agro chemical brand toxaphene 6e [8001-35-2] Vol. I: 1118

agrocide [608-73-1] Vol. I: 564

agrocit [17804-35-2] Vol. II: 148

agroforotox [52-68-6] Vol. III: 707

agronal [100-56-1] Vol. III: 699

agrosan [62-38-4] Vol. II: 535

agrosan-d [62-38-4] Vol. II: 535

agrosan-gn-5 [62-38-4] Vol. II: 535

agrosil-lr [1344-09-8] Vol. III: 868

agrosil-s [1344-09-8] Vol. III: 868

agrosol-s [12427-38-2] Vol. III: 531

agrosol-s [133-06-2] Vol. II: 276

agrothrin [52315-07-8] Vol. II: 251

agroxon [94-74-6] Vol. I: 25

agroxone [94-74-6] Vol. I: 25

agroxone [94-75-7] Vol. I: 30

agrox 2-way and 3-way [133-06-2] Vol. II: 276

agrypnal [50-06-6] Vol. I: 223

agsco toxaphene [8001-35-2] Vol. I: 1118

agway toxaphene-6e [8001-35-2] Vol. I: 1118

ai3 03700 [124-04-9] Vol. II: 48

ai-23024 [82-68-8] Vol. II: 132

ai-28,597 [98-87-3] Vol. III: 84

ai30005 [107-13-1] Vol. I: 123

ai3-0310 [65-85-0] Vol. I: 293

ai3-01955 (USDA) [111-15-9] Vol. II: 365

ai3-00046 [78-59-1] Vol. II: 272

ai3-00052 [79-01-6] Vol. II: 427

ai3-00053 [95-50-1] Vol. I: 250

ai3-00056 [1300-71-6] Vol. I: 1171

ai3-00075 [59-50-7] Vol. II: 258

ai3-00134 [87-86-5] Vol. I: 928

ai3-00140 [92-87-5] Vol. I: 288

ai3-00142 [88-06-2] **Vol. II:** 732

ai3-00262 [131-11-3] **Vol. II:** 784

ai3-00283 [84-74-2] **Vol. II:** 774

ai3-00404 [141-78-6] **Vol. II:** 19

ai3-00409 [67-56-1] **Vol. I:** 836

ai3-00553 [51-79-6] **Vol. II:** 224

ai3-00698 [86-30-6] **Vol. I:** 622

ai3-00808 [71-43-2] **Vol. I:** 239

ai3-01006 [1303-33-9] **Vol. II:** 99

ai3-01031 [55-21-0] **Vol. III:** 86

ai3-01202 [57-13-6] **Vol. I:** 1136

ai3-01229 [108-10-1] **Vol. II:** 692

ai3-01238 [67-64-1] **Vol. I:** 77

ai3-01240 [110-86-1] **Vol. II:** 874

ai3-01241 [91-22-5] **Vol. I:** 1022

ai3-01552 [95-65-8] **Vol. I:** 1169

ai3-01553 [108-68-9] **Vol. I:** 1165

ai3-01636 [67-63-0] **Vol. II:** 515

ai-3.01719- [118-74-1] **Vol. I:** 268

ai3-01860 [127-18-4] **Vol. II:** 421

ai3-01916 [74-83-9] **Vol. II:** 577

ai3-02061 [71-55-6] **Vol. I:** 684

ai3-02435 [65-30-5] **Vol. I:** 877

ai3-02583 [98-07-7] **Vol. II:** 139

ai3-02729 [628-63-7] **Vol. II:** 27

ai3-02738 (USDA) [21908-53-2] **Vol. II:** 559

ai3-03053 [62-53-3] **Vol. I:** 164

ai3-03582 [62-56-6] **Vol. I:** 1140

ai3-03967 [7733-02-0] **Vol. II:** 1103

ai3-03995 [120-80-9] **Vol. III:** 190

ai3-04273 [117-81-7] **Vol. II:** 769

ai3-04458 [1600-27-7] **Vol. II:** 24

AI3-04465 [557-34-6] **Vol. III:** 12

ai3-04470 [7646-85-7] **Vol. I:** 1178

ai3-07540 (USDA) [78-93-3] **Vol. I:** 347

ai3-07541 [75-56-9] **Vol. I:** 991

AI3-07775 [120-82-1] **Vol. III:** 92

ai3-08197 [95-47-6] **Vol. I:** 1155

ai3-08524 [576-26-1] **Vol. I:** 1167

ai3-08584 [90-04-0] **Vol. II:** 76

ai3-08765 [142-90-5] **Vol. III:** 391

ai3-08937 [12125-02-9] **Vol. I:** 141

ai3-10509 [1120-36-1] **Vol. III:** 904

ai3-15071 (USDA) [117-84-0] **Vol. II:** 789

ai3-15300 [100-39-0] **Vol. II:** 1019

ai3-15329 [111-92-2] **Vol. III:** 311

ai3-15425 [121-44-8] **Vol. I:** 1125

ai3-15514 [74-97-5] **Vol. II:** 583

ai3-15516 [87-61-6] **Vol. II:** 137

ai3-15518 [100-44-7] **Vol. I:** 1099

ai3-15639 [75-50-3] **Vol. II:** 573

ai3-15717 [79-10-7] **Vol. I:** 110

ai3-16115 [71-23-8] **Vol. II:** 856

AI3-16225 [60-57-1] **Vol. III:** 356

ai3-16292 [96-45-7] **Vol. I:** 757

ai3-16418 [131-52-2] **Vol. III:** 676

ai3-17034 [121-75-5] **Vol. II:** 950

ai3-17181 [60-00-4] **Vol. I:** 36

ai3-17612 [105-67-9] **Vol. I:** 1162

ai3-18303 (USDA) [4170-30-3] **Vol. I:** 547

ai3-18445 [96-12-8] **Vol. II:** 818

ai3-24037 [142-84-7] **Vol. III:** 378

ai3-24160 [107-02-8] **Vol. I:** 99

ai3-24215 [109-89-7] **Vol. I:** 597

ai3-24237 [64-18-6] **Vol. I:** 711

ai3-24474 [75-00-3] **Vol. II:** 321

AI3-24838 [7778-44-1] **Vol. III:** 65

ai3-25675 [2636-26-2] **Vol. III:** 271

ai3-25991 [950-10-7] **Vol. III:** 536

ai3-26110 [373-02-4] **Vol. II:** 648

ai3-26263 [75-21-8] **Vol. II:** 401

ai3-26435 [98-51-1] **Vol. III:** 169

ai3-27572 [22224-92-6] **Vol. II:** 743

ai3-28257 [556-61-6] **Vol. III:** 600

ai3-28579 [103-23-1] **Vol. II:** 291

ai3-28745 [544-92-3] **Vol. I:** 522

ai3-28748 [506-64-9] **Vol. II:** 915

ai3-28749 [151-50-8] **Vol. I:** 978

ai3-28752 [557-21-1] **Vol. I:** 1181

ai3-50325 [75-55-8] **Vol. I:** 219

ai3-51770 [303-34-4] **Vol. III:** 501

ai3-52118 [77-78-1] **Vol. II:** 976

ai3-52255 [106-42-3] **Vol. I:** 1152

ai3-52358 [87-62-7] **Vol. II:** 1092

ai3-52483 [139-13-9] **Vol. III:** 648

ai3-52649 [111-92-2] **Vol. III:** 311

ai3-62048 [15663-27-1] **Vol. III:** 237

AIM [7783-47-3] **Vol. III:** 881

ai3-02209-x [1330-20-7] **Vol. I:** 1159

ai3-15637-x [74-89-5] **Vol. I:** 840

ai3-15638-x [124-40-3] **Vol. I:** 613

ai3-25550-X [16919-19-0] **Vol. II:** 908

ai3-31100-x [74-90-8] **Vol. I:** 743

ai3-24178-x (USDA) [1300-73-8] **Vol. III:** 976

ak (adsorbent) [7440-44-0] **Vol. III:** 179

akar-50 [510-15-6] **Vol. III:** 108

akar-338 [510-15-6] **Vol. III:** 108

akar [510-15-6] **Vol. III:** 108

akaritox [116-29-0] **Vol. III:** 906

akroleina (Polish) [107-02-8] **Vol. I:** 99

akrolein (Czech) [107-02-8] Vol. I: 99

akrylamid (Czech) [79-06-1] Vol. I: 106

akrylonitryl (Polish) [107-13-1] Vol. I: 123

akticon [1912-24-9] Vol. II: 1050

aktikon [1912-24-9] Vol. II: 1050

aktikon-pk [1912-24-9] Vol. II: 1050

aktinit-a [1912-24-9] Vol. II: 1050

aktinit-pk [1912-24-9] Vol. II: 1050

aktinit-s [122-34-9] Vol. II: 1047

aktisal [144-62-7] Vol. I: 904

aktivex [123-03-5] Vol. III: 196

akzo-chemie-maneb [12427-38-2] Vol. III: 531

alachlore [15972-60-8] Vol. II: 9

alanex [15972-60-8] Vol. II: 9

alanox [15972-60-8] Vol. II: 9

alar [1596-84-5] Vol. III: 884

alar-85 [1596-84-5] Vol. III: 884

alatex [75-99-0] Vol. I: 1016

albacol [71-23-8] Vol. II: 856

albone [7722-84-1] Vol. II: 482

albone-ds [7722-84-1] Vol. II: 482

Alcoa-331 [21645-51-2] Vol. II: 57

alcoa-331 [21645-51-2] Vol. III: 35

alcoa-c-330 [21645-51-2] Vol. II: 57

alcoa-c-330 [21645-51-2] Vol. III: 35

alcoa-c-333 [21645-51-2] Vol. II: 57

Alcoa-c-333 [21645-51-2] Vol. III: 35

alcoa-c-30bf [21645-51-2] Vol. II: 57

alcoa-c-30bf [21645-51-2] Vol. III: 35

alcoa sodium fluoride [7681-49-4] Vol. II: 925

alcohol [64-17-5] Vol. II: 394

alcohol, anhydrous [64-17-5] Vol. II: 394

alcohol C-6 [111-27-3] Vol. I: 734

alcohol c-7 [111-70-6] Vol. III: 474

alcohol C-8 [111-87-5] Vol. I: 897

alcohol C-9 [143-08-8] Vol. I: 893

alcohol c-12 [112-53-8] Vol. II: 316

alcohol c-16 [36653-82-4] Vol. III: 193

alcohol, dehydrated [64-17-5] Vol. II: 394

alcohol, ethyl [64-17-5] Vol. II: 394

alcohol, methyl [67-56-1] Vol. I: 836

alcohol methyl amylique (French) [108-11-2] Vol. I: 913

alcohol, propyl [71-23-8] Vol. II: 856

alcool allilico (Italian) [107-18-6] Vol. II: 52

alcool allylique (French) [107-18-6] Vol. II: 52

alcool-,amilico-(Italian) [123-51-3] Vol. II: 190

alcool amylique (French) [71-41-0] Vol. II: 700

alcoolbenzylique [100-51-6] Vol. I: 313

alcool butylique (French) [71-36-3] Vol. I: 352

alcool butylique secondaire (French) [78-92-2] Vol. II: 194

alcool butylique tertiaire (French) [75-65-0] Vol. II: 197

alcool etilico (Italian) [64-17-5] Vol. II: 394

alcool,-isoamylique-(French) [123-51-3] Vol. II: 190

alcool isobutylique (French) [78-83-1] Vol. I: 765

alcool isopropilico (Italian) [67-63-0] Vol. II: 515

alcool isopropylique (French) [67-63-0] Vol. II: 515

alcool-methylique (French) [67-56-1] Vol. I: 836

alcool metilico (Italian) [67-56-1] Vol. I: 836

alcool propilico (Italian) [71-23-8] Vol. II: 856

alcool propylique (French) [71-23-8] Vol. II: 856

alcool-,thylique (French) [64-17-5] Vol. II: 394

alcophobin [97-77-8] Vol. III: 382

alcox-e-30 [25322-68-3] Vol. II: 811

aldacide [30525-89-4] Vol. I: 907

aldecarb [116-06-3] Vol. III: 19

c-10 aldehyde [112-31-2] Vol. III: 294

aldehyde acrylique (French) [107-02-8] Vol. I: 99

aldehyde butyrique (French) [123-72-8] Vol. I: 357

aldehyde-c10 [112-31-2] Vol. III: 294

aldehyde c-10 [112-31-2] Vol. III: 294

aldehyde c-6 [66-25-1] Vol. II: 456

aldehyde–collidine [104-90-5] Vol. III: 770

aldehyde crotonique (French) [123-73-9] Vol. III: 263

aldehyde propionique (French) [123-38-6] Vol. I: 1009

aldehydine [104-90-5] Vol. III: 770

aldeide acrilica (Italian) [107-02-8] Vol. I: 99

aldeide butirrica (Italian) [123-72-8] Vol. I: 357

aldesen [111-30-8] Vol. III: 468

aldicarbe (French) [116-06-3] Vol. III: 19

aldocit [309-00-2] Vol. III: 343

aldol-54 [36653-82-4] **Vol. III:** 193

aldrec [309-00-2] **Vol. III:** 343

aldrex [309-00-2] **Vol. III:** 343

aldrex-40 [309-00-2] **Vol. III:** 343

aldrine (France) [309-00-2] **Vol. III:** 343

aldron [309-00-2] **Vol. III:** 343

aldrosol [309-00-2] **Vol. III:** 343

alfamethrin [67375-30-8] **Vol. III:** 30

alfanaftilamina (Italian) [134-32-7] **Vol. III:** 637

alfa-naftyloamina (Polish) [134-32-7] **Vol. III:** 637

alfa-tox [333-41-5] **Vol. III:** 758

alfol-8- [111-87-5] **Vol. I:** 897

alfol-12 [112-53-8] **Vol. II:** 316

alfol-16 [36653-82-4] **Vol. III:** 193

algaedyn [7440-22-4] **Vol. I:** 1036

algae k [7722-64-7] **Vol. I:** 915

algeon-22 [75-45-6] **Vol. II:** 591

algimycin [62-38-4] **Vol. II:** 535

algistat [117-80-6] **Vol. II:** 645

algofrene-type-1 [75-69-4] **Vol. II:** 602

algofrene type-2- [75-71-8] **Vol. I:** 825

algofrene-type-5 [75-43-4] **Vol. II:** 595

algofrene-type-6 [75-45-61] **Vol. II:** 591

algofrene-type-67 [75-37-6] **Vol. III:** 411

algrain [64-17-5] **Vol. II:** 394

algran [309-00-2] **Vol. III:** 343

algylen [71-55-6] **Vol. I:** 684

algylen [79-01-6] **Vol. II:** 427

alkargen [75-60-5] **Vol. I:** 214

alk-aubs [97-77-8] **Vol. III:** 382

alkohol (German) [64-17-5] **Vol. II:** 394

alkron [56-38-2] **Vol. III:** 750

alkyl dimethylbenzyl ammonium chloride [8001-54-5] **Vol. II:** 67

alkyldimethylbenzylammoniumchloride [8001-54-5] **Vol. II:** 67

alkyldimethyl (phenylmethyl) quaternary ammonium chloride [8001-54-5] **Vol. II:** 67

allbri natural copper [7440-50-8] **Vol. I:** 519

all clear root destroyer [7758-98-7] **Vol. I:** 510

alleron [56-38-2] **Vol. III:** 750

allile, (cloruro di)-(Italian) [107-05-1] **Vol. II:** 830

allilowy alkohol (Polish) [107-18-6] **Vol. II:** 52

alltex [8001-35-2] **Vol. I:** 1118

alltox [8001-35-2] **Vol. I:** 1118

allyl al [107-18-6] **Vol. II:** 52

allyl aldehyde [107-02-8] **Vol. I:** 99

allylalkohol (German) [107-18-6] **Vol. II:** 52

allylchlorid (German) [107-05-1] **Vol. II:** 830

allyl chlorocarbonate [2937-50-0] **Vol. III:** 27

allyle, (chlorure d') (French) [107-05-1] **Vol. II:** 830

allylic alcohol [107-18-6] **Vol. II:** 52

allyl trichloride [96-18-4] **Vol. III:** 943

almond artificial essential oil [100-52-7] **Vol. I:** 237

alochlor [15972-60-8] **Vol. II:** 9

alotano [151-67-7] **Vol. III:** 141

alp [20859-73-8] **Vol. I:** 133

alphadrate [57-13-6] **Vol. I:** 1136

al phos [20859-73-8] **Vol. I:** 133

alugel [21645-51-2] **Vol. II:** 57

alugel [21645-51-2] **Vol. III:** 35

alum [10043-01-3] **Vol. III:** 39

alumigel [21645-51-2] **Vol. II:** 57

alumigel [21645-51-2] **Vol. III:** 35

alumina hydrate [21645-51-2] **Vol. II:** 57

alumina hydrate [21645-51-2] **Vol. III:** 35

alumina hydrated [21645-51-2] **Vol. II:** 57

alumina hydrated [21645-51-2] **Vol. III:** 35

alpha-alumina trihydrate [21645-51-2] **Vol. II:** 57

alumina trihydrate [21645-51-2] **Vol. II:** 57

alpha-alumina trihydrate [21645-51-2] **Vol. III:** 35

alumina trihydrate [21645-51-2] **Vol. III:** 35

aluminium fluorure (French) [7784-18-1] **Vol. I:** 131

aluminium fosfide (Dutch) [20859-73-8] **Vol. I:** 133

aluminium phosphide [20859-73-8] **Vol. I:** 133

aluminum alum [10043-01-3] **Vol. III:** 39

aluminum fluoride [7784-18-1] **Vol. I:** 131

aluminum hydrate [21645-51-2] **Vol. II:** 57

aluminum hydrate [21645-51-2] **Vol. III:** 35

aluminum (III) hydroxide [21645-51-2] **Vol. II:** 57

aluminum (III) hydroxide [21645-51-2] **Vol. III:** 35

aluminum monophosphide [20859-73-8] Vol. I: 133

aluminum oxide 3H20 [21645-51-2] Vol. II: 57

aluminum oxide 3H₂O [21645-51-2] Vol. III: 35

aluminum oxide hydrate [21645-51-2] Vol. II: 57

aluminum oxide hydrate [21645-51-2] Vol. III: 35

aluminum oxide trihydrate [21645-51-2] Vol. II: 57

aluminum oxide, trihydrate [21645-51-2] Vol. II: 57

aluminum oxide trihydrate [21645-51-2] Vol. III: 35

aluminum oxide, trihydrate [21645-51-2] Vol. III: 35

aluminum phosphide [20859-73-8] Vol. I: 133

aluminum sulfate (2:3) [10043-01-3] Vol. III: 39

aluminum sulfate (Al₂(SO₄)₃) [10043-01-3] Vol. III: 39

aluminum sulphate [10043-01-3] Vol. III: 39

aluminum, triethyl- [97-93-8] Vol. II: 59

aluminum trifluoride [7784-18-1] Vol. I: 131

aluminum trihydrate [21645-51-2] Vol. II: 57

aluminum trihydrate [21645-51-2] Vol. III: 35

aluminum trisulfate [10043-01-3] Vol. III: 39

alvit [60-57-1] Vol. III: 356

alzodef [156-62-7] Vol. II: 265

amalgum [7440-22-4] Vol. I: 1036

amatin [118-74-1] Vol. I: 268

ambiben [133-90-4] Vol. III: 44

ambiocide [608-73-1] Vol. I: 564

ambox [485-31-4] Vol. II: 261

ambush [52645-53-1] Vol. III: 282

ambush-c [52315-07-8] Vol. II: 251

amchem-grass-killer [76-03-9] Vol. II: 32

amchemical 68–250 [16672-87-0] Vol. III: 426

amchemical 3-CP [101-10-0] Vol. III: 806

amchem-2,4,5-tp [93-72-1] Vol. II: 845

amchem weed killer 650 [94-11-1] Vol. II: 15

amchlor [12125-02-9] Vol. I: 141

ameisensaeure (German) [64-18-6] Vol. I: 711

amephyt [834-12-8] Vol. II: 1054

amercide [133-06-2] Vol. II: 276

American Cyanamid 12,880 [60-51-5] Vol. III: 740

American Cyanamid-3422 [56-38-2] Vol. III: 750

American-Cyanamid-4,049 [121-75-5] Vol. II: 950

American Cyanamid 4,124 [2463-84-5] Vol. III: 748

American Cyanamid 4124 [2463-84-5] Vol. III: 748

American Cyanamid-5,223 [2439-10-3] Vol. III: 290

americium [14596-10-2] Vol. III: 42

ametrex [834-12-8] Vol. II: 1054

ametryn [834-12-8] Vol. II: 1054

ametycine Vol. III: [07-7], 1054

am fol [7664-41-7] Vol. I: 135

amibin [133-90-4] Vol. III: 44

amid-kyseliny-benzoove (Czech) [55-21-0] Vol. III: 86

amidocyanogen [420-04-2] Vol. II: 263

amidosulfonic acid [5329-14-6] Vol. III: 886

amidosulfuric acid [5329-14-6] Vol. III: 886

amine,diethyl, 2,2-dihydroxy [111-42-2] Vol. I: 658

amine, dipentyl [2050-92-2] Vol. III: 395

amine, triethyl, 2,2',2''-trihydroxy [102-71-6] Vol. II: 374

aminic acid [64-18-6] Vol. I: 711

2-aminoaethanol (German) [141-43-5] Vol. II: 356

m-aminoaniline [108-45-2] Vol. III: 98

2-aminoaniline [95-54-5] Vol. III: 697

3-aminoaniline [108-45-2] Vol. III: 98

2-aminoanisole [90-04-0] Vol. II: 76

o-aminoanisole [90-04-0] Vol. II: 76

4-aminoazobenzene [60-09-3] Vol. II: 74

p-aminoazobenzene [60-09-3] Vol. II: 74

4-aminoazobenzol [60-09-3] Vol. II: 74

p-aminoazobenzol [60-09-3] Vol. II: 74

o-aminoazotoluene [97-56-3] Vol. III: 46

aminobenzene [62-53-3] Vol. I: 164

4-(4-aminobenzyl) aniline [101-77-9] Vol. III: 304

1-amino-butaan (Dutch) [109-73-9] Vol. II: 203

1-aminobutane [109-73-9] Vol. II: 203

3-aminophenylmethane [108-44-1] Vol. III: 47

2-amino propaan (Dutch) [75-31-0] Vol. I: 787

2-aminopropane [75-31-0] Vol. I: 787

2-aminopropan (German) [75-31-0] Vol. I: 787

2-amino propano (Italian) [75-31-0] Vol. I: 787

1-aminopropan-2-ol [78-96-6] Vol. II: 821

3-amino-1-propene [107-11-9] Vol. III: 23

3-aminopropene [107-11-9] Vol. III: 23

3-aminopropylene [107-11-9] Vol. III: 23

3-amino-p-toluidine [95-80-7] Vol. II: 1016

amino-4 pyridine [504-24-5] Vol. II: 879

p-aminopyridine [504-24-5] Vol. II: 879

aminosulfonic acid [5329-14-6] Vol. III: 886

aminosulfuric acid [5329-14-6] Vol. III: 886

4-amino-6-tert-butyl-4,5-dihydro-3-methyl-thio-1,2,4-triazi n-5-one [21087-64-9] Vol. III: 626

4-amino-6-tert-butyl-3-(methylthio)-as-triazin-5 (4h)-one [21087-64-9] Vol. III: 626

4-amino-6-tert-butyl-3-(methylthio)as-triaz-in-S (4H)-one [21087-64-9] Vol. III: 626

3-aminotoluen (Czech) [108-44-1] Vol. III: 47

4-aminotoluen (Czech) [106-49-0] Vol. II: 1040

4-aminotoluen (Czech) [106-49-0] Vol. III: 927

alpha-aminotoluene [100-46-9] Vol. II: 170

aminotoluene [100-46-9] Vol. II: 170

p-aminotoluene [106-49-0] Vol. II: 1040

p-aminotoluene [106-49-0] Vol. III: 927

m-aminotoluene [108-44-1] Vol. III: 47

o-aminotoluene [95-53-4] Vol. I: 1114

aminotriacetic acid [139-13-9] Vol. III: 648

4-amino-3,5,6-trichloro-2-picolinic acid [1918-02-1] Vol. II: 791

4-amino-3,5,6-trichloropicolinic acid [1918-02-1] Vol. II: 791

4-aminotrichloropicolinic acid [1918-02-1] Vol. II: 791

4-Amino-3,5,6-Trichlorpicolinsaeure (German) [1918-02-1] Vol. II: 791

2-amino-1,3-xylene [87-62-7] Vol. II: 1092

aminozid [1596-84-5] Vol. III: 884

aminozide [1596-84-5] Vol. III: 884

ammo [52315-07-8] Vol. II: 251

ammo hypo [7783-18-8] Vol. I: 1088

ammon chlor [12125-02-9] Vol. I: 141

ammoneric [12125-02-9] Vol. I: 141

ammonia, anhydrous [7664-41-7] Vol. I: 135

ammonia aqueous [1336-21-6] Vol. I: 150

ammoniaca (Italian) [7664-41-7] Vol. I: 135

ammoniac (French) [7664-41-7] Vol. I: 135

ammonia gas [7664-41-7] Vol. I: 135

ammoniak (German) [7664-41-7] Vol. I: 135

ammonia, monohydrate [1336-21-6] Vol. I: 150

ammonia sesquicarbonate [506-87-6] Vol. I: 139

ammonia solution [1336-21-6] Vol. I: 150

ammoniated cupric sulfate [10380-29-7] Vol. III: 268

ammoniated mercuric chloride [10124-48-8] Vol. II: 533

ammoniated mercury chloride [10124-48-8] Vol. II: 533

ammoniated mercury chloride hydrargy-rum ammoniatum [10124-48-8] Vol. II: 533

ammonii chloridum [12125-02-9] Vol. I: 141

ammonio (bicromatodi) (Italian) [7789-09-5] Vol. I: 590

Ammonio cupric [10380-29-7] Vol. III: 268

ammonio (dicromatodi) (Italian) [7789-09-5] Vol. I: 590

ammonium acid fluoride [1341-49-7] Vol. II: 64

ammonium acid sulfite [10192-30-0] Vol. II: 985

ammonium, alkyldimethylbenzyl-, chloride [8001-54-5] Vol. II: 67

ammonium amidosulfate [7773-06-0] Vol. II: 959

ammonium amidosulfonate [7773-06-0] Vol. II: 959

ammonium amidosulphate [7773-06-0] Vol. II: 959

ammonium aminosulfonate [7773-06-0] Vol. II: 959

ammonium, benzyldimethyl (2-(2-(p-(1,1,3,3-tetramethylbutyl) phenoxy) eth-oxy) ethyl)-, chloride [121-54-0] Vol. III: 105

ammonium salz der amidosulfonsaeure (German) [7773-06-0] Vol. II: 959

ammonium silicon fluoride [16919-19-0] Vol. II: 908

ammonium sulfate (2:1) [7783-20-2] Vol. I: 159

ammonium sulfate (solution) [7783-20-2] Vol. I: 159

ammonium sulfide [12135-76-1] Vol. I: 161

ammonium sulfide (solution) [12135-76-1] Vol. I: 161

ammonium sulfite, hydrogen [10192-30-0] Vol. II: 985

ammonium sulfocyanate [1762-95-4] Vol. II: 1004

ammonium sulfocyanide [1762-95-4] Vol. II: 1004

ammonium sulphamate [7773-06-0] Vol. II: 959

ammonium sulphamidate [7773-06-0] Vol. II: 959

ammonium sulphate [7783-20-2] Vol. I: 159

ammonium tetrachlorozincate [52628-25-8] Vol. III: 982

ammonium tetrafluoroborate [13826-83-0] Vol. I: 144

ammonium tetrafluoroborate (1-) [13826-83-0] Vol. I: 144

ammonium thiosulfate [7783-18-8] Vol. I: 1088

ammonium trioxalate ferrate(III) [2944-67-4] Vol. III: 454

ammonium vanadate [7803-55-6] Vol. II: 1088

ammonium vanadate(V) [7803-55-6] Vol. II: 1088

ammonium zinc chloride [52628-25-8] Vol. III: 982

ammonyx [8001-54-5] Vol. II: 67

ammonyx-cpc [123-03-5] Vol. III: 196

amoben [133-90-4] Vol. III: 44

amokin [54-05-7] Vol. III: 229

amoniak (Polish) [7664-41-7] Vol. I: 135

amorphous silica dust [16871-71-9] Vol. II: 911

amorphous-silica-dust [7631-86-9] Vol. III: 846

amphojel [21645-51-2] Vol. II: 57

amphojel [21645-51-2] Vol. III: 35

amprolene [75-21-8] Vol. II: 401

amresco acryl-40 [79-06-1] Vol. I: 106

ams [7773-06-0] Vol. II: 959

ams [98-83-9] Vol. II: 947

ams [salt] [7773-06-0] Vol. II: 959

amthio [7783-18-8] Vol. I: 1088

amudane [126-07-8], Vol. III: 472

amyl acetate [628-63-7] Vol. II: 27

amylacetic ester [123-92-2] Vol. I: 780

amyl acetic ester [628-63-7] Vol. II: 27

amyl acetic ether [628-63-7] Vol. II: 27

amyl alcohol [71-41-0] Vol. II: 700

n-amyl alcohol [71-41-0] Vol. II: 700

amyl alcohol, normal [71-41-0] Vol. II: 700

n-amylalkohol (Czech) [71-41-0] Vol. II: 700

amylazetat (German) [628-63-7] Vol. II: 27

amyl carbinol [111-27-3] Vol. I: 734

amylcarbinol [111-27-3] Vol. I: 734

amylester kyseliny octove (Czech) [628-63-7] Vol. II: 27

amyl hydride [109-66-0] Vol. II: 682

amyl-hydrosulfide [110-66-7] Vol. III: 53

n-amyl mercaptan [110-66-7] Vol. III: 53

amyl methyl alcohol [105-30-6] Vol. I: 912

amyl methyl cetone (French) [110-43-0] Vol. II: 454

n amyl methyl ketone [110-43-0] Vol. II: 454

amylofene [50-06-6] Vol. I: 223

amylol [71-41-0] Vol. II: 700

amylowy alkohol (Polish) [123-51-3] Vol. II: 190

amyl-sulfhydrate [110-66-7] Vol. III: 53

amyl-thioalcohol [110-66-7] Vol. III: 53

amyltrichlorosilane [107-72-2] Vol. III: 941

an [107-13-1] Vol. I: 123

anac-110- [7440-50-8] Vol. I: 519

anadonis green [1308-38-9] Vol. III: 234

anaesthetic ether [60-29-7bf] Vol. III: 412

analgit [119-36-8] Vol. III: 622

anamenth [79-01-6] Vol. II: 427

anatase [13463-67-7] Vol. III: 921

anatox [8001-35-2] Vol. I: 1118

aneldazine [2008-41-5] Vol. II: 216

anestan [151-67-7] Vol. III: 141

anesthenyl [109-87-5] Vol. I: 829

anesthesia ether [60-29-7bf] Vol. III: 412

anesthetic ether [60-29-7bf] Vol. III: 412

anethol [104-46-1] Vol. III: 56

antetan [97-77-8] Vol. III: 382

antethyl [97-77-8] Vol. III: 382

anteyl [97-77-8] Vol. III: 382

anthon [52-68-6] Vol. III: 707

anthracen (German) [120-12-7] Vol. II: 79

anthracin [120-12-7] Vol. II: 79

anthrapole-73 [90-43-7] Vol. II: 176

anthrasorb [7440-44-0] Vol. III: 179

anthydride arsenique (French) [1303-28-2] Vol. I: 199

antiaethan [97-77-8] Vol. III: 382

antibulit [7681-49-4] Vol. II: 925

anticarie [118-74-1] Vol. I: 268

antietanol [97-77-8] Vol. III: 382

anti-ethyl [97-77-8] Vol. III: 382

antietil [97-77-8] Vol. III: 382

antifebrin [103-84-4] Vol. III: 3

antiformin [7681-52-9] Vol. II: 498

antigal [333-41-5] Vol. III: 758

anti-germ 77 [121-54-0] Vol. III: 105

antikol [97-77-8] Vol. III: 382

antimoine fluorure (French) [7783-56-4] Vol. I: 185

antimoine (pentachlorure d') (French) [7647-18-9] Vol. II: 83

antimoine (trichlorure d') (French) [10025-91-9] Vol. I: 182

antimonio (pentacloruro di) (Italian) [7647-18-9] Vol. II: 83

antimonio (tricloruro di) (Italian) [10025-91-9] Vol. I: 182

antimonious oxide [1309-64-4] Vol. II: 86

antimonous bromide [7789-61-9] Vol. I: 179

antimonous chloride [10025-91-9] Vol. I: 182

antimonous fluoride [7783-56-4] Vol. I: 185

antimonpentachlorid (German) [7647-18-9] Vol. II: 83

antimontrichlorid (German) [10025-91-9] Vol. I: 182

antimony black [7440-36-0] Vol. I: 177

antimony bromide [7789-61-9] Vol. I: 179

antimony bromide (Sbbr3) [7789-61-9] Vol. I: 179

antimony-butter [10025-91-9] Vol. I: 182

antimony chloride [10025-91-9] Vol. I: 182

antimony chloride (SbCl15) [7647-18-9], Vol. II: 83

antimony fluoride [7783-56-4] Vol. I: 185

antimony (iii) bromide [7789-61-9] Vol. I: 179

antimony (iii) chloride [10025-91-9] Vol. I: 182

antimony (iii) fluoride (1:3) [7783-56-4] Vol. I: 185

antimonyl potassium tartrate [28300-74-5] Vol. I: 188

antimony oxide [1309-64-4] Vol. II: 86

antimony (3) oxide [1309-64-4] Vol. II: 86

antimony oxide (O3Sb2) [1309-64-4] Vol. II: 86

antimony oxide (Sb2O3) [1309-64-4] Vol. II: 86

antimony perchloride [7647-18-9] Vol. II: 83

antimony peroxide [1309-64-4] Vol. II: 86

antimony, regulus [7440-36-0] Vol. I: 177

antimony sesquioxide [1309-64-4] Vol. II: 86

antimony trichloride [10025-91-9] Vol. I: 182

antimony(V) chloride [7647-18-9] Vol. II: 83

antimony white [1309-64-4] Vol. II: 86

antimoonpentachloride (Dutch) [7647-18-9] Vol. II: 83

antimoontrichlride (Dutch) [10025-91-9] Vol. I: 182

antimucin-wdr [62-38-4] Vol. II: 535

antinonin [534-52-1] Vol. I: 540

antinonnin [534-52-1] Vol. I: 540

antioxidant-116 [135-88-6] Vol. III: 640

antioxidant-4k [128-37-0] Vol. III: 317

antioxidant-kb [128-37-0] Vol. III: 317

antioxidant-pbn [135-88-6] Vol. III: 640

antipyonin [1303-96-4] Vol. III: 134

antiren [110-85-0] Vol. III: 776

anti rust [7632-00-0] Vol. I: 890

antisal-1- [127-18-4] Vol. II: 421

antisal 4 [7601-54-9] Vol. II: 751

antisal-1a [108-88-3] Vol. I: 1109

antisal-2b [7664-39-3] Vol. I: 750

antiseptol [121-54-0] Vol. III: 105

antisol-1- [127-18-4] Vol. II: 421

antitrombosin [66-76-2] Vol. III: 336

antivitium (Spain) [97-77-8] Vol. III: 382

antox [1309-64-4] Vol. II: 86

antrancine [128-37-0] Vol. III: 317

antymon (Polish) [7440-36-0] Vol. I: 177

antyrost [123-33-1] Vol. II: 870

anyvim [62-53-3] Vol. I: 164

ao-4k [128-37-0] Vol. III: 317

a-15881p [1309-64-4] Vol. II: 86

4-ap [504-24-5] Vol. II: 879

ap 50 [1309-64-4] Vol. II: 86

apachlor [470-90-6] Vol. III: 716

apamidon [13171-21-6] Vol. III: 704

apavap [62-73-7] Vol. I: 946

APCO-2330 [108-45-2] Vol. III: 98

aperochemical (etu-22) [96-45-7] Vol. I: 757

apex-4 [123-95-5] Vol. III: 167

aphalon [330-55-2] Vol. II: 1084

aphamite [56-38-2] Vol. III: 750

API no. 2 fuel oil [68476-30-2] Vol. III: 465

appa [732-11-6] Vol. III: 737

aprocarb [114-26-1] Vol. II: 231

aptal [59-50-7] Vol. II: 258

aqua ammonia [1336-21-6] Vol. I: 150

aqua care [57-13-6] Vol. I: 1136

aqua care hp [57-13-6] Vol. I: 1136

aquachloral [302-17-0] Vol. III: 197

aquadrate [57-13-6] Vol. I: 1136

aquaffin [25322-68-3] Vol. II: 811

aqua fortis [7697-37-2] Vol. I: 879

aqua-kleen [1929-73-3] Vol. III: 155

aqua maid permanent algaecide [7758-98-7] Vol. I: 510

aqua-nuchar [7440-44-0] Vol. III: 179

aquarex-me [151-21-3] Vol. II: 981

aquarex-methyl [151-21-3] Vol. II: 981

aquathol [145-73-3] Vol. I: 900

aquathol plus [145-73-3] Vol. I: 900

aquathol plus granular [145-73-3] Vol. I: 900

aquatronicssnarl a-cide dri pac snail powder [7758-98-7] Vol. I: 510

aqua-vex [93-72-1] Vol. II: 845

aqueous hydrogen chloride [7647-01-0] Vol. I: 738

aquilite [7704-34-9] Vol. II: 965

aquisal [144-62-7] Vol. I: 904

ar-3 [7440-44-0] Vol. III: 179

aracide [140-57-8] Vol. III: 61

aracnol-K [116-29-0] Vol. III: 906

araldite hardener hy-951- [112-24-3] Vol. I: 1129

araldite hy-951 [112-24-3] Vol. I: 1129

aralen [54-05-7] Vol. III: 229

aralo [56-38-2] Vol. III: 750

aramit [140-57-8] Vol. III: 61

aramiteararamite-15W [140-57-8] Vol. III: 61

aramite-15W [140-57-8] Vol. III: 61

arasan [137-26-8] Vol. I: 632

arasan m [137-26-8] Vol. I: 632

arasan sf [137-26-8] Vol. I: 632

arasan sf x [137-26-8] Vol. I: 632

aratron [140-57-8] Vol. III: 61

arboricid [93-79-8] Vol. II: 34

arborol [534-52-1] Vol. I: 540

arbortrine [17804-35-2] Vol. II: 148

arcton-4 [75-45-6] Vol. II: 591

arcton-7 [75-43-4] Vol. II: 595

arcton-9 [75-69-4] Vol. II: 602

arcton-11 [75-69-4] Vol. II: 602

arcton-12 [75-71-8] Vol. I: 825

arcton-33 [76-14-2] Vol. III: 408

arcton-63 [76-13-1] Vol. II: 353

arcton-114 [76-14-2] Vol. III: 408

arctuvin [123-31-9] Vol. I: 753

aredion [116-29-0] Vol. III: 906

argentate (I-), bis (cyano-c)-, potassium [506-61-6] Vol. II: 813

argentate (I-), dicyano-, potassium [506-61-6] Vol. II: 813

argentates(I-) (sol), dicyano [506-61-6] Vol. II: 813

argenti nitras [7761-88-8] Vol. I: 1038

argentum [7440-22-4] Vol. I: 1036

argerol [7761-88-8] Vol. I: 1038

argezin [1912-24-9] Vol. II: 1050

argucide [7722-64-7] Vol. I: 915

arilate [17804-35-2] Vol. II: 148

arklone-p [76-13-1] Vol. II: 353

Armco iron [7439-89-6] Vol. III: 487

armofos [7758-29-4] Vol. II: 1061

arochlor-1260- [11096-82-5] Vol. II: 808

aroclor-1254 [11097-69-1] Vol. II: 806

aroclor [1336-36-3] Vol. I: 974

arosol [122-99-6] Vol. III: 695

arprocarb [114-26-1] Vol. II: 231

arquad-b-100 [8001-54-5] Vol. II: 67

arrivo [52315-07-8] Vol. II: 251

arsan [75-60-5] Vol. I: 214

arsenate [7778-39-4] Vol. II: 91

arsenenous acid, sodium salt [7784-46-5] Vol. I: 208

arseniate de calcium (French) [7778-44-1] Vol. III: 65

arsenic-75 [7440-38-2] Vol. I: 190

arsenic acid anhydride [1303-28-2] Vol. I: 199

arsenic acid, calcium salt (2:3) [7778-44-1] Vol. III: 65

arsenic acid, disodium salt [7778-43-0] Vol. II: 93

arsenic acid (HaAsO₄) [7778-39-4] Vol. II: 91

arsenic acid (H₃AsO₄), calcium salt (2:3) [7778-44-1] Vol. III: 65

arsenic acid (H₃AsO₄), monopotassium salt [7784-41-0] Vol. III: 71

arsenic acid, lead salt [3687-31-8] Vol. III: 69

arsenic acid, liquid [7778-39-4] Vol. II: 91

arsenic acid, monopotassium salt [7784-41-0] Vol. III: 71

arsenic anhydride [1303-28-2] Vol. I: 199

arsenic bisulfide [1303-32-8] Vol. III: 75

arsenic black [7440-38-2] Vol. I: 190

arsenic blanc (French) [1327-53-3] Vol. I: 203

arsenic bromide (AsBr3) [7784-33-0] Vol. II: 96

arsenic butter [7784-34-1] Vol. I: 196

arsenic chloride [7784-34-1] Vol. I: 196

arsenic hydrid [7784-42-1] Vol. I: 212

arsenic hydride [7784-42-1] Vol. I: 212

arsenic(II) bromide [7784-33-0] Vol. II: 96

arsenic (iii) chloride [7784-34-1] Vol. I: 196

arsenic (iii) oxide [1327-53-3] Vol. I: 203

arsenic (iii) trichloride [7784-34-1] Vol. I: 196

arsenic oxide [1303-28-2] Vol. I: 199

arsenic oxide [1327-52-2] Vol. III: 64

arsenic oxide [1327-53-3] Vol. I: 203

arsenic pentaoxide [1303-28-2] Vol. I: 199

arsenic pentoxide [1327-52-2] Vol. III: 64

arsenic red [1303-33-9] Vol. II: 99

arsenic sesquioxide [1327-53-3] Vol. I: 203

arsenic sesquisulfide [1303-33-9] Vol. II: 99

arsenic sesquisulphide [1303-33-9] Vol. II: 99

arsenic sulfide [1303-32-8] Vol. III: 75

arsenic sulfide [As2S3] [1303-33-9] Vol. II: 99

arsenic sulfide red [1303-32-8] Vol. III: 75

arsenic sulfide yellow [1303-33-9] Vol. II: 99

arsenic sulphide [1303-33-9] Vol. II: 99

arsenic tersulfide [1303-33-9] Vol. II: 99

arsenic tersulphide [1303-33-9] Vol. II: 99

arsenic trihydride [7784-42-1] Vol. I: 212

arsenic trisulphide [1303-33-9] Vol. II: 99

arsenicum album [1327-53-3] Vol. I: 203

arsenic (v) oxide [1303-28-2] Vol. I: 199

arsenic yellow [1303-33-9] Vol. II: 99

arsenigen saure (German) [1327-53-3] Vol. I: 203

arsenious acid [1327-53-3] Vol. I: 203

arsenious acid [H3AsO3], potassium salt [10124-50-2] Vol. II: 104

arsenious acid anthydride [1327-53-3] Vol. I: 203

arsenious acid, monosodium salt [7784-46-5] Vol. I: 208

arsenious acid, potassium salt [10124-50-2] Vol. II: 104

arsenious acid, sodium salt [7784-46-5] Vol. I: 208

arsenious bromide [7784-33-0] Vol. II: 96

arsenious chloride [7784-34-1] Vol. I: 196

arsenious oxide [1327-53-3] Vol. I: 203

arsenious sulfide [1303-33-9] Vol. II: 99

arsenious sulphide [1303-33-9] Vol. II: 99

arsenious trioxide [1327-53-3] Vol. I: 203

arsenite de potassium (French) [10124-50-2] Vol. II: 104

arsenite de sodium (French) [7784-46-5] Vol. I: 208

arseniuretted hydrogen [7784-42-1] Vol. I: 212

arsenolite [1327-53-3] Vol. I: 203

arsenous acid anhydride [1327-53-3] Vol. I: 203

arsenous acid, calcium salt [52740-16-6] Vol. II: 109

arsenous acid, potassium salt [10124-50-2] Vol. II: 104

arsenous anhydride [1327-53-3] Vol. I: 203

arsenous bromide [7784-33-0] Vol. II: 96

arsenous chloride [7784-34-1] Vol. I: 196

arsenous hydride [7784-42-1] Vol. I: 212

arsenous oxide [1327-53-3] Vol. I: 203

arsenous oxide anhydride [1327-53-3] Vol. I: 203

arsenous sulfide [1303-33-9] Vol. II: 99

arsenous tribromide [7784-33-0] Vol. II: 96

arsenous trichloride [7784-34-1] Vol. I: 196

arsenowodor (Polish) [7784-42-1] Vol. I: 212

Arsenwasserstoff (German) [7784-42-1] **Vol. I: 212**

arsine, dichlorophenyl [696-28-6] **Vol. II: 107**

arsine, dichloro(phenyl)- [696-28-6] **Vol. II: 107**

arsineoxide, hydroxydimethyl [75-60-5] **Vol. I: 214**

arsodent [1327-53-3] **Vol. I: 203**

arsonate [2163-80-6] **Vol. II: 610**

arsonic acid, calcium salt (1:1) [52740-16-6] **Vol. II: 109**

arsonic acid, methyl-, monosodium salt [2163-80-6] **Vol. II: 610**

arsonous dichloride, phenyl [696-28-6] **Vol. II: 107**

art-2 [7440-44-0] **Vol. III: 179**

arthrochin [54-05-7] **Vol. III: 229**

artificial almond oil [100-52-7] **Vol. I: 237**

artone [108-94-1] **Vol. II: 268**

artrichin [54-05-7] **Vol. III: 229**

arwood copper [7440-50-8] **Vol. I: 519**

arylamine [62-53-3] **Vol. I: 164**

asana [66230-04-4] **Vol. III: 400**

asarco 1-15 [7440-66-6] **Vol. I: 1175**

asazol [2163-80-6] **Vol. II: 610**

asbestos dust [1332-21-4] **Vol. I: 217**

Asbestose (German) [1332-21-4] **Vol. I: 217**

asbestos fiber [1332-21-4] **Vol. I: 217**

asbestos fibre [1332-21-4] **Vol. I: 217**

as-bond-1001 [1344-09-8] **Vol. III: 868**

Ascabin [120-51-4] **Vol. III: 117**

ascabiol [120-51-4] **Vol. III: 117**

ascarite [1332-21-4] **Vol. I: 217**

a-150 [silane] [75-94-5] **Vol. III: 945**

as-dimethylhydrazine [57-14-7] **Vol. II: 470**

as-methylphenylethylene [98-83-9] **Vol. II: 947**

Asmofug-e [7681-11-0] **Vol. III: 785**

as m-xylenol [105-67-9] **Vol. I: 1162**

as o-xylenol [95-65-8] **Vol. I: 1169**

asp-47 [3689-24-5] **Vol. II: 1010**

asphalt cements [8052-42-4] **Vol. II: 114**

asphalt (cut) [8052-42-4] **Vol. II: 114**

asphaltic bitumen [8052-42-4] **Vol. II: 114**

asphalt, liquid medium curing [8052-42-4] **Vol. II: 114**

asphalt, liquid rapid curing [8052-42-4] **Vol. II: 114**

asphalt, liquid slow curing [8052-42-4] **Vol. II: 114**

asphaltum [8052-42-4] **Vol. II: 114**

aspon-chlordane [57-74-9] **Vol. II: 619**

astm-d458 [497-19-8] **Vol. II: 922**

astonex [35367-38-5] **Vol. III: 339**

as trimethylbenzene [95-63-6] **Vol. I: 281**

astrobot [62-73-7] **Vol. I: 946**

asulfa-supra [7704-34-9] **Vol. II: 965**

asym-dichloroethylene [75-35-4] **Vol. II: 409**

asym-dimethylhydrazine [57-14-7] **Vol. II: 470**

asymmetrical trimethylbenzene [95-63-6] **Vol. I: 281**

asymmetric-dimethylhydrazine [57-14-7] **Vol. II: 470**

o-at [97-56-3] **Vol. III: 46**

atalco-c [36653-82-4] **Vol. III: 193**

atazinax [1912-24-9] **Vol. II: 1050**

ate [97-93-8] **Vol. II: 59**

atgard [62-73-7] **Vol. I: 946**

atgard c [62-73-7] **Vol. I: 946**

Athylen (German) [74-85-1] **Vol. I: 691**

Athylenglykol (German) [107-21-1] **Vol. I: 653**

athylenglykol-monoathylather (German) [110-80-5] **Vol. II: 362**

atlas-'a'- [7784-46-5] **Vol. I: 208**

atlas white titanium dioxide [13463-67-7] **Vol. III: 921**

atomic sulfur [7704-34-9] **Vol. II: 965**

atranex [1912-24-9] **Vol. II: 1050**

atrasine [1912-24-9] **Vol. II: 1050**

atrataf [1912-24-9] **Vol. II: 1050**

atratol-a [1912-24-9] **Vol. II: 1050**

atrazin [1912-24-9] **Vol. II: 1050**

atred [1912-24-9] **Vol. II: 1050**

atrex [1912-24-9] **Vol. II: 1050**

atrivyl [443-48-1] **Vol. III: 628**

attac 4-2 [8001-35-2] **Vol. I: 1118**

attac 4-4 [8001-35-2] **Vol. I: 1118**

attac 6-3 [8001-35-2] **Vol. I: 1118**

attac-6- [8001-35-2] **Vol. I: 1118**

attac-8- [8001-35-2] **Vol. I: 1118**

attack [137-26-8] **Vol. I: 632**

au-3 [7440-44-0] **Vol. III: 179**

aules [137-26-8] **Vol. I: 632**

auripigment [1303-33-9] **Vol. II: 99**

austiox r-cr 3 [13463-67-7] **Vol. III: 921**

austrostab 110 e [7428-48-0] **Vol. III: 882**

avadex [2303-16-4] **Vol. II: 218**

barbenyl [50-06-6] **Vol. I:** 223

Barber's weed killer (ester formulation) [94-11-1] **Vol. II:** 15

barbiphenyl [50-06-6] **Vol. I:** 223

barbipil [50-06-6] **Vol. I:** 223

barbiris [50-06-6] **Vol. I:** 223

barbita [50-06-6] **Vol. I:** 223

barbituric acid, 5-ethyl-5-phenyl [50-06-6] **Vol. I:** 223

bario (perossido di) (Italian) [1304-29-6] **Vol. III:** 82

bario (Spanish) [7440-39-3] **Vol. I:** 225

barium binoxide [1304-29-6] **Vol. III:** 82

barium carbonate (1:1) [513-77-9] **Vol. I:** 227

barium carbonate [513-77-9] **Vol. I:** 227

barium cyanide [542-62-1] **Vol. I:** 230

barium dichloride [10361-37-2] **Vol. II:** 119

barium dicyanid [542-62-1] **Vol. I:** 230

barium dihydroxide [17194-00-2] **Vol. II:** 122

barium dinitrate [10022-31-8] **Vol. I:** 233

barium dioxide [1304-29-6] **Vol. III:** 82

barium hydroxide (Ba(OH)2) [17194-00-2] **Vol. II:** 122

barium (ii) nitrate (1:2) [10022-31-8] **Vol. I:** 233

barium nitrate [10022-31-8] **Vol. I:** 233

barium oxide, per- [1304-29-6] **Vol. III:** 82

barium peroxide (Ba(O₂)) [1304-29-6] **Vol. III:** 82

Bariumperoxid (German) [1304-29-6] **Vol. III:** 82

bariumperoxyde (Dutch) [1304-29-6] **Vol. III:** 82

barium superoxide [1304-29-6] **Vol. III:** 82

barquat mb-50 [8001-54-5] **Vol. II:** 67

barquat mb-80 [8001-54-5] **Vol. II:** 67

barricade [52315-07-8] **Vol. II:** 251

baryum (French) [7440-39-3] **Vol. I:** 225

basaklor [76-44-8] **Vol. II:** 628

basamid [533-74-4] **Vol. III:** 293

basamid-fluid [137-42-8] **Vol. III:** 861

basamid-g [533-74-4] **Vol. III:** 293

basamid-granular [533-74-4] **Vol. III:** 293

basamid-p [533-74-4] **Vol. III:** 293

basamid-puder [533-74-4] **Vol. III:** 293

basanite [88-85-7] **Vol. II:** 704

basfapon [75-99-0] **Vol. I:** 1016

basfapon b [75-99-0] **Vol. I:** 1016

basfapon/basfapon-N [75-99-0] **Vol. I:** 1016

basic mercuric sulfate [1312-03-4] **Vol. III:** 563

basic zinc chromate [13530-65-9] **Vol. III:** 984

basic zirconium chloride [7699-43-6] **Vol. II:** 1110

basinex [75-99-0] **Vol. I:** 1016

basinex p [75-99-0] **Vol. I:** 1016

basudin [333-41-5] **Vol. III:** 758

basudin-10-g [333-41-5] **Vol. III:** 758

batazina [122-34-9] **Vol. II:** 1047

batrilex [82-68-8] **Vol. II:** 132

battery acid [7664-93-9] **Vol. II:** 969

bau [7440-44-0] **Vol. III:** 179

Bay-5024 [2032-65-7] **Vol. II:** 308

Bay-5122 [114-26-1] **Vol. II:** 231

Bay-7162 [10265-92-6] **Vol. II:** 744

Bay-9010 [114-26-1] **Vol. II:** 231

Bay-9026 [2032-65-7] **Vol. II:** 308

bay-9027 [86-50-0] **Vol. III:** 744

Bay-10756 [8065-48-3] **Vol. III:** 755

bay-14981 [2463-84-5] **Vol. III:** 748

bay-17147 [86-50-0] **Vol. III:** 744

Bay-19149 [62-73-7] **Vol. I:** 946

bay-21097 [301-12-2] **Vol. III:** 661

Bay-30130 [26638-19-7] **Vol. II:** 298

bay-34727 [2636-26-2] **Vol. III:** 271

bay-36205 [2439-01-2] **Vol. III:** 663

Bay-37344 [2032-65-7] **Vol. II:** 308

Bay-39007 [114-26-1] **Vol. II:** 231

bay-45432 [1113-02-6] **Vol. III:** 657

Bay-68138 [22224-92-6] **Vol. II:** 743

Bay-70143 [1563-66-2] **Vol. I:** 420

Bay-94337 [21087-64-9] **Vol. III:** 626

baychrom a [10101-53-8] **Vol. I:** 482

baychrom f [10101-53-8] **Vol. I:** 482

bayclean [8001-54-5] **Vol. II:** 67

bay-dic-1468 [21087-64-9] **Vol. III:** 626

Bay-e-393 [3689-24-5] **Vol. II:** 1010

Bay e-605 [56-38-2] **Vol. III:** 750

Bayer-15, 922 [52-68-6] **Vol. III:** 707

bayer-4964 [2439-01-2] **Vol. III:** 663

Bayer-5546 [10265-92-6] **Vol. II:** 744

Bayer-8169 [8065-48-3] **Vol. III:** 755

Bayer-9027 [86-50-0] **Vol. III:** 744

Bayer-15-922 [52-68-6] **Vol. III:** 707

Bayer-17147 [86-50-0] **Vol. III:** 744

bayer-21097 [301-12-2] **Vol. III:** 661

Bayer-22/190 [2463-84-5] **Vol. III:** 748

Bayer-30-130 [26638-19-7] **Vol. II:** 298

benzene, 1-methyl-3-nitro- [99-08-1] **Vol. II:** 1033

benzene, 1-methyl-4-nitro [99-99-0] **Vol. I:** 1105

benzene, 2-methyl-1,3,5-trinitro [118-96-7] **Vol. I:** 1093

benzene, 1,2-(1,8-naphthalenediyl)- [206-44-0] **Vol. II:** 435

benzene, 1,2-(1,8-naphthylene)- [206-44-0] **Vol. II:** 435

benzenenitrile [100-47-0] **Vol. I:** 296

benzene-o-dicarboxylic acid di-n-butyl ester [84-74-2] **Vol. II:** 774

benzene, o dichloro [95-50-1] **Vol. I:** 250

benzene, o-dihydroxy [120-80-9] **Vol. III:** 190

benzene, o dinitro [528-29-0] **Vol. I:** 258

benzene, 1,1'-oxybis- [101-84-8] **Vol. II:** 392

benzene, p dichloro [106-46-7] **Vol. I:** 254

benzene, p-dinitro- [100-25-4] **Vol. II:** 128

benzene, pentachloro [608-93-5] **Vol. I:** 277

benzene, pentachloronitro- [82-68-8] **Vol. II:** 132

benzene, phenoxy [101-84-8] **Vol. II:** 392

benzene, (phenylamino) [122-39-4] **Vol. I:** 623

benzenephosphonic acid, thiono-, ethyl-p-nitrophenyl ester [2104-64-5] **Vol. II:** 738

benzene phosphothionic acid, ethyl4-nitro pheny lester [2104-64-5] **Vol. II:** 738

benzenephosphothionic acid, ethyl 4-nitrophenyl ester [2104-64-5] **Vol. II:** 738

benzene, propyl- [103-65-1] **Vol. II:** 135

benzene sulfochloride [98-09-9] **Vol. III:** 103

benzenesulfonamide, N-(2-mercaptoethyl) -, S-ester with o,o-diisopropylphosphorodithioate [741-58-2] **Vol. III:** 731

S-2-benzenesulfonamidoethyl o,o-di-isopropyl phosphorodithioate [741-58-2] **Vol. III:** 731

benzenesulfon chloride [98-09-9] **Vol. III:** 103

benzene sulfonechloride [98-09-9] **Vol. III:** 103

benzenesulfonic (acid) chloride [98-09-9] **Vol. III:** 103

benzenesulfonic acid, dodecyl [27176-87-0] **Vol. I:** 282

benzenesulfonic acid, dodecyl-, calcium salt [26264-06-2] **Vol. II:** 209

benzenesulfonic acid, dodecyl-, sodium salt [25155-30-0] **Vol. III:** 101

benzenesulfonic acid, 4-hydroxy-, zinc salt (2:1) [127-82-2] **Vol. II:** 142

benzenesulfonic acid, p-hydroxy, zinc salt (2:1) [127-82-2] **Vol. II:** 142

benzenesulfonic chloride [98-09-9] **Vol. III:** 103

S-2-benzenesulphonamidoethyl o,o-di-isopropyl phosphorodithioate [741-58-2] **Vol. III:** 731

s-2-benzenesulphonamidoethyl o,o-diisopropyl phosphorodithioate [741-58-2] **Vol. III:** 731

benzenesulphonyl chloride [98-09-9] **Vol. III:** 103

benzene tetrachloride [95-94-3] **Vol. I:** 279

benzene, 1,2,3,4-tetrachloro [634-66-2] **Vol. III:** 899

benzene, 1,2,3,5-tetrachloro [634-90-2] **Vol. III:** 901

benzene, 1,2,4,5-tetrachloro [95-94-3] **Vol. I:** 279

benzenethiol [108-98-5] **Vol. III:** 917

benzene, 1,2,3-trichloro- [87-61-6] **Vol. II:** 137

benzene, 1,2,4-trichloro- [120-82-1] **Vol. III:** 92

benzene, 1,2,4-trichloro-5-((4-chlorophenyl)sulfonyl)- [116-29-0] **Vol. III:** 906

benzene, 1,1'-(2,2,2-trichloroethylidene) bis (4-methoxy [72-43-5] **Vol. II:** 345

benzene, 1,2,3-trihydroxy- [87-66-1] **Vol. II:** 882

benzene, 1,2,4-trimethyl [95-63-6] **Vol. I:** 281

benzene, 1,2,5-trimethyl [95-63-6] **Vol. I:** 281

benzene, 1,3,5-trinitro- [99-35-4] **Vol. III:** 948

1,2,3-benzenetriol [87-66-1] **Vol. II:** 882

benzene, vinyl [100-42-5] **Vol. I:** 1060

benzenol [108-95-2] **Vol. I:** 920

benzenosulfochlorek (Polish) [98-09-9] **Vol. III:** 103

benzenosulphochloride [98-09-9] **Vol. III:** 103

benzen (Polish) [71-43-2] **Vol. I:** 239

benzenyl chloride [98-07-7] **Vol. II:** 139

benzenyl trichloride [98-07-7] **Vol. II:** 139

benzethonium [121-54-0] **Vol. III:** 105

benzethoniumchloride [121-54-0] Vol. III: 105

benzethonium chloride 1622 [121-54-0] Vol. III: 105

benzetonium chloride [121-54-0] Vol. III: 105

benzex [608-73-1] Vol. I: 564

8,9-benzfluoranthene [207-08-9] Vol. II: 151

benzhydrol, 4,4'-dichloro-alpha(trichloromethyl) [115-32-2] Vol. II: 144

benzidam [62-53-3] Vol. I: 164

benzidina (Italian) [92-87-5] Vol. I: 288

benzidin (Czech) [92-87-5] Vol. I: 288

p benzidine [92-87-5] Vol. I: 288

benzidine base [92-87-5] Vol. I: 288

benzidine, 3,3'-dichloro [91-94-1] Vol. I: 285

benzidine, 3,3'-dichloro-, dihydrochloride [612-83-9] Vol. III: 326

benzidine, 3,3'-dimethyl [119-93-7] Vol. III: 364

benzilan [510-15-6] Vol. III: 108

benzile (clorurodi) (Italian) [100-44-7] Vol. I: 1099

benzilic acid, 4,4'-dichloro-, ethyl ester [510-15-6] Vol. III: 108

2-benzimidazolecarbamic acid, 1-(butylcarbamoyl)-, methyl ester [17804-35-2] Vol. II: 148

1-benzine [91-22-5] Vol. I: 1022

benzinoform [56-23-5] Vol. I: 435

benzinol [79-01-6] Vol. II: 427

1,2-benzisothiazol-3 (2h)-one, 1,1-dioxide [81-07-2] Vol. III: 824

1,2-benzisothiazolinone, 1,1-dioxide [81-07-2] Vol. III: 824

1,2-benzisothiazolin-3-one, 1,1-dioxide [81-07-2] Vol. III: 824

3-benzisothiazolinone 1,1-dioxide [81-07-2] Vol. III: 824

benzite [99-35-4] Vol. III: 948

benz (j) aceanthrylene, 1,2-dihydro-3-methyl [56-49-5] Vol. III: 604

benzo (a)anthracene [56-55-3] Vol. I: 235

1,2-benzoanthracene [56-55-3] Vol. I: 235

benzoanthracene [56-55-3] Vol. I: 235

benzo (a)phenanthrene [218-01-9] Vol. I: 489

benzoate [65-85-0] Vol. I: 293

benzo (b)phenanthrene [56-55-3] Vol. I: 235

benzo (b)pyridine [91-22-5] Vol. I: 1022

benzo (b)quinoline [260-94-6] Vol. I: 98

benzo chinon (German) [106-51-4] Vol. I: 303

benz-o-chlor [510-15-6] Vol. III: 108

benzo (def) phenanthrene [129-00-0] Vol. III: 811

benzo-(d,e,f)-phenanthrene [129-00-0] Vol. III: 811

1,3-benzodioxole, 5-((2-(2-butoxyethoxy) ethoxy) methyl)-6-propyl [51-03-6] Vol. III: 779

benzoepin [115-29-7] Vol. II: 661

benzoesaeure (German) [65-85-0] Vol. I: 293

10,11-benzofluoranthene [205-82-3] Vol. III: 110

11,12-benzofluoranthene [207-08-9] Vol. II: 151

benzo-12,13-fluoranthene [205-82-3] Vol. III: 110

7,8-benzofluoranthene [205-82-3] Vol. III: 110

8,9-benzofluoranthene [207-08-9] Vol. II: 151

7-benzofuranol, 2,3-dihydro-2,2-dimethyl-, methylcarbamate [1563-66-2] Vol. I: 420

benzofuroline [10453-86-8] Vol. III: 285

benzohydroquinone [123-31-9] Vol. I: 753

benzoic-acid amide [55-21-0] Vol. III: 86

benzoic acid, 3-amino-2,5-dichloro [133-90-4] Vol. III: 44

benzoic acid, ammonium salt [1863-63-4] Vol. II: 157

benzoic acid, benzyl ester [120-51-4] Vol. III: 117

benzoic acid, chloride [98-88-4] Vol. I: 307

benzoic acid, 3,6-dichloro-2-methoxy [1918-00-9] Vol. II: 159

benzoic acid, 2-hydroxy-, methyl ester [119-36-8] Vol. III: 622

benzoic-acid, mercury(2) salt [583-15-3] Vol. III: 551

benzoic acid nitrile [100-47-0] Vol. I: 296

benzoic acid, peroxide [94-36-0] Vol. I: 310

benzoic acid, phenylmethyl ester [120-51-4] Vol. III: 117

benzoic aldehyde [100-52-7] Vol. I: 237

benzoic-sulfimide [81-07-2] Vol. III: 824

o-benzoic sulfimide [81-07-2] Vol. III: 824

benzoic trichloride [98-07-7] Vol. II: 139

benzo (jk) fluorene [206-44-0] Vol. II: 435

11,12-benzo (k) fluoranthene [207-08-9] Vol. II: 151

benzol [71-43-2] Vol. I: 239

benzol-90 [71-43-2] Vol. I: 239

benzol-black [7440-44-0] Vol. III: 179

benzole [71-43-2] Vol. I: 239

benzolo (Italian) [71-43-2] Vol. I: 239

benzonitrile, 3,5-dibromo-4-hydroxy [1689-84-5] Vol. II: 163

benzonitrile, 2,6-dichloro [1194-65-6] Vol. I: 299

benzoperoxide [94-36-0] Vol. I: 310

1,12-benzoperylene [191-24-2] Vol. II: 156

1,2-benzophenanthrene [218-01-9] Vol. I: 489

2,3-benzophenanthrene [56-55-3] Vol. I: 235

(1) benzopyrano (3,4-b) furo (2,3-h) (1) benzopyran-6 (6aalpha)-one, 1,2,12,12a alpha-tetrahydro-2alpha-isopropenyl-8,9-dimethoxy [83-79-4] Vol. II: 166

(1) benzopyrano (3,4-b) furo (2,3-h) (1) benzopyran-6 (6ah)-one, 1,2,12,12a-tetrahydro-2-alpha-isopropenyl-8,9-dimethoxy [83-79-4] Vol. II: 166

(1) benzopyrano (3,4-b) furo (2,3-h) (1) benzopyran-6 (6ah)-one, 1,2,12,12a-tetrahydro-8,9-dimethoxy-2-(1-methyl ethenyl)-, (2r-(2alpha, 6aalpha, 12aalpha))- [83-79-4] Vol. II: 166

2h-1-benzopyran-2-one, 4-hydroxy-3-(3-oxo-1-phenylbutyl) [81-81-2] Vol. I: 525

2H-1-benzopyran-2-one, 3,3'-methylenebis (4-hydroxy [66-76-2] Vol. III: 336

benzopyridine [91-22-5] Vol. I: 1022

benzopyrrole [120-72-9] Vol. I: 760

2,3-benzopyrrole [120-72-9] Vol. I: 760

1,4-benzoquine [106-51-4] Vol. I: 303

benzoquinol [123-31-9] Vol. I: 753

2,3-benzoquinoline [260-94-6] Vol. I: 98

benzoquinone [106-51-4] Vol. I: 303

p benzoquinone [106-51-4] Vol. I: 303

benzo (rst) pentaphene [189-55-9] Vol. III: 308

o-benzosulfimide [81-07-2] Vol. III: 824

benzosulfinide [81-07-2] Vol. III: 824

benzo-2-sulphimide [81-07-2] Vol. III: 824

benzosulphimide [81-07-2] Vol. III: 824

o-benzosulphimide [81-07-2] Vol. III: 824

benzo-sulphinide [81-07-2] Vol. III: 824

2-benzothiazolethiol [149-30-4] Vol. III: 540

2-benzothiazolethione [149-30-4] Vol. III: 540

benzothiazole-2-thione [149-30-4] Vol. III: 540

2 (3h)-benzothiazolethione [149-30-4] Vol. III: 540

2-benzothiazolinethione [149-30-4] Vol. III: 540

2-benzothiazolyl mercaptan [149-30-4] Vol. III: 540

benzotriazine derivative of a methyl dithiophosphate [86-50-0] Vol. III: 744

benzotriazinedithiophosphoric acid, dimethoxy ester [86-50-0] Vol. III: 744

1,2,3-benzotriazin-4 (3H)-one, 3-(mercaptomethyl)-, O,O-dimethyl phosphorodithioate [86-50-0] Vol. III: 744

benzotricloruro (Spanish) [98-07-7] Vol. II: 139

benzoylamide [55-21-0] Vol. III: 86

benzoylbenzene [119-61-9] Vol. III: 114

benzoyl methide [98-86-2] Vol. I: 86

benzoylperoxid (German) [94-36-0] Vol. I: 310

benzoylperoxyde (Dutch) [94-36-0] Vol. I: 310

o-benzoyl sulfimide [81-07-2] Vol. III: 824

benzoylsulfonic imide [81-07-2] Vol. III: 824

o-benzoyl sulphimide [81-07-2] Vol. III: 824

benzoyl superoxide [94-36-0] Vol. I: 310

1,12-benzperylene [191-24-2] Vol. II: 156

1,2-benzphenanthrene [218-01-9] Vol. I: 489

benzulfide [741-58-2] Vol. III: 731

benzydyna (Polish) [92-87-5] Vol. I: 288

benzyfuroline [10453-86-8] Vol. III: 285

benzyl alcohol benzoic ester [120-51-4] Vol. III: 117

benzyl alcohol, 2,4-dichloro-alpha-(chloromethylene)-, diethyl phosphate [470-90-6] Vol. III: 716

benzyl benzenecarboxylate [120-51-4] Vol. III: 117

benzyl butylphthalate [85-68-7] Vol. II: 767

benzylcarbonochloridate [501-53-1] Vol. III: 120

benzylcarbonyl chloride [501-53-1] Vol. III: 120

benzyl chloride (DOT) [100-44-7] Vol. I: 1099

beta, beta'-dichlorodiisopropyl ether [108-60-1] Vol. II: 386

beta, beta-dichlorodiisopropyl ether [108-60-1] Vol. II: 386

beta, beta'-dichloroethyl ether [111-44-4] Vol. II: 379

beta, beta'-dichloroethyl sulfide [505-60-2] Vol. II: 962

beta, beta'-dichloroethyl sulphide [505-60-2] Vol. II: 962

beta, beta-dicyano-o-chlorostyrene [2698-41-1] Vol. III: 218

beta, beta'-dihydroxydiethyl ether [111-46-6] Vol. II: 306

beta-butoxy-beta'-thiocyanodiethyl ether [112-56-1] Vol. III: 521

beta-butylene glycol [107-88-0] Vol. III: 148

beta-2-chloro-1-(2',4'-dichlorophenyl) vinyl diethylphosphate [470-90-6] Vol. III: 716

beta-chloroethanol [107-07-3] Vol. III: 220

beta-chloroethyl alcohol [107-07-3] Vol. III: 220

beta-chloroethyl-beta'-(p-t-butylphenoxy)-alpha'-methylethyl sulfite [140-57-8] Vol. III: 61

beta-chloroethyl-beta-(p-t-butylphenoxy)-alpha-methylethyl sulphite [140-57-8] Vol. III: 61

beta-chloroethyl-beta-(p-tert-butylphenoxy)-alpha-methylethyl sulphite [140-57-8] Vol. III: 61

beta chloroprene [126-99-8] Vol. I: 336

beta-cyanoethanol [109-78-4] Vol. II: 466

beta dichloroethane [107-06-2] Vol. I: 643

beta-(2-(3,5-dimethyl-2-oxocyclo hexyl)-2-hydroxyethyl) glutarimide [66-81-9] Vol. II: 447

beta-dinitrophenol [573-56-8] Vol. II: 717

beta-di-p-hydroxyphenylpropane [80-05-7] Vol. II: 720

beta-ethanolamine [141-43-5] Vol. II: 356

beta-ethoxyethanol [110-80-5] Vol. II: 362

beta-ethoxyethyl acetate [111-15-9] Vol. II: 365

beta-ethylmercaptoethyl diethyl thionophosphate [8065-48-3] Vol. III: 755

1-(beta-ethylol)-2-methyl-5-nitro- 3-azapyrrole [443-48-1] Vol. III: 628

beta-hydroxyethyl acrylate [818-61-1] Vol. III: 484

beta-hydroxyethylamine [141-43-5] Vol. II: 356

beta-hydroxyethyl chloride [107-07-3] Vol. III: 220

1-(beta-hydroxyethyl)- 2-methyl-5-nitroimidazole [443-48-1] Vol. III: 628

beta-hydroxyethyl phenyl ether [122-99-6] Vol. III: 695

beta-hydroxypropionitrile [109-78-4] Vol. II: 466

beta hydroxytricarballylic acid [77-92-9] Vol. I: 494

beta hydroxy tricarboxylic acid [77-92-9] Vol. I: 494

beta ketopropane [67-64-1] Vol. I: 77

beta-mercuribenzoate [583-15-3] Vol. III: 551

beta-methallyl chloride [563-47-3] Vol. III: 202

beta-methoxyethanol [109-86-4] Vol. II: 370

beta-methyl acrolein [123-73-9] Vol. III: 263

beta methylacrolein [4170-30-3] Vol. I: 547

beta methyl acrolein [4170-30-3] Vol. I: 547

beta-methylallyl chloride [563-47-3] Vol. III: 202

beta methylbivinyl [78-79-5] Vol. I: 784

beta methylbutyl acetate [123-92-2] Vol. I: 780

beta-methylnaphthalene [91-57-6] Vol. III: 611

beta methylpropyl ethanoate [110-19-0] Vol. I: 42

beta methylpyridine [108-99-6] Vol. I: 964

beta naftilamina (Italian) [91-59-8] Vol. I: 860

beta naftylamin (Czech) [91-59-8] Vol. I: 860

beta naftyloamina (Polish) [91-59-8] Vol. I: 860

beta naphthylamine [91-59-8] Vol. I: 860

beta naphthylamin (German) [91-59-8] Vol. I: 860

beta-naphthyl- phenylamine [135-88-6] Vol. III: 640

beta-octyl alcohol [123-96-6] Vol. III: 655

beta-phenoxyethanol [122-99-6] Vol. III: 695

beta-phenoxyethyl alcohol [122-99-6] Vol. III: 695

beta-phenylpropene [98-83-9] Vol. II: 947

beta-phenylpropylene [98-83-9] Vol. II: 947

beta picoline [108-99-6] Vol. I: 964

betaprone [57-57-8] Vol. II: 672

beta-propionolactone [57-57-8] Vol. II: 672

beta-proprolactone [57-57-8] Vol. II: 672

beta-pyrene [129-00-0] Vol. III: 811

beta pyridyl alpha n-methylpyrrolidine [54-11-5] Vol. I: 874

beta quinol [123-31-9] Vol. I: 753

5'beta-rotenone [83-79-4] Vol. II: 166

betasan [741-58-2] Vol. III: 731

betasan-e [741-58-2] Vol. III: 731

betasan-g [741-58-2] Vol. III: 731

beta-t [79-00-5] Vol. II: 341

beta thiopseudourea [62-56-6] Vol. I: 1140

beta-trichloroethane [79-00-5] Vol. II: 341

bethrodine [1861-40-1] Vol. II: 1043

betula [119-36-8] Vol. III: 622

betula oil [119-36-8] Vol. III: 622

bexane [94-81-5] Vol. III: 174

bexon [443-48-1] Vol. III: 628

bexone [94-81-5] Vol. III: 174

bexton-4l [1918-16-7] Vol. III: 789

bfv [50-00-0] Vol. I: 706

bhc [608-73-1] Vol. I: 564

bhc [66-76-2] Vol. III: 336

bh dalapon [75-99-0] Vol. I: 1016

bht [128-37-0] Vol. III: 317

bht (food grade) [128-37-0] Vol. III: 317

bi-58 [60-51-5] Vol. III: 740

bi-3411 [302-17-0] Vol. III: 197

biacetyl [431-03-8] Vol. III: 299

N, N'-bianiline [122-66-7] Vol. II: 479

4,4'-bianiline [92-87-5] Vol. I: 288

p,p'-bianiline [92-87-5] Vol. I: 288

bianisidine [119-93-7] Vol. III: 364

bibenzene [92-52-4] Vol. I: 326

bibesol [62-73-7] Vol. I: 946

bicarburet of hydrogen [71-43-2] Vol. I: 239

bicarburretted hydrogen [74-85-1] Vol. I: 691

bichlorendo [2385-85-5] Vol. III: 583

bichloride of mercury [7487-94-7] Vol. II: 539

1,2-bichloroethane [107-06-2] Vol. I: 643

bichlorure de mercure (French) [7487-94-7] Vol. II: 539

bichlorure de propylene (French) [78-87-5] Vol. I: 987

bichlorure d'ethylene (French) [107-06-2] Vol. I: 643

bichromate d'ammonium (French) [7789-09-5] Vol. I: 590

bichromate de sodium (French) [10588-01-9] Vol. II: 300

bichromate of soda [10588-01-9] Vol. II: 300

bicyclo (2.2.1)heptane, 2,2-dimethyl-3-methylene [79-92-5] Vol. I: 406

bicyclo (2,2,1) heptan-2-one, 1,7,7-tri-methyl [76-22-2] Vol. II: 211

bicyclo (2.2.1) heptene-2-dicarboxylic acid, 2-ethylhexylimide [113-48-4] Vol. III: 631

bicyclo (2.2.1) hept-5-ene-2,3-dicarboxylic acid, 1,4,5,6,7,7-hexachloro [115-28-6] Vol. III: 200

bicyclo (2.2.1) hept-2-ene, 5-ethylidene [16219-75-3] Vol. III: 440

bicyclo (3.1.1) hept-2-ene, 2,6,6-trimethyl [80-56-8] Vol. III: 32

bicyclopentadiene [77-73-6] Vol. II: 635

bicyelopentadiene [77-73-6] Vol. II: 635

bidrin [141-66-2] Vol. III: 720

bifenthrine [82657-04-3] Vol. III: 131

bifex [114-26-1] Vol. II: 231

bifluoriden (Dutch) [7782-41-4] Vol. I: 703

biformal [107-22-2] Vol. I: 728

biformyl [107-22-2] Vol. I: 728

bi-k [57-13-6] Vol. I: 1136

bilarcil [52-68-6] Vol. III: 707

bilorin [64-18-6] Vol. I: 711

bimethyl [74-84-0] Vol. III: 401

2,3,1',8'-binaphthylene [207-08-9] Vol. II: 151

binitrobenzene [99-65-0] Vol. III: 89

binitrotoluene [25321-14-6] Vol. II: 1021

binnell [1861-40-1] Vol. II: 1043

bio-5,462 [115-29-7] Vol. II: 661

bio-clave [120-32-1] Vol. III: 122

biogas [74-82-8] Vol. III: 579

biogrisin-fp [126-07-8], Vol. III: 472

bionol [8001-54-5] Vol. II: 67

biosept [123-03-5] Vol. III: 196

bioxyde d'azote (French) [10102-43-9] Vol. I: 887

biphenate [82657-04-3] Vol. III: 131

biphenthrin [82657-04-3] Vol. III: 131

biphentrin [82657-04-3] Vol. III: 131

1,1'-biphenyl [92-52-4] Vol. I: 326

biphenyl [92-52-4] Vol. I: 326

p,p'-bis (dimethylamino) diphenylmethane [101-61-1] **Vol. I:** 173

4,4'-bis (dimethylamino)diphenylmethane [101-61-1] **Vol. I:** 173

bis((4-dimethylamino)phenyl)methane [101-61-1] **Vol. I:** 173

bis(4-dimethylaminophenyl)methane [101-61-1] **Vol. I:** 173

bis(4-(dimethylamino)phenyl)methane [101-61-1] **Vol. I:** 173

bis (dimethylamino) phosphonous anhydride [152-16-9] **Vol. II:** 884

bis (dimethylamino) phosphoric anhydride [152-16-9] **Vol. II:** 884

2,6-bis (1,1-dimethylethyl)-4-methylpheno [128-37-0] **Vol. III:** 317

bis (dimethylthiocarbamoyl) disulfide [137-26-8] **Vol. I:** 632

bis (dimethyl thiocarbamoyl) disulfid (German) [137-26-8] **Vol. I:** 632

bis (dimethylthiocarbamoyl) disulphide [137-26-8] **Vol. I:** 632

bis (dimethylthiocarbamyl) disulfide [137-26-8] **Vol. I:** 632

bis (dithiophosphate de O,O-diethyle) de S-S'(1,4-dioxanne-2,3-diyle) [French] [78-34-2] **Vol. II:** 756

4,4'-bis (2,3-epoxypropoxy) diphenyldimethylmethane [1675-54-3] **Vol. III:** 132

S-(1,2-bis (ethox-carbonyl)-ethyl)-o,o-dimethyl-dithiofosfaat (Dutch) [121-75-5] **Vol. II:** 950

s-1, 2-bis (ethox-carbonyl)-ethyl-o,o-dimethyl thiophosphate [121-75-5] **Vol. II:** 950

S-(1,2-bis (ethoxycarbonyl) ethyl) o,o-dimethyl phosphorodithioate [121-75-5] **Vol. II:** 950

2,4-bis (ethylamino)-6-chloro-s-triazine [122-34-9] **Vol. II:** 1047

bis (ethylenediamine) copper(2) [13426-91-0] **Vol. III:** 270

bis (ethylenediamine) copper ion [13426-91-0] **Vol. III:** 270

bis (ethylenediamine) copper(2) ion [13426-91-0] **Vol. III:** 270

bis (2-ethylhexyl) 1,2-benzenedicarboxylate [117-81-7] **Vol. II:** 769

bis (2-ethylhexyl)-1,2-benzenedicarboxylate [117-81-7] **Vol. II:** 769

bis-(2-ethylhexyl) ester kyseliny adipove (Czech) [103-23-1] **Vol. II:** 291

bis (2-ethylhexyl) hexanedioate [103-23-1] **Vol. II:** 291

1,4-bis(2-ethylhexyl) sodium sulfosuccinate [577-11-7] **Vol. III:** 859

bis(2-ethylhexyl) sodium sulfosuccinate [577-11-7] **Vol. III:** 859

bis(2-ethylhexyl) s-sodium sulfosuccinate [577-11-7] **Vol. III:** 859

bis(2-ethylhexyl) sulfosuccinate sodium salt [577-11-7] **Vol. III:** 859

S-(1,2-bis (etossi-carbonil)-etil)-o,o-dimetil-ditiofosfato (Italian) [121-75-5] **Vol. II:** 950

bisferol-a (Czech) [80-05-7] **Vol. II:** 720

bis (4-glycidyloxyphenyl) dimethylmethane [1675-54-3] **Vol. III:** 132

8014-bis-hc [60-51-5] **Vol. III:** 740

bishydroxycoumarin [66-76-2] **Vol. III:** 336

bis-3,3'-(4-hydroxycoumarinyl) methane [66-76-2] **Vol. III:** 336

bis (4-hydroxycoumarin-3-yl) methane [66-76-2] **Vol. III:** 336

bis (2-hydroxyethoxyethane) [112-27-6] **Vol. II:** 1058

1,2-bis (2-hydroxyethoxy) ethane [112-27-6] **Vol. II:** 1058

bis (hydroxyethyl)amine [111-42-2] **Vol. I:** 658

bis (2-hydroxyethyl)amine [111-42-2] **Vol. I:** 658

bis (2-hydroxyethyl) ether [111-46-6] **Vol. II:** 306

bis (hydroxymethyl) acetylene [110-65-6] **Vol. III:** 172

bis (4-hydroxyphenyl) dimethylmethane [80-05-7] **Vol. II:** 720

bis (4-hydroxyphenyl) dimethylmethane diglycidyl ether [1675-54-3] **Vol. III:** 132

bis (4-hydroxyphenyl) propane [80-05-7] **Vol. II:** 720

bis (2-hydroxypropyl) amine [110-97-4] **Vol. II:** 823

bis (isodecyl) phthalate [26761-40-0] **Vol. III:** 340

2,4-bis (isopropylamino) -6-chloro-s-triazine [139-40-2] **Vol. III:** 802

2,4-bis(isopropylamino)-6-methoxy-s-triazine [1610-18-0] **Vol. III:** 787

2,4-bis(isopropylamino)-6-methoxy-1,3,5-triazine [1610-18-0] **Vol. III:** 787

2,4-bis (isopropylamino)-6-methyl mercapto-S-triazine [7287-19-6] **Vol. II:** 1056

2,4-bis (isopropylamino)-6-methyl thio-S-triazine [7287-19-6] **Vol. II:** 1056

2,4-bis (isopropylamino)-6-methyl thio-1,3,5-triazine [7287-19-6] **Vol. II:** 1056

blue copper as [7758-99-8] Vol. II: 255
blue oil [62-53-3] Vol. I: 164
blue-ox [1314-84-7] Vol. II: 1100
blue powder [7440-66-6] Vol. I: 1175
bluestone [7758-99-8] Vol. II: 255
blue-vicking [7758-99-8] Vol. II: 255
blue vitriol [7758-99-8] Vol. II: 255
blulan [1861-40-1] Vol. II: 1043
blu phen [50-06-6] Vol. I: 223
bna [91-59-8] Vol. I: 860
b-nine [1596-84-5] Vol. III: 884
b nitropropane [79-46-9] Vol. I: 999
bnm [17804-35-2] Vol. II: 148
bnp-20 [88-85-7] Vol. II: 704
bnp-30 [88-85-7] Vol. II: 704
bolero [28249-77-6] Vol. II: 213
bolfo [114-26-1] Vol. II: 231
bolls eye [75-60-5] Vol. I: 214
bonalan [1861-40-1] Vol. II: 1043
bondelane-a [126-33-0] Vol. II: 1007
bondolane-a [126-33-0] Vol. II: 1007
bonibal [97-77-8] Vol. III: 382
bonide blue death rat killer [7723-14-0] Vol. II: 764
Bonide blue death rat killer [7723-14-0] Vol. III: 973
bonide root destroyer [7758-98-7] Vol. I: 510
Bonoform [79-34-5] Vol. III: 414
boracic acid [10043-35-3] Vol. I: 327
borascu [1303-96-4] Vol. III: 134
borate (1-), tetrafluoro-, ammonium [13826-83-0] Vol. I: 144
borate(1–), tetrafluoro-, hydrogen [16872-11-0] Vol. III: 457
borate(1-), tetrafluoro-, lead(2) (2:1) [13814-96-5] Vol. III: 507
borate(1-), tetrafluoro-, lead(2) [13814-96-5] Vol. III: 507
borax-2335 [1332-07-6] Vol. III: 980
borax decahydrate [1303-96-4] Vol. III: 134
borax, dehydrated [1330-43-4] Vol. III: 874
borax glass [1330-43-4] Vol. III: 874
borax (Na₂ (B₄O₇) [1303-96-4] Vol. III: 134
bordermaster [94-74-6] Vol. I: 25
borer sol [107-06-2] Vol. I: 643
boric acid (h3bo3) [10043-35-3] Vol. I: 327
Boric acid (HBO₂), anhydride [1303-86-2] Vol. III: 138

boric acid (H₂B₄O₇) disodium salt [1330-43-4] Vol. III: 874
boric acid (H₂B₄O₇), disodium salt, decahydrate [1303-96-4] Vol. III: 134
boric acid (H₂B₄O₇) sodium salt [1330-43-4] Vol. III: 874
boric acid, zinc salt [1332-07-6] Vol. III: 980
boric anhydride [1303-86-2] Vol. III: 138
boricin [1303-96-4] Vol. III: 134
boric oxide [1303-86-2] Vol. III: 138
boric oxide (B₂O₃) [1303-86-2] Vol. III: 138
2-bornanone [76-22-2] Vol. II: 211
bornate [115-31-1] Vol. III: 491
boroethane [19287-45-7] Vol. II: 292
borofax [10043-35-3] Vol. I: 327
borofluoric acid [16872-11-0] Vol. III: 457
boron hydride [17702-41-9] Vol. I: 584
boron hydride [19287-45-7] Vol. II: 292
boron oxide (B₂O₃) [1303-86-2] Vol. III: 138
boron sesquioxide [1303-86-2] Vol. III: 138
boron sodium oxide (B₄Na₂O₇) [1330-43-4] Vol. III: 874
boron sodium oxide (B₄Na₂O₇), decahydrate [1303-96-4] Vol. III: 134
boron trihydroxide [10043-35-3] Vol. I: 327
boron trioxide [1303-86-2] Vol. III: 138
borsaure (German) [10043-35-3] Vol. I: 327
boruho [114-26-1] Vol. II: 231
botrilex [82-68-8] Vol. II: 132
bov [7664-93-9] Vol. II: 969
bovinox [52-68-6] Vol. III: 707
bowl cleaner [7647-01-0] Vol. I: 738
bp [50-32-8] Vol. II: 153
bpl [57-57-8] Vol. II: 672
bpps [2312-35-8] Vol. II: 983
bradsyn-peg [25322-68-3] Vol. II: 811
101-brand-pcnb-75-wettable [82-68-8] Vol. II: 132
brasilazina-oil-yellow-g [60-09-3] Vol. II: 74
brasilazina oil yellow-r [97-56-3] Vol. III: 46
brassicol-75 [82-68-8] Vol. II: 132
brassicol [82-68-8] Vol. II: 132
brassicol-super [82-68-8] Vol. II: 132
bravo [1897-45-6] Vol. II: 513
bravo-500 [1897-45-6] Vol. II: 513

butoxide [51-03-6] Vol. III: 779

Butoxone [94-82-6] Vol. III: 176

Butoxone ester [94-82-6] Vol. III: 176

1-butoxy-alpha-(2-thiocyanoethoxy) ethane [112-56-1] Vol. III: 521

1-butoxybutane [142-96-1] Vol. III: 313

butoxy-d 3 [1929-73-3] Vol. III: 155

butoxydiethylene glycol [112-34-5] Vol. III: 421

butoxydiglycol [112-34-5] Vol. III: 421

butoxy d 3: 1 liquid emulsifiable brushkiller lv96 (Canada) [94-75-7] Vol. I: 30

2-butoxyethanol, acetate [112-07-2] Vol. II: 360

butoxyethanol ester of (2,4-dichlorophenoxy) acetic acid [1929-73-3] Vol. III: 155

butoxyethoxyethanol [112-34-5] Vol. III: 421

2-(2-butoxyethoxy) ethanol [112-34-5] Vol. III: 421

2-(2-butoxyethoxy) ethanol acetate [124-17-4] Vol. III: 425

alpha-(2-(2-n-butoxyethoxy)-ethoxy) -4,5-methylenedioxy-2-propyltoluene [51-03-6] Vol. III: 779

5-((2-(2-butoxyethoxy) ethoxy) methyl-6-propyl-1,3-benzodioxole [51-03-6] Vol. III: 779

butoxyethoxyethyl acetate [124-17-4] Vol. III: 425

2-(2-butoxyethoxy) ethyl acetate [124-17-4] Vol. III: 425

2-(2-butoxy) ethoxy) ethyl ester of thiocyanic acid [112-56-1] Vol. III: 521

2-(2-butoxyethoxy) ethyl thiocyanate [112-56-1] Vol. III: 521

2-(2-(butoxy) ethoxy) ethyl thiocyanic acid ester [112-56-1] Vol. III: 521

2-[2-(butoxy) ethoxy] ester of thiocyanic acid [112-56-1] Vol. III: 521

2-[2-(butoxy) ethoxy]ethyl ether of thiocyanic acid [112-56-1] Vol. III: 521

butoxyethyl acetate [112-07-2] Vol. II: 360

2-butoxyethyl acetate [112-07-2] Vol. II: 360

2,4-d butoxyethyl ester [1929-73-3] Vol. III: 155

2,4-d 2-butoxyethyl ester [1929-73-3] Vol. III: 155

2-butoxyethylester kyseliny octove (Czech) [112-07-2] Vol. II: 360

butoxyrhodanodiethyl-ether [112-56-1] Vol. III: 521

2-butoxy-2'-thiocyanodiethyl ether [112-56-1] Vol. III: 521

1-butoxy-2-(2-thiocyanoethoxy) ethane [112-56-1] Vol. III: 521

buttercup yellow [13530-65-9] Vol. III: 984

butter of antimony [10025-91-9] Vol. I: 182

butter of arsenic [7784-34-1] Vol. I: 196

butter of zinc [7646-85-7] Vol. I: 1178

buttersaeure (German) [107-92-6] Vol. I: 360

butter-yellow [97-56-3] Vol. III: 46

butyl acetate [123-86-4] Vol. I: 18

1-butyl acetate [123-86-4] Vol. I: 18

2-butyl acetate [105-46-4] Vol. I: 64

t butyl acetate [540-88-5] Vol. I: 66

butylacetaten (Dutch) [123-86-4] Vol. I: 18

butylacetat (German) [123-86-4] Vol. I: 18

butylacetone [110-43-0] Vol. II: 454

n-butyl acrylate [141-32-2] Vol. II: 842

2-butylalcohol [78-92-2] Vol. II: 194

2-butyl alcohol [78-92-2] Vol. II: 194

butyl alcohol [71-36-3] Vol. I: 352

butylaldehyde [123-72-8] Vol. I: 357

butyl aldehyde [123-72-8] Vol. I: 357

n butyl aldehyde [123-72-8] Vol. I: 357

butylamine [109-73-9] Vol. II: 203

t-butylamine [75-64-9] Vol. II: 199

butylamine, N-nitrosodi [924-16-3] Vol. II: 187

butylamine, tertiary [75-64-9] Vol. II: 199

n-Butylamin (German) [109-73-9] Vol. II: 203

butylated-hydroxytoluen [128-37-0] Vol. III: 317

butylbenzyl phthalate [85-68-7] Vol. II: 767

butyl benzylphthalate [85-68-7] Vol. II: 767

N-butyl benzyl phthalate [85-68-7] Vol. II: 767

2-butylbutanoic acid [149-57-5] Vol. II: 461

1-(butylcarbamoyl)-2-benzimidazolec arbamic acid, methyl ester [17804-35-2] Vol. II: 148

1-(n-butylcarbamoyl)-2-(methoxy-carboxamido)-benzimidazol (German) [17804-35-2] Vol. II: 148

butyl-carbinol [71-41-0] Vol. II: 700

n-butyl octadecanoate [123-95-5] **Vol. III:** 167

butyl octadecanoate [123-95-5] **Vol. III:** 167

butyl octadecylate [123-95-5] **Vol. III:** 167

butylowy alkohol (Polish) [71-36-3] **Vol. I:** 352

butyl oxide [142-96-1] **Vol. III:** 313

2-(p-t-butylphenoxy) cyclohexyl propargyl sulfite [2312-35-8] **Vol. II:** 983

butylphenoxyisopropyl-chloroethyl-sulfite [140-57-8] **Vol. III:** 61

2-(4-t-butylphenoxy) isopropyl-2-chloroethyl sulfite [140-57-8] **Vol. III:** 61

2-(p-butylphenoxy) isopropyl-2-chloroethyl sulfite [140-57-8] **Vol. III:** 61

2-(p-t-butylphenoxy) isopropyl-2'-chloroethyl sulphite [140-57-8] **Vol. III:** 61

2-(p-t-butylphenoxy)-1-methylethyl 2-chloroethyl ester of sulphurous acid [140-57-8] **Vol. III:** 61

2-(p-butylphenoxy)-1-methylethyl-2-chloroethyl sulfite [140-57-8] **Vol. III:** 61

2-(p-butylphenoxy)-1-methylethyl 2-chloroethyl sulfite [140-57-8] **Vol. III:** 61

2-(p-t-butylphenoxy)-1-methylethyl 2'-chloroethyl sulphite [140-57-8] **Vol. III:** 61

2-(p-t-butylphenoxy)-1-methylethyl sulphite of 2-chloroethanol [140-57-8] **Vol. III:** 61

butyl phenylmethyl 1,2-benzene carboxylate [85-68-7] **Vol. II:** 767

butyl phosphate [126-73-8] **Vol. I:** 951

butyl phosphate, tri [126-73-8] **Vol. I:** 951

butyl-phosphorotrithioate [78-48-8] **Vol. III:** 768

butyl-phosphorotrithioate ((BuS)₃ PO) [78-48-8] **Vol. III:** 768

butyl phosphorotrithioite [(bus) 3p] [150-50-5] **Vol. III:** 574

butyl-phthalate [84-74-2] **Vol. II:** 774

n-butyl phthalate [84-74-2] **Vol. II:** 774

butyl 2-propenoate [141-32-2] **Vol. II:** 842

butyl stearate [123-95-5] **Vol. III:** 167

butyl 2,4,5-t [93-79-8] **Vol. II:** 34

butyl 2,4,5-trichlorophenoxyacetate [93-79-8] **Vol. II:** 34

n-butyl (2,4,5-trichlorophenoxy) acetate [93-79-8] **Vol. II:** 34

butyl (2,4,5-trichlorophenoxy) acetate [93-79-8] **Vol. II:** 34

butynediol [110-65-6] **Vol. III:** 172

2-butyne-1,4-diol [110-65-6] **Vol. III:** 172

2-butynediol [110-65-6] **Vol. III:** 172

butynediol-1,4 (French) [110-65-6] **Vol. III:** 172

Butyrac-200 [94-82-6] **Vol. III:** 176

Butyrac [94-82-6] **Vol. III:** 176

Butyrac ester [94-82-6] **Vol. III:** 176

butyral [123-72-8] **Vol. I:** 357

n butyraldehyde [123-72-8] **Vol. I:** 357

butyraldehyde (Czech) [123-72-8] **Vol. I:** 357

butyraldehyd (German) [123-72-8] **Vol. I:** 357

butyrate [107-92-6] **Vol. I:** 360

n butyric acid [107-92-6] **Vol. I:** 360

2,4-D butyric acid [94-82-6] **Vol. III:** 176

butyric acid, 2-amino-4-(methylthio)- [63-68-3] **Vol. III:** 593

butyric acid, 4-((4-chloro-o-tolyl) oxy)- [94-81-5] **Vol. III:** 174

butyric acid, ethyl ester [105-54-4] **Vol. III:** 438

butyric acid nitrile [109-74-0] **Vol. III:** 152

butyric alcohol [71-36-3] **Vol. I:** 352

butyric aldehyde [123-72-8] **Vol. I:** 357

butyric ether [105-54-4] **Vol. III:** 438

n-butyronitrile [109-74-0] **Vol. III:** 152

butyronitrile [109-74-0] **Vol. III:** 152

butyrylaldehyde [123-72-8] **Vol. I:** 357

butyrylonitrile [109-74-0] **Vol. III:** 152

bye-bugs [16919-19-0] **Vol. II:** 908

bygran [101-21-3] **Vol. II:** 234

bzcf [501-53-1] **Vol. III:** 120

bzt [121-54-0] **Vol. III:** 105

c-31 [21645-51-2] **Vol. II:** 57

c-31 [21645-51-2] **Vol. III:** 35

c-33 [21645-51-2] **Vol. II:** 57

c-33 [21645-51-2] **Vol. III:** 35

C-56 [77-47-4] **Vol. II:** 283

c-570 [13171-21-6] **Vol. III:** 704

c-709 [141-66-2] **Vol. III:** 720

c-1933 [1982-47-4] **Vol. III:** 232

c-1983 [1982-47-4] **Vol. III:** 232

c-2059- [2164-17-2] **Vol. II:** 1086

c8949 [470-90-6] **Vol. III:** 716

c-8949 [470-90-6] **Vol. III:** 716

cab-o-sil m-5 [16871-71-9] **Vol. II:** 911

cab-o-sil m-5 [7631-86-9] **Vol. III:** 846

cabosil-N-5 [16871-71-9] **Vol. II:** 911

cabosil-N-5 [7631-86-9] **Vol. III:** 846

cabosil st-1 [16871-71-9] **Vol. II:** 911

cabosil st-1 [7631-86-9] **Vol. III:** 846

cachalot 1-50 [112-53-8] **Vol. II:** 316

cachalot 1-90 [112-53-8] **Vol. II:** 316

cachalot c-51 [36653-82-4] **Vol. III:** 193

cacodylic acid [75-60-5] **Vol. I:** 214

cacp [15663-27-1] **Vol. III:** 237

cadmium bromide dimer [7789-42-6] **Vol. I:** 373

cadmium dibromide [7789-42-6] **Vol. I:** 373

cadmium dichloride [10108-64-2] **Vol. I:** 376

cadmium monosulfate [10124-36-4] **Vol. I:** 364

cadmium sulfate (1:1) [10124-36-4] **Vol. I:** 364

cadmium sulphate [10124-36-4] **Vol. I:** 364

cadox-ps [94-17-7] **Vol. III:** 685

cadox-tbh [75-91-2] **Vol. III:** 896

CaF [532-27-4] **Vol. I:** 89

cairox [7722-64-7] **Vol. I:** 915

cake alum [10043-01-3] **Vol. III:** 39

calcanthite [7758-99-8] **Vol. II:** 255

calcia [1305-78-8] **Vol. I:** 402

calcid [592-01-8] **Vol. I:** 395

calcium acetylide [75-20-7] **Vol. I:** 391

calcium alkylaromatic sulfonate [26264-06-2] **Vol. II:** 209

calcium alkylbenzenesulfonate [26264-06-2] **Vol. II:** 209

calciumarsenat [7778-44-1] **Vol. III:** 65

calcium arsenate (Ca₃(AsO₄)₂) [7778-44-1] **Vol. III:** 65

calcium carbide [75-20-7] **Vol. I:** 391

calcium carbimide [156-62-7] **Vol. II:** 265

calcium chlorohydrochlorite [7778-54-3] **Vol. II:** 494

calcium chlorohypochloride [7778-54-3] **Vol. II:** 494

calcium cyanamid [156-62-7] **Vol. II:** 265

calcium cyanamide (DOT) [156-62-7] **Vol. II:** 265

calcium dicarbide [75-20-7] **Vol. I:** 391

calcium dinitrate [10124-37-5] **Vol. I:** 399

calcium dodecylbenzensulfonate [26264-06-2] **Vol. II:** 209

calcium hypochloride [7778-54-3] **Vol. II:** 494

calcium (ii) nitrate (1:2) [10124-37-5] **Vol. I:** 399

calcium meta arsenite [52740-16-6] **Vol. II:** 109

calcium monoxide [1305-78-8] **Vol. I:** 402

calcium n-dodecylbenzenesulfonate [26264-06-2] **Vol. II:** 209

calcium o-arsenate [7778-44-1] **Vol. III:** 65

calcium orthoarsenate [7778-44-1] **Vol. III:** 65

calcium oxychloride [7778-54-3] **Vol. II:** 494

calcium phosphide [1305-99-3] **Vol. I:** 404

calcium photophor [1305-99-3] **Vol. I:** 404

calcotone-black [7440-44-0] **Vol. III:** 179

calcotone-white-t [13463-67-7] **Vol. III:** 921

calcyan [592-01-8] **Vol. I:** 395

calcyanide [592-01-8] **Vol. I:** 395

caldon [88-85-7] **Vol. II:** 704

calmogastrin [21645-51-2] **Vol. II:** 57

calmogastrin [21645-51-2] **Vol. III:** 35

calochlor [7487-94-7] **Vol. II:** 539

caloxol w-3 [1305-78-8] **Vol. I:** 402

calx [1305-78-8] **Vol. I:** 402

calx usta [1305-78-8] **Vol. I:** 402

calxyl [1305-78-8] **Vol. I:** 402

camathion [121-75-5] **Vol. II:** 950

campaign [1702-17-6] **Vol. III:** 239

2-camphanone [76-22-2] **Vol. II:** 211

camphechlor [8001-35-2] **Vol. I:** 1118

camphene (2,2-dimethyl-3-methylene nor-bornane) [79-92-5] **Vol. I:** 406

camphochlor [8001-35-2] **Vol. I:** 1118

camphofene huileux [8001-35-2] **Vol. I:** 1118

camphor-natural [76-22-2] **Vol. II:** 211

campilit [506-68-3] **Vol. I:** 558

camposan [16672-87-0] **Vol. III:** 426

candaseptic [59-50-7] **Vol. II:** 258

candex [1912-24-9] **Vol. II:** 1050

canogard [62-73-7] **Vol. I:** 946

cao-1 [128-37-0] **Vol. III:** 317

cao-3 [128-37-0] **Vol. III:** 317

CaP [532-27-4] **Vol. I:** 89

caparol [7287-19-6] **Vol. II:** 1056

caparol-80w [7287-19-6] **Vol. II:** 1056

capitol [8001-54-5] **Vol. II:** 67

caporit [7778-54-3] **Vol. II:** 494

Capquin [54-05-7] **Vol. III:** 229

capraldehyde [112-31-2] **Vol. III:** 294

capric aldehyde [112-31-2] **Vol. III:** 294

caprinaldehyde [112-31-2] **Vol. III:** 294

caprinic aldehyde [112-31-2] **Vol. III:** 294

caproaldehyde [66-25-1] **Vol. II:** 456

N-caproaldehyde [66-25-1] **Vol. II:** 456

caproic acid [142-62-1] **Vol. III:** 479

caproic aldehyde [66-25-1] **Vol. II:** 456

N-caproic aldehyde [66-25-1] **Vol. II:** 456

carbamothioic acid, bis (1-methylethyl)-, S-(2,3-dichloro-2-propenyl) ester [2303-16-4] **Vol. II:** 218

carbamothioic acid, bis (1-methylethyl) -, S-(2,3,3-trichloro-2-propenyl) ester [2303-17-5] **Vol. III:** 931

carbamothioic acid, bis (2-methylpropyl)-, S-ethyl ester [2008-41-5] **Vol. II:** 216

carbamothioic acid, dipropyl-, S-propyl ester [1929-77-7] **Vol. II:** 221

carbamyl [116-06-3] **Vol. III:** 19

carbanilic acid, isopropyl ester [122-42-9] **Vol. I:** 426

carbanilic acid, m-chloro-, isopropyl ester [101-21-3] **Vol. II:** 234

carbanilic acid, m-hydroxy, methyl ester, m-methylcarbanilate (ester) [13684-63-4] **Vol. III:** 130

carbathiin [5234-68-4] **Vol. II:** 670

carbathion [137-42-8] **Vol. III:** 861

carbathione [137-42-8] **Vol. III:** 861

carbation [137-42-8] **Vol. III:** 861

carbax [115-32-2] **Vol. II:** 144

carbazotic acid [88-89-1] **Vol. II:** 793

carbetovur [121-75-5] **Vol. II:** 950

carbetox [121-75-5] **Vol. II:** 950

carbicron [141-66-2] **Vol. III:** 720

carbimide [420-04-2] **Vol. II:** 263

carbinamine [74-89-5] **Vol. I:** 840

carbinol [67-56-1] **Vol. I:** 836

carbobenzoxy chloride [501-53-1] **Vol. III:** 120

carbobenzyloxy chloride [501-53-1] **Vol. III:** 120

carbodiamide [420-04-2] **Vol. II:** 263

carbofos [121-75-5] **Vol. II:** 950

carbolac [7440-44-0] **Vol. III:** 179

carbolic acid [108-95-2] **Vol. I:** 920

Carbolsaure (German) [108-95-2] **Vol. I:** 920

3-(carbomethoxyamino) phenyl 3-methyl-carbanilate [13684-63-4] **Vol. III:** 130

carbomicron [141-66-2] **Vol. III:** 720

carbon-12 [7440-44-0] **Vol. III:** 179

carbona [56-23-5] **Vol. I:** 435

carbon bisulfide [75-15-0] **Vol. I:** 428

carbon bisulphide [75-15-0] **Vol. I:** 428

carbon chloride [56-23-5] **Vol. I:** 435

carbon chlorosulfide [463-71-8] **Vol. I:** 1086

carbon d [142-59-6] **Vol. I:** 411

carbon dichloride oxide [75-44-5] **Vol. I:** 939

carbon difluoride oxide [353-50-4] **Vol. III:** 186

carbon dioxide (CO₂) [124-38-9] **Vol. III:** 182

carbon disulphide [75-15-0] **Vol. I:** 428

carbone (oxychlorurede) (French) [75-44-5] **Vol. I:** 939

carbone (oxydede) (French) [630-08-0] **Vol. I:** 433

carbone (sulfurede) (French) [75-15-0] **Vol. I:** 428

carbon fluoride oxide (COF₂) [353-50-4] **Vol. III:** 186

carbon hydride nitride [74-90-8] **Vol. I:** 743

Carbonica [124-38-9] **Vol. III:** 182

carbonic acid, ammonium salt [506-87-6] **Vol. I:** 139

carbonic acid, barium salt (1:1) [513-77-9] **Vol. I:** 227

carbonic acid, cobalt(2) salt (1: 1) [513-79-1] **Vol. III:** 244

carbonic acid, diammonium salt [506-87-6] **Vol. I:** 139

carbonic acid dichloride [75-44-5] **Vol. I:** 939

carbonic acid, disodium salt [497-19-8] **Vol. II:** 922

carbonic acid, dithallium (1) salt [6533-73-9] **Vol. I:** 1076

carbonic acid, dithio-, cyclic s, s-(6-methyl-2,3-quinoxalinediyl) ester [2439-01-2] **Vol. III:** 663

carbonic acid gas [124-38-9] **Vol. III:** 182

carbonic acid, monoammonium salt [1066-33-7] **Vol. II:** 62

carbonic acid, sodium salt [497-19-8] **Vol. II:** 922

carbonic anhydride [124-38-9] **Vol. III:** 182

carbonic dichloride [75-44-5] **Vol. I:** 939

carbonic dichloride, thio [463-71-8] **Vol. I:** 1086

carbonic difluoride [353-50-4] **Vol. III:** 186

carbonic oxide [630-08-0] **Vol. I:** 433

carbonio (ossiclorurodi) (Italian) [75-44-5] **Vol. I:** 939

carbonio (ossidodi) (Italian) [630-08-0] **Vol. I:** 433

carbonio (solfurodi) (Italian) [75-15-0] **Vol. I:** 428

carbon monobromide trifluoride [75-63-8] **Vol. III:** 143

carbon monoxide (DOT) [630-08-0] Vol. I: 433

Carbon monoxide monosulfide [463-58-1] Vol. III: 188

carbon nitride [460-19-5] Vol. I: 555

carbonochloridic acid, ethyl ester [541-41-3] Vol. III: 433

carbonochloridic acid, methyl ester [79-22-1] Vol. III: 597

carbonochloridic acid, phenylmethyl ester [501-53-1] Vol. III: 120

carbonochloridic acid, 2-propenyl ester [2937-50-0] Vol. III: 27

carbonochloridic acid, trichloromethyl ester [503-38-8] Vol. III: 376

carbonothioic dichloride [463-71-8] Vol. I: 1086

carbon oxide [630-08-0] Vol. I: 433

carbon oxide (CO₂) [124-38-9] Vol. III: 182

carbon oxide, di- [124-38-9] Vol. III: 182

carbon oxide sulfide [463-58-1] Vol. III: 188

carbon oxychloride [75-44-5] Vol. I: 939

carbon oxyfluoride [353-50-4] Vol. III: 186

carbon oxyfluoride (COF₂) [353-50-4] Vol. III: 186

carbon oxysulfide [463-58-1] Vol. III: 188

carbon sulfide [75-15-0] Vol. I: 428

carbon tet [56-23-5] Vol. I: 435

carbonyl chloride [75-44-5] Vol. I: 939

carbonylchloride, thio [463-71-8] Vol. I: 1086

carbonylchlorid (German) [75-44-5] Vol. I: 939

carbonyl diamide [57-13-6] Vol. I: 1136

carbonyldiamine [57-13-6] Vol. I: 1136

carbonyl dichloride [75-44-5] Vol. I: 939

carbonyl fluoride [353-50-4] Vol. III: 186

carbonyl nickel powder [7440-02-0] Vol. I: 862

carbophos [121-75-5] Vol. II: 950

carbopol-extra [7440-44-0] Vol. III: 179

carbopol-m [7440-44-0] Vol. III: 179

carbopol-z-4 [7440-44-0] Vol. III: 179

carbopol-z-extra [7440-44-0] Vol. III: 179

carbothialdin [533-74-4] Vol. III: 293

carbothialdine [533-74-4] Vol. III: 293

carbowax-6000 [25322-68-3] Vol. II: 811

5-carboxanilido-2,3-dihydro-6-methyl1,4-oxathiin [5234-68-4] Vol. II: 670

carboxine [5234-68-4] Vol. II: 670

carboxin-sulfone [5259-88-1] Vol. III: 659

carboxybenzene [65-85-0] Vol. I: 293

carboxyethane [79-09-4] Vol. I: 1012

carbure de calcium (French) [75-20-7] Vol. I: 391

carburo calcico (Spanish) [75-20-7] Vol. I: 391

carfene [86-50-0] Vol. III: 744

carpen [2439-10-3] Vol. III: 290

carpene [2439-10-3] Vol. III: 290

carpidor [1861-40-1] Vol. II: 1043

carplex [16871-71-9] Vol. II: 911

carplex-30 [16871-71-9] Vol. II: 911

carplex-30 [7631-86-9] Vol. III: 846

carplex [7631-86-9] Vol. III: 846

carplex-80 [16871-71-9] Vol. II: 911

carplex-80 [7631-86-9] Vol. III: 846

Carrel-Dakin solution [7681-52-9] Vol. II: 498

carsil [1344-09-8] Vol. III: 868

carsil (silicate) [1344-09-8] Vol. III: 868

carsonol-sls [151-21-3] Vol. II: 981

carsoron [1194-65-6] Vol. I: 299

carzol [23422-53-9] Vol. III: 463

carzol-sp [23422-53-9] Vol. III: 463

casalis green [1308-38-9] Vol. III: 234

casaron [1194-65-6] Vol. I: 299

casoron [1194-65-6] Vol. I: 299

casoron-133- [1194-65-6] Vol. I: 299

casoron g-10 [1194-65-6] Vol. I: 299

casoron g-4 [1194-65-6] Vol. I: 299

casoron gsr [1194-65-6] Vol. I: 299

casoron g20-sr [1194-65-6] Vol. I: 299

casoron w-50 [1194-65-6] Vol. I: 299

casul-70hf [26264-06-2] Vol. II: 209

Caswell no. 506- [78-59-1] Vol. II: 272

Caswell no. 043 [16919-19-0] Vol. II: 908

Caswell no. 049A [628-63-7] Vol. II: 27

Caswell no. 056 [7778-39-4] Vol. II: 91

Caswell no. 057- [1303-28-2] Vol. I: 199

Caswell no. 062 [8052-42-4] Vol. II: 114

Caswell no. 063 [1912-24-9] Vol. II: 1050

Caswell no. 093 [115-32-2] Vol. II: 144

Caswell no. 097 [7287-19-6] Vol. II: 1056

Caswell no. 106 [8052-42-4] Vol. II: 114

Caswell no. 109B [1303-86-2] Vol. III: 138

Caswell no. 125-d [886-50-0] Vol. II: 1045

Caswell no. 137 [7778-44-1] Vol. III: 65

Caswell no. 145- [7778-54-3] Vol. II: 494

Caswell no. 184 [139-40-2] Vol. III: 802

Caswell no. 185a [59-50-7] Vol. II: 258

Caswell No. 206 [101-10-0] Vol. III: 806

chemcolox-340 [60-00-4] **Vol. I:** 36

chemetron fire shield [1309-64-4] **Vol. II:** 86

chem-fish synergized [83-79-4] **Vol. II:** 166

chemform [123-33-1] **Vol. II:** 870

chemform [72-43-5] **Vol. II:** 345

chemical insecticide's isopropyl ester of 2,4-d liquid concentrate [94-11-1] **Vol. II:** 15

chemical mace [532-27-4] **Vol. I:** 89

chemichlon-g [7778-54-3] **Vol. II:** 494

chemifluor [7681-49-4] **Vol. II:** 925

chem-mite [83-79-4] **Vol. II:** 166

chem-neb [12427-38-2] **Vol. III:** 531

chemox [88-85-7] **Vol. II:** 704

chemox-pe [88-85-7] **Vol. II:** 704

chem pels c [7784-46-5] **Vol. I:** 208

chem phene [8001-35-2] **Vol. I:** 1118

chemrat [83-26-1] **Vol. III:** 774

chem-rice [26638-19-7] **Vol. II:** 298

chemsect dnoc [534-52-1] **Vol. I:** 540

chem sen 56 [7784-46-5] **Vol. I:** 208

chevron-9006 [10265-92-6] **Vol. II:** 744

Chevron-orthene [30560-19-1] **Vol. III:** 1

chevron-ortho-9006 [10265-92-6] **Vol. II:** 744

chexmate [75-60-5] **Vol. I:** 214

childion [116-29-0] **Vol. III:** 906

Chile saltpeter [7631-99-4] **Vol. III:** 864

chimiclor [15972-60-8] **Vol. II:** 9

chingamin [54-05-7] **Vol. III:** 229

chinine [130-95-0] **Vol. III:** 821

Chinin (German) [130-95-0] **Vol. III:** 821

chinoleine [91-22-5] **Vol. I:** 1022

chinolin (Czech) [91-22-5] **Vol. I:** 1022

chinoline [91-22-5] **Vol. I:** 1022

chinomethionat [2439-01-2] **Vol. III:** 663

chinomethionate [2439-01-2] **Vol. III:** 663

chinon (Dutch,German) [106-51-4] **Vol. I:** 303

chinone [106-51-4] **Vol. I:** 303

p Chinon (German) [106-51-4] **Vol. I:** 303

chinorto [311-45-5] **Vol. III:** 669

chinozan [82-68-8] **Vol. II:** 132

chinufur [1563-66-2] **Vol. I:** 420

Chiolite [15096-52-3] **Vol. III:** 37

chip cal [7778-44-1] **Vol. III:** 65

chip cal granular [7778-44-1] **Vol. III:** 65

chipco thiram-75 [137-26-8] **Vol. I:** 632

chipro-buctril [1689-84-5] **Vol. II:** 163

chiptox [94-74-6] **Vol. I:** 25

chlofenvinphos [470-90-6] **Vol. III:** 716

chlon [87-86-5] **Vol. I:** 928

chloorbenzeen (Dutch) [108-90-7] **Vol. I:** 245

2-chloor-1,3-butadieen (Dutch) [126-99-8] **Vol. I:** 336

chloordaan (Dutch) [57-74-9] **Vol. II:** 619

o-2-chloor-1-(2,4-dichloor-fenyl)-vinyl-o,o-diethylfosfaat (Dutch) [470-90-6] **Vol. III:** 716

(2-chloor-3-diethylamino-1-methyl-3-oxo-prop-1-en-yl)-dimethyl-fosfaat [Dutch] [13171-21-6] **Vol. III:** 704

chloor (Dutch) [7782-50-5] **Vol. I:** 446

1-chloor-2,3-epoxy propaan (Dutch) [106-89-8] **Vol. I:** 981

chloorethaan (Dutch) [75-00-3] **Vol. II:** 321

2-chloorethanol (Dutch) [107-07-3] **Vol. III:** 220

3-(4-(4-chloor-fenoxy)-fenoxy)-fenyl)-1,1-dimethylureum (Dutch) [1982-47-4] **Vol. III:** 232

3-(4-chloor-fenyl)-1,1-dimethylureum (Dutch) [150-68-5] **Vol. II:** 1075

N-(3-chloor-fenyl)-isopropyl carbamaat (Dutch) [101-21-3] **Vol. II:** 234

chloor-methaan (Dutch) [74-87-3] **Vol. II:** 585

1-chloor-4-nitrobenzeen (Dutch) [100-00-5] **Vol. III:** 210

O-(4-chloor-3-nitro-fenyl)-O, O-dimethylmo-nothiofosfaat (Dutch) [2463-84-5] **Vol. III:** 748

chloorpikrine (Dutch) [76-06-2] **Vol. II:** 606

chloorwaterstof (Dutch) [7647-01-0] **Vol. I:** 738

chloracetic acid [79-11-8] **Vol. I:** 22

chloracetone (French) [78-95-5] **Vol. II:** 828

chloracetyl chloride [79-04-9] **Vol. III:** 214

2-Chloraethanol (German) [107-07-3] **Vol. III:** 220

chloraethylen,-tri (German) [79-01-6] **Vol. II:** 427

2-Chloraethyl-phosphonsaeure (German) [16672-87-0] **Vol. III:** 426

chlorak [52-68-6] **Vol. III:** 707

chloraldural [302-17-0] **Vol. III:** 197

Chloraldurat (German) [302-17-0] **Vol. III:** 197

chloralex [302-17-0] **Vol. III:** 197

chlorali-hydras [302-17-0] **Vol. III:** 197

chlorallylene [107-05-1] Vol. II: 830

chloral,-monohydrate [302-17-0] Vol. III: 197

chloralvan [302-17-0] Vol. III: 197

chlorambed [133-90-4] Vol. III: 44

chloramben [133-90-4] Vol. III: 44

chlorambene [133-90-4] Vol. III: 44

Chlorameisensaeureaethylester (German) [541-41-3] Vol. III: 433

Chlorameisensaeure-methylester (German) [79-22-1] Vol. III: 597

chlorammonic (France) [12125-02-9] Vol. I: 141

4-chloranilin (Czech) [106-47-8] Vol. I: 169

p chloraniline [106-47-8] Vol. I: 169

o-chloraniline [95-51-2] Vol. III: 216

chloraquine [54-05-7] Vol. III: 229

chlorbenzene [108-90-7] Vol. I: 245

chlorbenzilat [510-15-6] Vol. III: 108

chlorbenzilate [510-15-6] Vol. III: 108

Chlorbenzol [108-90-7] Vol. I: 245

chlorbenzylate [510-15-6] Vol. III: 108

2-Chlor-1,3-butadien (German) [126-99-8] Vol. I: 336

chlorcyan [506-77-4] Vol. I: 561

chlordantoin [628-63-7] Vol. II: 27

(2-Chlor-3-diaethylamino-1-methyl-3-ossido-prop-1-en-yl)-dimethyl-phosphat [German] [13171-21-6] Vol. III: 704

o-2-Chlor-1-(2,4-dichlor-phenyl)-vinyl-o,o-diaethylphosphat (German) [470-90-6] Vol. III: 716

chlorea [150-68-5] Vol. II: 1075

chloreal [87-90-1] Vol. III: 936

chlore (French) [7782-50-5] Vol. I: 446

chlorene [75-00-3] Vol. II: 321

1-Chlor-2,3-epoxy propan (German) [106-89-8] Vol. I: 981

chlorethene [75-01-4] Vol. II: 416

chlorethephon [16672-87-0] Vol. III: 426

chlorethyl [75-00-3] Vol. II: 321

chlorethylene [75-01-4] Vol. II: 416

chlorex [111-44-4] Vol. II: 379

chlorextol [1336-36-3] Vol. I: 974

chlorfenidim [150-68-5] Vol. II: 1075

p-chlorfenol (Czech) [106-48-9] Vol. II: 710

p-chlorfenol (Czech) [95-57-8] Vol. II: 707

chlorfenvinfos [470-90-6] Vol. III: 716

chlorfenvinphos [470-90-6] Vol. III: 716

chlorfos [52-68-6] Vol. III: 707

Chlor (German) [7782-50-5] Vol. I: 446

chlorhydrate de nicotine (French) [2820-51-1] Vol. II: 652

chlorid ammonia (Czech) [12125-02-9] Vol. I: 141

chlorid amonny (Czech) [12125-02-9] Vol. I: 141

chlorid antimonity (Czech) [10025-91-9] Vol. I: 182

chlorid-fenylrtutnaty (Czech) [100-56-1] Vol. III: 699

chlorid-rtutnaty (Czech) [7487-94-7] Vol. II: 539

chlorierte-biphenyle, chlorgehalt-54% (German) [11097-69-1] Vol. II: 806

chlor-ifc [101-21-3] Vol. II: 234

chlorilen [79-01-6] Vol. II: 427

chlorinated biphenyl [1336-36-3] Vol. I: 974

chlorinated camphene [8001-35-2] Vol. I: 1118

chlorinated diphenyl [1336-36-3] Vol. I: 974

chlorinated diphenylene [1336-36-3] Vol. I: 974

chlorindan [57-74-9] Vol. II: 619

chlorine cyanide [506-77-4] Vol. I: 561

chlorine mol [7782-50-5] Vol. I: 446

chlorine of lime [7778-54-3] Vol. II: 494

chlor-ipc [101-21-3] Vol. II: 234

chlorisol [7722-64-7] Vol. I: 915

p-chlor-m-cresol [59-50-7] Vol. II: 258

chlor-methan (German) [74-87-3] Vol. II: 585

4-(4-Chlor-2-methyl-phenoxy)-buttersaeure [German] [94-81-5] Vol. III: 174

4-(4-Chlor-2-methylphenoxy)-buttersaeure [German] [94-81-5] Vol. III: 174

3-Chlor-2-methyl-prop-1-en (German) [563-47-3] Vol. III: 202

1-Chlor-4-nitrobenzol (German) [100-00-5] Vol. III: 210

O-(4-Chlor-3-nitro-phenyl) -O,O-dimethyl-monothiophosphat (German) [2463-84-5] Vol. III: 748

2-chloro- [95-57-8] Vol. II: 707

4-chloro- [106-48-9] Vol. II: 710

o-chloro- [95-57-8] Vol. II: 707

2-chloroacetaldehyde [107-20-0] Vol. II: 1

chloroacetaldehyde monomer [107-20-0] Vol. II: 1

alpha chloroaceticacid [79-11-8] Vol. I: 22

chloroacetic acid chloride [79-04-9] Vol. III: 214

chloroacetic acid, solid (DOT) [79-11-8]
Vol. I: 22

chloroacetic chloride [79-04-9] Vol. III:
214

1-chloroacetone [78-95-5] Vol. II: 828

chloroacetone [78-95-5] Vol. II: 828

alpha-chloroacetone [78-95-5] Vol. II: 828

1-chloroacetophenone [532-27-4] Vol. I: 89

chloroacetophenone [532-27-4] Vol. I: 89

alpha chloroacetophenone [532-27-4] Vol.
I: 89

chloroaethan (German) [75-00-3] Vol. II:
321

chloroaldehyde [107-20-0] Vol. II: 1

alpha chloroallylchloride [26952-23-8] Vol.
I: 1007

2-chloroallyl chloride [78-88-6] Vol. III:
805

chloroalonil [1897-45-6] Vol. II: 513

4-chloro-alpha-(4-chlorophenyl)-alpha-hy-
droxybenzeneacetic acid ethyl ester
[510-15-6] Vol. III: 108

4-chloro-alpha-(4-chlorophenyl)-alpha-(trichloromethyl) benzenemethanol [115-32-
2] Vol. II: 144

4-chloro-alpha-(4-chlorophenyl)-alpha-(trichloromethyl) benzyl alcohol [115-32-2]
Vol. II: 144

4-chloro-alpha-(1-methylethyl) benzeneacetic acid cyano (3-phenoxyphenyl)
methyl ester [51630-58-1] Vol. III: 95

4-chloro-alpha-phenyl-o-cresol [120-32-1]
Vol. III: 122

4-chloro-alpha-phenyl-ortho-cresol [120-
32-1] Vol. III: 122

p chloroaniline [106-47-8] Vol. I: 169

o-chloroaniline [95-51-2] Vol. III: 216

chloroben [95-50-1] Vol. I: 250

chlorobenzal [98-87-3] Vol. III: 84

alpha chlorobenzaldehyde [98-88-4] Vol. I:
307

(o-chlorobenzal) malononitrile [2698-41-1]
Vol. III: 218

4-chlorobenzamine [106-47-8] Vol. I: 169

4-chlorobenzenamine [106-47-8] Vol. I:
169

chlorobenzen (Polish) [108-90-7] Vol. I:
245

chlorobenzol [108-90-7] Vol. I: 245

4-chlorobenzoyl peroxide [94-17-7] Vol. III:
685

p-chlorobenzoyl peroxide [94-17-7] Vol. III:
685

S-(p-chlorobenzyl) diethylthiocarbamate
[28249-77-6] Vol. II: 213

S-4-chlorobenzyl diethyl (thiocarbamate)
[28249-77-6] Vol. II: 213

S-4-chlorobenzyl diethylthiocarbamate
[28249-77-6] Vol. II: 213

o-chlorobenzylidene malonitrile [2698-41-
1] Vol. III: 218

2-chlorobenzylidene malononitrile [2698-
41-1] Vol. III: 218

S-(4-chlorobenzyl) N,N-diethylthiocarba-
mate [28249-77-6] Vol. II: 213

(S-(4-chlorobenzyl) N,N-diethylthiolcarba-
mate) [28249-77-6] Vol. II: 213

4-chloro-2-benzylphenol [120-32-1] Vol.
III: 122

1-chloro-2,-(beta-chloroethoxy) ethane
[111-44-4] Vol. II: 379

1-chloro-2-(beta-chloroethylthio) ethane
[505-60-2] Vol. II: 962

chloro 1,1-biphenyl [1336-36-3] Vol. I:
974

chloro biphenyl [1336-36-3] Vol. I: 974

2-chloro-4,6-bis (ethylamino)-s-triazine
[122-34-9] Vol. II: 1047

2-chloro-4,6-bis (ethylamino)-1,3,5-triazine
[122-34-9] Vol. II: 1047

1-chloro, 3,5-bisethylamino-2,4,6-triazine
[122-34-9] Vol. II: 1047

2-chloro-4,6-bis (isopropylamino) -s-tria-
zine [139-40-2] Vol. III: 802

chloroble-m [12427-38-2] Vol. III: 531

2-chlorobmn [2698-41-1] Vol. III: 218

chlorobrom [74-97-5] Vol. II: 583

chlorobutadiene [126-99-8] Vol. I: 336

2-chlorobutadiene-1,3 [126-99-8] Vol. I:
336

2-chlorobuta-1,3-diene [126-99-8] Vol. I:
336

2-chlorobutadiene [126-99-8] Vol. I: 336

1-chlorobutane [109-69-3] Vol. I: 344

chlorocamphene [8001-35-2] Vol. I: 1118

m-chlorocarbanilic acid, isopropyl ester
[101-21-3] Vol. II: 234

3-chlorocarbanilic acid, isopropyl ester
[101-21-3] Vol. II: 234

chlorocarbonate de methyle (French) [79-
22-1] Vol. III: 597

chlorocarbonate d'ethyle (French) [541-41-
3] Vol. III: 433

chlorocarbonic acid ethyl ester [541-41-3]
Vol. III: 433

chlorocarbonic acid methyl ester [79-22-1] Vol. III: 597

chlorochin [54-05-7] Vol. III: 229

3-chlorochlordene [76-44-8] Vol. II: 628

chloro (chloromethoxy) methane [542-88-1] Vol. II: 384

chlorochromic anhydride [14977-61-8] Vol. II: 247

chloro-3-cresol [59-50-7] Vol. II: 258

chlorocresol [59-50-7] Vol. II: 258

p-chlorocresol [59-50-7] Vol. II: 258

chlorocresolo [59-50-7] Vol. II: 258

chlorocresolum (latin) [59-50-7] Vol. II: 258

chlorocyan [506-77-4] Vol. I: 561

chlorocyanide [506-77-4] Vol. I: 561

chlorocyanogen [506-77-4] Vol. I: 561

2-chloro-4-(1-cyano-1-methylethyl amino)-6-ethylamine-1,3,5-triazine [21725-46-2] Vol. II: 854

2-chloro-4-(1-cyano-1-methylethyl amino)-6-ethylamino-s-triazine [21725-46-2] Vol. II: 854

3-chloro-1,2-dibromopropane [96-12-8] Vol. II: 818

1-chloro-2,3-dibromopropane [96-12-8] Vol. II: 818

1-chloro-2,2-dichloroethylene [79-01-6] Vol. II: 427

2-chloro-1-(2,4-dichlorophenyl)-ethenyl diethyl phosphate [470-90-6] Vol. III: 716

2-chloro-1-(2,4-dichlorophenyl) vinyldiethylphosphate [470-90-6] Vol. III: 716

2-chloro-1-(2,4-dichlorophenyl) vinyl diethyl phosphate [470-90-6] Vol. III: 716

N-(2-chloro-1-(diethoxyphosphinothioylthio)ethyl)- phthalimide [10311-84-9] Vol. III: 728

7-chloro-4-(4-diethylamino-1- methylbutylamino) quinoline [54-05-7] Vol. III: 229

(7-chloro-4-(4-diethylamino-1-methylbutylamino) quinoline [54-05-7] Vol. III: 229

2-chloro-3-(diethylamino) -1-methyl-3-oxo-1-propenyl dimethyl phosphate [13171-21-6] Vol. III: 704

2-chloro-2-diethylcarbamoyl-1-methylvinyl dimethyl phosphate [13171-21-6] Vol. III: 704

1-chloro-diethylcarbamoyl-1-propen-2-yl dimethyl phosphate [13171-21-6] Vol. III: 704

2-chloro-2-diethylcarbamyl-1-methylvinyl-dimethyl phosphate [13171-21-6] Vol. III: 704

2-chloro-2',6'-diethyl-N-(methoxy methyl) acetanilide [15972-60-8] Vol. II: 9

alpha-chloro-2', 6'-diethyl-N-methoxymethylacetanilide [15972-60-8] Vol. II: 9

2-chloro-2', 6'-diethyl-N-methoxy methylacetanilide [15972-60-8] Vol. II: 9

S-(2-chloro-1-(1,3-dihydro-1,3-dioxo-2h-isoindol-2-yl)ethyl)o,o-diethyl phosphorodithioate [10311-84-9] Vol. III: 728

S-(2-chloro-1-(1,3-dihydro-1,3-dioxy-2H-isoindol-2-yl)ethyl)o,o-diethyl phosphorodithioate [10311-84-9] Vol. III: 728

2-chloro-4,6-di(isopropylamino) -1,3,5-triazine [139-40-2] Vol. III: 802

7-chloro-4,6-dimethoxycoumaran-3-one-2-spiro-1'-(2'-methoxy-6'-methylcyclohex-2'-en-4'-one) Vol. III: [126-07-8], 472

chlorodimethyl ether [107-30-2] Vol. II: 390

chlorodimethylether (Czech) [107-30-2] Vol. II: 390

5-chloro-3-(1,1-dimethylethyl)-6-methyl-2,4 (1H, 3H)-pyrimidinedione [5902-51-2] Vol. III: 894

chlorodiphenyl (60%Cl) [11096-82-5] Vol. II: 808

3-chloro-1,2-epoxypropane [106-89-8] Vol. I: 981

1-chloro-2,3-epoxypropane [106-89-8] Vol. I: 981

chloroethanal [107-20-0] Vol. II: 1

2-chloro-1-ethanal [107-20-0] Vol. II: 1

chloroethane [75-00-3] Vol. II: 321

chloroethane-NU [71-55-6] Vol. I: 684

2-chloroethanephosphonic acid [16672-87-0] Vol. III: 426

chloroethanoic acid [79-11-8] Vol. I: 22

chloroethanol [107-07-3] Vol. III: 220

2-chloro-1-ethanol [107-07-3] Vol. III: 220

chloroethene [71-55-6] Vol. I: 684

chloroethene [75-01-4] Vol. II: 416

2-chloroethyl alcohol [107-07-3] Vol. III: 220

2-chloro-4-ethylamineisopropyl amine-s-triazine [1912-24-9] Vol. II: 1050

1-chloro-3-ethylamino-5-isopropyl amino-s-triazine [1912-24-9] Vol. II: 1050

2-chloro-4-ethylamino-6-isopropyl amino-s-triazine [1912-24-9] Vol. II: 1050

2-chloro-4-(ethylamino)-6-(isopropyl amino)-s-triazine [1912-24-9] Vol. II: 1050

2-chloro-4-ethylamino-6-isopropyl amino-1,3,5-triazine [1912-24-9] Vol. II: 1050

p-chlorophenylamine [106-47-8] **Vol. I:** 169

2-chlorophenylamine [95-51-2] **Vol. III:** 216

4-chlorophenylamine [106-47-8] **Vol. I:** 169

(N-((4-chlorophenyl) amino) carbonyl)-)2,6-difluorobenzamide [35367-38-5] **Vol. III:** 339

N-(((4-chlorophenyl) amino) carbonyl)-)2,6-difluorobenzamide [35367-38-5] **Vol. III:** 339

N-((4-chlorophenyl) amino)-)2,6-difluoro-benzamide [35367-38-5] **Vol. III:** 339

N-(3-chloro phenyl) carbamate d'isopropyle (French) [101-21-3] **Vol. II:** 234

N-(3-chlorophenyl) carbamic acid, isopropyl ester [101-21-3] **Vol. II:** 234

(3-chlorophenyl) carbamic acid, 1-methylethyl ester [101-21-3] **Vol. II:** 234

N-(4-chlorophenylcarbamoyl)-)2,6-difluoro-benzamide [35367-38-5] **Vol. III:** 339

p chlorophenylchloride [106-46-7] **Vol. I:** 254

2,2-bis (4-chlorophenyl)-1,1-dichloroethane [72-54-8] **Vol. III:** 404

2,2-bis (p-chlorophenyl)-1,1- dichloroethane [72-54-8] **Vol. III:** 404

1,1-bis (4-chlorophenyl)-2,2-dichloroethane [72-54-8] **Vol. III:** 404

1,1-bis (p-chlorophenyl) -2,2-dichloroethane [72-54-8] **Vol. III:** 404

2,2-bis (4-chlorophenyl) 1,1-dichloroethene [72-55-9] **Vol. II:** 414

2,2-bis (4-chlorophenyl)-1,1-dichloro ethylene [72-55-9] **Vol. II:** 414

2,2-bis (p-chlorophenyl)-1,1-dichloro ethylene [72-55-9] **Vol. II:** 414

1,1-bis (p-chlorophenyl)-2,2-dichloro ethylene [72-55-9] **Vol. II:** 414

1-(4-chlorophenyl)-)3-(2,6- difluorobenzoyl) urea [35367-38-5] **Vol. III:** 339

1-(4-chlorophenyl)-)3-(2,6-difluorobenzoyl) urea [35367-38-5] **Vol. III:** 339

1-(p-chlorophenyl)-)3-(2,6-difluorobenzoyl) urea [35367-38-5] **Vol. III:** 339

3'-(4'-chlorophenyl)-1,1-dimethylurea [150-68-5] **Vol. II:** 1075

3-(4-chlorophenyl)-1,1-dimethylurea [150-68-5] **Vol. II:** 1075

3-(p-chlorophenyl)-1,1-dimethylurea [150-68-5] **Vol. II:** 1075

1-(4-chlorophenyl)-3,3-dimethylurea [150-68-5] **Vol. II:** 1075

1-(p-chlorophenyl)-3,3-dimethylurea [150-68-5] **Vol. II:** 1075

1-(4-chloro phenyl)-3,3-dimethyluree (French) [150-68-5] **Vol. II:** 1075

N-3-chlorophenylisopropylcarbamate [101-21-3] **Vol. II:** 234

chlorophenylmercury [100-56-1] **Vol. III:** 699

chlorophenylmethane [100-44-7] **Vol. I:** 1099

S-((4-chlorophenyl) methyl) diethyl carbamothioate [28249-77-6] **Vol. II:** 213

S((4-chlorophenyl) methyl) diethylcarbamothioate [28249-77-6] **Vol. II:** 213

((2-chloro-phenyl) methylene) propanenitrile [2698-41-1] **Vol. III:** 218

4-chloro-2-(phenylmethyl) phenol [120-32-1] **Vol. III:** 122

N-(p-chlorophenyl)-N', N'-dimethyl urea [150-68-5] **Vol. II:** 1075

N'-(4-chlorophenyl)-N, N-dimethyl urea [150-68-5] **Vol. II:** 1075

N-(4-chlorophenyl)-N', N'-dimethylurea [150-68-5] **Vol. II:** 1075

1,1-bis (p-chlorophenyl)-2,2,2-trichloro ethanol [115-32-2] **Vol. II:** 144

1,1-bis (4-chlorophenyl)-2,2,2-trichloro ethanol [115-32-2] **Vol. II:** 144

p-chlorophenyl 2,4,5-trichlorophenyl sulfone [116-29-0] **Vol. III:** 906

4-chlorophenyl, 2,4,5-trichlorophenyl sulfone [116-29-0] **Vol. III:** 906

chlorophos [52-68-6] **Vol. III:** 707

S-(2-chloro-1-phthalimidoethyl) o,o-diethyl phosphorodithioate [10311-84-9] **Vol. III:** 728

S-2-chloro-1-phthalimidoethyl o,o-diethyl phosphorodithioate [10311-84-9] **Vol. III:** 728

chlorophthalm [52-68-6] **Vol. III:** 707

S-(2-chloro-1-phthalmido- ethyl)-o,o-diethylphosphoro- thionate [10311-84-9] **Vol. III:** 728

chlor-o-pic [76-06-2] **Vol. II:** 606

chloropicrine (French) [76-06-2] **Vol. II:** 606

chloropicrin, liquid (DOT) [76-06-2] **Vol. II:** 606

chloropreen (Dutch) [126-99-8] **Vol. I:** 336

chloroprene [126-99-8] **Vol. I:** 336

3-chloroprene [107-05-1] **Vol. II:** 830

chloropren (German, Polish) [126-99-8] **Vol. I:** 336

chlorure de butyle (French) [109-69-3] **Vol. I:** 344

chlorure de chloracetyle (French) [79-04-9] **Vol. III:** 214

chlorure de chromyle (French) [14977-61-8] **Vol. II:** 247

chlorure de cyanogene (French) [506-77-4] **Vol. I:** 561

chlorure de methallyle (French) [563-47-3] **Vol. III:** 202

chlorure de methyle (French) [74-87-3] **Vol. II:** 585

chlorure de methylene (French) [75-09-2] **Vol. I:** 820

chlorure d'ethyle (French) [75-00-3] **Vol. II:** 321

chlorure d'ethylene (French) [107-06-2] **Vol. I:** 643

chlorure d'ethylidene (French) [75-34-3] **Vol. II:** 325

chlorure de vinyle (French) [75-01-4] **Vol. II:** 416

chlorure-de-vinylidene (French) [75-35-4] **Vol. II:** 409

chlorure de zinc (French) [7646-85-7] **Vol. I:** 1178

chlorure d'hydrogene anhydre (French) [7647-01-0] **Vol. I:** 738

chlorure d'hydrogene (French) [7647-01-0] **Vol. I:** 738

chlorure ferrique (French) [7705-08-0] **Vol. I:** 699

chlorure mercurique (French) [7487-94-7] **Vol. II:** 539

chloruro de hidrogeno (Spanish) [7647-01-0] **Vol. I:** 738

chlorvinphos [62-73-7] **Vol. I:** 946

Chlorwasserstoff (German) [7647-01-0] **Vol. I:** 738

chloryl [75-00-3] **Vol. II:** 321

chloryl anesthetic [75-00-3] **Vol. II:** 321

chlorylea [79-01-6] **Vol. II:** 427

chlorylen [71-55-6] **Vol. I:** 684

chorylen [79-01-6] **Vol. II:** 427

chp [80-15-9] **Vol. II:** 490

chromate of soda [7775-11-3] **Vol. I:** 469

chrome [7440-47-3] **Vol. I:** 476

chrome (French) [7440-47-3] **Vol. I:** 476

chrome green [1308-38-9] **Vol. III:** 234

chrome ocher [1308-38-9] **Vol. III:** 234

chrome oxide [1308-38-9] **Vol. III:** 234

chrome oxide green gn-m [1308-38-9] **Vol. III:** 234

chrome (trioxydede) (French) [1333-82-0] **Vol. I:** 493

Chrom (German) [7440-47-3] **Vol. I:** 476

chromia [1308-38-9] **Vol. III:** 234

chromic acetate [1066-30-4] **Vol. I:** 26

chromic acetate (iii) [1066-30-4] **Vol. I:** 26

chromic acid, diammonium salt [7789-09-5] **Vol. I:** 590

chromic acid, dipotassium salt [7789-00-6] **Vol. I:** 467

chromic acid, disodium salt [7775-11-3] **Vol. I:** 469

chromic acid green [1308-38-9] **Vol. III:** 234

chromic acid (H2CrO4), dilithium salt [14307-35-8] **Vol. II:** 242

chromic acid (H2Cr2O7),-disodium salt [10588-01-9] **Vol. II:** 300

chromic acid (H_2CrO_4), zinc salt (1:1) [13530-65-9] **Vol. III:** 984

chromic acid, solid [1333-82-0] **Vol. I:** 493

chromic acid, strontium salt (1:1) [7789-06-2] **Vol. I:** 473

chromic acid, zinc salt [13530-65-9] **Vol. III:** 984

chromic anhydride [7738-94-5] **Vol. II:** 239

chromic arthydride [1333-82-0] **Vol. I:** 493

chromic oxide [1333-82-0] **Vol. I:** 493

chromic oxychloride [14977-61-8] **Vol. II:** 247

chromic sulfate [10101-53-8] **Vol. I:** 482

chromic sulphate [10101-53-8] **Vol. I:** 482

chromic (vi) acid [1333-82-0] **Vol. I:** 493

chromic(VI) acid [7738-94-5] **Vol. II:** 239

chromitan b [10101-53-8] **Vol. I:** 482

chromitan ms [10101-53-8] **Vol. I:** 482

chromitan Na [10101-53-8] **Vol. I:** 482

chromium acetate [1066-30-4] **Vol. I:** 26

chromium chloride [10049-05-5] **Vol. I:** 479

chromium chloride oxide [14977-61-8] **Vol. II:** 247

chromium chloride oxide (CRO2C12) [14977-61-8] **Vol. II:** 247

chromium dichloride [10049-05-5] **Vol. I:** 479

chromium dichloride dioxide [14977-61-8] **Vol. II:** 247

chromium, dichlorodioxo- [14977-61-8] **Vol. II:** 247

chromium dioxide dichloride [14977-61-8] Vol. II: 247

chromium dioxychloride [14977-61-8] Vol. II: 247

chromium disodium oxide [7775-11-3] Vol. I: 469

chromium (ii) chloride [10049-05-5] Vol. I: 479

chromium (ii) chloride (1:2) [10049-05-5] Vol. I: 479

chromium (III) oxide (2: 3) [1308-38-9] Vol. III: 234

chromium iii sulfate [10101-53-8] Vol. I: 482

chromium (iii) sulfate (2:3) [10101-53-8] Vol. I: 482

chromium lithium oxide (CrLi2O4) [14307-35-8] Vol. II: 242

chromium oxide [1308-38-9] Vol. III: 234

chromium oxide [1333-82-0] Vol. I: 493

chromium (3) oxide [1308-38-9] Vol. III: 234

chromium oxide (Cr2O3) [1308-38-9] Vol. III: 234

chromium oxide green [1308-38-9] Vol. III: 234

chromium oxide greens [1308-38-9] Vol. III: 234

chromium oxide pigment [1308-38-9] Vol. III: 234

chromium oxide-X1134 [1308-38-9] Vol. III: 234

chromium oxychloride (CRO2C12) [14977-61-8] Vol. II: 247

chromium sesquioxide [1308-38-9] Vol. III: 234

chromium sodium oxide [7775-11-3] Vol. I: 469

chromium sodium oxide (Cr3Na2O7) [10588-01-9] Vol. II: 300

chromium sulfate [10101-53-8] Vol. I: 482

chromium sulphate [10101-53-8] Vol. I: 482

chromium sulphate (2:3) [10101-53-8] Vol. I: 482

chromium triacetate [1066-30-4] Vol. I: 26

chromium trioxide [1333-82-0] Vol. I: 493

chromium (3) trioxide [1308-38-9] Vol. III: 234

chromium (6) trioxide [1333-82-0] Vol. I: 493

chromium trioxide [7738-94-5] Vol. II: 239

chromium(VI) dioxychloride [14977-61-8] Vol. II: 247

chromium (vi) oxide [1333-82-0] Vol. I: 493

chromoxychlorid (German) [14977-61-8] Vol. II: 247

chromozin [1912-24-9] Vol. II: 1050

chromsaeureanhydrid (German) [1333-82-0] Vol. I: 493

Chromtrioxid (German) [1333-82-0] Vol. I: 493

chromyl chloride (CRO2C12) [14977-61-8] Vol. II: 247

chromylchlorid (German) [14977-61-8] Vol. II: 247

chroomoxychloride (Dutch) [14977-61-8] Vol. II: 247

chroomtrioxyde (Dutch) [1333-82-0] Vol. I: 493

chroomzuuranhydride (Dutch) [1333-82-0] Vol. I: 493

N-(chrysanthemoxymethyl)-1-cyclohexene-1,2-dicarboximide [7696-12-0] Vol. III: 912

chrysanthemumdicarboxylic acid monomethyl ester pyrethrolone ester [121-29-9] Vol. III: 817

chryson [10453-86-8] Vol. III: 285

chwastox [94-74-6] Vol. I: 25

ci-7811 [16871-71-9] Vol. II: 911

ci-7811 [7631-86-9] Vol. III: 846

ci-10305 [88-89-1] Vol. II: 793

ci-11000 [60-09-3] Vol. II: 74

ci-11160 [97-56-3] Vol. III: 46

ci-23060- [91-94-1] Vol. I: 285

ci-37035 [100-01-6] Vol. II: 70

ci-37107 [106-49-0] Vol. II: 1040

ci-37107 [106-49-0] Vol. III: 927

ci-37230- [119-93-7] Vol. III: 364

ci-37265 [134-32-7] Vol. III: 637

ci-76010 [95-54-5] Vol. III: 697

ci-76025- [108-45-2] Vol. III: 98

ci-76035 [95-80-7] Vol. II: 1016

ci-76500 [120-80-9] Vol. III: 190

ci-76515 [87-66-1] Vol. II: 882

ci-77002 [21645-51-2] Vol. II: 57

ci-77002 [21645-51-2] Vol. III: 35

ci-77052 [1309-64-4] Vol. II: 86

C.I. 77085 [1303-32-8] Vol. III: 75

ci-77086 [1303-33-9] Vol. II: 99

ci-77180- [7440-43-9] Vol. I: 370

ci-77288 [1308-38-9] Vol. III: 234

cis-9, cis-12-octadecadienoic acid [60-33-3] Vol. II: 667

cis, cis-9, 12-oetadeeadienole acid [60-33-3] Vol. II: 667

cis-ddp [15663-27-1] Vol. III: 237

cis delta (9) octadecenoic acid [112-80-1] Vol. I: 895

cis delta (sup(9)octadecenoic acid [112-80-1] Vol. I: 895

cis-diaminodichloroplatinum [15663-27-1] Vol. III: 237

cis-diaminodichloroplatinum(II) [15663-27-1] Vol. III: 237

cis-diamminedichloroplatinum(II) [15663-27-1] Vol. III: 237

cis-diammineplatinum(II) chloride [15663-27-1] Vol. III: 237

cis-dichlorodiaminoplatinum [15663-27-1] Vol. III: 237

cis-dichlorodiaminoplatinum(II) [15663-27-1] Vol. III: 237

cis-dichlorodiammineplatinum [15663-27-1] Vol. III: 237

cis-dichlorodiammine platinum(II) [15663-27-1] Vol. III: 237

cis-dichlorodiammineplatinum(II) [15663-27-1] Vol. III: 237

cis-2-dimethylcarbamoyl-1-methylvinyl dimethylphosphate [141-66-2] Vol. III: 720

cis-1,2-ethylenedicarboxylic acid [110-16-7] Vol. II: 528

cislin [52918-63-5] Vol. III: 296

cis-N-triehloromethylthio-4-cyclo hexene-1,2-diearboximide [133-06-2] Vol. II: 276

cis-9-octadecenoicacid [112-80-1] Vol. I: 895

cis octadec-9-enoicacid [112-80-1] Vol. I: 895

cis oleicacid [112-80-1] Vol. I: 895

ci-solvent-yellow-1 [60-09-3] Vol. II: 74

ci-solvent-yellow-3- [97-56-3] Vol. III: 46

cis-2-(1-phenylethoxy) carbonyl-1-methylvinyl dimethylphosphate [7700-17-6] Vol. III: 266

cis-platin [15663-27-1] Vol. III: 237

cisplatin [15663-27-1] Vol. III: 237

cisplatine [15663-27-1] Vol. III: 237

cisplatino (Spanish) [15663-27-1] Vol. III: 237

cis-platinous diaminodichloride [15663-27-1] Vol. III: 237

cis-platinous diammine dichloride [15663-27-1] Vol. III: 237

cis-platinum [15663-27-1] Vol. III: 237

cis-platinum diaminodichloride [15663-27-1] Vol. III: 237

cis-platinum diamminedichloride [15663-27-1] Vol. III: 237

cis-platinum(II) [15663-27-1] Vol. III: 237

cis-platinum(II) diaminedichloride [15663-27-1] Vol. III: 237

cis-platinum(II) diaminodichloride [15663-27-1] Vol. III: 237

cis-platinum(II) diamminedichloride [15663-27-1] Vol. III: 237

cisplatyl [15663-27-1] Vol. III: 237

cispt(II) [15663-27-1] Vol. III: 237

cis, trans-1,3-dichloropropene [542-75-6] Vol. II: 834

citretten [77-92-9] Vol. I: 494

citric acid, ammonium iron (3) salt [1185-57-5] Vol. I: 491

citric acid, anhydrous [77-92-9] Vol. I: 494

citrine ointment [10045-94-0] Vol. II: 554

citro [77-92-9] Vol. I: 494

ckb-1220 [10265-92-6] Vol. II: 744

cl-12880 [60-51-5] Vol. III: 740

cl-47470 [950-10-7] Vol. III: 536

clairsit [594-42-3] Vol. II: 613

clarosan [886-50-0] Vol. II: 1045

claudelite [1327-53-3] Vol. I: 203

claudetite [1327-53-3] Vol. I: 203

clear-eyes [8001-54-5] Vol. II: 67

clf-II [7440-44-0] Vol. III: 179

cl-ifk [101-21-3] Vol. II: 234

4,6-clinitrocresol (Dutch) [534-52-1] Vol. I: 540

clofenvinfos [470-90-6] Vol. III: 716

clophen [1336-36-3] Vol. I: 974

cloprop [101-10-0] Vol. III: 806

clopropane carboxylic acid, 3-(2,2-dichloroethenyl)-2,2-dimethyl, cyano (3-phenoxyphenyl) methylester, (+-)- [52315-07-8] Vol. II: 251

cloralio (Italian) [75-87-6] Vol. I: 442

clordan (Italian) [57-74-9] Vol. II: 619

clordecone [143-50-0] Vol. III: 587

cloretilo [75-00-3] Vol. II: 321

clorex [111-44-4] Vol. II: 379

cloroben [95-50-1] Vol. I: 250

clorobenzene (Italian) [108-90-7] Vol. I: 245

2-cloro-1,3-butadiene (Italian) [126-99-8] Vol. I: 336

clorocresol (Spanish) [59-50-7] **Vol. II:** 258

o-2-cloro-1-(2,4-dicloro-fenil)-vinil-o,o-dietil-fosfato (Italian) [470-90-6] **Vol. III:** 716

(2-cloro-3-dietilamino-1-metil-3-oxo-prop-1-en-il)-dimetil-fosfato [Italian] [13171-21-6] **Vol. III:** 704

clorodifenili, cloro-54%-(Italian) [11097-69-1] **Vol. II:** 806

1-cloro-2,3-epossipropano (Italian) [106-89-8] **Vol. I:** 981

cloroetano (Italian) [75-00-3] **Vol. II:** 321

2-cloroetanolo [Italian] [107-07-3] **Vol. III:** 220

clorofene [120-32-1] **Vol. III:** 122

3-(4-cloro-fenil)-1,1-dimetil-urea (Italian) [150-68-5] **Vol. II:** 1075

N-(3-cloro-fenil)-isopropil-carbammato (Italian) [101-21-3] **Vol. II:** 234

3-(4-(4-cloro-fenossil)-1,1-dimetil-urea (Italian) [1982-47-4] **Vol. III:** 232

cloroformiato de etilo (Spanish) [541-41-3] **Vol. III:** 433

cloroformio (Italian) [67-66-3] **Vol. I:** 453

cloro (Italian and Spanish) [7782-50-5] **Vol. I:** 446

clorometano (Italian) [74-87-3] **Vol. II:** 585

3-cloro-2-metil-prop-1-ene (Italian) [563-47-3] **Vol. III:** 202

1-cloro-4-nitrobenzene (Italian) [100-00-5] **Vol. III:** 210

O-(4-cloro-3-nitro-fenil) -O, O-dimetil-mo-notiofosfato (Italian) [2463-84-5] **Vol. III:** 748

clorophene [120-32-1] **Vol. III:** 122

cloropicrina (Italian) [76-06-2] **Vol. II:** 606

cloropool [7681-52-9] **Vol. II:** 498

cloroprene (Italian) [126-99-8] **Vol. I:** 336

clorox [7681-52-9] **Vol. II:** 498

clorox liquid bleach [7681-52-9] **Vol. II:** 498

clorthalonil (German) [1897-45-6] **Vol. II:** 513

clortosip [1897-45-6] **Vol. II:** 513

cloruro de amonio [12125-02-9] **Vol. I:** 141

cloruro de bencilideno (Spanish) [98-87-3] **Vol. III:** 84

cloruro de hidrogeno anhidro (Spanish) [7647-01-0] **Vol. I:** 738

cloruro di ethene (Italian) [107-06-2] **Vol. I:** 643

cloruro di etile (Italian) [75-00-3] **Vol. II:** 321

cloruro di etilidene (Italian) [75-34-3] **Vol. II:** 325

cloruro di metallile (Italian) [563-47-3] **Vol. III:** 202

cloruro di metile (Italian) [74-87-3] **Vol. II:** 585

cloruro di vinile (Italian) [75-01-4] **Vol. II:** 416

cmb-200 [7440-44-0] **Vol. III:** 179

cmb-50 [7440-44-0] **Vol. III:** 179

cmu [150-68-5] **Vol. II:** 1075

cmu-weedkiller [150-68-5] **Vol. II:** 1075

CN [532-27-4] **Vol. I:** 89

cnv-weed-killer [150-68-5] **Vol. II:** 1075

co-12- [112-53-8] **Vol. II:** 316

co-1214 [112-53-8] **Vol. II:** 316

coalite ntp [25155-23-1] **Vol. I:** 1170

coal naphtha [71-43-2] **Vol. I:** 239

coal oil [8008-20-6] **Vol. II:** 522

coal oil fuel oil no, 1 [8008-20-6] **Vol. II:** 522

coal tar oil [8001-58-9] **Vol. I:** 495

coat b1400 [67-56-1] **Vol. I:** 836

cobalt(2) acetate [71-48-7] **Vol. III:** 5

cobalt acetate [71-48-7] **Vol. III:** 5

cobalt acetate (Co(OAc)₂) [71-48-7] **Vol. III:** 5

cobalt black [1307-96-6] **Vol. III:** 252

cobalt carbonate (1: 1) [513-79-1] **Vol. III:** 244

cobalt(2) carbonate (CoCO₃) [513-79-1] **Vol. III:** 244

cobalt carbonate (CoCO₃) [513-79-1] **Vol. III:** 244

cobalt chloride [7646-79-9] **Vol. I:** 501

cobalt diacetate [71-48-7] **Vol. III:** 5

cobalt dichloride [7646-79-9] **Vol. I:** 501

cobalt diformate [544-18-3] **Vol. III:** 248

cobalt dinitrate [10141-05-6] **Vol. I:** 504

cobalt formate [544-18-3] **Vol. III:** 248

cobalt(II) acetate [71-48-7] **Vol. III:** 5

cobalt (ii) chloride [7646-79-9] **Vol. I:** 501

cobalt(II) formate [544-18-3] **Vol. III:** 248

cobalt (ii) nitrate [10141-05-6] **Vol. I:** 504

cobalt (ii) nitrate (1:2) [10141-05-6] **Vol. I:** 504

cobalt(II) oxide [1307-96-6] **Vol. III:** 252

cobalt monooxide [1307-96-6] **Vol. III:** 252

cobalt monoxide [1307-96-6] **Vol. III:** 252

cobalt muriate [7646-79-9] **Vol. I:** 501

cobalt nitrate [10141-05-6] **Vol. I:** 504

cobalt (2) nitrate [10141-05-6] **Vol. I:** 504

copper, metallic powder [7440-50-8] **Vol. I:** 519

copper milled [7440-50-8] **Vol. I:** 519

copper m-l- [7440-50-8] **Vol. I:** 519

copper monosulfate [7758-98-7] **Vol. I:** 510

copper oxalate [5893-66-3] **Vol. III:** 259

copper powder [7440-50-8] **Vol. I:** 519

copperslag airborne [7440-50-8] **Vol. I:** 519

copperslag milled [7440-50-8] **Vol. I:** 519

copper (2) sulfate (1:1) [7758-98-7] **Vol. I:** 510

copper sulfate (1:1) [7758-98-7] **Vol. I:** 510

copper (2) sulfate [7758-98-7] **Vol. I:** 510

copper sulfate [7758-99-8] **Vol. II:** 255

copper sulfate (CuSO4)-pentahydrate [7758-99-8] **Vol. II:** 255

copper(2) sulfate pentahydrate [7758-99-8] **Vol. II:** 255

copper sulfate pentahydrate [7758-99-8] **Vol. II:** 255

copper sulfate powder [7758-98-7] **Vol. I:** 510

copper sulphate [7758-99-8] **Vol. II:** 255

copper tetraammine sulfate monohydrate [10380-29-7] **Vol. III:** 268

copper(2+), tetraammine sulfate(1:1) monohydrate [10380-29-7] **Vol. III:** 268

copper(2+), tetraammine-, sulfate, monohydrate [10380-29-7] **Vol. III:** 268

coprol [577-11-7] **Vol. III:** 859

corasil-ii [16871-71-9] **Vol. II:** 911

corasil-ii [7631-86-9] **Vol. III:** 846

cormate [14484-64-1] **Vol. III:** 489

cornox m [94-74-6] **Vol. I:** 25

corosul-d-and-S [7704-34-9] **Vol. II:** 965

corothion [56-38-2] **Vol. III:** 750

corrosive mercury chloride [7487-94-7] **Vol. II:** 539

corrosive sublimate [7487-94-7] **Vol. II:** 539

corsair [52645-53-1] **Vol. III:** 282

cortilan-neu [57-74-9] **Vol. II:** 619

co-1214s [112-53-8] **Vol. II:** 316

cosair [52645-53-1] **Vol. III:** 282

cosan [7704-34-9] **Vol. II:** 965

cosan-80- [7704-34-9] **Vol. II:** 965

cosan-I [133-07-3] **Vol. III:** 461

cosmetic hydrophobic green-9409 [1308-38-9] **Vol. III:** 234

cosmetic-hydrophobic-tio2-9428 [13463-67-7] **Vol. III:** 921

cosmetic micro-blend chrome oxide 9229 [1308-38-9] **Vol. III:** 234

cosmetic-micro-blend-tio2-9228 [13463-67-7] **Vol. III:** 921

cosmetic white c47-5175 [13463-67-7] **Vol. III:** 921

cosolv [79-01-6] **Vol. II:** 427

cotneon [86-50-0] **Vol. III:** 744

cotnion [86-50-0] **Vol. III:** 744

cotnion-methyl [86-50-0] **Vol. III:** 744

cotogard [2164-17-2] **Vol. II:** 1086

cotoran [2164-17-2] **Vol. II:** 1086

cotoron [2164-17-2] **Vol. II:** 1086

cottonex [2164-17-2] **Vol. II:** 1086

cotton tox mp 82 [8001-35-2] **Vol. I:** 1118

coumafene (French) [81-81-2] **Vol. I:** 525

coumarin,3-(alpha acetonylbenzyl) 4-hydroxy [81-81-2] **Vol. I:** 525

coumarin,4-hydroxy-3 (1-phenyl-3-oxo butyl) [81-81-2] **Vol. I:** 525

coumarin, 3,3'-methylenebis (4-hydroxy) [66-76-2] **Vol. III:** 336

cp-25 [108-88-3] **Vol. I:** 1109

cp-27 [108-90-7] **Vol. I:** 245

cp-32- [110-86-1] **Vol. II:** 874

cp-105 [106-91-2] **Vol. III:** 470

cp-15336 [2303-16-4] **Vol. II:** 218

cp-23426- [2303-17-5] **Vol. III:** 931

cp-31393 [1918-16-7] **Vol. III:** 789

cp-50144- [15972-60-8] **Vol. II:** 9

cpca [115-32-2] **Vol. II:** 144

1,3-cpd [77-73-6] **Vol. II:** 635

cpdc [15663-27-1] **Vol. III:** 237

cpdd [15663-27-1] **Vol. III:** 237

cr-3029 [12427-38-2] **Vol. III:** 531

crackdown [52918-63-5] **Vol. III:** 296

crag-1 [136-78-7] **Vol. III:** 855

crag-974 [533-74-4] **Vol. III:** 293

crag fungicide-974 [533-74-4] **Vol. III:** 293

crag herbicide [136-78-7] **Vol. III:** 855

crag herbicide-1- [136-78-7] **Vol. III:** 855

crag nemacide [533-74-4] **Vol. III:** 293

crag-ses [136-78-7] **Vol. III:** 855

crag sesone [136-78-7] **Vol. III:** 855

crag-85w [533-74-4] **Vol. III:** 293

cratecil [50-06-6] **Vol. I:** 223

crawhaspol [79-01-6] **Vol. II:** 427

credo [7681-49-4] **Vol. II:** 925

creme [107-30-2] **Vol. II:** 390

creosote [8001-58-9] **Vol. I:** 495

creosote, from coal tar [8001-58-9] **Vol. I:** 495

creosote oil [8001-58-9] Vol. I: 495

creosote pl [8001-58-9] Vol. I: 495

creosotum [8001-58-9] Vol. I: 495

2-cresol [95-48-7] Vol. I: 536

3-cresol [108-39-4] Vol. III: 261

m-cresol (ACGIH, DOT, OSHA) [108-39-4]
Vol. III: 261

m-cresol, 4-chloro [59-50-7] Vol. II: 258

m-cresol, 4-chloro- [59-50-7] Vol. II: 258

o-cresol, 4-chloro-alpha-phenyl [120-32-1]
Vol. III: 122

o cresol,4,6, dinitro [534-52-1] Vol. I: 540

cresol diphenyl phosphate [26444-49-5]
Vol. III: 723

m-cresole [108-39-4] Vol. III: 261

Crest [7783-47-3] Vol. III: 881

crestoxo [8001-35-2] Vol. I: 1118

cresyl diphenyl phosphate [26444-49-5]
Vol. III: 723

cresyldiphenyl phosphate [26444-49-5]
Vol. III: 723

m-cresylic acid [108-39-4] Vol. III: 261

o cresylicacid [95-48-7] Vol. I: 536

cresylic creosote [8001-58-9] Vol. I: 495

o-cresyl phosphate [78-30-8] Vol. III: 724

criofluorano (Spanish) [76-14-2] Vol. III:
408

crisapon [75-99-0] Vol. I: 1016

crisatrina [1912-24-9] Vol. II: 1050

crisatrine [834-12-8] Vol. II: 1054

crisazine [1912-24-9] Vol. II: 1050

crisfuran [1563-66-2] Vol. I: 420

cristobalite [7631-86-9] Vol. III: 846

cristoxo-90 [8001-35-2] Vol. I: 1118

crisuffan [115-29-7] Vol. II: 661

crisuron [330-54-1] Vol. II: 1078

crodacol-c [36653-82-4] Vol. III: 193

crodacol-cas [36653-82-4] Vol. III: 193

crodacol-cat [36653-82-4] Vol. III: 193

crolean [107-02-8] Vol. I: 99

cromile, cloruro di (Italian) [14977-61-8]
Vol. II: 247

cromo, ossicloruro di (Italian) [14977-61-8]
Vol. II: 247

cromo (triossidodi) (Italian) [1333-82-0]
Vol. I: 493

croprider [94-75-7] Vol. I: 30

crop rider 3-34d-2 [94-11-1] Vol. II: 15

crop rider 3,34d [94-11-1] Vol. II: 15

crop rider 6d-os weed killer [93-79-8] Vol.
II: 34

crop rider 6d weed-killer [93-79-8] Vol. II:
34

croptex-chrome [101-42-8] Vol. III: 452

croptex-ruby [101-42-8] Vol. III: 452

crotenaldehyde [123-73-9] Vol. III: 263

crotonal [4170-30-3] Vol. I: 547

crotonaldehyde [123-73-9] Vol. III: 263

crotonaldehyde, (e)- [123-73-9] Vol. III:
263

crotonamide, 2-chloro-N, N-diethyl-3-hy-
droxy-, dimethyl phosphate [13171-21-6]
Vol. III: 704

crotonamide, 3-hydroxy-N,N-dimethyl-, cis-
, dimethyl phosphate [141-66-2] Vol. III:
720

crotonamide, 3-hydroxy-N-N-dimethyl-, dimethyl phosphate, cis [141-66-2] Vol.
III: 720

crotonamide, 3-hydroxy-N-N-dimethyl-, dimethyl phosphate, (e) [141-66-2] Vol. III:
720

crotonic acid, 3-hydroxy-, alpha-methylbenzyl ester, dimethyl phosphate, (e)-
[7700-17-6] Vol. III: 266

crotonic acid, 3-methyl-, 2-sec-butyl-4,6-
dinitrophenyl ester [485-31-4] Vol. II:
261

crotonic aldehyde [4170-30-3] Vol. I: 547

crotoxyfos [7700-17-6] Vol. III: 266

crotoxyphos [7700-17-6] Vol. III: 266

crotylaldehyde [4170-30-3] Vol. I: 547

crude arsenic [1327-53-3] Vol. I: 203

Crump leather-lasting dressing [554-84-7]
Vol. III: 693

crusader-s [1689-84-5] Vol. II: 163

crusader-s [1702-17-6] Vol. III: 239

cryofluoran [76-14-2] Vol. III: 408

cryofluorane [76-14-2] Vol. III: 408

cryofluoranum (Latin) [76-14-2] Vol. III:
408

cryolite [15096-52-3] Vol. III: 37

cryolite (AlNa₃F₆) [15096-52-3] Vol. III: 37

cryolite (Na₃(AlF₆)) [15096-52-3] Vol. III:
37

crystal ammonia [506-87-6] Vol. I: 139

crystallized verdigris [142-71-2] Vol. II: 11

crystal propantl-4 [26638-19-7] Vol. II:
298

crystals of venus [142-71-2] Vol. II: 11

crystex [7704-34-9] Vol. II: 965

crysthion-21 [86-50-0] Vol. III: 744

crysthyon [86-50-0] Vol. III: 744

crystol carbonate [497-19-8] Vol. II: 922

cs [2698-41-1] Vol. III: 218

csac [111-15-9] Vol. II: 365

csf giftweizen [7446-18-6] Vol. I: 1079

cs (lacrimator) [2698-41-1] Vol. III: 218

csp [7758-99-8] Vol. II: 255

cube [83-79-4] Vol. II: 166

cube extract [83-79-4] Vol. II: 166

cube-pulver [83-79-4] Vol. II: 166

cube-root [83-79-4] Vol. II: 166

cubic niter [7631-99-4] Vol. III: 864

cubor [83-79-4] Vol. II: 166

cucumber dust [7778-44-1] Vol. III: 65

Cu m3 [7440-50-8] Vol. I: 519

cuma [66-76-2] Vol. III: 336

cumeen (Dutch) [98-82-8] Vol. I: 551

cumeenhydroperoxyde (Dutch) [80-15-9]
Vol. II: 490

alpha-cumene hydroperoxide [80-15-9] Vol.
II: 490

cument hydroperoxide [80-15-9] Vol. II:
490

cumenyl hydroperoxide [80-15-9] Vol. II:
490

cumid [66-76-2] Vol. III: 336

cumol [98-82-8] Vol. I: 551

cumolhydroperoxid (German) [80-15-9] Vol.
II: 490

7-cumyl hydroperoxide [80-15-9] Vol. II:
490

cumyl hydroperoxide [80-15-9] Vol. II: 490

alpha-cumyl hydroperoxide [80-15-9] Vol.
II: 490

cunitex [137-26-8] Vol. I: 632

cupric diacetate [142-71-2] Vol. II: 11

cupricin [544-92-3] Vol. I: 522

cupric sulfate [7758-98-7] Vol. I: 510

cupric sulfate, pentahydrate [7758-99-8]
Vol. II: 255

cupric sulphate [7758-98-7] Vol. I: 510

cupriethylene diamine [13426-91-0] Vol.
III: 270

cupriethylenediamine, bis (ethylenediamine)-, ion [13426-91-0] Vol. III: 270

cuprous cyanide [544-92-3] Vol. I: 522

cuprtic acetate [142-71-2] Vol. II: 11

cuprtic sulfate [7758-99-8] Vol. II: 255

cuprum (Latin) [7440-50-8] Vol. I: 519

cuptic sulfate anhydrous [7758-98-7] Vol.
I: 510

curafume [74-83-9] Vol. II: 577

curaterr [1563-66-2] Vol. I: 420

curetard a [86-30-6] Vol. I: 622

curex flea duster [83-79-4] Vol. II: 166

curitan [2439-10-3] Vol. III: 290

curling-factor Vol. III: [126-07-8], 472

curpric tetraammine sulfate monohydrate
[10380-29-7] Vol. III: 268

c urzate-m [12427-38-2] Vol. III: 531

cutaval [7439-96-5] Vol. III: 534

cvp [470-90-6] Vol. III: 716

cvp [pesticide] [470-90-6] Vol. III: 716

C-weiss 7 (German) [13463-67-7] Vol. III:
921

cwn-2 [7440-44-0] Vol. III: 179

cy-1 500 [156-62-7] Vol. II: 265

cyaanwaterstof (Dutch) [74-90-8] Vol. I:
743

cyanamid [156-62-7] Vol. II: 265

cyanamide [156-62-7] Vol. II: 265

cyanamide calcique (French) [156-62-7]
Vol. II: 265

cyanamide, calcium salt (1:1) [156-62-7]
Vol. II: 265

cyanamide, calcium salt [156-62-7] Vol. II:
265

cyanamid, granular [156-62-7] Vol. II: 265

cyanamid, special grade [156-62-7] Vol. II:
265

cyanhydrine d'acetone (French) [75-86-5]
Vol. I: 794

cyanide of calcium [592-01-8] Vol. I: 395

cyanide of potassium [151-50-8] Vol. I:
978

cyanide of sodium [143-33-9] Vol. I: 1045

cyanoacetonitrile [109-77-3] Vol. II: 531

N-cyanoamine [420-04-2] Vol. II: 263

cyanoamine [420-04-2] Vol. II: 263

cyanobenzene [100-47-0] Vol. I: 296

cyanobromide- [506-68-3] Vol. I: 558

4-cyano-2,6-dibromophenol [1689-84-5]
Vol. II: 163

cyanoethane [107-12-0] Vol. II: 851

2-cyanoethanol [109-78-4] Vol. II: 466

2-cyanoethyl alcohol [109-78-4] Vol. II:
466

cyanoethylene [107-13-1] Vol. I: 123

(r,s)-alpha-cyano-4-fluoro-3-phenoxybenzyl-
(1r,s)-cis, trans-3-(2,2- dichlorovinyl)-2,2-
dimethylcyclopropanecarboxylate [68359-
37-5] Vol. III: 288

(rs)-alpha-cyano-4-fluoro-3-phenoxybenzyl
(1rs, 3rs: 1rs, 3sr)-3-(2,2-dichlorovinyl)-
2,2-dimethylcyclopropanecarboxylate
[68359-37-5] Vol. III: 288

cyano (4-fluoro-3-phenoxyphenyl) methyl 3-(2,2-dichloroethenyl-2,2-dimethyl cyclopropanecarboxylate [68359-37-5] **Vol. III:** 288

cyanogen [460-19-5] **Vol. I:** 555

cyanogenamide [420-04-2] **Vol. II:** 263

cyanogen bromide [506-68-3] **Vol. I:** 558

cyanogene (French) [460-19-5] **Vol. I:** 555

cyanogen monobromide [506-68-3] **Vol. I:** 558

cyanogen nitride [420-04-2] **Vol. II:** 263

cyanogran [143-33-9] **Vol. I:** 1045

cyanol [62-53-3] **Vol. I:** 164

cyanomethane [75-05-8] **Vol. I:** 81

1-cyano-1-methylethane [78-82-0] **Vol. III:** 798

alpha-cyano-3-phenoxy-benzyl alpha-(4-chlorophenyl) isovalerate [51630-58-1] **Vol. III:** 95

alpha-cyano-3-phenoxybenzyl-alpha-(4-chlorophenyl) isovalerate [51630-58-1] **Vol. III:** 95

alpha-cyano-3-phenoxy-benzyl alpha-isopropyl-4-chlorophenylacetate [51630-58-1] **Vol. III:** 95

alpha-cyano-3-phenoxybenzyl 2-(4-chlorophenyl)-3-methylbutyrate [51630-58-1] **Vol. III:** 95

(s)-alpha-cyano-3-phenoxybenzyl (s)-2-(4-chlorophenyl)-3-methylbutyrate [66230-04-4] **Vol. III:** 400

alpha-cyano-3-phenoxybenzyl-0 cis, trans-2,2-dichlorovinyl 2,2-dimethyl-cyclopropanecarboxylate [52315-07-8] **Vol. II:** 251

(-)-alpha-cyano-3-phenoxybenzyl-(-)-cis, trans-3-(2,2-dichlorovinyl)-2,2-dimethylcyclopropane carboxylate [52315-07-8] **Vol. II:** 251

(rs)-alpha-cyano-3-phenoxybenzyl (s)-2-(4-difluoromethoxyphenyl)-3-methylbutyrate **Vol. III:** [70124-77-5], 455

alpha-cyano-3-phenoxybenzyl isopropyl-4-chlorophenylacetate [51630-58-1] **Vol. III:** 95

(rs)-alpha-cyano-3-phenoxybenzyl N-[2-chloro-alpha, alpha, alpha-trifluoro-p tolyl)-d-valinate. [69409-94-5] **Vol. III:** 459

(s)-alpha-cyano-3-phenoxybenzyl-(1r)-cis-3-(2,2-dibromovinyl)-2,2-dimethyl cyclopropane carboxyate [52918-63-5] **Vol. III:** 296

(S)-alpha-cyano-3- phenoxybenzyl (1R) - cis-2,2-dimethyl- 3-[(RS)-1,2,2,2-tetrabromoethyl]- cyclopropanecarboxylate [66841-25-6] **Vol. III:** 929

(s)-alpha-cyano-3-phenoxybenzyl (1r, 3r)-3-(2,2-dibromovinyl)-2,2- dimethyl cyclopropan-1-carboxylate [52918-63-5] **Vol. III:** 296

(rs)-alpha-cyano-3-phenoxybenzyl (r)-2-[2-chloro-4-(trifluoromethyl) anilino-3-methyl butanoate] [69409-94-5] **Vol. III:** 459

(rs)-alpha-cyano-3-phenoxybenzyl(rs)-2-(4-chlorophenyl)-3-methylbutyrate [51630-58-1] **Vol. III:** 95

(rs)-alpha-cyano-3-phenoxybenzyl (1rs)-cis, trans-3-(2,2-dichlorovinyl)-2,2-dimethylcyclopropanecarboxylate [52315-07-8] **Vol. II:** 251

(S)-alpha-cyano-3-phenoxybenzyl (1R,3S)-2,2-dimethyl-3-[(RS)-1,2,2,2- tetrabromoethyl]-cyclopropane- carboxylate. [66841-25-6] **Vol. III:** 929

(rs)-alpha-cyano-3-phenoxybenzyl (z)-(1rs)-cis-3-(2-chloro-3,3,3-trifluoropropenyl)-2,2-dimethylcyclopropane carboxylate [68085-85-8] **Vol. II:** 288

(rs)-alpha-cyano-3-phenoxybenzyl (z)-(1rs, 3rs)-(2-chloro-3,3,3-trifluoropropenyl)-2,2-dimethylcyclopropane carboxylate [68085-85-8] **Vol. II:** 288

cyano (3-phenoxyphenyl) methyl 4-chloro-alpha-(1-methylethyl) benzene acetate [51630-58-1] **Vol. III:** 95

cyano (3-phenoxyphenyl) methyl 4-chloro-alpha-(1-methylethyl) benzeneacetate [51630-58-1] **Vol. III:** 95

(cyano (3-phenoxyphenyl) methyl-4-chloro-alpha-(1-methylethyl) phenylacetate) [51630-58-1] **Vol. III:** 95

[1alpha, 3alpha (z)] 1-(-)-cyano-(3-phenoxy phenyl) methyl 3-(2-chloro-3,3,3-trifluoro-1-propenyl)-2,2-dimethylcyclopropanecarboxylate, [68085-85-8] **Vol. II:** 288

(1r-(1-alpha (s),3-alpha))-cyano-(3-phenoxyphenyl) methyl-3-(2,2-dibromovinyl)-2,2-dimethylcyclopropanecarboxylate) [52918-63-5] **Vol. III:** 296

cyano (3-phenoxyphenyl) methyl 3-(2,2-dichloroethenyl)-2,2-dimethylcycloptopane carboxylate, [52315-07-8] **Vol. II:** 251

[1alpha (S),3alpha]-(-)-cyano (3-phenoxyphenyl) methyl 3-(2,2-dichloroethenyl)— 2,2-dimethylcyclopropanecarboxylate [67375-30-8] **Vol. III:** 30

(cyano (3-phenoxyphenyl) methyl 3-(2,2-di-chlorovinyl-2,2-dimethylcyclo propane carboxylate [52315-07-8] **Vol. II**: 251

(+ −)-cyano-(3-phenoxyphenyl) methyl (+)-4-(difluoromethoxy)-alpha-(methylethyl) benzeneacetate **Vol. III**: [70124-77-5], 455

(rs)-cyano-(3-phenoxyphenyl) methyl (s)-4-(difluoromethoxy)-alpha-(1-methylethyl)-benzeneacetate **Vol. III**: [70124-77-5], 455

cyano (3-phenoxyphenyl) methyl 2,2-di-methyl-3-(1,2,2,2-tetrabromoethyl) cyclo-propanecarboxylate [66841-25-6] **Vol. III**: 929

cyano (3-phenoxyphenyl) methyl N-[2-chloro-4-(trifluoromethyl) phenyl]-d-valinate [69409-94-5] **Vol. III**: 459

(o-p-cyanophenyl-o,o-dimethylphosphoroth-ioate) [2636-26-2] **Vol. III**: 271

o-(4-cyanophenyl) o,o-dimethyl phospho-rothioate [2636-26-2] **Vol. III**: 271

cyanophos [62-73-7] **Vol. I**: 946

1-cyanopropane [109-74-0] **Vol. III**: 152

2-cyanopropane [78-82-0] **Vol. III**: 798

2-cyano-2-propanol [75-86-5] **Vol. I**: 794

2-cyano-1-propene [126-98-7] **Vol. II**: 839

2-cyanopropene-1 [126-98-7] **Vol. II**: 839

2-cyanopropene [126-98-7] **Vol. II**: 839

cyanox [2636-26-2] **Vol. III**: 271

cyanure d'argent (French) [506-64-9] **Vol. II**: 915

cyanure de baryum (French) [542-62-1] **Vol. I**: 230

cyanure de calcium (French) [592-01-8] **Vol. I**: 395

cyanure de mercure (French) [592-04-1] **Vol. II**: 544

cyanure de methyl (French) [75-05-8] **Vol. I**: 81

cyanure de potassium (French) [151-50-8] **Vol. I**: 978

cyanure de sodium (French) [143-33-9] **Vol. I**: 1045

cyanure de vinyle (French) [107-13-1] **Vol. I**: 123

cyanure de zinc (French) [557-21-1] **Vol. I**: 1181

Cyanwasserstoff (German) [74-90-8] **Vol. I**: 743

cyap [2636-26-2] **Vol. III**: 271

cyazin [1912-24-9] **Vol. II**: 1050

cybolt [70124-77-5], **Vol. III**: 455

cyclic propylene (diethoxy-phosphinyl) di-thioimidocarbonate [950-10-7] **Vol. III**: 536

cyclic propylene (diethoxyphosphinyl) di-thiomidocarbonate [950-10-7] **Vol. III**: 536

cyclic propylene p,p-diethyl phosphonodi-thioimidocarbonate [950-10-7] **Vol. III**: 536

cyclic s, s-(6-methyl-2,3-quinoxalinediyl) dithiocarbonate [2439-01-2] **Vol. III**: 663

cyclic tetramethylene sulfone [126-33-0] **Vol. II**: 1007

cyclite [100-39-0] **Vol. II**: 1019

cyclodan [115-29-7] **Vol. II**: 661

cycloheptane, 1-aza [111-49-9] **Vol. III**: 477

cyclohexadienedione [106-51-4] **Vol. I**: 303

1,4-cyclohexadienedione [106-51-4] **Vol. I**: 202

2,5-cyclohexadiene-1,4-dione [106-51-4] **Vol. I**: 303

1,4-cyclohexadienedioxide [106-51-4] **Vol. I**: 303

cyclohexamethylenimine [111-49-9] **Vol. III**: 477

cyclohexanamine [108-91-8] **Vol. II**: 280

1,2-cyclohexanedicarboxylicacid, 3,6-endo epoxy [145-73-3] **Vol. I**: 900

cyclohexane, 1,2,3,4,5,6-hexachloro [608-73-1] **Vol. I**: 564

cyclohexane, 1,2,3,4,5,6-hexachloro-, (mixed isomers) [608-73-1] **Vol. I**: 564

cyclohexane, methyl [108-87-2] **Vol. III**: 607

1-cyclohexanol [108-93-0] **Vol. I**: 580

cyclohexanon (Dutch) [108-94-1] **Vol. II**: 268

cyclohexanone iso-oxime [105-60-2] **Vol. III**: 79

cyclohexatriene [71-43-2] **Vol. I**: 239

4-cyclohexene-1,2-dicarboximide, N(trichloromethyl)thio- [133-06-2] **Vol. II**: 276

4-cyclohexene-1,2-dicarboximide, N((trichloromethyl) thio)- [133-06-2] **Vol. II**: 276

1-cyclohexene-1,2-dicarboximidomethyl-2,2-dimethyl-3-(2-met hylpropenyl) cyclo-propane carboxylate [7696-12-0] **Vol. III**: 912

degesch calcium cyanide a dust [592-01-8] Vol. I: 395

degrassan [534-52-1] Vol. I: 540

de-green [78-48-8] Vol. III: 768

deh-24- [112-24-3] Vol. I: 1129

deha [103-23-1] Vol. II: 291

dehp [117-81-7] Vol. II: 769

dehydrag-sulfate-gl-emulsion [151-21-3] Vol. II: 981

dek [96-22-0] Vol. III: 679

dekrysil [534-52-1] Vol. I: 540

delac j [86-30-6] Vol. I: 622

delagil [54-05-7] Vol. III: 229

deleaf-defoliant [150-50-5] Vol. III: 574

delicia [20859-73-8] Vol. I: 133

delicia gastoxin [20859-73-8] Vol. I: 133

delmofulvina [126-07-8], Vol. III: 472

delnatex [78-34-2] Vol. II: 756

delnav [78-34-2] Vol. II: 756

delsene-m [12427-38-2] Vol. III: 531

delta (3)-2-butenone [78-94-4] Vol. III: 624

delta-chloroethanol [107-07-3] Vol. III: 220

delta (9) cis oleic acid [112-80-1] Vol. I: 895

deltamethrine [52918-63-5] Vol. III: 296

deltan [67-68-5] Vol. II: 641

deltic [78-34-2] Vol. II: 756

delusal [1314-84-7] Vol. II: 1100

demasorb [67-68-5] Vol. II: 641

dematon [8065-48-3] Vol. III: 755

demavet [67-68-5] Vol. II: 641

demeso [67-68-5] Vol. II: 641

demeton [8065-48-3] Vol. III: 755

demeton-methyl sulphoxide [301-12-2] Vol. III: 661

demeton-o and demeton-s [8065-48-3] Vol. III: 755

demeton-o-methyl sulfoxide [301-12-2] Vol. III: 661

demeton-s methyl sulfoxide [301-12-2] Vol. III: 661

Demeton-s-methyl sulfoxid [German] [301-12-2] Vol. III: 661

demon [52315-07-8] Vol. II: 251

demos-140 [60-51-5] Vol. III: 740

demox [8065-48-3] Vol. III: 755

demsodrox [67-68-5] Vol. II: 641

den [109-89-7] Vol. I: 597

den [55-18-5] Vol. I: 601

dena [55-18-5] Vol. I: 601

denkaphon [52-68-6] Vol. III: 707

densinfluat [79-01-6] Vol. II: 427

deosan [7681-52-9] Vol. II: 498

deosan green label steriliser [7681-52-9] Vol. II: 498

dep [52-68-6] Vol. III: 707

de pester ded weed lv-2 [94-75-7] Vol. I: 30

d-e-r-332- [1675-54-3] Vol. III: 132

deriban [62-73-7] Vol. I: 946

deril [83-79-4] Vol. II: 166

dermacid [149-30-4] Vol. III: 540

dermasorb [67-68-5] Vol. II: 641

dermaton [470-90-6] Vol. III: 716

derribante [62-73-7] Vol. I: 946

derrin [83-79-4] Vol. II: 166

derris [83-79-4] Vol. II: 166

des [62-73-7] Vol. I: 946

des i-cate [145-73-3] Vol. I: 900

desiccant 1-10 [7778-39-4] Vol. II: 91

desicoil [88-85-7] Vol. II: 704

desitin-dabaways [8001-54-5] Vol. II: 67

desitin skin care lotions [8001-54-5] Vol. II: 67

desmodur-t80 [584-84-9] Vol. II: 125

2,4-des-Na [136-78-7] Vol. III: 855

2,4-des-na [136-78-7] Vol. III: 855

2,4-Desnatrium (German) [136-78-7] Vol. III: 855

desormone [94-75-7] Vol. I: 30

desoxon-1 [79-21-0] Vol. III: 682

desprout [123-33-1] Vol. II: 870

de-sprout [123-33-1] Vol. II: 870

dessicant 1-10 [7778-39-4] Vol. II: 91

2,4-des sodium [136-78-7] Vol. III: 855

destruxol borer sol [107-06-2] Vol. I: 643

destruxol orchid spray [54-11-5] Vol. I: 874

deta [111-40-0] Vol. I: 606

detal [534-52-1] Vol. I: 540

detf [52-68-6] Vol. III: 707

detia [20859-73-8] Vol. I: 133

detia ex b [20859-73-8] Vol. I: 133

detia gas exb [20859-73-8] Vol. I: 133

detia gas ex-m [74-83-9] Vol. II: 577

detmol-ua [2921-88-2] Vol. III: 762

developer-11- [108-45-2] Vol. III: 98

developer-b [95-80-7] Vol. II: 1016

developer-c [108-45-2] Vol. III: 98

developer-db [95-80-7] Vol. II: 1016

developer-dbj [95-80-7] Vol. II: 1016

developer-h [108-45-2] Vol. III: 98

developer-m [108-45-2] Vol. III: 98

developer-mc [95-80-7] Vol. II: 1016

developer-mt [95-80-7] Vol. II: 1016

developer mt-cf [95-80-7] Vol. II: 1016

developer-mtd [95-80-7] Vol. II: 1016

developer p [100-01-6] Vol. II: 70

developer-t [95-80-7] Vol. II: 1016

devigon [60-51-5] Vol. III: 740

devikol [62-73-7] Vol. I: 946

devisulphan [115-29-7] Vol. II: 661

devol orange b [88-74-4] Vol. I: 173

devol orange salt b [88-74-4] Vol. I: 173

devol red gg [100-01-6] Vol. II: 70

dextrone [4685-14-7] Vol. III: 670

dextrone-x [4685-14-7] Vol. III: 670

dezodorator [91-20-3] Vol. I: 854

diacetonalcohol (Dutch) [123-42-2] Vol. II: 688

diacetonalcool (Italian) [123-42-2] Vol. II: 688

diacetonalkohol (German) [123-42-2] Vol. II: 688

diacetone alcohol [123-42-2] Vol. II: 688

diacetone alcool (French) [123-42-2] Vol. II: 688

diacetonyl alcohol [123-42-2] Vol. II: 688

1,2-diacetoxyethane [111-55-7] Vol. II: 398

diacetoxymercury [1600-27-7] Vol. II: 24

diacetylmethane [123-54-6] Vol. I: 910

Diaethanolamin (German) [111-42-2] Vol. I: 658

diaethanolnitrosamin (German) [1116-54-7] Vol. II: 377

Diaethylamin (German) [109-89-7] Vol. I: 597

Diaethylnitrosamin (German) [55-18-5] Vol. I: 601

O,O-Diaethyl-o-(2-isopropyl-4-methyl-pyrimidin-6-yl)-monot hiophosphat (German) [333-41-5] Vol. III: 758

O,O-Diaethyl-o-3,5,6-trichlor-2-pyridylmonothiophosphat (German) [2921-88-2] Vol. III: 762

o, o'-Diaethyl-p-nitrophenylphosphat (German) [311-45-5] Vol. III: 669

diak-5- [123-31-9] Vol. I: 753

diakon [80-62-6] Vol. II: 567

dialifos [10311-84-9] Vol. III: 728

dialiphor [10311-84-9] Vol. III: 728

dialiphos- [10311-84-9] Vol. III: 728

diallaat (Dutch) [2303-16-4] Vol. II: 218

di-allate [2303-16-4] Vol. II: 218

diallat (German) [2303-16-4] Vol. II: 218

dialuminum sulfate [10043-01-3] Vol. III: 39

dialuminum sulphate [10043-01-3] Vol. III: 39

dialuminum trisulfate [10043-01-3] Vol. III: 39

diamide [957-51-7] Vol. II: 6

diamine [302-01-2] Vol. I: 735

m-diaminobenzene [108-45-2] Vol. III: 98

1,2-diaminobenzene [95-54-5] Vol. III: 697

1,3-diaminobenzene [108-45-2] Vol. III: 98

o-diaminobenzene [95-54-5] Vol. III: 697

4,4'-diaminobiphenyl [92-87-5] Vol. I: 288

p,p'-diaminobiphenyl [92-87-5] Vol. I: 288

1,8-diamino-3,6-diazaoctane [112-24-3] Vol. I: 1129

4,4'-diamino-3,3'-dichlorobiphenyl [91-94-1] Vol. I: 285

4,4'-diamino-3,3'-dichlorodiphenyl [91-94-1] Vol. I: 285

2,2'-diaminodiethylamine [111-40-0] Vol. I: 606

p, p'-Diaminodifenylmethan (Czech) [101-77-9] Vol. III: 304

4,4'-diamino-3,3'-dimethylbiphenyl [119-93-7] Vol. III: 364

4,4'-diamino-3,3'-dimethyldiphenyl [119-93-7] Vol. III: 364

4,4'-diaminodiphenyl [92-87-5] Vol. I: 288

p diaminodiphenyl [92-87-5] Vol. I: 288

diaminodiphenylmethane [101-77-9] Vol. III: 304

p,p'-diaminodiphenylmethane [101-77-9] Vol. III: 304

4,4'-Diaminodiphenylmethan (German) [101-77-9] Vol. III: 304

4,4'-diamino ditan [101-77-9] Vol. III: 304

diaminoditolyl [119-93-7] Vol. III: 364

1,6-diaminohexane [124-09-4] Vol. I: 732

4,4'-diamino-1, 1'-biphenyl [92-87-5] Vol. I: 288

2,4-diamino-1-methylbenzene [95-80-7] Vol. II: 1016

1,3-diamino-4-methylbenzene [95-80-7] Vol. II: 1016

1,6-diamino n-hexane [124-09-4] Vol. I: 732

di-(4-aminophenyl) methane [101-77-9] Vol. III: 304

2,4-diaminotoluen (Czech) [95-80-7] Vol. II: 1016

2,4-diamino-1-toluene [95-80-7] Vol. II: 1016

di (2-chloroethyl) ether [111-44-4] **Vol. II:** 379

dichloroethyl formal [111-91-1] **Vol. I:** 815

2,2-dichloroethyl formal [111-91-1] **Vol. I:** 815

di-2-chloroethyl formal [111-91-1] **Vol. I:** 815

dichloroethyl oxide [111-44-4] **Vol. II:** 379

2,2'-dichloroethyl sulfide [505-60-2] **Vol. II:** 962

di-2-chloroethyl sulfide [505-60-2] **Vol. II:** 962

dichloroethyl-sulfide [505-60-2] **Vol. II:** 962

2,2'-dichloroethyl sulphide [505-60-2] **Vol. II:** 962

di-2-chloroethyl sulphide [505-60-2] **Vol. II:** 962

dichlorofluoromethane [75-43-4] **Vol. II:** 595

dichlorohydridomethylsilicon [75-54-7] **Vol. III:** 609

dichloroisopropyl ether [108-60-1] **Vol. II:** 386

2,2-dichloroisopropyl ether [108-60-1] **Vol. II:** 386

dichlorokelthane [115-32-2] **Vol. II:** 144

dichloromercury [7487-94-7] **Vol. II:** 539

3,6-dichloro-2-methoxybenzoic acid [1918-00-9] **Vol. II:** 159

2,5-dichloro-6-methoxybenzoic acid [1918-00-9] **Vol. II:** 159

(dichloromethyl) benzene [98-87-3] **Vol. III:** 84

dichloromethyl ether [542-88-1] **Vol. II:** 384

dichloromethylsilane [75-54-7] **Vol. III:** 609

dichloromonofluoromethane [75-43-4] **Vol. II:** 595

2,3-dichloro-1,4-naphthalenedione [117-80-6] **Vol. II:** 645

2,3-dichloro-1,4-naphthaquinone [117-80-6] **Vol. II:** 645

dichloronaphthoquinone [117-80-6] **Vol. II:** 645

2,3-dichloro-1,4-naphthoquinone [117-80-6] **Vol. II:** 645

2,3-dichloronaphthoquinone-1,4 [117-80-6] **Vol. II:** 645

2,3-dichloronaphthoquinone [117-80-6] **Vol. II:** 645

3,6-dichloro-o-anisic acid [1918-00-9] **Vol. II:** 159

1,5-dichloro-3-oxapentane [111-44-4] **Vol. II:** 379

dichlorooxozirconium [7699-43-6] **Vol. II:** 1110

(2,4-dichlorophenoxy)acetic acid [94-75-7] **Vol. I:** 30

2,4-dichlorophenoxyacetic acid [94-75-7] **Vol. I:** 30

2,4-dichlorophenoxyacetic acid, butoxyethyl ester [1929-73-3] **Vol. III:** 155

2,4-dichlorophenoxyacetic acid isopropyl ester [94-11-1] **Vol. II:** 15

4-(2,4-dichlorophenoxy) butanoic acid [94-82-6] **Vol. III:** 176

(2,4-dichlorophenoxy) butyric acid [94-82-6] **Vol. III:** 176

2-(2,4-dichlorophenoxy) ethanol hydrogen sulfate sodium salt [136-78-7] **Vol. III:** 855

2,4-dichlorophenoxyethyl hydrogen sulfate sodium salt [136-78-7] **Vol. III:** 855

2,4-dichlorophenoxyethyl sulfate, sodium salt [136-78-7] **Vol. III:** 855

dichlorophenylarsine [696-28-6] **Vol. II:** 107

2,6-dichlorophenyl cyanide [1194-65-6] **Vol. I:** 299

3-(3,4-dichlorophenyl)-1,1-dimethyl urea [330-54-1] **Vol. II:** 1078

1-(3,4-dichlorophenyl)-3,3-dimethyl urea [330-54-1] **Vol. II:** 1078

1-(3,4-dichlorophenyl)-3,3-dimethyl uree (French) [330-54-1] **Vol. II:** 1078

dichlorophenylmethane [98-87-3] **Vol. III:** 84

3-(3',4'-dichlorophenyl)-1-methoxy-1-methylurea [330-55-2] **Vol. II:** 1084

3-(3,4-dichlorophenyl)-1-methoxy-1-met hylurea [330-55-2] **Vol. II:** 1084

3-(3,4-dichlorophenyl)-1-methoxy methylurea [330-55-2] **Vol. II:** 1084

1-(3,4-dichlorophenyl) 3-methoxy-3-methyluree (French) [330-55-2] **Vol. II:** 1084

N'-(3,4-dichlorophenyl)-N-methoxy-N-methylurea [330-55-2] **Vol. II:** 1084

N-(3,4-dichlorophenyl)-N'-methyl-N'-methoxyurea [330-55-2] **Vol. II:** 1084

N'-(3,4-dichlorophenyl)-N, N-dimethylurea [330-54-1] **Vol. II:** 1078

N-(3,4-dichlorophenyl)-N',N'-dimethyl urea [330-54-1] **Vol. II:** 1078

N-(3,4-dichlorophenyl) propanamide [26638-19-7] Vol. II: 298

3,6-dichloropicolinic acid [1702-17-6] Vol. III: 239

dichloro-1,2-propane [78-87-5] Vol. I: 987

alpha,beta dichloropropane [78-87-5] Vol. I: 987

alpha, alpha-dichloropropane [78-99-9] Vol. III: 796

dichloro-1,3-propene [542-75-6] Vol. II: 834

1,3-dichloro-1-propene [542-75-6] Vol. II: 834

1,3-dichloropropene-1 [542-75-6] Vol. II: 834

2,3-dichloropropene [78-88-6] Vol. III: 805

1,3-dichloro-2-propene [542-75-6] Vol. II: 834

1,3-dichloropropene and 1,2-dichloropropane mixture [8003-19-8] Vol. III: 792

2,3-dichloro-2-propene-1-thiol, diisopropyl-carbamate [2303-16-4] Vol. II: 218

S-(2,3-dichloro-2-propenyl) bis (1-methylethyl) carbamothioate [2303-16-4] Vol. II: 218

3',4'-dichloropropionanilide [26638-19-7] Vol. II: 298

3,4-dichloropropionanilide [26638-19-7] Vol. II: 298

2,2-dichloropropionic acid [75-99-0] Vol. I: 1016

alpha dichloropropionicacid [75-99-0] Vol. I: 1016

alpha,alpha dichloropropionic acid [75-99-0] Vol. I: 1016

1,3-dichloropropylene [542-75-6] Vol. II: 834

1,3-dichloro-1-propylene [542-75-6] Vol. II: 834

2,3-dichloropropylene [78-88-6] Vol. III: 805

3,6-dichloro-2-pyridinecarboxylic acid [1702-17-6] Vol. III: 239

3,6-dichloropyridine-2-carboxylic acid [1702-17-6] Vol. III: 239

1,2-dichlorotetrafluoroethane [76-14-2] Vol. III: 408

dichlorotetrafluoroethane [76-14-2] Vol. III: 408

dichlorothiocarbonyl [463-71-8] Vol. I: 1086

dichlorotin [7772-99-8] Vol. II: 1013

alpha, alpha-dichlorotoluene [98-87-3] Vol. III: 84

dichlorovas [62-73-7] Vol. I: 946

(2,2-dichloro vinil)dimetil fosfato (Italian) [62-73-7] Vol. I: 946

2,2-dichlorovinyl dimethyl phosphate [62-73-7] Vol. I: 946

dichlorovos [62-73-7] Vol. I: 946

2,4-Dichlor phenoxy essigsaeure (German) [94-75-7] Vol. I: 30

3-(3,4-Dichlor-phenyl)-1,1-dimethyl-harnstoff (German) [330-54-1] Vol. II: 1078

3-(3,4-Dichlor-phenyl)-1-methoxy-1-methyl-harnstoff (German) [330-55-2] Vol. II: 1084

dichlorphos [62-73-7] Vol. I: 946

dichlorpropan [26638-19-7] Vol. II: 298

Dichlorpropan-dichlorpropengemisch (German) [8003-19-8] Vol. III: 792

o-(2,2-Dichlorvinyl) o,o dimethylphosphat (German) [62-73-7] Vol. I: 946

dichromic acid, diammonium salt [7789-09-5] Vol. I: 590

dichromic acid, disodium salt [10588-01-9] Vol. II: 300

dichromic acid (H2Cr2O7), disodium salt [10588-01-9] Vol. II: 300

dichromium sulphate [10101-53-8] Vol. I: 482

dichromium trioxide [1308-38-9] Vol. III: 234

dichromium trisulphate [10101-53-8] Vol. I: 482

dicid [333-41-5] Vol. III: 758

diclone [117-80-6] Vol. II: 645

S-(2,3-dicloro-allil)-N, N-diisopropilmonotio-carbammato (Italian) [2303-16-4] Vol. II: 218

p diclorobenzene (Italian) [106-46-7] Vol. I: 254

1,4-diclorobenzene (Italian) [106-46-7] Vol. I: 254

1,1-dicloro-2,2-bis (4-chloro-fenil)-etano (Italian) [72-54-8] Vol. III: 404

diclorodifluometano (Spanish) [75-71-8] Vol. I: 825

1,1-dicloroetano (Italian) [75-34-3] Vol. II: 325

1,2-dicloroetano (Italian) [107-06-2] Vol. I: 643

2,2'-dicloroetiletere (Italian) bcee [111-44-4] Vol. II: 379

3-(3,4-dicloro-fenil)-1-metossi-1-metilurea (Italian) [330-55-2] Vol. II: 1084

di (2-hydroxyethyl) amine [111-42-2] **Vol. I:** 658

di-(2-hydroxyethyl) nitrosamine [1116-54-7] **Vol. II:** 377

4,4'-dihydroxy-3,3'-methylene bis coumarin [66-76-2] **Vol. III:** 336

di-4-hydroxy-3,3'-methylenedicoumarin [66-76-2] **Vol. III:** 336

2,4-dihy- droxy-2-methylpentane [107-41-5] **Vol. III:** 620

2,2-di (4-hydroxyphenyl) propane [80-05-7] **Vol. II:** 720

1,2-dihydroxypropane [57-55-6] **Vol. I:** 1003

3,6-dihydroxypyridazine [123-33-1] **Vol. II:** 870

1,4-diidrobenzene (Italian) [123-31-9] **Vol. I:** 753

diisobutilchetone (Italian) [108-83-8] **Vol. I:** 730

diisobutylcarbinol [108-82-7] **Vol. II:** 453

di isobutylcetone (French) [108-83-8] **Vol. I:** 730

diisobutylketon (Dutch,German) [108-83-8] **Vol. I:** 730

diisobutylphenoxyethoxyethyldimethyl benzyl ammonium-chloride [121-54-0] **Vol. III:** 105

diisobutylphenoxyethoxyethyldimethyl benzyl ammonium chloride [121-54-0] **Vol. III:** 105

(diisobutylphenoxyethoxyethyl) dimethyl-benzylammonium chloride [121-54-0] **Vol. III:** 105

(2-(2-(4-diisobutylphenoxy) ethoxy) ethyl) dimethylbenzylammonium chloride [121-54-0] **Vol. III:** 105

p-diisobutyl phenoxyethoxyethyl dimethyl benzylammonium chloride [121-54-0] **Vol. III:** 105

diisobutylthiocarbamic acid, S-ethyl ester [2008-41-5] **Vol. II:** 216

diisocarb [2008-41-5] **Vol. II:** 216

di-isocyanate de toluylene (French) [584-84-9] **Vol. II:** 125

di-iso-cyanatoluene [584-84-9] **Vol. II:** 125

2,4-diisocyanato-1-methylbenzene [584-84-9] **Vol. II:** 125

2,4-diisocyanatotoluene [584-84-9] **Vol. II:** 125

diisocyanatotoluene [26471-62-5] **Vol. II:** 509

Diisocyanat-toluol (German) [584-84-9] **Vol. II:** 125

2,6-diisopropylamino-4-methoxytriazine [1610-18-0] **Vol. III:** 787

o,o-diisopropyl 2-(benzenesulfonamido) ethyl dithiophosphate [741-58-2] **Vol. III:** 731

o,o-diisopropyl dithiophosphate S-ester with N-(2-mercaptoethyl) benzenesulphonamide [741-58-2] **Vol. III:** 731

N-(2-(o,o-diisopropyldithiophosphoryl) ethyl) benzenesulfonamide [741-58-2] **Vol. III:** 731

diisopropyl ether [108-20-3] **Vol. I:** 996

diisopropyl methanephosphonate [1445-75-6] **Vol. III:** 342

N,N'-diisopropyl-6-methoxy-1,3,5-triazine-2,4-diamine [1610-18-0] **Vol. III:** 787

N, N-di-isopropyl-6-methylthio-1,3,5-triazine-2,4-diamine [7287-19-6] **Vol. II:** 1056

diisopropyl oxide [108-20-3] **Vol. I:** 996

S-(o,o-diisopropyl phosphorodithioate) ester of N-(2-mercaptoethyl) benzenesulfonamide [741-58-2] **Vol. III:** 731

S-(o, o-diisopropylphosphorodithioate) of N-(2-mercaptoethyl) benzenesulfonamide [741-58-2] **Vol. III:** 731

o,o-diisopropyl phosphorodithioate S-ester with N-(2-mercaptoethyl) benzenesulfonamide [741-58-2] **Vol. III:** 731

o,o-diisopropyl S-(2-benzenesulfonylamino-ethyl) phosphorodithioate [741-58-2] **Vol. III:** 731

o,o-di-isopropyl S-2-phenylsulphonylamino-ethyl phosphorodithioate [741-58-2] **Vol. III:** 731

diisopropylthiocarbamic acid, S-(2,3-dichloroallyl) ester [2303-16-4] **Vol. II:** 218

diisopropylthiocarbamic acid S-(2,3,3-trichloroallyl) ester [2303-17-5] **Vol. III:** 931

N-diisopropylthiocarbamic acid s-2,3,3-trichloro-2-propenyl ester [2303-17-5] **Vol. III:** 931

di-isopropylthiolocarbamate de S-(2,3-dichloro allyle) (French) [2303-16-4] **Vol. II:** 218

N,N-Diisopropyl-2,3,3-trichlorallyl-thiolcarbamat (German) [2303-17-5] **Vol. III:** 931

2,3-diketobutane [431-03-8] **Vol. III:** 299

diketone alcohol [123-42-2] **Vol. II:** 688

dikotes [94-74-6] **Vol. I:** 25

dikotex [94-74-6] **Vol. I:** 25

dilantin db [95-50-1] **Vol. I:** 250

1,1-dimethyl-3-(alpha, alpha, alphatrifluoro-m-tolyl) urea [2164-17-2] **Vol. II:** 1086

N, N-dimethyl-alpha, alpha-diphenyl acetamide [957-51-7] **Vol. II:** 6

dimethyl 2-(alpha-methylbenzyloxycarbonyl)-1-methylvinyl phosphate [7700-17-6] **Vol. III:** 266

N, N-dimethyl-alpha-phenyl-benzeneacetamide [957-51-7] **Vol. II:** 6

dimethylamidoethoxy- phosphoryl cyanide [77-81-6] **Vol. III:** 891

dimethylamine, nnitroso- [62-75-9] **Vol. I:** 610

dimethylamine, n nitroso- [62-75-9] **Vol. I:** 610

4-(dimethylamine)-3,5-xylyl N-methylcarbamate [315-18-4] **Vol. II:** 227

2',3-dimethyl-4-aminoazobenzene [97-56-3] **Vol. III:** 46

(dimethylamino)benzene [121-69-7] **Vol. I:** 171

N-dimethyl amino-beta-carbamyl propionic acid [1596-84-5] **Vol. III:** 884

4-dimethylamino-3,5-dimethylphenyl N-methylcarbamate [315-18-4] **Vol. II:** 227

p,p dimethylaminodiphenylmethane [101-61-1] **Vol. I:** 173

4,4'-(dimethylamino) diphenylmethane [101-61-1] **Vol. I:** 173

m-[((dimethylamino) methylene) amino) phenyl methylcarbamate hydrochloride [23422-53-9] **Vol. III:** 463

m-[((dimethylamino) methylene) amino] phenyl methylcarbamate hydrochloride [23422-53-9] **Vol. III:** 463

3-dimethylaminomethyleneiminophenyl- N-methylcarbamate, hydrochloride [23422-53-9] **Vol. III:** 463

3-(dimethylamino)-1-methyl-3-oxo-1-propenyl dimethyl phosphate (e)-isomer [141-66-2] **Vol. III:** 720

dimethylaminosuccinamic acid [1596-84-5] **Vol. III:** 884

n-(dimethylamino) succinamic acid [1596-84-5] **Vol. III:** 884

N-Dimethylamino-succinamidsaeure (German) [1596-84-5] **Vol. III:** 884

N-Dimethylamino, succinamidsaeure [German] [1596-84-5] **Vol. III:** 884

4-dimethylamino-3,5-xylyl methylcarbamate [315-18-4] **Vol. II:** 227

4-dimethylamino-3,5-xylyl-N-methyl carbamate [315-18-4] **Vol. II:** 227

4-dimethylamino-3,5-xylyl N-methylcarbamate [315-18-4] **Vol. II:** 227

dimethylanfiline [121-69-7] **Vol. I:** 171

dimethylaniline [1300-73-8] **Vol. III:** 976

2,6-dimethylaniline [87-62-7] **Vol. II:** 1092

dimethylarnideacetate [127-19-5] **Vol. I:** 10

dimethylarsinic acid [75-60-5] **Vol. I:** 214

dimethylbenz (a) anthracene [57-97-6] **Vol. III:** 88

7,12-dimethyl-1,2-benzanthracene [57-97-6] **Vol. III:** 88

dimethylbenzanthracene [57-97-6] **Vol. III:** 88

7,12-dimethylbenzanthrascene [57-97-6] **Vol. III:** 88

2,6-dimethylbenzenamine [87-62-7] **Vol. II:** 1092

p dimethylbenzene [106-42-3] **Vol. I:** 1152

1,2-dimethylbenzene [95-47-6] **Vol. I:** 1155

dimethylbenzene [1330-20-7] **Vol. I:** 1159

1,3-dimethylbenzene [108-38-3] **Vol. III:** 975

1,4-dimethylbenzene [106-42-3] **Vol. I:** 1152

o dimethylbenzene [95-47-6] **Vol. I:** 1155

n,n dimethylbenzeneamine [121-69-7] **Vol. I:** 171

dimethyl 1,2-benzenedicarboxylate [131-11-3] **Vol. II:** 784

dimethyl 1,4-benzenedicarboxylate [120-61-6] **Vol. I:** 1076

dimethyl benzeneorthodicarboxylate [131-11-3] **Vol. II:** 784

3,3'-dimethylbenzidin [119-93-7] **Vol. III:** 364

7,12-dimethylbenzo (a) anthracene [57-97-6] **Vol. III:** 88

1,4-dimethyl-2,3-benzphenanthrene [57-97-6] **Vol. III:** 88

alpha, alpha-dimethylbenzyl hydro peroxide [80-15-9] **Vol. II:** 490

3,3'-dimethyl-(1,1'-biphenyl)-4,4'-diamine [119-93-7] **Vol. III:** 364

3,3'-dimethylbiphenyl-4,4'-diamine [119-93-7] **Vol. III:** 364

3,3'-dimethyl-4,4'-biphenyldiamine [119-93-7] **Vol. III:** 364

1,1'-dimethyl-4,4'-bipyridinium [4685-14-7] **Vol. III:** 670

N,N'-dimethyl-4,4'-bipyridinium [4685-14-7] **Vol. III:** 670

1,1'-dimethyl-4,4'-bipyridinium cation [4685-14-7] **Vol. III:** 670

dimilin-g1 [35367-38-5] **Vol. III:** 339

dimilin-g4 [35367-38-5] **Vol. III:** 339

dimilin odc-45 [35367-38-5] **Vol. III:** 339

dimilin wp-25 [35367-38-5] **Vol. III:** 339

dimit [957-51-7] **Vol. II:** 6

dimp [1445-75-6] **Vol. III:** 342

dimpylat [333-41-5] **Vol. III:** 758

dimpylate [333-41-5] **Vol. III:** 758

di-n-amylamine [2050-92-2] **Vol. III:** 395

dinapacryl [485-31-4] **Vol. II:** 261

Di natrium aethylenbisdithiocarbamat (German) [142-59-6] **Vol. I:** 411

dinatrium-(n,n'-aethylen bis (dithiocarbamat)) (German) [142-59-6] **Vol. I:** 411

dinatrium-(n,n'-ethyleen bis (dithiocarbamaat)) (Dutch) [142-59-6] **Vol. I:** 411

di-n-butylamine [111-92-2] **Vol. III:** 311

di-n-butyl ether [142-96-1] **Vol. III:** 313

di-n-butylnitrosamine [924-16-3] **Vol. II:** 187

n, N-di-N-butylnitrosamine [924-16-3] **Vol. II:** 187

Di-n-butylnitrosamin (German) [924-16-3] **Vol. II:** 187

di-n-butyl phthalate [84-74-2] **Vol. II:** 774

dinex [131-89-5] **Vol. II:** 713

dinitrall [88-85-7] **Vol. II:** 704

2,4-dinitraniline [97-02-9] **Vol. III:** 368

dinitratodioxouranium [10102-06-4] **Vol. II:** 1071

dinitro [534-52-1] **Vol. I:** 540

2,4-dinitroanilina (Italian) [97-02-9] **Vol. III:** 368

2,4-Dinitroanilin (German) [97-02-9] **Vol. III:** 368

2,4-dinitrobenzenamine [97-02-9] **Vol. III:** 368

p-dinitrobenzene [100-25-4] **Vol. II:** 128

2,4-dinitrobenzene [99-65-0] **Vol. III:** 89

o dinitrobenzene [528-29-0] **Vol. I:** 258

dinitrobenzene [99-65-0] **Vol. III:** 89

m-dinitrobenzene [99-65-0] **Vol. III:** 89

dinitrobenzene, para- [100-25-4] **Vol. II:** 128

1,3-dinitrobenzol [99-65-0] **Vol. III:** 89

dinitrobenzol [25154-54-5] **Vol. I:** 261

dinitrobutylphenol [88-85-7] **Vol. II:** 704

dinitrocresol [534-52-1] **Vol. I:** 540

dinitrocyclohexylphenol [131-89-5] **Vol. II:** 713

2,4-dinitro-6-cyclohexyl phenol [131-89-5] **Vol. II:** 713

dinitrodendtroxal [534-52-1] **Vol. I:** 540

dinitrogen monoxide [10024-97-2] **Vol. I:** 889

dinitrogen oxide [10024-97-2] **Vol. I:** 889

3,5-dinitro-2-hydroxytoluene [534-52-1] **Vol. I:** 540

2,4-dinitro-6-methylphenol [534-52-1] **Vol. I:** 540

4,6-dinitro-2-(1-methyl-propyl) phenol [88-85-7] **Vol. II:** 704

2,4-dinitro-6-(1-methylpropyl) phenol [88-85-7] **Vol. II:** 704

2,4-dinitro-6-(1-methyl-propyl) phenol (French) [88-85-7] **Vol. II:** 704

2,4-dinitro o-cresol [534-52-1] **Vol. I:** 540

dinitro o-cresol [534-52-1] **Vol. I:** 540

4,6-dinitro o-cresolo (Italian) [534-52-1] **Vol. I:** 540

dinitro-o-cyclohexylphenol [131-89-5] **Vol. II:** 713

4,6-dinitro-o-cyclohexylphenol [131-89-5] **Vol. II:** 713

4,6-dinitro o-kresol (Czech) [534-52-1] **Vol. I:** 540

dinitro-ortho-cyclohexylphenol [131-89-5] **Vol. II:** 713

dinitro-ortho-sec-butyl phenol [88-85-7] **Vol. II:** 704

4,6-dinitro-o-sec-butylphenol [88-85-7] **Vol. II:** 704

2,4-dinitrophenylamine [97-02-9] **Vol. III:** 368

dinitrophenylmethane [25321-14-6] **Vol. II:** 1021

4,6-dinitrophenyl-2-sec-butyl-3-methyl-2-butenonate [485-31-4] **Vol. II:** 261

4,6-dinitro-2-sec, butylfenol (Czech) [88-85-7] **Vol. II:** 704

4,6-dinitro-2-sec butylfenol (Czech) [88-85-7] **Vol. II:** 704

4,6-dinitro-2-sec-butylphenol [88-85-7] **Vol. II:** 704

2,4-dinitro-6-sec-butylphenol [88-85-7] **Vol. II:** 704

4,6-dinitro-2-sec-butylphenyl beta, beta-dimethylacrylate [485-31-4] **Vol. II:** 261

2,4-dinitro-6-sec-butylphenyl 2-methylcrotonate [485-31-4] **Vol. II:** 261

dinitrosol [534-52-1] **Vol. I:** 540

2,4-dinitrotoluol [121-14-2] **Vol. II:** 1023

dinitrotoluol [25321-14-6] **Vol. II:** 1021

dinoc [534-52-1] **Vol. I:** 540

dinopol-nop [117-84-0] **Vol. II:** 789

dinoseb, 3,3-dimethylacryl ester [485-31-4] Vol. II: 261

dinosebe (French) [88-85-7] Vol. II: 704

dinoseb-methaerylate [485-31-4] Vol. II: 261

dinoxol [93-76-5] Vol. I: 59

di-n-propylamine [142-84-7] Vol. III: 378

dinurania [534-52-1] Vol. I: 540

dioctlyn [577-11-7] Vol. III: 859

dioctylal [577-11-7] Vol. III: 859

dioctyl 1,2-benzenedicarboxylate [117-84-0] Vol. II: 789

dioctyl-medo forte [577-11-7] Vol. III: 859

dioctyl o-benzenedicarboxylate [117-84-0] Vol. II: 789

dioctyl phthalate [117-81-7] Vol. II: 769

dioctyl-phthalate [117-84-0] Vol. II: 789

dioctyl sodium sulfosuccinate [577-11-7] Vol. III: 859

dioform [540-59-0] Vol. II: 406

dioksyny (Polish) [10192-30-0] Vol. II: 985

dioktylester-kyseliny-ftalove (Czech) [117-84-0] Vol. II: 789

diolamine [111-42-2] Vol. I: 658

diolane [107-41-5] Vol. III: 620

diol-14b [110-63-4] Vol. III: 150

diomedicone [577-11-7] Vol. III: 859

diomethane diglycidyl ether [1675-54-3] Vol. III: 132

di-on [330-54-1] Vol. II: 1078

1,4-diossan-2,3-diil-bis (O, O-dietil-ditiofosfato) (Italian) [78-34-2] Vol. II: 756

1,4-diossibenzene (Italian) [106-51-4] Vol. I: 303

diosuccin [577-11-7] Vol. III: 859

diotilan [577-11-7] Vol. III: 859

4,4'-di-o-toluidine [119-93-7] Vol. III: 364

diovac [577-11-7] Vol. III: 859

diox [577-11-7] Vol. III: 859

1,4-dioxaan-2,3-diyl-bis (O, O-diethyldithiofosfaat) (Dutch) [78-34-2] Vol. II: 756

1,4-dioxan [123-91-1] Vol. I: 618

1,4-Dioxan-2,3-diyl-bis (O, O-diaethyl-dithiophosphat) (German) [78-34-2] Vol. II: 756

1,4-dioxan-2,3-diyl bis (O, O-diethyl phosphorothiolothionate) [78-34-2] Vol. II: 756

1,4-dioxan-2,3-diyl OOO'O'-tetraethyl di (phosphorodithioate) [78-34-2] Vol. II: 756

p dioxane [123-91-1] Vol. I: 618

2,3-p-dioxanedithiol S,S-bis (O, O-diethyl phosphorodithioate) [78-34-2] Vol. II: 756

p-dioxane-2,3-dithiol, s,s-diester with O,O-diethyl phosphorodithioate [78-34-2] Vol. II: 756

p-dioxane-2,3-diyl ethyl phosphorodithioate [78-34-2] Vol. II: 756

2,3-p-dioxane S,S-bis (O, O-diethyl phosphorodithioate) [78-34-2] Vol. II: 756

1,3-dioxan-4-ol, 2-6-dimethyl-, acetate [828-00-2] Vol. III: 363

m-dioxan-4-ol, 2,6-dimethyl-, acetate [828-00-2] Vol. III: 363

1,4-dioxan, tetrahydro [123-91-1] Vol. I: 618

3,6-dioxaoctane-1,8-diol [112-27-6] Vol. II: 1058

dioxation [78-34-2] Vol. II: 756

dioxide of vitavax [5259-88-1] Vol. III: 659

1,1-dioxidetetrahydrothiofuran [126-33-0] Vol. II: 1007

1,1-dioxidetetrahydrothiophene [126-33-0] Vol. II: 1007

dioxido de carbono (Spanish) [124-38-9] Vol. III: 182

dioxin [10192-30-0] Vol. II: 985

dioxine [10192-30-0] Vol. II: 985

p dioxobenzene [123-31-9] Vol. I: 753

dioxodichlorochromium [14977-61-8] Vol. II: 247

S-((1,3-dioxo-2h-isoindol-2-yl) methyl) O,O-dimethyl phosphorodithioate [732-11-6] Vol. III: 737

2,4-dioxopentane [123-54-6] Vol. I: 910

1,3-dioxophthalan [85-44-9] Vol. I: 960

3,6-dioxopyridazine [123-33-1] Vol. II: 870

dioxosulfatouranium [1314-64-3] Vol. III: 956

1,1-dioxothiolan [126-33-0] Vol. II: 1007

dioxothiolan [126-33-0] Vol. II: 1007

dioxothion [78-34-2] Vol. II: 756

m dioxybenzene [108-46-3] Vol. I: 1026

o-dioxybenzene [120-80-9] Vol. III: 190

1,4-dioxybenzene [106-51-4] Vol. I: 303

1,4-Dioxy benzol (German) [106-51-4] Vol. I: 303

dioxyde de baryum (French) [1304-29-6] Vol. III: 82

dioxyde de carbone (French) [124-38-9] Vol. III: 182

dioxyethylene ether [123-91-1] Vol. I: 618

dipropylthiocarbamic acid S-propyl ester [1929-77-7] Vol. II: 221

dipterex [52-68-6] Vol. III: 707

dipthal [2303-17-5] Vol. III: 931

direct-brown-br [108-45-2] Vol. III: 98

direct-brown-gg [108-45-2] Vol. III: 98

direx-41 [330-54-1] Vol. II: 1078

dirnethylnitrosoamine- [62-75-9] Vol. I: 610

dirurol [330-54-1] Vol. II: 1078

disan [741-58-2] Vol. III: 731

disan (pesticide) [741-58-2] Vol. III: 731

di-sec-octyl phthalat [117-81-7] Vol. II: 769

disetil [97-77-8] Vol. III: 382

disflamoll-dpk [26444-49-5] Vol. III: 723

disflamoll tb [126-73-8] Vol. I: 951

disflamoll-tp [115-86-6] Vol. III: 951

disilyn [121-54-0] Vol. III: 105

disinall [8001-54-5] Vol. II: 67

disodium acid orthophosphate [7558-79-4] Vol. II: 933

disodium acid phosphate [7558-79-4] Vol. II: 933

disodium arsenate [7778-43-0] Vol. II: 93

disodium carbonate [497-19-8] Vol. II: 922

disodium chromate [7775-11-3] Vol. I: 469

disodium dichromate [10588-01-9] Vol. II: 300

disodium difluoride [7681-49-4] Vol. II: 925

disodium-dithionite [7775-14-6] Vol. III: 858

disodium 1,2-ethanediylbis (carbamodithioate) [142-59-6] Vol. I: 411

disodium ethylene-1,2-bisdithiocarbamate [142-59-6] Vol. I: 411

disodium ethylenebis (dithio carbamate) [142-59-6] Vol. I: 411

disodium hydrogen phosphate [7558-79-4] Vol. II: 933

disodium hydrophosphate [7558-79-4] Vol. II: 933

disodium hydrosulfite [7775-14-6] Vol. III: 858

disodium monohydrogen phosphate [7558-79-4] Vol. II: 933

disodium monosulfate [7757-82-6] Vol. III: 871

disodium monosulfide [1313-82-2] Vol. II: 935

disodium orthophosphate [7558-79-4] Vol. II: 933

disodium phosphoric acid [7558-79-4] Vol. II: 933

disodium selenite [10102-18-8] Vol. II: 903

disodium selenite [10102-18-8] Vol. III: 828

disodium selenium trioxide [10102-18-8] Vol. II: 903

disodium selenium trioxide [10102-18-8] Vol. III: 828

disodium sulfate [7757-82-6] Vol. III: 871

disodium sulfide [1313-82-2] Vol. II: 935

disodium sulphate [7757-82-6] Vol. III: 871

disodium tetraborate [1330-43-4] Vol. III: 874

disodium tetraborate decahydrate [1303-96-4] Vol. III: 134

disonate [577-11-7] Vol. III: 859

dispermine [110-85-0] Vol. III: 776

dissolvant-apv [111-46-6] Vol. II: 306

distilled mustard [505-60-2] Vol. II: 962

distokal [67-72-1] Vol. II: 330

distopan [67-72-1] Vol. II: 330

distopin [67-72-1] Vol. II: 330

disul [136-78-7] Vol. III: 855

disulfan [97-77-8] Vol. III: 382

disulfatozirconic acid [14644-61-2] Vol. I: 1193

disulfide, bis (dimethylthiocarbamoyl) [137-26-8] Vol. I: 632

disulfuram [97-77-8] Vol. III: 382

disulfure de tetramethylthiourame (French) [137-26-8] Vol. I: 632

disulfuro di tetrametiltiourame (Italian) [137-26-8] Vol. I: 632

disul-na [136-78-7] Vol. III: 855

disul-Na [136-78-7] Vol. III: 855

disul-sodium [136-78-7] Vol. III: 855

2,6-di-terc. butyl-p-kresol (Czech) [128-37-0] Vol. III: 317

di-tert-butylcresol [128-37-0] Vol. III: 317

2,6-di-tert-butylcresol [128-37-0] Vol. III: 317

2,6-di-tert-butyl-1-hydroxy-4-methylbenzene [128-37-0] Vol. III: 317

3,5-di-tert-butyl-4-hydroxytoluene [128-37-0] Vol. III: 317

2,6-di-tert-butyl-4-methylphenol [128-37-0] Vol. III: 317

di-tert-butyl-p-creso [128-37-0] Vol. III: 317

dmp [131-11-3] Vol. II: 784

2,6-dmp [576-26-1] Vol. I: 1167

3,4-dmp [95-65-8] Vol. I: 1169

3,5-dmp [108-68-9] Vol. I: 1165

dms-70 [67-68-5] Vol. II: 641

dms [75-18-3] Vol. I: 846

dms [77-78-1] Vol. II: 976

dms-90 [67-68-5] Vol. II: 641

dmsa [1596-84-5] Vol. III: 884

dms (methylsulfate) [77-78-1] Vol. II: 976

dmso [67-68-5] Vol. II: 641

dmt [120-61-6] Vol. I: 1076

dmtp [950-37-8] Vol. III: 734

dmtt [533-74-4] Vol. III: 293

dmu [330-54-1] Vol. II: 1078

dn-111 [131-89-5] Vol. II: 713

dn [131-89-5] Vol. II: 713

dn-1 [131-89-5] Vol. II: 713

dn-289 [88-85-7] Vol. II: 704

dn [534-52-1] Vol. I: 540

m-dnb [99-65-0] Vol. III: 89

dnbp [88-85-7] Vol. II: 704

dn-dry-mix-no-1 [131-89-5] Vol. II: 713

dnoc [534-52-1] Vol. I: 540

dnochp [131-89-5] Vol. II: 713

dnok (Czech) [534-52-1] Vol. I: 540

dnop [117-84-0] Vol. II: 789

dnosbp [88-85-7] Vol. II: 704

dnp [25550-58-7] Vol. III: 371

2,5-dnp [329-71-5] Vol. III: 374

dn (pesticide) [131-89-5] Vol. II: 713

dnsbp [88-85-7] Vol. II: 704

2,4-dnt [121-14-2] Vol. II: 1023

dnt [25321-14-6] Vol. II: 1021

2,6-dnt [606-20-2] Vol. I: 1096

3,4-dnt [610-39-9] Vol. II: 1028

do-14 [2312-35-8] Vol. II: 983

doa [103-23-1] Vol. II: 291

dobendan [123-03-5] Vol. III: 196

1,2,3,4,5,5,6,7,8,9,10,10-dodeca- chlorooctahydro-1,3,4-metheno-2-cycl obuta-(c,d)pentalone [143-50-0] Vol. III: 587

1,1a, 2,2,3,3a, 4,5,5,5a, 5b, 6-dodecachlorooctahydro-1,3,4-metheno-1- H-cyclobuta(cd) decane [2385-85-5] Vol. III: 583

1,1a, 2,2,3,3a, 4,5,5a, 5b, 6-dodecachlorooctahydro-1,3,4-metheno-1H-cyclobuta(cd) pentalene [2385-85-5] Vol. III: 583

dodeca- chlorooctahydro-1,3,4-metheno-2H-cyclobuta (c, d) pentalene [2385-85-5] Vol. III: 583

dodecachlorooctahydro-1,3,4-metheno-2h-cyclobuta(cd) pentalene [2385-85-5] Vol. III: 583

dodecachloropentacyclodecane [2385-85-5] Vol. III: 583

dodecachloro- pentacyclo (5.2.1.0 (2,6).0 (3,9).0 (5,8))decane [2385-85-5] Vol. III: 583

dodecachloropentacyclo (3.2.2.0 (sup2,6),0 (sup3,9),0 (sup5,10))decane [2385-85-5] Vol. III: 583

n-dodecane [112-40-3] Vol. III: 386

n-dodecanethiol [112-55-0] Vol. III: 388

n-Dodecan (German) [112-40-3] Vol. III: 386

n-dodecanol [112-53-8] Vol. II: 316

n-dodecan-1-ol [112-53-8] Vol. II: 316

dodecanoyl peroxide [105-74-8] Vol. I: 798

alpha-dodecene [112-41-4] Vol. III: 390

n-dodec-1-ene [112-41-4] Vol. III: 390

dodecyl [112-53-8] Vol. II: 316

dodecyl alcohol [112-53-8] Vol. II: 316

n-dodecyl alcohol [112-53-8] Vol. II: 316

1-dodecyl alcohol [112-53-8] Vol. II: 316

dodecyl alcohol, hydrogen sulfate, sodium salt [151-21-3] Vol. II: 981

dodecylbenzene-sodium-sulfonate [25155-30-0] Vol. III: 101

dodecylbenzenesuffonic acid, calcium salt [26264-06-2] Vol. II: 209

dodecylbenzenesulfonic acid [27176-87-0] Vol. I: 282

n dodecylbenzenesulfonic acid [27176-87-0] Vol. I: 282

dodecylbenzenesulfonic acid, sodium salt [25155-30-0] Vol. III: 101

dodecylbenzenesulphonic acid [27176-87-0] Vol. I: 282

dodecylbenzensulfonan-sodny (Czech) [25155-30-0] Vol. III: 101

dodecylbenzensulfonic acid, calcium salt [26264-06-2] Vol. II: 209

alpha-dodecylene [112-41-4] Vol. III: 390

dodecylguanidine acetate [2439-10-3] Vol. III: 290

n-dodecylguanidine acetate [2439-10-3] Vol. III: 290

dodecylguanidine monoacetate [2439-10-3] Vol. III: 290

dodecyl mercaptan [112-55-0] Vol. III: 388

n-dodecyl mercaptan [112-55-0] Vol. III: 388

1-dodecyl mercaptan [112-55-0] Vol. III: 388

dodecyl methacrylate [142-90-5] Vol. III: 391

dodecyl 2-methyl-2-propenoate [142-90-5] Vol. III: 391

dodecyl sodium sulfate [151-21-3] Vol. II: 981

dodecyl sulfate, sodium salt [151-21-3] Vol. II: 981

dodigen-226 [8001-54-5] Vol. II: 67

dodin [2439-10-3] Vol. III: 290

dodine [2439-10-3] Vol. III: 290

dodine acetate [2439-10-3] Vol. III: 290

dodine, mixture with glyodin [2439-10-3] Vol. III: 290

dof (Russian plasticizer) [117-81-7] Vol. II: 769

doguadine [2439-10-3] Vol. III: 290

dojyopicrin [76-06-2] Vol. II: 606

dol [608-73-1] Vol. I: 564

dolamin [7783-20-2] Vol. I: 159

dolen pur [87-68-3] Vol. I: 340

dolicur [67-68-5] Vol. II: 641

doligur [67-68-5] Vol. II: 641

dolmix [608-73-1] Vol. I: 564

dolochlor [76-06-2] Vol. II: 606

dominex [67375-30-8] Vol. III: 30

domoso [67-68-5] Vol. II: 641

dop [117-81-7] Vol. II: 769

doquadine [2439-10-3] Vol. III: 290

dorlone [542-75-6] Vol. II: 834

dormal [302-17-0] Vol. III: 197

dormiral [50-06-6] Vol. I: 223

dormol [50-00-0] Vol. I: 706

dormon [94-75-7] Vol. I: 30

dormone [94-75-7] Vol. I: 30

dorthion [121-75-5] Vol. II: 950

doruplant [834-12-8] Vol. II: 1054

dorvert [116-29-0] Vol. III: 906

doscalun [50-06-6] Vol. I: 223

doubledown [944-22-9] Vol. II: 735

doubledown [944-22-9] Vol. III: 714

double-strength [93-72-1] Vol. II: 845

dowanol-33b [107-98-2] Vol. II: 825

dowanol-50b [34590-94-8] Vol. III: 381

dowanol-db [112-34-5] Vol. III: 421

dowanol-dpm [34590-94-8] Vol. III: 381

dowanol-ee [110-80-5] Vol. II: 362

dowanol-ep [122-99-6] Vol. III: 695

dowanol-pm [107-98-2] Vol. II: 825

dowanol-te [112-50-5] Vol. II: 369

dowchlor [57-74-9] Vol. II: 619

dowcide-2s [88-06-2] Vol. II: 732

dowclene ls [71-55-6] Vol. I: 684

dowco-139 [315-18-4] Vol. II: 227

dowco-179 [2921-88-2] Vol. III: 762

dowco-186 [76-87-9] Vol. II: 941

dowco-290 [1702-17-6] Vol. III: 239

dow dormant fungicide [131-52-2] Vol. III: 676

dowfroth-250 [37286-64-9] Vol. II: 812

dowfume mc-2 [74-83-9] Vol. II: 577

dowfume mc-2 fumigant [74-83-9] Vol. II: 577

dowfume mc-2r [74-83-9] Vol. II: 577

dowfume mc-2 soil fumigant [74-83-9] Vol. II: 577

dowfume-n [8003-19-8] Vol. III: 792

Dow general weed killer [88-85-7] Vol. II: 704

dowicide-1 [90-43-7] Vol. II: 176

dowicide-2- [95-95-4] Vol. I: 936

dowicide-6 [58-90-2] Vol. I: 934

dowicide-7- [87-86-5] Vol. I: 928

'dowicide-a' [90-43-7] Vol. II: 176

dowicide-1-antimicrobial [90-43-7] Vol. II: 176

dowicide-7-antimicrobial [87-86-5] Vol. I: 928

dowicide ec-7 [87-86-5] Vol. I: 928

dowicide-g [131-52-2] Vol. III: 676

dowicide g-st [131-52-2] Vol. III: 676

dowicide-2s [88-06-2] Vol. II: 732

Dow-Per [127-18-4] Vol. II: 421

dowpon [75-99-0] Vol. I: 1016

dowpon m [75-99-0] Vol. I: 1016

Dow selective weed killer [88-85-7] Vol. II: 704

dowspray-17 [131-89-5] Vol. II: 713

dowtherm [107-21-1] Vol. I: 653

dowtherm-209- [107-98-2] Vol. II: 825

dowtherm e [95-50-1] Vol. I: 250

dowtherm sr-1- [107-21-1] Vol. I: 653

Dow-Tri, dukeron [79-01-6] Vol. II: 427

doxarfume mc-33 [74-83-9] Vol. II: 577

doxinate [577-11-7] Vol. III: 859

doxol [577-11-7] Vol. III: 859

dp-222 [1344-09-8] Vol. III: 868

2,2-dpa [75-99-0] Vol. I: 1016

dpa [26638-19-7] Vol. II: 298

dpa [75-99-0] Vol. I: 1016

DPH [122-66-7] Vol. II: 479

dpp [56-38-2] Vol. III: 750

DPX-3674 [51235-04-2], Vol. III: 481

dracylic acid [65-85-0] Vol. I: 293

dragnet [52645-53-1] Vol. III: 282

dragon [52645-53-1] Vol. III: 282

drapolene [8001-54-5] Vol. II: 67

drapolex [8001-54-5] Vol. II: 67

draza [2032-65-7] Vol. II: 308

dreft [151-21-3] Vol. II: 981

drest [8001-54-5] Vol. II: 67

drexar [2163-80-6] Vol. II: 610

drexel [330-54-1] Vol. II: 1078

drexel-parathion-8e [56-38-2] Vol. III: 750

drexel plant bed gas [74-83-9] Vol. II: 577

drexel prop-job [26638-19-7] Vol. II: 298

drexel-super p [123-33-1] Vol. II: 870

dri-kil [83-79-4] Vol. II: 166

dri tri [7601-54-9] Vol. II: 751

dromisol [67-68-5] Vol. II: 641

dr roger's tox ene [8001-35-2] Vol. I: 1118

dry and clear [94-36-0] Vol. I: 310

dry ice [124-38-9] Vol. III: 182

dry mix no, 1 [131-89-5] Vol. II: 713

dryseq [1344-09-8] Vol. III: 868

dse [142-59-6] Vol. I: 411

dsp [7558-79-4] Vol. II: 933

dss [577-11-7] Vol. III: 859

dtmc [115-32-2] Vol. II: 144

du-112307 [35367-38-5] Vol. III: 339

dual-murganic-rpb [5234-68-4] Vol. II: 670

dublofix [75-00-3] Vol. II: 321

du cor concentrated fly insecticide [105-67-9] Vol. I: 1162

dufalone [66-76-2] Vol. III: 336

dulsivac [577-11-7] Vol. III: 859

duneryl [50-06-6] Vol. I: 223

duodecyl alcohol [112-53-8] Vol. II: 316

duo kill [62-73-7] Vol. I: 946

duo-kill [7700-17-6] Vol. III: 266

duosol [577-11-7] Vol. III: 859

duphacid [35367-38-5] Vol. III: 339

Duphar [116-29-0] Vol. III: 906

duponal [151-21-3] Vol. II: 981

duponal-waqe [151-21-3] Vol. II: 981

duponol-c [151-21-3] Vol. II: 981

duponol-me [151-21-3] Vol. II: 981

duponol-methyl [151-21-3] Vol. II: 981

duponol-wa [151-21-3] Vol. II: 981

duponol-waq [151-21-3] Vol. II: 981

dupont-26 [1344-09-8] Vol. III: 868

du-pont-326 [330-55-2] Vol. II: 1084

Du Pont-732 [5902-51-2] Vol. III: 894

dupont 1179 [16752-77-5] Vol. II: 38

du-Pont-1991 [17804-35-2] Vol. II: 148

du-Pont-herbicide-326 [330-55-2] Vol. II: 1084

du Pont herbicide-326 [330-55-2] Vol. II: 1084

DuPont Herbicide-732 [5902-51-2] Vol. III: 894

du Pont insecticide 1179 [16752-77-5] Vol. II: 38

durafur developer c [120-80-9] Vol. III: 190

duran [330-54-1] Vol. II: 1078

durasorb [67-68-5] Vol. II: 641

dura treet ii [87-86-5] Vol. I: 928

duravos [62-73-7] Vol. I: 946

duravos [7700-17-6] Vol. III: 266

Durham-nematicode-em-17,1 [96-12-8] Vol. II: 818

dursban [2921-88-2] Vol. III: 762

dursban-4e [2921-88-2] Vol. III: 762

dursban-f [2921-88-2] Vol. III: 762

dusicnan barnaty (Czech) [10022-31-8] Vol. I: 233

dusicnan zirkonicity (Czech) [13746-89-9] Vol. I: 1191

dusitan sodny (Czech) [7632-00-0] Vol. I: 890

du sprex [1194-65-6] Vol. I: 299

Dutch liquid [107-06-2] Vol. I: 643

Dutch oil [107-06-2] Vol. I: 643

du-ter [76-87-9] Vol. II: 941

duter [76-87-9] Vol. II: 941

du-ter fungicide [76-87-9] Vol. II: 941

du-ter fungicide wettable powder [76-87-9] Vol. II: 941

du-ter pb-47 fungicide [76-87-9] Vol. II: 941

du-ter w-50 [76-87-9] Vol. II: 941

du-tur flowable-30 [76-87-9] Vol. II: 941

dvb-100 [1321-74-0] Vol. III: 967

dvb [1321-74-0] Vol. III: 967

dvb-22 [1321-74-0] Vol. III: 967

dvb-27 [1321-74-0] Vol. III: 967

dvb-55 [1321-74-0] Vol. III: 967

dvb-80 [1321-74-0] Vol. III: 967

dw-3418 [21725-46-2] Vol. II: 854

dwell [2593-15-9] Vol. III: 898

dwuchlorodwuetylowy-eter (Polish) [111-44-4] Vol. II: 379

dwuchlorodwufluorometan (Polish) [75-71-8] Vol. I: 825

2,4-dwuchlorofenoksyoctowy kwas (Polish) [94-75-7] Vol. I: 30

dwuchlorofluorometan (Polish) [75-43-4] Vol. II: 595

dwuchloropropan (Polish) [78-87-5] Vol. I: 987

dwuetyloamlna (Polish) [109-89-7] Vol. I: 597

dwumetyloanilina (Polish) [121-69-7] Vol. I: 171

dwumetylowy-siarczan (Polish) [77-78-1] Vol. II: 976

dwunitrobenzen (Polish) [99-65-0] Vol. III: 89

dwunitro o-krezol (Polish) [534-52-1] Vol. I: 540

dyak [1596-84-5] Vol. III: 884

dyanacide [62-38-4] Vol. II: 535

dybar [101-42-8] Vol. III: 452

dyclomec [1194-65-6] Vol. I: 299

dyfonat [944-22-9] Vol. II: 735

dyfonat [944-22-9] Vol. III: 714

dyfonate [944-22-9] Vol. II: 735

dyfonate [944-22-9] Vol. III: 714

dyfonate tillam 1-4e [944-22-9] Vol. II: 735

dyfonate tillam 1-4e [944-22-9] Vol. III: 714

dykanol [1336-36-3] Vol. I: 974

dylox [52-68-6] Vol. III: 707

dymel-22 [75-45-6] Vol. II: 591

dymid [957-51-7] Vol. II: 6

dymon SWH wasp and hornet spray [628-63-7] Vol. II: 27

dynasil-a [78-10-4] Vol. III: 447

dynex [330-54-1] Vol. II: 1078

dyprin [63-68-3] Vol. III: 593

dyrex [52-68-6] Vol. III: 707

dysect [52315-07-8] Vol. II: 251

dytol j-68 [112-53-8] Vol. II: 316

dytol m-83 [111-87-5] Vol. I: 897

dytop [88-85-7] Vol. II: 704

dyvon [52-68-6] Vol. III: 707

dyzol [333-41-5] Vol. III: 758

e393 [3689-24-5] Vol. II: 1010

e-600 [311-45-5] Vol. III: 669

e-605 [56-38-2] Vol. III: 750

e-1059 [8065-48-3] Vol. III: 755

e-3314- [76-44-8] Vol. II: 628

earthcide [82-68-8] Vol. II: 132

easygel [7783-47-3] Vol. III: 881

easy off-d [150-50-5] Vol. III: 574

EB [100-41-4] Vol. I: 264

ecatox [56-38-2] Vol. III: 750

eccothal [7446-18-6] Vol. I: 1079

echlomezol [2593-15-9] Vol. III: 898

echlomezole [2593-15-9] Vol. III: 898

ectiban [52645-53-1] Vol. III: 282

ectrin [51630-58-1] Vol. III: 95

edathamil [60-00-4] Vol. I: 36

edco [74-83-9] Vol. II: 577

edetic [60-00-4] Vol. I: 36

edetic acid [60-00-4] Vol. I: 36

edta [60-00-4] Vol. I: 36

edta acid [60-00-4] Vol. I: 36

edta (chelating agent) [60-00-4] Vol. I: 36

effomoll-doa [103-23-1] Vol. II: 291

effusan-3436- [534-52-1] Vol. I: 540

effusan [534-52-1] Vol. I: 540

efloran [443-48-1] Vol. III: 628

e-605-forte [56-38-2] Vol. III: 750

egitol [67-72-1] Vol. II: 330

egm [109-86-4] Vol. II: 370

egme [109-86-4] Vol. II: 370

ei [151-56-4] Vol. I: 671

ei-47-470 [950-10-7] Vol. III: 536

ei-47470 [950-10-7] Vol. III: 536

Eisendimethyldithiocarbamat (German) [14484-64-1] Vol. III: 489

Eisen(III)-tris (N,N-dimethyldithiocarbamat) (German) [14484-64-1] Vol. III: 489

EK-1700 [95-54-5] Vol. III: 697

ekagom-g [149-30-4] Vol. III: 540

ekagom tb [137-26-8] Vol. I: 632

e-kaprolaktam (Czech) [105-60-2] Vol. III: 79

ekatin-wf-and-wf-ulv [56-38-2] Vol. III: 750

ekatox [56-38-2] Vol. III: 750

eksmin [52645-53-1] Vol. III: 282

ektafos [141-66-2] Vol. III: 720

ektasolve-DB [112-34-5] Vol. III: 421

ektasolve-eb [112-07-2] Vol. II: 360

ektasolve eb acetate [112-07-2] Vol. II: 360

ektasolve-ee [110-80-5] Vol. II: 362

el-103 [34014-18-1] Vol. III: 892

elaic acid [112-80-1] Vol. I: 895

elaldehyde [123-63-7] Vol. III: 665

elaol [84-74-2] Vol. II: 774

elayl [74-85-1] Vol. I: 691

elbanil [101-21-3] Vol. II: 234

EPA pesticide chemical code 010501 [115-32-2] Vol. II: 144

EPA pesticide chemical code 011301 [96-12-8] Vol. II: 818

EPA pesticide chemical code 013501 [7778-44-1] Vol. III: 65

EPA pesticide chemical code 013905 [7647-14-5] Vol. III: 851

EPA pesticide chemical code 014701 [7778-54-3] Vol. II: 494

EPA pesticide chemical code 014703 [7681-52-9] Vol. II: 498

EPA pesticide chemical code 021201 [101-10-0] Vol. III: 806

EPA pesticide chemical code 022001 [8052-42-4] Vol. II: 114

EPA pesticide chemical code 022002 [8052-42-4] Vol. II: 114

EPA pesticide chemical code 025902 [108-94-1] Vol. II: 268

EPA pesticide chemical code 028001 [84-74-2] Vol. II: 774

EPA pesticide chemical code 028002 [131-11-3] Vol. II: 784

EPA pesticide chemical code 029001 [542-75-6] Vol. II: 834

EPA pesticide chemical code 042301 [75-21-8] Vol. II: 401

EPA pesticide chemical code 044003 [141-78-6] Vol. II: 19

EPA pesticide chemical code 044105 [108-10-1] Vol. II: 692

EPA pesticide chemical code 045001 [60-57-1] Vol. III: 356

EPA pesticide chemical code 045201 [67-72-1] Vol. II: 330

EPA pesticide chemical code 047501 [67-63-0] Vol. II: 515

EPA pesticide chemical code 047502 [71-23-8] Vol. II: 856

EPA pesticide chemical code 052001 [7487-94-7] Vol. II: 539

EPA pesticide chemical code 052102 [21908-53-2] Vol. II: 559

EPA pesticide chemical code 052104 [1600-27-7] Vol. II: 24

EPA pesticide chemical code 053201 [74-83-9] Vol. II: 577

EPA pesticide chemical code 056502 [82-68-8] Vol. II: 132

EPA pesticide chemical code 057701 [121-75-5] Vol. II: 950

EPA pesticide chemical code 063003 [131-52-2] Vol. III: 676

EPA pesticide chemical code 063201 [79-21-0] Vol. III: 682

EPA pesticide chemical code 064206 [59-50-7] Vol. II: 258

EPA pesticide chemical code 064212 [88-06-2] Vol. II: 732

EPA pesticide chemical code 068103 [556-61-6] Vol. III: 600

EPA pesticide chemical code 068304 [10588-01-9] Vol. II: 300

EPA pesticide chemical code 072003- [7488-56-4] Vol. III: 833

EPA pesticide chemical code 072603 [1344-09-8] Vol. III: 868

EPA pesticide chemical code 075301 [16919-19-0] Vol. II: 908

EPA pesticide chemical code 075307 [16871-71-9] Vol. II: 911

EPA pesticide chemical code 077501 [7704-34-9] Vol. II: 965

EPA pesticide chemical code 078001 [7664-93-9] Vol. II: 969

EPA pesticide chemical code 078501 [127-18-4] Vol. II: 421

EPA pesticide chemical code 078601 [79-34-5] Vol. III: 414

EPA pesticide chemical code 081202 [79-01-6] Vol. II: 427

EPA pesticide chemical code 083001 [56-35-9] Vol. II: 313

EPA pesticide chemical code 088601 [1314-84-7] Vol. II: 1100

EPA pesticide chemical code 089002 [127-82-2] Vol. II: 142

EPA pesticide chemical code 08901 [7733-02-0] Vol. II: 1103

EPA pesticide chemical code 100601 [22224-92-6] Vol. II: 743

EPA pesticide chemical code 268200 [2636-26-2] Vol. III: 271

EPA pesticide chemical code 326200 [76-14-2] Vol. III: 408

EPA pesticide chemical code 598300 [76-01-7] Vol. II: 336

EPA pesticide Code 011102 [1303-96-4] Vol. III: 134

EPA pesticide code-028801 [510-15-6] Vol. III: 108

EPA pesticide code 029101 [72-54-8] Vol. III: 404

EPA pesticide code 034001 [72-43-5] Vol. II: 345

EPA pesticide code 035001 [60-51-5] Vol. III: 740

EPA pesticide code 053101 [142-90-5] **Vol. III: 391**

EPA pesticide code-057501 [56-38-2] **Vol. III: 750**

EPA pesticide code-059101 [2921-88-2] **Vol. III: 762**

EPA pesticide code 066003 [62-38-4] **Vol. II: 535**

EPA pesticide code 079401 [115-29-7] **Vol. II: 661**

EPA pesticide code 080801 [834-12-8] **Vol. II: 1054**

EPA pesticide code 080803 [1912-24-9] **Vol. II: 1050**

EPA pesticide code 080805 [7287-19-6] **Vol. II: 1056**

EPA pesticide code 080807 [122-34-9] **Vol. II: 1047**

EPA pesticide code 080808. [139-40-2] **Vol. III: 802**

EPA pesticide code 080813 [886-50-0] **Vol. II: 1045**

EPA-Shaughnessy-#058001 [86-50-0] **Vol. III: 744**

EPA-Shaughnessy code 057701 [56-38-2] **Vol. III: 750**

EPA Shaughnessy code: 109301 [51630-58-1] **Vol. III: 95**

e-600 (pesticide) [311-45-5] **Vol. III: 669**

ephorran [97-77-8] **Vol. III: 382**

epichloorhydrine (Dutch) [106-89-8] **Vol. I: 981**

epichlorhydrin [106-89-8] **Vol. I: 981**

epichlorhydrine (French) [106-89-8] **Vol. I: 981**

alpha epichlorohydrin [106-89-8] **Vol. I: 981**

epichlorohydrin (DOT) [106-89-8] **Vol. I: 981**

epichlorohydryna (Polish) [106-89-8] **Vol. I: 981**

epichlorophydrin [106-89-8] **Vol. I: 981**

epi clear [94-36-0] **Vol. I: 310**

epicloridrina (Italian) [106-89-8] **Vol. I: 981**

epicure-ddm [101-77-9] **Vol. III: 304**

epifrin ophthalmic solution [8001-54-5] **Vol. II: 67**

epihydrinaldehyde [765-34-4] **Vol. II: 449**

epihydrine-aldehyde [765-34-4] **Vol. II: 449**

epikure-ddm [101-77-9] **Vol. III: 304**

e-pilo ophthalmic solution [8001-54-5] **Vol. II: 67**

epinal [8001-54-5] **Vol. II: 67**

epi-rez 510 [1675-54-3] **Vol. III: 132**

epn-300 [2104-64-5] **Vol. II: 738**

epoxide-a [1675-54-3] **Vol. III: 132**

epoxybutane [106-88-7] **Vol. II: 185**

1,2-epoxybutane [106-88-7] **Vol. II: 185**

1,4-epoxybutane [109-99-9] **Vol. I: 720**

1,2-epoxy-3-chloropropane [106-89-8] **Vol. I: 981**

3,6-epoxycyclohexane-1,2-dicarboxylicacid [145-73-3] **Vol. I: 900**

1,2-epoxyethane [75-21-8] **Vol. II: 401**

epoxyheptachlor [1024-57-3] **Vol. II: 626**

2,3-epoxy-1-propanal [765-34-4] **Vol. II: 449**

2,3-epoxypropanal [765-34-4] **Vol. II: 449**

1,2-epoxypropane [75-56-9] **Vol. I: 991**

epoxypropane [75-56-9] **Vol. I: 991**

2,3-epoxypropionaldehyde [765-34-4] **Vol. II: 449**

2,3-epoxypropylchloride [106-89-8] **Vol. I: 981**

2,3-epoxypropyl methacrylate [106-91-2] **Vol. III: 470**

2,2-bis (4-(2,3-epoxypropyloxy) phenyl) propane [1675-54-3] **Vol. III: 132**

eppy/n [8001-54-5] **Vol. II: 67**

epsilon-caprolactam [105-60-2] **Vol. III: 79**

epsylon-kaprolaktam (Polish) [105-60-2] **Vol. III: 79**

1,4,5,6,7,8,8-eptacloro-3a,4,7,7a-tetraldro-4,7-endo-metano-indene (Italian) [76-44-8] **Vol. II: 628**

eptacloro (Italian) [76-44-8] **Vol. II: 628**

equigel [62-73-7] **Vol. I: 946**

equino-aid [52-68-6] **Vol. III: 707**

erade [2439-01-2] **Vol. III: 663**

eradex [2921-88-2] **Vol. III: 762**

erase [75-60-5] **Vol. I: 214**

eraverm [110-85-0] **Vol. III: 776**

erazidon [2439-01-2] **Vol. III: 663**

erban [26638-19-7] **Vol. II: 298**

erbanil [26638-19-7] **Vol. II: 298**

ergoplast-addo [103-23-1] **Vol. II: 291**

ergoplast-fdb [84-74-2] **Vol. II: 774**

ergoplast-fdo [117-81-7] **Vol. II: 769**

ergoplast fdo-s [117-81-7] **Vol. II: 769**

erinitrit [7632-00-0] **Vol. I: 890**

erithane [76-87-9] **Vol. II: 941**

erl-2774 [1675-54-3] **Vol. III: 132**

ersoplast-FDA [84-74-2] Vol. II: 774

es-100 [78-10-4] Vol. III: 447

es-28 [78-10-4] Vol. III: 447

esaclorobenzene (Italian) [118-74-1] Vol. I: 268

esaidro-1,3,5-trinitro-1,3,5-triazina (Italian) [121-82-4] Vol. III: 277

esani (Italian) [110-54-3] Vol. II: 459

esbecythrin [52918-63-5] Vol. III: 296

escort [1702-17-6] Vol. III: 239

escre [302-17-0] Vol. III: 197

esen [85-44-9] Vol. I: 960

es-28 (ester) [78-10-4] Vol. III: 447

esgram [4685-14-7] Vol. III: 670

eskabarb [50-06-6] Vol. I: 223

eskimon-11 [75-69-4] Vol. II: 602

eskimon-12- [75-71-8] Vol. I: 825

eskimon-22 [75-45-6] Vol. II: 591

esperal (France) [97-77-8] Vol. III: 382

essence of mirbane- [98-95-3] Vol. I: 273

Essigester (German) [141-78-6] Vol. II: 19

essigsaeureanhydrid (German) [108-24-7] Vol. I: 73

essigsaeure (German) [64-19-7] Vol. I: 12

Esso fungicide 406 [133-06-2] Vol. II: 276

ester-25 [311-45-5] Vol. III: 669

ester of 2-chloroethanol with 2-(p-tert-butylphenoxy)-methyl sulphite [140-57-8] Vol. III: 61

esteron 44 [94-11-1] Vol. II: 15

esteron [93-76-5] Vol. I: 59

esteron-99-concentrate [1929-73-3] Vol. III: 155

esterone [93-76-5] Vol. I: 59

estol-1550 [84-66-2] Vol. II: 779

estonox [8001-35-2] Vol. I: 1118

estosteril [79-21-0] Vol. III: 682

estrex-1b-54, -1b-55 [123-95-5] Vol. III: 167

estricnina [57-24-9] Vol. I: 1054

estron-245 [93-76-5] Vol. I: 59

estrosel [62-73-7] Vol. I: 946

estrosol [62-73-7] Vol. I: 946

etain (tetrachlorure d') (French) [7646-78-8] Vol. III: 876

etanolamina (Italian) [141-43-5] Vol. II: 356

etanolo (Italian) [64-17-5] Vol. II: 394

etantiolo (Italian) [75-08-1] Vol. III: 419

etcmtb [2593-15-9] Vol. III: 898

eter monoetilico del etilenglicol (Spanish) [110-80-5] Vol. II: 362

ethaanthiol (Dutch) [75-08-1] Vol. III: 419

ethal [36653-82-4] Vol. III: 193

ethanal [75-07-0] Vol. I: 4

ethanamide [60-35-5] Vol. II: 4

ethanamine, n ethyl [109-89-7] Vol. I: 597

ethanamine, n ethyl n-nitroso [55-18-5] Vol. I: 601

ethanamine, n,n diethyl [121-44-8] Vol. I: 1125

ethanamine, N,N-dimethyl- [75-50-3] Vol. II: 573

ethandial [107-22-2] Vol. I: 728

1,2-ethandiol [107-21-1] Vol. I: 653

ethane, 2,2-bis (p-anisyl)-1,1,1-trichloro [72-43-5] Vol. II: 345

ethane, 2,2-bis (p-ethylphenyl)-1,1-dichloro [72-56-0] Vol. III: 687

ethane, bromo [74-96-4] Vol. III: 431

ethane, 1-butoxy-2-(2-thiocyanatoethoxy)- [112-56-1] Vol. III: 521

ethanecarboxylic acid [79-09-4] Vol. I: 1012

ethane, chloro- [75-00-3] Vol. II: 321

ethane, chloropentafluoro [76-15-3] Vol. III: 226

ethanedial [107-22-2] Vol. I: 728

1,2-ethanediamine, n-(2-aminoethyl) [111-40-0] Vol. I: 606

1,2-ethanediamine, n,n'-bis (2-aminoethyl)- [112-24-3] Vol. I: 1129

ethane, 1,2-diamino-, copper complex [13426-91-0] Vol. III: 270

ethane dichloride [107-06-2] Vol. I: 643

ethane, 1,1-dichloro [75-34-3] Vol. II: 325

ethane, 1,2-dichloro [107-06-2] Vol. I: 643

ethane, 1,1-dichloro-2,2-bis (p-chlorophenyl)- [72-54-8] Vol. III: 404

ethane, 1,1-dichloro-2,2-bis (p-ethylphenyl)- [72-56-0] Vol. III: 687

ethane, 1,2-dichloro-1,1,2,2-tetrafluoro- [76-14-2] Vol. III: 408

ethane, 1,2-dichlorotetrafluoro- [76-14-2] Vol. III: 408

ethane, 1,1-difluoro [75-37-6] Vol. III: 411

ethane, 1,2-difluoro-1,1,2,2-tetrachloro- [76-12-0] Vol. II: 328

ethanedinitrile [460-19-5] Vol. I: 555

ethanedioic acid [144-62-7] Vol. I: 904

ethanedioic acid, ammonium iron salt [2944-67-4] Vol. III: 454

ethanedioic acid, ammonium iron(3+) salt (3:3:1) [2944-67-4] Vol. III: 454

2-ethoxyethyle, acetate de (French) [111-15-9] Vol. II: 365

2-ethoxyethylester kyseliny octove (Czech) [111-15-9] Vol. II: 365

ethoxy-4-nitrophenoxyphenyl phosphine sulfide [2104-64-5] Vol. II: 738

ethoxy-(((4-nitrophenoxy) phenyl) phosphine) sulfide [2104-64-5] Vol. II: 738

5-ethoxy-3-trichloromethyl-1,2,4-thiadiazole [2593-15-9] Vol. III: 898

ethoxytriethylene glycol [112-50-5] Vol. II: 369

ethoxytriglycol [112-50-5] Vol. II: 369

ethrel [16672-87-0] Vol. III: 426

ethrel-c [16672-87-0] Vol. III: 426

ethylacetaat (Dutch) [141-78-6] Vol. II: 19

ethylacetic acid [107-92-6] Vol. I: 360

ethyl acetic ester [141-78-6] Vol. II: 19

ethylacrylaat (Dutch) [140-88-5] Vol. I: 114

ethylakrylat (Czech) [140-88-5] Vol. I: 114

ethyl alcohol [64-17-5] Vol. II: 394

ethyl alcohol, anhydrous [64-17-5] Vol. II: 394

ethylalcohol (Dutch) [64-17-5] Vol. II: 394

ethyl aldehyde [75-07-0] Vol. I: 4

ethyl alpha-methyl acrylate [97-63-2] Vol. III: 575

5–ethyl–alpha–picoline [104-90-5] Vol. III: 770

ethylamine, 2,2'-iminobis ethylene diamine, n-(2-aminoethyl) [111-40-0] Vol. I: 606

2-(ethylamino)-4-(isopropylamino)-6-(methylthio)-s-triazine [834-12-8] Vol. II: 1054

2-ethylamino-4-isopropylamino-6-methylthio-s-triazine [834-12-8] Vol. II: 1054

2-ethylamino-4-isopropylamino-6-methylthio-1,3,5-triazine [834-12-8] Vol. II: 1054

ethylan [72-56-0] Vol. III: 687

S-ethyl azepane-1-carbothioate [2212-67-1] Vol. III: 77

ethylbenzeen (Dutch) [100-41-4] Vol. I: 264

ethyl benzene [100-41-4] Vol. I: 264

ethylbenzol [100-41-4] Vol. I: 264

alpha-ethyl-beta-n-propylacrolein [645-62-5] Vol. II: 464

alpha-ethyl-beta-propylacrolein [645-62-5] Vol. II: 464

S-ethyl bis (2-methylpropyl) carbamothioate [2008-41-5] Vol. II: 216

ethyl butanoate [105-54-4] Vol. III: 438

2-ethyl-1-butanol [97-95-0] Vol. I: 343

2-ethylbutanol-1 [97-95-0] Vol. I: 343

2-ethylbutanol [97-95-0] Vol. I: 343

ethylbutylacetaldehyde [123-05-7] Vol. II: 457

2-ethylbutyl alcohol [97-95-0] Vol. I: 343

ethyl butyrate [105-54-4] Vol. III: 438

alpha-ethylcaproaldehyde [123-05-7] Vol. II: 457

2-ethylcaproaldehyde [123-05-7] Vol. II: 457

alpha-ethylcaproic acid [149-57-5] Vol. II: 461

2-ethylcaproic acid [149-57-5] Vol. II: 461

ethyl carbinol [71-23-8] Vol. II: 856

ethyl carbonochloridate [541-41-3] Vol. III: 433

ethyl-cellosolve [110-80-5] Vol. II: 362

ethyl cellosolve acetaat (Dutch) [111-15-9] Vol. II: 365

ethyl-cellosolve-acetate [111-15-9] Vol. II: 365

ethylchloorformiaat (Dutch) [541-41-3] Vol. III: 433

ethyl 4-chloro-alpha-(4-chlorophenyl)-alpha-hydroxybenzeneacetate [510-15-6] Vol. III: 108

ethyl chlorocarbonate [541-41-3] Vol. III: 433

ethylchlorohydrin [107-07-3] Vol. III: 220

ethyl chloromethanoate [541-41-3] Vol. III: 433

ethyl cyanide [107-12-0] Vol. II: 851

p,p'-ethyl-ddd [72-56-0] Vol. III: 687

p,p-ethyl ddd [72-56-0] Vol. III: 687

ethyl 4,4'-dichlorobenzilate [510-15-6] Vol. III: 108

ethyldichlorobenzilate [510-15-6] Vol. III: 108

ethyl 4,4'-dichlorodiphenyl glycollate [510-15-6] Vol. III: 108

ethyl 4,4'-dichlorophenyl glycollate [510-15-6] Vol. III: 108

S-ethyldiisobutyl thiocarbamate [2008-41-5] Vol. II: 216

S-ethyl diisobutylthiocarbamate [2008-41-5] Vol. II: 216

S-ethyl di-isobutyl (thiocarbamate) [2008-41-5] Vol. II: 216

ethylene glycol monoethyl ether [110-80-5] Vol. II: 362

ethylene glycol monoethyl ether acetylated [111-15-9] Vol. II: 365

ethylene glycol monomethyl ether [109-86-4] Vol. II: 370

ethylene glycol, monomethyl ether [109-86-4] Vol. II: 370

ethylene glycol monophenyl ether [122-99-6] Vol. III: 695

ethylene glycol phenyl ether [122-99-6] Vol. III: 695

ethylene glycol polymer [25322-68-3] Vol. II: 811

ethylene hexachloride [67-72-1] Vol. II: 330

ethylene monochloride [75-01-4] Vol. II: 416

ethylene oxide, homopolymer [25322-68-3] Vol. II: 811

ethylene oxide, methyl [75-56-9] Vol. I: 991

ethylene oxide polymer [25322-68-3] Vol. II: 811

ethylene, phenyl [100-42-5] Vol. I: 1060

ethylene polyoxide [25322-68-3] Vol. II: 811

ethylene tetrachloride [127-18-4] Vol. II: 421

ethylene, tetrachloro [127-18-4] Vol. II: 421

1,3-ethylenethiourea [96-45-7] Vol. I: 757

1,3-ethylene-2-thiourea [96-45-7] Vol. I: 757

ethylenethiourea [96-45-7] Vol. I: 757

n,n'-ethylenethiourea [96-45-7] Vol. I: 757

n,n ethylene thiourea [96-45-7] Vol. I: 757

ethylene trichloride [79-01-6] Vol. II: 427

ethylene, trichloro- [79-01-6] Vol. II: 427

ethylenimine [151-56-4] Vol. I: 671

ethyl ester [141-78-6] Vol. II: 19

S-ethyl ester hexahydro-1H-azepine- 1-carbothioic acid [2212-67-1] Vol. III: 77

ethylester-kyseliny-karbaminove (Czech) [51-79-6] Vol. II: 224

ethylester kyseliny octove (Czech) [141-78-6] Vol. II: 19

ethyl ester of 4,4'-dichlorobenzilic acid [510-15-6] Vol. III: 108

n ethylethanamine [109-89-7] Vol. I: 597

ethyl ethanoate [141-78-6] Vol. II: 19

ethyl ethylene glycol [110-80-5] Vol. II: 362

ethylethylene oxide [106-88-7] Vol. II: 185

ethyl formate (DOT) [109-94-4] Vol. II: 438

ethylformiaat (Dutch) [109-94-4] Vol. II: 438

ethylformic acid [79-09-4] Vol. I: 1012

ethyl formic ester [109-94-4] Vol. II: 438

ethyl glycol [110-80-5] Vol. II: 362

ethyl glycol acetate [111-15-9] Vol. II: 365

ethylglykolacetat (German) [111-15-9] Vol. II: 365

S-ethyl hexahydroazepine-1-carbothioate [2212-67-1] Vol. III: 77

S-ethyl hexahydro-1H-azepine-1-carbothioate [2212-67-1] Vol. III: 77

ethylhexaldehyde [123-05-7] Vol. II: 457

ethyl 1-hexamethyleneiminecarbothiolate [2212-67-1] Vol. III: 77

S-ethyl 1-hexamethyleneiminothiocarbamate [2212-67-1] Vol. III: 77

2-ethylhexanal [123-05-7] Vol. II: 457

ethylhexanoic acid [149-57-5] Vol. II: 461

alpha-ethylhexanoic acid [149-57-5] Vol. II: 461

2-ethyl-2-hexen-1-al [645-62-5] Vol. II: 464

2-ethyl-2-hexenal [645-62-5] Vol. II: 464

2-ethylhexenal [645-62-5] Vol. II: 464

alpha-ethyl-2-hexenal [645-62-5] Vol. II: 464

ethylhexoic acid [149-57-5] Vol. II: 461

2-ethylhexoic acid [149-57-5] Vol. II: 461

2-(2-ethylhexyl)-3a, 4,7,7a-tetrahydro-4,7-methano-1h-isoindole-1,3 (2h)-dione [113-48-4] Vol. III: 631

2-ethylhexylaldehyde [123-05-7] Vol. II: 457

n-(2-ethylhexyl) bicyclo-(2,2,1)-hept-5-ene-2,3-dicarboximide [113-48-4] Vol. III: 631

n-2-ethylhexylimide endomethylenetetrahydrophthalic acid [113-48-4] Vol. III: 631

n-(2-ethylhexyl)-5-norbornene-2,3-dicarboximide [113-48-4] Vol. III: 631

ethylhexyl phthalate [117-81-7] Vol. II: 769

2-ethylhexyl phthalate [117-81-7] Vol. II: 769

2-ethylhexyl, 2-propenoate [103-11-7] Vol. II: 45

n-(2-ethylhexyl) -8,9,10-trinorborn-5-ene-2,3-dicarboximide [113-48-4] Vol. III: 631

ethyl hydrate [64-17-5] Vol. II: 394

ethyl hydride [74-84-0] Vol. III: 401

ethyl hydrosulfide [75-08-1] Vol. III: 419

ethyl hydroxide [64-17-5] Vol. II: 394

ethyl 2-hydroxy-2,2-bis (4-chlorophenyl) acetate [510-15-6] Vol. III: 108

ethyl 2-hydroxy-2,2-di (p-chlorophenyl) acetate [510-15-6] Vol. III: 108

5-ethylidenebicyclo (2,2,1) hep-2-ene [16219-75-3] Vol. III: 440

5-ethylidenebicyclo (2.2.1) hept-2-ene [16219-75-3] Vol. III: 440

ethylidene-chloride [75-34-3] Vol. II: 325

ethylidene dichloride [75-34-3] Vol. II: 325

ethylidene difluoride [75-37-6] Vol. III: 411

ethylidene-fluoride [75-37-6] Vol. III: 411

ethylidenelactic acid [50-21-5] Vol. I: 792

ethylidenenorbornene [16219-75-3] Vol. III: 440

ethylimine [151-56-4] Vol. I: 671

ethyl ketone [96-22-0] Vol. III: 679

ethyl 'm [77-81-6] Vol. III: 891

ethylmercaptaan (Dutch) [75-08-1] Vol. III: 419

ethyl mercuric phosphate [2235-25-8] Vol. III: 441

ethylmercury phosphate [2235-25-8] Vol. III: 441

ethyl mercury phosphate [2235-25-8] Vol. III: 441

ethylmerkaptan (Czech) [75-08-1] Vol. III: 419

ethyl 2-methacrylate [97-63-2] Vol. III: 575

ethyl methacrylate [97-63-2] Vol. III: 575

ethyl methanoate [109-94-4] Vol. II: 438

ethylmethyl carbinol [78-92-2] Vol. II: 194

ethylmethylcetone (French) [78-93-3] Vol. I: 347

ethyl methyl cetone (French) [78-93-3] Vol. I: 347

ethyl-methylene-phosphorodithioate [563-12-2] Vol. II: 759

ethylmethylketon (Dutch) [78-93-3] Vol. I: 347

ethyl methyl ketone [78-93-3] Vol. I: 347

ethyl-methyl-ketone-peroxide [1338-23-4] Vol. II: 192

ethyl 3-methyl-4-(methylthio) phenyl iso-propylphoshoramidate [22224-92-6] Vol. II: 743

ethyl 3-methyl-4-(methylthio) phenyl (1-methylethyl) phosphoramidate [22224-92-6] Vol. II: 743

ethyl 2-methyl-2-propenoate [97-63-2] Vol. III: 575

5–ethyl–2–methyl pyridine [104-90-5] Vol. III: 770

ethyl 4-methylthio-m-tolyl isopropylphos-phoramidate [22224-92-6] Vol. II: 743

ethyl 4-(methylthio)-m-tolyl isopropylphos-phoramidate [22224-92-6] Vol. II: 743

ethyl N-dimethylphos- phoramidocyanidate [77-81-6] Vol. III: 891

S-ethyl-N-hexamethylenethiocarbamate [2212-67-1] Vol. III: 77

N-ethyl-N-isopropyl-6-methylthio-1,3,5-tria-zine-2,4-diamine [834-12-8] Vol. II: 1054

ethyl nitrile [75-05-8] Vol. I: 81

N-ethyl-N'-(1-methylethyl)-6-(methyl-thio)-1,3,5-triazine-2,4-diamine [834-12-8] Vol. II: 1054

ethyl-N, N-diisobutylthiocarbamate [2008-41-5] Vol. II: 216

ethyl N,N-diisobutylthiocarbamate [2008-41-5] Vol. II: 216

S-ethyl N,N-diisobutyl thiocarbamate [2008-41-5] Vol. II: 216

S-ethyl N,N-diisobutylthiocarbamate [2008-41-5] Vol. II: 216

ethyl-N, N-diisobutyl thiolcarbamate [2008-41-5] Vol. II: 216

S-ethyl N,N-diisobutylthiolcarbamate [2008-41-5] Vol. II: 216

S-ethyl N,N-diisobutyl thiolcarbamate [2008-41-5] Vol. II: 216

O-ethyl N,N-dimethylphos- phoramidocya-nidate [77-81-6] Vol. III: 891

S-ethyl N,N-hexamethylenethiocarbamate [2212-67-1] Vol. III: 77

n ethyl n-nitroso ethanamine [55-18-5] Vol. I: 601

n ethyl n-nitrosoethanamine [55-18-5] Vol. I: 601

ethylolamine [141-43-5] Vol. II: 356

o-ethyl-o-((4-nitro-fenyl)-fenyl)-monothiofos-fonaat (Dutch) [2104-64-5] Vol. II: 738

o-ethyl o-(4-nitrophenyl) benzenethiono-phosphonate [2104-64-5] Vol. II: 738

o-ethyl o-(4-nitrophenyl) phenylphospho-nothioate [2104-64-5] Vol. II: 738

o-ethyl o-4-nitrophenyl phenyl-phospho-
nothioate [2104-64-5] **Vol. II:** 738

o-ethyl o-4-nitrophenyl phenylphospho-
nothioate [2104-64-5] **Vol. II:** 738

o-ethyl o-p-nitrophenyl benzenethiophos-
phonate [2104-64-5] **Vol. II:** 738

o-ethyl o-(p-nitrophenyl) phenylphospho-
nothioate [2104-64-5] **Vol. II:** 738

o-ethyl-o-p-nitrophenylphenyl phospho-
nothioate [2104-64-5] **Vol. II:** 738

o-ethyl o-p-nitrophenyl phenylphospho-
nothioate [2104-64-5] **Vol. II:** 738

o-ethyl o-p-nitrophenyl phenylphosphono-
thiolate [2104-64-5] **Vol. II:** 738

o-ethyl o-(p-nitrophenyl) phenylphosphono-
thionate [2104-64-5] **Vol. II:** 738

o-ethyl o-p-nitrophenyl phenylphospho-
rothioate [2104-64-5] **Vol. II:** 738

o-ethyl-o, p-nitrophenyl phenylphospho-
rothioate [2104-64-5] **Vol. II:** 738

ethyl orthosilicate [78-10-4] **Vol. III:** 447

ethyl oxide [60-29-7bf] **Vol. III:** 412

ethyloxirane [106-88-7] **Vol. II:** 185

2-ethyloxirane [106-88-7] **Vol. II:** 185

ethyl-paraoxon [311-45-5] **Vol. III:** 669

S-ethyl perhydroazepin-1-carbothioate
[2212-67-1] **Vol. III:** 77

S-ethyl perhydroazepine-1-thiocarboxylate
[2212-67-1] **Vol. III:** 77

5-ethyl-5-phenylbarbituric acid [50-06-6]
Vol. I: 223

2,2-bis (p-ethylphenyl)-1,1-dichloroethane
[72-56-0] **Vol. III:** 687

1,1-bis (p-ethylphenyl)-2,2-dichloroethane
[72-56-0] **Vol. III:** 687

o-ethyl phenylphosphonothioic acid o-(4-
nitrophenyl) ester [2104-64-5] **Vol. II:**
738

o-ethyl phenyl (p-nitrophenyl) thiophos-
phonate [2104-64-5] **Vol. II:** 738

o-ethyl phenyl p-nitrophenyl thiophospho-
nate [2104-64-5] **Vol. II:** 738

ethylphosphonodithioic acid o-ethyl S-phenyl ester [944-22-9] **Vol. III:** 714

ethylphosphonodithioic acid o-ethyl s-phenyl ester [944-22-9] **Vol. II:** 735

ethyl phosphorodimethylamidocyanidate
[77-81-6] **Vol. III:** 891

ethyl-phthalate [84-66-2] **Vol. II:** 779

5–ethyl–2–picoline [104-90-5] **Vol. III:** 770

5–ethyl–4–picoline [104-90-5] **Vol. III:** 770

ethyl p-nitrophenyl benzenethionophos-
phonate [2104-64-5] **Vol. II:** 738

ethyl p-nitrophenyl benzenethiophosphate
[2104-64-5] **Vol. II:** 738

ethyl p-nitrophenyl benzenethiophospho-
nate [2104-64-5] **Vol. II:** 738

ethyl p-nitrophenyl phenylphosphonoth-
ioate [2104-64-5] **Vol. II:** 738

ethyl p-nitrophenyl thionobenzene phosphate [2104-64-5] **Vol. II:** 738

ethyl-p-nitrophenylthionobenzene phosphate [2104-64-5] **Vol. II:** 738

ethyl (p-nitrophenyl) thionobenzene phosphonate [2104-64-5] **Vol. II:** 738

ethyl p, p′-dichlorobenzilate [510-15-6]
Vol. III: 108

ethyl propenoate [140-88-5] **Vol. I:** 114

ethyl 2-propenoate [140-88-5] **Vol. I:** 114

ethyl-propionyl [96-22-0] **Vol. III:** 679

2-ethyl-3-propylacrylaldehyde [645-62-5]
Vol. II: 464

ethyl pyrophosphate (Et4p207) [107-49-3]
Vol. II: 886

ethyl pyrophosphate, tetra [107-49-3] **Vol.
II:** 886

ethyl silicon trichloride [115-21-9] **Vol. III:**
933

o-ethyl s-phenyl ethyldithio phosphonate
[944-22-9] **Vol. II:** 735

o-ethyl S-phenyl ethyldithiophosphonate
[944-22-9] **Vol. III:** 714

(°2°)-o-ethyl s-phenyl ethylphosphonodithioate [944-22-9] **Vol. II:** 735

o-ethyl s-phenyl ethylphosphonodithioate
[944-22-9] **Vol. II:** 735

(−)-o-ethyl S-phenyl ethylphosphonodithioate [944-22-9] **Vol. III:** 714

o-ethyl s-phenyl ethylphosphono- dithioate
[944-22-9] **Vol. III:** 714

o-ethyl s-phenyl ethylphosphono thiolo-
thionate [944-22-9] **Vol. II:** 735

o-ethyl S-phenyl ethylphosphonothiolothionate [944-22-9] **Vol. III:** 714

o-ethyl s-phenyl(rs)-ethylphosphonodith-
ioate [944-22-9] **Vol. II:** 735

o-ethyl S-phenyl(rs)-ethylphosphonodith-
ioate [944-22-9] **Vol. III:** 714

o-ethyl s,s-dipropyl dithiophosphate
[13194-48-4] **Vol. III:** 427

o-ethyl s,s-dipropyl phosphorodithioate
[13194-48-4] **Vol. III:** 427

o-ethyl s,s-dipropylphosphorodithioate
[13194-48-4] **Vol. III:** 427

o-ethyl-s, s-dipropyl phosphorodithionate
[13194-48-4] **Vol. III:** 427

ethyl sulfhydrate [75-08-1] Vol. III: 419

s-(2-(ethylsulfinyl) ethyl) o,o-dimethl phosphorothioate [301-12-2] Vol. III: 661

ethyl-systox [8065-48-3] Vol. III: 755

ethyl-tetraphosphate [757-58-4] Vol. II: 992

ethyl tetraphosphate, hexa- [757-58-4] Vol. II: 992

ethyl thioalcohol [75-08-1] Vol. III: 419

(N-(2-ethylthio) benzene sulphonamide-S, o,o-diisopropylphosphorodithioate [741-58-2] Vol. III: 731

ethyl thiopyrophosphate [3689-24-5] Vol. II: 1010

ethyl thiudad [97-77-8] Vol. III: 382

ethyltrichlorosilane [115-21-9] Vol. III: 933

ethyltriglycol [112-50-5] Vol. II: 369

ethyl-tuads [97-77-8] Vol. III: 382

ethyl urethan [51-79-6] Vol. II: 224

ethylurethan [51-79-6] Vol. II: 224

ethyl urethane [51-79-6] Vol. II: 224

o-ethylurethane [51-79-6] Vol. II: 224

ethyne [74-86-2] Vol. I: 95

ethyne, calcium deriv [75-20-7] Vol. I: 391

ethynylcarbinol [107-19-7] Vol. II: 864

eticol [311-45-5] Vol. III: 669

etil acrilato (Italian) [140-88-5] Vol. I: 114

etilacrilatului (Roumanian) [140-88-5] Vol. I: 114

etilbenzene (Italian) [100-41-4] Vol. I: 264

etil clorocarbonato (Italian) [541-41-3] Vol. III: 433

etil-cloroformiato (Italian) [541-41-3] Vol. III: 433

etile (acetato di) (Italian) [141-78-6] Vol. II: 19

etile (formiato di)-(Italian) [109-94-4] Vol. II: 438

N,N'-etilen-bis (ditiocarbammato) di manganese (Italian) [12427-38-2] Vol. III: 531

n,n'-etilen bis (ditiocarbammato) di sodio (Italian) [142-59-6] Vol. I: 411

etilenimina (Italian) [151-56-4] Vol. I: 671

etileno [74-85-1] Vol. I: 691

etilfen [50-06-6] Vol. I: 223

etilmercaptano (Italian) [75-08-1] Vol. III: 419

etilon [56-38-2] Vol. III: 750

o-etil-o-((4-nitro-fenil)-fenil)-monotio fosfonato (Italian) [2104-64-5] Vol. II: 738

etiol [121-75-5] Vol. II: 950

etmt [2593-15-9] Vol. III: 898

eto [75-21-8] Vol. II: 401

etoksyetylowy alkohol [110-80-5] Vol. II: 362

2-etossietil-acetato (Italian) [111-15-9] Vol. II: 365

etridiazol [2593-15-9] Vol. III: 898

etridiazole [2593-15-9] Vol. III: 898

etu [96-45-7] Vol. I: 757

etylenu-tlenek (Polish) [75-21-8] Vol. II: 401

etylobenzen (Polish) [100-41-4] Vol. I: 264

etylowy alkohol (Polish) [64-17-5] Vol. II: 394

etylu-bromek (Polish) [74-96-4] Vol. III: 431

etylu-chlorek (Polish) [75-00-3] Vol. II: 321

etylu-krzemian (Polish) [78-10-4] Vol. III: 447

eucanine-gb [95-80-7] Vol. II: 1016

eumin [443-48-1] Vol. III: 628

eunatrol [143-19-1] Vol. I: 899

euneryl [50-06-6] Vol. I: 223

Eureka products, criosine [67-56-1] Vol. I: 836

Eureka products criosine disinfectant [67-56-1] Vol. I: 836

euxyl k-100- [100-51-6] Vol. I: 313

evercyn [74-90-8] Vol. I: 743

evik [834-12-8] Vol. II: 1054

eviplast-80 [117-81-7] Vol. II: 769

eviplast-81 [117-81-7] Vol. II: 769

evits [7664-38-2] Vol. I: 942

evola [106-46-7] Vol. I: 254

exact s [75-18-3] Vol. I: 846

exagien [119-36-8] Vol. III: 622

exhoran [97-77-8] Vol. III: 382

exitelite [1309-64-4] Vol. II: 86

exo-dieldrin [60-57-1] Vol. III: 356

exodin [333-41-5] Vol. III: 758

exolit lpkn [7723-14-0] Vol. II: 764

exolit vpk-n 361 [7723-14-0] Vol. II: 764

exotherm [1897-45-6] Vol. II: 513

exotherm-termil [1897-45-6] Vol. II: 513

exo-1,7,7-trimethylbicyclo[2.2.1]hept-2-yl thiocyanatoacetate [115-31-1] Vol. III: 491

expar [52645-53-1] Vol. III: 282

experimental fungicide 5223 [2439-10-3] Vol. III: 290

experimental herbicide-1 [136-78-7] Vol. III: 855

experimental herbicide 732 [5902-51-2]
Vol. III: 894
experimental insecticide 12,880 [60-51-5]
Vol. III: 740
experimental insecticide-269 [72-20-8] Vol.
III: 349
experimental insecticide 4124 [2463-84-5]
Vol. III: 748
explosive d [131-74-8] Vol. I: 156
exporsan [741-58-2] Vol. III: 731
exsel [7488-56-4] Vol. III: 833
exsiccated sodium phosphate [7558-79-4]
Vol. II: 933
extermathion [121-75-5] Vol. II: 950
extrar [534-52-1] Vol. I: 540
extrax [83-79-4] Vol. II: 166
extrema [1309-64-4] Vol. II: 86
extrusil [16871-71-9] Vol. II: 911
extrusil [7631-86-9] Vol. III: 846
e-z-off d [78-48-8] Vol. III: 768
f-11 [75-69-4] Vol. II: 602
f-12- [75-71-8] Vol. I: 825
f-21 [75-43-4] Vol. II: 595
f-22 [75-45-6] Vol. II: 591
f-40 [14484-64-1] Vol. III: 489
f-112 [76-12-0] Vol. II: 328
f-113 [76-13-1] Vol. II: 353
f-114 [76-14-2] Vol. III: 408
f-115 [76-15-3] Vol. III: 226
f-461 [5259-88-1] Vol. III: 659
f461 [5259-88-1] Vol. III: 659
f-735- [5234-68-4] Vol. II: 670
f1991 [17804-35-2] Vol. II: 148
faber [1897-45-6] Vol. II: 513
fac [1185-57-5] Vol. I: 491
factitious air [10024-97-2] Vol. I: 889
fair-2 [123-33-1] Vol. II: 870
fair-30 [123-33-1] Vol. II: 870
fair-plus [123-33-1] Vol. II: 870
fair-ps [123-33-1] Vol. II: 870
falisilvan [101-42-8] Vol. III: 452
falitiram [137-26-8] Vol. I: 632
falkitol [67-72-1] Vol. II: 330
faltan [133-07-3] Vol. III: 461
faltex [133-07-3] Vol. III: 461
famfos [13171-21-6] Vol. III: 704
fannoform [50-00-0] Vol. I: 706
fargo [2303-17-5] Vol. III: 931
far-go [2303-17-5] Vol. III: 931
farmaneb [12427-38-2] Vol. III: 531
farmco-atrazine [1912-24-9] Vol. II: 1050
farmco-diuron [330-54-1] Vol. II: 1078

farmco fence rider [93-76-5] Vol. I: 59
farmco-propanil [26638-19-7] Vol. II: 298
fartox [82-68-8] Vol. II: 132
fasciolin [56-23-5] Vol. I: 435
fasciolin [67-72-1] Vol. II: 330
fasco terpene [8001-35-2] Vol. I: 1118
fasco wy-hoe [101-21-3] Vol. II: 234
fastac [52315-07-8] Vol. II: 251
fastac [67375-30-8] Vol. III: 30
fast corinth base b [92-87-5] Vol. I: 288
fast dark-blue base-r [119-93-7] Vol. III:
364
fast-garnet-base-b [134-32-7] Vol. III: 637
fast-garnet-b-base [134-32-7] Vol. III: 637
fast orange base gr [88-74-4] Vol. I: 173
fast orange base jr [88-74-4] Vol. I: 173
fast orange gr base [88-74-4] Vol. I: 173
fast orange gr salt [88-74-4] Vol. I: 173
fast orange o-base [88-74-4] Vol. I: 173
fast orange o-salt [88-74-4] Vol. I: 173
fast orange salt gr [88-74-4] Vol. I: 173
fast orange salt jr [88-74-4] Vol. I: 173
fast red base gg [100-01-6] Vol. II: 70
fast red base 2j [100-01-6] Vol. II: 70
fast red 2g base [100-01-6] Vol. II: 70
fast red gg base [100-01-6] Vol. II: 70
fast red gg salt [100-01-6] Vol. II: 70
fast red 2g salt [100-01-6] Vol. II: 70
fast red mp base [100-01-6] Vol. II: 70
fast red p base [100-01-6] Vol. II: 70
fast red p salt [100-01-6] Vol. II: 70
fast red salt gg [100-01-6] Vol. II: 70
fast red salt 2j [100-01-6] Vol. II: 70
fast-spirit-yellow [60-09-3] Vol. II: 74
fast-white [7446-14-2] Vol. III: 517
fast-yellow-at [97-56-3] Vol. III: 46
fast-yellow-b [97-56-3] Vol. III: 46
fast-yellow-gc-base [95-51-2] Vol. III: 216
fat-yellow-aab [60-09-3] Vol. II: 74
fazor [123-33-1] Vol. II: 870
f-13b1 [75-63-8] Vol. III: 143
fbhc [608-73-1] Vol. I: 564
fc-11 [75-69-4] Vol. II: 602
fc-12- [75-71-8] Vol. I: 825
fc-21 [75-43-4] Vol. II: 595
fc-22 [75-45-6] Vol. II: 591
fc-112 [76-12-0] Vol. II: 328
fc-113 [76-13-1] Vol. II: 353
fc-114 [76-14-2] Vol. III: 408
fc-115 [76-15-3] Vol. III: 226
fc-152a [75-37-6] Vol. III: 411

flordimex [16672-87-0] Vol. III: 426
florel [16672-87-0] Vol. III: 426
flores martis [7705-08-0] Vol. I: 699
floridine [7681-49-4] Vol. II: 925
florocid [7681-49-4] Vol. II: 925
flo-tin-41 [76-87-9] Vol. II: 941
flour-sulphur [7704-34-9] Vol. II: 965
flowers of antimony [1309-64-4] Vol. II: 86
flowers of benjamin [65-85-0] Vol. I: 293
flowers of benzoin [65-85-0] Vol. I: 293
flozenges [7681-49-4] Vol. II: 925
fl-tabs [7681-49-4] Vol. II: 925
fluate [79-01-6] Vol. II: 427
flubalex [1861-40-1] Vol. II: 1043
flucarmit [119-36-8] Vol. III: 622
flue gas [630-08-0] Vol. I: 433
flugene-113 [76-13-1] Vol. II: 353
flugex-13b1 [75-63-8] Vol. III: 143
flukoids [56-23-5] Vol. I: 435
fluktan [151-67-7] Vol. III: 141
fluoboric acid [16872-11-0] Vol. III: 457
fluoboric acid (H(BF₄)) [16872-11-0] Vol. III: 457
fluophosgene [353-50-4] Vol. III: 186
fluor [7782-41-4] Vol. I: 703
fluoraday [7681-49-4] Vol. II: 925
fluoral [7681-49-4] Vol. II: 925
fluor (Dutch) [7782-41-4] Vol. I: 703
Fluoressigsaeure (German) [62-74-8] Vol. I: 39
fluorhydric acid [7664-39-3] Vol. I: 750
t-fluoride [7681-49-4] Vol. II: 925
fluorident [7681-49-4] Vol. II: 925
fluorid-sodny (Czech) [7681-49-4] Vol. II: 925
fluorigard [7681-49-4] Vol. II: 925
fluorine-19 [7782-41-4] Vol. I: 703
fluorineed [7681-49-4] Vol. II: 925
fluorinse [7681-49-4] Vol. II: 925
fluoristan [7783-47-3] Vol. III: 881
fluoritab [7681-49-4] Vol. II: 925
fluoroacetic acid, sodium salt [62-74-8] Vol. I: 39
fluorocarbon-114 [76-14-2] Vol. III: 408
fluorocarbon-115 [76-15-3] Vol. III: 226
fluorocarbon-11 [75-69-4] Vol. II: 602
fluorocarbon-12 [75-71-8] Vol. I: 825
fluorocarbon-1301 [75-63-8] Vol. III: 143
fluorocarbon-22 [75-45-6] Vol. II: 591
fluorochloroform [75-69-4] Vol. II: 602
fluorocid [7681-49-4] Vol. II: 925

fluorodichloromethane [75-43-4] Vol. II: 595
fluoroethene [75-02-5] Vol. III: 966
fluoroethylene [75-02-5] Vol. III: 966
fluoroformyl fluoride [353-50-4] Vol. III: 186
fluoro (Italian) [7782-41-4] Vol. I: 703
fluor-o-kote [7681-49-4] Vol. II: 925
fluorol [7681-49-4] Vol. II: 925
fluorophosgene [353-50-4] Vol. III: 186
fluorosilicic acid, ammonium salt [16919-19-0] Vol. II: 908
fluorotane [151-67-7] Vol. III: 141
fluorotrichloromethane [75-69-4] Vol. II: 602
fluorotrojchlorometan (Polish) [75-69-4] Vol. II: 602
fluorowodor (Polish) [7664-39-3] Vol. I: 750
fluorufo amonico (Spanish) [12125-01-8] Vol. I: 146
fluorure-acide-d'ammonium (French) [1341-49-7] Vol. II: 64
fluorure d'ammonium (French) [12125-01-8] Vol. I: 146
fluorure de sodium (French) [7681-49-4] Vol. II: 925
fluorure d'hydrogene anhydre (French) [7664-39-3] Vol. I: 750
fluoruro de hidrogeno anhidro (Spanish) [7664-39-3] Vol. I: 750
fluorwasserstoff (German) [7664-39-3] Vol. I: 750
fluorwaterstof (Dutch) [7664-39-3] Vol. I: 750
fluotane [151-67-7] Vol. III: 141
fluothane [151-67-7] Vol. III: 141
flura-drops [7681-49-4] Vol. II: 925
flurcare [7681-49-4] Vol. II: 925
flursol [7681-49-4] Vol. II: 925
flux maag [54-11-5] Vol. I: 874
flytrol [333-41-5] Vol. III: 758
fma [62-38-4] Vol. II: 535
FMC-1240 [563-12-2] Vol. II: 759
fmc-2070- [137-26-8] Vol. I: 632
FMC-5273 [51-03-6] Vol. III: 779
fmc-5488 [116-29-0] Vol. III: 906
FMC-9044 [485-31-4] Vol. II: 261
fmc-9260 [7696-12-0] Vol. III: 912
fmc-10242- [1563-66-2] Vol. I: 420
FMC-17370 [10453-86-8] Vol. III: 285
FMC-30980 [52315-07-8] Vol. II: 251

gallex [105-67-9] Vol. I: 1162
gallotannic acid [1401-55-4] Vol. II: 237
gallotannin [1401-55-4] Vol. II: 237
gallotox [62-38-4] Vol. II: 535
gamasol-90 [67-68-5] Vol. II: 641
gamaspra [608-73-1] Vol. I: 564
gamma-aminopyridine [504-24-5] Vol. II: 879
alpha,gamma butadiene [106-99-0] Vol. I: 333
gamma-chloroallyl chloride [542-75-6] Vol. II: 834
gamma-chloroisobutylene [563-47-3] Vol. III: 202
gamma-(4-chloro-2-methylphenoxy) butyric acid [94-81-5] Vol. III: 174
gamma chloropropylene oxide [106-89-8] Vol. I: 981
gammacide [608-73-1] Vol. I: 564
gammacold [608-73-1] Vol. I: 564
gamma-(2,4-dichlorophenoxy) butyric acid [94-82-6] Vol. III: 176
alpha, gamma-dichloropropylene [542-75-6] Vol. II: 834
gamma-mcpb [94-81-5] Vol. III: 174
gamma-2-methyl-4-chlorophenoxybutyric acid [94-81-5] Vol. III: 174
gamma methylpyridine [108-89-4] Vol. I: 1019
gamma-methylthio-alpha-aminobutyric acid [63-68-3] Vol. III: 593
gamma-oxo-alpha-butylene [78-94-4] Vol. III: 624
gamma picoline [108-89-4] Vol. I: 1019
gammexane [608-73-1] Vol. I: 564
gamtox [608-73-1] Vol. I: 564
garantose [81-07-2] Vol. III: 824
gardenal [50-06-6] Vol. I: 223
garden-tox [333-41-5] Vol. III: 758
gardentox [333-41-5] Vol. III: 758
gardepanyl [50-06-6] Vol. I: 223
gardopax [834-12-8] Vol. II: 1054
garlon [93-72-1] Vol. II: 845
garnitan [330-55-2] Vol. II: 1084
t-Gas [75-21-8] Vol. II: 401
gas oil [68476-30-2] Vol. III: 465
gas oil (DOT) [68476-30-2] Vol. III: 465
gaultheria oil [119-36-8] Vol. III: 622
gaultheria oil, artificial [119-36-8] Vol. III: 622
GB [107-44-8] Vol. III: 826
gc-1189 [143-50-0] Vol. III: 587

gc-1283 [2385-85-5] Vol. III: 583
gc 3944-3-4 [82-68-8] Vol. II: 132
gc-4072 [470-90-6] Vol. III: 716
gearphos [56-38-2] Vol. III: 750
gebrannter kalk [1305-78-8] Vol. I: 402
gebutox [88-85-7] Vol. II: 704
Geigy-338 [510-15-6] Vol. III: 108
Geigy-13005 [950-37-8] Vol. III: 734
geigy-24480 [333-41-5] Vol. III: 758
geigy-27,692 [122-34-9] Vol. II: 1047
geigy-30,027 [1912-24-9] Vol. II: 1050
geigy-30,028 [139-40-2] Vol. III: 802
Geigy gs-13005 [950-37-8] Vol. III: 734
gelber phosphor (Crerman) [7723-14-0] Vol. II: 764
gelber Phosphor (German) [7723-14-0] Vol. III: 973
gelbes-quecksilberoxyd [21908-53-2] Vol. II: 559
gelbkreuz [505-60-2] Vol. II: 962
gel-kam [7783-47-3] Vol. III: 881
gel-tin [7783-47-3] Vol. III: 881
gemalgene [71-55-6] Vol. I: 684
ge-materials-d4d5 [497-19-8] Vol. II: 922
genate [2008-41-5] Vol. II: 216
genate-plus [2008-41-5] Vol. II: 216
gen diur (Spain) [12125-02-9] Vol. I: 141
General chemicals 1189 [143-50-0] Vol. III: 587
genesolv-D [76-13-1] Vol. II: 353
genetron-12- [75-71-8] Vol. I: 825
genetron-21 [75-43-4] Vol. II: 595
genetron-22 [75-45-6] Vol. II: 591
genetron-100 [75-37-6] Vol. III: 411
genetron-112 [76-12-0] Vol. II: 328
genetron-113- [76-13-1] Vol. II: 353
genetron-114 [76-14-2] Vol. III: 408
genetron-115 [76-15-3] Vol. III: 226
genetron-316 [76-14-2] Vol. III: 408
genetron-152a [75-37-6] Vol. III: 411
geniphene [8001-35-2] Vol. I: 1118
genithion [56-38-2] Vol. III: 750
genklene [71-55-6] Vol. I: 684
genoplast-b [84-74-2] Vol. II: 774
gentanol [111-70-6] Vol. III: 474
Geonter [5902-51-2] Vol. III: 894
gepiron [2163-80-6] Vol. II: 610
geranium crystals [101-84-8] Vol. II: 392
germalgene [79-01-6] Vol. II: 427
German saltpeter [6484-52-2] Vol. I: 153
Germany: pigment 2 [7440-22-4] Vol. I: 1036

glycol monophenyl ether [122-99-6] Vol. III: 695

glycols, polyethylene [25322-68-3] Vol. II: 811

glycomonochlorhydrin [107-07-3] Vol. III: 220

glycon ro [112-80-1] Vol. I: 895

glycon wo [112-80-1] Vol. I: 895

glyodex 3722 [133-06-2] Vol. II: 276

glyodex 37-22 [133-06-2] Vol. II: 276

glyoxal aldehyde [107-22-2] Vol. I: 728

glyoxal, dimethyl [431-03-8] Vol. III: 299

glyoxylaldehyde [107-22-2] Vol. I: 728

ppg-2 methyl ether [34590-94-8] Vol. III: 381

1721-gold [7440-50-8] Vol. I: 519

gold bronze [7440-50-8] Vol. I: 519

gold crest h-60, termide [76-44-8] Vol. II: 628

gontochin [54-05-7] Vol. III: 229

good-rite gp 264 [117-81-7] Vol. II: 769

gothnion [86-50-0] Vol. III: 744

gp-40-66:120 [87-68-3] Vol. I: 340

gpkh [76-44-8] Vol. II: 628

grain alcohol [64-17-5] Vol. II: 394

gramerin [75-99-0] Vol. I: 1016

gramoxone [4685-14-7] Vol. III: 670

granol-nm [12427-38-2] Vol. III: 531

granosan [107-06-2] Vol. I: 643

granox [118-74-1] Vol. I: 268

granox nm [118-74-1] Vol. I: 268

granox-nm [12427-38-2] Vol. III: 531

granox-pfm [12427-38-2] Vol. III: 531

granox-pfm [133-06-2] Vol. II: 276

granular crystals copper sulfate [7758-98-7] Vol. I: 510

granular zinc [7440-66-6] Vol. I: 1175

grape seed oil [60-33-3] Vol. II: 667

graphlox [77-47-4] Vol. II: 283

grascide [26638-19-7] Vol. II: 298

grasex [75-87-6] Vol. I: 442

graslan [34014-18-1] Vol. III: 892

gray arsenic [7440-38-2] Vol. I: 190

gray selenium [7782-49-2] Vol. I: 1030

11661-green [1308-38-9] Vol. III: 234

green chrome oxide [1308-38-9] Vol. III: 234

green chromic oxide [1308-38-9] Vol. III: 234

green chromium oxide [1308-38-9] Vol. III: 234

green cinnabar [1308-38-9] Vol. III: 234

green-cross-warble-powder [83-79-4] Vol. II: 166

green cross weed no more-80 [94-75-7] Vol. I: 30

Greenland spar [15096-52-3] Vol. III: 37

green nickel oxide [1313-99-1] Vol. III: 645

green oil [120-12-7] Vol. II: 79

green oxide of chromium [1308-38-9] Vol. III: 234

green oxide of chromium oc-31 [1308-38-9] Vol. III: 234

green rouge [1308-38-9] Vol. III: 234

grenade [68085-85-8] Vol. II: 288

greosin [126-07-8], Vol. III: 472

gresfeed [126-07-8], Vol. III: 472

gricin [126-07-8], Vol. III: 472

griffex [1912-24-9] Vol. II: 1050

griffin-manex [12427-38-2] Vol. III: 531

grifulin [126-07-8], Vol. III: 472

grifulvin [126-07-8], Vol. III: 472

grisactin [126-07-8], Vol. III: 472

griscofulvin [126-07-8], Vol. III: 472

grisefuline [126-07-8], Vol. III: 472

griseofulvin-forte [126-07-8], Vol. III: 472

griseofulvinum [126-07-8], Vol. III: 472

griseofulvin, (+)- [126-07-8], Vol. III: 472

grisetin [126-07-8], Vol. III: 472

grisofulvin [126-07-8], Vol. III: 472

grisovin [126-07-8], Vol. III: 472

gris-peg [126-07-8], Vol. III: 472

grizeofulvin [126-07-8], Vol. III: 472

groco-2- [112-80-1] Vol. I: 895

groco-4- [112-80-1] Vol. I: 895

groco-51 [112-80-1] Vol. I: 895

groco-5810 [123-95-5] Vol. III: 167

groco-6- [112-80-1] Vol. I: 895

grosafe [7440-44-0] Vol. III: 179

gro-tone nematode granular [96-12-8] Vol. II: 818

ground-vocle-sulphur [7704-34-9] Vol. II: 965

grower service toxaphene-6e [8001-35-2] Vol. I: 1118

grower service toxaphene mp [8001-35-2] Vol. I: 1118

gr 48-11ps [131-52-2] Vol. III: 676

gr 48-32s [131-52-2] Vol. III: 676

grundier arbezol [87-86-5] Vol. I: 928

grysio [126-07-8], Vol. III: 472

gs-13005 [950-37-8] Vol. III: 734

gs-13-005 [950-37-8] Vol. III: 734

gs-13005 [950-37-8] **Vol. III:** 734

gs-14260 [886-50-0] **Vol. II:** 1045

guanidine, dodecyl-, monoacetate [2439-10-3] **Vol. III:** 290

guardian [70124-77-5], **Vol III:** 455

gusathion-20 [86-50-0] **Vol. III:** 744

gusathion-25 [86-50-0] **Vol. III:** 744

gusathion [86-50-0] **Vol. III:** 744

gusathion-k [86-50-0] **Vol. III:** 744

gusathion-m [86-50-0] **Vol. III:** 744

gusathion-methyl [86-50-0] **Vol. III:** 744

guservin [126-07-8], **Vol. III:** 472

gustafson captan 30-dd [133-06-2] **Vol. II:** 276

gustafson-terraclor-80%-dust-concentrate [82-68-8] **Vol. II:** 132

guthion [86-50-0] **Vol. III:** 744

gyben [608-73-1] **Vol. I:** 564

gy phene [8001-35-2] **Vol. I:** 1118

h-34 [76-44-8] **Vol. II:** 628

h-46 [21645-51-2] **Vol. II:** 57

h-46 [21645-51-2] **Vol. III:** 35

h-133- [1194-65-6] **Vol. I:** 299

h-321 [2032-65-7] **Vol. II:** 308

h-1803 [122-34-9] **Vol. II:** 1047

ha-1- [65-85-0] **Vol. I:** 293

hag-107 [66841-25-6] **Vol. III:** 929

haiari [83-79-4] **Vol. II:** 166

haitin [76-87-9] **Vol. II:** 941

haitin-wp-20-(fentinhydroxide 20%) [76-87-9] **Vol. II:** 941

haitin-wp-60 (fentinhydroxide 60%) [76-87-9] **Vol. II:** 941

halan [151-67-7] **Vol. III:** 141

halane [118-52-5] **Vol. III:** 329

halite [7647-14-5] **Vol. III:** 851

halocarbon-112 [76-12-0] **Vol. II:** 328

halocarbon-115 [76-15-3] **Vol. III:** 226

halocarbon-13B1 [75-63-8] **Vol. III:** 143

halon-11 [75-69-4] **Vol. II:** 602

halon-104- [56-23-5] **Vol. I:** 435

halon-112 [75-43-4] **Vol. II:** 595

halon-122- [75-71-8] **Vol. I:** 825

halon-1001 [74-83-9] **Vol. II:** 577

halon-1011 [74-97-5] **Vol. II:** 583

halon-1301 [75-63-8] **Vol. III:** 143

halon-2001 [74-96-4] **Vol. III:** 431

halon-242 [76-14-2] **Vol. III:** 408

halon [75-71-8] **Vol. I:** 825

halotan [151-67-7] **Vol. III:** 141

halotano (Spanish) [151-67-7] **Vol. III:** 141

halothan [151-67-7] **Vol. III:** 141

halothane [151-67-7] **Vol. III:** 141

halothanum (Latin) [151-67-7] **Vol. III:** 141

Halowax 1014 [1335-87-1] **Vol. III:** 475

halowax [1335-88-2] **Vol. III:** 902

halsan [151-67-7] **Vol. III:** 141

haltox [74-83-9] **Vol. II:** 577

haltron-22 [75-45-6] **Vol. II:** 591

hamidop [10265-92-6] **Vol. II:** 744

hampshire-nta-acid [139-13-9] **Vol. III:** 648

haptarax [470-90-6] **Vol. III:** 716

haptasol [470-90-6] **Vol. III:** 716

hartshorn [506-87-6] **Vol. I:** 139

hatcol-dop [117-81-7] **Vol. II:** 769

havidote [60-00-4] **Vol. I:** 36

hc-2072 [311-45-5] **Vol. III:** 669

hcb [118-74-1] **Vol. I:** 268

hcbd [87-68-3] **Vol. I:** 340

hcch [608-73-1] **Vol. I:** 564

hccp [77-47-4] **Vol. II:** 283

hccpd [77-47-4] **Vol. II:** 283

hce [1024-57-3] **Vol. II:** 626

HCFC-124a [63938-10-3] **Vol. III:** 230

hch [608-73-1] **Vol. I:** 564

hcn [74-90-8] **Vol. I:** 743

hcs-3260 [57-74-9] **Vol. II:** 619

hd [505-60-2] **Vol. II:** 962

he-5- [123-31-9] **Vol. I:** 753

hea [818-61-1] **Vol. III:** 484

heavy oil [8001-58-9] **Vol. I:** 495

hedapur m-52- [94-74-6] **Vol. I:** 25

hedarex m [94-74-6] **Vol. I:** 25

hedolit [534-52-1] **Vol. I:** 540

hedonal m [94-74-6] **Vol. I:** 25

heksan (Polish) [110-54-3] **Vol. II:** 459

heksogen (Polish) [121-82-4] **Vol. III:** 277

hel-fire [88-85-7] **Vol. II:** 704

heliotridine ester with lasiocarpum and angelic acid [303-34-4] **Vol. III:** 501

hemostyp [71-36-3] **Vol. I:** 352

p-(phenylazo) aniline [60-09-3] **Vol. II:** 74

p-phenylazonaniline [60-09-3] **Vol. II:** 74

p-phenyl-nitrobenzene [92-93-3] **Vol. III:** 650

heod [60-57-1] **Vol. III:** 356

hepta [76-44-8] **Vol. II:** 628

1,4,5,6,7,8,8-heptachloor-3a,4,7,7a-tetra-hydro-4,7-endo-methano-indeen (Dutch) [76-44-8] **Vol. II:** 628

heptachloor (Dutch) [76-44-8] **Vol. II:** 628

1,4,5,6,7,8,8-Heptachlor-3a,4,7,7,7a-tetra-hydro-4,7-endo-methano-inden (German) [76-44-8] Vol. II: 628

heptachlorane [76-44-8] Vol. II: 628

heptachlore (French) [76-44-8] Vol. II: 628

1,4,5,6,7,8,8-heptachloro-2,3-epoxy-3a,4,7,7a-tetrahydro-4,7-methano indan [1024-57-3] Vol. II: 626

1-heptadecanecarboxylic acid [57-11-4] Vol. I: 1052

heptagran [76-44-8] Vol. II: 628

heptagranox [76-44-8] Vol. II: 628

heptamak [76-44-8] Vol. II: 628

heptamul [76-44-8] Vol. II: 628

n-heptane [142-82-5] Vol. II: 450

3-heptanecarboxylic acid [149-57-5] Vol. II: 461

n-heptanol [111-70-6] Vol. III: 474

n-heptan-1-ol [111-70-6] Vol. III: 474

1-heptanol [111-70-6] Vol. III: 474

4-heptanol, 2,6-dimethyl- [108-82-7] Vol. II: 453

4-heptanone, 2,6-dimethyl [108-83-8] Vol. I: 730

heptasol [76-44-8] Vol. II: 628

heptox [76-44-8] Vol. II: 628

heptyl alcohol [111-70-6] Vol. III: 474

n-heptyl alcohol [111-70-6] Vol. III: 474

heptyl carbinol [111-87-5] Vol. I: 897

herb-all [2163-80-6] Vol. II: 610

herban-m [2163-80-6] Vol. II: 610

herbatox [330-54-1] Vol. II: 1078

herbax [26638-19-7] Vol. II: 298

herbax-3e [26638-19-7] Vol. II: 298

herbax-4e [26638-19-7] Vol. II: 298

herbax lv-30 [26638-19-7] Vol. II: 298

herbax-technical [26638-19-7] Vol. II: 298

herbazin [122-34-9] Vol. II: 1047

herbazin-50 [122-34-9] Vol. II: 1047

herbex [122-34-9] Vol. II: 1047

herbicide-273 [145-73-3] Vol. I: 900

herbicide-326 [330-55-2] Vol. II: 1084

s-95 (herbicide) [75-99-0] Vol. I: 1016

herbicide m [94-74-6] Vol. I: 25

herbicides, silvex [93-72-1] Vol. II: 845

herbon pennout [145-73-3] Vol. I: 900

herbon-yellow [101-42-8] Vol. III: 452

herboxy [122-34-9] Vol. II: 1047

hercoflex-260 [117-81-7] Vol. II: 769

herco prills [6484-52-2] Vol. I: 153

Hercules-14-503 [10311-84-9] Vol. III: 728

Hercules-14503- [10311-84-9] Vol. III: 728

hercules-3956- [8001-35-2] Vol. I: 1118

hercules-528 [78-34-2] Vol. II: 756

hercules-ac528 [78-34-2] Vol. II: 756

hercules toxaphene emulsifiable concentrate [8001-35-2] Vol. I: 1118

herkal [62-73-7] Vol. I: 946

herkol [62-73-7] Vol. I: 946

hermal [137-26-8] Vol. I: 632

hermat tmt [137-26-8] Vol. I: 632

hermesetas [81-07-2] Vol. III: 824

heryl [137-26-8] Vol. I: 632

het [757-58-4] Vol. II: 992

het acid [115-28-6] Vol. III: 200

hetp [757-58-4] Vol. II: 992

hex [77-47-4] Vol. II: 283

hexablanc [608-73-1] Vol. I: 564

hexabutyldistannoxane [56-35-9] Vol. II: 313

hexacap [133-06-2] Vol. II: 276

hexa cb [118-74-1] Vol. I: 268

hexachlor [608-73-1] Vol. I: 564

hexachlor-aethan (German) [67-72-1] Vol. II: 330

hexachloran [608-73-1] Vol. I: 564

hexachlorane [608-73-1] Vol. I: 564

Hexachlorbenzol (German) [118-74-1] Vol. I: 268

hexachlor-1,3-butadien (Czech) [87-68-3] Vol. I: 340

hexachlorcyklopentadien (Czech) [77-47-4] Vol. II: 283

hexachlornaftalen (Czech) [1335-87-1] Vol. III: 475

3,4,5,6,9,9-hexachloro-1a,2,2a,3,6,6a,7,7a-octahydro-2,7:3,6-dimethanonaphth (2,3-b) oxirene [60-57-1] Vol. III: 356

3,4,5,6,9,9-hexachloro-1a,2,2a,3,6,6a,7,7a-octahydro-2,7:3,6-dimethanonaphth[2,3-b] oxirene [60-57-1] Vol. III: 356

(1aalpha, 2beta, 2aalpha, 3beta, 6beta, 6aalpha, 7beta, 7aalpha)-3,4,5,6,9,9-hexachloro- 1a,2,2a,3,6,6a,7,7a-octahydro-2,7:3,6-dimethanonaphth[2,3-b]oxirene [60-57-1] Vol. III: 356

3,4,5,6,9,9-hexachloro-1a,2,2a,3,6,6a,7,7a-octahydro-2,7:3,6- dimethanonaphth[2,3-b]oxirene [72-20-8] Vol. III: 349

(1aalpha, 2beta, 2abeta, 3alpha, 6alpha, 6abeta, 7beta, 7aalpha)-3,4,5,6,9,9-hexachloro- 1a,2,2a,3,6,6a,7,7a-octahydro-2,7:3,6- dimethanonaphth[2,3-b]oxirene [72-20-8] Vol. III: 349

hoe-02-810 [330-55-2] Vol. II: 1084

hoe-2671 [115-29-7] Vol. II: 661

hoe-2784 [485-31-4] Vol. II: 261

hoe-2810 [330-55-2] Vol. II: 1084

hoe-2784-oa [485-31-4] Vol. II: 261

hokmate [14484-64-1] Vol. III: 489

holiday pet repellent [628-63-7] Vol. II: 27

holiday repellent dust [628-63-7] Vol. II: 27

hombitan [13463-67-7] Vol. III: 921

home heating oil no. 2 [68476-30-2] Vol. III: 465

#2 home heating oils [68476-30-2] Vol. III: 465

homopiperidine [111-49-9] Vol. III: 477

hong-nien [62-38-4] Vol. II: 535

hornotuho [94-74-6] Vol. I: 25

horse head a-420 [13463-67-7] Vol. III: 921

hospex [111-30-8] Vol. III: 468

hospital milton [7681-52-9] Vol. II: 498

hostaquick [62-38-4] Vol. II: 535

hostaquik [62-38-4] Vol. II: 535

Hostetex l-pec [120-82-1] Vol. III: 92

hrs-1276 [2385-85-5] Vol. III: 583

hrs-1655- [77-47-4] Vol. II: 283

hs-14260 [886-50-0] Vol. II: 1045

hs [302-17-0] Vol. III: 197

ht-972 [101-77-9] Vol. III: 304

hth [7778-54-3] Vol. II: 494

htp [757-58-4] Vol. II: 992

huile d'aniline (French) [62-53-3] Vol. I: 164

huile-de-camphre (French) [76-22-2] Vol. II: 211

humko industrene r [57-11-4] Vol. I: 1052

hungazin [1912-24-9] Vol. II: 1050

hungazin-dt [122-34-9] Vol. II: 1047

hungazin-pk [1912-24-9] Vol. II: 1050

hw-920 [330-54-1] Vol. II: 1078

hy-951- [112-24-3] Vol. I: 1129

hyadur [67-68-5] Vol. II: 641

hyamine [121-54-0] Vol. III: 105

hyamine-1622 [121-54-0] Vol. III: 105

hyamine-3500 [8001-54-5] Vol. II: 67

hy-chlor [7778-54-3] Vol. II: 494

hychol-705 [21645-51-2] Vol. II: 57

hychol-705 [21645-51-2] Vol. III: 35

hyclorite [7681-52-9] Vol. II: 498

hydan [118-52-5] Vol. III: 329

hydantoin, dichlorodimethyl [118-52-5] Vol. III: 329

hydantoin, 1,3-dichloro-5,5-dimethyl [118-52-5] Vol. III: 329

hydout [145-73-3] Vol. I: 900

hydracrylic acid, beta-lactone [57-57-8] Vol. II: 672

hydracrylic acid beta-lactone [57-57-8] Vol. II: 672

hydracrylonitrile [109-78-4] Vol. II: 466

hydrafil [21645-51-2] Vol. II: 57

hydrafil [21645-51-2] Vol. III: 35

hydral [302-17-0] Vol. III: 197

hydral-705 [21645-51-2] Vol. II: 57

hydral-705 [21645-51-2] Vol. III: 35

hydral-710 [21645-51-2] Vol. II: 57

hydral-710 [21645-51-2] Vol. III: 35

hydralin [108-93-0] Vol. I: 580

hydrargyrum [7439-97-6] Vol. II: 549

hydrargyrum ammoniatum [10124-48-8] Vol. II: 533

Hydrargyrum-bijodatum (German) [7774-29-0] Vol. II: 547

hydrargyrum-oxid-flav [21908-53-2] Vol. II: 559

hydrargyrum precipitatum album [10124-48-8] Vol. II: 533

hydrated-alumina [21645-51-2] Vol. II: 57

hydrated alumina [21645-51-2] Vol. III: 35

hydrated aluminum oxide [21645-51-2] Vol. II: 57

hydrated aluminum oxide [21645-51-2] Vol. III: 35

hydrazide maleique [123-33-1] Vol. II: 870

hydrazine anhydrous [302-01-2] Vol. I: 735

hydrazine base [302-01-2] Vol. I: 735

hydrazine, 1,1-dimethyl [57-14-7] Vol. II: 470

hydrazine, 1,2-dimethyl [540-73-8] Vol. II: 468

hydrazine, 1,1-diphenyl- [530-50-7] Vol. II: 473

hydrazine, 1,2-diphenyl- [122-66-7] Vol. II: 479

hydrazine, methyl- [60-34-4] Vol. II: 475

hydrazobenzen (Czech) [122-66-7] Vol. II: 479

hydrazobenzene [122-66-7] Vol. II: 479

1,1'-hydrazodibenzene [122-66-7] Vol. II: 479

hydrazoic acid, sodium salt.- [26628-22-8] Vol. I: 1043

hydrazomethane [540-73-8] Vol. II: 468

4-hydroxy-4-methyl pentan-2-one [123-42-2] Vol. II: 688

4-hydroxy-4-methyl-pentan-2-on (German, Dutch) [123-42-2] Vol. II: 688

1-hydroxymethylpropane [78-83-1] Vol. I: 765

2-hydroxy-2-methylpropionitrile [75-86-5] Vol. I: 794

p-hydroxynitrobenzene [100-02-7] Vol. II: 726

2-hydroxynitrobenzene [88-75-5] Vol. II: 723

3-hydroxynitrobenzene [554-84-7] Vol. III: 693

4-hydroxynitrobenzene [100-02-7] Vol. II: 726

m-hydroxynitrobenzene [554-84-7] Vol. III: 693

o-hydroxynitrobenzene [88-75-5] Vol. II: 723

3-hydroxy-N,N-dimethyl-cis-crotonamide dimethyl phosphate [141-66-2] Vol. III: 720

2-hydroxy- n-octane [123-96-6] Vol. III: 655

1-hydroxyoctane [111-87-5] Vol. I: 897

2-hydroxyoctane [123-96-6] Vol. III: 655

4-hydroxy-3-(3-oxo-1-fenyl butyl) cumarine (Dutch) [81-81-2] Vol. I: 525

4-hydroxy-3-(3-oxo-1-phenyl butyl) cumarin (German) [81-81-2] Vol. I: 525

4-hydroxy-3-(3-oxo-1-phenyl butyl) 2h-1-benzopyran-2-one [81-81-2] Vol. I: 525

m hydroxyphenol [108-46-3] Vol. I: 1026

o-hydroxyphenol [120-80-9] Vol. III: 190

p hydroxyphenol [123-31-9] Vol. I: 753

2-hydroxyphenol [120-80-9] Vol. III: 190

3-hydroxyphenol [108-46-3] Vol. I: 1026

4-hydroxyphenol [123-31-9] Vol. I: 753

1-hydroxy-2-phenoxyethane [122-99-6] Vol. III: 695

2,2-bis (p-hydroxyphenyl) propane [80-05-7] Vol. II: 720

2,2-bis (p-hydroxyphenyl) propane, diglycidyl ether [1675-54-3] Vol. III: 132

2,2-bis (4-hydroxyphenyl) propane diglycidyl ether [1675-54-3] Vol. III: 132

2-hydroxypropanamine [78-96-6] Vol. II: 821

1-hydroxypropane [71-23-8] Vol. II: 856

3-hydroxypropanenitrile [109-78-4] Vol. II: 466

2-hydroxy-1,2,3-propanetricarboxylic acid [77-92-9] Vol. I: 494

2-hydroxypropanoic acid [50-21-5] Vol. I: 792

2-hydroxypropanol [57-55-6] Vol. I: 1003

3-hydroxypropene [107-18-6] Vol. II: 52

alpha hydroxypropionicacid [50-21-5] Vol. I: 792

3-hydroxypropionic acid, beta-lactone [57-57-8] Vol. II: 672

3-hydroxypropionic acid lactone [57-57-8] Vol. II: 672

3-hydroxypropionitrile [109-78-4] Vol. II: 466

2-hydroxy-1-propylamine [78-96-6] Vol. II: 821

2-hydroxypropylamine [78-96-6] Vol. II: 821

3-hydroxy-1-propyne [107-19-7] Vol. II: 864

1-hydroxy-2-propyne [107-19-7] Vol. II: 864

hydroxytoluene [100-51-6] Vol. I: 313

alpha hydroxytoluene [100-51-6] Vol. I: 313

m-hydroxytoluene [108-39-4] Vol. III: 261

2-hydroxytoluene [95-48-7] Vol. I: 536

3-hydroxytoluene [108-39-4] Vol. III: 261

o hydroxytoluene [95-48-7] Vol. I: 536

1-hydroxy-2,2,2-trichloroethylphosphonic acid, dimethyl ester [52-68-6] Vol. III: 707

2-hydroxy-1,3,5-trinitrobenzene, [88-89-1] Vol. II: 793

hydroxytriphenylstannane [76-87-9] Vol. II: 941

hydroxytriphenyltin [76-87-9] Vol. II: 941

hydrozobenzene [530-50-7] Vol. II: 473

hydrure de lithium [French] [7580-67-8] Vol. III: 525

hyfatol [36653-82-4] Vol. III: 193

hygeia creme magic bowl cleaner [7647-01-0] Vol. I: 738

hylene-t [584-84-9] Vol. II: 125

hyperoxia [7782-44-7] Vol. II: 675

hypersal [7647-14-5] Vol. III: 851

hy phi 1055 [112-80-1] Vol. I: 895

hy phi 1088 [112-80-1] Vol. I: 895

hy phi2066 [112-80-1] Vol. I: 895

hy phi2088 [112-80-1] Vol. I: 895

hy phi2102 [112-80-1] Vol. I: 895

hypnon [98-86-2] Vol. I: 86

hypnone [98-86-2] Vol. I: 86

hypochlorite de calcium (French) [7778-54-3] Vol. II: 494

hypochlorous acid, calcium salt [7778-54-3] Vol. II: 494

hypochlorous acid, calcium-salt, (dry mixture) [7778-54-3] Vol. II: 494

hypochlorous acid, sodium salt [7681-52-9] Vol. II: 498

hyponitrous acid anhydride [10024-97-2] Vol. I: 889

hyponitrous ether [109-95-5] Vol. III: 444

hyposan and voxsan [7681-52-9] Vol. II: 498

hysteps [50-06-6] Vol. I: 223

hystrene-80- [57-11-4] Vol. I: 1052

hystrene s-97 [57-11-4] Vol. I: 1052

hystrene t-70 [57-11-4] Vol. I: 1052

hytrol-o [108-94-1] Vol. II: 268

hy vic [137-26-8] Vol. I: 632

i amyl acetate [123-92-2] Vol. I: 780

i-butylamine [78-81-9] Vol. III: 493

icc-23 7-disinfectant, sanitizer, destainer, and deodorizer [7722-64-7] Vol. I: 915

ice-spar [15096-52-3] Vol. III: 37

icetone [15096-52-3] Vol. III: 37

ici-146-814 [68085-85-8] Vol. II: 288

ici cf 2 [71-55-6] Vol. I: 684

p picoline [108-89-4] Vol. I: 1019

ida, imc flo tin 41 [76-87-9] Vol. II: 941

ideal concentrated wood preservative [67-56-1] Vol. I: 836

ideno (1,2,3-cd)pyrene [193-39-5] Vol. I: 759

idrochinone (Italian) [123-31-9] Vol. I: 753

idrogeno solforato (Italian) [7783-06-4] Vol. II: 485

idroperossido di cumene (Italian) [80-15-9] Vol. II: 490

idroperossido di cumolo (Italian) [80-15-9] Vol. II: 490

idrossido di stagno trifenile (Italian) [76-87-9] Vol. II: 941

4-idrossi-4-metil-pentan-2-one (Italian) [123-42-2] Vol. II: 688

4-idrossi-3-(3-oxo-) fenil butil) cumarine (Italian) [81-81-2] Vol. I: 525

idryl [206-44-0] Vol. II: 435

ifo-13140- [121-75-5] Vol. II: 950

igran-500- [886-50-0] Vol. II: 1045

igran-50- [886-50-0] Vol. II: 1045

igran [886-50-0] Vol. II: 1045

illoxol [60-57-1] Vol. III: 356

imidan [732-11-6] Vol. III: 737

imidan-5-dust [732-11-6] Vol. III: 737

imidan-wp [732-11-6] Vol. III: 737

imidathion [732-11-6] Vol. III: 737

imidazole-2 (3h) thione, 4,5-dihydro [96-45-7] Vol. I: 757

2,4-imidazolidinedione, 1,3-dichloro-5,5-dimethyl [118-52-5] Vol. III: 329

2-imidazolidine thione [96-45-7] Vol. I: 757

2-imidazolidinethione [96-45-7] Vol. I: 757

imidazolidinethione [96-45-7] Vol. I: 757

imidazoline-2 (3h) thione [96-45-7] Vol. I: 757

2-imidazoline-2-thiol [96-45-7] Vol. I: 757

imidazoline-2-thiol [96-45-7] Vol. I: 757

imidocarbonic acid, phosphonodithio-, cyclic propylene p,p-diethyl ester [950-10-7] Vol. III: 536

2,2'-iminobis (ethanol) [111-42-2] Vol. I: 658

2,2'-iminobisethanol [111-42-2] Vol. I: 658

1,1'-iminobis (2-propanol) [110-97-4] Vol. II: 823

iminodiethanol [111-42-2] Vol. I: 658

2,2'-iminodi-1-ethanol [111-42-2] Vol. I: 658

2,2'-iminodiethanol [111-42-2] Vol. I: 658

2,2'-iminodi-N-nitrosoethanol [1116-54-7] Vol. II: 377

1,1'-iminodi-2-propanol [110-97-4] Vol. II: 823

o-impc [114-26-1] Vol. II: 231

imperator [52315-07-8] Vol. II: 251

imperator [52645-53-1] Vol. III: 282

impruval [128-37-0] Vol. III: 317

impruvo [128-37-0] Vol. III: 317

imsol-a [67-63-0] Vol. II: 515

inactisol [121-54-0] Vol. III: 105

inakor [1912-24-9] Vol. II: 1050

incracide-10a [7758-98-7] Vol. I: 510

incracide e-51- [7758-98-7] Vol. I: 510

1,3-indandione, 2-pivaloyl [83-26-1] Vol. III: 774

1h-indene-1,3 (2h)-dione, 2-(2,2-dimethyl-1-oxopropyl)- [83-26-1] Vol. III: 774

indenopyrene [193-39-5] Vol. I: 759

1h indole [120-72-9] Vol. I: 760

indole-4,7-dione, 6-amino-8-(((aminocarbonyl)oxy)methyl)-1,1a, 2,8,8a,8b-hexahydro-8-a-methoxy-5-methyl-, (1aS-(1a-alpha, 8beta, 8a-alpha, 8balpha))-(9CI) Vol. III: [07-7], 633

indol (German) [120-72-9] Vol. I: 760

isopropene-cyanide [126-98-7] **Vol. II:** 839

isopropenil-benzolo (Italian) [98-83-9] **Vol. II:** 947

isopropenyl-benzeen (Dutch) [98-83-9] **Vol. II:** 947

isopropenylbenzene [98-83-9] **Vol. II:** 947

isopropenyl-benzol (German) [98-83-9] **Vol. II:** 947

1,2,12,12aalpha-tetrahydro-2a-iso propenyl-8,9-dimethoxy[1] benzo pyrano[3,4-b]furo[2,3-h][1]benzo pyran-6 (6ah)-one [83-79-4] **Vol. II:** 166

isopropenylnitrile [126-98-7] **Vol. II:** 839

isopropilamina (Italian) [75-31-0] **Vol. I:** 787

isopropilbenzene (Italian) [98-82-8] **Vol. I:** 551

isopropile (acetatodi) (Italian) [108-21-4] **Vol. I:** 46

isopropil n-fenil carbammato (Italian) [122-42-9] **Vol. I:** 426

isopropoxymethylphoshoryl fluoride [107-44-8] **Vol. III:** 826

o-isopropoxyphenyl methylcarbamate [114-26-1] **Vol. II:** 231

o-isopropoxyphenyl N-methylcarbamate [114-26-1] **Vol. II:** 231

o-(2-isopropoxyphenyl) N-methylcarbamate [114-26-1] **Vol. II:** 231

2-isopropoxyphenyl N-methylcarbamate [114-26-1] **Vol. II:** 231

2-isopropoxypropane [108-20-3] **Vol. I:** 996

isopropylacetaat (Dutch) [108-21-4] **Vol. I:** 46

Isopropylacetat (German) [108-21-4] **Vol. I:** 46

isopropylacetone [108-10-1] **Vol. II:** 692

isopropyl acetone [108-10-1] **Vol. II:** 692

isopropyl alcohol [67-63-0] **Vol. II:** 515

iso-Propylalkohol (German) [67-63-0] **Vol. II:** 515

N-Isopropyl-alpha-chloroacetanilide [1918-16-7] **Vol. III:** 789

isopropylbenzeen (Dutch) [98-82-8] **Vol. I:** 551

isopropylbenzene [98-82-8] **Vol. I:** 551

isopropyl benzene [98-82-8] **Vol. I:** 551

isopropylbenzene hydroperoxide [80-15-9] **Vol. II:** 490

isopropylbenzol [98-82-8] **Vol. I:** 551

Isopropyl benzol (German) [98-82-8] **Vol. I:** 551

N-isopropyl-2-chloroacetanilide [1918-16-7] **Vol. III:** 789

isopropyl 3-chlorocarbanilate [101-21-3] **Vol. II:** 234

isopropyl 3-chlorophenylcarbamate [101-21-3] **Vol. II:** 234

isopropyl cyanide [78-82-0] **Vol. III:** 798

isopropyl 2,4-d ester [94-11-1] **Vol. II:** 15

isopropyl (2,4-dichlorophenoxy) acetate [94-11-1] **Vol. II:** 15

isopropyl dimethytcarbinol [105-30-6] **Vol. I:** 912

isopropyle (acetated') (French) [108-21-4] **Vol. I:** 46

2,4-d, isopropyl ester [94-11-1] **Vol. II:** 15

isopropylester-kyseliny-methylfluorfosfonove (Czech) [107-44-8] **Vol. III:** 826

isopropylformaldehyde [78-84-2] **Vol. I:** 769

isopropylformic acid [79-31-2] **Vol. I:** 787

isopropylideneacetone [141-79-7] **Vol. II:** 696

4,4'-isopropylidenebis (1-(2,3-epoxypropoxy) benzene) [1675-54-3] **Vol. III:** 132

4,4'-isopropylidenebisphenol [80-05-7] **Vol. II:** 720

4,4'-isopropylidenebis (phenol) [80-05-7] **Vol. II:** 720

p,p'-isopropylidenebisphenol [80-05-7] **Vol. II:** 720

4,4'-isopropylidenediphenol [80-05-7] **Vol. II:** 720

p,p'-isopropylidenediphenol [80-05-7] **Vol. II:** 720

4,4'-isopropylidenediphenol diglycidyl ether [1675-54-3] **Vol. III:** 132

isopropyl m-chlorocarbanilate [101-21-3] **Vol. II:** 234

isopropyl meta-chlorocarbanilate [101-21-3] **Vol. II:** 234

Isopropyl-methylphosphonofluoridate [107-44-8] **Vol. III:** 826

o-isopropyl methylphosphonofluoridate [107-44-8] **Vol. III:** 826

isopropylmethylpyrimidyl-diethyl-thiophosphate [333-41-5] **Vol. III:** 758

O-2-isopropyl-4-methylpyrimidyl-O,O-diethyl phosphorothioate [333-41-5] **Vol. III:** 758

isopropyl N-chlorophenylcarbamate [101-21-3] **Vol. II:** 234

isopropyl N-(3-chlorophenyl) carbamate [101-21-3] **Vol. II:** 234

lead oxide (PbO) [1317-36-8] Vol. III: 513
lead oxide yellow [1317-36-8] Vol. III: 513
lead protoxide [1317-36-8] Vol. III: 513
lead s2- [7439-92-1] Vol. I: 800
lead (2) sulfide [1314-87-0] Vol. I: 807
lead sulfide (pbs) [1314-87-0] Vol. I: 807
lead tetraethide [78-00-2] Vol. I: 966
lead,tetraethyl [78-00-2] Vol. I: 966
lead tetrafluoroborate [13814-96-5] Vol.
   III: 507
lead tetramethyl- [75-74-1] Vol. II: 796
lead, tetramethyl- [75-74-1] Vol. II: 796
leaf green [1308-38-9] Vol. III: 234
le captane (French) [133-06-2] Vol. II: 276
leda-benzalkonium-chloride [8001-54-5]
   Vol. II: 67
ledinitrocresol-4,6 (French) [534-52-1] Vol.
   I: 540
ledon-114 [76-14-2] Vol. III: 408
ledon-11 [75-69-4] Vol. II: 602
ledon-12- [75-71-8] Vol. I: 825
legumex [94-81-5] Vol. III: 174
leinoleic acid [60-33-3] Vol. II: 667
leivasom [52-68-6] Vol. III: 707
lemoflur [7681-49-4] Vol. II: 925
lemonene [92-52-4] Vol. I: 326
lepinal [50-06-6] Vol. I: 223
lepinaletten [50-06-6] Vol. I: 223
lerophyn [62-38-4] Vol. II: 535
lethalaire g-54 [56-38-2] Vol. III: 750
lethalaire g-57 [3689-24-5] Vol. II: 1010
lethane-384-regular [112-56-1] Vol. III:
   521
lethurin [79-01-6] Vol. II: 427
l'ethylene thiouree (French) [96-45-7] Vol.
   I: 757
leucethane [51-79-6] Vol. II: 224
leucogen [7631-90-5] Vol. II: 920
leucol [91-22-5] Vol. I: 1022
leucoline [91-22-5] Vol. I: 1022
leucothane [51-79-6] Vol. II: 224
leukol [91-22-5] Vol. I: 1022
leuna m [94-74-6] Vol. I: 25
levanox green-ga [1308-38-9] Vol. III: 234
levanox-white-rkb [13463-67-7] Vol. III:
   921
levoxine [302-01-2] Vol. I: 735
lexone [21087-64-9] Vol. III: 626
lexone [608-73-1] Vol. I: 564
lexone-df [21087-64-9] Vol. III: 626
lexone-4l [21087-64-9] Vol. III: 626
leytosan [62-38-4] Vol. II: 535

l-gamma-methylthio-alpha-aminobutyric
   acid [63-68-3] Vol. III: 593
libavius fuming spirit [7646-78-8] Vol. III:
   876
lignasan [2235-25-8] Vol. III: 441
likuden [126-07-8], Vol. III: 472
Lilly-34, 314 [957-51-7] Vol. II: 6
lime [1305-78-8] Vol. I: 402
lime, burned [1305-78-8] Vol. I: 402
lime chloride [7778-54-3] Vol. II: 494
lime nitrate [10124-37-5] Vol. I: 399
lime-nitrogen [156-62-7] Vol. II: 265
lime saltpeter [10124-37-5] Vol. I: 399
line rider [93-76-5] Vol. I: 59
linex [330-55-2] Vol. II: 1084
9, 12-linoleic acid [60-33-3] Vol. II: 667
linolic acid [60-33-3] Vol. II: 667
linormone [94-74-6] Vol. I: 25
linorox [330-55-2] Vol. II: 1084
lintox [608-73-1] Vol. I: 564
linurex [330-55-2] Vol. II: 1084
linuron (herbicide) [330-55-2] Vol. II: 1084
liqueified petroleum gas [106-97-8] Vol. II:
   183
liquid ammonia [7664-41-7] Vol. I: 135
liquid-derris [83-79-4] Vol. II: 166
liquidolefiant gas [74-85-1] Vol. I: 691
liquid oxygen [7782-44-7] Vol. II: 675
liquid pitch oil [8001-58-9] Vol. I: 495
liquid silver [7439-97-6] Vol. II: 549
liquiphene [62-38-4] Vol. II: 535
liquital [50-06-6] Vol. I: 223
lirobetarex [150-68-5] Vol. II: 1075
liro-cipc [101-21-3] Vol. II: 234
liromate [14484-64-1] Vol. III: 489
liropon [75-99-0] Vol. I: 1016
liroprem [87-86-5] Vol. I: 928
lirothion [56-38-2] Vol. III: 750
listab 28 [7428-48-0] Vol. III: 882
litarol-m [1689-84-5] Vol. II: 163
litharge [1317-36-8] Vol. III: 513
litharge-pure [1317-36-8] Vol. III: 513
litharge yellow L-28 [1317-36-8] Vol. III:
   513
lithium chromate (Li2CrO4) [14307-35-8]
   Vol. II: 242
lithium chromate(vi) [14307-35-8] Vol. II:
   242
lithium hydride (LiH) [7580-67-8] Vol. III:
   525
lithium, metallic [7439-93-2] Vol. I: 810

lithium monohydride [7580-67-8] **Vol. III:** 525

lixophen [50-06-6] **Vol. I:** 223

l-3-(1-methyl-2-pyrrolidyl)pyridine [54-11-5] **Vol. I:** 874

l nicotine [54-11-5] **Vol. I:** 874

lobamine [63-68-3] **Vol. III:** 593

lo-estasol [1929-73-3] **Vol. III:** 155

loha [7439-89-6] **Vol. III:** 487

loisol [52-68-6] **Vol. III:** 707

lonocol-m [12427-38-2] **Vol. III:** 531

lontrel [1702-17-6] **Vol. III:** 239

lontrel-sf-100 [1702-17-6] **Vol. III:** 239

lorex [330-55-2] **Vol. II:** 1084

lorinal [302-17-0] **Vol. III:** 197

lorol [112-53-8] **Vol. II:** 316

lorol-11 [112-53-8] **Vol. II:** 316

lorol-20- [111-87-5] **Vol. I:** 897

lorol-24 [36653-82-4] **Vol. III:** 193

lorol-5- [112-53-8] **Vol. II:** 316

lorol-7 [112-53-8] **Vol. II:** 316

lorox [330-55-2] **Vol. II:** 1084

lorox-linuron-weed-killer [330-55-2] **Vol. II:** 1084

lorox-weed-killer [330-55-2] **Vol. II:** 1084

lorox-41-weed-killer [330-55-2] **Vol. II:** 1084

lorsban [2921-88-2] **Vol. III:** 762

losantin [7778-54-3] **Vol. II:** 494

lost [505-60-2] **Vol. II:** 962

s-lost [505-60-2] **Vol. II:** 962

lo vol [93-76-5] **Vol. I:** 59

lox [7782-44-7] **Vol. II:** 675

loxanol-k [36653-82-4] **Vol. III:** 193

loxanol-k-extra [36653-82-4] **Vol. III:** 193

loxanwachs-sk [36653-82-4] **Vol. III:** 193

lpg-ethyl-mercaptan 1010 [75-08-1] **Vol. III:** 419

l-96 (salt) [1344-09-8] **Vol. III:** 868

lsobutyl aldehyde [78-84-2] **Vol. I:** 769

lsobutyryl aldehyde [78-84-2] **Vol. I:** 769

lsopropylidenebis (4-hydroxybenzene) [80-05-7] **Vol. II:** 720

l-tartaric acid, ammonium salt [14307-43-8] **Vol. III:** 50

lubergal [50-06-6] **Vol. I:** 223

lubrokal [50-06-6] **Vol. I:** 223

lucidol-70 [94-36-0] **Vol. I:** 310

lucidol [94-36-0] **Vol. I:** 310

lucidol b-50- [94-36-0] **Vol. I:** 310

lucidol g-20- [94-36-0] **Vol. I:** 310

lucidol kl-50- [94-36-0] **Vol. I:** 310

lucidol-50p [94-36-0] **Vol. I:** 310

ludox-hs-40 [16871-71-9] **Vol. II:** 911

ludox-hs-40 [7631-86-9] **Vol. III:** 846

lumbrical [110-85-0] **Vol. III:** 776

luminal [50-06-6] **Vol. I:** 223

lunar caustic [7761-88-8] **Vol. I:** 1038

luperco ast [94-36-0] **Vol. I:** 310

luperco-bdb [94-17-7] **Vol. III:** 685

lupersol-DDM [1338-23-4] **Vol. II:** 192

lupersol-DEL [1338-23-4] **Vol. II:** 192

lupersol delta-X [1338-23-4] **Vol. II:** 192

luprosil [79-09-4] **Vol. I:** 1012

lurgo [60-51-5] **Vol. III:** 740

luride [7681-49-4] **Vol. II:** 925

luride-SF [7681-49-4] **Vol. II:** 925

lutosol [67-63-0] **Vol. II:** 515

lutrol [25322-68-3] **Vol. II:** 811

lutrol-9 [107-21-1] **Vol. I:** 653

lye [1310-58-3] **Vol. III:** 782

lysoform [50-00-0] **Vol. I:** 706

lysol brand disinfectant [105-67-9] **Vol. I:** 1162

m-1- [7440-50-8] **Vol. I:** 519

m-3- [7440-50-8] **Vol. I:** 519

m-4- [7440-50-8] **Vol. I:** 519

m-5055- [8001-35-2] **Vol. I:** 1118

maac [108-84-9] **Vol. II:** 686

mace [532-27-4] **Vol. I:** 89

mace (lacrimator) [532-27-4] **Vol. I:** 89

mach nic [54-11-5] **Vol. I:** 874

macondray [94-75-7] **Vol. I:** 30

Macquer's salt [7784-41-0] **Vol. III:** 71

macrogol [25322-68-3] **Vol. II:** 811

macrogol-400-bpc [107-21-1] **Vol. I:** 653

mad [2163-80-6] **Vol. II:** 610

mafu [62-73-7] **Vol. I:** 946

mafu strip [62-73-7] **Vol. I:** 946

magnetic-6 [7704-34-9] **Vol. II:** 965

magnetic 70, 90, and 95 [7704-34-9] **Vol. II:** 965

mah [123-33-1] **Vol. II:** 870

maintain-3 [123-33-1] **Vol. II:** 870

malacide [121-75-5] **Vol. II:** 950

malafor [121-75-5] **Vol. II:** 950

malagran [121-75-5] **Vol. II:** 950

malakill [121-75-5] **Vol. II:** 950

malamar-50 [121-75-5] **Vol. II:** 950

malasol [121-75-5] **Vol. II:** 950

malaspray [121-75-5] **Vol. II:** 950

malataf [121-75-5] **Vol. II:** 950

malathiazol [121-75-5] **Vol. II:** 950

malathion-e50- [121-75-5] Vol. II: 950

malathion-lv-concentrate [121-75-5] Vol. II: 950

malathon [121-75-5] Vol. II: 950

malathyl [121-75-5] Vol. II: 950

malation (Polish) [121-75-5] Vol. II: 950

malatol [121-75-5] Vol. II: 950

malatox [121-75-5] Vol. II: 950

maleic acid anhydride [108-31-6] Vol. I: 812

maleic hydrazine [123-33-1] Vol. II: 870

maleic maleic acid hydrazide [123-33-1] Vol. II: 870

malein-30 [123-33-1] Vol. II: 870

maleinic acid [110-16-7] Vol. II: 528

Maleinsaeurehydrazid (German) [123-33-1] Vol. II: 870

malenic acid [110-16-7] Vol. II: 528

N, N-maleoylhydrazine [123-33-1] Vol. II: 870

malerbane-giavoni-l [2212-67-1] Vol. III: 77

maliazide [123-33-1] Vol. II: 870

malipur [133-06-2] Vol. II: 276

malix [115-29-7] Vol. II: 661

malmed [121-75-5] Vol. II: 950

malonic acid dinitrile [109-77-3] Vol. II: 531

malonic acid mononitrile [372-09-8] Vol. I: 29

malonic dinitrile [109-77-3] Vol. II: 531

malonic mononitrile [372-09-8] Vol. I: 29

malononitrile [109-77-3] Vol. II: 531

malononitrile, (o-chlorobenzylidene)- [2698-41-1] Vol. III: 218

malphos [121-75-5] Vol. II: 950

malurane [330-55-2] Vol. II: 1084

malzid [123-33-1] Vol. II: 870

manam [12427-38-2] Vol. III: 531

mandb-10064 [1689-84-5] Vol. II: 163

maneb-80 [12427-38-2] Vol. III: 531

maneba [12427-38-2] Vol. III: 531

manebe (French) [12427-38-2] Vol. III: 531

manebgan [12427-38-2] Vol. III: 531

maneb-r [12427-38-2] Vol. III: 531

maneb-zl4 [12427-38-2] Vol. III: 531

manesan [12427-38-2] Vol. III: 531

mangaan(II) -(N,N'-ethyleen-bis(dithiocarbamaat)) (Dutch) [12427-38-2] Vol. III: 531

manganese, ((1,2-ethanediyl-bis(carbamodithioato))(2-)) [12427-38-2] Vol. III: 531

manganeseethylene-bis-dithiocarbamate [12427-38-2] Vol. III: 531

manganese ethylenebis-dithiocarbamate [12427-38-2] Vol. III: 531

manganese ethylene-1,2-bisdithiocarbamate [12427-38-2] Vol. III: 531

manganese, (ethylenebis (dithiocarbamato))- [12427-38-2] Vol. III: 531

manganese(II)-ethylene-di (dithiocarbamate) [12427-38-2] Vol. III: 531

Mangan(II)-(N,N'-aethylen-bis (dithiocarbamate)) (German) [12427-38-2] Vol. III: 531

manganous-ethylenebis (dithiocarbamate) [12427-38-2] Vol. III: 531

mangan (Polish) [7439-96-5] Vol. III: 534

manosil-vn-3 [16871-71-9] Vol. II: 911

manosil-vn-3 [7631-86-9] Vol. III: 846

manoxol-ot [577-11-7] Vol. III: 859

manzate [12427-38-2] Vol. III: 531

manzate-200 [12427-38-2] Vol. III: 531

manzate-d [12427-38-2] Vol. III: 531

manzate-maneb fungicide [12427-38-2] Vol. III: 531

manzeb [12427-38-2] Vol. III: 531

manzin [12427-38-2] Vol. III: 531

maoh [108-11-2] Vol. I: 913

mapasol [137-42-8] Vol. III: 861

maposol [137-42-8] Vol. III: 861

maprofix-563 [151-21-3] Vol. II: 981

maprofix-neu [151-21-3] Vol. II: 981

maprofix-wac [151-21-3] Vol. II: 981

maprofix wac-la [151-21-3] Vol. II: 981

maralate [72-43-5] Vol. II: 345

marinol [8001-54-5] Vol. II: 67

marisan-forte [82-68-8] Vol. II: 132

marlate [72-43-5] Vol. II: 345

marmer [330-54-1] Vol. II: 1078

marsh gas [74-82-8] Vol. III: 579

martinal [21645-51-2] Vol. II: 57

martinal [21645-51-2] Vol. III: 35

martinal a [21645-51-2] Vol. II: 57

martinal a [21645-51-2] Vol. III: 35

martinal-a/s [21645-51-2] Vol. II: 57

martinal-a/s [21645-51-2] Vol. III: 35

martinal f-a [21645-51-2] Vol. II: 57

martinal f-a [21645-51-2] Vol. III: 35

marvex [62-73-7] Vol. I: 946

marvik [69409-94-5] Vol. III: 459

mascagnite [7783-20-2] Vol. I: 159

ma-100 [carbon] [7440-44-0] Vol. III: 179

masoten [52-68-6] Vol. III: 707

masposol [137-42-8] Vol. III: 861

massicot [1317-36-8] Vol. III: 513

massicotite [1317-36-8] Vol. III: 513

matrigon [1702-17-6] Vol. III: 239

mattling acid [7664-93-9] Vol. II: 969

Maxx-90- [139-40-2] Vol. III: 802

Mayer's reagent [7783-33-7] Vol. III: 542

maz [315-18-4] Vol. II: 227

mazide [123-33-1] Vol. II: 870

mb-10,064 [1689-84-5] Vol. II: 163

mb-10064 [1689-84-5] Vol. II: 163

mb [74-83-9] Vol. II: 577

mbc [17804-35-2] Vol. II: 148

m-b-c fumigant [74-83-9] Vol. II: 577

mbc-33 soft fumigant [74-83-9] Vol. II: 577

mbc-soil-fumigant [74-83-9] Vol. II: 577

mbk [591-78-6] Vol. II: 462

m-b-r 98 [74-83-9] Vol. II: 577

mbt [149-30-4] Vol. III: 540

2-mbt [149-30-4] Vol. III: 540

mbx [74-83-9] Vol. II: 577

20-mc [56-49-5] Vol. III: 604

3-MC [56-49-5] Vol. III: 604

m-44 capsules (potassium cyanide) [151-50-8] Vol. I: 978

mcb [108-90-7] Vol. I: 245

4-(MCB) [94-81-5] Vol. III: 174

mcf [79-22-1] Vol. III: 597

3-MCh [56-49-5] Vol. III: 604

2m-4ch [94-74-6] Vol. I: 25

mcp [94-74-6] Vol. I: 25

mcpa [94-74-6] Vol. I: 25

(2,4-mcpb) [94-81-5] Vol. III: 174

2,4-mcpb [94-81-5] Vol. III: 174

4mcpb [94-81-5] Vol. III: 174

MCPB [94-81-5] Vol. III: 174

MCP-butyric [94-81-5] Vol. III: 174

m-44 cyanide capsules [143-33-9] Vol. I: 1045

mda [101-77-9] Vol. III: 304

mdba [1918-00-9] Vol. II: 159

Me-700 [72-54-8] Vol. III: 404

mea [141-43-5] Vol. II: 356

mea [151-38-2] Vol. III: 594

mea (alcohol) [141-43-5] Vol. II: 356

meb [12427-38-2] Vol. III: 531

mebetizole [149-30-4] Vol. III: 540

mebithizol [149-30-4] Vol. III: 540

mebr [74-83-9] Vol. II: 577

3-meca [56-49-5] Vol. III: 604

mediben [1918-00-9] Vol. II: 159

medilave [123-03-5] Vol. III: 196

medizinc [7733-02-0] Vol. II: 1103

meetco [78-93-3] Vol. I: 347

meg [107-21-1] Vol. I: 653

meidon-15-dust [2104-64-5] Vol. II: 738

mek [78-93-3] Vol. I: 347

mek-peroxide [1338-23-4] Vol. II: 192

melinite [88-89-1] Vol. II: 793

melipax [8001-35-2] Vol. I: 1118

melitoxin [66-76-2] Vol. III: 336

melprex [2439-10-3] Vol. III: 290

melprex-65 [2439-10-3] Vol. III: 290

mema [151-38-2] Vol. III: 594

mendrin [72-20-8] Vol. III: 349

meonine [63-68-3] Vol. III: 593

mep [104-90-5] Vol. III: 770

mephanac [94-74-6] Vol. I: 25

mephospholan (French) [950-10-7] Vol. III: 536

mercaptan-amilique (French) [110-66-7] Vol. III: 53

mercaptan methylique (French) [74-93-1] Vol. II: 615

mercaptan-methylique-perchlore (French) [594-42-3] Vol. II: 613

2-mercaptoacetic acid [68-11-1] Vol. III: 538

alpha-mercaptoacetic acid [68-11-1] Vol. III: 538

mercaptobenzene [108-98-5] Vol. III: 917

2-mercaptobenzothiazole [149-30-4] Vol. III: 540

mercaptobenzthiazole [149-30-4] Vol. III: 540

2-mercaptobenzthiazole [149-30-4] Vol. III: 540

mercaptodimethur [2032-65-7] Vol. II: 308

1-mercaptododecane [112-55-0] Vol. III: 388

mercaptoethane [75-08-1] Vol. III: 419

(N-(2-mercaptoethyl) benzenesulfonamide) [741-58-2] Vol. III: 731

N-(2-mercaptoethylbenzene) sulfonamide S-(o,o-diisopropylphosphorodithioate) m. N-(beta-o,o-diisopropyldithio- phosphoryle-thyl) benzenesulfonamide [741-58-2] Vol. III: 731

mercaptofos [8065-48-3] Vol. III: 755

2-mercapto-2-imidazoline [96-45-7] Vol. I: 757

mercury biniodide [7774-29-0] Vol. II: 547

mercury, bis(fulminato)- [628-86-4] Vol. III: 548

mercury, bis (thiocyanate [592-85-8] Vol. I: 1082

mercury bisulfate [7783-35-9] Vol. II: 563

mercury chloride (HgC12) [7487-94-7] Vol. II: 539

mercury cyanide (Hg(CN-)2) [592-04-1] Vol. II: 544

mercury diacetate [1600-27-7] Vol. II: 24

mercury dichloride [7487-94-7] Vol. II: 539

mercury dicyanide [592-04-1] Vol. II: 544

mercury diiodide [15385-57-6] Vol. III: 559

mercury dinitrate [10045-94-0] Vol. II: 554

mercury dithiocyanate [592-85-8] Vol. I: 1082

mercury fulminate (dry) [628-86-4] Vol. III: 548

mercury fulminate [Hg(ONC)₂] [628-86-4] Vol. III: 548

mercury fulminate (wet) [628-86-4] Vol. III: 548

mercury(II) acetate [1600-27-7] Vol. II: 24

mercury(II) acetate, phenyl [62-38-4] Vol. II: 535

mercury(II) benzoate [583-15-3] Vol. III: 551

mercury(II) chloride [7487-94-7] Vol. II: 539

mercury(II) chloride ammonobasic [10124-48-8] Vol. II: 533

mercury(II) cyanide [592-04-1] Vol. II: 544

mercury(II) fulminate [628-86-4] Vol. III: 548

alpha-mercury(II) iodide [7774-29-0] Vol. II: 547

mercury(II) nitrate [10045-94-0] Vol. II: 554

mercury(II) nitrate (1:2) [10045-94-0] Vol. II: 554

mercury(II)-oxide [21908-53-2] Vol. II: 559

mercury(II) potassium iodide [7783-33-7] Vol. III: 542

mercury(II) sulfate (1:1) [7783-35-9] Vol. II: 563

mercury(II) sulfate [7783-35-9] Vol. II: 563

mercury (ii) thiocyanate [592-85-8] Vol. I: 1082

mercury(I) nitrate [10415-75-5] Vol. III: 570

mercury iodide (Hg₂I₂) [15385-57-6] Vol. III: 559

mercury iodide (HgI2) [7774-29-0] Vol. II: 547

mercury iodide (HgI) [7774-29-0] Vol. II: 547

mercury(I) oxide [15829-53-5] Vol. III: 561

mercuryl acetate [1600-27-7] Vol. II: 24

mercury, metallic [7439-97-6] Vol. II: 549

mercury(1+), methyl [22967-92-6] Vol. III: 566

mercury(1+), methyl-, ion [22967-92-6] Vol. III: 566

mercury monoxide [21908-53-2] Vol. II: 559

mercury nitrate [10045-94-0] Vol. II: 554

mercury(1) nitrate [10415-75-5] Vol. III: 570

mercury(2) nitrate [10045-94-0] Vol. II: 554

mercury nitrate monohydrate [10045-94-0] Vol. II: 554

mercury, (nitrato-o) phenyl [55-68-5] Vol. III: 702

mercury, nitratophenyl [55-68-5] Vol. III: 702

mercury oleate [1191-80-6] Vol. III: 553

mercury(2) oxide [21908-53-2] Vol. II: 559

mercury oxide (HgO) [21908-53-2] Vol. II: 559

mercury oxide (Hg₂O) [15829-53-5] Vol. III: 561

mercury oxide sulfate [Hg₃O₂(SO₄)] [1312-03-4] Vol. III: 563

mercury oxonium sulfate [1312-03-4] Vol. III: 563

mercury perchloride [7487-94-7] Vol. II: 539

mercury pernitrate [10045-94-0] Vol. II: 554

mercury potassium cyanide [591-89-9] Vol. III: 556

mercury potassium iodide [7783-33-7] Vol. III: 542

mercury potassium iodide (K₂HgI₄) [7783-33-7] Vol. III: 542

mercury protoiodide [15385-57-6] Vol. III: 559

mercury protonitrate [10415-75-5] Vol. III: 570

mercury sulfate, basic [1312-03-4] Vol. III: 563

mercury sulfate (HgSO4) [7783-35-9] Vol. II: 563

mercury thiocyanate hg (scn) [592-85-8] Vol. I: 1082

merex [143-50-0] Vol. III: 587

merfazin [100-56-1] Vol. III: 699

mergal-a-25 [62-38-4] Vol. II: 535

merge [2163-80-6] Vol. II: 610

merge-823 [2163-80-6] Vol. II: 610

merit [1689-84-5] Vol. II: 163

2-merkaptobenzotiazol (Polish) [149-30-4] Vol. III: 540

merkazin [7287-19-6] Vol. II: 1056

merkon [13171-21-6] Vol. III: 704

merocet [123-03-5] Vol. III: 196

meronidal [443-48-1] Vol. III: 628

merpan [133-06-2] Vol. II: 276

merpectogel [55-68-5] Vol. III: 702

merphenyl nitrate [55-68-5] Vol. III: 702

merpol-oj [25322-68-3] Vol. II: 811

merrillite [7440-66-6] Vol. I: 1175

mersolite-2 [100-56-1] Vol. III: 699

mersolite [62-38-4] Vol. II: 535

mersolite-7 [55-68-5] Vol. III: 702

mersolite-8 [62-38-4] Vol. II: 535

mertax [149-30-4] Vol. III: 540

mertionin [63-68-3] Vol. III: 593

mesamate [2163-80-6] Vol. II: 610

Mesityloxid (German) [141-79-7] Vol. II: 696

mesityloxyde (Dutch) [141-79-7] Vol. II: 696

mesomile [16752-77-5] Vol. II: 38

mesurol [2032-65-7] Vol. II: 308

mesylith [54-05-7] Vol. III: 229

met [63-68-3] Vol. III: 593

meta-aminoaniline [108-45-2] Vol. III: 98

meta-benzenediamine [108-45-2] Vol. III: 98

metabrom [74-83-9] Vol. II: 577

metacetone [96-22-0] Vol. III: 679

metacetonic acid [79-09-4] Vol. I: 1012

metachlor [15972-60-8] Vol. II: 9

metachloronitrobenzene [121-73-3] Vol. III: 207

metachlorphenprop [101-10-0] Vol. III: 806

meta-diaminobenzene [108-45-2] Vol. III: 98

metadichlorobenzene [25321-22-6] Vol. III: 325

meta-dichlorobenzene [541-73-1] Vol. III: 321

meta-dinitrobenzene [99-65-0] Vol. III: 89

metaisosystox sulfoxide [301-12-2] Vol. III: 661

metakrylan-metylu (Polish) [80-62-6] Vol. II: 567

metallic arsenic [7440-38-2] Vol. I: 190

metam [137-42-8] Vol. III: 861

metam-fluid basf [137-42-8] Vol. III: 861

metamidofos-estrella [10265-92-6] Vol. II: 744

metamidophos [10265-92-6] Vol. II: 744

metam-sodium [137-42-8] Vol. III: 861

metam-sodium (Dutch, French, German, Italian) [137-42-8] Vol. III: 861

metanolo (Italian) [67-56-1] Vol. I: 836

metanol (Spanish) [67-56-1] Vol. I: 836

metaphenylenediamine [108-45-2] Vol. III: 98

metaquest a [60-00-4] Vol. I: 36

metasol-30 [62-38-4] Vol. II: 535

metasystemox [301-12-2] Vol. III: 661

metasystox-r [301-12-2] Vol. III: 661

meta-tcpn [1897-45-6] Vol. II: 513

meta-tetrachlorophthalodinitrile [1897-45-6] Vol. II: 513

meta-toluene diisocyanate [584-84-9] Vol. II: 125

meta-toluylene-diamine [95-80-7] Vol. II: 1016

meta-toluylenediamine [95-80-7] Vol. II: 1016

meta-tolylenediamine [95-80-7] Vol. II: 1016

N-(meta-trifluoromethylphenyl)-N, N'-dimethylurea [2164-17-2] Vol. II: 1086

metaupon [112-80-1] Vol. I: 895

metaxon [94-74-6] Vol. I: 25

metazene [142-90-5] Vol. III: 391

methaanthiol (Dutch) [74-93-1] Vol. II: 615

methacetone [96-22-0] Vol. III: 679

methachlor [15972-60-8] Vol. II: 9

methacide [108-88-3] Vol. I: 1109

methacrylate de butyle (French) [97-88-1] Vol. III: 164

methacrylate de methyle (French) [80-62-6] Vol. II: 567

methacrylic acid, butyl ester [97-88-1] Vol. III: 164

6,9-methano-2,4,3-benzodioxathiepin, 6,7,8,9,10,10-hexachloro-1,5,5a,6,9,9a hexahydro-, 3,3-dioxide [1031-07-8] Vol. III: 398

6,9-methano-2,4,3-benzodioxathiepin, 6,7,8,9,10,10-hexachloro-1,5,5a,6,9,9a hexahydro-, 3-oxide [115-29-7] Vol. II: 661

4,7-methano-1h-indene, 1,4,5,6,7,8,8-heptachloro-3a,4,7,7a-tetrahydro [76-44-8] Vol. II: 628

4,7-methano-1h-indene, 1,4,5,6,7,8,8-heptaehloro-3a,4,7,7a-tetrahydro [76-44-8] Vol. II: 628

4,7-methano-1h-indene, 1,2,4,5,6,7,8,8-octaehloro-2,3,3a,4,7,7a-hexahydro- [57-74-9] Vol. II: 619

2,5-methano-2h-indeno (1,2-b) oxirene, 2,3,4,5,6,7,7-heptaehloro-1a, 1b, 5,5a,6,6a-hexahydro-, (laalpha, lbbeta, 2alpha, 5alpha, 5abeta, 6beta, 6aalpha)- [1024-57-3] Vol. II: 626

4,7-methano-1h-isoindole-1,3 (2h)-dione, 2-(2-ethylhexyl)-3a, 4,7,7a-tetrahydro [113-48-4] Vol. III: 631

4,7-methano-1h-indene, 1,2,4,5,6,7,8,8-oetachloro-2,3,3a,4,7,7a-hexahydro- [57-74-9] Vol. II: 619

2,5-methano-2h-indeno (1,2-b) oxirene, 2,3,4,5,6,7,7-heptachloro-1a, 1b, 5,5a,6,6a-hexahydro-, (laalpha, lbbeta, 2alpha, 5alpha, 5abeta, 6beta, 6aalpha)- [1024-57-3] Vol. II: 626

2,5-methano-2H-oxireno (a) indene, 2,3,4,5,6,7,7-heptachlor-la, 1b, 5,5a,6,6a hexahydro [1024-57-3] Vol. II: 626

methanoic acid [64-18-6] Vol. I: 711

4,7-methanoindan, 1,4,5,6,7,8,8-heptachloro-2,3-epoxy-3a,4,7,7a-tetrahydro [1024-57-3] Vol. II: 626

4,7-methanoindan, 1,2,4,5,6,7,8,8-octachloro-3a,4,7,7a-tetrahydro- [57-74-9] Vol. II: 619

4,7-methanoindene, 3a,4,7,7a-tetrahydro- [77-73-6] Vol. II: 635

4,7-methanoindene, 1,4,5,6,7,8,8-heptachloro-3a,4,7,7a-tetrahydro [76-44-8] Vol. II: 628

methanolacetonitrile [109-78-4] Vol. II: 466

methanol, ethynyl [107-19-7] Vol. II: 864

methanol, (2-furyl) [98-00-0] Vol. I: 723

4,7-methano-lh-indene, 3a,4,7,7a-tetrahydro- [77-73-6] Vol. II: 635

methanol, phenyl [100-51-6] Vol. I: 313

methanol, sodium salt [124-41-4] Vol. II: 639

methanol, trimethyl [75-65-0] Vol. II: 197

methanthiol (German) [74-93-1] Vol. II: 615

methaphenamiphos [22224-92-6] Vol. II: 743

1,2,4-methenecyclopenta (c, d) pentalene-r-carboxaldehyde, 2,2a,3,3,4,7-hexachlorodecahydro [7421-93-4] Vol. II: 319

1,2,4-methenocyclopenta(cd) pentalene-5-carboxaldehyde, 2,2a,3,3,4,7-hexachlorodecahydro-, (1alpha, 2beta, 2abeta, 4beta, 4abeta, 5beta, 6abeta, 6bbeta, 7r)- [7421-93-4] Vol. II: 319

1,3,4-metheno-1H-cyclobuta(cd) pentalene, 1,1a, 2,2,3,3a, 4,5,5,5a, 5b, 6-dodecachlorooctahydro- [2385-85-5] Vol. III: 583

1,3,4-metheno-2H-cyclobuta(cd) pentalen-2-one, 1,1a,3,3a,4,5,5,5a,5b, 6-decachlorooctahydro- [143-50-0] Vol. III: 587

methenyl chloride [67-66-3] Vol. I: 453

methenyl tribromide [75-25-2] Vol. I: 832

methenyl trichloride [67-66-3] Vol. I: 453

methidathion [950-37-8] Vol. III: 734

methilanin [63-68-3] Vol. III: 593

methionine [63-68-3] Vol. III: 593

s-methionine [63-68-3] Vol. III: 593

methionine, 1 [63-68-3] Vol. III: 593

methlyfluorophosphonic acid isopropyl ester [107-44-8] Vol. III: 826

meth-o-gas [74-83-9] Vol. II: 577

methogas [74-83-9] Vol. II: 577

methoxcide [72-43-5] Vol. II: 345

methoxo [72-43-5] Vol. II: 345

methoxone [94-74-6] Vol. I: 25

2-Methoxy-aethanol (German) [109-86-4] Vol. II: 370

6-methoxy- alpha-(5-vinyl-2-quin-uclidinyl)-4-quinolinemethanol [130-95-0] Vol. III: 821

2-methoxy-1-aminobenzene [90-04-0] Vol. II: 76

2-methoxyaniline [90-04-0] Vol. II: 76

o-methoxyaniline [90-04-0] Vol. II: 76

2-methoxybenzenamine [90-04-0] Vol. II: 76

2-methoxy-4,6-bis (isopropylamino)-s-triazine [1610-18-0] Vol. III: 787

2-methoxy-4,6-bis (isopropylamino)-1,3,5-triazine [1610-18-0] Vol. III: 787

methylacrylaat (Dutch) [96-33-3] Vol. I: 119

methyl acrylate, monomer [96-33-3] Vol. I: 119

methyl acrylat (German) [96-33-3] Vol. I: 119

alpha-methylacrylonitrile [126-98-7] Vol. II: 839

2-methylacrylonitrile [126-98-7] Vol. II: 839

methyl alcohol [67-56-1] Vol. I: 836

methyl aldehyde [50-00-0] Vol. I: 706

methylalkohol (German) [67-56-1] Vol. I: 836

methylallene [106-99-0] Vol. I: 333

1-methylallene [106-99-0] Vol. I: 333

2-methylallyl chloride [563-47-3] Vol. III: 202

methylallyl chloride [563-47-3] Vol. III: 202

2-Methyl-allylchlorid (German) [563-47-3] Vol. III: 202

methyl alpha-methylacrylate [80-62-6] Vol. II: 567

N-methyl-alpha-pyrrolidinone [872-50-4] Vol. II: 893

N-methyl-alpha-pyrrolidone [872-50-4] Vol. II: 893

methylaminen (Dutch) [74-89-5] Vol. I: 840

N-methyl- aminobenzene [100-61-8] Vol. III: 602

(methylamino) benzene [100-61-8] Vol. III: 602

2-methyl-1-aminobenzene [95-53-4] Vol. I: 1114

1-methyl-2-aminobenzene [95-53-4] Vol. I: 1114

n-((methylamino) carbonyl) oxy) ethanimidothioic acid methyl ester [16752-77-5] Vol. II: 38

N-methylaminodithioformic acid˙sodium salt [137-42-8] Vol. III: 861

N-methylaminodithioformicacid, sodium salt [137-42-8] Vol. III: 861

1-methyl-2-aminoethanol [78-96-6] Vol. II: 821

N-methyl- aminomethanethionothiolicacid, sodium salt [137-42-8] Vol. III: 861

N-methylamino- methanethionothiolic acid sodium salt [137-42-8] Vol. III: 861

2-methyl-2-aminopropane [75-64-9] Vol. II: 199

methylamyl acetate [108-84-9] Vol. II: 686

methylamyl alcohol [105-30-6] Vol. I: 912

methylamyl alcohol [108-11-2] Vol. I: 913

methyl amyl cetone (French) [110-43-0] Vol. II: 454

methyl amyl ketone [110-43-0] Vol. II: 454

methylaniline [100-61-8] Vol. III: 602

p-methylaniline [106-49-0] Vol. II: 1040

p-methylaniline [106-49-0] Vol. III: 927

m-methylaniline [108-44-1] Vol. III: 47

2-methylaniline [95-53-4] Vol. I: 1114

3-methylaniline [108-44-1] Vol. III: 47

4-methylaniline [106-49-0] Vol. II: 1040

4-methylaniline [106-49-0] Vol. III: 927

o methylaniline [95-53-4] Vol. I: 1114

methylarsenic acid, sodium salt [2163-80-6] Vol. II: 610

methylate de sodium (French) [124-41-4] Vol. II: 639

1-methylazacyclopentan-2-one [872-50-4] Vol. II: 893

2-methylazacyclopropane [75-55-8] Vol. I: 219

methylazinphos [86-50-0] Vol. III: 744

2-methylaziridine [75-55-8] Vol. I: 219

2-methylbenzamine [95-53-4] Vol. I: 1114

N-methylbenzazimide, dimethyldithiophosphoric acid ester [86-50-0] Vol. III: 744

N-methylben- zenamine [100-61-8] Vol. III: 602

p-methylbenzenamine [106-49-0] Vol. II: 1040

p-methylbenzenamine [106-49-0] Vol. III: 927

m-methylbenzenamine [108-44-1] Vol. III: 47

2-methylbenzenamine [95-53-4] Vol. I: 1114

3-methylbenzenamine [108-44-1] Vol. III: 47

4-methylbenzenamine [106-49-0] Vol. II: 1040

4-methylbenzenamine [106-49-0] Vol. III: 927

o methylbenzenamine [95-53-4] Vol. I: 1114

methylbenzene [108-88-3] Vol. I: 1109

4-methyl-1,3-benzenediamine [95-80-7] Vol. II: 1016

3-methylbenz (j) aceanthrene [56-49-5] Vol. III: 604

methylbenzol [108-88-3] Vol. I: 1109

methyl isobutyl carbinol [105-30-6] Vol. I: 912

methylisobutyl carbinol [108-11-2] Vol. I: 913

methylisobutyl carbinol acetate [108-84-9] Vol. II: 686

methyl isobutyl cetone (French) [108-10-1] Vol. II: 692

methylisobutylketon (Dutch, German) [108-10-1] Vol. II: 692

methylisocyanaat (Dutch) [624-83-9] Vol. II: 507

methyl isocyanat (German) [624-83-9] Vol. II: 507

methyl- isothiocyanaat (Netherlands) [556-61-6] Vol. III: 600

methylisothiocyanat (Denmark) [556-61-6] Vol. III: 600

Methyl-isothiocyanat (German) [556-61-6] Vol. III: 600

methylisothiokyanat (Czech) [556-61-6] Vol. III: 600

methyljodide (Dutch) [74-88-4] Vol. III: 581

Methyljodid (German) [74-88-4] Vol. III: 581

methyl ketone [67-64-1] Vol. I: 77

2-methyllactonitrile [75-86-5] Vol. I: 794

methylisoamyl acetate [108-84-9] Vol. II: 686

methylmercaptaan (Dutch) [74-93-1] Vol. II: 615

methylmercaptan [74-93-1] Vol. II: 615

2-methylmercapto-4,6-bis (isopropylamino)-s-triazine [7287-19-6] Vol. II: 1056

4-methylmercapto-3,5-dimethylphenyl N-methylcarbamate [2032-65-7] Vol. II: 308

2-methylmercapto-4-ethylamino-6-isopropylamino-s-triazine [834-12-8] Vol. II: 1054

4-methylmercapto-3,5-xylyl methylcarbamate [2032-65-7] Vol. II: 308

methylmercury(1+) [22967-92-6] Vol. III: 566

methylmercury(I)-cation [22967-92-6] Vol. III: 566

methylmercury ion (1+) [22967-92-6] Vol. III: 566

methylmercury ion [22967-92-6] Vol. III: 566

4-methyl-meta-phenylene diisocyanate [584-84-9] Vol. II: 125

methylmethacrylaat (Dutch) [80-62-6] Vol. II: 567

methyl-methacrylat (German) [80-62-6] Vol. II: 567

n methylmethanamine [124-40-3] Vol. I: 613

methylmethane [74-84-0] Vol. III: 401

methyl methanoate [107-31-3] Vol. II: 441

2-methyl-2-methoxypropane [1634-04-4] Vol. II: 388

methyl-methylacrylate [80-62-6] Vol. II: 567

methyl 2-methyl-l-propenyl ketone [141-79-7] Vol. II: 696

methyl 2-methyl-2-propenoate [80-62-6] Vol. II: 567

methyl 2-methylpropenoate [80-62-6] Vol. II: 567

2-methyl-2-(methylthio) propanal, o-((methylamino) carbonyl) oxime [116-06-3] Vol. III: 19

2-methyl-2-(methylthio) propionaldehyde o-(methylcarbamoyl) oxime [116-06-3] Vol. III: 19

2-Methyl-2-methylthio-propionaldehyd-o-(N-methyl-carbamoyl)-oxim (German) [116-06-3] Vol. III: 19

methyl m-hydroxycarbanilate, m-methylcarbanilate [13684-63-4] Vol. III: 130

methyl monosulfide [75-18-3] Vol. I: 846

4-methyl-m-phenylenediamine [95-80-7] Vol. II: 1016

4-methyl-m-phenylene diisocyanate [584-84-9] Vol. II: 125

methyl-m-phenylene isocyanate [26471-62-5] Vol. II: 509

4-methyl-m-phenylene isocyanate [584-84-9] Vol. II: 125

methyl-3-m-tolycarbamoloxyphenyl carbamate [13684-63-4] Vol. III: 130

methyl mustard [556-61-6] Vol. III: 600

methyl mustard oil [556-61-6] Vol. III: 600

methyl n-amyl ketone [110-43-0] Vol. II: 454

methyl n amylketone [110-43-0] Vol. II: 454

alpha methylnaphthalene [90-12-0] Vol. I: 858

methyl n-butyl ketone [591-78-6] Vol. II: 462

2-methyl-1-nitrobenzene [88-72-2] Vol. II: 1036

methyl pentyl ketone [110-43-0] **Vol. II:** 454

m-methylphenol [108-39-4] **Vol. III:** 261

2-methylphenol [95-48-7] **Vol. I:** 536

3-methylphenol [108-39-4] **Vol. III:** 261

o methylphenol [95-48-7] **Vol. I:** 536

methylphenylamine [100-61-8] **Vol. III:** 602

N-methylphenylamine [100-61-8] **Vol. III:** 602

methylphenyl diphenyl phosphate [26444-49-5] **Vol. III:** 723

4-methyl-phenylene diisocyanate [584-84-9] **Vol. II:** 125

methylphenylene isocyanate [26471-62-5] **Vol. II:** 509

4-methyl-phenylene isocyanate [584-84-9] **Vol. II:** 125

1-methyl-1-phenylethene [98-83-9] **Vol. II:** 947

methyl phenyl ketone [98-86-2] **Vol. I:** 86

o methylphenylol [95-48-7] **Vol. I:** 536

methylphosphonofluoride acid, isopropyl ester [107-44-8] **Vol. III:** 826

methylphosphonofluoridic acid isopropyl ester [107-44-8] **Vol. III:** 826

methylphosphonofluoridic acid 1-methyl-ethyl ester [107-44-8] **Vol. III:** 826

methyl phthalate [131-11-3] **Vol. II:** 784

2-methyl-1-propanal [78-84-2] **Vol. I:** 769

2-methylpropanal [78-84-2] **Vol. I:** 769

methylpropanal [78-84-2] **Vol. I:** 769

methyl propanal [78-84-2] **Vol. I:** 769

1-methyl-1,3-propanediol [107-88-0] **Vol. III:** 148

2-methylpropane nitrile [78-82-0] **Vol. III:** 798

2-methyl-2-propanethiol [75-66-1] **Vol. III:** 163

alpha methylpropanoicacid [79-31-2] **Vol. I:** 787

1-methyl-1-propanol [78-92-2] **Vol. II:** 194

2-methylpropanol-1 [78-83-1] **Vol. I:** 765

2-methylpropanol-2 [75-65-0] **Vol. II:** 197

2-methyl propanol [78-83-1] **Vol. I:** 765

methyl propenate [96-33-3] **Vol. I:** 119

2-methylpropenenitrile [126-98-7] **Vol. II:** 839

2-methyl-2-propenenitrile [126-98-7] **Vol. II:** 839

methyl-2-propenoate [96-33-3] **Vol. I:** 119

methyl prop-2-enoate [96-33-3] **Vol. I:** 119

methyl propenoate [96-33-3] **Vol. I:** 119

2-methyl-2-propenyl chloride [563-47-3] **Vol. III:** 202

2-methylpropionaldehyde [78-84-2] **Vol. I:** 769

2-methylpropionic acid [79-31-2] **Vol. I:** 787

alpha methylpropionicacid [79-31-2] **Vol. I:** 787

2-methylpropionitrile [78-82-0] **Vol. III:** 798

alpha methylpropionaldehyde [78-84-2] **Vol. I:** 769

1-methylpropyl acetate [105-46-4] **Vol. I:** 64

2-methyl-1-propylacetate [110-19-0] **Vol. I:** 42

2-methylpropyl acetate [110-19-0] **Vol. I:** 42

2-methylpropyl acrylate [106-63-8] **Vol. III:** 15

1-methylpropyl alcohol [78-92-2] **Vol. II:** 194

2-methylpropyl alcohol [78-83-1] **Vol. I:** 765

1-methylpropylamine [13952-84-6] **Vol. III:** 159

3-methyl-2-propylamine [78-81-9] **Vol. III:** 493

2-methylpropylamine [78-81-9] **Vol. III:** 493

6-(1-methyl-propyl)-2,4-dinitrofenol (Dutch) [88-85-7] **Vol. II:** 704

(6-(1-methyl-propyl)-2,4-dinitrofenyl)-3,3-di-methyl-acrylaat (Dutch) [485-31-4] **Vol. II:** 261

2-(1-methylpropyl)-4,6-dinitrophenol [88-85-7] **Vol. II:** 704

2-(1-methylpropyl)-4,6-dinitrophenyl beta, beta-dimethacrylate [485-31-4] **Vol. II:** 261

2-(1-methylpropyl)-4,6-dinitrophenyl 3,3-di-methylacrylate [485-31-4] **Vol. II:** 261

(6-(1-Methyl-propyl)-2,4-dinitrophenyl)-3,3-dimethyl-acrylat (German) [485-31-4] **Vol. II:** 261

2-(1-methylpropyl)-4,6-dinitrophenyl 3-methyl-2-butenoate [485-31-4] **Vol. II:** 261

2-methylpropyl methyl ketone [108-10-1] **Vol. II:** 692

p methylpyridine [108-89-4] **Vol. I:** 1019

1-methyl-2-(3-pyridyl)pyrrolidine [54-11-5] **Vol. I:** 874

mondur td-80 [26471-62-5] Vol. II: 509

mondur-tds [584-84-9] Vol. II: 125

monex [2163-80-6] Vol. II: 610

monitor [10265-92-6] Vol. II: 744

monitor [insecticide] [10265-92-6] Vol. II: 744

Monoaethanolamin (German) [141-43-5] Vol. II: 356

monoallylamine [107-11-9] Vol. III: 23

monoammonium carbonate [1066-33-7] Vol. II: 62

monoammonium sulfamate [7773-06-0] Vol. II: 959

monoammonium sulfite [10192-30-0] Vol. II: 985

monoban [2163-80-6] Vol. II: 610

monobenzylamine [100-46-9] Vol. II: 170

monobromoacetone [598-31-2] Vol. III: 800

monobromodichloromethane [75-27-4] Vol. I: 817

monobromoethane [74-96-4] Vol. III: 431

monobromomethane [74-83-9] Vol. II: 577

monobromotrifluoromethane [75-63-8] Vol. III: 143

monobutylamine [109-73-9] Vol. II: 203

monocalcium arsenite [52740-16-6] Vol. II: 109

monochloorazijnzuur (Dutch) [79-11-8] Vol. I: 22

monochloorbenzeen (Dutch) [108-90-7] Vol. I: 245

monochloracetic acid [79-11-8] Vol. I: 22

monochloracetone [78-95-5] Vol. II: 828

monochlorbenzene [108-90-7] Vol. I: 245

Monochlorbenzol (German) [108-90-7] Vol. I: 245

Monochloressigsaeure (German) [79-11-8] Vol. I: 22

monochlorethane [75-00-3] Vol. II: 321

monochlorhydrine du glycol [French] [107-07-3] Vol. III: 220

monochloroacetaldehyde [107-20-0] Vol. II: 1

monochloroacetic acid [79-11-8] Vol. I: 22

monochloroacetic acid solution (DOT) [79-11-8] Vol. I: 22

monochloroacetone [78-95-5] Vol. II: 828

monochloroacetyl chloride [79-04-9] Vol. III: 214

monochlorobenzene [108-90-7] Vol. I: 245

monochlorodifluoromethane [75-45-6] Vol. II: 591

monochlorodimethyl ether [107-30-2] Vol. II: 390

monochloroethane [75-00-3] Vol. II: 321

monochloroethanoic acid [79-11-8] Vol. I: 22

2-monochloroethanol [107-07-3] Vol. III: 220

monochloroethene [75-01-4] Vol. II: 416

monochloroethylene [75-01-4] Vol. II: 416

monochloromethane [74-87-3] Vol. II: 585

monochloromethyl methyl ether [107-30-2] Vol. II: 390

mono-chloro-mono-bromo-methane [74-97-5] Vol. II: 583

monochloromonobromomethane [74-97-5] Vol. II: 583

monochloropentafluoroethane [76-15-3] Vol. III: 226

monochlorosulfuricacid [7790-94-5] Vol. I: 457

monochlorotrimethylsilicon [75-77-4] Vol. III: 946

monochromium trioxide [1333-82-0] Vol. I: 493

monocite methacrylate monomer [80-62-6] Vol. II: 567

monoclorobenzene (Italian) [108-90-7] Vol. I: 245

monocobalt oxide [1307-96-6] Vol. III: 252

monocresyl diphenyl phosphate [26444-49-5] Vol. III: 723

monocyanoacetic acid [372-09-8] Vol. I: 29

monoethanolamine [141-43-5] Vol. II: 356

monoethylene glycol [107-21-1] Vol. I: 653

Monofluoressigsaures natrium (German) [62-74-8] Vol. I: 39

monofluoroethylene [75-02-5] Vol. III: 966

monofluorotrichloromethane [75-69-4] Vol. II: 602

monohydrated selenium dioxide [7783-00-8] Vol. II: 901

monohydroxybenzene [108-95-2] Vol. I: 920

monohydroxymethane [67-56-1] Vol. I: 836

monoiodomethane [74-88-4] Vol. III: 581

monoioduro di metile (Italian) [74-88-4] Vol. III: 581

monoisobutylamine [78-81-9] Vol. III: 493

monoisopropanolamine [78-96-6] Vol. II: 821

monoisopropylamine [75-31-0] Vol. I: 787

N-monomethylamide of O,O-
dimethyldithiophosphorylacetic acid [60-
51-5] **Vol. III**: 740

monomethylamine [74-89-5] **Vol. I**: 840

N-monomethylaniline [100-61-8] **Vol. III**:
602

monomethylaniline [100-61-8] **Vol. III**:
602

monomethyldichlorosilane [75-54-7] **Vol.
III**: 609

monomethyl ether of ethylene glycol [109-
86-4] **Vol. II**: 370

monomethylhydrazine [60-34-4] **Vol. II**:
475

mono-n-butylamine [109-73-9] **Vol. II**: 203

mononickel oxide [1313-99-1] **Vol. III**: 645

mononitrogen monoxide [10102-43-9] **Vol.
I**: 887

mono-n-propylamine [107-10-8] **Vol. II**:
861

monoperacetic acid [79-21-0] **Vol. III**: 682

monophenol [108-95-2] **Vol. I**: 920

monoplex-doa [103-23-1] **Vol. II**: 291

monopotassium arsenate [7784-41-0] **Vol.
III**: 71

monopotassium dihydrogen arsenate
[7784-41-0] **Vol. III**: 71

monoprop [79-09-4] **Vol. I**: 1012

monopropylamine [107-10-8] **Vol. II**: 861

monopropyleneglycol [57-55-6] **Vol. I**: 1003

monosodium acid methanearsonate [2163-
80-6] **Vol. II**: 610

monosodium acid metharsonate [2163-80-
6] **Vol. II**: 610

monosodium methanearsonate [2163-80-6]
Vol. II: 610

monosodium methanearsonic acid [2163-
80-6] **Vol. II**: 610

monosodium methylarsonate [2163-80-6]
Vol. II: 610

monosodium sulfite [7631-90-5] **Vol. II**:
920

monovinyl chloride (MVC) [75-01-4] **Vol.
II**: 416

monsanto 2,4-d isopropyl ester [94-11-1]
Vol. II: 15

montar [1336-36-3] **Vol. I**: 974

monter [1336-36-3] **Vol. I**: 974

montrose-propanil [26638-19-7] **Vol. II**:
298

monuroun [150-68-5] **Vol. II**: 1075

monurox [150-68-5] **Vol. II**: 1075

monuuron [150-68-5] **Vol. II**: 1075

morbicid [50-00-0] **Vol. I**: 706

morestan [2439-01-2] **Vol. III**: 663

morestan-2 [2439-01-2] **Vol. III**: 663

morestane [2439-01-2] **Vol. III**: 663

moringine [100-46-9] **Vol. II**: 170

morocide [485-31-4] **Vol. II**: 261

morpholine, 4-nitroso [59-89-2] **Vol. III**:
634

morrocid [485-31-4] **Vol. II**: 261

morton-ep-332- [23422-53-9] **Vol. III**: 463

morton ep-161e [556-61-6] **Vol. III**: 600

moscarda [121-75-5] **Vol. II**: 950

mo-1202t [1313-27-5] **Vol. I**: 848

mota maskros [94-75-7] **Vol. I**: 30

moth balls [91-20-3] **Vol. I**: 854

moth flakes [91-20-3] **Vol. I**: 854

motox [8001-35-2] **Vol. I**: 1118

mottenhexe [67-72-1] **Vol. II**: 330

mous-con [1314-84-7] **Vol. II**: 1100

mouse nots [57-24-9] **Vol. I**: 1054

mouse rid [57-24-9] **Vol. I**: 1054

mouse tox [57-24-9] **Vol. I**: 1054

moxie [72-43-5] **Vol. II**: 345

moxon [94-75-7] **Vol. I**: 30

mqd [2439-01-2] **Vol. III**: 663

m3r [7440-50-8] **Vol. I**: 519

mrowczan-etylu (Polish) [109-94-4] **Vol. II**:
438

mrowkozol [114-26-1] **Vol. II**: 231

m3s [7440-50-8] **Vol. I**: 519

msma [2163-80-6] **Vol. II**: 610

mtbe [1634-04-4] **Vol. II**: 388

mtd [10265-92-6] **Vol. II**: 744

mtd [95-80-7] **Vol. II**: 1016

mudekan [330-55-2] **Vol. II**: 1084

mulhouse-white [7446-14-2] **Vol. III**: 517

murfos [56-38-2] **Vol. III**: 750

murfulvin [126-07-8], **Vol. III**: 472

murganic [5234-68-4] **Vol. II**: 670

muriatic acid [7647-01-0] **Vol. I**: 738

muriatic ether [75-00-3] **Vol. II**: 321

murine for the eyes [8001-54-5] **Vol. II**:
67

murite of ammonia [12125-02-9] **Vol. I**:
141

murocoll-2 ophthalmic solution [8001-54-
5] **Vol. II**: 67

murocoll-19 opthalmic solution [8001-54-
5] **Vol. II**: 67

Muro's opcon a ophthalmic solution
[8001-54-5] **Vol. II**: 67

Muro's opcon ophthalmic solution [8001-54-5] Vol. II: 67

muro tears ophthalmic solution [8001-54-5] Vol. II: 67

Murphy's rose fungicide [133-07-3] Vol. III: 461

s-mustard [505-60-2] Vol. II: 962

mustard-gas [505-60-2] Vol. II: 962

mustard-hd [505-60-2] Vol. II: 962

mustard, sulfur [505-60-2] Vol. II: 962

mustard-vapor [505-60-2] Vol. II: 962

mutamycin [07-7], Vol. III: 633

m xylenol [105-67-9] Vol. I: 1162

mylone [533-74-4] Vol. III: 293

mylone-85 [533-74-4] Vol. III: 293

mylon [Czech] [533-74-4] Vol. III: 293

myrmicyl [64-18-6] Vol. I: 711

mystox-d [131-52-2] Vol. III: 676

n1030 [16871-71-9] Vol. II: 911

n1030 [7631-86-9] Vol. III: 846

n-2790 [944-22-9] Vol. III: 714

n-38 [1344-09-8] Vol. III: 868

n-521 [533-74-4] Vol. III: 293

na-22 [96-45-7] Vol. I: 757

na-22- [96-45-7] Vol. I: 757

naa-173- [57-11-4] Vol. I: 1052

nabame (French) [142-59-6] Vol. I: 411

nabasan (discontinued) [142-59-6] Vol. I: 411

na-22-d [96-45-7] Vol. I: 757

nadone [108-94-1] Vol. II: 268

nafpak [7681-49-4] Vol. II: 925

naftalen (Polish) [91-20-3] Vol. I: 854

1-naftilamina (Spanish) [134-32-7] Vol. III: 637

alpha-naftylamin (Czech) [134-32-7] Vol. III: 637

1-naftylamine (Dutch) [134-32-7] Vol. III: 637

2-naftylamine (dutch) [91-59-8] Vol. I: 860

nafun ipo [142-59-6] Vol. I: 411

nako-tmt [95-80-7] Vol. II: 1016

nalco-1050 [16871-71-9] Vol. II: 911

nalco-1050 [7631-86-9] Vol. III: 846

nalcon-243 [533-74-4] Vol. III: 293

nalfloc-N-1050 [16871-71-9] Vol. II: 911

nalfloc-N-1050 [7631-86-9] Vol. III: 846

nalox [443-48-1] Vol. III: 628

napclor-g [131-52-2] Vol. III: 676

1-naphthalamine [134-32-7] Vol. III: 637

1-naphthalenamine [134-32-7] Vol. III: 637

2-naphthalenamine [91-59-8] Vol. I: 860

2-naphthalenamine, N-phenyl [135-88-6] Vol. III: 640

naphthalene, alpha methyl [90-12-0] Vol. I: 858

naphthalene, beta-methyl [91-57-6] Vol. III: 611

1,4-naphthalenedione, 2,3-dichloro [117-80-6] Vol. II: 645

naphthalene, hexachloro [1335-87-1] Vol. III: 475

naphthalene, 1-methyl [90-12-0] Vol. I: 858

naphthalene, 2-methyl [91-57-6] Vol. III: 611

naphthalene, octachloro- [2234-13-1] Vol. III: 653

naphthalene oil [8001-58-9] Vol. I: 495

naphthalene, tetrachloro [1335-88-2] Vol. III: 902

naphthalene, trichloro [1321-65-9] Vol. III: 939

naphthalidam [134-32-7] Vol. III: 637

naphthalidine [134-32-7] Vol. III: 637

naphthalin [91-20-3] Vol. I: 854

naphthaline [91-20-3] Vol. I: 854

naphthene [91-20-3] Vol. I: 854

1,4-naphthoquinone, 2,3-dichloro [117-80-6] Vol. II: 645

alpha-naphthylamine [134-32-7] Vol. III: 637

1-Naphthylamin (German) [134-32-7] Vol. III: 637

2-naphthylamin (German) [91-59-8] Vol. I: 860

N-(2-naphthyl) aniline [135-88-6] Vol. III: 640

1,2-(1,8-naphthylene) benzene [206-44-0] Vol. II: 435

naphthyleneethylene [83-32-9] Vol. I: 1

2-naphthylphenylamine [135-88-6] Vol. III: 640

naphtoelan red gg base [100-01-6] Vol. II: 70

naphtol as-kg [106-49-0] Vol. II: 1040

naphtol as-kg [106-49-0] Vol. III: 927

naphtol as-kgll [106-49-0] Vol. II: 1040

naphtol as-kgll [106-49-0] Vol. III: 927

napthalene-1, 2, 3, 4-tetrahydride [119-64-2] Vol. III: 910

narcogen [79-01-6] Vol. II: 427

narcotan [151-67-7] Vol. III: 141

narcotane [151-67-7] Vol. III: 141

narcotann ne-spofa (Russian) [151-67-7] **Vol. III:** 141

narcotile [75-00-3] **Vol. II:** 321

narcylen [74-86-2] **Vol. I:** 95

narkosoid [79-01-6] **Vol. II:** 427

narkotan [151-67-7] **Vol. III:** 141

nata [76-03-9] **Vol. II:** 32

natal [76-03-9] **Vol. II:** 32

natasol fast orange gr salt [88-74-4] **Vol. I:** 173

natreen [81-07-2] **Vol. III:** 824

natrium [7440-23-5] **Vol. II:** 918

Natriumacetat (German) [127-09-3] **Vol. III:** 849

natrium aceticum [127-09-3] **Vol. III:** 849

Natriumaluminiumfluorid (German) [15096-52-3] **Vol. III:** 37

natriumbichromaat (Dutch) [10588-01-9] **Vol. II:** 300

natrium-carbonicum-calcinatum [497-19-8] **Vol. II:** 922

natrium-carbonicum-siccatum [497-19-8] **Vol. II:** 922

Natriumchlorid (German) [7647-14-5] **Vol. III:** 851

Natrium-2,4-dichlorphenoxyaethylsulfat (German) [136-78-7] **Vol. III:** 855

natriumdichromat (German) [10588-01-9] **Vol. II:** 300

natriumfluoracetat (German) [62-74-8] **Vol. I:** 39

natrium fluoride [7681-49-4] **Vol. II:** 925

natriumfluotscetaat (Dutch) [62-74-8] **Vol. I:** 39

Natriumhexafluoroaluminate (German) [15096-52-3] **Vol. III:** 37

natriumhydroxid (German) [1310-73-2] **Vol. II:** 929

natriumhydroxyde (Dutch) [1310-73-2] **Vol. II:** 929

natriumnitrit (German) [7632-00-0] **Vol. I:** 890

natrium-N-methyl-dithiocarbamaat (Dutch) [137-42-8] **Vol. III:** 861

Natrium-N-methyl-dithiocarbamat (German) [137-42-8] **Vol. III:** 861

natriumphosphat (German) [7558-79-4] **Vol. II:** 933

natriumselenit (German) [10102-18-8] **Vol. II:** 903

Natriumselenit (German) [10102-18-8] **Vol. III:** 828

Natriumsulfat (German) [7757-82-6] **Vol. III:** 871

natural anglesite [7446-14-2] **Vol. III:** 517

natural bromellite [1304-56-9] **Vol. III:** 123

natural galena [1314-87-0] **Vol. I:** 807

natural lead sulfide [1314-87-0] **Vol. I:** 807

natural molysite [7705-08-0] **Vol. I:** 699

natural montroydite [21908-53-2] **Vol. II:** 559

natural orpiment [1303-33-9] **Vol. II:** 99

natural wintergreen oil [119-36-8] **Vol. III:** 622

naugard tjb [86-30-6] **Vol. I:** 622

Naugatuck d-014 [2312-35-8] **Vol. II:** 983

nauli "gum" [104-46-1] **Vol. III:** 56

navadel [78-34-2] **Vol. II:** 756

Na-X [497-19-8] **Vol. II:** 922

naxol [108-93-0] **Vol. I:** 580

nayper bo [94-36-0] **Vol. I:** 310

nb- [98-95-3] **Vol. I:** 273

nba [71-36-3] **Vol. I:** 352

N4-(7-chloro-4-quinolinyl)-N1,N1-diethyl-1,4-pentanediamine [54-05-7] **Vol. III:** 229

NCI-9579- [99-99-0] **Vol. I:** 1105

NCI-554813 [74-96-4] **Vol. III:** 431

NCI-C00044 [309-00-2] **Vol. III:** 343

NCI-C00055 [133-90-4] **Vol. III:** 44

NCI-C00066 [86-50-0] **Vol. III:** 744

NCI-C00077 [133-06-2] **Vol. II:** 276

NCI-C00099 [57-74-9] **Vol. II:** 619

NCI-C00102 [1897-45-6] **Vol. II:** 513

NCI C00113 [62-73-7] **Vol. I:** 946

NCI-C00124 [60-57-1] **Vol. III:** 356

nci-c00135 [60-51-5] **Vol. III:** 740

NCI-C00157 [72-20-8] **Vol. III:** 349

NCI-C00180 [76-44-8] **Vol. II:** 628

NCI-C00191 [143-50-0] **Vol. III:** 587

NCI-C00215 [121-75-5] **Vol. II:** 950

NCI C00259 [8001-35-2] **Vol. I:** 1118

NCI-C00260 [76-87-9] **Vol. II:** 941

NCI-C00395 [78-34-2] **Vol. II:** 756

nci-c00408 [510-15-6] **Vol. III:** 108

NCI-C00419 [82-68-8] **Vol. II:** 132

NCI-C00475 [72-54-8] **Vol. III:** 404

NCI-C00486 [115-32-2] **Vol. II:** 144

NCI-C00497 [72-43-5] **Vol. II:** 345

NCI-C00500 [96-12-8] **Vol. II:** 818

NCI C00511 [107-06-2] **Vol. I:** 643

NCI-C00533 [76-06-2] **Vol. II:** 606

neopentanoic acid [75-98-9] **Vol. III:** 781

neoplatin [15663-27-1] **Vol. III:** 237

neopynamin [7696-12-0] **Vol. III:** 912

neosabenyl [120-32-1] **Vol. III:** 122

neosone-d [135-88-6] **Vol. III:** 640

neosyl [16871-71-9] **Vol. II:** 911

neosyl [7631-86-9] **Vol. III:** 846

neo-tric [443-48-1] **Vol. III:** 628

neozin [7733-02-0] **Vol. II:** 1103

neozon-d [135-88-6] **Vol. III:** 640

neozone [135-88-6] **Vol. III:** 640

neozone-d [135-88-6] **Vol. III:** 640

neracid [133-06-2] **Vol. II:** 276

nereb [12427-38-2] **Vol. III:** 531

nerkol [62-73-7] **Vol. I:** 946

nervanaid b-acid [60-00-4] **Vol. I:** 36

nespor [12427-38-2] **Vol. III:** 531

Nessler reagent [7783-33-7] **Vol. III:** 542

neston [63-68-3] **Vol. III:** 593

netagrone [94-75-7] **Vol. I:** 30

netazol [94-74-6] **Vol. I:** 25

netzschwefel [7704-34-9] **Vol. II:** 965

neurobarb [50-06-6] **Vol. I:** 223

neutral ammonium chromate [7788-98-9] **Vol. I:** 464

neutral ammonium fluoride- [12125-01-8] **Vol. I:** 146

neutral lead stearate [7428-48-0] **Vol. III:** 882

neutral sodium chromate [7775-11-3] **Vol. I:** 469

neutra-verdigris [142-71-2] **Vol. II:** 11

neutrazyme [151-21-3] **Vol. II:** 981

nevax [577-11-7] **Vol. III:** 859

nevugon [52-68-6] **Vol. III:** 707

new improved ceresan [2235-25-8] **Vol. III:** 441

new improved granosan [2235-25-8] **Vol. III:** 441

nexoval [101-21-3] **Vol. II:** 234

ni-270- [7440-02-0] **Vol. I:** 862

NIA-1240 [563-12-2] **Vol. II:** 759

NIA-5273 [51-03-6] **Vol. III:** 779

NIA-5462 [115-29-7] **Vol. II:** 661

nia-5488 [116-29-0] **Vol. III:** 906

nia-5996- [1194-65-6] **Vol. I:** 299

NIA-9044 [485-31-4] **Vol. II:** 261

nia-10242- [1563-66-2] **Vol. I:** 420

NIA-17370 [10453-86-8] **Vol. III:** 285

NIA-33297 [52645-53-1] **Vol. III:** 282

niacide [14484-64-1] **Vol. III:** 489

Niagara-1240 [563-12-2] **Vol. II:** 759

niagara-5006- [1194-65-6] **Vol. I:** 299

Niagara-5,462 [115-29-7] **Vol. II:** 661

niagara-5,996- [1194-65-6] **Vol. I:** 299

Niagara-9044 [485-31-4] **Vol. II:** 261

niagara-10242- [1563-66-2] **Vol. I:** 420

Niagara estasol [94-11-1] **Vol. II:** 15

niagaramite [140-57-8] **Vol. III:** 61

niagara nia-10242 [1563-66-2] **Vol. I:** 420

niagara p.a.-dust [54-11-5] **Vol. I:** 874

nialk [79-01-6] **Vol. II:** 427

nibren [1321-65-9] **Vol. III:** 939

nibren-wax [1321-65-9] **Vol. III:** 939

nichel (Italian) [7440-02-0] **Vol. I:** 862

nichel tetracarbonile (Italian) [13463-39-3] **Vol. I:** 865

nickel-200- [7440-02-0] **Vol. I:** 862

nickel-201- [7440-02-0] **Vol. I:** 862

nickel-205- [7440-02-0] **Vol. I:** 862

nickel-207- [7440-02-0] **Vol. I:** 862

nickel-270- [7440-02-0] **Vol. I:** 862

nickel ammonium sulfate (Ni(NH4)2(SO4)2) [15699-18-0] **Vol. II:** 973

nickel carbonyle (French) [13463-39-3] **Vol. I:** 865

nickel carbonyl (ni (co)4) [13463-39-3] **Vol. I:** 865

nickelcarbonyl, (t-4) [13463-39-3] **Vol. I:** 865

nickel (2) chloride [7718-54-9] **Vol. I:** 867

nickel chloride [7718-54-9] **Vol. I:** 867

nickel diacetate [373-02-4] **Vol. II:** 648

nickel dichloride [7718-54-9] **Vol. I:** 867

nickel dicyanide [557-19-7] **Vol. II:** 650

nickel dinitrate [13138-45-9] **Vol. III:** 641

nickel(II) acetate [373-02-4] **Vol. II:** 648

nickel (ii) chloride [7718-54-9] **Vol. I:** 867

nickel(II) cyanide [557-19-7] **Vol. II:** 650

nickel(II) nitrate (1: 2) [13138-45-9] **Vol. III:** 641

nickel(II) oxide (1:1) [1313-99-1] **Vol. III:** 645

nickel(II) oxide [1313-99-1] **Vol. III:** 645

nickel (ii) sulfate [7786-81-4] **Vol. I:** 870

nickel (it) chloride (1:2) [7718-54-9] **Vol. I:** 867

nickel(2) nitrate [13138-45-9] **Vol. III:** 641

nickel nitrate [Ni(NO3)2] [13138-45-9] **Vol. III:** 641

nickelous acetate [373-02-4] **Vol. II:** 648

nickelous chloride [7718-54-9] **Vol. I:** 867

nickelous nitrate [13138-45-9] **Vol. III:** 641

nitric acid, cobalt (2) salt [10141-05-6] Vol. I: 504

nitric acid, lead salt [18256-98-9] Vol. III: 511

nitric acid, mercury(I) salt [10415-75-5] Vol. III: 570

nitric acid, mercury(1) salt [10415-75-5] Vol. III: 570

nitric acid, mercury(2) salt [10045-94-0] Vol. II: 554

nitric acid, nickel(II) salt [13138-45-9] Vol. III: 641

nitric acid, nickel(2) salt [13138-45-9] Vol. III: 641

nitric acid, phenylmercury salt [55-68-5] Vol. III: 702

nitric acid, silver(I) salt [7761-88-8] Vol. I: 1038

nitric acid silver (i) salt [7761-88-8] Vol. I: 1038

nitric acid silver (1) salt [7761-88-8] Vol. I: 1038

nitric acid sodium salt [7631-99-4] Vol. III: 864

nitric acid, sodium salt [7631-99-4] Vol. III: 864

nitric acid, thallium(1) salt [10102-45-1] Vol. II: 1001

nitric acid, zinc salt [7779-88-6] Vol. I: 1185

nitric acid, zirconium (4) salt [13746-89-9] Vol. I: 1191

nitric acid, iron (3) salt [10421-48-4] Vol. I: 883

nitrileacrilico (Italian) [107-13-1] Vol. I: 123

nitrileacrylique (French) [107-13-1] Vol. I: 123

nitrile adipico (Italian) [111-69-3] Vol. I: 128

nitrile kyseliny malonove (Czech) [109-77-3] Vol. II: 531

nitril kyseliny malonove (Czech) [109-77-3] Vol. II: 531

nitril kyseliny malonoye (Czech) [109-77-3] Vol. II: 531

nitriloacetonitrile [460-19-5] Vol. I: 555

nitrilo-2,2',2''-triacetic acid [139-13-9] Vol. III: 648

nitrilotriethanol [102-71-6] Vol. II: 374

2,2',2''-nitrilotriethanol [102-71-6] Vol. II: 374

nitrilo-2,2',2''-triethanol [102-71-6] Vol. II: 374

2,2',2''-nitrilotris(ethanol) [102-71-6] Vol. II: 374

nitrite de sodium (French) [7632-00-0] Vol. I: 890

nitrito [10102-44-0] Vol. I: 885

nitrito sodico (Spanish) [7632-00-0] Vol. I: 890

nitro [10102-44-0] Vol. I: 885

p-nitroanilina (Polish) [100-01-6] Vol. II: 70

p-nitroaniline [100-01-6] Vol. II: 70

p-nitroaniline [100-01-6] Vol. III: 58

o nitroaniline [88-74-4] Vol. I: 173

nitrobarite [10022-31-8] Vol. I: 233

2-nitrobenzenamine [88-74-4] Vol. I: 173

4-nitrobenzenamine [100-01-6] Vol. II: 70

nitrobenzol [98-95-3] Vol. I: 273

4-nitrobiphenyl [92-93-3] Vol. III: 650

nitrocalcite [10124-37-5] Vol. I: 399

nitrocarbol [75-52-5] Vol. II: 597

p-nitrochloorbenzeen (Dutch) [100-00-5] Vol. III: 210

p-nitrochlorobenzene [100-00-5] Vol. III: 210

m-nitrochlorobenzene [121-73-3] Vol. III: 207

4-nitro-1-chlorobenzene [100-00-5] Vol. III: 210

4-nitrochlorobenzene [100-00-5] Vol. III: 210

1-nitro-4-chlorobenzene [100-00-5] Vol. III: 210

o-nitrochlorobenzene [88-73-3] Vol. III: 204

nitrochlorobenzene, para-, solid (DOT) [100-00-5] Vol. III: 210

p-Nitrochlorobenzol (German) [100-00-5] Vol. III: 210

nitrochloroform [76-06-2] Vol. II: 606

p-nitroclorobenzene (Italian) [100-00-5] Vol. III: 210

4-nitrodiphenyl [92-93-3] Vol. III: 650

p-nitrodiphenyl [92-93-3] Vol. III: 650

nitroetan (Polish) [79-24-3] Vol. II: 334

nitrofan [534-52-1] Vol. I: 540

4-nitrofenol (Dutch) [100-02-7] Vol. II: 726

nitrogen-14 [7727-37-9] Vol. II: 657

nitrogen dioxide (liquid) [10102-44-0] Vol. I: 885

nitrogen dioxide (no2) [10102-44-0] Vol. I: 885

nitrogen gas [7727-37-9] Vol. II: 657

nitrogen hypoxide [10024-97-2] Vol. I: 889

niuif-100 [56-38-2] Vol. III: 750

nivaquine-B [54-05-7] Vol. III: 229

nix [52645-53-1] Vol. III: 282

3-(n methylpyrrollidino)pyridine [54-11-5] Vol. I: 874

nmor [59-89-2] Vol. III: 634

nmp [872-50-4] Vol. II: 893

nmt [75-52-5] Vol. II: 597

nobecutan [137-26-8] Vol. I: 632

nobunt [118-74-1] Vol. I: 268

no bunt-40- [118-74-1] Vol. I: 268

nobunt-80- [118-74-1] Vol. I: 268

no bunt liquid [118-74-1] Vol. I: 268

nocbin [97-77-8] Vol. III: 382

nocceler-22- [96-45-7] Vol. I: 757

noctec [302-17-0] Vol. III: 197

noflamol [1336-36-3] Vol. I: 974

nogos [62-73-7] Vol. I: 946

nomersan [137-26-8] Vol. I: 632

nonalol [143-08-8] Vol. I: 893

n nonan-l-ol [143-08-8] Vol. I: 893

nonan-l-ol [143-08-8] Vol. I: 893

1-nonanol [143-08-8] Vol. I: 893

nonox-d [135-88-6] Vol. III: 640

nonox-dn [135-88-6] Vol. III: 640

nonox-tbc [128-37-0] Vol. III: 317

n nonylalcohol [143-08-8] Vol. I: 893

nonyl hydride [111-84-2] Vol. II: 659

nonyl phenol [25154-52-3] Vol. I: 926

nonyl phenol (mixed isomers) [25154-52-3] Vol. I: 926

nopcocide [1897-45-6] Vol. II: 513

nopcocide n-96 [1897-45-6] Vol. II: 513

nopcocide-N-96- [1897-45-6] Vol. II: 513

nopcocide-n40d-and-n96 [1897-45-6] Vol. II: 513

nopeocide-n40d-and-n96 [1897-45-6] Vol. II: 513

no pest [62-73-7] Vol. I: 946

no pest strip [62-73-7] Vol. I: 946

noptil [50-06-6] Vol. I: 223

no-pyr [930-55-2] Vol. II: 891

5-norbornene-2,3-dicarboximide, N-(2-ethylhexyl)- [113-48-4] Vol. III: 631

5-norbornene-2,3-dicarboxylic acid, 1,4,5,6,7,7-hexachloro [115-28-6] Vol. III: 200

5-norbornene-2,3-dimethanol, 1,4,5,6,7,7-hexachloro-, cyclic sulfate [1031-07-8] Vol. III: 398

5-norbornene-2,3-dimethanol, 1,4,5,6,7,7-hexachloro-, cyclic sulfite [115-29-7] Vol. II: 661

2-norbornene, 5-ethylidene [16219-75-3] Vol. III: 440

norcamphor, 1,7,7-trimethyl [76-22-2] Vol. II: 211

norex [1982-47-4] Vol. III: 232

norforms [62-38-4] Vol. II: 535

norit [7440-44-0] Vol. III: 179

norkool [107-21-1] Vol. I: 653

normal primary butyl alcohol [71-36-3] Vol. I: 352

normal primary hexadecyl alcohol [36653-82-4] Vol. III: 193

normersan [137-26-8] Vol. I: 632

norosac [1194-65-6] Vol. I: 299

norval [577-11-7] Vol. III: 859

norway saltpeter [10124-37-5] Vol. I: 399

norway saltpeter [6484-52-2] Vol. I: 153

norwegian saltpeter [10124-37-5] Vol. I: 399

nosilen [25322-68-3] Vol. II: 811

nourithion [56-38-2] Vol. III: 750

novacorn [1689-84-5] Vol. II: 163

novochlorhydrate [302-17-0] Vol. III: 197

novonidazol [443-48-1] Vol. III: 628

novoscabin [120-51-4] Vol. III: 117

nowsouth safti sol brand concentrated bowl cleanser with magic action [7647-01-0] Vol. I: 738

noxfire [83-79-4] Vol. II: 166

noxfish [83-79-4] Vol. II: 166

2-np [79-46-9] Vol. I: 999

npyr [930-55-2] Vol. II: 891

NRDC-104 [10453-86-8] Vol. III: 285

NRDC-143 [52645-53-1] Vol. III: 282

NRDC-149 [52315-07-8] Vol. II: 251

NRDC-161 [52918-63-5] Vol. III: 296

ns-02- [17804-35-2] Vol. II: 148

nsc-185 [66-81-9] Vol. II: 447

nsc-389 [79-01-6] Vol. II: 427

nsc-746- [51-79-6] Vol. II: 224

nsc-1573 [120-80-9] Vol. III: 190

nsc-1771- [137-26-8] Vol. I: 632

nsc-3122 [90-04-0] Vol. II: 76

nsc-3809 [100-25-4] Vol. II: 128

NSC-5354 [95-54-5] Vol. III: 697

NSC-7189 [99-65-0] Vol. III: 89

nsc-8819 [107-02-8] Vol. I: 99

nsc-9577 [88-72-2] Vol. II: 1036

nsc-9578 [99-08-1] Vol. II: 1033

octa-klor [57-74-9] Vol. II: 619

octalene [309-00-2] Vol. III: 343

octalox [60-57-1] Vol. III: 356

octamethyl-difosforzuur-tetramide (Dutch) [152-16-9] Vol. II: 884

octamethyldiphosphoramide [152-16-9] Vol. II: 884

Octamethyl-diphosphorsaeuretetramid (German) [152-16-9] Vol. II: 884

octamethylpyrophosphoramide [152-16-9] Vol. II: 884

octamethylpyrophosphoric acid tetramide [152-16-9] Vol. II: 884

octamethyl pyrophosphortetramide [152-16-9] Vol. II: 884

octamethyl tetramido-pyrophosphate [152-16-9] Vol. II: 884

octan amylu (Polish) [628-63-7] Vol. II: 27

octane [111-65-9] Vol. II: 668

octan etoksyetlu (Polish) [111-15-9] Vol. II: 365

octan etylu (Polish) [141-78-6] Vol. II: 19

octan-fenylrtutnaty (Czech) [62-38-4] Vol. II: 535

octan mednaty (Czech) [142-71-2] Vol. II: 11

octann butylu (Polish) [123-86-4] Vol. I: 18

octanol [111-87-5] Vol. I: 897

1-n octanol [111-87-5] Vol. I: 897

n octanol [111-87-5] Vol. I: 897

n octan-1-ol [111-87-5] Vol. I: 897

n-octan-2-ol [123-96-6] Vol. III: 655

octan propylu (Polish) [109-60-4] Vol. I: 55

octan winylu (Polish) [108-05-4] Vol. I: 69

octilin [111-87-5] Vol. I: 897

octoil [117-81-7] Vol. II: 769

octowy bezwodnik (Polish) [108-24-7] Vol. I: 73

octowy kwas (Polish) [64-19-7] Vol. I: 12

octyl adipate [103-23-1] Vol. II: 291

octyl alcohol [111-87-5] Vol. I: 897

n octyl alcohol [111-87-5] Vol. I: 897

s-octyl alcohol [123-96-6] Vol. III: 655

2-octyl alcohol [123-96-6] Vol. III: 655

octyl alcohol, normal primary [111-87-5] Vol. I: 897

octylaldehyde [123-05-7] Vol. II: 457

n-octyl bicycloheptene dicarboximide [113-48-4] Vol. III: 631

n-octylbicyclo-(2.2.1)-5-heptene-2,3-dicarboximide [113-48-4] Vol. III: 631

octyl carbinol [143-08-8] Vol. I: 893

octyl phthalate [117-81-7] Vol. II: 769

N-octyl phthalate [117-84-0] Vol. II: 789

octyl-phthalate [117-84-0] Vol. II: 789

ofhc cu [7440-50-8] Vol. I: 519

og-25 [76-06-2] Vol. II: 606

oil green [1308-38-9] Vol. III: 234

oil of aniseed [104-46-1] Vol. III: 56

oil of mirbane- [98-95-3] Vol. I: 273

oil of turpentine [8006-64-2] Vol. II: 1065

oil of turpentine, rectified [8006-64-2] Vol. II: 1065

oil of vitriol [7664-93-9] Vol. II: 969

oil of wintergreen [119-36-8] Vol. III: 622

oil-soluble-aniline-yellow [60-09-3] Vol. II: 74

oilstop, halowax [30525-89-4] Vol. I: 907

oil-yellow-21 [97-56-3] Vol. III: 46

oil-yellow-2681 [97-56-3] Vol. III: 46

oil-yellow-aab [60-09-3] Vol. II: 74

oil-yellow-ab [60-09-3] Vol. II: 74

oil-yellow-an [60-09-3] Vol. II: 74

oil-yellow-at [97-56-3] Vol. III: 46

oil-yellow-b [60-09-3] Vol. II: 74

oil-yellow-c [97-56-3] Vol. III: 46

oil-yellow-i [97-56-3] Vol. III: 46

oil-yellow-2r [97-56-3] Vol. III: 46

oktanen (Dutch) [111-65-9] Vol. II: 668

oktanex [72-20-8] Vol. III: 349

oktan (Polish) [111-65-9] Vol. II: 668

oktaterr [57-74-9] Vol. II: 619

okultin m [94-74-6] Vol. I: 25

olamine [141-43-5] Vol. II: 356

olate flakes [143-19-1] Vol. I: 899

oleate of mercury [1191-80-6] Vol. III: 553

oleic acid, mercury(2) salt [1191-80-6] Vol. III: 553

oleic acid, sodium salt [143-19-1] Vol. I: 899

oleine-7503- [112-80-1] Vol. I: 895

oleinic acid [112-80-1] Vol. I: 895

oleobidrin [141-66-2] Vol. III: 720

oleodiazinon [333-41-5] Vol. III: 758

oleofos-20 [56-38-2] Vol. III: 750

oleogesaprim [1912-24-9] Vol. II: 1050

oleoparaphene [56-38-2] Vol. III: 750

oleophosphothion [121-75-5] Vol. II: 950

Olin-Mathieson-2,424 [2593-15-9] Vol. III: 898

olin-terraclor-90%-dust-concentrate [82-68-8] Vol. II: 132

olin-terraclor-technical-grade-pcnb99%-soil-fungicide [82-68-8] Vol. II: 132

olin-terraclor-75%-wettable-powder [82-68-8] Vol. II: 132

olow (Polish) [7439-92-1] Vol. I: 800

olpisan [82-68-8] Vol. II: 132

om-2424 [2593-15-9] Vol. III: 898

omal [88-06-2] Vol. II: 732

omchlor [118-52-5] Vol. III: 329

omega-aminotoluene [100-46-9] Vol. II: 170

omega-bromotoluene [100-39-0] Vol. II: 1019

omega-caprolactam [105-60-2] Vol. III: 79

omega chloroacetophenone [532-27-4] Vol. I: 89

omega chlorotoluene [100-44-7] Vol. I: 1099

omega, omega, omega-trichlorotoluene [98-07-7] Vol. II: 139

omite-57e [2312-35-8] Vol. II: 983

omite-85e [2312-35-8] Vol. II: 983

Omnii-gel [7783-47-3] Vol. III: 881

oms-14- [62-73-7] Vol. I: 946

oms-33 [114-26-1] Vol. II: 231

oms-47 [315-18-4] Vol. II: 227

oms-93 [2032-65-7] Vol. II: 308

oms-197 [72-20-8] Vol. III: 349

oms-214 [2463-84-5] Vol. III: 748

oms-219 [2104-64-5] Vol. II: 738

oms-570 [115-29-7] Vol. II: 661

oms-639 [315-18-4] Vol. II: 227

oms 771- [116-06-3] Vol. III: 19

oms-800 [52-68-6] Vol. III: 707

oms-844 [950-37-8] Vol. III: 734

oms-864- [1563-66-2] Vol. I: 420

oms-0971 [2921-88-2] Vol. III: 762

oms-1206 [10453-86-8] Vol. III: 285

oms-1328 [470-90-6] Vol. III: 716

oms-1804 [35367-38-5] Vol. III: 339

oms-3023 [66230-04-4] Vol. III: 400

ona [88-74-4] Vol. I: 173

n-nonane [111-84-2] Vol. II: 659

oncb [88-73-3] Vol. III: 204

122Sb [14374-79-9] Vol. III: 60

ontracic-800 [1610-18-0] Vol. III: 787

ontrack [1610-18-0] Vol. III: 787

ontrack-we-2 [1610-18-0] Vol. III: 787

ontrack we herbicide [87-86-5] Vol. I: 928

onyx-btc (Onyx Oil and Chem Co,) [8001-54-5] Vol. II: 67

opal [7631-86-9] Vol. III: 846

opp [90-43-7] Vol. II: 176

optal [71-23-8] Vol. II: 856

op-thal-zin [7733-02-0] Vol. II: 1103

optimum [79-06-1] Vol. I: 106

optised [7733-02-0] Vol. II: 1103

optraex [7733-02-0] Vol. II: 1103

Oral-B rinsing solution, concentrate [7783-47-3] Vol. III: 881

orange base ciba ii [88-74-4] Vol. I: 173

orange base irga ii [88-74-4] Vol. I: 173

orange grs salt [88-74-4] Vol. I: 173

orange salt ciba ii [88-74-4] Vol. I: 173

orange salt irga ii [88-74-4] Vol. I: 173

orazinc [7733-02-0] Vol. II: 1103

ordam [2212-67-1] Vol. III: 77

ordinary lactic acid [50-21-5] Vol. I: 792

ordram [2212-67-1] Vol. III: 77

organol-yellow-2a [60-09-3] Vol. II: 74

organol-yellow-2t [97-56-3] Vol. III: 46

orgasol-1002d-white-10-extra-cos [13463-67-7] Vol. III: 921

ornamental-weeder [133-90-4] Vol. III: 44

orpiment [1303-33-9] Vol. II: 99

orthamine [95-54-5] Vol. III: 697

orthene [30560-19-1] Vol. III: 1

Ortho-124120 [30560-19-1] Vol. III: 1

ortho-4355 [300-76-5] Vol. II: 746

ortho-9006 [10265-92-6] Vol. II: 744

ortho-aminoanisole [90-04-0] Vol. II: 76

orthoarsenic acid [1327-52-2] Vol. III: 64

orthoarsenic acid [7778-39-4] Vol. II: 91

ortho-benzenediol [120-80-9] Vol. III: 190

ortho-benzyl-para-chlorophenol [120-32-1] Vol. III: 122

orthobenzylparachlorophenol [120-32-1] Vol. III: 122

orthobenzyl-p-chlorophenol [120-32-1] Vol. III: 122

orthoboric acid [10043-35-3] Vol. I: 327

ortho-chlorobenzylidene malononitrile [2698-41-1] Vol. III: 218

ortho chlorotoluene [95-49-8] Vol. I: 1103

orthocide [133-06-2] Vol. II: 276

orthocide 406 [133-06-2] Vol. II: 276

orthocide 50 [133-06-2] Vol. II: 276

orthocide 75 [133-06-2] Vol. II: 276

orthocide 7,5 [133-06-2] Vol. II: 276

orthocide 83 [133-06-2] Vol. II: 276

orthocresol [95-48-7] Vol. I: 536

orthodichlorobenzene [95-50-1] Vol. I: 250

orthodichlorobenzol [95-50-1] Vol. I: 250

pac [56-38-2] Vol. III: 750

pacol [56-38-2] Vol. III: 750

paladin [121-75-5] Vol. II: 950

palatinol-a [84-66-2] Vol. II: 779

palatinol-ah [117-81-7] Vol. II: 769

palatinol-bb [85-68-7] Vol. II: 767

palatinol-e [84-74-2] Vol. II: 774

palatinol-m [131-11-3] Vol. II: 784

palatinol-z [26761-40-0] Vol. III: 340

palmityl alcohol [36653-82-4] Vol. III: 193

pamisan [62-38-4] Vol. II: 535

pamolyn-100- [112-80-1] Vol. I: 895

pan [85-44-9] Vol. I: 960

panogen-m [151-38-2] Vol. III: 594

panogen-metox [151-38-2] Vol. III: 594

panomatic [62-38-4] Vol. II: 535

panoram-75- [137-26-8] Vol. I: 632

panoram d-31 [60-57-1] Vol. III: 356

panoxyl [94-36-0] Vol. I: 310

pansoil [2593-15-9] Vol. III: 898

panthion [56-38-2] Vol. III: 750

pantozol-1 [7700-17-6] Vol. III: 266

papermaker's alum [10043-01-3] Vol. III: 39

papite [107-02-8] Vol. I: 99

paraacetaldehyde [123-63-7] Vol. III: 665

para-aminonitrobenzene [100-01-6] Vol. II: 70

parabar-441 [128-37-0] Vol. III: 317

para benzoquinone [106-51-4] Vol. I: 303

parabis-a [80-05-7] Vol. II: 720

paracetaldehyde [123-63-7] Vol. III: 665

paracetate [108-21-4] Vol. I: 46

parachlorometacresol [59-50-7] Vol. II: 258

parachlorophenol [106-48-9] Vol. II: 710

3-para-chlorophenyl-1,1-dimethyl urea [150-68-5] Vol. II: 1075

1-para-chlorophenyl-3,3-dimethyl urea [150-68-5] Vol. II: 1075

N-para-chlorophenyl-N', N'-dimethyl urea [150-68-5] Vol. II: 1075

paraderil [83-79-4] Vol. II: 166

paradi [106-46-7] Vol. I: 254

paradichlorobenzene [106-46-7] Vol. I: 254

paradichlorobenzol [106-46-7] Vol. I: 254

para dihydroxybenzene [123-31-9] Vol. I: 753

para-dinitrobenzene [100-25-4] Vol. II: 128

para-dioxan [123-91-1] Vol. I: 618

para dioxane [123-91-1] Vol. I: 618

para dioxybenzene [123-31-9] Vol. I: 753

paradow [106-46-7] Vol. I: 254

paradust [56-38-2] Vol. III: 750

paraform [30525-89-4] Vol. I: 907

paraformaldehydum [30525-89-4] Vol. I: 907

para formic aldehyde [30525-89-4] Vol. I: 907

para hydroquinone [123-31-9] Vol. I: 753

paral [123-63-7] Vol. III: 665

paraldehyde-draught (BPC1973) [123-63-7] Vol. III: 665

paraldehyde-enema (BPC1973) [123-63-7] Vol. III: 665

paralkan [8001-54-5] Vol. II: 67

paramar-50 [56-38-2] Vol. III: 750

paramar [56-38-2] Vol. III: 750

paramex [2385-85-5] Vol. III: 583

paramoth [106-46-7] Vol. I: 254

para-nitroaniline [100-01-6] Vol. II: 70

paranitrofenol (Dutch) [100-02-7] Vol. II: 726

paranitrofenolo (Italian) [100-02-7] Vol. II: 726

paranitrophenol [100-02-7] Vol. II: 726

paranox-441 [128-37-0] Vol. III: 317

para-oxon [311-45-5] Vol. III: 669

paraoxone [311-45-5] Vol. III: 669

para, para,-kelthane [115-32-2] Vol. II: 144

paraphos [56-38-2] Vol. III: 750

paraquat-dication [4685-14-7] Vol. III: 670

paraquat-ion [4685-14-7] Vol. III: 670

parathene [56-38-2] Vol. III: 750

parathion-ethyl [56-38-2] Vol. III: 750

parawet [56-38-2] Vol. III: 750

parazene [106-46-7] Vol. I: 254

pardner [1689-84-5] Vol. II: 163

parenaphthalene [120-12-7] Vol. II: 79

parmetol [59-50-7] Vol. II: 258

parol [59-50-7] Vol. II: 258

paroxan [311-45-5] Vol. III: 669

parozone [7681-52-9] Vol. II: 498

parsons 2,4-d weed killer isopropyl ester [94-11-1] Vol. II: 15

parzate [142-59-6] Vol. I: 411

parzate liquid [142-59-6] Vol. I: 411

patrole [10265-92-6] Vol. II: 744

pay-off [70124-77-5], Vol. III: 455

PB [51-03-6] Vol. III: 779

p-30bf [21645-51-2] Vol. II: 57

p-30bf [21645-51-2] Vol. III: 35

pbna [135-88-6] Vol. III: 640

pb-nox [83-79-4] Vol. II: 166

pb s 100 [7439-92-1] Vol. I: 800

pbx(af)-108 [121-82-4] Vol. III: 277

p Ca [106-47-8] Vol. I: 169

pcb-1016 [12674-11-2] Vol. I: 970

pcb-1242 [53469-21-9] Vol. II: 803

pcb-1248 [12672-29-6] Vol. I: 972

pcb-1254 [11097-69-1] Vol. II: 806

pcb-1260 [11096-82-5] Vol. II: 808

pcb [1336-36-3] Vol. I: 974

pcc [8001-35-2] Vol. I: 1118

pce [127-18-4] Vol. II: 421

pchk [8001-35-2] Vol. I: 1118

pcl [77-47-4] Vol. II: 283

pcm [594-42-3] Vol. II: 613

pcmc [59-50-7] Vol. II: 258

pcnb-100 [82-68-8] Vol. II: 132

pcnb [82-68-8] Vol. II: 132

pcnb-technical-material-for manufacturing purposes only [82-68-8] Vol. II: 132

pcp [87-86-5] Vol. I: 928

pcp sodium [131-52-2] Vol. III: 676

pcp sodium salt [131-52-2] Vol. III: 676

pd-185- [57-11-4] Vol. I: 1052

pdb [106-46-7] Vol. I: 254

pdd [15663-27-1] Vol. III: 237

pdq [94-81-5] Vol. III: 174

PDU [101-42-8] Vol. III: 452

pearl alum [10043-01-3] Vol. III: 39

pear off [628-63-7] Vol. II: 27

pear oil [123-92-2] Vol. I: 780

pedinex (French) [131-89-5] Vol. II: 713

peg [25322-68-3] Vol. II: 811

peg-400 [25322-68-3] Vol. II: 811

pegalan [80-62-6] Vol. II: 567

peg-6000ds [25322-68-3] Vol. II: 811

pei-75 [60-51-5] Vol. III: 740

pelagol grey-c [120-80-9] Vol. III: 190

pelagol-grey-j [95-80-7] Vol. II: 1016

pelagol-J [95-80-7] Vol. II: 1016

pelargonic alcohol [143-08-8] Vol. I: 893

pelikan-c-11/1431a [7440-44-0] Vol. III: 179

pencal [7778-44-1] Vol. III: 65

penchlorol [87-86-5] Vol. I: 928

pennac cra [96-45-7] Vol. I: 757

pennamine d [94-75-7] Vol. I: 30

penncap-e [56-38-2] Vol. III: 750

penphene [8001-35-2] Vol. I: 1118

pentaborane (9) [19624-22-7] Vol. II: 679

pentaborane (B5 H9) [19624-22-7] Vol. II: 679

(9)-pentaboron nonahydride [19624-22-7] Vol. II: 679

pentaboron nonahydride [19624-22-7] Vol. II: 679

pentacetate 28 [628-63-7] Vol. II: 27

pentacetate [628-63-7] Vol. II: 27

pentachloorethaan (Dutch) [76-01-7] Vol. II: 336

pentachloraethan (German) [76-01-7] Vol. II: 336

pentachlorethane (French) [76-01-7] Vol. II: 336

pentachlornitrobenzol (German) [82-68-8] Vol. II: 132

pentachloroa°ntimony [7647-18-9] Vol. II: 83

1,2,3,4,5-pentachlorobenzene [608-93-5] Vol. I: 277

pentachlorophenate sodium [131-52-2] Vol. III: 676

pentachlorophenol sodium salt [131-52-2] Vol. III: 676

pentachlorophenoxy-sodium [131-52-2] Vol. III: 676

pentachlorophenyl chloride [118-74-1] Vol. I: 268

Pentachlorphenol (German) [87-86-5] Vol. I: 928

pentacloroetano (Italian) [76-01-7] Vol. II: 336

penta concentrate [87-86-5] Vol. I: 928

2,4-pentadione [123-54-6] Vol. I: 910

pentagen [82-68-8] Vol. II: 132

pentalin [76-01-7] Vol. II: 336

pentamethylene [287-92-3] Vol. III: 280

1-pentanamine, n-pentyl [2050-92-2] Vol. III: 395

pentan-2,4-dione [123-54-6] Vol. I: 910

N-pentane [109-66-0] Vol. II: 682

pentanedial [111-30-8] Vol. III: 468

1,5-pentanedial [111-30-8] Vol. III: 468

1,4-pentanediamine, N(4)-(7-chloro-4-quinolinyl)-N (1),N (1)-diethyl [54-05-7] Vol. III: 229

2,4-pentane- diol, 2-methyl [107-41-5] Vol. III: 620

1,5-pentanedione [111-30-8] Vol. III: 468

2,4-pentanedione [123-54-6] Vol. I: 910

pentane, 2-methyl- [107-83-5] Vol. III: 613

pentanen (Dutch) [109-66-0] Vol. II: 682

pentanethiol [110-66-7] Vol. III: 53

1-pentanethiol [110-66-7] Vol. III: 53

pentani (Italian) [109-66-0] Vol. II: 682

pentanol-1 [71-41-0] Vol. II: 700

1-pentanol [71-41-0] Vol. II: 700

n-pentan-1-ol [71-41-0] Vol. II: 700

n-pentanol [71-41-0] Vol. II: 700

pentanol [71-41-0] Vol. II: 700

1-pentanol acetate [628-63-7] Vol. II: 27

1-pentanol, acetate [628-63-7] Vol. II: 27

2-pentanol,acetate [626-38-0] Vol. I: 52

2-pentanol,4-methyl [108-11-2] Vol. I: 913

2-pentanol, 4-methyl-, acetate [108-84-9] Vol. II: 686

pentanone-3 [96-22-0] Vol. III: 679

2-pentanone, 4-hydroxy-4-methyl [123-42-2] Vol. II: 688

2-pentanone, 4-methyl [108-10-1] Vol. II: 692

pentan (Polish) [109-66-0] Vol. II: 682

pentaphenate [131-52-2] Vol. III: 676

penta ready [87-86-5] Vol. I: 928

pentasodium triphosphate [7758-29-4] Vol. II: 1061

pentasol [71-41-0] Vol. II: 700

pentasulfure de phosphore (French) [1314-80-3] Vol. I: 955

penta wr [87-86-5] Vol. I: 928

3-penten-2-one, 4-methyl- [141-79-7] Vol. II: 696

1-pentyl acetate [628-63-7] Vol. II: 27

pentyl acetate [628-63-7] Vol. II: 27

n-pentyl acetate [628-63-7] Vol. II: 27

1-pentyl alcohol [71-41-0] Vol. II: 700

pentyl alcohol [71-41-0] Vol. II: 700

pentylamine, pentyl [2050-92-2] Vol. III: 395

pentylcarbinol [111-27-3] Vol. I: 734

3-pentylcarbinol [97-95-0] Vol. I: 343

pentyl mercaptan [110-66-7] Vol. III: 53

n-pentylmercaptan [110-66-7] Vol. III: 53

n pentyl methyl ketone [110-43-0] Vol. II: 454

pentylsilicon trichloride [107-72-2] Vol. III: 941

pentyltrichlorosilane [107-72-2] Vol. III: 941

peo-18 [25322-68-3] Vol. II: 811

2,2-bis (p-(2,3-epoxypropoxy) phenyl) propane [1675-54-3] Vol. III: 132

per [127-18-4] Vol. II: 421

per-a-clor [79-01-6] Vol. II: 427

perawin [127-18-4] Vol. II: 421

perbulate [1929-77-7] Vol. II: 221

perbutyl-h [75-91-2] Vol. III: 896

perc [127-18-4] Vol. II: 421

perchloorethyleen,-per (Dutch) [127-18-4] Vol. II: 421

perchlor [127-18-4] Vol. II: 421

Perchloraethylen,-per (German) [127-18-4] Vol. II: 421

perchlordecone [2385-85-5] Vol. III: 583

perchlorethylene [127-18-4] Vol. II: 421

perchlorethylene,-per (French) [127-18-4] Vol. II: 421

perchloric acid, ammonium salt [7790-98-9] Vol. III: 684

perchloride of mercury [7487-94-7] Vol. II: 539

perchlormethylmerkaptan (Czech) [594-42-3] Vol. II: 613

perchlorobenzene [118-74-1] Vol. I: 268

perchlorobutadiene [87-68-3] Vol. I: 340

perchlorocyclopentadiene [77-47-4] Vol. II: 283

perchlorodihomocubane [2385-85-5] Vol. III: 583

perchloroethane [67-72-1] Vol. II: 330

perchloroethylene [127-18-4] Vol. II: 421

perchloromethane [56-23-5] Vol. I: 435

perchloromethanethiol [594-42-3] Vol. II: 613

perchloromethylformate [503-38-8] Vol. III: 376

perchloromethyl formate [503-38-8] Vol. III: 376

perchloromethylmercaptan [594-42-3] Vol. II: 613

perchloromethyl-mercaptan [594-42-3] Vol. II: 613

perchloromethylmercaptan [594-42-3] Vol. II: 613

perchloronaphthalene [2234-13-1] Vol. III: 653

perchloropentacyclodecane [2385-85-5] Vol. III: 583

perchloropentacyclo (5.3.0.0 (2,6).0 (3,9).0 (4,8))decane [2385-85-5] Vol. III: 583

perchloropentacyclo (5.2.1.0 (2,6).0 (3,9).0 (5,8))decane [2385-85-5] Vol. III: 583

perchloropentacyclo[5.3.0.0 (2,6).0 (3,9).0 (4,8) decan-5-one [143-50-0] Vol. III: 587

ph-6040 [35367-38-5] Vol. III: 339

phaldrone [302-17-0] Vol. III: 197

phaltan [133-07-3] Vol. III: 461

phaltane [133-07-3] Vol. III: 461

phc [114-26-1] Vol. II: 231

Phelps triangle brand copper sulfate [7758-98-7] Vol. I: 510

phemeride [121-54-0] Vol. III: 105

phe-mer-nite [55-68-5] Vol. III: 702

phemerol [121-54-0] Vol. III: 105

phemerol chloride [121-54-0] Vol. III: 105

phemersol chloride [121-54-0] Vol. III: 105

phemithyn [121-54-0] Vol. III: 105

phenachlor [88-06-2] Vol. II: 732

phenacide [8001-35-2] Vol. I: 1118

phenaclor [88-06-2] Vol. II: 732

phenacylchloride [532-27-4] Vol. I: 89

phenacyl chloride [532-27-4] Vol. I: 89

phenador x [92-52-4] Vol. I: 326

phenalco [55-68-5] Vol. III: 702

phenalgene [103-84-4] Vol. III: 3

phenalgin [103-84-4] Vol. III: 3

phenamiphos [22224-92-6] Vol. II: 743

Phenanthren (German) [85-01-8] Vol. I: 919

phenatox [8001-35-2] Vol. I: 1118

phene [71-43-2] Vol. I: 239

pheneene germicidal solution and tincture [8001-54-5] Vol. II: 67

phenemal [50-06-6] Vol. I: 223

phenemalum [50-06-6] Vol. I: 223

phenethylene [100-42-5] Vol. I: 1060

phenic acid [108-95-2] Vol. I: 920

phenitol [55-68-5] Vol. III: 702

phenmad [62-38-4] Vol. II: 535

phenmedipham [13684-63-4] Vol. III: 130

phenmerzyl nitrate [55-68-5] Vol. III: 702

phenobal [50-06-6] Vol. I: 223

phenobarb [50-06-6] Vol. I: 223

phenobarbitalum [50-06-6] Vol. I: 223

phenobarbitone [50-06-6] Vol. I: 223

phenobarbituric acid [50-06-6] Vol. I: 223

phenoclor [1336-36-3] Vol. I: 974

phenohep [67-72-1] Vol. II: 330

phenol, beta-dinitro- [573-56-8] Vol. II: 717

phenol, 2,6-bis (1,1-dimethylethyl)-4-methyl [128-37-0] Vol. III: 317

phenolcarbinol [100-51-6] Vol. I: 313

phenol, 3-chloro- [108-43-0] Vol. III: 689

phenol, 2-chloro-4-nitro-, O-ester with O,O-dimethyl phosphorothioate [2463-84-5] Vol. III: 748

phenol, 4-chloro-2-(phenylmethyl)- [120-32-1] Vol. III: 122

phenol, 6-cyclohexyl-2,4-dinitro [131-89-5] Vol. II: 713

phenol, 2-cyclohexyl-4,6-dinitro [131-89-5] Vol. II: 713

phenol, 2,3-dichloro [576-24-9] Vol. III: 332

phenol, 2,6-dichloro- [87-65-0] Vol. II: 715

phenol, 3,4-dichloro- [95-77-2] Vol. III: 691

phenol, dimethyl [1300-71-6] Vol. I: 1171

phenol, 2,4-dimethyl [105-67-9] Vol. I: 1162

phenol, 2,6-dimethyl [576-26-1] Vol. I: 1167

phenol, 3,4-dimethyl [95-65-8] Vol. I: 1169

phenol, 3,5-dimethyl [108-68-9] Vol. I: 1165

phenol, 4-(dimethylamino)-3,5-dimethyl-, methylcarbamate (ester) [315-18-4] Vol. II: 227

phenol, 4,4'-dimethylmethylenedi [80-05-7] Vol. II: 720

phenol, 3,5-dimethyl-4-(methylthio)-, methylcarbamate [2032-65-7] Vol. II: 308

phenol, dimethyl-, phosphate (3:1) [25155-23-1] Vol. I: 1170

phenol, dinitro [25550-58-7] Vol. III: 371

phenol, 2,5-dinitro [329-71-5] Vol. III: 374

phenol, 2,6-dinitro- [573-56-8] Vol. II: 717

phenole (German) [108-95-2] Vol. I: 920

phenol, gamma-dinitro [329-71-5] Vol. III: 374

phenol, hexahydro [108-93-0] Vol. I: 580

phenol, 4,4'-isopropylidenedi [80-05-7] Vol. II: 720

phenol, m-chloro- [108-43-0] Vol. III: 689

phenol, 2-methyl [95-48-7] Vol. I: 536

phenol, 3-methyl (9CI) [108-39-4] Vol. III: 261

phenol, 2-(1-methylethoxy)-, methylcarbamate [114-26-1] Vol. II: 231

phenol, 4,4'-(1-methylethylidene) bis [80-05-7] Vol. II: 720

phenol, 2-(1-methylpropyl)-4,6-dinitro [88-85-7] Vol. II: 704

phenol, m-nitro- [554-84-7] Vol. III: 693

phenol, 2-nitro- [88-75-5] Vol. II: 723

phosphorodithioic acid, O,O-dimethyl ester, S-ester with 4-(mercaptomethyl)-2-methoxy-delta (sup (2) -1,3,4-thiadiazolin-5-one [950-37-8] **Vol. III: 734**

phosphorodithioic acid, O,O-dimethyl ester, S-ester with 3-(mercaptomethyl)-1,2,3-benzotriazin-4 (3H)-one [86-50-0] **Vol. III: 744**

phosphorodithioic acid, O,O-dimethyl ester, S-ester with 4-(mercaptomethyl)-2-methoxy-delta (2)-1,3,4-thiadiazolin-5-one [950-37-8] **Vol. III: 734**

phosphorodithioic acid, O,O-dimethyl ester, S-ester with 2-mercapto-N-methylacetamide [60-51-5] **Vol. III: 740**

phosphorodithioic acid O,O-dimethyl ester S-ester with N-(mercaptomethyl) phthalimide [732-11-6] **Vol. III: 737**

phosphorodithioic acid, O,O-dimethyl ester, S-ester with N-(mercaptomethyl) phthalimide [732-11-6] **Vol. III: 737**

phosphorodithioic acid, O,O-dimethyl s-(2-(methylamino)-2-oxoethyl) ester [60-51-5] **Vol. III: 740**

phosphorodithioic acid, O,O-dimethyl S-((4-oxo-1,2,3-benzotriazin-3 (4h) -yl) methyl) ester [86-50-0] **Vol. III: 744**

phosphorodithioic acid, S-(2-chloro-1-(1,3-dihydro-1,3-dioxo-2H-isoindol-2-yl)ethyl) o,o-diethyl ester [10311-84-9] **Vol. III: 728**

phosphorodithioic acid S-(2-chloro-1-(1,3-dihydro-1,3-dioxo-2H-isoindol-2-yl) ethyl) o,o-diethyl ester [10311-84-9] **Vol. III: 728**

phosphorodithioic acid, S-(2-chloro-1-phthalimidoethyl) o,o-diethyl ester [10311-84-9] **Vol. III: 728**

phosphorodithioic acid, S-(2-chloro-1-phthalimidoethyl)-o,o-diethyl ester [10311-84-9] **Vol. III: 728**

phosphorodithioic acid S-((1,3-dihydro-1,3-dioxo-2h-isoindol-2-yl) methyl) O,O-dimethyl ester [732-11-6] **Vol. III: 737**

phosphorodithioic acid, s-((1,3-dihydro-1,3-dioxo-2h-isoindol-2-yl) methyl) O,O-dimethyl ester [732-11-6] **Vol. III: 737**

phosphorodithioic acid, S-((1,3-dihydro-1,3-dioxo-isoindol-2-yl) methyl) O,O-dimethyl ester [732-11-6] **Vol. III: 737**

phosphorodithioic acid, S-(5-methoxy-2-oxo-1,3,4-thiadiazol-3 (2h)-yl) methyl) O,O-dimethyl ester [950-37-8] **Vol. III: 734**

phosphorodithioic acid S-[(5-methoxy-2-oxo-1,3,4-thiadiazol-3 (2h) -yl) methyl] O,O-dimethyl ester [950-37-8] **Vol. III: 734**

phosphorodithioic acid, S,S'-I, 4-dioxane-2,3-diyl O,O,O',O'-tetraethyl ester [78-34-2] **Vol. II: 756**

phosphorodithioic acid, S,S'-methylene o o,o',o'-tetraethyl ester [563-12-2] **Vol. II: 759**

phosphorodithioic acid, S,S'-p-dioxane-2,3-diyl O,O,O',O'-tetraethyl ester [78-34-2] **Vol. II: 756**

phosphorothioate, O,O-diethyl O-6-(2-isopropyl-4-methylpyrimidyl) [333-41-5] **Vol. III: 758**

phosphorothioic acid, O-(2-chloro-4-nitrophenyl) O, O-dimethyl ester [2463-84-5] **Vol. III: 748**

phosphorothioic acid O-(2-chloro-4-nitrophenyl) O,O-dimethyl ester [2463-84-5] **Vol. III: 748**

phosphorothioic acid, o-(4-cyanophenyl) o,o-dimethyl ester [2636-26-2] **Vol. III: 271**

phosphorothioic acid o-(4-cyanophenyl) o,o-dimethyl ester [2636-26-2] **Vol. III: 271**

phosphorothioic acid, O,O-diethyl O-(2-(ethylthio) ethyl) ester, mixed with O,O-diethyl S-(2-(ethylthio) ethyl) ester (7:3) [8065-48-3] **Vol. III: 755**

phosphorothioic acid, O,O-diethyl O-(2-(ethylthio) ethyl) ester, mixture with O,O-diethyl S-(2-(ethylthio) ethyl) phosphorothioate [8065-48-3] **Vol. III: 755**

phosphorothioic acid, O,O-diethyl O-(isopropylmethylpyrimidinyl) ester [333-41-5] **Vol. III: 758**

phosphorothioic acid, O,O-diethyl O-(2-isopropyl-6-methyl-4-pyrimidinyl) ester [333-41-5] **Vol. III: 758**

phosphorothioic acid, O,O-diethyl O-(6-methyl-2-(1-methylethyl)-4-pyrimidinyl) ester [333-41-5] **Vol. III: 758**

phosphorothioic acid, O,O-diethyl O-(4-nitrophenyl) ester [56-38-2] **Vol. III: 750**

phosphorothioic acid, O,O-diethyl O-(3,5,6-trichloro-2-pyridinyl) ester [2921-88-2] **Vol. III: 762**

phosphorothioic acid, O,O-diethyl O-(3,5,6-trichloro-2-pyridyl) ester [2921-88-2] **Vol. III: 762** phosphorothioic acid o,o-dimethyl ester, o-ester with p-hydroxybenzonitrile [2636-26-2] **Vol. III:** 271

phosphorothioic acid, o,o-dimethyl ester, o-ester with p-hydroxybenzonitrile [2636-26-2] **Vol. III:** 271

phosphorothioic acid, s-(2-(ethylsulfinyl)ethyl) o,o-dimethyl ester [301-12-2] **Vol. III:** 661

phosphorotrithioic acid, S,S,S-tributyl ester [78-48-8] **Vol. III:** 768

phosphorotrithious acid, s,s,s-tributyl ester [150-50-5] **Vol. III:** 574

phosphorotrithious-acid,-tributyl-ester [150-50-5] **Vol. III:** 574

phosphorous chloride [7719-12-2] **Vol. I:** 952

phosphoroxychloride [10025-87-3] **Vol. I:** 957

Phosphorsaeureloesungen (German) [7664-38-2] **Vol. I:** 942

Phosphortrichlorid (German) [7719-12-2] **Vol. I:** 952

phosphorus-31 [7723-14-0] **Vol. II:** 764

phosphorus chloride [10025-87-3] **Vol. I:** 957

phosphorus chloride (c16p2) [7719-12-2] **Vol. I:** 952

phosphorus chloride oxide (poc13) [10025-87-3] **Vol. I:** 957

phosphorus chloride (pc13) [7719-12-2] **Vol. I:** 952

phosphorus oxide trichloride [10025-87-3] **Vol. I:** 957

phosphorus oxytrichloride [10025-87-3] **Vol. I:** 957

phosphorus pentasulfide [1314-80-3] **Vol. I:** 955

phosphorus pentasulfide (p4s10) [1314-80-3] **Vol. I:** 955

phosphorus persulfide [1314-80-3] **Vol. I:** 955

phosphorus (red) [7723-14-0] **Vol. II:** 764

phosphorus sulfide (p2s5) [1314-80-3] **Vol. I:** 955

phosphorus (white) [7723-14-0] **Vol. II:** 764

phosphorus (white) [7723-14-0] **Vol. III:** 973

phosphorus, white, molten (dry) [7723-14-0] **Vol. II:** 764

phosphorus white, molten (UN2447) (DOT) [7723-14-0] **Vol. III:** 973

phosphorus, white or yellow, dry or under water or in solution (UN1381) (DOT) [7723-14-0] **Vol. III:** 973

phosphorus yellow [7723-14-0] **Vol. III:** 973

phosphorus, yellow (ACGIH: OSHA) [7723-14-0] **Vol. III:** 973

phosphoryl chloride [10025-87-3] **Vol. I:** 957

phosphoryl trichloride [10025-87-3] **Vol. I:** 957

phosphothion [121-75-5] **Vol. II:** 950

phosphure de zinc (French) [1314-84-7] **Vol. II:** 1100

phosphures d'alumium (French) [20859-73-8] **Vol. I:** 133

phostoxin [20859-73-8] **Vol. I:** 133

phostoxin a [20859-73-8] **Vol. I:** 133

phosvin [1314-84-7] **Vol. II:** 1100

phosvit [62-73-7] **Vol. I:** 946

photophor [1305-99-3] **Vol. I:** 404

phph [92-52-4] **Vol. I:** 326

phtalofos [732-11-6] **Vol. III:** 737

1,3-phthalandione [85-44-9] **Vol. I:** 960

phthalandione [85-44-9] **Vol. I:** 960

phthalhydroquinone [120-80-9] **Vol. III:** 190

phthalic acid, benzyl butyl ester [85-68-7] **Vol. II:** 767

phthalic acid, bis (2-ethylhexyl) ester [117-81-7] **Vol. II:** 769

phthalic acid, bis (8-methylnonyl) ester [26761-40-0] **Vol. III:** 340

phthalic-acid,-dibutyl-ester [84-74-2] **Vol. II:** 774

phthalic-acid-dibutyl-ester [84-74-2] **Vol. II:** 774

phthalic acid, diethyl ester [84-66-2] **Vol. II:** 779

phthalic acid, diisodecyl ester [26761-40-0] **Vol. III:** 340

phthalic acid, dimethyl ester [131-11-3] **Vol. II:** 784

phthalic acid, dioctyl ester [117-81-7] **Vol. II:** 769

phthalic acid, dioctyl ester [117-84-0] **Vol. II:** 789

phthalic acid, hexahydro-3,6-endo oxy [145-73-3] **Vol. I:** 900

phthalic acid, methyl ester [131-11-3] **Vol. II:** 784

phthalimide, N-(mercaptomethyl)-, S-ester with O,O-dimethyl phosphorodithioate [732-11-6] **Vol. III: 737**

phthalimide, N-((trichloromethyl) thio)- [133-07-3] **Vol. III: 461**

phthalimidomethyl O,O-dimethyl phosphorodithioate [732-11-6] **Vol. III: 737**

phthalimido O,O-dimethyl phosphorodithioate [732-11-6] **Vol. III: 737**

phthalic acid anhydride [85-44-9] **Vol. I: 960**

phthalol [84-66-2] **Vol. II: 779**

phthalophos [732-11-6] **Vol. III: 737**

Phthalsaeureanhydrid (German) [85-44-9] **Vol. I: 960**

Phthalsaeuredimethylester (German) [131-11-3] **Vol. II: 784**

phthaltan [133-07-3] **Vol. III: 461**

phthalthrin [7696-12-0] **Vol. III: 912**

phthorothanum [151-67-7] **Vol. III: 141**

phyban [2163-80-6] **Vol. II: 610**

phygon [117-80-6] **Vol. II: 645**

phygon-xl [117-80-6] **Vol. II: 645**

phytar-138- [75-60-5] **Vol. I: 214**

phytar-560- [75-60-5] **Vol. I: 214**

phytar [75-60-5] **Vol. I: 214**

pic-clor [76-06-2] **Vol. II: 606**

picfume [76-06-2] **Vol. II: 606**

picket [52645-53-1] **Vol. III: 282**

pickle alum [10043-01-3] **Vol. III: 39**

m picoline [108-99-6] **Vol. I: 964**

3-picoline [108-99-6] **Vol. I: 964**

4-picoline [108-89-4] **Vol. I: 1019**

picolinic acid, 4-amino-3,5,6-trichloro- [1918-02-1] **Vol. II: 791**

picral [88-89-1] **Vol. II: 793**

picratol [131-74-8] **Vol. I: 156**

picric acid, ammonium salt [131-74-8] **Vol. I: 156**

picride [76-06-2] **Vol. II: 606**

picronitric acid [88-89-1] **Vol. II: 793**

pielik [94-75-7] **Vol. I: 30**

pigment-white-27 [16871-71-9] **Vol. II: 911**

pigment-white-27 [7631-86-9] **Vol. III: 846**

pigment yellow 36 [13530-65-9] **Vol. III: 984**

pikrinezuur (Dutch) [88-89-1] **Vol. II: 793**

pikrinsacure (German) [88-89-1] **Vol. II: 793**

pikrynowy-kwas (Polish) [88-89-1] **Vol. II: 793**

pillarfuran [1563-66-2] **Vol. I: 420**

pillarzo [15972-60-8] **Vol. II: 9**

pimelic ketone [108-94-1] **Vol. II: 268**

pimelin ketone [108-94-1] **Vol. II: 268**

pin [2104-64-5] **Vol. II: 738**

pinakon [107-41-5] **Vol. III: 620**

pindon [Dutch] [83-26-1] **Vol. III: 774**

2-pinene [80-56-8] **Vol. III: 32**

pinene [80-56-8] **Vol. III: 32**

pine-o disinfectant [95-57-8] **Vol. II: 707**

pink [93-79-8] **Vol. II: 34**

piombotetra etile (Italian) [78-00-2] **Vol. I: 966**

piombo tetra-metile (Italian) [75-74-1] **Vol. II: 796**

piperazidine [110-85-0] **Vol. III: 776**

1,4-piperazine [110-85-0] **Vol. III: 776**

piperazine, anhydrous [110-85-0] **Vol. III: 776**

Piperazin (German) [110-85-0] **Vol. III: 776**

2,6-piperidinedione, 4-(2-(3,5-dimethyl-2-oxocyclohexyl)-2-hydroxy ethyl)-, (1s-(1alpha(s),3alpha,5beta)) [66-81-9] **Vol. II: 447**

pipersol [110-85-0] **Vol. III: 776**

piretrina-1 (Portuguese) [121-21-1] **Vol. III: 814**

piridina (Italian) [110-86-1] **Vol. II: 874**

pirofos [3689-24-5] **Vol. II: 1010**

pisol [112-53-8] **Vol. II: 316**

pitezin [1912-24-9] **Vol. II: 1050**

pittcide [7778-54-3] **Vol. II: 494**

pittsburgh px-138 [117-81-7] **Vol. II: 769**

pivacin [83-26-1] **Vol. III: 774**

pival [83-26-1] **Vol. III: 774**

pivaldione [French] [83-26-1] **Vol. III: 774**

pivaldion [Italian] [83-26-1] **Vol. III: 774**

pivalic acid [75-98-9] **Vol. III: 781**

2-pivaloyl-indaan-1,3-dion [Dutch] [83-26-1] **Vol. III: 774**

2-pivaloylindan-1,3-dione [83-26-1] **Vol. III: 774**

2-pivaloyl-1,3-indandione [83-26-1] **Vol. III: 774**

2-Pivaloyl-indan-1,3-dion [German] [83-26-1] **Vol. III: 774**

2-pivaloylindane-1,3-dione [83-26-1] **Vol. III: 774**

pivalyl [83-26-1] **Vol. III: 774**

2-pivalyl-1,3-indandione [83-26-1] **Vol. III: 774**

pivalyl indan-1,3-dione [83-26-1] Vol. III: 774

pivalyl-indandione [83-26-1] Vol. III: 774

pivalyl-valone [83-26-1] Vol. III: 774

pivalyn [83-26-1] Vol. III: 774

pkha [7790-98-9] Vol. III: 684

PKhFN [131-52-2] Vol. III: 676

pkhnb [82-68-8] Vol. II: 132

placidol-e [84-66-2] Vol. II: 779

planotox [1929-73-3] Vol. III: 155

plant-dithio-acrosol [3689-24-5] Vol. II: 1010

plantfume-103-smoke-generator [3689-24-5] Vol. II: 1010

plantgard [94-75-7] Vol. I: 30

plantifog-160m [12427-38-2] Vol. III: 531

plantulin [139-40-2] Vol. III: 802

plantvax-20 [5259-88-1] Vol. III: 659

plant-vax [5259-88-1] Vol. III: 659

plantvax [5259-88-1] Vol. III: 659

Plastiazan-41 (Russian) [122-99-6] Vol. III: 695

plasticized-ddp [26761-40-0] Vol. III: 340

plastomoll-doa [103-23-1] Vol. II: 291

platiblastin [15663-27-1] Vol. III: 237

platinex [15663-27-1] Vol. III: 237

platinol [15663-27-1] Vol. III: 237

platinol-ah [117-81-7] Vol. II: 769

platinol-dop [117-81-7] Vol. II: 769

platinum, diamminedichlo-, cis [15663-27-1] Vol. III: 237

platinum diamminodichloride [15663-27-1] Vol. III: 237

plomb fluorure (French) [7783-46-2] Vol. III: 502

plumbane,tetraethyl [78-00-2] Vol. I: 966

plumbane, tetramethyl [75-74-1] Vol. II: 796

plumbous chloride [7758-95-4] Vol. I: 803

plumbous fluoborate-tetrafluoroborate (Pb(BF₄)₂) [13814-96-5] Vol. III: 507

plumbous fluoride [7783-46-2] Vol. III: 502

plumbous oxide [1317-36-8] Vol. III: 513

plumbous sulfide [1314-87-0] Vol. I: 807

plumbum [7439-92-1] Vol. I: 800

pluracol-245 [80-05-7] Vol. II: 720

pluracol-e [25322-68-3] Vol. II: 811

pluriol-e-200 [25322-68-3] Vol. II: 811

pma [62-38-4] Vol. II: 535

pmac [62-38-4] Vol. II: 535

pmacetate [62-38-4] Vol. II: 535

pmal [62-38-4] Vol. II: 535

pmas [62-38-4] Vol. II: 535

pmc [100-56-1] Vol. III: 699

p-methoxy-beta-methylstyrene [104-46-1] Vol. III: 56

pmm [594-42-3] Vol. II: 613

pmp [732-11-6] Vol. III: 737

pmp (pesticide) [732-11-6] Vol. III: 737

pna [100-01-6] Vol. II: 70

pnb [92-93-3] Vol. III: 650

pncb [100-00-5] Vol. III: 210

pneumax-mbt [149-30-4] Vol. III: 540

pnp [100-02-7] Vol. II: 726

polacaritox [116-29-0] Vol. III: 906

polisin [7287-19-6] Vol. II: 1056

polsulkol-extra [7704-34-9] Vol. II: 965

pol thiuram [137-26-8] Vol. I: 632

poly [7758-29-4] Vol. II: 1061

polybor [1303-96-4] Vol. III: 134

polybor 3 [1303-96-4] Vol. III: 134

polychlorcamphene (former USSR) [8001-35-2] Vol. I: 1118

polychlorobiphenyl [1336-36-3] Vol. I: 974

polychlorocamphene [8001-35-2] Vol. I: 1118

polycizer-162 [117-84-0] Vol. II: 789

polycizer-332 [123-95-5] Vol. III: 167

polycizer-dbp [84-74-2] Vol. II: 774

polydiol-200 [25322-68-3] Vol. II: 811

poly(ethyleneether) glycol [25322-68-3] Vol. II: 811

polyethylene glycol-400 [25322-68-3] Vol. II: 811

polyethylene oxide [25322-68-3] Vol. II: 811

polyformaldehyde [30525-89-4] Vol. I: 907

poly-g [25322-68-3] Vol. II: 811

poly-g600 [25322-68-3] Vol. II: 811

polyglycol-6000 [25322-68-3] Vol. II: 811

polygon [7758-29-4] Vol. II: 1061

polyhydroxyethylene [25322-68-3] Vol. II: 811

polylin no 515 [60-33-3] Vol. II: 667

polymerised formaldehyde [30525-89-4] Vol. I: 907

polymine-d [121-54-0] Vol. III: 105

polyox-wsr-301 [25322-68-3] Vol. II: 811

poly(oxy-1,2-ethanediyl), alpha-hydroome-ga-hydroxy [25322-68-3] Vol. II: 811

polyoxyethylenediol [25322-68-3] Vol. II: 811

potassium mercuric iodide [7783-33-7] Vol. III: 542

potassium mercuriiodide [7783-33-7] Vol. III: 542

potassium metaarsenite [10124-50-2] Vol. II: 104

potassium metaarsenite, acid [10124-50-2] Vol. II: 104

potassium monoiodide [7681-11-0] Vol. III: 785

potassium (permanganatede) (French) [7722-64-7] Vol. I: 915

potassium silver cyanide (KAg(CN)2) [506-61-6] Vol. II: 813

potassium tetracyanomercurate(II) [591-89-9] Vol. III: 556

potassium tetracyanomercurate (K₂Hg(CN)₄) [591-89-9] Vol. III: 556

potassium tetraiodomercurate(II) [7783-33-7] Vol. III: 542

potassium zirconifluoride [16923-95-8] Vol. II: 1107

potassium zirconium fluoride [16923-95-8] Vol. II: 1107

potassium zirconium hexafluoride [16923-95-8] Vol. II: 1107

potato alcohol [64-17-5] Vol. II: 394

pounce [52645-53-1] Vol. III: 282

powder-and-root [83-79-4] Vol. II: 166

p-25- (oxide) [13463-67-7] Vol. III: 921

pp383 [52315-07-8] Vol. II: 251

pp-383 [52315-07-8] Vol. II: 251

pp557 [52645-53-1] Vol. III: 282

pp-557 [52645-53-1] Vol. III: 282

pp-563 [68085-85-8] Vol. II: 288

pptc [1929-77-7] Vol. II: 221

pracarbamin [51-79-6] Vol. II: 224

pracarbamine [51-79-6] Vol. II: 224

pramex [52645-53-1] Vol. III: 282

pramitol-5p [1610-18-0] Vol. III: 787

prebane [886-50-0] Vol. II: 1045

precipitated sulfur [7704-34-9] Vol. II: 965

prefar [741-58-2] Vol. III: 731

prefar-e [741-58-2] Vol. III: 731

prefix d [1194-65-6] Vol. I: 299

prefrin-z [7733-02-0] Vol. II: 1103

premalox [101-42-8] Vol. III: 452

premazine [122-34-9] Vol. II: 1047

premerge [88-85-7] Vol. II: 704

prenfish [83-79-4] Vol. II: 166

prentox [83-79-4] Vol. II: 166

prentox synpren-fish [83-79-4] Vol. II: 166

prep [16672-87-0] Vol. III: 426

pre-san [741-58-2] Vol. III: 731

preserv o-sote [8001-58-9] Vol. I: 495

prespersion,-75-urea [57-13-6] Vol. I: 1136

prevenol [101-21-3] Vol. II: 234

preventoil [95-95-4] Vol. I: 936

preventol [101-21-3] Vol. II: 234

preventol-56 [101-21-3] Vol. II: 234

preventol-cmk [59-50-7] Vol. II: 258

preventol-o-extra [90-43-7] Vol. II: 176

prevostselvon-100 [25154-52-3] Vol. I: 926

preweed [101-21-3] Vol. II: 234

prezervit [533-74-4] Vol. III: 293

prim amyl acetate [628-63-7] Vol. II: 27

primary amyl acetate [628-63-7] Vol. II: 27

primary amyl alcohol [71-41-0] Vol. II: 700

primary isoamyl alcohol [123-51-3] Vol. II: 190

primary octyl alcohol [111-87-5] Vol. I: 897

primatol [1610-18-0] Vol. III: 787

primatol [1912-24-9] Vol. II: 1050

primatol [7287-19-6] Vol. II: 1056

primatol-a [1912-24-9] Vol. II: 1050

primatol-p [139-40-2] Vol. III: 802

primatol-q [7287-19-6] Vol. II: 1056

primatol-s [122-34-9] Vol. II: 1047

primatol-z-80 [834-12-8] Vol. II: 1054

primaze [1912-24-9] Vol. II: 1050

princep [122-34-9] Vol. II: 1047

pri-n-heptyl alcohol [111-70-6] Vol. III: 474

printop [122-34-9] Vol. II: 1047

prioderm [121-75-5] Vol. II: 950

pristacin [123-03-5] Vol. III: 196

pro [67-63-0] Vol. II: 515

prodalumnol [7784-46-5] Vol. I: 208

prodalumnol double [7784-46-5] Vol. I: 208

product 308 [36653-82-4] Vol. III: 193

profalon [330-55-2] Vol. II: 1084

profume-a [76-06-2] Vol. II: 606

programin [62-38-4] Vol. II: 535

pro-gro [5234-68-4] Vol. II: 670

prokarbol [534-52-1] Vol. I: 540

prolate [732-11-6] Vol. III: 737

prolate-5-dust [732-11-6] Vol. III: 737

prolate-e [732-11-6] Vol. III: 737

prolate 8-os [732-11-6] Vol. III: 737

prolex [1918-16-7] Vol. III: 789

1-propen-3-ol [107-18-6] **Vol. II: 52**
propenyl alcohol [107-18-6] **Vol. II: 52**
2-propenyl alcohol [107-18-6] **Vol. II: 52**
2-propenylamine [107-11-9] **Vol. III: 23**
p-propenylanisole [104-46-1] **Vol. III: 56**
p-1-propenylanisole [104-46-1] **Vol. III: 56**
4-propenylanisole [104-46-1] **Vol. III: 56**
2-propenyl bromide [106-95-6] **Vol. III: 25**
2-propenyl chloride [107-05-1] **Vol. II: 830**
2-propenyl chloroformate [2937-50-0] **Vol. III: 27**
propenyl)-2,2-dimethyl-, cyano (3-phenoxyphenyl) methyl ester2-isopropoxyphenyl methylcarbamate [114-26-1] **Vol. II: 231**
p-propenylphenyl methyl ether [104-46-1] **Vol. III: 56**
prophos [13194-48-4] **Vol. III: 427**
propilenimina (Spanish) [75-55-8] **Vol. I: 219**
1,3-propiolactone [57-57-8] **Vol. II: 672**
3-propiolactone [57-57-8] **Vol. II: 672**
propiolactone [57-57-8] **Vol. II: 672**
propiolic alcohol [107-19-7] **Vol. II: 864**
propional [123-38-6] **Vol. I: 1009**
propionaldehyde, 2,3-epoxy- [765-34-4] **Vol. II: 449**
propionaldehyde,2-methyl [78-84-2] **Vol. I: 769**
propionaldehyde, 2-methyl-2-(methylthio)-, o-(methylcarbamoyl) oxime [116-06-3] **Vol. III: 19**
propionanilide, 3',4'-dichloro- [26638-19-7] **Vol. II: 298**
propione [96-22-0] **Vol. III: 679**
propionic acid anhydride [123-62-6] **Vol. II: 849**
propionicacid, 2,2-dichloro [75-99-0] **Vol. I: 1016**
propionic acid-3,4-dichloroanilide [26638-19-7] **Vol. II: 298**
propionic acid 3,4-dichloroanilide [26638-19-7] **Vol. II: 298**
propionic acid grain preserver [79-09-4] **Vol. I: 1012**
propionic acid 3-hydroxy-betalactone [57-57-8] **Vol. II: 672**
propionic acid, 3-hydroxy-betalactone [57-57-8] **Vol. II: 672**
propionicacid, 2-methyl [79-31-2] **Vol. I: 787**
propionic acid, 2-(2,4,5-trichlorophenoxy)- [93-72-1] **Vol. II: 845**
propionic aldehyde [123-38-6] **Vol. I: 1009**

propionic nitrile [107-12-0] **Vol. II: 851**
propionitrile, 3-hydroxy- [109-78-4] **Vol. II: 466**
propiononitrile [107-12-0] **Vol. II: 851**
propionyl oxide [123-62-6] **Vol. II: 849**
prop-job [26638-19-7] **Vol. II: 298**
propoksuru (Polish) [114-26-1] **Vol. II: 231**
propon [93-72-1] **Vol. II: 845**
propotox [114-26-1] **Vol. II: 231**
propoxure [114-26-1] **Vol. II: 231**
proprop [75-99-0] **Vol. I: 1016**
propyl acetate [109-60-4] **Vol. I: 55**
1-propylacetate [109-60-4] **Vol. I: 55**
2-propylacetate [108-21-4] **Vol. I: 46**
propylacetone [591-78-6] **Vol. II: 462**
1-propyl alcohol [71-23-8] **Vol. II: 856**
propyl alcohol [71-23-8] **Vol. II: 856**
n-propyl alcohol [71-23-8] **Vol. II: 856**
propyl aldehyde [123-38-6] **Vol. I: 1009**
propylaldehyde [123-38-6] **Vol. I: 1009**
n-Propyl alkohol (German) [71-23-8] **Vol. II: 856**
n-propylamine [107-10-8] **Vol. II: 861**
propylamine [107-10-8] **Vol. II: 861**
1-propylamine [107-10-8] **Vol. II: 861**
2-propylamine [75-31-0] **Vol. I: 787**
propylamine, 1-methyl [13952-84-6] **Vol. III: 159**
propylbenzene [103-65-1] **Vol. II: 135**
1-propylbenzene [103-65-1] **Vol. II: 135**
propyl carbinol [71-36-3] **Vol. I: 352**
propylcarbinol [71-36-3] **Vol. I: 352**
n propylcarbinylchloride [109-69-3] **Vol. I: 344**
propyl cyanide [109-74-0] **Vol. III: 152**
n-propyl cyanide [109-74-0] **Vol. III: 152**
propyldihydride [74-98-6] **Vol. II: 815**
n-propyl-di-N-propylthiolcarbamate [1929-77-7] **Vol. II: 221**
s-propyl dipropylcarbamothioate [1929-77-7] **Vol. II: 221**
s-propyl dipropyl (thiocarbamate) [1929-77-7] **Vol. II: 221**
s-propyldipropylthiocarbamate [1929-77-7] **Vol. II: 221**
S-propyl dipropylthiocarbamate [1929-77-7] **Vol. II: 221**
propyl-dipropylthiolcarbamate [1929-77-7] **Vol. II: 221**
1-propylene [115-07-1] **Vol. I: 1004**
propylene aldehyde [4170-30-3] **Vol. I: 547**
propylene dichloride [78-87-5] **Vol. I: 987**

propylenedichloride [78-87-5] Vol. I: 987

alpha,beta propylenedichloride [78-87-5] Vol. I: 987

alpha, alpha-propylene dichloride [78-99-9] Vol. III: 796

propylene, 2,3-dichloro- [78-88-6] Vol. III: 805

propylene epoxide [75-56-9] Vol. I: 991

1,2-propyleneglycol [57-55-6] Vol. I: 1003

alpha propyleneglycol [57-55-6] Vol. I: 1003

. alpha-propylene glycol monomethyl ether [107-98-2] Vol. II: 825

propylene glycol monomethyl ether [107-98-2] Vol. II: 825

1,2-propyleneimine [75-55-8] Vol. I: 219

propylene imine [75-55-8] Vol. I: 219

propyleneoxide [75-56-9] Vol. I: 991

propylene oxide [75-56-9] Vol. I: 991

propylene oxide methanol adduct [37286-64-9] Vol. II: 812

Propylenglykol-monomethylaether (German) [107-98-2] Vol. II: 825

1,2-propylenimine [75-55-8] Vol. I: 219

propylenimine [75-55-8] Vol. I: 219

propylformic acid [107-92-6] Vol. I: 360

propyl-hydride [74-98-6] Vol. II: 815

propylic alcohol [71-23-8] Vol. II: 856

propylic aldehyde [123-38-6] Vol. I: 1009

propylidene chloride [78-99-9] Vol. III: 796

n-propyl mercaptan [107-03-9] Vol. III: 808

1-propylmercaptan [107-03-9] Vol. III: 808

propyl methanol [71-36-3] Vol. I: 352

propylmethanol [71-36-3] Vol. I: 352

propylnitrile [107-12-0] Vol. II: 851

propyl N,N-dipropylthiolcarbamate [1929-77-7] Vol. II: 221

propyl-N, N-dipropylthiolcarbamate [1929-77-7] Vol. II: 221

propylowy-alkohol (Polish) [71-23-8] Vol. II: 856

propyl oxirane [106-88-7] Vol. II: 185

6-(propylpiperonyl)-butyl carbityl ether [51-03-6] Vol. III: 779

6-propylpiperonyl butyl diethylene glycol ether [51-03-6] Vol. III: 779

n-propyl-1-propanamine [142-84-7] Vol. III: 378

5-Propyl-4-(2,5,8-trioxa-dodecyl)-1,3-benzo-dioxol (German) [51-03-6] Vol. III: 779

2-propyn-1-ol [107-19-7] Vol. II: 864

2-propynol [107-19-7] Vol. II: 864

3-propynol [107-19-7] Vol. II: 864

1-propyn-3-ol [107-19-7] Vol. II: 864

propynyl alcohol [107-19-7] Vol. II: 864

2-propynyl alcohol [107-19-7] Vol. II: 864

propyon [114-26-1] Vol. II: 231

proxitane-4002 [79-21-0] Vol. III: 682

proxol [52-68-6] Vol. III: 707

prozinex [139-40-2] Vol. III: 802

prozoin [79-09-4] Vol. I: 1012

prussic acid [74-90-8] Vol. I: 743

prussite [460-19-5] Vol. I: 555

ps [76-06-2] Vol. II: 606

pseudoacetic acid [79-09-4] Vol. I: 1012

pseudocumene [95-63-6] Vol. I: 281

pseudocumol [95-63-6] Vol. I: 281

pseudocyanuric acid [108-80-5] Vol. III: 275

pseudohexyl alcohol [97-95-0] Vol. I: 343

pseudothiourea [62-56-6] Vol. I: 1140

pseudourea [57-13-6] Vol. I: 1136

psi cumene [95-63-6] Vol. I: 281

psoriacid-s-stift [69-72-7] Vol. II: 898

pt-01 [15663-27-1] Vol. III: 237

P-tbt [98-51-1] Vol. III: 169

ptbt [98-51-1] Vol. III: 169

puralin [137-26-8] Vol. I: 632

purasan-sc-10 [62-38-4] Vol. II: 535

puratronic chromium trioxide [1333-82-0] Vol. I: 493

puraturf-10 [62-38-4] Vol. II: 535

pure chromium oxide green-59 [1308-38-9] Vol. III: 234

pure oxygen [7782-44-7] Vol. II: 675

px-104 [84-74-2] Vol. II: 774

px-120 [26761-40-0] Vol. III: 340

px-138 [117-84-0] Vol. II: 789

px-238 [103-23-1] Vol. II: 291

pydrin [51630-58-1] Vol. III: 95

py-kill [7696-12-0] Vol. III: 912

pynosect [10453-86-8] Vol. III: 285

pynosect [52645-53-1] Vol. III: 282

N-pyr [930-55-2] Vol. II: 891

N-N-pyr [930-55-2] Vol. II: 891

pyradex [2303-16-4] Vol. II: 218

pyralene [1336-36-3] Vol. I: 974

pyramid-1 [1344-09-8] Vol. III: 868

pyramid-8 [1344-09-8] Vol. III: 868

pyranol [1336-36-3] Vol. I: 974

pyranton [123-42-2] Vol. II: 688

pyranton-a [123-42-2] Vol. II: 688

pyrazine-hexahydride [110-85-0] Vol. III: 776

pyrazine, hexahydro [110-85-0] Vol. III: 776

pyrenone-606 [51-03-6] Vol. III: 779

(+)-pyrethrate [121-29-9] Vol. III: 817

(+)-pyrethronyl [121-29-9] Vol. III: 817

(+)-pyrethronyl (+)-trans-chrysanthemate [121-21-1] Vol. III: 814

3,6-pyridazinediol [123-33-1] Vol. II: 870

3,6-pyridazinedione, 1,2-dihydro [123-33-1] Vol. II: 870

pyridin [51630-58-1] Vol. III: 95

4-pyridinamine [504-24-5] Vol. II: 879

pyridine, 4-amino [504-24-5] Vol. II: 879

2-pyridinecarboxylic acid, 4-amino3,5,6-trichloro- [1918-02-1] Vol. II: 791

pyridine,3-methyl [108-99-6] Vol. I: 964

pyridine,4-methyl [108-89-4] Vol. I: 1019

pyridine,3-(1-methyl-2-pyrrolidinyl) , (s) [54-11-5] Vol. I: 874

pyridine, 3-(1-methyl-2-pyrrolidinyl)-, hydrochloride, (s)- [2820-51-1] Vol. II: 652

pyridine, 3-(1-methyl-2-pyrrolidinyl)-, (s)-, (r-(r, r))-2,3-dihydroxybutane dioate (1:2) [65-31-6] Vol. II: 654

pyridine, 3-(1-methyl-2-pyrrolidinyl) , (s), sulfate (2:1) [65-30-5] Vol. I: 877

pyridine,3-(tetrahydro-1-methylpyrrol-2-yl) [54-11-5] Vol. I: 874

pyridin (German) [110-86-1] Vol. II: 874

2-pyridinol, 3,5,6-trichloro-, o-ester with O,O-diethyl phosphorothioate [2921-88-2] Vol. III: 762

4-pyridylamine [504-24-5] Vol. II: 879

pyrinex [2921-88-2] Vol. III: 762

pyrisept [123-03-5] Vol. III: 196

pyro [87-66-1] Vol. II: 882

pyrobenzol [71-43-2] Vol. I: 239

pyrobenzole [71-43-2] Vol. I: 239

pyrocatechin [120-80-9] Vol. III: 190

pyrocatechine [120-80-9] Vol. III: 190

pyrocatechinic-acid [120-80-9] Vol. III: 190

pyrocatechol [120-80-9] Vol. III: 190

pyrocatechuic acid [120-80-9] Vol. III: 190

pyrodone [113-48-4] Vol. III: 631

pyrofax [106-97-8] Vol. II: 183

pyrogallol [87-66-1] Vol. II: 882

Pyrokatechin (Czech) [120-80-9] Vol. III: 190

Pyrokatechol (Czech) [120-80-9] Vol. III: 190

m-pyrol [872-50-4] Vol. II: 893

pyroligneous acid [64-19-7] Vol. I: 12

pyrophosphate de tetraethyle (French) [107-49-3] Vol. II: 886

pyrophosphoramide, octamethyl [152-16-9] Vol. II: 884

pyrophosphoric acid, dithiono, tetraethyl ester [3689-24-5] Vol. II: 1010

pyrophosphoric acid octamethyl tetraamide [152-16-9] Vol. II: 884

pyrophosphoric acid, tetraethyl ester [107-49-3] Vol. II: 886

pyrophosphorodithioic acid, o,o,o,o-tetraethyl ester [3689-24-5] Vol. II: 1010

pyrophosphorodithioic acid, tetraethyl ester [3689-24-5] Vol. II: 1010

pyrophosphoryltetrakisdimethylamide [152-16-9] Vol. II: 884

pyroxylic spirits [67-56-1] Vol. I: 836

pyrrole, tetrahydro-N-nitroso [930-55-2] Vol. II: 891

1-pyrrolidinamine, N-nitroso- [930-55-2] Vol. II: 891

pyrrolidine, 1-methyl-2-(3-pyridal) [54-11-5] Vol. I: 874

pyrrolidine, 1-methyl-2-(3-pyridyl) , surfate [65-30-5] Vol. I: 877

pyrrolidine, 1-nitroso- [930-55-2] Vol. II: 891

2-pyrrolidinone, 1-methyl [872-50-4] Vol. II: 893

pzh-2 [7439-89-6] Vol. III: 487

pzh3- [7439-89-6] Vol. III: 487

pzh1m1- [7439-89-6] Vol. III: 487

pzh2m1- [7439-89-6] Vol. III: 487

pzh2m2- [7439-89-6] Vol. III: 487

pzh2m [7439-89-6] Vol. III: 487

pzh-1m3 [7439-89-6] Vol. III: 487

pzh3m [7439-89-6] Vol. III: 487

pzh4m [7439-89-6] Vol. III: 487

q-70 [1344-09-8] Vol. III: 868

q-137 [72-56-0] Vol. III: 687

qac [121-54-0] Vol. III: 105

qcb [608-93-5] Vol. I: 277

quakeral [98-01-1] Vol. II: 443

quamlin [52645-53-1] Vol. III: 282

quartz [7631-86-9] Vol. III: 846

quaternario-cpc [123-03-5] Vol. III: 196

quaternary ammonium compounds, alkylbenzyldimethyl, chloride [8001-54-5] Vol. II: 67

quaternium-1 [8001-54-5] Vol. II: 67

quatrachlor [121-54-0] Vol. III: 105

Quecksilber-chlorid (German) [7487-94-7] Vol. II: 539

Quecksilber (German) [7439-97-6] Vol. II: 549

questuran [2439-10-3] Vol. III: 290

quicklime [1305-78-8] Vol. I: 402

quickphos [20859-73-8] Vol. I: 133

quicksan-20 [62-38-4] Vol. II: 535

quicksan [62-38-4] Vol. II: 535

quicksilver [7439-97-6] Vol. II: 549

quilan [1861-40-1] Vol. II: 1043

quinagamine [54-05-7] Vol. III: 229

(-)-quinine [130-95-0] Vol. III: 821

quinol [123-31-9] Vol. I: 753

quinolin [91-22-5] Vol. I: 1022

quinoline, 7-chloro-4-((4-(diethylamino)-1-methylbutyl) amino)- [54-05-7] Vol. III: 229

quinomethionate [2439-01-2] Vol. III: 663

p-quinone [106-51-4] Vol. I: 303

quinone [106-51-4] Vol. I: 303

quinosan [82-68-8] Vol. II: 132

2,3-quinoxalinedithiol, 6-methyl-, cyclic carbonate [2439-01-2] Vol. III: 663

2,3-quinoxalinedithiol, 6-methyl-, cyclic dithiocarbonate (ester) [2439-01-2] Vol. III: 663

quintar [117-80-6] Vol. II: 645

quintar-540f [117-80-6] Vol. II: 645

quintex [101-42-8] Vol. III: 452

quintocene [82-68-8] Vol. II: 132

quint ox [60-57-1] Vol. III: 356

quintozen [82-68-8] Vol. II: 132

quintozene [82-68-8] Vol. II: 132

2-quinuclidine- methanol, alpha-(6-methoxy-4- quinolyl)-5-vinyl [130-95-0] Vol. III: 821

quolac ex-ub [151-21-3] Vol. II: 981

quso-51 [16871-71-9] Vol. II: 911

quso-51 [7631-86-9] Vol. III: 846

quso-g-30 [16871-71-9] Vol. II: 911

quso-g-30 [7631-86-9] Vol. III: 846

r-10- [56-23-5] Vol. I: 435

r-11 [126-15-8] Vol. III: 909

r-20- [67-66-3] Vol. I: 453

r-22 [75-45-6] Vol. II: 591

r-40 [74-87-3] Vol. II: 585

88r [140-57-8] Vol. III: 61

88-r [140-57-8] Vol. III: 61

r-113 [76-13-1] Vol. II: 353

r-114 [76-14-2] Vol. III: 408

r-115 [76-15-3] Vol. III: 226

r-290- [74-98-6] Vol. II: 815

r-600 [106-97-8] Vol. II: 183

r-680 [13463-67-7] Vol. III: 921

r-717- [7664-41-7] Vol. I: 135

r1270 [115-07-1] Vol. I: 1004

r-1504 [732-11-6] Vol. III: 737

r-1607 [1929-77-7] Vol. II: 221

r-1910 [2008-41-5] Vol. II: 216

r-2170 [301-12-2] Vol. III: 661

r-4461 [741-58-2] Vol. III: 731

r-4572 [2212-67-1] Vol. III: 77

R-152a [75-37-6] Vol. III: 411

racemic lactic acid [50-21-5] Vol. I: 792

racusan [60-51-5] Vol. III: 740

radapon [75-99-0] Vol. I: 1016

radazin [1912-24-9] Vol. II: 1050

rade-cate 25 [75-60-5] Vol. I: 214

radioactive antimony [14374-79-9] Vol. III: 60

radioactive antimony isotope 122 [14374-79-9] Vol. III: 60

radizine [1912-24-9] Vol. II: 1050

radocon [122-34-9] Vol. II: 1047

rafex [534-52-1] Vol. I: 540

rafex-5- [534-52-1] Vol. I: 540

[1R-[1 alpha, 3 alpha (E)]]-[5-(phenylmethyl)-3-furanyl]methyl 3-[(dihydro-2-oxo-3 (2H)-thienylidene) methyl]-2,2-dimethylcyclopropanecarboxylate [58769-20-3] Vol. III: 499

[1r-[1alpha (s),3alpha]]-cyano (3-phenoxyphenyl) methyl 3-[2,2-dibromoethenyl]-2, 2-dimethyl-cyclopropanecarboxylate [52918-63-5] Vol. III: 296

ramor [7440-28-0] Vol. II: 996

ramp [107-21-1] Vol. I: 653

ramrod [1918-16-7] Vol. III: 789

ramrod-65 [1918-16-7] Vol. III: 789

raney alloy [7440-02-0] Vol. I: 862

raney copper [7440-50-8] Vol. I: 519

raney nickel [7440-02-0] Vol. I: 862

range-oil [8008-20-6] Vol. II: 522

raphalox [534-52-1] Vol. I: 540

raphone [94-74-6] Vol. I: 25

raschit [59-50-7] Vol. II: 258

raschit-k [59-50-7] Vol. II: 258

rasen-anicon [59-50-7] Vol. II: 258

rasorite-65 [1330-43-4] Vol. III: 874

ratbane-1080- [62-74-8] Vol. I: 39

rat death liquid [7784-46-5] Vol. I: 208

rat-nip [7723-14-0] Vol. II: 764

Rat-Nip [7723-14-0] Vol. III: 973

ratol [1314-84-7] **Vol. II:** 1100

rattengfftkonserve [7446-18-6] **Vol. I:** 1079

rayox [13463-67-7] **Vol. III:** 921

r-13b1 [75-63-8] **Vol. III:** 143

rb [56-38-2] **Vol. III:** 750

rch-55/5- [7440-02-0] **Vol. I:** 862

rc plasticizer b-17 [123-95-5] **Vol. III:** 167

rc-plasticizer-dbp [84-74-2] **Vol. II:** 774

rc-plasticizer-dop [117-81-7] **Vol. II:** 769

RCRA waste number U052 [108-39-4] **Vol. III:** 261

rc-schwefel extra [7704-34-9] **Vol. II:** 965

rdx [121-82-4] **Vol. III:** 277

re-4355 [300-76-5] **Vol. II:** 746

re-9006 [10265-92-6] **Vol. II:** 744

realgar [1303-32-8] **Vol. III:** 75

rebelate [60-51-5] **Vol. III:** 740

reclaim [1702-17-6] **Vol. III:** 239

rectules [302-17-0] **Vol. III:** 197

red [101-42-8] **Vol. III:** 452

red arsenic [1303-32-8] **Vol. III:** 75

red arsenic glass [1303-32-8] **Vol. III:** 75

red arsenic sulfide [1303-32-8] **Vol. III:** 75

redax [86-30-6] **Vol. I:** 622

red devil dry weed killer [94-75-7] **Vol. I:** 30

reddon [93-76-5] **Vol. I:** 59

reddox [93-76-5] **Vol. I:** 59

red 2g base [100-01-6] **Vol. II:** 70

red mercuric oxide [21908-53-2] **Vol. II:** 559

red off [112-80-1] **Vol. I:** 895

red orpiment [1303-32-8] **Vol. III:** 75

red oxide of mercury [21908-53-2] **Vol. II:** 559

red phosphorus [7723-14-0] **Vol. II:** 764

red precipitate [21908-53-2] **Vol. II:** 559

red shield [60-57-1] **Vol. III:** 356

red top toxaphene-8-spray [8001-35-2] **Vol. I:** 1118

reduced michler's ketone [101-61-1] **Vol. I:** 173

refrigerant-114 [76-14-2] **Vol. III:** 408

refrigerant-115 [76-15-3] **Vol. III:** 226

refrigerant-11 [75-69-4] **Vol. II:** 602

refrigerant-12- [75-71-8] **Vol. I:** 825

refrigerant-22 [75-45-6] **Vol. II:** 591

refrigerant-152A [75-37-6] **Vol. III:** 411

refrigerant-13B1 [75-63-8] **Vol. III:** 143

refrigerent-112 [76-12-0] **Vol. II:** 328

Refusal (Netherlands) [97-77-8] **Vol. III:** 382

regulox [123-33-1] **Vol. II:** 870

regulox-36 [123-33-1] **Vol. II:** 870

regulox-50w [123-33-1] **Vol. II:** 870

regulox-50-w [123-33-1] **Vol. II:** 870

regulus of antimony [7440-36-0] **Vol. I:** 177

regutol [577-11-7] **Vol. III:** 859

reheis-f-1000 [21645-51-2] **Vol. II:** 57

reheis-f-1000 [21645-51-2] **Vol. III:** 35

remasan-chloroble-m [12427-38-2] **Vol. III:** 531

remol-trf [90-43-7] **Vol. II:** 176

renal-md [95-80-7] **Vol. II:** 1016

reofos-95- [25155-23-1] **Vol. I:** 1170

reomol-doa [103-23-1] **Vol. II:** 291

reomol-dop [117-81-7] **Vol. II:** 769

reomol-d-79p [117-81-7] **Vol. II:** 769

repeftal [131-11-3] **Vol. II:** 784

repulse [1897-45-6] **Vol. II:** 513

resmethrine [10453-86-8] **Vol. III:** 285

resmetrina (Portuguese) [10453-86-8] **Vol. III:** 285

resoquine [54-05-7] **Vol. III:** 229

resorcin [108-46-3] **Vol. I:** 1026

resorcine [108-46-3] **Vol. I:** 1026

responsar [68359-37-5] **Vol. III:** 288

retard [123-33-1] **Vol. II:** 870

retarder ak [85-44-9] **Vol. I:** 960

retarder ba [65-85-0] **Vol. I:** 293

retarder esen [85-44-9] **Vol. I:** 960

retarderj [86-30-6] **Vol. I:** 622

retarder pd [85-44-9] **Vol. I:** 960

retarder-w [69-72-7] **Vol. II:** 898

retardex [65-85-0] **Vol. I:** 293

reumachlor [54-05-7] **Vol. III:** 229

revac [577-11-7] **Vol. III:** 859

revenge [75-99-0] **Vol. I:** 1016

rezifilm [137-26-8] **Vol. I:** 632

rhenogran etu [96-45-7] **Vol. I:** 757

rhenosorb c [1305-78-8] **Vol. I:** 402

rhenosorb f [1305-78-8] **Vol. I:** 402

rhodanin s-62- [96-45-7] **Vol. I:** 757

rhoden [114-26-1] **Vol. II:** 231

rhodiachlor [76-44-8] **Vol. II:** 628

rhodialothan [151-67-7] **Vol. III:** 141

rhodianebe [12427-38-2] **Vol. III:** 531

rhodiasol [56-38-2] **Vol. III:** 750

rhodiatox [56-38-2] **Vol. III:** 750

rhodiatrox [56-38-2] **Vol. III:** 750

rhomenc [94-74-6] **Vol. I:** 25

rhomene [94-74-6] **Vol. I:** 25

rhonox [94-74-6] **Vol. I:** 25

rhoplex ac-33 [97-63-2] Vol. III: 575

Rhothane [72-54-8] Vol. III: 404

Rhothane d-3 [72-54-8] Vol. III: 404

r hweed rhap 20 [94-75-7] Vol. I: 30

ridall-zinc [1314-84-7] Vol. II: 1100

ridect pour-on [52645-53-1] Vol. III: 282

rigo toxaphene-8- [8001-35-2] Vol. I: 1118

rikabanol [80-05-7] Vol. II: 720

rimso-5 [67-68-5] Vol. II: 641

riogen [62-38-4] Vol. II: 535

ripcord [52315-07-8] Vol. II: 251

riselect [26638-19-7] Vol. II: 298

r-ll [75-69-4] Vol. II: 602

ro-2 [13463-67-7] Vol. III: 921

roach-salt [7681-49-4] Vol. II: 925

road asphalt [8052-42-4] Vol. II: 114

road tar [8052-42-4] Vol. II: 114

robac-22- [96-45-7] Vol. I: 757

roccal [8001-54-5] Vol. II: 67

rock salt [7647-14-5] Vol. III: 851

rodalon [8001-54-5] Vol. II: 67

rodanins-62 [96-45-7] Vol. I: 757

ro dex [57-24-9] Vol. I: 1054

Rogor-40 [60-51-5] Vol. III: 740

Rogor [60-51-5] Vol. III: 740

Rogor-20l [60-51-5] Vol. III: 740

Rogor-l [60-51-5] Vol. III: 740

Rogor-p [60-51-5] Vol. III: 740

rogue [26638-19-7] Vol. II: 298

ro-ko [83-79-4] Vol. II: 166

roll-fruct [16672-87-0] Vol. III: 426

roman vitriol [7758-98-7] Vol. I: 510

roman vitriol [7758-99-8] Vol. II: 255

romergal-cb [8001-54-5] Vol. II: 67

ronone [83-79-4] Vol. II: 166

rosanil [26638-19-7] Vol. II: 298

ro-sulfram-500 (USA) [97-77-8] Vol. III: 382

rosuran [150-68-5] Vol. II: 1075

rotacide-e, c,- [83-79-4] Vol. II: 166

rotalin [330-55-2] Vol. II: 1084

rotax [149-30-4] Vol. III: 540

rotefive [83-79-4] Vol. II: 166

rotefour [83-79-4] Vol. II: 166

rotenon [83-79-4] Vol. II: 166

rotenona (Spanish) [83-79-4] Vol. II: 166

(-)-rotenone [83-79-4] Vol. II: 166

rotessenol [83-79-4] Vol. II: 166

Rothane [72-54-8] Vol. III: 404

rotocide [83-79-4] Vol. II: 166

rotox [74-83-9] Vol. II: 577

rovokil [13194-48-4] Vol. III: 427

roxion [60-51-5] Vol. III: 740

roxion-ua [60-51-5] Vol. III: 740

royal brand bean tox-82- [8001-35-2] Vol. I: 1118

royal-mbt [149-30-4] Vol. III: 540

royal mh-30 [123-33-1] Vol. II: 870

royal slo-gro [123-33-1] Vol. II: 870

roztoczol [116-29-0] Vol. III: 906

roztoczol-extra [116-29-0] Vol. III: 906

roztozol [116-29-0] Vol. III: 906

RP-3377 [54-05-7] Vol. III: 229

s-(r,r)-cyano (3-phenoxyphenyl) methyl 4-chloro-2-(1-methylethyl) benzene- acetate [66230-04-4] Vol. III: 400

r-10-refrigerant) [56-23-5] Vol. I: 435

r-12(refrigerant) [75-71-8] Vol. I: 825

r-20(refrigerant) [67-66-3] Vol. I: 453

r-21 (refrigerant) [75-43-4] Vol. II: 595

R-50 (refrigerant) [74-82-8] Vol. III: 579

r-11 [insect repellent] [126-15-8] Vol. III: 909

rtu-1010 [82-68-8] Vol. II: 132

rtu [5234-68-4] Vol. II: 670

rtv-vitavax [5234-68-4] Vol. II: 670

RU-15525 [58769-20-3] Vol. III: 499

RU-22974 [52918-63-5] Vol. III: 296

ru-25474 [66841-25-6] Vol. III: 929

ruberon [62-38-4] Vol. II: 535

ruby arsenic [1303-32-8] Vol. III: 75

rucoflex-plasticizer-doa [103-23-1] Vol. II: 291

rumetan [1314-84-7] Vol. II: 1100

runa-arh-200 [13463-67-7] Vol. III: 921

runa-rh20 [13463-67-7] Vol. III: 921

ruphos [78-34-2] Vol. II: 756

rutile [13463-67-7] Vol. III: 921

rutiox-cr [13463-67-7] Vol. III: 921

rutranex [69-72-7] Vol. II: 898

rycopel [52315-07-8] Vol. II: 251

rydichloro-1,2-ethane [107-06-2] Vol. I: 643

s-10165 [26638-19-7] Vol. II: 298

s-1298 [112-53-8] Vol. II: 316

s-1315- [75-99-0] Vol. I: 1016

s-1844 [66230-04-4] Vol. III: 400

s-1- [76-06-2] Vol. II: 606

S-3151 [52645-53-1] Vol. III: 282

s-400 [7758-29-4] Vol. II: 1061

s-4084 [2636-26-2] Vol. III: 271

s-5602 [51630-58-1] Vol. III: 95

s-5602alpha [66230-04-4] Vol. III: 400

solution potassium iodohydragyrate [7783-33-7] Vol. III: 542

solvanol [84-66-2] Vol. II: 779

solvanom [131-11-3] Vol. II: 784

solvarone [131-11-3] Vol. II: 784

solvay-soda [497-19-8] Vol. II: 922

solvent-111- [71-55-6] Vol. I: 684

solvent ether [60-29-7bf] Vol. III: 412

solvent-yellow-1 [60-09-3] Vol. II: 74

solvezinc [7733-02-0] Vol. II: 1103

solvezink [7733-02-0] Vol. II: 1103

solvo powder [65-85-0] Vol. I: 293

solvulose [110-80-5] Vol. II: 362

somalia-yellow-2g [60-09-3] Vol. II: 74

somalia-yellow-r [97-56-3] Vol. III: 46

sometam [137-42-8] Vol. III: 861

somipront [67-68-5] Vol. II: 641

somnos [302-17-0] Vol. III: 197

somonal [50-06-6] Vol. I: 223

somonil [950-37-8] Vol. III: 734

sonac [7664-38-2] Vol. I: 942

sonacide [111-30-8] Vol. III: 468

sontec [302-17-0] Vol. III: 197

sopranebe [12427-38-2] Vol. III: 531

soprathion [56-38-2] Vol. III: 750

soprocide [608-73-1] Vol. I: 564

soprofor-s-70 [26264-06-2] Vol. II: 209

soprothion [56-38-2] Vol. III: 750

sovol [1336-36-3] Vol. I: 974

soxinol-22- [96-45-7] Vol. I: 757

soxinol-m [149-30-4] Vol. III: 540

SP-1103 [7696-12-0] Vol. III: 912

spectracide [333-41-5] Vol. III: 758

spectracide-25ec [333-41-5] Vol. III: 758

sperlox-S [7704-34-9] Vol. II: 965

spersul [7704-34-9] Vol. II: 965

sphaerocobaltite [513-79-1] Vol. III: 244

spin-aid [13684-63-4] Vol. III: 130

spirit of ethyl nitrite [109-95-5] Vol. III: 444

spirit of hartshorn [7664-41-7] Vol. I: 135

spirit of turpentine [8006-64-2] Vol. II: 1065

spirits of salt [7647-01-0] Vol. I: 738

spirits of turpentine [8006-64-2] Vol. II: 1065

spirits of wine [64-17-5] Vol. II: 394

spiro (benzofuran-2 (3h),1'-(2)cyclohexene)-3,4'-dione, 7-chloro-2',4,6-trimethoxy-6'-beta-methyl [126-07-8], Vol. III: 472

spiro (benzofuran-2 (3h),1'-(2) cyclohexene)-3,4'-dione, 7-chloro-2',4,6-trimethoxy-6'-methyl-, (2s-trans)- [126-07-8], Vol. III: 472

spirofulvin [126-07-8], Vol. III: 472

spolacid [133-07-3] Vol. III: 461

spontox [93-76-5] Vol. I: 59

spor-kil [62-38-4] Vol. II: 535

sporostatin [126-07-8], Vol. III: 472

spotfete f [137-26-8] Vol. I: 632

spotrete [137-26-8] Vol. I: 632

spotrete f [137-26-8] Vol. I: 632

spra cal [7778-44-1] Vol. III: 65

spracal [7778-44-1] Vol. III: 65

Spray-Tox [58769-20-3] Vol. III: 499

spring bak [142-59-6] Vol. I: 411

spritz hormin [94-75-7] Vol. I: 30

sprout-nip [101-21-3] Vol. II: 234

sprout/off [123-33-1] Vol. II: 870

sprout-stop [123-33-1] Vol. II: 870

spruce-seal [62-38-4] Vol. II: 535

spud-nic [101-21-3] Vol. II: 234

spud-nie [101-21-3] Vol. II: 234

spur [69409-94-5] Vol. III: 459

sq-1489- [137-26-8] Vol. I: 632

SQ-15500 [95-54-5] Vol. III: 697

sq-9453- [67-68-5] Vol. II: 641

sr-406 [133-06-2] Vol. II: 276

sr406 [133-06-2] Vol. II: 276

sr-999- [7440-22-4] Vol. I: 1036

sra-5172 [10265-92-6] Vol. II: 744

sranan sfx [137-26-8] Vol. I: 632

ss-2074 [2439-01-2] Vol. III: 663

stabilizator-ar [135-88-6] Vol. III: 640

stabilized ethyl-parathion [56-38-2] Vol. III: 750

sta-fast [93-72-1] Vol. II: 845

staflex-dbp [84-74-2] Vol. II: 774

staflex-doa [103-23-1] Vol. II: 291

staflex-dop [117-81-7] Vol. II: 769

stagno (tetracloruro di) (Italian) [7646-78-8] Vol. III: 876

stam [26638-19-7] Vol. II: 298

stam-f-34 [26638-19-7] Vol. II: 298

stam f-34 [26638-19-7] Vol. II: 298

stam-lv-10 [26638-19-7] Vol. II: 298

stam m-4 [26638-19-7] Vol. II: 298

stampede [26638-19-7] Vol. II: 298

stampede-3e [26638-19-7] Vol. II: 298

stam-supernox [26638-19-7] Vol. II: 298

Stancare [7783-47-3] Vol. III: 881

Stanide [7783-47-3] Vol. III: 881

stannane, hydroxytriphenyl [76-87-9] Vol. II: 941

stannane, tetrachloro [7646-78-8] Vol. III: 876

stannane, tributyl [688-73-3] Vol. II: 945

stannic chloride, anhydrous [7646-78-8] Vol. III: 876

stannic tetrachloride [7646-78-8] Vol. III: 876

stannous dichloride [7772-99-8] Vol. II: 1013

stannous fluoride (SnF$_2$) [7783-47-3] Vol. III: 881

star [1344-09-8] Vol. III: 868

starfol bs-100 [123-95-5] Vol. III: 167

stathion [56-38-2] Vol. III: 750

stauffer-captan [133-06-2] Vol. II: 276

Stauffer-ferbam [14484-64-1] Vol. III: 489

stauffer-N-2790- [944-22-9] Vol. II: 735

Stauffer-N-2790 [944-22-9] Vol. III: 714

stauffer-N-521- [533-74-4] Vol. III: 293

stauffer-r-1504 [732-11-6] Vol. III: 737

stauffer r-1910 [2008-41-5] Vol. II: 216

Stauffer r-4,572 [2212-67-1] Vol. III: 77

stavox [128-37-0] Vol. III: 317

steara pbq [106-51-4] Vol. I: 303

stearex beads [57-11-4] Vol. I: 1052

stearic acid, butyl ester [123-95-5] Vol. III: 167

stearic acid, lead salt [7428-48-0] Vol. III: 882

stearix-brown-4r [60-09-3] Vol. II: 74

stearophanic acid [57-11-4] Vol. I: 1052

steladone [470-90-6] Vol. III: 716

stental extentabs [50-06-6] Vol. I: 223

stepanol-me [151-21-3] Vol. II: 981

stepanol-me-dry-aw [151-21-3] Vol. II: 981

stepanol-methyl [151-21-3] Vol. II: 981

stepanol-methyl-dry-aw [151-21-3] Vol. II: 981

stepanol-t-28 [151-21-3] Vol. II: 981

sterolamide [102-71-6] Vol. II: 374

s tetrachlorobenzene [95-94-3] Vol. I: 279

stibine,tribromo [7789-61-9] Vol. I: 179

stibine,trichloro [10025-91-9] Vol. I: 182

stibine,trifluoro [7783-56-4] Vol. I: 185

stibium [7440-36-0] Vol. I: 177

Stickdioxyd (German) [10024-97-2] Vol. I: 889

Stickmonoxyd (German) [10102-43-9] Vol. I: 887

Stickstoffdioxid (German) [10102-44-0] Vol. I: 885

stikstofdioxyde (Dutch) [10102-44-0] Vol. I: 885

sting-kill [102-71-6] Vol. II: 374

stink-damp [7783-06-4] Vol. II: 485

stirolo (Italian) [100-42-5] Vol. I: 1060

stock-guard [70124-77-5], Vol. III: 455

stomoxin [52645-53-1] Vol. III: 282

stopaethyl [97-77-8] Vol. III: 382

stopethyl [97-77-8] Vol. III: 382

stopgerme-s [101-21-3] Vol. II: 234

Stop-home-treatment [7783-47-3] Vol. III: 881

stopspot [100-56-1] Vol. III: 699

stpp [7758-29-4] Vol. II: 1061

strathion [56-38-2] Vol. III: 750

strazine [1912-24-9] Vol. II: 1050

strel [26638-19-7] Vol. II: 298

stricnina (Italian) [57-24-9] Vol. I: 1054

strobane t [8001-35-2] Vol. I: 1118

strontium chromate (1:1) [7789-06-2] Vol. I: 473

strontium chromate [7789-06-2] Vol. I: 473

strychinos [57-24-9] Vol. I: 1054

strychnidin-10-oe [57-24-9] Vol. I: 1054

strychnidin-10-one, 2,3-dimethoxy [357-57-3] Vol. II: 180

strychnine, 2,3-dimethoxy- [357-57-3] Vol. II: 180

Strychnin (German) [57-24-9] Vol. I: 1054

stuntman [123-33-1] Vol. II: 870

stunt-man [123-33-1] Vol. II: 870

stutox [1314-84-7] Vol. II: 1100

styreen (Dutch) [100-42-5] Vol. I: 1060

styren (Czech) [100-42-5] Vol. I: 1060

alpha,beta styrene [25013-15-4] Vol. I: 1057

styrene,armethyl [25013-15-4] Vol. I: 1057

styrene,methyl [25013-15-4] Vol. I: 1057

styrene monomer [100-42-5] Vol. I: 1060

styrol [100-42-5] Vol. I: 1060

styrole [100-42-5] Vol. I: 1060

styrolene [100-42-5] Vol. I: 1060

styron [100-42-5] Vol. I: 1060

styropol [100-42-5] Vol. I: 1060

styropor [100-42-5] Vol. I: 1060

su2000 [7440-44-0] Vol. III: 179

suazin [2008-41-5] Vol. II: 216

subitex [88-85-7] Vol. II: 704

sublimat (Czech) [7487-94-7] Vol. II: 539

sublimate [7487-94-7] Vol. II: 539

sublimed sulfur [7704-34-9] Vol. II: 965

sublimed sulphur [7704-34-9] Vol. II: 965

submar [608-73-1] Vol. I: 564

succinic acid 2,2-dimethylhydrazide [1596-84-5] Vol. III: 884

succinic acid, mercapto-, diethyl ester, s-ester with o,o-dimethyl phosphorodithioate [121-75-5] Vol. II: 950

succinic acid, mono (2,2-dimethylhydrazide) [1596-84-5] Vol. III: 884

succinic acid N,N-dimethylhydrazide [1596-84-5] Vol. III: 884

succinic acid, sulfo-, 1,4-bis(2-ethylhexyl) ester, sodium salt [577-11-7] Vol. III: 859

succinic 1,1-dimethyl hydrazide [1596-84-5] Vol. III: 884

succinic N',N'-dimethylhydrazide [1596-84-5] Vol. III: 884

suchar-681 [7440-44-0] Vol. III: 179

sucker-stuff [123-33-1] Vol. II: 870

sucol-b [110-63-4] Vol. III: 150

sucre edulcor [81-07-2] Vol. III: 824

sucrette [81-07-2] Vol. III: 824

sudan-yellow-'r [60-09-3] Vol. II: 74

sudan-yellow-ra [60-09-3] Vol. II: 74

sudan-yellow-rra [97-56-3] Vol. III: 46

sudol [1300-71-6] Vol. I: 1171

suffa [7704-34-9] Vol. II: 965

suffolan [126-33-0] Vol. II: 1007

suffureted hydrogen [7783-06-4] Vol. II: 485

sufran [7704-34-9] Vol. II: 965

sufran-d [7704-34-9] Vol. II: 965

sulem [7487-94-7] Vol. II: 539

sulfamate [7773-06-0] Vol. II: 959

sulfamate d'ammonium [7773-06-0] Vol. II: 959

sulfamic acid, cobalt(2) salt (2:1) [14017-41-5] Vol. III: 256

sulfamic acid monoammonium salt [7773-06-0] Vol. II: 959

sulfamidic acid [5329-14-6] Vol. III: 886

sulfaminsaeure (German) [7773-06-0] Vol. II: 959

sulfate de cuivre (French) [7758-98-7] Vol. I: 510

sulfate-de-dimethyle (French) [77-78-1] Vol. II: 976

sulfate-de-methyle (French) [77-78-1] Vol. II: 976

sulfate de nicotine (French) [65-30-5] Vol. I: 877

sulfate de plomb (French) [7446-14-2] Vol. III: 517

sulfate-dimethylique (French) [77-78-1] Vol. II: 976

sulfate mercurique (French) [7783-35-9] Vol. II: 563

sulfatep [3689-24-5] Vol. II: 1010

sulfato-de-dimetilo (Spanish) [77-78-1] Vol. II: 976

sulfatom ammoniya (Russian) [7783-20-2] Vol. I: 159

sulfex [7704-34-9] Vol. II: 965

sulfidal [7704-34-9] Vol. II: 965

sulfide, bis (2-chloroethyl) [505-60-2] Vol. II: 962

sulfide (Na(SH)) [16721-80-5] Vol. II: 937

sulfinylbis (methane) [67-68-5] Vol. II: 641

sulfinylbismethane [67-68-5] Vol. II: 641

o-sulfobenzimide [81-07-2] Vol. III: 824

o-sulfobenzoic acid imide [81-07-2] Vol. III: 824

sulfone-aldoxycarb [116-06-3] Vol. III: 19

sulfone, p-chlorophenyl 2,4,5-trichlorophenyl [116-29-0] Vol. III: 906

sulfonic acid, monochloride [7790-94-5] Vol. I: 457

sulforon [7704-34-9] Vol. II: 965

sulfospor [7704-34-9] Vol. II: 965

sulfosuccinic acid bis(2-ethylhexyl) ester sodium salt [577-11-7] Vol. III: 859

sulfotep [3689-24-5] Vol. II: 1010

sulfotepp [3689-24-5] Vol. II: 1010

sulfur atom [7704-34-9] Vol. II: 965

sulfur dioxide [7446-09-5] Vol. I: 1064

sulfure de methyle (French) [75-18-3] Vol. I: 846

sulfur hydride [7783-06-4] Vol. II: 485

sulfuric acid, aluminum salt (3:2) [10043-01-3] Vol. III: 39

sulfuric acid aluminum(3) salt (3:2) [10043-01-3] Vol. III: 39

sulfuric acid, ammonium iron (2) salt (2:2:1) [10045-89-3] Vol. I: 1068

sulfuric acid, beryllium salt (1:1) [13510-49-1] Vol. III: 126

sulfuric acid, cadmium salt (1:1) [10124-36-4] Vol. I: 364

sulfuric acid, cadmium (2) salt [10124-36-4] Vol. I: 364

sulfuric acid, copper (2) salt (1:1) [7758-98-7] Vol. I: 510

sulfuric acid, copper (2) salt (1:1), pentahydrate [7758-99-8] Vol. II: 255

sulfuric acid, copper(2) salt, pentahydrate [7758-99-8] Vol. II: 255

sulfuric acid, diammonium salt [7783-20-2] Vol. I: 159

sulfuric acid, dimercury(1) salt [7783-36-0] Vol. III: 545

sulfuric-acid,-dimethyl-ester [77-78-1] Vol. II: 976

sulfuric acid disodium salt [7757-82-6] Vol. III: 871

sulfuric acid, disodium salt [7757-82-6] Vol. III: 871

sulfuric acid, dithallium (1) salt [7446-18-6] Vol. I: 1079

sulfuric acid, lead(2) salt (1:1) [7446-14-2] Vol. III: 517

sulfuric acid, mercury(2) salt (1:1) [7783-35-9] Vol. II: 563

sulfuric acid, monododecyl ester, sodium salt [151-21-3] Vol. II: 981

sulfuric acid, nickel (2) salt (1:1) [7786-81-4] Vol. I: 870

sulfuric acid nickel (2) salt [7786-81-4] Vol. I: 870

sulfuric acid, thallium (1) salt [7446-18-6] Vol. I: 1079

sulfuric acid, zinc salt [7733-02-0] Vol. II: 1103

sulfuric acid zinc salt [7733-02-0] Vol. II: 1103

sulfuric acid, zirconium (4) salt (2:1) [14644-61-2] Vol. I: 1193

sulfuric chlorohydrin [7790-94-5] Vol. I: 457

sulfuric ether [60-29-7bf] Vol. III: 412

sulfur-mustard [505-60-2] Vol. II: 962

sulfur-mustard-gas [505-60-2] Vol. II: 962

sulfur ointment [7704-34-9] Vol. II: 965

sulfurous acid arthydride [7446-09-5] Vol. I: 1064

sulfurous acid, 2-chloroethyl 2-(4-(1,1-dimethylethyl) phenoxy)-1-methylethyl ester [140-57-8] Vol. III: 61

sulfurous acid, 2-(4-(1,1-dimethyl ethyl) phenoxy) cyclohexyl 2-propynyl ester [2312-35-8] Vol. II: 983

sulfurous acid, monoammonium salt [10192-30-0] Vol. II: 985

sulfurous acid, monosodium salt [7631-90-5] Vol. II: 920

sulfurous acid, 2-(p-t-butylphenoxy)-1-methylethyl-2-chloroethyl ester [140-57-8] Vol. III: 61

sulfurous acid, 2-(p-tert-butyl phenoxy) cyclohexyl 2-propynyl ester [2312-35-8] Vol. II: 983

sulfurous acid, 2-(p-tert-butylphenoxy)-1-methylethyl 2-chloroethyl ester [140-57-8] Vol. III: 61

sulfurous anhydride [7446-09-5] Vol. I: 1064

sulfurous oxide [7446-09-5] Vol. I: 1064

sulfur oxide [7446-09-5] Vol. I: 1064

sulfur, pharmaceutical [7704-34-9] Vol. II: 965

sulfur phosphide [1314-80-3] Vol. I: 955

sulfur selenide [7488-56-4] Vol. III: 833

sulfur selenide (SSe) [7446-34-6] Vol. III: 841

sulikol [7704-34-9] Vol. II: 965

sulkol [7704-34-9] Vol. II: 965

sulourea [62-56-6] Vol. I: 1140

sulphamic acid [5329-14-6] Vol. III: 886

sulphatep [3689-24-5] Vol. II: 1010

sulphatepp [3689-24-5] Vol. II: 1010

2-sulphobenzoic imide [81-07-2] Vol. III: 824

sulphocarbonicanhydride [75-15-0] Vol. I: 428

sulpholane [126-33-0] Vol. II: 1007

sulphos [56-38-2] Vol. III: 750

sulphoxaline [126-33-0] Vol. II: 1007

sulphur dioxide [7446-09-5] Vol. I: 1064

sulphuret of carbon [75-15-0] Vol. I: 428

sulphuric acid [7664-93-9] Vol. II: 969

sulphuric acid, nickel (ii) salt [7786-81-4] Vol. I: 870

sulphur-mustard [505-60-2] Vol. II: 962

sulphur-mustard-gas [505-60-2] Vol. II: 962

sultaf [7704-34-9] Vol. II: 965

sumi-alfa [66230-04-4] Vol. III: 400

sumi-alpha [66230-04-4] Vol. III: 400

sumibac [51630-58-1] Vol. III: 95

sumicidin [51630-58-1] Vol. III: 95

sumifleece [51630-58-1] Vol. III: 95

sumifly [51630-58-1] Vol. III: 95

sumipower [51630-58-1] Vol. III: 95

sumitick [51630-58-1] Vol. III: 95

sumitomo-s-4084 [2636-26-2] Vol. III: 271

sumitox [121-75-5] Vol. II: 950

sumkidin [51630-58-1] Vol. III: 95

suncide [114-26-1] Vol. II: 231

sundaram-1975 [13171-21-6] Vol. III: 704

super-arsonate [2163-80-6] Vol. II: 610

superarsonate [2163-80-6] Vol. II: 610

super-arsonate [2163-80-6] Vol. II: 610

superarsonate [2163-80-6] Vol. II: 610

super-cosan [7704-34-9] Vol. II: 965

super-de-sprout [123-33-1] Vol. II: 870

super de-sprout [123-33-1] Vol. II: 870

super d-weedone [93-76-5] Vol. I: 59

superfloss [16871-71-9] Vol. II: 911

superfloss [7631-86-9] Vol. III: 846

superior dri-die [16919-19-0] Vol. II: 908

superior methyl bromide-2 [74-83-9] Vol. II: 577

superlysoform [50-00-0] Vol. I: 706

supernox [26638-19-7] Vol. II: 298

superoxol [7722-84-1] Vol. II: 482

superpalite [503-38-8] Vol. III: 376

superpalite-phosgene [503-38-8] Vol. III: 376

super-rodiatox [56-38-2] Vol. III: 750

super-six [7704-34-9] Vol. II: 965

supersorbon-iv [7440-44-0] Vol. III: 179

supersorbon-sl [7440-44-0] Vol. III: 179

super-sprout-stop [123-33-1] Vol. II: 870

super-sucker-stuff [123-33-1] Vol. II: 870

super sucker-stuff [123-33-1] Vol. II: 870

super sucker-stuff hc [123-33-1] Vol. II: 870

super-tin-41-gardian-flowable-fungicide [76-87-9] Vol. II: 941

supona [470-90-6] Vol. III: 716

supone [470-90-6] Vol. III: 716

supracid [950-37-8] Vol. III: 734

supracide-ulvair [950-37-8] Vol. III: 734

suprathion [950-37-8] Vol. III: 734

sup-'r-flo [12427-38-2] Vol. III: 531

sup'r-flo ferbam flowable [14484-64-1] Vol. III: 489

sup'r-rio [330-54-1] Vol. II: 1078

surchlor [7681-52-9] Vol. II: 498

surcopur [26638-19-7] Vol. II: 298

sure death 2,4,5-t concentrated #4 [93-79-8] Vol. II: 34

surfalone [126-33-0] Vol. II: 1007

surfate de zinc (French) [7733-02-0] Vol. II: 1103

surflo b17 [67-56-1] Vol. I: 836

surfuric acid, ammonium nickel (2) salt (2:2:1) [15699-18-0] Vol. II: 973

surfuric acid, chromium (3) salt (3:2) [10101-53-8] Vol. I: 482

surpass [1929-77-7] Vol. II: 221

surpass-e [1929-77-7] Vol. II: 221

surpur [26638-19-7] Vol. II: 298

suscon [2921-88-2] Vol. III: 762

sustan [128-37-0] Vol. III: 317

sustane-bht [128-37-0] Vol. III: 317

sutan [2008-41-5] Vol. II: 216

sutan-6e [2008-41-5] Vol. II: 216

sutar'-85-e [2008-41-5] Vol. II: 216

suy b-2 [7439-89-6] Vol. III: 487

suzu-h [76-87-9] Vol. II: 941

sv-102 [577-11-7] Vol. III: 859

svovl [7704-34-9] Vol. II: 965

swasconol d-60 [25322-68-3] Vol. II: 811

sweep [1897-45-6] Vol. II: 513

sweet birch oil [119-36-8] Vol. III: 622

sweet spirit of niter [109-95-5] Vol. III: 444

sweet spirit of nitre [109-95-5] Vol. III: 444

Swift's gold bear 44 ester [94-11-1] Vol. II: 15

sycoporite [1317-42-6] Vol. III: 241

sykose [81-07-2] Vol. III: 824

syllit [2439-10-3] Vol. III: 290

syllit-65 [2439-10-3] Vol. III: 290

symazine [122-34-9] Vol. II: 1047

symclosen [87-90-1] Vol. III: 936

symclosene [87-90-1] Vol. III: 936

sym dichloroethane [107-06-2] Vol. I: 643

sym-dichloro-dimethyl ether [542-88-1] Vol. II: 384

sym-dichloroethylene [540-59-0] Vol. II: 406

sym-dichloroethyl ether [111-44-4] Vol. II: 379

sym-dichloromethyl ether [542-88-1] Vol. II: 384

sym-dichlorotetrafluoroethane [76-14-2] Vol. III: 408

sym diisopropylacetone [108-83-8] Vol. I: 730

sym-dimethylhydrazine [540-73-8] Vol. II: 468

symetryczna-dwumetylohydrazyna (Polish) [540-73-8] Vol. II: 468

symmetrical-dimethylhydrazine [540-73-8] Vol. II: 468

symmetrical diphenyl hydrazine [122-66-7] Vol. II: 479

1,2,12,12a-tetrahydro-8,9-dimethoxy2-(1-methylethenyl)-[1]benzopyrano [3,4-b]furo[2,3-h][1]benzopyran-6 (6ah)-one [83-79-4] Vol. II: 166

[2r-(2alpha, 6aalpha, 12aalpha)]-1,2,12,12a-tetrahydro-8,9-dimethoxy 2-(1-methylethenyl)[1]benzopyrano[3,4-b]furo[2,3-h]benzopyran-6 (6ah)-one [83-79-4] Vol. II: 166

tetrahydro-3,5-dimethyl-2h-1,3,5-thiadiazine-2-thione [533-74-4] Vol. III: 293

1,2,3,6-tetrahydro-3,6-dioxopyridazine [123-33-1] Vol. II: 870

tetrahydrofuraan (Dutch) [109-99-9] Vol. I: 720

tetrahydrofuranne (French) [109-99-9] Vol. I: 720

tetrahydro-2h-3,5-dimethyl-1,3,5-thiadiazine-2-thione [533-74-4] Vol. III: 293

tetrahydro-2h imidazole-2-thione [96-45-7] Vol. I: 757

2,3,5,7alphabeta-tetrahydro-1-hydroxy-1h-pyrrolizine-7-methanol 1-angelate-7-(2,3-dihydroxy-2 (1-methoxyethyl))-3-methyl-butyrate [303-34-4] Vol. III: 501

3a,4,7,7a-tetrahydro-4,7-methanoindene [77-73-6] Vol. II: 635

1,2, 3, 4-tetrahydronaphthalene [119-64-2] Vol. III: 910

tetrahydronicotyrine, dl [54-11-5] Vol. I: 874

3a,4,7,7a-tetrahydro-n-(trichloromethane-sulphenyl) phthalimide [133-06-2] Vol. II: 276

1,2,3,6-tetrahydro-N-(trichloro methylthio) phthalimide [133-06-2] Vol. II: 276

tetrahydro pain dioxin [123-91-1] Vol. I: 618

tetrahydroparadioxin [123-91-1] Vol. I: 618

N-(3,4,5,6-tetrahydrophthalimide) methyl-cis, trans-chrysanthemate [7696-12-0] Vol. III: 912

3,4,5,6-tetrahydrophthalimidomethyl(-)-cis, trans-chrysanthemate [7696-12-0] Vol. III: 912

tetrahydrothiophene 1,1-dioxide [126-33-0] Vol. II: 1007

2,3,4,5-tetrahydrothiophene-1,1-dioxide [126-33-0] Vol. II: 1007

tetrahydrothiophene-dioxide [126-33-0] Vol. II: 1007

tetrahydrothiophene 1-dioxide [126-33-0] Vol. II: 1007

3a,4,7,7a-tetrahydro-2-((trichloro methyl) thio)-1h-isoindole-1,3(2h)-dione [133-06-2] Vol. II: 276

tetraidrofurano (Italian) [109-99-9] Vol. I: 720

tetrakisdimethylaminophosphonous anhydride [152-16-9] Vol. II: 884

tetraleno [127-18-4] Vol. II: 421

tetralex [127-18-4] Vol. II: 421

tetralin [119-64-2] Vol. III: 910

tetramethrine [7696-12-0] Vol. III: 912

tetramethyldiaminodiphenylmethan [101-61-1] Vol. I: 173

n,n'-tetramethyldiaminodiphenyl methane [101-61-1] Vol. I: 173

(p,p'-tetramethyl)diaminodiphenyl methane [101-61-1] Vol. I: 173

p,p'-tetramethyldiaminodiphenyl methane [101-61-1] Vol. I: 173

p,p tetramethyldiaminodiphenylmethane [101-61-1] Vol. I: 173

(4,4'-tetramethyl) diaminodiphenylme thane [101-61-1] Vol. I: 173

4,4'-tetramethyldiaminodiphenyl methane [101-61-1] Vol. I: 173

n,n,n',n'-tetramethyl-4,4'-diaminodiphenyl-methane [101-61-1] Vol. I: 173

tetramethyldiurane sulphite [137-26-8] Vol. I: 632

tetramethylene cyanide [111-69-3] Vol. I: 128

tetramethylene dicyanide [111-69-3] Vol. I: 128

tetramethylene glycol [110-63-4] Vol. III: 150

1,4-tetramethylene glycol [110-63-4] Vol. III: 150

tetramethylene oxide [109-99-9] Vol. I: 720

tetramethylene-sulfone [126-33-0] Vol. II: 1007

tetramethyllead [75-74-1] Vol. II: 796

p-1',1',4',4'-tetramethyloktylbenzensulfonan sodny (Czech) [25155-30-0] Vol. III: 101

tetramethylplumbane [75-74-1] Vol. II: 796

n,n,n',n'-tetramethyl p,p'-diaminodiphenyl-methane [101-61-1] Vol. I: 173

tetramethylthioperoxydicarbonic diamide [137-26-8] Vol. I: 632

Tetramethyl thiramdisulfid (German) [137-26-8] Vol. I: 632

thiophosphate de o,o-dimethyle et de s-2-ethylsulfinylethyle [French] [301-12-2] **Vol. III:** 661

thiophosphoric acid 2-isopropyl-4-methyl-6-pyrimidyl diethyl ester [333-41-5] **Vol. III:** 758

thiophosphoric anhydride [1314-80-3] **Vol. I:** 955

Thiophosphorsaeure-O, S-dimethyl ester-amid (German) [10265-92-6] **Vol. II:** 744

2-thiopropane [75-18-3] **Vol. I:** 846

n-thiopropyl alcohol [107-03-9] **Vol. III:** 808

thiopyrophosphoric acid (((HO)2PS) 20), tetraethyl ester [3689-24-5] **Vol. II:** 1010

thiopyrophosphoric acid, tetraethyl ester [3689-24-5] **Vol. II:** 1010

thiosan [137-26-8] **Vol. I:** 632

thiosan [97-77-8] **Vol. III:** 382

thioscabin [97-77-8] **Vol. III:** 382

thiosuffan [115-29-7] **Vol. II:** 661

thiosulfan-tionel [115-29-7] **Vol. II:** 661

thiosulfuric acid, diammonium salt [7783-18-8] **Vol. I:** 1088

thiosulfuric acid (H2S203), lead salt [26265-65-6] **Vol. II:** 525

thiosulfuric acid, lead salt [26265-65-6] **Vol. II:** 525

thiotax [149-30-4] **Vol. III:** 540

thiotep [3689-24-5] **Vol. II:** 1010

thiotepp [3689-24-5] **Vol. II:** 1010

thiotex [137-26-8] **Vol. I:** 632

thiotox [115-29-7] **Vol. II:** 661

thiotox [137-26-8] **Vol. I:** 632

thiotox (insecticide) [115-29-7] **Vol. II:** 661

2-thiourea [62-56-6] **Vol. I:** 1140

thiourea, n,n'-(1,2-ethanediyl) [96-45-7] **Vol. I:** 757

thiovanic acid [68-11-1] **Vol. III:** 538

thiovit [7704-34-9] **Vol. II:** 965

thiram-75- [137-26-8] **Vol. I:** 632

thiramad [137-26-8] **Vol. I:** 632

thirame [137-26-8] **Vol. I:** 632

thirampa [137-26-8] **Vol. I:** 632

thirasan [137-26-8] **Vol. I:** 632

thireranide [97-77-8] **Vol. III:** 382

thitrol [94-81-5] **Vol. III:** 174

thiulix [137-26-8] **Vol. I:** 632

thiurad [137-26-8] **Vol. I:** 632

thiuram [137-26-8] **Vol. I:** 632

thiuramin [137-26-8] **Vol. I:** 632

thiuram m-rubber accelerator [137-26-8] **Vol. I:** 632

thiuramyl [137-26-8] **Vol. I:** 632

Thompson-Hayward-6040 [35367-38-5] **Vol. III:** 339

thorium-X [872-50-4] **Vol. II:** 893

threamine [78-96-6] **Vol. II:** 821

three elephant [10043-35-3] **Vol. I:** 327

threthylen [79-01-6] **Vol. II:** 427

threthylene [79-01-6] **Vol. II:** 427

thu [62-56-6] **Vol. I:** 1140

thylate [137-26-8] **Vol. I:** 632

tiazon [533-74-4] **Vol. III:** 293

tilcarex [82-68-8] **Vol. II:** 132

tillram [97-77-8] **Vol. III:** 382

timonox [1309-64-4] **Vol. II:** 86

tin bifluoride [7783-47-3] **Vol. III:** 881

tin chloride [7647-18-9] **Vol. II:** 83

tin chloride (SnC12) [7772-99-8] **Vol. II:** 1013

tin chloride (SnCl₄) [7646-78-8] **Vol. III:** 876

tin dichloride [7772-99-8] **Vol. II:** 1013

tin difluoride [7783-47-3] **Vol. III:** 881

tin fluoride [7783-47-3] **Vol. III:** 881

tin fluoride (SnF₂) [7783-47-3] **Vol. III:** 881

tin, hydroxytriphenyl [76-87-9] **Vol. II:** 941

tin(II) chloride (1:2) [7772-99-8] **Vol. II:** 1013

tin(II) chloride [7772-99-8] **Vol. II:** 1013

tin(IV) chloride [7646-78-8] **Vol. III:** 876

tin(IV) tetrachloride [7646-78-8] **Vol. III:** 876

tinotepp [3689-24-5] **Vol. II:** 1010

tin perchloride [7646-78-8] **Vol. III:** 876

tin protochloride [7772-99-8] **Vol. II:** 1013

tin tetrachloride [7646-78-8] **Vol. III:** 876

tin tetrachloride, anhydrous [7646-78-8] **Vol. III:** 876

tintetrachloride (Dutch) [7646-78-8] **Vol. III:** 876

tiofine [13463-67-7] **Vol. III:** 921

tiofos [56-38-2] **Vol. III:** 750

tiona-td [13463-67-7] **Vol. III:** 921

tionel [115-29-7] **Vol. II:** 661

tionex [115-29-7] **Vol. II:** 661

tiovel [115-29-7] **Vol. II:** 661

tioxide [13463-67-7] **Vol. III:** 921

tioxide ad-m [13463-67-7] **Vol. III:** 921

tioxide r-cr [13463-67-7] **Vol. III:** 921

3-(3-trifluoromethylphenyl)-1,1-di methylurea [2164-17-2] Vol. II: 1086

N-(3-trifluoromethyl) phenyl-1,1-dimethylurea [2164-17-2] Vol. II: 1086

3-(m-trifluoromethylphenyl)-1,1-dimethylurea [2164-17-2] Vol. II: 1086

N-(m-trifluoromethylphenyl)-N', N'-dimethylurea [2164-17-2] Vol. II: 1086

N-(3-trifluoromethylphenyl)-N, N'dimethylurea [2164-17-2] Vol. II: 1086

N'-(3 trifluoromethylphenyl)-N, N-dimethylurea [2164-17-2] Vol. II: 1086

N-(3-trifluoromethylphenyl)-1T-N'-di methylurea [2164-17-2] Vol. II: 1086

trifluoromonobromomethane [75-63-8] Vol. III: 143

trifluorostibine [7783-56-4] Vol. I: 185

trifocide [534-52-1] Vol. I: 540

trifolex [94-81-5] Vol. III: 174

trifrina [534-52-1] Vol. I: 540

trifungol [14484-64-1] Vol. III: 489

trigen [112-27-6] Vol. II: 1058

triglycine [139-13-9] Vol. III: 648

triglycol [112-27-6] Vol. II: 1058

triglycollamic acid [139-13-9] Vol. III: 648

triglycol monoethyl ether [112-50-5] Vol. II: 369

trigonox a-75 (Czech) [75-91-2] Vol. III: 896

trigonox-M-50 [1338-23-4] Vol. II: 192

trigosan [62-38-4] Vol. II: 535

triherbicide-cipc [101-21-3] Vol. II: 234

trihydrated alumina [21645-51-2] Vol. II: 57

trihydrated alumina [21645-51-2] Vol. III: 35

trihydroxyaluminum [21645-51-2] Vol. II: 57

trihydroxyaluminum [21645-51-2] Vol. III: 35

1,2,3-trihydroxybenzen (Czech) [87-66-1] Vol. II: 882

1,2,3-trihydroxybenzene [87-66-1] Vol. II: 882

3,4,5-trihydroxybenzoicacid [149-91-7] Vol. I: 726

trihydroxycyanidine [108-80-5] Vol. III: 275

trihydroxytriazine [108-80-5] Vol. III: 275

2,4,6-trihydroxy-1,3,5-triazine [108-80-5] Vol. III: 275

trihydroxytriethylamine [102-71-6] Vol. II: 374

trikacide [443-48-1] Vol. III: 628

trikamon [443-48-1] Vol. III: 628

triklone [79-01-6] Vol. II: 427

trikojol [443-48-1] Vol. III: 628

trikozol [443-48-1] Vol. III: 628

o-Trikresylphosphate (German) [78-30-8] Vol. III: 724

trilene [79-01-6] Vol. II: 427

trilit [118-96-7] Vol. I: 1093

trilon-a [139-13-9] Vol. III: 648

trilon bw [60-00-4] Vol. I: 36

trimangol [12427-38-2] Vol. III: 531

trimangol-80 [12427-38-2] Vol. III: 531

trimar [79-01-6] Vol. II: 427

trimaton [137-42-8] Vol. III: 861

trimatron [137-42-8] Vol. III: 861

trimegol [133-06-2] Vol. II: 276

trimeks [443-48-1] Vol. III: 628

trimethylacetic acid [75-98-9] Vol. III: 781

2-(trimethylacetyl)-1,3-indandione [83-26-1] Vol. III: 774

alpha,alpha',alpha''-trimethyl- aminetricarboxylic acid [139-13-9] Vol. III: 648

trimethylaminomethane [75-64-9] Vol. II: 199

1,2,5-trimethylbenzene [95-63-6] Vol. I: 281

1,3,4-trimethylbenzene [95-63-6] Vol. I: 281

1,7,7-trimethylbicyclo (2,2,1)-2-heptanone [76-22-2] Vol. II: 211

2,6,6-trimethylbicyclo (3.1.1) hept-2-ene [80-56-8] Vol. III: 32

2,6,6-trimethyl- bicyclo-(3,1,1)-2-heptene [80-56-8] Vol. III: 32

1,7,7-trimethylbicyclo (2,2,1) hept-2-yl-thiocyanatoacetate [115-31-1] Vol. III: 491

trimethyl-carbinol [75-65-0] Vol. II: 197

trimethyl chlorosilane [75-77-4] Vol. III: 946

3,5,5-trimethyl-2-cyclohexene-1-one [78-59-1] Vol. II: 272

1,1,3-trimethyl-3-cyclohexene-5-one [78-59-1] Vol. II: 272

3,5,5-trimethyl-2-cyclohexen-1-one [78-59-1] Vol. II: 272

3,5,5-trimethyl-2-cyclohexenone [78-59-1] Vol. II: 272

3,5,5-Trimethyl-2-cyclohexen-1-on (German, Dutch) [78-59-1] Vol. II: 272

trimethyleentrinitramine (Dutch) [121-82-4] Vol. III: 277

u-5954 [127-19-5] Vol. I: 10

uantax-SBS [7631-90-5] Vol. II: 920

uc-21149 [116-06-3] Vol. III: 19

ucar-17- [107-21-1] Vol. I: 653

ucar-bisphenol-hp [80-05-7] Vol. II: 720

ucar-4c [25322-68-3] Vol. II: 811

ucar solvent lm [107-98-2] Vol. II: 825

UCAR-solvent-2lm [34590-94-8] Vol. III: 381

ucc-974 [533-74-4] Vol. III: 293

ucon-12- [75-71-8] Vol. I: 825

ucon-22 [75-45-6] Vol. II: 591

ucon-112 [76-12-0] Vol. II: 328

ucon-114 [76-14-2] Vol. III: 408

ucon-fluorocarbon-113 [76-13-1] Vol. II: 353

ucon-12/halocarbon-12- [75-71-8] Vol. I: 825

ucon-lb-1715 [37286-64-9] Vol. II: 812

ucon-refrigerant-11 [75-69-4] Vol. II: 602

u-46-d [94-75-7] Vol. I: 30

u-dimethylhydrazine [57-14-7] Vol. II: 470

udmh [57-14-7] Vol. II: 470

u-46dp [94-75-7] Vol. I: 30

u-46KW [93-79-8] Vol. II: 34

u46KW [93-79-8] Vol. II: 34

ultracid-40 [950-37-8] Vol. III: 734

ultracid [950-37-8] Vol. III: 734

ultracide [950-37-8] Vol. III: 734

ultracide Ciba-Geigy [950-37-8] Vol. III: 734

ultracide-ulvair-250 [950-37-8] Vol. III: 734

ultramarine green [1308-38-9] Vol. III: 234

ultramide [57-13-6] Vol. I: 1136

ultrasil-vh-3 [16871-71-9] Vol. II: 911

ultrasil-vh-3 [7631-86-9] Vol. III: 846

ultrasil-vn-3 [16871-71-9] Vol. II: 911

ultrasil-vn-3 [7631-86-9] Vol. III: 846

ultra-sulfur [7704-34-9] Vol. II: 965

ulv [8065-48-3] Vol. III: 755

u46-mcpb [94-81-5] Vol. III: 174

UN2194 [7783-79-1] Vol. III: 839

unden [114-26-1] Vol. II: 231

UN1381 (DOT) [7723-14-0] Vol. III: 973

un-1578 (DOT) [100-00-5] Vol. III: 210

UN2076 (DOT) [108-39-4] Vol. III: 261

UN2447 (DOT) [7723-14-0] Vol. III: 973

unh [10102-06-4] Vol. II: 1071

unicrop-cipc [101-21-3] Vol. II: 234

unidron [330-54-1] Vol. II: 1078

unifac-6550 [60-33-3] Vol. II: 667

uniflex-dbp [84-74-2] Vol. II: 774

unifos [62-73-7] Vol. I: 946

unimoll-bb [85-68-7] Vol. II: 767

unimoll-da [84-66-2] Vol. II: 779

unimoll-db [84-74-2] Vol. II: 774

Union Carbide-21149- [116-06-3] Vol. III: 19

union carbide a-150 [75-94-5] Vol. III: 945

Union Carbide UC-21149 [116-06-3] Vol. III: 19

unipon [75-99-0] Vol. I: 1016

uniroyal [117-80-6] Vol. II: 645

uniroyal-d014 [2312-35-8] Vol. II: 983

unisept bza [65-85-0] Vol. I: 293

uniston cr-ht 200 [7772-99-8] Vol. II: 1013

unitane 0-110 [13463-67-7] Vol. III: 921

unitane 0-220 [13463-67-7] Vol. III: 921

unitane [13463-67-7] Vol. III: 921

unitane or-150 [13463-67-7] Vol. III: 921

unitane or-340 [13463-67-7] Vol. III: 921

unitane or-342 [13463-67-7] Vol. III: 921

unitane or-350 [13463-67-7] Vol. III: 921

unitane or-540 [13463-67-7] Vol. III: 921

unitane or-640 [13463-67-7] Vol. III: 921

univerm [56-23-5] Vol. I: 435

universal-ZPS [127-82-2] Vol. II: 142

universne acid [60-00-4] Vol. I: 36

uniwhite-ao [13463-67-7] Vol. III: 921

uniwhite-ko [13463-67-7] Vol. III: 921

uniwhite-or-450 [13463-67-7] Vol. III: 921

uniwhite-or-650 [13463-67-7] Vol. III: 921

uns-dimethylhydrazine [57-14-7] Vol. II: 470

unslaked lime [1305-78-8] Vol. I: 402

uns trimethylbenzene [95-63-6] Vol. I: 281

unsym-dimethylhydrazine [57-14-7] Vol. II: 470

unsymmetrical-dimethylhydrazine [57-14-7] Vol. II: 470

unsym-trichlorobenzene [120-82-1] Vol. III: 92

ur [57-13-6] Vol. I: 1136

uracil, 3-tert-butyl-5-chloro-6-methyl [5902-51-2] Vol. III: 894

uranium acetate [541-09-3] Vol. I: 1134

uranium,bis (acetato) dioxo [541-09-3] Vol. I: 1134

uranium,bis (acetato o) dioxo [541-09-3] Vol. I: 1134

uranium,bis (aceto) dioxo [541-09-3] Vol. I: 1134

uranium, bis (nitrato-O) dioxo [10102-06-4] Vol. II: 1071

uranium, bis (nitrato-O, 0') dioxo(solid) [10102-06-4] Vol. II: 1071

uranium diacetate dioxide [541-09-3] Vol. I: 1134

uranium, difluorodioxo [13536-84-0] Vol. III: 954

uranium, dinitratodioxo- [10102-06-4] Vol. II: 1071

uranium, dioxosulfato [1314-64-3] Vol. III: 956

uranium, dioxo (sulfato(²-)-O)- [1314-64-3] Vol. III: 956

uranium fluoride oxide [13536-84-0] Vol. III: 954

uranium fluoride oxide (UO₂F₂) [13536-84-0] Vol. III: 954

uraniumi [7440-61-1] Vol. I: 1132

uranium nitrate oxide (UO2(NO3)2) [10102-06-4] Vol. II: 1071

uranium oxide peroxide (UO2(O2)) [19525-15-6] Vol. II: 1069

uranium oxide sulfate [1314-64-3] Vol. III: 956

uranium oxide sulfate (UO₂SO₄) [1314-64-3] Vol. III: 956

uranium oxide sulfate (UO₂(SO₄)) [1314-64-3] Vol. III: 956

uranium oxide (UO4) [19525-15-6] Vol. II: 1069

uranium oxyacetate [541-09-3] Vol. I: 1134

uranium oxyfluoride [13536-84-0] Vol. III: 954

uranium oxynitrate [10102-06-4] Vol. II: 1071

uranium oxysulfate [1314-64-3] Vol. III: 956

uranium peroxide (UO4) [19525-15-6] Vol. II: 1069

uranyl (2) acetate [541-09-3] Vol. I: 1134

uranyl acetate (uo2 (oac)2) [541-09-3] Vol. I: 1134

uranyl diacetate [541-09-3] Vol. I: 1134

uranyl dinitrate [10102-06-4] Vol. II: 1071

uranyl-fluoride (UO₂F₂) [13536-84-0] Vol. III: 954

uranyl nitrate (UO2(NO3)2) [10102-06-4] Vol. II: 1071

uranyl sulfate (UO₂ (SO₄)) [1314-64-3] Vol. III: 956

uranyl sulfate (UO₂SO₄) [1314-64-3] Vol. III: 956

urea ammonium nitrate solution [57-13-6] Vol. I: 1136

urea-13c [57-13-6] Vol. I: 1136

urea, 3-(4-chlorophenyl)-1,1-dimethyl [150-68-5] Vol. II: 1075

ureacin-20 [57-13-6] Vol. I: 1136

ureacin-40creme [57-13-6] Vol. I: 1136

ureacin-10lotion [57-13-6] Vol. I: 1136

urea, 3-(3,4-dichlorophenyl)-1,1-dimethyl [330-54-1] Vol. II: 1078

urea, 3-(3,4-dichlorophenyl)-1-methoxy-1-methyl [330-55-2] Vol. II: 1084

urea, 1,1-dimethyl-3-(alpha, alpha, alpha-trifluoro-m-tolyl)- [2164-17-2] Vol. II: 1086

urea, 1,1-dimethyl-3-phenyl [101-42-8] Vol. III: 452

urea, 1,3-ethylene-2-thio [96-45-7] Vol. I: 757

urea, N'-(4-(4-chlorophenoxy) phenyl)-N, N-dimethyl [1982-47-4] Vol. III: 232

urea, N'-(4-chlorophenyl)-N, N-dimethyl [150-68-5] Vol. II: 1075

urea, N'-(3,4-dichlorophenyl)-N-methoxy-N-methyl [330-55-2] Vol. II: 1084

urea, N'-(3,4-dichlorophenyl)-N, N-dimethyl [330-54-1] Vol. II: 1078

urea, N,N-dimethyl-N'-(3-(trifluoromethyl) phenyl)- [2164-17-2] Vol. II: 1086

urea, 1-(p-chlorophenyl)-3-(2,6-difluorobenzoyl)- [35367-38-5] Vol. III: 339

urea, 3-(p-chlorophenyl)-1,1-dimethyl [150-68-5] Vol. II: 1075

urea, 3-(p-(p-chlorophenoxy) phenyl)-1,1-dimethyl [1982-47-4] Vol. III: 232

urea, 2-seleno [630-10-4] Vol. III: 844

urea, seleno [630-10-4] Vol. III: 844

urea, 1-(5-(t-butyl) -1,3,4-thiadiazol-2-yl)-1,3-dimethyl [34014-18-1] Vol. III: 892

urea,2-thio [62-56-6] Vol. I: 1140

urea,thio [62-56-6] Vol. I: 1140

ureophil [57-13-6] Vol. I: 1136

urepearl [57-13-6] Vol. I: 1136

uretan-etylowy (Polish) [51-79-6] Vol. II: 224

uretano [51-79-6] Vol. II: 224

urethan [51-79-6] Vol. II: 224

urethane [51-79-6] Vol. II: 224

urethanum (INN-Latin) [51-79-6] Vol. II: 224

urevert [57-13-6] Vol. I: 1136

urox-d [330-54-1] Vol. II: 1078

USAFA-4600 [109-77-3] Vol. II: 531

usafcb-22 [91-59-8] Vol. I: 860

USAF CB-35 [68-11-1] Vol. III: 538

USAF CY-2 [156-62-7] Vol. II: 265

USAF EK-1375 [60-09-3] Vol. II: 74

USAFEK-1597 [141-43-5] Vol. II: 356

USAF EK-1995 [420-04-2] Vol. II: 263

usafek-218 [91-22-5] Vol. I: 1022

USAF EK-2219 [90-43-7] Vol. II: 176

usafek-356 [123-31-9] Vol. I: 753

USAF EK-3 [103-84-4] Vol. III: 3

usafek-496 [98-86-2] Vol. I: 86

usafek-497 [62-56-6] Vol. I: 1140

usafel-62 [96-45-7] Vol. I: 757

USAF gy-3 [149-30-4] Vol. III: 540

usaf kf-11 [2698-41-1] Vol. III: 218

usafkf-17 [372-09-8] Vol. I: 29

usafp-220 [106-51-4] Vol. I: 303

USAF P-7 [330-54-1] Vol. II: 1078

USAF P-8 [150-68-5] Vol. II: 1075

USAF RH-7 [109-78-4] Vol. II: 466

usafrh-8 [75-86-5] Vol. I: 794

usaf sc-2 [126-07-8], Vol. III: 472

USAF ST-40 [126-98-7] Vol. II: 839

usaf xr-29 [149-30-4] Vol. III: 540

USAFXR-41 [150-68-5] Vol. II: 1075

USAFXR-42 [330·54-1] Vol. II: 1078

USR-604 [117-80-6] Vol. II: 645

US-rubber-604 [117-80-6] Vol. II: 645

ustaad [52315-07-8] Vol. II: 251

uts silverator water treatment unit [7761-88-8] Vol. I: 1038

uvilon [110-85-0] Vol. III: 776

uvon [7287-19-6] Vol. II: 1056

uzgen [17804-35-2] Vol. II: 148

v-9- [7440-22-4] Vol. I: 1036

vac [108-05-4] Vol. I: 69

vagilen [443-48-1] Vol. III: 628

valentinite [1309-64-4] Vol. II: 86

valetone [108-83-8] Vol. I: 730

valine aldehyde [78-84-2] Vol. I: 769

vam [108-05-4] Vol. I: 69

vanadate (VO3(1-)), ammonium [7803-55-6] Vol. II: 1088

vanadic acid anhydride [1314-62-1] Vol. I: 1145

vanadic acid (HVOa), ammonium salt [7803-55-6] Vol. II: 1088

vanadic anhydride [1314-62-1] Vol. I: 1145

vanadic oxide [1314-34-7] Vol. III: 384

vanadic sulfate [27774-13-6] Vol. II: 1090

vanadio, pentossido di (Italian) [1314-62-1] Vol. I: 1145

vanadium chloride [7632-51-1] Vol. III: 962

vanadium chloride oxide (vocl3) [7727-18-6] Vol. I: 1148

vanadium chloride (VCl₄) [7632-51-1] Vol. III: 962

vanadium chloride (VCl₄), (t-4) [7632-51-1] Vol. III: 962

vanadium(IV) chloride [7632-51-1] Vol. III: 962

vanadium monoxide trichloride [7727-18-6] Vol. I: 1148

vanadium oxide [1314-34-7] Vol. III: 384

vanadium oxide [1314-62-1] Vol. I: 1145

vanadium(3) oxide [1314-34-7] Vol. III: 384

vanadium oxide, sesqui [1314-34-7] Vol. III: 384

vanadium oxide sulfate (VO(SO4)) [27774-13-6] Vol. II: 1090

vanadium oxide trichloride [7727-18-6] Vol. I: 1148

vanadium oxide (V₂O₃) [1314-34-7] Vol. III: 384

vanadium oxosulfa [27774-13-6] Vol. II: 1090

vanadium, oxo (sulfato (2-)-o)- [27774-13-6] Vol. II: 1090

vanadium oxychloride [7727-18-6] Vol. I: 1148

vanadium oxysulfate [27774-13-6] Vol. II: 1090

vanadium pentaoxide [1314-62-1] Vol. I: 1145

vanadiumpentoxid (German) [1314-62-1] Vol. I: 1145

vanadium, pentoxyde de (French) [1314-62-1] Vol. I: 1145

vanadiumpentoxyde (Dutch) [1314-62-1] Vol. I: 1145

vanadium sesquioxide [1314-34-7] Vol. III: 384

vanadium sulfate [27774-13-6] Vol. II: 1090

vanadium trichloride monooxide [7727-18-6] Vol. I: 1148

vestinol-oa [103-23-1] **Vol. II:** 291

vestrol [79-01-6] **Vol. II:** 427

vetiol [121-75-5] **Vol. II:** 950

vetron-2t [93-76-5] **Vol. I:** 59

vianol [128-37-0] **Vol. III:** 317

vicad [10108-64-2] **Vol. I:** 376

vic m-xylenol [576-26-1] **Vol. I:** 1167

vic-trichlorobenzene [87-61-6] **Vol. II:** 137

vidden-d [542-75-6] **Vol. II:** 834

villiaumite [15096-52-3] **Vol. III:** 37

villiaumite [7681-49-4] **Vol. II:** 925

vinegar acid [64-19-7] **Vol. I:** 12

vinegar naphtha [141-78-6] **Vol. II:** 19

vinfle, (cloruro di)-(Italian) [75-01-4] **Vol. II:** 416

vinicizer-80 [117-81-7] **Vol. II:** 769

vinicizer-85 [117-84-0] **Vol. II:** 789

vinicoll [133-07-3] **Vol. III:** 461

vinile (acetatodi) (Italian) [108-05-4] **Vol. I:** 69

vinkeil-100- [60-00-4] **Vol. I:** 36

vinylacetaat (Dutch) [108-05-4] **Vol. I:** 69

vinyl acetate hq [108-05-4] **Vol. I:** 69

vinyl acetate monomer [108-05-4] **Vol. I:** 69

vinylacetat (German) [108-05-4] **Vol. I:** 69

vinylamine [151-56-4] **Vol. I:** 671

vinyl a-monomer [108-05-4] **Vol. I:** 69

vinylbenzen (Czech) [100-42-5] **Vol. I:** 1060

vinylbenzen (Dutch) [100-42-5] **Vol. I:** 1060

vinylbenzene [100-42-5] **Vol. I:** 1060

vinylbenzol [100-42-5] **Vol. I:** 1060

vinyl beta-chloroethyl ether [110-75-8] **Vol. III:** 225

vinyl carbinol [107-18-6] **Vol. II:** 52

vinylcarbinol [107-18-6] **Vol. II:** 52

vinyl chloride monomer [75-01-4] **Vol. II:** 416

vinylchlorid (German) [75-01-4] **Vol. II:** 416

vinyl 2-chloroethyl ether [110-75-8] **Vol. III:** 225

vinyl-c-monomer [75-01-4] **Vol. II:** 416

vinylcyanide [107-13-1] **Vol. I:** 123

vinyle (acetatede) (French) [108-05-4] **Vol. I:** 69

vinyle, (chlorure de)-(French) [75-01-4] **Vol. II:** 416

vinylethylene [106-99-0] **Vol. I:** 333

vinylformic acid [79-10-7] **Vol. I:** 110

vinylidene chloride (inhibited) [75-35-4] **Vol. II:** 409

vinylidene chloride, monomer [75-35-4] **Vol. II:** 409

vinylidene dichloride [75-35-4] **Vol. II:** 409

vinylidine chloride [75-35-4] **Vol. II:** 409

vinylidene chloride(I, I) [75-35-4] **Vol. II:** 409

vinyl methyl ketone [78-94-4] **Vol. III:** 624

vinylofos [62-73-7] **Vol. I:** 946

2-vinyloxyethyl chloride [110-75-8] **Vol. III:** 225

vinylphate [470-90-6] **Vol. III:** 716

vinylsilicon trichloride [75-94-5] **Vol. III:** 945

vinylstyrene [1321-74-0] **Vol. III:** 967

vinyltoluene [25013-15-4] **Vol. I:** 1057

p-vinyltoluene [622-97-9] **Vol. III:** 970

vinyl trichloride [79-00-5] **Vol. II:** 341

vinyltrichlorosilane [75-94-5] **Vol. III:** 945

vinyphate [470-90-6] **Vol. III:** 716

violet-3 [1330-20-7] **Vol. I:** 1159

violet phosphorus [7723-14-0] **Vol. II:** 764

Virginia-Carolina vc 9–104 [13194-48-4] **Vol. III:** 427

visalens soaking/cleaning solutions [8001-54-5] **Vol. II:** 67

visalens-wetting-solution [8001-54-5] **Vol. II:** 67

visco-1152 [67-63-0] **Vol. II:** 515

visine-ac [7733-02-0] **Vol. II:** 1103

visine-ac [8001-54-5] **Vol. II:** 67

visko rhap [93-76-5] **Vol. I:** 59

vitaflow [5234-68-4] **Vol. II:** 670

vitasil-220 [16871-71-9] **Vol. II:** 911

vitasil-220 [7631-86-9] **Vol. III:** 846

vitavax-100- [5234-68-4] **Vol. II:** 670

vitavax-200 [5234-68-4] **Vol. II:** 670

vitavax-300 [5234-68-4] **Vol. II:** 670

vitavax-34- [5234-68-4] **Vol. II:** 670

vitavax [5234-68-4] **Vol. II:** 670

vitavax 30-c, [5234-68-4] **Vol. II:** 670

vitavax-735d [5234-68-4] **Vol. II:** 670

vitavax-pcnb [5234-68-4] **Vol. II:** 670

vitavaxsulfone [5259-88-1] **Vol. III:** 659

vitavax-thimm-ltndane [5234-68-4] **Vol. II:** 670

vitavax-thiram [5234-68-4] **Vol. II:** 670

vitavax-75w [5234-68-4] **Vol. II:** 670

vitavex [5259-88-1] **Vol. III:** 659

vitawax [5234-68-4] **Vol. II:** 670

vito-spot-fungicide [76-87-9] **Vol. II:** 941